ENCYCLOPEDIA
OF THE
EARLY CHURCH

ENCYCLOPEDIA OF THE EARLY CHURCH

Produced by the Institutum Patristicum Augustinianum
and edited by
Angelo Di Berardino

Translated from the Italian by
Adrian Walford

With a foreword and bibliographic amendments by
W.H.C. Frend

Volume II

New York
OXFORD UNIVERSITY PRESS
1992

Translated from the Italian
Dizionario Patristico e di Antichità Cristiane
© Casa Editrice Marietti, S.p.A.

English Version
Copyright © 1992 by James Clarke & Co. Ltd

First published in Great Britain in 1992 by
James Clarke & Co. Ltd.
P.O. Box 60
Cambridge CB1 2NT

First published in the United States in 1992 by
Oxford University Press, Inc.
200 Madison Avenue
New York, New York 10016

Oxford is a registered trademark of Oxford University Press

All rights reserved. No part of this publication may be reproduced,
stored in a retrieval system, or transmitted, in any form or by any means,
electronic, photocopying, recording, or otherwise,
without the prior permission of Oxford University Press.

Library of Congress Cataloging-in-Publication Data

Dizionario patristico e di antichità cristiane. English
Encyclopedia of the early church / edited by Angelo Di Berardino;
translated from the Italian by Adrian Walford; with a foreword and bibliographic
amendments by W.H.C. Frend.
p.   cm.
Translation of: Dizionario patristico e di antichità cristiane.
Includes bibliographical references and index.
ISBN 0-19-520892-7
1. Christian literature, Early — Dictionaries — Italian. 2. Church history — Primitive and early church, ca. 30–600 — Dictionaries — Italian.
I. Di Berardino, Angelo. II. Title.
BR66.5.D5813 1992
270.1'03—dc20   91-23934
                                                              CIP

Printing (last digit): 9 8 7 6 5 4 3 2 1

Printed in Hong Kong

Contents

Volume I

Foreword to the English Translation by W.H.C. Frend	ix
Note on Bibliographical Amendments	ix
List of Contributors	xi
Biblical Abbreviations	xv
Bibliographical Abbreviations	xvii
Entries A–M	1–578

Volume II

Entries N–Z	579–886
Synoptic Table	887
Maps	919
Illustrations	949
Index	1095

N

NABOR and FELIX. Milanese *martyrs. Feast celebrated at *Milan 12 July, in agreement with their Passion. In *Mart. hier.* they are also mentioned on 10 and 11, but these are mistaken repetitions. The notice of the 10th contains distorted elements taken from the Passion. Acc. to this they were of African origin, martyred at Lodi (*Laus Pompeia*) and thence translated to Milan. In their Milanese basilica were discovered the bodies of Sts *Gervasius and Protasius, whose cult eventually eclipsed their own.

Delehaye *OC*, 335-37; *LTK* 7, 755; *Vies des SS.* 7, 268; *BS* 9, 689-93.

V. Saxer

NAG HAMMADI Writings. A Library of *Coptic writings, discovered by Muhammad 'Ali al-Samman Muhammad Khalifah in Dec 1945 near Nag Hammadi, *Egypt (ancient Chenoboskion, between Siut and Luxor), N of Hamra Dom, near the city of Farshut.

It comprises 12 codices, plus eight leaves removed in *late antiquity and folded into the cover of the sixth, which have given us a complete text. Each codex, except the tenth, consists of a collection of relatively short works, totalling 52 treatises. Of these, six are duplicates (the *Apocryphon of John* is present in three redactions: II, 1; III, 1; IV, 1; there are two redactions of the *Gospel of the Egyptians*: III, 2 and IV, 2; two of *Eugnostos*: III, 3 and V, 1; two of the *Gospel of Truth*: I, 3 and XII, 2; two of the *Untitled work*: II, 5 and XIII, 2). Six were already known before the Library's discovery, either in the original Greek (VI, 5: a passage of Plato, *Rep.* 588b-589b; a Hermetic thanksgiving prayer; XII, 1: the *Sentences of Sextus*) or in Latin tr. (VI, 8=*Asclepius* 21-29) or Coptic tr. (II, 1 and par.: *Apocryphon of John*; III, 4: *Sophia of Jesus Christ*). Of the forty new treatises that thus remain, some (VIII, 1; IX, 1, 2, 3; X, 1; XI, 1, 2, 3, 4; XII, 2) are very fragmentary.

The contents of the Library are not specifically *gnostic. Alongside the passage of Plato and the *Sentences of *Sextus*, we can place the *Teachings of Silvanus* (VII, 3), an example of Christian wisdom literature, probably from a monastic milieu. The *Acts of Peter and the twelve apostles* (VI, 1) do not contain specifically gnostic themes, though certain motifs (the stranger, the journey, the hidden pearl) easily lend themselves to gnostic interpretation. Of the group of *Hermetic works, VI, 6 (*Discourse on the *Ogdoad and Ennead*) is closely related to treatise XIII of the *Corpus Hermeticum* on regeneration. The more properly gnostic works show the use of a great variety of *literary genres: gospels, acts, epistles, apocalypses, *dialogues, treatises, revelatory addresses, prayers, etc. They must belong to different sectarian groups. Some come within the ambit of the *Valentinian school (I, 3: *Gospel of truth*; I, 4: *Treatise on the *resurrection*; I, 5: *Tripartite treatise*; II, 3: *Gospel of Philip*) or show signs of some such influence (I, 1: *Prayer of the apostle Paul*; I, 2: *Apocryphon of James*). Others can be ascribed to groups of *Sethian type or are outside the classifications of the *heresiologists.

Many hypotheses have been advanced as to the Library's origin and nature, made more difficult by the fact that, to this day, archaeological excavations *in loco* have found no further clarifying elements. The collection may have been the library of a group or sect of gnostics or even of one gnostic; or it may have been the evidence collected by a controversialist or heresiologist for purposes of refutation; or, more simply, it may be the remains of a larger *library. Different elements, including names of persons, titles and addresses, obtainable from an analysis of the material in which the codices were wrapped, suggest a mid 4th-c. monastic milieu of Pachomian type.

Although the texts are written in *Coptic, their original language was Greek. Linguistic elements suggest that we have to do with a plurality of translators (which does not necessarily mean a plurality of scribes as authors of the copies that have come down to us). As the existence of duplicates or works with several versions confirms, there must have been different translators who worked on divergent Greek texts. Attempts so far to study the copyists' different "hands" lead to the conclusion that they are translations and transcriptions made all through the 4th c. The originals are undoubtedly older, even if their dating problems are far from being resolved.

Study of the Library has been made difficult by the complex vicissitudes that marked their acquisition by private individuals (the first codex or *Jung Codex*) and the Egyptian government. Today the Library is preserved in its entirety in the Cairo Coptic Museum. A facsimile edition, prepared between 1972 and 1977 makes it accessible to all interested scholars.

Editions: CPG 1, 1156-1222; *The Facsimile Edition of the Nag Hammadi Codices*. Published under the auspices of the Department of Antiquities of the Arab Republic of Egypt in conjunction with the UNESCO, Leiden 1972-1977; J.W.B. Barns - G.M. Browne - J.C. Shelton, *Nag Hammadi Codices, Greek and Coptic Papyri from the Cartonnage of the Covers*, NHS 16, Leiden 1981. (Other N.H. Studies listed by J.M. Robinson, *The Roots of Egyptian Christianity*, Philadelphia 1986, Preface). At present three projects of edition and translation are in progress:
- the *Coptic Gnostic Project* of the Institute for Antiquity and Christianity at Claremont (Cal.) under the direction of James M. Robinson (cf. J.M. Robinson, The Coptic Gnostic Library Today, *NovT* 14 [1967/68] 356-401). Edited so far are: *Nag Hammadi Codices III, 2 and IV, 2. The Gospel of Egyptians* (The Holy Book of the Great Invisible Spirit) edited with Translation and Commentary by A. Boehlig and F. Wisse, in Cooperation with P. Labib, Leiden 1975; Douglas M. Parrott (ed.), *Nag Hammadi Codices V, 2-5 and VI with Papyrus Berolinensis 8502, 1 and 4*, Leiden 1979.
- the Berliner Arbeitskreise für koptisch-gnostische Schriften under the direction of H.M. Schenke. It has so far published the translation, with commentary, of some fifteen works in *ThLZ* in preparation for a general edition.
- the *team*, directed by J.E. Ménard, of the universities of Strasburg and Laval. So far they have edited 12 vols. of texts (plus one of studies), including J.E. Ménard, *La lettre de Pierre à Philippe*, Quebec 1977; id. *L'Authentikos Logos*, Quebec 1977; J.P. Mahé, *Hermès en Haute-Egypte*, Quebec 1978; Y. Janssens, *La Protennoia trimorphe*, Quebec 1978; B. Barc, *L'Hypostase des archontes, suivi de Norea*, par M. Roberge, Quebec-Louvain 1980; L. Painchaud, *Le IIe Traité du Grand Seth*, Quebec-Louvain 1982. The group directed by J.M. Robinson has also published the first entire translation of the Library: *The Nag Hammadi Library*, translated into English under the editorship of J.M. Robinson, Leiden 1977 (rev. 1988). The introduction also contains an account of the discovery and various mishaps undergone by the Library before its publication.
Systematic bibliography, with annual supplements in *NovT*, by D.M. Scholer, *Nag Hammadi Bibliography 1948-1969*, Leiden 1971. Recent books: M. Krause, *Die Texte von Nag Hammadi: Gnosis. Festschrift für H. Jonas*, Göttingen 1978, 216-243; K.W. Troeger, Zum gegenwärtigen Stand der Gnostic- und Nag-Hammadi-Forschung, in id. (Hrsg.), *Altes Testament-Frühjudentum-Gnosis*, Gütersloh 1980, 11-33; J.M. Robinson-B. van Elderen, The Nag Hammadi Excavations 1972-1980, *The Institute for Antiquity and Christianity Report* (ed. M.W. Meyer), Claremont 1981, 37-44.
Articles: in "Gnostica" *SP* 18.1 (1985); C. Scholten, Die Nag-Hammadi-Texte als Buchbesitz der Pachomioner, *JbAC* 31 (1988) 144-172; J. Dart, *The Laughing Savior*, New York 1976 (account of discovery and publication of N.H. Document); C. Colpe, Heidinsche, jüdische und christliche Uberlieferungen, in den Schriften aus N.H. I-IX, *JbAC* 15-25 (1972-1982); P. Cherix, *La Concept de Notre Grande Puissance* (CG. VI.4), Göttingen 1982 (bibl.); W. Myszor, Antonius-Briefe und Nag-Hammadi Texte, *JbAC* 32 (1989) 72-88.

G. Filoramo

NAIN - widow's son: *iconography. Infrequent in early Christian art, limited for now to *sculpture, of which we have some 20 examples in *Gaul (e.g. *Ws* 26, 2/3; 227, 1/2) and at *Rome (e.g. *Ws* 214, 2; 8, 4). The episode, which in Lk 7, 11-15 seems very rich in characters, is summarily described: Jesus, accompanied by at least one *apostle, touches with the **virga* the body of the widow's son, already prepared for burial (e.g. *Ws* 214, 2), emphasizing by his act the message of *faith inherent in the composition. The mummy lies either in a sarcophagus, sometimes spirally fluted and set on a Leonine base, or on the ground (*Ws* 128, 1). On some reliefs the mother also appears (e.g. *Ws* 8, 4; 225, 2; 26, 2/3). *Sarcophagi with this scene date from c.300-330 (*Ws* 8, 4).

DACL 12, 1, 582 ff; *LCI* 4, 542 ff.; *Ws* 304 f.; *Rep.*

D. Calcagnini Carletti

NAME (nomen, ὄνομα). The n. is not a word that refers to something or some person in a purely external relationship: it expresses the essence of a being and the knowledge of a higher reality than man, or of a deity who ensures a certain power over the being whose n. has come into someone's possession, i.e. it makes possible, acc. to the uniform conviction of the ancients, a *magic, a manipulation of the n. itself. In the OT a certain cult developed round the n. of God, though the *Bible always forbade manipulation of the divine n. The n. of God in the OT is represented by the so-called tetragram JHWH, whose Greek translation, going back to the 4th c. BC, is Κύριος, later referred by the NT to Christ. The fact that God has only one n. made possible the distinction of the Israelite religion from other Middle Eastern religions, since the other nn. attributed to God such as "'El Shaddai" and "'El 'Elyon" (Ex 6, 2; Gen 14, 22) were only substitute nn. The holy n. of God was entrusted to the priests, but the OT never knew a "hidden", "secret" n. For the Jews, invocation of the n. made it possible to enter and become part of God's community or to sanctify the place in which the n. was invoked. According to Israelite conviction, the n. dwelt in the temple of Jerusalem; it sometimes appears almost hypostatised (Ex 23, 21; Is 30, 27; Ps 20, 1; 54, 1; 89, 24; Mal 1, 11). The *epiclesis of the n. on some person assumed the meaning of taking possession, while the formula "in the n." (1 Sam 25, 9) expressed recognition, legitimation of someone. Change of n., a phenomenon known in the OT, is also found in the NT. The n. was given during *circumcision by the Jews, during *baptism by the Christians, since for the latter it represented the moment at which the baptized began to serve Christ (Mk 3, 16; Mt 16, 18; Acts 13, 9). The formula "in God's n." was replaced in the NT by the formula "in Jesus' n." (Mt 21, 9; Jn 5, 43; Acts 4, 17; 5, 40). Another important formula is "to baptize in the n." (εἰς τὸ ὄνομα τοῦ... Mt 28, 19; Acts 2, 38; 8, 16; 19, 5) with the meaning of "to become Jesus' property". The multiplication of nn. expressed abundance, growth of power, and the divine nn. had considerable importance in *blessing or cursing: the interpretation of the divine nn. was an obligation in the OT and then became so for patristic *theology. The ὄνομα, acc. to the church *Fathers, being a human word, cannot express the divine essence; no word fully satisfies this purpose (Just., *I Apol.* 61, 11) except the divine Logos, since we do not know the word expressing the divine nature (Greg. Nyss., *Contra Eunom.*, ed. W. Jaeger, II, 169 ff.), to us inexpressible and transcendent.

The Fathers stressed that the word θεός is not the n. of the essence, a n. impossible to find (John Chrys., *Hom. in Hbr.* 2, 2), that the divine nn. derive from the divine actions (Just., *II Apol.* 6, 2) and the various nn. correspond to the various divine energies (Basil, *C. Eunom.* I, 7: PG 29, 525A), in accordance with *God's beneficent actions towards men (Greg. Nyss., *C. Eunom.*, ed. W. Jaeger, II, 242, 10). This profound modification of the theology of the divine nn. was a consequence of negative theology on the basis of so-called "Hellenistic acosmic theology". This current of theological thought appeared in the 4th c. in the Cappadocians and then, in a complete form, in *Dionysius the Areopagite in the 5th c. (God is ἀνώνυμος, πολυώνυμος, ὑπερώνυμος: PG 3, 593-596).

Greek personal *names in antiquity were usually composed of two elements (e.g. Ἄριστ-ιππος); they often indicated personal (Νικίας) or circumstantial (Δίδυμος) characteristics. The choice of n. was not based on precise laws, but there were customs that regulated it: e.g., the son inherited his father's name (Ἰουστῖνος Πρίσκου τοῦ Βακχείου), *slaves' nn. alluded to their place of origin (e.g. Θετταλή), their features, external appearance or some quality (e.g. Ὀνήσιμος). The Romans gave two, three or even more nn. to boys and girls, but two nn. were always fundamental: the *praenomen* (personal name), the n. which represented that of the *gens*, and the *cognomen* (additional name), indicating some physical peculiarity. *Praenomines* were always written in abbreviated form, except in cases where someone subscribed his *praenomen* only. In the *family circle the *praenomen* was used, among *friends the *cognomen*, in official nomenclature always the complete name. Generally an adoptive son could take the whole name of his adoptive father; it was possible to take also a *cognomen* formed from the diminutive suffix *-anus, -inus*, or else to keep the adoptive father's *cognomen* unaltered.

DB 4, 1669-77; *LTK* 7, 780-83; *ODC* 720 f.; *KLP* 4, 657-661; *RGG*³ 4, 1298-1306; *Evangelisches Kirchenlexikon*, 2, 1499-1502; J. Fichter-Ch. de Morè-Pontgibaud, Sur l'analogie des noms divins, *RSR* 19 (1929) 481-512; 20 (1930) 193-223; 38 (1952) 161-188; 42 (1954) 321-360; C. Mazzantini, La questione dei "nomi divini", *Giornale di Metafisica* 9 (1954) 113-124; E. Sittig, *De graecorum nominibus theophoris*, Diss. Philol. Halenses, vol. XX. pars I, 1911.

L. Vanyó

NAMES, PERSONAL. While in the first two centuries AD the name-system based on three elements (*praenomen, gentilicium* and *cognomen*) prevails everywhere in Latin *pagan inscriptions, from the 3rd c. first the *praenomen* and then the *gentilicium* fell into disuse. Most Christian titles, both men and *women, contain only the *cognomen*: presumably it sufficed to identify people. In more conservative provincial or social circles, the *tria nomina* persist sporadically, while the *duo nomina* are quite frequent in the 4th c. and become rare only in the 5th and 6th. In Christian inscriptions the *cognomen* sometimes functions as a *praenomen* or *gentilicium*, or *vice versa* (cf. e.g. CIL V, 1661, from *Aquileia: *Furia Firminia Gaudentia*; Diehl, 286, from *Rome: *Valerius Victor Paternus*). In Greek there is generally only one name-element, often followed by the patronymic in the simple genitive, or preceded by υἱός, παῖς, something very rare in Latin Christian dedications. That Greek examples of two or three names are imitations of the Roman use is proved by transliteration of *praenomina* and *gentilicia* (e.g. Μάρκος Αὐρέλιος, Κλ. Ἀκείλιος, Τίτος Φλάβιος).

Christians (unlike Jews) attached no special importance to the etymology of p. nn. They normally had names of pagan or specifically mythological origin: Aphrodites, Galatea, Herculius, Isidorus, Liber, Mercurius, Phoebe, Martialis, Saturus, Silvanus, Socrates, Ptolomaeus, etc. Biblical names were not very common before the 4th/5th c. (the commonest were Maria, Susanna, Iohannes, Petrus) while typically Christian or Christianized *cognomina* like Agape, Anastasius, Benedictus, Martyrius, Redemptus, Spes or Renatus seem to have become more common as infant *baptism became established. Note also the so-called theophorous names, widespread in other forms even in the pre-Christian era, as if to place oneself under a special tutelage of the deity. Many of them were used by Africans: Quodvultdeus, Habetdeus, Deogratias, Deusdedit, Adeodatus, Theodulos. Frequently enough we encounter, in inscriptions, names of humiliation (or vituperative names), originally peculiar to Christians, but soon adopted by pagans too, no longer paying any attention to their meaning, but simply as a fashion: Asellus, Calumniosus, Contumeliosus, Importunus, Onager, Stercorius, Coprion, Proiectus and Proiecticius (the last two given to abandoned babies).

Surnames, or *signa*, may be introduced by various expressions (*qui et, quae et, ὃς καί, signum, sive, id est*) and follow the *cognomen*, or be used on their own. Sometimes, despite ending in *-ius*, they refer to *women (Amantius, Simplicius, Toribius). A number of female *cognomina* have an irregular lengthening in the genitive and dative, with weak forms prevailing over strong forms and with Greek endings: Agnes *(-etis, -eti)*, Aproniane *(- enis, -eni* or *-etis, -eti)*, Cyriace *(-etis, -eti)*, Iuliane *(-enis, -eni* or *-etis, -eti)*. These are certainly not diminutives, as was once hypothesized. Barbarian p. nn. appear in the 4th c. and increase considerably in the following centuries.

We cannot generally tell people's origins from their names, since Greek and Oriental p. nn. were commonly used by Westerners, and *vice versa*, out of exoticism or fashion. The function and importance of apparently unusual *cognomina*, which were only used locally has not yet been sufficiently pointed out.

DACL 12, 1481-1553; *EC* 5, 435-436; W. Schulze, *Zur Geschichte lateinischer Eigennamen*, Berlin 1904 (= Berlin-Zürich-Dublin 1966); M. Schoenfeld, *Wörterbuch der altgermanischen Personen - und Völkernamen*, Heidelberg 1911 (= *ibid.*, 1965); F. Grossi Gondi, *Trattato di epigrafia cristiana latina e greca del mondo romano occidentale*, Rome 1920 (= *ibid.* 1968), 71-91; P. Testini, *Archeologia cristiana*, Rome 1958 (= Bari 1980), 367-371; I. Kajanto, On the Problem of Names of Humility in Early Christian Epigraphy, *Arctos* 3 (1962) 45-53; id., Les noms, in *SyllInscrLatVetMusVat* 2, Helsinki 1963; 40-72; id., *The Latin Cognomina*, Helsinki 1965; id., *Supernomina. A Study in Latin Epigraphy*, Helsinki 1966; A. Ferrua, I nomi degli antichi cristiani, *La Civiltà Cattolica* 117 (1966) 92-98; D. Mazzoleni, Nomi di barbari nelle iscrizioni paleocristiane della Venetia et Histria, *RomBarb* 1 (1976) 159-180; I. Kajanto, The Emergence of late single Name System, in *L'onomastique latine*, Colloques internationaux du CNRS 564, Paris 1977, 421-430; H.I. Marrou, Problèmes généraux de l'onomastique chrétienne, *ibid.*, 431-435; Ch. Pietri, Remarques sur l'onomastique chrétienne de Rome, *ibid.*, 439-445; N. Duval, Observations sur l'onomastique dans les inscriptions chrétiennes d'Afrique du nord, *ibid.*, 447-456; C. Carletti, Appendice bibliografica: Epigrafia, in P. Testini, *Archeologia cristiana*, Bari ²1980, 817-818.

D. Mazzoleni

NAPLES

I. Christian origins - II. Archaeology.

I. Christian origins. The initial period of the history of the church of N. is still partly obscure. Our sources are late or legendary, and the information in them must be evaluated with care. Acc. to the account handed down by the Life of St Asprenus (in ASS, *Augusti* I, 200-202) and his *biography by Alberic (in F. Ughelli, *Italia Sacra sive de episcopis Italiae*, vol. VI, Venice ²1720, coll. 18-25), the Christian community of N. was founded by St *Peter who, on his way from *Antioch, stopped at N., where he healed and converted the young Asprenus, who was consecrated bishop of N. in AD 44 and martyred in 79.

The facts given in these accounts (both post-9th c.) do not seem to clash with those of more authoritative sources like the *Chronicon episcoporum Neapolitanorum*, preserved in the *Liber Pontificalis (cod. Vat. Lat. 5007), the Bianchinian Catalogue (cod. laur. 604, summary of cod. Vat. Lat. 5007), the Marble Calendar, which names Asprenus (in the variants *Aspren* and *Asprenus*) first in the list of bishops of N., but with no date. The dating of Maximus, the tenth bishop, who died between 355 and 362, far from lowering the terms of Asprenus's pontificate, may just be a sign of the incompleteness of the catalogues. Other factors seem to confirm the presence of a Christian community at N. in the 1st c.: 2nd-c. frescos in the catacomb of S. Gennaro; the frequency of the name Asprenus in the 1st c.; the probable cultural influences of Puteoli, where a Christian community was active in the second half of the 1st c.

DACL 12, 691-776; EC 8, 1631-44; BHL I, 117, nn. 724-726; D. Mallardo, *Storia antica della Chiesa di Napoli, Le fonti*, Naples 1943; id., Le origini della Chiesa di Napoli, *Miscellanea P. Paschini*, I, Rome 1948, 27-68.

M.L. Angrisani Sanfilippo

II. *Archaeology. Despite the greater antiquity of cemeterial evidence, any description of the city's layout must necessarily start from the episcopal complex. Recent excavations allow us to outline more clearly the buildings composing this. Acc. to the picture proposed by Farioli, we can recognize: the Constantinian cathedral, variously described by ancient sources as dedicated to the Saviour and to the Apostles, now the church of S. Restituta, its original five aisles still traceable in the present changed plan; a basilica built by bishop Stephen at the end of the 5th c., dedicated to the Saviour and called from its builder the Stephania, seemingly oriented on an axis parallel to the former, from which it was separated by an alley. The recently discovered *mosaic pavement has been ascribed to the Stephania, but nothing is known of its architecture. From the 5th c., the presence of two *church buildings posed liturgical and functional problems common to "double cathedral" situations; strips of mosaic found under the Stephania may mean that even in the 4th c. there was a hall used as a *catechumeneum* beside the cathedral. Between the two churches was the baptistery of S. Giovanni in Fonte, ascribed to bishop *Severus (362-408): square plan, cupola with octagonal drum, richly decorated with mosaics. Besides these three buildings, there must have been in the episcopal *insula* a second baptistery ascribed to bishop Vincent and an *accubitorium*, i.e. the episcopal *triclinium*, to which has been referred an apse with mosaic paving found in the aforementioned excavations.

In the vicinity of the *insula episcopalis* - in the area of the old Greek *agorà* and Roman forum - was discovered in 1954 the basilica of S. Lorenzo Maggiore, ascribed to the episcopate of *John II (535-555): three aisles, semicircular apse flanked by *pastophoria*, narthex: much of its mosaic paving remains. Finally the church of S. Giorgio, at the crossing of the present Via Vicario Vecchio and Via Duomo: three aisles, apsed transept, going back to bishop Severus, but radically rebuilt in the 17th c.; the church of S. Maria Maggiore - now S. Maria Maggiore alla Pietra Santa - ascribed to the episcopate of Pomponius (514-532), completely rebuilt in the 17th c.; the basilica of the Holy Apostles, completely restored in the first half of the 17th c.; the basilica of St John - now S. Giovanni Maggiore - on the hill of that name, ascribed to bishop Vincent (555-578) and rebuilt in the 17th c.; and a group of *diaconiae*: St Andrew, which certainly existed at the start of the 8th c., St Januarius (now S. Gennaro all'Olmo), attested from the end of the 7th c., Sts John and Paul, completely destroyed, but known from the start of the 8th c. in the area of the present University, and S. Maria ad Cosmedin (now S. Maria di Porta Nova) which seems to go back to the 8th c. As for *monasteries - a good 14 mentioned by the sources - traces remain only of that of St Severinus, mentioned at the time of bishop Victor (492-496), in the Castello dell'Ovo, with remains of the church and the monks' underground *cells dug in the rock.

In the suburbs, we know four underground *cemetery complexes: S. Gennaro, S. Gaudioso, S. Eufebio and S. Severo. The cemetery of S. Gennaro, opening in the hill of Capodimonte, consists of two superimposed levels, each developed from primitive nuclei. The lower and older level originates in a non-Christian family complex with a large trapezoid room - the so-called "vestibule" - with benches along the richly decorated walls, off which open smaller rooms, and a small community nucleus where bishop Agrippinus was laid in the mid 4th c. This tomb gave rise, perhaps at the start of the 4th c., to a small oratory and, later but still in the 4th c., a larger basilica. The development of the cemetery, roughly in the mid 4th c., centres round three main galleries of exceptional width, from which smaller galleries lead off. There is considerable variety of types of burials and decorative schemes. The upper level also developed from a primitive family nucleus, Christian from the start, as the *paintings decorating it prove, consisting of two adjacent and communicating rooms, later enlarged with galleries which, in their progress, meet and incorporate other small underground rooms. But the most obvious transformation followed the translation to a cubicle of this cemetery - seemingly at the start of the 5th c. - of the relics of St *Januarius. This involved the reconstruction of the surrounding area and the formation of a sort of sanctuary decorated in time with several superimposed layers of pre-9th-c. frescos, incl. the earliest image of St Januarius (early 6th c.). Also related to the transformations caused by the *martyr's deposition must be the so-called *basilica minor*, also known as the basilica of the bishops since, starting with *John I, it housed the tombs of the Neapolitan prelates, of whom some mosaic images remain. Immediately W of the catacomb is the great surface basilica also dedicated to Januarius, with three aisles and apse pierced by great arcades. Of the later life of the complex, we must mention the so-called emergency bishop's house of Paul II, consisting of various buildings incl. a *triclinium* and a *marmoreus baptismatis fons*, identified in the lower vestibule; and a church dedicated to St *Stephen, built at the end of the 5th c., of which only part of the apse foundations remain. Some would see this church and that of St Januarius as examples of twin basilicas originating in a suburban villa.

Of the cemetery of S. Gaudioso - which takes its name from the bishop of Abitina exiled at N. by *Genseric in 439 and buried there - it is hard to reconstruct the original ground-plan because of the radical 17th-c. transformations. Of what remains, the hypogeum consists essentially of a wide ambulatory onto which open cubicles decorated with frescos and mosaics.

Still less can be said of the two small hypogea of S. Eufebia and S. Severo. The former, which opens in the hill of Capodichino, consists of an ambulatory with cubicles on both sides and an oratory at the end which gave rise to the present church of S. Eframo Vecchio or the Immaculate Conception. Chronologically, the earliest evidence is a fresco dating from the 5th c., though sources speak of an earlier time. Of the latter, which opens in the hill of Capodimonte, the one surviving trace is a cubicle with traces of fresco datable to the 5th c., reached from the present church of S. Severo della Sanità.

Still in the suburbs, sources mention the churches of St Euphemia (5th c.) between the catacomb of S. Gennaro and the city walls; Sts Fortunatus and Maximus (mid 4th c.) opposite the catacomb of S. Gaudioso; St Sossius (first half of 8th c.), seemingly on the hill of Capodimonte; of the latter no monumental traces remain. However, though profoundly modified, the oratory of St Asprenus, in the old port area of the city, is composed of two small superimposed halls; the lower one, which reuses part of an old thermal establishment, may perhaps be referred to the 6th c. and the service of a community of fishermen; the upper one, T-shaped, is dated by its marble remains to the early Middle Ages.

The map of the city is completed by a crown of monasteries mentioned by the sources, but of which no monumental traces remain, except that of St Severinus already mentioned. [Fig: 231]

S. D'Aloe, Catalogo di tutti gli edifici sacri della città di Napoli e suoi sobborghi, *Archivio Storico per le Province Napoletane* 8 (1883) 111 ff., 287 ff., 499 ff. and 670 ff.; B. Capasso, *Topografia della città di Napoli nell'XI secolo*, Naples 1895; A. Venditti, *L'architettura dell'Alto Medioevo*, in *Storia di Napoli*, II, 2, Naples 1969, 774 ff.; R. Di Stefano, *La cattedrale di Napoli. Storia, restauro, scoperte, ritrovamenti*, Naples 1874; U.M. Fasola, *Le catacombe di S. Gennaro a Capodimonte*, Rome 1974; R. Farioli, Gli scavi nell' "insula episcopalis" di Napoli paleocristiana: tentativo di lettura, *Atti IX Congr. Intern. di Archeolog. Crist.*, Rome 1978, 275 ff.; id., Aggiornamento a E. Bertaux, *L'art dans l'Italie méridionale*, IV, Rome 1978, 153-162, 189-193; P. Testini et al., La Cattedrale in Italia: Schede, Napoli, *Actes XIᵉ Cong. int. d'arch chrét*, CEFR 123, Rome 1989, 95-97.

L. Pani Ermini

NARBONNE

I. Christian origins - II. The Visigothic (462-719) and Arab (719-759) eras.

Narbo Martius (later Narbona) was the first Roman colony in *Gaul, founded in 118 BC by Caesar's favourite legion, the 10th Decumana, under the direction of Tiberius Claudius Nero. Capital of the province of Narbonensis which extended from the Alps to the Pyrenees, situated at the crossing of the Aquitaine road with the Via Domitia which led from the Rhône to *Spain, it had an active port from which maritime commerce left for the *East, Sicily and *Rome; at Ostia the *navicularii* of N. had their free port. No ancient monuments are preserved apart from a few inscriptions, but the city had a forum, a theatre, *baths, an amphitheatre and the capitol. In the 3rd c. the barbarian threat on the Rhine shifted the centre of gravity of interests, and in the 4th c. the centre of provincial administration towards the Rhône and the N, and the late 4th-c. administrative reforms confirmed the eclipse of N., reducing the territory of its province to Narbonensis I with N., *Toulouse (*Tolosa*), *Béziers (*Biterris*), *Nîmes (*Nemausus*) and Lodève (*Luteva*) as main centres. Its ecclesiastical *organization was modelled on the civil administration.

I. Christian origins. The hypothesis has been advanced that St *Paul stopped at N. on his way to Spain, but both the *journey and the route are suppositions and doubly gratuitous.

The first known bishop, *Paul, seems to have been absolutely the first. He was buried *ad Albolas*, on the Via Domitia, where a basilica later rose in his name in the middle of the S necropolis. His name is given by *Prudentius (*Perist*. 4, 34), *Mart. hier*. (22. 3; Italic recension) and *Gregory of Tours (*Hist. Franc*. I, 30). *Caesarius calls him a disciple of *Trophimus of Arles, Gregory lists him among the seven missionaries sent from Rome to *Gaul in c.250. A *Life* (BHL 6589-90), from which the *Passio SS. Dionysii, Rustici* *et Eleutherii* seems to derive, may go back to the 8th c. when the *Franks reconquered N. (759). Ado of Vienne confuses the saint with the Sergius Paulus of Acts 13, 7-8.

Nothing is known of the church of N. in the 4th c., but its history in the 5th is well known thanks to its difficult relations with *Arles on the subject of metropolitan rights during *Hilary's episcopate (417-427). At the time of the Visigoths, numerous *churches were built under bishop *Rusticus (9 Oct 427-26 Oct 461), as inscriptions attest: their chief promoters were Rusticus and a priest named Othia to whom we owe the cathedral, the church of St Felix, built in the N necropolis on the road to Béziers, and the churches of Minerve and the nearby Ensérune. Rusticus, perhaps because of his Marseillais origin, also played a role in the religious life of the province and of the West in general, over the consecration of *Ravennius of Arles, the council of *Chalcedon, ecclesiastical discipline and the conflict between Fréjus (*Forum Iulii*) and *Lérins. His successor Hermes, previously appointed bishop of Arles but not accepted by the inhabitants of that diocese, was uncanonically elected bishop of N. Pope *Hilarus (461-68) intervened and accepted his election for N., but took away his metropolitan *authority for life.

II. The Visigothic (462-719) and Arab (719-759) eras. From the late 5th to the early 8th c., Narbonensis I, under the name *Septimania*, was a Visigothic province under the religious influence of *Spain, *Arian until 589 and then *Catholic. For the council of 589 cf. *Merovingian councils. The *archbishops of N. took part (very rarely) in the councils of *Toledo, celebrated the *Hispanic or Visigothic liturgy (the oldest *Liber commicus*, preserved fragmentarily in palimpsest *Paris. lat.* 2269, is that of St Nazarius of Carcassonne, 8th/9th cc.) and received from Spain relics of *martyrs (*Vincent of Saragossa, Felix of Gerona, *Eulalia of Mérida, Justus and Pastor of Complutum), canonical legislation and patristic inspiration. During this period the systematic *evangelization of the countryside was continued and completed through the creation of rural parishes.

Few archbishops are known: Caprarius, present at the council of *Agde (506); Migetius, at the councils of Toledo and N. (589) and Toledo (597); Sergius, attested in 610; Selva, at two councils of Toledo (633, 638); Argebaudus, mentioned by *Julian of Toledo in 673; Sunifred, represented at two councils of Toledo (683, 684) and present at that of 688, recipient of a *letter of Idalius of Barcelona *c*.689; Daniel (788-789). The series of Carolingian archbishops opens with Nebridius (799-822): he was in touch with Benedict of Aniane, duke William of Toulouse and king, later emperor, Louis the Pious. With him ended the ancient era of the church of N.: Rusticus's cathedral was destroyed and rebuilt.

DACL 12, 791-878; L. Duchesne, *Fastes* I, 302-9; E. Griffe, *Histoire religieuse des anciens pays de l'Aude*, t. 1, Paris 1933; A. Mundó, *El commicus palimpsest Paris. Lat.* 2269, Montserrat 1956, 151-275; H.I. Marrou, Le dossier épigraphique de l'évêque Rusticus de Narbonne, *RAC* 46 (1970) 331-349; D. Mazzoleni, Vescovi e Cattedrali nella documentazione epigrafica in occidente (Gallia, Iberia, Africa), *Actes XI*ᵉ *Cong. int. d'arch. chrét.*, CEFR 123, Rome 1989, 779-794.

V. Saxer

NARCISSUS of Neronias. Supporter of *Arius from the start, for which he was condemned with *Eusebius of Caesarea and *Theodotus of Laodicea at the *council held at Antioch just before that of *Nicaea (325). At Nicaea he agreed to subscribe the anti-Arian formula of faith, but immediately became one of the protagonists of the anti-Nicene reaction. He took part in the councils of *Antioch (327) and *Tyre (335), which respectively condemned *Eustathius and *Athanasius. One of a *Eusebian delegation to *Constans at *Milan in 342, he took part in the council of *Sardica and was condemned by the Westerners as an Arian. His participation in the council of *Sirmium (351) against *Photinus is uncertain. In 358-9 he worked with *Constantius against the *homoeousians in favour of a moderate Arian *theology.

DCB 4, 3-4; Fliche-Martin 3, *passim*; Simonetti 593.

M. Simonetti

NARCISUS of Jerusalem. Between 180 and 192, under the emp. *Commodus, N. succeeded Dolichianus as bishop of *Jerusalem. In 195 he presided (with *Theophilus, bishop of *Caesarea) over the synod of Palestine, called by pope *Victor (189-199) to deliberate the problem of celebration of *Easter. Subjected to calumny, N. abandoned his see and withdrew to the *desert. Later, for unknown reasons, he resumed his office: because of his advanced age, he had *Alexander of Cappadocia with him as coadjutor. Died after 212, aged 116 or more. *Epiphanius (*Pan*. II, 2, 66, n. 20: PG 42, 61) puts his death under *Alexander Severus (222-235).

Tillemont III, 177-180; ASS *Oct.* XII, 782-790; *BS* 9, 719-721; *Vies des SS.* 10, 971-973; Euseb., *HE* V, 12, 1-2; 22; 23, 3; VI, 8, 7; 9, 1-8; 10; 11, 1-3; CPG 1340.

M. Spinelli

NARRATIO. In a broad sense, the Latin term n. means the *rei gestae expositio*, the *actus narrandi*; in a specialized sense it is a technical term used in *rhetoric, where three kinds of *narrationes* are distinguished. The first consists of the exposition of the facts that relate directly to the case in dispute. The second of an account in the manner of *digressio*, made again in the context of *oratio* and linked to it (in such cases, n. is one of the six parts of which the classical *oratio* is composed acc. to the rules of *inventio*; cf. e.g. *Rhetorica ad Herennium* I, 3, 4; Cic., *Part. orat.* 31 f.; *De orat.* II, 326-330). The third, which has nothing to do with *causae civiles*, embraces the whole world of letters: it is the *pars in negotiis posita* which is in turn divided into *fabula, historia, argumentum*: these are so many required passages for the pupil who progresses under the guidance first of the *grammaticus*, then of the rhetor (cf. *Rhet. ad Herenn.* I, 8, 12; Cic., *De invent.* I, 19, 27). Christian writers often use the term and concept of n. with regard to the *literary genre of *historia*, comprised in the *pars in negotiis posita*: for them the content of n. is sacred history, *historia salutis*, which extends from *creation up to the present time of the church (cf. Aug., *De cat. rud.* 3, 5; 6, 10).

For the rhetorical notion, cf. K. Barwick, Die Gliederung der narratio in der rhetorischen Theorie und ihre Bedeutung für die Geschichte des antiken Romans, *Hermes* 63 (1928) 261-288; H. Lausberg, *Handbuch der literarischen Rhetorik. Eine Grundlegung der Literaturwissenschaft*, I, Munich 1960, 147 ff. and *passim*. Notes on the use of n. in Christian authors in: V. Grossi, Regula veritatis e narratio battesimale in Sant'Ireneo, *Augustinianum* 12 (1972) 437-463; P. Siniscalco, Christum narrare et dilectionem monere. Osservazioni sulla narratio del "De catechizandis rudibus" di S. Agostino, *Augustinianum* 14 (1974) 605-623.

P. Siniscalco

NARRATIO DE REBUS ARMENIAE (Διήγησις): history of the church of *Armenia in its relations with the Byzantine church from the time of the council of *Nicaea (325) to the end of the 7th c., written in a way favourable to the council of *Chalcedon. The Armenian original is lost: only a pre-13th-c. Greek version survives.

G. Garitte, *La Narratio de rebus Armeniae. Edition critique et commentaire*, CSCO 132/subs. 4, Louvain 1952.

S.J. Voicu

NARRATIO DE REBUS PERSICIS (Ἐξήγησις τῶν πραχθέντων ἐν Περσίδι). A "romance with dogmatic-apologetic aims" (Bardenhewer), in the form of a conference on true religion held at the court of the *Sassanid king Arrinatus by Christians, *Jews, *pagans and a Persian mage. A central role is played in the conference by the official Aphrodizianus, head of the bodyguard, a pagan very well-disposed towards Christianity. The author presents himself as one of the Christians invited to the debate. The main part of the *Narratio* is the first, where Christianity and *Hellenism, or paganism, are debated.

Here it is shown that even the pagan gods witnessed to Christ: among arguments illustrating this thesis, Aphrodizianus relates the marvels manifested in the temple of Hera or Fons in the Persian capital at the moment of Christ's birth, culminating in the apparition of a star and a shining diadem above the goddess's statue. This was the same star that led the *Magi, sent by the king, to *Jerusalem. There they found *Mary and the Babe, then returned to Persia with the *image of Mary and Jesus. In the rest of the *Narratio*, some "archimandrites" make accusations against Aphrodizianus for his pro-Christian sentiments, but the king vindicates him. Then there is a trial of *magic, in which bishop Casteleus defeats the Persian Oricatus. Next comes a dispute with the Jews, in which they are criticized for their unbelief.

The aim of the whole account is perhaps not so much the triumph of Christianity over the other religions, as the demonstration of the thesis that paganism has made a valid contribution to the success of Christianity, as appears from the centrality of Aphrodizianus in the whole account.

The work, of Syrian or Asiatic origin, is later than the *Historia Christiana* of *Philip of Side (434-439), often cited as source, and before the Sassanid collapse in the first half of the 7th c.: the editor, E. Bratke, put it in the second half of the 5th c., but most think it late 6th-c. Its attribution to patriarch *Anastasius I of Antioch or *Anastasius the Sinaite, attested in some MSS, seems invalid. Indeed we may doubt whether the author was a Christian, as he claims, given his scanty knowledge of church *organization and institutions: it is even supposed that he was a pagan who wrote at a time when pagans risked being persecuted and suppressed in the Roman empire.

CPG 6968; BHG 802-805 g; PG 10, 97-108 (just Aphrodizianus's account); E.

Bratke, *Das sogennante Religiongespräch am Hof der Sasaniden*, TU 19, N.F. 4, 3, 1899; Bardenhewer 5, 151; U. Monneret de Villard, *Le leggende orientali sui Magi evangelici*, ST 163, Vatican City 1952, 107-111; E. Honigmann, Philippus of Side and his Christian History, in *Patristic Studies*, ST 173, Vatican City 1953, 82-91; Beck 381; Erbetta I, 1, 224.

F. Scorza Barcellona

NARRATIONES DE CAEDE MONACHORUM IN MONTE SINAI. Account of a barbarian raid on *Sinai, written by an anon. hermit, ex-court official of *Theodosius II, with interesting details on animal sacrifices among the Arabs. Despite the title of the MSS, the author cannot be *Nilus of Ancyra; even if he really was a 5th-c. Sinaite monk, what he says cannot be taken as authentic history. F. Conca is preparing a crit. ed.

CPG 6044 (with bibl.); BHG 1301-1307; PG 79, 589-693; K. Heussi, Untersuchungen zu Nilus dem Asketen, in TU 42/2, Leipzig 1917, 123-159; V. Christides, Once again the "Narrationes" of Nilus Sinaiticus, *Byzantion* 43 (1973) 39-59; F. Conca, Per una edizione critica di [Nilo], "Narrationes" (PG 79, 589-694), *Acme* 31 (1978) 37-57.

J. Gribomont

NARSAI or Narses (399-502). N., surnamed the "Leper", theologian *par excellence* of the *Nestorian church and founder of the School of *Nisibis. Born 399 in a village N-E of Mosul, he was distinguished by intellectual precocity. Fleeing from Mazdaist persecution and orphaned, he was brought up in a *monastery of which his uncle was *abbat. He subsequently studied for ten years at the School of *Edessa. His uncle tried more than once to keep him in his monastery and make him his successor, but his love of teaching brought him back to Edessa in *c*.435. On the death of Qiyore (Cyrus), director of the school, in 437, N. was unanimously elected to succeed him. He directed the School for 20 years, propagating the doctrine of *Theodore of Mopsuestia and maintaining a *friendship with *Ibas of Edessa. Ibas died in 457, and his successor Nonnus dismissed N. on account of his adherence to Nestorianism. Suspected of plotting with the Persians and condemned by Nonnus to be burnt alive, N. fled from Edessa and, giving way to the pressure of *Barsauma, metropolitan of Nisibis, founded the School of Nisibis, which he directed from 471 and which replaced that of Edessa, closed in 489. Died 502, aged 103.

N.'s literary activity was displayed in three fields: *exegesis, preaching and *liturgy. The authenticity of his exegetical works has been doubted because of the silence of later exegetes, and gaps in the *manuscript tradition, which goes back no further than the 13th c. Some liturgical compositions are also attrib. to other authors, e.g. *Ephrem and *Balai. But his 82 (or 84) verse *homilies (*memre*) in twelve-syllable verse, on liturgical, theological, exegetical and moral subjects, of which only half have been published, are certainly authentic.

N.'s *theology is not original: it is that of Theodore of Mopsuestia, who was his true teacher. In lesser manner we can discern the influence of *Theophilus of Antioch, doubtless through the mediation of *Aphraates; Ephrem's influence seems still less. N.'s originality consists in his having "fixed, as a true exegete and theologian ... a whole group of traditions that from then on would be transmitted as a whole, until the rebirth of Nestorian exegesis in the 9th and 10th cc." (Ph. Gignoux: PO 34, 515).

It is not useless, following Gignoux, to mention the influence that N. may have exercised on an "authentic representative of Nestorian exegesis, the famous traveller *Cosmas Indicopleustes" (*ibid.*, 510). Cosmas certainly depends primarily on Theodore of Mopsuestia and on Mar *Aba, "*catholicós" of *Persia, but, if we consider that certain of his developments are simplified expositions of Theodore's complex *theology, we may ask whether Cosmas may not have been inspired in this work of simplification by certain passages of N.'s work - e.g. the analogies he uses to clarify the doctrine of the *Trinity, the *mystery of the generation of the Son, the cosmic duties of the *angels, etc. (cf. Gignoux, *op. cit*., 511-513).

A. Mingana, *Narsai Doctoris Syri Homiliae et Carmina*, 2 vols., Mosul 1905; Ph. Gignoux, *Homélies de Narsai sur la création*, PO 34, 419-716 (with Fr. tr.); F.G. McLeod, *Narsai's metrical Homilies on the Nativity, Epiphany, Passion, Resurrection and Ascension*, PO 40, 1-193 (with Eng. tr.); Ortiz de Urbina 115-118; A. Vööbus, *History of the School of Nisibis*, CSCO 266, subs. 26, 57-65; *DSp* 11, 40-41; F.G. McLeod, Man as the Image of God: its meaning and theological significance in Narsai, *ThS* 42 (1981) 458-467.

R. Lavenant

NATALE, Dies Natalis. The custom of celebrating the anniversary of death, rather than of *birth as did the *pagans, was a peculiarity of Christianity: *oblationes pro defunctis*, says *Tertullian, *pro natalitiis, annua die facimus* (*Cor*. 3, 3). The custom is attested for *martyrs in 167 (*Mart. Polyc*. 18, 3). *Cyprian recommends keeping note of the death-day of the martyrs (*Ep*. 12, 2) so that, on the anniversary, the *sacrifice of the *Eucharist may be celebrated in their memory (*Ep*. 39, 3).

Delehaye OC 33-36; V. Saxer, *Morts, martyrs, reliques en Afrique chrétienne aux premiers siècles*, Paris 1980, 69-73, 105-7, 157-58.

V. Saxer

NATALIUS. Known only through *Eusebius, who cites extracts of an anon. work against the *adoptianists (*HE* V, 28, 8-12). Persuaded by *Theodotus the Tanner and *Theodotus the Banker, N. agreed to be ordained as their bishop in *Rome, for which he received a monthly stipend. On his repenting, bishop *Zephyrinus accepted him back into *communion, but only as a *layman. The episode is very important for the question of *penitence, since it indicates that even the *sin of *heresy was pardoned in the Roman church at the start of the 3rd c. N. was a *confessor in the *persecution (perhaps of 202/203).

A. D'Alès, *L'édit de Calliste*, Paris 1914, 124 f.; P. Galtier, *Aux origines du sacrement de la pénitence*, Rome 1951, 152 f.

A. Di Berardino

NATIVITY: *iconography. The representation arose to celebrate not the poverty of the new-born Christ, but his majesty (Testini, *Giuseppe*, 313 ff.). In the first half of the 4th c., in the atmosphere of triumph created by the religious peace, the older prediction scenes relating to the OT prophecies of the Messiah's *birth from a Virgin, i.e. scenes that only indirectly alluded to the *incarnation, were replaced by realistic depictions of the N., aimed at glorifying the incarnation of the Logos and emphasizing the reality of his earthly journey and the annunciation of the coming of the Messiah to the Shepherds, i.e. to the Jewish people.

Worked out very probably in marble-workers' workshops - the oldest example is a *sarcophagus from 343 (*Ws* 2, 182 f., figs. 175-176) -, the fixed components of the N. scene are the Babe in swaddling-bands in the manger, and the ox and ass recorded in the prophecies of Isaiah (1, 3) and Habakkuk (3, 2) and then by the *Fathers (Ambr., *In Luc*. 2, 7: PL 15, 2649; Prud., *Cathemerinon* XI, 78: PL 59, 896). While ox and ass are nearly always present, *Mary and esp. *Joseph appear much less frequently. They may even both be missing, e.g. in a late 4th-c. fresco in the hypogeum of S. Maria in Stelle at *Verona (Testini, *Giuseppe*, 314) or on the contemporary lid of St *Ambrose's sarcophagus at *Milan (*Ws*, pl. 189, 2). However, the presence of just the ox and ass at the Virgin's feet on the mid 4th-c. sarc. of St *Trophimus at *Arles suggests a shortened formulation of the N. scene and a fusion of it with that of the Adoration of the *Magi (*Ws*, pl. 242, 1). More often than Mary and Joseph, esp. on 4th-c. sarcophagi, are one or more shepherds, in poor clothing, holding a *baculus* or an instrument of work (*Ws*, pl. 243, 6), in the act of venerating the Babe or pointing to the star. In older depictions, Mary appears sitting beside the manger (cf. e.g. *Ws*, pls. 198, 1; 201, 5; 226, 1; 249, 11); from the 6th c. she is often shown lying on a pallet, a position perhaps intended to emphasize the fact that Christ really put on our humanity and to make her suffering more evident (cf. e.g. Volbach, *Elfenbeinarbeiten*, n. 169, pl. 54). In later compositions, esp. from the 10th c., Mary sometimes appears bending over the cradle (cf. miniature of the *Codex Egberti*, *c*.980: Kraus, *Miniaturen*, pl. 12). Joseph however, who appears in depictions of the N. only from the first half of the 5th c., is generally shown seated (e.g. Volbach, *Elfenbeinarbeiten*, nn. 118, 119, pls. 36, 37: sole examples holding a saw), more rarely standing beside the Babe (e.g. a panel of *Maximian's *cathedra at *Ravenna: Cecchelli, *Cattedra*, p. 160, pl. 25; a miniature on f.4b of the Syriac evangelary of Rabbula: Cecchelli, *Rabb. Gosp.*, 54 ff.). For the background of the scene, in the West, in early representations, the crib appears under a cabin or a roof; while in the *East they preferred to set it in the open or in a grotto. **[Fig: 232]**

RBK 2, 637 ff.; *LCI* 2, 86-120; M. Schmid, *Die Darstellung der Geburt Christi in der bildenden Kunst*, Stuttgart 1890; F. Noak, *Die Geburt Christi in der bildenden Kunst bis zur Renaissance*, Darmstadt 1894; E.B. Smith, *Early Christian Iconography and a school of ivory carvers in Provence*, Princeton 1918, 13 ff.; H. Cornell, *The Iconography of the Nativity of Christ*, Uppsala 1924; L. Réau, *Iconographie de l'art chrétien*, III, 2, Paris 1958, 752 ff.; G.A. Wellen, *Theotokos. Eine ikonographische Abhandlung über das Gottesmutterbild in früchristlicher Zeit*, Utrecht-Antwerp 1961, 20 ff., 49 ff.; G. Ristow, *Die Geburt Christi in der frühchristlichen und byzantinisch-ostkirchlichen Kunst*, Recklinghausen 1963; P. Testini, Alle origini dell'iconografia di Giuseppe di Nazareth, *RAC* 48 (1972) in part. pp. 273 f., 313 ff.; B. Bagatti, La "luce" nell'iconografia della Natività, *SBF* 30 (1980) 233-250. For examples cited cf. also: F.X. Kraus, *Die Miniaturen des Codex Egberti*, Freiburg i. Br. 1884; C. Cecchelli, *La cattedra di Massimiano ed altri avorii romano-orientali*, Rome 1936; W.F. Volbach, *Elfenbeinarbeiten des Spätantike und des frühen Mittelalters*, Mainz 1952; C. Cecchelli - G. Forlani - M. Salmi, *The Rabbula Gospels*, Olten-Lausanne 1959; K. Michalowski, *Faras, die Kathedrale aus dem Wüstensand*, Zürich 1967.

M. Marinone

NAUCRATIUS (c.320-c.356). Elder brother of *Basil of Caesarea. Before Basil, he finished his studies at *Athens, withdrew to a family property at Annesi with his mother and sister, and sacrificed his worldly career for the *ascetic life (cf. Greg. Nyss., *Life of St *Macrina*). That this was a life in which hunting and fishing, with servants become his companions, played a considerable part, appears from the fact that he died in an accident during one such excursion.

M.M. Hauser-Meury, *Prosopographie zu den Schriften Gregors von Nazianz*, Bonn 1960, 125-126.

J. Gribomont

NAVIGIUS. Brother of *Augustine and son of *Patricius and *Monica, native of Thagaste in *Africa. Of delicate health (Aug., *De b. vita* 2, 14). Present at Cassiciacum, where he appears in the dialogues *Contra Academicos* (I, 2, 5-6), *De beata vita* (I, 6; I, 7; 2, 14) and *De ordine* (I, 2, 5). Present at Monica's death at Ostia in 387 (*Confess.* IX, 11, 17). It seems certain that he had religious daughters (Possid., *Vita Aug.* 26, 1). We do not know whether Patricius, a cleric of the church of *Hippo, and his brother, a cleric of bishop *Severus at Milevis, were his sons (Aug., *Serm.* 356, 3: PL 39, 1575).

W.H.C. Frend, The Family of Augustine: a microcosm of religious change in North Africa, *Atti del Congresso internaz. su S. Agostino nel XVI Centenario della conversione*, Studia Ephemeridis Augustinium 24, Rome 1987, I, 135-151.

A. Di Berardino

NAZARETH. Small town in Galilee, mentioned in the NT as home of *Mary, Jesus' mother, and of Jesus himself after the return from Egypt. In the Byzantine period it had no bishop, since Christians of Jewish stock still lived there up to the 7th c. Excavations on the site of the Annunciation have uncovered a first building phase with religious grottos, a square baptistery and a *martyrium*; a second phase with a synagogue-type church, facing the grottos; finally in the 5th c. a three-aisled church facing E, while the grottos on its N side at a lower level continued to be venerated. In the Middle Ages a great basilica 70 m. long was built. Inside the baptismal font and then under the foundations of the Byzantine church were found many graffiti, in various languages, which documented the frequency of the *faithful and, at the same time, the Marian cult with invocations to Mary, even before this got under way after the council of *Ephesus. In the traditional "house of *Joseph" there was a similar building development. **[Fig: 233]**

B. Bagatti, *Gli scavi di Nazaret*, I, Jerusalem 1967; id., Scavo presso la chiesa di S. Giuseppe a Nazaret, *SBF* 21 (1971) 5-32; E. Testa, *Nazaret Giudeo-cristiana*, Jerusalem 1969; G. Kroll, *Auf den Spuren Jesu*, Innsbruck 1973, 110-123; G. Cornfeld, *Archeology of the Bible. Book by book*, New York-London 1976.

B. Bagatti

NAZARIUS and CELSUS, Milanese martyrs. Acc. to his biographer *Paulinus, *Ambrose, after *Theodosius's death (395), discovered the bodies of Sts N. and C. and transferred that of N. to the basilica of the Apostles (*Vita Ambr.* 32-4), which later took his name. In the 10th c., archbishop Landulf transferred C.'s body there too. Their Passion (mid 5th c.) has no historical value. Many Latin (BHL 6034-50) and Greek versions (BHG 1323-24) of it were made. From Ambrose's time their relics were scattered all over the empire. *Mart. hier.* records them with *Gervasius and Protasius on 19 June, 28 July and 30 Oct; Florus and Usward on 12 June as well.

Vies des SS. 7, 676-78; *Cath* 2, 776-77; *LTK* 7, 853-54; *BS* 9, 780-85.

V. Saxer

NAZORAEI. *Judaeo-Christian sect first mentioned by *Epiphanius (*Pan.* 29). He and *Jerome (*De vir. ill.* 3) agree that its members lived in Beroea. The former thought them descendants of the *Jerusalem Christians who fled to Pella after the fall of the city in AD 70 (Euseb., *HE* III, 5, 2-3). They spoke Aramaic and had a gospel of their own originally written in Aramaic, presumably known to Jerome (*Ep.* 112, 13 and other passages). Acc. to Epiphanius, the N. may be called *orthodox, though he did not know if they accepted Christ's virgin *birth. This is obviously the group cursed thrice a day by the Jews (*Pan.* 29, 9, 2, often cited in early Christian literature: cf. Just., *Dialogue with Trypho*, ch. 16). Jerome may have been in contact with some of its members, but gives no precise information on the sect.

P. Vielhauer, *Jewish-Christian Gospel*, in W. Schneemelcher, *N.T. Ap.*, Cambridge 1991-92; A.F.J. Klijn, Jerome's Quotations from a Nazoraean Interpretation of Esaiah, *RecSR* 60 (1972) 241-255; A.F.J. Klijn-G.J. Reinink, *Patristic Evidence for Jewish-Christian Sects*, NTS 36, Leiden 1973, 44-52.

A.F.J. Klijn

NEBRIDIUS. Born near *Carthage, a close and valued *friend of *Augustine, whom he followed to *Milan (Aug., *Confess.* VI, 10, 17; VI, 2, 3). He was not present at the Cassiciacum discussions, when Augustine was converted with his friends. But he helped convince Augustine to reject divination, since "no art exists to foresee the future" (*op. cit.* VII, 6, 8), and abandon *Manichaeism (*op. cit.* VII, 2, 3). N. was converted soon after Augustine and, returning to *Africa, converted his family (*op. cit.* IX, 3, 6). Died before 391. Meanwhile he had kept up a close friendship with Augustine, esp. through exchange of *letters. Three of his letters (among Augustine's: *Epp.* 5, 6, 8) and nine of Augustine's remain.

DCB 4, 9-10; *PLRE* 1, 620 n. 4; *PAC* 1, 774-776; P. Brown, *Augustine of Hippo, a biography*, London 1967, 67-68, 133-137.

A. Di Berardino

NECTARIUS. Saint, *archbishop of *Constantinople (381-397). Native of *Tarsus in *Cilicia, senator and praetor, he was in the capital in June 381, when the emp. *Theodosius I unexpectedly proposed him to the bishops of the 2nd ecumenical *council to succeed the outgoing *Gregory Nazianzen. Though still a *catechumen, the imperial candidate received the conciliar suffrage. The formula of faith recited by him as a baptismal profession seems to have become the Nicene-Constantinopolitan *creed. His first acts were connected with the conciliar decrees (dogmatic definition, *letter to the emperor [9 July 381] and synods of 382, 383, 384), and he modestly inaugurated, in virtue of the council's can. 3, a new step in the Constantinopolitan church's progress towards the patriarchate. Died 27 Sept 397, when the Byzantine *Synaxarion* commemorates him and *Flavian of Antioch. In another liturgical commemoration (11 Oct), his name precedes that of four archbishops of Constantinople, incl. his brother *Arsacius. As well as some patriarchal acts, we have an Armenian version of his *Eulogy* of St *Stephen. The *homily on St Theodore of Tyre handed down under his name is considered spurious.

CPG 4300-4301; PG 39, 1821-1840; *Cath* 9, 93-94; *BS* 9, 831-832.

D. Stiernon

NEMESIANUS and Companions. N. and his companions - bishops, priests, deacons and *laymen - were *martyrs of *Valerian's *persecution. Among them we can distinguish a first group of martyrs from *Numidia (Cypr., *Epp.* 76-79); of this group, the notes of the council of *Carthage (256) make Iader of Midilis and Littaeus of Gemellae *confessors, and Lucius of Theveste (Tebessa, Algeria) a martyr and confessor. Also identified are Nemesianus of Thubunae (Tobnae, Tobna, Algeria), Felix of Bagai (Algeria), Felix of Bamacorra, Dativus of Bades, Lucius of Castra Galbae, Polianus of Milevis, Victor of Octavu. *Mart. hier.* puts Dativus on 14 March and Lucius of Theveste on 18 Jan; *Cal. Carth.* puts Nemesianus on 18/24 Dec. The notes of the council of 256 also mention Clarus of Mascula as a confessor and Salvianus of Gazaufascula as a martyr. A second group of martyrs were from Proconsular *Africa: Successus of Abbir Germaniciana (Cypr., *Ep.* 80; *Mart. hier.* 18 Jan; council of 256: marginal notes), Faustus of Thimida Regia, Lucius of Membressa, Quietus of Uruc, Saturninus of Thugga, Terapius of Bulla, Venantius of Thinissa, Victor of Assuras, all of whom the notes designate as confessors; Peter of Hippo Diarrhytus, Verulus of Rusiccade, Libosus of Vaga, described by the same document as martyrs. Paul of Obba is inserted in *Mart. hier.* (18 Jan).

BS 9, 798-800; V. Saxer, *Saints anciens d'Afrique du Nord*, Vatican City 1979, 16-17, 104-116.

V. Saxer

NEMESIUS of Emesa. Late 4th-c. bishop of Emesa in *Syria. His *De natura hominis*, written at the turn of the 5th c. (as now seems certain, not the mid 5th c. as E. Zeller and E. Amman maintained; cf. W. Vanhamel: *DSp* 11, 93), is important more from the viewpoint of cultural history than that of dogmatic *theology (though the emphasis in ch. 3 on the total absence of mixture between Christ's human and divine natures, expressed by the term ἀσυγχύτως, anticipates the formulation of the council of *Chalcedon; cf. A. Grillmeier, Vorbereitung der Formel von Chalkedon, *CGG* I, 158-159; S. Lilla: *Augustinianum* 22 [1982] 553 f.; W. Vanhamel: *DSp* 11, 95). As has been shown esp. by German *Quellenforschung* (cf. bibl.), N.'s work uses the most disparate philosophical sources, from peripatetic *Aristotelianism to *Stoicism and *neoplatonism, from *Galen to *Origen's *Commentary on Genesis*; indeed, the work is fundamental for our knowledge of *Porphyry's *Symmikta Zetemata* and, as W. Jaeger has pointed out, also of various Poseidonian doctrines. The first complete treatise on Christian *anthropology, it deals with the various problems concerning the nature of man and his *soul; an exposition of its contents is in *DSp* 11, 94-98.

Editions: CPG 3550; Greek text: *Nemesius Emesenus De natura hominis*, ed. C.F. Matthaei, Halae 1802 (= PG 40, 504-817); on the Latin version of Nicola Alfano, archbp. of Salerno, died 1085: C.J. Burkhard, Leipzig 1917; on Riccardus Burgundius of Pisa's Latin version (1165), dedicated to the emp. Frederick Barbarossa: *Némésius d'Emèse*, ed. G. Verbeke-J.R. Moncho, Corpus Comm. in Aristotelem Graecorum Supp. 1, Leiden 1975.
Studies: Bardenhewer 4, 275-280; *PWK* Suppl. 7, 562-566; *DSp* 11, 92-99; A. Gercke, Eine platonische Quelle des Neoplatonismus, *RhM* 41 (1886) 266-291; D. Bender, *Untersuchungen zu Nemesios von Emesa*, Leipzig 1898; B. Domanski, *Die Psychologie des Nemesius*, Münster 1900; W. Jaeger *Nemesios von Emesa. Quellenforschungen zum Neuplatonismus und seinen Anfängen bei Poseidonios*, Berlin 1914; H.A. Koch, *Quellenforschungen zu Nemesios von Emesa*, Berlin 1921; H. Schöne, Verschiedenes, *RhM* 73 (1920-1924) 154-156; E. Skard, Nemesiosstudien, I, Nemesios und die Genesisexegese, *SO* 15/16 (1936) 23-43; id., Nemesiosstudien, II, Nemesios und Galenos, *SO* 17 (1937) 9-25; id., Nemesiosstudien, III, Nemesios und die Elementenlehre des Galenos, *SO* 18 (1938) 31-41; E. Dobler, *Nemesios von Emesa und die Psychologie des menschlichen Aktes bei Thomas von Aquin*, Werthenstein 1950; A. Grillmeier, Vorbereitung der Formel von Chalkedon, *CGG* I, 158-159; H. Dörrie, *Porphyrios' Symmikta Zetemata*, Munich 1959; M. Morani, La versione armena del trattato Περὶ φύσεως ἀνθρώπου di Nemesio di Emesa, *Memorie Ist. Lombardo di Sc. e Lett.* 31/2 (1970) 105-193; id., Il manoscritto Chigiano di Nemesio, *Rendiconti dell'Ist. Lombardo*, Classe di Lett. 105 (1971) 621-635; id., Un commento armeno inedito al De natura hominis di Nemesio, *ibid.* 106 (1972) 407-410; id., Contributo per un'edizione critica della versione armena di Nemesio, *Memorie Ist. Lombardo di Sc. e Lett.* 33/3 (1973) 195-135; id., *La tradizione manoscritta del De natura hominis di Nemesio*, Milan 1981; R.W. Sharples, Nemesius of Emesa and some theories of Divine Providence, *VChr* 37 (1983) 141-156.

S. Lilla

NEOCAESAREA. City of N *Syria, now Athis, on the Euphrates (Theodoret, *HE* I, 7; *Notitia dign.* 33, 4), whose bishop was present at the council of *Nicaea. In it have been found two 4th-c. basilicas and a 5th-c. *baths.

H. Harper, *Antiquités de l'Euphrate*, Aleppo 1974, 103-106.

B. Bagatti

NEOCAESAREA, *Council of. At N., a city of *Pontus Polemoniacus, between 314 and 319, Vitalis of Antioch presided over a council of 17 bishops, six of them from *Galatia. The council was certainly held before Vitalis's death in 319 but, since it did not deal with the problem of the *lapsi*, the *persecution must have been over for some years. The Greek text is transmitted by *John Scholasticus (V. Benesevic, *Joannis Scholastici Synagoga L. titolorum*, Munich 1937) and the Latin by *Dionysius Exiguus (A. Strewe, *Die Canonessammlung des Dionysius Exiguus*, Berlin 1931). The council forbade *marriage to priests on pain of deposition (can. 1), set norms on sexual morals and laid down liturgical laws. Other canons concerned marriage (in particular, those who remarried had to submit to *penitence). Instruction of *catechumens was regulated. It was specified that the *baptism of a pregnant woman did not communicate *grace to the foetus. The canons of the council of N. became part of the *canonical collections of the Eastern church.

Mansi 2, 539; J.G. Herbst, Die Synoden von Ancyra und Neucaesarea, *ThQ* 3 (1821) 399-447; C. Busioc, *Sinodul local din Neacesarea*, Bucharest 1915; Joannou I, 2, 74-83; Palazzini 3, 175 f.

C. Nardi

NEOCHALCEDONIANISM. By this term modern scholars define that approach to *christology which integrated the formula of *Chalcedon (two natures united in one *hypostasis) with the *Unus ex Trinitate passus est* of the Scythian monks and *Cyril's 12 *anathemata, to give more meaning to the unity of the two natures and lessen the hostility of the *monophysites. This theological tendency prevailed in the first half of the 6th c., through the influence of the doctrinal rethinking favoured by the policy of *Zeno's *Henoticon*, and was actively favoured by *Justinian. Its main spokesmen were *Nephalius, *John of Scythopolis, *John of Caesarea, *John Maxentius, *Ephrem of Antioch, Justinian himself and *Leontius of Jerusalem.

M. Richard, Le Néo-chalcédonisme, *MSR* 3 (1946) 156-161; Ch. Moeller, *Le chalcédonisme et la néo-chalcédonisme en Orient de 451 à la fin du VIe siècle*, CGG 1, 637-720; S. Helmer, *Der Neuchalkedonismus*, Bonn 1962; L. Perrone, *La chiesa di Palestina e le controversie cristologiche*, Brescia 1980, 224-285; J. Meyendorff, *Christ in Eastern Christian Thought*, New York 1975.

M. Simonetti

NEONAS of Seleucia (in Isauria). *Homoeousian, deposed at the council of *Constantinople of 360, where the moderate *Arians and *homoeans prevailed. In *c.* 365 he was one of the homoeousian bishops who subscribed *letters to be sent to *Liberius of Rome to establish relations with him.

DCB 4, 175; Simonetti, 340, 395.

M. Simonetti

NEONICENISM. Ever since T. Zahn's study (*Marcellus von Ancyra*, published in 1867), scholars, esp. Protestants, have characterized the interpretation of the Nicene faith upheld by the Cappadocians and their friends and sanctioned by the council of *Constantinople of 381, as N., speaking of "*Jungnizäner*", "*neue Orthodoxie*", "*kappadokische Neugläubigkeit*", etc. The novelty of this Neo-Nicene position is made to consist in the fact that it understood the *homoousios of *Nicaea as *homoiousios*, unity of substance (*unius substantiae*, imposed at Nicaea by the Latins) as equality of substance. This theory, however, starts from undemonstrated premises (unsustainable interpretations of Nicaea, *Athanasius, the Cappadocians, etc.) and so remains unacceptable. But it is licit to speak of a reinterpretation of the Nicene faith in the sense that the Cappadocians replaced ὁμοούσιος by the more complete formula of μία οὐσία - τρεῖς ὑποστάσεις and above all opposed not just *Arianism, but also *Sabellianism, as their formula precisely expresses.

R. Arnou, Unité numérique et unité de nature chez les Pères après le concile de Nicée, *Gregorianum* 15 (1934) 242-254; A.M. Ritter, *Das Konzil von Konstantinopel und sein Symbol*, Göttingen 1965, 270-293 (fundamental, but must be supplemented by the same author's article, Zum Homoousios von Nizäa und Konstantinopel, *Festschr. C. Andresen*, Göttingen 1979, 404-423).

B. Studer

NEOPHYTE. Greek term for a new shoot, still fragile. *Neophytos* appears once in 1 Tim 3, 6 to indicate a new *convert. Transliterated by the Latins, it was used by *Tertullian to mean a convert (*Praesc.* 16; *Adv. Marc.* 1, 20. It is not at all frequent in the first Christian centuries. The Greeks preferred to speak of the newly-illuminated. The term prevailed particularly in *Augustine and in *epigraphy to designate the newly baptized.

DACL 12, 1103-1110; Lampe 905.

A. Hamman

NEOPLATONISM

I. General characteristics - II. Plotinus - III. Porphyry - IV. 4th-c. Neoplatonism: Iamblichus and Julian the Apostate - V. The school of Athens and exegesis of the *Parmenides* - VI. Proclus and Damascius - VII. Neoplatonism in the Latin West.

I. General characteristics. N. was the philosophical current which dominated *late antique thought from the 3rd c. AD on, and which, both on the metaphysical and on the ethical level, provided the framework for the thought of many *Fathers, both Greek- and Latin-speaking (cf. *Platonism and the Fathers). Usually it is closely associated with the name of *Plotinus, its creator and most illustrious exponent. But without detracting from the greatness and originality of Plotinian thought, we must bear in mind that n. had behind itself a long preparation lasting several centuries: starting from the Old Academy which immediately succeeded Plato, it passed through Aristotle and his commentator *Alexander of Aphrodisias, *Stoicism (esp. the "Middle Stoa" of Poseidonius), the neopythagoreanism of the first two centuries AD (Moderatus, Ps.-Brutinus, *Numenius), the "New Academy" of Antiochus of Ascalon, *middle Platonism, the *Judaeo-Hellenism represented by *Philo of Alexandria, down to *Ammonius Saccas, Plotinus's Alexandrian teacher. As two great modern scholars, W. Theiler (Plotinius und die antike Philosophie, *MH* [1944] 209-225=*Forschungen zum Neuplatonismus*, Berlin 1966, 140-159) and E.R. Dodds (Tradition and personal Achievement in the Philosophy of Plotinus, *JRS* 50 [1960] 2) have pointed out, one of the most distinctive characteristics of n. was its marked cultural *syncretism, into which flowed not just Platonic, but also Academic, Aristotelian-peripatetic, Stoic, neopythagorean, middle Platonic and Philonian doctrines, to form a new and original philosophical system (E.R. Dodds, *CQ* 22 [1928] 139-140 and 140 n. 1, rightly rules out any direct influence of Philo on Plotinus; but some Philonian doctrines probably reached Plotinus through Ammonius Saccas, who had originally been a Christian and who, like all educated Alexandrian Christians, certainly knew the writings of this Alexandrian Jew; this could explain some surprising analogies between Philo and Plotinus, such as, e.g., the doctrine of the dual ethical phase of *metriopátheia-*apátheia*, or the doctrine of the second *hypostasis as *image of the first). Like subsequent exponents of n., Plotinus wanted primarily to be an exegete and systematizer of Plato: his main intention was to "explain" and "order" into a coherent system the various doctrines which the Athenian philosopher had left scattered through his dialogues and had often covered with veils, which had to be lifted (*Enn.* V, 1, 8; E.R. Dodds, *JRS* 50 [1960] 1). But to carry out this intention Plotinus did not disdain, indeed he held it necessary, to make use of nearly all earlier Greek thought: *Porphyry (*Vita Plot.*, ch. 14) gives us an idea of the vast range of authors read and discussed in his school. His example would be followed by his successors.

II. *Plotinus. While *Ammonius Saccas, faithful to Aristotle's thought and to the prevailing tendency in *Philo and *middle Platonism, had considered *nous* as the highest principle - in which he would be followed by *Origen the Neoplatonist and later, in the 5th c., by Hierocles - Plotinus introduced into his master's teaching a correction that would remain basic to the whole of later "orthodox" n.: at the summit of his edifice he put no longer *nous*, but the absolutely negative One of the first hypothesis of Plato's *Parmenides*, which he identified with the absolute good of *Rep.* VI, 509b (*Enn.* V, 1, 8). The adoption of the negative One as supreme principle, the clear formulation of the doctrine of its superiority to mind (*nous*) and to being, and its identification with the good have precedents in the history of Greek thought: firstly, as E.R. Dodds observed in his fundamental article (The Parmenides of Plato and the Origin of the Neoplatonic One, *CQ* 22 [1928] 129-142), the points of contact between the absolutely negative One of the first hypothesis of the *Parmenides* and the Plotinian doctrine of the One (cf. Dodds, *art. cit.*, 132-133; H.R. Schwyzer: *PWK* 21, 1, 553-554) presuppose the metaphysico-theological interpretation of the One discussed in the first three hypotheses of the *Parmenides* that had been given above all by the neopythagorean Moderatus who, acc. to Simplicius (*In Phys.* 230, 34-231, 2 Diels), had identified the One of the first hypothesis with the metaphysical One above being, that of the second hypothesis with the intelligible and the ideas, and that of the third hypothesis with the *soul (this same interpretation is also found in Plotinus, *Enn.* V, 1, 8); secondly, Plato himself (*Rep.* VI, 509b) had stated that the good "is not being, but excels it in dignity and power" (this passage, along with the first hypothesis of the *Parmenides*, represents the supreme "Platonic" authority on which, in all "orthodox" n., the doctrine of the first hypostasis rests); thirdly, as Dodds (*art. cit.*, 140) and P. Merlan (*The Cambridge History of Later Greek and Early Medieval Philosophy*, Cambridge 1967, 30-32) have pointed out, even in the Old Academy, Speusippus had reached the point of distinguishing the One from mind and being; fourthly, Aristotle (*On prayer* fr. 46 Rose) and two Neopythagoreans, Ps.-Brutinus and Ps.-Architas, had asserted the superiority of the One to mind (on Ps.-Brutinus and Ps.-Architas cf. esp. J. Whittaker: *VChr* 23 [1969] 95 and 102-103); fifthly, the identification between the One and the good had in all probability been maintained by Plato himself in his *Lecture on the good* (cf. Aristoxenus, *Harm.* 30, p. 44, 11; Aristotle *Met.* N 1091b 13-14; Numenius fr. 19 Des Places p. 59, 12-13) and taken up by the Neopythagorean Ps.-Brutinus (cf. Syrianus, *In Met.* 183, 1-2; Ps.-Alexander, *In Met.* 821, 33-822, 1 and Whittaker, *art. cit.*, p. 95).

The principal motifs which characterize the Plotinian doctrine of the first *hypostasis (one-good) are the following: *1)* the One is the first and highest principle (V, 1, 1; V, 1, 7; V, 3, 11; V, 4, 1; V, 5, 10; VI, 9, 3); *2)* it is absolutely simple (II, 9, 1; V, 1, 5; V, 1, 9; V, 3, 11; V, 4, 1; V, 5, 10); *3)* it is absolutely pure, without quantity or quality (V, 1, 9; V, 5, 10; VI, 9, 3); *4)* it is free from any mixture (V, 1, 9; V, 4, 1; V, 5, 10; V, 5, 13; VI, 7, 36); *5)* it is prior to all and above everything (V, 3, 11; V, 3, 13; V, 4, 1; V, 5, 6; V, 5, 12; V, 5, 13); *6)* it is different from all other things and cannot be identified with anything (V, 1, 7; V, 3, 11; V, 4, 1; V, 5, 6; V, 5, 13); *7)* but at the same time it is all, in that it contains all things potentially in itself in an undifferentiated state (V, 2, 1; V, 3, 15); *8)* it is completely self-sufficient (II, 9, 1; V, 3, 12; V, 3, 13; V, 4, 1; VI, 9, 6); *9)* it is above virtue, in that it does not possess it (I, 2, 3); *10)* it is desired by all beings (I, 6, 7; V, 5, 12; V, 1, 6); *11)* though generating *nous*, it remains motionless, in a state of absolute rest and always identical with itself (V, 1, 6; V, 3, 12; V, 4, 2; V, 5, 5; V, 5, 12) (on this idea is based the concept of *moné*, which, with those of *próodos* and *epistrophé*, would be further developed by *Proclus); *12)* in the generation of *nous*, its power is not subject to diminution or exhaustion (VI, 7, 36; VI, 9, 5); *13)* though remaining motionless in the generation of *nous*, we cannot say that it moves nor that it stays still, since it is above the state of rest and movement (V, 5, 10; VI, 9, 13); *14)* it is not *nous* and it is above and prior to it, in that it generates it and is the cause of its existence and of the multiplicity that it contains (V, 1, 7; V, 3, 11; V, 3, 16; V, 1, 5; V, 1, 8; V, 3, 12; V, 3, 16; V, 4, 2); *15)* it is above being (V, 1, 8; V, 1, 10; V, 4, 1; V, 5, 6; V, 3, 17; V, 4, 2); *16)* it "surpasses being", in that it is at the summit of the intelligible world and reigns over it (I, 1, 8; V, 5, 3; V, 3, 12); *17)* not being "being", it is a non-being (V, 2, 1; V, 5, 6; VI, 9, 3; VI, 9, 5); *18)* on the one hand, it is identified with the absolute good; on the other, the good that it transmits remains beneath its true nature (II, 9, 1; V, 1, 8; V, 5, 13; V, 3, 11; VI, 9, 6); *19)* it is without form or shape (V, 1, 7; V, 5, 6; V, 5, 11; VI, 7, 32); *20)* it is indivisible (V, 1, 7; VI, 9, 2); *21)* from the point of view of physical size, it is neither limited nor infinite (V, 5, 10); *22)* it is, however, infinite as regards its inexhaustible generative power (V, 5, 10; VI, 9, 6); *23)* by itself it radiates its *light, which is *nous* and which also shines in man's intellect, esp. when he draws near to it (V, 1, 6; V, 3, 12; V, 3, 15; V, 3, 17; VI, 7, 36; VI, 9, 4; VI, 9, 7; VI, 9, 9); *24)* it is everywhere, and at the same time it is not in any place (V, 5, 9; VI, 8, 16; VI, 9, 3); *25)* it is outside *time (VI, 9, 3); *26)* it is above beauty, since it is its source (I, 6, 9; V, 5, 12); *27)* it is totally without will or thought, and does not even think itself (V, 1, 6; V, 3, 11; V, 3, 13; VI, 7, 35; VI, 7, 40; VI, 9, 6); *28)* but, on the other hand, its act is one with its will (VI, 8, 21); *29)* it does not carry on any cognitive activity (V, 3, 12); *30)* it is unknown, in that it is not the object of thought or consciousness (V, 3, 14; V, 4, 1; VI, 7, 35; VI, 9, 4); *31)* it is above speech, hence nameless and ineffable (V, 3, 13; V, 3, 14; V, 4, 1; V, 5, 6; V, 9, 4; V, 9, 5); *32)* it flees from those who seek it (V, 5, 10); *33)* the only means the human mind possesses, to draw near to it, is the *via negativa*, that process which consists in denying it any attribute whatever and which makes known, not what it is, but what it is not (V, 3, 14; V, 3, 17; V, 5, 4; V, 5, 6; V, 5, 13; VI, 8, 21).

**Nous*, the second *hypostasis, is for its part distinguished by the following characteristics: *1)* it is generated by the One *ab aeterno* as the result of an overflowing of its superabundant power and an irradiation of its light (V, 1, 6; V, 2, 1; V, 3, 16; V, 4, 1; V, 5, 5) (apropos of this last, cf. point 23 of the doctrine of the One); *2)* it is *nous* in that, once generated by the One, it turns towards it and contemplates it (V, 1, 6; V, 1, 7; V, 2, 1; V, 4, 2); *3)* it is the act, image and imitation of the One, but remains below it, though it is above all other beings (V, 1, 6; V, 1, 7; V, 3, 15; V, 4, 2; V, 5, 5); *4)* it is the one-many which contains in itself an indistinct multiplicity, and in that sense corresponds to the One of the second hypothesis of the *Parmenides* (V, 3, 15; V, 4, 1); *5)* it contains in itself all the intelligible beings or ideas which are the product of its noetic activity (III, 8, 11; V, 1, 4; V, 1, 7; V, 3, 5; V, 3, 11; V, 3, 15; V, 4, 2; V, 5, 2; V, 5, 3; V, 5, 6; V, 6, 6; V, 9, 5; V, 9, 6; V, 9, 8); *6)* it is identical with the intelligible beings or ideas (there is total identity between being, *nous*, the ideas or forms, and the act of being) (III, 9, 1; V, 1, 4; V, 1, 8; V, 1, 10; V, 3, 5; V, 4, 2; V, 5, 6; V, 9, 3; V, 9, 5; V, 9, 6; V, 9, 8); *7)* consequently it is identical with the intelligible world, model of the sensible world, i.e. with the "living intelligible being" of Plato's *Timaeus* (III, 8, 11; V, 1, 4; V, 3, 16; V, 5, 4; V, 9, 9); *8)* it is identical with absolute beauty (I, 6, 9); *9)* it is identical with the "cause" of the *Philebus* and *Timaeus*, and the demiurge of the *Timaeus* (V, 1, 8; V, 9, 3); *10)* in the first moment of its existence, i.e. before turning to the one and contemplating it (cf. above, n. 2), it is identical with *apeiría* and with the "indeterminate dyad" (II, 4, 15; V, 1, 5; V, 4, 2; VI, 5, 11; VI, 5, 12); *11)* in this first moment, before the formation of the ideas, it is identical with intelligible formless matter, which is the model of sensible matter just as the intelligible universe is the model of the sensible universe (II, 4, 3; II, 4, 15); *12)* before turning towards the one and receiving from it the imprint of limit or unity (cf. *Enn.* II, 4, 5), it is an indeterminate being (II, 4, 5), prior to mind proper (V, 2, 1; V, 9, 8) (on points 10-12 cf. J.M. Rist, The Indefinite Dyad and Intelligible Matter in Plotinus, *CQ* n.s. 12 [1962] 99-107; A.H. Armstrong, *Spiritual or Intelligible Matter in Plotinus and St. Augustine: Plotinian and Christian Studies*, London 1979, 277-283; and E.R. Dodds, *Proclus. The Elements of Theology*, Oxford 1933, 246-247; the motif of the priority of being to mind - which in Plotinus represent two moments of the same *hypostasis, distinguishable only logically - would be further developed by later n.); *13)* it contains life in itself (V, 6, 6); *14)* it thinks itself (V, 3, 13; V, 9, 5); *15)* the ideas considered individually, existing within *nous*, are intelligent beings like *nous* (V, 1, 4; V, 9, 8).

The third hypostasis is represented by the **anima mundi* of which Plato had spoken above all in the *Timaeus, Philebus* and book X of the *Laws*. Its characteristics are: *1)* it is born from *nous* as a result of the overflowing of the latter's power, just as *nous* is born from the One (cf. *nous*, n. 1) (V, 2, 1); *2)* it contemplates *nous* just as *nous* contemplates the one, and from this *contemplation it draws its capacity to act (cf. *nous*, n. 2) (VI, 3, 6; V, 1, 3; V, 1, 6; V, 2, 1); *3)* it is the manifestation, act and image of *nous*, just as *nous* is of the One (cf. *nous*, n. 3) (V, 1, 3; V, 1, 6; V, 2, 1); *4)* indeterminate at the moment when it is generated by *nous*, it receives the limit and form of its generative principle, analogously to what happens with *nous* (cf. *nous*, n. 12) (V, 1, 7); *5)* it participates in both the intelligible world and the sensible world (IV, 8, 7; V, 1, 7; V, 3, 7); *6)* it generates the sensible world (V, 1, 2; V, 1, 7; V, 2, 1); *7)* it receives the rational principles (*logoi*) from *nous* and thence transmits them to the sensible world (IV, 3, 10; V, 9, 3); *8)* it contains in itself all the *logoi*, and is itself the universal *logos* and a multiple unity (II, 3, 16; IV, 3, 8; V, 3, 8; V, 2, 5); *9)* it awakens the universe and is the cause of its life (V, 1, 2); *10)* it rules, holds together and governs the sensible universe, which is permeated by it in all its parts and which, thanks to it, is a god (IV, 3, 9; IV, 8, 2; V, 1, 2); *11)* it governs and adorns the universe in accordance with the *logos* which is its law (II, 3, 16; IV, 3, 10; IV, 3, 39); *12)* it is infused in the sensible universe *ab aeterno* (IV, 3, 9).

Nearly all the points listed above are traceable back to earlier Greek thought. Without proceeding to a detailed investigation of the various sources, we will confine ourselves here to mentioning that, as regards the one, points 13, 17, 19, 24, 25, 30 and 31 go back to the first hypothesis of the *Parmenides*; points 2, 3, 4 and 9 are of Aristotelian-peripatetic derivation; and point 16 has a precedent in *Numenius; that as regards *nous*, points 4 and 5 are traceable respectively to the second hypothesis of the *Parmenides* and to the middle Platonic doctrine (also present in the Old Academy, Antiochus of Ascalon and Philo) of ideas as identical to thoughts contained in the mind of God; points 6, 13 and 14 are of Aristotelian origin; and point 15 goes back to both Plato (*Sophist* 248e-249a) and *Alexander of Aphrodisias (*Mantissa* 108, 15-18); and that as regards the *anima mundi*, points 2 and 9 have precedents in Numenius and Alcinous, while points 7, 8, 10 and 11 are of Stoic derivation (cf. R.E. Witt, The Plotinian Logos and its Stoic Basis, *CQ* 25 [1931] 103-111).

Plotinus's views on the matter which is the basis of sensible bodies depend in part on earlier Platonic-peripatetic tradition, which had identified the "receptacle" of Plato's *Timaeus* with primordial matter, absolutely without form, shape or quality, and ready to receive in itself the impressions of the forms, producers of qualities: for Plotinus - as for Aristotle, Antiochus of Ascalon, middle Platonism and Numenius - sensible matter is the absolute negation of form (I, 8, 9), it is without quality, shape or magnitude (I, 8, 10; II, 4, 8; VI, 9, 7), and receives in itself the forms and qualities (I, 8, 8; I, 8, 10; II, 4, 8; VI, 9, 7). But Plotinus goes beyond the classical position of *Middle Platonism, for which matter, together with God and the ideas, was one of the "three principles", and draws closer to the views of some neopythagoreans: in substantial agreement with Numenius (fr. 50 Des Places), he asserts that there is an incompatibility between being and matter, which is incapable of combining with being (III, 6, 13); that what enters into matter is not a true being, but only an image (III, 6, 13); and, like Moderatus in Simplicius (*In Phys.* 231, 4-5), he maintains that matter, in itself without forms - which are identical with beings - can be considered a non-being, without this implying a denial of its existence (I, 8, 5; II, 4, 10; II, 4, 16; II, 5, 4; III, 6, 7; III, 6, 13).

As regards the origin of matter, Plotinus (IV, 8, 6) seems to consider both hypotheses: either it exists *ab aeterno* (the classical thesis of middle Platonism and *Numenius) or it derives from a higher principle which has granted it existence as a grace: this is the thesis of Eudorus, *Ammonius Saccas and some neopythagorean circles (cf. Calcidius, *In Tim.* 235; Diogenes Laertius, VIII, 25; and Simplicius, *In Phys.* 181, 18-19).

To the problem of the origin and nature of *evil, Plotinus gives, in the eighth treatise of the first *Ennead*, two replies: on the one hand evil consists of total absence, privation and lack of good, it is not found in beings - which, as products of good, participate in it - and is a form of non-being (I, 8, 1; I, 8, 3; I, 8, 5): this thesis would be taken up by *Proclus and, among the church Fathers, *Origen, *Basil the Great, *Gregory of Nyssa, *Augustine and Ps.-*Dionysius the Areopagite; on the other hand its essence - allowing that one can speak of an "essence" of evil - must be identified with matter (I, 8, 3): this thesis would be rejected by *Proclus and, following him, by Ps.-Dionysius. In *souls, evil consists primarily in lack of measure, the main characteristic of matter (I, 8, 4). For Plotinus, the two solutions are perfectly consistent with each other, in that they can be derived from his theory of matter: evil, total absence of good and non-being, is identical with matter in that this last, of itself, does not in any way participate in good, is totally without it and is a non-being (I, 8, 5). All intelligible beings and all sensible objects originally contained in the One in an undifferentiated state (cf. point 7 of the doctrine of the One) proceed from it by a process of emanation (*próodos*), which is the result of the overflowing of its inexhaustible power. The One, on the one hand, in this way does not remain alone in itself; but on the other, despite the generation of the beings, it remains always in the same state (cf. points 11 and 12 of the doctrine of the One). For their part, the beings generated by the One imitate their principle by generating in turn the lower beings. These are the ideas expounded in IV, 8, 6 (223, 1-224, 18). They are motifs which, though undergoing further accentuation in later n., esp. in Proclus and Ps.-Dionysius, are already clearly formulated here (the term *próodos* appears in 223, 5, while the term *ménontos* [223, 10] is the exact equivalent of the Proclian *moné*).

Despite the presence of matter, the sensible universe is very beautiful since it imitates the intelligible nature of whose excellence it is the revelation (IV, 8, 6 [224, 23-28]). Plotinus decisively rejects the contempt shown by the *gnostics towards the created *world, whose beauty, symmetry and order he praises, thus conforming to all previous Greek thought (II, 9, 16) (cf. also *Cosmos). In fact, the sensible universe is characterized by the "sympathy" of the various beings among themselves (their great variety is ruled by the unique and multiform universal *logos*, and each part collaborates with the universal soul and with the other parts to the life of the whole, from which it cannot be excised [IV, 3, 8; IV, 4, 35; IV, 4, 40; IV,

7, 3]) and by the harmony which reigns supreme in it (IV, 4, 35). That we are here in the presence of Stoic, in particular Poseidonian, motifs has been shown by R.E. Witt (Plotinus and Posidonius, *CQ* 24 [1930] 198-207, esp. 202-203).

To the treatment of ethical problems Plotinus dedicates two treatises in particular, that *On the virtues* and that *On felicity*, respectively the second and fourth of the first *Ennead*. If the evil of the *soul consists in lack of measure (cf. above), the task of the first class of *virtues - the political virtues - is to moderate and regulate the passions, imposing a measure on them (I, 2, 2): so the preliminary ethical state is represented by that *metriopátheia* which Plato had already touched on in the *Republic* (IV, 423e, 431c; X, 619a) and which had become the ethical norm *par excellence* in the Old Academy, the Peripatos, middle Stoicism and middle Platonism (cf. references in S. Lilla, *Clement of Alexandria*, Oxford 1971, 99-103). Man's supreme end, which Plato, in the famous passage of *Theaetetus* 176b, had made to consist in "likeness to God as far as possible" and in *flight from this world towards the ideal world, cannot be achieved for Plotinus by the mere moderation of the passions produced by the political virtues, but by their total destruction or *apátheia*, product of the higher class of virtues, the "cathartic virtues", and virtually identical with that *catharsis* or detachment of the soul from the *body on which Plato had so insisted in the *Phaedo* (I, 2, 3). Once completely purified, the soul is able to turn towards mind, the second hypostasis, and contemplate it, thus realizing in itself the third and highest class of virtues, the "contemplative virtues" (I, 2, 4; I, 2, 6). Whoever attains this third state is a real god (I, 2, 6); his intellect then returns to that state of perfection in which it was at the moment of his descent from the ideal world (I, 2, 6) (this motif would be at the basis of *Gregory of Nyssa's doctrine of "*images"; cf. *Nous and *Platonism and the Fathers). The perfect man is absolutely imperturbable: Plotinus adopts both the ideal of *apátheia*, characteristic of early Stoicism, and the motifs used by the Stoics to characterize the wise man, which reappear in the portrait of the wise man drawn in the fourth treatise of the first *Ennead*. In the adoption of the dual ethical state of *metriopátheia-apátheia*, Plotinus has a clear precursor in *Philo of Alexandria; and since the identical scheme is found also in *Clement, this motif can be considered one of the most salient characteristics of the Alexandrian cultural milieu in which n. arose (for more details cf. S. Lilla, *op. cit.*, 103-105, 227-228). But a state exists which is still more elevated than that of *contemplation of the intelligible: such a condition is reached when the human intellect attains the vision of the source of the intelligible, the one-good, and, carried away by loving impulse, aspires to "mingle" with it (I, 6, 7). It is then that union comes about between the human mind and the One, which become one thing (VI, 7, 34). And since the One is above mind and does not think (cf. points 14, 27 and 29 of the doctrine of the One), so too the human intellect which is identified with it automatically ceases from all noetic activity (VI, 7, 35). This motif of union with the first principle characterized by the extinction of all thought becomes a constant of all n. and receives particular emphasis in Ps.-Dionysius the Areopagite (cf. S. Lilla, *Augustinianum* 22 [1982] 564-565).

III. *Porphyry. Plotinus's greatest disciple continued his syncretistic approach: following the example of Ammonius, who had shown the agreement between Plato and Aristotle and united their doctrines in a single system (cf. Photius, *Bibl.*, cod. 214, 172a; cod. 251, 461a), Porphyry also wrote a work *On the identity between the school of Plato and that of Aristotle* (no. 20 of Beutler's list: *PWK* 22/1, 285). But Porphyry's *syncretism was still broader than Plotinus's, since it also comprised *oracles, above all the "Chaldaean *oracles" (whose fragments have been edited by G. Kroll, *De oraculis chaldaicis*, Bratislava 1894, and more recently by E. Des Places, *Oracles chaldaïques*, Paris 1971), and an interest in magical practices and *theurgy: this is attested by the juvenile work *De philosophia ex oraculis haurienda* (no. 44 of Beutler's list: *PWK* 22/1, 295) and by the *Commentary on the Chaldaean oracles*, composed prob. after Plotinus's death (nos. 48 and 48a in Beutler: *PWK* 22/1, 296-297; on these two works and their related problems cf. P. Hadot's article, *REAug* 6 [1969] 205-247). It was precisely his contact with the "Chaldaean oracles" which led to some modifications in Porphyry's theologico-metaphysical views with respect to his "Plotinian" phase, which was characterized by greater fidelity to his teacher's thought: these concerned above all the nature of the first *hypostasis, its relations with the other hypostases, and the "triadic" structure of the Porphyrian metaphysic. In fact Porphyry laid the foundations of a metaphysical system which recurred in its main lines, though with some corrections, in later n. and esp. in Iamblichus and *Proclus.

The *History of philosophy* and the *Sententiae ad intelligibilia ducentes*, dating in all probability from his Plotinian phase, reproduce both the general scheme of the three hypostases, fixed by Plotinus, and certain Plotinian doctrines relating to each of them. The three hypostases are represented by the good, the demiurge and the *anima mundi (Hist. philos. fr. 16 Nauck, Sent. 31). The first hypostasis generates the second by a process of emanation which does not take place in *time (Hist. philos. fr. 18), and which is due to the first principle's superabundance of power (Macrobius, Comm. in Somn. Scip. I, 14, 5; cf. Beutler: PWK 22/1, 303) (cf. point 1 of Plotinus's doctrine of *nous); similarly, nous generates the anima mundi (Macrobius, ibid.) (cf. point 1 of Plotinus's doctrine of the anima mundi). The first hypostasis is the first and highest principle of all (Macrobius, ibid.) (cf. point 1 of Plotinus's doctrine of the One); it is the one-good, absolutely simple, self-sufficient, without parts; it is the object neither of names nor of knowledge, and cannot be defined by any terms, which remain always beneath it (Hist. philos. fr. 18) (cf. points 18, 2, 8, 20, 30, 31 and 33 of Plotinus's doctrine of the One); it is above mind and being, and is a non-being (Sent. 25 [15, 1-2 Lamberz], 26 [15, 9. 12]) (cf. points 14, 15 and 17 of Plotinus's doctrine of the One); it cannot be compared to any other being, and no being is on its level (Hist. philos. fr. 18) (cf. points 5 and 6 of Plotinus's doctrine of the One); it is everywhere and in no place (Sent. 31) (cf. point 24 of Plotinus's doctrine of the One); it can be contemplated only when one has attained that absence of all noetic activity which is superior to thought itself (Sent. 25 [15, 1-2]) (cf. what we said of "union" with the one in Plotinus). The second hypostasis is mind, in that it contemplates the first (Macrobius, ibid.) (cf. point 2 of Plotinus's doctrine of nous); it is similar to the first (Macrobius, ibid.) (cf. point 3 of Plotinus's doctrine of nous); it is identical with the totality of beings and with their essence (Hist. philos. fr. 18) (cf. points 5, 6 and 7 of Plotinus's doctrine of nous); and it is also identical with absolute beauty (Hist. philos. fr. 18) (cf. point 8 of Plotinus's doctrine of nous). Before becoming familiar with Plotinus's teaching, Porphyry, who still shared the theories of his previous teacher Longinus, had criticized the doctrine of the presence of intelligible ideas in the divine nous, a doctrine maintained by Plotinus in the fifth book of the fifth Ennead, and had written a treatise on it; only after two refutations of his thesis by Amelius did he change his opinion and embrace Plotinus's doctrine, to which he remained faithful from then on (cf. Vita Plot. 18, 8-22). So the middle Platonic and Plotinian doctrine of ideas contained in the divine intellect is also present in Porphyry (cf. Macrobius, Comm. in Somn. Scip. I, 2, 14).

While, in his "Plotinian" phase, Porphyry's first hypostasis appears identical to the Plotinian One, his adherence to the theology of the "Chaldaean oracles" led him to identify this principle with the "*Father" named in those oracles (cf. e.g. fr. 3, 7, 22, 79 Des Places) and with the first member (identical with the "Father") of the intelligible triad celebrated by them (cf. e.g. fr. 31 Des Places). Both *Proclus (In Parm. 1070, 15 Cousin) (cf. W. Theiler, Die chaldäischen Orakel und die Hymnen des Synesios, in Forschungen zum Neuplat., Berlin 1966, 260-261, and P. Hadot, La métaphysique de Porphyre, in Entr. sur l'ant. class. XII, Geneva 1966, 132) and Damascius (Dub. et sol. 43 [I 86, 8-10 Ruelle]) (cf. Hadot, ibid.) mention this identification of Porphyry's, while taking a definite stand against it. From these witnesses it is clear that Porphyry's first principle does not in any definite way transcend the intelligible triad, but is an integral part of it although it is its "Father", which substantially differentiates it both from the Plotinian One and from the first hypostasis as it appears in the History of philosophy and the Sententiae. As P. Hadot has shown (op. cit., in Entr. sur l'ant. class. XII, 127-157), the way which leads to a better understanding of Porphyry's thought on this matter, and which shows a connection, however partial, with the History of philosophy and the Sententiae, is through the anon. commentary on the Parmenides attrib. by Hadot with good reason to Porphyry himself (REG 74 [1961] 410-438) and published by him in the second volume of his Porphyre et Victorinus (pp. 64-112). In this commentary, the first principle - identified with the One of the first hypothesis of the Parmenides - is, like the One of Plotinus, the History of philosophy and the Sententiae, the negative One, beyond plurality, act, thought, simplicity and the One itself (I, 31-35; II, 10, 12; on this last point cf. Plato, Parm. 141e), beyond diversity, identity, likeness and unlikeness (III, 33-34), approachable only through a negative procedure, which better corresponds to its nature (II, 4-5; II, 15-16; IV, 17-18; VI, 23-28; IX, 27-28; X, 9-11; X, 23-25; XIII, 16-23), above all beings and not comparable with them (I, 4-5; I, 18-19; III, 7-8; III, 11; III, 35; IV, 1; IV, 11; IV, 28-30; VI, 20), simple (I, 8), without parts (VI, 17-18), different from nous (II, 1-3), above word and thought, and knowable only by unknowing (II, 15-17; IX, 24-26), ineffable and nameless (I, 3-5; II, 20; II, 24), superessential (II, 11; XII, 23-24), prior to essence (X, 25), without essence (XII, 5), outside *time (VIII, 8-9), above and prior to knowledge and ignorance (V, 10-11; V, 15; V, 28-30; VI, 10-11); and it is endowed with infinite power, thanks to which it is the cause and principle of everything (I, 25-27; cf. also Macrobius, Comm. in Somn. Scip. I, 14, 6 apropos of Porphyry, and point 22 of Plotinus's doctrine of the One). But on the other hand, despite its radical difference from all beings, this One is not the one non-being of Sent. 25 and 26 and of Plotinus (cf. point 17 of Plotinus's doctrine of the One), nor is it without the attribute of existence like the One of the first hypothesis of the Parmenides (cf. 141e); on the contrary - and this fact has been justly brought out by J.M. Rist (Hermes 92 [1964] 220-225) - Porphyry is at pains to specify that the attribute par excellence of the first One is existence: it is inseparable from the existence with which it is identified (IV, 7-8; XII, 26; XIV, 6-7; XIV, 22-23; XIV, 25-26) and is real being in comparison with subsequent beings (IV, 27-28; note in XIV, 6-7, 22-23, 25-26 the application of the term hyparxis to the first principle; cf. also Rist, art. cit.). By virtue of its infinite and overwhelming power, the first One is endowed with an activity of its own, enérgeia (XIV, 6; XIV, 21-26), which leads it to emanate from itself a second One (XI, 19-23): this is the One of the second hypothesis of the Parmenides, different from the first, though deriving from it (XI, 23-28; XI, 33-XII, 4), one-being (XI, 30-31; XII, 5-10), the unity of all beings (XII, 4), the One which possesses being at a lower level, since it participates in the existence of the first One (XII, 27-29) (so for Porphyry there are two levels of existence: that of the first One, prior to genuine being and definable as "the idea of being", and that of the second One, participating in the first [XII, 29-35]).

The infinite power emanating from the first One and the expression of the activity (enérgeia) of its existence is life, which is thus also infinite (XIV, 17-21; XIV, 25-26); the turning back of this power - destined to form the second One - towards the original principle, the first One, and the act by which it contemplates it in trying to return to it (it sees, in the first principle, itself and an intelligible being, noetón) produce mind (XIV, 16-21, 23-25) (cf. point 2 of Plotinus's doctrine of nous). The outlines of the three members of the highest triad are thus delineated with precision: the first member - the Father of the "Chaldaean oracles" - is on the one hand almost completely identical with the negative One of the first hypothesis of the Parmenides, but on the other - and in this it departs from the One of Plotinus and later neoplatonists - it is identified with absolute existence (hyparxis) and is thus being par excellence: the second member - produced by the emanation of the original source's infinite power - is life; the third member - identical with the patrikòs nous of the Chaldaean oracles (cf. fr. 37, 1; 39, 1 Des Places) and called that by Porphyry too (De regr. an. fr. 8 Bidez) (= Augustine, De civ. Dei X 23) - is mind, destined to form, as in Plotinus, the composite unity or one-being of the second hypothesis of the Parmenides. The characteristics of each member of the triad - the first is characterized by simple existence (hyparxis) which remains in itself, the second by the enérgeia which is set free by hyparxis, the third by the turning of the emanation towards its own source (XIV, 5-8, 21-26) - clearly adumbrate the three moments of moné, próodos and epistrophé which, as we have seen, are already present in Plotinus (cf. point 11 of the doctrine of the One and points 1 and 2 of the doctrine of nous) and which would become still more marked in later n. Though conceptually distinguishable from each other, the three members of the triad form a higher unity (IX, 3-4) (even the first member, at the moment when it produces the emanation destined to turn towards its source, can be defined as nous and life [XIV, 16-21]). This explains why we can simultaneously apply to this unity thus conceived the opposite conclusions - negative and positive - of the first two hypotheses of the Parmenides (XIV, 26-34) (in this aspect Porphyry foreshadows the theology of Ps.-*Dionysius the Areopagite: cf. Nom. div. V, 10: PG 3, 825B; VII, 3: PG 3, 872A).

It must be said that the nature and function of the triad's intermediate member, life, appear rather obscure to *Augustine, who hesitates to identify it with the *anima mundi which, in the Plotinian scheme and also in Porphyry's History of philosophy, occupies not the second, but the third place (De civ. Dei X, 23=De regr. an. fr. 8 Bidez); in fact the Porphyrian "life" cannot be identified with the anima mundi, since the "intelligible triad", of which "life" is an integral part, corresponds essentially to the Plotinian nous (the Plotinian nous also includes "life" in itself: cf. point 13 of Plotinus's doctrine of nous). Acc. to the evidence of John of Lydia (De mens. IV, 122 [159, 5-8 Wuensch]), the summit of the Porphyrian metaphysical system has a henadic structure: in other words, acc. to Porphyry - who would have based it on his own interpretation of the "Chaldaean oracles" - it is formed not of one, but of three intelligible triads. This means that the inexhaustible power of the first principle is not limited to producing the first triad but, by an analogous process, emanates a second triad from the first and a third from the second. We can certainly accept Hadot's reconstruction (in Entr. sur l'ant. class. XII, 138-140) acc. to which both the second and the third triad would be composed of members analogous to those of the first, i.e. of being, life or power, and intelligence; but while in the first triad the dominant character would be provided by the first member, being or the Father, in the second it would be provided by the second, life or power, and in the third by the third, intelligence. So being, life and intelligence would not just enter into each triad as constitutive members, but would also respectively dominate the first, second and third triads (cf. Hadot, 138). Already clearly delineated in the Porphyrian scheme is the hierarchical system of later n., particularly that of Iamblichus and *Proclus (though these would clearly separate the first principle from the first member of the first triad).

Two other important aspects of the Porphyrian metaphysic, which would be taken up esp. in Proclus's *Elements of theology*, are represented by the two following laws: *1)* the production (or "generation") of lower beings by higher (on this idea cf. Plotinus, *Enn.* IV, 8, 6) necessarily involves a subdivision or multiplication of the original unity and also a diminution of its power as it descends down the scale of beings: the being generated is hence inferior to its producer (cf. *Sent.* 11 and 13; W. Theiler, Porph. u. Augustin, in *Forschungen zum Neuplat.* 174; R. Beutler: *PWK* 22/1, 305); *2)* each being occupies its place in the scale of beings on the basis of its dignity (*axía*); the ordered arrangement (*táxis*) of beings depends on their relationship of superiority to the lower and subordination to the higher, and produces a harmony (*syntaxis*) which is an image of the original unity (cf. Theiler, *op. cit.*, 172, 180-184: *PWK* 22/1, 305).

For Porphyry, as for Plotinus, matter is without form, it is true non-being and absence of being (*Sent.* 20, 3; 20, 10); but while, as we have seen, Plotinus in *Enn.* IV, 8, 6 had considered the origin of matter from the first principle solely as a hypothesis, Porphyry, adopting the views of *Ammonius Saccas and the "Chaldaean *oracles" (cf. fr. 34 Des Places and nn. 1-2 on p. 129), transmuted his teacher's hypothesis into certainty: acc. to *Aeneas of Gaza (*Theophr.* 45, 4-9 Colonna), both in his *Commentary on the Chaldaean oracles* and in his exegesis of *Enn.* I, 8 Porphyry maintained that matter can be considered "atemporal" (*áchronos*) but not ungenerate, and rejected as impious the middle Platonic doctrine which made matter one of the three principles; this passage of Aeneas of Gaza is echoed by that of John of Lydia (*De mens.* IV, 159 [175, 5-9]) where it is affirmed that matter, "atemporal" but not "without beginning", received its existence by the will of the Father (cf. Theiler, *op. cit.*, 179-180). Proclus (*In Tim.* I, 396, 5-7, 21-26) (cf. Theiler, 177), citing Porphyry's thought, says that the demiurge "by his thought alone" gives reality to the sensible world, having no need of matter and producing everything from himself (the passage of John of Lydia and these two passages of Proclus, which stress the self-sufficiency of God's will and thought in *creation, show a close analogy with two parallel passages of *Photius, *Bibl.*, cod. 214. 172a, 251. 461b, reflecting the views of Hierocles and *Ammonius Saccas; cf. S. Lilla, *Clement of Alexandria*, 224). Porphyry shares the Plotinian theory of *evil as non-being: it is clear from *Sent.* 26 that this non-being is the opposite of the non-being represented by the first principle. Man's departure from the first principle also brings about his separation from real being, and generates in him that "fallacious passion" which is non-being, the extreme degradation of being.

The Porphyrian ethic does not differ substantially from that of Plotinus: par. 32 of the *Sententiae* closely follows what Plotinus had said in his treatise *On the virtues* (*Enn.* I, 2). Like Plotinus, Porphyry distinguishes the preliminary ethical state of *metriopátheia*, realized through the political virtues, from *apátheia, identified with *homoíosis theó and achieved completely by the third class of man's virtues, the theoretical or contemplative virtues (*Sent.* 32: pp. 25, 6-9; 28, 3-4). The only difference from Plotinus is that while the latter in *Enn.* I 2, 6 (57, 14-17) had simply spoken of the presence in the divine *nous* of "paradigms" of the higher human virtues - "paradigms" superior to any virtue and not definable by the term "virtue" - Porphyry in *Sent.* 32 (28, 6-29, 7) calls such models "paradigmatic virtues" and hence postulates the existence of a fourth and highest class of virtues, those present in the divine *nous*. For Porphyry too, separation of the *soul from the *body, or "catharsis", is the indispensable premise for the *contemplation of the intelligible and the return of the human intellect to *God: he insists on this not just in *Sent.* 32 where, like Plotinus, he dwells on the second class of virtues, the cathartic virtues, but also in his lost work *De regressu animae* (cf. e.g. fr. 11, 2; 11, 5 Bidez) and in his work *De abstinentia*. In *De regressu animae* (cf. in particular fr. 2 Bidez=Augustine, *De civ. Dei* X, 9), alongside a firm condemnation of the lower magical practices we can observe a partial re-evaluation of *theurgy, for which Porphyry no longer feels the admiration manifested in the juvenile work *De philosophia ex oraculis haurienda* (cf. E. Des Places, *Oracles chaldaïques*, Paris 1971, 18-19; 23-24): theurgy can be useful for the *purification of the lower (or "spiritual") part of the human soul, by which sensible objects are perceived, but has no effect on its intellectual part, whose conversion to God takes place without it.

Finally, for Porphyry too the highest knowledge of the supreme principle presupposes the cessation - above thought - of all noetic activity (*anoesía*)

(*Sent.* 25 [15, 2]). This *anoesía* is accompanied by absolute ignorance (*agnosía*) of the first unknowable principle (*Comm. in Parm.* IX, 24-26). This total identity between ignorance and knowledge of God foreshadows similar formulations of Ps.-Dionysius the Areopagite (cf. S. Lilla, *Augustinianum* 22 [1982] 563-564). On the decisive influence exercised by Porphyry on authors like *Ambrose, *Marius Victorinus and *Augustine, cf. *Platonism and the Fathers.

IV. 4th-century Neoplatonism: Iamblichus and *Julian the Apostate. In Iamblichus, who may be considered the author of the final phase of n., since the Proclian system is largely foreshadowed in his, the tendency to *syncretism, already considerable in *Porphyry, underwent further development: from the philosophical point of view it embraced *Aristotelianism and the neopythagorean tradition of the 1st and 2nd cc. AD (one of his major works in ten books - of which five remain under different titles: *The life of Pythagoras, Protrepticus to philosophy, Mathematical science, The arithmetic of Nicomachus, Arithmetical theology* - was entitled *Collection of Pythagorean doctrines*; cf. W. Kroll: *PWK* 9/1, 646-647); from the religious point of view, it embraced the "Chaldaean *oracles" and *theurgy, which Iamblichus reappraised much more than Porphyry did (cf. Dalsgaard Larsen, in *Entr. sur l'ant. class.* XXI, Geneva 1975, 1-4; E. Des Places, *ibid.*, 69-94; J. M. Dillon, *Iamblichi Chalcidensis in Platonis dialogos commentariorum fragmenta*, Leiden 1973, 26-29). Iamblichus also began the systematic exegesis of the Platonic dialogues that would be characteristic of later n.: they were interpreted from the ethical, physical, mathematical and metaphysical points of view (cf. esp. K. Praechter, in F. Überweg, *Grundriss der Gesch. der Philos.*, I, Berlin 1926, 615-616). J. Dillon (*op. cit.*, 27) has observed: "Plato is brought into agreement not just with Aristotle and Pythagoras, but also with Homer, Hesiod, Orpheus and the Chaldaean oracles . . . each dialogue, in its entirety, is pervaded by a higher meaning . . . to each dialogue is assigned a purpose, to which the introductory and apparently accessory parts must also conform."

While inheriting from Porphyry both the triadic scheme and the components of each triad (being, life, intelligence), Iamblichus does not hesitate to take more than one stand against his predecessor, whom he accuses of not being sufficiently "Platonic" or "Plotinian" (cf. e.g. *Comm. in Tim.* fr. 16, 34, 70 Dillon, and Dillon, *op. cit.*, 28). In fact a first, great difference between the two springs immediately to mind: while Porphyry had identified the first principle with the first member of the intelligible triad and with the "Father" of the Chaldaean oracles, Iamblichus sets above the first intelligible triad: a) the two principles destined to compose both the first and the subsequent triads, i.e. *péras* and *apeiría* (cf. Plato, *Philebus* 23c, 24a); b) the One prior to these two principles, simple, not co-ordinated with the first triad, identical with the monad, prior to essence (*proousios*), father of essence (*ousiopátor*), principle of the intelligible (*noetárches*), pre-existent being (*proóntos ón*), good, first god and first king: this One occupies an intermediate position between *péras* and *apeiría* on the one hand and the absolute first principle on the other; c) the absolute first principle (*prótistos*), completely ineffable, prior to the first god and first king, motionless, solitary, free from any contact with the intelligible, model (*parádeigma*) of the first god, who is born from it thanks to a process of irradiation (cf. esp. *De myst.* VIII, 2; Damascius, *Dub. et sol.* 43, 50, 51 and the scheme in Dillon, *op. cit.*, 32). Even if Iamblichus's second One (cf. "b") is not co-ordinated with the first triad (cf. Damascius, *Dub. et sol.* 43), while Porphyry's first principle is an integral part of the first intelligible triad, still these two principles show some close analogies: both are "prior to essence", "principle of essence", "father", "pre-existent being" and "principle of the intelligible" (cf. *De myst.* VIII, 2; Porphyry, *Comm. in Parm.* IV, 27-28; X, 24-25; XII, 26-27; XII, 29-30; and Damascius, *Dub. et sol.* 43). Given Iamblichus's great openness towards 2nd-c. neopythagoreanism, it is no wonder if his second One also has great affinity with *Numenius's first god, defined as "good", prior to *ousía*, its cause, its principle and absolute being (cf. fr. 16 and 17 Des Places).

From the combination of *péras* and *apeiría* originates one-being (cf. Dillon, 32), the third member of the first intelligible triad formed by *péras, apeiría* and the one-being (*hèn òn*) which characterizes it. This one-being or absolute being is identified by Iamblichus with the eternal being (*tò òn aeí*) of *Timaeus* 27d, with the One of the second hypothesis of the *Parmenides* (142b), with the *èn òn* mentioned several times in *Sophist* 244d, 245 a-e (*In Tim.* fr. 29 Dillon, 132), with eternity (*aión* - cf. *Aeon*) (*In Tim.* fr. 64) and with the model (*parádeigma*) of *Timaeus* 28a (*In Tim.* fr. 35) (cf. Dillon, 33-36). And since Plato (*Sophist* 248e) closely associates absolute being with life and intelligence (cf. also Dillon, 36), this absolute being is also identical with life and intelligence in their highest form; intelligence in this state is the *katharòs* or *patrikòs nous* which contains in itself the "monads" of the forms in an undifferentiated state and which is characterized by symmetry, truth and beauty, but especially symmetry (*In Phil.* fr. 4 and 7; Proclus, *Theol. Plat.* III, 13; cf. Dillon, 37, 260, 262-263). The third member of Iamblichus's first intelligible triad corresponds fully to the third member of the Porphyrian first intelligible triad, also represented by *patrikòs nous* and one-being (cf. above); and it also reflects the Plotinian doctrine of the presence of being, life and intelligence in the second *hypostasis. That Iamblichus, like Porphyry, supposed the existence of two more intelligible triads after the first is clear from Proclus, *In Tim.* I, 308, 20-21 and *Theol. Plat.* III, 13: in the latter passage *Proclus ascribes to Iamblichus the doctrine of the three mixed beings - the three intelligible triads - characterized respectively by symmetry, truth and beauty. As earlier for Porphyry and later for Proclus, so for Iamblichus the second intelligible triad must have been characterized to a greater degree by life (*noetè zoè*), identical to the *zoon* of *Timaeus* 37d and 39e, and the third by intelligence (*noetòs nous*); this last intelligence - which must not be confused with the *katharòs* or *patrikòs nous* of the first triad, identical with being (cf. *In Phil.* fr. 7) - produces the forms proper (*In Phil.* fr. 4; cf. Dillon, 37 and 260). But Iamblichus goes beyond Porphyry, anticipating Proclus's scheme: from the last intelligible triad originate three triads of "intelligible and intelligent" beings (Proclus, *In Tim.* I, 380, 21=fr. 230 Dalsgaard Larsen; Festugière in his translation rightly adds *noèton te kaì* before *noeròn*; cf. Dalsgaard Larsen, in *Entr. sur l'ant. class.* XXI, 16, and also Dillon, 308 n. 1); from the last triad of "intelligible and intelligent" beings originates a series of seven intelligent beings (*noerà hebdomàs*), which corresponds in all probability to that of the Chaldaean oracles (cf. Dillon, 308-309, and the list in Dalsgaard Larsen, in *Entr. sur l'ant. class.* XXI, 17) and in which the demiurge of the *Timaeus* occupies third place (Proclus, *In Tim.* I, 308, 22-23=fr. 230 Dalsgaard Larsen). Hence Proclus's statement (*In Tim.* I, 307, 17-18=*In Tim.* fr. 34 Dillon), that Iamblichus identified the demiurge with the whole intelligible universe placed, as we have seen, well above the intelligent beings, is not exact; in any case, Iamblichus's demiurge, in order to make the sensible *world, "gathers into himself" and controls the intelligible models present in the *noetón* (Proclus, *In Tim.* I, 307, 20-25=*In Tim.* fr. 34 Dillon; cf. also Dillon, 38): thus Iamblichus must have interpreted this passage of *Timaeus* 39e relating to the *contemplation of the ideas by the demiurgic *nous*. So in Iamblichus's system too there re-emerges, at various levels, the doctrine of the presence of the model-ideas in the divine mind which was formulated by the Old Academy and taken up by Antiochus of Ascalon, *Philo, *middle Platonism, *Plotinus and *Porphyry. Descending further down the scale, acc. to Iamblichus we must distinguish the non-participated or separate intellect from the intellect which is the object of participation by the hypercosmic soul and the *anima mundi* and which is inseparable from them (*In Tim.* fr. 56; cf. Dillon, 39); and the hypercosmic, transcendent, absolute, non-participated and monadic soul - with which the Porphyry of the post-Plotinian phase had identified the demiurge (cf. Proclus, *In Tim.* I, 306, 32-307, 2) and to which is referred the soul mentioned in *Timaeus* 34b (*In Tim.* fr. 50, 19-27) - from the *anima mundi* proper, object of participation by the sensible world (*In Tim.* fr. 54, 9-10); both the *anima mundi* and individual *souls are produced by the hypercosmic soul (*ibid.*; cf. Dillon, 39). Also present in Iamblichus, as in Porphyry and Plotinus, is the law of the three moments represented by *moné*, *próodos* and *epistrophé*, as is clear from Proclus (*In Tim.* II, 215, 5-15=text 252 Dalsgaard Larsen, *In Tim.* fr. 53 Dillon; cf. also Dalsgaard Larsen, in *Entr. sur l'ant. class.* XXI, 19); in this passage, which contains an allegorical interpretation of the first four numbers, *moné* is related to the number one, *próodos* to two, *epistrophé* to three; and more, we can observe in it the close association between *hénosis* and *tautótes* (both identical to *moné*) and between *próodos* and *diákrisis* (= multiplication) which would later be characteristic of Proclus and Ps.-*Dionysius the Areopagite.

Like Plotinus, Iamblichus took over the Poseidonian idea of "sympathy" and "*communion" between the various parts and the various elements of the sensible universe, as we see from *De myst.* I, 5, 17. As Porphyry had done and as Proclus would later do, he upheld the origin of matter from the higher principle (cf. Proclus, *In Tim.* I, 386, 10-16). The main characteristic of matter is otherness, going back to the dyad (*apeiría*) that is found after the second One (*In Tim.* fr. 7) (cf. Dillon, 47-48).

Also in Iamblichus we find the Plotinian and Porphyrian scheme of the various classes of *virtues (political, cathartic, theoretical, paradigmatic), as *Ammonius (*In Categ.* 135, 12-32) attests. To these virtues Iamblichus adds the "hieratic virtues", also called "unitary" since they bring about union with the first principle (cf. Olympiodorus, *In Phaed.* II, 138 ff.; Praechter, in Überweg, *Grundriss der Gesch. der Philos.*, I, 617).

Closely dependent on Iamblichus (cf. esp. R.E. Witt, Iamblichus as a Forerunner of Julian, in *Entr. sur l'ant. class.* XXI, 35-63) is *Julian the Apostate. In his *Address to the Sun King* (ed. Ch. Lacombrade, *L'empereur Julien*, II, 2, Paris 1964), in which he makes great use of a lost work of Iamblichus (cf. W. Kroll: *PWK* 9/1, 646), he adopts the distinction, characteristic of Iamblichus, between "intelligible" (*noetón*) and "intelligent" (*noerón*): beyond the sensible sun (the third sun) are the sun of intelligent beings (the second sun) and the sun of the intelligible world (the first sun). While the first sun is the principle and cause of everything, beyond mind, and is virtually identified with Plotinus's one-good (*In Reg. sol.* II, 2), the second sun (*noerós*) is born from the first, receives intelligence from it and functions as an intermediary between the intelligible and sensible worlds. This work of Julian represents an attempt at reconciliation between n. and the cult of the *sun, a tendency characteristic of the 3rd and 4th cc. AD.

V. The school of *Athens and exegesis of the *Parmenides*. We have already had a chance to observe that the correspondence established by *Plotinus in *Enn.* V, 1, 8 (26, 23-27) between his three *hypostases and the first three hypotheses of Plato's *Parmenides* presupposes the metaphysico-theological interpretation of the *Parmenides* that had been given by Neopythagoreanism, esp. by the neopythagorean Moderatus; and that *Porphyry too was the author of a *Commentary on the Parmenides*, which explains the agreement between the post-Plotinian phase of his metaphysic with the theology of the Chaldaean *oracles (cf. above). In later n. too, the *Parmenides* continued to be an object of exegesis: in the eyes of the Neoplatonists, it was the dialogue that expounded what they held to be the true Platonic theology, turning on the hierarchical system of the various beings which, starting from the absolutely transcendent superessential One, finally arrived at matter. As H.D. Saffrey and L.G. Westerink have shown in the preface of the first volume of their edition of the *Theologia platonica* (*Théologie platonicienne*, I, Paris, 1968, in particular LXXV-LXXXIX), *Proclus, in the sixth book of his *Commentary on the Parmenides* (1051, 34-1064, 12 Cousin), provides a detailed review of the interpretations of the *Parmenides* that were given in n. after Plotinus: he expounds and criticizes the interpretations of Amelius and Porphyry (the damaged text edited by Hadot is only the beginning of the latter's commentary), that of Iamblichus and that of the so-called "philosopher of Rhodes"; and finally reaches his own immediate predecessors in the school of Athens: Plutarch of Athens and Syrianus. It is not possible to proceed here to a detailed examination of the interpretations of Amelius, Porphyry, Iamblichus and the "philosopher of Rhodes" and of Proclus's criticisms of them, something that has already been done by Saffrey and Westerink; so we will confine ourselves to saying that the interpretations of Plutarch of Athens (cf. Proclus, *In Parm.* 1058, 21-1061, 20) and Syrianus (*ibid.* 1061, 20-1064, 12) agree with each other to a great extent: acc. to Proclus, Plutarch of Athens identified the first hypothesis of the *Parmenides* with the first principle, the second with mind, the third with the soul, the fourth with the forms united to matter and the fifth with matter; while Syrianus made the first hypothesis correspond to the absolutely transcendent One, the second to all the classes of gods (intelligible, intelligible-intelligent, intelligent, hypercosmic, encosmic), the third to the *souls derived from the divine soul, the fourth to the forms united with matter, the fifth to matter. Still acc. to Proclus, Syrianus's interpretation, though not differing substantially from that of Plutarch of Athens, showed some progress from it in that it included in the second hypothesis all the hierarchies of gods, introduced - still in the second hypothesis - the concept of henads, and limited the third hypothesis to the souls participating in the divine soul (cf. Saffrey Westerink, *op. cit.*, LXXXIV LXXXVIII).

Syrianus's interpretation of the *Parmenides* coincides with that of Proclus and is thus basic for an understanding of the whole Proclian system.

VI. Proclus and Damascius. *Proclus too, like his teacher Syrianus, refers the first hypothesis of the *Parmenides* to the absolutely transcendent One, the second to all the classes of beings (or gods) including the lowest, the third to particular souls, the fourth to forms united with matter and the fifth to matter itself (cf. *Theol. plat.* I, 12). In the Proclian system, the process of "hierarchization" of beings is pushed to its greatest degree; and the relations between the various classes of beings arranged according to this rigid hierarchical criterion appear to be regulated by laws that, deriving in large measure from earlier n., receive their systematic codification in the *Elements of theology*. Another great work, the *Platonic Theology*, whose first six books survive, represents the *summa* of the whole Proclian system, since it aims to be a detailed treatment both of the first principle and of the various subsequent classes of gods. In his commentaries on the Platonic dialogues (*On Parmenides, On the Republic, On Timaeus, On Cratylus*), Proclus, following the example of Iamblichus and of his teachers of the school of *Athens, interprets Plato in accordance with the patterns of his own philosophy: i.e. he aims to demonstrate that his own system is already prefigured in the philosophy of Plato (cf. e.g. *Theol. Plat.* I, 2, p. 9, 16-18 Saffrey-Westerink). Our treatment will briefly examine the following points: 1) the more important laws that regulate Proclus's hierarchical edifice; 2) the first principle; 3) the henads; 4) the various classes of beings or gods; 5) the origin of matter and the characteristics of the universe; 6) the nature of *evil; 7) ethics and union with the supreme principle.

1. Of the laws that regulate the relations between the various classes of hierarchically-arranged beings, the most important are the following: a) the three moments of *moné, próodos* and *epistrophé* (*El. theol.* 29-35): this law is already present in *Plotinus, *Porphyry, Iamblichus (cf. above) and Syrianus (*In Met.* 127, 9-10); b) the likeness of the produced being to the productive principle (*El. theol.* 28. 29. 32; cf. Syrianus, *In Met.* 106, 26-29); c) the inferiority of the produced being to the productive principle (*El. theol.* 28. 36); d) the unalterability of the productive principle despite the production of the lower being (*El. theol.* 27: cf. points 11 and 12 of Plotinus's doctrine of the One); e) production due to the superabundance and overflowing of the power of the productive principle (*El. theol.* 27: cf. point 1 of Plotinus's doctrine of *nous*); f) the close connection between the production of lower beings and the subdivision (or multiplication) of the unity and power of the higher being (*El. theol.* 27. 35. 64. 95: cf. above, Porphyry); g) the greater or lesser participation of beings in the original perfection and unity, according to the higher or lower point they occupy in the hierarchical scale (*El. theol.* 36. 62. 86: cf. Plotinus I, 2, 2); h) the close connection between the place occupied by a being in the hierarchical scale, its dignity (*axía*) and its aptitude to receive illuminations from the higher being (*El. theol.* 122. 142: cf. above, Porphyry); i) the presence in higher beings of all the properties of lower beings, and the presence in lower beings of some of the properties of higher beings (*El. theol.* 18. 97. 150: cf. Plotinus I, 2, 7); j) the relations between higher beings and lower ones, represented by "transmission" (*metádosis*) (*El. theol.* 18), and those between lower beings and higher ones, represented by "participation" (*méthexis, metoché, metousía*) and "conversion" (*epistrophé*) (*El. theol.* 1. 2. 3. 12. 148); k) the function of the central member of a series, which extends towards the higher member and the lower member, joining them to each other (*El. theol.* 148). Some of these "laws" would be taken up in the *Celestial hierarchy* of Ps.-*Dionysius the Areopagite (cf. S. Lilla, *Augustinianum* 22 [1982] 554-557).

2. Unlike Iamblichus, who had placed two "ones" above absolute being or one-being, Proclus, like Theodore of Asine, Plutarch of Athens and Syrianus, returned to the Plotinian conception of the single One, superior to being, to the intelligible and to mind, and identical with the negative One of the first hypothesis of the *Parmenides*. The major points characterizing the Proclian conception of the first principle are the following: a) the multiple participates *in some way* in the transcendent One (*Theol. Plat.* II, 1; *El. theol.* 1); b) the One transcends multiplicity (*Theol. Plat.* II, 1; *El. theol.* 5); c) the One is the cause of the multiple (*Theol. Plat.* II, 1); d) the One is unique (*Theol. Plat.* II, 2); e) the One is beyond being (*Theol. Plat.* II, 2); f) the One is a non-being (*Theol. Plat.* II, 2; II, 5) (cf. point 17 of Plotinus's doctrine of the One); g) the One is the cause of all that exists (*Theol. Plat.* II, 3); h) the One is above mind (*Theol. Plat.* II, 4; *El. theol.* 20); i) the One is identical with the absolute good (*Theol. Plat.* II, 6; *El. theol.* 20); j) though identified with the absolute good, the One is above the idea of the good (*Theol. Plat.* II, 7); k) the One is beyond knowledge, i.e. it is absolutely unknown and nameless (*Theol. Plat.* II, 7); l) the One generates all the divine orders (*Theol. Plat.* II, 7); m) the One is the source of the goodness that fills the whole universe with itself (*Theol. Plat.* II, 7); n) the unity of the One is fecund, but its generative activity does not involve divisions, movements or multiplications in it (*Theol. Plat.* II, 7) (cf. point 11 of Plotinus's doctrine of the One); o) the One is desired by all beings (*El. theol.* 12); p) the *via negativa*, characteristic of the first hypothesis of the *Parmenides*, is that which is most suitable to the One (*Theol. Plat.* II, 10).

3. As Saffrey and Westerink have shown in the introduction to the first volume of their edition of the *Platonic Theology* (p. LXIII), the Proclian system, from the One down to the lowest deities, is divided into nine classes of deities, which are treated in the various books of the *Platonic Theology* in the following order: 1) the One (book II); 2) the henads (book III); 3) the intelligible gods (book III); 4) the intelligible-intelligent gods (book IV); 5) the intelligent gods (book V); 6) the hypercosmic gods (book VI); 7) the encosmic gods; 8) the universal souls; 9) the *angels, *demons and heroes (those books of the *Platonic Theology* dealing with classes 7-9 of the gods are lost). The absolutely transcendent One produces firstly a multitude of henads, like though not wholly identical with itself (*Theol. Plat.* III, 2-3). Apropos of these henads we must mention: a) their generation takes place in a way that is fitting to the One, i.e. in a unitary way (*Theol. Plat.* III, 3); b) the henads, like the One, are superessential, i.e. superior to all beings (*Theol. Plat.* III, 3); c) unlike the One, the henads are objects of participation by the lower beings (*Theol. Plat.* III, 4); d) the henads function as intermediaries between the One and the beings, linking the latter to the One and making them turn towards themselves and towards the One (*Theol. Plat.* III, 3; III, 4); e) the henads are hierarchically ordered: those closer to the One are participated in by the more elevated beings, those further from the One are participated in by beings of inferior grade (*Theol. Plat.* III, 5).

4. Proclus inherits from Iamblichus both the partition of the beings or gods into the three great classes of intelligible (*noetoí*), intelligible-intelligent (*noetoí te kaì noeroì*) and intelligent (*noeroì*) gods; and the triadic scheme which is found in the two classes of the intelligible and intelligible-intelligent gods and which is characterized by the presence, at various levels, of being, life and intelligence; and the manner of generation of the various classes (the highest class originates from the One, while each of the subsequent classes is originated by the class immediately above it). From this point of view Iamblichus may be considered the most important source of Proclus, who regularly calls him "divine".

For Proclus, as for Porphyry and Iamblichus, being, life and intelligence form three descending steps (*El. theol.* 101; *Theol. Plat.* III, 6; IV, 1); while being characterizes the class of the intelligible gods, life characterizes that of the intelligible-intelligent gods and intelligence that of the intelligent gods (*Theol. Plat.* IV, 1; IV, 3). As for Iamblichus before him, so too for Proclus the triad being-life-intelligence is not limited to determining the difference between these three great classes of gods, but is found within the class of the intelligible gods and that of the intelligible-intelligent gods, in that it characterizes the three triads that compose each of these two classes.

Like Iamblichus, Proclus also postulates the existence of two constitutive principles of the various triads, of which Plato had spoken in *Philebus*, i.e. the *próton péras* (or second One, deriving directly from the first principle) and the *próte apeiría* which is its infinite generative power, which occupies an intermediate position between the *próton péras* and absolute being, and which is diffused everywhere producing the other beings by virtue of a process of multiplication (*Theol. Plat.* III, 8; *El. theol.* 89. 90). While the *próton péras* corresponds to *moné*, the *proté apeiría* expresses the moment of divine *próodos* (*Theol. Plat.* III, 8). *Péras* and *apeiría*, combining together, generate a product which, participating in both, is mixed, *miktón* (cf. *Philebus* 27b): this is absolute being *par excellence* (*autóon*), which, together with its two constitutive elements, *péras* and *ápeiron*, forms the first and highest triad of the intelligible gods, distinguished by being. Since it participates in *péras*, this absolute being is a unity (*èn òn*); since it participates in *apeiría*, it contains in itself, in a still hidden and indistinct way, the multiplicity of beings destined to take on ever clearer and sharper outlines in the subsequent classes (*Theol. Plat.* III, 9). *Péras* and *ápeiron* also produce the other two subsequent triads: the second triad of the intelligible gods is formed by *péras, ápeiron* and intelligible life, the third by *péras, ápeiron* and intelligible intelligence (*Theol. Plat.* III, 12; III, 14); the second and third triad, though also resulting from the *péras-ápeiron* combination, proceed respectively from the first and second triads (*Theol. Plat.* III, 12; III, 13). Apropos of the three intelligible triads it must be noted: a) the first triad is distinguished by *péras*, the second by *apeiría*, the third by mixture (*Theol. Plat.* III, 13); b) the first triad reflects the moment of *moné*, the second that of *próodos*, the third that of *epistrophé* (*Theol. Plat.* III, 14); c) the first triad participates more in symmetry, the second in truth, the third in beauty (*Theol. Plat.* III, 13); d) proceeding from the first to the third intelligible triad, we see a gradual process of unfolding and revelation of beings (*Theol. Plat.* III, 13; III, 14).

From the class of the intelligible gods originates the second great class, that of the intelligible-intelligent gods, which is on the level of life (*Theol. Plat.* IV, 1; IV, 3), just as the class of the intelligible gods is on the level of being and that of the intelligent gods is on the level of intelligence. Its main characteristics are the following: a) the intelligible-intelligent gods function as intermediate link and point of union between the superior class (the intelligible gods) and the inferior class (the intelligent gods) (*Theol. Plat.* IV, 1; IV, 2; IV, 3); b) in them we note a progressive multiplication and revelation of the *dynámeis* which in the higher class appear still united (*Theol. Plat.* IV, 1; IV, 2; IV, 3); c) like the intelligible gods, they are divided into three triads, each of which is formed of being, life and intelligence. In the higher triad the intelligible component dominates, in the intermediate there is a balance between the intelligible and the intelligent elements, while in the third the intelligent element dominates (*Theol. Plat.* IV, 3).

Classes 5-7 of deities (the intelligent gods, hypercosmic gods and encosmic gods) are based on allegorical interpretations of some deities of Greek *mythology and of the "Chaldaean *oracles" of which Proclus, like Iamblichus, was a great admirer: a detailed examination of them, which cannot be conducted here, can be found in Saffrey's and Westerink's introduction (*Proclus. Théologie platonicienne*, I, LXVI-LXXV). In his treatment of these inferior classes too, Iamblichus must have been Proclus's principal source (for some doctrines of Iamblichus inherited by Proclus cf. J.M. Dillon, Iamblichi Chalcidensis, in *Platonis dialogos commentariorum fragmenta*, Leiden 1973, 52-53.

5. Like Porphyry and Iamblichus, Proclus maintains the origin of matter from the divine principle: taking to its extreme consequences the Plotinian identification between *apeiría* and intelligible matter (cf. points 10 and 11 of Plotinus's doctrine of *nous*), Proclus identifies sensible matter with the *apeiría* of the sensible world deriving from *proté apeiría* (*In Tim*. I, 385, 7-8); indeed sensible matter proceeds from *proté apeiría* and one-being (*In Tim*. I, 385, 12-14). Just as the supreme god originates *apeiría*, so it originates the matter that is its extreme degradation (*In Tim*. I, 384, 30 - 385, 2); and as it generates all beings, so it also generates matter (*In Tim*. II, 102, 6-11). The demiurge, who occupies a rather low rung on the scale of the gods, since he belongs to the class of the intelligent gods (*In Tim*. I, 394, 8-9), does not produce matter himself but finds it already formed by the higher principle and is limited to imposing order on it (*In Tim*. I, 300, 6-8; I, 384, 16-24). The imposition of order on matter by the demiurge takes place, however, *ab aeterno*, since the demiurge is good by nature, and there has never been a moment when he has not manifested his goodness (*In Tim*. I, 394, 23-25); hence the "wicked and disordered" movement of corporeal nature (cf. *In Tim*. 30a) is not real and prior to the intervention of the demiurge, still less is it due to the existence of a wicked *anima mundi* as Plutarch of Chaeronea had supposed (*In Tim*. I, 394, 9-11), but only hypothetical: i.e. it would have taken place if the demiurge's intervention had not occurred (*In Tim*. I, 394, 27-31). In Proclus too, as in Plotinus and Iamblichus, recurs the Poseidonian doctrine of the close *communion of beings among themselves (*El. theol.* 148 [130, 10-11]).

6. Proclus inherits from Plotinus the doctrine of *evil as a lack of good, which he deals with at length in his *De malorum subsistentia*, surviving in the Latin translation of William of Moerbeke, archbishop of Corinth in 1280 (ed. H. Boese, *Procli Diadochi opuscula De providentia, libertate, malo*, Berlin 1960). The points that distinguish the Proclian theory of evil are four: a) absolute evil cannot be numbered among beings, since all beings, by nature, aspire to the good (ch. 2, 24-27, 29-32); b) while absolute evil does not exist at all, relative evil, to be able to exist, must in some way participate in the good, as does every being (ch. 9, 5-18); c) evil can be considered only as a relative, not a total, absence of good (ch. 10, 12-13); d) evil is not in matter (chs. 32, 1-4; 35, 12-14): in this last point Proclus clearly distances himself from Plotinus (cf. above). On the use of this Proclian work by Ps.-*Dionysius the Areopagite in ch. IV of the *Divine names* cf. S. Lilla, *Augustinianum* 22 (1982) 557-558, and the bibliography cited there.

7. In his *Life of Proclus*, Marinus draws an ideal picture not just of the master, but also of the way of life he achieved; into his description in chs. 21-26 come Platonic (esp. from *Phaedo* and *Phaedrus*), Plotinian, Porphyrian and Iamblichean motifs. In ch. 21 appears the motif of the *soul's total detachment from the *body and its withdrawal into itself, formulated in the *Phaedo* and taken up by Plotinus in his treatise *On the virtues* and by Porphyry esp. in *De regressu animae*; still in ch. 21, in the footsteps of Plotinus and Porphyry, he brings out the difference between *metriopátheia* and *apátheia*, identified with the soul's absolute independence of the body and with the "cathartic virtues". Ch. 22 celebrates the master's progress beyond the cathartic virtues to the vision of the ideas - whose description, through Plotinus and Porphyry, goes back in the final analysis to the *Phaedrus* - and to the attainment of the "paradigmatic virtues" present in the divine *nous (this last is a Porphyrian motif). Ch. 25, in the footsteps of Plotinus and Porphyry, emphasizes the difference between the inferior life of the "good man", achieved through the political virtues, and the superior life of the gods attained by Proclus, as well as the identity betwen *apátheia* and *homoíosis theó*. Ch. 26 explicitly names Iamblichus in celebrating Proclus's ascent to the highest class of virtues, those which Iamblichus had called "theurgic". In all probability these were the "hieratic" virtues (cf. above apropos of Iamblichus).

Like Plotinus, Proclus too was an upholder of union with the supreme principle, a union characterized by the cessation and surpassing of all noetic activity (cf. e.g. *Exc. chald.* IV; *Theol. Plat.* I, 3; I, 25: Ps.-Dionysius the Areopagite had this last passage in mind in *Nom. div.* I, 1: PG 3, 585B-592A; cf. S. Lilla, *Augustinianum* 22 [1982] 565 n. 189).

Of Damascius, pupil of Proclus and last scholarch of the school of *Athens at the time when *Justinian decreed its closure in 529, there remain the works *Dubitationes et solutiones de primis principiis*, *Dubitationes et solutiones in Platonis Parmenidem* (beginning incomplete; edited with the former by C.E. Ruelle, Paris, 1889), the *Lectures on Philebus* (ed. L.G. Westerink, Amsterdam 1959) and the *Life of Isidore* (*Vitae Isidori reliquiae*, ed. C. Zinten, Hildesheim 1967). His thought marks no substantial progress from that of Proclus. His main work, *Dubitationes et solutiones de primis principiis*, is a minute examination of the Proclian system and the problems relating to it. But Damascius returns to Iamblichus's conception of the two higher "ones" (cf. R.T. Wallis, *Neoplatonism*, London 1972, 158); and he insists on the unknowability not just of the first principle, but also of the way in which the various classes of beings originate: the origin of multiplicity from the one, and also the fundamental Proclian law of *moné, próodos* and *epistrophé* are just supports on which the weak human mind leans so as to have an idea of facts which remain incomprehensible (cf. Wallis, *ibid.*). Recent studies have shown his influence on Ps.-Dionysius (cf. for more details S. Lilla, *Augustinianum* 22 [1982] 542).

VII. Neoplatonism in the Latin West. The increased research into the spread of n. in the Latin culture of the West is due above all to P. Henry and P. Courcelle (cf. bibl.): while the former has shown the presence of Neoplatonic motifs in *Firmicus Maternus (pp. 25-43), Macrobius (pp. 146-192), Servius, *Ammianus Marcellinus and *Sidonius Apollinaris (pp. 192-202), the latter has drawn attention to Macrobius (pp. 16-34), Marcianus Capella (pp. 198-205), *Claudianus Mamertus (pp. 223-235) and Sidonius Apollinaris (pp. 235-246). On the dependence on n. of various Latin patristic authors, such as *Hilary of Poitiers, *Marius Victorinus, St *Ambrose, St *Augustine and *Boethius, cf. *Platonism and the Fathers with relative bibliography. On Calcidius, the studies of J.C.M. Winden and J.H. Waszink (cf. bibl.) are still fundamental.

Studies: *1.* Preparation of Neoplatonism: E.R. Dodds, The Parmenides of Plato and the Origin of the Neoplatonic One, *CQ* 22 (1928) 129-142; W. Theiler, *Die Vorbereitung des Neuplatonismus*, Berlin 1930; C.J. De Vogel, A la recherche des étapes précises entre Platon et le Néoplatonisme, *Mnemosyne*, ser. IV, 7 (1957) 111-122; P. Merlan, *From Platonism to Neoplatonism*, The Hague 1960; id., Greek Philosophy from Plato to Plotinus, in *The Cambridge History of Later Greek and Early Medieval Philosophy*, Cambridge 1967, 14-132; H.J. Krämer, *Der Ursprung der Geistmetaphysik, Untersuchungen zur Geschichte des Neuplatonismus zwischen Platon und Plotin*, Amsterdam 1964; R.T. Wallis, *Neoplatonism*, London 1972, 16-36; (ed.) H.J. Blumenthal and R.A. Markus, *Neo-Platonism and Early Christian Thought*, London 1981; R.B. Harris (ed.), *The Structure of Being: a Neoplatonic approach*, Norfolk, Va. 1982; S. Gersh, *Middle Platonism and Neoplatonism: the Latin tradition*, 2 vols. Notre Dame 1986.
2. On Plotinus and his sources: H.R. Schwyzer: *PWK* 21/1, 471-592; W.R. Inge, *The Philosophy of Plotinus*, London-New York 1918; R.E. Witt, The Plotinian Logos and its Stoic Basis, *CQ* 25 (1931) 103-111; id., Plotinus and Posidonius, *CQ* 24 (1930) 198-207; P. Henry, *Les états du texte de Plotin*, Paris 1938; id., *Les manuscrits des Ennéades*, Paris 1941; A.H. Armstrong, *The Architecture of the intelligible Universe in the Philosophy of Plotinus*, Cambridge 1940 (reprint Amsterdam 1967); id., Plotinus, in *The Cambridge History of Later Greek and Early Medieval Philosophy*, Cambridge 1967, 195-268; id., *Plotinian and Christian Studies*, London 1979 (reprint of various important articles that appeared earlier); W. Theiler, Plotinus und die antike Philosophie, *MH* (1944) 209-225 (reprinted in *Forschungen zum Neuplatonismus*, Berlin 1966, 140-159); id., Plotin zwischen Platon und Stoa, in Les sources de Plotin, in *Entr. sur l'ant. class.* V, Geneva 1960, 65-86 (reprint in *Forschungen zum Neuplatonismus*, 124-139); L. Sweeney, Infinity in Plotinus, *Gregorianum* 38 (1957) 515-535, 713-732; W.N. Clarke, Infinity in Plotinus: a Reply, *Gregorianum* 40 (1959) 75-98; P. Hadot, Etre, vie, pensée chez Plotin et avant Plotin, in *Entr. sur l'ant. class.* V, Geneva 1960, 107-141; E.R. Dodds, Tradition and personal Achievement in the Philosophy of Plotinus, *JRS* 50 (1960) 1-7; J.M. Rist, The indefinite Dyad and intelligible Matter in Plotinus, *CQ* n.s. 12 (1962) 99-107; ibid., *Plotinus: the Road to Reality*, Cambridge 1967; P. Merlan, *Monopsychism, Mysticism, Metaconsciousness*, The Hague 1963; R. Arnou, *Le désir de Dieu dans la philosophie de Plotin*, Rome 1967; A. Graeser, *Plotinus and the Stoics*, Leiden 1972; R.T. Wallis, *Neoplatonism*, London 1972, 37-93; See also the numerous contributions which appeared in the three volumes: Les sources de Plotin, in *Entr. sur l'ant. class.* V, Geneva 1960, *Le Néoplatonisme. Royaumont 9-13 juin 1969*, Paris 1971 and *Plotino e il Neoplatonismo in Oriente e in Occidente*, Accademia Naz. dei Lincei, Rome 1974.
3. On Porphyry: R. Beutler: *PWK* 22/1, 275-313; H. Dörrie, *Porphyrios' Symmikta Zetemata*, Zetemata 20, Münster 1959; id., Das fünffach gestufte Mysterium: der Austieg der Seele bei Porphyrios und Ambrosius, in *Mullus*, Festschr. Th. Clauser, JbAC Ergbd. 1, Münster 1964, 79-92; id., Die Schultradition im Mittelplatonismus und Porphyrios, in *Entr. sur l'ant. class.* XII, Geneva 1966, 3-32; id., Die Lehre von der Seele, *ibid.*, 165-187; P. Hadot, Fragments d'un commentaire de Porphyre sur le Parménide, *REG* 74 (1961) 410-438; id., La métaphysique de Porphyre, in *Entr. sur l'ant. class.* XII, Geneva 1966, 87-123; J.M. Rist, Mysticism and Transcendence in Later Neoplatonism, *Hermes* 92 (1964) 220-225; A.R. Sodano, Porfirio commentatore di Platone, in *Entr. sur l'ant. class.* XII, 195-228; J.H. Waszink, Porphyrios und Numenios, *ibid.*, 35-83; A.C. Lloyd, *The Cambridge History of Later Greek and Early Medieval Philosophy*, Cambridge 1967, 272-295; R.T. Wallis, *Neoplatonism*, London 1972, 94-137; J. Whittaker, Neopythagoreanism and the transcendent Absolute, *SO* 48 (1973) 77-86; A. Smith,

Porphyry's Place in the Neoplatonic Tradition. A Study in postplotinian Neoplatonism, The Hague 1974; W. Deuse, Der Demiurg bei Porphyrios und Jamblich, in *Die Philos. des Neuplatonismus herausgeg. von C. Zintzen*, Darmstadt 1977, 238-278; F. Romano, *Porfirio di Tiro*, Catania 1979.
4. On theurgy and Iamblichus: W. Kroll: *PWK* 9/1, 645-651; J. Bidez, Le philosophe Jamblique et son école, *REG* 32 (1919) 29-40; E. Eitrem, La théurgie chez les Néoplatoniciens et dans les papyrus magiques, *SO* 22 (1942) 49-79; E.R. Dodds, Theurgy and its Relationship to Neoplatonism, *JRS* 37 (1947) 55-69; H. Lewy, *Chaldean Oracles and Theurgy*, Cairo 1956, Paris ²1978; B. Dalsgaard Larsen, *Jamblique de Chalcis exégète et philosophe*, Aarhus 1972; id., La place de Jamblique dans la philosophie antique tardive, in *Entr. sur l'ant. class.* XXI, Geneva 1975, 1-26; R.T. Wallis, *Neoplatonism*, London 1972, 118-137; H. Dillon, *Iamblichi Chalcidensis in Platonis dialogos commentariorum fragmenta*, Leiden 1973; E. Des Places, La religion de Jamblique, in *Entr. sur l'ant. class.* XXI, Geneva 1975, 69-94; R.E. Witt, Iamblichus as a Forerunner of Julian, *ibid.*, 35-64.
5. On the school of Athens: H.D. Saffrey-L.G. Westerink, *Proclus, Théologie platonicienne*, I, Paris 1968, XXVI-XLVIII, LXXV-LXXXIX; R.T. Wallis, *Neoplatonism*, London 1972, 138 ff.
6. On Proclus: R. Beutler: *PWK* 23/1, 186-247; E.R. Dodds, *Proclus. The Elements of Theology*, Oxford 1933; W. Beierwaltes, *Proklos. Grundzüge seiner Metaphysik*, Frankfurt/M. 1965; A.C. Lloyd, *The Cambridge History of Later Greek and Early Medieval Philosophy*, Cambridge 1967, 302-314; H.D. Saffrey-L.G. Westerink, *Proclus. Théologie platonicienne*, I-IV, Paris 1968-1981 (introduction to the individual volumes); R.T. Wallis, *Neoplatonism*, London 1972, 138-158; S.E. Gersh, *Kínesis Akínetos*, Leiden 1973, G. Martano, *Proclo di Atene*, Naples 1974; D.J. O'Meara (ed.), *Neoplatonism and Christian Thought*, Albany 1981; C. Andresen, The Integration of Platonism into early Christian Theology, *SP* 15.1, (1984) 399-413.
7. On Neoplatonism in the Latin West: P. Henry, *Plotin et l'Occident. Firmicus Maternus, Marius Victorinus, Saint Augustin et Macrobe*, Louvain 1934; P. Courcelle, *Les lettres grecques en Occident. De Macrobe à Cassiodore*, Paris 1948; J.C.M. van Winden, *Calcidius on Matter. His Doctrine and Sources*, Leiden 1959; J.H. Waszink, *Timaeus a Calcidio translatus commentarioque instructus*, London-Leiden ²1975; AA. VV., Néoplatonisme, in *Mél. offerts à J. Trouillard*, Fontanay aux Roses 1981.

S. Lilla

NEPHALIUS. Alexandrian monk. He had a considerable role in the religious life of the years 482-507. At first he was a *monophysite of schismatic tendency, as *Zacharias the Rhetor (*Hist. Eccl.; Vit. Sev.*) and other sources testify. The opinion of J. Lebon (*Le monophysisme séverien*, Louvain 1909, 33 n. 3; 43, n. 4) who, misinterpreting some passages of *Evagrius and *Liberatus, considers him as always a Chalcedonian, is unbelievable. In 482 he took part in the rebellion of the 30,000 monks who, led by Theodore of Antinoë, claimed an explicitly anti-Chalcedonian interpretation of the *Henoticon*: N. was among the more intransigent and turbulent (Fliche-Martin 4, 362-367). His intolerance was based on considerations of church politics rather than on doctrinal extremism: thus in 507 we find him leading a campaign of violent agitation against the Severian monophysite monks. In *Palestine, on this occasion, he preached an address in defence of the two natures and the council of *Chalcedon: this *apologia* (συνηγορία) is lost. Fragments are preserved in the two *Orationes ad Nephalium* (of the first we have only the ending, the second survives entire in a *Syriac version) composed in reply by *Severus of Antioch: N., who uses patristic arguments, appears to be the source of the original *theology of *John Grammaticus and precursor of "*Neochalcedonianism", a *christology aimed at reconciling, with the help of Aristotelian categories, the thought of *Cyril of Alexandria and the conclusions of Chalcedon.

CPG 3, 6825; CSCO 119, 1-69 (text); CSCO 120, 1-50 (Fr. tr.); W. Ensslin, Nephalios, *PWK* 16, 2489-2490; Ch. Moeller, Un représentant de la Christologie néochalcédonienne au début du sixième siècle en Orient: Nephalius d'Alexandrie, *RHE* 40 (1944-1945) 73-140; Beck 284-285; S. Helmer, *Der Neuchalkedonismus*, Bonn 1962, 151-159.

A. Labate

NEPOS. (First?) bishop of Arsinoë (=Medinet el Fayyum), *c.*240. A *millenarian, fervent defender of the *Apocalypse, he opposed *Origen in a lost *Refutation of the allegorists*; *Eusebius (*HE* VII, 24, 1-3) seems to know him through *Dionysius of Alexandria, his adversary. Nevertheless he constitutes an important aspect of Egyptian *tradition, whose evidence has not been made use of, but who conditioned the history of *Origenism. N. also contributed to the writing of liturgical poetry (in Greek?).

Quasten I, 371.

J. Gribomont

NEPOTIANUS. Nephew of *Heliodorus, bishop of Altinus, and *friend of *Jerome: second half of 4th c. After initially embracing military life, he dedicated himself to *God; at first he thought to withdraw to an *ascetic life in the *East or on a Dalmatian island, but Heliodorus influenced him into staying at home. He became a priest and his uncle's coadjutor; died 396. Jerome, who dedicated to him *Ep.* 52 on the life of clerics and monks, gave his funeral eulogy in *Ep.* 60 to Heliodorus, in which the description of N.'s *virtues and qualities becomes a portrait of the ideal priest.

F. Cavallera, *Saint Jérôme. Sa vie et son oeuvre*, Louvain-Paris 1922, I, 182-184.

S. Zincone

NEREUS and ACHILLEUS. Roman *martyrs. The earliest evidence of their cult comes from the 4th c.: *Damasus's poem (Ferrua, *Epigr. Damas.*, no. 8) and the colonnade of the ciborium that covered the primitive altar of their cemeterial basilica. Acc. to these sources, they were two soldiers who deserted for religious reasons and were condemned to death. Their bodies were buried in the *cemetery of Domitilla, and over their tombs were erected first a chapel, by Damasus (366-84), then in 390 a semi-underground basilica, by *Siricius. Feast 12 May in *Mart. hier.* The stone *cathedra*, from which *Gregory the Great preached his *Hom. in Ev.* 28 (PL 76, 1210-13), was transferred by Baronius from the cemeterial basilica to his *titulus* of Sts N. and A., near the Baths of Caracalla. The cemeterial basilica was destroyed by an earthquake in 817. G. B. Rossi excavated it in 1873-74.

A. Achelis, *Acta SS. Nerei et Achillei*, TU 11/2; P. Franchi de' Cavalieri, *I SS. Nereo ed Achilleo nell'epigramma damasiano*, ST 22, 43-55; U. Fasola, *La basilica dei SS. Nereo e Achilleo e la catacomba di Domitilla*, Rome ²1967.

V. Saxer

NERO. Lucius Domitius Ahenobarbus, Roman emperor 54-68 under the name N., was placed by *Claudius, under pressure from Agrippina and moderate circles, ahead of Britannicus in succession. A complex and controversial personality, N. reigned with moderation until 62, but was then influenced by extreme monarchist elements who pushed him towards ferocious purges of the opponents of absolute power: in 65-66, on the pretext of the Pisonian plot, *Seneca, Lucan, Petronius and others were eliminated. In 68 N. committed suicide, the senate and soldiers of many Western provinces having rebelled. Galba, acclaimed by the Praetorians, succeeded him.

With regard to the Christians, the fierce *persecution of 64, caused, acc. to *Tacitus (*Annal.* XV, 44), by popular anger over the burning of *Rome perhaps ordered by the emperor himself, and whose victims included the apostles *Peter and *Paul, was not due to a specifically Neronian law but, acc. to *Tertullian, to the application of a *senatus-consultum* of *Tiberius. For Tertullian (*Apol.* V, 1 ff.) N. was the initiator of the *institutum* of the persecutions. But acc. to Suetonius (*Nero* 16, 38, 39), the Christians were condemned for *superstitio illicita* and not for responsibility for the fire. Acc. to M. Sordi the anti-Christian change in N.'s policy, even before 64, is attested in 1 Peter, written at Rome before 64 (Sordi, *Il cristianesimo*, 85). Scholars researching the legal bases of the persecutions have long debated the existence of an "*institutum Neronianum*", i.e. a specific persecutory law.

L. Dieu, La persécution au IIe siècle. Une loi fantôme, *RHE* 38 (1942) 5 ff.; M.A. Levi, *Nerone e i suoi tempi*, Milan 1949; J. Zeiller, Institutum Neronianum, *RHE* 50 (1955) 393-399; V. Monachino, *Il fondamento giuridico delle persecuzioni nei primi due secoli*, Rome 1955; J. Beaujeu, L'incendie de Rome en 64 et les chrétiens, *Latomus* 19 (1960) 47 ff.; M. Sordi, *Il cristianesimo e Roma*, Bologna 1965, 79-94 and 421-423 (bibl.); E. Demongeot, A propos de la persécution de 64 contre les chrétiens et de l'Institutum Neronianum, *Recueil de mémoires et travaux de la Soc. d'hist. du droit écrit de Montpellier* VII (1970) 145-155; W.H.C. Frend, *Martyrdom and Persecution in the Early Church*, Oxford 1965; G.E.M. de Ste Croix, Why were the early Christians persecuted? *P&P* 26 (1963) 6-38; W. Rordorf, Die neronische Christenverfolgunt im Spiegel der apokryphen Paulusakten, *NTS* 28 (1982) 365-374.

L. Navarra

NERSETES (or Narsetes or Narsai or Nerses): name of various Armenian catholicoi.

- N. I (cath. 353/5- 372) is mentioned, with many semi-legendary details, by *Faustus of Byzantium. He spent some years as chamberlain at the court of king Arsak. His episcopal consecration revealed a man full of energy, concerned above all to reorganize the charitable structures and canonical rules of the church of *Armenia. Even as *catholicós he continued to carry out political missions for the king.

J. Markwart, *Die Entstehung der armenischen Bistümer*, Rome 1932, 223 [87]-233 [97]; *BS* 9, 742-746.

- N. II Astarekec'i (cath. 548-57): called the council of *Dvin (554-5) against the *Nestorians (or *Manichees?). Some canonical responses are attrib. to him.

EC 8, 1768.

S.J. Voicu

NESTORIUS - NESTORIANISM. In April 428 N. was consecrated bishop of *Constantinople. Born 381 (but the date is uncertain) at Germanicia (*Syria), educated at *Antioch, perhaps a pupil of *Theodore of Mopsuestia, he was monk and then priest at Antioch. His election as patriarch was warmly supported by court circles, who knew his gifts of virtue and eloquence. As part of a series of initiatives aimed at restoring purity of faith at Constantinople, N. publicly disapproved the custom, by now widespread among the people, of calling *Mary Mother of God (*Theotokos): in fact Antiochene *christology distinguished with the greatest precision the divine from the human properties in Christ, so that Mary strictu sensu must be considered only the mother of the man Jesus. N., however, preferred the more comprehensive title Christotokos. His attitude at once aroused discontent and protests, which caused *Cyril of Alexandria to intervene. Political and doctrinal motives were interwoven in Cyril's action: in fact, in view of the traditional rivalry between *Alexandria and Antioch, Cyril saw with dislike a prestigious Antiochene, who evoked for him the ghost of *John Chrysostom, on the episcopal see of Constantinople, now the principal see of the *East; and his Alexandrian christological position, putting maximum stress on the subordination of Christ's humanity to his divinity, had a better grasp on His unity and was suspicious of a too sharp distinction between the human and divine properties in Him. Cyril's second *letter to N. (the first was just a general request for information) and N.'s reply (430), both doctrinal in character, precisely defined the divergences between the Alexandrian and Antiochene christologies.

As part of their respective campaigns of propaganda, both Cyril and N. informed pope *Celestine of the dispute: he, without going into it too deeply, decided in Cyril's favour and, following a *council held at *Rome in August 430, invited N. to confess his errors and instructed Cyril to deliver the Roman diktat to him. Not until November did Cyril transmit it to N., adding to it a series of 12 *anathemata which presented the Alexandrian christology in its most radical form: they spoke, among other things, of unity of (human and divine) nature (hénosis physiké) in Christ. No Antiochene could have subscribed to them. But meanwhile N. had asked the emperor to convene a council, which was granted. The council that began at *Ephesus on 22 June 431 developed in a very irregular way, thanks to Cyril, whose supporters condemned and deposed N., while some days later N.'s supporters condemned and deposed Cyril. Faced with such irregularities, *Theodosius II approved both condemnations and depositions; but that of Cyril, who had returned to *Egypt, remained unoperative, while N. spontaneously renounced any further defence and retired to a *monastery at Antioch. At Constantinople, *Atticus was elected in his place. Subsequent negotiations between Cyril and *John of Antioch led to a reconciliation in April 433: the Antiochenes renounced N., whose condemnation they approved; Cyril renounced the 12 anathemata. The formula of faith (Act of union) approved by both parties stated that in the one Christ, Son and Lord, occurred the union of the two natures, human and divine, without confusion, so that he is consubstantial with the *Father according to his divinity and consubstantial with us according to his humanity, and Mary is defined as mother of God. N. was expelled from Antioch and exiled, first to *Petra and then to the Great Oasis, in the desert of *Libya. He survived until the council of *Chalcedon (451) and, in his late apologia, the Book of Heraclides, affirmed the congruence of his doctrine with that of *Leo's *Tomus ad Flavianum.

N. was accused of dividing Christ, affirming two Christs and two Sons, i.e. man and *God, and reviving the *adoptianism of *Paul of Samosata, in that he had conceived the union of man and God in Christ as solely external. N. always denied the validity of these accusations and accused Cyril of being an *Apollinarist. It is not easy to specify his doctrine in all its details on the basis of what remains of his writings: some *letters and *homilies; fragments, esp. of homilies, extracted by his enemies, going back to the beginning of the controversy; and the late Book of Heraclides, surviving in *Syriac tr., whose ideas do not necessarily correspond in toto with those expressed so many years before. On this question the opinions of modern scholars have not yet reached agreement: some speak of the tragedy of N., accused of errors he did not commit, others support the validity of the accusations. We will sum up the main points of his doctrine.

N.'s concern, as a good Antiochene, was to safeguard, against Apollinarists and *Arians, the integrity of Christ's human nature understood as a complete personality, capable of free initiative, where the Alexandrians reduced it to a mere passive instrument of the Logos. Therefore he carefully distinguished the properties of the two natures and the names which refer to them: that is why he preferred for Mary the name Christotokos (= mother of Jesus in his union with the Logos) to that of *Theotokos. But despite the distinction, he rejected the charge of preaching two Christs, constantly confirming the indivisibility and unity of Christ, the incarnate Logos. To indicate the union of the two natures, he sometimes spoke of ineffable unity (hénosis), but he preferred synápheia (= conjunction), to avoid the union being considered a mixture. He adopted the traditional terminology of Antioch, and spoke of man assumed by the Logos, of a temple which the Logos made its dwelling, i.e. a terminology which brought out the distinction in Christ between man and God.

Given the priority of this interest, N. could not accept the Cyrillian formulae of union of nature, union of *hypostasis in Christ: for him indeed a nature had no subsistence, no concrete reality if it were not also a hypostasis, so that the union of natures and of hypostases appeared to him as a confusion of the two natures and the two hypostases in the manner of *Apollinaris. So he preferred to speak of union kat'eudokían (= out of courtesy, condescension), which was charged by his adversaries with *adoptianism, but which should be understood as a voluntary union of the Logos with the man. To render this concept more concrete, N. spoke repeatedly of a single *prosopon in which the two natures are united, but it is not easy to clarify this concept, given the vagueness of meaning of prosopon, which corresponds to the Latin *persona, but in which the connotation of appearance or external aspect is always present. The expression had already been used by *Theodore of Mopsuestia, but Cyril rejected it as insufficient from the first moment of the controversy: it has been thought that it may signify the psychological or moral personality into which come together unitarily the very distinct properties of the two natures, each one provided with its own hypostasis and its own prosopon (= mode of being, mode of appearance of each real entity). In the Book of Heraclides, N. speaks directly of an exchange of prosopa in Christ, in the sense that one of the two natures uses the prosopon of the other as if it were its own, which suggests a late acceptance by him of the communicatio idiomatum (= exchange of the predicates of humanity and divinity in Christ by virtue of the union), rejected years before in his reply to Cyril's second letter.

We may conclude that N., like Theodore before him, sensed the danger of compromising Christ's unity inherent in the divisionist christology of the Antiochenes, and sought to obviate it, but without encapsulating his good intention in a theologically unexceptionable concept. This lack was magnified by the convergence of a series of adverse circumstances of disparate character (political, personal, etc.), which, favoured also by N.'s malleability and lack of prudence, set in movement a machine that crushed him without pity, turning the good monk into the impious divider of Christ.

In Antiochene circles, however, not everyone accepted his condemnation. His supporters won over to the Nestorian faith (two natures, two hypostases, one prosopon of Christ) the Christian church of *Persia which, amidst difficulties and adversity, was destined to survive into the modern age.

CPG 3, 5665-5766; Quasten II, 518-523; F. Loofs, Nestoriana, Halle 1905; DTC 11, 76-187; L. Scipioni, Nestorio e il concilio di Efeso, Milan 1974 (bibl.); W. de Vries, Die syrisch-nestorianische Haltung zu Chalkedon, CCG 1, 603-635; R.A. Greer, The Antiochene Christology of Diodore of Tarsus, JThS n.s. 17 (1966) 327-341; H.E.W. Turner, Nestorius Reconsidered, SP 13 (1975) 306-321; R.C. Chesnut, The two Prosopa in Nestorius' Bazaar of Heracleides, JThS n.s. 29 (1978) 392-409; S. Gero, Barsauma of Nisibis and Persian Christianity in the Fifth Century, CSCO 426, Subs. 63, Leuven 1981; S.P. Brock, The Christology of the Church in the East in the Synods in the Fifth to Early Seventh Centuries, Aksum-Thyateira (Festsch. Archbishop Methodius), Athens 1985, 125-142.

M. Simonetti

NICAEA

I. The city - II. The council of 325.

I. The city. A city in *Bithynia (now Iznik), legendarily founded by Dionysus, it became Roman in 72 BC at the conclusion of the Mithridatic war (App., Bell. civ. V, 139, 1). In *Claudius's time it contended with *Nicomedia for the seat of the provincial governor. Ruined by an earthquake in AD 112, it was restored under *Hadrian, who visited it in 123 and encouraged the foundation of the walls, finished at the time of *Claudius Gothicus. Damaged by the *Goths in 258, it underwent various restorations in subsequent centuries; again in the 5th c. the walls were rebuilt. Christianized early, a series of *martyrs is recorded there (cf. E. Josi, Nicea, EC 8, 1827). Bishops: *Theognis (Arian, exiled in 325, bishop again in 328, †344), Chrestus (325-328), Eugenius (Arian, †370), Hypatius (Arian, †379), Dorotheus (381), Anastasius (at *Chalcedon, 451), Peter, Appius, Anastasius II, Stephen, Theophilus, Photius, George, Anastasius III.

PWK 17, 226-243; DACL 12, 1179-1211; EC 8, 1827 f.; EAA 5, 452-453; J. Sölch, Bitinische Städte im Altertum, Klio 19 (1925) 140 ff.; for the bishops cf. M. Le Quien, Oriens Christianus, I, Paris 1740, 640-662; P.B. Gams, Series Episc., Graz 1957 (repr.) 443.

M. Forlin Patrucco

II. The *council of 325. After his victory over *Licinius, emperor of the *East, in Sept 324, *Constantine made every effort to settle the disputes between the Eastern bishops, as he had already sought to do for the West over the *Donatist schism, promoting the synods of *Rome (311) and *Arles (314). So from autumn 324 he called the bishops to a synod, similar to the *comitia* of the civil administration of the empire, at first at *Antioch and then, for reasons of convenience, at N. in the immediate vicinity of his imperial residence of *Nicomedia. The *cursus publicus* was put at the disposal of the conciliar *Fathers. The essential aims of the synod were two: to smooth out the *Arian question and to resolve the question of *Easter. A first *journey of bishop *Ossius of Cordova, principal agent of Constantine's religious policy, to *Alexander of Alexandria, to reconcile him with *Arius, was a complete failure. On the other hand, a synod held at Antioch during the winter of 324/5 confirmed Alexander's condemnation of the group of early Arians, put out by the local synod which had brought together a hundred bishops from *Egypt and *Libya. Constantine inaugurated his council on 20 May 325, the day after the festivities for his victory over Licinius, celebrated at Nicomedia. All the synodal Fathers had been present at N. for some days. After a speech of welcome by the emperor, spoken in Latin, Arius's sympathizers spoke first, proposing a formula of faith read by *Eusebius of Nicomedia: but this was rejected. So *Eusebius of Caesarea put forward his formula of faith in a personal capacity, to free himself from a suspicion of *heresy that had been weighing on him for some months due to a censure by the synod of Antioch, but also to help the conciliar Fathers find a formula of faith convenient for all. As *Athanasius of Alexandria related 25 years later in his *letter *De decretis*, the discussions of the council of N. were long and laborious. The traditional formulae of the *Creed were better defined in an anti-Arian sense by subsequent additions, to the point of inserting the attribute *homoousios*, "consubstantial", to qualify the unity of essence of the *Father and the Son. We do not know who took the initiative in proposing this insertion, which became the touchstone of Nicene *orthodoxy. At the end of the debates, only Arius and two bishops, Secundus of Ptolemais and Theonas of Marmarica, refused to subscribe. They were *excommunicated, the two bishops deposed, and all three exiled to Illyricum. Just three months after the council, Eusebius of Nicomedia would also be exiled.

The polemical originality of the "faith of N." lies in these words: "from the substance of the *Father" and "true *God from true God, generated, not created, consubstantial with the Father". This way of defining Christ's divinity was particularly close to the position of *Alexander of Alexandria, but without reproducing *à la lettre* the formulations of his *theology. Neither Alexander, nor *Athanasius after him, seems to have used on his own account the word *homoousios*, which for the great majority of the Eastern bishops had become a source of doctrinal difficulty, even if they condemned *Arius. The political developments of the so-called "Arian" crisis brought into this episcopate an opposition to N., which lasted openly until the beginning of the reign of *Theodosius I, and until the celebration of the first council of *Constantinople (381).

Among the disciplinary decrees of N., the most important concerns the date of celebration of *Easter, according to the Roman custom and according to the Alexandrian custom, i.e. on the *Sunday immediately after the first full moon after the spring equinox. The task of the necessary astronomical calculations was entrusted to the see of *Alexandria. The canons of N., 20 in number, speak of the structures of the church (cann. 4-7, 15, 16); the *clergy (cann. 1-3, 9, 10, 17); public *penitence (can. 11-14); readmission of *schismatics and *heretics (can. 19); and finally give norms on *liturgy (cann. 18 and 20).

The emperor closed the council, whose presidency of honour he had personally assumed during the main sessions, with a banquet and an offer of *gifts to the Fathers; this, no doubt, as part of the festivities organized from 25 July 325 on the occasion of his *vicennalia* (he had become Caesar, aged 18, on 25 July 306).

In the course of the 4th c., the full ecclesial significance of this first ecumenical *council gradually became clear in the minds of those responsible for the church. The "faith of N." would remain the dogmatic rule invoked by all the other ecumenical councils of the ancient church.

Sources: CPG 4, 8511-8527; G.L. Dossetti, *Il simbolo di Nicea e di Costantinopoli*, crit. ed., Rome 1967.
Recent studies: I. Ortiz de Urbina, *Nicée et Constantinople, Hist. des Conciles oecum.*, 1, Paris 1963; M. Aubineau, Les 318 serviteurs d'Abraham et le nombre des Pères au concile de Nicée (325), *RHE* 61 (1966) 5-43; H. Chadwick, Les 318 Pères de Nicée: *ibid.*, 808-811; H.J. Sieben, Zur Entwicklung der Konzilsidee, I-II, *TheolPhilos* 45 (1970) 353-389; 46 (1971) 40-70; D.L. Holland, Die Synode von Antiochien (324/25) und ihr Bedeutung für Eusebius von Cäsarea und das Konzil von Nizäa, *ZKG* 81 (1970) 163-181; M. Simonetti, *La crisi ariana nel IV secolo*, Rome 1975; L. Abramowki, Die Synode von Antiochien 324/25 und ihr Symbol, *ZKG* 86 (1975) 356-366; C. Kannengiesser, Nicée 325 dans l'histoire du christianisme, *Concilium* 138 (1978) 39-47; B.M. Weischer, Die ursprüngliche nikänische Form des ersten Glaubenssymbol im ankyratus des Epiphanios von Salamis, *TheolPhilos* 53 (1978) 407-414; R. Lorenz, Das Problem der Nachsynode von Nizäa (327), *ZKG* 90 (1979) 22-40; J.N.D. Kelly, The Nicene Creed: A Turning Point, *SJT* 36 (1983) 23-39; C. Lubhéid, The Alleged Second Session of the Council of Nicaea, *JEH* 34 (1983) 165-174; A. de Halleux, La Réception du symbole Oecuménique de Nicée à Chalcédoine, *EThL* 61 (1985) 1-47; R.P.C. Hanson, *The Search for the Christian doctrine of God*, Edinburgh 1988, chs. 5 and 6; H. Chadwick, *History and Thought of the Early Church*, London 1982; R. Gregg (ed.), *Arianism*, Cambridge, Mass. 1985; F. Young, *From Nicaea to Chalcedon*, London 1983; O. Skarsaune, A neglected detail in the Creed of Nicaea (325), *VChr* 41 (1987) 34-54.

C. Kannengiesser

NICETAS of Remesiana. After having been confused, up to the end of the 19th c., with the homonymous bishop of *Aquileia and with *Nicetius of Trier, N. is now, after the studies of Morin and Burn, almost certainly identifiable with the N. mentioned by *Gennadius (*Vir. ill.* 22), with *Paulinus of Nola's guest (*Carm.* 17; 27; *Ep.* 29, 14), with the author of a book on the faith recommended by *Cassiodorus (*Inst.* 16), with the recipient of a *letter, in 414, from pope *Innocent I (*Ep.* 17), who also names him in another letter (*Ep.* 16) of *c*.409. So N. was bishop of Remesiana (now Bela Palanka in Serbia) in the second half of the 4th c. He went to *Italy at least twice, in 398 and 402, and was a guest of the bishop of Nola who, on his departure, dedicated his *Propempticon* to him: from this poem (17) we seem to deduce that N. exercised his *ministry even in areas beyond the borders of his episcopal territory, which was in *Dacia Mediterranea, politically joined to the *East by *Theodosius I in 379, but ecclesiastically dependent on the patriarchate of *Rome. If Paulinus's work helps us to reconstruct N.'s personality (besides *Carm.* 17 see *Carm.* 27, composed on the occasion of N.'s visit to Nola for the feast of St *Felix in 402), Gennadius's constitutes our source of knowledge of his writings. These were: *Instructio ad competentes*, whose six books survive fragmentarily, dedicated to *catechumens and written with pastoral rather than theoretical aims. Fragments remain of books I and II, respectively dealing with the conduct of candidates for *baptism and the errors of the *pagans. Two surviving treatises which affirm, against the *Arians, the divinity and consubstantiality of the Son and, against the *Macedonians, the divinity of the *Holy Spirit, must be part of book III. But book IV, which was concerned with the practice of *astrology, is completely lost. Book V, which survives entire, is an *Explanatio symboli* having considerable affinities with *Cyril of Jerusalem's *catecheses: in it the formula "communion of saints" is proposed, for the first time in the West. It is still uncertain whether book VI, which dealt with the Paschal victim, should be identified with the *De ratione Paschae* variously attrib. to *Athanasius (PG 28, 1605) and *Martin of Braga (PL 72, 49). Gennadius also mentions the *Libellum ad lapsam virginem*, recently identified by Gamber with the short recension of the pseudo-Ambrosian *De lapsu virginis* (PL 16, 383-400). Among *Jerome's apocryphal letters, but attrib. to N., is *De vigiliis servorum Dei* (PL 30, 240-246), while under the name of Nicetius of Trier is *De psalmodiae bono* (PL 68, 371-376), a *sermon which champions the singing of *hymns and *psalmody and which attributes the *Magnificat* (Lk 1, 46) to Elizabeth, in accordance with some Latin MSS. Finally *De diversis apellationibus* (PL 52, 863-866), on the various christological titles, may be a juvenile work. Attribution to N. of the *Te Deum* is uncertain and disputed.

PL 52, 837-876; PLS 3, 189-202; CPL 646-652; A.E. Burn, *Niceta of Remesiana, His Life and Works*, Cambridge 1905; K. Gamber, *Textus patristici et liturgici*, 1, 2, 5, 7: Regensburg 1964, 1965, 1966, 1969; M.G. Mara, Niceta di Remesiana, in *Patrologia* III, 180-183; G.G. Walsh, *FOTS* 7 (1949) 1-76; H.C. Brennecke, Nicetas, *TRE* (forthcoming).

M.G. Mara

NICETIUS of Trier. The main source for his life is the *biography by *Gregory of Tours (*De vita* pp. 17; *Gl. conf.* 92). To this we may add two *letters by him and five to him (MGH, *Epp.* 3, 116-122, 126-127, 133-134, 137-138) and two poems by *Venantius Fortunatus (MGH, *Auct. ant.* 4/1, 63-65). Native of the Limousin, where he lived as a monk, he was appointed bishop of *Trier (*Treverus*) in 525-6 by Theodoric I of Austrasia. He exercised great influence over his contemporaries (he led Aredius to become a monk) and over kings. He brought Italian workmen to Trier to restore the cathedral of St John, the church of St Peter of Neumagen and the abbey of St Maximinus. None knew better than this Roman how to revive the old capital of the Western empire, grafting onto it the *Romanitas* of his native Aquitaine. We may call the 6th c. at Trier "the century of N.". Died perhaps 569 and was buried in the basilica of St Maximinus. His local cult seems to have been immediate (Greg. of Tours, *Gl. conf.* 92), but the city's liturgical books preserve traces of it only from the 10th c. (*AB* 69, 285). Feast 5 Dec (Ado of Vienne), 5 Dec or 3 Oct (13th-c. *calendars of Trier) or 1 Oct (addition to the *Corbeiensis* [11th c.] of the *Mart. hier.*).

LTK 7, 941; *Vies des SS.* 12, 182-189; M. Rouche, *L'Aquitaine des Wisigoths aux Arabes*, Paris 1979, 427-428; N. Gauthier, *L'évangélisation des pays de la Moselle*, Paris 1980, 172-204.

V. Saxer

NICHOLAS of Ancyra. Presbyter of *Ancyra. We know nothing certain of him or when he lived: perhaps the 5th or 6th c. Useful clues can be drawn from the dates of the MSS of the *catenae that preserve fragments of works attrib. to him and which call him N., N. the presbyter, N. presbyter of Ancyra, or N. monk and presbyter.

He may be the same as the *ascetic of Ancyra whom the hermit *Mark addressed with his work Πρὸς Νικόλαον (CPG 3, 6095 - PG 65, 1027-1050) and who replied with a *letter of thanks (PG 65, 1051-1054). From some references present in scholia of catenae on the Acts of the Apostles, it seems that he was prob. the author of *commentaries on biblical books: *In prophetam Ioel* and *In prophetam Amos*. Perhaps we may also attribute to him a commentary *In Psalmos* and one *In Ionam* of which we possess two fragments with interpretations of literal type.

CPG 3, 6104; J.A. Cramer, *Catenae graecorum patrum in N.T.*, t. 3, Oxford 1838, 35, 37, 38, 124; G. Karo-I. Lietzmann, *Catenarum graecarum catalogus*, Göttingen 1902, 35; K. Staab, *Die Pauluskatenen nach den handschriftlichen Quellen untersucht*, Rome 1926, 16-17; Y.-M. Duval, *Le livre de Jonas dans la littérature chrétienne grecque et latine*, Paris 1973, 454-456, 663-665; Bardenhewer 4, 181; R. Devreesse, Chaînes exégétiques grecques, *DBS* 1, 1146-1147, 1209; W. Ensslin, Nikolaos, *PWK* 17, 361.

A. Labate

NICHOLAS of Myra. Among the signatories of the council of *Nicaea (325), as bishop of Myra in *Lycia (*Asia Minor). Died 6 Dec 343 and was buried outside the city, towards the sea, where his remains stayed until transferred to Bari on 9 May 1087. The mausoleum has the aspect of a 4th-c. chapel, and its marble *sarcophagus, decorated with pastoral scenes, was under an arcosolium. It seems to have been inserted into a basilica of *ad corpus* type, square plan with inscribed cross, surmounted by vaults and domes (6th c.), last restored in the second quarter of the 11th c. His cult had enormous and universal popularity, to which his translation to Bari gave a new lease of life. But his legend (BHG 1347-64; BHO 10; BHL 6104-21, Suppl. 235-237) has no historical value. Feasts 6 Dec and 9 May.

BS 9, 923-948; *Vies des SS.* 12, 200-213; G. Anrich, *Hagios Nikolaos*, 2 vols., TU 2, Leipzig 1913-1917; F. Darsy, Il sepolcro di S. Nicola di Mira, in *Mél. Tisserant*, ST 232, Vatican City 1964, 29-40; E. Martin, *Saint Nicholas. Life and Legend*, New York 1975; Ch. W. Jones, *S. Nicola. Biografia di una leggenda*, Bari 1983.

V. Saxer

NICOLAÏTANS. The data of Rev 2, 6. 14. 16. 20 do not allow us to identify the sect, or to fix its origins even approximately. It was prob. a doctrinal and ethical movement. If the N. are to be recognized in the "false apostles" of Rev 2, 2, they would correspond to those itinerant preachers who passed for *prophets and *apostles without being so (cf. 1 Th 5, 20-21; 1 Jn 4, 1), who have left traces even at the start of the 2nd c. (Ign., *Eph*. 9, 1; *Did.* 11, 8-10). We can infer their doctrine by examining their conduct. The N. had no scruples about taking part in *pagan sacred banquets (Rev 2, 14-15), which were often accompanied by immoral practices. Archetypes of such conduct are *Balaam (Num 31, 16 ff.) who, acc. to Rabbinic exegesis, advised king Balak of Moab to offer Moabite women to the Israelites in order to make them deviate from monotheism and turn to *idolatry, making them eat the meat of unclean animals; and Jezebel, certainly a symbolic name corresponding to a real person, which is linked with the homonymous Phoenician queen of Israel who led her husband Ahab, followed by a large part of the people, into idolatrous cults (1 Kgs 16, 31; 2 Kgs 9, 22). This woman, passing herself off as a prophetess, gathered around herself a large group of followers whom she led astray with prophecies and seductions (Rev 2, 20-23). The indications of the *Apocalypse would suggest that we have here a species of pre-*gnostics who, in the name of a "higher wisdom" (Rev 2, 24), professed and practised a certain theoretical and practical laxism, pushing *Paul's teachings on Christian freedom (1 Cor 8; Rom 14) to their extreme consequences. Such consequences would be the participation in pagan sacred banquets, since "an *idol is nothing", and a certain complacency about the cult of the *emperor since, in the act of worship, not the emperor's person but his protective genius was adored. The symbolic language of the Apocalypse (*idolothytes - *fornication) seems to conceal some such doctrine. Apart from the identity of name, there is no proof that the proselyte of *Antioch, the deacon Nicolaus (Acts 6, 5), was the movement's founder. Patristic *tradition is divided: *Irenaeus (*Adv. haer.* I, 26, 3; III, 11, 1), *Hippolytus (*Refut.* VII, 36) and *Epiphanius (*Pan.* 25) repeat and amplify the data of the Apocalypse. Their credibility is slight as regards the deacon Nicolaus. *Clement of Alexandria (*Strom.* II, 20; III, 4), followed by *Eusebius (*HE* III, 29) and *Theodoret (*Haer. fab.* III, 1), reject the identification with the deacon of Acts 6, 5 and explain that perverse men corrupted his saying "you must disregard the *flesh" from its authentic meaning to make it a principle of libertinage. In the Middle Ages all who opposed ecclesiastical *celibacy were classed as N.

DB 4, 1616-1619; *DTC* 11, 499-506; *EC* 8, 1859; *LTK* 7, 976; M. Goguel, Les Nicolaïtes, *THR* 115 (1937) 5-36; *EB* 5, 129-131.

E. Peretto

NICOMEDIA

I. City and Christianity - II. The 1003 martyrs of N.

I. City and Christianity. N., civil and religious capital of *Bithynia, is now Izmit in Turkey, at the end of the homonymous gulf on the Sea of Marmora (ancient Propontis). Founded in 264 BC, its name derives from Nicomedes I of Bithynia. With *Rome, *Alexandria, *Antioch and *Athens, it was called one of the greatest cities of the Graeco-Roman world, and was seat of the prefect (perhaps from there *Pliny the Younger wrote his *letter to *Trajan on the Christians) and residence of *Diocletian, who in 303 unleashed a tremendous *persecution whose main victim was the Christian community of N. which, already flourishing at the time of *Dionysius of Corinth (c.175), by that time made up half the population. In this way the martyrology of N. is one of the richest, with individual commemorations (above all of St *Anthimus and *Lucian of Antioch) and collective *memoriae*: 18 March (10,000), 22 June, 23 Dec (20), 25 Dec, etc.

By his solemn entry into N. in 324, *Constantine marked his final victory over his rivals and the establishment of religious *peace. Even after the foundation of *Constantinople (330), N. remained the preferred residence of the monarch, who, in the suburban villa of Ancyrona, received *baptism in *articulo mortis* (May 337) from bishop *Eusebius, who was linked to the cause of *Arius, as were Eudoxius and Demophilus, his successors at Constantinople. In 358 *Constantius set up a new diocese, with N. as capital, called *Pietas* in honour of his wife Eusebia. Weakened, in favour of its rival *Nicaea, by the pro-*Nestorian option of its bishop Himerius at the council of *Ephesus (in Sept 431 *Theodosius II called a synod at N. to attain agreement between *Cyril of Alexandria and *John of Antioch), N.'s precedence in Bithynia was restored at the 4th ecumenical *council (451) within the limits allowed by *Chalcedon's can. 28 on the primacy of *Constantinople. On the metropolis cf. *Bithynia.

Invaded by *Goths, shaken by earthquakes (24 Aug 358, Oct 359, Nov-Dec 362, 368, 15 Aug 554, 26 Oct 740 - this last date liturgically commemorated), the city was more than once rebuilt, in particular the renowned *martyrion* of St Anthimus, which *Justinian made all of marble and gold, the ideal scenario to welcome pope Constantine I in 711, under Justinian II.

The passage of the Persians and other invaders has cancelled out the ancient civic splendour still attested by the city walls.

PWK 17, 468-492; *CE* 9, 70-71; *EC* 8, 1864; *EAA* 5, 455-457; V. Laurent, *Le corpus des sceaux de l'empire byzantin*, V, 1, Paris 1963, 268-281; Beck 164-165; R. Janin, *Les Eglises et les monastère des grands centres byzantins*, Paris 1975; J. Darrouzès, *Notitiae episcopatuum Ecclesiae byzantinae*, Paris 1981, 208, nn. 171-179; 221, nn. 191-201.

D. Stiernon

II. The 1003 martyrs of N. The epic Passion BHG 1219 tells of four imperial bodyguards who, in the first days of *Diocletian's *persecution, presented themselves before the emperor with their *families, *slaves and clients to proclaim themselves Christians. They were massacred by soldiers on 13 Mechir (= 7 Feb). This account has no historical value; the particular date of the *martyrdom suggests an Egyptian origin for the Passion, which omits nothing of the craziest eccentricities of the *hagiography of that country.

Their cult presents as many problems. The Synaxaria put their feast, not on the same date as the Passion, but on 12 Feb; and the most trustworthy ancient sources (Euseb., *HE* VIII, 5-6; *Mart. hier.*; *Mart. syr.*) are completely silent about them. The cult has the air of having arisen from the Passion and having the same grounds.

BS 9, 982-985.

V. Saxer

NILUS of Ancyra (†430?). Ascetic writer, who presents himself as a disciple of *John Chrysostom. Since he does not mention the christological disputes, he must be placed in the early 5th c.; the signs confirm this hypothesis. More than 20 people named Nilus are listed by J.A. Fabricius - G.C. Harles (*Bibliotheca Graeca*, X, Hamburg 1807, 1-40: cf. Index, Leipzig 1838, 70), and the works handed down under that name (CPG 6043-6084; PG 79) pose thorny problems of authenticity. N.'s links with the emp. *Theodosius II and with *Sinai are supposed only by the *Narrationes de caede monachorum in monte Sinai*, a legendary account which it seems we must attribute to a different author. There are no old Latin translations and few Eastern versions.

K. Heussi began the study of the *corpus* of *letters; the order of the MSS, badly dealt with by the editors, eliminates a number of false problems. Even if, as A. Cameron has shown, we must fear touching-up in the recipients and even if there are *interpolations of some passages - like e.g. the emperor's intervention in Chrysostom's favour - these letters still remain valuable evidence on a monk of *Ancyra. Linked with the letters is the *Logos askètikos* (= *De monastica exercitatione*), a vigorous call for monks to return to *poverty, which is the foundation of their "philosophy", with wide use of the OT interpreted allegorically (in a way not at all Origenian). N. uses *Philo, the *Vita Antonii*, *Basil and the *Apophthegmata*. A *Syriac version of this work exists. The same themes are found in the work addressed to Magna, a *deaconess of Ancyra (early 5th c.), *On voluntary poverty*, which explicitly alludes to the previous work; but it also combats the contrary *vice, refusal to work (chs. 21-27) and names in this context *Adelphios*, one of the great *Messalians of Mesopotamia, and *Alexander the Acoemete "recently arrived at Constantinople" (PG 79, 997A), an episode which takes us up to 427.

De monachorum praestantia, a defence of life far from cities, lived in recollection, has points of contact with the previous treatises, though the critique of *monasticism is less apparent on account of new preoccupations. Perhaps we should add the *Panegyric of Albianus*, since this monk, who had been at the school of Leontius of Ancyra (who became bishop *c.*400) and then a *pilgrim at *Jerusalem and hermit in the *desert of Nitria, is close to our author; but the *literary genre is completely different.

The *commentaries on Ecclesiastes and *Song of Songs are known only through the *catenae, of which there is still no systematic edition. H. Ringshausen, who is preparing the text of that on the Song of Songs, thinks that its author is the same as that of the *Peristeria*, an Alexandrian work whose author is generally considered ill-known. Ringshausen identifies him with the author of the Sinaitic *Narrationes*; he also believes, with great likelihood, that this N. is the author of the treatise *On voluntary poverty*, an argument that seems to me to lead to an identification with the monk studied above. Ringshausen shows a great (exaggerated!) mistrust of the letters, and makes no mention of the *Logos askètikos*. It is hard to judge the reasons that he adduces from a still unpublished commentary, but he may be right, at least in part. This would leave us with two authors both called N.

Under the same name there also exist collections of sentences (there is a *Syriac version): they are *antitheses (on the theme corruptible/incorruptible) without great depth.

Numerous works of *Evagrius have been put under N.'s name, doubtless at the time of the council of *Constantinople of 553 and the condemnations of *Origenism: *De oratione, De octo spiritibus malitiae, Ad Eulogium*. His *De octo vitiosis cogitationibus* is a compilation from the Greek *Cassian. The *Enchiridion* adapts Epictetus. Research into the MSS and florilegia would be necessary to disentangle this history.

CPG 6043-6084; P. Bettiolo, *Gli scritti siriaci di Nilo il Solitario*, intr., ed. and tr., Louvain-La-Neuve 1983; *DIP* 6, 296-298; *DSp* 11, 345-354; H. Ringshausen, *Zur Verfasserschaft und Chronologie der dem Nilus Ancyranus zugeschriebene Werke*, Frankfurt a.M. 1967; J. Gribomont, La tradition manuscrite de s. Nil, *StudMon* 11 (1969) 231-267; A. Cameron, The Authenticity of the Letters of St. Nilus, *GRBS* 17 (1976) 181-196; P. Bettiolo, Le Sententiae di Nilo, *Cristianesimo nella Storia* 1 (1980) 155-184; V. Messana, Praxis et Theoria chez Nil d'Ancyre, *SP* 18.2 (1989) 235-242.

J. Gribomont

NIMBUS: *iconography. Luminous disc set around the head of deities, divinized persons or allegorical figures. Particularly widespread in the Graeco-Roman and Oriental world as an attribute of gods, emperors and symbolic personifications (J. Strzygowski, *Calenderbilder*, pls. IX, XIII, XIV), it was taken up by Christian art, marking a continuity in the translation of an idea into an *image, in symbolic and religious evocation. On the vault of the mausoleum of the Julii in the *Vatican necropolis, Christ-Helios is depicted with a radiate n. (O. Perler, *Die Mosaiken der Juliergruft*). In catacomb frescos it was initially attributed to ornamental and symbolic figures, to the sun (*Wp* 2, 1-2), in accordance with classical tradition. The seasons in the catacomb of SS. Marcellino e Pietro (*Wp* 100), the *phoenix in the Greek chapel at Priscilla (P. Testini, *Catacombe e ant. cimiteri*, fig. 115), the figure of *Tellus*, also held to be Cleopatra, in the hypogeum of Via Latina (A. Ferrua, *Via Latina*, pl. 102) all possess a n. From the mid 4th c. we find it attributed to Christ, both his real figure and the related *symbol of the *Agnus Dei* (*Wp* 252), and to the dove. But it was applied without regularity: its absence does not have the same value for chronology as its presence.

In the frescos of the catacombs of Domitilla (*Wp* 115, 2; 225, 1; 181, 1), Callisto (*Wp* 243, 1), Marco e Marcellino (*Wp* 153, 1), Ciriaca (*Wp* 205; 206, 1), Christ alone among the *apostles and saints has a n. Only later did *Peter and *Paul themselves (Commodilla, *Traditio Clavum, WMM*, 148-149; U. Fasola, *La cat. di S. Gennaro*, pl. VI, fig. 71), and the numerous figures of saints that crowd the catacombs, obtain the n. (*Wp* 255, 1-2; 256; 259, 1; 260; 261; 264, 2; M. Marinone, *Cat. di Albano*, fig. 16). That of Christ, from the 5th c. on, often has a cross and is distinguished by the *monogram, with or without *alpha and omega (e.g. Marcellino e Pietro, *Wp* 252; sarcophagi at *Arles, *Tarragona, *Ravenna). More rare, but still present, on *sarcophagi, where it must often have been painted. The form varies little and the colour on the frescos goes mostly from brown to dark blue, to dark green, to yellow. Very common from the second half of the 4th c. on *mosaics, which allow better preservation of colours and on which the translucent, gilded and jewelled n. rapidly became popular. At S. Vitale, Ravenna, the classical tradition of the n. as an honorific sign is still present: *Justinian and *Theodora, among their courts, are nimbate. From the 6th c. on, the n. was constantly extended to *Mary, the *angels, the saints, the symbols of the *evangelists.

The *tabula circa verticem*, a rectangular or square n. or "pseudo-nimbus", considered by Paul the Deacon, in his Life of *Gregory the Great (PL 75, 231), as a *signum viventis*, surrounds the head of living persons in frescos, mosaics, miniatures. Certainly the expression of a particular dignity or, acc. to some, of a particular role (e.g. founder), intended to give the impression of a genuine portrait, it took on considerable importance in mediaeval portraiture. [**Fig: 234**]

DACL 12, 1272-1312; *EC* 8, 1884-1887; *EAA* 5, 493-497; *LCI* 3, 324-332; G. Ladner, The so called square nimbus, *Medieval Studies* 3 (1941) 15-45; M. Collinet Guérin, *Histoire du nimbe des origines aux temps modernes*, Paris 1961 (with bibl.); J. Strzygowski, *Die Calenderbilder der Chronographen vom Jahre 354*, JDAI Supp. 1, Berlin 1888; O Perler, *Die Mosaiken der Juliergruft in Vatikan*, Freiburg/Sviz. 1953; P. Testini, *Le catacombe e gli antichi cimiteri cristiani in Roma*, Bologna 1966; M. Marinone, La decorazione pittorica della catacomba di Albano, *RIA* 19/20 (1972-1973) 103-138; U.M. Fasola, *La catacomba di S. Gennaro a Capodimonte*, Rome 1975.

A.M. Giuntella

NIMES, *Council of. In 394 or 396, some 20 bishops, all followers of *Felix of Trier, met at N. (*Nemausus*) in *Gaul, to compose the doctrinal conflicts linked to *Priscillianism, and the disputes between the bishops of Aix and *Arles and those of *Marseilles and *Vienne. The absence of the opposition impeded the solution of these problems, but to make up for this the council issued many disciplinary canons, particularly against *ordination of *women as deacons (can. 2) and against vagrant clerics (can. 1).

CCL 148, 49-51; Gaudemet 124-131; Hfl-Lecl 2, 79-81; Palazzini 3, 198-199.

Ch. Munier

NINIAN. Our earliest information on this Scottish saint comes from *Bede (*HE* III, 4), from two late 8th-c. poems (*Miracula* and *Hymnus Nynie ep.*) and from a late Life by Aelred of Rievaulx (BHL 6239-40). A 4th/5th-c. missionary, he evangelized the Picts and founded a church which later took the name of St Martin, at Candida Casa, now identified with Whithorn or Withern in Galloway. Feast (in Scotland) 16 Sept.

BS 9, 1012-14; C. Thomas, *The Early Christian Archaeology of North Britain*, Oxford 1971.

V. Saxer

NINO. Georgian saint. *Rufinus (*HE* 10, 11) tells the story of an anonymous slave-girl of the king of *Georgia (*Armenia), who converted her masters to Christianity. In this form the account appears in the Byzantine Synaxaria (*Syn. eccl. CP.* 27. 10). In the 8th c., Ps.-*Moses of Khorene gave the anon. girl the name Nounè, and a *biography of the saint was composed on this basis. Feast 14 Jan in the Georgian Synaxaria, 29 Oct in the Armenian. In the West, Baronius inserted a mention of the "Christian prisoner", "the saintly Christian handmaid" who converted Georgia, into the *Mart. rom.* on 15 Dec.

P. Peeters, Les débuts du christianisme en Géorgie d'après les sources hagiographiques, *AB* 50 (1932) 5-58; *BS* 8, 1018-1021.

V. Saxer

NISIBIS

I. City and Christianity - II. Archaeology.

I. City and Christianity. N. (now Nusaybin in Turkey), transit-point of the caravans between *East and West, was also in the path of the many invaders who took possession of it at various times. Populated with Greeks by Seleucus I, at the start of the Christian era it was bitterly disputed between Persians and Romans. The latter ruled it from 297, when it enters history as a Christian city, though it was evangelized much earlier, since the presence of Mari, disciple of *Addai - apostle of *Edessa - is attested further S, at Ctesiphon, between 79 and 116. But the earliest historical evidence of Christianity at N. is the epitaph of Abercius of Hierapolis (cf. *Epigraphy). The *memoriae* of the *martyrs of N. in the old *Breviarium Syriacum* testify that N.'s *evangelization took place at an early date.

The city's first bishop was *Jacob of N. (Mar Yaqub), celebrated by *Ephrem in his *Carmina Nisibena*. His episcopate lasted some 30 years (308/9-c.338); he took part in the council of *Nicaea (325). Jacob also founded the School of N., appointing Ephrem as "Interpreter", i.e. commentator on the *Scriptures. But the School of N. only began properly with *Narsai (399-502), after the closure of the School of the Persians at *Edessa in 489. While Jacob was alive, N. knew a period of peace, but shortly before his death in 338, and again in 346 and 350, the Persians returned to besiege the city. After the death of Jacob's successor Vologeses II, the emp. *Jovian (363) ceded N. to the Persians. After the exodus of the Hellenistic population, Shapur II transplanted 12,000 Persians to the city to replace the inhabitants who had left with the Roman troops. And so, after having been Greek for six centuries, N. became Persian: this change would have a considerable influence on the orientation of theological crises in the following centuries. With N., the five provinces beyond the Tigris passed under Persian rule, becoming the five dioceses dependent on N. Among the East Syrian metropolises, N. held second place after Elam, i.e. Beth Lapat-Gundi-Shapur, summer residence of the *Sassanid royal family.

The most famous metropolitan of N., in the course of the 5th c., was *Barsauma, who ensured the triumph of *Nestorianism in *Persia. Decisive for this was his having attracted to the city the great teacher Narsai, who had to flee Edessa in 457 or 471. Pressed by Barsauma, he came to N. and founded the School of N., nursery of bishops and missionaries, intellectual and spiritual centre of Nestorianism. Barsauma also founded a hospital, in which were trained the doctors of N., who would have great influence at the court of the king of Persia and whose tradition would be perpetuated at the court of the Caliphs of Baghdad (9th c.).

At the start of the 6th c. the *monophysites appeared at N., though the city remained the stronghold of Nestorianism, whose influence was felt as far as Roman *Syria. Meanwhile, contacts and exchanges of ideas with the *Byzantines were not interrupted, and metropolitan *Paul of N. was one of the theologians sent by *Chosroes I to *Justinian for a theological contest. In *c.*567 theological controversies broke out in the School of N. itself, with *Henana of Adiabene, who was expelled from the School for having taught things opposed to the Interpreter (*Theodore of Mopsuestia, †428).

From 565, with the emp. *Justin II, a new period began in Persian-Byzantine relations. The Byzantines demanded the restoration of N., and hostilities resumed in 572; but the Persians remained rulers of the city. At the death of metropolitan Paul in 573, Henana returned as director of the School and resumed teaching his errors. The new metropolitan Gregory was powerless against the rebel: in 601 Chosroes, circumvented by the doctors and the patriarch, sent Gregory into exile where he died in 611/12.

The struggle between Persians and Byzantines resumed in 602. N. rebelled against Chosroes, and the king took it by a stratagem and sacked it. Nor was the church at *peace, disturbed as it was by doctrinal controversies. Metropolitan Cyriac of N. came to an agreement with numerous other metropolitans and appointed *Babai the Great, supervisor of the convents, to fight the *Messalians and the followers of Henana and *Joseph Hazzaya. For more than ten years, until his death soon after 628, Babai worked to consolidate the Nestorian faith. Meanwhile the struggle between Persians and Byzantines continued, coming to a head under *Heraclius, who expelled the Persians and advanced as far as the royal residential cities, which he captured in 629. N. at this time was in Byzantine hands. In the same period (631) is mentioned the presence in the city of a West Syrian bishop named Abraham: we do not know if he had any successors. During the Byzantine occupation N. suffered a terrible famine. In 639 Heraclius had to abandon it, and the city surrendered to the Muslim Arabs, with whom a new era began. In this period the Nestorian church was mainly occupied with the struggle against the monophysites, who sought to get a foothold in the city: in 707 this design was brought to an end. In the same year a monk of Qartamin, Simon of Zayté, built the great church of the *orthodox and a hospice for strangers. But the uninterrupted series of West Syrian bishops of N. began only in the 9th c. The city's School, despite Henana's doctrinal innovations, which helped discredit it, continued to exist until 832, when it was replaced by the School of Baghdad.

DTC 11, 157-323; *CE* 10, 473-474; Labourt, *passim*; Ortiz de Urbina, 15-17; 120-122; A Vööbus, *History of the School of Nisibis*, CSCO 266, Subs. 26, Louvain 1965; J.-M. Fiey, *Jalons pour une histoire de l'Eglise en Iraq*, CSCO 310, Subs. 36, Louvain 1970; id., *Nisibe, métropole syriaque orientale et ses suffragants, des origines à nos jours*, CSCO 388, Subs. 54, Louvain 1977.

R. Lavenant

II. *Archaeology. The church of Mar Yaqub, built over the crypt, is thought to have housed the saint. Built in 359 and transformed in the 8th and 9th cc., it is a square-plan building with apse to E and atrium in front. **[Figs: 235, 236]**

J. Strzygowski, *L'ancien art chrétien de Syrie*, Paris 1936, 9 and 147; U. Monneret de Villard, *Le chiese della Mesopotamia*, Rome 1940; M.F. Castelfranchi, Le Sepolture di Vescovi e Monaci in Mesopotamia (IV-VIII secolo), *Actes XI*[e] *Cong. arch. chrét.*, CEFR 123, Rome 1989, 1267-1279.

B. Bagatti

NOAH

I. In the Fathers - II. Noah's ark: iconography.

I. In the Fathers. Mentioned in Babylonian texts, N. is the name of the patriarch who lived at the time of the *flood (Gen 5, 29). In Hebrew it means "consoled". In biblical tradition, N. is a cultivator of the earth, inventor of viticulture, who condemns Ham to serve Shem (Gen 9, 18-27), but above all he is the central figure of the flood (Gen 6, 5-9; 6, 17), the "saviour" of mankind with whom God concludes the new alliance after the fall of *Adam. The figure of N. also appears elsewhere in the OT (Ez 14, 12-23; Is 54, 9; Sir 44, 17) and NT (Mt 24, 37; Heb 11, 7; 1 Pet 3, 20; 2 Pet 2, 5). Rabbinic tradition attributed to him the function of "Paraclete", intercessor for righteous Israelites. Continuing this tradition, the church *Fathers developed a typology around the figure of N.: he is the antitype of Adam, since he renews mankind corrupted by *sin (Athan., *Oratio c. arian.* 2, 51: PG 26, 256A) and has saved, in himself, *God's *image and rule over the lower creatures (Basil Sel., *Or.* 5, 2: PG 85, 84B; Cosmas Ind., *Topogr. Christ.* 5: PG 88, 236C; John Chrys., *Hom. in Gen.* 25, 5: PG 53, 224-25); the vine produced by N., naked and drunk, may be compared to the fruit of the tree of knowledge of good and *evil, since the vine too can become an occasion of sin; but N. sinned through ignorance (Orig., *Hom. in Gen.* 9, 20: PG 12, 109C). The just N. prefigures Christ, the eight persons saved with him in the ark represent the "*ogdoad" accomplished by Christ who saved us through the wood of the *cross, as N. saved the just from catastrophe by the wood of the ark (Just., *Dial.* 138, 1-2). The ark became a *symbol of the church in the controversy between *Hippolytus and pope *Callistus, which generated the following question: does the ark-church symbolize a Christian community containing sinners and just in equal measure, or must it be considered only a means of salvation for the pure? N. and the dove bearing an olive-branch in its beak prefigures Christ in the Jordan, when the *Holy Spirit rose above him in the form of a dove (Cyril of Jer., *Cat.* 17, 7). For *Asterius the Sophist, the patriarch N. in the ark during the flood represents Christ in the tomb (*Hom. in Ps.* 6: PG 40, 448C). The name N., acc. to *Origen, means ἀνάπαυσις, δικαιοσύνη (*Hom. in Gen.* 2, 3: PG 12, 169 AB); for the Fathers, the figure of N. is an example of morality and *faith (Greg. Naz., *Or.* 43, 70: PG 36, 592B) and is a preacher of *penitence (*1 Clem.* 7, 6; Theoph. Ant., *Ad Autol.* 3, 19).

RGG[3] 4, 1501 f.; *LTK* 7, 1016-17; *DB* 4, 1661-1667; *BS* 9, 1027-1028; J. Daniélou, *Sacramentum Futuri*, Paris 1950, 60 ff.; *Evangelisches Kirchenlexikon* 2, 1613-15; *DB* 4, 1661-1667; J.P. Lewis, *A study of the interpretation of Noah and the Flood in the Jewish and Christian Literature*, Leiden 1968.

L. Vanyó

II. Noah's ark: *iconography. N.'s ark is linked to a very important symbolism (*Wp* 316): it is in fact the *allegory of the church, which saves the just and outside which none can find salvation. St *Peter relates it to *baptism. Acc. to the biblical source (Gen 6, 3-22), God gave N. precise instructions for building the ark, which was a three-floored rectangular structure: a sort of enormous box. And this is how it appears in early Christian representations. Probably this iconography derives from the classical repertoire, in particular the mythical episode of Deucalion and Pyrrha, cast into the sea in a chest. The oldest representation of the ark is held to be a *painting in the hypogeum of the Flavii in Domitilla, dating from the first 20 years of the 3rd c. Nearly 50 examples are preserved in the Roman catacombs. The scene is also common in plastic art. The oldest examples are the so-called *Jonah sarcophagus (*Ws* 9, 3=*Rep.* 35) and the sarc. ex-Lateranense 236 (*Ws* 57=*Rep.* 46), both late 3rd-c. In the latter the dead man, as an *orans, appears in the ark in N.'s place, since in sepulchral symbology the figure of N. has been assimilated with the concept of salvation of the *soul, just as the dove, when it appears, normally alludes to the Lord's saving intervention. A painted arcosolium, mid 4th c., in the *cemetery of Panfilo at *Rome shows a variant: the ark closed (it is normally open), without N., and the dove to the left in flight, while on the right it returns with the olive-branch. In the 5th c., at El-Bagawat in *Egypt, the ark (uniquely) is shown as a gondola. In *mosaics, a damaged depiction of N.'s ark remains in the dome of the mausoleum of Centcelles in *Spain (mid 4th c.). In the pavement mosaic of the basilica of *Mopsuestia (first half of 5th c.) in *Cilicia, the ark is shown with feet, and the animals are arranged in a circle around it: it bears the caption NΩEP. In *Palestine, finally, is known a mosaic pavement, *c.*530, in the synagogue discovered under the church of Sts Cosmas and Damian at Gerasa, where N. is leaving the ark after the flood.

In plastic art, N. turning towards the olive-bearing dove and the raven is attested in a fragment of baptismal font (4th c.) in the Bardo Museum, coming from Sufetula (Sbeitla). In a late 4th-c. sarc. in the Velletri Museum (*Rep.* 4, 3), the ark appears with feet. N. is always alone in the ark, except in the Trier sarc., of disputed date (4th-5th c.), where he appears with all his family.

The same variant is found in the Tours Pentateuch (early 7th c.) which also shows the ark during the flood. Still in miniature, the Cotton Codex and the Vienna Genesis (era of *Justinian) show the ark with several floors. In the sumptuary arts, a series of scenes relating to N. is attested in gold-glass and on clay *lamps. A glass of Constantinian date at Cologne shows N. in the ark, together with other OT episodes. On the reverse of some coins from *Apamea, from the reigns of *Septimius Severus, Macrinus and *Philip, the ark appears with two male figures. On it is the word NΩE. **[Fig: 237]**

DACL 12, 1397-1400; *EC* 8, 1909-1910; *LCI* 3, 611-620; R.P.J. Hooyman, Die Noe Darstellungen in der frühchristliche Kunst, *VChr* 12 (1958) 113-135; J. Fink, Bemerkungen zur frühchristlichen Noe-ikonographie, *RAC* 49 (1973) 171; R. Pillinger, Noe zwischen zwei Tauben, *RAC* 54 (1978) 97-102; R. Garrucci, *Storia dell'arte cristiana*, III, Prato 1873-80, pll. 169-202; W. von Hartel-S. Wichkoff, *Der Wiener Genesis*, Vienna 1895, pl. 4; W. DeBock, *Matériaux pour servir l'archéologie chrétienne de l'Egypte*, St Petersburg 1901, 22, pl. 12; M.A. Merlin, *Catalogue des Musées et Collections Archéologiques de l'Algérie et de la Tunisie. Musée Aloui*, III, Paris 1921, *c.*1444, p. 73; E. Josi, Il cimitero di Panfilo, *RAC* 3 (1926) 191, fig. 76; F.M. Biebel, *Mosaics in Gerasa City of Decapolis*, New Haven 1938, 297-351; J. Daniélou, *Les figures du Christ dans l'Ancien Testament*, Paris 1950, 55; A. Grabar, *L'età d'oro di Giustiniano*, Milan 1966, 202; H. Laag, *Der Trierer Noe-Sarkophage*: Festschrift A. Thomas, Trier 1967, 233-238; J. Porcher, in J. Porcher-J. Hubert-W.F. Volbach, *L'Europa delle invasioni barbariche*, Milan 1968, 128; L. Budde, *Antike Mosaiken in Kilikien*, Recklinghausen 1969, pls. 26-44; L. Pani Ermini, L'ipogeo detto dei Flavi a Domitilla, osservazioni sulla sua origine e sul carattere della decorazione, *RAC* 45 (1969) 138, fig. 9, p. 173; G. Gutmann, Noah's Raven in Early Christian and Byzantine Art, *CArch* 26 (1977) 63-75; H. Schlunk-T. Hauschild, *Die Denkmäler der frühchristlichen und westgotischen Zeit*, Mainz a.R. 1978, 25-27, pls. 12-19; J.A. Maritz, Noah's Ark and the Animals in Early Christian Art, *Akroterion* 28 (1983), 102-108; J. Fink, *Noe der Gerechte in der frühchristliche Kunst*, Archiv für Kulturgeschichte, Heft 4, Münster/Cologne 1955.

A.M. Di Nino

NOETUS of Smyrna. The first to spread the *Patripassian doctrine, at the end of the 2nd c., for which he was condemned by the presbyters of Smyrna. *Hippolytus (*Noët.* 1) attests that N. claimed that he was *Moses and his brother was *Aaron.

A. Hilgenfeld, *Die Ketzergeschichte des Urchristentum*, Leipzig 1884, 616-618.

M. Simonetti

NOMINA SACRA. In this way Ludwig Traube defined a certain number of particular terms linked to the three persons of the *Trinity, to biblical characters, or relating to attributes of worship or reverence, as well as to grades of the ecclesiastical hierarchy, which appear abbreviated by contraction in Latin and Greek biblical MSS from the 4th c. Many of these abbreviations occur frequently, even in early Christian *epigraphy, but today the term N. S. is no longer in common use in epigraphical criticism.

It was hypothesized that the first N. S. were transcribed in abbreviated form to stress their sacral character; it was also thought that the overlining was likewise connected with this *symbolism, but these theories do not seem to be acceptable. Apart from any other consideration, abbreviated letters and numerals with overlining, which has a purely distinctive value, occur normally in epigraphy.

Among the N. S. most commonly used in Greek *palaeography, we should mention at least $\overline{\text{KC}}$ (Κύριος), $\overline{\text{IC}}$ $\overline{\text{XC}}$ ('Ιησοῦς Χριστός), $\overline{\text{YC}}$ (Υἱός), $\overline{\text{MP}}$ (Μήτηρ), $\overline{\text{CHP}}$ (Σωτήρ), $\overline{\text{ΔAΔ}}$ (Δαυίδ); in Latin $\overline{\text{DS}}$ (*Deus*), $\overline{\text{IHS}}$ (*Jesus*), $\overline{\text{XPS}}$ (*Christus*), and - later - $\overline{\text{EPS}}$ (*episcopus*), $\overline{\text{PBR}}$ (*presbyter*), $\overline{\text{REVS}}$ (*reverendissimus*). Abbreviations by contraction were widely adopted from the 6th c. also in secular texts.

DACL 12, 1478-1481; *EC* 8, 1926-1927; L. Traube, *Nomina sacra*: Quellen und Untersuchungen zur lateinischen Philologie des Mittelalters 2, Munich 1907; C.H. Turner, The nomina sacra in early latin Christian manuscripts, *Miscellanea F. Ehrle*, IV, Rome 1924, 62-74; C.H. Roberts, *Manuscript, Society and Belief in Early Christian Egypt*, Oxford 1979.

D. Mazzoleni

NONNA (†*c.*374). Mother of *Gregory Nazianzen, born of an old Christian family. Gregory's *Or.* 18 describes her *virtues and how she converted her husband.

F.J. Dölger, Nonna. Ein Kapitel über christliche Volksfrömmigkeit des IV. Jhrh., *AC* 5 (1936) 44-75; M.M. Hauser-Meury, *Prosopographie zu den Schriften Gregors von Nazianz*, Bonn 1960, 134-135.

J. Gribomont

NONNUS, Pseudo-. The earliest (6th c.) of the scholiasts on *Gregory Nazianzen, various of whose *Orations* he commented on, taking particular interest in allusions to *mythology. The work had an enormous circulation: 139 MSS have been counted. It was used by numerous other Byzantine scholiasts and philologers, and translated into *Syriac (the existing version dates from 624), *Armenian and *Georgian. The comments on *Or.* 4 and 5, with extracts relating to *Or.* 39 and 43, are published in PG 36, 985-1072 under the title *Collectio et expositio historiarum*.

CPG 3011; K. Weitzmann, *Greek Mythology in Byzantine Art*, Princeton 1951; F. Lefherz, *Studien zu Gregor von Nazianz*, Bonn 1958; J. Declerck, Les commentaires mythologiques du Ps. Nonnos sur l'hom. XLIIIe de Grégoire de Nazianze, *Byzantion* 47 (1977) 92-112.

J. Gribomont

NONNUS of Panopolis (*c.*400-470). A native of Panopolis in Upper *Egypt (modern Akhmim), N. is a character with imprecise outlines. He composed an epic poem in 48 cantos, the longest to survive from classical antiquity, entitled *Dionysiakà*, which describes Dionysius's adventurous journeys and warlike and amorous exploits in India and Greece. The poem, absolutely *pagan in character, rich in mythological information and baroque amplifications, shows precise stylistic and linguistic affinities with a hexametrical *Paraphrasis (Metabolè)* of *John's gospel, commonly attributed for this reason to the same author. We do not know if the Christian poem was composed after N.'s *conversion to Christianity, or if the mythological poem is an intentional ironical demolition of pagan religion or, finally, if the Egyptian poet's dual literary output does not attest a situation of cultural and religious *syncretism quite normal in his environment at that time: think of *Claudian! The presence in the *Paraphrasis* of the definition of *Mary as *Theotokos offers no sure chronological reference-point, since this must have been widespread even before the declaration of *Ephesus in 431. The Nonnian hexameter is characterized by predominance of dactyls and by its evidence of particular attention to the intensive nature of stress.

CPG 3, 5641-5642; PG 43, 665-1284; A. Scheindler, *Nonni Panopolitani Paraphrasis S. Evangelii Ioannei* (Bibliotheca Teubneriana), Leipzig 1881; R. Keydell, *Nonni Panopolitani Dionysiaca*, 2 vols., Berlin 1959; F. Vian, *Les Dionysiaques*, I, Paris 1976; P. Chuvin, *Les Dionysiaques*, II, Paris 1976. Studies: R. Janssen, *Das Johannesevangelium nach der Paraphrase des Nonnos* (TU 8/4), Leipzig 1903; J. Golega, *Studies über die Evangeliendichtung des Nonnos von Panopolis. Ein Beitrag zur Geschichte der Bibeldichtung im Altertum*, Breslauer Studien zur historischen Theologie 15, Breslau 1930; G. D'Ippolito, *Studi noniani. L'epillio nelle "Dionisiache"*, Palermo 1964; J. Golega, Zum Text der Johannesmetabole des Nonnos, *ByzZ* 59 (1966) 9-36; Quasten II, 116-118; W. Peek, *Kritische und erklärende Beiträge zu den Dionysiakà des Nonnos*, Berlin 1969; id., *Lexicon zu den Dionysiakà des Nonnos*, 4 Lief., Berlin-Hildesheim 1968-1975; P. Volpe Cacciatore, Osservazioni sulla Parafrasi del Vangelo di Giovanni di Nonno di Panopoli, *Annali Fac. Lett. Napoli* 32 (1979-1980) 41-50; W. Fauth, *Eidos poikilon. Zur Thematik der Metamorphose und zum Prinzip der Wandlung aus dem Gregensatz in den Dionysiakà des Nonnos von Panopolis*, Hypomnemata 66, Göttingen 1981.

P.F. Beatrice

NORICUM

I. Christianity - II. Archaeology.

I. Christianity. N. was a Roman province corresponding, in the period that concerns us, to much of modern Austria (excluding Vienna and the left bank of the Danube). To S it bordered the Carnian Alps, to W the river Inn, to N the Danube, to E *Pannonia, along that ideal line which joins the rivers Danube, Sava and Drava and comprises the cities of Celeia (Celje in Slovenia) and Flavia Solva (near Leibnitz). In the tetrarchical period N. was divided into two parts: N. Ripense (between the Danube and the Taurus mountains) and N. Mediterraneum (between the Taurus and the Carnian Alps), under the rule of a *praes*.

Information on N.'s first *evangelization is scarce and uncertain. One probable way of penetration of Christianity, apart from the colonial and military, could be that of commercial links with *Aquileia. Important centres like *Milan and *Sirmium also had considerable influence. Early evidence of the spread of Christianity into N. at the time of *Diocletian comes from the *Passio* - surviving in two recensions (ASS, *Mai* I, 466-472) - of *Florian of Lauriacum (Lorch) and 40 companions, whose historicity, at least in its essential nucleus, is supported by the *Mart. hier.* (IV *Non. Mai*). Other evidence comes from *Athanasius who, without specifying names, sees or number, mentions the presence at the synod of *Sardica (343) of representatives from N. supporting the cause of *orthodoxy (*Apol. sec.* 1. 36: PG 25, 312A; *H. Ar.* 28: PG 25, 725B). The late 4th and early 5th cc. probably saw a more extensive spread of Christianity, if it is true that various *castella* (fortified retreats that arose on high ground at the time of the barbarian incursions) reveal, in the light of archaeological excavation, the church and baptistery almost always at their centre or in their highest part.

But by far the most important historical evidence for the life and *organization of the church of N., at a particularly difficult period (453-486) for this province on the border of *Romanitas*, is *Eugippius's *Life* of St *Severinus. For that time we have knowledge, among other things, of two episcopal sees, Lauriacum and Tiburnia or Teurnia (St Peter im Holz). Other Christian centres mentioned by the *Life* are: Asturis (Klosterneuburg), Boiotrum (Innstadt), Comagenis (Tulln), Favianis (Mautern?), Ioviacum (Shlögen) and Iuvavum (Salzburg).

J. Zeller, *Les origines chrétiennes dans les provinces danubiennes de l'empire romain*, Paris 1918 (repr. Rome 1967); R. Noll, *Frühes Christentum im Oesterreich von den Anfängen bis zum 600 n. Chr.*, Vienna 1954; K. Gamber, Die Severins-Vita als Quelle für das gottesdienstliche Leben in Norikum während des 5. Jh., *RQA* 65 (1970) 145-157; M. Pavan, Stato romano e cumunità cristiane nel Norico, *Clio* 9 (1973) 453-496; V. Pavan, Note sul monachesimo di s. Severino e sulla cura pastorale nel Norico, *VetChr* 15 (1978) 347-360; G. Winkler, Noricum und Rom, *ANRW* II.6. 183-262; H. Wolff, Über die Rolle der christlichen kirche in den administrationsfernen Gebieten von Noricum in 5 jh. n. Chr., in (ed.) W. Eck, *Religion und Gesellschaft in der römischen Kaiserzeit*, Cologne-Vienna 1989, 265-293.

V. Pavan

II. *Archaeology. Among the *church buildings so far found in N., the most numerous are basilicas of the type characteristic of the Upper Adriatic, consisting of a longitudinal hall in whose E section, or raised presbytery, is a semicircular bench for the *clergy, accessible from all parts. Churches like this are at Lauriacum, Aguntum, Lavant, Grazerkogel and church no. 1 at Hemmaberg. Of them, the most interesting is that of Lavant, built on a hill and fortified early in the 5th c., since it is made up of two similar halls, on an E-W line, both with clergy bench and *cathedra*. The presence of a baptismal font in the E hall shows that it was an episcopal basilica.

The second type of church found in N. has a common plan with semicircular apse and clergy bench of the same form. Church no. 2 at Hemmaberg belongs to this type, which may have rooms along the side walls (as at Duel), but also *pastophoria* beside the apse, as at Hoischügel. At Celeia the church was decorated with pavement *mosaics of some quality. At Ulrichsberg the hall appears to be strengthened with pilaster strips on the outside, while the section of pavement belonging to the clergy's semicircular *subsellia* was decorated with mosaics. These churches sometimes also have a narthex.

Elsewhere there were different and more developed plans. At Teurnia (St Peter im Holz), the hall has a cruciform plan with narrower transverse arms. The W arm contains the clergy's semicircular bench; the E one is flanked by apsidal rooms. The first has an apse with pilaster strips, the other a pavement mosaic showing animals among decorative motifs. All these rooms are connected to one transverse arm by a door and to the E arm by another. The type-plan of the basilica of Teurnia is probably derived from churches of similar type in N *Italy. The church of Säben seems to be a simpler variant of that at Teurnia. Of particular interest is a religious complex built on a fortified hill recently excavated at Vranje near Sevnica in Slovenia and consisting of three buildings: one of basilica type situated on the upper terrace, and two others on the lower terrace (one of the two has a square apse), doubtless with baptismal functions. The presence of the *cathedra* between the clergy seats shows that it was an episcopal complex. The complex of Rifnik in Slovenia was also built on a fortified hill.

At Iuvavum (Salzburg), above St Peter's *cemetery, religious rooms are excavated in the precipitous rock with *arcosolia* similar to those of the Roman catacombs. From the form of the arcosolia, and acc. to ancient tradition, these rooms can prob. be attrib. to eremitical monks in the first centuries of Christianity.

Among small objects we must mention a fragment of *sarcophagus showing a *shepherd with a *lamb on his shoulders. There are also inscriptions, gems, *lamps, fragments of *plutei* from Teurnia, etc. **[Fig: 238]**

R. Egger, *Frühchristliche Kirchenbauten im südlichen Noricum*, Vienna 1916; R. Noll, *Frühes Christentum in Österreich von den Anfängen bis um 600 nach Chr.*, Vienna 1954; A. Alföldi, *Noricum*, London 1974, 198 ff.; P. Petru-Th. Ulbert, *Vranje bei Sevnica. Frühchristliche Kirchenanlagen auf dem Ajdovski Gradec*, Ljubljana 1975; R. Pillinger, Neue Ausgrabungen und Befunde frühchristlicher Denkmäler in Österreich (1974-1986), *Actes XIe Cong. int. d'arch. chrét.*, CEFR 123, Rome 1989, 2089-2125.

N. Cambi

NOTAE DE TEXTU EVANGELIORUM. MS St Gall 1935 (Lowe, *Codices Latini Antiquiores*, 984), a palimpsest of which many leaves contain the gospels according to the Vulgate, is so old that it may have been written while *Jerome was alive. The margins contain first-hand *glosses, which specify, e.g., the case of a noun or the sense of a verb whose form is ambiguous; often they transcribe the Greek term translated and its various equivalents. These glosses recur in some later MSS, going back to the 9th-c. school of Rheims. Their most probable author is Jerome.

CPL 590a; B. Bischoff, Zur Rekonstruktion der Sangallensis (Σ) und der Vorlage seiner Marginalien, *Biblica* 29 (1941) 153-158.

J. Gribomont

NOTARIUS. The noun is derived from *nota, notare* and means one who takes notes with the help of abbreviation marks. Therefore he is called a tachygrapher. The sign is more of a symbol than a letter. There existed schools of *notarii*, where stenography was learned. Nn. were used for public affairs, conferences, trials. They noted down speeches, depositions in lawsuits, as we see from the Acts of the *martyrs, esp. the *Scillitan martyrs and *Cyprian. Nn. were present at the conference of *Carthage in 411. Their acts were carefully preserved in the archives, as Cyprian (*Ep.* 68, 6) and *Augustine (*C. Cresc.* III, 70) attest. Above all from the 4th c. on, the church had its own nn., who drew up formal acts: e.g. the enfranchisement of a *slave, a deed of *donation, etc. Normally they were *clerics. At *Rome, the regional nn. were people of importance.

DACL 12, 1623-1640; A. Mentz, *Antike Stenographie*, Berlin 1944; H. Hagendahl, Die Bedeutung der Stenographie für die spätlatein. christ. Literatur, *JbAC* 14 (1971) 24-38 (bibl.).

A. Hamman

NOTITIA DIGNITATUM. Official document of the imperial chancellery no earlier than 395, whose full title must have been: *Notitia omnium dignitatum et administrationum tam civilium quam militarium*. In its present state, it takes the form of a sort of yearbook with quite evident chronological "stratifications". The copy in our possession was that of the *primicerius of the West. The text is divided in two parts, i.e. relating to the two *partes* of the Roman empire, of which it lists the functionaries and their official titles, their territorial competence, their offices or dependent troops, and the composition of their *officium*. Discussion of the genesis of the work, i.e. whether it is derived directly from an official document or whether, from an original short nucleus, it attained its present form through interpolations and additions by private compilers, is still open. Many dates have been proposed, oscillating from the end of the 4th c. (L.A. Constans and F.S. Salisbury) to the fourth decade of the 5th c. (O. Seek, Th. Mommsen, J.B. Bury). Recently G. Clement has demonstrated that the document was "the product of an experienced bureaucratic tradition, with aims of information, kept constantly up to date so as to be usable for a long period of time". The textual anomalies and its own development are linked to the discontinuity of the imperial chancelleries.

E. Böcking, Bonn 1839-1853, still useful for its full commentary; O. Seek, Berlin 1876.
Studies: E. Polaschek: *PWK* 17, 1, 1077 ff.; A. Lippold: *KLP* 4, 166-167; P. Lot, La "Notitia Dignitatum Utriusque Imperii". Ses Tares, Sa Date de Composition, Sa Valeur, *REA* 38 (1936) 285-338; G. Clemente, *La "Notitia dignitatum"*, Cagliari 1968; D. Hoffmann, *Das spätrömische Bewegungsheer und die Notitia dignitatum*, 2 vols., Düsseldorf 1969; AA. VV., *Aspects of the Notitia Dignitatum*, Oxford 1976.

M.L. Angrisani Sanfilippo

NOTITIA GALLIARUM. Document of Roman administrative origin, drawn up between the end of the 4th c. and the first decades of the 5th, with additions in the Frankish era. Its complete title is *Notitia provinciarum et civitatum Galliarum*. It lists the 17 provinces - a list corresponding to that of the *Notitia dignitatum* - and the cities of *Gaul. It is very useful for knowing the structure of ecclesiastical *organization in 5th-c. Gaul, since it gives to all intents and purposes the list of ecclesiastical provinces, the main metropolitan cities and the other episcopal cities.

CPL 2342; MGH, *Auct. ant.* IX, Chron. min. I, 552-612 (text: 584-612); Schanz IV, 2, 128-130; E. Griffe, *La Gaule chrétienne à l'époque romaine*, II, Paris ²1966, 111-125; P.-M. Duval, *La Gaule jusqu'au milieu du V⁵ siècle*, Paris 1971, 681-682; J. Harries, Church and State in the "Notitia Galliarum", *JRS* 68 (1978) 26-48.

A. Di Berardino

NOTITIA PROVINCIARUM ET CIVITATUM AFRICAE. This document has come down to us through only two MSS, the second of which, Laon 113 (9th c.) (*Cat. gén. mss. Dép.*4°, I [1849] 97), is now lost. The other was, in the 16th c., the property of L. Haller of Hallerstein (Ortellius, *Thesaurus geographicus*, Anversa 1596). The main editions are those of Ruinart, used by Migne (PL 58, 269-76), C. Halm in 1878 (MGH, *Auct. ant.* 3/1, 63-71) and M. Petschenig in 1881 (CSEL 7, 117-34).

The document lists, by provinces (*Africa Proconsularis, *Numidia, *Byzacena, *Mauretania Caesariana and Sitifiana, Tripolitania, *Baleari), the African bishops, with their sees, called to *Carthage in 484 by the *Vandal king *Huneric, who hoped the meeting would result in an accommodation between *Arians and *Catholics - an attempt foiled by the firmness of the latter, who drew up an *orthodox profession of faith. In reprisal, the Catholic churches were closed and their goods confiscated. 46 bishops were exiled to *Corsica and 302 to the borders of Africa: there were one *martyr, one *confessor, some *apostates and many dead. Marginal annotations appear in the document, one of which, abbreviated to *prbt*, is prob. short for *peribat*, not in the sense of apostasy, but of physical death. Besides its information on Huneric's persecution, the *Notitia* reflects the situation of the African church at the end of the 5th c.: Proconsularis, 54 bishops; Numidia, 125; Byzacena, 115; Mauretania Caesariana, 120, and 3 vacant sees; Mauretania Sitifiana, 42; Tripolitania, 5; Baleari, 8; in all, 469 bishops and 472 bishoprics. Mauretania Tingitana does not appear, being dependent on *Spain.

Potthast, *Bibliotheca historica medii aevi*, Berlin ²1896, 2, 869; L. Duchesne, *Histoire ancienne de l'Eglise*, 3, Paris 1929, 645, n. 1.

V. Saxer

NOUS

I. In Greek philosophy - II. In patristics - III. In gnosticism.

I. In Greek philosophy. We owe to Anaxagoras the first elaboration of the doctrine of the metaphysical n., which subsequent Greek thought could not exclude: for this presocratic philosopher, n. is the ordering principle and the cause of everything (texts 42, 47, fr. 12 Diels, *Die Fragm. der Vors.* II ⁶1952; cf. also Plato, *Cratylus* 400a 8-10); it is simple, without mixture, pure (texts 55, 56 and 61, fr. 12 and 13), free from passions (text 56), identical with the One (text 61), infinite (fr. 12) and at the same time above infinity (cf. Clem. Al., *Protr.* 66, 1: GCS I, 50, 15-17), principle of movement (texts 42, 55, 56, fr. 13), unmoving (text 46) and at the same time pervading everything (text 55). For Plato, the demiurge, maker of the *anima mundi* and the sensible universe in accordance with the model of the ideas, is himself a n. (*Tim.* 39e) and infuses n. into the soul of the universe (*Tim.* 30b); the latter, in consequence, is a being endowed with soul and intelligence (*Tim.* 30b; cf. also *Cosmos*). In *Philebus* 28c-d, n. is presented as the supreme ruler of the universe; in 28e its ordering function is emphasized (cf. also *Phaedo* 97d=Anax. text 47; *Cratylus* 413c=Anax. text 55e; *Cratylus* 400a 8-10, and other passages in which Anaxagoras is named); and in 30c it is identified with the supreme cause and wisdom, above limit and infinity. In *Laws* X 897b, n. is the divine principle which, infusing itself into the *anima mundi* (cf. *Tim.* 30b, *Philebus* 30d), enables it to rule the whole universe rightly and successfully. Of the various types of movement, the circular is that which most resembles the movement of n. (*Laws* X 898a) (on the Platonic conception of n. as a transcendent metaphysical principle, cf. R. Hackforth's still important article, Plato's Theism, *CQ* 30 [1936] 4-9). Human intellect, the divine element destined to govern man, is "sown" in him directly by the demiurge (*Tim.* 41c). Only it, the pilot of the *soul, is able to contemplate the essence - colourless, formless and untouchable - represented by the totality of the ideas (*Phaedrus* 247c); but only the thought of the philosopher has *wings that carry him to the heights, beyond the heavenly vault, and initiate him into a vision and a higher knowledge (249c, 250c). In the Old Academy, Xenocrates (fr. 15 Heinze) considered the first god a n. identical with the monad (it is quite probable that Xenocrates had already formulated the doctrine - which would become characteristic of *middle Platonism, *neoplatonism, *Philo and some *Fathers - of the ideas as identical with the thoughts contained in the divine mind: having identified n. with the monad and the ideas with numbers [fr. 15 and 34 Heinze], he cannot have found it hard to make the ideas derive from n. just as numbers derive from the monad; on this question and related studies, cf. S. Lilla, *Clement of Alexandria*, Oxford 1971, 202-203, n. 3, 246). In another representative of the Old Academy, Speusippus (fr. 38 Lang, 89 Isnardi-Parente=Aetius in Stobaeus, *Ecl.* I 29, vol. I, 35, 3-4 Wachsmuth), we find already formulated a doctrine which would become basic in orthodox neoplatonism, that of the distinction of mind from the One and from the good (cf. E.R. Dodds, *CQ* 22 [1928] 140 and P. Merlan, *The Cambridge History of Later Greek and Early Medieval Philosophy*, Cambridge 1967, 32).

Both with regard to the metaphysical n. and to the human n., Aristotle worked out a doctrine, taken in part from Anaxagoras, which left profound traces on subsequent thought. In *On prayer* fr. 46 Rose (Bekker V, 1483a 27-28), he seems, resembling Speusippus and anticipating orthodox Neoplatonism, to consider the hypothesis of the supreme deity's transcendence even over mind; but in ch. 7 of book Λ of the *Metaphysics* (1072a-1073a), he retains the identification of the first principle with mind. The first principle (1072b 11, 13-14) is a n. which always thinks itself and which is hence in perennial activity and totally identical with the object of its thought (1072b 18-23); it is identical with life, since its activity is life (1072b 26-28); it is without magnitude or parts and is indivisible (1073a 5-6); it is absolutely without passions and unalterable (1073a 11); it is eternal, totally separate from sensible things (1073a 4); it is unmoving (1073a 4) and, remaining motionless and changeless, it moves the first heaven (1072a 24-25, b 7-8), since it is absolute beauty, the object of desire, thought and love (1072a 26, 27-28, b3); it is simple being *par excellence* (1072a 32); and it is endowed with an infinite power, since it causes motion for an infinite *time (1073a 8-9; cf. W.D. Ross, *Aristotle's Metaphysics*, II, Oxford 1948, 372-374). The human n. is a divine principle infused from outside, *De gen. an.* B 736b 28, and the most divine element that man possesses: by its activity it is the cause of perfect happiness (*Eth. Nic.* X, 1177a 13-17), since this activity, man's highest, is the same thing as *theoria (*Eth. Nic.* X, 1177a 19-22); in the *soul nothing is superior to it (*De an.* I, 410b 13-14). But it is above all in chs. 4-5 of book III of *De anima* (429a-430a) that Aristotle develops his theory of human n.: n. in activity (cf. *De an.* III, 430a 17), distinct from passive or potential n. (*De an.* III, ch. 5, 430a 10-14), is without passions (429a 15, b 23, 430a 18, a 34), free from any mixture whatever (429a 18-19, b 23-24 - Anaxagoras is named in these two passages - 430a 18), simple (429b 23), not confused with the *body, indeed separate from it (429a 24-25, b 5, 430a 17, a 22-23); it is the seat of the ideas (429a 27-28); because it carries on its noetic activity, it is identical with the intelligible objects which it thinks (429b 30-31; 430a 3-4), and is intelligible like them (430a 2-3); it is immortal and eternal (430a 23). This conception of the metaphysical n. which thinks itself and is identical with the object of its own thought was taken up by the 2nd-c. (AD) philosopher *Alexander of Aphrodisias (*Mantissa* 108, 7-8; 109, 4-7; *De an.* 87, 89-88, 2; 88, 5-6) to whom, as R. Walzer has shown (*Plotino ed il Neoplatonismo in Oriente e in Occidente*, Rome 1974, 430), we owe the complete elaboration of the doctrine of the *nous poietikós*, which consists of the application to the metaphysical n. described by Aristotle in *Metaphysics*, book Λ, ch. 7, of the various properties of the "active" human intellect illustrated in Aristotle's *De anima*, book III, ch. 5 (cf. in particular Alexander, *De an.* 88, 17-89, 12 and *Alexander of Aphrodisias). The theory of the metaphysical n. worked out by Aristotle and Alexander of Aphrodisias would be adopted entire into the Plotinian conception of n.

For the Stoics the metaphysical n. was identical with the rational principle immanent in the universe, which it pervades, permeates with itself, administers and rules; it also takes the names of *logos*, spirit, soul, fire, Zeus, law, ether, nature, fate and providence (SVF I 102, 154, 160, 162, 172, 176, 530, 534; II 580, 634, 937, 1026-1027, 1151), and contains in itself all the rational or spermatic principles, which, combining with the passive element, primordial matter, give rise to the various elements and to all sensible beings (SVF I 98, 1202; II 1027). As regards man's n., unlike Plato who had always stressed the difference between the rational part of the soul and the two irrational parts, *Stoicism asserted the substantial unity of the human *soul: the seven faculties of the soul distinct from the rational faculty or *hegemonikón* (the five senses, the reproductive faculty and the vocal faculty) are in fact emanations of the *hegemonikón*, whose seat is in the heart rather than in the head (SVF II 826, 827, 836-838, 839, 910); so there is a close interdependence between irrational faculties and rational faculty, between sensations and thought: rather than faculties differing by nature, it would be more correct to speak of a single force which, according to the case, carries out distinct functions (SVF II 849). On the close parallelism established by the Stoics between man governed by reason and the universe ruled by the *logos*, cf. *Macrocosm and Microcosm.

The *syncretistic, but basically Stoic, character of the thought of Antiochus of Ascalon - who, though professing to be an "Academic" and a follower of the *antiqui*, i.e. the Old Academy and the Peripatos, was in fact a Platonizing Stoic - is clearly seen in his conception of the metaphysical n. and the human n. Cicero's passage (*Academ. post.* I, 29): *quam vim animum esse dicunt mundi, eandemque esse mentem sapientiamque perfectam, quam deum appellant . . . quam interdum necessitatem appellant* denotes a mixing of Platonic and Stoic elements: the term *vim* is both Platonic and Stoic (cf. *Sophist* 265b; SVF I 176; II 1025, 1151); the *animum mundi* is found both in Plato (cf. e.g. *Tim.* 30b; *Laws* X, 897 b) and in the Stoics (cf. e.g. SVF I 495, 530; II 1026); the words *mentem sapientiamque* go back both to *Philebus* 28d-e, 30c and to the Stoics (cf. SVF I, 102, 112); and the identification of the divine intelligent principle with *necessitas* is of Stoic origin (cf. e.g. SVF I, 160, 176; II 580). As J. Dillon has rightly noted (*The middle Platonists*, London 1977, 83 and 95), this Ciceronian passage reproducing Antiochus's point of view presupposes an identification between the *nous*-demiurge, the *anima mundi* of the *Timaeus*, and the Stoic *logos*. W. Theiler (*Die Vorbereitung des Neuplatonismus*, Berlin 1930, 18-19 and 39-40) and J. Dillon (*op. cit.*, 93-96), examining those passages that depend on Antiochus in Cicero (*Orator* 8-10) and in *Augustine (*De civ. Dei* VII, 28, on Varro's allegorical interpretation of Jove, Minerva and Juno: Jove is God's mind, his daughter Minerva his thoughts, and Juno matter), reach the conclusion that Antiochus too championed the doctrine of the ideas as thoughts contained in the mind of God; but, as Dillon observes (*op. cit.*, 95), we must bear in mind that Antiochus, precisely because he identified the Platonic *nous*-demiurge with the Stoic *logos*, was led to identify the Platonic ideas contained in the *nous*-demiurge with the *logoi* or spermatic principles contained in the universal *logos* of the Stoics. Antiochus's doctrine on the human n. (Cicero, *Academ. Pr.* II, 30: *mens enim ipsa quae sensuum fons est atque ipse sensus est, naturalem vim habet quam intendit ad ea, quibus movetur*) is nothing but a reproduction of the Stoic doctrine (cf. SVF II 826, 827, 836).

In *Philo of Alexandria's conception of the metaphysical n. there are elements of the Platonic, peripatetic and Stoic traditions. The relationship he establishes between n. and the first principle coincides with what we saw in Aristotle, *On prayer* fr. 46 Rose: on one hand, as J. Whittaker noticed (*VChr* 23 [1969] 102), in *Leg. Alleg.* II 46 he puts God above n.; but on the other - and this is the dominant tendency in his work - he is led to identify God with the n. of the universe: *De migr. Abr.* 192 (II 306, 3-4), *De gig.* 41 (II, 50, 5), *De opif. m.* 8 (I, 2, 19). In *De opif. m.* 8, this n. is on one hand identified with the active principle, in accordance with the teaching of the Stoics (cf. SVF I 85); on the other, it is given the attributes εἰλικρινέστατος and ἀκραιφνέστατος which bring it close to the Anaxagorean-Aristotelian conception (but on εἰλικρινής cf. also Plato, *Sympos.* 211e) and it is presented, in conformity with the peripatetic tradition, as "above virtue" (cf. *Magna mor.* B 1200 b 14) and "above science" (cf. *Eth. Eud.* H 1248a 28). It directs the universe like the ship's pilot and the charioteer: *De ebr.* 86 (II 186, 14), *De conf. ling.* 98 (II 247, 22-23), *De aet. m.* 83 (VI, 98, 20), an image that recurs in the pseudo-Aristotelian *De mundo*, 400b 6-7, and which ultimately goes back to Plato (*Phaedrus* 246a-b, 247c, 253c, 253e), where, however, it is applied not to the metaphysical n., but to the human n. Like Aristotle's metaphysical n., Philo's God causes movement while remaining unmoved: *De post. C.* 28 (II 7, 14). Of Platonic-Aristotelian derivation (cf. *Philebus* 20d, *Eth. Nic.* I 1097b 7-8, *Eth. Eud.* H 1244b 8, 1249b 16, *Met.* N 1091b 16-17) is the doctrine of the absolute self-sufficiency of n.: *De Cher.* 44 (I 181, 4), 46 (I 181, 12), 123 (I 199, 16), *De vita Mos.* 111 (IV 145, 15), *De spec. leg.* II 38 (V, 95, 1). In Philo too recurs the doctrine of the ideas as identical with the thoughts contained in the mind of God: in *De opif. m.* 17-20 he makes the comparison - like that present in Cicero (*Orator*. 8-10) and going back to Antiochus - between the architect who plans the city in his mind, and God who, before originating the sensible universe, thinks it in his mind; and in *De Cher.* 49 (I 182, 3) God is defined as "incorporeal seat of the incorporeal ideas", a passage closely parallel to *De opif. m.* 20 (I 6, 9-10): both passages represent an application to the divine n. of what Aristotle (*De an.* III 429a 27-28) had said apropos of the thinking human n. Another application to the divine n. of the Aristotelian doctrine of the human n. is represented by the total absence of passions in God: Philo's divine n. is completely impassible: *Quod D. sit imm.* 52 (II 68, 10-12), just as the active human n. is for Aristotle: *De an.* III 430a 18 (but cf. also *Met.* Λ 1073a 11). For Philo, the human n. is a "divine fragment" "breathed" from on high: *De somm.* I 34 (III 212, 7), *Quis. rer. div. her.* 64 (III 15, 19-20), since it is "an image of the image", i.e. an *image of the divine *logos* (thus Philo interprets the "second image" of Gen 1, 26-27; cf. S. Lilla, *Clement of Alexandria*, Oxford 1971, 21). Philo also takes over the Stoic doctrine (cf. SVF I 826, 827, 836) of n. as source of sensations: *Leg. alleg.* I, 28 (I 68, 11-12), *De post. C.* 126 (II 27, 24-26). The term πηγή present in these two passages must be compared with the term *fons* in Cicero: *Academ. Pr.* II, 30 (cf. above).

For Plutarch of Chaeronea, the supreme god is identical with the intellect and with the good *anima mundi* of book X of Plato's *Laws*: *De Is. et Os.* 371A (II 523, 12-13 Bernardakis), 370F (II 522, 16-20); with the good: *De Is. et Os.* 372F (II 527, 20-21); with the One and with true being: *De E apud Delphos* 392E (III 22, 9-11), 393A-B (III 23, 6-13), 393B (III 23, 15-19), *De def. or.* 428F (III 117, 3); with the intelligible: *De Is. et Os.* 373A (II 528, 9-10), 382D (II 552, 21-22); with the beautiful: *De Is. et Os.* 383A (II 554, 5-6); with the ideas and with the demiurge: *De Is. et Os.* 373F (II 590, 9-10). The identification of the first principle with absolute being and with the intelligible is derived not just from the application to God of the Platonic doctrine of the immutability of the ideas, but also from Aristotle: *Met.* Λ 1072b 10, 1072a 26; the identification of the first principle with the intellect is based both on *Timaeus* 39e and *Philebus* 30c, and on *Metaphysics*, book Λ; the close connection between n. and the *anima mundi* goes back to *Philebus* 30c, *Laws* X 897b; the identification of the model-idea (cf. *Tim.* 28a, 28c, 30c, 31a, 39e) with the πατήρ or *nous*-demiurge (cf. *Tim.* 28c, 39e) is explained by the fact that Plutarch interpreted in an Aristotelian-peripatetic sense the passage of *Tim.* 39e in which the *nous*-demiurge contemplates the ideas present in the living intelligible being (acc. to this interpretation, in this passage the intellect contemplates itself, i.e. its own thoughts). This proves that Plutarch too, following the footsteps of the Old Academy and Antiochus, considered the Platonic model-ideas as thoughts contained in the mind of God; this hypothesis finds full confirmation in *De plac. philos.*, 882D (V 287, 5-7).

In Alcinous's doctrine of n. (that this, not Albinus, is the exact spelling of the author of the *Didaskalikós* has been demonstrated by J. Whittaker, *Phoenix* 28 [1974] 450-456), the presence of Aristotelian motifs is still more marked. Like Aristotle (*De an.* III ch. 5, 430a 10-14), Alcinous (*Did.* 164, 16-20) distinguishes n. in potency from n. in act, applying to the heavenly n. (the second n. which governs the *anima mundi*) the distinction which Aristotle had reserved to the human n. The further distinction, present in this passage, between the n. of heaven and the n. its cause, identical with the first god, is based on *Timaeus* 30b - where the *nous*-demiurge (cf. *Tim.* 39e) infuses n. into the *anima mundi* and the latter into the sensible body of the universe - and on *Laws* X, 897b. Of peripatetic derivation (cf. *Magna Mor.* B 1200a 14) is the idea - present, as we have seen, in Philo too - of the first n.'s superiority to virtue (*Did.* 181, 36-37). Like the n. of Aristotle and Alexander of Aphrodisias, Alcinous's n. thinks itself (*Did.* 164, 24-27) and remains unmoving and is at the same time an object of desire and active, in that it moves the intellect of the heavens (*Did.* 164, 20-24). For Alcinous too, the ideas are an integral part of the mind of God, in that they are his thoughts and his activity (*Did.* 163, 12-13, 27-30; 164, 26-27). The identification of the ideas with the thoughts of God is also present in Atticus, fr. 9 Des Places (p. 69, 40-41).

In neopythagoreanism we can observe two tendencies apropos of n., considered now as the supreme principle, now as subordinate to the One (this last tendency, which is connected with Speusippus and Aristotle [*On prayer* fr. 46 Rose], is that which would prevail in orthodox neoplatonism). Like Alcinous, *Numenius of Apamea asserts the existence of two intellects, which are for him the first and second god: the "first god" is the supreme intellect (fr. 16 Des Places), absolute being and good (fr. 16 and 17), the living intelligible being of the *Timaeus* (fr. 22), above and prior to *ousia as its cause and principle (fr. 2 and 16) and at the same time innate to it (fr. 16); the "second god", identical with the Platonic demiurge (fr. 16 and 21), is the second intellect (fr. 17 and 22), directs the heavens (fr. 12) and is the cause of the harmony of matter (fr. 18; cf. *Numenius). As H.J. Krämer has pointed out *TheolPhilos* 44 [1969] 486), Numenius's fr. 22 Des Places (= text 25 Leemans), in which the first n. is identified with the living intelligible being of *Tim.* 39e, proves that Numenius too considered the ideas as present in the mind of God (in fr. 16 the first n. is considered the principle and cause of the ideas, because they are the result of its noetic activity, and "innate" to them because, once produced, they are contained in n.). On the other hand both Pseudo-Brutinus (Syrianus, *In Metaph.* 166, 5-6) and Pseudo-Architas (Stobaeus, *Anth.* I, 41, vol I, 280, 15-17) place

n. below the One (cf. Whittaker: *VChr* 23 [1969] 95 and 102-103).

In *neoplatonism, as in Neopythagoreanism, there are two different approaches to n.: on the one hand *Ammonius Saccas, his faithful disciple *Origen the Neoplatonist and later, in the 5th c., Hierocles continue to consider the *nous*-demiurge as the highest principle (on Ammonius Saccas cf. K.O. Weber, *Origenes der Neuplatoniker*, Zetemata 27, Munich 1962, 160; W. Theiler, *Forschungen zum Neuplatonismus*, Berlin 1966, 10 and 41; S. Lilla, *Clement of Alexandria*, 223-224; on Origen the Neoplatonist cf. Proclus, *Theol. Plat.* II 4=fr. 7 Weber pp. 5-6; on Hierocles cf. Photius, *Bibl.*, cod. 214, 172a; cod. 251, 461b, where "will" appears as one of God's attributes and *In aureum carmen* 125, 6-7 Mullach, apropos of the presence of the ideas in the mind of God); on the other, the so-called "orthodox neoplatonism", inaugurated by *Plotinus and concluded by *Proclus and Damascius, considers n. not as the first principle, but as a subsequent *hypostasis, since the first principle is represented by the absolutely transcendent and negative One of the first hypothesis of Plato's *Parmenides*. For a more detailed examination of the doctrine of n. in "orthodox neoplatonism", see *neoplatonism: suffice here to say that for Plotinus, Iamblichus, Proclus and Damascius, n., independently of its hierarchical grade with respect to the first principle (for Plotinus it represents the second *hypostasis, while for *Porphyry, Iamblichus, Proclus and Damascius it is the third member of the triad formed of being, life and intelligence), is the seat of the model-ideas and is in general identified with the demiurge; and that the Plotinian doctrine of n., like that of Philo, Plutarch and Alcinous, accepts important Aristotelian motifs.

II. In *patristics. Patristic thought was inevitably influenced by the speculations on the relationship between n. and the first principle (oscillating, as we have seen, between total identity and subordination) that were characteristic of contemporary Platonism and neopythagoreanism. For *Athenagoras, *God is an eternal intellect (*Legat.* 10) comprehensible only by the mind (*Legat.* 4) (cf. *Platonism and the Fathers). *Clement of Alexandria takes over the Philonian conception of God as a n. which is the seat of ideas: the passage of *Strom.* IV, 155, 2 (II, 317, 11) ("mind is the seat of ideas, mind is God") is indebted to Philo, *De Cher.* 49 (I 182, 9) ("God . . . is the incorporeal seat of incorporeal ideas"). That Clement considers God as a n. is also clear from the emphasis he puts on his will (fr. 48 Stählin [III 224]), a conception that he has in common with his teacher *Pantaenus and with *Ammonius Saccas (cf. also S. Lilla, *Clement of Alexandria*, 222-224). For Clement as for *Philo, the human mind is a divine element (*Strom.* I, 94 [II 60, 24-25]) and a most pure essence, "breathed" into man at the moment of his *creation (*Strom.* V, 87, 4 [II 383, 20-384, 3]), since it is "the *image of the image", i.e. of the *logos* (*Strom.* V, 94, 4-5 [II 388, 13-16]). *Origen, if on one hand (*De princ.* I, 1, 6 [21, 13]) he considers God an intellect, on the other (*C. Celsum* VII, 38 [II 188, 11]) openly asserts that God can be conceived both as n. and as something superior to n., agreeing with Aristotle (*On prayer* fr. 46 Rose) and showing himself fully aware of the debate taking place in the heart of contemporary neoplatonism (cf. J. Whittaker: *VChr* 23 [1969] 92-93). Certain aspects of his conception of God (God thinks himself and is completely self-sufficient) show the influence of the Aristotelian-peripatetic doctrine of the metaphysical n. (cf. *Aristotelianism, under Origen, nos. 3-4). Furthermore, his conception of the *logos*'s origins from God and of the relationship between them coincide substantially with the Plotinian conception of the relationship between n. and the One (cf. *Platonism and the Fathers apropos of Origen's *logos*, nos. 1-9). Motifs of Aristotelian origin can be observed in the Origenian conception of the human n. (cf. *Aristotelianism, under Origen, nos. 6-10). *Gregory of Nyssa, in full agreement with *Plotinus, states that God is above n. (*In cant. cantic. hom.* V, 157, 15) and reserves to his wisdom or *logos* - the equivalent of Plotinus's n. - the presence of the thoughts, prior to the *creation of the sensible universe (the same idea as in Origen and Philo; cf. *Platonism and the Fathers under Gregory of Nyssa). Gregory sees in the human n., as created by God, the expression of man's original purity, beauty and perfection (thus he interprets Gen 1, 26: "in our image, after our likeness"); obscured and disfigured by his predilection for matter, this beauty can be restored by separating the n. as far as possible from the *body (on the agreement between Gregory and Plotinus on this point too, demonstrated by Merki, cf. Platonism and the Fathers under Gregory of Nyssa). Quite frequent in Gregory is the motif of the winged intellect which, free from every material bond, flies towards the divine (cf. e.g. *De virg.* 294, 8-10 and Plato, *Phaedrus* 246d). Gregory accepts the Stoic doctrine - present, as we have also seen, in Antiochus of Ascalon and Philo - of the close dependence of sensations on the intellect (*De an. et res*: PG 46, 29B 2-4, 32A 8-11, 60B 4-6; cf. also *De hom. opif.*: PG 46, 140A 1-9). For *Eusebius, *Cyril of Alexandria, St *Augustine and *Theodoret, the second person of the *Trinity (the Johannine *logos*) corresponds to Plotinus's and Porphyry's n. (cf. *Platonism and the Fathers). For Ps.-*Dionysius the Areopagite, n., like the other "divine names", does not designate the first principle considered in its μονή, above n. and *ousia*, but refers only to its πρόοδος; and in that sense too it is above any n. (cf. *Nom. div.* I, 5: PG 3, 583A 7-8, C 9; VII, 1: PG 3, 868A 5; VII, 2: PG 3, 868C 14; 869A 7); in the divine n., considered in this sense, all the model-ideas are present (*Nom. div.* V, 8: PG 3, 824C 7-8). The angelic creatures are commonly called by him "intellects" (νόες).

III. In *gnosticism. In the *Corpus Hermeticum* too (cf. *Hermetism) we can note the oscillation between identification of the n. with God and the tendency to consider it subordinate to the first principle: in *Poimandres* 6 (I 8, 16-18) the highest god is the n. that generates the *logos*; and also in *Corp. Herm.* V, 11 (I 65, 4), the supreme god is presented as n.; but, as J. Whittaker has noted (*VChr* 23 [1969] 95), in *Corp. Herm.* XII, 1 (I 174, 3) n. appears as the product of God's essence. In the gnostic system of *Basilides, n. is born of the *Father; cf. Irenaeus, *Adv. haer.* I, 24, 3 (SCh 264, 324, 40-326, 43): *Basilides autem . . . ostendens Nun primo ab innato natum Patre* (cf. also Whittaker, *art. cit.*, 103). The *Valentinian school distinguishes *Ennoia* (also called *Cháris* or *Sigé*) - which, with *Bythós* or *Propátor*, forms the first pair of *aeons - from n., which is the *Monoghenés* born together with *Alétheia* from the first pair: it is the principle of everything and the only one able fully to know the Father (Iren., *Adv. haer.* I, 1, 1: SCh 264, 28, 74-30, 90). In the *Apocryphon of John* (ed. W. Till, TU 60, Berlin 1955), *Ennoia* is identical with Barbelos and *Pronoia*, the first image of God (27, 5-18 p. 95 Till), while n. is produced subsequently to *Monogenés*, is generated by Barbelos and is identical with Christ (30-31, 9 pp. 101-103 Till).

DSp 11, 461-465; A.H. Armstrong, The Background of the Doctrine "That the Intelligibles are not outside the Intellect", Les sources de Plotin, in *Entr. sur l'ant. class.* V, Geneva 1960, 393-425; P. Hadot, Etre, vie, et pensée chez Plotin et avant Plotin, *ibid.*, 107-157; H.J. Krämer, *Der Ursprung der Geistmetaphysik*, Amsterdam 1967; id., Grundfragen der aristotelischen Theologie, II, *TheolPhilos* 44 (1969) 486-496; J. Whittaker, Ἐπέκεινα νοῦ καὶ οὐσίας, *VChr* 23 (1969) 91-104; R. Walzer, Aristotle's Active Intellect νοῦς ποιητικός in Greek and early Islamic Philosophy, in *Plotino e il Neoplatonismo in Oriente e in Occidente*, Acc. Naz. dei Lincei, Quaderno 198, Rome 1974, 423-446; A.H. Armstrong, The Negative Theology of *Nous* in Later Platonism, in (eds.) H.D. Blume-F. Mann, *Platonismus und Christentum* (Festschr. Dörrie), JbAC Ergbd. 10, Münster 1983, 31-37; W. Foerster, *Gnosis* (tr. R. McL. Wilson), 2 vols., Oxford 1974.

S. Lilla

NOVATIAN. Probably Roman by birth, proud of belonging to the community of *Rome (cf. Cypr., *Ep.* 30, 32), N. did not go through all the grades of the ecclesiastical hierarchy (*Cornelius to *Fabius, in Euseb., *HE* VI, 43, 13). His presbyterial *ordination by *Fabian of Rome was opposed by the *clergy and the community, perhaps because of the fact that he had received *baptism when gravely ill. Pope Fabian, Cornelius's predecessor, had been impressed by N.'s outstanding intellectual gifts; indeed, there was nothing in Latin *theology comparable to N.'s *De Trinitate* until at least the 4th c. N.'s *letters (30 and 36 in Cyprian's corpus) greatly excel other Roman letters, and not just in style. For this reason Cornelius himself, *Cyprian (*Ep.* 55, 24) and the anon. author of *Ad Novatianum* manifest a certain esteem for N. A passage in *De Trinitate* (29, 169) seems to show that he was cured of a grave illness and began writing the work immediately after his baptism *in extremis*. For this and for *ascetic reasons, he seems to have lived as a hermit up to the time of *Decius's *persecution; only in summer 250 did he once more put himself at the community's complete disposal, immediately taking a leading role, and also exercise his *authority in the pastoral work of *penitence. While the first Roman letter envisages reconciliation, in case of illness, for those who gave way (the *lapsi*) in the persecution, *Ep.* 30, 8 permits "prudent help" only when there is no human hope of a cure. Moreover Cyprian, whose *flight is much criticized in *Ep.* 8, is now called *gloriosissimus* and thus put on the same plane as the *confessors; N. now clearly sees Cyprian, whose rigid attitude over penitential practice had meanwhile become known at Rome, as a natural ally.

When, in 251 at Rome, Cornelius rather than the eminent N. was elected *pope, N. got himself ordained by three Italian bishops and sought to win bishops of other cities over to his side or to install alongside them new bishops favourable to himself, so that a true new church was set up. N. can thus be considered the first antipope who fully deserves the title. His community allowed no pardon for grave *sins, nor did it lead to *penitence, as *Tertullian had done even after he became a *Montanist. N.'s *schism arose over the desire for a pure and holy church. But this image of the church did not derive from the genuine heritage of primitive Christianity, but was an expression of the admission of a rigorous *Stoicism into Christian life (Cypr., *Ep.* 55, 16).

Late sources maintain that N. became a *martyr under *Valerian. In 1932, in a small catacomb near S. Lorenzo *fuori le mura*, an inscription was found: *Novatiano beatissimo martyri Gaudentius diaconus fecit*, but the identification of this martyr with the founder of the sect is not demonstrable, indeed is rather improbable.

N.'s dogmatic masterpiece, *De Trinitate*, was not conceived as a treatise on the *Trinity. The author makes no use of the term *trinitas*, though it had appeared 40 years earlier in Tertullian (*De pud.* 21). He calls the *Holy Spirit neither *God nor a *persona*, and does not even include it in his discussion of the unity of God. The Spirit indeed, for N., is something divine and is third alongside the *Father and the Son (not just *fidei auctoritas*, i.e. the weight of the baptismal confession, but even the rational order demands faith in the Spirit). In fact, in his *theology of the Holy Spirit, N. remains a long way behind Tertullian. Since Tertullian's clearest assertions are not just connected in a casual way with the defence of *Montanism and of the new prophecy of the Holy Spirit (*Adv. Praxean* combats the opponents of Montanism), it may be that N.'s reticence has its basis in his lasting rejection of Montanism.

N. introduced into Christian *Latin the expressions *incarnari* (*Trin.* 138) and *praedestinatio* (*Trin.* 94). This latter term too appears in a christological context. *Christology is the basic question for N., though he clarifies in first place the concept of *God (accepting his immutability: *Trin.* 23) and takes on the defence of the God of the *Bible against those who accused Him of being envious (*Trin.* 7, 28). Christ came on earth in a true *body (*Trin.* 56); he is not just man, but also God (*ibid.*). This is clear from the fact that he is invoked everywhere (*Trin.* 74); and he is *concordia utriusque substantiae* (*Trin.* 140). As God, he is not to be identified with the *Father, since he is a second person alongside him (*Trin.* 146). Only in that he came in our *flesh does he have the capacity to save us (*Trin.* 53). To tell the truth, nowhere does the discussion touch on liberation from *sin. Moreover, no expiatory efficacy is attributed to Christ's death; it simply proves Christ's true human nature. God's word (*deus, *sermo*) assumed at the *incarnation a *body which he put off at death, like a tunic, and which he resumed for ever at the *resurrection (*Trin.* 124). Thus was made definitive what had happened at the start through the incarnation. Because the son of God became the son of man, and the son of man the son of God, mankind can be led to eternal salvation (*Trin.* 134). Christ operates this as a teacher, whose word confers immortality and divinity on those who believe it (*Trin.* 84). *Faith, however, seems to be identified with morality; the fault of the *Jews was that they despised the law, while the merit of Christians coming from *paganism lies in the fact of their having come to the law of the gospel, becoming worthy of the help of the Holy Spirit (*Trin.* 164). For N., the burden of man's *sins consists above all in mortality, in being subject to the passions and weaknesses of the flesh. Salvation is obtained through faith and austerity of life. Morality and doctrine take first place, sacramentality and *grace only second place.

N. wrote the wholly allegorical work *De cibis iudaicis* before Decius's persecution; he did it for one of the partial communities that were being formed in the 3rd c., to which he had been assigned. He also composed treatises, now lost, in which he clarified the true meaning of the *Sabbath and *circumcision (*Cib.* I, 6). In *De cibis* he shows how the unclean animals, prohibited in *Scripture, signify wicked human actions (*Cib.* 3, 12); e.g. it is not the eating of raptors, but theft, that is prohibited (*Cib.* 3, 20). *De spectaculis* combats those who tried to justify the licitness of the theatre and their participation in public *entertainments by using arguments from Scripture, adducing the episode of *David dancing (*Spect.* 3, 2), or who wished to reconcile Christianity and the culture of the theatre, the *church and the Roman state. Enough to think of the time before the Decian persecution, under the emp. *Philip the Arab, who quite favoured the Christians. N. sought to call attention to the immoral contents of pagan entertainments. For Christians the only decent entertainments are narrations of Holy Scripture (*Spect.* 10, 1).

De bono pudicitiae was written by N. some time after his episcopal *ordination: in it he mentions that he preached almost daily on the gospel (*Pud.* 1, 1). The basic reason for the practice of *chastity is the church herself, virgin and bride (*Pud.* 2, 2), as well as the desire to spare oneself the shame of sin and *paenitentia*, which is nothing but "shameful evidence of sin committed" (*Pud.* 13, 4). N.'s virgin mother church does not heed sins, does not lead back to *penitence, it cares only that its members remain "the Lord's temple, member of Christ, habitation of the Holy Spirit, chosen for *hope, consecrated in faith, destined for holiness, sons of God, brothers of Christ, consorts of the Holy Spirit" (*Pud.* 2, 1).

CPL 68-76; PL 3, 911-1000 and part of PL 4, among Cyprian's works; CCL 4; H. Weyer, *Novatianus. De Trinitate. Über den dreieinigen Gott*, Text, Übersetzung, Einleitung, Kommentar (Testimonia II), Düsseldorf 1962; H.J. Vogt, *Coetus Sanctorum. Der Kirchenbegriff des Novatian und die Geschichte seiner Sonderkirche*, Theophaneia 20, Bonn 1968; R.J. De Simone, *The Treatise of Novatian the Roman Presbyter on the Trinity*, Studia Ephem. Augustinianum 4, Rome 1970; H. Gülzow, *Cyprian und Novatian. Der Briefwechsel zwischen den Gemeinden Rom und Kartago zur Zeit der Verfolgung des Kaisers Decius*, Beiträge zur historischen Theologie 48, Tübingen 1975; G.W. Clarke, *The Letters of St. Cyprian of Carthage*, ACW 49, New York 1984; P. Mattei, *Novatianus, Oeuvres*, ed. and Fr. tr., SCh (1990); A. Wlosok, Novatianus, in new ed. of Schanz-Hosius-Krüger, *Handbuch der Altertumswissenschaft*; H. Gülzow, *Cyprian und Novatian*, Tübingen 1975.

H.J. Vogt

NOVATIANISTS. The schismatic church founded by *Novatian, the 3rd-c. Roman priest who then became antipope, began in *Rome after the election of pope *Cornelius (March 251), over the question of the *lapsi* at the time of the *persecution of *Decius (249-250); it continued to exist uninterruptedly until the time of *Innocent I (401-17), who closed some churches of the N., and *Celestine I (422-32), who expelled them from Rome. Novatian accused Cornelius of weakness for having readmitted *lapsi* to the church. To avoid contaminating his church of "the pure: καθαροί" (Euseb., *HE* VI, 43, 1), Novatian denied reconciliation to *lapsi* under any condition, proclaiming that no-one but *God had the power to pardon the *lapsi* their faults (Socr., *HE* IV, 28; Cypr., *Ep.* 55, 26-27). The church, for him, was the community of the saved under the guidance of the *Holy Spirit (*De Trin.* 29, 26), a select group of *prophets and *martyrs, the immaculate virgin (*ibid.*). This idea of a Messianic church, heroic and in conflict with the *world, resumed in some way the conflict that opposed *Hippolytus to *Callistus.

To obtain followers, Novatian sent to *Carthage the priest Maximus, the deacon Augendus and the Roman Novatianists Machaeus and Longinus (Cypr., *Ep.* 44, 1). Later, Evaristus, the bishop who had ordained Novatian, *Novatus of Carthage (cf. *Ep.* 14, 10), Primus and Dionysius, organized a Novatianist church in Carthage by consecrating Maximus, who had been rejected by *Cyprian's community, as bishop (Cypr., *Epp.* 50 and 57; 59, 9). Cyprian probably wrote his *De unitate* (May 251) against Novatian and his schismatic church. From Carthage the movement spread to the rest of *Africa. At the time of *Leo the Great (*Ep.* 12, 6) there were still N. in Africa. The Novatianist churches of Africa used to rebaptize (Cypr., *Ep.* 73, 1). W.H.C. Frend (129-130) has observed: "The N. of Africa are shown to be in the same line of *tradition as the *Donatists. The factors which had led to the Donatist *schism were still active". The Latin term "*mundi-*καθαροί" has been found on Donatist inscriptions (CIL VIII, 10656, Henchir el Guiz).

For the Novatianist movement: a) in *Gaul (3rd-5th cc.): 1. *Arles, see Cyprian (*Ep.* 68); 2. *Autun, cf. Jerome (*De vir. ill.* 82); 3. Rouen, cf. Innocent I (*Ep.* 2, 11: PL 20, 475); - b) in *Spain (late 4th c.): *Barcelona, cf. Pacianus (*Epp. tres ad Sympron. Novat.*: PL 13, 1051-1082) - c) in the *East: 1. *Alexandria (5th-7th cc.), cf. Socrates (*HE* VII, 7), Photius (*Bibl.*, cod. 182, 208, 280); 2. *Syria (3rd-4th cc.), cf. Eusebius (*HE* VI, 44, 1; 46, 5), Jerome (*De vir. ill.* 91); 3. in *Asia Minor (3rd-7th cc.) the old prophetic church was succeeded by the *Montanists and they by the Novatianist church (Frend 59); cf. *In Symb. Apost.* 39: PL 21, 376; Socrates, *HE* IV, 28. *Socrates had some sympathy for the N.: cf. what he says of the Novatianist bishops of *Constantinople (V, 21; VII, 6, 12, 17), *Scythia (VII, 46) and Nicaea (VII, 25), their churches in *Phrygia and Paphlagonia (VI, 28), Lydia (VI, 19), Cyzicus (II, 38; III, 11), Nicaea (VII, 12), *Nicomedia and Cotyaeum (IV, 28), the N. and the *Easter controversy (V, 22) and the *martyrdom of Novatian (IV, 28). Cf. Sozomen (*HE* IV, 20; VII, 18; VIII, 1). At the council of *Nicaea, the N. accepted the *homoousios (Socr., *HE* I, 10; V, 10): the *Fathers required from them a profession of faith, the repudiation of errors, *laying-on of hands, but not a new *baptism (can. 8: Mansi 2, 671). In *c.*390 *Ambrose wrote his *De paenitentia* against the N. of *Milan and N *Italy.

Chr. W.F. Walch, *Entwurf einer vollständigen Historie der Ketzereien, Spaltungen und Religionstreitigkeiten, bis auf die Zeiten der Reformation*, II, Leipzig 1764, 185-288; A. Harnack, Novatian. Novatianisches Schisma. Kirche der Katharer, *RE* 14, 223-242; *DTC* 11, 815-849; R. Janin, Les Novatiens orientaux, *Echos d'Orient* (1929) 385-397; A. Casamassa, *Novaziano*, Rome 1947/48, 92-101; E. Peterson, Novaziano e Novazianismo, *EC* 8 (1952) 1976-80; W.H.C. Frend, *The Donatist Church*, Oxford 1952; H.J. Vogt, *Coetus Sanctorum. Der Kirchenbegriff des Novatian und die Geschichte seiner Sonderkirche*, Bonn 1968; H. Crouzel, *L'Eglise primitive face au divorce*, Paris 1971, 221-229; Ch. Pietri, *Roma Christiana*, BEFAR 224, 2 vols., Rome 1976, *passim*; M. Bévenot, Cyprian and his Recognition of Cornelius, *JThS* n.s. 28 (1977) 346-359; G. Cereti, *Divorzio, nuove nozze e penitenza nella chiesa primitiva*, Bologna 1977, 287-319.

R.J. De Simone

NOVATUS. Bishop of Sitifis (now Sétif in Algeria) for 37 years, as his epitaph (Diehl 1101) recites. Ordained in 403, he seems to be the bishop N. to whom *Augustine's *Ep.* 84 is addressed. Indeed he seems not just a *friend of Augustine, whose brother was with N. at *Hippo, but one who made him known to others (Aug., *Epp.* 229, 1 and 230, 1-2); a *letter of his to Augustine is lost (Aug., *Ep.* 229, 1). In 416 he was present at the *council of Milevis (Aug., *Ep.* 176, *intest.*); and at that of *Carthage in 418 (cf. CCL 149, p. 158, 54-56) and esp. that of 419, still more important since it had to deal with the affair of *Apiarius, as one of the legates of *Mauretania Sitifiensis (CCL 149, p. 92 and p. 151). In that year *Galla Placidia invited him to a council to be held at Spoleto - but never held - to deal with the dispute between Boniface, bishop of *Rome, and his rival Eulalius (*Coll. Avel., ep.* 28: CSEL 35, p. 73). In 413 he was expelled from his see by the *Vandal *Genseric (Prosp. Aq., *Chron. ad an.* 437: MGH, *Auct. ant.* IX, 475).

S.A. Morcelli, *Africa christiana*, I, Brescia 1816, 283-284; *PWK* 3 A, 393-394; J.-L. Maier, *L'épiscopat de l'Afrique romaine, vandale et byzantine*, Rome 1973, 369; *PAC* 1, 783-784; A. Mandouze, Novatus in *Prosopographie chrétienne du Bas-Empire*, 1, *Afrique (303-533)*, Paris 1982, 783; *Letter* 28* (Divjak), newly discovered Augustinian Letters publ. J. Divjak, CSEL 88, Vienna 1981; A. Schindler, Vermitteln die neuentdeckten Augustin-Briefe auch neue Erkenntnisse über den Donatismus?, in (ed.) C. Lepelley, *Les Lettres de Saint Augustin découverts par Johannes Divjak*, Paris 1983, 117-121.

A. Di Berardino

NOVATUS of Carthage. Priest of *Carthage (mid 3rd c.). With *Fortunatus and Gordius, he opposed *Cyprian's election (Cypr., *Ep.* 24, 4). A follower of *Felicissimus, greedy, malign and always in search of novelty (*Ep.* 50, 52, 2), he separated from the church (*Ep.* 47, 52, 1) because of his obstinacy. *Eusebius of Caesarea (*HE* VI, 43, 1-2. 21; VI, 45. 46, 1. 3; VII, 4, 8) and all the Greek writers who follow him confuse N. with *Novatian.

E. Romero Pose

NOVATUS the Catholic. Monk, prob. 4th-c. (but, acc. to B. Fischer, later), author of the *Sententia de humilitate et oboedientia et de calcanda superbia*, a brief exhortation to charity and humility.

CPL 639.1154; PL 18, 67-70; PLS 1, 672.1751; *DSp* 11, 477-478.

E. Romero Pose

NOVIODUNUM (Isaccea, district of Tulcea, Romania). Celtic place-name: Roman, then Byzantine, fortress and port in N Dobruja, on the right bank of the Danube; rebuilt at the start of the 4th c. and again under *Justinian (527-565). Excavations in 1955-56 identified part of the wall and the buildings on the N-E side of the city, mostly destroyed by the river. Here were found ruins of a Christian basilica backing onto the city wall, with a wide apse to N-E, three aisles and a simple narthex. A cemeterial basilica was identified in the great necropolis of the S zone of the city (not excavated).

The most important early Christian monument, found in 1971 some 10 km S-E of the city in the centre of the village of Niculitel, is a crypt with martyrs' tombs (*martyrium*), above which rose a Christian *church. The *martyrium*, preserved entire, is a square-plan building with sides of *c.* 3.50 m., height 2.25 to 2.30 m., with a brick cupola 3 m. in diameter. On the W side is a small entrance (0.70 x 0.69 m.), closed at the time of its discovery by a limestone slab. Inside lay a great coffin of fir planks with four human skeletons. On the upper part of the N and S walls, each under a monogrammatic *cross, were *epigraphs incised in the raw plaster and later coloured red: Μάρτυρες Χριστοῦ (N wall); Μάρτυρες Ζώτικος, Ἄτταλος, Καμάσις, Φίλιππος, (S wall). The names of these four *martyrs, Zotikos, Attalos, Kamasi(o)s and Philippos, with "25 others", are recorded by *Mart. hier.* for 4 June. Further digging in 1973 and 1975 revealed beneath this crypt two small rooms, on whose entrance was the epigraph: Ὧδε καὶ ὧδε ἰχὼρ μαρτύρων ("Here and there [is] martyrs' *blood"). Within were two ceramic vases and a great quantity of burnt human bones belonging to two bodies. It is supposed that the six martyrs fell during *Diocletian's great *persecution of 303-304. Placed initially or later in a *martyrium*-type building, traces of whose foundations were identified W of the surviving *martyrium*, the relics of the six martyrs were transferred to the latter, prob. in the 5th/6th c., when a basilica, whose foundations have been partly uncovered, was also built over them. **[Fig: 215]**

V.H. Baumann, *Dacia N.S. La basilique découverte à Niculitel*, 16 (1972) 189-202; E. Popescu, *Inscriptiile grecesti si latine din. secolele IV-XIII descoperite in România*, Bucharest 1976; I. Barnea, *Les monuments paléochrétiens de Roumanie*, Vatican City 1977; id., *Christian Art in Romania*, I, Bucharest 1979.

I. Barnea

NUBIA

I. Christianity - II. Ancient Nubian.

I. Christianity. N. is a region situated along the Nile between N Sudan and S *Egypt, reaching more or less from the confluence of the Blue and White Niles (Khartoum) to the First Cataract (Aswan). Christianity probably arrived sporadically in the 3rd c. The earliest archaeological evidence dates from *c.* 400. *John of Ephesus's *History* attributes the *evangelization of N. to rival missions sent by *Justinian and *Theodora in *c.* 540, which converted the region's three kingdoms: Nobadia (capital Faras/Pachoras), Makuria (capital Old Dongola) and Alodia (capital Soba). Relations with *Byzantium were interrupted in *c.* 640 by the Arab conquest of Egypt, which ushered in a permanent state of war which ended in 1323 with the invasion of the kingdom of Dongola (united with Nobadia *c.* 700). The kingdom of Alodia was destroyed at the end of the 15th c. by the Fung, a desert people.

Apart from the possible presence of missionaries of Chalcedonian obedience in the 6th c., ecclesiastically N. emerges as an anti-Chalcedonian region organized around the episcopal sees of Faras (later transferred to Qasr Ibrim/Primis) and Djebel Adda/Daw). Recent archaeological research has revealed a Christian culture with original traits, flourishing in the 9th-11th cc. both architecturally (*churches with characteristic ground-plans, sometimes frescoed) and literarily (*epigraphy and literary remains in Greek, Sahidic *Coptic, Nubian and *Arabic).

U. Monneret de Villard, *Storia della Nubia cristiana*, OCA 118, Rome 1938; id., *La Nubia medioevale*, I-IV, Cairo 1935-57; W.B. Emery, *Egypt in Nubia*, London 1965; S. Jakobielski, *A History of the Bishopric of Pachoras on the Basis of Coptic Inscriptions*, Faras 3, Warsaw 1972; K. Michalowski (ed.), *Nubia. Récentes recherche*, Warsaw 1975; J.M. Plumley, *The Scrolls of Bishop Timotheo*, London 1975; W.H.C. Frend-I.A. Muirhead, The Greek manuscripts from the Cathedral of Q'asr Ibrim, *Muséon* 89 (1976) 43-49; *Nubia christiana*, I, Warsaw 1982; Arts. by J.M. Plumley et al., *JEA* 1964 and following on Egypt Exploration Excavations at Q'asr Ibrim 1963-; W.H.C. Frend et al., Fragments of an Acta Martyrum from Qasr Ibrim, *JbAC* 19 (1986) 66-70; id., A Eucharistic Sequence from Q'asr Ibrim, *ibid.* 30 (1987) 90-98; id., Fragments of the Acta S. Georgii from Q'asr Ibrim, *ibid.* 32 (1989) 89-104; id., The Cult of Military Saints in Christian Nubia, in (ed.) C. Andresen, *Theologia Crucis-Signum Crucis* (Fetschr. E. Dinkler), Tübingen 1979, 155-163; P. van Moorsel et al., *The Central Church of Abdallah Nirgi*, Leiden 1975; W.Y. Adams, *Nubia: Corridor to Africa*, London 1977 (indispensable work); W.Y. Adams, The Architectural Evolution of the Nubian Church, 500-1400 AD, *JARCE* (1965) 87-139; P.L. Shinnie-H.N. Chithick, Ghazali - a monastery in the Northern Sudan, and Excavations at Soba, *Sudan Antiquities Service Occasional Papers* 3 and 5 (1955 and 1961); N.B. Millet, *Gebel Adda*, Reports in *JARCE* 1963, 1964 and 1967; W.H.C. Frend, The Exploration of Christian Nubia; retrospect and prospects, *Proceedings of PMR Conference* 1981, Villanova 1985, 51-53.

II. Ancient Nubian. A language of the Nilo-Cyadic family, in which some works of Christian interest were written or translated in the Middle Ages, using an alphabet derived from ancient *Coptic: we possess parts of the *Bible (esp. a lectionary for the Christmas period), a collection of the miracles of St *Mennas, the canons of *Nicaea, a *homily on the *cross and numerous minor or fragmentary texts, some still unidentified.

B.M. Metzger, *The Early Versions of the New Testament, Their Origin, Transmission and Limitations*, Oxford 1977, 268-274; G.M. Browne, An Old Nubian Fragment of Revelation, *Studia Papyrologica* 20 (1981) 73-82; E.A.W. Budge, *Texts Relating to Saint Mena of Egypt and Canons of Nicaea in a Nubian Dialect*, London 1909; F.L. Griffith, *The Nubian Texts of the Christian Period* (= Abh-Berlin 1913, 8), Berlin 1913; F.L. Griffith, Christian Documents from Nubia, *PBA* 14 (1928) 1-3; E. Zyhlarz, Neue Sprachdenkmäler des Altnubischen, *Studies Presented to F.L. Griffith*, London 1932, 187-195; J. Barns, A Text of the Benedicite in Greek and Old Nubian from Kasr El-Wizz, *JEA* 60 (1974) 206-211; C.D.G. Müller, Deutsche Textfunde in Nubien, in E. Dinkler, *Kunst und Geschichte Nubiens in christlicher Zeit*, Recklinghausen 1970, 245-258; J.M. Plumley-G.M. Browne, *Old Nubian Texts from Q'asr Ibrim*, Texts from Excavations 9, London 1988.

S.J. Voicu

NUMBERS, Symbolism of. In the symbolism of n., for the ancient world and the *Fathers, we must go back to the *flexio digitorum* which was their way of counting and to grammatry and gematry (psephy and isopsephy). The term *digitus*, "finger", also meant "skill in counting" and so came under the rule of *actio*, which played so great a part in *rhetoric. The first chapter, *De loquela digitorum*, of *Bede's *De temporum ratione* gives a clear picture of all the n. as expressed by the fingers. The flexing of the fingers produced images which attributed definite values to biblical n. A typical example is the 100 (*martyrs), 60 (*widows) and 30 (married people) of the *parable of the sower. The digital representation of 100 (a circle formed by the right thumb and index finger) developed images which become a possession of the spirit, i.e. it was the crown of martyrdom. The representation of 100 signified perfection. To the 99 *sheep one more must be added: the lost one. To express 30, the index finger was set down gently on the thumb as if to indicate the conjugal embrace. So 30 was attributed to faithful spouses. 60, from the uncomfortable position of the thumb against the index finger, was for widows. In many texts we have the expression: (s)he has obtained 100, or 60, or 30 - incomprehensible unless we understand the meanings given to these numbers. The same goes for the Kingdom of Heaven expressed by the *ogdoad, indicated by the sign of eight: thumb, index finger and middle finger raised, ring finger and little finger folded on the palm of the hand. This was not three, as it would be today: that was made with index finger and thumb raised, and the other fingers folded on the palm. Much incomprehension has been caused by the gesture of eight, which has not been understood in our times. It has been confused with the gesture of *prayer partly because of the dead who, however, hope in the ogdoad: the Kingdom of Heaven. Arabic numerals have universalized the signs, so we must go back to methods and ways different from our own in order to understand the past. The symbology of n. as recognized by the early Christians may seem affected or artificial, but we must enter into the spirit of their culture in order to read and understand them. The actio of n., which lay outside any common procedure and did not arise spontaneously, had no consistency. This was one subject of *Irenaeus's polemic (Adv. haer. I, 16, 2) against the *Valentinians. For St *Augustine, without knowing the language of n. we cannot understand the metaphorical and mystical senses of *Scripture (Doct. christ. 2, 16, 25). The Lord has arranged all things by measure, number and weight (Wis 11, 21). Augustine would have liked a dictionary compiled, giving the mystical senses of the n. mentioned in the *Bible (Doct. christ. 2, 39, 59). He thought it stupid to imagine that n. appeared in *Scripture without reason (Trin. 4, 6). Besides Augustine and Bede, *Isidore of Seville wrote a short treatise: Liber numerorum qui in Sanctis Scripturis occurrunt. Under the title Liber de computo we have a vast mass of disordered material on the factors relating to the Easter cycle. The term *computus was, by antonomasia, that which concerned *Easter.

There is no early Christian author who does not, directly or indirectly, mention the symbolism of n. *Methodius of Olympius's Symposium (8, 11, 199-203 Bonwetsch) gives the values attrib. to certain n.: 1000 (100 x 10) symbolizes the *Father; 200, the sum of two perfect n., the *Holy Spirit; 60 (6 x 10), Christ. Even earlier, *Clement of Alexandria gives the numerology of the culture of his time. For him, 6, 7 and 8 are the three n. which link man most directly to the deity (Strom. VI, 16, 138, 5-6; VI, 16, 140, 1-2; VI, 16, 141, 3 Stählin).

We now come to grammatry and gematry or, as some say, psephy and isopsephy. The relationship between the letters composing a word and the n. signified by those letters, also had symbolic value. The early Christians thought that every letter of the Hebrew, Greek and Latin alphabets had a mysterious nature. *Origen, referring to the 22 inspired books of the Bible, discerned in the 22 letters of the Hebrew alphabet an introduction to the wisdom and divine teachings imprinted in men (Select. in Ps. 1: PG 12, 1084). Even signatures often consisted of the number that corresponded to the *name, as in the graffiti of Pompeii. But each number could correspond to many different names and subjects. Irenaeus (Adv. haer. 5, 30, 3 Rousseau) says that many names can have the number 666. The number 801, given by the letters A (1) and Ω (800), is attrib. to Christ as fullness and perfection, the beginning and the end, the first and the last of the *Apocalypse (21, 6; 22, 12) (cf. *Alpha and Omega). The symbolism of n. is a vital part of *patristics and must be studied in order to understand a method of research based on the culture of the time. The *theology of the Fathers had no wish to detach itself from the popular mind.

A. Quacquarelli, Il triplice frutto della vita cristiana 100, 60 and 30 (Mt. 13, 8 nelle diverse interpretazioni), Rome 1953; id., Ai margini dell'actio: La loquela digitorum (La rappresentazione dei numeri con la flessione delle dita in un prontuario trasmesso dal Beda), VetChr 7 (1970) 199-224; id., La fortuna di Archimede nei retori e negli autori cristiani antichi, in Saggi patristici-Retorica ed esegesi biblica, Quaderni di VetChr 5, Bari 1971, 381-424; id., L'ogdoade patristica e i suoi riflessi nella liturgia e nei monumenti, Quaderni di VetChr 7, Bari 1973; A.Y. Collins, Numerical Symbolism in Jewish and Early Christian Apocalyptic Literature, ANRW II.21.1 (in press).

A. Quacquarelli

NUMENIUS of Apamea. Platonizing neopythagorean philosopher (2nd c. AD) of great interest for the religions and cultures of the Oriental peoples and the history of philosophy, Greek and other. His theologico-metaphysical system based on the "three gods" - the first god or absolute good, the second god or demiurge, the third god or *world (fr. 21 Des Places, p. 60) - clearly foreshadows, though with some differences, the Plotinian system of three *hypostases. His works - the most important was On the good - are lost, surviving only in fragments and accounts of his doctrines by later authors (*Origen the Christian, *Eusebius of Caesarea, *Porphyry, Iamblichus, *Nemesius of Emesa, *Olympiodorus, *Aeneas of Gaza, Calcidius, *Proclus, Macrobius); these have been collected and edited first by E. A. Leemans and then by E. Des Places (see final bibl.). The "first god" (on this expression cf. frr. 11, 15 and 21 Des Places) is the Pythagorean monad (fr. 52, p. 95, 5-6 Des Places), the absolute good of Plato's Rep. VI 509b (fr. 16, p. 57, 4, 14; fr. 20, p. 60), the supreme intellect (fr. 16, p. 57), the completely unknown absolute being (fr. 17, p. 58): on the one hand it "surpasses essence" (fr. 2, p. 44, 16), is prior to it and is its principle (fr. 16, p. 57); on the other, it is "inherent in essence" (fr. 16, p. 57, 10) and possesses an essence, though different from that of the second god (fr. 16, p. 57, 15-16); it is absolutely inactive and motionless (fr. 12, p. 54; fr. 15, p. 56), but at the same time it "sows" the seminal principles (fr. 13, p. 55).

The "second god" is the "maker" and "demiurge" of Plato's Timaeus (fr. 21, p. 60; fr. 16, p. 57), the second intellect (fr. 17, p. 58), concerned with intelligible as well as sensible things (fr. 15, p. 56), "good" (cf. Tim. 29e) in that it participates in the absolute good represented by the first god (fr. 19, p. 59; fr. 20, p. 60); it watches over matter, takes care of it and is associated with it, remaining separate from it and raising it to its own character (fr. 11, p. 53); it moves (fr. 15, p. 56), proceeds across the heaven and directs it (fr. 12, p. 54); it is the cause of the harmony that reigns in matter, and guides it by contemplating the ideas (fr. 18, pp. 58-59). Matter is ungenerate and originally without forms or qualities, an opinion which N. shares with contemporary *middle Platonism (fr. 52, p. 95); but once it has received forms, ornaments and harmony, it becomes the sensible universe and in this sense may be considered generate (fr. 52, p. 95, 10-12).

The sensible universe, the "third god" (cf. fr. 21), is beautiful in that it is an imitation of the essence of the "second god" and receives its ornaments from its participation in the beautiful (fr. 16); along with the "second god" - which, as we have seen, is united with matter - it forms a single entity (fr. 11, p. 53, 14).

Like so many upholders of the "Oriental mirage", N. sees Platonic doctrines already adumbrated in other older Eastern peoples, including the Jewish (fr. 1, p. 42), and sees in Plato "an Attic-speaking *Moses" (fr. 8, p. 51). *Plotinus read and commented on N.'s works in his lectures, as he did those of other philosophers (Porphyry, Vita Plot., ch. 14, vol. I, p. 15, 12 Bréhier); he was even accused of having plagiarized N. (Porph., ibid. ch. 17, p. 17, 1-2). Indeed there is no lack of analogies, even terminological ones, between N. and Plotinus, brought out by E.R. Dodds (Les sources de Plotin, in Entr. sur l'ant. class. V, 16-20: here we will merely note that the expression ἐποχούμενον ἐπὶ τῇ οὐσίᾳ, referring to the first god (fr. 2, p. 44, 16), is echoed by Plotinus's ἐποχούμενον τῇ νοητῇ φύσει καὶ τῇ οὐσίᾳ (Enn. I 1, 8 [I 44, 9-10]). Another important point in which N. anticipates *neoplatonism, also brought out by Dodds (op. cit. pp. 23-24), is the idea that the conferring of benefits by higher principles to lower beings causes no impoverishment in the former, since the benefits, however profuse, continue to remain in their source (fr. 14, pp. 55-56). The three Plotinian *hypostases, however, represent a more coherent system than N.'s system turning on the "three gods": Plotinus's first hypostasis, the one good, would be unequivocally raised above intellect and οὐσία; his second hypostasis, or intellect, though remaining identical with the demiurge, would be clearly separated from matter and the sensible *world; the third hypostasis or *anima mundi - not the second - would have the task of guiding and directing the sensible universe, just as Plato had said in the Timaeus, the Philebus and book X of the Laws. N. was also appreciated by various Greek *Fathers: the phrase "What is Plato but an Attic-speaking Moses?" is cited by *Clement, *Eusebius and *Theodoret (cf. refs. in Des Places, p. 52, n. 4); numerous fragments of N. are preserved by Eusebius; some points of the *theology of *Origen - who, acc. to *Porphyry (Euseb., HE VI, 19, 8), was an assiduous reader of N. - show clear dependence on his doctrines (cf. *Platonism and the Fathers, apropos of Origen); and *Nemesius of Emesa (De natura hominis, ch. 2) invokes the authority of *Ammonius Saccas and N. (fr. 4b, pp. 46-47) to support the theory of the incorporeality of the *soul.

PWK Suppl. 7, 664-678; KLP 4, 192-194; F. Thedinga, De Numenio philosopho platonico, (Diss.) Bonn 1875; H.Ch. Puech, Numénius d'Apamée et les théologies orientales au second siècle, in Mélanges Bidez, 2, 1934, 745-778 (repr. in En quête de la Gnose, I, Paris 1978, 25-54); A.J. Festugière, La révélation d'Hermès Trismégiste, III, Paris 1953, 42-47; IV, Paris 1954, 123-132; G. Martano, Numenio d'Apamea, Naples 1960; E.R. Dodds, Numenius and Ammonius, in Entr. sur l'ant. class., V, Geneva 1960, 3-24; P. Merlan, The Cambridge History of Later Greek and Early Medieval Philosophy, Cambridge 1967, 96-106; J.H. Waszink, Porphyrios und Numenios, in Entr. sur l'ant. class. XII, Geneva 1966, 35-78; E. Des Places, Numénius Fragments, Paris 1973.

S. Lilla

NUMIDIA. Annexed to Roman *Africa by *Augustus (AD 25); established as a military territory under *Claudius (37) and as a province under *Septimius Severus (197-213), under the authority of the legate of the 3rd Legion Augusta stationed at *Lambaesis; divided by *Diocletian into two, with capitals at *Cirta in the N and Lambaesis in the S, with four centres linked in an autonomous confederation: Cirta-*Constantine; Chullu-Collo; Rusicade-Skikda, Milevis-Mila; conquered by the *Vandals (429-456), by *Justinian (535) and by the Arabs (c.700). N. was a turbulent province, which ill supported foreign rule, with periodic revolts breaking out starting from Aurès; an essentially agricultural province (grain, olives), where urbanization (old cities: *Hippo Regius, Cirta, Calama; Roman-founded cities: Lambaesis, Thamugadi-*Timgad, Mascula, Khenchéla, Cuicul-Djémila, etc.) favoured Roman occupation before the Christian penetration. The economic disorganization at the time of the Vandals compromised the Roman and Christian presence in the long run.

Christian presence in N. is ascertained firstly by the existence of *martyrs in the early 3rd c., cited by Maximus of Madaura: Miggin, Sanaem, Namphamus, Lucita (Aug., *Ep.* 16, 2). In *Cyprian's time, the functioning of the institution of *councils supposes some 30 bishoprics in N., of which 21 have been identified thanks to the *Sententiae episcoporum* of 256: Badia-Bades, 15; Bagai, 12; Bamaccora, 33; Castra Galbae, 6; Cedias-Hr Oum Kif, 11; *Cirta, 8; Cuicul-Djémila, 71; Gadiaufala-Ksar Sbahi, 76; Gemellae-Sidi Aich, 82; *Hippo Regius-Annaba, 14; Lamasba-Hr Merouane, 75; *Lambaesis, 6; Madaura-Mdaourouch, 73; Mascula-Khenchéla, 79; Midili-Hr Bou Rebia, 45; Milevis, 13; Nova-Hr Ben Khelifi or Hr el Ateuch, 60; Octavu, 78; Rusicade-Skikda, 70; Thamugadi-*Timgad, 4; Thubunae-Tobna, 5.

*Valerian's *persecution (259) produced *martyrs, mentioned in the *Passio Mariani* (BHL 131): Agapius, an otherwise unknown bishop; Secundinus, bishop of Cedias; *Marianus and James, clerics of Cirta; Aemilianus, Tertulla, Antonia. Other martyrs of this persecution are annotated in the *Sententiae episcoporum* of 256, like Nemesianus and his companions. In 303 the *Gesta apud Zenophilum* (proceedings of the sequestration of the goods of the church of Cirta) give a concrete idea of the community's economic importance and *liturgical furniture. *Donatism, born in the aftermath of Diocletian's persecution, was deeply rooted in N., where it dictated laws. At the conference of *Carthage of 411, N. was represented by 55+9 Catholic bishops and 67+4 Donatist ones, i.e. 64 against 71. This figure gives a picture, though very imperfect, of the predominance of Donatists in the province. The development of the *schism seems to have benefited from missionary efforts in the countryside after the *peace of the church (313) and the lack of strong personalities among the *Catholics (except *Optatus of Milevis), in contrast to the Donatists with their warrior-bishops like Donatus of Bagai and *Optatus of Thamugadi and their great controversialists like *Parmenian or *Petilian. This situation would be reversed by the activities of *Augustine. Further confirmation is possible thanks to the *Notitia provinciarum et civitatum Africae*, which mentions 125 Numidian bishops.

PWK 17, 1343-57; Delehaye *OC*, 376-99, *passim*; V. Saxer, *Vie liturgique et quotidienne à Carthage vers le milieu du IIIe s.*, Vatican City 1969, *passim*; id., *Saints anciens d'Afrique du Nord*, *passim*, Vatican City 1979; Y.-M. Duval, *Loca sanctorum Africae*, CEFR 58, 2 vols., Rome 1982; A. Berthier et al., *Les Vestiges du christianisme antique dans la Numidie centrale*, Algiers 1942 (indispensable); W.H.C. Frend, A Note on Religion and Life in a Numidian Village in the Later Roman Empire in *Archaeology and the Study of Early Christianity*, London 1988; id., The End of Byzantine North Africa: Some evidence of Transitions, *ibid.*; C. Lepelley, *Les Cités de L'Afrique romaine au Bas-Empire*, 2 vols., Paris 1979 and 1981; W.H.C Frend, Fussala, Augustine's Crisis of Credibility (Divjak Letter 20*), in (ed.) C. Lepelley, *Les Lettres de Saint Augustin découvertes par Johannes Divjak*, Paris 1983; S. Lancel, L'affaire d'Antoninus de Fussala, *ibid.*, 267-285.

V. Saxer

NUMISMATICS

1. Preconstantinian period. Two Aquileian bronze coins from the reign of *Maxentius bear a small *cross incised on the tympanum of the temple of the goddess Roma. The sign, considered the first Christian mark on Roman coinage (Laffranchi, Il problematico, 45 ff.), predating *Constantine, has in fact been shown to be the decussate form of the numeral X, alluding to the tenth year of the emperor's reign (Picozzi, *I "folles"*, 75 ff.).

2. Constantinian period. Even after the edict of 303, Christian *symbolism does not immediately appear on the coinage (Pautasso, *I segni*, 501). The first symbol is the *monogram which Constantine had engraved on his soldiers' shields on the eve of the battle of the Milvian Bridge in 312, acc. to the accounts of *Lactantius (*De mort. persec.* 45, 5) and *Eusebius of Caesarea (*Vita Const.* 28-29). The *coeleste signum*, an interwoven X and P - initials of Χριστός - appear in a medallion of 315 coined at *Ticinum*, engraved on Constantine's helmet, the whole accompanied by the legend IMP. CONSTANTINVS P.F.AVG. (Alföldi, The helmet, 9 ff.). In 316 some *folles* from Ticinum bore a Greek cross alongside the figure of Mars or the *Sun (Bruun, The Christian signs, 5 ff.); in 317 and 318 similar *folles* were coined at *Rome and Londinium (Daniélou, *Les Symboles*, 143 ff.). The Greek cross, variously situated, appears again in *folles* from Ticinum, *Trier and Lugdunum in 318 and 319, while in the same years some *folles* from Siscia show Constantine with the christological monogram on his helmet with the legend VICTORIAE LAETAE PRINC. PERP., accrediting the evidence of Eusebius (*Vita Const.* I, 13), who relates that the emperor bore Christ's initials on his helmet. In 320 *folles* were issued at Londinium, Ticinum, *Aquileia, Siscia and *Thessalonica with the cross or the monogram between two conquered barbarians and the legend VIRTUS EXERCITVS or VIRTVS PRINCIPI. A gold medallion, coined at Siscia in 326, shows on the obverse Constantine, his face stretched out towards the deity and his arms in the *gesture of *expansibus manibus*, to signify the new relationship between the emperor and *God and, at the same time, the *pietas* of the sovereign (Breglia, *L'arte romana*, 218 ff.). On the reverse is the emperor again, holding the *labarum* with the christogram, the whole accompanied by the legend GLORIA SAECVLI. In 327-8 the Christian *labarum* appears again, crushing a serpent - an allusion to the defeated *Licinius (Franchi de' Cavalieri, Il labaro, 61 ff.) - on a *follis* of *Constantinople with the legend SPES PVBLICA, which expresses the empire's hope in the three sons of Constantine, founder of the second Flavian dynasty.

3. The era of the Constantinids. From 334 to 336 *folles* were issued at Aquileia, Trier, *Arles and *Lyons with christograms variously situated and the legend VIRTVS EXERCITVS (Pautasso, I segni, 508). In 336 at *Antioch, *solidi* were minted with the legend VICTORIA CONSTANTINI AVG. and the monogrammatic cross ☧, also mentioned by Eusebius (*Vita Const.* 2, 9). On the *pecunia maiorina* - a new denomination instituted between 346 and 348 to revalue the bronze coinage, with the legend FEL. TEMP. REPARATIO - the emperor appears with the *labarum*, or between two standards, or on a ship guided by a winged genius between christograms and the legend CONCORDIA MILITVM. The conflicts (350-353) between *Magnentius and *Constantius II made their way into numismatic *iconography (Kellner, *Libertas und Christogram*, *passim*) with personifications of Liberty and Victory and the legends VICTORIA AVG. LIB. ROMANOR., RESTITVTOR LIBERTATIS, LIBERATOR REIPUBLICAE. Maxentius's bronze coins have great free-standing christograms enriched by the apocalyptic letters *Alpha and Omega, i.e. taking up an anti-*Arian position in order to gain the support of the *Catholics, while Constantius put on his coins, as well as the *labarum* with the Christian emblem, the legend HOC SIGNO VICTOR ERIS, reclaiming the famous Constantinian auspice on the eve of the Milvian Bridge.

4. From Theodosius to Justinian. With *Theodosius I (379-395) some *solidi* represent the emperor with the *labarum* and a processional cross, Victory enthroned with a shield blazoned with a cross, the statues of Rome and Constantinople with the cross-surmounted globe. In the *solidi* and *maiorinae* of Theodosius's wife Aelia *Flaccilla, Victory engraves a christogram on a shield. *Centenionales* of *Arcadius (394-408) and *solidi* of *Honorius (395-423) depict the emperor being crowned by the *manus Dei*, and the gold coinage of *Galla Placidia presents the legend SALUS REI PUBLICAE and a christogram inside a laurel crown. In the following years the iconography of the emperor bearing a cross is enriched by the element of the *volumen* (Panvini Rosati, Il rotolo, 557-566), and in a *solidus* of *Theodosius II the emperor crushes a human-headed serpent, referring to the usurper John (423-425), defeated in 425 by *Valentinian III and recalling Ps 90 [91], 13: *super aspidem et basiliscum ambulabis, leonem et draconem calpestabis*. Later, legends and reliefs remain unchanged apart for rare marginal elements until *Zeno (479-491) and Phocas (602-610), after which the Latin cross appears on its own until the end of the 10th c., sometimes accompanied by the suggestive saying LVX MVNDI. But the christogram too endures on the *solidi* of *Eudoxia, Arcadius and *Justinian, with the comment SALVS ORIENTIS FELICITAS OCCIDENTIS.

H. Cohen, *Description des monnaies frappées dans l'empire romain*, vols. VII-VIII, Paris 1888-1892; J. Maurice, *Numismatique Constantinienne*, 3 vols., Paris 1908-1912; P. Franchi de' Cavalieri, Il labaro descritto da Eusebio, *Studi Romani* 50 (1914) 161; L. Laffranchi, Il problematico segno della croce nelle monete precostantiniane di Aquileia, *Aquileia Nostra* 3 (1932) 45-52; A. Alföldi, The helmet of Constantine with the Christian Monogram, *JRS* 21 (1932) 9-23; J. Daniélou, *Les symboles chrétiens primitifs*, Paris 1961, 143 ff.; P. Bruun, The Christian signs on the coins of Constantine, *Arctos* 3 (1962) 5-35; F. Panvini Rosati, Il rotolo come attributo dell'Imperatore sulle monete romane, in *Atti del VI Cong. Internaz. di Arch. Crist.*, Vatican City 1965, 557-566; P. Bruun, *The Roman*

Imperial Coins, VII, London 1966; L. Breglia, *L'arte romana nelle monete di età imperiale*, Milan 1968; W. Kellner, *Libertas und Christogram*, Karlsruhe 1969; S. De Caro Balbi, Comparsa di simboli cristiani sulle monete dell'impero in età costantiniana, *Annali* 16-17 (1969-1970) 143-169; F. Panvini Rosati, La zecca di Aquileia, in *Aquileia e Ravenna*, Udine 1973, 289-298; V. Picozzi, I "folles" con leggenda CONSERV VRB SVAE coniati nella zecca di Aquileia sotto Massenzio, *Rivista Italiana di Numismatica* (1979) 175 ff.; A. Pautasso, I segni del Cristianesimo nella monetazione romana, in *Atti del V Congresso Nazionale di Archeologia Cristiana*, Rome 1982, 491-525; C.H.V. Sutherland et al., *Roman Imperial Coinage*, London 1967-1981, vols. 6-9.

F. Bisconti

NYSSA. City in *Cappadocia, in the Strategia Morimene (cf. Ptol., V, 6, 23) on the banks of the Halys: E of *Caesarea, on the road to *Ancyra and not far from the border with *Galatia (cf. Greg. Nyss., *Ep.* 6). Its site is not definitely identified (cf. Ramsay, *Historical Geography*, 287-288). Episcopal see of *Gregory, younger brother of *Basil of Caesarea, who consecrated him in 372 when N. became important in ecclesiastical politics because it was part of Cappadocia II, a province created by *Valens's division of Cappadocia into two parts (371-2). Basil aimed to create bishops linked to himself in order to challenge the rule of the bishop of *Tyana - capital of the new province - who opposed the Nicene front. Bishops: Gregory (372-94); Heraclidas (431); Musonius (449-51), present at *Chalcedon (Mansi 6, 981 ff.); John, at *Constantinople (553): (Mansi 9, 175, 192 and 391); John II (680); Paul (692).

W.M. Ramsay, *The Historical Geography of Asia Minor*, London 1890; W. Ruge, Nyssa, *PWK* 17, 2, 1662. For the bishops: M. Le Quien, *Oriens Christianus*, I, Paris 1740, 391 ff., and P.B. Gams, *Series Episc.*, Graz 1957 (repr.), 440.

M. Forlin Patrucco

O

OAK, Synod of the. In this place not far from *Constantinople, a *council met in 403 to settle the dispute between *John Chrysostom, patriarch of Constantinople, and *Theophilus of Alexandria. Certain monks expelled from *Egypt (401) by Theophilus, who accused them of *Origenism, had found refuge at Constantinople. The conflict between them and Theophilus flaring up again, the emp. *Arcadius had called Theophilus to Constantinople to clear himself before a tribunal presided over by Chrysostom. Reaching the capital in 403, the able Theophilus, exploiting support at court and among members of the local *clergy hostile to their bishop, quickly managed to reverse the situation, so that it was he who accused John at the council which met at the O. Taking part in it were some Egyptian bishops who had come with Theophilus and a few others from various regions hostile to John, including the Syrians *Acacius of Beroea, *Severian of Gabala and *Antiochus of Ptolemais. John was accused of having lent help to the Origenists and of various administrative irregularities. He refused to be judged by his enemies and protested by *letter against the procedure, which had various irregularities. He refused to appear even when enjoined to do so by an imperial rescript, and the participants in the council used this refusal to condemn and depose him. The emperor immediately had him transported to a place in *Bithynia.

Hfl-Lecl 2, 137-150; C. Baur, *Der heilige Johannes Chrysostomus und seine Zeit*, II, Munich 1930, 202-220.

M. Simonetti

OBLATIONS OF THE FAITHFUL. The NT uses two terms: *eleemosyna*, often associated with *prayer and *fasting, as forms of worship (cf. *Did.* 1, 6; 15, 4; *2 Clem.* 16, 4), and *prosphora*, dear to St *Paul, who emphasizes the liturgical character of the offering, whether of Christians offering themselves (Rom 15, 16; *1 Clem.* 36, 1; *Mart. Pol.* 14, 1) or of offerings brought for worship. In the Jewish context, the *Didachè* requires the offering of firstfruits and the tithe to the *ministers of the gospel. The same principle is in *Irenaeus (*Adv. haer.* IV, 18, 1-2) and esp. in *Origen (*Hom. Num.* 11, 2; *C. Cels.* VII, 34-35; Hamman, 255-256). *Hermas knows offerings for bishops and for the *poor (*Sim.* V, 2, 9-11). *Justin mentions the offerings twice: they are spontaneous and linked to the Sunday *assembly, for the poor (*I Apol.* 1, 17 and 67, 1-6). The *Sibylline* *oracles describe the custom of placing clothes for the needy on the altar, as well as *bread and *wine (VIII, 483-486; 496-500). Sometimes the bishop appealed to the generosity of the faithful (Tertull., *De ieiunio* 13; *Hom. Clem.* III, 71, 5; for *Africa, cf. Hamman, 257-259). The liturgical and *canonical collections also regulate the offerings and specify their organization and the distribution of the gifts (*Didasc.* VIII-IX; *Trad. Ap.* 5, 1-2; 28, 1-8; *Const. Ap.* II, 34, 5; 36, 3-8; VII, 29, 1-3; VIII, 18, 30; *Can. Ap.* 2-4; *Can. Hipp.* 159-163; *Test. D. II*, 11; in the following *councils: *Gangra, cann. 7-8; IV *Carthage, can. 95; *Vaison, can. 4; II *Arles, can. 4; Orléans, cann. 14-16; Mâcon, can. 4 [581]; and Mâcon [585], which renews the old prescription of the tithe on pain of *excommunication, can. 5).

*John Chrysostom mentions the offering of the faithful at every liturgical assembly (*Adv. Jud.* 3, 4; *Hom. Hebr.* 11, 3), which the cleric must distribute, without storing up, for *works of assistance and esp. for the poor. Benefactors' names are mentioned in the litanical prayer (*In Act. hom.* 18, 5; on which v. Hamman, 198-200). In the West, *Cyprian (*De op. et el.* 15) and later *Ambrose protest against the faithful who forget their offering at the liturgical assembly (*In Ps.* 118, prol. 2). *Augustine specifies that the offering remits a great number of *sins; in this sense liturgical offering and *almsgiving are intimately connected (*Ench.* 110). *Caesarius of Arles, especially, preaches on the importance of the offering; the avaricious man is a homicide, he makes himself guilty of the death of the poor. His teaching recurs in the councils of *Gaul.

A. Hamman, L'offrande des fidèles, in *Vie liturgique et vie sociale*, Paris 1968, 229-295; R.M. Grant, The Organization of Alms, in *Early Christianity and Society*, London 1978, 124-145; Libéralités des fidèles, *DACL* 9, 489-497; J. Fasiori, La dîme du début de deuxième siècle jusqu'à l'édit de Milan, *Lateranum* 49 (1983) 5-24.

A. Hamman

OCEANUS. Of noble Roman family, relative of *Fabiola and *friend of *Jerome (*Epp.* 69 and 77), whom he visited in *Palestine and who valued him for his conduct and theological knowledge. Indeed he wrote: "our holy brother Oceanus . . . is so worthy from every point of view and so erudite in the law of the Lord that he has no need for me to ask him before he can instruct you and illustrate my thought on all questions of *Scripture" (*Ep.* 126, 3). Though a *layman, O. was concerned with church discipline (cf. Jerome, *Ep.* 69) and *orthodoxy. For this reason, alarmed by *Rufinus's translation of *Origen's *De principiis*, he and *Pammachius, another layman, asked Jerome for a complete translation of the work so as to condemn its errors (*letter among Jerome's *Epp.*, 83): Jerome made the translation (*Ep.* 84, 12). O. also corresponded with *Augustine (cf. Aug., *Ep.* 180).

DCB 4, 63; *PLRE* 1, 636.

A. Di Berardino

OCTATEUCH of Clement. To distinguish the *Syriac canonical *Clementine tradition from the *Apostolic Constitutions* (ed. Funk) in eight books, P. de Lagarde introduced the convenient expression *O. of C.* The Syriac form is very different from the Greek tradition. It contains, in books I and II, the *Testament of Our Lord Jesus Christ*. These documents, cited by *Severus of Antioch in 512-518, were published by E. Rahmani at Mainz in 1899. They were then discovered at the head of *canonical collections by A. Vööbus, who re-edited them in 1979 on the basis of all the MSS. Book III is the *Doctrina XII apostolorum* (cf. *Ecclesiastical canons), a combination of texts already published in Greek by Bickell in 1843, and then by Funk. This is the "Apostolische Kirchenordnung" of the Germans. Book IV is entitled *Ordinance of the apostles*. Clement addressed it to the nations: it concerned *charisms, *ordinations and ecclesiastical canons. This title actually corresponds to the contents of books IV-VII. It is the material of book VIII of the Greek *Apostolic Constitutions*, in a different order. Book VIII is entitled: eighth book addressed by Clement to the peoples; it concerns ordinances of the holy *apostles and canons. These are the 85 or 84 canons preserved in Greek and printed apart in all canonical collections. The *Testament* itself parallels certain parts of book VIII of the Greek Constitutions. The whole of the Octateuch was recopied by *Jacob of Edessa in the 7th c. In the *Arabic, *Coptic and *Ethiopic traditions, the Octateuch has taken a different form, containing other *apocrypha.

F. Nau, *La version syriaque de l'Octateuque de Clément*, Paris 1913; A. Vööbus, *The Didascalia Apostolorum in Syriac*, in CSCO 401-402, 406-407, *Script. syri* 175-176 and 179-180, Louvain 1979 bibl. 402, pp. 50-220); G. Graf, *Geschichte der christlichen arabischen Literatur*, V/1 Rome 1944, 283-292, *DDC* 6, 1065-1066.

M. van Esbroeck

ODES OF SOLOMON. A collection of 42 poetic compositions, well known in the early church. *Lactantius (*Inst.* IV, 12, 3) cites *Ode* 19, 6 in Latin. The *Pistis Sophia* (3 and ff.) cites *Odes* 1, 5, 6 and 25 in *Coptic. Ps.-Athanasius (*Synopsis of Holy Scripture*, 6 ff.) and Nicephorus (*Stichometria*, 9 ff.) mention it.

The *Odes* were discovered by J. R. Harris, in *Syriac, in 1907: 56 leaves in fascicles of six, containing 40 *Odes* and 18 *Psalms of Solomon*: hence the title *Odes of S.* They were published in English: 1909 (²1911). A facsimile edition, with translation and commentary, was published by J. R. Harris and A. Mingana at Cambridge in 1916-20. A second Syriac MS, incomplete, was discovered by F. C. Burkitt in the British Museum in 1912 (Add. 14583=*Odes* 17, 7-end). The 11th *Ode*, in Greek, is in *Bodmer Papyrus* XI (Cologne-Geneva 1959).

It seems certain that the 42 *Odes* were originally written in Greek in the second half of the 2nd c.; they comment on the *liturgy, esp. that of *baptism and *Easter, of a *Judaeo-Christian community in *Syria.

The *Odes* show no trace of the *dualism characteristic of the principal *gnostic systems. The OT is venerated and paraphrased in their very title. Their *theology is that of the trinitarian *God, creator and saviour, precisely that of the great ecclesiastical *tradition. Their mystical doctrine is based on the concrete accomplishment of salvation, acquired in Jesus Christ, dead and risen. All the spiritual attitudes expressed in the *Odes* presuppose the historical reality of this accomplishment; they bring out its significance in praise, thanksgiving and profession of faith. There are still unresolved enigmas concerning the correct interpretation of the *Odes* in the history of 2nd-c. Christian doctrines: particularly the gnosticizing form of their lyricism and the use of certain expressions which may be of Essene origin.

CPG 1350; Quasten I, 145-150; J.H. Charlesworth, *The Odes of Solomon*, Oxford 1978 (with Syriac text, Eng. tr. and notes); M. Lattke, *Die Oden Salomos in ihrer Bedeutung für N.T. und Gnosis*, 2 vols., Fribourg-Göttingen 1979 (cf. *RecSR* 1981, 455 f.; Lattke announces a sequel: complete, state-of-research bibliography, and commentàry); A. Hamman, *Les Odes de Salomon*, Paris 1981; *DSp* 11, 602-608; Erbetta I, 1, 608-658 (It. tr.); H.J.W. Dryvers, under "Odes of Solomon" (vii-x) in *East of Antioch*, London 1984.

C. Kannengiesser

ODILIA. The earliest evidence of a cult of this saint are the invocations in the litanies: *St Emmeram of Ratisbon, 9th c.; Freising, Münstereifel, Utrecht, 10th c.; Tegernsee, 11th c.* Then follow the *calendars and *martyrologies: *Echternach, Eichstätt, 12th c.; Metz, Paris, St Maur of Verdun, 13th c.* In the same century appear the offices and *masses: *Metz, St Maur of Verdun*. From the 13th c. the evidence multiplies. In the *monastery of Hohenburg, which claims O. as its *abbess and foundress, the saint's name appears in documents of the 10th c. In the 11th c. it is attested in the monastery of Niedermünster. In the same century a cleric of Hohenburg wrote, on the model of the *Vita Salabergae*, the unhistorical Life of St O. (BHL 6271). From it we can learn only that O. founded Hohenburg, where she is buried. Feast 13 Dec.

M. Barth, *Die hl. Odilia. Ihr Kult in Volk und Kirche*, Strasbourg 1938, 2 vols.; *Vies des SS.* 12, 413-17; *BS* 9, 1110-16.

V. Saxer

ODOACER (Odovacar). Scythian by origin, O. was born in 434. He soon entered the Palatine guard. In 476 the barbarian soldiers of which the Roman army was largely composed acclaimed him king, at the same time claiming a third of the lands of *Italy. To legitimize his power, O. sent the imperial standards to the Eastern emperor *Zeno and requested the title of Roman patrician, though he never assumed it. He occupied *Dalmatia (481), defeated the Rugians and freed Sicily from the *Vandals. In 489 *Theodoric marched on Italy at the head of the Ostrogoths; O., repeatedly defeated (Isonzo, Verona, Adda), shut himself up in *Ravenna and sustained a two years' siege ending in the extermination of himself and his people (15 March 493).

O.'s policy was essentially preservative of Roman prerogatives; distribution of lands to the barbarians did not affect small landed properties, since it was limited to the *latifundiae*. Thus began a period of mutual adjustment between Romans and Germans. *Arian by faith, he tolerated the *Catholic church, whose bishops he respected; the *schism between *Rome and *Constantinople, provoked by the "unitive" edict (*Henotikon*), favoured the cohabitation of the two ethnic groups on Italian soil. He issued a decree against the transference of church *property and intervened directly in the election of pope Felix III.

PWK 18, 1888-1896; *EC* 11, 74 f.; L. Schmidt, *Die Ostgermanen*, Munich 1934; A. Chastagnol, *Le sénat romain sous le règne d'Odoacre. Recherches sur l'épigraphie du Colisée au Ve siècle*, Bonn 1966; M. McCormick, Odoacer, Emperor Zeno and the Rugian Victory Legation, *Byzantion* 47 (1977) 212-222; A.H.M. Jones, The Constitutional Position of Odoacer and Theodoric, *JRS* 52 (1962) 126-130; W. Goffart, *Barbarians and Romans AD 415-584*, Princeton 1980; J.M. Wallace-Hadrill, *The Barbarian West 400-1000 AD*, London ³1967; C. Pietri, Aristocratie et société clericale en Italie au temps d'Odoacre et de Théodoric, *MEFR Ant.* 93 (1981) 417-467; B. Mac Bain, Odovacer the Hun?, *CPh* 78 (1983) 323-327.

M.L. Angrisani Sanfilippo

OECUMENIUS. 6th-c. philosopher and rhetor, traditionally called *scholasticus*: identified by Pétridès in 1903 as the *comes* of Isauria who corresponded with the *monophysite *Severus of Antioch (†538). Previously placed as late as the 11th c., Donato Veronese (16th c.) first attributed to him three large late 8th-c. *commentaries in the form of *catenae: *In Acta Apostolorum, In epistulas catholicas, In Pauli epistulas* (PG 118-119). Unconfirmed even by *manuscript tradition, this false hypothesis probably originated from the presence of some scholia cited under the name *Oikoumenion* (K. Staab, *Die Pauluskatenen*, Rome 1926, 93-99) and drawn from his lost works (*Commentarii in Pauli epistulas, Scholia in Iohannem Chrysostomum*). The *Commentarius in Apocalypsin*, known by the 7th c. in a *Syriac edition and edited by Hoskier (the most important codex used is *Messan. gr. 99*), remains firmly linked with O.'s name; it must be distinguished from that published under his name by J.A. Cramer (*Catenae graecorum Patrum in Novum Testamentum* VIII, Oxford 1840, 497-582), which in fact only contains *excerpta* from a similar commentary by *Andrew of Caesarea. In a limited number of MSS containing a *synopsis scholike* of the commentary on the *Apocalypse, O. is indicated as bishop of Trikka (Trikkala) in *Thessaly: this error can be ascribed, as Schmid has shown, to the compiler of the *synopsis*, a 10th-c. contemporary of a bishop of Trikka, Oecumenius, perhaps the same celebrated as a thaumaturge, on 3 May, in the Greek church.

CPG 3, 7470-7475; H.G. Hoskier, *The complete Commentary of Oecumenius to the Apocalypse for the first time from manuscripts at Messina, Rome, Salonika, and Athos edited*, Ann Arbor, 1928; K. Staab, *Pauluskommentare aus der griechischen Kirche*, Münster i. W. 1933, 423-49; Krumbacher, 131-133; S. Pétridès, Oecumenius de Trikka, ses oeuvres, son culte, *Echos d'Orient* 6 (1903) 307-310; W. Ensslin, Oikumenios, *PWK* 17, 2174; J. Schmid, Ökumenios der Apokalypsen-Ausleger und Ökumenios der Bischof von Trikka, *Byz. neugr. Jahrb.* 14 (1938) 322-330; Beck 417-418; J. Schmid, Oikumenios, *LTK* 7, 1122-1125; J.M. Sauget, Ecumenio, *BS* 4, 899; Altaner 554.

A. Labate

OGDOAS - OGDOAD (ὀγδοάς).

*1. In *gnosticism.* a) In the *Valentinian system the O. is represented firstly by the first four pairs of *aeons: Βυθός - Σιγή, Νοῦς -'Αλήθεια, Λόγος - Ζωή, ῎Ανθρωπος - ῾Εκκλησία; these form the πρώτη ὀγδοάς: cf. Irenaeus, *Adv. haer.* I, 1, 1 (ed. Rousseau-Doutreleau, SCh 264, Paris 1979, 28, 74-32, 105), I, 11, 1 (168, 1205). This O. is the first of the three parts into which the *pleroma of thirty aeons is divided: the O., the *Decad* and the *Dodecad*: Irenaeus, *Adv. haer.* I, 1, 2 (34, 123). The O. is also identified with Σοφία -' Αχαμώθ (on the identity between Σοφία and 'Αχαμώθ cf. Irenaeus, *Adv. haer.* I, 4, 1, p. 62, 354-355), the heavenly *Jerusalem and universal mother who, expelled from the pleroma, generates the demiurge of the sensible *world, in turn related to the *Hebdomas*: cf. Irenaeus, *Adv. haer.* I, 5, 2 (80, 502-504), I, 5, 3 (82, 522-523); Hippolytus, *Ref.* VI, 31 (159, 16-18 Wendland), VI, 34 (165, 20-21). Finally, the O. is also the place where Σοφία -'Αχαμώθ resides, i.e. the eighth heaven of the fixed stars: Hippolytus, *Ref.* VI, 32 (161, 14-15), VI, 33 (162, 4). Considered under this last point of view, the O. is the seat of *life, the heavenly Jerusalem, the place of rest to which the elect are destined before entering the pleroma of the aeons, the day of the Lord (as the Lord's day is the eighth day that follows the *Sabbath, the seventh day of the Jews, so on the cosmological plane the O. comes after the *Hebdomas*, the group of seven lower heavens or the seventh heaven where the biblical demiurge resides): cf. Hippolytus, *Ref.* VI, 32 (161, 16-18); Clement, *Exc. ex Theod.* 63, 1 (III 128, 9-12), 80, 1 (III 131, 25). b) For the school of Secundus, the original O. is formed of προαρχή, ἀνεννόητον, ἄρρητον and ἀόρατον, which respectively generate ἀρχή, ἀκατάληπτον, ἀνονόμαστον and ἀγέννητον: cf. Hippolytus, *Ref.* VI, 38 (169, 2-9). c) The gnostic *Mark identifies the O. with Jesus: Hippolytus, *Ref.* VI, 47 (179, 4). d) *Basilides identifies the O. both with the μέγας ἄρχων who creates and rules the sensible world: Hippol., *Ref.* VII, 25 (203, 9), VII, 27 (207, 14-15), and with the heaven of the fixed stars in which he and his sons reside: *Ref.* VII, 23 (201, 15-16), X, 14 (275, 22-23). e) For the *Barbelognostics, the O. is the eighth heaven in which Barbelo, the universal mother, corresponding to the Σοφία -' Αχαμώθ of the Valentinians, resides: Epiphanius, *Pan.* 6, 10 (I 287, 9-10 Holl). f) The same doctrine recurs in the *Arcontici*, for whom the "resplendent mother", corresponding to Barbelo and Σοφία -' Αχαμώθ, resides in the eighth heaven: Epiphanius, *Pan.* 40, 2 (II, 82, 16-20). g) For the *Wisdom of Jesus Christ*, as for the Valentinians, the O. is the eighth heaven, the seat of the elect (124, p. 289, 6-9 Till, TU 60, Berlin 1955).

*2. *Hermetism.* In *Poimandres*, the O., called ὀγδοατική φύσις, is the eighth heaven where the elect *soul arrives after having laid down all its *vices in the lower heavens and where, before being deified and entering into divinity, it resides with the other powers singing hymns in praise of the *Father (*Corp. Herm.* I, 26 vol. I, 16, 4-13 Nock-Festugière); the correspondence with the Valentinian doctrine found in *Clement (*Exc. ex Theod.* 63) is very close. The same motif of the hymn in praise of the deity sung by the powers present in the O. also recurs in CH XIII, 15 (II, 206, 16-18). In Leida Papyrus W, O. is the sovereign name of the deity to which all the hierarchies of *angels and *demons and all created beings are subject (cf. R. Reitzenstein, *Poimandres*, Leipzig 1904, 54; A.D. Nock-A.J. Festugière, *Corp. Herm.* II, Paris 1945, 216 n. 66). In the ancient Egyptian religion, O. indicates the group of eight primal deities, a doctrine closely approaching that of the schools of *Valentinus and Secundus (cf. Reitzenstein, op. cit. 54).

*3. *Clement of Alexandria* was profoundly influenced by gnostic speculations on the O. For him the O. is: a) the eighth heaven of the fixed stars, the nearest to the intelligible universe: *Strom.* IV, 159, 2 (II, 318, 31-319, 1), V, 36, 3 (II, 350, 18); b) the place where "gnostic" souls reside in *contemplation after having overcome the *Hebdomas* and, at the same time, the eighth day of the Lord: *Strom.* VI, 108, 1 (II, 486, 6-9), VII, 57, 5 (III, 42, 11-14); c) Jesus Christ himself: *Strom.* VI, 140, 3 (II, 503, 15-16). While doctrines a) and b) show very close analogies with the Valentinian school, what is evident about doctrine c) is its dependence on *Mark the gnostic, brought out by A. Delatte, *Etudes sur la littérature néopythagoricienne*, Paris 1915, 231-245.

R. Reitzenstein, *Poimandres*, Leipzig 1904, 53-55; W. Bousset, *Hauptprobleme der Gnosis*, Göttingen 1907, 12-15-17-19, 164, 340, 354; J. Kroll, *Die Lehren des Hermes Trismegistos*, Münster 1914, 304, 308, 363, 412; A. Delatte, *Etudes sur la littérature néopythagoricienne*, Paris 1915, 231-245; A.D. Nock-A.J. Festugière, *Corpus Hermeticum*, I, Paris 1945, 25 n. 64; II, Paris 1945, 215-216 nn. 65-66 (where further bibliography can be found); A.J. Festugière, *La révélation d'Hermès Trismégiste*, III, Paris 1953, 130-133; F. Sagnard, *La gnose valentinienne et le témoignage de saint Irénée*, Paris 1947, 301-304; id., *Clément d'Alexandrie: extraits de Théodote*, SCh 23, Paris 1948, 185 n. 2; P.A. Recheis, *Engel Tod und Seelenreise*, Rome 1958, 150-151; S. Lilla, *Clement of Alexandria*, Oxford 1971, 184-186; A. Quacquarelli, *L'ogdoade patristica e suoi riflessi nella liturgia e nei monumenti*, Bari 1973.

S. Lilla

OIL. In the ancient Mediterranean, esp. in the sphere of Judaeo-Hellenistic culture, oil and the olive were of great importance. "With *bread, cheese, salt and *wine, olives and olive oil were the primary sources of sustenance" (*PWK* 17, 2012). "Like the palm and the laurel, the olive too had an important symbolic function in the field of the arts and religion" (*PWK* 17, 2020). The same is true in the OT world. We often read of *anointing with oil: it was poured on religious objects, for the consecration of priests and kings. Metaphorically, anointing indicates the outpouring of the Spirit of God. So the eschatological Saviour would be called the "Anointed" (*Christòs*). All this was fulfilled in Jesus the Christ; the Spirit of God rested on him, and this act was his anointing (Lk 4, 18-21; Acts 10, 38; cf. also 2 Cor 1, 21 f.; 1 Jn 2, 20; Rom 11, 17. 24; Heb 1, 9; Rev 11, 4; also Mt 25, 3-8). But alongside these important allegorical uses, the NT also knows the actual use of oil in the Lord's anointings (Lk 7, 38; Mk 14, 3-9) and for the healing of the sick (Mk 6, 13 and, already in a "liturgical" way, Jas 5, 14-16). All this is accepted by the *Fathers in ever-growing measure. *Ephrem the Syrian celebrates the significance of oil widely in his *hymns (*De virginitate* IV-VII and XXXVII: CSCO 224, 12-29; 116). "Oil possesses a great efficacy ... like Christ, who is all in all ... the name of oil is a *symbol, and from it is derived the name 'Christ' ... And if they (the disciples) have been anointed and consecrated, mysteriously present in the oil was Christ, who has cast out all ills ... The name of oil, then, is a symbol and shadow of the name 'Christ'" (IV, 6-8). *Hymn* V exalts the oil of the candlestick and the lamp. *Hymn* VI celebrates the "oil, worthy of the Highest, giver of royal dignity"; it was used for the anointing of priests and kings; now "the multitudes are anointed and purified (in Christ)". In *Hymn* VII, 6, he says of *baptism: "The *Holy Spirit impresses his seal with oil on the *sheep of his flock ... The invisible imprint of the Spirit is also impressed with oil in bodies... (14) ... Oil is a real mirror. And from whatever side I may observe oil, I see the gaze of Christ who shines in it." In the post-apostolic period, our earliest source is *Tertullian. He attests the use of oil for the healing of a sick man (*Ad Scap.* 4: PL 1, 703 B) and explains its meaning in detail in the ceremony of *initiation: *Exinde, egressi de lavacro, perungimur benedicta unctione de pristina disciplina, qua ungi oleo de cornu in sacerdotium solebant. Ex quo Aaron a Moyse unctus est, unde Christus dicitur, a chrismate quod est unctio, ... facta spiritualis, quia spiritu unctus est a Deo Patre* (*De bapt.* 7: PL 1, 1206C-1207A). Not long after, *Hippolytus of Rome's *Apostolic Tradition* gives us, in concrete liturgical settings, the foundation of all further functions and uses of oil in the church. The oil consecrated by the bishop, during the baptismal rite, with a "prayer of thanksgiving", is hence called *oleum gratiarum actionis* (*eucharistia*); whereas the oil that is exorcized is called *oleum exorcismi*. Baptizands are exorcized with the latter, those who have been baptized in *water are anointed with the former (ch. 21: ed. Botte, 46, 50, 52). Ch. V hands down a formulary for the consecration of oil, "with which you have anointed priests and prophets" and by means of which "you give to all who receive it, strength and health" (p. 18). What is expressed here summarily is put forward in richer formularies in the *Euchologion* of *Serapion of Thmuis: consecration of oil for healing of the sick (chs. 17 and 29), for baptismal exorcism (ch. 22), for anointing of the baptized (ch. 25) (Funk II, 178-192). That the oil, to be thus corporeally and spiritually efficacious, must be consecrated, is abundantly attested e.g. by *Cyprian (*Ep.* 70, 2: CSEL III/2, 768) and in a particularly penetrating way in the *Mystagogical catecheses* attrib. to *Cyril of Jerusalem: in II, 3 (PG 33, 1080 AB), above all in no. III, which is entirely dedicated to *chrisma* (1088A-1091A). In the church of *Rome, the consecration of these three types of oil has been fixed, from the earliest times, on feria V of Holy Week. The *Sacramentarium Gelasianum Vetus* gives us in the *Missa Chrismatis* the formularies of consecration, still valid today in the Roman rite (Mohlberg, nn. 378-390). We must at least mention the important passage on the consecration of the *chrisma*: *... Te igitur deprecamur ... ut huius creaturae pinguedinem sanctificare ... digneris ... per potentiam Christi tui a cuius sancto nomine chrisma nomen accepit, unde unxisti sacerdotes reges prophetas et martyres tuos, ut sit his qui renati fuerint ... chrisma salutis* (n. 388). But the use of holy oil was not just limited to these strictly sacramental circumstances. Not only was oil taken to sick people at home, as medicine, but the custom became established of taking oil from *lamps lit on tombs of saints or in holy cities, in small containers (*ampullae*), as a sort of physically and spiritually miraculous ointment (e.g. John Chrys., *In mart. hom.*: PG 50, 664 to end).

Texts: F.X. Funk, *Didascalia et Const. Apostol.* I-II, Paderborn 1905; L.C. Mohlberg, *Liber Sacr. R. Aeccl.* (= Gelasianum vetus) (RED Fontes IV), Rome 1960; E. Beck, *Des hl. Ephraem d. Syrers Hymnen, De virginitate*, CSCO 224, Louvain 1962; B. Botte, *La Tradition Apostolique de S. Hippolyte*, LQF 39, Münster i.W. ³1966.
Studies: Dictionaries: F.X. Kraus, *Realenc. d. class. Altertümer* 2 (1886), 522-525, 710-714; *DACL* 1, 1722-1747; 6, 2777-2791; *TWNT* 1, 230-232 (*GLNT* 1, 617-626); 2, 468-478 (*GLNT* 3, 381-388); 4, 807-809 (*GLNT* 7, 639-646); 9, 482-576; *PWK* 17/2, 1998-2022; 2454-2474; *LTK* 7, 1143-1145. P. Hofmeister, *Die Öle in der morgenländischen und abendländischen Kirche*, Würzburg 1948; *DSp* 11, 788-819; *Cath* 5, 1056-1063; *DTC*, Tables génér. 2135-2143; W. Nagel, Exorzismus, *TRE* 9, 747-756; L.L. Mitchell, *Baptismal Anointing*, London 1966; (ed.) M.J. O'Connell, *Temple of the Holy Spirit: Sickness and Death of the Christian in the Liturgy*, New York 1983.

B. Neunheuser

OLYMPIA(S), deaconess. Various ancient documents mention her: *Palladius's *Dialogue on the Life of St John Chrysostom* and his *Lausiac History*; *Sozomen's *Ecclesiastical History*; 17 *letters sent to her by Chrysostom; and an anon. *Life* (second half of 5th c.).

Born between 361 and 368 at *Constantinople, of a family belonging to the circle of the imperial palace, she was brought up by Theodosia, sister of St *Basil's friend *Amphilochius of Iconium. Sources attribute to her friendly and familiar relations with *Gregory Nazianzen and *Gregory of Nyssa. She married Nebridius, prefect of Constantinople, in 386, but was soon widowed. It is related in tones of sublime spirituality how she refused to obey the emp. *Theodosius and marry a high imperial official. Her refusal resulted in the sequestration of all her goods and the start of her life of *poverty and *ascesis. She used her goods - restored because of her edifying life - for great charitable activities, for which *Nectarius, archbishop of Constantinople 381-397, made her a *deaconess when aged just 30. O. lived at Constantinople, sharing her life of service with numerous *women who joined her. When *John Chrysostom arrived there as *archbishop (398), he highly esteemed and supported O.'s community; O. and her companions in turn collaborated with him; various of her companions were ordained deaconesses. After the emp. *Arcadius's decree of 404 had exiled Chrysostom to Cucusa in *Armenia, O. also suffered various persecutions, being finally exiled to *Nicomedia, where she died in 410. We possess 17 letters written to her by Chrysostom from his exile in Armenia.

BHL 2, 154, nn. 1374-76; Palladius, *Dialogus* X, 16 (PG 47, 35-61); id., *La Storia Lausiaca*, 56, 1; 61, 3, Milano 1975; Sozomen *HE* VIII, 9, 24.27 (PG 67, 1538); Jean Chrysostome, *Lettres à Olympias et à tous les fidèles*, SCh 103, Paris 1964; *EC* 9, 96; *BS* 9, 1154-58.

E. Cavalcanti

OLYMPIODORUS. 6th-c. deacon of *Alexandria, exegete. Ordained under patriarch John III Niciota (505-515/6), he is known for the works handed down under his name, essentially moralizing *Commentaries* on the *Wisdom books and *Prophets, part of the scriptural *Catenae: Commentaries on *Job (a work taken up by Nicetas of Heraclea), Ecclesiastes, Jeremiah (and letter of Jeremiah), Lamentations, Baruch and (partly preserved) *Psalms. A brief fragment of his work against *Severus of Antioch survives (published under St Athanasius's name). Fragments of a commentary on Proverbs and on the gospel passage "Rejoice in that day" are considered spurious.

CPG 7453-7465; PG 93, 13-780; 89, 1189; *DB* 4, 1796; (ed.) U. and D. Hagerdorn, *Olympiodorus von Alexandria, Kommentar zu Hiob*, Berlin 1984.

D. Stiernon

OLYMPIUS. 4th-c. bishop, some say of *Toledo, some say of *Barcelona. Acc. to *Gennadius (*De vir. ill.* 23), he wrote a (lost) *Librum Fidei adv. eos qui naturam et non arbitrium in culpam vocant* (on original *sin). *Augustine cites him together with *Irenaeus, *Cyprian, *Reticius, *Hilary, *Ambrose and Gregory (*C. Iul.* I, 3, 8; II, 10, 33; III, 17, 32) and in the same work (*C. Iul.* I, 2, 8: PL 44, 644-645) alludes to a work - *in quodam sermone ecclesiastico* - which may be the same one cited by Gennadius.

E. Flórez, *ES* 29, 77-81.

E. Romero Pose

OLYMPIUS. *Honorius's *Magister officiorum*, native of *Pontus. He succeeded *Stilicho, whom he ruined, obtaining his arrest (408) and persecuting his followers. Maybe O. inspired the law of 14 Nov 408, which excluded *pagans from the palace administration. Upholder of a last-ditch anti-barbarian policy, he prevented any agreement with the invader *Alaric. He was finally disgraced and put to death by *Constantius III (Zosimus, *Historia nova* V, 32-46; Photius, *Bibl.*, cod. 80; *CT* XVI, 5, 42). *Augustine wrote two *letters to O. (*Epp.* 96 and 97), one asking help for a bishop in a question of property, the other requesting action against the *heretics.

PLRE 2, 801-802.

E. Prinzivalli

ONUPHRIUS. Egyptian *anchorite, prob. mid 4th c. His life, and that of other monks, is described in a *Coptic text attrib. to one Papnutis (Paphnutius). The part referring to O. also exists in Greek and other languages. The Coptic is not necessarily the original. O. lived quite alone, naked, clothed only in his hair, which reached his knees. A palm-tree gave him one fruit per month for food. For reasons that are not clear, his figure became popular outside *Egypt and gave rise to various legends. His *iconography too is particularly rich, inside and outside Egypt. Feast 12 (or 10) June.

BHG 1378-1382; 2330; *Auctarium* p. 143; E.A.T. Wallis Budge, *Coptic Martyrdoms in the Dialect of Upper Egypt*, London 1914, LIV-LXVII, 205-224, 455-473; A.G. Elliott, *Roads to Paradise: reading the Lives of the early saints*, London 1987, chs. 3 and 6.

T. Orlandi

OPERA OMNIA. In the age of the manuscript, the idea of bringing together the works of a single author had to contend with two difficulties: one financial, the other critical. The first was superable, but the second required a critical sense which appeared only in the Renaissance, thanks to the efforts of the great editors, mainly Johann Froben (Frobenius) assisted by Erasmus. So, with regard to the *Fathers, we know only one attempt to bring together St *Augustine's works, at Clairvaux (now *Troyes BM* 40). For him, apart from the *Retractationes*, there was *Possidius's *Indiculus* (PL 46, 5-22), which simplified the collecting.

The publication of *O. O.* began at Basle with the publisher J. Froben (1460-1527), who worked with Johann Petri (1441-1511) and Johann Amerbach (1441-1513). This brought improvements to the correctness of the texts; above all it required a critical examination of authenticity, and discarded spurious works.

From 1505-06 the *editio princeps* of St Augustine's *O. O.* appeared, but without critical apparatus. In 1513 Erasmus's collaboration brought a critical surge: *O. O.* of *Jerome, in 10 vols., 1516; of *John Chrysostom (in Latin), 1517; of *Cyprian, 1520; of *Tertullian, 1521 (ed. Renanus); of *Arnobius, 1522; of *Hilary, 1523; of *Irenaeus, 1526; of *Ambrose, 1527. A new 10-vol. edition of Augustine's works by Erasmus appeared in 1528-9, a landmark for choice of MSS, correctness of texts and elimination of apocrypha. All *Origen's works, edited by Erasmus, were published after his death by Renanus.

Erasmus's activities were carried on by his disciple Beat Bild, surnamed Renanus. The impetus had been given. In the following centuries the *Maurists brought out exemplary editions of *O. O.*, at first the Latin Fathers, then the Greeks (published with the Greek text for the first time). They had the good sense not to discard the *spuria* completely, but to put them in appendixes.

J. de Ghellinck, *Patristique et moyen âge*, II and III, Brussels 1947-1949; *DHGE* 15, 667-679; *DHGE* 19, 126-129. For Renanus cf. A. Horawitz, *SAW* 70 (1872) 189-244, 323-378.

A. Hamman

OPHITES - NAASSENES. A 2nd/3rd-c. *gnostic group, in whose religious doctrines and practices the serpent (Gk ὄφις; Heb. *nahas*) occupied a prominent place. The use of the serpent must be seen against the background of their polemical attitude, characteristic of various gnostic systems, towards the God of the OT: indeed, the corruptor of *Adam and Eve, presented in the Genesis account as a symbol of *evil and origin of *sin, was venerated by the O. as a spreader of "*gnosis*", i.e. that knowledge of good and evil which the OT Demiurge-God had forbidden to man. His power was further manifested by *Moses in the episode of the bronze serpent (Num 21) and recognized by Jesus himself (Jn 3, 14) (cf. Iren., *Adv. haer.* I, 30, 7; Ps.-Tertull., *Adv. omn. haer.* 2). Besides their polemical value, they concurred in expounding the good luck of the serpent, within gnostic speculation, the great importance and fundamental function of this animal in so many aspects of *pagan religiosity as a symbol of the regenerative force of nature, of divination and prophecy, and as an animal linked to the cult of the dead and to *astrology (ref. in Leisegang).

Within gnostic speculation, the symbolism of the serpent took on values and functions that varied considerably according to context. Most likely, the facts relating to the O. or N., documented by the *heresiologists, should be understood not as referring to a single or precise gnostic group, but to various schools which, though starting from some common doctrinal elements, then developed diversified theological systems.

*Irenaeus, who identifies the O. with the *Sethians, attributes to them a cosmogonic and anthropogonic mythical account which in many aspects anticipates, in a more simplified form, the *Valentinian myth (*Adv. haer.* I, 30). In the so-called "Diagram" of the O. (described in Orig., *C. Cels.* VI, 24 ff.), a design schematically representing the universe as these sectarians imagined it, the serpent becomes a cosmic symbol: it is the monster (the biblical Leviathan) that surrounds the terrestrial *world, whose lord and master it is. Acc. to *Epiphanius (*Pan.* 37), who describes some of their religious practices, the O. used a live serpent for the consecration of the *Eucharist. Particularly important is the evidence of *Hippolytus (*Ref.* V, 6, 3-11, 1; X, 9, 1-3) who quotes, citing literally as well as paraphrasing and summarizing, a long passage taken from a Naassene work and an anapaestic psalm also used among them. The former (also known as the "Preaching of the Naassenes") recounts the myth of man's origin, following a scheme that shows some analogies with the doctrine of *Saturnilus. Acc. to the author of this work, the theme of the Primordial Man and his descent to the world, which makes up the nucleus of the gnostic myth, is present, though in confused and obscure form, in the religious heritage of various peoples, as attested by the rich review of pagan myths (Greek, Phrygian, Assyrian, Egyptian, etc.) that accompanies the anthropogonic account. The psalm develops the theme of the *Soul which wanders in the labyrinths of the world until Jesus descends to free it by revealing *gnosis* to it. A *painting, found in the catacomb of Via Manzoni, *Rome, seems to refer to this Naassene doctrine of the Soul. Acc. to Hippolytus, the Naassene group was closely related to the Perati, the Sethians and the gnostic *Justin.

CPG 1155 f.; W. Völker, *Quellen zur Geschichte der christlichen Gnosis*, Tübingen 1932, 11-33; M. Simonetti, *Testi gnostici cristiani*, Bari 1970, 25-71; A. Hilgenfeld, *Die Ketzergeschichte des Urchristentums urkundlich dargestellt*, Leipzig 1884, repr. Hildesheim 1963, 249 ff., 277 ff.; A. Hönig, *Die Ophiten*, Berlin 1899; R. Reitzenstein, *Poimandres*, Leipzig 1904, 83-102; H. Leisegang, *Die Gnosis*, Leipzig 1924, 111-185; *EC* 9, 80-81; *LTK* 7, 1178-79, M. Simonetti, Qualche osservazione su presunte interpolazioni nella predica dei Naasseni, *VetChr* 7 (1970) 115-124; J.D. Kaestli, L'interprétation du serpent de Genèse 3 dans quelques textes gnostiques et la question de la gnose "ophite", in ed. J. Ries, *Gnosticisme et monde hellénistique*. Actes Coll. de Louvain-La-Neuve (11-14 mars 1980), Louvain-La-Neuve 1982, 116-130; A.J. Welburn, Reconstructing the Ophite Diagram, *NT* 23 (1981) 261-287; J. Frickel, *Hellenistische Erlösung in christlicher Deutung: Die gnostische Naassenerschrift*, Leiden 1984; J. Frickel, *Hellenistische Erlösung in christlicher Deutung: Die gnostische Naasenerschrift*, NHS 19, Leiden 1984. See also under *Gnosis.

C. Gianotto

OPTATIAN: Publilius Optatianus Porphyrius. Born 260-270, proconsul of *Achaia before 306, exiled 315 or shortly before, pardoned by the emp. *Constantine in 325, *praefectus urbi* in 329 and 333. His famous epistolary exchange with Constantine must be dated Nov/Dec 311. This versifier was pardoned as a result of a collection of 20 poems in praise of Constantine, later enriched with seven more. He also wrote anacyclic verses, i.e. verses whose letters, read back to front, form other verses; they are handed down by the *Codex Salmasianus* (*Anthologia Latina*, vol. I). No. 81 (ed. Riese) - if we must find a meaning in the artifice - sings of the triumph of love (Venus) over death, Jove, Bacchus, Hercules, the Sun, the Moon, Mars and finally over Venus herself, conquered by Adonis. O. is also a master in the technique of so-called "figured poetry". Alongside elements of ancient origin, we also find Christian elements.

Ed. G. Polara, *Carmina*, I. *Textus adiecto indice verborum*. II. *Commentarium criticum et exegeticum*, Corpus Scr. Lat. Paravianum, Turin 1973; G. Polara, Naples 1976; Schanz 4/1, 11-14; G. Polara, *Ricerche sulla tradizione manoscritta di Publilio Optaziano Porfirio*, Salerno 1971; id., Cinquant'anni di studi su Optaziano (1922-1973), *Vichiana* 3 (1974) 110-124; 282-301, with later articles in the same periodical; T.D. Barnes, Publilius Optatianus Porfirius, *AJPh* 96 (1975) 173-186; J. Fontaine, *Naissance de la poésie dans l'occident chrétien. Les IIIe-VIe siècles*, Paris 1981.

I. Opelt

OPTATUS of Milevis. 4th-c. bishop of Milevis in *Numidia. Of him we know only, from *Jerome (*Vir. ill.* 110), that he wrote his six books against *Donatism, *Valentiniano et Valente principibus*, i.e. some time between 364 and 367 (prob. 366-7, acc. to some scholars). The work's original title is not known, either from Jerome or from *manuscript tradition: those proposed by modern editors and scholars (the best-known being *Adversus Parmenianum Donatistam* and *De schismate Donatistarum*) have a merely indicative value. Contrary to Jerome's indication and to the plan laid down by O. himself in book I, ch. 6, the work as handed down consists of seven books: passages of book VII are interpolated into book III. It is now accepted that, around 385, O. had wanted to prepare a second edition (witness the Catholic and Donatist *episcopal lists of *Rome cited in II, 3-4, brought up to date in some codices with the names of pope *Siricius, elected 384, and the Donatist bishops Lucian and Claudian), but was unable to complete it: someone in his entourage would have been instructed to complete the new edition, using the material collected by O. which appears in book VII, not without making transpositions and interventions of his own.

The work is a reply to the anti-Catholic treatise written by the Donatist *Parmenianus in 362. In the first book O., who addresses Parmenian as "brother", recalls the points faced by the Donatist in his work and explains the order he intends to follow in his treatment: then he corrects some statements of Parmenian on Christ's *flesh, and distinguishes the concept of *heresy from that of *schism, retorting on *Donatus the accusation of schism that Parmenian had thrown at the Catholics. The main argument of book I is the history of the origins of the schism up to the council of *Rome held under pope *Miltiades: in this context O. wishes to demonstrate that some of the bishops who had accused *Caecilian's consecrator *Felix of Apthugni of *traditio* should themselves be considered *traditores*. The theme of book II is the unicity of the *Catholic church, characterized by its universality: in *Africa, only the Catholic church is in *communion with the universal church and only the Catholic church is in possession of the qualities that, acc. to *Scripture, are proper to it. In book III, O. defends the imperial intervention in Africa, reminding the Donatists that they had been the first to appeal to the emperor's authority: the severe measures adopted against them had been made necessary by their own violence, and their claim to style themselves martyrs in a persecution unleashed by the Catholic church was vain. In book IV O. rejects Parmenian's thesis that, in the sinners of Scripture, we should see the Catholics: it is rather the Donatists, responsible for so many crimes, who are sinners. Book V is dedicated to the theme of the validity of *baptism; against the Donatists who, following ancient African practice, considered invalid the baptism conferred by heretics and therefore rebaptized Catholics who (spontaneously or otherwise) came over to their church, O. defends the baptismal doctrine of the church of Rome accepted by the African Catholics at the council of *Arles in 314, by which the efficacy of the *sacrament does not depend on the person of the *minister, but on the action of *God. Book VI describes the violence of the Donatists, esp. at the time of *Julian the Apostate: but above all O. laments the spiritual harm worked by the Donatists with their propaganda. Book VII, added to the second edition, attempts to reply to the objections raised among Donatists by the treatise: after an invitation to unity in the Catholic church, it examines the interpretation of some biblical passages to which the Donatists had appealed in their controversy, and gives a defence of Macarius, the architect of unity in 347. O.'s work was also to have provided a collection of documents on the origin of the schism (cf. I, 22. 26). Only one of our MSS, *Parisinus* 1711, cites some documents (*Gesta apud Zenophilum, Purgatio Felicis* [cf. *Gesta Purgatione*]) which make up part of the original documentation collected by O., as well as eight *letters (six by *Constantine) added to the collection at a later date. O.'s work has great value, both historical and theological. He collected a great number of documents to reconstruct the events of the schism and rebut the charges of the Donatists, anticipating what *Augustine would do esp. on the occasion of the council of *Carthage of 411. Theologically too, anticipating Augustine, O. thoroughly explores the ecclesiastical and sacramental doctrine in the points contested by the Donatists (unity of the church, efficacy of the *sacraments in relation to who administers them). Also characteristic of O. is the tendency to attenuate the reasons for disagreement between the parties, presenting his opponents as extremists with whom he can hold a dialogue and demonstrate the groundlessness of their positions, rather than as enemies.

A *sermon, *In natali sanctorum Innocentium* (PLS 1, 288-294: in reality a *Christmas sermon adapted to the feast of the Holy *Innocents) features in a MS under the name of O. of Milevis, and has been attrib. to him since A. Wilmart's edition of it: the prevalent opinion now is that formulated earlier by A. Pincherle, who saw it as the work of a Donatist, perhaps of *Optatus of Thamugadi: precise points of contact have been seen between it and the work of *Tichonius (F. Scorza Barcellona, E. Romero Pose). We must certainly reject the attribution to O., upheld by G. Morin, of three pseudo-Augustinian sermons on *Epiphany: *Serm.* 131 and 132 of the *Maurists' *Appendix* (PL 39, 2005-2007 and 2007-2008) and the *Sermo sancti Augustini de Epiphania* (PLS 1, 297-300). O.'s name has been proposed, on the basis of stylistic arguments which are finally unconvincing, for two *Easter sermons (PLS 1, 295-296; PLS 4, 665-667).

CPL 244: PL 11, 883-1104 (and Ellies du Pin's edition, Anversa 1702); CSEL 26 (ed. C. Ziwsa) 1893; CPL 245-249 and 940 (spurious sermons of doubtful attribution); *DTC* 11, 1077-1084; *EC* 9, 449-451; *BS* 9, 1307-1311; *PWK* 18, 1, 765-772; E. Michaud, La théologie d'Optat de Milève d'après le "De schismate Donatistarum", *Revue Intern. de Théol.* 16 (1908) 208 ff.: O.R. Vassall-Phillips, *The Work of St. Optatus against the Donatists*, London 1917; P. Batiffol, *Le catholicisme de saint Augustin*, I, Paris 1920, 77-108; A. Wilmart, Un sermon de saint Optat pour la fête de Noël, *RSR* 2 (1922) 271-302; G. Morin, Deux sermons africains du Ve-VIe siècle avec un texte inédit du symbole, *RBen* 35 (1923) 233-236; A. Pincherle, Un sermone donatista attribuito a S. Ottato di Milevi, *Bilychnis* 22 (1923) 134-148; id., L'ecclesiologia nella controversia donatista, *Ricerche R* 1 (1925) 5-55; id., Noterelle ottaziane: *ibid.* (1927) 440-445 (= *Cristianesimo antico e moderno*, Rome 1956 70-76); U. Moricca, *Storia della letteratura latina cristiana*, II, Turin 1928, 698-722; A. Wilmart, Un prétendu sermon pascal de S. Augustin, *RBen* 1 (1929) 197-203; A. Pincherle, Due postille sul donatismo, *Ricerche R* 18 (1947) 160-164 (cf. *Cristianesimo antico e moderno*, cit., 76-79); S. Blomgren, *Eine Echtheitsfrage bei Optatus von Mileve*, Stockholm 1959; id., Spicilegium Optatianum, *Eranos* 58 (1960) 132-141; H.D. Altendorf, ed. S. Blomgren, Eine Echtheitsfrage, cit., *ThLZ* 85 (1960) 598-600; A.C. de Veer, A propos de l'authenticité du livre VIIe d'Optat de Milève, *REAug* 7 (1961) 389-391; H. Silvestre, Trois sermons à retirer définitivement de l'héritage d'Optat de Milève, *PACA* 1964, 61-62; V. Saxer, Un sermon médiéval sur la Madeleine. Reprise d'une homélie antique pour Pâques attribuable à Optat de Milève (392), *RBen* 80 (1970) 17-50; R.B. Eno, The Work of Optatus as a Turning Point in the African Ecclesiology, *Thomist* 37 (1973) 668-685; F. Scorza Barcellona, L'interpretazione dei doni dei Magi nel sermone natalizio di (Ps.) Ottato di Milevi, *SSR* 2 (1978) 129-149; E. Romero Pose, Ticonio y el sermón In natali sanctorum Innocentium (Exégesis de Mt 2), *Gregorianum* 60 (1979) 513-544; *Patrologia* III, 112-116; *PAC* 1, 795-801; *DSp* 11, 824-829; B. Kriegbaum, *Optatus von Mileve, Historiker und Theologe* (forthcoming); J.E. Merdinger, Optatus Reconsidered, *SP* 22 (1989) 294-299.

F. Scorza Barcellona

OPTATUS of Thamugadi. *Donatist bishop of Thamugadi (*Timgad), 388-398. A tyrannical personality who exercised enormous power in *Numidia during his ten years' episcopate (Aug., *C. Ep. Parmen.* II, 2, 4: *Optatum Gildonianum decennalem totius Africae gemitum*). Revered by the Donatist population, who celebrated the anniversary of his consecration with enthusiastic displays (Aug., *C. litt. Petil.* II, 23, 53), O. was accused by the *Catholics of having abused his *authority against minors and others, whose property had been entrusted to him, and against landed proprietors who got in his way (*ibid.* II, 35, 82); he also sought to suppress by force the *schism of the *Maximianists within the ranks of Donatism (Aug., *C. Cresc.* IV, 25, 32; cf. also III, 54, 65). In *Augustine's mind, O. was associated with the plundering, increasingly widespread (*C. litt. Petil.* I, 24, 26; *Ep.* 43, 8, 24) and increasingly organized, of the *circumcellions (*C. Ep. Parmen.* I, 11, 17: *[Circumcelliones] qui primo tantummodo fustibus, nunc etiam ferro se armare coeperunt*). In 397/8, O. threw in his lot with count *Gildo in the last rebellion against the authority of the emp. *Honorius (Aug., *C. Cresc.* III, 12, 15: *inter Gildonis satellites praecipuus haberetur teneretur moreretur [Optatus])* and, with him, was defeated, captured and executed (Aug., *C. litt. Petil.* II, 92, 209). Though perhaps not himself a revolutionary (cf. E. Tengström, *Donatisten und Katholiken*, 74-78), his violent actions and his despotic ways (Aug., *C. Cresc.* IV, 25, 32) gave an opening to the revolutionary social elements in the Donatist church. His power is symbolized by the large cathedral and dependent buildings, built to glorify his name, which rose on a hill overlooking Timgad (cf. E. Albertini, *CRAI* [1939] 100-102; Ch. Courtois, *Timgad, l'antique Thamugadi*, Alger 1951, 72 ff.).

Sources: Augustine, texts cited (ed. Petschenig, CSEL 51-52).
Studies: H.J. Diesner, Die Periodisierung des Circumcellionentums, *Wiss. Zeitsch. Univ. Halle* 10 (1962) 1329-1338; E. Tengström, *Donatisten und Katholiken*, Göttenborg 1964, 74-78 and 84-90; W.H.C. Frend, *The Donatist Church*, Oxford ²1971, ch. 15; *PAC* 1, 797-801.

W.H.C. Frend

OPUS IMPERFECTUM IN MATTHAEUM. Name given to a lengthy anon. Latin *commentary on *Matthew's gospel, which ends at ch. 25 and shows *lacunae* here and there. The available text is insufficient in many ways: only a modern crit. ed. can solve the many problems raised by this work. Though handed down under the name of *John Chrysostom, its author was an unknown early 5th-c. *Arian. The earlier hypothesis, making it a *Latin translation of a lost Greek original, has recently been revived on a new basis. The *exegesis is mainly allegorizing, with technical procedures closely recalling *Origen, touching on themes of a strongly existential character: man seen in the struggle between good and *evil, between the devil who has enslaved his *flesh and *God who succours his *soul, which is free in its decisions, but insufficient to achieve salvation without divine help. The author's Arian faith transpires in various polemical and doctrinal passages and esp. in the recurrent theme of persecution: he is head of a small Arian community which sees its members daily falling away, due to the pressure of the dominant *Catholics; so he insists on the theme of *temptation, which represents the sieve that awaits all the best. In the *eschatological perspective of the last *judgment, which will see the defeat of the persecutors, the army of *Antichrist is represented by the Catholic church, and the beginning of the end-time is made to commence with *Constantine and *Theodosius, the emperors who persecuted the Arians.

CPL 707; PG 56, 611-946; *Patrologia* III, 95-97 (bibl.); R. Etaix, Fragments inédits de l'Opus imperfectum in Matthaeum, *RBen* 84 (1974) 271-300; J. Banning, The Critical Edition of the Opus imp. in Matthaeum: An Arian Source, *SP* 17, Oxford 1982, 382-387.

M. Simonetti

ORACLES

I. In general - II. Chaldaean oracles - III. Sibylline oracles.

I. In general. A form of divination consisting of a reply, expressed in various ways and means, by a deity to a query put by the believer on personal or communal questions. In ancient Greece, Dodona and Delphi were seats of famous O. dedicated respectively to Zeus and to Apollo. At Dodona, divination was by observing the movement of the leaves of the sacred oak and the waters of a nearby spring; at Delphi the Pythia, inspired by the god, uttered enigmatic replies which the priests then interpreted. Christian polemic against *astrology and its practices also covered the oracular art, whose veracity was sometimes not wholly denied, but was attributed to the actions of *demons. Thus, e.g., *Tertullian claims that these, knowing some truths and borrowing others from the biblical prophecies, deceive men through O. to subject them to their power (*Apol*. 22). *Tatian too recognizes in O. a certain curative capacity inspired by evil spirits (*Orat*. 18). Meanwhile the *pagans objected to the use of oracular writings by the Christians themselves in order to confirm their own doctrines. *Lactantius, in support of the doctrine of the immortality of the *soul, appeals - apart from the *Bible - to the "prophecies of the Sibyl" and the replies of Milesian Apollo. One document attesting Jewish and then Christian use of the pagan oracular genre is the Sibylline oracles (cf. below). *Augustine's treatise *De divinatione daemonum*, while repeating the polemical arguments against oracular practices worked out in earlier patristic tradition, confirms the persistence and strength of such practices, which penetrated widely even into the Christian population, despite imperial decrees against *haruspices* and diviners.

A. Bouché-Leclercq, *Histoire de la divination dans l'antiquité*, I-IV, Paris 1878-1881; *DACL* 2, 1198-1212; P. Amandry, *La mantique apollonienne à Delphes*, Paris 1950; H.W. Parke, *The Oracles of Zeus*, Oxford 1957; R. Lane Fox, *Pagans and Christians*, London 1986, ch. 5; R.P.C. Hanson, The Christian attitude towards Pagan religions, in *Studies in Christian Antiquity*, Edinburgh 1985, ch. 9; H. Chadwick, Oracles of the End in the Conflict of Paganism and Christianity in the 4th c., in (eds.) E. Lucchesi-H.D. Saffrey, *Mémorial André-Jean Festugière: Antiquité païenne et chrétienne*, Geneva 1984, 125-129.

G. Sfameni Gasparro

II. Chaldaean oracles. Attributed to an author named Julian, perhaps Julian the Theurge (2nd c. AD), and highly esteemed in *neoplatonist circles, this collection of brief theosophical and cosmosophical hexametrical compositions survives only partly and indirectly, through the citations of *pagan and Christian authors. The first Christian to mention it is *Arnobius; later, *Marius Victorinus and *Augustine. Among pagans, besides *Porphyry, Iamblichus was particularly inspired by the *theology of the Chaldaean O., to which he dedicated a (lost) commentary in 28 books. Some extracts of a similar exegetical work by *Proclus are preserved by Michael Psellus (11th c.), who wrote a "Commentary on the Chaldaean Oracles". The doctrine expressed by the collection, Platonic in inspiration, contemplates the notion of a supreme transcendent *God (the *Father) from whom derive an Intellect with demiurgic function and a third principle (Hecate), intermediate between the two and identifiable with the *anima mundi*. Numerous *angels and *demons people the higher world, exercising their action on men positively or negatively. They play a fundamental role in the practice of *theurgy, which consists of establishing contact with the gods through ritual acts of initiation. The Chaldaean O. affirm a *dualistic *anthropology, with the notion of the *soul fallen from the divine world, imprisoned in the *body and subject to cosmic destiny. Salvation is possible through theurgic rites and rigorous abstention, aimed at dominion over the bodily passions.

E. Des Places, *Oracles Chaldaïques. Avec un choix de commentaires anciens*, Paris 1971.

G. Sfameni Gasparro

III. Sibylline oracles.

*1. The *pagan Sibyl.* Among the ancients, the Sibyl was a woman who went into *ecstasy and prophesied, very often announcing catastrophes. Her words were considered O. Her origin should perhaps be sought in Persia, whence she spread particularly to the Greek colonies of Asia Minor. She was considered now a demoniac, now a deity (*Thea Sibylla*). Various places of worship were dedicated to her, so that her name ended by becoming a common name, the Sibyls, to whom a collection of O. was attributed. The Greeks knew the Sibyl of Marpessos; more important was that of Erythrai, who settled at Delphi, where she was considered sometimes the wife, sometimes the daughter of Apollo. *Clement of Alexandria cites the opening of a *Carmen Sibyllinum* by Heraclius the Sophist (*Strom*. I, 108). Then the Sibyl came to Cumae, where she uttered O. in a grotto. Virgil evokes her in *Aeneid* VI, 98 ff. The collection of her O., preserved in the Capitol, was destroyed when the temple of Jupiter caught fire. The most famous oracle of Cumae was the announcement of the saviour of the world, immortalized in Virgil's 4th Eclogue, which *Constantine in his *Oratio ad sanctos* applies entirely to Christ (19-21). There we read the acrostic of Christ ΙΧΘΥΣ ΣΤΑΥΡΟΣ (*Or. Sib*. VIII, 217-250). *Augustine gives it in Lat. tr. (*De civ. Dei* 18, 23).

2. The Sibylline oracles. The Jews of *Alexandria used the collections of Sibylline O. and their authority in the Greek world for their own religious propaganda. To this end they adapted, transformed, interpolated and added new compositions. These O. usually announced, in the spirit of the prophets, catastrophes and the end of the world, as does the mediaeval hymn *Dies Irae*. The primitive Jewish nucleus is book III of the Sibylline O., in the first version, written in AD *c*.140. Books IV and V ooze hatred towards the Rome of the time of *Vespasian and *Titus, the destroyer of the temple. The Christians in turn operated in the same way as the *Jews, exploiting the Jewish source, interpolating, esp. in books I-III, and rewriting specifically Christian parts, like books VI-VII-VIII, prob. in the 2nd c. The *Shepherd* of *Hermas makes reference to them (*Vis*. II, 4). They rise in turn against Roman *persecution in book VIII, and announce the fall of *Rome and of the emp. *Nero.

3. Structure. Books I-II form a whole. A Jewish redaction (I, 1-223), with Christian *interpolations (224-400) of later redaction. In book II it is hard to distinguish what is Jewish from what is Christian. Book III announces various catastrophes; it seems to be entirely Jewish with no Christian interpolation. Books VI-VIII are totally Christian. Book VII shows *gnostic conceptions (late 2nd c.), like the baptism of fire (VII, 84). Book VIII - 500 verses - is much more important, since it refers to Christian *life and *liturgy (VIII, 402-411). *Lactantius takes 30 citations from it. The first part could have been written by a Jew. It announces God's chastisement and the fall of Rome. The second part is a triumphal *hymn to Christ, as judge and lord of the *world. The third is a hymn to *God the creator and to the Logos, his incarnate son. The book concludes with norms for Christian life. There seems to be no gap between this book VIII and the three books discovered by A. Mai (XI-XIV, a numbering that follows the *manuscript tradition of the family of codices [*Ambrosianus* E 64 sup., Vatican 1120 and 743, *Monacensis* 312], which divides book VIII into three).

4. Survival of the collection. Already known to Hermas, the O. are cited by *Justin (*I Apol*., 20, 1; 42, 2), *Athenagoras, *Theophilus, Ps.-*Melito, *Tertullian, *Clement of Alexandria, *Commodian, *Lactantius, *Eusebius, the *Apostolic Constitutions*, *Gregory Nazianzen, *Sozomen and *Augustine. Augustine knew a Lat. tr. of VIII, 217-43 (see also Kurfess's ed., 222-264). Its influence continued to make itself felt over the centuries: *Dies Irae*, Dante, Calderón, Giotto, Michelangelo. A *Prophetia Sibillae magae* exists in Latin, published by B. Bischoff (*Mél. de Ghellinck*, Gembloux 1951, 121-147).

Texts: CPG 1352; *Oracula Sibyllina*, ed. A. Rzach, Vienna 1891 (the best); ed. Geffcken, GCS 1902; ed. A. Kurfess, Munich 1951 (text with Ger. tr., bibl.); It. tr. Erbetta III, 497-529.
Studies: A. Rzach, *PWK* 2A, 2073-2183 (bibl. up to 1924; after which, see J. Quasten, *Patrologia*, I, Turin 1967, 151 f.); H. Fuchs, *Der geistige Widerstand gegen Rom*, Berlin 1938; H. Jeanmaire, *La Sibylle et le retour de l'âge d'or*, Paris 1939; A. Peretti, *La Sibilla nella propaganda ellenistica*, Florence 1943; B. Altaner, Augustinus u.die nt. Apokryphen, *AB* 67 (1949) 244-247; E. Demougeot, St. Jérome, les O.S. et Stilicon, *REA* 54 (1952) 83-92; A. Kurfess, Kaiser Konstantin u.die Erythr. Sibylle, *ZRGG* 3 (1951) 353-357; id., Zur V. Buch der O.S., *RhM* 29 (1956) 225-241; id., Juvenal und die Sibylle, *HJ* 76 (1957) 79-83;

id., *Dies irae, ibid.*, 77 (1958) 328-338; J.B. Bauer, Die Gottesmutter in den Or. Sib., *Marianum* 18 (1956) 118-224; F. Paschoud, *Roma aeterna*, Rome 1967; V. Nikiprowetzky, *La troisième Sibylle*, Paris 1970.

A. Hamman

ORANGE, *Council of.* Met 8 Nov 441 at Orange (Lat. *Arausio*) in *Gaul; *Hilary of Arles presided, *Eucherius of Lyons took part. It issued 30 disciplinary canons, which show clearly the pastoral problems of the time and the customs of the *Gallican liturgy. For the council of 529, cf. *Merovingian councils.

CCL 148, 77-93; Hefl-Lecl 2, 430-454; Griffe, *La Gaule chrétienne*, Paris ²1966, II, 121-122; III, *passim*; Palazzini 3, 237.

Ch. Munier

ORANS: *iconography. This term is used to indicate a male or female figure, usually in front view, in the *gesture of *expansis manibus*. This attitude has its justification if we consider it proper to someone praying or asking for help: at any rate it expresses a dialogue, following a custom indicated more than once in the OT: "Every day I call upon thee, O Lord; I spread out my hands to thee" (Ps 88, 9 [10]; see also Ex 17, 1; Lam 3, 41). In the past this *symbol was considered exclusively Christian, though with various interpretations (the *soul, *Mary, the *ecclesia*), but the studies of W. Neuss and T. Klauser have insisted on the existence of not dissimilar representations outside the Christian figurative context. Besides monumental evidence (the famous "statue of Livia" [1st c. AD] which substantially repeats the type of the Artemisia of Halicarnassus [4th c. BC], the stuccos of the Pythagorean basilica of Porta Maggiore, Rome, to cite only the best-known), literary sources document the familiarity of this gesture in the Hellenistic and Roman artistic cultures (Pliny, *Nat. Hist.* XXXIV, 73; 78; 90).

In a quite numerous series of coins, going from *Trajan to Maximianus Herculeus, a female figure recurs in the indisputable gesture of the o., indicated by the legends as *Pietas, Pietas publica, Pietas augusta, Pietas Augusti* or *Augustae, Augustorum*, etc., i.e. a personification of *Pietas erga homines* and *adversus deos*. The *pietas* of the deceased *erga deos* appears in some 3rd-c. *sarcophagi next to the figure of the shepherd carrying a ram, symbol of φιλανθρωπία, and generalized figures of philosophers (*Ws* 1, 3; 2, 3; 19, 1). Certainly belonging to the same iconographical tradition are some clearly Christian-inspired monuments, in which the figure of the o. is combined, for its christological significance, with that of the Good *Shepherd. These two symbolic images are repeated on the oldest Christian sarcophagi (*Ws* 1, 2; 3, 1; 19, 3-6; 178, 2, etc.) and on the vaults of cubicles in the catacombs (*Wp* 25; 61), figuratively expressing the two poles of the work of salvation: the Saviour and the saved. The introduction of biblical scenes and the evident allusion to the means of salvation help to clarify the precise sense of these symbolic figures, taken from the pre-Christian repertoire (*Ws* 127, 1; 127, 2). The iconographical type of *Pietas* or Εὐσέβεια was taken up by the Christians with no formal change. The isolated *orantes* are depicted among the trees which characterize *Paradise (*Wp* 110; *Ws* 19, 6) or with the *capsa* of the *volumina*, an allusion to doctrine as means of salvation, or between two *apostles who introduce the deceased into Paradise (*Ws* 111, 4; 215, 7), according to the indispensable formulae for expressing the state of being "saved".

When the artist endows the image with individual elements (*name, facial characteristics, age, etc.), he fixes in the portrait of the deceased the general characteristics of the type: e.g., in the *cemetery of Domitilla the deceased Veneranda, in the act of praying, is introduced by St *Petronilla into Paradise (*Wp* 213). Further evidence of the desire to represent the deceased in heavenly *peace is provided by numerous funerary slabs, on which the text of the *epigraph is accompanied by the image of the dead o. (e.g. ICUR V, 14385; 15137). Finally, some well-known biblical characters are shown as *orantes*: *Noah, *Jonah, *Daniel, *Susanna, the three young men in the *fiery furnace. The presence of the dove alongside Noah (*Ws* 9, 3) or of the *angel alongside Daniel (*Ws* 96; 218, 1-2) and the three young men (*Wp* 137, 1; 23, 1; *Ws* 161, 1; 201, 4), indicates the divine intervention that has taken place. *Mary, the *martyrs and saints also appear *expansis manibus*, again testifying to the state of eternal *beatitude: e.g. St *Agnes on a 5th-c. *pluteus* slab (U. Broccoli, *Corpus Scultura*, pl. 106) and on gold-glass (C.R. Morey, *Gold Glass*, pl. XIV, 83-85), St *Cecilia in the homonymous crypt in Callisto (*Wp* 260, 2), St *Januarius (U.M. Fasola, *La cat. di s. Gennaro*, pl. VII) and St *Apollinaris in his basilica at *Ravenna (F.W. Deichmann, *Mosaiken von Ravenna*, pl. 38). This gesture, variously interpreted but certainly familiar to the first Christians in its most simple and spontaneous meaning, in harmony with the words of St *Paul (1 Tim 2, 8) and of the *Fathers (Clem. Rom., *I ad Cor.* 39, 1; Tertull., *De orat.* 14, 29: PL 1, 1272, 1286-88; Orig., *De orat.* 21: PG 11, 479-482), has been exhaustively clarified in De Bruyne's proposed hermeneutic: the biblical characters do not invoke salvation, they are already saved. Thus the dead o. is a symbol of joy because (s)he is in divine *peace; the isolated images are nothing but a stereotyping of this symbol. Primitive Christian iconography knew no doubt and no pessimism, but was permeated by the constant certainty of Christ's saving intervention. **[Fig: 239]**

DACL 12, 2298 ff.; *EC* 9, 179-181; *EAA* 5, 704-708; *LCI* 3, 352-354; W. Neuss, Die Oranten in der altchristlichen Kunst, *Festchrift P. Clemen*, Bonn 1926, 130 ff.; H. von Schoenebeck, Die Christliche Sarkophagplastik unter Konstantin, *MDAI (R)* 51 (1936) 275-320; A. Grabar, *Martyrium*, II, Paris 1946, 24, 292-295; A. Stuiber, *Refrigerium interim. Die Vorstellungen vom Zwischenzustand und die frühchristliche Grabeskunst*, Bonn 1957; T. Klauser, Studien zur Entstehungsgeschichte der christlichen Kunst, *JbAC* 2 (1959) 115-131; 3 (1960) 112 ff.; L. De Bruyne, Les Lois de l'art paléochrétien comme instrument herméneutique, *RAC* 39 (1963) 12 ff.; E. Sauser, *Frühchristliche Kunst*, Innsbruck-Vienna-Munich 1966, 368-377; P. Testini, *Le catacombe e gli antichi cimiteri in Roma*, Bologna 1966, 269 ff. (with bibl.); V. Saxer, "Il étendit les mains à l'heure de sa Passion": le thème de l'orant dans la littérature chrétienne des IIe et IIIe siècles, *Augustinianum* 20 (1980) 335-365. For the examples cited in the text, see: F.W. Deichmann, *Frühchristliche Bauten und Mosaiken von Ravenna*, Baden Baden 1958; C.R. Morey, *The Gold Glass in the Vatican Library with Additional Catalogue of other Gold Glass Collections*, Vatican City 1959; P. Testini, *op. cit.*, 266 ff.; C. Carletti, *I tre giovani Ebrei di Babilonia nell'arte cristiana antica*, Brescia 1975; U.M. Fasola, *La catacomba di s. Gennaro a Capodimonte*, Rome 1975; U. Broccoli, *La diocesi di Roma: Corpus della scultura altomedievale*, VII, 5, *Il Suburbio*, 1, Spoleto 1981; G.W. Meats, *The Roman Villa at Lullingstone*, 2 vols., Kent Archaeological Society, 1987, vol. 2, *The Wall Paintings*.

A.M. Giuntella

ORATIO CYPRIANI. Two works exist with this title, wrongly attrib. to *Cyprian of Carthage and to the legendary *Cyprian of Antioch. They are two *prayers preserved in Latin. The Greek original was published by T. Schermann (Die griech. Kyprianosgebete, *OC* 3 [1903] 303-23). The first bears the title *Oratio I Cypriani antiocheni pro martyribus*, and begins with the angelic *trisagion *Hagios*: the whole thing is at once a profession of faith and an invocation. Biblical names appear in it (*Tobias, *Daniel, etc.). It has recently been recognized as a source of inspiration for symbolic figures painted in the catacombs (H. Leclercq, *DACL* 12, 2230). The *Oratio II Cypriani quam dixit sub die passionis suae* is full of examples of answered prayer. It gives an impression of being a series of exorcisms invoking *God against the infernal powers.

CPL 67 (p. 14); PL 4, 985-99; 101, 567-69; CSEL 3, 3, 144; *DACL* 12, 2, 2324-45; A. Wilmart, *RBen* 48 (1936) 281 n. 1; A. Di Berardino, *Patrologia* III, 298; G. Philippart, Orationes Cypriani, *AB* 91 (1973) 298. For the value of this type of prayer in hagiographical literature, cf. B. de Gaiffier, *Etudes critiques d'hagiographie et d'iconologie*, Subsidia Hagiographica 43, Brussels 1967, 58-60.

L. Dattrino

ORDERS - ORDINATION. Basing themselves on the apostolic writings, the ecclesiastical writers consider Christ's priesthood as the one priesthood, in which ordained *ministers and *faithful participate. Many authors speak in this way, incl. *Origen (*Hom. in Lev.* IX, 1: GCS VI, 417-419), *Clement of Alexandria (*Adumbratio in I Pt.*: PG 9, 730), *Ambrose of Milan (*Expos. in Lc.* V, 33; VIII, 52: PL 15, 1645, 1781), *Augustine of Hippo (*De civ. Dei* XX, 10: PL 41, 676; CCL 48, 717 f.), *Leo the Great (*Serm.* IV, 1: CCL 138, 14). At the same time they make a clear distinction between ordained ministers and people. *Justin distinguishes ἀδελφούς from προεστῶς and διάκονοι (*I Apol.*, 65: PG 6, 428). *Tertullian makes the same distinction (*De baptismo* 27: CCL 1, 291-292; *De praescr.* 32, 41: CCL 1, 212-213, 221-222; *De virginibus velandis*, 9: CSEL 2, 1218-1219). *Hippolytus of Rome, in his ritual for ordination, states it with the utmost clarity (*Trad. Apost.*, ed. B. Botte 1963, 7-20), and Ps.-*Dionysius repeats it in his *De eccles. Hier.* (V, 1-6: PG 3, 500-505; SCh 58, 10-103).

The same distinction appears very early within the hierarchy itself; e.g. in *Hermas (*Vis.* III, 5, 1: PG 2, 103; SCh 53, 110-112) and still more in *Ignatius of Antioch (*Magn.* 3, 7: PG 5, 664-665; Funk 2, 116; *Trall.* 2, 3: PG 5, 676-677; Funk 2, 96; *Philad.* 3, 7: PG 5, 700-701; Funk 2, 170, 174, 184), who also specifies that the people are subject to the bishop, the priest and the deacon (*Ephes.* 2, 4, 20: PG 5, 733-734, 753-756; Funk 2, 236, 240, 256; *Magn.* 2, 3, 6, 13: PG 5, 757, 764, 766; Funk 2, 114, 130; *Trall.* 2, 13: PG 5, 676-677, 684-685; Funk 2, 94, 112, etc.). This distinction became more and more firmly established.

On the other hand, we know that in the West one controversy tended to see the bishop as simply a priest to whom a power had been added. Thus *Ambrosiaster (*In epist. ad Tim. primam* 13, 8-10: PL 17, 496, CSEL 81, 267; *In epist. ad Ephes.* 14, 11-12: PL 17, 409-410, CSEL 81, 98-101).

*Jerome in particular defends this thesis (*Epist. ad Evangelum* 116: PL 22, 1192-1195, CSEL 56, 308-312; *Com. in epist. ad Titum* 1, 5: PL 26, 562-563). Tertullian, in various places, takes up the distinctions mentioned (*De bapt.* 17: PL 1, 1326, 1329, CCL 1, 291-292; *De praescr.* 32: PL 21, 44, CCL 1, 1218-1219), and *Cyprian does not fail to mention them in his *letters (*Ep.* 5, 1: PL 4, 325, CSEL 3, 478; *Ep.* 17: PL 4, 276, CSEL 3, 521; *Ep.* 18: PL 4, 278, CSEL 3, 523; *Ep.* 23: PL 4, 325-326, CSEL 3, 568). Though from the beginning the *Fathers distinguish an ordained hierarchy from the people, they are not so clear when they distinguish between ἐπίσκοπος and πρεσβύτερος. E.g. *Clement of Rome calls those who hold power at *Corinth πρεσβύτερος (*1 Clement* 47, 6; 44, 2; 47, 1: PG 1, 308, 317, Funk 1, 160, 170-171). *Theodore of Mopsuestia seeks to explain the progressive distinction between ἐπίσκοπος and πρεσβύτερος, esp. in his *Commentary on 1 Tim*. In small communities the πρεσβύτερος did a bit of everything and for this reason was called ἐπίσκοπος. Later, only those would be called bishops who had the power of *laying-on of hands (χειροτονεῖν), and they were indicated by this name. They were responsible for a province and bore the name of *apostle.

After the deaths of the apostles, their successors bore the name of ἐπίσκοποι (Theod. Mops., *In Tim.* III, 8: ed. Swete 2, 117-126). It is interesting to note that it was apparently *Irenaeus of Lyons who, for the first time in the West, called a local bishop ἐπίσκοπος (*Adv. haer.* III, 3, 1: PG 7, 348, 349; SCh 211, 31). *Hippolytus shows us above all the "great priest", the bishop: this name clearly indicates that he constitutes the first order, and the priest shares a part of these powers. The place Hippolytus assigns to the deacon - who was ordained to serve the bishop not just in performing the *liturgy, but for every concrete need of the life of the ecclesial community - offers us a *theology of the *diaconate whose richness would later, and quite quickly, be lost from sight; already by the 5th c. the deacon had only, and almost exclusively, a liturgical function. We have proof of this decline in the formularies for ordination contained in the Verona sacramentary (*Sacr. Veronense*, ed. L.C. Mohlberg, 120-121, nn. 948-951). If Hippolytus leaves the priest little to do, it is because it is the bishop who habitually celebrates the *Eucharist and the other *sacraments. The priest celebrates as a delegate of the bishop, as is openly shown by the rite of *fermentum* in *Ordo Romanus* I (M. Andrieu, *Les ordines romani du Haut moyen âge*, II, Louvain ²1960, 61-64). *Innocent I's well-known *letter to *Decentius of Gubbio speaks very clearly and distinctly of the rite of *fermentum* (*Ad Decentium Eugubinum* 5: PL 56, 516-517; ed. R. Cabié, *La lettre d'Innocent I à Decentius de Gubbio*, Louvain 1973; *Ordo Romanus* II, ed. M. Andrieu, 6). *Cyprian describes the custom in *Africa, where the priest celebrates in the bishop's absence (*Ep.* 13, 2: CSEL 3/2, 510-511). The offering of the Eucharist is proper to the bishop or the priest, and is never entrusted to the deacon. *Athanasius clarifies this idea, mentioning a *layman in a small village where there was no priest, who had been permitted to celebrate the Eucharist (*Apol. c. arian.*: PG 25, 269).

A certain theological exploration of the matter unfortunately issued in a form of legalism, to such a point that the council of *Sardica (343) required anyone being consecrated bishop to be first, for a certain length of time, lector, deacon or presbyter (Mansi 3, 4). Luckily, as we know, these legal prescriptions were not truly observed until later: St Ambrose's consecration as bishop is a case in point. As for rites of ordination, from Hippolytus of Rome through the Roman sacramentaries to the 10th-c. Romano-Germanic pontifical we find only *laying-on of hands and a consecratory *prayer (*Pontificale romano-germanico del X secolo*, ed. 1972, C. Vogel - R. Elze, ST 269, Vatican City, I, 1972, 20-36; 205-226), while in the *East, as we know from the writings of e.g. *John Chrysostom, the rite was very different. Laying-on of hands was considered the essential rite of ordination, and it was frequently indicated as conferring the *grace of the *Holy Spirit and the power that the bishop received to communicate it to others (Basil, *Ep.* 188, 1: PG 32, 668); the gospel-book was also laid on the bishop's head to indicate that he received *Aaron's tiara and *authority in the church, with a particular obligation to make known the gospel and to observe himself what the gospel required (for laying-on of hands, *In Tim. III, 8 hom.* II: PG 62, 553; for the gospel-book, the *homily is published in John Chrysostom's *Spuria*: PG 56, 397, author *Severian of Gabala). The same text is found in *Photius, *bibl.*, cod. 277: PG 104, 276 (cf. J. Zellinger, *Studien zu Severian von Gabala*, Münster i. W. 1926, 60-64). *Palladius too (*Dialogus de vita et conversatione beati Joannis Chrysostomi*, 14: PG 43, 43) records the rite of laying on the gospel-book. *Jerome, expounding his thought on the episcopate, makes the ordination of a bishop consist solely in the choice of a mere priest who is made to sit on a higher seat (*Ep. 146 ad Evangelum*: PL 22, 1194; CSEL 56, 316). The Latin Fathers, then, give no indication of other rites than that of laying-on of hands; the other rites appear only in the 10th c. with the Romano-Germanic pontifical.

Though St *Augustine says much about the grades of the hierarchy, on the rites he repeats what his predecessors had written; more than once, however, esp. in his *letters, he insists on the need for irreproachable moral conduct in those who have received orders (*Ep.* 21: PL 33, 38; CSEL 88, 112; *Ep.* 22: PL 33, 90; CSEL 88, 113-119, etc). But he does not deny the validity of orders conferred by an unworthy, heretical or schismatic minister. His anti-*Donatist works try to explain the respective roles of the *minister, the *sacrament as such and the *faith of the subject. What is said of *baptism is also applicable to the other sacraments, which must be considered valid even if conferred outside the church (*De baptismo* III, 17, 22: PL 43, 149; CSEL 51, 213-214; *ibid.* III, 17, 23: PL 43, 150; CSEL 51, 214-215). An expression in *De baptismo* sums up Augustine's theological opinion on the validity of the sacrament: *non cogitandum quis det, sed quid det* (*De bapt.* IV, 10, 16: PL 43, 164; CSEL 51, 240). St *Leo, and with him many other ecclesiastical writers, followed this *Augustinian doctrine on ordinations conferred outside the church (Leo Gt, *Ep.* 12, 6: PL 54, 653). Augustine's *De catechizandis rudibus* and *De doctrina christiana* give the same doctrine.

We should not forget the influence exercised by *Gregory the Great. His *Liber regulae pastoralis*, which had considerable success and was even translated into Greek, takes up the Augustinian doctrine on ordination conferred on a subject whose moral conduct before ordination left something to be desired, referring to the *theology of baptism: *sicut baptizatus semel baptizari iterum non debet, ita qui consecratus est semel in eodem ordine iterum non valet consecrari* (*Ep.* 46: PL 77, 585). For the doctrine pertaining to the validity of the sacrament of orders conferred by a *heretic, we cannot here go into the controversies of the 8th c. and later, but in general they relate to the Augustinian doctrine. On the other hand, the Roman *council of 769 invalidated all ordinations made by the usurping pope Constantine (Mansi 12, 701). The same case was repeated with Hincmar of Rheims, resolved through the intervention of Nicholas I, for the ordinations conferred by Ebo, bishop of the same city. The decisions of the council of Soissons (853) on this matter followed the Augustinian line, as did the theological position adopted to resolve the problem of the ordinations conferred by pope Formosus. On this occasion Auxilius, in *c.*911, provided a patristic document: *De ordinationibus papae Formosi*, and a dialogue entitled *Infensor et defensor*: small works summing up the Augustinian doctrine for the unworthy bishops or the usurping *pope (*De ordinat.* 19. 20. 25. 26: PL 129, 1060-1074; *Inf. et def.* 5-6: PL 129, 1082). These problems returned to the fore in various minor episodes, and they were usually resolved by following the Augustinian doctrine on baptism and ordination.

A. Michel, Ordre, DTC 11, 1275-1286; AA. VV., *Etudes sur le sacrement de l'Ordre*, Paris 1957; J. Lécuyer, Episcopat et presbytérat dans les écrits d'Hippolyte de Rome, RecSR 41 (1953) 30-50; id., *Le sacerdoce dans le mystère du Christ*, Paris 1967 (ch. 13, Les successeurs des Apôtres et le sacrement de l'Ordre, 339-392; Evêque et prêtre, 393-410); J. Colson, *Ministre de Jésus Christ ou Sacerdoce de l'évangile*, Paris 1966 (awareness in the subapostolic period 211-346; patristic index 367-374); A. Lemaire, *Les ministères aux origines de l'Eglise*, Paris 1971; R.D. Dupuy, *Mysterium salutis*, 8, Brescia 1975, 618-647; A. Santantoni, *L'ordinazione episcopale*, SA 69, Rome 1976; J. Lécuyer, *Le sacrement de l'Ordre*, Paris 1983; P.F. Bradshaw, The Participation of other Bishops in the Ordination of a Bishop in the Apostolic Tradition of Hippolytus, SP 18.2 (1989) 335-339.

A. Nocent

ORDINES ROMANI

1. Significance and history. While the *liber sacramentorum* (cf. *Liturgy: IX) contains the euchological formulary of the Christian rites, the *liber ordinum* gives a detailed exposition of them and indispensably complements the former. There is also a difference in use between the two *books: the sacramentary is the text of the celebrant, while the *ordines* serve as a guide to the Master of Ceremonies. Before being brought together in collections, the *ordines* existed as separate booklets, each one intended for a specific function. This is the case of the *O. R*. They were preserved thanks to the religious policy of the Carolingian kings Pippin the Short and Charlemagne, who introduced the Roman *liturgy into their state, but in order to achieve this, had to reach a compromise with *Gallican customs. The oldest MSS of *ordines* go back to this time (early 9th c.): among them we can distinguish those in which the Roman rites have kept their original purity from those in which they were adapted to Frankish customs.

2. Collections of "ordines". The collection that has preserved the Roman usages intact is called A: it was formed in all probability in N France, in *c.*750, on the basis of isolated *opuscula* coming from *Rome in the late 7th and early 8th cc. It comprises *ordines* I, XI, XXVII, XLII, XXXIV, XIIIA concerning the Papal Mass, *baptism, Holy Week, deposition of relics in new churches, *ordinations from acolytes to bishops, and liturgical readings through the year.

Collection B comprises the same *ordines* (the *ordo* of *Veronensis* 92 has XIIIB, I, XI, XXVIII, XLI, XXXVIIB), but according to a Gallican recension and with Gallican additions (ordinations, episcopal *blessings). It was drawn up between 790 and 814/18: its purpose was to serve as a complement to the 8th-c. Gelasian sacramentary.

Alongside these wide-ranging collections, we distinguish the collection of St Amandus (9th c., N France), the *Capitulare ecclesiastici ordinis* (Austrasia-Burgundy, late 8th c.) and the collection of *Sangallensis* 614 (mid century), an important step in the formation of the Romano-Germanic pontifical.

3. Studies and editions. After the old editions of Hittorp (*De div. cath. Eccl. officiis*), Mabillon (*De liturg. Gall. libri III*) and Martène (*De antiq. Eccl. ritibus*), it was M. Andrieu who took the decisive step in the study of the Roman *ordines*, and his edition has become a classic. He gave the *O. R.* a new numbering from I to L, making a selection between MSS and excluding certain texts from the collection. This is evident when we look at the funerary *ordines*: while Andrieu published only one, XLIX of his series, D. Sicard has recently published seven more. R. Elze has done the same for coronation ceremonies: for that of the emperor, Andrieu published *ordines* XLV-XLVIII, while Elze knows 22 and four more for the empress; for that of kings we know at least 20, of which Andrieu makes no mention. So although Andrieu has opened up a vast field of research, with a magisterial sureness of method, there is still much work to be done in his wake.

M. Andrieu, *Les ordines romani du haut moyen âge*, 5 vols., (unfortunately without indexes), Louvain 1931-61; R. Elze, *Ordines coronationis imperialis*, Hanover 1960; C. Vogel, *Introduction aux sources de l'histoire du cult chrétien au moyen âge*, Turin ²1976, 101-181; D. Sicard, *La liturgie de la mort dans l'église latine des origines à la réforme carolingienne*, LQF 63, Münster 1978.

V. Saxer

ORGANIZATION, ECCLESIASTICAL

I. Dioceses - II. Parishes - III. Ecclesiastical provinces - IV. Patriarchates.

I. Dioceses. The term diocese is derived from Roman public law: a *dioecesis* was a vast area of several provinces, ruled by a *vicarius*. In the ecclesiastical sphere the term diocese (διοίκησις) was reached by a transition, in Christian antiquity, from the term church (ἐκκλησία) extending to the universal church, the particular church and the *church building, to that of parish (παροικία) and only later exclusively to that of diocese. In the West, besides *paroecia*, the terms *ecclesia, territorium, fines episcopatus* and *dioecesis* were also used. In 417 pope *Zosimus referred to rural communities, previously called *dioceses*, as *paroeciae*. From the end of the 6th c., with *Sidonius Apollinaris, the two terms were used equivalently. From the 7th c. a bishop's territory was called *dioecesis* and the particular community led by him was the *paroecia*. The origin of the dioceses coincided with the beginnings of Christian preaching; it is visible in the context of St *Paul's *journeys. In the territories of the Graeco-Roman world (the Palestinian Jewish sphere was ruled by the apostolic college), Paul founded Christian communities in the great cities, to each of which he appointed a college of presbyters, supported by deacons. Paul is seen assisted by other hierarchical members with subordinate powers, whom the community members obey (1 Cor 16, 15 f.; 1 Th 5, 12; Rom 12, 6 ff.). They are called elders, presbyters (Acts 14, 23); they are overseers (ἐπίσκοποι) to rule God's church as pastors (Acts 20, 17. 28): for now, presbyters and bishops designate holders of the same functions. Deacons have different tasks from those of presbyters-bishops (1 Tim 1, 1-10; 5, 17-19; Tit 1, 5. 11). Alongside the hierarchical members, there are the charismatics with gifts, such as *prophecy and tongues, aiming to keep the new faith alive. As for Paul, his having been directly called as an Apostle to the Gentiles justifies his unique place in the ordering of his communities (2 Cor 10, 8; 13, 10; 1 Cor 4, 21), where he appears as teacher, judge and legislator (1 Cor 7, 17; Tit 1, 5). The same with *Peter, beginning from *Antioch. The disciples, sent into these territories, inherited the burden of government from their respective *apostles, establishing an uninterrupted apostolic *succession. In the urban centres, which were missionary strongpoints and more populated by Christians, bishops were installed. Following the propagation of Christianity in various cities, a presbyter was put with the bishop. Each of these bishops was equal in *authority to the other bishops, but any could excel in personal gifts. The setting-up of a diocese was more and more linked to the number of believers and to the rise of new cities (council of *Sardica, can. 6; *Chalcedon, can. 17).

Territories with many cities usually had small but numerous bishoprics, as in N *Africa and S-central *Italy, while territories with few cities had vast bishoprics, as in N Italy, *Gaul, *Spain, the Danubian and Balkan provinces.

The diocesan area usually coincided with the urban-administrative area and, through an intrinsic mechanism, the lesser dioceses usually gravitated around the dioceses corresponding to the provincial capitals, with which they already had economic and political links. Residence was obligatory for bishops, though often disregarded, esp. in the 4th c., by Eastern courtier-bishops (council of *Nicaea, can. 16; *Antioch, can. 3; *Sardica, cann. 7-8, 11-12; *Carthage [390], can. 7; [397], can. 56). In the *East, the care of the rural population of a local church was entrusted to a χωρεπίσκοπος or *chorbishop, who was completely dependent on the urban bishop: thus the synods of *Ancyra, *Neocaesarea, *Nicaea (4th c.). They later tended to be replaced by *periodeuti (visitatores, circuitores)*, i.e. simple itinerant priests (synod of *Laodicea [c.380], can. 57). The West, all civil order having been overturned by the barbarian invasions, turned to the *pope for the founding of dioceses and the setting-up of the church hierarchy. Otherwise, the bishop of *Rome did not as a rule intervene in the setting-up of new local churches. If the bishops of *Africa, in the West, moved in a rather autonomous way in bringing local *Donatist churches within the orbit of the *Catholic churches, in the East the state authority intervened or was drawn into the matter in cases of conflict. For *Italy, 16 dioceses are documented at the end of the 3rd c.; 55 more arose in the 4th c.; another 155 in the 5th c.; in the 6th c., 57 more; and at the start of the 7th c., another 13. In all, at the start of the 7th c. peninsular and insular Italy counted 258 dioceses. The reasons for this increase include the freedom of the church (*edict of Milan, 313), eminent episcopal figures such as St *Ambrose at *Milan, St *Eusebius at Vercelli and St *Maximus at *Turin, and the determining role of the bishop of Rome.

EC 4, 1651-1653; F. Lanzoni, *Le diocesi d'Italia dalle origini al principio del secolo VII*, Faenza 1927; S. Mochi Onory, *Vescovi e città*, Bologna 1933; A. Schürmann, Diözese, *RACh* 3, 1053-63; J. Gaudemet, *L'Eglise dans l'Empire Romain (IV-V siècles)*, Paris 1958, 322-377; K. Baus-E. Ewig, *Storia della Chiesa. I. Le origini (I-IV sec.)*, Milan 1977, 132-135; K. Baus-E. Ewig, *Storia della Chiesa. II. L'epoca dei concili (IV-V sec.)*, Milan 1977, 253-255; Fliche-Martin, *The Church in the Christian Roman Empire*, II, London 1952, 600-606; F. Dvornik, *Byzantium and the Roman Primacy*, 1966; H. Chadwick, Faith and Order at the Council of Nicaea, *HThR* 53 (1960) 171-195; id., The origin of the title "Oecumenical Council", *JThS* n.s. 23 (1972) 132-135; id., *Episcopacy in the New Testament and Early Church*, (ed.) J. Howe, Lambeth Conference preparatory papers, London 1977, 206-214; A. Tuillier, La Doctrine des Apôtres et la hiérarchie dans l'Eglise primitive, *SP* 18. 3 (1989) 229-262; W.H.C. Frend, *The Rise of Christianity*, London 1984; Various authors, *Cristianizazzione e organizzazione ecclesiastica delle campagne nell 'alto medioevo*, Settimane di Studio 28, 2 vols., Spoleto 1982.

O. Pasquato

II. Parishes. Trustworthy historical information on communities subject to the bishop, but not directly led by him, comes from the councils of *Antioch (341) for the *East and *Sardica (343) for the West. The bishop led not just the urban community, but also that of the village (*chora*). If there were bishops in the villages (*chorepiskopoi*), they should only administer the churches dependent on them (Antioch, can. 10). From this we deduce that not all country churches had their bishop, and that country parishes existed. At Sardica, bishops were forbidden to interfere in the parishes of another bishop (can. 18). Pope *Innocent (402-417) said that all his churches were inside the city walls; but he supposed that the other bishops administered more distant parishes. *Rome seems to have had no country parishes. Only those presbyters in charge of the *cemeteries outside the city walls were comparable to the parish priests of other episcopal areas. Innocent calls the churches in Rome *Tituli* (*Ep. ad Decent.* 5). The Roman parishes must have arisen long before Innocent, but the accounts of them in the *Liber Pontificalis* were written after his time, at the end of the 5th c. By the mid 3rd c., *Cornelius lists 46 presbyters, 7 deacons, 7 sub-deacons, 42 acolytes, 52 exorcists, lectors and doorkeepers in the Roman community, as well as 1500 *widows and needy people fed by the community (Euseb., *HE* VI, 43, 11). The Roman community must have numbered tens of thousands, who could not all have come together at once to celebrate the *Eucharist. Acc. to the *LP*, Cletus (*c.*80-90) ordained 25 presbyters and Evaristus (after 100) allotted the Roman *Tituli* to the presbyters (and ordained 7 deacons). Urban (†230) acquired 25 silver *patenae* (prob. for 25 Tituli). *Fabian (236-250) allotted the municipal regions to the deacons and installed 7 sub-deacons (the areas of the *diaconiae* were not identical to those of the Tituli). That *Dionysius (260-268) entrusted the churches to the presbyters is something that we must certainly put at an earlier time; he also raised cemeteries (and their parish churches) to the status of *dioeceses*, i.e. he elevated the cemetery churches outside the walls into parish churches. *Marcellus (*c.*300) elevated the 25 Tituli into *dioeceses*, for the numerous *baptisms (*neophytes) and penitents.

The author of the *LP*, then, gives the name *dioeceses* to autonomous presbyterial communities with the right to confer baptism. This linguistic use is first attested by *Sulpicius Severus who relates of *Martin of Tours (*Dial.* II, 9, 6; *Ep.* 1, 10; 3, 6) that in the vast region of his episcopal city he founded about 10 country parishes (*dioeceses*). In the 5th and 6th cc. the parochial organization of the countryside was completed everywhere: as we see esp. in *Gaul, where only in *c.*600 did *dioecesis* become the common term for an episcopal community, while presbyterial communities were now called parishes. The urban parishes may have been developed in the other great cities, as they were in Rome, by the 3rd c.

Duchesne *LP*; K. Müller, Parochie und Diözese im Abendland in spätrömischer und merowingischer Zeit, *ZNTW* 1933, 149-185; E. Griffe, Les premières paroisses de la Gaule, *BLE* 50 (1949) 229-239; id., A travers les paroisses rurales de la Gaule au VIᵉ siècle, *BLE* 75 (1975) 3-26; W. Geerlings, Augustinus und sein Bistum, *ThQ* 158 (1978) 27-35.

H.J. Vogt

III. Ecclesiastical provinces (with metropolitan centre). From the 4th c. a co-ordination appeared, which gave rise to a more complex and hierarchized structure: the ecclesiastical province (ἐπαρχία). The fact is explained along the lines of missionary practice which, spreading from the capital to the other centres of the province, generated communities, which were filiations of the central episcopal or mother-church. The bishop of the mother-church (μητρόπολις) was, from the 4th c., called the metropolitan (ἐπίσκοπος μητροπολίτης), later ἀρχιεπίσκοπος, and had jurisdiction over the other bishops (ἐπαρχιῶται), who from the 8th c. were also called "suffragans". Another factor in the formation and cohesion of ecclesiastical provinces, from the end of the 2nd c., was the synods, set up with the aim of bringing together the bishops, not just of one political province, but also of a wider territory (Euseb., *HE* V, 23, 25) to debate important ecclesiastical affairs going beyond the local sphere of the churches. The provincial *council was called and presided over by the metropolitan (council of Antioch [341], cann. 14, 16, 20). It had competence over all questions of the province (Constantinople I [381], can. 2).

The ecclesiastical provinces, increasing in number, went on conforming to the boundaries of the political provinces, which were multiplied by *Diocletian and his successors until, at the start of the 5th c., there were 120 of them. All the bishops of a province, with the metropolitan's confirmation, could install a bishop in a diocese (council of Nicaea [325], can. 4).

The vague imprecision of the territorial boundaries was at the root of jurisdictional problems, later regulated by conciliar canons. But bishops' interventions outside their own territory were more properly "the consequence of a universal ecclesial sense that the bishops of previous centuries had exercised. The most typical example of this universal interest is given by the correspondence the bishops held with many other sees with the aim of resolving questions and difficulties" (J.-R. Palanque, in Fliche-Martin). But in the West a more complex type of grouping later arose: the episcopal sees of N *Africa met under *Carthage and those of S-central *Italy under *Rome, whose prestige and power went beyond the metropolitan sphere. There was a similar configuration in the *East: for *Antioch, with respect to *Syria and the Eastern provinces of *Asia Minor as regarded participation in the synods of Antioch (Euseb., *HE* VI, 46, 3; 7, 5, 1-2), and with respect to *Cilicia and Osrhoene as regarded missionary activity (*ibid.*, VI, 12, 2); and for the see of *Alexandria, on which all the sees of *Egypt, *Libya and Pentapolis depended (council of Nicaea, can. 6). The two sees appeared as super-metropolises: the way to the patriarchates lay open. Then the ecclesiastical province had its own metropolitan centre in the civil capital of the province, where the metropolitan resided, i.e. close to the civil governor and more able to help his colleagues in their relations with him. Synods (4th-5th cc.) established his powers: as well as the presidency of the provincial synod, he had a certain role watching over the religious-ecclesiastical life of the province and over his own bishops. The bishop, though autonomous in his own territory, had to obtain approval from the metropolitan and the other bishops for extra-territorial undertakings. The metropolitan institution prevailed first in the Greek East, except for Egypt (and its dependants Libya and Pentapolis) where the bishops depended on the bishop of Alexandria. In the Latin West there was marked heterogeneity. The bishops of suburbicarian Italy depended directly on the bishop of Rome. The bishop of *Milan, residence of the emperor and the *vicarius annonarius*, one of the few large cities of N Italy, was the main metropolitan of an ecclesiastical territory comprising several civil provinces. From *c.*425, the see of *Aquileia was metropolis of the provinces of Venice and Istria, *Rhaetia and *Noricum, as was *Sirmium of *Pannonia. Aquileia and *Ravenna were pre-eminent at the time of *Honorius and *Arcadius (395-423; 395-408). In the provinces of N Africa, the metropolitan (primate) was the oldest bishop, who did not necessarily reside in the metropolis. In Proconsularis, however, the metropolitan was the bishop of Carthage, who also had a certain influence in the other provinces. In *Gaul, from the 4th c. the metropolitan sees coincided with the provincial capitals. The *popes set up the apostolic vicariate of *Arles, a see founded by St *Trophimus, a disciple of St *Peter. For *Spain, we have no evidence of the existence of the metropolitan institution until after the mid 5th c. Only in the 6th c. do we see the great provincial synods there. Everywhere, in general, ecclesiastical provinces corresponded to civil administrative units, but the churches reserved the right to keep their own areas despite the changeability of the civil ones (Innocent I, *Ep.* 24, 2; council of Chalcedon, can. 12). For individual metropolises, Eastern and Western, cf. J.-R. Palanque, in Fliche-Martin, 2.

DACL 9, 786-790; *EC* 8, 914-915; J. Gaudemet, *L'Eglise dans l'Empire Romain (IVᵉ-Vᵉ siècles)*, Paris 1958, 380-407; Fliche-Martin, *The Church in the Christian Roman Empire*, 2, London 1952; K. Baus-E. Ewig, *Storia della Chiesa*, II, Milan 1977, 450-452.

IV. Patriarchates. The legal bases for their foundation were laid at the council of *Nicaea ([325], can. 6), which made the bishops of *Egypt, *Libya and Pentapolis dependent on the bishop of *Alexandria and recognized the privileged situation of *Antioch, whose bishop had metropolitan powers over the 22 bishops of Coele-Syria as well as those of *Cilicia, Mesopotamia, *Palestine and *Cyprus. These positions were recognized by the councils of *Constantinople ([381], can. 2) and *Ephesus (431). Can. 2 of Constantinople also named the political dioceses of the Eastern part of the empire: Oriens, *Asia, *Pontus and *Thrace, without alluding to a church to whose bishop the government of their respective dioceses belonged: we can already foresee the institution of the Eastern patriarchate, the more so if we take into account can. 3: "The bishop of Constantinople will have primacy of honour (τὰ πρεσβεῖα τῆς τιμῆς) immediately after the bishop of Rome, since that is a new Rome (νέαν 'Ρώμην)": the *thronos* of Alexandria, the principal one in the *East, had to take second place. Ambitions of prelates and imperial *caesaropapism, as well as the fact of having been founded by *Constantine (330) and made capital of the empire (so that its bishop had to be raised in dignity), took *Constantinople in just 70 years from a suffragan diocese of Heraclea to first patriarchate of the whole East. Its effective power was consolidated through the appeals of the Eastern *clergy to the emperors, transmitted from them to the bishop of Constantinople, who presided over the permanent synod (σύνοδος ἐνδημοῦσα) composed of visiting bishops. As for *Jerusalem, the council of Constantinople (can. 7) recognized its bishop's "traditional honorific position" (ἀκολουθία τῆς τιμῆς), but was mindful of the metropolitan rights of *Caesarea. The primary factor in the origin of the three patriarchates of Alexandria, Antioch and Rome was their geopolitical and economic role, a factor that had partly guided the Christian missionaries. This motive, though considered completely inadequate by bishops of Rome such as *Leo I and *Gelasius I, was the only reason for the institution of the patriarchate of Constantinople. Then there was the linguistic-cultural factor of the inhabitants of the areas where they arose: there was, in fact, a patriarchate respectively in the territories of Egyptian culture (Alexandria), Syrian culture (Antioch), Greek culture (Constantinople), Latin culture (Rome, though Carthage's dominance in N Africa was greater than Roman influence elsewhere in the West) and Jewish culture (Jerusalem). Thirdly there was the element of the apostolic origin of the leading episcopal sees (Petrine principle), which went on acquiring more and more weight, partly through Roman pressure. Only at the council of *Chalcedon (451) was Jerusalem recognized as a patriarchal see with jurisdiction over *Palestine, through the good offices of its bishop *Juvenal (422-458), and made independent of Antioch. Following the finding of the presumed tomb of the apostle *Barnabas (488) in *Cyprus, the Cypriot church obtained equality with that of Antioch; from definitive independence, obtained for it by the emp. *Zeno, it then passed to "autocephaly". As for *Rome, popes Leo I (440-461) and Gelasius I (492-496) added to the Petrine principle the peculiar link between *Peter and the Roman church, claiming that this was to be considered the apostolic see *par excellence* with the *popes as heirs of Peter's primatial powers.

After having dealt with the origin of the dioceses and seen that a group of them constituted an ecclesiastical province and several provinces a patriarchate, we must now follow the development of the five patriarchates: *Alexandria, *Antioch, *Rome, *Constantinople and *Jerusalem. The patriarchate of Alexandria: many, rejecting Leo I's *Tomus ad Flavianum* and the decisions of the council of Chalcedon, detached themselves from the *Catholic church and formed a new "*Coptic" church, calling the Catholics "*Melchites" or emperor's men. The Greek patriarchate, its numbers greatly reduced, remained faithful to the council of Chalcedon. The Melchite patriarchs include three saints: St Proterius (452-457), St *Eulogius (580-607) and St *John III, the Almsgiver (609-619). *Maximus

Confessor and *Sophronius defended Catholic *orthodoxy at the time of the Arab invasion (638), on whose eve the Greek church of Alexandria counted only 200,000 *faithful, mostly foreigners, against the 5/6,000,000 of the indigenous Coptic church. The patriarchate of Antioch was displeased by the rise of the patriarchate of Constantinople, at the council of Chalcedon (451), to first see of the East. Meanwhile it was infiltrated by *heresy, and the patriarchal see alternated between heretics and orthodox until, under the *monophysite patriarch *Severus (512-518), *Jacob Baradaeus organized the "*Jacobite" Syrian church: by the mid 6th c. about half the inhabitants of *Syria adhered to this. The situation deteriorated during the Arab occupation (638-969). The patriarchate of Jerusalem, to defend itself from Antioch and Alexandria, sought help from Constantinople, whose faith it followed. The Persian invasion (614) put an end to a brief period of splendour. The histories of the patriarchates of Constantinople and Rome are interwoven. The bishops of Constantinople aimed at "second place in the hierarchy after the *those of Rome and (at) privileges equal to his, though without demanding universal primacy, still acknowledged by them to the Roman see, though passed over in silence as far as possible" (M. Jugie, *Le schisme byzantin. Aperçu historique et doctrinal*, Paris 1941, 37).

The matter appeared more logical on the death of *Theodosius I (395) and the division of the Roman empire into two parts. Now, in can. 28 of the council of Chalcedon (Mansi 7, 369), which modified can. 3 of Constantinople (381) (Mansi 2, 560), the privileges of the see of Constantinople were made equal to those of Rome, while conceding first place to the latter: the bishop of Constantinople had the right to ordain the metropolitans of the (civil) dioceses of *Pontus, *Asia and *Thrace. The patriarchate of Constantinople, comprising these three exarchates, was now founded *de jure*.

Though without legal value, can. 28 served the bishop of Constantinople for his patriarchal jurisdiction. In the mid 7th c., the patriarchate counted 33 metropolises with 352 suffragan bishoprics and 24 autocephalous *archbishoprics. From the 7th c. the see of Constantinople was also called apostolic; indeed in the 9th c. it was declared to be of apostolic origin (heir of *Ephesus, therefore of St *John the apostle; linked to the legend of St *Andrew the apostle, evangelizer of *Byzantium): we thus arrive at the theory of the "pentarchy", installed by the *Holy Spirit, of apostolic *succession and composed of superior leaders of identical dignity. The Roman see was agreeable to the pluralistic structure as long as the patriarchates were shown to be based on apostolic origin. Conflicts arose through Rome vaunting her pre-eminent apostolic origin, claiming to be the universal leader of the church. The *Decretum Gelasianum*, the case of *John Chrysostom under *Innocent I (402-417), *Leo I's (440-461) opposition to can. 28 of *Chalcedon, the patriarch of Constantinople's assumption, against Roman protest, of the title "ecumenical" (= imperial) patriarch (6th c.), etc., are so many steps in a gradual tension and distancing between Rome and Constantinople. But the primacy of the bishops of Constantinople among the churches of the *East was uncontested, all the more so since the other Eastern patriarchates went over either to the *Nestorian (Antioch) or to the *monophysite (Alexandrian) heresy, and ended up under Muslim rule (7th c.). Relationships with the emperor were regulated by mutual exchanges of synodal *letters with a profession of faith by the patriarch elect; with the insertion of the patriarch's name into the *diptychs and its commemoration during the canon of the *mass, read by the deacon from the ambo; with the diplomatic representation of the pope or patriarch to the emperor through permanent ambassadors or special nuncios (*apocrisiarii*). This diplomatic network, though necessary, failed to prevent the rift between the sees of Constantinople and Rome from deepening further and further and culminating in the *schism of 1054.

EC 6, 759-763: patriarchate of Al.; *ibid.*, 1461-1463: p. of Ant.; *EC* 4, 732-737: p. of Const.; *EC* 6, 201-202: p. of Jer.; J. Gaudemet, *L'Eglise dans l'Empire Romain (IV-V siècles)*, Paris 1958, 389-407; P.P. Joannou, *Pape, Concile et patriarches dans la tradition de l'Eglise orientale jusqu'au IXe siècle*, Grottaferrata 1962; W. de Vries, *Rom und die Patriarchate des Ostens*, Freiburg i Br. 1963; K. Baus-E. Ewig, *Storia della Chiesa. II. L'epoca dei concili (IV-V sec.)*, Milan 1977, 258-263; AA. VV., *I Patriarcati orientali nel primo millennio*, OCA 181, Rome 1968; V. Monachino, *Il canone 28 di Calcedonia. Genesi storica*, L'Aquila 1979.

O. Pasquato

ORIENTATION. See **EAST**.

ORIENTIUS. Author of a poetic composition called by its first editor, Delrio, in 1600, *Commonitorium*, divided into two books of 518 distychs (309+209). The author's name, otherwise unknown, is given at the end (*Comm.* 2, 417). The first to name him is *Venantius Fortunatus in the 6th c. (*Vita s. Martini* 1, 17: PL 88, 366), who places him between *Prudentius and *Sedulius. O. certainly lived in *Gaul (*Comm.* 2, 184), devastated by the many barbarian invasions of the early 5th c., of which he speaks at length without naming the invaders (*Comm.* 2, 165-202). We know three *biographies of an O., bishop of *Augusta Ausciorum* (Auch in Gascony): everything leads to his being our author. The work is a long exhortation to live a fully Christian life. He intends to instruct on the frailty of human things, the importance of worship of *God and of a worthy conduct of life, in view of the heavenly kingdom. He denounces the various *vices prevalent around him. The moral aspect is preponderant, indeed we may call it a *sermon in verse. It abounds in descriptions. In the sole MS of the *Commonitorium* we also find various other poetic compositions, almost certainly not by O.: *De nativitate Domini, De epithetis Salvatoris, de Trinitate, Explanatio nominum Domini, Laudatio* (the last three considered to form a literary unity) and two *Orationes*.

Editions: CPL 1465-68; PL 61, 977-1006; R. Ellis: CSEL 16, 1 (1881) 191-261; L. Bellanger, Paris-Toulouse 1903 (with Fr. tr.); M.D. Tobin: PSt 74 (1945) (with Eng. tr.); C.A. Rapisarda, Catania 1958, ^{2}1970; id., It. tr. in *NDid* 10 (1960) XII-36, and Catania 1970. Studies: Bibl. in *Patrologia* III, 310; L. Bellanger, *Etudes sur le poème d'O.*, Toulouse 1902; id., *Le poème d'O.*, Paris-Toulouse 1903 (revision and enlargement of the previous); G. Brugnoli, L'oltretomba in O., *Orpheus* 4 (1957) 131-137; C.A. Rapisarda, Introduzione critica ad O. con bibliografia, *NDid* 8 (1958) 1-78; *DSp* 11, 903-906; P.-M. Duval, *Les sources de l'histoire de France*, I, Paris 1974, 701-702.

A. Di Berardino

ORIGEN

I. Life - II. Works - III. Character of O.'s thought - IV. The exegete - V. The spiritual man - VI. The speculative theologian.

O. is, with *Cyprian, the pre-Nicene author about whom we possess the greatest amount of biographical data: scarce in his works, but abundant in book VI of *Eusebius's *Ecclesiastical History*, to which we can add the information provided by *Jerome and *Photius from lost works of Eusebius. His relations with his pupils and his teaching programmes are described in *Gregory the Thaumaturge's *Address of thanks to O.*

I. Life. O., surnamed Adamantius, man of steel, or of diamond, was born prob. at *Alexandria of a Christian family. His father *Leonides gave him a particularly thorough education, Greek and biblical. In 202 Leonides suffered *martyrdom during *Septimius Severus's *persecution and the family's goods were confiscated. Some months later O. opened a *school of grammar (i.e. literature) so as to maintain his mother and six younger brothers; bishop *Demetrius put him in charge of instructing *catechumens, while the prefects of *Egypt continued the persecution. For some time he kept up this dual teaching role; then, probably when his family no longer needed his help, he gave up teaching secular culture to dedicate himself entirely to *catechesis. Urged by the radicalism which characterized his youth, he sold all the manuscripts in his possession for a petty sum, and this gesture seems to indicate a renunciation of all that was not knowledge of *God: but the apostolic demands made him return to what he had abandoned and, to deepen his philosophical knowledge, he followed the courses of *Ammonius Saccas, the father of *neoplatonism. He lived a life of extreme austerity and, taking Mt 19, 12 literally, castrated himself. Having obtained great success in his teaching, he put *Heraclas in charge of catechesis proper, reserving for himself the students most anxious to make progress and keeping in touch with *heretics and *pagans. After the age of 30 he began to write, urged by *Ambrose, a wealthy Alexandrian whom he had brought back from *Valentinianism to *orthodoxy. Ambrose had been drawn to *gnosis by not finding in the Great Church the intellectual nourishment he craved. Converted by O., he persuaded his teacher to produce what he lacked, putting considerable means at his disposal. During this first period living in Alexandria, O. made many *journeys: to *Rome, where he seems to have heard *Hippolytus preach; to Palestinian *Caesarea, where bishop *Theoctistus, like *Alexander of Jerusalem, allowed him to preach although a *layman, provoking a protest from Demetrius; to Roman *Arabia (Jordan), called there by the governor; finally to *Antioch with a military escort, summoned by Julia Mammaea - mother of the emp. *Alexander Severus - who wished to be informed about Christianity.

In *c.*231, the bishops of *Achaia invited O. to *Athens to dispute with groups of heretics. On his way through Caesarea in Palestine, Theoctistus and Alexander ordained him priest. On his return to Alexandria, Demetrius, annoyed by an *ordination carried out without his consent, called a *council of bishops and priests who exiled O. from Egypt; then Demetrius, supported by some bishops, declared him deposed from the priesthood. Saddened, O. retired to Caesarea, where he was welcomed by his Palestinian *friends and where, as in many other provinces of the *East, Demetrius's sentence was ignored. O. resumed teaching: from the *Address of thanks* to him by a pupil - who, despite recent challenges, we persist in identifying with *Gregory the Thaumaturge (cf. *Gregorianum* 60 [1979] 287-320) - we know something of his teaching in that city: a type of *school that was more a sort of "missionary" course of study for young pagan sympathizers, developed with the aim of presenting the Christian version of philosophical problems, without yet teaching Christian doctrine proper. O. preached frequently, and we have many of his *homilies. His literary activities remained considerable, since Ambrose joined him at Caesarea with stenographers and copyists. He travelled often: to Athens, where he began his *Commentary on the Song of Songs*; to Arabia (Jordan) where, during a synod, he brought bishop *Beryllus of Bostra back to orthodoxy and where he disputed with a group of Christians who asserted that the *soul died with the *body and was raised with it (the *Arabian heresy); to *Nicomedia, where he dictated his *letter to *Julius Africanus; to *Cappadocia, called there by bishop *Firmilian. Perhaps it was in Arabia that a small *council met around bishop Heraclides: its proceedings were discovered in 1941 at Tura in Egypt. In 250, this overflowing and multiform activity was abruptly ended by *Decius's *persecution: imprisoned and tortured, O. courageously proclaimed his faith. What they desired was not his death but his *apostasy, which, as the most outstanding Christian figure of his time, would have had a considerable effect. The emperor's death restored him to freedom within a few months, but in broken health. He died soon after, aged 69, prob. in 254. His tomb was still visible in the 13th c. at Tyre, in the church of the Holy Sepulchre.

II. Works. O. was prob. the most prolific author of antiquity, pagan or Christian: the list of his works given by *Jerome (*Ep.* 33, to *Paula), though incomplete, is astonishing. His *Hexapla* set down the whole text of the OT in six columns: Hebrew, in Hebrew and Greek characters; the four Greek versions of Aquila, *Symmachus, the Septuagint and *Theodotion; for some books he had supplementary versions, called Fifth, Sixth, Seventh. With graphic signs borrowed from Alexandrian grammarians, he represented the additions and *lacunae* of each text. He also commented, by different methods, on much of the *Bible. Of his learned *commentaries we possess: in Greek, nine books on *John and eight on *Matthew; in *Rufinus's Latin version, ten books on Romans and four on *Song of Songs; an anon. Lat. tr. gives us the whole second half of the Commentary on Matthew. About 300 *homilies are preserved: in Greek, on Jeremiah and on *Saul and the Witch of Endor*; in Rufinus's Latin version, on *Genesis, *Exodus, Leviticus, Numbers, Joshua, Judges, the birth of Samuel, Psalms 36, 37 and 38; in Jerome's Latin version, on Isaiah, Jeremiah, *Ezekiel, *Song of Songs, *Luke, as well as homilies on various *psalms, until now attrib. to Jerome himself, but which he only translated. O. also wrote many *Scholia*: brief exegetical notes on scriptural passages; such *Scholia* were often made into collections. They are hard to identify today in the mass of innumerable short or long fragments from lost commentaries or homilies. These fragments have come down to us in one of three ways. We possess two collections of excerpts: the *Apologia for Origen* in six books (only book I remains, in Rufinus's Lat. tr.), composed by the martyr *Pamphilus of Caesarea with the help of the historian *Eusebius, defends O. from the accusations made against him in the 3rd and 4th cc. and cites numerous other texts of his, many cited nowhere else; while the *Philocalia of Origen*, surviving in Greek, contains extracts collected by *Gregory Nazianzen and *Basil of Caesarea for a veiled apologetic motive. We also find many fragments attrib. to O. in the exegetical *Catenae, compilations of patristic interpretations on a book of Scripture, which is thus commented on verse by verse. Finally, O. is very often cited by later authors accusing or defending him. In addition, *Jerome tells us that his own commentaries on the Pauline epistles Gal, Eph, Tit and Philem closely follow O.'s corresponding commentaries, now lost save for fragments.

A second group of works, though not declaredly exegetic, give plenty of room to *Scripture. *De principiis*, or *Perì Archôn*, the source of posthumous accusations against O., is the first Christian attempt at theological reflection to start from the *regula fidei* and lean on Scripture and reason: it survives entire only in Rufinus's Lat. tr.; one seventh of it is preserved in Greek in the *Philocalia*, and some fragments are cited by Jerome, *Justinian and other authors. The other works survive in Greek: *De oratione* (*Perì Euchès*) with a commentary on the *Lord's Prayer; *Exhortatio ad martyrium*, addressed to *Ambrose when he was in danger of being arrested during the persecution of *Maximinus Thrax; the *Dispute with Heraclides*, rediscovered at Tura (cf. above); *Contra Celsum*, a refutation in eight books of the attack by the philosopher *Celsus in his anti-Christian *True doctrine*, and the most important *apologetic work of Christian antiquity. Fragments survive, Greek and Latin, of other lost works such as *De resurrectione* and *Stromata*. Of O.'s abundant correspondence, two complete *letters remain, to *Julius Africanus and *Gregory the Thaumaturge, and fragments of some others.

The greater part of O.'s immense work has not escaped the ravages of time and the persecuting violence of the emperor *Justinian. What remains still has considerable breadth, but the greater part consists only of *Latin translations and fragments: this raises important critical problems, since his translators Rufinus and Jerome did not just translate, but paraphrased: their translations aimed to fit the texts into the Latin thought-world; and the fragments in the *Catenae often consist of summaries made by the catenists, who sought to retain the main ideas. The result of this procedure was certainly not satisfactory, esp. in *De principiis*, since Rufinus on one side and Jerome and Justinian on the other translated or chose passages according to the oppposed positions they took up in the *Origenist dispute: even the texts preserved in the *Philocalia* underwent cuts.

III. Character of O.'s thought. Later we will look at the accusations, mostly unfounded, made against O.'s thought after his death. His accusers have some excuse. Firstly, the historical mentality which tries to place a writer in his own time is a relatively recent acquirement - even today we cannot say it is all that widespread among theologians; the same with the idea of a development of *dogma or, rather, of the church's awareness of *revelation. Secondly, O. himself never troubles to "define" his theological thought clearly. He depends heavily on the biblical text he is commenting on and follows it step by step, not bothering to counterbalance it immediately by complementary statements, which will be found elsewhere, in the interpretation of another text. In the theorising contained in *De principiis*, it often happens that he examines two or even three different solutions, which he seeks to expound in all their clarity, often leaving the conclusion to the reader. For this reason, while studying him we must keep in mind the whole of his work, at least what remains of it: we can then see that the different passages balance each other, making up a *theology full of tensions and nuances which generally turns out to be *orthodox. He must never be judged on an isolated text, since this expresses only a partial aspect which, taken on its own, is unilateral and hence heterodox. But his detractors from the 4th c. on had neither the time nor the patience nor even the will to study him in this way. The need to "define", which appears in the many *councils provoked by the *Arian crisis, from *Nicaea to *Constantinople, with their various formulae of faith, arose both from the influence of *Roman law, which, from the time of *Constantine, assimilated the rule of faith more and more to the civil law, and from the need, against *heretics, to enunciate doctrine in the clearest and most complete terms possible, so as to leave the enemy no loophole. The resulting mentality created the most adverse possible conditions for judging and understanding O. fairly. We must also draw attention to another essential point: in all his works, O. is simultaneously an exegete, a spiritual man and a speculative theologian. We cannot understand his *exegesis and his speculation if we forget that they are the product of a spiritual man, since they are informed by the pre-eminence of the spiritual element, as is the cosmology of *De principiis*. And *vice versa*: his *spirituality and his *theology are those of an exegete; he has made theology of his exegesis and his spirituality. Many distortions of his thought by modern criticism come from ignorance of this spiritual element. Also, O. knew the *philosophy and science of his time very well and taught it, as we see in the programme illustrated by *Gregory the Thaumaturge. But he himself was not a philosopher: he used philosophy as a theologian. As for science, he used it to explain the literal meaning of *Scripture; as he did with philology, which he knew through his Alexandrian education.

IV. The exegete. O. is, together with *Jerome, the greatest critical exegete (the *Hexapla*) and the greatest literal exegete of antiquity. He possessed inexhaustible curiosity about the different readings found in MSS of both OT and NT, and pointed them out in his *commentaries and even in his *homilies: but, for him as for all the *Fathers before Jerome, the Greek text took precedence over the Hebrew, since it was the one the *apostles had given to the church. He always explains the literal meaning accurately, helped by philology and by all the disciplines of his time and using his knowledge of Hebrew customs and exegesis, which he gained by cultivating relationships with Rabbis. Many have been scandalized by the statement in *De principiis* IV, 2, 5 that some texts have no valid literal sense. But for O. the literal sense means the material meaning of the letter, not, as for us, the meaning which the sacred author wished to express: consequently, when the *Bible speaks in figurative language, which it does frequently, it has no valid literal meaning in O.'s sense. If we examine the cases of this sort which we find in the *Homilies on the Hexateuch*, we will see that they arise either from errors of interpretation of the Septuagint (which was, as we said, the text of the church, excluding the Hebrew text even when that was known) or from the fact that, in relatively rare cases, O. failed to put himself in the context, whether literal or literary, psychological or historical. But these objections develop from details of little importance, and O. believed much more in the historicity of scriptural passages than the most conservative of contemporary exegetes: cf. his defence of the historicity of *Noah's ark against the objections of the *Marcionite *Apelles, who rightly said that the dimensions given would not allow room for so many animals (*Hom. Gen.* II, 2; *C. Cels.* IV, 31). In some NT events, such as the theophany of Jesus' *baptism, Jesus' *temptation, etc., he rightly sees a sensible expression of a spiritual vision.

But if it was for us Christians that *Scripture, incl. the OT, was composed (1 Cor 10, 11), the *Holy Spirit did not do so just in relation to the literal meaning: otherwise the legal and ceremonial prescriptions of the old covenant abolished by Christ ought to be suppressed from the Bible, as ought those narratives which, in the pastoral perspective in which the OT is set, are of interest only for the teaching that can be drawn from them. The meaning intended by the Spirit is the spiritual one, which O. finds in the NT and in earlier *tradition, but of which he is the great theorist. The OT is only *revelation in that it is, in its entirety, a prophecy of Christ, and *exegesis must show that Christ is the key to it. NT exegesis applies what is said of Christ to every Christian and shows, in Christ's actions, the prophecy and anticipated possession of the *eschatological blessings. Christ's first coming is both the fulfilment of the time which preceded it and the prophecy of his second *parousia*; the difference between the "temporal gospel" as lived here on earth and the "eternal gospel" (Rev 14, 16) of *beatitude does not lie in ὑπόστασις, substance, since there is only one gospel, but in ἐπίνοια, in the human manner of considering things, since we see only "through a glass darkly" what we will see "face to face": yet even down here we possess the "true goods", and all the sacramentalism that characterizes the time of the church is found in germ in this distinction.

Despite the incomprehension it would meet from the 16th c. on, this type of *exegesis is essential to Christian *dogma, since, if *revelation is Christ, the OT is not everywhere revelation unless it speaks everywhere of Christ. We can certainly criticize O.'s exegesis for not taking the human author sufficiently into account, in the persuasion that the *Holy Spirit is the author of *Scripture and, since the exegesis of Scripture must be worthy of *God, for seeking a meaning in every detail, which often results in an arbitrary procedure, compensated by the beauty and spiritual wealth of many of his expositions. But around the pattern we have expounded are interwoven many influences, biblical, rabbinic, Philonian, Hellenic, *gnostic, etc., which give Origenian exegesis an extremely complex character. There has been no lack of attempts at classification. Above all, the theory of triple meaning, going back to O. himself (*De principiis* IV, 2, 4), in which the meanings correspond to the divisions of his trichotomic *anthropology: the corporeal or literal meaning, the psychical or moral meaning and the spiritual or mystical meaning. P.H. de Lubac (*Exégèse Médiévale* I/1, 198-211) traces the doctrine of the fourfold meaning, first enunciated by *Cassian, back to O.'s practice: literal meaning, allegorical meaning (affirmation of Christ as the centre of history), tropological meaning (concerning the Christian's behaviour after Christ's coming), anagogical meaning (giving a presentiment of the eschatological blessings). Some modern authors distinguish "typology" from "*allegory", understanding the latter in a different sense from that of the doctrine of the fourfold meaning: interpretation which conforms to the horizontal, progressive and linear conception of Christian *time is "typological"; exegesis which supposes above the actual *world a divine or angelic world which it reflects in some way is "allegorical". This distinction is legitimate if it refers to different literary forms: what is not acceptable is the value-judgment on "allegory", which is accused of not being Christian: in fact, "allegorical" exegeses of this type are not rare in the NT, and if Christian time is sacramental it implies, as well as the horizontal dimension, a vertical dimension which manifests the fact that the *eschatological realities are already present in the actual world.

V. The spiritual man. The spiritual doctrine is present throughout in every exegetical work. O.'s ascetic and moral conceptions are still largely unstudied. *Spiritual combat dominates his *anthropology and his angelology. Man is at once spirit (πνεῦμα), *soul and *body: *pneuma* is the gift *God gives to each one to guide him in knowledge, *prayer and *virtue. But the soul is dual: its higher part, mind (νοῦς) or heart, is educated by *pneuma* and is the faculty of the soul which receives *grace; the lower part, "*flesh" or "thought of the flesh" (Rom 8, 6-7), corresponds in some measure to the concupiscence which attracts man towards the body. Moreover the soul is solicited by *angels and *demons who stand behind their military commanders, Christ and Satan. We find in O. a fairly complete doctrine, dispersed here and there in his exegetical pages, in those on *martyrdom, on *virginity or *chastity, on virtue and the *virtues, on *sin and on all that Christian *ascesis implies.

O. was the creator of many mystical themes that would be taken up by his successors. To the traditional, collective exegesis of the bride of the *Song of Songs as representing the church, he added an individual exegesis which applied to the Christian soul. He compared Is 49, 2 to Song 2, 5, thus creating the image of the arrow and wound of *love. Because Christ's *incarnation produces its effects in everyone, he must also be born and grow in every soul. Spiritual ascent, through prayer and virtue, is represented by the apostles' ascent of the mountain, on whose summit Jesus appeared to them in his divinity, which appeared through his humanity: the transfiguration is the symbol of the highest knowledge that man below can have of *God in his Son. The five spiritual senses which develop in him who progresses, allow him, in analogy with the five bodily senses, to know intuitively and innately the divine realities to which he adhered before by *faith. This innateness comes about with the development of his participation in God's *image, the Word, which man received at the moment of his *creation and which he makes grow in himself with Christian *life and with the practice of the *virtues - since only like knows like - until he reaches the "likeness" of *beatitude which coincides with perfect vision.

Knowledge has for object the *mysteries, those of visible and invisible beings, those of the relations within the *Trinity, all unified in the person of the Son, Mystery *par excellence*, Image of the Father, intelligible World: in Him are found the plans and germs of creation in that he is God's wisdom. The mystery is understood in the *light communicated by the divine *Persons, where the Son, adapting himself in various ways according to each one's capacity, becomes food which nourishes man with the divine nature, which the Son receives constantly from the *Father; he is a vine which rejoices the soul, filling it with sweetness, delight, "enthusiasm", i.e. the feeling experienced through the divine presence. Knowledge is the meeting between the divine freedom that gives itself and the human freedom that receives: God does not take possession of man against his will, in an *ecstasy that would take away his fundamental conscience and freedom, since only the devil acts in such a way in those possessed and those in the grip of a passion. Knowledge has as its usual starting-point the exegesis of Scripture, meditated in renunciation of *sin and the *world, in purity of heart: *faith is its necessary beginning, but its object becomes ever more present to the five spiritual senses and, perfectly understood, knowledge is indistinguishable from *love, in union: "*Adam knew Eve, his wife" (Gen 4, 1). The esotericism of which Origenian *spirituality has been wrongly accused is really a matter of not giving a *soul something which it could not support and which would damage it: a rule common to all spiritual directors. Similarly, to accuse him of an aristocratic vision would be to forget that he expounded all these arguments in *homilies preached to all the Christians of *Caesarea, exhorting his hearers to go forward in order to know. Remember too that O. often manifests a profoundly affective *devotion to Christ, and that in his works we find some evidences, few but clear, of a personal mystical experience.

VI. The speculative theologian. O.'s *theology is inseparable from his *exegesis and his spiritual doctrine and is inspired by them. He remains faithful to the *regula fidei* of his time, which he sets out in the preface to *De principiis*, and, starting from this, with the help of *Scripture, reason and his own spiritual and pastoral experience, he conducts his research with all modesty, without claiming to dogmatize: *De principiis* is a theology ἐν γυμνασίᾳ, "in exercise", i.e. in research. In large measure, it is developed in reaction to the *heresies of his time. Against the *Marcionites, O. affirms the goodness of the Creator and his identity with Jesus' *Father, as well as the agreement of the two Testaments and the value of the OT; against the *Valentinians, he affirms free will, personal responsibility and the rejection of *predestination seen as a law of nature; against *Docetism, he affirms the authentic humanity assumed by Christ as a condition of *redemption. Similarly he opposes two trinitarian heresies: against the modalists, he proclaims the distinct personality of each *Person, and against the *adoptionists, the eternal generation of the Word. Finally, the tendencies towards *anthropomorphism, *millenarism and literalism present in the Great Church furnish him with the occasion to profess the incorporeality of *God, of the *soul and of the final *beatitude, and the abolition, in Christ, of the Jewish law in its ceremonial and legal precepts. It would evidently be unjust to criticize him, as his detractors frequently do, for not having foreseen later heresies and for sometimes using in an *orthodox sense (as comparison with other texts shows) expressions which later received a heterodox meaning. The philosophy which inspires his theology is *middle Platonism, an eclectic *Platonism mixed with a lot of *Stoicism and a little *Aristotelianism, but O. is not a philosopher proper: he uses his philosophical knowledge as a theologian.

As regards the *Trinity, O. seeks to express, in a way that is more dynamic than ontological, the unity of the *Persons and the distinct personality of each one. The *Father is the source of the divinity that he communicates to the Son in his eternal and continuous generation and to the Spirit, in such a way that the Son is and remains in the Father, as the Father in the Son: the Son remains in the Father even when, in the *incarnation, he is on earth with his human *soul. We find in O., for the first time, apropos of the eternity of the Son's generation, the formula οὐκ ἦν ὅτε οὐκ ἦν, "there was no moment when (the Son) was not" (*Perì Archôn* I, 2,9; IV, 4, 1, confirmed by Athanasius, *Com. Rom.* I, 5). On the other hand, he opposes in the Valentinians their representation of the generation of the Son as *probolè* or *prolatio*, i.e. a representation implying, like human or animal generation, a division of substance. The Father is the first since he is the beginning and since it is he who sends the Son and the Spirit on their mission: this is the essence of O.'s "*subordinationism" which, save for a few clumsy expressions, does not involve an inferiority of power for the other two Persons. The Son possesses, in the unity of his *person, a multiplicity of titles (ἐπίνοιαι): the different *names attrib. to him by *Scripture and which express the different aspects of his relations with the Father and with men. As Wisdom, he is the intelligible World which contains the principles and germs of beings and all knowledge of *mysteries; as Logos, he reveals them and it is he who performs the *creation. The union of the Word with human nature is, in accordance with the doctrine of pre-existence of souls, prior to the incarnation since the Word's human soul was created together with the other souls in pre-existence: through its union with the Word it was "in the form of God", impeccable, and so Christ, in his humanity, is the bridegroom of the church, which, in pre-existence, was formed of the totality of the other souls. To redeem his fallen bride, he was incarnate in *Mary's womb, and with him the Word. He revealed his divinity to men by translating it into a human person. This soul, in the Passion, was abandoned as a ransom to Satan and *descended into Hell, where he freed the souls of the just dead and led them back with him in his ascent. O. teaches, in his way, the *hypostatic union and, for the first time, the *communicatio idiomatum*. The *Holy Spirit comes from the Father through the Son, who communicates his titles to it: it is the Sanctifier and constitutes the "matter" or "nature" of the *charisms which correspond to the "actual *graces" of scholastic theology.

O. works out a doctrine of *angels and *demons: guardians of individuals or peoples, or overseers of the various realms of nature. Man has been created, like the angel, in the *image of God, the Word: he partakes of the existence and divinity of the Father and the filiation and rationality of the Word, understanding all these terms in a sense that is more supernatural than natural. But *sin covers God's image in man with diabolical or bestial images which only the Redeemer can suppress. O.'s speculations on raised *bodies tend to affirm simultaneously their identity with and their otherness from the earthly body, according to the Pauline image of the seed and the plant (1 Cor 15, 35-44). If O. does not see *Mary as completely exempt from sin, he is the first to affirm clearly her perpetual virginity, he sees in her one of the great "types" of the spiritual man and, acc. to *Socrates (*HE* VII, 32), he called her *Theotokos. He also works out a precise doctrine on *baptism, the *Eucharist, *penitence, *orders and *marriage.

What is called *Origenism is taken not from the whole of O.'s theology, which would constitute the nourishment of the 4th-c. *Fathers, but from some speculations in *De principiis*, deprived of their hypothetical and antithetical character, and systematized by successors. The one clearly demonstrable point is the hypothesis of the pre-existence of *souls, Christ's included. In the beginning, God created the "minds" all equal, clothed in an ethereal body and immersed in divine *contemplation. Slackening of fervour constituted the primitive fault which divided them into *angels, men and *demons: at that moment the "minds" cooled down into souls, given that ψυχή is connected with ψῦχος, cold. This doctrine, inherited from Plato, provided O. with a reply to the *Marcionite attacks on God the creator, since it allowed him to trace men's inequality at birth to a consequence of free will: it also allowed him to avoid the difficulties of the two other replies to the problem of the soul's origin, *traducianism and creationism. In reference to his time, he cannot be called a *heretic, since the church had no doctrine on the origin of souls, apart from their *creation by God.

As for *apocatastasis, the restoration at the end of *time, it is derived from 1 Cor 15, 23-26: it has no pantheistic character, and if certain texts seem to affirm the final salvation of the demon and the damned, others have a contrary sense and the *Letter to his friends in Alexandria* expressly denies it. O. has no clear vision of this problem: in fact it is a great *hope of his, and to make it into a rigorous theory is incompatible with his great stress on free will. The other errors he was accused of are often contradicted by reliable Greek texts and arose from the incomprehension of his accusers or from a subsequent specialization of vocabulary: it was not understood that the world created from all eternity by God is the intelligible World of the Platonic "ideas" or of the Stoic "reasons", projects or germs of beings, world contained in the Word, hence created from all eternity by the Father in the generation of the Son; it was not understood that the assertion that the Son does not see the Father is directed against *anthropomorphism and that "see" is used here in its ocular sense: because in fact O. frequently speculated on the Son's knowledge of the Father; that *Methodius's accusations apropos of resuscitated bodies arose from a misunderstanding, since he took the word εἶδος in its common meaning of external appearance, while in O. it indicates a metaphysical principle; that O. cannot have claimed in *De principiis* to believe in metempsychosis, which he judges absurd in his remaining Greek commentaries, judging it incompatible with the teaching of the church; nor in the renewal of Christ's *sacrifice in heaven, since in book I of *Com. Jo.* (35 [40], 255), contemporary with *De principiis*, he clearly affirms its uniqueness; it was not understood that to apply to Christ the term γενητός, not distinct from γεννητός, and κτίσις, which does not have precisely the meaning of creature, does not make the Word a creature; that his "*subordinationism", motivated by his conception of origin and of "*economy", remains *orthodox, despite some clumsy expressions; finally, it was not understood that by attributing to him the absurd theory of spherical glorious bodies, *Justinian interpreted what O. said of the stars in *De Oratione* XXXI, 3 as referring to risen bodies; and that the doctrine of "successive *worlds" is nothing but a passing hypothesis to be reconciled with the doctrine of *apocatastasis.

Almost complete bibliography up to 1969 in H. Crouzel, *Bibliographie critique d'Origène*, La Haye-Steenbrugge 1971: supplement up to 1980: *ibid.* 1982.
Editions: In the GCS collection of various authors in 12 volumes; numerous works with text and Fr. tr. in SCh; Ch. Delarue's edition in J.P. Migne's re-edition (PG 11 to 17) with the *Hexapla* and some fragments, published by A. Galland and A. Mai. For the *Philocalia* ed. Robinson, Cambridge 1893. Fragments on the Pauline epistles in *JThS* 3 (1902); 9-10 (1908-1909); 13-14 (1912-1913). Homilies on Psalms in CCL 78 (cf. V. Peri, *Omelie Origeniane sui Salmi*, Vatican City 1980); H. Crouzel-M. Simonetti, *Traité des Principes* T.iii, Bks iii and iv, SCh 268 (1980) and T.iv: Bks iii and Commentaire, SCh 269 (1980); M. Borret, *Homélies sur le Lévitique* T.i. Introd. and Homilies i-vii, SCh 286 (1981) and viii-xvi, Index, SCh 287 (1981); C. Blanc *Commentaire sur S. Jean*, T.iv, Bks xix-xx, SCh 290 (1982); M. Harl-N. de Lange, *Philocalie 1-20 et Lettre à Africanus*, SCh 302 (1983); (ed.) H.J. Vogt, *Der Kommentar zum Evang. nach Matthaus*, Stuttgart 1985; H. Crouzel - M. Simonetti, *Traité des Principes* T.v., SCh 312 (1984).
General works: DTC 12, 1489-1565; R. Cadiou, *La jeunesse d'Origène*, Paris 1935; J. Daniélou, *Origène*, Paris 1948; P. Nautin, *Origène, sa vie et son oeuvre*, Paris 1977 (a work to be used with caution, until it is confirmed by further studies).
Exegesis: H. de Lubac, *Histoire et Esprit*, Paris 1950; R.P.C. Hanson, *Allegory and Event*, London 1959; R. Gögler, *Zur Theologie des biblischen Wortes bei Origenes*, Düsseldorf 1963; A. Zöllig, *Die Inspirationslehre des Origenes*, Freiburg i. Br. 1902; E. Molland, *The Conception of the Gospel in the Alexandrian Theology*, Oslo 1938; G. Lomiento, *L'esegesi origeniana del Vangelo di Luca*, Bari 1966; N. de Lange, *Origen and the Jews*, Cambridge 1976; H. Crouzel, Pourquoi Origène refuse-t-il parfois le sens littéral dans ses Homélies sur l'Hexateuque?, *BLE* 70 (1969) 241-263.
Spirituality: W. Völker, *Das Vollkommenheitsideal des Origenes*, Tübingen 1931; A. Lieske *Die Theologie der Logosmystik bei Origenes*, Münster 1938; P. Bertrand, *Mystique de Jésus chez Origène*, Paris 1951; H. Crouzel, *Théologie de l'image de Dieu chez Origène*, Paris 1956; H.U. von Balthasar, *Parole et Mystère chez Origène*, Paris 1957; F. Fässler, *Der Hagios-Begriff bei Origenes*, Fribourg (Switz.) 1958; G. Teichtweier, *Die Sündenlehre des Origenes*, Regensburg 1958; F. Hartmann, Origène et la théologie du martyre, *EThL* 34 (1958) 773-824; H. Crouzel, *Origène et la "connaissance mystique"*, Bruges/Paris 1961; G. Gruber, ZΩH. *Wesen Stufen und Mitteilung des wahren Lebens bei Origenes*, Munich 1962; M. Martínez Pastor, *Teología de la Luz en Orígenes*, Comillas 1963; H. Crouzel, *Virginité et Mariage chez Origène*, Bruges/Paris 1963; J. Dupuis, *L'esprit de l'homme*, Bruges 1967; M. Eichinger, *Die Verklärung Christi bei Origenes*, Vienna 1969; W. Gessel, *Die Theologie des Gebetes nach "De Oratione" von Origenes*, Munich 1975.
Theology: C. Verfaillie, *La doctrine de la justification chez Origène*, Strasbourg 1926; C. Vagaggini, *Maria nelle opere di Origene*, Rome 1942; K. Rahner, La doctrine d'Origène sur la pénitence, *RecSR* 37 (1950) 47-97, 252-286, 422-456; M. Harl, *Origène et la fonction révélatrice du Verbe Incarné*, Paris 1958; P. Nemeshegyi, *La paternité de Dieu chez Origène*, Paris 1960; J. Chênevert, *L'Eglise dans le commentaire d'Origène sur le Cantique des Cantiques*, Brussels 1969; J. Rius-Camps, *El dinamismo trinitario en la divinización de los seres racionales según Orígenes*, Rome 1970; J.A. Alcain, *Cautiverio y redención del hombre en Orígenes*, Bilbao 1973; H.J. Vogt, *Das Kirchenverständnis des Origenes*, Cologne 1974; L. Lies, *Wort und Eucharistie bei Origenes*, Innsbruck 1978. The Acts of the two international Conferences on Origen, Montserrat 1973 and Bari 1977: *Origeniana*, Bari 1975.
Conferences (since 1980): H. Crouzel-A. Quacquarelli (eds.), *Origeniana Secunda*, Rome 1980; R.P.C. Hanson-H. Crouzel (eds.), *Origeniana Tertia*, Rome 1985; L. Lies (ed.), *Origeniana Quarta*, Innsbruck-Vienna 1987.
Studies: J.W. Trigg, *Origen: The Bible and Philosophy in the Third-Century Church*, Atlanta 1983; G. af Hällström, *Fides Simpliciorum according to Origen of*

Alexandria, Commentationes Humanarum Litterarum 76, Helsinki 1984; C. Rabinowitz, Personal and Cosmic Salvation in Origen, *VChr* 38 (1984) 319-329; J.N. Rowe, *Origen's doctrine of Subordination*, 23/27, Berne 1987; C. Kannengiesser-W.L. Petersen, *Origen of Alexandra: his world and his legacy*, Notre Dame 1988; H. Crouzel, *Origen*, San Francisco 1989; id., Theolgical construction and research. Origen on free-will, in (eds.) R. Bauckham - B. Drewery, *Scripture, Tradition and Reason* (Festsch. R.P.C. Hanson), Edinburgh 1988, 239-265; id., The Literature on Origen 1970-1988, *ThS* 49 (1988) 499-516; A. Labate, Origene e la guerra giusta (*C. Celsum* viii 73), *CCC* 9 (1988) 257-282; W. Biernert, Der Streit im Origens, *Oikonomia* 25 (1989) 93-106, 159 ff; E. dal Covolo, Morte e Martirio in Origene, *Filosofia e teologia* 4 (1990); C.P. Bammel, Adam in Origen, in (ed.) R. Williams, *The Making of Orthodoxy* (Essays in Honour of Henry Chadwick), Cambridge 1989, 62-93; W. Schütz, *Der christliche Gottesdienst bei Origenes*, Stuttgart 1984.

H. Crouzel

ORIGENISM. We can distinguish in "Origenism" six successive moments. The first was that of *Origen, that is to say the whole of the speculations which, through the incomprehension of his successors, constituted the basis of later O. The second was that of O. as understood by his 3rd- and 4th-c. detractors - *Methodius, *Peter of Alexandria, *Eustathius of Antioch: these were answered by *Pamphilus's *Apology for Origen*. Besides the pre-existence of the *soul and *apocatastasis, they contested, through a series of misunderstandings, the doctrine of the resurrected *body and of eternal *creation. The third was the O. of the Egyptian and Palestinian monks (second half of 4th c.): it was expounded mainly by *Evagrius of Pontus in the *Kephalaia Gnostica* - acc. to the unexpurgated *Syriac tr. published by A. Guillaumont in PO 28/1 (1958) - and in the *Letter to *Melania*, ed. in the Syriac tr. and a Greek retroversion by W. Frankenberg (*Evagrius Ponticus*, Berlin 1912). Evagrius "scholasticized" Origen's thought, suppressing its internal tensions and leaving out a great part of his doctrine so as to construct a system with what remained; this was the surest way to make it heretical, since *heresy is the suppression and fragmentation of the *antitheses that characterize Christian doctrine. The fourth moment, the most important, was O. as the 4th- and 5th-c. anti-Origenists - *Epiphanius, *Jerome and *Theophilus of Alexandria - supposed him (while Origen was defended by *John of Jerusalem and *Rufinus of Aquileia). Their propositions must pass through the critical sieve, since, as was quite normal for their time, they lacked a historical sense: they had no notion of the development of *dogma - something of which awareness has been attained quite recently - and did not judge Origen starting from the situation of his time. Moreover they excelled neither in philosophical nor in theological understanding. They totally failed to understand the change of mentality which separated the persecuted minority church of Origen's time and the triumphant church of their own time, especially as regarded the importance of a Christianization of *philosophy for the guiding of the intellectual world, and the need for a *theology "in exercise" (γυμνασία), i.e. in research. They accused Origen in view of the heresies of their own time, esp. *Arianism, without asking what were those that he had to face and which determined his particular problems. They had absolutely no awareness of the doctrinal progress brought about in the church by the reaction to Arianism in comparison with the succinct *regula fidei* of the 3rd c., which Origen had expounded in the preface to *De principiis*; nor of the evolution of vocabulary, and thus they understood the terms used by Origen in the sense which such terms had taken on in the 4th c. and which, in some theologically important cases, was much more precise than that of the 3rd c. They read Origen projecting onto him the O. of their own time, that of the third moment, since it was that which they actually had in their sights. They never made systematic studies of Origen's work and they based their accusations on isolated texts, taking no account of the explanations often found in other passages in the same book, sometimes only a few lines away. Whatever Origen wrote in the form of exercise (γυμναστικῶς), they understood as said in the form of doctrine (δογματικῶς), with a conception of *orthodoxy and *regula fidei* which was being increasingly modelled on that of civil law and expressed in "definitions" in the struggle against heresy.

The battle began with *Epiphanius, metropolitan of Salamis or Constantia in *Cyprus: he classified Origen's "heresy" together with those that filled his *Ancoratos* and his *Panarion*, and insisted in obtaining a condemnation of Origen from bishop *John of Jerusalem. In 393 a certain Atarbius, by what right we do not know, made a round of the convents of *Palestine gathering signatures for Origen's condemnation. Ill received by *Rufinus in his convent on the Mount of Olives, against all expectation he was welcomed by *Jerome, until then an ardent defender of Origen, in his *monastery at *Bethlehem. The battle grew more bitter, with Rufinus and John against Jerome and Epiphanius. A reconciliation was reached between Rufinus and Jerome, but the dispute was revived when Rufinus, back in *Rome, translated book I of Pamphilus's *Apologia*, followed by the *Perí Archôn*, a manuscript which, purloined by *Eusebius of Cremona, a monk and *friend of Jerome, scandalized Jerome's Roman friends. They obliged Jerome to make a new translation of the *Perí Archôn* which, with the intention of being literal, highlighted Origen's heresies and Rufinus's inexactitudes, and did everything to embitter thoughts. Meanwhile the patriarch of *Alexandria, *Theophilus, was chosen to arbitrate between the two contending parties. At first favourable to Origen, in the interests of the politics of the patriarchate he changed sides, expelled the auxiliary bishop *Isidore and the "Long Brothers", and obtained the deposition of *John Chrysostom who had given them asylum in *Constantinople. He condemned Origen at a regional synod in 400: these events had immediate repercussions in the West, thanks to Jerome, and are echoed in two *letters of pope Anastasius. This first dispute terminated in 402 with Rufinus's silence.

The fifth moment was the O., or rather the Evagrianism, of the Palestinian monks (first half of 6th c.) who lived in convents under the obedience of St *Sabas, the Great *Laura and the New Laura. The main expression of their doctrine is the *Book of St Hierotheus*, the work of the Syrian monk *Stephen bar Sudayle, who aggravated Evagrius's Origenist "scholasticism" into a radical *pantheism. Between *Justinian's first and second interventions, these Origenists divided into two groups. The extremists were called *isochristi* (ἰσόχριστοι) since they held that both at the beginning and at the end all the "minds" are equal to Christ: his superiority over them is only provisional; he had no part in the original *sin. The moderates, whose tardy alliance with the anti-Origenists led to the condemnation of the *isochristi*, were called *protoctisti* (πρωτόκτιστοι) since they attributed to Christ a superiority over the other minds: their opponents inflicted the surname *tetraditi* on them, accusing them of transforming the *Trinity into a tetrad by introducing Christ's humanity into it.

The sixth moment consisted of the presumed O. against which the emperor *Justinian's condemnatory documents were directed: we must not confuse those of 543 with those of 553. The former, occasioned by an appeal by anti-Origenist Palestinian monks, presented to the emperor by the papal *apocrisarius Pelagius, consist of the *Liber adversus Origenem* or *Letter to *Menas*, followed by extracts from *De principiis* and *anathemata (DS 403-411): approved by the emperor's domestic synod, they were presented for signature to the *pope and the patriarchs. They take aim at Origen himself, seen in the perspective of the O. of the time and with some gross misunderstandings, both in the Letter and in the anathemata. The other condemnatory documents come from the 5th ecumenical *council, *Constantinople II, compiled before and after the official opening. While Justinian sought to obtain the assent of pope *Vigilius, whom he had brought to the capital by force, before declaring the council open, acc. to the hypothesis of Fr. Diekamp, he addressed a *letter to the assembled bishops, which is the rough draft of 15 anathemata which do not figure in the official *Acts of Constantinople II and which are thus not legally the product of an ecumenical council: they take aim explicitly at the *isochristi*, and Origen is cited only as the standard-bearer to whom they appeal; moreover, some of these anathemata reproduce texts of Evagrius word for word. In the council's official Acts - the emperor, having given up trying to obtain the *pope's consent, had declared the council open on his own authority - the anathemata directed against the "*Three Chapters" contain Origen's name, last, in a list of heretics, in anathema no. 11. The same list, but without Origen's name, appears in the emperor's *Homonoia*, which is the rough draft of these anathemata. Very probably Origen's name was added following the discussions on the Origenists which preceded the official opening, as the symbol of the *isochristi*. Pope Vigilius's documents, approving, after its close, a council that was held against his will, do not mention Origen. From a legal point of view, the presence of Origen's name in the list in question is not particularly important: it does not mean that he was formally a *heretic - the conciliar *Fathers were prob. persuaded of it by certain rumours of an alleged *apostasy, spread by Epiphanius - but it prob. means that in his writings there are some errors, taking into account the way in which they were read at that time; but it is more exact to say, perhaps, that under his name it was really the *isochristi* who were condemned. A consequence of this condemnation was the loss of the greater part of his works in the original language. This greatly held up the propagation of Origen's thought in the Byzantine world, or at least its direct propagation, since, through the 4th-c. Fathers, who were nearly all his disciples, Origen's *theology invisibly marked the whole Christian *tradition.

If Origen seems to have been generally rejected by the Byzantine Middle Ages, the Latin Middle Ages, up to the 13th c., read him passionately in the *Latin translations: Origen influenced Bernard of Clairvaux, William of Saint-Thierry and, through them, the Cistercian tradition. On the other hand, recent attempts to make him the source of Bulgarian Bogomilism and of Catharism do not seem worthy of consideration. Between the 13th and 14th cc., his *Platonism was hard to reconcile with the ruling Aristotelianism: so in this period he was remembered only for his *exegesis. In the Renaissance he was defended by Pico della Mirandola and much admired by Erasmus. In the 16th c. appeared the first editions that would be perfected in the following centuries, as would studies. At the start of the 20th c., Origen, confused with later O., judged almost exclusively on a rather one-sided reading of the *De principiis*, was seen more as a Greek philosopher than as a Christian theologian: his *spirituality was not noticed at all and his *exegesis was considered arbitrary and absurd (H. von Harnack, E. de Faye, P. Koetschau, H. Koch). Origen's spirituality was rediscovered by W. Völker (1931), and our understanding of his exegesis is the work of H. de Lubac (1950): his personality has recovered its essential dimensions. At present, Origen is, after *Augustine, the most widely read of the ecclesiastical writers of antiquity.

F. Diekamp, *Die origenistischen Streitigkeiten im sechsten Jahrhundert und das fünfte allgemeine Concil*, Münster 1899; F. Cavallera, *Saint Jérôme*, Louvain 1922; Origénisme, DTC 11, 1565-1588; A. Guillaumont, *Les "Kephalaia Gnostica" d'Evagre le Pontique et l'histoire de l'origénisme chez les Grecs et les Syriens*, Paris 1962; H. Crouzel, Qu'a voulu faire Origène en composant le Traité des Principes?, BLE 76 (1975) 161-186, 241-260. Origen's influence on the 4th-c. Fathers is frequently emphasized in studies of them. For the Latin Middle Ages: H. de Lubac, *Exégèse Médiévale*, I/1, Paris 1959, 198-304; J. Deroy, *Bernardus en Origenes*, Haarlem 1964; P. Verdeyen, *La théologie mystique de Guillaume de Saint-Thierry*: in various contributions, in *Ons geestelijk erf* (Anversa) from 1977 to 1979; H. Crouzel, Origène est-il la source du catharisme?, BLE 80 (1979) 3-28. For the Renaissance: H. Crouzel, *Une controverse sur Origène à la Renaissance: Jean Pic de la Mirandole et Pierre Garcia*, Paris 1977; M. Schär, *Das Nachleben des Origenes im Zeitalter des Humanismus*, Basel 1979; A. Godin, *Erasme lecteur d'Origène*, Geneva 1982; J.G. Bunge, Origenismus Gnosticismus: Zum geistesgeschichtlichen Standort des Evagrios Pontikos, VChr 40 (1986) 24-54; J. Meyendorff, *Christ in Eastern Christian Thought*, New York 1975, ch. 3; Various authors, Section xii, "Origenism", SP 14 (1979).

H. Crouzel

ORIGEN the Neoplatonist. Nearly all modern scholars (Daniélou, *Origène*, Paris 1948, 89; Schwyzer, PWK 21, 1, 480; Dörrie, *Hermes* 83 [1955] 468-472; Dodds, *Les sources de Plotin*, Geneva 1960, 26 and 31, n. 1; H. Chadwick: Numenius and Ammonius, in *Early Christian Thought and the Classical Tradition*, Oxford 1966, 68 ff.; Theiler, *Forschungen zum Neuplatonismus*, Berlin 1966, 3) agree that the O. mentioned by *Porphyry in *Vita Plot*. ch. 3 (I 3, 24-4, 32 Bréhier) and ch. 14 (I 15, 20-16, 23), and by Longinus, cited by Porphyry in *Vita Plot*. ch. 20 (I 22, 37), and who was a disciple of *Ammonius Saccas, with *Plotinus and Herennius (fr. 2, 5 and 6 Weber, 4-5), must be distinguished from the other much more famous Christian *Origen - also a disciple of Ammonius - whom Porphyry refers to in Eusebius, HE VI 19, 6-7 (II 558, 26-560, 7 Schwartz: but Dörrie, *Hermes* 83 [1955] 469-470, is wrong in holding that Porphyry, in this passage of *Eusebius, confused the Christian Origen with O. the Neoplatonist). The attempt to identify the two and deny the independent existence of O. the Neoplatonist, made by R. Cadiou (*La jeunesse d'Origène*, Paris 1935, 231-262) and recently repeated, with no new arguments, by F.H. Kettler (*Origenes, Ammonius Sakkas und Porphyrius, Kerygma und Logos, Festschrift C. Andresen*, Göttingen 1979, 325-327), does not seem tenable. A deeper examination of the facts which make the distinction necessary, a collection of the evidence relating to O. the Neoplatonist (preserved by Porphyry, *Eunapius, *Photius, *Proclus and *Nemesius of Emesa) and a reconstruction and interpretation of his philosophical system can be found in K.O. Weber's monograph (*Origenes de Neuplatoniker*, Zetemata 27, Munich 1962), which remains the classical work on which any study of him must be based. Here we will simply say that, as Porphyry relates, he was at *Alexandria, with Plotinus and Herennius, when Ammonius Saccas died in 242 (*Vita Plot*. ch. 3=fr. 2 Weber); that he visited Plotinus when the latter held his lectures in *Rome after 244 (*Vita Plot*. ch. 4=fr. 3 Weber; on the date, cf. Cadiou, 235 and Dörrie, 471); that he wrote two treatises, Περὶ τῶν δαιμόνων and "Ὅτι μόνος ποιητὴς ὁ βασιλεύς, the latter during the reign of *Gallienus, which began in 253 (*Vita Plot*. ch. 3=fr. 2 Weber); that he also left notes for a commentary on the *Timaeus*, mentioned by Proclus (fr. 8-16 Weber, pp. 6-11); and that, as Proclus relates (*Theol. plat*. II, 4 [= fr. 7 Weber]), his metaphysical system differed from that of Plotinus in that he did not put the One above mind and being, but considered mind (equivalent to absolute being) as the supreme principle, remaining more faithful to the teaching of his master Ammonius Saccas (cf. Schwyzer, PWK 21, 1, 480; Weber, 160).

R. Beutler, PWK 18, 1, 1033-1036; H.R. Schwyzer, PWK 21, 1, 479-480; K.O. Weber, *Origenes der Neuplatoniker*, Zetemata 27, Munich 1962; H. Crouzel, *Origène et Plotin. Comparaisons doctrinales*, 1990; arts. Various authors in *Actes, V Colloquium Oregenianum*, Boston 1989.

S. Lilla

OROSIUS. We first hear of the Spanish priest (Paul?) O. in 414 when, still a young man, he arrived at *Hippo and made contact with *Augustine. In 415, on Augustine's advice, he went to *Palestine, where he met *Jerome and where his imprudence involved him in the start of the *Pelagian controversy. In 416 he returned to the West with various messages and the claimed relics of *Stephen the protomartyr, which contributed to the *conversion of numerous Jews at Minorca. Returning to Hippo, he wrote his historical work in 416 and 417. We know nothing of his later life (details, discussion of controverted points and *testimonia* are in Schanz IV, 1, 483-485).

We possess three works by O.: 1) *Commonitorium de errore Priscillianistarum et Origenistarum*. In this very brief memoir, sent by O. to Augustine after his first stay at Hippo, he provided a guide to these heretical doctrines, widespread in *Spain and Portugal, and invited the bishop to refute them. 2) *Liber apologeticus contra Pelagianos*. This is our sole source on the conference of *Jerusalem of 415 which, under the presidency of bishop *John, discussed the ideas of *Pelagius, who was present. O. railed against Pelagius's doctrine, but his Latin was badly translated into Greek by the interpreter, and John later accused him of having maintained that, even with *God's help, man cannot be without *sin (*Apol*. 9, 2). In this treatise O. gives an account of the conference, denies having pronounced the heretical opinion for which he was rebuked, and above all refutes *Pelagianism (for these two works cf. *Patrologia* III, 467-469). 3) *Historiarum adversus paganos libri VII*. O.'s main work, and the earliest universal Christian history. The material is divided thus: Bk I: geographical introduction, history from the *flood to the foundation of Rome. Bk II: Rome, from its foundation to the capture of the *Urbs* by the Gauls (390); history of Persia and Greece up to the battle of Cunassa (401). Bk III: Roman history up to the eve of the war against Pyrrhus; Greek history up to the disintegration of Alexander's empire. Bk IV: history of Rome up to the destruction of Carthage (146). Bk V: Rome from 146 to 70 BC. Bk VI: Rome from 70 BC to the accession of *Augustus. Bk VII: history of the empire up to AD 417.

The writing of this work was, like Augustine's *City of God*, a consequence of the capture of *Rome by *Alaric in 410. The *pagans claimed that this catastrophe was the consequence of the abandonment of the traditional cults, finally prohibited by *Theodosius in 392, and many Christians were shaken: for nearly a century in fact, the doctors of the church had proclaimed that the Roman empire had been established by divine *Providence in order to spread and bring about the triumph of Christianity, whose lot seemed so bound up with Rome. Augustine, in the *City of God*, refuted this political theology. In the prologue to his *History*, O. states that Augustine, while writing book XI of the *City*, had asked him to make a list of the innumerable ills that had afflicted humanity in the past (evidently to attenuate the gravity of the catastrophe of 410); O. concludes by declaring that his research had led him to discover that the evils of the past were infinitely more terrible than those of the present, and that the kingdom of death was paralyzed after Christianity had triumphed (*Prol*. 10-14). His *History* is not the serene exposition of objective research, but an impassioned demonstration. Against the pagans - as the title makes clear - and with *apologetic aims, O. seeks to demonstrate the coherence of the actions of Providence in history: for him God's designs are not inscrutable, but evident.

Like many ancient historians, O. does not directly use the authors he names (esp. Plato, Polybius, Sallust, Trogus Pompeius) and does not cite his true sources, whose identification is still disputed (cf. Goetz, 25-29) but among whose number are certainly the *Bible, *Justin (for events not belonging to Roman history), a summary of Livy, Florus, Eutropius, *Eusebius-*Jerome's *Chronicle*, Eusebius's *Ecclesiastical History* in *Rufinus's Lat. tr., and *Augustine's *City of God*. O. is neither too erudite nor too accurate: he thinks that Suetonius wrote the *Gallic War* (VI, 7, 2) and that Sallust recounted the era of *Domitian (VII, 10, 4). From his sources he draws what is useful for his demonstration, i.e. catastrophes, defeats, great numbers of deaths; faced with two different versions, he chooses the most dramatic. Thus his narrative, at least up to *Augustus, is essentially a sequence of tragic reverses.

O. follows the scheme of the succession of four world empires (Dan 2 and 7), but with a geographical and chronological correction of his own: Babylon, Macedonia, Carthage, Rome (II, 1-3 and VII, 2). He is fascinated by synchronisms and the symbology of *numbers, to which he sacrifices exactness: some distortions allow him to work out an extraordinary construction of chronological parallels between the four world empires (*loc. cit. supra*): to emphasize the Christ-Augustus synchronism, he falsifies the date of three episodes that marked the start of the monarchy at Rome, putting them on the day of *Epiphany, manifestation of Christ's monarchy (VI, 20 and 22; VII, 1; on these passages cf. Paschoud, 115-119 and 125-131). O.'s *History* is divided into seven books, as the *week into seven days; the last book is dedicated to the imperial era, which thus appears as the *Sunday of the history of mankind. In fact, from the time of Augustus, acc. to O., good once more began to triumph over *evil and to predominate almost exclusively under the Christian empire. O. thus almost completely eliminates from history any *eschatological dimension: he counts ten *persecutions of the Christians, compared with the ten plagues of Egypt; after the tenth plague the Egyptians were engulfed in the Red Sea, which the Hebrews crossed safe and sound; just as, O. suggests, after the tenth persecution the Christians are safe from evil, they will escape the eleventh persecution, that of *Antichrist (VII, 26-27; cf. Paschoud, 119-125).

O. ends thus in order to demonstrate the contrary of what Augustine expounded in his *City of God*; no wonder Augustine never mentioned O. after the publication of his *History* and indeed refuted him without naming him (e.g. *De civ. Dei* XVIII, 52). We may ask, then, whether O. is telling the exact truth when he claims to have written his work at Augustine's request (cf. Corsini, 35-51). In the Middle Ages, O.'s interpretation of history was better understood than were Augustine's lofty speculations, his work was very popular and had considerable influence: only the "century of lights" forced O. off his pedestal. O.'s importance today is not in his value as a source: in fact he gives us little that we cannot find in older and more reliable historians. But he provides fundamental evidence of the history of ideas in his time, even if he fails to overcome his contradictory tendencies (Christian cosmopolitanism *v.* loyalty to the empire, hatred of the barbarian invader *v.* adhesion to the idea that the invasion makes possible the *evangelization of the Germans), and he created the universal Christian history which would be at the basis of mediaeval *historiography. Finally O., master of the resources of *rhetoric, vigorous and impassioned, is not without literary talent.

CPL 571-574; PL 31, 663-1216; CSEL 5 (Zagenmeister, 1882: *Liber apol., Hist.*) and 18, 149-157 (Schepss, 1889: *Comm.*). Serie Mondadori-Valla (text, It. tr. by A. Bartalucci, intro. and notes by A. Lippold), 2 vols., Milan 1976 (just *Hist.*); FC 50 (Eng. tr. by I.W. Raymond-R.J. Deferrari, 1964 (just *Hist.*); G. Fink, Recherches Bibliographiques sur Paul Orose, *RevArch BiblMus* 58 (1952) 271-322 (bibl. up to 1952); Schanz IV, 1, Munich 1920; J. Svennung, *Orosiana*, Uppsala 1922 (problems of language); B. Lacroix, *Orose et ses idées*, Montréal-Paris 1965; E. Corsini, *Introduzione alle "Storie" di Orosio*, Turin 1968; *Patrologia* III, 466-470; F. Fabbrini, *Paolo Orosio, uno storico*, Rome 1979; H.W. Goetz, *Die Geschichtstheologie des Orosius*, Darmstadt 1980; F. Paschoud, La polemica providenzialistica di Orosio, in *La storiografia ecclesiastica nella tarda antichità* (Atti del Convegno di Erice), Messina 1980, 113-133; W.H.C. Frend, Augustine and Orosius on the fall of the Roman Empire in the West, *AugStud* 20 (1989) 1-38. Editions: Budé, *Oeuvres* (forthcoming 1991); K.D. Daur, Orosius, *De errore Priscillianistarum et Origenistarum*, CCL 49 (1986); H. Daiber, Orosius *Historiae adversus paganos*, in arabischer Überlieferung, in (ed.) J.W. van Henten et al., *Tradition and re-interpretation in Jewish and Early Christian Literature*, Leiden 1986, 202-249.

F. Paschoud

ORPHEUS

I. In general - II. Iconography.

I. In general. Mythical singer and poet; Greek tradition usually linked him with Thrace: there were different versions of his parentage and life. Son of a Muse, sometimes Calliope, and of Apollo or of the river-god Oiagros, O. was connected with the voyage of the Argonauts and involved in an infernal *katabasis* in search of his wife Eurydice. A particular tradition tells of his violent death at the hands of the Maenads, followers of Dionysus, whose orgiastic cult O. had ignored in his special veneration of Apollo. But beyond such traditions on his "life", the fundamental fact remains the quality of "theologian" attrib. to him by virtue of the many religious writings which, from the 6th-5th cc. BC, circulated in Greece in O.'s name and in that of people variously connected with him (Musaeus, Linus). This literature, usually called *ta orphikà* ("the Orphic things"), was interested in a particular way in theo-cosmogonic and anthropogonic speculations and became very popular in the Late Hellenistic period. In it was expressed a particular religious vision with *dualistic leanings, based on the idea of the fragmentation of the original One, which, through a process of evolution, led up to the present situation implying the presence, in the human *body, of a divine substance (the *soul) which aspires to liberation. An abstemious way of life, sometimes accompanied by cathartic practices, guarantees such liberation, freeing the soul from the cycle of rebirth.

To the first generations of Christians the figure of O., like that of other *pagan "sages", appeared in an ambivalent light: rejected as a spokesman of the error of *idolatry, he yet appeared, on account of some aspects of the religious teaching connected with him, as one who had prefigured some fact of *revelation. Enough to recall the statement of *Augustine who, reflecting an opinion quite widespread among Christians, saw in O. one who, with the Sibyls (cf. *Oracles), "predicted" some truths about the *Father and the Son (*C. Faust.* XVII, 15). *Clement of Alexandria attests the use of the figure of the mythical Thracian singer, who could tame wild beasts with the sweetness of his music, as a "symbol" of Christ who draws men to himself by the fascination of the divine word, and leads them to the truth (*Protr.* 1, 1-10). This symbolism justifies the presence of O. in the Christian art of the first centuries, his image often being characterized by the attributes of the Good *Shepherd: the frescos of some Roman catacombs (3rd and 4th cc.) show O. sometimes acc. to classical iconographical schemes, sometimes in the type of the Good Shepherd, as a figure of Christ.

O. Kern, *Orphicorum Fragmenta*, Dublin-Zürich ³1972; W.K.C. Guthrie, *Orpheus and the Greek Religion*, 1935; *DACL* 12, 2735-2755.

G. Sfameni Gasparro

II. *Iconography. In early Christian figurative art, the myth of O. is reduced to his figure, in Oriental *dress, seated among wild beasts or other animals, which he soothes with the sound of the lyre. But there is no reason to consider this as evidence of the presence of an Orphic component in Christianity from the start. The figure of O. thus rendered - in the intentional likeness of the Good *Shepherd - recalls, in a unique transference of images and hence only for purposes of communication, the figure of Christ who fascinates and softens even the most restive hearts; but precisely because of the danger of confusion between the two, the scene had little popularity among Christians (Clem. Al., *Protr.* 1, 3; Euseb., *Praep. ev.* XII, 12; *Vita Const.* 14; Ps.-Justin, *Cohort. ad Graec.* 14; Cyr. Al., *Contra Jul.* 1, 26: PG 76, 541; Aug., *C. Faust.* XVII, 5). O. appears however, acc. to a well-defined scheme, in some Roman cemetery *paintings (first half of 3rd - end of 4th c.): in Callisto (*Wp* 37), Domitilla (*Wp* 229), SS. Marcellino e Pietro (*Wp* 98 and Nestori 62, n. 79), Priscilla (*ibid.*, 26, n. 29). The Christian character of some figures of O. sculpted on two *sarcophagus fronts from Ostia (*Rep.* 70 and 1022) and one from Porto Torres (Pesce, *Sarcofagi*, n. 57) seems less probable.

Acc. to the most recent findings of *iconological research (Murray, The Christian Orpheus, 19 ff.), the identification of O. with Christ involves the other, already made in Jewish art, between O. and *David in his role as lyrist and psalmist. There is clear evidence of this in a painting in the synagogue of *Dura Europos, a *mosaic in that of Gaza, one in *Jerusalem and a painting - now lost - in the Roman *cemetery of Vigna Randanini (Corby-Finney, Orpheus-David, 7 ff.). A painted ceiling in the cemetery of Domitilla (*Wp* 55), which includes the figure of O. in a central *clipeus* and that of David in a lateral scene, constitutes, for the declining 3rd c., an important proof and a salient phase of this process of assimilation. **[Fig: 240]**

A. Boulanger, *Orphée: rapports de l'orphisme et du christianisme*, Paris 1925; M. de Fraipont, *Orphée aux catacombes*, Paris-Tournai 1935; B. Bagatti, Il mosaico dell'Orfeo a Gerusalemme, *RAC* 28 (1952) 145-160; G. Pesce, *Sarcofagi romani di Sardegna*, Rome 1957; J.B. Friedman, Syncretism and Allegory in the Jerusalem Orpheus mosaic, *Traditio* 23 (1967) 1-13; A. Ferrua, Una nuova regione della catacomba dei SS. Pietro e Marcellino, *RAC* 44 (1968) 29-68; F.M. Schoeller, *Darstellungen des Orpheus in der Antike*, Freiburg i. Br. 1969; H. Stern, Orphée dans l'art paléochrétien, *CArch* 23 (1974) 1-16; C. Murray, The Christian Orpheus, *ibid.* 26 (1977) 19-27; P. Corby-Finney, Orpheus-David: a Connection in Iconography between Graeco-Roman Judaism and Early Christianity?, *Journal of Jewish Art* 5 (1978) 6-16; P. Testini, Arte mitraica e arte cristiana, in *Mysteria Mithrae*, Rome 1979, 435-437; C. Murray, *Rebirth and Afterlife*, Oxford 1981, 37 ff.

F. Bisconti

ORTHODOXY

I. Orthodoxy-heterodoxy - II. Orthodox church.

I. Orthodoxy-heterodoxy. In ancient Christianity the term orthodoxy, which means "right or correct opinion" (*doxa orthe*), connotes Christian doctrine and *life in conformity with the original truth about Jesus of Nazareth and his teachings. It is therefore opposed to heterodoxy, whose error it shows up, in the synonymous sense of *pseudodoxy* (false opinion). In this development there is a departure from the classical philosophical meaning of heterodox (having different opinions), a meaning opposed to those who share the same opinion (*homodoxoi*). The pastoral epistles speak of heterodox teaching (1 Tim 1, 3 and 6, 3), and 2nd-c. authors speak of heterodox doctrines (Ign., *Magn.* 8, 1; *Smyrn.* 6, 2) and heterodox teachers (Polyc., 3). Just as the term *heresy, from its original meaning of "choice", acquired a pejorative sense, so too did heterodoxy, though this was perhaps a parallel phenomenon, and only later did they come to be identified as synonyms of deviation from Christian doctrine and *disciplina. This fusion is clear in *Eusebius, where the *quartodeciman churches, excommunicated by pope *Victor, are called "heterodox" (*HE* V, 24, 9) and false teachers in the church "heterodidaskaloi" (*HE* III, 28, 8; VII, 7, 4). These are opposed not to *homodoxia* but to *orthodoxia*, while orthodox teachers, like *Irenaeus and *Clement, are called "ambassadors of the church's orthodoxy" (*HE* III, 23, 2), in whom there are no heresies (Athan., *C. Arian.* II, 42-43). The affirmation of orthodoxy involves the same set of problems as the dialectic orthodoxy/heresy.

J.F. McCue, Orthodoxy and Heresy: Walter Bauer and the Valentinians, *VChr* 33 (1979) 118-130; M. Simon, From Greek Hairesis to Christian Heresy, in *Early Christian Literature and the Classical Intellectual Tradition*, Paris 1979, 101-116; A. Meredith, Orthodoxy, Heresy and Philosophy in the Later Half of the Fourth Century, *The Heythrop Journal* 16 (1975) 5-21; J.M. Leroux, Le surgissement d'une orthodoxie au IVe siècle, *VSp*, Suppl. 118 (1976) 294-309; (ed.) R. Williams, *The Making of the Orthodoxy* (Essays in Honour of Henry Chadwick), Cambridge 1989; R.P.C. Hanson, The Achievement of Orthodoxy in the Fourth Century, *ibid.*, 142-156.

II. Orthodox church. In subsequent eras the term o. was also used to indicate the Eastern churches in *communion with *Constantinople, which, altogether, were called "the holy, orthodox, *catholic, apostolic Eastern church", distinguishing themselves from the other communities that were not in communion with them, e.g. the *Nestorian, *Jacobite and *Coptic churches. The first *Sunday in Lent is celebrated as the feast of orthodoxy (going back to AD 824) to commemorate the triumph over iconoclasm and heresy in general.

S. Bulgakov, *The Orthodox Church*, London 1935; W. de Vries, *Der christliche Osten in Geschichte und Gegenwart*, Würzburg 1951; P. Evdokimov, *L'orthodoxie*, Paris 1959; J.D. Zizioulas, La continuité avec les origines apostoliques dans la conscience théologique des églises orthodoxes, *Istina* 19 (1974) 65-94; J. M. Hussey, *The Orthodox Church in the Byzantine Empire*, (eds. Hand-Chadwick, *Oxford History of the Christian Church*), Oxford 1986; V. Lossky, *Mystical Theology of the Eastern Church*, London 1965.

V. Grossi

OSSIUS. Born c.256, bishop of Cordova (*Corduba*) c.300, *confessor during *Diocletian's *persecution. *Friend of *Constantine, who asked him to make *peace in the *Arian controversy (c.324). In this role he presided over the council of *Antioch early in 325 and played a considerable part in that of *Nicaea. He also played a prominent part in that of *Sardica (343) where, with the local bishop Protogenes, he headed the Western bishops. Later, when the West too passed under *Constantius's administration, he refused to subscribe the condemnation of *Athanasius in 356. But in 357, aged over 100, he was forced to subscribe the pro-Arian formula of *Sirmium. He died soon after. His yielding, due to old age, made a great impression and provoked violent reactions in the West, from *Hilary, *Phoebadius and the *Luciferians. O.'s *De laude virginitatis* and *De interpretatione vestium sacerdotalim*, mentioned by *Isidore (*Vir. ill.* 5), are lost. But we have some canons approved at Sardica, a *letter written with Protogenes to inform *Julius of Rome of some decisions of the council of Sardica, and one written in 356 to Constantius, who was putting pressure on O. to subscribe Athanasius's condemnation: here, for the first time after the start of the Constantinian policy of collaboration between *church and empire, a bishop points out how advantageous it would be for the emperor to abstain from meddling in the internal problems of the church, as the Arian controversy was. On the basis of Mt 22, 21, O. affirms the separation of the two powers.

CPL 537-539; PL 8, 1317-1332; 10, 557-564; 632-848; PLS 1, 184-196; V.C. De Clercq, *Ossius of Cordova*, Washington 1954; *Patrologia* III, 58-59; H. Chadwick, Ossius of Cordoba and the Presidency of the Council of Antioch in 325, *JThS* n.s. 9 (1958) 292-304; R.P.C. Hanson, *The Search for the Christian Doctrine of God*, Edinburgh 1988.

M. Simonetti

OUSIA. The term *ousia* is derived from the Greek verb *einai*, "to be". It is usually translated "substance" or "essence" (Lat. *substantia, essentia*), but carries an ample series of meanings not completely expressed by these words. *Einai*, "to be", was normally opposed to *genesthai*, to "come to be" or "become"; so that *ousia* represented a thing considered as persisting, or that which was permanent in something, beyond its changeable states. The problem remained of how this was to be explained or with what it should be identified: on this there was little agreement, and various theories were developed.

Plato was the first great thinker who used the word in philosophical discussion. He showed Socrates asking the question: "What is" - e.g. - "justice?", as distinct from any specific just act. For Plato the *ousia* of a thing was its "Idea", its eternal and perfect archetype; *ousia* could also be used in a collective sense for "reality", i.e. the entirety of Ideas, as opposed to "appearance", changeable phenomena.

Aristotle made limited use of this conception, holding that things were composed of their form (*eidos, idea*) and their matter: the form of things was permanent and characteristic of the species, but did not subsist outside its realization in matter: in this way the Platonic conception of Ideas was rejected. Aristotle explained that *ousia* could refer both to the form of things and to their matter, or to their union, i.e. to the thing itself. In the *Categories* he distinguished between the first *ousia*, the individual person or thing, and the second *ousia*, the species or genus to which it belonged. Considering the importance it had subsequently, this nomenclature had surprisingly little diffusion until about AD 200: in any case it failed to clarify the distinction between the species understood collectively (e.g. the human race) and its specific form (e.g. human nature).

*Stoicism saw the world as a process in evolution, whose development was controlled not by a permanent system of Ideas or Forms, but by a single rational principle operating within it: formless matter and informing reason were thus two aspects of what was really a single and entire complex. Against this, however, *Platonism considered the Stoics materialists for whom the permanent reality (*ousia*) of the world was its matter, rather than the intelligible and spiritual realities (*noetai ousiai*), as they themselves held.

In the *Bible, *ousia* is not used as a philosophical term, but only in the sense of "property" or "possession". The Christian writers quickly became familiar with the Platonic distinction between material *ousia* and immaterial *ousia*, the latter including not just intelligible realities (e.g. moral ideas), but also rational beings, such as *demons or *angels (as in some old Platonizing writers, e.g. *Philo). The discussion soon turned to *God's *ousia*: it was commonly held that we could be sure of God's existence (one possible sense of *ousia*), but that we could never comprehend his nature (which could also be called *ousia*). Thus God's *ousia* was declared incomprehensible, though in fact it was often discussed, having recourse to biblical texts such as "God is ... fire ... *light ... *love". It was often asserted that God could be classified among "immaterial realities", though with the important proviso that God, for Christians, was the *sole* creator of all other reality.

Ousia played an important role in the *theology of the *Trinity, in which the term - a composite of *ousia* - *homoousios* ("of the same *ousia*", "*consubstantial*", "*coessential*") was decisive. Discussion centred on the Logos or Son of God and it was soon admitted that God's Word could not be understood simply as an activity, like a pronounced word, but must be a totally independent reality: whence it seemed to follow that it must be a distinct *ousia*. Applying the same principle to the *Holy Spirit, it might seem that there were three *ousiai* in God, a formulation which can be found in some Greek writers immediately before the council of *Nicaea. The formulation was criticized as suggesting three different natures or degrees of divinity. Yet Greek authors, from *Origen on, had expressed God's triple reality with the alternative expression "three hypostases", using a term that had come to be more clearly associated with individual beings. But this formulation too was opposed by some Latin authors, for whom *hypostasis* corresponded to their term *substantia*, which was its etymological equivalent. Thus, according to the Latins, the Origenian Greeks believed in three *types* of divinity, while the Latin expression *una substantia* (and its equivalent "one hypostasis", used by some Greeks) risked being understood as "one single *Persona*".

The council of Nicaea, opposing the teaching of *Arius, established that there was a single *ousia* in God and that the Son came from the *Father's *ousia*; but the doctrine of the three hypostases was not directly attacked, and the Eastern theologians continued to insist on the distinct reality of Father, Son and Spirit. Modern discussions on Nicaea often hold that the conciliar bishops must have presupposed a definition of *ousia* which implied the Aristotelian distinction between first and second *ousia*: either "a thing", or "a nature". This is mistaken: the *Categories* had little influence on the Christian writers of that period and in any case the Nicene formula was conceived in order to ensure wide agreement, while it condemned *Arianism (cf. Kelly, *Doctrines*, 233-237; Stead, *Divine Substance*, 223-266). The solution was clarified in the course of the 4th c., as discussions proceeded between those who (like *Athanasius) spoke of "one hypostasis" in three Persons, and those who continued to defend the "three hypostases". An important reference-point was the council of *Alexandria of 362, in which Athanasius decided for tolerance: the expression "one hypostasis" need not be understood as a fusion of the three Persons, and the expression "three hypostases" was not to be condemned as if a division were being made. Later the Cappadocians opted for the formula "three hypostases, but one *ousia*", which became the normal *orthodox expression. *Ousia* thus came to indicate the essential attributes common to all three Persons: but even among the Cappadocians themselves we can find traces of their earlier and more flexible use (cf. Dörrie et al. (eds.), *Gregor von Nyssa*, 107-119). Both they and later orthodox theologians emphasized the incomprehensibility of God's *ousia*: all that we can understand are his activities or "energies".

S. Gonzales, *La fórmula* Μία οὐσία τρεῖς ὑποστάσεις *en San Gregorio de Nisa*, Rome 1939; I. Ortiz de Urbina, *El Símbolo Niceno*, Madrid 1947; id., *Nicée et Constantinople*, Paris 1963; J. Whittaker, Ἐπέκεινα νοῦ καὶ οὐσίας, *VChr* 23 (1969) 91-104; J.N.D. Kelly, *Early Christian Creeds*, London ³1972; G.C. Stead, The Concept of Divine Substance, *VChr* 29 (1975) 1-14; H. Dörrie et al. (eds.), *Gregor von Nyssa und die Philosophie*, Leiden 1976; J.N.D. Kelly, *Early Christian Doctrines*, London ⁵1977; G.C. Stead, *Divine Substance*, Oxford 1977 (bibl.); R.P.C. Hanson, *The Search for the Christian Doctrine of God*, Edinburgh 1988.

G.C. Stead

PACATUS. A character hard to identify. Harnack sought to recognize his identity with Latinius Drepanius P., a rhetor of Aquitaine (4th/5th c.) (Sitzungsb. der preuss. Ak. Berlin, 1921, 266-284; 834-835). Another identification proposed by Harnack (*art. cit.*) was the name of P., *apologist and author of a treatise *Contra Porphyrium* of which only a few fragments remain (PL 68, 359; PG 5, 1025-28). Ruling out these hypotheses (W. Bährens, Pacatus, *Hermes* 56 [1921] 443-45), it seems preferable to refer to a certain P., a personal *friend of *Paulinus of Nola. To him was addressed a *letter on Paulinus's death, by the presbyter *Uranius (*Epistula de obitu Paulini ad Pacatum*). It seems certain, as we can presume from the letter itself, that this P. had the intention of celebrating Paulinus's *biography in verse (ch. 1: *Nunc autem veniamus ad ea quae tibi, qui vitam eius versibus illustrare disponis, dicendi materiam subministrent*).

CPL 1152a; *Pacatus (Ps. Polycarpi Smyrnensis) Contra Porphyrium*, ed. F.X. Funk-F.D. Diekamp, *Patres Apostolici*, II, Tübingen 1913, 397-400; *Epistula de obitu Paulini*: PL 53, 859-66; E. Galletier, *Panégyriques latins*, III, Paris 1955, 49, n. 4; J.F. Matthews, Gallic Supporters of Theodosius, *Latomus* 30 (1971) 1073-99.

L. Dattrino

PACHOMIUS (*c.*292-347). Founder of *coenobitism. Of a *pagan family from the far S of *Egypt, he was converted by the sight of Christian charity towards imperial conscripts. From *baptism (313), he felt attracted to a monastic life in the service of his *brothers, and submitted to the austere school of the solitary Palamon. After an unfruitful experience of common life, he understood the need to impose strict *poverty and rigid discipline on his monks. He had the gift of judging men, gaining their loyalty and organizing their collaboration. At times he had a heavy hand: this did not please everyone equally. Some even accused him of acting on the basis of spiritual visions and intuitions, since he was certainly no abstract intellect. He brought together in various *monasteries, esp. Tabennesi and Pbou, thousands of monks and nuns, maintained by a thriving economy. Shortly before his death in 345, he was subjected to the judgment of a synod of local bishops, meeting at Latopolis, but the energy of his disciples saved him. *Athanasius of Alexandria constantly supported him. If at the last moment he rejected his disciple *Theodore, who seemed destined to succeed him, the latter long kept the most lively, respectful, but human memories of him: these are the basis of the Coptic Lives and the Greek *First life* (drawn up perhaps before the *Coptic text was written down, since the interpreting brothers were more at their ease with Greek, but on the basis of memories better preserved in the Coptic; the Greek recension thinks of a distant public, omits various details and attenuates the founder's *visions and charismatic authority). The other Greek Lives, and the Latin Life that depends on it, are already far removed from the certain authenticity of the older documents. As for the *Rules, it is hard to know whether they were put into writing before the founder's death. In *Jerome's translation they appear in four collections which do not tally, and it was this Latin text that exercised a great influence in the West. In Coptic, only fragments have been found (the same goes for the Lives); in Greek we have only extracts. P. also wrote *letters in cryptic language, translated by Jerome, recently discovered in Coptic and Greek. Some Coptic *catecheses have also been found.

The Lives make P. an adversary of *Origen; nevertheless it seems to be established that the "*gnostic" books found at *Nag Hammadi were collected in his community and perhaps hidden on the occasion of some canonical visit.

A. Boon, *Pachomiana latina*, Louvain 1932; F. Halkin, *S. Pachomii Vitae Grecae*, Brussels 1932; L. T. Lefort, *Les vies coptes de s. Pachôme*, Louvain 1943, 1966; id., *Oeuvres de s. Pachôme et de ses disciples*, CSCO 159-160, Louvain 1956; A.J. Festugière, *La première Vie grecque de S. Pachôme: Les moines d'Orient*, IV, 2, Paris 1965; A. Veilleux, *La liturgie dans le cénobitisme pachômien*, Rome 1968; F. Ruppert, *Das pachomianische Mönchtum*, Münsterschwarzach 1971; *DIP* 6, 1067-1073; A. de Vogüé, Saint Pachôme et son oeuvre d'après plusieurs études récentes, *RHE* 1974, 425-453; F. Moscatelli, *Vita copta di S. Pacomio*, Padua 1981; A. Veilleux, *Pachomian Koinonia*, I-III, Kalamazoo 1980-1981; H. Quecke, *Die Briefe Pachoms*, Regensburg 1975; P. Rousseau, *Ascetics, Authority and the Church in the Age of Jerome and Cassian*, Oxford 1978; id., *Pachomius: the making of a community in fourth century Egypt*, Berkeley 1985 (bibl. of primary sources and secondary works); J.E. Goehring, New frontiers in Pachomian Studies, in (eds.) B.A. Pearson-J.E. Goehring, *The Roots of Egyptian Christianity*, Philadelphia 1986, 236-257.

J. Gribomont

PACIANUS. A late 4th-c. bishop of *Barcelona (*Barcino*), who died under *Theodosius. He wrote various *opuscula* (Jerome, *Vir. ill.* 106), incl. *Cervus* (or *Cervulus*), to dissuade the *faithful from feasts in honour of the new year. We have three *letters to the *Novatianist *Simpronianus, opposing those heretics' rigorist *ecclesiology, contrasting the plurality of the *heresies with the unity of the *Catholic church, recalling that *Novatian himself had approved the pardon of *lapsi* before becoming a *schismatic, and showing from numerous NT passages that even after *baptism the church allows repentance to the sinner, wanting not his death but his *conversion. P. deals with *penitence in *Paraenesis ad paenitentiam*, exhorting sinners to public penance, which many Christians then rejected, both for the penalties it imposed and out of human respect. We also have a baptismal *sermon. The *Liber ad Iustinum Manichaeum* and *De similitudine carnis peccati*, though attrib. to him by Morin, are not his. The latter has been claimed for *Eutropius.

CPL 561-563; PL 13, 1051-1094; Ed. L. Rubio Fernández, Barcelona 1958; *Patrologia* III, 124-126 (bibl.).

M. Simonetti

PAGAN - PAGANISM. The Latin term *paganus, -a, -um* (from *pagus, -i*: village) designated what was rustic, of the countryside or village; its secondary meaning was that of "civilian, bourgeois", as opposed to military. There has been much discussion of the process that gave rise to the religious meaning of the word *paganus*, whose corresponding adjective became equivalent to "who" (or "which") "belongs to the old *polytheism, gentile". Some derive its religious meaning from the secondary meaning (T. Zahn); others, with more reason, from the original meaning (J. Zeiller). In any case we must observe that *paganus* in the sense of idolatrous prevailed only in the 4th c. AD and that, if we want to go back to the history of the idea of paganism as religion with no relation to Judaism or Christianity, we must bear other notions in mind. From the 2nd c., the Greek *apologists used to divide mankind into three groups: Greeks (Ἕλληνες), *Jews (Ἰουδαῖοι) and Christians (Χριστιανοί); this derived from the fact that the Christians identified themselves neither with *Hellenism nor with Jewish tradition. Hence their definition of themselves as τρίτον γένος or καινὸν γένος with the intention of defending the novelty and originality of Christianity as against paganism and Judaism; therefore their apologies were called either πρὸς Ἰουδαίους or πρὸς Ἕλληνας.

When *Tertullian, the first or one of the very first Christians to write in Latin, wrote a work against the pagans, he entitled it *Ad nationes*. In so doing he probably recognized that the concept *Romanus* retained primarily its ethnic or political meaning and did not deprive the Christians of the title of Roman citizens, nor intend to exclude them from the *res publica*. The term *nationes* (or *gentes*) translates the Greek ἔθνη and, as has been suggested (cf. A. Schneider, *Le premier livre*, 13 ff.), is perhaps influenced by the OT - the LXX designates the holy people as λαός and the pagan "nations" as ἔθνη, the latter a term taken up by the NT, with its derivative ἐθνικοί - and by Roman "nationalism", which contrasted the *populus Romanus* with the *gentes* or *nationes*: foreign peoples, provincials, hardly differing from barbarians.

At the end of the 3rd c. *Arnobius would also entitle his apologetic work *Adversus nationes*. Later the term *paganus* would prevail: think of Paulus *Orosius's work, *Historiarum adversus paganos libri VII*.

The pejorative sense which the notions of "pagan" and "paganism" assumed among Christians, is clear: they represented beliefs in false gods, around which negative and blameworthy rites, practices, customs and usages gathered.

T. Zahn, Paganus, *Neue Kirchl. Zeitschr.* 10 (1899) 18 ff.; J. Zeiller, *Paganus. Etude de terminologie historique*, Paris-Fribourg 1917; E. Demougeot, Remarques sur l'emploi de Paganus, *Studi in onore di A. Calderini and R. Paribeni*, I, Milan 1956, 337-350; H. Grégoire, *Les persécutions dans l'Empire Romain*, Brussels ²1964, 188-220; I. Opelt, Griechische und lateinische Bezeichnungen der Nichtchristen. Ein terminologischer Versuch, *VChr* 19 (1965) 1-22; A. Schneider, *Le premier livre "Ad nationes" de Tertullien*, Geneva-Rome 1968, 10 ff., (Christians as *tertium genus*) and *passim*; J.H.W.G. Liebeschütz, *Continuity and Change in Roman Religion*, Oxford 1979; E.R. Dodds, *Pagan and Christian in an Age of Anxiety*, Cambridge 1965; R. MacMullen, *Paganism in the Roman Empire*, New Haven 1981; R.P.C. Hanson, The Christian attitude to Pagan Religions, in *Studies in Christian Antiquity*, Edinburgh 1985, 199-232; R. Lane Fox, *Pagans and Christians*, London 1986.

P. Siniscalco

PAHLAVI. Dead language of the Western Iranian group, into which many works of Christian interest seem to have been translated, generally from *Syriac. Only some *Manichaean texts and fragments of a psalter survive.

B.M. Metzger, *The Early Versions of the New Testament*, Oxford 1977, 276-277; W. Sundermann, Christliche Evangelientexte in der Überlieferung der iranischmanichäischen Literatur, *MIOF* 14 (1968) 388-415; P. Gignoux, L'auteur de la version pehlevie du psautier serait-il nestorien?, *Mémorial Mgr G. Khouri-Sarkis*, Louvain 1969, 233-244.

S.J. Voicu

PAIDEIA. The p. that bordered on *rhetoric and was not rhetoric, was directed towards a general vision of Christian culture. Its elements were compared, one by one, with the biblical unity manifested in the *incarnation of the Word. The basic *book was Holy *Scripture. Scriptural words and phrases entered the daily language of the *faithful. The *Bible became the irreplaceable means of man's growth. But reconstruction had to be carried out without destroying values, wherever found. The vital points of classical civilization were cast into new *literary genres which acquired their own features, into *liturgy which became christological and into biblical *exegesis. The great problem that p. discussed was chiefly that of knowledge: self-knowledge as awareness of the motives of our actions, and knowledge of principles and things. *Irenaeus maintained that the Christian cannot act impassionedly from states of mind. Even when he refutes *heretics he cannot ignore persons, intentions and writings. He himself was an example. Besides studying the works of *Valentinus's disciples, he had frequented them in order to understand their thought (*Adv. haer., praef.* 1, 2). *Hippolytus (*Refut.* 10, 5) did not disenchant the labyrinth of *heresies except through the strength of truth. We can see how Christian sensibility was wounded by facile gossip and *a priori* condemnation. *Tertullian (*Apol.* 1, 2) cried out that truth must be known before being condemned. Christian p. denounced the insufficiency of the *philosophy of the time, which was accustomed, on the educational level, to make judgments without examining not just the recondite aspects, but even the more obvious ones. For *Gregory the Thaumaturge (*Or. Pan.* 7, 95-96), research, in order to form consciences, had to be objective; otherwise, by vain and useless disputes and subtleties, it led to *impiety. Philosophy - maintained *Clement of Alexandria (*Strom.* 1, 7, 37) - must be rooted in the teaching of *justice and *piety in order to be a preparatory education for spiritual knowledge. Otherwise, as *Arnobius said (1, 58), it was reduced to vacuous syllogisms, enthymemes and definitions which sought to gain belief and assent, never to discover the lineaments of truth. In this way a close relationship grew up between philosophers, who, because of their contradictions, were always more inclined to argue than to seek the truth, and heretics. Tertullian's diatribe against philosophy (*Praescr.* 7, 1-13) had its foundation here. Secular knowledge could not but suborn heresies. *Eusebius (*HE* V, 28, 13-18) noted the impossibility of putting in order, for the sake of comparison, the biblical exemplars followed by each heretic: they all differed from each other. Consequently their interpretations could not but be wholly personal.

In three centuries, day by day, Christian p. constructed the new culture as *peace of spirit. *Origen succeeded in generating in the souls of his followers a state of security and tranquillity. But this depended, as Gregory the Thaumaturge said (*Or. Pan.* 7, 98-103), on his concept of philosophy as a school of *virtue for the sake of peace of mind. A spiritual doctrine, then, in which each discipline was not for its own sake, but was coordinated with the others for a single end. Origen's moral judgments had a metaphysical origin and sprang from an attentive and minute examination that urged him to daring conquests. In scriptural exegesis, the golden strands that constituted the basis of Christian p. had to be sought. The elements of this p. developed and pervaded the Eastern and Western culture that revealed the true face of the era which followed *Constantine. The mind was prepared by the trivium (grammar, *rhetoric and *philosophy) and quadrivium (arithmetic, geometry, astronomy and *music) to perceive higher things. It was manifestly a p. whose end was not earthly ethicity, like that of the classical world (so well studied by Jaeger), but a first step to the *life of the other world, through the moral freedom that came from the biblical word, which was the word of salvation. For other aspects cf. *School.

H.I. Marrou, *Saint Augustin et la fin de la culture antique*, Paris 1958; A. Quacquarelli, *Le fonti della paideia antenicena (Renovatio mundi)*, Brescia 1967; id., *Scuola e cultura nei primi secoli*, Brescia 1974; W. Jaeger, *Cristianesimo primitivo e paideia greca*, Florence 1974 (Eng. tr. *Early Christianity and Greek Paideia*, Cambridge, Mass. 1962); R. Dostalova, Christentum und Hellenismus, *ByzS* 44 (1983) 1-12; H.I. Marrou, *Histoire de l'Education dans l'Antiquité*, Paris 1965, 6th enlarged ed.; id., L'École de l'Antiquité tardive, in *La Scuola nell' Occidente latino dell'alto Medioevo*, Settimane di Studio 19, Spoleto 1972, 127-143.

A. Quacquarelli

PAINTING

I. Wall decoration. Style and content - II. Miniature painting - III. Icons.

I. Wall decoration. Style and content. Now that we have finally abandoned the 1st- and 2nd-c. dating proposed by the first students of early Christian p., it seems that the earliest pictorial manifestations of Christianity (after investigations conducted comparatively with contemporary *pagan products and the triumph of a line of research which, alongside purely stylistic elements, also examines antiquarian and archaeological-topographical data, decisive for chronology) cannot go back beyond the first years of the 3rd c. From the formal point of view, it is likewise acknowledged that primitive Christian p., like early Christian art in general, fits perfectly into *late antique artistic tendency (an expression of the profound economic, political, social and spiritual crisis that rocked the Roman world in the 3rd c.), a tendency from which it took the most fluent characteristics of the popularizing strand, which favoured the content and expressive value of the image at the expense of the formal rigour and organic correctness of classical tradition. This tendency, moreover, was perfectly in line with the very essence of early Christian art, an art which, beyond the search for formal perfection, aimed to evoke the fundamentals of its own faith through a figurative "language" capable of "expressing" its concepts in a way that was clear and easily understandable to all.

At least until the religious *peace, early Christian p. is documented for us almost exclusively through cemeterial frescos (of the Roman catacombs, in particular), where the custom of decorating the tomb as a second *domus*, already widespread in the classical world, was perpetuated in order to evoke, by means of *images, the means by which *faith ensured the believer's salvation. An incomplete documentation, then, which leaves the question of whether, in this period, alongside funerary decoration there also developed an ornamentation of *church buildings, as the - so far unique - example of the *domus ecclesiae* of *Dura Europos, a caravan city on the banks of the Euphrates, destroyed by the Persians soon after 256 (C. H. Kraeling - C. Bradford Welles, *The Christian Building. The Excavations at Dura Europos. Final Report VIII, Part II*, New Haven 1967), would lead us to believe. Here, in a room of the house used as a baptistery, images from the OT and NT, evoking the *neophyte's salvation through the death-*resurrection effected in *baptism by Christ, the new *Adam, are painted in an extremely diagrammatic and sketchy style (but also monumental, as in the scene of the *Pious Women at the sepulchre), in which the frontality of the figures, the paratactical composition and the loss of naturalistic proportions have led us to see strong Iranian influences.

The appearance of these paintings in the baptismal chamber and not in the meeting-room of the *domus*, has been held by some as evidence of a resolute aversion on the part of the church authorities, in the early centuries, to decorating the walls of a church (an aversion also attested by can. 36 of the council of *Elvira in 306: Mansi 2, col. 11), whereas portrayals in a baptistery could carry out the pure task of illustrating for neophytes the sacramental means of regeneration of the *soul.

At *Rome, the first pictorial manifestations - visible, as we have said, only in a cemeterial context - date from the first decades of the 3rd c. The wall decoration of the cubicles and galleries of the catacombs - obtained by the technique of fresco or, more rarely, of tempera - is marked, in this period, by the typical illusionistic ornamental style of red and green lines on a white ground, a style which in composing - on walls and ceilings - niches, panels, medallions, lunettes, etc., in various patterns, connected by linear segments, enlivens the extreme simplification of the old decorative architectural partitions of Pompeian tradition. Within this linear syntax (as we can observe, e.g., in the crypts of Lucina in the *cemetery of S. Callisto [*Wp* 24-25; 26, 1; 27, 1; 28; 29, 1=Nestori, 99, nos. 1-2], in the cubicle of the *coronatio* in the catacomb of Pretestato [*Wp* 17-20=Nestori, 87, no. 3] or in the so-called cubicles of the sacraments in S. Callisto [*Wp* 15, 2; 26, 2-3; 27, 2-3; 29, 2; 38-39; 40, 3; 41; 46-48, 1=Nestori, 102 f., nos. 21-25]), the figures (the ornamental ones often taken from the common classical repertoire) are treated with a rapid, casual, impressionistic technique which, by means of lively and essential matching of colours, destroys the naturalistic image, annihilating its volumes and details.

The subject matter, in these first depictions, is marked by a powerful *symbolism, centred on the concept of salvation, alluded to by symbolic images (e.g. the Good *Shepherd and the *Orans*, alternately repeated, as in the ceiling of the cubicle of the crypts of Lucina, to signify the two poles of salvation, the Saviour and the saved) and by the first biblical portrayals, such as those of *Daniel in the lions' den, the story of *Jonah, Jesus' *baptism, the miracle of the rock, the raising of *Lazarus, the healing of the *paralytic, etc. Sometimes, alongside scenes that would not be repeated in subsequent eras (e.g. the "fishes of Lucina", perhaps allusive of the

*Eucharist [*Wp* 28=Nestori, 99, no. 2] or the so-called "crowning with thorns" in the cited homonymous cubicle [*Wp* 18=Nestori, 87, no. 3]), we find, in this period, iconographical formulations which have no exact parallels in later paintings (e.g. the scene of the Samaritan woman or the raising of Lazarus in cubicles A3 and A6 of the sacraments [*Wp* 39, 1; 46, 2=Nestori, 102 f., nos. 22, 25]) and which are evidence of what we may call the initial or formative phase of the Christian iconographical repertoire, a phase characterized by the elaboration of figurative formulae not yet codified into those fixed schemes which, typical of subsequent eras, would make their subjects easily recognizable and comprehensible to all.

Outside Rome, but still in a funerary context, we can also assign to the first decades of the 3rd c. the paintings of a mausoleum in the necropolis of Cimitile, near Nola, with scenes of *Jonah and of *Adam and Eve (P. Testini, Cimitile: l'antichità cristiana, in *L'art dans l'Italie méridionale*. A revision of Emile Bertaux's work under the direction of Adriano Prandi, Rome 1978, 167 f.), thematically and stylistically, as well as chronologically, very close to the paintings in the vestibule of the upper floor of the catacomb of S. Gennaro at *Naples (U.M. Fasola, *Le catacombe di S. Gennaro a Capodimonte*, Rome 1975, 26 ff.).

Back in Rome, around the mid 3rd c., the spacious red-green linear style of the first decades of the century tends to become more and more illusionistic and to lose compositional coherence through the progressive insertion of suspended elements and infilling which tend to break up the harmonious linear arrangement and destroy the free spaces (e.g. in the cubicle of the Good Shepherd at Domitilla: *Wp* 6, 2; 7, 2; 9-12, 1=Nestori, 121, no. 23). These characteristics can be grasped, halfway through the 3rd c., in the hypogeum of the Aurelii in Viale Manzoni (Nestori, 44-47, nos. 1-4), associated with a resurgence of the classicizing tendency, obvious esp. in the monumental figures of people recurring on the walls of one of the two underground cubicles. The disputed interpretation of some scenes has led to the monument being attrib. to a heterodox Christianity, perhaps of *gnostic inspiration (M. Chicoteau, *Glanures au viale Manzoni. Le Monumentum Aureliorum à Rome*, Brisbane 1976, with bibl.).

Great chronological problems - which can be extended to much of the Christian pictorial output of the early centuries - have been created by the frescos depicting OT and NT episodes (as well as a banquet scene, considered by some a veiled allusion to the Eucharist) in the so-called "Greek Chapel" in Priscilla (*Wp* 13-15, 1; 16=Nestori 27 f., no. 39), for which have been proposed (and, mind you, in studies conducted more or less at the same time) dates oscillating between 170-180 and the first decades of the 4th c. (cf. Reekmans, *La cronologie*, 284 f.). The megalographic dimensions of the *Susanna cycle (unique in its formulation) and the undeniable classicism that pervades the "chapel" have recently led to a date after *Gallienus (✝268) being proposed (A. Recio, La "Cappella Greca" vista y diseñada entre los años 1783 y 1786 por Seroux d'Agincourt, *RAC* 56 [1980] 49-94), i.e. within the sphere of that classical renaissance whose signs can be seen clearly, in the same catacomb of Priscilla, in the so-called cubicle of the "Velatio" (*Wp* 66, 1; 78-81; 82, 2=Nestori, 23, no. 7), where the portrayals on the end arcosolium, perhaps representing three moments in the life of a young woman (C. Dagens, A propos du cubiculum de la "Velatio", *RAC* 47 [1971] 119-129), are painted with intensity of colour and fine modelling.

The period going from the last decades of the 3rd c. to the Constantinian era is well documented by the numerous paintings in the cemetery of SS. Marcellino e Pietro. The oldest group of frescos, still characterized by the linear wall pattern and by a solid, well-constructed placing of the figures (e.g. the so-called cubicle of *Nicerus* with the unique scene of the lame woman: G.P. Kirsch, Pitture inedite di un arcosolio dei SS. Marcellino e Pietro, *RAC* 7 [1930] 210 ff.=Nestori, 58 f., no. 65) is followed, in some cubicles on so-called staircase Y and in the region of the *agapai* (*Wp* 97, 1; 98; 99, 1; 100=Nestori, 58, no. 64; *Wp* 93-94; 95, 2, 3; 96; 97, 2=Nestori, 57, no. 58), by compositions marked by sharp chromatic contrasts which attain great expressive intensity by means of Pointillistic optical effects. A singular variety of floral and geometrical ornaments recurs in these paintings, together with a gradual transformation (starting esp. from the tetrarchical era) of the lines of the 3rd-c. ornamental scheme into bands, like frames to the figured scenes, by now an essential element of the decoration. From the thematic point of view, we should note the introduction of new OT and NT scenes and the frequent representation of *fossores* at work and of banquets, sometimes perhaps allusive of the heavenly feast, sometimes realistically portraying the funeral banquets offered by the living to the *dead.

With the religious *peace, triumphalistic visions were introduced into *iconography to express the triumph attained and the advent of a new era. Images of the heavenly court abounded, with Christ as sovereign, surrounded by his dignitaries the *apostles, a theme that, already known in abbreviated formulations from the start of the 4th c. (e.g. in the so-called *Orpheus cubicle in the catacomb of SS. Marcellino e Pietro: *Wp* 96=Nestori, 57, no. 58), was reproposed in more solemn terms in the years that followed the religious peace, to express, in polemic with the imperial power (which was interfering more and more in religious matters), the divine origin of the *imperium*. At the same time new scenes were introduced celebrating episodes that marked Christ's earthly life and victory. From the formal point of view, the Constantinian era marks a recovery of plasticism and a reconsolidation of forms contained in strong outlines; the figures appear in panels bordered by ever thicker frames, with frontal, roundish faces with large, dilated eyes, framed by dark, compact hair (see, e.g., the well-known depiction of the "Madonna and Child" in Maius: *Wp* 163, 1; 207-209=Nestori, 36, no. 22).

At Rome, the most significant document of 4th-c. p. - dateable somewhere between the Constantinian era and the last decades - is certainly the catacomb of Via Dino Compagni (A. Ferrua, *Le pitture della nuova catacomba di via Latina*, Vatican City 1960), a *cemetery of private character, almost entirely frescoed, in which, alongside products of unusual artistic quality, we find themes formulated with unusual iconographies and, in some cases, totally new scenes (e.g. the now famous "*Medical lecture": Ferrua, *op. cit.*, pl. 107=Nestori, 77, no. 9), singularities all to be connected with the monument's private character, in which could work, evidently, more talented painters, not tied to those traditional iconographical schemes that were invariably repeated by the modest artisans who worked in the *cemeteries administered by the *clergy. If the pictorial quality makes it reasonable to think of a particularly wealthy client, some would relate the extraordinarily rich iconographical repertoire (esp. of OT scenes) to the existence of miniature bibles from which the painters drew themes and iconographies. Then the existence, among the various paintings, of subjects of pagan inspiration is certain evidence of the presence, in the hypogeum, in the second half of the 4th c., alongside totally Christian groups, of family nuclei not yet converted to the new faith. From the formal point of view, the characteristics of the Constantinian era, visible in the first cubicle, are followed by portrayals marked by a more masterly classicism and a formal composure, in the style of the decades immediately following the middle of the 4th c.

The same classicist manner, but characterized by a more marked linearism, is visible in the figured compositions of the late 4th and early 5th c. Burial in underground cemeteries declining, pictorial activity in this period lost its specifically funerary character and turned, often, to glorifying the *martyrs and the kingship of the Teacher, inspired by the grandiose exemplars of *church buildings. In the vault fresco of a cubicle of Marcellino e Pietro (*Wp* 252-254=Nestori, 48, no. 3), the derivation from lost basilical models is evident in the characteristic overlapping of two compartments, in which unfolds the exaltation of Christ's royal magisterium, between *Peter and *Paul, and the *acclamatio* by the four eponymous martyrs of the *Agnus Dei* on the mount of *paradise. The stylistic characteristics of the time are evident in the elongated figures, the hieratic faces, the almond-shaped eyes, the "helmet" hairstyles, the marked linearity of the forms, dematerialized in an abstract atmosphere. Similar characteristics can also be encountered in a painting, depicting Christ between six people, in the catacomb of Albano near Rome, dateable to the early 5th c., to which we may add, in the same catacomb, at the end of the century, a fresco with Christ between four saints, characterized by greater linearity and abstraction (M. Marinone, La decorazione pittorica della catacomba di Albano, *RIA* n.s. 19-20 [1972-73] 103-138).

Outside the funerary sphere, we find at Rome, in the second half of the 4th c., the frescos of a small religious room beneath the church of SS. Giovanni e Paolo on the Coelian (M. Cecchelli, Osservazioni sul complesso della "domus" celimontana dei SS. Giovanni e Paolo, *Atti IX Congr. Int. Arch. Crist.*, I, Vatican City 1978, 551-562) and of a private oratory near the Lateran basilica (V. Santa Maria Scrinari, Documenti paleocristiani nell'ambito dello storico ospedale di S. Giovanni al Laterano, *Rend. PARA* 48 [1975/76] 377-91). The two monuments show the stylistic characteristics of the cemeterial p. of the time (in particular, for the Lateran oratory, resemblances have been seen with the frescos of the catacomb of Via D. Compagni) and both present scenes of considerable interpretative difficulty.

Problems of interpretation - not so much for the paintings, as for the function of the monument - are also presented by the hypogeum of S. Maria in Stelle, at *Verona (W. Dorigo, L'ipogeo di S. Maria in Stelle in Val Pantena, *Saggi e memorie di Storia dell'Arte* 6 [1968] 9-31), by some thought to be funerary, by others religious: the depictions, taken from the OT and NT, that decorate the walls of one of the two underground rooms, drawn up in monumental forms and characterized by elongated figures, contained in strongly marked outlines, are dated between the late 4th and early 6th cc.

The representation of Christ with the apostolic college, however, marked by a greater linearity, must prob. be dated to a later era.

Leaving *Italy, important pictorial evidence, still assignable to the 4th c., has been found, always in a funerary setting, in the Balkans, at Nis (L. Mirkovic, La nécropole paléochrétienne de Nis, *Archaeologica Iugosl.* 2 [1956] 85-100), at Pécs (F. Fülep, *Sopianae. Die Stadt Pécs zur Römerzeit*, Budapest 1975, 24 and 26) and esp. at *Thessalonica, where an uninterrupted pictorial activity is documented from the end of the 3rd c. (tomb of the E necropolis, decorated in the red-green linear scheme) all through the 4th (tomb of the Good *Shepherd, so-called tomb of Eustorgius, on the Via S. Demetrio, in the E necropolis, with various biblical portrayals), up to the 7th c. (frescos found in the area of the Agorà [D. Pallas, *Les monuments paléochrétiens de Grèce découverts de 1959 à 1973*, Vatican City 1977, 65-73]). Particularly important are the frescos of Pécs, come to light in 1975-76, which decorate the lower room of a great mausoleum with depictions of a person seated on a throne in *paradise and the biblical episodes of *Adam and Eve and of *Daniel in the lions' den, compositions presenting iconographical and stylistic characteristics comparable to Roman cemeterial pictorial output of the second half of the 4th c. (F. Fülep, Excavations of the early Christian mausoleum of Pécs, *Archaeologiai Ertesitö* 104 [1977] 246-257). In *Thrace, in the village of Osenovo near Varna, more or less in the same period were executed the paintings of a chamber tomb (picturing the deceased, members of his family, birds and other animal and plant motifs, under a starry vault with images of the Sun and Moon) of considerable interest for their characteristic style, marked by a singularly primitive taste which constitutes important evidence of late Roman provincial artistic trends (A. Mincev - P. Georgiev, The tomb at Osenovo. A monument of the late antique paintings on the Thracian coast of Pontos, in *Actes du II*e *Congrès International de Thracologie* (Bucharest, 4-10 Sept 1976), II. "Histoire et Archéologie", Bucharest 1980, 411-423).

In *Egypt, great interest has been aroused by the paintings, dateable from the mid 4th to the mid 5th c., that decorated the cupola of a tomb in the necropolis of El-Bagawat, esp. because of the presence, alongside the usual depictions inspired by soteric paradigms, of a scene of the *Exodus of the Israelites, formulated in a singularly diagrammatic and summary manner, leading some to think of a strong Jewish influence (M.L. Thérel, La composition et le symbolisme de l'iconographie du mausolée de l'Exode à El-Bagawat, *RAC* 45 [1969] 223-270). Still in *Africa, in the second quarter of the 4th c. an important funerary hypogeum found at Gargaresh (Tripolitania) was frescoed with scenes of *Adam and Eve, Jesus' *entry into Jerusalem and figures of torch-bearers, paintings showing close contact with the Roman frescos of Via D. Compagni and the *mosaics of Piazza Armerina (A. Di Vita, L'Ipogeo di Adamo e Eva a Gargaresc, *Atti IX Congr. Int. Arch. Crist.*, II, Vatican City 1978, 199-256).

With the 6th c., at Rome, alongside the persistence of the classical tradition, we begin to be aware of the incipient penetration of Byzantine taste, which stiffens the images into an abstract immobility, dematerializing them in essential chromatic schemes. These tendencies are clearly visible in, besides various cemeterial paintings, e.g. the famous fresco in the catacomb of Commodilla, representing the Virgin and Child between Sts *Felix and Adauctus, the latter depicted in the act of presenting the dead Turtura (*WMM* 136=Nestori, 138, no. 3). Alongside the flat, diagrammatic style which depicts figures with linear outlines and fixed, abstract faces, we find, in the image of Turtura, modes that are still clearly Western in the admirable realism of the face. Also inspired by the characteristics of the Byzantine fashion are the earliest frescos (6th-early 7th cc.) of the *diaconia* of S. Maria Antiqua al Foro (J. Nordhagen, S. Maria Antiqua: the frescos of the seventh century, *AAAH* 8 [1978] 89-142), like the "Maria Regina" and the *Annunciation with the "Pompeian *angel", the latter characterized by a marked formal classicism in the Hellenistic tradition, visible in the gentle transition of the planes and the use of dense, warm colours. In the catacombs of S. Gennaro at *Naples, an early 6th-c. fresco, with portraits of the dead Theotecnus, his wife and little daughter, formulated in a linear way without plasticity, is close in style to that of Turtura (Fasola, Le catacombe, *op. cit.*, 96).

Considerable evidence of the p. of the 6th and 7th cc. comes to us from *Egypt, thanks to the frescos of Bawit and Saqqara which decorated the walls of the small chapels that the Coptic monks built in large numbers in their convents. At Bawit, we can see in the majority of the works an advanced process of geometrization of forms, esp. in the frequent apsidal depictions of Christ in majesty and of *Mary enthroned among *apostles and saints, frontally and hieratically rendered, with clear colours and strong outlines (J. Clédat, Le monastère et la nécropole de Baouît, *Mémoires publiés par les membres de l'Institut français d'archéologie orientale du Caire* 12 [1904]). Still more schematic are the Saqqara paintings: on the apse vaults, the images of Christ enthroned and the Madonna *lactans* are uncompromisingly two-dimensional and linear (J. Quibell, *Excavations at Saqqara [1906-1907]*, Cairo 1908, pl. XL ff.; J. Maspéro - E. Drioton, *Fouilles exécutées à Baouît, Mémoires op. cit.*, 59 [1931]).

Returning again to Rome, a vibrant style, animated by a return to the concise manner, characterizes the layer of frescos in S. Maria Antiqua with *Solomon and the seven Maccabaean brothers, attributable to the mid 7th c., i.e. earlier than the grandiose composition of Christ on Golgotha commissioned by John VII (705-707) in the presbytery of the church (J. Nordhagen, S. Maria Antiqua, *op. cit.*, 114-120; id., The Frescoes of John VII [A.D. 705-707] in S. Maria Antiqua in Rome, *AAAH* 3 [1968]). The oldest frescos of the church of S. Saba, with a procession of saints, also prob. date from John VII, while the paintings of the same subject in the basilica of S. Crisogono (Matthiae, *Pittura*, I, 151 and 188) must be considered slightly later. We meet ever more decorative and calligraphic styles in the last depictions that decorate the historical nuclei of the catacombs, as e.g. in the coronation of Sts *Abdon and Sennen in the Pontian cemetery (*Wp* 258=Nestori, 146, no. 6), in the saints painted near the tomb of pope *Cornelius in S. Callisto (*Wp* 256=Nestori, 100, no. 5), in the martyrdom of St *Callistus depicted in the cemetery of Calepodio (A. Nestori, La catacomba di Calepodio al III miglio dell'Aurelia Vetus e i sepolcri dei papi Callisto I e Giulio I, *RAC* 47 [1971] 203-211=Nestori, 150, no. 2) or in the faded figure of St *Luke in the basilica of Commodilla (*WMM* 147=Nestori, 138, no. 3), dated by the inscription below it to the late 7th c.

To the 7th c. and to an Eastern milieu - we must finally mention - some would attribute the now famous frescos found in 1944 in S. Maria Foris Portas at Castelseprio, with scenes of the life of Christ whose extraordinary classicist reviviscence - unique in its formulation - has divided scholars over their date (from 7th to 10th c.) and their school (G.P. Bognetti, *Castelseprio. Guida storico-artistica*, Vicenza 1974, 27 ff.). **[Figs: 241, 242, 243]**

II. Miniature painting. In the wake of the late antique miniatures inspired by Hellenistic-Roman tradition (*Vergilius Romanus, Vergilius Vaticanus, Ilias Ambrosiana, Vienna Dioscoris*, etc.: Weitzmann, *Manuscrits*, 32-72), from the 5th c. the earliest Christian miniatures appear, ornamenting (usually full-page, in tempera) biblical MS texts on the parchment *codices* that had gradually replaced the old *rotuli* (cf. *Book). To the early 5th c. and to a Roman *scriptorium* we must prob. attribute the miniatures of the *Itala of Quedlimburg* (Weitzmann, *Manuscrits*, 40 f.), whose rendering of figures and still illusionistic landscape (as we can observe in the best-preserved scenes of the life of Samuel) have led to comparisons with the *Vergilius Vaticanus*. A strong classical influence is visible in the few fragments of the "*Cotton*" *Genesis*, attributable to a 5th/6th-c. Alexandrian milieu (Weitzmann, *Manuscrits*, 73-75), and esp. in the famous *Vienna Genesis*, the most sumptuous surviving early Christian manuscript, in which, alongside typically Byzantine characteristics (denial of spatiality, no attempt to create a precise relationship between figure and surroundings, abstract backgrounds, stylization of some figurative elements), we find strong reminiscences of Romano-Hellenistic naturalistic art, esp. in the *Flood scene, still marked by great spatiality and a fully perspective vision (R. Bianchi Bandinelli, La composizione del diluvio nella Genesi di Vienna, *MDAI(R)* 62 [1955] 66-77). For the Vienna miniatures - copies of an older model and executed by more than one artist - the 6th c. and *Syria or *Palestine seem the most probable time and place. Similar in style and date to the *Vienna Genesis* are the illustrations of St Matthew's Gospel in the *Codex Sinopensis* in the Bibliothèque Nationale, Paris, and those of Matthew's and Mark's Gospels in Rossano cathedral (Weitzmann, *Manuscrits*, 14, 89-90), all attributable to 6th-c. Syria. The Rossano miniatures in particular - grouped outside the text, at the beginning of the work - show a highly refined style, by now fully Byzantinized in the flattening and tendency to abstraction of the figures and backgrounds. We depart even further from classical naturalism with the miniatures of the *Codex of Rabbula* (Weitzmann, *Manuscrits*, 101-108), executed in Syria in 586 and preserved in the Biblioteca Laurentiana, Florence: eight full-page miniatures and a series of canonical tables ornamented with NT scenes and portraits. The style is rather flowing and schematic, but there are realistic accents in the lively and expressive faces and the striving after spacious scene-painting in some compositions - like the Crucifixion scene - which, even in the late 6th c., hark back to the classical tradition. To the late 6th c. should perhaps be assigned the *Paris Bible* and the *Gospel of St Augustine* (Weitzmann, *Manuscrits*, 109-115), the latter containing two notable miniatures, one showing scenes of the Passion, the other the *evangelist Luke enthroned. To the 7th-c. and N *Africa, finally, we must attribute the miniatures of OT stories in the "Ashburnham" Pentateuch in the Bibliothèque Nationale, Paris (Weitzmann, *Manuscrits*, 118-125), in which the absence of any trace of spatial composition, the dramatic sense, the abundance of fantastic architecture and the taste for a richer colouring seem to announce the mediaeval aesthetic.

III. Icons. The oldest sacred *images painted with encaustic technique on wood for devotional ends cannot be older than the 6th c. The majority of them come from St Catherine's Monastery on Mt *Sinai. To three artisans of great quality who worked in the second half of the 6th and first half of the 7th c. at *Constantinople, where the classical tradition was preserved with particular purity, we owe the three magnificent icons representing Christ, the Virgin enthroned between two saints, and St *Peter. In the first (Weitzmann, *The Monastery*, no. B1, 13-15, pls. I-II, XXXIX-XLI), the great refinement of the modelling builds up, with gentle transitions of planes, Christ's face, whose rigidity is counterpointed by the evident asymmetry of the great eyes, the moustache and the beard; the same contrasting rhythms recur in the Madonna enthroned with the little Jesus (Weitzmann, *The Monastery*, no. B3, 18-21, pls. IV-VI, XLIII-XLVI), whose relative mobility is set against the two saints George and Theodore, rigidly frontal and hieratic; above them, the heads of the archangels looking up at the divine hand are rendered in an impressionistic technique. Finally, inspired by the compositions of ivory consular *diptychs, the icon of St Peter (Weitzmann, *The Monastery*, no. B5, 23-26, pls. VIII-X, XLVIII-LI), whose expressive and spiritual face is the focal point of the whole composition. More abstract and calligraphic are the faces of Sts *Sergius and Bacchus in one 7th-c. icon, again from Mt Sinai, now in the Kiev Museum (Weitzmann, *The Monastery*, no. B9, pp. 28-30, pls. XII, LII-LIII).

The two images of the Virgin in S. Maria Nova (P. Cellini, Una Madonna molto antica, *Proporzioni* 3 [1950] 1-6) and S. Maria in Trastevere (C. Bertelli, *La Madonna di Santa Maria in Trastevere*, Rome 1961), Byzantine in inspiration but with a linear harmony typical of the Western tradition, must date from 7th/8th-c. Rome.

I. In general on early Christian painting: G. Wilpert, *Le pitture delle catacombe romane*, Rome 1903; id., *Die römischen Mosaiken und Malereien der kirchlichen Bauten vom IV. bis XIII. Jahrhundert*, Freiburg i.Br. 1916; F. Wirth, *Römischen Wandmalerei von Untergang Pompejis bis Ende des dritten Jahrhunderts*, Berlin 1934; J. De Wit, *Spätromischen Bildnismalerei*, Berlin 1938; R. Bianchi Bandinelli, Continuità ellenistica nella pittura dell'età medio e tardo romana, *RIA* (1953) 1-85 (cf. id., *Archeologia e cultura*, Rome ²1979, 344-423); A. Grabar-C. Nordenfalk, *Early Medieval Painting from the fourth to the eleventh century*, Lausanne 1957; M. Borda, *La pittura romana*, Milan 1958; R. Farioli, *Pitture di epoca tarda nelle catacombe romane*, Ravenna 1963; M. Chatzidakis-A. Grabar, *La pittura bizantina e dell'alto medioevo*, Milan 1965; J. Kollwitz, Die Malerei der konstantinischen Zeit, *Akten VII. Intern. Kong. für christliche Archäologie*. Trier 5-11 September 1965, Vatican City 1969, 29-158; L. De Bruyne, La peinture cémétériale constantinienne, *ibid.*, 159-214; G. Matthiae, *Pittura romana del Medioevo*, I-II, Rome 1965; P. Du Bourguet, *La peinture paléochrétienne*, Paris 1965; W. Dorigo, *Pittura tardo-romana*, Milan 1966; A. Grabar, *Le premier art chrétien (200-395)*, Paris 1966; id., *L'âge d'or de Justinien*, Paris 1966; P. Testini, *Le catacombe e gli antichi cimiteri cristiani in Roma*, Bologna 1966, 255-309 (with full bibl. on pp. 357-368); F. Gerke, *Spätantike und frühes Christentum*, Baden-Baden 1967; C. Delvoye, *L'art byzantin*, Paris 1967; V. Lazarev, *Storia della pittura bizantina*, Turin 1967; T. Hubert-J. Porcher-W.F. Volbach, L'ipogeo detto dei Flavi in Domitilla I. Osservazioni sulla sua origine e sul carattere della decorazione, *RAC* 45 (1969) 119-173; L. De Bruyne, La "Cappella greca" di Priscilla, *RAC* 46 (1970) 291-330; R. Bianchi Bandinelli, *Rome. La fin de l'art antique*, Paris 1970; L. Pani Ermini, L'ipogeo detto dei Flavi in Domitilla II. Gli ambienti esterni, *RAC* 48 (1972) 235-269; L. Reekmans, La chronologie de la peinture paléochrétienne. Notes et réflexions, *RAC* 49 (1973) 271-291; M. Gough, *The origins of Christian Art*, Geneva 1973; A. Nestori, *Repertorio topografico delle pitture delle catacombe romane*, Vatican City 1975; V. Fazzo, *Le giustificazioni delle immagini religiose dalla tarda antichità al Cristianesimo I. La tarda antichità (con un'Appendice sull'Iconoclasmo Bizantino)*, Naples 1977; H. Mielsch, Zur Stadtrömischen Malerei des 4. Jahrhunderts n. Chr., *MDAI(R)* 85 (1978) 151-207; J. Fink, *Die römischen Katakomben*, Antike Welt. Zeitschrift für Archäologie und Kulturgeschichte 9, Munich 1978; R. Bianchi Bandinelli, *La pittura antica*. A cura di Filippo Coarelli e Luisa Franchi dell'Orto, Rome 1980; K. Michalowski, *Faras, die Kathedrale der Wüstensand*, Zürich 1967; id., *Faras*, Warsaw 1974; J.M. Plumley, Some examples of Nubian Christian art from the excavations at Q'asr Ibrim, in (ed.) E. Dinkler, *Kunst und Geschichte Nubiens in Christlicher Zeit*, Recklinghausen 1970, 129-140 (Nubian Christian Art); (eds.) J.P.C. Kent - K.S. Painter, *The Wealth of the Roman World, 300-700*, London 1977; E. Kitzinger, *Byzantine Art in the Making*, Cambridge, Mass. 1977; id., The Cleveland Marbles, *Atti IX Congresso Internaz. di Arch. crist.*, Rome 1978, 653-675; H. Brandenburg, Ars Humilis: Zur Frage eines christlichen Stils in der Kunst des 4 Jahrhunderts nach Christus, *JbAC* 24 (1981) 71-84; J. Deckers - H.R. Seeliger, *Die Katalcombe "Santi Marcello e Pietro", Repertorium der Malereien*, Münster 1987; R.L. Milburn, *Early Christian Art and Architecture*, Berkeley 1988; Sister C. Murray, Artistic Idiom and doctrinal development, in (ed.) R. Williams, *The Making of Orthodoxy* (Essays in Honour of Henry Chadwick), Cambridge 1989; id., Art and the Early Church, *JThS* n.s. 28 (1977) 303ff.
II. On miniature painting: K. Weitzmann, *Ancient Book Illumination*, Cambridge, Mass. 1959; id., *Illustrations in Roll and Codex. A Study of the Origin and Method of Text Illustration*, Princeton ²1970; id., *Studies in Classical and Byzantine Manuscript Illumination*, Chicago-London 1971; id., *Manuscrits gréco-romains et paléochrétiens*, New York 1977; id., *Byzantine Book Illumination and Ivories*, London 1980.
III. On icons: K. Weitzmann-M. Chatzidakis-K. Miatev-S. Radojcic, *Frühe Ikonen. Sinai, Griechenland, Bulgarien, Jugoslawien*, Vienna-Munich 1966; D. and T. Talbot Rice, *Icons and Their Dating*, London 1974; K. Weitzmann, *The Monastery of Saint Catherine at Mount Sinai. The Icons*. Vol I: *From the Sixth to the Tenth Century*, Princeton 1976; id., *The Icons. Holy Images. Sixth to Fourteenth Century*, New York 1978.

V. Fiocchi Nicolai

PALAEOGRAPHY

I. Greek script from the 1st to the 8th c. AD - II. Latin script from the 1st to the 8th c.

P. seeks to reconstruct the history of writing, in all its manifestations, placing it in its socio-cultural setting. For much of our period (1st-8th cc. AD), Greek and Latin script evolved in the same cultural setting: but a true Greco-Latin p. is still to be constructed. The great basic notions are common to them. The usual script, that of private documents and notes, tended for reasons of rapidity and economy to reduce the number of strokes and join them up: it was a cursive script. Writing intended for books aimed above all at legibility and beauty: the strokes were distinct, the execution slower (steadier) and more accurate, the aesthetic effect obtained through deliberate stylization. But the forms and transformations of book-script, though autonomous up to a point, can be understood only by bearing in mind the changes in everyday script, the normal scene of evolution of script.

I. Greek script from the 1st to the 8th c. AD. Even before the 1st c. there was a differentiation between ordinary script and book-script. Both were still in capital letters, i.e. contained between two ruled lines; but, while the former stressed its cursive character to the point of compromising its legibility, the latter developed, sometimes in rather banal forms close to common usage, but sometimes in genuine calligraphic styles, two of which, very characteristic, mark the Roman era of the history of Greek script (30 BC - AD 324). The "severe style" or "Bacchylidean majuscule" is typically represented by Bacchylides's Papyrus B.M. 733: without ornamental strokes, it alternates wide and narrow letters, an opposition that would return more than once in the history of Greek script; it flourished in the period between the early 2nd and late 3rd c. One of the finest examples of the "Roman uncial" or "round majuscule" style is the Hawara Papyrus of the Iliad: the letters, in a rounded stylization, are sober and extremely regular; the golden age of this script coincided with that of the Antonines. The end of the Roman era saw the birth of scripts which would be established, fixed and perpetuated in "canonized" form in the Byzantine era. This era saw, with the triumph of the codex over the *rotulus* and of parchment over *papyrus, the establishment and spread of "Biblical majuscule".

Though born in a *pagan environment, this script reached perfection in the biblical codices of the full 4th c. (*Vaticanus* and *Sinaiticus*) and, with the victory of Christianity, spread everywhere. Its appearance is extremely calligraphic, there is a clear tendency to geometrical forms (nearly all the letters, even the round ones, can be inscribed in squares); the layout is poised, the contrast between light upstrokes and heavy downstrokes is balanced; ornamental strokes are absent, at least until the period of perfection. "Canonized", the Biblical majuscule style lasted from the 5th to the 8th c.: sharper contrasts and the presence of more and more artificial ornamental "peaks" marked its gradual decadence. Alongside the Biblical majuscule, another script became established particularly at *Alexandria: the "Alexandrian majuscule" (a name preferable to "Coptic uncial"). A more flowing script (the strokes are connected by hooks), the commonest type has the abovementioned contrast between wide letters and narrow letters (see the famous *codex Marchalianus* of the Prophets, *Vat. Gr.* 2125). In use until the 8th c., its characteristics became exaggerated with time. Biblical majuscule and Alexandrian majuscule are "vertical-axis" scripts. In its earliest type, whose origin may perhaps be sought in a form of Bacchylidean majuscule, the "ogival majuscule" is a script leaning to the right, showing the same contrast between wide and narrow letters, and whose curved strokes gradually tended, as time passed, to fracture. This script, which reached perfection in the *Freer codex* of the Gospels (5th c.), spread esp. from the 6th to the 9th c., gradually replacing the biblical and Alexandrian majuscules (a vertical type was also created) and then disappearing from use with the advent of the minuscule. If the majuscule kept its monopoly of book-script until the 8th c., ordinary script had for some time been evolving towards the minuscule, a script contained between four lines. From the 6th c. the script of documents was completely minuscule. Its evolution was quite distinct from that of book-majuscule, by now totally artificial, but whose canonized forms put up a stiff resistance. While the script of the first Christian *books, themselves of modest workmanship, was often close to that of common usage, the script of *codices*, esp. biblical ones, became more and more hieratic, the sign of a change in the function of the book. It is probable that from the 8th c., and maybe even from the 7th, there were attempts to use a form of cursive minuscule in unpretentious books (e.g. personal copies on papyrus), but evidence of such attempts is nearly all lost.

II. Latin script from the 1st to the 8th c. Until the fall of the Western Roman empire, the evolution of Latin script runs closely parallel to that of Greek. The first written documents on papyrus go back to the 1st c. BC; the first literary papyri to the start of the Christian era. The book-script of this time was a majuscule, the *capitalis rustica*: the use of a pen with a wide, flexible point created a sharp contrast between light upstrokes and heavy downstrokes, the layout was poised and accurate. The *capitalis cursiva*, used for documents, was quite different: here the hard, fine pen generated equal strokes, which tended to merge and, at times, to be elongated. The use of the *capitalis rustica* is documented up to the 6th c.; but in some MSS it was replaced by a more majestic and solemn script, modelled on the square majuscule of the monuments, the *capitalis elegans* (e.g. in the "Augustan" Virgil). Between the 2nd and 4th cc., technical factors (use of parchment, change of the pen's position *vis-à-vis* the page) and cultural factors (spread of the Christian book), which are still disputed, favoured evolution (from majuscule to minuscule) and innovations, which would lead in the 6th c. to the rise and establishment of new scripts: uncial and semi-uncial in books, cursive minuscule in documents. The uncial, whose original significance remains obscure, was characterized by the rounded stylization of the letters, some of which drew inspiration from the cursive; the most typical were A, D, E, M. Quite accurate in the 4th and 5th cc., the uncial then evolved into much less careful forms, surviving until the 9th c.; from the 4th to the beginning of the 7th c. it was the most widespread book-script. The origin of the semi-uncial is controversial; it was, in any case, a mixed script, in which majuscule forms mingled with many minuscules of cursive origin: particularly characteristic were a, b, d, g, m, r, s. In documents too, a new form of cursive script appeared in the 4th c.: this was the cursive minuscule, characterized by the decisive adoption of the four-line system and by a series of graphic forms which, whatever their origin, were freely adapted and transformed in the play of the many ligatures. More disciplined forms of minuscule and semi-minuscule cursive were also used in books, esp. in 7th- and 8th-c. *Italy. With the decadence and subsequent disintegration of the Western Roman empire, local cultures and particular scripts, called national scripts, arose and became established. From the 5th to the 8th c. the *monasteries and episcopal *schools, which replaced the workshops of professional copyists, adopted the majuscule scripts and developed, on the basis of the cursive minuscule, particular book-scripts. Thus in N and central Italy we distinguish various types of pre-Caroline minuscule: in S Italy, which suffered a long cultural eclipse, the Beneventan minuscule acquired its characteristic traits in the course of the 8th c.; in *Spain the Visigothic script, both book- and cursive, was distinguished by a great number of very particular forms (e.g. a and g). In *Gaul and *Germany under the *Franks, the Merovingian script, Chancery - and then book-script, had a brief but notable development from the 6th to the 8th c.; we can distinguish several types, associated with the centres of Luxeuil, Laon (*Laudunum*) and Corbie and those of Germany and Switzerland. But the most characteristic area was undoubtedly that of *Britain and *Ireland, which exercised a powerful influence even on the continent, thanks to the foundations of the Irish missionary monks, e.g. Bobbio. The scripts used in this area, called insular, were two: one, called insular majuscule or semi-uncial, was used in the more solemn and calligraphical manuscripts, like the *Lindisfarne Gospels*; the other, with narrow, pointed forms and characterized by the use of particular ligatures, is called insular minuscule. It was the latter that spread to the *monasteries of the continent, before the Carolingian minuscule, a script connected with the formation of the new Roman-German empire and with the cultural renaissance of Charlemagne, put an end to the age of graphical particularism. But by now the patristic era was over.

Retrospective bibliography: J. and M.D. Mateu Ibars, *Bibliografía palaeografica*, Barcelona 1974 (use with caution). Current bibliography: Scriptorium. Revue internationale des études relatives aux manuscrits, from 1946 (with annexed "Bulletin codicologique"); excellent retrospective for the Greek sector: J. Irigoin, Les manuscrits grecs 1931-1960, *Lustrum* 7 (1962) 5-93, 332-335; recent selected bibliographies: B. Bischoff, *Paläographie des römischen Altertums und des abendländischen Mittelalters*, Grundlagen der Germanistik 24, Berlin 1979, 299-323; J. Irigoin, Bibliographie, in A. Dain, *Les manuscrits*, Paris 1975, 191-205; D. Harlfinger, *Griechische Kodikologie und Textüberlieferung*, Darmstadt 1980, 657-678. History of Latin script, general expositions: G. Battelli, *Lezioni di paleografia*, Vatican City ³1949; B. Bischoff, cf. above; C. Cencetti, *Lineamenti di storia della scrittura latina*, Bologna 1954; id., *Compendio di paleografia latina*, Naples 1963; basic documentation for the patristic era: E.A. Lowe, *Codices latini antiquiores*, 1-11 and suppl., Oxford 1972; A. Bruckner-A. Marichal, *Chartae latinae antiquiores*, Olten-Lausanne 1954. History of Greek script, basic handbook: V. Gardthausen, *Griechische Paläographie*, I-II, Leipzig 1911-1913 (anast. repr. Leipzig 1978); for the period of the majuscule: E.G. Turner, *Greek Manuscripts of the Ancient World*, Oxford 1971 (collection of facsimiles with important introduction); G. Cavallo, *Ricerche sulla maiuscola biblica*, Florence 1967; AA. VV., *La paléographie grecque et byzantine*, Paris 21-25 octobre 1974 (C.N.R.S. Colloque international no. 559), Paris 1977; C.H. Roberts, The Codex, *PBA* 40 (1954), 169-204; id., *Manuscript, Society and Belief in Early Christian Egypt*, London 1979; J.W. Barns et al., *Nag Hammadi Codices: Greek and Coptic Papyri from the Cartonnage of the Covers*, NHS 16, Leiden 1981.

P. Canart

PALAEOSLAVONIC, Translations in. Methodius, one of the creators of the Slav alphabet and, at the same time, one of the promotors of P. literature, worked mainly as a translator from Greek, both with his brother Constantine the Philosopher (= Cyril) and on his own, particularly after Cyril's death (Feb 869). His biographer - prob. his disciple Clement of Ochrida - says in the *Vita Methodii* (ch. 15) that he translated, among other things, *oc'skija knigy*, i.e. "books of the Fathers" a statement whose interpretation has aroused real controversy among specialists. It seems hardly likely that he meant the *Paterikòn, which describes the life and deeds of the desert Fathers, since that sort of work would not have been indispensable to Slavs newly converted to Christianity. It seems much more probable that he meant works by the "Church *Fathers", i.e. patristic works in the strict sense, since this type of work was necessary for any people that had just embraced the Christian faith. The *conversion to Christianity of a new people required the translation into their language of the Holy *Scriptures and the basic liturgical books, as well as of patristic texts that could offer material for further education. The knowledge of such works was the best way to consolidate and deepen Christian doctrine in the newly-converted: they constituted a common literary wealth, no less precious than the *books used in the *liturgy, and all the more necessary since heretical doctrines that threatened the purity of the true faith soon began to be propagated among the new converts. Important evidence of this is found in the famous *Responsa Nicolai I Papae ad consulta Bulgarorum* of 866 (ch. 106): the *pope instructs the newly converted Bulgars in these words: *de his autem, quos in patriam vestram de diversis locis advenisse perhibetis varia et diversa docentes*. Other historical sources tell us that preachers of various heretical doctrines reached the newly converted country: *Paulicians, *Manichees, dualists, even Mohammedans. The elements of *dualism, thus awakened, later crystallized into the heresy of the Bogomils.

Alongside the few original works, P. literature, in particular the Palaeobulgar, consisted mainly of translations from the Greek, i.e. patristic works rather than works by Byzantine authors proper. Bulgaria, the first of the southern and eastern countries to be converted, became the centre of translation and diffusion of patristic texts among the *orthodox Slavs. This activity was later taken up and continued by the Serbs and Russians: during the Middle Ages nearly all the works of the major Fathers, and even some by less well-known authors, were translated. Translations of patristic works became an essential component of mediaeval literature. Yet it is hard to fix their chronology, particularly for the centuries after the conversion, when many centres of literary activity were formed in these Slav countries, though for some of them it is relatively easy to establish the precise chronology of translations, done in Bulgaria in the second half of the 9th and the 10th. c. Some of these translations bear a specific date: that of others can be determined from palaeographic, linguistic or historical evidence. The choice of works to translate was not casual, but was determined by the needs of spiritual life: works were chosen in order to answer doubts or queries which arose in the religious life.

The names of great Fathers like *Basil of Caesarea, *John Chrysostom and *Gregory Nazianzen became widely known through the translation of their liturgies: they were venerated as saints and their *images regularly painted in places of honour on altars. St *John Damascene, that illustrious representative of Greek *patristics, was also translated. In 893/4 his famous *De orthodoxa fide* was translated into Palaeobulgar, but only 48 of its 100 chapters - those particularly useful to new converts and perhaps more accessible in their content: on the earthly and heavenly *world, the human organism, certain natural phenomena, etc. His *Dialectics* were translated rather later. Despite the difficulties the translator had to overcome in rendering the complicated theological and philosophical terminology, the translation was done with laudable accuracy.

In the Early Middle Ages, dualistic tendencies among the S Slavs were partially identified with the doctrine of *Arius, and this chiefly because of its denial of the *Trinity. In the Late Middle Ages the doctrine of the Bogomils, Patarini and Cathars was called *ariana haeresis*. Certainly it was not "authentic" *Arianism, by now long vanished from history, that the Bulgars intended to combat when, at the request of prince Simeon (893-927), bishop Constantine of Preslav (then the Bulgarian capital) prepared to translate *Athanasius of Alexandria's famous sermons against the Arians; his translation was finished in 906. Overcoming numerous difficulties to render this text, particularly complicated on account of an abstract terminology, in a language still with no literary tradition of that kind, he gave his contemporaries an effective weapon with which to fight the dualistic *heresy that was crystallizing at the time.

*Gregory Nazianzen was among the authors best known to the Slavs from very early on. Constantine the Philosopher (=Cyril) knew his works in his youth and had a particular veneration for him, which lasted all his life. The *Vita Constantini-Cyrilli* (ch. 3) tells how after suffering a disappointment he, the younger son of a wealthy family, locked himself up at home and learned Gregory's works by heart. It was under this influence that Constantine composed his first literary work: a poem in seven strophes in praise and honour of his first "teacher". Doubtless Constantine read and learned "by heart" not just Gregory's major works, i.e. his *sermons and beautiful poetry, but also his correspondence: elsewhere in the *Vita* (ch. 5), a citation from one of Gregory's *letters is attributed to the young scholar. The first Slav authors who composed the *Vita Constantini-Cyrilli*, the *Vita Methodii* and the eulogies in their honour, demonstrate, among other things, that they had been influenced by the Cappadocian Father.

The Bulgarian writer John the Exarch (*c*.850-*c*.930) wrote his voluminous *Hexaemeron* shortly before 918 - a vast compilation based mainly on *Basil of Caesarea's *Hexaemeron*, *Severian of Gabala, *John Chrysostom and other less well-identified sources. A *Hexaemeron* was a vast *commentary on *Genesis ch. 1 with wide digressions on connected problems. In reply to the dualistic heretics who generally denied the importance of the OT as a sacred book for the Christian church, it stressed its fundamental value while insisting on the intrinsic bond between the two Testaments. It offered its Slav readers an authoritative explanation of the problems of *creation, according to the conceptual categories of *Moses. Its vast canvas contained information on the heavenly and earthly worlds, on flora and fauna, often in a truly realistic form with details derived from careful observation of nature. While using primarily patristic writings, John the Exarch also exploited other sources and personal observations, thus creating a real dogmatic and naturalistic encyclopaedia. Following Basil's basic lead and that of some other Fathers (e.g. *Theodoret of Cyrrhus, long passages of whose *Graecarum affectionum curatio* he cites), John made his compilation with a deep sense of divine Wisdom and of the admirable harmony of the creation of the universe and of man. In this too was concealed a heartfelt reply to the pessimistic ideas of the dualistic heretics, who claimed that the visible, material world and man were created by the *evil principle. He also, still following his sources, criticized the astrological beliefs of the Protobulgars.

In the late 9th or early 10th c., the so-called *Dialogi quattuor* (CPG 7482) of Ps.-*Caesarius, the pretended brother of Gregory Nazianzen (actually the work of an unknown mid 6th-c. *Acoemete author), were translated in Bulgaria. Here too we read, as well as very interesting ethnographical information, criticisms of *astrology. The anon. author has used *Bardesanes's (†222) *Book of the Laws of the Countries*, citing copious and very interesting extracts. The P. translation, more complete and more correct than the surviving Greek text, is a source to supplement and correct that text, preserved in relatively later MSS.

Among the panoply of writings used in the struggle against Arian tendencies was the *Visio S. Petri Alexandrini*, linked with the name of the Alexandrian bishop and martyr *Peter (†311), and translated at some early date. The popularity of this text of *hagiography was such that the scene of the *vision was reproduced on wall-paintings and icons: the bishop is shown before an *image of Christ with torn clothing, and the inscription gives the bishop's question: "Who tore it?", and the Lord's reply: "The impious Arius". Since dualistic tendencies long persisted, these depictions, inspired by hagiographical sources, continued until the Late Middle Ages.

In answer to the philosophical determinism of the dualists, who denied individual responsibility for actions, the *resurrection of the dead or retribution for *sins, *Methodius of Olympius's *De autexusio (De libero arbitrio)* was translated in the first decades of the 10th c. The Greek original being partly lost, the P. translation - edited in 1930 by A. Vaillant (PO 22, 5, 631-889) with attempted Greek retroversion and French translation - has considerable value for patristic studies. In the same period the *Topographia Christiana* of *Cosmas Indicopleustes, a merchant and then monk at *Alexandria, was translated into Slav. Besides refuting Ptolemy's cosmographical system, it offered Slav readers an extremely rich mine of geographical, ethnographical and naturalistic information on *Ethiopia, India, Ceylon, etc., with information on flora and fauna and with the author's own painted illustrations, faithfully reproduced in the copies of the Slav translation. The work was a precise reply to ideas, widespread in mid 10th-c. Greater Moravia, about the sphericity of the earth and the existence of the antipodes, ideas bitterly fought by Constantine the Philosopher during his stay in the region and also by the official church.

Numerous other patristic works, prose and poetry, were subsequently translated: we cannot give a detailed list, only the most important. We know that when the *monastery of Rila, the greatest monastic centre in S-W Bulgaria, was founded in the third decade of the 10th c., *Ephrem the Syrian's basic work, the *Paraineseis (Admonitiones)*, was translated here for the edification of the monks. We find fragments of this work and allusions to it in the *Spiritual Testament* of St John of Rila, the monastery's founder, written *c*.940; other fragments remain in a late 10th-c. Glagolithic MS, and a complete copy of the P. translation in a 14th-c. Cyrillic MS. *Hippolytus's *Commentary on Daniel*, whose Greek text is incomplete, also survives entire in Slav translation. Thanks to the translations, other patristic works of edification, for both monks and *laymen, were accessible to Slav readers in their original version: e.g. *John Moschus's *Pratum spirituale* (6th c.) and *John Climacus's famous and very popular *Scala Paradisi* (6th/7th c.), which was copied, read and frequently cited all through the mediaeval period and even later. A fragment of *Palladius of Helenopolis's *Dialogue on the life of St John Chrysostom*, modelled on Plato's *Phaedo* and containing information on the "peoples of India and the Brahmins", attracted particular interest and was circulated as an independent narrative. *Isidore of Pelusium (†*c*.435), some of whose *letters were translated and circulated in the *orthodox Slav world, enjoyed great fame. *Cyril of Jerusalem's (387) *Catecheses* were translated into Slav in the early mediaeval period; the second of them has been magisterially analysed and edited by the late Slavicist A. Vaillant (1932). We also know some works of *Epiphanius of Salamis (†403) in mediaeval Slav translation.

Translators used to take works by famous Fathers, e.g. *Gregory Nazianzen or *John Chrysostom, and make them into voluminous collections (*sborniki*), which usually circulated widely. One of these collections, that of John Chrysostom, is known by the eloquent title *Zlatostruj (Chrysorrhemon)*; composed early in the 10th c. at the suggestion of king Simeon, it had particular renown. *Hesychius of Jerusalem's *Commentary on the psalms* (5th c.) was translated in the first decades of the 10th c. and placed alongside the text of the *psalms themselves: the two texts appear thus in the famous Bologna University Library codex known as the *Psalterium Bononiense* (first half of 13th c.). In reply to certain Renaissance naturalistic interpretations of the *Song of Songs, *Theodoret of Cyrrhus's *Interpretatio* of it (5th c.) was translated, prob. before the end of the 14th c.: it is cited, sometimes tacitly, by Slav authors of that time. The enigmatic author Ps.-*Dionysius the Areopagite was known to the Slavs in the 9th/10th c. and cited by them more than once, but a complete translation of his works seems to have been made only in the second half of the 14th c. (preserved in, among others, a copy of 1371).

To conclude, we may say that the translation of so many works by the Fathers constituted the most solid possible basis for the birth and development of orthodox Slav literature. The Slav translations are of considerable interest for our knowledge of Greek patristic literature and must be preserved with gratitude and care.

CPG 2-3 (useful but incomplete); A.S. Archangelskij, *Tvorenija otcov cerkvi v drevnerusskoj pis'mennosti. Izvlecenija iz rukopisej i opyty istoriko literaturnykh izocenj*, I-IV, Kazan 1889-1890 (a fundamental work, still not superseded); G.H. Bonwetsch, *Hippolyt's Kommentar zum Buche Daniel und die Fragmente des Kommentars zum Hohenliede*, Leipzig 1897 (cf. Dujcev: *Medioevo bizantinoslavo*, II, Rome 1968, 10, n. 2); bishop Mikhail, Agrafy v tvorenijach sv. Metodija, *Zhurnal Moskovskoj patriarchii* 6 (1954) 43-50; M. Heppell, Slavonic Translations of Early Byzantine Ascetical Literature. A Bibliographical Note, *JEH* 5 (1954) 86-100; V.V. Danilov, Pis'ma Isidora Pelusiota v "Izbornike Svjatoslava 1073 goda", *Trudy Otdela drevnerusskoj literatury* 11 (1955) 335-341; M. Heppell, Some Slavonic Manuscripts of the "Scala Paradisi" (Lestvica), *ByzS* 18 (1957) 233-270; A. Vaillant, Le saint Ephrem slave, *ibid.* 19 (1958) 279-286; A. de Santos Otero, *Los Evangelios apócrifos*, Madrid ²1963; D. Bogdanovic, *Jovan Lestvicnik u vizantijskoj i staroj srbskoj knjizevnosti*, Beograd 1968 (cf. Dujcev: *ByzZ* 66 [1973] 102-103); I. Dujcev, *Medioevo bizantino-slavo*, II, Rome 1968, 3-27, 589-594 (for the Slav version of Ps.-Caesarius's *Dialogues*: 191-205, 604-605); R. Riedinger, *Pseudo-Kaisarios. Überlieferungsgeschichte und Verfasserfrage*, Munich 1969 (cf. Dujcev: *ByzZ* 67 (1974) 403-406); G.M. Prochorov, Korpus socinenij s imenem Dionisija Areopagita, *Trudy Otdela drevnerusskoj literatury* 31 (1976) 351-361; A. de Santos Otero, *Die handschriftliche Überlieferung der altslavischen Apokryphen*, I-II, Berlin-New York 1978-1981; I. Dujcev, La letteratura greca patristica nella Bulgaria medievale, Παρνασσός 25 (1983) 173-178.

I. Dujcev

PALESTINE

I. Palestine I - II. Palestine II - III. Palestine III.

Pompey conquered *Syria in 64/63 BC, making it a Roman province with ill-defined borders: *Cappadocia, *Egypt, the Arabian desert and the Euphrates. In general, the Romans left client kings or set up independent cities in the various territories. After the fall of Jerusalem in AD 70, Judaea became a separate imperial province under a *legatus*, and after the war of 130/131 it took the name Syria-Palestine, which from 295 was extended to some Arabian territories. From 358, the province was subdivided into Palaestina and Palaestina Salutaris. In about 400, Palestine was subdivided into three parts: two comprised Palestine proper (Palaestina I and Palaestina II), while Palaestina III or Salutaris also included some territories E of the Dead Sea. To N of this extended the province of *Arabia, bordering on *Syria. The capital of P. I was *Jerusalem, that of P. II Scythopolis, and that of P. III *Petra. (In identifying dioceses I follow F.M. Abel, *Géographie de la Palestine*, II, Paris 1938, 200-202; B. Bagatti, *The Church from the Gentiles in Palestine*, Jerusalem 1971, 92-95.)

I. Palestine I. Along the coast, the most northerly diocese was Dor. Occasional excavations have revealed the church *mosaic, whose inscription records bishop Acacius (*Enc. Arc. Excavations*, 334-337). A stone bears an authentication of a fragment of rock from Calvary exposed to the veneration of the faithful (Bagatti, *Villaggi di Galilea*, 111).

Ascalon and Maiuma of Ascalon. Occasional excavations at Barnea, N of the present (mediaeval) wall of Ascalon, now a public park, have shown that this area was largely inhabited in the Roman-Byzantine period. Two churches have been found: one with three aisles; in the floor of the other, which seems an annexe of other sacred buildings, is a citation of Ps 23 (22), 1, and the name of bishop Anastasius with the date 493. This suggests that in the Christian period Ascalon was here, and that modern Ascalon was actually Maiuma, since it was built on the port (Bagatti, Ascalon, *SBF* 24 [1974] 227-264). The bishop of Ascalonian Maiuma is recorded only in the 6th c. In the park are capitals from churches and other sacred objects, but there are no sure traces of an early Christian church in this area (*Enc. Arc. Excavations*, 125-130).

*Gaza and Maiuma of Gaza. Though the city had many churches ornamented with mosaics (Abel, Gaza, *RBi* 40 [1931] 5-31), no sure trace of them remains. But we know many, mainly 6th-c., sepulchral inscriptions recording the *faithful, e.g. two deacons Alexander and Patricius, and the characteristic phrase wishing the deceased "rest in Christ among the saints". Some *crosses surmount Calvary, represented by three bows. A Latin inscription mentions *Juvenal, prob. the first patriarch of Jerusalem (Clermont-Ganneau, *Archeological Researches*, 397-429). Of the many *monasteries that arose in the region, there is no sure trace, or none of any importance; but establishments not recorded in literary texts have been found, some with mosaic floors. Some are of artistic interest: the mosaics of the church of Umm Gerar, with birds; of Kissifin, with human figures: a horseman and portraits, among the finest in Palestine; of St Cyriac's church at Magen, with birds, grapes and portraits; of the church of Shellal, with quadrupeds, birds, fish and fruit, executed in 561-562. At Maiuma was found a synagogue mosaic with *David playing the harp and various animals (Bagatti, *Villaggi di Giudea e Neghev*, 150-189).

Eleutheropolis-Beit Gebrin. Two *church buildings are preserved from the late Roman and early Christian period: the church of Muhatt el-Urdi, with mosaics with animals and scenes of *Jonah (now covered up) and one with the seasons and hunting scenes with a chapel dedicated to Christ, universal king, originally on the hill of el-Muqarqah, now at Jerusalem (Bagatti, Eleuteropoli, *SBF* 22 [1972] 109-129). Further S are the remains of a great church called Sanhanna, studied in the 19th c. but deteriorating rapidly (Ovadiah, *Corpus*, 135-137). Near it were found Christian tombs decorated with painted birds and crosses.

*Diospolis-Lydda. With the Arab foundation of Ramle, Lydda was almost abandoned. Nothing remains of buildings of the Roman period and the old church of St *George, except perhaps some remains incorporated in the later crusaders' church. A half-obliterated inscription on a reused column in the mosque which has replaced the sanctuary of St George says that "the most worthy pastors" took care to "ornament this most renowned temple" which might, perhaps, be made better known by excavation (Bagatti, *Villaggi di Samaria*, 160-169). Being a *cemetery church, it was prob. at the edge of the city.

Nicopolis-Emmaus, previously Amwas. After the complete destruction of the village, nothing remains but three apses built with great unmortared blocks in the Roman technique and a group of Christian buildings, all in ruins, with a three-aisled church, a baptistery with quadrilobate pool and mosaic scenes with wild animals biting domestic ones. The 2nd-c. date proposed in the past for the mosaic has been shown to be unacceptable by a recent excavation which revealed the existence of another white mosaic beneath it (Bagatti, *Villaggi di Samaria*, 148-152). So the chronology must be changed, and the Christian character of the apses is in doubt.

Jericho. Remains of various churches: on the Kelt, that of St Andrew with 6th-c. inscriptions; in the modern built-up area, that of St George, a small oratory built by the priest George (†566); the church at Tell Hassan, thought to be the cathedral; and, to N towards the fountain, that belonging to the monk Anthimos, whose floors have birds, quadrupeds, fishes and the inscription of three deacons: Daniel, Macarius and John. The church built on the spring, partly visible in earlier centuries, is no longer so (Augustinovic, *Gerico*, 65-92).

Flavia Neapolis, now Nablus, preserves no remains of ancient churches, but prob. has the *proseuché* built as a forum for the company of *Messalians, with inscriptions of their members on the stone sedilia (Bagatti, *Villaggi di Samaria*, 52). Further S, at the Well of the Samaritan Woman, remains of mosaics allow us to reconstruct the cruciform church, built in the 4th c. when the well was already a baptistery, and a *sarcophagus with the shield of the Amazons and a cross reveals Christian worship on the spot before the erection of the church. Material taken from the well during cleaning shows that it was in use from the Iron Age, centuries before Christ.

Sebaste had the temple of *Augustus, partly destroyed by excavators searching for earlier levels, the prototype of Christian churches with apse flanked by two *pastophoria* between a rectilinear wall. The access stair was also imitated, e.g. at Et-Taiybe (Bagatti, *Villaggi di Samaria*, 31-36). In the E side of the hill is the traditional tomb of *John the Baptist, later enclosed in the 4th-c. church (partly found) and then in that of the crusaders. S of the city, near the fountain which provided water, almost at the start of the conduit, is an underground chapel dedicated to *Aaron, which has Christian graffiti (Ovadiah, *Corpus*, 157-159).

Cf. also under *Jerusalem and *Caesarea. **[Fig: 244]**

II. Palestine II. At *Diocaesarea*, now *Ziporis*, a rock-cut church near the brow of the hill, where the city was built, was prob. transformed into a church by bishop Marcellinus in 518 (Ovadiah, *Corpus*, 182 f.). Lower down and to N, a Hebrew inscription seems to be from a synagogue, where the crusaders built the church of St Anne.

Capernaum. The excavation of the traditional "house of Peter" has revealed, under the floor of the octagonal church built in the 5th c. by gentile Christians, a primitive stairway supported by a great arch, where *pilgrims scratched names and invocations in various languages: Hebrew, Aramaic, Greek and Latin, thus attesting the worship practised there. Lower down are small houses joined together near a courtyard going back to the Hellenistic period. The synagogue built with white conchs and ornamented with friezes and symbols is no earlier than the 4th c.

At *Et-Tabga*, the traditional site where Jesus used to talk to the crowd and work miracles, are preserved remains of 4th- and 5th-c. churches; one records the multiplication of loaves and fishes: in its mosaic floor are Nile motifs.

Scythopolis-Beth-Sh'an was the metropolis. On the brow of the hill the cathedral, an apsed circular plan with atrium and *cells around it, was destroyed by excavators in order to explore earlier buildings (Ovadiah, *Corpus*, 38-40). Not all the details of construction are certain (e.g., we do not know if the conduits served the baptistery found there). The church of St Procopius had geometrical mosaics. There are mosaics with human figures, zodiac, months and birds in the monastery of the Lady Mary, on the rocky side N of the *wadi*, where was also a funerary chapel with mosaic representations of the months, now in the Rockefeller Museum at Jerusalem (*Enc. Arc. Excavations*, 221-228), and a mosaic with an inscription recording the lepers' bath-house, not yet found.

Tiberias. During the building of the aqueduct that carries away the salt waters from the region of Et-Tabga, a complex was found consisting of a small church lower down and a 6th-c. church higher up. The former is architecturally singular, perhaps an adaptation of a *pagan room, supposed to be the work of Count Joseph, a *Jew converted at the time of *Constantine (Ovadiah, *Corpus*, 180). Of the *Judaeo-Christian bishop's residence recorded by St *Epiphanius, nothing is yet known. Christian fragments include a piece of architrave with crosses, reused in the small mosque, now a museum, and a Greek inscription, now lost, variously transcribed (Bagatti, *Villaggi di Galilea*, 48-61).

Beth-Yerak. Further S of Tiberias, on the same shore of the lake, is the ruin of Beth-Yerak, comprising a synagogue and a church with baptistery which has had several phases. An inscription mentions the presbyters Elias and Basil and the year 528/9 (Ovadiah, *Corpus*, 40 f.; *Enc. Arc. Excavations*, 260-262).

Pella, refuge of the Christians in the war of AD 70, has, to W, a church with atrium resting on older remains where was found a *sarcophagus ornamented with bunches of grapes and a *cross. Other Christian elements have long been known and other Christian *church buildings are still being excavated.

Helenopolis: tentatively identified with the village of Daburieh, at the foot of Tabor on the W slope, whose bishopric was then transferred to the mountain. No ancient Christian remains have been found here, in contrast to Kefer Kama, a Circassian village at the foot of the mountain but on the E side. Here, in two adjacent and communicating chapels, with superimposed levels and mosaic inscriptions, are recorded a deacon, a presbyter and a bishop called Eustathius. Birds, fish and geometrical designs are represented in the pavement mosaic. There is even a reliquary. Thanks to this wealth of finds, we may believe that the episcopal see was in Kefer Kama (Bagatti, *Villaggi di Galilea*, 274-280).

Hippos-Sussita is situated on the E shore of the lake, on a hill whose name derives from its position. The remains of four churches have been found, none yet completely excavated. One has a baptistery on the N side and floor in *opus sectile*; from the inscriptions we know that it was dedicated to Sts *Cosmas and Damian. The baptismal font is in the centre of the apse. A pagan inscription, found *in situ*, suggests that the life of the Christian city was continuous with the pagan settlement (Ovadiah, *Corpus*, 174-178).

Gadara, now Umm Qeis, once a diocese, preserves the Roman town-plan with classical buildings and a fine colonnade. Excavations have revealed a *bath of the Byzantine period with mosaic floors with geometrical designs and Greek inscriptions. At Es-Salt, part of the diocesan territory, has been found a *monastery with a church in which are preserved a mosaic floor with Nile scenes and a figure of a boy, ruined during the 8th-c. iconoclastic movement.

Capitolias, now Beit Ras, has Christian remains (Saller-Bagatti, *Nebo*, 222), incl. ruins of a church at Weli El-Khader. A building with central plan, apse and *opus sectile* floor, it was excavated in 1976-78 near the little western theatre. The city has fine classical monuments.

Abila, now El-Qoweilbe, has remains of two churches and a *cemetery (Saller-Bagatti, *Nebo*, 221; Piccirillo, *Chiese e mosaici della Giordania settentrionale*, 5-34). [Figs: 245, 246]

III. Palestine III. The main city of the Negev was *Elusa*, or *Khalasa*, but it was pillaged in antiquity for building materials. At present all its monuments are buried under the sand, but recent excavations still in progress have revealed much of a large basilica, originally with one apse, then with three, with a singular "throne" in the central apse, formed of a stair of seven steps, not next to the wall, quite hard to climb. It was covered in marble, as was the presbytery. Other monuments have been identified, but not yet excavated (*Enc. Arc. Excavations*, 359 f.). Happier finds in other towns of the Negev have revealed important ruins of churches, two or more in each centre, with many inscriptions in *mosaic or stone. We should mention Eboda or Audat, Nitzana, previously *Haugia el Hafi*, where liturgical and secular manuscripts were also found, incl. a little Greek-Virgilian dictionary, Shivta or Sbaita and Mefis or Kornub. The churches began to be built in the 5th/6th c.: common to all of them are niches in the side apses, intended for relics.

Charachmoba, now Kerak, preserves no plan of its Byzantine church, but a good 200 Greek inscriptions, many sepulchral, often reused for paving. The dated ones are from between 375 and 661. In the surroundings, various villages still preserve Christian remains, including some hermitages, e.g. El-Habis or another on the Wadi ed-Defali with remains of religious *paintings (Saller-Bagatti, *Nebo*, 229 f.).

Areopolis, now Rabba, has remains of a church, inscriptions with the name of bishop John (597-8) and a metropolitan named Stephen, and in a niche an invocation to the *Theotokos (R. Canova, *Iscrizioni*, Rome 1954, 198-208).

Zoara, now Ghor es-Safi, recorded in the map of Madaba with *Lot's church (Saller-Bagatti, *Nebo*, 230), has preserved capitals with crosses, inscriptions and ceramics. Similar materials are found at Arindella, now Gharandal.

Feinan, a place well-known to those condemned to forced labour during the later *persecutions, has various remains of churches, inscriptions (one with a dedication of the time of bishop Theodore, 587-8) and sepulchral titles.

Aila, now Aqaba, preserves Christian remains, incl. two capitals: one with saints Longinus and Theodore and two *angels holding the globe; the other with saints George and Isidore, all named in the inscriptions (Saller-Bagatti, *Nebo*, 233).

Pharan is on *Sinai on Djebel Meharret. Capitals with crosses and a man praying, perhaps in memory of *Moses, have been seen in the Orchard of the monks of St Catherine, where a new chapel has been built. The well-known monastery of St Catherine has recently been written about (cf. bibl.).

Cf. also under *Petra. [Fig: 247]

Encyclopedia of Archaeological Excavations in Holy Land, Jerusalem 1970; C. Clermont-Ganneau, *Archaeological Researches in Palestine*, II, London 1896; F.M. Abel, Gaza au VIe siècle d'après le rhéteur Chorikios, *RBi* 40 (1931) 5-31; id., *Géographie de la Palestine*, II, Paris 1938; S. Saller-B. Bagatti, *The town of Nebo*, Jerusalem 1949; A. Augustinovic, *Gerico e dintorni*, Jerusalem 1951; G.H. Forsyth-K. Weitzmann, *The Monastery of S. Catharina*, Ann Arbor 1965; A. Ovadiah, *Corpus of the Byzantine Churches in the Holy Land*, Berlin 1970; B. Bagatti, *Antichi villaggi cristiani di Galilea*, Jerusalem 1971; id., *The Church from the Gentiles in Palestine*, Jerusalem 1971; id., Il Cristianesimo ad Eleuteropoli (Beit Gebrin), *SBF* 22 (1972) 109-129; id., Ascalon e Maiuma di Ascalon nel VI secolo, *SBF* 24 (1974) 227-264; id., *Antichi villaggi cristiani di Samaria*, Jerusalem 1979; id., *Antichi villaggi di Giudea e Neghev*, Jerusalem 1983; V. Corbo-S. Loffreda-E. Testa-A. Spijkerman, *Cafarnao*, 4 vols., Jerusalem 1972-1975; R.H. Smith, *Pella of Decapolis*, Vooster 1973; K. Weitzmann, *The Icons*, Princeton 1976; D. Baldi-B. Bagatti, *Saint Jean-Baptiste dans les souvenirs de sa patrie*, Jerusalem 1980; M. Piccirillo, *Chiese e mosaici della Giordania settentrionale*, Jerusalem 1981; G. Stemberger, *Juden und Christen im Heiligen Land: Palästina unter Konstantin und Theodosius*, Munich 1987; P.W.L. Walker, *Holy City, Holy Places? Christian attitudes to Jerusalem and the Holy Land in the Fourth Century*, Oxford 1990; Y. Tsafrir, Christian Archaeology in Israel in Recent Years, *Actes XIe Cong. internat d'arch. chrét*, CEFR 123, Rome 1989, 1737-1770; P. Figueras, Découvertes récentes d'épigraphie chrétienne en Israël, *ibid.*, 1771-1785; M. Fischer, An Early Byzantine Settlement at Kh. Zikrin (Israel), *ibid.*, 1787-1807; G. Foerster, Decorated marble chancel screens in sixth century synagogues in Palestine and their relation to Christian art and architecture, *ibid.*, 1808-1820.

B. Bagatti

PALLADIUS. The name of this evanescent personality is linked to the start of the *evangelization of *Ireland. Acc. to *Prosper of Aquitaine (*Chron.*, s.a. 404), in 431 pope *Celestine sent P. to the Irish (*Scotti*) as their first bishop. On the news of his death (c.432), *Patrick left from *Gaul to replace him.

DCB 4, 176-177.

M. Simonetti

PALLADIUS of Helenopolis. Born in *Galatia c.363/4; seems to have received a good classical education. Became a monk c.386, lived for some time in *Palestine, then went to *Egypt, where he was introduced to the life of the Egyptian *ascetics. In 390 he went to Nitria, thence to the desert of Cellae, where he lived nine years and met *Evagrius of Pontus, who had a profound influence on him (cf. *Hist. Laus.* 38). In c.400 he became bishop of Helenopolis in *Bithynia; in 403, at the time of the synod of the *Oak, he was involved in the *Origenist controversy.

When *John Chrysostom was sent into exile, P. pleaded his cause at *Rome before pope *Innocent I. Returning to *Constantinople in 406, he was arrested and exiled to Egypt. When he was allowed to return, he went home to Galatia and, acc. to *Socrates (*HE* VII, 36), became bishop of Aspuna. He must have died before 431, when Eusebius appears as bishop of Aspuna.

During his exile, c.408, P. composed the *Dialogus de vita S. Iohannis Chrysostomi*, an important biographical source esp. for the last years of Chrysostom, whose *friend and great admirer he was. This work, formally modelled on Plato's *Phaedo*, takes the form of a dialogue at Rome between an Eastern bishop and the Roman deacon Theodore. In it, P. defends Chrysostom from accusations made against him by *Theophilus of Alexandria in a defamatory *libellus* (cf. Facund., *Defens.* VI, 5), now lost.

Around 419-20, P. wrote his *Historia Lausiaca*, dedicated to Lausus, *Theodosius II's chamberlain. It consists of a collection of profiles of various ascetics, men and *women, mainly Egyptian and, to a lesser extent, Palestinian.

P. used personal memories of ascetics he knew directly as well as accounts learned from others and containing many legendary elements. Writing essentially a work of edification, he wished to stress the spiritual value of the life of the *desert, which he knew well; he drew this world with vivacity and observed its merits and its defects, e.g., above all, pride. Textually, there are three recensions of the *Historia Lausiaca*: a short one (G), which seems to be the original, a long one (B) composed, acc. to E. Honigmann, by *Heraclidas of Nyssa, and another (A), which appears to be a contamination of the other two and is fused with the *Historia monachorum in Aegypto*. The *Lausiac History*, which remains an important source for the history of the beginnings of *monasticism, had great popularity; it was soon translated into Latin and many Eastern languages.

The treatise *De gentibus Indiae et Bragmanibus* is preserved under P.'s name, but only the first part, containing some stories of India, seems to be his; the second part, attrib. by the author himself to Arrian, is an amalgam of diverse tendencies and influences.

CPG 3, 6036-6038; *Historia Lausiaca*: PG 34, 995-1260; C. Butler, TSt VI, 2, Cambridge 1904; G.J.M. Bartelink, Milan 1974. *Dialogus de vita S. Iohannis Chrysostomi*: PG 47, 5-82; P.R. Coleman-Norton, Cambridge ²1958. *De gentibus Indiae et Bragmanibus*: J.D.M. Derrett: *CM* 21 (1960) 100-135; W. Berghoff, Meisenheim am Glan 1967. The numerous modern-language translations of the *Historia Lausiaca* include that into French by A. Lucot, Paris 1912; German by St. Krottenthaler, *BKV*² 5, Kempten-Munich 1912, 315-440; English by R.T. Meyer, ACW 34, London 1965; Italian by G. Gottardi, Sienna 1961, and by M. Barchiesi, Milan 1974, with introduction by C. Mohrmann and text and commentary by G.J.M. Bartelink. Translations of the *Dialogus* include that in English by H. Moore, New York-London 1921; Spanish by D. Ruiz Bueno, *BAC* 169, Madrid 1958, 125-296; German by L. Schläpfer, Düsseldorf 1966. Full bibliography on P. in Quasten II, 178-182. The many studies include: C. Butler, *The Lausiac History of Palladius*, TSt VI, 1, Cambridge 1898; P. Ubaldi, Appunti sul "Dialogo storico" di Palladio, *Mem. Acc. Sc. Tor.* 56 (1906) 217-296; C. Butler, Palladiana, *JThS* 22 (1921) 21-35, 138-155, 222-238; F. Halkin, L'Histoire Lausiaque et les Vies grecques de S. Pachôme, *AB* 48 (1930) 257-301; E. Schwartz, Palladiana, *ZNTW* 36 (1937) 161-204; S. Linnér, *Syntaktische und lexikalische Studien zur Historia Lausiaca des Palladios*, Uppsala-Leipzig 1943; R. Draguet, L'Histoire Lausiaque. Une oeuvre écrite dans l'esprit d'Évagre, *RHE* 41 (1946) 321-364; 42 (1947) 5-49; J. Dumortier, La valeur historique du Dialogue de Palladius et la chronologie de S. Jean Chrysostome, *MSR* 8 (1951) 51-56; E. Honigmann, *Heraclidas of Nyssa (about 440 A.D.)*, ST 173, Vatican City 1953, 104-122; J.D.M. Derrett, The History of "Palladius on the Races of India and the Brahmans", *CM* 21 (1960) 64-99; B. Berg, Dandamis: an early christian portrait of Indian ascetism, *CM* 31 (1970) 269-305; F. van Ommeslaeghe, Que vaut le témoignage de Pallade sur le procès de saint Jean Chrysostome?, *AB* 95 (1977) 389-413; E.D. Hunt, Palladius of Helenopolis: a Party and its Supporters in the Church of the Late Fourth Century, *JThS* n.s. 24 (1973) 456-480.

S. Zincone

PALLADIUS of Ratiaria. Leader of the Arians of Illyricum, in 380 he asked *Gratian to call a *council that included the Easterners. But the council of *Aquileia (381), restricted to Italian bishops inspired by *Ambrose, condemned and deposed him for *Arianism together with *Secundianus of Singidunum. In 383 he went with *Ulfila to *Constantinople to ask *Theodosius to abrogate the Aquileian sentence, but without success. The *Dissertatio Maximini contra Ambrosium* has a long polemical passage by him on the council and on Ambrose. Attribution to him of the *Sermo Arrianorum* and *Fragmenta Arriana* has no basis.

M. Meslin, *Les Ariens d'Occident*, Paris 1967, 85-92, 111-134; *Patrologia* III, 89. 97-98.

M. Simonetti

PALLADIUS of Saintes. Bishop of Saintes (*Santonae*), first half of 6th c. Of wealthy family (Greg. Tours, *Gl. conf.* 60), he built basilicas dedicated to St *Martin of Tours, St *Eutropius (first bishop of Saintes), St *Stephen, Sts *Peter, *Paul, *Lawrence and *Pancras. *Gregory the Great sent him relics for the latter (*Ep.* VI, 48). He took part in the 4th council of Paris (573). In 584, by order of metropolitan Bertcrann of Bordeaux, he consecrated Faustianus, favourite of the usurper Gundovald, as bishop of Dax (Greg. Tours, *Hist. Franc.* VII, 31; VIII, 2). He took part in the 2nd council of Mâcon, called by king Guntram in 585; it deposed Faustianus and ordered P., Bertcrann and another prelate, Orestes, to pay Faustianus an annual pension (*ibid.* VIII, 20). A piece of land was later extorted from him for having compromised himself by giving hospitality to messengers of Fredegund, Guntram's enemy (*ibid.* VIII, 43). Feast remembered locally 7 Oct.

BS 10, 59-60; DCB 4, 177.

E. Malaspina

PALLADIUS of Suedri. Of this region of Pamphilia, which after *Constantine had its own governor once more, P. was a *politeuómenos*, i.e. a citizen (*Bürger in*, acc. to K. Holl) or a magistrate (*curialis* or senator, acc. to Petau), certainly an exemplary citizen, having sold all to help the *poor and the church, "burning with zeal and holy living, admirable for his *Catholic faith and perfect *communion", acc. to *Epiphanius of Salamis. He wrote that bishop of *Cyprus a *letter, invoking him as prince of his soul and of the church, to ask him to intervene in the pneumatological questions which agitated the community of Suedri, defining the faith through a detailed exposition of the trinitarian *creed according to the *orthodox belief. Pamphilia had been in communion with Cyprus from time immemorial: even its dialect was akin to the Cypriot. To the Christian magistrate and others who had requested him, Epiphanius replied with the *Ancoratus*, in July of the 90th year (374) of the era of *Diocletian.

CPG 3744; PG 43, 13-17; GCS 25, 3-5; C. Riggi, *Epifanio. L'ancora della fede*, Rome 1977, 35-37.

C. Riggi

PALMAS. Late 2nd-c. bishop of *Amastris, in *Pontus. *Eusebius of Caesarea informs us that in the time of pope *Victor (189-199), as well as the Roman, Palestinian and Gallic *councils, the bishops of Pontus too called a council, whose president was P. "as the oldest" (*HE* V, 23, 3). The *letters of these local councils were known in Eusebius's time, and all celebrated the feast of *Easter according to the Roman tradition, i.e. the *Sunday after 14 Nisan. Eusebius also refers to P.'s *martyrdom: *Dionysius of Corinth wrote a letter to the believers of Amastris in which he mentions P., their martyr bishop (*HE* IV, 23, 6).

G. Ladocsi

PALMYRA

I. History - II. Archaeology.

I. History. Modern Tadmor (a small village inhabited by settled Bedouin) in Syria, P. was an extremely ancient caravan city, whose notoriety is recorded in the Bible (*Tadmur*: 2 Chr 8, 4). A large oasis in the Syro-Arabian desert, half-way between the Mediterranean and the Euphrates (300 km from that river), it was traversed by the great commercial trade-routes from the *East and favoured by abundant springs. Involved in the long Roman-Parthian wars (in 41 BC it risked sack by Anthony, but the inhabitants escaped, fleeing with their belongings beyond the Euphrates), it owed its extraordinary fortune to its character as a neutral semi-independent city, in which goods could be exchanged between the two great hostile powers, Rome and the Parthian empire. It entered the Roman sphere of influence after *Trajan's victory over the Parthians, when Mesopotamia became a Roman province. *Septimius Severus tried to make it a base of operations against the East, but the rise of the *Sassanid dynasty favoured the city's autonomy: after the disastrous Roman defeat by Shapur in 258, the Palmyrene Odenathus, of the powerful family of the Iulii Aurelii Septimii, defeated the Persian king in battle and received from the emp. *Gallienus the honorific title of *imperator (restitutor, corrector) totius Orientis*. Odenathus then extended his authority over *Syria and much of the East; he died in a plot in 266-7 and was succeeded by his son Vaballatus, under the tutelage of his mother *Zenobia, who continued the policy of conquest, to a point where she refused to recognize Roman sovereignty: P.'s great adventure ended in 272 with its defeat by *Aurelian, who besieged the city and captured Zenobia. Christianity spread when the city was already in full decline: a bishop, Marinus, was present at *Nicaea.

J.G. Février, *Essai sur l'histoire politique et économique de Palmyre*, Paris 1931; M. Rostovtsev, *Caravan Cities*, Eng. tr. Oxford 1932.

M. Forlin Patrucco

II. *Archaeology. Within the great circuit of the walls built on the Roman plan, W of the great temple, are preserved remains of two three-aisled churches, one of which has an apse with radiating buttresses, as in the church of the Ascension on the Mount of Olives. Two more churches are outside the city wall (Lassus, *Sanctuaires*, 168-231). More recent excavations have revealed *pagan monuments, except for a portrait of Christ. In the Polish dig of 1960, jewels and pendant ear-rings, one in the form of a *cross, were found in a small jar; in the dig of 1962, a seal with a Maltese cross and Greek inscription. Various sepulchral inscriptions mention Christians. For Jewish and Christian influence, see L. D. Merino's study.

J. Lassus, *Sanctuaires chrétiens de Syrie*, Paris 1947; K. Michalowski, *Palmyre. Fouilles polonaises 1960*, Paris 1962; L.D. Merino, Influencias judía y cristiana en los signos e inscripciones palmirenas, *SBF* 21 (1971) 76-148.

B. Bagatti

PALUT. Legendary disciple of *Addai, with whom he evangelized the city of *Edessa. In fact, he must have lived during the late 2nd and early 3rd c., and was ordained bishop at *Antioch by Serapion. The episode shows the dependence of the Edessene community on that of Antioch. The *orthodox Christians of Edessa, after P.'s death, were called Palutiani.

J.B. Segal, *Edessa "The Blessed City"*, Oxford 1970, 79-81; E. Peretto, Il problema degli inizi del cristianesimo in Siria, *Augustinianum* 19 (1979) 197-214.

A. Di Berardino

PAMMACHIUS. Senator of the family of Furii, fellow-student and *friend of *Jerome, who dedicated some works to him and wrote him many *letters (*Epp.* 48; 49; 57; 66; 83; 84; 97). *Augustine also wrote to him (*Ep.* 58). He married Paulina, daughter of the Roman matron *Paula, and, widowed in 397, entered monastic life; he spent his fortune to assist the *poor and built a hospice for *pilgrims at Porto and the church of SS. Giovanni e Paolo (*titulus Pammacchii*) at *Rome. He took an interest in the controversies over *Jovinian and *Origen. Died 410 during the invasion of the *Goths.

LTK 8, 16 (bibl.), *BS* 10, 72-74 (bibl.); *PLRE* 1, 663.

A. Pollastri

PAMPHILUS of Caesarea. Of a noble family of Berytus (Beirut), he held public office and then became a disciple of *Pierius, called "the younger Origen", at the *Didaskaleion* of *Alexandria (Photius, *Bibl.*, cod. 118-119). He moved to *Caesarea to reanimate the *school founded by *Origen, and was there ordained priest by bishop Agapius. His teaching, like Origen's, involved a spiritual and scriptural approach. He restored and developed the *library attached to the school and organized a workshop of copyists. Arrested in Nov 307, he spent two years in prison and was beheaded on 16 Feb 310 under *Maximinus Daia. Two lost *biographies were written by his teacher Pierius and by his pupil *Eusebius, who called himself "Eusebius Pamphili" and spoke of him in *HE* and in *De Mart. Pal.* In prison, with Eusebius's help, he wrote an *Apology for Origen* in six books (book VI was written after his death by Eusebius alone: cf. Photius, *Bibl.*, cod. 118). Only book I remains, in *Rufinus's *Latin translation. The introduction, addressed to the *martyrs of *Palestine condemned *ad metalla*, illustrates the method to follow in order to read and judge Origen fairly, emphasizing the hypothetical and antithetical character of his speculations, which should not be understood as dogmatic assertions, but often contradict each other, as well as his attachment to *orthodoxy. Citing many texts of Origen, some otherwise unknown, P. refutes accusations concerning Origen's thought on the *Trinity, the *incarnation, the historicity of *Scripture, the *resurrection, punishment, the *soul, metempsychosis. To deprive this defence of Origen of the too glorious heritage of a martyr, *Jerome would have us believe that it was the exclusive work of the semi-Arian Eusebius, but the other evidence refutes this.

Apologia di Origene: PG 17, 521-616.

H. Crouzel

PAMPHILUS of Jerusalem. Two works survive under his name, one dogmatic, the other a work of *hagiography. *Capitulorum diversorum seu dubitationum solutio* (A. Mai, *Nova Patrum Bibliotheca*, II, Rome 1844, 597-693) deals with questions of *christology and refutes *Severus of Antioch, the *acephali and the *tritheist *monophysites. Long citations of it appear in the *Doctrina Patrum* (ed. F. Diekamp, Münster 1907, repr. ²1982). The author has used, sometimes literally, *Leontius of Byzantium (particularly the *Epilysis*) and the emp. *Justinian's *Confessio rectae fidei*. The work was composed between 560 and 630. To celebrate the holy *virgin and *martyr Soteris, also venerated at *Jerusalem in the 6th c., P. wrote the *Encomium s. Soteridis* (P. Franchi de' Cavalieri, *Hagiographica*, ST 19, Rome 1908, 113-120). The saint is made a companion of St *Pancras.

CPG 3, 6920 f.; Bardenhewer 5, 18 f.; Beck 379 and 465; Altaner 549; M. Richard, Léonce et Pamphile, *RSPh* 27 (1938) 27-52 (=*Opera Minora* III. Turnhout 1977, no. 58); id., Pamphile de Jérusalem, *Muséon* 90 (1977) 277-280. Edition: J. Declerck (ed.), *Pamphili opera Capitulorum diversorum seu dubitationum solutio (CPG 6920), necron Soteridis (CPG 6921)*, CCG 1990.

A. De Nicola

PANCRAS, martyr. P. is inscribed in *Mart. hier.* after Sts *Nereus and Achilleus, on 12 May: *et natale s. Pancrati m.* Pope *Symmachus (498-514) built or enlarged the basilica over his tomb, outside the homonymous gate on the Via Aurelia (Duchesne *LP* I, 262). *Gregory the Great preached *Hom. in Ev.* 27 (PL 76, 1204-10) there and founded a *monastery nearby (*Reg. epp.* 4, 18). Honorius I (625-38) had a semi-circular crypt and an altar *ad corpus* built in the choir of the church (De Rossi, ICUR 2, 24 and 156). *Gregory of Tours (*Gl. mart.* 39) mentions the custom by which accused people could, by an oath sworn in that church, be absolved of charges against them. The Gelasian and Gregorian sacramentaries cite a *mass in honour of the *martyr on his feast-day. His Passion (BHL 6420-27) is of epic type and goes back to the 6th or 7th c.

Vies des SS. 5, 238-40; *LTK* 8, 22; *BS* 10, 82-89; A. Amore, *I martiri di Roma*, Rome 1975, 251-53; G.N. Verrando, Le iscrizioni di S. Pancrazio, *RIA* 59 (1983) 151-209.

V. Saxer

PANNONIA

I. History - II. Archaeology.

I. History. Corresponding to part of modern Hungary, between the Danube, the Drava, the Sava and the frontier of *Noricum, during the Roman empire P. was divided into the administrative dioceses of P. Prima which included, among others, the territories of Vindobona, Carnuntum, Scarbantia and Savaria - St *Martin's birthplace (Sulp. Sev., *Mart.* 2, 1) - and P. Secunda, between the Danube and the lower Drava and Sava, whose main centre was *Sirmium: there was also a civitas Valeria, birthplace, acc. to *Ennodius, of Anthony of Lérins (*Opusc.* 4, 7).

A first sure evidence of Christianity in P. is the *martyrdom (304) of *Victorinus, bishop of Petovium (Ptuj) on the border with Noricum (Jerome, *De vir. ill.* 74). The presence of this able Christian author - so *Jerome judges him - testifies to a Christianity rooted in P. for some time and already at a mature level. Something also signified by some authentic elements in the accounts of other *passiones*, such as that of Quirinus of Siscia (ASS *Jun.* I, 372 ff.) and, for P. Secunda, of Pollio, *primicerius lectorum* of the church of Cibali (ASS *Apr.* III, 571-573), and of *Irenaeus of Sirmium (ASS *Martii* III, 553-555), on whom, however, cf. M. Simonetti, *Studi*, 55-79. If we remember also the so-called Four Crowned Martyrs and their companion Simplicius (ASS *Nov.* III, 748 ff.), an ample and vital picture of the church of P. is already evident by the late 3rd c. Further confirmation of this, in the subsequent period, is the repeated presence of various spokesmen of that church at various *councils and synods.

J. Zeiller, *Les origines chrétiennes dans les provinces danubiennes de l'empire romain*, Paris 1918 (repr. Rome 1967); H. Leclercq, Illyricum, *DACL* 7, 1, 99-104. 114; id., Pannonie, *DACL* 13, 1, 1046-1063; M. Simonetti, *Studi Agiografici*, Rome 1955; A. Lippold-E. Kirsten, Donauprovinzen, *RACh* 4, 166 ff.; L. Várady, *Das letzte Jahrhundert Pannoniens (376-476)*, Amsterdam 1969.

V. Pavan

II. *Archaeology.

1. Pannonia Prima was established as a province on the former territory of Pannonia Superior in 293 when *Diocletian reformed the system of provinces; its borders were the Danube to N, the old frontier with Pannonia Inferior to E, *Noricum (i.e. a line between the Drava and Sava a little to E of the Norican cities of Celeia and Flavia Sola) to W, and the river Drava to S. Its administrative centre was Savaria (Sombathely). The dioceses of P. Prima are not known: Savaria was perhaps the most important. Very few early Christian remains are preserved. At Vindobona (Vienna), two basilicas with semicircular projecting apse were excavated. At Kekkut, near Lake Balaton, two churches were discovered: one of common type, the other with three apses and a seat for the *clergy. At Fénékpuszta, near Balaton, two churches were excavated, the second built over an older building. Besides basilicas, some smaller *church buildings are preserved (*Kisdiòspùszta*). At Savaria was excavated the imperial residence, once thought to be a basilica in honour of St Quiricus, bishop and *martyr of Siscia (*Savia*), until excavations showed this supposition to be groundless. The hall had very fine pavement *mosaics. From Savaria comes an altar pillar decorated with rosettes, ivy, the *monogram of Christ, a dolphin and a cuttlefish.

J. Zeiller, *les origines chrétiennes dans les provinces danubiennes de l'empire romain*, Paris 1918 (repr. Rome 1967); T. Nagy, *La storia del Cristianesimo in Pannonia sino alla caduta della copertura romana di confine*, Budapest 1939 (in Hungarian, with German résumé); Z. Kádár, Lineamenti dell'arte romana della Pannonia nell'epoca dell'antichità tarda e paleocristiana, *CCAB* 16, 179 ff.; A. Mócsy, *Pannonia and Upper Moesia*, London 1974, 297 ff.

2. Pannonia Secunda, established as a province on the former territory of Pannonia Inferior in 293, extended to S as far as the border with *Dalmatia, i.e. to the Dinaric Alps; to E as far as the Danube; to N it went slightly beyond the Drava. Its W border coincided with *Savia, i.e. the old border with P. Superior.

Its episcopal centres were the great cities of *Sirmium (the provincial capital), Mursa (now Osijek) and Cibalae (now Vinkovci). Except at Sirmium, no religious buildings have yet been found in P. Secunda. At Cibalae, a building on the necropolis called Kamenice, 1.5 km from the city, is not yet excavated. Here, very probably, the *martyr Pollio was laid. From Cibalae comes a *sarcophagus with *tabella ansata* without inscriptions, but with figures of fish and *bread. At Mursa were found a *monogram of Christ, some inscriptions, *lamps and other small objects. At Strbinci, near Dakovo, was discovered a gold-glass with figures of spouses and the inscription FLORENTIS.

J. Zeiller, *Les origines chrétiennes dans les provinces danubiennes de l'empire romain*, Paris 1918 (repr. Rome 1967); T. Nagy, *La storia del Cristianesimo in*

Pannonia sino alla caduta della copertura romana di confine, Budapest 1939 (in Hungarian, with German résumé); A. Mócsy, *Pannonia and Upper Moesia*, London 1974, 325 ff.; B. Vikc, Elementi ranog krsc'anstva u sjevernoj Hrvatskoj, *Archeoloski vestnik* 29 (1978) 588 ff; N. Duval-V. Popovic, Urbanisme et topographie chrétienne dans les provinces septentrionales d'Illyricum, *Actes Xe Congrès internat. d'arch. chrét*, I, Vatican City-Thessalonica 1984 (publ. 1985) 541-579; N. Duval, L'architecture religieuse de Tsaritchin Grad dans le cadre de l'Illyricum oriental au VIe siècle, *Villes et peuplement dans l'Illyricum protobyzantin*, CEFR 77, Rome 1984, 399-481.

N. Cambi

PANODORUS. Egyptian monk, flourished between 395 and 408, author of a lost *Chronography*, continuator of Sixtus *Julius Africanus and *Eusebius, known only through extracts cited by Syncellus, whose main source it is. P.'s work, completed in 412 by another Egyptian monk, Ammianus (CPG 3, 5537), began from the *creation of the world, which he dated 5493 years before Christ's birth, and was wholly set in a chronological scheme based on biblical tradition, into which he inserted secular history from time to time. It was he who introduced into Christian *chronology Ptolemy's Κανὼν Βασιλειῶν, Manetho's canon, the book of Soti, and the "list of the Theban kings" of Eratosthenes and Apollodorus. His main sources were Dexippus, Julius Africanus and Eusebius, whom Syncellus knew through him. He had a decisive influence on George Syncellus. H. Gelzer (*Sextus Iulius Africanus und die byzantinische Chronographie*, 3 vols., Leipzig 1880-1898) denied the 9th-c. chronicler any originality, and attributed any merit to the Egyptian monk, linked with the official church, but also part of the *neoplatonist culture of the 4th/5th c., which facilitated the harmonization between Christian and *pagan history: this thesis was partly reappraised by R. Laqueur (Synkellos, *PWK* 2, 4, 1388-1410).

CPG 3, 5535; G. Dindorf, *Georgius Syncellus et Nicephorus Constantinopolitanus* (CSHB), I, Bonn 1829, 61[3-5] - 62[2-5] - 63[11] - 66[3]; Van Der Hagen, *De cyclo magno paschali seu de periodo DXXXII annorum Panodori, monachi Aegyptii*: Dissertationes de cyclis paschalibus, Amsterdam 1736, 93-106; Krumbacher 340-342; O. Seel, Panodoros, *PWK* 18, 632-635; Altaner 236.

A. Labate

PANTAENUS. All the relevant evidence (preserved by *Eusebius, *Jerome, *Clement, *Origen, *Alexander of Jerusalem, *Pamphilus, *Anastasius the Sinaite, *Maximus Confessor) has been collected and cited entire by A. von Harnack in *Geschichte der altchristlichen Literatur bis Eusebius*, I, Leipzig 1893, 291-296. Eusebius (*HE* V, 11, 1 [II 452, 10-12 Schwartz]) rightly held that Clement (*Strom*. I, 11, 2 [II 8, 22-9, 3]), speaking of the teacher whom he found "hidden in Egypt", was alluding to P.; the same passage of Clement tells us of P.'s Sicilian origin: Σικελικὴ τῷ ὄντι ἦν μέλιττα (cf. T. Zahn, *Forschungen zur Geschichte des neutestamentlichen Kanons*, III, Erlangen 1884, 161); and the words δυνάμει δὲ οὗτος πρῶτος ἦν (*Strom*. I, 11, 2 [II 8, 23-24]) prove that Clement held P. in higher esteem than all his other teachers, the "presbyters" and guardians of apostolic *tradition, whose oral teachings Clement claims to have listened to (*Strom*. I, 11, 1 and 11, 3). Eusebius (*HE* V, 10, 1-4 [II 450, 12-452, 5]) tells us that P., after having been a *Stoic philosopher, directed the catechetical *school of *Alexandria with great fervour during the reign of *Commodus, and even travelled to India to announce the Christian message to that distant people. Acc. to Zahn (p. 176), his death must be put around AD 200. Clement, in his exegetical works and esp. in the 'Υποτυπώσεις, appeals regularly to the teaching of the πρεσβύτεροι - the most authoritative of whom was undoubtedly P. - or simply to the πρεσβύτερος *par excellence*, who must certainly be identified with P. (cf. the various evidence in Harnack, 292-293, and Zahn, 157-161). But P.'s name is explicitly mentioned only in two passages, cited in full as the only two sure fragments by M.J. Routh (*Reliquiae sacrae*, I, Oxford 1846, 378-379): Clement, *Ecl. proph*. 56, 2 (III 152, 28-153, 4=fr. I Routh 378), where apropos of Ps 18, 6 [19, 6] it is said that the prophets were accustomed to use the present tense in place of the future or the past (but it is arbitrary to try to trace back to P. a much wider context comprising much of the *Eclogae propheticae* and the *Excerpta ex Theodoto*, as W. Bousset did in *Jüdisch-christlicher Schulbetrieb in Alexandria und Rom*, Göttingen 1914, 156 ff., 174 ff., 263); and Maximus Confessor, *De variis difficilibus locis Dionysii et Gregorii*: PG 91, 1085 A10-B12 (= fr. II Routh, 378-379; Clement, fr. 48 Stählin, III, 224), where *God's will is stressed as the creative principle). This last passage, to which H. Langerbeck has rightly drawn attention (*JHS* 77 [1957] 77 and *AAWG* phil. hist. Kl. 69 [1967] 157), shows a close analogy with *Photius's two passages (*Bibl*., cod. 214, 172a [III 126, 25-26 Henry] and 251, 461b [VII 192, 8-9]) relating to Hierocles and in all probability reflecting doctrines of *Ammonius Saccas.

The description of Stoic philosopher attrib. to P. by Eusebius (*HE* V, 10, 1 [II 450, 18-19]) is taken in too strict a sense by Pohlenz (*NAWG* phil. hist. Kl. 1943, p. 166, n. 3), given the *syncretistic character of Greek philosophy at the end of the 2nd c. It is not hazardous to suppose that Clement inherited from P. his tendency to combine in a single system, reflecting the truth, the best doctrines of the various philosophical systems (cf. *Strom*. I, 37, 6 and I, 57, 6): the image of the bee used by Clement (in *Strom*. I, 33, 6 and IV, 9, 2) in support of πολυμαθία, is always applied by Clement to P. too (*Strom*. I, 11, 2); the same "eclectic" tendency reappears in Ammonius Saccas, who taught at Alexandria in the same period and aimed to harmonize Plato and Aristotle (cf. no. 7 in *Ammonius Saccas). Also quite likely is the hypothesis formulated by R.E. Witt (*CQ* 25 [1931] 195) on the existence of very close links between Ammonius Saccas and the Alexandrian Christians at the close of the 2nd c.: these seem to be confirmed not just by the correspondence (cf. above) between P.'s and Ammonius's doctrines of God's will as creative principle, but also by the parallels between Clement and *Plotinus noted by Witt (*CQ* 25 [1931] 195-204); to these we may add the parallel, on the infinity of God, between *Strom*. V, 81, 6 (II 380, 22-23) οὐ κατὰ τὸ ἀδιεξίτητον νοούμενον and *Enn*. VI, IX, 6 (VI[2]179, 10 Bréhier) οὐ τῷ ἀδιεξιτήτῳ (cf. S. Lilla, *JThS* n.s. 31 [1980] 100 n.2). Finally, since Clement (*Strom*. I, 11, 2-3) presents P. as one of the repositories of the esoteric *tradition going back to the *apostles, it is legitimate to hold that P. too, like Clement, maintained the existence of a secret tradition, an idea characteristic of the Christian *gnosticism which flourished in *Egypt in the 2nd c. AD (cf. S. Lilla, *Clement of Alexandria*, Oxford 1971, 234).

M.J. Routh, *Reliquiae Sacrae*, I, Oxford 1846, 375-379; T. Zahn, *Forschungen zur Geschichte des neutestamentlichen Kanons und der altkirchlichen Literatur*, III, Supplementum Clementinum, Erlangen 1884, 156-176; A. von Harnack, *Geschichte der altchristlichen Literatur bis Eusebius*, I, Leipzig 1893, 291-296; G. Bardy, Aux origines de l'école d'Alexandrie, *RecSR* 27 (1937) 65-90; S. Lilla, *Clement of Alexandria*, Oxford 1971, 233-234.

S. Lilla

PANTALEON, martyr. P. is the Latinized form of *Panteleimon* "all-merciful". The saint's name is inscribed in *Mart. hier.* on 27 July. It has no equivalent in *Mart. syr.*, incomplete at this point. The Eastern synaxaria cite P.'s name on various dates. Acc. to his Passion (BHG 1412-14m; BHO 835-37; BHL 6429-42) and the Panegyrics, this *martyr of *Nicomedia was a doctor, able to cure all diseases: an aetiological legend derived from his name. The many versions of the legend also attest the popularity of his cult. The *martyrology of Florus of Lyons follows the Latin version of the legend. At *Rome, three churches were built in his honour in the Middle Ages.

Vies des SS. 7, 651-52; *LTK* 8, 24-25; *BS* 10, 108-18.

V. Saxer

PANTHEISM. Pantheistic conceptions of reality can be observed in Heraclitus (divinity is represented by universal fire, which transforms itself into all things, text 8 and fr. 90 Diels, *Die Fragm. der Vors*. I[6], 1951, 145 and 171), Diogenes of Apollonia (God is identical to air, which reaches everywhere, fr. 5 Diels, II[6], 1952, 61), Aeschylus (Zeus is identical to all things found in nature, fr. 70 Nauck, *Trag. graec. fragm*., Leipzig 1889, 24) and Euripides (Zeus is identified with the "chariot" of the world and with fate, *Trojan Women* 884-886).

"Pantheistic" tendencies also appear in the young Aristotle, in the peripatetic school, in book X of Plato's *Laws* (though here, above the *anima mundi* is *nous*, a principle distinct from it: cf. *Laws* X 897b 1) and in the *Epinomis*: for the young Aristotle cf. Περὶ φιλοσοφίας, fr. 26 Walzer (= Cicero, *De nat. deor*. I, 13, 33: *Aristotelesque ... modo mundum ipsum deum dicit esse*), Clement, *Protr*. 66, 4 (I 51, 1-2) and Plutarch, *Plac. philos*. V 20, 908F (cited by R. Walzer, *Aristotelis dialogorum fragmenta*, Florence 1934, 92, n. 2); for the peripatetic school cf. Cicero, *De nat. deor*. I 13, 35: *nec audiendus eius* (Theophrastus) *auditor Strato ... qui omnem vim divinam in natura sitam esse censet*, and Clement, *Protr*. 66, 5 (I 51, 6-7); on book X of the *Laws* and the *Epinomis* cf. the two passages cited by R. Walzer, *op. cit*., 92, nn. 2 and 3.

But the pantheistic conception is above all characteristic of Stoic philosophy, in which Heraclitean influence is particularly evident. For it, the divine principle is none other than spirit (πνεῦμα), identical with ether, fire, the *logos*, fate and *providence; it is immanent in the universe, permeates it with itself, spans it from one end to the other, proceeding from the centre to the periphery and from the periphery to the centre, and governs and administers it as its supreme law: cf. SVF I 153-177, II 1021-1056 (so for *Stoicism there is a close affinity between the concepts of φύσις, νόμος, λόγος, κόσμος, πνεῦμα, and πρόνοια). Patristic authors, while accepting in general the Stoic and Philonian conception of the divine *logos* immanent in the universe, certainly could not share - like *Philo and contemporary Platonic philosophers - the Stoic identification between the sensible universe and the first creative principle, which in this way came to be material, corporeal and corruptible, and they constantly argued against it (indeed for them, as for Philo, the *logos* was no longer the first principle, but the son of *God generated by him, his power, his instrument in the creation of the sensible universe, and the soul and law of the latter). The rejection of Stoic p. can be observed, e.g., in *Justin Martyr (*I Apol.* 20 [I 180, 7-11 Otto], II, 7 [I 300, 6-7 and 9-12]), in *Clement of Alexandria (*Protr.* 66, 3 [I 50, 24-27=SVF II 1039], *Strom.* I, 51, 1 [II 33, 12-14=SVF II 1040], V, 89, 2 [II 384, 18-19=SVF II 1035] and V, 89, 3 [II 384, 22-385, 1]) and in some passages of *Origen's *Contra Celsum* and *Commentary on John*, cited in SVF II 1051, 1052, 1053 and 1054. Passages of *Theodoret, *Hippolytus, *Eusebius, *Tertullian, *Lactantius and *Salvianus in which this Stoic doctrine is cited and often rejected can be found in SVF II 1029-1043.

S. Lilla

PANTOLEON. Presbyter of a *monastery τῶν Βυζαντιῶν prob. at *Jerusalem and author of the *Homilia in exaltationem crucis* (BHG 430). This used to be dated between the 5th and 12th cc. This last date was refuted by the discovery of two 9th-c. Greek MSS (*Marc.* II 17 and *Vat. Palat.* 205) containing the *Homilia*, and a codex (*Berolin. Syr.* 28) with a *Syriac translation of it, dateable to the 8th-9th c. Recently Honigmann, after having clarified the time of the institution of the feast of ὕψωσις and established that the monastery τῶν Βυζαντιῶν was prob. that founded in Jerusalem by *Abraham of Ephesus, has dated the composition of P.'s *Homilia* some time between 650 and 750. This rules out the identification of our man with the deacon and cartophylax of the Great Church of *Constantinople, Pantoleon (or Pantaleon), author of some *homilies (V. Grumel, Pantaléon, *DTC* 11, 1855; H.G. Beck, Pantoleon, *LTK* 8, 31). Perhaps the presbyter P. may be the same P. to whom pope *Martin I addressed a *letter (Jaffé 2068) on the morrow of the Lateran *council, rebuking him for his opposition to its decisions. There is no evidence to support the attribution to him of the unpublished *Homilia de exaltatione crucis* (BHG 427 p), present in Messan. Gr. S. Salv. 4 (13th c.); and he is certainly not the author of the *Tractatus contra errores Graecorum* (PG 140, 487-574), written in 1252 by a homonym.

CPG 3, 7915-7918; PG 98, 1265-1269; Krumbacher 167; Baumstark 264; E. Honigmann, La date de l'homélie du prêtre Pantoléon sur la fête de l'Exaltation de la Croix (VIIe siècle) et l'origine des collections homiliaires, *Bull. de l'Acad. royale de Belgique* 36 (1950) 547-559; Beck 457-458.

A. Labate

PAPACY

I. The Christian community of Rome in the first three centuries - II. The emergence of the papacy in the 4th-5th cc. - III. The transition to the mediaeval papacy (6th-8th cc.) - IV. Elements of historical evaluation - V. Historiography of the papacy - VI. The popes - VII. Letters of the popes.

I. The Christian community of Rome in the first three centuries. For the period preceding the Constantinian era there is little exact information on the bishops of *Rome. The evidence, mostly in *Eusebius of Caesarea's *Hist. Eccl.*, is scanty and unreliable. Thus the catalogues of *popes, composed for an apologetic demonstration of apostolic *succession, are problematical. *Irenaeus's *episcopal list (*Adv. haer.* III, 3, 3) deserves some credit for its names, taken from credible traditions. But the list does not allow us to ascertain the moment when the collegial direction of the Roman community became a monarchical episcopate. Still more uncertain is the chronology which Eusebius (†339), and after him the so-called *catalogus liberianus* (336 or 354), tried to fix, synchronizing individual pontificates with the dates of imperial governments. Despite the fragmentary nature of the historically ascertained information, we can hold that almost from the beginning Christianity spread rapidly to the capital of the empire (cf. Acts 2, 10). The first historical data on a Christian presence in Rome are in Suetonius's *Life of *Claudius* (41-54), written in the 2nd c. (Mirbt n. 3; cf. Acts 18, 2). The Christian community in Rome cannot have been founded either by *Peter or by *Paul, as the tradition first related by Irenaeus (*Adv. haer.* III, 1, 1; III, 2, 3) would have it.

The Church in the imperial capital was founded by unknown *Judaeo-Christians. But its importance is soon clearly attested by Paul's letter to the Romans, written in *c.*57/8. Much clearer is the evidence of *1 Clement* (96), with which the Roman community intervened on the occasion of disagreements emerging in the community of *Corinth. The way in which this refers to the *martyrdom of Peter and Paul (ch. 5) and insists on the principle of succession, guarantee of order in the community (chs. 40-44), heralds the care that the Roman church would later claim over the other churches. Besides the evidence of 1 Peter (after 94), that of *Ignatius's letter to the Romans (*c.*110), with its very solemn address and its reference to Peter and Paul (ch. 4, 2a), is important since it already contains some elements of later papal theory: the importance of the capital, the presence of the two *apostles, primacy in *faith and charity (*praef.*). Towards the end of the 2nd c. we encounter the most famous evidence, that of Irenaeus, which not only traces the Roman community back to Peter and Paul, but considers *communion with this community as the surest demonstration of being within the doctrinal *tradition derived from the apostles (*Adv. haer.* III, 3, 2). This evidence is confirmed by the similar, though less explicit, positions of *Tertullian (*Praescr.* 32. 36) and *Origen (Euseb., *HE* VI, 14, 10). And above all it finds valid confirmation in the attraction which the community of Rome exercised over the earliest Christians. As esp. Eusebius relates, not a few people went to the capital to find greater certainty in their faith, indeed to join such a famous community: *Polycarp of Smyrna, *Justin the philosopher and martyr, *Hegesippus, *Marcion (139), *Montanus, Irenaeus, Origen. This attraction cannot be explained just by the fact that the Roman community was at the centre of the empire. Moreover, certain facts concerning the history of the so-called apostolic *creed, the form of the NT canon and the first elaborations of liturgical traditions in *Hippolytus's *Apostolic Tradition*, show incontrovertibly that in *c.*200 the community of Rome, whatever the exact explanation, possessed in effect an *authority which no other community had or was even comparable to. This importance of the Roman community cannot be understood unless we bear in mind the concrete initiatives taken by some popes. Around 200, we begin to have precise information not just on the monarchical organization, previously unclear, but also on individual bishops. *Cornelius (251-253) gives the first description of the Roman *clergy, led by a single bishop (Euseb., *HE* VI, 43, 11 f.; cf. Hippolytus of Rome's slightly earlier evidence). We do not know whether the Roman bishop of that time had a permanent residence. But we do know that, from *Callistus on (217-22), the popes were buried in a community *cemetery. The first biographical fact whose exact historical date we know is *Pontianus's renunciation (28 Sept 235). Before this date we know of the action of *Victor I, who in *c.*190 *excommunicated *Polycrates of Ephesus and other bishops of *Asia Minor because they would not allow the celebration of *Easter on the *Sunday after 14 Nisan: an action criticized, though accepted, by Irenaeus (Euseb., *HE* V, 23 ff.). Victor also expelled *Theodotus, who sought to spread *adoptionism at Rome. Acc. to Hippolytus's polemical information, *Zephyrinus (199-217?) and Callistus propagated a too unitarian doctrine of the *Trinity. Callistus was also accused of tolerating moral laxity. After 250, Cornelius was involved, with *Cyprian of Carthage (†258), in the controversy over the readmission of those (the *lapsi*) who had in some way denied the Christian faith during *Decius's *persecution and whom *Novatian and *Novatus thought should be excluded for ever from the community, considered a church of saints. On this occasion Cyprian wrote his most famous work, *De unitate catholicae ecclesiae*, in which he presented Peter's confession of faith (Mt 16) as a symbol of the unity of the church. Later, opposing the position of *Stephen I (254-257) on the validity of *baptism administered by *heretics, Cyprian made it clear that in his opinion Peter's primacy did not extend to his successors. Indeed, he explicitly rejected the Roman claims, which Stephen himself seems to have based on Mt 16. The dispute, in which *Firmilian of Caesarea was also involved, was ended by a new persecution in which both protagonists died as *martyrs. After *Gallienus revoked his father's anti-Christian edict, the churches enjoyed some 40 years of *peace. *Dionysius (260-268) profited from this to reorganize the community, tested by nearly ten years of persecutions. Appealed to by Alexandrian presbyters over the trinitarian controversy in *Egypt, at a synod he took a stand against the *Sabellian and *tritheist tendency and communicated to his homonym of *Alexandria, *Dionysius, the charges made against him. The latter cleared himself, as Eusebius (*HE* VII, 26) and *Athanasius relate. Dionysius of Rome also helped the community of *Caesarea in Cappadocia, suffering as a result of barbarian invasions (cf. Basil, *Ep.* 70). We lack precise information on the following pontificates. *Marcellus I (307/8) and Eusebius I (308-310?) died in exile.

From this rather scanty and not always reliable information on the Roman community and its bishops, it emerges that during the first centuries the church of Rome sometimes took initiatives in order to maintain harmony in the communities or to restore the unity of Christians. In particular it claimed to defend the *apostolicity of the faith, recording the succession of those who were responsible for it. But its *theology was hardly original. It would be the same later. Rome would continue to claim *custodia fidei et disciplinae*.

Other churches sometimes accepted these interventions for the sake of harmonious communion in the one apostolic faith. Thus *I Clement* continued to enjoy great authority, as is attested by its almost canonical transmission and by a letter of *Dionysius of Corinth (Euseb., *HE* IV, 23). Irenaeus admonished Victor I, but nevertheless acknowledged his *authority (*HE* V, 24). Dionysius of Alexandria, defending himself against accusations of *tritheism, did not question the intervention of the Roman synod under Dionysius, whom, in an earlier letter, he considered a learned and admirable man (*HE* VII, 6). Finally, we should note the homage paid by Ignatius, and later by *Basil, to Rome's charitable activity.

II. The emergence of the papacy in the 4th-5th cc. With *Galerius's edict of 311, a new political situation officially began for Christianity. This period, called the time of the imperial church or Constantinian era, was decisive for the history of the p. At the start of these *tempora christiana*, the p. hardly emerges. *Constantine gave *Miltiades (310-314) the Lateran palace as residence and also instructed him to resolve the *Donatist problem, which arose after the persecutions, when a minority at *Carthage opposed bishop *Caecilian, whom they considered a *traditor*. But Miltiades's arbitration was unsuccessful. Constantine involved himself in the affair, convoking the synod of *Arles (314), in which neither Miltiades nor *Sylvester (314-335) had any part. Sylvester had no part in the council of *Nicaea (325), called by Constantine to end the *Arian crisis. In recognizing the privileges of the see of *Alexandria, on the model of those of *Rome, this *council laid the foundations of the ecclesiastical hierarchy in accordance with the civil divisions of the empire. This decision (cann. 6 f.) became the more important when in 330 Constantine transferred his residence to *Constantinople, whose episcopal see came into competition not just with the main sees of the *East, but also with that of of Rome. In the Arian controversy, which got worse after Constantine's death (337), *Julius I (337-352), with his synod of 341, asssumed the defence of *Athanasius and *Marcellus of Ancyra, claiming a priority of judgment in such affairs (DS 132). The Western *Fathers of the synod of *Sardica (343) vindicated this judgment and approved the right of appeal to the Roman see, but the Eastern bishops did not agree (DS 133 ff.). Ten years later, *Liberius (352-366) obtained even less. Under political pressure he subscribed the so-called third formula of *Sirmium (DS 139 f.), compromising his Nicene loyalty, though without breaking with Athanasius, and provoking a grave crisis in the community of Rome.

Only with *Damasus (366-384) did papal policy start to be crowned by success. This "turning-point" (Pietri) was certainly due to the personal ability of Damasus and his successors. But it was nevertheless favoured by the political and ecclesiastical circumstances and obtained esp. thanks to the *conversion of the Roman aristocracy, which was by now supporting the church economically and which made a particular contribution to the formation of the ideology of *Roma aeterna*, without which the petrine claims would not have carried the weight they did in the 5th c. After overcoming the difficulties arising out of his election, not without the help of *Gratian (375-383) who approved his jurisdiction over all the metropolitan sees of the West (Mirbt n. 305), Damasus intervened with his synods in the dogmatic controversies of the time (*Tomus Damasi* of 377: DS 152-161). Appealed to simultaneously by the Eastern bishops on the discussions about the *schism of *Antioch, in 380 Damasus received a solemn confirmation of his primatial policy from *Theodosius I in the so-called edict of the three emperors (Mirbt n. 310). One year later however, the council of *Constantinople (381) and its decision on the second rank pertaining to the new Rome (can. 2: Mirbt n. 312) showed that Theodosius had no more need of Roman support. In his protest against the synodal decision (Mirbt n. 313) and on other occasions, Damasus showed that behind his interventions lay very clear ideas on the role of the *Sedes Apostolica*. Referring to Mt 16, 18 f., already used by the African tradition to emphasize the unity of the church, symbolized by the *cathedra Petri*, and leaning primarily on the *authority of *Peter and *Paul, venerated at Rome as princes of the *apostles, he founded a theory of Roman primacy which his successors needed only to extend.

*Siricius (384-399) first applied Paul's *sollicitudo omnium ecclesiarum* (2 Cor 11, 28) to the apostolic ministry of the popes. Anastasius I (399-402) and *Innocent I (402-417) extended it in their "decretal" *letters, in which they replied to disciplinary questions put by bishops in mission fields such as *Gaul and *Spain, or communicated the decisions of synods to bishops who could not attend them. If we consider not just the recipients, mostly bishops of *Italy or of lands in which the metropolitan regime was not yet sufficiently established, but also the relatively restricted fields (*causae maiores*, interventions in contested episcopal elections), we see a considerable contrast between jurisdiction - limited to the Western regions, from which N *Africa was still excepted, and mainly concerning disciplinary questions - and theory, which increasingly seems to have extended the *cura*, the *onera* and the *principatus* to the churches of the whole Roman empire. No less noteworthy is the fact that these popes, while invoking the apostolic origin of their universal authority, sought to confirm it not just by *consuetudo*, but also by synodal law, apropos of which Siricius seems to have been the first to refer to the so-called *corpus romanum*, i.e. to that tradition which had confused the canons of *Nicaea with those of *Sardica (*Ep.* 5, 2). Although this reference to the canons is explained mainly by the pressing need to back up his interventions in the field of civil law (Spiegl), it remains significant that these popes were concerned to emphasize the conformity of decretal law with synodal law. The contrast between the relatively limited practice and the apparently universal theory becomes even more manifest if we consider that the pope's interlocutors did not always give his interventions the same weight, indeed contested them outright. This goes particularly for the position taken by *Zosimus (417-418) in favour of *Pelagius and *Caelestius (417), but also in the case of *Apiarius, a presbyter of *Sicca who had appealed to Zosimus and later to *Celestine I (422-432). Even Boniface I (418-422), who so vigorously expressed the need for *communion with the Roman church, head of all the churches (*Ep.* 14, 1), had to recognize the limits of Roman jurisdiction, ignored by an African synod (419) and contested by the pretensions of Constantinople as expressed in the edict of *Theodosius II (421) (*Epp.* 13 ff.). His successor Celestine, appealed to by both the Constantinopolitan and Alexandrian sees, pronounced with his synod on the question of Christ, *God born of the virgin, and charged Cyril as his deputy to take the necessary measures against *Nestorius (*Epp.* 11-14). But he could not help seeing not just that the council of *Ephesus, convoked by Theodosius II (431), resumed discussion of the question which he believed he had resolved, but also that Cyril cared little for the legal form with which the Roman church pursued its interests. It remains to note that Celestine, like his predecessors, did not engage overmuch in theological disputes. Instead, he too was content to exclude authoritatively the errors of Nestorius, as he had previously confirmed the condemnation of the *Pelagians (*Ep.* 21). *Sixtus III (432-440), informed of the agreement reached in 433 between *John of Antioch and *Cyril of Alexandria, congratulated the two reconciled bishops on their restored *peace, subsequently approving the creed of union and attributing the success to the presence of Peter, guarantor of the true faith (*Epp.* 5 f.). He had undoubtedly worked from the start for a reconciliation of the two principal sees of the *East (*Epp.* 1 f.). However it is obvious that this union was an Eastern affair.

This whole history of the p., with its steps forward and steps backward, culminated in the pontificate of *Leo the Great (440-461). Well prepared for his task, at a politically and ecclesiastically difficult time, not just by a solid literary training but also by his activities in the service of the Roman church, Leo deserves to be called "the Great" especially for his contribution to the development of the Roman see, which brought to an end the work begun by Damasus and his successors. Thanks to his numerous *letters, in part also to his *sermons, we know quite well the *ministry he performed in the Roman community, his interventions in favour of the true faith and of discipline in the West, esp. in *Italy and *Gaul, but also his relations with the Eastern churches, in particular his contribution to the definition and defence of the faith of *Chalcedon. Working for the unity of all the churches solicitude, but also with a moderation remarkable in a *custodia fidei et canonum* (*Ep.* 115, 1), he always recognized the political implications of his pastoral activity, as the situation of the "imperial church" required at that time. He worked for unity of faith, the basis of the political unity of the empire. He also put himself at the disposal of the peoples of *Italy, invaded by Huns and *Vandals. Nor did he hesitate to recognize a certain priestly dignity in the emperor, i.e. the duty of working for the harmony of the churches in the empire's own interests. But this very open attitude to the *salus rei publicae*, still marked by a certain Roman patriotism, did not prevent him rejecting can. 28 of *Chalcedon, in which the order of the great churches was settled according to one-sidedly political criteria. He reminded the emp. *Marcion of the distinction between *church and empire, demanding freedom of action for the church and in this way vigorously affirming the Roman primacy, though without reciprocity from the court of Constantinople. In all this pastoral activity, more or less universal in extent, Leo followed very clear ideas on Roman primacy, expressed in his letters, esp. in their introductions and conclusions, as well as in the sermons preached on the anniversary of his episcopal *ordination. These can be summed up in three words: Christ-Peter-Pope. The primatial role of the Apostolic See is founded on two facts: the intimate union between Christ and *Peter, prince of the *apostles; and the apostolic legacy that survives in the bishops of Rome, successors and vicars of Peter, always present in his church. Like his predecessors, Leo leaned principally on the Petrine succession, established in his opinion at the Lord's own command (Mt 16, 18 ff.; Lk 22, 32; Jn 21, 15-19). He applied the same principle to the sees of *Alexandria and *Antioch, which should therefore have precedence over that of *Constantinople (*Ep.* 106, 5). But his reinterpretation of apostolic *authority, granted by Christ to Peter and his heirs, reflects still more clearly than that of his predecessors a certain political orientation. Leo, in justifying his claims, did not just, like them, use legal-political categories like *principatus, dignitas, haeres, vices, ius potestatis, consortium potestatis*: more than they had done, he placed them in the perspective of *Roma aeterna, caput orbis* and source of peace. In his sermon on Peter and Paul (*Serm.* 82), he did not just take up the traditional theme of *pagan Rome's providential role in the *evangelization of the world: he insisted, with a vigour not yet encountered, on the pacific, even more glorious, activity of Christian Rome, now renewed on the foundation of the apostles. It is true that the contrast between the theory of universal right and the practice, obviously limited by the ecclesiastico-political situation of the time, recurs in Leo too, especially if we compare the pyramidal presentation of church hierarchy in a letter to *Anastasius of Thessalonica (*Ep.* 14, 11) with the concrete exercise of *sollicitudo omnium ecclesiarum*, esp. with respect to the Eastern churches. It is notable with what delicacy Leo respects *consuetudo*, the canons of the synods and, above all, the rights of the other bishops, who with him constitute the *collegium caritatis* (*Epp.* 5, 2; 6, 1; 12, 2). But the fact remains that he formulated, with a theological precision unknown until then, the Roman doctrine of the primacy of Peter's successors with regard to *custodia fidei et disciplinae*.

While the activity of *Hilarus (461-468) was limited to some interventions in jurisdictional controversies in *Gaul and *Spain, *Simplicius (468-483) and Felix III (483-492) confronted the religous policy of the Constantinopolitan court which, in the post-Chalcedonian disputes, favoured the *monophysite movement. This conflict was aggravated by the fact that during this time the Western part of the empire was passsing under the control of Germanic peoples of *Arian faith. Felix III's protests against the so-called *Henotikon*, promulgated in 482 by *Basiliscus, led to the break between him and *Acacius, bishop of Constantinople. This first great *schism between East and West, called the Acacian schism, lasted until 519; one of its effects was that in 490 all the patriarchal sees of the East were occupied by monophysites. Meanwhile, *Gelasius I (492-496), also engaged in controversy with the new Rome, formulated in his letter to the emp. *Anastasius his doctrine of the two powers, which would determine the political thought of the mediaeval West for a millennium (DS 347). Taking up the Leonine distinction between the *auctoritas* of the bishops and the imperial *potestas*, he affirmed the independence of the two spheres, civil and ecclesiastical, though without wanting to separate them. In his letters and treatises, he further defended his position energetically against any imperial interference in church affairs. At the same time he sought to impose his primatial rights on the Latin churches.

III. The transition to the mediaeval papacy (6th-8th cc.). In the pontificates of the second half of the 5th c. were heralded the vicissitudes that would characterize the Byzantine era, the period when the p. still depended more or less on the imperial court of *Constantinople, represented in part by the exarch of *Ravenna, up to the close collaboration of the popes with the kingdom of the *Franks: the rivalry between the sees of *Rome and Constantinople, the struggle of the popes for the freedom of the church, the Apostolic See's growing orientation towards the Germanic kingdoms. After the brief pontificate of Anastasius II (496-498), considered too favourable to Byzantine policy and even to *monophysism, the majority reacted against this pro-Byzantine attitude and chose *Symmachus (498-514), to whom the minority opposed *Lawrence (498-506). Though supported by *Theodoric, king of the *Goths, and then also by the synod which met in 501, Symmachus had difficulty in imposing his authority. This situation explains the aspirations that lay behind the so-called Symmachian falsifications in which, on the basis of forged documents, the principle was claimed that would be invoked so many times during the Middle Ages: *prima sedes a nemine iudicatur* (Mirbt no. 468). Still tied to "Roman patriotism", Symmachus took little interest in the spread of *Catholic Christianity in the kingdoms of the *Franks and *Burgundians. Only from *Hormisdas (514-523) did reconciliation between West and *East become possible, because *Justin I (518-527) and his nephew *Justinian I (527-565) followed a pro-Chalcedonian policy in the interests of the unity of the empire. In 519 the imperial court accepted the *Regula fidei Hormisdae*, which not only insisted on the faith of *Chalcedon, but also claimed that the Roman See was the one guarantor of the true Christian religion. Although Justin and Justinian understood this claim of primacy, based on Mt 16, 18 ff., in a different way, having another concept of the coexistence of church and state, the formula entered Western ecclesiastical law as one of the bases of the development of papal primacy. But the *rapprochement* between the Apostolic See and the court of Constantinople created difficulties, both in East and West. Theodoric watched it with growing mistrust. Together with *Boethius, *John I (523-526) became a victim of the political conflicts between Germans and Byzantines: a sign of the future situation in which the p. would suffer on account of political events. Popes quite favourable to the Goths, Felix III (526-530) and Boniface II (530-532), contested by Dioscorus (530), were followed by popes open to Byzantine policy, *John II (533-535) and Agapitus I (535-536). After *Silverius (536-537), supported by the Goths and consequently condemned by court-martial by the *Byzantines, who took Rome in 536, *Vigilius (537-555) was completely engulfed by Byzantine ecclesiastical policy. In 548, and again after some resistance in 554, he accepted the condemnation of the so-called *Three Chapters, imposed by Justinian in 543/4 in the interests of reconciliation with the *monophysites, and even recognized the council of *Constantinople II (553), in which he had wanted no part and which had excommunicated him. The result was that much of the Western episcopate, particularly the sees of *Milan and *Aquileia, separated from Rome. While confessing himself faithful to Chalcedon, Vigilius's successor *Pelagius I (556-561) was also dependent on Justinian and so was not able, even by helping the Roman population oppressed by economic need, to overcome the *schism in the West. John III (561-574), finally approved by Justinian, restored peace in great measure. After Justinian's death (565), the invasions of the *pagan *Longobards created new difficulties against which Benedict I (575-579) and *Pelagius II (579-590) were powerless. A new policy was needed, an opening-up towards the Germanic peoples who for some time had been more open to *Catholicism.

This turning-point, which for the West included the decisive transition from *Late Antiquity to the Middle Ages, took place under *Gregory the Great (590-604). Born in 540 into a senatorial family, he was profoundly Roman. But his vision was to convert the peoples and integrate them into the universal church. His many *letters (854 preserved), show not only his attractive personality but also his political and pastoral ideals. While accepting the imperial power of *Byzantium, he sought an understanding with the Longobards and even began to approach the *Franks. But Gregory's main claim to greatness is not this new policy, nor even the defence of the Roman primacy which was part of it. It emerges much more in his charitable activity, for which he reorganized the *patrimony of St Peter (the *properties of the Roman church), in his pastoral *ministry (training of the *clergy and tireless preaching), in his interest in monastic life, understood in *Augustine's sense as *actio* and *contemplatio*, and in his missionary undertaking, stimulated by a certain eschatological outlook, to the Anglo-Saxons, i.e. beyond the frontiers of Roman civilization. His successors, most of whom had short reigns, failed to maintain the p. on this level. Despite the increasing separation between East and West, they still remained oriented towards the Byzantine court, so that one may say that at Rome the Western Middle Ages began only half-way through the 8th c. In the long series of these "Byzantine" popes, there are few important events to record. In 607 a Roman synod reformed the rules for electing the bishop of Rome. The pontificate of *Honorius I (625-638) became famous for the "question of Honorius", much discussed in modern controversies on papal infallibility. Honorius's ambiguous position on the problem of *monoenergism led the council of *Constantinople III (680) to include him in its condemnation of *monothelism. Leo II (682-683) accepted this decision, accusing his predecessor of having betrayed the apostolic position, always preserved entire by the Roman See. Honorius, who was certainly not equal to the theological demands of the time, also stands out for his strict administration of ecclesiastical *property which allowed him to exercise great influence in the political situation of *Italy, and for his missionary activity among the Anglo-Saxons and Longobards. His successors Severinus (640), *John IV (640-642) and esp. *Theodore I (642-649) took a clear stand against monophysism. In the empire's difficult position, this question of faith became a political matter. *Martin I (649-653) who, with the help of *Maximus Confessor, had got monothelism condemned by his synod in 649, was deported in 653 to Constantinople, where he was condemned and exiled. He died in 655, a martyr for the freedom of the church. His successors Eugenius I (654-657) and *Vitalian (657-672) were under Byzantine influence. In 666 the emperor elevated *Ravenna into an autocephalous church, a brutal interference in the rights of the Western patriarchate. Since Byzantine power was felt as foreign to Italy and as at the same time the links between the Apostolic See and the West, especially *Britain, were becoming closer, the definitive turn towards the Germanic kingdoms was only a question of time. This time it was not impeded by the resumption of ecclesiastical *communion between Rome and *Byzantium at the council of *Constantinople IV (680/81), under Agatho (678-681), nor by the rather favourable attitude of the popes of Sicilian, Greek and Syrian origin who occupied the Apostolic See between 678 and 752. Rome's growing separation from *Byzantium and ever closer union with the West, which was passing under Frankish rule, were furthered by the imperial action against *Sergius (687-701), whom the Roman people protected against Byzantine soldiers (692: after the so-called "2nd Trullan council"), by the political and military difficulties of the Byzantine empire itself, esp. in an Italy dominated by the *Longobards, by the iconoclast controversy which divided the East (from 726 on), by Roman interest in the missions, undertaken particularly by *Boniface, in Friesland and Saxony (719-754), and finally by the reconciliation of the p. and the Longobards under Zacharias (741-752), who also instructed a Frankish metropolitan to consecrate Pippin the Short as king of the the Franks. Under Stephen II (752-757), the union between Apostolic See and Frankish kingdom, conditioned by the hostile attitude of both the Byzantines and the Longobards, definitively opened a new era in papal history.

IV. Elements of historical evaluation. There is no doubt that the Roman bishops' awareness of being charged with the *cura ecclesiae universalis*, as successors of *Peter, grew only slowly during the first Christian centuries. It is no less certain that their ever more decided commitment to the *sollecitudo omnium ecclesiarum*, a decisive contribution to the defence of ecclesiastical freedom, constituted a complex history of claims on one hand and rejections on the other, moments of advance and moments of stagnation, or of rapid progress (*Stephen I, *Damasus, *Leo I, *Gelasius, *Gregory I); at times the process was slow, but a harbinger of more incisive times (*Liberius, *Zosimus, *Vigilius). It is notable that papal theory preceded the effective exercise of universal primacy (distinction of zones of influence). In this context Leo I's famous phrase, *Si quid itaque a nobis recte agitur recteque decernitur, si quid a misericordia Dei cotidianis supplicationibus obtinetur, illius est operum atque meritorum, cuius in sede sua vivit potestas, excellit auctoritas* points out at once the ideal extent and the concrete limits of the Petrine ministry (*Serm.* 3, 3). But we cannot understand its history except as a long and complex evolution. Opposing factors, favourable and restraining, interacted in this historical development. Yet judgment is not always easy. Facts which appear as obstacles are revealed at a distance as causes of progress, and *vice versa*.

On the civil level it was an advantage for the p. to be linked to the historical destiny of the Urbs, governing centre of the empire until the end of the third century. But the privileged position which it owed to this political situation was called into question when imperial residences were set up in other cities. Yet this political demotion of *Rome, particularly in favour of *Constantinople, became in the *tempora christiana* a positive point once more, since the p. profited from it for greater independence from the emperors, who were taking more and more interest, indeed too much, in ecclesiastical affairs, but who now resided far from the Urbs.

On the level of ecclesiastical politics the emergence of great episcopal sees, later called patriarchates, undoubtedly favoured the development of the Rome. On one hand, the p.'s jurisdiction over the West was recognized without too many difficulties by the Eastern sees (sometimes more than by those of the West), indeed it was used at *Nicaea as a model for *Alexandria's jurisdiction over *Egypt; on the other hand, the rivalry between the Egyptian see and that of "New Rome" allowed the Apostolic See to play a mediating role between the two rival sees and thus reinforce its own position. In the Western part of the empire, *Rome gradually overcame all the resistance of the other important sees. *Milan, above all *Carthage, later even *Ravenna were forced to admit, besides the always acknowledged rank of *caput et origo*, a much wider jurisdiction of the Apostolic See once those respective cities lost their political importance.

On the theoretical level we meet the same opposition of political ideas and theological premises. The idea of the *Urbs aeterna*, after the *conversion of the Roman aristocracy taken up in a Christian spirit by the popes of the 5th c., certainly gave an ideological foundation to the papal claims. But a similar aspiration could also animate the outlook of the bishops of New Rome, especially after *Romanitas*, to say nothing of *Graecitas*, had been compromised in the West (cf. the title of imperial ecumenical patriarch). The principle of *apostolicity particularly favoured Rome, the foundation of *Peter and *Paul, princes of the apostles. But Alexandria and *Antioch could also boast of being apostolic sees, as the popes themselves stressed against the pretensions of Constantinople, which itself later claimed to be the see of the apostle *Andrew. Indeed, as *Tertullian and *Cyprian attest, the principle of the apostolicity of all bishops was older. A whole series of other factors were needed before the Roman see could establish itself as the one *Sedes Apostolica*. Among them, the development of the cult of Peter and Paul and the visiting of their tombs merits particular consideration, though we must not totally overlook the fact that the same phenomenon, i.e. the veneration of relics and *pilgrimages to other sanctuaries, contributed not a little to a certain "Christian nationalism", which accentuated the centrifugal forces, the diverse languages and mentalities. Finally, much more important was what we may call a conflict of *ecclesiology. The Roman see had the advantage of representing the *Una Sancta*. This role must have been especially felt from the time Christianity was recognized as the official religion of the empire, and the Christians became so identified with it that the idea of *extra Ecclesiam nulla salus* meant in practice *extra Ecclesiam imperii nulla salus* (cf. the attitude of *Ambrose and many others to the barbarians).

Origo unitatis, symbolized by the singular vocation of *Peter, also included, esp. with regard to the Western churches, the privilege of *antiquitas*, in the chronological sense too. Against the Eastern churches, this prerogative could only apply in the sense that the church of Rome claimed always to have remained faithful to the apostolic *tradition, in particular that it had never abandoned the faith of *Nicaea. But the principle of the one apostolic church found itself in conflict with that of *catholica*, the *koinonia of all the apostolic churches, a principle no less felt in the historical context of the imperial church. This conflict was all the more crucial since catholicity included the idea of consensus, no less dear to the ancients than that of *antiquitas*. It was undoubtedly simpler to affirm one's agreement with all the churches by demonstrating *communion with that church which possessed the *principalitas* of all the apostolic churches (cf. *Irenaeus). Yet was not union with all the churches more convincing?

Finally, the personal level. There is no doubt that the church of Rome owed its rise, in great part, to the strong personalities who led it: *Victor I, *Stephen I, *Dionysius, *Damasus, *Innocent I, *Leo I, *Gelasius and *Gregory I. Yet this rise was sometimes contested by equally important personalities: *Cyprian, *Basil of Caesarea, *Ambrose, *Patroclus of Arles, *Cyril of Alexandria. On the other hand certain backward steps, indeed eclipses, of Roman primacy were due above all to weak, imprudent or over-hasty *popes: *Liberius, *Zosimus, *Vigilius, *Honorius, Eugenius I. However it is also true that the needs and difficulties of certain bishops, often caused by the failings of others, civil or ecclesiastical, gave the Apostolic See occasion to intervene in favour of oppressed bishops and thus to increase its own prestige, as the appeals of *Athanasius, *John Chrysostom, *Nestorius, *Flavian and in a certain sense even Basil demonstrate. In view of all these diverse and often conflicting factors, it is not easy to reach a just and balanced explanation of the ancient history of the p. (see below: V). To consider all these human, and even too-human, phenomena does not mean, for a Roman Catholic, casting doubt on the church's need for the continuous ministry of Peter. It simply means taking seriously the historicity, i.e. the incarnational structure, of the church of Christ.

Bibliographies: cf. in manuals of Church History, studies of individual popes, and above all the *Archivum hist. pont.*, Rome 1963 ff.
Sources: cf. eds. of the works of the popes and of *Epp.* addressed to them, indicated by CPL and CPG; P. Jaffé-G. Wattenbach, *Regesta Pontificum Romanorum*, I (-1143) Leipzig 1885; C. Vogel (L. Duchesne), *Le Liber Pontificalis*, Paris ²1955; H. Foerster, *Liber diurnus Romanorum Pontificum*, Bern 1958; C. Mirbt-K. Aland, *Quellen zur Geschichte des Papsttums und des römischen Katholizismus* I: *Von den Anfängen bis zum Tridentinum*, Tübingen ⁶1967.
General studies: cf. the manuals of Church History ed. by A. Fliche-V. Martin, H. Jedin, L.J. Rogier, etc., C. Andresen, K.D. Schmidt-E. Wolf; E. Caspar, *Geschichte des Papsttums von den Anfängen bis zur Höhe der Weltherrschaft*, Tübingen 1930-1933; P. Batiffol, *Cathedra Petri*, Paris 1938; J. Haller, *Das Papsttum*, I, Stuttgart ²1950; F.X. Seppelt, *Geschichte der Päpste*, I-II, Munich ²1954 f.; V. Monachino, *I papi nella storia* I, Rome 1961; G. Schwaiger, *Geschichte der Päpste*, Munich 1964; C. Falconi, *Storia del papato e dei papi*, Rome 1967 ff.; A. Franzen-R. Bäumer, *Papstgeschichte*, Freiburg ²1978; W. Ullmann, Leo I and the theme of Papal primacy, *JThS* n.s. 11 (1960) 25-51; D.E. Lanne, L'Eglise de Rome "a gloriosissimis duobus apostolis Petro et Paulo Romae fundatae et constitutae ecclesiae", *Irenikon* 49 (1976) 275-322; J. Richards, *The Popes and the Papacy in the Early Middle Ages, 476-752*, London 1979; M. Woytowytsch, *Papstum und Konzile von den Anfängen bis zu Leo I (440-461)*, Stuttgart 1981; C. Munier, La question des appels à Rome d'après la Lettre 20* (Divjak) d'Augustin, in (ed.) C. Lepelley, *Les Lettres de Saint Augustin découvertes par Johannes Divjak*, Paris 1983, 287-299; P. Lampe, *Die stadtrömischen Christen in den ersten beiden Jahrhunderten*, Tübingen 1987 (bibl.); J.E. Merdinger, The politics of persuasion: Augustine's tactics towards the Papacy, (Divjak Letters 22, 23 and 23A), *Atti Cong. Internatz. su S. Agostino nel XVI centenario della conversione*, Studia Ephemerides Augustinianum 24, 3 vols., Rome 1987.
Studies of individual periods or particular aspects: P. Batiffol, *Le siège apostolique*, Paris ³1924; G.B. Ladner, *Die Papstbildnisse des Altertums und des Mittelalters*, I, Vatican City 1941; O. Bertolini, *Roma di fronte a Bisanzio e ai Longobardi*, Bologna 1941; J. Ludwig, *Die Primatsworte Mt 16, 18-19 in der*

altkirchlichen Exegese, Münster 1952; M. Maccarrone, *Vicarius Christi. Storia del titolo papale*, Rome 1952; G. Langgärtner, *Die Gallienpolitik der Päpste im V. und VI. Jahrhundert*, Bonn 1964; M. Maccarrone, S. Pietro in rapporto a Cristo nelle più antiche testimonianze, *StudRom* 15 (1967) 397-420; W. Ullmann, *The growth of Papal Government in the Middle Ages*, London ³1970; W. Marschall, *Karthago und Rom. Die Stellung der nordafrikanischen Kirche zum apostolischen Stuhl in Rom*, Stuttgart 1971; P.P. Joannou, *Die Ostkirche und die Cathedra Petri im 4. Jahrhundert*, Stuttgart 1972; L. Magi, *Le sede romana nell corrispondenza degli imperatori e patriarchi bizantini (VI-VII sec.)*, Rome 1972; J. Hennig, Zur Stellung der Päpste in der martyrologischen Tradition, *Archivum hist. pont.* 12 (1974) 7-32; O. Wermelinger, *Rom und Pelagius*, Stuttgart 1975; J. Speigl, Das entstehende Papsttum, die Kanones v. Nizäa und die Bischofseinsetzungen in Gallien, *Festschrift H. Tüchle*, Paderborn 1975, 43-61; Ch. Pietri, *Roma christiana (311-440)*, BEFAR 224, 2 vols., Rome 1976; P. Stockmeier, Römische Kirche und Petrusamt im Licht frühchristlicher Zeugnisse, *Archivum hist. pont.* 14 (1976) 357-372; P. Conte, Il significato del primato papale nei padri del VI concilio ecumenico, *Archivum hist. pont.* 15 (1977) 7-111; V. Monachino, *Il canone 28 di Calcedonia, Genesi storica*. L'Aquila 1979; J.M.R. Tillard, *The Bishop of Rome*, London 1983.

B. Studer

V. *Historiography of the papacy. Historical interest in the p. appears first in the catalogues of Roman bishops. To stress the *apostolicity of the authentically Christian *tradition, *Hegesippus and *Irenaeus of Lyons in the 2nd c., and *Eusebius of Caesarea (†339) soon after, between them drew up these chronological *episcopal lists. An exclusive concern with *chronography led the so-called "Chronographer of 354" to harmonize the "Liberian" catalogue with the Roman consular *fasti*. Some 5th-c. chronicles as well as *Theodoret and *Socrates, both continuators of Eusebius's *Historia Ecclesiastica*, complete the catalogues up to their own time. The series of papal portraits made for St Paul's basilica attests the interest of that time in the history of the p. The chronographical undertakings of the 4th and 5th cc. were continued in the *Liber Pontificalis*, whose first two recensions (up to 530 and up to the end of the 7th c.) concern the patristic period. It is a collection of *biographies of Roman bishops, those before 400 brief, the later ones more extensive, all in stereotyped form (name, origin, length of pontificate, decrees, liturgical institutions, ordinations, funeral, vacancy of the see). The first recension contains few reliable facts, but the second includes precise information. While in the Middle Ages papal historiography was part of the ecclesiastical historiography of the time, distinguished by its rather edifying pattern of the ages of the world, that of the 16th c. was excessively dominated by the confessional controversies of that period, although Valla's work on the *Donation of Constantine was an early study in objective history. But from the start of the 17th c. a more critical and objective historiography commenced, thanks to the efforts of Panvinio, Baronius, Tillemont and others. The new methods, worked out by the *Maurists (Coustant) and Bollandists and finally by the historical criticism of the 19th c., contributed to an enormous progress in research on the p. The opening-up of the Vatican archives and the founding of the national institutes of History in Rome, while leading to great progress in the mediaeval and modern historiography of the p., could not help also stimulating studies of its beginnings.

Thus Duchesne, editor of the *Liber Pontificalis* (1886-92), Caspar (1930-37), Haller (1934-45), Seppelt (1931-41) and various Italian scholars, particularly of the history of Rome, such as Brezzi, dedicated themselves to the study of the origins of the p. However, as is plain from the most recent work on the subject, that of Pietri, unfortunately limited to the 4th and first half of the 5th cc., pontifical historiography must reach out to wider dimensions, i.e. it must consider not just papal documents and ancient writings, but also the development of the city of Rome, the forms of pastoral *ministry, social conditions, including what emerges from the data of *archaeology. Yet it is obvious that such undertakings still suffer from the poverty of preparatory monographs, as is shown, e.g., by the restricted case of the list of the popes.

Sources: besides the letters of the popes (cf. CPL 1627-1744) and the historical works of the ancient authors mentioned, cf. in particular the *Lib. Pontif.*, ed. L. Duchesne, Paris 1886-92, repr. 1955.
Studies: A. Mercati, La serie dei papi nell'Annuario pontificio per l'anno 1947, *Oss. Rom.* 19 genn. 1947 (Eng. vers. *Mediaeval Stud.* 9 [1947] 71-80); A. Frutaz, Papa, IV. La cronotassi papale, *EC* 9, 756-765; H. Tüchle, Papstgeschichts-schreibung, *LTK* 8, 49-53 (bibl.); R. Bäumer, Papstliste, *LTK* 8, (*331-440*) 54-59; Ch. Pietri, *Roma christiana*, BEFAR 224, 2 vols., Rome 1976; M. Wojtowytsch, *Papsttum und Konzile von den Anfängen bis zu Leo I (440-461)*, Stuttgart 1981; W. Ullmann, *Gelasius I (492-496)*, Stuttgart 1981.

B. Studer

VI. The *popes. Up to 752, the lists give 92 bishops of *Rome, including *Peter. Many of these have their own specific treatment elsewhere in this Dictionary. Here we will give brief biographical details of the others.

Linus (67?-79?). Nothing certain is known of L., but acc. to *LP* he was from Tuscia. *Irenaeus and *Eusebius identify him with the L. named in 2 Tim 4, 21, and consider him the first successor of the apostles (Iren., *Adv. haer.* III, 3, 3; Euseb., *HE* III, 2; III, 4, 8). His *martyrdom is uncertain.

Anacletus or *Cletus* (79?-88?). Chronology again uncertain, but Irenaeus and Eusebius make him Linus's successor (Iren., *Adv. haer.* III, 3, 3; Euseb., *HE* 13; III, 21; V, 6, 1). A Cletus placed before him in *LP* I, 22 may just be an abbreviation of Anacletus (*LP* I, 2-3; cf. *Carmen c. Marc.* III, 278: CSEL 2, 1441).

*Clement I (88?-97? [92-101 acc. to Eusebius's chronology]).

Evaristus (97?-106?). Certainly Clement's successor (Iren., *Adv. haer.* III, 3, 3; Euseb., *HE* III, 34; IV, 1). Of Greek origin, he died a *martyr; the writings attrib. to him are apocryphal (PG 5, 1047-1058; Jaffé 20-23).

Alexander I (105?-115?). Evaristus's successor; seemingly not martyred (Iren., *Adv. haer.* III, 3, 3; Euseb., *HE* IV, 1; V, 6, 4). The writings attrib. to him are apocryphal (Jaffé 24-30; cf. *DTC* 1, 709).

Sixtus I (115-125?). *LP* gives him a Roman origin and attributes liturgical and disciplinary regulations to him (I, XCI and 128; cf. Jaffé 31-33). The sixth in Irenaeus's list (*Adv. haer.* III, 3, 3; Euseb., *HE* IV, 4; V, 6, 4). He did not expel from ecclesiastical *communion those who celebrated *Easter on a date different from his own (Euseb., *HE* V, 24, 4).

Telesphorus (125?-136?). Seventh bishop of Rome; died a martyr (Iren., *Adv. haer.* III, 3, 3). The decretal attrib. to him is not genuine (Jaffé 34).

Hyginus (136?-140?). Length of his pontificate very uncertain. In his time the gnostics *Valentinus and *Cerdo arrived in Rome; Cerdo, after a profession of faith, was received into the Roman community, but later expelled (Iren., *Adv. haer.* III, 4, 2).

Pius I (140?-155?). Chronology uncertain, but ninth bishop of Rome (Iren., *Adv. haer.* III, 3, 3; Euseb., *HE* IV, 22, 3). The Muratorian Fragment makes him brother to *Hermas, author of the *Shepherd*. During his pontificate various *heretics propagated their doctrines at Rome: *Valentinus (Iren., *Adv. haer.* III, 4, 3) and *Marcion (Tertull., *Adv. Marc.* I, 19). *Justin also taught there. For the numerous writings attrib. to him, cf. Jaffé 43-56.

Anicetus (155?-166?). In his time the Roman church must have been enlivened by the presence or visits of various personalities: Justin, *Tatian, *Hegesippus (Euseb., *HE* IV, 11, 7; IV, 22, 3). *Polycarp was given such a welcome there that, despite differences over the date of Easter, he and A. parted in *peace (Euseb., *HE* IV, 15; V, 24, 14-17). Polycarp also met various heretics there, whom he brought back to *orthodoxy (Iren., *Adv. haer.* III, 3, 4): they were numerous in the capital.

Soter (166?-175?). We know more of him than of his predecessors. He wrote to the community of *Corinth; his *letter (lost) was well received, and *Dionysius of Corinth replied praising the generosity of the Roman community and of S. (Euseb., *HE* IV, 23, 9-11). Harnack thought *2 Clement* could be S.'s letter, an improbable opinion. Acc. to *Praedestinatus, he wrote against the Marcionites (I, 26: PL 53, 506); *Marcionism did indeed start to spread in his time and might have come to his knowledge.

Eleutherius (175?-189). Of Greek origin, Anicetus's deacon, he succeeded Soter (Euseb., *HE* IV, 22, 1); last on Irenaeus's list (*Adv. haer.* III, 3, 3): *Irenaeus, recommended by his community of *Lyons, brought him, from the Lyonais Christians in prison, a letter on *Montanism (Euseb., *HE* V, 3, 4; V, 4, 1-2), on which he was prob. still ill-informed.

*Victor (189-199).

*Zephyrinus (199-217).

*Callistus I (217-222).

Urban I (222-230). Bishop in a period of religious *peace, but the Christian community of Rome was still disturbed by the *schism of *Hippolytus. Buried in the *cemetery of S. Callisto, where an inscription has been recovered. Legend anachronistically associates him with the martyr *Cecilia (*LP* I, 63).

*Pontianus (230-235).

Antherus (235-236). Elected after Pontianus was exiled to *Sardinia; seemingly not a martyr. Buried in the cemetery of S. Callisto.

*Fabian (236-250).

*Cornelius (251-253).

Lucius I (253-254). Hardly elected when he was exiled, but soon allowed back to Rome; corresponded with *Cyprian (*Epp.* 61; 68, 5). Perhaps not a martyr; buried in Callisto. His pontificate lasted about eight months (Euseb., *HE* VII, 2). For his apocrypha cf. Jaffa 122 f.

*Stephen I (254-257).

*Sixtus II (257-258).

*Dionysius (259-268).

Felix I (269-274). Not a martyr, though *Cyril of Alexandria calls him one. The council of *Antioch, having condemned *Paul of Samosata, wrote a *letter to his predecessor (Euseb., *HE* VII, 30, 2-17); Dionysius being dead, F. received and answered it. A sentence cited by Cyril may well be his, though probably altered (ACO I, I, 7 p. 45; Quasten I, 488).

Eutychianus (275-283). The facts given by *LP* are rather fantastic; not a martyr. For his spurious writings, cf. Jaffé 145-156.

Caius or *Gaius* (283-296). Of Dalmatian origin, pope for 12 years (15, acc. to Euseb., *HE* VIII, 22, 1). Not a martyr. *LP* attributes to him regulations on admittance to *orders (I, 161). Named in the *Acts* of Sebastian and those of Susanna. Buried in Callisto, where his epitaph is preserved.

*Marcellinus (296-304).

*Marcellus I (308-309).

Eusebius (309-310). A Greek doctor, acc. to *LP*. A poem of Damasus tells how the disorders that broke out under Marcellus I over the readmission of *lapsi continued under him (Ferrua 129-136; 181). Exiled by the emp. *Maxentius to Sicily, where he died. For his spurious writings, cf. Jaffé 163-170.

*Miltiades (311-314).

*Sylvester I (314-335).

Marcus (18/I-7/X/336). Roman, Sylvester's deacon. Laid down that the pope should be consecrated by the bishop of Ostia (*LP* I, 202; cf. Aug., *Brev. coll. cum don.* III, 16, 29: PL 43, 641). Built churches in Rome, including S. Marco, now part of Palazzo Venezia. Buried in the cemetery of S. Balbina. Cf. Jaffé 181.

*Julius I (337-352).

*Liberius (352-366).

*Damasus I (366-384).

*Siricius (384-399).

Anastasius I (399-401). Roman, esteemed by *Paulinus of Nola (*Ep.* 20, 2) and *Jerome (*Epp.* 127, 10; 130, 16). He condemned some *Origenist propositions (Jaffé 276; 282) and was very severe with *Rufinus of Aquileia, who sent him an *Apologia* to vindicate his *orthodoxy (PL 21, 623-629). He encouraged the African bishops against *Donatism (Jaffé 283; Mansi 3, 1023) and opposed the *Manichees (*LP* I, 218).

*Innocent I (401-417).

*Zosimus (417-418).

Boniface I (418-422). On Zosimus's death, part of the *clergy elected Eulalius (27/XII/418); next day, the other part nominated the presbyter B. Eulalius was supported and defended at the court of *Ravenna by the prefect *Symmachus, who at first called a *council at Spoleto but then, because of Eulalius's attitude, recognized B., who laid down that future contested cases should be decided by a new election (Jaffé 353). He got the emp. *Honorius to condemn the *Pelagians (cf. Prosper of Aq., *Contra coll.* 21, 57: PL 45, 1831); *Augustine dedicated *Contra duas epist. Pelagianorum* to him. He intervened in the affair of *Apiarius of Sicca and in *Gaul (Jaffé 349; 362), and defended the rights of the Roman see over Illyricum (Jaffa 350 f.; 363 f.; 368). Cf. CPL 1648 f.; PL 20, 749-792; *LP* I, 227-229.

*Celestine I (422-432).

*Sixtus III (432-440).

*Leo I, the Great (440-461).

*Hilarus or Hilarius (461-468).

*Simplicius (468-483).

Felix III (II) (483-492). Roman, son of a presbyter (*LP* I, 252) and ancestor of Gregory the Great (*Dialogues* IV, 16), he assumed the pontificate at the time of *Odoacer. No sooner elected than he sent a delegation to *Constantinople with a series of requests about *Africa, *Peter Mongus and patriarch *Acacius (Jaffé 591 f.); then he invited Acacius to Rome to give an account of his conduct (Jaffé 593). But, ill-informed, on his legates' return he *excommunicated them and Acacius at a Roman synod in 384 (Jaffé 599-604). Acacius reacted by erasing F.'s name from the *diptychs: thus began the Acacian schism which, despite various attempts in the following years, lasted until 519 (cf. Acacius and Hormisdas). He intervened in *Dalmatia on Pelagian questions (CSEL 35, 398-439 among Gelasius's letters).

*Gelasius I (492-496).

Anastasius II (496-498). Son of a Roman priest; he sought to end the Acacian schism (cf. above), sending a delegation to the emp. *Anastasius and being very conciliatory (Jaffé 748). He addressed a letter to the bishops of Gaul condemning the opinion that the human *soul is generated by the parents (Jaffé 751). His pro-Byzantine policy displeased many clerics (*LP* I, 258) and, on his death, caused the Lawrentian schism (cf. *Lawrence and *Symmachus).

*Symmachus (498-514).

*Hormisdas (514-523).

*John I (523-526).

Felix IV (III) (526-530). Imposed by *Theodoric; through *Caesarius of Arles, he supported the condemnation of the *semi-Pelagians by the council of Orange (529).

Dioscorus (22/IX-14/X/530). Alexandrian, elected pope by the majority of the Roman clergy, though Felix IV had designated Boniface as his successor.

Boniface II (530-532). *Archdeacon at Rome, designated by Felix IV in order to avoid rivalries, but this choice was ratified by only part of the clergy, the others choosing Dioscorus, who died immediately; then B. claimed the submission of his opposers. At first he too wanted to name a successor in the person of Vigilius, but he later retracted (*LP* I, 281). He confirmed the canons of the council of Orange (Jaffé 881).

*John II (533-535).

Agapitus I (535-536). Roman, son of the presbyter Gordian (*LP* I, 287) who died in the riots of the Lawrentian schism (501/502). He accepted the appeal of Contumeliosus of Riez, deposed by *Caesarius of Arles, to whom he refused the right to alienate *property (Jaffé 890 f.; Mansi 8, 855 f.). Wrote to the African bishops on the treatment of Arian converts (Jaffé 892); sent a letter to the emp. *Justinian on the same question (Jaffé 883). The Ostrogothic king Theodahat sent him to negotiate with Justinian, from whom he obtained the deposition of patriarch *Anthimus, a *monophysite, previously bishop of Trebizond. Died at *Constantinople; his body was taken to Rome; venerated as a saint in the Latin and Greek churches.

*Silverius (536-537).

*Vigilius (537-555).

*Pelagius I (556-561).

John III (561-574). At a time of disorder in *Italy, suppressed by general Narses (*LP* I, 305), and of religious struggles over the *Three Chapters, he managed to restore unity with *Africa and N Italy, but not with *Aquileia. He also intervened in *Gaul to restore bishops Salonius and Sagittarius to their sees (Jaffé 1040).

Benedict I (575-579). Consecrated after a long vacancy of the Roman see. During his pontificate, bands of *Longobards sacked Italy. Died in the besieged city.

*Pelagius II (579-590).

*Gregory I, the Great (590-604).

Sabinianus (604-606). Born at Blera in Tuscany; before succeeding Gregory, he had been *apocrisiary at Constantinople.

Boniface III (19/II-12/XI/607). Roman; apocrisiary at Constantinople in 603; persuaded the emp. Phocas to recognize the Roman see as *caput* of all the churches (*LP* I, 316); held a *council at Rome which laid down that, during the life of the pope or of a bishop, no preparations could be made for the election of his successor (*LP* I, 316).

Boniface IV (608-615). Native of Marsica. Continued good relations with Phocas (*LP* I, 317). Turned the Pantheon into a Christian church; favoured the introduction of Roman, rather than *Celtic, customs into the English church.

Adeodatus I (Deusdedit) (615-618). Of Roman origin.

Boniface V (619-625). Neapolitan. Shared his predecessors' concern with the English church (Jaffé 2005-2009).

*Honorius I (625-638).

Severinus (28/V-2/VIII/640). Roman. Had to wait nearly 20 months for the emp. *Heraclius, whose *Ekthesis* he did not sign, to confirm his election. During his pontificate the papal treasury was sacked by the *chartularius* Maurice (*LP* I, 238).

*John IV (640-642).

*Theodore I (642-649).

*Martin I (649-655).

Eugenius I (654-657). Roman presbyter, elected while Martin was still alive but exiled in the Crimea. By diplomatic means, he continued negotiations to reconcile the Roman church and the emp. *Constans II (641-668), who supported *monothelism.

*Vitalian I (657-672).

Adeodatus II (672-676). Roman monk, continued to oppose monothelism. Granted the privilege of exemption to the abbey of Saint-Martin of Tours (Jaffé 2105).

Donus (676-678). Roman. Seems to have obtained recognition of the apostolic see's authority, from the *archbishop of *Ravenna. Promoted the restoration and decoration of many Roman churches.

Agatho (678-681). Of Sicilian origin. With him the autocephaly of the see of Ravenna ended. He intervened in England on matters concerning subdivision of dioceses (Mansi 11, 179). He accepted the invitation of emp. Constantine Pogonatus to put an end to the disputes on monothelism; to this end he called a council at Rome in 680, which condemned the *heresy. Thence he sent a delegation to a council at Constantinople (the 6th ecumenical, 680/681), which ended after his death.

Leo II (682-683). Also Sicilian. After his election he had a long wait before the emperor allowed his consecration. He ratified the acts of the 6th ecumenical council and the condemnation of Honorius (Jaffé). But in his letters to the bishops of *Spain he toned down this condemnation (Jaffé 219-2122). An educated man who knew Greek.

Benedict II (684-685). Roman. Also had to wait long for his consecration to be authorized; persuaded the emperor to delegate the authorization to the exarch of Ravenna. Persuaded the 16th council of *Toledo (684) to accept the 6th ecumenical council.

John V (685-686). Of Syrian origin - most popes of this period were Oriental - and previously involved in the 6th ecumenical council.

Conon (686-687). On John V's death (2 Aug 686), the clergy proposed the *archpriest Peter, and the army the presbyter Theodore. Finally they agreed on the old man C.; brought up in Sicily, his father was a soldier in the Thracian corps (*LP* I, 368-370).

*Sergius I (687-701).

John VI (701-705). Of Greek origin, elected against the wishes of the emp. Tiberius III (698-705), who tried to get rid of him. During the Longobard invasion, he ransomed Campanian prisoners with church treasures.

John VII (705-707). Also Greek. Refused, like Sergius I, to approve the synod of Constantinople of 692 (called *Quinisext* or *in Trullo*).

Sisinnius (15/I-4/II/708). Of Syrian origin, he lasted only 20 days.

Constantine (708-715).

Gregory II (715-731). Roman and well prepared to perform his office. He withstood *Leo III the Isaurian (717-741) and condemned iconoclasm at the Roman synod of 729. Rebuilt the abbey of Montecassino and restored S. Paulo *fuori le mura*. Encouraged the apostolate of St *Boniface - even changing his name - in *Germany, giving him directives on moral, liturgical and disciplinary questions.

Gregory III (731-741). Syrian. Condemned the iconoclasts again at the council of 731, definitively removing the church from the influence of the exarch of Ravenna, *longa manus* of the emperor (732).

Zacharias (741-752). A Calabrian Greek. He brought the ancient papacy to an end and began the mediaeval papacy. As well as supporting St Boniface's German missions, he recognized Pippin the Short as king, laying the basis of that alliance between the Holy See and the *Franks which culminated in the coronation of Charlemagne in 800.

The ancient church also records the following antipopes: *Hippolytus (217-235); *Novatian (251-258?), both Roman; Heraclius (309 or 310); Felix II (355-365), also Roman; Ursinus (366?-381?), a Roman deacon; Eulalius (418-419), Roman *archdeacon; *Lawrence (498-501/507), Roman archpriest; Paschal (687-692), Roman archdeacon. But Peter and Theodore, involved in the events connected with the papal succession in 686 and 687, should not be considered antipopes.

J.N.D. Kelly, *The Oxford Dictionary of Popes*, Oxford 1986.

M. Spinelli

VII. Letters of the popes. Under the general term "Letters of the Popes" (*epistulae pontificiae*), church historians include all documents attributed to bishops of *Rome, without distinction of particular forms, such as the decretals going back to the end of the 4th c., the encyclicals, first attested under *Martin I (649-653), chirographs, apostolic letters, etc., except for the so-called papal privileges. Among these documents, undoubtedly primordial sources for the history of the papacy and of the whole church, we must distinguish those of the first three centuries from later ones.

For the first period, documents preserved are very scarce. Except for *1 Clement* which presents itself as a *letter from the community of Rome, we possess entire only two letters of *Cornelius I (251-253), included among the letters of *Cyprian (CPL 50). The fragments of the letters of some other popes (Soter, Eleutherius, *Victor I, *Pontianus, Lucius I, *Stephen I, *Dionysius) have been handed down, like so many other writings of the early centuries, in *Eusebius of Caesarea's *Historia Ecclesiastica*. Among these documents, the most important are those of Victor I on the question of *Easter, Stephen I on *baptism administered by *heretics (DS 110 f.), and Dionysius on the *theology of the *Trinity (DS 112-115). Other letters are mentioned or partly preserved in other ecclesiastical authors, like the two letters of *Julius I (337-352) in *Athanasius (CPL 1627), and the letters of *Liberius (352-366) in *Hilary and *Socrates (CPL 1628 ff.).

With *Damasus I (366-384) the second, much better documented, period begins. Thanks to papal registers (*regesta, registri*) in which, from the 4th c. on, following Roman custom, copies of documents issued by the apostolic chancellery were placed, though handed down only in a fragmentary state, and above all thanks to the ecclesiastico-political and *canonical collections (decretal law), of which the *Avellana* is the most famous, we know the papal correspondence of this second period quite well, esp. those of *Innocent I (401-417), (CPL 1641), *Leo I (440-461) (CPL 1656: more than 170 letters), *Hilarus (461-468) (CPL 1662 f.), *Gelasius I (492-496) (CPL 1667 f.), *Symmachus (498-514) (CPL 1678) and *Gregory I (590-604) (CPL 1714). Even if these writings must be attributed more to the anonymous activity of the apostolic chancellery than to the literary initiative of individual authors, except perhaps some letters of Leo and Gregory, nonetheless they deserve to be considered not just as sources of the history of the Roman church, but also as evidence for the history of ancient Christian literature. The papal letters, indeed, esp. those written in the more ardent periods of papal history, have great importance for the history of Christian doctrines, canon law and *liturgy, in particular for the *Augustinian theology of *grace, some sacramental questions, *christology, the *ecclesiological positions of the Roman curia in relation to those of the imperial court and of other bishops. At the same time, they allow us to understand more deeply the theological and ecclesial situation of various *Fathers of the church: *Ambrose, *John Chrysostom, *Augustine, John *Cassian, *Cyril of Alexandria, *Theodoret and, naturally, those popes who, like Leo and Gregory, are numbered among the *Fathers of the church. Even from a literary point of view they are not without interest, allowing us to follow the development of rhythmical prose, i.e. the transition from quantitative rhythm to stressed rhythm in the final *clausulae*. They also demonstrate the influence of the Roman legal-political mentality on Christian *Latin, as we see particularly in Leo and his immediate successors. They constitute a typical case of Christian adaptation to a secular *literary genre, i.e. the resumption of forms of imperial legislation in the decretal legislation of the Apostolic See. Finally, the later letters, of Gregory and others, attest the transition from the ancient to the mediaeval mentality.

A study of all this papal documentation, which is of interest not just for the history of the church of Rome, but also for *patrology, the history of later Latin literature and even the history of the end of the Roman empire and the beginnings of the Byzantine empire and the Germanic kingdoms, requires a vaster knowledge of editions, subsidiary works and studies than that normally presupposed for patristic research.

On editions: CPL 1568-1774, and in particular P. Jaffé-G. Wattenbach, *Regesta Romanorum Pontificum ab condita Ecclesia ad a. 1198*, Leipzig ²1885-1888; C. Silva-Tarouca, Le antichi lettere dei Papi e le loro edizioni (sec. IV-VI), *CivCatt* 72 (1921) 13-22, 323-336; id., Nuovi studi sulle antiche lettere dei Papi, *Gregorianum* 12 (1931) 3-56, 349-425, 547-498.
Editions: besides PL, MGH, CSEL, CCL, cf. A. Thiel, *Epistulae Romanorum Pontificum genuinae et quae ad eos scriptae sunt*, Braunsberg 1867/8 (repr. Hildesheim 1974); C. Mirbt-K. Aland, *Quellen zur Geschichte des Papsttums und des römischen Katholizismus*, I, Tübingen ⁷1967.
Studies: besides Church Histories, cf. in particular E. Caspar, *Geschichte des Papsttums von den Anfängen bis zur Höhe der Weltherrschaft*, I-II, Tübingen 1930, 1933; F. Di Capua, *Il ritmo prosaico nelle lettere dei Papi e nei documenti della cancelleria romana dal IV al XIV secolo*, Rome 1937/9; H. Erharter, Papstbriefe, *LTK* 8, 48 f. (bibl.); C. Andresen, *Die Kirchen der alten Christenheit*, Stuttgart 1971, esp. 579-601; Ch. Pietri, *Roma christiana (311-440)*, BEFAR 224, 2 vols., Rome 1976; B. Studer, *Patrologia* III, Turin 1978, 546 ff. (bibl.); P.A. McShane, *La Romanitas et le Pape Léon le Grand*, Paris 1979 (bibl.).

B. Studer

PAPAS BAR AGGAI. In the first three decades of the 4th c. he was *archbishop of *Seleucia-Ctesiphon and worked to create the centre of the Christian community under the *Sassanids in the capital itself. He came into conflict with many bishops, more than one of whom were to die as *martyrs in the subsequent persecutions. He also tried to obtain recognition from the Eastern bishops of the Roman empire. The *letter on the church hierarchy as mirror of the heavenly hierarchy, addressed to his disciple Aggor, is a later Nestorian *pseudoepigraph.

EC 9, 777; Baumstark 29-30; *DTC* 11, 164 f.

M. van Esbroeck

PAPHNUTIUS. Name of various Egyptian *ascetics or hermits. The first, mentioned by *Rufinus and *Sozomen, was a bishop and *confessor, present at the council of *Nicaea (325), where he sided against ecclesiastical *celibacy (he seems legendary). The second († before 394) is known from the *Historia monachorum in Aegypto*. A homonymous contemporary, surnamed Bubalos (wild ox), figures in the *Apophthegmata of *Cassian and *Palladius. The best-known (and wholly legendary) was the disciple and presumed biographer of St *Onuphrius.

BS 10, 35-37 and 26-28.

J. Gribomont

PAPIAS of Hierapolis. Bishop of Hierapolis in *Phrygia, his birthplace, and author of the *Explanation of the Lord's sentences*, published c.130/140, a work that considerably influenced *Irenaeus, *Hippolytus and *Victorinus of Petovium. Only 13 fragments remain, preserved by Irenaeus and *Eusebius of Caesarea (Iren., *Adv. haer.* V, 33, 4; Euseb., *HE* II, 15, 2; III, 36, 2; 39, 13-17). P. dedicated his work to the *exegesis of the Lord's words and deeds. The introduction (Euseb., *HE* III, 39, 15 ff.) wishes to reassure the reader of the historical authenticity of the book's contents, since the author refers to John's seven disciples, who belonged to the group of the presbyters *Aristion and *John, though only the last two should be considered historical figures (the seven disciples perhaps represent a literary pattern quite widespread at the time). Acc. to Irenaeus, P. was a "very archaic" person, a follower of John and a contemporary of *Polycarp (Iren., *Adv. haer.* V, 33, 4); but acc. to Eusebius (*HE* III, 39, 13), he was a weak thinker and held many *Judaeo-Christian ideas. The fragments of P. mention only the gospels of *Mark and *Matthew, they say nothing of *Luke, the Pauline letters or *John's gospel. The most famous fragment, on which there were opposing opinions (Euseb., *HE* III, 39, 15), refers to the tradition of a presbyter named by P., acc. to whom *Mark's gospel was nothing but the interpretation of the Petrine *kerygma adapted to the various circumstances. Acc. to P., *Matthew composed his gospel in "Hebrew dialect", words that have given rise to numerous debates among scholars.

LTK 8, 34-36; *RGG* 5, 47-48; A. v. Harnack, *Geschichte der altchristlichen Literatur*, I, 1893, 65 ff.; II, 1, 1897, 333 ff.; W. Lahrenfeld, Ein verhängnisvoller Schreibfehler bei Eusebius, *Byzantinisch-neugriechische Jahrbücher* 3 (1922) 282-85; J.F. Bligh, The Prolog of Papias, *ThS* 12 (1952) 234-40; F.X. Funk-K. Bihlmeyer, *Die Apostolischen Väter, Sammlung Ausgewählten Dogmen- und Kirchengeschichtlichen Quellenschriften*, Tübingen ²1956, 133 ff.; J. Munck, Presbyters and Disciples of the Lord in P., *HThR* 52 (1959) 223-43; W. Bauer, *Rechtgläubigkeit und Ketzerei im ältesten Christentum*, Tübingen ²1968, 187 ff.; Altaner 54 f.; J. Kürzinger, *Papias von Hierapolis und die Evangelien des NT*, Regensburg 1983; U.H.J. Körtner, *Papias von Hierapolis: ein Beitrag zur Geschichte des frühen Christentums*, Gottingen 1983; R.M. Grant, Papias in Eusebius' Church History, *Mél. H.C. Puech*, Paris 1974, 209-213.

L. Vanyó

PAPYRUS - PAPYROLOGY. Papyrus as a writing material, obtained from the plant of that name (*cyperus papyrus*) by careful treatment of the stem, originated and was extraordinarily widespread in *Egypt, where it was in common use by the 3rd millenium BC. The technique used in making papyraceous paper is described in detail, though with some inexactness, by Pliny (*Nat. Hist.* XIII, 74-77; 81-82). Technical and statistical data, as well as socio-economic information, have helped us refute two inexact suppositions, i.e. that papyrus was expensive and fragile. The causes that led, from the 4th c., to the dominance of the parchment codex over that of papyrus, must have been of a different kind. Parchment too, obtained by treating the skins of small quadrupeds, became an excellent writing material: it too comes, together with other materials, under the heading of papyrology. But its origin is less ancient than that of papyrus, with which it came into competition esp. because of the use of the codex.

The original form of the *book was the scroll or *volumen*, consisting of a number of sheets pasted together to form a strip of any length, which was then rolled up so that the writing was on the inner surface, usually smoother as well as more protected. But by the 1st c. BC we sense something new arising, with the substitution by the Romans of pieces of wood joined in a group (*codex*), with parchment sheets bound together to form rudimentary books (*membranae*), to which St *Paul refers (2 Tim 4, 13). But while the *pagan world was almost indifferent to this innovation, remaining attached to the traditional form, the Christians realized its usefulness and advantages and perfected its composition, and very soon the codex "prevailed as the only possible format for the Christian Scriptures". It has been suggested that such decisive innovations may have been the work of a dominant figure in the primitive church who introduced, at the same time as the adoption of the papyrus codex, the use of the *nomina sacra*, "stereotyped" abbreviations of characteristic Christian words, such as "Lord", "Father", "Son", "Spirit", "Cross", "Saviour", etc. The papyrus codex thus became the distinctive book-form of Christian manuscripts in the early church, differing "from the parchment scroll of Judaism and the papyrus scroll of the pagan world" (Skeat).

By the term papyrology we mean the science which arose at the start of the 20th c. and concerned itself with the Greek, and the very few Latin, documents of antiquity, written mainly on papyrus but also on parchment or other writing-materials: from the oldest Greek papyri, going back to the 4th c. BC, to documents of the 7th c. AD (Arab conquest of Egypt) and even as far as the 9th c. The field of papyrology includes specializations such as Egyptian papyrology (Egyptian languages) and Arabic papyrology.

The 20th c. has truly been, in the "prophetic" expression of Theodore Mommsen, the century of papyri, as the 19th was that of *epigraphy. Papyrological discoveries, sometimes sensational, made as the result of excavations or antiquarian acquisitions, provide an exceptional quantity and variety of evidence, concerning all fields of the sciences of antiquity from political history to literature, from law and economics to religious history. It has particularly favoured the documentation of early Christianity, primarily in its most specific sector, that of biblical evidence.

The quantity and details of texts, particularly of the NT, which have come down to us on papyrus or parchment confirm and enrich the literary-historical evidence of the daily use and diffusion at all levels of the Holy *Scriptures within the primitive communities. Of this evidence, the oldest and most valuable are Yale P. I, 1 (late 1st c.), a fragment of papyrus codex of Genesis; the well-known Ryl. P. III 457, a fragment of papyrus codex, perhaps from el-Faiyum or *Oxyrhynchus, dateable by *palaeography to the first decades of the 2nd c. and containing Jn 18, 31-32, 37-38; and the important group of Chester-Beatty PP., the oldest of which are dateable to the 2nd c. Next in antiquity and importance is Bodmer P. II (2nd-3rd c.), prob. found in Upper Egypt (Panopolis): a papyrus codex with an accurate edition of the whole of *John's gospel, from which we deduce that exemplars of the Fourth Gospel circulating in Egypt at the end of the 2nd c. must have been many and varied. To the 4th c. belong the two most famous codices of the *Bible, which mark the triumph of the parchment over the papyrus codex: the *codex Sinaiticus* (Brit. Mus. Add. MS 43725) thought to be from *Palestine, and the *codex Vaticanus* (Vat. Gr. 1209), prob. from Egypt (*Alexandria?). Some have thought to link the two codices with the well-known *letter sent by *Constantine the Great to bishop *Eusebius of Caesarea, ordering him to provide 50 parchment Bibles written "by professionals with precise knowledge of their art". Among NT *apocrypha, the best-attested is the *Protogospel of James*, of which Bodmer P. V (3rd c.), from upper Egypt, entitled Γένεσις Μαρίας, gives the oldest version.

The contributions of papyrology in the strictly patristic field, whose basis and constant reference-point is the NT text, are considerable, though proportionally inferior to biblical and secular Greek finds. *Patristics, besides extending its interests beyond its traditional borders, has nowadays increasingly taken historical and philological aspects into its properly theological sphere (K. Treu); and papyrology itself, thanks both to the distinctiveness of its evidence and to this fruitful convergence, has left a mark on the contact of specifically philological data with the contents of theological thought, and been influenced by it in turn: a kind of thoughtful convergence capable of pointing to a way of overcoming that philological-theological dualism which has been the subject of long and lively debate. Into this perspective come some of the most notable examples among the many that are registered in the respective specialized catalogues.

Two singular readings of OxP XV 1782 (late 4th c.) relating to *Didachè* 1, 4 have suggested that, in the compilation of the text transmitted by the original small parchment codex to which the Oxyryhnchus fragment belongs, educational intentions and ethical interests were deliberately stressed.

The main "papyrological" evidence of the *Pastor Hermae* is in Mich. P. 130, a fragment of papyrus scroll with *Mand.* II, 6-III, 1 (late 2nd c.: el Faiyum), and in Mich. P. 129 (second half of 3rd c.: Theadelphia or Asyut?), long fragments of papyrus codex that give considerable parts of the "Precepts" and "Similitudes" and which confirm the hypothesis that the whole text was not published at the same time and that for some time the work began with the 5th Vision.

*Melito of Sardis's *De Pascha* is handed down in two series of papyrus codex sheets: Chester-Beatty P. XII and Bodmer P. XIII, dateable respectively to the 4th and 3rd-4th cc. Contacts between this celebrated work and the apocryphal *Acta Ioannis* (2nd c.) have led to the supposition of a common theological *tradition on the central theme of trinitarian and christological doctrine.

The oldest and most interesting direct evidence of *Irenaeus's *Adversus haereses* is OxP III 405, seven fragments of a papyrus scroll almost contemporary with the author, who, judging from a citation of Mt 3, 6-7 corresponding to MS D, must have known the text now represented by the Beza Codex.

Numerous papyrus finds that have enriched our knowledge of *Origen's texts include the well-known Tura Codex (6th-7th c.), published by Scherer in 1949: this gives the interesting "christological" debate between Origen and Heraclides, which shows the active participation of *laymen in synodal debate. Also from Tura, and also by Origen, is a papyrus codex with *De Pascha*, recently published by Guéraud and Nautin and dateable to the 6th-

7th c.: a text with useful comparisons in the exegetical *catenae. A curious, though not rare, occurrence is represented by a group of fragments which ended up in the Florentine collection (Pubblic. Soc. Ital. per la ricerca dei papyri: PSI inv. 2101), but have been shown to come from the same codex as various fragments now preserved in the British Museum (Egerton P. 3), all dateable to the mid 3rd c. and almost certainly belonging to a work of Origen, perhaps to an otherwise missing or mutilated section of his *Commentary on the 4th Gospel* (or on *Genesis?).

Various commentaries of *Didymus the Blind, originating from the monastery of St Arsenius (*Origenist in tendency), come from the papyrological casket of the Tura cave, an outstanding source of precious contributions to patristic research.

There is no lack of contributions to the works of *Basil the Great, *Gregory Nazianzen, *John Chrysostom, *Cyril of Alexandria and others. The one example of *Gregory of Nyssa is in BKT VI, 4 (Berliner Klassikertexte) (Berlin P. inv. 5863), six double sheets of a 5th-c. codex (el Faiyum) with extracts of *De vita Moysis*. This papyrus decisively confirms the authority of cod. Venetus 67 on Gregory's use of the term ἀποκατάστασις. Also important is the group of unidentified or anonymous patristic fragments (more than 100 in van Haelst's catalogue), which constitute a considerable literary enrichment and allow us to glimpse, by comprehensive study, further contributions to patristic thought. Nor, finally, can we ignore the contribution of papyrology in the field of hagiographical and liturgical literature, and in that of so-called minor or popular literature. *Coptic-language papyri constitute a fertile and continually growing field, whose remarkable documentation extends from the biblical sphere to patristic literature, particularly *hagiography. Among the main publications, three series stand out: that of Leiden on the gnostic library of *Nag Hammadi, the Coptic series edited by T. Orlandi in the Milanese series "Testi e Documenti per lo studio dell'antichità" and the Sahidic texts from the collection of the Bodmer papyri.

Among the few Latin witnesses, we should mention four 4th-c. parchment sheets with sections of *Augustine's sermons 34 and 38 (van Haelst, *Catalogue*, n. 1209), and a *Psalmus responsorius* (Barc. P. inv. 149-153; van Haelst, *Catalogue*, n. 1210), wrongly named by its editor Roca-Puig *Hymn to the Virgin Mary*, whose main source is the *Protogospel of James*.

Editions: C. Wessely, *Les plus anciens monuments du Christianisme écrits sur papyrus, I-II*: PO 4, 2; 18, 3, Paris 1906, 1924; A. Deissmann, *Licht vom Osten*, Tübingen 1923. The two major collections containing Christian papyri: *Greek Papyri in the British Museum*, 1893, and the *Oxyrhynchus Papyri*, London 1898-. Also the *Kölner Papyri*, Opladen 1976-.
Studies: W. Derouaux, Littérature chrétienne antique et papyrologie, NRTh 62 (1935) 810-843; A. Bataille, *Les papyrus*, Traité d'Etudes Byzantines, II, Paris 1955; A. Calderini, *Papyri*, Milan 1962; K. Aland, *Kurzgefasste Liste der griechischen Handschriften des Neuen Testament*, Berlin 1963; G. Cavallo, *Ricerche sulla maiuscola biblica*. Florence 1967; E.G. Turner, *Greek Papyri. An Introduction*, Oxford 1968; T.C. Skeat, Early Christian Book Production: papyri and manuscripts, in *The Cambridge History of the Bible*, II, *The West from the Fathers to the Reformation*, ed. G.W.H. Lampe, Cambridge 1969, 54-79, 512-513; AA. VV., *La produzione libraria cristiana delle origini: Papiri e Manoscritti*, tr. M. Manfredi, Florence 1976; M. Naldini, La letteratura cristiana antica e i papiri. Note e osservazioni, in *Proceedings of the Twelfth International Congress of Papyrology*, Toronto 1970, 379-384; O. Montevecchi, *La papirologia*, Turin 1973; K. Treu, Papyri und Patristik, *Kairos* 16, 2 (1974) 97-114; G. Cavallo, Libro e pubblico alla fine del mondo antico, in *Libri, editori e pubblico nel mondo antico. Guida storica e critica*, ed. G. Cavallo, Bari 1975, 83-132; K. Aland, *Repertorium der griechischen Papyri*, I, *Biblische Papyri*, Berlin-New York 1976; V. Bartoletti, *Papiri e papirologia* (Istituto Papirologico "G. Vitelli"), Florence 1976, (M. Manfredi); J. van Haelst, *Catalogue des Papyrus Littéraires Juifs et Chrétiens*, Paris 1976; E.G. Turner, *The Typology of the Early Codex*, Philadelphia 1977; E.A. Judge-S.R. Pickering, Papyrus documentation of Church and Community in Egypt to the mid-fourth century, *JbAC* 20 (1977) 47-71; C.H. Roberts, *Manuscript, Society and Belief in early Christian Egypt*, Oxford 1979; E.A. Judge, The magical use of Scripture in the Papyri, in (eds.) W.E.W. Conrad-E.G. Ewing, *Perspectives on Language and Text*, Eisenbrams 1987; A. Pietersma (ed.), *The Acts of Phileas, Bishop of Thmuis: P. Chester Beatty XV*, Geneva 1984. For Coptic-language papyri: W. Kammerer, *A Coptic Bibliography*, Ann Arbor 1969 (notifications up to 1950). Bibliographical updating from 1949 on in the review *Orientalia*. D.M. Scholer, *Nag Hammadi Bibliography 1948-1969*, Leiden 1971; R. McL. Wilson, *The Future of Coptic Studies*, Leiden 1978; (eds.) A. Atiya et al., *The Coptic Encyclopaedia*, New York 1990.

M. Naldini

PARABALANI. Acc. to the *Codex Theodosianus, the p. were *qui ad curanda debilium agra corpora deputantur* (*CT* XVI, 2, 43) at *Alexandria: i.e. they formed the nursing personel of the Alexandrian church. We know neither the exact nature of their work nor when they originated, but by the start of the 5th c. they were a powerful corporation which at times could even cause disorders. For this reason, after the death of Hypatia - we do not know if they were involved in this - in 416 *Theodosius II issued a law restricting their privileges and any activities not specifically assistential: he also prescribed that they should be of *poor background - *honorati* and *curiales* could not be p. - and that their number could not exceed 500 (*CT* XVI, 2, 42), increased two years later to 600 (*CT* XVI, 2, 43). Acc. to this law, the p. were under the control of the bishop of Alexandria; on certain occasions they could be his right arm, as at the *Latrocinium* of *Ephesus in 449 (Mansi 6, 828). The p. enjoyed certain clerical privileges, but though in the *CT* they come under the heading *de clericis*, we are not sure if they belonged to the *clergy. Part of the *CT* legislation passed into the *Codex Justinianus* (*CT* XVI, 2, 42=*CJ* I, 3, 17; *CT* XVI, 2, 43=*CJ* I, 3, 18).

DACL 13, 1574-1578. F. Martroye, Les parabalani d'Alexandrie, *Bulletin de la Société nationale des Ant. de France* (1923) 275-281; H. Grégoire, Sur le personnel hospitalier des Eglises. "Parabolans" et "Privataires", *Byzantion* 13 (1938) 283-285; A. Philipsborn, La compagnie d'ambulanciers, "parabalani" d'Alexandrie, *Byzantion* 20 (1950) 185-190; W. Schubart, Parabalani, *JEA* 40 (1954) 97-101.

A. Di Berardino

PARABLE. In the technical language of ancient *rhetoric, the term παραβολή (Lat. *conlatio, similitudo*; cf. Cic., *De invent*. I, 30, 49) meant a more or less broad and detailed comparison between two terms belonging to different spheres. Aristotle (*Rhet*. II, 20) distinguished p. from fable, which he placed outside reality, while he numbered p. among examples that furnished the orator with generally accepted proofs. In the NT, besides Heb 9, 9; 11, 19, where it takes on the meaning of *symbol, the term παραβολή occurs many times in the synoptic gospels and is used in the whole vast gamut of meanings given by the Hebrew *mashal* (Aram. *mathla*): at different times it may designate a solemn, enigmatic sentence, advice on *savoir vivre*, a typical case, a metaphor, a brief image, a comparison, an analogy, a narrative. In other places in the NT we find the term παροιμία, whose sense oscillates between simile (Jn 10, 6), proverb (2 Pet 2, 22) and enigmatic discourse (Jn 16, 25. 29). As Dodd observes (*The parables of the Kingdom*, 19-21), in its simplest form the p. is a metaphor or simile drawn from daily life which strikes the hearer and stimulates reflection; the simple metaphor can be enriched by details and become a complete image and eventually a story and a genuine narrative. In the *Apostolic Fathers* the term παραβολή occurs in the so-called *Epistle of *Barnabas* and in the *Shepherd* of *Hermas. In *Barn*. 6, 10, apropos of Ex 33, 3, it means an enigmatic assertion which must be interpreted allegorically (cf. also *Barn*. 17, 2). The third part of Hermas's *Shepherd* is designated παραβολαί; the first five of these resemble the synoptic pp.: pp. needing an explanation (cf. *Sim*. V, 3, 1; 4, 1), in accordance with a conception analogous to that in Mk 4, 34 (where the same term, ἐπιλύω, appears with reference to the explanation of the p.).

Acc. to *Clement of Alexandria, a p. is a narrative that leads the hearer from what resembles the proper subject to what is true and proper, or an expression that shows with vigour, by means of other circumstances, what the proper subject is (*Strom*. VI, 15, 126). *Origen, who also bears in mind this type of definition of p., adds another: p. is a discourse relating to a fact presented as if it had happened, while it has not actually happened, but could happen; such a discourse indicates realities, in a figurative sense, by virtue of a transference of meaning of what is said in the p. (*In Ps.* 77, 2; *Analecta Sacra* III, p. 111; cf. also *Fragm. in Pr.* 1, 6: PG 13, 20C). He distinguishes p. from αἴνιγμα, which consists in the exposition of events that have not been experienced, cannot happen and express a secret meaning in a hidden manner (*Fragm. in Pr.* 1, 6: PG 13, 25B; cf. Jerome, *Tract. de Ps.* 77, 2: CCL 78, 65); the fact recounted by the enigma is thus unlikely, unlike that of the p. Origen also shows how to distinguish the p. from the simile, on the basis of Mk 4, 30, emphasizing that the former is specific and the latter is generic (*Comm. in Mt*. X, 4), but later he seems to question this distinction (*ibid*. 16 and also 11). At any rate, for Origen the veiled teaching expressed by the p. has the aim of pushing the human spirit into searching, a necessary condition for any discovery of truth (cf. M. Harl, *Origène et la fonction révélatrice du Verbe incarné*, Paris 1958, 245); the Logos, to exercise his hearers' understanding, pointed to some truths under the form of enigmas, others in obscure discourses, others in pp., others in problematic questions (*Contra Cels*. III, 45). *John Chrysostom, who seems to associate the term παραβολή with the term αἴνιγμα (*In Jo. hom.* 47, 1), observes apropos of Mt 13, 15 that if the Lord had not desired the salvation of his hearers, he would have kept quiet and not spoken in pp., whereas by means of this parabolic language, this veiled mode of expression, he solicits their attention (*In Mt. hom.* 45, 2). For *Cyril of Alexandria, pp. are images of things intelligible and spiritual rather than visible; what cannot be seen with the eyes of the body, the p. shows to the eyes of the mind, giving a beautiful form to the subtleties of intelligible things through sensible, almost tangible things (*Comm. in Lc.*, ch. VIII: PG 72, 624 CD).

Among Latin Christian authors, *Tertullian, referring to the gospel pp., uses the term *parabola* alternately with *similitudo* (cf. *De praescr. haer*. 26, 3; *De paen*. 8, 4); *Hilary commonly calls the NT pp. *similitudines* (*In Mt*. 13, 7; 21, 8) or *comparationes* (*In Mt*. 6, 6; 20, 5; 21, 13). On *parabola=comparatio*, cf. also *Cassiodorus (*Exp. in Ps*. 56, 2 apropos of Mt 23, 37). The association of the concept of p. with those of *similitudo* and *comparatio* recurs in *Gaudentius of Brescia, who stresses that it is proper for the perfect teacher to use parabolic language to instruct his disciples and urge them to action (*Tract*. 18: CSEL 68, 153). *Jerome, noting that it is the custom of the Syrians, particularly in *Palestine, to include pp. in everything they say, since what cannot be understood by the hearers through simple teaching may be understood through likenesses and examples (*Comm. in Mt*. 18, 23), considers the p. as a likeness "*quae ab eo vocatur, quod alteri* παραβάλλεται, *hoc est adsimilatur, et quasi umbra prooemiumve veritatis est*" (*Ep*. 121, 6: CSEL 56, 23. For the contrast *parabola-veritas*, see *Ep*. 49, 13: CSEL 54, 370; cf. also Theodoret, *In Ez*. 20, 49: PG 81, 1008 A). He stresses that the Lord spoke many pp. to the crowd, so that they would grasp the different teachings according to each man's inner disposition, and joined clear things with obscure ones so as to lead his audience to the knowledge of what it did not understand by means of things it did understand (*Comm. in Mt*. 13, 3). *Ambrose, for whom p. is a figure requiring solution (*Expl. Ps*. 43, 56), distinguishes *narratio* from *parabola*, as in the p. of the rich man and Lazarus, where the name of one of the characters appears (*Expos. ev. sec. Lc*. VIII, 13; cf. also Jerome, *Hom. de Lazaro et divite*: CCL 78, 507). In *Augustine we find the concept of *parabola* associated with those of *similitudo* and *aenigma* (cf. *En. in Ps*. 68, Serm. 1, 15; *In Ps*. 48, Serm. 1, 5, where he says: "*Aenigma est obscura parabola quae difficile intelligitur*"; Serm. 93, 1); in pp. and in figures, which should not be understood in the proper sense, "*aliud ex alio est intellegendum*" (*Contra mend*. 24). Augustine divides the gospel pp. into two groups, of which the first is depicted "*secundum similitudinem aliquam*", e.g. Mt 18, 23-35; Lk 7, 41-43; 15, 11-32, and the second "*ex ipsa dissimilitudine*", e.g. Mt 6, 30; 7, 7-11; Lk 16, 1-13: in the one case, the distinctive characteristic of the p.'s two terms of comparison is likeness, in the other case, unlikeness (*Quaest. evang. lib*. II, 45).

In ancient Christian writers, even in the variety of their outlooks and modes of expression, we can grasp one common fact: the parabolic language expressed by Jesus tends substantially to elicit reflection and understanding in its hearers so that, when they are moved by right intention, they can more easily understand the Lord's saving message and consequently translate it into action.

GLNT 9, 519-568; *DBS* 6, 1149-1177; J. Jeremias, *The parables of Jesus*, London 1963; C.H. Dodd, *The parables of the Kingdom*, London ²1961; M. Marin, *Ricerche sull'esegesi agostiniana della parabola delle dieci vergini (Mt 25, 1-13)*, Bari 1981, 49-64.

S. Zincone

PARADISE. The term is derived from the Avestic *pairi-daeza*, "(enclosed) park", whence the Late Babylonian *pardisu* and the Hebrew *pardes*. This last term appears in the OT in Neh 2, 8; Song 4, 13, with the meaning of "orchard, forest". The LXX version translates it παράδεισος (which in, e.g., Xenophon meant the park of the Persian kings), attributing a religious meaning to it to the same extent as does the account of Gen 2, 3. Hence, finally, the *paradisus* of the Vulgate. In inter-testamentary Judaism, the garden of Eden became the dwelling of the souls of the blessed, patriarchs and visionaries. Thus the idea arose that the original p., now hidden, would return visibly at the end of time, an idea paralleled in certain NT passages (Lk 23, 43; 2 Cor 12, 4; Rev 2, 7).

In the *Fathers the term assumed a plurality of meanings, to indicate the definitive place where the eschatological destiny of the blessed would be fulfilled, *Abraham's bosom where they would rest, the heavenly *Jerusalem destined to welcome them, the blessed life and its location, heaven, often identified with the Edenic p. Thus *Cyprian (*Ad Fort*. XII) recalls, in times of *persecution, that if death takes us away from this deceitful earth it is to take us to p., the kingdom of heaven. *Basil uses the two names - heaven and p. - indifferently. He is the first to establish, using contemporary cosmology, the particular origin and nature of this paradisiacal heaven. It is an empyrean place, which evades any definition, possessing a radiant *light, created to receive the angelic natures and make God's friends happy (*Hom. I in Hex*. V). *Cyril of Jerusalem, however, maintains that heaven is made of *water, the most beautiful element (*Cat*. III; *De bapt*. 5). We must also - he maintains - recognize the plurality of the heavens: only the third, to which St *Paul ascended, is the same as p. (*Cat*. XIV; *De res*. 26).

*Augustine places this seat of the blessed in the highest heaven (*illud summum coelum*: *De Gen. ad litt*. XII, 35), which coincides with Abraham's bosom or p. (*In Jo*. 91; *Serm*. 280, 5), though it must not be confused with the heaven of the *angels (*Serm*. 26, 5; *Enchir*. 73). Besides, if p. were distinct from heaven, where else could it be situated (*Enarr. in Ps*. XXX, 3, 8)? At bottom, the different names conceal the same reality: the seat of the blessed (*De Gen. ad litt*. XII, 34, 65); faced with more pressing requests to locate it, he says that after this life *God himself will be the place where our *souls will rest (*Enarr. in Ps*. XXX, 3, 8).

Augustine's difficulties reflect the Fathers' inevitable oscillations on the problem of locating p. *Irenaeus, e.g., had distinguished the heaven of p. from that of the holy city (*Adv. haer*. V, 36, 1): because unequal merits deserve unequal rewards. *Ephrem the Syrian specifies that the saints will be admitted to the kingdom of heaven only after the *resurrection. Meanwhile, the souls of the blessed will wait in heaven, not in the supreme part of glory, but in Eden, i.e. the heaven of p. (*Carmina nisibena* LXXIII, ed. Bickell, Leipzig 1866, 222-223). But these variations do not affect depictions of the content of the paradisiacal life of the last days, of which we also find important evidence in *epigraphy and *iconography. The Apostolic Fathers already mention the prize that awaits the heirs of the kingdom: they will be allowed to see Christ and live in fellowship with him (*Barn*. VII, 11), an infinite happiness will be their lot (*1 Clem*., XXXIV, 7-8). The writings of *Ignatius are animated by this idea: to have God as one's lot (θεοῦ ἐπιτυχεῖν: *Rom*. IV, 1; *Eph*. X, 1), enjoyment of God which will be mediated by Christ. These themes are also at the centre of the Passions and Martyrdoms. The *martyrs of *Lyons yearn to hasten to Christ to receive the crown of immortality (Euseb., *HE* V, 1). Saturus (*Passio Perp. et Fel*. XI) has a *vision, in the heights of heaven, to the east, of an immense space like a flower-garden. Heaven thus appears to the martyrs as the promised p., the luminous place of rest and *peace.

Later theologians enlarged on these images. *Origen's picture is important. He says that the seat of the blessed life, however situated, is above the firmament and is a world of true *light in which it will be possible to live thanks to the glorious *bodies we will possess at the resurrection (*Contra Cels*. III, 42). When the redeemed reach this heaven, they will learn from God himself the true causes of phenomena and will consequently know invisible and ineffable things. At the end of this process, they will be pure intellects, able to contemplate rational substances directly (*De princ*. II, 11 7). The problem of the beatification of the blessed, the nature of the spiritual body and its relation to the earthly *body, the tasks and functions of the redeemed, are a favourite subject of reflection. *Lactantius, e.g., emphasizes the transformation that will take place in our faculties, adapted to the new conditions of the heavenly dwelling (*Div. Inst*. VII, 26). *Ambrose favours the mystical theme of the union of the elect with each other and with God (*De ob. Th*. XXIX). *Eusebius of Caesarea shows *Constantine's soul being reunited with God clothed in the splendour of light, his gaze fixed on the celestial vault (*De vita Const*. I, 2). The Cappadocians describe the paradisiacal life in vivid images. *Basil says that, after the *resurrection, the elect will contemplate *God directly (*Hom. in Ps*. XXXIII, 11). He compares the calm and infinite joy of this *contemplation with the sudden raptures of *ecstasy that can sometimes be experienced in this life (*ibid*. XXXII, 1). For *Gregory Nazianzen, our joy will consist in contemplating the *Trinity: this will be possible because we will become truly sons of God (*Or*. XXIV, 9). This theme returns in the thought of *Gregory of Nyssa: besides immortality, our nature will be clothed in divine qualities such as *glory, power, perfection (*De an. et res*.: PG 46, 156 ff.). The descriptions of some Oriental Fathers are rich in detail. *Aphraates, e.g., maintains that the air of those sublime regions will be infinitely sweet, fit for the glorious bodies clothed in heavenly light and nourished on divinity. An eternal spring is destined to bring to bloom the marvellous trees planted by the Lord in a garden without limits or borders (*Dem*. VIII, *De res. mort*. 22; PS 1, 402).

For Augustine, finally, our moral perfection and our final happiness will consist in knowing and loving the divine *Trinity: this is the *beatitude that awaits *angels and redeemed (*De Trin*. VIII, 4-8). This will be possible thanks to the eyes of the risen and transfigured *body (*De civ. Dei* XXII, 29, 3-6). The greatest joy will consist in the praise of God. There will be degrees of honour based on merit, but no jealousy. The saints will continue to make use of their free will, finally deprived of the possibility of sinning. Eternal *life will be, for the redeemed, a perpetual Sabbath, in which they will be filled with God's blessing and sanctification. Thus will the psalmist's words be fulfilled: "Be in peace and know that I am God" (*De civ. Dei* XXII, 30, 1-4). **[Fig: 248]**

DTC 2, 2474-2511 (Ciel); *RGG*³ 5, 96-98 (Paradies. II im AT); *MySal* XI, Brescia 1978; J.L. Ruiz de la Peña, *La otra dimensión*, Madrid ²1975 (ch. VII: *La vida* *eterna*); D.A. Stuiber, *Refrigeriun interim: Die Vorstellungen vom Zwischenzustand und die frühchristliche Grabeskunst*, Theophaneia 11, Bonn 1957; E. Pagels, The Politics of Paradise: Augustine's exegesis of Gen i-iii versus that of John Chrysostom, *HThR* 88 (1985) 67-99.

G. Filoramo

PARALYTIC, Healing of the: *iconography. The gospels describe two healings of paralytics by Christ, in different places and conditions: one at Capernaum (Mt 9, 1-8; Mk 2, 3-12; Lk 5, 18-26), the other near the pool of Bethesda (Jn 5, 1-9). Very common even in the earliest repertoire, that of catacomb *painting, the gospel account is generally summed up in a single image: the p. walks away with his bed on his shoulders in obedience to Christ's command: "Take up your bed and walk" (cf. e.g. the fresco in the Chapel of the Sacraments A3 in the *cemetery of Callisto, second quarter of 3rd c., one of the oldest examples; *Wp* 27, 3). Often it appears next to a *Baptism scene, to emphasize the symbolic nexus between the healing of the p. and the remission of *sin. In few paintings does Christ appear together with the p. (the most important is in the baptistery of *Dura Europos, second quarter of 3rd c.: cf. C.H. Kraeling, *The Christian building. The excavation at Dura-Europos. Final report VIII.* 2, New Haven 1967, pl. 18), whereas in funerary reliefs the Redeemer, in a *gesture of speech or *impositio manus*, and frequently one or two disciples, become an essential part of the composition. Neither in paintings nor in *sarcophagi is there the least suggestion of place or circumstance that would allow us to identify which miracle is intended. Interest is centred on the figure of the p. as paradigm of Christ's saving power and symbol of trust in that hidden power. Exceptions are the so-called Bethesdà-type sarcophagi, particularly rich in details including many of the elements of *John's text: the episode is shown in two images divided by a wavy ledge indicating the water of the pool: below, the p. surrounded by other sick people lies on his bed, approached by Christ; above, next to the Lord in a gesture of command, the p. departs with his litter on his shoulders (cf. eg., among the completest examples, sarc. 125 of the Museo Pio Cristiano [last quarter of 4th c.]=*Rep.* n. 63).

*Mosaics include the splendid one in the 6th-c. basilica of S. Apollinare Nuovo at *Ravenna (F.W. Deichmann, *Frühchristliche Bauten und Mosaiken von Ravenna*, Baden-Baden 1958, pls. 175-176). Though less frequently than in paintings or funerary reliefs, the miracle of the p. occurs in the repertoire of the so-called minor arts: gold-glass (Morey, *The gold-glass*, nn. 159, 347, 366, 448; L. Kötzsche, *Age of Spirituality*, 442, n. 401); ceramics (Salomonson, *Late Roman*, 72 f.; L. Kötzsche, *op. cit.*, 443, n. 402 [with other examples]); ivories (Volbach, *Elfenbeinarbeiten*, nn. 113, 119, 125, 142, 152, 170, 179, 180, 181, 185, 221, 233, 234) and metalwork (Volbach, *Metallarbeiten*, 38, pl. 7). **[Fig: 249]**

DACL 13, 1615-1626; *EC* 9, 806-808; *Wp*, 201 ff.; *Ws* 2, 293 ff.; E. Baldwin Smith, *Early christian iconography and school of ivory carvers in Provence*, Princeton 1918, 102 ff.; W.F. Volbach, *Metallarbeiten des christlichen Kultus in der Spätantike und frühen Mittelalter*, Mainz 1921; M. Simon, Sur l'origine des sarcophages chrétiens du type Béthesda, *MEFR Ant.* 55 (1938) 107 ff.; F. Gerke, *Die christlichen Sarkophage der vorkonstantinischen Zeit*, Berlin 1940, 216 ff.; L. De Bruyne, L'imposition des mains dans l'art chrétien ancien, *RAC* 20 (1943) 113-266; W.F. Volbach, *Elfenbeinarbeiten der Spätantike und frühen Mittelalters*, Mainz ²1952; F.W. Deichmann, *Frühchristliche Bauten und Mosaiken von Ravenna*, Baden-Baden 1958; C.R. Morey, *The gold-glass collection of the Vatican Library with additional catalogues of other gold-glass collections*, Vatican City 1959; J.W. Salomonson, Late Roman earthenware with relief decoration found in Northern Africa and Egypt, *Oudheidkundige Mededelingen* 43 (1962) 72 f.; R. Giordani, Di un singolare rilievo funerario cristiano del Museo Archeologico Nazionale di Cagliari, *RAC* 52 (1976) 171-184; L. Kötzsche, in *Age of spirituality. Late Antique and Early Christian Art. Third to seventh century*, New York 1979.

M. Marinone

PARAPETASMA. The Greek term παραπέτασμα, corresponding to Latin *cortinae, vela* or *antependia*, is generally used to indicate the drapery that, from the 3rd c., is found with some frequency on *pagan and Christian *sarcophagi, at the shoulders of busts of the dead or of *orantes depicted in the centre of the front, on the side panels or on the lid. Sometimes this drapery is held up at each end by two small figures (erotes, genii, victories), as e.g. the lid of the so-called Albani sarc. from S. Sebastiano (*Ws* 40=*Rep.* 241), but in most cases it is just suspended, with showy knots at the upper angles (cf. e.g. *Ws* 98, 1=*Rep.* 73 and *Rep.* 76). The significance of the p. is not completely clear, but it is thought to be a *symbol alluding vaguely to the next world. Acc. to Bovini, it may be related to the veils used in funeral ceremonies, but its true function was probably lost with time. In this way the p. became a mere ornament to enrich and give movement to the background behind some characters (in the so-called *Plotinus sarc., the p. occupies the whole front). Cumont's theory, that the p. reflected some apocryphal Hebrew texts, acc. to which good spirits wrap the dead in a radiant garment (rendered by the p.) to carry it to heaven, has had no takers. The veils that appear in *mosaic depictions of public and private buildings (e.g. *Theodoric's *palatium* in S. Apollinare Nuovo at *Ravenna) and in the so-called minor arts (cf. the scene of Christ's teaching in the Brescia Reliquary) have a completely different meaning, connected with a real practical use. **[Fig: 250]**

DAGR 9, 670-677; G. Rodenwaldt, Cortinae. Ein Beitrag zur Datierung der antiken Vorlage der mittelalterlichen Terenzillustrationen, *NGWG* (1925) 33-49; F. De Ruyt, Etudes de symbolisme funéraire, *Bulletin de l'Institut historique Belge de Rome* 17 (1936) 143-185; F. Cumont, *Lux perpetua. Recherches sur le symbolisme funéraire des Romains*, Paris 1942, 476; G. Bovini, *I sarcofagi paleocristiani*, Vatican City 1949, 57-58; Volbach-Hirmer, figs. 85-86, 75; F.W. Deichmann, *Frühchristliche Bauten und Mosaiken von Ravenna*, Baden-Baden 1958, figs. 107-109.

D. Mazzoleni

PARAPHRASE, BIBLICAL. "Paraphrase" which, in modern literature, has outlasted rival terms in ancient treatises on *rhetoric such as "periphrase", "metaphrase", "*metabolè*", continues to designate any type of composition involving *translatio* either from one language to another or, within the same language, from one style to another, one metre to another, from poetry to prose and *vice versa*. Fundamental classical texts on p. are Aelius Theo (*progymn.* 1=*Rhet. Gr.* II, 62-64 Spengel) and Quintilian (1, 9, 2 and 10, 5, 4-5). For the former, p. is not confined to the slavish substitution of synonyms or the rendering of a text through periphrase, but includes the taking up and reworking by an author of a concept or image of his model. So it is not a distinct type of literary composition, but a particular technique of mimesis. In Roman education, however, p. was the scholastic exercise that preceded that of literary composition. Alongside translation from Greek to Latin, Quintilian points out the usefulness of the p. of Latin poets into prose, especially if, not confining oneself to interpretation pure and simple, one engages in a genuine contest (*certamen atque aemulationem*) with one's model (*Inst. Or.* 10, 5, 5). Paraphrastic exercise - on whose utility in the field of eloquence Cicero expresses some reservations (*De or.* 1, 34, 154) - was practised mainly in prose. Mythological and historical themes (*progymnasmata*), often taken from poetry, were treated in prose, as *Augustine informs us (*Confess.* 1, 17, 27) and as appears later in *Ennodius's *dictiones*. Less common was metrical p. which, in the guise of *hypothesis*, usually prefaced prose p.

Indebted to ancient rhetorical p. (in prose) was biblical p., composed in the epic verse of Homer and Virgil, but also - esp. in the Byzantine world - in iambics. To follow the lines of development and delineate the characteristics of paraphrastic Christian poetry, which reached its greatest splendour in the 4th-6th cc., would be in large part to trace the history of biblical epic poetry, which presupposes both an *interpretatio christiana* of the classical epic and an *interpretatio epica* of the *Bible (Thraede), and, in short, to revisit ancient Christian poetry *tout court* (esp. the Latin).

The *certamen* and *aemulatio* (of which Quintilian speaks) mark, with greater or lesser intensity, the various moments of Christian p., which however does not exhaust its function in the learned *aemulatio* of the poetical rewriting of the Bible. Indeed, as a prolongation of the *lectio divina* of *Scripture, it, no less than that, feeds the souls of the poet and his public, hungry for the Word of God. Poetic dimension and spiritual dimension are thus conjoined in p., which is often also a devout exercise.

Pre-Christian examples of hexametrical p. of the OT are attested in the *Judaeo-Hellenistic world by *Eusebius, who preserves three fragments of Philo the Elder's *De Jerusalem* (3rd c. BC), which imitates Lycophron and Euphorion (*Praep. ev.* IX, 20, 24, 37), and one of Theodotus's *De Iudaeis* (2nd c. BC), which Homerizes Gen 33 and 34 (*Praep. ev.* IX, 22).

The first Christian biblical p. is the Greek one of *Gregory the Thaumaturge (3rd c.), who composed a prose p. (*Metaphrasis*) of Ecclesiastes, following the LXX text. Still in the Greek world, leaving aside Byzantine literature which frequently falls into this *literary genre, we will confine ourselves to mentioning the hexametrical pp. of the psalter (*Metaphrasis*) attrib. to *Apollinaris of Laodicea (4th c.) and of *John's Gospel (*Metabolè*) by *Nonnus of Panopolis (5th c.). In the Latin world, pp. of the Bible, almost exclusively hexametrical, assumed an astonishing variety of characters and forms: now faithful transcription, respecting the order of facts in their chronological succession and averse to rhetorical amplification; now free *retractatio*, enlivened by digressions or embellished with passages of doctrinal *exegesis. Using Virgil in a more or less emulatory way, p. conferred an epic colouring on the sacred text and Romanized it. In *Epos*, Thraede has tried to expound a homogeneous development of Latin paraphrastic poetry, esp. of the OT, and to label its different components. Although these definitions emphasize sometimes only one - and not always the most important - aspect of generally quite complex works, Thraede's remains a remarkable attempt at systematization. In the sphere of OT p.,

Thraede distinguishes: a) historico-grammatical p. (*Cyprian the Poet's *Heptateuchos*, the anon. *De Sodoma* and *De Iona*, etc.); b) rhetorico-didactic p. (Claudius *Marius Victorinus's *Metrum in Genesim* and *Alethia*); c) elegiac-hymnic p. (*Dracontius's *Laudes Dei*); d) lyrico-dramatic p. (*Avitus's *De spiritalis historiae gestis*); e) biographical and autobiographical p. (*Paulinus of Périgueux's *Vita Martini* and *Paulinus of Pella's *Eucharisticos*); f) allegorical p. (*Prudentius's *Hamartigenia*). This summary omits the paraphrastic activity of *Paulinus of Nola, who, after proving himself on the reduction of Suetonius's *De regibus* (of which *Carmen* III is a fragment), paraphrased Luke 1 in *Carmen* VI, treating the figure of John the Precursor with free invention, and Psalms 1, 2 and 136 in *Carmines* VII (in iambic trimeters), VIII and IX respectively. Poetic p. of the NT is represented by *Juvencus's *Evangeliorum libri IV* which, stimulated by *Constantine's religious political programme, inaugurated this literary genre in the West; *Sedulius's *Carmen Paschale* and his prose p. (*Opus Paschale*), which allows itself a certain freedom from its models; and *Arator's *De Actibus Apostolorum*.

S. Gamber, *Le livre de la Genèse dans la poésie latine au Ve siècle*, Paris 1899 (repr. Geneva 1977); G. Krüger, *Die Bibeldichtung zur Ausgang des Altertums*, Giessen 1919; J. Golega, *Der homerische Psalter*, Ettal 1960; K. Thraede, Epos, *RACh* 5, 983-1042; R. Herzog, *Die Bibelepik der lateinischen Spätantike*, I, Munich 1975; D. Kartschoke, *Bibeldichtung, Studien zur Geschichte der epischen Bibelparaphrase von Juvencus bis Otfrid von Weissenburg*, Munich 1975; A. Pignani, Parafrasi o metafrasi?, *AttAccPont Napoli* 24 (1976) 219-25; P. Volpe Cacciatore, *Sulla "Parafrasi di Giovanni" di Nonno di Panopoli*, Salerno 1980; J. Fontaine, *Naissance de la poésie dans l'Occident chrétien*, Paris 1981; A. Pignani, La parafrasi come forma d'uso strumentale, *JOEByz* 32 (1982) 21-32; A.V. Nazzaro, La parafrasi salmica di Paolino di Nola, in *Atti del Convegno XXXI Cinquantenario della morte di S. Paolino (431-1981)*, Rome 1983, 93-119.

A.V. Nazzaro

PARENTIUM. We know little or nothing of Parentium (Porec-Parenzo) before the Romans decided to use the convenient position of the Parentine peninsula, halfway between the two colonies of Tergeste and Pola and at the natural convergence of all the internal roads of the territory. Then, in relation to the moving of *Italy's E frontier from Formio (Risano) to Arsa in 16 BC, the supposed pre-Roman centre must have received an exact boundary and an urban organization still recognizeable in the traces of the two main roads, in the road-network that divides the city-plan into rectangular *insulae* and in the forum whose archaeological layer has been explored beneath the modern Piazza Marafor. This was fronted by the largest Roman temple in Istria, rebuilt by T. Abudius Verus, previously vice-admiral of the Ravenna squadron stationed at P. This city is the one place where archaeological data confirm a sure episcopal constitution predating the edict of tolerance: the protobishop *Maurus, whom constant tradition venerates as a *martyr from at least the 6th c., is attested by the famous *epigraph (4th-6th c.) which records his translation from the *cemetery outside the walls to the urban basilica, *ubi episcopus et confessor est factus*. Also, to N of the splendid basilica built by bishop Euphrasius in the mid 6th c. have been uncovered the remains of a probable *ecclesia domestica* adapted from the hall of a Roman house later enlarged into the first *church buildings. The *mosaic inscriptions here dated to the end of the 4th c. document a varied community composed of men of all sorts of social conditions. After Maurus, no other bishops' names are known until Euphrasius, but the apse vault of his basilica reveals a point of arrival in the process of Christianization which took place in Istria over two and a half centuries. [Fig: 251]

G. Cuscito, *Parenzo dalle origini all'età di Giustiniano*, Padua 1976.

G. Cuscito

PARIS, *Council of. Although Christianity must have reached the Roman city of Lutetia Parisiorum during the 2nd c., its first known bishop is *Dionysius (Denis) in the second half of the 3rd c. In summer 360, Gallic bishops met here for a council, at which they disavowed their adherence to the acts of the pro-*Arian council of *Rimini (359) and subscribed the definitions of *Nicaea. *Saturninus of Arles and Paternus of Périgueux were declared deposed from the episcopate, as supporters of that *heresy.

For the councils of 552, 556-73, 573, 577 and 614, cf. *Merovingian councils.

CCL 148, 32-34; Gaudemet 90-99; Palazzini 3, 295; Hilary of Poitiers, *Fragm. Hist.* XI.

Ch. Munier

PARMENIAN. *Donatist bishop of *Carthage 362-391/2. Gallic or Spanish by birth, not N African (Opt. of M., *De schismate* II, 7), he prob. adhered to *Donatus's cause during the latter's exile and emerged as leader after his death in c.355. He became bishop of Carthage after the triumphant return of the Donatist leaders to N *Africa, supported by the emp. *Julian, in 362 (Opt. of M., *op. cit.* II, 16; Aug., *Contra litteras Petiliani* II, 97, 224). Julian's death on 26 June 363 did not halt the Donatist revival; in c.364 P. wrote a treatise, *Adversus ecclesiam traditorum*, vindicating his church's legitimacy and asserting the irredeemable unworthiness of the *Catholics ((Opt. of M., *op. cit.* I, 4-6). The treatise's five books were on *baptism, church unity, the unworthiness of the Catholics as *traditores* and instigators of the persecution of Paul and Macarius (imperial commissioners in 347/8) and finally on the reason for rejecting the Catholic *sacraments, "the *oil of the wicked" (cf. Ps 141, 5). P. is heavily influenced by the arguments and language of *Cyprian's *De unitate Ecclesiae catholicae*, though his definition of the church by the possession of the gifts (*dotes*) attributed to the bride in the Song of Songs makes an original contribution to Western *ecclesiology. He claimed for the Donatists the **cathedra*, sign of the *authority and unity of the episcopate, the *angelus*, the *angel who hovered over the *waters of baptism, the *fons*, i.e. the baptismal font, the *sigillum* of baptism, and the altar. He showed again the importance the Donatists attached to the liturgy and symbolism of baptism. His forceful denial of the emperor's right to interfere in church affairs ((Opt. of M., *op. cit.* I, 22) followed Donatus's views. His treatise was answered in c.365 by *Optatus of Milevis's *De schismate Donatistarum*.

In his thirty years as a bishop, P. kept firm control over the Donatist church. In *Mauretania, *Rogatus's schism was controlled, as were the activities of the *circumcellions, though P. failed to suppress the Claudianist schism that struck the Donatists of Rome in c.385 (Aug., *Serm. II in Ps.* 36, 20). Even the higher imperial officials, Donatists like Flavian, *Vicarius Africae* in 377, were subjected to church discipline (Aug., *Ep.* 87, 8). P.'s *Psalms*, written in a popular style, spread Donatist teaching across N Africa (*Praedestinatus, De haer.* 43; Aug., *Ep.* 55, 18, 34). His prestige almost rivalled that of his predecessor Donatus. In c.380, however, P. found himself in conflict with the lay Donatist theologian *Tichonius, who seemed averse to the justification of the *schism and the Donatist insistence on rebaptizing converts (Aug., *Ep.* 93, 10, 43-44, and *Contra Ep. Parmeniani* I, 1, 1). At first P. tried to convince Tichonius that the universality of the church was acceptable only in the context of its integrity (*Ep. ad Tyconium*); not succeeding, he had Tichonius condemned by a *council: c.385 (Aug., *C. Ep. Parmen.* I, 1, 1). His firmness succeeded, since Tichonius seems to have listened to his appeal and not left the Donatist church (Aug., *C. Ep. Parmen.* III, 6, 29), and there was no schism in the Donatist ranks.

*Augustine respected P. for his ability as a theologian and administrator: "He strengthened the Donatist sect" (*Serm.* 46, 8, 17); Augustine and Optatus considered him an able speaker, with a caustic tongue (Opt. of M., *op. cit.* II, 14). His acceptance of Donatism also shows the existence of a point of view similar to the Donatist one in other Western Christian communities. On his death he left a church which was indisputably that of the majority of N African Christians (Opt. of M., *op. cit.* VII, 1 and Aug., *Ep.* 22, 2).

Some extracts from P.'s works can be found in Optatus of Milevis, *De schismate Donatistarum* (ed. C. Ziwsa, CSEL 26) and Augustine, *Contra Epistolam Parmeniani* (ed. Petschenig, CSEL 51). PAC 1, 816-821; P. Monceaux, Parmenianus, primat donatiste de Carthage, *JS* IV Ser. 7 (1909) 19-26, 157-169; id., *Histoire litt. de l'Afrique chrét*, V, 220 ff.; A. Pincherle, L'ecclesiologia nella controversia donatista, *Ricerche Religiose* 1 (1925) 35-55; W.H.C. Frend, *The Donatist Church*, Oxford ²1971, ch. 13.

W.H.C. Frend

PAROUSIA. Greek term (= presence, coming) present in the NT, where it means Christ's second coming in *glory, announced more than once by himself. Uncertainty as to its date led to various speculations, both Christian and heterodox, aimed at establishing the precise moment: these included the doctrine of *millenarism. For the *Fathers, p. means the *eschatological *judgment (*Diogn.* VII, 6), *God's presence beside the *martyrs (*Diogn.* VII, 9), Christ's coming in the *incarnation (Ign., *Philad.* 9, 2; Just., *Dial.* 88, 2; *I Apol.* 48, 2; 54, 7; Clem. Al., *Strom.* I, 18; II, 16). Christ's p. at the end of *time will be such that each will know him and thus will come about the separation of the good from the bad (Origen, *Comm. in Mt.* 70). Related to Christ's first p. in the *flesh, but contrasted to it in external character, the eschatological p. is an element of *catechesis in connection with the themes of judgment, *resurrection of men and things, vigilant expectation but free from curious inquiry (Cyr. of Jer., *Catech.* XV, 1-4). It will occur in two moments: first will appear the *cross, to the condemnation of the reprobate, then the *angels to the glorification of the just (John Chrys., *In Mt. hom.* 76, 3-4).

R. Trevijano Etcheverría, Ἐπιδημία y παρουσία en Orígenes, *Scriptorium Vict.* 16 (1969) 313-337; Lampe 1043 f.; A. Brontesi, *La soteria in Clemente Alessandrino*, Rome 1972, cf. index 725 f; M. Werner, *The Formation of Christian Dogma*, London 1957, Section 1; R.M. Grant, The Coming of the Kingdom, *JBL* 67 (1948) 297-303; D.G. Dunbar, The delay of the Parousia in Hippolytus, *VChr* 37 (1983) 313-327.

F. Cocchini

PARRHESIA. In the *polis* of the ancient Greeks, παρρησία meant freedom of speech, freedom of the citizen. In patristic literature the term is used in the pejorative sense of excessive casualness, offensive familiarity, fruit of vanity (Doroth., *Doctrinae* IV, 5-6: SCh 92, 252). In a positive sense, p. characterizes man's relations with *God before the fall. *Gregory of Nyssa identifies it with ἐλευθερία ("structural freedom") (J. Gaïth, *La conception*, 65 f.), familiarity conceded by God to his elect in the OT: *Abel, *Moses (John Chrys., *De sac.* 6, 4: PG 48, 681a). Various authors join p. with some special *virtue, most often with *prayer (Nilus, *Peristeria*, 4, 2: PG 79, 828ab), *sophia* and truth (the contemplatives), humility, *martyrdom. It is a prerogative of the *Theotokos (Epiph., *Hom. 5 in laudes s. Mariae Deiparae*: PG 43, 501b). Finally, p. with God is reflected in brotherly contact with men, esp. in the trusting relationship with a spiritual father.

J. Gaïth, *La conception de la liberté chez Grégoire de Nysse*, Paris 1953, 65 f.; H. Jaeger, Παρρησία et fiducia, *SP* 1 (= TU 63) (1957) 221-39.

T. Špidlík

PARTHEMIUS. Unknown 5th- or 6th-c. presbyter, perhaps African, whose reply to a *letter of the *Vandal count Sigistheus is preserved, in a deliberately artificial style, ending in 12 hexameters and an elegiac distich.

CPL 804-805; PLS 3, 448-449; *PAC* 1, 821.

F. Scorza Barcellona

PARTHENIUS of Constantinople. Presbyter and *archimandrite of *Constantinople, supporter of *Nestorius. A *Letter* of his, preserved in Latin (*Synodicum*, 153, cf. 152: PG 84, 767 f.), addressed to metropolitan *Alexander of Hierapolis (Syria), an intransigent follower of Nestorius and opponent of *Cyril of Alexandria (*Synodicum* 152: PG 84, 767b), calls Nestorius *sanctissimo illo et deo honorabili archiepiscopo nostro et teste Christi domino Nestorio*. So we must interpret in a Nestorian sense his profession of faith: *Veritas vero est confiteri Dominum nostrum Iesum Christum, Filium Dei vivi, Deum perfectum hominemque perfectum; et damus passiones quidem humanitati Christi, divinitati vero miracula, unum Christum, unum Dominum praedicantes*.

CPG 3, 5780; ACO I, 4, 175 f.; Bardenhewer 4, 212.

A. De Nicola

PASCENTIUS the Arian. *Arian count of the imperial court (4th-5th c.). In 406 he challenged *Augustine to a debate at *Carthage (Possid., *Vita Aug.* 17) and boasted of having beaten him. Augustine professed his faith in the three divine *persons (*Ep.* 238, 2, 10), expounded the *mystery of the *Trinity (*Ep.* 238, 2, 11; 238, 4, 25; 238, 5, 28) and the word *homoousios (*Ep.* 238, 1, 4; 238, 5, 27), and exhorted P. to expound his own faith (*Ep.* 238, 5, 26) so that he might be happy believing, without deforming the truth (*Epp.* 239, 1; 238, 5, 29; 239, 2-3). P. bitterly contested Augustine (*Ep.* 240), who replied to his calumnies in *Ep.* 241.

CPL 703: *Ep. ad Augustinum*: PL 33, 1051; CSEL 42, 559-560; CPL 366: *Altercatio cum Pascentio Ariano*: PL 33, 1156-1162; *PAC* 1, 827-830.

E. Romero Pose

PASCHASINUS of Lilybaeum. Bishop of Lilybaeum (Marsala) in Sicily, P. enjoyed the confidence of pope *Leo I (440-461) who, in 444, consulted him on the question of the date of *Easter. P. replied (Leo, *Epp.* 3; 16, 7) giving preference to the Alexandrian solution over the Roman. Seven years later Leo again addressed P. over the date of Easter 455. We do not know P.'s answer. But at the same time the *pope sent him as head of the papal delegation to *Constantinople, to represent him at the council of *Chalcedon (cf. *Epp.* 88; 89; 91; 92). As we see from the synodal *acts, P. was indeed the Roman spokesman at the *council's debates.

CPL 1656; Leo, *Ep.* 3: PL 54, 606-610; T. Jalland, *The Life and Times of St. Leo the Great*, London 1941, *passim*; cf. Ballerini's notes in PL.

B. Studer

PASCHASIUS of Dumium. Monk of the *monastery of *Dumium in *Gallaecia and disciple of the bishop of that city, St *Martin of Braga (*c.*515-*c.*580). The foundation of that monastery, decided on by Martin even before his election to the episcopate, was inspired by the *Rules professed by the old monks of the *East. To this end, Martin charged the learned monk P. to translate from Greek the sentences of the Desert Fathers (*Prologus* [PL 73, 1024]: *Vitas Patrum graecorum [ut caetera] facundia studiose conscriptas, iussus a te, sanctissime pater, in latinum transferre sermonem*). The version attrib. to P. is contained in book VII of the great collection of *Vitae Patrum first published by the Jesuit H. Rosweyde (Anversa, 1615), and is part of the collection known as *Verba seniorum*. P. of Dumium and Martin of Braga are thus the greatest humanists of the early Portuguese Middle Ages.

CPL 1079 c.; PL 73, 1025-1066; 74, 381-394; C.W. Barlow, *Martini episcopi Bracarensis opera omnia*, New Haven, 1950, 30-51; M. Martins, Pascasio Dumiense traductor dos Padres do deserto, *Broteria* 51 (1950) 295-304; J.H. Waszink, *VChr* 6 (1952) 60; A. Kurfess, Weitere Textkritische Bemerkungen zu Martin ep. Bracarensis opera, *Athenaeum* 33 (1955) 60-63; C.W. Barlow, *Martin of Braga. Paschasius of Dumium*, Washington 1969, 113-171; Quasten, *Patrologia* II, 189-191; J.G. Freire, *A versão latina por Pascasio de Dume dos Apophthegmata Patrum*, 2 vols., Coimbra 1971.

L. Dattrino

PASCHASIUS of Rome. Venerated at *Rome after his death (*c.*514). During the *schism that followed the election of pope *Symmachus, the deacon P. declared with obstinate persistence in favour of the antipope *Lawrence (498-505). His loyalty can only be explained by P.'s probable *ordination by Lawrence. *Gregory the Great, at a distance of time and perhaps in view of his penitent life, judged him favourably: *Mirae sanctitatis vir, eleemosinarum maxime operibus vacans, cultor pauperum et contemptor sui* (*Dial.* IV, 40) and attributed his choice to ignorance rather than malice. He seems to have written a book on the *Holy Spirit, now lost. We have a *letter he wrote in 513 to *abbat *Eugippus, author of the life of St *Severinus.

CPL 678 and 962; *Commemoratorium de vita S. Severini cum epistulis amoeboeis Eugippii et Paschasii diaconi*: PL 62, 39 ff.; CSEL 9, 2, 1 ff.; R. Cessi, Lo scisma laurenziano e le origini della dottrina politica della Chiesa di Roma, *Archiv. Soc. Romana di Storia patria* 42 (1919) 5-229; M. Pellegrino, *Comm. Vitae s. Severini, RSCI* 12 (1958) 1-26; *LTK* 8, 131; *BS* 10, 347; Fliche-Martin 4, 435.

L. Dattrino

PASTOR. Mid 5th-c. anti-Priscillianist bishop, of Galician origin, seemingly bishop of Palencia. Nothing is known of his life before his consecration as bishop (431). Mentioned by *Idatius (*Chron.* 102) as a *friend of *Agrestius and *Syagrius, and by *Gennadius (*De vir. ill.* 65. 76) who also mentions a *Libellum in modum symboli* written by P. Died at Orléans in 457, a prisoner of the *Goths. He was the first to use the term *filioque*. His *creed was directed against the *Priscillianists. Morin, following Gennadius, identified the creed of the 1st council of *Toledo with P.'s *Libellum*: an opinion accepted with reservations. Acc. to J. A. Aldama, the Toledan profession of faith had two versions: short (Toledo, 400) and long (Toledo, 447). The latter would have been P.'s.

Libellus in modum symboli: CPL 559; G. Morin, Pastor et Syagrius, deux écrivains perdus du cinquième siècle, *RBen* 10 (1893) 385-390; A. Künstle, *Antipriscilliana*, Freiburg 1902, 40-45; J.A. de Aldama, *El símbolo Toledano I. Su texto, su origen, su posición en la historia de los símbolos*, AG 7, Rome 1934, 29-37; A. Tranoy, *Hydace. Chronique*, SCh 219, 68-69; A.C. Vega, Un poema inédito titulado "De fide" de Agrestio, obispo de Lugo, siglo V, *Bol.R.Acad. Hist.* 159 (1966) 167-209.

E. Romero Pose

PATERIKON. Greek term (*sc. biblion*=book of the *Fathers; pl.: *paterikà*) which indicates, in monastic literature, those collections that bring together the anecdotes and thoughts of the *desert Fathers: cf. *Apophthegmata Patrum*.

A. Di Berardino

PATERIUS. Disciple and *notarius secundicerius* of *Gregory the Great (6th-7th cc.; *Epp.* I, 37; V, 26; VI, 12; IX, 97; XI, 15; John Diac., *Vita Greg.* II, 11). He projected a continuous *commentary on *Scripture, with the interpretations given by his teacher. He divided the work into three books: two for the OT (historical books, *Psalms, Proverbs, *Song of Songs, Wisdom, Eccl., Prophets) and one for the NT. We do not know if he finished the whole work, or just that part from *Genesis to Song of Songs. Many mediaeval authors cite Gregory through P.

CPL 1718: *Liber Testimoniorum veteris Testamenti quem Paterius ex opusculis S. Gregorii excerpi curavit*: PL 79, 683-916; A. Wilmart, Le recueil grégorien de P. et les fragments wisigothiques de Paris, *RBen* 39 (1927) 81-104; R. Etaix, Le Liber Testimoniorum de P., *RSR* 32 (1958) 66-78.

E. Romero Pose

PATIENS of Lyons. Mentioned several times by *Sidonius Apollinaris (*Epp.* II, 10; IV, 25; VI, 12; IX, 3). After his election as bishop in 469/70, P. was elevated to the see of *Lyons when that city fell to the *Burgundians. A welcome guest at the table of their kings, esp. of Chilperic, he had a good influence on them and their nobles, converting esp. the *women to *Catholicism. *Clothilde, future wife of the Frankish king *Clovis, was born of a princely Catholic family in *c*.475. Known for his building activity - basilicas built in his time include the cathedral, at whose *dedication *Faustus of Riez delivered a *homily - P. also stood out for his charity, distributing free grain during the Visigothic raid of 471. He took part in the council of *Arles (471) and himself called another, soon after, at *Lyons, to examine the case of the priest *Lucidus, accused of predestinationism. The Bollandists think he died *c*.480. Anniversary 11 Sept (*Mart. hier.*).

Duchesne, *Fastes* II, 163; E. Griffe, *La Gaule chrétienne à l'époque romaine*, II, Paris 1947, 99-101, 288-290; *Mart. Hier.*, 502; ASS Sept. III, 791-797; *BS* 10, 426 f.

V. Saxer

PATRICIANI. Acc. to *Filaster (*De haer.* 62: CSEL 38, 32), the P. were derived from one Patricius, who lived in *Rome and taught that the human *body was created not by *God, but by the devil. So the P. despised the body, even to the point of suicide (cf. Aug., *De haer.* 61). Thus they seem a *gnostic sect (cf. Aug., *Contra adv. legis* II, 12, 40: PL 42, 664). *Praedestinatus (61: PL 53, 608) adds that they sometimes asked strangers to kill them and that the sect was widespread in *Mauretania and *Numidia, followed subsequently by the *Donatists (cf. also Ambrosiaster, *In I Ep. Tim.* 4: PL 17, 499).

A. Di Berardino

PATRICIUS. Husband of *Monica and *curialis* of Thagaste in N *Africa (Aug., *Confess.* II, 3, 5), father of three children: St *Augustine, *Navigius and a daughter whose name is unknown (Possid., *Vita Aug.* 26, 1). Though a *pagan, he respected Monica's faith and the upbringing she gave her *children (*Confess.* I, 11, 17); but in agreement with his wife, although a modest landowner, he wanted his son Augustine to rise socially through higher education at *Carthage: at that time study was one of the main means of social ascent (cf. *Confess.* II, 3, 5 and II, 3, 8). Though inclined to anger and conjugal infidelity, P. had affection and respect for his wife, who in turn served with tenderness and affection and sought to convert him to Christianity (*Confess.* IX, 9, 19), managing to win even the sympathy and devotion of her mother-in-law (*Confess.* IX, 9, 20). P. was converted (*Confess.* IX, 9, 22) and died in 370/71, when Augustine was seventeen.

W.H.C. Frend, The Family of Augustine; a microcosm of religious change in North Africa, *Cong. internaz. su S. Agostino nel XVI centenario della conversione*, Studia Ephemerides Augustiniana 24, 3 vols., Rome 1987, I, 135-151.

A. Di Berardino

PATRICK. Saint (†*c*.492?), apostle of *Ireland. The chronology of his life is much disputed. From his works we learn that he was born in *Britain of the deacon Calpornius, a decurion, son of the presbyter Potitus and proprietor of a holding at Bannaventa Berniae (?), unidentified, but perhaps near Carlisle. Kidnapped by pirates at 16, he worked as a shepherd in Ireland. Back with his family, he felt a call to the apostolate and followed it despite opposition. He prepared by a period of training which lasted, by his own account (*Confess.* 23), several years and took place - acc. to the *biographies of Muirchú and Tirechán - in *Gaul, with *Germanus of Auxerre and at *Lérins (something hardly probable). In fact this training may have been, at least partly, in Britain. Acc. to tradition, P., consecrated bishop, landed in 432 in N Ireland, where he propagated Christianity, renewing in himself the experience of St Paul, whom he continually cites. Feast 17 March. From his see of Armagh, his cult spread to the continent from the 8th c. through the Irish missionaries, and many legends flourished about him.

Two of his works remain, both in vulgar Latin (P. prob. used Gaelic in his oral preaching): *Epistola ad milites Corotici*, against a British king, probably of S-W *Scotland (Strathclyde), and the *Confessio*, written in old age. P. "confesses" the mercy of *God who called him to evangelize Ireland; presenting himself as a *peccator rusticissimus*, P. contrasts himself with the learned and powerful British *Pelagians: they were linked to *Romanitas* (now departed from Britain) both by ascetic interests - with nostalgia for the *virtutes* of ancient Rome, betrayed by the corruption of the more recent Roman administration - and by politico-economic interests, threatened by Picts, Scots and Saxons. From this complex situation, P. emerges, with his powerful experience of *grace, choosing the *evangelization of a "barbarous" land (cf., *Ep.* 1), Ireland, while remaining deeply attached to his roots as a Romanized Celt and therefore also to the Christians of Roman Gaul (*Ep.* 14; *Confess.* 43).

Works of doubtful authenticity are the *Epistola ad episcopos in Campo hAi* (frag.) and the *Dicta Patricii*. As for the *acts of the first *Synodus episcoporum, id est Patricii, Auxilii et Isernini*, we think they may go back in part to P.'s own thought.

CPL 1099-1100; 1102-1104; PL 53, 801-826; SCh 249 (R.P.C. Hanson); L. Bieler, *Libri epistolarum Sancti Patricii Episcopi*, *CM* 11 (1950) 1-150; 12 (1951) 81-214; L. Bieler, *The Irish Penitentials*, Script. lat. Hib. 5, Dublin 1963, 54-59; M.J. Faris (ed.), *The Bishops' Synod*, Liverpool 1976; L. Bieler, *Four Latin lives of St. Patrick*, Script. lat. Hib. 8, Dublin 1971; id., *The Patrician Texts in the Book of Armagh*, Script. lat. Hib. 10, Dublin 1979; L. Bieler, *The Life and Legend of St. Patrick*, Dublin 1949; R.P.C. Hanson, *Saint Patrick, His origins and career*, Oxford 1968; C.E. Stancliffe, Kings and Conversion: some comparisons between the Roman mission to England and Patrick's to Ireland, *FMS* 14 (1980) 59-94; C. Thomas, *Christianity in Roman Britain to AD 500*, London 1981, 295-346; R.P.C. Hanson, *Life and Writings of the Historical St. Patrick*, San Francisco 1983; J.F. Kelly, The Escape of St. Patrick from Ireland, *SP* 18 (1985) 41-45; E.A. Thompson, *Who was St. Patrick?*, Woodbridge 1985.

E. Malaspina

PATRIMONY OF ST PETER. The problem of *property and properties did not arise for the church of *Rome until after the era of the *persecutions, with the emp. *Constantine. He recognized the church's right to possess (*CT* XVI, 2, 4). (The so-called "Donation of Constantine", however, is apocryphal, and was shown as a forgery in the 15th c).

During the following centuries, the generosity of princes and the *faithful enriched the the *papacy with *donations, goods, lands which made up what was called from the 6th c. the *Patrimonium S. Petri*. These lands were situated first in Sicily, then in Illyria, *Gaul, Corsica, *Sardinia, *Africa (*Hippo), and administered by a central authority, the *rector*. The revenue maintained not just the pontifical administration, but also social works, the *poor, prisoners; later they were also used for defence. The Holy See thus defended the regions around Rome, the exarchate of *Ravenna and the duchy of Rome. When the castle of Sutrium was destroyed in 707, *Liutprand restored it "as a gift for the holy apostles Peter and Paul".

In 753, pope Stephen II invoked the protection of Pippin the Short, king of the *Franks, against the attacks of the *Longobards. Pippin promised the *pope to defend his interests and "the goods of St Peter". After the victory of 755, the Longobards ceded the cities of the exarchate of Ravenna. In 773 and 774 Charlemagne in turn defeated the Longobards, who had to cede other territories, which made up the Papal States and consolidated the pope's temporal power. Acc. to Saltet and Griffe, the donation of Quierzy to pope Stephen in 754 is an *interpolation of the *Vita Hadriani* (*LP* I, 498), as is the promise made by Charlemagne at Rome in 774.

Liber Censuum, ed. Fabre, Paris 1899; *Liber Pontificalis*, ed. Duchesne-Vogel, Paris ²1957; P. Fabre, *De patrimoniis Rom. Eccl. usque ad aetatem Carolinorum*, Paris ¹1892; L. Duchesne, *Les premiers temps de l'état pontifical*, Paris ²1904; E. Spearing, *The Patrimony of the Roman Church in the time of Gregory the Great*, Cambridge 1918; R. Macaigne, *L'Eglise mérovingienne et l'état pontifical*, Paris 1929; L. Saltet, La lecture d'un texte et la critique contemporaine, *BLE* 41 (1940) 176-206; 42 (1941) 61-85; E. Griffe, Aux origines de l'état pontifical, *BLE* 53 (1952) 216-231; 55 (1954) 65-89; *EC* 9, 957-960; P. De Leo, *Il "Constitutum Constantini"*, Reggio Calabria 1974; V. Recchia, *Gregorio M. e la società agricola*, Bari 1978 (bibl.).

A. Hamman

PATRIPASSIANS. Thus the Latins called those *monarchians who were known in the *East from the mid 3rd c. as *Sabellians, and whom the moderns call modalists since they make the Son a mode of being and manifestation of the *Father. The founder of this doctrine was *Noetus of Smyrna (late 2nd c.). He taught that if *God is one, Christ, who is God, is identical with the Father, in the sense that the Father, appearing as Son, was generated as man, suffered and raised himself. To support this doctrine he adduced Scriptural passages on the unicity of God (Ex 3, 6; 20, 3; Is 44, 6; *et al.*), from which he inferred the Son's identity with the Father (Jn 10, 30; 14, 9-10), and again Is 44, 6; Bar 3, 36-38; Rom 9, 5. He interpreted allegorically Jn 1, 1, on whose basis the Logos-theologians considered the Son to be God distinct from the Father.

Noetus was condemned by the presbyters of Smyrna, but *Epigonus introduced his doctrine to *Rome, where *Cleomenes and *Sabellius developed it, while *Praxeas spread it at *Carthage. The doctrine of these early 3rd-c. p. was that of Noetus, but also knew a variant; in the incarnate Christ the divine component represented the Father, the human component the Son; only the latter suffered on the *cross, since the Father did not suffer, but commiserated with the Son. Indeed Lk 1, 35 calls *Mary's child the Son of God. Moreover *Hippolytus (*Philos*. IX, 10, 11) puts forward the patripassian doctrine at Rome in the sense that one and the same God is manifested now as Father, invisible, immortal, etc., now as Son, visible, mortal, etc.

This doctrine was opposed by *Tertullian at Carthage and by Hippolytus at Rome. *Callistus, c.220, had both Sabellius and Hippolytus condemned at Rome, so as to hold a *via media* between the two extremes; but he himself professed a doctrine of monarchian type: Father and Son are distinguished by name, but not by *ousia* (substance), they are a single indivisible Spirit; their *prosopon* is one, distinguished only by name. In the *incarnation the man is the Son, the divine Spirit that descended into him is the Father; thus the Father suffered together with the Son. Callistus, like the p., accused the Logos-theologians, i.e. Hippolytus, of being ditheists.

A. Hilgenfeld, *Die Ketzergeschichte des Urchristentums*, Leipzig 1884, 615-626; J.N.D. Kelly, *Early Christian Doctrines*, London 1958, 119-123; M. Slusser, The Scope of Patripassianism, *SP* 17, Oxford 1982, 169-175; A. Grillmeier, *Christ in Christian Tradition* (Eng. tr. J.S. Bowden), London 1975, 144 ff.

M. Simonetti

PATRISTICS. See PATROLOGY.

PATROCLUS of Arles.

In 412 P. replaced Heros, entangled in the fall of the usurper Constantine (408-414); despite the irregularity of his advancement, he got on well in the time of pope *Zosimus (417-418), at whose *ordination he was present. He immediately obtained from the *pope the parishes of Ceyreste and Garguier, on the pretext that they had been usurped by Proculus of Marseilles, as well as metropolitan rights over Viennensis and Narbonensis I and II, to the detriment of *Vienne, *Narbonne and Aix (*Aquae*). He made use of the apostolate of St *Trophimus, supposedly sent by the apostle *Peter to occupy the see of *Arles (*Arelatus*), to act as metropolitan and to lord it over and traffic in the bishoprics. But Zosimus's successor Boniface (418-422) did not continue his predecessor's policy and did not renew these exorbitant privileges of P.'s. Entangled in turn in the political storm which followed the emp. *Honorius's death (423), P. was assassinated (426). Duchesne speaks of him as an "ecclesiastical adventurer".

Duchesne, *Fastes* I, 95-112; id., *Histoire ancienne de l'Eglise*, III, Paris 1929, 228-31.

V. Saxer

PATROLOGY - PATRISTICS

I. Meaning of the terms - II. Literary history of the Fathers - III. The transmission of patristic texts - IV. Printed editions of the Fathers - V. Patristics today.

It was the title *Pater* ("*Father") that fashioned the term "patrology", then the term "patristics"; two close terms, but which tend to be distinguished. The creator of the term "patrology" was the Lutheran J. Gerhard (†1637) in his posthumous study "*Patrologia sive de primitivae ecclesiae christianae doctorum vita ac lucubrationibus opusculum*", which appeared at Jena in 1653; the book goes from *Hermas to Bellarmine.

I. Meaning of the terms.

1. The term "patrology" tends to express mainly the historical and literary study (life and works) of the ancient writers. A certain number of historians, to include all the authors of the church, both *orthodox and heterodox, prefer to speak of "History of Christian literature" (Harnack) or "of ecclesiastical literature" (Bardenhewer), a title shared by numerous contemporary works, of diverse origin and tendency (Batiffol, Puech, Labriolle, Bardy, Moricca, Pellegrino). "Patristic" was originally an adjective, the noun "*theology" being understood. It apeared in the 17th c. among Lutheran and Roman Catholic theologians, who divided theology into "biblical, patristic, scholastic, symbolic and speculative". Those who now give this term their preference favour the study of ideas and doctrines over the philological and literary aspect. Enough to glance through the this change.

2. The ancients traced no rigid frontier between Christian antiquity and the Middle Ages; they easily gave the name "Fathers" to later writers, as in the case of the old *Bibliothecae Patrum*, which go up to the 15th and 16th cc. Even Mabillon still considered St Bernard the "last of the Fathers". Migne followed this example for his *Patrologia* (to the great despair of J.B. Pitra). The moderns fix more precise limits: for them, in general, the *Fathers end with *Gregory the Great or *Isidore of Seville for the Latins, and *John Damascene for the Greeks. Some would include the Venerable *Bede and Byzantinism.

II. Literary history of the Fathers.

*Eusebius of Caesarea may be considered the father of Patrology. In his *Hist. Eccl.*, he provided irreplaceable information on writers and their works, citing long extracts. In *De vir. ill.*, *Jerome drew up 135 brief items, more or less in chronological order, going from Simon *Peter to *Sophronius and ending modestly with himself. He was inspired by Eusebius. The best items are those on his contemporaries. Jerome's work was continued by *Gennadius of Marseilles, who added 97 (100) new items; it is valuable for the 5th c. The continuation by *Isidore of Seville and *Ildefonsus of Toledo is enlarged mainly by the addition of Spanish writers (PL 83, 1084-1106).

An anon. author introduced the Christian writers into *Hesychius of Miletus's *Onomasticon*, whence they passed into *Suidas and *Photius's *Myriobiblon sive Bibliotheca*. In 1317, Ebedjesus bar Berika established the Catalogue of writers of *Syriac literature.

In the West, the historians of the Middle Ages (Sigebert of Gembloux [†1112], Honorius of Autun, Anonymous of Melk, falsely identified with Henry of Gand, continued by John of Tritenheim, called Trithemius [†1516]), were content to provide summaries (not without errors) of Jerome, Gennadius or Isidore. Documentation will be found in J.A. Fabricius, *Bibliotheca ecclesiastica*, Hamburg 1718.

Renaissance and Humanism brought classical and Christian antiquity back into favour, on the level of texts and history, and there was a genuine return to the sources. Knowledge of ancient languages, esp. Greek, led to the discovery of the Eastern and Western heritage. Now the Fathers were invoked by both sides in the controversies of the Reformation. This is the case of the case of the *Centuries of Magdeburg* (1559), Lutheran historical research which tried to demonstrate the Roman Catholic deviations. They provoked the monumental work of C. Baronius (1538-1607), *Annali Ecclesiastici* in 12 vols.

The 17th c. saw the appearance of epoch-making works: R. Bellarmine, *Liber de scriptoribus ecclesiasticis*, Rome 1613, continued by P. Labbé and C. Oudin. Then follow three historians of great stature: Louis Ellies du Pin, *Nouvelle bibliothèque des auteurs ecclésiastiques*, Paris 1686-1711, 47 vols.; L.S. Le Nain de Tillemont, *Mémoires pour servir à l'histoire des 6 premiers siècles*, Paris 1693-1712, 16 vols., a work still reissued today and valued by scholars; then R. Ceillier, *Histoire générale des auteurs sacrés et ecclésiastiques*, Paris 1729-63, 23 vols.

The Protestant and Anglican churches furnished the works of Dodwell (†1711), Cave (†1713), C. Oudin (†1717), Fabricius and C.T.G. Schönemann; these passed into Migne. The theological studies of Petau (†1652) and Thomassin (†1695) led to the emergence of a serious knowledge of patristics, as did the various researches on liturgical sources; for the Eastern church those of Jean Morin (†1659) and Renaudot (†1720), whose *Liturgiarum Orientalium Collectio* is still reissued today. To these we must add J. Goar's (†1654) *Rituale Graecorum* and Martène's (†1739) *Ancient rites of the Church.*

Exploration of the rich libraries was intensified in the 17th and 18th cc., allowing the *publication of numerous unpublished works, which went to swell the partial collections of the Fathers. Here we must cite L.A. Muratori (1672-1750) and Mansi (1692-1769) whose *Amplissima collectio* of the *councils, preceded by that of P. Labbé (†1667), is still reissued today. For its part the Assemani family, mainly Giuseppe Simeone, revealed the riches of *Syriac literature to the West.

The early 19th c. was content to continue this exploration, thanks to the sagacity of Angelo Mai (1782-1854), who published 38 volumes in four successive collections. J.B. Pitra (1812-1889), who had already deciphered the inscription of Pectorius of Autun, continued Mai's work with less rigour. To these researchers we must add the Norwegian C.P. Caspari (1814-1892) for the publication of unpublished *creeds (3 vols.). Patristic research intensified in the second half of the century. This was the time of J.H. Newman in England. Universities created seats of patrology at Louvain (J.B. Malou, B. Jungmann), Tübingen (J.A. Mühler, then F.X. Funk) above all Berlin, which became a centre of publication and research with A. Harnack (†1930) and Th. Zahn (†1933), and Munich (Bardenhewer). France came to the fore with L. Duchesne (1843-1922) and P. Batiffol (1861-1929). All these authors of the patristic "revival" smoothed the path for contemporary research. *Archaeology in turn entered by force to enrich historical research on Christian antiquity. In this field, the precursor was Antonio Basio (†1629), who from 1620 described subterranean *Rome. In the 19th c., G.B. De Rossi (†1894) shed light on Christian, esp. pre-

Constantinian, Rome, with its *epigraphy and with the *iconography of the catacombs.

III. The transmission of patristic texts. From the 4th c., some monks dedicated themselves to the transcription of Christian as well as *pagan works. To this end, *Cassiodorus founded the *monastery of *Vivarium, where Greek texts were translated. Both in *East and West the monasteries preserved and transmitted the patristic, esp. the spiritual, heritage. There were rich *libraries of works of the Fathers esp. in the great abbeys (Bobbio, St Gall, Reichenau, Corbie). In the West, the Greeks were known only in *Latin translation.

From the end of the patristic era, texts were transmitted through collections (Eclogae, Philocaliae, *Catenae, dogmatic and monastic Florilegia, *Apophthegmata), more convenient and less costly.

The first anthology of *Origen (*Philocalia) goes back to the Cappadocians. In the West, such selections were largely dominated by *Augustine. *Prosper, *Eugippius, later Florus of Lyons and the Venerable *Bede composed Sentences taken from Augustine's works. *Paterius composed the *Liber testimoniorum* (extracts from the works of *Gregory the Great).

*Isidore inaugurated the *literary genre of Sentences and composed a theological *summa* with citations of Augustine and Gregory I. In the 7th c., *Defensor of Ligugé's *Liber scintillarum*, one of the most read books, enlarged his collection esp. with moral and ascetic sentences; besides the Latins, the Greeks and even *Ephrem figured in it. To it we must add the *Testimonia divinae Scripturae et Patrum* of a contemporary. The East and later the West conceived dogmatic and spiritual florilegia, which ransacked patristic literature. The dogmatic florilegia spring from the importance accorded to argument from authority (cf. Patristic *argumentation) in theological controversies and in the *councils; *Scripture was replaced by the *Fathers. *Basil (*Treatise on the Holy Spirit*, ch. 2 f.), *Vincent of Lérins (*Commonitorium* 29-31), *Augustine (*Contra Julianum* I, 6-34), *Cyril of Alexandria, *Theodoret (*Eranistes*), *Severus of Antioch (*Contra Johannem Grammaticum*, and his sources), constituted patristic *dossiers*. The same would happen with *Ephrem of Antioch and *Sophronius of Jerusalem (lost), with those formed by *Maximus Confessor, with the *Doctrina Patrum de incarnatione Verbi* by one Anastasius (7th-8th c.), with the *Sacra Parallela* and the work of Florus of Lyons. The *Armenian tradition transmitted *Timothy Aelurus's florilegium against the council of *Chalcedon, and composed original ones (*Seal of faith*, etc.).

Spiritual and monastic florilegia collected the sayings of the Fathers, esp. of *John Chrysostom and the Byzantine Fathers. Thus the *Pandectae* of Holy *Scripture (7th c.), the *Quaestiones et Responsiones* of *Anastasius the Sinaite (later augmented) and the *Evergetinon* of Paul of Constantinople (11th c.). In the West, the *Liber exhortationis* of Paulinus of Aquileia (†802), Alcuin's *Book of vices and virtues*, the *Diadem* of the monks of Smaragdus (11th c.). Many of these florilegia are still unpublished. Among the *Gnomai* (Sentences) and *Loci communes*, besides the famous *Sentences* of *Sextus, the *Precepts* of the deacon *Agapetus of Constantinople were circulated in all the *libraries of Europe and inaugurated the *literary genre of so-called *Mirrors of princes*; the best-known are Ps.-Maximus's *Loci communes*, in which the Fathers are cited at length. The *homiliaries also drew on the patristic heritage and popularized "selected passages" of preaching, but also biblical *commentaries and theological and spiritual works of the Fathers. A minute exploration has allowed us to reconstruct in part the works of *Chromatius. For centuries such readings nourished the *prayer of monks and of the whole church.

Finally we must add that the *canonical collections (also open to theological questions, like the *sacraments and *ministries) were vehicles of some patristic texts. Gratian's *Decretum*, in vogue in the Middle Ages, cites 1200, of which 1022 are authentic, though attribution is not always correct. The Latins predominate, esp. *Augustine; the Greeks are modestly represented. Of 33 texts attrib. to *John Chrysostom, only 14 are authentic. It consists of often long extracts, sometimes covering several columns of Migne. It was through these media that the Fathers were known in the Scholastic era. Only rarely did the Schoolmen have direct knowledge of their works; usually they had recourse to the collections. Even Peter Lombard, who provided in his *Sentences* the theological manual of the Schoolmen, presented his work as a compilation of the Fathers "*in quo maiorum exempla doctrinamque invenies . . . brevi volumini complicans Patrum sententias*" (Prologue). In it the Fathers figured as authorities to whom appeal must be made. The Schoolmen consulted them through the *Tabulae originalium*, the *Libri auctoritatum*.

We must not pass over in silence the more intensive study of the Fathers in the great *monasteries, mediators of their texts during the Middle Ages, whether in the Benedictine order or in the Cluniac monasteries, in Cistercian abbeys or Augustinian communities, or in various Charterhouses. In the Fathers the monks sought not so much *auctoritates* as nourishment for their faith and for their spiritual life (J. Leclercq, *L'amour des lettres et le désir de Dieu*, Paris 1957, 87-107).

Within the Universities themselves, Jean Gerson (†1429) prepared the era of the Humanists, setting off, against the tendencies of the age, a return to *Gregory the Great, *Augustine, Ps.-*Dionysius, the *Vitae Patrum*.

IV. Printed editions of the Fathers. The Renaissance and the invention of printing favoured the spread of the patristic writings. At first, the tendency prevailed to print those books most in demand. The first text was an apocryphon of *Augustine, *De vita christiana*, printed at Mainz in 1461; then *De arte praedicandi* (book IV of *De doctrina christiana*). The first Christian works printed in Italy, at Subiaco in 1465, were the works of *Lactantius, followed by the *City of God*. The first Greek text, *Gregory Nazianzen's *Orations*, was published at Venice in 1516 by Aldo Manuzio. Previously overlooked texts like Lactantius, *Tertullian and *Jerome were honoured once more.

The Humanists gradually improved the printed text, clearing up and eliminating the copyists' errors that had accumulated over the centuries. They were the technical collaborators of the publishers: Andrea de Bossi collaborated in the publications of Schweynheim and Pannartz, first at Subiaco, then at Rome. At Basle, Amerbach and then Froben were surrounded by a *pléiade* of Humanists; Johann Heylin, Beatus Rhenanus, Johann Pellikan and above all Erasmus, the prince of Humanists.

*Publication tended toward the "*Opera Omnia", which imposed a critical examination to discern the authentic from the apocryphal, after centuries of fanciful attributions. *Ambrose's complete works were issued first at Milan in 1492, then at Basle, where they were followed by Augustine's in 1506 and Jerome's in 1508. No critical selection had yet been made: this would be the initiative of Erasmus, who prepared for publication the works of *Cyprian (1521), *Arnobius (1522), *Hilary (1523), Jerome (1524-26), *Irenaeus (1526), Ambrose (1527) and finally Augustine (1527-29). *Origen, in Latin, appeared only in 1536, after Erasmus's death. Erasmus had already started to collate, with friends, various manuscripts, to establish the "censorship" of the works, to clear the text of the slips and errors of the copyists. Thanks to him, publication began to be established on a more critical basis.

Erasmus's efforts were carried on by important publishers, often flanked by philologists like Robert Estienne (†1559) and his son Henri. Here we must cite the Franciscan Feuardent (†1610), author of a series of patristic editions; the Jesuit Fronton du Duc (†1624), publisher of *John Chrysostom, *Gregory of Nyssa and *Basil; J. Sirmond (†1651), historian and publisher of the Fathers, *Fulgentius, *Theodoret of Cyrrhus, *Eusebius and *Rufinus, and of the collection *Concilia antiqua Galliae*; the Dominican F. Combefis (†1878), who published *Maximus Confessor and a vast patristic *Homiliary; the historian Etienne Baluze (†1718) who, among other things, prepared a *Conciliorum nova collectio*; J. B. Cotelier (†1686) who, at Paris, established the catalogue of Greek manuscripts and published the three volumes of *Ecclesiae Graecae Monumenta*.

In England, Henry Savile (†1622), founder of what became the Bodleian Library, furnished a notable edition of the works of John Chrysostom, collated on the best manuscripts; and in Germany, J. A. Fabricius (†1736), philologist and publisher of *Hippolytus and *Filaster (1721). To them we must add the teamwork of the *Maurists and Bollandists.

In the 19th c., Jaques Paul Migne (1800-1875) collected in a single *Patrologia* in two series, Latin and Greek, all the partial publications, the great editions of the Fathers, bringing together *Bibliothecae, Spicilegia, Analecta, Anecdota, Miscellanea, Monumenta* and *Reliquiae*, published since the invention of the press. To them he added the historical, critical and theological dissertations of past centuries, to which today's scholar can still go with profit, and which would be hard to find elsewhere.

From the mid 19th c., the critical edition was based on more rigorous criteria, formulated and applied by K. Lachmann; from these the patristic texts and the new *corpora*, primarily CSEL and GCS, then SCh and CCL, now benefit. The East finds its own place in PO, PS and CSCO.

V. Patristics today. The field of research has been enormously enlarged; monographs and specialized reviews have accumulated. The bibliography of *Origen and *Augustine requires large works of its own. Today patristics is enriched thanks to the discovery of new Latin, Greek and Oriental texts which ceaselessly augment the *dossier*. The *Patrologia Latina* alone has needed four supplementary volumes.

Numerous Greek texts are proposed to us in various versions, esp. in *Syriac, which allow us bit by bit to reconstruct the work of a hitherto misunderstood *Theodore of Mopsuestia. Geerard's *Clavis Patrum* has established a first inventory, in which we meet many unpublished works.

*Papyrology and *epigraphy do not cease to provide unpublished texts like Origen's *Debate with Heraclides* or *Didymus's commentaries on *Genesis, *Job and Zechariah, discovered at Tura in 1941. The *Nag Hammadi library has opened a new phase in the study of *gnosticism. More modest *papyri permit us to know better the daily life of the Christians of the first centuries. *Archaeology enriches patristics, which explains the place we have given it in this Dictionary.

Papyri, versions, fragments, precise inventories of manuscripts favour the ever more rigorous publication of ancient texts, which also takes advantage of a deeper knowledge of Christian *Latin and Christian *Greek. To measure the distance we have come, we need only analyse the latest edition of *Adversus haereses* in SCh.

The contribution of philology has allowed real progress in the determining and study of ancient texts. Not by chance have Harnack's works been anastatically republished, despite their unequal historical value. After the school of Uppsala, that of Nijmegen has taught us to discern better the originality of Christian Latin. Questions of authenticity and *chronology have been studied with new rigour, author's sources carefully and patiently analysed. The *Sitz im Leben*, the cultural, religious, political, social, literary, philosophical *milieu* are meticulously explored. The way opened by F. J. Dölger on the comparison between Antiquity and Christianity is bearing abundant fruit. A rereading carried out with more existential preoccupations gives greater coherence to interpretation, a truer dimension to analysis.

Over the centuries, the texts of the Fathers have been invoked mainly as a norm of *orthodoxy, and they nourished the opposed positions at the time of the Reformation and Counter-reformation. This had two consequences which still weigh on patristics: the hypertrophy of dogmatics to the detriment of other spheres, and the embitterment of orthodoxy, which led to the ostracism of heresiarchs and even of an Origen, who in his own time was a judge of orthodoxy in an assembly of bishops.

Patristics has too often favoured the Latins rather than the Greeks, *Augustine rather than *Hilary of Poitiers. Fortunately *Irenaeus, *Origen and *Gregory of Nyssa are resuming their due place, while Hilary and many others do not yet occupy theirs. All the *apocryphal literature, long the Cinderella of patristics, is only now starting to be better studied and edited.

Since the appearance of *Dogmengeschichten*, patristics has fled almost exclusively to the early history of *dogma. Recent studies begin to show us all the other fields that could be explored: moral and ascetic teaching, theological and monastic life, political and theological doctrine of history, spiritual and mystical life, social and economic ideas. We may add *liturgy, *catechesis, homiletic. These are subjects that would be sought in vain in the patristic scholars of the early 19th c., or in many modern works, and that have allowed us to bring to light the original contribution of a *Salvian of Marseilles.

To all these fields, still not completely explored, we can add the influence exercised by the Fathers on subsequent centuries, whence to measure how far *Tertullian, *Origen, *Lactantius or *Augustine have been able to inform with themselves the whole of European thought.

Patristics, once (like *theology) a domain reserved for the clergy, has now returned to the public domain. The *Fathers have found their own place in general literature (M. Schanz). They interest researchers, religious and otherwise, Christians and non-Christians, who come together in the study of the Fathers with their philological and philosophical culture. But to be a patrologist, it is not sufficient just to want to; besides, the term *patristic* is nothing but the adjective of *theology*, in the most complete and authentic sense of the term.

Select bibliography: *I.* E. Amann, Pères de l'Eglise, *DTC* 12, 1192-1215 (summary of discussions on the term).
II. Patristik, *RE* 15, 1013 (still a useful exposition); A. Benoit, *L'actualité des Pères de l'Eglise*, Paris 1961, 10-52. Cf. also the manuals of Altaner and Quasten. H. Hurter-Fr. Paugerl, *Nomenclator lit. theol. cath.*, Innsbruck ⁴1926, 5 vols.; J. de Ghellinck, *Patristique et Moyen Age*, II and III, Brussels 1947/48; E. Peterson, Patrologia, *EC* 973-976; B. Loth-A. Michel, *Tavole generali del DTC* 3, 3592-94 (documentation).
III. A. Siegmund, *Die Überlieferung der griech. christ. Literatur in der latein. Kirche bis zum 12. Jh.*, Munich 1949; T. Gottlieb, *Über die mittelalterlichen Bibliotheken*, Graz 1955; C. Munier, *Les sources patristiques du droit de l'Eglise du VIIIe au XIIIe s.*, Strasbourg 1957; Articles: Bibliothèque, Florilège, Homéliaire, in *DSp* 1, 1589-1606; 5, 435-512; 7, 597-616 (bibl.).
IV. T. Ittig, *De bibliothecis et catenis Patrum*, Leipzig 1707 (collections classified topographically [Basle, Paris], detailed works: from 1528 to 1707). Continued by J.G. Dowling, *Notitia scriptorum ss. Patrum aliorumque veterum Eccl. Monumentorum*, Oxford 1839 (collections from 1700 to 1839); C.T.G. Schönemann, *Bibliotheca historico-litteraria Patrum Latinorum*, Leipzig 1792/94, 2 vols. (gives the publications of the apocrypha and from Tertullian to Isidore: life, works, indications in chronological order of editions and translations, usually taken up by PL). For certain Greek Fathers, information is in J.A. Fabricius, *Bibliotheca Graeca*, Hamburg 1712/14, books V and VI (from the origins to the fall of Constantinople. There is a restricted number of Fathers, less complete than Ittig and Schönemann). For modern editions and translations, *De studio theologiae patristicae et historicae*, Seminarium 17 (1977) (brought up to date by specialists). Cf. the manuals: Quasten I, 17-21; Altaner 33 ff. For bibliography, consult *L'Année Philologique* (much room is given to the Fathers) and *Bibliographia Patristica*.
V. B. Altaner, Stand der patrologischen Wissenschaft, *Misc. Mercati* I, ST 121, Rome 1946, 483-520; J. de Ghellinck, *Patristique et Moyen Age*, Brussels 1947/48, II-III (dense with information); id., Les recherches patristiques, progrès et problèmes, *Mélanges Cavallera*, Toulouse 1948, 65-85; W. Schneemelcher, Wesen und Aufgaber der Patristik innerhalb der ev. Theologie, *EvangTheol* 10 (1950) 207-222; M. Pellegrino, Un cinquantennio di studi patristici in Italia, *Scuola Cattolica* 80 (1952) 424-52; A. Benoit, *L'actualité des Pères de l'Eglise*, Neuchâtel-Paris 1961; A.G. Hamman, Pour un aggiornamento des manuels de patrologie et de patristique, TU 107 (1970) 95-99; id., *Jacques Paul Migne. Le retour aux Pères de l'Eglise*, Paris 1975; id., Pour une lecture concrète des textes, TU 125 (1982) 285-292; id., *L'épopée du livre. La transmission des textes anciens, du scribe à l'imprimerie* (forthcoming); G. May, Lateinische Patristik (Literaturbericht) *ThRund* 53 (1988) 250-276; S.P. Brock, Syriac Studies 1981-1985. A classified bibliography, *PdO* 14 (1987); C. Nardi, Il primo quinquennio della "Bibliotheca Patristica", *Homo Vivens*, Annali dell'Istituto superiore di scienze e religiose Galantini di Firenze 1 (1989).

A. Hamman

PATRON. The custom of putting a *church building, a person or an association under the protection of a saint does not appear in primitive times. Originally, a church was the κυριακόν, the *dominicum*, before becoming the basilica or church. In the same way, in the beginning, Christians gave their *children the same *names as the *pagans used. The later custom of giving a saint's name to an association dates from the Middle Ages and is outside our scope.

1. Churches. The custom of giving a saint's name to a church seems to have begun at *Constantinople with *Constantine's construction of a "basilica of the Apostles", in which he was buried and to which his son *Constantius translated some relics of the *apostles. It was also called the *apostoleion*. The custom travelled west and is attested at about the same time at *Rome and *Milan. *Ambrose played a decisive role in its development with the discoveries of Sts *Gervasius and Protasius, *Nazarius and Celsus, Vitalis and Agricola (386-394). At Rome, the *cemetery churches took the name of the *martyr buried there: St *Peter *in Vaticano*, St *Paul without the walls, St *Lawrence without the walls, St *Agnes without the walls, etc. From the cemetery churches where it first appeared, the custom was extended to the city churches or *tituli*. Proof of this development is contained, as regards Rome, in the synods of 499 and 595: in the former, the titular priests of the Roman churches still sign, as usual, as incumbents of the pastoral cure of the *tituli Pammachi, Iulii, Vestinae, Damasi, Eusebi, Tigridae, Aequitii*, etc.; in the latter, of the *tituli SS. Iohannis et Pauli, S. Mariae in Trastevere, S. Vitali, S. Eusebii, S. Balbinae, SS. Silvestri et Martini*, etc. Other churches bore a saint's name from the start: St Clement, St Matthew, St Mary Major, the Holy Apostles. *Pilgrimages and translations of relics (real or representative) multiplied the number of churches dedicated to foreign saints: Roman saints outside *Italy, Eastern saints to the West or *vice versa*. In the end, no new church was built without being dedicated to a saint, whose name it took, and without placing his relics there.

*2. Personal *names*. *Cyprian names, among his colleagues, bishops called Peter, Paul, Moses (*Epp.* 27-28; 31-32; 67; 71, etc.), a *pope was called Stephen, some *women Mary. *Dionysius of Alexandria points out the habit of some Christians naming themselves after the *apostles. *Eusebius of Caesarea cites the case of five *martyrs who left their *pagan names for names taken from the *Bible (*Mart. Pal.* 11, 8). *John Chrysostom (*Hom. in Gen.* 21, 3) and *Ambrose (*Exh. virg.* 3) exhort Christian parents to give their *children the names of virtuous men and martyrs, to place them under their protection and use them as examples. Pope *John II (533-35) was originally named Mercurius, and numerous inscriptions in the church of S. Clemente mention him with this name when he was still a priest, while the *plutei* of the *schola cantorum* bear the name Johannes, in *monogram. We think he ceased to bear the name connected with paganism in the act of assuming the papal throne. The custom of calling oneself by Christian names became general in this period, while the bearing of pagan names was felt as inconvenient.

LTK 8, 187-192; H. Delehaye, Loca sanctorum, *AB* 48 (1930) 5-64; *AB* 48 (1930) 5-64; A. Orselli, *L'idea e il culto del santo patrono cittadino nella letteratura latina cristiana*, Bologna 1965; ead., Il santo patrono cittadino fra Tardo Antico e Alto Medioevo, in *La cultura in Italia fra Tardo Antico e Alto Medioevo*, Convegno del 1979, II, Rome 1981, 771-784.

V. Saxer

PATROPHILUS of Scythopolis. One of *Arius's supporters from the start, he tried to support his cause at the council of *Nicaea. He worked for the success of the subsequent anti-Nicene reaction and was one of the bishops who condemned *Athanasius at *Tyre (335). *Eusebius of Vercelli, exiled to Scythopolis after the council of *Milan in 355, suffered ill-treatment from him. In 358-59, P. worked to oppose the anti-Arian activity of the *homoeousians, who condemned him (without effect) at the council of *Seleucia in 359. In 360, at the start of the controversy over the *Holy Spirit, he expressed himself negatively on its divinity, together with *Acacius.

DCB 4, 216-217; Simonetti 595.

M. Simonetti

PAUL

I. The man and his evangelizing work - II. Paulinism - III. Commentaries on the Pauline epistles - IV. Iconography.

I. The man and his evangelizing work. The *letters written by P. to the communities he founded and in which he lived and worked are the primary source for our knowledge of his personality and evangelizing work; but despite the autobiographical details in them (cf. Gal 1, 11-2, 14; 1 Cor 15, 9; 2 Cor 11, 22-23; 12, 1-10; Rom 11, 1; Phil 3, 4-6; Philem 9), the letters do not provide sufficient elements to draw a complete biographical picture; so, despite problems over the authenticity of the Acts of the Apostles as a historical work composed by P.'s disciple *Luke, NT historians tend to use the narrative elements of that work, for three-quarters of which P. is the protagonist, to supplement the information given by the Epistles. The Apostle calls himself Paulos (cf. 1 Th 1, 1; Rom 1, 1), the Greek form of the Roman surname Paulus ("Small"), originally usual in the Gens Aemilia. But in the Acts he is called Saul until his meeting with Sergius Paulus, governor of *Cyprus (Acts 13, 7). From Acts 13, 9 on, as if to emphasize the Apostle's missionary insertion into the Romano-Hellenistic world, he is called Paulos. In fact, in the diaspora it was usual to use two names, Semitic and Hellenic, (cf. Acts 1, 23; 12, 25). In Phil 3, 5, P. declares himself "circumcized on the eighth day, of the people of Israel, of the tribe of Benjamin, a Hebrew born of Hebrews, as to the law a Pharisee" (cf. also Rom 11, 1). Acc. to Acts 22, 3, his native city was *Tarsus, the great commercial emporium of *Cilicia, not far from the mouth of the navigable river Cydnus. His birth must be placed at the start of our era (between AD 5 and 10). Again acc. to Acts 22, 3 and 26, 5, P. spent his youth at *Jerusalem where he was a disciple of Rabbi Gamaliel (Acts 22, 3). As belonging to the group of Pharisees, P. stood out in zeal for the Law and the ancestral traditions (cf. Gal 1, 13-14). Moved by zeal for the Law and the Mosaic tradition, he opposed the nascent Christian religion, harshly persecuting the church (cf. Gal 1, 13-16; 1 Cor 15, 10-11; Phil 3, 5). But, through divine calling and election (cf. Gal 1, 13-16; 1 Cor 15, 9-10; Phil 3, 6. 12), he became a follower and *apostle of Jesus who appeared to him risen (cf. 1 Cor 15, 8-9; 9, 1). Acc. to Acts 9, 3; 22, 6; 26, 12, Christ's appearance to P. took place on the road to *Damascus, but this is not explicitly confirmed in the Pauline epistles. What is certain is that, acc. to 1 Cor 15, 8-9, the vision of the risen Christ was for P. a guarantee of his election to the dignity and mission of an apostle. After his acceptance of Christ, he withdrew to Arabia, where he remained about three years; then he was at Damascus (cf. Gal 1, 17): P. himself (cf. 2 Cor 11, 32-33) tells how he escaped the governor of king Aretas, which offers the possibility of a date, since king Aretas died in AD 40.

From Damascus he went for some days to Jerusalem where he met *James and *Peter (cf. Gal 1, 18 ff.); then he stayed in Cilicia (Gal 1, 21). For some years he did no evangelistic work; in all likelihood he remained at Tarsus (cf. Acts 11, 25). Acc. to Acts 11, 25, it was *Barnabas who invited him to *Antioch and included him in his mission (cf. Acts 13, 2). From the Pauline epistles we can identify some important stages of his evangelizing mission in *Asia Minor and Greece, but we cannot reconstruct a complete picture of his missionary affairs. The Acts describe at length three missionary *journeys by P. through the regions of Anatolia, Asia Minor, *Macedonia and Greece. At the end of the third journey, he went to Jerusalem to offer the Judaean *brothers the collection held in the communities he had founded (cf. Acts 21, 17 ff.). At Jerusalem, P. was arrested by Roman soldiers on account of a popular riot that broke out on his account in the Temple area (cf. Acts 21, 27 ff.). Faced with the danger of his violent elimination at the hands of the *Jews, the Roman tribune had him taken before the procurator Felix (procurator AD 52-58 [final date disputed]) at *Caesarea. Since in Felix's view there was no reason to proceed according to Roman Law, P. would have been sent back to the religious judgment of the Sanhedrin; but P., a Roman citizen from birth (cf. Acts 22, 28), appealed to the imperial tribunal. Felix temporized for so long that in the end it was his successor *Festus (AD 58-62) who sent P. to *Rome. The Acts dwell on the account of the sea-journey with its various misadventures ending in the disembarkation at Puteoli, and conclude by saying that P. spent two years in forced residence at Rome, able to receive visitors and preach the Kingdom of God and the Lord Jesus Christ. Almost certainly, P. must have resumed his missionary activity after this Roman imprisonment, since we learn from his Epistles that he founded the communities of Colossae and Laodicea in Asia Minor and, acc. to Tit 1, 5, evangelized the island of *Crete. An old tradition, based on Rom 15, 24. 28 and implicitly confirmed by *Clement of Rome (1 Clem. 5, 7), claims that P. made a missionary journey to *Spain. Clement's letter (chs. 5-7) also seems to confirm the old tradition of the Roman church that *Peter and P. were *martyrs at Rome during *Nero's *persecution (between 64 and 68); *Ignatius seems to allude to this tradition in his Epistle from Smyrna to the community of Rome (Ign., Rom. 4, 3). At the end of the 2nd c., the presbyter *Gaius knew the tropaion of P., martyr on the Via Ostiense (cf. Euseb., HE II, 25, 7).

V. Loi

II. Paulinism. With his *theology, P. was from the start an element of stimulus and disquiet in the church's theological debates. The Acts, a good half of which is dedicated to the Apostle, guard his memory, exalting his figure and presenting him as a great thaumaturge and powerful preacher, while saying nothing about his aspiration to be considered an *apostle with the same title as the Twelve. Other letters composed (edited), in the closest circle of Pauline preaching, by P.'s collaborators and disciples, touch on and explore single aspects (Ephesians: the church, una sancta, even if composed of *Jews and *pagans; Colossians: the cosmic, universal character of Christ and the *redemption). 1 Peter and, with less evidence, Hebrews are influenced by Pauline thought. But still more problematic is the relationship between P. and *John. How common to both is the insistence on certain fundamental concepts (law and *grace; *faith; *love), which distinguishes them from the synoptic gospels; and yet how independent is the Johannine exposition. In some late NT letters (Jas, 2 Pet), we already note some critical tendencies towards P. and some hints of a deformation of the Pauline doctrine of *justification; such tendencies and hints, in the *canonical writings as well as in the later writings of the *orthodox church, never quite reach an open rejection and condemnation of the Apostle. Such an attitude was limited to some scattered groups of heretical *Judaeo-Christians (cf. Ps.-*Clementines). The writings of the Apostolic Fathers (1 Clem., *Ignatius and *Polycarp) presuppose knowledge of some Pauline letters, esp. of 1 Corinthians, but they do not let it transpire clearly on the strength of what basic concept they are concerned to keep P.'s theological heritage alive. The same goes for the Pastoral Epistles (1 and 2 Timothy; Titus) which, at the end of the NT era, with their concern for the integral preservation of genuine doctrine, appeal once more and forcefully to the sole *authority of the Apostle, but without embarking upon debates with their opponents on the content of the Pauline expressions. Common to the Apostolic Fathers and the Pastoral Epistles is their great esteem for P. as missionary and apostle who has suffered much for Christ.

1. Controversies over Paul in the 2nd c. Many of the church writings (*Barnabas, 2 Clem., the majority of the *apologists, *Papias) disregard P. without declaring any specific intention of doing so and without enabling us clearly to guess the reason. But as a result of *Marcion's extremist manipulation of P.'s letters, the church writers too were obliged to take up a position. We cannot be sure whether the predilection for P. of the presbyter of Asia Minor who wrote the apocryphal Acts of Paul (Tertull., De bapt. 17) was determined by an anti-Marcionite attitude. Due to the unfavourable conditions of the sources, we cannot give a precise idea of the knowledge that *Justin or *Melito of Sardis had of P., nor of their opinion of the Apostle's doctrine. An exceptional case is the author of the Letter to *Diognetus who, without being at all Marcionite, shows a real disdain for the OT and a clear predilection for P. For the *gnostics, P. was not a particularly authoritative witness, though he was called by *Tertullian "the apostle of the *heretics" (Adv. Marc. 3, 5, 4). The *Valentinians above all sought to make use of him in their system, on account of his allegorical *exegesis, despite some fundamentally important Pauline ideas (*God as *Father of Jesus Christ, the universality of God's saving plan, the *resurrection of the *flesh as an irrelinquishable moment of the eschata) being irreconcilable with their conceptions. The writings in which we note real consonances with the Pauline doctrine of the resurrection, like e.g. the "letter to Reginus", present doubts on the genuineness of their gnostic character. At the end of the 2nd c., *Irenaeus (Adv. haer.) and Tertullian (Adv. Marc.) not only affirmed the integration of the Pauline epistles in the revealed NT writings, but also demonstrated P. From then on the place of the corpus paulinum in the NT canon that was being formed was no longer contested.

2. *The Pauline tradition in *East and West*. In no age has P. been used in his entirety on the theological level; preference has always been given to now one, now another aspect of his thought. With the 3rd c. it happened that the Pauline tradition followed divergent paths in the Eastern and in the Western church. The former decidedly contested with P. the doctrine of *justification, to which the West gave the greatest importance. The East propagated the P. of I and II Corinthians, the West that of Romans (Benz 291). *Clement of Alexandria cites Romans, yes, but he does not accept the doctrine of justification; *Origen even comments on Romans, but its basic concept is absent from his theological system. *Athanasius and *Gregory of Nyssa comment only on 1-2 Corinthians, remaining silent on Romans. In *Didymus the Blind, who deals with the whole *corpus paulinum*, only Romans is missing. Greek *commentaries on Romans certainly exist - above all that of *John Chrysostom, P.'s greatest admirer among the Greek *Fathers - but they give the impression of having been compiled for a scruple of literary completeness than for the investigation of their theological content (Benz 297). The mystical-sacramental aspect of Pauline preaching, with the stress on conforming to Christ, new *birth, new creation and sanctification of men, appears very marked in all the Greek theologians.

In Western theology, for a long time P. played no role that could be called, in an overall view, determinant. A serious consideration of the Pauline letters began not before the end of the 4th c., primarily - and this is significant - at the margins of the church's official theology. The commentaries on P. of so-called *Ambrosiaster and the Aquileia MS are of unknown authorship, while those of *Marius Victorinus are by a rhetor recently come to the faith from *neoplatonism. But *Ambrose of Milan was also seriously concerned with P., above all in some late *letters addressed to his fellow-bishops (cf. *Epp.* 73-76). A particular knowledge of P. is attributed to *Priscillian, who was condemned as a *heretic. In Galatians and then more completely in Romans, P. gave form to his *theology of *grace and *justification, which was accepted above all in the West, inflaming the controversies between *Pelagius and *Augustine on the cooperation between grace and free will, and which was developed to its extreme consequences (doctrine of original *sin and *predestination). Luther and the Reformation, putting the stress on Romans and the doctrine of justification, further deepened the rift between Western and Eastern theology. P., who wished to be "all things to all men" (1 Cor 9, 22), wrote to Greeks and Romans adapting himself to their different bent. The Apostle's intention is turned into its contrary when we forget that he wished to adapt himself to particular recipients and situations, and we pretend to adopt the P. of particular letters or particular theological assertions as Pauline doctrine in its totality.

A. Merzagora, Giovanni Crisostomo commentatore di S. P., *Didaskal.* 10 (1931) 1-73; E. Aleith, *Paulusverständnis in der Alten Kirche*, Beihefte ZNTW 18, Berlin 1937; E. Benz, Das Paulus-Verständnis in der morgenländischen und abendländischen Kirche, *ZRGG* 3 (1951) 289-309; M.F. Wiles, *The Divine Apostle*, Cambridge 1967; W. Schneemelcher, Paulus in der griechischen Kirche des 2. Jh., *Gesammelte Aufsätze zum Neuen Testament und zur Patristik*, AV 22, Thessalonica 1974, 145-181; A. Lindemann, *Paulus im ältesten Christentum*, BHTh 58, Tübingen 1979; E. Dassmann, *Der Stachel im Fleisch. Paulus in der frühchristlichen Literatur bis Irenäus*, Münster 1979; B. Lohse, Beobachtungen zum Paulus-Kommentar des Marius Victorinus und zur Wiederentdeckung des Paulus in der lateinischen Theologie des 4. Jh., *Kerygma und Logos* (Festschrift Andresen), Göttingen 1979, 351-366; W. Erdt, *Marius Victorinus Afer, der erste lateinische Paulus-Kommentator*. Europäische Hochschulschriften 23, 135, Frankfurt a.M. 1980; W. Geerlings, Hiob und Paulus. Theodizee und Paulinismus in der lateinischen Theologie am Ausgang des 4. Jahrhunderts, *JbAC* 24 (1981) 56-66; E. Dassmann, *Paulus in frühchristlicher Frömmigkeit und Kunst*, Opladen 1982.

E. Dassmann

III. *Commentaries on the Pauline epistles. It has been repeated many times and by various scholars (Harnack, Lagrange, Wiles, Lyonnet, Godsey) that the *corpus paulinum*, and in particular the letter to the Romans, marked, by the reflections and commentaries to which it gave rise, the most salient moments of the history of Christianity. Ample information on the Greek commentaries will be found in *Jerome: in *Ep.* 33 he gives a catalogue of *Origen's commentaries on the *corpus paulinum*, again present in *Vir. ill.* 54; in *Ep.* 49 he mentions various patristic commentaries on 1 Corinthians and, in *Ep.* 119, those on 1 Thessalonians. Altogether from Jerome we know of some 20 Greek commentaries dedicated to various Pauline epistles: the list of *Fathers cited includes *Origen, Dionysius, *Pierius, *Eusebius of Caesarea, *Didymus, *Diodore of Tarsus, *Theodore of Heraclea and *Apollinaris of Laodicea. Information on a *Comm. Rom.* by *Eunomius is in *Socrates (*HE* IV, 7). Such references indicate that Pauline commentaries of the 3rd and first half of the 4th c., the majority of them unknown to us, were still circulating at the end of the 4th c. Knowledge of other commentaries can be deduced, however fragmentarily, from those scriptural commentaries of the Byzantine era known as *catenae*. If we leave out the uncertain information of a possible Marcionite commentary on the Pauline letters and another attrib. now to *Pantaenus (Jerome, *Vir. ill.* 36), now to *Heraclitus (Euseb., *HE* V, 27), the earliest and most important commentator on P. seems to be *Origen. Of all the commentaries attrib. to him, besides *Rufinus's *Latin translation of the *Comm. Rom.* and chs. 3-5 of that work in the original Greek cited by a Tura *papyrus, there remain only fragments handed down by *Pamphilus, *Eusebius, *Epiphanius or the *catenae*. Of *Eusebea of Emesa's *Comm. Rom.* and *Comm. Gal.*, *Acacius of Caesarea's *Comm. Rom.* and Didymus's *Comm. Rom.* and *Comm. 1-2 Cor.*, only catenic fragments survive. Of the commentaries on five Pauline letters attrib. by Jerome to *Apollinaris, we know only important fragments of *Comm. Rom.* Of *Diodore's commentaries too, only catenic fragments of *Comm. Rom.* remain. However, we know *John Chrysostom's 250 *homilies on the whole *corpus paulinum* and his systematic commentary on Galatians. In these homilies and this commentary, the pastoral character of Chrysostom's works is evident from the continual practical applications to which the text gives rise; but a first literal approach is not normally overlooked. Again in the *catenae* are preserved numerous fragments of *Severian of Gabala's Pauline commentaries and *Theodore of Mopsuestia's commentaries on Romans, 1 and 2 Corinthians and Hebrews. Theodore's commentary on P.'s ten minor letters survives in a 5th-c. Latin version. The loss of the greater part of Theodore's works is connected with the charges of *Nestorianism made against him and his consequent condemnation a century after his death. Of *Cyril of Alexandria's Pauline commentaries, we have fragments on Romans, 1 and 2 Corinthians and Hebrews. But we do have *Theodoret of Cyrrhus's systematic commentaries on the 14 epistles of the *corpus paulinum*. These commentaries (first half of 5th c.) are characterized by brevity and unaffected originality. In the later 5th c., *Gennadius of Constantinople wrote commentaries on P.'s letters, of which we have only extensive catenic fragments of *Comm. Rom.* and brief passages of other epistles.

As for Latin commentaries on the *corpus paulinum*, the fact that many of them have survived entire allows us to give a more exact picture of them. The first commentator is *Marius Victorinus, of whom we have part of the commentaries on Galatians, Ephesians and Philippians, dominated by attention to the text, which he interprets literally. Slightly later is *Ambrosiaster, whose systematic commentary on all P.'s letters survives, dominated by literal-historical *exegesis, but not without typological interpretations. Thanks to Ambrosiaster, we possess the pre-Vulgate Latin text of P.'s letters as used in the second half of the 4th c. An anonymous Pauline commentary was recently discovered in a codex at Budapest. *Jerome's commentaries on Philippians, Galatians, Ephesians and Titus, written soon after 386 while he was at *Bethlehem, make wide use of earlier commentaries by Origen, Didymus and Apollinaris, esp. for Galatians and Ephesians. *Augustine first attempted to comment on Romans in 394; he confined himself to a brief, literal exegesis of just a few passages (*Expositio quarundam propositionum ex Epist. ad Rom.*). His complete commentary on Galatians is of the same date. Soon after, he returned to Romans, intending to comment on it systematically, but got no further than Rom 1, 1-7, (*Epistulae ad Romanos inchoata expositio*). In 397 he expounded chs. 7-9 for *Simplicianus (*Quaest. ad Simplicianum*). Other chapters of Romans are commented on in *De peccatorum meritis et remissione* and *De spiritu et littera*. *Pelagius had a lively interest in the *corpus paulinum*, and between 406 and 409 he commented on 13 of P.'s letters. His exegesis is literal, sober, with rare digressions; the interlinear explanation that characterizes it makes Pelagius's work more like the *scholia* than a true commentary. Souter has restored to *Cassiodorus the *Comm. Rom.* published in the 16th c. under the name of *Primasius. In the mid 6th c. indeed, Cassiodorus and his school gave us those Pauline commentaries that modify Pelagius's commentaries in an anti-Pelagian sense (*Expositio s. Pauli epistulae ad Romanos, una cum complexionibus in XII sequentes s. Pauli epistulas a quodam Cassiodori discipulo anonymo concinnatis*).

The limited presence of the *corpus paulinum* in the 2nd c. and the few commentaries, largely lost, of the 3rd c. has been explained as a consequence of the attitude taken by the Great Church in the face of the use the *Marcionite and *gnostic *heresies made of P., and in the face of the strong anti-Paulinism that arose particularly in *Judaeo-Christian circles. But the profusion of Pauline commentaries in the 4th and 5th cc. does not seem fortuitous if we think that authors like M. Victorinus, Ambrosiaster and Pelagius - who were not interested in commenting on the whole *Bible - dedicated much of their exegetical work almost exclusively to P.'s epistles. It seems insufficient to explain the presence of these commentaries by the abating of the danger represented in the 2nd c. by Marcionites and gnostics, by *Ebionites and in general by *Judaizers. Several things are not irrelevant to the phenomenon: *1)* the doctrinal maturity that the Christian community must have attained and which is witnessed by the fact that now, in antiheretical polemic, they had recourse to the *corpus paulinum* (cf. Ambrosiaster, *Comm. Rom., praef.*); *2)* the rise of a marked interest by *Jews and Christians in mutual proselytism: once more a rereading of P., particularly of Romans, could clarify the problems born from this circumstance; *3)* the changed political situation, which saw the proscription of *paganism by the empire and the consequent *conversion to Christianity of multitudes of pagans: P.'s epistles became a reference-point both for those who stressed the sufficiency of *faith and those who dwelt on mankind's *sin and man's consequent responsibility from the ethical point of view; *4)* the monastic *ascesis that institutionalized in the new historical situation the longing for witness of life to be borne, with emphases that varied according to the intensity with which preference was given to grace or free will, mercy or merits; all themes that found full development in P.

J.A. Cramer, *Catenae Graecorum Patrum in Novum Testamentum*, Oxford 1844 (for the Pauline fragments, vols. IV-VII); C.H. Turner, Greek Patristic Commentaries on the Pauline Epistles, in *Dictionary of the Bible* by J. Hastings, New York 1906, 485-531 (extra vol. 484-531, Edinburgh 1947); A. Souter, *The Earliest Latin Commentaries on the Epistles of St. Paul*, Oxford 1927; K. Staab, *Pauluskommentare aus der Griechischen Kirche*, Münster 1933; K.H. Schelkle, *Paulus Lehrer der Väter*, Düsseldorf 1959; M.F. Wiles, *The Divine Apostle. The interpretation of St. Paul's Epistles in the Early Church*, Cambridge 1967; H.J. Frede, *Ein neuer Paulus Text und Kommentar*, I-II, Freiburg 1973-74; W.A. Meeks, *The first urban Christians: The social world of the Apostle Paul*, New Haven 1983; E.P. Sanders, *Paul, the Law and the Jewish People*, Philadelphia 1983; J. Zeisler, *Pauline Christianity*, Oxford 1983; D.R. MacDonald, *The legend and the Apostle: the Battle for Paul in Story and Canon*, Philadelphia 1983; W.H.C. Frend, *The Rise of Christianity*, London 1984, ch. 3; W. Rordorf, *Pastoralbriefe und Paulusakten*, (*Mél E.E. Ellis*); Various authors, *The Church Fathers and the New Testament*, Patristica Nordica 2, Lund 1987; C.P. Bammel, *Origenis in Epistolam Pauli ad Romanos, Aus der Geschichte der lateinischen Bibel*, Freiburg 1990; F. Cocchini, Il Paolo di Origene nel periodo alessandrino, *Actes du V Colloquium Origenianum*, Boston 1989; Various authors, *Le epistole Paoline nei Manichai i Donatisti e il prima Agostino*, Sussidi Patristici 5, Rome 1989; J.S. Pobee, *Persecution and Martyrdom in the theology of St. Paul*, JSNT Suppl. 6, Sheffield 1985.

M.G. Mara

IV. *Iconography. "Short of stature, bald head, bandy legs, well-formed body, joined eyebrows, rather prominent nose, full of kindness. Sometimes he seemed a man, at others the face of an angel". Thus P. is described by the anon. author of the *Acta Pauli et Theclae*, a Pauline apocryphon dating from before AD 200 (Erbetta 2, 259). Acc. to Nicephorus (*HE* I, 2, 37), the Apostle's physical characteristics were a small frame, curved shoulders, high forehead, aquiline nose and long face.

The problems of the origin of the Pauline iconography in early Christian art and why it appeared only relatively late are not yet completely cleared up. P.'s *portrait was a reconstruction, as were those of many other biblical characters. The first surviving depictions of the Apostle are referrable to the end of the 3rd c., when some Roman catacomb *paintings (e.g. in Domitilla: *Wp* 126; 148, 2) depict the apostolic college with Christ in the centre, but P.'s features, like those of *Peter, are not yet well-determined as they would be some decades later. The same phenomenon recurs in some *sarcophagi of the same period or slightly later, in which the deceased appears flanked by two *apostles with still variable physical characteristics, but who must in all probability be identified as Peter and P., representatives *par excellence* of the Twelve (cf. e.g. *Ws* 81, 4=*Rep*. 855; *Ws* 269, 3=*Rep*. 565). From the full Constantinian era however, representations of the Princes of the apostles take on new peculiarities and become increasingly widespread in the early Christian figurative repertoire, in *painting, *sculpture, *mosaic and the so-called minor arts (ivories, gold-glass, metalwork). The phenomenon can be explained by the great expansion of the cult of the *martyrs, and by their increasingly frequent depiction in the guise of companions and defenders of the *souls of the deceased. In the mid 4th c. the Pauline iconography (analogously to the Petrine) was now fixed and returns in innumerable examples of varied date and provenance, with variants few and unimportant: abundantly bald head, beard usually long and pointed, generally rather fine features. This physiognomy contrasts with that of Peter, which - probably by the law of contrast - became established with different characteristics: short and curly hair, wavy and flowing beard, strong features. Even if the depictions (mainly pictorial) appear sketchy in some cases, the identification of the two apostles is certain and is sometimes confirmed by captions with their names alongside the figures. P. often appears beside Christ (opposite Peter) in the scene of so-called *traditio legis*, or, in the group of sarcophagi (mainly Roman and Gallic) known as "Passion" sarcophagi, in episodes of his arrest (tied between two soldiers, or with a soldier placing a rope round his neck) and *martyrdom (about to be beheaded by a soldier, who draws his sword). The best-known example, in which both depictions appear together, is the front of the sarcophagus of Junius Bassus (*Ws* 13=*Rep*. 680, 1), dated 359 and preserved in the Vatican Grottos. Among so many, we may mention the columned sarcophagus preserved in the Vatican basilica (*Ws* 154, 4=*Rep*. 675, 3) and a late 4th-c. sarcophagus front in the Museo Pio Cristiano (*Ws* 142, 3=*Rep*. 61).

In *painting, the characteristic Pauline features appear particularly clear in the paintings in the cubicle of the **fossor* Diogenes at Domitilla (*Wp* 182, 2), or in room I of the catacomb of Via Latina (Ferrua, *Via Latina*, pl. CVIII and p. 69); later examples include the painting in the little basilica of the *cemetery of Commodilla, certainly mid 6th-c. (*WMM* 146). Among mosaic scenes are that in *Constantia's mausoleum in Rome (*WMM* 4) with *traditio legis* (with considerably later restorations) and the triumphal arch of S. Maria Maggiore (5th c.) where St Peter appears beside the throne (cf. C. Cecchelli, *I mosaici della basilica di S. Maria Maggiore*, Turin 1956, pl. XLVII). Among sculpted reliefs, we can recognize the profiles of Peter and P. in a piece from the Museo Paleocristiano di Monastero at *Aquileia, dateable to the 4th or early 5th c. (B. Forlati Tamaro - L. Bertacchi, *Aquileia. Il Museo Paleocristiano*, Padua 1962, fig. on p. 37). The Apostle is also shown on ivory objects, e.g. the *traditio legis* on the ivory *capsella* of Samagher, now in Venice Arch. Mus. (M. Guarducci, La capsella eburnea di Samagher. Un cimelio d'arte paleocristiana del tardo Impero, *Atti e Memorie della Soc. Istriana di storia patria*, n.s. 26 [1978] 1-146). P.'s bust, alone or with Peter, other saints or the dead, is often recognizable in gold-glasses. Very rare is the depiction of P. being stoned at Lystra (Acts 14, 19), which appears on an ivory casket made in the West in the third decade of the 5th c. and preserved in the British Museum (Testini, L'iconografia degli Apostoli, 293, n. 16). **[Fig: 252]**

DACL 2, 2694-2699; *EC* 9, 720-722; *LCI* 8, 128-147; L. De Bruyne, L'iconographie des Apôtres Pierre et Paul dans une lumière nouvelle, in *Saecularia Petri et Pauli*, Vatican City 1969, 35-84; P. Testini, L'iconografia degli Apostoli Pietro e Paolo nelle cosiddette "arti minori", *ibid.*, 241-323; id., L'apostolo Paolo nell'iconografia cristiana fino al VI secolo, in *Studi Paolini*, Rome 1969, 61-93; id., La lapide di Anagni con la "Traditio legis", *Archeologia Clas.* 25-26 (1973-1974) 718-740; Y. Christe, Apocalypse et "Traditio legis", *RQA* 71 (1976) 42-55; E. Dassmann, *Paulus in frühchristlicher Frömmigkeit und Kunst*, Opladen 1982; J.M. Huskinson, *Concordia Apostolorum. Christian Propaganda at Rome in Fourth and Fifth Centuries. A Study of Early Christian Iconography and Iconology*, London 1982.

D. Mazzoleni

PAUL II. Patriarch of *Constantinople (641-653); previously *oikonomos* treasurer of Hagia Sophia, he was elected patriarch on 1 Oct 641 to succeed the retiring *monophysite *Pyrrhus. Pope *Theodore rejected his synodal letter as too vague, requiring from him a synodical condemnation of his predecessor. To the *pope's repeated demands for a guarantee of *orthodoxy, P. replied in 646/7 with a *monothelite profession of faith, while claiming not to be in disagreement with the bishop of *Rome. Theodore replied with a sentence of deposition, and P. took reprisals against the altar of the chapel of the papal palace and the Roman *apocrisiaries. Persuaded by P., in 648 *Constans II issued a new monothelite edict, the *Typos*, which was rejected by the Roman synod presided over by *Martin I, who renewed the *anathema against P. (Oct 649), while the Byzantine patriarch sent the Armenians a dogmatic document (*c.*650) aiming at union. Shortly before dying (27 Dec 653), he intervened in favour of Martin, who was tried at Constantinople in Dec 653. The sixth ecumenical *council (680) anathematized him together with Pyrrhus, after reading three of his *letters and other writings on the one will of Christ.

CPG 7620-21; Grumel, *Regestes*[2], 299-301; L. Maggi, *La Sede Romana nella corrispondenza degli imperatori e patriarchi bizantini (VI-VII sec.)*, Rome-Louvain 1972, 212-223; J.L.D. van Dieten, *Geschichte der Patriarchen von Sergios I. bis Johannes VI. (610-715)*, Amsterdam 1972, 76-103; P. Conte, *Chiesa e primato nelle lettere dei papi del sec. VII*, Rome 1971, *passim* (cf. index 582).

D. Stiernon

PAULA. A noble Roman matron, she adopted the *ascetic life on her husband's death and was one of the group of Roman noblewomen who met on the Aventine under the spiritual guidance of *Jerome. In 385 she and her daughter *Eustochium followed Jerome to *Palestine and, after visiting the monks in the *desert of Nitria, settled in 386 at *Bethlehem, where she founded a hospice for *pilgrims, a male *monastery and a female monastery in which she lived until her death (404).

Jerome, *Epp.* 46; 108; Palladius, *Hist. Laus.* 36; 41; *BS* 10, 123-136 (bibl.).

A. Pollastri

PAUL Helladicus. 5th-c. hagiographer. Of Greek origin, adopted the Hesychastic life (solitude and *contemplation) at Elusa in Idumaea (Arabia Petraea). He wrote, prob. 526, an *Encomium* of St Theognius, bishop of Betelia near Gaza (†522), resumed and amplified by *Cyril of Scythopolis. A treatise by him on luxury survives in the form of a *letter.

CPG 7530-7531; Beck 199.

D. Stiernon

PAULICIANS. Heterodox mediaeval sect originating in *Asia Minor. If we disregard Armenian sources, far from clear, it is first attested in 655 (preaching of Constantine-Silvanus) in Byzantine Mesopotamia. After various ups and downs (imperial persecutions, migrations into Arab territory), by the first half of the 9th c. it was widespread in the eastern part of the empire and had its own "state" at Tefrik, crushed by the *Byzantines in 878. Until the end of the 12th c. the P. are attested sporadically in the West and the Near East and with some continuity in the Balkans (where they converged with the Bogomils) and at *Constantinople. The exact nature of their doctrine is still disputed. Byzantine sources wrongly call them "*Manichees", but they considered themselves Christian and willingly appealed to the NT, esp. the Pauline writings. Disregarding possible doctrinal evolutions, two traits seem characteristic of the sect: a certain *dualism and, as regards the *incarnation, *docetism. Their worship had no place for the saints and their relics, nor even for the *cross or the *sacraments. Their affinities with the iconoclasts seem to be only superficial. Lemerle has questioned the theories of Conybeare (and Garsoian) that they were of *adoptianist origin and that their doctrines were contained in the Armenian treatise called the *Key of faith*.

F.C. Conybeare, *The Key of Truth: A Manual of the Paulician Church of Armenia*, Oxford 1898; N. Garsoian, *The Paulician Heresy*, The Hague-Paris 1964; id., Byzantine Heresy: A Reinterpretation, *DOP* 25 (1971) 85-113; C. Astruc et al., Les sources grecques pour l'histoire des Pauliciens d'Asie Mineure, *Travaux et Mémoires* 4 (1970) 1-227; P. Lemerle, L'histoire des Pauliciens d'Asie Mineure d'après les sources grecques, *ibid.* 5 (1973) 1-144.

S.J. Voicu

PAULINIANS. Followers of the doctrines professed by *Paul of Samosata (Aug., *De haer.* 44). Can. 19 of *Nicaea requires them to be rebaptized if they wished to return to the *Catholic church, and attests the existence within this group of clerics, deacons and *deaconesses (Athan., *Or. c. Ar. sec.* 48).

F. Cocchini

PAULINIANUS. Younger brother of *Jerome whom he followed, with their sister, into monastic life. With *Vincent, Jerome's *friend and benefactor, he left *Rome for *Palestine. In 394, not yet 30, despite their resistance he was ordained priest by *Epiphanius of Salamis (*Cyprus), violating the rights of the ordinary of the place, *John of Jerusalem, with whom Jerome was in dispute over the *Origenist question. P., who wished only to help his brother in the *monastery, exercised his priestly *ministry in it. He remained always on good terms with Jerome, despite *Palladius's insinuations (*Hist. laus.* 79), and may have survived him, but we know nothing of his later life.

DCB 4, 230 f.

M.G. Bianco

PAULINUS. Biographer of *Ambrose of Milan. The scanty information on which first Bouvy, then Palanque sought to reconstruct his life is based on his own work (*Vita s. Ambrosii*) and the facts given by *Marius Mercator (*Commonitorium*, ACO I, 5, 1, 66. 4), *Augustine and *Isidore of Seville (*De vir. ill.* 17). Eye-witness of the last period of Ambrose's life (his earliest explicit fact dates from 395), we do not know when he became familiar with Ambrose, whose *notarius* he later became. After Ambrose's death (397) he remained at *Milan and was there when Mascezel arrived from *Africa, victorious over his brother *Gildo, and again when the bodies of the *martyrs Sisimus, Martyrius and Alexander were transferred there. We do not know when *Simplicianus sent him to Africa to administer the *property of the Milanese church there. But it was prob. there that, at Augustine's request, he wrote the *Vita s. Ambrosii*: 422 seems a more likely date than 412 (cf. M. Pellegrino, *Paolino di Milano*). The historical value of this has been much disputed. Today it is considered substantially reliable (M. Pellegrino, *op. cit.*, 23) since the encomiastic intention, the moral preaching, the presence of the marvellous and the interest in the ecclesial dimension of Ambrose's life take nothing away from the reliability of the information he gives. Of little literary value, with no *rhetoric, the work follows Ambrose's life chronologically. The *Libellus adversus Caelestium Zosimo episcopo datus* is also attrib. to P. Baronius and Tillemont have maintained, and Palanque has confirmed, the identity between P. the biographer of Ambrose and P. the opponent of *Caelestius. In this brief work, P., who had opposed the Pelagian theories in the assembly of bishops meeting at *Carthage in 412 (Aug., *De gratia Christi et de pecc. orig.* II 3. 8; *Contra duas epist. Pel.* II 6), refuses to appear, as pope *Zosimus wished, before the Roman tribunal to renew his accusations; he demonstrates the errors of *Pelagius and Caelestius and asks for their condemnation to be ratified. We know nothing further of P. after this episode.

CPL 169; *Vita s. Ambrosii*: PL 14, 28-50; *Libellus adversus Caelestium Zosimo episcopo datus*: PL 20, 711-716; L. Tillemont, *Mémoires pour servir à l'histoire ecclésiastique des six premiers siècles*, Venice 1732, X, 81; E. Bouvy, Paulin de Milan, *Revue Augustinienne* 1 (1902) 497-514; J.R. Palanque, La Vita Ambrosii de Paulin. Etude critique, *RSR* 4 (1924) 26-42; 401-420; id., *Saint Ambroise et l'Empire Romain*, Paris 1933, 400-416. M. Pellegrino, *Paolino di Milano. Vita di S. Ambrogio*, Rome 1961; E. Lamirande, *Paulin de Milan et la "Vita Ambrosii"*, Paris 1983.

M.G. Mara

PAULINUS of Antioch. Antiochene priest, around 350 he was head of the small Nicene community at *Antioch, which had separated from the mass of the church, since this, after the deposition of *Eustathius (327), had always been led by pro-*Arian bishops. The dispute continued even when, between 360 and 363, the local bishop *Meletius came close to the Nicene *theology. In 362 *Lucifer of Cagliari ordained P. bishop and in 363 *Athanasius recognized him as bishop of Antioch in Meletius's place: these facts aggravated the Antiochene *schism, since *Egypt and the West supported P. and the whole anti-Arian *East supported Meletius. *Basil of Caesarea was particularly hostile to P., accusing him of *Sabellian and Marcellian tendencies. The *Apollinarist controversy further afflicted the life of the small community, but P. rejected, *c.*380, a proposal of reconciliation presented to him, on what terms we do not know, by Meletius. When Meletius died during the council of *Constantinople (381), P., despite the support of the West, failed to get himself recognized as sole bishop of Antioch, and *Flavian was elected in Meletius's place. The council of *Rome (382) again recognized P. as sole bishop of Antioch, but without effect. P. died soon after and was replaced at the head of the schismatic community by *Evagrius.

DCB 4, 232-233; F. Cavallera, *Le schisme d'Antioche*, Paris 1905.

M. Simonetti

PAULINUS of Bordeaux. Contemporary of *Faustus of Riez, perhaps the same P. named by *Gennadius as author of works on *penitence, the beginning of Lent, *Easter, obedience, *neophytes (*Vir. ill.* 68). A fragment of his *De paenitentia* remains, wrongly attrib. to Faustus. He may also be the P. who wrote *Ep.* 4 among Faustus's *letters (CSEL 21, 181-183); attribution to him of the *Tractatus duo de initio quadragesime* (A. Mai, *Spicilegium Romanum*, IV, Rome 1840, 309-313) is doubtful.

CPL 981-982; PL 58, 875-876 (= PL 103, 699-702); B. Poschmann, *Die abendländische Kirchenbusse im Ausgang des christlichen Altertums*, Munich 1928, 128-129, n. 4.

S. Zincone

PAULINUS of Nola (355-431). Christian poet: we possess a collection of his *Epistolae* and one of his *Carmina* (ed. Hartel, CSEL 29-30). Born 355 (or earlier) at Burdigala (Bordeaux) in Aquitaine, of a rich, noble senatorial family related to the most illustrious in the empire, he was educated in circles connected with the teaching of *Ausonius, earning forensic success and the poetic laurel. Appointed senator in 378, he consequently obtained a curule magistrature, prob. the governorship of Campania, which he exercised with humanity. In 383, after *Gratian's death and the rise to power of the *Arian *Valentinian II, who persecuted *orthodox magistrates, esp. those who had supported the usurper *Maximus in *Gaul, P., returning to Aquitaine and perhaps passing through *Milan, met St *Ambrose, with whom he seems to have had subsequent encounters and who gave him his first instruction for *baptism, which he then received at home from bishop *Delphinus in 389. In the same year - perhaps to escape the persecution his family was then suffering from - he fled with his wife Terasia to the recesses of the Pyrenees, breaking his links with his old circles and even with his friend and teacher Ausonius. In *Spain he was finally converted to wholehearted Christianity and gave away his immense wealth, causing great scandal everywhere, but approved and praised by Ambrose, *Augustine, *Jerome and Severus. At *Barcelona in 394 the people acclaimed him priest (the procedure is unclear). He left Spain in 395 and took refuge at Nola, where he still had property, attracted there by the call to put himself at the service of St *Felix, to whom he had dedicated himself since the years of his governorship: he lived here until 409 or 410, leading an *ascetic life with his wife and other companions, next to St Felix's sanctuary at Cimitile near Nola, keeping in touch by letter with influential bishops and learned Christians and receiving *friends and visitors, who came from all parts to talk to him. In 409/10 he left the retirement of Cimitile for the episcopal see of Nola, vacant after the death of bishop Paul. Terasia had died in 409. He himself died 22 June 431.

P. wrote most of his surviving *letters and poems at Cimitile. The letters (51, acc. to Hartel), addressed to Augustine, Jerome, *Rufinus, *Jovius, *Pammachius, *Desiderius, *Sulpicius Severus, *Amandus, Delphinus and others, show sentiments of warm *friendship, renewed by the new Christian *spirituality and by a sociability made purer, through *ascesis, by the bonds of brotherhood in Christ. Though neither an acute and original theologian nor a profound exegete when faced with problems of *theology or biblical *exegesis, P.'s letters reveal a pure and intensely felt *faith and a deep and rich humanity. His poetry (33 poems, acc. to the Hartel ed., but Carm. I, IV, V, XXXII and XXXIII are wrongly attrib. to him) was on secular subjects until 389, after which it was all directed to singing the praises of Christ and St Felix: in Spain, between 389 and 394, he composed Carm. VI, VII, VIII and IX; in 393-4, Carm. X and XI to Ausonius; in 395 Carm. XII, the first natalicium in honour of St Felix; at Nola, in 396, he composed Carm. XIII, the second natalicium and, from then on, every year, one for St Felix's *natale: 12 more in all, from 397 to 408 or 409 (XIV in 397, XV in 398, XVI in 399, XVIII in 400, XXIII in 401, XXVI in 402, XXVII in 403, XXVIII in 404, XIX in 405, XX in 406, XXI in 407, XXIX in 408 or 409); he composed other poems for three occasions: XVII, a propempticon for *Nicetas, and XXIV, an epistle to Citherius, in 400; XXV, an epithalium for the *wedding of *Julian of Eclanum and Titia, between 401 and 404; XXXI, a consolatorium for the death of Celsus, some time between 393 and 408, prob. before 401. In all these poems, except Carm. VI in honour of *John the Baptist and Carm. VII, VIII and IX, which are paraphrases of Psalms 1, 2 and 136, P. uses the *literary genres already consecrated by classical tradition (propempticon, epithalamium, consolatorium, natalicium), but gives them a profoundly new and Christian content.

Like *Prudentius, whom he resembles in education, participation in public life and very similar religious experience, P. renewed, in accordance with the requirements of a comprehensive conception of Christian life, the poetic conceptions not just of the *pagan poets, but also of his Christian predecessors, and by consecrating poetry to Christ, made it into a spiritual exercise of the type which, on the basis of *lectio divina, raises the creature to the Creator and obtains salvation for it. Judgment of P.'s poetry has been led astray by comparison with that of Prudentius, which is more varied, richer in imagery and so more communicative. It is true that P.'s poems are sometimes excessively prolix and too oratorical, but even the digressions often have a precise poetic function. His poetry, though it may seem simple, in fact makes difficult reading and is not limited to sympathy for his neighbour (many scenes of events, realistically narrated; characterization, now of holy and cultivated visitors, now of humble peasants and pilgrims) or descriptions of feasts, represented in a lively way (but in harmony with the ascetic's love for St Felix and, through him, of Christ), but reveals a constant poetic tension, always at work, even when weakened by mannerisms and excesses.

Mildness is commonly considered the principal note of P.'s character. This judgment is derived from the friendly tone in which he addresses his friends and his sympathy for the humble. But this mildness must not be taken for weakness: P. knows how to be severe and decisive, as when he rejects Ausonius's rebukes for having abandoned secular poetry or the criticisms of his detractors, or when he has to rebuke the citizens of Nola who had denied him water for his sanctuary. What is really dominant in P.'s character is the fervour of his religious faith, which conditions all his actions and his literary activity itself, turning it, in *love for his neighbour, towards the love and praise of Christ.

Editions: PL 61; CSEL 29-30 (Hartel).
Translations: Eng.: P.G. Walsh, Letters and Poems, 3 vols., ACW 35, 36, 40, London-Westminster 1966, 1967, 1975.
For an exhaustive list of studies of P. (basic among which are P. Fabre's two, Essai sur la chronologie de l'oeuvre de St. Paulin de Nole, Paris 1949, and St. Paulin de Nole et l'amitié chrétienne, Paris 1949) cf. S. Costanza, Meropio Ponzio Paolino: Antologia di carmi, Messina 1971, 17-36; J.T. Lienhard, Paulinus of Nola and Early Western Monasticism, Bonn 1977, 11-16 and 192-204. To the works indicated there we must now add: W. Erdt, Christentum und heidnisch-antike Bildung bei Paulinus von Nola, mit Kommentar und Übersetzung des 16 Briefes, Heisenheim a.G. 1976; S. Costanza, Cristianesimo e Romanità in Paolino di Nola (Studi in on. di S. Pugliatti), V, Messina-Milan 1979; K. Kohlwes, Christliche Dichtung und stilistische Form bei Paulinus von Nola, Bonn 1979; A. Ruggiero, Nola crocevia dello spirito (with tr. of Carm. XXI and XXVII); and finally, A. Ruggiero-H. Crouzel-G. Santaniello, Paolino di Nola, momenti della sua vita e delle sue opere, Nola 1983, with tr. of Carm. XV, XVI, XVIII (Ruggiero) and Ep. XXIII (Santaniello), and the studies collected in the Atti del Convegno su Paolino di Nola (Nola 20-21 marzo 1982), Rome 1984; W.H.C. Frend, Paulinus of Nola and the last century of the Western Empire, JRS 59 (1969) 1-11; id., The two worlds of Paulinus of Nola, in (ed.) J.W. Binns, Latin Literature of the Fourth Century, London 1974, 100-133; J. Desmulliez, Paulin de Nole, Etudes chronologique (393-397), Paris 1985; Y.-M. Duval, Les premiers rapports entre Paulin de Nole et Jérôme, Studi in onore di S. Costanza, 1987; J.T. Lienhard, Friendship in Paulinus of Nola and Augustine, Mél. van Bavel, Leuven 1990.

S. Costanza

PAULINUS of Pella. Born late 376 or early 377 at Pella in *Macedonia, where his father Thalassius was vicarius; from the age of three he lived in *Gaul, his parents' land. At home he spoke Greek and found it hard to learn Latin. At 20 he married. The Gothic invasions of the early 5th c., with which he collaborated in some way, and the subsequent vengeance of the Gallo-Romans, brought him considerable suffering and the loss of much of his property. His "*conversion", i.e. return to the faith of his infancy, is placed in 421/22. He spent his last years at *Marseilles, where he had a property, and died after 459. P. was a nephew of the poet *Ausonius, whose sister married Thalassius; but it is possible that he was born of a previous marriage, since Ausonius never mentions him. P. wrote a short autobiographical poem, Eucharistos (thanks to God), published in 459 when he was 83 (v. Euch. 12 f.). As he says in the preface, he is recounting his life not for the curiosity of others, but to give thanks to *God, who has always guided and guarded him in the vicissitudes of life. For this reason, the 616-verse poem is a reflection on his past life, in the light of *faith in God. The poetry is decadent, both in form and content. *Paulinus of Nola's Carm. IV, a prayer to God for a serene life with his family without injuring others, is also attrib. to P. (Hartel, CSEL 30, 3).

CPL 1472-73; PLS 3, 1115-1128; CSEL 16 (1888) 291-318; P.H. Evely White, London 1921 (among Ausonius's works and with Eng. tr.); C. Moussy, SCh, Paris 1974 (with Fr. tr.).
Studies: G. Funaioli, De Paulini Pellaei carminis "eucharisticos" fontibus, Le Musée belge 9 (1905) 159-179; P. Courcelle, Un nouveau poème de Paulin de Pella, VChr 1 (1947) 101-113 (also in Hist. Litt. des grandes inv. germ., Paris ³1964, 293-302); P. Tordeur, Concordance de P.P., Brussels 1964; Patrologia III, 311-313.

A. Di Berardino

PAULINUS of Périgueux. We know little of the life of this 5th-c. poet, often confused even in antiquity with the better-known *Paulinus of Nola. His birth must be placed early in the 5th c.: cf. his reference to his gravis senecta in De visitatione nepotuli sui, written between 470 and 473. There is much doubt whether he was bishop of Périgueux (Petrocoricum). His best-known work is the hagiographical De Vita S. Martini, a poetic paraphrase in six books of *Sulpicius Severus's work, composed, perhaps to be read in public, at the request of bishop *Perpetuus of Tours, who suggested that he also turn into verse an opusculum of his own on the miracles worked by St *Martin after his death. P. amplified Sulpicius's account with sometimes tedious pomposity, giving way with excessive frequency to moral and apologetic digressions and minute descriptions, some of them not without fine observation and lively representation. To the extolling of St Martin's miracles we owe two minor compositions of little literary value, an inscription of 25 hexameters De orantibus, intended to embellish a new basilica of St Martin, and the 80 hexameters De visitatione nepotuli sui.

CPL 1474-1477; PL 61; M. Petschenig, CSEL 16, I, 1888.
Studies: A. Huber, Die poetische Bearbeitung der Vita S. Martini des Sulpicius Severus durch Paulinus von Périgueux, Kempten 1901; A.H. Chase, The Metrical Lives of St. Martin of Tours of Paulinus and Fortunatus and the Prose Life by Sulpicius Severus, HSCPh 43 (1932) 51-76; Helm, PWK 32 (1949) 2355-2359; F. Châtillon, Paulin de Périgueux, auteur de la Vita Martini, et Sidoine Apollinaire, panégyriste des empereurs, RMAL 23 (1976) 5-12; J. Fontaine, Hagiographie et politique de Sulpice Sévère à Venance Fortunat: La christianisation du pays entre Loire et Rhin (Actes du Colloque de Nanterre), RHEF 62 (1975) 113-140.

S. Costanza

PAULINUS of Trier. At the council of *Arles (353) he was the only Gallic bishop to refuse to subscribe *Athanasius's condemnation, imposed by *Constantius. Deposed and exiled to *Phrygia, where he died soon after.

DCB 4, 232.

M. Simonetti

PAULINUS of Tyre. Supporter of *Arius from the start (c.320), he shared his doctrine and was the recipient of an important doctrinal *letter from

*Eusebius of Nicomedia. At *Nicaea he subscribed the anti-Arian formula of faith, at *Constantine's request, but took part in the subsequent anti-Nicene reaction. Elected bishop of *Antioch, 327, in place of the deposed *Eustathius, he died some months later.

CPG 2, 2065; DCB 4, 231-232.

M. Simonetti

PAULINUS of York. Sent to Kent by *Gregory the Great in 601 (Bede, *Hist. eccl. gent. angl.*, I, 29). When Ethelburga, sister of king Eadbald of Kent and a *Catholic, went to *York to marry Edwin, *pagan king of Northumbria (*op. cit.*, II, 9), P. was consecrated bishop of York (21 July 625) and went with her. The aim of his transference was to *convert the king and court to Christianity, an aim he embraced zealously, baptizing not just the king but also numerous nobles (*op. cit.*, II, 12. 14). He was forced to return to Kent (*op. cit.*, II, 20) after King Edwin's defeat in 632 by the Briton Cadwallon, who brought paganism back to the Christianized areas. Subsequently named bishop of Rochester, P. died 10 Oct 644.

P. Hunter Blair, *The World of Bede*, London 1970, ch. 10.

F. Cocchini

PAUL of Antioch. *Monophysite patriarch of *Antioch, P. (of Beth Ukkame, or the Black) was born in *c*.500 (?) at *Alexandria. He was *syncellus* (secretary) to the monophysite patriarch *Theodosius of Alexandria, in exile at *Constantinople, who appointed him to succeed Sergius of Tella, founder of the dissident monophysite hierarchy of Antioch, who died early in 561. P. was quite moderate towards the imperial church, ready to retire if it condemned *Chalcedon, and even entered *communion with it for a time, convinced that it was on the point of announcing that condemnation. He more than once took part in peace talks in the capital. He was opposed by more extremist elements, including *Jacob Baradaeus, the great organizer of the monophysite hierarchy in the *East, who had certainly hoped to become patriarch himself. In *Egypt in 575, P. supported the moderate candidate for the see of Alexandria, *Theodore, who was defeated. Within the monophysite ranks, P. opposed the so-called "*tritheist" tendency and was accused of *Sabellianism. During these disputes, when his authority was reduced to very little, he always found support among the Ghassanid Bedouin, and stayed for a long time with their Sheikh Al-Hareth bar Gabala. The *letters he wrote to justify himself are published in *Syriac in J. B. Chabot, *Documenta ad origines monophysitarum illustrandas* CSCO 17, Louvain ²1952).

CPG 3, 7203-7214; E. Honigmann, *Evêques et évêchés monophysites d'Asie antérieure au VIᵉ siècle*, Louvain 1951, 195-205; W.H.C. Frend, *The Rise of the Monophysite Movement*, Cambridge 1972, 291-293, 318-328.

J. Gribomont

PAUL of Apamea. Bishop of *Apamea and metropolitan of *Syria Secunda, after the deposition of the *monophysite *Peter by the emp. *Justin in 513. A Chalcedonian, he attended the synod of *Constantinople of 536 and wrote to *Justinian, in the name of the suffragan bishops, a *letter containing an *orthodox profession of faith and a condemnation of *Anthimus, *Severus and his predecessor Peter, monophysites (Mansi 8, 980-984).

E. Prinzivalli

PAUL of Aphrodisia. Native of *Asia Minor (*Ephesus?), ordained metropolitan of Aphrodisia in Caria in 558 by *Jacob Baradaeus assisted by two other *monophysite bishops. In 566, at *Constantinople, he subscribed a *letter to his Eastern colleagues. In 571, at the start of the persecution of the monophysites, he was removed from the Carian *monastery to which he had retired some time before, taken in chains to Constantinople and imprisoned in the patriarchal palace. Patriarch *John Scholasticus forced him to withdraw his assent to monophysism and sent him to Aphrodisia to be deposed and reordained by the Chalcedonian metropolitan, thus creating a singular case of episcopal reordination. He died in poverty before 576/7.

CPG 7234; E. Honigmann, *Evêques et évêchés monophysites d'Asie Antérieure au VIᵉ siècle*, Louvain 1951, 218-231.

D. Stiernon

PAUL of Callinicum. We know few details of P., bishop of Callinicum (now Rakkah) in Osrhoene. His declared *monophysite tendencies had led to his elevation to the episcopate in 503, at the time of the emp. *Anastasius, but were also the cause of his deposition, together with some 50 anti-Chalcedonian bishops, soon after the accession of the emp. *Justin. P. fled to *Edessa, where he dedicated himself to translating from Greek to *Syriac most of the works of patriarch *Severus of Antioch. This enabled the survival (despite the *damnatio memoriae* of the Greek works) of the polemical correspondence between Severus and *Julian of Halicarnassus on the incorruptibility of Christ's *body before the *resurrection, and a discourse against Julian. We also prob. owe P. the translation of the correspondence between Severus and *Sergius Grammaticus, the patriarch's treatises *Contra impium Grammaticum*, i.e. *John of Caesarea, and a treatise by the same author entitled *Philalethes* (cf. CSCO 133; 134). Taking into account the similarity in style between these translations and an old version of the *Cathedral Homilies*, again by Severus of Antioch, this last version has also been attrib., following W. Wright, to P., who rightly deserves the title "Interpreter (translator) of Books" traditionally conferred on him by the *Jacobite church. We must not confuse P. with the homonymous bishop of Edessa who was exiled to Euchaita in 522, restored to his see in 526 (Duval 316, n. 2) and died in 527. We must also distinguish him from bishop Paul of Edessa (Chabot 79) who, fleeing the Persian invasion in 619, i.e. a century later, took refuge in *Cyprus where he translated Severus of Antioch's *Octoechus* (a collection of some 300 *hymns for the feasts of the *liturgical year). This Paul of Edessa must in turn be distinguished (against F. Nau: *ROC* 7 [1902] 100, n. 5) from another homonym, the *abbat Paul, who at about the same time (624) and also in Cyprus, translated *Gregory Nazianzen's works into Syriac. The translations of the *homilies and that of the *Octoechus* were revised in the 7th c. by the celebrated *Jacob of Edessa.

Duval 359; Baumstark 174; Chabot 70-71; Ortiz de Urbina 163, 245-246.

J.-M. Sauget

PAUL of Canopus. Patriarch of *Alexandria (538-541). Born at *Tarsus, monk at Tabennesi and *abbat of a *monastery at Canopus. The emp. *Justinian made him patriarch of Alexandria in place of *Theodosius, deposed for anti-Chalcedonianism. Consecrated by patriarch *Mennas of Constantinople in the presence of *Pelagius, *apocrisiarius of pope *Vigilius and future *pope, P. set out for *Egypt with military support to suppress the *monophysite party. After two years he was deposed by the council of *Gaza, prompted by the empress *Theodora, a secret monophysite. A revolt provided the pretext for deposing P., but the circumstances of his fall are rather obscure. His successor Zoilus was an insignificant figure.

L. Duchesne, *L'Eglise du VIᵉ siècle*, Paris 1925, 103-5, 169-70; *DCB* 4, 250 f.; *DTC* 12, 660.

G. Ladocsi

PAUL of Concordia. From *Jerome's correspondence (*Ep*. 5, 2) it seems that this P. was old and was a *friend of his. *Ep*. 10 was sent directly to him (*ad Paulum senem Concordiae*): Jerome asks P. to lend him some *books and, in exchange, sends the *Life* of *Paul of Thebes, just written (the *letter is perhaps of 375). P.'s having passed his hundredth year induces Jerome to congratulate him on his green old age, so singular as to seem almost a prize from *God for his *virtues, as happened to the patriarchs of old (*Tu semper Domini praecepta custodiens*). Jerome mentions P. again in *De vir. ill.* 53, for a testimony given to *Tertullian's celebrity.

J. Labourt, *Saint Jérôme, Lettres*, Paris 1949, I, 17 and 27; P. Zovatto, Paolo di Concordia, *AAAd* 5 (1974) 165-180.

L. Dattrino

PAUL of Constantinople. Bishop of *Constantinople in 332, he took part in the council of *Tyre (335) and subscribed *Athanasius's condemnation; but soon after he too was deposed and exiled to *Pontus. Subsequently he gained credit and was held a martyr of the anti-*Arian cause, but his anti-Arianism is far from certain: he was deposed as an instigator of seditions, and this would always be the charge subsequently brought against him. He returned to Constantinople on *Constantine's death (337), but was immediately thrown out and *Eusebius of Nicomedia elected in his place. P. fled to *Rome, but on Eusebius's death (*c*.342) he reappeared in Constantinople. The *Eusebians opposed *Macedonius to him, and the riots degenerated into a genuine insurrection, following which P. was once more expelled from the city and exiled, first to Sinagra near the Persian border, then to *Emesa. But he did not give up, and reappeared in Constantinople in 344, still strong in the favour of part of the population, but was secretly arrested and put ashore at *Thessalonica, in the part of the empire administered by *Constans. Under pressure from Constans, *Constantius allowed him to return in 346 to Constantinople, where Macedonius, who had been elected in his place, was reduced to exercising his functions in a single church. When Constans was killed by the usurper

*Magnentius and Constantius made war on Magnentius (350), P. was arrrested again. The fact that Magnentius had sent an embassy to P. and *Athanasius could favour the charge of treason; and the fact that some of his partisans, the so-called *Sancti Notarii*, were executed shows that the charge brought against them was not a religious one. P. was transferred to Cucusa in *Cappadocia, where he died soon after. Claims by ancient sources that he was strangled seem unlikely.

W. Telfer, Paul of Constantinople, *HThR* 43 (1950) 31-92; Simonetti 395.

M. Simonetti

PAUL of Emesa (first half of 5th c.). Bishop of *Emesa (now Homs) in Phoenicia after 410. At the council of *Ephesus (431) he was one of the synod headed by *John of Antioch which *excommunicated *Cyril of Alexandria and *Memnon of Ephesus. In 432, John made him his ambassador in *Alexandria in charge of peace talks with Cyril, giving him what would later be the Creed of Union. At Alexandria, having agreed to excommunicate *Nestorius and recognize *Maximian as bishop of *Constantinople, P. was received into *communion by Cyril and several times invited to preach in the cathedral. He visited *Egypt again in 433, returning to *Antioch with Cyril's *letter *Laetentur coeli*, which happily crowned his peace mission. Died before 445, when another bishop appears in the see of Emesa. We have part of his correspondence over the negotiations and fragments of his three Alexandrian *homilies.

CPG 6365-6369; PG 77, 165-168, 1433-1444; Hfl-Lecl 2, 1, 393-402.

D. Stiernon

PAUL of León. Paulus Aurelianus is known from a late *biography written by Wrmonoc (884): among many legendary elements, it gives a few reliable facts, e.g. his birth in *Britain (*c*.480). Crossing the Channel with some companions, he stopped at the isle of Batz (Brittany), where he founded a *monastery and, in the fortress facing it on the mainland, a missionary centre for priests (Kastel-Paul or Saint-Pol-de-Léon). Childebert, king of the *Franks, made him accept the episcopal see of Kastel-Paul. He died at Batz in 572/575, but was buried at Kastel-Paul. A town developed round his tomb and took its name from him.

ASS *Martii* II, 107-119; *Vies des SS.* 3, 260-262; *LTK* 8, 229-230; *BS* 10, 296-299.

E. Romero Pose

PAUL of Mérida. Deacon of *Mérida (*Spain), early in the 7th c. he wrote the *Vitae sanctorum patrum Emeritensium*, in declared imitation of *Gregory the Great's *Dialogues* and in a simple form for the sake of the *imperiti*. An account of the miracles of the monks and early *Fathers, as it approaches P.'s own time it is rich in important historical information, esp. concerning bishops Fidelis and *Massona. He narrates at length the wanderings of Massona, who was of Gothic origin, during *Leovigild's "persecution".

CPL 2069; PL 8, 115-164; Díaz 214.

M. Simonetti

PAUL of Narbonne. The first known bishop of *Narbonne was named P., and seems to have been the first bishop. Buried at Albolas on the Via Domitia, where a basilica in his name was later built, at the centre of the city's S necropolis. Mentioned by *Prudentius (*Perist.* 4, 34), *Mart. hier.* (22, 3, Italic recension), *Caesarius of Arles (*De myst. S. Trin.*) and *Gregory of Tours (*Hist. Franc.* 1, 28). Caesarius calls him a disciple of *Trophimus of Arles, Gregory numbers him among the seven missionaries sent from *Rome to *Gaul in *c*.250. The 8th c., when the *Franks reconquered Narbonne (759), may be the date of a *Life* (BHL 6589-90), from which the *Passio SS. Dionysii, Rustici et Eleutherii* seems to be derived. Ado of Vienne confuses him with Sergius Paulus (Acts 18, 7-8). Nothing else is known of him.

Duchesne, *Fastes* I, 302-303; E. Griffe, *Histoire religieuse des anciens pays de l'Aude*, I, Paris 1933, 16-22; id., *La Gaule chrétienne à l'époque romaine*, I, Paris ²1964, 164-167 and *passim*.

V. Saxer

PAUL of Nisibis (or Paul the Persian). *Nestorian exegete and theologian, erudite and endowed with great judgment. Director for 30 years, at the time of Mar *Aba I, of the *school of Hadiab; bishop of *Nisibis from 551; died 571. At the head of a strong delegation, he took part in a debate at *Constantinople with *Justinian's theologians (532-3), where he defended (for the first time?) the doctrine of two *hypostases in Christ. He can undoubtedly be identified with Paul the Persian, very active at Constantinople around 527 and Christian spokesman, at Justinian's request, at an official debate with Photinus the Manichee (cf. PG 88, 529-574). His Introduction to the *Bible served as basis for the *Instituta regularia divinae legis*, written in Latin at Constantinople at that time by the *quaestor sacri palatii* *Junilius Africanus and dedicated by him to *Primasius of Hadrumetum (PL 68, 15-54; crit. ed. by H. Kihn). This lucid manual, *Aristotelian in spirit, sums up the doctrine of the school of Nisibis, i.e. of *Theodore of Mopsuestia; it contains teaching on the *Trinity, the *Incarnation, the Law. It was very popular at Saint Gall and in S Germany in the 9th c.

CPG 7010-7015; H. Kihn, *Theodor von Mopsuestia und Junilius Africanus*, Fribourg 1880; G. Mercati, *Per la vita e gli scritti di Paolo il Persiano*, ST 5, Vatican City 1901, 180-206; A. Vööbus, *History of the School of Nisibis*, CSCO 266, Louvain 1965, 170-172; A. Guillaumont, Un colloque entre orthodoxes et théologiens nestoriens de Perse sous Justinien, *CRAI* 1970, 201-207; id., Justinien et l'Eglise de Perse, *DOP* 23-24 (1969-70) 39-66; M. Richard, *Iohannis Caesariensis Opera*, CCG 1 (1977) XXXIX-XLI.

J. Gribomont

PAUL of Samosata. Native of Samosata (*Syria), of non-Greek stock; between 260 and 270, when Syria was ruled by *Zenobia, he was a high official of the queen's financial administration (*decenarius*) as well as bishop of the church of *Antioch. His conduct, more in keeping with his public than his ecclesiastical office, and his doctrine aroused unfavourable reactions in the church: between 264 and 268 there were many attempts to bring him back to a way of life and belief more consonant with the norms of the church, but in vain; a *council (of maybe two) of bishops from nearby cities had no success. In 268 a new council at Antioch managed to accuse and convict him of *heresy, following a debate between P. and the priest *Malchion, a teacher of *rhetoric well versed in *theology. P. was deposed from his office, but refused to abandon the *church buildings he occupied. When Antioch returned to Roman administration, an appeal was made to the emp. *Aurelian and P. was expelled by the public authorities. His followers formed the sect of *Paulinians, still alive at the time of the council of *Nicaea (325), but of little importance.

*Eusebius and other ancient authors had access to the *acts of the council of 268, and Eusebius (*HE* VII, 27-30) uses them at length, but he is very vague on P.'s doctrine, and the surviving fragments of the debate between P. and Malchion are not exhaustive and not all certainly authentic. But they seem to confirm the unanimous judgment of the ancients that P. was a *monarchian of *adoptianist type. P.'s adoptianism is more highly evolved than that of the late 2nd / early 3rd c.: it takes account of the Logos-theology developed by the *apologists and *Origen. But for P. the divine Logos (= Wisdom) is not personal, i.e. endowed with a distinct *hypostasis, but only a *dynamis*, an operative faculty of *God: it is the order and command by which God operates in the *world. P. gives the name "Son of God" not to the Logos, but to the man Jesus in whom the Logos dwelt as in a temple and was clothed as in a garment. This union raised the man Jesus to a higher rank than the patriarchs and prophets, and gave him a constitution somehow different from that of other men, but it was still always an extrinsic, moral union, by which Jesus, while being Son of God, remained a mortal man.

We are unable to verify the grounds and significance of the information (from a *homoeousian source) that the council of 268 also accused P. of considering the Logos consubstantial (*homoousios) with the *Father: perhaps P. meant to say that the Logos, as a divine impersonal power, had no *ousia (essence, substance) of its own, distinct from that of the Father.

P.'s monarchianism was clearly opposed to the Logos-theology worked out by Origen, and many of the bishops who variously worked against him were of Origenian filiation (*Gregory the Thaumaturge, *Firmilian of Caesarea, *Dionysius of Alexandria, *Theotecnus of Caesarea, etc.). The episode of P. must therefore be interpreted as a critical moment in the spread of Alexandrian culture and theology into an area of *Asiatic culture, where monarchianism in its various forms had originated and remained vital. P. had radicalized this monarchianism in a way that allowed the Origenians to lay charges against it, but because his doctrine had its roots in the city's tradition, cleansed from its radical points it survived its condemnation. More than 50 years after P.'s condemnation, monarchianism was still represented at Antioch in a moderate form by *Eustathius, and precise terminological similarities between the two authors (indwelling of the Logos in Jesus, image of the temple, etc.) confirm the continuity of the monarchian tradition at Antioch.

Quasten I, 401-402 (bibl.); F. Loofs, *Paulus von Samosata*, TU 44, 5, Leipzig 1924; G. Bardy, *Paul de Samosate*, Louvain 1929; H. De Riedmatten, *Les Actes du procès de Paul de Samosate*, Fribourg 1952; F.W. Norris, Paul of Samosata: *Procurator Ducenarius*, *JThS* n.s. 35 (1984) 50-70; H.C. Brennecke, Zum Prozess gegen Paul von Samosata, *ZNTW* 35 (1984) 270-290; J.A. Fischer, Die antiochenischen Synoden gegen Paul von Samosata, *AHC* 18 (1986) 9-30.

M. Simonetti

PAUL of Thebes. Tradition has designated him "the first hermit" of the Christian world. Born 228, under the emp. *Alexander Severus, in the Lower *Thebaïd, of a very rich family, he was able to receive an uncommon education, becoming "exceptionally versed in Greek and Egyptian culture" (*Jerome). At 16, he lost his parents and inherited a huge fortune, shared with an elder sister. To escape *Decius's *persecution (249) he withdrew to a remote house in the country; but learning that his brother-in-law, driven by the "cursed lust for gold", was thinking of handing him over to the authorities, he fled to an even more deserted place, where "he came to transform that state of necessity into a voluntary choice". Entering a cave in a rocky mountain, he found it suitable for living, as if by divine *providence: an ancient palm whose long branches extended over the cave entrance gave food and clothing; a hidden spring gave limpid water. Here he decided to pass his life in recollection and *prayer; a crow miraculously brought him a daily ration of bread. Aged 113, P. was visited by the great monk *Anthony, in his nineties, whom he could charge with being neither the only nor the first hermit in the *desert. That day the crow brought a double ration of bread for the two monks immersed in prayer and heavenly conversation.

Even P.'s end was surrounded by the miraculous. From his monastery, Anthony returned, this time with special care, to the grotto; but on the way he saw the holy *anchorite's soul ascend to heaven amid a choir of *angels and saints singing "Hosanna"; finding his body inert in the grotto, still on its knees, he wrapped it in the cloak given him by the saintly bishop *Athanasius and buried it in a grave dug by two great lions (AD 341).

These and other prodigious events narrated around the figure of P. obviously arouse perplexity and require critical discernment; but the facts in the first chapters of *Jerome's *Vita Pauli* seem to be substantially trustworthy.

BS 10, 269-280; Jerome, *Vita Pauli*: PL 23, 17-30; P. de Labriolle, *Vie de Paul de Thèbes et vie de St. Hilarion*, Paris 1907; S. Girolamo, *Vite di Paolo, Ilarione e Malco*, It. tr. Milan 1975.

M. Naldini

PAUL of Verdun (†648/9). Native of the region of Autun (Augustodunum), after a short time at the court of Clotaire II (584-629) he retired to a mountain near *Trier (Paulusberg or Bulisberg), but not to organized monastic life. From 625/30 to 648/9 he was bishop of Verdun, elected by the *clergy and people and appointed by king Dagobert (early 7th c.-639). He worked hard for his church, which he found in miserable conditions: he restored divine worship, gave new fervour to canonical life and trained the clergy. He stood out for his cult of the *Eucharist, devotion to *Mary, goodness to the *poor. We have two gracious and affectionate *letters from P. to *Desiderius of Cahors in *c.*641, and a note from Desiderius inviting P. to take part in the *dedication of a basilica.

CPL 1303⁰; PL 87, 260, 261; CCL 117, 333, 334; LTK 8, 234; BS 8, 280-281; DCB 4, 271.

M.G. Bianco

PAUL Silentiarius. Byzantine poet, *primicerius* of the *silentiarii* (chamberlains who kept order at imperial audiences and *processions) at the court of *Justinian (527-565). *Friend of the historian and poet Agathias (*Historiae* 5, 9). In the Alexandrian tradition, he composed *epigrams and descriptions of works of art in verse. His *Descriptio (ekphrasis) sanctae Sophiae* in 887 hexameters (in the manner of *Nonnus) was preceded by two introductions, one addressed to the emperor, the other to patriarch *Eutychius. His second didactic composition was the *Descriptio ambonis* in 275 hexameters, with introduction in iambic trimeters. This pulpit of the cantors was under the dome of Hagia Sophia, raised on eight marble columns. The church, inaugurated by Justinian in 537, had been gravely damaged by the earthquake of 558. It was restored in five years, and the solemnities of the second "consecration" began on Christmas Eve 562. P.'s first composition was prob. read on 6 Jan 563 and the second a few days later, both in the presence of the emperor and the patriarch. The two poems, of literary and historical importance, are of great interest for the history of Byzantine art: the author lingeringly describes the magnificence of the temple, rich in polychrome marbles, tall columns, *mosaics, precious stuff and gold and silver vessels (calendars, *lamps of various forms, etc.). The *Anthologia Palatina* contains 78 of P.'s *Epigrams*, mainly amorous. The poem in 190 choliambics *In thermas pythicas* (in *Bithynia) is not his.

CPG 3, 7513-7516; PG 86, 2119-2268; P. Friedländer, *Johannes von Gaza und Paulus Silentiarius*, Leipzig-Berlin 1912; G. Viassino, *Paolo Silenziario, Epigrammi* (text, It. tr. and comm.), Turin 1963; *Anthologia Palatina* (ed. P. Waltz, "Les belles lettres"); Bardenhewer 5, 24 f.; PWK 3, 2366-2377; Impellizzeri 247 f. and 434 f. (bibl.).

A. De Nicola

PAUL the Simple, saint (4th c.). Mentioned by *Rufinus (*Hist. monach.* 31 [PL 21, 457-459]) and *Palladius (*Hist. Laus.* 28: PG 34, 1076-1084; cf. *Appendix ad Palladium*: PG 65, 381-385; *De vitis Patrum seu verba seniorum*, 167: PL 73, 795 f.; Soz., *HE* 1, 13: PG 67, 900 BC). An Egyptian peasant, he peacefully left his wife after surprising her in *adultery and went to the *Thebaïd, to St *Anthony Abbot who, before accepting him as a disciple (aged 60), put his monastic vocation to a hard test. Humility, simplicity, obedience and great *faith characterized his life; he had great power over *demons.

ASS *Maii* I, 643-647; LTK 8, 214; BS 10, 264 f.

A. De Nicola

PEACE. The term p. was an important word in ancient Christianity. Its meanings were many: we will give the main ones.

1. P. in relation to Christ, the p. of every man (Eph 2, 14). Reference of the concept of p. to Christ's person later developed, esp. in *Augustine, into an understanding of the term in relation to the human person. This may have been Christianity's most specific contribution to the understanding of the term p., normally understood in relation to *war and as its absence. *2.* The p. of Christ as fruit of the Spirit (Gal 5, 22; Rom 14, 17), given to whoever believes in Christ ("I leave you peace . . . my peace I give you": Jn 14, 27). This is the p. of believers, distinct from that which reigns in other groups. The *Fathers put great emphasis on the reception by Christians of this p. which is proper to them (comments of *Origen and Augustine on Jn 14, 27). *3.* P. in relation to civil and religious institutions, and to relationships between peoples contending for the supremacy of economic goods and of utopian and political cultural models.

The first meaning of p., referred to Christ and man, finds ample space in the *liturgy and in *anthropological reflection, esp. *Augustinian; the second, relative to the p. "of believers" as a religious group, is very much present in the internal *orthodoxy-*heresy dialectic and in relations between the various churches (cf. *Cyprian's work and Augustine's anti-*Donatist writings); the third was subject to the evolution of the Christian understanding of its own place in the *world. Before *Constantine we see an objection to war and, in general, to all that could endanger human life (the clearest evidence is in *Traditio Apostolica*, 16, ed. B. Botte 1963, p. 36, forbidding *catechumens and *faithful to kill in war or to enlist; and the judge to use *ius gladii*; and Origen, *Contra Celsum*, book 5 and 7-8). This attitude is summed up by the *Vita Martini* (ch. 4) in the confession: "I am not allowed to fight" (*pugnare mihi non licet*). After Constantine, the same line is taken by Augustine (*Epp.* 220; 229; 230; 231) in the years 426-427 before the *Vandal invasion. His principle that wars should be ended at the negotiating table and p. reached with the weapons of p., not those of war (*ipsa bella verbo occidere . . . et acquirere vel obtinere pacem pace, non bello*: *Ep.* 229, 2), remains famous. The situation of Christians had changed by Augustine's time. We see no *conscientious objection, but the soldier's duty to fight; it is insisted that leaders, those responsible, should not make war and should use every means to put it off and mitigate it. What, in the pre-Constantinian church, was a necessity for all (against any aggression on human life) became after Constantine a distinction between life as a citizen, which brought with it a duty to defend the state, and life as a Christian, which advised against making war. Augustine summed up the new situation thus: "If you ask me for advice conforming to the maxims of this world one cannot give certain advice on uncertain things. But if you ask me for advice conforming to the law of God, I know perfectly well what I should tell you . . . the soldiers of Christ fight not to kill men, but to defeat the princes, powers and spirits of *evil (Eph 6, 12) . . . conjugal duties do not prevent you . . . from seeking peace even in wars, if it is ever necessary for you to take part in them" (*Ep.* 220, 9 and 12). Augustine's passage in *De civ. Dei* (4, 15) on wars justified by a neighbour's wickedness, in accordance with the model applied by Livy to the expansion of the Roman empire, must be read in the same perspective.

In an overall view, we observe again that the term p. is used to indicate now the harmonious relationship between church and state, now the state of man's reconciliation with *God, eternal rest, the Christian community founded on *orthodoxy and harmony (Tertull., *De or.* 18; Orig., *In Jer. hom.* 9, 2). It is therefore synonymous with *communio* (Ursacius, *Ep. ad Athanasium*). Like war, p., in the gospel, is not linked to a political condition but to the coming of Christ, who came down to earth to restore man's p. with God (never the other way round, since tension comes from *sin). *Paul calls Christ "our peace" (Eph 2, 14). So Christ's disciple must be a "peacemaker" (Mt 5, 9).

*Clement of Rome attaches great importance to p., a term which recurs seven times in his *letter. The p. of *creation, according to the *Stoic model, serves as a paradigm to exhort the community of *Corinth to live in p. For

*Ignatius, p. is an eschatological good which God concedes to the *assembly (*Phil.* 7, 2). *Cyril of Jerusalem asks for *prayers for the authorities to promote p. (*Cat.* 15, 6). Among the other Fathers we can distinguish various attitudes. All develop the soteriological and spiritual aspect of p., in accordance with the gospel. *Tertullian remains faithful to the evangelical line (*Adv. Marc.* 3, 14; 4, 28; 5, 14. 17). Reconciliation with God concludes p. (*Pud.* 2, 12, 15). In his *Montanist period, he distinguishes *pax humana, pax ecclesiastica* and *pax divina*. So does *Clement of Alexandria (*Strom.* I, 29, 4). *Gregory of Nyssa interprets the *beatitude of p. in a psychological sense, while looking at the harm done by war (*De beat. hom.* 7, 2-4). Interpretation of p. on a level of a political theology already appears in *Melito. It prevails esp. from the 4th c., in the sense of an identification of *pax Christi* and *pax Constantini*, in *Eusebius, *Ambrose, *Lactantius and *Orosius. *Augustine gradually distances himself from a political vision, to see in p. an interior and eschatological good.

P. appears in the funerary acclamation *in pace*, never used by *pagans but very frequent in Christian *epigraphy. In *Africa we find: *fidelis in pace*. *Schism gave rise to the epigraphical formula: *pax et unitas*. *Liturgy knew the *kiss of p. from the time of *Justin (*I Apol.* 65, 2), for fraternal reconciliation. P. recurs in liturgical salutations (*Const. Ap.* VIII, 13, 1; 15, 10). We find a prayer for p. (*Const. Ap.* VIII, 10, 3; 15, 4; John Chrys., *In Ac. hom.* 37, 3).

RACh 8, 434-505; M. Viano, Contributo alla storia semantica della famiglia latina di pax, *AAT* 88 (1953) 168-183; C. Tibiletti, Il senso escatologico di pax e di refrigerium in un passo di Tertulliano, *Maia* 10 (1958) 209-219; J. Laufs, *Der Friedensgedanke bei Augustinus. Untersuchungen zum 19. Buch des Werkes "De civitate Dei"*, Wiesbaden 1973; A. Portolano, *L'etica della pace nei primi secoli del cristianesimo*, Naples 1974; A.M. Papes, Il concetto di pace in Tertulliano, *Salesianum* 42 (1980) 341-350; M. Toschi, *Pace e vangelo. La tradizione cristiana di fronte alla guerra*, Brescia 1980; V. Grossi, I nodi della pace nella storia del cristianesimo, in *La pace del Regno*, Turin 1983, 73-95; E. Dinkler, *Eirene: Der urchristliche Friedensgedanke*, SHAW (Phil.-Hist. Klasse.), Heidelberg 1973.

A. Hamman

PELAGIA. St P. of Antioch is commemorated by *Mart. syr.* and *Mart. hier.* on 8 Oct. *John Chrysostom preached a (definitely authentic) *sermon in her honour (PG 50, 579-84). *Ambrose mentions her twice (*De virgin.* 3, 7, 33; *Ep.* 27). The first to allude to her may have been *Eusebius of Caesarea (*HE* VIII, 12, 2). These authors all make her a voluntary *martyr who threw herself from the roof of her house to escape the outrages of the soldiers who came to arrest her. Ambrose wrongly calls her the sister of two other Antiochene martyrs, Berenice and Prosdoce. The error passed into Florus's *martyrology and thence through the other historical martyrologies into *Mart. rom.*

Vies des SS. 10, 227-30; *LTK* 8, 245; *BS* 10, 430-32; P. Franchi de' Cavalieri, La "Homilia II in s. Pelagiam" è veramente di s. Giovanni Crisostomo?, ST 65, 281-303; Séminaire d'Histoire des Textes, Les Vies latines de S. Pélagie, *REAug* 12 (1977) 3-29; 15 (1980) 265-304; *Pélagie la Pénitente. Métamorphose d'une légende*, 2 vols., Paris 1981-1984.

V. Saxer

PELAGIUS. Meletian bishop of Laodicea in *Syria (*c*.360-*c*.382). *Theodoret of Cyrrhus describes his life before his episcopate. He married very young, but lived in perfect *chastity from the first day (*HE* IV, 12; V, 8). *Acacius of Caesarea consecrated him bishop as part of the reinforcement of the *homoousian party. At the councils of *Antioch (363) and *Tyre (367), he was an ardent defender of *orthodoxy, for which the emp. *Valens exiled him to *Arabia (372). After Valens's death he returned to his see; in the *schism of Antioch he sided with *Meletius, and at the council of *Constantinople (381) he protected *Gregory Nazianzen. Date of death unknown.

Simonetti, cf. index.

G. Ladocsi

PELAGIUS - PELAGIANS - PELAGIANISM

I. Historical outline - II. The polemical context - III. Writings and thought.

These three terms had a considerable historical import, relating to the time, the persons and the ideas they incarnated, and constitute a symbol in relation to what they represented later in the history of Christianity. Benefiting from the progress of studies in this field, we will try to shed light on the historical importance of the terms and "redefine" them in a way that owes more to the sources than to the symbolic identity they developed from *Augustine's death (430) on. Over the centuries, the Pelagian controversy has been read mostly from the viewpoint of Augustine's writings, which present it as a "new *heresy" in Christianity (*Retract.* II, 33). This has had a negative influence which extends to our understanding of Augustine's own thought, leading us to blame him for the anthropological attitudes of others, within which he replied in order to show their shallowness. Studies of Pelagianism, from Plinval (1943) to the present, take us beyond the orthodox judgment of it and allow us to glimpse, more widely than before, the whole movement which, for some 20 years (from 410), stirred up the waters of Christianity, esp. in the West. The terms "Pelagians - Pelagianism" derive from Pelagius, born *c*.354 in *Britain and baptised *c*. 380-84 in *Rome, where he long lived and was one of the most influential voices of his time. Of this movement of thought we will give the historical outline, the polemical context, the writings and the thought.

I. Historical outline. We distinguish three quite clearly-defined moments: the Pelagian movement before 411; from 411 to 418; and after 418.

1. Before 411. To this period we ascribe the work *De induratione cordis Pharaonis*, which was Pelagius's position on the understanding of Christianity, then debated in Italy in intellectual Christian circles, perhaps in polemic with *Ambrosiaster and with Augustine's *Quaestiones 83*. In this work, rejecting any kind of *Manichaean predestinationism, he insists on the meriting of a destiny by observing, through the freedom innate in nature (*Indur.* 46 and 51), *God's precepts. This first period, if the most suggestive, is also the hardest to approach, given the scarcity of explicit concrete elements at our disposal.

2. From 411 to 418. In this period the Pelagian controversy began and ended publicly. There was the synod of *Carthage of 411 against *Caelestius, followed by a condemnation which was always a reference-point for the rest of the debate; the synod of *Diospolis of 415 against Pelagius, whom it absolved; the African reaction which led to *Innocent I's condemnation of Pelagius and Caelestius in 417 (27 Jan); their temporary rehabilitation by pope *Zosimus (Sept 417); the *concilium plenarium* of Carthage of May 418 which condemned Pelagianism in 9 canons; pope Zosimus's *Tractoria* condemning Pelagius and Caelestius, sent to all the bishops to subscribe, and their condemnation by the emp. *Honorius (there are no exact dates to establish priority between Honorius and Zosimus). This period also saw Pelagius's principal works (*Ep. ad Demetriadem; De natura; De libero arbitrio*).

3. After 418. This last period was characterized by *Augustine's controversy with *Julian of Eclanum on how to reconcile the goodness of *marriage with the transmission of original *sin, and by the transformation of the Pelagian controversy into a question of *anthropology. From this context came Augustine's great theological works on human freedom (*De gratia et libero arbitrio, De correptione et gratia, De praedestinatione sanctorum, De dono perseverantiae*). In this period the problems of *De induratione cordis Pharaonis* were revived (given the similarity between many questions in that work and those that emerged between 425 and 427, many questions hang over its authorship and date). The Pelagian controversy found asylum in the African (*Hadrumetum) and Provençal convents, but no longer as a *heresy, rather as a debate.

The distinction made by Julian of Eclanum to Zosimus between heresy and *quaestio* was accepted as a fact once the great Pelagian teachers had disappeared (Pelagius died, perhaps in Egypt, *c*.427; Caelestius and Julian were in exile). It dragged on in questions about *grace and freedom, approached *more pelagiano*, i.e. one against the other (for Augustine, grace is an aid to freedom, not a contender with it), issuing in the questions of *initium fidei* (*semi-Pelagianism) and predestinationism, condemned at the council of *Arles of 473. The interpretation of Pelagianism and Augustine that prevailed after 431 usually only remembered this third period. On the imperial side, there were Honorius's rescript against defaulters (9 June 419) and that of *Valentinian III against the Pelagianism of S *Gaul (9 July 425).

II. The polemical context. The Pelagian controversy has gone down in history as a question about the right understanding of Christian *anthropology. In fact this was the doctrinal synthesis to which it led in the years after the condemnation of the Pelagianists, esp. from 425 on. In reality, there was a whole movement of ideas spanning the Christian world in the first half of the 5th c. It was put forward as a plea for consistency with the gospel, which required *ascesis and commitment in every area of human life, not just in convents. It was a matter of imitating Jesus Christ in all seriousness. But what might have been just a stimulus to do better was investigated on the theoretical level, and the argument led to the possibility and merit of human freedom, the nature of *God's *grace, the value of the *sacraments in Christian life. Just when the post-Constantinian understanding of Christian life seemed to have prevailed, i.e. the Christian's presence within the state and not in religious opposition to it, everything was once more under discussion. In Roman intellectual circles, divided between *Origenists (a non-anthropomorphic reading of the *Bible) and anti-Origenists (a literal reading of *Scripture for which they were accused of *anthropomorphism), there was a very bitter controversy, which was spread and radicalized by fugitives from *Rome after its fall to *Alaric in 410.

At Rome, *Jovinian's interpretation of Christianity was further investigated. He, resting on the common grace of *baptism, had denied any special merit and hence any special reward to the choice of *virginity and continence, so encouraged at that time by authoritative voices like *Ambrose, *Jerome, etc. The circle of Roman Origenists, linked to powerful families and led by Pelagius, had already produced *Rufinus's anti-traducianist *Liber de fide*. Now it gathered together many scattered voices, thanks to Pelagius's moral stature and *Caelestius's dialectical capacity, in a new unitary reading of Christianity: every Christian is called to follow Christ in the choices of virginity and *chastity, and the possibility of doing so is within everyone's freedom (*Ep. ad Demetriadem*). This ideal, radically opposed to Jovinian, found some favourable ground everywhere, according to circumstance: at *Syracuse, in the social world; in the noble family of the Anicii, who felt honoured to support the *monasteries; in the monasteries themselves, which found Pelagian ascetic ideas a stimulus to living and promoting the choice of that *propositum* (embracing *monasticism) so disparaged by Jovinian. Augustine attests that the following Pelagian adage circulated in the monastery of *Hadrumetum: "It is in my power to do good, it is I who direct my own freedom" (*Ep.* 216, 5). The ambiguities which the Pelagian synthesis brought with it concerned not so much the affirmation of the possibility of human freedom, as the way of understanding that freedom and the consequences deriving from it. The Pelagians thus denied the birth of mankind in "original" *sin, so that it was not in need of Christ's *redemption from *birth, and consequently opposed the practice of baptizing *children for the remission of sins. Finally, they understood God's grace reductively, making it simply an external aid to freedom. The Pelagian ambiguities, carried over to *Africa where they came into contact with *Augustine's ecclesiological-sacramental solutions to the *Donatist question, now concluded (411), and with the question of *traducianism, connected with the problem of the origin of the *soul, were received as "a new scandal in the church" (*Ep.* 177, 15), a "new heresy" (Aug., *Retr.* II, 33). Pelagius and Caelestius were identified as the heads of this movement, men of great persuasive capacity (*Ep.* 175, 1), but "pernicious authors of a new heresy" (*Ep.* 175, 1 and 182, 3) and definitively condemned as such at the council of Carthage of 418, in the circular letter (the *Tractoria*) of pope *Zosimus (418) and at the council of *Ephesus of 431.

The Pelagian movement must be seen as an attempt by Christianity to understand itself, coming to light in the first decade of the 5th c. In fact Jovinian, Pelagius and Augustine reflect the fundamental ideas on the different way of thinking about God's grace and man's freedom, that is to say the man-God relationship, the anthropological problem at the heart of Christianity. For Jovinian, God's grace is given equally to all, it cannot be appropriated or enlarged by any personal merit due to one's own efforts. For Pelagius, God's grace is just an external aid to freedom, an aid of the same importance as *creation, *revelation, remission of sins, not an aid to freedom itself to be itself and to operate on the level of a good that merits eternal *life. In fact freedom has, from creation, its own radical autonomy in decisions regarding its own destiny. To its choices, and to them alone, are due the merits of wanting them and of actualizing them and the corresponding reward.

For Augustine, God's grace is itself the good of human freedom: through it, freedom can be what it is and operate on the level of a good that merits eternal life. Left to itself, not having the support which is God, it wanders adrift (*Ep.* 194, 2, 3).

The help that freedom receives from grace does not mean that it is replaced in its decisions; it is simply put in a position of being able to express itself at a level of freedom and not of conditioning. Grace helps freedom as one friend helps another, adapting to possibilities. Freedom, in this view, is not an absolute which can be realized at the first desire. It has degrees of growth: a state of infancy when it knows what it wants, but does not know how and has not the strength to do it; and an adult state when it is able to carry out what it desires.

III. Writings and thought. The Pelagian writings, all aimed at exhorting men to perform what is set out in the examples of *Scripture, must be set in the cultural-historical context of the controversy over the understanding of Christianity which occurred at the start of the 5th c. In the past they were seen as a single *corpus* attributed to Pelagius, but modern studies take into account the differences of content and style in many works, as well as their homogeneity of thought. In fact, the anonymous propagation of many ideas, slogans and writings made it very hard even at the time to formalize the real thought of Pelagius and other disciples. The *Liber Testimoniorum* (a composite of 160 titles) from which six propositions were taken during Pelagius's trial at *Diospolis in 415, remains the largest collection of the Pelagian ideas which circulated with no precise paternity. At present the *corpus pelagianum* is divided into three groups: works of Pelagius, dubious works of Pelagius, works of other authors. The major contributions to this clarification have been made by: C.P. Caspari (*Briefe, Abhandlungen und Predigten*, Oslo 1890), who ascribes to bishop *Fastidius works previously attrib. to Pelagius; G. Plinval (cf. in particular Vue d'ensemble sur la littérature pélagienne, *REL* 29 [1951] 284-294: an approach followed by Hamman in PLS 1, 1101 f.); R.F. Evans (Pelagius, Fastidius and the pseudo-Augustinian *De vita christiana*, *JThS* 13 [1962] 72-98); P. Courcelle (*Histoire litt. des grandes invasions germaniques*, Paris ³1964, 303-317); J. Morris (Pelagian Literature, *JThS* 16 [1965] 25-60); the specifications of CPL 728-766. Finally it should be observed that the majority of Pelagius's works have reached us through Ps.-Jerome.

The thought of Pelagius's exegetical, theological and ascetico-moral writings can be summed up as a call to imitate Christ as set forth in the gospel. All Scripture must be understood as the *revelation of a *Lex* to be observed and examples to be imitated, in particular that of Jesus Christ (the outlook of his *commentary on *Paul: *Expositiones XIII ep. Pauli*, ed. A. Souter, Cambridge 1926). The theological and ascetic writings can be grouped around the *De natura* of 414, which upholds the fundamental possibility of man, innate in his own nature, being able to direct his choices according to God's commandments, living without sin. Pelagius explored this leading idea of his thought, based on the doctrine of *creation, again in the *De libero arbitrio* of 415. Freedom is implanted in man at the moment of creation like a root, whose fruits will be determined by the choices he makes. God's grace is added to help this primary decision, in the guise of an incitement to the will to follow Christ's example. His other writings, esp. those on *chastity and *virginity, set out this idea concretely.

In conclusion, Pelagianism took its name from Pelagius: at the time of the controversy, the group headed by *Caelestius was also prominent. *Julian of Eclanum did not really represent a group, but much writing after him (e.g. *Praedestinatus) referred to his thought, esp. that on concupiscence, as a Pelagian doctrine. From Augustine's death on, Pelagianism was a summary term to indicate all those who, appealing to human freedom, were suspected of being "enemies of God's grace". Pelagius was understood symbolically, either as the heretic who was against the grace of God, or as the defender of human freedom; just as Augustine, in the Pelagian counter-light, was considered the defender of God's grace, but suspected of using it to supplant human freedom (the heresy of predestinationism, i.e. God determines freedom to good and *evil; *predestination itself was later understood this way by many, projecting onto Augustine questions which arose after him). A greater knowledge of the Pelagian movement is also rendering historical justice to the real content of Augustine's own writings.

PLS 1, 1101f.; CPL 728-766; R.F. Evans, *Four Letters of Pelagius*, London 1968; G. Plinval, *Pélage. Ses écrits, sa vie et sa réforme. Etude d'histoire littéraire et religieuse*, Lausanne 1943; R.F. Evans, *Pelagius. Inquiries and Reappraisals*, London 1968; G. Greshake, *Gnade als konkrete Freiheit. Eine Untersuchung zur Gnadenlehre des Pelagius*, Mainz 1972; O. Wermelinger, *Rom und Pelagius*, Stuttgart 1975; V. Grossi, in *Patrologia* III, 437-458; id., La crisi antropologica nel monastero di Adrumeto, *Augustinianum* 19 (1979) 103-33; id., Pelagio, *DIP* 6, 1327-1330; R.F. Evans, Pelagius: Fastidius and the pseudo-Augustinian *De Vita Christiana*, *JThS* n.s. 13 (1962) 72-94; J. Morris, Pelagian Literature, *JThS* n.s. 16 (1965) 26-60; P. Brown, Pelagius and his supporters: Aims and Environment, *JThS* n.s. 19 (1968) 93-114; id., The Patrons of Pelagius: the Roman Aristocracy between East and West, *JThS* n.s. 21 (1970) 56-72; G. Bonner, *Augustine and Modern Research on Pelagianism*, 1972; E. TeSalle, Rufinus the Syrian, Caelestius, Pelagius: Explorations in the prehistory of the Pelagian Controversy, *Philippian Studies* 22 (1974) 26-48; J.P. Burns, Augustine's Role in the Imperial Action against Pelagius, *JThS* n.s. 30 (1979) 67-83; G. Bonner, Some Remarks on Letters 4* and 6*, in (ed.) C. Lepelley, *Les Lettres de Saint Augustin découvertes par Johannes Divjak*, Paris 1983, 155-164; R.A. Markus, Pelagianism: Britain and the Continent, *JEH* 37 (1986) 191-204; B.R. Rees, *Pelagius: A reluctant Heretic*, Woodbridge 1988.

V. Grossi

PELAGIUS I, pope (556-561). Of the Roman aristocracy, P. distinguished himself as *Vigilius's deacon and *apocrisiary. Under his influence, *Justinian condemned *Origenism (PG 86, 945-994). Defending the Latin position, he pronounced against the condemnation of the *Three Chapters. While Vigilius was detained at *Constantinople, P. held an important position in *Rome. He actively helped the population of the Urbs during the conflict between *Totila and the *Byzantines. Back in Constantinople, he supported Vigilius in his resistance to the condemnation of the Three Chapters, even writing a treatise on it. Though he rebuked Vigilius for giving way, after the *pope's death he too recognized the condemnation and even the *council of Constantinople II (553). The emperor had him consecrated bishop of Rome against the resistance of the *clergy and people. In the context of imperial policy, P. restored the discipline and finances of the Roman church. While his charitable activities succeeded in winning over the people, he continued to meet resistance in the Western churches. In N *Italy a long *schism began. From P. on, the Roman bishops were obliged to ask the emperor to approve their election. The *papacy seemed no more than the Western patriarchy of the imperial church.

CPL 1698-1703; PLS 4, 1277-1396; P.M. Gassó-C.M. Batlle, *Pelagii I Papae epistulae quae supersunt*, Montserrat 1956 (PLS 4, 1284-1312): R. Devreesse, *Pelagii in defensionem trium capitulorum*, Rome 1932 (PLS IV, 1313-1369); B. Rubin, *Das Zeitalter Justinians*, I, Berlin 1960; *LTK* 8, 249 f.; K. Baus et al., *Storia della Chiesa*, III, *passim*.

B. Studer

PELAGIUS II, pope (579-590). Elected while the - *Arian - *Longobards were besieging *Rome, P. could not obtain the emperor's approval. He asked help from the *Franks (Jaffé 1048), and from the *Byzantines through his *apocrisiary in *Constantinople, the future *Gregory I (Jaffé 1052). The political situation in *Italy led to the creation of the Byzantine exarchate of *Ravenna (584). When calm returned, P. worked to end the *schism of the bishops dependent on *Aquileia (Jaffé 1054-1056), which had arisen under *Pelagius I over the *Three Chapters; but despite the collaboration of Gregory and the exarch of Ravenna, he failed. In a long *letter (ACO V, 2, 112-132) in which we may also see Gregory's hand, he confronted all the theological problems raised by the *dossier* sent by those bishops, and distinguished conciliar decisions on *custodia illibatae fidei* from others, as having a different binding value; he also examined in detail the case of each of the three condemned men (*Theodore of Mopsuestia, *Ibas of Edessa and *Theodoret of Cyrrhus). P. protested because *John IV of Constantinople had given himself the title "ecumenical patriarch", and refused to accept the *Acts of the *council held by John in 587, indeed he forbade his apocrisiary to have any relations with John; but his protests were useless (Greg. Gt: MGH, *Epp.* IV, 32, 34, 38; V, 44; VI, 41). The *Liber Pontificalis* and some inscriptions attest that he carried out many works at Rome; a *mosaic in the basilica of S. Lorenzo *fuori le mura* represents him as a restorer of the church. He died of plague and was buried in St Peter's.

CPL 1705-1707; *Verzeichnis* 482; PL 72, 701-760; PLS 2, 1413 f.; LP 1, 309-311; Jaffé 1047-1065; ACO V, 2, XXIII and 105-132; *DTC* 12, 669-675; *EC* 9, 1078 f.; O. Bertolini, *Roma di fronte a Bisanzio e ai Longobardi*, Bologna 1941, 225-300; M.J. Higgins, Two notes, in *Polychronion, Fest. Dölger*, Heidelberg 1966, 238-243; J. Orlandis, Sobre el origen de la "lex in confirmatione Concilii", *Anuario de Historia del Derecho Esp.* 41 (1971) 113-126; L. Magi, *La sede romana nella corrispondenza degli imperatori e patriarchi bizantini*, Rome-Louvain 1972, 162-165 (cf. index).

A. Di Berardino

PENITENCE

I. Penitence and reconciliation - II. Iconography - III. Penitential books.

I. Penitence and reconciliation. The first and only reconciliation is procured by *baptism, through which *sins committed before Christian *initiation are cancelled. After baptism, the Christian should not sin any more: so the problems of (second) penitence (and reconciliation) are, as to their effects, baptismal problems (penance=second baptism).

1. Pre-Nicene period. The earliest record of a reconciliation allowed to faults committed after baptism appears in the works of *Hermas (Rome, c.140), *Tertullian († after 220) and *Cyprian (†258) (both from N *Africa). For Hermas, the only reconciliation is that received in baptism; exceptionally, for the generation contemporary with him, a "second" penance and reconciliation is possible, but only once in a lifetime: "He who has obtained remission of sins (in baptism) should sin no longer.... For those who were called before these last times, the Lord has instituted a penance.... If anyone... falls into sin, he can do penance, but only once" (Hermas, *Shepherd, Mand.* IV, 1, 8).

Tertullian (c.204) definitively formulates the fundamental principle of the non-repeatability of penance in his *De paenitentia* (7, 9, 10): "(A second penitence is possible), but only once, since it is the second time (the first penitence being that of baptism), and never again". In *De paenitentia*, apropos of the readmission of adulterers and fornicators to *communion (against which Tertullian himself argued after becoming a *Montanist), he describes for the first time the penitential process: a life of hard mortifications, ragged clothing, prolonged *fasting, *prayers and lamentations, prostrations before the presbyters and servants of God, appeal to the intercession of the community, final pardon (by *God alone? through the community's mediation?).

Cyprian (†258), in agreement with pope *Cornelius (253-255), admitted *apostates, numerous during the *persecutions of *Decius and *Valerian, to reconciliation, not without opposition from rigorist circles.

It seems then that penitential discipline was worked out starting from concrete cases, like those mentioned, and that it was hence variable according to community.

2. Early Christian post-Nicene period (from 325 to early 7th c.). All the documents (*Augustine, *Caesarius of Arles, *councils, decretals, correspondence of bishops, esp. from S *Gaul) agree in describing the organization of official penitence in the following way. The process developed in three chronologically separate moments: 1. Entry into penitence, with the bishop's consent, in the course of a community ceremony (*petere, accipere paenitentiam*). 2. Remaining for a shorter or longer time in the order of penitents (*ordo, status paenitentium*). 3. Readmission to the community through *laying on of hands by the bishop, ordinarily on Maundy Thursday, before the whole community (*reconciliatio*).

Since this process was not repeatable, sinners who were still young were excluded from penitence because of a possible fall back into sin, in which case the official church could do nothing for them. Since the penitent state required perfect and final continence, as well as the giving up of all public duties, it followed that married people could with difficulty be allowed to become penitents except in old age; magistrates and soldiers were excluded. Even after reconciliation, the sinner was a marked man for the rest of his life: entry into penitence was equivalent to entry into religion, with the added mark of infamy, an end to civil and *family life. The entire process (but not the confession of faults) was public, in the sense that penitents occupied a special place in church, the community interceded for them and they were publicly reconciled. But penitents were not excommunicate: penance must not be confused with *excommunication. The rigour and consequences of the state of penitence had the effect that the *faithful, with the bishop's explicit consent, deferred reconciliation to the end of their lives, as in the pre-Nicene period they had deferred baptism *in extremis*. The result was a complete "penitential void": the great majority of believers of the early Christian period could not benefit from reconciliation except at the point of death. During their life they "did" penance as best they could, given that they did not "receive" penance, before being allowed, without ecclesiastical absolution, to receive the *Eucharist. Recourse to penitence was forbidden to bishops, presbyters and deacons, not as a privilege but because of the incompatibility between the state of penitent and the clerical state (in both senses).

3. Tariffed penance (early 7th to 12th c.). At the end of the early Christian period there appeared on the continent, brought there by Anglo-Saxon and Celtic monks, a penitential system representing a break with the earlier discipline: tariffed penance or insular penance, so called because of its origin (Irish and Anglo-Saxon *monasteries). Its fundamental principle was this: to each fault a precise penance was applied, consisting of different mortifications, particularly *fasts (days, months, years). These tariffs or taxes were contained in the Penitential Books. Tariffed penance was administered by the presbyter (no longer just the bishop); the sinner made a detailed confession of his faults and had imposed on him a period of fasting that varied with the gravity of the acts committed. His fasts completed, the sinner was considered absolved; only from the 9th c. is an explicit absolution by the confessor attested (declaratory formulae; the formula *Ego te absolvo* is late). Unlike the older penitence, the sinner could go to the confessor every time he sinned. Major *clerics were not excluded from tariffed penance. Days, months and years of fasting mounted up with the number of sins, often exceeding a human lifetime. "Redemptions" or ransoms provided possible commutations to obviate these situations:

- A long fast was redeemed by a shorter but more rigorous fast, or by beatings administered by the confessor (sometimes flagellation) or other bizarre mortifications (e.g. spending nights in the tombs).

- The fast owed by a sinner could be redeemed by fasts carried out by third parties, esp. monks, in exchange for economic compensation (money, legacies, gifts in kind, land or rivers and ponds).

- A fast of several years could be redeemed by *masses celebrated for the sinner by monks (who, at this period, became priests) in exchange for *donations, esp. money. This was the most frequent penitential ransom, at least for rich sinners. Hence the equations: sin=fast=mass=sum of money given to confessor or monastery.

The Carolingian reformers who tried to fight against tariffed penance and the penitential books in order to restore the earlier discipline were only half successful, which led to a penitential dichotomy: a grave hidden fault required tariffed penance, a grave public fault required penance according to the earlier practice. Sin was now evaluated not by its intrinsic gravity, but by the publicity surrounding it.

4. Penitential reorganization (early 13th c.). At the end of the 12th c. the penitential system evolved: the Penitential Books disappeared and gave place to the Confessors' manuals. A tripartite system was set up:

- Solemn public penance: this was the old penitential system in all its rigour and consequences, imposed on *laymen for particularly scandalous faults (parricide, infanticide, regicide, burning of cities, etc.). Major clerics were excluded from it.

- Public, but not solemn, penance: the penitential *pilgrimage. The confessor sent the sinner to more or less famous sanctuaries of Christendom, giving him, after a blessing, a pilgrim's staff, purse and hat, and a letter attesting that he was travelling to expiate his sins. This generally assured the sinner-pilgrim lodging, bread and water. Penitential pilgrimage was imposed on laymen for less scandalous public faults, and on clerics for any grave public fault whatever (wandering clerics).

- Tariffed penance: imposed for all grave hidden faults committed by clerics or *laymen. This type of penance rapidly evolved into modern sacramental confession: confession and subsequent immediate absolution (the expiation being emptied of any moral significance by the game of commutations).

*5. Penitential discipline in the *East.* The penitential system in the East is, in its organization and evolution, much less well-known than the Western system. Sinners were divided into various classes on the basis of the gravity of their faults. These classes were fixed by the Eastern *councils and the penitential letters of the *Fathers, esp. those of St *Basil:

The implorers (*ploratio*) stood on foot outside the church, supplicating the people on their way to the *Eucharist.

The auditors (*auditio*) took part in the *liturgy of the word and were sent away with the *catechumens with no special prayer; their place was in the narthex.

The prostrate (*prostratio*) left the service together with the catechumens after having received *laying on of hands, *ante absidem* (in front of the apse, i.e. in front of the sanctuary).

The participants (*assistentia*) took part in the whole Eucharistic office, but without communicating.

This process was necessarily public, but confession of sins was not public. Sinners went through the different grades, constrained to one or other of them according to the gravity of their faults.

At a date hard to determine, but some time in the 7th c., the penitential classes disappeared to give way to a penitential system in which confession, as in the Western tariff system, had a preponderant place, with the difference that the penalties (*epitimia*) imposed on sinners were not left to the judgment of the confessor, but were determined in conformity with the conciliar penitential canons, with no possibility of redemption. This extremely harsh discipline led to the more or less general abandonment of any individual form of penance.

B. Poschmann, Busse und letzte Ölung, *Handbuch der Dogmengeschichte*, IV, 3, Freiburg i. Br. 1951; C. Vogel, *Le pécheur et la pénitence dans l'Eglise ancienne*, Paris 1966; id., *Le pécheur et la pénitence au moyen âge*, Paris 1969; id., *Les Libri paenitentiales*, Typologie des sources du moyen âge occidental 27, Turnhout 1978; H.E.W. Turner, *The Patristic Doctrine of Redemption*, London 1952; E. Langstadt, Tertullian's Doctrine of Sin and the power of Absolution in *de pudicitia*, SP 2 (1957) 251-257; H. Karpp, *Die Busse*, Zürich 1969; L.M. White, Transactionalism in the penitential thought of Gregory the Great, *Restoration Quarterly*, 21 (1978) 33-51; B. Ward, *Harlots of the Desert: a study of repentance in early monastic sources*, Kalamazoo 1987.

C. Vogel

II. *Iconography.* We have no direct representations of penance from the early Christian era that depict the liturgical-sacramental act of confession or reconciliation. J. Wilpert would interpret two *sarcophagus fragments, largely reconstructed, as representing such a sacramental event (*La fede*, 256-260), but, quite apart from doubts on the correctness of the reconstruction, it must be said that the *laying of hands on *orantes is capable of many interpretations. Achelis (7-11) has tried to link a vault fresco in the catacomb of Pretestato in Rome (*WK* 51, 1) with the theme of penitence. It shows, acc. to scholars, a *shepherd placing his right hand on a *sheep approaching him, while his left hand pushes back a pig and an ass. Achelis explains this by reference to *Novatianist rigorism, which denied any return of sinners (pigs) and *heretics (asses) to the church. The fresco would have been commissioned by someone who wished to expound his personal convictions on the subject of penitence, which differed from those of the community of Rome. If however the animals on his left were not a pig and an ass but wolves, it could be interpreted as the biblical shepherd scene. In the sphere of the iconography of *baptism, a scene in the region of Lucina in the catacomb of S. Callisto (*WK* 29, 1) is open to, among others, a penitential interpretation. Fink (106 ff.) explains this fresco by reference to a passage in the *Syriac *Didascalia* (2, 15, 8), where the bishop is urged not to allow the sinner to sink into despair, but to hold out his hand in reconciliation. That the question of the possibility and feasibility of the remission of post-baptismal *sins was, besides the few examples so far known, a theme of early Christian iconography, can be accepted as certain. Not by chance was Christian art born during those same decades when, in pastoral and theological discussion, the problem of the *lapsi* and of the return of public sinners to the church broke out. Particularly on the point of death and in view of passing away, the question of the verification, but also of the recovery, of baptismal *grace acquired a character of extreme urgency. Referring to Ezek 14, 13-20, which says that *God, inflicting punishment on the earth, would save only the just men *Noah, *Daniel and *Job, but not their sons and daughters, *Cyprian (*De lapsis* 19) cites these three in a call to personal repentance, which cannot be surrogated by the intercession of confessors. Fink (61-73) points out that these three, whom he holds must be quite frequently intended in the contextual nexus of catacomb *paintings, can be interpreted as representations of penitence. Even if the connection of the scene with Noah, Daniel and Job is sometimes so uncertain that their inclusion in other iconographical contexts seems possible, it is still indisputable that numerous OT and NT scenes, often recurrent in early Christian burial art, are connected, in contemporary *homily and *catechesis, with an exhortation to repentance and the *hope of remission of sins. This is so of the figure of the *shepherd which Christians of the Great Church, opposed by *Tertullian, used as a decorative element on offering *chalices to express their hope in the remission of sins (*De pudicitia* 10, 12). Many *Fathers emphasize the saviour Shepherd's love for sinners (Dassmann, 322-340).

It must be noted that, in pre-Constantinian patristic *exegesis, the image of *Jonah too possessed an exclusively penitential rather than paschal symbolism, as we could argue on the basis of Mt 12, 39 ff. From a formal point of view, the scene of Jonah's repose under the booth - connected with a scene of him thrown into the sea - could be, esp. in sarcophagus reliefs, a survival of the *pagan myth of Endymion, at the same time symbolizing the blessed repose of the dead. But when, in some sarcophagi (e.g. those of Belgrade and of Julia Julianete [Dassmann, 387]), the repose scene (open to mythological interpretation) is absent, and, next to Jonah thrown into the sea or cast out by the fish, there appears the shepherd saving the sheep or *Noah after the *flood, then there is no doubt that the figurative programme is that of the Christian symbolism of salvation, and is therefore open to the theme of remission of sins. In some catacomb paintings, the normally tripartite cycle (Jonah thrown in the sea, cast out by the fish, at rest) is supplemented by the so-called *Jonas irritatus* scene, where the prophet sits sadly under the dried-up plant (*WK* 61; 96; 100). Jonah must learn not to be angry at God's pity on the city of Nineveh which repented.

Depictions of the shepherd and those of Jonah are often linked in early Christian burial art. If NT motifs also appear, e.g. the healing of the *paralytic, the woman with an *issue of blood or analogous miracles, the penitential meaning of the entire figurative programme becomes still more evident (e.g. cubicle 5 of the catacombs of SS. Pietro e Marcellino [*WK* 98]). In the patristic exegesis of the 3rd and 4th cc., the exegesis of the original sin of *Adam and Eve also stresses hope in *redemption.

The relationship that exists between images of *martyrs and intercessors of any type and hope in remission of sins is still little studied. We know that the cult of the martyrs, which began in the 3rd c. and flourished in the 4th, was rooted in the hope of the saints' intercession at the last *judgment. But we cannot be sure whether such an intercessory function can be applied to OT scenes of *Daniel in the lions' den or the young men in the *fiery furnace, since until the time of *Constantine we know no iconographical depictions of martyrs or scenes of *martyrdom, and whether those and other scenes do not just allude to a general soteric symbolism, such as is known from the prayers of Ps.-Cyprian, is still an open question.

H. Achelis, Altchristliche Kunst, 4, *ZNTW* 16 (1915) 1-23; G. Wilpert, Un'antica rappresentazione della penitenza, *Ill. Vat.* 6 (1935) 15-16; id., *La fede della chiesa nascente secondo i monumenti dell'arte funeraria antica*: Collezione "Amici delle catacombe", Vatican City 1938; J. Fink, *Noe der Gerechte in der Frühchristlichen Kunst*, Beiträge zum Archiv für Kulturgeschichte 4, Münster-Cologne 1955; E. Dassmann, *Sündenvergebung durch Taufe, Busse und Martyrerfürbitte in den Zeugnissen frühchristlicher Frömmigkeit und Kunst*, Münsterische Beiträge zur Theologie 36, Münster 1973.

E. Dassmann

III. Penitential books. The Penitentials are *books (or opuscula) that originated in the British Isles and spread over the continent thanks to Irish and Scots monks who came to evangelize the regions that had remained *pagan or reverted to paganism after the great Germanic invasions. They presented a system of penance characterized essentially by the application of tariffs to the penalties fixed for each type of *sin and their number. Historians have long given this type of penance the name "tariffed penance". It competed with and finally supplanted the old canonical penance of primitive times, and was the prelude to the private penance of the late mediaeval and modern periods. Unlike the old *canonical penance, it was imposed not just by the bishop but by any priest, and could be approached not just once in a post-baptismal life, but every time the need was felt. On every occasion it presupposed confession of sins and an act of penance, but it consisted essentially of a more or less severe and prolonged *fast, to the point that the two terms *paenitere* and *ieiunare* became synonymous. If, at least in theory, the penalties imposed for sins confessed on a single occasion could exceed a human lifetime, in practice and from the beginning, penalties could be commuted and replaced by compensations that made them shorter. Estimation of penalties was not left to individual initiative, but was carefully regulated in writing.

Since the list of Penitential Books is far from complete, we can only trace their history. Up to *c*.650, the books are all of insular origin: Penitential of *Gildas, *Canones Wallici*, Irish and Celtic Penitentials and those of *Columbanus. From the mid 7th c. to the era of the Carolingian reform, they were at their height. Some continued to come from the Isles, whether of the old type (Penitential of Cummean or *Cummineus, *Iudicia Theodori, Discipulus Umbrensium*) or of the new (Penitentials of *Bede, Egbert, Alberes, Bigot, Gwynn, etc.). Others came from the continent, but defy classification on account of their complexity and lack of consistency. Attempts have been made to remedy this disorder by distinguishing the Penitentials in which the influence of Columbanus, Cummean or Theodore predominates, and separating this section from books of Spanish origin. With the Carolingian reform a third period opened, characterized by a claim to be reviving the old canonical penance, but ending in a compromise based on the following principle: if the sin was public, penance must be public (i.e. canonical penance); if private, penance must be private (i.e. "tariffed penance"). From this period date, among others, the Penitential of Alitgarius of Cambrai (825-830) and the two of Rabanus Maurus, at Otgar (841-842) and at Heribald (852-854). The great *canonical collections were also compiled and propagated in the Carolingian era: the *Dionysio-Hadriana*, the *Hispana*, the *Dacheriana* and the *Quadripartita*, which set themselves up as *summae* of the old canonical penance, even that which failed to establish itself and whose failure shows the limits of the Carolingian reform. Among the Penitentials of the last period (late 9th to 11th cc.) are three groups with original characteristics: post-Carolingian Penitentials, which pass themselves off as Roman, but are really Frankish; Penitentials in which the Anglo-Saxon character predominates; and mixed extracts, in which ancient conciliary canons and citations taken from the Penitentials appear side by side. The penitential collection of Burchardt of Worms, *Corrector sive Medicus*, if not chronologically the last, is the most representative of the genre and the most used in the later Middle Ages. Not until the Lateran council of 1215 did "tariffed" penance in turn fall into disuse, to be replaced by the present system of private penance, so far as that still survives after Vatican II.

DTC 12, 1160-1179. For editions, cf. *Verzeichnis* 469 f. P. Fournier-G. Le Bras, *Histoire des collections canoniques en Occident depuis les Fausses Décrétales jusqu'au Décret de Gratien*, Paris 1931; J.A. Jungmann, *Die lateinischen Bussriten in ihrer geschichtlichen Entwicklung*, Innsbruck 1932; R. Lasczcz, *Organisation de la pénitence "tarifée" d'après les "Ordines" des "Libri paenitentiales" jusqu'au "Corrector sive Medicus" de Burchard de Worms (1008-1072)*, Strasbourg 1971 (typed thesis); C. Vogel, *Les "Libri paenitentiales"*, Turnhout 1978; R. Kottje, *Die Bussbücher Halitgars von Cambrai und des Hrabanus Maurus*, Berlin 1980.

V. Saxer

PENTECOST. The Christian origin of the feast of P. goes back to the event described in Acts 2, 1-11. In the first centuries the feast was closely linked to *Easter, so much as to be considered its conclusion. It appears more as a period of time, the "seven *weeks", than as a single feast-day, and was characterized by joy (Tertull., *De bapt.* XIX, 2; *De coron.* III, 4; Basil, *De Spir.* XXVII, 66), a joy expressed in a ban on *fasting (Tertull., *De coron.* III, 4; Euseb., *Pasch.* 5; Athan., *Ep. ad Orsisium* 2; Ambr., *Apol. Dav.* 8, 41; Ps.-Just., *Quaest. et resp.* CXV); during P. the Alleluia was sung, there was continuous reading of the Acts of the Apostles and the wonders worked by the Lord in his church were remembered.

It was during the 4th c. that the custom was fixed of celebrating P. to make it coincide with the last of the 50 days of Easter, of which it was the solemn close: this is apparently the meaning of can. 43 of the council of *Elvira (early 4th c.). Can. 20 of *Nicaea (325) established that this celebration must be held standing up, without genuflections as on other *Sundays. During the final third of the 4th c. the feast's meaning was specified, at *Constantinople, as a reminder of the gift of the *Holy Spirit to the *apostles and the church; at *Rome, as the gift of the New Law and the descent of the Spirit; at *Milan, as the ending of Eastertide. But in *Spain, the preaching of *Gregory of Elvira (PLS I, 46-72) does not allow the same certainty, since the reading of Acts 2, 1 ff. was not evidently linked to the day of P. Finally, the tendency to historicize liturgical feasts, evident everywhere in the 4th c., brought with it a separation of the two primitive aspects contained in a single celebration: the commemoration of the Lord's *glory, at first fixed on the 50th day, was transferred to the 40th, giving rise to the feast of *Ascension, while *meditation on the church, linked to the outpouring of the Holy Spirit and the start of the apostolic mission, became the essential object of the feast on the 50th day.

This evolution was complete by the 5th c., as we see from texts of *Augustine, *Paulinus of Nola, *Peter Chrysologus, *Leo the Great, *Faustus of Riez, *Caesarius of Arles. The meaning of the feast was further modified: its character as the end of Eastertide became less clear, while the process of subdividing the 50 days of Easter was accentuated. In the end P. became a feast like the others, whose essential object was fixed by Acts 2, 1-41, but its link with Easter was no longer discerned.

In theologically investigating the feast's specific content, the *Fathers did not forget its Jewish origin: P., the "feast of weeks", was one of the three great feasts when all Israel went up to Jerusalem to honour God; its importance is attested by various biblical calendars which name and characterize it (Ex 23, 14-17; 34, 18-23; Lev 23, 15-21; Num 28, 26-30; Dt 16, 9-12). At first, patristic typology brought out the agricultural character of the OT P.: the firstfruits of grain were a figure of Christ, firstborn from the dead who returns to the *Father (Epiph., *Haer.* II, 51, 31); of the coming of the Holy Spirit (Orig., *Hom. in Lev.* II, 2); of the calling of the gentiles (Euseb., *De Sol. pasch.* 4); of the first evangelical preaching (John Chrys., *Hom. in Act.* IV, 1). Then there was a typology that compared the Christian P. with the Jewish P. via the theme of the Law and the covenant (Ambrosiaster, *Quaest. V. et N.T.* XCV; John Chrys., *In Mt. Hom.* I). A parallel with the event of *Sinai was made possible by its date: 50 days after Easter (Jerome, *Ep.* 78; Aug., *Ep.* 55, 16; *De catech. rud.* 23, 41). The meaning of P. as a memorial of the covenant was not always present in Judaism, or at least not everywhere. At Qumran in the 2nd c. BC the memorial of the Pact was already being celebrated at P., but it was only in AD 270 that the Talmud made a connection between P. and the gift of the Torah (*Pes.* 68b). Acts ch. 2 contains various references to the event of Sinai (Cypr., *Testim.* III, 101), but not until the 4th c. did the typology of the Fathers notice the gift of the Law as an element of P., Jewish as well as Christian. Some have hypothesized an Essenic influence on Acts, followed by a Christian influence on Rabbinism, thus historicizing a feast which could not have been celebrated after the destruction of the Temple if it had retained its primitive agricultural character. A tradition that goes from *Tertullian (*De bapt.* XIX, 2) to pope *Siricius (*Ep. et Decr.*) allowed for celebration of *baptisms at P.-tide.

DBS 7, 858-879; O. Casel, Art und Sinn der ältesten christlichen Osterfeier Jahrbuch für Liturgie Wissenschaft, *JLW* 14 (1938) 1-78, G. Kretschmar, Himmelfahrt und Pfingsten, *ZKG* 66 (1954-55) 209-53; B. Noack, *The Day of Pentecost in Jubilees. Qumran and Acts*, Leiden 1962; R. Cabié, *La Pentecôte. L'évolution de la cinquantaine pascale au cours des cinq premiers siècles*, Tournai 1965; J. Potin, *La fête juive de la Pentecôte*, Paris 1971, F. Cocchini, L'evoluzione storico-religiosa della festa di pentecoste, *RivBib* 25 (1977) 297-326, M.G. Durand, Pentecôte johannique et Pentecôte lucanienne chez certains Pères, *BLE* 79 (1978) 37-126; J. Gunstone, *The Feast of Pentecost: the great Fifty Days in the Liturgy*, Westminster 1967.

V. Saxer - F. Cocchini

PEREGRINUS. A foreigner, not a Roman citizen. In Republican times, most of the population of *Rome consisted of *peregrini* from conquered countries. In general they had no political rights (they took no part in popular assemblies, could not hold magistratures, did no military service). In the sphere of private law, they could not give evidence in court, draw up a will in the form of a Roman citizen or inherit a citizen's property. Their *marriages were considered legal (*iustae nuptiae*) only if they enjoyed the *ius conubii* granted either personally or - more often - to their *civitates* of origin, which had obtained them from Rome (cf. *Citizenship).

M. Forlin Patrucco

PEREGRINUS. A person dedicated to biblical studies. He made a new edition of the *Wisdom books and, perhaps, also of others, based on *Jerome. He revised and ordered *Priscillian's canons for the Pauline epistles. It seems certain that he was a bishop, and quite likely that he was from *Baetica. Attempts have recently been made to prove that a complete recension of the *Bible is his, but the problem remains open. When he lived is uncertain: prob. 5th c., but perhaps first half of 6th.

T. Ayuso Marazuela, *La Vetus Latina Hispana*, I, Madrid 1953, 520-522; B. Fischer, Algunas observaciones sobre el "Codes Gothicus" de la R.c. de S. Isidoro en León, *Archivos Leoneses* 15 (1961) 5-47.

M. Díaz y Díaz

PEREGRINUS of Auxerre. Cited in *Mart. hier.* (16 May), with a development of information from one family of MSS to another. Between the two comes the 7th-c. Passion (BHL 6623), which makes P. a Roman priest whom *Sixtus II (257-8) consecrated bishop and sent to evangelize the Puisaye. He preached there on a *pagan feast-day, was imprisoned at Bouhy (*vico Baiaço*) and beheaded. His relics were transferred to Saint-Denis between the 7th and 9th cc.; the church was near the hospice for *pilgrims who came from France (Duchesne *LP* II, 47).

Duchesne, *Fastes* I, 50; *Vies des SS.* 5, 327-29; *LTK* 8, 269-70; *BS* 10, 460-62.

V. Saxer

PERFUME. P. is solar energy, sign of divine *life. The gods breathe it to preserve their immortality. To mortals it gives *joie de vivre* and forgetfulness of care. It is the ambassador of the spirit: it foretells the arrival of the beloved and evokes presence in absence. It is perceived by the sense of smell, which releases arcane affective charms in man. In Ancient Egypt, temples were drowned in incenses and aromas to transport the faithful to the heavenly kingdom, and embalming gave life to the dead. Greeks and Romans saw p. as the sign of theophany, and ointment as life-giving medicine. For the Persians, p. and foul smells manifested good and *evil.

The Hebrews saw the "tree of life" planted by God on the earth as the source of the "perfume of the Law", which leads via the "perfume of Wisdom" to the "garden of justice" (Gen 2, 9-12; Sir 24, 15; *1 Enoch* 24, 4-32, 6). The "sweet-smelling sacrifice" (Gen 8, 21) is human life offered to God and to one's brothers: it is theophany, since it makes man a collaborator in God's manifestation in the world (*Ap. Baruch* 67, 6; *Jubil.* 2, 22). God's love for the community, his bride, is described in the *Song of Songs in a triumph of aromas in the countryside and in the nuptial chamber. Acc. to Josephus, "the altar of incense with its thirteen perfumes . . . signified that all things are of God and for God" (*Bell. Iud.* V, 5, 5).

Jesus at his *birth received as a gift "incense and myrrh", announcing sweetness and bitterness for mankind (Mt 2, 11); "anointed" with the *Holy Spirit in the Jordan and with ointment at Bethany, he foretold the p. of his death and *resurrection, which begot the church (Jn 1, 32-34 and 12, 1-8). *Paul smells in *sin an "odour of death" and in *justice a "perfume of life" (2 Cor 2, 16). The breath given by *God to *Adam is called by *Philo "a light breath", "a breeze and a perfumed exhalation, mild and sweet" (*Leg. all.* 1, 42). Adam, on receiving it, opened his eyes to life; so said the *Mandaeans, acc. to whom "when the baby is in its mother's womb, it breathes the odour of life" (*Joh.* I, in G. Furlani, Il buon odore e il cattivo odore nella religione dei Mandei, *RAL* VIII, VII [1952] 321).

God is "the perfect perfume", say the Christians (Athenag., *Suppl.* 13). He possesses "every sweet odour and all the fragrance of spices" (Iren., *Adv. haer.* IV, 14, 3). The *Father gives the Son his immortal p. through *anointing in the Holy Spirit; the incarnate Word carries the breath of the life of the *Trinity to the world, inaugurating the outpouring of love on all mankind (Hipp., *Co. Ct.* 2, 1-30; Orig., *Co. Ct.* I, 101, 8-102, 8 and 107, 17-108, 8). The *Gnostics emphasize the p. of knowledge: Sophia possesses "an aroma of immortality left to her by Christ and by the Holy Spirit" (Iren., *Adv. haer.* I, 4, 1), and her sons on earth "strive to maintain in themselves the ray and spark of *light with the perfume of the Spirit" (Hipp., *Ref.* V, 19, 6 and X, 11, 4). Various *Fathers recognize in the suffering Jesus the sacrificial offering acceptable to God for its "sweet fragrance" (Eph 5, 2): Christ's *cross is the perfumed tree that conquers the smell of death and fills mankind with new immortal life (Hipp., *Co. Ct.* 12, 1-13, 4; Orig., *Co. Ct.* I, 102, 10-20; Greg. Nyss., *Hom. Ct.* III; Ambr., *In Ps.* 40, 15).

Christ, giving his life, renews *creation, in which God had communicated to man "the perfumed breath of his divinity" (Ps.-Clem., *Rec.* IV, 9, 1). The memory of the p. of the earthly *paradise, enlivened by expectation of the p. of the eternal paradise, gives depth to the involvement of Christians in history. The heralds of God's justice "breathe incorruptibility everywhere and give life to men" (Iren., *Adv. haer.* III, 11, 8), revealing that "a sweet smell to the Lord is the heart that praises its creator" (Barn. 2, 10; cf. Greg. Nyss., *Hom. Ct.* XI; Ambr., *De virgin.* 61-67 and *In Lc.* 6, 13-34; Aug., *In Jo.* 50, 1-10). The outpouring of the p. of Christ reaches its height in "*martyrdom"; he who dies for the Lord's name spreads in the world the bouquet of his *resurrection and, through the fragrance of his *sacrifice, draws the generations of believers to God: "wherever the relics of the martyrs are found, we run quickly towards the p. of the ointments of Christ" (Phil. Carpas., *In Ct.*: PG 40, 41 C; cf. *Mart. Polyc.* 15, 2). *Ascesis of life is an everyday martyrdom: the fragrance of the gifts of the Spirit becomes the p. of believers who, in *contemplation, attain the height of God's p. (Greg. Gt, *Co. Ct.* 12-30).

E. Lohmeyer, Vom göttlichen Wohlgeruch, *SHAW* 9, 1-52; A. Stumpff, εὐωδία, *TWNT* 2, 808-810; W. Deonna, Euodia. Croyances antiques et modernes: L'odeur suave des dieux et des élus, *Geneva* 17 (1939) 167-262; H. C. Puech, Parfums sacrés, odeur de sainteté, effluves paradisiaques, *L'Amour de l'Art*, Paris 1950, 36-40; G. Delling, ὀσμή, *TWNT* 5, 492-495; E. Cothenet, Parfums, *DBS* 4, 1653-1660; A. Orbe, *La teología del Espíritu Santo*, Rome 1966, 377-390; M. Detienne, *Les jardins d'Adonis*, Paris 1972; id., Aromi e seduzione, *Religione e Civiltà* 1 (1972) 529-537; G. Piccaluga, Adonis e i profumi di un certo strutturalismo, *Maia* 26 (1974) 33-51; P. Meloni, *Il profumo dell'immortalità. L'interpretazione patristica di Cantico 1, 3*, Rome 1975 (qv. for remaining bibliography); M. Detienne. *Dionysos mis à mort*, Paris 1977; P. Rovesti, *Alla ricerca dei profumi perduti*, Venice 1980.

P. Meloni

PERICHORESIS (Gk περιχώρησις). The term which, in *neoplatonist anthropology, served to explain how the *soul is intimately united to the *body without being confused with it, was applied by *Gregory Nazianzen, in a similar sense, to the union of the two natures in Christ (*Ep.* 101; *Or.* 38, 13). In this sense it was taken up by the Byzantine authors who saw in the human composition an analogy with the *incarnation (Lampe 1077 ff.). Following Ps.-Cyril (PG 77, 1172D), *John Chrysostom used p. in an analogous sense for the union, inseparable but not confused, of the three divine *persons.

G.L. Prestige, *God in Patristic Thought*, London ²1952, 297-305; H.A. Wolfson, *The Philosophy of the Church Fathers*, Cambridge, Mass. 1964.

B. Studer

PERIGENES. Bishop of *Corinth (419-after 435). Born at Corinth, where he was a priest. His bishop Rufus ordained him bishop of Patrae (Patrai) in *Achaia in 419. The Christians of Patrae never accepted him; the reasons are unknown. Soon after Rufus's death, he was elected bishop of Corinth. The Corinthians asked for pope Boniface's (418-422) approval: after receiving information from his legate, bishop Rufus of Thessalonica, he sent it the same year (Socr., *HE* VII, 36). P. recognized the supremacy of the *pope's legate (*Ep. Bonif.* 4: PL 20, 760), but, after Rufus's death, refused to acknowledge that of his successor *Anastasius, who was confirmed in that office by pope *Sixtus III (432-440). Anastasius called a synod at *Thessalonica, the pope wrote a *letter to the Eastern bishops and another (435) to P. asking him to recognize Anastasius's *authority (*Ep. Sixt.* 7, 8, 10: PL 1, 610). His reply and the date of his death are unknown. P. took part in the council of *Ephesus among the *orthodox *Fathers.

G. Ladocsi

PERIODEUTA. Priest in charge of visiting and caring for village and rural Christians, dependent on the city bishop. This institution goes back to the 3rd c. and is met in the Eastern provinces: *Asia Minor, *Syria and esp. *Egypt, where the bishop was not easily accessible since he resided only in district capitals, or for other reasons, e.g. *persecution (cf. Phileas, *Ep.*: PG 10, 1566B). The late 4th-c. council of *Laodicea (can. 57) laid down that bishops should not be consecrated for villages and countryside, but only a p., who had no fixed residence and cared for the Christians of several villages (cf. Athan., *Apol.* 85: PG 25, 400).

DACL 14, 369-379; Lampe, s.v.

A. Di Berardino

PERPETUA AND FELICITY

I. The *Passio* - II. Iconography.

I. The Passio. Archetype of Acts of Christian *martyrs, this *Passio* relates the arrest of Vibia Perpetua together with other *catechumens (Revocatus, Felicity, Saturninus and Secundinus), the events that occurred in prison, the judgment, the execution (ch. 3-10) and the *visions of Saturus, who had spontaneously hidden among the detainees (chs. 11-13). On the basis of the first (1-2) and last (14-21) chapters and other information, some would attribute the *Passio Perpetuae et Felicitatis* to *Tertullian or one of his circle. Written by an anon. member of the community, it contains historically authentic elements, related by the martyrs themselves (Perpetua's three visions and that of Saturus), and reflects the first *persecution in *Carthage following an edict (202). Dated 203, it is mentioned by Tertullian (*De* *anima* 55) and *Augustine (*Sermo* 280: PL 38, 1281). The text, preserved in Greek and Latin versions, is evidence of the Christianity lived by the people and, more than an apologetic document, was an instrument of edification during the *liturgy. It contains typically *apocalyptic *Judaeo-Christian elements.

CPL 32; PL 3, 13-18 (Ruinart); P. Franchi de' Cavalieri, *RQ* 5 (1896) 104-148 (ST 221, 1962, 41-155); H. Musurillo, *The Acts of Christian Martyrs*, Oxford 1972, 107-131; Quasten I, 163-164; E. Corsini, Proposte per una lettura della "Passio Perpetuae", in *Forma Futuri* (Misc. M. Pellegrino), Turin 1975, 481-541; V. Lomanto, Rapporti fra "Passio Perpetuae" e "passiones" africane, *ibid.*, 542-565; C. Mazzucco, Il significato cristiano della "libertas" proclamata dai martiri della "Passio Perpetuae", *ibid.*, 556-586; J. Daniélou, *Les origines du christianisme latin*, Paris 1978, 61-64; S. Prete, Il motivo onirico della "scala". Note su alcuni Atti di martiri africani, *Augustinianum* 19 (1979) 521-526; M.R. Lejkorvitz, The Motivations for St. Perpetua's Martyrdom, *Journal of the American Acad. of Religion* 44 (1976) 417-421; W.H.C. Frend, Blandina and Perpetua: two early Christian Martyrs, *Les Martyrs de Lyon (177)*, Paris 1978, 167-177; M.A. Rossi, The Passion of St. Perpetua: Everywoman of Late Antiquity, in (eds.) R.C. Smith-J. Lounibos, *Pagan and Christian anxiety: A Response to E.R. Dodds*, Lanham 1984, 53-86; A. Pettersen, Perpetua, Prisoner of Conscience, *VChr* 41 (1987) 139-153.

E. Romero Pose

II. *Iconography. Martyrs in 202 at Carthage, P., F. and their companions were greatly venerated in that city; over the place of their burial, acc. to *Victor of Vita (*De persec. Vand.* 1, 3), was built the *basilica maiorum*. This was identified, in 1907, thanks to an inscription naming the martyrs, and, in 1929, the *confessio* of the basilica. The oldest iconographical reference to the two martyrs is at *Ravenna in the (now archiepiscopal) chapel of S. Andrea (*c.*519) and in the procession of martyrs in the church of St Martin *in Caelo aureo*, now S. Apollinare Nuovo (6th c.). They also appear on the underside of the apsidal arch in the Euphrasian basilica of *Parentium (543-554). The scenes on the rear face of a *sarcophagus in Burgos Arch. Mus., variously dated 5th-8th c., include a ladder (=heavenly ladder), which some would recognize as one of the visions of the *Passio Perpetuae*. **[Fig: 253]**

E. Josi: *EC* 9, 1191 ff.; L. Schütz: *LCI* 8, 155-156; *Wp* 445 ff.; A.L. Delattre, *Les martyrs de Carthage*: Comptes-rendus de l'Académie des Inscriptions et Belles Lettres 1907, 176-177 and 193-195; id., Nuove scoperte di monumenti cristiani antichi a Cartagine. La "confessio" della Basilica Maiorum, *RAC* 7 (1930) 303; Y.-M. Duval, *Loca Sanctorum Africae*, CEFR 58, 2 vols., Rome 1982, I, 13-16, 682-684.

G. Santagata

PERPETUUS of Tours. 6th bishop of *Tours (*Turones*). Received two *letters from *Sidonius Apollinaris (*Epp.* 4, 18; 7, 5, 8-9). Presided over two *councils, at *Tours (18 Nov 461) and *Vannes (before 491) (CCL 148, pp. 147 and 150). *Gregory of Tours wrote a biographical sketch, and mentions him twice more (*Hist. Franc.* 10, 31; 2, 14, 23; *Vit. S. Mart.* 1, 2, 6). We possess a regulation by P. on vigils and *fasting, still in force in Gregory's time: fasts were fixed on Wednesday and Friday, from *Pentecost to the feast of St John in June, from 1 Sept to 1 Oct, from St Martin (11 Nov) to *Christmas, from 13 Jan to 15 Feb; vigils at Christmas, *Epiphany, the *Cathedra of St Peter, Resurrection of Christ (27 March), *Easter, *Ascension, anniversaries of St *Martin (4 July and 11 Nov), St Symphorian (22 Aug), St Lictorius (13 Sept), St Brize (13 Nov), St *Hilary (13 Jan). Thanks to him, we thus possess the late 5th-c. liturgical *calendar of Tours. P. built a basilica at Tours in honour of St *Peter, one at Mont-Louis in honour of St *Lawrence, and the parish churches of Esvres, Mougon, Barron, Ballau, Vernon (Indre-et-Loire). He left all his property to various churches in his diocese and was buried in St Martin's on 30 Dec. The will of 475 is a *falsification. The regulation on fasting and vigils is not contained in the canons of Tours and Vannes, but at the end of Gregory of Tours's *Historia Francorum*.

Duchesne, *Fastes* II, 304; *DACL* 15, 2619-20; *LTK* 8, 281; *BS* 10, 502-3; *Vies des SS.* 4, 182-88.

V. Saxer

PERSECUTIONS. May be defined as actions of a violent nature directed against the Christian church and its *members by their enemies.

Jesus' message of salvation entailed his confrontation with the Jewish authorities and eventual crucifixion by order of Pontius *Pilate, Roman procurator in Judaea. The gospels warn the disciples that their lot may imply suffering and p.; they may be "delivered up to councils and flogged in synagogues" (Mt 10, 17). Jesus himself said that the *blood of his covenant was shed for many (Mk 14, 24; Mt 26, 28, cf. Is 53, 12) and that the blood of the Christians too would be shed for his Name (Mk 10, 39).

The primitive community of *Jerusalem soon felt the truth of these words. In AD *c.*35, *Stephen became the first *martyr (Acts 7, 60). In 44, Herod *Agrippa's zeal for precise observance of the law, during the brief period of his servile government (41-44), culminated in the *martyrdom of *James, brother of *John, and the arrest of *Peter (Acts 12, 2-3). Though this p. was not pushed to its limits, *Paul preferred to preach in the synagogues of the diaspora (AD 47-57), bitterly opposed by many Jews, and was persecuted from city to city (cf. Acts 17, 10-13). His opponents depicted him as a "fomentor of continuous revolts among all the Jews throughout the world" (Acts 24, 5) and tried to have him condemned and executed by the Roman authorities, first at *Corinth (Acts 18, 12-17), then at Jerusalem (Acts 24, 1-6 and 25, 2). *Luke (Acts) shows that the Roman authorities looked on Christianity as an affair within Judaism (Acts 18, 14-17) and were not hostile to Paul and his preaching (Acts 13, 7-12). So what was the cause of *Nero's p. of AD 64?

We know little of the Christian community in *Rome at this time (but cf. Acts 18, 10 and 28, 22; Rom 1, 8 and Suet., *Claudius* 25, 4; cf. Frend, *Martyrdom and Persecution*, 160-161). Three factors seem important: *a)* Nero's desperate attempt to find a way out of the fire of 19 July 64, which he was suspected of having caused; *b)* official and popular opinion in Rome, which disapproved of any threat to the majesty of the Roman gods by foreign cults, Judaism included, and the ease with which the Jews were suspected of misanthropy (Tac., *Hist.* V, 5, 1) and incendiarism (Josephus, *Bell. Iud.* VII, 3, 3); *c)* the hostility of the *Jews towards the Christans.

*Tacitus's account (*Annales* XV, 44) of the p., written some 60 years later, may have been modelled on Livy's description of the plot of the Bacchanali and its suppression in 186 BC (Liv., XXXIX, 14, 10). Moreover, the Christians were seen as Jewish sectarians and, as such, guilty of "hostility to the human race" (*odium generis humani*), which manifested itself in the adoption of a perverse religion (*prava religio*) and conspiracy to set fire to Rome. Thus those who "confessed" (Christianity) were killed in a cruel and theatrical way, intended to placate the anger of the gods. A generation later, the author of *1 Clement* (ch. 5) seems to blame the catastrophe on the "envy and jealousy" of the church's internal enemies, i.e. the Jews.

The p., though not extended to *Italy and the provinces, had the result of putting the Christians in a legally difficult position. Tacitus leaves no doubt that for him Jesus' execution by Pilate was justified and that Christianity was a "mortal *superstition" whose followers deserved punishment. His contemporary Suetonius illustrates the repression of the Christians by Nero's police operations (*coercitio*), which he approved (*Nero* 16, 2). Acc. to him, the Christians were guilty of *magic practices and of having introduced "a new and dangerous religion", but he does not connect the p. with the fire of Rome.

In the 2nd c., *Melito of Sardis (Euseb., *HE* IV, 26, 8) and *Tertullian (*Apol.* 5, 4) mention *Domitian as the second "persecuting emperor". But it seems that Domitian's repressive measures were aimed at forcefully discouraging members of the Roman nobility from adopting Jewish customs (Dio Cassius, *Hist. Rom.* LXVII, 14; cf. Suet., *Domit.* 18). It is possible that *Domitilla, the emperor's niece (*neptis*) who was exiled to Pandataria (Pantelleria), was a Christian (cf. Euseb., *HE* III, 18, 4 and *Chron. ad ann. Abraham* 2111).

From the end of the 1st c., however, the Roman authorities distinguished between Jews and Christians (cf. Suet., *Domit.* 12) and Christianity was considered illegal (cf. 1 Pet 4, 15-16). In the province of *Asia, the *Apocalypse points to savage p. by Jews, local people and authorities (Rev 2, 9. 13; 17, 6 and 13, 16-17: an evident commercial boycott). The correspondence between *Trajan and *Pliny, his special legate (*legatus pro praetor*) in *Bithynia in 112-113, shows that Christians were exposed to summary execution if denounced to the authorities (*ad me tamquam christiani deferebantur*: Plin., *Ep.* X, 96, 2). Pliny asked the emperor if the mere profession of the name Christian deserved punishment or if only the crimes associated with it should be punished. In reply, Trajan explained that, though Christians should not be sought out (like common malefactors) - *conquirendi non sunt* - they should be punished if they refused to retract and "worship our gods". If they retracted, however, they should be freed. In another rescript, Trajan's successor *Hadrian instructed the proconsul of Asia, C. Minucius Fundanus, in 124-25 to condemn Christians only if found guilty of infractions of the law by a tribunal. They must not be victims of popular clamour and denunciations, and they have the right to make an accusation of defamation against their accusers (procedure of *calumnia*).

These two decisions established the authorities' line of approach to the Christians up to the end of the century. They had the effect of discouraging *accusations (cf. A.N. Sherwin White, The Early Persecutions and Roman Law Again, *JThS* n.s. 3 [1952] 199-213), and the Christians enjoyed a period of relative tranquility until the reign of *Marcus Aurelius (161-180). But from then on, official reluctance to act against the Christians was overcome by the growth of popular anger. Besides personal rancour, which cost *Justin martyr his life at Rome in 165 (cf. Just., *II Apol.* 3 and 11), there were *accusations, fervently believed, of incest ("Oedipal relations"), cannibalism ("Thyestean banquets") and atheism (rejection of the Greek gods). The Christians were held responsible for natural catastrophes which demonstrated the anger of the gods (cf. Tertull., *Apol.* 40, 2 and Aug., *De civ. Dei* II, 3, which cites the popular saying: *Pluvia desit causa Christiani sunt*). Again, at this time they were still sufficiently close to Judaism (cf. Orig., *C. Cels.* II, 1 and 4; III, 1) to inherit the unpopularity earlier reserved for the Jews. The result was a series of sporadic but savage p. inspired by popular resentment: the best-known is that which led to the martyrdom of *Polycarp of Smyrna, prob. *c.*165 (or perhaps between 156 and 160; cf. *Mart. Polyc.*, ed. H. Musurillo, *The Acts of the Christian Martyrs*, 5-20, and cf. Euseb., *HE* IV, 15) and the pogrom of *Lyons of 177, recorded in a *letter sent by the survivors to the churches of Asia and Phrygia (Euseb., *HE* V, 1 and 2, 1-8). At Smyrna, the Jews joined the *pagans in demanding Polycarp's death (*Mart. Polyc.* 13 and 18, and Euseb., *HE* IV, 15, 26. 29), and the authorities consented.

New light is shed on the reasons for the unpopularity of the Christians in this period by the Platonist *Celsus, whose *Alethès Logos*, written *c.*178, was a complete attack on every aspect of Christian *theology, *anthropology and conduct. For Celsus, the Christians formed an "illegal association" whose members were bound by secret oaths (Orig., *C. Cels.* I, 1; cf. Caecilius in *Minucius Felix's *Octavius*, 8, 4: "*profanae conjurationis*"): not only were they unable to undertake normal public duties, but with their *proselytism (*C. Cels.* III, 52 and 55) they risked undermining the traditional social and religious structure of the empire. The p., including the active searching out of Christians, were justified (*C. Cels.* VIII, 69).

The crisis of the reign of Marcus Aurelius was overcome. In the last quarter of the 2nd c., Christians at Rome began to be more respected and, thanks to their conduct, considered as persons comparable to "genuine philosophers" (*Galen, *Commentary on Plato's "Republic"*, cited by R. Walzer, *Galen on Jews and Christians*, Oxford 1949, 15; cf. also Euseb., *HE* V, 21). Sporadic p. nevertheless continued, involving the martyrdom at Rome, *c.*183, of the senator (?) *Apollonius (*Acta Apollonii*; cf. Musurillo, *op. cit.*, XXV and 90-104; G. Lanata, *Gli Atti dei martiri*, 145-157, with abundant bibl.).

The accession of the dynasty of the Severi in 193 opened a new era in relations between the Christians and the empire. There were fewer popular accusations of black *magic and promiscuous relations, but more of atheism and opposition to the deities that protected the Roman empire. Furthermore, Christianity now rediscovered its old missionary ideal. At *Alexandria, *Clement preached the duty of spreading the gospel (*Strom.* I, 1, 3; 3, 1 ff.). At *Carthage, *c.*197, *Tertullian boasted of the ever-growing number of Christians (*Adv. Nat.* 1, 14) and of *conversions to Christianity (*fiunt non nascuntur christiani*: *Apol.* 18, 4). This progress provoked a strong popular reaction, supported this time by the authorities. P., in the form of lynching (at Alexandria, cf. Clem., *Strom.* II, 20, 125 and II, 283, 18) or mass attacks (at Rome, cf. Hipp., *In Dan.* 1, 23) or judicial executions (at Carthage, *Passio Perpetuae*), took place over a wide area between 197 and 212, with particular violence in 202-203. *Eusebius (*HE* VI, 1, 1) associated the p. with the emp. *Septimius Severus, but it seems more likely that the rescript attributed by Spartianus (*Historia Augusta, Vita Severi* 17, 1) to that emperor in 202, which forbade conversions to Judaism or Christianity, was his reply to a provincial governor who had asked for instructions about popular movements against the Christians, particularly against converts (cf. Frend, Open Questions, 349, against K.H. Schwarte, Das angebliche Christengesetz des Septimius Severus, *Historia* 12 [1963] 185-208). It is noteworthy that, in contrast with all other periods of p., Christian converts rather than leaders were the target of pagan attacks (so *Passio Perpetuae* and, for Alexandria, Euseb., *HE* VI, 3).

The wave of resentment receded again. Between 212 and 235, the church enjoyed a further period of quasi-tolerance. Whether Lampridius's statement that *Alexander Severus (222-235) "let the Christians be" (*Hist. Aug.*: Lampridius, *Vita Alex. Severi* 22, 6: *christianos esse passus est*) is true or not, the church could now maintain *property (the oldest Roman catacombs) and erect buildings for prayer in some cities (*Dura Europos).

The revolution of 22 March 235 put an end to the dynasty of the Severi. Alexander Severus's successor *Maximinus Thrax (235-238) liquidated the Christian *slaves and officials of his predecessor's court (Euseb., *HE* VI, 28) and struck at the Christian leaders (*ibid.*). Pope *Pontianus and *Hippolytus of Rome were among the exiles (they died in *Sardinia at the end of 235), and *Origen reveals that in 236 there were important Christians in Palestinian *Caesarea who feared for their lives (*Exhort. ad mart.*, addressed to *Ambrose). Maximinus fell to another revolution inspired by the landed proprietors of N *Africa (the Gordiani), supported by the Roman senate. During the reigns of Gordian III (238-244) and *Philip the Arab (244-249), the church went through another period of prosperity (Euseb., *HE* VI, 36; cf. Orig., *C. Celsum* I, 43). It was too good to last. Writing his *Commentary on Matthew* in 238, Origen foresaw that p. against the Christians would take on a world-wide dimension (*In Matth.* 24, 9; *Serm.* 39).

He was not mistaken; a new change in the political wheel of fortune brought with it a new emperor and a new policy towards the Christians. C. Quintus Messius *Decius, who took the surname "Trajan" (249-251), was a good general and believed passionately in the traditional virtues of Rome. Like Maximinus before him, he thought the Christians should be made to take responsibility for the catastrophes that had struck the empire under his predecessor. *Eusebius (*HE* VI, 39, 1) writes that "because of his hatred for Philip, he launched a persecution against the churches". In January 250, Decius ordained by edict that the annual *sacrifice made on the Capitol to the Roman gods should be repeated in the provincial cities, and at the same time had many eminent Christians arrested. On 21 Jan, pope *Fabian was tried before him and condemned. A similar fate befell *Babylas, bishop of *Antioch (Euseb., *HE* VI, 39, 4). At Carthage, *Cyprian had to go into hiding, and at Alexandria, *Dionysius had an eventful escape from arrest (Euseb., *HE* VI, 40, 2 ff.). This first phase was followed by the establishment of commissions in the provinces to supervise sacrifices to the gods of the empire and the genius of the *emperor (cf. Cypr., *Epp.* 43, 3; 67, 4; also Euseb., *HE* VI, 42, 1). Trials were held from February to March at Carthage and Smyrna, and until June-July in some parts of *Egypt. Egyptian *papyri have preserved 43 *libelli (certificates) given to those who sacrificed (cf. esp. J.R. Knipfing: *HThR* 16 [1923] 345-390). It was presumed that all would sacrifice, since refusal meant death (*Acta Pionii* 7, 4). Decius's measures had an initial success. His intention was more to oblige the Christians to conform to the worship of the Roman gods than to ban the practice of Christianity (thus, in the *Acts of Conon* 4, 4 [Musurillo, *op. cit.*, 189], the judge says to *Conon: "If you have recognized Christ, recognize our gods also"). Some Christians resisted; some important bishops like Euthemon of Smyrna openly *apostatized (*Acta Pionii* 15, 2), while *confessors like *Pionius of Smyrna were objects of scorn and commiseration (*Acta Pionii* 10, 16 and 20). *Paganism was once more at its height, while Christianity remained an urban religion diffused in minority communities in the principal cities of the empire. It had never been in greater danger.

But Decius fell in battle against the *Goths (June 251). An attempt by his successors Gallus and Volusianus to resume the p. (Cypr., *Ep.* 59, 8) resulted in the exile and death of pope *Cornelius at Centumcellae in June 253, but little else. The church quickly recovered lost ground. The 87 bishops who met at *Carthage on 1 Sept 256 to support Cyprian's stand against pope *Stephen in the dispute over the *baptism of *heretics gave evidence of its durability and adaptability.

In summer 257, the emp. *Valerian (253-260), hard pressed by the Persians on the empire's eastern frontier, made a new attempt to force the Christians to recognize the Roman gods. The text of his edict is lost, but we seem to have a paraphrase in the sentence preserved by the *Acta Proconsularia* relative to Cyprian's exile (*Sacratissimi imperatores Valerianus et Gallienus litteras ad me dari dignati sunt, quibus praeceperunt eos, qui Romanam religionem non colunt, debere Romanas caerimonias recognoscere*: Acta Proc. 1, ed. W. Hartel, CSEL 3, 3, p. CX). So the authorities accepted that the Christians would not worship the gods; nevertheless they demanded that they recognize formally, in some way, their saving power for the peoples of the empire. It was this that Aemilianus, vice-prefect of Egypt, had expressed in his talk (hardly a "trial") with bishop *Dionysius of Alexandria in 257 (cf. Euseb., *HE* VII, 11, 9-11, which gives Dionysius's account of it, contained in a letter to bishop Erammon). Both Cyprian and Dionysius rejected compromise and were interned, Cyprian in a comfortable villa he owned at Curubis on the gulf of Hammamet; the less fortunate Dionysius in the oasis of Kuphra.

Next year, Valerian gave much harsher instructions to the Senate, to be transmitted to the provincial governors. Acc. to Cyprian (*Ep.* 80), bishops, presbyters and deacons were to be punished immediately (by death), Roman senators (*viri egregii*) and knights (*equites*) were to lose their dignity and property, and even *matronae* were to lose their property and be banished. Civil servants (*caesariani*) would be reduced to *slavery and sent in chains to work on the imperial estates. This is the harshest and most detailed list of punishments ever issued against the Christians and, in the West, the p. lasted far longer and was probably more costly in human lives, especially among the leaders, than even the "great persecution". On 13 Sept, Cyprian was taken from Curubis to Carthage and brought the next day before the proconsul, Galerius Maximus. He was condemned to death after a brief audience and executed on the evening of 14 Sept. Galerius's reasons for condemning him to death are interesting: the Christians were "irreligious" (*sacrilega mente*), they "met together as an illegal association" (*nefariae conspirationis*), and Cyprian, as their "standard-bearer", had "openly professed himself an enemy of the gods and the religion of Rome" (*Acta Proconsularia* 4, pp. CXII-CXIII). The accusations seem to be those that Celsus had formulated eighty years earlier, but the "conspiracy" now directly challenged the primacy of the Roman gods.

Once more, external events helped the Christians. Valerian was taken prisoner by the Persians at *Edessa (June 260) and his son and successor *Gallienus (253-268) ordered the restoration of the church's places of worship and allowed freedom of existence (and hence of worship). A paraphrase of the text sent in c.262 to Dionysius and his dependent bishops in Egypt after the end of the civil war has been preserved by Eusebius (*HE* VII, 13: "I have given orders that the benefits of my generosity should expand through the whole world, so that they [the authorities] should withdraw from the places of worship and thus you also may profit from the directions contained in my rescript, so that no-one may molest you").

Over the next 40 years, a *modus vivendi* was established between *church and empire. The church obtained something like an intermediate situation, that of an autonomous body, governed by its own officials according to its own laws, but subject to the overall authority of the imperial governing power. Although, curiously, no great Christian leaders emerged in this period, the church penetrated every sphere of society, most importantly among the indigenous peoples of the important territories of *Egypt, *Syria and N *Africa, and also beyond the frontiers of the empire as far as the kingdom of *Armenia. It was no longer concentrated exclusively in the cities, a relatively easy target for the authorities.

The great persecution of 303-305, in the West, and 303-312, in the *East, can today be seen as the culmination of the measures of *Diocletian and his fellow-tetrarchs, aimed at imposing on the empire a strictly uniform administration inspired by the cult of the Roman gods and the practice of the so-called traditional virtues of Rome. One by one, the army, provincial administration, taxes and coinage were reformed with a view to providing an administrative and economic system that was simple and uniform, but inflexible, over the whole empire.

The motto "*genio populi romani*" on the reverse of silvered bronze coins, coined in tens of thousands of pieces to be used in the empire, summed up the imperial message, just as the emperor's edict on prices (301) reflected his paternalistic search for uniformity. Inevitably, uniformity was extended to religion. In March 297 (302, acc. to some), Diocletian ordered the prefect of Carthage to put the *Manichees to death and burn their books (*Coll. legum mosaicarum et romanorum, in Iurisprud. anteiustin.*, ed. Seckel-Kübler, II, 2, p. 381, and discussion in F. Décret, *L'Afrique manichéenne, IVe et Ve siècles*, I, Paris 1978, 162-165). They could be seen as "enemy agents" in the pay of Rome's adversary, Persia, but from the moment of Persia's final defeat in 298, attention turned to the Christians.

The p. developed gradually - so it seemed to *Eusebius, as a contemporary (*Chron. ad ann.* 301: *Veturius magister militae Christianos milites persequitur, paulatim ex illo iam tempore persecutione adversum nos incipiente*). The army and civil administration were purged of Christians, but during the latter part of 302 Diocletian hesitated. His wife and daughter may have been favourably disposed towards Christianity. In the end, the arguments of his Caesar, *Galerius, and a visit to the oracle of Milesian Apollo at Didyma, got the upper hand. The Christian challenge had to be accepted, but there must be no bloodshed (Lact., *De mortibus persecutorum*: SCh 39, 11, 8). The day chosen to put an end to the Christian menace (*ibid.*, 12) was 23 Feb 303, the feast of *terminalia*.

The text of the first edict has not survived, but it seems to have been along the lines of Valerian's measures, though without the death penalty. Christian churches were to be destroyed, their religious functions prohibited and the *Scriptures handed over to the authorities to be burnt (cf. Lact., *De mort.* 13). Christians belonging to the upper classes were to lose their privileges. Those holding civil office (*oiketiae*, in Euseb., *HE* VIII, 2, 4) were to be reduced to slavery and Christians were forbidden to claim or defend their rights in court. The edict was enforced without problems by the provincial and local authorities. After the revolts in Melitene and Syria, which, it was claimed, were inspired by Christians, a second edict condemned the *clergy to imprisonment (Euseb., *HE* VIII, 6, 9), but the prisons were unable to hold so many people; so in summer 303, a third edict ordered that they should be made to sacrifice and then freed. Every effort was made, sometimes with grotesque results, to ensure their consent (Euseb., *Mart. Pal.* 1, 3-5). Diocletian then left for Rome to celebrate his *Vicennalia* (20 Nov 303).

Up to this point, the persecution had had a certain success (Euseb., *HE* VIII, 2, 1). The churches were destroyed, many bishops had capitulated (like Paul, bishop of *Cirta in *Numidia), others had equivocated or fled (like *Peter of Alexandria) and the communities had been dispersed. Only a small number of unwavering men and women had been put to death (cf. Euseb., *Mart. Pal.* 1, 5). During his visit to the West, however, Diocletian fell ill and Galerius took advantage of the situation. In spring 304, a fourth edict was issued calling on everyone, men, women and children, to sacrifice and offer a libation under pain of death. From then on, there were so many martyrs that in N Africa the *dies thurificationis* was remembered a century later (CIL VIII, 6700) alongside the *dies traditionis*.

Apart from N Africa (Euseb., *HE* VIII, 6, 10), the great p. in the West was more violent in later fame than in reality; it virtually ended at the time of the abdication of Diocletian and *Maximian (1 May 305), and was not renewed. In the East, however, things were different. Galerius was now Senior Augustus and, after a pause lasting until Easter 306, the p. began again. *Maximinus Daia, Galerius's Caesar, issued an edict ordering all provincial governors to oblige everyone to sacrifice to the gods. At Palestinian *Caesarea, officials summoned everyone individually by name on the basis of the census lists (*Mart. Pal.* 4, 8). Eusebius records many deaths in *Palestine in the next two years (*Mart. Pal.* 5-7). Between July 308 and November 309, there was another interval centred on the meeting between Diocletian, Maximian and Galerius at *Carnuntum* (Nov 308). From this moment, however, the pagan effort began to decline. When the p. was resumed in Palestine in November 309, every article bought or sold in the market had to be contaminated with libations and sprinkled with the *blood of sacrifice (*Mart. Pal.* 9, 2). The pagans themselves were sceptical of the efficacy of this measure; Eusebius comments that they "judged it out of place" (*Mart. Pal.* 9, 3); though one illustrious victim, the presbyter *Pamphilus, suffered *martyrdom (*Mart. Pal.* 11) in March 310.

The same fate awaited Maximian's attempts to restore pagan worship and temples. Even before Galerius fell ill in spring 311, it was clear that Christianity had become too strong to be suppressed by violent means.

Galerius's recantation (30 April 311) is rightly considered the turning point in the struggle between paganism and Christianity in the Roman empire. Though drawn up in the conviction that everything "must be reordered according to the old laws and public order (*disciplinam*) of the Romans", it admitted that too many Christians had refused to obey. Now the Christian God, deprived of worship, was angry, so "let the Christians once more exist (*denuo christiani sint*) and rebuild their churches" and "pray to their God for our well-being, for that of the state and for themselves, so that in every place the state may be preserved intact and they themselves live in safety in their own homes" (text in Euseb., *HE* VIII, 17, 3-10). But all was in vain: Galerius died in agony six days later, 5 May 311.

This edict restored Christianity to the same situation it had enjoyed before the start of the p.: but it could not be the end of the struggle. There were now two divinities responsible for the safety of the empire: the "immortal gods" and the "Christian god". For the first time in the history of Rome, a power different from the old one was accepted as influencing, even though in a negative way, the destiny of the empire.

But Maximinus Daia did not accept the spirit of Galerius's edict. The Christians "could exist", but what would happen if their pagan neighbours objected? Petitions against the "atheists" were organized, thanks to the civic *curiae* of *Nicomedia, *Tyre and *Antioch (Euseb., *HE* IX, 7, 10-11 and IX, 9, 4) and the provincial senates of *Lycia and Caria (Diehl 1a and b). A final atrocious tremor of p. shook Egypt in 311 and cost the lives of bishop *Peter of Alexandria, 25 Nov 311, and innumerable Copts in the *Thebaïd (Euseb., *HE* IX, 9, 4-5). This time, the Christians knew that death meant victory; they would never forget it. Diocletian's rise to power initiated the "age of the martyrs", which marked the point of departure of the *computus of the *Coptic Christians.

The last victim of the p. was *Lucian of Antioch, martyred at Nicomedia on 7 Jan 312. On 28 Oct 312 *Constantine defeated *Maxentius at the Milvian Bridge; in 313 he met *Licinius at *Milan: this solemnly ended the era of p. (*Edict of Milan).

Why did the Christians suffer two and a half centuries of suspicion, attacks and official repression? It is impossible to give a single reply. Rivers of ink have been poured out, esp. at the beginning of the 20th c., seeking to define the nature of the imputations against the Christians. Most likely the p. were the final result of a complex of political, social and legal factors, with different emphases at different times. In the first place, there was the conviction, deeply rooted in the governing class of Rome, that the salvation of the empire depended on an appropriate acknowledgment of the gods. There could be no toleration of any attack on their pre-eminence. In AD 64 the Christians were accused, rightly or wrongly, of having outraged this sentiment and, as *Tacitus shows, this was not forgiven them; theirs could never have become a *religio licita*. Secondarily, to the 2nd-c. inhabitants of the provinces, the links between *Jews and Christians could appear closer than they actually were, and the Christians gradually inherited the unpopularity previously experienced by Jews in the Greek cities of the Eastern Mediterranean. Thirdly, their apparent secrecy and close *organization made them objects of popular fears and resentments. *Celsus's dual rebuke, "illegal organization" and "apostate Judaism", sums up this attitude. They deserved repression as a *latebrosa et lucifuga natio* (Caecilius's opinion in Minucius Felix, *Octavius* 8, 4). The authorities were in agreement. To these various *accusations was added, in the 3rd c., that of fighting directly against the gods and hence against the political stability of the empire. This was behind the p. from Decius to Diocletian. It needed the genius of Constantine to understand that the victory of the Christian *God could be accepted without a political disaster. The organization of the church and the conviction of personal salvation suggested to him that the pagan world had crumbled and that its defeat could be placed at the service of a new, Christian empire.

Acta martyrum: R. Knopf-G. Krüger, *Ausgewählte Märtyrerakten*, 4. Auflage, with Appendix by G. Ruhbach, Tübingen 1965; H. Musurillo, *The Acts of the Christian Martyrs*, texts with Eng. tr. (Oxford Early Christian Texts), Oxford 1972; G. Lanata, *Gli Atti dei martiri come documenti processuali*, Milan 1973.
Studies: up to 1963, bibl. in W.H.C. Frend, *Martyrdom and Persecution in the Early Church*, Oxford 1965, 572-604; M. Sordi, *Il cristianesimo e Roma*, Bologna 1965; G.E.M. de St-Croix, Why were the early Christians persecuted?, *P&P* 26 (1963) 6-38; C. Delvoye, *Les Persécutions contre les chrétiens dans l'Empire romain*, Brussels 1967; R. Freudenberger, Christenreskript. Ein umstrittenes Reskript des Antoninus Pius, *ZKG* 78 (1967) 1-14; id., *Das Verhalten der römischen Behöorden gegen die Christen in 2. Jahrhundert*, Munich 1967; T.D. Barnes, Legislation against the Christians, *JRS* 58 (1968) 32-50; id., Pre-Decian Acta Martyrum, *JThS* n.s. 19 (1968) 509-531; E. Bickermann, Trajan, Hadrian and the Christians, *RFIC* 96 (1968) 290-315; P. Keresztes, Marcus Aurelius a Persecutor?, *HThR* 61 (1968) 321-341; G.W. Clarke, Some observations on the Persecution of Decius, *Antichton* 3 (1969) 63-76; J. Molthagen, *Der römische Staat und die Christen im zweiten und dritten Jahrhundert*, Göttingen 1970; A. Wlosok, *Rom und die Christen*, Stuttgart 1970; id., Die Rechtsgrundlagen der Christenverfolgungen der ersten zwei Jahrhunderte, in Klein (ed.), *Das frühe Christentum im römischen Staat*, Darmstadt 1971, 275-301; G.W. Clarke, Two measures in the Persecution of Decius, *BICS* 20 (1973) 118-124; W.H.C. Frend, Open questions concerning the Christians and the Roman Empire in the Age of the Severi, *JThS* n.s. 25 (1974) 333-351; P. Keresztes, The Peace of Gallienus, *WS* N.F. 9 (1975) 174-185; L.F. Jannsen, "Superstitio", and the Persecution of the Christians, *VChr* 33 (1979) 131-159; B.W. Workman, *Persecution in the Early Church*, Oxford ²1980; K.H. Schwarte, Die Religionsgesetze Valerians, in *Religion und Gesellschaft in der römischen Kaiserzeit*, Cologne-Vienna 1989, 103-163; P.S. Davis, Origin and Purpose of the Persecution of 303 AD, *JTS* n.s. 40 (1989) 66-94; P. Keresztes, *Imperial Rome and the Christians*, 2 vols., Lanham-New York-London 1989.

W.H.C. Frend

PERSIA. Persian territory, which extended from Mesopotamia to the borders of India and from *Armenia to *Arabia, a fief of the Arsacid (Parthian) and *Sassanid (224-632) empires and then of the caliphates of Damascus (637-750) and Baghdad, was the cradle of an Aramean Christian church, very lively from the beginning. Developing on the margins of the Roman, later the Byzantine empire, in an area at all times subject to religious *syncretism, it preserved important aspects of Eastern ecclesiastical *tradition.

1. Origins. It is on the question of origins that our knowledge has made most progress in recent years. Thanks to discoveries made and work done in the field of *gnosis (*Nag Hammadi, *Manichaeism), heterodox Judaism (Qumrân) and particularly *Judaeo-Christianity (*Elkesaïtes), we can say that P., even before Osrhoene, was the first bastion of Arameo-Syrian Christianity. The important Jewish communities that had long resided in Mesopotamia, esp. those of "Palestinian obedience" - starting with the small realm of *Adiabene, whose state religion from AD 36 was Judaism - explain a rapid *evangelization which reached the edge of Ctesiphon between 76 and 116 (ancient tradition relating to Mari preserved in Arabic in the *Book of the Tower*. The apocryphal literature on Judas *Thomas, *Addai, Mari and Aggai is quite late and tries to make everything dependent on *Edessa, its own place of origin). The characteristics of the earliest Persian Christianity (which can be reconstructed with the help of *Aphraates, the Persian sage, and *Ephrem of Nisibis) - "*ascesis", *encratism, *Targumic and *Midrashic traditions, *liturgy (Eucharistic blessings, music, architecture), theological school - take us back to a Judaeo-Christianity often contaminated by an underlying sectarian Judaism, as we find it in the *Odes of Solomon*, the *Gospel* and *Acts of Thomas*, the Pseudo-*Clementines*, the *Manichaean psalter*, etc. (The evangelization of India (*Malabar) from apostolic times, though backed up by no scientific proof, is historically admissible since India did have commercial relations with the Roman empire: cf. *ANRW* II.9.2 [1978]).

2. 2nd-4th centuries. Evidence for the period immediately following the beginnings of evangelization is still rare. The stele of Abercius (cf. *RACh* 1, 12-17) speaks of a Christianity beyond the Euphrates, *c.*150, and the *Book of the laws of the countries*, attributed to a disciple of *Bardesanes, briefly describes the life of the communities of "Parthia, Media and Hatra" *c.*200-220. It is clear that Christianity must have developed in the 3rd and 4th cc. in the Persian empire, as we can deduce from the following phenomena: a) the appearance of *heresies like *Marcionism and esp. *Manichaeism; b) the welcome given to Christian exiles and prisoners, deported during the campaigns successfully conducted in the West (as far as *Cilicia) by the first Sassanids; c) existence of the earliest *martyrs; d) echoes of struggles over the see of *Seleucia-Ctesiphon starting from 310 (cf. *Papas bar Aggai); e) foundation of a theological school at *Nisibis. The Sassanid court followed a policy of tolerance, in contrast to their Roman enemies who were ousted at the time of *Gallienus (260-268). In fact it was during the "little peace" that the party of the Magi, instigated by their powerful leader Kartîr, put very heavy pressure on the Shahanshâh Bahrâm II (276-296) to begin a first persecution against foreign religions (in which *Mani, among others, was killed), and it was after the restored peace that, under *Constantius II, Shapur II (309-379) perpetrated the "great massacre" against those who now had to be considered "allies of the Caesars". On the other hand, in their internal policy the kings of kings always had to come to terms with the very influential national religion, Mazdaism, but also with the Jewish, Manichaean and sometimes even the "Greek" Christian communities, who were useful to the development of their economy.

3. 5th-7th centuries. Persian Christianity benefited, from *Yezdegerd I (399-420) to the coming of Islam, from the protection of the Shahs. These, according to their policy, could sometimes make use of the Christians, even as their ambassadors to *Byzantium, and favour them against the Magi, and sometimes profit from their internal struggles or put very violent pressure on them. At any rate, this period was characterized by the reorganization of Christianity and its consolidation with the help of the West. This is evidenced particularly by the explosion of *monasticism, the struggles against *Messalianism and *Origenism (cf. *Henana of Adiabene, †610), the theological work (translation of Greek *Fathers, foundation of the school of *Seleucia-Ctesiphon in 541 by Mar *Aba), but also the divisions, and all this under the direction of great leaders, a good many of whom were doctors of the Sassanid court. The first of these, *Maruta of Majferqat, got the faith and ecclesiastical discipline of *Nicaea adopted by the synod of 410. This synod also attests the extent of the Persian church after this period (from the isles of the Persian Gulf to the borders of Armenia, and from Persis [Fars] to the region of Nisibis). The expulsion of the Persians from *Edessa, the foundation of the new school of *Nisibis by *Narsai (†c.502) and the strong personality of *Barsauma (†491/6) were the principal instruments of the Persian church's adoption of *Nestorianism (486). But the young Nestorian church was not organized sufficiently quickly and strongly to oppose the infiltration of the *monophysites. These, profiting from the conquest of the north of the country by *Byzantium, succeeded in installing their own hierarchy at Takrit from 629.

4. After the 7th century. The tolerance of the Ummayads (637-750) and esp. the Abassids (750-1258) (the patriarchal see was transferred to Baghdad, the new capital) must have been favourable to the *Jacobite and Nestorian Persian churches. The latter, though beginning to disintegrate, dedicated itself in a peculiar way to new missions, esp. in China, and, by intense labour, transmitted its Aramean and Greek heritage to the new Muslim culture.

Sources: Besides secular Roman, Byzantine, Persian and Arab sources, to reconstruct the history of the church in P. we must go above all to ecclesiastical, more specifically Syrian, sources (listed in Ortiz de Urbina 206-212 and esp. in J.-M. Fiey, *Jalons pour une histoire de l'Eglise en Iraq*, CSCO 310 [subs. 36] Louvain 1970, 8-31) and, preferably, to the *Chronicle of Seert* (in Arabic), the *Oriental synodicon* (ed. Chabot), to Bedjan and various authors, starting with Aphraates and Ephrem (cf. Ortiz de urbina 115-153).
Studies: J. Labourt, *Le Christianisme dans l'empire perse sous la dynastie sassanide (224-632)*, Paris 1904; E. Tisserant, Nestorienne (Eglise), in *DTC* 11, 157 ff.; S.P. Brock, A classified bibliography, *PdO* 4 (1973) 393-465; *ANRW* II, 8 and 9; J. Neusner, *History of the Jews in Babylonia*, Leiden 1965-1970; R. Murray, *Symbols of Church and Kingdom: A Study in Early Syriac Tradition*, Cambridge 1975, esp. 1-38; J. Ries, Mani et le Manichéisme, *DSp* 10, 198-216; F. Decret, Les conséquences sur le christianisme en Perse de l'affrontement des empires romain et sassanide de Shâpûr 1er à Yazdgard 1er, *RecAug* 14 (1979) 91-152; G. Wiessner, *Zur Märtyrerüberlieferung aus der Christenverfolgung Shapurs*

II, AKW Göttingen, Phil.-hist.Kl., Folge 67, Göttingen 1967; J.-M. Fiey, *Assyrie chrétienne*, 3 vols., Beirut 1965-68; id., *Nisibe*, CSCO 388 (sub. 54) Louvain 1977; id., *Chrétiens syriaques sous les abassides*, CSCO 420 (subs. 59) Louvain 1980. Cf. also: *Symposium Syriacum 1980*, OCA (forthcoming) and the imminent new bibliography of Syriac studies (1970-1980) by S.P. Brock; W.G. Young, *Patriarch, Shah and Caliph*, Rawalpindi 1974; S.P. Brock, Christians in the Sasanid Empire: a case of Divided Loyalties, Religion and National Identity, in (ed.) S. Mews, *Religion and National Identity*, SCH 18, Oxford 1982, 1-19; id., A Martyr at the Sasanid Court under Vahram II: Cantida, *AB* 91 (1978) 167-181.

F. Rilliet

PERSONA, Person. Besides the classical meanings of the term p. (role, character, individuality), Latin *theology knew from the start a technical meaning. The refutation of the *monarchians gave *Tertullian occasion to develop its trinitarian use and suggest a christological meaning (*Adv. Prax.*). Having certainly taken it from secular exegesis, which distinguished *personae* both grammatically and aesthetically (dignity of personality), under the influence of legal tradition he invested it with a realistic note (*persona=res*). In the 4th c., the anti-*Arian authors clarified its trinitarian meaning, excluding the purely functional sense that the equivalent πρόσωπον (*prosopon*) seemed to suggest (cf. *Hilary). *Augustine, while avoiding the term p. in his theology of the *Trinity, introduced it in a positive sense into *christology (from 400: *unitas personae*; from 411: *una persona*). As its parallel use in *ecclesiology demonstrates (*ecclesia quasi una persona*), the exegetical echoes remained strong even in Augustine. But, by applying the expression *una persona* also to man, composed of *soul and *body, Augustine showed the degree to which, for him, p. meant not just a logical subject of attribution, but also an ontological principle of the one Christ (though not yet explicitly in the sense of a *hypostatic union). Under the influence of the 4th-c. philosophical and christological disputes, *Boethius finally defined p. as *naturae rationalis individua substantia* (*De duab. naturis* 3), a fundamental definition for scholastic theology.

M. Nédoncelle, Prosopon et persona dans l'antiquité classique, *RSR* 22 (1948) 277-299; T. van Bavel, *Recherches sur la christologie de s. Augustin*, Fribourg 1954; C. Andresen, Zur Entstehung und Geschichte des trinitarischen Personbegriffs, *ZNTW* 52 (1961) 1-39; J. Moingt, *Théologie trinitaire de Tertullien*, II, Paris 1966; S. Otto, *Person und Subsistenz*, Munich 1968; R.V. Sellers, *Two ancient Christologies*, London 1954; L.I. Scipioni, *Nestorio e il concilio di Efeso*, Milan 1974, esp. 388 ff; H.R. Drobner, *Person-Exegese und Christologie bei Augustinus Zur Herkunft der Formel una persona*, Philosophia Patrum 8, Leiden 1986.

B. Studer

PERSONIFICATIONS

1. P. of concepts. The earliest p. in Christian figurative art may be said to be the figures of the *orans and the Good *Shepherd, esp. if on their own, as symbolic expressions of *pietas* and *philanthropia* (Klauser, *Studien*, II, 397 and III, 112). A woman with a palm branch in the pavement *mosaic of the S hall of the Theodorian complex at *Aquileia (4th c.) is thought to represent *victoria eucharistica* (Mian, *La "vittoria"*, 131 ff.).

2. P. of the church. P. of the church appear on the front of a late 4th-c. *sarcophagus from Perugia showing a matron beside Christ, facing St *Peter and replacing St *Paul (*Ws*, pl. 28, 3), and in the mosaics of the Roman basilicas of S. Pudenziana (*WMM* 3, 42-44) and S. Sabina (5th c.), where two female figures are captioned *ecclesia ex gentibus* and *ecclesia ex circumcisione* (ibid. 3, 47, 1 and 2). On a panel of the wooden door of S. *Sabina, more or less contemporary with the mosaic cited, the church - or possibly *Mary - is shown in the form of a woman between Sts Peter and Paul, in a triumph scene (Jeremias, *Die Holztür*, pl. 68).

3. P. of the physical elements. P. taken from the common repertoire persist in ancient Christian art, such as those of the seasons, months, heaven, frequently repeated in *paintings and sarcophagi, but not taking on any new or renewed values. In the pavement mosaic of the martyrial basilica of Gasr Elbia (ancient Olbia) in Cyrenaica (6th c.), a series of panels includes a number of p., among them the *lighthouse of Alexandria (Guarducci, *Epigrafia*, IV, fig. 148) and the four rivers of *Paradise (*ibid.*, figs. 144, 145). A painted figure in the *cemetery of the Jordani is probably not the Tigris (*Wp* 212); but figures in scenes of Christ's *baptism, in the mosaics of the Baptisteries of the Orthodox (Bovini, *Mosaici*, pl. 2) and of the Arians (ibid., pl. 19) at *Ravenna, probably represent the Jordan. Mountains, rivers, geographical locations and cities are expressed as persons in codex miniatures and *itineraria picta*, as in the *Joshua Scroll*, the *Chronographer of 354* and the *Tabula peutingeriana*.

4. P. of the sun and moon. Portrait or full-length p. of the sun and moon, on the front of the late 3rd-c. "*Jonah" sarcophagus (*Ws* 9, 3), sarcophagus-lids of the Constantinian era (*Ws* 2, 300) and the painting of *Joseph's Two Dreams in the 4th-c. hypogeum of Via D. Compagni (Ferrua, *Le pitture*, pl. 96), indicate the celestial vault, alluding perhaps to the other world or simply to the vicissitudes of human life. The symbolic component seems more explicit in 5th- and 6th-c. ivories, where the sun and moon are shown in person, in scenes of Christ's crucifixion, to signify the end of his earthly life and foretell his *resurrection, and of his *baptism, to indicate the start of his human journey (Testini, *Arte mitraica*, 433-452). To the phenomenon of p. is added that of symbolic assimilation, in the representation of Christ-Helios, as in the mosaic in the vault of the mausoleum of the Julii in the *Vatican necropolis (Sear, *Roman wall*, 53, 2-3).

5. Zoomorphic substitutions. As final manifestation of the personificatory process we must mention a phenomenon collateral to p. itself: zoomorphic substitution for the human figure. On the painted front of the "Celerina arcosolium" at Pretestato, the episode of *Susanna molested by the elders is shown as a *lamb threatened by two wolves, captioned *Susanna* and *senioris* (Dagens, Autour du pape libère, 327 ff.). The cubicle of Leo (4th c.) at Commodilla shows a lamb multiplying the loaves in place of Christ (Ferrua, *Scoperta*, 35), the *Agnus Dei* between two lambs (Peter and Paul) (*ibid.*, 37), a dove with monogrammatic *cross accompanied by **alpha* and *omega* (*ibid.*, 36) and a dove with *nimbus on a mountain (*ibid.*, 10) facing 12 other doves (the *apostles) (*ibid.*, 11). The 12 doves surrounding a triple *monogram on the vault of the baptistery of Albenga also allude to the apostolic college (Sciarretta, *Il battistero*, fig. 10).

LCI 3, 394-407; H. Stern, *Le calendrier de 354*, Paris 1953; G. Bovini, *Mosaici di Ravenna*, Milan 1957: A. Ferrua, Scoperta di una nuova regione della catacomba di Commodilla, *RAC* 34 (1958) 10 ff.; id., *Le pitture della nuova catacomba di via Latina*, Vatican City 1960; T. Klauser, Studien zur Entstehungsgeschichte der christlichen Kunst, *JbAC* 2 (1959) 115-145 and 3 (1960) 112-133; F. Darsy, *Santa Sabina*, Rome 1961, 81; M.C. Dagens, Autour du pape Libère. L'iconographie de Suzanne et des martyrs romains sur l'arcosolium de Celerina, *MEFR Ant.* 78 (1966) 327 ff.; F. Mian, La "vittoria" di Aquileia, *AAAd* 8 (1975) 131-153; R. Fairoli, *Ravenna roman e bizantina*, Ravenna 1977; V. Sciarretta, *Il battistero di Albenga*, Ravenna 1977; F.B. Sear, *Roman wall and vault mosaics*, Heidelberg 1977, 127-128; M. Guarducci, *Epigrafia greca*, IV, Rome 1978, 476-481; P. Testini, *Arte mitraica ed arte cristiana: Mysteria Mithrae*, Rome 1979, 443-452; G. Jeremias, *Die Holztür der Basilika S. Sabina in Rom*, Tübingen 1980; C. Murray, *Rebirth and afterlife*, Oxford 1981, 77-84; B. Domagalski, Der Hirsch in spätantiker Literatur und Kunst, in JbAC Ergbd. 15, 1990.

F. Bisconti

PETER

I. The ancient evidence - II. Legends of Peter - III. Veneration of Peter at Rome - IV. Veneration of Peter in the ancient world - V. Iconography.

The cult of the *apostle P. rests on presuppositions whose truthfulness must be tested: his coming to *Rome, his death and burial on the Vatican hill. Critical examination of these presuppositions clears the way for a demonstration of the different forms his cult assumed in the course of time.

I. The ancient evidence. From the chronological point of view, we must list the first evidence referring to P. as follows: *Clement of Rome, *1 Clem.* 5-6: AD 95-96; *Tacitus, *Ann.* 15, 44: 98-117; Jn 21, 18: before 100; *Ascension of Isaiah* 4, 25, 8: 100-125; *Ignatius of Antioch, *Rom.* 4, 3: 24 Aug 107; Suetonius, *Vita Claudii*, 25, 4: c.120; *Dionysius of Corinth (Euseb., *HE* II, 25, 8): c.166-174; the Roman priest *Gaius (*ibid.* II, 25, 6-7): 199-217. This leaves out *Papias of Hierapolis, whose evidence is ambiguous and disputable. From all the others we can draw the following conclusions: *1)* in his Epistle to the Romans *Paul says nothing about P.'s stay at Rome, which means that when he wrote, P. was not there at that time. None of the other texts oblige us to affirm or deny any such sojourn; *2)* on P.'s *martyrdom there is a whole *tradition going back to the end of the 1st c., through John, Clement of Rome, the *Ascension of Isaiah* and Dionysius of Corinth; *3)* on the circumstances of his martyrdom we must distinguish: the place, Rome, is apparent from Clement and the *Ascension of Isaiah*; as for date, the reign of *Nero. Clement and Tacitus agree on the great number of the first Christian victims, and seem to refer to the same event. The manner of martyrdom, crucifixion, is indicated by John; *4)* only one text, Gaius as related by *Eusebius, specifies his burial. This text supports the "red wall" in the *Vatican necropolis as the place where P.'s body rested, and where the *faithful periodically went to carry out the traditional commemorative rites in his honour. These explain how it was possible to preserve the *memory of P.'s burial-place.

II. Legends of Peter. From AD *c.*200 texts appear, anonymously or under assumed names, claiming to supplement the *canonical writings by accounts of P.'s teaching, actions and career: *1)* the *Doctrine of P.* (early 2nd c.); *2)* the *Kerygma of P.* (2nd c.): these first two texts confine themselves to P.'s preaching in the *East, and mention no connection with Rome; *3)* the *Acts of P.*: the primitive form is known only from fragments and some citations, which suggest a date of *c.*200. A first part refers to P.'s acts at *Jerusalem, and contains the two episodes of the apostle's daughter and the gardener. The second contains the struggle against *Simon Magus and P.'s crucifixion, head down. This primitive recension was used in the Ps.-*Clementine writings (early 3rd c.); the definitive form is contained in the 4th-c. Latin *Acts* of Vercelli. This last redaction is of Roman origin and offers some points of comparison with the topography and iconography of *Rome. The *Quo vadis?* story appears, but neither the place of martyrdom nor that of burial are specified; *4)* the so-called *Passion* of Ps.-Linus (BHL 6655) (4th c.). This takes the *Quo vadis?* from the earlier text, but adds P.'s imprisonment in the Mamertine prison, the episode of the *titulus Fasciolae*, and locates the apostle's death "in the *naumachia* near Nero's obelisk on the hill". It too is based on 2nd/3rd-c. Vatican traditions, but it has a rather unrealistic vision of things and says absolutely nothing about P.'s tomb; *5)* the *Acts of P. and Paul* by Ps.-Marcellus (BHL 6657, chs. 22-88) (*c*.400). This is a further development of the legend, still based on the earlier traditions. New elements are Pontius *Pilate's letter to *Claudius, the localizing of Simon's flight, the print left by the apostle's knees when praying on the Via Sacra *iuxta templum Romae*, P.'s burial "under the terebinth near the Vatican *naumachia*"; *6)* the *Passion of P. and Paul* by Ps.-Hegesippus or Ps.-Ambrose (BHL 6648), compiled *c*.580. It adds nothing new. The martyrdom is dated 29 June 57 on the basis of the *Latin Chronicle* of 553; *7)* the *Martyrdom of P.* by Ps.-Abdias (BHL 6663) (6th c.): another compilation made using numerous pre-existent elements, e.g. Ps.-Hegesippus. In conclusion, in their ever more frequent and ever less trustworthy references (from no. 3 on) to Roman and then Vatican topography, these legendary texts reveal the continuity of a local tradition connected with Gaius's monument and P.'s burial as well as the ever more fantastic and inexact sayings which circulated about the circumstances of P.'s sojourn and martyrdom at Rome.

III. Veneration of Peter at *Rome. In the *Depositio martyrum*, on 29 June, we read: *III kal. iul. Petri in Catacumbas et Pauli Ostense Tusco et Basso cons.* From this it appears that from 258 the cult of P. at Rome was celebrated in the catacomb of S. Sebastiano, and that of *Paul on the Via Ostia, on 29 June. Graffiti on the walls of the *triclia* (cf. **Memoria apostolorum*) of the catacomb, dating from the second half of the 3rd c., associate Paul with the cult of P. and specify the nature of that cult: the celebration of **refrigeria* in honour of the two *apostles (cf. **Cathedra: C. Petri*). Neither from the graffiti nor from the *Depositio* can we discover the cause of this cult at S. Sebastiano, but two different hypotheses have been suggested. Duchesne proposed that the bodies were translated to the catacomb at the time of *Valerian's *persecution. This theory is back in fashion and found apparent confirmation in the excavations in the Vatican necropolis, but subsequent discussions showed that the translation of P.'s relics to S. Sebastiano remains a hypothesis. The second theory was H. Delehaye's: he explained the creation of this place of worship by the impossibility of celebrating P.'s *memoria* at the Vatican during Valerian's persecution (257-259); acc. to Delehaye, only the cult, not the relics, was transferred: since *refrigeria* near the tomb had become impossible, they were moved to S. Sebastiano, near a cenotaph. This hypothesis was recently backed up by M. Guarducci. At any rate, it is at the Vatican, not at S. Sebastiano, that *Constantine built P.'s great basilica, apparently begun in 324; the reasons are obvious. However, *c*.350, a *basilica apostolorum* was raised at S. Sebastiano, dedicated to the conjoint worship of the two apostles. Dedicated specifically to P. was the basilica on the Esquiline where the apostle's chains were venerated from at least 431, when *Sixtus III tried, in vain, to join Paul's *patronage to that of P. The second family of MSS of *Mart. hier.* consider the basilica as the first church founded and dedicated by P., but this novelty cannot go back to that *martyrology's Italic recension. Another ancient church of P. was the chapel marking the place where P. and Paul prayed, when *Simon Magus crashed onto the Via Sacra. Paul I (757-67) restored it.

To these buildings, the work of emperors and popes, we can add the evidence of private citizens preserved in the *iconography of the catacombs and *sarcophagi, in gold-glass, medallions, and other objects of the "minor arts". Usually popular in character, they come mostly from Rome. Their subjects obey two fundamental tendencies: either P. and Paul appear together, with Christ (personified or symbolized) or without him, with or without the apostolic college, their attitude similar (both either sitting or standing) or differentiated by their attributes or their relation to Christ (only P. appears in scenes of *Traditio Legis et Clavium*); or else P. is represented alone, at various moments of his life and legend (e.g. the cock scene, the water springing from the rock, P. reading or being led to execution).

In this manner was expressed a faith to which the liturgical formularies give a more theological and somehow official expression. The formularies too invoke P. and Paul together: when they speak of the "apostles" without further specification, they always mean just the two *apostles of Rome: together they founded the church of Rome with their *blood, and they continue to instruct it and rule it in the person of their successor, the *pope; their conjoint *authority and their care extend to the universal church, of which they are the foundation, the leaders and the coping-stone. Paul as the apostle to the nations (in parallel with the progress of Christianization, the primitive meaning of *gentes*, = *pagan, disappears in favour of its universal meaning), P. as shepherd of the flock. But the two titles also express each one's different relationship to Christ: although they form "a pair to whom equal veneration is due", *par mundo venerabilis*, Paul remains "the last of the apostles", since he was the last of them to be called to the faith; and P. "the first", since he was the first to confess it and to receive from Christ, with the primacy, the office of universal pastor, which the popes inherited. Wherever we look, then, in the religious topography of the city, in iconography, under the most varied forms, in liturgical *prayer itself, in every field appears this fundamental dual tendency in Rome's veneration of P.: on one hand he is the inseparable companion of Paul, with whom he symbolizes the inner unity of the Roman church born *ex circumcisione et ex gentibus*; on the other, he is the head of the apostolic college and, consequently, of the church of Rome and the universal church.

IV. Veneration of Peter in the ancient world. Among the Roman feasts, only that of 29 June spread everywhere, but with the peculiarity that in the *East the date was adapted to local circumstances. In *Mart. syr.* it is on 28 Dec, in the Coptic *calendar of Alexandria on 5 Epiphis (11 July). In both cases P. is considered the *coryphaeus* or leader of the *apostles. The feast of 22 Feb (**cathedra Petri*) was unknown in *Africa (*Cal. Carth.*), and in *Gaul it competed with that of 18 Jan. This gave rise to the compromise of the Hieronymian and Gallican *Martyrologies: 18 Jan, cathedra of St Peter at *Rome; 22 Feb, cathedra of St Peter at *Antioch. The Calendar of *Perpetuus of Tours (†490) is ambiguous on the date of this feast, and the Hispanic calendars always accepted only that of 22 Feb.

The movements of *pilgrims crystallized round these feasts: esp. 29 June drew them to Rome to venerate the apostle's tomb, and they came from everywhere: as witness *Ambrose (hymn *Apostolorum passio*), *Prudentius (*Perist.* 11, 1-2, all 12), *Paulinus of Nola (*Epp.* 17-18, 29, 35), *Gregory the Great (*Hom. in Ev.* 37, 9). This movement of people was accompanied by the circulation of ideas. The Roman ideology of P. is echoed by *Asterius of Amasea (PG 40, 2640), *John Chrysostom (PG 60, 678-679; 62, 57), *Maximus of Turin (CCL 23, 1-8, 30-33, 426-28). Among the most copious preachers on St P. is *Augustine (*Serm.* 295-99, 381; PLS 2, 462 ff., 598-608, 756-758); among the most important is *Theodoret of Cyrrhus, who took up the ideas of *Leo the Great (PG 83, 1311-1313).

It comes as no surprise to see that pilgrims and others spread the apostle's relics and set up *churches in his name in their own lands. They were usually representative relics, or filings from his chains. In Africa we find them from 359 at Kherbet-Oum-el-Ahdam, near *Sétif (Alg., CIL 8, 20600), in *Spain from the 4th-7th cc. (J. Vives, *Inscr. crist.* 199, 316, 335, 374, 389, 513). Churches of the apostle existed in *Gaul, at *Tours at the time of Perpetuus (†490), at *Vienne in the 5th c., at *Arles at the time of *Caesarius (†543), St-Pierre-de-Mouleyrès and St-Pierre-de-Gallègue. At Paris, the basilica of Sts P. and Paul was the burial-place of *Clovis (†511), *Clothilde (†545) and St *Genoveffa (†*c*.500), who ended by giving the church her own name; at the end of the Merovingian era, there were three more churches at Paris dedicated to the apostle: St-Pierre d'Arcis, St-Pierre-des-Boeufs and St-Pierre-des-Bois. A basilica of St P. existed at Corbie, thanks to Clothilde and then Clotaire (†561). At Nantes (*Namnetes*), the church of Sts P. and Paul existed in 567; at Rouen (*Rotomagus*), the church of St P. took the name of St Ouen on the latter's death (†684). The Roman missionaries sent by *Gregory the Great to *Britain spread the cult of the apostle: the church of Sts P. and Paul at London was founded by *Augustine of Canterbury (end of 6th c.); St P. of Bradwell on Sea was built after 653, St P. of Wearmouth in 674. *Boniface in turn planted the cult of the apostle in *Germany, but that takes us beyond the scope of this book.

L. Duchesne, La memoria apostolorum della via Appia, *Mem. Pont. Acc. Rom. Arch.* 1 (1923) 1-22; H. Delehaye, Le sanctuaire des apôtres sur la voie Appienne, *AB* 45 (1927) 294-304; H. Lietzmann, *Petrus und Paulus in Rom*, Bonn 1927; AA. VV., *Esplorazioni sotto la confessione di S. Pietro in Vaticano*, 2 vols., Vatican City 1951; O. Cullmann, *S. Pierre, disciple, apôtre, martyr*, Neuchâtel 1952; A. Prandi, *La zona archeologica della confessione Vaticana*, Vatican City 1956; M. Sotomayor, *S. Pedro en la iconografia paleocristiana*, Granada 1962; Erbetta, II; AA. VV., *Saecularia Petri et Pauli*, coll. Studi di antichità cristiana 28, Vatican City 1969; D. O'Connor, *Peter in Rome. The Literary, Liturgical and Archaeological Evidence*, New York 1969; A.G. Martimort, Vingt-cinq ans de travaux et recherches sur la mort de S. Pierre et sur sa sépulture, 1946-1971, *BLE* 73 (1972) 73-101; E. Kirschbaum, *Die Gräber der Apostelfürsten*, Frankfurt/M.

²1974; M. Maccarrone, *Apostolicità, episcopato e primato di Pietro*, Rome 1976; U. Fasola, *Pietro e Paolo a Roma*, Rome 1980; M. Guarducci, *Pietro in Vaticano*, Rome 1983; J.M.C. Toynbee-J. Ward Perkins, *The Shrine of St. Peter and the Vatican Excavations*, London 1956; J.D. Kingsbury, The figure of Peter in Matthew's Gospel as a theological problem, *JBL* 98 (1979), 68-83; M. Hengel, *Studies in the Gospel of Mark*, Philadelphia 1985, esp. 50-63; T.V. Smith, *Petrine controversies in early Christianity*, Tübingen 1985.

V. Saxer

V. *Iconography. P. is certainly the character most frequently represented in early Christian iconography, a fact that corresponds perfectly to the equally numerous citations of him in the gospels. Going beyond the useful statistical surveys of the number of times P. appears on figurative monuments (fully summed up by E. Josi), thanks to recent studies we can say that the physiognomical definition of P. in the figurative arts does not appear before the mid 4th c.

Before that date, P. was depicted in a few scenes. Among the oldest examples in which he can certainly be identified is the baptistery of *Dura Europos (256), where the symbolic value linked to the *hope of salvation clearly appears in the episode of walking on the water.

With the gradual spread of scenes showing Christ among the *apostles, P., though not yet clearly individualized by precise facial traits, certainly had a role of primary importance. At the end of the 3rd c. this composition developed almost in antithesis to the image of the emperor surrounded by *apparitores*, a theme widely used later.

The characteristics which would later distinguish P.'s *portrait (abundant hair, short beard, broad forehead) took shape in the period immediately following, from the religious *peace (313) through the whole first half of the 4th c., in parallel with the birth and development of numerous other new scenes that entered the figurative repertoire of Christian compositions.

The so-called Petrine cycle was formed, with episodes relating to P.'s life and works: the moment of the denial (occurring at least 70 times), where P. generally occupies the centre with the cock, key to the reading of the episode; equally common is P. between two soldiers arresting him; also recurrent is the miracle of the spring, where P. is assimilated to *Moses, confirming what Augustine wrote at more or less the same time (*Serm*. 351, 4: [*Moses*] *figura fuit Petri*). These three moments of P.'s career are frequently shown together as parts of a single trilogy.

With facial characteristics now defined, P. is often associated with *Paul on either side of Christ, in *sculpture, *painting, *mosaic (cf. also *Traditio legis et clavium*): Christ, now *imperator* as well as *magister*, is with his court, and both are symbolically contrasted with the *imperatores* and their earthly courts. Scenes of this kind mark the evolutionary steps of Christian representations in the 4th c., at which time they became vehicles of ideological opposition to the power of certain social categories. From the depiction of the apostle with no precise iconography in scenes containing a message of hope in salvation, we come to the iconographical definition of P. in a strongly symbolic context containing, alongside the primitive message, traits of the the one or another of the ideologies of the moment.

Though numerous pictorial or mosaic examples exist, a good many of the momuments through which we can read the Petrine cycle are *sarcophagi (a fact which reveals a certain clientèle). P.'s career, his miracles or simply himself and Paul are common in the "minor arts": there is an enormous amount of incised glass or gold-glass on which the old heroes of classcal *mythology are perfectly replaced by the new heroes of the ideology of the ruling class, which set the fashion. P. and the miracle of the spring are on a glass cup from Podgoritza: P. and Paul wish health to the users of gilded glasses or cups. **[Fig: 254]**

DACL 14, 822-891 (esp. 935 ff.); *EC* 9, 1417-1420; *EAA* 6, 162-167; *LCI* 8, 158-174; C. Cecchelli, *Iconografia dei Papi*, I, *S. Pietro*, Rome 1937; G. Ladner, *I ritratti dei Papi nell'antichità e nel medioevo*, Vatican City 1941; A. Grabar, Le portrait en iconographie paléochrétienne, *RSR* 36 (1962) 87-109; M. Sotomayor, *S. Pedro en la iconografía paleocristiana*, Granada 1962; P. Testini, Gli apostoli Pietro e Paolo nella più antica iconografia cristiana, in *Studi Petriani*, Rome 1968, 105-130. For sarcophagi: G. Wilpert, *I sarcofagi cristiani antichi*, Rome 1929-1936; F.W. Deichmann-G. Bovini-H. Brandenburg, *Repertorium der christlich-antiken Sarkophage*, I, *Rom und Ostia*, Wiesbaden 1967. For paintings in the Roman catacombs: A. Nestori, *Repertorio topografico delle pitture delle catacombe romane*, Roma Sotterranea Cristiana V, Vatican City 1975; J.M. Huskinson, *Concordia Apostolorum. Christian Propaganda at Rome in Fourth and Fifth Centuries. A study in Early Christian Iconography and Iconology*, London 1982; W. Wischmeyer, *Die Tafeldeckel der christliche Sarkophage konstantinischer Zeit in Rom*, Freiburg i. Br. 1982; C. Pietri, *Roma christiana*, BEFAR 224, 2 vols., Rome 1976, esp. 227-282, 327-336 and 1413-1450; E. Dassmann, Die Szene Christus-Petrus mit dem Hahn, in *Pietas* (Festsch. B. Kötting), JbAC Ergbd. 8, Münster 1980, 509-527.

U. Broccoli

PETER. First bishop of Parembolae, a military station in *Palestine († before 451). In his *Life of St Euthymius*, *Cyril of Scythopolis tells the legendary story of P.'s life up to his episcopate (*Vita s. Eut.* 18-21). Of Greek stock, P. was a senior officer of the Persian empire and bore the name *Aspebetus*. During the last persecution (late 4th c.) inspired by the Magan party, he was ordered to occupy the mountain passes so as to capture Christians fleeing to the Roman empire. He soon started helping the Christians and was himself forced to flee, with his son Terebus and his parents. Anatolius, prefect of the *East, welcomed him and made him governor of the Bedouin vassal tribes. Terebus, who was paralytic, was told in a dream that St *Euthymius could cure him. After the miraculous cure, *Aspebetus* was baptized with all his family, took the homophonous name *Petrus*, and dedicated himself to the religious life. At the wish of the Bedouin converts, patriarch *Juvenal of Jerusalem ordained him their bishop, against the jurisdictional rights of the metropolitan see of Palestinian *Caesarea. The precise date of his *ordination is unknown, but was before 428 (Tillemont, *Mem. Eccl.* XV, 196). P. took part in the council of *Ephesus where, following Euthymius's advice, he sided with *Cyril of Alexandria. He was one of the four bishops who summoned *Nestorius to appear before the *council to justify himself (Mansi 4, 1132; ACO I, 1, 2, 9), and one of the three who informed the latecomer *John of Antioch of the decisions taken (ACO I, 1, 2, 18). P. condemned Nestorius (Mansi 4, 1149; ACO I, 1, 2, 20; Mansi 4, 1176; ACO I, 1, 2, 34; Mansi 4, 1219). On his way back, he went to St Euthymius to inform him of the decisions taken at the council. We know nothing of the rest of his life.

DCB 4, 344, n. 35; ACO IV, 3, 2, index 388; L. Perrone, *La chiesa di Palestina e le controversie cristologiche*, Brescia 1980, 75.

G. Ladocsi

PETER. Bishop of Myra, third quarter of 5th c. One of the bishops to whom the emp. *Leo I (457-474) sent an encyclical *letter in 458 to restore *peace and *orthodoxy in *Egypt after the council of *Chalcedon. We know two works by him: *Contra Apollinarem* (Diekamp, *An. Patr.*, p. 50) and a letter *Ad Leonem imperatorem* (ACO II, 5, 60-63).

CPG 3, 6157; F. Diekamp, *Analecta patristica*, OCA 117, Rome 1938, 50-53; *DCB* 4, 344.

G. Ladocsi

PETER. Patriarch of *Jerusalem 524-552, succeeding *John II. A great devotee of the ascetic St *Sabas, whom he sent to *Constantinople to plead with *Justinian to suspend the taxes of the Christians of *Palestine, reduced to misery by the insurrection of the *Samaritans. When pope Agapitus deposed *Anthimus, the *monophysite patriarch of Constantinople, and replaced him with *Menas (536), P. subscribed the condemnation. Because of disorders caused by the *Origenist monks of Palestine, supported by *Domitian of Ancyra and *Theodore Ascidas of Caesarea in Cappadocia, P. appealed to Justinian, who issued an edict obliging bishops and *abbats to subscribe *Origen's condemnation (543). Later, however, P. and abbat Gelasius of Mar Saba refused to subscribe Justinian's edict condemning the *Three Chapters: only after being summoned to Constantinople did he consent to sign, and was forced to accept two Origenist *syncelli*. He died in 552. A Christmas *homily of his survives in a *Georgian version, and a fragment on *fasting. One of the main sources for his life is *Cyril of Scythopolis's *Vita Sabae*.

CPG 7071 (Christmas homily): I. Abuladaze, *Mravalthavi*, Bulletin de l'Institut Marr de langue, d'histoire et de culture matérielle 14, Tiflis 1944, 307-316; CPG 7018 (fragment on fasts): PG 95, 76B (in Jo. Damasc., *De sacris ieiuniis*); *DCB* 4, 343 f.; Hfl-Lecl 3, 1, p. 17; L. Duchesne, *L'Eglise au VIᵉ siècle*, Paris 1925, 207-208; P. Batiffol, Justinien et le Siège apostolique, *RecSR* 16 (1926) 236-237; Fliche-Martin 4, 584, 603; L. Perrone, *La chiesa di Palestina e le controversie cristologiche*, Brescia 1980, *passim*.

F. Scorza Barcellona

PETER I of Alexandria. Elected bishop of *Alexandria in 300, after having directed the *didaskaleion*. Imprisoned during the great *persecution, freed around 306, rearrested and beheaded *c*.311. During the years of persecution he had to cope with *Meletius of Lycopolis who, not sharing P.'s moderation towards *lapsi* who wished to return to the church, replaced imprisoned bishops with adherents of his own ideas. P. had Meletius condemned and deposed by a *council held in the interval between his two imprisonments, thus beginning the Meletian *schism. We have a brief *letter of P. from prison, putting his congregation on guard against the prevarications of Meletius. Of his various works we know the titles and some fragments: *On divinity*, which dealt with Christ's divinity; *On the coming of our Saviour*, perhaps the same as the previous work; *On *the soul*, against the Platonist theory, taken from *Origen, of the preexistence of the *soul; *On the *resurrection*, perhaps also a polemic agenst Origen, whose doctrine on the spiritual condition of the risen human *body he opposed; a letter, prob. festal, *On penitence*, of which 14 disciplinary canons survive relating to the sanctions to be inflicted on *lapsi* who requested readmission to the church. Perhaps the treatise *On Easter* dedicated to Tricenius was also a festal letter.

P.'s polemic against some of Origen's doctrines has led some modern scholars to postulate a wide anti-Origenian reaction in the Alexandria of his time, which even involved Origen's exegetical principles. In fact it was simply a stand against Origen's most disputed doctrines, since Origen's *exegesis and *theology had a decisive influence at Alexandria even after P.

PG 18, 449-522; Quasten I, 378-382 (bibl.); *DCB* 4, 331-334; L.B. Radford, *Three Teachers of Alexandria: Theognostus, Pierius and Peter*, Cambridge 1908; M. Simonetti, Le origini dell'arianesimo, *RSLR* 7 (1971) 317-330; T. Vivian, *St. Peter of Alexandria, Bishop and Martyr*, Philadelphia 1988; C.W. Griggs, *Early Egyptian Christianity*, Leiden 1990.

M. Simonetti

PETER II of Alexandria. Priest of *Alexandria, designated as his successor by the dying *Athanasius (373), but the emp. *Valens imprisoned him and substituted the *Arian *Lucius. P. managed to escape and found shelter with pope *Damasus at *Rome, where he worked in favour of *Paulinus in the schism of *Antioch and against *Basil's attempts to unify the anti-Arians in the *East. He returned to Alexandria shortly before Valens's death (379) and in 380 tried unsuccessfully to install *Maximus instead of *Gregory Nazianzen as bishop of the anti-Arian community of Constantinople. Died before the start of the council of *Constantinople of 381.

CPG 2, 2515-2517; *DCB* 4, 334-336.

M. Simonetti

PETER II of Sebaste (*c*.344-*c*.394). Brother of *Basil the Great, born while his father was dying, received an ascetic education from his sister *Macrina (Greg. Nyss., *Vita Macr.*, chs. 12-13). A priest, he prob. lived at Annesi near *Neocaesarea (Basil, *Ep*. 216) and acted as intermediary with the bishops of *Pontus (*Ep*. 203, end). He was elected to the see of Sebaste before May 381, in opposition to the recently-dead *Eustathius. *Gregory of Nyssa dedicated to P. his openly *Origenist *De opificio hominis*, and wrote *Ep*. 29 to him; *Ep*. 30 is P.'s reply (PG 45, 241-244; ed. G. Pasquali, Leiden ²1952, 89-91). He figures in a legendary *letter on Basil's miracles (Ps.-*Amphilochius, CPG 3253).

P. Devos, S. Pierre I évêque de Sébastée dans une lettre de Grégoire de Nazianze, *AB* 79 (1961) 346-390; *BS* 10, 771-772.

J. Gribomont

PETER Apselamos or Balsamos. Native of Anea (Judaea), acc. to *Eusebius (*Mart. Pal*. 10, 2). He refused, despite his youth, to *sacrifice to the gods and was burnt with the *Marcionite bishop Asclepius. A Latin Passion (BHL 6702), which reproduces a lost Greek original, gives him the surname Balsamos and says he was crucified at Aulana (Samaria). In fact he died at *Caesarea in Palestine, 11 Jan 309.

Vies des SS. 1, 50-53; *LTK* 8, 347; *BS* 10, 791-92.

V. Saxer

PETER Chrysologus. The life of P., *archbishop of *Ravenna, known from the 9th c. as Chrysologus, is still obscure. From the *Liber Pontificalis* and a 9th-c. *biography, as well as certain passages of the works attrib. to him, we learn that he was born at Imola (*Forum Cornelii*) *c*.380. Between 425 and 429, certainly before 431, he became metropolitan of Ravenna. In 445 he was present at the death of St *Germanus of Auxerre. Three or four years later he wrote to *Eutyches, presbyter of *Constantinople, who had appealed to him after his condemnation by *Flavian. P. advised him to submit to the decisions of *Leo, bishop of *Rome and successor of *Peter. He died between 449 and 458 (date of a letter of Leo to his successor, Neones), prob. 3 Dec 450, perhaps at Imola.

Today we recognize as authentic works of P. a *letter, 168 *sermons of the 8th-c. *Collectio Feliciana* and 15 other sermons. Other writings, like the famous Ravenna Scroll, a 7th-c. collection of *prayers, cannot be considered authentic. The preaching that made P. famous is marked by the careful preparation of a well-educated rhetor, by human warmth and the divine fervour of a holy man. By its examples, it reflects the particular situation of Ravenna, an imperial residence, important port and agricultural centre. It consists mostly of *homilies on gospel passages, but also on *Paul's epistles, the *psalms, the baptismal *creed, the *Lord's Prayer and the saints, as well as exhortations to repentance.

Chrysologus, commenting on the *Bible and taking cues from the liturgical celebrations, is an authoritative witness to the theological concerns of his time: the Latin doctrine of the *incarnation, *Catholic positions on *grace and Christian living, even recognition of the primacy of the bishop of Rome. His vast preaching activity has left us, above all, an inestimable documentation of the *liturgy and culture of Ravenna, situated between Rome and N *Italy. No other bishop of the time gives us such a complete picture of the *liturgical year. Opposing the resistance of moribund *paganism and polemizing against the *Jews of his city, P. represents the pastoral attitude of the bishops of the imperial church of his day.

Editions: CPL 227-237; PLS 3, 153-183; 5, 399 f.; PL 52, 183-666: *Coll. Feliciana*; A. Olivar, crit. ed.: CCL 24 (1975), I: *Serm*. 1-62a; *Ep. ad Eutychen*: inter ep. Leonis M. (= CPL 229) 25; PL 54, 739-744=ACO II/3 1 (1935) 6-7 and ACO II, 1, 2 (1933) 45 f. (Greek text); Italian ed. A. Pasini, Siena 1953; German ed. G. Böhmer, *BKV*² 43, Kempten 1923.
Studies: A. Olivar, *Los sermones de san Pedro Crisólogo de Ravena*, Diss. Greg., Rome 1969; G. Lucchesi, Stato attuale degli studi sui santi della antica provincia ravennate, *Atti dei convegni di Cesena e di Ravenna* I, Cesena 1969, 51-80; F. Sottocornola, *L'anno liturgico nei sermoni di Pietro Crisologo*, Cesena 1973 (bibl.); M. Spinelli, *Pier Crisologo. Omelie per la vita quotidiana*, Rome 1978; id., Il ruolo sociale del digiuno in Pier Crisologo, *VetChr* 18 (1981) 143-156.

B. Studer

PETER MONGUS. Patriarch of *Alexandria 477-490. Elected by the anti-Chalcedonian party on the death of *Timothy Aelurus, at first he was not recognized by the patriarchs of *Constantinople (*Acacius) and *Rome (*Gelasius), who both favoured *Timothy Salofaciolus. But on the latter's death, since his successor *John Talaias gave no political guarantees, Acacius convinced the emp. *Zeno to send a conciliatory *letter (the *Henoticon*) to Alexandria. The result was a reconciliation between Acacius and P., though differences remained over the theological interpretation of the text of the *Henoticon*. This reconciliation displeased many of the anti-Chalcedonian Egyptian *clergy (*Acephali), as well as Rome, which separated from Acacius (*Acacian schism). But P.'s work enabled his anti-Chalcedonian successors to rule the see of Alexandria without rivals, until the time of *Justinian and patriarch *Theodosius.

W.H.C. Frend, *The Rise of the Monophysite Movement*, Cambridge 1972, esp. 174-183.

T. Orlandi

PETER of Altinum. Bishop of Altinum, sent by *Theodoric in 501 to administer the church of *Rome while awaiting the outcome of the dispute between *Symmachus and *Lawrence. P. discharged his office in a way unfavourable to Symmachus and was later condemned by him in a synod of 115 bishops (Mansi 8, 245-248; 273-290; Anast. Libr., *Hist. di vitis rom. pont*. 53: PL 128, 451).

E. Prinzivalli

PETER of Apamea. Follower of the monophysite *Severus, patriarch of *Antioch (512-518, †537/8), around 510 (at the time of the emp. *Anastasius) he became bishop of *Apamea (metropolis of Syria Secunda) against the *canonical rules. He cancelled the names of his *orthodox predecessors from the *diptychs and led a scandalous and violent life (*turpem ac omni abominatione plenam*: *Libellus monachorum*, Mansi 8, 1129). On the restoration of orthodoxy by *Justin (518-527), P. was *excommunicated by the council of *Tyre (16 Sept 518; Mansi 8, 577 and 1093 ff.; Hfl-Lecl 2, 1049 f.) and had to leave his see. *Justinian, at the start of his reign, tried to bring the *monophysites into the unity of the church. They were protected by *Theodora, who also managed to bring the monophysite-leaning *Anthimus into the see of *Constantinople. Only by the intervention of pope Agapitus in 536 was Anthimus removed and replaced by patriarch *Menas. A *council was immediately called, and Severus, Peter and Zoara, whose intense pro-monophysite propaganda during their stay at Constantinople had caused a scandal, summoned to appear. They were condemned by the council and declared deposed (Mansi 8, 975-1156); a second condemnation followed at the council of *Jerusalem of 536 (Mansi 8, 1163-1176: HFL-Lecl 2, 1142-1155). Justinian ratified the conciliar decisions with *Novella* 42.

DBC 4, 340 f. (n. 12); Bardenhewer 5, 22; *DHGE* 3, 920; Fliche-Martin 4; Beck 54 and 379.

A. De Nicola

PETER of Callinicum. Also known as Petrus minor; *Jacobite patriarch of *Antioch 581-591. Famed in the history of *dogma for his christological dispute with *Damian, patriarch of *Alexandria (578-605), against whom he wrote a long treatise in four books, each of 25 chapters. Certainly written in Greek, it survives only in *Syriac tr., and not entire. If we may believe the Syriac MSS, P. also wrote an *Anaphora, a treatise against *tritheists, *letters and a pentasyllabic metrical *homily on the Crucifixion.

Duval 365-366; Baumstark 177; Chabot 78-79; Ortiz de Urbina 165; R.Y. Ebied-L.R. Wickham, The discourse of Mar Peter Callinicus on the Crucifixion, *JThS* n.s. 26 (1975) 23-37; R.Y. Ebied, Peter of Antioch and Damian of Alexandria: the end of a friendship, *A Tribute to A. Vööbus*, Chicago 1977, 277-282; A. van Roey, Une controverse christologique sous le patriarcat de Pierre de Callinique, in *Symp. Syriacum* 1976, OCA 205, Rome 1978, 349-357; R.Y. Ebied-A. von Roey-L.R. Wickham, *Peter of Callinicum, Anti-Tritheist Dossier*, Leuven 1981.

J.-M. Sauget

PETER of Trajanopolis. Bishop of Trajanopolis (*Thrace). Took part in the council of *Ephesus in 431, supporting the positions of *John of Antioch. Condemned, with John and his supporters, by the *council (ACO I, 1, 3, pp. 13, 25, 26), he dedicated to it a *libellus* in which, rejecting his earlier behaviour, he made a professon of *orthodoxy and repudiated *Nestorius.

CPG 3, 5799; ACO I, 1, 7, 139 (text of the *libellus*); PWK 19, 1327.

J. Irmscher

PETER the Deacon. Cardinal deacon of *Rome, disciple of St *Gregory the Great, under whom he led a monastic life, after giving his goods to the *poor. Gregory made him administrator of church *property in Sicily, then in Campania: their *letters, frequently exchanged (cf. Greg., *Regist. Epist.*: PL 77 442-1327 *passim*), show the *pope's concern that the profits should go to the poor, esp. of Rome. Returning to the city, after Gregory's death P. became the most tireless defender of that pontiff, accused by some ill-wishers of having squandered the goods of the church of Rome, even if on charity. He appears as protagonist in Gregory's *Dialogues* (PL 77, 150-430). There is a letter in his name entitled *Liber Petri diaconi et aliorum qui in causa fidei a Graecis ex Oriente Romam missi fuerunt* (it affirms the doctrine of two natures and one *person in Christ, following the belief of *Chalcedon [451] and *Cyril of Alexandria).

CPL 663 and 817; PL 45, 1772-76; 62, 83-92; 65, 442-451 (among the *Letters* of St Fulgentius of Ruspe, no. 16); John the deacon, *Vita S. Gregorii*, II, 11: PL 75, 92; A. Grillmeier, in *Das Konzil von Chalkedon*, II, Würzburg 1953, 799-800, nn. 13-14; A. Mancone, *EC* 9, 1435; G. Lucchesi, *LTK* 8, 360; Quasten II, 228.

L. Dattrino

PETER the Fuller. First *monophysite patriarch of *Antioch (†488). The sources do not allow a sure reconstruction of his life, but it seems that P., originally an *Acoemete monk at *Constantinople, left the *monastery, perhaps because of theological disputes with his Chalcedonian brethren. On good terms with *Zeno the Isaurian, son-in-law of the emp. *Leo I, he followed him to Antioch when Zeno was made *magister militum per Orientem*. Here P. stirred up opposition to patriarch *Martyrius, whom he briefly succeeded when he was forced out (471). Deposed and exiled by Leo, he returned a second time to the throne of Antioch on the occasion of *Basiliscus's revolt (475-477). Expelled again, he finally returned (485-488) thanks to the emp. Zeno's policy of conciliating the monophysites (the *Henoticon*).

A controversial figure, hard to evaluate precisely, P. left important traces of himself in some liturgical innovations. The most important and disputed was his addition to the *Trisagion* of the *theopaschite formula ὁ σταυρωθεὶς δι' ἡμᾶς during his first episcopate. It was the object of lively controversy in the 6th c., as apocryphal *letters to P. preserved in the *Collectio Sabbaitica* attest. In 476 he began to recite the *Creed during *mass, and he also introduced the rite of consecration of ointment (μύρον) and *blessing of *water on *Epiphany night. It is above all some of these liturgical aspects which have suggested P.'s identification with the author of the *Corpus Areopagiticum* where, among other things, the consecration of the μύρον is raised to the rank of a *sacrament (*Ecc. h.* IV: PG 3, 472 D-485 B). The theory, recently put forward again by Riedinger, has not been accepted, but it is admitted that the writings of Ps.-*Dionysius reflect a liturgical context close to that of P.

CPG 6522-6525; Zacharias the Rhetor, *HE* V, 10: CSCO 83 (tr.) 233-235, CSCO 87 (tr.) 161-162; ACO III, 6-25, 217-231; E. Schwartz, Publizistische Sammlungen zum Acacianischen Schisma, *ABAW* NF 10 (1934) 182-183, 192-193, 210, 287-300; E. Honigmann, *Evêques et évêchés monophysites d'Asie Antérieure au VI* siècle* (CSCO 127), Louvain 1951; U. Riedinger, Pseudo-Dionysios Areopagites. Pseudo-Kaisarios und die Akoimeten, *ByzZ* 52 (1959) 276-296; id., Akoimeten, *TRE* 2, 148-153; W. Strothmann, *Das Sakrament der Myron-Weihe in der Schrift De Ecclesiastica Hierarchia des Pseudo-Dionysios Areopagita in syrischen Übersetzungen und Kommentaren*, Wiesbaden 1978; G. O'Daly, Dionysios Areopagita, *TRE* 8, 772-780; W.H.C. Frend, *The Rise of the Monophysite Movement*, Cambridge ²1979, 167-170, 188-190.

L. Perrone

PETER the Iberian (†491). So called as a native of *Georgia († *Iberia). One of the outstanding figures of the first *monophysite generation. Son of king Bosmyrios (his real name was Nabarnugi), he was sent aged 12 as a hostage to *Constantinople. Seized by ascetic zeal, in 437-8 he fled to *Jerusalem, where *Melania junior and *Gerontius welcomed him, clothed him in the monastic *habit and gave him the Christian name P. In 445 he settled at Maiuma near *Gaza, where he was ordained priest. Bishop of that city during the Palestinian rebellion at the council of *Chalcedon, in 453 he was exiled to *Egypt, where he later consecrated *Timothy Aelurus as patriarch of *Alexandria (457). Back in *Palestine, he consolidated the monophysite presence there. Linked with *Isaiah of Gaza, with him he avoided demands to adhere to the *Henoticon*. Despite his intransigent rejection of Chalcedon, P. was not a radical monophysite, as his condemnation of the Eutychian theories of John the Rhetor shows. He exercised great attraction over the intellectual circles from which *Severus of Antioch issued. His qualities as a visionary, attested esp. by the *Plerophoriae* of *John of Maiuma (Rufus), have led some to identify him with Ps.-*Dionysius the Areopagite.

R. Raabe, *Petrus der Iberer, ein Charakterbild zur Kirchen- und Sittengeschichte des 5. Jahrhunderts*, Leipzig 1895; F. Nau, *Jean Rufus, évêque de Maiouma. Plérophories*: PO VIII, 1, Paris 1912; E. Schwartz, *Johannes Rufus, ein monophysitischer Schriftsteller*, SHAW 16, Heidelberg 1912; E. Honigmann, *Pierre l'Ibérien et les écrits du Pseudo-Denys l'Aréopagite*: Mém. de l'Ac. Roy. de Belg., Cl. des lettr. 47, 3, Brussels 1952; P. Devos, Quand Pierre l'Ibère vint-il à Jérusalem?, *AB* 86 (1968) 337-350; W.H.C. Frend, *The Rise of the Monophysite Movement*, Cambridge ²1979, chs. 4 and 5.

L. Perrone

PETER the Patrician, also called Peter the Rhetor, Byzantine historiographer. Born at *Thessalonica, *c*.500, he practised as an advocate at *Constantinople and became the empress *Theodora's court *curator*. In 539/40 he became *magister officiorum* and, before 550, a patrician. In 535 he had conducted negotiations in *Italy with Theodahat, king of the *Goths; on that occasion he was imprisoned for three years. In 552 he was sent to pope *Vigilius who, in the course of the dispute of the *Three Chapters, had fled from Constantinople to *Chalcedon. In 561/2 he advantageously negotiated peace with the Persians. His diplomatic ability, vast culture and rhetorical talent aroused the admiration of his contemporaries. He died soon after his return from *Persia. Fragments of his *Histories* (Ἱστορίαι) are preserved in the collections of *excerpta de legationibus* and *de sententiis*, compiled by Constantine VII Porphyrogenitus (first half of 10th c.). The work was prob. divided by emperor's reigns and went from the 2nd triumvirate (43 BC) to at least the time of *Julian. His main source, cited textually in some places, is Dio Cassius. A second work was called *On the organization of the state* (Περὶ πολιτικῆς καταστάσεως). To it certainly belong chs. 84-85 of the first book and perhaps also chh. 86-95 of Constantine Porphyrogenitus's book on ceremonial. Attribution to P. of the palimpsest fragment Περὶ πολιτικῆς ἐπιστήμης, discovered by A. Mai (*Scriptorum veterum nova collectio*, 2, Rome 1827, 571-609), is disputed. In his reflections, P. manifestly wished to offer practical guidance for the development of a model of constitutional monarchy based on philosophical principles. A third work, completely lost, dealt, in popular language (a thing of considerable interest), with the Byzantine/Persian peace negotiations of 560/562.

PG 113, 663-676 (very incomplete); C. Müller, *Fragmenta historicorum Graecorum*, 4, Paris 1868, 181-199 (without the Constantinian *excerpta*); A. Mai, *loc. cit*., 197-246; L. Dindorf's edition of Dio Cassius, 5, Leipzig 1865, 181-232; PWK 19, 1296-1304; H. Hunger, *Die hochsprachliche profane Literatur der Byzantiner*, 1, Munich 1978, 300-303.

J. Irmscher

PETILIAN. Donatist bishop of *Constantine (*c*.395-412) and opponent of *Augustine. A jurist by training and a *Catholic, he was converted, at first unwillingly, to *Donatism (Aug., *Sermo ad Caesariensis ecclesiae plebem* 8); by Augustine's time he had become the main spokesman of the Donatist church. In 400-401 he wrote an encyclical, *Epistola ad presbyteros*, to his clergy, trying to prove that *Donatus's church was the true Catholic church in N *Africa. Augustine replied in 401 with his first two books *Contra*

Litteras Petiliani, to which P. replied with a work addressed personally to Augustine, *Ad Augustinum*, in which he claimed that his adversary had never ceased to be fundamentally a *Manichee. Augustine replied again with the last book *Contra Litteras Petiliani*. Later, in 410, P. wrote *De unico baptismo* in defence of the Donatist practice of rebaptizing converts; Augustine replied in *De unico baptismo contra Petilianum*. At the conference of *Carthage of 411, P. quickly emerged as the champion of Donatism: he tirelessly defended the cause with subtle oratory and, still more effectively, with brief and pungent comments on the arguments of the Catholics. He spoke nearly 150 times; towards the end, when the Donatist cause was clearly weakening, he tried to obtain an adjournment by claiming that a sore throat prevented him speaking (*Gesta Collationis* III, 541-542: CCL 159, 50). Deposed from his see after the proscription of Donatism (*CT* XVI, 5, 52), he still tried, where possible, to reunite the remaining Donatists and even convened a *council of bishops in *Numidia (Aug., *Contra Gaudentium* I, 37, 47-48). He disappeared from the scene after 415.

Sources: for a reconstruction of the *Epistola ad presbyteros* cf. Monceaux 5, 311-318. Augustine's works against Petilian are edited by CSEL 52-53.
Studies: Monceaux 6, Paris 1922, 1 ff.; S. Lancel, *Actes de la conférence de Carthage en 411*, I, SCh 194, Paris 1972, 231-238; *PAC* 1, 855-868; W.H.C. Frend, *The Donatist Church*, Oxford 1952, ch. 15; id., Manichaeism in the struggle between Saint Augustine and Petilian of Constantine, *AugM* 2, Paris 1954, 859-866.

W.H.C. Frend

PETITIONES ARRIANORUM. When *Athanasius, after his return to *Alexandria on *Julian's death, went in 363 to *Antioch to meet *Jovian, some *Arians presented the emperor with repeated requests not to permit Athanasius to take possession of his see again, but to appoint someone else. These unsuccessful requests have survived in the form of a dialogue between the emperor, the Arians and the crowd, as an appendix to Athanasius's *letter to Jovian (PG 26, 820 ff.). Among the postulants was *Lucius, Arian bishop of Alexandria.

Simonetti 376.

M. Simonetti

PETRA, now Selah, a city in the mountains of Idumaea, 300 km. S of Amman, capital of the Nabataeans and an important caravan city, was annexed by the Romans in 105. Obtained by cutting and embellishing the rock, it was rich in fantastic colours. One of the major temples, that "of the urns", was turned into a *church in 447 by bishop Jason, as we learn from the inscription placed there (Saller-Bagatti, *Nebo*, 233; Brünnow-Domaszewski, *Provincia Arabia*, I, 393). Another church has been found S-E of Qasr el-Biut. Some inscriptions and graffiti bear *crosses, as do a *lamp and various steles (Dalman, *Petra-Forschungen*, 26). With the Arab occupation, the city lost its strategic value.

R.E. Brünnow-A.V. Domaszewski, *Die Provincia Arabia*, I, Strassburg 1904; G. Dalman, *Neue Petra-Forschungen unter der Heilige Felsen*, Leipzig 1912; S. Saller-B. Bagatti, *The Town of Nebo*, Jerusalem 1949; A. Negev, The Nabateans and Provincia Arabia, *ANRW* II.8, 520-686.

B. Bagatti

PETRONILLA. The oldest evidence of this *martyr is archaeological: the 4th-c. "Veneranda fresco", in an arcosolium of the *retro sanctos* of the basilica of the catacomb of Domitilla, shows the dead woman being introduced into heaven by *Petronella mart*. Her name later appears in the *Index oleorum*, in the oldest catalogue of the *cemeteries of *Rome, in the *Itineraries. The latter attest the popularity of the saint, whose name has replaced those of Sts *Nereus and Achilleus as titular of the cemeterial basilica. She was inserted into the legendary Passion of Sts Nereus and Achilleus (BHL 6061): not only do these, from soldiers, become eunuchs and palace guards of *Domitilla, but P. becomes the daughter of the apostle *Peter, miraculously cured by her father, as we see in the *Acts of Peter*. The identification does not seem to be derived from the fresco, but from the *sarcophagus, inscribed: *Aur. Petronillae filiae dulcissimae*. The sarc. was considered that of Peter's daughter, the inscription written by the *apostle's own hand, the whole transported by Paul I (757-67) to a chapel which took the saint's name in the Vatican basilica. The translation was made to reunite in burial those believed to be father and daughter and to please Pippin the Short, king of the *Franks, eldest son of the church. The *mass of St P. is still celebrated for France every year in her chapel on her feast-day, 31 May.

DACL 4, 1409-17; *Vies des SS*. 5, 608-12; *LTK* 8, 327-28; *BS* 10, 514-21; P. Testini, *Le catacombe e gli antichi cimiteri cristiani di Roma*, Bologna 1966, 149, 183, 241, 269, 296; A. Amore, *I martiri di Roma*, Rome 1975, 201; U. Fasola, *Pietro e Paolo a Roma*, Rome 1980, 71-74.

V. Saxer

PETRONIUS of Bologna. Acc. to the evidence of *Eucherius of Lyons (PL 50, 719) and *Gennadius (PL 58, 1082 f.), P., after holding high civil office, was bishop of Bologna (*Bononia*) 431/2-450. His name is linked to the building of the church of S. Stefano, modelled on the Constantinian basilicas of *Jerusalem. P. cannot be the author of the *Vita patrum monachorum Aegypti*, attrib. to him by Gennadius, but may be the author of two *sermons, in honour of St *Zeno and for the anniversary of his episcopal consecration. From the Middle Ages, he was much venerated at Bologna.

CPL 210 f.; G. Morin, *RBen* 14 (1897) 3-8; *LTK* 8, 328; *BS* 10, 521-530.

B. Studer

PHILEAS of Thmuis. Early 4th-c. bishop of Thmuis, a city on the Nile delta, and archon of *Alexandria. We possess first-hand evidence of the circumstances of his *martyrdom: *1)* the *letter he sent to his diocesans about his arrest, his imprisonment and the tortures inflicted on the martyrs of Alexandria (Euseb., *HE* VIII, 10, 2-10); *2)* his defence, i.e. the account of the proceedings of his fifth and last audience, at which Culcianus, prefect of *Egypt, condemned him to death, 4 Feb 306 (Papyrus Bodmer XX). *3)* The Latin Acts of his and Philoromus's martyrdom (BHL 6799). *1)* and *2)* are very close to the facts, *3)* dates from the end of the 4th c.

F. Halkin, L'Apologie du martyr Philéas de Thmuis et les Actes latins de Philéas et Philorome, *AB* 81 (1963) 5-30; V. Martin, *Papyrus Bodmer XX. Apologie de Philéas, évêque de Thmuis*, Geneva 1964; *BS* 5, 686-7; H. Musurillo (ed. and tr.), *The Acts of the Christian Martyrs*, Oxford 1972, xlvi-xlviii and 320-353; A. Pietersma, *The Acts of Phileas Bishop of Thmuis*, Geneva 1984.

V. Saxer

PHILIP, apostle. Named between *John and *Bartholomew in Mt 10, 3 / Lk 6, 14; after *Andrew in Mk 3, 18 / Acts 1, 13; with Andrew in Jn 12, 22. The two names are Greek. Acc. to Jn 1, 44, P. was born in Bethsaida like Andrew and *Peter. In Jn 6, 5-7, Jesus questions him before the multiplication of the loaves. In Jn 12, 20-23, the Gentiles who wish to be presented to Jesus apply to him. Acc. to *Clement of Alexandria (*Strom.* III, 25, 3), he was the anonymous disciple to whom Jesus addressed the words of Mt 8, 22 / Lk 9, 60. *Papias, *Polycrates of Ephesus and the apocryphal *Act. Andr.* and *Act. Phil.* confuse the *apostle P. with *Philip "the deacon" (Acts 6, 8 and 21, 8). Clement (*Strom.* III, 52, 5-cf. Euseb., *HE* III, 30, 1) also mentions the daughters of the apostle P., but only to relate that their father married them off. He is one of the 11 *apostles presented as authors of the *Ep. Apostolorum* (2, 13) and one of those who interrogate Jesus in the course of *gnostic debates (*Sophia J. C., Pistis Sophia, Books of Jeu*). *Pistis Sophia* 42-44 presents P. as author of the sayings and deeds of Jesus, like *Thomas and *Matthew. So he is the presumed author of one of the three main *apocryphal gospels, so highly thought of in some gnostic circles. The *Ev. Phil.* was known only through a brief citation of *Epiphanius (*Pan.* 26, 13, 2-3). It was known to have been used by the *Manichees (Tim. of Const.: PG 86, 21 C; Ps.-Leontius of Byz.: PG 86, 1213 C). The *Coptic document discovered at *Nag Hammadi (NHC II, 51, 29-86, 19) is a collection of sentences of Jesus, offering an important contribution to our very limited knowledge of gnostic theology and sacramental practice. Also from Nag Hammadi (NHC VIII, 132, 10-140, 27) comes a *Letter of Peter to Philip* (a gnostic dialogue of the risen Saviour with his disciples), in which P. is mentioned only at the beginning. Acc. to the oldest *tradition (*Heracleon, in Clem., *Strom.* IV, 71, 3), P. did not die a martyr's death.

R. Trevijano

PHILIP. Disciple of *Bardesanes, of the Eastern branch of *Valentinianism. Some scholars (e.g. F. Nau, J. Quasten) make him the author of the *Book of the laws of the nations*, a work in *dialogue form traditionally attrib. to Bardesanes, where the master answers the disciples's queries on the characters of men and the influence of the planets. This text, mentioned by *Eusebius (*HE* IV, 30, 1-3) and preserved in the original language, is one of the oldest documents of *Syriac literature (if we except the translations of the *Bible).

Syriac text of the *Liber legum regionum* with Fr. tr.: F. Nau, PS, pars I, tomus 2, Paris 1907, 490-658; It. tr.: G. Levi della Vida: *Il dialogo delle leggi dei paesi*, Rome 1921.

C. Gianotto

PHILIP, priest (†455/6): disciple of *Jerome (Gennadius, *De vir. ill.* 63). His *commentary on *Job, important evidence of the Vulgate, exists in a long and a short recension (CPL 643 and 757; PL 26, 619-802 and 23, 1407-1470). The forwarding *letter and some fragments are in A. Wilmart, *Analecta Reginensia*, Vatican City 1933, 316-322.

J. Bauer, Corpora orbiculata. Eine verschollene Origenesexegese bei Pseudo-Hieronymus, *ZKTh* 82 (1960) 333-341; *Patrologia* III, Sp. ed. (by A. Di Berardino) Madrid 1981, 290.

J. Gribomont

PHILIP of Gortyna. Bishop of Gortyna (*Crete), c.170. *Dionysius of Corinth, writing to the church of Gortyna, praises P. for the *faith and courage of his church and exhorts it to be on guard against *heretics, perhaps Marcionites (Euseb., *HE* IV, 23, 5). P. wrote a work against *Marcion (Euseb., *HE* IV, 25).

E. Prinzivalli

PHILIP of Side. Disciple of Rhodo in *Side (Pamphilia), ordained deacon and presbyter by *John Chrysostom, whose close *friend he was (*Ep.* 213: PG 52, 729). Ran three times for the patriarchate of *Constantinople (426 *Sisinnius; 428 *Nestorius; 431 *Maximian). Refuted *Julian the Apostate's work *Against the Galilaeans*. His voluminous *Christian History* in 36 books of *c.*1000 chapters went from the *creation of the world to at least AD 426. Published in 434-39, its chronological order was often defective and it made continual digressions of all kinds (geometrical, astronomical, musical, geographical, botanical, etc.); the style was verbose (Asian) and undistinguished. Fragments and extracts remain; those that complete *Eusebius's *HE* (e.g. on *Papias of Hierapolis, on the *didaskaleion* of *Alexandria) are important.

CPG 3, 6026; Socrates, *HE* 7, 26-27. 29. 35; Photius, *Bibl.* 35; Bardenhewer 4, 135-137; Quasten II, 533-535; Altaner 228.

A. De Nicola

PHILIPPI. "The city of Philippi was once called Datos and, earlier still, Krenides, because of the many springs of water (Krenai) that rise in its hills. Philip fortified it, since it was the most flourishing city of *Thrace, and called it Philippi after himself" (App., *Bell. civ.* 105, 439). Late in 49/50, the *apostle *Paul and his followers came from the Troad to this Greek city of *Macedonia, following the vision that disclosed new ways and opened new horizons to his apostolic mission (Acts 16, 8-10). Nothing is preserved in P. which can shed light on the organization, worship and activity of the early Christian community. The small, poor buildings that were probably turned into domestic churches have been destroyed. The church recently discovered is of the second half of the 2nd c. *Late antiquity was the period of greatest splendour for Philippian Christianity. But we should not logically expect grandiose Christian *church buildings to appear immediately after the *persecutions. In the city centre, E of the *agorà*, has been found an important "Macedonian" tomb of the 2nd c. BC with a wealth of gold objects (crown of oak leaves, finely worked diadem, etc.). Above the funerary chamber was a building in the form of a *naos*, whose foundations remain. It was a monument (as the *epigraph of the *sarcophagus says) dedicated to the dead heroized *Euephenes Exekestou* who, as we deduce from some discoveries, being priest of the Cabiri, was also considered, in accordance with a later legend, as the founder of the city and venerated with a special cult. Beside this late Hellenistic tomb was founded the Christian community's first house of prayer: a one-aisled hall with square apse, whose form seems dictated by a road linking the Via Egnatia with the market street, which formed the S boundary of the Roman forum. The building, 10.50 m. up to the boundary of the altar, was later enlarged to 18 m. The rebuilding allowed the inclusion of the tomb. More or less at the centre of the oratory a transverse wall, in which arcades or passages must have opened, divided the hall into two rooms: its function was probably to support the roof. The *mosaic pavement decoration corresponds to this layout. The E part has six sections, side by side in pairs, bordered by frames, with intervals enlivened by trees, peacocks and small birds pecking fruit, peacocks beside vases, foliage spilling out of vases, etc. On the E side of this composition, at the centre of the *tabula ansata*, is a three-line inscription composed of gold, red, blue and white stones: ΠΟΡΦΥΡΙΟΣ ΕΠΙΣΚΟΠΟΣ ΤΗΝ ΚΕΝΤΗΣΙΝ ΒΑΣΙΛΙΚΗΣ ΠΑΥΛΟΥ ΕΠΟΙΗΣΕΝ ΕΝ ΧΡΙΣΤΩ.

The Porphyrius of the inscription is the known bishop of P. who subscribed at the synod of *Sardica (342-3). In the years of his episcopal *ministry, Porphyrius replaced an older, cruder, less valuable pavement mosaic with the present one. So the church must have been built some decades before him, but after the *edict of Milan (313). The mosaic in the S part of the church is divided into four large sections, with animal and floral themes surrounded by geometrically decorated frames. At the centre of these sections is an inscription in black letters on a white ground: ΧΡΙΣΤΕ ΒΟΗΘΙ ΤΟΥ ΔΟΥΛΟΥ ΣΟΥ ΠΡΙΣΚΟΥ ΣΥΝ ΠΑΝΤΙ ΤΟΥ ΟΙΚΟΥ ΑΥΤΟΥ. About the end of the 4th c., P. began to lose its Roman town plan with its sumptuous public buildings and regular streets. With the religious *peace, there began to develop Christian church architecture, whose principal monuments we will record.

The octagonal basilica: Starting as a small church dedicated to St Paul, it became the city cathedral. It is an octagonal building with an inner *peribolos* of columns forming an ambulatory within the central nucleus: to E is the shallow apse with *subsellia*; to W, a narthex slightly wider than the *naos*. To W of the narthex and of the same width is the triporticoed atrium with the font occupying the space of the W portico. To S opens a semicircular vestibule (*propylaeum*) from which a monumental entrance with three flights of steps connects the temple with the market street. The pavement was of marble slabs; the perimeter walls of the hall and atrium had rich marble and *opus sectile* decorations with various motifs (plants, fish, etc.). About the mid 5th c., the octagonal structure was transformed by adding an inscribed square, widening the apse, providing new *subsellia* and adding galleries. In the 1st half of the 6th c., a very few structural alterations were carried out to the original inscribed octagon, limited to reinforcing with plinths the corner pillars of the temple which formed four corner apses, which therefore underwent alterations. The now thicker corner pillars altered the original proportions of the building; the atrium which connected octagon A and B was eliminated, with the font and the splendid vestibule giving onto the market street, and replaced by a modest courtyard open to the sky. Centred around the octagon were the buildings indispensable for liturgical life: to N-W the font, to N-E the baptistery, perhaps of its genre the best-preserved in all its parts (vestibule, dressing-room, *catechumeneum*, baptistery hall with central cruciform pool and *chrismarium* linked to the E section of the *naos*). Further N, separated from the baptistery complex by a dividing wall, was found the structure of a *baths in use from the late Hellenistic age to the destruction of the city. To N-E of the octagon was found an imposing two-storied bishop's palace, whose architectural phases mirror those of the octagonal basilica. The octagon complex forms a pattern of early Christian church architecture rare, if not unique, in the E Mediterranean basin, *Syria included.

The basilica outside the walls: 500 m. E of the walls are the ruins of an early Christian basilica excavated in 1956-7: 33 by 15.50 m., it has three aisles, galleries and semicircular apse to E with presbyterial bench. In the place of the altar, nothing remains except the altar of consecration and the *thalassidion*. The *naos* proper shows two architectural periods; it was preceded by a narthex and a quadriportico, of which only the E side could be excavated. Backing onto the N perimeter wall were five rooms, the functions of three of which have been identified. The smallest room, to N-W, contained the stairs leading to the galleries; the other two, as finds demonstrate, were the place of offerings (*prothesis*) and the deacons' room (*diaconicon*). In the 4th c. the basilica had a transept with stylobate columns and a pavement *mosaic with geometrical, animal or plant motifs framed by various-shaped polychrome borders. Under the floor of the basilica, both nave and narthex, were found many vaulted or chest tombs, mostly of *clergy, and inscriptions now preserved in the local museum. The foundation of the basilica goes back to the mid 4th c., i.e. the reign of *Constantius II (337-361). Coins found in the later building layers belong to the time of *Theodosius II (408-450). In 473 the *Goths reached P. but failed to take it: "they only made raids outside the city wall and committed no other cruelties" (Malchus, ed. Bonn 1829, 234 ff.). This basilica, being outside the wall, "facing the city", was evidently partly devastated during this raid. It is impossible to say whether repairs were carried out: the church may have remained abandoned for several decades or been used as best it could until the time of *Justinian, when rebuilding modified it as mentioned. The basilica's second period ended shortly before the time of the emp. Leo VI (886-912). In the 5th c. in particular, but also in the 6th, other splendid churches were built, but the Octagon kept its old status as cathedral.

Basilica A: built about the end of the 5th c. at the foot of the acropolis, on a natural terrace N of the Roman forum, it stands out by its dimensions (130 by 50 m.). The basilica was linked with the Via Egnatia by a monumental staircase which reached into the atrium or quadriportico, whose S side was occupied by a fountain for washing, a two-storied structure. Five niches opened into the thickness of the wall of the pool: the central one, larger than the others, served as a dressing-room and for other liturgical necessities; the others, alternately square and semicircular, were closed by parapets which formed small basins. From the atrium three entrances led to the narthex, and from here other doors led into the basilica, which had three aisles and a transept. In the altar were discovered the hole for the consecration stone and the series of *subsellia* on the N and S sides. On the N side of the basilica is the stair leading to the galleries and two other rooms of unknown function.

Basilica C: a short distance to W of basilica A was found a three-aisled basilica with transept, a wealth of *sculpture and splendid marble decoration in the lateral sectors of the transept. Excavation is still in progress.

Basilica B: a domed basilica, built *c*.550 in the centre of the Roman city, S of the forum, on the market street. It has three aisles separated by colonnades of six columns, large altar and projecting semicircular apse to E, narthex to W, rooms backing along the perimeter wall, including washing fountain, and prothesis to N. The sculptures that decorated capitals, bases, *abaci* and *cancelli* are of exceptional quality, comparable to those of Hagia Sophia in *Constantinople. From the 7th c. P. lost the importance it had held for so many centuries. The monuments slowly went to ruin and the city was finally reduced to insignificance. Despite this, the cult of the apostle Paul was kept alive even after the city's radical destruction under Turkish rule. [Fig: 255]

P. Lemerle, Les inscriptions latines et grecques de Philippes, *BCH* 59 (1935) 126-164; P. Collart, *Philippes, Ville de Macédoine*, Paris 1937; P. Lemerle, *Philippes et la Macédoine orientale*, Paris 1945; St. Pelekanidis, ἀσεβής, ἀσέβεια, ἀσεβέω, *AE* 1955 (1959) 114-179; *PAAH* 1958-1965, 1967-1979 (for a general reconnaissance of the excavations of the Octagon complex at Philippi); R.F. Hoddinot, *Early Byzantine Churches in Macedonia and Southern Serbia*, London 1963, 99-116, 169-173, 188-193; St. Pelekanidis, Kultprobleme in Apostel-Paulus-Oktagon von Philippi im Zusammenhang mit einem älteren Heroenkult, *Atti del IX Congr. intern. di Archeologia cristiana* (Rome 21-27 Sept. 1975), Vatican City 1978, 393-397; id., Συμπεράσματα ἀπὸ τὴν ἀνασκαφὴ τοῦ 'Οκταγώνου τῶν Φιλίππων in 'Η καβάλα καὶ ἡ περιοχὴ τῆς Α' Τοπικὸ Συμπόσιο„ Thessalonica 1980, 149-158. E. Pelekanidou, 'Η κατὰ παράδοση φυλακὴ τοῦ Ἀποστόλου Παύλου στοὺς Φιλίππους, *ibid.*, 427-435, figs. 1-5, pl. 1; W. Müller-Wiener, Bischopsresidenzen im ostlichen Mittelmeer-raam, *Actes XIe Congrès internat. d'arch. chrét.*, CEFR 123, Rome 1989, 651-667.

E. Pelekanidou

PHILIP the Arab (244-259). Acc. to *Eusebius (*HE* VI, 34) - who prob. based it on *Origen's *letters to P. and his wife, mentioned in VI, 36, 3: information supplemented by two authors independent of him and in the Antiochene tradition, *John Chrysostom in a *sermon on *Babylas of Antioch (PG 50, 539-544) and the *Chronicon Paschale* (ed. Dindorf I, 503, Bonn 1832) - the emp. P. and the empress Otacilia Severa were obliged by bishop Babylas of Antioch to do public penance during an *Easter vigil for the murder of Gordian III's young son, entrusted to them by his father. P. would thus be the first Christian emperor, and if Eusebius does not mention him in his *Life of Constantine* it is because that emperor associated P.'s memory with the *damnatio memoriae* of his claimed descendant *Licinius. The objections constantly made by historians to P.'s Christianity can thus be answered.

J.M. York Jr., The image of Philip the Arab, *Historia* 21 (1972) 320-332; H. Crouzel, Le Christianisme de l'empereur Philippe l'Arabe, *Gregorianum* 56 (1975) 545-550.

H. Crouzel

PHILIP "the Deacon". Acts 6, 5; 8, 4. 13. 26-40 speak of "one of the Seven", an "evangelist" (Acts 21, 8) named Philip. The identifircation of the *ministry of the "Seven" in the church of *Jerusalem (Acts 6) with that of the deacons, is highly disputable and does not appear until *Irenaeus (*Adv. haer.* I, 26, 3; III, 12, 10; IV, 15, 1) and, later, in *Cyprian (*Ep.* 3, 3). Some writers confused this "evangelist" with the *apostle *Philip, one of the Twelve. They spoke of the apostle Philip, father of daughters (*Papias, in Euseb., *HE* III, 39, 9), *virgins (*Polycrates of Ephesus, in Euseb., *HE* III, 31, 3), prophetesses (the Montanist *Proclus, in Euseb., *HE* III, 31, 4), buried at Hierapolis. These are the daughters of the Philip of Acts 21, 8-9. In exchange, Philip's house and those of his daughters were pointed out to *Jerome (*Ep.* 108, 8) at *Caesarea in 385. The same confusion appears in the *martyrdom of *Andrew (Lipsius-Bonnet II, 1, p. 47), where the apostle Philip ends up as a missionary in Samaria. In the *Act. Phil.* (*ibid.* pp. 1-98), the apostle and the "deacon" are the same person, the main character. The work, which depends on older legends, was prob. composed in the 5th c.

R.A. Lipsius-M. Bonnet, *Acta Apost. Apoc.* I, 1-2, Leipzig 1898, repr. Darmstadt 1959; Erbetta I, 213-243 (gospel), II, 451-490 (acts); R.M. Wilson, *The Gospel of Philip*, London 1962; J.-E. Ménard, *L'Evangile de Philipe*, Strasbourg 1967; A.H.C. van Eijk, The Gospel of Philip and Clement of Alexandria, *VChr* 25 (1971) 94-120; *BS* 5, 719-721; (ed.) W.W. Isenberg, The Gospel of Philip, in J.M. Robinson, *The Nag-Hammadi Library*, New York 1977, 131-151.

R. Trevijano

PHILO, historiographer (4th-7th c.). As we learn from two citations by *Anastasius the Sinaite (8th c.), P. wrote an ecclesiastical history (Ἐκκλησιαστικὴ ἱστορία). He should not be identified with bishop *Philo of Carpasia (4th c.; cf. *PWK* 15, 362 ff.).

CPG 3, 7512; G. Mercati, Un preteso scritto di san Pietro vescovo d'Alessandria e martire sulla bestemmia e Filone l'istoriografo, in *Opere minori*, 2, Vatican City 1937, 426-439 (for the fragments: 429-431 and 437 f.).

J. Irmscher

PHILOCALIA. *Gregory Nazianzen sent his metropolitan *Theodore of Tyana an Origenian anthology in memory of *Basil, recently dead. There is reason to think it was composed by himself and his *friend, prob. *c*.360, at the start of their theological studies: not surprising if we consider that from that time both possessed exceptional maturity. Preserving some stupendous pages of *Origen in the original, the work is essentially interested in exegetical method and the doctrine of freedom, passing over controversial theological subjects in silence.

J. Robinson, *The Philocalia of Origen*, Cambridge 1893; E. Junod, *Origène. Philocalie 21-27*, SCh 226, Paris 1976 (the other chapters are published in SCh among the relevant works of Origen); (eds.) M. Hall-N. de Lange, *Origene: Philocalie, 1-20. Sur les Ecritures et La Lettre à Africanus*, SCh 302, Paris 1983.

J. Gribomont

PHILOCALUS (or Filocalus), FURIUS DIONYSIUS. Author of the *calendar known as the *Chronograph of 354*, whose frontispiece is signed *Furius Dionysius Filocalus titulavit*. P.'s name is linked to the work of pope *Damasus, for whom he incised the famous *epigrams celebrating the deeds of the *martyrs. The characters P. invented for these epigraphs, new in their proportions and in their apexes, represent an important step in the history of *palaeography and *epigraphy. Died *c*.382.

CPL 2249 ff.; *EAA* 679 f.; A. Ferrua, Filocalo l'amante della bella lettera, *CivCatt* 1 (1939) 35 ff.; id., *Epigrammata Damasiana*, Vatican City 1942, 21 ff.; J. Vives, Las inscripciones Damasianas, in *La tumba de San Pedro y las catacumbas Romanas*, Madrid 1954, 426 ff.

U. Dionisi

PHILOGONIUS of Antioch. Bishop of *Antioch when the Arian controversy began, he was among *Arius's opponents; in his *letter to *Eusebius of Nicomedia, Arius mentions him as one of those who conceived the Son's generation from the *Father in a way unacceptable to him. Died 324/5, a few months before the council of *Nicaea.

DCB 4, 389-390.

M. Simonetti

PHILOMENA. The hagiographical romance of P. was built up in the 19th c. on the basis of a Christian inscription (Diehl 2251 A). The inscription was on a tomb in the catacomb of Priscilla, and the bones found there (2 May 1802) were held to belong to a *martyr. Transferred to Mugnano (diocese of Nola), they immediately led to miracles and became an object of pilgrimage: P.'s fame was universal. The Curé d'Ars attributed his own miracles to her mediation. P. became "the saint of the 19th c."

DACL 5, 1600-5; *Vies des SS.* 8, 172-73; *LTK* 8, 469; *BS* 5, 796-800.

V. Saxer

PHILO of Alexandria. The main representative of *Judaeo-Hellenism. He greatly influenced the *exegesis, *theology and *spirituality of the *Fathers through the mediation of *Clement of Alexandria, *Origen, *Gregory of Nyssa and *Ambrose, who all read him directly. *Eusebius and *Jerome treat him almost as a Christian, and it is to the Christians that we seem to owe the preservation of his works. Some years older than Jesus, he was still alive in AD 41. He led an ascetic and contemplative life and seems to have done the work of a rabbi. He belonged to a very rich Alexandrian business family: his brother Gaius Julius Alexander, called the Alabarch, administered the property of many members of the imperial family and the Herodian dynasty; his nephew Tiberius apostatized from Judaism, became procurator of Judaea, prefect of Egypt, then *Titus's lieutenant at the siege of *Jerusalem; his nephew Marcus was the first husband of the famous Berenice. P. received a thorough Greek education, but remained sincerely attached to the Jewish faith. His work is, to our knowledge, the first to show on a grand scale the meeting of two cultures. But he also defended the Jewish community of *Alexandria against the pogrom carried out in AD 38 by the prefect Flaccus (cf. *In Flaccum*) and led the embassy sent to the emp. *Caligula (cf. *Legatio ad Caium*): only the assassination of the mad emperor prevented a disaster.

P.'s work is mostly exegetical. A part of his treatises consist of literal and moral exegesis: *De opificio Mundi, De Abrahamo, De Josepho, De vita Mosis, De Decalogo, De Specialibus Legibus, De Virtutibus, De Praemiis et Poenis, De Exsecrationibus.* Another part consists of allegorical explanations of certain passages of *Genesis: *Legum Allegoriae, De*

Cherubim, De Sacrificiis Abelis et Caini, Quod Deterius Potiori insidiari soleat, De posteritate Caini, De Gigantibus, Quod Deus sit immutabilis, De Agricultura, De Plantatione, De Ebrietate, De Sobrietate, De Confusione linguarum, De Migratione Abrahami, Quis Rerum Divinarum Haeres sit, De Congressu Eruditionis et gratiae, De Fuga et Inventione, De Mutatione nominum, De Somniis. The two exegetical methods meet in the *Quaestiones et Solutiones in Genesim et in Exodum* preserved in *Armenian. Then there are his philosophical works: *Quod Omnis Probus Liber sit, De Vita contemplativa* (describes the life of Judaeo-Hellenistic monastic communities, the *Therapeutae*), *De Aeternitate Mundi, De Providentia, De Animalibus*, and the topical works cited above, *In Flaccum* and the *Legatio ad Caium*.

P.'s interpretation of *Scripture is multiform. The literal sense is the basis of the spiritual, and he explains it with all the science of his time. The sources of his *paideia* are the Jewish law, esp. the Pentateuch, and Hellenistic culture with its encyclopaedic education (grammar, *rhetoric, dialectic, *music, geometry, astronomy, physics, etc.), handmaid of philosophy, itself subordinate to Wisdom, which is spiritual understanding. He explains the text and its anomalies, makes wide use of symbolic arithmetic inspired by the Pythagoreans, and knows law and jurisprudence, both Jewish and Greek. The spiritual interpretation is developed on various levels. In the end it is cosmological: the Temple is the symbol of the *world, and its different parts regions of the *cosmos; the high priest's garments are the same. From the *macrocosm he passes to the microcosm, from the world to man: psychological exegesis like that which shows us, in the couples *Adam/Eve or *Abraham/Sarah, intelligence and sensation. From here it is an easy transition to moral *allegory, like that of the animals representing the passions. But the essential interpretation is the mystical: the reading of the text becomes the growing awareness of a spiritual experience which awakes and expresses itself: thus the exegesis of the three patriarchs, developed in *De Congressu* and repeated everywhere, as three types of acquisition of Wisdom: *Abraham by study, *Jacob by practice (*ascesis) and Isaac, the greatest, who possesses it by nature. We cannot dwell on his *theology: *God, the Logos, the Powers, *angels and men, the world.

P. uses *philosophy for the sake of theology in an eclectic way: his inspiration is mainly Stoic and Platonic, following the amalgam inaugurated by Poseidonius of Apamea and middle Stoicism, but the *Aristotelianism of other schools is not absent from his works. His biblical exegesis can be found elsewhere in the various kinds of rabbinic interpretation, but is not unrelated to that which the Stoics Cornutus and Chaeremon applied to Greek mythology or Egyptian traditions. He influenced patristic *exegesis, despite differing in one essential aspect, the central role of Christ. Though fully Greek, P. remained a Jew in his basic convictions, a Jew who is often, in his theology and *spirituality, close to a Christian: note the declaration that God alone works virtuous actions in the *soul and that man must give him thanks; the distinction of creator and creature replaces the more Greek one between incorporeal and corporeal; finally the importance given to charity.

L. Cohn - P. Wendland - J. Reiter, Berlin 1896-1930, 8 volumes; F.H. Colson, London 1928-1952, 12 vols. (with Eng. tr.); R. Arnaldez - J. Poilloux - C. Mondésert, 36 vols., Paris from 1961 (with Fr. tr.). For a general introduction see J. Daniélou, *Philon d'Alexandrie*, Paris 1958, and R. Arnaldez's extensive introduction in the first volume of the French edition; H. Chadwick, Philo and the Beginning of Christian Thought, in (ed.) A.H. Armstrong, *Later Greek and Early Mediaeval Philosophy*, Cambridge 1967, 137-157; S. Sandmel, *Philo of Alexandria: an introduction*, New York, Oxford 1979; (eds.) D. Winston-J. Dillon, *Two Treatises of Philo of Alexandria*, Chico 1983; R.M.Berchman, *From Philo to Origen: Middle Platonism in Transition*, Chico 1984; D.T. Runia, *Philo of Alexandria and the Timaeus of Plato*, Leiden 1986; id., The Structure of Philo's Allegorical Treatises, VChr 38 (1984) 209 ff; J.C.M. van Winden, The World of Ideas in Philo of Alexandria, An Interpretation of *De opificio mundi* 24-25, VChr 37 (1983) 209-217; A. van den Hoek, *Clement of Alexandria and his use of Philo*, Leiden 1988; P. Radice - D.T. Runia, *Philo of Alexandria: an annotated Bibliography, 1937-1986*, Leiden 1988; G. Downing, Ontological Asymmetry in Philo, JThS n.s. 41 (1990) 423-440; D.T. Runia, *Exegesis and Philosophy, Studies in Philo of Alexandria*, London 1991.

H. Crouzel

PHILO of Carpasia. Early 5th-c. bishop of Carpasia (*Cyprus). A deacon expert in interpreting *Scripture, contemporary of *Epiphanius of Salamis, who ordained him bishop (*Vita S. Epiph.* II, 49). Of his *Commentary on the *Song of Songs*, we have a summary of the Gk text, published by Giacomelli in 1772, a Lat. tr., important because closer to the original, by *Cassiodorus's friend *Epiphanius Scholasticus (but Cassiodorus [*Inst.* I, 5, 4] attributed P.'s work to the better-known Epiphanius of Salamis) and extracts in *Procopius's and Ps.-*Eusebius's *catenae on the Song of Songs and in *Cosmas Indicopleustes's *Topographia christiana*. P.'s *exegesis, based on the traditional identification Christ-bridegroom and Church-bride, is modest and compilative.

PG 40, 27-154 (M. Giacomelli); A. Ceresa Gastaldo, *Il commento al "Contico dei Cantici" di Filone di Carpasia*, Turin 1979 (CP 6) (ed. of the Latin version and It. tr.).

E. Prinzivalli

PHILOSOPHY and the Fathers. Christianity existed for *c*. 150 years before its initial confrontation with Greek philosophy. The earliest Christian writers, known as the Apostolic Fathers (*c*.90-160), were concerned almost exclusively with pastoral questions, with the result that their writings show little interest in the philosophical schools of their time. But a decisive confrontation between Christianity and Greek thought was inevitable, since both claimed exclusive possession of a wisdom able on its own to provide a truthful account of man's nature and destiny. Historically, the oldest "dialogue" between the two was occasioned by the *conversion to Christianity of intellectuals who had frequented the philosophical schools and, in some cases, continued to consider themselves philosophers even after their conversion (*Justin, *Aristides). But clear and irreconcilable difficulties quickly arose over the proper attitude of the Christian thinker towards a philosophical world that could not be ignored. These differences are more marked than ever in the first two *apologists, Justin martyr and his first disciple, *Tatian. For the former, philosophy is a divine gift that should bring man closer to *God (*Dial.* 2. 1). Far from being mutually opposed, in Justin's conception philosophy and Christianity represent respectively the partial *revelation of the truth through the power of the divine Logos seminally present in all *souls, and the full and final revelation of the same Logos in the *incarnation. On the other hand, Tatian's profound aversion to philosophy was just one aspect of his total hostility to Greek culture in general. His harsh criticisms of philosophy in particular were taken up with characteristic incisiveness 50 years later by *Tertullian, in antithetical formulae that have become classics: "What has Athens to do with Jerusalem, and the Academy with the Church?" (*Praes*. 7). But in the end it was Justin's position that found general acceptance among later church writers, in both *East and West. Of the Easterners - including *Origen, *Gregory of Nyssa and *Eusebius - it was *Clement of Alexandria who showed, in theory and practice, how philosophy could contribute richly to a deeper understanding and scientific exposition of the revealed truth. Among the Westerners, from *Marius Victorinus to Anselm and the Middle Ages, the motif of "faith in search of understanding" (*fides quaerens intellectum*) guaranteed philosophy a permanent role in the theological development of the Latin church. Finally, the imposing doctrinal synthesis achieved by St *Augustine through the fusion of revealed truth with *neoplatonism was the demonstration *par excellence* of the fact that *faith and reason could be considered natural and compatible partners.

Many specific articles in this book confront in greater detail the complex relationship between Christian and Greek thought: in particular *Aristotelianism, *Hellenism and Christianity, *Judaeo-Hellenism, *Pantheism, *Platonism and the Fathers, *Stoicism, besides other articles of a theological, spiritual and artistic character.

A.H. Armstrong-R.A. Markus, *Christian Faith and Greek Philosophy*, London 1964; H.A. Wolfson, *The Philosophy of the Church Fathers*, I, Cambridge, Mass. ²1964; R. Harvanek, Patristic Philosophy, CE 10, 1103-1107; (ed.) A.H. Armstrong, *Later Greek and Early Mediaeval Philosophy*, Cambridge 1967; id., *St. Augustine and Christian Platonism*, Villanova 1967; W. Pannenberg, The Appropriation of the Philosophical Concept of God as a Dogmatic Problem of Early Christian Thought, in *Basic Questions of Theology*, II, Philadelphia-London 1971, 119-189; E.P. Meijering, *Orthodoxy and Platonism in Athanasius: Synthesis or Antithesis?*, Leiden 1974; A.H. Armstrong, *Plotinian and Christian Philosophy*, Cambridge 1981; J.M. Rist, Basil's neoplatonism: its Background and Nature, in (ed.) P.J. Fedwick, *Basil of Caesarea, 1600th anniversary symposium*, Toronto 1981; A.M. Ritter, Platonismus und Christentum in der Spätantike, ThRu 49 (1984) 31-56; S. Gersh, *Middle Platonism and Neoplatonism: the Latin Tradition*, 2 vols., Paris 1986; D.L. Balás, Philosophy, in (ed.) E. Ferguson, *Encyclopaedia of Early Christianity*, New York-London 1990, 727-731 (with bibl.).

R.J. De Simone

PHILOSTORGIUS, writer of ecclesiastical history. Born *c*.368 in the village of Borissus in *Cappadocia, he died after 425 (his works mention events of that year). A *layman, we know nothing of his profession. He went to *Constantinople aged 20 and lived there, making occasional *journeys (to *Palestine and *Antioch, among others) which broadened his education: he shows a knowledge of geography, astronomy and natural science. The direction of his life was determined by a meeting with his compatriot *Eunomius, a disciple of *Aetius and leader of the intransigent *Arians: from then on he shared the destinies of the *anomoean community and his work was extraordinarily influenced by them. Two works (ἐγκώμιον of

Eunomius; ἀγῶνες against *Porphyry), mentioned by him in the *Historia Ecclesiastica*, are lost. Nor has the complete text of the latter survived, but we have extensive extracts of it (crit. ed. by Bidez; earlier edd. by J. Gothofredus [Geneva 1643] and H. Valesius [Paris 1673]). The work, conceived as a continuation of *Eusebius's, comprised events between 320 (start of the Arian controversy) and 425; it was divided into two tomes, each introduced by an *epigram (*Anth. Pal.* I, 193-194): twelve books in all, whose initial letters formed an acrostic on the author's name. *Photius (*Bibl.*, cod. 40) knew the whole work and made a summary of it, preserved by the *manuscript tradition (cod. *Barocc.* 142 and its descendents), not always exact and sometimes tendentious, since he omitted what clashed with *orthodoxy. Many other fragments have been found in the *Artemii Passio* (PG 96, 1251-1320), written before the 10th c. by John of Rhodes, and in *Suidas (in character with this Lexicon, these are histories of famous people like the martyr *Babylas, Agapetus of Synnada, Auxentius of Mopsuestia, *Leontius of Tripoli). There are also *excerpta* in Nicetas Acominatus's *Thesaurus orthodoxae fidei* and esp. in the *Vita Constantini* of cod. Angel. 22, ed. H. G. Opitz (*Byzantion* 9 [1934] 535-593): here the author summarizes a compilation, made by others from the *HE*, in which all heretical excesses were removed. P.'s sources are unknown. He certainly used *letters of emperors, *acts of *councils and *martyrs and perhaps also the work of *Eunapius.

*Paganism and *orthodoxy were his enemies and the *persecutions of the *heretics, as Batiffol has shown (*Quaestiones*, p. 12), led him into *apocalyptic views: thus he recognized in the misfortunes of the Roman empire the fulfilment of the divine wrath and the prophecies of *Daniel and the gospels. His *apologetic intent towards the Arianism of Aetius and Eunomius led him above all to cite Arian sources: it is this partiality that makes the *HE* valuable to supplement our knowledge of a historical period for which the majority of sources are orthodox. Photius (*Bibl.*, cod. 40) allows P. some elegance of style.

CPG 3, 6032; PG 65, 460-637; GCS 21, Berlin 1972; Translation: E. Walford, *The Ecclesiastical History of Sozomen ... also the Ecclesiastical History of Philostorgius as Epitomized by Photius*, London 1855; P. Batiffol, *Quaestiones Philostorgianae*, Paris 1891; L. Jeep, Zur Überlieferung des Philostorgius, TU 17, 3b, 2; Schmid-Stählin, *Geschichte* VII, 2, 2, 1433-1434; P. Heseler, Neues zur Vita Constantini des codex Angelicus 22, *Byzantion* 10 (1935) 399-402; J. Bidez, Fragments nouveaux de Philostorge sur la vie de Constantin, *Byzantion* 10 (1935) 403-442; G. Fritz, Philostorge, *DTC* 12, 1495-1498; G. Geutz, Philostorgios, *PWK* 20, 119-122; G. Moravcsik, *Byzantinoturcica* I, Berlin 1958, 473-474; H. Rahner, Philostorgios, *LTK* 8, 468; Quasten II, 535-537; Altaner[7], 229.

A. Labate

PHILOTHEUS. Monk of Our Lady of the Burning Bush on *Sinai; in line with the *hesychasm of *John Climacus and *Hesychius, he wrote some spiritual works (CPG 7864-7866) published partly in the *Philocalia* of Nicodemus the Hagiorite (Athens 1958, II, 274-286), partly in PG 98, 1369-1372, with a further fragment under the name of Philotheus Kokkinos, PG 154, 729-745 (cf. col. 717, nn. 21 and 22). See the MSS, hesychastic collections written on Mt Athos in the 14th c.: *Vat. gr.* 658, 703, 730; *Marc.* II 73; *Vindob. theol. gr.* CLVI, CCXXIV, CCXXVIII. Another collection, not registered in CPG, is in the Italo-Greek MS *Paris. Suppl. gr.* 1277. Acc. to Beck, 453-454, these works predate Simeon the New Theologian; they certainly do not belong to antiquity.

J. Gribomont

PHILOXENUS of Mabbug (†523). By his bitter political and doctrinal struggles and his assiduous activity with the monks of *Syria, P. was the main architect of the installation of the Severan regime at *Antioch and deserves the title of pioneer of Syrian *monophysism. Though his memory has been eclipsed by that of *Severus, what remains of his works reveals an original theological synthesis, expressed in very beautiful language, which is at the meeting-place of the Syrian (*Ephrem, John of Apamea) and Alexandrian *traditions (*Athanasius, *Cyril, *Evagrius). A native of the province of Beit Garmai, born *c.*440 and named Xenaias (perhaps Joseph), he was prob. very soon sent to the theological school of the Persians at *Edessa. At the end of a sound training he embraced the Cyrillian ideology, abandoning *Nestorian *theology in order to be able to stay in the convents of Syria-Mesopotamia and W Syria. Towards 470 he went to Antioch (at this time he must have assumed the Grecized form of his name) and gained the esteem of patriarch *Peter the Fuller. When Peter was driven out of the see (476-484), P. too was expelled by patriarch Calendio (482-484). Sent to *Constantinople by the monks of his party, from the emp. *Zeno he obtained Calendio's deposition and the return of Peter, who ordained him to the metropolitan see of Mabbug (Hierapolis) on 16 Aug 485. P. dedicated his episcopate, at least up to Severus's ascent of the patriarchal see of Antioch (512), to the struggle against the council of *Chalcedon and pope *Leo's *Tomus ad Flavianum*, to the point of accepting Zeno's *Henotikon*. He deployed all his strength, zeal, eloquence and his pen to convince the *clergy and monks of his province, rather hostile to his ideas, to support the *monophysites of *Persia and neutralize his adversaries, the chief of whom was *Flavian of Antioch who, after ten years of bitter fighting, had him deposed. During the all too few years of Severus's patriarchate, P. sought to calm the extremist spirits of his party. The accession of the emp. *Justin I (518) put an end to monophysism at Antioch; P., refusing to abjure, was deported. He died very old and tired at Philippopolis in *Thrace.

P.'s works can be divided into two literary groups: monastic and dogmatic. For the former, P. inherited the *Messalian and Evagrian terminology, from which he broke away, opposing to it the Syriac tradition. An evolution has been observed in his attitude towards *Evagrius's "learned *spirituality": this allows us to establish a relative chronology of his works, otherwise hard to date. Thus the *Epistle to Abraham and Orestes* follows that to *Patricius*, which in turn comes after the 13 paraenetic *memre*. To the second group of works belong treatises, *commentaries and *letters, which, being more polemical, are more easily datable. To a first pre-episcopal period belong the ten *memre* against Habib (*De uno e sancta trinitate incorporato et passo*). The initial period of his episcopate saw the great *Evangelical commentaries*, intended as a companion to the similar works of *Diodore of Tarsus and *Theodore of Mopsuestia. To carry out this work, P. commissioned from his suffragan, bishop Polycarp, a new translation of the NT more faithful to the Greek; this version, now lost, is known as the *Philoxeniana*. During the third period we see a P. hardening his positions, anti-Chalcedonian even in words, but then a little more conciliatory from the moment of Severus's accession, as the *Book of the sentences* attests. Finally, the last letters echo Justin's anti-monophysite persecution. (For a complete list of works, *spuria* and references, see bibliography below and esp. A. de Halleux's fundamental work).

P.'s "verbal" and archaizing monophysism has its roots both in the theocentric Alexandrian tradition and the apophatic Syrian tradition, so that his *theology is not just a simple reaction against the council of Chalcedon; it is characterized by a dual divine becoming: without changing, *God is born, suffers and dies (*theopaschism); with the destruction of *sin, the Son makes man into a new creature (filiation: the state of the simple believer) and unites him with God (restoration to the primordial state: the monk). Closely connected with his doctrine on the *Trinity (the Saviour assumes no other nature, nor second *hypostasis, than that of the God-Word), his *christology rests above all on soteriological bases: "our author's monophysism conceives the union less as a principle of singularity and personal unity than as a principle of mediation by the individual humanity of the Word, between the divine person and all mankind" (de Halleux, 514). This *soteriology is not intellectualist (like e.g. Evagrius), but sacramental: its roots are in the *mysteries of Christ's saving *economy.

Feast celebrated among the Syrian Orthodox on several occasions: 18 Aug (*ordination), 10 Dec (death), 18 Feb (translation).

M. Brière-F. Graffin, *De uno e sancta trinitate incorporato et passo, diss.* 3, 4, 5, PO 38 (1977); *diss.* 6, 7, 8, PO 39 (1979); *diss.* 9-10: PO 40 (1980); A. de Halleux, *Commentaire du prologue johannique*, CSCO 380-81, Syr. 165-66 (1977); J.W. Watt, *Fragments of the Commentary on Matthew and Luke*, CSCO 292-293, Syr. 172 and 174 (1978); cf. A. de Halleux: *Muséon* 93 (1980) 5-35; S. Brock, A classified bibliography, *PdO* 4 (1973) 343-465; id., for the years 1970-1980: PdO in press; Baumstark: 141-144; Ortiz: par. 107; A. de Halleux, *Philoxène de Mabbog, sa vie, ses écrits, sa théologie*, Louvain 1963 (with bibl.); P. Harb, Les origines de la doctrine de "la-hasusuta" (apatheia) chez Ph. de M., *PdO* 5 (1974) 227-241; F. Graffin, Le Florilège patristique de Ph. de M., *OCA* 197 (1974) 267-290; A. Vööbus, La biographie de Ph. de M., tradition de manuscrits, *AB* 93 (1975) 111-114; A. de Halleux, La Philoxénienne du Symbole, *OCA* 205 (1976) 295-315; A. Grillmeier, Die Taufe Christi und die Taufe der Christen. Zur Tauftheologie des Ph. von M. und ihrer Bedeutung für die christliche Spiritualität, in *Fides sacramenti, sacramentum fidei* (Studies in honor of P. Smulders), Assen 1981, 137-175 (with bibl.); R.C. Chestnut, *Three Monophysite Christologies: Severus of Antioch, Philoxenus of Mabbug and Jacob of Sarug*, Oxford 1972; D.J. Fox, *The Matthew-Luke Commentary of Philoxenus*, Atlanta 1979.

F. Rilliet

PHINEHAS: *iconography. To date, early Christian art has handed down a single depiction of the biblical episode of the Israelite P.'s killing of the Hebrew Zimri, guilty of having united himself to the idolatrous Midianite woman Cozbi (Num 25). It occurs at *Rome in a *painting in the catacomb of Via D. Compagni (Nestori, 72, no. 2), datable to the mid 4th c. P. appears dressed as a Roman officer: in his right hand is a lance on which are impaled the bodies of Zimri and Cozbi. The identification of the scene must be considered sure, on the basis of close iconographical resemblances to mediaeval Bible miniatures and a 15th-c. Serbian psalter (Kötzsche

Breitenbruch, *Die neue Katakombe*, 85-87, pl. 21), in which the episode is formulated in a wholly analogous way. But the scene, as represented in the painting and in these late miniatures, seems to have drawn less on the biblical account than on old Rabbinic traditions which had amplified it and were evidently widespread in the Christian world. Even from the formal point of view, a Jewish archetype has been hypothesized; but from the 9th c. on, illustrated versions of the Octateuch and psalters seem closer to the biblical passage (Kötzsche Breitenbruch, *Die neue Katakombe*, 86 f.; Ferrua, *Una scena nuova*, 110).

A. Ferrua, Una scena nuova nella pittura catacombale, *RPAA* 30-31 (1957-58, 1958-59) 107-116; id., *Le pitture della nuova catacomba di via Latina*, Vatican City 1960, 48, pl. 92; L. Kötzsche Breitenbruch, *Die neue Katakombe an der via Latina in Rom. Untersuchungen zur Ikonographie der alttestamentlichen Wandmalereien*, JbAC Ergbd. 4, Münster 1976, 85-87, pl. 21.

V. Fiocchi Nicolai

PHOCAS of Sinope. The sole historically ascertained saint of this name was a gardener, a *martyr at Sinope, whose Panegyric *Asterius of Amasea wrote in c.400. At this time his cult was already widespread in *Pontus, the Cyclades, Sicily, *Rome, *Antioch and *Tyre. His feast was 22 Sept. Starting from this authentic character, and using his name, were created a bishop P. (feast same day), a P. son of a ship-owner (*id.*) and a martyr of Antioch (5 March). All three are pure invention.

Vies des SS. 9, 449-51; *LTK* 8, 481; *BS* 5, 948-50.

V. Saxer

PHOEBADIUS. Bishop of Agen (*Aginnum*) in *Gaul, he had a prominent (anti-Arian) part in the council of *Rimini (359); he was the last to submit to subscribe the pro-Arian formula and was allowed to add some clarifications to it. *Jerome (*Vir. ill.* 108) calls him very old in 392. He also calls him the author of several *opuscula*, but names only *Contra arianos*. This short work, refuting the (pro-Arian) formula of *Sirmium of 357 and composed immediately after that date, is P.'s only surviving work. In it, for lack of adequate literature, P. exploits *Tertullian's *Adv. Praxean*, adapting its anti-*Monarchian arguments to the requirements of anti-Arian polemic. While never using the term *homoousios*, characteristic of Nicene *theology, P. defends that theological approach and considers the formula of 357, pro-Arian but put forward as a compromise formula, as *Arianism pure and simple.

CPL 473; PL 20, 11-30; *Patrologia* III, 77-78; A. Durengues, *Le livre de S. Fébade contre les Ariens*, Agen 1927.

M. Simonetti

PHOENIX: *iconography. Originating in the ancient East and Egypt as a symbolic expression of the exceptional or everyday events of the *cosmos, the myth of the p., as introduced to the West by Herodotus (*Hist.* II, 73), has it that a sacred bird, similar to an eagle and a peacock, dies every 500 years to rise immediately from its ashes. Latin authors use it to exemplify the concept of eternity understood as *palingenesis*, continual cyclical recurrence; the Church *Fathers use it to communicate the *mystery of the *resurrection of the *flesh (*1 Clem.* 1, 25; Tertull., *De res. carn.* III; Commod., *Carmen apolog.* vv. 139-142; Zen. Ver., *Tract.* I, 2, 9; Ambr., *Hexaem.* V, 23, 79; *Const. Apost.* V, 7, 15; Aug., *De natura et orig. anim.* IV, 20, 33; Ps.-Epiph., *Phys.* XI) and the concepts of *virginity, *chastity (Ruf., *Expos. symb.* IX, 10) and filial *piety (Orig., *C. Cels.* IV, 98, 25). Its most extensive and definite literary expression is Ps.-Lactantius's *De ave phoenice*. In early Christian art the myth is reduced to a *symbol and the bird is usually depicted with *nimbus and rays - typical solar attributes - in the fiery and majestic aspect of its accomplished regeneration: thus on an incised slab from S. Callisto (ICUR IV, 10785) and, perhaps, on a *painting on a masonry block in the catacomb of Pretestato (Nestori, 88, n. 11). The mythical bird also appears associated with the palm, whose symbolic significance and Greek name it shares, on some lead seals, on an incised slab from Priscilla (Bisconti, *Lastra incisa*, 45) and on a bronze candelabrum from *Aquileia (Bertacchi, *Il grande lampadario*, 341).

In the scene of the *traditio legis*, the p. is shown on the palm, corresponding to St *Paul, greatest assertor of the founding aspect of the resurrection (1 Cor 15, 14). This scheme is followed by a group of late 4th-c. *sarcophagi (*Ws* 150, 2; 154, 5; 151, 1; 12, 4; 39, 2), a painting in the catacomb of Grottaferrata (*WMM* 1, 132), an incised slab preserved at Anagni (Testini, *La lapide de Anagni*, 178) and some gold-glass. To designate the place of *paradise, the p. also appears in later paintings and *mosaics which use the scheme and setting of the *traditio*, e.g. the painting in the little basilica of the *cemetery of Felicita (8th c.) (De Rossi, Scoperta de una cripta, 149) or the apse mosaic of the basilica of SS. Cosma e Damiano (*WMM* 3, 102). Depictions of the bird at the moment of its death in the flames are more rare. We know a painted example in the so-called Greek chapel of Priscilla (Ferrua, Tre note di iconografia, 273) and one in mosaic in a room attached to the post-Theodorian basilica of Aquileia (Menis, *I mosaici*, 33). Mosaics and coins, esp. of the Constantinid dynasty, also attest the scheme of the p. on the mount and on the globe (Van den Broek, *The myth*, pls. II, III, VIII, XX, XXX). **[Fig: 256]**

G.B. de Rossi, Scoperta d'una cripta storica nel cimitero di Massimo "ad sanctam Felicitatem" sulla via Salaria Nuova, *BACr* 3 (1885) 149-184; L. Charbonneau-Lassay, *Le bestiaire du Christ*, Bruges 1940, 402-422; A. Ferrua, Tre note di iconografia paleocristiana: La fenice sul rogo, *Miscell. G. Belvederi*, Vatican City 1954-1955, 237-277; G.C. Menis, *I mosaici cristiani di Aquileia*, Udine 1965, 33-35; R. Van den Broek, *The myth of phoenix according to classical and early Christian tradition*, Leiden 1972; P. Testini, La lastra di Anagni con la "Traditio legis". Nota sull'origine del tema, *Archeologia Classica* 25-26 (1973-1974) 718-740; L. Bertacchi, Il grande lampadario di Aquileia, *Aquileia Nostra* 50 (1979) 341-350; F. Bisconti, Aspetti e significati del simbolo della fenice nella letteratura e nell'arte del cristianesimo primitivo, *VetChr* 16 (1979) 21-40; id., Lastra incisa inedita dalla catacomba di Priscilla, *RAC* 57 (1981) 43-67.

F. Bisconti

PHOS HILARON. The *hymn *Phos Hilaron*, "radiant light", acc. to *Basil the Great (*De Sp. Sancto* 29, 73), is of the greatest antiquity. It was an evening *chant, when the lamps were lit (Greg. Nyss., *De vit. Macr.* 25). In it, *light is seen as a *symbol of Christ, sent by the *Father. It may represent a reaction to the cult of the *sun, dear to the *pagans. The hymn is addressed to Christ and closes with a trinitarian *doxology, in which the three divine *persons are co-ordinated. We find an analogous structure in the *Gloria and in a 3rd-c. *papyrus (PO 18, 506). From Basil's arguments we can deduce that the hymn predates the *Arian controversy. Dölger thinks the 2nd c.

Ch. Pelargus, *Enchiridion Graeco-Latinum Hymnorum*, Frankfurt 1594 (editio princeps); M.J. Routh, *Reliquiae sacrae*, 3, Oxford 1815, 299, = ed. Usher; R.R. Smothers, Phos hilaron, *RecSR* 19 (1929) 266-283 (text and analysis). F.J. Dölger, Der christ. Abendhymnus, Phos hilaron, *AC* 5, 11-26; A. Hamman, *La prière*, II, Paris-Tournai 1962, 251.

A. Hamman

PHOTINUS of Constantinople. Presbyter and *defensor* of the church of *Constantinople, he wrote a *Life of *John (IV) the Faster*, the patriarch of Constantinople (589-95) against whose assumption of the title ecumenical patriarch (i.e. of the Byzantine empire) popes *Pelagius II (579-90) and *Gregory the Great (590-604) remonstrated in vain. A fragment of the *Life* survives, cited in the *Acts of Nicaea II* (787), *Actio* IV (Mansi 13, 79-80): it tells of a miracle obtained through the *image of Our Lady. From the account it appears that P. knew the patriarch personally and wrote his life after the death of the emp. Maurice (602), who is called martyr and, almost certainly, after that of the emp. Phocas (610), called 'serpent of the abyss and tyrant'.

CPG 3, 7971; Bardenhewer 5, 75 and 131; Beck 459.

A. De Nicola

PHOTINUS of Sirmium. Native of *Galatia, deacon and disciple of *Marcellus of Ancyra, he was bishop of *Sirmium (*Pannonia) when the Easterners first associated him with his teacher, condemning him in the *Ekthesis makrostichos*, and the Westerners subscribed his condemnation at *Milan (345). After an action against him in 347, of which we know little (Milan? Sirmium?), a *council of Eastern bishops met at Sirmium in 351 and deposed him. He returned to Sirmium at *Julian's accession (362), but was later expelled again. He had a certain following in the West, as far as *Africa, but never represented a real danger to *orthodoxy. *Jerome (*Vir. ill.* 107) says that P. wrote much and names *Contra gentes* and *Ad Valentinianum*. His works and those written against him are lost. Among the little surviving evidence, the *anathemata of the council of Sirmium of 351 and the information of *Epiphanius (*Pan.* 71) are important.

Though linked by the ancients with *Paul of Samosata as a representative of *adoptianism, P.'s closest points of contact were with Marcellus. He affirmed a rigid *monarchianism in which the Logos was conceived as a mere impersonal *dynamis* of the *Father, now enfolded in Him at rest, now manifest and exteriorized in the divine action inherent in the *world (*Logopator*). In this way the OT theophanies, which a unanimous ancient *tradition ascribed to the Logos conceived personally, were ascribed by P. to the Father operating through the Logos, since he considered this tradition ditheist. P., like Marcellus, made the Son of God be born of *Mary, in the sense that the Logos became Son only by being incarnate in the man Jesus and taking up his dwelling in him. Compared with his teacher, P. was more clear in affirming that the humanity assumed by the Logos was complete even in *soul, against the Logos/*sarx* *christology typical of the Alexandrians. As *Adam had sinned with soul and *body, so the Logos, to redeem man, was incarnate in a humanity complete in soul and body.

M. Simonetti, *Studi sull'arianesimo*, Rome 1965, 135-159; Simonetti 590; R.P.C. Hanson, *The Search for the Christian Doctrine of God*, Edinburgh 1988, 235-238.

M. Simonetti

PHOTIUS. Born at *Constantinople in c.820 of a noble iconodule family, he had an excellent education. Under the regency of Theodora (843-58), he was professor of philosophy at the university and then head of the imperial chancellery and member of the senate. Bardas called him to be patriarch (858) in place of Ignatius; Basil I deposed him (867) and recalled Ignatius; but on Ignatius's death, he was reinstated (878) and remained in office until 886, when he was finally removed by Leo VI. He died in 891. P. has importance both in the history of the church and as a theologian and humanist. This last aspect (which alone interests us) is revealed above all by the work commonly called the *Bibliotheca* or *Myrionbiblon*, but whose real title seems to be *Description and list of the volumes read by us*. As is said in the *Letter* of preface, addressed to his brother Tarasius, and repeated at the end of the work, this was put together and published immediately before P. left on an embassy to the Caliph Mutawakkil (855), and was intended as evidence of affection for his brother during their separation. So the work was compiled when the author was not yet 40 and was still a *layman holding office at court. It brings together the readings made and summarized in the preceding years, from the time when the desire to become cultured first arose in him. In it are reviewed, in no pre-established order, 279 works, which attest the variety and breadth of the author's interests: works by philologists, grammarians, lexicographers, metrists, orators, sacred and secular historians, doctors, philosophers, romancers, theologians. Each work forms a chapter, commonly called *Codex*; their length varies considerably, from simple annotations of a few lines (cf. Cod. 2. 4. 5 etc.) to real summaries of works with anthologies of extracts (cf. Cod. 220. 223. 230. 242. 245 etc.). Some authors are treated more than once; very often P. gives a judgment on the utility and literary value of the work under examination. Well over half the books cited in the *Bibliotheca* have been lost: hence the importance of the Photian work.

Surprising in P., still a layman, is the great place given to ecclesiastical literature, both in number (20 historians; 55 theologians) and in length (more than half the work). The authors or texts reviewed are: *a*) historians: Basilius Cilix, Charinus, Chrysippus presb., *Evagrius, *Eusebius, Eustratius, Georgius Alex. episc., Hermias *Sozomenus, *Hesychius ill., Ioannes presb. Aegeates, Ioannes (*John) Moschus, Fl. Iosephus iud., *Metrodorus, Nicephorus, Philippus Sid. (*Philip of Side), Philo Iud. (*Philo of Alex.), *Philostorgius, Sergius, *Theodoretus, Theophanes Byz.; *b*) theologians: Adriani alex. Andronici Isagoge, Andronicianus, Apollinarius (*Apollinaris), *Athanasius, *Asterius Amasiae ep., Basilius M. (*Basil the Great), *Basilius Seleuc., *Cassianus, *Caesarius, Iohannes (*John) Chrys., Clemens (*Clement) Alex., *Diadochus Phot., *Diodorus Tars., Ephraem (*Ephrem) Theop., Ephraim (*Ephrem) Syrus, *Epiphanius, Eugenius, *Eulogius, *Eusebius ep. Thess., *Gelasius, *Germanus patr., *Gregorius Nyss., *Heracleanus, *Hesychii duo, *Hippolytus ep., *Hippolytus Irenaei discip., Iobius (*Jobius) mon., Ioannes (*John) Carpathius, Ioannes (*John) Scytop., *Irenaeus ep., Iustinus (*Justin) martyr, *Leontius ep., Marcus mon., *Maximus Conf., *Methodius ep., *Modestus ep. Hier., Nicias mon., *Nilus mon., Origenes (*Origen), *Pamphilus martyr, *Pierius presb., *Polycarpus martyr, *Procopius, *Sophronius patr. Hier., Stephanus (*Stephen) Gobarus, *Synesius Cyr., Synodi variae, Theodorus antioch., Theodorus mon., Theodorus presb., Theodosius mon., *Theognostus.

E. Bekker, Berlin 1824 (in PG 103 f.); R. Henry, Paris 1959 ff.; *PWK* 20, 1 (1941) 667-737; S. Impellizzeri, *La letteratura bizantina*, Florence 1975, 340-365; *LTK* 8, 484-488.

A. De Nicola

PHRYGIA. Included from the start in the province of *Asia (list of cities, later all episcopal sees, in Pliny, *Nat. Hist.* V, 95 and 145), *Diocletian's reorganization divided it into two lesser provinces, Phrygia I (cap. *Laodicea) and Phrygia II (cap. Eukarpia). From the doctrinal point of view, a distinction can be made in the region on the basis of the vast amount of *epigraphy preserved: the *Montanist inscriptions of the Tembris valley, in N-E Phrygia, the *orthodox ones found in the central and southern urban areas. The latter attest the Christianization carried out through the Pauline churches of Laodicea, Colossae and Hierapolis; the former attest the spread of the Montanist *heresy in the 3rd c. - driven out of the Hellenized cities of S Phrygia - in the villages of the north, among rustic classes untouched by *Hellenism, where even in the 4th c. we know of the presence of a heretical Christianity (cf. e.g. Socr., *HE* IV, 28, on the *Novatianist bishop of Cotiaeum, metropolis of the upper Tembris valley). A special place in the Christianization of the region is occupied by the city of Laodicea Combusta, at the crossing of the roads from *Cappadocia to *Cilicia, a lively centre of commercial traffic: always absent from the cities represented at the 4th-c. *councils (first mentioned at *Chalcedon in 451), acc. to *Epiphanius (*Haer.* 47) it owed the name κεκαυμένης (Combusta) to the burning of the *encratite heresy in its territory; cf. also Basil, *Ep.* 188, reply to a query by *Amphilochius of Iconium on the admission of members of heretical sects into his church: no academic query, but doubtless based on the concrete problem represented by the presence in his district of a heretical city like Laodicea, as also shown by the epigraphs of Laodicea which mention the same sects dealt with in *Basil's *letter (Cathars, Saccophori, Apotactites, Encratites).

W.M. Ramsay, *The Cities and Bishoprics of Phrygia*, Oxford 1895-6; W.M. Calder, The Epigraphy of the Anatolian Heresies, in *Anatolian Studies Presented to Sir W.M. Ramsay*, Manchester 1923, 59-91; J. Schmidt, Phrygia, *PWK* 20, 1, 781-891; U. Peschlow, Bericht über Constantinopel und Kleinasien (Phrygien), in *Actes XIe Congrès internat. d'archéologie chrétienne*, CEFR 123, Rome 1989, 1612-1615. See also under *Montanism.

M. Forlin Patrucco

PHYSIOLOGUS (Gk *physiologos*): name of many Christian collections of questions and answers on *mirabilia* of natural science, circulated in Greek, Latin and the Oriental languages. The replies are given by a person called P., i.e. Aristotle, but the work was substantially anonymous, though some late revisions go under the names of famous *Fathers (*Epiphanius, *Basil, *John Chrysostom, *Jerome). The oldest redaction is Gk, prob. 3rd-c. (Treu's and Riedinger's attempts to take it back to the 2nd c., and Peterson's to bring it forward to the 4th, are not convincing); it consists of 48 "natures", mostly of animals, but also of plants and minerals. Their sources are very varied: alongside biblical material we observe classical, Hellenistic and even traditional Egyptian beliefs. The whole has been revisited in a Christian sense in accordance with the rules of allegorical *exegesis, which draws out *christological (sometimes with archaic *adoptianist tendencies) and moral (often *encratite) applications. An Alexandrian origin has sometimes been deduced from the use of *allegory and some allusions to *Egypt; but the question remains disputed. The work's popular character favoured its diffusion in the *East and in mediaeval Europe, in numerous languages and various redactions, variously interpolated from MS to MS - which obstructs research on its literary history. It had a considerable influence on Christian *symbolism and *iconography.

CPG 3766; *PWK* 20, 1 (1941) 1074-1129; F. Sbordone, Rassegna di studi sul Physiologus (1936-1976), *RFIC* 105 (1976) 494-500; F. Sbordone, *Physiologus*, Milan 1936; D. Offermanns, *Der Physiologus nach den Handschriften G und M*, Meisenheim am Glan 1966; D. Kaimakis, *Der Physiologus nach der ersten Redaktion*, Meisenheim am Glan 1974; U. Treu, Zur Datierung des Physiologus, *ZNTW* 57 (1966) 101-104; R. Riedinger, Der Physiologus und Klemens von Alexandria, *ByzZ* 66 (1973) 273-307; E. Peterson, *Frühkirche, Judentum und Gnosis. Studien und Untersuchungen*, Rome 1959, 236-253 (cf. *ByzZ* 47 [1954] 60-72); E. Peters, *Der griechische Physiologus und seine orientalischen Übersetzungen*, Berlin 1898; M. Goldstaub, *Der Physiologus und seine Weiterbildung, besonders in der lateinischen und in der byzantinischen Literatur*, Leipzig 1899-1901; F. Lauchert, *Geschichte des Physiologus*, Strasburg 1899; E. Brunner-Traut, Spitzmaus und Ichneumon als Tiere des Sonnengottes, *NGWG* 7 (1965) 123-163; E. Brunner-Traut, Ägyptische Mythen im Physiologus (zu Kapitel 26, 25 und 11), *Festschrift für S. Schott*, Wiesbaden 1968, 13-44; G. Graf, *Geschichte der christlichen arabischen Literatur*, v. 1 (ST 118), Vatican City 1944, 548-549; A. van Lantschoot, A propos du Physiologus, *Coptic Studies in Honor of W.E. Crum*, Boston 1950, 339-363; C. Conti Rossini, Il Fisiologo etiopico, *Rassegna di studi etiopici* 10 (1951) 5-51; A. van Lantschoot, Fragments syriaques du Physiologus, *Muséon* 72 (1952) 37-51; N. Henkel, *Studien zum Physiologus im Mittelalter*, Tübingen 1966; E. Brunner-Traut, Altägyptische Mythen im Physiologus, *Antaios* 10 (1969) 184-198 [repr.: id., *Gelebte Mythen. Beiträge zum altägyptischen Mythus*, Darmstadt 1981, 99-113]; F.J. Carmody, *Physiologus latinus. Etudes préliminaires, versio B*, Paris 1939; K. Grubmüller, Überlegungen zum Wahrheits-anspruch des Physiologus im Mittelalter, *FMS* 12 (1978) 160-177; J.H. Declerck, Remarques sur la tradition du Physiologus grec, *Byzantion* 51 (1981) 148-158; *LCI* 3 (1971) 432-436 and *passim*; R. van den Broek, *The myth of the phoenix according to classical and early Christian Traditions*, Leiden 1971; U. Treu, Otterngezücht. Ein patristischer Beitrag zur Quellenkunde des Physiologus, *ZNTW* 50 (1969) 113-122; E. Otto, Das Pelikan-Motiv in der altaegyptischen Literatur, *Studies presented to D.M. Robinson*, I, Washington 1971, 215-222.

S.J. Voicu

PIERIUS of Alexandria. Priest of *Alexandria, preacher and *ascetic during the episcopate of *Theonas, c.281-300 (Euseb., *HE* VII, 32, 26-30; Jerome, *Ep.* 70, 4). Head of the *didaskaleion*, acc. to *Philip of Side and *Photius (*Bibl.*, cod. 119), he was nicknamed "Origen junior" (Jerome, *De vir. ill.* 76). His writings were widely known. Ammonius, an *anchorite of the *desert of Nitria, recited numerous passages from *memory; Sylvia, a noble Roman lady on *pilgrimage to *Palestine, read some of his *homilies on the *Bible (Pall., *Hist. laus.* 12 and 14; Jerome, *De vir. ill.* 76). Photius knew personally, much later, ten of these homilies; he praises their style - clear, easy and free from artifice - the propriety of their *argumentation and the originality of their ideas. Unfortunately, he copies only two fragments: one, on *Luke's gospel, where he says that honour and dishonour done to the *image (*eikôn*) redounds on the "prototype", is on the *theology of the *Trinity; the other, an extract of an Easter homily on Hosea, concerns the cherubim placed by *Moses on the ark of the covenant, and *Jacob's pillar. Two more fragments appear in what remains of Philip of Side's *HE* (ed. de Boor, *TU* 5, 2). A *Vita S. Pamphili* and a treatise on the Mother of God, mentioned by Philip of Side, are lost. Either because of *Diocletian's *persecution, or rather because his life was made impossible by the anti-*Origenism of Theonas's successor, *Peter I of Alexandria, P. fled to *Rome, where he spent his last years. Speaking of some MSS of the NT, *Jerome calls them *exemplaria Pierii* (*Com. Mt.* 24, 36). It is possible that *Pamphilus, P.'s disciple, brought these copies to *Caesarea, where Jerome could have consulted them.

M.J. Routh, *Reliquiae Sacrae*, Oxford 1846, III, 423-435; PG 20, 733 ff.; A. v. Harnack, *Geschichte der altchr. Literatur bis Eusebius*, Leipzig 1893, I, 439 ff.; L.B. Radford, *Three teachers of Alexandria: Theognostus, Pierius and Peter. A study in the early history of Origenism and Anti-origenism*, Cambridge 1908; *DTC* 12, 1744-46.

C. Kannengiesser

PIETY. P. (εὐσέβεια) means the application of all one's being to *God's service. Its opposite is *impiety. Since classical antiquity it has meant the profound religious sentiment of respect and *love for the deity (cf. Isocr. 258; Plat., *Symp.* 193) and for one's parents (Plat., *Rep.* 615). The OT presupposes the "fear of God", not so much "fright" as sovereign respect for the divine majesty and transcendence, and as a whole suggests the idea of religion or p. towards God. On man's part, p. takes concrete forms in *Abraham's heroic obedience (Gen 22, 12), *Jacob's faith (Gen 48, 15), *Joseph's flight from offence to God (Gen 39, 9), Josiah's (Sir 49, 3) and Onias III's (2 Mac 3, 1) fidelity to God and in the theological-Messianic definition of *Jerusalem: "glory of piety" (Bar 5, 4). In the NT, the word appears only in the Pastoral epistles and 2 Peter. Its original meaning is enriched by a stress on the filial love of God, who manifested himself uniquely and sublimely in the *incarnation of the Word. The "teaching which accords with p." is a euphemism for the religion of Christ (1 Tim 6, 3). The universalism of p., which has the promises of eternal *life (1 Tim 4, 7; cf. 1 Cor 9, 24-26; Eph 6, 12; Phil 3, 12-14), presupposes a close link with Christ and with religious knowledge. It is the way to acquire imperishable goods (1 Tim 6, 6-7) and is part of the *virtues of the man of God (1 Tim 6, 11). The proof of its authenticity is in effective *love of neighbour (Jas 1, 26-27).

Primitive Christian literature accepted and adopted the secular meaning of love and respect for the gods (religion) as well as the biblical meaning, making the word mean the sense of duty (Justin, *I Apol.* 3, 2), the p. of *devotion in its various manifestations (Euseb., *Praep. ev.* I, 1). To this meaning, which seems predominant, are added others variously defined: religion (or worship?) of the *pagans, *Jews and Christians, according to specification (Justin, *Dial.* 14, 2; Athenag., *Leg.* 13, 1; Euseb., *Praep. ev.* XI, 15), *orthodox faith as opposed to *heresy (Athan., *C. Arian.* II, 45; Epiph., *Haer.* XXXIV, 21), correct moral behaviour (Justin, *Dial.* 4, 7), religious observances of the pagans (Athenag., *Leg.* 30, 3) and Christians (Euseb., *HE* IV, 24), worship of the deity (Clem. Al., *Strom.* I, 24). In *late antiquity it becomes the honorific title of monks, bishops, emperors (Athan., *Ad Monacos Epist.* I; *Epist. ad Rufinianum; Apol.* II, 44).

The concept of p. also appears in a derivative and secondary way in the Greek word εὐλάβεια. In the LXX, the meaning of fear-reverence opens the way to the preferred meaning of piety-devotion (Ex 3, 6; 1 Sam 18, 15. 29; Jer 5, 22; Mal 3, 16). In the NT, we can only examine Heb 12, 28 concerning worship, which is offered with p. and fear in continuity by the church-community to God, a consuming fire (a theological title: Dt 9, 3). In the religious vocabulary of Christian writers, the meaning of reverence and p. is prevalent (Polyc., *Ad Phil.* 6, 3; Evagr. Pont., *De orat.* 42; Cyr. Al., *Ps.* 5, 8; Euseb., *HE* IV, 15, 8; Clem. Al., *Strom.* VI, 17; Hipp., *In Dan.* 3, 22, 1; etc.). It had more success as an honorific title of monks, bishops, priests and emperors (Alex. of Al., *Epist. ad Alex.* I; Greg. Naz., *Epist.* 102; Athan., *Epist. ad Jovianum* 1).

EC 9, 1388-1389; *GLNT* 9, 1458-1486; *Enc. Bib.* 5, 728-729.

E. Peretto

PILATE

I. Pilate in tradition - II. Iconography.

I. Pilate in tradition. The canonical gospels agree on the fact that, although P. recognized Jesus' innocence, he committed a judicial crime to satisfy the people. This seems to contradict what we otherwise know of him. Procurator of Judaea from AD 26, many stories could be told of his administration, full of provocations, violence, rapine, executions without trial (Phil., *Leg. Gai* 301-2; Josephus, *De bello iud.*, II, 169-177; *Antiq. iud.*, XVIII, 55-59, 62). Lk 13, 1 also alludes to a cruel act committed by P. The Christian description of P.'s conduct may be due in part to apologetic concern to present Jesus (and primitive Christianity) as innocent before the Roman authorities, attributing all hostility against him to the malice and intrigues of the Jewish leaders; but this explanation is insufficient. The fall and execution of Sejanus put P.'s political career in danger. If Jesus' trial is placed in the period following Sejanus's fall, AD 32 or 33, we can understand why P. weakly gave way to the pressures against Jesus. Later, realizing that his career was safe, P. returned to his old ways (Josephus, *Antiq. iud.* XVIII, 85-87). In 36 he was deposed and sent to Rome to give an account of his actions before the emperor. Christian legend has him commit suicide (Euseb., *HE* II, 7), or even executed.

The apologetic tendencies of the *Ev. Petri* also concern the figure of P., who seeks to free himself from all responsibility in Jesus' condemnation, to blame it on *Herod and the *Jews. *Justin (*I Apol.* 35. 48) mentions the Acts of Jesus' trial before P.; but this is prob. just a supposition of their existence. *Tertullian (*Apol.*) speaks of a *letter from P. to *Tiberius in which P. narrates Jesus' miracles in detail; as much as to say that its author was a Christian. *Eusebius (*HE* II, 2) also mentions this apocryphal letter. But he does not mention the Christian *Acta Pilati*; yet he knows well anti-Christian Acts, which had to be learnt by *memory in schools under the persecutor *Maximus (*HE* I, 9, 3-4; IX, 5, 1; IX, 7, 1-2). It is possible that the Christian *Act. Pil.* were written as a reply to them. The surviving literature concerning P. is much later. There is an apocryphal correspondence between P. and Tiberius; another between P. and Herod; and a letter from P. to *Claudius. Relatively old is the account in *Paradosis Pilati*, where P., judged before the senate, proclaims his innocence of Jesus' death and throws the blame on the Jews. Condemned to be beheaded, he prays to the Lord asking for pardon. A voice from heaven proclaims him blessed and announces that he will be a witness of the Lord at his second coming. This explains and justifies the St P. venerated in the *Coptic church. A *Martyrium Pilati* survives in *Arabic and *Ethiopic. At the other end of the spectrum, we have three Western accounts (*Mors Pilati, Cura sanitatis Tiberii, Vindicta Salvatoris*) in which Tiberius, cured by Veronica's veil, orders P. to kill himself (or exiles him); evil spirits then take possession of his corpse. The evidence of *Orosius (*Hist.* VII, 4) in the 5th c., and of *Gregory of Tours (*Hist. Franc.* I, 21-24) in the 6th, confirm the hypothesis of a further revision of the primitive *Act. Pil.*, which survive under the title *Evangelium Nicodemi* and had such influence on Christian *iconography from the 5th c.; they assert both P.'s testimony and that of the Jews themselves in favour of Jesus' innocence and divinity.

EC 9, 1472-1476; K. von Tischendorf, *Evangelia Apocrypha*, Leipzig ²1876; Hennecke-Schneemelcher I, 330 ff.; Erbetta 3, 119-135; A. de Santos Otero, *Los Evangelios apócrifos*, Madrid 1956, cf. index p. 739; E. Bammel, Φίλος τοῦ Καίσαρος, *ThLZ* 77 (1952) 206-210; J. Blinzler, Die Niedermetzelung von Galiläern durch Pilatus, *NovT.* 2 (1957) 24-49; O.V. Volkoff, Un saint oublié: Ponce Pilate, *Bulletin Arch. Copte* 20 (1969/70) 165-175; E. Cerulli, Tiberius and Pontius Pilate in Ethiopian Tradition and Poetry, *PBA* 59 (1973) 141-158; J. Lemonon, *Pilate et le gouvernement de la Judée: Textes et Monuments*, Paris 1981.

R. Trevijano

II. *Iconography. The depiction of P.'s judgment appears in Christian iconography in the course of the 4th c. Completely absent from the figurative repertoire of catacomb *painting, it had considerable diffusion and success in that of *sarcophagi, in the so-called "passion" contexts elaborated after the religious *peace and in the celebrative atmosphere of victory.

The scene is easily identifiable by its many characteristic elements: the *suggestus* and *sella curulis* on which P. sits, surrounded by a varying number of advisers; the *clepsydra* to measure the time taken by the trial; the *camillus* bearing the basin in which P. washes his hands; finally

Christ, alone or flanked by soldiers, brought in or taken out by them.

Well-known sarcophagi, datable to the mid 4th c., reproduce the scene: e.g. Lat. 55, "of the two brothers" (*Rep.* 45), Lat. 174 (*Rep.* 677, 1), that of Junius Bassus, dated 359 (*Rep.* 680, 1). A *unicum* (but Wilpert's guess is accepted by all recent scholars) would be the case of the Constantinian sarc. of Berja (Spain), where P. is replaced by *Nero, in the guise of the persecutor Nebuchadnezzar, and Jesus by the two apostles *Peter and *Paul in a symbolic context with clear ecclesiological allusion.

From the 4th c. on, the scene carries on with variations in the attitudes of the protagonists and in other kinds of monuments: e.g. the ivory *diptych of Milan Cathedral Treasury and the British Museum ivory, both 5th-c.; in the 6th c. appears the famous ciborium of San Marco, Venice.

If evidence is lacking in 4th/5th-c. painting, in the period immediately following and in line with the examples of court art so far examined, the scene returns in the *mosaics of S. Apollinare Nuovo, *Ravenna: finally, the recently re-examined *Codex Purpureus Rossanensis* deserves special mention. [**Fig: 257**]

Exhaustive bibliography collected by C. Carletti, s.v. Pilato, *EC* 9, 1476-1477; W.F. Volbach, *Elfenbeinarbeiten der Spätantike und des frühen Mittelalters*, Mainz 1952; F. Zuliani, I marmi di S. Marco, in *Alto Medio Evo*, Venice (undated); A. Saggiorato, *I sarcofagi paleocristiani con scene di passione*, Bologna 1968, 83-89; M. Sotomayor, *Sarcófagos romano-cristianos de España. Estudio iconográfico*, Granada 1975, 105 f.; F. De Maffei, *Il codice purpureo di Rossano Calabro: la sua problematica e alcuni risultati di ricerca*, Rossano 1978.

U. Broccoli

PILGRIM - PILGRIMAGE (Peregrinatio). The first known Christian pilgrim, a Cappadocian bishop named *Alexander, went to *Jerusalem soon after 200 "to pray and visit the sites there" (Euseb., *HE* VI, 11, 2). One hundred years later *Eusebius (*Dem. ev.* VI, 18, 23) declared, apropos of Zech 14, 4, that "all who believe in Christ come here (to Jerusalem) from every part of the world, not, as in the past, to admire the splendour of the city or to pray in the ancient Temple, but . . . to wonder at the effects of the conquest and destruction of Jerusalem . . . and to pray on the Mount of Olives opposite the city . . . where the Saviour's feet rested". Naturally, not all Christians went there and the Mount of Olives was not, at least at the start of the 4th c., the only place where Christians prayed; Golgotha and Christ's sepulchre were inaccessible since *Hadrian's Temple was on top of them (Jerome, *Ep.* 58, 3), but on the hill S-W of the city, on the Christian Sion, before *Constantine's basilica there was already a church. Constantine's many buildings at Jerusalem and in the Holy Land attested the already lively interest in the Holy Places and conferred a new impulse on pilgrimages. Not only Constantine's mother *Helena went as a pilgrim to the Holy Land (Euseb., *Vita Const.* 3, 42), but also Eutropia, mother of the emperor's wife Fausta (Soz., *HE* 2, 4). For both women the motive was probably the blood shed in their family, so that the idea of expiation was added to the two classical motives (*prayer and visiting the sites), an idea that later left a decisive mark on pilgrimage.

The earliest surviving Christian account of a pilgrimage, that of the Bordeaux pilgrim of 333, shows us that Christians in the Holy Land visited not only the sites of Jesus' life, but also those of the OT. It cites the NT only 9 times, but the OT 22, and mentions not just the sites of the saving events but, e.g., the tombs of the patriarchs at Hebron. Christian pilgrimages ideally continued those of the *Jews, though Eusebius radically stresses the difference. In the 4th and 5th cc. a peculiar stimulus was given by the fact that some Roman ladies had moved to *Palestine: after 372 *Melania senior (Pall., *Hist. Laus.* 46), in 385 *Paula and her daughter *Eustochium (Jerome, *Ep.* 108), in 417 *Melania junior (*Gerontius's *Life*) and Paulina. *Jerome settled at *Bethlehem with Paula, who created a hospice for pilgrims. In this way there arose in the *East Latin bases which increasingly reinforced, through *peregrinatio*, awareness of the unity of the church. To the three pilgrims named above must be added the Spanish nun *Egeria, whose surviving account constitutes a new stage. Not just the sites of memories and tombs - including those of Christian *martyrs - were visited, but also the dwellings of living *ascetics. Indeed Melania went first to the monks of *Egypt. Besides visiting the memorials of the patriarchs, Egeria was very interested in the many monks and nuns of Mesopotamia, whom she considered holy people (*Peregr.* 20 and 21). In particular the *stylites (monks living on a column), like *Simeon the Elder at Kalat Siman in *Syria (†459), *Simeon the Younger near *Antioch (†592), *Daniel near *Constantinople (†493) attracted many pilgrims. Even after their deaths their columns remained places of pilgrimage, as were the famous tombs of the martyrs at Resafa in Syria (*Sergius and Bacchus), *Seleucia in Isauria (*Thecla), *Chalcedon (*Euphemia), the city of *Mennas near *Alexandria, etc. The guide to the *journey was and remained the *Bible, even when special *itineraries of the Holy Land were compiled in the 5th c. (e.g. that of *Eucherius of Lyons). From it were taken not just itineraries, but suitable passages were read on the spot, a psalm fitting the circumstance was sung and prayers were said (*Peregr.* 3, 6; 4, 3; 14, 1). Where there was a church, the holy *sacrifice was celebrated (*Peregr.* 3, 6 and 4, 4 ff.), so that it was sometimes celebrated twice in one day, though those who were not priests communicated only once (*Peregr.* 4, 4). Pilgrims took part in local religious functions and festivals and, once home, recounted them (e.g. *Peregr.* 24-29, on the sumptuous *liturgies of Jerusalem). They also tried to bring back relics, e.g. some fragment of Christ's *cross, venerated at Jerusalem from c.350. At *Rome, however, in the most important centre of pilgrimage in the West because of the tombs of the apostles *Peter and *Paul, which conferred on it a prestige greater than that of *Carthage, Valencia or *Saragossa, they gave only "contact relics", i.e. pieces of material that had been lowered into Peter's tomb. Elsewhere Rome offered sufficient relics of martyrs, and thus arose a new form of pilgrimage with the aim of procuring relics (Avit., *Ep.* 27; Greg. of Tours, *In glor. mart.* 28).

Yet pilgrimages were also criticized by great doctors: *Gregory of Nyssa exhorted men to leave the body were it was and make pilgrimage to the Lord, and not from Cappadocia to Palestine (*Ep.* 2, 18); Jerome, though he had led Paula's pilgrimage, insisted that: "*Anthony and all the ranks of monks . . . did not visit Jerusalem, and yet the doors of *paradise were thrown open to them" (*Ep.* 58, 2). But among monks we find another form of pilgrimage, which developed in parallel with *peregrinatio*, i.e. ascetic detachment from one's country, in imitation of *Abraham who left his own land. Thus it is better, acc. to *abbat Jacob, to live as a stranger than to give *hospitality to strangers (*Apophth. Patr. Jacob.* 1). This does not mean that the monk must undertake a pilgrimage with a fixed destination, but that he must detach himself from his house and *family. For abbat Titoe, only he who reins in his tongue truly lives in a foreign land (*ibid., Titoe* 2; cf. also the reply of Lucius and Longinus).

B. Kötting, *Peregrinatio religiosa. Wallfahrten in der Antike und das Pilgerwesen der alten Kirche*, Münster 1950; J. Wilkinson, *Egeria's Travels*, London 1971; H. Donner, *Pilgerfahrt ins Heilige Land. Die ältesten Berichte christlicher Palästinapilger (IV.-VII. Jh.)*, Stuttgart 1979; P. Maraval, *Lieux saintes el pélérinages d'Orient*, Paris 1985; P.W.L. Walker, *Holy City, Holy Places? Christian attitudes to Jerusalem and the Holy Land in the Fourth Century*, Oxford 1990; L. Reedimans, Siedlungsbildung bei spätantiken Wallfahrtsstätten, in (eds.) E. Dassman et al., *Pietas* (Festsch. B. Kötting), JbAC Erg vol. 8, Münster 1980, 325-358.

H.J. Vogt

PIMENIUS. The *Mart. hier.* names this *martyr on 18 Feb and 2 Dec. The true date seems the former, which agrees with the Passion and the marble *calendar of Naples. The other date is explained by that of *Bibiana, into whose legendary cycle P. was inserted. His body rested in the *cemetery of Pontianus ad ursum pileatum on the Via Portensis: the information also appears in the *Itineraries. In the cemetery his image can be seen with those of two other martyrs, accompanied by an inscription: *SCS. Milix. SCS. Pimenius. SCS. Pollion*. His name sometimes appears in the Passion of Bibiana (BHL 1322), the legend of Crescentius (BHL 1986) and that of Donatus (BHL 2289), where he is called a priest. His relics were transferred to S. Silvestro *in capite*.

Vies des SS. 3, 518; *BS* 10, 871, 73.

V. Saxer

PINYTUS. *Eusebius of Caesarea (*HE* IV, 21 and 23, 7-8) is the only source to preserve the name of P., bishop of Knossos (*Crete) under the emp. *Marcus Aurelius (161-180), apropos of an exchange of *letters with bishop *Dionysius of Corinth, who put his colleague on guard against certain manifestly *encratite tendencies. P. replied insisting on his duty to give his flock a progressive spiritual nourishment that would lead them to the stature of adults. Eusebius cannot check his admiration for P.: "In this letter, as in a complete picture, are manifested Pinytus's *orthodoxy in the faith, his care for all that could help his flock, his learning and his understanding of divine things" (ed. Bardy, p. 204). On the basis of Eusebius's testimony, *Jerome included P. in his *De vir. ill.*, 27 (PL 23, 679-680).

Eusebius, *HE* (SCh 31, 199, 204); *BS* 10, 876-877; P. Nautin, *Lettres et ecrivains chrétiens des IIe et IIIe siècles*, Paris 1961.

J.-M. Sauget

PIONIUS. *Eusebius (*HE* IV, 15, 47) puts the *martyrdom of P., a priest of Smyrna, at the time of that of *Polycarp, but acc. to the *Acts* of P. (ch. 23) he was martyred under *Decius. Scholars are divided between the two: on one hand, Eusebius may have been wrongly led to make the two Smyrnan martyrs contemporary by the fact of reading P.'s *Acts* in a collection which also contained those of Polycarp; on the other hand, hagiographical tradition has an undue tendency to link some Acts of martyrs to the time of Decius, the first empire-wide *persecution. Eusebius praises P.'s outspokenness, his confession of faith before the people and magistrates, his exhortations to the *brothers who flocked to him in prison. The *Acts* of P., whose degree of historical credibility varies from one part to another, but which certainly contains a trustworthy nucleus, has been preserved in the Greek text together with a Latin and other Eastern versions.

Delehaye *PM*, 26-33; *BS* 10, 919-921; *RACh* 2, 1175; M. Simonetti, *Studi agiografici*, Rome 1955, 9-51; Sist. Cyrilla, Pionius of Smyrna, *SP* 10 (1970) 281-284; G. Lanata, *Gli Atti dei martiri come documenti processuali*, Milan 1973, 162-177; *Acta Mart.* see (ed.) H. Musurillo, *Acts of the Christian Martyrs*, Oxford 1972, xxviii and xxix; R. Lane Fox, *Pagans and Christians*, London 1986, 460-492.

S. Zincone

PIOUS WOMEN: *iconography. The episode of the finding of the Holy Sepulchre empty by the p. w. (varying from one to three), related in Mt 28, 10, Mk 15, 46-16, 8, Lk 23, 55-24, 8 and Jn 20, 1-18, is so far absent from Roman cemetery *painting, but is depicted in the baptistery of *Dura Europos (pre-256), where the p. w. advance from right to left towards the closed tomb with ointments and torches; above the sepulchre two stars are interpreted as *symbols of two *angels. The fact that the scene appears first in a baptistery is evidently related to its clear eschatological symbolism. In plastic art, however, it appears at *Rome (*Ws* 325) and *Milan (*Ws* 243, 6), where we find the angel and the two Maries near the tomb, elements repeated in a fragment from Aix (Le Blant, *Sarc.*, no. 208), while in the Servanne *sarcophagus (*Ws* 15, 2) the third woman is added. Only two *women, then, are shown on two of the wooden door-panels of S. *Sabina, Rome (*c*.430): in the first the angel appears to them, in the second the risen Jesus stands before them. In the minor arts, the scene is attested on the ivory *diptych of Munich (4th/5th c.) (three women, two soldiers, wingless angel), that of Milan (5th c.) and another in London (two women, two soldiers): in the marble *capsella* of St John the Baptist at *Ravenna (5th c.), two women face the cross-bearing Christ, who ascends to heaven, drawn by the divine hand (two different moments of the gospel narratives fused synoptically). In the *ampullae* of Monza and Bobbio (6th c.), two or three p. w. appear, with a winged angel with *nimbus. The best-known *mosaic example is in S. Apollinare Nuovo, Ravenna (early 6th c.), with two women and an angel; in MSS the episode recurs in the Evangelarium of Rabbula (586), f. 13a, with three guards on the ground, angel, two women; next to it, the same women at Jesus' feet as he ascends to heaven. **[Fig: 258]**

DACL 15, 1, 570-573; *EC* 7, 141; *LCI* 2, 55-57; R. Bartoccini, La capsella marmorea di Ravenna, *RAC* 7 (1930) 299-302. W.F. Volbach, *Elfenbeinarbeiten der Spätantike und des frühen Mittelalters*, Mainz 1952, no. 116, pl. 35 (London diptych); F.W. Deichmann, *Frühchristlichen Bauten und Mosaiken von Ravenna*, Baden Baden 1958, pl. 206; id., *Ravenna. Geschichte und Monumente*, Wiesbaden 1969, 187-188; A. Grabar, Le fresque des Saintes Femmes au tombeau à Doura, *CArch* 8 (1956) 9-26; A. Grabar, *Ampoules de Terre Sainte*, Paris 1958, pls. 11, 2; 19; 28; 34/39; Cecchelli, *Rabb. Gosp.*, 70-71, f. 13a; Volbach-Hirmer, *Arte*, 75, fig. 92 (Milan diptych), 76, fig. 93 (Munich diptych); E. Dyggve, *Sepulcrum Domini*, Festschrift F. Gerke, Baden Baden 1962, 11-20; C.H. Kraeling, *The excavations at Dura Europos*. II. *The Christian Building*, New Haven 1967, 213 ff., pls. 20, 26/28; J. Engemann, Palästinensische Pilgerampollen im F.J. Dölger Institut in Bonn, *JbAC* 6 (1973) 5-27; E. Dassmann, *Sündenvergebung durch Taufe, Busse und Martyrerfürbitte in den Zeugnissen frühchristlicher Frömmigkeit und Kunst*, Münster 1973, 43-44 and 377.

M. Perraymond

PIRMINIUS. Perhaps of Spanish origin, he carried on an intense missionary activity in S *Germany in the first half of the 8th c. He founded the *monastery of Reichenau on an island in Lake Constance. In 727 he went to Alsace, where he founded Murbach and other monasteries. He wrote *De singulis libris canonicis scarapsus* (=*excerptus*), almost a florilegium of scriptural passages arranged so as to illustrate the catechetical theme of the history of the *world seen as a work of divine *providence for man's salvation. Since the text had to serve for the training of missionaries in a *pagan land, it also developed themes against *idolatry, taken from *Martin of Braga's *De correctione rusticorum*.

PL 89, 1029-1050; G. Jecker, *Die Heimat des heiligen Pirmin, des Apostels der Alamannen*, Münster 1927 (with ed. of the *Scarapsus*).

M. Simonetti

PISENTHIUS of Kepht. Person of considerable importance in the *Coptic church (late 6th-early 7th c.). A monk in the *monastery of Apa Phoebammon at Gemes near Thebes, he was consecrated bishop of Kepht (Koptos) by patriarch *Damian. We have two Coptic *biographies, highly romanticized, and the original archive of his correspondence, found near the monastery of Epiphanius, also near Thebes, where he long resided. His sole surviving literary work is an *Encomium of St *Onuphrius*, well-written, but with no particular characteristics.

E.A.T. Wallis Budge, *Coptic Apocrypha*, London 1913, 75-127, 258-334; E. De Lacy O'Leary, The Arabic Life of S. Pisenthius, *PO* 22 (1930) 313-488; M. Krause, *Revue d'Egyptologie* 24 (1972) 101-107; W.E. Crum, *ROC* 20 (1915-17) 38-67; L.S.B. MacCoull, *Dioscorus of Aphrodito: his work and his world*, Berkeley 1988.

T. Orlandi

PLATONISM AND THE FATHERS

I. General considerations - II. The second century - III. Clement of Alexandria and Origen - IV. The Greek Fathers - V. The Latin Fathers.

I. General considerations. Of all the Greek philosophers, Plato was undoubtedly the one most venerated by the *Fathers. If we except figures like *Tatian, the author of the *Cohortatio ad Graecos*, *Hermias, *Hippolytus, *Epiphanius and *Tertullian, who inclined to a radical condemnation of *all* Greek *philosophy, what strikes us when we examine the attitude of those Fathers who were open to Greek thought is that the often violent arguments over Aristotle and the Stoics (cf. *Aristotelianism and *Pantheism) - not to mention Epicurus, considered ===== *tout court* - do not touch Plato, who was considered the author of the highest and most inspired philosophy, a genuine "*theology" able to provide the Christian who was not content with "simple *faith" with an adequate conception of the deity and the necessary means to reach it, hence worthy of being compared with Christianity and even of being used in the search for its highest truths. The rejection by some Fathers of certain doctrines of Plato or of the Platonic school - e.g. metempsychosis, matter coeternal with *God, demonology (cf. *Demon) - did not diminish their admiration for the man who remained for them the greatest philosopher of all time. But the Fathers' veneration of Plato was not an isolated phenomenon: Platonic philosophy was an object of veneration and study by the exponents of the Academy and, in a wider setting, by the *pagan intelligentsia, and continued as a current of thought in the pagan world until the start of the 6th c. In its initial phase, that of the Old Academy of Xenocrates, Crantor and Speusippus (4th-3rd c. BC), this current was limited to Academic circles and was only one of many philosophical schools; but in the imperial age it became the dominant philosophy, replacing *Stoicism, the "philosophy" *par excellence* of the Hellenistic period, quite a few of whose doctrines it adopted. Syncretistic Platonism (cf. *Syncretism), which takes the name *middle Platonism in the first two centuries AD and *neoplatonism from the start of the 3rd c. and was also related to the *Judaeo-Hellenism of *Philo of Alexandria, ran parallel to the development of patristic thought; the period between the 2nd and 6th cc. saw an uninterrupted series of very close connections between the two currents, which gave rise to a continual osmosis. In studying *Justin martyr we cannot ignore the middle Platonism of the 2nd c.; in the case of *Clement and *Origen we must take into account both middle Platonism and the initial phase of neoplatonism; nor can the three great Cappadocians be explained without the neoplatonism of the 3rd and 4th cc., or *Dionysius the Areopagite be separated from the neoplatonism of *Proclus and Damascius; and it is now well-known that from *Marius Victorinus and St *Augustine we must go back to *Plotinus and *Porphyry. So the study of the "Platonism of the Fathers", though we must take into account some direct knowledge of Plato's writings, is mainly the study of their relationship with the Platonic current of thought that prevailed from the 1st to the 6th c.: it must examine the way in which the Fathers' interpretation of Platonic thought was influenced by contemporary Platonism, as well as the doctrines arising from this Platonism, appropriated by the Fathers and used by them to interpret their religious creed. In this enquiry it would be quite sterile to adopt the method followed in J. Meifort's study of Clement's Platonism: to an abstract Platonic philosophy drawn from the Platonism of the first two centuries AD, Meifort opposes an equally abstract Christianity, thus emphasizing their essential incommunicability. Nor do the positions of W. Völker and E. von Ivanka seem much more promising: Völker considers Greek thought and Christianity as two irreconcileable forces between which there can be no real synthesis, and is thus inclined to reduce the presence of Platonism in the Greek Fathers that he examines to a terminology without any real significance, incapable of altering the original essence of their Christianity; von Ivanka, though admitting the adoption of Platonic doctrines by the Fathers (*Übernahme*), insists on their transformation in a Christian sense (*Umgestaltung*); this last point of view, though acceptable in some particular cases, cannot be generalized to the point of becoming a universally valid canon of interpretation.

II. The second century. While *Tatian seems to treat Plato no better than the other Greek philosophers (accusing him of γαστριμαργία, *Ad Graecos* 2, p. 2, 22-23 Schwartz, 25, p. 26, 23-24, and of imitating Pythagoras and Pherecydes, 3, p. 4, 7-8), *Athenagoras shows great respect for him as well as a detailed knowledge of contemporary Platonism, which left a profound impression on his thought. In *Pro Christ.* 6 (7, 5-7 and 9-11 Schwartz), to demonstrate that Plato also believed in one *God, eternal and not subject to becoming, he cites with approval the two passages of *Timaeus* 28c - fundamental in the theology of *middle Platonism - and 41a (the latter also in ch. 23, p. 29, 19-21); and in ch. 23 (pp. 30, 23-31, 2) he cites *Phaedrus* 246e (cf. also Hippolytus, *Ref.* I, 19, 8, p. 20, 17-18), asserting that Plato called the supreme deity μέγας Ζεύς for reasons of clarity, εἰς σαφήνειαν, and adding (p. 31, 6-7) ὅτι μὴ δυνατὸν εἰς πάντας φέρειν τὸν θεόν: these last words echo *Timaeus* 28c; the expression ὁ τοῦδε τοῦ παντὸς ποιητὴς in *Pro Christ.* 8 (8, 32) also derives from *Timaeus* 28c. The influence of middle Platonism on Athenagoras is evident in the following points: *1)* God is above all beings (*Pro Christ.* 8 [9, 5-6]): cf. Plutarch, *De def. or.* 426b; *De Is. et Os.* 382f; Alcinous, *Did.* 164, 15-16; *2)* God is immortal, immobile, external, not subject to change and becoming (*Pro Christ.* 22 [27, 24-25, 28-29], 4 [5, 10], 6 [7, 7]); cf. Diogenes Laertius, III, 77; Hippolytus, *Ref.* I, 19, 6 (20, 12); Alcinous, *Did.* 164, 21 and 28; Maximus of Tyre, *Or.* XI, 59a; Philo, *De Cher.* 90; *3)* God is self-sufficient and above desire (*Pro Christ.* 29 [39-40]): cf. Alcinous, *Did.* 164, 28; Philo, *De virt.* 9; *4)* God is an eternal mind (*Pro Christ.* 10 [11, 9-10]): cf. Plato, *Tim.* 39e, *Philebus* 28c-e, 30c; Alcinous, *Did.* 164, 24; Philo, *Leg. Alleg.* III, 29; *5)* God can be understood only with the mind (*Pro Christ.* 4 [5, 10-11]): cf. Alcinous, *Did.* 165, 4-5; *6)* the doctrine of the three principles (God, matter, ideas) formulated in *Pro Christ.* 7 (8, 14-15) is typically middle Platonic (cf. Alcinous, *Did.* 163, 10-12; Hippolytus, *Ref.* I, 19, 1; Diogenes Laertius III, 76); *7)* the association between the λόγος and the idea in *Pro Christ.* 10 (11, 2-3) is Philonian (cf. e.g. *De opif. m.* 20; *De sacr. Ab. et C.* 83) and would be taken up by Clement, *Strom.* V, 16, 3 (II, 336, 8-9), who referred it to the "barbarians", i.e. to *Philo; behind this passage of Athenagoras lies the middle Platonic doctrine of the ideas as thoughts contained in the mind of God (cf. below apropos of Clement); *8)* the εἶδος is the idea immanent in matter, which thanks to its presence assumes a form and becomes a definite sensible object, *Pro Christ.* 22 (28, 7 and 21), 24 (32, 7 and 20): this originally Aristotelian doctrine is common in middle Platonism, as we see from the passages of Plutarch, Alcinous and Philo collected by E. Schwartz (index p. 105 TU IV, 2, Leipzig 1888); *9)* God originated the *world by impressing forms and order on originally formless and disordered matter, *Pro Christ.* 15 (16, 11-15); *De res. mort.* 3 (51, 16-17): this is the classical middle Platonic view of the origin of the world, which we find, e.g., in Plutarch's *De animae procr. in Timaeo*, in Alcinous (*Did.* ch. VIII, pp. 162-163), in Philo's *De opificio mundi*, in Hippolytus, *Ref.* I, 19, 1-3 (19, 7; 19, 4-20, 1) and in Diogenes Laertius III, 76-77; *10)* primordial matter is πανδεχές, *Pro Christ.* 15 (16, 14): this epithet, which goes back to Plato, *Tim.* 51 a 7 (thus he designates the formless receptacle, which subsequent philosophy would identify with primordial matter), is found, e.g., in Alcinous, *Did.* 162, 26 (cf. also the δεξαμενή of Hippolytus, *Ref.* I, 19, p. 19, 6-7); *11)* Athenagoras appropriates the contrast - originating with Plato and present in all subsequent Platonism - between the intelligible world, the world of being, eternal and not subject to becoming, and the sensible world, subject to becoming, having a beginning and an end: cf. *Pro Christ.* 15 (16, 2-4) and 19 (21, 15-19) and Plato, *Tim.* 27d-28a; *12)* the motif of the wonder produced by the vision of the beauty of the universe, which leads men back to its maker, is found in *Pro Christ.* 16 (17, 13-14): present in Plato (*Philebus* 28e; *Laws* XII, 966d-e), it constantly returns in subsequent Platonism and in various patristic authors (cf. *Cosmos).

*Justin Martyr describes Plato and Pythagoras as two "wise men, who were the wall and support of our philosophy" (*Dial. c. Tryph.* ch. 5, p. 28, 16-17 Otto), thus conforming to the contemporary tendency in philosophy (cf. C. Andresen, *Justin und der mittlere Platonismus*, *ZNTW* 44 [1952-53] 162; id., *Logos und Nomos*, Berlin 1955, 124 and 239-240; S. Lilla, *Clement of Alexandria*, Oxford 1971, 44); and, referring to the period before his *conversion, he says of the scholastic Platonism of the 2nd c. AD which attracted him more than all the other philosophical schools: "The thought of the incorporeal essences took possession of me, and the *contemplation of the ideas gave my mind *wings; in a short time I believed I had become wise, and in my simplicity I hoped soon to see God: because this is the end of the Platonic philosophy" (*Dial.* ch. 2, pp. 10, 23-12, 2). At the beginning of the *Dialogue with Trypho*, Justin speaks like any Platonist of his time (cf. also C. Andresen, *ZNTW* 44 [1952-53] 163): nearly every sentence contains echoes of the Platonic dialogues and of the scholastic Platonism that was the basis of his training. The term δεξαμενή (*Dial.* 2 [10, 24]) is linked to *Phaedrus* 246 c 1, 249 c 4, 249 d 6, 251 b 7, and foreshadows the motif - so popular with subsequent Fathers, esp. *Gregory of Nyssa - of the winged mind flying towards the heights. The phrase "that which tears the mind away from sensible things and makes it apt to intelligible ones, so that it can see absolute beauty and truth" (*Dial.* 2 [10, 11-13]) takes us back to the motifs of the *soul's departure from the *body and the sensations, present e.g. in *Phaedo* 65e, 67c, and of the contemplation of the beautiful (*Sympos.* 211d-212a). The motif of philosophy as the one source of happiness (*Dial.* 3 [14, 13-15]) follows the Stoic and Platonic view, taken over by middle Platonism, of the self-sufficiency of virtue (cf. S. Lilla, *op. cit.*, 68-72). The definition of philosophy as the "science of being and knowledge of the true" (*Dial.* 3 [14, 18-19]), besides being based on a terminology very common in Plato, recalls Alcinous's definition of the contemplative life, "which consists in the knowledge of the truth" (*Did.* 152, 28-29). His definition of *God (*Dial.* 3 [16, 1-2]) recalls the two Platonic doctrines of the unalterability of the ideas (*Tim.* 28a) and of the absolute good on which the existence of all beings depends (*Rep.* VI 509b): cf. also *Timaeus* 29a and Alcinous, *Did.* 164, 35-36. His definition of science as knowledge of human and divine things (*Dial.* 3 [16, 8-10]) takes up the definition of wisdom that was formulated by *Stoicism and became common in *middle Platonism (cf. C. Andresen, *ZNTW* 44 [1952-53] 162 n. 20). The motif of God's knowability only by the mind (*Dial.* 3 [18, 1-2]) is the same as is found in Athenagoras and Alcinous (cf. above, Athenagoras no. 5, and C. Andresen, *ZNTW* 44 [1952-53] 166) and goes back finally to the Platonic concept of the knowability of the ideas by means of the intellect (*Phaedo* 66 a 1; *Phaedrus* 247c; *Tim.* 28 a 1). Justin's words (*Dial.* ch. 4, p. 18, 7-13): "the mind's eye was given us for this, so that with this pure instrument we could see the being that is the cause of all intelligible beings, which has neither colour, form nor magnitude visible to the eyes, but is something that is above all essence, unspeakable and inexpressible, which alone is the beautiful and the good, and which manifests itself unexpectedly in well-prepared souls thanks to their affinity with it and their desire to see it", are reminiscent of various Platonic dialogues: the expression εἰλικρινεῖ αὐτῷ ἐκείνῳ has its exact correspondent in *Phaedo* 66a; "the being that is the cause of all intelligible beings" is the absolute good of *Rep.* VI 509b; the absence of colour, form and magnitude in God goes back to *Phaedrus* 247c, *Parmen.* 137d, *Sympos.* 211e, and, as C. Andresen has shown (*ZNTW* 44 [1952-53] 166), also appears in *Celsus (fr. VI, 64 Bader) (this is the famous *via negativa*, common in middle Platonism and *neoplatonism and also adopted by subsequent Fathers: on its presence in middle Platonism and neoplatonism cf. S. Lilla, *Clement of Alexandria*, 221 and 222 n. 5); the supreme being's superiority to essence is based on the famous passage of *Rep.* VI, 509b, which would become fundamental in the theology of neoplatonism; its ineffability goes back to *Sympos.* 211a, *Parmen.* 142a, *Tim.* 28c and is also common in middle Platonism (cf. S. Lilla, *op. cit.*, 220-221); the identification of absolute Beauty and absolute Good (i.e. between the Beautiful of *Sympos.* 210e-212a and the Good of *Rep.* VI, 509b) occurs in the Platonism of the 2nd c. AD (cf. e.g. Alcinous, *Did.* 164, 30-33; 165, 27), but not in neoplatonism (where the Good would be identified with the first *hypostasis, the Beautiful with the second); the words "which manifests itself unexpectedly in the soul" recall *Epist.* VII, 341 c 6-d 1; and the motif of the desire (ἔρως) for *contemplation is found in *Phaedo* 66e and *Sympos.* 210e. The phrase "it is above all when we are torn away from the body and remain alone with ourselves that we obtain what we love" (*Dial.* 4 [22, 2-3]) is indebted to *Phaedo* 66 a 3-5, 66 e 2-3, 67 c 6-d 2, 68 a 6-7. The doctrine of the ingenerate and immortal *soul, explicitly attributed by Justin to "some Platonists" (*Dial.* 5 [24, 9-10]), has a precise counterpart in middle Platonism (cf. Hippolytus, *Ref.* I, 19, 10, p. 21, 5). The theory of the ingenerate universe, which Justin rejects (*Dial.* 5, p. 24, 11-12), belongs not just to the Peripatetic school but also to those middle Platonists - such as Alcinous, Apuleius, Celsus, Taurus - who did not interpret the *Timaeus* literally (cf. S. Lilla, *op. cit.*, 197-198). To consider the universe's materiality and subjection to becoming as proof of its generation by a higher cause - a thesis accepted by both Justin and Trypho (*Dial.* 5, p. 24, 13-16) - goes back directly to *Timaeus* 28b (in middle Platonism the generation of the universe was maintained by Plutarch and Atticus; cf. also Hippolytus, *Ref.* I, 19, 4, p. 20, 4-6; C. Andresen, *ZNTW* 44 [1952-53] 163-164; id., *Logos u. Nomos*, 280-283 and S. Lilla, *op. cit.*, 197). Justin's words on the universe, which, though of itself subject to corruption, is made immortal by the will of God (*Dial.* 5 [26, 8-10]), refer explicitly to *Timaeus* 41 a-b (on the agreement between Justin and middle Platonism in the interpretation of this passage of the *Timaeus*, cf. C. Andresen, *ZNTW* 44 [1952-53] 163). The attributes "ingenerate and incorrupt" ascribed to God (*Dial*. 5 [28, 3-4]), already observed in Athenagoras, are common in middle Platonism (with a precise correspondence in Hippolytus [*Ref*. I, 19, 6, p. 20, 12]). The motif of the dialectical process terminating in the absolute principle, *Dial*. 5 (p. 28, 14-15), is the same as is found at the end of the sixth book of the *Republic* (511b) and in Alcinous (*Did*. 162, 8-10). Even after his conversion, Justin continued to speak in the language of contemporary Platonism: his conception of the supreme deity remained that characteristic of this Platonism, as is proved by his citation of the famous passage of *Timaeus* 28c in *II Apol*. 10 (1304, 22-24), based, as C. Andresen has demonstrated (*ZNTW* 44 [1952-53] 167-168), not on the original text of the Platonic dialogue, but on a middle Platonic scholastic tradition. Also dependent on middle Platonism, like those of Athenagoras, are Justin's views on formless primordial matter and the origin of the world, which is due to God's ordering action on matter (*Apol*. I, 10 [I, 156, 6-7] and I, 50 [I, 252, 26-254, 1]): cf. C. Andresen, *ZNTW* 44 [1952-53] 164-165. And in his criticism of the Peripatetics, Stoics and Epicurus, we see the influence of the scholastic Platonism of the 2nd c. (cf. S. Lilla, *op. cit.*, 45-51). An echo of the Pythagorean maxim ἕπου θεῷ, widespread also in middle Platonism, appears in *Apol*. I, 14 (I, 164, 18-19): C. Andresen, *ZNTW* 44 [1952-53] 163 n. 20), has shown that this passage has an exact correspondence in Plutarch, *De recta rat. aud*. 1, 37 d.

The author of the *Cohortatio ad Graecos* rebukes Plato and Aristotle for contradicting each other and even themselves (chs. 6-7, pp. 28-32 Otto). While acknowledging that Plato had a correct opinion on the true God (ch. 20, p. 62, 2-3), he does not hesitate to claim that he owed it to *Moses: like *Orpheus, Homer, Solon and Pythagoras, Plato too was able during his stay in Egypt to get to know the Mosaic books and thus reach the high conception of the ineffability of God (ch. 14, p. 48, 25-29; ch. 20, p. 60, 6-8; p. 62, 3-6). This assumption provides Pseudo-Justin with an occasion to accuse Plato of cowardice: despite having learnt the "true theology" from Moses, Plato did not have the courage to profess it openly at Athens for fear of sharing the fate of Socrates; in introducing into the *Timaeus* (41a) the speech addressed by the demiurge to the astral gods, he wished to protect himself from any possible accusation of *impiety towards the traditional religion and to give the impression of professing polytheism (ch. 20, pp. 60-62; ch. 22, p. 64, 27-30). Beyond these criticisms, the knowledge of the Platonic tradition showed by Ps.-Justin betrays the unmistakable imprint of middle Platonism: his doctrine of the three principles (God, ideas, matter) (ch. 6 [28, 10-14]) is that in vogue in the Platonism of the 1st and 2nd cc. (cf. above: Athenagoras, no. 6); his connection of the ideas with God (ch. 6 [28, 17-18]) suggests the middle Platonic doctrine of the ideas as thoughts of God, explicitly mentioned in the next chapter (7 [30, 14-15]): cf. below, apropos of Clement; numerous other expressions and ideas go back to Plato and Platonic authors. Here we will simply mention that, in ch. 22 (66, 5-9), the words of *Tim*. 27d on the immutable being are connected with the ὁ ὤν of Ex 3, 14: this tendency to refer to God those words which in *Tim*. 27d-28a concerned being is middle Platonic in origin (cf. Alcinous, *Did*. 165, 4-5).

The few words given to Plato by *Hermias (*Irrisio gent. philos*. 11 [Diels, *Dox. gr*. 653, 27-30]) derive from a middle Platonic source. Also middle Platonic in origin is the chapter dedicated by *Hippolytus to the exposition of Plato's system, considered as one of the *heresies (*Ref*. I, 19 [19-23 Wendland; cf. Diels, *Dox. gr*. 567-570]); it is one of our most important sources of knowledge not just of the scholastic Platonism of the first two centuries AD in general, but also of its various currents (cf. also K. Praechter, in F. Überweg, *Grundriss der Gesch. der Philos*. I, Berlin 1926, 556).

III. Clement of Alexandria and Origen. In *Clement of Alexandria, veneration of Plato and the influence of contemporary Platonism take on wider dimensions and richer developments than in *Justin. In *Protr*. 68, 1 (I, 51, 25-27), Clement prays Plato to accompany him in his search for *God; immediately after, he cites the two famous passages of *Tim*. 28c and *Epist*. VII, 341c on the ineffability of God and addresses Plato again, approving his words and renewing the invitation of a few lines earlier (*Protr*. 68, 2 [I, 52, 1-2]). In *Strom*. I, 42, 1 (II, 28, 2-4), Plato is presented as "the friend of truth . . . as if carried away by God"; in *Strom*. II, 100, 3 (II, 167, 23-168, 4), he is praised for his definition of human destiny as ὁμοίωσις θεῷ (cf. *Theaetetus* 176b); and in *Strom*. V, 29, 4 (II, 345, 3), he is celebrated together with Pythagoras as he who "succeeded in catching the reflections of the correct meaning of truth". The theory of the use of the OT by the Greeks, which in *Tatian and the author of the *Cohortatio ad Graecos* produces only an attitude of disdain and self-sufficiency towards Greek *philosophy, in Clement, even more than in Justin, increases his admiration for the Athenian philosopher. While affirming, like Justin, Plato's dependence on *Scripture (cf. the passages collected in S. Lilla, *op. cit.*, 42 n. 4), Clement, like Justin and even more than him, sees this not as a reason to condemn Plato's philosophy, but as a further occasion to celebrate its divine origin and invite Christians to follow it. The most evident proofs are in *Stromata*, book I, ch. 25 and the two passages of *Strom*. II, 100, 2 and V, 29, 3-4: in these three cases, the agreement between Plato and *Moses enables Clement to bring out the sublime character of the Platonic conception of the contemplative life, the formula *homoiosis theô* and the Platonic-Pythagorean theology in general. So we should not marvel at the great number of Platonic citations and echoes in the writings of Clement, who in all probability had direct knowledge of Plato's dialogues (besides the *Register* of Stählin's ed. of Clement, cf. also F.L. Clark, Citations of Plato in Clement of Alexandria, *TAPhA* 33 [1902] XII-XX). Among Clement's many passages it will suffice here to call attention to *Strom*. V, 83, 1-4: the terms πτεροῦται and βρῖθον used to refer to the soul's leap up to the heights (cf. above: Justin) go back to *Phaedrus* 246c, 255 c-d, 247b; to indicate the divine origin of virtue, he cites two passages: *Meno* 100b and 99e, which contain the expression θείᾳ μοίρᾳ, also used by *Philo and contemporary Platonism (cf. S. Lilla, *op. cit.*, 66, n. 1). One important "Platonic" motif taken over by Clement - pointed out esp. by J. Wytzes (*VChr* 9 [1955] 148-158; 11 [1957] 226-245; 14 [1960] 129-153) - is that of the educational and providential function of the punishments inflicted by God on men: this would recur in *Origen and *Gregory of Nyssa. The *middle Platonism of the 2nd c. and of *Philo, and early *neoplatonism (Clement and *Ammonius Saccas lived in the same city, *Alexandria, in the late 2nd and early 3rd cc.), are the source of many of Clement's doctrines and confer an unmistakable imprint on his Christianity. Clement's attitude to the various Greek philosophical schools (veneration of Plato and Pythagoras; condemnation of Stoic materialism, of Epicurus and of the *Aristotelian doctrine of *providence) conforms faithfully to the canons of middle Platonism; and his conception of philosophy as "cultivation of wisdom", a wisdom defined in Stoic terms as "knowledge of divine and human things and their causes", and of the subordination of philosophy to *theology (cf. S. Lilla, *op. cit.*, 41-59), are derived from Philo and middle Platonism. Examination of Clement's system of *ethics shows the triple influence of middle Platonism, Philo and neoplatonism. Of middle Platonic and Philonian derivation - though in the final analysis, of Stoic, Platonic or Aristotelian origin - are the following doctrines: virtue as a disposition of the well-knit *soul which conforms to the *logos* and as a harmony of the soul itself; its self-sufficiency; its derivation from natural disposition, exercise and learning; the definition of the four cardinal *virtues and their application to the three parts of the human soul; the close connection between the various virtues; the origin and nature of the passions, irrational movements produced not by erroneous judgments of reason, but by the two lower parts of the soul (the irascible and the concupiscent) and by bodily conditions; regulation of the passions by the *logos* - the charioteer described in *Phaedrus* 246b - and obedience to the law of nature laid down by the universal *logos*, present also in human reason (ethic of μετριοπάθεια). The adoption of *apátheia* and its identification with *homoiosis theô* link Clement closely to the Alexandrian tradition, represented by Philo and neoplatonism (cf. S. Lilla, *op. cit.*, 60-117).

Nearly all the fundamental elements of the Clementine conception of *gnosis* also go back to Platonic tradition and Philo. The use of mystery-terms, the strong esoteric stamp given to higher knowledge, the ideal of the contemplative life and *communion with the intelligible realities, whose precondition is *catharsis* or detachment from the *body and the sensible world in general, the role of the auxiliary disciplines: all these motifs are present in Plato, Philo, middle Platonism and neoplatonism (cf. S. Lilla, *op. cit.*, 144-173). The same can be said of his cosmology and his theology. From the *Timaeus* comes the distinction - taken over by Philo and Platonic tradition - between the intelligible world or model and the sensible world or image, present in *Strom*. V, 93, 4; the doctrine of matter as ἄχρονος, which, acc. to *Photius, Clement adopted in his *Hypotyposes* (ed. Stählin, III, p. 202, 10-11), is that of middle Platonism; equally middle Platonic (we have already noted its presence in Athenagoras and Justin) is the conception of matter as without qualities or form (cf. S. Lilla, *op. cit.*, 191-194). In one particular point Clement departs from scholastic Platonism and approaches neoplatonism: in *Strom*. V, 89, 5-6 (II, 385, 5-9), partly anticipating the attitude of *Gregory of Nyssa, he rejects the middle Platonic doctrine of matter as a principle and retorts that it was defined by Plato not just as without qualities or form, but as a non-being (μὴ ὄν); this latter expression, referred to matter, appears in *Plotinus (*Enn*. II, 4, 10 [II, 64, 34-35]), who inherited it from Moderatus the neopythagorean (cf. Simplicius, *In phys*. 231, 4-5; cf. also S. Lilla, *op. cit.*, 197-199).

The doctrine of ideas as thoughts contained in the mind of God, present in *Strom*. IV, 155, 2 (II, 317, 11); V, 16, 3 (II, 336, 8-9); V, 73, 3 (II, 375,

18-19), is of clear middle Platonic and "Philonian" derivation. One doctrine in which Clement foreshadows neoplatonism is his definition of the *logos* as a unity comprising everything, πάντα ἕν, *Strom*. IV, 156, 2: the same idea appears in the Plotinian conception of *nous (cf. S. Lilla, *op. cit.*, 205-206) and in St *Augustine (cf. below). The various characteristic motifs of the Clementine conception of God's transcendence (distance from beings, superiority to space, *time and virtue, ineffability, *via negativa*) are those of Philo, middle Platonism and neoplatonism (cf. S. Lilla, *op. cit.*, 212-224). On the close parallel between *Strom*. V, 81, 6 (II, 380, 22-23) and *Enn*. VI, 9, 6 (VI², 170, 10-11), on the infinity of God, cf. S. Lilla, *JThS* n.s. 31 [1980] 100 n. 2; cf. also *Pantaenus).

*Origen does not show the same enthusiasm and admiration for Plato as does Clement. His lack of explicit veneration for Plato is part of his colder and more detached attitude towards Greek philosophy, pointed out by H. Koch (*Pronoia und Paideusis*, Berlin-Leipzig 1932, 176-180, 307), J. Quasten (*Patrologia*, I, Turin 1967, 318) and H. Chadwick (*Early Christian Thought*, 101-103). In his *Contra Celsum*, however, there are some positive appreciations of the Athenian philosopher: in III, 63 (I, 257, 4-5) he rebukes *Celsus for not having followed the Platonic teaching on humility and composure (cf. *Laws* IV, 716a); in III, 80 (I, 270, 16-22) he asserts the essential identity between the Christian conception of the blessed life and of communion with the divine and the Platonic doctrine of the soul's ascent to the heavenly vault and contemplation of the hyper-Uranian world (cf. *Phaedrus* 247a-c, 250b-c); in IV, 39 (I, 312, 26-29) he takes up Clement's theory (cf. *Strom*. II, 100, 3; II, 167, 24-168, 4) that Plato was in agreement with the Mosaic Law either by chance or because he got to know the texts directly during his stay in Egypt (this last point repeats the theory of the author of the *Cohortatio ad Graecos*); in VI, 3 (II, 72, 20) he fully approves the passage of *Epist*. VII, 341 c-d; in VI, 5 (II, 74-75) he connects the *light discussed in this passage with the light of Jn 1, 9 and Mt 5, 14 and renews his approval (p. 75, 16-17); in VII, 43 (II 194, 6), repeating the words of *Timaeus* 28c, he shows that he fully shares them (on Origen's attitude to Plato's philosophy cf. esp. H. Crouzel, *Origène et la philosophie*, Paris 1962, 49-65). As H. Chadwick observes (*op. cit.*, 102), despite his reservations about Greek philosophy, Origen makes an even deeper synthesis between Christianity and Platonism than does Clement. *Porphyry tells us that he was a pupil of *Ammonius Saccas, the founder of *neoplatonism (cf. Eusebius, *HE* VI, 19, 6), and that he lived always with Plato and the most important neopythagorean and Platonic philosophers (Eusebius, *HE* VI, 19, 8). In fact all his thought reveals the deep imprint of both middle Platonism and neoplatonism. H. Koch (*op. cit.*, 17, 206-207, 237-238, 268-274, 286-291) has shown that Origen's judgments on the various Greek philosophical schools (substantially analogous to those of Clement and Justin, cf. e.g. *C. Celsum* I, 21 and III, 75: I, 266, 20-267, 6) agree with those of the exponents of middle Platonism, which was thus no less important for his cultural formation than it had been for his two predecessors. Clearly dependent on middle Platonism - as well as on *Phaedrus* 247c - is the passage of *De principiis* IV, 3, 15 (347, 9-12) on incorporeal creatures: *si sit aliqua substantia in qua neque color neque habitus neque tactus neque magnitudo intellegenda sit, mente sola conspicabilis*, which has two exact parallels in Maximus of Tyre (*Or.* XI, 60a [139, 20-140, 5]) and in Alcinous (*Did.* 164, 14); in particular, the words *mente sola conspicabilis*, which refer not to God but to spiritual creatures, should be compared with *Phaedrus* 247 c 7-8, *Tim.* 28 a 1, Maximus of Tyre (*Or.* XI, 60a [140, 4-5]), Alcinous (*Did.* 165, 4-5), Ps.-Justin (*Coh. ad Gr.* 66, 19-20), Justin (*Dial.* 3 [18, 1-2]) and Athenagoras (*Legat.* 4 [5, 10-11]). Yet, as we have just noted, the middle Platonic authors and the *apologists refer them directly to God. The Origenian conception of *God depends both on the scholastic Platonism of the 2nd c. and on neoplatonism. Like the supreme deity of the middle Platonists, Origen's God is ingenerate, uncorrupt, above all beings (*C. Cels.* VI, 66: II, 136, 26; IV, 14: I, 284, 27; cf. e.g. Hippolytus, *Ref.* I, 19, 6; Alcinous, *Did.* 164, 15-16, 18-19; 166, 12), immutable and unalterable (*De or.* 24, 2, p. 354, 8-9; cf. e.g. Philo, *De Cher.* 90; Maximus of Tyre, *Or.* XI, 59a; Alcinous, *Did.* 166, 12), self-sufficient (*C. Cels.* VII, 65: II, 215, 6; *Comm. in Jo.* XIII, 34, p. 259, 20; cf. Celsus in *C. Cels.* VIII, 21: II, 238, 16; Alcinous, *Did.* 164, 28; Philo, *De spec. leg.* II, 38), without form, magnitude or colour (*C. Cels.* VI, 64: II, 134, 15; *De princ.* I, 1, 6, p. 21, 16; cf. Plato, *Phaedrus* 247c; *Parmen.* 137d; Hippolytus, *Ref.* I, 19, 3; Plutarch, *De Is. et Os.* 382b: II, 552, 2-4; Maximus of Tyre, *Or.* XI 60a), above any spatial location (*De or.* 23, 1, p. 349, 25-27; 23, 3, p. 351, 2-4; *De princ.* I, 1, p. 21, 15, p. 22, 4; cf. Plato, *Parmen.* 138a; *Sympos.* 211a; Philo, *De post. C.* 14; *De conf. ling.* 136; Apuleius, *Apol.* 64, 7), ineffable and nameless (*C. Cels.* VI, 65; VII, 43; cf. Plato, *Parmen.* 142a; *Sympos.* 211a; *Tim.* 28c; *Epist.* VII, 341c; Alcinous, *Did.* 164, 7; 164, 28; 165, 4; Celsus in *C. Cels.* VI, 65, p. 135, 18 and 26; VII, 42, pp. 192, 30; 193, 1; Maximus of Tyre, *Or.* XI, 60a; Apuleius, *De Plat.* I, 190; *Apol.* 64, 7), absolutely devoid of passions (*C. Cels.* VI, 65, p. 136, 6-7: where Celsus's position is accepted; *De princ.* II, 4, 4, p. 131, 27; cf. Philo, *De Abr.* 202: IV, 45, 8-9; Celsus in *C. Cels.* VI, 65, p. 136, 6-7). Origen also applies respectively to the *Father and to the *logos* the expressions πρῶτος θεός and δεύτερος θεός, characteristic of *Numenius and Alcinous (*De martyr.* 46 [42, 18 and 23]; *C. Cels.* VI, 47 [II, 119, 2]; V, 39 [II, 43, 22-26]; cf. Numenius, frs. 11 and 21 Des Places [pp. 53, 3-4; 60, 4-5] and Alcinous, *Did.* 164, 19). Nearly all these doctrines turn up again in *Plotinus; as for their affinities with neoplatonism, the following points must be noted: *1)* God is not composite, but simple and indivisible (*C. Cels.* IV, 14 [I, 284, 27-28]): cf. Plato, *Rep.* II, 380d, *Parmen.* 137d; Plotinus, *Enn.* IV, 2, 1 (IV, 7, 17-8, 30; 30, 41-42); V, 2, 1 (V, 33, 3), and also Philo, *Leg. Alleg.* II, 2 (I, 90, 10); Alcinous, *Did.* 166, 6; Clement, *Strom.* V, 81, 6 (II, 280, 21-22): this whole passage of Clement, 380, 20-25, is based on a neopythagorean interpretation of the first hypothesis of Plato's *Parmenides*; cf. S. Lilla, *JThS* n.s. 31 (1980) 97 n. 4. *2)* As well as *nous, God can be defined as a something which is above *nous* and *ousia (*C. Cels.* VI, 64 [II, 134, 24-135, 5]; VII, 38 [II, 188, 11]; *Comm. in Jo.* 19, 6 [305, 16-17]). This doctrine is typical of neoplatonism (cf. e.g. Plotinus, *Enn.* V, 1, 8 [V, 26, 7-8]; V, 5, 6 [V, 98, 11]; V, 6, 6 [V, 118, 30-31]) and goes back on one hand to Plato, *Rep.* VI 509b, on the other to Aristotle, Περὶ εὐχῆς fr. 46 Rose (as J. Whittaker points out, VChr 23 [1969] 92, it is present in Celsus as well; cf. *C. Cels.* VII, 45 [II 197, 2-3]). *3)* God remains absolutely unknown to human reason (*C. Cels.* VI, 65 [II, 135, 18-20]; *De princ.* I, 1, 5 [20, 7; 20, 22-23]); the same doctrine is present in Plato (*Parmen.* 142a; *Sympos.* 211a), Philo, Clement, *Numenius and Plotinus (cf. S. Lilla, *op. cit.*, 217-219, 222) and also in Celsus (*C. Cels.* VI, 65 [II, 135, 17-18, 24-25]). *4)* God creates the various beings by making them emanate from his own unity, and in this sense he can be considered their source (cf. *De princ.* II, 1, 1 [107, 3-5]; I, 1, 6 [21, 13]; I, 3, 8 [61, 14]). This is a typically neoplatonist idea, which in the patristic field would later be taken up and developed by Ps.-*Dionysius the Areopagite: for Plotinus, the *one* generates *nous* (in which is contained the totality of beings) since it "overflows" and releases its own superabundant energy (*Enn.* V, 2, 1 [V, 33, 7-8]; V, 3, 12 [V, 66, 40-41]).

The Platonic tradition has also left evident traces in the Origenian conception of the *logos*: *1)* in a surviving fragment of the *Commentary on the Epistle to the Hebrews* (PG 14, 1308C), Origen, describing the generation of the *logos* by the *Father, appropriates the doctrine of the Wisdom of Solomon (7, 25), acc. to which σοφία is an ἀπόρροια τῆς τοῦ παντοκράτορος δόξης εἰλικρινής; Plotinus uses the same idea of emanation from the originating principle to describe the generation of *nous* from the One (*Enn.* V, 2, 1 [V, 33, 7-8]; V, 3, 12 [V, 66, 40-41]). *2)* For Origen, the *logos* is the splendour that proceeds from the *light which is the Father (*De princ.* I, 2, 4 [33, 1-2]; I, 2, 7 [37, 7-8]; IV, 4, 1 [349, 17-19]; *Comm. in Jo.* XIII, 25 [249, 29-30]; *In Ier. hom.* IX [70, 19]): this doctrine too, while going back to Wis 7, 26, shows an analogy with Plotinus (*Enn.* V, 1, 6 [V, 22, 28-29]; V, 2, 12 [V, 66, 40-42]), who compares the *nous* generated by the One with light emanating from the sun (cf. also W. Theiler, *Forschungen zum Neuplatonismus*, Berlin 1966, 24). *3)* Just as Origen's *logos* is generated by the Father *ab aeterno* (*De princ.* I 2, 4 [33, 1]; *In Ier. hom.* IX [70, 16-22]), so Plotinus's *nous* is generated *ab aeterno* by the One (*Enn.* V 1, 6 [V, 22, 29-30]) (cf. Theiler, *op. cit.*, 24). *4)* Just as Origen's *logos* receives its existence and divinity from its *contemplation of the Father (*Comm. in Jo.* II, 2 [55, 5-8]; cf. H. Crouzel, *Origène et la connaissance mystique*, Bruges 1961, 497), so Plotinus's *nous* exists as *nous* in that it contemplates the One (*Enn.* V, 2, 1 [V, 33, 9-11]; V, 4, 2 [V, 81, 24-26]). *5)* Going back to Ps 35, 10 and Jn 8, 12, Origen (*Comm. in Jo.* II, 23 [80, 16]) calls both the Father and the Son "lights"; Plotinus, speaking of the one and of the *nous*, expresses himself in the same way (*Enn.* V, 3, 12 [V, 66, 44]). *6)* The *logos*, the second god generated by the first god (cf. above), is also called by Origen "demiurge" (*C. Cels.* VI, 47 [II, 119, 2]), exactly as *Numenius calls his second god "demiurge" (cf. 20-21 Des Places, p. 60) and as Plotinus calls *nous*, the second *hypostasis, "demiurge" (*Enn.* II, 3, 18 [II, 45, 15]; V, 9, 3 [V, 163, 26]). *7)* While the Father is the absolute Good, the *logos* is the *image of his "goodness", but not the absolute good (*In Matth.* XV, 10 [375, 12-376, 13]; *De princ.* I, 2, 13 [46, 13-47, 9]; *Comm. in Jo.* XIII, 34 [261, 27-28]): Numenius uses exactly the same terms (fr. 20 Des Places [p. 60]): referring to Plato (*Rep.* VI 509b), he calls the first god αὐτοαγαθόν (cf. the αὐτοαγαθόν of Origen, *De princ.* I, 2, 13, p. 47, 4) and model (ἰδέα) of the second god, who is "good" in that he participates in the Good which is the first god (cf. also Theiler, *op. cit.*, 24). *8)* To emphasize the superiority of the *logos* over all other beings, Origen (*Comm. in Jo.* XIII, 25 [249, 27]) uses the words ὑπερέχων οὐσίᾳ καὶ πρεσβείᾳ καὶ δυνάμει, derived from Plato (*Rep.* VI, 509b). *9)* Present in the divine wisdom, which is identical to the *logos*, are all the forms (τύποι), which are at once the models of sensible things and the "thoughts" or "rational principles" conceived in anticipation by God (*Comm. in Jo.* I, 19 [I, 24, 1-2; 7-8]; cf. *Comm. in Jo.* I, 34 [I, 43, 20-22]; cf. also Theiler, *op. cit.*, 21-22). *10)* The Origenian doctrine of the presence in the *sophia* or *logos* of the various *logoi* or rational principles has an exact correspondence in the Plotinian conception of the *anima mundi, which contains in itself the totality of the *logoi* (cf. on one hand *Comm. in Jo.* I, 19 [I, 24, 7-8]; I, 34 [I, 43, 20-22]; *De princ.* I, 2, 2 [30, 7-8]; *C. Cels.* V, 39 [II, 43, 22-27]; on the other, Plotinus, *Enn.* VI, 2, 5 [VI¹, 104, 12-14]): this doctrine, Stoic in origin (cf. e.g. SVF I, 98; I, 102; II, 1027), would be taken up by *Gregory of Nyssa (*De anima et res.*: PG 46, 2802-2941) and Ps.-*Dionysius the Areopagite (*De div. nom.* V, 8: PG 3, 824C). *11)* For Origen, the *logos* is also God's power which embraces and holds together the whole universe (*De or.* 23, 1 [349, 28-350, 1]): this doctrine, also present in Philo and Clement (cf. S. Lilla, *op. cit.*, 210-211), goes back on one hand to the Stoic conception of the *logos* (cf. e.g. SVF I, 530; III, 439) and on the other to the *anima mundi* of Plato's *Timaeus*, which also surrounds and embraces the sensible universe (*Tim.* 34 b 3-4; 36 e 2-3); it is also present in middle Platonism (cf. S. Lilla, *op. cit.*, 211-212, and *JThS* n.s. 31 [1980] 90 n. 2). As Theiler (*op. cit.*, 23 n. 37) has shown, we also find in Origen the Plotinian doctrine of "intelligible matter", the basic element of the ideas (or forms) which make up *nous*; cf. Calcidius, ch. 278. Plotinus dedicates the fourth treatise of the second *Ennead* to this problem.

While decisively rejecting (cf. Chadwick, 115-116; Theiler, 18-19; M. Simonetti, *I principi di Origene*, Turin 1968, 68 n. 35, 228-229 nn. 29-30 apropos of *De princ.* I, 8, 4, and also *C. Cels.* VIII, 30 [II, 245, 24-25]) the Platonic and Orphic-Pythagorean doctrine of the transmigration of the *soul from the human *body to that of an animal (cf. Plato, *Phaedo* 81e-82a; *Phaedrus* 249 b 2-3; *Tim.* 41d-42d), Origen considers the presence of the soul in the human body to be the result of its fall from heaven, produced by an inclination towards the corporeal and material and the "cooling" of its love for God; the differences between different men reflect the differences between the original condition of their souls (*De princ.* II, 8, 3, pp. 157, 14-158, 2; 158, 17-159, 1; 159, 4-14; 160, 1-7, 19-20; 161, 2-3; IV, 2, 7,p. 319, 10-14); the same idea was expounded in Plato's *Phaedrus* (246c-248e) and also taken up by Plotinus (*Enn.* VI, 9, 9 [VI², 184, 22-24]). Apart from the rejection of transmigration to animal bodies, the correspondence between the cyclical rhythm described in the *Phaedrus* and that in Origen (*De princ.* IV, 3, 11) is perfect: in the *Phaedrus* the souls fall from heaven into human bodies and, if they remain attached to corporeal things, transmigrate from them into irrational animals, thence to trace the same route back again; Origen establishes a parallel between the descent of the soul to this world and its subsequent descent from this world to the underworld (*De princ.* IV, 3, 11, p. 339, 7-9), and between the soul's reascent from *hell to earthly life and its ascent from earth to the firmament (*De princ.* IV, 3, 11, p. 339, 10-15). Theiler (*op.cit.*, 29) has noted that the same motif of the soul's fall into a human body by way of punishment and its subsequent rise after a longer or shorter period recurs in Hierocles.

Another Platonic motif in Origen, to which H. Koch (pp. 13-144) has called attention, is the idea of the value of punishments: these are not inflicted on the sinner by God in retaliation, but as a real cure and medicine (*De princ.* II, 5, 3, pp. 135, 30-136, 2); they lead to *purification (*De orat.* 29, 15, p. 390, 12-15) and are thus an important element of the divine *paideia* (*C. Cels.* V, 31 [II, 33, 9-10]). Clement had expressed himself, and Gregory of Nyssa would express himself, in similar terms. The idea of the educative and purificatory function of punishment goes back to Plato, as is proved by passages like *Phaedo* 113d and *Gorgias* 525b; it is also present in middle Platonism: for Plutarch, divine punishment is a medicine and a cure for the soul (*De sera num. vind.* 549f-550a [III, 421, 22-422, 8]); the terms *iatreia* and *pharmacon* used by Plutarch (pp. 421, 24; 422, 2) should be compared with the words *curari... medicamentis* in *Rufinus's *Latin translation (*De princ.* II, 5, 3 [135, 32]). The punitive function, acc. to Origen, is often exercised by *angels; similarly, for Hierocles, it is rational beings who punish men (on this correspondence cf. Theiler, p. 29).

In the doctrine of *apocatastasis too, the correspondences between Origen and Hierocles pointed out by Theiler (*op. cit.*, 27-28) are very close: for both, apocatastasis consists of a return to the original condition, involves the consequence of *divinization and union with the deity (on this coincidence cf. also Koch, *op. cit.*, 297) and presupposes man's spontaneous submission to God.

In *ethics, Origen is linked to the Alexandrian tradition represented by *Philo, *Clement and *Plotinus. For him, as for Plato (*Rep.* IV, 439 d 1-2; 444 d 13-e 2; *Tim.* 86 b 5-7), Chrysippus (cited in *C. Cels.* VIII, 51 [II, 266, 18-30]), Poseidonius, *Galen, Philo and Clement (cf. S. Lilla, *op. cit.*, 96 and 98), the passions are a real sickness of the soul (*Comm. in Matth.* XIII, 16 [I, 220, 12-14]); he too favours, to start with, their control and moderation to make them conform to nature (*Hom. XXII in lib. Jesu Nave*, 4 [436, 17-20]), but he too sees their total destruction as the highest ethical ideal (*Comm. in Matth.* XIII, 16 [I, 219, 28-29]; XV, 4 [II, 358, 22-23, 32-33]; XV, 17 [II, 398, 27]; fr. 64 [III, 41]); *Hom. I in lib. Jesu Nave* [294, 18-20]; *Comm. in Jo.* XX, 36 [376, 27]). So H. Koch's claim (p. 267) that Origen, like Alcinous, rejects *apátheia does not correspond to the truth. Origen too sees "likeness to God" (Plato, *Theaetetus* 176b) as man's final destiny: *De princ.* III, 6, 1 (280, 2-281, 5), *Comm. in Jo.* XX, 17 (349, 26-27).

As for higher knowledge or *gnosis*, H. Crouzel (*Origène et la connaissance mystique*, Bruges 1961) has emphasized and examined in detail the use made by Origen of terms taken from *mystery language or technical terms such as, e.g., μυστήριον, μυστικός, ἀπόρρητος, ἄρρητος, κρυπτός, ἀόρατος, (pp. 25-46), θέα, θεωρία, (*theoría*), θεᾶσθαι, θεωρεῖν, θέαμα, (pp. 375-379), ὁρᾶν, διορατικός, (pp. 380-381), νοεῖν, ἐννοεῖν, (pp. 383-389), χωρεῖν, χωρητικός, (pp. 392-395), γνῶσις, γνωστικός (pp. 395-398). A good many of these terms are present in Plato and were taken up by the whole subsequent Platonic tradition, as well as being widely used by Philo and Clement (cf. S. Lilla, *op. cit.*, 146-150). Subsequent patristic authors like Gregory of Nyssa and Ps.-Dionysius the Areopagite would also make regular use of them.

Finally, the Platonic tradition left an indelible mark on the Origenian conception of the *resurrection of the body: Origen argues vigorously against those who, interpreting *Scripture to the letter, believe in the resurrection of the material body, subject to the same physical needs as the body which lives on earth (*De princ.* II, 11, 2 [184-186]); and emphasizes the fact that the risen body ends by taking on a completely spiritual nature (*De princ.* III, 6, 6 [288, 6-7; 289, 7-8]): *iam tum corpus quasi spiritui ministrans in statum qualitatemque proficiat spiritalem*. This Origenian view was criticized by *Methodius (*De res.* III, 16, 9 [413, 5-8]).

*Gregory the Thaumaturge's *Thanks to Origen* is rich in Platonic echoes, both direct and filtered through later authors: they have been pointed out by A. Brinkmann (Gregors des Thaumaturgen Panegyricus auf Origenes, *RhM* 56 [1901] 56-58), who lays particular stress on the definitions of the four cardinal *virtues.

IV. The Greek Fathers. *Epiphanius considers Platonic philosophy a *heresy derived, like the other Greek philosophical schools, from the mystery religions in vogue among the Egyptians, Phrygians, Phoenicians and Babylonians (*Pan.* 4, 6-9 [182, 13-183, 10 Holl]). The brief sketch he gives of it (*Pan.* 6, 1-3, [185, 13-25]) depends both on *middle Platonism and on the neopythagoreanism of the 1st and 3rd cc. AD: while the doctrine of the three causes and of matter coeternal with God (p. 185, 19-20, 24-25) reveals a clear middle Platonic imprint and should be compared with *Hippolytus's information (cf. also Holl, *App. font.*, p. 185), the mention of the opposite theory, in which matter is not coeternal with God but is generated by him (p. 185, 21-22), goes back to the speculations of some Neopythagoreans on matter, as we see from Calcidius (*In Tim.* 295 [297, 16-19 Waszink]).

*Eusebius of Caesarea dedicates the whole of books XI-XIII and part of book XIV of his *Praeparatio evangelica* to Plato. Book XI aims to demonstrate exhaustively the thesis - present in the *Cohortatio ad Graecos*, in *Justin and *Clement - of the dependence of Plato's metaphysico-theological system on Mosaic wisdom (cf. esp. the title of ch. 2, vol. II, p. 6). Eusebius's attitude is not the hostile one of *Tatian and the author of the *Cohortatio ad Graecos*, but closer to that of Clement and marked by deep respect. Plato is presented as "the greatest of the Greek philosophers" (XI, 1, 3, vol. II, 5, 17) and "he who usually spoke the truth, though he did not always express it happily" (XI, 1, 5, vol II, 6, 6-8) (on Eusebius's judgments of Plato cf. E. Des Places, Eusèbe de Césarée juge de Platon dans la Préparation évangélique, in *Mélanges A. Diès*, Paris 1966, 69-77). Careful reading of book IX of the *Praeparatio* shows that what really matters to Eusebius is not proving Plato's dependence on *Moses, but rediscovering Platonic philosophy in the OT books, interpreted according to the canons of middle Platonism and *neoplatonism: he cites copious extracts from Platonic dialogues and from works by later authors like Atticus, Aristocles, *Numenius, Plutarch, *Philo, *Plotinus, Amelius and Clement. The interpretation of Plato followed by Eusebius is, in fact, strongly conditioned by Platonic tradition and also by earlier patristic authors: in ch. 10 (II, pp. 24-25) he adopts the Platonic distinction, taken over by all subsequent Platonism, between the sensible, changeable *world and the intelligible, unchanging world, connecting the famous passage of *Tim.* 27d-28a not just with Eccl 1, 9 but also with Ex 3, 14 (cf. above, apropos of the *Cohortatio ad Graecos*, and below, apropos of St Augustine); in chs. 15-19, like the middle Platonic and neoplatonic philosophers whose authority he invokes, he affirms the existence of the "second cause" in the Platonic system; in ch. 21 (II, p. 46), he sees in the famous passage of *Epist.* II, 312 d-e a proof of the existence of the three *hypostases in the Platonic theology: Plotinus - cited by Eusebius in ch. 18 (II, 39, 22-40, 8) - uses the same passage to give his three hypostases a "Platonic" basis (*Enn.* V, 1, 8); the very title of ch. 21, *Perì tòn archikòn hypostáseon*, betrays an unmistakable Plotinian imprint (cf. the title of the first treatise of the fifth *Ennead*). Book XII (II, 83-162) is wholly dedicated to demonstrating the agreement between Platonic philosophy and Mosaic wisdom. Book XIII (II, 163-256) on one hand praises Plato's rejection of Greek religion and puts forward Aristobulus and Clement as the principal authorities able to confirm the theory of the Greeks' dependence on and agreement with the OT, on the other it criticizes some points of Platonic philosophy as not being in agreement with the teaching of Moses. Book XIV, ch. 4 (II, 263-268) mentions with approval Plato's criticisms of previous physical theories. Chs. 5-9 give, on the basis of Numenius, a history of the Platonic Academy up to Carneades. Book XV contains a coherent criticism of the *Aristotelian philosophy from the point of view of the middle Platonist philosopher Atticus. These books of the *Praeparatio* reveal Eusebius's great interest not just in Plato, but also in all later Platonism: they are our main source of knowledge of Numenius and Atticus. The *excerptum* on the *soul taken from the seventh treatise of the fourth *Ennead* and preserved in book XV, ch. 22, of the *Praeparatio* (II, 387-399) is considered by P. Henry as authoritative evidence of the history of Plotinus's text (*Les états du texte de Plotin, Etudes plotiniennes*, I, Paris 1938, ch. 4, 68-71, 77-124).

The Platonic echoes in the works of *Basil the Great have been pointed out by K. Gronau in his still important study, *De Basilio Gregorio Nazianzeno Nyssenoque Platonis imitatoribus*, Göttingen 1908. Basil's use above all of the *Timaeus* is evident in his nine *Homilies on the *Hexaemeron* (cf. S. Giet, SCh 26a, Paris 1968, 57-60, the notes at the end of the edition and the index of authors, 537), though, as Gronau demonstrated (*Posidonius, Eine Quelle für Basilius' Hexaemeros*, Braunschweig 1912, and *Posidonius und die jüdisch-christliche Genesisexegese*, Leipzig-Berlin 1914, 7-112), it is based not just on a direct reading, but also and particularly on an intermediate source, the commentary on the *Timaeus* by the Stoic Poseidonius, who played a decisive role in all late antique and patristic thought. Among the many echoes of the *Timaeus*, we will mention here only the expression *áphthonos agathótes* referred to *God (*Hom.* 1, 2 [96, 12 Giet]), which is indebted to *Tim.* 29 e 1-2. The motif of the *contemplation of that which is beyond the heavens, present in *De fide*, ch. 1 (PG 31, 465B 7-10), goes back to *Phaedrus* 247 b-d; the image of the chariot drawn by two horses, symbol of the intellect overcome by the passions (*Ad iuv.* IX [57, 74-76 Boulenger]), is also from *Phaedrus* 246b, 247b; the motif of the *body as an obstacle and prison to the *soul, which finds liberation in philosophy (*Ad iuv.* IX [54, 4-55, 6; 57, 87]), goes back to *Phaedo* 65a, 67b, 83a. Still more important is Basil's direct dependence on neoplatonism, particularly Plotinus, which, though rather undervalued by Giet as regards the *Homilies on the Hexaemeron* (*op. cit.*, 60-61), has been given its proper weight by other scholars. A. Jahn, in his work *Basilius Magnus plotinizans*, Bern 1838, has shown the close dependence of *De spiritu* (PG 29, 768-774) on *Enn.* V, 1, 1-5 and of ch. 9 of *De Spiritu sancto* (PG 32, 108-109) on various treatises of the *Enneads*. P. Henry (*Les états du texte de Plotin, Etudes plotiniennes*, I, Paris 1938, ch. 5, 159-196), by means of numerous precise synoptic comparisons, has called attention to the correspondences, even terminological, between the *Enneads* and various writings of Basil (*Epist.* II; *In hexaem.*; *De fide*; *De Spir. sancto*; *De spir.*) and has upheld with strong arguments the authenticity of *De spiritu*, denied by Garnier. H. Dörries too (*De Spiritu sancto, Der Beitrag des Basilius zum Abschluss des trinitarischen Dogma*, Göttingen 1956, p. 53) has admitted, though with some reservations and limitations, the presence of Plotinian motifs in ch. IX of *De Spiritu sancto*. Finally H. Dehnhard (*Das Problem der Abhängigkeit des Basilius von Plotin* [PTS 3], Berlin 1964) has taken up the question of Basil's use of Plotinus particularly in *De spiritu*, bringing out the numerous coincidences between the two authors (cf. esp. 6-13, 57).

For *Gregory of Nyssa too, K. Gronau (*op. cit.*) has listed numerous Platonic echoes. Even if Gregory cleanly rejects some conceptions of Plato or later scholastic Platonism (in particular the doctrine of metempsychosis: *De an. et res.*: PG 46, 108B-121A; that of the fall of souls from heaven, characteristic of the *Phaedrus* and taken over by *Origen: *De an. et res.*: PG 46, 112B, 116B-117C; that of matter as an ingenerate principle coeternal with God, adopted, as we have seen, by the *apologists: *De an. et res.*: PG 46, 121C 8-124A 5), we can say that nearly every sentence he wrote is a reworking of motifs taken either from Plato or from later Platonic tradition and expressed in their characteristic terms. Among his numerous ideas also present in Plato, a few will suffice here: the passage on absolute beauty (*De virg.* ch. 11 [296, 15-20 Cavarnos]), which goes back to *Sympos.* 211a; the refusal to attribute feelings of envy to the deity (*De Sp. sancto* 99, 23-24 Müller; cf. Plato, *Tim.* 29e; *Phaedrus* 247a; *Epin.* 988b and W. Jaeger, *Gregor von Nyssa's Lehre vom heiligen Geist*, Leiden 1966, p. 38); the motif - frequent in Gregory and deriving from *Phaedrus* 246 d 6-7 - of the *winged intellect which, free from any material bond, flies towards the divine (cf. e.g. *De virg.* ch. 11, 294, 8-10); his views on the nature of the *soul, deriving from *Phaedo* (the soul is an intelligible and invisible essence [*De an. et res.*: PG 46, 36A; cf. *Phaedo* 66b-67b, 67d]; it must gather itself to itself and separate itself from the *body before it can contemplate the intelligible [PG 46, 88A, 89C; cf. *Phaedo* 65c, 65d, 67a, 79d, 80e, 83b]; and if it shows excessive attachment to the body, it is condemned to roam like a ghost around the tombs [PG 46, 88B-C; cf. *Phaedo* 81c-d]: on these correspondences cf. also M. Pellegrino, Il platonismo di Gregorio Nisseno nel Dialogo intorno all'anima ed alla risurrezione, *Riv. fil. neosc.* 30 [1938] 437-474); and the idea of the cathartic and educational value of punishment, also adopted, as we have seen, by Clement and Origen (*De an. et res.*: PG 46, 88A, 89B, 97C-100A, 100C, 157B-C, C-D, 160C; cf. *Phaedo* 113d; *Gorgias* 525b). The Platonic tradition, and neoplatonism in particular, left evident traces in Gregory's *theology: the supreme deity is above colour, form, magnitude and all corporeal properties (*De virg.* 10, 290, 23-291, 2; *In cant. cantic. hom.* V, 157, 17-18 Langerbeck; cf. e.g. Plato, *Sympos.* 211e; *Phaedo* 247b; *Parmen.* 137b; Alcinous, *Did.* 164, 14; Plotinus, *Enn.* V, 5, 6 [V, 97, 4-5]), above good itself (*In eccl. hom.* VII, p. 425, 8-13 Alexander; cf. Philo, *De opif. m.* 9 [I, 3, 21]; Iamblichus, *De myst.* VIII, 2 [196, 12-14 Des Places]), above mind (*In cant. cantic. hom.* V, 157, 15) (for references to neoplatonism and Aristotle cf. above, apropos of Origen, and Langerbeck, *op. cit.*, App. font., p. 157), above all beings (*De perf.* 188, 15-16 Jaeger; cf. Plotinus, *Enn.* V, 5, 6 [V, 97, 9-10]; V, 1, 6 [V, 22, 13]), admits only the *via negativa* (*De an. et res.*: PG 46, 40C; *In cant. cantic. hom.* V, 157, 16-19) (for the presence of this motif in middle Platonism and neoplatonism cf. above, apropos of Justin), is infinite (*Contra Eunom.* II [I, 246, 16-17 Jaeger]; *In cant. cantic. hom.* V, 157, 20-21; *De an. et res.*: PG 46, 97A; cf. E. Mühlenberg, *Die Unendlichkeit Gottes bei Gregor von Nyssa*, Göttingen 1966; Plato, *Parmen.* 137d; Plotinus, *Enn.* IV, 3, 8 [74, 38]), absolutely simple (*De an. et res.*: PG 46, 93C) (for references to the whole Platonic tradition cf. above, apropos of Origen), indefinable and so incomprehensible and ineffable (*De virg.* 10, 288, 21-22, 25-26, 290, 21-23; *De profess. Christ.* 134, 7-8 Jaeger; *De perf.* 194, 14-26; *C. Eunom.* II [II, 58, 26-28]; *Quod non sint tres dii* 52, 15-20 Müller) (for parallels in the Platonic tradition cf. above, apropos of Clement and Origen), flees from whoever seeks it (*Contra Eunom.* II [I, 246, 16-22]; *De perf.* 194, 12-14; cf. Philo, *De post. C.* 13 and 18; Plotinus, *Enn.* V, 5, 10 [V, 102, 9]), is self-sufficient, loves itself and desires nothing outside itself (*De an. et res.*: PG 46, 92C, 93A-B; cf. Plotinus, *Enn.* III, 8, 11 [III, 167, 9-10; 168, 42-43]; VI, 8, 15 [VI², 153, 11; 152, 1-153, 6]) and is the absolute beauty that beautifies all things which participate in it (*De virg.* 11, p. 292, 14-15; cf. Plato, *Sympos.* 211b; Plotinus, *Enn.* I, 6-7 [I, 103, 25-30]; I, 6, 8 [I, 104, 6-8]; but for Plotinus the beautiful is identified not with the first principle, but with the second *hypostasis). Two other important characteristic motifs of the Platonic tradition appropriated by Gregory are that of God's noetic activity or, more precisely, the presence of thoughts in the divine wisdom (before creating the sensible universe, the divine wisdom thinks it [*De perf.* 182, 14-15; 183, 3-4]; cf. Philo, *De opif. m.* 19, and above, apropos of Origen's *logos*, no. 9); and that of God's power which embraces, holds together and pervades the whole universe (*De profess. Christ.* 138, 27-139, 4; *De an. et. res.*: PG 46, 24C, 28A; cf. above, apropos of Origen's *logos*, no. 11). Gregory too, like his predecessors, adopts the Platonic ideal of *homoíosis theô (cf. *Theaetetus* 176b), which he combines with the expression *kat'eikóna kaì homoíosin* of Gen 1, 26; but while Clement and Origen, following Philo, see in the *eikón* of Gen 1, 26 a reflection of the divine *logos* in human reason, and in *homoíosis* a perfection understood as the crown of man's ethical endeavours and hence attainable only later, Gregory considers the two expressions of Gen 1, 6 as synonymous and is hence led to consider *eikón-homoíosis* as the expression of man's original perfection, which he lost as a result of *sin - which obscured the *image of the primitive beauty - and which he must strive to restore in himself by breaking, as far as possible, his links with the body. To H. Merki (*Homoiosis Theo. Von der platonischen Angleichung an Gott zur Gottähnlichkeit bei Gregor von Nyssa*, Freiburg in der Schweiz 1952, 92-164) is due the credit of having thoroughly analysed the meaning of *homoíosis* in Gregory and pointed out its analogies with neoplatonism: Plotinus too relates *homoíosis* closely to *apátheia, katharótes (purity) and the soul's return to its original beauty, obscured by its links with the body (cf. parallels in Merki, 98-99, 115-117). A further link between Gregory and Platonic tradition is the motif of higher knowledge of divine things (*theoría*): in him too, as in Clement and Origen, it takes on a strongly esoteric imprint, is expressed in terms taken from mystery-language and finds its Scriptural *symbols in images of cloud and darkness (on all these points cf. J. Lemaître, R. Roques, M. Villier, the section in *Contemplation* dedicated to Gregory in *DSp* 3, 1772-1775; and esp. J. Daniélou, *Platonisme et théologie mystique*, Paris 1944, and the section *Mystique de la ténèbre chez Grégoire de Nysse*, *DSp* 3, 1872-1885). As regards the motif of the soul's union with God, examined esp. by Daniélou (cf. the studies cited above), we will simply say here that some ideas in *De anima et resurrectione* recall analogous Plotinian motifs: for Gregory, as for Plotinus, the pure, beautiful soul is joined and mixed with the divine beauty to which it is akin (*De an. et res.*: PG 46, 89B, 93C; cf. *Enn.* I, 6, 7 [I, 103, 13]; I, 6, 9 [I, 105, 12; 106, 29-30]), sees in its own beauty the image and reflection of the original beauty which is its model (PG 46, 89C; cf. *Enn.* I, 6, 7 [I, 103, 25-30]), becomes like God by seeking to imitate that transcendent beauty (PG 46, 89D-92A; cf. *Enn.* I, 6, 7 [I, 103, 27-28], I, 6, 9 [I, 106, 29-30, 32-33]) and now possesses what it hoped to possess (PG 46, 93B; cf. *Enn.* VI, 9, 9 [VI², 185, 46]). On the relationship between Gregory of Nyssa and the Platonic tradition cf. also H.F. Cherniss's book, *The Platonism of Gregory of Nyssa*, New York 1971 (reprint of the 1930 edition).

In *Gregory Nazianzen too, K. Gronau (*op. cit.*) has found numerous Platonic echoes (cf. also R. Gottwald, *De Gregorio Nazianzeno platonico*, Diss. Breslau 1906; H. Pinault, *Le platonisme de saint Grégoire de Nazianze*, La Roche-sur-Yon 1925; E. Fleury, *Saint Grégoire de Nazianze et son temps*, Paris 1930, 79-80; and C. Moreschini, Luce e purificazione nella dottrina di Gregorio Nazianzeno, *Augustinianum* 13 (1973) 535-549 and Il platonismo cristiano di Gregorio Nazianzeno, *ASNP* serie III, 4 [1974] 1347-1392). Of the various Platonic ideas we will mention only four here, present in the second *Theological discourse* (*Orat.* 28): the allusion to the famous passage of *Timaeus* 28c (PG 36, 29C), also mentioned, as we have seen, by *Athenagoras, *Clement, *Justin and *Origen; the denial of any feeling of envy in God, met also in Gregory of Nyssa (PG 36, 40B); the motif of the winged thought flying to the heights, going back to *Phaedrus* 246e (PG 36, 65B); and the comparison between *God and the sun of the sensible world, which goes back to *Rep.* VI, 508 b 12-c 2 (PG 36, 69A); other passages of other *homilies are cited by C. Moreschini (*Augustinianum* 13 [1973] 538). We also find in Nazianzen's *theology the motifs characteristic of the whole Platonic tradition, already observed in previous Fathers (cf. esp. Origen and Gregory of Nyssa): God is ingenerate, unalterable, incorruptible (*Orat. 28 theol.* 2: PG 36, 36C), infinite, formless and untouchable (*Orat. 28 theol.* 2: PG 36, 33B), simple (*Orat. 30 theol.* 4: PG 36, 105B), not composite (*Orat. 28 theol.* 2: PG 36, 44A: *Orat. 31 theol.* 5: PG 36, 169B), not found in any place (*Orat. 28 theol.* 2: PG 36, 37C), far away (*Orat. 28 theol.* 2: PG 36, 29B), incomprehensible and unreachable (*Orat. 28 theol.* 2: PG 36, 29C, 32B, 40B, 53B-C), ineffable and nameless (*Orat. 28 theol.* 2: PG 36, 29C; *Orat. 30 theol.* 4: PG 36, 125B), desired by all rational beings (*Orat. 28 theol.* 2: PG 36, 44A; cf. Plotinus, *Enn.* I, 6, 7 [I, 103, 1-5], and also Aristotle, *Met.* = 1072a 26b 3), admits only the *via negativa* (*Orat. 28 theol.* 2: PG 36, 44A 4-6; *Orat. 29 theol.* 3: PG 36, 76B 15), pervades everything (*Orat. 28 theol.* 2: PG 36, 33D) and holds it together (*Orat. 28 theol.* 2: PG 36, 32C). Two further neoplatonic motifs adopted by Nazianzen are those of the mutual *communion of beings (*Orat. 28 theol.* 2: PG 36, 48A) (cf. e.g. Iamblichus, *De myst.* I, 5, 17 [47, 12-13 Des Places]) and the use of the term *parágein* to indicate the *creation of beings by the deity (*Orat. 28 theol.* 2: PG 36, 48A 15-16) (cf. e.g. Plotinus, *Enn.* VI, 8, 20 [VI², 159, 21]). On one point, however, Gregory Nazianzen takes a stand against neoplatonism: in *Orat. 29 theol.* 3 (PG 36, 76C), he openly criticizes the Plotinian conception of the production of the second *hypostasis by the automatic and involuntary overflowing of the power present in the One (cf. *Enn.* V, 2, 1 [V, 33, 8-9]).

Nor does *Nemesius of Emesa manage to escape the influence of neoplatonism, as chs. 2 and 3 of *De natura hominis* demonstrate: in ch. 2 (pp. 59, 12-70, 2 Matthaei), to refute the thesis of the corporeity of the human *soul, he appeals to the authority of *Ammonius Saccas and the neopythagorean *Numenius, one of the philosophers most read and admired by *Plotinus; and in ch. 3 he resolves the problem of the relationship between the divine and human natures in Christ's *person (cf. pp. 140, 6-141, 1) by referring to what Ammonius Saccas and *Porphyry, in his *Symmikta Zetemata*, had maintained of the soul's union with the body: this union does not involve any confusion between the two elements (cf. esp. pp. 129, 9-14; 139, 4; 140, 5).

As we see from his work *Adversus Iulianum*, the trinitarian theology of *Cyril of Alexandria is based on the neoplatonic doctrine of the three *hypostases. In the ten surviving books of this work, intended as a refutation of the emperor *Julian's three books *Against the Galilaeans*, Cyril aims to demonstrate that the Christian doctrine of the *Trinity is substantially shared by various Greek philosophers, and to this end he quotes Plato, Numenius, Hermes Trismegistus, Plotinus and Porphyry, citing longer or shorter passages of all these authors. As with Eusebius, the work most useful for Cyril's thesis is the Plotinian treatise *Perì tòn archikòn hypostáseon* (the first of the fifth *Ennead*), from which he approvingly cites selections, esp. from book VIII and also from books IV and II: cf. esp. PG 76, 921B-C, 921D-924A, 924B, 929C, 917D-920A, 920C for book VIII; PG 76, 724A-B for book IV; PG 76, 604B for book II (all these citations are pointed out by P. Henry, *Les états du texte de Plotin, Etudes plotiniennes*, I, Paris 1938, 123-140).

In his *Graecarum affectionum curatio* (PG 83, 784-1152; P. Canivet, SCh 57, 1-2, Paris 1958), *Theodoret of Cyrrhus more than once names both Plato and the exponents of the subsequent neopythagorean, middle Platonic and neoplatonic tradition (cf. Canivet's very accurate indexes under Amelius, Ammonius Saccas, Atticus, neoplatonists, Numenius, Plato, Platonists, Plotinus, Plutarch of Chaeronea and Porphyry, SCh 57, 2, pp. 473, 475, 482, 483-486). He also gives numerous citations of works by all these authors (cf. the relevant index, pp. 451-466). Plato has a magnificent style (I, 9, p. 105, 25), the beauty of his language puts all other philosophers in the shade (I, 12, p. 106, 21-23), he is the best of the philosophers (II, 6, p. 138, 15-16), like Pythagoras and Anaxagoras he went to Egypt to learn true wisdom from the Hebrews (II, 23-26, pp. 144, 19-145, 9), he agrees with Pythagoras esp. in his doctrine of the soul, an essence of divine origin which falls into bodies by way of punishment (V, 13, p. 229, 18-20; V, 28, p. 234, 9-15), he sees purity of soul as the precondition for receiving divine teachings (I, 85, p. 126, 4-7, citing *Phaedo* 67b), he speaks of a multiplicity of gods only for fear of the Athenians (II, 38, p. 149, 111-13; II, 42, p. 150, 10-11; III, 74, p. 192, 4-6), he believes in one God (II, 38, p. 149, 9-11; II, 42, p. 150, 6-9, citing the famous passage of *Tim.* 28c), he admirably distinguishes supersensible and changeless being from sensible and changeable becoming (II, 33-36, pp. 147, 14-148, 19, citing *Tim.* 27c-28a, 37e-38a), he identifies the supreme ethical ideal with detachment from the things of this world (XII, 22, p. 424, 19-21) and with likeness to God as far as is possible (XI, 9, p. 394, 3-7; XII, 22-23, pp. 424, 21-425, 6, citing *Theaetetus* 176 a-b), he studies Scripture (VI, 61, p. 276, 5-6), follows and appropriates the teaching of Moses and the prophets (II, 26, p. 145, 8-9; II, 43, p. 150, 13-18; VI, 31, p. 265, 14-15; VI, 33, p. 266, 10-11; XI, 27, p. 40, 13-22) and is, in Numenius's words, "an Attic-speaking Moses" (II, 114, p. 169, 17-18). His philosophy is way above that of Aristotle, who did not pay him due respect and who opposed him particularly in the two doctrines of the mortality of the soul and divine *providence (V, 46-47, p. 242, 13-15; cf. *Aristotelianism). Particularly worthy of admiration and praise are his metaphysico-theological ideas: the universe is born from a cause represented by the demiurge, good by nature and without any feeling of envy; everything owes its existence to the good, superior to essence in dignity and power; the best proof of the derivation of the universe from a cause is its sensible and corporeal nature, subject to birth and becoming; it is hard to find its maker and demiurge, and harder still to express him; the sensible universe is the most beautiful of born things; the demiurge is the most beautiful of causes; changelessness is proper to more divine things, change is proper to bodily things (IV, 32-45, pp. 212, 16-217, 5, citing *Timaeus*, the *Republic* and the *Statesman*); the universe was formed according to a model-idea, to be identified with God's thought and with the intelligible heaven (IV, 49, p. 218, 13-19). But some aspects of Plato's teaching and conduct cannot be approved: the doctrine of matter as a principle coeternal with God, also shared by Pythagoras, the Stoics and Aristotle (IV, 46, p. 217, 10-11, cf. IV, 37, p. 214, 7-8); the doctrine that places the origin of *evil in corporeality (IV, 46-48, pp. 217, 12-218, 12); the doctrine of metempsychosis, taken from Pythagoras (XI, 34, p. 403, 11-13); and his predilection for the sumptuous table of the Syracusan tyrant (XII, 70, p. 439, 1-6, citing Xenophon's apocryphal letter to Aeschines). These judgments on Plato and his philosophy depend as much on the Platonic tradition of the first centuries AD as on earlier patristic authors. From middle Platonism are derived, in the last analysis, his veneration for Plato, the comparison of Plato's philosophy with that of Pythagoras, the distinction between the intelligible and sensible worlds (cf. also above, apropos of Athenagoras, no. 11), the extolling of likeness to God as the supreme ethical ideal, the contrast between Platonic and Aristotelian teaching on the soul and on *providence (think of the middle Platonist philosopher Atticus), the citation of the well-known passage of *Rep.* VI, 509b on the superiority of the Good (cf. above, apropos of Justin), the citation of the two famous passages of *Tim.* 28 b-c and 28c on the derivation of the physical universe from a higher cause and on the demiurge whom it is hard to find and impossible to express (cf. above, apropos of Justin); the doctrine of matter as a principle coeternal with God and of the ideas as thoughts of God and model of the sensible world. The beauty of Plato's language and style is mentioned in the *Cohortatio ad Graecos* (cf. also *Aristotelianism); the two theories that Plato went to Egypt to study the wisdom of the Hebrews and that he expressed polytheistic ideas only for fear of the Athenians are also present in the *Cohortatio*; the theory of his dependence on Moses and the prophets and his thefts from them is the classic thesis of Justin, the author of the *Cohortatio*, Clement and Eusebius; the thesis of his love of sumptuous tables is present in Tatian.

In Ps.-*Dionysius the Areopagite, the metaphysico-theological system characteristic of the final phase of *neoplatonism - that of *Proclus and Damascius, whose "auditor" at *Athens he probably was - becomes both the basis of *theology and the interpretative canon of *Scripture and the *liturgy, as a rich series of studies has now demonstrated beyond any doubt (cf. S. Lilla, Introduzione allo studio dello Ps. Dionigi l'Areopagita, *Augustinianum* 22 [1982] 533-577; for the list of relevant studies cf. esp. 541-542, 571-573). Of the Proclian motifs appropriated by Ps.-Dionysius, the most important are: the law of *moné, próodos* and *epistropé*; the stress on the transcendence of the first principle, defined as a non-being; the application to *moné* of the negative concepts of the first hypothesis of Plato's *Parmenides* and to *próodos* of the positive concepts of the second hypothesis; the definition of the Son and the *Holy Spirit as "flowers" and "superessential lights" of the *Father; the view that everything is potentially contained in the unity of the first principle and emanates from it in *próodos* through the effect of the overflowing of its superabundant power, though without exhausting the original source, which remains unchangeable; the simultaneous adoption of positive theology, which descends with *próodos* from the first principle to beings, and of negative theology, which seeks to ascend from beings to the first principle; the preference accorded to negative theology; certain fundamental laws of the angelic hierarchy (higher beings transmit divine illuminations to lower ones and possess all the properties of lower ones, while lower beings do not possess all the properties of higher ones; the *angels are the revealers of the deity's *arcanum*); the idea of the "mystical tradition"; and the union of the human mind with the supreme principle, which takes place without intermediaries and presupposes the stilling and overcoming of all knowledge and noetic activity (these and other motifs are examined in detail, together with their dependence on the Platonic tradition and on Proclus in particular, in S. Lilla, *art. cit.*, where attention is also called to the more important contributions of previous scholars). We must not forget that Ps.-Dionysius's dependence on Proclus is not confined to the adoption of doctrines, but goes as far as the use of his technical terminology. There are places in the *Corpus Dionysiacum* where passages from Plato's dialogues emerge: e.g. *De div. nom.* I, 5 (PG, 3, 593 A 10-B2), which goes back to *Parmen.* 142a, and *De div. nom.* IV, 7 (PG 3, 701 D3-704 A 2), which depends on *Sympos.* 210 a-b.

Not a few neoplatonic - esp. late neoplatonic - motifs and terms made their way into the work of *Maximus Confessor through Ps.-Dionysius the Areopagite. Maximus read and studied the *Corpus Dionysiacum* deeply: he wrote the *Ambiguorum liber* (or *De variis difficilibus locis SS. Dionysii et Gregorii*: PG 91, 1032-1417) and at least part of the scholia to the *Corpus* handed down under his name; in his works he often refers more or less openly to Ps.-Dionysius (cf. the list of relevant passages in P. Sherwood, *DSp* 3, 296-297). Maximus's dependence on Ps.-Dionysius has been studied by Sherwood (*DSp* 3, 295-300) and also in three works by W. Völker (*Festschrift für Bischof Dr. A. Stohr*, Mainz 1960, 243-254; *Der Einfluss des Pseudo-Dionysius Areopagita auf Maximus Confessor*, in *Studien zum neuen Testament und Patristik E. Klostermann zum 90 Geburtstag dargebracht*, TU 77, Berlin 1961, 331-350; and *Maximus Confessor als Meister geistlichen Lebens*, Wiesbaden 1964).

The same may be said of *De Trinitate*, a work falsely attrib. to Cyril of Alexandria, but actually composed in the 7th c. (cf. B. Fraigneau-Julien, *RecSR* 49 [1961] 188), and of *John Damascene's *Expositio fidei*. But Damascene's use of Ps.-Dionysius (esp. *De divinis nominibus*) is not always direct: as B. Kotter (*Die Schriften des Johannes von Damaskos* II) has shown, various chapters of the *Expositio fidei* reproduce sections of *De Trinitate* which in turn depend on Ps.-Dionysius's *De divinis nominibus*. Two detailed lists of parallels between *De divinis nominibus* and the *Expositio fidei* can be found in S. Lilla, *Studi in memoria di C. Ascheri*, Differenze 9, Urbino 1970, 176-177, and *Augustinianum* 13 (1973) 611-623; the latter study also contains the references to *De Trinitate*.

V. The Latin Fathers. Of the Latin Christian authors in whom Platonism left evident traces, we must particularly mention *Minucius Felix, *Lactantius, *Hilary of Poitiers, *Marius Victorinus, St *Ambrose, St *Augustine and *Boethius. Minucius Felix in *Octavius* XIX, 14 (29, 31-30, 4 Waltzing) mentions with approval the famous passage of *Timaeus* 28c, a classic passage not just in *middle Platonism but also among the Greek *Fathers: *Platoni itaque in Timaeo deus et ipso suo nomine mundi parens, artifex animae, caelestium terrenorumque fabricator, quem et invenire difficile prae nimia et incredibili potestate et cum inveneris in publicum dicere impossibile praefatur*. Cf. also XXVI, 12 (45, 13-14): *Plato, qui invenire deum negotium credidit*. In XXVI, 12 (45, 15-46, 3), naming the *Symposium*, he calls attention to the Platonic opinion on *demons (cf. esp. *Sympos.* 202 d-e). The direct dependence of *Octavius* XIV, 6, on the requirement not to hate true reasoning, on *Phaedo* 88c-91a was discovered by J.P. Waltzing (Minucius Félix et Platon, in *Mélanges Boissier*, Paris 1903, 445-460; cf. esp. 459). P. Shorey (Plato and Minucius Felix, *Classical Review* 18 [1904] 302-303) has brought out the dependence of *Octavius* XVI, 1 (19, 30-32): *ut conviciorum amarissimam labem verborum veracium flumine diluamus*, on *Phaedrus* 243d.

A. Kurfess (Lactantius und Plato, *Philologus* 78 [1923] 381-392) has listed various points of *Lactantius's *Divinae institutiones, Epitome* and *De ira dei* in which Lactantius either refers explicitly to Plato or clearly depends on him (though we cannot, acc. to Kurfess, infer from these passages any direct knowledge of Plato by Lactantius, whose sources would have been Cicero - esp. the *Tusculanae* - anthologies used by the *apologists, digests of philosophy *ad usum christianorum* and *Arnobius). Here we will simply mention the reminiscence of the famous passage of *Tim.* 28c in *Div. inst.* I 8, 1, the definition of death as separation of the *soul from the *body, taken from *Phaedo* 64c and 67d and present in *Div. inst.* II 12, 9, and the reference in *Div. inst.* II 14, 9 to the doctrine of demons expounded by Plato in *Sympos.* 202e (cf. Kurfess, pp. 383-384; echoes of *Tim.* 28c and *Sympos.* 202e also occur, as we have seen, in Minucius Felix). A. Wlosok (*Laktanz und die philosophische Gnosis*, Heidelberg 1960, 252-253) has rightly insisted on the decisive importance of *Hermetism and middle Platonism in Lactantius's interpretation of the passage of *Tim.* 28c; his study also has the merit of stressing the close connections linking Lactantius's *theology and his ideas on *revelation, knowledge of *God and *contemplation with the Hermetic-Platonic tradition of the 2nd c. AD (cf. esp. the paragraph *Das Verhältnis des Laktanz zur philosophischen Gnosis*, 222-231).

*Hilary of Poitiers's work is the first great example in the Latin Christian world of the affirmation of that marked negative theology so much in vogue, as we have seen, among the Greek Fathers. L. Longobardo (*Il linguaggio negativo della trascendenza di Dio in Ilario di Poitiers*, Naples 1982) has dedicated a careful and detailed analysis to the concepts of ingeneracy, infinity, incomprehensibility, ineffability, incorporeity, immutability, incorruptibility and impassibility that characterize Hilary's theology, showing their connections with Platonic and patristic tradition: these are virtually the same concepts noted in *Clement, *Origen and the Cappadocians, and are of clear Platonic, middle Platonic, neoplatonic and Philonian derivation. Here we will simply mention that the idea that the *Father's generation of the Son does not involve any change in the Father's nature, expressed by Hilary in *De Trin.* V, 37 (PL 10, 155B; cf. Longobardo, 55), closely follows the neoplatonic idea of the absolute unalterability of the first principle, which is not prejudiced by the emanation from it of *nous and the beings (cf. Plotinus, *Enn.* V, 3, 12 [V, 66, 34-36]; VI, 7, 36 [VI² 111, 24-25]; VI, 9, 5 [VI² 178, 36-37]).

In his still important study *Plotin et l'Occident* (Louvain 1934, 1-24), P. Henry, taking *Porphyry's *Vita Plotini* as his starting-point, has investigated the diffusion of Plotinian philosophy in the Latin West. To this diffusion contributed not just Porphyry and *Plotinus's Latin-speaking disciples (like the Etruscan Amelius), but also, in the 4th c., the African rhetor *Marius Victorinus, converted to Christianity in old age (cf. Henry, 44): as Augustine records in *Conf.* VIII, 2, 3 (cf. P. Hadot, *Marius Victorinus. Recherches sur sa vie et ses oeuvres*, Paris 1971, 201), he was the author of a *Latin translation - read with great avidity by Augustine (cf. the passages from *Contra Academicos* and *De beata vita* cited by Hadot, 202) - of certain *libri platonicorum* which, in all probability, comprised some treatises of *Plotinus collected and commented on by Porphyry (this is the conclusion reached by Hadot, 202 and 207, sharing the thesis of W. Theiler, *Porphyrios und Augustin, Forschungen zum Neuplatonismus*, Berlin 1966, 163; but Henry, *op. cit.*, 46, sees the *libri platonicorum* simply as Plotinus's *Enneads*). But Marius Victorinus's interest in neoplatonic philosophy was not just confined to translation: the investigations of P. Henry and esp. of P. Hadot have shown the extent of the dependence on *neoplatonism of Victorinus's dogmatic-theological works. P. Henry (*op. cit.*, 47-62) has called attention to the presence of Plotinian motifs in the

Liber de generatione divini verbi ad Candidum arianum and esp. in the *Adversus Arium* (in the latter [IV, 22] we can observe a direct textual citation of *Enn.* V, 2, 1 [V, 34, 1-2 Bréhier], on the One which is both all things and nothing in particular; cf. *op. cit.*, 49-54); while P. Hadot, in his fundamental work (*Porphyre et Victorinus*, I-II, Paris 1968), has dedicated a comprehensive and minute investigation to the presence of "Porphyrian" motifs in Victorinus's theological works (vol. I analyses all the parallels between Porphyry's theology and that of Victorinus; cf. esp. 102-461: among "Porphyrian" motifs present in Victorinus, we should specially mention that of the first principle conceived as non-being and that of the intelligible triad formed of being, life and thought; vol. II gives the text of those passages of *Adversus Arianum* and *Ad Candidum* that depend on Porphyry, the text of Porphyry's commentary on the *Parmenides* and a section of *Proclus's commentary on the *Parmenides*, containing "important elements for the understanding of some Porphyrian themes of Victorinus"; cf. Hadot, vol. II, preface). The fundamentally "Porphyrian" character of Victorinus's theology was confirmed by Hadot in his subsequent work, *Marius Victorinus. Recherches sur sa vie et ses oeuvres*, Paris 1971, 203.

The researches of A. Solignac, J.H. Waszink and P. Courcelle (cf. the relevant references in P. Hadot, *Marius Victorinus. Recherches sur sa vie et ses oeuvres*, Paris 1971, 204 n. 10) have called attention to the existence at *Milan in c.386 of a circle of Latin cultivators of neoplatonism, both Christian and *pagan. They included St *Ambrose and his teacher *Simplicianus, an old friend of Marius Victorinus (cf. P. Courcelle, *Rev. Philol.* 76 (1950) 56, and Hadot, *op. cit.*, 204). Indeed St Ambrose deeply assimilated certain Platonic doctrines and the neoplatonism of Plotinus and Porphyry, as the studies of P. Courcelle, L. Taormina, P. Hadot and A. Solignac have demonstrated by precise textual comparisons. P. Courcelle (Plotin et Saint Ambroise, *Rev. Philol.* 76 [1950] 29-56; cf. also his book *Recherches sur les Confessions de Saint Augustin*, Paris 1950, 106-132) has pointed out numerous Plotinian echoes in Ambrose's *De Isaac vel anima* and *De bono mortis*, and emphasizes the dependence of his sermon *In Isaiam* on Porphyry's *De regressu animae* (on this last point cf. esp. *Recherches sur les Confessions de Saint Augustin*, Paris 1950, 133-137). L. Taormina (Sant'Ambrogio e Plotino, *MSLC* 4 [1954] 41-85) has dwelt on numerous parallels between Plotinus and the bishop of Milan, esp. as regards the doctrine of God's transcendence and some problems of the soul. P. Hadot (Platon et Plotin dans trois sermons de Saint Ambroise, *REL* 34 [1956] 202-220), taking the direction pointed out by Courcelle, has discovered further parallels between Plato and Plotinus on one hand and Ambrose's *De Isaac vel anima* and *De bono mortis* on the other. Echoes of *Phaedrus* 246a, 247b appear in *De Isaac* VIII, 64-65 (cf. Hadot, *op. cit.*, 208), while the Platonic dialogue most used in *De bono mortis* is *Phaedo*, though the famous passage of *Theaetetus* 176b concerning *flight from the *world understood as likeness to God emerges in *De bono mortis* V, 17 (cf. Hadot, *op. cit.*, 213). Ambrose's direct source, acc. to Hadot, could be either Porphyry's *De regressu animae* (*art. cit.*, 220) or some work of the Cappadocian Fathers, the only Fathers in whom we find the fusion between Philonian, Origenian and Plotinian elements which is characteristic of *De Isaac* (on this last hypothesis cf. *Marius Victorinus. Recherches sur sa vie et ses oeuvres*, 206). Courcelle's example was also followed by A. Solignac, Nouveaux parallèles entre Saint Ambroise et Plotin: le "De Jacob et vita beata" et le Perì eudaimonías (Enn. I, 4), *Archives de Phil.* 20 (1956) 148-156. Finally, P. Courcelle's second study (Nouveaux aspects du platonisme chez Saint Ambroise, *REL* 34 [1956] 220-239) returned to the question of Ambrose's dependence on neoplatonism: the first section (*Ambroise lecteur de Plotin et Porphyre*, 221-226) brings out further parallels between the *Enneads* on one hand and Ambrose's *De Jacob* and *De officiis* on the other, and gives a favourable reception to Hadot's hypothesis (*REL* 34 [1956] 220) that Porphyry's *De regressu animae* was the direct source of Ambrose's *De bono mortis* (the various echoes of Plato's *Phaedo* would thus have reached Ambrose through the Porphyrian work); the second section (*Ambroise lecteur du Phèdre de Platon*, 226-232), taking up and partly modifying W. Wilbrand's hypothesis (Ambrosius und Plato, *RQA* 25 [1911] 42-49) that the direct source of the echoes of the *Phaedrus* present in Ambrose's *De Abraham*, *De Isaac* and *De virginitate* was *Origen, brings out further parallels between *Phaedrus* 246d-247e, 248b, 254a and *De virg.* XV, 96 and XVII, 109: while not ruling out that Ambrose's direct source may have been Origen's *Commentary on the *Song of Songs*, Courcelle is inclined to think that Ambrose read *Phaedrus* 246-254 directly (*art. cit.*, 231-232); the third section (*Ambroise lecteur de Macrobe*, 232-239) shows some parallels between Macrobius's commentary on Cicero's *Somnium Scipionis* and some of Ambrose's writings (*Hexaemeron, De Cain et Abel, De bono mortis, De Noe, De Isaac et anima, De Nabuthae historia*).

St *Augustine, like *Justin, *Clement, *Eusebius and *Theodoret, shows great admiration for Plato: his eulogy of him in *De civ. Dei*, book VIII, ch. 4, where he presents him as the greatest theologian among the Greek philosophers (CCL 47, p. 219, 1-3: *non quidem immerito excellentissima gloria claruit, qua omnino ceteros obscuraret, Plato*), smacks of the veneration of Plato by middle Platonism (Augustine knew of and admired one of its most illustrious exponents, Apuleius, author of *De Platone et eius dogmate: Apuleius Afer extitit Platonicus nobilis* [*De civ. Dei* VIII, 12: CCL 47, 229, 23-28]). In *De civ. Dei* XXII, 22 (CCL 48, 845, 123-124) he applaudingly cites Cicero's paraphrase of the passage in *Tim.* 47b on the divine origin of philosophy, *nec hominibus, inquit, ab his aut datum est donum maius aut potuit ullum dari*; and in *De civ. Dei* VIII, 11 (CCL 47, 227-228) he does not hesitate to make a close connection between Platonic doctrines and the books of *Moses, with the aim of emphasizing the substantial identity between Platonic philosophy and Mosaic wisdom, as Justin, Clement and Eusebius had done, and as Theodoret would soon do: the beginning of the book of *Genesis is interpreted on the basis of the *Timaeus*; and the famous passage of Ex 3, 14, *ego sum qui sum*, is related to the Platonic or, more precisely, middle Platonic doctrine of the immutability of the first principle: *qui vere est quia incommutabilis est ... vehementer hoc Plato tenuit et diligentissime commendavit* (we have already observed that the same connection between Ex 3, 14 and the passage in *Tim.* 27d on the immutable ideas is made by the author of the *Cohortatio ad Graecos*, ch. 22 - which, influenced by middle Platonism, relates, like Augustine, the passage of *Tim.* 27d not to the ideas but to the first principle - and by Eusebius, *Praep. ev.* XI, 10; cf. above). It is this coincidence between Plato and Moses that induces Augustine to take up the theory of Plato's knowledge of the OT, *ut paene assentiar Platonem illorum librorum expertem non fuisse* (*De civ. Dei* VIII, 11 [CCL 47, 228, 41-42]), though with a circumspection not shown by the Greek Fathers: if Plato really got to know the Mosaic wisdom during his stay in Egypt, he must have done so through oral teaching rather than a Greek translation of the Hebrew books, which did not yet exist: *qua propter in illa peregrinatione sua Plato nec Hieremiam videre potuit tanto ante defunctum nec easdem scripturas legere, quae nondum fuerant in Graecam linguam translatae, qua ille pollebat; nisi forte, quia fuit acerrimi studii, sicut Aegyptias, ita et istas per interpretem didicit, non ut scribendo transferret... sed ut conloquendo quid continerent, quantum capere posset, addisceret* (*De civ. Dei* VIII, 11 [CCL 47, 228, 15-23]). But Plato's dependence on the OT does not exclude direct inspiration or *revelation from God, this latter a hypothesis that Augustine seems to prefer: *sed undecemque ille ista didicerit, sive praecedentibus eum veterum libris sive potius, quo modo ait apostolus, quia quod notum est Dei manifestum est in illis; Deus enim illis manifestavit (Rom 1, 19)* (*De civ. Dei* VIII, 12 [CCL 47, 229, 1-4]). Augustine's putting forward the hypothesis of the divine inspiration of Platonic philosophy alongside the hypothesis of its dependence on the Mosaic books shows a surprising analogy with what Clement says of Plato and Pythagoras in *Strom.* II, 100, 3 (II, 167, 23-168, 4) and *Strom.* V, 29, 3-4 (II, 344, 23-345, 3). Not just Plato, but also the exponents of subsequent Platonism, esp. those of *middle Platonism and *neoplatonism, receive high praise from Augustine, for whom their teachings coincide in great part with Christian beliefs: *elegimus enim Platonicos omnium philosophorum merito nobilissimos, propterea quia sapere potuerunt licet immortalem ac rationalem vel intellectualem hominis animam nisi participato lumine illius Dei, a quo et ipsa et mundus factus est, beatam esse non posse; ita illud, quod omnes homines appetunt, id est vitam beatam, quemquam isti assecuturum negant, qui non illi uni optimo, quod est incommutabilis Deus, puritate casti amoris adhaeserit* (*De civ. Dei* X, 1 [CCL 47, 271, 13-272, 21]); *recentiores tamen philosophi nobilissimi, quibus Plato sectandus placuit, noluerint se dici Peripateticos aut Academicos, sed Platonicos ex quibus valde sunt nobilitati Graeci Plotinus, Iamblichus, Porphyrius; in utraque autem lingua, id est Graeca et Latina, Apuleius Afer extitit Platonicus nobilis* (*De civ. Dei* VIII, 12 [CCL 47, 229, 23-28]). But the light of Platonic philosophy shines again especially in *Plotinus, who is to be considered a Plato *redivivus*: *os illud Platonis, quod in philosophia purgatissimum est et lucidissimum, dimotis nubibus erroris emicuit maxime in Plotino, qui Platonicus philosophus ita eius similis diiudicatus est, ut simul vixisse, tantum autem interest temporis, ut in hoc ille revixisse putandus sit* (*Contra Acad.* III, 18, 41 [CCL 29, 59, 41-60, 46]). Reading Marius Victorinus's Latin version of some of Plotinus's books - very probably with comments by *Porphyry; cf. above, apropos of Marius Victorinus - at Milan in 386 lit a vehement fire in Augustine's heart: *libri quidam pleni ... etiam mihi ipsi de me ipso incredibile incendium concitarunt* (*Contra Acad.* II, 2, 5 [CCL 29, 20, 51-57]); *lectis autem Plotini paucissimis libris ... conlataque cum eis, quantum potui, etiam illorum auctoritate, qui divina mysteria tradiderunt, sic exarsi, ut omnes illas vellem ancoras rumpere* (*De beata vita* I, 4 [CCL 29, 67, 98-102]). This reading convinced Augustine, then just over 30 years old, of the essential identity between the Plotinian metaphysic and Johannine theology, which, according to him, is also present in the *Enneads*: *procurasti mihi quosdam platonicorum libros ex graeca lingua in latinam versos et ibi legi non quidem his verbis, sed hoc idem omnino multis et multiplicibus suadere rationibus quod in principio erat verbum et verbum erat apud Deum et Deus erat verbum: hoc erat in principio apud Deum . . . et quia hominis anima, quamvis testimonium perhibeat de lumine, non est tamen ipsa lumen, sed verbum, Deus, est lumen verum . . . et quia in hoc mundo erat, et mundus per eum factus est (Confess.* VII, 9, 13 [CCL 27, 101, 4-16]); *quod enim ante omnia tempora et supra omnia tempora incommutabiliter manet unigenitus filius tuus coaeternus tibi et quia de plenitudine eius accipiunt animae, ut beatae sint, et quia participatione manentis in se sapientiae renovantur, ut sapientes sint, est ibi (Confess.* VII, 9, 14 [CCL 27, 102, 32-36]); *gratulatus est mihi* (Simplicianus) *quod . . . in istis autem omnibus modis insinuari deum et eius verbum (Confess.* VIII, 2, 3 [CCL 27, 114, 6-9]). The identification of the *god* and *logos* of the prologue of John's gospel respectively with Plotinus's first and second *hypostases is the same interpretation of the Plotinian metaphysic adopted, as we have seen (cf. above), by *Eusebius, *Cyril of Alexandria and *Theodoret. Indeed neoplatonic philosophy has a right, acc. to Augustine, to be considered the truest philosophy and the crown of human thought (cf. W. Theiler, *Porphyrios und Augustin. Forschungen zum Neuplatonismus*, Berlin 1966, 160). His enthusiasm for Plotinian philosophy followed the bishop of *Hippo all his life: as his biographer *Possidius relates (*Vita Aug.* ch. 28), on his deathbed he repeated the words of a *sapiens*: *non erit magnus magnum putans quod cadunt ligna et lapides et moriuntur mortales*. To P. Henry (*Plotin et l'Occident*, Louvain 1934, 137-138) goes the credit for noticing that the *sapiens* in question was Plotinus: the phrase cited is a translation of *Enn.* I, 4, 7 (I, 77, 23-24 Bréhier). No wonder the connections between Augustine and neoplatonism have been the object of numerous studies: here we will mention only those of L. Grandgeorge (*Saint Augustin et le Néoplatonisme*, Paris 1896; cf. in particular the list of the various Plotinian and Porphyrian treatises cited and borne in mind by Augustine, 39-41); P. Henry (*Plotin et l'Occident*, Louvain 1934; cf. the two chapters: Saint Augustin le converti, 63-119, and Saint Augustin l'évêque, 120-145, containing numerous synoptic comparisons between Augustine and Plotinus; and the important collection of passages from Augustine's works in which neoplatonism is mentioned, 79-82; on the use of Porphyry's writings, cf. in particular 121); P. Courcelle (*Recherches sur les Confessions de Saint Augustin*, Paris 1950, 153-174); W. Theiler (*Porphyrios und Augustin. Forschungen zum Neuplatonismus*, Berlin 1966, 160-251), who not only thoroughly examines the multiple analogies between Plotinus's pupil and the bishop of Hippo, but also provides (250-251) an exhaustive list of all the Augustinian passages containing doctrines traceable to Porphyry; A.H. Armstrong (St. Augustine and Christian Platonism, *Plotinian and Christian Studies*, London 1979, XI, 1-66, and Spiritual or intelligible Matter in Plotinus and St. Augustine, *ibid.*, VII, 277-283); R. Russell (The Role of neoplatonism in St. Augustine's *De civitate Dei*, in *Neoplatonism and Early Christian Thought. Essays in Honour of A.H. Armstrong*, London 1981); and J.J. O'Meara (*The Neoplatonism of Saint Augustine. Neoplatonism and Christian Thought*, New York 1981). Nor is it possible here to cite even a small part of the many analogies between Augustine, Plotinus and Porphyry discovered by these and other scholars: we can only mention the parallels betwen the Augustinian conception of God and the Plotinian conception of the One, brought out by Grandgeorge (*op. cit.*, 57-84), though this subject really needs a more precise and thorough investigation; the doctrine of intelligible matter, also present, as we have seen (cf. above), in Origen (A.H. Armstrong's *Plotinian and Christian Studies*, London 1979, VII, 277-283, has called attention to its presence in Augustine); the motif of man's recollection into himself, condition of his union with the first principle, pointed out by Henry (*op. cit.*, 112); the Platonic ideal (cf. *Theaetetus* 176b) of likeness to God by flight from this world, which Augustine appropriates, reproducing Plotinus's own words (cf. Henry, 107). Particular attention is due to the Augustinian idea of the "Word" or divine wisdom which contains in itself the rational principles of all things: *veritatem fixam, stabilem, indeclinabilem, ubi sunt omnes rationes rerum omnium creaturarum* (*Sermo* 141: PL 38, 776); *spiritus sapientiae multiplex, eo quod multa in sese habeat; sed quae habeat, hoc et est, et ea omnia unus est, neque enim multae sed una sapientia est . . . in quibus sunt omnes invisibiles atque incommutabiles rationes rerum etiam visibilium et mutabilium, quae per ipsam factae sunt* (*De civ. Dei* XI, 10 [CCL 48, 331, 72-332, 78]). There is a surprising similarity not just with the Origenian idea of the *logos* (cf. above, apropos of Origen's *logos*, no. 9); but also with the Plotinian doctrine of *nous which contains all the ideas in itself and which is a composite unity, identical to this multiplicity or totality of ideas. Augustine's words *eo quod multa in sese habeat; sed quae habeat, haec et est, et ea omnia unus est* can be explained by bearing in mind the ἓν ἄρα πολλά of *Enn.* V, 3, 15 (V, 69, 11), the ἓν πάντα of *Enn.* V, 3, 15 (V, 69, 23) and the ἔχει οὖν ἐν ἑαυτῷ πάντα of *Enn.* V, 1, 4 (V, 20, 21); cf. also the expression πάντα ἕν with which Clement describes the *logos* (*Strom.* IV, 156, 2 [II, 318, 1]). The Plotinian *anima mundi*, the third *hypostasis, also contains in itself the rational principles received from *nous* (cf. e.g. *Enn.* II, 3, 16 [II, 42, 19-20] and VI, 2, 5 [VI1, 104, 12-13]). On the correspondences between Plotinus's second hypostasis and Augustine's "Word"-wisdom, cf. also O. Perler's dissertation, *Der Nus bei Plotin und das Verbum bei Augustinus als vorbildliche Ursache der Welt*, Freiburg Schweiz 1931. Finally we should mention that the famous Augustinian doctrine of *evil as a privation of good (cf. e.g. *De civ. Dei* XI, 22 [CCL 48, 341, 22-23]; XII, 3 [CCL 48, 357, 26] and the references in Courcelle, *Recherches sur les Confessions de Saint Augustin*, Paris 1950, 124 n. 4) was already formulated by Plotinus in the eighth treatise of the first *Ennead* (cf. *Enn.* I, 8, 1 [I, 115, 11-12; I, 115, 19]). Acc. to Courcelle (*op. cit.*, 124), Augustine was very probably led to embrace this Plotinian doctrine by St Ambrose's sermon *De Isaac*, itself dependent on Plotinus (cf. Courcelle, *op. cit.*, 106-107).

The role played by Plato and neoplatonism - esp. by *Proclus - in some aspects of *Boethius's thought has been brought out by P. Courcelle (*La consolation de la philosophie dans la tradition littéraire*, Paris 1967, *passim*; for Proclus cf. 203-231) and H. Chadwick in *Boethius: The Consolations of Music, Logic, Theology and Philosophy*, Oxford 1981 (cf. in particular 16-22 and the whole of chs. 4 and 5, 174-253).

Besides the studies cited here, R. Arnou's article, Platonisme des Pères, *DTC* 12, 2294-2392 remains fundamental. There is a useful anthology of texts in R. Arnou's other work, *De Platonismo patrum*, Rome 1935. Cf. also: J. Daniélou, *Message évangélique et culture hellénistique*, Tournai 1961; E. von Ivanka, *Plato Christianus. Ubernahme und Umgestaltung des Platonismus durch die Väter*, Einsiedeln 1964; J.H. Waszink, *Der Platonismus und die altchristliche Gedankenwelt*, *Entr. sur l'ant. class.*, III, Geneva 1957, 139-169; G. Vlastos, *Platonic Studies*, Princeton 1981; J.M. Rist, *Platonism and its Christian Heritage*, London 1985; H. Crouzel, Idees Platoniciennes et raisons Stoïciennes dans la Théologie d'Origène, *SP* 18.3 (1989) 365-383; R.M. Berchman, *From Philo to Origen: Middle Platonism in Transition*, Chico 1984; C. Fabricius, Zu den Aussagen der greichischen Kirchenväter über Platon, *VChr* 42 (1988) 179-187; E. Osborn, *The Beginning of Christian Philosophy*, Cambridge 1981; H. Dörrie, *Der Platonismus in der Antike: Grundlagen, System, Entwicklung*, I, Stuttgart 1987.

S. Lilla

PLEROMA (πλήρωμα). Term signifying fullness, totality, perfection; used by the translators of the LXX, it then became a term of theological content coloured with Stoic elements by *Philo, and was taken up and applied in various senses by Hellenistic Christian writers, *orthodox and *gnostic. In the gnostic schools, the p. meant the entire divine, celestial dimension in its multiplicity and unity, which contains in itself the invisible, transcendent beings and is their point of origin and goal of return as a place of rest in the state of salvation. The opposite of p. is *kenoma*, meaning for the gnostics the present *world where *souls exist outside the p. Besides this "spatial" sense, the p. assumed a temporal meaning: esp. in non-gnostic Christian texts, it means "the totality of *aeons", "the fullness of *time" (*pleroma tôn aionôn*), the perfect state of the world in the Son and with the Son of God (*Odes of Sol.* 17, 7; 19, 5; 26, 7; 36, 6; 41, 13; Ign., *Ephes.* proem.; *Trall.* proem.; Hipp., *Philosoph.* 8, 10, 3; *Acta Thom.* 148). The p. makes possible perfect knowledge and that knowledge where, after long searching, rest is found (*Odes of Sol.* 7, 13; 35, 6 and 36).

The term is used in *John's gospel (1, 16) and esp. in the so-called deutero-Pauline epistles (Eph 1, 10 and 23; 2, 17; 3, 19; 4, 10 and 13; Col 1, 19; 2, 19) to indicate the perfect presence of divinity in Christ, who is the place where *God dwells bodily, but also to indicate Christ as the whole measure of divinity, knowable through the church, his body. Non-gnostic Christian authors used the term in the singular, while gnostic authors preferred the plural. For *Origen, the fullness of divine *revelation is Christ (*In Io.* VI, 3: PG 14, 209BC). Acc. to the gnostics, the temple of *Jerusalem symbolized various *pleromata*: the sanctuary the p. accessible only to the high priest, the pronaus the p. accessible to the levites, i.e. the *psychici* (Orig., *In Io.* 10, 19: PG 14, 365CD). *Tertullian translates the term by *universitas*, using it to mean the totality of spiritual gifts given to Christ or the full belonging of creatures to God, who admits no other gods (*Adv. Marc.* I, 10: PL 2, 283 AB).

LTK 8, 560-561; *TWNT* 6, 283-309; Lampe 1094-95; P. Benoit, Corps, Tête et Plérôme dans les épîtres de la captivité, *RBi* 63 (1956) 6-44; A. Feuillet, L'Eglise, plérôme du Christ, d'après Ephés. 1, 22, *NRTh* 78 (1956) 449-472, 593-610; K. Koschorke, Eine gnostische Paraphrase des johanneischen Prologs, *VChr* 34 (1979) 383-392.

L. Vanyó

PLEROPHORIAE. From Greek πληροφορεῖν (to give evidence). Thus were called the collections of anecdotes attrib. to various persons living at the time of the council of *Chalcedon (451), who proved the soundness of the anti-Chalcedonian position particularly by means of miraculous events. The most famous of these collections, in which the same episodes often appeared, was drawn up by John, a monk at the *monastery of Rufinus (Beth Rufina) and bishop of Maiuma (*Palestine), hence known as John Rufus or *John of Maiuma. The name P. is therefore commonly used by antonomasia to designate this work, which survives in a *Syriac tr. from the Greek. But fragments of similar works survive in *Coptic, and works which now bear a different title (e.g. the Life of *Peter the Iberian or the Memories of *Dioscorus) must be considered as belonging to this *literary genre.

F. Nau, *Jean Rufus évêque de Maiouma, Plérophories*, PO 8, 1; T. Orlandi, Un frammento delle Pleroforie in copto, *Studi e Ricerche sull'Oriente Cristiano* 2 (1979) 3-12.

T. Orlandi

PLINY the Younger. Gaius Plinius Caecilius Secundus, nephew and adoptive son of Pliny the Elder. Rhetor and letter-writer, Roman senator, consul in AD 100 and legate of *Bithynia and *Pontus 111-113. Died soon after.

Of his abundant correspondence comprising 10 books, one of the most important *letters is that addressed to *Trajan on the Christians (X, 96). He asks the emperor what policy he should adopt towards these, who are becoming ever more numerous and concerning whom he has received anonymous denunciations. Trajan's reply firmly diapproves of the latter.

P.'s letter is a document of incontestable authenticity and capital interest for 2nd-c. Christianity. P. says that the Christians form "a considerable crowd, of every age, condition and sex". Their communities are widely spread beyond the cities, invading "villages and countryside". The letter provides details of Christian liturgical and social life. The *synaxis appears as a solid and respected institution, indispensable to the life of the community. There are two meetings, one in the morning, one in the evening. The first is held *stato die* (certainly on *Sunday), *ante lucem*, and a *carmen Christo quasi deo* is sung. Two things are clear: at the *assembly there is praying and singing; and Christ is at the centre of worship. In the evening there is a *promiscuum et innoxium* meal, doubtless the *agapè. He also mentions two *ancillae*, perhaps *deaconesses, put to the torture.

This is the first official text concerning Christians to adduce the legal arguments for *persecution. *Tertullian criticises the validity of the arguments in *Apolog.* 2, 6.

M. Durry, *Lettres de Pline*, IV, Paris 1947 (Coll. Budé); M. Schuster-R. Hanslik, Leipzig 1958; *RACh* 2, 1170-1173; *KLP* 4, 937-38; J. Moreau, *La persécution du christianisme dans l'empire romain*, Paris 1956; L. Rusca, *Saggio sulle persecuzioni. Plinio il Giovane. Carteggio con Traiano*, Milan 1963; Plinio il Giovane, *Lettere scelte*, ed. R. Scarcia, Rome 1967 (full bibl. 29-35); A. Hamman, Chrétiens et christianisme vus et jugés, par Suétone, Tacite et Pline le Jeune, in *Forma Futuri*, Turin 1975, 91-109; Ch. Munier, *L'Eglise dans l'empire romain*, Paris 1979; A.N. Sherwin White, *The Letters of Pliny: a historical and social Commentary*, Oxford 1966 (esp. 691-712); W.H.C. Frend, *Martyrdom and Persectuion in the early Church*, Oxford 1965, ch. 8; G.E.M. de Ste Croix, Why were the Early Christians persecuted?, *P&P* 26 (1963) 6-38; R.L. Wilken, *The Christians as the Romans saw them*, New Haven 1984, esp. 1-30; G.J. Johnson, *De conspiratione delatorum*, Pliny and the Christians revisited, *Latomus* 47 (1988) 417-422.

A. Hamman

PLOTINUS. His biographical details are given by *Porphyry, whose *Vita Plotini*, composed between 301 and 305, is our sole primary source (as H.R. Schwyzer has noted [*PWK* 21/1, 476], *Firmicus Maternus, *Eunapius, the *Lexicon Suidae* and Ps.-Eudocia cannot be considered independent of Porphyry; the one detail given by Eunapius but not in Porphyry is P.'s birthplace, Lykon, identifiable either with Lycopolis in Upper *Egypt or Lykon on the Nile Delta). Born in the 13th year of the reign of *Septimius Severus, i.e. 204-205 (a date calculated by Porphyry [*Vita Plot.* 2, vol. I, 2, 34-3, 37 Bréhier] on the basis of information provided by his doctor Eustochius, acc. to whom P. was 66 at the time of his death in the second year of *Claudius II [cf. *Vita Plot.* 2, p. 2, 29-31]). Aged 28, i.e. in 232-33, he attended various teachers of philosophy at *Alexandria, but was deeply disappointed by them all until he met *Ammonius Saccas, who was a revelation for him and with whom he remained for 11 years. Early in 243 - Ammonius had probably died shortly before - wishing to know the Persian religion, he joined the emp. Gordian III's expedition against Persia. In the first months of 244, following the failure of the campaign and the death of the emperor, he went to *Antioch; that same year he arrived in *Rome, where he finally settled and began his teaching career. For ten years, following Ammonius's example, he confined himself to oral teaching; only in the first year of *Gallienus (253) did he begin to write some treatises. When in summer 263 - tenth year of Gallienus - Porphyry came to Rome and met him, he had already composed 21 treatises. Between 263 and 268, during the period when Porphyry remained with him before leaving Rome, he wrote another 24. The last five treatises were composed during the first year of Claudius II (268-69). In 269, after 26 years at Rome, now increasingly ill, he retired to the property of his deceased friend Zetus near Minturno in Campania, where he died in 270 aged 66. His project of founding Platonopolis, the ideal city ruled according to the model fixed by Plato in the *Laws*, originally approved by the emp. Gallienus but then dropped, was never put into practice.

For his doctrines, cf. *Aeon, *Apatheia, *Cosmos, *Nous and esp. *neoplatonism; for his influence on the Fathers, both Greek and Latin, cf. *Platonism and the Fathers and *neoplatonism.

H. Oppermann, *Plotinos' Leben*, Heidelberg 1929; H.R. Schwyzer, *PWK* 21/1, 472-477;
Editions: P. Henry and H.R. Schwyzer, *Opera*, 2 vols., Paris-Brussels, 1951-1959; A.H. Armstrong (tr.) *Plotinus*, 7 vols. LCL, 1966-1988.
Studies: A.H. Armstrong (ed.), *The Cambridge History of Later Greek and Early Medieval Philosophy*, Cambridge 1967, 195-268; H.J. Blumenthal, Plotinus in the Light of Twenty Years Scholarship, 1951-1971, *ANRW* II.36.1 (1988) 528-570; K. Corrigan-P. O'Cleirigh, The Course of Plotinian Scholarship 1971 to 1986, *ibid.*, 571-623; L. de Blois, Plotinus and Gallienus, in (eds.) A.A. Bastiaensen et al., *Fructus Centesimus* (Mel. G.J.M. Bartelink), Steensbrugge, 1989, 69-82; M. Sells, Apophasis in Plotinus, a critical approach, *HJR* 78 (1985) 47-65.

S. Lilla

PNEUMA (πνεῦμα). A term meaning wind, breath. In Hellenistic thought and biblical tradition it took on many nuances, and its various meanings were inherited by the *Fathers of the church. In medical schools, p. was always a material reality, a living and lifegiving fluid, e.g. the system of circulation of the *blood, the breath of man or of any living creature. The concept of p. played a very important role in ancient philosophical currents, starting with *Stoicism. Acc. to the Stoics, the p. was the *anima mundi*, an impersonal wind-fire which permeated and unified the universe, participating in the divine nature itself. Through the LXX, the term was transmitted to the vocabulary of the Fathers; in the LXX, p. is the Greek translation of the Hebrew *ruah* which means "wind", "air", "breath of God" (Gen 3, 8). It gives life to man, but sometimes appears as a force which tends to abolish and suppress the creature (Ex 15, 8; Ps 33, 6; Is 11, 4). In an immaterial sense, it is an invisible, mysterious power; it represents the nature of *God, omnipotent, spiritual (Ps 139, 7), brooding over the waters (Gen 1, 2); it inspires the prophets, stimulates men to prophecy (Num 11, 25-29); it is the spirit of wisdom that enables some people to execute works of art (Ex 28, 3; 31, 3).

In prophetic language there is a close connection between divine *logos* and p. (Dan 5, 11; Wis 7, 12), which serves as a basis, in primitive Christianity, for logos-*christology, pneuma-christology and angelochristology. The divine p. as a permanent gift is a promise for the Messiah (Is 11, 2) and for all the just. The plural *pneumata* is a collective notion applied to various spirits, good or bad, the spirit of zeal, of licence (Num 5, 14; Hos 4, 12; 5, 4; Zech 13, 2), and then has the meaning of *angels and *demons; the good spirits, faithful to God, are considered as guardian angels of men, intercessors, "paracletes", who bring the prayers of the just before God and remind him of the merits of believers. The liturgical service of the angels remains a characteristic idea in Christian thought. In the NT, p. in the singular means the *Holy Spirit, while the plural *pneumata* is a collective concept referring to various spirits. The importance of the Holy Spirit in patristic *theology is incomparable to that of other spirits, whether angels or demons: it permeates not just the Messianic activity of Jesus and his disciples, but the whole life of the church.

In the vocabulary of the Fathers, the singular p. is reserved for the Holy Spirit and the plural *pneumata* for the angels, the *soul, the wind, *virtue, the sinful act, the adverse demon (Cyr. of Jer., *Cat.* 16, 13). *Gregory of Nyssa and *Nemesius hold that the term p. was inherited from Greek medical writers, since it was the technical term used for the circulation of the blood and for the liquids of the human organism (PG 44, 240C; PG 40, 697A); but it also designates man's *nous*, his rational and spiritual element (Ign. of Ant., *Ep. Polyc.* 5, 1; Just., *Dial.* 135, 6) as opposed to the soul (*anima*), principle of lower animal life (Tat., *Or.* 12), or represents a spiritual being superior to the *anima mundi* (Tat., *Or.* 4; Athenag., *Legatio* 22, 3; Theoph. of Al., *Ad Autol.* 2, 4). The p. is also a suggestion which determines man's moral or intellectual attitude (Orig., *In Jo.* 10, 39; Hermas, *Mand.* 5, 2, 8), the spirit of a person (Just., *Dial.* 49, 7), a supernatural power of the Christian soul, redeemed and capable of resisting *idols (*Hom. Clem.* 9, 15), the immaterial substance of the angels (Basil, *De Sp. sancto* 38: PG 32, 137A) or the divine nature itself (Athenag., *Legatio* 16, 8) or the Logos and the pre-existent Son of God (Iren., *Adv. haer.* III, 10, 3; V, 1, 2; Orig., *Dial. Her.* ed. J. Scherer, SCh 67, p. 60).

Characteristic of *gnostic vocabulary is the use of the plural *pneumata* to refer to the duality rooted in the world of spirits. In Syriac Christian theology, the feminine gender (*ruha'*) caused some difficulties, since it allowed feminine images to be associated with the concept of the Spirit. Besides this, traces of the theory of mediating Spirits also appear in the *prayer of the *Roman canon, which says: *Supplices te rogamus: iube haec perferri per manus sancti Angeli tui in sublime altare tuum, in conspectu divinae maiestatis tuae*.

F. Mussner, *LTK* 8, 568-576; F. Kamlah, *Begriffslexikon zum NT*, I, 479-487; H. Winel, *Die Wirkungen des Geistes und der Geister im nachapostolischen Zeitalter bis auf Irenaeus*, Freiburg i. Br.-Leipzig-Tübingen 1899; J. Barbel, *Christos Angelos*, Theophaneia 3, Bonn 1941; A. Recheis, *Engel, Tod und Seelenreise, Das Wirken der Geister beim Heimgang des Menschen in der Lehre der Alexandrinischen und Kappadokischen Väter*, Temi e Testi 4, Rome 1958; H. Dörries, *De Spiritu Sancto*, Göttingen 1956; H. Opitz, *Der heilige Geist nach den Auffassungen der römischen Gemeinde bis ca. 150*, Berlin 1960; O. Benz, *Der Paraklet, Fürsprecher im häretischen Spätjudentum, im Johannes-Evangelium und in neu gefundenen gnostischen Schriften*, Leiden-Cologne 1963; A. Miklósházy, *East-Syrian Eucharistic Pneumatology*, Rome 1968; C.K. Barett, *The Holy Spirit and the Gospel Tradition*, London 1975; R.G. Tanner, "Pneuma" in Saint Ignatius, TU 115, Berlin 1975, 265-270; H.J. Jaschke, *Der heilige Geist im Bekenntnis der Kirche*, Münster 1976; H. Soake, Pneuma, *PWK* Suppl. 14, 399 ff.

L. Vanyó

PNEUMATOMACHI. This term (πνευματομαχοῦντες, πνευματομάχοι) was used initially by *Athanasius and others to describe those who, without being *Arians, did not accept the divinity of the *Holy Spirit. They were later called *Macedonians.

P. Meinhold, Pneumatomachoi, *PWK* 41, 1066-1101; W.-D. Hauschild, *Die Pneumatomachen*, Hamburg 1967.

M. Simonetti

POITIERS. Ancient Limonum (Lemonum) Pictonum, then Pictavis, in the 6th c. Pictavium; at the time of *Augustus it belonged to the province of Aquitania and, after *Diocletian, to Aquitania Secunda. It was one of the greatest cities of Roman *Gaul: its amphitheatre with 40,000 seats was the largest. Destroyed at the end of the 3rd c., as attested by a layer with traces of fire (Bagaudi? Germans?), it was almost immediately rebuilt on a smaller scale but more luxuriously, and surrounded with the strongest system of fortifications in Gaul. Even the temples were all rebuilt, which proves that the city was still completely *pagan. Its territory corresponded to the three departments of Vienne, Vendée and Deux Sèvres.

Christianity put down roots and developed at P. in the 4th c.: from this era date the first episcopal church (beneath the modern cathedral), the baptistery and the *domus ecclesiae* (bishop's house); from the 4th or 5th c. date the churches of St Martin, St Hilaire-entre-deux-Eglises, St Hilaire-de-la-Celle (above St *Hilary's house?), St Savin (above Savinus's dwelling?) and perhaps St Paul (the future priory). Outside the city, Notre-Dame-la-Grande was built on the ruins of a temple, while St Hilaire-le-Grand was built over the funerary *cella* of Hilary's wife and daughter. The 5th c. was an obscure period in P.'s history: but we can say that it was then that paganism totally disappeared. From this era we know the names of bishops *Hilary (356-13 Jan 367/8), Adelphus and Elaphus.

The battle of Vouillé (507) forced P. to submit to *Clovis. Under his successors, it changed hands 20 times in the 6th c., while the 7th was a period of peace; the 8th was again marked by the struggle of the dukes of Aquitaine, Hunaud and Gaiffier, against the *Franks Charles Martel and Pippin the Short, with an Arab interlude in 732. This era is marked esp. by the building of the abbey of the Holy Cross by Queen *Radegund, who lodged 200 nuns there and built St Mary's abbey for monks alongside it. The new churches of St Luke (hospital and oratory) and St *Leodegarius were built at P.

Outside the city St Hilary's developed, becoming the episcopal necropolis (tomb of *Venantius Fortunatus), and St Saviour's (tomb of St *Porcarius) and the oratories of Elidius and Leonius, Martin and Gelasius were built. E of the city were built the *churches of St Hilaire-de-la-Celle, St Pierre-le-Puellier, Notre-Dame-l'Ancienne; W of the city, Notre-Dame-la-Grande, St Michel, St Radegund (enlarged in the 8th c.).

From this period we know bishops Pientius, who helped St Radegund found the abbey in *c*.544; Pascentius, to whom Venantius Fortunatus dedicated his *Life* of St Hilary (BHL 3885-87), after 565; Maroveus, a contemporary of *Gregory of Tours (*Hist. Franc.* 7, 24; 9, 33, 39-41, 43; 10, 15-16; *Virt. S. Mart* 2, 44; *Gl. Conf.* 104); Plato, Gregory of Tours's *archdeacon (*Hist. Franc.* 5, 49; *Virt. S. Mart* 4, 32); Venantius Fortunatus (Baudonivia, *Vita S. Radeg.* 2, prol.; Paulus Diaconus, *Hist. Lang.* 2, 13); Caregisilus; Ennoald (614); John (627); Diddus, uncle of St *Leodegarius of Autun (628/9-669/70); Ansoald (676-696/7); Eparchius (794); Maximinus.

Duchesne, *Fastes* II, 75-87; *DACL* 14, 1252-1340; M. Vieillard Troiekouroff, Les monuments religieux de Poitiers d'après Grégoire de Tours, in *Etudes mérovingiennes*, Poitiers 1952, 285-92; id., *Les monuments religieux de la Gaule d'après les oeuvres de Grégoire de Tours*, Paris 1976, 218-230; D. Claude, *Topographie und Verfassung der Städte Bourges und Poitiers bis in das 11. Jh.*, Lübeck-Hamburg 1960.

V. Saxer

POLEMIUS SILVIUS. A *calendar surviving under the title *Polemii Silvii laterculus* contains information of mainly Christian interest, with supplementary information on matters of general culture, collected under the following titles: *Nomina omnium principum Romanorum, Nomina provinciarum, Nomina cunctorum spirantium atque quadrupedum, Quae sint Romae, breviarium temporum, Voces variae animantium, Nomina ponderum vel mensurarum*. The compilation is explicitly dated *Postumio et Zenone viris clarissimis consulibus*, i.e. 448; it is dedicated *Domino beatissimo Eucherio episcopo*, i.e. to *Eucherius of Lyons. P. must be identified with Silvius, *friend of *Hilary of Arles, in whose *Life* (ch. 11) he is numbered among *eiusdem temporis qui suis scriptis meriti summi claruere*; a brief item on him also survives in the *Chronica Gallica a.* 452, relating to the year 438: *Silvius turbatae admodum mentis post militiae in palatio exacta munera aliqua de religione conscribit*.

CPL 2256; PL 13, 676; T. Mommsen, *Chronica minora*, I, 518-551; edition of *Quae sint Romae* in R. Valentini-G. Zucchetti, *Codice topografico della città di Roma*, I, Rome 1940, 294-301, 308-310. Information in Schanz IV/2, 130; *PWK* 21, 1960-1963.

V. Loi

POLEMON the Apollinarist. Leader of the more extreme group of *Apollinaris's disciples. *Theodoret of Cyrrhus (*Haer. fab. comp.* IV, 8), describing the division of Apollinaris's disciples into parties, mentions the "Polemians" in ch. 9. When this happened is not very clear to us; at any rate P. emerged as leader of the opposition to the "Great Church"; a fragment of his, often cited by later sources (cf. Lietzmann, fr. 174, pp. 274-275), contains very harsh expressions against the "novelties" of the Cappadocians and the "idle fancies" of the Italians (cf. also a fragment of his *Antirrheticus*: Lietzmann, fr. 173, p. 274). P. warns against the danger of "divisions" in Christ, which would lead to the idea of two wills in the incarnate Logos (cf. Lietzmann, fr. 175, p. 275). His position was shared by Eunomius of Beroea in *Thrace and by *Julian the Apollinarist, to whom Apollinaris himself dedicated a treatise (frs. 150-152 Lietzmann) replying to the problem of the two wills in Christ. Of a *letter from P. to Julian, two fragments remain (176-177 Lietzmann, pp. 275-276). P.'s pupil *Timothy, bishop of Berytus, separated from this group and in 381 subscribed the canons of the council of *Constantinople, which among other *heresies condemned the Apollinarists (can. 1). It was prob. then that P. wrote the work against Timothy of which his best-known fragment is a part (fr. 174 Lietzmann, *cit.*).

CPG 3710-3713; H. Lietzmann, *Apollinaris von Laodicea und seine Schule*, Tübingen 1904, 40 f., 153; frs. pp. 274-276; *DCB* 4, 422.

E. Cavalcanti

POLITICS (Christian political thought). The following observations are no more than a very brief introduction to the political thought of the Christians of the first centuries, with reference not so much to the history of *dogma as to the history of political ideas, in the meaning given to these terms by J. Touchard (*Histoire des idées politiques*). In fact ancient Christian literature did not include treatments of political theory like those the Greek world had known, but put forward new experiences and problems that were destined to influence later centuries considerably. The first and essential interest was religious (and manifested mostly starting from the biblical text, esp. the NT, and its interpretation): any words of a political nature addressed by bishops to those responsible for the *civitas* or to their congregations must be understood in this light, just as the evidence provided in the first three centuries by the *martyrs must be assessed in this light. This does not diminish, but establishes and explains, the political importance of the primitive Christian communities, which lived and worked in different social situations.

Rome had long superseded the idea and reality of the old Greek city-states and had united peoples of different origins, languages and traditions in the same political organization; and to maintain unity, it had worked out a series of rules of common life and given the inhabitants of the Roman *oikoumene* a legal status (cf. *Citizenship). The Christians generally recognized the positive aspect of this situation, which allowed the gospel to be proclaimed to the peoples with greater ease (not all, however: alongside the "providential" line, there was an "apocalyptic" line, which tended to consider the empire negatively, seeing it as evil personified).

Historically, the former idea prevailed over the latter; after all, it conformed to Paul's instructions (cf. Rom 13, 1 ff.) to obey the appointed authorities because they were instituted by God in that they did good, instructions followed from apostolic times and even in the more critical times of the persecuted church (2nd-3rd cc.). Peter's first epistle (2, 13 ff.) takes the same line. From this arose a Christian attitude of full loyalty to the Roman empire, more and more often declared and proved by specific arguments: *God being held as the foundation and source of all good, imperial authority was conceived as being descended from him, its exercise being necessary after *Adam's *sin or, as some writers say (cf. e.g. Irenaeus, *Adv. haer.* 5, 24, 1 ff.), after Cain's fratricide, when fear of God disappeared.

Christianity, then, showed men the way of simultaneous obedience to God and to those responsible for public affairs, provided these did not attribute absolute prerogatives to themselves or to the institutions they represented, nor require idolatrous acts, i.e. not acknowledging either of these things as receiving its justification from on high. Here is shown the originality of Christian political thought. Starting with Jesus' teaching (cf. Mt 22, 15), it made a distinction between the "political" and the "religious", of great importance in a world in which history, tradition and mentality took for granted the close, indissoluble bond between the one and the other, and which had raised the edifice of the *civitas* on precisely this bond. Thus was born a conception of the *res publica* - in modern terms, let us say the state - considered as legitimate in its order, but without religious value as such, a conception that differed not just from the sacred character of the ancient city, but also from the theocratic ideal of that Judaism which identified the Messianic kingdom with earthly political systems.

Investigation of the problem of the nature of politics was allowed, indeed in a certain way required, by the recognition of God, one and triune, as Lord of all men and all things and by the belief that he had revealed to mankind a message of salvation, containing some imperative norms which everyone had to observe, whatever his social position: whence arose the sense of a personal autonomy, claimed even in the face of the empire; freedom of *conscience was discovered, and the concept of individual freedom gradually transformed. During the very first centuries these ideas became concrete especially in the behaviour and words of the martyrs.

Thus Christian thought poses with clarity the question of the connection between politics and morals and, at the same time, it seems to have been well aware from the start how difficult it was to maintain the right balance between the two spheres, how hard it was, in the actuality of historical situations, to preserve the reciprocal autonomy between church and state (cf. *Church and empire), since: "the correlation of both powers has its root in the personal unity of the Creator and Redeemer, but must not become a confusion that issues in the state's power over the church (*Caesaropapism) or the church's power over the state (papal theocracy)" (H. Rahner, *Chiesa e struttura politica nel cristianesimo primitivo*, Milan 1979, 19).

R.W.-A.J. Carlyle, *A History of Mediaval Political Theory in the West*, I, Edinburgh-London ²1927; G. Combès, *La doctrine politique de Saint Augustin*, Paris 1927; J. Touchard, *Histoire des idées politiques*, Paris 1959; H. Rahner, *Kirche und Staat im frühen Christentum. Dokumente aus acht Jahrhunderten und ihre Deutung*, Munich ²1961; (ed.) G. Barbero, *Il pensiero politico cristiano*, 2 vols., Turin 1962-65; W.H.C. Frend, The Roman Empire in the eyes of the Western Schismatics during the Fourth Century AD, *Miscellanea Historiae Eccles.*, Louvain 1961, 9-22; id., The Roman Empire in Eastern and Western Historiography, *PCPhS* 194 (n.s. 14) (1968) 19-32; R.A. Markus, *Saeculum: History and Society in the Theology of St. Augustine*, Cambridge 1976 (repr. 1989); P. Brown, *Religion and Society in the Age of Saint Augustine*, London 1972, esp. Part iii; W.H.C. Frend, Augustine and State Authority: the example of the Donatists, *Agostino d'Ippona, "Quastiones Disputatae"*, Palermo 1989; id., Augustine and Orosius on the End of the Ancient World, *AugStud* 20 (1989) 1-38.

P. Siniscalco

POLLENTIUS. *Friend of *Augustine, known to us through *De coniugiis adulterinis* (PL 40, 451-486), the first treatise in the whole of the first five centuries dedicated exclusively to *divorce and second marriage. Augustine calls P. "*brother" and P. addresses Augustine as "father". We do not know if he was a priest or a bishop. P. had sent Augustine a *letter about matrimonial problems. Thus began a discussion between the two, as a result of which Augustine wrote *De coniugiis adulterinis* and referred P. to the explanations he had given in *De Serm. in Monte*, I, 14, 39 on the *exegesis of 1 Cor 7 (Pauline privilege, *adultery and *marriage) and Mt 5, 31-32; 19, 9 (marriage in OT and NT). In *Retract.* II, 57, Augustine acknowledges the difficulties of this polemical discussion.

H. Crouzel, *L'Eglise primitive face au divorce*, Paris 1970, 337-353; *PAC* 1, 878-880.

E. Romero Pose

POLYCARP. Bishop of Smyrna, P. died a *martyr on 23 Feb 167. *Irenaeus, who came from Smyrna, claims to have known him in childhood and heard him speak of his relations with "John", whom he identifies with the *apostle, but who was prob. the "*John the presbyter" whom *Papias, a contemporary and compatriot of P., expressly distinguishes from the apostle (cf. Euseb., *HE* III, 39, 4). We possess a *letter from *Ignatius of Antioch to P. and, slightly later, a letter from P. to the Christians of *Philippi. Soon after, *c.* 160, P. went to *Rome to confer with pope Anicetus (155-166) "on a question concerning *Easter" (Euseb., *HE* IV, 14, 1). The church of Rome always celebrated this feast on a *Sunday, while the churches of *Asia celebrated it according to "ancient custom", i.e. on 14 Nisan, following the Hebrew calendar, on whatever day of the *week it fell. The bishop of Smyrna and the bishop of Rome remained firm in their respective positions, but "in *peace". In 167 (date provided by Euseb., *Chron.* ed Helm, GCS 47, p. 205 and *HE* IV, 14, 10), there was a *persecution at Smyrna. P., who was hidden outside the city, was betrayed by a young *slave and arrested. He died at the stake on 23 Feb, which, acc. to the account of his *martyrdom, was "the day of the great *sabbath", i.e. a Sunday (for the meaning of this expression cf. Epiph., *Pan. Expos. fidei* 24; in 167, 23 Feb did in fact fall on a Sunday). He was 86 years old.

1. The letter to the Philippians. This has been preserved partly in Greek and completely in a *Latin translation; the hypothesis that the present text is the result of a fusion of two primitive letters has no sufficient basis. We see from this letter that some Philippians had written to P. complaining of a certain Valens who "had been ordained presbyter" (probably bishop) among them and whom they accused of having committed wrongs for love of money. At the same time they asked P. to send them a copy of Ignatius's letters in his possession. In his reply, P. insisted on the need to abstain from any kind of cupidity, but also asked them not to treat Valens and his wife as enemies. To his letter he added the collection of Ignatius's letters that we possess.

2. The martyrdom of P. This work, composed soon after his death in the form of a letter addressed to the church of Philomelium, is the first Christian work exclusively dedicated to describing the passion of a martyr and also the first to use this title of "martyr" to designate a Christian who died for the faith. It was influenced by similar Jewish accounts (III and IV Maccabees) and in turn influenced the development of this *literary genre among the Christians.

Altaner 52-54; Quasten I, 202-208. Editions: SCh 10, Paris 1969; F.X. Funk-K. Bihlmeyer-W. Schneemelcher, Tübingen 1970; J.A. Fischer, Darmstadt 1976, with Ger. tr.; It. tr. A. Quacquarelli, *I Padri Apostolici*, Rome 1976, 151-172.
Studies: T. Baumeister, *Die Anfänge der theologie des Martyriums*, Münster 1980, 289-306 (with bibl.); P. Brind' Amour, La date du martyre de Polycarpe (le 23 fév. 167), *AB* 98 (1980) 456-462; W.R. Schoedel, *Polycarp, Martyrdom of Polycarp. Fragments of Papias*, London 1966; L.W. Barnard, The Problem of St. Polycarp's Epistle to the Philippians, *Studies in the Apostolic Fathers and their background*, Oxford 1966, 31-40; T.D. Barnes, A Note on Polycarp, *JThS* n.s. 18 (1967) 433-437; B. Dehandschutter, *Martyrium Polycarpi*, Leuven 1979; H. Koester, *Introduction to the New Testament*, Philadelphia 1982, II, 306-308.

P. Nautin

POLYCRATES of Ephesus. Late 2nd-c. bishop of *Ephesus. We have his *letter written to pope *Victor on the occasion of the *Easter controversy (Euseb., *HE* V, 25). At that time synods were celebrated in which many bishops declared against the *quartodeciman practice and in favour of celebrating Easter on a *Sunday. P. too, exhorted by Victor, held a synod in *Asia, where all the participants declared themselves followers of the quartodeciman *tradition. In his letter he also gives some personal information: "Even I, Polycrates, the least of you all, observe the tradition of my ancestors, some of whom were also my predecessors: seven of my ancestors, indeed, were bishops; I am the eighth. They always celebrated Easter day when the Jewish people abstain from leavened bread. My *brothers, I have lived 65 years in the Lord; I have been in touch with the brothers of the whole world; I have read all *Scripture and I will not let myself be frightened by scarecrows, since greater men than me have said that we must 'obey the Lord rather than men' (Acts 5, 29)" (Euseb., *HE* V, 24, 6 f.).

A. Di Berardino

POLYCHRONIUS of Apamea. Bishop of *Apamea (*Syria) and brother of *Theodore of Mopsuestia, died 430, important exegete of the Antiochene school. Wrote *commentaries on *Job, *Daniel, *Ezekiel, perhaps also Jeremiah, fragments of which, in variously revised forms, are in the *catenae.

CPG 2, 3878-3880; A. Mai, *Scriptorum veterum nova collectio*, 12, Rome 1825, 105-160 (comm. on Daniel); A. Mai, *Nova Patrum bibliotheca*, VII 2, Rome 1854, 92-127 (comm. on Ezekiel); PG 64, 739-1038 (the disputed comm. on Jeremiah); *RE*³ 15, 528; *PWK* 21, 1598; O. Bardenhewer, *Polychronius, Bruder Theodors von Mopsuestia und Bischof von Apamea*, Freiburg i. Br. 1879; M. Faulhaber, *Die Propheten-Catenen nach römischen Handschriften*, Freiburg i. Br. 1899 (with references to unpublished material).

J. Irmscher

POMERIUS Julianus. Native of *Mauretania, he settled, perhaps in consequence of the difficulties that followed the *Vandal invasion, in S *Gaul and lived (second half of 5th c.) at *Arles, where he gained fame for his doctrine and his rhetorical knowledge. He had links with illustrious persons of his time, such as *Ennodius of Pavia, *Caesarius of Arles (whose teacher he was when, in 498, Caesarius left the *coenobium of *Lérins to settle at Arles) and *Ruricius of Limoges. Ruricius (*Epist*. II, 10) calls him *abbat, while Ps.-*Gennadius (*Vir. ill.* 109) calls him a presbyter of the Gallic church. Ps.-Gennadius and *Isidore of Seville (*Vir. ill.* 25) attribute numerous writings to him. All that survives is the treatise in three books, *De vita contemplativa*, once attrib. to *Prosper of Aquitaine, now accepted as P.'s. The title is better suited to book I, while books II and III deal with the wider theme of the Christian *virtues and the means to practise them.

CPL 998-998a; PL 59, 415-520; *DTC* 12, 2537-2543; Schanz IV, 2, 554-556; F. Degenhart, *Studien zu Iulianus Pomerius*, Eichstätt 1903; J.C. Plumpe, Pomeriana, *VChr* 1 (1947) 227-239.

S. Pricoco

POMPONIA GRAECINA. Roman noblewoman, wife of Aulus Plautius, lived in the reign of *Nero (54-68). Acc. to *Tacitus, she was accused of *superstitionis externae rei* (*Ann*. XIII, 32). Her husband examined her conduct and the gossip circulating about her before her *family, and declared her innocent. Tacitus adds that P. lived for 40 years in continual sadness, without wearing luxurious clothes. A Roman tradition calls her a Christian, but Tacitus's words cannot be taken as proof of this. Scholars think her "foreign *superstition" was either the Jewish or the Christian religion, which were detestable to the *pagans, and rule out the Jewish, since the *Jews were favourites at Nero's court. To reinforce this opinion they mention certain fragments of early Christian inscriptions which contain forms of this name (De Rossi, *La Roma sotteranea*, II, 282). But these arguments are not enough to confirm P.'s Christianity. She may have been a follower of one of the superstitious and despised religions of the time, which were numerous at *Rome, or have suffered from a mental illness which manifested itself in strange or superstitious religiosity.

EC 9, 1734; *PWK* 21, 2351, n. 83 f.; G.B. De Rossi, *La Roma sotterranea cristiana*, II, Rome 1867.

G. Ladocsi

PONTIANUS, pope (230-235). Succeeded Urban in 230. During the *persecution of *Maximinus Thrax, in 235, he was deported with the presbyter (?) *Hippolytus to *Sardinia, renouncing his office on 28 Sept (*Chron. of 354*: MGH *Chron. min*. 1, 75) and being succeeded by Antherus. He died on the island. His body was brought back to *Rome by pope *Fabian, who buried him in the *cemetery of Callisto. His epitaph was found in 1909 (Diehl 953).

LP I, XCIV f.; 145 f.; 3, 74; *DTC* 12, 2553 f.; *BS* 10, 1013-1015.

A. Di Berardino

PONTIANUS. African bishop of unknown see (perhaps to be identified with the Pontianus or Potentianus, bishop of Thenae in *Byzacena, to whom *Fulgentius of Ruspe appeared in a vision to announce the choice of his successor: cf. Ferrandus, *Vita S. Fulgentii* 29); author of a *letter, 544-45, to *Justinian in favour of the authors involved in the condemnation of the *Three Chapters, who, being dead, he thought could not be subject to condemnation. We cannot rule out P.'s possible identity with Ponticianus, Catholic bishop of an unspecified see of Byzacena, who took part in the synod of Iunca in 523 (*Concilia Africae*: CCL 149, 227 f.). There is little basis for the theory that chs. 1-12 of *Isidore of Seville's *De viris illustribus* were taken from a similar work by P., written to supplement *Gennadius's *De viris illustribus*.

CPL 864; PL 67, 995-998; F. Schütte, *Studien über den Schriftstellerkatalog (de viris ill.) des hl. Isidor von Sevilla*, Kirchengeschichtliche Abhandlungen, 1, Breslau 1902, 102; Schanz IV, 2, 583 f.; Moricca III, 2, 1479 f.; *PAC* 1, 883 f.

F. Scorza Barcellona

PONTICIANUS. African Christian, member of the imperial administration at *Milan. In 386 he visited *Augustine and *Alypius. Finding them reading *Paul's epistles, he told them the story of the hermit *Anthony, the Egyptian monks and two imperial officials at *Trier who abandoned everything to become monks. The visit and P.'s account gave the decisive push to the *conversion of Alypius and Augustine (Aug., *Confess*. VIII, 6-7).

DCB 4, 439; *PAC* 1, 884.

A. Di Berardino

PONTIFEX MAXIMUS, (pontiff). Head of the most prestigious Roman religous institution, the *collegium pontificium*, the P. M. supervised all religious activities and the college's complex disciplinary system; thus he gradually acquired considerable political weight. The origins of this magistrature are lost in the mists of time, while the earliest known election of a P. M. (Livy, 25, 5, 2) dates from the 3rd c. BC. At the end of the republican era, on the death of the last P. M. *Augustus had himself acclaimed P. M. (AD 12), as did his successors in the empire. From Nerva on, the dignity of P. M. was included among the attributes of the *princeps*, without any electoral procedure. *Constantine and his sons kept the title; the first to renounce it, demonstratively, was *Gratian under pressure from *Ambrose in 382-83 (Zosimus, *Hist. nova* 4, 36, 5). With this the state abandoned all interest in the *pagan cults; they continued to exist only as private convictions. The episode symbolizes the sunset of the old pagan spirit of Rome. The title of P. M. remained as an attribute of the highest religious authority and was assumed from the 5th c. by the *popes (cf. *Leo I). In the Middle Ages, even certain bishops bore the title (but these were sporadic cases), while from the Renaissance on it has been the pope's exclusive prerogative.

G. Wissowa, *Religion und Kultus der Römer*, Munich ²1912, 501-523; K. Latte, *Römische Religionsgeschichte*, Munich 1960, 195-200, 400-407; K. Ziegler, *KLP* 4, 1046-1048; L. Schumacher, Die vier hohen römischen Priesterkollegien unter den Flaviern, den Antoninen und den Severern (69-235 n. Chr.), *ANRW* II.16.1 (1978) 737-768; I. Kajanto, Pontifex maximus as title of the Pope, *Arctos* 15 (1981) 37-52; G. Gottlieb, *Ambrosius von Mailand und Kaiser Gratian*, Hypomnemata 40, Göttingen 1973; W. Eck, Religion und Religiositat in der soziopolitischen Führungsschicht der Hohen Kaiserzeit, in (ed.) W. Eck, *Religion und Gesellschaft in der römischen Kaiserzeit* (Colloquium . . . Vittinghoff), Cologne-Vienna 1989, 15-52.

B. Aland

PONTIUS. Deacon of *Carthage. Known only as the biographer of *Cyprian, whom in 257 he accompanied to exile at Curubis (Korfa, Tunisia). Soon after his teacher's death, he wrote his life, panegyrical in type but containing some precious information on Cyprian's *conversion, his attitude during the plague of 252 and his literary work. Date of P.'s death unknown. Ado introduced P.'s name into his *martyrology on 8 March.

LTK 8, 616; Monceaux 2, 179-97; A. Harnack, *Das Leben Cyprians von Pontius*, TU 39/3.

V. Saxer

PONTUS. Region of N *Asia Minor, a flourishing kingdom in Hellenistic and Roman times. Its most famous king, Mithridates VI Eupator, extended his power to the N coast of the Black Sea, annexing the Greek colonies and occupying Armenia Minor, Paphlagonia and S Cappadocia; he fought against Rome, but was finally defeated by Pompey after a war of 20 years (88-67 BC). At his death, part of the region remained independent, while Pompey joined the areas on the Black Sea with *Bithynia into a single Roman province (64 BC). The territory underwent various repartitions and attributions to different provinces, while some regions of limited extent - esp. in the 1st c. AD - were assigned to vassal kings. The final arrangement, which lasted until the reign of *Justinian, was part of *Diocletian's division of the provinces into smaller units: the territory was then divided into the three provinces of Diospontis (capital *Amasea), Pontus Polemoniacus (capital *Neocaesarea) and Armenia Minor (capital Sebaste).

P. was also the name of one of the Eastern dioceses among which Diocletian's reorganization distributed the provinces and which, with the resolutions of the council of *Constantinople of 381, were recognized as ecclesiastical areas: seat of the *vicarius* of the diocese, as well as of the metropolitan bishop, was *Caesarea, formerly capital of the province of *Cappadocia. Christian origins in P. go back to apostolic times: *Peter's first epistle is addressed, among others, to the believers in that region, showing that he knew that Christianity had reached a certain degree of diffusion there. We know the names of bishops in various cities even before the time of *Constantine: that of Comana Pontica in 248; Nicetius, Phoedimus and Athenodorus (between 250 and 275) of Amasea; *Phocas of Sinope, martyred in c.300; the best-known, through his own writings and those of *Basil of Caesarea and *Gregory of Nyssa, is *Gregory called the Thaumaturge, bishop of Neocaesarea in Pontus Polemoniacus between 240 and 264.

E. Olshausen, Pontus und Rom (63 v. Chr.-64 n. Chr.), *ANRW* II.7.2, 903-912; R.D. Sullivan, Dynasts in Pontus, *ibid.*, 913-930, both with rich bibliography; bishops in P.B. Gams, *Series episcoporum ecclesiae catholicae*, Regensburg 1873, 441-448.

M. Forlin Patrucco

POOR - POVERTY. The theme poor-poverty, in the *Fathers, has its roots in the *Bible where it occupies a considerable place: think of the rich vocabulary used to describe it in the OT. Faced with the fundamental needs of existence, the biblical point of view is the most realistic possible: there is no exaltation of poverty for its own sake; its legislation does not allow pauperism, the poverty that involves a condition of dependence, recalling the life of slavery in Egypt from which Israel was freed. The Exodus model is important from this standpoint: during the wandering in the desert, division into social classes was impossible since everyone was poor and equal in practice. Mosaic legislation on the theme of poverty is composed of different layers and, in some points (like the redemption of slaves at the end of seven years and the institution of the Jubilee: cf. Ex 21, 1-11; 23, 11; Lev 25, 3-9; Dt 15, 1-18), is perhaps more an ideal project than an effectual norm. Some texts introduce original initiatives of Israel (Lev 19, 9; 25, 3-6. 35; Ex 22, 24-26; Ruth 2, 1-3). Behind these norms is an idea of real parity, not just an affirmation of principle. If temporal blessings are the concrete sign of Israel's fidelity to God, poverty is seen as a curse that strikes the faithless or lazy. But experience shows that poverty is not always a punishment, but often accompanies virtue: esp. in the Psalms, the poor are often ranked together with the just. Thus we find many texts showing how wisdom must tend towards a "just mean" (Pr 30, 8-9) and others telling a different tale: many poor people are so because they are victims of man's injustice.

We see a change of perspective with the transition from nomadic life to settlement in Canaan, with consequent division of the land and onset of those social inequalities that increased in the period of the monarchy. This period sees the intense activity of the prophets in defence of the poor (Amos 4, 1; 5, 11-12; Is 3, 14-15; 10, 1-2; Jer 21, 11-12), with whose cause God comes to identify his own (Is 11, 4; 14, 30-32; Amos 2, 6-7; Bar 2, 18). This identification overturns the concept of poverty, which is no longer economic or sociological, but spiritual (Zeph 2, 3): the mediator of salvation is no longer the sovereign, but the Servant of Jahweh, "Jahweh's poor man" who rises to a messianic figure. The theme of "Jahweh's poor" - a technical expression which prevailed after the exile in wisdom texts (the identification of these "poor" with a socially organized group is much disputed) - touches on a mass of data: poverty, political power, the earth, *impiety, *justice, which are part of the specific history of Israel, a history that would become a model for the Christian communities of the first centuries.

NT and Christian *tradition discovered in the figure of the "servant of Jahweh" the announcement of Christ, suffering and redeemer. The NT texts present a dual evaluation of the theme of poverty: not only is it not a pure *evil (cf. the theology of the *beatitudes), but it can become a sign of eschatological fullness. A page of *Matthew (25, 31-46) on the last *judgment provides the reason for this acquisition by the poor of a privileged position in the Kingdom: the poor man is a universal sign of Christ's presence among men. The *Judaeo-Christian community of *Jerusalem (Acts 2, 44-46; 4, 32-37) required an equality that excluded misery, recalling the parity of condition proposed by Jewish legislation as an ideal aim: but communion of goods was, in contrast to the Essenes, a free choice of poverty (cf. P. Brezzi and M. Del Verme) which, being essentially a lack of guarantees, recalled the freedom from the Law proclaimed by *Paul. Communion of goods was not just a sign of mutual acceptance among believers, but expressed, as a manifestation of poverty, the fact that every Christian was, like Christ, a "servant of Jahweh". In the NT, discussion of poverty starts from Christ's poverty (Mt 8, 20; Lk 2, 7. 12. 16. 24; Jn 7, 15; 19, 25-27; 2 Cor 8, 9) and his friendship with the poor (Lk 4, 16; 7, 22) and ends by requiring poverty of those who wish to follow him (Mt 19, 16 ff.; Mk 1, 16; Lk 9, 3; 10, 4), not imposing the same renunciation on all, but teaching freedom from money, mutual help in need, the need to work for a living (2 Thess 3, 10), to all.

In the Christianity of the first centuries, the theme of poor-poverty is given different emphases according to geographical and cultural milieux. Against *encratite and *dualistic tendencies, Platonic and *Manichaean in origin, which, seeing an irresolvable fracture between the material and spiritual worlds, end by condemning any contact with earthly goods, even to prohibiting *labour (cf. the *heresy of the *apotactites* or *apostolici*: Epiph., *Haer.* 61), the Christians proposed a fundamental continuity between the two worlds on the basis of *faith in the one creator *God. *Didachè* IV, 8 stresses the obligation to share goods, the fruit of labour; *Aristides (*Apol.* 15, 7) and *Tertullian (*Apol.* 39, 10) consider this sharing a specific duty of Christians. For *Clement of Alexandria, communion of goods among men is commanded (*Quis dives salv.* 13). He tells us who is the poor man according to the gospel: he whose renunciation of goods is not an end in itself, but is founded on the Kingdom; in this perspective we may not judge the genuineness of a state of poverty, but must give to whoever asks, since the basis of giving without discrimination is human dignity (*Quis div. salv.* 33). For *Minucius Felix, poverty is not an infamy (*Oct.* 36); Christ is present in the poor and needy (Tertull., *Adv. Marc.* IV, 14, 9; Cypr., *De op. et elem.* 15; Greg. Naz., *De paup. amand.* 36; Aug., *Serm.* 86). *Cyprian, pastorally concerned with the presence of the poor in the Christian community, stresses the need for *almsgiving (*De op. et elem.* 5) to restore a situation of equality between *rich and poor. For *Basil, the social condition of the poor is not willed by God, but is the fruit of *sin (*Hom.* VI, 7; Greg. Naz., *Or.* 14, 25); so in helping the poor we cancel original sin (Basil, *Hom.* VIII, 7). The superfluity of the rich belongs to the poor (id., *Hom.* VI, 7), but relationships of generosity are possible even among the poor themselves (id., *Hom.* VIII, 6). Not all the poor are blessed, since poverty is not in itself a positive thing: poverty without freedom does not give *life (Clem. Al., *Quis div. salv.* 11); those poor are blessed who are poor through imitating Christ (Ambr., *Exp. Ev. sec. Luc.* V, 53; John Chrys., *In Mt. hom.* 90, 4); they are poor because they have not sinned, they have no *vices, they have nothing belonging to the prince of this world (Ambr., *Exp. Ev. sec. Luc.* V, 53). One who understands poverty not just in a spiritual sense is *Gregory of Nyssa: for him there must be no difference between actual poverty and spiritual poverty because, for the Christian, there is only one true poverty, which is the fruit of the Spirit and does not allow the possession of earthly riches (PG 44, 1208). The poor are seen as models of life in supporting affliction: for this reason they are stationed at church doors, to be an example and to encourage pity (John Chrys., *In I Thess. hom.* 11, 4); they are free to talk: for this reason the disciples sent by Christ were poor (id., *In Hebr. hom.* 18, 2); they are our true defenders before God (Hermas, *Past. alleg.* 2; Paul. Nol., *Epist.* 13, 11; Fulg. of Ruspe, *Serm.* 1, 7-8). *Augustine warns the poor against the danger of falling into covetousness (*Serm.* 85), since the truly poor man is he who not only does not possess, but does not desire to possess (Aug., *Enarr. in Ps. 103, ser.* 3, 16; John Chrys., *In Hebr. hom.* 11, 3). Poverty is the mother of all *virtues (Theodoret, *Prov. div.* 6), but not the poverty which is the fruit of dissipation or laziness (John Chrys., *In I Cor. hom.* 34, 6). A complementarity between poor and rich is indicated by *Hermas in the *allegory of the vine and the elm (*Past., alleg.* 2). A certain complementarity is also demonstrated by *John Chrysostom in the image of the two cities, that of the rich and that of the poor (*In I Cor. hom.* 34, 5), where by "poor" is meant those who work laboriously. Constant in patristic writings are: a negative evaluation of indigence and of poverty suffered for injustice, not chosen freely as *sequela Christi*; exhortation to Christian communities to share, to give generously and create concrete structures (centres of collection, funds for the needy, centres of voluntary work) capable of remedying situations of need. In the ancient church, the social problem was not the primary object of teaching, which was all aimed at evangelizing the poor, i.e. individuals, since it is acceptance of the good news, acc. to the *Fathers, that leads to the recognition of man's common origin and of the goodness of things, whence flows the rich man's duty to share until he changes the poor man's condition.

A. Causse, *Les pauvres d'Israël*, Paris 1922; *DBS* 7, 387-406; G. Barbieri, Le dottrine economiche nel pensiero cristiano, *Grande Antol. Filos.*, Milan 1954, V, 1089-1174; P. Brezzi, *Fonti e studi di storia della Chiesa*, Milan 1972; various articles in *Augustinianum* 17 (1977); M. Del Verme, *Comunione e condivisione dei beni*, Brescia 1977; V. Grossi, La chiesa precostantiniana di fronte alla povertà, in *L'annuncio del Regno ai poveri*, Turin 1978, 69-101; M.G. Mara, *Ricchezza e povertà nel cristianesimo primitivo*, Rome 1980; R.M. Grant, *Early Christianity and Society*, New York 1977; various authors, *Per Foramen Acus, Il cristianesimo antico di fronte alla pericope evangelica del "giovane ricco"*, Milan 1986.

M.G. Mara

POPE. The name "pope", from Greek πάππας, πάπας (father) was, until nearly the 9th c., but esp. in the 3rd-5th cc., given to *abbats and bishops as an expression of affectionate veneration. P. was sometimes given to mere priests. It is the normal term for Greek priests today. In *Egypt the πάπας *par excellence* was the bishop of *Alexandria. The bishop of *Rome is first described as p. only in a late 3rd-c. inscription. Towards the end of the 4th c. it tended to become his specific title (v. Ambr., *Ep.* 42; synod of Toledo [400]: Mansi 3, 1006 f.; Vincent of Lérins, *Comm.* 32 f.). During the 5th c. the name was still often accompanied by certain qualifications, such as *papa Urbis aeternae, papa Urbis Romae*. With the 6th c. it became, in official documents, save for ever rarer exceptions, a distinctive title of the bishop of Rome (v. *Lib. Pont.*, from the life of Agapitus I [535 f.] on). See also under *Papacy.

EC 9, 752 f.; *LTK* 8, 36 f.; B. Labanca, Del nome di "papa" nelle Chiese cristiane di Oriente ed Occidente, in *Actes du XIIe Congrès des Orientalistes*, III/2, Rome 1899, 47-101 (repr. Rome 1978); Ch. Pietri, *Roma christiana*, BEFAR 224, Rome 1976, 1609-1612 (bibl.); J. Moorhead, "Papa" as Bishop of Rome, *JEH* 36 (1985) 337-350.

B. Studer

PORCARIUS of Lérins. P.'s legend belongs to the hagiographical cycle of *Lérins, a cycle still awaiting study and on which A. Dufourcq's book sheds no light. At the time of Charles Martel (*c.*732), the Saracens (another subject on which B. Luppi's book sheds only false lights) devastated Provence: P., warned by a dream, saved treasure, relics and children. Ten days later, the Saracens arrived and massacred *abbat P. and 500 monks. Only Eleutherius escaped, buried the dead and, 20 years later, rebuilt the *monastery (BHL 6899-6901). At Lérins a supposedly 10th-c. chapel marks the traditional site of the massacre.

Vies des SS. 8, 204; *BS* 10, 1035-36; A. Dufourcq, *Etude sur les Gesta martyrum romains*, 2: *Le mouvement légendaire lérinien*, Paris 1907; B. Luppi, *I Saraceni in Provenza, in Liguria e nelle Alpi occidentali*, Bordighera 1952.

V. Saxer

PORPHYRIUS. Bishop of *Gaza. P.'s life was written by the deacon *Mark, his faithful disciple. The publication of this document (ed. H. Grégoire and M.A. Kugener, Paris 1930) has led to disagreements about its historical value. But on the succession and chronology of the facts related, it seems almost completely trustworthy.

P. was born in 347 at *Thessalonica of a rich family. Aged *c.* 30, he left the *world to dedicate himself to *God. He spent five years in *Egypt, as a monk in the *desert of *Scete, then five more in *Palestine, in a cave on the banks of the Jordan. His excessive austerities reduced him to a very precarious state of health, but he wished to visit the holy places of *Jerusalem, where Mark, a copyist by profession, met him and gradually became attached to him, not leaving him until his death. In the holy city, P., who had given away his inheritance to the *poor, set up as a cobbler to provide for his modest means. The fame of his sanctity reached bishop *John, who ordained him priest and made him guardian of the relics of the Holy Cross (392, aged 45). In 395, on the death of bishop *Aeneas of Gaza, *archbishop John of Caesarea, as metropolitan, appointed P. to succeed him. His rule was at first disturbed, esp. by the hostile resistance of the *pagans. At least twice P. had to request imperial support: once through the intervention of *John Chrysostom, the second time going to *Constantinople himself (402). P.'s episcopate lasted 25 years: he died 26 Feb 420. On this day the Byzantine Synaxaria put his *memoria* and Baronius introduced him into the *Mart. rom.*

BS 10, 1039-1043; Altaner 243; H. Leclercq, *DACL* 14.1 (1939) 1464-1504; H. Grégoire-M-A. Kugener, *Marc de Diaers, Vie de Porphyre*, Paris 1930; P. Peeters, La Vie géorgienne de S. Porphyre de Gaza, *AB* 59 (1941) 65-216; R.M. Grant, *Early Christianity and Society*, New York 1977; R. MacMullen, *Christianizing the Roman Empire*, New Haven 1984, ch. 10.

J.-M. Sauget

PORPHYRIUS of Antioch. Elected bishop of *Antioch at *Flavian's death (404), he is presented differently by our two main sources: *Palladius's *Dialogue on the life of St John Chrysostom* and *Theodoret of Cyrrhus's *Ecclesiastical History*. Acc. to Palladius, P. was *John Chrysostom's enemy and a brutal man, dissolute and a political adventurer, persecutor of Constantius, who was Chrysostom's *friend and a deacon of Antioch, the people's choice for the see; P. was consecrated with summary rites while the city was empty on account of the games at Daphne. Disorders then broke out, but were suppressed by the head of the army, Valentinus. The emp. *Arcadius intervened to oblige all to enter into *communion with P. (*CT* XVI, 4 6: 8 Nov 404; cf. PG 47, 13 and 37). But *Theophilus of Alexandria and pope *Innocent would not recognize him.

Theodoret however presents him as a man who "left many signs of his gentleness and prudence" (*HE* III, 35), for which "his memory is holy" (*Ep.* 83). A surviving fragment of a *letter of Theophilus informs P. of the calling of a *council against the new followers of *Paul of Samosata.

DCB 4, 443, n. 4; Fliche-Martin 4, 178 f.

G. Ladocsi

PORPHYRY, philosopher. Of Syrian origin, Malchus (=*Basileus*), called Porphyrius (232/3-305?), was one of the most prestigious and original exponents of *neoplatonism. A disciple of the philologist Longinus at *Athens and of *Plotinus at *Rome (263-268), where he revised his earlier religious opinions in the light of Plotinus's teaching; after a brief stay in Sicily for reasons of health (268-271), he returned to Rome after Plotinus's death (†270) and took on the direction of his School until his own death. Author of numerous strictly philosophical works, P. also interests Christian history and patristic literature for various reasons.

In his youth he probably knew *Origen at *Caesarea (cf. Euseb., *HE* VI, 19, 3 ff.) and we cannot rule out his having connections with Syrian Christian circles. In his *Letter to Marcella* there are surprising affinities with the Pythagorean-*encratite *Sentences of *Sextus*, of which Origen thought highly, and in his work *De abstinentia* P. uses the biblical expression "*coats of skins" (Gen 3, 21).

In any case, his deep knowledge of both Judaism and Christianity allowed him to make the most ferocious attack on Christianity that antiquity records: fifteen books *Katà Christianôn* (*Against the Christians*), composed after 270, which aroused a long series of Christian apologetic replies such as those of *Methodius of Olympius, *Eusebius of Caesarea, *Apollinaris of Laodicea and, at least indirectly, *Jerome and *Macarius Magnetes. Destroyed by order of *Theodosius II and *Valentinian III, the work is known only through fragments and more or less explicit allusions traceable in ancient church authors, but knowledge of it has been enlarged in recent decades and may yet be further enlarged by new discoveries. P.'s criticisms are aimed esp. against the Christian use of the *Scriptures, showing with considerable philological accuracy the inauthenticity of the book of the prophet *Daniel, dated to the 2nd c. BC, and various inexactitudes in the gospel accounts. Through the work of the African rhetor *Marius Victorinus and the preaching of *Ambrose of Milan, P. exercised a profound influence on *Augustine's neoplatonist formation. P.'s work *De regressu animae* is confronted at length in book X of *De civitate Dei*.

J. Bidez, *Vie de Porphyre, le philosophe néoplatonicien*, Gand-Leipzig 1913; A. Harnack, Porphyrius' "Gegen die Christen" 15 Bücher, Zeugnisse, Fragmente und Referate, *Abhandlungen der Königlichen Akademie der Wissenschaften*, Phil.-hist. Klasse, Berlin 1916, 1-116; J.R. Laurin, *Orientations maîtresses des apologistes chrétiens de 270 à 361*, AG 61, Rome 1954; H. Chadwick, *The Sentences of Sextus*, TSt n.s. 5, Cambridge 1959; J.J. O'Meara, *Porphyry's Philosophy from Oracles in Augustine*, Paris 1959; P. Hadot, Citations de Porphyre chez Saint Augustin, *REAug* 6 (1960) 205-244; H. Dörrie, Porphyrios als Mittler zwischen Plotin und Augustin, *Antike und Orient im Mittelalter*, Miscellanea Medievalia 1, Berlin 1962, 26-47; id., Das fünffach gestufte Mysterium. Der Aufstieg der Seele bei Porphyrios und Ambrosius, *Mullus* (Festschrift T. Klauser), JbAC Ergbd. 1, Münster i. W. 1964, 79-92; AA. VV., *Porphyre*, Entretiens sur l'antiquité classique 12, Vandoeuvres-Geneva 1966; W. Theiler, *Forschungen zum Neuplatonismus*, Berlin 1966; P. Courcelle, *Recherches sur les Confessions de Saint Augustin*, Paris ²1968, 311-382; P. Hadot, *Porphyre et Victorinus*, 2 vols., Paris 1968; E.L. Fortin, The "Viri novi" of Arnobius and the Conflict between Faith and Reason in the Early Christian Centuries, *The Heritage of the Early Church* (Essays in Honor of G.V. Florovsky), OCA 195, Rome 1973, 197-226; E. Des Places, *Platonismo e tradizione cristiana*, Milan 1976, 176-187; J. Pépin, *Ex platonicorum persona. Les lectures philosophiques de Saint Augustin*, Amsterdam 1977; A. Benoit, Le "Contra Christianos" de Porphyre: où en est la collecte des fragments?, *Paganisme, Judaïsme, Christianisme*, (Mél. M. Simon), Paris 1978, 261-275; F.H. Kettler, Origenes, Ammonius Sakkas und Porphyrius, *Kerygma und Logos* (Festschrift C. Andresen), Göttingen 1979, 322-328; M. Stern, *Greek and Latin Authors on Jews and Judaism*, II, Jerusalem 1980, 423-483; A. Meredith, Porphyry and Julian against the Christians, *ANRW* II.23.2 (1980) 1119-1149; G. Rinaldi, L'Antico Testamento nella polemica anticristiana di Porfirio di Tiro, *Augustinianum* 22 (1982) 97-111; T.D. Barnes, Porphyry *Against the Christians*: Date and Attribution of Fragments, *JThS* n.s. 24 (1973) 424-442; B. Croke, Porphyry's anti-Christian Chronology, *JThS* n.s. 34 (1983) 168-185; R.L. Wilken, *The Christians as the Romans saw them*, New Haven 1984, 126-163; W.H.C. Frend, Prelude to the Great Persecution: the Propaganda War, *JEH* 38 (1987) 1-18; id., *The Rise of Christianity*, London 1984, 441-444; A. Smith, Porphyrian Studies since 1913, *ANRW* II.36.2 (1988) 717-773.

P.F. Beatrice

PORTRAIT. By p. is meant the reproduction from life of an individual's *image, seen in its contingent reality. In a true p., imitation of physical likeness is completed by a search for psychological traits; in this way we pass from the facial, merely realistic, p. to the more complete, physiognomical, p. When it is impossible to model a person from life, we have the reconstructed p., which seeks imaginatively to do justice to the subject's characteristics on the basis of elements drawn from literary sources or *tradition. It follows, however, that in this way we can create different images of the same individual. Of this type, in early Christian art, are the pp. of Christ, shown in various forms; of Sts *Peter and *Paul, whose *iconography was fixed only from the 4th c. Also reconstructed, with variable characteristics, are the pp. of many *martyrs, saints and biblical characters, and of the college of *apostles, whose components often show similar or stereotyped characteristics. The phenomenon is certainly connected to the persistence in the first centuries of an iconoclastic tradition inherited from Judaism, which did not allow people's physical peculiarities to be fixed. By the time the need was felt, much later, these elements had been lost and so it was necessary to use reconstructed pp.

Egyptian pp., many coming from El-Fayum (1st-4th cc. AD), painted on wooden tablets that covered mummys' faces, while important in the history of the Roman p., show no signs of Christianity and so can only be borne in mind for general stylistic comparisons with other pp. of the dead from cemetery *painting, esp. Roman. Such pp., characterized by a certain sketchiness, but with considerable facial essentiality, can be recognized esp. in figures of *orantes, often with very marked outlines and large eyes, as in the examples in Callisto (*Wp* 88), Domitilla (*Wp* 138, 2) or the cemetery of the Jordani (*Wp* 120, 2). Many of the small figures of *orantes* incised on gravestones at *Rome and, in large number, in the Museo Paleocristiano at Aquileia are probably pp. Pp., presumably of deceased, can also be made out in scenes of life or *métiers* (as in the hypogeum of Trebius Justus at Rome), or again of symbolic type (e.g. St *Petronilla's introduction of Veneranda into *paradise, in the catacomb of Domitilla: *Wp* 213).

Among later paintings, from the 6th c. dates that of Turtura in the Roman catacomb of Commodilla which, by its expressive intensity and the essential simplification of its features, recalls the empress *Theodora in the mosaic of S. Vitale at *Ravenna.

In *mosaics, besides the African funerary mosaics esp. of *Tabarka (Bardo Museum), in the 14 busts of donors (men and *women) in the pavement of the S Theodorian hall at *Aquileia (second decade of 4th c.) we can recognize characteristics of the period of transition from the expressionism of the tetrarchy to that of the Constantinian era, esp. in the marked outlines of the faces, the women's hairstyles, the *dress. Other pp. can be identified in the figures of clients or clerics (often with the square *nimbus of the living) who appear in mosaic compositions, mostly apsidal: thus in some Roman basilicas, in the Euphrasian basilica of *Parentium, in St Demetrius's at *Thessalonica or in S. Vitale at Ravenna, where the long tradition of imperial iconography is renewed, often in new ways, in the mosaic panels of *Justinian and Theodora. Numerous pp. can also be recognized in gold-glass, following a fashion particularly common in the 4th c. In the late imperial age, imperial pp. (or pp. of other dignitaries) occur on ivory *diptychs.

The *imagines clipeatae* of *sarcophagi are in the tradition of Roman pp., but sometimes the faces are only roughed out. This is probably due to lack of time, though some scholars suggest neglect by heirs or an intention to apply a stucco mask with the deceased's features, or clients' superstitious fear of death. In the curious case of the "sarc. of the two brothers" in the Museo Pio Cristiano (*Ws* 91=*Rep.* 45), two males resembling each other have taken the place of the two spouses for whom the busts had originally been prepared. Sometimes the deceased are shown standing, holding a *rotulus*, against a background of *parapetasma. As for portraits of *popes, some would recognize these in 4th-c. gold-glass, but this theory is disputed. The oldest paintings would be some of the medallions showing the series of *pontiffs in the basilica of S. Paolo, Rome, not all preserved. The *tondi* with the popes from *Innocent I to *Leo the Great would originally have been pp., though considerably modified and retouched in the mediaeval and baroque periods. [Fig: 259]

DACL 14, 1, 1543-1573; EC 10, 1008-1009; EAA 6, 695-738; LCI 3, 446-455; H.P. L'Orange, *Studien zur Geschichte des spätantiken Porträts*, Oslo 1933; J. Bolten, *Die "imago clipeata". Ein Beitrag zur Portrait-und Typengeschichte*, Paderborn 1937; G.B. Ladner, *I ritratti dei papi nell'antichità e nel Medioevo*, I, Vatican City 1941; G. Bovini, *I sarcofagi paleocristiani*, Vatican City 1949; C. Casalone, Note sulle pitture dell'ipogeo di Trebio Giusto a Roma, *CArch* 11 (1962) 53-64; A. Grabar, Le portrait en iconographie paléochrétienne, *RSR* 36 (1962) 87-109; P. Testini, *Le catacombe e gli antichi cimiteri cristiani in Roma*, Bologna 1966; K. Parlasca, *Mumienportraits und verwandte Denkmäler*, Wiesbaden 1966; F. Zanchi Roppo, *Vetri paleocristiani a figure d'oro conservati in Italia*, Bologna 1969; B. Forlati Tamaro, Epigrafi cristiane sepolcrali con graffiti di Aquileia, *Archeologia Classica* 25-26 (1973-1974) 280-296; G.C. Menis, I ritratti nei mosaici pavimentali di Aquileia, *AAAd* 8 (1975) 73-92.

D. Mazzoleni

POSSIDIUS. Among our main sources of knowledge of P. are his own *Vita Augustini*, *Augustine's *letters and *Prosper's *Chronicon*. From 391 he lived in the monastic community directed by Augustine at *Hippo; soon after 397 he became bishop of Calama in *Numidia. As such he took part in the anti-*Donatist *councils of *Carthage of 403 and 407, played an important role in the conference of Carthage of 411 and took part in the anti-*Pelagian councils of Milevis (416) and Carthage (419). He undertook two official missions to *Italy (409 and 410) to ask the emperor to renew the laws against *pagans and *heretics. Abandoning Calama, invaded by *Vandals in 428, P. fled with other bishops to Hippo, where he remained until its fall, attending Augustine in his final illness. He returned to Calama, but was expelled by the *Arian *Genseric in 437. After this date we know no more of him. Between 432 (death of Boniface) and 437 (exile of the *Catholic bishops) P. composed his *biography of Augustine, to which he added a complete list of his writings (*Indiculus*), for fear that in the violence of the invasion the bishop and his works should be condemned to oblivion. The *Vita* fits into the old Suetonian biographical tradition. After the prologue, the matter is distributed not in four (Weiskotten, Diesner), but in three parts (Pellegrino): 1) the *vita*, a chronological account of Augustine's actions and vicissitudes starting from his *conversion (2-18). To avoid competing with the *Confessions*, earlier material is given only one chapter; 2) *mores*, understood as an exposition of the hero's conduct in daily life, public and private (19-27, 5); 3) final period, illness and death (27, 6-31). This last section (30) contains Augustine's letter to bishop *Honoratus on the behaviour of the *clergy during the enemy invasion. The biography is substantially truthful and objective, within the limits of this *literary genre: P. uses his knowledge of Augustine's deeds and writings, from which he often takes images and expressions. Though a biography of Suetonian type, it has characteristics typical of Christian *hagiography. In the prologue, P. emphasizes his own personal experience as a guarantee of truthfulness. Compared with *Paulinus of Milan, P. follows chronological order more faithfully (errors are few); he is able to appreciate his subject's literary activity; he gives little room to the marvellous or the demonic (only two miracles are recorded); and in his presentation of the *vita* and *mores*, he reduces *apologetics to a minimum.

Editions: CPL 358-59; PL 32, 33-66; H.T. Weiskotten, *Sancti Augustini Vita scripta a Possidio episcopo*. Edited with Revised Text, Introduction, Notes and an English Version, Princeton 1919; M. Pellegrino, *Possidio: Vita di S. Agostino*, introd., crit. text, version and notes, Alba 1955; *Vita di Cipriano. Vita di Ambrogio. Vita di Agostino* (Introd. by C. Mohrmann, Crit. text and comm. by A.A.R. Bastiaensen. Trs. by L. Canali and C. Carena, Milan 1975. It. version by M. Simonetti, Rome 1977).
Studies: A. v. Harnack, *Das Leben Augustin's von Possidius*, Berlin 1930 (introd. and tr.; Sp. tr.: *Revista de Occidente* 9 (1931) 52-100); H.J. Diesner, Possidius und Augustinus, *SP* 6, Berlin 1962, 350-365; *PAC* 1, 890-896.

A.V. Nazzaro

POTAMIAENA and BASILIDES, *martyrs. *Eusebius (*HE* VI, 5, 1-6) sums up, in a form already contaminated by legend, the story of P., a young martyr of *Alexandria: on her way to execution, she managed to convert the soldier B. who was escorting her and had earlier heard *Origen's lectures. With her mother, P. was smeared with boiling pitch and burned. A few days later, B. proclaimed himself a Christian and was arrested. In a dream he saw P. place the crown of martyrdom on his head; he was baptized in prison and beheaded three days later. The three martyrs are commem. on 28 June in *Mart. hier.*, 30 in *Mart. rom.*

Vies des SS. 6, 517; *LTK* 8, 645; *DHGE* 6, 1175-76; *BS* 2, 902-6.

V. Saxer

POTAMION of Heraclea. Appears among the participants of the council of *Nicaea (325) as bishop of Heraclea. Ten years later, at the council of *Tyre organized by *Eusebius of Nicomedia in favour of *Arius's theses and against St *Athanasius, P. took a clear anti-Arian position, defended the *orthodox position and cast doubts on the conduct of Eusebius, who had survived the *persecution of *Maximinus Daia in unclear circumstances, while P. had, at the same time, lost an eye and a foot. P. died as a result of maltreatment by the Arians. *Mart. rom.* records him on 18 May as a bishop, killed during the Arian controversy at the time of the emp. *Constantius II. We cannot completely trust *Mart. hier.*, acc. to which P. suffered *martyrdom on 18 May during the anti-Christian persecutions, and was only a presbyter.

EC 9, 1848; *BS* 10, 1060; Athanasius, *Hist. Arian* 73; id., *Ap. ad Const.* 28; Epiphanius, *Panarion* 68; ASS *Mai* IV, 165 ff.

M. Perraymond

POTAMIUS. Bishop of Lisbon (*Olisipo*) c.350, joined the ranks of the *Arians c.357, *praemio fundi fiscalis* (Faustinus, *Lib. prec.* 32). He played a prominent part in upholding the pro-Arian formula of *Sirmium in 357 and took part in the council of *Rimini (359) in the Arian ranks. We learn from some of his writings that he later rejoined the *Catholics.

From P.'s Arian period come an epistolary fragment handed down by *Phoebadius (*C. Arrian.* 5) and an item of Alcuin (PL 101, 113). From the Catholic period, prob. after 359, come the anti-Arian *Epistula ad Athanasium* and *Epistula de substantia*. In the latter, ample text, P. presents the union of *Father and Son (at times also the *Holy Spirit) not just as unity of substance common to the three *Persons in a general way, but as mutual and total interpenetration (chs. 18; 19). In the final part, on the basis of Gen 1, 26, he seeks the *image of the *Trinity in the features of the material man.

Two brief *homilies, *De Lazaro* and *De martyrio Esaiae prophetae* also survive: his interest is not in interpreting biblical passages, but is purely descriptive, tending to represent with maximum crudity and most marked realism the most macabre and repellent details of the two accounts (the smell emanating from *Lazarus's corpse, the rasping of the saw as it quarters Isaiah's body). Elsewhere P. shows the same predilection for strong tones and garish effects: a heavily baroque style, in fact, characteristic of the author.

CPL 541-545; PL 8, 1409-1418; PLS 1, 202-216; *Patrologia* III, 75-77 (bibl.); A. Montes Moreira, *Potamius de Lisbonne et la controverse arienne*, Louvain 1967.

M. Simonetti

POTHINUS of Lyons. One of the martyrs of *Lyons (*Lugdunum*) of 177: bishop of that city. His advanced age (over 90) did not prevent him being dragged before the tribunal and imprisoned, or abused by the over-excited crowd: "Though his body was giving way due to old age and disease, his soul was kept in him, so that Christ might triumph through him". He died in prison two days later (Euseb., *HE* V, 1, 29-31). With his companions, he is inscribed in *Mart. hier.* on 2 June.

DACL 10, 72-121; *Vies des SS.* 6, 26-40; *LTK* 8, 648; *BS* 8, 61-65.

V. Saxer

POVERTY. See **POOR.**

PRAEDESTINATUS. Under this title P.J. Sirmond published in 1643 a work, anon. in the MSS, which scholars have variously attributed: to *Arnobius the Younger, to *Primasius or, preferably, to someone in the circle of *Julian of Eclanum. It is in three books: *1)* a catalogue of *heresies, following *Augustine's *De haer.* and adding interesting information on the Tertullianists at *Rome, the *Nestorians and the predestinarians; *2)* a *sermon, attrib. to Augustine, but prob. by the author himself, with very radical formulae on *predestination; *3)* a refutation, phrase by phrase, of this theory of predestination. The work dates from 432/440 and deserves to be better studied.

Editions: CPL 243; PL 53, 587-672.
Studies: *LTK* 8, 672 (bibl.); H. von Schubert, *Der sogenannte "Praedestinatus"*, TU 24, 4, Leipzig 1903; M. Abel, Le "Praedestinatus" et le pélagianisme, *RTAM* 35 (1968) 5-25; *Patrologia* III, 464 and 539.

B. Studer

PRAETEXTATUS Vettius Agorius. Born, perhaps *Rome, c.310; present at the inaugural ceremonies of the new capital of *Constantinople in 330, if a passage of John of Lydia (*De mens.* IV, 2) refers to him. Had a particularly rich *cursus honorem*, as an inscription (CIL VI, 1779=ILS 1259) and, in part, two others (CIL VI, 1777=ILS 1258; CIL VI, 1778) attest. Among other things he was *consularis* of *Lusitania, proconsul of *Achaia (362), urban prefect (367), praetorian prefect, prefect of *Italy, Illyricum and *Africa (384). In 384 he was designated consul and died. In 344 he had married Aconia Fabia Paulina (Jerome, *Ep.* 49, 3), by whom he had a son. Two epigraphs (CIL XV, 7563; CIL VI, 1777) attest his ownership of *domus* on the Esquiline and the Aventine. Praised by *Ammianus for his moral integrity (*Hist.* XXVII, 9), he was a stout defender of *paganism, like his friend *Symmachus and *Nicomachus Flavianus. At his intervention *Valentinian I abrogated the law of 364 forbidding nocturnal *sacrifices. P. was close to several mystery cults: among other things he was augur, *pontifex* of Vesta and the *Sun, *curialis* of Hercules, *tauroboliatus* (cf. CIL VI, 1778, 1779). The senate and vestals raised statues in his honour. Prefect of Rome at the time of pope *Damasus (366-384), who failed to convert him to Christianity. In the literary field, he translated Greek lyric poets and prose writers and had some success as an orator.

EC 9, 1985-1986; R. Paribeni, *Da Diocleziano alla caduta dell'Impero d'Occidente*, Bologna 1942, 205 ff.; *PLRE* 1, 722-724; L. Storoni Mazzolani, *Sul mare della vita*, Milan 1969, 61-136; L. Cracco Ruggini, Il paganesimo romano tra religione e politica (384-394 d. C.). Per una reinterpretazione del Carmen contra paganos, *Atti Acc. Lincei, Memorie Sc. mor. stor.*, Rome 1979; J.F. Matthews, The historical setting of the *Carmen contra paganos* (Cod. Par. Lat. 80 84), *Historia* 19 (1970) 464-479.

M. Perraymond

PRAXEAS. We know this *patripassian *monarchian only through *Tertullian's *Adversus Praxean*. This tells us that P. introduced the monarchian *heresy to *Rome and then to *Carthage. Here, under accusation, he had to retract and subscribe a document (*libellus*) to that effect; but since he continued to propagate his heresy, Tertullian refuted it in his work, written in 213. Since *Hippolytus and other sources know nothing of P., while naming other monarchians, and since *Praxeas* may be taken as a nickname (= mischief-maker, swindler), some have thought that under this name Tertullian is aiming at one of the monarchians known to us: *Epigonus, who introduced *Noetus's monarchian heresy to Rome from *Asia, or even *Callistus. Today we tend to consider P. as the real name of an otherwise unknown heretic distinct from the others.

R. Cantalamessa, Prassea e l'eresia monarchiana, *Scuola Cattolica* 90 (1962) 28-50; A. Grillmeier, *Christ in Christian Tradition*, London 1965, 144 ff; J.N.D. Kelly, *Early Christian Doctrines*, London 1978, 122-126.

M. Simonetti

PRAXEDIS. The church of S. Prassede in *Rome was originally the *titulus Praxidae* (Roman *council of 499). During the 6th c. it took the name of St Praxedis (Roman council of 595). This change of title is an example of the transformation into saints of the founders of the churches of Rome. It was then that the legend must have been born which made P. the sister of Pudentiana, titular of the nearby church (example of the formation of a topographical legendary cycle), and daughter of the senator *Pudens. She was also identified with her homonym in the cemetery of *Priscilla. Paschal I (817-24) rebuilt her church, reversing its orientation. It was doubtless at this time that P. began to enjoy a liturgical cult.

DACL 14, 1691-1700; *Vies des SS.* 7, 516; *LTK* 8, 702; *BS* 10, 1062-72; A. Amore, *I martiri di Roma*, Rome 1975, 68-69.

V. Saxer

PRAYER

I. The first three centuries - II. The East: 4th and 5th centuries - III. The Latin West in the 4th and 5th centuries.

I. The first three centuries. P. is an integral part of any religion, its heart and its essence. Christianity rejected neither the terminology nor the *gestures of *pagan p., but expressed a systematic reserve about everything that savoured of *idolatry. To εὐχή, the common term of antiquity, the Christians, esp. St *Paul, preferred προσευχή. The Latin gospels usually translate this by *orare*, to distance themselves from the classical vocabulary.

The *apostle Paul transposed into his personal experience the same reality taught in the synoptic gospels and declared by *John when he spoke of worship "in spirit and in truth". P. immerses the Christian in the heart of the *mystery of the *Trinity, as the *Lord's Prayer* shows. Its roots are in the Lord who has come, who comes and who will come, and it culminates in *contemplation of God's gift, received from Christ in the *Holy Spirit. It has, therefore, a triple dimension: ecclesial, existential and eschatological (Hamman, *Prière*, I, 423-434).

The first two centuries of spiritual experience were a preparation for the golden age of the *Fathers: Christian p. started from the gospel, esp. from the Lord's Prayer. *Judaeo-Christianity borrowed the forms and customs of p. from Judaism. It continued to pray three times a day, using the *Oratio dominica* (*Did.* 8), which ended with the *doxology dear to contemporary Judaism. The references scattered through the *Epistle of *Barnabas* make up a small moral treatise on p. - it must be humble, joyful, confident and vigilant. The importance given to *eschatology directs p. itself towards expectation of the promised benefits (*Did.* 16; *Barn.* 21, 3; Hermas, *Sim.* II, 9) and towards the Lord's *parousia*, which is both reward and punishment. The church born of the mission to the pagans shows greater autonomy from the Jewish heritage. *Clement's p., the first explicit formulation outside the NT, expresses the belief of the community. Addressed to the *Father, creator and father of the ages, it is praise and thanksgiving for the Son, and gathers together the whole church's intentions through the mediation of Christ, our elder brother (*1 Clem.* 59, 3-61, 3). Though *Ignatius gives no examples of p., his *letters are full of allusions to it. The bishop's soul is turned continually towards *God the Father and towards Jesus Christ, from whom he expects all blessings. Father and Son are united in one worship and one *faith. His invocations to Christ are sighs, cries of the heart, which pours out its most personal aspirations without losing sight of the community (Hamman, *Prière*, II, 97). *Irenaeus turns spontaneously to the prayers of the NT (*Adv. haer.* III, 6, 4; 25, 2). P. is an existential attitude of the Christian: "Every day he worships God in God's temple, his *body, practising *justice at all times" (*Dem.* 96). In this temple, the Spirit cries: *Abba, Father!* and models man in God's likeness (*Dem.* 6). A sorrowful *kenosis*, which wrings from man a cry of expectation and an awareness of incompleteness.

The Acts of the *martyrs offer us the witness of a sober and committed p. The *confessor of the faith, like Christ on the cross and the protomartyr *Stephen, expresses in his p. the sense of his last rite and his total and universal gift. Martyrological literature describes with satisfaction the long prayers during detention and the continual dialogue with Christ even under torture. Christ accompanies and strengthens the martyr, who is visibly transformed by his presence. This helps us understand the ecclesial dimension of his p. and his confession (*Mart. Polyc.* 8; *Mart. Fructuosi* 3). The soul of the martyrs' p., at once individual and ecclesial, is the Holy Spirit: it prays in each of the martyrs, as it does in all Christians, esp. when they confess their faith before the *world. It guides the martyr like a *Eucharist to his *teléiosis*, in *Clement of Alexandria's expression, i.e. to his completeness and perfection.

The *apocrypha show a little-studied aspect of ancient p. They are rich in prayers and *hymns, though some seem interpolated. Sometimes quite disordered prayers, in which lyricism inspires a religious fervour corresponding to the popular tastes, impregnated with pathos and poetry, whence they arose (Hamman, *Prière*, II, 176). The p. of the apocrypha shows a rather uncertain, even *gnostic, *theology. Usually addressed to Christ, it is continually centred on the *Kyrios* in *glory, who manifests his presence and power, and allows us to discover the cosmic dimension of salvation. It leads to a consciousness of the cosmic importance of *faith. A christocentric, paschal, cosmic and eschatological p.

The 3rd c. saw the first treatises on p.: they turned on the *Lord's prayer, which was thus at the centre of daily and liturgical p. It was solemnly entrusted (*traditio*) to the *catechumen as the expression of his new *birth. It was by now the norm of Christian p. All the first treatises on p. comment on the Lord's prayer (*Tertullian, *Cyprian, *Origen), as if all teaching on this matter necessarily rested on it.

Tertullian stresses first of all the absolute novelty of Christian p. Christ is "the spirit that prevails, the word that teaches, the reason that gives life" (*De or.* 1). Properly understood, the Lord's prayer "sums up the whole gospel" (*ibid.*). Tertullian joins to p. the *gestures that accompany it (genuflection, prostration), but also *fasting and *hospitality (*De or.* 25). P. is the offering of the new covenant, which makes us "perfect worshippers and true priests" (*De or.* 28). His *commentary stresses the eschatological aspect: p. must hasten the coming of the Kingdom (*De or.* 5; 29).

Cyprian's treatise possesses neither the ascetic character nor the expressive brilliance of Tertullian's, but it is more pastoral: he is more concerned with seeing his congregation pray fruitfully. For this reason he is mainly interested in the ecclesial dimension of p., suggested by the plural form of the *Pater noster*. He has a missionary concern quite rare in the first centuries. Daily p. stimulates perseverance and disposes towards *martyrdom (*De or. dom.* 18).

*Clement of Alexandria gives no commentary on the *Pater* but, to make up for this, book VII of the *Stromata* offers "the first Christian and theological exposition of prayer" (E. von der Goltz). Two prayers occur in the *Pedagogue* (III, 12, 101; cf. I, 6, 42, 1-2): they are addressed to Christ, but associated with the Father and the Holy Spirit. Book VII of the *Stromata* describes p. as "an intimate and silent communication of the *soul with *God" (Pourrat, *Spiritualité*, 113). Spiritual progress is also progress in p., because this completes the perfect gnostic and leads him to the mystical heights (*Strom.* VII, 49, 6). This doctrine exercised a lasting influence on *Origen, *Cassian and *Diadochus.

Origen's treatise *On Prayer*, which Quasten justly calls "a real jewel", gives a profound and substantial textual *commentary on the *Pater* and a *theology of p., which must be biblical, ascetic and eschatological. P. marks out the itinerary of God's people and its stages: return to the divine likeness, elimination of the sensible through *purification, awakening of the spiritual senses leading to perfect union. The Father is the terminus of all p., the Holy Spirit is its soul (*De or.* 2). Note the pedagogical way in which Origen describes the preparation, beginning and conditions for every p. (*De or.* 31-32; 9).

The various treatises also tell us much about practices of p. The ternary system of daily p. is already mentioned in the *Didachè*. Morning and evening prayers had a more solemn character and were often linked to an *assembly (Tertull., *De or.* 9; *Trad. Ap.* 41). The Christians kept up the habit of p. before meals. Prayers made at the right time, but which must not lead us to forget that "the Christian's whole life is one long celebration" (Clem. Al., *Strom.* VII, 7, 35, 6).

*Polycarp prayed on his feet, facing the *East (*Mart. Polyc.* 5, 1), like John in the apocryphal Acts, either in expectation of the *Kyrios* or in the direction of *Paradise (Tertullian, Clement of Alexandria and Origen attest the same custom). The protagonist of the *Acts of Hipparchus* painted a *cross on the east wall of his house, where he prayed seven times a day.

The Christian prayed standing in memory of the *resurrection, or kneeling or prostrate to adore God or confess his *sins. He prayed with arms outstretched like Christ on the cross, or with hands open like *orantes* (V. Saxer, *Augustinianum* 20 [1980] 335-365).

Christian p. is never detached from existence, but extends into concrete situations "of a charity without limits": *agapè*, *hospitality, sharing of goods are the manifestations of a fraternity discovered in Christ and lived in the church. Origen repeats Clement's maxim drawn from *Scripture: "Prayer accepted by God is a good deed" (*Ped.* III, 69, 3).

Rooted in biblical *revelation and in Christ, Christian p. is at one and the same time thanksgiving and expectation, recapitulation and *eschatology.

E. von der Goltz, *Das Gebet in der ältesten Christenheit*, Leipzig 1901; G. Békés, *De continua oratione Clem. Alex. doctrina*, Rome 1942; A. Hamman, *Prières des premiers chrétiens*, Paris 1952, ²1981; id., *Le Pater expliqué par les Pères*, Paris 1952; id., *La Prière*, I, *Les origines chrétiennes*; II, *Les trois premiers siècles*, Paris 1959-1963; F.J. Dölger, Beiträge zur Geschichte des Kreuzzeichens, *JbAC* 1 (1958) 5-19; 2 (1959) 15-29; 4 (1961) 5-17; Gebet, *RACh* 8, 1134-1258; *La preghiera*, 3 vols., Rome-Milan 1966; *Ecclesia orans*, Mélanges A. Hamman, Rome 1980 (= *Augustinianum* 20 [1980]); A. Hamman, La prière chrétienne et la prière païenne, formes et différences, *ANRW* II.23, 1190-1247; R.L. Simpson, *The Interpretation of Prayer in the Early Church*, Philadelphia 1965; P. Bradshaw, *Daily Prayer in the Early Church*, London 1981; R.P.C. Hanson, The liberty of the bishop to improvise prayer in the Eucharist, in *Studies in Christian Antiquity*, Edinburgh 1985, 113-116; H.B. Green, Mt 28 19, Eusebius and the Lex orandi, in (ed.) R. Williams, *The Making of Orthodoxy* (essays in honour of Henry Chadwick), Cambridge 1989, 124-141; L.E. Phillips, Daily Prayer in the *Apostolic Tradition* of Hippolytus, *JThS* n.s. 40 (1989) 389-400.

A. Hamman

II. The East: 4th and 5th centuries. From the 4th-5th c., spiritual literature in the *East became ever more monastic. It was in monastic circles that works of teaching and all but exhaustive treatises on the subject of p. flourished (cf. Spidlík, *La spiritualité de l'Orient chrétien*, 409 ff.).

1. Definitions and divisions. For the monks, p. was the science *par excellence*, which combined everything in itself: *faith, *life according to faith, salvation. So the various definitions of p. proposed by the *Fathers certainly do not claim to be such in the strict sense of the word, but are focused on one or another aspect of that vital act which is praying. Man's conversation with *God, p. is primarily a "request to God for convenient goods" (Basil, *Hom. in mart. Julittam* 3: PG 31, 244A). The ancients themselves spoke of "raising the spirit to God" (a definition taken over by *Evagrius [*De orat.* 35: PG 79, 1173]). As a raising up to God our *Father, this *ascesis is not just a "vision" in the Platonic sense, but a "dialogue of the spirit with God" (*ibid.* 3, col. 1168). *John Damascene (*De fide orth.* 3, 24: PG 94, 1089C) joins the two aspects in the formula later used by so many others: "Prayer is a raising up of the spirit to God and a request to God for convenient goods". To avoid the danger of Platonic intellectualism, it was necessary to give new meanings and to adapt the term νοῦς, instrument of this ascesis, to indicate the totality of the person vivified by the Spirit.

To indicate progress, degrees of p. were listed and explained by the classical authors on *contemplation and *mysticism. The traditional order of these degrees goes back in general to the trichotomic structure of human composition: thus bodily or oral p. is distinguished from mental or interior p. and again from spiritual p., which takes place in the depths of the spirit.

In commenting on the four types of p. present in St *Paul (1 Tim 2, 1): δέησις, προσευχή, ἔντευξις, εὐχαριστία, the Fathers follow *Origen's exposition (*De orat.* 14: PG 11, 460; Cassian, *Collat.* 9, 9: PL 49, 780).

2. Prayer of petition. In opposition to the aristocraticism of some, the Fathers defend the p. of petition for "all the needs" of believers, though it is certainly a sign of higher perfection to ask for heavenly rather than earthly and temporal goods (Evagr., *De orat.* 37-38: PG 79, 1176).

To the question why our prayers are not always heard, the Fathers often reply that it is because we pray too little and are sinners (John Chrys., *In dimissione Chananaeae* 10: PG 52, 457). This is why, in the penitential tradition of the Eastern monks, one request is repeated insistently: for the remission of *sins - *Kyrie eleison*.

3. Uninterrupted prayer. In full agreement with all Eastern spiritual teachers, *Maximus Confessor writes that "Holy Scripture commands nothing impossible" when it prescribes: "Pray unceasingly" (1 Th 5, 17) (*Liber asc.* 25: PG 90, 929D, 932A). The *Messalians took the command literally: for them praying was equivalent to saying prayers, refusing any profane work, esp. manual *labour. In opposition to such utopias, *Augustine wrote *De opere monachorum* (PL 40, 547-582).

The *acoemetes believed it possible to achieve uninterrupted p. by collaboration, employing all the community in turn, through the succession of the offices of different groups of monks. The classic solution is in Origen: "the whole life of a saint is one great prayer, of which what we call prayer is only a part" (*De orat.* 12: PG II, 452). This doctrine would become that of Augustine (*De haeres.* 57: PL 42, 40) and *Aphraates (*Demonstr.* IV, 14-17). Though the monks always sought to augment their practice of p., their aim can be summed up in *Cassian's expression: *orationis status* (*Collat.* 10, 14: SCh 54, p. 95), a habitual disposition of the heart, a κατάστασις, a state of life.

4. *The victory over iconoclasm.* The contest over *images unleashed a grave crisis in the *Byzantine empire, which, with alternating phases, occupied part of the 8th and 9th cc. The significance of the anti-iconoclast reaction can only be understood by comparing it with the long series of *christological controversies. An idea dear to the Greek Fathers was that the sanctification of the *world culminates, and the union of the whole universe is brought about, in and through Christ. The Byzantine East placed the image of the *Pantocrator* at the centre of its cupolas, where it was the central motif of all *iconography and the pivot of the *cosmos. The aim of iconography is to testify to God's presence in the visible world. "Once God, the Incorporeal and Invisible, was never depicted. But now that he has been manifested in the *flesh and lived among men, I represent the "visible" of God: I do not worship matter, but the creator of matter" (John Dam., *Imag.* I, 16: PG 94, 1245A).

5. *Liturgical prayer.* The Greek Fathers have some fine pages on the ecclesial character of public p. The victory over iconoclasm was also the victory of the monks. The essential points of the "Studite reform" were the development of the *liturgy and the stabilization of rites. From the beginning of the 4th c., more precisely from the time the church was accepted and favoured by the imperial authorities, freedom to improvise was reduced. Texts and rites gradually became fixed, and the same period saw a transition from the oral liturgy of many communities to a written liturgy. In the Fathers there are five groups of writings on the theme of liturgical p.: mystagogical *catecheses; explicative treatises (addressed to *faithful, *clergy or monks, which multiplied esp. after the decline of the discipline of the *catechumenate); *homilies pronounced on the occasion of liturgical feasts; "Paschal letters"; works of a more general theological character. The Fathers saw the liturgy as an authoritative argument, a "*locus theologicus*" *ante litteram*. The most celebrated example is in *Prosper of Aquitaine, who, between 435 and 442, in his *Indiculus de gratia Dei* (8: PL 51, 209F) expounded the classic formula: *legem credendi statuat lex supplicandi*. Later, among the monks, liturgical contemplation arrived at the patristic conception of *theology as wisdom and gnosis of life.

6. *Hesychasm.* Nowadays, when we speak of hesychasm, we generally think of a certain method of p. codified in the monasteries of Mount Athos. In fact hesychasm, in its primitive meaning, was a system of *spirituality so old as to coincide with the origins of Eastern *monasticism.

Tranquillity, ἡσυχία, being without anxiety, ἀμεριμνία, safekeeping of the heart and spirit, vigilance, προσοχή constitute the climate of p. and derive from it. "Attentive practice of the *hesychia* of the heart will reveal a vertiginous abyss, and the ear of *hesychia* will hear marvellous things of God" (Hesychius of Batos, *Cent*, II, 30: PG 93, 1521A). In the history of this movement we can distinguish five periods: 1) the hesychasm of the *desert Fathers: insistence on solitude; 2) the Sinaites (*John Climacus, *Hesychius the Sinaite, Philotheus the Sinaite): *spiritual combat leads to *nepsis*, sobriety, which helps cast out diabolical suggestions; 3) Simeon the New Theologian (949-1022): direct experience of the Spirit; 4) the hesychasm of Mount Athos in the second half of the 13th and all of the 14th c.: the "method" of p. and the Palamite disputes; 5) the era of the *Philocalia* (end of 18th c.): return to patristic sources.

The *Philocalia of the neptic saints* (Venice 1782) is a collection of the most important texts on the subject of p. in the tradition of the Fathers of the Christian East.

DSp 1, 169-175; 7, 381-399; 1224-39; (bibl.); *DIP* 3, 1306-13; I. Hausherr, *Penthos. La doctrine de la componction dans l'Orient chrétien*, OCA 132, Rome 1944; id., Comment priaient les Pères, *RevAscMyst* 32 (1956) 38-58, 284-296; id., *Les leçons d'un contemplatif. Le Traité de l'Oraison d'Evagre le Pontique*, Paris 1960; id., *Noms du Christ et voies d'Oraison*, OCA 157, Rome 1960; id., *La prière perpétuelle du chrétien: Laïcat et Sainteté*, II, Rome 1965, 111-166, (repr. in *Hésychasme et prière*, OCA 176, Rome 1966, 255-306); J.-M. Sauget, *Bibliographie des Liturgies orientales (1900-1960)*, Rome 1962; F. Vandenbroucke, *Théologie des Gebetes nach "De oratione" von Origenes*, Munich-Paderborn-Vienna 1975; C. von Schönborn, *L'icône du Christ. Fondements théologiques élaborés entre le Ier et le IIe Concile de Nicée (325-787)*, Paradosis 24, Fribourg Suisse 1976; T. Spidlík, *La spiritualité de l'Orient Chrétien. Manuel systématique*, OCA 206, Rome 1978, 293 ff., 382 ff. (bibl.); S.P. Brock, *The Syriac Fathers on Prayer and the Spiritual Life*, Kalamazoo 1987; W.H.C. Frend, Some Greek Liturgical Fragments from Qasr Ibrim, *SP* 15 (1984) 545-553; and with G. Dragas, A Eucharistic Sequence from Qasr Ibrim, *JbAC* 30 (1987) 90-99.

T. Spidlík

III. The Latin West in the 4th and 5th centuries. The Latin Fathers of the Golden Age all commented on the book of *Psalms, "*vox Christi and Patrem*" (Ambr., *Ps.* 61). Thus Hilary, Ambrose, Jerome, Augustine. It is from these *Enarrationes* that any analysis of their teaching on Christian p. must start. In all of them, it is rooted in *Scripture. The majority of the Westerners are heavily influenced by *Origen. This is particularly obvious in Hilary, Ambrose and Jerome.

In *Hilary, the *Tractatus in Psalmos* are not just a *commentary on the psalter, but the tracing of a spiritual itinerary to *God. Christian p. is not just a formula, but a stimulus to *life: not so much *oratio verborum* as *affectus cordis* (*In Ps.* 60, 2). It is primarily an attitude of *faith (*ibid.*) and a motion of the heart (*Op. Hist. fr.* VIII). It requires interior assent (*In Ps.* 63, 5-6).

The *Treatise on the *Trinity* begins and ends with a p. (I, 38; VI, 19, 21; XII, 52-57). This, better than long explanations, demonstrates that p. leads to *faith (I, 37-38). God is a *mystery that we must accept in p. and with p. To find God, silent adoration is better than speculation (*De Trin.* I, 6; II, 7; X, 50, 54, 55; XI, 9, 23, 47).

*Ambrose is perhaps the Latin author most influenced by Origen. It is particularly visible in his *devotion to Christ. A preacher of *Scripture, Ambrose naturally draws on Origen's works, above all on the long *commentary on Ps 118, esp. vv. 22, 27-34. Like that of his teacher, his preaching willingly turns into explicit p. (*De fide* I, 20; V, 19; *Sp. s.* II, prol. 8; I, 14; *De paen.* II, 67 and 73; *Or. in Valent.* 80; *Or. in Theodos.* 36). He prefers to adore and praise God, like the Seraphim, rather than to discuss his nature (*De fide* II, 106; *Sp. s.* III, 21). Spontaneous prayers are habitually addressed to Christ (*De fide* I, 137; *Sp. s.* prol. 13. Cf. Baus, *Das Gebet beim h. Ambrosius*, Trier 1952). Dogmatic expositions end in confession, confession in *oratio* and *adoratio* (Dassmann, 91). Like Origen, he expresses his devotion to Christ in affective p. (Baus. *RQA* 44-46).

*Jerome is equally dependent on Origen in his p. to Christ. He invokes Christ and the *Holy Spirit mainly when he comments on Scripture (*Vita Hilar.* 1), e.g. *In Os. comm.* prol.; *In Hierem. comm.* 4. 6, praef.; *De vir. ill.* praef. It is not possible to accept God's word except "*Domino revelante*" (*Ep.* 65, 22, 4), "*nunc adiuvante, imo inspirante nobis Christo*" (*In Is. comm.* 8, *in fine*). In this p. to God, Jerome depends not just on Origen, but equally on martyrological *piety and literature (Baus, Das Gebet zu Christus beim h. Hieronymus, Trier, *ThZ* 60 [1951] 188).

*Augustine, in his *catechesis, his preaching and his correspondence, accords as important a place to p., esp. the *Lord's Prayer, which he calls a "*sacramentum*" (*Serm.* 228, 3). He comments on it incessantly (cf. A. Hamman, *Le Pater expliqué par les Pères*, Paris 1952, 154-169). In his *Enarrationes*, the psalter traces the Christian's road to God. See also his commentary *in I Johannis*. This is particularly obvious in *En. in Ps.* 41, in the psalms which evoke the heavenly *Jerusalem (*En. in Ps.* 61; 85; 86; 98), and in the gradual psalms (*En. in Ps.* 124; 125; 136; 147; 149). In the *psalms, Augustine sees the p. and *mystery of Christ and the church, the Head and the body. Christ's p. is our own; our own p. becomes Christ's. "Our prayers are to him, for him and in him" (*En. in Ps.* 85, 1). And he adds: "So there is one man who lasts until the end of days, and it is always his members who cry out" (*ibid.*, 4). The *Enarrationes* prolong the p. of the *Confessions* and manifest the spiritual and mystical experience of Augustine, the man of p. He had already expounded this in *De quant. an.* 33, 73-76. Augustine summed up his thought in *Ep.* 130 to *Proba. In the *Sermon on the mount*, he connects the petitions of the Lord's Prayer with the *Beatitudes and the Gifts of the Holy Spirit, which mark out the way to God.

After Augustine we find p. in preaching, particularly in commentaries on the Lord's prayer, esp. in *Peter Chrysologus (*Serm.* 67-72, ch. 39; 43), as well as in monastic literature, as in the East: John *Cassian, *Coll.* IX and X; *Martin of Tours, where p. is rooted in *lectio divina* (*Vita Mart.* 26, 2-4). Monastic *Rules insist on continual p. (Aug., *Ep.* 211, 7; Cass., *Coll.* IX, 3, 6, 7; X, 14), a recommendation which made its way into the *Rule of* St *Benedict.

Ch. Morel, La vie de prière de S. Augustin d'après sa correspondance, *RAM* 23 (1947) 222-258; A. de Bovis, Le Christ et la prière, *RAM* 25 (1949) 180-193; K. Baus, Das Gebet zu Christus, beim hl. Hieronymus, Trier. *ThZ* 60 (1951) 178-188; J. Delamare, La prière à l'école de saint Augustin, *VSp* 86 (1952) 477-493; K. Baus, Die Stellung Christi im Gebet des hl. Augustinus, Trier. *ThZ* 63 (1954) 321-339; id., Das Nachwirken des Origenes in der Christusfrömmigkeit des hl. Ambrosius, *RQA* 49 (1954) 21-55; A.M. Besnard, Les grandes lois de la prière. Saint Augustin, maître de prière, *VSp* 101 (1959) 237-280; T.A. Hand, *St. Augustin on Prayer*, Dublin 1964; M. Abad, *La oración misionera y sus fuentes según san A.*, Madrid 1964; E. Dassmann, *Die Frömmigkeit des Kirchenvaters Ambrosius v. M.*, Münster 1965, 86-94 and *passim*; G. García Montana, *La eficacia de la oración según la doctrina de San Agustín*, Madrid 1966; E. von Severus, Gebet, *RACh* 8, 1242-1247; A.G. Hamman, *Saint Augustin prie les Psaumes*, Paris 1981; L. Verheijen, La prière dans la Règle d'Augustin, *S. Augustin et la Bible*, Bible de tous les temps 3, Paris 1986, 167-179; S. Poque, La prière du catechumène Augustin en septembre 386, (*Conf* ix.4), in *Congresso Internazionale su S. Agostino nell XVI centenario della conversione*, Studia Ephemeridis Augustinianum 28, 3 vols., Rome 1987.

A. Hamman

PREDESTINATION

I. Before St Augustine - II. St Augustine.

St Augustine defined p. as "divine prescience and preparation of God's benefits by which those who are saved are most certainly saved" (*De dono persev.* 14, 35). The various supernatural gifts (call to *faith, *grace, *justification, etc.) are acts of divine benevolence, which wants to save us. In the polemical context of the *Fathers, problems emerge when we seek to connect p. with other data: existence of *evil in the *world, human freedom, retribution in the next life, condemnation after death. When this happens, we can always be tempted to play down one of its elements, denying either *God's intervention in history (*Pelagianism) or human freedom (*Manichaeism) and casting doubt on the unity of the saving plan (gnostic *dualism). Here we will emphasize the most important names in the history of the doctrine of p. in the first centuries.

I. Before St Augustine. Despite the infrequency of this theme in the earliest Christian writers, *Clement of Rome puts forward the following principles: salvation depends on God's initiative (*1 Clem.* 29, 1); salvation cannot be attained without good works (*ibid.*, 33, 1); these salutary works are themselves a gift of God (*ibid.*, 32, 3-4; 33, 1; 33, 8). Against *gnosticism and *Manichaeism, the *apologists and their followers affirm the unity of the saving plan, God's universal saving will and man's necessary co-operation under the influence of *grace. *Tatian (*Orat. ad Graecos* 13) and, with greater clarity, *Justin (*I Apol.* 12; 42) and *Irenaeus (*Adv. haer.* IV, 29, 2) give this clear expression, even if by p. they mean God's prescience rather than His decision. *Clement of Alexandria speaks of the "just whom God has predestined, having known them as such before the foundation of the world" (*Strom.* VII, 17). This fact strengthens the exercise of freedom, since "the just man's free will particularly obeys the divine will" (*Strom.* VI, 17). Salvation by grace requires the co-operation of good works (*Strom.* V, 1). *Origen, who deals quite fully with this subject, determined the later rhythm of Eastern *theology. He discusses the problem of p. particularly in his *Commentary on Romans* ch. 8 (cf. *Comm. in Rom.* VII, 7-8). He distinguishes between a general divine prescience and a prescience of *love. The latter prescience, a knowledge that approves and not just one of prevision, is a courtesy that only reaches the good (*Comm. in Rom.* VII, 7). Prescience and p., then, have the same range and the same object (*ibid.*, VII, 8). P. happens *post praevisa merita* and not as a result of prescience, which in no way annuls man's freedom and responsibility (*ibid.*, VII, 16; *De orat.* 6). Salvation depends on both causes, but primarily on the initiative of God, who calls all men to salvation. *Gregory Nazianzen (*Oratio* XXXVII, 13), *John Chrysostom (*In Ep. ad Romanos hom.* XVI, 5) - also *Cyril of Alexandria (?) (*In Ep. ad Romanos* 8, 30) - are more precise than Origen when speaking of God's universal saving will and of p.: God always gives his grace to the predestined and to those whom he has called from the beginning. Human complicity is necessary for salvation, though everything - even willing - must be attributed to God (Greg. Naz., *Orat.* XXXVII, 13). John Chrysostom, bearing in mind the effects, *ex consequenti*, distinguishes a double will of God: one of benevolence and another of punishment (*In Ep. ad Ephes. hom.* I, 2). And "though God works in us, both to will and to act" (*In Ep. ad Philipp. hom.* 8, 1), man, to be saved, has need of good works (*In Mt. hom.* 19, 2). *Commentaries of Latin writers on the Scriptural passages which mention this *mystery contain innumerable texts on p. It is clearly affirmed by *Hilary of Poitiers (*In ps.* LXIV, 5) and in even more detail by *Ambrose (*De fide* V, 6, 83).

II. St *Augustine. For this as for other problems, Augustine is the apogee of *patristics, and his importance for its history is decisive. Of places where the subject of p. is studied, we could cite the following: *Expositio quorundam propositionum ex Epist. ad Romanos; De diversis quaestionibus ad Simplicianum, II; Contra Iulianum; Contra duas epistolas Pelagianorum; Enchiridion; De gratia et libero arbitrio; De correptione et gratia; De praedestinatione sanctorum; De dono perseverantiae; Contra Iulianum opus imperfectum; Epistolae*: 194; 214; 215; 217; 225; 226. Interpretations of his doctrine are many and sometimes contradictory. To understand the *Augustinian theology of p., we must take into account both *God's universal saving will and the gratuitous character of man's supernatural *life. Bearing these two postulates in mind, Augustine looks continually at the mystery of the p. and predilection of God, who, *ante praevisa merita*, elects many and gives them his *grace so that they may win salvation. And, since all men form a *massa damnationis* in consequence of original *sin, we must also explain why God chooses some from this community of sinners, predesting them to salvation and not others. Acc. to St Augustine, prescience and the preparation of the efficacious means of salvation are the two necessary elements of p., fruit of divine benevolence (*De dono persev.* 14, 35). On God's part, p. is free and infallible. This, however, does not exclude or annul man's freedom, indeed it helps and strengthens it. The difficulties raised over Augustine's doctrine by the *semi-Pelagians of S *Gaul were partly caused by a confusion between prescience and p.; in reality, since they are not identical, the one can exist without the other (*De praedest. sanct.* 10, 19). P. means giving benefits; prescience means previous knowledge of good and *evil. In saying this, there is not the least allusion to a p. to damnation, though there are certainly prescience and permission of a penalty consequent on demerits. "He that keeps a knowledge of the causes of all things, cannot leave men's wills out of that knowledge, knowing them to be the causes of their actions" (*De civ. Dei* V, 9). Contrary to any form of *Pelagianism, God's prescience does not at all annul the free responsibility of man's will, which co-operates, or does not, in the accomplishment of the saving plan. *Fulgentius of Ruspe and *Caesarius of Arles carried on the line of Augustinian thought, which in one way or another would inspire Western *theology over the centuries.

DTC 12, 2815-2896; 1, 2398-2408 (Augustine); J. Saint-Martin, *La pensée de Saint Augustin sur la prédestination gratuite et infaillible des élus à la gloire, d'après ses derniers écrits (426-430)*, Paris 1930; L. Pelland, *S. Prosperi Aquitani doctrina de praedestinatione et voluntate Dei salvifica*, Montreal 1936; G. Nygren, *Das Prädestinationsproblem in der Theologie Augustins*, Göttingen 1956; J. Chéné, *La théologie de Saint Augustin. Grâce et prédestination*, Le Puy-Lyon 1961; V. Boublík, *La predestinazione: S. Paolo e S. Agostino*, Rome 1961; A. Trapè, A proposito di predestinazione. S. Agostino e i suoi critici moderni, *Divinitas* 7 (1963) 243-284; R. Bernard, La prédestination du Christ total selon saint Augustin, *RecAug* 3 (1965) 1-58; *MySal* IV/2, 776-800; P. Brown, *Augustine of Hippo*, London 1967, 398-407; J.M. Rist, Augustine on Free Will and Predestination, *JThS* n.s. 20 (1969) 420-447; J.P. Burns, *The Development of Augustine's doctrine of Operative Grace*, Paris 1980.

S. Folgado Flórez

PRESENTATION IN THE TEMPLE

I. Liturgy - II. Iconography.

I. *Liturgy. The first attestation of a liturgical memorial of the P. in the T. is at *Jerusalem, where *Egeria's *Peregrinatio* vaguely indicates it under the title *Quadragesima de Epiphania*, since it was fixed not on 2 Feb, but, in dependence on *Epiphany (the day when Christ's *birth was solemnized in the *East), on 14 Feb. The feast was celebrated *cum summa laetitia ac si per Pascha* (SCh 21, 36, p. 206). Egeria makes no mention of candles carried in procession. It was the Roman matron Ikalia, at the time of the emp. *Marcian (450-457), who suggested this custom. From Jerusalem the solemnity, named *hypapante=occursus (Domini)*, i.e. the Saviour's meeting with Simeon, spread beyond *Palestine: it is attested by, among others, *Gregory of Nyssa's *sermon for the feast of the P. (PG 46, 1151-1182) and that of *Cyril of Alexandria (PG 77, 1039-1050). An edict of *Justinian in 542 prescribed its annual solemnization at *Constantinople (Theophanes, *Chronographia*: PG 108, 487-488). Following the example of the Eastern churches, *Rome accepted the feast of the P., but naturally put it on 2 Feb. It was pope *Sergius I (687-701) who introduced it together with the three other Marian feasts of the Annunciation, Assumption and Nativity (*LP* I, 376). At Rome, as in the East, the feast of the P. was celebrated with great solemnity marked by *penitence, with a great procession from S. Adriano to S. Maria Maggiore, as witnessed by the 7th-c. *Ordo* XX (M. Andrieu, *Les Ordines Romani du Haut Moyen Age*, III, Louvain 1951, 235), which makes no mention of any rite of *blessing of candles, attested only in the late 9th or early 10th c. by some formulae added to the Padua Sacramentary (A. Ebner, *Quellen und Forschungen zur Geschichte und*

Kunstgeschichte des Missale Romanum im Mittelalter, Freiburg i. Br. 1896, 130). In the other churches of the West, the feast of the P. was introduced rather late. In *Spain, for example, we do not find it until the 11th c.

II. *Iconography. The only known depiction of the P. in the T. is in the *mosaic on the right side of the triumphal arch in S. Maria Maggiore, *Rome, executed under *Sixtus III (432-440) (C. Cecchelli, *I mosaici della basilica di S. Maria Maggiore*, Turin 1956, 219, pls. 53-58). The scene solemnly unfolds before a portico and concludes at the Temple. Escorted by two *angels, *Mary, in royal garments and statuary pose, advances with the Babe in her arms. Next to *Joseph, who points to them, are the prophetess Anna and the old Simeon, approaching full of reverence with his hands veiled by the *pallium* to receive the Babe *in pallio suo*, as specified by Ps.-Matthew (XV), who elsewhere adheres to Luke's *canonical narrative. Behind Simeon is the stupendous group of the Jewish priests, and below them four birds, two white and two dark, whose number recalls Mary's offering acc. to Ps.-Matthew, who here alters the canonical source.

A P. in the T. was among the subjects depicted in the mosaics of La Daurade de Toulouse (5th-6th c.), of which we have only vague knowledge (H. Woodruff, The iconography and date of the mosaics of La Daurade, *The Art Bulletin* 13 [1931] 80 ff.). The same subject appeared in the decoration of the church of St Sergius at *Gaza (first half of 6th c.) (Coricio, *Laud. Marc.* I, 48: ed. Foerster p. 14).

On the feast: P. Batiffol, *Etudes de liturgie et d'archéologie chrétienne*, Paris 1919, 193-215; L. Duchesne, *Origines du culte chrétien. Etude sur la liturgie latine avant Charlemagne*, Paris ⁵1925, 287-288; G. Löw, Purificazione, *EC* 10, 341-345; M. Righetti, *Manuale di storia liturgica*, II, Milan 1955, 90-95. On the iconography: C. Cecchelli, *I mosaici della basilica di S. Maria Maggiore*, Turin 1956, 219 f.; G.A. Wellen, *Theotokos. Eine ikonographische Abhandlung über das Gottesmutterbild in frühchristlicher Zeit*, Utrecht-Antwerp 1961, 60 f.; K. Wessel, Darstellung Christi im Tempel, *RBK* 1, 1134-1135; E. Lucchesi Palli, Darbringung Jesu im Tempel, *LCI* 1, 473-477; P. Testini, Alle origini dell'iconografia di Giuseppe di Nazaret, *RAC* 48 (1972) 334.

M. Marinone

PRIESTHOOD OF BELIEVERS. Since the Reformation, debate over the p. of b. has emerged and become a particularly disputed question in Christianity. In the context of an unequal church society (Bellarmine's thesis against the Protestants), the p. of b. was understood, when it was considered, as something markedly distinct from the priestly *ministry of the *clergy, hence limited to the *laity. Developing Roman Catholic theology of the laity after the 2nd World War led to Pius XII's formulation of the difference between the p. of b. and the ministerial priesthood as *essentia et non gradu tantum* (*Acta Apost. Sedis* 46 [1954] 669). The formula, today under linguistic and cultural examination (cf. *Rassegna di teol.* 1980, 409-412; 1981, 471-473), has been taken up again by Roman Catholic theologians. In the Roman Church, though not in other denominations, from the Reformation onwards, the theological picture of the expression "priesthood of believers" has usually reflected a preoccupation with asserting the superiority of the ministerial priesthood over that of the people; semantically, when the ordained ministry has been discussed, it has been expressed in terms and categories of "priesthood".

Having established this premise, useful for grasping the different understanding of Christian priesthood today from that of antiquity, we will give a few clues to it. The scriptural texts used in *patristics for the priesthood of Christians can be grouped in two series: 1) those connected with human life, lived along the religious lines of Jesus, in particular Jn 4, 23 ("the true worshippers of the *Father"); Rom 15, 16 and Phil 2, 17 (apostolic life and *conversion to Jesus Christ as *liturgy or worship offered to *God); Rom 12, 1 (the spiritual worship of offering oneself to God as a living *sacrifice); 2) passages on the Christian people endowed with *regale sacerdotium*: 1 Pet 2, 4-10; Rev 1, 6; 5, 10; 20, 6. The first series is usually found in apologetic contexts, against *Jews or *pagans, on the nature of their religions (in Christian eyes: a worship no longer accepted by God, that of the Jews; and a worship limited to offerings of things rather than of a holy life, that of the pagans). Christian worship is different: it is a new way of understanding religion. This is expressed in the holiness of a life lived in relation to God, not in mere ritual gestures or religious offerings independent of such a life. Explaining their worship as a new way of understanding the "religious" bond, Christians described themselves as "priests", indeed as the only true priests in the whole of mankind. This response was the watermark of Christian *apologetic literature until the gaining of religious freedom (*Galerius 311; *Licinius and *Constantine 313). A text of *Tertullian may be used as a hermeneutical key to similar passages in him and in other authors. "This" he writes in *De oratione* (28, 1-2) "is the spiritual offering that has abolished all the old sacrifices. . . . It is written: 'The hour will come when the true worshippers will worship the Father in spirit and in truth' and these worshippers he seeks. We are the true worshippers and the true priests because, praying in spirit, in spirit we offer our *prayer to God, as a sacrifice due to Him and accepted by Him. This is the offering He has asked for and the offering that has His approval" (see also Minuc. Fel., *Octav.* 32; Cypr., *Or. dom.* 2; *Ep.* 77, 3). In this sense, i.e. in the link that exists between living and worship, Christians considered themselves "priests of *peace" because they opposed the violence of the circus (Tertull., *Spect.* 16), and spoke of the "priesthood of *widowhood" (Tertull., *Ad uxorem* 1, 7) and the priesthood of *martyrdom in witness to Christ (Cypr., *Ep.* 77, 3).

The second series of texts, relating to *regale sacerdotium*, appears in two contexts: that of the *sacraments, esp. *baptism; and sometimes that of dispute with the hierarchical *ministry, which seemed to be reserving the title of "priesthood" for itself.

The NT expression is derived from the LXX which, in the original Hebrew text, in Ex 19, 6 has *regnum sacerdotum* for *regale sacerdotium*, applied to the entire Jewish nation: this will one day be "a kingdom of priests". In the NT, those baptized in Christ's name are seen as fulfilling the ancient promise, i.e. Christians are all kings and priests. In antiquity, the sacred was the place of access to an encounter, mediated by a priest, who was the man of separation and of mediation with the divine. The Christian is seen as one who approaches *God directly, no longer in need of further mediation; in this sense he is a priest, he is endowed with *regale sacerdotium*. The baptismal context particularly brought out this aspect. It was seen as "consecration to God" (Just., *I Apol.* 61), bearing in itself the *anointing of the Spirit, which, on the lines of Christ's anointing, permits approach to God, i.e. priesthood (Tertull., *Bapt.* 7; Iren., *Adv. haer.* IV, 8, 3; IV, 18, 2; Ephrem, *In festum Epiphaniae, hym.* 3; Ambr., *Myst.* 6, 30; *In ps.* 118, 6, 34; Aug., *En. in ps.* 26, II; Leo Gt., *Serm.* 4, 1). In the semantic context of this meaning, the Alexandrian school developed the concept of a priesthood of the just, which reaches perfection after the *resurrection. Since priesthood means being close to God, it is linked to the subject's holiness and will be perfect in the kingdom of heaven (Clem. Al., *Strom.* VI, 8, 107; Orig., *In ps.* 38; *Hom.* 1, 1; *Exhort. ad mart.*, 30 and 39; *In Num. hom.* 10, 2; Ruf., *De ben. patriarch.* 2-3; Cyr. Al., *De ador. in spir. et verit.* 9, etc.). In the same context, *Augustine defines *sacrifice as "that work which puts us in holy fellowship with God" (*De civ. Dei* X, 5).

In all this, naturally, the link was brought out between the priesthood and Christ's priesthood, which had one of its main fulfilments in the offering of the *Eucharist, as well as in holiness of living (Iren., *Adv. haer.* IV, 18, 1 ff.; Cyr. Al., *De ador. in spir. et verit.* 12 and 16; Ambr., *Sacr.* 4, 1; *In ps.* 118, prol. 2; *In Lc. ev.* 8, 52; Aug., *Serm.* 82, 5). The application of the concept of priestly dignity to all Christians, in line with Christ (the Anointed), was also an apologetic response both to the pagan mediations with the deity proposed to the initiate of the mystery religions (only in Christ is such a mediation possible) and to Greek *philosophy, which held God to be absolutely inaccessible (in Christ, however, everyone is offered the possibility of approaching God). Within a priesthood understood as a possibility of relating to God in line with Jesus Christ, Christian antiquity knew a diversity of *ministries that, on the concrete plane, mirrored the church's hierarchized structure: in particular the deacon, the presbyter and the bishop. In Augustine's time, all Christians were already classified into three well-defined groups: the *clergy, the *continentes* and the *faithful, a foretaste of Gratian's later *ordo clericalis et laicalis* (*Decr.* IV, p. 1, c. 2). Though first generation Christians knew no distinction between community leaders and others by reason of a different priesthood, they were moving in that direction, particularly in the West. *Clement of Rome, intervening in the affairs of the church of *Corinth which wished to dismiss its presbyters, by making a comparison between *episcopé* and priesthood, a comparison alien to classical and biblical usage, was the first to interpret the ministry of those in charge of the Christian community in terms of the OT Jewish (Aaronic) priesthood (*1 Clem.* 43-44). The Montanist *Tertullian contested this approach because, for him, the community leaders were not the same as the "spiritual men" (who alone had true *authority) and, besides, for such problems, we must get away from the general notion of "church" and from general sacral distinctions (*De exhort. cast.* 7; *De pudic.* 21. In *De praescr.* 41, he had stigmatized as heretical the practice of allowing primacy to spiritual men, the very solution he later put forward as a *Montanist).

Several things contributed to the ever more marked distinction between common priesthood and ministerial priesthood and the interpreting of the latter in sacerdotal categories: the application of the OT title of "shepherd" (Ezek 34) by Jahweh to Christ and by Him to the community leaders, in the perspective of the Aaronic priesthood (*1 Clem.* 43-44); the "sacral" character enjoyed by community leaders in ancient society, particularly

Roman society, which was heavily institutionalized (which explains many transformations and identifications in the "head" of Latin Christendom); the equation, after *Constantine, of church ministries with civil dignities, which also led to the assimilation of court *dress into public *assemblies, esp. liturgical ones. In an inquiry into the p. of b. on the evidence of the patristic period (1st-6th cc.) there remains, in conclusion, the starting-point of a priesthood, common to all Christians, on the lines of Jesus Christ's religious relationship, which, in attaining the order of ends (approach to God), transcends any order of concrete ministries; and these ministries, however transforming they may be, do not overstep their functional character.

B. Botte, Secundi meriti munus, *Questions Lit.Par.* 21 (1936) 84-88; L. Cerfaux, Regale Sacerdotium, *RSPh* 28 (1939) 5-39; J. Lécuyer, Essai sur le sacerdoce des fidèles chez les Pères, *La MaisonD.* 27 (1951) 7-50; P. Dabin, *Le sacerdoce royal des fidèles dans la tradition ancienne et moderne*, Paris 1950 (Westerners 71-168; Easterners 509-577); J. Lécuyer, Le sacerdoce royal des chrétiens selon saint Hilaire de Poitiers, *AnThA* (1949) 302-325; G. Folliet, Les trois catégories des chrétiens, *AugM* 2, Paris 1954, 631-644; J. Pintard, *Le sacerdoce selon s. Augustin. Le prêtre dans la Cité de Dieu*, Paris 1960; L. Ryan, Patristic Teaching on the Priesthood of the Faithful, *The Irish Theol. Quarterly* 29 (1962) 25-51; R. Jacob, Le martyr, épanouissement du sacerdoce chrétien dans la littérature patristique jusqu'en 258, *MSR* 24 (1967) 57-83, 153-172; A. Quacquarelli, L'epiteto sacerdote (iereus) ai cristiani in Giustino martire (Dial 116, 3), *VetChr* 7 (1970) 5-19; G. Otranto, Nonne et laici sumus? (Exh. cast. 7, 3), *VetChr* 8 (1971) 27-47; M. Jourjon, Remarques sur le vocabulaire sacerdotal de la 1 Clementis, in *Epektasis* (Mél. J. Daniélou), Paris 1972, 107-110; M. Bévenot, Tertullian's Thoughts about the Christian "Priesthood", in *Corona Gratiarum* (Mél. E. Dekkers), I, Brugge 1975, 125-137; M. Bévenot, "Sacerdos", as understood by Cyprian, *JThS* n.s. 30 (1979) 421-423; R.R. Noll, The Search for a Ministerial Priesthood in I Clement, *SP* 13 (1979) 250-254; (eds.) Rauch-Imhof, Das Priestertum in der einen Kirche, *Koinonia* 4 (1987) 45-89; J.T. Lienhard, *Ministry*, Wilmington 1984; S.G. Hall, Ministry, Worship and Christian Life, in (ed.) I. Hazlett, *Early Christianity* (Festsch. W.H.C. Frend), London 1991, 101-111.

V. Grossi

PRIMASIUS of Hadrumetum. Bishop of *Hadrumetum (*Africa) between 550 and 560, one of the few Africans to support the condemnation of the *Three Chapters. In his *Commentary on the *Apocalypse*, he claims to be inspired by *Augustine and *Tichonius, whose commentary he has purified from anti-*Catholic ideas. Tichonian influence is very strong in his allegorizing interpretation, which tends to see the prophetic text as referring to the history of the church, not just in the last days, but from the beginning, in such a way as to relieve the eschatological tension. The transparent anti-Roman ideas in John's work are totally eliminated.

Moricca III, 2, 1485-1487; CPL 873; PL 68, 793-936.

M. Simonetti

PRIMIAN of Carthage. *Donatist bishop of *Carthage c.392-412. A man of violent temperament and intransigent views; he represented the views of the people of Carthage and *Numidia. Only *recens ordinatus* when he became primate of the Donatist church (Aug., *Serm. 2 in Ps.* 36, 20), P. was opposed by the Donatist *clergy of Carthage under the deacon *Maximian (a descendant of *Donatus of Carthage). The *schism exploded at the end of 392: P. was condemned for tyrannical acts, first by a *council of 43 bishops at Carthage, then, 24 June 393, by a council of 100 bishops at Cebarsussa, presided over by the Donatist primate of Byzacena, Victorinus of Munatiana (Aug., *op. cit.* and *C. Cresc.* II, 6, 7). P. reciprocated: on 24 April 394 he was absolved by a council of 310 bishops, mainly from Numidia and *Mauretania, at Bagai in S Numidia. The Maximianists were in turn condemned (*C. Cresc.* III, 15, 18; 16, 19; 19, 22; 56, 62; IV, 4, 5; 10, 12; 28, 35; 31, 38; 38, 45; 39, 46). Between 395 and 397 P. instituted a series of lawsuits before the proconsul of Africa and the city magistrates of Carthage to regain possession of the *property held by the Maximianist bishops (Aug., *op. cit.*, III, 59, 65; IV, 47, 57). He tenaciously supported *Optatus of Thamugadi, in whose anniversary celebrations he took part (Aug., *Ep.* 108, 2, 5). In Sept 403 he scornfully refused an invitation from the *Catholics to discuss ending the schism (Aug., *Ad Donatistas post Collationem* I, 31, 53), and sought to discredit *Augustine on the basis of his equivocal past, accusing him of crypto-*Manichaeism (Aug., *En. Ps. 36, serm.* 3). At the conference of *Carthage of 411, P. led the Donatist delegation but said little. Though he must have been deposed at the time of the proscription of the Donatist church in 412 (*CT* XVI, 5, 52), we know nothing more of him.

Monceaux 4, 157 ff. (severe judgment); W.H.C. Frend, *The Donatist Church*, Oxford ²1971, ch. 15; *PAC* 1, 905-913.

W.H.C. Frend

PRIMICERIUS. Title (*primus in cera*, i.e. in the *tabula cerata*) given in the late empire to one holding the first rank in duties or functions of various kinds, as e.g. the *primicerius sacri cubiculi*, the *primicerius notariorum* or of other *officia* in the imperial chancellery. In particular, the *comites sacrarum largitionum* (imperial ministers of finance: the title varies greatly in the sources) had under them a general *primicerius* and particular *primicerii* for each of the various *scrinia* (offices). A similar organization existed in other *officia*, for some of which a *secundicerius* and *tertiocerius* are also mentioned (cf. *CT* VI, 24, 7 ff.; *CJ* I, 27; XII, 7. 28; CIL 13, 2385; Cassiod., *Variar.* XI, 25 and 30-32: MGH, *Auct. ant.* 12, 346 ff.). The title was also used in ecclesiastical circles for various offices: *primicerius scholae, primicerius defensorum*, etc. *Isidore of Seville describes the office of p. of the lower ecclesiastical *ministries (*Ep.* 1, 3: PL 83, 896). The *primicerius notariorum*, head of the secretariat, had particular importance in the great churches (cf. Diehl 1287; 3768; council of Ephesus: ACO I, I, 2 p. 7; Mansi 7, 377; Nilus, *Ep.* 2, 238: PG 79, 321B; Evagr., *HE* II, 18: PG 86, 2560A; Anast. Apocr., *Acta S. Max.* 3: PG 90, 113D; Eugipp., *Vita Sev.* 46: CSEL 9, 2 pp. 66; Greg. of Tours, *Hist. Franc.* 2, 37). The p. of the *notarii of the church of *Rome played an important role in the life of the community (cf. PL 68, 55; Roman council of 649: Mansi 10, 891; *Ordo Rom.* I, ed. Andrieu II, 70; *Liber diurnus, passim*).

DACL 14, 1171-1181; *DAGR* 4, 647; *EC* 10, 20-22; Lampe, *s.v.*; Niermeyer, *Mediae Latinitatis Lexicon Minus*, Leiden 1976, 848.

M. Forlin Patrucco

PRISCIAN. Latin (Christian) grammarian. Of his life we know little but that he was born at *Caesarea in Mauretania and lived largely at *Byzantium under the emp. *Anastasius (491-518). In the manner of *Pliny's *Panegyric to Trajan*, P. exalted Anastasius's deeds in a *Panegyric* of just over 300 verses, an uninspired work. His verse translation of the Greek Dionysius Periegetes's geographical poem (*Periegesis of the Earth*) is of modest value. P.'s fame rests on his grammatical work: his *Institutio de arte grammatica* was known and studied throughout the Middle Ages. Of its 18 books, the first 16 (*Priscianus maior*) elaborate the principles and rules of grammar; the last two (*Priscianus minor*), on syntax, come to a sudden end, either intentionally or through the loss of the final books. While recognizing the superiority of the Greeks, P. offers a very organic and comprehensive work on the whole traditional technique of the Latin language. His theoretical rules are enriched by a wealth of citations of lost Latin authors. In particular, in his *De metris fabularum Terentii*, he was able to establish the metric characteristics of the ancient comedians, by now almost lost.

Editions: CPL 1546-1554. For the *Institutiones* and other minor works, M. Hertz, *apud* H. Keil, *Grammatici Latini*, vols. II and III, Leipzig, 1855-1923. For the *Carmen in laudem Anastasii imp.*, E. Baehrens, *Poetae latini minores*, V, Leipzig 1883, 264-74. For the *Periegesis*, P. van de Woestijne, Bruges, 1953. Studies: Schanz IV/2, 221-238; *PWK* 44 (1954) 2328-2346; *KLP* 4, 1141 f.; G. Mazzini, Il codice vat. 3313 di Prisciano, *ALMA* 1 (1924) 213-222; 2 (1925) 5-14; M. Passalacqua, *I codici di Prisciano*, Rome 1978. For studies of recent years cf. *Medioevo latino*, Spoleto 1980, 220 ff.

L. Dattrino

PRISCILLA. Innumerable legends have been built around P., both anciently and recently. Hagiographers made her the mother of the senator *Pudens, grandmother of *Praxedis and Pudentiana and hostess of the apostle *Peter in a villa on the Via Salaria, beneath which the homonymous catacomb was later dug. At this point the archaeologists intervened, led by G.B. De Rossi, and brought all these people, as well as Aquila and Prisca or Priscilla (Acts 18, 2-4; Rom 16, 3; 1 Cor 16, 19; 2 Tim 4, 19), into the family of the *Acilii Glabriones*; Marucchi went so far as to identify, in the catacomb, the font in which Peter had baptized. All this is mere romance, with no more value than the hagiographical romance. What is certain is that the name P. appears several times in inscriptions (*DACL* 6, 1264-66; 1274) and that we must distinguish at least three women with this name: Vera Priscilla, wife of M. *Acilius Glabrio; Plaria Vera Priscilla, her daughter-in-law; Priscilla, wife of one M. Acilius V. It follows that one of the nuclei of the homonymous catacomb doubtless owes its existence to them.

DACL 6, 1269-74; *Vies des SS.* 1, 365-66; *BS* 10, 1113-14; P. Testini, *Le catacombe e gli antichi cimiteri cristiani di Roma*, Bologna 1966, 50, 166; A. Amore, *I martiri di Roma*, Rome 1975, 66-67.

V. Saxer

PRISCILLIAN - PRISCILLIANISM. Priscillian, a *layman of high position and great capacity, began, c.370-375, to preach in *Spain a very rigid ascetic doctrine, which had great success but attracted the hostility of bishops *Idatius of Mérida and *Ithacius of Ossonuba. In 380 a *council at

*Saragossa condemned Priscillian's ideas, but took no measures against persons. Then bishops *Instantius and Salvian, to give Priscillian's preaching greater *authority, consecrated him bishop of Avila (*Abela*). Meanwhile Idatius and Ithacius obtained from the emp. *Gratian a decree against the *Manichees in terms that could also include the Priscillianists. Then Priscillian, who had extended his activities to S *Gaul where the rich lady Eucrotia joined his group, went to *Italy to seek support from *Ambrose and *Damasus; though unsuccessful, he managed to get Gratian's decree annulled. After Gratian's murder, Ithacius accused Priscillian before the usurper *Maximus at *Trier. Salvian being dead, Priscillian and Instantius were deferred by him to a council at *Bordeaux (384). Instantius was deposed; Priscillian did not turn up and appealed directly to Maximus. Idatius and Ithacius also went to Trier and worked on Maximus so well that they obtained Priscillian's condemnation, despite the contrary efforts of *Martin of Tours. Priscillian and Eucrotia were beheaded on a charge of black *magic, Instantius and others were exiled. It was the first time a *heretic had been condemned to death as such, and the unfavourable impression was great: even Ambrose, who had refused Priscillian his support, protested strongly. In reaction, Ithacius was deposed; Idatius resigned spontaneously. Priscillian having gained the martyr's crown, his movement was active in N Spain for much of the 5th c.

Priscillian's *multa opuscula* (Jerome, *Vir. ill.* 121) being lost, until the end of the 19th c. the Priscillianist doctrine had to be reconstructed on the basis of the writings of opponents and the decrees of Spanish councils, which presented it as *gnostic and *encratite in content: distinction between the *God of the OT and the God of the NT, divine nature of the *soul, Christ's humanity not real, condemnation of *marriage, practise of magic and *astrology. In 1889 the publication of a *corpus* of Priscillianist writings from a Würzburg MS disappointed those who expected decisive evidence, since, apart from a few passages, these writings are completely *orthodox. As a *dossier* of texts collected for defensive purposes, it avoids the movement's most characteristic ideas. It was long ruled out that these texts could be Priscillian's direct work, but this theory now finds authoritative support. The *Liber apologeticus* and *Liber ad Damasum* condemn various *heresies and astrological doctrines. The *Liber de fide et apochryphis* maintains that not all inspired books are contained in the Canon and that the *apocrypha should not all be indiscriminately rejected just because heretics have interpolated them with their errors: they should be read, though with due caution; the divine Spirit is not confined to the *canonical books, because where Christ is, there is freedom. Another eight treatises are homiletic or liturgical: the *homilies deal with OT episodes, following the traditional typological reference to Christ and the church. They were attached to the *dossier* in order to refute accusations of rejecting the OT, as did the gnostics and Manichees. Priscillian's *Canones epistularum Pauli apostoli*, a florilegium of 90 propositions taken from the Pauline epistles, have reached us separately.

Nothing remains of the Priscillianist writers mentioned by *Jerome (*Vir. ill.* 122; 123) or others: *Latronianus, *Tiberianus, *Asarbus. In 1913 G. Morin published an anon. *De Trinitate fidei catholicae*, perhaps by Priscillian himself, almost certainly by a Priscillianist: it gives an archaizing doctrine of the *Trinity, tending to *monarchianism, which confirms an accusation made against the Priscillianists by *Orosius. The fundamental themes of their preaching cannot be specified, beyond a certain rigorous *encratism nourished on a negative conception of the material *world, with perhaps some astrological infiltration.

CPL 785-796; CSEL 18; PLS 2, 1389-1542; *Patrologia* III, 126-131 (bibl.); A. Orbe, Doctrina trinitaria del anónimo prisciliantista De Trinitate fidei catholicae, *Gregorianum* 49 (1968) 510-562; H. Chadwick, *Priscillian of Avila*, Oxford 1976; J. Fontaine, Panorama espiritual del Occidente peninsular en los siglos IV y V, in *Primera Reunión gallega de estudios clásicos* (Pontevedra 2-4 julio 1979), Santiago de Compostela 1981, 185-209; J.M. Blázquez Martínez, Prisciliano introductor del ascetismo en Gallaecia, *ibid.*, 210-236; A.B.J.M. Goosen, Algunas observaciones sobre la pneumatología de Prisciliano, *ibid.*, 237-242; A. Van Dam, *Leadership and Community in Late Antique Gaul*, Berkeley 1985; A.M. La Bonnardière, Du Nouveau sur le Priscillianisme, (Ep 11*), in (ed.) C. Lepelley, *Les Lettres de Saint Augustin découvertes par Johannes Divjak*, Paris 1983, 205-214.

M. Simonetti

PRIVATUS of Lambaesis. This name appears in *Cyprian's correspondence (*Epp.* 36, 4; 59, 10). From *Ep.* 36, circulated by the Roman *clergy at the most crucial moment of *Decius's *persecution, we learn that Cyprian had warned the church of *Rome about P.'s situation, a situation that we know from *Ep.* 59. P., bishop of Lambaesis, deposed "for numerous and grave errors, some years before, by a sentence of 90 bishops", had returned to plead his cause at the *council of spring 251, but his request was rejected. He had then thrown in his lot with a small group of Carthaginians who had separated from the official church and irregularly ordained a certain *Fortunatus as bishop in opposition to Cyprian. After this *letter (summer 252), we lose sight of P.

Monceaux 2, 5; *DCB* 4, 479 f; G.W. Clarke, *The Letters of St. Cyprian of Carthage*, vol. 3 (letters 45-66), ACW 46, New York 1986.

V. Saxer

PROBA. P. Faltonia, wife of Clodius Celsinus Adelphius, *praefectus urbi* in 351, mother of Olybrius, consul in 379, and of the *praefectus urbi* of 391, was of the higher nobility. She became a Christian and wrote a historical *epos*, now lost, on the struggle between *Magnentius and *Constantius II, conceived as a *bellum civile* in the manner of Lucan. Between 360 and 370, she put together a *cento of 694 hexameters from pieces of Virgil: a *carmen sacrum* (v. 9), in which she hymned the Christian *revelation in 14 OT and 18 NT scenes. Though much read, it was condemned by the *Decretum Gelasianum*. Contrary to older negative evaluations, today we see in it the fascination of a perfect piece of Virgilian art. Though P. could not use biblical names, she represented concrete events with extraordinary skill: the creation of sea-creatures after the model of Aeneas's shield, that of plants in the style of the Eclogues, the serpent in the earthly *Paradise by the one that killed Laocoön, the expulsion from Paradise in the words of the Cumaean Sibyl, the *flood, the crossing of the Red Sea, the birth and *baptism of Christ, when *John welcomes Jesus in the words of Anchises, the *Sermon on the Mount with threats of punishment in the Virgilian inferno, the storm on the lake using marine elements from books III and V of the Aeneid, Jesus' *entry into Jerusalem (on Silenus's ass), the crucifixion, with Christ in a threatening attitude, the *resurrection and *ascension.

Editions: CPL 1480; PL 19, 805-818; Schenkl, CSEL 16, 1, 568-609.
Studies: Schanz IV, 1, 219-221; R.H. Schelkle, Cento, *RACh* 2, 972-973; I. Opelt, Der zürnende Christus im Cento der Proba, *JbAC* 7 (1964) 106-111; M.R. Cacioli, Adattamenti semantici e sintattici nel centone virgiliano di Proba, *Studi italiani di filol. classica* 41 (1969) 188-246 (on the language); C. Cariddi, *Il centone di Proba Petronia*, Naples 1971; *Patrologia* III, 257 f.; J. Fontaine, *Naissance de la poésie dans l'occident chrétien*, Paris 1981; E.A. Clarke-D.F. Hatch, Jesus as hero in the Vergilian Cento of Faltonia Betitia Proba, *Vergilianus* 27 (1981) 31-39.

I. Opelt

PROCESSION. The existence of pp. throughout history seems to be derived from man's psychological and sociological needs. We find pp. in the OT (cf. *La MaisonD.* 43 [1955] 5-22) and the NT (*ibid.* 22-28). The Christian meaning of p. is that of a people on *pilgrimage accompanied by *God (in an OT sense) or walking with Christ (towards death and *resurrection). The main examples of liturgical pp. are those connected with the celebration of the *Eucharist, i.e. the entrance of the celebrant, who advances with the gospel book and the offerings (cf. *Euchologium Serapionis* 19: A. Hänggi-I. Pahl, *Prex eucharistica*, Fribourg 1968, 133; Aug., *Retract.* 2, 11: PL 32, 634; *Confess.* 5, 9, 17: PL 59, 55); holy week, with the p. of palms, which flourished at the end of the 4th c. in the church of *Jerusalem (*Peregr. Etheriae*: CSEL 39, 31) before spreading to *Spain (6th c.) and *Gaul (9th c.) (cf. Férotin, 178-183); stational churches, which developed from the feasts of the *martyrs from the 5th c. on; Marian devotions, like those of *Christmas, the *Annunciation, Purification and Dormition, which flourished at *Rome at the end of the 7th c. and seem to have originated in the Byzantine *East; penitential practices, e.g. those of the major litanies, which originated in 6th-c. Rome and replaced the *pagan Robigalia, and the minor litanies (Rogations) which began in *Vienne (Gaul) and were introduced to Rome at the end of the 8th c., none of which had a wholly penitential character; translations of the relics of saints and funeral processions.

Monumenta Ecclesiae Liturgica 5, ed. M. Férotin, Paris 1904; *Les Ordines Romani*, II, ed. M. Andrieu, Louvain 1960; *DACL* 14, 1895-96; *LTK* 8, 843-844; *CE* 11, 819-821; *RACh* 2, 422-429; W. Rötzger, *Des Heiligen Augustinus Schriften als liturgie-geschichtliche Quelle*, Munich 1930; *La Maison D.* 43 (1955); M. Righetti, *Manuale di storia liturgica*, I, Milan 1955; F. Van der Meer, *Augustine the Bishop*, London 1961; A.G. Martimort, Processioni, pellegrinaggi, giubilei, in (ed.) A.G. Martimort, *La Chiesa in preghiera*, Rome 1966, 680-690.

P. Fahey

PROCLUS. P. (*Proculus*) was a second-generation Montanist, who wrote on behalf of *Montanism at *Rome in the opening years of the 3rd c., as did *Tertullian at *Carthage at much the same time. His works are lost. Our information comes from the *dialogue against Proclus, attrib. by *Eusebius (*HE* III, 31, 4) to *Gaius, a learned church writer (not a presbyter) and opponent of John's *Apocalypse, author of anti-Montanist works at the time of pope *Zephyrinus (198-217) (Euseb., *HE* II, 25, 6; 6, 20, 3). P.

headed a Montanist group named after him, distinct from another named after Aeschines, a modalist *monarchian (Ps.-Tertull., *Adv. omn. haer.* 7, 2). From this we can deduce that P., unlike the Montanism of his time and the common opinions of the movement (recognition of *Montanus as Paraclete), professed no *christology that could be charged with *heresy. Tertullian (in his Montanist period) therefore mentions him among the most renowned anti-heretical church writers (*Justin, *Miltiades and *Irenaeus) and pays tribute to his ascetic lifestyle and literary qualities. P.'s identification with the "Christian Proculus" cited by Tertullian (*Ad Scap.* 4, 5) is taken for granted.

P. de Labriolle, *La crise montaniste*, Paris 1913, 277-280; A. v. Harnack, *Geschichte der altchristlichen Literatur bis Eusebius*, Leipzig ²1958 (1904) 2, 2, p. 206.

B. Aland

PROCLUS Lycius Diadochus. A biography of him was left by his scholar Marinus (ed. F. Boissonade, Leipzig 1814): encomiastic in character, its models are Philostratus's *Life of *Apollonius of Tyana* and *Porphyry's *Life of *Plotinus* (cf. R. Beutler, *PWK* 23/1, 186). Born at *Byzantium between 409 and 412 of well-to-do parents of Lycian origin, who took him back to *Lycia soon after his birth. As a boy he studied under the rhetor Leo of Isauria in *Alexandria, where he remained several years, studying grammar, Latin, *rhetoric and philosophy (esp. Aristotle). Not yet 20, he transferred to the neoplatonic school of *Athens, directed at the time by Plutarch of Athens and Syrianus, both of whom had a decisive influence on him. It was they - esp. Syrianus - who initiated him into *neoplatonism through the study and exegesis of Plato and Aristotle. After the brief scholarchate of Domninus of Larissa, who succeeded Syrianus on his death - exact year unknown - P., already famous and respected, took on the direction of the school, which he kept up until his death in 484/5. As a result of political disturbances, not unrelated to his attempts to restore the worship of the *pagan deities, his stay at Athens was interrupted for a year, which he had to spend in Lydia. On his doctrines cf. *Nous*, and neoplatonism; on his influence on Ps.-*Dionysius the Areopagite cf. also *Platonism and the Fathers.

R. Beutler, *PWK* 23/1, 186-190.

S. Lilla

PROCLUS of Constantinople. Born prob. at *Constantinople before 390, deacon and priest there during the episcopate of *Atticus, he was a tenacious pretender to that see. His hopes were disappointed in 426 (election of *Sisinnius), 428 (*Nestorius) and 431 (*Maximian); on Maximian's death (434), he crowned his long dream. Sisinnius had earlier designated him bishop of Cyzicus, but the population did not accept the designation and P. remained in the capital, where, on 23 Dec 428, in a *homily preached in Nestorius's presence, he defended the divine motherhood of *Mary and the title of *Theotokos*, which the patriarch disliked.

During his episcopate (434-446), P. worked in various ways to extend the sphere of Constantinople's patriarchal *authority, from *Cappadocia to Illyricum; in January 437 he brought the relics of *John Chrysostom, who had died in exile, back to the capital. The most important episode of his episcopate occurred in the wake of the Nestorian controversy. In 435, Nestorius's enemies having turned their attention to the works of *Diodore of Tarsus and *Theodore of Mopsuestia, whom they considered Nestorius's teachers, the bishops of *Armenia asked P.'s opinion on the matter. In his reply, *Tomus ad Armenios*, P. rebutted the more divisionistic aspects of Theodore's *christology and, following *Cyril of Alexandria, particularly stressed the unity of Christ, man and *God, in the formula "God the Word, one of the *Trinity, became incarnate". This later developed into the formula *Unus ex Trinitate passus est*, which was spread by the Scythian monks at the start of the 6th c. and was known as P.'s. P. claimed that *John of Antioch and other bishops of the Antiochene tendency had subscribed the *Tomus* and condemned the adjoined series of extracts from the works of Theodore of Mopsuestia, whose name he concealed. The refusal of the Antiochenes to condemn the passages of Theodore, who had died at *peace with the church decades before, provoked a controversy in which Cyril also took part; but faced with the firm resistance of the Antiochenes, P., Cyril and the emp. *Theodosius II abandoned their attempt to get the Theodorian extracts condemned (*c.*437).

P. was a brilliant preacher: more than 20 of his orations survive, three only in *Syriac tr. The authenticity of some is doubted. They are nearly all to do with the christological controversy, on the anti-Nestorian side. The same argument dominates his seven surviving *letters (all only in Lat. tr., except the second, the *Tomus ad Armenios*, of which we also have the Greek original). But the fragment of a letter in which P. calls Christ "one of the Trinity crucified in the *flesh", i.e. the *theopaschite formula, is probably false.

CPG 3, 5800-5915; PG 65, 680-850; *DTC* 13, 662-670; M. Richard, Proclus de Constantinople et le théopaschisme, *RHE* 38 (1942) 303-331; F.J. Leroy, *L'homilétique de Proclus de Constantinople*, ST 247, Vatican City 1967; M. Aubineau, Bilan d'une enquête sur les homélies de Proclus de Constantinople, *REG* 85 (1972) 572-596; L. van Rompay, Proclus of Constantinople, "Tomus ad Armenios" in the post-Chalcedonian tradition, in (ed.) C. Laga et al., *After Chalcedon: studies in theology and Church history* (Festsch. van Roey), OLA 18, Louvain 1985, 425-449.

M. Simonetti

PROCOPIUS of Gaza. Christian sophist. Born 465 at *Gaza in *Palestine, educated there and at *Alexandria. Soon called home to fill the teaching chair which he held until he died (soon after 530). His pupil and successor Choricius attests in his commemorative address that P. enjoyed great esteem for his teaching and publications, but also for his openness and simplicity. His theological training was so sound that he was considered and honoured as a bishop. His teaching shows a linguistic training founded on the classical writers and the unmistakable personality of a rhetor. An important example is his panegyric to *Anastasius I, spoken at the opening of a monument. Only two of his many rhetorical exercises (*ekphráseis*) are published: the description of the artistic clock in the market of Gaza (ed. H. Diels, *Über die von Prokop beschriebene Kunstuhr von Gaza*, Berlin 1917) and that of a series of late antique paintings at Gaza (ed. P. Friedländer, Vatican City 1939). P.'s correspondence, published in his own lifetime as an exemplary collection of masterpieces of Atticist stylistic purity, consists of 163 *letters; despite their sins of abstractness, they bring out the spiritual links between Gaza and the cultural centres of the Mediterranean. P. also left a great mass of theological writings, much of which still awaits an editor. It is undoubtedly wrong to try to distinguish between P. the sophist and P. the theologian. P. is the first of the series of authors of *catenae*. An adherent of Alexandrian *theology, P. made full use of *Origen, repudiating the Antiochene *Theodoret. To his great catenae on the Octateuch and the other historical books of the OT, which were entire in *Photius's time, he gave the title Ἐκλογαὶ ἐξηγητικαί; what remains are the salient parts of an extract prepared by P. himself under the title Ἐπιτομὴ ἐκλογῶν. In P.'s name we have two *commentaries on the *Song of Songs, one authentic, the other a fragment of the catena of the three Fathers, based on *Nilus, *Gregory of Nyssa and *Maximus Confessor. The catenae on the *prophet Isaiah and on Ecclesiastes survive, rather incomplete. The catena on Proverbs under P.'s name is spurious. The fragments (ed. A. Mai, *Nova Patrum bibliotheca*, IV 2, Rome 1847, 155-201 and VII, 2, Rome 1854, 1-81) of a genuine work of P. suggest a dual recension, as for the commentary on the Octateuch. A polemic against the neoplatonist *Proclus, of which fragments remain (A. Mai, *Classici auctores e Vaticanis codicibus editi*, 4, Rome 1831, 274 ff.), and a passage of Christian *apologetic developed with dialectical method are shown to be genuine by the most recent studies, which have removed earlier doubts. But a description of Hagia Sophia is more probably by Procopius of Caesarea; a monody on the destruction of that church by an earthquake is probably by Michael Psellus. Both works are assigned to P. in the earliest editions.

CPG 3, 7430-7448; PG 87, 21-2842 (incl. *spuria*); A. Garzya-R.-L. Loenertz, *Procopii Gazaei epistolae et declamationes*, Ettal 1963; S. Leanza, *Procopii Gazaei catena in Ecclesiasten*, CCG 4, Turnhout 1978, 1-50; *PWK* 23, 259-272; *EC* 10, 85; Beck 414-416.

J. Irmscher

PROCULIANUS. *Donatist bishop of *Hippo at the time of the *Catholics Valerius and *Augustine. Augustine recognized P.'s peaceful sentiments (*Ep.* 33, 1) and willingly and unconditionally accepted his invitation to meet him to sort out the difficulties between Catholics and Donatists (in 396) (*Ep.* 33, 4). But the two do not seem to have met, partly because meanwhile a Catholic, in dispute with his mother, had joined P.'s group, an action that led to civil consequences (*Ep.* 34). Their relations became tenser because of other Catholics joining the Donatists (*Ep.* 35, 1-2), Donatists joining the Catholics (*Ep.* 78, 8) and violence done to the latter (*Ep.* 105, 2, 3). Date of P.'s death unknown, but in 410 we find *Macrobius as his successor at Hippo.

DCB 4, 489; PAC 1, 924-926.

A. Di Berardino

PROFUTURUS. In the lay community of *Hippo there was a P. (Aug., *Ep.* 158, 9 [of *Evodius]), who must be the same P. whom *Augustine called *frater* and charged to go to *Bethlehem and give *Jerome a *letter and some writings (Aug., *Epp.* 28, 1, 1; 71, 1). But at the moment of departure P. was made a bishop (*Ep.* 71, 2; cf. Paul. Nol., *Ep.* 7, 1). His see may have been *Cirta (=*Constantine), where we find a bishop of that name (cf. Aug., *De unico bapt.* 16, 29), recipient of a letter from Augustine, who considered him an intimate *friend (*Ep.* 38).

PAC 1, 928-929.

A. Di Berardino

PROFUTURUS of Braga. Metropolitan of *Gallaecia in 538, he fought hard against *Priscillianist groups. Interested in theological and disciplinary problems, in a (lost) *letter he submitted the custom of *baptism by triple immersion, the celebrations of *Easter and the *Eucharist and other similar subjects to the examination of *Rome, for correct instructions. His letter led to a Decretal of pope *Vigilius (538) - included in the *Hispana* *Canonical Collection - in consequence of which, from the 1st council of *Braga (561), even before the liturgical unification under the Visigoths at the start of the 7th c. (cf. Hispanic liturgy), the Suevian liturgy resembled the Roman.

DHEE 3, 2030, Z. García Villada, *Historia eclesiástica de España*, II/2, Madrid 1933.

M. Díaz y Díaz

PROHAERESIUS. Christian sophist, perhaps from an Armenian family of Kukussus, born 276 in *Cappadocia. Pupil of Ulpian of Antioch and Julian the Cappadocian in *Athens. From 340 to 347 *Constans sent him to *Gaul as *magister militum*. On the way he passed through *Rome, where the senate dedicated a statue and inscription to him on account of his great oratorical gifts. *Basil and *Gregory Nazianzen were his auditors; *Julian the Apostate, during his stay at Athens, esteemed his eloquence. Following Julian's decree forbidding Christians to teach, P., with the African *Marius Victorinus, was obliged to abandon his *school despite his good relations with the emperor. At Athens his most feared rival was *Himerius. P. prob. held the prefecture in the first half of 356 and his oratorical struggle with Himerius must have occurred at the end of the proconsulate of Musonius.

PWK 23, 1, 30-32; J. Bidez, *La vie de l'empereur Julien*, Paris 1930, 55 ff. and 133 ff.; Fliche-Martin 3, 234; *PLRE* 1, 731.

M. Perraymond

PROLOGUES TO THE BIBLICAL BOOKS. The *Bible has been handed down through a succession of editions (and translations) sometimes popular, sometimes erudite, which have left traces not just in the text itself, but also in prologues, *glosses and more or less extensive footnotes, providing what seems necessary to introduce us to the text and sometimes to combat certain interpretations. Before the books, lists of numbered *Capitula* can provide an analysis and help trace the desired passage. In *DBS* 8 (1969) 688-692, B. Botte has given a good overall view, with bibliography; in the *Bulletin de la Bible latine* annexed to the *Revue Bénédictine*, P.M. Bogaert analyses recent works (cf. indexes, under *Préfaces et prologues*). In fact, the Latin tradition is much richer on this point. For Greek MSS, see R. Devreesse, *Introduction à l'étude des manuscrits grecs*, Paris 1954, 102-121 (*Origen), 139-141, 159-171, which escaped Botte. There has often been a tendency to attribute these anonymous texts to great *heretics (or to polemic against them): *Marcion, the *monarchians, *Priscillian, *Pelagius. A more prudent inquiry is more reserved in the face of these hypotheses and tends slightly to lower the date of the most important texts. Cf. M.E. Schild, *Abendländische Bibelvorreden*, Gütersloh 1970; J. Regul, *Die antimarcionitischen Evangelienprologe*, Freiburg 1969; and the introductions to the volumes of Beuron's *Vetus Latina*. The majority of the Latin texts are now published with critical editions of the *Vulgate*.

J. Gribomont

PROPERTY, ECCLESIASTICAL

1. Origins. The evidence of the Pauline epistles and Acts is that the first communities possessed no particular buildings reserved exclusively for their meetings. The Christians met in private houses (Acts 1, 14; 12, 12; 20, 7-12; 1 Cor 16, 19; Rom 16, 5; etc.). As far as we know, the construction of special *church buildings reserved for liturgical meetings began to be considered only in the 3rd c. (*Dura Europos [c.232]; cf. Tertull., *Adv. Valentin.* 3, 1; Clem. Al., *Strom.* VII, 29, 2), when Christian *cemeteries also began to be thought of. In fact, before organizing *areae* of their own, the Christians buried their dead as best they could in Jewish or *pagan burial-places, on the basis of their own family ties, social relationships or mere neighbourhood. From the 3rd c. on, *Rome had cemeteries administered by the local communities and integrated into their possessions. The fact that the first places for meetings and burials belonged to private people had appreciable advantages: the communities were free from the material problems inherent in the possession and management of real estate. But, with the progress of *evangelization, the usefulness of having proper church possessions was felt ever more strongly; at the same time, the inconveniences involved in private property forced church leaders to look for a safer system of possessing goods. In fact, at the start of the 3rd c., in various places and particularly at Rome, a property of corporative, properly ecclesiastical, type began to exist (Hipp., *Ref.* IX, 7; Tertull., *Scap.* 3, 1; *Historia Augusta, Alex. Sev.* 49). Evidence multiplies in the second half of the 3rd c. (rescript of *Gallienus: Euseb., *HE* VII, 13; 30, 19). In 311, *Galerius's edict authorized the Christians to re-establish their meeting-places.

2. Legal status. The legal existence of a church estate is attested even before the peace of *Constantine. The *Edict of Milan of 313 clearly speaks of the owner of church property: this is the *corpus* of Christians, i.e. of the various local communities, presided over by their respective heads. Contrary to the theories worked out in the 19th c. to justify these facts (G. B. De Rossi: *collegia tenuiorum*; Duchesne: *de facto* tolerance), modern criticism has recognized that each community could acquire, from then on, and administer a corporative estate by private right, within the common law (Munier, 274). It is clear that this state of things was confirmed after the peace of the church. Though the legislative texts of the late empire recognize church property, they do not define its nature. But it appears that the estate was that of the local community, under the direction of its bishop; we cannot say that a property of the church in its universality (in the sense of can. 1495, par. 1, of the Code of Canon Law of 1917) existed at that time. The idea of a foundation, in the modern sense of this word (i.e. an institution, itself the owner of the estate by which it exists), corresponds only to the system installed in the *East at the end of the 5th c. At that time the West, where the bishop remained the administrator of the whole of the property (Gelasius, fr. 24: Thiel, 498), does not seem to have known this system. The tendency to fragmentation, further favoured by other factors (rapid growth of estates; increasing autonomy of churches, hospices, *monasteries; *donations bound by precise stipulations; claims by *laymen to retain rights over goods given [cf. Gaudemet, 302-306]), finally prevailed.

3. Administration. Belonging to the local community, goods were administered by its head, the bishop. The oldest documents describe this type of management as completely discretional: the bishop was responsible only to God. The *Didascalia* thinks it out of place that clerics or laymen should ask the bishop to be accountable. As long as the allotment of resources was limited to the immediate distribution of daily offerings, this state of affairs involved no grave risk of abuse: the bishop could not allow open injustice. But there is nothing in common between the direct, public, patriarchal management described by the apostolic *Apocrypha and that set up at *Rome under pope *Cornelius (Euseb., *HE* VI, 43) or at *Carthage under bishop *Cyprian. The business ability of the Roman church appears above all in the organization of funeral services and the management of the catacombs; we can well believe that it was capable of dealing with the needs of *brothers who were *poor, sick or in trouble. Much evidence allows us to illustrate the administration of church goods during the late empire: a) by and large, the bishop continued to enjoy total freedom. Then, there supervened and prevailed an ever stronger tendency to associate the local *clergy (council of Antioch, can. 25; *Canons of the Apostles*, can. 40; Gelasius, *Ep.* 17, 1). In the East, the creation of bursars was made obligatory by the council of *Chalcedon (451); in the West, the *Statuta ecclesiae antiqua* extol the same institution. Pope *Gelasius established strict control over the temporal management of the Italian bishops under his jurisdiction (Gaudemet, 309); b) on the other hand, the bishop had to respect the destination of goods as wished by the donor. If the donation involved no special stipulations, the bishop had to employ the income for pious or charitable purposes (poor, imprisoned, wayfarers, etc.). Until the end of the 5th c., the bishop enjoyed total freedom to distribute the revenues among these various recipients. But in *Italy a more rigid rule prevailed with *Simplicius (*Ep.* 1: Thiel 176) and Gelasius. Indeed, Gelasius imposed division into four parts (bishop, poor, clerics, buildings). *Gaul and *Spain adopted other types of division; c) though in principle church goods were inalienable, some exceptions were admitted, in particular for helping the poor, redeeming prisoners and helping those in great need.

G. Bovini, *La proprietà ecclesiastica e la condizione giuridica della Chiesa in età preocostantiniana*, Milan 1949; J. Gaudemet, *L'Eglise dans l'Empire romain (IVe-V^e f.)*, Paris 1958, 288-311; Ch. Pietri, *Roma christiana*, BEFAR 224, 2 vols., Rome 1976, I, 1-96; Ch. Munier, *L'Eglise dans l'Empire romain (IIe-IIIe f.)*, Paris 1979, 264-274; J. Guyon, *La cimitière Aux Deux Lauriers*, BEFAR 264, Rome 1987.

Ch. Munier

PROPHET - PROPHECY

I. Prophecy in the primitive church - II. Iconography - III. Commentaries on the prophetic books.

I. Prophecy in the primitive church. The division *apostle-prophet-teacher, evoked by the *Didachè (11, 17), seems to be of Jewish origin and corresponds to law-prophecy-wisdom. It is already present in 1 Cor 12, 28 and Acts 13, 1. In Acts 11, 27, some prophets of *Jerusalem go to *Antioch. In Didachè 11, 17 and 12, 3, apostles and prophets are cited together; in 15, 2, prophets and teachers. *Hermas's Shepherd similarly links prophets and teachers.

We are in the presence of a primitive community *organization and a "variety of *ministries" (1 Cor 12, 4). This organization would be replaced by deacons and bishops in Didachè 15, 1.

The Didachè presents the prophet as endowed with a *charism of the *Holy Spirit, not subject to the judgment of the community. This will judge him not by his words, but by his conduct and disinterestedness (11, 7-12). The prophet is an itinerant who may settle among a community. The prophets can give thanks freely after meals (10, 7) not celebrate the *Eucharist, as some have said. They can ask for food for the *poor (11, 9). Didachè 13, 3 says: "The prophets are your high priests"; but Audet considers this passsage a late redaction of the interpolator (Didachè, Paris 1958, p. 105).

Hermas (Precept 11) provides the criteria by which to distinguish true from false prophets. The latter are recognized by their venality (J. Reiling). *Justin knows false prophets (Dial. 35, 3; 51, 2; 69, 1; 82, 1-2). Cf. also *Eusebius (HE IV, 26; V, 17-19). *Montanism claimed to prolong the era of the prophets (Tertull., De an. 9, 4; Cypr., Ep. 75, 10; Euseb., HE V, 16, 3, 7, 9; 17, 2-4).

E. Fascher, Prophetes, Giessen 1927; J.P. Audet, Didachè, Paris 1958, 203-206, 439-441, 447-453; A. Lemaire, Les ministères aux origines de l'Eglise, Paris 1971; J. Reiling, Hermas and Christ. Prophecy, Leiden 1973; W. Rordorf-A. Tuilier, La doctrine des Douze Apôtres, SCh 248, Paris 1978; D.E. Aune, Prophecy in early Christianity and the Ancient Mediterranean World, Grand Rapids 1983; H.K. Stander, Prophets in the early Christian Church, Ekklesiastikos Pharos 66-67 (1984-85) 113-122.

A. Hamman

II. *Iconography. Images of prophets appear in the earliest repertoire of Christian art, either as vague evocations of the messianic prophecy or as historical representations linked to narrations of the biblical text or in figures of groups.

1. Evocations of the messianic prophecy. A vague allusion to the foretelling of the Messiah's *birth from a Virgin rather than the historical identification with a prophet, Balaam, Isaiah or Micah, is discernible in those isolated depictions of palliate male figures in the act of pointing to a sign in the sky: a star in some frescos, first half of 4th c., of the catacomb of SS. Marcellino e Pietro (Wp 158, 2; 159, 2; 165) and in that of the *cemetery of Via D. Compagni (Ferrua, pl. 86, 1); a christogram in a *painting, second half of 4th c., in the cemetery of Ciriaca (Wp 241). Recently, some would see the fusion of a historical narrative intention and a symbolic expression of the evocation of the messianic prophecy in a disputed painting preserved in the cemetery of S. Tecla, mid 4th c. (Santagata, S. Tecla, fig. 2), in which are two male figures pointing to the star. Alongside these isolated depictions, again as vague allusions to the messianic prophecy must be understood the images of palliate men pointing to a star that appear alongside the figure of the Virgin and Child (cf. *Mary), as in the fresco of a catacomb gallery in the cemetery of Priscilla, first years of 3rd c. (Wp 21-22) and, at times, in scenes of the Adoration of the *Magi, as e.g. in the inscription of one Severa in the Museo Pio Cristiano (Kirschbaum, Balaam, pl. 3).

2. Historical representations. The prevailing of a historical narrative intention is seen, however, in those depictions more or less closely dependent on the biblical text, linked to the vissicitudes of individual prophets (many of whom we deal with under their own names). Of *Elijah, we find depictions of his meeting with the widow of Sarepta, the raising of the widow's son and the holocaust on Mt Carmel in the frescos of the synagogue of *Dura Europos, mid 3rd c. (Kraeling, Synagogue, pls. 31, 62-63) and that of his ascent to heaven in a chariot of fire, elaborated on the classical model of Helios's ascent to heaven on a quadriga, in the paintings of the catacombs of Domitilla (Wp 230, 2) and of Via D. Compagni (Ferrua, pl. XXIII) and in some sarcophagi (Ws 189, 2 and 190, 3). Of Jeremiah, acc. to the description of the illustrative distich (Forcella, Iscrizioni, no. 227), the vision of the *angel in the form of a *lamb was depicted in the mosaic biblical cycle of the basilica of St Ambrose at *Milan. Of *Ezekiel, the vision of the field of dry bones repopulated by the living appears frequently on sarcophagi (Ws 112, 2; 123, 3; 151; 184, 1; 206, 7; 215, 7; 219, 1); but the earliest depiction of an Ezekiel cycle is provided by the frescos of the synagogue of Dura Europos (Kraeling, Synagogue, pls. 69-72). The fresco of the catacomb of Domitilla, second half of 4th c., with a bearded figure pointing to two turreted buildings (allusion to Bethlehem?), before which are the Madonna and Child (Wp 229), may perhaps be linked to Micah and his annunciation of the Messiah's birth at *Bethlehem.

3. Group images. Among the first monumental representations of the series of prophets must be mentioned that, in stucco, of the Baptistery of the Orthodox at *Ravenna, mid 5th c. (Deichmann, Ravenna, pls. 72-79, 81). In the same city, some 50 years later, a series of 16 prophets, shown as majestic old men holding scrolls in their hands (a constant attribute in later iconography), appears in the *mosaics of the basilica of S. Apollinare Nuovo (Deichmann, Ravenna, pls. 136-153). Still at Ravenna, mid 6th c., images of prophets appear in the mosaic decoration of S. Vitale (Deichmann, Ravenna, pls. 312-321). [Fig: 260]

EC 10, 101-103; LCI 3, 461-462; cf. articles on individual prophets in: DACL, LCI, RBK, RACh; G.B. De Rossi, Altri monumenti di sacre vergini nell'Agro Verano, BACt 1 (1863) 76 ff.; Garrucci, Storia, II, 63 ff.; Wp 174 ff., 183 ff.; G. Wilpert, La divina maternità di Maria Vergine e il profeta Isaia, RAC 11 (1934) 151 ff.; E. Kirschbaum, Der Prophet Balaam und die Anbetung der Weisen, RQA 49 (1954) 129 ff.; L. Réau, Iconographie de l'art chrétien, II/1, Paris 1956, 343-419; L. De Bruyne, L'Initiation chrétienne et ses reflets dans l'art paléochrétien, RSR 36 (1962) 27-85; id., Les lois de l'art paléochrétien comme instrument herméneutique, RAC 39 (1963) 7-91; P. Testini, Alle origini dell'iconografia di Giuseppe di Nazareth, RAC 48 (1972), esp. 281 ff.

For the monuments cited in the text, cf.: V. Forcella, Iscrizioni di Milano anteriori al sec. IX, Codogno 1897; C.H. Kraeling, The Excavations at Dura-Europos. Final Report VIII. Part I: The Synagogue, New Haven 1956; F.W. Deichmann, Frühchristliche Bauten und Mosaiken von Ravenna, Baden Baden 1958; A. Ferrua, Le pitture della nuova catacomba di via Latina, Vatican City 1960; G. Santagata, Su due discusse figurazioni conservate nel cimitero di S. Tecla, in Esercizi. Arte, Musica, Spettacolo 3 (1980) 7-14; M.L. Kessler, Prophetic Portraits in the Dura Synagogue, JbAC 30 (1987) 149-156. See also *Montanism.

M. Marinone

III. Commentaries on the prophetic books. Fundamental above all for our knowledge of the exegetical method used by the *Fathers in interpreting the OT, the presence of patristic *commentaries on the prophetic books marks the history of Christianity from the first centuries. Before the systematic commentaries, the prophetic texts had undoubtedly been the object of *homilies, as witness the fact that prophetic passages were read and commented on during the liturgical *assembly (Just., I Apoll, 67). The reading, alongside an extract from the Torah, and commentary on a prophetic text (haftara) was already in use in the *Jewish liturgy (cf. Lk 4, 16-21). It is probable that the liturgical and catechetical context helped emphasize the messianic character of the prophetic preaching (cf. Iren., Dem. 30). Undoubtedly it was in the works of the *apologists that the presence of prophetic passages began to make itself felt: the prophets were presented to the *pagan world as witnesses of *God's truth and of his *revelation (Just., I Apol. 53; Dial. 1-9; Tat., Orat. 29; Theoph. Ant., Autol. 1, 14), while to the Jewish world they were reproposed in accordance with the messianic interpretation already consecrated in the NT (Just., Dial. 43; Ign., Philad. 5, 2; 9, 2; Tertull., Adv. Jud. 8; cf. Euseb., Dem. ev. 8, 2; Athan., De inc. 11; Cyr. of Jer., Catech. 12, 19). With *Irenaeus, the prophets became the irrefutable proof to demonstrate both the dynamism and the unity of the divine *economy, thanks to the centrality of Christ who was seen as both the announcer and the fulfiller of all prophecy: of him the prophets spoke, it was he who fulfilled what they revealed (Iren., Adv. haer. IV, 33-34; cf. also Just., I Apol. 30).

Among the works expressly dedicated to commenting on the prophetic books, but surviving only in a few fragments or known only because recorded by other authors, we will mention: the commentaries on Isaiah, *Ezekiel and Zechariah attrib. to *Hippolytus; the systematic commentaries of *Origen, who wrote 30 books on Isaiah (Euseb., HE VI, 24, 2), 21 on Ezekiel and at least 26 on the 12 minor prophets (Euseb., HE VI, 32, 2); *Victorinus of Petovium's commentaries on Isaiah, Ezekiel and Habakkuk and *Pierius of Alexandria's homily (rather than treatise) on Hosea (Jerome, Vir. ill. 74 and 76). More or less ample *glosses on various passages of Jeremiah are all that remain of a commentary, surely complete, on all the prophetic books, composed by *Ephrem, following a christocentric and ecclesiological interpretation, though always faithful also to historical sense and philological analysis. We must also mention *Apollinaris of Laodicea's commentaries on Isaiah, Hosea and Malachi (cited by Jerome in the prologues to his commentaries on the same books); *Cyril of Alexandria's on Ezekiel and *Daniel; *Theodore of Mopsuestia's on Isaiah, Ezekiel, Jeremiah and Daniel, strongly anti-*Origenist in their condemnation of any christological interpretation of the text; those of Ps.-*John Chrysostom on Jeremiah, Ezekiel and Daniel, marked by a strong moralizing interest; the fragments on Jeremiah by *Victor of Antioch and *Olympiodorus of Alexandria, in the form of glosses surviving in the *catenae, and those on Isaiah by *Procopius of Gaza. A mainly historical and rational interpretation is prominent in *Julian of Eclanum's fragments on Hosea, Joel and Amos.

The surviving patristic works that comment on whole, or parts of, prophetic books are in the form of *homilies or treatises (*tractatus). The earliest surviving homilies are the nine on Isaiah composed by *Origen, but which we have only in *Jerome's *Latin translation (but Baehrens considers homily IX inauthentic: GCS VIII). Again by Origen are 14 homilies on Jeremiah translated by Jerome: of these, 12 (on Jer 1, 1-20, 12) are also preserved in the Greek text and constitute a model of balance between respect for the historical meaning and spiritual topicality, very recurrent given the perhaps catechetical character of these homilies. They dwell particularly on the problems concerning God's *providence, *evil and suffering. Again by Origen, but translated by Jerome, are 14 homilies on Ezekiel which offer a mainly moral and typological interpretation. Of *John Chrysostom there remain six homilies on Isaiah, strongly moralizing in tendency, and two homilies *De Prophetarum obscuritate*, dealing with the prophetic books in general. Of *Gregory the Great we have 22 homilies on Ezekiel where, however, the text is often used just as a cue to develop a wider spiritual discourse.

As for *commentaries proper, the first that survives - which is also the first exegetical treatise handed down by the Fathers - is *Hippolytus's *Comm. Dan.*, dateable to the early 3rd c. It consists of four books on the prophetic text of Daniel, in *Theodotion's Greek version, but also comprising the deuterocanonical sections. The author proposes an interpretation aimed mainly at calming down those who expected the imminent end of the *world. His *exegesis is historical: so he holds all the information that Daniel gives about himself and his visions to be real, but he also makes use of *allegory, esp. in a christological sense. Jerome mentions (*Vir. ill.* 81) *Eusebius of Caesarea's *Comm. Is.* in ten books (or 15, as he says in the prologue of his own *Comm. Is.*). But this work (recently reconstructed and published in 1975 in GCS 9, *Eusebius Werke*, IX) which, according to the author's promise, was to have given a historical interpretation of the prophecies, depends largely on the spiritual and allegorical reading of them given by Origen. The text of Isaiah is taken from the LXX according to the *Hexapla*. Still on Eusebius, we can consider books 2-9 and 15 of his *Demonstratio Evangelica* as genuine commentaries on prophetic passages. A long commentary on Is 1-16 seems falsely attrib. to *Basil of Caesarea, despite some signs to the contrary. *Ambrose's *Expositio Isaiae prophetae* has been partly reconstructed, thanks to a myriad of fragments found in other works of his or in Augustinian references (P.A. Ballerini: CCL 14, 1957, 403-408). A systematic commentary on all the prophetic books is offered by *Jerome (*Vir. ill.* 135) who, after having given the texts in both the Hebrew and the LXX version, comments on them first literally and then spiritually. His spiritual interpretation depends considerably on Origen and, for the *Comm. Zach.*, on Didymus. A commentary by *John Chrysostom on Is 1, 1-8, 10 surviving in *Armenian version is considered probably authentic. Of *Didymus we have 18 books on Is 40-60 (the one part that he considers original to the prophet) and the whole *Comm. Zach.*, yielded by the Tura *papyri. In this work, composed at Jerome's request (*Vir.ill.* 109), Didymus uses the literal interpretation for a first approach to the text, but passes immediately to a spiritual reading in accordance with anagogy, *allegory, tropology and theory, though he often uses these terms without too many distinctions. Of his *Comm. Os.* only catenic fragments remain. *Ephrem's *Comm. Zach.*, strongly christocentric and composed in a sober and terse way so as to appear almost in the form of scholia, survives in *Syriac and in an 18th-c. Latin version. The books of the 12 so-called minor prophets were commented on systematically by *Cyril of Alexandria and *Theodore of Mopsuestia: while Cyril makes wide use of typology and interprets the majority of the prophecies in a christological sense, Theodore refers to the history of the Jewish people - thanks to a sound historical and philological research - even those prophecies held by the great tradition to be messianic. *Glosses on the 12 minor prophets, still nearly all unpublished, were written by *Hesychius of Jerusalem, of whom we also have a *Comm. Is.* in the form of very brief glosses on individual verses. Of *Theodoret of Cyrrhus, we have commentaries on all the prophetic books. The first, on *Daniel, strongly anti-Jewish, then that on Jeremiah in ten books, the most extensive of all to survive on this prophet, also comprising Baruch and Lamentations; on *Ezekiel (often reduced to a mere paraphrase of the text, though at the start it seems to echo something of the interpretation offered by Didymus) and on the 12 minor prophets. That on Isaiah, in 20 books, shows clearly how Theodoret knew how to make himself independent of the rigid literalism of his predecessor Theodore. In fact, alongside the literal sense, he makes wide use of the allegorical and typological method, defending the messianic interpretation of numerous prophecies. In this sense his commentary on Isaiah is close to that of Cyril of Alexandria who, in five books (respectively concerning Is 1-10; 10-24; 25-42; 42-51; 52-66), interprets the prophetic text first in a literal sense, then in the spiritual. A list (not exhaustive) of patristic commentaries on the prophets can be found in *Cassiodorus (*Inst. div.* 3).

J. Harvey, Ezéchiel, *DSp* 4, 2217-2220; C. Kannengiesser, Jérémie (chez les Pères de l'Eglise), *DSp* 8, 889-901; M. Simonetti, Note sull'esegesi veterotestamentaria di Teodoro di Mopsuestia, *VetChr* 14 (1977) 69-102; id., Note sul commento di Cirillo d'Alessandria ai Profeti minori, *VetChr* 14 (1977) 301-330; M. Delcor, La prophétie de Daniel (Ch. 2 et 7), dans la littérature apocalyptique juive et chrétienne en référence spécial à L'Empire romain, *Popoli e spazio Romano tra diritto e profezia*, Da Roma alla Terza Roma, Studi 3, Naples 1986.

M.G. Mara

PROPHETIAE EX OMNIBUS LIBRIS COLLECTAE. Anon. book (CPL 84; PLS I, 177-180, 1738-1741), prob. of *Donatist origin, late 4th c. A "very brief disputation" on the seven ways of prophecy is followed by a list of the *prophets, from *Adam to Zachariah father of *John the Baptist, and some NT passages concerning prophecy. After a respectful mention of *Cyprian, the work closes with a condemnation of *Montanist prophecy.

J. Gribomont

PROSELYTE. From Gk "foreigner", it translates the Hebrew *ger*. It indicates converts to Judaism. They are named together with the "God-fearers" in Acts 2, 11; 13, 43; 10, 2, from whom they may have been distinguished by *circumcision. Proselytism was intense in the NT period (Mt 23, 15); after some easing off, concurrent with the Jewish revolts of AD 70 and 135, it remained lively until the 5th c., despite the fact that, from the 4th c., imperial edicts had sometimes forbidden it under pain of death. It is evidence of the vitality of Judaism in the Christian era and of a universalistic mentality; in fact it declined when Judaism tended to turn in on itself. The Talmud attests the attitudes, sometimes conflicting, of the rabbis towards pp., the requisites and the ritual to become one; in this, the bath (*tebhila*), whose religious value even *John Chrysostom acknowledged, had particular importance.

DBS 8, 1353-1356; *EC* 10, 158 f.; Lampe 1171; M. Simon, *Verus Israel*, Paris 1958, cf. index.

S. Cavalletti

PROSOPON. While keeping the various meanings of non-Christian Greek, the term p. became a technical term, first in the *theology of the *Trinity and then in *christology. The specific trinitarian use, first attested by *C. Noetum* (7; 14) and also indirectly by *Tertullian (*Prax.*), is explained primarily by three factors: the scriptural use of p., esp. in 2 Cor 4, 6: "the *glory of God in the face of Christ", a meaning that remained important in all *patristics (Lampe 1186); the "prosopic" *exegesis prepared by *Philo, in which, against the *Jews and *monarchians, the divine *persons and their names were distinguished (cf. Justin, *I Apol.* 36, 1 f.); the *Stoics' way of opposing p., as individuality, to the generic (cf. Iren., *Adv. haer.* III, 11, 9). The term p., very rare in Eastern authors before the 4th c., is initially found, as a trinitarian term, principally in texts that rebuke the *Sabellians for speaking of one p., i.e. a single divine reality, or for speaking of three *prosopa*, but in a transitory sense (Lampe 1187). Subsequently the anti-Sabellians too took up the term p., identifying it more or less with *hypostasis (Lampe 1187 f.): as does the *Epistula synodica* of the council of *Constantinople of 381 (*COeD* 24). In the second half of the 4th c., p. was also used as a key term in christology, initially in the form of a negation of two persons in Christ, then as affirmation of a single person (Lampe 1188 ff.). Two particulars must be stressed: the expression *one person* is often characterized by an exegetical nuance, referring to the single subject of the biblical attributes, human and divine, in Christ; equally important is the very complicated theology of *Nestorius, who distinguished two natural *prosopa* from the *prosopon* of union (Grillmeier, *Jesus*, 707-726; Lampe 1188 ff.).

G.L. Prestige, *God in Patristic Thought*, London 1948; R. Braun, *Deus Christianorum*, Paris 1962, 207-242; C.J. de Vogel, The Concept of Personality in Greek and Christian Thought, in *Studies in Philosophy and History of Philosophy*, Washington 1963, 20-60; A. Grillmeier, *Jesus der Christus im Glauben der Kirche*, Freiburg 1979 (Eng. tr. J.S. Bowden, *Christ in Christian Tradition*, London 1989); R.V. Sellers, *Two Ancient Christologies*, London 1940; R.A. Norris, *Manhood and Christ: A Study of the Christology of Theodore of Mopsuestia*, Oxford 1963; L.I. Scipioni, *Nestorio e il Concilio di Efeso*, Milan 1974.

B. Studer

PROSPER of Aquitaine. Acc. to *Gennadius (*De vir. ill.* 85 [86]), P. was born in Aquitaine at the end of the 4th c. and was at *Marseilles in *c.*426, attracted by the monastic and theological environment of St Victor and *Lérins. There he had numerous contacts with the *monasteries, while remaining a *layman, and was involved in the *semipelagian controversy. In 428 he wrote to *Augustine who sent him the treatises *De praedestinatione sanctorum* and *De dono perseverantiae*. After Augustine's death he went to *Rome to obtain the condemnation of the ideas circulating in the monastic communities of Provence, for which *Celestine I sent a *letter to the bishops of *Gaul (PL 50, 528-530). P. returned to Marseilles where, between 432 and 436, he composed his most important works. Then he went back to Rome where he entered the service of *Leo I. He died there after 455. Feast 25 June.

A lettered man, P. put his *classical culture at the service of *theology. A letter of his to *Rufinus is an excellent exposition of the Augustinian doctrine of *grace. Two verse works, *De ingratis carmen* and *Epigrammata*, inspired by the *Pelagian controversy, defend Augustine's theological thought. The works of theology composed in Provence maintain the Augustinian theses in the semipelagian controversy: *Pro Augustino responsiones*, *De gratia Dei et libero voluntatis arbitrio contra collatorem* (= *Cassian) and the *Capitula* added to Celestine I's letter. At Rome he wrote the *Expositio psalmorum*, inspired by Augustine, the *Liber sententiarum ex operibus S. Augustini* and *De vocatione omnium gentium* which critics have restored to him (first work in Christian literature to be concerned with the salvation of infidels). To these we must add the *Epitome Chronicae*, history of mankind from the beginning of the *world to 433, then continued up to 444 (*Chronicon vulgatum*) and to 446-455 (*Chronicon integrum*). The *Epistula ad Demetriadem de humilitate* (PLS 3, 149) is probably his.

An unconditional Augustinian, P. passionately defended his teacher's theology of grace. His Roman sojourn moderated this defence somewhat, teaching him not to be more Augustinian than the pope. The *Capitula* have a more controlled tone. On the same lines, *De vocatione* upholds *God's saving will for individuals and for all mankind, teaching man's responsibility in history and manifesting missionary concerns.

The *Liber sententiarum* is the first Augustinian florilegium. It inaugurated a *literary genre that flourished above all in the Middle Ages. Its influence would be exercised at the council of *Orange in 529. P. is "the first representative of mediaeval *Augustinianism" (Cappuyns).

Editions: CPL 516-535; PL 51 and PLS 3, 147-149; MGH, *Auct. ant.* 9, 341-499 (*Epitome*); CCL 68A (*Liber sententiarum* and *Expositio psalmorum*); PL 45, 1756-1760 and 1833-1850 (*Pro Augustino responsiones* and *Capitula*); PL 55, 161-180 (*Epist. ad Demetriadem*). Eng. tr. (partial): ACW 14 and 32.
Studies: L. Valentin, *S. Prosper d'Aquitaine*, Toulouse-Paris 1900 (still useful for literary questions, surpassed for critical and theological problems); G. Bardy: *DTC* 13, 846-850; J. Gaidioz, *La christologie de S. Prosper d'A.*, Lyon 1947; G. de Plinval, P. interprète de s. Augustin, *REAug* 1 (1958) 339-355; C. Bartnik, L'universalisme de l'histoire du salut dans le De voc., *RHE* 68 (1973) 731-758; *Patrologia* III, 522-528.

A. Hamman

PROTOLOGY

I. General notions: religious-historical typology - II. History of Christianity.

I. General notions: religious-historical typology. We speak of "protology" and "protological explanation" in analogy with and sometimes coupled with "*eschatology" and "eschatological explanation". The "beginnings", the *prota*, the *arché*, are thus seen in relation to the "last things", the *eschata*, the *telos*. If we look at the religions and philosophies of the ancient world, this relationship can take various forms. Sometimes the beginnings are just inchoate, implicit, in comparison with the fulfilment; but sometimes the fulfilment is presented as a restoration of the beginnings; or again the two viewpoints are combined in various ways and proportions. Extreme cases of this phenomenology are the positions, alien to patristic writers but familiar to pre-Socratic Greek thought, of an evolutionism-hylozoism that derives the variety of the world from a simple, elementary *arché*, or a repetitive-cyclical conception based on the contrast - or dialectic - of the One and the Many, as e.g. in the dualistic monism of Empedocles and of certain Orphic formulations. Another protological subject, which directly interests the student of Christian thought, concerns a motif widespread in the world of religions: the character of the *arché* as determining the actual state of the universe and of man. The primordial persons and events determine or at least condition the present, substantially as well as existentially. Here lies the importance of the study of "protological explanation". We can easily understand the importance assumed, outside and inside Christian thought, by the characterization of the entities and agents to whom creative, emanative or demiurgic power is attributed, and by the idea of a fault, fall (or accident) that occurred in the beginning and impressed on the *world and on man a mark that conditions them and gives urgency to a prospect of salvation, a *soteriology. Which, in turn, is made to depend on eschatology, but also, in various ways, on the restoration of the beginning, i.e. of the condition of primordial integrity. In this context, still on a phenomenological level - but a phenomenology constructed with an eye on concrete historical research - alternative religious-historical categories emerge: a) that of a unique creative act, though this may be divided into phases or moments (not motivated by faults) or into incidents set at the level of the first creatures; or b) that of a "*double creation" which is itself expressed in alternative forms: *1)* a creator *God who uses the collaboration of other inferior creative "agents", who render Him "innocent" of any possible contact with the *evil that may emerge from the ontological imperfection of creatures (Plat., *Timaeus* 41A-43A: the Demiurge and the "young" gods, his collaborators in the creation of man; Philo of Al., *Op. mundi* 72-75; *Fug.* 69 f., etc., partly taken from the *Timaeus*; late Jewish conceptions of a *deuteros theos* or of demiurgic *angels not hostile to God); *2)* a unique creator who, however, "adds" to the first *creation a second creative act (second in an ontological, not necessarily a chronological sense), motivated by a fall of the first creatures (*Origen, *Gregory of Nyssa, etc.); *3)* the *gnostic theory of a proud and ignorant demiurge who transmits to the beings whom he forms his own worst characteristics, though these are not such as to abolish that pneumatic "self" which, to the gnostic, is derived from an original outpouring of divine substance, however unconsciously conveyed by the demiurge himself; *4)* finally, in the *pagan world, an alternative to the gnostic position is the Plotinian position of radical monism (the One, whence *Nous*, whence the *Anima mundi*) which, through a sort of rationality-necessity intrinsic to the system, implies the manifestation, at emanative levels below the One (*souls, but also, in its own way, *Nous*), of a *tolma*, a "courage", not just ethical but also ontological, which marks with itself the whole "devolutive" process, down to the incorporation of souls and the final shore of pure matter; a devolutive conception which is also implicit in gnostic systems like that of *Valentinus, in which the final *aeon Sophia, peripheral and female, has almost written into her qualities the ontological premise of a crisis that will give rise to the demiurge and to the psychic and hylic substance in the world and in man.

This being said, we can understand how the notions - varied but specific - of "double creation" may imply dualistic positions, varying from an ontological *dualism of monistic type (*Plotinus and, in their way, the gnostics) to a dualism in a broad theological sense (God and the demiurge in the gnostics, but also, in another sense which may or may not be compatible with biblical monotheism, the doctrines of Plato's *Timaeus* and of *Philo of Alexandria: Philo obviously does not talk about "gods", but about powers or angels with demiurgic duties), or to an anthropological dualism (though even this is ontological in its way, like any dualism properly so called) like that implicit (inherited, mediated in various ways, from Plato) in doctrines of Origenian type concerning the "heavy" incorporation of souls (*noes*) or of the primordial *Adam, "*image of God" and therefore - according to this line of interpretation - not destined in the creator's first intentions (i.e. in the "first creation") for the physiological function of *sexuality (a similar position is in the radical *encratism of a *Tatian or a *Julius Cassianus, with the essential difference that these infer from it the absolute incompatibility of this function with God's intentions for man, or at least with the New Covenant régime); there are intermediate positions in the *Liber Graduum* and in some *Fathers, though never in the sense of a condemnation of *marriage, nor even, as far as these Fathers are concerned, as part of an overall position on the subject, but rather in the context of ascetic aims which (while referring to protological explanations like those just indicated) intend to express the novelty of evangelical *spirituality.

Returning to the distinction between a unique creative act (even if divided, but not as a consequence of sin) and "double creation" (in its different forms), we see how different in the two cases is the protological explanation of *evil, as well as the protological explanation of the actual constitution of the *world and of man. In the light of a unique creative act, the fault of the first-formed certainly remains in some way determinative (or explanatory) with respect to actual man in the qualifying conditions of his existence (death, fatigue, *labour, need of salvation, hereditary sin as expressed in the conception of "original *sin"), but does not imply anthropological dualism. The conception of "double creation" does imply such dualism, in a way that fits the different theorizations of it (Platonic, gnostic, Origenian, and perhaps some positions of apocalyptic Judaism: interpretations of the biblical account of the "sons of God" in Enoch, etc.). In all these cases, the phrase "antecedent fault" is preferred to that of "original sin".

Lastly we will show how protological subject-matter (and explanation) becomes a type, in various ways, of eschatological subject-matter (and explanation). In every form of Christian thought, an original integrity, lost by the first-formed, is conceived as requiring a final restoration or *apocatastasis*; but when the idea of "double creation" is involved, in the different meanings shown above, a specific position follows. In this case the restoration will involve the elimination of the significance of the "second creation", i.e. that of the psychic-hylic components (for gnosticism) or of "heavy" corporeity implying the sexed structure of mankind (for the Origenian position). Thus in *Origen, on the basis of the correspondence (even if not total in every aspect) between *arché* and *telos*, *arché* will be described on the basis of *telos*: i.e., *Adam's original situation will be conceived - on the model of the eschatological situation - as "angelic", not sexed or at least not destined for the exercise of the sexual function, sexual procreation. With this projection into the origins, the Origenian position is dependent, phenomenologically if not historically, on the position of *encratism - in its varied and graduated meanings (from *Tatian and *Julius Cassianus to the *Odes of Solomon*, Ps.-*Macarius, the *Gospel of *Thomas* [in its encratite component, as far as that is distinguishable from the gnostic component]) - naturally with the essential difference of the admission, quite clear in the Origenian position, of the lawfulness and propriety of the institution of marriage in the present *economy. To this we can add that, having safeguarded the recognition (sometimes undervalued or reduced to terms of vague Platonizing "language") of the actual ontological and anthropological depth of protological explanation when this is expressed in notions of "double creation" and "antecedent fault", it would be wrong to forget, for the Fathers of the Origenian position and for non-gnostic encratism, the other dimensions and inspirations of their spirituality and the biblical sources from which they start, themselves often of a protological nature (Adam's primordial innocence, the original splendour of the angelic creatures before their fall, the "new creation" introduced by Christ and by evangelical spirituality); just as it would be wrong to undervalue the Fathers' constant concern with safeguarding *orthodoxy, not just concerning *God the creator, the ontological goodness of creatures and the doctrine of the *resurrection of the *body, but also on the lawfulness and divine origin of the institution of *marriage, contested with varying emphases by radical encratism.

Thus on *virginity, justifications of great protological depth (Adam's original nature not made for sexual procreation) are lined up with others, "protological" but not implying "double creation" (innocence and non-concupiscence of the primordial couple and of their eventual *synousia*), eschatological (on the basis of Mt 22, 30, Mk 12, 15 and particularly Lk 20, 34-36), *ascetic, for the sake of the Kingdom (Mt 19, 12), etc. We must also consider the different distribution of these justifications, taking account of *chronology, *literary genre, etc., as well as their concrete mutual interference. Under this last aspect, we should not forget the frequent association, as a pair, between protological and eschatological explanation. Nor, finally, should we overlook the possibility that, among the Fathers themselves, protological explanations of the "double creation" type, while remaining important for the reconstruction of their thought and its underpinnings, may have been used by them as conventional means of expressing frames of mind that were more broadly ascetic or spiritual or more specifically linked to personal spiritual experiences.

We have left to the end a conception which, though fundamentally gnostic, is typical of protological explanation. This is the *Manichaean conception, which implies the original separation and coeternity of the two substances, *light and darkness. This original situation is contrasted with the state of mixture that characterizes the mid-time and will be followed by the final, restored separation of the two substances, characteristic of Manichaean eschatology and of the sexual and alimentary encratism theorized by that religion. The religious-historical presupposition of Manichaean protological explanation, i.e the formula of radical *dualism (implying coeternity of the two principles, unlike the "mitigated" dualism of other gnostics and the dualistic monism of the *neoplatonists), can be found in the radical dualism of Zoroastrianism, which also admits a darkness coeternal with light (but not, as for the gnostics, identical with matter). Zoroastrianism, in fact, also combines protology and eschatology on the basis of the idea of the original and final separation of light and darkness: a cyclical but not repetitive procedure, which, both in Manichaeism and in Zoroastrianism, remains alien to the Greek idea of an eternal return or, at least, a pure and simple final restoration of the primordial condition. In fact, the darkness at the end will be devoid of that aggressive potential that characterized it at the beginning.

U. Bianchi

II. History of Christianity. In the Christian *tradition, the basis of any *anthropological formulation is provided by Gen 1-3 which, after the dual account of the *creation of the first couple, illustrates the situation of the first-formed in *Paradise, followed by the transgression of the precept, the expulsion from the garden and the definition of man's state as subject to death, toil and, for *woman, submission to her husband and the pains of childbearing.

In the other books of the OT, references to this protological picture are few and allusive; the only important uses of it are in the deuterocanonical books. Besides the brief mention in Tob 8, 5-6 of the union of *Adam and Eve, ben Sirach mentions the transgression, attributing the responsibility to the woman (Sir 25, 24), while Adam is extolled as being "above every living creature" (Sir 49, 16). The *Wisdom of Solomon* defines the first-formed as "Father of the world", endowed with the "power to rule all things", Wisdom having redeemed him from his "false step" (Wis 10, 1-2). Ps.-Solomon also knows another agent responsible for death's entry into the world: "God created man for incorruption, and made him in the image of his own eternity, but through the devil's envy death entered the world" (Wis 2, 23 f.). The motif of *phthonos* towards man, evidently connected with his special dignity, the implicit identification of the serpent of *Genesis with the "adversary" Satan of other biblical contexts, the contrast between the original *aphtharsia* and the death that entered after transgression: all these elements enrich the outlines of the protological picture that emerges from Gen 1-3. They are further amplified and complicated in the rich apocryphal literature which flourished in the Judaism of the turn of the Christian era, characterized by strong *apocalyptic and eschatological interests and by a specific attention to the beginnings of human history, decisive in many ways for determining the present state. The literary cycle turning on the figure of *Adam, whose earliest expression is the *Life of Adam* in its two versions, Latin (*Vita Adae*) and Greek (*Apocalypse of Moses*), shows the original state of the first-formed pair in terms of community of life with the *angels, living with them in *Paradise and eating "angelic food", as opposed to the "animal" food which they find on earth when they are expelled from their first home (*Vita Adae* IV, 1-2). Of particular interest in characterizing the protological picture seen in this work is the theme of the "poison of iniquity" injected by the serpent-Satan into the forbidden fruit, identified with "concupiscence" (*epithymia*), the beginning and root of every *evil (*Apoc. Mos.* 19, 3). Another tradition, quite solid and widely diffused, attributes a decisive weight in determining human destiny to the equally primordial attitude of a group of angels, the Watchmen, identified with the "sons of God" of Gen 6, 1-4, come on earth to couple with the daughters of men and teach them forbidden knowledge. This theme, whose earliest and most elaborate exemplification is in the Ethiopic *Enoch*, runs through a large part of the Jewish apocryphal literature from the *Book of Jubilees* to the *Testaments of the XII Patriarchs* and *2 Baruch*, and is known to the Qumrân writers, to *Philo of Alexandria and to Josephus. Its appearance in patristic tradition from *Justin to *Ambrose, *Jerome and others reveals the interest it aroused within primitive Christianity, which, however, attributed decisive importance rather to the Adamic protology, taken up into the Pauline scheme of the first and second Adam as a paradigmatic sign of salvation-history. Without pretending to exhaust the pregnancy of that scheme, it will suffice to note how *Paul's concentration of attention on the redemptive universality of Christ's work leads to an equally exclusive concentration on the efficacy of the contrary sign of the founder of mankind, who, as anti-type of the Heavenly man, "the first man . . . from the earth" (1 Cor 15, 45-49), is depicted as him "in whom all die, as also in Christ all will be made alive", in the context of the invitation to put off the earthly image of Adam and clothe ourselves in that of the *epouranios*. The theme of Adam *typos tou mellontos* in Rom 5, 12-21 turns on the contrast between the *paraptoma* of the first-formed and the saving *charisma* (*charis/dorea*) of Jesus Christ, the one presaging death and condemnation, the other *life and *justification, under the respective signs of disobedience and total obedience.

We cannot here follow the various lines of development of protological themes in patristic tradition. We will only say that, alongside the Pauline position which strongly stresses the negativity of Adam's role and is preoccupied with his status before his sin, at least two main strands can be distinguished, one characterized by the idea of the inchoative character of the privileges of the original condition, the other by considerable insistence on Adam's perfection and integrity, sometimes also put forward as the objective of the eschatological expectation. The first of these attitudes finds its most organic and articulate expression in *Irenaeus, who while exalting the very special privileges of the pre-lapsarian Adam, favours the idea of his state of "*childhood", which implies "simplicity" and absence of malice (cf. *Dem.* 14) but also expresses that fundamental incompleteness and immaturity which made his fault possible, while the divine work of *redemption unfolding along the course of history is described as both education and guide to man's complete maturity, in the setting of the "*recapitulation" of all creation in Christ. *Clement of Alexandria is part of the tradition that attributes to the first-formed the innocence, but also the incompleteness, of childhood: in the *Protrepticus* (XI, 111, 1) he shows him, "free from everything", playing in Paradise. His sin then appears as submission to pleasure: Clement, following Philo, points to the biblical serpent as the figure of "pleasure which goes on its belly, earthly vice which clings to matter". Here we are face to face with a notion which, variously elaborated, would characterize the protological picture of the interpretative position which is concerned to exalt the exceptional privileges of the pre-lapsarian Adam. Man's original state is defined as free from any material care, entirely directed to the *contemplation of *God, while his transgression consists of a decline to the sensual level, clearly qualified as *hedoné*. On the other hand Clement, alone among the Fathers except for *Zeno of Verona, explicitly identifies the first sin with the practice of *marriage. With that he makes the maximum concession to his *encratite adversaries against whom he argues strongly in book III of the *Stromata*, accepting their fundamental postulate of the identification of Adam's sin with physical union. But Clement contradicts the essence of the encratite position by affirming the intrinsic goodness and lawfulness of marriage, planned by God for man from the beginning, but unduly "anticipated" by him, since - the first couple being still "young" (*neoi*) - they should have waited until the proper time (*Strom.* III, XIV, 93, 3 p. 239, 16-21 Stählin; cf. *ibid.* XVII, 103, 1 p. 243, 23-25). Though he sees the first sin in a nuptial light, marriage for Clement is part of the divine plan for man, who undergoes no substantial modifications or "additions" as a result of the primordial transgression.

The encratite view, on the contrary, implies the idea that physical union and consequent generation are not just the matter of the first sin, but activities essentially alien to man's nature, performed through diabolic instigation and in imitation of the beasts. The doctrine of *Tatian, *Julius Cassianus and the anonymous encratites who, already denounced by Irenaeus for their radical abstentionism, are the object of Clement's polemic, follows the biblical position by admitting a single divine demiurge, but defines marriage and procreation as "animal" physical activity, object and sign of the degeneration initiated by Adam. With this it attributes to the first sin a specific founding efficacy, having introduced into human life, against the Creator's will, a constitutive element of its present dimensions. These peculiar protological justifications of the radical *enkrateia* that condemned marriage as "*fornication and corruption" were accompanied in encratite circles by a whole series of other justifications only partly held in common with the *orthodox, essential among which, together with the motif of *imitatio Christi*, is eschatological tension, in the dual sense that total continence procures final *resurrection and, at the same time, anticipates it and makes it already present in this life. Clement accuses his hearers of considering themselves already risen (probably on the basis of identifying *baptism with *anastasis*) and for this reason rejecting marriage and generation. At the same time there is attested in encratite circles a particular *exegesis of the Lucan pericope on the condition of the risen (Lk 20, 34-36), according to which those who do not marry show themselves to be the "sons of that age", since even in the present age they live "like angels", who take neither wife nor husband. In the complexity of the themes that come together in the encratite view to form the basis of their radical rejection of marriage, the peculiar protological scheme underlying this view provides one of the essential justifications; in fact, given the idea of an original state of perfection (*symphonia* of *soul and spirit for Tatian, integrity of the soul free from *epithymia* in Julius Cassianus, unity preceding sexual differentiations in the *Gospel of the Egyptians*), and with Christ's saving action being depicted as a restoration of pre-lapsarian conditions, the specific connection between the Adamic transgression and the custom of marriage leads the encratites to hold that abandonment of this custom, identified by antonomasia with sin, is indispensably necessary for salvation itself.

In patristic *tradition, holding firm to the principle of the legitimacy of marriage and generation, we can identify a variegated but essentially homogeneous line of interpretation which, insisting on Adam's exceptional prerogatives and at the same time extolling the values of Christian continence and *virginity, leaves room in the full picture of ethico-ascetic, eschatological and ecclesiological explanations on which those values repose for a typical protological explanation. This occurs when, presenting the perfect Adam in Paradise as a virgin and making marriage and physical generation a "remedy" necessitated by the state of post-lapsarian decadence and weakness, it depicts *enkrateia* and *virginity not as a "sign" of the condition of the risen, but as an anticipated realization of it and at the same time a restoration of the original perfection. This line of interpretation includes authors of differing doctrinal presuppositions, from *Tertullian and *Origen to *Athanasius, *Didymus the Blind, *Basil of Ancyra, *Gregory of Nyssa, *Gregory Nazianzen, *John Chrysostom and, in the West, *Ambrose, *Jerome and, in an early phase of his position on this subject, *Augustine.

The strong eschatological tension of Tertullian's thought needs no particular exemplification here; if the totality of Christian *virtues, and above all continence, are finalized on reaching that blessed *telos* in which the saints, according to the Lucan text, will be *translati . . . in angelicam qualitatem et sanctitatem* (*Ad uxor.* I, I, 5), virgins, who *iam in terris non nubendo de familia angelica deputantur* (*ibid.* I, IV, 4), achieve this eschatological condition in anticipation. This *status* is at the same time a return to the pre-lapsarian Adamic condition, being characterized by the complete extirpation of concupiscence, which is alien to *Paradise. This was a place of perfect integrity (cf. *De anima* 38, 2; *De monog.* 17, 5), so that those "who wish to be welcomed into Paradise must finally cease from that activity (i.e. marriage) by which Paradise was contaminated" (*De exhor. cast.* 13, 4). We observe how the pre-eminent eschatological justification of Tertullian's ethic and its strong ascetic components cannot be separated from the particular protological vision which, based on the notion of a virgin Adam in Paradise and identifying marriage and generation as clear signs of the post-lapsarian situation, characterizes these activities - legitimate, in that they are allowed by God - as "added" elements qualifying a fallen human existence, neither useful nor necessary in the state of original integrity.

In *Origen, protology falls into two moments: the first creation of free and equal intellects united to God and to each other in perfect charity, and the "second" creation (second in the dual sense of chronological posteriority and ontological inferiority) of the material substance which provides a support, both place of fall and means of ransom, to those intelligent creatures who have departed in various ways, through "satiety" and ineptitude, from the initial goodness and perfection. Human history is thus characterized by two beginnings, one metahistorical, on the level of the pre-existence of the *noes* (intellects), and the other specifically connected with the events of Adam, where Origen upholds the biblical assumption of a first man created by God, protagonist of the events of Eden. A fragment of the *Commentary on the Letter to the Corinthians* contrasts this "historical" Adam's original state of perfect integrity, in total dedication of soul to the contemplation of God, with the post-lapsarian condition, in which marriage, for which his body was not intended in the original divine plan, is practised. We see from this that the Origenian themes of virginity and continence, so central to his *spirituality, and his whole *anthropological vision, in the general cosmological and eschatological picture, are profoundly conditioned by his protology, worked out according to the scheme of *double creation.

A further typical example of mutual, irrevocable connection between anthropological, ethico-ascetic and eschatological themes on the one hand, and protology on the other, is offered by *Gregory of Nyssa. In his *De virginitate*, Gregory forcefully proclaims the post-lapsarian character of the institution of marriage which, as remedy and "consolation" for the death established by sin, prolongs the present régime of *genesis-phthorà* by the sequence of births and deaths, while *parthenia*, breaking through this unstoppable cycle, sets a barrier to death itself, restores the integrity of Paradise and anticipates the blessed *eschaton*; his *De opificio hominis*, however, presents a more complex and ontologically qualified protological picture. The perfect *creation of "total" humanity, in *God's *image and likeness, envisages the absence of sexual distinctions, the multiplication of historical man being entrusted to a mysterious form of generation in the "angelic" manner. The prevision of Adam's future sin, through which man will decline towards matter and sensual pleasure, leads the Creator to add to the "image" the animal element of *sexuality which, through physical union, permits the multiplication of the fallen human race until it reaches the "plenitude" for which it is destined. This results in a scheme of "double creation" which, discarding the Origenian idea of pre-existence and implying the chronological simultaneity of the "first" (element of "image") and "second" creations ("addition" consisting of sexual distinctions), depicts a protology equally graduated on two levels, which conditions Gregory's views of *anthropology, *ethics and *eschatology.

*Augustine, after a first period of accepting the now consolidated protological scheme in which marriage is incompatible with Adam's integral situation, rejects that scheme in order to affirm, with ever clearer determination in the course of his anti-*Pelagian polemic, that marriage and physical generation are part of God's original project for man. He thus affirms a different protology, already present in a broad patristic tradition to which belong, in their various ways, *Irenaeus, *Clement of Alexandria and, later, *Theodore of Mopsuestia and other representatives of the Eastern and Western church. In Augustine, it acquires a specific consciousness and articulation in connection with his doctrine of "original *sin" which, while it brings out the multiple, dramatic and devastating effects of Adam's transgression on the whole of mankind, attributes to them an exclusively ethico-religious importance, rejecting the explicit or implicit anthropological implications of a protology that had seen in Adam's sin the "founding" event of a typical human activity and existential dimension, the sexual, expressed in marriage and generation.

J. Bonsirven, *Le Judaïsme palestinien au temps de Jésus-Christ*, vols. I-II, Paris 1934; C. Tibiletti, Verginità e matrimonio in antichi scrittori cristiani, in *Annali FacLettFilos Università di Macerata* II, 1969, 1-227; U. Bianchi (ed.), *La "doppia creazione" dell'uomo negli Alessandrini, nei Cappadoci e nella gnosi*, Rome 1978; id. (ed.), *Arché e Telos. L'antropologia di Origene e di Gregorio di Nissa. Analisi storico-religiosa*, Atti del Coll. di Milano 17-19 maggio 1979, Milan 1981; G. Sfameni Gasparro, Protologia ed encratismo: esempi di esegesi encratita di Gen. 1-3, *Augustinianum* 22 (1982) 75-81; P. Pisi, *Genesis e phthorà. Le motivazioni protologiche della verginità in Gregorio di Nissa e nella tradizione dell'enkrateia*, Rome 1981; U. Bianchi (ed.), *Atti del Colloquio internazionale su "La tradizione dell'enkrateia: motivazioni ontologiche e protologiche"*, Milan 20-23 aprile 1982; G. Sfameni Gasparro, *Enkrateia e antropologia. Le motivazioni protologiche della "continenza" nel cristianesimo dei primi secoli e nello gnosticismo*, Rome 1984.

G. Sfameni Gasparro

PROTOPASCHITES. According to Roman and Alexandrian liturgical *tradition, the Christian *Easter should not be celebrated before the spring equinox; but the church of *Antioch and other churches of *Syria, *Cilicia and Mesopotamia, as well as *Novatianist churches in *Phrygia (cf. Socr., *HE* IV, 28; V, 21), following the *computus used among the Jewish communities, who from the 3rd c. did not keep to the observance of the *terminus post quem* of the spring equinox, sometimes celebrated Easter ahead of the other Christian churches: from this anticipation comes the conventional name of p. given to the followers of this liturgical practice. The council of *Nicaea recalled the Eastern churches to the observance of the date of Easter indicated by the Roman and Alexandrian ecclesiastical computi.

V. Grumel, *Traité d'études byzantines*, I, *La Chronologie*, Paris 1958, 42; R. Cantalamessa, *La Pasqua della nostra salvezza*, Casale Monf. 1984 (repr.), 134.

V. Loi

PROVERBIA GRAECORUM. A collection of 74 sentences, translated from Greek, known to us through Sedulius Scotus and through references in other sources, such as a letter of Cathwulf to Charlemagne (MGH, *Epist.* IV, 503). This collection is generally considered an adaptation, made in 6th-c. *Ireland, from the original Greek.

CPL 1130; S. Hellmann, *Sedulius Scottus*, Munich 1906, 121-135; J.F. Kenney, *The Sources for the Early History of Ireland*, I, New York 1929, 566.

S. Zincone

PROVIDENCE. Patristic teaching on p. as an active divine attribute is a product of two traditions, both of which it develops and investigates theologically: that of classical theodicy, for which πρόνοια or *providentia* is the divine rationality that harmoniously organizes and rules the course of the *cosmos; and the biblical tradition for which p. is the active and loving will of a personal creator *God on behalf of the beings created by him, particularly in view of the fulfilment of his saving plan for man.

1. The doctrine of a divine p. directing the life of the cosmos was formulated by Plato (*Tim.* 30 and 44) and became the common heritage of the main classical philosophical schools in the Hellenistic-Roman age, with the sole exception of the Epicurean; but it was particularly important in the philosophical-religious tradition inspired by *Stoicism: Panaetius wrote a περὶ προνοίας, used by Cicero in the second book of *De natura deorum*; Seneca wrote a *De providentia* (cf. also Diog. Laert., VII, 147; Cic., *Nat. deor.* II, 22, 58; Sen., *Nat. Quaest.* II, 45, 2). Christian writers, particularly the *apologists, often refer to the classical teaching on p., esp. when trying to vindicate Christian monotheism (cf. Just., *Dial.* 1; Athenag., *Leg.* 19, 3; Theoph. Ant., *Ad Autol.* II, 8; Tat., *Orat.* 2; Iren., *Adv. haer.* III, 25; Hipp., *Refut.* I, 21, 1; Orig., *C. Cels.* IV, 74; Tertull., *An.* 20, 5; Min. Fel., *Oct.* 19, 10; Lact., *Div. inst.* I, 2, 2-3; VII, 3, 6; *Ira Dei* 9). In celebrating divine p. in the created *world, the teaching of Christian writers follows classical theodicy and does not substantially depart from it; p. is the divine reason that contrives, arranges and governs the universe (cf. Athenag., *Leg.* 8; Clem. Al., *Strom.* VI, 148, 2; VII, 9, 1-4; Novat., *Trin.* VIII, 43; Lact., *Ira Dei*, chs. 9-10; Aug., *De civ. Dei* XXII, 24). P. extends to the whole cosmos as well as to its details (cf. Athenag., *Res.* 18, 5; *Leg.* 24, 3; Clem. Al., *Strom.* VI, 153, 4; Orig., *De princ.* IV, 1, 7; *C. Cels.* V, 71; Min. Fel., *Oct.* 18, 3; Novat., *Trin.* VIII, 43). P. is manifested particularly evidently in the functionality, beauty and harmony of the members of the human *body (cf. Min. Fel., *Oct.* 18, 1 ff.; Lact., *De opif. Dei* 1, 16; 2, 10; 3, 4; 4, 23-24; 6, 6; 8, 15; Aug., *De civ. Dei* XXII, 24; Greg. Nyss., *Hom. opif.*, chs. 8-9). P. directs and governs the life of all mankind and of individual men (cf. Just., *I Apol.* 44, 11; Clem. Al., *Protr.* 91, 3; *Strom.* VII, 9, 1-4; Novat., *Trin.* VIII, 43; Lact., *Ira Dei* 19, 6). But at this point Christian writers stress the profound difference between the Christian conception and the classical one of εἱμαρμένα or *fatum*, which denies man's freedom (cf. Just., *II Apol.* 7, 4; Theoph. Ant., *Ad Autol.* III, 7; Hipp., *Refut.* I, 21, 2; Clem. Al., *Exc. Theod.* 74; *Orig.*, *C. Cels.* I, 10; Euseb., *Praep. ev.* VI, 6; Aug., *De civ. Dei* V, ch. 1 ff.). P. governs and directs the destinies of peoples and of human kingdoms. *Origen (*C. Cels.* II, 30) presents as providential the coincidence between the *pax Augusta* in Rome's universal empire and the birth of Christianity; a similar idea is developed by *Eusebius of Caesarea (*Praep. ev.* I, 4, 2-5; *Dem. ev.* VII, 30-33. 35). *Lactantius celebrates the intervention of divine p. in the collapse of the persecutors of the church (cf. *De mortibus persecutorum*, ch. 1) and in the raising of *Constantine to the purple (*Div. inst.* VII, 26, 12). The celebration of p. in the unfolding of human history is the idea that inspires *Augustine's *De civitate Dei*; P. directs the destinies of human empires (*De civ. Dei* V, 12; V, 13; V, 15); it willed to reward the moral virtues of the ancient Romans with the power of empire (*De civ. Dei* V, 12; V, 13; V, 15). Not wholly alien to the classical philosophico-religious tradition (cf. *Asclepius* 39) is the conception of divine p. exercising its cosmological functions through intermediary "powers"; in the Christian conception, God exercises p. through the *angels (cf. Athenag., *Leg.* 24, 3; Clem. Al., *Strom.* VI, 148, 2; Novat., *Trin.* VIII, 45; Method., *Res.* I, 37). We can also link to the classical philosophico-religious tradition the conception of the divine Logos as a rational power that organizes and directs the universe, so that the divine providentiality that pervades the cosmos can be referred immediately to the Logos: this doctrine can be found in the *Letter to *Diognetus* (7, 2) and recurs in *Clement of Alexandria (*Strom.* VII, ch. 2), *Irenaeus (*Adv. haer.* I, 22, 1-2) and Ps.-Athanasius (*Annunt.* 5). But it is particularly dear to Eusebius, for whom the Logos is the same divine πρόνοια that creates and directs the universe (cf. *Laud. Const.* XII, 7); the Logos cares for men with paternal love (*Theoph. syr.* I, 79), is responsible for the government of kings and their legislation (*Dem. ev.* V, 1, 6) and educates mankind, leading it to the heights of Christian civilization (*Theoph. syr.* II, 93).

2. In the patristic *tradition, esp. the Greek, the classical theodicy of the cosmological function of p. is undoubtedly a predominant influence. But since Christian thinkers explicitly refer to the conception of p. as an active attribute of *God, creator and father of biblical *revelation, there is no lack of patristic sayings specifically inspired by the biblical teaching, for which p. is manifested above all in salvation-history. Indeed, in some Christian thinkers the conception of p. coincides with the divine will unfolding in the saving "*economy", so that the whole overall picture of the saving "dispensation" and its individual characteristics are a manifestation of divine p. Divine p. is manifested in human history in a teaching activity that leads to a progressive and ever deeper knowledge of *God and of his religion (cf. Clem. Al., *Strom.* VI, 17, 153, 1; Aug., *De civ. Dei* X, 14). It stirred up the OT prophets to foretell Christ (cf. Just., *Dial.* 118, 3; Clem. Al., *Strom.* V, 6, 2; VI, 128, 3; Euseb., *Theoph. syr.* II, 93); it illuminates mankind through Holy *Scripture (cf. Orig., *De princ.* IV, 7; *Sel. in Ps.* 1). The gradual education of mankind and its development in *grace to attain *contemplation of God is brought about by divine p. (or the divine "economy") through the *incarnation of the Word, dispenser of paternal *glory for the good of men (cf. Iren., *Adv. haer.* III, 20, 2; IV, 20, 6-7; Novat., *Trin.* XVIII, 100). Acc. to *Eusebius of Caesarea, the Word is the dispenser of divine p., which it does first by liberation from the worship of *demons, then by the call to know and worship the true God; by means of his laws and precepts, by means of the *prophets and other just men, it educates in the fear of God; finally, instilling love of truth and goodness, it raises mankind to civilization and the encounter with Christian truth (cf. *Theoph. syr.* II, 93). Divine p. arranged the calling of the Gentiles to be God's people instead of the *Jews (cf. Just., *Dial.* 118, 3); it disposes men to receive the divine call to salvation (cf. Clem. Al., *Strom.* I, 2, 3); it created the church (cf. Hermas, *Vis.* I, 3, 4) and builds it up with Holy *Scripture (cf. Orig., *De princ.* IV, 7). The conception of a divine p. which permits *evil in the *world but is not indifferent to it, indeed which uses it as an occasion of merit for the good (cf. Iren., *Adv. haer.* IV, 38, 4; Clem. Al., *Strom.* I, 17, 86, 1 ff.; Aug., *Gen. c. Man.* II, 28; *De civ. Dei* XIV, 11; *Enchir.* 11), is inspired by biblical *revelation. Particularly inspired by biblical revelation is the *eschatological view of the doctrine of p., which prepares rewards or punishments according to merit. This thesis is developed by *Lactantius in *De ira Dei*, but also recurs in *De mortibus persecutorum*. Lactantius categorically rejects the Stoic conception of an exclusively benign p.; because he is provident, God cannot fail to be angry with those who transgress his law: *Si est providens, ut oportet Deum, consulit utique generi humani, quo sit vita nostra et copiosior et melior et tutior. Si est pater ac dominus universorum, certe et virtutibus hominum delectatur et vitiis commovetur. Ergo et iustos diligit et impios odit* (*De ira Dei* 19, 6); because he cannot be indifferent to moral evil, God judges and condemns sinners (*De ira Dei* 22, 3; 24, 9). The final *judgment on the good and evil done by men is consistent with divine p.: this teaching frequently recurs in the *Fathers (cf. Athenag., *Res.* 18; Iren., *Adv. haer.* IV, 36, 6; Clem. Al., *Strom.* IV, 12, 87, 2; Orig., *De princ.* II, 9, 8); p. is clearly revealed in the variety and proportionality of divine penalties against sinners (cf. Aug., *De civ. Dei* XXI, 13; John Chrys., *Hom. in Mt.* XIII; *In Ps.* IV, 10).

1) H. Koch, *Pronoia und Paideusis. Studien zu Origenes und sein Verhältnis zum Platonismus*, Berlin 1932; G.L. Prestige, *Dieu dans la pensée patristique*, Paris 1955; M. Spanneut, *Le Stoïcisme des Pères de l'Eglise*, Paris 1957, 325-331; V. Loi, *Lattanzio nella storia del linguaggio e del pensiero teologico pre-niceno*, Zürich 1970, 66-69. 2) H.D. Simonin, Providence selon les Pères grecs, *DTC* 13, 941-960; A. Rascol, Providence selon S. Augustin, *ibid.*, 961-983; J. Behm, πρόνοια, in *GLNT* 7, 1217-1220; Lampe, πρόνοια, 1157-1158; J. and P.G. Walsh, *Divine Providence and Human Suffering*, Wilmington 1985.

V. Loi

PRUDENTIUS. Born prob. at Calagurris (Calahorra on the Upper Ebro) in 348, Aurelius Prudentius Clemens belonged to that provincial Hispano-Roman aristocracy that built its career on the emp. *Theodosius (379-395) of Cauca (doubtless modern Coca in Old Castile). Full of a still Horatian melancholy, the verse introduction with which P. prefaced his collection of poems published in 404/5 is a sort of miniature *Confessions*, an autobiography followed by an exposition of his poetic project. We see the student in love with *rhetoric, the ambitious lawyer, the provincial governor who ends his career as private adviser to the emperor (*proximus*), perhaps until the latter's death in 395. For this Christian man of letters, retirement is the occasion of a dual *vita nova*: his *conversion to the perfect life leads him to asceticism and to great poetry. The latter expresses and upholds the aspirations of the former; or rather, it becomes a spiritual exercise integrated with *ascesis, praise of *God and service to the church, a singular monument of *faith and poetic culture, in an extremely diversified output.

Did P. plan all his work in advance, like a vast basilica, a sort of Christian *archipoème*, aiming to emulate the traditions of ancient poetry in every genre (W. Ludwig)? Or, more likely, did he organize bit by bit, and mainly after the event, an output spread over the years and perhaps begun before his retirement from high office to his lands in *Tarraconensis (at Calahorra? or at *Saragossa or *Tarragona? three cities whose *martyrs he particularly celebrates)? The fact is that the summary outlined in the *praefatio* does not rigorously correspond to the works preserved under P.'s name. "That my sinful soul may compose hymns all my days and not pass a night without singing to the Lord; that it may combat *heresies and manifest the *Catholic faith": this strophe alludes precisely to the lyrical collection of the *Cathemerinon* and its title, and then, less precisely, seems to evoke the didactic and theological poems, in hexameters, of the *Apotheosis* and the *Hamartigenia*. "Trampling on the *pagan rites, it brings ruin, O Rome, to your *idols" is a good description of the anti-pagan polemic of the two epic books *Contra Symmachum*. The celebration in *hymns of the passion and cult of the (esp. Roman and Spanish) martyrs appears in the next verse: "it consecrates a cult to the martyrs; it celebrates the apostles" (cf. *Peristeph.* 12, on the martyrdom and Roman sanctuaries of *Peter and *Paul). But here the presentation stops; nothing leads us to suppose the existence of the allegorical epos of the *Psychomachia* (the "combat of the soul" - for the *soul and within the soul - of personified *virtues and vices), nor of the *epigrams that serve as captions to figurative representations of biblical scenes: the *Dittochaeon* ("double nourishment" - of the two Testaments?). Were these two poems perhaps written after the publication of the 404/5 edition?

Collections of poetry in various lyrical metres and long poems in hexameters contrast with each other, in their form, as the two sides of P.'s poetry: on one hand the lyric, on the other the epic and didactic. But in fact the whole project draws its unity from a properly hymnodic design, even if this only afterwards appeared to P. as the unifying principle of his poetry: "Being unable to do so by his merits, he celebrates God at least with his voice": this whole poetic output is felt, thought, willed as a "lyrical oblation", the literary form of a "spiritual *sacrifice". From this source wells a specifically religious poetry which, through hymnody, is directly linked to *liturgy: it is a celebration (*concelebret*). As such, and even if gradually distinguished from it in a very personal manner (we observe this in the course of the *Cathemerinon*: cf. J.L. Charlet), it is rooted in the rich experience of his predecessors, the poet-bishops *Hilary of Poitiers and *Ambrose of Milan.

This cultivated poetry, work of a *layman of wide and refined culture, produced a difficult synthesis of various currents of Latin poetic tradition, as renewed and perpetuated in the course of the 4th c. This is reflected in P.'s poetry like a sort of vast panorama. In the background is the classical training and taste of a "man of the muses", who knows not only Virgil and Horace, but also Lucretius and Catullus, as well as Ovid, Lucan and the poets of the Flavian age. Further to the fore, the rebirth of an innovative poetry in 4th-c. Latin letters: the hermeneutical research of *Optatian, the glittering preciosity of *Ausonius (J.L. Charlet has recently shown that their influence was considerable), the use in literate society of *epigrams, wordplay and the *cento. But right to the fore, P. reaped the harvest of three quarters of a century of Latin Christian poetry. After an epic and meditative "re-presentation" of *Scripture (*Juvencus - a Hispano-Roman of the time of *Constantine), P.'s own generation saw the epigrams of *Damasus, the hymns of Hilary and Ambrose, *Paulinus of Nola's search for a great Christian poetry (but the relative chronology of Paulinus and P. and their works, and therefore their mutual influence, remains hard to establish).

P.'s genius lay in absorbing, in the service of the project we have tried to define above, these three different styles and conceptions of poetry, trying at once to unify them and follow their inflections in poetry of a quite different genre and tone. We can no longer reduce him to a "neoclassicist", conscious of his means, demanding in his technique, an ardent *retractor* of his great forebears of the Augustan age and the early empire. He is a poet deeply impregnated with the poetic aesthetic of *late antiquity: whence his exaggerations and his baroque animation, the violence of his passion and his imagination, his alternating taste for the pathetic and the graceful, the too Hellenistic sparkle (*poikilia!*) of his tones and his standards of vocabulary and style, in short a mixture of genres characteristic of the new "late" Alexandrianism. All this is based on the religious inspiration of the Christian poet, who is led to express in his verses the symbolic perception of the world and of existence that is common to the *homo religiosus* and the poet. To avoid misjudging P.'s talent, we must not reduce him to one of his three characters, nor eliminate one of them, but see that the first two are at the service of the third.

P.'s lyrical collections are the most transparent evidence of his spiritual project. With a concentric (or "mandala") composition of clear Hellenistic derivation, each poem gravitates round a centre which is always the meditative paraphrase of one or more episodes of sacred history: this is thus brought up to date, in the lived time of a "ruminated" *lectio divina* - as Ambrose and the mediaeval monks would say. Thus each poem is like the *libretto* of a little liturgy of the Christian imagination.

The pious ballads of the *Peristephanon* end by sublimating this literary cult of the *martyrs, which had already been extolled in prose through the ever more romantic literature of the *Acts* and above all the *Passions*. P. displays an art of lyrical and dramatic narration which is sometimes coloured by a popular liveliness, following oral legends whose key is often lost to us (but composition 6 is dedicated to the martyrs of *Tarragona whose *Acts* are preserved and we can, in this privileged case, check the poetic work against a prose source).

The *iconography of the frescos or *mosaics placed in the sanctuaries of the martyrs also played a role which helps us ascertain the poet's precise allusions: thus for *Cassian at Imola, or for *Peter and *Paul at *Rome. The geography of the *martyria* he celebrates traces the itinerary followed by the poet during a mysterious *journey to *Rome, to whose cause he makes only veiled allusions.

The poetic works in hexameters are more austere, but no less original. Here the poet laboriously becomes a verse theologian: but this layman does not have the solid theological ability of a Hilary or an Ambrose. In the structures of his demonstrations and in the forms and vocabulary that express them, he tries to be a Lucretius of Christian doctrine. The rhetorical and polemical epos (reminiscent of Lucan and Juvenal, imitation of whom is demonstrable here) of the two songs *Contra Symmachum* juxtaposes a rhetoric violently satirical of pagan religions and a celebration of the ideology of the Christian empire, in the *chiaroscuro* of an *antithesis between barbarians, rural and *pagan, and Romans, citizens and Christians.

The poem that was most read and most depicted in mediaeval iconography deserves still greater attention. The invention of the *Psychomachia* disconcerts us by its *allegory. It is a Homeric series of duels between personified vices and *virtues, dressed as heroes and heroines of an *Aeneid* which reminds us of opera, or *auto sacramental*, before its time. *Ekphrasis* (poetic description of works of art, in Alexandrian poetry) reappears, in this temple of the soul, as heir of the allegorical temples of antique poetry (cf. the prologue of the third *Georgic*) and of the prophetic visions of the heavenly *Jerusalem, made concrete here in a splendour of precious materials and colours.

Latin Christian poetry of the great generation of *Ambrose, *Augustine and *Jerome reached its highest expression in P. He immediately became a model who was reread by 5th-c. Western poets and, above all, by mediaeval scholars.

This "sweet P. with the unequalled mouth, so great and so famous for his diverse poetic compositions", as *Isidore of Seville wrote at the beginning of the 7th c. on his *library wall, where there may well have been a portrait of P., may be considered the prince of Christian poets.

Editions: CPL 1437-1446; PL 59, 767-1078; 60, 11-594; CSEL 61; CCL 126. Text and translation: ed. M. Lavarenne in 4 vols., Paris ²1955-1963; LCL, 2 vols., London 1949-1963; BAC 58, Madrid 1950.

Studies: I. Rodríguez Herrera, *Poeta Christianus, Prudentius' Auffassung von der Aufgabe des christlichen Dichters*, Diss. Munich, Speyer, 1936 (Sp. tr. Salamanca 1981); B.M. Peebles, *The poet Prudentius*, New York 1951; I. Lana, *Due capitoli prudenziani*, Rome 1962; Chr. Gnilka, *Studien zur Psychomachie des Prudentius*, Wiesbaden 1963; Kl. Thraede, *Studien zur Sprache und Stil des Prudentius*, Göttingen 1965; R. Herzog, *Die allegorische Dichtkunst des Prudentius*, Munich 1966; W. Ludwig, Die christliche Dichtkunst des Prudentius und die Transformation der klassischen Gattungen, *Entretiens de la Fondation Hardt* 23 (1977) 303-372; *Patrologia* III, 267-281 (with bibl.); W. Evenepoel, *Studies in the Liber cathemerinon of Aurelius Prudentius Clemens* (in Dutch, résumé in English), Brussels 1979; M. Sotomayor, La Iglesia en la España romana, *Historia de la Iglesia en España*, I, Madrid 1979, 312 ff. and 318-333; J.L. Charlet, *L'influence d'Ausone sur la poésie de Prudence*, Aix-Paris 1980; J. Fontaine, *Naissance de la poésie dans l'occident chrétien*, Paris 1981 (a quarter of the work is dedicated to P.); J.L. Charlet, *La création poétique dans le Cathemerinon de Prudence*, Paris 1982; V. Edden, Prudentius, in (ed.) J.W. Binns, *Latin Literature of the Fourth Century*, London 1974, 160-182; E. Paratore, Prudenzio fra antico e nuovo, *Passaggio dal Mondo antico al Medioevo da Teodosio a san Gregorio magno*, Atti dei Convegni Lincei 45, Rome 1980, 51-86; R.Palla (ed.), *Prudenzio, Hamartigenia*, Pisa 1981; J. Harries, Prudentius and Theodosius, *Latomus* 43 (1984) 69-85; V. Buchheit, *Resurrectio Carnis* bei Prudentius, *VChr* 40 (1986) 261-285; id., Prudentius contra Symmachum, eine Einheit?, *ibid.* 66-82; R. Henke, Der Romanushymnus des Prudentius und die griechische Prosapassio, *JbAC* 29 (1986) 59-65; A.M. Palmer, *Prudentius on the Martyrs*, Oxford 1989; M. Kah, *Die religiosität des Prudentius zwischen "pietas christiana" und "pietas Romana"*, Bonn 1990.

J. Fontaine

PSALMODY. Everywhere in the Psalter we find allusions to acts of worship: *sacrifices, *oracles, *blessings, *processions, feasts; if assistance is required, the psalmist sings and mentions musical instruments. Titles and rubrics (*selah, alleluia*) show that the collection was made for the temple liturgy. In synagogue worship, the Psalter was not part of the readings (i.e. the Law and the Prophets) and seems to have had no fixed place, but no doubt it was used. It is hard to say how much of these varied and adaptable traditions of worship the first Christian *assemblies kept; we must doubtless insist on the charismatic, inspired character of the *hymns and *psalms joined to the readings, the *Eucharist and the exhortations. There was soon great interest in the biblical canticles, analogous to the psalms and certainly added to the canonical collection (if necessary, one could calculate the references to these canticles and those to the psalms in order to evaluate the comparative influence of groups of biblical writings on different *Fathers). We can certainly link to Judaism the custom of having a psalm read, on a slightly embellished melody, by a soloist, joined by the community for acclamations in the form of a refrain ("responsorial psalm").

The well-prepared "choir" which knew the Psalter by heart, after long consisting of consecrated *virgins, pious persons and *clerics, was enlarged on the appearance of *monasticism. In *Egypt, the communities of monks tended rather to listen in silence to the chanting of a lector, accompanied from time to time by a silent prostration and a final *prayer. Traces are preserved of the introduction of more active psalmodic practices, introduced at *Antioch or *Milan in the second half of the 4th c. to reinforce the Nicene community against the conservative and less popular customs of the *Arians. The spread of this antiphonal *chant was rapid and introduced an abundant p. into the divine office. We can recognize that one of the temptations of the richer and more artistic *liturgy was rapidly to limit the monastic use of the Psalter, to give more time to the melodies or to free liturgical poetry (much more banal and less rich in profound *spirituality). Each local church had its own evolution, with its exceptional values and its weaknesses. This does not alter the fact that p. remained the fundamental structure of the liturgical prayer of the canonical hours.

H. Schneider, *Die altlat.bibl.Cantica*, Beuron 1938; J. Gelineau, Les formes de la psalmodie chrétienne, *La MaisonD.* 33 (1953) 134-172; P. Salmon, in A.G. Martimort, *L'Eglise en prière*, Paris 1961, 789-824; E. Ferguson, The Patristic Interpretation of some Musical Terms, *SP* 16.2 (1988) 15-23; A.G. Hamman, L'Utilisation des Psaumes dans les deux premiers siècles Chrétiens, *SP* 18.2 (1989) 363-374.

J. Gribomont

PSALMS, Book of. The NT is full of allusions to the p., as also to Isaiah (cf. Nestle-Aland, *Novum Testamentum Graece*, Stuttgart ²⁶1979, 752-757), in their christological interpretation (songs of Messianic triumph, prayers of suffering innocence, interpretation of salvation-history, the kingdom, the holy city). Perhaps the absence of heavy historical context in these ever-topical prayers particularly facilitated their evangelical reading (see e.g. J. Dupont, *Etudes sur les Actes des Apôtres*, Paris 1967, 245-390). During the first three centuries, the p. remain (with Isaiah) the prophetic book most used for *apologetic and pastoral purposes (cf. *Biblia Patristica* I-III, Paris 1975-1981: respectively, 24 pp. out of 175; 18 out of 177; and 45 out of 157 in the OT index). The main desire was to understand Christ's prayer and to participate in it as a church. The LXX translation, often obscure, but the expression of a more universal and more prophetic prayer than that of the Hebrew Psalter, lent itself particularly well to this adaptation (cf. B. Fischer, Le Christ dans les Psaumes. La dévotion aux Psaumes dans l'Eglise des martyrs, *La MaisonD.* 27 [1957] 86-108; A. Rose, *Psaumes et prière chrétienne*, Bruges 1965).

*Commentaries on the Psalter were so numerous and copious that it was impossible to preserve them all in their original form; they are mainly handed down by exegetical *catenae, composite works that are only now beginning to be studied in their entirety. The fragments that numerous scholars have taken from isolated manuscripts are so mutilated and uncertain that we still have no idea of the history and development of this *literary genre. The richness of *Origen's works was such that an immense number of later commentators (*Eusebius, *Hilary, *Ambrose, *Jerome and many others) pillaged them: this was bad for their survival, but favoured their influence. At present, M. Harl, at the head of a group of Hellenists, and E. Mühlenberg are working systematically on the *publication of the catenae; R. Devreesse, who had opened the way, has in old age provided an overall analysis of what is valuable in the material already published. So we can attempt a systematic presentation of the principal commentaries.

Let us start with *Origen. Virtually his first attempt at *exegesis, at *Alexandria in 222, was a commentary on the p., which went no further than Ps 15 (16) (or 25 [26]) and of which P. Nautin (*Origène*, I, Paris 1977, 262-275) has identified a dozen important fragments. A second commentary was recommenced at *Caesarea in 247, after the *homilies and the volumes on the *Prophets and *Song of Songs; it reached Ps 72 (73) and also included Ps 118 (119). Finally, unable to complete this monumental work, Origen made brief notes, *Excerpta*, on the whole Psalter. According to V. Peri's fine discovery (*Omelie origeniane sui Salmi*, Vatican City 1980), these *Excerpta* are identical with the *Tractatus* translated by *Jerome (CCL 76) (and perhaps with that already used by Jerome in his *Epp.* 20, 23, 25, 26, 28, 30, 34 and 37). The fragments of Origen published in various places are brought to our attention by Devreesse (1-88). We must hope that they are soon published correctly, like those of Ps 118 (119) edited by M. Harl.

Nautin suggests that Origen's second commentary was due to his discovery of the *Hexaplar versions *Quinta* and *Sexta*, still unknown in the Alexandrian commentary, and also to the publication of a homily by *Hippolytus (printed by Nautin in his *Dossier d'Hippolyte et de Méliton*, Paris 1953, 161-183) which claimed Davidic authorship of the whole Psalter.

In the 4th c., the first commentator of the whole Psalter, in 330, was *Eusebius of Caesarea (CPG 3467). He had Origen's *library at his disposal and followed his tracks. His comments on Ps 37 (38) (PG 30, 81-104) and 51-95 (52-96) (PG 23, 441-1221) are well preserved; for the rest, see Devreesse (89-146). Eusebius inherited Origen's erudition, not his religious genius.

To the Arian *Asterius the Sophist, who died after 341, Jerome attributes a commentary on the whole Psalter. These *homilies, rhetorical preaching aimed at exhortation, are far from being a historical-philological commentary, but they try to insert the texts into *David's life, following the traditional titles. 31 homilies are preserved (all authentic? CPG 2815): PG 40, 389-477; better, M. Richard, *Asterii Sophistae commentariorum in Psalmos quae supersunt*, Oslo 1956.

The catenae allow us to reconstruct *Athanasius's scholia to the psalter (CPG 2140), also preserved in *Syriac and *Coptic. We do not know when he wrote them, nor are we even sure of their authenticity. Very elementary, they bear no traces of the subjects dear to Athanasius, but show clear dependence on Eusebius. They can be read with safety, taking account of PG 27 and G. M. Vian (*Testi inediti dal commento ai Salmi di Atanasio*, Rome 1978). But his *letter to Marcellinus (CPG 2097, PG 27, 12-45) is certainly authentic.

Chronologically, perhaps we should cite here *Hilary's Treatises on the p. (CPL 428, PL 9), strongly inspired by Origen.

The fragments of *Didymus and *Apollinaris (CPG 2551 and 3681) provided by the catenae have been admirably published by E. Mühlenberg. Didymus's great commentary handed down by the Tura *papyri (CPG 2550) was published at Bonn by M. Gronewald (with L. Doutreleau and A. Gesché) from 1968 to 1970.

*Basil left a series of homilies, many Origenian, on selected p. (CPG 2836, PG 29, 209-494; CPG 2910, PG 30, 104-116). The commentaries attrib. to *Diodore of Tarsus, certainly very Antiochene, must be studied again when they are published entire (CPG 3818, CCG 6). Those of *Theodore of Mopsuestia (CPG 3833) have been published by R. Devreesse (*Le commentaire de Théodore de Mopsueste sur les Psaumes*, Vatican City 1939); for *Julian of Eclanum's Latin version see CCL 88A, Turnhout 1977.

Meanwhile the West had produced *Ambrose's *Explanationes* on Twelve Psalms, then on Ps 118 (119) (CPL 140-141, PL 14 and 15), inspired by Origen, as were the *Commentarioli in psalmos* (CPL 582, CCL 72, 163-245) of *Jerome, whose *Tractatus* (CPL 592-593) can be considered, as we have said, a translation of Origen's *Selecta*.

Still following Origen's line, but this time in the context of a more refined *spirituality, we may cite *Gregory of Nyssa (*In inscriptiones psalmorum*: CPG 3155, PG 44, 432-608) and above all *Evagrius of Pontus (CPG 2455; we await an edition from M.J. Rondeau, who has already published excellent studies).

On the Antiochene side, the homilies of *John Chrysostom (CPG 4413, PG 55) and the commentary of *Theodoret (CPG 6202, PG 80) had great popularity, and the published text still awaits improvements. Of *Cyril of Alexandria, only fragments remain in the catenae (CPG 5202; Devreesse, 224-233). Of *Hesychius of Jerusalem, on the other hand, we know three distinct commentaries (CPG 6552-6554), which also still await careful study.

In the West, the most masterly and profound work is obviously *Augustine's highly personal *Enarrationes* (CPL 283, PL 36, CCL 38-40). *Late antiquity would add *Cassiodorus's *Expositio* (CPL 900, PL 70, CCL 97-98).

This rapid glance at ancient *exegesis must confine itself to these names. The best of this exegesis has been put together, verse by verse, in the two anon. volumes *La tradition médite le Psautier chrétien*, brought together by E. de Solms (Paris 1973). This reading felicitously supplements the historical commentaries on the Psalter and goes deeper into liturgical understanding and *prayer.

R. Devreesse, *Les anciens commentateurs grecs des Psaumes*, Vatican City 1970; M. Harl, *La chaîne palestinienne sur le Ps. 118*, I-II, SCh 189-190, Paris 1972; E. Mühlenberg, *Psalmenkommentare aus der Katenenüberlieferung*, I-III, Berlin 1975-1978; M.-J. Rondeau, *Les commentaires patristiques du Psautier*, OCA 219, Rome 1982; E. Ferguson, Athanasius Epistola ad Marcellinum in interpretationem Psalmorum, *SP* 16.2 (1985) 295-308.

J. Gribomont

PSALMUS RESPONSORIUS. A composition published in 1965 from a 4th-c. *papyrus, initialled 149¹-153. The first 12 strophes survive, telling the story of Jesus, starting from *David and cut short at the wedding-feast of *Cana (Jn 2, 1-11). We do not know what other gospel episodes were included, since the last two leaves allow no hypothesis. The composition obeys the poetic rules of the "*responsio*" with its *cursus* (cf. *clausula*), harmony of sounds and use of similar terms. It is a flexible structure, the free combination of whose verses and strophes is guided by poetic inspiration and mastery of the Latin language. The sources are the *Protogospel of James*, followed as far as the coming of the *Magi, *Matthew's Gospel for the Magi episode and *John's Gospel for the Cana episode. Signs of *Luke's Gospel are few and disputable. Distinguishing points of doctrine are *Mary's virginity, divine motherhood and intercession. Among the oldest Latin liturgical texts, the presence of some Greek consonants reminds us that the Latin alphabet had not yet completely replaced the Greek alphabet in rhythmic compositions for liturgical use.

R. Roca Puig, *Himne a la Virge Maria. "Psalmus Responsorius". Papir llatí del sigle IV*. Barcelona ²1965; E. Peretto, "Psalmus Responsorius". Un inno alla Vergine Maria di un papiro del IV secolo, *Marianum* 29 (1967) 255-265; *Patrologia* III, 320-321; W.H.C. Frend et al., Further Liturgical Documents, to be published in *JbAC* (1992).

E. Peretto

PSEUDOEPIGRAPHY. P., in the strict sense, is a literary falsification regarding the author of a work, to be distinguished from *falsification of a text. The term is mostly used for the OT and NT apocryphal books, but is also extended to other works, Christian or secular. We must distinguish two cases: a) the real author intentionally hides under a false name or behind a well-known name, the fraud being the act of the author who thus hides; or b) texts circulate under a name which is not the author's, without him being responsible. The error may arise from a confusion of names, a false attribution or a copyist's error. The fundamental task of criticism is to rectify the error.

First there are the gospels, acts and apocalypses, either anon. or falsely attrib. to biblical characters (studied under *Apocrypha). Then: the *Epistola apostolorum*, which in the *Ethiopic version bears the title *Testament of our Lord and Saviour Jesus Christ*; Paul's letter to the Laodiceans and third letter to the Corinthians; the letter of *Barnabas and that of Titus and, if we wish, also Ps.-*Dionysius the Areopagite. The authors of all these are unknown. *Tertullian says the *Acts of Paul* were the work of a priest who composed them out of devotion to the Apostle (*De bapt.* 4). A taste for the marvellous is evident, as in *hagiography.

To the *apostles, and even to Christ, are attrib. a legislative collection which goes from the 2nd to the 5th c. and whose authors are neither named nor known, except for the *Apostolic Tradition. The main contributions are: the *Didachè*, the Catholic *Didascalia of the twelve apostles*, the *Apostolic Constitutions, the *Ecclesiastical canons (or constitution) of the apostles* (a title often modified: cf. A. Hamman, *Vie liturgique*, 114, n. 1), the *Testament of our Lord Jesus Christ*. With all these collections, the difficulty is to discover their successive layers, the value and origin of their elements, and their influences, with the aim of fixing their place, date and sometimes motives. The *Apostolic Constitutions*, e.g., incontestably have a Jewish source in the "Clementine liturgy". Their intention is often polemical: the *Didascalia* rails against *Judaizers, the *Apostolic Constitutions* defend the hierarchy against *monasticism.

Apart from NT characters, some authors hide under a pseudonym. The best-known cases are Ps.-*Dionysius and *Macarius. The author who hides under the pseudonym of Dionysius the Areopagite seeks to strengthen his doctrine through NT authorship, but most of all he wishes, at a time of violent polemic over *christology, to act as a conciliator. A moderate *monophysite, in this way he avoids having to cite *Chalcedon and the monophysites, as well as the *Henoticon*. A certain Macarius figures as the author of some hundred spiritual discourses or *homilies: the pseudonym is not intentional or imputable to the author. It would seem that the work was anon. and was attrib. to a holy man, a *makarios*, like the *Liber Graduum*; similarly the author of the *Commonitorium* was called *Peregrinus. In time the adjective came to be understood as a proper name. So it was easy to confuse this "Macarius" with another *Macarius, the Great, also called the Egyptian. It is also possible that, since the author professed *Messalian ideas condemned by the *councils, the editors held it prudent to have recourse to subterfuge, putting the homilies under the guarantee of the Egyptian *ascetic or of *Simeon Stylites.

There is no author of any calibre who does not drag a train of spurious works behind him. Those of the most illustrious (*John Chrysostom [cf. *John Chrysostom, Pseudo-], *Ephrem, *Augustine) attain a considerable amount (CPG 4500-5079, against 4305-4495 for Chrysostom's authentic works; for Augustine, cf. *Verzeichnis*, 158-190). The reasons are many. They are connected not with the author's responsibility, but with his success and with the transmission of texts. The point of departure may have been the imitation of famous writers. At one time the whole West "Augustinized". Indeed it did so right up to the Middle Ages. The celebrated *letter or homily *Cogitis me*, which circulated under *Jerome's name, was equally a fraud, perpetrated by Paschasius Radbertus and unmasked by modern criticism.

A great many errors are due to copyists, who tended to turn every "John" into John Chrysostom (Latin texts in CPL 916-945). A whole collection of homilies attrib. to Chrysostom seems to be the work of *John Mediocris, bishop of *Naples. A Latin text going under the name of Gregory was assigned to *Gregory Nazianzen, when it was really by *Gregory of Elvira. Augustine himself fell into the error. A copyist, coming across the abbreviation CHR, read Chrysostom instead of *Chromatius. In the same way, all writings in a codex were assigned to the author of the first text or of the immediately preceding text. The old catalogues did not always examine with due accuracy the content of each manuscript. Through lack of control, errors were handed down on the authority of others. The ancients themselves asked questions about authenticity. To resolve them, *Clement of Alexandria and *Origen examined language and style. We find the beginnings of hermeneutics in the correspondence between Sextus *Julius Africanus and Origen, in *Tertullian (*Adv. Marc.* IV), *Rufinus (*De adult. libr. Origenis*), *Jerome (*Adv. Rufinum*), *Amphilochius of Iconium (*Pseudepigrapha haereticorum*: PG 39, 115-117), the anon. author of the *Adv. Fraud. Apollon.* (PG 86, 1947-1976) and Nicephorus (*Contra Eus. et Epiphanium* 2).

For the printed editions of the humanists, two fundamental questions were asked: was a work authentic? if not, who was its author? From the first they tried to separate genuine from spurious works. To answer these questions and especially to discover the true authors, today we have perfected instruments - indexes, *Claves*, PLS - which allow us to venture into this immense forest. A more perfect and complete inventory of manuscripts, a better knowledge of the chronological problems, vocabulary and doctrines of Christian authors and of currents of ideas allow us to identify more clearly the authors of anonymous writings and pseudoepigrapha.

T. Birt, *Kritik und Hermeneutik*, Munich 1913; J. de Ghellinck, *Patristique et Moyen Age*, Brussels 1947, II, 80-92, 200-214; W. Speyer, Fälschung, *RACh* 7, 236-277 (bibl.); id., Religiöse Pseudoepigraphie und literarische Fälschung im Altertum, *JbAC* 8/9 (1965-1966) 88-125; id., *Die literarische Fälschung im heidnischen und christlichen Altertum*, Munich 1971; N. Brox (ed.), *Pseudoepigraphie in der heidnischen und jüdisch-christlichen Antike*, Darmstadt 1977.

A. Hamman

PSOTES of Psoi. Egyptian bishop, a *martyr under *Diocletian. His *Passion* (of the type that seems genuine, or at least not "epic") is handed down in *Coptic, Latin and *Ethiopic. Some traits relate it to that of *Peter of Alexandria. Many other Coptic texts mention him. A speech, supposedly pronounced by him before execution and consisting mostly of recommendations to believers, is a later *falsification. Nothing is known for certain of P. and his activities, but he was certainly venerated.

H. Delehaye, Les martyrs d'Egypte, *AB* 40 (1922) 5-154 and 299-364; esp. 316-19 and 343-352; T. Orlandi, *Il dossier copto del martire Psote*, Milan 1978.

T. Orlandi

PTOLEMY and LUCIUS. These Roman *martyrs are known to us from the philosopher *Justin, who used their punishment as an occasion to address his second *Apology* to *Antoninus Pius. P. had brought to the faith a woman who, with her husband, had led a dissolute life: the emperor allowed her to divorce him. In reprisal the ex-husband accused P. of being a Christian, and he was condemned and executed. L., present at the trial, protested against the unjust sentence, and was himself condemned and executed.

The two martyrs had no cult at *Rome until Baronius inscribed them in the *Mart. rom.* He followed Florus of Lyons, who had put them on 23 Aug: Ado and Usward preferred 19 Oct, which date Baronius kept.

Vies des SS. 10, 628-29; *LTK* 8, 894; *BS* 12, 526-27.

V. Saxer

PTOLEMY the Gnostic. Gnostic teacher, second half of 2nd c. A disciple of *Valentinus, he belonged with *Heracleon to the Italic or Western branch of his school. His *Letter to Flora*, whose text is cited entire by *Epiphanius (*Haer*. I, 33, 3-8), is about whether or not to accept the OT. P. distinguishes three parts: the first is of divine origin, the second comes from *Moses, the third from the Jewish elders. Jesus came to perfect, not to abolish, the first part. But he abolished the second (e.g. the *lex talionis*). The third part is the ritual law, which the Saviour, opening the pneumatic age, spiritualized.

T. also wrote a brief exegetical *commentary to the Prologue of *John's gospel (Iren., *Adv. haer*. I, 8, 5-6), which he interpreted allegorically so as to find in it the first eight fundamental *aeons of the Valentinian *pleroma. He was also the author of the system on which *Irenaeus's Great Notice (*Adv. haer*. 1, 1, 1-8, 4), of fundamental importance for the study and understanding of Valentinian *gnosis, is based.

SCh 24; W. Völker, *Quellen zur Geschichte der christlichen Gnosis*, Tübingen 1932; G. Quispel, *Ptolémée, Lettre à Flora*, SCh 24, Paris 1949; F.M. Sagnard, *La gnose valentinienne et le témoignage de saint Irénée*, Paris 1947; A. Orbe, *La teología del Espíritu Santo*, Est. Valentinianos 4, Rome 1966; G. Quispel, La Lettre de Ptolémée à Flora, *VChr* 2 (1948) 17-54; K. Rudolph, *Gnosis* (tr. R. McL. Wilson), Edinburgh 1983, 259 ff; W.H.C. Frend, *Rise of Christianity*, London 1984, 208-217; R.M. Grant, A Woman of Rome, The Matron in Justin Apology 2, *ChHist* 54 (1985), 461-472; Text of Ptolemy's Letter to Flora (Eng. tr. W. Foerster-R. McL. Wilson), *Gnosis, a selection of Gnostic Texts*, Oxford 1972, I, 121-161.

G. Filoramo

PUBLICATION, Technique of. The sources worked on by the science of Greco-Roman antiquity, and hence also by *patristics, do not come down to us in the original, but only in later MS copies. While the first editions (*editiones principes*) of the humanistic era and of subsequent centuries were content to reproduce the one or the few MSS accessible to the editor, from the middle of the 19th c. rigorous rules were worked out for the scientific edition (*editio critica*), which could provide information on the *manuscript tradition and on the constitution of the text which was based on it; this was of considerable advantage for understanding the texts. Furthermore, from the standard edition for the expert (*editio maior*), popular editions (*editiones minores*) were often made, providing just the restored text without critical apparatus. A critical edition must be so conceived, in principle, as to render its predecessors superfluous, i.e. it should combine, in the most correct way, all the evidence of any value contained in them. An indispensable element is the introduction (*Praefatio*) to the edition. This gives information about the sources of the text edited: existing MSS, if necessary *papyri, more rarely inscriptions, secondary tradition in the form of translations and citations, earlier printed editions, esp. those based on lost MSS. All these bearers of tradition are called *lege artis*; their mutual relations of dependence are made clear by a *stemma* ("family tree"). They are distinguished by initials, if possible the same ones already used in the specialist literature; these are then collected in an appropriate list of abbreviations. If the research on the *recensio* of the text, involving the determination of the archetype and of the common copies, which often have to be reconstructed, and of the available bearers of variants, turns out to be vast, then it should be published separately, while only the results are given in the *Praefatio*. The *Praefatio* is usually, and conveniently, written in Latin. Then follows the main part, i.e. the *editio critica* of the text thus restored, accompanied by the *apparatus criticus*. When the genuine external form of the archetype cannot be reconstructed by the means available, the editor himself must decide on its articulation, orthography, punctuation and diacritical marks, explaining in the *Praefatio* the principles he has followed to this end. As far as is useful, an apparatus of textual witnesses (*testes*), on whom places in the text are based from time to time, precedes the *apparatus criticus*. This last gives information about the variants of the tradition and the conjectures of modern scholars as well as an account of *lacunae*, integrations, additions, expunctions, displacements, solutions of abbreviations, corruption. To point out all these phenomena, there is a system of critical marks and parentheses, already worked out in a rudimentary way in antiquity. Recently the so-called Leiden system, worked out by the papyrological section of the International Congress of Orientalists, held at Leiden in 1931, has spread widely in papyrological studies, while in editions of *epigraphical or literary texts it is used sporadically. Texts based on a single witness, like inscriptions, papyri and palimpsests, usually have a faithful *palaeographical edition. In the case of texts with radically different redactions - e.g. *hagiography and other popular literature - the individual redactions can be published together in different parallel columns. For the edition of texts handed down only fragmentarily, appropriate models have been worked out. The *apparatus criticus* is preceded or followed by the apparatus of witnesses, which gives the sources, citations, *excerpta*, parallels, translations, commentaries and so on. In historical texts the deciphering of the data is a great help to the reader. Finally, precious help is offered by the indexes, to whose compilation great importance should be attributed. The indexes allow us to study the linguistic and grammatical peculiarities of the text, record names and things, places of other works (e.g. biblical references). It must also be established, case by case, whether the edition should be provided with a specialist bibliography; this is a practice which has met with success after being followed by the Bibliotheca Teubneriana.

O. Stählin, *Editionstechnik. Ratschläge für die Anlage textkritischer Ausgaben*, Leipzig 1909; G. Pasquali, *Storia della tradizione e critica del testo*, Florence ²1952; M.L. West, *Textual Criticism and Editorial Technique applicable to Greek and Latin Texts*, Stuttgart 1973.

J. Irmscher

PUBLISHING. For antiquity, *edere* in the etymological sense of the word meant "to put forth" and was used for persons as well as things, the voice as well as the *book. In reality, between private transcription and publication things followed not one well-fixed pattern, but circumstances and situations. It seems that the beginning must be put at the level of intention: to accept or permit the copying of a text, a letter or a sermon. This entailed the risk of seeing further reproductions of it. This is the evident case of the apostolic and extra-apostolic letters of *Clement and *Ignatius, and of the *Martyrdom of *Polycarp*. For the latter we possess the names of the two copyists who in turn transcribed the text from the manuscript of a certain Irenaeus. The *literary genres must be distinguished: a *Sermo* was initially something spoken, not written; a *letter did not seek publicity, except in the case of professional writers who, like *Cyprian, *Gregory Nazianzen or *Sidonius Apollinaris, intended to preserve its text, make collections of them and often encourage their circulation. Things are clearer in the case of an *opus*. In general this was written for others. Once the text was fixed, the author kept one copy and entrusted the archetype to a *librarius*, who was both copyist and seller. He worked to order and looked after circulation. Transcription was gradually transformed into commercial publication. It was always an artisan's job. The author was not at all bound to one copyist-publisher. Once on sale, the book was in the public domain: there was no copyright. Sometimes we have several recensions of a single text: e.g. *Tertullian's Adversus Marcionem* and his *Apologeticum*. This may be the result of circulation before the work was complete (as happened for *Augustine's *De Trinitate*), of modifications, alterations or *interpolations introduced by copyists or later publishers (e.g. *Eucherius's *Instructiones* were triplicated in the course of time).

Th. Birt, *Das antike Buchwesen*, Berlin 1892; F.M. Beare, Books and Publication in the ancient world, *Un. Toronto Quart.* 14 (1945) 150-167; J. de Ghellinck, *Patristique et moyen âge*, II, Brussels 1947; H.I. Marrou, La technique de l'édition à l'époque patristique, *VChr* 3 (1949) 208-224; *RACh* 2, 664-688 (bibl.).

A. Hamman

PUBLIUS. Bishop of *Athens, *martyr early in the reign of *Marcus Aurelius (161-180). *Eusebius of Caesarea mentions him in connection with the *letter of *Dionysius of Corinth (*c.*160-170) to the Athenians. In the letter, Dionysius rebukes the Athenian *faithful for having cooled in the faith "after their leader P. had been martyred during the *persecutions of those times". Eusebius adds: "Quadratus (to be distinguished from the apologist *Quadratus!) became their bishop after P.'s martyrdom" (*HE* IV, 23, 2).

EC 10, 293; *BS* 10, 1237.

G. Ladocsi

PUDENS and PUDENTIANA. The church of S. Pudenziana appears as *titulus Pudentis* in texts, as *ecclesia Pudentiana* in the inscription of the apse *mosaic. Excavations ending in 1962 showed the existence of a church dating from 390-398, inserted into a pre-existing *baths, to which an apse with the mosaic and inscription was added early in the 5th c. The founder, of whom we have no historical knowledge, was transformed into a saint in the course of the 6th c. At about the same time he appears in the legend of *Peter as the *apostle's host, father of Pudentiana and *Praxedis. His identification with the homonymous figure of 2 Tim 4, 21 savours of romance. His daughter Pudentiana, born of a misinterpretation of the mosaic and its inscription, appears in texts from the 7th c. at the same time as her sister. In the *itineraries, *Martyrologies and *liturgy, she is also called Potentiana.

A. Dufourcq, *Etude sur les Gesta martyrum romains*, 1, Paris 1900, 127-30; *DACL* 14, 1967-73; *LTK* 8, 897; *Vies des SS.* 5, 364-66; *BS* 10, 1062-72; A. Amore, *I martiri di Roma*, Rome 1975, 68-69.

V. Saxer

PULCHERIA (Aelia P. Augusta). Eastern empress, born 19 Jan 399 at *Constantinople, daughter of the emp. *Arcadius and *Eudoxia; died July 453. Becoming Augusta on 4 July 414, P. consecrated her *virginity, with her younger sisters Arcadia and Marina. She exercised considerable influence on the upbringing and education, as well as the political and ecclesiastical activities, of her brother the emp. *Theodosius II (408-450). This influence was reduced after 439 by the eunuch Chrysaphius, whom P. executed in 450. After her brother's death she ascended the throne, contracting a mystical *marriage with the older *Marcian. The council of *Chalcedon, which was to end the disputes over *christology, but which unleashed new conflicts by the disputed terminology of its definition (*horos*), was prepared and organized wholly by P.; can. 28, which fixed the position of the patriarch of New Rome, led to opposition from pope *Leo I. P., who left her fortune to the *poor and promoted humanitarian institutions and the building of *churches, was proclaimed a saint (10/11 Sept).

PWK 23, 1954-1963; *BS* 10, 1245-1256.

J. Irmscher

PURGATORY. The term p. in its technical meaning does not appear until the 11th c., in Hildebert (PL 171, 741). Before that, there were efforts to depict the next life before the last *judgment; but since *eschatology developed only slowly, there is some fluctuation in the *vision of the afterlife. *Martyrs were thought to reach *beatitude directly; non-martyrs had to expiate their post-baptismal *sins in a sort of *sheol* (Tert., *De an.* 58). So p. was a medicine rather than a punishment. For *Origen (and *Clement of Alexandria before him), the faults of the just will be purified by passing through a river of fire (*Num. hom.* 15; cf. Edsman, 1-14). The existence of a purifying punishment is affirmed by *Cyprian (*Ep.* 55, 22), *Ambrose (*In Ps.* 36, 26), *Gregory of Nyssa (*Or. de mort.*), *Caesarius of Arles (*Serm.* 167 and 179). *Augustine bases his own teaching on *purification in the afterlife on the church's prayer for the dead (*De cura mort., De octo Dul. quaest.* II, *De civ. Dei* 21, 13 and 24). Eastern and Western *epigraphy attests that the *gens Christiana* prayed for its *dead. *Tertullian even knows a *Eucharist for the dead. Likewise Cyprian and *Cyril of Jerusalem. A "memento" of the dead is incorporated in the Eastern *liturgies, and the Latin Sacramentaries contain *masses *pro defunctis*.

DACL 4, 427-456; 14, 1878-1951; H. Rondet, *Le Purgatoire*, Paris 1948; J.J. Gavigan, S. Augustini doctrina de purgatorio, *Ciudad de Dios* 167 (1954) 283-296; C.M. Edsman, *Le baptême de feu*, Uppsala 1940; P. Jay, Le purgatoire dans la prédication de Césaire d'Arles, *RecSR* 24 (1957) 5-14; A. Stuiber, *Refrigerium interim*, Bonn 1957; J. Le Goff, *La naissance du purgatoire*, Paris 1981; R.R. Attwell, From Augustine to Gregory the Great, an evaluation of the emergence of the doctrine of Purgatory, *JEH* 38 (1987) 173-186.

A. Hamman

PURIFICATION. Purity is a common concept in all religions: it is the disposition required before one may approach sacred things. According to the primitive conception, p. is obtained through rites, while it is lost by contact with certain material things. In the *Bible, the notion of purity becomes interior and moral.

Greek *philosophy insists on p. for the sake of *theoria*: only the pure may win the Pure, says Plato in a generally accepted aphorism (*Phaedo* 66d, 67bc, 70a). The theme is taken up by the *Fathers, but in a wider context. They present perfection as an *apocatastasis*, i.e. a restoration of the primitive condition. In this context, perfection and catharsis coincide (cf. Greg. Naz., *Or.* 2, 20: PG 35, 812c), the negative term expresses positively the fullness of eternal *life. The Mother of God is "most pure" or "purified" by the coming of the Spirit, divinized (cf. M. Gordillo, *Mariologia orientalis*, OCA 141, Rome 1954, 111 ff.). But in the language of *ascesis, the term p. is reserved for the first degree of the spiritual life, *praxis* (as opposed to *theoria*, *contemplation), defined by *Evagrius (*Praktikós* 78: SCh 171, 666) as "the spiritual method that purifies the passional part of the *soul".

To determine the various elements of p., we must determine what is considered as *evil and an obstacle to perfection. For Christians, the first and essential p. is that of *sin, which is done by *penitence. The second p. concerns the consequences and occasions of sin, esp. the passions. The result of this ascetic p. is *apatheia*. The more radical purifications are described by the classics of contemplation: the soul must renounce everything that can in any way come between it and the vision of *God. *Isaac of Nineveh establishes three degrees of this p.: *1)* the bodily phase (*dûborô pagranô*): to oppose the passions of the *body; *2)* the psychical (*napsonô*) phase: the struggle against dissipating thoughts; *3)* the spiritual (*ruhonûtô*) phase: to let oneself be governed totally by the spirit (cf. *DSp* 7, 2043 ff.). Isaac also returns to Evagrius's idea that perfect p. requires "nakedness" of intellect, despoiled not only of passions and imaginary images, but also of the multiplicity of rational, and therefore partial, concepts in order to reach the intuitive vision of "pure *light" (I. Hausherr, Ignorance infinie, *OCP* 2 [1936] 351-362; repr. in *Hésychasme et prière*, OCA 176, Rome 1966, 38-49).

TWNT 3, 416-434; *DSp* 8, 1664-1883; R. Arnou, *Le désir de Dieu dans la philosophie de Plotin*, Paris 1967, 153 ff.; M. Simon, Souillure morale et souillure rituelle dans le christianisme primitif, *SMSR* 38 (1967) 498-511; T. Spidlík, *Grégoire de Nazianze. Introduction à l'étude de sa spiritualité*, OCA 189, Rome 1971 26 ff.; id., *La spiritualité de l'Orient chrétien. Manuel systématique*, OCA 206, Rome 1978, 180 ff.

T. Spidlík

PYRRHUS. Patriarch of *Constantinople (638-641; 654). Presbyter of Hagia Sophia and *hegumen of the *monastery of the *Theotokos* at Chrysopolis on the Bosphorus, elected patriarch 20 Dec 638. In late 638 or early 639, he issued a dogmatic synodal decree approving the *Ekthesis* and *excommunicating those who would not subscribe it, a subscription twice requested (639 and 641) from pope *John IV. The accession of the emp. *Constans II overturned his situation and forced him to renounce the patriarchate (night of 28/29 Oct 641). He retired to monastic life in *Africa, convinced of his errors in a public debate at *Carthage with St *Maximus Confessor (July 645). At *Rome, he repudiated the *monothelite heresy, but at *Ravenna, soon after, he abjured *orthodoxy, bringing on himself pope *Theodore's synodal sentence of *anathema and deposition. After the death of *Paul II, 8 or 9 Jan 654, he reoccupied his see for six months, during which time he approved the *Typos*. He died on the day of *Pentecost (1 July) 654. The 3rd council of *Constantinople (680) condemned him as a *heretic.

CPG 7615-7618; Grumel, *Regestes*² 294-298; P. Conte, *Chiesa e primato nelle lettere dei papi del secolo VII*, Rome 1971, *passim* (index p. 583); L. Magi, *La Sede Romana nella corrispondenza degli imperatori e patriarchi bizantini (VI-VII sec.)*, Rome-Louvain 1982, 205-215; J.L. van Dieten, *Geschichte der patriarchen von Sergios I. bis Johanes VI (610-715)*, Amsterdam 1972, 57-78, 104-105.

D. Stiernon

Q

QUADRATUS. Mentioned by *Eusebius (*HE* IV, 3, 1-2) as author of a work in defence of the Christian religion, presented to the emp. *Hadrian (117-138) and so the earliest example of *apologetic known to us. Eusebius read it and noted its author's intelligence and *orthodoxy. It is now lost, save for a fragment cited by Eusebius as evidence of Q.'s archaicity. In it Q. speaks of men healed or revived by Christ as still living in his day. The writer's identity was, and is, disputed since the name Q. recurs elsewhere in *HE* in relation to persons whose identity with the apologist seems dubious. Scholars now agree that *Jerome's identification (*Vir. ill.* 19; *Ep.* 70, 4) of the apologist with the homonymous bishop of *Athens (cf. *HE* IV, 23, 3) is unfounded, since the latter lived at the time of *Marcus Aurelius (161-180), when people cured by Christ could not possibly still be alive. As for the date and place of presentation, some scholars maintain, on the basis of Eusebius's *Chronicle* (cf. *vers. Armen.*, ed. Schoene, II, 166; Jerome's version, GCS III, 199), that it was in Athens during Hadrian's visit there, i.e. in 125/6, or else in 129. Others think it was composed in *Asia Minor and given to Hadrian there in 123/4 or 129. Also disputed is the apologist's identification with the homonymous *prophet whom Eusebius's evidence (*HE* V, 17, 2), taken from the Anonymous anti-Montanist (cf. also *HE* III, 37, 1, prob. from the same source), seems to number among the direct descendants of the *apostles and to indicate as working in Asia Minor.

PG 5, 1261-1266; J.C. T. Otto, *Corpus Apologetarum Christianorum*, IX, Jena 1872, 333-341; A. Harnack, *Die Überlieferung der Griechischen Apologeten*, TU, I, 1-2, Leipzig 1882, 100-114; id., *Geschichte der altchristlichen Literatur bis Eusebius*, I, Leipzig 1893, 95-96; II, 1, Leipzig 1897, 269-271; T. Zahn, *Forschungen zur Geschichte des neutestamentlichen Kanons*, VI, Leipzig 1900, 41-53; Bardenhewer 1, 183-187; *DTC* 13, 2, 1429-1431; G. Bardy, Sur l'apologiste Quadratus, *Mélanges Henri Grégoire*, Brussels 1949, 75-86; Quasten I, 170 f; R.M. Grant, *Greek Apologists of the Second Century*, London 1988, ch. 4.

V. Zangara

QUAESTIONES ET RESPONSIONES on Holy Scripture. Among the exegetical genres of the *Bible - scholia, *homilies and *commentaries - "Q. et R." are related to scholia. Like those, they give no complete systematic interpretation of a whole sacred book (the task of commentaries), but consist of explanatory notes, usually brief, on given passages, verses or even biblical terms. Unlike scholia, they use the generally artificial and fictitious method of question and answer, introduced (esp. the question) by stereotyped formulae repeated with tiresome monotony. Before being applied to *Scripture, the genre was widely used in secular literature. Aristotle wrote a work entitled Ζητήματα καὶ λύσεις and, before him, sophists and rhetors had used it in schools to spur disciples to resolve the numerous questions raised by texts, particularly the Homeric poems.

The first to compose "Quaestiones" on the Bible seems to have been *Philo of Alexandria, whose *Problems and solutions on Genesis and Exodus* (Euseb., *HE* II, 18, 1) commented on some chapters of those texts. We do not know which Christian writer first had the idea of using this exegetical genre. None of *Origen's works comes quite into this category, though he sometimes, in homilies and commentaries, introduces an imaginary interlocutor to propound a difficulty which is then given its solution. To *Eusebius we owe a specific work, *Evangelical questions and solutions*, of which we possess some Greek and *Syriac fragments and, fortunately, a later summary which allows us quite a precise idea of the original and its contents. The work is in two parts: books I and II dealt with problems relating to Jesus' infancy; book III with the resurrection accounts. Of the *Diverse questions* of *Acacius, Eusebius's disciple and successor in the see of *Caesarea, thanks to *Jerome (*Ep.* 119, 6) we have a long fragment on 1 Cor 15, 51; from it we deduce that Acacius's work also dealt with biblical questions.

Of the *Questions on the Old and New Testaments* handed down by some MSS under Augustine's name, we have three recensions respectively comprising 151, 127 and 115 questions. On the basis of linguistic, stylistic and conceptual arguments, the work has been claimed for the author of the commentary on St *Paul's epistles known as *Ambrosiaster. The *Questions* are of various natures: some are real treatises developing all the aspects of a problem, some are limited to clarifying a particular problem rapidly, some are aimed at enemies of the church or believers holding dangerous opinions, some are like *sermons dedicated to the interpretation of a *psalm or a biblical character. The author shows a good knowledge of classical and Christian writers: Souter, the greatest expert on the *Questions*, numbers among them Cicero, Sallust, Livy, *Irenaeus, *Tertullian, *Cyprian, *Victorinus of Petovium, *Hilary of Poitiers, etc.

Various *letters of *Jerome, dedicated to resolving scriptural difficulties on which that learned *Father had been consulted, also come into the genre of "Quaestiones". Such are *Epp.* 35, resolving five questions on *Genesis raised by pope *Damasus, 36, addressed to the same *pope, 120 and 121. At the beginning of his exegetical activity, Jerome also composed a specific work, *Quaestiones Hebraicae in Genesim*: in it he compares the Hebrew and LXX versions through 220 questions, which constitute a real commentary on the whole book in the form of scholia.

Several of *Augustine's works are in this genre: *Quaestionum in Heptateuchum libri 7; De octo quaestionibus ex Veteri Testamento; Quaestionum evangeliorum libri 2; Quaestiones 17 in Ev. secundum Matthaeum* (on whose authenticity there are some doubts); *Expositio 84 propositionum ex epist. ad Romanos; De diversis quaestionibus 83 liber unus; De diversis quaestionibus ad Simplicium libri duo; De octo Dulcitii quaestionibus liber unus*. Some scholars question the placing of these works in the genre of "Quaestiones" on the rather specious grounds that they resolve not imaginary difficulties, but real problems that occurred to Augustine and his correspondents. As for character and content, some of them are quite short, some of considerable length; some are a sort of systematic inventory of the difficulties presented by the study of a sacred book, others are confined to resolving a few problems.

The first of *Eucherius of Lyons's two books of *Instructiones* are dedicated to the interpretation of selected biblical passages, from *Genesis to *Revelation, in dialogue form. The *Instructiones* are dedicated to his son *Salonius, bishop of Geneva, to whom have been attributed - but not indisputably (cf. Salonius) - four *opuscula* resolving, through a close-packed concatenation of questions and answers, questions on selected passages of Proverbs, Ecclesiastes, *Matthew and *John. The *Instructiones* and Salonius's *opuscula* offer few personal thoughts, but use, indeed sometimes plagiarize, earlier works. Lack of originality also characterize *Isidore of Seville's *Expositiones mysticorum sacramentorum seu Quaestiones in Vetus Testamentum* (which belong more to the genre of scholia than to "Quaestiones" proper) and the *De Veteri et Novo Testamento quaestiones*, convincingly claimed for the same Isidore. The Venerable *Bede's *Quaestiones in librum Regum* show the same lack of originality.

Three writings that appear among *Justin's works, *Quaestiones et responsiones ad orthodoxos, Quaestiones christianorum ad gentiles* and *Quaestiones gentilium ad christianos*, are in the genre of "Quaestiones". Their Justinic paternity being rightly discarded, both *Diodore of Tarsus and *Theodoret of Cyrrhus have been suggested, though with no real agreement for either, as author of these and of the *Confutatio quorundam Aristotelis dogmatum*. At present we can only say that the four works are all from the same hand and prob. from the mid 5th c. We possess two recensions of the *Quaestiones et responsiones ad orthodoxos*, one of 146 questions, the other - prob. the original - of 161. The questions are extremely varied and only partly on *Scripture; other subjects include *apologetics, *dogma, *ethics, *exegesis, *liturgy and natural science. The interlocutors are equally varied: *catechumens, educated *pagans, disciples. Ps.-Justin's other two opuscula, *Quaestiones christianorum ad gentiles* and *Quaestiones gentilium ad christianos* deal with philosophical problems and are outside our field.

*Theodoret of Cyrrhus also did work in this genre: *Quaestiones in Octateuchum* and *In libros Regnorum et Paralipomenon*. Both are addressed to Hypatius, who had insistently asked the author for explanations of difficult passages of the *Bible. Rightfully, only the *Quaestiones in Octateuchum* belong to this genre, since the other work - a sequel of the first - esp. the questions on the *Chronicles*, is really a commentary in the form of scholia.

*Hesychius of Jerusalem is the probable author of the *Collection of objections and solutions*, which seems to be a summary of his lost *Evangelical concordance*. The *Collection* is a sort of concordance intended to illustrate, through questions and answers, 61 evangelical problems on Jesus' public life, passion and death.

To *Maximus Confessor we owe at least two collections of scriptural questions: *Quaestiones ad Thalassium* (PG 90, 243-786) and *Quaestiones et dubia* (PG 90, 785-856). In the first, the author explains 65 questions to

*Thalassius, at his explicit request. In the second, some of the problems posed are unrelated to Scripture. On the whole, this second work is simpler than the other and corresponds better to the classical type of scriptural questions. But the questions to Thalassius are characterized by long considerations, undoubtedly more important, though less conformable to the genre.

Of the works of *Anastasius the Sinaite, the only one that interests us is entitled *Quaestiones et responsiones* (PG 89, 312-824). In current editions it comprises 154 questions; but the MSS do not all attest the same number, and it is certain that some of the 154 published do not belong to Anastasius. Apart from scriptural problems, they deal with dogmatic, moral and liturgical subjects.

G. Bardy, La littérature patristique des "Quaestiones et Responsiones" sur l'Ecriture Sainte, *RBi* 41 (1932) 210-236, 341-369, 515-537; 42 (1933) 14-30, 211-229, 328-352.

C. Curti

QUAILS, Miracle of the: *iconography. Judging by surviving monuments, the biblical episode of Ex 16, 13, repeated in Num 11, 13-14, was not common in the repertoire of ancient Christian art. The best-known example is on a *sarcophagus front in the Camposanto Monumentale at Pisa, where the scene appears in a group of men, women and children, with arms raised to catch the birds or kneeling towards great birds in the foreground. In other sculpted monuments - front of a sarc. lid in the Musée Lapidaire at Avignon, side of a sarc. from *Arles, now in the Musée de Beaux-Arts at Aix-en-Provence, a fragment of sarc. lid found at Trinquetaille with three groups of people divided by trees, perhaps a sarc. front in the Civic Museum of Carcassonne - the scene is essentially represented by the same iconography and, as at Pisa, shows associations with the depiction of the Crossing of the Red Sea. The episode is present in the *mosaic decoration of S. Maria Maggiore in *Rome; recently, some would recognize it in a fresco of the Roman *cemetery of the Jordani, where a person, apparently naked, holds in his left arm a bag in which to put the birds towards which he advances. The monuments span the period from *Constantine (Pisa sarc.) to *Sixtus III (mosaic of S. Maria Maggiore). **[Fig: 261]**

For the Pisa sarcophagus cf. P. Arias, in *Camposanto Monumentale di Pisa. Le antichità*, Pisa 1977, 168, figs. 244-245 (the list is not exhaustive and contains errors of interpretation); for sculptures: F. Benoit, Fragments de sarcophages inédits, *RAC* 17 (1940) 266 ff.; for the mosaic, in particular C. Cecchelli, *I mosaici della basilica di S. Maria Maggiore*, Turin 1956, 157 f., pl. 33; G. Matthiae, *Mosaici medievali delle chiese di Roma*, Rome 1967, 106; J. Wilpert-W.N. Schumacher, *Die römischen Mosaiken der Kirchlichen Bauten vom IV-XIII. Jahrhundert*, Freiburg-Basel-Vienna 1976, pl. 40; U. Fasola, Le recenti scoperte nelle catacombe sotto Villa Savoia. Il "coemeterium Iordanorum ad S. Alexandrum", *Actas VIII Congreso Int. Arqueol. Crist.*, Vatican City-Barcelona 1972, 283, pl. CXII, fig. 11.

L. Pani Ermini

QUARTODECIMANS. The ancient *heresiologists called Q. those Christians who, following the Johannine *chronology of the Passion, celebrated *Easter on the 14th day of the first moon of spring, i.e. the date of the Jewish Passover, 14th Nisan in the Jewish calendar, and broke their penitential *fast on that date. In the 2nd c. the Quartodeciman practice was followed by the churches of *Asia Minor which followed the Johannine tradition (cf. Euseb., *HE* V, 24, 2-8). Among the best-known exponents of this practice are *Melito of Sardis, *Apollinaris of Hierapolis, *Polycarp and *Polycrates of Ephesus. The Quartodeciman Easter was centred on the celebration of Christ immolated as the true Paschal *lamb; this liturgical conception inspired the etymological interpretation of πάσχα from πάσχειν "to suffer", dear to Quartodeciman tradition; particularly important in the *liturgy of the Easter vigil was the reading of and typological commenting on Ex 12: its typological interpretation must have been open to eschatological views. At the end of the 2nd c., a presbyter named Blastus tried to establish the Quartodeciman practice at *Rome (cf. Euseb., *HE* V, 15 and V, 20, 1; Ps.-Tert., *Adv. omn. haer.* 8, 1), arousing the reaction of pope *Victor, who entered into open conflict with the Eastern churches which observed this practice. Victor threatened to *excommunicate the Q. (cf. Euseb., *HE* V, 23, 3): we do not know if he carried out his threat. What is certain is that the Q. are listed among the *heretics by *Hippolytus of Rome (*Refut.* VIII, 18) and Ps.-Tertullian (*Adv. omn. haer.* 8, 1) and recur among the heretical groups cited by *Epiphanius (*Pan.* 50, 1), *Theodoret (*Haer. fab. comp.* III, 4) and *Filaster (*Haer.* 30). But after the council of *Nicaea's decision on the *Sunday celebration of the Christian Easter, the Quartodeciman observance was restricted to marginal groups of Middle Eastern Christians (cf. Epiph., *Pan.* 50, 1, 5-8).

F.E. Brightman, The quartodeciman question, *JThS* 15 (1924) 254-270; C.C. Richardson, The Quartodecimans and the synoptic chronology, *HThR* 23 (1940) 177-190; B. Lohse, *Das Passafest der Quartadecimaner*, Gütersloh 1953; R. Cantalamessa, *L'Omelia "In S. Pascha" dello Pseudo-Ippolito di Roma*, Milan 1967, 67-95; W. Huber, *Passa und Ostern. Untersuchungen zur Osterfeier der alten Kirche*, Berlin 1969; R. Cantalamessa, *La Pasqua della nostra salvezza*, Casale Monf. 1984 (repr.), 147-156; A. Strobel, *Texte zur Geschichte des frühchristlichen Osterkalenders*, Münster 1984; S.G. Hall, The Origins of Easter, *SP* 15.1 (1984) 554-567.

V. Loi

QUINTASIUS of Carales. The first bishop of Carales (Cagliari) whose name is known: he subscribed the canons of the council of *Arles in 314 with his presbyter Ammonius (Mansi 2, 477).

A. Pollastri

QUIRICUS of Barcelona. Bishop of *Barcelona (*Barcino*) c.654-c.666. Took part in the 10th council of *Toledo (656) and probably, as an *abbat, in the 8th (653). Maintained friendly relations with *Taio of Saragossa, exhorting him to publish his books of *Sentences*. Wrote two *letters to *Ildefonsus of Toledo: the first to thank him for a copy of the treatise *On *Mary's Virginity*, which he considers rich in edification and spiritual consolation. A *hymn in honour of St *Eulalia of Barcelona is attrib. to him.

CPL 1271-1273; Díaz 211-213; PL 80 729-730; 96, 193-196; J. Tarré, Santa Eulalia de Barcelona, *Vida cristiana* 2 (1915) 122; A. Fábrega, *Santa Eulalia de Barcelona*, Rome 1958, 45-53.

M. Díaz y Díaz

QUODVULTDEUS. Succeeded *Capreolus as bishop of *Carthage: perhaps the same deacon of that city who asked and obtained from *Augustine the treatise *On heresies* (PL 42, 21-50). He had been bishop for hardly two years when *Genseric conquered Carthage, 19 Oct 439: he saw the *Catholic churches confiscated and he himself was exiled. He found refuge with bishop Nostrianus of *Naples, where he took part in the anti-*Pelagian struggle. His *depositio* appears in the *Cal. Carth.* on 8 Jan, in the Marble Calendar of Naples on 19 Feb. A tomb with the portrait of an African bishop, found in the catacomb of S. Gennaro, may be Q.'s. He had been replaced at Carthage by 454, so must have died shortly before that. A collection of the works that we can attribute to him has recently been published: *1) *De promissionibus Dei*; 2)* 13 *sermons against *Jews, *pagans and *heretics, to *catechumens and candidates for *baptism; the two *letters of the deacon Q. to Augustine.

SCh 101 (intro.); CCL 60; *LTK* 8, 856; *PWK* 24, 1396-98 (n. 1); V. Saxer, *Saints anciens d'Afrique du Nord*, Vatican City 1979, 184-86.

V. Saxer

R

RABBULA of Edessa. The outlines of R.'s life and work can be traced from his *biography, written in *Syriac by one of his clerics (name unknown). Born at Qennesrin (Calcis), not far from Aleppo, his father was a *pagan priest, his mother a Christian. He studied Greek and Syriac and began a career in civil administration. Attracted to Christianity, his religious search was helped by Eusebius, bishop of Qennesrin, and Acacius, bishop of Aleppo. One tradition has him baptized in the Jordan during a *pilgrimage to *Palestine. He left his wife and children, sold his goods for the *poor and embraced monastic life in the convent of a certain Abraham, later withdrawing to eremitical solitude (through mere homonymity, the episode of R.'s *conversion has been interpolated into the life of *Alexander the Acoemete).

On the death of bishop Diogenes of *Edessa, R. was designated his successor. His ardent faith and intrepid zeal made him an energetic fighter against *heresies. He initially favoured the doctrine of *Theodore of Mopsuestia, widely spread by *Nestorius, but at the council of *Ephesus (431) - if he was there himself - he at first supported *John of Antioch, the declared enemy of *Cyril of Alexandria, but soon changed sides and became Cyril's supporter and *friend, *excommunicating Theodore of Mopsuestia; John of Antioch did the same to him. At *Constantinople, in the presence of the emp. *Theodosius II, R. preached a *homily clearly aimed against Nestorius. This partly survives in a Syriac tr. made by R.'s biographer, but of the 46 *letters mentioned by the biographer, which R. wrote "to priests, princes, notables and monks", only part are so far known. Among others, he corresponded with Cyril of Alexandria and *Andrew of Samosata.

His austerity and rigour, to others as to himself, and his ardour in fighting Nestorianism, made him few friends. He was more feared than loved, even by his congregation and *clergy, headed by his future successor *Ibas. But he was of unbounded charity, esp. to the poor. He died 7 Aug 435 and was buried next day.

His literary and doctrinal activities include a Syriac tr. of Cyril's *De recta fide* to the emp. Theodosius, of which the author had given him a copy. He prob. translated other works of Cyril. He appears to be the author of a still unpublished homily on the legitimacy of the cult of the *dead. His activity as a reformer and legislator is known from a series of disciplinary canons for priests and monks.

R.'s possible role in the replacement of the Syriac *Diatessaron* by translations of the separate gospels (the simple version or Pesitta) is debated. It may be worth pointing out, finally, that R. has absolutely nothing to do with the homonymous copyist of the famous Syriac evangelary in the Bibliotheca Laurentiana at Florence, made in 586 (the oldest known decorated Syriac MS), known as the *Codex of Rabbula*.

Duval 339-341 (and *passim*); Baumstark 71-73; Chabot 45-47; Ortiz de Urbina 89-91; G.G. Blum; *Rabbula von Edessa, der Christ, der Bischof, der Theolog*, CSCO 300, subs. 34, Louvain 1969.

J.-M. Sauget

RADEGUND. Daughter of king Berthar of Thuringia, R. was taken as a hostage to France after the defeat of the Thuringians in 531. It is uncertain whether she was a Christian then, but she certainly was when she married Lothar I in 538. When her husband had her brother Lothacar murdered in 555, she fled first to St Médard at Noyon (*Noviomagus*), who consecrated her a nun (*diacona*), then to her possessions at Saix (*Vienne), finally to *Poitiers (*Pictavium*) where she founded a female *monastery, which took the name of the Holy Cross when she obtained a relic of the Cross from *Justin II in 569. On that occasion *Venantius Fortunatus composed the *hymns *Pange lingua* (*Repertorium Hymn.* 14, 475) and *Vexilla Regis* (*op. cit.*, 21, 481). The monastery followed the *Rule* of St *Caesarius of Arles. R. died 13 Aug 587 and was buried in Notre-Dame, which later took the name of St Radegund. Venantius Fortunatus composed a first *biography of her (BHL 7048), the nun Baudonivia a second (BHL 7049) and Hildebert of Lavardin a third, in the 12th c. (BHL 7051).

DACL 14, 2044-55; *Vies des SS.* 8, 227-34; *LTK* 8, 963; *BS* 10, 1347-52; J. Leclercq, *Sainte Rodegonde de Venance Fortunatus et celle de Baudovinie, Essai d'hagiographie comparée*, *Fructus Centesimus* (Festsch. Bartelink), Steenbrugge 1989, 207-216.

V. Saxer

RATIONES SEMINALES. We first meet the concept of r. s. (λόγοι σπερματικοί) among the *Stoics, who thus explained the origin and development of the *world. Broken into so many particles, the Logos is present everywhere, organizing the *cosmos and creating a sympathy between all things. This gives an understanding of prophecy, *providence and life in conformity with man's nature. In patristic literature (Lampe 1248 f.), the theme is taken up in various forms: by *Justin, to show that all men, participating in the Logos, know at least part of the truth (*II Apol.*, 10); by *Origen, to make us understand the struggle of opposed forces in man (*Com. Jo.* 20, 2-5); by *Methodius, to demonstrate the *resurrection (*Res.* 1, 24), by *Eusebius of Caesarea, to define the notion of Logos (*Ecc. theol.* 2, 13); by *Augustine, to emphasize the nexus between natural causality and the divine will that created all things in a single instant (*Gen. Litt.*).

M. Pohlenz, *Die Stoa*, Göttingen ³1964; G. Faggin, Rationes seminales, *Encicl. Filosof.* 2 (1967) 549 f.; *BA* 48 (1972) 653-668, 685-690; *Historisch. Wörterbuch Phil.* 5 (1980) 484-489.

B. Studer

RAVENNA

I. City and archaeology - II. Councils.

I. City and *archaeology. Rising on a group of sandy knolls (the old shoreline), R. was in immediate contact with the open sea and defended by swamps that cut it off from *terra firma*. Because of its singular geographical position, which rendered it well-nigh impregnable, the city, already a flourishing commercial emporium, became, after the Roman conquest of the Po Valley, a strategic bulwark of considerable importance. To improve the defence of the Adriatic and E Mediterranean, *Augustus built S-E of the city the famous military port of Classis, base of a praetorian fleet of 250 ships, and joined it to the S branch of the Po by a canal that crossed R.: the *fossa Augusta* thus assured the city's function as a nexus between the Adriatic routes and the internal navigation of the lagoons and the Po. From then on, the busy life of the shipyards also developed, as we see from the sepulchral stele of Publius Longidienus, carpenter of the fleet (*faber navalis*) shown, axe in hand, near a ship under construction. Early in the 2nd c. *Trajan built an aqueduct that conveyed to R., poor in water resources, the waters of the nearby Appenines. The city continued to grow in importance and population, extending beyond the old municipal *oppidum* into the area later called Regio Caesarum. Considerable growth of the urban centre, enlargement of the circuit of the walls and vast building activity began in 402 when, faced with the danger of *Alaric's invasion, *Honorius, for strategic safety, transferred the capital of the Western empire from *Milan to R.: then the city started to assume the magnificent aspect of an imperial residence. In the new additions to the built-up area, superb *church buildings arose, decorated inside with splendid *mosaics, like the great Ursian cathedral pulled down in the 18th c., the adjacent baptistery decorated with stucco and magnificent mosaics by bishop Neon (451-475), the so-called Mausoleum of *Galla Placidia attached to the church of S. Croce, the Placidian basilica of St John the Evangelist, built by the Augusta to fulfil a vow made during a storm at sea while she was returning to R. from the *East (424), and that dedicated to the *Apostles by bishop Neon - now called S. Francesco, near Dante's tomb.

During this period the Ravennate church, whose history is often linked with the city's fortunes, was assuming a role of great importance. At the end of the 4th c. bishop Ursus transferred the episcopal see from Classis to the urban centre, where he built the first intramural basilica, mentioned above.

7th- and 8th-c. traditions, which place the origins of Christianity and the first see of the Ravennate bishops in the Roman port of Classis, have been confirmed by the discovery in that area of some cemeterial *areae*, partly used by Christians, as we see from the few but important *epigraphs found there, together with many inscriptions of sailors, perhaps reused. From the "Probian" *area* come the epigraphs of Flavius Anastasius, Valeria Maria and Antifons, now in the Museo Arcivescovile; particularly the last two, if we accept their Christian character - as seems reasonable - date the first Christian settlement of the area of Classis and all Aemilia to the end of the 2nd c. or the first decades of the 3rd, among the inhabitants of the *oppidum Classis*, mainly merchants and seamen.

Little is known of St *Apollinaris, the first bishop and sole *martyr of the

Ravennate church, almost surrounded by a halo of legend, beyond the scanty information handed down by *Serm.* 128 of *Peter Chrysologus, R.'s oldest ecclesiastical writer and the first bishop of R. to be invested with metropolitan *authority, *c.*430. Some information also comes from the old *episcopal list, which survives authentic and entire: between Apollinaris and *Severus, present at the council of *Sardica (343), ten bishops succeeded each other in the see of Classis and were buried in those sepulchral *areae* outside the walls of the *castrum* where Probus I or, acc. to others, Probus II built the first disputed cathedral and *Maximian later consecrated the majestic basilica of St Apollinaris (549).

After the fall of the Western empire in 476 and the tragic end of the experiment of *Odoacer, the first of the Romano-barbarian kings in *Italy, R. knew its second moment of political greatness and building activity during the Gothic kingdom of *Theodoric (493-526). A wise and enlightened monarch, in the first decades of his reign he tried to follow a policy of peaceful coexistence between victors and vanquished. He restored Trajan's aqueduct; he erected his residence, the *Palatium* depicted in mosaic on the right wall of S. Apollinare Nuovo, the palatine basilica originally dedicated to the Saviour and built by Theodoric next to his court; for his *Arian people he built their own churches, such as the *Anastasis Gothorum* (today Santo Spirito), which functioned as a cathedral with its own baptistery.

Following *Justinian's conquest of Italy, the city, now the seat of the exarch who held civil and military power in the name of the Eastern emperor, had its third period of political power and civic splendour. The *Byzantines vigorously reaffirmed the dogmas of *orthodoxy, reconciled the Arian temples to *Catholic worship and brought into the city the resources of the *East, adorning buildings with marbles from Proconnesus and *mosaics executed by workmen perhaps trained at *Constantinople. Linked with the more important monuments of this phase, among them the basilica of S. Vitale and the imposing basilica of S. Apollinare in Classe, are the names of Julianus Argentarius, probably Justinian's emissary to R. to reconcile political opinion to that prince, and *Maximian of Pola who, destined to fight tenaciously against the insurgent *schism of the *Three Chapters, was the first to assume the title of *archbishop. A tireless builder and reorganizer of his church, he must be considered, with Chrysologus, one of Apollinaris's most prestigious successors and the founder of that power of the Ravennate church whose struggles and claims against the Roman pontificate were a prelude to its proclamation of autocephaly. It was bishop Maurus, in 666, who obtained recognition from the Eastern emperor of his hierarchical independence from *Rome on the basis of his see's presumed *apostolicity, derived from a legendary *passio* which made St Apollinaris a disciple of St *Peter. [Fig: 262]

EAA 6, 608-641; G. Bovini, *Saggio di bibliografia su Ravenna*, Bologna 1968; other bibl.: *FR* 107-108 (1974) 253-257; *CCAB* 1965 ff.; F. Lanzoni, *Le diocesi d'Italia dalle origini al principio del secolo VII*, II, Faenza 1927, 723-767; F. Farioli, Chiarificazione sulla topografia delle necropoli pagane e delle aree cimiteriali cristiane nella zona di Classe (Ravenna), *Atti I Congr. inter. archeol. dell'Italia sett.*, Turin 1963, 79-92; G. Bovini, Il problema della ricognizione archeologica del "portus Augusti" di Ravenna e del "castrum Classis", *op. cit.*, 69-78; id., Memorie cristiane scomparse dell'antica città di Classe, *CCAB* 12 (1965) 105-113; G. Cortesi, *Il porto e la città di Classe*, Alfonsine 1967; G. Bovini, *Edifici di culto di età paleocristiana nel territorio ravennate di Classe*, Bologna 1969; F.W. Deichmann, *Ravenna. Hauptstadt des spätantiken Abendlandes*, Wiesbaden 1969-1974, ii.3, *Geschichte, Topographie Kunst und Kultur* 1989; *Corpus della scultura paleocristiana, bizantina e altomedievale di Ravenna*, 3 fasc., ed. G. Bovini, Rome 1968-1969; R. Budriesi, *Le origini del cristianesimo a Ravenna*, Ravenna 1970; G. Zattoni, *Il valore storico della "Passio" di S. Apollinare e la fondazione dell'episcopato a Ravenna e in Romagna, Scritti storici ravennati*, Ravenna ²1975; G. Bovini, *Eglises de Ravenne, Churches of Ravenna*, Novara 1960; A. Guillou, *Régionalisme et indépendance dans l'empire Byzantinia an VIIe siècle. L'example de l'Exarchat et de la Pentapole d'Italie*, Rome 1969; J. Richards, *The Popes and the Papacy in the Early Middle Ages, 476-752*, London 1979, 176-213; T.S. Brown, *Gentlemen and Officers; imperial administration and aristocratic power in Byzantine Italy*, London 1984; O.G. von Simson, *Sacred Fortress; Byzantine Art and Statecraft in Ravenna*, Princeton 1987; M. Cecchelli-G. Bertelli, Edifici di Culto ariano in Italia, *Actes XIe Cong. Int. d'arch. chrét.*, CEFR 123, Rome 1989, 233-247; P. Testini et al., La cattedrale in Italia: Schede, Ravenna, *ibid.*, 140-142 (bibl.); C. Rizzardi, Note sull'antico Episcopo di Ravenna; formazione e sviluppo, *ibid.*, 711-732; J.-C. Picard, L'Atrium dans les Eglises paléochrétiennes d'Occident, *ibid.*, 549-551 (bibl.); E. Dinkler, *Das Apsismosaik von S. Apollinare in Classe*, Cologne 1964.

G. Cuscito

II. *Councils. 419.* After the death of pope *Zosimus, a small group at *Rome elected Eulalius, while the majority elected Boniface (418-422). The emp. *Honorius, then residing at R., convened a council early in February 419 to resolve the question. Both contenders were present, but the council failed to reach a solution because of discord between the bishops taking part. The emperor then charged *Achilleus of Spoleto to celebrate the *Easter rites at Rome and, soon after, due to the insubordinate attitude of Eulalius, recognized Boniface.

Mansi 4, 399; Hfl-Lecl 2, 211; Palazzini 4, 42 f.; *DTC* 2, 988 f.; *EC* 2, 1863 f.; Fliche-Martin 4, 315-317; A. Simoncini, *La Chiesa Ravennate*, Ravenna 1964, 165-169.

499. In 498, on the death of pope Anastasius II, his successor *Symmachus was opposed by *Lawrence, supported by the senate and a minority of the people. Both pretenders were then called by *Theodoric to R., where archbishop Peter convened a council at which Symmachus was recognized. Lawrence submitting, Symmachus made him bishop of Nocera (*Nuceria Alfaterna*).

EC 11, 629 f.; Fliche-Martin 4, 424 f.; A. Simoncini, *La Chiesa Ravennate*, Ravenna 1964, 170-174.

A. Di Berardino

RAVENNIUS of Arles. *Hilary's successor in the see of *Arles, whose cause he had defended at *Rome before *Leo I while still a priest. Elected unanimously by *clergy and people after Hilary's designation (Leo, *Ep.* 40), the *pope sent him his *Tomus ad Flavianum* to hand on to the bishops of *Gaul (*Ep.* 67). He presided over the *council of Arles in 453. Died 461.

E. Griffe, *La Gaule chrétienne à l'époque romaine*, II, Paris ²1966, 145 ff. and *passim*; Y.-M. Duval, *La Gaule jusqu'au milieu du Ve siècle*, Paris 1971, 775.

A. Hamman

REBECCA: *iconography. The only known scene of R., Isaac's wife (Gen 24 ff.), is thought to be that in the Mausoleum of the Exodus at El-Bagawat in *Egypt (early 5th c.). It shows R. and Eliezer at Nahor's well (Gen 24, 11 ff.). R. (captioned PENBEKA), facing the well, gives a drink to Eliezer, next to whom are two camels and a servant, who leads a laden ass. However, R. often appears in miniatures. Among the oldest is the 6th-c. Vienna Genesis, where she appears on three pages: in VII, 13, R. comes from Nahor with a pitcher on her shoulders and, in the lower part, gives a drink to *Abraham's servant, who has ten camels. In VII, 14, the servant gives R. a gift of gold bracelets; in the lower register, to the left he questions her and, inset on the right, he talks to Laban. Finally in VIII, 16, in the scene of Isaac and Abimelech, R. stands next to Isaac, who has the two Gerarenes behind him; below, she embraces Isaac beneath Abimelech's palace. In the Latin Tours (Ashburnham) Pentateuch in the Bib. Nat., Paris (7th c., inspired by a 6th-c. model), f. 22v shows the episode of R. giving birth; beside her stand three figures.

EC 10, 601; *LCI* 3, 503-504; Garrucci, *Storia* 3, pl. 115, 1, 2, 4; G. von Gerhardt, *The Miniatures of the Ashburnham Pentateuch*, London 1883; W. von Hartel-F. Wickhoff, *Die Wiener Genesis*, Vienna 1895, 150-151: pll. XIII, XIV, XVI; H. Stern, Les peintures du Mausolée de l'Exode à El'Bagawat, *CArch* 11 (1960) 93-119; A. Grabar, *Christian Iconography*, Princeton 1968, fig. 308; M.L. Thérel, La composition et le symbolisme de l'iconographie de l'Exode à El'Bagawat, *RAC* 45 (1969) 228-270.

M. Perraymond

RECAPITULATION. Following the classical and biblical use of ἀνακεφαλαίωσις (Rom 13, 10; Eph 1, 10) and keeping closely to the Pauline theme of the two Adams, *Irenaeus worked out his famous doctrine of r. in the framework of his presentation of salvation-history, further characterized by the concepts of οἰκονομία (*economy), mutual adaptation of *God and man, progress and education. By ἀνακεφαλαίωσις he understands the *incarnation as it sums up and fulfils the whole of man's earlier history, the institution of Christ as head of the whole universe, the fact that Christ and *Mary, by their obedience, have repaired the disobedience of *Adam and Eve. Though later authors did not take up this theory of *soteriology entire, they took inspiration from it, esp. the idea of the ineffable exchange between sinful man and God the saviour.

RACh 1, 411-414; J. Daniélou, *Message évangélique et culture hellénistique*, Paris 1961, 156-169.

B. Studer

RECARED. King of the Visigoths of *Spain, he succeeded his father *Leovigild who, dying, entrusted him to bishop *Leander of Seville to be instructed in the *Catholic faith. Already shaken by the violent death of his brother *Hermenegild, who had rebelled against their *Arian father, he abjured *heresy, by his example led his subjects to embrace *orthodoxy and called the 3rd council of *Toledo (589). The addresses he gave there were probably written by Leander. His *conversion marked a decisive step in the political and religious history of Spain. Died at Toledo in 601.

Historia de la Iglesia en España, directed by R. García Villoslada, I, Madrid 1979, 401-413; J. Fontaine, Conversion et Culture chez les Wisigoths d'Espagne, in *La Conversione al Cristianesimo nell Europa dell'Alto Medioevo*, Settimane di Studio 14, Spoleto 1967, 87-147; E.A. Thomson, *The Goths in Spain*, Oxford 1969, ch. 4.

L. Navarra

REDEMPTION. In modern languages, under the terms either derived from the Latin *redemptio* or literally translating it (Ger. *Erlösung*), we commonly understand the whole of Christ's saving work, effected especially by his death (*DTC* 13/2, 1912). In a strict sense, however, r. is equivalent to the ransom paid by Christ to free men from the forces of *evil (devil, *demons, cosmic powers). But since behind the metaphor of ransom - taken from military language (prisoners of war) rather than that of mercantile law (slavery) (Elert) - lies the idea of liberation, we must also take into consideration the other categories and images that evoke the same idea of liberation, such as victory over demons, tricking the devil, abuse of power by the devil, man's "return match" on equal terms, etc. This is all the more legitimate in that, in patristic literature itself, all these motifs are continually interwoven, not just with each other, but with other soteriological themes (cf. *Soteriology, *Expiation, *Divinization, *Illumination). It is also significant that in Latin *redemptio* and *redimere* do not include the price of ransom in texts before the 4th c. (Braun, *Deus*, 506-511).

As the question of Christ's victorious supremacy is quite often mentioned in the NT itself, in the gospels (Lk 10, 18; Jn 12, 31; 14, 30) or in the *corpus paulinum* (1 Cor 2, 8; Eph 2, 2; Col 2, 15), it is not surprising that the theme soon appears in patristic writers. In the 2nd c., so dominated by dualistic tendencies and so impressed by Judaeo-Greek demonology, *Justin attaches great importance to stressing the victory obtained by Christ over the *demons, both in his earthly existence, primarily the *temptation and passion, and in the persecuted church (Studer, *Soteriologie*, 69 f.). After him and in part even before him, Christian authors, like Christian *iconography, turn continually to the *mysteries of Jesus, understood as manifestations of his superiority to the demons (Turner, *Christ*, 55-59). This is true principally with regard to his death which put an end to death (Iren., *Adv. haer.* III, 18, 6. 8; Aug., *Tract. Jo.* 12, 11; Leo, *Serm.* 22, 4), his *descent into hell in which the devil's last forces were routed (Grillmeier) and his *ascension as a triumphal return through the heavenly spheres (Daniélou). In the 2nd c., polemic against *Marcion, who opposed the vengeful *God of the OT to the good and merciful God of the NT, led to the combining of the biblical themes of ransom (Mk 10, 45; 2 Pet 2, 1) and the precious *blood (1 Pet 1, 19) (Braun, *Deus*, 506 f.) so as to elaborate the redemptive theory by which God, or rather Christ, paid the just price to the devil to redeem men, who were his prisoners, indeed were sold to him as the price of *sin. This theory, which allows the devil a certain right of acquisition, i.e. which attributes to God, always just, though good, a certain "fair play", is already present in *Irenaeus (*Adv. haer.* V, 1, 1). Joined to the motif of the devil's ignorance (1 Cor 2, 8; Ign., *Eph.* 19, 1-3) and that of his deception (bait, poison), it was fully developed by *Origen (*Hom. Ex.* 6, 9; *Com. Mt.* 16, 8) and *Gregory of Nyssa (*Cat.* 22 f.; Barbel, *Gregor*, 146-149 and Studer, *Soteriologie*, 71 f.). But the idea that God, while recognizing a certain right of the devil over men, tricked him into claiming rights over an innocent man, Christ, who was without sin, and thus losing his dominion over men, was criticized even in the 4th c. by an anon. writer (*Adam. dial.*: PG 11, 1756 f.) and by *Gregory Nazianzen (*Or.* 45, 22). Both in the *East (*DTC* 13/2, 1939 f.) and esp. in the West, a similar theory, that of *abusus potestatis*, is found in patristic literature, esp. in *homilies (Kessler, *Bedeutung*, 76 f.). Saying that the devil's ignorance was mainly ignorance of the *mystery of Jesus' virginal origin, it emphasizes still more strongly the devil's abuse of power over Christ, who was pure from original sin, though fully man and hence in solidarity with the whole human race (Hilary, *Psalm.* 68, 8; Aug., *Trin.* 13, 12, 16-15, 19; Leo, *Serm.* 22, 3 f., etc.). On the other hand, it is significant that *Augustine, followed by *Leo and others, plays with the term *redemptio* (with *pretium, mercator*, etc.), making clear that for him this is a metaphor which, at bottom, means nothing more than the gospel idea of the "stronger" (Mt 12, 29) (*Serm.* 130, 2; *En. Ps.* 147, 16). In the polemical perspective of the 2nd c. (anti-*gnostic and anti-Marcionite), in which it was necessary to defend the unity of the two Testaments, the Pauline *antithesis of the first and second Adam (Rom 5, 12-19) was taken up in the sense of a "return match" (cf. *Recapitulation). He who was conquered in *Paradise must himself, in Christ, conquer his conqueror. Thus Irenaeus (*Adv. haer.* III, 18, 7) and, later, *Augustinian theology (Aug., *Ench.* 108; *Trin.* 13, 17, 22 f.; Leo, *Serm.* 22, 3; 42, 3; 63, 1) and the *liturgy inspired by it (cf. *Miss. Rom., praef. De cruce*: Rivière, *Dogme* 98, n. 1). When in the 4th c. Christianity had finally shown itself victorious over antique *paganism, Christian authors, esp. in their *Easter preaching, expressed Christ's supremacy over all adverse forces in the language of the Roman triumph. Already sketched out in the NT (Col 2, 15; 2 Cor 2, 14; Rev 19, 11-16) and always present in the writings of the first three centuries, esp. in the Acts of the *martyrs, the manner of comparing Jesus' death and *resurrection to a military victory now follows, even in its details, the description of the Roman *triumphus*. Thus in *Ambrose (*Fid.* IV, 1-2; *Luc.* 10, 107-112) and *Leo (*Serm.* 69, 4; 74, 1), the *via crucis* and the ascension are presented as a triumph of the *rex gloriae* (cf. also John Chrys., *Hom. 2 Cor.* 5, 1 and other texts in Lampe 654 f. and Blaise 830 f.). From ancient criticisms of the theory of r., but also from certain reservations advanced by the very authors who had taken it up (Aug., *Trin.* 13, 13, 17), it seems that in some way the categories and images used should be judged as concessions, often excessive ones, both to anti-Marcionite polemic and to popular taste. So we can understand that later *theology largely eliminated this theory, replacing it with that of *substitutio vicaria* (e.g. Anselm). And yet, exception being made for a certain onesidedness of this last theory, we cannot deny that the theology of r., particularly in its essential idea of Christ's victory over the forces of *evil, includes a fundamental evangelical truth, i.e. that God alone, in Christ, leads man, free and created by Himself, but exposed to many external harmful influences, to a free dialogue with himself.

J. Rivière, *Le dogme de la rédemption après s. Augustin*, Paris 1930; *DTC* 13, 1912-2004; W. Elert, Redemptio ab hostibus, *ThLZ* 72 (1947) 265-270; J. Daniélou, *Bible et Liturgie*, Paris 1951, 409-428; R. Braun, *Deus Christianorum*, Paris 1962; J. Barbel, *Gregor v. N., Die grosse katechetische Rede*, Stuttgart 1971; H. Kessler, *Die theologische Bedeutung des Todes Jesu*, Paderborn 1970; A. Grillmeier, *Der Gottessohn im Totenreich: Mit ihm und in ihm*, Freiburg 1975, 76-174; B. Studer, *Soteriologie der Kirchenväter*, HDG 3/2a, Freiburg 1978; H.E.W. Turner, *The Patristic Doctrine of Redemption*, London 1952; id., *Christ the Saviour*, 1965; M.F. Wiles, Soteriological Argument in the Fathers, *SP* 9 (1966) 321-325; J. Roldanus, *Le Christ et l'homme dans la théologie d'Athanase d'Alexandrie*, Leiden 1968; R. Greer, *Captains of our Salvation*, Tübingen 1973; R.C. Gregg - D.E. Groh, *Early Arianism; a view of Salvation*, London 1981; B. Studer, *Dieu Sauveur, La rédemption dans la foi de l'Eglise*, Paris 1988.

B. Studer

REFRIGERIUM. The Latin terms *refrigerium, refrigerare, refrigeratio* occupy a much larger place in the Latin field than does their Greek equivalent ἀναψύξις. Their *Sitz im Leben* must be sought above all in the Latin world; and this renders rather fragile the hypothesis of borrowings from Eastern, esp. Egyptian, vocabulary and religions, advanced by Cumont, van der Leeuw and Parrot. The primary meaning of *refrigerare*, employed by Cicero (*De sen.* 16, 57) and Pliny (*Natur. Hist.* 21, 46), is "to cool", speaking of water and shade. The term is sometimes used in a moral sense: "to cool the ardour of passion". More technical is its use in the cult of the *dead, in which the verb means, acc. to some, "to revive the memory" or, better, to celebrate a banquet "in memory" of the dead (Feltre Inscription). If the verb, and above all the noun, are rare in classical Latin, the Christians make great use of them (in particular *refrigerium; refrigeratio* rarely occurs). We find the word r. in Christian writings and *epigraphy. At the start, it keeps its primitive meaning of refreshing and cooling, in a physical and moral sense. Then it moves over to meaning rest and relief. Later still, it expresses future *beatitude, either provisional, "in *Abraham's bosom" (Gelasian and Gregorian sacramentaries, Roman and Mozarabic liturgies), or final, in *Paradise (*Tertullian, Christian inscriptions: examples in *DACL* 14, 2184-2188). R. finally becomes the technical term for funeral meals, connected with the dead or with their *natale*, originally *pagan but to which Christians remained deeply attached, esp. in *Africa. To this end they built real *cellae*, hypogea for funeral services and banquets in honour of the dead (Tertull., *De monog.* 10; Zen. of Ver., *Tract.* 1, 15, 6).

Before forbidding them, the bishops made efforts to turn these banquets in two directions: banquets for the *poor, and the cult of *martyrs. The word sometimes indicated the charitable meal served to the poor and those on *assistance, and ended by becoming synonymous with *agapè* (with which term it should not be confused). In Africa, at *Tipasa, *mensae* are still visible near the churches of the martyrs. Abuses led to the suppression of the r., first at *Milan, then, but here with greater resistance, in Africa.

P. de Labriolle, Refrigerium, *BALAC* 2 (1912) 214-219; F. Cumont, *Les religions orientales dans le paganisme romain*, Paris 1924; A.M. Schneider, *Refrigerium*, I, Freiburg 1928; E. Buonaiuti, Refrigerio pagano e refrigerio cristiano, *Ricerche religiose* 5 (1929) 60-67; G. van der Leeuw, Refrigerium, *Mnemosyne* 3 (1935) 125-148; A. Parrot, *Le Refrigerium dans l'au-delà*, Paris 1937; J. Quasten, *Vetus superstitio et nova religio*: The Problem of Refrigerium in the Ancient Church of North Africa, *HThR* 33 (1940) 253-266; E. Josi, Refrigerium, *EC* 10, 627-631 (archaeological point of view); C. Mohrmann, *Etudes sur le latin des chrétiens*, II, Rome 1961, 81-92; A. Hamman, *Vie liturgique et vie sociale*, Paris 1968, 202-218; J. Stevenson, *The Catacombs*, London 1978.

A. Hamman

REGINUS. Bishop of Constantia (Salamis) in *Cyprus, he took part in the council of *Ephesus in 431. With his colleagues Zeno of Kurion and Evagrius of Soli he presented the *council with a *libellus that accused the patriarch of *Antioch of having violated the autocephaly of the Cypriot church. In a *homily preached on the same occasion, also surviving in an *Ethiopic version, he favoured the deposition of *Nestorius from the patriarchate of *Constantinople.

CPG 3, 6485 f.: the *Libellus* in Greek: ACO I, 1, 7, pp. 118-119; in Latin: ACO I, 5, pp. 357-358; the *Sermo* in Greek: ACO I, 1, 2, pp. 70-71; in Latin: ACO I, 3, pp. 168-169.

J. Irmscher

REGULA FIDEI. This expression indicated the norm that bounded the limits of the Christian search for truth. It was much used in the first period of theological reflection, when the Christian message was being translated into the categories of Greek thought. The *regula fidei* constituted the methodological rule of research: in the reading of *Scripture (Ps.-Athan., *De Trin.* VII, 5), in what language to use (Aug., *De civ. Dei* X, 23), what *traditions to respect, such as that relating to the date of *Easter (Euseb., *HE* V, 24, 6), or what trinitarian formula to use when administering *baptism (Ps.-Athan., *De Trin.* VII, 10). *Irenaeus calls it *regula veritatis* in *Adv. haer.* (I, 1, 20) and *regula fidei* in *Demonstr. apost.* (ch. 3); *Tertullian mentions it in *De virgin. vel.* 1 (*regula fidei credendi scilicet*), *Adv. Prax.* 3 and particularly in *De praescript.*, where in ch. 13 he enumerates its articles. The problems inherent in the *regula fidei* are: *1)* the precise identification of its contents, which are generally recognized to have a substrate that borders on the baptismal *creeds, since it embraces not just a confession of faith, but also the observance of customs affirmed in the church (*Clement of Alexandria calls it "ecclesiastical canon" or "rule of tradition" [*Strom.* I, 1, 15; VI, 18, 165]; *Origen calls it "preaching of the church" [*In Jo. com.* V, 8]); *2)* a second problem is related to its origin (how it was formed and under what directive norm) and function (a rule of *faith, which is at the same time a summary of it?). The presence of such norms in church tradition also raises the question of the possibility of having *regulae fidei* brought up to date in accordance with transitions of culture and *regulae fidei* peculiar to individual Christian cultures; *3)* finally we must note how the use of *regulae fidei* was increasingly limited to expressing a negative judgment on heterodox Christian ideas, rather than being a regulative rule of faith in the *orthodoxy/*heresy dialectic, i.e. a guiding rule for faith which must always submit to the impact of acculturation and is thus at the service of faith and theological research. A recovery of this meaning of the *regulae fidei* would bring out their positive function rather than the negative one of discrimination (though this is inherent in them), help to create a more appropriate dialectic in the relationship between magisterium and theological research, and propose formulations more meaningful to all believers, not comprehensible only to theologians. The *regulae fidei* of Christian antiquity are not just a backward-looking relic of the safeguarding of the Christian faith, but, as a part of Christian history, have transmitted an essential nucleus of Christian truths, indicating a method to orient research in its continual process of assimilating culture to faith.

C.H. Dodd, *The Apostolic Preaching and its Development*, New York 1944; L.W. Countrymen, Tertullian and the *Regula Fidei, Second Century* 2 (1982) 208-277; W.R. Farmer, Galatians and the Second Century Development of the *Regula Fidei, Second Century* 4 (1984) 143-170; R.P.C. Hanson, The Rule of Faith of Victorianus and Patrick, in *Studies in Christian Antiquity*, Edinburgh 1985.

V. Grossi

REMIGIUS of Rheims. R. is one of the most famous bishops of *Rheims (*Remis*) but, though the fact of having baptized *Clovis has given him a certain celebrity, we know little of his life. *Sidonius Apollinaris congratulated him on his talent as a writer (*Ep.* 9, 7); R. wrote to Clovis to congratulate him on his accession to the throne as king of Tournai (*Turnacum*) in 481 (MGH, *Epp. mer. aev.* I, 113). He catechized him and baptized him on 25 Dec sometime between 496 and 506, in a city generally identified as Rheims. R. also wrote to numerous bishops (*ibid.*, 113-116). Acc. to *Gregory of Tours (*Gl. conf.* 78), he was bishop for 70 years. He appears in *Mart. hier.* on 15 Jan, the day of his *depositio*, and 1 Oct, the day of his *translatio*. At Rheims his *natale was celebrated on 13 Jan. We have two *Lives* of him, one wrongly attrib. to *Venantius Fortunatus, but truthful (BHL 7150), the other written by Hincmar, but of no historical value (BHL 7152-59). His will has been preserved in two versions: one long with *interpolations, the other short with alterations, but substantially trustworthy.

DACL 14, 2331-37; LTK 8, 1226-27; Vies des SS. 10, 13-17; BS 11, 104-13; Duchesne, *Fastes* III, 81-82.

V. Saxer

RESTITUTUS (or RESTUTUS). *Catholic bishop of *Carthage, he succeeded *Gratus before 359, in which year he took part in (and perhaps presided over) the council of *Rimini. An opponent of the formula of *Sirmium of 22 May 359 (the dated *creed), he was designated as head of the delegation of the Nicene majority that went to *Constantinople to give the emp. *Constantius the council documents and a *letter. This delegation, preceded by the pro-*Arian one, was not received but had to stop at Nike in *Thrace, where on 10 Oct 359 it was forced to give way and sign a new formula of faith (formula of Nike).

It was then that R. dictated a text justifying his rejection, and that of the delegation over which he presided, of the Nicene formula of Rimini and his acceptance of *communion with the pro-Arian party he had earlier condemned (cf. Hilary of Poitiers, *Frag. hist.* 8, 5-6: PL 10, 702). R.'s attitude was criticized in *Africa (cf. PLS 1, 303-304), but the episode must soon have been forgotten. We know no more of R.'s life, except that by 390 Geneclius was bishop of Carthage.

Simonetti 314-321; *PAC* 1, 968-969.

A. Di Berardino

RESURRECTION OF THE DEAD. The affirmation of the r. of the dead or of the *flesh, contained in the oldest *creeds, e.g. the *Traditio Hippolyti* (early 3rd c.), the *Dêr-Balyzeh* papyrus (mid 4th c., but from an older tradition), and in professions of faith, emerges as a constant factor of the belief of the early church and generally contains three components: *1)* it is an *eschatological event that will take place "on the last day" at Christ's second *parousia*; *2)* it is universal: everyone will rise; *3)* it simultaneously includes the concepts of bodily identity and bodily newness ("*Resurget non aliud corpus, quamvis in aliud*": Hilary of Poitiers, *In ps.* 2, 41).

This belief was the benchmark of ancient Christianity esp. against Greek thought which, anchored to spiritualism, could not accept the bodily and sensible as part of man's good. Once past the phase of defending the given fact of r., the question became one of specifying the form of the risen *body and how it could be imagined.

In the Apostolic Fathers, the r. of the flesh is often pointed to as the true *life, of which Christ's r. is the beginning, the pledge, the model. *Clement of Rome tries to make it credible by appealing to the rhythms of nature and the story of the *phoenix and basing it on OT citations. *2 Clement* uses it to exhort to *chastity. The main traditional assertions are often contained in passing allusions.

The *Apologists take the same approach, further developed in *Tatian (*Or. ad Graecos* 6 and 13), in *Justin's *Apologies* and *Dialogue*, which contains a millenarian development (80-83), in the fragment of a *De resurrectione* attrib. to Justin, in *Theophilus of Antioch and in *Melito of Sardis.

The first important treatise on the r. of the dead is *Athenagoras's *De resurrectione*: its authenticity, rejected by R. Grant who sees it - we think wrongly - as a reply to *Origen, is upheld by L. Barnard, E. Bellini and J. Vermander, with good reasons. As a *lectio publica*, a conference preached prob. before a *pagan audience to show that *faith in the r. is not repugnant to reason, it is not based on *revelation and contains no scriptural citations, except a few allusions. Constructed on the rules of *rhetoric, it first discards the objections - *God will not or cannot - and then goes on to the positive arguments. The first part shows a rather crude conception of r., its great problem being: if an animal eats a man and then another man eats the animal, to whom will the elements that passed from the first to the second body belong at the r.? The arguments of the second part are more valid, being essentially based on the fact that man is both *body and *soul.

*Irenaeus deals specifically with the r. in book V of *Adversus haereses*. His is a traditional and quite complete doctrine, upheld by apologetic arguments, imbued with a *millenarism expressed above all in a surprising citation of *Papias (V, 33, 3-4), which he fully approves. *Clement of Alexandria does not deal with the r. But *Tertullian dedicates an important treatise to it, *De resurrectione*, written when he was under the influence of *Montanism but had not yet broken with the Great Church. He reacts against the allegorical interpretations of r. given by the *gnostics, particularly by *Marcion and his disciple *Apelles, by *Basilides and *Valentinus. After an introduction attacking the opponents of r. and extolling the dignity of the *flesh (1-17), the first part magisterially expounds the scriptural arguments (18-55) and the second (56-62) explains how that same flesh that suffered will be, by God's working, no longer subject to death and suffering, and examines the problems arising out of this.

*Origen's doctrine of r. is present in nearly all his works: his juvenile *De resurrectione* is lost, apart from a few fragments. This doctrine contains many important and traditional elements, but it was mainly his conception of the risen body, caricatured by later scholars, that caused scandal. This tried to show that between the earthly body and the glorious body there is both an identity and an otherness, using the comparison of the seed and the plant in 1 Cor 15, 34-44. It also observes a fact that seems to have escaped his predecessors, i.e. the fluid character of the earthly body, which cannot be defined by its constantly changing material elements. The glorious body that will be endowed with an ethereal quality is the same earthly body with its coarse quality: the same in substance, but different in quality. The continuity of the earthly body in the different moments of its existence and the continuity between the earthly body and the glorious body are expressed both by the Stoic idea of *rationes seminales - expressing the Pauline comparison in philosophical language - and by the Platonic term *eidos*, "form", the metaphysical principle of the unity and identity of the body.

This term, understood by *Methodius in his *De resurrectione* in its vulgar meaning of external appearance, caused a fundamental error of interpretation whose repercussions would echo down the centuries. Methodius's *millenarism and *anthropomorphism, though refined compared with those of his predecessors, could certainly not accept a doctrine that had been constructed in reaction against millenarism and anthropocentrism.

Among the Valentinian gnostics, the *cosmos and history are the moment of recovery of the "scattered divine seeds", their liberation through *gnosis of the cosmos (=the visible, the material), in a salvation that will consist of the reintegration of all that is divine but has fallen into matter, the moment of *apocatastasis or restoration of everything to its original state: the reconstitution of the scattered members of God (*Ev. veritatis* 41, 28-29). In the Letter to Reginus, r. is gnosis, i.e. the awakening in the gnostic of the spiritual self: the moment of rising from death to ascend to the kingdom of *light, the original state of existence in the *pleroma (in Erbetta, I, 1, 244-248). The anon. author rejects psychic r., i.e. the r. of the *soul, and r. of the *flesh as biological continuity. The sole reality is the spiritual r. of the live members, i.e. the thought, the mind and the *pneuma. The gnostic vision is on the level of the inner man, the spiritual man; that of Origen is similar, but saves the r. of the body; Irenaeus's vision and that of Justin and Hippolytus are set within the vision of apocalyptic Judaism: belief in the millenium, end of the *world as cosmic tragedy, end of the empire and appearance of *Antichrist. Origen goes beyond this *tradition and tends to interiorize the end of the millennium (*Princ*. 2, 11, 2) and the end of the world as a breaking out of cosmic events. Apocalyptic themes are of marginal interest to him. He even speaks of the heavenly *Jerusalem as a provisional place until the number of the just is completed, but it is not the place of millenarian beliefs where the just, risen with Christ, will live for a thousand years before passing into the definitive eternity, so much as a chance for the just to progress in the knowledge of *God. The book of the *Apocalypse, in the different reading of it given by the Asiatics (Irenaeus) and Romans (esp. *Hippolytus) on one hand and the *Origenists on the other, marks the two great approaches to the understanding of the r. of the dead (or of the flesh) within ancient Christianity. In the *East it was read without any concern to see it as a unitary text, and it was oriented towards an allegorical reading of the images and verses; but in the West it was read in the light of the theory of *recapitulation as a succession of real events (*Victorinus of Petovium and *Tichonius). After Origen, Christ's second coming and the r. of the flesh became one of the truths that were handed down without constituting a problem. With the peace of *Constantine, the next life lost its interest and historical *time acquired a place in reflection on the faith (*Eusebius of Caesarea). In the Cappadocians the future has particular value for its ethical interest, given the moment of *judgment that awaits everyone. The West also stressed the problem of *eschatology as a moral problem, linked in particular to the book of *Daniel, where the fourth kingdom is spoken of (2, 40).

Out of all this, *Augustine's contribution emerges as a synthesis between *Asiatic and Alexandrian Christianity. Understanding the *eschaton* as that which is perfect, he sees man's ultimate destiny in terms of his body-soul, a unity that must be saved entire if it is to be happy. He rejects the Platonic theory of man's *beatitude as involving freedom from the *body, which therefore sees a possible r. of the *flesh as an evil (*Ep*. 166, 9, 27), and the ironies of *Porphyry and his followers over Christian belief in the r. (*De civ. Dei* 13, 16, 1). His thesis leans on two arguments: *1)* the body, as such, is no obstacle to felicity, but only the body fallen into mortality and corruptibility as the result of ancestral *sin (*De civ. Dei* 13, 16, 1 and 24; *En. in Ps*. 141, 17-19); *2)* the *soul, the vital principle of the body, does not fall into the body as the result of a fault (*Gen. ad litt*. 7, 27, 38; *De imm. animae* 15, 24), but naturally tends towards it. In fact even after death there remains in it "the natural appetite to rule the body, which in some way retards it and impedes it from tending with all its strength towards the highest heaven" (*Gen. ad litt*. 12, 36, 68). Augustine overcomes the Platonic stumbling-block which sees the flesh as a prison of the soul by setting up the category of the hierarchy between body-soul-God. The soul can fulfil itself only by contact with the body, which, in the scale of beings, has a dignity of its own. Therefore it cannot fulfil itself by repudiating the body, but by restoring its destroyed order. This will take place with the r. of the flesh (*De musica* 6, 5, 13). The risen body will again have its health, but its real state will be known only when it is experienced. So Augustine wastes no time theorizing on the condition of the risen body; for him the *eschaton* is a fact of faith. The drama, for Augustine, is the time of history, the state of human existence before the *eschaton*.

DTC 13, 2504-2572; A. Fierro, Las controversias sobre la resurrección en los siglos II-V, *RET* 28 (1968) 3-21; T. van Eijk, *La Résurrection des morts chez les Pères Apostoliques*, Paris 1974; P. Siniscalco, *Ricerche sul De resurrectione di Tertulliano*, Rome 1966; H. Crouzel, La doctrine origénienne du corps ressuscité, *BLE* 81 (1980) 175-200, 241-266; J.G. Davis, Factors leading to the emergence of belief in the Resurrection of the Flesh, *JThS* n.s. 23 (1972) 448-455; A.H.C. van Eijk, Resurrection-language. Its various meanings in early Christian Literature, *SP* 12 (1975) 271-276; J.E. McWilliam Dewart, *Death and Resurrection*, Wilmington 1986; M. Dimura, The Resurrection of the Body and Soul, Origen, *Contra Celsum*, *SP* 18.3 (1989) 385-392; C.R. Hennessy, The Scriptural and Classical Roots of Origen's theology of Death, *ibid*., 403-4.

H. Crouzel - V. Grossi

RETICIUS of Autun. R. was certainly not the first bishop of *Autun (*Augustodunum*), but he appears as such in the mediaeval *episcopal lists. An educated man of noble family (Greg. of Tours, *Gl. conf*. 74), he made a *commentary on the *Song of Songs and wrote against *Novatian. In 357-77, *Jerome asked for an exemplar of his "sublime" commentary on the Song to make a copy of it (PL 22, 337). Having obtained it, in 385 he found it full of "*ineptiae*"; the author, an "eloquent man", spoke with the magnificence typical of the Gauls, but did not try to make the reader understand what the sacred writer wished to say (PL 22, 461-63). In *De viris illustribus* (82), he judged him more serenely: "Reticius, in the land of the Edui, bishop of Autun, was very celebrated in *Gaul under Constantine. We possess his commentaries on the Song of Songs, and another large volume, but apart from these works I have found nothing of his". His commentary on the Song was still known to Berengarius Scholasticus in the 12th c. (PL 178, 1864-66). The second treatise against Novatian is cited by St *Augustine, who speaks of "the great *authority in the church" of the bishop of Autun, and praises him for publicly admitting original *sin (PL 44, 644; 45, 1078). The two treatises seem to be lost. R. also enjoyed the esteem of *Constantine, who sent him to the councils of *Rome (313) and *Arles (314), which deliberated on *Donatism. He seems to have died *c*.334. *Memoria* in *Mart. hier*. on 15 July; at Autun 20 July. The Bollandists put it on the 19.

Duchesne, *Fastes* II, 174-176-77; *DTC* 13, 2571-72; *Vies des SS*. 7, 442-44; *LTK* 8, 1258; *BS* 11, 140; E. Griffe, *La Gaule chrétienne à l'époque romaine*, I, Paris ²1964, 188-190; J. Gaudemet, *Conciles gaulois du IVᵉ siècle*, SCh 241, Paris 1977.

V. Saxer

REVELATION. It is rather difficult to present patristic ideas on God's r. Perhaps too conditioned by the debates provoked by the critical and rationalistic mentality of the last two centuries, which on the Catholic side found a dogmatic solution in Vatican I (DS 3004-3007), we easily risk introducing into the *Fathers a set of problems that were alien to them. In fact, they never summed up the whole of *God's saving action or the content of the *Bible in the overall concept of r. In particular, they never discussed the possibility of a gratuitous manifestation of the divine *mystery accessible only to supernatural *faith. Nor did they fix the end of the so-called public r. at the moment of the last *apostle's death (DS 2421), owning rather that the fullness of time in Jesus Christ was known through faith only by virtue of a continuous r. by Christ himself, always present in his church. Moreover, research into this question is made difficult by the complexity of the biblical premises. The viewpoint within which the writings of the OT, and then of the NT, speak of the dialogue in which the living God communicates himself to men - first to the people of Israel, then to all those who believe in Jesus - is understood differently by the various authors. It may be nomistic (covenant or law of the people freed from Egypt), prophetic (word recording God's faithfulness), sapiential (inquiring into the meaning of human existence), rabbinic (interpretation of the Torah, unique manifestation of the divine will), or apocalyptic (explanation of the present moment in view of the future). Finally, patristic *theology itself is demonstrably very complex. Its vocabulary is far from uniform. So it is not enough to refer to the terms ἀποκάλυψις, φανέρωσις, *revelatio* and the like. We must also consider εὐαγγέλιον, οἰκονομία (*economy), μυστήριον (*mystery), γνῶσις (*gnosis), παιδεία, παράδοσις, etc., as well as common images like word, divine voice, vision, inspiration and others. The theological positions of the individual authors also diverge, since they correspond to the diverse conditions of their communities, the multiform objections of *Jews, *pagans and *heretics, and their own varied personal situations (*conversion experiences, culture, etc.). Furthermore, monographs on the subject are rather scarce (Stockmeier, *Offenbarung*, 29 f.).

In this as in other fields, the Apostolic Fathers had not yet abandoned the primitive viewpoints. Expectation of the imminent *parousia did not cease to dominate their interests. And they were also faced with the question of the *secundum scripturas* of Jesus' destiny. So they do not investigate the biblical data very much. However they do require in a particular way that Christian *life should conform to Jesus' commandments, as well as to his example as true teacher. *Clement even goes so far as to compare the *oracles and divine manifestations of *paganism to the biblical evidence (I, 25 f.; 55). Above all he sees in the cosmic order the model of community life (60). *Ignatius innovates, for his part, considering the mysteries of Jesus' life, i.e. the content of the gospel, as a document superior to that of the Jewish Bible (*Philad.* 8, 2; 9, 2). Finally, the *Shepherd* of *Hermas delivers heavenly revelations to the church, symbolized by an old Woman (*Vis.* 1-3).

The confrontation with the *Jews who doubted Jesus' messianic status (Just., *Dial.* 48-108) and with the pagans who ruled out any real *incarnation of a son of God (Just., *II Apol.* 13; 54 f.; Orig., *C. Cels.* 5, 2), still more the universal significance of an event that had taken place only a few years before (Just., *I Apol.* 46; *Diogn.* 9), led the 2nd-c. *Apologists, particularly *Justin, to the first thematization of divine r. They claimed to possess the whole truth in Jesus Christ who, after having appeared to the ancient Fathers and spoken to them through the *Prophets (Just., *I Apol.* 63, 15 f.; *Dial.* 127, 4; Theoph., *Autol.* II, 22), and after having established all knowledge and all just legislation in the pagan world (Just., *II Apol.* 10), had by his *incarnation made himself God's definitive and complete word to the Christians (Just., *II Apol.* 7 [8], 3).

This idea of the incarnation of the Logos, supreme manifestation of *God, was further investigated by the anti-gnostic authors. Faced with pretensions to a perfect *gnosis, reserved only for the "spirituals" who recognized their kinship with God, but without ever seeing the transcendent God, as well as with the rejection of the goodness of the visible *creation and so of Israel's history as well, *Irenaeus stresses the unity of the saving *economy, making it culminate in the incarnation of the Son of God, which, together with all the other wholly gratuitous divine manifestations, will lead men in the second *resurrection to the vision of the immortal God (*Adv. haer.* IV, 20, 5). He also demonstrates that Holy *Scripture, including the quadriform gospel and the apostolic epistles, though only in the interpretation in accordance with the *regula fidei* maintained in the *apostolic churches, is the authentic testimony of this r., initiated in the reign of the Spirit, extended in the reign of the Son, but to be definitively fulfilled in the reign of the *Father (*Adv. haer.* IV, 20, 1-2).

In conditions that were similar, but also partly new, *Tertullian took up Irenaeus's *theology, though introducing it in the Latin category of the concept of truth and of its ancient, i.e. divine, origin (*Praescr.* 6, 4; 32, 1), and insisting more on the possibility of knowing *God from his creation (*Adv. Marc.* I, 18, 2: *natura ex operibus - doctrina ex praedicationibus*) as well as from the testimony of the *soul (*Test.* 2).

Soon afterwards the Alexandrians *Clement and *Origen followed the same anti-gnostic line, though condescending much more to the legitimate aspirations of the *gnostics and thus elaborating a Christian gnosis. Clement, adapting to the biblical doctrine not just the Platonic vision of the *world, but also the mystery-language of Greek *philosophy, develops his theory of divine pedagogy. Acc. to him, the Logos is the mystagogue who enlightens the believer, esp. in *baptism, leading him in a continual education to the knowledge of *God (*Protr.* 15; *Paed.* I, 6). Origen, for his part, more dependent on the Bible and more faithful to the *regula fidei*, outlines in a grandiose picture the gnoseological mediation of the Logos. The Logos - Wisdom, in that it turns to the Father, i.e. God's knowledge of himself, and Word, as facing the world, i.e. manifestation of God - being present in all things created through him and also acting all through history, reflects God's *glory everywhere (*Princ.* I, 2, 2, f.). Adapting himself to the capacity of man, creature and sinner, he has laid down in the incarnation in which he has announced himself (*Jo.* 13, 28) the basis of the *faith without which no-one can rise to the perfect vision of God (*C. Cels.* 6, 68). Identifying the written word with the Logos, Origen sees in the *Bible a coming of the latter which heralds the eternal gospel, though without reducing all r. to *Scripture. While thus impressing a very intellectual character on his theology of divine r., he does not wholly overlook its historical dimensions, understanding it esp. as a dialogue between God and man (Stockmeier, *Offenbarung*, 65).

This Origenian doctrine of the *mystery of the Logos survived in particular in the theology of *Eusebius of Caesarea (†339), though Eusebius conferred a political orientation on it, emphasizing the triumph of divine r. attained by *Constantine, earthly vicar of the eternal Word, in the victory over paganism (Farina, *L'impero*).

In the 4th c. the dogmatic controversies, caused primarily by Origen's misunderstood opinions on the mediation of the Logos, led to a further philosophical elaboration of the biblical premises in question. While *Arius put the Logos-Christ on the side of the creatures, even denying him full knowledge of the *Father (Athan., *Ar.* II, 22), the Nicene theologians recognized Jesus Christ as the son of God, basing on his true divinity the possibility of an authentic r. of the Father. Thus *Athanasius considers the incarnation of Christ, true God, as guarantee of the divinizing gnosis, now definitively restored to mankind, all the more so because, acc. to him, the corruption of death, the almost insurmountable obstacle to any knowledge of God, has been overcome at the root in the incarnation itself (*De incarn.* 9-14, 54).

*Apollinaris of Laodicea, in his doctrine of the λόγος ἔνσαρκος, proposes the same theory of *divinization in an even more stringent way (Mühlenberg, *Unendlichkeit*). But *Hilary, reinterpreting the Origenian idea of the Logos, *gloria patris*, in the light of the Nicene faith, declares concisely: *Summa dispensationis est Filio, ut noveris Patrem* (*De Trin.* III, 22). In the second phase of the *Arian controversy, Athanasius and, still more, *Basil and the other Cappadocians brought out the role of the *Holy Spirit, also put on God's side, in man's *illumination (Basil, *De Spir. S.* 24, 55 ff.; Greg. Naz., *Orat.* 31, 26; Greg. Nyss., *Vita Mos.* 2, 158 f.). Moreover Basil, very expert in the doctrine of the knowledge of God through his *creation (*Hexaem.* 1, 6), even stated that knowledge of divine existence precedes *faith in God (*Ep.* 235, 1; cf. Euseb., *Praep. ev.* I, 1, 11). Finally the debate on Christ's true humanity, compromised by the Origenian tradition, contributed to the definition of the way in which the intimate union between divinity and complete humanity in Christ constitutes the supreme r. of the divine *mystery. This was attested in particular by *Cyril of Alexandria who, constKrained by the Antiochene tradition, certainly recognized the integrity of the assumed humanity, but nevertheless held forcefully that the incarnation is the manifestation in the *flesh (1 Tim 3, 16) of Christ himself who, according to his nature, is invisible (*Chr. un.*: SCh 97, 324), and that therefore the illumination of the knowledge of God the Father shines forth in Christ's very person (*De recta fide* 32: PG 76, 1181B).

In the West, at the same time, *Marius Victorinus with his neoplatonist approach to the mediation of the Logos (*Cand.* 14; *Ar.* I, 19), *Ambrose with his *lex - evangelium* *antithesis (*Psal.* 61, 33; *Off.* I, 48, 238), *Jerome with his philological analyses of *revelatio*, considered as a Christian word (Gal 1, 1), and *Augustine with his extensive renewal of the patristic heritage and esp. with his linguistic finesse (De Veer, *Revelare*), did not fail to put new emphases on the *theology of r., exalting the absolute value of God's manifestation in Christ. The bishop of *Hippo in particular, caught up in a vast debate with the African *Manichees and faced with Milanese *neoplatonism, considerably developed the principal data of the Christian *tradition. On one hand, he stresses even more that Christ reveals the true meaning hidden in the OT books (Wieland, *Offenbarung*, 263 ff.). On the other, he bases all knowledge of truth on Christ, *magister interior* (*Mag.* 37 f.; *Trin.* IV, 2, 4), though without renouncing Christ, *auctoritas fidei*, who, present in the Bible, the incarnation and the church, leads us to that *faith without which *illumination of the heart is not possible (*Confess.* XI, 8; *Trin.* VII, 4; *De civ. Dei* X, 32). Though Augustine, in the setting of his Platonic vision of the world, may not give great importance to God's historical r., we should not undervalue the extent to which, partly under the influence of Latin *rhetoric, and even more under the impress of the *regula fidei*, he succeeded in overcoming the risks of his philosophical approach, reducing *scientia* and *sapientia* to the one Christ, God and man (*Trin.* XIII, 19, 24).

While the *Augustinian doctrine of divine r. dominated the West until the scholastic period, in the *East two theologians manifested new accents. Under the influence of *neoplatonism, Ps.-*Dionysius, closely uniting *creation and r., presented the universe as a gradual communication of God's intimate secret (*Cael. hier.* IV, 1 f.), confirmed however by Scripture, the norm of all knowledge of truth (*Eccl. hier.* V, 3, 7), and guaranteed above all by the theandric life of Jesus Christ (*Ep.* 4). Correcting this too unitarist viewpoint, thanks to his fidelity to the *indivise-inconfuse* of the creed of *Chalcedon, *Maximus Confessor distinguished clearly not just between the creator and the creation, but also between the latter's symbolic function and the divine reality symbolized by it (*Quaest. Thal.* 59), as appears above all in the incarnation, which reveals God's *love in the most striking way and at the same time attests the full freedom of the human response.

Observing all the patristic developments of the complex biblical doctrine of r., we notice above all the great openness of the ancient theologians towards the various phenomena of divine manifestation and the wholly natural supposition that God communicates himself to man. Exceptions to this, in their way, were the *gnostics who denied the possibility of seeing the transcendent deity, and *Arius who extended his negative theology even to Christ. So the difficulties against which the Fathers had to struggle did not, except in these two cases, touch on the possibility of a gratuitous r. of God's inner life, but rather on its concrete form, i.e. God's manifestation by means of the incarnation of the true Son of God, contested by *Jews, gnostics and non-Christian philosophers. So *Justin, *Irenaeus, *Athanasius and *Maximus Confessor defend the *incarnation as supreme theophany. But not even *Origen, *Augustine and Ps.-*Dionysius fail to emphasize that the Word, incarnate from *Mary, has revealed itself and so also God the *Father. But this Christo- or Logo-centrism takes various forms. Some, if not all, insist on Christ's revealing presence in the church, as above all does Justin, who brings out the Logos's conquering strength against all error, or Augustine, who holds so firmly to the *magister interior*. Some, like Origen, emphasize the action performed by the Logos in the past, becoming not just flesh, but even Scripture, i.e. a word accessible to every believer. But all look at the incarnation in a future perspective, considering it as the basis of the spiritual ascent towards the eternal vision of God, perhaps recalling the central affirmation of the gospel that the Son of man will appear as judge of the living and the dead to definitively reveal God's *glory.

M. Harl, *Origène et la fonction révélatrice du Verbe incarné*, Paris 1958; A.C. De Veer, Revelare-Revelatio, *RecAug* 2 (1962) 331-357; W. Pannenberg (ed.), *Offenbarung als Geschichte*, Göttingen ³1965; R. Farina, *L'impero e l'imperatore cristiano in Eusebio di Cesarea*, Zürich 1966; E. Mühlenberg, *Die Unendlichkeit Gottes bei Gregor von Nyssa*, Göttingen 1966; V. Hahn, *Das wahre Gesetz. Eine Untersuchung der Auffassung des Ambrosius von Mailand vom Verhältnis der beiden Testamente*, Münster 1969; P. Stockmeier, *Offenbarung in der frühchristlichen Kirche*, HDG I, 1a, Freiburg 1971, 27-87 (fundamental bibl.); M. Seybold, *Die Offenbarungsthematik in der Spätpatristik und Frühscholastik, ibid.*, 88-115; B. Studer, *Soteriologie der Kirchenväter*, HDG III, 2a, Freiburg 1978, 55-225; R. Tremblay, *La manifestation et la vision di Dieu selon s. Irénée de Lyon*, Münster 1978; W. Wieland, *Offenbarung bei Augustinus*, Mainz 1978 (bibl.); H. von Campenhausen, *The Formation of the Christian Bible* (Eng. tr. J.A. Baker) London 1972, esp. 211-243; J. Barr, Revelation in History, *The Interpreter's Dictionary of the Bible*, Suppl. vol., Nashville 1976.

B. Studer

RHAETIA. Roman province between the Tridentine Alps and the Danube, corresponding in part to present Switzerland and the Bavarian highlands. Under *Diocletian's reorganization of the empire, R. was included among the *regiones* of *Italy and divided into *R. prima* and *R. secunda*. We have little knowledge of the spread of Christianity, which must have reached it from the neighbouring *Noricum; but we must think in terms of a difficult propagation, if it is true that archaeological finds in the region suggest that attachment to indigenous or, at the most, Oriental cults was more important and widespread, esp. in *R. prima* (but on this cf. recently R. Degen, Liechtenstein, 202-222).

So far three bishoprics are known for certain: *1)* Augusta Vindelicorum (Augsburg), where St Aphra suffered *martyrdom under Diocletian (MGH, *Scr. rer. mer.* 2, 41-64; 7, 192-204; and cf. evidence of Ven. Fort., *Vita Martini*, vv. 640-644); *2)* Curia (Chur) if, at the synod of *Milan (451), bishop *Abundantius of Como subscribed in the name of the absent Asinius or Asimus, *episcopus ecclesiae Curiensis* (Mansi 6, 144); *3)* Sabiona (Säben), whose bishop, Ingenuus or Ingenuinus, took part in 590 in the synod of Marianus (MGH, *Scr. rer. lang. et it. Saec.* VI-IX, 393, and cf. Paulus Diac., *Hist. Lang.* III, 26). Apart from this, a general idea of the conditions of *R. secunda* in the mid 5th c. can be gathered from *Eugippius's *Vita s. Severini*, esp. chs. 15. 19. 20. 22. 27, whence we learn of the existence of some Christian communities at Quintanis (Künzing) and Batavis (Passau), whose *basilica*, however, was in nearby Boiotrum (Innstadt), beyond the river Inn and therefore in Noricum. Some doubts have been expressed over another fact in the *Vita* (ch. 41, 1), about a bishop of R., Valentinus (M. Heuwieser, *Geschichte des Bistums Passau*, 37 ff.).

R. Heuberger, *Rätien im Altertum und Frühmittelalter*, Innsbruck 1932; A. Bigelmair, Bavière, *DHGE* 6, 1524-1528; M. Heuwieser, *Geschichte des Bistums Passau*, I, Passau 1939; *EC* 10, 842-843; R. Degen, Liechtenstein zwischen Spätantike und Mittelalter, *HA* 9 (1978) 202-222; K. Gamber, *Sarmannina. Studien zum Christentum in Bayern und Österreich während der Römerzeit*, Regensburg 1982, 7-57.

E. Pavan

RHEIMS. Durocorturum, in the late empire Urbs Remensis, became the metropolis of Belgica II under *Diocletian and an active economic and military centre. The first reliably attested bishop, Imbetausius, preceded in the *episcopal list by three unknown titulars (Sixtus, Sinicius, Amausius), appears in 314 (council of *Arles). *Gregory of Tours's evidence on the two local *martyrs Timothy and Apollinaris (*Glor. Mart.* 55) is not enough to guarantee these two saints' historicity, as hypothetical as that of a founder Sixtus, first in the list of bishops (Flodoard, *Hist. Rem. Eccles.* I, 2; Duchesne, *Fastes* 3, 77). Imbetausus is followed in the list by Aprus, Maternianus, Donatianus, Viventius, Severus, Nicasius (a victim either of the *Vandals in 407 or of *Attila in 451), Baruch (*Barucius*), Barnabas, Bennagius, all of little importance until *Remigius (bishop from at least 457, died 518). In 482 Remigius hailed the accession of *Clovis, whom he baptized between 496 and 508 at R., by now one of the metropolis's of *Catholic *Gaul for the southern churches previously under the rule of the *Arian kings. But his successors Romanus, Flavius (...535...), Mappinius (...549) seem to be blotted out; in the late 6th c., Egidius served the political interests of king Sigebert and of Fredegund, but was deposed by Childebert in 590. He is a sign that the bishops were now playing the role of palace prelates.

For the 7th-c. bishops we know nearly nothing but the evidence of Flodoard, who recounts their *donations: Romulf, Sonnatius (...614-627), Laudigesilus, Angelbert, Lando, Nivo, Reolo (an ex-count of Champagne); for the 8th c., Rigobert (deposed in 717 by Charles Martel, who imposed Milo, bishop of *Trier, on R.), then an Irish monk, Abel, in 743, succeeded by another monk (of Saint-Denis), Tilpin (748-794).

Of the little-known Christian community's *church buildings, nothing remains but the evidence provided by texts: *intra muros*, a cathedral (initially dedicated to the *apostles), doubled by a second building (attested from the 6th c. and dedicated to St *Mary), a baptistery and a church of St *Peter (7th c., acc. to Flodoard); outside the walls, the mausoleum of Jovinus, *magister militum*, which later bore the title of St Nicasius, three *basilicae* dedicated to Timothy and Apollinaris, St Julian and St Remigius, attested by Gregory of Tours, and, from the 7th c., an *ecclesia Sancti Sixti et Sinicii*: these buildings are grouped S of the *castrum* in a cemeterial zone that later became the *bourg* of St-Rémy.

J. Leflon, *Histoire de l'Eglise de Reims du Ier au Ve s.*, Travaux de l'Académie de Reims 152, 1941; *DACL* 14, 2213-2290; H. Reinhardt, La cathédrale du VIe s. à Genève et l'Eglise du baptême de Clovis à Reims, *Geneva* 11 (1963) 127-139; L. Pietri, *Topographie chrétienne des cités de la Gaule (IV-VII s.)*, I, Paris 1974, 73-83.

Ch. Pietri

RHETORIC. Both in the classical and in the Christian world, r. involved the same conception of life. The Greeks called "rhetors" those whom the Latins called "orators". Nowadays the term r. is used pejoratively, mostly to mean an ornate form of speech, without reference to its content. Recently, however, r. has got past the negative phase of being identified with its vices and has been rediscovered as the nexus of form and content, decisive for evaluating artistic, literary and figurative expression. Well-known German scholars, from Curtius to L. Arbusow (*Colores rhetorici*, Göttingen 1963) and H. Lausberg (*Elemente...; Handbuch...*), have stressed the literary importance of theological and liturgical texts and the importance of linguistic phenomena in the public life of antiquity. V. Florescu has emphasized the dilemma that appeared at the time of the first *patristics, "when God's Word was considered so rich in intrinsic penetration that it could do without the subtle and complicated outward expressions of verbal debate" (*La retorica nel suo sviluppo storico*, 7).

For the ancients, besides the schooling of that name which immediately followed that of grammar, r. was a constant factor of relations between content and form. Common to classico-pagan and ancient Christian r. was the condemnation of formal values without moral values. There was a universal r. from which man could not depart and which did not obey the rigid rules of theorists. There was now a general tendency to see in the figurative arts the norms of universal r. which underlay the various modes of artistic expression. This was particularly so in *iconography which, like a literary page, could not be seen in isolation from the natural patterns of semantics; i.e. of signs that revealed the spontaneity of certain states of mind. This was the line taken by the church *Fathers, well described by *Paulinus of Nola and *John Damascene to explain how there came to be a wide-ranging participation in the *Bible; the so-called Bible of the poor.

The relation between word and action is intimately close, indissoluble for the Christian. According to the gospel (Mt 12, 36), on *judgment day men will render account for every careless word they utter. The Fathers seem to say that if the truth of things were dependent on the speaker's ability, the world would end in the triumph of lies. The OT and NT continually state that perverse words express perverse morals, and insist on the truthful word of the humble, simple mind. The central fulcrum of all patristic Christianity was Holy *Scripture, which from the spiritual point of view replaced Homer and Virgil. Ancient Christian authors thought and wrote biblically, a dimension that should not escape students of ancient r. It was no longer a decadent Christian *Greek and *Latin, but a fertile culture generating new forms and new contents.

Through biblical *exegesis, the Fathers followed the ideals of *sapientia* more than of *eloquentia*. They used fine metaphors to indicate the immense *sapientia* present in the Bible. With Christ, a new world had opened; the *studia liberalia* were neither excluded nor required. All roads led to Christ, both manual *labour and study. The humble *slave, the peasant, the fisherman, the artisan, young and old, men and *women, people of any social condition could live Christ, which was understanding the Scriptures. Biblical penetration by the Fathers and the community exercised a great influence on language, which was transformed to express another culture. New forms and elements unknown to the *pagan world entered speech. In rhetorical terms, we could say that there was an *elocutio* dependent on biblical *inventio*. The language of the biblical versions, of *liturgy and *catechesis, the language of the community, had to be understood. Literary forms evolved with language. Christian homiletic had to speak in everyday language. The characteristics of the various *genera dicendi* whose roots were in evangelical preaching prevailed. Biblical *commentaries are philologically richer than their pagan counterparts. Servius is poorer than *Jerome in elements of comparison and in vocabulary, and he has no historical sense of language, nor Jerome's knowledge of Hebrew. In every Christian author we can see the exercise learnt in the rhetor's school moving in a biblical direction in search of elements of development. Declamations in the form of *controversia*, debate both in lawsuits and on *suasoria*, discussion on matters of history, *mythology and culture were presented as preparatory exercises for moral life. Many points of declamation became themes of Christian authors. Even the Christian *lectorate* should be connected to the custom of declamation and reading aloud so that the Sacred Texts could be understood before being practised. This transformation of the Greek and Latin languages included many new elements of metric that concerned quantity and stress. *Augustine examined the phenomenon and explained many things about the modification of stresses in relation to quantity, which was growing weaker, and about *melos*. He would have written six books on *melos* if circumstances had permitted, but was prevented by "the burden of ecclesiastical cares" which as a bishop he had to bear (*Ep.* 101, 3). The *trivium* (grammar, rhetoric and philosophy) was undergoing profound modifications. To get a clear idea of them we should read book IV of Augustine's *De doctrina christiana*. The Fathers were the first to discern the elements common to all men *naturaliter rhetorici*. And these were patterns such as *antithesis, parallelism, *gradatio*, *auxesis*, etc., which even became an instrument of scriptural *exegesis. The *quadrivium* too (arithmetic, geometry, *music, astronomy) was enlarged and developed in the ancient Christian authors. The encyclopaedism of *Isidore of Seville was something quite new. Bishop *Anatolius of Laodicea, who attained the highest degree of mathematical knowledge, lived in the 3rd c. (Euseb., *HE* VII, 32, 6 Schwartz). Few fragments of his works survive. He was in touch with Pappus, to whom we owe our knowledge of Euclid's and Apollonius's works on higher geometry. Pappus was a convert, as we deduce from the pledge in which he affirms his belief in *God, author of the OT and NT. Augustine, anticipating the direction of analytical geometry, considered number the foundation of mathematics and the natural sciences. The mathematicians of the 5th and 6th cc. were incorporated into ancient Christianity. *Boethius the rhetor was induced to write on arithmetic, geometry and music; his last book was the *Consolatio Philosophiae*: and we have reached the *quadrivium*. Studies of Christian r. still need to be deepened by a surer knowledge of the evolution of the exact sciences, which found their application here.

H. Lausberg, *Elemente der literarischen Rhetorik*, Munich ²1967; id., *Handbuch der literarischen Rhetorik*, 2 vols., Munich ²1973; V. Florescu, *La Retorica nel suo sviluppo storico*, Bologna 1971; G.A. Kennedy, *Classical Rhetoric and its Christian and Secular Tradition from Ancient to Modern Times*, London 1980 (rich bibl.); A. Quacquarelli, *Retorica e iconologia*, Bari 1982; H.I. Marrou, *Histoire de l'Education dans l'antiquité*, Paris 1965⁶; G. Kennedy, *Greek Rhetoric under Christian emperors*, Princeton 1983; J.L. Kinneavy, *Greek Rhetorical Origins of Christian Faith: an enquiry*, Oxford 1987; R.A. Kaster, *Guardians of Language; The Grammarian and Society in Late Antiquity*, Berkeley 1988; A. Spira, The Impact of Christianity on Ancient Rhetoric, *SP* 18.2 (1989) 137-154; C. Smith, Christian Rhetoric in Eusebius's Panegyric at Tyre, *VChr* 43 (1989) 226-247; F.W. Norris, *Faith gives fullness to Reason: a commentary on Gregory of Nazianzen's Theological Orations*, Leiden 1990.

A. Quacquarelli

RHODANIUS of Toulouse. Sources disagree on the identification of the *council at which bishop R. of *Toulouse (*Tolosa*) resisted the pro-Arian aims of the emp. *Constantius II, for which he was exiled to *Phrygia, where he died. *Jerome seems to make it that of *Arles of 353, *Rufinus and *Sozomen that of *Milan of 355 - a list of signatories suggests that R. subscribed *Athanasius's condemnation then - *Hilary of Poitiers and *Sulpicius Severus that of *Béziers, where R., of a conciliatory nature, was drawn along by Hilary's example. These last witnesses seem more credible. Hilary describes the sufferings of the church of Toulouse on that occasion, echoed in the *mass of the translation of the martyr *Saturninus (1 Nov) preserved in the Mozarabic sacramentary. Despite his death in exile, equivalent to *martyrdom, R. does not appear as a saint in the *episcopal list of Toulouse: the church of that city, which 50 years later became the capital of the Arian Visigothic kings, could not celebrate an opponent of *Arianism as a saint, and so R. was forgotten by his compatriots.

H. Crouzel, Un "résistant" toulousain à la politique de l'empereur Constance II: l'évêque Rhodanius, *BLE* 77 (1976) 173-190.

H. Crouzel

RHODO. Our knowledge of the Christian writer R. (second half of 2nd c.) depends on *Eusebius (*HE* V, 13). There seems no basis in what we infer from *Jerome (*Vir. ill.* 37; 39) who, otherwise faithful to Eusebius, differs in attributing to R. a work against the *Cataphrygians, which would imply his identity with the homonymous anti-*Montanist mentioned by Eusebius (*HE* V, 16). Acc. to Eusebius, R. wrote works of various kinds. Of Asiatic origin, he was one of *Tatian's disciples at *Rome, as Tatian himself stated in an anti-*Marcionite work. His relations with his old teacher are further illustrated by the fact, found by Eusebius in that same work, that R. promised to write a book resolving the obscurities that Tatian thought he had found in *Scripture, and which he had expounded in a work entitled *Problems*. Eusebius does not say whether the promise was kept, but he does mention another work by R., a *commentary on the *Hexaemeron*. As for the anti-Marcionite work, in which the author addressed an otherwise unknown Callistio, it was to illustrate and refute the various doctrinal positions developed within the Marcionite *heresy. Eusebius cites an ample fragment of it, followed by two more relating to a dispute between *Apelles, now old, and R., who exposes and derides the Marcionite's claims.

PG 5, 1331-1338; A. Harnack, *Geschichte der altchristlichen Literatur bis Eusebius*, I, Leipzig 1893, 599; II, 1, Leipzig 1897, 313 f.; Bardenhewer 1, 392-394; P. de Labriolle, *Les sources de l'histoire du montanisme*, Fribourg-Paris 1913, XX f.; A. Harnack, Rhodon und Apelles, in *Geschichtliche Studien f. A. Hauck*, Leipzig 1916, 39-51; id., *Marcion. Das Evangelium vom fremden Gott*, TU 45, Leipzig ²1924, 163 f., 180-196.

V. Zangara

RICH - RICHES - PROPERTY. In the *Bible, riches are a good *per se* and are praised because, like all things, they belong to God and man's possession of them can be interpreted as a sign of divine protection and blessing (Gen 13, 2; 26, 12 ff.; Dt 8, 7-10; 28, 1-11). If man is blessed by God for his faithfulness and this blessing or reward consists in having descendants and riches, health and prestige, the link between reward and faithfulness has a long development in the history of Israel until it attains the insight that it is not always true; indeed, often the contrary is true: it is the impious man who habitually does well (Jer 16, 1-4). The wisdom books reflect that, although a good, wealth is never the best good (Pr 15, 15; 16, 8; 22, 1; Sir 30, 14 ff.; Job 28, 15-19; Wis 7, 8-11). The difficult existential balance between *poverty and riches and the advantages of one and the other are summed up in Pr 30, 7 ff. The close link that was established, with the development of the monarchy, between riches and *idolatry is obvious to the prophets (Hos 10, 1; Is 2, 7-8; 7, 20; Ezek 7, 19-20). The king became the country's main landowner, and the weak were forced to sell their inherited land, an inheritance which visibly expressed man's part in God's people. In these terms, the owning of large estates was not just a social injustice, but led directly to a desacralization of Israel and severed the people from the land, where people and land were the indissociable terms of the promise to Abraham (Mic 2, 1-5; Hos 5, 10; Is 5, 8; 1 Kgs 21, 2).

In the NT we find all the OT's reservations about riches, but the link between *justice and earthly reward is absent. Even if goods are not considered bad in themselves, the acquisition of riches is associated with the idea of injustice (Lk 16, 9-11) and the debate moves from goods to their possessor, from riches to the rich (Lk 6, 24-25; Jas 5, 1-5; 1 Tim 6, 17): it is hard for them to enter the Kingdom (Mt 19, 23). Christ has friendly relations with rich men (Mt 27, 57; Mk 2, 3-17; 14, 3; Lk 19, 1-10; 14, 1-12), but always with a view to their salvation, which is gained by giving their wealth away to the *poor (Mt 19, 21; Lk 12, 33). *Luke especially insists on the dangers deriving from riches (1, 52-53; 6, 20. 24. 25; 8, 14; 12, 13-21. 32-34; 14, 12-14. 15-24; 16, 12-13. 19-31; 18, 18-30; 19, 1-10; 21, 1-4).

Patristic reflection on the theme of rich-riches starts from *Scripture. The first text to deal fully and directly with the relationship between riches and the Christian life is *Clement of Alexandria's *Quis dives salvetur?* Bearing in mind Mk 10, 17-31, which taken literally seems to be a categorical condemnation of the rich, the author, through an allegorico-spiritual *exegesis, explains to rich men who have become Christians the deepest meaning of the Marcan passage: the invitation to detach the heart from riches, without completely depriving oneself of them (*Quis div.* 11; 14; 18); the debate thus moves from possession to use (*Quis div.* 14; 26), since Clement's opposition to *dualism leads him to affirm the goodness of created things, which he commends because they allow us to come to the help of those in need (*Quis div.* 13). But riches held on their own account are iniquitous (*Quis div.* 31) and the rich man who does not share is a murderer (*Quis div.* 7). Riches are compared to a serpent: those who cannot take hold of it in the right way are easily bitten and poisoned (Clem. Al., *Paedag.* III, 6, 37). The remedy for cupidity is parsimony (*ibid.*); true wealth is to possess the immortal Logos (*Paedag.* II, 3, 35). The Jewish wisdom tradition (goods are good and a sign of blessing), the *Stoic ethic (riches are a danger), and the evangelical announcement (circulation of goods is a sign of brotherly *love) meet in Clement's thought. Material riches must be abandoned by those who wish to become followers of Christ (Orig., *In Mt.* XV, 25). Attachment of Christians to wealth is given as one of the main causes of denial of the faith during *Decius's *persecution (Cypr., *De lapsis* 11); the dangerous nature of riches (id., *De op. et elem.* 13), all the more if accumulated, is shown by comments on Luke 12, 16-21 (Cypr., *De dom. orat.* 20; Ambr., *Exp. evang. sec. Luc.* VII). But riches have a positive aspect: they can be distributed (Hermas, *Shepherd, Sim.* I, 6, 8), handed out as *alms (Cypr., *De op. et elem.* 8-11), donated and shared (Lact., *Div. Instit.* 5, 15; Greg. Naz., *De pauper. amand.* 22). *Basil's homily VI (*De avaritia*) is an extensive commentary on the Lucan parable of the rich fool. The rich man's proper position is constantly pointed out in all Basil's works: he is an administrator of divine goods (*De avar.* VI, 1-2; cf. also Ambr., *De Nabut.* VI-VIII). In homily VII (*In divites*), Basil comments on Mt 19, 16-26, showing how the rich man's possession of goods must be resolved in the sharing which changes the social condition of the poor or restores the justice willed by God in giving to all. The relationship between *right to possess* and *right to private property* in Ambrosian thought is one of the subjects in which scholars are most interested, giving different interpretations (cf. M.G. Mara, *Ricchezza e povertà*, 70 n. 121). *Ambrose does not appear to deny the right to private property, provided this right is available to everyone (*De Nabut.* 2; 11; 29). At the basis of Ambrose's thought are motifs common to *patristics: the goods of the earth were given by *God for men to enjoy; the rich man is only their administrator, not their owner; monopoly of goods by the few is not allowable; refusal to give and share them is a rejection of eternal *life; not only does death contradict the lasting possession of earthly goods, but the eschatological view of life reveals their provisionality. In his *homily* 63 on Matthew, *John Chrysostom shows the rich the necessity of leaving their riches to follow Christ; in *hom.* 85 he invites church leaders to give the church's own wealth to the poor; in *hom.* 34 he shows how the rich man who is so by God's gift should be useful to his neighbour. For *Augustine the judgment is continually transferred from goods to their use: the rich man is in self-contradiction if, wishing to possess good things, he does not wish to be good himself by sharing them (*Serm.* 85).

Patristic thought on rich-riches (goods) is in agreement on: the goodness of things, the equality of men, the duty of the rich to share to the point of changing the condition of the poor, the role of the rich as administrators rather than owners of goods, the transfer of any moral judgment from possession of riches to the use made of them, condemnation of accumulation of riches, the obligation to order to man's service all that relates to the economic order (riches, goods, acquisition, use, etc.). The are nuances, due to historical (*persecutions, failure of political authorities to tackle social problems, decline of morals and hence of *faith in the Christian community, pestilence, famine), environmental (different economic situations of the various cities and regions) and personal (different economic situation of individual *Fathers) circumstances over inherited wealth (Clem. Al., *Quis div.*), the question of whether wealth originates in injustice (John Chrys., *In I Tim. hom.* 12), or *usury; whether selling everything and giving it as *alms is perfect *virtue, not required of all (Cyr. Al., *De adorat. et cult. in spir. et ver.* XIII, 462), not commanded (John Chrys.), or a necessary condition for eternal *life (Cypr., Greg. Naz.), a perfect duty of charity (Ambr.).

DB 5, 1093-1094; P. Christophe, *Les devoirs moraux des riches. L'usage du droit de propriété dans l'Ecriture et la Tradition patristique*, Paris 1964; L. Orabona, *Cristianesimo e proprietà. Saggio sulle fonti antiche*, Rome 1964; G. Barberi, I nuovi principi economico-sociali nella elaborazione dei Padri greci e latini, *Grande Ant. Filos.*, Milan 1954, V, 1129-1174; *Augustinianum* 17 (1977); M.G. Mara, *Ricchezza e povertà nel cristianesimo primitivo*, Rome 1980; Autori vari, *Per foramen acus; il cristianesimo antico di fronte alle pericope evangelica del "giovano ricco"*, Milan 1986; H. Gülzow, *Armut und Reichtum als Problem der frühen Kirche*.

M.G. Mara

RIEZ, *Council of. In this Alpine city (*Reii*) of Upper Provence in *Gaul, a council met on 29 Nov 439 under the presidency of *Hilary of Arles to resolve problems arising from the irregular *ordination of the *layman Armentarius to the episcopal see of Embrun (*Ebrodunum*).

CCL 148, 61-75; Hfl-Lecl 2, 423-430; Palazzini 4, 109-110.

Ch. Munier

RIMINI, *Council of. Called by the emp. *Constantius in parallel with the Eastern council of *Seleucia to bring *peace to the church, deeply disturbed by the events of 357-358 relating to the *Arian controversy, the council of Ariminum opened early in June 359 in the presence of *c.*400 bishops, presided over by *Restitutus of Carthage. *Liberius of Rome was absent: he may have sent representatives. The pro-Arians *Valens, Ursacius and *Germinius, at the head of *c.*80 bishops, insisted on ratifying the preparatory formula of faith published at *Sirmium on 22 May 359, and confirmed its condemnation of the Nicene *homoousios*. But the great majority of the council firmly opposed them, and, after long drawn out discussions, reaffirmed the validity of the Nicene formula of 325, including the *homoousios*. Valens and his party left the *council and were condemned on 21 July 359, when the main Arian propositions were condemned together with other *Sabellian and Photinian ones.

At this point, a delegation led by Restitutus went at Constantius's order to report to him at *Constantinople, where Valens had preceded them. Constantius, who would not accept the *homoousios* and demanded a theological formula of vague compromise, ordered the delegation to be stopped near Constantinople and transferred to Nike, a postal station in *Thrace. Here Valens and his collaborators submitted the Westerners to a long series of debates mixed with threats and enticements, until their resistance broke and they were persuaded to subscribe a formula of faith in which the Son was defined only as "like the *Father according to the *Scriptures".

Valens and the delegation returned to R., where the Western bishops were obliged to remain by order of the emperor, in whose name the prefect Taurus announced that no-one could leave the city without first subscribing, as had the delegates. After initial resistance, mistrust and discouragement gradually prevailed and one by one the delegates subscribed the so-called formula of R., a brief text in which the Son is defined as "like the Father according to the Scriptures", while the term *ousia* (and so the two composite terms *homoousios* and *homoiousios*) was proscribed. The last to sign, late in November, was *Phoebadius of Agen, who was allowed to add to the formula, on his own account, some *anathemata condemning the main Arian propositions.

The formula of R. was so vague that it was capable of an interpretation consonant with Arian doctrine. For this reason the council was held all over Christendom as a great victory for *Arianism, and Arians would always later appeal to it as the solemn conciliar sanction of their doctrine.

Hfl-Lecl 1, 929-955; Simonetti 313-325; Y.M. Duval, La "manoeuvre frauduleuse" de Rimini, in *Hilaire et son temps*, Paris 1969, 51-103; K.M. Girardet, *Kaiser Konstantius II als Episcopus Episcoporum, und das Herrscherbild des kirchlichen Widerstandes*, *Historia* 26 (1977) 95-128; R. Klein, *Constantius II und die christliche kirche*, Darmstadt 1977; W.H.C. Frend, *The Rise of Christianity*, Philadelphia-London 1984, 534-543.

M. Simonetti

RING. The Romans borrowed from the Greeks the use of a r. as reward for success in war. From an iron r. they went on to a gold r., first for ambassadorial senators, then for all senators, finally for knights. Over the centuries various laws regulated its use, which was always reserved for free men. The emperor could also allow a gold r. to freedmen; finally Novella 78, ch. 1 (539) declared that with his freedom a *slave bought the right to wear a gold r. With growing luxury, the use of gold rr. multiplied until *Tertullian (*De cultu fem.*, 9) could reprove *women for wearing one on each finger. *Clement of Alexandria (*Paed.* 3, 11) urged Christians to engrave Christian *symbols on their signet-rings. It is impossible to determine when, in the Church, the r. was no longer just a seal, but came to indicate *authority and its use to be reserved for bishops, thus virtually continuing the custom by which the *flamines Dialis*, dedicated to the cult of Jupiter and considered equal to senators, also had the right to bear a gold r. as a sign of honour. In the 7th c. the r. is numbered among episcopal insignia. From *Ambrose, we can infer the custom in his time, at *Rome and *Milan, of giving Christian *virgins a r. on the day of their consecration, but not until the 10th-c. Romano-Germanic Pontifical do we know for certain of a r. for virgins, blessed with a formula like that used for blessing the nuptial r. The custom of giving an iron r. in pledge of *marriage goes back to the Romans and is attested by Plautus.

J. Braun, *I paramenti sacri, loro uso, storia e simbolismo*, It. tr. Turin 1914; P. Salmon, *Etude sur les insignes du pontife dans le rite romain*, Rome 1955; P. Jounel, La liturgie romaine du mariage, *La MaisonD*. 50 (1957) 30-57; I.H. Dalmais, La liturgie du mariage dans les Eglises orientales, *La MaisonD*. 50 (1957) 58-69. A. Chavasse, *Le sacramentaire gélasien*, Tournai 1958, 5-27; V. Noè, Il matrimonio nella liturgia, *Enciclopedia del matrimonio*, Brescia 1959, 684-732; R. Metz, Le nouveau rituel de consécration des vierges, sa place dans l'histoire, *La MaisonD*. 110 (1972) 88-115.

M.G. Bianco

ROGATUS. 4th-c. *Donatist bishop of Cartenna (Tenés) (Aug., *Ep*. 93, 1, 1). In *c*.370, at the time of *Parmenian, he broke with his party and founded a sect, the Rogatists (*Ep*. 93, 3, 11; 93, 8, 24). These appealed to the emp. *Julian (*Ep*. 93, 4, 12), claiming their rights - they considered themselves the authentic heirs of Christ and *Donatus - while Parmenian's followers, supported by Firmus (*C. litt. Pet*. 2, 133, 184; *C. ep. Parm*. I, 10, 16; *Ep*. 87, 10), tried to eliminate R. and his partisans and take possession of their churches. R. defended himself from the attacks of Firmus's followers (= Donatists). The lawsuits between Primianists and *Maximianists were imminent (*Gesta cum Em*. 9; *C. Cresc*. III, 62, 68; IV, 70, 73; *C. ep. Parm*. I, 11, 17). R. allied himself with the latter. The Rogatists and the various parties that adhered to them were centred on *Mauretania (*In Ioh. ev*. 10, 6). *Augustine judged them harshly (*Ep*. 93, 6, 20-21. 7, 23. 8, 24-25. 10, 36. 11, 49).

Monceaux 4, 253-254; 6, 316-317; E. Tengström, *Donatisten und Katholiken*, Göteborg 1964, 79-82Z; *PAC* 1, 990 f.

E. Romero Pose

ROMAN CANON. The Roman eucharistic prayer is called "*canon actionis*" by the old Gelasian sacramentary (Mohlberg, *Liber sacramentorum*, 183). The *Eucharist is an action regulated by a canon. The R. C. has preserved the expression "*actionem*" for the parts known as "*Communicantes*" and "*Hanc igitur*". The idea of observing a canon as norm of action in the celebration of the Eucharist is formulated by St *Augustine: "*Totum illum agendi ordinem quem universa per orbem servat Ecclesia*" (*Ep*. 18: PL 33, 203); Pope *Vigilius, writing to *Profuturus of Braga, gives the name "*canonica prex*" to the immutable part of the Roman eucharistic prayer.

In the 6th c., *Ordo Romanus I* mentions the canon without preface (Andrieu, *Les Ordines Romani*, 95), a way of doing things which contradicts the older tradition. For *Cyprian, the preface and the rest of the eucharistic prayer form a unity: "*Ideo et sacerdos ante orationem praefatione praemissa parat fratrum mentes dicendo: sursum corda, ut dum respondet plebs: habemus ad Dominum, admoneatur nihil aliud se quam Dominum cogitare debere*" (*De dom. orat*. 31: CSEL 3, 289, 14-17). The change may have been caused by the fact that from the 4th c. only the prefaces became interchangeable, while the rest of the canon, apart from certain embolisms, remained fixed. In 538 pope Vigilius wrote about this to Profuturus of Braga: "*Ordinem quoque precum in celebritate missarum nullo nos tempore, nulla festivitate significamus habere divisum, sed semper eodem tenore oblata Deo munera consecrare. Quoties vero paschalis, aut ascensionis domini, vel pentecostes, et epiphaniae, sanctorumque Dei fuerit agenda festivitas, singula capitula diebus apta subjungimus, quibus commemorationem sanctae solemnitatis, aut eorum facimus, quorum natalitia celebramus: cetera vero ordine consueto prosequimur*" (PL 69, 18).

The origin of the R. C. is an open question, though scholars incline to date it to the time of pope *Damasus I. The language of the canon shows the rhythm and language of Roman speech before the 4th c., without the *cursus* that we find in later orations (cf. *clausula*), esp. at the time of pope *Leo I. The primitive *epiclesis too shows the antiquity of the canon's origin, i.e. before the Eastern pneumatological definitions. The earliest traces of the text are found in a partial citation made by St *Ambrose in *De Sacramentis* (IV, 21-27). It starts with "*Fac nobis hanc oblationem scriptam*" and ends with "*quod tibi obtulit summus sacerdos Melchisedech*". We may presume that Ambrose was also aware of a form of preface and petition, since in *De Sacram*. (IV, 14) he declares: "*Laus Deo defertur, oratio petitur pro populo, pro regibus, pro caeteris*". The present text of the canon is virtually that of St *Gregory the Great (Botte-Mohrmann, *L'ordinaire de la messe*, 15-25).

Patristic citations of some parts of the R. C. shed light on its history and *theology. At the "*Te igitur*", acc. to the evidence of St Leo, the church and its leaders are remembered: "*De fratribus vero quos et epistolis tuis et legatorum nostrorum relatione communionis nostrae cupidos esse cognovimus . . . pacis et communionis nostrae unitate laetentur . . . De nominibus autem Dioscori, Juvenalis, et Eustathii ad sacrum altare recitandis dilectionem tuam hoc decet custodire, quod nostri ibidem constituti faciendum esse dixerunt, quodque honorandae S. Flaviani memoriae non repugnet, et a gratia tua Christianae plebis animos non avertat*" (*Ep*. 80: PL 54, 913-915). In his *letter to *Justinian (*Ep. 4 ad Just*., 2: PL 69, 22), pope Vigilius cites the "*Te igitur*" with the words "*pro Ecclesia quam adunare, regere, custodire digneris*". The practice of naming bishops in this part of the *mass must be seen in the context of the doctrinal controversies of that time, when *communion with a group of bishops assured *orthodoxy of faith.

Also in the 4th c. arose the practice of reading the names of the offerers during the canon. The deacon read the names (*diptychs) aloud and the priest then continued with "*quorum tibi fides cognita estt*". St *Jerome criticized this practice in his works: *In Ezech*. 6, 18: "*Publiceque diaconus in Ecclesiis recitet offerentium nomina: tantum offert illa, tantum ille pollicitus est, placentque sibi ad plausum populi, torquente eos conscientia*" (PL 25, 175) and *In Hierem*. 2, 108: "*At nunc publice recitantur offerentium nomina et redemptio peccatorum mutatur in laudem: nec meminerunt viduae illius in evangelio, quae in gazophylacium duo aera mittendo omnium divitum vicit donaria*" (PL 24, 784). But pope *Innocent I approved this custom and ordered bishop *Decentius of Gubbio, who followed the practice of reading the diptychs before the canon, to observe the Roman custom "*inter sacra mysteria*" (*Ep*. 25: PL 20, 553).

The addition of saints' names among the *Communicantes* was made by pope Vigilius. In *Ep*. 2 (PL 69, 18), already cited, he speaks of special *capitula* for *Easter, *Ascension, *Pentecost and *Epiphany: however, those for *Christmas, Epiphany and Ascension seem to be of Leonine origin. The memoria of the Blessed Virgin, introduced in the Eastern *anaphorae after the council of *Ephesus, was in the R. C. from the time of pope *Gelasius.

The "*Quam oblationem*" presents a problem. St Ambrose concludes it with the words "*quod est figura Corporis et Sanguinis Domini nostri Iesu Christi*" (*De Sacr*. IV, 21: CSEL 73, 55, 2), an expression that recalls the ancient expressions "*imago*" and "*similitudo*" used by Gelasius in his letter to *Eutyches (*Tract. III, De duabus naturis in Christo*, 14: Thiel I, 451). The Gregorian text has dropped the term "*figura*" and introduced the causative particle "*ut*", so as to indicate the epiclesis's independence of the account of institution. In the Ambrosian text, on the other hand, it is necessary to enter, by virtue of the "*quod est figura*", into the account of institution, which acc. to Ambrose is actually consecratory: "*Consecratio igitur quibus verbis est et cuius sermonibus? Domini Iesu. Nam reliqua omnia, quae dicuntur in superioribus, a sacerdote dicuntur. . . . Ubi venitur, ut conficiatur venerabile sacramentum, iam non suis sermonibus utitur sacerdos, sed utitur sermonibus Christi. Ergo sermo Christi hoc conficit sacramentum*" (*De Sacram*. IV, 4, 14: CSEL 73, 52, 10-16; the question of the *epiclesis is dealt with in its own article).

The commemoration of the dead must have been introduced into the R. C. from some Christian funerary prayer, around the 3rd c., as we see from the text whose Latin recalls that of the catacombs. St Augustine attests the practice of reading the names of the deceased in the eucharistic prayer: thus he writes in *Serm*. 172, 2: "*ut pro eis qui in corporis et sanguinis Christi communione defuncti sunt, cum ad ipsum sacrificium loco suo commemorantur, oretur, ac pro illis quoque id offerri commemoretur*" (PL 38, 936).

The final *doxology of the R. C., with the formula "*in unitate Spiritus Sancti*", seems to be of later date, since it uses a fully developed doctrine of the *Trinity. Ambrose's doxology, on the other hand, seems to be much older (*De Sacram*. VI, 5, 24: SCh 25a, 152). Both lack *Hippolytus's viewpoint, of the church in which the doxology is offered to the *Father and the Son with the *Holy Spirit (*Trad. Ap*.: LQF 39, 55).

L. Eizenhöfer, *Canon Missae Romanae* (Rer. Eccl. Doc., s. minor, subs. stud. I, VII): Pars prior: Traditio textus (Rome 1954), pars altera: Textus propinqui (Rome 1966); A. Hänggi-I. Pahl, *Prex eucharistica*, Fribourg (Switz.) 1968, 423-477 (L. Eizenhöfer-I. Pahl edition, with bibl.); F. Cabrol, Canon, *DACL* 2, 1847-1905; E. Bishop, *On the early texts of the Roman Canon: Liturgia historica*, Oxford 1918,

77-115; B. Botte, *Le canon de la messe romaine, éd. critique, introduction et notes*, Louvain 1935; A. Baumstark, Das Problem des römischen Messkanon, *EphemLiturg* 53 (1939) 204-244; M. Andrieu, *Les Ordines Romani du haut moyen-âge. 2. Les textes*, Spicil. Sacr. Lovan. 23, Louvain 1948; B. Capelle, Innocent Ier et le canon de la messe, *RecTh* 19 (1952) 236-247 (=*Travaux Liturgique* , 2, 236-247); B. Botte - C. Mohrmann, *L'ordinaire de la messe*, Paris 1953; B. Capelle, Problèmes du "Communicantes" de la messe, *Riv. Lit.* 40 (1953) 187-195 (=*Travaux liturgique*, 2, 269-275); id., L'évolution du qui pridie de la messe romaine, *RecTh* 22 (1955) 5-16 (= *Travaux liturgique*, 2, 276-286); id., L'intercession dans la messe romaine, *RBen* 65 (1955) 181-191 (= *Travaux liturgique*, 2, 248-257); P. Borella, Il canone della messa romana nella sua evoluzione storica, *Ambrosius* 35 (1959), suppl., 26-50; A.G. Martimort (ed.), *L'Eglise en prière*, Tournai, 1961; L.C. Mohlberg, *Liber sacramentorum romanae aecclesiae ordinis anni circuli*, Rer. Eccl. Doc. 4, Rome 1960, ²1968; A.G. Martimort (ed.), *The Church at Prayer: the Eucharist*, Shannon 1973; J. Jungmann, *The Mass*, Collegeville 1976; A. Keifer Ralph, The Unity of the Roman Canon: An Examination of its unique structure, *Studia Liturgica* 11 (1976) 39-58.

A. Chupungco

ROMANIANUS. A wealthy native of Thagaste, fellow-citizen and benefactor of *Augustine, whom, when his father died (371), R. enabled to go and study at *Carthage (*C. Acad.* 2, 2, 3 f.; a work dedicated to R.: *ibid.* 1, 1). Finding themselves together at *Milan (385), "it was [R.] who insisted more than anyone on putting into practice the project" (*Conf.* 6, 14, 24) of common life among friends (386). R. had a son, *Licentius (cf. *De ord.* 1, 2, 5), who defended the Academic doctrine in *De ord.* (1, 4, 10). R. was related to *Alypius (*Ep.* 27, 5). Augustine wrote *Contra Academicos* to induce R. to study *philosophy (*C. Acad.* 2, 8; *Conf.* 6, 24). A few years later (390) he addressed *De vera religione* to him, to convince him to embrace the Christian faith as the true religion (*C. Acad.* 2, 8; *De vera rel.* 12; *Epp.* 15, 1; 27, 4). R. became a Christian (396) and remained a follower of Augustine (*Ep.* 27, 4; *Ep.* 32: *Paulinus of Nola to R. and Licentius). The name R. appears on an inscription at Thagaste (CIL 8 *suppl.* 17226).

DCB 4, 550-51; *PAC* 1, 994-997; W.H.C. Frend, *The Rise of Christianity*, Philadelphia-London 1984, 664-5, 723-24.

R. De Simone

ROMAN LAW AND CHRISTIANITY

I. Christian doctrines and Roman law - II. Roman law and Christian institutions.

The problems that emerged from the impact of Christianity on Roman law are many and complex and have had very varied answers, some tending to magnify the influence of Christian doctrines on Roman law in the Late Empire (I), others to reduce it to truer proportions. Here, we will evaluate the influence of Roman law on the evolution of Christianity in general, and on ecclesiastical institutions in particular (II).

I. Christian doctrines and Roman law. Christian influence on Roman law derived primarily from the will of the *princeps*; its most direct expression was imperial legislation: *1)* the emperors accepted and imposed the new religion; they multiplied the institutions that favoured the church; they assured the triumph of the Christian faith, as defined by the church's own *authorities, threatening all sorts of sanctions against dissent (exclusion from civil functions, incapacity to succeed, etc.). An abundant legislation suppressed *heresy, *paganism and Judaism (*CT* XVI). The emperors also regulated the status of clerics and bishops, as well as the conditions necessary to proceed to *orders; they were lavish with privileges and *immunities favouring the *clergy. *Valentinian III formally recognized the Roman primacy (Novella 18 of 19 June 448); *2)* we must also note a Christian influence in legislation on the *family and on society. The church fought against *abortion, abandonment of newborn infants, trade in *children and abuses of *patria potestas*, and would have liked an official restriction on freedom of *divorce (council of *Carthage of 13 June 407). It penalized those who broke off betrothals without good reason (*CT* III, 5, 2; cf. council of *Elvira, can. 54) and obtained an interdiction on consanguineous marriage (*CT* III, 12, 2-4; cf. Elvira, can. 61). It supported *slavery, but demanded the abolition of the custom of branding slaves on the forehead and of keeping servants' families apart; it favoured enfranchisement and obtained civil recognition for *manumissio in ecclesia*. It fought against gladiatorial games and managed to obtain their suppression. It even went so far as to modify the festivals of the civil calendar and substitute the *Lord's day* for the *dies solis* (cf. *Sunday); it aimed to keep intact the sacred character of days reserved for God, getting theatrical and circus games forbidden on them; *3)* less incisive was its influence in the economic field. Some have wished to see such influence in the legislation on injury or limitations on rates of interest, in the enlargement of the notion of criminal intent, in the lessening of formalism in favour of the obligatory character of simple pledge, in sanctions against unjust enrichment; but it is hard to assess it in each of these cases, though we need not reject it altogether; *4)* criminal legislation in the Late Empire was marked by increased severity, almost cruelty. Christianity had very little influence on it, and that only sporadically and wholly fortuitously. The church obtained the suppression of the penalty of crucifixion, and the opportunity to take part in supervising prisons, so as to be able to comfort prisoners; *5)* historians have managed to discover further traces of Christian influence on Roman legal procedure, but the comparisons shown are very insignificant (Gaudemet, 512).

II. Roman law and Christian institutions. The influence of Roman law on the church appears mainly in terminology, in legal procedure and in institutions: in fact, in the transposition or adaptation of Roman rules to the church's own needs. *1)* among terminological borrowings, that of *ordo* to designate members of the church hierarchy is among the most significant, since this term, from the start, was used to describe the ruling groups of the *civitas*, particularly the municipal senate. We must also mention that the powers of the high priest, from the start, were defined by the technical terms *auctoritas* and *potestas*. As for the terminology of decretals, this was greatly indebted to that of imperial decrees; *2)* the influence of Roman law was particularly felt in the area of legal procedure. The *pope's legislative power was certainly not a mere borrowing from Roman institutions (it remained based on *Scripture and grew out of the primacy), but this made it no less indebted to contemporary models than was the elaboration of decretals. On the other hand, the ways of appealing to the pope were clearly inspired by imperial law (*relatio, provocatio*). The same was true of *councils. This institution was not simply borrowed from Roman public law, and in its organization it greatly benefited from contemporary models (ways of holding sessions, voting procedure, drawing up of *acts, etc.). But canon law drew as widely as possible on Roman models in the procedural sphere: it took over the technique of *praescriptiones (de persona, de mandato, de tempore)* and that of *contestatio litis*; it also adopted details of means of proof and conditions relating to witnesses (e.g. exclusion of *infames personae*). It goes without saying, then, that canon law owes the entire context of its procedure of appeal to Roman law; *3)* the influence of Roman law on Christian institutions was multiform; most clear in the matter of *marriage. The church took over, in essentials, the Roman notion of *matrimonium* and the conditions for fulfilling it (*dos, tabulae nuptiales*); it adopted its doctrine on marriage by mutual consent, though immediately perfecting it (Munier, 17); it retained betrothals according to the Roman laws, but soon distinguished them clearly from marriage proper. It was unyielding in its principles on matters of fidelity and indissolubility. Many of the norms relating to the status of clerics were inspired by those regulating civil careers (prohibition of moving from one post to another, promotion by degrees and at intervals). Finally, the outward signs of power (*dress and insignia) of Roman officials were taken over to signify the various ranks of ecclesiastical dignitaries (*pontificalia*, dalmatic, pallium). Lastly, we must note the clear parallel between the civil territorial divisions and the ecclesiastical ones.

Biondo Biondi, *Il diritto romano cristiano*, 3 vols., Milan 1952-1954; J. Gaudemet, *La formation du droit séculier et du droit de l'Eglise aux IVe et Ve siècles*, Paris 1957; *L'Eglise dans l'Empire romain (IVe-Ve siècles)*, Paris 1958; Ch. Munier, *L'église dans l'Empire romain (IIe-IIIe siècles)*, Paris 1979.

Ch. Munier

ROMANUS. A member of the *clergy of *Caesarea, passing through *Antioch at the time of *Diocletian's *persecution, he was arrested on account of his zeal in exhorting Christians brought before the judge. He was condemned to have his tongue torn out and was strangled in prison. We owe these facts to *Eusebius of Caesarea (*Mart. Pal.* 2). Among later panegyrics, that of *John Chrysostom (PG 50, 605-612) is basically faithful to Eusebius, adding only the doctor's intervention, while that of *Eusebius of Emesa (PG 24, 1097, where it is wrongly attrib. to Eusebius of Caesarea), *Prudentius's poem (*Perist.* 10) and *Severus of Antioch's three *homilies (*Hom.* 1, 35 and 80) indulge in legend. From the Middle Ages, Passions of the saint multiplied (BHG 1600yz; BHL 7298-7304, 7299b; BHO 1028); there is also an unpublished *Syriac version. Commem. 17 Feb *in Caesarea*, 18 Feb *in Antiochia* by *Mart. hier.*, 18 only in the Byzantine Synaxaria. Passions and *martyrologies give him a companion, Barula, a name which is a deformation of *Barlaam.

M. Simonetti, Nuovi studi agiografici. I. Su S. Romano, *RAC* 31 (1955) 223-33; *Vies des SS.* 11, 604-7; *LTK* 9, 15; *BS* 11, 338-42.

V. Saxer

ROMANUS Melodus. The question of whether R., the greatest of the Byzantine hymnographers, lived at the time of *Anastasius I (491-518) or of Anastasius II (713-716) was long disputed; but it was finally resolved in favour of the former by the discovery in 1905 of the original Greek text of the miracles of St Arthemius, which speaks of a miracle worked at the time of the emp. *Heraclius (610-641) in favour of a young man who sang R.'s *hymns: if Melodus's poems were sung at that time, he could not have lived at the time of Anastasius II. Little is known of R.'s life. Born at *Emesa in *Syria in the late 5th c., of a family of Jewish origin, he was a deacon when he settled at *Constantinople at the end of the reign of Anastasius I. His death seems to have been before that of *Justinian (565), but after 555.

R. composed a great number of sacred *hymns, the *kontakia*; we have c. 100 of them, but not all are authentic. The *kontakion*, whose origin is much disputed, may be considered a metrical *homily; introduced by the *proemium*, it consists of several strophes called *troparia*, of which the first, the *irmon*, is the model for the following strophes. At the end of each *troparion* is a refrain, called *ephymnion*; the various strophes of the *kontakion* follow the laws of isosyllaby and homotony and are linked to each other by the acrostic which generally gives the author's name, sometimes also the subject, more rarely the subject without the author's name. The *kontakion*'s metric has no connection with that of classical poetry, since it is purely tonic and syllabic. The melodic recitation of the *kontakion* is indicated in the superscript by the determination of the tone. The content of R.'s hymns is connected with the various themes of the liturgical celebrations; they have as object the main feasts of the church, OT and NT characters and episodes, *martyrs and saints, and various subjects. Their sources are in *Scripture, the old Passions of the martyrs, Greek homiletic and, in particular, *Ephrem. The poetry of R., called not without exaggeration the "Christian Pindar", often attains moments of high lyricism and intense participation: as e.g. the hymns on *Christmas or on *Mary's lament at the Cross, in which the dialogue between the Virgin and her suffering Son is rendered in tones of acute, vibrant dramatic force and manifests the depth of Melodus's inspiration.

CPG 3, 7570. The most complete crit. ed. of R.'s hymns is that of P. Maas-C.A. Trypanis, *Sancti Romani Melodi Cantica. Cantica genuina*, Oxford 1963; *Cantica dubia*, Berlin 1970.
Partial editions: J.B. Pitra, *Analecta Sacra*, I, Paris 1876, 1-241 (29 hymns); G. Cammelli, *Romano il Melode. Inni*, Florence 1930 (8 hymns, with It. tr.); N.B. Tomadakis, Athens, v. I, 1952 (14 hymns), v. II, 1954 (14 hymns), v. III, 1957 (8 hymns), v. IV, 1959-1961 (13 hymns); J. Grosdidier de Matons, SCh 99 (Hymns I-VIII), 110 (Hymns IX-XX), 114 (Hymns XXI-XXXI), 128 (Hymns XXXII-XLV), 283 (Hymns XLVI-LVI) (with Fr. tr.).
Other translations: French: R.R. Khawam, *Romanos le Mélode. Le Christ Rédempteur. Célébrations liturgiques*, Paris 1956; id., *Hymne sur la résurrection*, in *Le grec chrétien*, Paris 1965, 175-185; German: G.H. Bultmann, *Romanus der Melode. Festgesänge*, Zürich-Munich-Paderborn-Vienna 1960; English: M. Carpenter, *Kontakia of Romanos, Byzantine Melodist. I: on the Person of Christ*, Columbia 1970; *II: on Christian Life*, 1973; It. tr. by G. Gharib, Rome 1981 (63 hymns).
Studies: E. Mioni, *Romano il Melode. Saggio critico e dieci inni inediti*, Turin 1937; C. Chevalier, Mariologie de Romanos (490-550 environ), le roi des Mélodes, *RecSR* 28 (1938) 48-71; G. Zuntz, Probleme des Romanos-Textes, *Byzantion* 34 (1964) 469-534; C.A. Trypanis, The Metres of Romanos, *Byzantion* 36 (1966) 560-623; K. Mitsakis, *The language of Romanos the Melodist*, Munich 1967; E. Salvaneschi, Adattamento interlinguistico come mezzo espressivo in Romano Melodo, *AATC* 39 (1974) 21-68; J. Grosdidier de Matons, *Romanos le Mélode et les origines de la poésie religieuse à Byzance*, Paris 1977.

S. Zincone

ROMANUS of Roso. Bishop of Roso (or Rosi on the gulf of Issus, N of *Antioch and Seleucia Pieria, in Syria Secunda), contemporary and opponent of *Severus of Antioch (512-518); known only from references and citations. In 516 Severus, in a *Letter to Antiochus*, *abbat of the *monastery of Beth Mar Bassos, rebuked the abbat for having defended R.'s *orthodoxy. From the *Letter* it seems that R. was already dead. In the *Letter* and in his *Cathedral homilies* 119 and 124, Severus cites several passages of a work by R. called the *Ladder* (*Seblatha*, i.e. κλῖμαξ) and severely refutes their absurd *exegesis. R. interpreted *Mary at the wedding of *Cana as a symbol of concupiscence, and in this context distinguished *sins into natural and voluntary sins and sins against nature (*Hom.* 119, 390 ff.); he saw in the variety of the synoptic account of *Peter's confession an image of future *heresies (*Hom.* 124, 222 f.). Severus had known the *Ladder* at *Constantinople (503-511) and had refuted it in a book. The letter also cites a phrase of a work by R. against the *Theopaschites and states that R. had been condemned by a *council of bishops, first at Antioch and then (after 518) at *Alexandria: R. was compared to *Nestorius, the *Manichees and the *borboriani*.

CPG 3, 7117-7121; PO 26, 375-379; 29, 222 f.; Cod. Harvard (Houghton Library) Syr. 22, f. 68a. 75a. 76a.; R. Draguet, *Julien d'Halicarnasse et sa controverse avec Sévère d'Antioche sur l'incorruptibilité du corps du Christ*, Louvain 1924, 80 f.; E. Honigmann, *Evêques et évêchés monophysites d'Asie Antérieure au VIe siècle*, CSCO 127, Subsidia 12, Louvain 1951, 82 f.; S.P. Brock, Some New Letters of the Patriarch Severus, *SP* 12 (= TU 115), Berlin 1975, 22 ff.

A. De Nicola

ROME

I. History and archaeology - II. Councils - III. Liturgy.

I. History and *archaeology.

1. a) The existence at R. of a small community before the arrival of *Paul (and doubtless of *Peter) is attested in the letter to the Romans (AD 57-58), addressed by the Apostle to those who preceded him there: this small group came from a Jewish background, as the letter explicitly attests (2; 4, 1), and as the names of believers suggest (Rom 16, whose *manuscript tradition is less sure). These first converts who met in private houses (Prisca: Rom 16, 5) received from Paul a *deaconess who bore the apostolic *letter, but we do not know how this first community, founded by missionaries unknown to us, was organized. Some signs suggest a state of tension between these *Judaeo-Christians and the powerful urban synagogues: at *Corinth Paul met Aquila, a Jew from *Pontus, who had come from Italy with his wife Priscilla (Acts 18, 3). We must doubtless connect this exile with Suetonius's statement (*Vita Claudii* 18, 2) that in AD 49 *Claudius expelled from R. some Jews stirred up "by a certain *Chrestus*"; this may allude to Christians ill-identified by the police. In any case Paul, who knew another Judaeo-Christian, Epaenetus, who had come to R. from *Asia, certainly arrived at R. some years later, after appealing to Caesar, and was kept for two years in an imprisonment that did not interfere with his missionary zeal, since he had contact with Judaeo-Christians (*Mark, Aristarchus) or *converts from *paganism (like perhaps *Timothy: Acts 28, 30). We cannot date Paul's *martyrdom, which was certainly under *Nero; but no-one has contested his Roman mission. In the case of Peter, there has been a whole series of controversies over whether the *apostle came to Rome, but, as Lietzmann has demonstrated, this debate has been overtaken: the evidence of patristic literature (1 Pet 5, 13; Clem. Rom., *Cor.*; Irenaeus; etc.), the evidence suggested by archaeology (Fasola) and the local cult of the two *apostles are enough to establish that Peter came to R. and was martyred there. One *tradition insists - for symbolic and ecclesiological reasons - on putting the martyrdoms of Peter and Paul on the same day (29 June) and the same year (often 67). It is hard to define concretely the scope of this dual mission, though later traditions sometimes attributed to the community evangelized by the two apostles a particular importance in the transformation of the Good News into written gospels (*Mark, *Luke); in any case some evidence, in Clement, suggests the influence of Judaeo-Christian themes for the elaboration at R. of a whole *theology of apostolic tradition.

b) This church is hardly known for the first two centuries: *Tacitus brings it into Roman history (*Annali* 15, 44) when he records the massacre of Christians martyred after the fire of R. in 64 by Nero, looking for culprits to placate public indignation after a disaster that had devastated the centre of the city (from that moment, the Christians were marked out by the police and by Roman public opinion). But we must await the evidence of *Clement of Rome in the late 1st c. (*Ep. Cor.*) to catch a glimpse of the *organization of this community: it is aware of already belonging to a post-apostolic age; some permanent *ministries are established, and the author invokes them using the vocabulary of Jewish priestly tradition as well as the image of the *militia Christi*; at the head of the church, which is led by a college of ministers who guarantee the episcopate and *diaconate (as a still imprecise vocabulary indicates), is Clement, first of the college, representative of and responsible for the mission. In the mid 2nd c., a little apocalypse, written by *Hermas, an ex-slave, brother of bishop Pius, allows us to suppose the persistence at R. of charismatic functions, but he also mentions the priests and their college, attributing the duty of *assistance to the deacons and bishops. This evidence must be supplemented by that of *Irenaeus of Lyons, who demonstrates the existence of a unitary organization entrusted to a responsible figure, though he does not, in the first local use, clearly bear the name of bishop. Irenaeus lists those who, after Peter and Paul, transmitted the apostolic *tradition: Linus, Cletus, Clement to the end of the 1st c., Evaristus, Alexander, Sixtus, Telesphorus, Hyginus, Pius, Anicetus, Soter, Eleutherius and *Victor (189-199). He describes an *episcopal list *ante litteram*, i.e. the succession of those who from time to time held the ministry of local unity. From the 2nd c. we see, more clearly than in Clement's time, the personal intervention of the Roman bishops in the name of their church: e.g. those of Soter, solicited in the mid-century by bishop *Dionysius of Corinth, Anicetus, who received *Polycarp of Smyrna at R., Eleutherius, and finally Victor on the *Easter question. This divided Christians within the city itself; then it opposed the Asiatic churches to those of R. and *Alexandria on a question of *computus, but above all on the choice of the festive day: that of the Passion or the *Sunday of the Resurrection? These interventions (esp. that of Victor, who broke with the Asiatic episcopate) reflect the *authority, "in the Latin sense of the term", of this doubly apostolic church, to which Irenaeus assigns the concrete function of being, in the West, the privileged centre of reference for church unity (*Adv. haer.* III, 3, 2). The spread of R.'s role as political and economic capital of the Roman world facilitated this particular mission, but the Roman church increasingly claimed that it arose because of its dual *apostolicity. At the same time it was the missionary metropolis, probably for *Gaul and *Africa, certainly for *Italy. Besides, this church provided numerous examples (cf. *Justin for the organization of the *liturgy; the Easter festival); it welcomed debates on *apologetics and on a *theology that began, with Justin, to use the conceptual instrument of *philosophy. This explains the attraction exercised by R. on many itinerant preachers and theologians who sought to establish themselves or, at least, to make disciples there: *Marcion and the gnostic *Valentinus, but also *Theodotus of Byzantium, an *adoptianist expelled from the Roman church in 198; *Praxeas, who defended a *monarchian theology and was condemned by Eleutherius; and the disciples of *Montanus, who had a similar lack of success in 177.

c) The 3rd c. was a period of great organization dominated by some prestigious bishops: after *Zephyrinus (199-217), we must cite *Callistus, an ex-slave (217-222), followed by Urban (221-230), *Pontian (230-235) and Antherus (235-236), these last two victims of *persecution. *Fabian (236-250) enjoyed great spiritual prestige, acc. to *Eusebius (*HE* VI, 29); *Cornelius (251-253), who died in exile, showed great energy when *Novatian's *schism threatened; after Lucius (253-254), *Stephen (254-257) claimed primacy with authoritarian decisiveness; *Sixtus II (257-258) was overwhelmed by *Valerian's persecution, but *Dionysius occupied the see from 259-268, and after him Felix (269-274), Eutychian (275-283) and Caius (283-296), of whom we know little. From then until *Miltiades, the *chronology is uncertain because of *Diocletian's great persecution and the void caused by it: *Marcellinus until 304, then *Marcellus and Eusebius (309 or 310). For the whole of this period the evidence of the *Apostolic Tradition*, a small treatise of missionary organization and liturgy composed by a vigorous theologian, *Hippolytus of Rome, illustrates the organization of the *synaxis, over which, in the course of the century, the priests could preside in the bishop's absence and as his delegates. The precautions taken in accepting candidates for *baptism, and the project of a *catechumenate, attest the progress of *conversion. At the same time Hippolytus describes the system of *assistance, the meals of the *agapè*, the particular forms of a collective charity for burial (the deacon Callistus was in charge of the church *cemetery, a catacomb near the Appian Way). This was the time when the seven deacons were given the duty, for the collection and multiple uses of charity, of surveying each of the seven zones into which the city was approximately divided. This community probably had no *churches (in our sense of the word, as a specialized meeting-place permanently set aside for *liturgy); at any rate we do not know of any; but for these multiple services R. had, more than elsewhere, the use of a specialized *clergy: in 251 the clergy counted at least 46 priests and seven subdeacons with as many deacons, 42 acolytes, 56 lectors, exorcists and door-keepers and, acc. to the same witness (Euseb., *HE* VI, 43, 11), the church fed 1500 *widows and *poor people. This last figure shows the important following of *converts, who amounted to a considerable minority of the great city (2-5 per cent?). Pope Callistus showed pastoral intelligence in this new situation, organizing with benevolence the *penitence of believers who had fallen into *sin after *baptism and making efforts to facilitate the union of Christian aristocrats with spouses who shared their faith, even if they were not of the same rank. This last arrangement demonstrates the sociology of this primitive church: the aristocracy, despite some notable examples (*Domitilla at the end of the 1st c., *Apollonius in the 2nd c.), was still little penetrated by Christianity, but conversion affected a variety of circles: members of the Palatine service (e.g. Prosenes, whose epitaph, of 217, is one of the oldest Latin Christian inscriptions) and a whole population of Greek speakers, soldiers and artisans. Of this first Christian community we know mainly the *cemeteries and the epitaphs (see *Epigraphy), which attest the appearance, in the 3rd c., of a Christian formulary which, as in Jewish epigraphy, expresses the hope of victory over death with a conviction completely unknown to *pagan texts. At the same time the first *images of Christian figurative art appear in the decorations of the catacombs. Certainly, many trials ravaged this community: *persecution struck from the reign of the emp. *Maximinus Thrax in 235 (deportation of bishop *Pontian and of *Hippolytus); but it became general with *Decius's edict of 250 (*martyrdom of bishop *Fabian), the laws of *Valerian (257 and 258), which led to the martyrdom of pope *Sixtus and the deacon *Lawrence, and finally the edicts of *Diocletian (303-304), which raged ferociously (e.g. *Agnes). After these trials, the church was split on how to judge the numerous *lapsi*; in particular, in 251 the priest *Novatian gathered around himself a rigorist party which he led into *schism against bishop *Cornelius, elected by a more moderate majority. Cornelius called against his rival, now antipope, a *council of some 60 Italian bishops. This assembly witnesses to the expansion of R.'s primacy in the peninsula; in 254 bishop *Stephen received the appeal of two Spanish bishops; *Cyprian of Carthage invited him to depose a bishop of *Arles. The Roman bishop made known to Christian *Africa the rules of his discipline, which forbade the rebaptism of converted *heretics: even if this discipline met intransigent resistance from Cyprian, the episode is significant, exactly like that of bishop *Dionysius who took a stand, at the request of the Alexandrian churches, against the too *subordinationist preaching of bishop *Dionysius of Alexandria. These interventions made explicit the primatial pretensions of a church which claimed to guard and maintain the apostolic tradition, the church of the *cathedra Petri.

2. *The 4th and 5th cc.* After *Constantine's victory, Oct 312, at the Milvian bridge N of R., the attitude of the authorities towards the church changed radically: we pass from the privileged support at the time of the first "Christian" emperor (312-337) to an exclusive backing under *Theodosius (379-395). The evolution of the political situation, but also of social and economic conditions, certainly played a considerable role in the development of the Roman church at the time of the Christian empire (4th and 5th cc.).

a) Above all in the great movement of *conversion, which particularly in this period affected urban populations, R. gives an example of an extremely fruitful missionary experience, to which imperial favour gave only indirect and secondary help. Concrete proof of this are the numerous *churches (now understood as specialized buildings) erected for Christian *assemblies inside the walls and in the suburbs of the city. In the first half of the 4th c., R. owed to Constantine the foundation of the episcopal church with its baptistery (*basilica constantiniana*, then St John Lateran): a grandiose basilica in which, for the first time, the bishop could bring all his people together; and to the imperial benevolence of Constantine and his successors R. also owed the construction of basilicas near the tombs of the Roman *martyrs: St Peter's (cf. *Vatican), St Lawrence's near the Tiburtina, a building near St Paul's tomb, enlarged at the end of the century, a great basilica near the Via Labicana where Constantine originally thought to put his own tomb, and other *martyria* near the Via Appia (St *Sebastian's) or the Via Nomentana (St Agnes's). In all this activity, few urban foundations were financed by the emperor (except the Palatine church near the residence of the empress *Helena: now S. Croce *in Gerusalemme*); the building of the urban churches was carried out, particularly in this period, by the bishops (*Sylvester, Mark, *Julius, *Liberius), while the private Maecenatism that dominated from the mid 4th to the mid 5th c. was beginning to manifest itself (Lucina): in some decades we owe more than 20 edifices to Christian building activity (see *Tituli*); built according to the size of *donations, they ended by creating a tight network throughout the city. In the 5th c. the bishops always took a munificent part in building activities, as *Sixtus III's foundation of the great basilica on the Esquiline (S. Maria Maggiore) indicates; but after this *pope, who died in 440, the rhythm of new constructions slowed down visibly (note only pope *Simplicius's S. Stefano on the Caelian) and work was directed to embellishment and restoration, necessary after the two sacks of R., devastated for three days in 410 by the *Goths and three weeks in 455 by the *Vandals. Meanwhile bishops, *clergy and even *laymen looked after the catacombs, for the cult of their martyrs and the visits of pilgrims. Begun by the start of the 4th c., *pilgrimages developed particularly at the time of *Damasus, *Siricius and, in the 5th c., *Innocent and Boniface. Under *Leo I, an aristocratic lady financed the building of a basilica in honour of St *Stephen on the Via Latina. Meanwhile, the church was enriched by *donations which, added to existing imperial and private ones, allowed for the maintenance of the buildings and their clergy: if in the mid 4th c. this patrimony was still quite a small amount, the more so since the bishops seem to have been tied to the foundation deeds (which constituted the *tituli*), in the 5th c. the church's economic power asserted itself when the pope conceded, despite the resistance of donors, the power of capitalizing on the revenue and organizing a common *property. This material institution allowed a more efficient organization of the mission; the episcopal *liturgy was enriched by great community assemblies in the great basilicas (the episcopal church, S. Maria Maggiore, St Peter's, etc.), used by the bishop, in the course of the year, from station to station. Thus the period of Lent, extended from three to six weeks, united catechumens and penitents until *Easter in meetings of instruction or reconciliation held in the presence of the whole people. But we must also take into account the celebration of *Christmas, the institution of periods for *fasting and for the offertory (see *Liturgical year). Local pastoral activity was organized in the titular churches which began, at the end of the 5th c., to be given baptisteries. Entrusted to the priests (c.100 in the 5th c.) who celebrated in unity with the bishop thanks to the rite of *fermentum*, these churches had a fixed *clergy (acolytes, lectors).

In the 5th c. they became centres of local life, which gave hospitality to meetings for offertories (from Leo's time) and to the (now public) ceremonies of Christian *marriage. The system of *assistance, still entrusted to the deacons divided among the seven sectors, received the support of the "titles" and also of charitable foundations helped by generous believers (*xenodochia*), while the **fossores* looked after burial services in the catacombs. Meanwhile the organization of the sanctoral (**Depositio martyrum* of 336, then "Jerome's" *Martyrology for later evidence) established, in rivalry with pagan festivities, the celebration of *martyrs buried in the suburbs as well as the celebration of relics present in the city (St Peter's chains in S. Pietro in Vincoli). Monastic communities settled near great *martyria* (*Ad catacumbas*; St Peter's; S. Lorenzo; etc.). The organization of the *clergy now had a discipline rigorously fixed (see the decretals of *Siricius, *Innocent, etc.) and with controlled recruitment: custom marked out two types of career, that which led from exorcist or lector to acolyte, and finally to the priesthood; and that which led to the diaconate, from which the bishop usually came. In this clergy we distinguish a small educated élite (of priests and deacons) often recruited from Christian families connected with the service of the church: this was the case, e.g., of pope Damasus. From this élite came the bishops, including some with strong personalities: after *Miltiades (311-314) came *Sylvester (314-335), Mark (336), *Julius (337-352), *Liberius (352-366) and *Damasus (366-384). *Siricius (384-399) seems a colourless personality compared with his predecessor, a poet and diplomat, but he was an inflexible administrator. After Anastasius (399-401), the political situation unsettled the episcopate of a great pope, *Innocent (401-417), who was succeeded by the despotic and maladroit *Zosimus (417-418), then by Boniface (418-422), *Celestine (422-432), an outstanding spirit, and finally *Sixtus III (432-440). The pontificate of *Leo the Great (440-461) marked a decisive moment, all the more since R. had a bishop who was on the level of the great Eastern theologians. After him, *Hilarus or Hilarius (461-468) and, even more, *Simplicius (468-483) seem more colourless. Felix II (III) (483-492) was the first bishop produced by the Roman aristocracy. *Gelasius (492-496), inflexible by nature and duty, already belongs to another age, when the Western empire had finally disappeared. We can measure the effectiveness of this pastoral work by the success of *conversions. If at the start of the 4th c. only a few aristocratic families were Christian (the Anicii, the Probi), and if at the end of the same century a pagan resistance was organized around *Symmachus and Nicomachus *Flavianus, at the end of the 5th c. there were no more open defenders of the old *idols, but simply believers disposed to maintain, out of tradition or precaution, some old rituals like the Lupercalia. The crises that rent the Roman community tended increasingly to rend Christian factions: e.g. the strife whipped up by *Ursinus against the election of Damasus (366-370) or by Eulalius against Boniface (418).

b) In this new period for the church, the Roman see began to extend its influence more concretely, in reply to many solicitations: *i)* in traditional dependence on R., confined to suburbicarian *Italy after the mid 4th c., the *pope exercised a metropolitan *authority which is cited as an example at the council of *Nicaea of 325 (can. 6). He controlled episcopal elections, judged church affairs and rigorously imposed the canons of Roman discipline. In the northern provinces of the peninsula this activity came up against the influence of the bishop of the imperial residence, *Milan, esp. at the time of *Ambrose (374-397), and, in the 5th c., though less importantly, that of *Ravenna; *ii)* in the mission's western provinces, *Spain and still more *Gaul, where church *organization was rudimentary at the start of the 4th c., R. exercised a sort of *patriarchate. From these regions came the appeals of bishops: that of *Himerius of Tarragona to *Siricius, those of *Exuperius of Toulouse and *Victricius of Rouen to *Innocent. These mostly requested advice on organizing their clergy and their *liturgy; they received decretals, i.e. small disciplinary treatises that later circulated in collections and thus constituted a code of religious law. These solicitations more rarely involved the pope in theological conflicts: Damasus rejected *Priscillian's appeal, but Innocent replied to the council of *Toledo which had consulted him on the resistance of that heresiarch's partisans (404) and Leo, on the same problem of Priscillianism, gave advice to the bishop of *Gallaecia, *Turibius. In short, the Roman see dealt with *causa maiores* and intervened in episcopal affairs; when communications between R. and Gaul risked being broken by the barbarian invasions, Zosimus entrusted the functions of a vicar to the bishop of *Arles (418). But this system did not work well: Boniface restricted its importance, and Leo finally reduced the rights of the Arlesian *Hilary to those of a simple metropolitan; *iii)* the same system was adopted with the bishops of the Illyrian prefecture, which passed in the 5th c. under the authority of the Eastern emperor. Preoccupied with escaping the influence of a nascent patriarchate in the "Second Rome", *Constantinople, the episcopate accepted Innocent's establishment of a vicariate at *Thessalonica: this disappeared in the second half of the 5th c.; *iv)* *Africa, with a fulcrum in the great Christian metropolis of *Carthage and an organized episcopal college, did not tolerate this sort of experience: in 323 *Constantine had instructed the Roman *Miltiades to arbitrate in the controversy between *Caecilian of Carthage and his rigorist adversary. The failure of this intervention, at the start of the *Donatist schism, long discouraged Roman intervention. But in the 5th c. *Aurelius of Carthage and *Augustine of Hippo sought the authority of *communion with R. to defeat the Pelagian menace (417): they extolled Innocent's sentence all the more since his successor seemed, for a time, determined to reopen the question. With imperial support, they managed to make Zosimus give way: in summer 418 he decided to condemn *Pelagius and his disciple *Caelestius without half measures. Thus Christian Africa recognized the prestige of the Roman primacy and gave way more and more to the authority of its communion, esp. when the invasion of the Arian *Vandals created the danger of a persecution; *v)* for the Christian *East, Roman primacy more and more represented an authority that guaranteed a certain ecclesiastical freedom, since the pope was not directly involved in the conflicts of the Eastern parties: to the great *councils called by the emperor at *Nicaea (325), *Sardica (343), *Ephesus (431 and 447) and *Chalcedon (451), he sent his delegates. Analogously, bishops persecuted by imperial policy sought the support of the apostolic see: this was the case of *Athanasius, expelled from *Alexandria by the Arians in 339 and vindicated by a Roman council in 340. When the emp. *Valens once more favoured *Arianism, *Basil of Caesarea sought to obtain Roman support (372-379), despite the incomprehension of theological vocabulary and differences of ecclesiastical policy that separated him from Damasus. These solicitations induced the Roman bishop to intervene both in disciplinary questions and in problems of faith. In a council held at R. in 377, Damasus vigorously formulated a condemnation of the *pneumatomachi, which was adopted in 381 by the council of *Constantinople. Against bishop *Nestorius of Constantinople, *Cyril of Alexandria sought an alliance with the Roman see, which associated itself, in the person of *Celestine, in the decisions of the council of Ephesus, Celestine being hailed "as a new Paul" (431). With *Leo, Roman intervention in theological debate became really decisive, since the papal legates succeeded in getting the *christology expounded by the pope in his **Tomus ad Flavianum* (451) adopted in the conciliar definition of *Chalcedon. The letters of the popes (see *Papacy) illustrate the *ecclesiology that explains this exercise of the primacy: Damasus is the first to use the pericope of Mt 16, 18 to justify the powers of *Peter's successor. Leo and Gelasius work out a definitive theological explanation of the role of the see, where the *apostle's successor and vicar, invested for discipline and *faith with a special responsibility and an outstanding *authority, exercises his *ministry in a city renewed by the acts of two new founders, Peter and Paul.

3. The 6th and 7th cc. The political context changed when R. passed for half a century under the rule of the *Goths; the *Byzantine reconquest, completed in 554, set up a *vicarius in urbe* to represent the emperor.

a) The heritage of buildings was enriched not just by churches that restricted the old network of **tituli*: SS. Apostoli (mid 6th c.), S. Agata dei Goti, taken over from the *Arian Goths by St *Gregory, and on the outskirts of the city S. Pancrazio, built under Symmachus, and S. Giovanni at the Lateran gate; a new quarter was also occupied, the Forum, where the Christians ended by installing themselves in some public buildings (in the 7th c., the curia, under *Honorius I?), in annexes of the imperial palace (S. Maria Antica); in the 6th c. Felix III (IV) founded a church in honour of *Cosmas and Damian in an outbuilding of the Forum of Peace. Two new types of building multiplied: *monasteries (some fifteen foundations for the 6th c., including St Gregory's monastery on the Caelian; some ten in the 7th c., including S. Saba) and *diaconiae*, which served as liturgical annexes to charitable institutions created in the 7th c. The clergy's links with the aristocracy were consolidated from the Gothic era, even if the relations of popes with senators often degenerated into conflict (over the administration of the *patrimony or the regulation of episcopal succession: e.g. at the time of *Symmachus [498-514]). At any rate we more often see prelates from great families: after the humble Symmachus, the rich *Hormisdas (514-523), perhaps *John I (523-526); then, after Felix IV (526-530), Boniface

II (530-532) and *John II (533-535), Agapitus who belonged to a clerical family (535-536) and, above all, after *Silverius (536-537), *Vigilius, son of a prefect (537-555). We do not know the origins of *Pelagius (556-561), John III (561-574), Benedict (575-579) or *Pelagius II (579-590), but it is well-known that *Gregory the Great (590-604) came from a noble family. The list of the 7th c. includes some great bishops: after Sabinianus (604-606) and the two Bonifaces (III, 607, and IV, 608-615), Adeodatus, son of a subdeacon (615-618), Boniface V (619-625), *Honorius (625-638); after Severinus and *John IV, *Theodore (642-649), an Oriental, and *Martin (649-655); after Eugenius and *Vitalian, Adeodatus II and Donus, Agatho (678-681), Leo II, Benedict II, John V, an Antiochene (685-686), who preceded the Thracian Conon and another Antiochene, *Sergius (687-701).

b) Against the barbarians, the bishop became the defender of the city, the protector who sought to dispel the threat, as *Leo had done at the time of *Attila. Gregory particularly illustrated this role when there was a real threat from the *Longobards, and he received in his epitaph the surname "God's consul". *i)* Over suburbicarian *Italy the Roman bishop still exercised the jurisdiction of direct administration: Gregory regulated episcopal successions (e.g. that of *Naples), organized provincial *councils (for *Sicily) and conferred real authority on the Roman clerics who administered the *patrimony of St Peter. The northern provinces had long been disturbed by the *schism organized by an episcopate headed by the see of *Aquileia, hostile to the condemnation of the *Three Chapters, which had been ratified by pope *Vigilius in 553. But in 649 the Lateran council brought together a considerable number of Italian bishops to condemn the *monothelism proclaimed by a *typos* of Constantinople and rejected by pope *Martin; *ii)* relations with *Africa, renewed at the time of the Byzantine reconquest, slackened after the Muslim conquest: in the 5th c., R. gave refuge to bishops persecuted by the *Vandals; in the 6th c. *Gregory intervened in *Numidia against a Donatist revival, and the last manifestations of an organized African episcopate supported R. against monothelism; *iii)* with *Spain, on the contrary, the political situation was made more favourable by the *conversion of the Visigothic kings and by the links between Gregory and *Leander of Seville. But, to fix the rules of its pastoral activity and its mission, the Iberian peninsula had at its disposal a conciliar organization that made Roman intervention less justifiable: *Braulio of Saragossa noted this after an intervention, judged untimely, by pope Honorius (638) on the Jewish question; *iv)* *Gaul had an increasingly important role, especially after the establishment of the *Catholic monarchy of the *Franks; the episcopate generally supported the Roman positions in the theological conflicts that opposed the popes to the emperors (the *Three Chapters, *monothelism). Certainly the delegacy conceded to the bishop of *Arles at the time of *Caesarius (503-543) lost all efficacy in the course of the 6th c.; but, despite the organization of provincial councils, the apostolic see remained a model often imitated; *v)* after *Celestine, who supported *Patrick's mission to the Scotti (431), *Gregory organized a mission to *Britain headed by a monk from his Caelian monastery, *Augustine (597); in 668, Vitalian designated a bishop for *Canterbury; *vi)* as regards the *East, the policy of the emperors *Zeno and *Anastasius (476-518) had caused the final break, with the *excommunication of *Acacius of Constantinople by Felix III in 487; but in 518 the emp. *Justinian obliged bishop *John to submit and accept the "Second Rome's" *communion with R. Rome's primacy seemed better established when pope Agapitus, present on an embassy, deposed the patriarch and installed *Menas in his place. Yet, from the time of Justinian, the pope clashed with the system of the imperial church: *Vigilius, after a clumsy resistance, had to accept the condemnation of the Three Chapters willed by his sovereign (553); with *Gregory, who contested the bishop of Constantinople's title of ecumenical patriarch, the Roman see regained authority. Despite the submission of pope *Honorius, it fought, under *John IV, *Theodore and *Martin (Lateran council, 649), against the monothelite theology that the emperor wanted to impose; and despite reprisals (execution and martyrdom of pope Martin, 655), the Roman positions prevailed at the council of *Constantinople (681).

Situated at the margins of an empire that was crumbling in *Italy and preoccupied with finding close and effective protection against the direct threats of the *Longobards, R. supported with ever greater difficulty the vexatious interference of the *Byzantine authorities in papal elections. The iconoclastic policy of the emp. *Leo the Isaurian increased the alienation and pushed the bishop of R. ever closer to France. **[Figs: 263-269]**

H. Lietzmann, *Petrus und Paulus*, Berlin 1927; E. Caspar, *Geschichte des Papsttums*, I and II, Tübingen 1930; H. Grisar, *Roma alla fine del mondo antico*, Rome 1930; R. Krautheimer, *Corpus basilicarum christianarum Romae*, 5 vols., Rome 1937-80 (Eng. and It. ed.); P. Batiffol, *Cathedra Petri*, Paris 1938; B.M. Appollonj Ghetti-A. Ferrua-E. Josi-E. Kirschbaum, *Esplorazioni sotto la confessione di S. Pietro in Vaticano*, 2 vols., Rome 1951; L. Duchesne, *Le Liber Pontificalis*, text, introduction and commentary, I and II (III, additions and appendix by C. Vogel), Paris ²1957 (bibl.); G. Ferrari, *Early roman Monasteries*, Rome 1957; Y.M. Congar, *L'ecclésiologie du Haut Moyen Age*, Paris 1968; D.W.M. O'Connor, *Peter in Rome, The literary liturgical, and archaeological Evidence*, New York 1969; W. Marshall, *Kartago und Rom*, Päpste und Papsttum 1, Stuttgart 1971; P. Joannou, *Die Ostkirche und die Kathedra Petri*, Päpste und Papsttum 3, Stuttgart 1972; L. Duchesne, *Scripta minora. Etudes de topographie romaine et de géographie ecclésiastique*, Rome 1973; H.I. Marrou, *Patristique et Humanisme* (the Oriental origin of the diaconiae, 81 ff.), Paris 1975; C. Wermelinger, *Rom und Pelagiu*, Päpste und Papsttum 7, Stuttgart 1975; M. Maccarrone, *Apostolicità, episcopato e primato di Pietro*, Lateranum 42, Rome 1976; Ch. Petri, *Roma christiana, Recherches sur l'Eglise de Rome, son organisation, sa politique, son idéologie de Miltiade à Sixte III (311-340)*, BEFAR 224, 2 vols., Rome 1976 (bibl.); H.I. Marrou, *Christiana tempora*, Rome 1977; U.M. Fasola, *Pietro e Paolo a Roma*, Rome 1980; Ch. Pietri, Aristocratie et société cléricale en Italie au temps d'Odoacre et de Théodoric, *MEFR Ant.* 93 (1981) 418-473; J. Toynbee - J. Ward-Perkins, *The Shrine of St. Peter and the Vatican Excavations*, London 1956; J. Carcopino, *Les Reliques de Saint Pierre à Rome*, Paris 1965. E. Crueitti-Naber, *Rome: Pagan and Christian*, Rome 1975; J. Moorhead, Papa as Bishop of Rome, *JEH* 36 (1985) 337-350; P. Lampe, *Die stadtromischen Christen in den ersten beiden Jahrhunderten*, Tübingen 1989; R. Krautheimer, *Rome, Profile of a City, 312-1308*, Princeton 1979. Christianisation: L. Reekmans, L'Implantation monumentale chrétienne dans le paysage urbain de Rome, 300 à 850, *Actes XIe Cong. internat. d'arch. chrét*, CEFR 123, Rome 1989, 861-915. Organization: C. Pietri, Régions ecclésiastiques et paroisses romains, *ibid.*, 1035-1067. Topography and Excavations to 1986, see *ibid.*, index 2877-2888. General archaeology: P. Testini et al., La Cattedrale in Italia: Schede, Roma, *ibid.*, 14-26.

Ch. Pietri

II. *Councils

193. Apart from some councils mentioned by the *Libellus Synodicus*, a not wholly reliable source, the first Roman council known for certain is that held under pope *Victor I, which condemned the *quartodeciman way of determining the date of *Easter. The bishops of *Asia Minor, led by *Polycrates of Ephesus, declared themselves unwilling to follow the Roman practice. Victor intended to exclude them from *communion, but was dissuaded by the intervention of some bishops including *Irenaeus of Lyons.

231. Council held under pope *Pontian to ratify the measures taken by bishop *Alexander of Alexandria against *Origen because of his irregular *ordination and his doctrines. Origen later sent pope *Fabian a profession of faith (Euseb., *HE* VI, 36, 4).

251. Following the *schism of *Novatian, pope *Cornelius called a council which was attended by 60 bishops and a great many priests and deacons (Euseb., *HE* VI, 43, 2). They condemned Novatian and his followers, accepted the decisions on *lapsi* taken at the council of *Carthage (251) and further specified the line to be followed; they also examined the case of bishop *Trophimus.

260. Some members of the churches of the Libyan Pentapolis denounced bishop *Dionysius of Alexandria to R., accusing him of a *subordinationist position on the *Trinity and of saying that the Son was a creature (ποίημα). A council was held at R. in which *Dionysius of Rome condemned both *Sabellianism and *tritheism (cf. Athan., *Ep. de decr. Nic. syn.* 26). Dionysius of Alexandria retracted and later clarified some extreme points of his thought.

313. Council related to the rise of the *Donatist controversy. The African church had split into two parties: one led by *Donatus, the other by *Caecilian. The Donatists appealed to the emp. *Constantine, requesting the arbitration of the bishops of *Gaul. The emperor consented and a council was held at R.: *Miltiades presided over three Gallic and 15 Italian bishops and the representatives of the two parties, including Donatus and Caecilian. The council confirmed the legitimacy of Caecilian's election.

341. *Athanasius, deposed a second time from the see of *Alexandria (338), went to R. to plead his cause. Here, pope *Julius I thought to hold a council to try to resolve the problems raised by the council of *Nicaea. He also summoned the *Arians, led by *Eusebius of Nicomedia. But Eusebius delayed, not wishing to recognize the right of the bishop of R. to interfere in Eastern questions. Julius finally held the council in spring 341: he re-examined Athanasius's case and declared him the legitimate bishop of Alexandria; he re-examined the case of *Marcellus of Ancyra, accused of *Sabellianism, and, after hearing his profession of faith and judging it satisfactory, invalidated his deposition and *excommunication.

370. Council held by pope *Damasus and known to us through his synodal letter *Confidimus*, in which he confirms the validity of the Nicene *Creed and writes that the three *Persons of the *Trinity are *unus substantiae*. *Auxentius, Arian bishop of Milan, was condemned.

374. Pope Damasus condemned *Eustathius of Sebaste for his doctrine of the *Holy Spirit and *Apollinaris of Laodicea for his *christological doctrine.

380. The council examined the cause of *Maximus the Cynic: irregularly elected to the see of *Constantinople, he had been deposed and replaced by *Gregory Nazianzen. Maximus came to R. to plead his cause and the *Fathers of this council, ill-informed, declared him in the right.

382. Council attended by, among others, *Ambrose, *Jerome, *Epiphanius of Salamis, *Acholius of Thessalonica, *Paulinus of Antioch and three representatives of Constantinople. Jerome drew up a profession of faith for those Apollinarists who desired to return to *communion with the church; *Flavian of Antioch, *Acacius of Beroea and *Diodore of Tarsus were excluded from communion with R. The first three chapters of the *Decretum Gelasianum refer to the positions taken by this council against the anti-Roman and anti-Alexandrian deliberations of the council of *Constantinople (381) (Simonetti 548-552).

390. Pope *Siricius condemned the monk *Jovinian and his supporters and the doctrines apropos of impeccability consequent on baptism, denial of *Mary's perpetual virginity and rejection of penitential practices (cf. Jerome, *Ad Iovinianum*).

417. Pope *Innocent called a council to grant the requests of the African clergy to condemn *Pelagius and *Caelestius. The decisions taken against the two *heretics at the councils of *Carthage and Milevis (416) had been overturned by the council of *Diospolis, where *John of Jerusalem had welcomed Pelagius and Caelestius back to the bosom of the church. The African episcopate turned to the *pope as the only one with the right, in the last resort, to resolve the question. Innocent confirmed the condemnation of the two heretics, but left them an opening for repentance and reconciliation.

417. Innocent's successor *Zosimus re-examined the doctrines of Caelestius, who was present at R., and Pelagius, who sent a *libellus fidei*, convinced himself of their *orthodoxy and rehabilitated them. Next year (418) Zosimus, becoming aware of the haste with which he had absolved the heretics, and given the clear stand taken by the emp. *Honorius and the African clergy, put out the *Epistola tractoria* condemning *Pelagianism.

430. Following the *christological controversy that broke out between *Nestorius, bishop of Constantinople, and *Cyril of Alexandria, particularly about the attribute of *Theotokos* given to the Virgin, pope *Celestine intervened with a council. In 429 Nestorius had written to the pope clarifying his positions. Cyril also wrote to R., adding to his letter a *dossier* on the events: some *sermons of Nestorius, the *letters Cyril had sent him, a syllabus of Nestorian errors compiled by Cyril, and a collection of patristic testimonies on the *Incarnation, all translated into Latin - a very important detail and one that Nestorius had neglected. On the basis of this documentation, the pope condemned Nestorius, giving him ten days to retract his positions and entrusting the execution of the sentence to Cyril. The pope sent other letters to *John of Antioch, *Juvenal of Jerusalem, Rufus of Thessalonica, *Flavian of Philippi and the clergy and people of Constantinople (L.I. Scipioni, *Nestorio e il concilio di Efeso*, Milan 1974, 166-176).

443 or 444. Council celebrated by *Leo I at which the *Manichees and their doctrines were condemned.

445. The council examined the situation created in *Gaul, where bishop *Hilary of Arles had given proof of excessive zeal and excessive concentration of *authority in his own hands. Two controversial cases were examined and Leo I revoked the provisions taken by Hilary, who warmly defended his reasons in person. No specific provisions were taken against him, except the limitation of his authority to the diocese of *Arles. In his letter *Divinae cultum religionis* (*Ep.* 10), sent to the bishops of the province of *Vienne, Leo confirmed the primacy of R. over all the churches.

449. Condemnation of the decisions taken at the *latrocinium* of *Ephesus (Aug 449). Leo I sent letters to the emp. *Theodosius II, his sister *Pulcheria and the clergy and people of Constantinople. He asked the emperor to annul the decisions of the synod and call an ecumenical council to be celebrated in Italy.

484. In 482 the emp. *Zeno had promulgated the *Henotikon* or edict of union, in an attempt to compose the *monophysite crisis. Even while attempting to conciliate, the edict had a clear monophysite tendency. To clarify the situation, pope Felix III sent two legates to Constantinople, bishops Vitalis and Misenus, bearing letters to the emperor and to bishop *Acacius of Constantinople. The legates disregarded the pope's directives and managed to restore *communion with Acacius and with *Peter Mongus. In consequence of this, Felix called a council (July 484) in which he deposed Acacius and expelled him from communion with R. Acacius's reaction signalled the start of the so-called *Acacian schism, which lasted until 519.

499. Following the Lawrentian schism, which had paralysed the activity of the Roman see, *Symmachus called a council to take measures to prevent a repetition of similar events. It established a series of canons, including one enabling the pope to designate his successor. His antagonist *Lawrence, who also subscribed the council's decisions, was made bishop of Nocera in Campania.

501. Despite the legitimation of Symmachus's election, the party favourable to Lawrence, headed by the Roman senator Festus, continued to plot against the pope, accusing him of grave moral crimes, of having displaced the date of *Easter and dissipated the patrimony of the church. They appealed to *Theodoric, who called a council at R., to which he sent a visitor. The council was long and stormy; finally it was decided that only God could judge the accusations against Symmachus; for men he was acquitted and could continue to exercise his ministry.

531. Examined the appeal to R. of bishop *Stephen of Larissa, deposed by *Epiphanius of Constantinople. Stephen claimed that, since he was under the jurisdiction of the metropolitan of *Thessalonica, vicar of the apostolic see, only the pope, and not the bishop of Constantinople, could judge him. Pope Boniface II, having heard Stephen's suffragans and examined the evidence in his possession, confirmed the existence of the apostolic vicariate of Thessalonica. We do not know how Stephen's case ended.

Under *Gregory the Great a series of councils were celebrated (*595; 600; 601*), linked to that pope's reforming work.

Palazzini 4, 119-180; Hfl-Lecl 1-3; Mansi 1-9; DS; Fliche-Martin 1-4; P. Paschini-V. Monachino, *I Papi nella storia*, I, Rome 1961; Ch. Pietri, *Roma Christiana*, BEFAR 224, 2 vols., Rome 1976; W.H.C. Frend, Eastern Attitudes to Rome during the Acacian Schism, in (ed.) D. Baker, *The Orthodox Churches and the West*, SCH 13, Oxford 1976, 69-81; R.E. Brown-J.P. Maier, *Antioch and Rome: New Testament Cradles of Catholic Christianity*, New York 1983. Appeals to Rome: C. Munier, La Question des Appels à Rome d'après la Lettre 20* (*Divjak) d'Augustin, in (ed.) C. Lepelley, *Les Lettres de Saint Augustin découvertes par Johannes Divjak*, Paris 1983, 287-299.

U. Dionisi

III. *Liturgy. At R., as elsewhere, we must distinguish two periods in the history of the liturgy: before and after the recognition of the church by the empire (314).

1. For the first three centuries, we know the liturgy celebrated at R. through two main documents: *Justin's *Apologies* (*c*.150) and the priest *Hippolytus's *Apostolic Tradition* (*c*.215). Every *Sunday the Christians met for *mass, in the course of which "they read publicly the memories of the apostles (the gospels) and the writings of the prophets, as much as time permits". After the president's *homily and a universal *prayer, "they bring *bread and *wine with *water. Whoever presides raises to heaven the prayers and the *Eucharist, so far as he can, and all the people respond with the *acclamation *Amen*" (*I Apol*. 67). Afterwards came the distribution and division of these eucharistic elements, which were also taken to the absent by the deacons (*I Apol*. 67). We note that it was up to the president to improvise the eucharistic prayer, "as far as he can", keeping to a pattern provided by *tradition. Sixty years later, Hippolytus attests the same freedom; after giving an example of eucharistic prayer, he adds this observation "that the bishop gives thanks, as we have said above. But it is not necessary for him to pronounce the same words as we did, as if making an effort to say them from *memory ... but each one prays according to his own capacity" (*Trad. Ap*. 10).

Besides this Sunday Eucharist, every year the Christians celebrated the feast of *Easter: a memorial of the Lord's Passion and *Resurrection, but also the birth of a new generation of believers, who received *baptism at dawn on Easter Sunday. Normally baptism was conferred on adults, after a long *catechumenate, but it could happen that adolescents or even *children were baptized with their parents (*ibid*. 20-21). For the whole community the feast of Easter was preceded by two days of *fasting and followed by 50 days of joy (*ibid*. 33). The *liturgical year comprised no other feasts. But every year the community celebrated the anniversaries (**natale*) of its *martyrs: this would later be the origin of the sanctoral. Acc. to Hippolytus's evidence, the Christians also met during the *week, to pray together and listen to "instructions in the word" given by deacons, priests or some passing teacher (*ibid*., 39, 41). Such a liturgical practice corresponded to a situation in which the Christians of R. were a small community: they met by districts, in the house of one of them (*domus ecclesiae*) and teaching, like worship, was carried out by qualified *ministers.

2. From the 4th c. the transformation of the relations between *church and empire had numerous repercussions on the liturgy. From the time of *Constantine, great basilicas were built (Lateran, Vatican, St Paul's outside the walls), as well as other *churches, which took in increasing numbers of

Christians. So the *Eucharist was no longer celebrated in the domestic atmosphere of a house, but in the sacred atmosphere of a church; Christians no longer met around a table, but around an altar, which emphasized the sacrificial character of the *mass; the importance of the *assembly obliged greater organization; the growing number of rapidly trained ministers necessitated the publication of handbooks for worship and preaching. Oral liturgy thus gave way to written liturgy, flexible liturgy to fixed liturgy.

The collections of *prayers (*sacramentarii*) and *homilies (*homiliaries*) were written in Latin, since between 360 and 380 the church of R. stopped celebrating the liturgy in Greek, a language that many believers no longer understood, and adapted the language of the people, Latin.

The *Roman canon of the mass, used up to our own day, goes back to the pontificate of St *Damasus (366-384); it laid great stress on the sacrificial character of the mass, the presentation of the offerings by the priest and the assembly, and the acceptance of these sacrificial offerings by *God.

From the 4th to the 6th cc. numerous formularies were composed at R. for the mass and the *sacraments, brought together first in anonymous *libelli* and then in sacramentaries attributed to popes, who were not really their authors, but who had an important role in working out the Roman liturgy. We may cite the Leonian Sacramentary, compiled in the 6th c. (generally called the Verona Sacramentary, since handed down by a manuscript of that city), the Gelasian Sacramentary, compiled at the end of the 7th c., which gathers formularies used in the *tituli* or parish churches of R.; the Gregorian Sacramentary, whose basis was the personal Sacramentary of pope *Gregory the Great (590-604), but further enriched and then imposed by Charlemagne on the whole Western empire.

The 6th and 7th cc. also saw the composition of lectionaries for the mass: epistolaries, evangelaries and *comes* (*books indicating the references of readings), books of *chants for mass and offices: antiphonaries; finally directories for celebration of *mass, *baptism, *ordination, reconciliation of penitents, funerals: *ordines Romani*.

As for the *sacraments, here too ecclesial and social evolution caused more or less rapid transformations. In the 4th and 5th cc., candidates for baptism were still mostly adults converted from *paganism: they prepared for *initiation through a *catechumenate of three years completed by a time of more intense doctrinal and liturgical preparation during Lent, which preceded their baptism. At the end of the 5th c., the situation changed due to the almost total Christianization of Roman society: *catechumens now came from a background already imbued with Christianity, so that preparation for baptism was reduced, for them, to the scrutinies and rites of Lent; adult baptism became the exception and baptism of *children of Christian parents became the rule: the catechumenate had no more *raison d'être* and the rites of initiation and baptism were simplified and adapted, well or ill, to the conditions of those candidates, incapable of affirming their *faith. But R. resisted this transformation of the baptismal ritual longer than other churches: only in 726, at the time of Gregory II, do we find evidence of the replacement of the traditional triple interrogation: "*Credis in Deum*" by the formula "*Ego te baptizo*" pronounced by the priest (*Ep. 15 ad Bonifacium* 8: PL 89, 929).

Closely linked to baptism was the sacrament of *penitence or reconciliation, which the ancients considered as a second baptism. At R., as elsewhere, penance was accorded only once in a lifetime. From the 4th to the 6th c., the Roman church lived with a penitential legislation whose severity, justified at the time of the *persecutions, no longer corresponded to the situation of large and only moderately fervent communities. Christians guilty of grave *sins tended to put off as long as possible their request for reconciliation. To obtain it, they had to confess their failings to the bishop and then enter the *ordo poenitentium*; on Sundays they attended mass but did not communicate; they were reinstated in the course of a public rite of reconciliation, which, from the 4th c., was celebrated on Maundy Thursday to allow them to take Easter *communion (cf. *Sacram. Gelas.* 352-363).

The rites surrounding Christian death and burial are precisely described in *Ordo Romanus* 49 (ed. Andrieu IV, 523-530), which dates from the late 7th c. The Christian's *transitus* is presented as a paschal exodus, an entry into the promised land where he will be welcomed by *angels and saints.

The Roman liturgy was thus formed between about 350 and 680: popes *Leo I, *Gelasius I and *Gregory I were its inspirers and principal organizers. Altogether, this liturgy gave plenty of room to the word: biblical readings, *psalmody, *prayers and prefaces. The language of prayer combined Roman sobriety with a controlled but rich lyricism. *Chant occupied an important place in the celebrations, but musical instruments were not allowed. The language of *gesture was symbolic and very discreet: attitudes of prayer, gestures of offering or of *laying-on of hands, *processions, all were characterized by a sobriety which it was impossible to preserve when, in Charlemagne's time, the Roman liturgy was diffused and imposed all over the Western world: then arose a Romano-Frankish or *Gallican liturgy, whose heirs we are.

Sources: Justin, *Apol.* I, II: PG 6, 327-470 (Fr. tr. L. Pautigny, Paris 1904; It. tr. I. Giordani, Rome 1962); Hippolytus, *Trad. Apost.* (ed. and Fr. tr.: SCh 11a, Paris 1968); *Sacram Veronense* (ed. L.C. Mohlberg, Rome 1966); *Liber Sacramentorum Romanae Ecclesiae - Sacram. Gelas.* (ed. L.C. Mohlberg, Rome ²1968); *Le Sacramentaire Grégorien* (ed. J. Deshusses, Fribourg 1971); *Les ordines romani du haut moyen age*, 5 vols. (ed. M. Andrieu, Louvain 1931-61).
Studies: T. Klauser, *Das römische Capitulare evangeliorum*, LQF 28, Münster 1935; E. Bourque, *Etudes sur les sacramentaires romains*, I, Rome 1949; A. Chavasse, Les plus anciens types du lectionnaire et de l'antiphonaire romains de la messe, *RBen* 62 (1952) 3-94; E. Dekkers, Autour de l'oeuvre liturgique de S. Léon le Grand, *SEJG* 10 (1958) 363-398; A. Chavasse, *Le Sacramentaire gélasien*, Tournai 1958; J.M. Hanssens, *La liturgie d'Hippolyte*, OCA 155, Rome 1959; G. Pomarès, *Gélase Ier. Lettre contre les Lupercales et dix-huit messes du Sacramentaire léonien*, SCh 65, Paris 1959 (introduction); H. Ashworth, The liturgical prayer of Gregory the Great, *Traditio* 15 (1959) 107-161; J.A. Jungmann, *La liturgie des premiers siècles*, Paris 1962; K. Gamber, *Codices liturgici latini antiquiores*, Fribourg 1963; E. Cattaneo, *Introduzione alla storia della liturgia occidentale*, Rome ²1969; Quasten I, 177-180, 190-194 (studies and bibliography on St Justin's *Apologies*); 437-448 (studies and bibliography on the *Trad. Apost.*); A. Nocent, *Storia dei libri liturgici romani*, Anamnesis II, Turin 1978, 147-165; J.H. Walker, Reservation Vessels in the earliest Roman Liturgy, *SP* 16.1 (1984) 568-572; M.R. Moreton, Roman Sacramentaries and ancient Prayer-tradition, *ibid.*, 577-580; V. Saxer, L'Utilisation par la liturgie de l'espace urbain et suburbain: l'example de Rome dans l'Antiquité et le Haut Moyen Age, *ibid.*, 917-1031.

P. Rouillard

ROTHARI. King of the *Longobards (636-656), he stands out for the restoration of his people, rent by the intestine struggles of the dukes and the continuation of the war against the Greeks. Under R.'s rule the Longobards extended their conquests at the expense of *Byzantine territory. R.'s most important achievement was the promulgation of the edict that bears his name, the foundation of later Longobard legislation. The legal material, previously transmitted orally, is substantially Germanic. The edict, in ungrammatical Latin, is divided into 348 chapters dealing mainly with the supression of crimes against the state, security of persons and things, hereditary right, family law, royal rights, law of obligation, of responsibility for *slaves, damage, obligations.

P. Delogu-A. Guillou-G. Ortalli, Longobardi e Bizantini, in *Storia d'Italia*, ed. P. Delogu, Turin 1980; N. Tamassia, *Le fonti dell'Editto di Rotari*, Pisa 1889, now in *Scritti di storia giuridica*, II, Padua 1967, 181-260; A. Cavanna, Nuovi problemi intorno alle fonti dell'Editto di Rotari, *SDHI* 34 (1968) 269-361; B. Paradisi, Il prologo e l'epilogo dell'Editto di Rotari, *ibid.*, 1-31; C.G. Mor, Il diritto romano nel sistema giuridico longobardo del secolo VIII, *RAL* 27 (1973) 1-16.

M.L. Angrisani Sanfilippo

RUFINUS. *Friend of *Prosper of Aquitaine. From Prosper's *letter *Ad Rufinum de gratia et libero arbitrio* (PL 51, 77-79; PL 45, 1793-1802), we learn that R., becoming aware that some, enemies of the *Augustinian doctrine of *grace, were accusing his friend Prosper too of having fallen into *heresy, wrote him a letter. Prosper replied expounding and defending Augustinian thought and refuting the *Pelagian theories: this reply seems to have been written *c.*429, since it considers *Augustine as alive, but in the final period of his life.

CPL 516; DCB 4, 562.

A. Pollastri

RUFINUS FLAVIUS. The information on R. F. in *Ambrose (*Epist.* 52, 1), *Symmachus (*Epist.* III, 81-91), *Jerome (*Epist.* 60, 16), *Zosimus (*Hist. nova* IV, 57, 4; V, 1, 1. 3) and *Theodoret (*Hist. eccl.* V, 18, 6-12) makes him an illustrious person of great education. Acc. to Jerome and Theodoret, R. F. was an astute, cruel and able politician, a persecutor of *heretics. Born *c.*335 in S *Gaul, he found ways to exercise his talent for intrigue and political initiative under *Theodosius I and then under *Arcadius. Head of the imperial household under Theodosius, he became consul in 390 and, soon after, praetorian prefect of the *East, which he remained into Arcadius's reign, up to 395. Acc. to Theodoret's rather confusing account, the responsibility for the massacre of *Thessalonica fell on R. F. and he was given the job of persuading *Ambrose to withdraw his condemnation of Theodosius. After Theodosius's death, rivalry between *Stilicho and R. F. ended with the latter's murder at *Constantinople in 395.

A. de Broglie, *L'Eglise et l'Empire romain au IVe siècle, 3e partie: Valentinien et Théodose*, 2 vols., Paris 1869; G. Rauschen, *Jahrbücher der christlichen Kirche unter dem Kaiser Theodosius dem Grossen*, Freiburg i. Br. 1897, 439 ff.; J.R. Palanque, *Saint Ambroise et l'Empire romain*, Paris 1933; PLRE 1, 778-781.

M.G. Mara

RUFINUS of Aquileia. Tyrannius R., student of letters at *Rome until c.386 and *friend of *Jerome: the two formed part of an ascetic group who in the years around 370 sought to recreate in R.'s home town of Concordia the monastic and intellectual life of the *East. After a long stay in *Egypt (373-380), where he frequented *Didymus, he lived with *Melania in the *monastery on the Mount of Olives at *Jerusalem, very close to bishop *John and *Origenist circles (*Palladius, *Evagrius). After unhappy disputes with Jerome over the translation of *Origen (and of the *Bible - from the Hebrew or from the LXX), he returned to the West in 397, pursued at Rome and then at *Aquileia by the animosity of his old friend. Fleeing the *Goths, he went to die in Sicily. He was the friend and confidant of the best spirits of Christian Italy. His work, modest, is mainly *Latin translation: it is he who has preserved most of what we have of Origen. His aim being not to make known the Eastern texts which were unfortunately to be lost for ever, but to illuminate a Latin world suffering from great cultural retardation, his translations do not aim to give the detail of the original, but to be useful to 4th-c. readers. He has more *piety than critical sense. His original works (in particular the prefaces to his translations) are published in CCL 20.

His first translation was of *Basil's *Asceticon* (397). In 398 he published Origen's *De principiis*, which sparked off a controversy. R. then translated or summarized the *Apologia for Origen* by the martyr *Pamphilus and *Eusebius of Caesarea; he justified himself to pope Anastasius in a personal *Apologia* of four pages, then published two books against Jerome, but refrained from prosecuting the dispute. He dedicated himself to translating Origen's magnificent *homilies on *Genesis, *Exodus, Leviticus, Numbers, *Song of Songs and Romans; various homilies of Basil and *Gregory Nazianzen; the ascetic Sentences of Evagrius and *Sextus (a Pythagorean whom he confused with pope Sixtus); the *Dialogue* of *Adamantius, confused with Origen; Eusebius of Caesarea's *Historia ecclesiastica*, brought up to the death of *Theodosius the Great; the *Historia monachorum in Aegypto*; and the Ps.-*Clementine *Recognitiones*. Two smaller personal compositions, the *Commentary on the Apostles' Creed* and that on the *Blessings of the Patriarchs*, were also inspired by Eastern models. This work made a decisive contribution to the culture, biblical and otherwise, of the Latin Middle Ages.

CPL 195 ff.; PL 21; CCL 20 (1961); J. Gribomont, in *Patrologia* III, 234-240; F.X. Murphy, *Rufinus of Aquileia*, Washington 1945; C.P. Hammond, The Last Ten Years of Rufinus' Life, *JThS* n.s. 28 (1977) 372-429; T. Christensen, *Rufinus of Aquileia and the Historia Ecclesiastica, Lib VIII-IX of Eusebius*, Copenhagen 1989; C.P. Bammel, Products of fifth-century scriptoria preserving conventions used by Rufinus, *JThS* n.s. 35 (1984) 347-393; F. Thélanson, *Païens et chrétiens au IV siècle: l'apport de l'histoire ecclesiastique de Rufin d'Aquilée*, Paris 1981.

J. Gribomont

RUFINUS (the Syrian). We know a R., disciple of *Jerome, priest of his *monastery at *Bethlehem, sent to *Rome in 399 with Jerome's *Epp.* 81 and 84 and his translation of *Origen's *De principiis*. Acc. to B. Fischer (in K. Aland, *Die alten Übersetzungen des NT*, Berlin 1972, 73) and the editors of Beuron's *Vetus Latina*, R. corrected *Paul's epistles and the later parts of the NT in the text that became the *Vulgate*. This R. is identified by many scholars with Rufinus the Syrian, who, acc. to *Marius Mercator (ACO I, 5, p. 5), introduced Pelagian ideas to Rome at the time of pope Anastasius I (399-402) and whose teaching was spread by *Caelestius and *Pelagius. At *Carthage in 411, Caelestius cited the priest R., whom he had heard at Rome in *Pammachius's house, as the source of his doctrine. The *Liber de fide*, which defends Pelagian ideas and is attributed in MSS to R. of Palestine, is considered his work. It shows familiarity with Greek theological writings (esp. the Cappadocians) and, as regards *anthropology, with the school of *Antioch. A *Libellus de fide* of 12 *anathemata is sometimes attrib. to the same R.

CPL 199 f.; PL 21, 1123-1154; PL 48, 451-491; M.W. Miller, *Rufini presbyteri Liber de Fide*, Washington 1964; Il *Libellus*: PL 21, 1123-24; PL 48, 239-254; ACO I, 5, 4-5; B. Altaner, Der Liber de fide, ein Werk des Pelagianers Rufinus des "Syrers", *ThQ* 130 (1950) 432-449 (= TU 83, 467-482); F. Refoulé, Datation du premier concile de Carthage contre les Pélagiens et du libellus fidei de Rufin, *REAug* 10 (1963) 41-49; H.I. Marrou, Les attaches orientales du Pélagianisme, *CRAI* (1968) 459-472; G. Bonner, Rufinus of Syria and African Pelagianism, *AugStud* 1 (1970) 31-47; E. TeSelle, Rufinus the Syrian, Caelestius, *ibid.* 3 (1972) 61-95; L.J. van der Lop, The Man in the Shadow behind Pelagius, *SP* 15.1 (1984) 247-254; M.W. Miller (ed.), *Rufini presbyteri: Liber de Fide; critical text and translation with introd. and commentary*, Washington DC 1964.

J. Gribomont

RUFUS of Shotep. Author of vast *Coptic *commentaries on the NT (lived late 6th-early 7th c.). From the scanty sources on his life we deduce that, after a period as a monk and a *pilgrimage to the Holy Land, he was consecrated bishop of Shotep (Hypsele, in Upper *Egypt: 7 km. S-E of Lycopolis-Siout) by patriarch *Damian (578-604). Damian inspired a considerable literary renaissance, of which R. was one of the most eminent figures, in the Coptic church. Long fragments survive of four codices containing a Commentary on *Matthew and one on *Luke. The texts are still unpublished and await proper evaluation, but they testify to a theological education that was vast for his time, and make their author a worthy descendant of the Alexandrian exegetical tradition. The *exegesis is naturally allegorical, but does not exclude a philological look at the text. R. insists on the harmony of the OT and NT within the *economy of salvation; here and there he inserts polemic against heretical doctrines (notably *Marcion and the *Manichees); but his interest is mainly in the moral teaching taken from *Scripture, which he addresses to a multifarious public, but esp. to monks.

G. Garitte, Rufus évêque de Sotep et ses commentaires des Evangiles, *Muséon* 69 (1956) 11-33.

T. Orlandi

RULES, MONASTIC. The Rule, i.e. the constitutional document of a particular religious order, is a notion that arose in the West among the jurists and curia of the 12th c. as the result of a long evolution. In the early centuries the *virgins who, inscribed on the lists (*canon*) of the local church, received subsidies from it and collaborated in its *liturgy and works of *assistance, were naturally under a certain unwritten discipline. Male ascetics too, freer since not dependent on subsidies, were accountable to their bishops for their activities. When contemporary documents wish to justify the *traditions of the church, they generally appeal to the *authority of the *apostles; this pseudo-*apostolicity was an immemorial custom.

When *monasticism appeared, at the moment when the Constantinian church was becoming secularized and toning down its moral demands, disputes and distortions were not unusual. The best monks served as a model; when *Athanasius wrote his *Life of *Anthony* (†356), he knew what points to insist on and proposed a living rule for all, naturally without obliging anyone to reproduce the saint's miracles or even his prodigies of austerity. *Coenobite communities imposed certain conditions so that *prayer, *ascesis and *labour could be exercised in harmony, while providing guarantees against abuses of power by the strongest: episcopal *authority intervened here only through a tacit *nihil obstat*. *Pachomius's *Precepts*, written perhaps after his death (347) to maintain the customs he had set up, based on experience and not at all on a theoretical system, were intended to preserve the *canon* of the primitive church; they were so sensible that, through *Jerome's translation, they exercised great influence in the West. They willingly used the standard of right *measure*, but still more that of the common good. In *Cappadocia, where a revolutionary ascetic social movement came up against a court episcopate, *Basil managed to impose a mediation thanks to his *Moral rules* (ὅροι), drawn literally from the NT, which defined *God's will for all disciples of the gospel. A commentary on these Rules, the *Asceticon*, answered the *brothers' queries and descended to concrete details; this is what the Latins, in a more evolved perspective, called *Instituta* (*Rufinus) or *Regula* (*Benedict). All these documents were translated into Latin (and other languages) and were fundamental for the insertion of monasticism into the episcopal church.

In the *East, *councils and emperors gradually fixed canons or laws to remedy abuses and to discipline relations between the monastic movement and the *world. As for the inner life of the *monasteries, the *Typica*, drawn up by the founders, made clear what they required from beneficiaries of their *donations. But the living tradition, the example of the saints, remained essential, esp. in the case of solitaries, who were often the most characteristic phenomenon.

In the West, besides the *Rules* of Basil (translated in 397) and Pachomius (translated in 404) and the *Institutiones* of *Cassian (402-424) - a long text that related, for the reform of the monks of *Gaul, the (pseudo) traditions of *Egypt - not to speak of the *Life of *Martin*, we must point to the *libellus* (the title "Rule" is late) that St *Augustine gave, c.397, to the (lay) African monks, which was based on an earlier very short *ordo monasterii*, perhaps by *Alypius. This later became the basis of *canonical life, organically integrated into basilical worship and pastoral *ministry, following a tendency typical in the West, which was also found, *mutatis mutandis*, in its monastic communities.

A second generation produced the *Rule of the Four Fathers*, prob. at *Lérins, c.400-410; a second *Rule of the Fathers*, at Lérins, c.427 (?), supplemented the first, and both left traces in later texts. Still in S Gaul, the *Oriental Rule* and that of *Macarius appeared before 500; in 534 that of *Caesarius of Arles for *virgins and a similar one for monks; in c.550 the two *Rules* of Aurelian. This whole Arlesian tradition bears a strong *Augustinian imprint.

Meanwhile in *Italy the *Rule of the Master* (before 530) vindicated its charismatic authority. The cento-*Rule* of *Eugippius, at Lucullanum (near Naples) c.530, copied Augustine, the Master, Basil, etc. Towards 550 at Montecassino, *Benedict's *Rule* kept the doctrinal part of the Master (summarized and adapted to the needs of the church of *Rome), but adapted the practices to some more important and more laborious monasteries.

From the second half of the century, in Gaul, come the third *Rule of the Fathers*, that of Tarnant and that of St *Ferreolus; in Italy that of Paul and of Stephen. The *Regula monachorum* and *Regula coenobialis* of St *Columbanus (†615) reached Gaul from *Ireland; in *Spain we know, besides the (lost) Rule of *John of Biclaro, *Leander of Seville's *De institutione virginium*, the Rules of *Isidore and *Fructuosus, the *Regula communis*, the *Pactum s. Fructuosi* and the *Consensoria monachorum*. But St Benedict's *Rule*, strong in its own moderation and the recommendations of St *Gregory the Great, was adopted in Gaul and *Britain, first in a mixed form (with St Columbanus), then alone. All these texts, which have much in common, demonstrate that each shaped his rule according to his own possibilities (even if they adopted without modifications a document composed elsewhere). Only in the Carolingian era, to correct abuses, did the imperial authority (or that of synods of *abbats who depended on it) see in the *Rule* (of Benedict or, for canons, of Chrodegang) a precise law, whose observance was controlled by suitable visitors.

CPL 1838-1876; PL 103; A. de Vogüé, *La Règle du Maître*, 3 vols., Paris 1964-65; id., *La Règle de s. Benoît*, 7 vols., Paris 1972-77; G. Turbessi, *Regole monastiche antiche*, Rome 1974; V. Desprez, *Règles monastiques d'Occident*, Bellefontaine 1980; id., *Les Règles des saints Pères*, 2 vols., Paris 1982; *DIP* 7, 1410-1617.

J. Gribomont

RUPERT of Salzburg. Born near Worms (*Wormatia*), related to the Carolingians, R. may have been bishop of Worms. Trained in the methods of Irish missionaries, he set out to evangelize, acc. to legend, Ratisbon and Lorch, certainly Salzburg. He founded St Peter's abbey (the oldest in Austria) at ancient Iuvavum and another for women at Nonnberg. Duke Theodo made him a gift of the city, which gave him a fixed base from which to set out on his missions. R. was a bishop-*abbat in the manner of the Scots (Irish). The foundations of Seekirchen (Wallersee) and Maximilianszelle (Pongau) remain as evidence of his missionary activity. He died c.720. His cult began with the translation of his relics to the new cathedral by his successor Virgilius (24 Sept 774), who may have written or commissioned the first Life of R., substantially preserved in the 9th-c. *Gesta Hrodberti* (BHL 7390).

MGH, *SS.* 6, 157 ff.; *Vies des SS.* 3, 587; *LTK* 9, 106; *BS* 11, 506-7.

V. Saxer

RURICIUS of Limoges. Of high family position, a younger *friend of *Sidonius and of *Faustus of Riez, active in the second half of the 5th and early 6th c. Like Sidonius he entered church life at a mature age, being elected bishop of Limoges (*Lemovicum*) in 485. He left an important collection of 80 *letters. Those written as a *layman deal with the usual, mostly futile, subjects typical of this *literary genre. Such subjects are not absent even after his transition to religious life; but now they are mostly moral and ecclesial, and scriptural citations are more frequent. In both, the form is highly elaborate.

CPL 985; PL 58, 57-123; CSEL 21, 351-450; *DTC* 14, 205-206; H. Hagendahl, *La correspondance de Ruricius*, Göteborg 1952.

M. Simonetti

RUSTICUS, poet. A. Wilmart (*MSCA* II, 271-272) has given us four distichs entitled *Versus Rustici defensoris sancti Augustini*. The first of these distichs goes like this: *Ter quinos animo suadente per ardua libros / Augustinus trahens nobile condis opus*. The allusion to *Augustine's treatise *De Trinitate* is evident. The author's name, almost unknown, has often been confused with that of the poet Rusticus *Helpidius, author of the *Carmen de Iesu Christi beneficiis*, and the *Tristicha historiarum Testamenti Veteris et Novi* (CPL 1506-1507; cf. W. Baeherens, *RhM* 31 [1876] 94, n. 1). R. may also have written a short note of thanks, addressed to *Eucherius of Lyons and sent with two of his works (see *Rusticus, presbyter).

CPL 1506; *Anthologia latina* (ed. Riese) I, 2, p. 264, no. 785 (defective ed., corrected by Wilmart, cit.); (for the *Epistola Rustici ad Eucherium*: PLS 2, 46-47); S. Cavallin, Le poète Domnulus. Etude prosopographique, *SEJG* 7 (1955) 49-66, esp. 50, n. 2; *Patrologia* III, 499.

L. Dattrino

RUSTICUS, presbyter. We possess a *letter written by this obscure 5th-c. person to *Eucherius of Lyons to thank him for having sent copies of two of his books (perhaps *Instructionum ad Salonium libri duo* or the *Formulae*). Prob. identical to the R. to whom *Sidonius Apollinaris wrote (*Ep*. II, 11). Some have suggested an R. of Bordeaux (cf. Sidon., *Ep*. VIII, 11, 3, vers. 36) or *Rusticus, bishop of Narbonne from 427, but the latter hypothesis presents difficulties.

CPL 496; PL 58, 489 f.; CSEL 31, 198-199; PLS 3, 46-47; C. Wotke, CSEL 31, XXIII-XXIV; Bardenhewer 4, 570-571; *Patrologia* III, 499.

A. Pollastri

RUSTICUS. Roman deacon, nephew of pope *Vigilius, he accompanied his uncle to *Constantinople on the occasion of the question of the *Three Chapters, but distanced himself when Vigilius published the *Iudicatum*. For his defence of the Three Chapters and his aversion to the council of *Constantinople of 553, he was deposed and exiled to the *Thebaïd. After the emperor's death, he returned to Constantinople and spent some time in the *monastery of the *Acoemetes. He wrote a *Synodicon*, a collection of official documents of the councils of *Ephesus (431) and *Chalcedon, which he himself translated into Latin with the help of earlier translations, and, c.560, the *Disputatio contra Acephalos*, from which we learn that R. had written an earlier anti-monophysite work, now lost. The *Disputatio*, despite the poor state in which it survives, is an important work because its refutation of *monophysite positions is based on a good knowledge of monophysism, esp. that of *Severus. R. keeps to a strict Chalcedonianism, with no concessions to *Cyril's *anathemata and the *Unus ex Trinitate passus est*. He knows *Boethius's definition of *persona* as *individua et rationalis natura*, but does not accept it since it can easily be adapted to a monophysite meaning; for his definition of divine person, he goes back to *Basil's definition: *individuum naturae communis* (PL 67, 1196). He also exploits the distinction introduced by *Leontius of Byzantium between hypostatic and enhypostatic, to explain how Christ's human nature does not correspond to a second person (PL 67, 1198. 1235. 1240).

CPL 946-947; PL 67, 1167-1254; ACO II, 3, 1-3; DTC 14, 371-72; M. Simonetti, La Disputatio contra Acephalos del diacono Rustico, *Augustinianum* 21 (1981) 259-289.

M. Simonetti

RUSTICUS of Narbonne. Seemingly the same Marseillais monk who corresponded with *Jerome (*Ep*. 125). He lost his father when young and his mother saw to his education, completed at *Rome. In 427 he became bishop of *Narbonne (*Narbona*), whose cathedral, destroyed by fire, he rebuilt. A *letter of *Ravennius of Arles puts him among "the most expert doctors" to be consulted in case of crisis (*Inter Leonis ep*. 99). In 450 he called an assembly of bishops and notables to examine a grave accusation. In 458/9 he sent his *archdeacon Hermes to put a series of disciplinary questions to pope *Leo I (PL 54, 1199-1209). The *pope encouraged him to remain at his post and not to shut himself up "in silence". R. consecrated the same Hermes bishop of Béziers, but that city did not accept him. B. seems to have died 26 Oct 461; Hermes succeeded him.

E. Griffe, *La Gaule chrétienne à l'époque romaine*, II, Paris ²1966, 265 f.; *BS* 11, 513-514.

A. Hamman

S

SABAS. Founder of the Great *Laura. *Cyril of Scythopolis's *Life of St Sabas* allows us to know this monk, famous in the monastic history of *Palestine, with some precision. Born 439 in the village of Mutalaska, diocese of *Caesarea in Cappadocia. After a first experience in the nearby *monastery (*Flavianae*), he left for Palestine aged 17, wishing to see the Holy Places. At *Jerusalem he decided to live out, in the nearby *desert, the monastic ideal he had tasted in his own land. After two brief attempts in the monasteries of Passarion and *Euthymius, he stopped in that of Theoctistus, where he remained until 473. After five years of eremitical life, he settled in a cave on the left bank of the Cedron. This was the start of a foundation called to play an important role in the development of *monasticism in that region, both by the quality of the ascetic life led there and by the number of monks: it was soon called the Great Laura. S.'s severity caused discontent and quarrels. Despite the support of the patriarchs of Jerusalem, he had considerable difficulties, including exile. Far from his monastery, he founded another at Gadara, E of Lake Tiberias. But before long, patriarch *Elias ordered him to return to lead the Great Laura, around which other monasteries and laurae continued to grow. S. combined the government of his own monastery with that of all the hermits of Palestine. The Great Laura actively supported *orthodoxy at the time of the *Origenist and *monophysite questions. In 530, following a bloody revolt of the *Samaritans, S., despite his advanced age, did not hesitate - for the second time - to travel to *Constantinople to plead the Christians' cause before the emp. *Justinian. S. died Sunday 5 Dec 532. The Byzantine Synaxaria have long commemorated him on that day: his memoria was introduced in the West more recently.

Without going into the history of the Great Laura after S.'s death, we should at least mention that in the 8th and 9th cc. it produced a series of monks famous in the Byzantine church for their literary activity (*theology, liturgical and ascetic poetry, in Greek): *John Damascene, Cosmas Melodus, Stephen Melodus, Stephen the Thaumaturge, Michael Syncellus, the two *Graptos* brothers Theodore and Theophanes, Theodore of Edessa. After the Arab invasion S.'s laura soon became an active centre of translation from Greek to *Arabic, whose influence extended beyond Palestine as far as Mt *Sinai.

A.-J. Festugière, *Cyrille de Scytopolis: Vie de S. Sabas*, Les moines d'Orient III. Les moines de Palestine 2, Paris 1962 (this is the Fr. tr. of E. Schwartz's crit. ed.); S. Vailhé, Répertoire alphabétique des monastères de Palestine, *ROC* 5 (1900) 274-276; *DACL* 15, 189-211; *Vies des SS.* 12, 169-182; *BS* 11, 533-535; D.J. Chitty, *The Desert City*, Oxford 1966. Eng. tr. and ed.: R.M. Price, *Cyril of Scythopolis. The Lives of the Monks of Palestine*, Kalamazoo 1989.

J.-M. Sauget

SABBATH. The Christian churches are repeatedly asked (by *Jews: Trypho in Just., *Dial.* 10, 3; by Seventh-Day Adventists: S. Bacchiocchi, *Divine Rest for Human Restlessness*, Rome 1980) why they no longer celebrate the s., whose sanctification is required by the *decalogue and which has at all times played, and still plays, a central role in Jewish tradition. How, historically, did this abandonment of the s. come about?

Jesus, through curing the chronically sick on the s. (Mk 3, 1-6 par.; Lk 13, 10-17; Jn 5, 2-9) and his provocative transgressions of the s. (Mk 2, 23-28 par.; Jn 5, 10-11), so offended Pharisaic circles as to convince them that his behaviour merited the death penalty (Mk 3, 6 par.; Jn 5, 16-18). Jesus' attitude was less an expression of his philanthropy than an indirect indication of his Messianicity (Mk 2, 28 par.): with his coming the definitive s. had arrived (Lk 4, 16-21), since he offered men not just physical healing but also the "sabbath of the heart", forgiveness of sins (Mt 11, 28-30).

Early Christianity at first hesitated to claim Jesus' freedom for itself in the face of the commandment of the s. (Mt 24, 20 f.; Lk 6, 5, *cod. Bezae*). Sources all agree that the *Judaeo-Christians still observed the s. (Just., *Dial.* 47, 1-4; Iren., *Adv. haer.* I, 26, 2; Euseb., *HE* III, 27, 2-5). Probably the Hellenists around *Stephen were the first not to observe, among other things, the s., for which they drew down a bloody *persecution (Acts 6, 13-14).

Certainly the gentile Christian communities founded by *Paul no longer observed the s. (Gal 4, 8-11; Col 2, 8. 16-17). This became a general *tradition for the Great Church from the 2nd c. (Ign., *Magn.* 9, 1; *Barn.* 15, 1-9; Just., *Dial.* 12, 3; 18, 2; 19, 5-6; 27, 5; 29, 3; Iren., *Adv. haer.* IV, 8, 1-3; 16, 1). Naturally, here and there, later on, even the Great Church had to be on guard against "*judaizers" (Tertull., *Orat.* 23, 1-2; Cyr. of Jer., *Cat.* 4, 37; council of Laodicea, can. 29; John Chrys., *Adv. Iudaeos*; Ps.-Ign., *Magn.* 9 [PG 5, 768]; Ps.-Athan., *De semente hom.*).

What was the theological motivation for the Great Church's attitude? The most frequent motives are three: *a)* the s. was unknown to the OT patriarchs; *Moses gave it - with other ceremonial laws - to the Jewish people as a punishment for disobedience or as a preparation (cf. the hidden *creed of Col 2, 17) for its fulfilment to be realized in Christ (cf. the texts of Justin and Irenaeus cited above; also: Ptolemy, *Epist. ad Floram* 5, 8-9. 12; Ambrosiast., *Com. Col.* 2, 16-17; Ambr., *Exp. Ev. Lc.* V, 31-33; 39-40; VII, 173-175; etc.); *b)* with the coming of Christ, remaining without external work for 24 hours had become secondary; what mattered was to cease to serve *sin, throughout the course of life; he observes the true s. who, following Christ, honours *God with a priestly service and loves his neighbour (Iren., *Adv. haer.* IV, 16, 1 and *Dem.* 96; Tertull., *Adv. Iud.* 4, 1-5 and *Adv. Marc.* IV, 12, 9-11; Orig., *In Num. hom.* 23, 4 and *In Ev. Mt. comm.* XII, 36; Euseb., *In Ps.* 91 [92]; Epiph., *Pan.* 30, 32, 6-9; Aug., *Ep.* 55, 18-19, 22); *c)* Christians await the final s., in which they will become partakers of perfect *peace in God (Heb 3-4; *Barn.* 15, 1-9; Just., *Dial.* 80, 1-2. 5; Iren., *Adv. haer.* V, 28, 3; 30, 4; Hipp., *In Dan. comm.* IV, 23, 1-24, 6; Tertull., *Adv. Marc.* III, 24, 3-6; Vict. Petov., *Fabr. mundi* 5; Lact., *Div. inst. epit.* 67, 3-5 [72, 13-15]; Aug., *De civ. Dei* XX, 7, 9). Often - as the cited texts show - before *Augustine, this *hope of the final s. is connected to that of the millenium. *Sunday, in this context, is understood as an "eighth day".

From the 3rd c. there was a difference between the Western and Eastern churches over the sabbath *fast: since all Christians fasted on Holy Saturday, here and there the Latin churches began, in analogy with the Holy Saturday fast, to fast every Saturday (Tertull., *Ieiun.* 14, 3; Hipp., *In Dan. comm.* IV, 20, 3; *Liber Pont.* 17, 2; council of Elvira, can. 26; Aug., *Ep.* 36, 31). This custom was always rejected by the Eastern churches (*Const. Apost.* VIII, 47, 67; Ps.-Ign., *Phil.* 13 [PG 5, 938]), even during Lent (Athan., *Ep. fest.* 6, 13; Egeria, *Peregr.* 27). This divergence later became one of the reasons for the *schism between East and West.

Individual Eastern churches - at first, it seems, monastic circles - introduced a celebration of the *Eucharist on Saturday in memory of the fullness of creation (*Vita Pachomii* 28; Pall., *Hist. Laus.* 32, 3; Greg. Nyss., *Adv. eos qui castigationes aegre ferunt*; Epiph., *De fide* 24, 7; *Const. Apost.* II, 59, 3; VII, 23, 3-4; VIII, 33, 1-2; Tim. of Al., *Responsa can.* 13). A direct relationship between these Saturday celebrations and the Judaeo-Christian practice cannot, in my view, be proved (C.S. Mosna and R.A. Kraft think otherwise). After *Constantine, Sunday became more and more a "substitute" for the s.

C.W. Dugmore, *The Influence of the Synagogue upon the Divine Office*, London 1944, 26-37; H. Huber, *Geist und Buchstabe der Sonntagsruhe. Eine historischtheologische Untersuchung über das Verbot der knechtlichen Arbeit von der Urkirche bis auf Thomas von Aquin*, Studia Theologiae moralis et pastoralis 5, Salzburg 1958; K. Hruby, La célébration du sabbat d'après les sources juives, *OrSyr* 7 (1962) 435-463; 8 (1963) 55-86; Y.B. Tremel, Du sabbat au jour du Seigneur, *Lumière et Vie* 11 (1962) 29-49; C. Floristán, El sábado judío. Del sábado al domingo. El domingo, día del Señor, *Salmanticensis* 10 (1964) 429-444; E. Lohse, Sabbaton, *TWNT* 7, 1-30; R.A. Kraft, Some Notes on Sabbath Observance in Early Christianity, *Andrews University Seminary Studies* 3 (1965) 18-33; P. Delhaye-J.L. Leclat, Dimanche et Sabbat, *MSR* 23, 2 (1966) 3-14, 73-93; C.S. Mosna, *Storia della domenica dalle origini fino agli inizi del V secolo. Problema delle origini e sviluppo. Culto e riposo. Aspetti pastorali e liturgici*, AG 170, Rome 1969; A. Verheul, Du sabbat au jour du Seigneur, *Quest.Lit. et Par.* 51 (1970) 3-27; F. Mathys, Sabbatruhe und Sabbatfest, *ThZ* 28 (1972) 242-262; W. Rordorf, *Sabbat und Sonntag in der Alten Kirche*, Traditio Christiana 2, Zürich 1972; K.H. Strand, *Essays on the Sabbath in Early Christianity*, Ann Arbor (Mich.) 1972; P. Grelot, Du sabbat juif au dimanche chrétien, *La MaisonD.* 123 (1975) 79-104; (ed.) D.A. Carson, *From Sabbath to Lord's Day: a Biblical Historical and Theological Investigation*, Grand Rapids 1982.

W. Rordorf

SABELLIUS - SABELLIANISM. All we know of Sabellius is that in c.220 he was condemned at *Rome by *Callistus as an exponent of *patripassian *monarchianism. His Libyan origin is uncertain. After the condemnation, either he or his disciples spread the monarchian doctrine in *Libya and *Egypt, and developed it in opposition to the Logos-theology of *Origen and his school. They extended the original patripassian doctrine to take in the *Holy Spirit: one sole *God is manifested as *Father in the OT,

Son in the *Incarnation, Holy Spirit poured out upon the *apostles at *Pentecost. In this way they avoided, at least formally, *Noetus's statement, which had met such opposition, that the Father himself had been incarnate and had suffered. They also maintained, against Origen's doctrine of three distinct *hypostases in the *Trinity, that Father, Son and Holy Spirit constitute a single *prosopon and a single hypostasis. The slender evidence that the Sabellians affirmed three prosopa in the Trinity must be referred to monarchians of more moderate type. In the course of the 4th c. the Logos-theologians described every form of monarchianism as Sabellianism: even *Marcellus of Ancyra, whose monarchianism was radical but different from that of Sabellius, was called a Sabellian. For lack of evidence we cannot specify the diffusion and duration of Sabellianism: what is certain is that it was a considerable obstacle to the Logos-theologians in the second half of the 3rd and early 4th cc. *Epiphanius (Pan. 62, 1) says that Sabellians were still active in c.375 at Rome and in Mesopotamia.

M. Simonetti, Sabellio e il sabellianismo, SSR 4 (1980) 7-28; G.L. Prestige, God in Patristic Thought, London 1952, and also Fathers and Heretics, London 1954, ch. 2; H.J. Carpenter, Popular Christianity and the Theologians, JThS n.s. 14 (1963) 294-310.

M. Simonetti

SABINA. S. can serve as an example of how, in the course of the 6th c., the founders and foundresses of the churches of *Rome became saints. In the Roman synod of 499, the church on the Aventine which bears her name was called titulus Sabinae; in that of 595, titulus S. Sabinae. S. appears neither in Mart. hier. (except for some second-hand additions in manuscript S), nor in the Sacr. Leon. Her feast, on 29 Aug, appears only in the Capitulare evangeliorum of Würzburg, c.740, the so-called Gelasian sacramentaries (8th c.), the Gregorian sacramentary and the historical *martyrologies, starting with *Bede's second version. These seem to depend on a Passion (BHL 7407 and 7586) created in the 6th c. on a basis of pure fantasy. The basilica is a classic example of an early Christian *church building (5th c.).

DACL 15, 218-38; Vies des SS. 8, 580-82; LTK 9, 195; BS 11, 540-43; F. Darsy, Santa Sabina, Rome 1961; id., Recherches archéologiques à Sainte-Sabine sur l'Aventin, Rome 1968; R. Krautheimer, Corpus basilicarum christianarum Romae, IV, Rome 1976, 69-94.

V. Saxer

SABINUS of Canosa. Bishop of Canosa (Canusium) in Apulia, prob. 514-566. The certain facts of his life are: his presence at the synod of Rome of *531, in whose *Acts he appears next to Boniface II in favour of *Stephen of Larissa; his leadership of the embassy sent by pope Agapitus to *Justinian to remove *Anthimus, *monophysite patriarch of Constantinople; what *Gregory the Great says in his Dialogues (II, 15; III, 5) about his familiarity with *Benedict at Montecassino and meeting with *Totila at Canosa. S. prob. accompanied the unfortunate pope *John I in the delegation sent by *Theodoric to *Byzantium (525-6) to plead, naturally without success, the *Arian cause. The anon. Acts of S. of Canosa (ASS Febr. II, 323-328; 3rd ed. 324-329) speak of his sanctity and call him restaurator ecclesiarum - a detail confirmed by recent excavations (1967) of the baptistery of S. Giovanni at Canosa.

BHL 7443-7445; A.A. Tortora, Relatio status sanctae primatialis ecclesiae Canusinae, Rome 1758; R. Cessi, Un vescovo pugliese del sec. VI (S. Sabino di Canosa), Atti del R. Istituto Veneto di Scienze Lettere ed Arti 73 (1912-1913) 1141-1157; A. Lentini, Due legati papali a Costantinopoli nel secolo VI: Germano di Capua e Sabino di Canosa, Atti del IV Congresso Nazionale di Studi Romani, Rome 1938, 385-393; V. Recchia, Reminiscenze bibliche e "topoi" agiografici negli "Atti" anonimi di S. Sabino Vescovo di Canosa, VetChr 4 (1967) 151-184; R. Moreno Cassano, Il Battistero di S. Giovanni a Canosa, VetChr 5 (1968) 163-204.

V. Recchia

SABINUS of Heraclea. Bishop in *Macedonia, in c.375 he made a collection (now lost) of the *acts of the various *councils held during the *Arian controversy, with anti-Nicene intentions. It was widely used by *Socrates and still more by *Sozomen, who got very important information from it.

W.D. Hauschild, Die antinizänische Synodalaktensammlung des Sabinus von Heraklea, VChr 24 (1970) 105-126.

M. Simonetti

SABINUS of Piacenza. Saint: feast 11 Dec. We know a Milanese deacon S. (Mansi 3, 460), pope *Damasus's emissary to the *East over the *schism of *Antioch, esteemed by *Basil (Epp. 89 and 92). He may be the bishop of Piacenza (Placentia), elected c.376, who took part in the councils of *Aquileia (381), *Rome (390) and *Milan (392). A highly esteemed *friend of *Ambrose, whose Epp. 45-49 and 58 are addressed to him and who submitted his works to him for judgment (Ep. 48, 1 and 3). None of S.'s *letters to Ambrose survive. He built the basilica Duodecim apostolorum at Piacenza.

EC 10, 1521; BS 11, 701-704; G.F. Rossi, S. Savino vescovo di Piacenza, Roma 1955; id., S. Savino diacono milanese, poi vescovo di Piacenza, Divus Thomas (Piacenza) 59 (1956) 125-142.

A. Di Berardino

SACRAMENTS

I. A liturgico-typological approach - II. A liturgico-symbolic approach - III. Exploration in the Fathers.

We must not expect to find in the Greek or Latin *Fathers, and in general esp. before St *Augustine, a theology of the ss. constructed according to modern criteria. To organize their doctrine starting from our own categories would be gravely to falsify their thought. The brevity of this article justifies choice from among very abundant material.

I. A liturgico-typological approach. Before St Augustine, who tried to develop a more conceptual *theology, the Fathers did not even think in terms of a logical demonstration of what the ss. are. They related them to salvation-history: they were, in the present, "a step" along the way to *eschatology. In this way they avoided what could have led to a sort of sacramental "*magic", and showed what was meant by Christ's institution of the ss.: a new content in pre-existing forms prepared since millennia. Starting from the tangible and visible side of sacramental celebration (*water, *oil, *bread, *wine, rites), they ascended to the announcement of them in the OT and NT (τύπος), to arrive at the s. itself, which thus found its true root (ἀντίτυπος). From this they then deduced the consequences for the concrete *life of the Christian.

1. In doing this, the Fathers follow the method of the NT itself. 1 Corinthians (10, 1-5) and 1 Peter (3, 19-21) take the *flood and the rock of Horeb as types of *baptism. St *Paul writes that *Jewish worship is the shadow of the true worship still to come, Christ being the reality (Col 2, 17). The vocabulary: heavenly reality (Jn 3, 12; 1 Cor 15, 40-48; etc.), shadow, *image (Rom 1, 23; 8, 29; 1 Cor 11, 7; 15, 49; etc.), serve them as oft-used examples.

2. The NT itself is constantly used in so far as it offers important types, like the pool of Bethesda (Jn 5), in which the Fathers see one of the specific types of baptism. Thus *Tertullian (De bapt. 5: CCL 1, 281), who insists on the action of the *angel. He is followed, e.g., by *Ambrose of Milan (De sacramentis II, 3, 7: SCh 25a, 115). The NT is also used with regard to the *Eucharist, as in Mt 22, 3: e.g. the *parable of the feast and the guests, commented on by *Cyril of Jerusalem in his Procatechesis (PG 33, 336-341). In the same way the parable of the virgins (Mt 25, 1) is applied, with its eschatological meaning, to the ss. of Christian *initiation. *Gregory Nazianzen sees in it the beginning and the way towards the eschatological banquet (Oratio XL in sanctum baptismum 46: PG 36, 426). The wedding-feast of *Cana (Jn 3) is used as a type of the Eucharist by Tertullian (De bapt. 9: CCL 1, 284), *Cyprian (Ep. 63, 12: CSEL 3/2, 711) and Cyril of Jerusalem (Cat. Myst. IV, 2: SCh 76, 136).

3. It is mainly the OT which provides the Fathers with the types they need for their *catechesis. Tertullian offers us a list of OT types that, with those of the NT, would be used from then on. *a)* In particular the type of *water, used in many *commentaries; it is developed, from the liturgical point of view, in the *blessing of the baptismal water in the Gelasian Sacramentary, attrib. to *Peter Chrysologus (Sacram. Gel. 445-446). The water of *creation is a type used by Tertullian (De bapt. 3: CCL 1, 278). Ambrose of Milan fixes his attention on many aspects of it, such as that of the ἰχθύς and the pisciculi, i.e. the Christians (De sacramentis 3, 3: SCh 25a, 72). *b)* The *flood: the Fathers use this type on the basis of 1 Pet 3, 3-10. In the eight persons saved, *Justin sees the *ogdoad, the eighth day, day of *Easter and the *resurrection (Dial. 138: PG 6, 793). Tertullian develops the *symbolism of the dove and that of the ark-church (De bapt. 8: CCL 1, 283; De idol. 24: CCL 2, 1924), followed in this by Cyprian (De unit. 6: CCL 3, 253). *c)* The Red Sea, as a type, also attracts the Fathers' attention, backed up by 1 Cor 10, 2-6. Thus Tertullian (De bapt. 9: CCL 1, 283-284), Cyprian (Ep. 68, 15: CSEL 3/2, 764), *Basil (De Spir. Sancto 14: SCh 17a, 163), *Gregory of Nyssa (Vita Moysis 50, 2: PG 44, 361), Cyril of Jerusalem (Cat. Myst. I, 2-3: SCh 126, 83-84) and Ambrose, who frequently refers to this type (De sacr. I, 12: SCh 25a, 57-58; 61; De myst. 12: PL 16, 409; 13: PL 16, 410). *d)* The waters of the Jordan offer the Fathers another type, above all the crossing itself as a type of baptism: thus Gregory of Nyssa (De iis qui baptismum differunt: PG 46, 420 f.; In bapt. Christi: PG 46, 592 f.), Cyril of Jerusalem (*Cat. Myst.* 3, 3: SCh 126, 121. 123. 129). The type of the healing of Naaman in the Jordan also interests them: Gregory of Nyssa (*In bapt. Christi*: PG 46, 593), Ambrose (*Exp. in Lc.* IV, 50 f.: CSEL 32/4, 165-179; *De sacr.* I, 13 f.; SCh 25a, 164). *e)* They also use the types of *bread and of *Melchizedek: *Clement of Alexandria (*Strom.* IV, 25: SCh 30, 62), Cyprian (*Ep.* 63, 4: CSEL 3/2, 703), Ambrose (*De sacr.* V, 1: SCh 25a, 88; IV, 10: SCh 25a, 80). Naturally they do not forget the *manna; just look, e.g., at Ambrose (*De myst.* 46: SCh 25a, 182) and *Augustine (*Tr. in Io.* 26, 12: CCL 36, 265). *f)* The Fathers also seek a typology of the meal and find it in the OT (Is 55, 1-3) and the NT (Lk 22, 29): thus *Origen (*Com. in Cant. hom.* 2, 3: PG 13, 155), Gregory of Nyssa (*In Cant. Cant. hom.* 5: PG 44, 873), Ambrose (*De Cain et Abel* I, 5: CSEL 32/1, 356). *g)* Still based on 1 Cor 10, 4, the Fathers see in the rock of Horeb the type of the baptismal water or else of Christ's *blood: Cyprian (*Ep.* 63, 8: CSEL 3/2, 705 ff.), Ambrose (*De sacr.* V, 3-4: SCh 25a, 89; *De myst.* 48: SCh 25a, 123).

We have insisted abundantly, but not yet exhaustively, on this patristic method, the most characteristic method of the Fathers, who intend not to offer a true concept of s., but "to show it"; starting from the sign, they make plain its origin and its insertion into salvation-history, of which the ss. are, in the present, "a step" and, at the same time, a prefiguration of *eschatology.

II. A liturgico-symbolic approach. Yet not all the *Fathers follow this way in their *catechesis. *Theodore of Mopsuestia, e.g., hardly ever ties himself to types, but seeks rather to trace a parallel between the visible and invisible liturgies. In his homily XII, 2 (Tonneau, ST 145, 325), e.g., he bases himself on the epistle to the Hebrews (8, 5; 10, 1) and sees in the s. a ritual imitation of Christ's historical actions. Nor does he stop at generalities: he tries to relate the sacramental rites to the gospel accounts (*Hom.* 15, 25-27: ST 145, 503-509). The same line is followed by Ps.-*Dionysius, who is clearly oriented towards *symbolism: sensible realities are *images of intelligible realities (*De divinis nominibus*: PG 3, 525-1120). The Eastern *liturgy provides him with a vast symbolic material, on which he draws abundantly. This method is not without dangers and prepares a way for the false symbolism of Amalarius of Metz. As we can see, the Fathers are concerned mainly with the two fundamental ss.: *baptism and the *Eucharist; but they do not ignore the other ss. and they do deal with them: e.g. Ambrose and his treatise on *penitence.

III. Exploration in the Fathers. With all this, the Fathers, even before St Augustine, were not content with a presentation of the ss. limited to typology or symbolism. They also sought to express, in their way, what we may call the "mechanism" of the ss.

1. They were drawn to a pre-Christian terminology. The use of the word μυστήριον in St Paul certainly does not imply a link with the *pagan *mysteries; the term serves him to indicate the plan of salvation eternally hidden in *God and revealed in and by Christ. Yet the pagan term μυστήριον becomes increasingly the precise name of the ss. *Justin (*Apol.* I, 66: PG 4, 429), though he does not use the expression, and *Tertullian (*De praescr.* 40: PL 2, 66; CCL I, 221; *De bapt.* 2: PL 1, 1209; CCL I, 227), who, in these cases and to designate the Christian celebrations, prefers not to use the word *sacramentum*, see in the pagan mysteries an imitation of the Christian ss.: the mysteries try to provoke a presence of the activity of a deity. But the Fathers oppose the pagan mysteries, contrasting them with the true mysteries, those of Christianity. Thus, against the cults of the *sun, they see Christianity as true possession of the true sun. *Clement of Alexandria offers us a very precise and very important text in which he begs the adepts of the cults of the pagan gods to come and celebrate the true mysteries. He employs the current vocabulary of the pagan cults that he intends to reject, imitating them (*Protrepticus* XII: PG 9, 778). At a time when the mystery religions were in decline, from the 4th c. on, the Fathers had no fear of possible confusions and made abundant use of the vocabulary of contemporary cults; e.g. *Basil, in his treatise *De Spiritu Sancto* (27: SCh 17, 155). Typical in this respect is the title given by *Cyril of Jerusalem to his teachings: *Mystagogical catecheses* (cf. *Mystagogy). In Ps-*Dionysius we find many examples of these "borrowings", esp. with regard to Christian *initiation. This is not in the least a theological "borrowing". It could be thought that the mystery cults were a providential preparation for the Christian religion, and above all for the theology of the actualization of Christ's saving acts. But we should not think of the ἀνάμνησις of Hellenism or the *memoria* of Latin civilization, but rather of the Palestinian *zikkaron*: the bringing of the past into the present before God's (and not primarily men's) eyes, so that He may remember, act and make his people walk forward from the covenant towards the end of the ages.

In the African-Latin text of the *Bible, the word is habitually translated *sacramentum*; the *Itala* gives many similar examples. The Vulgate,

however, always prefers the word *mysterium*. And yet the epistles to the Ephesians and Colossians, e.g., use both words indifferently, without it being possible to determine the reason.

The military meaning of *sacramentum* - a soldier's oath of loyalty - did not apply to the term *mysterium*. Yet Tertullian, followed by many others, using it in a military sense, influenced the use of the term among Christians for the *Creed, the *regula fidei* for baptism. With Tertullian, therefore, this term ended by acquiring a particular meaning, which designated a precise content: the faith. This also enabled a better contrast to be made between the pagan and Christian religions: the pagan mysteries had only the "apparent matter" of the ss. (*Ad nationes* I, 16: CCL 1, 36), and pagan worship was called *non sacramenta*.

In *Ambrose we find a definition of the s. of baptism that is valid for any sacrament.: the *neophyte has seen all that can be seen with human eyes, but not all that has been produced. Now that which is not seen is much greater than that which is seen and which is temporal, because that which is not seen is eternal (*De sacr.* I, 4: SCh 25a, 66). In discussing the *Eucharist, he uses both words: *mysterium* and *sacramentum* (*Exp. in ps.* 118, 13, 20: CSEL 62, 294). And the better to specify his thought, Ambrose uses for the Eucharist the terms *mysteria* and *sacramenta* linked: *sacramenta mysteriorum* or *mysteria sacramentorum* (*Apologia David* 12, 58: CSEL 32/2, 339). Despite everything, the difficulty of specifying the meaning of the two terms remains. Sometimes *mysterium* indicates a sacramental rite (*De myst.* I, 1: SCh 25a, 156), but *prayer itself can also be called *mysterium* (*De fide* IV, 10, 24: CSEL 73, 90; *De sacr.* V, 3, 12: SCh 25a, 124). Specification is not entirely absent, however: *sacramentum* designates what is seen, i.e. the exterior; *mysterium*, the content, what is interior. The *sacramenta* introduce us to the *mysteria* and, in turn, the *mysteria* make us understand the outer sign, the *sacramenta* (*De myst.* 2: PL 16, 406).

We cannot pass over in silence the use of the word *sacramentum* in a sometimes precise and extensive sense in the writings of *Leo the Great. It occurs 128 times in his *Sermons and 93 in his *Letters. Of all the meanings of this term in St Leo, the two main ones are: 1) one reality as a sign of another, a figure or *symbol (e.g. *Serm.* 55, 3: CCL 138A, 325); 2) an efficacious sacred sign (examples are numerous: e.g. *Serm.* 3, 1: CCL 138, 10-11; *Serm.* 21, 3: CCL 138, 88; *Serm.* 70, 4: CCL 138A, 429). But St Leo also uses the word *mysterium* or *mysteria* with the same meaning, in the sense of a figure, one reality as a sign of another (*Serm.* 23, 4: CCL 138, 106; *Serm.* 51, 7: CCL 138A, 302), or in the sense of an efficacious sign (*Serm.* 16, 1: CCL 138, 61).

2. The Fathers also study how the ss. "operate". Tertullian strongly emphasizes the rite and its effect (*De resurr.* 8: CCL 2, 931; *De praescr.* 40: CCL 1, 220; *De bapt.* 1; 4; 9: CCL 1, 277. 279. 283). He applies the same procedure as that of *baptism to the other ss. (cf. under individual sacraments). As for Cyprian - note only that Tertullian does not acknowledge any efficacy for the sacramental signs practised outside the church (*De bapt.* 15: CCL 1, 290) - he applies this thesis not just to baptism, but to all the ss., e.g. *ordination (*Ep.* 69, 1: CSEL 3, 760). *Ambrose too is interested in this "action" of the ss., particularly those of initiation. In this respect, he is closely related to the Pauline doctrine of baptism in the *similitudo* of Christ's death and *resurrection (Rom 6, 5), and seeks to explain how the rite of baptism is a *similitudo* that encloses the reality (*De sacr.* II, 23: SCh 25a, 88). He does the same for the *Eucharist; in this context he uses the word *figura*, a sign or symbol that contains the reality (*De sacr.* IV, 5, 21: SCh 25a, 114). We see here a very close link with the way in which *Serapion of Thmuis expresses himself in the *Euchologion* attrib., perhaps wrongly, to him (*Euchologion of Ser.* 3: F.X. Funk, *Didascalia et Const. Apost.* II, *Sacramentarium Serapionis* 174). Ambrose also explains that there is a rite whose symbolism corresponds to man's nature (*In Lc.* I, 2, 79: PL 15, 1663). He also attributes considerable importance to the word; word and symbolic rite are for him two indispensable elements (*De sacr.* IV, 19: SCh 25a, 112). In the case of the Eucharist, Ambrose attributes to the word the expression *sermones Christi, verba caelestia*, etc. (*De sacr.* IV, 5, 21: SCh 25a, 114) and elsewhere calls it *sermo operatorius Christi* (*De sacr.* IV, 4, 1: SCh 25a, 110). The action of the *Holy Spirit is mentioned in Ambrose's theology (*De sacr.* I, 5, 15. 17: PL 16, 440; SCh 25a, 68; *De myst.* 4; 19; 20; 24; 59: SCh 25a, 164, 166, 168, 190-191; *De Spir. sancto* 3, 16: CSEL 79, 167). Nor does Ambrose forget the church's action in the s.: it is the church itself which is fertile in the Spirit (*De virginibus* I, 6, 31: PL 16, 197; *Exp. in ps.* 118, 19, 23: CSEL 62, 433-434). We have given particular attention to Ambrose, because his doctrine is very representative of the thought of the Latin Fathers.

Among the Eastern Fathers the doctrine is no less firm. *Clement of Alexandria makes the efficacy of the rite the basis of his sacramental theology. Of numerous possible texts, we will cite one concerning *confirmation, which he calls *sigillum* (*Quis dives* 42: PG 9, 48). The notion of an efficacious symbol is used by them all, particularly for baptism and the Eucharist. *Origen goes to *neoplatonist philosophy both for his preferred symbolism, sometimes endangering the typology by exaggerations, and for his sacramental doctrine. In the baptismal rite he sees a symbol which confers gifts (*In Io.* VI, 17: PG 14, 257). Even so, despite what is sometimes said, his theological symbolism applied to the Eucharist does not overshadow the real presence (*C. Celsum* VIII, 33: PG 11, 1565).

The baptismal ritual of the *Didascalia, like that of the *Apostolic Constitutions, takes up the Pauline symbolism. The theology of the seal impressed in baptism is present here (*Didascalia et Constitutiones Apostolorum* III, 12, 3; 16, 4; Funk 210, 211). *Sins are remitted and man enters a new life (*Didasc.* II, 39, 4: Funk 216; V, 9, 1: Funk 262; VI, 12, 2: Funk 326, etc.). If Origen saw the effect of baptism in the invocation of the *Trinity, the Greek Fathers see it more in the active presence of the Spirit (cf. Jn 3, 5). For *John Chrysostom, it is God who baptizes, not the minister (*In Mt. hom.* 50, 3: PG 57, 507; *In Act. apost. hom.* 14, 3: PG 60, 116). *Basil says that *water, in itself, produces nothing: the presence of the Holy Spirit is necessary (*De Spir. sancto* 15, 35: SCh 17, 171). For *Cyril of Jerusalem, the invocation of the Trinity consecrates the *oil and confers its efficacious power on it (*Cat. Myst.* III, 3-4: SCh 126, 124-126).

St *Augustine occupies a place apart in his exposition of the ss.: his *theology is by far the most developed. While using symbols, he prepares what would later be the theology of the 12th c. and the council of Trent. Augustine does not overlook typology, but he goes beyond it and studies the sign in itself, independently of its relation to this or that s. Like some Fathers who preceded him, he distinguishes in the s. the visible (*sacramentum*) from the invisible (*virtus sacramenti*). These distinctions occur in his commentaries on *John's gospel (*In Io. tract.* 26, 11: CCL 36, 263). Without being evidently conditioned by our notion of seven ss., Augustine states that everything that gives order to sacred things is a s. So in the rites of the *catechumenate he calls the blessed salt given to the *catechumen *sacramentum salis* (*De cat. rud.* 26, 50: CCL 46, 173-174; the possible variant *sane* for *salis* matters little here, since later he discusses the salt as "sacrament"). The Gelasian Sacramentary too applies the word "sacrament" to the blessed salt (*Sacram. Gel.*, ed. Mohlberg, 288, p. 43, 26) and it recurs in later rituals until the reform of Vatican II, which suppressed the rite. In his two treatises *De doctrina christiana* (PL 34, 15, 122) and *De magistro* (PL 32, 1021-1034), Augustine expounds a sort of philosophy of signs, in which it is easy to see a dependence on the Alexandrians. Ss. are material subjects which fall under the senses, but which indicate a spiritual content. E.g., in baptism, water is the visible element: it washes the *body, but this signifies what it produces in the *soul (*In I ep. Io.* VI, 11: SCh 75, 300-304). The signs are not just conventional: there is a relationship between the signifier and the signified (*Ep.* 98, 9: CSEL 34, 530-531). In the s. there is a force that produces the signified effect. On the other hand, it would be to betray Augustine's thought to try to reduce the *virtus sacramenti* to its effects alone: the *virtus* also signifies the *grace received (*In Io. tract.* 26, 11-13: CCL 36, 264-267). Christ acts through his church and his *minister, and the word is this *virtus* that brings about the gift of grace, independently of the dignity of the minister (*Ep.* 98, 2: CSEL 34, 521-523). It is the minister, however, who, through word and *gesture, makes the s., because he offers God the occasion to act and give His grace. "*Accedit verbum et fit sacramentum*" (*In Io. tract.* 53, 3: CCL 36, 529) thus echoes what we have seen in other Fathers, particularly in Ambrose of Milan. But we must not limit the *verbum* to what we now, following the scholastics, call "form". *Verbum*, in the baptismal interrogation, is also the reply given by the catechumen, questioned on his *faith during his immersion (*In Io. tract.* 53, 3: CCL 36, 529). Augustine applies all this to the different ss. After him some Fathers, such as *Maximus of Turin (*Serm.* 13: CCL 23, 44), take up the same thought and open the way for the theologians of the 11th and 12th cc.

DTC 14, 498-527; LTK 9, 218-225; J. de Ghellinck et al., *Pour l'histoire du mot "sacramentum"*, Louvain-Paris 1924; O. Casel, Zum Wort Sacramentum, *JLW* 8 (1928) 225-232; E. Neveut, La théologie sacramentelle de saint Augustin, *Divus Thomas* (Piacenza) 34 (1931) 3-27; J. Daniélou, *Bible et Liturgie*, Paris 1951; B. de Soos, *Le mystère liturgique d'après saint Léon le Grand*, LQF 34, Münster 1958; W. van Roo, *De sacramentis in genere*, Rome 1966; R. Johanny, *L'eucharistie, centre de l'histoire du salut chez saint Ambroise de Milan*, Paris 1968; R. Schulte, L'evento salvifico nella comunità di Gesù Cristo, *MySal* 8, Brescia 1975, 97-127; H.J. Auf der Maur et al., *Fides Sacramenti Sacramentum Fidei* (Festsch. P. Smulders), Assen 1981; M.B. Moreton, Roman Sacramentaries and Ancient Prayer Traditions, *SP* 15.1 (1984) 577-580; id., From the Sacrifice of Christ to the Sacrifice of the Church, *SP* 18.2 (1989) 385-390; D. Serra, The blessing of Baptismal water at the Paschal Vigil in the *Gelasianum Vetus*, *Ecclesia Orans* 6 (1989) 323-342; P. Tragan, Le radice bibliche dei sacramento della confirmazione, *Riv. lit.* 76 (1989) 214-231; A. Schmemann, *For the Life of the World: Sacrament and Orthodoxy*, New York ²1973; L. Weil, *Sacraments and Liturgy: the Outward Signs*, Oxford 1983; D.J. Sheerin, *The Eucharist*, Wilmington 1986; T.M. Finn, *Early Christian Baptism and the Catechumenate*, Wilmington 1989.

A. Nocent

SACRAMENTUM. A Latin term introduced into *theology, but which has no exact equivalent in Greek. Among other things, it translates the word *mysterion* in the variety of its meanings, esp. in the *Bible. Etymologically, *sacramentum* expresses a religious thing (*sacer*); later it took on a legal nuance. The two elements (sacred and legal bond) recur in *sacramentum*, which therefore meant "an initiation confirmed by an oath". It primarily designated initiation into military service, and military service itself.

Employed in a baptismal context, the term simultaneously expressed enlistment into Christ's service, the oath of initiation which sealed the commitment made, and finally the initiatory rite itself, where *sacramentum* replaced the Greek *mysterion*.

Schematically, the term developed around two centres: a baptismal centre, with meanings of pledge, profession of faith and bond contracted, and of sign of recognition, teaching and *regula fidei*; and a *mystery-centre (translation of *mysterion*): truths hidden and revealed, *economy of salvation, hence typology, the feasts of Christ and the liturgical feasts, the sacred or consecratory sign, and finally the sacramental rite, principally the rites of *initiation (*baptism, *confirmation, *Eucharist). *Augustine also uses it for *marriage (*Tract. in Io.* 9, 2), *laying-on of hands, *anointing of the sick and reconciliation of penitents (cf. Blaise-Chirat, *Dictionnaire latin-français des auteurs chrétiens*, Strasbourg 1954, 729 ff.).

AA. VV., *Pour l'histoire du mot sacramentum*, Louvain-Paris 1924; F. Dölger, Sacramentum militiae, *AC* 2 (1930) 268-280; C. Couturier, *Sacramentum et Mysterium*, Paris 1953, 161-274; C. Mohrmann, Sacramentum dans les plus anciens textes, in *Etudes sur le latin des chrétiens*, I, Rome 1958, 233-244; D. Michaélides, "*Sacramentum*" chez Tertullien, Paris 1970; B. Studer, "Sacramentum et exemplum" chez s. Augustin, *RecAug* 10 (1975) 87-141.

A. Hamman

SACRIFICE. By s. we mean the act of worship in which man offers a creatural element to the deity. *Augustine stresses that the visible s. is a *sacrament, i.e. a sacred sign of an invisible s.: for him "every work tending to effect our *beatitude by an holy conjunction with God" is a true s. (cf. G. de Broglie, La notion augustinienne du sacrifice "invisible" et "vrai", *RecSR* 48 [1960] 135-165). In every s. we must consider four aspects: to whom we offer (*God), by whom the offering is made (by a holy and righteous priest), what is offered (a substance taken from those for whom it is offered) and for whom (for defective beings in need of *purification, so the substance offered must be without defect): these aspects find full expression in the s. of Christ, the one true mediator (*De Trin.* IV, 14, 19).

On the question of pre-Christian s., Augustine observes that, from antiquity, worship of God was expressed in the form of s. (*De civ. Dei* X, 4). The ss. of the *pagans were in error because they were offered to *demons; they stressed the strength of creatures rather than the power of the Creator (Aug., *Ep.* 102, 16-21). The *Fathers saw the Jewish ss. as a remedy for *idolatry (Just., *Dial.* 19, 6; Greg. Naz., *Or.* 31, 25; Epiph., *Pan.* 66, 71). Ss. are pleasing to God not because they help him personally, but because they express a desire to honour God; but he rejects them when performed for personal advantage (Tertull., *Marc.* II, 22). The ss. of the Jews had a typological value (Clem. Al., *Strom.* VII, 6, 32, 7; Orig., *C. Cels.* IV, 31; John Chrys., *Hom. 17, 3 in Hebr.*; Aug., *In ps.* 39, 12; 74, 12; *C. Faust.* VI, 5; *C. adv. leg.* I, 18, 37) - in particular *Abraham's s. of Isaac (cf. J. Gribomont, Isaac le patriarche, *DSp* 7, 1922-2005) and that of *Melchizedek (cf. J.L. Ska, Melchisédech, *DSp* 10, 967-972) - and gave way to Christ's s. (John Chrys., *Hom. 18, 1 in Hebr.*; Aug., *De civ. Dei* X, 20).

The Fathers use sacrificial terminology to interpret Christ's death. They stress the voluntary character of his offering, which had expiatory value in bringing about man's reconciliation with God. In this offering, Christ is both victim and priest. For the interpretation of Christ's death in terms of s. offered to the *Father, present esp. in the 4th c. in the West, cf. *Expiation.

From the earliest time of the church, the *Eucharist was called a s. (*Did.* 14; Just., *Dial.* 41 and 117; Iren., *Adv. haer.* IV, 17, 5; Tertull., *De orat.* 19, 4; Hipp., *Trad. Apost.* 4 and 21), often in connection with Malachi's prediction (1, 10 ff.) of the "pure offering". For *Cyprian (*Ep.* 63, 17), in the Eucharist we mention Christ's passion because "our sacrifice consists of the Lord's passion". If the Eucharist is a s. (Ambr., *In psalm.* 38, 25; John Chrys., *In prod. Jud.* I, 6), true (Aug., *De spir. et litt.* 11, 18; *De civ. Dei* X, 20), spiritual and bloodless (Cyr. of Jer., *Cat.* 23, 8), and an immolation (Ambr., *Exp. in Lc.* I, 28), it is also a sacramental or "relative" s., i.e. one which is related to the s. of the *cross (Cypr., *Ep.* 63, 14; Ambr., *De fid.* IV, 10, 124; *Sacr.* V, 4, 25; Chromat., *Serm.* 17 A, 2; Greg. Gt., *Dial.* 4, 58). The identity between the s. of the cross and the s. offered on the altar, which is its memorial, is given by the fact that in both cases there is one priest and one victim: one and the same Christ is offered (John Chrys., *Hom. 17, 3 in Hebr.*), who, now as then, is the one eternal priest (Ambr., *In psalm.* 38, 25; Aug., *De div. quaest.* 61, 2). The eucharistic s. is an antitype (Greg. Naz., *Or.* 2, 95) or *homoioma* of Christ's death (Serap., *Euchol.* XIII, 13); a memorial of the one true oblation (Aug., *C. Faust.* VI, 5; XX, 18; Theodoret, *In Hebr.* 8, 4 f.), image of the eternal liturgy that Christ now celebrates in the heavens (Theod. Mops., *Hom. catech.* 15, 15 f.). It is the s. after the order of Melchizedek which has replaced the s. of *Aaron (Aug., *In ps.* 33, *serm.* 1, 5-6). The eucharistic s. is a s. of thanksgiving (Just., *Dial.* 41, 1; Iren., *Adv. haer.* IV, 17, 5) and praise (Aug., *C. adv. leg.* I, 20, 39), *intercession (Cyr. Jer., *Cat.* 23, 8) and *expiation (Ambr., *Sacr.* IV, 6, 28; Aug., *De anim.* II, 15, 21; *Quaest. Hept.* III, 57); it is also offered for the dead (Tertull., *Coron.* 3, 3; Cyr. Jer., *Cat.* 23, 9, 10).

Besides the cross and the Eucharist, the offerings of Christians who take part in the eucharistic s. are also called a s.: our offering and s. must correspond to the Lord's s. (Cypr., *Ep.* 63, 9); every day the church, Christ's body, offers itself through him (Aug., *De civ. Dei* X, 6; XIX, 23, 5) who is the Head of those who offer, and who, through the church, exercises his priesthood as man, while as God he receives what is offered (Aug., *De civ. Dei* X, 20; Theodoret, *In ps.* 109, 4).

DTC 10, 795-993, 1317-1332; 14, 662-692; *EC* 8, 757-778; 10, 1585-1597. For the S. of the Mass, cf.: M. Lepin, *L'idée du sacrifice de la Messe d'après les théologiens depuis l'origine jusqu'à nos jours*, Paris 1926; J. Solano, *Textos eucarísticos primitivos*, BAC 88 and 118, Madrid 1962 and 1964 (particularly the index of vol. II, 880-908); J. Betz, Sacrifice et action de graces, *La MaisonD.* 87 (1966) 78-96; id., Eucharistie in der Schrift und Patristik, HDG 4/4a. For the presence of the term s. (and related terms) in the Fathers, cf. A. Blaise-H. Chirat, *Dictionnaire latin-français des auteurs chrétiens*, Strasbourg 1954, 395, 409, 564, 731-732; Lampe 658-660; 671; 1184-1185; R.J. Daly, *Christian Sacrifice. The Judaeo-Christian Background before Origen*, Washington 1978; F.M. Young, *The Use of Sacrificial Ideas in Greek Christian Writers from NT to John Chrysostom*, Cambridge, Mass. 1979; E. Ferguson, Spiritual Sacrifice in Early Christianity and its Environment, *ANRW* II.23.1 (1980) 1151-1189; K. Stevenson, *Eucharist and Offering*, New York 1986.

A. Pollastri

SAECULUM. The oldest Latin translators of the *Bible normally adopted the term *saeculum* to translate the Greek term αἰών (*aeon*), itself equivalent to the Hebrew '*ôlam* in its various meanings of "duration of the world", "*world" or "human world", fallen and ephemeral compared to the divine world. But in the *Afra* tradition of the *Vetus Latina*, particularly in *John's gospel, *saeculum* also appears as the normal equivalent of the Greek κόσμος (*cosmos*), itself adopted by Judaeo-Hellenistic linguistic tradition as equivalent to '*ôlam*. It is probable that the Latin mentality resisted giving a negative semantic meaning to the term *mundus* (which etymologically recalled the idea of the world's order and beauty); but the preference accorded to *saeculum* is more likely to be explained by the fact that this term, even in secular Latin, could be employed with a negative ethical meaning, in reference to the decline of the generations of mankind (cf. Virg., *Georg.* I, 468; Horace, *Carm.* III, 6, 17; Cic., *Parad.* 50; Quintil., *Inst.* II, 5; Tac., *Germ.* 19, 9). This explains the clear preference accorded to *saeculum* rather than *mundus* by classically-educated writers like *Tertullian, *Cyprian and *Lactantius. In patristic language the term *saeculum*, continuing the classical meanings of "century" (a hundred years), "temporal world", "mankind living in *time", can therefore indicate "the present time", "the present life", without any specific negative meaning (cf. Tertull., *Spect.* 2, 9, acc. to which *God is *auctor saeculi*; *Val.* 15, 5; Cypr., *Dom. orat.* 13); analogously it can signify the temporal duration of the world (cf. Cypr., *Ep.* 58, 7; Hilary, *Myst.* I, 1; Aug., *Serm.* 239, 6). But more often, at the suggestion of the biblical linguistic tradition, it can have a negative semantic meaning, ethico-religious in inspiration, so that *saeculum* can mean the human world in its vanity and fallen state (cf. Tertull., *Res.* 19, 7; *Apol.* 21, 6; *An.* 1, 6; *Mart.* 2, 5; Cypr., *Mort.* 21; Lact., *Div. inst.* V, 1, 19). Sometimes indeed it emphasizes the sinful aspect of human life (cf. Tertull., *Spect.* 8, 10; 15, 8; Cypr., *Hab. virg.* 7; Commod., *Instr.* II, 18, 1; II, 25, 2). In the context of the *pagan world's persecutory attitude towards Christianity, *saeculum* can specifically indicate the persecuting pagan world (cf. Tertull., *Pud.* 1, 1; 1, 4; Cypr., *Laps.* 2; Lact., *Div. inst.* V, 1, 26; VI, 17, 25; *Epit.* 61, 6; *Mort. pers.* 16, 5). Finally, in the nexus *saeculum saeculorum* or even *saeculum saeculi*, biblical in inspiration and appearing in Christian *doxologies, *saeculum* serves to indicate indefinite duration, eternity.

H. von Soden, *Das lateinische Neue Testament in Afrika zur Zeit Cyprians*, Leipzig 1909, 63-71, 147; *DB* 5, 1707-1708; *GLNT* 1, 535-565; R.A. Markus, *Saeculum: History and Society in the theology of St. Augustine*, Cambridge ²1989.

V. Loi

SAHAK the Great (Isaac; Armenian *catholicós 387-438). Born *c.* 350 from the stock of *Gregory the Illuminator, whose last descendant he was. Becoming catholicós, he took an active part in the political life of *Armenia, at that period divided between the *Byzantine and Persian empires. Among other things, his intercession led to the end of the Persian king *Yezdegerd's persecution of the Christians. In *c.* 425 he was deposed for political reasons, but retained strong moral authority over the Armenian church. He played a very important role in the creation of the Armenian alphabet, giving full support to the work of *Mesrob and himself translating part of the *Scriptures into *Armenian (Pentateuch, Isaiah, *Wisdom books), as well as undertaking "explanations of Scripture" (prob. homiletic *commentaries). His literary work, if it ever existed, is all lost. Only a few *letters remain, addressed to *Theodosius II, *Atticus of Constantinople and the prefect Anatolius of Constantinople. Of particular interest is his correspondence with *Proclus of Constantinople, who wrote his *Tomus ad armenos* on this occasion. Armenian tradition attributes to S. a *liturgy, spurious, and canons, dubious.

BS 7, 916-918; *EC* 10, 1616; Bardenhewer 5, 195-197; K. Sarkissian, *The Council of Chalcedon and the Armenian Church*, London 1965.

S.J. Voicu

SAINTS, Intercession of. Belief in the intercession of ss. is a corollary of belief in *communion with them. From antiquity it took different forms according to whether it was exercised by or towards the living or the dead.

1. Intercession of the martyrs. The primitive church recognized in the *martyrs, because of their profession of faith, the power of intervening in favour of sinners to obtain an alleviation, if not remission, of the penalties incurred by their *sins and sanctioned by the church. The first example of this practice occurs with the martyrs of *Lyons of 177 (Euseb., *HE* V, 2, 3). *Tertullian describes penitents kneeling at the feet of "God's friends", who intervene with the bishop (*Ad mart.* 1; *De pud.* 1). *Cyprian regulated the intervention of the martyrs, making it subject to, not replacing, the *penitence of the guilty (*De laps.* 36; *Epp.* 15 f.; 18), since "only God can pardon, though we see that the merits of the martyrs have great weight before his tribunal" (*De laps.* 17).

2. Intercession of saints through prayer. The power of interceding with the ss. by a confession of faith presupposed that the *confessors were alive. This power disappeared with the end of the *persecutions, but the martyrs continued to exercise it after their death, through prayer. So now the i. of the ss. operated directly with *God to alleviate eternal punishments and obtain their remission. Although, from the start, martyrs had been considered partakers of eternal *beatitude after their death, the surviving archaic habits and forms of prayer to them were typical of those for the *dead. The *Eucharist was offered "for" (*pro*, ὑπέρ) the martyrs as "for" the dead (Cypr., *Epp.* 1, 2; 12, 2; 39, 3; Liturgy of St John Chrys., ed. Brightman, *Liturgies Eastern and Western*, Oxford 1896, 131, 12 ff.). There was also a general conviction that martyrs and saints prayed for their *brothers, living or dead. In 203 *Perpetua of Carthage prayed for her dead brother Dinocrates (*Pass. Perp.* 7-8). In the same period, at *Alexandria, *Potamiaena promised the soldier Basilides to pray for him when she was with God (Euseb., *HE* VI, 5, 3 and 6). At *Tarragona on 21 Jan 259, bishop *Fructuosus was prayed by his flock *ut illos in mente haberet*, and he replied that he must pray "for the whole Catholic church, from the East to the West" (BHL 3196, nn. 1. 7). At *Carthage some months later, the congregation commended themselves to the prayers of the martyr Lucius (BHL 6009, n. 13). At *Tyre in 310, Theodosia begged the *confessors awaiting condemnation to remember her when they received their heavenly reward (Euseb., *De mart. Pal.* 7, 1). Even non-martyr saints were recognized as having the same role as intercessors: *Sulpicius Severus saw *Martin of Tours "among the *apostles and *prophets" and close to the praying people, to give them God's blessing (*Epp.* 2, 4 and 16: SCh 135, 1196, 1251).

3. Prayer addressed to martyrs and saints. The invocation of the martyrs was a common practice in Christian antiquity, as attested primarily by *epigraphy, from the 3rd c. The characteristic expressions of prayer of petition return insistently to this concept: *in mente habete, orate, petite pro, subvenite, in orationes vestras nos in mente habete*. We read analogous formulae in literary texts (Hipp., *In Dan.* 2, 30; Orig., *Orat.* 14, 6; *Exh. mart.* 37). Examples could be multiplied, drawing on the Cappadocian Fathers, *John Chrysostom, *Augustine, *Theodoret of Cyrrhus and others. The i. of ss. was all the more effective if prayer was made near their tomb: the one imploring pardon touched the tomb directly or through a third party; the dead were buried near the saints (*ad sanctos) to benefit in the afterlife from their saving intercession; miracles took place over tombs as visible signs that prayer had been heard. In this way the saint, even if in a subordinate position, was associated with Christ, who interceded, as mediator, with God.

Delehaye *OC*, 100-140; id., *Sanctus. Essai sur le culte des saints dans l'antiquité*, Brussels 1927; 122-161; *DTC* 14, 886-939; E. Ferguson (ed.), *Encyclopedia of Early Christianity*, New York-London 1990, 823 (bibl.).

V. Saxer

SALONA (Solin-Spalato-Split). Acc. to Strabo, S. was the port of the Illyrian Delmatae who lived in the Dalmatian hinterland. In Caesar's time S. became capital of the province of Illyricum, and afterwards capital of *Dalmatia. Even in the 1st c. BC it was a quite large and developed city. Christianity may have reached it in the 1st c., but was not deeply rooted until the mid 3rd c., when, acc. to tradition, it had its organized church (cf. *Dalmatia). At the end of the 3rd c., bishop Domnius organized the Christian community and the ecclesiastical province; but the church must have existed earlier. In the first half of the 5th c., S. became the metropolis of Dalmatia under *archbishop *Hesychius. A century later, the see of the vicariate of W Illyricum was established at S. At the start of the 7th c., S. was sacked and destroyed by Avars and Slavs.

Numerous remains and monuments of the early Christian period have been found at S. The oldest *church building arose in the N-W corner of the so-called *urbs nova*, where the episcopal centre was later built. It has been held to be a *domus ecclesiae* set up in a private house, but it is not yet completely excavated. The principal room was to S: its presbytery, raised one step above the nave, had a semicircular bench for the *clergy, detached from the W wall in which was a fairly large door. The presbytery was probably divided from the rest of the building by a *cancellus*. Near by, to S, was another building with a square apse raised on several steps. Its Christian character is shown by two pillars ornamented with a *cross, flanking the apse. Soon after the mid 4th c., the first Christian basilica of S. was built near these two oratories. Its form is hard to determine, but it was certainly longitudinal. At the start of the 5th c., another basilica was built N of the first, with square baptistery and polygonal pool. The apse *mosaic records its construction. Thus a great Christian centre arose, consisting of two basilicas with accessory rooms and the bishop's palace. The cathedral was restored for the 1st and 2nd *councils of Salona (530 and 533): the S basilica was given a cruciform plan with equilateral arms, and the baptistery was given an octagonal form and a cruciform pool. Both basilicas were connected by a long narthex, at whose N end steps led up to the *catechumens' room. In front of the narthex an atrium, inaccessible from the basilicas or the narthex, ended on the N side in an oratory.

Within the walls of S. were other Christian buildings. Of particular importance is a basilica with three aisles and an octagonal baptistery with cruciform pool, held by E. Dyggve to be the cathedral of the *Arian *Goths and therefore correlated with the Arian cathedral of *Ravenna (but the relation has no substance). It should be observed that the plans of the basilica published by E. Dyggve (*History of Salonitan Christianity*) are not exact, because he has not excavated them completely, but only made small soundings.

Of Christian *cemeteries, the most important area is that of Manastirine, 100 m. N of the walls, developed amid a *pagan graveyard in which, at the start of the 4th c., the body of the *martyr Domnius was laid. Apsed memorial buildings rose around his tomb. The cemetery was destroyed at the end of the 4th c., but at the start of the 5th a three-aisled basilica was built, enclosing Domnius's memorial in its presbytery. In the 6th c. a narthex was added to the church. In the first decade of the 7th c., the basilica was destroyed and the Salonitans were no longer able to rebuild it entire.

A second important cemetery, that of Marusinac, *c*.1 km. N of the city, developed around the tomb of the martyr Anastasius, laid in the mausoleum of a matron named Asclepia. The mausoleum consisted of a barrel-vaulted crypt and an upper floor. The two floors had an inscribed apse and the walls of the building were reinforced with buttresses. At the start of the 5th c. the mausoleum was enclosed in an atrium with porticos; to E, a longitudinal basilica was built, floored with sumptuous *mosaics, into which Anastasius's relics were translated. N of the basilica was a courtyard, enclosed by porticos, ending to E in a group of three mausolea. Dyggve called this part of the complex a *basilica discoperta*; but his theory, much contested, has no credibility.

The third cemetery, at Kapljuc, near the amphitheatre, developed round the tombs of the martyr Asterius (presbyter) and four soldiers of *Diocletian's court named Paulinianus, Antiochianus, Gaianus and Telius. The martyrs were buried in a mausoleum which was then lengthened towards the W, transforming it into a three-aisled basilica, around which grew a vast cemetery. Numerous Christian monuments have been discovered at S.: figured *sarcophagi, *epigraphy, church decorations, *mensae*, *paintings, *mosaics, crockery, glass, *lamps, etc. **[Fig: 270]**

W. Gerber, *Forschungen in Salona*, I, Vienna 1917; R. Egger, *Forschungen in Salona*, II, Vienna 1926; R. Egger-E. Dyggve, *Forschungen in Salona*, III, Vienna 1939; J. Bronsted-Fr. Weilbach-E. Dyggve, *Recherches à Salone*, I, Copenhagen 1928; E. Dyggve, *History of Salonitan Christianity*, Oslo 1951; E. Marin, La Topographie de Salone, *Actes XIe Congès internat. d'arch. chrét.*, CEFR 123, Rome 1989, 1117-1131; id., Les Nécropoles de Salone, *ibid.*, 1227-1242.

N. Cambi

SALONIUS of Geneva. Son of bishop *Eucherius of Lyons, born *c*.400. He and his brother *Veranus received a good spiritual and doctrinal training at the monastery of *Lérins, where their father, following a custom common in *Gaul in the late 4th and early 5th cc., had retired with his wife Galla and children to escape the *world. S.'s teachers, besides his father, were *Honoratus, founder of the *monastery, *Hilary, *Salvian and *Vincent. *Gennadius gives S. no chapter of his own in *De viris illustribus*, but mentions him twice: in the chapter on Eucherius (*disseruit* [sc. Eucherius] *etiam ad personam filiorum Salonii et Verani postea episcoporum obscura quaeque sanctarum capitula scripturarum* [PL 58, 1097]), manifestly alluding to the two exegetical works, *Instructiones* and *Formulae spiritalis intelligentiae*, dedicated by Eucherius respectively to S. and Veranus; and again in the chapter on Salvian, where S. appears as dedicatee of a book of Salvian's *Epistolae* (*ad Salonium episcopum liber unus* [PL 58, 1099]). Salvian also dedicated *De gubernatione dei* to him; and in the 9th and last *letter of the aforementioned *Liber epistolarum*, he explains to his *decus atque subsidium* (as he calls S.) why he has published his *Ad ecclesiam* under the pseudonym of Timothy.

S. became bishop of Geneva (*Genava*) prob. *c*.441; in that year his name appears among the subscribers to the *acts of the council of *Orange. In 442 he took part in the council of *Vaison; in 450 he and Veranus, also a bishop, with bishop Cerethius of Grenoble (*Gratianopolis*), wrote to *Leo the Great (PL 54, 887-890) thanking him for sending them a copy of his *Tomus ad Flavianum*.

In 1532 J.A. Brassicanus published two works under S.'s name: *Interpretatio in Parabolas Salomonis* and *In Ecclesiasten*. We do not know his grounds for attributing these two exegetical opuscula to Eucherius's son. In codex 1278 of the National Library of Vienna, from which he took them, they are anonymous, identified only by the "inscriptiones" *Glose Parabolarum* and *De nominibus Salemonis* (the superscript giving S. as the author is in another hand, probably Brassicanus's own). We owe the first crit. ed. to C. Curti, who republished the two works in 1964, keeping the traditional attribution and using 12 more manuscripts besides the Viennese; in these too the works are anon., except in codex 2689 of the National Library of Munich, where the marginal annotation ascribing them to S. is quite recent. In 1968 Curti brought out the *editio princeps* of two other exegetical works, *De evangelio Iohannis* and *De evangelio Matthaei*, which are attested, also anonymously, by five of the 13 manuscripts containing the two former works. The editor has demonstrated that all four are by the same author, backing his demonstration with a series of internal considerations: structural, conceptual, linguistic and stylistic.

J.P. Weiss has more than once interested himself in the author of the four works, dwelling esp. on the commentary on the *Parabolae* (Proverbs), for which Curti had identified no source and had explained the agreements between it and *Bede's *Expositio in Parabolas Salomonis* by assuming S. to be Bede's source. But Curti did not rule out the possibility that S. had ransacked a (now lost) work for the *Parabolae* in the same way that, to interpret the other three texts, he had plagiarized *Jerome's commentaries on Ecclesiastes and *Matthew and *Augustine's on *John. Weiss's indisputable achievement is to have discovered dependences of this text on *Gregory the Great and Bede, so that the work on the *Parabolae*, and consequently the other three texts, cannot be attributed to S. but to an unknown author who lived between 800 and 1000; Bede died in 735 and the oldest manuscript containing the four treatises seems to date from the 10th c. Valerie I.J. Flint returned to the problem of the author and came to the conclusion that our author plagiarized Bede (for the *Parabolae*), Alcuin (for Ecclesiastes and John) and Rabanus Maurus (for Matthew). She proposes as the probable author Honorius of Autun, to whom, by a mechanical error, M. Denis had already assigned the opuscula in illustrating codex 807, one of the five that gives all four works, for a Catalogue of the Palatine Library of Vienna.

On examination, the writings belong to the genre of *quaestiones et responsiones* and, as such, none of them is a continuous *commentary systematically explaining the whole biblical work: only limited verses or lines are picked out. The text is divided into a dense concatenation of questions and answers. The INT. asks for clarification, the RESP. replies by interpreting the verse or line in question and, almost always, adduces other passages of *Scripture in support. The question is very brief: it is introduced by a few formulae recurring with tiresome monotony, common to this genre but even less varied than usual. The reply is drawn up with the same uniformity: either it begins with a causal conjunction (*quod, quia*, etc.) in immediate connection with the interrogative adverb (*cur, quare*, etc.) of the question or it axiomatically explains the passage or biblical term that is the object of *exegesis (INT. *Qui sunt isti pennati?* RESP. *Sancti et electi viri* . . .). But what is most disappointing about these works is the uniformity of structure and language, which leads to the clear conviction that they cannot but be by the same writer: a writer always equal to himself, as devoid of invention as he is incapable of formal variation, who cannot raise himself above the level of an elementary exegetical discourse, presented in a flat, unadorned, and downright shabby style. The exegetical criterion followed is that of allegorical interpretation, but it is artificial and constructed with truly singular obstinacy and effort and applied even to passages in which the literal sense seems to be more obvious and *allegory seems difficult to superimpose on it.

Parabolae and Ecclesiastes: J.A. Brassicanus, Haganoae 1532 (*editio princeps*) (=PL 53, 993-1012); C. Curti, Catania 1964; *Vangeli di Giovanni e di Matteo*: C. Curti, Turin 1968 (*editio principes*). M. Besson, Un évêque exégète de Genève au milieu du Ve siècle: Saint Salone, *Anzeiger für schweizerische Geschichte* 9 (1902-1905) 252-265; J.A. Endres, *Honorius Augustodunensis*, Kempten 1906, 73-75; G. Bardy, La littérature patristique des "Quaestiones et Responsiones" sur l'Ecriture sainte, *RBi* 42 (1933) 20-22; C. Curti, Vienna, Nationalbibliothek, Latino 807, ff. 66v-110r è di Salonio, non di Onorio d'Autun, *Orpheus* 11 (1964) 167-184; id., Osservazioni sul testo di Salonio, *Studi in memoria di Carmelo Sgroi*, Turin 1965, 549-559; id., *Due Commentarii inediti di Salonio ai Vangeli di Giovanni e di Matteo. Tradizione manoscritta, fonti, autore*, Turin 1968; J.P. Weiss, L'authenticité de l'oeuvre de Salonius de Genève, *SP* 10 (1970) 161-167; id., Essai de datation du Commentaire sur les Proverbes attribué abusivement à Salonius, *SEJG* 19 (1969-1970) 77-114; V.I.J. Flint, The True Author of the Salonii Commentarii in Parabolas Salomonis et in Ecclesiasten, *RTAM* 37 (1970) 174-186; J.P. Weiss, Les sources du Commentaire sur l'Ecclésiaste du Pseudo-Salonius, *SP* 12 (1975) 178-183.

C. Curti

SALVIAN of Marseilles. One of our most important authors for the attention he dedicates to the history of his own time, disturbed by the barbarian raids that preluded the dissolution of the Roman empire and ancient civilization, but unable to see the signs that heralded a new social order.

Biographical facts are scarce and vague. Besides *Gennadius of Marseilles (*Vir. ill.* 68), he is mentioned by *Hilary, bishop of Arles 428/9-449, in his *Sermo de vita s. Honorati* (ed. Cavallin, Lund 1952, ch. 19, p. 63) and by *Eucherius of Lyons (†c.450) in the preface to his *Instructiones* (CSEL 31, 1, p. 66). Other information, uncertain and incomplete, can be deduced from his works. Since Gennadius tells us that S. was still alive, old and in good health (*in senectute bona*) when he wrote (prob. 467-469), his birth can hardly postdate the first years of the 5th c. (on the basis of other factors, *c.*390). Hints in his *Ep.* V, 1 and in his main work *De gubernatione Dei* (VI, 68-81) suggest that he was born at *Trier or Cologne or thereabouts. His writings show a good education, with particular attention to legal studies. He married Palladia, daughter of Hypatius, a *pagan later converted to Christianity, and of Quieta. After the birth of a daughter, Auspiciola, he and his wife agreed on a life of continence, an idea decidedly uncongenial to his wife's parents (*Ep.* IV). We do not know when he moved to S *Gaul. From *Epp.* VIII, to Eucherius, and IX, to *Salonius, from Eucherius (*Instructiones*, loc. cit.), from Hilary of Arles (cit.) and from his relations with the most illustrious men who issued from the *coenobium on the isle of *Lérins, it is argued that S. belonged to that community, at least from 426. By 429 he was a presbyter, either at Lérins or at *Marseilles (*Massilia*), where monastic life had received a considerable impulse. Nothing is known of his priestly *ministry.

Hilary of Arles (cit.) records a thought expressed by S. "in his writings", taken from a work now lost. Some books mentioned by Gennadius are also lost: three books *On virginity*, the *Commentary on the last part of Ecclesiastes*, a verse *Hexaemeron*, *homilies "for the bishops" (preached to bishops or, more probably, composed to be preached by bishops), homilies on the *sacraments (on *prayer and liturgical rites). What remains are: *Ad Ecclesiam, De gubernatione Dei* and a collection of nine *Letters*. The *letters are addressed to monks (of Lérins) (I), to *Eucherius of Lyons (II and VIII), to *Agraecius of Sens (III), to his wife's parents (IV), to his "sister" Captura (V), to a *friend, Limenius (VI), to Aprus and Verus (VII) and to bishop *Salonius (IX). The correspondence, probably a collection compiled by the author himself and surviving incomplete, has little historical and autobiographical interest, but seems to be suggested more by literary intentions, following the custom of the time. In the four books *Ad ecclesiam* (*Adversus avaritiam*, acc. to Gennadius), S., deploring the avarice that dominated everyone, maintains that every believer, *layman or religious - save some exceptions - has a duty to dispose of his property, either in life or at least in death, for the good of the church. The singularity of this thesis (while bearing in mind that the recipients of the goods of the church were the *poor), on which the author insists with great emphasis, leads us to see in this work not so much a document of clerical fanaticism as the expression of a conviction matured in an atmosphere of *ascesis, attentive to a social reality marked by intolerable injustices, put forward in the highflown style learnt from the schools of *rhetoric that easily went beyond the bounds of one developing his arguments in tranquillity.

S.'s main work is *De gubernatione Dei* (in Gennadius, *De praesenti iudicio*), in eight books, of which the last remained (or is now) incomplete. After writing *Ad ecclesiam*, the author developed the theme of divine *providence, which included the "government" and "judgment" of the *world, referring to the situation of the empire, overturned by social calamities caused by the barbarian invasions. He addresses Christians of little *faith, who ask how to reconcile divine providence with the sad condition of the *Catholic Romans, conquered and oppressed by the heretical or pagan barbarians. S. first demonstrates *God's providence by reason and the *Bible, then turns to the problem posed by current events. His answer lies in the reprehensible conduct of the "Roman" Christians, who in their dissoluteness are much worse than the barbarians, whether *pagans or *heretics (whose grave *vices he does not conceal), something that S. seeks to demonstrate by examining the different behaviour of various countries: *Gaul, *Spain and *Africa. Together with its apologetic intention, the work follows the evident aim of blaming moral disorders and exhorting to *conversion. *De gubernatione Dei* is a historical source of considerable value both for individual events and for our knowledge of the West at that time, in various aspects: social-economic, religious and moral, less for political and cultural conditions, taking into account the intentions of the apologist, the concerns of the moralist and the emphasis of the writer easily carried away by love of his thesis into generalizing and giving his picture the darkest tints. His work reflects the mentality of at least a considerable part of the Christians and churchmen of that time who sought to understand, in the light of *faith, the tragic events in which they were caught up, and to draw from them the consequences for a behaviour consistent with their Christian profession. His style, which is not ineffective, shows the taste of his time and the habits of his school: emphasis, search for effect through recourse to the expedients of rhetoric, wordy prolixity.

Editions: CPL 485-498; PL 53, 9-238; MGH, *Auct. Ant.* I, 1 (1877); F. Pauly, CSEL 8 (1883); G. Lagarrigue, SCh 176 (1971); 220 (1975) (Fr. tr.: Lagarrigue, *op. cit.*; It.: E. Marotta, *Contro l'avarizia*, Rome 1978).
Studies: A. Hamman, *Patrologia* III, 500-509 (with bibl.); L. Rochus, *La latinité de Salvien*, Acad. Roy. de Belgique, Cl. lett. Mém., vol. 30/2, Brussels 1934; O. Janssen, *L'expressivité chez Salvien de Marseille*, I: *Les adverbes*, Nijmegen 1937; M. Pellegrino, *Salviano di Marsiglia*, Lateranum n.s. VI, 1-2, Rome 1940; M. Jannelli, *La caduta di un impero nel capolavoro di Salviano*, Naples 1948; Ph. Badot, La notice de Gennade relative à Salvien, *RBen* 84 (1974) 352-366; S. Pricoco, Una nota biografica su Salviano di M., *SicGymn* 39 (1976) 351-368; id., *L'isola dei santi. Il cenobio di Lerino e le origini del monachesimo gallico*, Rome 1978; H. Fischer, *Die Schrift des Salvian von M. "An die Kirche"*, Bern-Frankfurt/M. 1976; C. Leonardi, Alle origini della cristianità medievale: Giov. Cassiano e Salviano di M., *StudMed* 18/2 (1977) 491-608; A. Hamman, L'actualité de Salvien de M., *Augustinianum* 17 (1977) 378-393; J. Badewien, *Geschichtstheologie und Sozialkritik in Werk Salvians von Marseille*, Göttingen 1980; J.J. O'Donnell, Salvian and Augustine, *AugStud* 14 (1983) 25-34; J.M. Blasquez Martinez, La crisis del Bajo Imperio en Occidente en la Obra de Salviano de Marsella, *Gerión* 3 (1985) 157-182.

M. Pellegrino

SALVIUS. Bishop of unspecified see, perhaps Octodurus (Martigny), to whom one Eucherius, commonly identified with *Eucherius of Lyons (5th c.), wrote a *letter (*Ep. ad Salv.*: PL 50, 827-828; CSEL 31, 173; MGH, *scr. rer. mer.* III, 39-41) sending him the *Passio Acaunensium martyrum, S. Mauricii et sociorum eius*. Eucherius held that the information on the *martyrdom that it related came from Theodore, first bishop of Octodurus, in whose diocese Acaunum was. On S.'s identification with the *Polemius Silvius who dedicated his *Laterculus* to bishop Eucherius, cf. *DCB* 4, 581-582. We also find a letter addressed to an S. among the spurious writings of *Sulpicius Severus (PL 20, 243-244; CSEL 1, 254-256).

CPL 479 and 491.

A. Pollastri

SAMARITANS. Geographically, the term indicates the inhabitants of Samaria, a historical region of *Palestine occupying the central part of the highlands W of the Jordan; ethnically, it indicates the descendants of the Israelite inhabitants who remained in that territory after the destruction of the city of Samaria in 721 BC and the deportation of more than 27,000 persons by Sargon, then mixed and fused with colonists from E Syria, esp. from Kutah (Tell Ibrahim, 20 km. N of Babylon). There was tension between Jews and S. from the time of the rebuilding of the temple.

Persecuted by John Hyrcanus, left in peace by Pompey and *Herod, the S. suffered a bloodbath under Pontius *Pilate, who had forbidden them a meeting on Mt Gerizim called early in AD 36 by an impostor (Josephus, *Ant. Ius.* XVIII, 4, 1-2). During the war against the Romans (66-70), 11,000 were killed on Mt Gerizim, where they had organized a resistance (Josephus, *De bell. Iud.* III, 7, 32). They soon recovered and spread all over Palestine and Transjordan, building their own synagogues and cordially detesting both *Jews and Christians. The *evangelists' attitudes to them vary from *Luke's understanding (10, 30-37; 17, 16) and *John's conciliatory spirit (4, 4-42) to *Mark's silence and *Matthew's reservations (10, 5). *Philip was their *apostle (Acts 8, 5-25). At present they live concentrated at Nablus, at the foot of Gerizim and around their synagogue. They keep their ancient traditions alive: they accept only the Pentateuch; in the light of Dt 18, 15-19 they await a second non-Davidic Messiah who will restore their worship; they faithfully observe the *Sabbath and practise *circumcision; Mt Gerizim is still a place of worship.

G. Ricciotti, *Storia d'Israele*, II, Turin 1934, 175-184; R.T. Anderson, Le Pentateuque Samaritain CW 2473, *RBi* 77 (1970) 68-75, 550-560; R.J. Goggins, *Samaritans and Jews. The Origins of Samaritanism Reconsidered*, Growing Points in Theology, Oxford 1975; J. Macdonald, *The Theology of the Samaritans*, London 1964; J. Bowman, *Samaritenische Probleme*, Stuttgart 1967; G. Steinberger, *Jüden und Christen im Heiligen Land: Palästina unter Konstantin und Theodosius*, Munich 1987.

E. Peretto

SAMSON: *iconography. The figure of S. was compared by the church *Fathers (Paul. of Nola, *Ep.* 23: CSEL 28, 168 ff.) to Christ: his triumph over the Philistines became for Christians a *symbol of divine *redemption and victory over death. Despite this very important symbolism, the only known scenes of the S. cycle in cemeterial art come from the catacomb of Via D. Compagni, Rome. The cycle comprises: *1)* S. killing the Philistines with the jawbone of an ass (Jgs 15, 9-17) in cubicle F (Ferrua, *Via Latina*, pl. CV); *2)* S. setting the foxes among the corn (Jgs 15, 5) in cubicle B (*ibid.*, XXX, 1); *3)* S. killing the lion (Jgs 14, 6) occurs twice: in the vault of cubicle B (*ibid.*, pl. XVII) we see only S.'s feet and the lions's hind paws; in room L (*ibid.*, pl. CIX) S. is in tunic and pallium, and below him is the dead lion with bees making honey in its mouth. The hypogeum is dated between 320 and 360. The only *mosaic of S. is in the floor of the basilica of *Mopsuestia in *Cilicia (first half of 5th c.). Of the same period are fragmentary reliefs in the *martyrion* of Seleucia, near *Antioch. In the sumptuary arts, a silk cloth in the Vatican Museum (8th c., of Syrian origin) shows S.'s fight with the lion. Another scene of S. occurs among Christian *symbols on a bronze Roman medallion. **[Fig: 271]**

DACL 15, 740; *EC* 10, 1819-1820; *LCI* 4, 30-38; F. Buonarroti, *Osservazioni sopra alcuni frammenti di vasi antichi di vetri ornati di figure trovati nei cimiteri di Roma*, Florence 1716, 2, pl. I, i; P. Lauer, *Le trésor du Sancta Sanctorum*, Fondation E. Piot, Monuments et mémoires 15, 1906, pl. XVIII; K. Weitzmann, in (ed.) R. Stillwell, *Antioch-on-the-Orontes*, Princeton-London 1941, 137 f., pl. XVII; L. Budde, *Antike Mosaiken in Kilikien*, Recklinghausen 1969, 187, figs. 143-157.

A.M. Di Nino

SAMUEL of Qalamun. A prominent figure in *Coptic *monasticism at the time of the Arab invasion (7th c.). A native of the Delta, son of a priest, he became an *anchorite at *Scete. He suffered persecution by the *Melchite patriarch Cyrus Moqauqas in 630 for refusing to renounce the anti-Chalcedonian confession. He was then captured by the Mazici, who kept him prisoner for some years. Freed, he went to the Fayum and founded a new *monastery at Qalamun. As head of that community he gained great fame, remaining on good terms with the Arab conquerors (at this period, Copts and Arabs still coexisted largely in peace). The *Life* of S., written by one of his successors, Isaac of Qalamun, survives entire in the original Coptic and in a shortened *Ethiopic version, prob. transmitted through the *Arabic. It contains a great many miraculous episodes, but also much historical information of great interest and is on the whole a trustworthy historical source. Life in S.'s monastery was the normal life of those times, when monasticism was one of the components of Egyptian civil and religious life, with constant relations with the different authorities and the people, and the problems derived from this, spiritual and economic. *Ascesis did not as a rule have to be very rigorous.

Paul van Cauwenbergh, *Etude sur les moines d'Egypte*, Paris 1914 (repr. Milan 1973); A. Alcock, *The Life of Samuel of Kalamun*, Warminster 1983.

T. Orlandi

SARAGOSSA. The ancient Cantabrian city of Salduba, destroyed by the Romans in 45 BC, was rebuilt with the name Caesaraugusta in 24 BC, becoming the important centre of a *conventus iuridicus* of *Tarraconensis. Its first known bishop, Felix, was a contemporary of *Cyprian (*Ep.* 67,6). Its great bishops include *Braulio (631-651) and his successor *Taio. Three *councils were held here:

380. With 12 bishops from various Spanish provinces and some from Aquitaine, it was called to confront *Priscillianism. Its eight canons condemned the discipline of that heretical group, but not its persons, who were not present. Pope *Damasus had ordered them not to condemn anyone in their absence. Despite this, *Sulpicius Severus and the 1st council of *Toledo allude to the condemnation of some of the absent.

592. Provincial council: metropolitan Arthemius of *Tarragona presided over 11 bishops and two delegates, called to resolve some problems rising out of the *conversion of the Visigoths to *Catholicism. They also examined the authenticity of relics and the possibility of *Arian *clergy becoming Catholic clergy after a new *ordination.

691. Provincial council. The absence of signatures on the *acts prevents us knowing the number of participants. It promulgated five canons on ecclesiastical discipline.

Synoden, passim.

P. de Luis

SARCOPHAGI, EARLY CHRISTIAN. We consider as early Christian those ss. in whose decorations appear motifs or subjects related to Christianity. Many survive from the 3rd and 4th cc., from Roman workshops: the earliest were made in workshops whose output was mainly *pagan. Among pagan themes, that of the philosopher and bucolic themes, with the figures of the *orans* and the *shepherd, were common in the 3rd c. The first Christian ss. appeared halfway through the 3rd c. and had biblical scenes together with these pagan ones, which were acceptable to Christians and capable of a new and specifically Christian interpretation. The s. of S. Maria Antiqua (*Rep.* 747) has the *orans*, the shepherd and the philosopher, *Jonah and the *baptism of Christ; that of Baebia Hertophile (*Rep.* 778) has Jonah and a banquet on the lid, and a pastoral scene on the strigilate front; that of Via della Lungara (*Rep.* 777) has *orans*, fisher and shepherd on the strigilate front, Christ's baptism on one side.

Christian motifs became increasingly important towards the end of the 3rd c.: e.g. the decorated slab in the Capitoline Museum (*Rep.* 811), with its reader-philosopher and raising of *Lazarus; or the Velletri s. (*Ws* 4, 3), whose front, dominated by the great figures of the *orans* and two shepherds, also has small biblical scenes.

At the turn of the 4th c. this tendency settled down, and attention turned to *symbols and scenes expressing Christian *hope in the next life, concentrating particularly on the hero who made this hope possible, i.e. Christ, and his saving actions. Esp. from the time of the edicts of Christian toleration, the number of Christian ss. grew considerably. *Rome produced and exported all over the West many ss. whose style is, in many cases, clearly related to the friezes of the triumphal arch dedicated to *Constantine in 315. In them, generally compressed on both sides of the central *orans*, are OT and NT scenes, and scenes of St *Peter inspired by apocryphal narratives. Lat. 161 (*Rep.* 6), e.g., has on the front: Miracle of the spring and Arrest of Peter, Wedding-feast of *Cana, Orans, *Healing of the blind man, Multiplication of loaves and fishes, Raising of *Lazarus; on the sides: the three young men in the *fiery furnace, *Adam and Eve. In the Constantinian era, ss. with double friezes soon appear: e.g., the famous "dogmatic sarcophagus". This great s., incomplete (it lacks the last touches of chisel and drill, and the final cleaning of the marble), is like a second edition - corrected, reordered and with a much more consistent *iconography - of another s. with very similar characteristics, produced by the same Roman workshop and found at *Arles early in 1974 (J.-M. Rouquette, *Trois nouveaux sarcophages*). This confirms the theory that at this time, not always but in quite a few cases, the distribution and positions of the various scenes results from a clear aim of pursuing a certain uniformity of content in religious symbolism, and sometimes also a stylistic unity through symmetrical and rhythmical distribution of volumes and lines. But we have still to resolve some problems posed by this evident presence of a principle ordering the scenes, since we do not know precisely to whom to attribute any particular fixed order, nor whether the church hierarchy played any part in it. Nor is it at all clear who the religious message of symbolic decoration was aimed at. Along with frieze ss., others continued to exist with alternate fields of strigils and panels reserved for any of the above-mentioned scenes. In the late Constantinian period, we see s. fronts divided by columns supporting architraves or arches; sometimes columns and arches are replaced by trees, as in the Arles s. (Benoit, *Sarcophages*, 39). Among these columnar ss. we see a new theme that would embrace the four central decades of the 4th c.: the so-called "passion" ss., of which an important and unique example is Lat. 171 (*Rep.* 49): in the central field the victorious *cross; to the left, Simon of Cyrene and the Crowning with thorns; to the right, between two of the columns, Christ before *Pilate. Usually associated with Christ's passion are the *martyrdoms of *Peter and *Paul and the symbolic figures of *Abel and *Job (Lat. 164, *Rep.* 61).

In the mid 4th c. we see renewed influence of Greek artistic tendencies at Rome, with particular preference for the cult of form, plasticity and harmonious composition. This tendency includes, among others, two great double-frieze ss.: the continuous-frieze s. "of the two brothers" (Lat. 183A, *Rep.* 45) and the columnar s. of Junius Bassus (*Rep.* 680).

In the last third of the 4th c., s. *iconography shows the influence of solemn and triumphal themes familiar from *mosaics and *paintings in contemporary basilicas and baptisteries: Christ's triumph over death; Christ the Lord in the heavenly *Jerusalem surrounded by his *apostles or receiving crowns from their hands: e.g. the Borghese s. (*Ws* 82, 1), the Milan s. (*Ws* 188, 1) and others such as *Rep.* 65, 175, 208, 678 and 724. Also from this date are the ss. with scenes of the Crossing of the Red Sea and the so-called "Bethesda" ss. But the classical scenes of the Constantinian era do not disappear altogether. The style is manneristic, the taste for plastic forms is lost again and the uninhibited use of the drill returns. Roman workshops declined from the start of the 5th c. and then ceased altogether.

Early Christian ss. were produced in N *Italy, esp. at *Milan, esp. in the last decades of the 4th c.

*Arles, in S *Gaul, still preserves numerous early Christian ss. Quite a few were imported from Rome, but in the second half of the 4th c. others were worked in local workshops, though their style and subjects follow the known models of the Roman workshops. Very characteristic of local output is a series of strigilate ss. with small scenes on two levels (Benoit, *Sarcophages*, 73-74), particularly with Christ victorious in the central frame and apostles and saints at the extremities (Benoit, *Sarcophages*, 82-84).

Unfortunately we know only a small part of the ss. produced by the workshops of *Constantinople. The importance of this centre is known more by its influence on other regions of the empire, even in the Iberian peninsula and esp. at *Ravenna. The preferred iconography of Constantinopolitan ss. was symbolic, esp. from the 5th c.; but ss. decorated with figures and scenes were also produced. E.g., the s. of the little "prince" (Talbot Rice - Hirmer, 9), or the Bakirkoy relief (Talbot Rice - Hirmer, 8), both from the last third of the 4th c. New finds and facts confirm the existence of these ss. with figures and of ss. with columns.

Ravenna, as regards early Christian ss., was the temporal successor of Rome and inherited style and subject-matter from Constantinople. The earliest, e.g. that of the church of S. Francesco, can be dated from 380. Of these, that of Liberius (*CORPUS* 8) was probably imported from Constantinople. It has columns with *apostles and the *Traditio legis* to St *Paul. The plasticity of the figures, the adaptation of the *dress to the form of the body, the smooth background, etc., are characteristic of the Greek school. Ss. with figures and scenes predominate at Ravenna up to the mid 5th c., though in these same ss. the rear face is sometimes given a symbolic decoration, as in the "Pignatta" s. (*CORPUS* 11) or in that of Isacius (*CORPUS* 13). The same scene of the *Traditio legis* to St Paul appears on the front of the s. of S. Maria *in Porto Fuori* (*CORPUS* 12) and in that "of the Twelve Apostles" (*CORPUS* 16). In that of Rinaldo (*CORPUS* 15), Peter and Paul offer crowns to the enthroned Christ. In that of Barbatian (*CORPUS* 17), columnar with Christ, Peter and Paul, the decorative and symbolic element now predominates: this element asserted itself from the mid 5th c., dominated the output of the whole 6th c. and continued throughout the politico-cultural transformations of the 7th and 8th cc. Among ss. with symbolic decorations are, e.g., that of S. Apollinare in Classe (*CORPUS* 28), columned, with *crosses, palms, christogram and peacocks drinking water issuing from an amphora; and that of Theodore (*CORPUS* 24), with crowns of plants and a central christogram, flanked by peacocks. Characteristic of Ravenna are pilasters or columns at the extremities of the decorated front, and great semi-cylindrical lids.

In S France a centre of production existed at *Marseilles in the first half of the 5th c. From it seems to derive the abundant series of so-called Aquitanian ss., whose main centre of production was at *Toulouse from the end of the 5th c. and all through the 6th. Other centres were at *Bordeaux and Martres-Toulousane. The bodies of these ss. are trapezoid in form, and their lids, with rare exceptions, are double-sloping with rounded-off ends. Those decorated with figures have the front nearly always divided into fields by columns or pilasters (Briesenick, *Typologie*, 14-19). Symbolic and vegetable decoration abounds, as do strigils, which have very special characteristics ((Briesenick, *Typologie*, 20-33).

The one workshop of early Christian ss. known in *Spain is that of *Tarragona. The first ss. were imported from *Carthage, e.g. the strigilate front of the two *Orantes* (Schlunk, *Tarragona*, 29-32). Later on, some Carthaginian artisans must have moved to Tarragona. Frequently the façade is decorated with fields of strigils, sometimes in two rows, (Schlunk, *Tarragona*, 34-36). There are some ss. with figures of apostles or scenes such as that of *Abraham or of *Moses receiving the Law (Schlunk, *Tarragona*, 1 and 2). A local centre of production with Greek influence, perhaps through *Africa, is documented at Bureba (Burgos), as is a teaching tradition, also of Greek origin, in a broad area of Andalusia embracing Cordova, Ecija, Alcaudete and Antequara (Schlunk, *Sarcófagos*). **[Fig: 272, 273]**

Ws; *Rep.*; *CORPUS*; F. Gerke, *Die christlichen Sarkophage der vorkonstantinischen Zeit*, Berlin 1940; J. Kollwitz, *Oströmische Plastik der theodosianischen Zeit*, Berlin 1941; F. Gerke, *Christus in der spätantiken Plastik*, Mainz ³1948; H. Schlunk, Un taller der sarcófagos cristianos en Tarragona, *ArchEspArq* 24 (1951) 67-97; F. Benoit, *Sarcophages paléochrétiens d'Arles et de Marseille*, Paris 1954; G. de Francovich, Studi sulla scultura Ravennate I. I sarcofagi, *FR* 26/27 (1958); D. Talbot Rice - M. Hirmer, *Kunst aus Byzanz*, Munich 1959; H. Fournet-Pilipenko, Sarcophages romains de Tunisie, *Karthago* 11 (1961) 77-166; G. Bovini, Sarcofagi costantinopolitani dei secoli IV, V e VI d. C., *CCAB* 9 (1962) 179-192; B. Briesenick, Typologie und Chronologie der südwest-gallischen Sarkophage, *JRGZ Mainz* 9 (1962) 76-182; T. Klauser, *Frühchristliche Sarkophage in Bild und Wort*, Olten 1966; F.W. Deichmann, Konstantinopler und ravennatische Sarkophag-Probleme, *ByzZ* 62 (1969) 291-307; H. Schlunk, Sarcófagos paleocristianos labrados en Hispania, *Act. VIII Cong. Int. Arq. Crist.*, Barcelona 1972, 187-218; J. Engemann, *Untersuchungen zur Sepulkralsymbolik der späteren römischen Kaiserzeit*, JbAC Ergbd. 2, Münster 1973; J.-M. Rouquette, Trois nouveaux sarcophages chrétiens de Trinquetaille (Arles), *CRAI* 1974, 254-277; M. Sotomayor, *Sarcófagos romano-cristianos de España*, Granada 1975; H. Brandenburg, Stilprobleme der frühchristlichen Sarkophagkunst Roms im 4. Jahrhundert. Volkskunst, Klassizismus, spätantiker Stil, *MDAI (R)* 86 (1979) 439-471; P.-A. Février, Sarcophages d'Arles, *CongrArchPaysArles*, Paris 1979, 317-359; P. van Moorsel, Le Sarcophage de Marcia Romaina Celsa, in (ed.) E. Dassmann et al., *Pietas* (Festsch Kötting), JbAC Ergbd. 8, Münster 1980, 499-508; K. Eichner, Die Produktionsmethoden der stadtrömischen Sarkophagfabrik in der Blütezeit unter Konstantin, *JbAC* 24 (1981) 85-113; W. Wischmeyer, *Die Tafeldechel der christlichen Sarkophage konstantinischer Zeit in Rom*, RQA Suppl. 40, Rome-Freiburg-Vienna 1982 (bibl.); G. Koch-H. Sichtermann, *Römische Sarkophage*, Munich 1982. See also arts. in *PAC*.

M. Sotomayor

SARDICA

I. The city and its Christian origins - II. Council.

I. The city and its Christian origins. As with most of today's great Bulgarian cities, the origin of S. (Sofia, capital of Bulgaria) is lost in pre-Christian antiquity. As a town of little importance, it was the territorial centre of the Thracian tribe of Serdi or Sardi, whence the two forms of its name (Serdica or Sardica). According to the most likely hypothesis, the name is derived from a word whose meaning corresponds to the Late Slav *srêda, sreda*, "middle, centre"; after the 6th c. the Slavs changed it to *Srêdec (-tz)*, an exact translation. S. is in a vast, fertile plain, on the slopes of Mt Skomios (now Vitosa) and at the crossing of main roads from the S-W (*Macedonia), N-W (the famous *Via diagonalis*, going from Singidunum [Belgrade] on the Danube to *Byzantium-*Constantinople) and one going N along the river Oescus (Iskur). After conquering Macedonia (146 BC), the Romans pushed up into the Balkan peninsula; in 29/28 BC, M. Licinius Crassus brought these regions, including the Serdi, under Roman rule. After the conquest of *Thrace, this region was organized into the *strategia* of S., comprising a huge territory centred on S. The site was fortified, the ways of communication organized. During the war against the Dacians over the Danube, the emp. *Trajan (98-117) visited the Balkan peninsula and, among other things, contributed greatly to the development of S., which was organized as a *municipium* and fortified, receiving the name Ulpia Serdica. Popular memory has linked the famous emperor's name with the construction of various fortresses, roads, bridges, etc., and a section of the Via diagonalis W of S. still bears the name Via Traiana. Remains of the fortifications built in Trajan's time have been discovered by archaeologists. When *Aurelian (270-275) was obliged to abandon *Dacia (part of modern Transylvania and Wallachia), he organized the cis-Danubian region into Dacia ripensis (centre *Ratiaria*) and Dacia mediterranea (centre S.). Thus S. acquired particular importance as a (mainly Latin-speaking) military, administrative and cultural centre. Its greatest development came at the start of the 4th c., which led the emp. *Constantine I (306-337), born in the nearby city of Naissus (Nis), to favour for some time the idea of establishing his imperial residence at S. In the disturbed period of the 5th c., S. suffered several invasions and destructions: in 441 and 447 by *Attila, in the second half of the 5th c. by the *Goths. At the time of *Justinian I (527-565), the city experienced a new period of splendour: its fortifications were rebuilt and various public and private buildings put up. In 809 it was conquered by prince Krum (802-814) and became part of the Bulgar state (founded 681).

Christianity quickly reached S., which became one of the main centres of the new faith. A Roman *baths, with *hypocaustum* and *calidarium*, built in the 3rd c., was turned in the 5th c. into a Christian *church building and, in the 11th-12th and 14th cc., enlarged and embellished with murals. In the 15th c. it became the metropolitan church. In the N-E part of the city a small church was built at the beginning of the 4th c.: damaged by Goths in 376 and 382, restored in Justinian's reign and named Santa Sophia, embellished with mural *paintings and *mosaics, it still survives. By the start of the 14th c. the name of the church had become the official name of the city, as we see from written evidence dated 1329. From the necropolis next to this church and elsewhere in the city, some 30 Christian inscriptions, Greek and Latin, have been recovered, attesting a large Christian community up to the 6th-7th cc. The importance of S. as a Christian centre is confirmed, among other things, by the fact that in 343/44 the famous church *council of S. was held here. [Fig: 274]

B. Filov, *Sofijskata curkva Sv. Sofija*, Sofia 1913; K. Mijatev, *Dekorativnata zivopis na Sofijskija nekropol*, Sofia 1925; S.N. Bobcev, *Jubileina kniga na grad Sofija*, Sofia 1928 (with studies by A. Isirkov, G.I. Katzarov, Jord. Ivanov); B. Filov, *Sofijskata curkva Sv. Georgi*, Sofia 1933; S.N. Bobcev, *Serdica*, Sofia 1943; V. Besevliev, *Spätgriechische und spätlateinische Inschriften aus Bulgarien*, Berlin 1964, 1-20; EAA 7, 207; N. Tchanéva-Detchevska, Les édifices culturels sur le territoire Bulgare pendant la période paléochrétienne à la lumière des nouvelles données, *Actes XI° Congrès internat. d'arch. chrét.*, CEFR 123, Rome 1989, 2491-2509 (bibl.).

I. Dujcev

II. *Council. To resolve the disagreements between *East and West over the *Arian controversy (councils of *Rome and *Antioch, 341), at the wish of the Western bishops and the emp. *Constans a council was called at S., in the Western part of the empire but close to the border with the East. In autumn 343, *c.*100 Western bishops, headed by *Ossius of Cordova and Protogenes of S. (pope *Julius was represented by two priests and a deacon), met *c.*70 Eastern bishops, of whom the pro-Arians *Maris of Chalcedon, *Narcissus of Neronias and Stephen of Antioch stood out. The Westerners insisted that *Athanasius, *Marcellus of Ancyra and other Eastern bishops, deposed in the East as anti-Arians but restored by the Westerners at the council of Rome in 341, should take part in their labours. The Easterners refused to admit them; and the council was shipwrecked on this *impasse*. The Easterners left S. after writing a synodal letter in which, after various accusations, they *excommunicated the principal Western spokesmen and substantially confirmed the formula of faith of Antioch of 341. The Westerners continued their work alone: they once more restored Athanasius, Marcellus and *Asclepas of Gaza; approved a series of disciplinary canons, of which cann. 3 and 3b allowed a bishop condemned at a provincial council to appeal to the bishop of *Rome; put forward their version of the facts in a synodal letter in which they condemned the Eastern leaders; published a doctrinal document in which, without using the disputed term *homoousios*, they gave the Nicene *creed of 325 a *monarchian interpretation, affirming a single *hypostasis* of *Father and Son and condemning those who, like the Easterners, affirmed two distinct hypostases. The council sanctioned the break in relations between the pro-Nicene West and the mainly anti-Nicene East (the first in the history of the church) and initiated a phase of inactivity which lasted until Constans's death (350).

Hfl-Lecl 1, 737-823; Simonetti 161-187; H. Hess, *The Canons of Sardica*, Oxford 1958; L.W. Barnard, The Council of Serdica: Some problems re-assessed, *Annuarium Hist. Conc.* 12 (1980) 1-25; id., The Site of the Council of Serdica, *SP* 17, Oxford 1982, 9-13; Ch. Pietri, La Politique de Constance II: un premier Césaropapisme, ou *l'Imitatio Constantini*, in *L'Eglise et l'Empire au IV° siècle*, Geneva 1989, ch. 4; L.W. Barnard, *The Council of Sardica, 343 AD*, Sofia 1983; K.M. Girardet, Kaiser Konstantius II als "Episcopus Episcoporum" und das Herrscherbild des kirchlichen Widerstandes, *Historia* 26 (1977) 95-128; W.H.C. Frend, *The Rise of Christianity*, Philadelphia-London 1984, 528-532.

M. Simonetti

SARDINIA. S. and *Corsica seem to have constituted a single Roman province until *Diocletian separated them from each other and joined them to the Italician dioceses. S. was a place of deportation (cf. Tac., *Annal.* II, 85). Christians too were deported there: e.g. *Callistus in the reign of *Commodus; pope *Pontian and the priest *Hippolytus, Callistus's opponent, in 235. S. came under the *Vandal rule of *Genseric in 455 and the *Byzantine rule of *Justinian in 533. Literary sources tell us little about Christianity in S.; *archaeology tells us more (cf. under *Italy). The first historical bishop was *Quintasius of Carales (Cagliari), present at the council of *Arles in 314. The most famous bishop was *Lucifer of Cagliari, whom *Athanasius calls *Sardiniae metropolis episcopus* (PG 25, 650. 731); *Eusebius of Vercelli was a native of S. To find other episcopal sees we must wait until 484, when some Sardinian bishops took part in a *council at Carthage: Lucifer of Cagliari, Martinian of Forum Traiani (Fordongianus), Vitalis of Sulci (S. Antioco), Felix of Turris Libysonis (Porto Torres) and Boniface of Senafer (Cornus?) (Victor of Vita, *Hist. pers. Afr. prov.*: MGH, *Auct. ant.* 3, 71). The Vandal kings Genseric, *Huneric (484) and Thrasamund deported numerous African Christians to S.; among them *Fulgentius of Ruspe (467-533), who took St *Augustine's body with him and founded a *monastery near Cagliari. At the time of *Gregory the Great there were seven dioceses (not yet all identified), but the island was still impregnated with *paganism: Gregory sent two missionaries, Felix and Cyriac, and wrote: "From the account of bishop Felix, my brother, and of my son Cyriac, a monk, I knew that among you all the peasants scattered through your lands are given up to *idolatry" (*Ep.* IV, 23). The same was true of workers on church lands (*Ep.* IV, 26). This widespread paganism was the consequence of clerical indifference (*Epp.* V, 38; IX, 204). For this reason Gregory never tired of spurring on bishop Januarius of Cagliari, to whom he wrote a good 20 *letters. The results of the *pope's efforts were not slow in coming (*Ep.* V, 38). The Roman church possessed many lands in S., under the care of a **defensor ecclesiae* (*Epp.* IX, 203; XIV, 2). In this period the Sardinian church had close relations with the Roman, but transition to Byzantine rule brought Greek influence, also on customs and worship, and hence a certain autonomy from *Rome, at least after Gregory the Great.

DACL 15, 888-899; *CE* 12, 1087 f.; F. Lanzoni, *Le diocesi d'Italia dalle origini al principio del secolo VII (an. 604)*, Faenza 1927, 656-679; F. Cherchi Papa, *La repubblica teocratica sarda nell'Alto Medioevo*, Cagliari 1971; C. Bellieni, *La Sardegna e i sardi nella civiltà dell'Alto Medioevo*, I-II, Cagliari 1973; P. Meloni, *La Sardegna romana*, Sassari 1975; (ed.) F. Lo Schiavo, *Sardegna centro-orientale dal neolitico alla fine del mondo antico*, Sassari 1978; A. Boscolo, *La Sardegna bizantina e alto giudicale*, Sassari 1978; id., *La Sardegna dei giudicati*, Sassari 1979; E. Cau, Fulgenzio e la cultura scritta in Sardegna agli inizi del VI secolo, *Sandalion* 2 (1979) 221-229; L. Pani Ermini, Antichità cristiana e Alto Medioevo in Sardegna attraverso le più recenti scoperte archeologiche, in *La cultura in Italia fra Tardo Antico e Alto Medioevo*, II, Rome 1981, 903-911; P. Testini et al., La cattedrale in Italia: Schede, Provincia Sardinia, *Actes XI° Cong. Internat. d'arch. chrét.*, CEFR 123, Rome 1989, 133-138 (bibl.).

A. Di Berardino

SASSANIDS. Iranian dynasty who reigned in *Persia from the 3rd c. AD to 640, when it fell to the Arabs. The S. were always a thorn in the side of the Roman empire, esp. in 260 when Shapur I defeated and imprisoned the emp. *Valerian. The consequent deportation of Christians from *Syria favoured the spread of Christianity inside the Iranian empire. The Christians were always mistrusted because of their attachment, which increased after the peace of *Constantine, to the Roman empire. This sympathy aroused a violent reaction in Shapur II, who began a persecution (339-340) aimed mainly at the church hierarchy: three successive bishops of *Seleucia-Ctesiphon were martyred, and the see was vacant from 348 to 388. *Yezdegerd I (399-420), son of Shapur II, was more tolerant towards the Christians, and in 410 *Maruta of Maiferqat was able to call a *council of *c*.40 bishops at *Seleucia-Ctesiphon, which approved the decisions of *Nicaea and restored the Persian church.

J. Labourt, *Le Christianisme dans l'Empire Perse sous la dynastie Sassanide*, Paris 1904; J.M. Fiey, *Jalons pour une histoire de l'Eglise en Iraq*, CSCO 310, Subs. 36, Louvain 1970; G. Pugliese Carratelli, La Persia dei Sassanidi nella storiografia romana da Ammiano a Procopio, in *Scritti sul mondo antico*: Bibl. della Parola del Passato, Naples 1976, 35-46; O. Bucci, La posizione del cristianesimo occidentale e quella del cristianesimo orientale di fronte alla guerra tra Roma e l'impero dei Sasanidi, *Atti Accad. roman. costant.*, Perugia 1979, 99-139; S.P. Brock, A Martyr at the Sasanid Court under Vahran II: Candida, *AB* 96 (1978) 167-181; id., Christians in the Sasanid Empire: a case of divided loyalties, in (ed.) S. Mews, *Religion and National Identity*, SCH 18, Oxford 1982, 1-19. See also under *Manichaeism.

U. Dionisi

SATISFACTIO. The theological use of the term s., so characteristic of Western *patristics, can be understood only in a context of Latin law and *rhetoric - both in *Tertullian, who first used it, and in later *Fathers. Tertullian speaks of s. or *satisfacere Deo* mainly in penitential contexts, in texts concerning *baptism, the first *paenitentia* (*Bapt.* 20, 1), and in texts on *ascesis for all (*Orat.* 23, 4). In its use we distinguish two fundamental aspects. The first concerns the offering of s., *confessio*, in which wrong is admitted and reparation promised (cf. *Paen.* 9, 2: *satisfactio confessione disponitur*; 8, 9; 10, 2; *Pud.* 9, 16). The second considers the carrying out of s.; the promised reparation for the crime is made in order to avoid *poena* or to obtain *venia* and be restored to a state of righteousness (cf. *Paen.* 5, 2; 5, 9 ff.). In this second sense s. refers, in a general way, both to *paenitentia* as inner *conversion (*Paen.* 5, 9: *per delictorum paenitentiam domino satisfacere*) and to *prayer (*deprecatio*) and ascetic acts through which the penitent expresses the seriousness of his *penitence (cf. *Bapt.* 20, 1; *Pud.* 13, 14). In all this, Tertullian makes the following premises: the relationship between *God and man is based on *lex*. If man does not observe the law, he becomes God's debtor. His *sin, being a *violatio legis*, is a *culpa* that requires a *poena*. God could simply concede him the *venia*. But he does not do so without requiring *satisfactio paenitentiae* as a *compensatio poenae* justly inflicted on the debtor (cf. *Paen.* 6, 4). However, he is always disposed to accept this s. (cf. *Paen.* 7, 14).

This doctrine of s. recurs in *Cyprian, also in a penitential context but with some new nuances. Like the term *paenitentia*, s. much more often concerns the prayer and the penitential acts that the penitent must carry out before being readmitted to ecclesiastical *communion (*pax ecclesiae*) (cf. *Epp.* 59, 13; 16, 2; *Eleem.* 5). To be full, this s. must have the right measure (*Epp.* 64, 1; 43, 2; *Laps.* 16). Judgment of the legitimate measure belongs to the competent authorities (*Laps.* 29; *Ep.* 43, 3). Only thus can s. be *grata* to the Lord (*Laps.* 29), who *satisfactione placandus est* (*Laps.* 17). After Cyprian, Latin authors adopted substantially the same penitential terminology. We meet it in *Lactantius (*Ira*, 21, 9: *gratia-satisfactio*), *Ambrose (*Luc.* 7, 156 f.; 10, 88), *Sulpicius Severus (*Dial.* 2, 10: *poena-satisfactio*), *Augustine (*Serm.* 19, 2 f.; *Ench.* 65, 70 f.), *Innocent I (*Ep.* 25, 7), *Leo I (*Ep.* 108).

In the 4th c. this penitential language became part of *christology. Explaining Jesus' voluntary acceptance of his passion, *Hilary declares that this satisfied the penal obligation, though in relation to the suffering Christ it is not properly *poena* (*Psal.* 53, 12). Ambrose, following the same line, says that Christ accepted death so that the sentence should be fulfilled, the condemnation (of sinful *flesh) be satisfied (*Fug.* 7, 44). Though these are two rather isolated texts, they are of great importance. They do not just express, in legal terminology, a fundamental patristic theme, that of *expiation, but they also prepare for the theology of Anselm, which would be fully centred on the idea of vicarious satisfaction. At about the same time a theological meaning of s. appears which is partly new and partly very close to that first penitential use. In synodal *acts and papal correspondence, but also in preaching, s. is used in cases of readmitting *heretics to ecclesial communion or of acknowledging the *orthodoxy of bishops to be readmitted to the episcopal college. Then s. means principally confession, as we see particularly from the Greek versions, which render it ἀπολογία (cf. 1 Pet 3, 15 [vulg.]; Leo, *Ep.* 28, 6: PL 54, 778B. 780A; cf. Orig., *Hom. Iud.* 3, 2; *Princ.* III, 1, 16; Jerome, *Ep.* 52, 7). In this picture of confession of correct *faith, s. sometimes includes *justification, satisfactory expression of faith, but also repentance of error. In certain cases it is a question of written confessions (*satisfactio libellaris*: Leo, *Ep.* 31, 4) or of confessions to be signed (*Ep.* 30, 2). *Leo's examples are the most interesting (cf. *Serm.* 35, 5: *satisfactio legitima*, required from *Manichees; *Ep.* 89, 1: *satisfactio correctis*, required from *Dioscorus's supporters; *Ep.* 164, 5: *legitima satisfactione correctis paenitentiae remedium non negetur*; also *Ep.* 127, 1: *talia scripta*, without the term s.); other important ones are in *Priscillian (?) (*Tract.* 1, 40: CSEL 18, 33), *Augustine (*Gest. Pelag.* 11, 24; 20, 44) and *Sixtus II (*Ep.* 2, 3).

A. Deneffe, Das Wort "satisfactio", *ZKTh* 34 (1919) 158-175; J. Rivière, Sur les premières applications du terme satisfaction à l'oeuvre du Christ, *BLE* 25 (1924) 285-297; *DTC* 14/1, 1129-1210; A. Beck, *Römisches Recht bei Tertullian und Cyprian*, Aalen ²1967; J. Roussier, Satisfacere, *Studi in on. di P. Francisci*, II, Milan 1956, 113-157; M. Brueck, "Genugtuung" bei Tertullian, *VChr* 29 (1975) 276-290.

B. Studer

SATORNILUS (or SATURNINUS). The *gnostic S., a contemporary of *Basilides and pupil of *Menander, was active in *Syria. Acc. to *Irenaeus (*Adv. haer.* I, 24, 1) - on whose description the other sources largely depend - he claimed that the *world was created by seven *angels. They created man, but because of their incompetence he was unable to stand upright and crawled on the ground like a worm. Then the Higher Power (ἡ ἄνω δύναμις) sent a spark of *life which animated man and allowed him to stand up. This spark represents the spiritual and eternal element in man, which, after his death, is destined to return to its place of origin. The *God of the OT is one of the seven angels, whose revolt against the unknown God causes the coming of Christ, whose main purpose is to destroy the God of the Jews. The last part of Irenaeus's report is hard to reconcile with what precedes it. He says that the angels created two categories of men, one good, the other bad; the Saviour came to help the former by destroying the bad men and the *demons who help them. Acc. to S., generation and *marriage are the work of Satan, who, together with the angels who created the world, also inspired the prophecies of the OT.

Irenaeus, *Adv. haer*, I, 24; Hippolytus, *Ref.* VII, 28; Tertullian, *De anima* 23; Filaster, *Liber de haer*, 31; Epiphanius, *Pan.* 23; *EC* 10, 1964-1965; H. Schlier, Das Denken der frühchristlichen Gnosis (Irenäus, *Adv. Haer.* I, 23, 24), in *Neutestamentliche Studien für Rudolf Bultmann*, Berlin 1954, 67-82; K. Rudolph, *Gnosis* (Eng. tr. R. McL. Wilson), Edinburgh 1983; R.M. Grant, *Gnosticism and Early Christianity*, New York ²1966, 15-17, 100-119.

A. Monaci Castagno

SATURNINUS of Arles. We know of S. through *Sulpicius Severus (*Chron.* 2, 40, 45) and *Hilary of Poitiers (*De syn.* 2, 2; *Fragm. hist.* 2, 8; 11, 4; *Ad Constantium* 2, 2, 3; 3, 2; *Adv. Auxentium* 7). He was bishop of *Arles (*Arelate*) under *Constantius (337-61); we do not know if he was bishop at the time of the *council of Arles (353), but he took part in those of *Milan (355), *Béziers (356), *Rimini (359) and *Constantinople (360). He headed the *Arian faction in *Gaul and was deposed from the episcopate in the *orthodox reaction under *Julian (361-363). As a *heretic, regularly deposed, he was not inscribed in the episcopal *diptychs.

Duchesne, *Fastes* I, 254-55; E. Griffe, *La Gaule chrétienne à l'époque romaine*, I, Paris ²1964, 224-26, 243-46, 262-64.

V. Saxer

SATURNINUS of Toulouse. First bishop of *Toulouse (*Tolosa*) (mid 3rd c.). A 5th-c. Passion (BHL 7495-96) or rather panegyric, substantially trustworthy, relates his *martyrdom. During *Decius's *persecution (250), he refused to *sacrifice and the rioting people had him bound, with no regular trial, to a bull that was to be sacrificed, and dragged to death. The Passion tells how bishop Hilary (second half of 4th c.) built over S.'s tomb a wooden funerary chapel, replaced by his successor Silvius with a stone building. *Exuperius (. . . 405-411 . . .) transferred the relics, 30 Nov, some way away from the primitive tomb and built there the basilica of St Sernin, destroyed without trace by the Arabs in 721. Another basilica was built in the same place in the Carolingian era and, rebuilt in the 11th-12th cc., still stands.

The Passion does not give the day of S.'s death; *Mart. hier.* mentions him on 29 and 30 Nov. The 30th is the day of Exuperius's translation (1 Nov in the Mozarabic *calendars); the 29th must have been chosen as that of the Roman Saturninus. The data in *Mart. hier.* seem to go back to the Gallican recension. In the 6th c. the saint's legend developed. While *Caesarius of Arles (G. Morin, *S. Caes. Arel. Op. omn.* 2, Maredsous 1942, 179) makes him a disciple of the *apostles, *Venantius Fortunatus (PL 88, 99-101) and *Gregory of Tours (*Hist. Franc.* I, 30) simply say that he came from *Rome. Gregory adds "in 250", referring to the Passion. His cult spread among the Visigoths of *Spain from the 5th c. and in *Gaul, from the 6th c., in the region of Toulouse, along the Mediterranean coasts and the route of St James of Compostella, as far as the region of Paris.

Duchesne, *Fastes*, I, 25, 295; E. Griffe, *La Gaule chrétienne à l'époque romaine*, I, Paris ²1964, 148-52, 395-402.

V. Saxer

SATYRUS. *Ambrose's elder brother; born perhaps *Rome, 330, brought up at *Trier, studied at Rome, entered the civil service and became *consularis*. Retired to *Milan to live an intense Christian life with Ambrose. Returning from *Africa, where he had gone to administer family property, he fell gravely ill and died at Milan in 375 (acc. to Palanque; 377-8, acc. to other scholars). Buried in the basilica of S. Vittore in Ciel d'Oro, near the tomb of the martyr Victor. Our knowledge of him comes from Ambrose's two funeral orations in his honour (*De excessu fratris sui Saturi*), the first preached on the day of the obsequies, the second seven days later.

BS 11, 664-666; J.R. Palanque, *Saint Ambroise et l'Empire Romain*, Paris 1933, 488-493; A. Palestra, *San Satiro, fratello di s. Ambrogio e s. Marcellina*, Milan 1980.

M.G. Mara

SAVIA. S. was one of the provinces organized after 293 in what had been *Pannonia Superior. To N it bordered the Danube-Drava, to E it reached a line crossing the Pannonian plain and forming the border with Valeria. To W, it had a common border with *Noricum; to S, its border with *Dalmatia followed the line of the Sava. The administrative centre was Siscia (modern Sisak). Episcopal sees are known only at Siscia and Jovia (modern Ludbreg). The border city of Poetovium was left to Noricum in *Constantine's time. In the 6th c. Siscia must have been part of Dalmatia, since bishops Johannes and Constantius signed the conclusions of the councils of *Salona of 530 and 533.

Christianity began to develop in S. in the second half of the 3rd c. Early data are scarce: the one concrete fact is that in June 303 bishop Quirinus of Siscia drowned in the stream running through Savaria in Pannonia Prima. It seems that the *persecutions in S. and Pannonia Prima were less harsh than in Pannonia Secunda. A later inscription at *Aquileia mentions one Amantius, who was bishop of Jovia in the late 4th c. and died at Aquileia in 413.

Christian archaeological remains are very few. No *church buildings have yet been excavated. In Aquae Iasse (Varazdinske toplice), a thermal basilica was turned into a church; next to the hall, for other liturgical services, were added rooms decorated with *paintings, only small pieces of which survive. One fragment shows a bearded head with *nimbus. At Jovia too the *baths were turned into a church. Both transformations were carried out in the late 4th c. Small Christian objects, *sarcophagi, *lamps, *amulets, *fibulae, etc., have been found in S.

J. Zeiller, *Les origines chrétiennes dans les provinces danubiennes de l'Empire Romain*, Paris 1918; T. Nagy, *The History of Christianity in Pannonia up to the collapse of the Roman defence of the Border*, (in Hungarian), Budapest 1939; *PWK*, Suppl. 14, 739 ff.; A. Mócsy, *Pannonia and Upper Moesia*, London 1974, 325 ff.; B. Vikic, Elementi ranog krscanstva u sjevernoj Hrvatskoj, *Arheoloski vestink* 29 (1978) 588 ff.

N. Cambi

SCAPULA. Proconsul of *Africa 211-213, he instigated a harsh anti-Christian *persecution. *Tertullian wrote him a sort of open *letter, *Ad Scapulam*, pointing out the proconsul's personal responsibility for the persecution, while his predecessors had known how to reconcile their sense of humanity with their duty as magistrates (4, 1 and 3). Tertullian further points out that S. was going against his own received instructions to punish confessed offenders and torture those who denied their crimes, since he used torture to try to make confessed Christians deny their confession (4, 2). This S. should prob. be identified with a *Scapula Tertullus*, ordinary consul in 195.

J. Schmidt, Ein Beitrag zur Chronologie der Schriften Tertullians und der Proconsuln von Afrika, *RhM* 46 (1891) 77-98; *Prosopographia Imperii Romani saec. I. II. III*, pars III, eds. P. De Rohden - H. Dessau, n. 193; Monceaux 1, 244-246; T.D. Barnes, *Tertullian, a historical and literary study*, Oxford 1971.

S. Zincone

SCETE, Desert of. Place with several salt lakes, W of the Nile Delta, halfway between *Alexandria and Cairo, a little way into the Libyan *desert (now Wadi al-Natrun [pron. *Wadi 'n-Natrun*]); the Greek (and Latin) name is derived from the *Coptic *Shiet*. Known in classical times for its mineral resources. *Macarius the Egyptian settled there, *c.*330, and a monastic community formed around him. From then on it was a monastic centre of enormous importance, though subject to ups and downs (destruction by Berbers, persecution of *Origenists, etc.). Four *monasteries are still active today (of St Macarius, of St Pshoi, of the Silurians and of Abraham), from whose old *libraries have come many important Coptic and *Syriac codices. From the 9th to the 12th c., the patriarch of Alexandria was customarily chosen from the monastery of St Macarius. In modern times S. has been confused with Nitria, further E, within the Delta (now *el-Barnugi*, from Coptic *Pernug*).

D.J. Chitty, *The Desert a City*, London-Oxford ²1977; H.G. Evelyn White, *History of the Monasteries of Nitria and Scetis*, 3 vols., New York 1932.

T. Orlandi

SCHISM - SCHISMATIC. The term schism entered Christian *Latin usage in the 3rd c., meaning dissent leading to disunion in the community. The term schismatic did not mean a Christian separated from the community, but a dissenter. The difference between *heresy and schism, as well as their mutual relationship, are spelled out by *Jerome: "*Inter haeresim et schisma hoc esse arbitrantur, quod haeresis perversum dogma habeat, schisma propter episcopalem dissensionem ab ecclesia separetur . . . Ceterum nullum schisma non sibi aliquam confingit haeresim, ut recte ab ecclesia recississe videatur*" (*In ep. Tit.* 3, 10-11). So we find schismatics nearly always associated with heretics, and often, esp. in *Cyprian's *letters, not easily distinguishable (*Epp.* 33; 66, 5). In the Donatist controversy, *Augustine's writings make a clear distinction between heretic and schismatic (*De haeresibus* 50); yet Augustine is aware that a schism sooner or later leads to heresy, esp. when it becomes a *schisma inveteratum* (*C. Cresconium* II, 4). Indeed, after 411 he classes *Donatism no longer among the schisms but among the heresies (*Ep.* 93, 11, 46; *De haeresibus* 69). The Cappadocians, esp. *Basil, see heresy as separation from the belief of the community, but schism as dissent from the official community for ecclesiastical reasons and over remediable questions (Basil, *Ep.* 188, can. 1). Alongside schisms Basil puts parasynagogues, liturgical *synaxes celebrated by bishops, presbyters or *faithful who are not correctly or sufficiently instructed (*Ep.* 188, *can.* 1). Schism therefore is a disagreement with the hierarchy on, e.g., the times and ways of conceding *penitence to *lapsi*. The main cause of schism is *libido dominandi* or *philarchia* (Basil, *C. Eun.* 1, 13; John Chrys., *In Eph. 4 hom.* 11, 4-5; Theodoret, *In I Cor.* 11, 18).

Legislation on schism, fixed mainly at the *councils of *Nicaea (325), *Laodicea (345 or later), *Constantinople (381 and 382) and *Carthage (419), usually involved removal from office and *excommunication, sometimes even referral to the civil authorities for factiousness (synod of *Antioch [341], can. 5).

E. Buonaiuti, Scisma ed eresia nella primitiva letteratura cristiana, *Saggi sul cristianesimo primitivo*, Città di Castello 1923, 274-285; H. Pétré, Haeresis, schisma et leurs synonymes latins, *REL* 15 (1937) 316-319; M. Pontet, La notion de schisme d'après s. Augustin, *L'Eglise et les Eglises*, I, Chevetogne 1955, 163-180; M. Meinertz, Schisma und Hairesis im NT, *Biblische Zeitschr.* N.F. 1 (1957) 114-118; Joannou I, 1; S.L. Greenslade, *Schism in the Early Church*, London ²1964; E. Ferguson, Attitudes to Schism at the Council of Nicaea, in (ed.) D. Baker, *Schism, Heresy and Religious Protest*, Cambridge 1972, 57-63; A. Schindler, Die Unterscheidung von Schisma und Häresie in Gesetzgeburn und Polemik gegen den Donatismus, in (ed.) E. Dassmann et al., *Pietas* (Festsch. Kötting), JbAC Ergbd. 8, Münster 1980, 228-236; G. Bonner, Schism and Church Unity, in (ed.) I. Hazlett, *Early Christianity* (Festsch. W.H.C. Frend), Cambridge 1990, ch. 19.

V. Grossi

SCHOLASTICA. Our only knowledge of St S., whose very existence has recently been questioned, comes from book II of *Gregory the Great's *Dialogues*. Sister of St *Benedict, S. was consecrated to *God from childhood; once a year she would go with her brother to the neighbourhood of the *monastery of Montecassino and converse with him. Gregory's moving account of their last meeting on the eve of their deaths (*Dial.* II, 33) remains famous.

BS 11, 742-749; A. Pantoni, Sulla località del convegno annuale di S. Benedetto e S. Scolastica, e sul monastero di Piumarola, *Benedictina* 15 (1968) 206-228; A. de Vogüé, La rencontre de Benoît et de Scholastique. Essai d'interprétation, *RAM* 48 (1972) 257-273; P.A. Cusack, St. Scholastica: Myth or Real Person?, *DR* 92 (1974) 145-159.

S. Zincone

SCHOLASTICUS. In Christian antiquity the adjective s., mostly used as a noun, had a dual meaning. In some texts (cf. e.g. Ps.-Mac. Egyp., *Hom.* 15, 42: PG 34, 604D; *Hom.* 26, 17: PG 34, 685B) it means, in a rather vague way, the literate, instructed man, one who has at least profited from the teaching of the *grammaticus* (cf. F. Blättner, *Storia della pedagogia*, It. tr. Rome ⁶1972, 42-43). Other sources, however, lead us to identify the s. more specifically as the man of law, the advocate (cf. council of Sardica, can. 10; Socr., *HE* VI, 6, 36: PG 67, 681A; Cyr. of Scyth., *Vita Sabae* 61; John Mosch., *Leimon*. 131: PG 87, 2996B; *Chron. Pasch.*: PG 92, 980A), esp. in relation to legal affairs or controversies in some way involving the church community (cf. *Cod. Ecc. Afr.* 97).

Du Cange, *Glossarium mediae et infimae latinitatis*, VI, 351-52; *PWK* 21/1, 624-25; Lampe 121-23.

M. Spinelli

SCHOOL

I. Christianity and schooling - II. The thought of the Fathers - III. Christian teachers - IV. Christian schools of catechesis and theology - V. Monasticism - VI. The dissolution of the ancient school.

I. Christianity and schooling. Christianity did not modify the structures of the ancient s. It introduced no new method of teaching, nor, even when it prevailed as the state religion, did it impose forms and contents differing from traditional teaching. Both in the early centuries and in the 4th c. and later, *pagan and Christian boys frequented the same ss., read the same texts and received the same instruction, despite their parents' religious differences. The possession of secular culture was particularly necessary for the upper-middle-class boys who provided the nucleus of the administrative careers. But church dignitaries came from this same class. And among these, esp. from the mid 4th c. on, a good many reached high ecclesiastical office after having been state officials. Except for the first catechistic initiation, religious education was left to individual inquiry and to *families.

Ss. were Christianized later on, not through innovations and reorganizations by the church, but through the Christianization of society. The Roman scholastic system long survived the political end of the empire, the barbarian settlements, the ruins of war; it slowly disappeared as the old world slowly disintegrated, while new forms of economic, political and civil life took its place. In the West, church ss. started later and were less common than has usually been held, nor did they fill *tout court* the place and functions of the earlier lay ss. Recent studies (esp. by Riché) tend to correct the theory, still supported by Marrou, that church ss. arose solely to replace the ancient ss. when those disappeared. In some areas at least, the two types of s. coexisted for some time. Church ss. were born of the need to offer a new type of teaching, based on the *Bible and aimed at the training of clerics, while lay ss. continued to seek secular and classical instruction. Even in 535 *Cassiodorus, tracing the outline of a s. of higher religious studies, could lament that secular studies were brilliantly taught in public ss., while the Holy *Scriptures were totally neglected.

The organization of the ancient s. into three orders survived to the end: the *ludus litterarius* or primary s., in which *children from seven to twelve learnt to read, write and count; the s. of the *grammaticus*, dedicated to the study of the great classical writers; and finally, after the age of 18, higher education, dedicated to *rhetoric. As in classical antiquity, so in the Christian centuries, little information remains about the first grade of instruction: only the design of higher education can be made out to some extent. Among the few visible innovations are the growth of the study and importance of grammatical theory, *ars grammatica*: the greatest grammarians, *Donatus, Servius and *Priscian, flourished now.

Instruction, even of the upper classes, became less and less bilingual. Greek, from Republican times the language of culture, was gradually disappearing in the West; in the *East, Latin had never had literary or cultural prestige, but had been the language of law, administration and public acts; it remained so until the time of *Justinian.

Nor did Roman architecture create buildings specifically designed for teaching during these centuries. Rare exceptions seem to be the *Maenianae scholae* of *Autun, described by Eumenes in his speech addressed to the emperors early in 298 asking for its restoration (*Pan. Lat.* 5) or the *scholae* of Mactar in Tunisia and of Ostia, great public buildings intended as youth clubs, in which there were rooms for teaching.

The material position of teachers continued to be precarious. Except for the most celebrated rhetors, the others received modest stipends from the cities or, even worse, lived on the contributions of their students, who quite often - something all sources seem to agree on - either did not pay or haggled for the price or payed very late.

Methods of correction appear to have gone on unchanged, consisting mainly of corporal punishment. From the *plagosus* Orbilius recorded by Horace to the teacher with a rod on a Claudian bas-relief from Arlon in Luxembourg, to the implacable teachers remembered with anguish by *Augustine in the *Confessions* (I, 9, 14-15; I, 14, 23), the *ferula* and the rod remained the ineliminable instruments of the ancient s. Not even Christianity intervened to modify traditional pedagogical ideas, for which infancy and adolescence were a thankless age that should quickly be left behind. Only adulthood was the *aetas perfecta* or *legitima*, and the *puer* should be prepared for it with all severity.

Nor did Christianity bring substantial innovations in teaching methods, except perhaps in the growing use of *memory, derived from catechistical preparation, which was based on memorization rather than writing, and the growing custom of learning parts of Scripture, like the *psalms and gospels, by heart.

II. The thought of the *Fathers. The problem of schooling does not come into the picture of the primitive *kerygma*. The earliest church attended to the preaching of the gospel message and the organization of worship, and was not interested in the institutions of secular society. But by its nature as a "learned" religion, i.e. based on *Scripture and the "*tradition" that developed around it, Christianity required a clearly literary type of culture and never renounced the teaching function (Marrou). The early Greek-speaking Christian writers - like *Melito of Sardis, *Theophilus of Antioch, *Athenagoras - were formed in the ss. of the time and show clear signs of the rhetorical training they received there. But even those of later generations, and then the Latins, even when coming from Christian *families, show that they received the same education. In fact Christian writers, even those who in theory were less well-disposed to secular culture, never ceased to base their prose on classical models and to draw widely on them in every way. The presence of the classics in Christian writings very slowly diminished, not through ideological opposition, but through the degradation of the culture itself.

The need to reconcile secular culture with *faith was felt most vividly from the late 2nd and the 3rd c., the period in which Christians ceased to live in small groups and were gradually integrated into society. *Justin reached *conversion after passing from one philosophical s. to another and, satisfied in the possession of Christian truth, felt no incompatibility between Christianity and *Hellenism. Much more than in him, acceptance of Greek culture becomes a critical debate in *Clement of Alexandria and a genial reconsideration of Hellenic *philosophy in the light of Christian doctrines in *Origen, the greatest Oriental Greek Christian thinker. In the Latin West, however, opposition to traditional culture is clear in *Tertullian, *Arnobius and, though more dissemblingly, in *Lactantius: the Christian must find his every intellectual gratification in Scripture, which, with the wealth of its teaching, can constitute the liberal arts. *Basil the Great dedicated to his nephews an *Exhortation to the young on the way to draw profit from pagan literature*, which has wrongly been seen as a defence of Greek culture. In fact Basil proposes an attitude of caution towards these writers and exhorts his audience to draw from them only what does not clash with Christian morality. Later, *Theodoret of Cyrrhus, in a work *On the way to cure Hellenic maladies*, while admitting that he has approached Greek culture with some profit, crudely spells out its errors and harmful potential and contrasts its obscurity with the luminous truth of Scripture.

None of these writings were on the specific themes of education and schooling. Of such, old Christian culture is absolutely devoid. Even works like *De inani gloria et de liberis educandis*, now universally ascribed to *John Chrysostom, and *Jerome's *Ep.* 107 (*Ad Laetam de institutione filiae*: written in 403), both addressing the *ascetic life, provide points of Christian pedagogy, but not programmes that could be used concretely in ss.

In fact the scholastic problem was resolved empirically. Certainly the fact that Christian boys were educated by *pagan teachers, whose teaching was bound up with *idolatry, who taught *mythology, who observed pagan festivals, caused reservations. And yet there was no way in which boys could give up secular education, which was necessary for their maturation and was even a propaedeutic to religious studies. This was the position even of the intransigent Tertullian (*De idolatria* 10). An example of how a Christian boy could be educated in pagan ss. and at the same time directed towards Christian doctrine is that of Origen. *Eusebius (*HE* VI, 2, 8) tells how the young Origen was sent to go through the various stages of traditional studies, while at the same time his father instructed him in the Holy Scriptures and examined him on them daily.

III. Christian teachers. It was, however, considered inopportune for Christians to teach in secular ss. But this was not a peremptory prohibition. At the start of the 3rd c., the 18-year-old Origen could choose to be a teacher and teach grammar for some time at *Alexandria (Euseb., *HE* VI, 2, 15 f.); a little later *Anatolius, who later became bishop of Laodicea, taught *Aristotelian philosophy at Alexandria (Euseb., *HE* VII, 32, 6) and the presbyter *Malchion had a s. of *rhetoric at *Antioch (Euseb., *HE* VII, 29, 2). In the course of the 4th c. we find cases, even famous cases, of Christian teachers esteemed by pagans and *vice versa*. The pagan rhetor *Libanius numbered Christians among his disciples and had relations with *Basil and *John Chrysostom; Themistius, also a pagan, was called to educate *Arcadius, son of the emp. *Theodosius I; *Prohaeresius, a Christian rhetor famous at *Athens, was the teacher of *Gregory Nazianzen, but also of pagans like *Eunapius, who later celebrated him in his *Lives of sophists*. *Augustine (*Confess.* VIII, 2, 5) records the sensational *conversion of the famous rhetor *Marius Victorinus.

An attempt to marginalize Christian teachers was made by the emp. *Julian, who accused of immorality and insincerity those who taught Homer, Hesiod and the other great writers of the past without believing in their gods (*Ep.* 61). But it is wrong to talk, as is usually done, of a Julianine anti-Christian reform of ss.; his edict on teachers of June 362 (*CT* XIII 3, 5), which has been seen as the instrument of this, contains no measures against the Christians, either explicitly formulated or insidiously dissembled, but aims to control the professionality of teachers and is part of a general tendency in Julian's legislation towards the reorganization of the bureaucracy and the reappraisal of municipal interests and competences (Pricoco).

According to the 5th-c. historian *Socrates (*HE* III, 16), Julian's persecution of Christian teachers forced two Christian teachers, father and son, named *Apollinaris, one a grammarian, the other a rhetor, to an experiment in Christian schooling, based on the reduction of the OT to verse by the father, and of the NT to Platonic dialogues by the son. But, adds Socrates, Providence rendered the actions of Julian and the attempts of the two Apollinares equally useless.

IV. Christian schools of *catechesis and *theology. From the middle of the 2nd c., schools of religious studies flourished in various cities of the empire. *Justin, after teaching at *Ephesus, founded at *Rome, before 165, a s. in which *Tatian, later an implacable opponent of Greek culture, studied. A catechetical s. (*Didaskaleion) was founded at *Alexandria in the 2nd c. by *Pantaenus, the teacher of *Clement of Alexandria. In 189 Clement became director of this s. and impressed on it a more strongly intellectual and philosophical character. Of his writings, the *Pedagogue* appears to express most manifestly this pedagogical, but always religious, experience of education in truth through Christ, teacher of mankind. The s. of Alexandria reached its apogee under *Origen, who assumed its direction in 204, aged 18. He also taught secular literature (*enkyklia grammata*) and later expanded the study of philosophy more and more (Euseb., *HE* VI, 3, 3). He founded another s. at *Caesarea, where he remained until his death. *Gregory the Thaumaturge has left a record of this s. and the teaching methods followed there in his *Thanks to Origen*, composed in c.238. Natural science, geometry and astronomy were studied, but were always subordinated to the primary purpose, that of illustrating and investigating Christian doctrine. The teacher did not want his disciples to be ignorant of any principle of philosophy and "led them by hand as on a journey", preserving them from lies harmful to *piety (ch. 14, 170 f.) and exhorting them not to bind themselves to any philosophy, but only to *God and his *prophets (ch. 15, 173). Origen himself, in a *letter - perhaps a reply to Gregory - which survives as chapter 13 of the *Philocalia*, invites his disciples to "take from Greek philosophy all that can serve as encyclical teaching (*enkyklia mathemata*) or as a propaedeutic to Christianity, and to take from geometry and astronomy what can be useful for the interpretation of the Holy Scriptures".

These "higher schools of Christian theology" (as they have been called) were not followed up. The lectures of *Apollinaris at Antioch, *Didymus the Blind at Alexandria and *Gregory Nazianzen at Constantinople, which *Jerome refers to (*Epp.* 84, 3; 50, 1; 52, 8), were private initiatives with no institutional character (Marrou).

V. *Monasticism. From the middle of the 4th c. (some decades later in the West), renewed attitudes of opposition to secular culture were expressed within the nascent monastic movement, which contested many of the institutions of society, deplored the secularization of the church, reasserted the need for a uniquely Christian culture and proclaimed the *eremi philosophia* as the true philosophy. Between 380 and 385 *John Chrysostom dedicated to this theme a text entitled *Against the adversaries of the monastic life*. In it, books I and II discuss child education and propose to entrust boys not to secular ss., which educate them in bad morals and worldly things (*biotika*), but to monks, who instruct them in Scripture and spiritual things (*pneumatica*).

The presence of boys in *monasteries is already attested by the *Rules* of *Pachomius, before the mid 4th c., and then by those of *Basil. But it is difficult to maintain that these were real ss., intended to replace traditional teaching. It has been denied that monastic ss. existed in the convents of *Egypt and *Syria (Bardy). *Libraries existed in them; there were calligraphers among the monks; among those who came from the *world, there must also have been educated men, but there was no institutionalized teaching. Basilian monasteries took in boys who were orphaned or whom their parents intended for monastic life, and gave them instruction, but this was a purely religious instruction, based on the sacred books.

In the West too, monasteries admitted boys from the start. We know this with certainty for *Lérins where we find, soon after its foundation in the first years of the 5th c., the two sons, still *pueri*, of *Eucherius, future bishop of *Lyons. Eugendus was still a child when he was accepted into the monastery of Condat (*Vita Patrum Iurensium* III, 124). The *Rule* of *Aurelian and that of the Master allow for *infantuli* in monasteries; both these and other Western *rules, like those of *Caesarius of Arles and *Ferreolus, prescribe that all who wish to become monks should learn to read; the *Rule of the Master* lays down that boys should study their letters for three hours a day under the direction of a literate monk (*Reg. Mag.* 50); the *Rule* of *Benedict provides for the use of *books, pens and tablets for all (*Reg. Ben.* 55). The rules also mention the presence of libraries in convents and frequently also *scriptoria*. Yet, even in the West we cannot, at least until the end of the 6th c., speak of "monastic schools" as replacements or alternatives to the secular ones which followed an organized scholastic *cursus* and which were open even to boys who were not destined for the cloister.

VI. The dissolution of the ancient school. Roman ss. continued to exist in many regions even after the invasions. Pirenne's thesis, that a certain degree of literary culture among *laymen in Merovingian *Gaul must be explained not just by the fidelity of senatorial families to the classical tradition but also by the survival of some scholastic structures, is now established. The same can be said of Visigothic *Spain, which long kept its cities and, in them, its scholastic traditions. In N *Africa the *Vandals, after the destruction of the initial occupation, reduced their opposition to Latin culture to such a point that some have spoken of a "Vandal Renaissance" (Courtois). At *Carthage and in other centres, there were teachers with a good classical training, like Felicianus, celebrated by the poet *Dracontius, Faustus, whose pupil was the poet Luxorius, *Symphosius, author of riddles very popular in the mediaeval ss., and the poet *Corippus. These ss. were still alive at the time of the Byzantine reconquest and appear to have held out until the Arab conquest. In *Italy, ss. with a classical approach kept up their own lively vitality, partly through *Theodoric's favourable policy and the work of *Cassiodorus. *Ennodius testifies that *Rome remained a "city friendly to the liberal arts" (*Ep.* 6: MGH, *Auct. ant.* VII, 222). Provincial students went to Rome to complete their studies. Theodoric's successor Athalaric sent a letter to the senators asking them to ensure regular payment to all professors, "grammarians, rhetors, teachers of law . . ., who refine manners and bring forth eloquence in the palace" (Cassiod., *Variae* IX, 21: MGH, *Auct. ant.* XII, 286). *Justinian, in the *Pragmatic Sanction* of 554, set up provisions for teachers of grammar, *rhetoric, medicine and law. The basis of this last ancient s. remained classical culture; its main components were grammar, poetry and rhetoric. The texts read were, as in the 4th c., the great writers: Terence, Horace, Ovid, Tibullus, Lucan, Cicero, Sallust and esp. Virgil. But it was an essentially impoverished culture, from which Greek and philosophy had disappeared.

In the Greek *East the ancient s. was transformed, but never completely extinguished. Until the Turkish conquest, not only did cultural ideals remain profoundly attached to the classical tradition but, at least as regards higher education, traces of the public s. remained; in particular, the imperial university of *Constantinople "dominated the intellectual history of the Byzantine East" (L. Bréhier).

Alongside these lay forms of education, in the 6th c. there appeared "schools" founded and organized by the church. But, as we have already said apropos of monastic ss., these were "internal" initiatives, intended for the training of clerics. The *Vita Caesarii* (II, 5 and 6) attests that *Caesarius of Arles was surrounded by a group of clerics to whom he gave lessons on Holy Scripture. The council of *Toledo of 527 set up an episcopal s., frequented by boys whose parents intended them for ecclesiastical life. The council of *Vaison of 529 decreed that rural parish priests should keep with them and instruct in the sacred texts young men wishing to become clerics. This text is considered the birth certificate of the parochial ss., which, when the Roman ss. disappeared altogether, performed the functions of primary ss. and were open even to boys who were not going to embrace an ecclesiastical career. This took place in the course of the 7th c., when the process of transformation of the ancient world accelerated. The disintegration of urban life and the growing ruralization then finally removed from the ss. their juvenile public and their teaching cadres. In early mediaeval society, alongside a great mass of *illiterati* there survived an *élite* of clerics and monks instructed by the church and a lesser number of cultivated aristocrats, privately educated in their families or, more and more often, in the monasteries and the church ss.

M. Roger, *L'enseignement des lettres classiques d'Ausone à Alcuin*, Paris 1905; A. Müller, Studentenleben im 4. Jahrhundert n. Chr., *Philologus* 69 (1910) 292-317; C. Barbagallo, *Lo Stato e l'istruzione pubblica nell'Impero romano*, Catania 1911; H. Pirenne, De l'état de l'instruction des laïques à l'époque mérovingienne, *RBen* 46 (1934) 165-177; G. Bardy, L'Eglise et l'enseignement au IVe siècle, *RSR* 14 (1934) 525-549 and 15 (1935) 1-27; L. Bréhier, L'enseignement classique et l'enseignement religieux à Byzance, *RHPhR* 21 (1941) 34-69; A. Lorcin, La vie scolaire dans les monastères d'Irlande aux Ve-VIIe siècles, *Rev. du Moyen-Age Latin* 1 (1945) 221-236; P. Courcelle, *Les lettres grecques en Occident de Macrobe à Cassiodore*, Paris ²1948; M. Pavan, *La crisi della scuola nel IV sec. d.C.*, Bari 1952; G. Bardy, Les origines des écoles monastiques en Occident, *SEJG* 5 (1953) 86-104; id., Les origines des écoles monastiques en Orient, *Mélanges J. de Ghellinck*, Gembloux 1951, I, 293-309; Ch. Courtois, *Les Vandales et l'Afrique*, Paris 1955; G. Downey, Education in the Christian Roman Empire: Christian and Pagan Theories under Constantine and his Successors, *Speculum* 32 (1957) 48-61; J. Leclercq, *L'amour des lettres et le désir de Dieu*, Paris 1957; P. Riché, La survivance des écoles publiques en Gaule au Ve siècle, *MA* 43 (1957) 421-436; S. Dill, *Roman Society in the Last Century of the Western Empire*, New York ²1958; H.I. Marrou, *Saint Augustin et la fin de la culture antique*, Paris ²1958; J. Fontaine, *Isidore de Séville et la culture classique de l'Espagne wisigothique*, Paris 1959; P. Riché, *Education et Culture dans l'Occident barbare. VIe-VIIIe siècles*, Paris 1962; AA. VV., *Los monjes y los estudios*, IV Semana de estudios monásticos, Poblet 1963; H.I. Marrou, *Histoire de l'éducation dans l'antiquité*, Paris ⁶1964; *Centri e vie di irradiazione della civiltà nell'Alto Medioevo*, Settimane di Studio 11, Spoleto 1964; P. Blomenkamp, Erziehung, *RACh* 6, 502-559; *Erziehung und Bildung in der heidn. und christl. Antike*, hrg. H. T. Johann, Darmstadt 1976; V. Paladini, *L'istruzione nel mondo classico*, Naples 1968; *Arts libéraux et philosophie au Moyen Age*, Actes du IVe Congrès Int. de philos. médiév., Montréal 1969; P. Lemerle, *Le premier humanisme byzantin*, Paris 1971; *La scuola nell'Occident latino nell'Alto Medioevo*, Settimane di Studio 19, Spoleto 1972; A. Di Berardino, Maestri cristiani del III secolo nell'insegnamento classico, *Augustinianum* 12 (1972) 549-556; A. Quacquarelli, *Scuola e cultura dei primi secoli cristiani*, Brescia 1974; *La cultura antica nell'Occidente latino dal VII all'XI secolo*, Settimane di Studio 22, Spoleto 1975; J. Préaux, Securus Melior Felix, l'ultime "orator urbis Romae", *Corona Gratiarum Dekkers*, Bruges 1975, II, 101-121; *Eglise et enseignement: Actes du Colloque du Xe anniversaire de l'Inst. d'hist. du christ. de l'Univ. libre de Bruxelles*, ed. J. Préaux, Brussels 1977; J. Bowen, *Storia dell'educazione occidentale*, It. tr., I, Milan 1979; P. Riché, *Les écoles et l'enseignement dans l'Occident chrétien de la fin du Ve siècle au milieu du XIe siècle*, Paris 1979; S. Pricoco, L'editto di Giuliano sui maestri (CTh 13, 3, 5), *Orpheus* 1 (1980) 348-370; M. Pavan, La scuola nel Tardo Antico, in *La cultura in Italia fra Tardo Antico e Alto Medioevo*, Atti del Convegno di Roma 12-16 nov. 1979, Rome 1981, 553-560; P. Riché, L'école dans le Haut Moyen Age, ibid., 561-574; R.A. Koster, *Guardians of Language. The Grammarian and Society in Late Antiquity*, Berkeley 1988 (bibl.); S.F. Bonner, The Edict of Gratian on the Remuneration of Teachers, *AJPh* 86 (1965), 113-137; id., *Education in Ancient Rome*, Berkeley-Los Angeles 1977; A.D. Booth, The Academic Career of Ausonius, *Phoenix* 36 (1982) 329-343; A. Cameron, The Last Days of the Academy at Athens, *PCPhS* 195 (1969) 7-29; L. Holtz, *Donat et la tradition de l'enseignement grammatical*, Paris 1981; P. Riche, *Education and Culture in the Barbarian West from the Sixth through the Eight Century* (Eng. tr. J.J. Contressi), Columbia 1976; H.C. Youtie, Because they do not know letters, *ZPE* 19 (1975) 101-108; M. Gärner, *Die Familienerziehung in der Alten Kirche*, Cologne 1985.

S. Pricoco

SCILLITANI. Twelve *martyrs are designated under this name: Speratus, Nartzalus, Cittinus, Donata, Secunda, Vestia, Veturius, Felix, Aquilinus, Laetantius, Januaria and Generosa. They came from Scilli, an unidentified place in *Africa Proconsularis (not Cillium [Kassérine, Tunisia]). Apart from the fact of their *martyrdom, 17 July 180, we know nothing of their lives. The Acts of their martyrdom (BHL 7527) are among the best documents of African *hagiography. The different recensions (BHL 7528-34) excellently illustrate the changes an authentic text can undergo. A basilica was later built over their tomb: its name often appears in antiquity (Aug., *Serm.* 155 and perhaps *Serm.* 37, *Denis* 16, *Guelf.* 30; Vict. Vit., *Hist. pers. Afr. prov.* 1, 9). From a *sermon preached in their honour on their feast-day by *Augustine, we learn that their Passion was read on the occasion (*Sermm.* cited, and *Guelf.* 31, *Lambot* 9). Their feast is also recorded in *Mart. hier.* and *Calend. Carth.*; but only Speratus's name appears in the marble *Calendar of Naples.

We know that a basilica in their honour was built near the *monastery of Biguas in Africa (BHL 4906). Some of their relics were preserved at Kherbet Oum el Ahdam in 359 (CIL 8, 20600) and at Dermech, near *Carthage. Their bodies were transferred to France at the time of Charlemagne and placed in St John's basilica at *Lyons (BHL 2045).

DACL 15, 1014-21; *Vies des SS.* 7, 378-81; *LTK* 9, 551-52; V. Saxer, *Vie liturgique et quotidienne à Carthage vers le milieu du IIIes.*, Rome 1969, 309; id., *Saints anciens d'Afrique du Nord*, Vatican City 1979, 31-34; id., *Morts, martyrs, reliques en Afrique chrétienne aux premiers siècles*, Paris 1980, 185-6, 203, 213, 226, 317; G. Lanata, *Gli Atti dei martiri come documenti processuali*, Milan 1973, 136-144; H. Musurillo, *Acts of the Christian Martyrs*, Oxford 1972, 86-89; W.H.C. Frend, *The Donatist Church*, Oxford ³1985, ch. 7.

V. Saxer

SCOTLAND. The earliest tribe known to inhabit S. were the Picts, whom the Romans on their arrival found settled in the region. The earliest Scottish historians relate that in *c.*203 pope *Victor sent missionaries to S., but there is no sure proof of *evangelization until the time of St *Ninian. Born in S-W S. in *c.*360, he studied at *Rome and returned home as bishop *c.*400, building the first stone *church in S. at Candida Casa, now Whithorn. He died prob. 432.

In the 6th c. S. was invaded by the Scotti, a Celtic tribe from *Ireland, who had already been evangelized. They settled along the W coast of S. Among them was *Columba, who founded the *monastery of Iona and with his companions evangelized all the N of the country.

At the same time Kentigern was working in S-W S. and what is now the extreme N-W of England. The Celtic church was tribal and monastic in its organization, and its monks, many of whom lived as hermits, were known for their extreme austerity. Though in *communion with Rome, the Celtic Christians followed their *traditions on, e.g., the date of *Easter. This led to inevitable tensions when the influence of *Augustine's mission and its Roman customs began to be felt in N England. Northumbria oscillated between Celtic and Roman observances, and Scottish influence was exercised over the Northumbrians through the Celtic bishops of Lindisfarne. The synod of Whitby (664) and the definitive institution of a metropolitan see at *York considerably limited this influence, and even the see of Lindisfarne adopted Roman customs.

Gradually S. began to conform to the Roman rite, encouraged esp. by *Adamnan, bishop of Iona (†704). In 710 the Pictish king Nechtan imposed conformity to Roman tradition on his whole kingdom. The 8th c. saw the rise of the *Deicolae* or Culdees, hermits influenced by Celtic traditions who lived in small communities. For *archaeology, cf. *Britain.

W.C. Dickinson et al., *A Source Book of Scottish History*, 3 vols., London ²1958-1961; G. Donaldson, *Scotland: Church and Nation through sixteen centuries*, London 1960; W.C. Dickinson-G.S. Pryde, *A New History of Scotland*, 2 vols., New York 1962; *DACL* 4, 1889-1921; J.H.S. Burleigh, *A Church History of Scotland*, Oxford 1964; E.A. Thompson, The Origins of Christianity in Scotland, *Scot. Hist. Rev.* 37 (1958) 17-22; C. Thomas, Ardwall Isle: the excavation of an early Christian site of Irish type, *Trans. Dumfriesshire and Galloway Nat. Hist. and Antiq. Soc.* 43 (1966) 84-116; id., *The Early Christian Archaeology of North Britain*, Oxford 1971. See also *Britain and Ireland.

E. Harbert

SCRIBE - RABBI. Scribe, in the Jewish world, originally meant a secretary with civil (2 Sam 8, 17; 20, 25) or military (2 Kgs 25, 19) duties. From the time of Ezra, he was an expert in the Torah (Ez 7, 6. 11); in this role, Ezra read to the assembled people the Torah of Moses, given to Israel by the Lord.

In post-biblical times, experts in religious law and tradition were called Rabbis; their origin and authority were traced back to *Moses through an uninterrupted chain. The title was conferred in a religious ceremony (*semikhah* = χειροτονία), which was forbidden by the Romans and fell into disuse. With the predominance of the Pharisaic tendency after AD 70, the Rabbis, who were not priests, replaced the scribes as spiritual leaders.

Scribes in the Roman world (Gk *grammateus*) were generally secretaries of public officials and magistrates, and also of private individuals. The scribe should be distinguished from the *librarius* (also *scriba librarius*), who was a mere copyist. The emperors too had their scribes (*scriba ab epistolis*, etc.). The scribes were organized in corporations. The term was also used in the church (cf. **notarius*).

EC 11, 172 f.; *DACL* 15, 1027-1035; *PWK* 21, 848-857; *DAGR* 4, 1123-1124; J. Neusner, *Early Rabbinic Judaism*, Leiden 1975.

S. Cavalletti

SCRIPTURE, HOLY

I. The canon - II. Versions and manuscripts - III. Inspiration - IV. Function of the Bible in the church.

I. The canon. That *God's Word was handed down to his people as Holy Scripture was a fact that the first Christian community inherited from Judaism. Doctrinal reflection had not yet defined all its forms; and neither Jesus nor the *apostles subordinated *revelation to the scholastic conceptions of the *scribes and doctors of the law, conceptions that were reinforced when Judaism closed ranks after the fall of *Jerusalem. The Hebrew (in some part Aramaic) *Bible would then be definitively closed, and in it the Torah (Pentateuch) would have a normative priority, with rigorous traditions for its legal interpretation. With *Philo, but in quite another spirit, the Pentateuch alone was commented on. At Qumrân, however, Isaiah had a central role. The gospel was indebted to prophetic freedom: all through the NT, Isaiah and the *Psalms have preference. Apocryphal literature played a certain part, not so much as a new category of holy books, but as an eschatological and messianic climate conferring unity on the prophecies. And this Galilean mentality did not lack parallels in the Jewish diaspora, in which *Paul was formed.

The writings that would make up the NT were not conceived by their authors as a complementary collection to be added to the Hebrew Canon. The churches had taken care to preserve the apostolic *letters, though their editor's identity and activities remain in shadow: some historians assign him a considerable redactional role. As for Jesus' words and miracles, we possess three redactions of them, but the primitive documents to which *Luke alludes are lost; added to them is the Fourth Gospel, which comes from a very particular background.

From the start of the 2nd c., *Ignatius clearly knows many of these texts, but without citing them as authorities; in mid-century, *Justin mentions their liturgical reading; but it is only with *Irenaeus that the whole appears coherent, canonized, fixed. Polemic against the claims of the *gnostics to possess secret *traditions, and against *Marcion's claim to choose and correct the texts and to reject the Hebrew Scriptures, helped reinforce awareness of the privileged position reserved to the writings judged apostolic, on the strength of their acceptance by the main churches and taking account of internal criteria of seriousness and *orthodoxy. Since no central *authority had then defined the Canon, each church had formed particular customs: *Syria, e.g., read the gospel in the *Diatessaron. But the force of fraternal *communion slowly tended towards uniformity, while qualified intellectuals (*Origen, *Eusebius, *Jerome . . .) guided bishops and people in their choice. While rejecting the suspect *apocrypha, for a long time many admitted a category of disputed books, edifying but not normative. These perplexities, which in the 2nd c. had still affected such important texts as *John's gospel or, at the other extreme, some apocryphal Gospels or Acts, were confined in the 4th c. to certain writings that come near the end of our NT (Hebrews, Catholic epistles, *Apocalypse), for which the question of apostolic origin remained particularly thorny. As for the OT, *Fathers who cast doubts on the authority of the deuterocanonical books were rare; and those who, like Jerome, claimed to align themselves with the *Jews and reject those books, found no popular echo. Indeed the Christian art of the catacombs concedes an extraordinary priority to the deuterocanonical legends: the young men in the *fiery furnace, *Daniel in the lions' den, or even a late, though *canonical, tale like that of *Jonah. But to make up for this, it can be shown that the Fathers' citations show a predilection for more fundamental pericopes.

Some authors, like *Basil, who use nearly all the writings of our Canon when necessary, normally limit their biblical sources to a much more restricted list. So the history of the Canon is much less uniform than is sometimes said. Explicit lists, on which we could safely build, are rare. One of the most important documents for defining the *tradition, and one that ended by prevailing, was the list authoritatively compiled by *Athanasius in 367 (Paschal letter 39). The bishop thought less of the universal church than of his own flock, the monks especially, who continued to draw edification from pious apocrypha. But this list spread automatically. In the West, it is a striking fact that it was an anon. author who took pains to denounce, in the so-called *Decretum Gelasianum, a long and composite list of apocrypha. In fact, the Greek and Latin copyists allowed these apocrypha to perish, or at any rate transmitted them badly: while the *Coptic church and those of *Syria, *Arabia and esp. *Ethiopia preserved them much better.

Since *Palaeoslavonic literature (post 10th-c.) is also well provided in this sector, we must hold that some Greek circles went on transmitting this material for some time. These writings are rarely heterodox, but rather of a *Judaeo-Christian or *encratite *piety, popular rather than clerical.

II. Versions and manuscripts. Access to the Semitic originals of the OT was the privilege of few scholars. The NT, written in Greek, usually uses the classical version of *Judaeo-Hellenism, known by agreement as the Septuagint (LXX), though a certain number of citations, alien to this version, may go back to some Aramaic *Targum or, often, to *Testimonia rewritten and adapted to their Christian function. From the 1st c., Judaism, closed in on itself, rejected the LXX and produced new Greek versions, some elements of which must have contaminated a good many of the manuscripts of the LXX. The spread of Christianity then led to a gradual labour of translations into *Latin, *Coptic, *Syriac; their originals are obscure, because their first steps were hesitant, partial, imperfect. Yet some archaic texts exist that reflect a social milieu of quite unconventional culture. They are often linked to the textual forms known as Western, since first met in Latin (*Cyprian), but which also circulated in *Syria, and which are ill-preserved in Greek.

The importance of "selected passages", *Testimonia, in primitive Christian literature suggests that the complete *books were not easy to acquire or perhaps even to use. But one purely technical fact, the replacement of the volumen by the codex, in which pages followed each other as in our own books, is connected with the spread of the Christian *Bible and suggests the existence of a vast market of readers, detached from the old cultural conventions and wishing to consult, and even to take apart, the sacred books.

Study of the variants demonstrates that the texts were continually rewritten in recensions that cared for grammar and even for Attic purity, introduced echoes of parallel passages or of more or less historical traditions (the Western text), or, on the contrary, were inspired by secular philology and systematically revised on the purest models (Alexandrian text). Then, the versions were so full of corrections that their sense could no longer be made out; some adhered strictly to the original, others had pretensions to greater correctness and linguistic limpidity. So the critical editions must master a vast number of variants, in which accidental corruptions count for very little; they rather result from an age-long labour of *piety and culture.

J. van Haelst's Catalogue des papyrus littéraires juifs et chrétiens (Paris 1976) lists 567 biblical *papyri (plus 50 fragments of *apocrypha), most of them recently discovered, which materially attest the forms under which Scripture circulated. Textual criticism classifies these ancient fragments and the surviving manuscripts (in all languages) and the patristic citations; it must distinguish the regulative action of the great centres of culture (above all *Alexandria, then *Antioch, *Rome, *Edessa, *Jerusalem and so many others) and identify the rare documents that have escaped this levelling.

Many peculiarities, which would be attributed to caprice if one considered them in isolation, take on a precise significance when the picture is reconstructed. This operation becomes even easier if we are dealing with biblical versions, when the variants in translation come to increase considerably the really different readings, and characterize a citation.

III. Inspiration. No theory of this was worked out, nor was it reflected on: the *Bible's authority was an undisputed fact. The *gnostics, who contested it, by doing so put themselves outside the Christian faith. The idea of an ancient and mysterious wisdom hidden in the sacred books was so widespread, in an era that A. J. Festugière has characterized as one of declining rationalism and a fashion for Oriental prophets (La Révélation d'Hermès Trismégiste, I, Paris 1944, 1-44), that respect for the Bible easily took on excessive forms: a magic wand could draw from it all the *gnoses one wanted. Since the cultural milieu was wholly different from that in which the OT and even the NT had been written, the reader ran into many problems, all the more so if he had had an education based on classical humanism: the anthropomorphisms, the Jewish ritual, the moral examples, the historical references, the vocabulary and the syntax. The work of *Philo and his refined allegorism provided a precious model. In the mean time Jesus and *Paul had cast doubt on the value of the Law and the break with Judaism had occurred. Quite soon a *Barnabas (cf. letter to the Hebrews) would be led to deny the literal sense of the old precepts, in favour of a radically spiritual *exegesis. Even before *Irenaeus, an intelligent gnostic like *Ptolemy (Lettre à Flora, ed. G. Quispel, SCh 24a, Paris 1966) managed to propose distinctions that were often pertinent. The simple faith of the people, the weight of lived *tradition, the healthy framework of Christian *life and of the analogy of *faith, the responsibility of the bishops and the *charisms of the saints ensured the church a royal road, along which *Origen, *Jerome, *John Chrysostom, *Augustine and so many others walked and progressed, with the help of a sound literary education, the dialogue with Rabbinic science and the stimulus of preaching.

All the sacred authors of the OT and NT were tendentially assimilated to the *prophets; indeed for a long time the main characteristic of the *Holy Spirit was to have spoken through their mouth, and to Him was attributed the fullness of truth acknowledged to the text. In consequence, it was not easily believed that a single explanation could suffice to exhaust the divine sense. Little attention was paid to distinguishing between what the human author directly intended in the setting of his own time, and the interpretations of his intuition that could be given within other cultures, just as no distinction was made between the biblical statement of a fact of salvation-history and the innumerable metaphysical analyses that *theology laboured to give it. Origen (and others after him) loved to emphasize the absurdity of a literal explanation, based on Semitic ways of speaking, and deduced from this the need for a more subtle discussion.

If declarations of this sort may surprise a modern reader, they did not prevent the ancient church from taking into account the literary and historical context, familiarizing itself sufficiently with biblical language to the point of becoming aware of many philological phenomena, and understanding, in its intimate essence, the moral or mystical terminology of the *prophets and *apostles. The literary *schools of antiquity were of a level hard to equal. If, because of their exegetical training, the Fathers depended on one another (*Basil and *Gregory, e.g., discerningly brought together in the *Philocalia *Origen's best pages on the interpretation of the Bible), we would do them wrong if we tried to explain them by mere scholastic routines, since each one of them had a secular literary training of his own. When we observe a reaction by them against the abuses of allegorism, the cause often resides in hostility to an ambiguous *gnosis, which presumed to derive hazardous philosophies from the text.

So strong was the conviction that wisdom could spring from no source but biblical *revelation, that the majority of Platonizing theologians held that the masters of Greek religious philosophy had drawn on *Moses. Polemic against a *Celsus, a *Porphyry, a *Julian, defenders of *pagan culture, took this direction. On the contrary, the question lost its importance for those who used *philosophy, subordinating it entirely to *faith.

IV. Function of the *Bible in the church. The *liturgy of the Word, associated with a *psalmody of praise or with the *Eucharist, was a community activity whose absence from any church would be hard to conceive. Its primary purpose was to manifest Christ's presence in the believers. It spontaneously attracted a homiletic element, which expounded the literal sense and above all showed the text's relation to the gospel. The *prophets, in the style of *Corinth, also had to give salvation-history a lyrical tone, in the same way in which an *Ephrem would use Scripture; the rhythmical prose of a *Melito and so many others gives us an idea of this for the Greek world, in which the heavy encumbrance of classical versification long paralysed poetic inspiration (*Gregory Nazianzen, *Apollinaris . . .). *Apologetic (against *Jews, against Greeks) and the struggle against *heresy usually fell back on discussions of Scripture.

*Hippolytus was probably the first to attempt systematic *commentaries on some pericopes and on the entire book of *Daniel. *Origen, who in a highly cultivated milieu had been able to undertake a scientific study of the text (cf. *Hexapla), wrote cycles of *homilies on the entire content of one book, as well as prolix, systematic, written commentaries. He found models and stimuli in some gnostic works, in particular on the beginning of *Genesis.

If the greatest *Fathers were, evidently, familiar with the Bible and, apparently, knew it by *memory, we would also like to know how far private reading was practised. The resistance that a new version (*Vulgate*) could encounter suggests that numerous formulae had penetrated the popular mind. We also see how Bibles were hunted down during *Diocletian's *persecution and how the *martyrs, if their manuscript was taken away, gloried in preserving God's Word in their own heart.

Several thousand titles are carefully pointed out in the pages dedicated to *Bibel* (G. Wanke; E. Plümacher; W. Schneemelcher; H. Karpp; K. Aland; S.P. Brock; etc.) in *TRE* 6, 1-377. A. von Campenhausen, *Die Entstehung der christl. Bibel*, Göttingen 1968; P.F. Ackroyd-C.F. Evans, *The Cambridge History of the Bible*, I, Cambridge 1970. The *Biblia Patristica*, Paris 1975, gathers and puts in order all the patristic references to the Bible; the three vols. so far published, relating to the first three centuries, allow enormous progress in the study of the Fathers' treatment of Scripture; Y. Janssens, L'Evangile des Egyptians, *Muséon* 100 (1987); J.F. Kelly, *Why is there a New Testament?*, Wilmington 1986; A. Pietersma-C. Cox (eds.) *The Septuaginta* (Essays in Honour of J.W. Wevers), Toronto 1984; W.H.C. Frend, The Old Testament in the age of the Greek Apologists, *Scottish Journal of Theology* 26 (1973) 129-150.

J. Gribomont

SCULPTURE. Given the premise that the artistic product is, as we know, an expression on the figurative plane of the culture to which it belongs and that, consequently, we can speak of "Christian" s. only in so far as we refer to subject-matter and ideological and programmatic content, and not to the stylistic and formal characteristics which, every time, it necessarily shares with all contemporary work, it is worth pointing out how the cultural heritage of Judaism, the ever-present danger of idolatrous suggestions and practices, the repeatedly manifested aversion of the Christians to *idols (even precisely as material products), the non-existence of iconic statues in *church buildings, were, with others, all reasons that long impeded and retarded, in various ways, the formation and spread of a Christian plastic art in the round.

With all this, we naturally do not mean to claim that a s. in the round was completely absent in the figurative expressions of early Christianity: indeed, the existence of this artistic genre is documented, e.g., by some statuettes of the Good *Shepherd, referrable to the 3rd c., perserved at *Rome and elsewhere (outstanding example in Museo Pio Cristiano in Vaticano: cf. Wilpert, *Sarcofagi*, I, 71), or by a well-known statuette of Christ seated, of the young, beardless, so-called "Hellenistic" type, dateable to the 4th c. and preserved in the Museo Nazionale Romano; it is also attested by various passages drawn from literary sources - thus, e.g., *Eusebius of Caesarea mentions bronze *images of the Good Shepherd and *Daniel placed as ornaments on fountains erected in piazzas (*Vita Const.* III, 49), and the *Liber Pontificalis*, writing about pope *Sylvester, explicitly mentions gold and silver statues depicting Christ, *John the Baptist, deer and *lambs, which decorated the Lateran baptistery (*LP* I, 174) - but it is still beyond doubt that early Christian plastic art was expressed mainly in the field of the funerary relief (see *Sarcophagi).

Outside the funerary sphere, and still in the field of relief, s. with a Christian character was exercised prevalently in the production of *liturgical furniture such as *mensae, ambos, ciboria, *cathedrae, plutaea, presbyterial cancelli, etc.

From about the 6th c. - partly on account of suggestions exercised by the Byzantine artistic world - the almost exclusively geometrical motifs that characterized such work in Western artistic culture were enriched by flowers, *crosses, ribbons and garlands, variously composed and interwoven, exemplified in the decoration of the church of S. Clemente in Rome, and whose ornamental repertoire - though obviously with different conceptions and stylistic renderings - remained substantially unvaried all through that century. Yet we should note that the motifs of this repertoire were not exclusive to Christian work (however quantitatively prevalent), but were general in all contemporary s.

With the attenuation of such suggestions as a result of important political events in the course of the 7th c., s. seems to re-elaborate, in a local sense, the characteristics propagated in the previous century and, at least in the Western regions, shows an undeniable falling off even in quantity of output (the *schola* of S. Maria Antiqua is a masonry enclosure decorated with *paintings), destined to pick up, still characterized almost exclusively by ornamental and geometrical motifs, but with more self-awareness, in the following century.

F. Mazzanti, *La scultura ornamentale romana nei bassi tempi*, Rome 1896; G. Wilpert, *I sarcofagi cristiani antichi*, I, Rome 1929, 71 ff.; AA. VV., *Corpus della scultura paleocristiana, bizantina ed altomedioevale di Ravenna*, diretto da G. Bovini, vol. I and II, Rome 1968, vol. III, Rome 1969 (with bibl.); *Corpus della scultura altomedievale pubblicato a cura del Centro Italiano di Studi sull'alto Medioevo*, I, Spoleto 1959, in course of publication (with ample bibl.). Further bibliography under individual articles; N. Duval, Plastique chrétienne de Tunisie et d'Algérie, *BCTH* n.s. 8 1972 (publ. 1975) 55-146; R. Farioli, Osservaziani sulla scultura del V-VI secolo: problemi revennati, *Passagio dal mondo antico al Medioevo da Teodosio a San Gregorio Magno*, Atti dei Convegni Lincei 45, Rome 1980, 147-194; E. Dassmann, Die Szene Christus - Petrus mit dem Halen Zum Verhältnis von Komposition und Interpretation auf frühchristlichen Sarkophagen, in (eds.) E. Dassmann et al., *Pietas* (Festsch. B. Kötting), JbAC Ergbd. 8, Münster 1980, 509-527 (include also pl. under "Sarcophagi"); W. Wischmeyer, Die vorkonstantinische christliche kunst in neuem Licht: Die Cleveland-Statuetten, *VChr* 35 (1981) 253-287; C. Struibe, Die kapitelle von Qasr ibn Warden. Antiochia und Konstantinopel in 6 Jahrhundert, *JbAC* 26 (1983) 59-106.

R. Giordani

SCYTHIA MINOR. This name, given from Hellenistic times to the territory between the Danube and the Black Sea, corresponds almost wholly to the modern Romanian province of the Dobrudja. Strabo called this territory Μικρὰ Σκυθία (*Geogr.* VII, 4, 5) to distinguish it from Scythia Major, N of the Black Sea. Previously part of *Moesia Inferior, under *Diocletian it became an independent province called S. M. Tradition makes the apostle *Andrew the evangelizer of Scythia (Euseb., *HE* III, 1, 1). When we consider the intensive contacts of the Greek colonies on the W shore of the Euxine Sea with the metropolises of *Asia Minor and the great centres of the Aegean Isles and continental Greece, as well as the large number of soldiers from the already evangelized regions of the Near East camped on the Lower Danube border, we may admit the presence of some isolated Christians in S. M. during the first three Christian centuries. Diocletian's *persecution must have claimed many victims in S. M., esp. among soldiers. The *martyrologies record many *martyrs in centres such as *Tomi, the provincial capital, Halmyris, *Noviodunum, Dinogetia and *Axiopolis. Other martyrs were attributed to *Licinius's persecution. At Halmyris (*Salmorus*) in the Danube delta, on the right arm of the river, are recorded perhaps the first martyrs so far known in the Dobrudja: Epictetus and Astio, both natives of *Phrygia who fled here in c.290. Also mentioned in this period is Evangelicus, who seems to have been the first bishop of Tomi. A fragmentary epigraph found at Tomi (Constanta) mentions a bishop and martyr not yet identified, prob. from the time of Licinius.

Until the time of the emp. *Anastasius I (491-518), the church of S. M. was led by a single bishop, residing at Tomi: this is mentioned by the historians *Sozomen (*HE* VI, 21) and *Theodoret of Cyrrhus (*HE* IV, 35), by a law (c.480) of the emp. *Zeno (*CJ* I, 3, 35) and by other literary sources. The first bishop reliably attested by historical documents is St Betranius (Bretanius, Vetranius), followed by eight known bishops: Gerontios or more probably Terentius (381), Theotimos I (c.392-407), Timotheos (431), Ioannes (445-446), Alexandros (449), Theotimos II (458), Paternus (c.491-520) and Valentinianus (c.550).

During the final flowering of S. M., between the reigns of Anastasius and *Justinian I (527-565), more bishoprics were created in other centres, while that of Tomi became an *archbishopric headed by a metropolitan. The "Scythian monks", among whom *John Maxentius (c.520) stands out, played an important role in the religious, cultural and political life of S. M. Two other monks from S. M., though not active there, increased its fame: John *Cassian (c.360-430/5), born in a Dobrudjan village; and *Dionysius Exiguus, to whom we owe the *Latin translation of much fundamental Greek patristic *theology, as well as the *computus of the vulgar era from Christ's birth (the Dionysian era), a computus that is still in use, though incorrect.

Literary information on the church of S. M. before the collapse of the Lower Danube border (early 7th c.) is supplemented by the archaeological discoveries made on this late Roman territory, above all by *epigraphy, Christian *church buildings and their annexes.

About a hundred early Christian inscriptions (4th-6th cc.) have been found in the modern Dobrudja, three quarters in Greek and one quarter in Latin: one epigraph is bilingual (Graeco-Latin). Nearly half come from Tomi; the others from Callatis (10), *Histria (10), Salsovia (1), *Noviodunum and surroundings (4), Dinogetia (10), *Axiopolis (5), Sucidava (Izvoarele, district of Constanta) (1), Ulmetum (5), *Tropaeum Traiani (1). More than half of them are incised on local stone or good-quality marble imported from Proconnesus or various parts of Greece; the others are on metal (silver, bronze, lead), terracotta, bone, glass or other materials. The majority are funerary epigraphs; the most frequent object is the *cross or one of the christograms. On some also appear symbolic figures: e.g. fish, dove, palm-leaf, vine, crown, etc.

Some thirty *church buildings of the 4th-6th cc. have been found in the Dobrudja, all longitudinal-plan; often only the foundations or lower courses of the walls survive. They are distributed as follows: *Tomi, 6 (2 unpublished); Callatis, 1; *Histria, 4; Arganum, 2; *Noviodunum, 2 (including one cemeterial, not excavated); Niculitel, 1; Dinogetia, 1; Troesmis, 3; Beroe, 1, cemeterial (unpublished); Capidava, 1 (unpublished); *Axiopolis, 2 (one cemeterial); Sucidava (Izvoarele, district of Constanta), 1; *Tropaeum Traiani, 5 (one cemeterial); Ibida, 1 (partly excavated). Most have three aisles, with apse oriented to E. Only that of Tropaeum Traiani has a transept or T-pattern. The basilica of Callatis and the minor one of Troesmis seem to be inspired by the Syrian type; five of the basilicas of Tomi, three of Tropaeum Traiani, one of Histria and that of Niculitel had crypts for relics under the altar. The only baptistery so far attested is that found S of the atrium of the marble basilica of Tropaeum Traiani. Numerous architectural *sculptures include c.100 capitals, stones, balustrades (cancelli), ciboria and pulpits, mostly in Proconnesian marble. Outstanding among a great variety of small objects is the disc of bishop Paternus of Tomi and a whole series of little *crosses, small bronze, lead, silver and gold objects, numerous terracotta vases (amphorae, plates, *lamps) with crosses or Christian epigraphs.

E. Popescu, *Inscriptiile grecesti si latine din secolele IV-XIII descoperite in România*, Bucharest 1976; I. Barnea, *Les monuments paléochrétiens de Roumanie*, Vatican City 1977; id., *Christian Art in Romania*, I, Bucharest 1979; I. G. Coman, *Scriitori bisericesti din epoca straromâna*, Bucharest 1979; A. Radulescu-V. Lungu, Le Christianisme en Scythie Mineure à la lumière des dernières découvertes archéologiques, *Actes XIe Cong. internat. d'arch. chrét.*, CEFR 123, Rome 1989, 2561-2615; E. Norocel, La Vie ecclésiastique à Tomis du IVe au VIe siècle, *Miscell. Hist. Eccles.* 7 (1985) 16-24; N. Vornicescu, Propagation et developpement du Christianisme, (Carpathes), *ibid.*, 25-38.

I. Barnea

SEBASTIAN. The Roman *Depositio martyrum* records St S. *ad catacumbas* on 20 Jan. *Ambrose adds that he was of Milanese origin (*In Ps.* 118: PL 15, 1574): he lived at the time of *Maximian and *Diocletian, and was martyred at *Rome. His Passion belongs to the epic and cyclical genre: around the main hero are grouped, as actors, numerous Roman *martyrs, historical and otherwise, whom he helps in prison and whom *hagiography makes his kin. His cult, which originally developed in the shadow of that of the *apostles in the basilica named after them on the Appian Way, ended by overtaking theirs. The saint's body was the object of numerous translations, undoubtedly partial, by St Médard to Soissons, by St *Gregory to the Vatican, to Fulda, Farfa and other places. The relics worked many miracles everywhere.

A. Dufourcq, *Etudes sur les Gesta martyrum romains*, I, Paris 1900, 301-2; *Vies des SS.* 1, 395-400; *LTK* 9, 557; *BS* 11, 776-801.

V. Saxer

SECRECY, SACRAMENTAL. The development of the sacrament of *penitence in the primitive church soon led to an understanding of the need for secrecy in the confession of secret *sins. *Origen (early 3rd c.) speaks of a confession of secret sins not automatically requiring subsequent public confession (*Hom. 2 in Psalmum* 37: PG 12, 1386). The early 4th-c. Syrian ascetic *Aphraates, in his 7th Treatise, forbids priests to make known the faults revealed to them by penitents (*Dem.* 7, 3: PS I, 318). *Paulinus (*Vita s. Ambrosii* 39: PL 14, 43) says that *Ambrose left his successors a good example by speaking only to *God about sins confessed to him. It is clear from Ambrose himself (*De poen.* 2, 10: CSEL 73, 199) that not all faults should be revealed publicly, which implies a confession that remained secret. *Augustine attests the custom of the early 5th-c. church in N *Africa (*Serm.* 82, 11: PL 38, 511), saying that if he does not publicly blame some sinners, this is because he desires to care for them rather than to accuse them. *Sozomen (early 5th c.) states in his *Historia Ecclesiae* (7, 16: PG 67, 1458-63) that the office of penitentiary priest at *Constantinople was established to allow sinners to make a secret rather than a public confession, and that the penitentiary was chosen for his reserve and discretion. *Leo the Great (440-461) had occasion to rebuke the bishops of Campania because, contrary to the rule received from apostolic times, they published the sins of their penitents (*Ep.* 168: PL 54, 1211). The legislation of the *council of Tovina (*Armenia, c.527) against priests who violated the sacramental seal (Hfl-Lecl 2, 1079, 20) seems to have been the earliest legislation on this, indicating that from the start of the 6th c. the fact of the sacramental seal had now acquired the force of law. The subsequent development of the church's practice of private confession confirmed what had already become law.

DTC 3, 960-74; *EC* 11, 256-7; E. Vacandard, *La confession sacramentelle dans l'église primitive*, Paris 1903; L. Honoré, *Le secret de la confession. Etude historique-canonique*, Brussels 1924; C.F. Savio, *Ad sigillum sacramentale animadversiones*, Turin 1936.

P. Fahey

SECUNDIANUS of Singidunum. *Arian priest, he became bishop of Singidunum (Belgrade) on the death of Ursacius (c.370/75). He was condemned and deposed, with *Palladius of Ratiaria, at the council of *Aquileia (381).

M. Meslin, *Les Ariens d'Occident*, Paris 1967, 441.

M. Simonetti

SECUNDINUS. Manichaean auditor of Rome, late 4th c. *Augustine cites a *letter sent by him: *Secundini manichaei ad sanctum Augustinum epistola*, which he refuted in *Contra Secundinum manichaeum*, with which he claimed (*Retractiones*, 2, 10) to be particularly satisfied: "*Mea sententia omnibus quae adversus illam pestem scribere potui, facile praepono*". S. had written *velut amicus, honorifice obiurgans*, accusing him of understanding nothing of the struggle waged by the Majesty of the *Father, his Firstborn and the *Holy Spirit *adversus principes et potestates*, and so of never having been a true *Manichee. In his reply, Augustine constructs a sort of anti-Manichaean *summa* in 26 chapters, particularly profound in some parts, which reveal how truly meagre his Roman contradictor's talent was.

CPL 324, 325; PL 42, 571-602; CSEL 25, 2, 893-947; P. Alfaric, *L'évolution intellectuelle de saint Augustin*, I, Paris 1918, 88-89, 215-216; R. Jolivet-M. Jourjon, Six traités anti-manichéens, *BA* 17 (1961) 509-633; R. Décret, *L'Afrique manichéenne*, 2 vols., Paris 1978 (esp. II, 99-108).

C. Riggi

SECUNDINUS. Poet (†447). Bishop in *Ireland; named Sechnall. Perhaps one of the missionaries who came (from *Gaul?) to Ireland in the 5th c.: tradition makes him the son of the "Longobard" Restitutus and of a sister of St *Patrick. Author of an *alphabetical hymn in honour of St Patrick (incl. *Audite omnes*) preserved in the Bangor Antiphonary (ff. 13^v-15^v) under the title *Ymnum Sancti Patrici*: the first known *hymn composed in Ireland, and the beginning of the legend of Patrick. Despite its use of the present, it is thought to have been written after Patrick's death. In the Middle Ages S. was locally commemorated as a saint, 27 Nov.

J.F. Kenney, *The Sources for the Early History of Ireland*, New York 1929, I, no. 87, pp. 258-260; CPL 1101; PL 53, 837-840; 72, 590-592; W. Stokes, *The Tripartite Life of St. Patrick*, London 1887, II, 382-384; F.E. Warren, *The Antiphonary of Bangor*, London 1893-1895, II, 14-16; C. Blume, *Analecta hymnica Medii Aevi* 51, Leipzig 1908, 340-346; L. Bieler, The Hymn of St. Secundinus, *PRIA* 55 (1953) 117-127; *BS* 11, 805-807; L. Bieler, St. Secundinus and Armagh, *Seanchas Ardmhacha* 2 (1956) 21-27.

E. Malaspina

SECUNDUS of Tigisis. Early 4th-c. bishop of Tigisis (now Ain el Bordj, Algeria) and primate of *Numidia. After the *persecution of 303, *Mensurius of Carthage wrote a *letter asking S. to assure him that he had not been a *traditor* of the *Scriptures (Aug., *Brev. collat.* III, 13, 25: CSEL 53, 75). S. replied describing what had happened in Numidia. Thus there was still *communion between the two primates at this time. At the so-called council of *Cirta of 305 (Lancel in a recent article thinks 307) - really a meeting of 12 bishops to ordain *Silvanus there - Purpurius of Limata accused S., who presided, of having been a *traditor* (Opt. of Mil., *De schism.* I, 14: CSEL 26, 16; Aug., *Brev. collat.* III, 15, 27).

On the election of *Caecilian as bishop of *Carthage, a meeting was held there (311/2) of c.70 Numidian bishops presided over by S. Led by the primate, they supported the opponents of Caecilian, who was accused of having been ordained by the *traditor* *Felix of Apthugni, and ordained *Maiorinus in his place. Thus S., as *Augustine notes several times, had not only pardoned convicted *traditores* in the meeting at Cirta and failed to prove his own innocence before the other bishops, but had condemned an absent man (Caecilian) on an unproven charge and elected another bishop in his place, thus becoming responsible for the formal act of crystallizing the Carthaginian dispute into a Donatist ecclesial structure (Aug., *Contra Gaud.* I, 37, 47; *Contra ep. Parm.* I, 3, 5; *Ep.* 43, 2-9; Ep. 53, 2, 4). The *Donatists themselves considered S.'s consecration of Maiorinus as the birth of the Donatist *schism (cf. *Gesta apud Zen.*: CSEL 26, 185 and 189).

Monceaux 4, 7-10; *BA* 32, 738 f.; *PAC* 1, 1052-1054; S. Lancel, Les débuts du Donatisme: la date du "Protocole de Cirta" et l'élection épiscopale de Silvanus, *REAug* 24 (1979) 217-229; W.H.C. Frend, *The Donatist Church*, Oxford ³1985, ch. 1; B. Kriegbaum, *Kirche der Traditoren oder Kirche der Martyrer?*, Innsbrucker theol.-Studien 13, Tyrolia-Innsbruck-Vienna 1986.

A. Di Berardino

SEDATUS of Nîmes. Not to be confused (as in PL 72, 1849 and 1878) with the homonymous bishop of Béziers, who lived a century earlier. Bishop of *Nîmes (*Nemausus*) c.500, he had friendly relations with *Ruricius. Critics have restored to him three *letters and some *sermons. *Caesarius seems to have borrowed from the latter.

Editions: *Epistolae (Inter Faust. ep.)* 15, 16, 19; PL 58, 865; another letter (unpublished) is in codex A 24 Sup.Bibl.Ambr.: cf. G. Morin, Sermo de natale Domini, *RBen* 35 (1923) 12-14 and PLS 4, 1927-29. Reused by Caesarius (*Serm.* 190). The same sermon is found in codex *Mellicensis* 218 (E.8), where only the exordium is modified. G. Morin would see some of S.'s texts in Caesarius (*Serm.* 56, 4; 57, 4; 193; 194, pp. 445-446). Two sermons are doubtful: PL 39, 1977-1981 and CSEL 21, 247-252.
Studies: *RBen* 35 (1923) 5-16; CCL 103 and 104 *ad. loc.*

A. Hamman

SEDULIUS. 5th-c. Christian poet; usually given the *praenomen* Coelius, attested by late MSS. "On the life of Sedulius, on his complete name, on his condition and country, on the time and place of composition of his works, we possess only few and uncertain facts" (Mayr). Acc. to notations in codices, he flourished under *Theodosius II and *Valentinian II, between 425 and 450. The so-called *Decretum Gelasianum* (494) expresses a positive judgment on him and his works, and calls him *venerabilis*. Acc. to one tradition, he learnt philosophy in *Italy, was converted to *Catholicism, was baptized by the presbyter Macedonius, and wrote to edify. The facts of his "*conversion", which he mentions in a letter to Macedonius, show a singular affinity with those of *Prudentius (*Praef*): after secular studies, useless to his *soul, he turned to *God, to whom he dedicated his literary activity. We think he was a presbyter (citing vv. I, 23-25 of his poem, which make him assiduous in *psalmody in the choir: but the passage could suggest the cloister). He was sometimes confused with the 9th-c. poet Sedulius Scotus.

The Viennese edition contains the *Paschale Carmen*, the *Paschale Opus*, two *hymns, two *letters to Macedonius, and the *Carmina ad Sedulium pertinentia*. His main work is the poem in five books, *Paschale Carmen*, a title derived from 1 Cor 5, 5: *pascha nostrum immolatus est Christus*. Its subject is the *miracula Christi salutiferi*. Book I has 368 hexameters; II, 300; III, 339; IV, 308; and V, 438. Book I illustrates OT miracles; books II-V take their material from the four gospels, esp. *Matthew's. S. continues with more freedom the tradition begun by *Juvencus in the 4th c. The recent hypothesis that he used a harmonized gospel is sound. He recognizes the psychagogical value of poetry: he uses verse to get closer to the reader's soul. The poem, published only at the end of the 5th c. by Turcius Rufus *Asterius, consul in 494, is rich in moralizing observations: S. must have drawn on biblical exegetical literature (*commentaries, summaries) widespread in his time. He is sensitive to the influence of classical poetry: Virgil holds first place among his models. Then come others, like Lucan and *Claudian, and the Christians *Prudentius, *Paulinus of Nola and *Juvencus. Though the gospel account, rewritten and amplified, does not keep its original simplicity, the poem has a certain charm. In S. we can see the transition from classical poetry to a new art, which associates classical tradition, *Bible and *patristics, and announces the Middle Ages.

Book I is considered by all as an introduction: an accessory, artificial element, alien to the economy of the poem. But it ought to be considered essential: it echoes the polemic against the deniers of the OT (the *gnostics, *Marcion). The two Testaments have a single author, and are unthinkable apart from each other. Christ is already present in the Old Law, and governs the two Testaments (I, 145-146). In the OT the *Father works with the Son; in the NT the Son acts with the Father (I, 291-296). The very title of the poem refers to Ex 12, 1-19. The poem was widely read in the Middle Ages. S.

himself turned it into prose in the *Paschale Opus*. Verbose in style, this is useful for understanding the poem and the constitution of the poetic text. One of the two hymns, *A solis ortus cardine*, was used in the *liturgy.

CPL 1447-1454; PL 19, 433-752; Hymns: PL 19, 753-770; CSEL 10; It. tr.: F. Corsaro, *Sedulio poeta*, Catania 1956 (poem and hymns). Studies: Schanz IV/2, 368-374; Bardenhewer 4, 642-647; *Patrologia* III, 304-308 (bibl.); Th. Mayr, *Studien zu den Paschale Carmen des christlichen Dichters Sedulius*, Augsburg 1916; C. Weyman, *Beiträge zur Gesch. der christlich-lateinischen Poesie*, Munich 1926, 121-137; G. Moretti Pieri, Sulle fonti evangeliche di Sedulio, *AATC* 34 n.s. 20 (1969) 125-243; C. Tibiletti, Il Paschale Carmen di Sedulio, introductory notes, to be published in *ANRW*; A. Solignac, Sedulius, *DSp* 14, 510-514; C. Springer, *The gospel as epic in late antiquity: the Paschale carmen of Sedulius*, VChr Suppl. 2, Leiden 1988; C.P.E. Springer, Sedulius' *A solis ortu cardine*: the hymn and its tradition, *EphemLiturg* 101 (1987) 69-75.

C. Tibiletti

SELEUCIA-CTESIPHON

I. The city - II. Councils.

I. The city. S.-C. are two ancient twin cities on the Tigris, *c*.60 km N of Babylon. The former, S. (now Tel Umar), was founded by Seleucus I Nicator in *c*.312 BC on the right bank of the river with a mixed Greek-Babylonian population. It was destroyed by the Roman general Cassius in AD 164. The latter, C. (now Salman Pak), was founded by the Parthians on the opposite bank of the river, becoming the winter residence of the Arsacids and the capital of the *Sassanids until the Arab conquest of 636. The earliest Christian evidence comes from the hill of Koha, between the two cities, which was subsequently incorporated in the Sassanid Ardashir's new capital in *c*.230 (the Mahoza of Jewish and Christian sources, Veh Ardashir for the Iranians). The name of the twin cities is used to designate the centre of the *orthodox (later *Nestorian) Persian church, prob. from the 3rd c. and until the Abassids (750-1258) moved the centre to Baghdad. From the time of Mar *Isaac's synod in 410, the bishop residing at S.-C. received the title *catholicós and, later (between 450 and 550), that of catholicós-patriarch. The *acts of the synods celebrated there, collected prob. under Timothy I in *c*.800, allow us to reconstruct for nearly four centuries the expansion of the church in *Persia (Nestorian after the synod of 486), its *chronology, prosopography and the evolution of its doctrine and discipline. This important source was published by J.-B. Chabot, with a French tr., under the title *Synodicon orientale* (Paris 1902).

PWK 21/1, 1148-1184; *PWK* Suppl. 4, 1102-1118; *KLP* 2, 369; 5, 83-85; Ortiz de Urbina 121-122; *LTK* 9, 651-652; *CE* 13, 54; M. Streck, *Seleucia und Ktesiphon*, Leipzig 1917; W. de Vries, Antiochien und Seleucia-Ctesiphon, Patriarch und Katholikos?, in *Mélanges Tisserant*, III, ST 233, Rome 1964, 429-450; J.-M. Fiey, *Jalons pour une histoire de l'Eglise en Iraq*, CSCO 310, subs. 36, Louvain 1970, 40-44.

F. Rilliet

II. *Councils. S.-C., at least from the start of the 4th c., was the see of the *catholicós or archbishop of the "*East", i.e. the highest ecclesiastical *authority in the Persian empire. So it was the theatre of numerous synods (at a certain point expressly provided for at biennial intervals) of which the best-known are: that of 399-400, presided over by *Maruta of Maiferqat, which ratified the end of the imperial persecutions of the second half of the 4th c. and the nomination of *Isaac as catholicós; that of 410, presided over by Maruta and Isaac, which reorganized the ecclesiastical structure and legislative regulations (introducing the so-called Canons of Nicaea) and proclaimed (or rather confirmed) the autonomy of the Persian church from the "Western" church (prob. the patriarchate of *Antioch); this autonomy, already very great, became total independence with the synod of 422. That of 486, at the time of *Barsauma, attests the Persian church's transition to *Nestorianism and legitimated the marriage of bishops.

Palazzini 5, 41-45; J.B. Chabot, *Synodicon Orientale*, Paris 1902; J. Labourt, *Le christianisme dans l'empire perse sous la dynastie sassanide (224-632)*, Paris 1904; J.-M. Fiey, Les étapes de la prise de conscience de son identité patriarcale par l'église syrienne orientale, *OrSyr* 12 (1967) 3-22; J. Gribomont, Le symbole de foi de Séleucie-Ctésiphon (410), in *A Tribute to A. Vööbus*, Chicago 1977, 283-294; A. Vööbus, *The Canons Ascribed to Maruta Maipherqat and related sources*, CSCO 439-440/syr. 191-192, Louvain 1982; W.M. Macomber, The Authority of the Catholic Patriarch of Seleucia-Ctesiphon, in *I patriarchi orientali nel primo millenio*, Rome 1968, 179-200.

S.J. Voicu

SELEUCIA in Isauria, *Council of. Called by the emp. *Constantius, in parallel with the Western council of *Rimini, to bring *peace to a church profoundly unsettled by the events occurring in 357-8 over the *Arian controversy, the council of S. in Isauria (*Asia Minor) opened on 27 Sept 359 in the presence of c.100 bishops and lasted a few days. The anti-Arian majority, represented mainly by *homoeousians, headed by *Silvanus of Tarsus, *Eleusius of Cyzicus and *Eustathius of Sebaste, imposed the *creed of *Antioch of 341, rejecting after bitter arguments a compromise formula put forward by *Acacius of Caesarea in the name of a minority of *homoeans and pro-Arians: this formula rejected both the anti-Arian *homoousios and *homoiousios and the Arian *anomoion*, and proposed to affirm Christ only as "like the *Father". This formula being rejected, its supporters deserted the labours of the council, which *excommunicated and deposed them.

The council's commission communicated its results to Constantius at *Constantinople in October, but found the emperor, who favoured Acacius's formula, very hostile. Long discussions followed, during which the news arrived that the Westerners, meeting at Rimini, had bound themselves to the formula of faith supported by the pro-Arians and the emperor, which was similar to that of Acacius. Constantius then imposed subscription to this formula on the representatives of the council of S. who were with him, and obtained it after long insistence. The last subscriptions were given on the night of 31 Dec.

Hfl-Lecl 1, 929-955; Simonetti 314. 326-338; R.P.C. Hanson, *The Search for the Christian Doctrine of God*, Edinburgh 1988, 371-380.

M. Simonetti

SEMIARIANS. Thus *Epiphanius (*Pan.* 73) calls the *homoeousians: ἡμιάρειοι. *Filaster (*Haer.* 67) and *Augustine (*Haer.* 51) render it *semiarriani*. The term is tendentious and is derived from a too simplistic identification of anti-*Arian *orthodoxy with the Nicene *theology based on the *homoousios*, failing to appreciate the fundamental orthodoxy of the anti-Arian supporters of the *homoiousios*. If anything, the name s. could rightly characterize those homoeousians also called *Macedonians, who detached themselves from the group after 360, refusing to recognize the divinity of the *Holy Spirit or to accept the *homoousios*, which other homoeousians had accepted between 363 and 366.

R.P.C. Hanson, *The Search for the Christian Doctrine of God*, Edinburgh 1988, ch. 12.

M. Simonetti

SEMIPELAGIANS. Exponents of a theological movement that flourished in 5th-c. *Gaul. We incline to call them Provençal or Marseillais Teachers: the term "semi-Pelagian", which came into use at the start of the 17th c., implies collusion with *Pelagianism. Their main centres were the isle of *Lérins opposite Cannes, and *Marseilles. John *Cassian, who knew Eastern *theology and had acquired a wealth of *ascetic experience in the *deserts of *Egypt, founded two *monasteries at Marseilles (*Massilia*); he died c.435. At the start of the century, *Honoratus established a monastery at Lérins, which produced famous monks like *Vincent and bishops like *Hilary, *Lupus, *Faustus, *Eucherius and his sons *Veranus and *Salonius. The doctrine's main sources were: Cassian, *Conl.* XIII, *Prosper of Aquitaine's and Hilary's two *letters to *Augustine, and the treatise *De gratia* by *Faustus, who was bishop of *Riez and died at the end of the century.

The s. are no less opposed to *Pelagius than Augustine was. They admit original *sin, the necessity of *baptism and *grace for salvation. They part company with *Augustinianism on the question of the distribution of grace and the beginning of the saving act. For Augustine, from the *massa perditionis* of humanity *God chooses some, *ante constitutionem mundi*, by a free and merciful decree, without any merit of their own, to be elected and infallibly predestined to eternal *life. The others, objects of just judgment and not of mercy, are left (*relinquuntur*) in the state of damnation (*Persev.* 14, 35). Thus the possibility of saving himself is given to no-one. The s. teach that Christ died for all, and that all are called to salvation: there is no preclusion on God's part. Acc. to Faustus's theological *anthropology (*De gratia* II, 12), God created man and gave him not just free will, but a *bonum naturae*, an inclination towards good, a capacity to discern good from *evil, *piety towards God, abundance of grace, wisdom and prudence. Evil is extraneous to man's nature. The natural goodness that makes man God's *image is put forward in *Stoic guise, *bonae semina voluntatis* (*ibid.*, I, 12), derived from Cassian (*Conl.* XIII).

For Augustine, these gifts, not being natural, are totally lost through original sin (*Praed. sanct.* 5, 10). All that has to do with eternal life is derived not from nature, which is vitiated, but from grace, which restores nature (Prosp., *C. Coll.* 13, 5).

But for the s., who held them to be natural since given with human nature itself, as God conceived it and actualized it in *creation, these gifts, though gravely altered by the original sin, are not destroyed: what is part of nature can be diminished, but cannot cease completely. So there is present in each man a principle of good, operating on the level of salvation. With the gifts of nature, man can enter the *vestibula salutis*; with the subsequent grace of Christ, he is introduced *ad ipsa vitae penetralia* (Faust., *De gratia*, II, 9). By virtue of this initial *bona voluntas*, man has an incipient *faith (*ibid.*, II, 8) and can turn to God in *prayer (*ibid.*, II, 12); while for Augustine, both *bona voluntas* (*De civ. Dei* XIV, 11) and faith, in its beginnings and in its development (*Praed. sanct.* 11, 22), and the capacity to pray (*Persev.* 23, 64) are the work of God. Salvation, even in its beginnings, is for Augustine a work of *gratia Dei per Iesum Christum*; for the s., salvation can be initiated by *prima gratia* or gifts of creation, and must be brought to completion by *gratia Christi*: *dona praestita per naturam, sed confirmanda per gratiam* (Faust., *De gratia* II, 12). God the creator is no different from God the redeemer; the author of nature is the author of grace (*ibid.*, II, 10). This disposes of the objection that salvation is, in its beginnings, a work of man: man acts with the gifts received from God in creation. The gifts of *redemption are immensely superior to those of creation (*ibid.*, II, 10). For Faustus, man can never attribute to himself even the beginning of salvation (*ibid.*, II, 10): it must be referred entirely to God; *gratia Christi* is always a free gift and never a reward for merit. Man's initiative, made possible by the creator's gifts, far from constituting merit, is only the condition for *gratia Christi* to act (Faustus, *Ep.* 1). Both Cassian and Faustus recognize that often, by his grace, God anticipates human initiative. The s. thus claim to safeguard the fundamental characteristics of grace: necessity, gratuity, prevenience.

Faustus's treatise reflects the teaching of the councils of *Arles (473) and *Lyons (slightly later). On the other hand, the 2nd council of *Orange (529), presided over by *Caesarius of Arles, declared that to work, *ut expedit*, any *bonum* regarding eternal life is not possible *per naturae vigorem*, but only through the *illumination and inspiration of the *Holy Spirit (can. 7, DS 180). *Fides* is always a gift of God and is not derived from nature (DS 200). At the same time it condemned the doctrine of *predestination to evil.

Faustus of Riez and the s. of the 5th c. cannot formally be called *heretics: only in 529 was their doctrine condemned. The council of Orange, not accepting the entire Augustinian doctrine of grace, took no decisions on the main point of controversy and on the essential difference between Augustinianism and semi-Pelagianism (Koch, *Der heilige Faustus*, 194, 204).

Sources: J. Cassian: CPL 512-513; *Conlatio* 13: PL 49, 897-954; CSEL 13; SCh 54, 147-181; Faustus of Riez: CPL 961; PL 58, 783-836; CSEL 21; Prosper: CPL 523; letter to Augustine: PL 44, 947-954; BA 24, 392-412; Hilary, letter to Augustine: PL 44, 954-960; BA 24, 414-434.
Studies: A. Koch, *Der heilige Faustus. Bischof von Riez. Eine dogmengeschichtliche Monographie*, Stuttgart 1895; M. Simonetti, Il De gratia di Fausto di Riez, *SSR* I (1977) 125-144; C. Tibiletti, Giovanni Cassiano. Formazione e dottrina, *Augustinianum* 17 (1977) 355-380; S. Pricoco, *L'isola dei Santi. Il cenobio di Lerino e le origini del monachesimo gallico*, Rome 1978; C. Tibiletti, Libero arbitrio e grazia in Fausto di Riez, *Augustinianum* 19 (1979) 259-285; id., La salvezza umana in Fausto di Riez, *Orpheus* n.s. 1 (1980) 371-390; also bibl. in BA 24 and in *Patrologia* III, 486-496; B.R. Rees, *Pelagius, a Reluctant Heretic*, Woodbridge 1988, esp. 103 ff; A. Solignac, Semipelagiéns, *DSp* 14, 556-568.

C. Tibiletti

SENATOR of Milan. As a priest, he was one of the envoys sent by *Leo I to *Constantinople in summer 450 with three *letters condemning the doctrine of *Eutyches (*Ep.* 69 to the emp. *Theodosius II; *Ep.* 70 to the empress *Pulcheria; *Ep.* 71 to the *archimandrites). In 451 he and *Abundantius went to *Milan with a letter from Leo announcing the outcome of the council of *Chalcedon (council of Milan, 451). He was bishop of Milan for three years between 470 and 480. A poem by *Ennodius (*Carm.* II, 87) praises him for his learning.

A. Di Berardino

SENECA and PAUL, Correspondence of. A collection of 14 *letters, eight from Seneca to *Paul and six short replies from the *Apostle. These *apocrypha were known to *Jerome, who calls them "much read" (*De vir. ill.* 12), and to *Augustine (*Ep.* 153, 14). They are the work of Latin falsifiers from a 4th-c. school of *rhetoric, who forgot that Paul spoke and wrote Greek. Acc. to Liénard, they are stylistic exercises from the school of *Symmachus. The content is mediocre and the replies monotonous. The *apologetic intention is visible: to overcome the repugnance of literate *pagans, who accused early Christian writings of lacking literary elegance.

C.W. Barlow, *Epistolae Senecae ad Paulum et Pauli ad Senecam*, Roma 1938 (=PLS 1, 673-678); L. Bocciolini Palagi, *Il carteggio apocrifo di Seneca e Paolo*,

Florence 1978 (text, intro. and commentary; bibl.); E. Liénard, Sur la correspondance apocryphe de Sénèque et de saint Paul, *RBPh* 11 (1932) 5-23; P. Benoit, Sénèque et Paul, *RBi* 53 (1946) 7-35; A. Momigliano, Note sulla leggenda del cristianesimo di Seneca, *Rivista St. Italiana* 62 (1952) 325-344; A. Kurfess, Zum dem apokryphen Briefwechsel, *Aevum* 26 (1952) 42-48; J.N. Sevenster, *Paul und Seneca*, Leiden 1961.

A. Hamman

SEPTIMIUS SEVERUS, emperor. Born at *Lepcis Magna in 146, died at Eboracum (*York) in 211. In *c.*187 he married, as his second wife, Julia Domna, of the house of the priests of *Emesa, who had introduced her to the knowledge and practices of the Eastern religions. So in S. S.'s *family we meet tendencies and mentalities of two of the most eccentric provinces in the empire, *Africa and *Syria: which were reflected in his religious policy. Emperor 193-211, he lived in a period tormented by internal disturbances and external wars. In the first years of his reign, he was tolerant towards the Christians: as is proved by the synods and assemblies of bishops in many cities to examine the question of the date of *Easter. There were some violent episodes, but they seem isolated (a *persecution broke out at *Carthage in 197, prob. connected with the arrival of a new proconsul). At the start of the 3rd c., the situation seems to change totally (perhaps the need to maintain the unity of the empire might justify the new attitude). If we may believe the **Historia Augusta* (cf. Spartianus, *Septim. Sev.* 17, 1), in 202 an imperial edict forbade Jewish and Christian proselytism (*Iudaeos fieri sub gravi poena vetuit. Idem etiam de Christianis sanxit*). In this way, by a general legislative provision directed against the Christians (and *Jews), their propaganda was to be stopped, striking particularly at catechists. Recent critics are divided on the value of this passage: some consider it a product of imagination or misinformation, others see no good reason not to accept it. Those were important years for Christianity: the presence of important personalities - popes *Victor (189-199) and *Zephyrinus (199-217), *Tertullian and *Origen; and movements which by reviving apocalyptic hopes - *Montanism - weakened the unity of the empire. Undoubtedly the situation of the churches seems to have deteriorated in the tenth year of S. S.'s reign: *Eusebius of Caesarea (cf. *HE* VI, 1 ff.; VI, 7 ff.) talks of persecution, illustrious *martyrdoms in *Alexandria, where believers from all *Egypt and the *Thebaïd were taken and where, among others, Origen's father *Leonides was martyred; in *Africa, as told by the *Passio Perpetuae et Felicitatis* and confirmed by Tertullian; in *Cappadocia, and prob. at *Antioch and *Jerusalem. With *Caracalla (211-217), the persecution abated; if there really was an edict, he did not apply it.

M. Platnauer, *The Life and Reign of the Emperor L. Septimius Seuerus*, Oxford 1918; *PWK*, 2a, 1940-2002 s.v. Severus; S. Mazzarino, *L'Impero romano*, II, Bari 1973, 434 ff. (bibl. on 604 ff.). On the credibility of the item in the *Historia Augusta*, see, among others, against the thesis, T.D. Barnes, Legislation against the Christians, *JRS* 58 (1968) 32-50; R. Freudenberger, Das angebliche Christenedikt des Septimius Severus, *WS* 81 (1968) 206-17; in favour, W.H.C. Frend, A Severan Persecution? Evidence of the Historia Augusta, in AA. VV., *Forma Futuri*. Studi in onore del Card. M. Pellegrino, Turin 1975, 470-80; middle position, E. del Covolo, *I Severi e il cristianesimo*, Rome 1989.

P. Siniscalco

SERAPION of Thmuis. Disciple of *Anthony the Great, in 339 he became bishop of Thmuis in the Nile delta. He received a first letter from *Athanasius (PG 25, 685), then four (or three) more on the *Holy Spirit (PG 26, 529-676). He seems to have been present at the council of *Sardica in 343 and defended Athanasius (Athan., *Apol. c. A.* 50). In 536, with a delegation of four bishops, he refuted the accusations against Athanasius before *Constantius II (Soz., *HE* IV, 9). The Arian Ptolemy expelled him from his see; so *Jerome calls him "*confessor". Feast 23 March.

Of his considerable output (Jerome, *De vir. ill.* 93), there remain the treatise *Contra Manichaeos* and some *letters.

The *Euchologion* that bears his name is generally accepted as his (except for B. Botte). The *anaphora is explicitly attributed to him. It contains the *Sanctus*; the *epiclesis is addressed to the Logos, not to the Holy Spirit. A mixture of Egyptian *liturgy and personal creation.

CPG 2485-2495; PG 40, 895-942; *Contra Manichaeos*, ed. R.P. Casey, Cambridge, Mass. 1931 (only complete ed.); for new fragments of letters, cf. CPG 2493 f.; *Euchologion*: G. Wobbermin, TU 17/3b, Leipzig 1898; F.E. Brightman, *JThS* 1 (1899/1900) 88-113, 247-277; F.X. Funk, *Didascalia et Const. Apostol.*, II, Paderborn 1905, 158-195; Eng. tr.: J. Woodsworth, London 1899 (Hamden, Conn. ²1964); Ger. tr.: *BKV*² 5 (1912) 135-157; A. Hamman, *Prières des premiers chrétiens*, Paris 1952, 179-200 (It. tr. Milan 1954, 159-181). *Studies*: *DTC* 14, 1908-1912; *PWK* Suppl. 8 (1956) 1260-1267; *DACL* 11, 606-611; Quasten II, 81-87; A. Peters, Het Tractaat van Serapion, *SEJG* 2 (1949) 55-94; B. Capelle, L'anaphore de Sérapion, *Muséon* 59 (1946) 425-443; B. Botte, L'Euchologe de Sérapion est-il authentique?, *OC* 48 (1964) 50-56; K. Gamber, Die Serapion-Anaphora ihrem ältesten Bestand nach untersucht, *OS* 16 (1967) 33-42; J.H.

Srawley, *The Early History of the Liturgy*, Cambridge 1913; W.H.C. Frend, The Church in the Reign of Constantius II, in *L'Eglise et l'Empire au IVᵉ siècle*, Geneva 1989, 73-112 (esp. 107-111).

A. Hamman

SERGIA. *Hegumen of the *monastery of St *Olympias in *Constantinople and contemporary of patriarch *Sergius between 612 and 638. We have her *Narratio Sergiae de translatione Sanctae Olympiadis* (BHG 1376), the *deaconess commem. in the Byzantine church on 24/25/26 July and in *Mart. rom.* on 17 Dec, famous for the *letters addressed to her by the exiled *John Chrysostom (SCh 13a). S., who several times in the work refers to herself as ἐγὼ ἡ ἁμαρτωλὸς Σεργία, thus removing any doubts about attribution, used oral and written sources and the *Vita S. Olympiadis* (BHG 1374-1375), composed perhaps by *Heraclidas of Nyssa, re-edited by Malingrey and compiled in a totally different style. The *Narratio* was edited by Delehaye on the basis of a single MS, Paris. 1453 (11th c.).

CPG 3, 7981; H. Delehaye, Vita Sanctae Olympiadis et Narratio Sergiae de eiusdem translatione, *AB* 15 (1896) 400-423; 16 (1897) 44-51; J. Bousquet, Récit de Sergia sur Olympias, *ROC* 11 (1906) 255-268; Beck 461.

A. Labate

SERGIUS. Patriarch of *Constantinople 610-638. Of Syro-*Jacobite origin. Deacon and director of an asylum for the *poor at Constantinople. Ascended the patriarchal throne 18 April 610 in the reign of *Heraclius, whose wars against the Avars and Slavs he supported, esp. during their co-ordinated siege of Constantinople in 626, earning the title "saviour of the capital". With the intention of restoring the empire's religious unity, fractured by *monophysism, esp. in the Eastern regions, he thought out the compromise doctrine of one energy in Christ (*monoenergism), on which he wrote to the Syro-Arabic bishops George Arsas (618), *Theodore of Pharan (*c.*620), patriarch Paul of Antioch (622) and Cyrus of Phasis (626), who became patriarch of *Antioch in 633. Against the opposition of *Sophronius of Jerusalem, in August 633 he issued a dogmatic synodal decree imposing silence on the question of the singleness or duality of Christ's will, suggesting that the phrase "single operator" be used. He also sought support from pope *Honorius, who replied praising S.'s prudence. In Nov 638, another synodal decree approved the *Ekthesis* and condemned whoever rejected it. He died 9 Dec 638 and was condemned as a *heretic by the 6th ecumenical *council (*Constantinople 680).

CPG 7604-7608; Grumel, *Regestes* 278c-293; P. Conte, *Chiesa e primato nelle lettere dei papi del secolo VII*, Rome 1971, *passim* (index p. 584); L. Magi, *La Sede Romana nella corrispondenza degli imperatori e patriarchi bizantini (VI-VII sec.)*, Rome-Louvain 1972, *passim*, esp. 196-205; J.L. Van Dieten, *Geschichte der Patriarchen von Sergios I. bis Johannes VI. (610-715)*, Amsterdam 1972; Fliche-Martin 5, *passim*, esp. 103 f.

D. Stiernon

SERGIUS, pope (687-701). On the death of Conon (21 Sept 687), part of the Roman people elected Theodore and the other part Paschal, who enjoyed the support of *Ravenna; but immediately afterwards the majority of the *clergy and people nominated S., of an Antiochene family, but born in Sicily and at that time belonging to the Roman clergy (*LP* I, 371 f.). The exarch consented only after the payment of 100 gold *librae*, already promised by Paschal. In 692 the emp. Justinian II, who had called a *council at *Constantinople (*Quinisext* or *in Trullo* II), sent the canons to *Rome for S.'s signature to be put after that of the emperor. S. refused, since he judged some canons contrary to church *tradition (*LP* I, 373). Then Justinian thought to use force, as at the time of *Martin I, but the attempt failed (*LP* I, 373 f.). Like his predecessors, S. continued to take an interest in the religious problems of *Britain (Jaffé 2131-2133); he baptized Cadwalla, king of Wessex, on *pilgrimage to Rome, where he died (buried in St Peter's), and encouraged *Willibrord's *evangelization of the Frisians. He introduced the singing of the *Agnus Dei* into the *mass.

CPL 1740 f.; *Verzeichnis* 541; PLS 4, 2174; ASS *Sept*. III, 425-445; *LP* I, 371-382; *BS* 11, 873-875; P. Conte, *Chiesa e primato nelle lettere dei papi del secolo VII*, Milan 1971, 494-504 (cf. index); L. Magi, *La sede romana nella corrispondenza degli imperatori e patriarchi bizantini (VI-VII sec.)*, Rome-Louvain 1972, 162 f; J. Richards, *The Popes and the Papacy in the Early Middle Ages, 476-752*, London 1979.

A. Di Berardino

SERGIUS and BACCHUS. The Passion of these two saints (BHG 1624) was probably composed using some meagre historical details (date and place of burial) in order to explain the popularity of their cult, whose centre was Resafa, 200 km E of Aleppo, a celebrated place of *pilgrimage in the *East. There seems to have been a votive basilica in honour of Sergius at

Eitha (*Syria) from 354; during the 6th c., basilicas multiplied, at *Bostra, *Constantinople, *Ravenna, *Rome. Their depiction in S. Maria Antiqua, Rome, seems to reproduce a canonical model originating at Resafa. *Gregory of Tours knows the two saints (*Hist. Franc.* 7, 31; 10, 31; *G. mart.* 96). Their date is 7 Oct in *Mart. hier.* and the historical *martyrologies, beginning with Anonymous of Lyons.

Vies des SS. 10, 191-97; J. Kollwitz, *"Die Grabungen in Resafa", Neue deutsche Ausgrabungen im Mittlemeergebiet und im Vorderen Orient*, Berlin 1959, 45-70; *BS* 11, 876-82.

V. Saxer

SERGIUS Grammaticus (6th c.). We are completely ignorant of S.'s life. He earned the epithet *"grammaticus"* for his education, more literary than theological. His name is remembered thanks to an epistolary correspondence with patriarch *Severus of Antioch between 515 and 520, preserved only in *Syriac tr. in a 7th-c. MS in the British Library (*Addit.* 17154). The version's author is not mentioned, but we need not fear to attribute it to *Paul, bishop of Callinicum, translator of other works by Severus (cf. Baumstark 160). Their christological dispute turned on the unity of property and nature in Christ after the union of the *incarnation, a thesis upheld by S., following *Eutyches; he had previously submitted it to the judgment of a "priestly assembly" including Antoninus of Aleppo, but it had subsequently been rejected by the council of *Tyre in 515. We should distinguish S. from another *"grammaticus"* in contact with Severus, unanimously identified with *John Grammaticus.

J. Lebon, *Le monophysisme sévérien*, Louvain 1909, 163-172; *DTC* 14, 1996-1997; J. Lebon, *Severi Antiocheni orationes ad Nephalium: eiusdem ac Sergii grammatici epistulae mutuae*, CSCO, 119-120, Script. Syri 64-67, (Syriac text and Lat. tr.), Louvain 1949.

J.-M. Sauget

SERGIUS of Ris'ayna. Date and place of birth unknown. We know only that he studied at *Alexandria where he acquired a sound Hellenistic education. He subsequently worked as head doctor (ἀρχιατρός) at Ris'ayna (Theodosiopolis), but he was also a priest. Though officially part of the *monophysite church, he maintained very good relations with *Nestorians: e.g. Theodore, bishop of Merv, his disciple and *friend, to whom he dedicated some of his works. He seems to have remained outside the *christological disputes, but did not hesitate to go over to the *orthodox when necessary. Thus in 535 he went to see the orthodox patriarch *Ephrem at *Antioch to complain of his bishop Asilus. Ephrem, conquered by his manners, entrusted him with a mission to pope Agapitus. S. went to *Rome and brought the *pope to *Constantinople with the aim of obtaining the expulsion from the imperial city of the monophysites there, beginning with patriarch *Anthimus, as well as *Severus of Antioch and *Theodosius of Alexandria. S. died next year (536) at Constantinople. We can understand that his reputation in the monophysite community was not high. His treatise on *faith is lost, but various philosophical works survive: a *Treatise on logic* in seven books (incomplete); *On negation and affirmation; On the cause of the Universe according to Aristotle's principles; On genus, species and the individual*.

S. deserves particular attention as a translator of Greek into *Syriac. The literality and exactness of his versions are of great interest, esp. from a lexicographical point of view. Some of these translations later served as a basis for *Arabic versions.

Among the works translated by S. were *Porphyry's *Isagoge*, Aristotle's *Categories*, Περὶ κόσμου and a treatise on the *soul different from the Περὶ ψυχῆς preserved in Greek, *Galen's treatise on *Simples* and other works by Galen on medicine, nutrition and even *astrology, surviving only in fragments. He also translated the *Geoponics*. S. made known in Syriac-speaking circles all the works that circulated at the start of the 6th c. under the name of *Dionysius the Areopagite and which exercised an immense influence on mystical authors, both monophysite and Nestorian.

Duval 247-249, 266-273, 276-278, 315, 363-364; Baumstark 167-169; Chabot 71-72.

J.-M. Sauget

SERMO, Sermon. Some ideas on the sermon in the patristic and mediaeval eras, particularly on the transmission of patristic texts, are given under *Homily and *Homiliary because of the close relationship between the terms. It remains to present the concept and evolution of the term *sermo*.

In classical Latin, *s.* means a conversation, a common discourse (analogous terms are *oratio, locutio*), a familiar discussion (analogues: *colloquium, congressus*). It is distinguished from elaborate discourse, peroration and harangue, and from versified texts (poems, songs, etc.: exceptionally Horace's *Satyrae* are called *sermones*, because they come into the literary genre of satirical dialogue). The Greek equivalent is λόγος. To address a word is *sermonem proferre*; to hold a conversation with someone is *sermonem conferre cum aliquo*. Tender, unreserved language is, for Cicero (*Off.* I, 37, 134), *sermo lenis minimeque pertinax*.

In the *Bible, *s.* means a discourse (*sermonem istum proferre*: Josh 2, 20), an *oracle (*sermo Domini erat pretiosus in diebus illis*: 2 Sam 3, 1), a teaching (*Fuit vir propheta potens in opere et sermone*: Lk 24, 19), language (the whole earth had one language and the same words, *labii unius et sermonum*: Gen 11, 1). The old African biblical versions render λόγος as *s.*; the European versions prefer *verbum*; the Greek ῥῆμα is also translated by *s*. The *logos* of the LXX is not always translated *verbum*; in most cases the Vulgate prefers *s.*, perhaps because *verbum* had become "the word" by antonomasia, i.e. the creative and redemptive Word: *Omnipotens sermo tuus de caelo...*; Wis 18, 15. So the old theological associations of the Hebrew *dabar* disappeared from *s.*, but remained linked to *verbum* (in particular, creative and renovative efficacy, typical of the Hebrew concept and taken up by Christian *theology). This explains why the Word is sometimes Sermo: *Nos enim Sermonem Dei scimus indutum carnis substantiam*, writes *Novatian (*De Trin.* 21: PL 3, 928 C). The term is applied to the *aeon (Νοῦς) in *Valentinus's theory, acc. to *Tertullian: *Emittit et ipse semetipso Sermonem et Vitam* (*Adv. Valent.* 7: CCL 2, 758).

The notion of act implied in *s.*, in a precise historical temporal context, recurs in biblical literature: *Reliqua autem sermonum Jeroboam scripta sunt in libro sermonum dierum regum Israel* (2 Kgs 14, 28); *sermones*, i.e. the king's acts; *sermones dierum* are the annals. From the ordinary faculty of speech, from the natural capacity of self-expression and communicating a line of reasoning, we move on to a supernatural reality. The λόγος σοφίας (*s. sapientiae*) and the λόγος γνώσεως (*s. scientiae*; 1 Cor 12, 8) are given by the *Holy Spirit for the good of the community, with the necessary *interpretatio sermonum*, the explanation of the charismatic message. *God's Word is essentially *divinus s.* (Ambr., *Ep.* I, 7, 5: PL 16, 906 B), *s. caelestis* (Ambr., *Ep.* 30, 9: PL 16, 1063 B), *sacer s.* (*1 Clem.* 56, 3: PG 1, 321 A, ἅγιος λόγος). The old Latin version published by G. Morin (*Anecdota Maredsolana*, II, Maredsous 1894, p. 51) has *sanctus s*. For *Irenaeus, *s.* is also a rational force: *paratum fieri ad susceptionem perfectae rationis*, which corresponds to the Greek εἰς ὑποδοχὴν τοῦ τελείου λόγου (*Adv. haer.* II, 19, 4; cf. I, 5, 6). The Greek λόγου has fallen out of the *manuscript tradition, but can be restored on the basis of other texts of Irenaeus, and of Tertullian (*Adv. Valent.* 25, 2).

Thus *s.* becomes a sacred expressive entity; the word itself, now enriched by new elements, belongs to Christian *Latin. Primarily it means a way of communicating a religious teaching with simplicity, and supposes a reflective and redactional work. But a refined theological concept is also attested. Tertullian, in the context of his trinitarian theology, translating Jn 1, 1: *Sermo erat apud Deum et Deus erat sermo* (*Adv. Prax.* 5: CCL 2, 1163), distinguishes *s.* and *ratio*: *Rationalis enim Deus, et ratio in ipsum prius et ita ab ipso omnia. Quae ratio sensus ipsius est. Hanc Graeci λόγον dicunt, quo vocabulo etiam sermonem appellamus ideoque iam in usu est nostrorum per simplicitatem interpretationis sermonem dicere in primordio apud Deum fuisse, cum magis rationem competat antiquiorem haberi, quia non sermonalis a principio sed rationalis Deus etiam ante principium, et quia ipse quoque sermo ratione consistens priorem eam ut substantiam suam ostendat*. Tertullian distinguishes between *s.* and *ratio*, contrary to the African use (*in usu est nostrorum*); for him, in God, *ratio* is immanent, and becomes *s.* after the first creative deed: *Fiat lux* (cf. C. Mohrmann, *Etudes sur le latin des chrétiens*, III, Rome 1965, 110; Tertull., *Adv. Prax.* 5: CCL 2, 1163-1164). The shift from λόγος (also translated *ratio*) to *s.* and *verbum* concerns philology and linguistic criticism. *Lactantius equates the terms: *Melius Graeci λόγον dicunt quam nos verbum, sive sermonem: λόγος enim et sermonem significat et rationem* (*Div. inst.* IV, 9, 1: CSEL 19, 300). The Greek *Fathers also know διάλογος, sometimes translated *concertatio* (Orig., *C. Cels.* III, 1: GCS I, 203: a debate). Then we meet διαλογή (a word not translated directly into Latin: Epiph., *Pan.* 20, 3: PG 41, 273A, but rendered *vocabulum disceptatio*: ibid., 76, 9: PG 42, 532D); διαλογισμός, *disputatio* (John Chrys., *Hom. VIII in 1 Tim.*: PG 11, 590C). The verb διαλογίζομαι is translated *reputari*; the same translation is used for λογίζομαι (Athan., *Vita Antonii*, 42: PG 26, 905AB). It is always the same reality: discourse, public or private. But with the spread of Christian Latin, *s.* would be applied to a discourse aimed at the proclamation of God's Word and the edification of the hearers: it is an important element of Christian *tradition and teaching. It would become a *literary genre: the Fathers preached *sermones*, which were taken down by stenographers (in the case of *Augustine), and published their texts (e.g. *Leo the Great).

But at the same time *s.* kept its old meaning of saying, sentence, often qualified with an epithet: e.g. in the *Rule of St *Benedict: *sermo responsionis porrigatur bonus* (ch. 31: CSEL 75, 96); *auditus malae rei aut otiosi sermonis* (ch. 67, 172). The other meanings are: a) a text: *Quae enim pagina aut qui sermo divinae auctoritatis veteris ac novi Testamenti . . . (Rule of Benedict,* ch. 73, 180); b) an expression: *Multi super hoc sermone diversa finxerunt* (Jerome, *Ep.* 20, 1: CSEL 54, 104); c) a word: Δεός *sermo graecus est, qui latinae interpretatur "timor"* (Cassiod., *Exp. in ps.* 21, 1: CCL 97, 190); d) an interior discourse: *Quodcumque cogitaveris sermo est* (Tertull., *Adv. Prax.* 5). In the Middle Ages, *s.* is also equivalent to a legal action brought by a magistrate.

In the 4th c., *s.* occurs more often for any kind of preaching: catechetical, exegetical, paraenetic. *Ambrose informs us: *Hesterno sermo noster ac tractatus usque ad sancti altaris sacramenta deductus est* (*De sacr.* V, 1, 1: PL 16, 445C). Augustine also uses *s.* and *tractatus*: *Sermonem ad altare Dei debemus hodie infantibus de sacramento altaris. Tractavimus ad eos de sacramento symboli . . . Hodie illis de hac re sermo debetur* (*Serm.* 228, 3: PL 38, 1102). *Tractatus,* then, is a more consistent, more developed exposition than *s.*, which is usually more concise; its subject also seems more weighty than in the case of *s. Enarratio,* *commentary, is developed in the form of *s.* for the congregation: *Statui autem per sermones id agere, qui proferantur in populis, quas graeci* ὁμιλίας *vocant. Hoc enim iustius esse arbitror, ut conventus ecclesiastici non fraudentur etiam psalmi huius intelligentia, cuius, ut aliorum, delectari assolent cantilena*: so Augustine will get 32 *sermones* out of it (*Enarr. in ps. 118,* prooem.). The *s.* is brief and is heard standing: *Longam lectionem audivimus, brevis est dies: longo sermone etiam nos tenere vestram patientiam non debemus. Novimus quia patienter audistis, et diu stando et audiendo tanquam martyri compassi estis* (Aug., *Serm.* 274: *In natali martyris Vincentii*: PL 38, 1253).

The difference between the concepts is picked up by *Isidore: *Homiliae autem ad vulgus loquuntur, tomi vero, id est libri, maiores sunt disputationes. Dialogus est conlatio duorum vel plurimorum, quem Latini sermonem dicunt. Nam quos Graeci* dialogos *vocant, nos* sermones *vocamus. Sermo autem dictus quia inter utrumque seritur. Unde in Vergilio: Multa inter se serebant. Tractatus est unius rei multiplex expositio, eo quod trahat sensum in multa, sententiam contrectando secum. Differt autem* sermo, tractatus *et* verbum. Sermo *enim alteram eget personam*; tractatus *specialiter ad se ipsum est*; verbum *autem ad omnes. Unde et dicitur: Verbum fecit ad populum.* Commentaria *dicta quasi cum mente. Sunt enim interpretationes, ut commenta iuris, commenta Evangelii* (*Etymologiae* VI, 8, 2-5: PL 82, 237C-238B). For Isidore, *s.* is still a dialogue, while *verbum* is a discourse addressed to all. But it seems that, in pastoral reality, *s.* was applied to catechetical teaching, exegetical explanation, moral and disciplinary admonition, religious exhortation, any circumstantial address, independently of the number of hearers. It also seems that the dialogue aspect corresponded more to simplicity of content than to oratorical form: *homilia* was elaborated according to the methodology of wisdom theology, not according to a systematic or artificial criterion. *Caesarius of Arles put an anthology of sermons at the *clergy's disposal: *sermones istos . . . populo frequentius recitare* (*Serm.* 2: CCL 103, 19): they are easy sermons, *Augustinian in spirit, keeping to the human and spiritual situation of the congregation. Bishop Felix prepared a collection of *sermones* of *Peter Chrysologus (CCL 24, 24A, 24B): evidence of an editorial policy!

The reductive interpretation of *s.* into paraenetic discourse probably comes from the *Sermo Domini in monte,* interpreted by Augustine: *Inveniet in eo, quantum ad mores optimos pertinet, perfectum vitae christianae modum . . . Nam sic ipse sermo concluditur, ut appareat in eo praecepta esse omnia quae ad informandam vitam pertinent* (*De sermone Domini in monte* I, 1: CCL 35, 1). From the 4th c. on, the *sermon on the mount was the prototype of popular preaching, of a well-defined literary genre. *Homilia* was applied more exactly to an explanation of the biblical text proclaimed during the celebration of the *liturgy. *S.* became an argument developed with a precise methodology; and not necessarily in dependence on the biblical text. However, the distinction is very uncertain, partly because modern editors of patristic texts have sometimes used disputable criteria. *Leo the Great's *sermones* have become *tractatus* in the crit. ed. of A. Chavasse (CCL 138-138 A), on the basis of their entitlement in a 9th-c. MS. Uncertainty about terminology has often had repercussions on the description of the liturgical books of the Western and Eastern traditions. Mediaeval sermons are not necessarily introduced by a biblical passage (unlike *homilies), but their content easily distinguishes them from the texts of *hagiography, martyrological and historical in character (*martyrologies, *menaea* and *synaxaria,* passionaries and legendaries).

The *s.* was a liturgical act, inserted into the celebration: the *ministerium sermonis* is linked to the *ministerium altaris*: *. . . Qui iam ministrantes altari, quo vos accessuri estis, assistimus, nec ministerio sermonis vos fraudare debemus* (Aug., *Serm.* 214, 1: PL 38, 1065). The preacher participates in the "spirituality" of the sacred text: *Divinae lectiones omnes ita sibi connectuntur, tanquam una sit lectio; quia omnes ex uno ore procedunt. Multa sunt ora ministerium sermonis gerentium, sed unum est os ministros implentis* (Aug., *Serm.* 170, 1, 1: PL 38, 927). As the biblical texts come from the same mouth, so the preaching (the *s.*) of the different preachers finds its unity in the inspiration of the Holy Spirit: *Per ineuntem Spiritum officio oris nostri divini sensus sermo diffunditur* (Hilary, *Tract. in ps.* 118, Tau, 2: CSEL 22, 541).

*Egeria at *Jerusalem heard readings and sermons in *Syriac, Greek and Latin, translated by interpreters (*Itinerarium Egeriae,* 47: CCL 175, 89).

In the Middle Ages, the sermon was dissociated from the liturgy; it was a circumstantial oration heard in an appropriate place: *In sequenti dominica dicere matutinas et missarum ad usum praedictum et cum nota, et tali hora qua scholares post missam ire poterunt ad sermonem* (Statutes of the University of Paris, AD 1380; ed. G. A. Lobineau, *Histoire de la ville de Paris,* III, Paris 1725, 498). In the practice of the inquisition, the condemned had to listen to a *s. publicus* or *actio fidei* ("*auto da fé*"), also called *s. generalis de fide.*

In 1562 the council of Trent (sess. XXII, can. 8) prescribed preaching on biblical readings, *frequenter inter missarum celebrationem . . . , diebus praesertim dominicis et festis* (DS 1749). But liturgists considered the *s.* as an ordinary *interruptio* of the liturgical celebration (e.g. P. Radò, *Enchiridion liturgicum,* I, Rome 1961, 334). Only after Vatican II did the situation tend to change in a very general way; the homily was preferred to the sermon, and the typical sermon, in three points followed by a conclusion, was given up, as was the apposite oratorical technique, which was often hardly sacred (think of the "*Conférences*" of Notre-Dame at Paris, renowned Lenten sermons; or the English "sermon", a text declaimed without gestures and without vocal effects). In modern languages, the term "sermon" sometimes assumes an ironic note, suggesting moralizing contents, conceited and wearisome expression and length of address.

The main derivatory words, in patristic and mediaeval Latin, concern the same concepts:

1) *sermonarium, liber sermocinalis, sermologus*: collection of sermons. *Sermologus, sive liber sermonum,* distinct from *homeliarius* (John Beleth, *Rationale divinorum officiorum,* ch. 60: PL 202, 66AB; cf. ch. 42: PL 202, 69D-70A); 2) *sermunculus*: little discourse: *Huic sermunculo adnexui* (Jerome, *Ep.* 32, 1, *ad Marcellam*: CSEL 54, 252); *. . . ad compendium studiosae intelligentiae in brevem sermunculum . . . coartamus* (Hilary, *Instructio psalmorum,* 17: CSEL 22, 15); *prayer: *Quis sermunculis ocius terminatis domum egressus inter exercitus multitudinem . . . emigrat* (Adamnan of Iona, *De locis sanctis* III, 4: PL 88, 912); 3) *sermonari*: to discourse: *Pullus gratiarum sermonetur tibi* (Priscill., *Tractatus* I, 31: CSEL 18, 26); 4) *sermocinari*: to discourse: *Inductio est oratio qua rebus non dubiis captamus assensionem eius cum quo instituta est, sive inter philosophos, sive inter rhetores, sive inter sermocinantes* (the chattering classes! Cassiod., *De artibus ac disciplinis liberalium litteratum,* 2: PL 70, 1165A); 5) *sermocinator*: preacher: *Sermocinator dum veritatem quam meminit loquitur . . .* (Fulg. of Ruspe, *Contra sermonem Fastidiosi Ariani,* 17: PL 65, 523A); 6) *sermocinatio*: talk, conversation: *Tanti magistri non solum ad ipsos sermocinatio, sed etiam pro ipsis ad Patrem oratio* (Aug., *Tract. in Ioh.* 104, 2: CCL 36, 602).

M. Righetti, *Manuale di storia liturgica,* III, Milan 1949, 216-222; the sermon; C. Mohrmann, *Praedicare-Tractare-Sermo,* in *Etudes sur le latin des chrétiens,* II, Rome 1961, 63-72; J.B. Schneyer, *Geschichte der katholischen Predigt,* Freiburg i. Br. 1968; W. Schütz, *Geschichte der christliche Predigt,* Berlin-New York 1972; R. Grégoire, *Homéliaires liturgiques médiévaux. Analyse de manuscrits,* Spoleto 1981, 18-37; L. Wills, The Form of the Sermon in Hellenstic Judaism and early Christianity, *HThR* 77 (1984) 277-299.

R. Grégoire

SERMO ARRIANORUM. This Arian text was sent to *Augustine in 418 for him to refute: it is included in some MSS of his *Contra sermonem Arrianorum.* It must have circulated among the *Goths who entered *Italy with *Alaric and Athaulf; not a *homily, but a small complete compendium of the radical *Arianism characteristic, in the West, of the school of *Ulfila.

CPL 701; PL 48, 677-684; *Patrologia* III, 97-98; M. Meslin, *Les Ariens d'Occident,* Paris 1967, 129-134.

M. Simonetti

SERMON ON THE MOUNT: *iconography. The scheme of representation of the S. on the M. shows *Christus Magister* in a *gesture of *adlocutio* with his right hand, in his left an open scroll, at the moment of addressing his *audientes* (*apostles or followers). These characteristics appear in the scene represented on the polychrome *sarcophagus in the Museo Nazionale, Rome (late 3rd-early 4th c.) (*Rep.* 773b, p. 321, pl. 62): here the bearded figure of Christ is inspired by the iconography of *Zeus* seated and draped, but the scroll is not open. A sarc. fragment from Porto (second quarter of 4th c.) (*Rep.* 110, p. 80, pl. 29) shows the scroll open; the *audientes* are four young male figures facing him. Wilpert sees a precedent for this, the first in order of time, in the fresco in cubicle III of the hypogeum of the Aurelii (*c*.230-240): here a bearded figure in tunic and pallium, seated on a rise, holds an unrolled scroll between his hands, while *sheep and goats graze on the heights. Acc. to Cecchelli (1944) and more recently Himmelmann (1975), this is really an image of the *shepherd of the soul, inspired by the figure of the philosopher of life: the spiritual expression is typical of such representations and his audience (sheep and goats) are not facing him. The same interpretative uncertainty weighs on the depiction in cubicle A in the catacomb of Via D. Compagni, Rome (the frescos of this sector are dated *c*.320), in which Ferrua proposes to see the S. on the M. Christ, standing on a rock, has his arm raised in the act of speaking to a crowd of disciples below him. Unlike Ferrua, Klauser (1962) and recently Kötzsche-Breitenbruch (1976) prefer to interpret the image as that of *Moses speaking to the Israelites. Both Hempel (1961) and Klauser note in the same cubicle the depiction of *Christus Magister* speaking to 12 *apostles, a theme very common in the *mosaic decorations of basilicas from the 4th c. on. [Fig: 275]

For the sarc. fragments cited, cf. C. Cecchelli, *Monumenti cristiano-eretici di Roma*, Rome 1944, 14-15; F.W. Deichmann-G. Bovini-H. Brandenburg, *Repertorium der christlich-antiken Sarkophage*, I, *Rom und Ostia*, Wiesbaden 1967. For the hypogeum of the Aurelii, cf. earlier references in Cecchelli, *op. cit.*, 1-119, to which must now be added N. Himmelmann, *Das Hypogäum der Aurelier am Viale Manzoni. Ikonographische Beobachtungen* (*Abh. d. Wiss. und der Lit., Geistes-und Sozial-wiss. Klasse*, n. 7), Wiesbaden 1975. For the catacomb of Via D. Compagni, cf. A. Ferrua, *Le pitture della nuova catacomba di Via Latina*, Vatican City 1960, 47, pl. XIV; T. Klauser, *JbAC* 5 (1962) 179; H.L. Hempel, Zum Problem der Anfänge der AT Illustration, *Zeitsch. für die Alttestamentliche Wiss.* 73 (1961) 300; L. Kötzsche-Breitenbruch, Die neue Katakombe an der Via Latina in Rom, *JbAC* 19 (1976) 87-88.

L. Ungaro Testini

SERVANUS. A legendary Passion (BHL 7609-10) takes a Scottish bishop and saint named S., who flourished perhaps in the 8th c., through every sort of adventure. The centre of his activity, and later his cult, was the *monastery of Culross. Acc. to others, S. lived in the 5th c. and was the disciple of *Palladius and teacher of Kentigern. St-Servan is the name of the Breton town where L. Duchesne was born and is buried.

Vies des SS. 7, 27-28.

V. Saxer

SERVATIUS of Tongres. S. took part in the anti-*Arian struggles of the mid 4th c. as a zealous defender of *orthodoxy. *Athanasius cites him among his partisans (PG 25, 337-38) and relates that he was sent as ambasssador to *Constantius by the usurper Magnentius (*ibid.*, 606-7). He was at *Rimini in 359-60. Date of death unknown. He is made to speak at the council of Cologne in 346, but the conciliar document is an 8th-c. *falsification. *Gregory of Tours (*Hist. Franc.* 2, 4; *Gl. conf.* 71) relates that, miraculously warned of the invasion of the Huns, he went to die at Maastricht (*Traiectum*). His legend (BHL 7611-41) makes him a relative of Christ and a disciple of *Peter. His feast is on 13 May in the *cod. Wissemburgensis* of *Mart. hier.*

Vies des SS. 5, 253-55; *LTK* 9, 693; *BS* 11, 889-93.

V. Saxer

SERVUS DEI. This title may have originated from the heading of some canonical epistles (Rom 1, 1; Phil 1, 1; Jas 1, 1) in which the respective authors declare their position before *God, more precisely before Jesus Christ. Entering ecclesiastical jargon together with others of the same tenor (*ancilla Dei, famulus Dei*), in the curial style it became a common formula with no allusion to the *virtue of humility. *Damasus styled himself *beatissimorum martyrum cultor Damasus episcopus, servus Dei*. Other *popes were content with *famulus Dei*. From these simple formulae developed more complex ones like *Gregory I's *servus servorum Dei*, which became usual in the pontifical chancellory and was intermittently copied by individual bishops (cf. Bede, *Hist. eccl. gent. angl.* II, 4). When Gregory I used it in his correspondence with *John the Faster, self-proclaimed "ecumenical" patriarch of *Constantinople (John Diac., *Vita Gregorii* II, 1), he was merely employing a formula used by him and dear to him. The title *servus Dei* was given to bishops, priests and monks; in the Code of Canon Law it is reserved for Christians whose process of beatification has been introduced.

Gregory I, *Reg. Epist.*; *In Ezec., praef.*; *Mor. in Job*; A. Ferrua, *Epigrammata Damasiana*, Vatican City 1942, 189; *DACL* 15, 1360-1363; *EC* 11 420-422; I. Schuster, Il titolo di "Servus Dei" nell'epistolario di S. Gregorio M., *Scuola Catt.* 73 (1945) 137-138.

E. Peretto

SETHIANS. *Gnostic sectarians who derived their name from Seth, a figure who had a particularly important function in their *mythology. Our earliest heresiological evidence comes from Ps.-Tertullian (*Adv. omnes haer.* 2), who names the Sethoitae after the *Ophites and *Cainites; but his presentation of the sect and its doctrine is rather confused. *Epiphanius (*Pan.* XXXIX, 1-5) partly depends on Ps.-Tertullian, but is more precise and detailed. Later *heresiologists (*Filaster, *Isidore of Seville, *John Damascene, etc.) are all dependent on Epiphanius and add nothing new. Acc. to the heresiologists, there were other groups who appealed in some way to the figure of Seth. In particular *Irenaeus (*Adv. haer.* I, 7, 5) attributes to the *Valentinians a division of men into three groups: the *pneumatici*, the *psychici* and the *hylici*, corresponding respectively to Seth, *Abel and Cain. The S. mentioned by *Hippolytus (*Ref.* V, 19, 1-22, 1) have nothing to do with the homonymous *heretics referred to by Ps.-Tertullian and Epiphanius, and must rather be identified with the group mentioned by Irenaeus. Interest in Sethian gnostic groups has been reawoken by the discovery of the *Nag Hammadi texts, in many of which Seth is a central figure. Some scholars (e.g. H.M. Schenke) hold that on the basis of this new evidence we can make out the lines of an autonomous, coherent "Sethian" doctrinal system. The Nag Hammadi texts referring to them are: the *Apocryphon of John* (BG 8502; NHC II, 1; III, 1; VI, 1); the *Hypostasis of the Archons* (NHC II, 4); the *Gospel of the Egyptians* (NHC III, 2); the *Apocalypse of Adam* (NHC V, 5); the *Three steles of Seth* (NHC VII, 5); *Zostrianus* (NHC VIII, 1); *Melchizedek* (NHC IX, 1); *Norea* (NHC IX, 2); the *Triform Protennoia* (NHC XIII, 1). The adherents of this gnostic group were aware of their descendance, physical and spiritual, from Seth; as Seth's descendants, they were φύσει σωζόμενοι and Seth was their Saviour-Revealer. Other elements characteristic of this doctrinal system were: the divine triad (primordial *Father, Barbelo, αὐτογενής); the four illuminating *aeons (Harmozel, Oroiael, Daveithe, Eleleth); an attempt at a theology of history. The hypothesis of the existence of a sectarian group of "Sethians" or of a Sethian doctrinal system is still much disputed; various scholars have declared themselves sceptical (e.g. K. Rudolph; A. Klijn; F. Wisse). The problem remains open.

F. Wisse, The Sethians and the Nag Hammadi Library, *Soc. of Bibl. Lit. Seminar Papers* 1972, 601-607; H.M. Schenke, Das sethianische System nach Nag-Hammadi-Handschriften, *Studia Coptica* (ed.. P. Nagel), Berlin 1974, 165-73; A.F.J. Klijn, *Seth in Jewish, Christian and Gnostic Literature*, Leiden 1977 (with bibl.); B. Layton (ed.), *The Rediscovery of Gnosticism*, Proceedings of the International Conference on Gnosticism at Yale, New Haven (Connecticut), March 28-31, 1978, vol. II: *Sethian Gnosticism*, Leiden 1981.

C. Gianotto

SÉTIF, ancient Sitifis. Part of *Mauretania, it became a *colonia* under Nerva (96-98), capital of Mauretania Sitifiana under *Diocletian (late 3rd c.) and thus the most important city in the region. *Theodosius I, in his struggle against Firmus, who had rebelled in 372, established his headquarters there (Ammianus Marcellinus, 29, 5. 7. 50. 56). In *Augustine's time it suffered an earthquake which forced the inhabitants to leave the city and camp out for a fortnight in the countryside: on this occasion 2000 persons requested *baptism (Aug., *Serm.* 19, 6). During the *Vandal occupation, S. and its region reverted temporarily to the empire (442-455). Reoccupied by the *Byzantines, it became once more the capital of Mauretania I; at this time Solomon provided it with a new, much smaller, circuit of walls. It remained a flourishing city during the first two centuries of Arab occupation, and fell into ruin only because of Ketania.

The inscription of Sertoria (CIL 8, 8647) cannot be invoked as a convincing argument for Christianity at S. in the 3rd c.: the dove and palm are not specifically Christian *symbols, and the epigraph is not dated. Proofs of a Christian presence appear only in the late 4th c. with inscriptions, and in the 5th c. with bishops. The inscriptions, funerary, were discovered by P. A. Février in the basilicas he excavated. They form a continuous series from 378 to 429 (one is of 471), and are not characterized by any specific Christian symbol, except for the Constantinian *monogram which appears at a late date. As for *church buildings, Mesnage once counted half a dozen, and two have been published by Février.

We know the following bishops' names: Severus, early 5th c. (Aug., *Ep.* 111, 7); Novatus, his successor, 403-440 (*conc. Carth.* 419; Aug., *Ep.* 185, 220; CIL 8, 8634), and his *Donatist rival Martianus; Lawrence, who dedicated a church to his patron saint in 452 (CIL 8, 8630); Donatus, 484; Optatus, 525. S. is still mentioned as an episcopal see in the 8th-c. *Thronos alexandrinos* and in the List of Leo VI the Wise (883).

PWK 5, 393-94; *DACL* 15, 1363-84; P.A. Février, *Fouilles de Sétif. Les basiliques chrétiennes du quartier nordouest*, Paris 1965; id., *Fouilles de Sétif. Quartier nordouest. Rempart et cirque*, Alger 1970; A. Mohammedi-E. Fentress, Fouilles de Sétif 1978-1982, *IIe Colloque sur l'Histoire et l'Archéologie de l'Afrique du Nord, 1983*, BCTH 19, Paris 1985, 469-478.

V. Saxer

SEVEN SLEEPERS of Ephesus. At the time of *Decius's *persecution (250), seven young men of *Ephesus took refuge in a cave, which miraculously closed over them: they slept there and awoke in the reign of *Theodosius II (401-450). This legend arose at Ephesus at the time of bishop Stephen, known for his part in the famous "robber council" (449). It achieved extraordinary fame and was propagated in Greek (BHG 1593-99), various Oriental languages (BHO 1012-21) and Latin, by, among others, *Gregory of Tours (BHL 2313-19). Eastern authors, from *Jacob of Sarug (†521) to Michael of Antioch (1166-99), refer to it; Mohammed knew it (Koran, *sura* 18), as did Middle Latin writers from Paulus Diaconus to Pietro de' Natali. Versions exist in all the vulgar Western tongues.

The legend seems to have arisen from the misinterpretation of an archaeological fact: the underground necropolis of Panajir Dagh at Ephesus. This goes back to the 4th/5th c., precisely the time when *epigraphy names the seven saints. Other ancient sanctuaries are at Arabissos (Cappadocia), Paphos (Cyprus), Marmoutier near Tours and Stiffel in Brittany (France), and *Rome; their relics are at Stegaurach (Germany), Guadix (Spain), St. Victor at Marseilles (France). Feast: 22 or 24 Oct in the East; 27 June or 27 July in the West.

DACL 15, 1251-62; *Vies des SS.* 7, 649-51; *Cath* 3, 1038; *LTK* 9, 737-38; *BS* 11, 900-7.

V. Saxer

SEVERIAN of Gabala (†408/431). Of his life we know only, and briefly, the Constantinopolitan period. He arrived there *c.*400 and rapidly gained fame as a brilliant preacher, particularly appreciated at the imperial court. *John Chrysostom made him his vicar before his *journey to Ephesus (401). But S.'s scheming attitude provoked incidents which necessitated a reconciliation on Chrysostom's return. At the synod of the *Oak (403), he sided with John's accusers. Acc. to *Gennadius of Marseilles (*De vir. ill.* 21), his death occurred "*Theodosio imperante*", i.e. after 408; *terminus ante quem* is the council of *Ephesus (431) or perhaps even 425 (Altendorf).

After his death he continued to enjoy a certain notoriety, evident from the use of his works in the 5th- and 6th-c. florilegia and the existence of pseudo-Severianine *homilies in Greek and *Coptic. Prob. in the 6th c. (because of a confusion with *Severus of Antioch, condemned in 536?), his homilies were put under John Chrysostom's name. This reattribution became prevalent in the *manuscript tradition, making it difficult to recover his literary heritage, consisting of a good 50 homilies (partly unpublished or at least in need of critical editions) and fragments in *catenae of a *commentary on the Pauline epistles.

His homilies are in general long and disordered, with constant changes of subject. His preaching is dominated by anti-heretical concerns, mainly on a level of superficial, repetitive polemic, without systematic investigation. He has little interest in ethical exhortation. Occasionally he reveals a high level of philological knowledge, but often his "Hebrew" etymologies are explicable only as from the Aramaic. His exegetical positions, still insufficiently studied, come across as a contradictory mixture of tendencies. His cosmology and his preference for literal interpretations are typically Antiochene. But on occasion he does not disdain *allegory, sometimes very far-fetched. Sometimes his arguments are archaizing and worthy of note. Zelinger has demonstrated his very close contacts with *Ephrem.

CPG 4185-4295; *DTC* 14, 2000-2006; C. Baur, *Der heilige Johannes Chrysostomus und seine Zeit.* II: *Konstantinopel*, Munich 1930; PG 48-65 (*passim*); H. Savile (ed.), *Opera Chrysostomi* [?], Eton 1612-13, 5, 648-653, 898-906; K. Staab, *Pauluskommentare aus der griechischen Kirche aus Katenenhandschriften gesammelt und herausgegeben*, Neutestamentliche Abhandlungen 15, Münster i. W. 1933; J.B. Aucher, *Severiani sive Seberiani Gabalorum episcopi Emesensis homiliae* . . . Venice 1827; N. Akinian, Ewsebeay episkoposi Emesac'woy cark', *Handes Amsorya* 72 (1958) 161-182, 449-474; 73 (1959) 1-50, 161-182, 321-360; A. Wenger, Une homélie inédite de Sévérien de Gabala sur le lavement des pieds, *REByz* 25 (1967) 219-234; M. van Esbroeck, Deux homélies de Sévérien de Gabala (IVe-Ve siècle) conservées en géorgien;

Bedi Kartlisa 36 (1978) 71-91; B.M. Weischer, *Qerellos IV 3: Traktate des Severianos von Gabala, Gregorios Thaumaturgos und Kyrillos von Alexandrien*, Äthiopische Forschungen 7, Wiesbaden 1980, 19-67; J. Zellinger, *Die Genesishomilien des Bischofs Severian von Gabala*, Alttestamentliche Abhandlungen 7, 1, Münster i. W. 1926; B. Marx, Severiana unter den Spuria Chrysostomi bei Montfaucon-Migne, *OCP* 5 (1939) 281-367; H.D. Altendorf, *Untersuchungen zu Severian von Gabala*, Tübingen 1957 (diss.); H.J. Lehmann, Per Piscatores-Orsordawk': *Studies in the Armenian Version of a Collection of Homilies by Eusebius of Emesa and Severian of Gabala*, Århus 1975, 273-367; S.J. Voicu, Severiano di Gabala e Pseudocrisostomo, *Mélanges M. Geerard*, in the press; id., Nuove restituzioni a Severiano di Gabala, *Rivista Studi Biz. e Neoellenici* 30-31 (1983-1984); M. Aubineau, *La traité inédite de christologie de Sévérien de Gabala*, Cahiers d'Orientalisme 5, Geneva 1983, (rev. L.R. Wickham, *JThS* [1985] 235-236); P. Nautin, L'Homélie de Sévérien de Gabala, "Sur le Centurion, contre les Manichéens et Apollinaristes", *VChr* 38 (1984) 393-399.

S.J. Voicu

SEVERINUS of Cologne. Succeeded *Euphratas as bishop of Cologne (*Colonia*). *Gregory of Tours calls him a man "*honestae vitae et per cuncta laudabilis*" and tells how, at the moment of *Martin of Tours's death (397), it was revealed to S. by a vision of a choir of *angels (*De virt. S. Mart.* I, 4: MGH, *scr. rer. mer.* I, 590). So S.'s episcopate must be put in the mid 4th c. He built a church, dedicated to Sts Cornelius and Cyprian, near which he was buried; this church took his name in the 9th c. To S. is attributed the *Doctrina de sapientia*, a series of brief sentences. His *Life* (late 9th-early 10th c.) is without foundation.

CPL 1153; PL 74, 845-848; J. Schlecht, *Doctrina XII Apostolorum. Die Apostellehre in der Liturgie der katholischen Kirche*, Freiburg i. Br. 1901, 127-129; *DCB* 4, 627; *BS* 11, 963-965; *LTK* 9, 699; *DACL* 15, 1391-1393.

S. Zincone

SEVERINUS of Noricum. Saint, monk, "apostle" of *Noricum; known mainly through the *Vita s. Severini* (CPL 678) written in 511 by his young disciple *Eugippius. With it we must consider the two amoebaean *epistulae* of Eugippius and *Paschasius. Of a noble Latin, perhaps Italic, family, it is quite probable that, in the first half of the 5th c., he sought the solitude of the *desert somewhere in the *East (Eug., *Epp.* 8-10: CSEL 9, II, 4 f.), after which a divine inspiration (*Sev.* 9, 4) took him, *c.*454, to Noricum Ripensis on the Danube border, to bring help to those peoples, oppressed by continual barbarian raids. These few personal facts can be gathered from Eugippius, who tends to make S. a promotor of a new ideal of monastic life that more closely met the spiritual and temporal needs of the time. S. died in Noricum in 482 after having given life, in an apostolate of nearly 30 years, to a plan of moral and material uplift of those peoples, prior to consolidation of the Christian faith itself. As part of this spiritual awakening, a series of *monasteries arose along the Danube border, at once centres of organized resistance to the recurrent barbarian invasions and seats of charitable and socio-assistential programmes (agricultural economy, *labour, health, etc.). In 488, when the Roman population evacuated Noricum after the resumption of the barbarian offensive, S.'s body - says the *Life* - was taken from its primitive burial-place (the tomb under the Jakobskirche at Heiligenstadt?) and brought to *Italy. After a first stay at Montefeltro near Rimini at the invitation of a rich Roman matron, Barbaria, it was finally laid, *c.*495, in a mausoleum built at Castrum Lucullanum (Pizzofalcone), near *Naples. There the monastic community of Noricum founded a new monastery.

In appendix, we must mention a *Hymnus in laudem sancti Severini*, perhaps 9th-c. (CSEL 9, II, 71-73), and the deacon John's *Translatio s. Severini* (MGH, *scr. rer. lang. et ital. saec. VI-IX*, 452-459), written at the time of a further translation of his remains (with the rediscovered relics of the *martyr Sossus, companion of St *Januarius) to Naples. The *calendar of that church and the *Mart. rom.* commemorate him on 8 Jan.

ASS *Ian.*, I, 483-497; *BS* 11, 965-971; A. Baudrillart, *S. Séverin, apôtre du Norique*, Paris 1908; M. Pellegrino, Il Commemoratorium vitae sancti Severini, *RSCI* 12 (1958) 1-26; F. Lotter, *Severinus von Noricum. Legende und historische Wirklichkeit*, Stuttgart 1976; J. Haberl, *Favianis, Vindobona und Wien. Eine archäologisch-historische Illustration zur Vita S. Severini des Eugippius*, Leiden 1976; D. Norberg, *Notes critiques sur l'Hymnarius Severinianus*, Stockholm 1977; V. Pavan, Note sul monachesimo di s. Severino e sulla cura pastorale nel Norico, *VetChr* 15 (1978) 347-360; Translation of Eugippus *Vita Sancti Severini*, FOTC 55 (1965); R. Zumkeller and E. Widder, *Der heilige Severinus, sein Leben und seine Verehrung*, Linz 1982; H. Wolff, Uber die Rolle der christlichen Kirche in den administrationsfernen Gebieten von Noricum im 5 Jh. n. Chr., in (ed.) W. Eck, *Religion und Gesellschaft in der römischen Kaiserzeit* (Festsch. Wittinghoff), Cologne-Vienna 1989, 265-293 (bibl.).

V. Pavan

SEVERUS. *Encratite *heretic. *Eusebius puts him after *Tatian (*HE* IV, 29) and says that S. accepted the Holy *Scriptures except for *Paul and Acts: this reveals S.'s *Judaeo-Christian cultural milieu. His disciples were called Severiani. *Epiphanius (*Haer.* 45) gives them gnostic-encratite characteristics: they maintained the dualistic theory of the *creation of the *world by inferior powers and claimed that the seed of the devil-serpent gave rise to the vine and that the sexual organs and *woman were also his work: whence the need to abstain from *wine and *marriage.

E. Prinzivalli

SEVERUS of Antioch. Born at Sozopolis in Pisidia, he studied at *Alexandria and Berytus (Beirut), where he was converted to a more religious life. Baptized in 488, he entered a convent near *Gaza. His religious formation took place in a *monophysite setting and he embraced this confession from the start, quickly becoming its most representative member on a political and cultural level. He went to *Constantinople in 509 - during the pro-monophysite *Henoticon* régime - effectively propagated his ideas there and in 512 was consecrated bishop of *Antioch. After six years of intense activity on behalf of his cause, the anti-monophysite reaction caused by *Justin's accession to the throne (518) forced him to seek refuge in *Egypt, where he continued the anti-Chalcedonian struggle with every means at his disposal, esp. by his writings. In 535 S. accepted an invitation from *Justinian, who sought agreement with the monophysites, to come to Constantinople, but yet another change of course by the emperor led to his expulsion in 536, while a decree ordered the destruction of his books. He withdrew again to Egypt, where he died at Chois, 8 Feb 538.

S.'s immense literary activity defended his doctrinal line of moderate monophysism against both radical monophysism and Chalcedonianism. Justinian's condemnation and Chalcedonian hostility have caused the loss, save for fragments, of the Greek originals of his works; but they were immediately translated into *Syriac, in which language many of them survive. Doctrinal works, all polemical, include the anti-Chalcedonian *To *Nephalius* (c.508); *Philalethés* (=friend of truth), written during his first Constantinopolitan period, in which S. examines a florilegium of Cyrillian passages with a *dyophysite sense collected by a Chalcedonian, in order to vindicate the monophysite character of Cyril's *theology; *Against John Grammaticus* (in 3 books), in which S. refutes a defence of the council of *Chalcedon written by *John of Caesarea. Against monophysite tendencies not shared by S. are: *Against *Sergius Grammaticus*, four letters written soon after 515, in which S. opposes the radical monophysite thesis that, after the union, human and divine properties are confused together in Christ; the *corpus* of writings aimed against the *aphthartodocetism of *Julian of Halicarnassus, after 518: *Criticism of Julian's tome; Refutation of Julian's propositions; Against the additions to the tome; Against Julian's apology; Apology of Philalethès*.

There also survive: 125 *Cathedral *homilies*, preached at Antioch during the years of his episcopate (512-518) to celebrate festivities and saints, to illustrate scriptural readings, etc.; more than 300 complete or fragmentary *letters out of the 4000 he wrote, very important historically and doctrinally. A baptismal ritual and some religious poetry have also come down under S.'s name.

S. was a prolix writer who never tired of repeating *ad satietatem* the fundamental principles of his creed: he illustrated them by continual reference to *Scripture (which was ignored by some Chalcedonians like *Leontius) and patristic *tradition - following the custom of the time - among whom he favoured *Cyril of Alexandria. Though he programmatically rejected the use of Greek *philosophy, he was a very good dialectician. An implacable polemist, he presented his complex material with considerable clarity. So he rightly stood out among the writers of his time. He was the most qualified representative of "verbal" monophysism, which took the name "Severian" from him. S. wished above all to expound and defend Cyril's doctrine, but without that teacher's uncertainties and terminological oscillations, which could favour the Chalcedonians, and made rigid and organic in a monophysite sense on the basis of the famous expression "One sole nature of *God the incarnate Logos". S. understood *physis* as indicative of a concrete, subsistent individual creature, equivalent therefore to both *hypostasis and *prosopon. This *physis* of the Logos is incarnate with a union which is both according to nature and according to hypostasis (*katà physin, kath'hypóstasin*), where nature and hypostasis are obviously those of the Logos, united to humanity. This humanity is entire and complete: in the union its properties, while inhering in the nature and hypostasis of the Logos, remain distinct, without mixture or confusion with the divine properties (against Sergius); therefore Christ's *body before the *resurrection was corruptible and mortal (against Julian), and Christ is consubstantial with the *Father according to his divinity and with us according to his humanity. In the manner of Cyril, S. illustrated this union through the analogy of the union of the human *soul and *body. He could not admit that this entire and complete humanity is a nature, because for him to admit two natures in Christ would mean also admitting two hypostases, and so dividing Christ in the manner of *Nestorius: he therefore rejected *Chalcedon and *Leo's *Tomus ad Flavianum*. He was disposed to admit only, like Cyril, that it is possible to speak of two natures in Christ in theory: in this sense Christ is derived from two natures (*ek dúo physeon*) and from two hypostases, but in reality there ensues from the union a single nature and a single hypostasis, a composite nature (=hypostasis) because the human properties are added to the divine ones: therefore Christ cannot exist in two natures (*en dúo physeis*), as the Chalcedonians wished. Given the distinction of properties, S. distinguishes, in Christ, human acts, e.g. suffering and dying, and divine acts, e.g. working miracles and rising: but, by virtue of the subject's unity, it makes no sense to divide these actions between man and *God.

S.'s *theology was closer than he thought to the Chalcedonian, particularly the *Neochalcedonian; this is why his monophysism is called verbal. But he would never admit this affinity, because for him an affirmation of Christ in two natures was enough to evoke the *heresy of *Nestorius.

CPG 3, 7022-7080; *DTC* 14, 1988-2000; J. Lebon, *Le monophysisme sévérien*, Louvain 1909; id., La christologie du monophysisme syrien, *CGG* 1, 425-580; W.H.C. Frend, *The Rise of the Monophysite Movement*, Cambridge 1972, 403; W.H.C. Frend, Severus of Antioch and the Origins of the Monophysite Hierarchy, *OCA* 195, Rome 1973, 261-275; Id., Isauria, Severus of Antioch's Problem-child 512-518, *Festsch. Marcel Richard*, TU 125 1981, 209-216; G. Dorival, Nouveaux fragments grecs de Sévère d'Antioche, *Antidoron* 1, (Mél. Geerard), Brussels 1984, 101-121; A. Grillmeier, *Christ in Christian Tradition* (Eng. tr.), Atlanta 1987, II, 269-284; F. Graffin, Sévère d'Antioche, *DSp* 14, 748-751.

M. Simonetti

SEVERUS of Aquileia (586-607). Paulus Diaconus (*Hist. Langob.* III, 26) tells how, when patriarch Elias of *Aquileia died after 15 years as bishop, S. succeeded him. We do not know the details of the election, but it cannot have pleased the exarch, who would have preferred to intervene forcefully to put an end to the *schism of the *Three Chapters. Their condemnation (553) had met fierce opposition in the West, of which the Aquileian or Istrian schism was the most salient and lasting episode - it survived into the time of *Gregory the Great and occupied some of his *letters. Thus, after *Pelagius I's forceful but useless injunctions to the Byzantine authorities in *Italy to end the schism (spring 559), after *Pelagius II's heartfelt but vain *letters to the bishops of Istria and their metropolitan Elias to convince them of the indefectible doctrine of the Apostolic See and invite them to a meeting to clarify and remove the causes of the division, now that Elias was dead and S. elected, exarch Smaragdus judged it opportune to use more energetic methods than heretofore. He arrested S. and three of his suffragans (John of Parentium, Severus of Trieste, and Vindemius of Cyssa, probably at Grado to consecrate the metropolitan), who were taken to *Ravenna and persuaded to enter into *communion with bishop John, an opponent of the Three Chapters.

The mild and insinuating manners that John probably adopted, backed up by fear of imperial arms and the presence of the feared exarch, finally induced the four bishops to change their minds and resign themselves to the dual force of reason and threats; but once back in their sees, they went back to the schismatic party, as the sources attest. Rejected as renegades by the followers of the schism, they were constrained to retract their abjuration at a synod of ten bishops convoked at Marano (591). The bitterness and indignation of Gregory the Great, who had succeeded Pelagius II in 590, were expressed in a letter sent to S. at Aquileia ordering him to present himself and his suffragans at *Rome in order to settle the dispute at a synod. But the dissident bishops, rather than consent to the *pope's requests and the imperial *iussio* he had convoked, preferred to send a *libellus supplex* (MGH, *Epist.*, I, 17-21) to the emperor. The outcome of this appeal and the resistance to Rome's doctrinal line on the specific question of the Three Chapters represent one of those partial eclipses of the principle of Roman primacy that history records from time to time among the members of the Christian church and even in the consciences of some bishops. The emp. Maurice was shaken by this supplication and enjoined pope Gregory to adapt himself to the circumstances of the time. Meanwhile Grado, the new residence of the patriarch of Aquileia, had suffered an invasion of Slav pirates and was tormented by other natural calamities described by Paulus Diaconus (*Hist. Langob.* IV, 2, 4). Gregory assisted them in their misfortunes while continuing his attempts to end the schism, now on its way to extinction, as when he urged exarch Smaragdus not to leave undefended the distressed bishop Firminus of Trieste who, reconciled with Rome (602), was obliged to face a popular rebellion raised against him by S. (MGH, *Ep.*, II, 360, 399).

On S.'s death (607), Smaragdus once again had recourse to violence to impose on Grado a candidate favourable to Rome, Candidianus. The dissidents elected another schismatic patriarch, John, with the complicity of the *Longobard king *Agilulf and his duke of Friuli, Gisulf II. Thus began the double series of patriarchs of Aquileia, one residing at Grado with jurisdiction over the coastal strip and Istria, which were subject to the *Byzantines, the other residing in the fortified castle of Cormons with jurisdiction over the Longobard hinterland.

P. Paschini, *Storia del Friuli*, I, Udine 1934, 101-109; G. Cuscito, Aquileia e Bisanzio nella controversia dei Tre Capitoli, *AAAd* 12 (1977) 231-262; id., Testimonianze epigrafiche sullo scisma tricapitolino, *RAC* 53 (1977) 231-256; id., La politica religiosa della corte longobarda di fronte allo scisma dei Tre Capitoli, *Atti VI Congr. intern. di studi sull'Alto Medioevo*, Spoleto 1980, 373-381; id., *Cristianesimo antico ad Aquileia e in Istria*, Trieste 1977 (but of 1979), 293, 295-296, 305-306, 318.

G. Cuscito

SEVERUS of Malaga. *Friend and companion of *Licinianus of Cartagena and prob. also of *Eutropius of Valencia, with whom he may have been trained at the Servitan *monastery (near Valencia?). Bishop of a see subject to the *Byzantines, he took an active part in the anti-Arian struggles. After 580, he composed a *libellus* against bishop Vincent of Saragossa who, giving way to the pressure of king *Leovigild, had become an *Arian. He also wrote a treatise called *Anulus* (Isid., *Vir. ill.* 30) on *virginity, addressed to his sister. Nothing remains of these works. He collaborated with Licinianus in the writing and theological structure of his *letter to the deacon Epiphanius on the spiritual nature of *angels. He enjoyed a reputation as a good theologian. Died *c*.620.

J. Madoz, *Liciniano de Cartagena y sus cartas*, Madrid 1948; *DHEE* 4, 2446.

M. Díaz y Díaz

SEVERUS of Milevis. Early 5th-c. bishop of Milevis in *Numidia. A *friend of *Augustine from youth, he shared his studies and later, with *Alypius and *Evodius, his religious life. He corresponded with Augustine (cf. Aug., *Epp.* 62 and 63, on the *ordination of deacon Timothy; *Ep.* 109, from S., praising his friend's writings, and Augustine's reply, *Ep.* 110) and with *Paulinus of Nola (cf. Aug., *Ep.* 31, 9; Paul. Nol., *Ep.* 7, 1=Aug., *Ep.* 32, 1). S. went to the conference of *Carthage (411), but took no part, due to illness. His name appears among the signatories of the letter of the *council of Milevis, 416, on the *Pelagian question (*Ep.* 176 of Augustine's). S. indicated his own successor to the *clergy of his diocese, without informing the *laity; the problem thus created was peacefully resolved by Augustine's personal intervention (cf. Aug., *Ep.* 213, 1 [AD 426]).

PAC 1, 1070-1075.

F. Scorza Barcellona

SEVERUS of Minorca. Bishop of Minorca (*Baleari) in the first years of the 5th c. Baronius published and attributed to S. an encyclical *letter found in the Vatican Library, which he related to the report of an African work (*De miraculis S. Stephani*: PL 41, 833-854). The *Letter* tells of the tensions between *Jews and Christians and the conversion of the Jews of Mahón when Paul *Orosius translated St *Stephen's relics there; it also gives information on the island's liturgy, politics and social life. The text, published by Baronius, was used by the *Maurists and by Migne. G. Seguí Vidal critically edited the text and supported its authenticity and integrity, backed up by E. Dekkers, U. Domínguez del Val and F. Martí; but not by Blumenkranz or, more cautiously, by M. C. Díaz y Díaz, who date it to the 8th c., not to 418 (Baronius) or 417 (Seguí). S.'s disputed *Letter* cites a *Commonitorium* which G. Seguí and J. Hillgarth identify with the pseudo-Augustinian *Altercatio Ecclesiae et Synagoguae* (PL 42, 1131-1140). The *Altercatio* is no longer considered to be by *Augustine, though it can be dated to the 5th c. and not, as A. Oepke maintained, the 11th.

Editions: CPL 576-577; PL 20, 731-746; PL 41, 821-832.
Studies: G. Seguí Vidal, *La carta-encíclica del obispo Severo. Estudio crítico de su autenticidad e integridad con un bosquejo histórico del cristianismo balear anterior al s. VIII*, Palma de Mallorca 1937, 149-185 (CCL, SCh and the Fund. Univ. Española are preparing new editions); *Un prematur testimoni de la polèmica antijueva: La circular de Sever de Menorca (417)*, pres. i trad. per J. Amengual i Batle, *LLuc* 60 (1981); M.C. Díaz y Díaz, De patrística española, *RET* 17 (1957) 3-46; F. Martí, Severo, *DHEE* 4, 2445-2446; M. Sotomayor y Muro, *Historia de la Iglesia en España*, Madrid 1979, I, 313, 355-364 (bibl.); J. Amengual i Batle, Noves fonts per a la historia de les Balears dins el Baix Imperi, *Bolleti de la Societat Arqueològica Lulliana* 37 (1979) 99-111; *Patrología* III, (Span. ed. Madrid 1981), 683-685; E.D. Hunt, St. Stephen in Minorca. An Episode in Jewish-Christian Relations in the Early 5th Century A.D., *JThS* n.s. 33 (1982) 106-123.

E. Romero Pose

SEVERUS of Naples. Twelfth in the catalogue of bishops of *Naples (*Neapolis*), S. was bishop from Feb 363 to 29 April 409. He reinvigorated *orthodox Christianity after the spread of *Arianism and the last outbursts of *paganism. He is mentioned in a *letter sent in 397/8 by Q. Aurelius *Symmachus, senator and prefect of *Rome, to Decius, *consularis* of Campania. Symmachus expresses appreciation of S.'s work: he is favourably regarded not just by the *Catholics, but by "all sects" (Q. Aur. Symm., *Ep.* VII, 51: ed. O. Seeck, in MGH, *Auct. ant.* VI, I (1883) 191). A letter of *Ambrose (*Ep.* 59 [85]: PL 16, 1232-33) attests the good relations between S. and the bishop of *Milan. The *Gesta episcoporum Neapolitanorum* (ed. G. Waitz, in MGH, *Scrip. rer. lang. et ital.* [1878] 404-405) attribute considerable building activity to S., referring particularly to the foundation of numerous basilicas and the famous Baptistery of Naples (also called S. Giovanni in Fonte), the oldest in the West.

BS 11, 993-994.

M. Spinelli

SEVERUS of Ravenna. The one historically certain fact about S., bishop of *Ravenna, is his participation in the council of *Sardica (343) where he signed the decrees, the *letter to *pope *Julius II and the synodal letter to all the bishops (Mansi 3, 39. 42. 66). Other information comes from Agnellus, author of the *Liber Pontificalis* of Ravenna (*c*.830-832), the *Life* written by the presbyter Liutulf of Mainz soon after 856 (ASS *Febr.* I, 88-91), two *sermons of Peter Damian (*Serm.* IV-V), sources relating to his cult, and *iconography.

BHL 7679-7684, Suppl. p. 279; *BS* 11, 997-1004 (bibl.).

A. Pollastri

SEVERUS of Synnada. Bishop of Synnada in *Phrygia salutaris, he was present at the council of *Ephesus (431) where, in the first session, he subscribed the deposition of *Nestorius (Mansi 4, 1224c). A brief speech against the Nestorian position, given during the *council, is preserved in *Ethiopic.

CPG 3, 6141; A. Dillmann, *Chrestomathia Aethiopica*, Leipzig 1866, 99 f.; S. Euringer: *Orientalia* n.s. 12 (1943) 127-130 (German version); Bardenhewer 4, 200.

A. De Nicola

SEVILLE

I. Christian origins - II. Councils.

I. Christian origins. S. is the ancient Hispalis in Hispania Ulterior; with the division of this province by the emp. *Augustus, it became part of *Baetica, whose second city, after the capital Cordova (*Corduba*), it was. It gradually grew at Cordova's expense and finally took over its function. Destroyed by the *Vandal Guntheric on his way to *Africa (427), it later fell briefly under Suevian (441-456) and then Visigothic rule until the Muslim invasion (712), apart from the brief period of *Byzantine rule in the second half of the 6th c. S. was the scene of *Hermenegild's *conversion to *Catholicism and rebellion against his father, who had made him governor of Baetica. The Christian history of S., on the basis of the evidence we possess, starts for us at the time of *Diocletian, with the *martyrdom of Sts Justa and Rufina, during the episcopate of that Sabinus who took part in the council of *Elvira as titular of the see of S. The *cod. Emilianensis* mentions an earlier bishop, Marcellus, of whom nothing is known. But to judge from the intensive Christianization of Baetica, revealed by that council, Christianity must have arrived long before. In the religious sphere too, S. gained ground at the expense of Cordova. And at the end of the 5th c., pope *Simplicius (468-483), after eulogizing bishop Zeno of S., named him his vicar for a territory of unknown extent, with the task of watching over church discipline (*Ep.* 1: PL 58, 35). The same privilege was renewed by pope *Hormisdas (514-523) for bishop Sallustius, limited only to the provinces of Baetica and *Lusitania (*Ep.* 26: PL 63, 425-426). But S. attained its greatest prestige under bishops St *Leander (578-599) and St *Isidore (599-636). At a local level, we see the celebration of two provincial *councils, in 590 and 619, and perhaps a third. The episcopal *school, founded by Leander, became a centre of propagation of secular and theological culture; it was to spread all over *Spain and nourish the European Middle Ages, thanks esp. to Isidore's literary output. Also important for the national life was Leander's influence on Hermenegild's conversion to Catholicism, which paved the way for that of *Recared and the whole Visigothic people. The Sevillian church became the spiritual capital of Spain and had enormous influence through the 3rd (589) and 4th (633) councils of *Toledo, animated respectively by Leander and Isidore. After the death of these two bishops, Toledo took the lead, and the Baetican metropolitans never again acquired the same prestige.

Bishops: Marcellus (?), Sabinus (*c*.300), Evidius, Deodatus, Sempronianus, Geminius, Glauchius, Marcianus, Sabinus (expelled by the Suevian Rechila in 441 and reinstated with the arrival of the *Goths in 458), Epiphanius (intruded during Sabinus's exile), Zeno (contemporary with popes Simplicius and Felix), Asphalius, Maximian, Sallustius (contemporary of pope Hormisdas), Crispinus, Pigasius, Stephen, Theodulus, Hyacinth, Reparatus, Stephen II, *Leander (578-599), *Isidore (599-4 April 636), Honoratus (12/5/636-12/11/641), Anthony (646-653), Fugitivus (656), Bracarius (?), Julian (681), Floresindus (683-688), Felix (transferred to the see of Toledo in 693), Faustinus (693/694), Gabriel (?), Sisebert (?), Oppas (712, betrayer of king Roderic).

A. Morgado, *Prelados servillanos*, Sevill 1906; J. Fontaine, *Isidore de Séville et la culture classique dans l'Espagne visigotique*, Paris 1959; D. Mansilla, Orígenes de la organización metropolitana en la iglesia española, *Hispania Sacra* 12 (1959) 267-271; *ES* 9.

II. *Councils. *590*: provincial. *Leander presided over seven other bishops of *Baetica. We have the council's *letter communicating to bishop Pelagius of Ecija the decisions that concerned him directly. Its three canons deal with church *slaves freed by the bishop, whom his predecessor had given to his own relatives, and women living in clerics' houses. Probably the letter contained only what concerned the recipient.

619: provincial. *Isidore presided over eight other bishops. Of its 13 canons, the first two are concerned with the boundaries of some Baetican dioceses; those that follow deal with freed slaves, and treasurers of churches and *monasteries. Can. 12 speaks of a certain Syrian bishop, belonging to the "*Acephali", who denied two natures in Christ. The most important canon is no. 13, an ample dogmatic treatise refuting that *heresy.

624: on this council, cf. P. Séjourné, *Saint Isidore*, Paris 1929, 29-31.

Synoden, passim.

P. de Luis

SEXTUS, antignostic. Christian writer (late 2nd-early 3rd c.). Known exclusively through *Eusebius's mention (*HE* V, 27) of his name and of a work *On the resurrection*, perhaps antignostic in content.

E. Prinzivalli

SEXTUS, Sentences of. *Origen cites a collection, Pythagorean in tone, of sentences (451; 610 in an enlarged edition) propounding a devout *ascesis, and attributes it to the Christian S. It was used by *Basil (*De baptismo*) and *Evagrius. *Rufinus translated it into Latin: more than 50 MSS (cf. P.M. Bogaert: *RBen* 82 [1972] 26-46) and some citations of *Pelagius, the Master, St *Benedict, St *Columbanus and others attest its popularity. Rufinus attributed it to pope *Sixtus II. Two *Syriac versions, an *Armenian version and the *Sacra Parallela* ascribed to *John Damascene support the same attribution, which Chadwick considers not unacceptable; the sentences are obviously largely earlier, but the compilation was made in a Christian spirit, with Christian additions and corrections. The *Coptic fragments (nos. 158-180, 307-397) given by cod. XII of *Nag Hammadi (ff. 15-16 and 27-34) unfortunately have neither title nor colophon. The opusculum provides a typical example of the, sometimes unexpected, acceptance of the ascesis and *piety of *Hellenism in certain fervent circles. Against Rufinus, *Jerome attributes them to a Pythagorean S., but this could be his own invention, given that this author is known from no other source.

F. de Paola, *Le Sentenze di Sesto*, Milan 1937; H. Chadwick, *The Sentences of Sextus*, TSt II, 5, Cambridge 1959; F. Wisse, Die Sextus Sprüche, in *Zum Hellenismus in den Schriften von Nag Hammadi*, Wiesbaden 1975, 55-86; id., The Sentences of Sextus, *Nag Hammadi Library in English*, II, Leiden 1977, 454-459; P.-H. Poirier, *Les Sentences de Sextus (NH XII, 1). Fragments (NH XII, 3)*, Quebec 1983.

J. Gribomont

SEXUALITY. The evaluation of early Christianity on human s., its origin and meaning in the new *economy of salvation, was always the object of latent tensions which sometimes exploded into open and dramatic conflict. The reason for this must be sought in the spiritual and ecclesiological, as well as the properly theological and anthropological, implications of a decision in this field.

As far as it is allowable to schematize in this matter, we can say with sufficient precision that early Christian sources show a fundamental, though variously articulated, opposition between upholders of a rigoristically *ascetic attitude of *encratite type and supporters of a more tolerant and elastic position. Movements practising erotic libertinism, documented in the *gnostic area, lie outside a properly Christian discussion of s., being profoundly permeated with *pagan and *syncretistic motivations.

1. The encratite interpretation of s., whose main literary expression is in some apocryphal gospels (*Gospel of the Egyptians, Coptic gospel of Thomas*) and the apocryphal acts of the apostles (*Peter, *Paul, *Andrew, *John, *Thomas), appears widespread by the end of the 1st c. (1 Tim 4, 3) and was propagated in the 2nd c. by authors like the Syrian *Tatian (in his *On perfection according to the Saviour*) and *Julius Cassianus (in his *On continence or eunuch-hood*). The *Montanists also preached the dissolution of conjugal ties in view of the imminent eschatological catastrophe (here *Tertullian found nourishment for his sexual rigorism and condemnation of second *marriage), while a bishop like *Pinytus of Cnossus, in dispute with *Dionysius of Corinth, defended ascetic rigorism as the normal ideal of Christian perfection (Euseb., *HE* IV, 23, 6-8).

The conception of salvation-history underlying such positions was based on some firm points common to the various doctrinal elaborations. For the encratites, *God created man and *woman, forbidding them to touch the tree of knowledge of good and *evil, but *Adam and Eve, transgressing the commandment, coupled and gave rise, after the creation of the *bodies symbolized by the *coats of skins (Gen 3, 21) and their expulsion from the earthly *Paradise, to the chains of *sin and concupiscence, births and deaths (Gen 4, 1). The *revelation of Jesus Christ would then consist of the offer to believers of the means, including sacramental means like baptismal washing, to break the chain of corrupt generation and, once set in the steps of the Master, to inaugurate a new humanity, new since imitating the Lord's example of *chastity and *poverty. The ideal thus outlined is that of the *monachòs* (the word first appears in the *Gospel of Thomas*, log. 16), the celibate and solitary man, disciple and imitator of the Lord, whose entry into the Kingdom is guaranteed (*Gospel of Thomas*, log. 49 and 75) and who already in this *world partakes of the angelic life thanks to his anticipation of the *resurrection. In such a radically anticosmic thought-context there naturally developed the idea that man's *birth is something intrinsically sinful and that therefore the newly born come into the world polluted by the "sin of birth" or "original *sin"; to be free from the curse of Adam's sin, contracted by birth and consequent slavery to the *demon, man needs the *purification of *baptism (Clem. Al., *Strom.* III, 100, 5; Ps.-Cyprian, *De centesima* 2).

This complex of beliefs, for which sex, even within *marriage, must always and everywhere be considered negatively, gave rise to forms of solitary life which sometimes went as far as self-emasculation (cf. Justin, *Apol.* I, 29, 2; Orig., *Comm. in Mt.* XV, 3; Euseb., *HE* VI, 8, 1-4), but also to forms of ascetic cohabitation (*virgines subintroductae*, *agapetai*) which never ceased to cause scandal and reactions among bishops and polemists at various times, esp. in the Syrian area (*Didachè* 11, 11; Aphraates, *Dem.* VI, 4-5; John Chrys. in PG 47, 495-514), but also in the West (e.g. Ps.-Cypr., *De singul. cler.*).

2. The controversialists hostile to encratism dialectically elaborated a *theology of human s. which, while rejecting in no uncertain terms what it considered to be excesses foreign to the true Christian faith, did not hesitate to take into itself and translate into Christian terms ideas and motifs borrowed from the classical philosophical tradition, markedly *Stoic and *Aristotelian, primarily the theme of the allowability and goodness of a balanced use of s., the ethical ideal of *metriopátheia*.

For *Clement of Alexandria, who dedicated the whole of book III of the *Stromata* to refuting encratite and gnostic doctrines, marriage is as holy as, if not more so than, continence (*Strom.* VII, 12) and finds its specific *raison d'être* in the continuation of the species and the preservation of the world. Clement, like *Irenaeus of Lyons before him (*Adv. haer.* I, 28, 1), tends improperly to identify *encratites with *gnostics, attributing their evaluation of s. to the cosmological and anthropological *dualism of the gnostics, esp. the *Marcionites. This inexact and tendentious presentation of encratite *ascesis would reappear, as a common heresiological *topos*, in other ecclesiastical writers who ended by simply confusing *Messalians and *Priscillianists, heirs of the encratite tradition, with *Manichees, heirs of gnostic *dualism. This did not prevent many doctrines of encratite origin being accepted, after adequate reworking, into the wider synthesis of *orthodoxy, e.g. the idea that religious perfection consists of *virginity in imitation of Christ, the doctrine of original sin, the post-lapsarian interpretation of the *coats of skins, etc.

3. The debate on s. within the Christian communities was resumed with great vigour in the course of the 4th c. The tendency towards worldliness in the church, with its attendant risks, released new monastic energies and extremist reactions, like that of *Eustathius of Sebaste. The liveliness of encratite doctrines and of the ascetic ethos linked with them is shown by the resistance opposed by numerous itinerant groups to ecclesiastical pressures in the now Christian empire. Encratite spirituality, based on the principle that the true Christian is he who prays in a state of continual continence so as to be always united with *God, found convinced supporters, as we have said, in Eastern *Messalianism and Hispanic *Priscillianism, conditioning from close to, dialectically, the developments of orthodox monastic *theology and *spirituality.

On the other hand, the spokesmen of the new ranks recently converted to Christianity, often not for disinterested reasons, did not hesitate to impugn the Christian value of *virginity in order to justify an aristocratic conception of sexual and conjugal life, still profoundly linked to the most noble values of the pagan tradition, and to resist in this way the indiscriminate expansion of radical asceticism, which they opposed as an expression of the feared "Manichaeism".

Among these often bitter conflicts, orthodox authors, bishops responsible for the moral and spiritual life of the community, attempted the difficult way of a mediation which proposed both to condemn encratite and Manichaean excesses in no uncertain terms, even accepting the help of repressive imperial legislation, and to safeguard the supreme Christian ideal of virginity without having to express disapproval of *marriage and procreation. This highly conflictual situation gave rise to a vast literature on virginity, which engaged the most prestigious names in contemporary Greek and Latin *patristics: *Methodius of Olympus, *Athanasius, *Basil of Ancyra, *Gregory of Nyssa, *John Chrysostom, *Ambrose, *Jerome, *Augustine, *Pelagius; with the important difference, however, that while in the *East theological speculation on s. moved prevalently in the direction of a polemic against ascetic extremism, in the West the defence of Christian virginity continued to be elaborated against its detractors, firstly pagans and then Christians too. The result was that the Latin *Fathers appear on the whole more inclined than the Greeks to exalt virginity and continence. Not by chance was ecclesiastical *celibacy affirmed from the 4th c. in the West (council of *Elvira, can. 33). In the East, Chrysostom upheld the post-lapsarian character of s. (*De virg.* 15, 2; *Hom. Gen.* XVIII, 3-4), and an *Origenist like *Didymus the Blind interpreted the *coats of skins of Gen 3, 21 as the sexually differentiated *bodies created as a consequence of sin (cf. Tura papyrus *ad loc.* in SCh 233, 250). For *Gregory of Nyssa, God added to humanity the bestial and irrational way of propagation in prevision of sin and mortality (*De hom. opif.* 16-17), but *Theodore of Mopsuestia held that man was created sexed and mortal in his first *catastasis*, independently of the sin he may have committed from then on (*Fragm. in Gen.* III, 17). In the West, *Ambrose and *Jerome, renewing the African ascetic tradition of *Tertullian and *Cyprian, defended Christian virginity against the assaults of *Helvidius, *Bonosus, *Jovinian and *Vigilantius, while *Pelagius worked out criteria of ascetic life for Roman aristocratic circles. But the knots were combed out in the course of the polemic between *Augustine and *Julian of Eclanum over original *sin. For Julian, formed in the Antiochene school and influenced by Aristotelian philosophy, human s. is intrinsically good since it is created by God; only immoderate concupiscence is sinful. Consequently, *children are born innocent, without original sin. Acc. to Augustine, however, concupiscence is always and everywhere an *evil, even if used well in *marriage for the end of procreation, and therefore even the children of Christian parents are born with the stain of original sin. In Augustine's thought, human s. mirrors, in its post-lapsarian condition, the body's state of rupture and rebellion from the control of reason, which in turn reproduces the rebellion that originally led the spirit to evade God's commandment. Despite this, it was Augustine who first worked out a complete *theology of matrimonial good, fidelity, indissolubility and procreation. In this way Augustine, breaking definitively with earlier ascetic uncertainties and tensions, authoritatively introduced into Western Christianity the idea of the essential compatibility of generation with Christian marriage, elevating procreation to the dignity of one of the primary ends of the nuptial *sacrament.

H. Crouzel, *Virginité et mariage selon Origène*, Museum Lessianum, Sect. théol. 58, Paris-Bruges 1963; C. Tibiletti, Verginità e matrimonio in antichi scrittori cristiani, *Annali della Fac. di Lett. e Filos. dell'Univ. di Macerata* 2 (1969) 9-217; R. Gryson, *Les origines du célibat ecclésiastique*, Gembloux 1970; J.P. Broudéhoux, *Mariage et famille chez Clément d'Alexandrie*, Théologie historique 11, Paris 1970; Y.-M. Duval, L'originalité du "De virginibus" dans le mouvement ascétique occidental. Ambroise, Cyprien, Athanase, in *Ambroise de Milan 1974. XVIᵉ centénaire de son élection épiscopale*, Paris 1974, 9-66; K. Niederwimmer, *Askese und Mysterium. Über Ehe, Ehescheidung und Eheverzicht in den Anfägen des christlichen Glaubens*, Göttingen 1975; AA. VV., *Etica sessuale e matrimonio nel cristianesimo delle origine*, Milan 1976; P.F. Beatrice, *Tradux peccati. Alle fonti della dottrina agostiniana del peccato originale*, Milan 1978; G. Delling, Geschlechter, *RACh* 10, 780-803; id., Geschlechtstrieb, *ibid.*, 803-812; id., Geschlechtsverkehr, *ibid.*, 812-829; J. Gribomont, Askese IV, *TRE* 4, 204-225; G. Sfameni Gasparro, *Enkrateia e antropologia*, Rome 1984; *Atti del Colloquio Internazionale sulla tradizione dell'Enkràteia* (20-23 aprile 1982), a cura di U. Bianchi, Milan; J. van Oort, Augustine and Mani on Concupiscentia sexualis, *Augustiniana Traiectina*, Paris 1987, 137-152.

P.F. Beatrice

SHEEP. Present in the OT and NT as a symbolic figure, the s. of Lk 15, 4-7; Mt 18, 12-14 is variously interpreted by the *gnostics: for *Apelles it is the *angel who repents of having created the *world, thus confessing its negativity (Tertull., *De car.* 8, 3); for the *Simonians it is *Ennoia* who, having left the *Father's house, returns there as the redeemed church (Iren., *Adv. haer.* I, 23, 2; Tertull., *De anim.* 34, 4; Hipp., *Haer.* VI, 19, 2. 3); for the *Valentinians, the 99 sheep are the incomplete *pleroma, which must recover its unity (Iren., *Adv. haer.* I, 16, 1-2). Among Christians, the s. of Lk 15, 4-7 is all mankind created by *God, lost, sought and refound (Tertull., *Adv. Marc.* IV, 32, 1; Iren., *Adv. haer.* III, 19, 3; 23, 1; Orig., *Hom. IX, 3 in Gen.*; Hilary, *In Matth.* 18, 6; Cyr. of Jer., *Catech.* XV, 24; Jerome, *In Matth.* III; Greg. Nyss., *Hom. II in Cant.*); at other times it is the sinner (Tertull., *De paenit.* VIII, 3-5; Cypr., *Ep.* 55, 15; Jerome, *In Matth.* III) or one who has not sinned gravely and can therefore be pardoned (Tertull., *De pudic.* VII, 17-19) or the church set on the hill, which is Christ (Aug., *Serm.* 37, 2). It is a title of Christ (Cypr., *Testim.* II, 15; it is a figure of the *Jews or of the gentiles, on the basis of Mt 15, 24 or Jn 10, 16 (Aug., *in evang. Ioh.* 49, 27). The sheep of Mt 18, 12-14 are the poor and despised (John Chrys., *Hom. LIX in Mt.*), while those of Ps 94, 7 are the patriarchs and *prophets and finally *Mary (Melito: SCh 123, 176; Chromat., *Serm.* 23, 3).

A. Orbe, *Parábolas evangélicas en S. Ireneo*, II, Madrid 1972.

F. Cocchini

SHENOUTE (Gk *Sinouthios*), c.350-466. *Abbat of the *monastery at a place called Atripe, near Achmin (Upper *Egypt), now called the White Monastery (*Deir el Abiad*) or monastery of Shenoute (*Deir anba Shenudah*). Though known primarily as the greatest original author of *Coptic literature, his activities embraced many fields, spiritual and practical, and left deep traces on Egyptian *monasticism and church life. Unfortunately the *manuscript tradition of his works is in a deplorable state; the sources for his life are only in Coptic: no Greek source mentions him.

S. was primarily the head of his monastery, and he dictated a complete series of *rules of behaviour, inspired by the rules of the Pachomians, but adapted to new times and new needs created by the important role that monasticism had assumed. He preached the monks long and frequent *sermons, partly surviving, on moral and spiritual subjects. Outside the monastery, S. acted in concert with the bishops of *Alexandria (*Athanasius, *Theophilus, *Cyril and *Dioscorus) to spread and consolidate Christianity in Egypt (his attitude to moribund *paganism was particularly severe) and to work for the triumph of Alexandrian *theology (he seems to have accompanied Cyril to *Ephesus). He also had relations with high civil officials, who appealed to him on various questions, and provided help and refuge to the local people during barbarian raids (desert nomad tribes: Blemmi).

His literary work reflects these activities: *homilies and *catecheses on monastic subjects, *letters addressed to various churchmen and *laymen. But he also wrote theological treatises (esp. on *Nestorianism) and a work against gnosticizing and *Origenist tendencies.

J. Leipoldt, *Sinuthii archimandritae vita et opera omnia*, I. III. IV; CSCO 41, 42, 73; tr. H. Wiesmann, CSCO 129, 96, 108, Louvain 1906-1951; E.C. Amélineau, *Oeuvres de Schenoudi*, I-II, Paris 1907-1914; K.H. Kuhn, *Pseudo-Shenoute on Christian Behaviour*, CSCO 206-207, Louvain 1960; numerous fragments edited in various places: cf. W. Kammerer, *A Coptic Bibliography*, Ann Arbor 1950 (index); N. Bell, *Besa, the Life of Shenute*, Kalamazoo (intr., tr. and notes). The only monograph predates the edition: J. Leipoldt, *Schenute von Atripe und die Entstehung des national-ägyptischen Christentums*, TU 25, 1, Leipzig 1903; J. Limbi, The State of Research on the Career of Shenoute of Atripe, in (eds.) B. Pearson-J.E. Goehring, *The Roots of Egyptian Christianity*, Philadelphia 1986, 258-270.

T. Orlandi

SHEPHERD, The Good

I. In the Fathers - II. Iconography.

I. In the *Fathers. Jesus calls himself "the good Shepherd" in Jn 10, 1. 14 and shows us the traits of this figure: the S. enters the sheepfold by the door (1), calls his own *sheep by name, leads them out (3) and goes before them (4); the sheep listen to his voice (3. 27) and follow him (4. 27). The mutual knowledge of the S. and the sheep (4. 14. 27) is based on the mutual knowledge of the *Father and the S. (15). The G.S. gives his life for his sheep (11. 15), unlike the hireling, who flees before the wolf (12-13); he gives them *life eternal (28) and must also lead the sheep that are "not of this fold", so that there will be "one flock and one Shepherd" (16). The image of the G.S. abounds in patristic literature from the start (prob. in Abercius, *Epitaph* 3-6). Besides occasional references, it receives specific attention in systematic *commentaries on *John (John Chrys., *Hom. 59-60 in Jo.*; Theod. Mops., *Jo.*; Aug., *In evang. Ioh.* 45-48; Cyr. Al., *Jo.*; Nonnus, *Par. Jo.*; Ammonius, *Jo.*; Bede, *In evang. Ioh.*) and in some *homilies (cf. Ps.-Chrysost., *In mem. mart.*; Aug., *Serm.* 137 and 138; Peter Chrysol., *Serm.* 40; Basil of Sel., *Or.* 26; Greg. Gt, *In evang.* 14). At times the name "Shepherd" is included in a list of christological titles (Orig., *Comm. in Rom.* VII, 19; *Jo.* I, 126. 267; *Hom. in Jer.* I [III] 4 [SCh 238, 326]; Basil of Caes., *Eun.* I, 7; *De Spir.* VIII, 17; Greg. Naz., *Or.* 30, 21; Euch., *Formulae* 6); it reveals Jesus' divinity (Orig., *Jo.* I, 22; Cyr. of Jer., *Cat.* 10, 3), shows the need of an innate guide for those who still possess little *logos* (Orig., *Jo.* I, 122, 190. 198; XIX, 39), shows Christ's care for men (John Chrys., *Hom. 59 in Jo.* 2). The qualifying adjective "good" is emphasized (Jerome, *Tract. in psalm.* 96, 10; *In Am.* II on Amos 5, 14-15; *In Is.* on Is 5, 20; XV on Is 55, 3); it often demonstrates, in connection with Mt 19, 17 and parallel passages, the equality of Father and Son (Ephrem, *Diat.* XV, 9-12; Ambr., *De fid.* II, 2, 25-26; *In Luc.* VIII, 67; Jerome, *Tract. in psalm.* 142, 10; *In Matth.* 19, 17; Caes. of Arles, *Breviarium fidei adv. haeret.*). Christ is "Shepherd of the *Catholic church in the world" (*Mart. Polyc.* 19, 2); he, Logos of the Father, is "the solicitous Shepherd of infants", i.e. the pedagogue "who leads us *children to salvation" (Clem. Al., *Paed.* I, 7, 53, 2-3; 9, 83, 3-84, 2); he is not just the S. of believers, in a general sense, but in the *soul of every man he is the S. of those actions that are devoid of reason (Orig., *Hom. 5 in Jer.* 6; cf. also Philo, *De sacr. Ab. et Cain* 45). Perhaps in polemic with *Marcionites and *gnostics, it is stressed how the one G.S. is also the legislator (Clem. Al., *Strom.* I, 26, 169, 1-2). The G.S. is related to some OT figures: *Abel (Chromat., *Serm.* 23, 2), *Moses (Tertull., *De fuga* 11, 1; cf. also Jerome, *Ep.* 82, 3), the shepherd of Ps 22 [23] (cf. J. Daniélou, *Bibbia e liturgia*, Milan 1958, 235-253) and Ezek 34 (Clem. Al., *Paed.* I, 9, 2-3; Cypr., *Ep.* 8, 1-2; Didym., *In Zach.* II, 39; Aug., *Serm.* 46 and 47; Jerome, *In Ezech.* XI on Ezek 34, 1-31; Basil of Sel., *Or.* 26, 2), the paschal *lamb (Hipp., *Trad. Ap.* 41; Ps.-Chrysost., *Pasc.* 2, 2; Orig., *Hom. 14 in Gen.* 2).

The title and characteristics of the G.S. are also attributed to men: to the apostles *Peter (Aug., *Serm.* 137, 4, 4; 138, 4, 4) and *Paul (Ambr., *Spir.* II, 10, 108; Aug., *Serm.* 137, 9, 11; *In evang. Ioh.* 47, 3); to the *apostles, *martyr bishops and St *Cyprian (Aug., *Serm.* 138, 1, 1); to the person of the bishop (Ign., *Philad.* 2, 1; Tertull., *Pudic.* 13; *Const. App.* II, 20, 1; Leo Gt, *Ep.* 10, 5); to the leaders of the churches who are shepherds in the image of the G.S. (Clem. Al., *Paed.* I, 6, 37, 3; Orig., *Hom. 12 in Lc.* 2; Cypr., *Unit. eccl.* 8; *Ep.* 69, 5; Greg. Naz., *Or.* 33, 15-16; Ambr., *In Luc.* VII, 50; John Chrys., *Hom. 60 in Jo.* 1); to *Constantine, since he leads his flock to the knowledge and worship of *God (Euseb., *Laud. Const.* IV, 65), to whoever fights for the Kingdom and sacrifices himself even to death to serve God (Ps.-Mac. Egyp., *Hom.* 17, 1: TU 72, 94). *Augustine clarifies the relationship between the one S. and the good shepherds by placing Jn 10 and Jn 21, 15-17 in parallel (*Serm.* 46, 23. 30; 138, 4, 4) and declaring that the shepherds are members of the S. (*Serm.* 138, 5, 5; *In evang. Ioh.* 46, 7-8). From this relationship we deduce that the shepherds must imitate the G.S. (Basil of Caes., *Moral.* reg. 80, 16; reg. 70, 19; Aug., *In evang. Ioh.* 47, 2; 123, 5) and speak his words (Aug., *Serm.* 46, 30) and that, esp. in periods of *persecution, the leaders of the church must not flee from danger like hirelings (Tertull., *De fug.* 11, 2; Cypr., *Ep.* 8, 1-2); when the wolf appears, the difference between hireling and true S. comes to light (Greg. Gt, *In evang.* 14, 2).

Other NT texts in which the image of a S. is referred to God or Christ, but without the designation of G.S., are: Lk 15, 4-7; Mt 18, 12-14; Mt 15, 24; Mk 14, 27; Mt 26, 31; Mt 25, 32; Heb 13, 20; 1 Pet 2, 25; 5, 4; Rev 7, 17.

GLNT 10, 1193-1227. J. Knabenbauer, *Commentarius in quatuor s. Evangelia. IV ev. secundum Joannem*, Paris ²1906, 335-349 (for patristic references); J. Quasten, *Der gute Hirte in hellenist. und frühchristl. Theologie: Heilige Überlieferung*, Festschrift für I. Herwegen, Münster 1938, 51-58; I. de la Potterie, Le Bon Pasteur: Populus Dei. *Mélanges A. Card. Ottaviani*, Rome 1969, 927-968; P.R. Tragan, *La parabole du "Pasteur" et ses explications: Jean 10, 1-18 (La genèse, les milieux littéraires)*, SA 67, Rome 1980 (bibl.).

A. Pollastri

II. *Iconography. Bucolic subjects were particularly common in the artistic tradition of the late imperial age: they responded to a need to depict the afterlife as a place of peace and serenity, and to an idealization of rustic *otium*. Among the various figures, a particular role was attributed to the man carrying the *sheep or ram on his shoulders, long considered to represent Christ as G.S. The image originated in pastoral-idyllic representations and was an abbreviated depiction of a pastoral scene, an allegorical image, before becoming the G.S. of the Christian repertoire. *Symbol of φιλανθρωπία or *humanitas*, following the iconographical tradition of Ram-bearing Hermes, in Christian art this image draws attention to the saving action and should not be considered as a portrayal of Christ, but as an ideogram, so that it can be repeated many times in the same place or on the same object (*Wp* 25; *Ws* 178, 2; 269, 1-4). The animal borne on the shoulders is not the sacrificial victim, but the saved creature. Again assimilating contemporary iconography, Christ's saving power is translated into the image of the **orans*, a component of the fundamental dual concept of salvation-history: the Saviour and the saved (*Wp* 117, 1; *Ws* 1, 2). The *pagan repertoire thus suggested scenes to Christian art and responded to the need of Christian artists and clients to introduce symbolic elements agreeable to popular *spirituality and in tune with the literary tradition of the sacred texts. The theme of the G.S. comes from various gospel passages: the *parable of the lost *sheep (Lk 15, 4-7); Christ's own words: "I am the good Shepherd" (Jn 10, 11), in accordance with the tradition affirmed in the Old Testament as well as the New (Job 10, 1-19; Ezek 24, 11-12; 14-16; 23; 31; Is 44, 28; Ps 22, 1-5; 118, 176, to cite only the best-known) of the S. as leader, guide, logos for the Christian people, salvation for the flock whose head he is. Thus Abercius, bishop of Hierapolis, defines himself in his epitaph as ". . . disciple of the pious Shepherd . . . " (A. Ferrua, Epitaffio di Abercio, 279 ff.). Particularly dear to the Christians of the growing church, the image of the G.S. was extremely common in all figurative genres. To the common type of the ram-bearing S., now young and beardless, now bearded, now bald, are added iconographical variations that confer narrative intentions on a representation which is symbolic in its own right: he is shown milking the sheep (*Wp* 117, 1), guarding the flock, *pedum* in hand (*Wp* 117, 2), stroking his dog or holding a milk-pail (*Wp* 66, 2). His dress, with a few variations, is almost always the same: a short, belted tunic, either exomis (*Ws* 1, 2; *Wp* 66, 1) or sleeved, usually with *alicula* (*Wp* 116, 2) and sandals. One seeming exception is the so-called G.S. in "singular dress", Oriental in type, very rich, portrayed on the floor of an oratory at *Aquileia, come down to us through restorations and adaptations *ab antiquo* (L. Bertacchi, *Buon Pastore dall'abito singolare*, 429 ff.). This enigmatic figure is interpreted, following literary evidence, as a representation of Christ. But for this as for the figures of shepherds, one in the oratory of the "CAL" at Aquileia (I. Bortolotti, *Sacello paleocristiano*, 26 ff.), the other in the Roman villa at Desenzano (M. Mirabella Roberti, *Mosaico col B.P.*, 393 ff.), the problem of whether they represent the G.S. or should rather be interpreted as expressions of an aspiration to *peace, an idealization of the *felicitas temporum* that led people to chose bucolic themes to decorate small private rooms, still remains open. The arcosolium *painting in the *cemetery of Balbina on the Via Ardeatina, showing a person with *nimbus in pastoral dress surrounded by 12 fish in two rows, interpreted by the editor, in accordance with a well-known passage of *Tertullian (*sed nos pisciculi secundum* ἰχθύν *nostrum Christum aqua nascimur: De bapt.* I, 3: CSEL 1, 277), as an allegorical *image of Christ (A. Nestori, Pitture inedite, 156), is certainly singular and unusual. There is both descriptive and symbolic intention in a scene in a *cubiculus* of Pretestato (*Wp* 51) near the *spelunca magna*, where the S. defends the flock on the right, pushing back a wild ass and a pig, symbols of *evil, with a **virga*. The certainty of *resurrection and welcome into the sheepfold by the G.S. after death is clearly expressed in the slab of Veratius Nicatora (Diehl 4463), where the G.S. is shown between *Jonah thrown up by the sea-monster, and the lion, symbol of death. Outside the funerary sphere, the image was particularly common in the decoration of Christian baptisteries, e.g. *Dura Europos, *Naples (*WMM* 36-38), the Lateran (*WMM* 256 f., fig. 73), where it appears alongside OT and NT episodes clearly allusive of the theme of salvation. Probably inspired by *Matthew's metaphor of the *sheep and the goats and the solemn *judgment at the end of *time (Mt 25, 31-34) is a sarcophagus-lid relief in the Metropolitan Museum, New York (*Ws* 83), with a S. in philosophical dress fondling the flock to his right and dismissing the goats to his left. But a complete assimilation of the pastoral allegory to an undoubtedly christological iconography appears in the well-known *mosaic in the Mausoleum of *Galla Placidia at *Ravenna, where the nimbussed Christ, with processional *cross, sits on a hill surrounded by his flock. Frequent on small objects and utensils (e.g. C.R. Morey, *Gold Glass*, pls. III, 14; XX, 18; AA. VV., *Age of Spirituality*, 465), ram-bearing figures are also known in some statues in the round: although their date has led to them being interpreted as representations of the G.S. (*Ws* 52, 1; 52, 5; 52, 8), the intention of such objects is not yet clarified and so it is hard to be sure whether they are generalized pastoral depictions or images of the G.S.

After an extraordinary flowering in ancient Christian art, esp. in the West, this iconographical type disappears almost completely in the Middle Ages, to reappear in some bas-reliefs or codex miniatures. **[Figs: 276, 277, 278]**

DACL 13, 2272-2390; *EC* 9, 930-934; *RACh* 3, 11 ff.; *LCI* 2, 289-299; *EAA* 2, 223 f.; *Wp* I, 45 ff.; *Ws* I, 63 ff.; H.U. von Schönebeck, Die christlichen Paradeisos-Sarkophage, *RAC* 14 (1937) 289-343; J. Quasten, Der Guter Hirt in frühchristlicher Totenliturgie und Grabeskunst, in *Misc. G. Mercati*, ST 121, Vatican City 1946, 373-406; J. Fink, Mythologische und biblische Themen in der Sarkophagplastik des 3. Jahrhunderts, *RAC* 27 (1951) 167-190; L. de Bruyne, Le décoration des baptistères paléochrétiens, in *Actes du Ve Congres Int. d'Archéologie Chrét.*, Vatican City 1967, 341-369; A. Stuiber, *Regrigerium interim. Die Vorstellungen vom Zwischenzustand und die frühchristliche Grabeskunst*, Bonn 1957, 151 ff.; T. Klauser, Studien zur Entstehungsgeschichte der christlichen Kunst, *JbAC* 1-10, 1958-1967 (cf. note on vol. 10, 42, and index); L. de Bruyne, Les lois de l'art paléochrétien comme instrument herméneutique, *RAC* 39 (1963) 7-92; A. Grabar, *Christian iconography. A Study of its Origins*, Princeton 1968, 10 f.; id., *L'art de la fin de l'antiquité et du moyen âge*, III, Paris 1968; E. Dassmann, *Sündenvergebung durch Taufe, Busse und Martyrerfürbitte in den Zeugnissen frühchristlicher Frömmigkeit und Kunst*, Münster/W, 1973, 322 ff.; N. Himmelmann, Sarcofagi romani a rilievo. Problemi di cronologia e di iconografia, *ASNP* 4 (1974) 139-177 (cf. 156 ff.); P.-A. Février, Naissance d'un art chrétien, *Les Dossiers de l'archéologie* 18 (1976) 18 ff.; R. Giordani, Frammento di rilievo inedito con rappresentazione di Buon Pastore nella basilica di S. Marco a Roma, *RAL* 30 (1976) 342-360; W.N. Schumacher, *Hirt und "Guter Hirt"*, Rome-Freiburg-Vienna 1977; N. Himmelmann, *Über Hirten Genre in der antiken Kunst*, Opladen 1980. For the examples cited in the text: A. Ferrua, Nuove osservazioni sull'epitaffio di Abercio, *RAC* 20 (1943) 279-305; C.R. Morey, *The Gold Glass Collection of the Vatican Library*, Vatican City 1959; A. Nestori, Pitture inedite del cimitero della via Ardeatina, *Rend. PARA* 45 (1972/1973) 151-163; M. Guarducci, Il Buon Pastore tra i pesci, *ibid.*, 165-170; I. Bortolotti, *Il sacello paleocristiano della CAL ad Aquileia*, Udine 1973; L. Bertacchi, Il mosaico aquileiese del Buon Pastore dall'abito singolare, *AAAd* 12 (1977) 429-444; AA. VV., *Age of Spirituality*, New York 1978; M. Mirabella Roberti, Un mosaico con Buon Pastore a Desenzano, in *Atti del V Congr. Naz. Arch. Crist.*, Rome 1982, 393-405.

A.M. Giuntella

SICCA VENERIA. This city (now Le Kef, Tunisia) had various names in antiquity, but by preference Sicca or Veneria, or Sicca Veneria. It was part of *Africa Proconsularis. Perched on a mountain, Mt Eryx, where rose a temple to Venus famous for the voluntary prostitution of matrons (Valer. Max., *Fastorum* II, 6, 15), on the road from *Carthage to *Constantine, it sided early with the Romans. *Augustus made it a Roman *colonia*. In the 3rd c. the *curator civitatis* and *curator* of the temple of Venus were the same person (CIL 8, 15881); in the 6th c. *Justinian gave it a new line of fortifications (Procop., *Aedif.* 6, 7, 10).

Christianity is attested here from the 3rd c. Of its bishops we know Castus in 256 (*Sent. epp.* 28), Eparchius in 345-48 (*Conc. Carth. sub Grato ep.*), Paul the Donatist and *Fortunatian the Catholic in 411 (*Conf. of Carth.*). The latter had relations with *Augustine, as did his successor *Urbanus (Aug., *Epp.* 148, 229). *Victor of Vita mentions another Paul (*Hist. pers. Afr. prov.* 2, 6); Candidus was bishop in 646. S. V. was famous for *Arnobius the Elder, professor of *rhetoric there; his *conversion in 295/6 so stupefied the bishop's entourage that he was asked to provide tangible proof of his sincerity. He replied with books I and II of his treatise *Against the pagans*. In Augustine's time the priest *Apiarius of S. V. became famous by appealing, against African custom, to *Rome.

The city still preserves remains of early Christian *church buildings: *1)* Dar-el-Kuss, a basilica dedicated to St *Peter, as numerous surviving keys on the vault demonstrate; *2)* the episcopal complex, with baptistery; *3)* the building "*à auges*" attached to the great mosque.

DACL 8, 689-701; *PWK* 21/2, 2187-88; *LTK* 9, 730.

V. Saxer

SICILY, *Council of. The *homoeousian delegation that went to *Rome in 365/6 and was persuaded by pope *Liberius to subscribe the Nicene *homoousios*, stopped at Sicily on the return *journey. A council of local bishops met for the occasion and confirmed Liberius's position and the validity of the Nicene *creed of 325 as the parameter of *orthodoxy. A *letter from the Sicilian bishops was read out at the homoeousian council of *Tyana (*Cappadocia) in 366. For Sicilian *archaeology, see *Italy.

Hfl-Lecl 1, 976-979; Simonetti 397.

M. Simonetti

SIDE, *Council of. In this Pamphylian city a council of 25 bishops met between 383 and 394, presided over by *Amphilochius of Iconium. Its *Acts are lost, but *Photius claimed to have read them together with the evidence of various bishops about the *Messalians, some of whom, incl. their leader Adelphius, were present: failing to show sufficient evidence of repentance, they were condemned. The council sent a *letter to *Flavian of Antioch telling him what had happened. **[Fig: 279]**

Photius, *Bibl.*, cod. 52; Mansi 3, 651; Palazzini 4, 184; *DIP* 5, 1262 f.

A. Di Berardino

SIDONIUS APOLLINARIS. G. Sollius Apollinaris Sidonius (all these names are given at the beginning of *Carm.* 9) was born at *Lyons on 5 Nov (*Carm.* 20, 1), *c.*431 (as we deduce from *Ep.* VIII, 6, 5), of a rich aristocratic Gallo-Roman family. His father and grandfather had been prefects of *Gaul, the former under *Valentinian III, the latter under the usurper Constantine. His good education, first at Lyons, then at *Arles, included Greek. Aged 20 he married Papianilla, daughter of Avitus who was soon (455) to be elected emperor. In Jan 456 he accompanied his father-in-law to *Rome, where he celebrated his assumption of the *fasces* with a panegyric. When Avitus was defeated by the coalition of Maiorianus and Ricimer and stripped of the imperial attributes, S. managed to extricate himself from his father-in-law's ruin and in 458 welcomed the emp. Maiorianus to Lyons with a new panegyric. After Maiorianus's assassination by Ricimer, S. withdrew from public life for some years and retired - to read, write and cultivate *friendships - to his possessions at Avitacum in the Auvergne. Most of the poetry collected in the *nugae* (*Carm.* 9-24) and the first books of *letters dates from these years, 461-467. This period of retirement ended in autumn 467 when, invited by an official letter, S. went to Rome; here, on 1 Jan 468, he pronounced a panegyric on the emp. Anthemius. He was rewarded with the prefecture of the city for 468. He fulfilled this duty - not without difficulties (cf. *Ep.* I, 10) - and returned to Gaul. Here, while the Visigoths pressed, S., after another period in the solitude of his possessions, changed the course of his life: after being - probably - ordained priest, in 471 he was elected bishop of Arvernum (Clermont-Ferrand). This was no inner crisis, but a not infrequent choice among the aristocracy of the time for whom, faced with the barbarians, nothing remained but to "renounce either their country or their hair" (*Ep.* II, 1, 4). But S. took seriously to church life, renouncing his earlier life and *family and discharging his high *ministry with diligence. He looked to the needs of his city while the Visigothic advance spread out, and became the animator of the resistance to the barbarians. When the Auvergne was in *Euric's hands, S. was forced into exile at Livia, near Carcassonne. In late 476 or in 477 he was authorized to return to Clermont, but not before making an act of homage, including a verse composition, to the Visigothic king. He passed his remaining years in the cares of his episcopal office, composing other epistles and the rare poem. The last of the dateable letters is of 481, but on the strength of (not wholly certain) references in other letters to later events, S.'s death must be placed after 486-7. After his death he was honoured as a saint; his feast is still celebrated at Clermont on 23 Aug.

S.'s verse compositions have come down in a collection of 24 *carmina*, compiled by the author almost certainly in 469. It seems to be a union of two independent collections, that of the three hexametrical panegyrics, for Anthemius (584 vv.), Maiorianus (603 vv.) and Avitus (602 vv.), each preceded by a brief verse preface, and that of the *epigrammata* or *nugae*, in various metres. Originally separate and added later were *Carmina* 9 (*Ad Felicem*), seemingly the collection's dedication, 22 (description of *Burgus*), 23 (eulogy of *Consentius) and 16 (eulogy of *Faustus of Riez), longer and more committed in tone than the other compositions, which are, apart from the two *epithalamia* (*Carm.* 11 and 15), brief occasional notes. Thirteen more compositions in various metres are inserted among the letters: epitaphs, inscriptions for churches, eulogies of friends and their writings. S.'s poems are without originality, but technically well-constructed and generally metrically correct; they continually imitate Virgil, Ovid, Statius and *Claudian, and are dense with mythological references.

The 147 *letters are divided - in the manner of *Pliny the Younger and *Symmachus - into nine books. Their arrangement is not chronological, but obeys literary and artistic criteria, esp. the variation of succeeding subjects. Very probably the author initially published a single book; then followed an edition of the first seven books, *c.*477, to which an eighth and a ninth were subsequently added. The date of individual letters cannot always be determined. We owe the most credible dating of the majority of them mainly to Loyen. Composed in an ornate and extremely careful prose supported by a constant - sometimes even tedious - use of rhetorical expressions, S.'s letters are "artistic letters", written to be published and destined for posterity. In the stylistic ideals that inspire them, in their insistent professions of modesty - the more the author makes light (to the point of self-denigration) of his own works, the more he desires the applause of his friends and the acknowledgement of his readers -, in their insistence on formulae of courtesy, compliments and eulogies, in their lively sense of patriotism, these letters are an expression of the "precious" ways of a cultivated and refined aristocracy, lovers of hyperbole to the point of extravagance and all the more proud of their own Roman culture the more this seemed to be undermined by the presence of the barbarians. Of exceptional historical interest, the letters offer us an enormous mass of information, sometimes unique, on 5th-c. Gallic society, the situation of the empire, the aristocratic classes, the church and the barbarians.

S.'s translation of Philostratus's *Life of *Apollonius of Tyana*, made during his exile at Livia (*Ep*. VIII, 3, 1), the *contestatiunculae* sent to bishop Megetius (*Ep*. VII, 3, 1) and the *Rogationes* introduced to Clermont in winter 472-3 (*Ep*. V, 14, 1 and VII, 1, 2) are all lost.

Editions: CPL 986-987a; PL 58 (reproduces Sirmond's 1652 edition with commentary); C. Luetjohann, MGH, *Auct. ant.* 8 (1887); P. Mohr, Teubner (1895); W.B. Anderson, LCL (vol. 1: 1963; vol. 2: 1965); A. Loyen, Les Belles Lettres (vol. 1: 1960; vols. 2 and 3: 1970); Sidonius, *Poems and Letters* (tr. W.B. Anderson) LCL Vol. 1 (1936) and Vol. 2 (1965); id., O.M. Dalton, *The Letters of Sidonius*, Oxford 1915 (useful introd.).
Studies: L.A. Chaix, *St. Sidoine Apollinaire*, 2 vols., Clermont-Ferrand 1867-68; P. Allard, *St. Sidoine Apollinaire*, Paris 1910; C.E. Stevens, *Sidonius Apollinaris and His Age*, Oxford 1933; A. Loyen, *Recherches historiques sur les panégiriques de Sidoine Apollinaire*, Paris 1942; id., *Sidoine Appollinaire et l'esprit précieux en Gaule aux derniers jours de l'empire*, Paris 1943; K.F. Stroheker, *Der senatorische Adel im spätantiken Gallien*, Tübingen 1948, 217-219, no. 358; S. Dill, *Roman Society in the Last Century of the Western Empire*, New York ²1958, 187-223; S. Pricoco, Studi su Sidonio Apollinare, *NDid* 15 (1965) 71-150; W.H. Semple, Apollinaris Sidonius. A Gallo-Roman Seigneur, *BRL* (1967) 136-158; F.E. Consolino, Codice retorico e manierismo stilistico nella poesia di Sidonio Apollinare, *ASNP* 4 (1974) 423-460; I. Gualandri, *Furtiva lectio. Studi su Sidonio Apollinare*, Milan 1979; N.K. Chadwick, *Poetry and Letters in early Christian Gaul*, London 1955; R.P.C. Hanson, The Church in Fifth Century Gaul: the Evidence of Sidonius Apollinaris, *JEH* 21 (1970) 1-10.

S. Pricoco

SIGISMUND. King of the *Burgundians. Converted from *Arianism to *Catholicism by *Avitus of Vienne *c.*496-99, S. succeeded his father Gundobald in 516. He expiated the murder of his son Sigeric (522) by doing penance at the *monastery of St Maurice at Agaunum, which he had restored and in which he instituted the *laus perennis*. Defeated in 523 by the *Franks, he fled once more to St Maurice where he was made prisoner, taken to Orléans (*Aurelianum*) and drowned with his wife and children in a well at S.-Péravy-la-Colombe (Loiret). In 535-6 his remains were buried at St John's near Agaunum, while those of his sons Gundobald and Giscald were entombed at St Maurice itself. The relics were later dispersed to Matzenheim (Alsace), Prague, Freising and Plck. A *mass in his honour to cure fever occurs frequently in early sacramentaries. His name appears in *Mart. hier.* on 1 May. A *Life* of him was written (BHL 7717-19).

Vies des SS. 5, 20-23; *LTK* 9, 748-49; *BS* 11, 1043-47.

V. Saxer

SIGISTHEUS. *Vandal count living in *Africa presumably in the 5th or 6th c., author of a brief complimentary *letter to a presbyter named *Parthemius, full of deference and praise for his correspondent's literary qualities; the reply survives.

CPL 803: PLS 3, 447-448.

F. Scorza Barcellona

SILVANUS of Cirta. Subdeacon of bishop Paul at *Cirta (in the 4th c. *Constantina*: now *Constantine, Algeria). During *Diocletian's *persecution he collaborated fully with the Roman official who made an inventory of the community's goods. Some time later, late 304 or early 305 - Lancel, in a recent article, thinks spring 307 - he was called to succeed Paul in a very disputed election (cf. *Gesta apud Zenoph.*: CSEL 26, 192-196): his opponents called him a *traditor*, but the people supported him. His consecrator *Secundus of Tigisis, primate of *Numidia, was under the same accusation: yet they both contested the election of *Caecilian as bishop of *Carthage (cf. Aug., *Contra Cresc.* III, 27, 31: CSEL 52, 437) and elected *Maiorinus in his place, formally opening the *Donatist *schism, of which they became important leaders in Numidia. When S. excluded the deacon Nundinarius from *communion with him, the latter accused him before the *consularis* Zenophilus. At the trial, which took place at *Timgad on 8 Dec 320 and is known through the *Gesta apud Zenophilum* (CSEL 26, 185-197), S. was fully proved to be a *traditor*, a thief and a simoniac, and condemned to exile.

Monceaux 4, 4-13, 24-32; *PAC* 1, 788 f.; 1, 1078-1080; S. Lancel, Les débuts du Donatisme: la date du "Protocole de Cirta" et de l'élection de Silvanus, *REAug* 25 (1979) 217-229; W.H.C. Frend, *The Donatist Church*, Oxford ³1985, ch. 1; B. Kriegbaum, *Kirche der Traditoren oder Kirche der Märtyrer?*, Innsbrucker theol. Studien 13, Tyrolia-Cologne-Vienna 1986.

A. Di Berardino

SILVANUS of Tarsus. One of the main homoeousian spokesmen at the council of *Seleucia (359), he took part in the subsequent discussions with *Constantius at Constantinople. The council of *Constantinople (360) deposed him and many other *homoeousians; but after Constantius's death (362) he resumed his anti-Arian activities. In 365-6 he was part of the homoeousian delegation that went to *Rome to ask pope *Liberius for help and solidarity: on this occasion he agreed to subscribe the Nicene *homoousios*. He took part in the subsequent council of *Tyana at which the delegation reported the results of its mission to the West.

DCB 4, 669.

M. Simonetti

SILVERIUS, pope (536-537). On Agapitus's death the Ostrogothic king Theodahat imposed the election of the subdeacon S., son of pope *Hormisdas (†523). In the war between the Ostrogoths and the *Byzantines he sided with *Belisarius, by whom he was accused of being pro-Ostrogoth, deposed (21 March 537) and deported to *Lycia (Liberatus, *Breviar.* 22: PL 68, 1040). *Vigilius was immediately elected in his place (29 March). *Justinian sent S. back to *Rome for a more serious enquiry into his case but, as soon as he arrived, Vigilius's intrigues got him exiled to Ponza, where he renounced his office and died soon after (2 Dec 357). He is venerated as a *martyr for his defence of the Chalcedonian faith.

LP I, 290-295; *ASS Iun.* IV, 13-18; *BS* 11, 1069-1071; P. Hildebrandt, Die Absetzung des Papstes Silverius (537), *HJ* 42 (1922) 213-249; L. Duchesne, *L'Eglise au VIe siècle*, Paris 1925, 151-154; O. Bertolini, *Roma di fronte a Bisanzio e ai Longobardi*, Bologna 1941, 129 f., 145-176; L. Magi, *La sede romana nella corrispondenza dei patriarchi e imperatori bizantini (VI-VII sec.)*, Rome-Louvain 1972, 133.

A. Di Berardino

SIMEON. Bishop and *martyr (†*c.*107). Son of Clopas (Lk 24, 18; Jn 19, 25) and cousin of the Lord; he succeeded *James the Less as second bishop of *Jerusalem, but only after an interval, acc. to *Eusebius: "after the *martyrdom of James and the capture of Jerusalem" (*HE* III, 11). Under his episcopate the Jerusalem community fled to Pella. His election shows the importance given to blood-ties by the *Judaeo-Christian community. *Hegesippus says that *Vespasian (Euseb., *HE* III, 12) and *Domitian (Euseb., *HE* III, 20, 1-6) sought out *David's descendants. Probably S. managed to escape the searches. He was martyred under *Trajan (Euseb., *Chron.* a. 107); after long torments he was crucified. Many identify him with the *apostle *Simon the Cananean, but ancient *tradition distinguishes the two.

G. Ladocsi

SIMEON STYLITES the Elder (†459; feast: 27 July for the Syrians, 1 Sept for the Greeks). Born before 400 on the *Syria-*Cilicia border, a shepherd, he lived two years with some nearby *ascetics and ten years in the *monastery of Eusebonas at Teleda (later all the monasteries disputed the honour of having trained him). Outdoing them all in austerity, he was begged to go elsewhere. He adopted the life of a recluse, passing whole Lents without food. Finally, to escape the influx of *pilgrims, he began his life as a *stylite, living on a column, always in the open, on a narrow space. The columns rose ever higher (finally reaching 16 metres). A community below was at his service and that of the crowds of admirers. In *c.*430 bishops and *archimandrites, constrained by his sanctity, conceded their approval; *Theodoret's eulogy (*Histoire des moines de Syrie*: SCh 257, 158-218) is a convincing example; see also the *Syriac and Greek *biographies. He was admired by the whole world, even at Paris. The emp. *Marcian pressured him to proclaim his adherence to *Chalcedon. The sanctuary built around his column in the time of *Zeno was a second, incomparable marvel.

BS 11, 1116-1138; H. Delehaye, *Les saints stylites*, Brussels 1923; A.J. Festugière, *Antioche païenne et chrétienne*, Paris 1959, 388-401; A. Vööbus, *History of Asceticism in the Syrian Orient*, II, Louvain 1960, 207-223; G. Garitte, Compléments à l'édition de la vie géorgienne de S. Siméon Stylite l'Ancien, *Muséon* 75 (1963) 79-93. On the letters attrib. to the saint: CPG 6640-6650.

J. Gribomont

SIMEON STYLITES the Younger (521-592). The cult of a second S., a *stylite on Mons Mirabilis in the immediate vicinity of *Antioch, seems to have been an attempt to replace that of the elder *Simeon which had been appropriated by the *monophysites. From 541 the young *ascetic lived on a column; in 551 he moved to a higher one. He let himself be ordained priest in 554 to confirm his *orthodoxy. The memory of his mother St Martha is also preserved. Some ascetic discourses and *letters (CPG 7365-7370) may be authentic, but with later alterations. Characteristic of S. were the lead or wooden images which spread to the West in the 8th c.

P. van den Ven, Les Ecrits de s. Siméon Stylite le Jeune, *Muséon* 70 (1957) 1-57; BS 11, 1141-1157; P. van den Ven, *La vie ancienne de s. Syméon Stylite le Jeune*, I-II, Brussels 1962-1970.

J. Gribomont

SIMON, apostle. Surnamed the Cananean by Mk 3, 18 and Mt 10, 4, who place him 11th before *Judas Iscariot; Luke places him tenth before *Judas son of James and Judas Iscariot, and calls him the Zealot (Lk 6, 15; Acts 1, 13). The two epithets "Cananean" and "Zealot" have led to differing interpretations. Apart from the disputable identification with the Lord's homonymous cousin *Simeon or Simon (Mt 13, 55; Lk 6, 3), brother and successor of *James the Less as bishop of *Jerusalem 62-106, Copts and Byzantines hold that this name conceals Nathanael of Cana (cf. *Bartholomew) or even the steward of the wedding-feast (Jn 2, 8-9). This interpretation cannot be sustained, since "Cananean" is not derived from Cana, but from the Aramaic *qena`na`*, "Zealot". Some understand this in the religious sense of burning with zeal, enthusiast (cf. Gal 1, 14), others as a clear sign of his membership of that religious movement of fanatical integralists, widespread and active esp. in Galilee, transformed during the Jewish War into a political faction which allowed no distinctions between religious and political life. After *Pentecost he preached the gospel prob. in *Egypt and *Persia, where he witnessed to Christ with his *blood. Acc. to a 9th-c. source, his tomb and a *church dedicated to him were at Nicopius in the western Caucasus (Epiph., *Vita Andreae*: PG 120, 244).

BS 11, 1169-1173; O. Cullmann, *Gesù e i rivoluzionari del suo tempo*, Brescia ²1971, 20-22; M. Hengel, *Die Zeloten*, Leiden 1971, 72-76; M. Borg, The Currency of the Term "Zelot", *JThS* n.s. 22 (1971) 504-512; G.R. Edwards, *Jesus and the Politics of Violence*, New York 1972, 55-59; J.A. Morin, Les deux derniers des Douze: Simon le Zélote et Judas Iskariôt, *RBi* 80 (1973) 332-349; G. Jossa, *Gesù e i movimenti di liberazione della Palestina*, Brescia 1980, 61-77.

E. Peretto

SIMON MAGUS - SIMONIANS. Contemporary of the *apostles, born prob. at Gitton in Samaria, where he met the deacon *Philip who had gone there with the apostles' mandate to spread the gospel (Acts 8, 1-8). Simon captured the enthusiasm and acclamation of the people by the spectacular deeds he performed: "This man is that Power of God which is called great" (Acts 8, 10). Baptized with other *Samaritans, but dazed by the signs and wonders worked by Philip, when *Peter arrived to lay hands on the newly-baptized he offered him money to acquire the mysterious power he possessed. Peter roughly cast him out, threatening divine punishments, since he had reduced the *charisms of the Spirit to merchandise (Acts 8, 18-25). From Simon's act derives simony, the purchase of spiritual goods with money. The sect of Simonians, made to derive from Simon, was not markedly *gnostic. The most salient points of its teaching were: Simon was considered the supreme God, and Helena, a prostitute he had redeemed from a brothel at Tyre, was his Thought (*Ennoia*), sprung from his mind. Helena had created the intermediate powers (*angels and archangels) who in turn had created the *world. Envious and jealous, they had then enclosed Helena in a human *body and forced her to transmigrate from one body to another. Simon, to free Helena and all men from the power of the intermediate powers, descended to earth and made himself known as the Son in Judaea, the *Father in Samaria and the *Holy Spirit elsewhere. Salvation came through faith in Simon's liberating power.

These doctrines do not seem much in line with the gnostic systems: the most difficult problem they create is that of identifying their primitive central nucleus, since its Christian characteristics and its analogies with a certain type of preaching in Samaria can be considered either as primitive or as later additions. Gnostic in tendency are the hostility of the intermediate powers against *God and men, the consideration of the human body as a prison of the divine element, the disapproval of the OT. Other traits are indefinable, or at least cannot be justified as gnostic subject-matter: the divinization of Simon and Helena, their claimed immortality, the absence of any mention of a specific fault explaining Simon's descent or of any link between *redemption and knowledge of Simon's nature. One has the impression of being on the way towards *Judaeo-Christian gnosticism, but of not having yet entered the gnostic climate.

Justin: *I Apol* 26, 1-3; Irenaeus: *Adv. haer.* I, 23, 1-2; 24, 1-2; Hippolytus: *Refut.* VI, 3-18, 7; L. Cerfaux, La gnose Simonienne, *Recueil L. Cerfaux*, Gembloux 1954, 191-258; R. Grant, *La gnose et les origines chrétiennes*, Paris 1964, 63-82; J.M.A. Sables Dabadie, *Recherches sur Simon le Magicien*, Paris 1962; M. Simonetti, *Testi gnostici cristiani*, Bari 1970, 1-3; J. Daniélou, *La Teologia del giudeo-cristianesimo*, Bologna 1974, 99-106; K. Beyschlag, *Simon Magus und die christliche Gnosis*, Tübingen 1974; S. Arai, Simonianische Gnosis und die Exegese über die Seele, in (ed.) M. Krause, *Gnosis and Gnosticism*, NHS 8, Leiden 1977, 185-203.

E. Peretto

SIMPLICIANUS of Milan. Perhaps of Milanese origin, presbyter at *Milan (*Mediolanum*) and *Ambrose's successor as bishop from 397. An educated theologian, acquainted with *neoplatonism, S. played a part in the *conversions of *Marius Victorinus and *Augustine, and was close to Ambrose in his spiritual development (cf. *Confess.* VIII, 2, 3). Ambrose wrote him several *letters (37, 38, 65, 67=Maur.); S.'s letters to Ambrose and Augustine are lost. Augustine's *De diversis quaestionibus ad Simplicianum* discusses certain points of the books of Kings and the epistle to the Romans; Augustine's *Ep.* 37 was also addressed to S., who died 400-401.

F. Savio, *Gli antichi vescovi d'Italia dalle origini al 1300 descritti per regioni. La Lombardia*, Florence 1913, 145-150; *EC* 11, 648.

S. Zincone

SIMPLICIUS, pope (468-483). A native of Tivoli (*Tibur*), he succeeded *Hilarus at a time of disorder in *Italy, characterized by the revolt of Ricimer and the end of the Roman empire in the West with the deposition of Romulus Augustulus by the Arian *Odoacer (476), who forbade the alienation of ecclesiastical *property without his authorization (PL 62, 74). We know of few actions by S. in the West: he reprimanded John of Ravenna, who had consecrated the bishop of Modena, who refused the office (Jaffé 583); he made Zeno of Seville his vicar for *Spain (Jaffé 590). He turned his whole attention eastward, as did his successors for decades. He refused to recognize can. 28 of *Chalcedon (Jaffé 569). In the *East, *Basiliscus took power and condemned the council of Chalcedon in his *Encyclica* (475): a moment of triumph for *monophysism in many Eastern sees. S. corresponded intensely with the emp. *Zeno, who returned to power in 476, and with *Acacius, patriarch of *Constantinople, who did not properly inform him of the gravity of events for the *orthodox cause. However he died before becoming aware of Zeno's *Henoticon*. For the Roman basilicas (St Peter's, S. Lorenzo and St Paul's) S. established a parochial service for *penitence and *baptism (*LP* I, 249). He also restored many other Roman *churches and composed dedications.

CPL 1664; *Verzeichnis* 543; PL 58, 35-62; *Collectio Avell.*: CSEL 55, 124-155; Thiel 175-214; *LP* I, 92-93, 249-251; III, 86-87; *DTC* 14, 2161-2164; *CE* 13, 232; Fliche-Martin 4, 366 f., 421; F.X. Seppelt, *Storia dei Papi*, I, Rome 1962, 134-137; W.H.C. Frend, *The Rise of the Monophysite Movement*, Cambridge ²1979, 173-183.

A. Di Berardino

SIMPLICIUS of Vienne. Bishop of *Vienne (*Vienna*) (late 4th-early 5th c.: what Gams says, *Series esp.*, Regensburg 1873, is inexact); *Paulinus of Nola included him in the group of prelates "truly worthy of God". At the council of *Turin in 398 he defended his metropolitan rights. Pope *Zosimus accorded him the prerogatives of *Arles (PL 20, 665; Jaffé 334: note that no. 335 is not authentic). Feast 11 Feb.

P. Courcelle, Fragments historiques de Paulin de Nole, conservés par Grégoire de Tours, in *Mélanges Halphen*, Paris 1951, 146-148; *DACL* 15, 1351-1360; *Vies des SS.* 2, 245; *BS* 11, 1201.

A. Hamman

SIN

I. Personal sin - II. Original sin.

I. Personal sin. The theological and liturgical areas in which Christian thought on the subject of personal s. developed are essentially four: prebaptismal *catechesis for *catechumens; penitential remission of ss. committed after *baptism; anti-heretical polemic; and monastic experience.

The oldest lists of ss. appear in prebaptismal instructions addressed to *Jews and *pagans preparing to receive the *sacrament of baptismal remission of ss. (*áphesis tôn hamartiôn*): outside the NT (cf. Gal 5, 19 ff.; 1 Cor 6, 9 ff.), the texts of *Did.* 1-6 and *Ep. Barn.* 18-20 illustrate the two ways of *life and death, *virtue and salvation or s. and perdition. In order of importance, the gravest s. is always *idolatry, followed by *fornication, murder, fraud, *magic, *abortion and so on. But many professions followed by catechumens were also considered sinful, since intrinsically involved with the idolatry of paganism, and had to be abandoned (Hipp., *Trad. Ap.* 16).

Primitive Christianity was profoundly convinced that, before the Christian *revelation, s. reigned sovereign over the whole *world, producing as its most visible fruit the moral and physical death of mankind (Rom 5, 12; Mel. Sard., *De Pascha* 48 ff.). Only the coming of the Christian religion and its sacramental means of salvation introduced the possibility of opposing the universality of s. and death.

But daily experience soon led to the realization that, even after baptismal rebirth, Christians continued to sin, even gravely, losing the privilege of the indwelling *Holy Spirit and the lifegiving contact with divine *grace. The penitential remission of ss. committed after baptism must have triggered off a long and delicate debate on the nature of s., the possibility of remitting it once committed and the church's power to mediate *God's mercy to the human sinner. First of all it was established that there was a fundamental distinction between ss. that led to death through their intrinsic gravity and those that were lighter and somehow inevitable in the daily life of every Christian living in this world (1 Jn 5, 16). But if, in the conscience of the *Catholic communities, all ss. were considered equally remissible on condition of the sinner's sincere repentance, rigorist circles inspired by *Montanism believed that the church's *authority could, through public *penitence, remit only lighter ss. while graver ss. were reserved for God's inscrutable mercy and the violent purification of *martyrdom. It is interesting to follow the evolution of *Tertullian's thought on the subject, from the Catholic positions of *De paenitentia* to the Montanist ones of *De pudicitia*.

There were plenty of attempts to classify and list the graver ss.: e.g. Tertullian in *Adv. Marc.* IV, 9, 6 counts seven: *idolatry, blasphemy, murder, *adultery, rape, false witness, fraud; but in *De pudic.* 5, basing himself on Acts 15, he lists three; idolatry, *fornication, murder. *Augustine criticizes this reduction (*Speculum, de libro Act. apost.*: CSEL 12, 199 f.), reaffirming the distinction between grave or mortal ss. and light, daily, venial ss., whose remission is entrusted not to public penitence but to the works of *fasting, *almsgiving and esp. *prayer (Remit our debts as we remit those of our debtors, recites the *Lord's Prayer: cf. *Ep.* 167; *Serm.* 351 and 352; *In evang. Joh.* 56, 4, etc.). *Caesarius of Arles says the same (*Serm.* 19, 2; 202, 1). The theological terminology of the 4th c. distinguishes *peccatores* from *impii*: the former are Christians who have sinned although they have *faith, and will be judged for it; the latter do not believe, and consequently will be damned without *judgment (Hilary, *In psalm.* 1).

Greek church writers, without ignoring the problems of classifying personal ss., went deeper into the question of free will as the one efficient cause of man's morally negative behaviour. *Clement of Alexandria states that the beginning of s. is in free choice and desire (*Strom.* I, 84, 2); s. derives from not knowing how to judge what one should do (*ibid.* II, 62, 3) and is a matter of will, not of nature (*ibid.* IV, 93, 3). For *Origen, the whole of history was set in motion by the free disobedience of rational beings who thus gave rise to the variety of the world and the distinct categories of *angels, men and *demons (*De Princ.* II, 9, 2-6). The speculative tradition of Greek ethical intellectualism, mainly *Aristotelian and *Stoic, and the polemical need to oppose the moral determinism preached by the various *gnostic sects, both contributed to this interpretation. The gnostics divided mankind into three classes; the *hylici* or χοϊκοί, predestined to perdition, the *pneumatici*, predestined to salvation, and the *psychici*, the only ones from whom a decision of free will was required. We may say that for the gnostics good and *evil were in some way innate to individual categories of men. From the end of the 3rd c. the *Manichees inherited this predestinationist theology, replacing the gnostics of the 2nd and 3rd cc. as the target of *orthodox polemists, under whose condemnation fell not just the Manichees, but all those sects which, without being rigorously predestinationist and dualistic, preached that evil was innate to man's physical, bodily nature. In particular the *Messalians were repeatedly accused, over the 4th-6th cc., of maintaining the doctrine of "natural evil" and the "sin of birth".

Numerous 4th- and 5th-c. church authors were careful to emphasize vigorously that only freely committed ss. deserve punishment and eternal damnation, unlike the impurity contracted at *birth which needs to be purified (cf. Did. of Al., *Job* X, 15; Ambr., *In psalm.* 48, 8-9); the sole responsibility for *evil is borne by the individual's free will from the age of reason (Did. of Al., *Exp. in Ps.* 57, 5; Ambr., *Cain et Ab.* II, 7, 25; *Jac.* I, 3, 10). *Theodore of Mopsuestia wrote an entire work *Against those who maintain that man sins by nature and not by free will* (summarized by *Photius, *Bibl.*, cod. 177), while *Augustine, freed from the chains of Manichaeism, dedicated a treatise to the defence of free will. For Augustine, every man's s. follows the same mechanism as *Adam's: it consists essentially in an *aversio a Deo* moved by pride, at the same time as a *conversio ad creaturas* caused by the autonomous and rebellious use of the free will, not enlightened or helped by divine *grace (*Quaest. Simpl.* I, 2, 18). So s. is, in a well-known definition, the will to keep or obtain what *justice forbids and what one is free to abstain from (*De duab. anim.* 11, 15). Augustine's reply to Manichaeism starts from the idea, acquired from *neoplatonism, that *evil is not a substance, an autonomous nature opposed to good, but a privation, a deficiency of good itself, which sinful man suffers by distancing himself from *God and squandering himself in the multiplicity of creatures.

The existence of human freedom presupposes that ethical life must be commensurate with the dictate contained in the Law, i.e. with God's will manifested in the commandments of the *Decalogue. Only through abiding by God's saving prescriptions can we hope to attain profound liberation from slavery to s. Monastic *meditation in particular followed this road, often arriving at declaredly rigorist positions like those of *Basil or *Pelagius. According to these masters of the *ascetic life, obedience to God's will must always be prompt and total, since the least transgression could risk compromising all previous ascetic efforts and frustrating their results (Basil, *De jud.* 7; *Mor.* XII, 1; *Reg. fus. prooem.*, etc.). God did not condemn *Adam for eating an apple, but for transgressing his commandment (Pelag., *De div. leg.* 5; *De oper.* 13; *Virg. Laus* 6). From this point of view all ss. are equivalent, since all equally express the will to disobey God's precept, and attempts to classify ss. by human measurements of their relative gravity lose any meaning. The rigorism of this ethical approach seems to be a re-emergence of the well-known Stoic paradox of the equality of all faults, but is also influenced by Jas 2, 10: "Who sins against a single commandment is guilty of all". At any rate Christian authors reacted against such rigorism, which they judged theoretically mistaken and impossible to put forward to the mass of the faithful (Aug., *Ep.* 167; Jerome, *Adv. Iovin.* II, 18 ff.).

Ancient *monasticism also studied the psychology of *temptation and struggle against the *demon. To the learned monastic tradition of *Evagrius of Pontus (*Cap. Pract.* 6-33; *De octo vit. cog.*) and John *Cassian (*Instit.* V-XII; *Coll.* V) must be attributed the first elaboration of a genuine system of ss. or capital *vices over which, to the number of eight, specific demonic powers were thought to preside: gluttony, *fornication, avarice, sadness, anger, *accidia, vainglory, pride. The elements here systematically elaborated go back to a very old *Judaeo-Christian tradition present in the *Testaments of the XII Patriarchs*, the *Shepherd* of *Hermas and in *Origen, and which received a notable literary reinterpretation early in the 5th c. in the *Hamartigenia* and *Psychomachia* of the Spanish poet *Prudentius.

G. Teichtweier, *Die Sündenlehre des Origens*, Studien zur Geschichte der katholischen Moraltheologie 7, Regensburg 1958; AA. VV., *Théologie du péché*, Bibliothèque de théologie, série II, vol. VII, Paris-Tournai 1960; S. Visintainer, *La dottrina del peccato in S. Girolamo*, AG 117, Rome 1962; M. Huftier, *La tragique de la condition chrétienne chez S. Augustin* Paris-Tournai 1964; K. Rahner, *La penitenza della Chiesa. Saggi teologici e storici*, Biblioteca di cultura religiosa 60, It. tr. Rome ²1968, 311-876; M. Spanneut, *Tertullien et les premiers moralistes africains* (Recherches et synthèses-section de morale), Gembloux 1969; P.F. Beatrice, L'allegoria nella Psychomachia di Prudenzio, *Studia Patavina* 18 (1971) 25-73; A. and C. Guillaumont, *Evagre le Pontique. Traité pratique ou le moine*, SCh 170, Paris 1971, 38-112; W.E.G. Floyd, *Clement of Alexandria's Treatment of the Problem of Evil*, Oxford 1971; H. Karpp, *La penitenza. Fonti sull'origine della penitenza nella Chiesa antica*, Traditio christiana 1, It. tr. Turin 1975; AA. VV., *Commandements du Seigneur et libération évangélique. Etudes monastiques proposées et discutées à Saint Anselme 15-17 février 1976*, SA 70, Rome 1977; P.F. Beatrice, *Tradux peccati. Alle fonti della dottrina agostiniana del peccato originale*, Milan 1978; J. Pelikan, *The emergence of the Catholic Tradition (100-600)*, Chicago 1971, 278-331; P. Brown, *The Body and Society: men, women and sexual renunciation*, Columbia 1988.

P.F. Beatrice

II. Original sin. The term original s. can mean two things: the s. of the origins, i.e. *Adam's personal s.; and the s. deriving from the origin, i.e. transmitted to all men from *birth. But unless specifically stated, it takes the second meaning and is one of the most difficult and mysterious truths of Christian teaching.

In the patristic era the church's awareness of it rested on biblical (Gen 3, 1-20; Rom 5, 12-19; other texts cited are Ps 51, 7; Job 14, 4; Eph 2, 3), liturgical (infant *baptism), and above all dogmatic bases (necessity and universality of *redemption in Christ).

Even in the earliest *patristics we find the Pauline comparison between Adam and Christ, to which is added that between Eve and *Mary, respectively the causes of death and *life (cf. Just., *Dial.* 100, 4-6; Iren., *Adv. haer.* 3, 22, 4; 5, 19, 1). *Melito of Sardis develops this comparison, speaking of the "sin that stamped its mark on every *soul, and those on whom it was stamped were vowed to death". And again: "all *flesh fell under sin, every *body under death" (*On Pascha* 54-55, ed. R. Cantalamessa, *I più antichi testi pasquali della Chiesa*, Rome 1972). *Irenaeus refers to the principle of all men's solidarity with Adam and with Christ: "In the first Adam we have offended God by not observing his commandment; but in the second Adam we have been reconciled, becoming obedient unto death" (*Adv. haer.* 5, 16, 3). *Origen, drawing on the *tradition, which he calls apostolic, of baptizing *children, writes that they (the *Apostles) knew that there is in everyone the stain of s. which must be washed with *water and the *Holy Spirit (*Comm. ep. Rom.* 5, 9; cf. *Hom. in Lev.* 8, 3; 12, 4; *Hom. in Luc.* 14, 5).

In the West, *Cyprian maintains that children must not be kept from *baptism, giving this reason: although they have not sinned personally, by being born from Adam they contract "the contagion of the old death", and ss. "not their own but those of others" must be remitted (*Ep.* 64, 5). Even *Tertullian, though he follows material *traducianism and does not favour the baptism of the new-born (*De Bapt.* 18, 5), speaks of the *vitium originis* that, through the work of the *demon, became *naturale quodammodo* (*De an.* 40-41).

In the next century, testimonies in favour of original s. occur in the Syrian Fathers *Aphraates (*Tract.* 6, 14; 23, 3) and *Ephrem (*Hymn.* 4, 1) and, in the Alexandrian school, *Athanasius (*Orat. c. Ar.* 1, 51: "Adam sinning, sin passed to all men") and *Didymus the Blind. Didymus explicitly calls generation the means of transmission (κατὰ διαδοχήν) of Adam's "ancient sin". Only Christ, born of a virgin, is free from it (*C. Man.* 8: PG 39, 1095).

At this point we must remember that the notion of original s. includes both hereditary punishment (mortality) and hereditary blame. The anti-*Manichaean controversy may have led some *Fathers to skip the second aspect and insist only on the first: *John Chrysostom's thought has been argued over from the earliest times (cf. Aug., *C. Iulianum* 1, 6, 21-7, 30). But the *Pelagians denied both, reducing the effect of the first s. to bad example alone. This was the theory of imitation. The controversy thus caused led the Fathers to consider more diligently, understand more clearly and preach more insistently the doctrine of original s., so that the controversy itself became the occasion of progress in the understanding of *dogma (cf. Aug., *De civ. Dei* 16, 2, 1).

The church's first reaction occurred at the synod of *Carthage, where six propositions of *Pelagius's disciple *Caelestius, containing the outline of Pelagian thought, were condemned. Bishop *Aurelius of Carthage reduced them to two, one concerning death (whether or not it comes from Adam's sin), the other concerning blame (whether or not it is transmitted by birth). To both Caelestius answered no, and was considered to be outside the church (Aug., *De grat. Chr. et de pecc. orig.* 2. 11. 12; *Ep.* 157, 22).

*Augustine, who was not present at that synod, intervened later at the request of the tribune *Marcellinus, with *De peccatorum meritis et remissione*, a fundamental work for the controversy and for *theology. In it, and in many others written later, he explored the themes called into question by the Pelagian denial of original s., i.e. the theologies of death, (disordered) concupiscence, *redemption, child baptism, human solidarity with Adam and Christ, the evils of the *world. His conclusion was that original s. was not a secondary but a fundamental question, since linked essentially to that of Christ's redemption and Christian *justification.

The magisterial *authority of the church recognized this conclusion as legitimate and sanctioned the dogma of original s., including the hereditary nature of death and blame. Landmarks in this direction were: the synod of *Diospolis (Palestine) in 415, which absolved Pelagius, but only because he condemned the statements of his disciple Caelestius (Aug., *De gest. Pelag.*: cf. NBA 17/2, App., pp. 503 f.); the African synods of *Carthage and Milevis of 416 (*Epp.* 175-176 among Augustine's), approved by pope *Innocent (*Epp.* 181-183 among Augustine's; Aug., *Serm.* 131, 10; *C. duas epist. Pel.* 2, 3, 5); the "plenary *council of all Africa" of 418, which confirmed the doctrine of original s. (DS 222-223); pope *Zosimus's *Tractoria* of 418, which renewed the condemnation of the Pelagian doctrines and their authors (only a few fragments remain: cf. Aug., *Ep.* 190, 23; *De grat. Chr. et de pecc. orig.* 2, 21, 24; *De an. et eius orig.* 2, 12, 17; DS 231); the council of *Ephesus (431), which ratified the Roman *Acts* of this condemnation (Mansi 4, 1338; for pope *Celestine's letter cf. *ibid.* 4, 1026). A century later the council of *Orange (529) confirmed the doctrine of original s. (DS 371-372; for pope Boniface II's approval cf. *ibid.* 398-400).

Patristic *theology also explored through Augustine the notion of original s., whose obscurity he felt (cf. *De mor. eccl. cath.* 1, 22, 40: "nothing better known in preaching, nothing more obscure to the understanding"). He explored it using the notion of human solidarity with Adam and Christ, in whom the history and lot of all men is summed up, since "every man is Adam, every man is Christ" (*In ps.* 70 d. 2. 1). Their influence is exercised primarily not through imitation, but through propagation and the infusion of *grace, or generation and regeneration. Original s. therefore is also willed, but by the will of the first man (cf. *De nupt. et concup.* 2, 28, 42), "in whom we were all one" (*De pecc. mer. et rem.*, 1, 10, 11; *De civ. Dei* 13, 14), just as we are all one in Christ and participate even without personal co-operation (the case of the baptized child) in his justice.

Original s. must therefore be understood as a notion intermediate between that of personal s. and that of the penalty for s., being identifiable neither with the one nor with the other (*Retract.* 1, 13, 5; 15, 3; 16, 2; *Op. imp. c. Iul.* 1, 47, 105).

This theology would pass into scholastic theology and be substantially confirmed by the council of Trent.

A. Casamassa, *Il pensiero di S. Agostino nel 396-397*, Rome 1919; J. Tixeront, *Histoire des dogmes dans l'antiquité chrét.*, Paris 1924; J. de Blic, Le péché originel selon St. Augustin, *RecSR* 16 (1926) 97-119; 17 (1927) 414-433, 512-531; J. Mausbach, *Die Ethik des heil. Augustinus*, Freiburg i. Br. 1929, I, 139-416; N. Merlin, *St. Augustin et les dogmes du péché originel et de la grâce*, Paris 1931; F. Floëri, Le pape Zosime et la doctrine augustinienne du péché originel, *AugM*, II, Paris 1954, 755-761; J. Gross, *Entstehungsgeschichte des Erbsündendogmas*, 1: *Von der Bibel bis Augustin*, Munich-Basel 1960; Ch. Boyer, Le péché originel, in *Théologie du péché*, Tournai 1960, 243-291; A. Sage, Le péché originel. Naissance d'un dogme, *REAug* 13 (1967) 211-248; id., Le péché originel dans la pensée de St. Augustin de 412 à 430, *REAug* 15 (1969) 75-112; V. Grossi, *La liturgia battesimale in S. Agostino. Studio sulla catechisi del peccato originale negli anni 393-412*, Rome 1970; E. Testa, *Il peccato originale nella patristica (Gen III)*, Jerusalem 1970; M. Flick-Z. Alszeghy, *Il peccato originale*, Brescia 1972; J. Liébaert, La tradition patristique jusq'au Ve siècle, in AA. VV., *La culpabilité fondamentale. Péché originel*, Gembloux 1975, 34-55; O. Wermelinger, *Rom und Pelagius. Die theol. Position der römischen Bischöfe im pel. Streit in der Jahren 411-432*, Stuttgart 1975; P.F. Beatrice, *Tradux peccati. Alle fonti della dottrina agostiniana del peccato originale*, Milan 1978 (cf. A. Trapè, "Tradux Peccati". A proposito di un libro recente, *Augustinianum* 19 (1979) 531-538); A. Trapè, *Introduzione generale a S. Agostino, Natura e grazia*, NBA XVII/1, 7-215 (with more copious bibliography); M. Aflatt, The Development of the Idea of Original Sin in Saint Augustine, *REAug* 20 (1974) 113-134; W.S. Babcock, Augustine's Interpretation of Romans (AD 394-396), *AugStud* 10 (1979) 55-74; id., Augustine on Sin and Moral Agency, *Journal of Religious Ethics* 16 (1988) 28-55; G. Bonner, Augustine's Doctrine of Man as Image of God and Sinner, *Augustinianum* 24 (1984) 495-514; id., Les origines africaines de la doctrine augustinienne sur la chute et le péché originel, *Augustinus* 12 (1967) 97-116.

A. Trapè

SINAI. Today the term S. means the whole peninsula between the Red Sea and the Gulf of Aqaba, but in antiquity it referred only to the mountain range in the southern part. The peninsula was part of the Roman province of *Arabia established by *Trajan in 106. Acc. to Christian *tradition, well established by the 4th c., the facts narrated in the book of *Exodus concerning *Moses and the transmission of the Law (cf. Ex 3, 1 ff.; 17 and 19, 16-32, 20) occurred in these places. Indeed the mountain of the Law was identified with the present Djebel Mousa (Mount of Moses), 35 km from ancient Pharan (now Feirân) (cf. Egeria, *Itin.* chs. 1-5). Because of the veneration the Christians had for the place, by the 4th c. the neighbourhood of the mountain was peopled with hermits, visited by Julian Sabas (Theodoret, *Hist. rel.* 2, 13: SCh 234, 222 ff.), Silvanus and his disciple Zacharias (Soz., HE 6, 32, 8), Mark of Scete (PG 65, 296C and 312D), etc. The pilgrim *Egeria went there late in 383, encountering many hermits nearly all the way up the mountain (it was forbidden to live right on the summit) and four little churches for liturgical services (cf. also Anon. Piac., *Itin.* 37 ff.). In the 6th c., to protect the monks from Saracen raids, the emp. *Justinian built a fortified *monastery dedicated to the Mother of God on the N slope of Djebel Mousa. Only after AD 1000 did it take the name of St *Catherine, whose body, acc. to legend, was brought there by *angels. The monastery is still very rich in icons saved from iconoclasm (726-823). It also possesses an important collection of manuscripts: Greek, *Arabic, *Syriac, *Georgian, etc.; among them, the 4th-c. *Codex Sinaiticus* of the *Bible contains much of the LXX and the NT, the *Shepherd* of *Hermas and the *Epistle of *Barnabas*. **[Figs: 280, 281, 282]**

DACL 15, 1463-1490; G. Garitte, *Catalogue des manuscrits géorgiens littéraires du Mont Sinaï*, CSCO 165, subs. 9, Louvain 1956; H. Skrobucha, *Sinai*, Oxford 1966; F. Mian, Caput vallis al Sinai in Eteria, *SBF* 20 (1979) 209-223; ead., L'anonimo Piacentino al Sinai, *VetChr* 9 (1972) 267-301; B. Lifshitz, Inscriptions du Sinaï et de Palestine, *ZPE* 7 (1971) 151-163; P. Compagni-P. Acquistapace, *Il Sinai. Sulle orme dell'Esodo*, Milan 1975; K. Wietzmann, *The Monastery of St. Catherine at Mount Sinai. The Icons*, I: *From the sixth Century to the tenth Century*, Princeton 1976; H. Husmann, Die datierten griechischen Sinai-Handschriften des 9. bis 16. Jahrhunderts, Herkunft und Schreiber, *OS* 27 (1978) 143-168; K. Weitzmann, *Studies in the Arts at Sinai*, Princeton 1982; J. Galey, *Il Sinai e il Monastero di S. Caterina*, Florence 1982. For other bibl., cf. under *Arabia.

A. Di Berardino

SIRICIUS, pope (384-399). After a long career in the service of the church, S. was elected bishop of *Rome with the consent of *Valentinian II. Though *Ambrose's dominant role eclipsed his *authority over the Western churches, he made a notable contribution to the development of the *papacy. This is attested by his *letters, addressed to *Himerius of Tarragona (385), the bishops of *Italy and *Africa (386) and *Anisius of Thessalonica (392). They concern disciplinary questions, elucidate the meaning of religious *virginity and express awareness of the *sollicitudo omnium ecclesiarum* to which S. felt himself called (*Ep.* 6, 1).

CPL 1637; PLS 3, 567 ff.; *BS* 11, 1234-1237; PL 13, 1131-1178 (Coustant); Ch. Pietri, *Roma Christiana*, Rome 1976, esp. 888-909; *Patrologia* III, Rome 1978, 548 f.; D. Callam, Clerical Continence in the Fourth Century, *ThS* 41 (1980) 3-50.

B. Studer

SIRMIUM

I. Christian origins - II. Councils.

I. Christian origins. Situated on the right bank of the river Sava (modern Srijemska Mistrovica), S. arose in the 1st c. AD as a military *castrum* on the border of the Roman empire. In the 3rd c. urban expansion began and a fortified city wall was built. Its greatest development was between the start of the 4th c. and the death of *Valens. S. was then the residence of the emperors, whose palace arose near the Hippodrome.

In the period of the great *persecutions many Christians were martyred here. S. was then the main city of *Pannonia Secunda and the see of the most important diocese of the region. The first known bishop, Domnus, subscribed at the council of *Nicaea. In 441 S. passed into the hands of the Huns; in 567 it was occupied by the *Byzantine army. The church and the church hierarchy were then reorganized, but in 582 the city was again destroyed by the Avars led by Baian. The last bishop, St Sebastian, fled to *Rome.

The oldest Christian *church buildings were outside the city. The *martyrium* of St Sinerotis and, in the E necropolis, that of St *Irenaeus were discovered in the 19th c. A trilobate *martyrium* was found in the E necropolis. A fourth *martyrium*, in ruins, was excavated on the present right bank of the Sava. The cathedral has not yet been found.

At the end of the 4th c. began the constant decline of the city, which seems to have been destroyed by a great fire. Over the ruins rose a basilica, only the E part of which is well-preserved. Perhaps this was the new cathedral built in the second half of the 5th c. The building had three aisles and a projecting apse with *synthronon*. The cult of the martyr *Demetrius of Thessalonica was introduced to S. in the 5th c. **[Fig: 283]**

M. Mirkovic, Sirmium. Its History from the 1st Century A.D. to 528 A.D., *Sirmium* 1 (1971) 5 ff.; V. Popovic, A survey of the topography and urban organization of Sirmium in the Late Empire, *ibid.*, 119; J. Guyon, Sirmium chrétienne dans la Basse Antiquité et le Haut Moyen Age, *MEFR Ant.* 86 (1974) 621 ff.; N. Duval, Sirmium "ville impériale" ou "capitale", *CCAB* 26 (1979) 53 ff.

N. Cambi

II. *Councils

347. In this year *Hilary (*Frag. hist.* B II 9, 1) puts a council of Eastern bishops meeting at S. against *Photinus. But this seems not completely accurate: it conflicts with other data.

351. Various Eastern bishops, *Eusebians and moderate *Arians, met in the presence of *Constantius to accuse Photinus. *Basil of Ancyra held a debate with Photinus at the end of which he was condemned, deposed and exiled. The council published a formula of faith which repeated the Antiochene one of 342, supplemented by 27 *anathemata against the doctrine of *Marcellus of Ancyra and Photinus.

357. Midway through the year, *Germinius, *Valens, Ursacius and a few other pro-Arian bishops met and published a long formula of faith which, without repeating the more radical Arian expressions, clearly subordinated the Son to the *Father and proscribed the doctrinal use of the terms *homoousios* and *homoiousios*, while not doing the same for the Arian doctrine. The *Catholics, in both *East and West, considered it openly Arian (*blasphemia Sirmiensis*).

358-359. The publication of the Sirmian formula of 357 provoked a strong reaction by the *homoeousians led by Basil of Ancyra. Following the council of *Ancyra of 358, Basil and some of his followers went to S. and, profiting from Constantius's momentary favour, held a small council of several Eastern bishops present *in loco* (S. was an imperial residence), with the three Illyrian pro-Arians Valens, Ursacius and Germinius. The ideas of the homoeousians triumphed: they obtained the condemnation and deposition of many pro-Arians, incl. *Eudoxius and the *anomoeans *Aetius and *Eunomius. The three Illyrians subscribed the decisions of the council, which republished the (second) formula of *Antioch of 341, supplemented by the anathemata of the council of S. of 351 and by a document (now lost) on *homoousios* and *homoiousios*. The succession of continual *coups de théâtre* in 357 and 358 showed the need for an ecumenical council, which was called at *Rimini for the West and *Seleucia in Isauria for the East. In preparation for this council, some Eastern bishops with Valens, Ursacius and Germinius met at S. in spring 359. This time Basil was in a minority among a concentration of moderate Arians and so-called *homoeans: Eudoxius, *Acacius, *Mark of Arethusa. On 22 May a formula of faith was published which was to serve as a basis for the labours of the coming councils: it was vague in describing the Father/Son relationship, which it defined as a general likeness: the Son is like the Father in all things, according to the *Scriptures.

377/378. On the basis of information provided by *Theodoret (*HE* IV, 8. 9), we place at S. in this period a council of anti-Arian tendency. The bishops meeting there reaffirmed the Nicene faith, affirmed the consubstantiality of the whole *Trinity, condemned some Arian bishops and wrote an encyclical to the Asiatic bishops to inform them of their decisions and to strengthen them in the struggle against *Arianism. The data relative to this council provide some difficulties, and not all modern scholars are convinced even of its existence.

347.351: Hfl-Lecl 1, 852-862; Simonetti 202-203; *357*: Hfl-Lecl 1, 899-902; Simonetti 229-233; *358.359*: Hfl-Lecl 1, 903-933; Simonetti 241-247; *377/378*: J. Zeiller, *Les origines chrétiennes dans les provinces danubiennes*, Paris 1918, 310-327; M. Meslin, *Les Ariens d'Occident*, Paris 1967, 86-87; Simonetti 439-441; R.P.C. Hanson, *The Search for the Christian Doctrine of God*, Edinburgh 1988, 325-371.

M. Simonetti

SISEBUT of Toledo. King of the Visigoths of *Spain (612-621), remembered for the military successes of his generals Rechila and Suintila over the rebellious Asturians and Basques; for his own successful campaigns against the *Byzantines, whose presence in Spain was reduced in his reign to some fortresses in the region of the Algarvi (Isid., *Hist. Goth.* 61; Ps.-Fredeg., *Chron.* IV, 33); and for his fervent religious activity, displayed in the founding of churches such as that of St Leocadia at *Toledo (Isid., *Chron.* a. 516) and above all in the forced "conversion" of the *Jews (*Leg. Wisig.* XII, II, 13-14). His activity as a writer was important for the history of Christian literature in the West: some of his *letters survive: one to *Theodelinda (*Ep.* 9) on the trinitarian faith, another to his son Theudila (*Ep.* 8) who had fulfilled the vocation his father dreamed of, that of *conversio* to contemplative solitude; also his *Vita Desiderii ep. Viennensis*; and a poetic epistle, *Carmen de luna*, in 61 hexameters. The *hymn *De ratione temporum* is also attrib. to him. To his patronage we owe the *De natura rerum* and *Etymologiae* dedicated to him by *Isidore of Seville, who was his teacher and spiritual director.

Of particular importance for the number of editions (Baehrens, Goetz, Riese, Fontaine) and different readings is the *Carmen de luna*, S.'s reply to Isidore's dedicatory letter to *De natura rerum*. In it the king, after lamenting that his military, legislative and governmental duties prevent him following his teacher in *contemplation and the study of nature, describes in precise scientific language, according to the current ideas of astronomy, a lunar eclipse. The result is a poem in which the dry concision of the basically Lucretian tone is softened by polished Alexandrian workmanship: an "enlightened" response to Isidore's religious allegorism (Stach); a literary exercise by a king nostalgic for a life of study (Fontaine). In our opinion it is the regret of a man who, having nourished from youth a vocation for monastic solitude and seeing himself prevented by the *congeries turbida rerum* from following the spiritual flights of his teacher, describes his spiritual state using the symbol of the moon, onto which the cone of the earth's shadow projects, preventing him seeing the *light of the sun. *Nonnumquam vero*, says Isidore in the work to which the poem replies (*De nat. rer.* XVIII, 6), *eadem luna etiam ecclesia accipitur, pro eo quod sic ista a sole sicut ecclesia a Christo inluminatur*. There is no break in continuity between the symbolic interpretation of natural phenomena in terms of salvation-history, indicated by Isidore in *De natura rerum*, and the image used by S. to express his state of mind. In the Isidorian work commissioned by the Visigothic king, *natura rerum* and *causae astrorum* are seen, like biblical *littera*, as a means of seeking *God and of communication of religious experience in the Christian world. This is one of the most fundamental aspects of mediaeval culture.

Editions: CPL 1186, 1298-1301; PL 80, 363-384; 83, 1112-1114; 94, 605-606; 129, 1369-1372; MGH, *Scr.rer.mer.* III, 630-637, 662-675; MGH, *Poet.lat.* IV/2, 682-686; E. Baehrens, *Poetae Latini minores*, V, 357-360 (1883); G. Goetz, Jena 1887; A. Riese, *Anthologia Latina*, I, n. 483; J. Fontaine, *Isidore de Séville, Traité de la nature*, Bordeaux 1960, 329-335.
Studies: M. Méndez Bejarano, *El rey Sisebuto astrónomo*, Madrid 1919; W. Stach, Bemerkungen zu den Gedichten des Westgotenkönigs Sisebuts (612-621) in *Corona Quernea*, Leipzig 1941, 74-96; id., König Sisebut ein Mäzen des isidorischen Zeitalters, *Die Antike* 19 I (1943) 63-76; J. Fontaine, *Isidore de Séville: op. cit.*, 151-161 (introduction); V. Recchia, *Sisebuto di Toledo: il "Carmen de luna"*, Bari 1971; P.D. King, *Law and Society in the Visigothic Kingdom*, Cambridge 1972; E.A. Thompson, *The Goths in Spain*, Oxford 1969.

V. Recchia

SISINNIUS. *Archbishop of *Constantinople 426-427. He was a priest working in the suburb of Elaia when, on the death of *Atticus (10 Oct 425), the people of the capital, considering his *piety and love for the *poor, preferred him to the *clergy's candidates, *Proclus and *Philip of Side: in his *Christian History* Philip criticizes this popular election, which took place on 28 Feb 426. The same day or soon after, the new archbishop, together with archbishop *Theodotus of Antioch, promulgated a very severe synodal letter against the *Messalians. S.'s *ordination of Proclus, 426 or 427, as metropolitan of Cyzicus met with opposition from the Christians of that city, who elected *Dalmatius instead, telling the archbishop that his presumed right to interfere in the episcopal elections of Cyzicus had been a personal privilege of his predecessor Atticus. S. died 24 Dec 427 and was succeeded by *Nestorius.

PG 103, 89; Grumel, *Regestes*², 49-49a; G. Dagron, *Naissance d'une capitale. Constantinople et ses institutions de 370 à 451*, Paris 1974, 470, 492.

D. Stiernon

SISINNIUS, Novatianist. A contemporary of *John Chrysostom, he was instructed by the philosopher Maximus. He was a lector in the time of *Agelius, *Novatianist bishop of *Constantinople, who designated him his successor, but the Novatianist *laity preferred Marcian, whom S. eventually succeeded in 395. *Socrates (*HE* VI, 22) and *Sozomen (*HE* VIII, 1) call S. a temperate, affable, eloquent person, ready in debate, expert in the *Scriptures, versed in dialectic. Socrates (*ibid.*) says S. wrote various works manifesting affectation of expression and a taste for poetic speech; but he was more esteemed in speaking than in writing. His gifts earned him general good will, even from bishop *Atticus (cf. Socr., *ibid.*).

S. Zincone

SIXTUS II, pope (257-258). Succeeded pope *Stephen. The *Liber Pontificalis* makes him a Greek and a *philosophus*: this title was given him long after, because *Rufinus of Aquileia had thought him the author of the *Sentences* of *Sextus. *Pontius, author of the *Vita Cypriani*, calls him "a good and pacific priest" (CSEL 3/3, CV). *Cyprian describes his *martyrdom in a *cemetery with some of his deacons as a result of *Valerian's edict (*Ep.* 80: CSEL 3/2, 840), i.e. in 258, shortly before that of Cyprian himself. Some days after him, his protodeacon *Lawrence, the great Roman martyr, was killed. Harnack attributed the treatise *Ad Novatianum* to S. (cf. Quasten I, 598).

ASS *Aug.* II, 124-142; *LP* I, 155 f.; III, 75; *BS* 11, 1256-1262; 8, 108-129 (Lawrence); P. Franchi de' Cavalieri, Un recente studio sul luogo del martirio di Sisto II, in *Note agiografiche*, ST 33, Vatican City 1920, 145-178; H. Delehaye, Recherches sur le légendier romain, *AB* 51 (1933) 43-49; K.H. Schwarte, Die Christengesetze Valeriens, in (ed.) W. Eck, *Religion und Gesellschaft in der römischen Kaiserzeit* (Festsch. Vittinghoff), Cologne-Vienna 1989, 103-163.

A. Di Berardino

SIXTUS III, pope (432-440). Of Roman origin, a presbyter in 418 under *Zosimus. At that time he had sympathies with the *Pelagians, from whom he soon distanced himself: *Augustine wrote him two *letters on the subject (*Epp.* 191 and 194, *c.*418) (cf. also the Lateran baptistery inscription: *LP* I, 236). In 439 he refused to readmit *Julian of Eclanum to ecclesial *communion because he had not retracted his error (Prosp. Aq., *Chron. ad a. 439*: PL 51, 598). Present at his episcopal consecration in 432 were two Eastern bishops to whom he entrusted two conciliatory letters, one for *Cyril of Alexandria and the other for the Eastern bishops (Jaffé 389 f.: PL 50, 583-590), to straighten out the disputes caused by the council of *Ephesus; he also intervened in other ways (ACO I, 4, 93; PG 84, 640). Cyril and *John of Antioch came to an agreement in 433 and both communicated the fact to S. (ACO I, 1, 4, 33=Mansi 5, 286; cf. Sixtus's *Ep.* 5), who replied to both congratulating them on making *peace (*Epp.* 5 and 6). He also reaffirmed the rights of the Roman see over the bishops of Illyricum, who often turned to *Constantinople, which had an interest in extending its powers in that region (Jaffé 393-396), where the bishop of *Thessalonica fulfilled the function of papal vicar. In the West, however, the barbarian invasions prevented close contacts between *pope and bishops. During S.'s pontificate many sacred buildings were constructed at *Rome. In particular, S. Maria Maggiore - the first Roman church dedicated to the Madonna - was rebuilt, with its famous triumphal arch covered in *mosaics narrating the infancy of Jesus with *Mary, with the aim of magnifying Christ, Man-God and son of Mary (the Ephesian *Theotokos*) (Diehl 975a-976). This building fervour has led S.'s period to be called an "architectural renaissance", though this had begun in the 4th c. The *Gesta de Xysti purgatione* is a *falsification composed by the partisans of pope *Symmachus on the occasion of the latter's trial in 501 (PLS 3, 1249-1255).

CPL 1655; *Verzeichnis* 545; PL 50, 571-626; PLS 3, 21-23; *LP* I, 232, 237; III, 85; *Patrologia* III, 556 f.; *DTC* 14, 2196-2199 (cf. *Tables* 4050); *BS* 11, 1262-1264; G. Wilpert, La proclamazione efesina e i mosaici della basilica di S. Maria Maggiore, *AST* 7 (1931) 197-213; R. Krautheimer, The Architecture of Sixtus III: A Fifth-Century Renaissance?, in *Essays in Honor of E. Panofsky*, New York 1961, 291-302 (now in *Studies in Early Christian, Medieval and Renaissance Art*, London 1971, 181-196); id., *Corpus Basilicarum Christianarum Romae*, III, Vatican City 1971, 1-60; G. Bovini, I mosaici romani dall'epoca di Sisto III (432-440), *CCAB* 10 (1963) 67-80, 81-101; F. Magi, *Il calendario dipinto sotto Santa Maria Maggiore*, Vatican City 1972 (cf. *REL* 51 [1973] 41-48); N.A. Brodsky, *L'iconographie oubliée de l'arc éphésien de Sainte-Marie-Majeure à Rome*, Brussels 1976 (cf. *RAC* 46 [1971] 179-183); Ch. Pietri, *Roma christiana*, BEFAR 224, 2 vols., Rome 1976 (cf. index p. 1700); G. Zecchini, I "Gesta de Xysti purgatione" e le fazioni aristocratiche a Roma alla metà del V secolo, *RSCI* 34 (1980) 60-74.

A. Di Berardino

SLAVE - SLAVERY. Slavery was so much a part of the social structure of antiquity that some *Fathers admitted it to be an economic necessity. Yet the church made the institution evolve. It accepted slaves into its ranks on the same level and with the same prerogatives as free men. The slave took part in the *assembly, received the same *sacraments and held the same offices, including the episcopate. His *marriage was not considered a *contubernium* but a true *coniugium*, sacred and indissoluble. Christian *families allowed slaves in their family tombs. A martyred slave was inscribed in the *diptychs together with the most illustrious names. Pope *Callistus, an ex-slave, even authorized, against Roman law, the marriage of noble ladies with freedmen, even with slaves.

The earliest Christian writers recommended mutual respect between Christian masters and slaves (Philem; 1 Pet 2, 18-25; Ign., *Ad Polyc.* 4, 3; *Did.* 4, 10; Hermas, *Past., Praec.* 8, 10; *Sim.* 1, 8; 9, 28, 8). *Clement of Alexandria's *Pedagogue* is full of exhortations to reduce the number of slaves and to better the existence of those who remained (*Strom.* IV, 19). *Cyprian's advice to the rich Demetrian was no different (*Ad Dem.* 8).

From the 4th c. legal and canonical pressure for the defence and enfranchisement of slaves was felt more and more. The council of *Elvira *excommunicated *ad tempus* masters who punished their slaves with death. This attitude influenced the *Codex Theodosianus* (IX, 12, 1), which took multiple measures to protect the human and moral life of slaves and facilitate their *manumissio in ecclesia*, recognizing the church's competence in this regard (Aug., *Serm.* 21, 6; cf. also *Can. Hipp.* VI, 36).

Nearly all the *councils from the 6th to the 9th c., in *Gaul, *Britain, *Spain and *Italy, took an interest in the lot of slaves and made provisions in their favour to guarantee their marriage, moveable property, *Sunday rest, etc.

The Fathers were too much indebted to ancient structures to condemn slavery. Clerics and religious institutions possessed slaves. The church recognized the master's property rights. But the master should behave towards them as a good *paterfamilias* (Ambr., *Ep.* 2, 31), meet their needs and prevent any form of useless cruelty (Aug., *Exp. ep. ad Galatas* 64). *Salvian thought masters were often responsible for the *vices of their slaves (*De gub.* VIII, 14).

Above all the church encouraged the enfranchisement of slaves (John Chrys., *In I Cor. hom.* 40, 5; cf. Aug., *Serm.* 31, 6; 356, 3, 7). *John Chrysostom advised teaching them a trade before freeing them (*ibid.*). Their right of *asylum was recognized (*RACh* 1, 836-844). The council of *Chalcedon authorized slaves to become monks with the prior consent of their master (can. 4). Yet the Fathers considered slavery an evil originating from *sin (*Augustine) and from man's greed (John Chrys., *In ep. ad Eph. hom.* 22, 2), which had split mankind in two (Greg. Naz., *Poem. th.* II, 36). *Gregory of Nyssa went so far as to contest the master's rights over the slave (*In eccl. hom.* 4). The church contributed greatly to making men aware of how the institution of slavery wounded human dignity, mitigating its rigour and making its structure evolve towards its eventual suppression.

A. Steinmann, *Sklavenlos und alte Kirche*, Mönchen-Gladbach 1922; S. Talamo, *La schiavitù secondo i Padri della Chiesa*, Rome ²1927; J. Vogt, *Sklaverei und Humanität*, Wiesbaden 1965 (Eng. tr. Oxford 1974); H. Gülzow, *Christentum und Sklaverei in den drei ersten Jahrhunderten*, Bonn 1969; R. Sierra Bravo, *Doctrina social y económica de los Padres de la Iglesia*, Madrid 1967; P.A. Milani, *La schiavitù nel pensiero politico. I. Dai greci al Basso Medioevo*, Milan 1972; O. Robleda, *Il diritto degli schiavi nell'antica Roma*, Rome 1976; Ch. Munier, *L'Eglise dans l'empire romain, II-III° s.*, Paris 1979.

A. Hamman

SOCRATES. The figure of S. did not fail to impress numerous patristic authors: G. Giannantoni (cf. bibl.) has collected all the references to S. in the *Martyrium S. Apollonii* (19 and 41), the *Martyrium Pionii*, *Justin, *Tatian, *Athenagoras, *Theophilus of Antioch, *Clement of Alexandria, *Origen, the *pseudo-*Clementine homilies*, *Eusebius, *Tertullian, *Minucius Felix, *Cyprian, *Lactantius, *Jerome, *Augustine and *Basil. For quite a few of these authors, particularly the Greek *apologists and Clement and Origen, S. becomes the ideal emblem of the wise man, the teacher of morality who believes in one *God, for which he is unjustly persecuted and condemned. The comparison spontaneously arises between the death of S. and that of the Christian *martyrs, also guilty of believing in one God and rejecting official polytheism. Origen even finds it natural to establish a parallel between S. and Jesus (*C. Cels.* III, 41; VII, 56). For St Augustine, S. is the *magister omnium qui tunc maxime claruerunt tenens in ea parte quae moralis vel activa dicitur principatum* (*De civ. Dei* XVIII, 35). Among Christian authors of the first centuries, only Tertullian, consistent with his negative attitude towards *Hellenism and Greek *philosophy, shows no goodwill towards him.

G. Giannantoni, *Socrate. Tutte le testimonianze da Aristofane e Senofonte ai Padri cristiani*, Bari 1971; D. Jackson, Socrates and Christianity, *CF* 31 (1977) 189-206; I. Opelt, Das Bild des Sokrates in der christlichen Literatur, *JbAC* 10 (1983) 192-207.

S. Lilla

SOCRATES Scholasticus. Writer of ecclesiastical history. The little we know of S.'s life is taken from his works. Born *c*.380 at *Constantinople, died *c*.450. He frequented the *pagan teachers Helladius and Ammonius, who had fled to the capital from *Alexandria in *c*.390 on the outbreak of a revolt. He attended the biblical *exegesis of the Arian Timothy and studied law, as is shown by his knowledge of Latin, which in the *East was studied only by jurists, and confirmed by some MSS of his *Historia Ecclesiastica* which call him σχολαστικός. His *HE* is dedicated to Theodore, who induced him to write. Who Theodore was is uncertain: perhaps a monk or a priest. Recently Chesnut thought of the *comes* Theodore who as *magister memoriae* was part of the commission charged with drawing up the **Codex Theodosianus*; this theory is cautiously adopted by Mazza (*Sulla teoria*, 373-376).

S.'s work sets out to be a continuation of *Eusebius's account: it is divided into seven books and embraces the period from 306 (abdication of *Diocletian) to 439 (17th consulate of *Theodosius II). Each book corresponds to the reign of an emperor: it begins with his ascent to the throne and ends with his death. Book III joins the brief reign of *Jovian to that of *Julian. The historical material is organized in a chronological framework governed by the dating of the main events. The variety of S.'s sources is examined particularly by Jeep and Geppert. Esp. in the first books, he uses *Rufinus of Aquileia, prob. in a Latin text. Known and cited are the works of Eusebius (esp. the *Vita Constantini*, main source of the first book, then the *Historia Ecclesiastica*, the treatise *De ecclesiastica theologia* and some lost exegetical and polemical works), Eutropius's *Breviarium* and the works of *Athanasius used by S. as a basis for his description of the *Arian controversy. Sometimes he refers to *Evagrius Ponticus, *George of Laodicea, *Philip of Side, *Acacius of Caesarea and also *Gregory the Thaumaturge, *Gregory Nazianzen, *Palladius, *Libanius and Themistius. Timothy of Berytus seems to be used but not cited. As well as these authors, S. consults and carefully transmits *letters and lists of bishops and emperors, as well as synodal *acts, esp. the συναγωγή of *Sabinus of Heraclea, whom he criticizes for his Nicene tendency and accuses of suppressing inconvenient evidence. From book VI on, S. includes his own memories and uses oral evidence: of considerable importance is that of the Novatianist priest Auxanon, from whom he received that detailed information on the *Novatianist sect which led Nicephorus Callistus to advance the groundless hypothesis that S. was a Novatianist. He uses all his sources critically, seeking to compare them scrupulously in an honest search for the truth. Thus after completing the work, having, through reading Athanasius, discovered some errors taken from Rufinus, he decided, as he explains in the *HE* (II *Prooem*.), to make a second edition, which is the one we possess. He revised the text of the first two books, but the survival of a long passage from book VI in two different recensions suggests that he revised the whole work.

S. suits his style to the rigorous examination of evidence: it is deliberately plain and simple because he does not admit that the search for formal elegance may in any way be allowed to amplify or diminish the facts or impair the clarity that his readers, mostly *laymen, expect. The *HE* describes events of secular as well as church history. S. justifies this mixture by the need for completeness, the need not to appear monotonous by always expounding disputes between bishops, and especially by the inevitable correlation, in good and in *evil - evident in the figure of the emperors - that he supposes to exist between secular and religious events, both of which are determined by man's wickedness and *sin and the consequent divine punishment. The Stoic-Platonic concepts of καιρός and συμπάθεια thus assume an important role in S.'s historical vision (Mazza, Sulla teoria, 358-366).

The *HE* was first edited by R. Estienne (Paris 1544); other editions are those of H. Valois (Paris 1668) and R. Hussey (Oxford 1853, reprinted 1878 and 1893 by W. Bright) which used *Laurent.* 69 and *Laurent.* 70, the two best MSS of the whole *manuscript tradition. A new edition is urgently needed, because the archetype of S.'s text is itself suspected to be contaminated (Périchon, *Pour une édition*: 116-120). In any case we must take into account the Latin version of part of the work made in the 6th c. by the monk *Epiphanius under the direction of *Cassiodorus and the two *Armenian versions edited by M. Ter Mowsesean (Valarsapat 1897).

CPG 3, 6028; PG 67, 28-842; R. Hussey, *Socratis Scholastici Ecclesiastica Historia*, I-III, Oxford 1853; L. Jeep, *Quellenuntersuchungen zu den griechischen Kirchenhistorikern*, Leipzig 1884; F. Geppert, *Die Quellen des Kirchenhistorikers Socrates Scholasticus*, Leipzig 1898; W. Schmid-O. Stählin, *Geschichte der griechischen Literatur*, Munich 1924, II/2, 1434-1435; *PWK* 22/1, 893-901; *DTC* 14, 2334-2336; G. Moravcsik, *Byzantinoturcica*, I, Berlin 1958, 508-510; *LTK* 9, 859; W. Jacob-R. Hanslik, *Cassiodori-Epiphanii Historia Ecclesiastica tripartita*, CSEL 71, Vienna 1952; G. Downey, The Perspective of the Early Church Historians, *GRBS* 6 (1965) 59-63; P. Périchon, Pour une édition nouvelle de l'Historien Socrate: les manuscrits et les versions, *RecSR* 53 (1965) 112-120; G. Chr. Hansen, Prosarhytmus bei den Kirchenhistorikern Sozomenos und Sokrates, *ByzS* 26 (1965) 82-93; Quasten II, 537-539; G.F. Chesnut, Kairos and Cosmic Sympathy in the Church Historian Socrates Scholasticus, *ChHist* 44 (1975) 161-166; F. Winkelmann, Die Kirchengeschichtswerke im ostr̈om. Reich, *ByzS* 37 (1976) 172-175; Altaner 228-229; G.F. Chesnut, *The First Christian Historians: Eusebius, Socrates, Sozomen, Theodoret and Evagrius*, Paris 1977 (Mâcon ²1986); M. Mazza, Sulla teoria della storiografia cristiana: osservazione sui proemi degli storici ecclesiastici, in *La storiografia ecclesiastica della tarda antichità*, Messina 1980, 335-389; ed. of *HE* in preparation by G.C. Hansen, GCS, Berlin.

A. Labate

SODOM: *iconography. Episodes illustrating *Lot's flight from S. (Gen 19, 24-26) show, behind Lot's wife - now a pillar of salt - a depiction of the city destroyed by God to punish the perversion of its inhabitants. It usually appears surrounded by high walls and engulfed in flames. The oldest known scene is a *painting in the catacomb of Via Latina, Rome (Ferrua, *Via Latina*, pl. XXX) (*c*.320-360). Then follows the Lot *sarcophagus (a.340-*Rep.* 188) from S. Sebastiano, where S. is rendered by a door supported by a spiral post with trabeation and a hint of a tympanum. Prominent in the background are the city walls and rows of windows from which issue tongues of fire. The sole example of miniature is in the 7th-c. Tours Pentateuch (f. 18a, pl. VI).

L. De Bruyne, Il sarcofago di Lot scoperto a S. Sebastiano, *RAC* 27 (1951) 112; J. Porcher, in J. Hubert-J. Porcher-W.F. Volbach, *L'Europa delle invasioni barbariche*, Milan 1968, 128.

A.M. Di Nino

SOGDIAN. *Lingua franca* of the Iranian family, discontinuously attested from the 1st (?) to the 10th c. in central Asia (from Turkestan to Mongolia). Using an Aramaic alphabet, it was used to transcribe mainly Buddhist, but also *Manichaean and Christian, texts. Among the latter, generally translated from *Syriac, were parts of the NT, of *Hermas's *Shepherd*, of *Evagrius and some hagiographical works and apophthegms.

F.W.K. Müller, *Soghdische Texte*, I, Berlin 1913; O. Hansen, *Berliner Soghdische Texte*, I, *Bruchstücke einer soghdischen Version der Georgspassion (C 1)*, Berlin 1941; id., Berliner sogdische Texte. II. Bruchstücke der grossen Sammelhandschrift C 2, *Abhandlungen der Akademie der Wissenschaften in Mainz* 15 (1954) 819-918; O. Hansen, Die buddhistische und christliche Literatur, *Iranistik. Literatur.* 1, Handbuch der Orientalistik I, 4, 1, Leiden-Cologne 1968, 77-99; id., Die christliche Literatur der Soghdier. Eine Übersicht, *Jahrbuch der Wissenschaften und der Literatur in Mainz* (1951) 296-302; id., Über die verschiedenen Quellen der christlichen Literatur der Sogder, *Acta Orientalia* 30 (1966) 95-102; B.M. Metzger, *The Early Versions of the New Testament, Their Origin, Transmission and Limitations*, Oxford 1977, 279-281; E. Benveniste, Etudes sur quelques textes sogdiens chrétiens, *JA* 243 (1955) 297-337; 247 (1959) 115-136; D.N. MacKenzie, Christian Sogdian Notes, *Bulletin of the School of Oriental and African Studies* 33 (1970) 116-124; W.B. Henning, *Selected Papers*. I-II, Acta Iranica II, 14-15, Leiden-Teheran-Liège 1977; S. Sims-Williams, The Sogdian Fragments of Leningrad, *Bulletin of the School of Oriental and African Studies* 44 (1981) 231-240.

S.J. Voicu

SOLDIERS CASTING LOTS FOR JESUS'S CLOTHING: *iconography. The gospel episode of the soldiers who cast lots for Christ's tunic (Ps 22, 19; Mt 27, 35; Mk 15, 24; Lk 23, 34; Jn 19, 23 f.) does not appear in Christian art until after the religious peace, when, like the other passion scenes, it was introduced to evoke one of the steps that marked Christ's victorious earthly journey. It first appears - if we can accept A. Ferrua's interpretation (*Le pitture*, 74 f., pls. 72-73), recently contested (Kötzsche-Breitenbruch, *Die neue Katacombe*, 42-45) - in a *painting in the catacomb of Via D. Compagni, Rome (second half of 4th c.). Two soldiers, armed with shield and spear, are shown beside a gaming instrument consisting of a sort of framework supporting a rotating urn in the form of a vase, from which issue small disks. The iconographical scheme, identical to that which recurs in early mediaeval miniatures and ivories (E.T. De Wald, *The Illustrations of the Utrecht Psalter*, Princeton 1942, 13 f., pl. XIX; A. Goldschmidt, *Die Elfenbeinskulpturen aus der Zeit der karolingischen und sächsischen Kaiser. VIII-IX Jahrhundert*, I, Berlin 1914, 20, 66, nn. 31 and 132 a, pls. XV, LVII), makes Ferrua's proposed identification of the scene very probable. In the *East, the episode first occurs in a miniature of the Gospels of Rabbula, 586 (Cecchelli, *Rabb. Gosp.*, 69-71), whose iconography is very different from the Western: three soldiers, seated beneath the *cross, are in the act of playing for the tunic with the game of *morra*. The same figurative scheme - evidently characteristic of the East - recurs, substantially the same, in early and mediaeval Byzantine icons, one of which, preserved in St Catherine's monastery on Mt *Sinai, can be assigned to the 8th c. (K. Weitzmann-M. Chatzidakis-K. Miatev-S. Radojcic, *Frühe Ikonen. Sinai, Griecheland, Bulgarien, Jugoslawien*, Vienna-Munich 1966, LXXIX, no. 6). **[Fig: 284]**

A. Ferrua, *Le pitture della nuova catacomba di Via Latina*, Vatican City 1960, 74 f., pls. 72-74; E. Lucchesi Palli, Morraspiele, *LCI* 3, 281 f.; L. Kötzsche-Breitenbruch, *Die neue Katakombe an der via Latina in Rom. Untersuchungen zur Ikonographie der alttestamentlichen Wandmalereien*, JbAC Ergbd. 4, Münster 1976, 42-45.

V. Fiocchi Nicolai

SOLOMON: *iconography. The figure of S. is connected with a dual *symbolism. The best-known one sees S. as supreme judge and sage (a symbol of wisdom); the second, less well-known, makes him the victor in the struggle against malign forces. The oldest examples of the judgment of S. (1 Kgs 3, 16-28) are a parodistic fresco at Pompeii and, perhaps, a mutilated scene in the synagogue of *Dura Europos accompanied by an inscription relating to S. (first half of 3rd c.). The earliest Christian depiction is on a silver reliquary in San Nazaro, Milan (end of 4th c.), the latest a 7th-c. *painting in the underground church of Santa Maria in Via Lata, Rome (*WMM* 37, 1). Scenes relating to S. are quite frequent in the minor arts, like an incised agate, an ivory at Berlin and a glass medallion. The second symbolism, that of S. as hero of good triumphant over *evil, is illustrated in a series of small objects. In the Middle Ages S., who for Josephus (*Ant. Iud.* VIII, 2, 5) had exorcistic powers, was considered a real *magus*. A group of talismans and medallions is inspired by such beliefs. A series of medallions in the Beidun Museum, Jerusalem, related to the *Testamentum Salomonis*, show S. on horseback, intent on transfixing a female figure (the *demon Hormias) beneath his horse's belly, with a lance. This iconography inspired the Christian one of St *George and the dragon. In a *papyrus at Oslo, S. is invoked together with Christ.

DACL 15, 588-602; *EC* 10, 1693-1694; *LCI* 4, 15-24; A. De Longpérier, Une intaille antique inédite, *CRAI* 24 (1880) 275; O. Wulff, *Altchristliche und mittelalterliche byzantinische und italienische Bildwerke*, Berlin 1909, I, 237, pl. 56, no. 1152; L.A. Costans, Une amulette chrétienne sur papyrus, *JS* 20 (1922) 181-182; O. Wulff-W.F. Volbach, *Altchristliche und mittelalterliche byzantinische und italienische Bildwerke*, Ergänzungsband, Berlin 1923, II; C.H. Kraeling, *The Excavations at Dura-Europos, The Synagogue*, New Haven 1956, 349; Volbach-Hirmer, 80, pl. 115; B. Bagatti, Altre medaglie di Salomone cavaliere e la loro origine, *RAC* 47 (1971) 331-342; J. Gutmann, Was there Biblical Art at Pompei?, *Antike Kunst* 15 (1972) 122-124.

A.M. Di Nino

SONG OF SONGS. "The whole world is not worth the day on which the S. of S. was given to Israel. All the *Scriptures indeed are holy: but the S. of S. is the holy of holies" (Rabbi *Akiva: *Jadaim* III, 5). The Song reveals to Israel the mystery of love: the bridegroom's love for the bride, the immense love of *God who creates mankind and his people, the even greater love he will manifest in future history. Israel lives the *ascesis of expectation. The present announces the joy of eternal love. The Song is a paschal hymn uniting history and prophecy: God "has passed over" to visit the people he loves, God "will pass over" in the final visitation of *love for mankind. "Prophetic history" becomes musical prayer in the "psalms" and "canticles": at the summit is the "S. of S." *Solomon as prophet surpasses *David. Christ "is greater than Solomon" (Lk 11, 31).

God is a faithful lover to his bride, faithless but thirsting for his embrace. For a people who take God's word literally, this is not just *allegory. It is the discovery that God is "literally" in love with man: "For as a young man marries a virgin, so shall your creator marry you" (Is 62, 5). The echo of the Song resounds in the gospel: the bridegroom's friends feast because the bridegroom is present (Mt 9, 15). Joy will be complete in the "new earth", when the community will descend from heaven "prepared as a bride adorned for her husband" (Rev 21, 2). Christians sing the Song for the feast of *Easter: beginning of spring, time of *weddings, feast of the *resurrection. God's voice is clear: why dilute it in *exegesis? A reverential fear ensured that for more than a century no writer gave interpretations of the Song; only the fragrance of its perfumes, and the eternal question: who is the bride?

*Hippolytus, listening to *Paul (Eph 5, 22-33) and accepting all that was nuptial in the synagogue, replies: "the bride of Christ is the church". *Origen attains the idea of the bride-church as lived truth, and brings it to life by inviting each person to nuptial love: the bride-*soul is the awakening of personal awareness to the duty of perennially renewing the bride-church, preparing mankind to be Christ's bride. The Christian community, not yet thinking to draw from the Song a nuptial *theology bringing to life the mystery of *marriage, found the "fire" of its *spirituality in the *mystery of *God who has celebrated his *marriage with mankind. The mystery of love is explored, between *eros* and *agapè*, between human love and mystical love speaking the same language: "The lover seeks in the beloved object a fullness of happiness that it is incapable of giving him: thus he puts it in God's place and makes it a supreme end. God alone is the adequate and true object of human passion" (H. Crouzel, *Le thème du mariage mystique*, 37).

The *gnostics exploited the Song in order to create with its fragrances a mysterious atmosphere, in which the *theology of *life is suffused with ritual illuminations (Iren., *Haer.* I, 4, 1 and 21, 3; Hipp., *Ref.* V, 19, 3-4 and VII, 22, 13-16). The churchmen discovered in it the synthesis of all biblical history, and sing the event of liberation in the form of prophecy. Each image is a *symbol that guides us to life, and the whole Song has an existential import. The bride, who so far has lived in the earthly garden perfumed by the divine lover's presence, will be granted in the nuptial chamber the perfect vision of God. The history of the interpretation of the Song, which has still to be written, must be based on the allusions, citations and hermeneutical reflections scattered throughout the works of the *Fathers, starting with *Commentaries* on the Song. The diachronical reconstruction of the meaning of the individual verses is the necessary labour prior to focusing on general hermeneutical principles.

The earliest Christian commentator is *Hippolytus, whose commentary is preserved in a *Georgian version (Song 1, 1-3, 7: ed. G. Garitte, CSCO 263-264), as well as *Armenian, *Syriac, *Palaeoslavonic and Greek fragments. The typological *exegesis, which inaugurates the identification of the bridegroom with Christ and the bride with the church, follows the Asiatic *christology centred on Jesus' humanity, contemplating the *Trinity in its orientation towards *incarnation. God's wedding with mankind is accomplished on the *cross: consumed by love, the man-Christ acquires the divine sonship already possessed by the Word and draws men towards the *resurrection. Christ's gift is not lost, since the *perfume of life is poured out on his new body which is mankind. "In his other books, *Origen surpassed everyone; in the Song of Songs he surpassed himself" (Jerome, in Origen *Hom. Ct.*, prol.). We know Origen's two *homilies in *Jerome's Latin version (on Song 1, 1-2, 14: ed. W.A. Baehrens, GCS 33) and four books of his *Commentary* in *Rufinus's *Latin translation (on Song 1, 1-2, 15: ed. W.A. Baehrens, GCS 33), besides the Greek fragments in *Procopius's *Catena and in the *Philocalia* (PG 87, 1545-1554 and 13, 35-36). The *ecclesiological interpretation is enriched by the *anthropological, which leads the *soul towards union with God in the church. Love is the energy of the *Holy Spirit, which transfigures man's senses and makes him capable of penetrating the *mystery. The Song reveals God to the "perfect", who, with their spiritual nose, sense the bridegroom who approaches to awaken in men the fragrance of God's *image. *Gregory of Nyssa raises Origenian *exegesis to the heights of perfect *contemplation (on Song 1, 1-6, 8: ed. H. Langerbeck, Leiden 1960). The *soul is called to detach itself from ephemeral things in order to ascend to the royal bridal bed of the vision of God. The perfume of a life lived in *virtue and service is enlivened by the breath of the Spirit in the church through the inebriating *ecstasy of Christ's *bread and *wine, which assimilate the bride to the groom. *Philo of Carpasia's commentary offers for the first time the complete interpretation of the Song: we know a Greek redaction of it (ed. M.A. Giacomelli, Rome 1772 = PG 40, 27-154) and a Latin version (ed. A. Ceresa Gastaldo, Turin 1979; cf. P.F. Foggini, Rome 1750). *Nilus of Ancyra gives his reflections an ascetic-moral tone precious for the life of monks, tackling with moderation the theme of the relationship between the contemplative life and the active life (PG 87, 1545-1754). *Theodoret of Cyrrhus rejects the literalistic Antiochene interpretation, railing - without naming *Theodore of Mopsuestia - against those who interpret the Song "carnally"; he follows Origen in considering it a "spiritual book", never accepting the literal interpretation that Origen had set out before the allegorical, and seeking to avoid theological expressions which might endanger the outcome of the Chalcedonian controversy (PG 81, 27-214; cf. M. Simonetti, Teodoreto e Origene, 919-930). In the Latin world Jerome mentions the commentaries of *Victorinus of Petovium and *Reticius of Autun (Vir. ill. 74 and 82), now lost. To *Gregory of Elvira are attributed five homilies, Origenian in inspiration but rich in Hippolytian resonances, which stress the sacramental dimension of the nuptial theme (on Song 1, 1-3, 4: ed. J. Fraipont, CCL 69). *Aponius's commentary is a mine open to research for the originality of its language (ed. H. Bottino-J. Martini, Rome 1843=PLS 1, 799-1031), while *Justus of Urgel's is a compendium of modest value (PL 67, 963-994). *Ambrose is the Father who has most sown his works with flowers plucked from the Song, though he wrote no continuous commentary. He drank deeply at the springs of nuptial *spirituality and communicated his thirst to believers, esp. *virgins, presenting the Song as a *lighthouse that illuminates all the books of *Scripture. The mystical message for the soul's journey has an ecclesial savour: Christ, mankind's perfect bridegroom in the gift of the *cross, enables his bride to become a mother (cf. P. Meloni, L'influsso del commento al Cantico, 865-890). The figure of *Mary is present in the Song (G. Bardy, Marie et le Cantique chez les Pères, Bibl VieChr 7 [1954] 32-41). Among writers who loved the Song but made no continuous commentary on it, we will mention *Cyprian, *Methodius, *Jerome, *Augustine and *Cyril of Alexandria, while remembering that many pearls from the book were set into the *liturgy to beat out the biblical rhythms of Christian life (J. Daniélou, Bible et liturgie, Paris 1951, 259-280). *Gregory the Great's Expositio in Canticum (ed. P. Verbraken, CCL 144) closes the tradition of the Fathers and heralds the great flowering of mediaeval monastic commentaries. The Spiritalis intelligentia of the mysteria litterae is *light on the journey to perfection that man must traverse, wisely basing his *ascesis on the poetical symmetries of the Song. Life is a glorious career towards sublimitas contemplationis, which guides the heart of man and of the church to live in union with God. The Song is the history of the encounter between God and men from the moment of *creation to the "embrace of the presence" in the nuptial chamber of eternity; through its mystical atmosphere Christians feel the divine bridegroom penetrate the bride's human *flesh to transfuse into it the life-blood of its divinity (cf. also under *Wisdom books, II, 3, and *Sponsa Christi).

For basic bibliography up to 1975 see P. Meloni, Il profumo dell'immortalità. L'interpretazione patristica di Cantico 1, 3, Rome 1975. Also: F. Cavallera, Cantique des Cantiques. Histoire de l'interprétation spirituelle, DSp 2, 93-101; A.M. La Bonnardière, Le Cantique des Cantiques dans l'oeuvre de saint Augustin, REAug 1 (1955) 225-237; G. Nolli, Cantico dei cantici, Turin-Rome 1968; H. Crouzel, Origines patristiques d'un thème mystique: le trait et la blessure d'amour chez Origène, Kyriakon I, Münster 1970, 309-319; M. Harl, Cadeaux de fiançailles et contrat de mariage pour l'épouse du "Cantique des cantiques" selon quelques commentateurs grecs, Mélanges H. Ch. Puech, Paris 1974, 243-261; G. Chappuzeau, Die Auslegung des Hohenliedes durch Hippolyt von Rom, JbAC 19 (1976) 45-81; M. Simonetti, Origene. Commento al Cantico dei cantici, Rome 1976; H. Crouzel, Le thème du mariage mystique chez Origène, StMiss 26 (1977) 37-57; P. Meloni, Ippolito e il Cantico dei cantici, Ricerche su Ippolito, Rome 1977, 97-120; A. Ceresa Gastaldo, La dimensione dell'amore nell'interpretazione origeniana del "Cantico dei Cantici", Paradoxos Politeia, Milan 1979, 187-194; id., L'esegesi origeniana del "Cantico dei Cantici", Origeniana secunda, Bari-Rome 1980, 245-252; P. Meloni, L'influsso del Commento al Cantico di Ippolito sull'Expositio Psalmi CXVIII di Ambrogio, Letterature comparate. Problemi e metodo, Studi in onore di Ettore Paratore, Bologna 1981, 865-890; M. Simonetti, Teodoreto e Origene sul Cantico dei cantici, ibid., 919-930.

P. Meloni

SOPHRONIUS. *Friend of *Jerome, who mentions (Vir. ill. 134) some juvenile works written by S. in Greek, now lost (Laudes Bethlehem, a book on the destruction of the temple of Serapis at *Alexandria). S. translated many of Jerome's works into Greek: his versions of the *psalms and prophets, Ep. 22 to *Eustochium, the Life of *Hilarion (BHG 752 f.) and perhaps the Life of Malchus (BHG 1015 f.). The Gk tr. of De viris illustribus (TU 14/1b) is not his, but by a later homonym.

CPG 2, 3630-3636; Bardenhewer 1, 3 ff.; 3, 283; CCL 72, XIV, XLVIII (bibl.); H. Chadwick, John Moschus and his friend Sophronius the Sophist, JThS n.s. 25 (1974) 41-74; A. Cameron, The Epigrams of Sophronius, CQ 33 (1984) 284-292; D. Duffy, Observations on Sophronius' "Miracles of Cyrus and John", JThS n.s. 35 (1984) 71-90.

A. Pollastri

SOPHRONIUS of Jerusalem. Born at *Damascus c.550, died at *Jerusalem 11 March 638 (a year after its conquest by the caliph Omar). Prob. originally a teacher of *rhetoric, whence his title "Sophist" (for the debate over the identification of S. the Sophist and S. the patriarch, cf. S. Vailhé, Sophrone le Sophiste et Sophrone le Patriarche, ROC 7 [1902] 360-385; 8 [1903] 32-69, 356-387; G. Zuretti, Sofronio patriarca di Gerusalemme [634-638], Didaskal. n.s. 4 [1926] 19-68). He became a monk in the *monastery of St Theodosius near Jerusalem. With his teacher *John Moschus he went to *Egypt, where they dedicated themselves to converting the *monophysites. After further peregrinations they embarked in 615 for *Italy. Moschus died at *Rome in 619 (or perhaps 634): S. brought his remains back to the monastery of St Theodosius. In 633 S. was engaged in the struggle against *monotheism in Egypt and *Africa, and soon afterward at *Constantinople against patriarch *Sergius. Immediately after his election to the patriarchal see of Jerusalem (634), he published a synodal letter in which, without using the formula of two energies or operations, he inculcated that doctrine (Mansi 11, 461-510; PG 87, 3, 3125-3146; Epitome of the synodal letter, Νέα Σιων, Jerusalem 77 [1922] 178-86). Among his other works (PG 87, 3, 3147-4014) are works of *hagiography, incl. a panegyric of the *martyrs Cyrus and John (at *Alexandria under *Diocletian), presented with great formal perfection (PG 87, 3, 3379-3676, cf. BHG² 475/79). S. also left 11 *homilies and 23 anacreontic odes, composed for religious feasts (ed. M. Gigante, Rome 1957, text and It. tr.). The Commentarius liturgicus is not by S., nor is the Vita Mariae Aegyptiacae, but the Homily on Palm Sunday handed down under the name of patriarch *Eulogius of Alexandria (PG 86, 2, 2913-38), probably is. A work against monotheism is lost.

CPG 7635-7681; DTC 14, 2779-83; EC 11, 906 f.; BS 11, 1283-85; LTK 9, 888 f.; G. Cosmas, De oeconomia incarnationis secundum s. S. Hierosolymitanum, Rome 1940; P. Parente, Uso e significato del termine Θεοκίνητος nella controversia monotelitica, REByz 11 (1953) 241-51; C. von Schönborn, Sophrone de Jérusalem. Vie monastique et confession dogmatique, Paris 1972.

T. Špidlík

SOPHRONIUS of Pompeiopolis. One of the main *homoeousian spokesmen at the council of *Seleucia (359), for which he was condemned and deposed at the council of *Constantinople of 360. On *Constantius's death in 362, he and other homoeousians challenged the decisions of *Rimini and Constantinople and condemned their supporters. We know nothing further of him.

DCB 4, 717; Simonetti, cf. index.

M. Simonetti

SORTES SANGALLENSES. Anon. text, so called since contained in St Gall palimpsest 908 (Lowe 953). Lowe claims that the MS is from N *Italy and must have connections with Bobbio, but the editors put its origin in 4th-c. *Gaul. It consists of the replies of an *Oracle to questions with a Christian colouring. A number of questions are lost; those that remain turn on various problems of daily, moral and social life: life and health, love, *marriage and *family, *friends and enemies, hopes and fears, concerns for house, goods, choice of profession, various undertakings: *journeys, return home, financial, economic and professional worries, trials.

A. Dold-R. Meister-K. Mras, Die Orakelsprüche, SAW 225, 4-5 (1948-1951) 21-72; cf. A. Kurfess, Zu den S.S.: 5 (1953) 143-146 (emendations); R. Weister-L. Krestan, Die Orakelsprüche, SAW 225, 5 (1951); E. Schönbauer, Die Sortes Sang. als Erkenntnisquelle des röm. u. germ. Rechts, AAWW 90 (1953) 23-34.

A. Hamman

SOTER, *martyr. *Ambrose (De virg. 3, 7, 37-38; Exh. virg. 12, 82) mentions S. as noble, a virgin, martyr and member of his own family. As such he proposes him as an example to his sister *Marcellina. An inscription (De Rossi, ICUR I, 495) refers to his *natale, 11 Feb. The Mart. hier. gives the same date, and several others which are erroneous repetitions (6 and 10 Feb; 2 May). The Sacram. gelas., historical *martyrologies and Mart. rom. commemorate him on 10 Feb. The 8th-c. *Itineraries locate his tomb on the Appian Way, in the *cemetery of Callistus. Sergius II (844-47) transferred his relics to S. Martino ai Monti.

Vies des SS. 2, 222 f.; LTK 9, 897; BS 11, 1328 f.

V. Saxer

SOTERIOLOGY. Faith in salvation in Jesus Christ was one of the fundamental facts of patristic teaching and *theology. Though the Christians of the early centuries were content to confess in the baptismal *creed that the Lord Jesus died and rose for us, they did not fail to express, in their manner of living and praying, what had been preached to them about the one saviour of the *world. When theologians defined their ideas on salvation, their theological reflection was at heart nothing but a reply to the question: why was Christ made man? While failing to take any position on Christ's saving work, bishops and synods always included it in their preoccupations with *orthodoxy. When they defended sound belief, whether in the *resurrection of the *flesh, the divinity of the Son and the *Holy Spirit, the one Lord Jesus Christ *God and man, or the necessity of actual *grace, they always started from the premise that Jesus Christ was made man for us and that God glorified him for us, through his death and resurrection, in the strength of the Holy Spirit.

Because the doctrine of salvation was never dogmatically defined, but everywhere presupposed, it is not easy to describe it briefly. Indeed it is all the more difficult since not just the histories of *dogma and theology, but a whole series of other disciplines too are interested in clarifying the concepts of salvation and *redemption: the histories of Christian *spirituality, *liturgy and early Christian *archaeology, the history of ancient religions and cultures and even the history of post-classical languages. Only one who is familiar with all these fields of research will be able to gather up the often scattered results of so many scientific disciplines into an approximately complete synthesis. It is as well to set out the following premises: firstly, patristic s. was not confined only to reproducing biblical s. Conditioned by their culture and personal dispositions, the Church *Fathers reinterpreted the message of Jesus Christ by choosing certain aspects and setting them in the context of their time. This fact appears especially in early Christian art, which represents Christ in various ways: as good *shepherd, teacher, lord of the world, crucified. And it is certainly allowable to consider the highly complex s. of the Fathers in terms of subjects and models: *illumination, victory, *divinization, *sacrifice of reconciliation. But we must also bear in mind two things: firstly, a thematic consideration of patristic s. must never overlook its historical development: i.e. we must bear in mind how different ideas found a different interest in the individual periods of the patristic era. Secondly, we also need a theological evaluation; in other words we must show in what perspectives the Church Fathers spoke of the necessity, for *faith, of a theophany of *God, of the new Adam's contest with the devil, of the God-man's mediation between God and the *world, of the education imparted by the divine-human teacher or of the vicarious sacrifice of the *cross.

Bearing these facts in mind, we will point to the following aspects of the history of patristic s. In the subapostolic period (up to 150), preaching on salvation still moved almost entirely in the tracks of apostolic *tradition. However it did show some inquiry, in that it developed the biblical reflection by a wider use of the so-called *testimonia*, it described salvation in terms of quite Hellenistic categories and, particularly clearly in *Ignatius of Antioch, it related salvation to the true *incarnation of the true God. Towards the mid 2nd c., we meet a s. that may be labelled "the triumph of the heavenly saviour". It presents Christ as a heavenly being who comes down to earth to carry out his work of salvation (*illumination of all peoples, victory over the *demons) and then reascends to his *Father together with redeemed humanity. Towards the end of the 2nd c., the controversy over the resurrection of the flesh induced the anti-*gnostic authors *Irenaeus and *Tertullian to put the *incarnation at the centre of that history (*oikonomia* - *economy), in which man attains complete likeness to immortal God. *Clement of Alexandria and *Origen, more open to the authentic aspirations of the Hellenistic world, worked out a s. in which salvation was conceived as true gnosis. According to them, the eternal Word, mediator between God and the world, leads back all mankind without distinctions, via a history of universal pedagogy, to a knowledge of the transcendent *God. At a time of crisis for the Roman empire, *Cyprian of Carthage (†258) presented Christ as the teacher and example of true *salus*, while keeping within the perspective of Christ's actual presence in his church, and not forgetting the historical fact of the past *redemption and the expectation of the supreme judge's return. Another Latin writer, *Lactantius (after 300), was the author of a still more Roman doctrine. For him Christ must be at the same time God and man, since only as God is he able to teach us true *justice and only as man can he set us the example of a *virtue that perseveres until death.

With the turning-point of *Nicaea (325), first ecumenical *council and symbol of the *tempora christiana*, a new concept of Christ was proclaimed: *Eusebius of Caesarea, theologian to *Constantine the Great, glorified Christ as the true emperor, and both the *liturgy and the art of the now imperial church echoed him. Defending the Nicene belief in Christ's true divinity, *Athanasius (†373) and *Hilary of Poitiers (†367) understood redemption primarily as incarnation of the eternal Son of God, who divinizes, i.e. glorifies, all mankind. Under the influence of *neoplatonism, by now the Roman empire's only philosophy, *Gregory of Nyssa (†394) conceived Christ's death and resurrection as the decisive turning-point in the history of *evil, as well as the beginning of man's never perfect but always progressing *divinization. During the 4th c., the defence of the faith of Nicaea led to a christocentrism that characterized not just liturgy, esp. baptismal liturgy, now understood as a participation in Christ's death, but also Christian *spirituality, as we see in *Evagrius's (†399) doctrine of equality with Christ (isochristism) and *Ambrose's (†397) doctrine of *Christus omnia*. The patristic s. of the first centuries undoubtedly reached its culminating point in *Augustine (†430). For him, Christ's saving work consisted above all in the *revelation of the *love of the humble God. Jesus Christ, sole mediator between God and man, manifested throughout his existence, but especially in his teaching and his example, how man, fallen with *Adam into the misery of *sin, can rediscover his God. And with his death and resurrection, he took, in the name of all men, the first step on the way to divine justice, so that under his guidance the City of God may be achieved in his church.

Though, from the start, the development of patristic s. included the question of the Saviour's person, this set of problems was not clearly posed before the end of the 4th c. But at that point Christian thinkers understood more clearly that human salvation could be founded only on the one Christ, true God and true man. The faith of *Chalcedon (451), declaring that Jesus Christ's two natures existed in one *person, perhaps did not make all the soteriological implications of this dogma explicit. But the preaching and theology of those who prepared the way for it left no doubt. Acc. to *Theodore of Mopsuestia (†428), the most eminent representative of the Antiochene tradition, God has united all creation in Christ, because the divinity of the Word and the humanity of Jesus are for ever conjoined in him. This union in the *assumptus homo* will nevertheless reach its perfection only in the final *resurrection.

But Christ's paschal *mystery has already anticipated the Christian's transition to a state of immortality and freedom, since in this glorious event Christ has become the model of *hope in eternal salvation, as well as the pledge of the consummation of all things. Crowning the Alexandrian tradition that had insisted so much on the divine-human origin of redemption, *Cyril of Alexandria (†444) gave it its most coherent expression. As his doctrine of Christ's priesthood particularly demonstrates, Cyril was particularly interested in the eternal Word's ontic unity with the humanity assumed only so as to anchor Jesus' historical action more deeply in God himself. *Leo the Great (†461), as a good heir of the Latin tradition, founded his s. on the dogma of double consubstantiality. While emphasizing the distinction of the two natures, he did not overlook the fact that Jesus' obedience and abandonment had their deepest roots in his personal unity with the Son of God. How much, in this doctrinal position, the theme of man's salvation was depicted is apparent from the way in which Leo presents his doctrine of the God-man in the setting of the celebration of the liturgical mysteries, thanks to which, acc. to him, the risen Lord never ceases to live in his church. Finally *Maximus Confessor (†662), certainly the profoundest thinker of the later patristic era, affirmed in a truly grandiose way the soteriological significance of the faith of *Chalcedon. Acc. to him the *Easter of the Christians was prefigured in Christ's pasch. Indeed the whole universe has been recapitulated in the glorious Christ. Above all Maximus, more than any other Father, based the saving union of all things on Christ's divine-human freedom, thus overcoming in the most radical manner any Hellenization of the good news.

Looked at formally, all this s. of the Fathers constitutes an ever-renewed interpretation of the *Bible's main assertions on salvation. Responding to the pastoral needs of all time, the Church Fathers brought the apostolic message up to date, seeking always to remain faithful to the whole of the biblical doctrine. But many of their interpretations were closely linked to their own situation. So ideas like the *sacrifice of reconciliation, victory over the devil, return to *Paradise, must be considered as metaphors proposed for their own time, to illustrate a fundamental claim that is always valid. While adapting the gospel to the pastoral situation of the moment, the Church Fathers also show how, to interpret these metaphors properly, we must not allow ourselves to be led only by the aspirations of today, but must always begin anew from the facts of the gospel.

As for its content, patristic s. understands *Christus per nobis* primarily as the Saviour's actual presence in his church: Christ is saving "today" all those who believe in his name. But this updating of salvation will not be complete until the last moment, when Christ will return to submit all things, and finally also himself, to God his Father. But this "presential" understanding of Christ's work must not be isolated from the previous history of Israel, still less from the historical events of Jesus' life. Acc. to the Church Fathers, Christ is operating now (*nunc*) only inasfar as he once (*tunc*) came for our salvation.

This nexus between now and then is understood in two ways: on one hand, it is claimed that through the paschal mystery Christ has been enthroned as Saviour, so that he can save now, in the church, all who believe in him; on the other, we remember that, in his passing to the Father, Christ anticipated our own. These two main ways of explanation, which are not mutually exclusive, undoubtedly smack of the ancient ideas and categories by which they were surrounded. Thus today we certainly do not share their way of speaking of the Saviour's heavenly existence or of the inclusion of all the members in Christ, the suffering and risen head. But we are well aware that behind these perhaps antiquated presentations of *redemption stood the *sensus fidelium*: Christ lives and saves men and is doing so, since he truly died and rose for us. This ever-valid meaning of faith includes the fact that Christ was able to save us only because he is at the same time God and man. In the final analysis this means that Christ must be a free man in order to be able to orient himself towards God with full freedom in our place, and that equally this free choice of God could not be based on God's own freedom. So a Christian must believe with the Church Fathers that God himself was the one just man on this earth and that, in his righteousness until death, our salvation, i.e. our abandonment to God and to men, has already begun for all times.

C. Andresen, Erlösung, *RACh* 6, 54-219; B. Studer, *Soteriologie in der Schrift und Patristik*: HDG III/2a, Freiburg 1978 (ample bibl.); R. Lane Fox, *Pagans and Christians*, London 1986; R.C. Gregg-D. Groh, The centrality of soteriology in early Arianism, *SP* 15.1 (1984) 305-316.

B. Studer

SOUL, HUMAN

I. The contexts - II. The problems.

The human s., on which *Pamphilus said that *Origen would not have dared write a treatise (*Apologia* 8: PG 17, 604), was one of the main themes of classical and Christian antiquity, but is still without a comprehensive study. We will give its dialectical context and the problems that developed in the first five centuries of Christianity.

I. The contexts. The contexts of comparison for the Christians were mainly *Stoicism and *Platonism. Stoicism considered the human s. as a living principle, which originated from matter and dissolved at the end of each cyclical conflagration (*ekpyrôsis*); while Platonism held it pre-existent to each one's *birth and hence not created simultaneously with the *body, but fallen into it (Cic., *Tusculanae* 1, 10, 20 and 11, 22; Tertull., *De anima* 54, 1-4). Stoicism set Christianity the problem of the nature and destiny of the s.; Platonism that of its relationship with the body.

II. The problems. The problems that derived from this had as background the question of the concept of incorporeality linked with that of the spiritual, a link that seemed inherently contradictory; and that of the mutual *communion of substances, in this case the s. and the body, which constituted the major difficulty for the spirituality of the s.

1. The nature of the soul. *Augustine, in the 5th c., summed up the results of centuries of debate thus: "*Natura animae . . . res spiritualis est, res incorporea est, vicina est substantiae Dei*" (*En. in ps.* 145, 4, 40). This was a conclusion in line with *neoplatonism (Plot., *Enn.* IV, entirely dedicated to the s.), which held the s. to be a substance separate from the body and acting independently of it; spiritual and incorporeal by nature, hence without the *quantitas* proper to bodies; rational, distinct therefore from the irrational s. of animals; divine and eternal by nature (Plot., *Enn.* IV, 7, 10, 1); immortal, added many *Fathers, particularly Augustine. But on the s.'s divinity the Fathers, while admitting its spirituality, made clear its distance from the divine essence: it is only "image of the *image" (Greg. Nyss., *De anima*). Given, then, its possibility of deviating, it is far from divine perfection (Cassiod., *De anima*: PL 70, 1287). The s., specified Augustine, being incorporeal, is the substance closest to *God (*De civ. Dei* 11, 26, 1; *Quant. an.* 34, 77; *Gen. ad litt.* 10, 24); being a creature, it is changeable and inferior to Him (*In Jo. ev.* 38, 10, 19). Aristotle and the Stoics thought that the substance of the s. was corporeal, naturally different from that common to other bodies. The Platonic duality *psychè-nous* was brought to separation by Aristotle (*De anima* 2, 2, 413b), and *nous in particular, starting-point of the intelligible *world in neoplatonism (Plot., *Enn.* IV, 3, 1, 4), was understood as having a divine constitution since made of the *quinta essentia* or *natura* (the thesis of Eudaemos) or of ether (*De Philosophia*),

an idea taken up, via *Philo of Alexandria (*Quis haeres* 57), by many Christian authors (texts in Moraux, *PWK* 14/1, 1171-1266). To the s. was applied *endeléchia*, i.e. the circular movement of the heavens. So the s. was by nature astral and hence divine, implying immortality *per se* (Arist., *De anima* 1, 3, 407a), but endowed with corporeality, *corporales lineas* (Tertull., *De an.* 9, 1-3), though of a substance more subtle than the bodies known to us (Iren., *Adv. haer.* 2, 34). The theory of the s.'s corporeality arose, among Christian authors, for various reasons: to avoid identifying the s. with *God (Maximus Conf., *Ep.* 6: PG 91, 428); opposition to the *Platonism of considering man only by his s. (Iren., *Adv. haer.* 5, 6); the possibility of *resurrection, with the s. capable of recognizing and attaching itself to the *body (Iren., *Adv. haer.* 2, 33-34); finally, to give an explanation of the passions, that which is incorporeal not being subject to passibility, acc. to the Stoic adage (Tertull., *De an.* 6, 4), otherwise they would have had to follow the theory of those who ascribed the passions to the body alone. In this context the s.'s immortality was understood as a gift of God and not as an innate quality (Iren., *Adv. haer.* 2, 34, 1-4). In the 5th c., reflection on original *sin and its transmission led to the questions of the s.'s origin (creationism or *traducianism?) and its spirituality-immortality (Aug., *Epp.* 166 and 167 to Jerome; *De quantitate animae; De immortalitate animae; De anima et eius origine*; Claud. Mam., *De statu animae*).

2. Immortality of the soul. In ancient Christianity there was a gradual triumph of this idea, due above all to *Origen and *Augustine. In the 5th c. there were still some who taught that the s. was material and mortal like any creature. At that time *Claudianus Mamertus wrote the most complete study we possess on the s.'s spirituality, from which was deduced the proof of its immortality. *Paul's text on Col 1, 16 ("*in ipso condita sunt universa . . . visibilia et invisibilia*") was read by some in favour of immortality, likening "invisible" to "incorporeal" (*asômaton*) (Orig., *Princ.* 1, 7, 1; 4, 3, 15; Aug., *Ep.* 238, 15); others made a distinction between "invisible" and "incorporeal", by which the s. was invisible, but corporeal like all created beings (the thesis attached to *Stoicism: SVF II, 123, 16; Tertull., *De an.* 7, 3; *De carne Chr.* 11). Immortality was considered proper to *God: the s. was only invisible, not immortal, though it could become so by God's gift (the thesis of *Irenaeus and of Christians in general who believed in the *resurrection). *Tertullian's statement on the immortality of the s. ("Some truths are known to us directly from nature, such as the immortality of the soul": *De resurr.* 3, 1-3) gained ground from the 2nd c. on, absorbing the neoplatonist principle that the s. contains the body, a role possible only for an immaterial principle (Nemesius of Emesa, *De natura hominis*: PG 40, 540; Proclus, *In Timaeum* 90A: Diehl 1, 293, 22). I.e., the s. ensures its continuance in being, unlike the *body, being by nature imperishable, i.e. immortal. The Alexandrian school, developing *Philo's idea that immortality and incorruptibility are divine qualities that mutually imply each other (*De opif. hom.* 135), inquired into the relationship between *apatheia (the impassibility that man acquires through ethical effort, and which is a divine attribute) and *aphtharsia* (the incorruptibility that indicates a link with immortality: Clem. Al., *Protr.* 120, 3; *Strom.* 8, 3; Greg. Nyss., passim). *Origen puts the s. in the third degree of rational creatures, among the incorporeal, assigning man's corporeality to his diversity and not to the nature of the s., which is therefore immortal *per se* (*Princ.* 1, 4, 1; 1, 8, 3; 2, 1, 4 and 8, 3). *Athanasius deduces the s.'s "reasonable and immortal" nature from its activity: it thinks in such a way and "moves itself" independently of the body (*Contra Gentes* 33. Tertullian had already argued thus: *De an.* 43, 1-2; 49, 1-3; *De test. an.* 1, 5; 4, 1; 5, 1-2). *Augustine derived the s.'s immortality from the eternity of the truth perceived by it (*Solil.* 2, 19, 33: "Hence the soul is immortal"; *De trin.* 14, 4: "In the human soul, rational and intelligent, we must find the *image of the Creator, immortally incised in its immortality").

Tertullian, *De anima* (CCL 2, 779-869); *De testimonio animae* (CCL 1, 173-183); *De censu animae* (lost); Origen, *La disputa con Eraclide* (SCh 67, Paris 1960); Gregory of Nyssa, *De anima et resurrectione* (PG 46, 11-160); ed. H. Polack, *Gregorii Nyss. opera omnia* 2, Leiden 1966); Ambrose, *De Isaac et anima* (CSEL 32/1, 641-700); Augustine, *De quantitate animae* (PL 32, 1035-1080); *De immortalitate animae* (PL 32, 1021-1034); *De anima et eius origine* (CSEL 60, 423-470); Faustus of Riez, *Ep.* 3 (CSEL 11, 3-17); Claudianus Mamertus, *De statu animae* (CSEL 11, 18-197); Cassiodorus, *De anima* (PL 70, 1279-1308). *Trattati sull'anima dal V al IX secolo*, Milan 1979; E. Rhode, *Psyche. Seelenkult und Unsterblichkeitsglaube der Griechen*, Freiburg i. Br. 1890-1894 (It. tr. Bari 1914-1916; Fr. tr. Paris 1952); E. Bréhier, *La théorie des incorporels dans l'ancien stoïcisme*, Paris ²1928; H. Gomperz, Asômatos, *Hermes* 67 (1932) 155-167; J. Goldbrunner, *Das Leib-Seele Problem bei Augustinus*, Munich 1934; Ch. Josserand, L'âme-Dieu. A propos d'un passage du "Songe de Scipion", *AC* 4 (1935) 141-153; A.J. Festugière, *La révélation d'Hermès Trismégiste III. Les doctrines de l'âme*, Paris 1953; J. Coman, L'immortalité de l'âme dans le "Phédon" et la résurrection des morts dans la littérature chrétienne des deux premiers siècles, *Helikon* 3 (1963) 17-40; E.L. Fortin, *Christianisme et culture* *philosophique au Ve siècle. La querelle de l'âme humaine en Occident*, Paris 1959; P. Moraux, Quinta essentia, *PWK* 24 (1963) 1171-1266; A.H. Chroust, The Doctrine of the Soul in Aristotle's Last Dialogue "on Philosophy", *The New Scholasticism* 42 (1968) 364-373; G. Movia, *Anima e intelletto*, Padua 1968; C.G. Stead, *The Changing Self. A Study on the Soul in Later Neoplatonism: Iamblicus, Damascius and Priscianus*, Brussels 1978; C. Scanzillo, *L'anima nei Padri dei primi secoli*, in AA. VV., *Problemi di attualità*, Naples ²1979, 33-164. R. Iacoangeli, Anima e eternità nel "De Isaac" di S. Ambrogio, in *Morte e Immortalita nella catechese di Padre di III-IV secolo*, BSR 66, Rome, 1985, 103-138; R.J. O'Connell, *The Origin of the Soul in St. Augustine's Later Works*, New York 1987.

V. Grossi

SOZOMEN. Author of a *Historia Ecclesiastica*. Contemporary of *Socrates Scholasticus. We know little of him, and that from his own work. Years of birth and death unknown. Born prob. in the village of Bethelia near *Gaza in *Palestine, to a family of very committed Christians who founded parish churches and convents. He describes his grandfather, who lived at the time of *Julian and was respected for his learning and, after his *conversion, for his exposition of Holy *Scripture. Schooled by monks, he remained in contact with monastic circles, from whom he took information useful for his work. After a *journey to *Italy he finally settled at *Constantinople, where he followed the profession of σχολαστικός: this must have been after the death of *Atticus (425), since he claims not to have known that bishop. *Photius (*Bibl.*, cod. 30) says his full name was Salamon Hermias Sozomen, but the oldest MSS also give the variant Hermias Sozomen Salaminius.

We know of an earlier historical work by S. in two books, now lost, going from Christ's ascension to the destitution of *Licinius (323) and depending on Clement (not of Alexandria, but the author of the "Pseudo-*Clementine homilies"), *Hegesippus, *Julius Africanus and *Eusebius. The nine books (whose modern division into chapters is not the author's) of the *HE*, dedicated to Theodosius the Younger and written between 439 and 450, continue this. As S. announces, the work was to concern the period between 324 (3rd consulate of Crispus and *Constantine) and 439 (17th consulate of *Theodosius II). However, it stops at the year 421 (death of *Honorius), while later events are barely mentioned. It has been supposed that this mutilation was due to imperial censorship, but the hypothesis is hardly probable: however, the *manuscript tradition does not allow us to establish whether or not the first readers possessed a more complete text. Written at the same time as Socrates's work, the *HE* deals with the same period and depends on it even in details, though it avoids mentioning it. Jeep and Schoo in their studies have identified other sources: *Rufinus (not cited), *Eusebius of Caesarea's *Historia Ecclesiastica* and *Vita Constantini*, some works of St *Athanasius (not all cited), *Palladius, numerous documents, synodal *acts, *letters of emperors and bishops, dispersed in state and church archives and private *libraries. Occasionally the *HE* refers to *Gregory Nazianzen's *orationes*, to *Libanius and *Apollinaris, while echoes of the works of *John Chrysostom and *Jerome are uncertain. Probably S. knew the works of *Ephrem, *Eustathius of Antioch, *Eunomius and *Eutropius; book IX makes great use of Olympiodorus of Thebes. The collection of letters indicated in the *HE* as γραφαί should perhaps be identified with the συναγωγή of *Sabinus of Heraclea (Batiffol). Also present are Syrian sources and some other works unknown to Socrates, evident in the passages relating to the history of the Persian monks and martyrs in the reign of Shapur II. The problem of finding the evidence was S.'s primary concern, as we see from the *Prooemium* of the *HE*. He made every effort to uncover all the documentation possible to him. Unlike Socrates, he extended his researches to the *biographies of illustrious monks and used numerous sources concerning events in the West.

Faced with the problem of organizing the superabundant material he had managed to find, he gave priority to his literary requirements, linked to the canons of classical *historiography, and to his aim of addressing the work to upper-class, but not excessively competent, readers. This explains the style, plainer and simpler than that of Socrates (Photius, *Bibl.*, cod. 30), the work, hardly commendable, of fusion and synthesis of sources and documents present almost throughout the *HE*, the reluctance to go into theological questions and the tendency to anecdotal narrative rich in ascetic legends and miraculous events. In S.'s historical vision, divine *providence constantly intervenes to direct worldly events to the progress of Christianity: the emperors become the guardians of *orthodoxy and the history of the church takes on apologetic tones. Editions of the *HE* have been prepared by R. Estienne (Paris 1544), H. Valois (Paris 1668), R. Hussey (Oxford 1860) and recently by Bidez-Hansen. We have a Latin version of part of the work made in the 6th c. by the monk *Epiphanius under the direction of *Cassiodorus, but no Oriental translations.

Editions: CPG 3, 6030; PG 67, 843-1630; GCS 50; W. Jacob-R. Hanslik, *Cassiodori Epiphanii Historia Ecclesiastica tripartita*, CSEL 71, Vienna 1952. Studies: L. Jeep, *Quellenuntersuchungen zu den griechischen Kirchenhistorikern*, Leipzig 1884; P. Batiffol, Sozomène et Sabinos, *ByzZ* 7 (1898) 265-284; G. Schoo, *Die Quellen des Kirchenhistorikers Sozomenos*, Berlin 1911; W. Schmid-O. Stählin, *Geschichte der griechischen Literatur*, Munich 1924, II, 2, 1435-1436; *PWK* 22/1, 1240-1248; *DTC* 14, 2469-2471; G. Moravesik, *Byzantinoturcica*, I, Berlin 1958, 510-512; *LTK* 9, 933-934; G. Downey, The Perspective of the Early Church Historians, *GRBS* 6 (1965) 63-66; G.C. Hansen, Prosarhytmus bei den Kirchenhistorikern Sozomenos und Sokrates, *ByzS* 26 (1965) 82-93; Quasten II, 539-541; A. Primmer, Sozomenus, Kirchengeschichte (ed. J. Bidez), *Gnomon* 39 (1967) 350-358; F. Winkelmann, Die Kirchengeschichtswerke im östrom. Reich, *ByzS* 37 (1976) 175-177; G.F. Chesnut, *The First Christian Historians: Eusebius, Socrates, Sozomen, Theodoret and Evagrius*, Paris ²1986; Altaner 229; M. Mazza, Sulla teoria della storiografia cristiana: osservazione sui proemi degli storici ecclesiastici, in *La storiografia ecclesiastica della tarda antichità*, Messina 1980, 335-389.

A. Labate

SPAIN and PORTUGAL. The Roman conquest of Hispania, begun in 218 BC, ended only in 19 BC with full pacification. Yet its Romanization slowly proceeded and reached the various provinces at different levels: most of all, for rapidity and intensity, Baetica. From 197 BC Hispania was divided into two provinces: Citerior (NE) and Ulterior (SW). *Augustus divided Ulterior into two provinces: *Baetica (cap. Cordova [*Corduba*]) and *Lusitania (cap. *Mérida [*Emerita*]); *Diocletian divided Citerior into three: *Tarraconensis (cap. *Tarragona [*Tarraco*]), *Carthaginensis (cap. Cartagena [*Carthago Nova*]) and *Gallaecia (cap. *Braga [*Bracara*]). The five provinces, with *Mauretania Tingitana, made up the diocese of Hispania, which from the time of *Constantine was part of the Prefecture of the Gauls.

St *Paul left no trace in the peninsula; we do not know whether his wish to go to S. (Rom 15, 24) was fulfilled; it remains only a hypothesis, despite the abundance of patristic testimonies, some of them very old. The coming of *James and the so-called Seven Apostolic Men also collapses under the silence of the primitive sources (literature, *archaeology, *epigraphy, *liturgy). Bearing in mind the respective, and sometimes late, proofs of not a few scholars, largely Spanish, who incline towards an African origin for Christianity in the peninsula, we must conclude that Christianity arrived in S. through the usual channels of Romanization: army, commerce, missionaries sent by other Christian communities, particularly by *Rome. The earliest evidence of Christianity in S. comes from *Irenaeus (*Adv. haer.* I, 3) and *Tertullian (*Adv. Iud.* 7, 4), but is of little value, being too vague and perhaps also rhetorical. From the first valid evidence, *Cyprian's *Ep.* 67 (of 254), we know of the existence of bishops at Mérida and Astorga-León (*Asturica*), as at other nearby cities, besides *Saragossa (*Caesaraugusta*); of a *persecution during which Christians set an example to their pastors; of relations with Rome (and only reflexively with *Carthage). Through the authentic, but not proconsular, Acts of the *martyr St *Fructuosus, we know of the presence of Christianity in Tarragona and some of its customs from 259. The martyrs of Diocletian's persecution witnessed to their faith in Saragossa, Calahorra (*Calagurris*), *Gerona (*Gerunda*), Cordova, Mérida and Alcalá de Henares. But the document that tells us most about the extent, characteristics and problems of Spanish Christianity is the *Acts of the council of *Elvira (early 4th c.). From these we deduce a considerable penetration of the Christian religion into Baetica and neighbouring areas: a Christianity in some ways not wholly free from *paganism and in others rather rigid and traditional, in ways already abandoned by the church elsewhere. During the 4th c. the Spanish church remained in a sense outside the great crises of Christianity. Its relationship with the *Donatism of nearby *Africa was very minimal and can be seen in the presence of eight bishops at the council of *Arles (314) and the probable influence of *Ossius of Cordova on Constantine's decisions. Ossius was the representative of the Spanish church's participation in the *Arian crisis, as ecclesiastical adviser to the emperor and leader of the anti-Arian party in the West. Two more peninsular bishops took part in the conflict, though in different ways and with different outcomes: *Potamius of Lisbon, who signed the Arian formula of *Sirmium (357), and *Gregory of Elvira, anti-Arian to the bitter end. But the *peace of the Spanish church was disturbed at the end of the 4th c. by the Priscillianist movement, with wide civil and ecclesiastical repercussions inside and outside the peninsula. *Priscillianism stimulated a certain literary output, which we can add to that of Potamius and Gregory, that of the poets *Juvencus and *Prudentius, that of *Bachiarius, *Eutropius (Spanish?) and the works of St *Pacianus, who tells us about some subjects very topical in the Spanish church at the time: *penitence, *marriage-*virginity, *ascesis, *clergy, cult of the *martyrs, etc. The conciliar activity that now began reached its culmination in the Visigothic period.

Before the *Visigoths occupied Tarraconensis in 414, the Suevi had crossed the Pyrenees and occupied Gallaecia, along with the Asding *Vandals; the Alans had settled in Lusitania, and the Silings in Baetica. Pressed by Roman troops, these last two crossed immediately into Africa (429). After the fall of the last Roman emperor in the West (476), the Visigoths invaded the whole of S. except for the Basque and Suevian territories, which were eventually incorporated into the Visigothic kingdom under *Leovigild (574 and 585 respectively). But only in 621, under Suintila, was territorial unity established after the expulsion of the *Byzantines who, called in by Athanagild, had entered from Africa in 554 and occupied the coastal zone of Carthaginensis, part of Tarraconensis and much of Baetica. From the religious point of view, the Suevi were originally pagans who were then converted to *Catholicism (448); they later changed over to *Arianism (465) and, at the time of St *Martin of Dumium, back to Catholicism. The other groups remained Arians. With them the Hispano-Roman community had to live, sometimes persecuted but generally in peace. During this period an intense relationship with the Roman see was kept up: the Spanish bishops appealed to the *pope, and he nominated his vicars for the various provinces; but even in the 4th c. these relations were growing fewer, at the same time as conciliar activity was intensifying in cities like *Tarragona, *Toledo (*Toletum*), *Barcelona (*Barcino*), *Lérida (*Ilerda*) and Valencia (*Valentia*), as well as *Braga and *Mérida. The civil metropolises also became ecclesiastical metropolises. To the first period belong the historians Paul *Orosius and *Idatius. *Severus of Minorca, *Apringius of Beja, *Justinian of Valencia, *Licinianus of Cartagena, *Justus of Urgel and St *Martin of Dumium were also celebrated characters of this time. The incipient *monasticism is also worthy of mention.

After reaching almost full territorial unity, king *Leovigild (568-586), urged more by political than religious motives, wished also to attain religious unity under the Arian faith professed by the Visigothic people, and to this end he applied various kinds of pressures. This intention, besides provoking a civil war with the rebellion of his son *Hermenegild, who had previously been converted to Catholicism, was a real failure. Unity was achieved only under his son and successor *Recared who, for political as well as religious reasons, was converted to the Catholic faith in 587, followed by the majority of the Visigothic people. This *conversion was publicly proclaimed at the 3rd council of Toledo (589). From then on, the Spanish church began a century of splendour under the banner of unity. The Visigothic church, markedly national in character, came of age and operated with a considerable degree of autonomy (in fact relations with *Rome were reduced to a minimum and were not exempt from tension); it governed itself through national or provincial *councils, among which those of *Toledo stand out for their legislative activity, ecclesiastical and civil. The prestige of the episcopal sees was generally linked to celebrated pastors who governed them: *Seville (*Hispalis*) with St *Leander and St *Isidore; *Saragossa with St *Braulio and *Taio; *Mérida with *Massona; Palencia (*Pallantia*) with Conantius; and Toledo with St *Eugenius, St *Ildefonsus and St *Julian. Harmony between church and monarch produced unity of worship, which was confirmed at the 3rd council of Toledo (589). Culture flourished, though consisting more of recapitulation and transmission than of creativity: secular culture, but above all religious culture. There was no lack of people occupied with *Scripture (Isidore, Ildefonsus, Taio, Julian). *Theology was apologetic, anti-Arian (Leander, Isidore) and, above all, anti-Jewish (Isidore, Ildefonsus, Julian), since, in the civilly and religiously united kingdom, the presence of *Jews was both a religious and a political problem. But there was no lack of dogmatic theology (Isidore, Taio), particularly *eschatology (Julian).

In legal matters, the Hispania *canonical collection, probably by Isidore, remains famous. In the field of civil law, we must record Recceswinth's *Liber Iudiciorum*, elaborated by Braulio. Secular history had its scholars in *John of Biclaro, Isidore and Julian, and ecclesiastical history had the *De viris illustribus* of Isidore and that of Ildefonsus. There was room in this output for an abundant literature of *hagiography, particularly the legendary Acts of the martyrs. Then there were Isidore's labours to preserve *classical culture. The centres of this culture were episcopal *schools, among which Seville stood out, and parochial and monastic schools, under the impulse of the councils. *Monasticism flourished after its beginnings in the earlier period, assuming various forms, some of which worried the conciliar *Fathers; at the same time monastic *rules multiplied (of Leander, Isidore, Fructuosus, the *Common Rule*, etc.). Celebrated monks were Martin of Dumium, Isidore, *Valerius of the Bierzo, *Fructuosus of Braga. With the Muslim invasion of 711, a new and very different era began in Spanish history.

A great many *councils were celebrated by the church of S. The number of those known from the Roman and then the Visigothic period is over 40; to these must be added some more uncertain councils and one Arian one. They are traditionally divided into Hispano-Roman and Hispano-Visigothic, and both these into general and provincial. Those celebrated before the conversion of the Visigoths to Catholicism (589) are conventionally called Roman, though some of them were celebrated under Visigothic and Suevian rule. Provincial councils are those in which the bishops of a province took part, and where problems which concerned only that province were dealt with. General councils are those which dealt with problems of faith or which concerned the whole peninsular church, in which all the bishops in the kingdom took part, personally or through representatives. Despite the importance of some of the Hispano-Roman councils (e.g. that of *Elvira), the most famous remain those, much studied, of *Toledo during the Visigothic period, which present particular and uniform characteristics, among which emerge: *a)* the participation of the sovereign (who convoked them, proposed the subjects to be dealt with and confirmed their canons, which thus had the value of civil laws); *b)* the presence of *laymen, generally nobles, who sometimes signed the *Acts, thus giving the impression of being a manifestation of the courts of the kingdom; *c)* the simultaneous treatment of civil and ecclesiastical problems; *d)* their celebration, which was not regular, but determined by the needs of the moment. These councils were the expression of a singular symbiosis between church and state (for the councils, cf. individual entries).

DACL 5, 407-523; *EC* 11, 1042-1052; *DHGE* 15, 892-915; *CE* 13, 494-496; *DHEE* 2, 983-985; *ES*, Madrid 1747; A. González, *Epistolae, Decretales ac Rescripta Romanorum Pontificum*, Madrid 1808; P.B. Gams, *Die Kirchengeschichte von Spanien*, Regensburg 1882-1883; H. Leclercq, *L'Espagne chrétienne*, Paris 1906; Z. García Villada, *Historia eclesiástica de España*, I-III, Madrid 1935-1940; J. Vives, *Concilios visigóticos e hispano-romanos*, Barcelona-Madrid 1963; M. Blázquez, Posible origen africano del cristianismo español, *AEA* 40 (1967) 30-50; T.C. Akeley, *Christian Initiation in Spain 300-1100*, London 1967; E.A. Thompson, *The Goths in Spain*, London 1969; AA. VV., La patrología hispano-visigoda, in *XVII Semana española de Teología*, Madrid 1970; U. Domínguez del Val, Cultura y teología en la España visigoda, *Salmanticensis* 17 (1970) 581-612; R. Puertas Tricas, *Iglesias hispánicas (siglos IV al VIII). Testimonios literarios*, Madrid 1975; J. Orlandis, Iglesia, concilios y episcopado en la doctrina conciliar visigoda, in *La Iglesia en la España visigoda y medieval*, Pamplona 1976, 154-181; id., *Historia de España. La España visigótica*, Madrid 1977; R. García Villoslada (ed.), *Historia de la Iglesia en España*, I, Madrid 1979 (bibl.); J. Orlandis-D. Ramos Lissón, *Die Synoden auf der Iberischen Halbinsel bis zum Einbruch des Islams (711)*, Paderborn 1981 (bibl.); J.M. Blazquez Martinez, Origin del Cristianesimo Hispanio, and, El Cristianesimo durante el Siglo IV, in (ed.) J.M. Manfarres, *España Romana*, Madrid 1982, 415-483; J.E. Salisbury, *Iberian Popular religion 600 BC-700 AD: Celts, Romans and Visigoths*, New York 1985; J.N. Hillgarth, *Visigothic Spain, Byzantium and the Irish*, London 1985; J. Fontaine, *Culture et spiritualité en Espagne du IVᵉ au VIIᵉ siècle*, London 1986; M. Sotomayor, Legenda y Realidad en los origines del cristianesimo hispano, *Proyección* 36 (1989) 179-198; J. Fontaine, Hispania (Literargeschichtlich), *RAC* (forthcoming 1990).

P. de Luis

SPELUNCA THESAURORUM ("the cave of treasures"): an apocryphon composed in *Syriac between the 4th and 7th cc., which draws on Christianized Jewish traditions relating to the OT, putting forward a continuous history from *creation to Jesus Christ, dwelling particularly on the period from *Adam to *Abraham. The work, of which an *Arabic version also exists, shows evident affinities with two more *apocrypha: the *Struggle of Adam* and the *Ethiopic book of revelations* (Qalimentos); like them it is linked in some way to the pseudo-*Clementine *corpus*.

Ortiz de Urbina 95-96; C. Bezold, *Die Schatzhöhle, aus dem syrischen Texte*, Leipzig 1883; id., *Die Schatzhöhle nach dem syrischen Texte . . . nebst einer arabischen Version*, Leipzig 1888; A. Götze, Die Schatzhöhle. Überlieferung und Quellen, *SHAW* 4 (1922) 23-37; A. Battista-B. Bagatti, *La Caverna dei Tesori. Testo arabo con traduzione italiana e commento*, Jerusalem 1979.

S.J. Voicu

SPES. Bishop of Spoleto (*Spoletium*) (late 4th-early 5th c., normally put before *Achilleus); author of a 12-verse poem in honour of the *martyr Vitalis, whose body he had rediscovered. The poet recommends himself and the *virgin Claventia, almost certainly S.'s daughter, to the saint's intercession. He died after 32 years as a bishop (his epitaph is preserved). Some of his relics were taken by Charlemagne to Aquisgranae to ornament his chapel.

Diehl I, 364, n. 1851; *DACL* 15, 1639 f.; A.P. Frutaz, Spes e Achilleo vescovi di Spoleto, in *Atti del II Convegno di Studi Umbri*, Perugia 1964, 362-365; *Patrologia* III, 300; R. Grégoire, L'agiografia spoletana antica: tra storia e tipologia, in *Il Ducato di Spoleto*, Spoleto 1983, 335-366; C. D'Angelo, Il vescovo Spes e la basilica spoletina dei SS. Apostoli. Nota preliminare, *ibid.*, 851-858.

A. Di Berardino

SPIRITUAL COMBAT. Spiritual tradition, following straight from *Scripture (cf. 1 Cor 9, 24-27) and adding the Stoic ideal, frequently compares *ascesis to a battle against the enemies of the *soul. Texts to this effect are numerous: for *Clement of Alexandria, the true "gnostic" must ask *God to put him to the test (*Strom.* IV, 7, 55, 1: GCS 2, 273). The tested man *par excellence* is the monk: δόκιμος μόναχος (Pall., *Hist. Laus.* 18, ed. C. Butler, Cambridge 1904, 52). Combat is inevitable, whether one is perfect or not. If the *apatheia* of the *Messalians tends towards quietism (cf. Ps.-Macarius, *Hom.* 26, 14: PG 34, 684), *Evagrius on the contrary believes that *temptations increase progressively in the spiritual life (*Prakt.* 59: SCh 171, 679). The Scriptures unmask the enemies who stir up these wars: Satan, the *world, the *flesh (cf. Rom 7-8). Monastic writings distinguish between "visible" and "invisible" combats. Evagrius explains "visible combat" as concerning πράγματα, objects, contact with which gives birth to passions. The way to resist and overcome the *demon is by abstention, by renunciation of the objects themselves. "But with monks (the demons struggle) mostly by thoughts (λογισμοί)" (*Prakt.* 48, 609). The mechanics of *temptation, whose different moments are linked together, was deeply analysed by Eastern spiritual writers, esp. those of "Sinaitic" *spirituality (*Nilus, *John Climacus, Psychius, *Philotheus). Elimination of wicked thoughts (ἀπόθεσις νοημάτων), guarding of the heart, discernment of spirits, *antirrhesis*, opposing evil thoughts by Scriptural texts: all are frequent themes in the writings of the *Hesychasts.

DSp 2, 1135-1142; O. Chadwick, *John Cassian*, Cambridge 1950, 95 ff.; I. Malone, *The Monk and the Martyr*, Washington 1950, 91 ff.; H. Crouzel, L'anthropologie d'Origène dans la perspective du combat spirituel, *RAM* 31 (1955) 364-385; G.M. Colombás, *Paradis et vie spirituelle*, Paris 1961, 118 ff.; A. de Vogüé, *Règle du Maître*, I, Paris 1964, 89 ff.; A. Guillaumont, *Evagre le Pont: Traité pratique, introd.*, SCh 170, Paris 1971, 95; T. Spidlík, *Spiritualité de l'Orient chrétien. Manuel syst.*, OCA 206, Rome 1978, 225 ff.

T. Spidlík

SPIRITUALITY

I. The first two centuries - II. The third century - III. From Nicaea to Chalcedon.

The notion of s. is modern and is not encountered as such among the ancients, who preferred to speak of spiritual *theology, *ascesis and *mysticism or, simply, of Christian and evangelical life. We will confine ourselves to offering pointers on ascetic and spiritual *life.

I. The first two centuries. We possess a certain number of writings from the *Judaeo-Christian communities, such as the *Didachè, the *Odes of Solomon*, the *Letter of *Barnabas*, the *Shepherd* of *Hermas, which reflect the spiritual life they led. The *Didachè* describes the practices of Christian life in the framework of Jewish and evangelical *tradition: *fasting, *prayer, good works. The *Odes of Solomon* reflect a mystical exaltation, a spiritual fervour surprising for that time. The *Letter of Barnabas* develops a s. of *baptism, a typological reading of Holy *Scripture, in which the author depicts Christian perfection as the new temple indwelt by the *Holy Spirit (ch. 16). He gives this advice: the Christian does not work out his salvation in sorrow, but is a child of joy (4, 11). Hermas's *Shepherd* gets rid of the image of an idealized church. Hermas is aware of the demands of baptism, but also of the falls that will occur. A realist, he opposes the rigorists and is unwilling to drive sinners, who are also called to help build the tower, to desperation, provided they do *penitence (*Sim.* VI, 3, 4, 6). He asks that the indispensable *virtues be observed before the evangelical counsels (*Sim.* V, 1, 4-5). He insists on the importance of joy and trust (*Prec.* X, 1, 2) and on discernment of spirits (*Prec.* VI, 2, 3-5). All these Judaeo-Christian writings, particularly the *Didachè*, emphasize *eschatology, which polarizes the entire spiritual life: the communities reach out to Christ's return, which they consider imminent.

With *Ignatius of Antioch, s. takes two complementary directions: one ecclesial, the other individual. The bishop is primarily a man of the church. Spiritual life develops in church, in the *assembly, the preferred place of *prayer, in the *Eucharist, in obedience to and union with the bishop. The soul of all spiritual life is Christ, our *life, our eternal life, our inseparable life (*Eph.* 3, 2; *Magn.* 1, 2; *Smyrn.* 4, 1). Christ occupies all Ignatius's *letters. Spiritual life consists of putting on Christ, his passion, his death, to share his *resurrection. The Eucharist is at once this *mystery and this *hope. *Martyrdom is at once a *liturgy and the shortest way to reach Christ and the *Father. In Ignatius and *Polycarp we see the formation of a spiritual theology of martyrdom, confession of *faith, presence of Christ, prolongation of the passion and the Eucharist, ecclesial witness, affirmation of the future resurrection attested by the Acts and Passions of the martyrs.

The *Apologists are more concerned to present Christianity to the *pagans than to expound spiritual life. At least *Justin witnesses to their sacramental and community life. Christians, illumined by Christ, know the truth and possess the *grace to lead a virtuous life (*II Apol.* 10, 3). Justin's own life witnessed to this in martyrdom. The *Letter to *Diognetus* does not just illustrate the Christian life, but also shows its action in the *world: Christians are the soul of the world (*Diogn.* 6), provided they radiate God's *love for man, which is the *raison d'être* of *creation: "If we love God, we will imitate his tender affection" (10, 4-5).

*Irenaeus describes throughout his work the gradual ascent towards knowledge of *God. Slowly God prepares man to accept the Word. Man is born a baby and imperfect: "It was first of all necessary that he be created, grow and gain strength, then multiply, finally reach his full development to attain *glory and then see his Lord" (*Adv. haer.* IV, 38, 3). A slow ascent, which is a *purification and transformation of the whole man to receive incorruptibility (*Demonstr.* 7).

Apocryphal literature, though tendentious and sometimes manipulated by heterodox currents, reveals a singular spiritual fervour. There we meet incontestable religious values: irruption of *redemption into the world through Christ, universal rehabilitation of the *cosmos through the resurrection, unshakeable trust in Christ and in his power, loving *devotion to Jesus wherever present, faith in an eschatology to come, exaltation of *woman as *virgin. Here we have some essential elements of Christian *faith and life.

From the 2nd c. the church was forced to defend, against the excesses of the *encratites and sects, the legitimacy of *marriage and the invitation of everyone to a perfect life.

II. The third century. Two main centres stand out in the 3rd c.: *Alexandria and *Carthage. The church developed despite increasingly violent *persecution, preserving its spirit of vigilance and preparedness for *martyrdom. Charity, lived in community, manifested itself in concrete forms and initiatives. *Tertullian was a polemist and moralist rather than a teacher of s.; yet it would be wrong to undervalue his ascetic works, which develop the following spiritual themes: *baptism, *prayer, martyrdom, patience, *chastity, expressing an ideal disposed towards heroism and martyrdom.

*Cyprian's literary work, like his activity, was primarily pastoral and spiritual. His preferred themes are: the one church, prayer, martyrdom, vigilance. "The kingdom of God begins to be very close to us. The reward of *life, the joy of eternal salvation, perpetual *beatitude, possession of *paradise once lost: all this approaches as the world passes." Cyprian's writings were among those most read throughout the church's history.

*Clement's entire work, despite his metaphysical concerns, aims to guide the gnostic, i.e. the illuminated believer, to perfection. The perfect model to imitate is Christ (*Paed.* I, 2, 4). The Christian people are a people of *children, who must be led to a limpid *faith. "Incessant exercise is necessary for gnosis to become a habit; perfection, once reached through the mystical habit, remains stable thanks to charity" (*Strom.* VI, 9, 78).

The whole of *Origen's work is dominated by the spiritual life. His main point of reference is the divine indwelling, which makes the just man into a temple of God and the Spirit. Here we find the triple presence, dear to Origen, of God and Christ in *Scripture, in the church and in the Christian. More than presence, there is union, which is both the impulse and the goal of spiritual *ascesis. A union whose ultimate end is *ecstasy. Starting from biblical images particularly dear to him, Origen describes the *purifications that prepare the encounter, beyond exile and *sin, to find inner unity (*De or.* 21, 2) and reach God. Ascesis that is essentially eschatological, in that it prepares the eternal day when God will be all in all (Hamman, *La Prière*, II). A spiritual doctrine that, through *Gregory of Nyssa, *Evagrius of Pontus, Ps-*Dionysius and *Maximus Confessor, would exercise its influence in *East and West. "There is no-one quite so omnipresent as Origen", says Urs von Balthasar.

III. From Nicaea to Chalcedon. For the *Fathers of the golden age it is enough to trace the outlines of spiritual currents on the basis of subject-matter. Baptismal *catechesis, at once doctrinal, ascetic and liturgical, plays a considerable role. It is an invitation to *conversion and *faith through evangelical life (Cyr. Jer., *Cat.* 1). The great number of surviving baptismal *sermons demonstrates the pains the Fathers took to prepare a strong faith and make it comprehensible (cf. Aug., *De catechizandis rudibus*).

Lent was a time of retreat for the whole church: preparation for *catechumens, reconciliation for sinners, mobilization and deepening of faith for all. *Basil's *Discourses* and *Hexaemeron* show us the people mobilized. The preaching of the Fathers reveals the importance they attached to the perfection of the *laity: "Jesus preached the *Beatitudes to married people as well as to monks" (John Chrys., *In Mt hom.* 7, 7; cf. Aug., *De sermone in monte*).

Preaching, in both East and West, offered the elements of a spiritual doctrine for the Christian people. Nourished by *Scripture and *liturgy, it endeavoured to expound the sanctification of all states of life, esp. *marriage and the *family. The domestic hearth "is a small church", says *John Chrysostom. Modest potsherds with fragments taken from the *Bible and liturgy, inscriptions with verses of *Psalms allow us to follow the diffusion of this catechesis. Liturgical echoes on *ostraka* and *papyri, of *acclamations, *doxologies, the *trisagion, *maranatha, Kyrie eleison,* show the influence exercised by the liturgy on popular *piety. *Letters of guidance, by *Basil, *John Chrysostom, *Augustine, *Paulinus of Nola, intensified this spiritual work in their own way, mainly among influential laymen and laywomen. To these we must add writings for *virgins and *widows (Viller-Rahner, 41-42).

The fervour of the *martyrs was continued esp. by virgins and monks, who multiplied and diversified, esp. after the Constantinian *peace, taking on the most varied and radical forms. The writings inspired by them (the *Vita S. Antonii,* the *Apophthegmata) exercised considerable influence in both *East and West. John *Cassian fixed their spiritual teaching and adapted their way of life. *Sulpicius Severus's *Vita Martini* played a similar role in the West. The Constantinian era gave free rein to the cult of the martyrs, then of the saints, who now occupied an important place in the community's spiritual life. This cult in turn gave rise to a literature in which legend invaded history, a *hagiography that would provide nourishment for generations of Christians. A great movement of *pilgrimage began in the 4th c. and grew continually: *Egeria may be considered its patron and model.

Both East and West developed the spiritual *theology inherited from Origen. *Gregory of Nyssa is rightly called "the father of *mysticism". His influence left its mark esp. on the spiritual doctrine of *Diadochus of Photica and *Maximus Confessor. Ps-*Dionysius, translated early and often into Latin, introduced this mystical theology to the West, where it decisively influenced the Victorines and St Bonaventure.

For his part *Augustine, with his theology of *grace, his *Rule for religious, his journey towards God, had a lasting influence on the Middle Ages. One of his disciples, *Gregory the Great, provided, in his *Moralia, Homilies* and *Regula Pastoralis,* manna for later generations of Christians.

Articles in *DSp*; P. Pourrat, *La spiritualité chrétienne,* I, Paris 1926; M. Viller, *La spiritualité des premiers siècles chrétiens,* Paris 1930; M. Viller-K. Rahner, *Aszese und Mystik,* Freiburg 1939 (tr. of the preceding with wealth of bibl.); J. de Ghellinck, *Lectures spirituelles dans les écrits des Pères,* Paris 1935; M.J. Rouët de Journel-J. Dutilleul, *Enchiridion Asceticum,* Freiburg ³1958; L. Bouyer, *La spiritualité du N.T. et des Pères,* Paris 1960; A. Hamman, *La prière,* II: *Les trois premiers siècles,* Paris 1963; *Les chemins vers Dieu,* Paris 1967 (texts selected by A. Hamman and F. Quéré); G. Bardy-A. Hamman, *La vie spirituelle des trois premiers siècles,* Paris 1968. Various authors, *Spiritualità del lavoro e catechisi del Padri III-IV sec.* Rome 1985; A. Sotignar and P. Dupuy, Spiritualité *DSp* 14 (1989) 1142-1173; L. Bouyer-L. Dattrino, *Spiritualità dei Padri,* vol. 3/A (ed.) Dehoniane, Bologna 1984; J. Biarne, Maître spirituel et règle cénobitique en Occident IVᵉ-VIᵉ siècles, *La Vie monastique en Occident de saint Benoît à saint Bernard, Actes du Colloque tenu à Senanque, 19-21 September 1983,* Paris n.d..

A. Hamman

SPONSA CHRISTI. Following Hosea, the prophets presented Jahweh's union with Israel as a marriage, and the *Song of Songs was interpreted in this sense. In the NT this theme was transposed into the union of Christ and his church, and *Hippolytus's *Com. Ct.* understands the bride only in the collective sense. The individual interpretation, in which the bride indicates the Christian *soul, is suggested in 1 Cor 6, 15-17; 7, 32-34, and occurs three times in *Tertullian apropos of *widows and *virgins (*Uxor.* I, 4, 4; *Resur.* 61, 6; *Virg. Vel.* 16, 4). But it was *Origen above all, in his *Com. Ct., Hom. Ct.* and other works, who powerfully orchestrated this double meaning of the bride, church and soul, thus creating the theme of mystical *marriage: the Christian soul is a bride because it is part of the church-bride; the church as bride progresses through the progress of souls. *Sin then becomes *adultery with the diabolical lover, adultery of the soul or adultery of the Synagogue.

J. Chênevert, *L'Eglise dans le Commentaire d'Origène sur le Cantique des Cantiques,* Brussels 1969; H. Crouzel, Le thème du mariage mystique chez Origène et ses sources, *StMiss* 26 (1977) 37-57.

H. Crouzel

SPYRIDION. Bishop of Trimethus (*Cyprus). Took part in the council of *Nicaea (325). Among the signatories of *Sardica in 343 (Athan., *Apol.* II 50, 2). Acc. to *Rufinus (*HE* X, 35), *Socrates (*HE* I, 12) and *Sozomen (*HE* I, 11), even as a bishop he continued his calling as a shepherd and worked many miracles. The historical figure was soon engulfed by legend. The first hagiographical *Life,* an iambic poem falsely attrib. to Triphillius, is completely legendary: lost in the original, it is preserved in two prose paraphrases, one anon., in the *Codex Laurentianus* XI, 9, the other as part of the *Life* of S. written by the monk *Theodore of Paphos (second half of 7th c.), both edited by Van den Ven. Theodore of Paphos's *Life* is the most important, though of no historical credibility: it served as the basis of an *Anonymous metaphrase* (11th c.), a *Short life* (11th c.), published by Van den Ven, and the *Life* by Simeon Metaphrastes (BHG, 1648).

P. van den Ven, *La légende de S. Spyridon évêque de Trimithonte,* Louvain 1953.

E. Prinzivalli

STATUTA ECCLESIAE ANTIQUA. Among the documents that tell us about the life of the Gallo-Roman church in the 5th c., the S.E.A. occupy a special place. For a long time they appeared in *canonical collections and editions of *councils under the erroneous title of a 4th council of Carthage; but the studies of Ballerini (1757), Maassen (1870), Malnory (1888), G. Morin (1913) and B. Botte (1939) have led to a gradual clarification of their place of origin, tendencies and even author: maybe *Gennadius of Marseilles, c.476/485.

The S.E.A. begin with an enumeration of the qualities required in a bishop, and a profession of faith; then follow 89 disciplinary canons on various subjects; finally a brief ritual for *ordination. Each of these three parts follows the plan of an earlier work: the prologue's profession of faith depends directly on Gennadius of Marseilles's work entitled *De ecclesiasticis dogmatibus*; many of the propositions are taken textually from this work. For the disciplinary canons, the author seems to have followed the model of the Eastern pseudo-apostolic compilations (*Didascalia,* *Apostolic Constitutions*); and the ritual of ordination is clearly influenced by *Hippolytus's *Apostolic Tradition*; the same series of orders recurs in the solemn *prayers for Good Friday.

A complex work, whose author draws on very different sources, the S.E.A. are nevertheless a coherent whole, thanks to the tendencies that inspire them. The author wishes, first and immediately, to limit the powers of the monarchical episcopate, submitting it to the dual control of the provincial synod and the "diocesan" *council, the *presbyterium,* whose involvement he considers indispensable for all the more important affairs: ordinations (ch. 10), trials (ch. 14), administration of church *property (chs. 15. 50). Sensitive to the rights of priests (chs. 2. 12. 56), whose dignity vis-à-vis bishops and deacons he emphasises (chs. 57-61), he gives his whole work a clearly ascetic imprint. From the prologue on, he carefully enumerates the qualities necessary for every ordinand: prudence, moderation, decorous life, good reputation. Nourished on Holy *Scripture, firm in doctrine and able to transmit it to his flock in appropriate language, the bishop must be completely dedicated to *prayer, *meditation on the Holy Books, preaching of God's word (ch. 3), not wasting his time reading profane or heretical works. Without being affected in his *dress and tenor of life (ch. 4), he must be gentle with his *clergy (ch. 2) and full of good will to people of low condition, *humilibus affabilis* (prologue). He must know how to obtain help in the many temporal duties that assail him (chs. 6-10); in his government he will not follow his own impulses or will, but will remain faithful to the healthy *traditions of the church, *patrum definitionibus* (prologue). Rules concerning the conduct of clerics are equally rigorous. They must live with simplicity (chs. 25. 26) and dignity (chs. 27. 28. 34. 68), assiduous in the liturgical offices (ch. 35); they must faithfully observe the obligatory *fasts (chs. 76. 77), guard themselves from the dangers of a too solicitous ministry, from too merry parties, from familiarity with young *widows and consecrated *virgins (chs. 27. 68. 75). They are asked to be modest and grave (chs. 73. 74. 26) and not let themselves drift into blasphemy, superstitious practices or silly or salacious jokes. To flee the temptations of sloth, aimless wandering and suspect *friendships (chs. 28. 34), they must take care to earn a living by the work of their own hands: *artificiolo vel agricultura* (chs. 29. 45), and to enrich their free time by study (chs. 45. 79).

The author of the S.E.A. dedicates some 15 canons to recommending harmony, union, understanding and charity to all *members of the church, from bishops down to simple *laymen. He invites clerics to abstain from all slander, envy, interested adulation and gossip (chs. 42-44). He demands *excommunication for those who unjustly accuse their *brothers, and that the evidence of the captious should only be accepted with extreme prudence (chs. 17. 46). In the same way he reproves practitioners of *usury, fomentors of discord and those who implacably avenge real or supposed wrongs against themselves (ch. 55). He proposes very severe penalties against those who violate the divine law of charity (chs. 17. 43. 44. 47-49); finally he denounces the carelessness of the Christian people, negligent in its duties and enamoured of enjoyment and lucre (chs. 69. 86), love of profane *entertainments (ch. 33), religious indifference and lack of fervour (chs. 31. 80. 82. 20-24).

Many passages of the *Statuta* suppose the presence among the people of a heretical power, with its own judges, liturgical *assemblies and *ministries (chs. 30. 80-82). *Catholics were exposed to vexations and robberies, out of hatred to their faith (ch. 70). However, the churches do not seem to have suffered systematic confiscations, because crowds of *poor people continued to receive their service of charity and *assistance (chs. 7. 64. 70. 86).

The evidence of the S.E.A. provides a good picture of the period of transition between the patristic golden age and the early Middle Ages. The theological and mystical impetus no longer animated the Christian people, seemingly taken up with earthly cares and involved in material responsibilities of all kinds; the clergy carried out their duties with difficulty. Culture itself seems already about to flee to the *monasteries, which would prepare the bishops of tomorrow.

To reawaken the drowsy *faith of the Christian people, to win the barbarian over to the truth, to put the church's still relatively intact wealth to the service of all adversities: these were the directives offered by the S.E.A. to the Provençal episcopate. From the first years of the 6th c., the councils presided over by *Caesarius of Arles would make every effort to effect these in the people and the clergy.

CCL 148, 162-188; Ch. Munier, *Les Statuta ecclesiae antiqua*, Paris 1960; Hfl-Lecl 2, 102-121; *CE* 13, 682.

Ch. Munier

STENOGRAPHY. The ancients knew a procedure of rapid, abbreviated *writing called tachygraphy, perhaps Egyptian in origin. Tiro, Cicero's freedman and secretary, worked out a system of abbreviated writing called *Notae tironianae*. Tachygraphy was much in use in the 3rd and 4th cc. in *East and West. Public services and courts used stenographers called *notarii*. Christian communities and the *Fathers often employed clerics as stenographers to take down conversations and public debates.

The stenographer set down the text on a *chartula* (rough paper) or a wax tablet (sometimes wood, esp. boxwood). A pointed metal stylus traced furrows on the soft wax. The tablets could serve indefinitely. The text, subsequently transcribed in full, was revised by the author before receiving its definitive form: a wearisome labour, as *Jerome confesses (*In Is. comm.* 13, *praef.*; *In Zach. comm.* 2 and 3, *praef.*). *Augustine reproves himself for not having always verified his references and revised his works with due attention (*Retr.* II, 41, 2).

Cf. first of all the *Archiv für Stenographie*; E. Chatelain, *Introduction à la lecture des notes tironiennes*, Paris 1900; C. Johnen, *Geschichte der Stenographie*, Berlin 1911; A. Mentz, *Antike Stenographie*, Munich 1927; E.G. Turner, *Greek Papyri. An Introduction*, Oxford 1969; H.H. Boge, *Griechische Tachygraphie und Tironische Noten*, Berlin-Hildesheim 1974 (bibl.); H. Hagendhal, Die Bedeutung der Stenographie, *JbAC* 14 (1971) 24-38 (bibl.).

A. Hamman

STEPHEN I, pope (254-257). S.'s brief pontificate is known from *Cyprian's correspondence (*Epp.* 67-75) and some information in *Eusebius's *HE* (VII, 2-9). He lived during the *persecutions of *Decius and *Valerian and his pontificate was disturbed by the penitential and baptismal controversies. The fact that, on the former occasion, a Spanish bishop appealed to S. (*Ep.* 67, 5) and that, in the case of bishop *Marcian of Arles, Cyprian asked S. to intervene (*Ep.* 68) shows us the *authority the Roman church had obtained at that time. But when, acc. to the evidence of Cyprian and *Firmilian, S. rejected any repetition of *baptism, resting on an unchangeable *tradition (*Ep.* 74, 1-2) and on his own Petrine succession (*Ep.* 75, 17), he met with resistance from the African and Asiatic churches. This encounter shows still more clearly how aware the Roman church had become of its primatial position in the universal church. S. died at the start of the new persecution, though not as a martyr. His death put an end to the controversy with the African church.

ASS *Aug.* I, 112-146; K. Baus, HKG I (Freiburg ³1965) 379 f., 401-407; A. Amore, *BS* 12, 22 ff.; U. Wickert, Paulus, der erste Klemens und Stephan von Rom, *ZKG* 79 (1968) 145-158; M. Lauria, Infames ed altri esclusi dagli ordini sacri secondo un elenco probabilmente precostantiniano, *Iura* 21 (1970) 182-186; S.G. Hall, Stephen of Rome and the One Baptism, *SP* 17, Oxford 1982, 796-798; G.W. Clark, *The Letters of St. Cyprian of Carthage*, vol. 4 (letters 67-82), ACW 47, New York 1989.

B. Studer

STEPHEN II of Hierapolis-Mabbug (late 6th c.). Author, *c.*592-594, of a Passion of St Golindukh. The original *Syriac text is lost, but a *Georgian tr. and a Gk revision by *Eustrathius of Constantinople remain. A fragmentary Gk treatise against the *monophysite sect of the *agnoetae*, handed down under the name of S. of Hierapolis, may or may not be by the same author.

CPG 3, 7005-7006; *DSp* 4, 1493-94; *DHGE* 15, 1234-35; F. Diekamp, *Analecta patristica*, OCA 117, Rome 1938, 154-160; G. Garitte, La passion géorgienne de sainte Golindouch, *AB* 74 (1956) 405-440.

S.J. Voicu

STEPHEN BAR SUDAYLE. *Monophysite monk, born at *Edessa, prob. in the last quarter of the 5th c. The most detailed contemporary source of S.'s life is a *letter of *Philoxenus of Mabbug, written *c.*510 to two Edessan priests, Abraham and Orestes (ed. G. Frohtingam, Leiden 1886), to put them on guard against the exaggerations of S.'s teaching, which he accuses of pantheistic tendencies. We must see in this severe judgment a reflection of S.'s influence on the *Origenistic doctrine which he knew through the works of *Evagrius of Pontus. *Jacob of Sarug (†521), in a letter addressed to S., while praising his zeal and *piety, rebukes him for his eschatological ideas.

Acc. to Philoxenus, S. was a disciple of a monk who upheld *heresies, whom he calls John the Egyptian. This claim gave credit to a theory that S. visited *Alexandria in his youth. A. Guillaumont has advanced the completely acceptable hypothesis that S.'s teacher may have been *John of Apamea, whose sojourn in *Egypt is fully ascertained: he was in fact opposed by Philoxenus. What is certain is that at a certain moment S. had to leave Edessa on account of his ideas. He fled to *Jerusalem, or at least *Palestine, to the Origenist monks, where he found a better climate of understanding and from where he certainly continued to propagate his ideas and correspond with his disciples who remained at Edessa (for this cf. Philoxenus's letter).

Of S.'s literary output (mystical *commentaries on the *Bible, esp. the *Psalms, *letters and theological treatises), nothing survives. Critics now unanimously recognize S.'s paternity of the work surviving in *Syriac under the name *Book of Hierotheus* (ed. and Eng. tr. by F.S. Marsh, London-Oxford 1927) or "of Saint Hierotheus, on the mystery hidden in God's dwelling" in five parts: I: Cosmology; II-IV: Ascent of the intellect towards its perfection. V: *Eschatology. What incontestably made the fortune of this markedly esoteric composition was its attribution to Hierotheus who, in the pseudo-Dionysian writings, was *Dionysius's teacher. Recent studies have demonstrated that the *Book of Hierotheus*, at least in the form in which it has survived, post-dates the appearance of the pseudo-Dionysian writings.

Duval 356-358; Baumstark 167, 350; Chabot 66-67; Ortiz de Urbina 103-104; *DSp* 4, 1481-1488.

J.-M. Sauget

STEPHEN of Bostra. We cannot prove that S. was a bishop of *Bostra (province of *Arabia). He lived in the 7th and 8th cc., certainly before *John Damascene, who cites from his work *Against the *Jews* (Λόγος κατὰ τῶν Ἰουδαίων) - who, evidently, favoured the iconoclasts - two passages in defence of the cult of sacred *images (*De imag.* III, *testimonia*: PG 94, 1376). The work is also mentioned and cited by other writers. All the fragments are collected by A. Mai (*Opere minori*, ST 76, Rome 1937, 202-207).

CPG 3, 7790; Bardenhewer 5, 47; Beck 447 and 831 (index); *DHGE* 15, 1211.

A. De Nicola

STEPHEN GOBAR. *Monophysite, apparently 6th-c., adherent of the so-called *tritheism of *John Philoponos. *Photius (*Bibl.*, cod. 232) describes a work by S. dealing with 52 questions on various subjects (*theology, cosmology, *anthropology, *eschatology), citing, for each question, passages of authors who represented the church's official doctrine and passages of authors who maintained erroneous doctrines.

CPG 3, 7300; PG 103, 1092-1105 (fragments); R. Henry, in Photius, *Bibliothèque*, Paris 1967, V, 67 ff.; G. Bardy, Le florilège d'Etienne Gobar, *REByz* 5 (1947) 5-30; 7 (1949) 51-52.

M. Simonetti

STEPHEN of Larissa. Bishop of Larissa at the time of *Justinian (527-565). Known from the *Acts of the Roman *council that met *in consistorio B(eati) Andreae* on 7 and 9 Dec 531, presided over by pope Boniface II (530-532). S., though a *layman and a soldier, was called by the *clergy and people of Larissa to succeed the deceased bishop Proclus. The presbyter Anthony and two suffragan bishops, Demetrius and Probianus, though they had approved the election, appealed to patriarch *Epiphanius of Constantinople (520-535) against this *ordination, claiming that it had been made against canonical rules. Epiphanius immediately deposed S.

without hearing his defence and had him taken by force to *Constantinople, where he kept him prisoner. S., who from the start had appealed to *Rome since *Thessaly had long belonged to the jurisdiction of the bishop of Rome, managed from Constantinople to send the *pope two *Letters*, which affirm papal primacy over the whole church and attest Rome's competence over Illyricum. The second *Letter* also describes S.'s trial by Epiphanius. He had taken no account of S.'s appeal to the pope, indeed this had irritated him still more, so that S. had been accused of seeking to harm the rights of the patriarchal church. Epiphanius did everything to prevent the appeal reaching Rome, but failed. Theodosius, suffragan bishop of Larissa, came to Rome and handed over the *Letters* of S. and the other suffragan bishops of Larissa, accompanied by 26 documents (*letters), nearly all by popes and emperors and proving Roman jurisdiction over Illyricum. In the council the *Letters* were read and the authenticity of the documents verified by comparison with the originals preserved in the archives of the Roman church.

The sentence pronounced by the council has not survived, since the Acts lack an ending. The episode of S. must be seen against the background of attempts by Constantinople to remove Illyricum from Roman jurisdiction.

Mansi 8, 739-784; crit. ed.: C. Silva-Tarouca, *Epistolae Roman. Pontif. ad vicarios per Illyr . . . Collectio Thessalonic.*, Textus et documenta, Series Theologica 23, Rome 1937; Hfl-Lecl 2, 2, 1117-1119; DCB 4, 740, n. 22; Fliche-Martin 4, n. 809, p. 676.

A. De Nicola

STEPHEN the Martyr. The account of S.'s *martyrdom (Acts 6-7) is the first hagiographical passage in the NT. He was stoned outside *Jerusalem at a time when the office of Roman prefect was vacant, perhaps in AD 31 or 32. Stoning was in accordance with Jewish custom, but outside Roman procedure.

The saint's relics were rediscovered in 415 by *Lucian, priest of Caphargamala, who wrote an account of it (BHL 7850-56). On 26 Dec 415 they were transferred to the church of St Sion at Jerusalem. Even before that, the *Mart. syr.* had celebrated S.'s feast on 26 Dec. Bishop *Juvenal built a basilica on what was considered the site of the stoning, and got *Cyril of Alexandria to open it, 439 (PG 85, 469). The empress *Eudoxia had it enlarged, but the Persians destroyed it. Another basilica was built by *Melania (SCh 90, 255-258).

*Gregory of Nyssa (PG 56, 701-736), *Asterius of Amasea (PG 40, 337-52) and *Basil of Seleucia (PG 85, 461-473) had already written panegyrics of the saint. The feast-day of 26 Dec passed from the *Mart. syr.* to the *Mart. hier.* and the *Cal. carth.* But the rediscovery of the relics in 415 and their later diffusion gave a new impulse to the spread of S.'s cult. *Augustine's attitude was important for *Africa and all-important for Western Europe. From the 5th c. we find the saint's relics everywhere: in the *Baleari, Africa, *Constantinople, *Rome (where the church of S. Stefano Rotondo goes back to c.450), Bourges, etc. Numerous episcopal churches bore his name. The epistle of his feast (Acts 6, 8-11; 7, 54-8, 1) was translated into the Romance languages and these translations, together with the Latin text, continued to be sung into the 19th c.

Vies des SS. 12, 687-702; *Cath* 4, 571-574; *LTK* 9, 1050-52; *BS* 11, 1376-92; V. Saxer, L'épître farcie de la Saint-Etienne "Sesta Lesson", *Provence historique* 23 (1973) 318-26; 24 (1974) 423-67; id., *Morts, martyrs, reliques en Afrique chrétienne aux premiers siècles*, Paris 1980, 245-78; M. Simon, *St. Stephen and the Helenists in the Primitive Church*, New York 1958; Y.-M. Duval, *Loca Sanctorum Africae*, CEFR 58, 2 vols., Rome 1982, II, 624-632.

V. Saxer

STILICHO. General of the *Vandals. Born c.360, he made his name in the war against the Visigoths (391-392), reaching the rank of *magister militum*. He married *Theodosius's niece Serena, but met resistance from the Roman aristocratic classes. Despite his repeated military successes he soon had to abandon the Eastern part of the empire, becoming guardian of the emp. *Honorius in the West. He was victorious against *Gildo in *Africa (398); he defeated *Alaric, once during the invasion of 401 and again at the battle of Pollentium. In 406 he was victorious against Radagaisus. His successes in war failed to hold the emperor's sympathies. In 408 a revolt broke out against which he would not and could not defend himself and, handed over to the rebels, he was executed with his wife Serena and son Eucherius.

An ungrateful judgment weighs upon S.'s attempt to govern the state, now torn between particularistic propensities and the new requirements of the barbarians, in a traditional way. *Augustine's cautious silence contrasts with the accusations of *Jerome and *Orosius. The sole voice of opposition based on objective data is that of Rutilius Namatianus.

S. Mazzarino, *Stilicone. La crisi imperiale dopo Teodosio*, Rome 1942; L. Vàrady, Stilicho proditor arcani imperii, *Acta Antiqua Hung.* 16 (1968) 413-432; A.D.E. Cameron, Theodosius the Great and the regency of Stilicho, *HSCPh* 73 (1969) 247-280; M.R. Alföldi, Zum Datum der Aufgabe der Residenz Treviri unter Stilicho, *Jahrbuch f. Numismatik und Geldgeschichte* 20 (1970) 241-248; L. Cracco Ruggini, "De morte persecutorum" e polemica antibarbarica nella storiografia pagana e cristiana. A proposito della disgrazia di Stilicone, *RSLR* 4 (1968) 433-447; M. Miller, Stilicho's Pictish War, *Britannia* 6 (1975) 141-145; S. Cristo, The Relationship of Symmachus to Stilicho and Claudian, *Atene e Roma* 21 (1976) 53-59; J.L. Sebesta, On Stilicho's Consulship. Variations on a Theme by Claudian, *The Classical Bulletin* 54 (1977-78) 72-75.

M.L. Angrisani Sanfilippo

STOBI (Στόβοι). Ancient Yugoslavian city at the confluence of the rivers Crna (Erigon) and Vardar (Axios), c.160 km. N of *Thessalonica, an episcopal see in *late antiquity, when it was also capital of *Macedonia Salutaris and later of Macedonia Secunda. The city already existed as part of Paeonia in the late 4th-early 3rd c. BC; it became a *municipium* in the first years of the Roman empire, even coining its own money from at least AD 69 to 222. John Stobaeus (5th c.) was born there. The city was abandoned, then destroyed, perhaps by an earthquake, before the end of the 6th c., though the bishops continued to bear the name of the see at least until the end of the 7th c.

Joint American-Yugoslav excavations in 1970-80 have considerably increased our knowledge of the life, culture and art of S., particularly in *late antiquity. Earlier Yugoslav excavations (1924-1940) had already uncovered parts of the urban centre. The excavations were particularly important not just for our knowledge of the evolution of the urban structure in the first six centuries AD, but also for the well-preserved architecture and monumental decorative arts, esp. *mosaics and frescos. The 4th c. is represented by the imperial residences - which continued into the 5th c. - such as *Theodosius's palace and the adjacent house of Parthenius, and the house of Peristeria: all buildings with a rich variety of pavement mosaics. Theodosius's palace even had wall mosaics in one room and an open courtyard paved in *opus sectile*. Near the city centre, a 4th-c. synagogue - replacing the earlier one of Polycharmus - had pavement mosaics with both rough and polished *tesserae*. Both pavement mosaics and wall frescos had designs that were mainly imitations of marble *opus sectile* or of panelling. S of the synagogue, near the theatre, was the oldest and best-preserved *church building in the Balkan peninsula: a three-aisled basilica built in the 4th c., with pavement mosaics with geometrical designs in the nave and presbytery - here combined with *opus sectile* - red limestone flooring in the side-aisles and frescos on the walls and ceilings.

Both the synagogue and the primitive church were rebuilt in the 5th c. with larger Christian basilicas (respectively the central and episcopal basilicas); four more basilicas are known for this period: that to N with a small cruciform baptistery, the cemeterial basilica just outside the gate giving onto the road to Heraclea Lyncestis, the Palikura basilica c.2 km. to S and finally that beyond the Erigon. The episcopal basilica is of particular importance on account of the quantity of mosaics preserved in the S aisle, the narthex, a S room and the nave, and for the frescos and mosaics in the very well-preserved baptistery, a tetraconch contained in a square structure. Inside, the mosaic which surrounds the large baptismal font has four main scenes, two of deer and two of peacocks beside an overflowing *kantharos*. The life-size *paintings on the baptistery walls comprise scenes perhaps of Christ's life and pictures of saints addressing audiences. The architectural embellishments of the baptistery and episcopal basilica, with the decorations of the capitals, further increase the importance of this ancient city for the study of late antique art and architecture. [**Figs: 285, 286, 287**]

E. Kitzinger, A Survey of the Early Christian Town of Stobi, *DOP* 3 (1946) 81-161; R.E. Hoddinott, *Early Byzantine Churches in Macedonia and Southern Serbia*, London 1962; J.R. Wiseman, Stobi in Yugoslavian Macedonia: Excavations and Research, *Journal of Field Archaeology* 5 (1978) 391-491 (with bibl.); *Studies in the Antiquities of Stobi*, ed. J.R. Wiseman, I-II Beograd 1973-1975; ed. B. Aleksova-J. Wiseman, III, Titov Veles 1981; Princeton University Press has begun publication of a series of studies entitled *Stobi*, Princeton 1984 ff; D. Pallas, Le Baptistère dans l'Illyricum Oriental, *Actes XIe Cong. Internat. d'arch. chret.*, CEFR 123, Rome 1989, 2485-2490.

J.R. Wiseman

STOICISM and the FATHERS. The earliest preaching and exposition of Christianity had a Semitic structure. In the 2nd c. Christianity, spreading into areas of Hellenistic-Roman culture, clothed itself in the forms of Western thought. It appropriated forms of expression from Graeco-Roman *philosophy: S., impregnated sometimes with *middle Platonism, and then *neoplatonism. The presence of S. is considered dominant until *Clement of Alexandria, c.230, when Platonic influence prevailed. The doctrine that is referred to the Stoà also resulted in part from *Aristotelian and Platonic precedents: S. was an eclectic system that gathered elements of various origins and fused them together in unity, giving them characteristic features. "We owe to the Stoà the ideas which middle Platonism transmitted to the Fathers, when they bear the specific imprint of the Stoic school" (Spanneut, IV, 154). All the ideas present in the *Bible but developed with particular insistence by S. should also be considered Stoic. In the surviving texts of the Early and Middle Stoà we see no traces of Judaism (Spanneut, II, 130).

Stoic influence includes terminology, transferred to Christian thought, and concepts, which could be adapted to express Christian truths and which were sometimes gathered up and inserted into the texture of Christian doctrine. At other times reference to the Stoà consists in the general tone, echoing the spiritual orientation and optimism of S. We will confine ourselves here to pointing out some specific convergences between Christian writers and S.

1. The 2nd-c. *Fathers follow the Stoà in the concept of man as *animal rationale* (Sen., *Ep.* 41, 8), composite in unity, against Platonic *dualism (*body the prison of the *soul). Cf. SVF III, 96, p. 24, 2; 50, p. 251: the soul, which is corporeal, grows and develops with the body and follows its lot (*compati corpori*, Tertull., *De an.* 5, 5=SVF I, 518, p. 117, 3). For *Irenaeus the body is part of man, as is the soul; the perfect man is made of soul, *flesh and Spirit (*Adv. haer.* 5, 6, 1: SCh 153, p. 76). In *Tertullian, man has a unitary structure: soul and body form a stable unity, a *textura* (*De res.* 34, 10), a *collegium* (*De res.* 15, 3), a *societas* (*De res.* 15, 1-6). As in the Stoà (SVF II, 604, p. 185, 45; Sen., *De provid.* 6, 9), death is separation of the soul from the body (*De res.* 19, 3; *De an.* 51, 1). The Stoic doctrine allows Tertullian to defend - against the *gnostics - the *resurrection and recomposition of soul and body in view of the *judgment (*De res.* 34, 8): gnosticism was Platonizing.

2. The *soul for Tertullian is corporeal (*corporalis, De an.* 22, 2). It is born of the breath (*flatus*) of *God (*ibid.*). The breath of God is something other than the spirit (*Adv. Marc.* II, 9, 1). "Corporeal" is not equivalent to "material" (*De an.* 3, 4). For the Stoics all reality is corporeal; what has no *corpus* is unreal: *nihil si non corpus* (*De an.* 7, 3); *nihil est incorporale nisi quod non est* (*De carne Ch.* 11, 4). *Soma* and *ousía* are the same thing (SVF II, 359, 123). The soul (= *flatus*) is an idea of the Stoics, who present it as *pneuma (SVF I, 135, p. 38, 3), and it is also biblical. The Stoic-Tertullianean thesis is not to be understood as anti-spiritualistic. It was taken over by John *Cassian (*Col.* 7, 13: SCh 42, 257): souls are not to be understood as incorporeal; only God is incorporeal. In Irenaeus we read: "souls are incorporeal compared to mortal bodies" (*Adv. haer.* 5, 7, 1, p. 84). He seems to be affirming a certain corporeality (cf. A. Orbe, *Antropología de San Ireneo*, Madrid 1969, 440); the context is Stoicizing.

3. According to the S. of Cicero and Seneca, the soul possesses by nature an ontological and moral good, which persists even when enveloped and suffocated by prejudices and errors (*Tusc.* III, 1, 2; *De fin.* V, 15, 43 and 21, 59-60; Sen., *Ep.* 22, 15; 94, 31. 54. 55. 56; *Ep.* 108, 8). In Tertullian (*De an.* 41) we see an analogous *bonum naturae*; it is at the basis of the *testimonium* of the "*anima naturaliter christiana*", based on the *sanitas* of the soul (*Apol.* 17, 4-6).

The initiative of good, made possible by the *bonum* of nature, recurs in the 4th c. in *Basil of Caesarea in a Stoicizing context (*Regul. fusius tract.*: PG 31, 908. 909. 910, etc.); in *John Chrysostom (*In epist. II ad Cor.*: PG 61, 397; *In epist. ad Hbr*: PG 43, 99); in *Optatus of Milevis (2, 20: CSEL 26, p. 55); in *Jerome (*Dial. adv. Pelag.* III, 1: PL 23, 596). *Bonum naturae* is an obvious concept in *Pelagius (*Ep.* to *Demetrias, chs. 2. 3. 4. 7. 8. 9: PL 30, 15-45). Hence the ideal of a life in accordance with nature (SVF III, 65, p. 16, 28; Sen., *Ep.* 41, 8; M. Aurelius, *Ric.* II, 17).

4. *Bonum naturae* also finds expression in the concept - Stoic in origin - of "seeds of virtue" (*semina virtutum*), known to Cicero (*De fin.* V, 15, 43; V, 7, 18) and to Seneca (*Epp.* 73, 16; 94, 29; 108, 8; 120, 4). Such seeds are innate and originate from nature in John *Cassian (*Conl.* 13, 12: SCh 54, 166; 23, 11: SCh 64, 154) and in *Faustus of Riez (*De gratia* I, 12: CSEL 21, 40).

5. *God immanent in the *world is manifested through the order of the world (cf. Chrysippus in SVF II, 1009, p. 299; and Cicero, who develops Cleanthes's thought in *Nat. deor.* II, 5, 13-15). *Tertullian considers *God knowable *ex operibus* (*Apol.* 17, 4). The world was created so that God could be known through it (*Adv. Marc.* I, 10, 1). The same in *Aristides (*Apol.* 1, 1, p. 71 Vona). S. sings the beauty of the world. The *cosmos=Aristotle's *diakósmesis* (*Perì philosophías* 12 b, p. 81 Ross) recurs in *Marcus Aurelius (*Ric.* 9, 1; 12, 5). It is used in the same sense by *Aristides (*Apol.* 1, 1, p. 71 and 117 Vona) and by Basil of Caesarea (*Hexaem., hom.* 1, 1. 5. 6: SCh 24a, 104. 110). *Athenagoras uses four times, in the cosmological sense, the verb *diakosméo* (*Legat.* 7, 1; 10, 1; 22,

12; 24, 3). The term enjoyed exceptional favour among the Fathers (Spanneut, I, 364). The optimistic Stoic doctrine that saw the cosmos as permeated by divinity (Virgil, *Aen.* 6, 726-727) was antithetical to the cosmic pessimism of *gnosticism (Tertull., *Adv. Marc.* I, 13, 5; 14, 3-5).

6. In *Ad Diognetum* (early 3rd c.) the Christians fulfil the same function in the world that S. assigned to the *anima mundi*; to be a principle of cohesion: the action is expressed by *synéchein* (to hold together from within) (6, 7), proper to Stoic cosmology (SVF II, 439, p. 144, 26-28). It was taken from the Stoic tradition by Wis 1, 7, where the word designates the cosmic action of the Spirit (cf. Iren., *Adv. haer.* V, 2, 3: SCh 153, p. 36, and Athenag., *Legat.* 6, 2, still in the sense of cosmic action; from there it passed into *Ad Diognetum*).

7. S. knew God's *providence (*prónoia*) (SVF II, 1106-1126, pp. 322-327; M. Aurelius, *Ric.* 9, 1), which it identified with the *world itself, since divinity regulated individual things (SVF II, 528, p. 169, 35). God was immanent in the world (Cic., *Nat. deor.* II, 22, 58; Sen., *De providentia* 5; *Nat. Quaest.*, praef.). Aristides attributes God's coming into the world to his *prónoia* (p. 117 Vona). The Stoic concept was adapted to Christian *dogma (cf. also Athenag., *Legat.* 8, 4; 19, 3; 24, 3; 25, 2).

8. Among Christians, the Stoic distinction between *lógos endiáthetos* and *prophorikós* (SVF II, 135, p. 43, 18) designates the divine Word immanent in God's bosom or manifested externally (Theoph. Ant., *Ad Autol.* II, 10; II, 22; p. 122 and 154: SCh 20).

9. Stoic cosmopolitanism, expounded by Cicero (*Nat. deor.* II, 62, 154; SVF II, 1131, p. 328): *mundus quasi communis deorum atque hominum domus*, and by Seneca (*De otio* 4, 1): *respublica magna et vere publica, qua dii atque homines continentur*, whose borders were marked by the sun (cf. also M. Aurelius, *Ric.* 6, 44), finds a conscious echo in Tertullian (*Apol.* 38, 3): *unam omnium rem publicam agnoscimus, mundum*; and in *Minucius Felix (*Oct.* 17, 2): *hanc communem omnium mundi civitatem*.

10. The doctrine of cosmic anthropocentrism, which puts the cosmos at the service of man, is Stoic (SVF II, nos. 1152-1167). It was fully developed in Cicero (*Nat. deor.* II): *omnia quae sint in hoc mundo, quibus utantur homines, hominum causa facta esse et parata* (II, 61, 154, p. 112 Ax). God created the things of this world for man's use: so *Justin (*I Apol.* 10, 2; *II Apol.* 4, 2; 5, 2); *Theophilus of Antioch (*Ad Autol.* I, 6; II, 18); *Ad Diognetum* (4, 2; 10, 2); Tertullian (*De an.* 33, 9; *De pat.* 4, 3; *De spect.* 2, 4: *universitatem tam bonam quam homini mancipatam*, etc.); *Ambrose (*Off.* I, 132). Notwithstanding Gen 1, 26. 28-30, the theme in Christian authors seems to be derived from S. (Waszink, *Tertulliani De Anima*, Amsterdam 1947, 296), which agrees with the *Bible without depending on it.

11. A text of Seneca (*Benef.* IV, 7; SVF II, 1024, p. 305, 33) identifies nature with God and with the divine reason that pervades the world. It is echoed in Minucius Felix: *quid aliud Deus quam mens et ratio et spiritus* (*Oct.* 19, 2). In Tertullian creation is the work of the *verbum*, the *ratio*, the *spiritus* (*Apol.* 17, 1): *Dei ratio quia Deus omnium conditor nihil non ratione providit disposuit ordinavit* (*De paenit.* 1, 2). God and world are linked, while excluding Stoic monism. Tertullian identifies "natural" and "rational" (*De an.* 43, 7). Clement of Alexandria sings the rational order of the world in a way that evokes the Stoà (*Strom.* VI, 155, 3). Nature for Tertullian is the first teacher; natural law binds because God is its author (*Virg. vel.* 11, 6; *De cor.* 5, 1); what does not proceed from nature is derived from the devil (*De c. fem.* I, 8, 2). Living according to nature is the norm of the Stoà (Sen., *Ep.* 41, 8); the law of physical nature takes on moral value, since man is an integral part of the *cosmos. Sometimes we find the same in Tertullian (*De cor.* 5, 1; *De c. fem.* I, 8, 2; cf. Spanneut, IV, 164).

12. Stoic sources, esp. the more recent ones, assign to *marriage the sole aim of procreation: *gámos* and *paidopoiía* are inseparable. Pleasure isolated from procreation is ruled out. See Epictetus (*Diss.* III, 7, 2; III, 12, 7); Lucan (*Bell. civ.* II, 387-391); Seneca (*Consol. ad Elv.* 13, 3; SVF III, 686, p. 172, 19). The identical doctrine appears in *Justin (*I Apol.* 29, 1); *Athenagoras (*Legat.* 33, 1); *Minucius Felix (*Oct.* 31, 5); *Tertullian (*Ad ux.* I, 2, 1); *Clement of Alexandria (*Paed.* II, 10, 92, 2; 10, 95, 3). Rigorism in conjugal relations goes back to S. as well as to the Bible. Acc. to the Stoics, pleasure is an irrational exaltation of the soul (SVF III, 406, p. 98, 30) and, as such, reprehensible: only the *lógos* must triumph (cf. n. 404, p. 98, 15). The passions must be eradicated (Cic., *Tusc.* IV, 19, 43).

In *monasticism, Epictetus was particularly popular. The perfect monk was described with the features of the Stoic sage. There are considerable Stoic elements in *John Chrysostom (Pohlenz, *Stoa*, II, 334, 338).

13. S.'s popularity with the Fathers was based on its similarities, often only external, to Christianity: *Seneca saepe noster* (Tertull., *De an.* 20, 1); *Stoici nostro dogmati in plerisque concordant* (Jerome, *In Esaiam* IV, 11: CCL 73, 151, 42). Terms and categories of thought, as well as Stoic doctrinal elements, which concurred with the scientific formulation of Christianity, even when they were part of the heritage of a philosophical *koiné*, attest the vitality of the movement they arose from.

Sources: H. von Arnim, *Stoicorum Veterum Fragmenta* (= SVF) I-IV, Stuttgart 1905, repr. 1964.
General treatments: J. Stelzenberger, *Die Beziehungen der frühchristlichen Sittenlehre zur Ethik der Stoa*, Munich 1933; M. Pohlenz, *La Stoa. Storia di un movimento spirituale*, It. tr., Florence 1967 (for our theme: II, 261-400); M. Spanneut, *Le Stoïcisme des Pères de l'Eglise de Clément de Rome à Clément d'Alexandrie*, Paris ²1969 (= Spanneut, I); id., *Permanence du Stoïcisme de Zénon à Malraux*, Gembloux 1973 (= Spanneut, II).
Studies: Tertulliani, *De Testimonio animae*, intr., text and comm. by C. Tibiletti, Turin 1959; S. Otto, *Natura und Dispositio. Untersuchung zum Naturbegriff und zur Denkform Tertullians*, Munich 1960; C. Tibiletti, Verginità e matrimonio in antichi scrittori cristiani, *Ann. Fac. Lettere Univ. di Macerata* II (1969) 1-217: Rome ²1983; R. Joly, *Christianisme et Philosophie*, Brussels 1973; M. Spanneut, Le Stoïcisme et Saint Augustin, in *Forma Futuri*, Studi M. Pellegrino, Turin 1975, 896-914 (= Spanneut, III); C. Moreschini, Tertulliano tra Stoicismo e Platonismo, in *Kerygma und Logos* (Festschrift C. Andresen), Göttingen 1979, 367-379; M. Spanneut, Les normes morales du Stoïcisme chez les Pères de l'Eglise, *Studia Moralia* 19 (1981) 153-175 (= Spanneut, IV); J.M. Rist, *The Stoics*, Berkeley 1978; M.L. Colish, *The Stoic Tradition from Antiquity to the Early Middle Ages*, 2 vols., Leiden 1985; J. Mansfeld, Resurrection added: The Interpretatio Christiana of a Stoic Doctrine, *VChr* 37 (1983) 218-233.

C. Tibiletti

STRATEGIUS (or STRATEGUS). 7th-c. monk of the *laura of Mar Saba, author of a *De Persica captivitate* comprising a narration of the siege and capture of *Jerusalem by the Persians in 614 (615, acc. to A. Butler, *The Arab Conquest of Egypt*, Oxford 1902, 61), an account by a Jerusalem Christian named Thomas who, after the foe's departure, buried the dead and compiled a list of them, and, as epilogue, the return of the Holy Cross by the emp. *Heraclius. Only three fragments remain in Greek: 1) *Epistula Zachariae Hierosolymitani*; 2) *De Persica captivitate opusculum*; 3) *Narratio in apophthegmatibus* (BHG^a 1448 w; Marr, *Antioh Stratig.*, 42-44). But we have an *Arabic version and a *Georgian one (text and tr. published by Garitte in CSCO). The name Stratiki (Στρατηγός or Στρατήγιος) appears in the title of the Georgian version, while cod. Vat. Arab. 697, discovered by Peeters, gives as the author one E(u)stratios (no better identified): palaeographical error may account for the difference (Peeters, *La prise de Jérusalem*, 84-85). Marr's opinion (appearing in the title of his work and taken up by J. Phokylides: *Ecclesiastikós Pháros* 22 [1923] 188-205) that S. should be identified with *Antiochus (CPG 7842-7844), *abbat (not monk) of Mar Saba and author of an *Exomologesis* (PG 89, 1846-1856) which, though inspired by the same events, has a different character from the *De Persica captivitate*, is unacceptable. In our present state of knowledge it is also impossible, despite chronological coincidence, to identify S. of Mar Saba with the S., abbat of the monastery of St Theodosius, cited by *John Moschus (PG 87, 2961).

CPG 3, 7846; PG 86, 2, 3227-3268; N. Marr, *Antioh Stratig. Plènenie Ierusalima Persami V 614 g.*, St Petersburg 1909, 42-44; G. Garitte, *La prise de Jérusalem par les Perses en 614*, CSCO 202/203, Louvain 1960; G. Garitte, *Expugnationis Hierosolymae a.d. 614 recensiones arabicae*, I: *A et B*, CSCO 340/341, Louvain 1973; II: *C et V*, CSCO 347-348, Louvain 1974; P. Peeters, La prise de Jérusalem par les Perses, *Mél. de l'Univ. Saint-Joseph* 9 (1923-24) (= *Rech. d'hist. et de philol. orient.*, I, Brussels 1951), 78-116; Beck 449-450.

A. Labate

STYLITE - STYLITISM. Of the ascetic exercise of *stasis*, "standing", we find various examples in the 4th c. (*Peregrinatio Aetheriae* 20, 6: CSEL 39, 65); in the sense of standing on a column, non-Christian parallels have been sought (Lucian, *De Syria Dea* 28-29, in the *pagan ascetics of Hierapolis, 2nd c. AD). Among Christians this way of life was inaugurated by *Simeon Stylites the Elder (†459 or 458?: *BS* 11, 1116-1118) at Telanissos or Telnesin (now Der Sim'an) N-E of *Antioch, where he came in 413 or 415, yearning to "fly to heaven" and escape from the crowd. From the defence of stylitism sketched out by *Theodoret (*Hist. Religiosa* 26, 12), we may infer that this unusual form of *ascesis was contested in religious circles. But it found numerous imitators in *Syria, *Egypt, *Palestine, *Asia Minor, Greece: *Daniel of Constantinople (†493) (*BS* 4, 470-471), *Simeon Stylites the Younger, near Antioch (†592) (*BS* 11, 1141-1157), Alypius, near Adrianople in Paphlagonia (7th c.) (*BS* 1, 867-870), Luke the Stylite (†979) (*BS* 8, 225-226) and others. Eustathius of Salonica (†1193-98) wrote a eulogy of stylitism (*Ad stylitam quemdam Thessalonicensem*: PG 136, 217-264).

LTK 9, 1128-9; *EC* 11, 1337-8; *DACL* 15, 1697-1718; H. Delehaye, *Les saints stylites*, Brussels 1923; B. Kötting, *Peregrinatio religiosa*, Münster 1950, 113-118; P. Peeters, *Le tréfonds oriental de l'hagiographie byzantine*, Brussels 1950, 93-136; T. Spidlík, *CE* 13, 750-1; I. Peña-P. Castellana-R. Fernández, *Les stylites syriens*, Jerusalem-Milan 1975; P. Brown, Town, Village and the Holy Man, in *Society and the Holy in Late Antiquity*, London 1982, 153-165; S.A. Harvey, The Sense of a Stylite: Perspectives on Simeon the Elder, *VChr* 42 (1988) 376-404; R. Doran (tr.), *Simeon Stylites: the Biographies*, Kalamazoo 1988; J.M. Sansterre, Les saints stylites du Ve au XIe siècle, *Problèmes d'histoire du Christianisme* 19 (1989).

T. Spidlík

SUBORDINATIONISM. Thus we call the tendency, strong in the *theology of the 2nd and 3rd cc., to consider Christ, as Son of *God, inferior to the *Father. Behind this tendency were gospel statements in which Christ himself stressed this inferiority (Jn 14, 28; Mk 10, 18; 13, 32, etc.), and it was developed esp. by the Logos-*christology. This theology, partly under the influence of *middle Platonism, considered Christ, *logos* and divine wisdom, as the means of liaison and mediation between the Father's transcendent divinity and the *world, and therefore in a subordinate position to him. When the conception of the *Trinity was enlarged to include the *Holy Spirit, as in *Origen, this in turn was considered inferior to the Son. Subordinationist tendencies are evident esp. in theologians like *Justin, *Tertullian, Origen and *Novatian; but even an *Irenaeus, to whom trinitarian speculations are alien, commenting on Jn 14, 28, has no difficulty in considering Christ inferior to the Father.

In polemic with the *monarchians, s. tended to become radicalized, because the Son's inferiority was stressed for the purpose of bringing out his distinction from the Father, which they denied. This was so with *Dionysius of Alexandria and esp. with *Arius: while the moderate s. of his predecessors had not doubted that Christ was the real Son of God and a partaker of his divine nature, Arius considered him created rather than generated, extraneous to the Father's nature and therefore having a second-order divine nature. In reaction to this radical s. the anti-Arian theologians, esp. *Athanasius and later the Cappadocians, eliminated all trace of s. among the three divine *persons and considered them all equal in nature and dignity.

W. Marcus, *Der Subordinationismus als historisches Phänomen*, Munich 1963; R.P.C. Hanson, *The Search for the Christian Doctrine of God*, Edinburgh 1988.

M. Simonetti

SUB TUUM PRAESIDIUM. This *troparion*, known to the *Coptic, Greek, *Ambrosian and Roman liturgies, is one of the oldest *prayers addressed to the Virgin *Mary, a fact confirmed by Rylands Papyrus 470, discovered in Egypt in 1917 and, judging by the results of palaeographic examination, dating from the 3rd c. The text of the *papyrus allows the following literal reconstruction: "under your protection we take refuge, Mother of God; do not reject our supplications in our need, but free us from danger, you only chaste, only blessed one" (G. Giamberardini, Il "Sub tuum praesidium", *Marianum* 31 [1969] 330). The troparion has a dual value: on the level of liturgical history, it is the oldest known evidence of the veneration of the Mother of God in the pre-Ephesine period; on the theological level, it contains the technical term *Theotokos*, relating to the divine maternity; the two terms *only chaste, only blessed* indicate Mary's sanctity, eminence and moral integrity, and faith in her mediation is implied by the words "under your protection we take refuge, free us from danger". Present in the principal liturgies gravitating round the shores of the Mediterranean, only the *Byzantine rite has preserved the simplest inflection (and the closest to that of the papyrus). Despite its liturgical position, which differs from rite to rite, we see a common denominator, which consists, in today's rites, of considering it as part of the Marian liturgy and, in the ancient rites, of reproducing it with more specific instructions as a post-gospel antiphon for a Marian feast. It was prob. part of the *troparia* of the office of *Christmas, a feast celebrated with a Marian emphasis in *Egypt from the 3rd c. As for the 3rd-c. date, some critics have raised difficulties on account of the presence of the word *Theotokos*, which became current only after the council of *Ephesus, but this could have been adopted in the *liturgy before or during the disputes over its meaning and the measure of its acceptability.

C.H. Roberts, *Catalogue of the Greek and Latin Papyri in the John Rylands Library*, III, *Theological and Literary Texts*, Manchester 1938, 46-47; F. Mercenier, L'Antienne mariale grecque la plus ancienne, *Muséon* 52 (1939) 229-233; V. Buffon, L'antica preghiera mariana "Sub Tuum Praesidium", *Mater Ecclesiae* 2 (1966) 142-145; P.I. Cecchetti, Il fidente ricordo al materno cuore di Maria. Il "Sub Tuum praesidium", primo cenno al culto di Maria, *Lateranum* 33 (1967) 281-293; G. Giamberardini, Il "Sub tuum Praesidium" e il titolo "Theotokos" nella tradizione egiziana, *Marianum* 31 (1969) 324-362; A. Malo, La plus ancienne prière à Notre Dâme, in *De Primordiis cultus mariani*, II, Rome 1970, 475-485; H. Quecke, Das "Sub tuum Praesidium" in koptischen Horologion, *Enchoria* 1 (1971) 9-17.

E. Peretto

SUCCENSUS. Bishop of Diocaesarea (Ouzoundja-Bourdj) in Isauria (R. Devreesse, *Le patriarcat d'Antioche*, Paris 1945, 147). Died c.440. After the acceptance of the "creed of union" (433), a compromise over the 12 *anathemata proclaimed by *Cyril of Alexandria, perplexities manifested themselves among the Cyrillians, which Cyril himself sought to dissipate by epistolary means (A. Rehrmann, *Die Christologie des hl. Cyrillus von Alexandrien*, Hildesheim 1901, 131-132; Fliche-Martin 4, 247 ff.). Among his correspondents was S., fragments of whose *letters are preserved among Cyril's writings, as well as two extracts in *Syriac discovered by Van Roey in a MS in the British Museum (Add. 17197). The bishop manifests the dissatisfactions of those around him, but does not reveal his personal opinion. In the *catenae on *Genesis, three fragments survive under S.'s name: on 4, 7[b] (Cain and the Lord), 22, 5[b]-6 and 22, 7-9 (*Abraham's sacrifice of Isaac).

CPG 3, 6488-6489; ACO I, I, 6, pp. 158[8-10], 159[9-10], 160[14-17], 161[19-25]; ACO I, 4, pp. 119[23-33], 237[8-9]/[37-39], 238[20-30], 239[9-14]; ACO I, 5, pp. 299[37], 300[2]/[30-32], 301[20-23], 302[10-16]; A. van Roey, Deux fragments inédits des lettres de Succensus, évêque de Diocésarée à Saint Cyrille d'Alexandrie, *Muséon* 55 (1942) 87-92; R. Devreesse, *Les anciens commentateurs grecs de l'Octateuque et des Rois*, ST 201, 180.

A. Labate

SUCCESSION, APOSTOLIC. *1 Clem.* (42, 1-4; 44, 1-3) mentions the s. of presbyters to their predecessors in office in the local church. The weight of its argument is based on the principle of a hierarchy constituted by divine right. This idea of s. is hardly hinted at in *Ignatius, though for him too the divine mission is the foundation of all episcopal *authority. For Ignatius, the authority of the *Apostles finds its hinterland beyond the Christian structure of his time. In the interval between *Paul's death and the *Apocalypse and, at latest, in the time of Ignatius and *Polycarp, there was a transition from the missionary apostolate to the local episcopate. Each community had a body of local administrators and, consequently, it was only the celebration of the *Eucharist that made it, among other reasons, monarchical. If the office of the Apostles, as immediate witnesses of the time of the *incarnation and first executors of Christ's will, was unique, his charge to them contained a mission that went beyond the time of their earthly existence. With it the idea of apostolic s. was objectively founded. The *Catholic church was distinguished from most of the first *heresies precisely because it did not leave doctrinal and illuminative tradition in the hands of spontaneous spiritual guides, teachers or *prophets, but possessed a group of responsible men. In the 2nd c. the concepts of *tradition and succession took on new theological weight and acquired their technical meaning. We cannot prove that the idea of a doctrinal transmission appeared first in the milieu of *gnosis, with its free magisterium (*Ptolemy, in Epiph., *Haer.* 33, 7, 9); but it took the typically Catholic form when it was linked to the s. of monarchical bishops. *Hegesippus appears as the disseminator of this model. He provides a list of names, as do the *gnostics (*Basilides, in Hipp., *Philosoph.* 7, 20, 1; *Valentinus, in Clem., *Strom.* 7, 106, 4), but he does it in a linked s. determined by the monarchical episcopal office. In so doing, he seems to have conceived the history of the heresies as a pseudo-apostolic succession (Euseb., *HE* IV, 22, 5-6). In defence of the church's belief, he used the principle of s. (Euseb., *HE* IV, 22, 3) and on reaching *Rome he compiled a list of her bishops up to Anicetus (Epiph., *Haer.* 27, 6), though without describing the episcopal successions as "apostolic" (though *Eusebius did so indirectly, *HE* IV, 8, 2). *Irenaeus fought on the same antignostic front and may have taken the idea of s. from Hegesippus. His most original contribution concerns the role of the *Holy Spirit who, as source of truth and *life, poured out by Christ, flowed only in the *Catholic and hierarchical church. In *Adv. haer.* III, 3, 1, he claims that the truth could be known by everyone in all the churches through the s. of the bishops, guardians of the apostolic tradition. Since it would take too long to enumerate the s. of the bishops of all the churches, he chose that of Rome (III, 3, 2-3). It was necessary to remain united to those who did not separate themselves from the original s. (IV, 26, 5; 33, 8). So the continuity of the true faith was inseparable from that of the apostolics. For *Hippolytus too, doctrinally a disciple of Irenaeus, the creator of the church's life was the Spirit with his gifts; but he too emphasized that it was the bishops who, as successors to the Apostles, received it.

For *Trad. Ap.*, 3, the bishop was in continuity with the apostolic *ministry. For *Tertullian (*Praescr.* 32, 1; *Adv. Marc.* IV, 5, 3; IV, 29), the uninterrupted s. of bishops was the convincing sign of the churches' *apostolicity. On the whole he sees it as the s. of apostolic churches rather than of apostolic bishops (*Praescr.* 20, 5-6). On becoming a *Montanist, he abandoned the idea (*De Pudic.* 21, 7). After him, appeal to apostolic s. loses its force in Christian *apologetics. In *Cyprian, the theme becomes a commonplace: *Epp.* 66, 4; 75, 16 (*Firmilian); *Sent. Epp.* 79 (Clarus of Mascula). He brings out the order of bishops as successors (*Ep.* 45, 3) and the connection with the church as a whole (*Ep.* 33, 1). *Clement of Alexandria uses the concept of s. from the apostles, but applies it to teachers who taught apostolic doctrine (*Strom.* VI, 61, 3), because he is interested in the true gnostic, the perfect Christian (*Strom.* VI, 106, 2; VII, 77, 4). But, faced with *heretics, he also emphasizes the priority of church *tradition (*Strom.* VII, 104, 1). *Origen's vocabulary on apostolic s. seems limited (*Princ.* praef., 2; IV, 22; *In Mt. comm., ser.* 46); between the Apostles and present preaching, he interposes s. in the church. His point of view is an expression of parity between presbyters and teachers, whose functions, however, he distinguishes (*Hom. in Lev.* VI, 6). Apostolic s. is a term dear to *Eusebius (*HE* I, 1, 1; III, 3, 3; 4, 11; VII, 32, 32; VIII, praef.); it is the guarantee of tradition (V, 6, 5; 25; VI, 9, 1). He cites the s. up to Anicetus, fixed by Hegesippus (IV, 22, 3), and that up to Eleutherius, fixed by Irenaeus (V, 5, 9-VI, 5). He also applies the common term "succession" to the teachers of the *school of *Alexandria (V, 10, 1; 11, 2; VI, 6; 29, 4) and even to the Greek philosophical schools (VII, 32, 6). His interest is more historical than doctrinal; but what interests him most is the s. of the bishops from the Apostles. He often mentions the s. of bishops in the four main sees, on which most of the Christian information in his *Chronicon* is based. Neither he nor Irenaeus includes the *Apostles in the list of bishops. But Cyprian's identification of *episcopate* and *apostolate* (*Epp.* 3, 3; 67, 4) continued to appear sporadically in the *East (Socr., *HE* V, 8); in the West there was a common respect for the Roman see. *Leo the Great (*Serm.* 4, 2-3; 3, 3-4; *Ep.* 45, 2; *Serm.* 5, 4) expresses the crucial importance of Petrine s. in the Roman see: *Peter as prince of the Apostles and primate of the universal church. *Gregory the Great (*Epp.* 2, 48; 3, 30; 7, 40; 9, 59; etc.) summed up this whole doctrine in the principle of papal *authority, though without using Leo's legal language.

G. Dix, The Ministry in the Early Church, in K.E. Kirk, *The Apostolic Ministry*, London 1946, 183-303; H. von Campenhausen, *Kirchliches Amt und geistliches Vollmacht in den ersten drei Jahrhunderten*, Tübingen 1953 (Eng. tr. London 1969); E. Molland, Irenaeus of Lugdunum and the Apostolic Succession, *JEH* 1 (1950) 12-28; id., Le développement de l'idée de succession apostolique, *RHPhR* 34 (1954) 1-29; A. Ehrhard, *The Apostolic Succession in the First Two Centuries of the Church*, London 1953; B. Botte, A propos de l'Adversus Haereses III, 3, 2 de Saint Irénée, *Irénikon* 30 (1957) 156-163; R.P.C. Hanson, *Tradition in the Early Church*, London 1962; A.M. Javierre, *El tema literario de la sucesión*, Zürich 1963; G.G. Blum, *Tradition und Sukzession*, Berlin-Hamburg 1963; O. Perler, L'évêque, représentant du Christ, selon les documents des premiers siècles, in Y. Congar-B.D. Dupuy, *L'épiscopat et l'Eglise universelle*, Paris 1964, 31-66; J. Colson, *Ministre de J.C. ou le sacerdoce de l'Evangile*, Paris 1966; A. Lemaire, *Les ministères aux origines de l'Eglise*, Paris 1971; R.F. Evans, *One and Holy. The Church in Latin Patristic Thought*, London 1972; *Mysterium Salutis*, Brescia 1972, VII, 639-713; H. Küng, *Strutture della Chiesa*, Turin 1965; J. Finkezeller, Zur Diskussion über das Verständnis der apostolischen Sukzession, *Theol. Prakt. Quart.* 123 (1975) 321-340; M.M. Garijo, La sucesión apostólica en los tres primeros siglos, *Diálogo ecúmenico* 2 (1976) 179-231; *TRE* 4, 430-466.

R. Trevijano

SUIDAS (or SUDAS). Anon. late 10th-c. Byzantine lexicographical compilation; S. does not indicate the author's name, which is unknown, but the title of the monumental lexicon. This is written Suidas or Sudas according to historical or etymological preference. The compilation consists of some 30,000 headings, giving it more the character of an encyclopaedia than a lexicon. Under numerous headings it touches on problems of secular and sacred history and describes writers of classical and Christian antiquity. Its historical information is important and abundant for authors and facts either ignored or just touched on by others; the many citations allow us to reconstruct or supplement lost or fragmentary literary works. For this markedly compilatory work the author used, for the grammatical and etymological part, lexicons of his own time (one prob. by *Photius) containing scholia on Homer, Sophocles, Aristophanes, Herodotus, Thucydides, *Gregory Nazianzen; for the historical part, the Byzantine chronicles, esp. the collection of Constantine Porphyrogenitus; for the biographical part, various *biographies, the Epitome of *Hesychius of Miletus, the Lives of Diogenes Laertius, etc.; for the philosophical part, commentaries on Aristotle, Plato, etc. He took much from authors still alive in his time. The S.'s value is not uniform: not very important for the lexical and scientific parts, it becomes so in the literary-historical part. It lacks method: its information is arranged uncritically and in a rather confused order.

Sudas, PG 117, 1193-1424; A. Adler, *Lexicographi graeci*, 5 vols., Leipzig 1928-1938; P. Maas, Der Titel des Suidas, *ByzZ* 32 (1932) 1; F. Dölger, Der Titel des sogenannten Suidaslexikons, *SBAW* 6 (1936) 1-37; id., Zur Σούδα-Frage, *ByzZ* 3 (1938) 36-58; *KLP* 5, 407 f.; S.G. Mercati, Intorno ai titoli dei lessici Suida-Suda e di papia, *MAL* Ser. 8, X, 1, Rome 1960 (now also in: S.G. Mercati, *Collectanea Byzantina*, I, Bari 1970, 641-708); N. Walter, Suda. Ein Literaturbericht zum Titel des sogenannten Suidas-Lexikon, *Altertum* 8 (1962) 169-175.

E. Peretto

SULPICIUS SEVERUS

SULPICIUS SEVERUS. A lawyer of *Bordeaux (*Burdigala*), "illustrious by birth and for his literary works" (Genn., *De vir. ill.* 19); converted to Martinian *ascesis by his mother-in-law Bassula (of a rich consular family: Paul. Nol., *Ep.* 5, 5), this Aquitainian, born *c.*360, became its zealous and gifted propagandist: sometimes by personal action (he organized a "Martinian" community on his property of Primuliacum, prob. near "the Lauraguais road", not far from the road from *Toulouse to *Narbonne, but above all by his sharp and subtle pen. His *friend and correspondent *Paulinus of Nola (cf. Paul. Nol., *Epp.* 1, 5, 11, 17, 22-24, 27-32) addressed to S. his long *Ep.* 22, full of epigraphical poetry to adorn the religious buildings (baptistery and churches) raised by S. at Primuliacum. S. was a typical representative of that Gallo-Roman aristocracy who, in the *schools of Bordeaux, were disciples of the *professores* described in a series of *epigrams by *Ausonius (book V of his poems). But he belonged, more precisely, to that active minority converted to the radical and demanding gospel of monastic asceticism which contested, to the point of anti-clericalism, the "worldly" bishops of the Gallo-Roman church "established" in the empire.

The success of S.'s Martinian works contributed not a little to the literary, and consequently the religious, popularity of the personality of St *Martin, whom from the end of the 4th c. the ascetics of *Gaul and *Italy made their model and standard-bearer. S. ended his life as a priest and, contaminated by the *Pelagian heresy, in an old age silently dedicated to *penitence (if we can believe Genn., *De vir. ill.* 19).

S.'s essential work consists of a tryptich consecrated to St Martin, widely copied and circulated in the Middle Ages under the (recent) name of Martinellus ("the little Martin"). It comprises the *Vita Martini*, prob. finished in 397 before Martin's death; three *letters to that Bishop of Tours; and the two (or three) books of *Dialogues*, a dossier in *dialogue form annexed to the *Life*. S. also wrote a *Chronicle* in two books, embracing sacred history from the *creation of the world to his own times: starting with the universal, then the Gallo-Roman, background, and leading up to his exaltation of Martinian asceticism (it went up to the consulate of *Stilicho in 400).

A "masterpiece", in the dual sense of *chronology and literary quality, of ascetic and episcopal *biography in the Latin West, the *Vita Martini* is a brilliant manifesto of the earliest Latin *monasticism, through the life and deeds of a monk-bishop, thaumaturge and evangelist, spiritual teacher and confessor of the faith. For centuries to come, the *Vita Martini* fixed the traits of a certain cumulative model of Christian sanctity, integrated with the ancient models (apostolic, martyrological, episcopal, ascetic) as they had been presented by, in particular, the NT, the Acts and Passions, the *Vita Cypriani*, the *Latin translations of *Athanasius's Greek *Life of Anthony*. The wealth of S.'s early Christian literary culture was thus adapted to the ideal of life which was, historically, that of Martin: imperial guard turned *miles Christi*, apostle of the countryside, witness to the *orthodox faith and the demands of the gospel, from time to time persecuted by *Arians and worldly bishops, a pastor trained by *Hilary for the responsibilities of *evangelization, a mediator of the monastic ascesis he had known at *Poitiers and *Milan.

Criticized in radical and positivistic terms at the start of the 20th c., the historicity of the facts related by S. in his Martinian works requires a prudent literary analysis that follows the stages of perception of the facts: Martin's own reading of his own experience; creation of oral traditions in which the Gallo-Roman imagination (Celtic and Latin, popular and literary) played an important role, and one hard to decipher today; finally the personal virtuosity of a biographer writing for a certain public: literary, aristocratic, a bit élitist, even cliquish.

A fervent admirer of Sallust, S. ably imitated even his stylistic mannerisms. His lively imagination, disciplined by a formally rigorous work, offers an apparently objective account of miraculous, or sometimes frankly fantastic, facts. The new Alexandrianism of the 4th c., so well represented in poetry by an *Ausonius of Bordeaux, brought out in S. effects of dramatization and finesse in which are still reflected the Hellenistic values of pathos and grace. Hence the incontestable literary elegance of these narratives and consequently their effectiveness. S.'s Martinian dossier was exceptionally popular in the literary and religious culture of the Middle Ages, and not just in the already vast field of *hagiography: it was put into verse by *Paulinus of Périgueux in the mid 5th c. and by *Venantius Fortunatus of Ravenna (then "of Poitiers") in the course of the 6th.

But S.'s Sallustianism is too precise to remain merely formal. The Roman historian's grumbling, bitter, dramatic moralism has tinged his vision of sacred history - and still more the history of his time, in the *Chronicle* - with *chiaroscuro*. The second book is practically dedicated to the *Arian question in Gaul, to the *Priscillianist heresy and, still more, through this gloomy event, to the scandal caused by the worldly bishops who embittered the struggle against this *heresy, ending in the blood shed at *Trier. S. rewrites the events like a new *Conspiracy of Catiline*, but victims and butchers are confused in a single invective, violent and without charity. Here, more than in the Martinian dossier, S. is revealed as a passionate, enthusiastic and vengeful man, without whose strange personality, against *Priscillian's "negative", the figure of Martin would perhaps appear less radiant.

Editions: CPL 474-476; PL 20, 95-240; CSEL 1; P.-M. Duval, *La Gaule jusqu'au milieu du Ve siècle*, Paris 1971, 656-662; *Vita Martini*: ed. J. Fontaine, SCh 133-135, Paris 1967-1969 (includes the three letters; critical text, introduction and full commentary); ed. J.W. Smit (critical text and It. notes) and L. Canali (It. tr.), Coll. Vita dei Santi 4, Mondadori s.d. (1975), IX-XXX (intr. by C. Mohrmann), 1-67 (text and tr.) 245-290 (notes and bibl.); *Dialogues*: tr. by P. Monceaux, Paris 1926 (with the whole of the Martinian dossier); reprint of J. Fontaine's ed. of the *Life* and letters, and P. Monceaux's of the *Dialogues*, with introduction by L. Pietri, Ligugé 1978; *Chronica*: ed. A. Lavertujon, with annotation (disputable), Paris 1895-1899; ed. De Senneville (in preparation for SCh); Ed and trans. B.M. Peebles, *FOTC* 7 (1949)

Studies: Introductions by J. Fontaine and C. Mohrmann to the eds. cited above. On the *Life*, studies by J. Fontaine, Sulpice Sévère a-t-il travesti S. Martin de Tours en martyr militaire?, *AB* 81 (1963) 35-58; id., Une clé littéraire de la Vita Martini de Sulpice Sévère, la typologie prophétique, in *Mélanges Chr. Mohrmann*, Utrecht-Anvers 1963, 84-95; id., Alle fonti della agiografia europea Storia e leggenda nella vita di S. Martino di Tours, *RSLR* 2 (1966) 187-206; on the *Chronica*, J. Fontaine, L'affaire Priscillien ou l'ère des nouveaux Catiliana. Observations sur le sallustianisme de Sulpice Sévère, *Classica et Iberica*, (Festschrift in honor of Joseph M.F. Marique), Worcester, Mass. 1975, 355-391; K. van Andel, *The Christian Concept of History in the "Chronicle" of Sulpicius Severus*, Amsterdam 1976; F. Murru, La concezione della storia nei Chronica di Sulpicio Severo, alcune idee di studio, *Latomus* 38 (1979) 961-981. Important summary by F. Ghazzani, *Sulpicio Severo*, Parma 1983; C. Stancliffe, *St. Martin and his Hagiographer, History and Miracle in Sulpicius Severus*, Oxford 1983; for S.S.'s role in the development of the cult of St Martin: L. Pietri, *La ville de Tours du IVe siècle: Naissance d'une cité chrétienne*, Rome 1983; F. Ghizzoni, *Sulpicio Severo*, Rome 1983.

J. Fontaine

SUN, Cult of the

SUN, Cult of the. The existence of solar deities is a fact notoriously very widespread in the various religious contexts, esp. those of clearly polytheistic type, characterized by divine personalities connected with cosmic functions and "departments". Suffice to recall the Babylonian Shamash, the Egyptian Amon-Ra, god of Thebes, and Aton of Heliopolis, also a solar deity, who at the time of Amenophis IV (1375-1358 BC) was at the centre of a famous "enotheistic" religious reform. In the Greek world, if the importance of Helios was rather limited in the classical and Hellenistic period, under the Roman empire the cults of solar deities of Eastern, esp. Syrian, origin gained considerable popularity. The rise of the Syrian gods proceeded in parallel with the establishment of dynasties of Eastern origin on the imperial throne. Thus the time of *Caracalla saw the establishment of the cult of the solar god of *Emesa, home of the emperor's mother, that Julia Domna of priestly stock who played a fundamental role in the religious policy of the time. With *Heliogabalus's ascent to the throne this cult reached its acme, he being a priest of Helios of Emesa, whom he intended to make the empire's principal god. Rome was then host, in the purpose-built temple, to the stone fallen from heaven, seat and manifestation of the god, whose cult was celebrated in typical Oriental forms. Heliogabalus's fall led to the return of the sacred stone to Emesa and the cessation of worship in his temple at Rome. But there was no decline of interest in Oriental solar deities or deities qualified by important luminous and celestial connotations. We need only think of the Iranian *Mithras, titular of a mystery-cult, whose solar characteristics were so pronounced that he could be invoked as "Mithras, unconquered Sun", even though Mithraism saw the figure of Sol, with whom Mithras himself had relations, variously represented, as a distinct figure.

The Egyptian Serapis, a god with a complex personality connected with the underworld and fertility, also had quite marked solar characteristics. Interest in the Sun-god, also expressed on a level of philosophical and theological speculations in the works of *neoplatonist authors like *Porphyry, the emp. *Julian and later Macrobius, found its most lively expression in the solar cult installed by *Aurelian (270-275). He, victoriously entering Emesa, recognized the city's solar god as the protector of the empire, transferred its cult to Rome in a temple built at state expense, and established an official cult. Worship was entrusted to a college of priests, and a four-yearly contest was instituted on 25 Dec, *dies natalis* of the *Sol invictus*.

F. Cumont, *Les religions orientales dans le paganisme romain*, Paris ⁴1929; F. Altheim, *Der unbesiegte Gott*, Hamburg 1957; G. Sfameni Gasparro, Le religioni orientali nel mondo ellenistico-romano, in (ed.) G. Castellani, *Storia delle religioni*, III, Turin 1971, 423-564; U. Bianchi, *Mysteria Mithrae*, Rome 1979; H. Dörrie, Die Solar-Theologie in der Kaiserzeitliche Antike, in (ed.) H. Frohnes-U.W. Knorr, *Kirchengeschichte als Missiousgeschichte*, I, Munich 1974, 283-292.

G. Sfameni Gasparro

SUNDAY

I. Origin - II. The earliest form of Sunday celebration - III. Significance.

I. Origin. It has frequently been asserted that the Christian S. depends on a pre-existent non-Christian institution.

1. A day consecrated to the *sun, from which it took its name, occurs in Graeco-Roman antiquity from the 1st c. AD; it was the second day of the planetary *week. But we know of no S. celebration in non-Christian antiquity; not even in the cult of *Mithras did this day assume special importance so early. It is thus an ill-founded hypothesis to suppose an original influence of a solar cult on the Christian S. (against S. Bacchiocchi; though a secondary influence is perceptible (cf. below).

2. It was recently believed that a Jewish precursor of the Christian S. had been discovered: after the finds at Qumrân, the solar calendar used by this sect was reconstructed; in it, feasts recurred every year on the same day of the week, on Wednesday, Friday or S. (A. Jaubert). So far, however, no-one has proved that at Qumrân the weekly S. had any particular importance. So it is hazardous to talk of an influence of Qumrân on the Christian S. (against E. Hilgert), esp. if at the same time we suppose a genuinely Christian origin of the S. celebration (against C. S. Mosna).

3. So the S. celebration appears to be a creation of the Christian church (first mentioned in 1 Cor 16, 2). How was it arrived at? The usual reply that Christ's *resurrection on *Easter morning led to this emphasis on S. is only partly correct. For one thing, the first Christian S. celebrations took place not in the morning, but in the evening (Acts 20, 7); for another, up to the 2nd c. Jesus' resurrection on Easter morning is mentioned only fleetingly as the reason for the S. celebration (Barn. 15, 9; Just., *I Apol.*, 67, 7). So in scholarly circles two more exact, though differing, theories are put forward.

4. At first the Christians met on Saturdays (H. Riesenfeld; R. Staats). For practical reasons, it was the most natural thing to meet together to "break *bread" in connection with the Jewish *Sabbath (which the first Christians still celebrated). Only in the 2nd c. was the celebration moved to S. morning, in memory of Christ's resurrection.

5. Since the theory of a "displacement" of the celebration raises difficulties, it seems better-founded to suppose that the first Christians met for the *Eucharist on S. evening (Acts 20, 7; Pliny the Younger, *Ep.* X, 96, 7), in continuity with Christ's appearances on Easter evening (Jn 20, 19) and the following S. (Jn 20, 26), and in continuity with the risen Christ's taking food with his disciples (Lk 24, 30. 41-43; Acts 1, 3 f.). Thus e.g. C. Callewaert, O. Cullmann, H. Dumaine, W. Rordorf.

6. The Adventist theory (now revived by S. Bacchiocchi), that the S. celebration was introduced in the late 2nd c. by the church of *Rome as an anti-Jewish reaction and in dependence on the *pagan cult of the *sun, rests on a fragile basis (cf. *ZKG* 91 [1980] 112-116).

II. The earliest form of Sunday celebration. This may be deduced from descriptions in the *Didachè, in *Pliny's *letter*, in *Justin and in *Hippolytus's *Apostolic Tradition*.

1. To begin with, the *Eucharist was celebrated in the evening in the context of a complete meal (1 Cor 11, 25; *Did.* 10, 1; Pliny, *Ep.* X, 96, 7: second meeting). This celebration had an *eschatological character, but not exclusively in the sense of awaiting the end of the world, so much as in the fact that they were aware of the presence of the glorified Lord and were therefore concerned for the holiness of the community.

For this reason they submitted to a serious examination of conscience before *communion and forgave each other their mutual failings (1 Cor 11, 28 ff.; *Did.* 10, 6; 14, 1-2 [in ref. to Mt 5, 23 f.]; Tertull., *Apol.* 39). The communal evening meal was still continued along with the *agapè.

2. Besides this there was, perhaps very early, a Christian celebration in the morning before dawn, first recorded by Pliny (*Ep.* X, 96, 7) and certainly related to the conferring of *baptism. After the abolition, in the same period, of the evening celebration (Pliny gives a reason: the prohibition of *Hetairiai*), the Christians celebrated the liturgy only early on S. morning.

3. This is exactly the situation we find in Justin. In *I Apol.*, 65 he describes the celebration of baptism. But since this, like the S. morning celebration described in *Apol.* I, 67, ended with the Eucharist, we may suppose that the celebration of baptism in Justin's time was still linked with the S. morning liturgy. After the baptism, which took place in running *water (Acts 8, 38; *Did.* 7, 1), the newly-baptized were led among the asssembled community. Then began the liturgy of the word: following the model of *Jewish synagogue worship, this consisted of readings followed by an exposition given by whoever presided. After this, *prayers were said standing up (in memory of the *resurrection: Tertull., *Cor.* 3; Basil., *Spir. sanct.* 27, 67). Deacons took communion to the absent. There was a collection for the needy.

4. In the *Apostolic Tradition* (ch. 4), in the context of an episcopal *ordination (which took place on S.: ch. 3), is preserved for the first time a complete eucharistic *liturgy: not only was this the godfather, in antiquity, of the Eastern and Western eucharistic liturgies, but it has recently played a great role in all the churches' attempts at liturgical reform.

III. Significance. This will be described by examining the various names of S. in Christian *tradition, according to the theological importance of the individual names.

1. Lord's Day (*kyriake hemera*). This is the new Christian name of this day, and it is preserved in various languages (esp. the Slav and Romance). First attested in Rev 1, 10, then in *Did.* 14, 1; *Ev. Petr.* 35. 50; Dionysius of Corinth (= Euseb., *HE* IV, 23, 11); etc. "Lord's Day" is a form analagous to "Lord's Supper" (1 Cor 11, 20); S. is the day that recalls the Lord (Christ) (perhaps mainly because on it the "Lord's Supper" was celebrated). The central meaning of the *Eucharist emerges from the texts cited above (para. II). All Christians took part (Just., *Apol.* I, 67, 3. 5), even at the risk of their lives (Tertull., *Fuga* 14, 1; *Acta Saturnini, Dativi et al.* 9. 11). The rediscovery of the S. communion of the whole community is one of the happiest characteristics of our ecumenical age, which draws its greatest strength from a deepened attachment to *tradition.

2. Eighth day. This name (Barn. 15, 8-9; Just., *Dial.* 41, 4; 138, 1; Cypr., *Ep.* 64, 4; *Didascalia* VI, 18, 11-16; Ambrosiaster, *Liber quaest.* 95, 2; Basil, *Spir. sanct.* 27, 64) expresses the fact that the reality of S. "transcends" the *week and represents an open window onto eternity. Also, in Jewish and Greek antiquity the *Ogdoad was a symbol of perfection (Würfel), related by Christians to the outpouring of the *Holy Spirit in baptism (on S.) (1 Pet 3, 18-21; 2 Pet 2, 5; Ambr., *Ep.* 41, 6, 17 [44, 4, 15]; cf. octagonal baptisteries), and which in heretical (Iren., *Adv. haer.* I, 5, 3; Clem. Al., *Exc. ex Theod.* 63, 1) and ecclesiastical speculations (Clem. Al., *Strom.* V, 106, 2; VI, 108, 1; Orig., *Comm. in Ps.* 118, 4; Aug., *Ep.* 55, 23) became a symbol of approach to *God.

3. First day. Thus S. is called in Jewish tradition and in the gospels (Mk 16, 2 par.; Jn 20, 19). The Christian week therefore begins with S.; we would do well to remember this today, in an age absolutely dominated by economic viewpoints. The festival day opens the round of working days (K. Barth, *Kirchl. Dogm.* III, 4, 51-79). Acc. to biblical tradition, this is the day of the *creation of *light (Gen 1, 3; cf. Just., *I Apol.*, 67, 7; Euseb., *Comm. in Ps.* 91 [92]); this fact makes it possible for Christians to adopt without difficulty even the pagan name day of the sun (first in Just., *I Apol.*, 67, 3), which has prevailed in the Germanic languages; Christ too was compared to the sun (Ign. of Antioch, *Magn.* 9, 1; Just., *Dial.* 100, 4; Mel. of Sardis, *Bapt.* 4; Orig., *In Lev. hom.* 9, 10; Athan., *De sabbato et circumc.* 5; Jerome, *In die dom. Paschae hom.*). To be sure, this left open the danger of *syncretistic confusion (Tertull., *Nat.* I, 13, 1-5; *Apol.* 16, 9-11; Euseb., *Vita Const.* IV, 19-20; cf. the fixing of *Christmas on 25 Dec).

4. But the gravest confusion was the misunderstanding of S. as if it were a Christian *Sabbath. This false step was taken by the church after *Constantine the Great had, in 321, proclaimed S. a public day of rest in the Roman empire (*CJ* III, 12, 2; *CT* II, 8, 1). To give a meaning to the obligatory S. rest (previously Christians had gone on working: Jerome, *Ep.* 108, 20, 3; Pall., *Hist. Laus.* 59, 2; Benedict of Nurcia, *Reg.* 48, 22 f.), the unhappy idea emerged of motivating the S. rest by the Sabbath commandment (Euseb., *Comm. in Ps.* 91 [92]; Ephrem Syr., *Sermo ad noct. dom. resurrectionis* 4; John Chrys., *De bapt. Christi hom.* 1; id., *In Gen. hom.* 10, 7; Euseb. Al., *Sermo* 16); in the 6th c. the equation of S. with the Sabbath was completely established (Caes. Arel., *Sermo* 10, 3, 5; council of Orléans, can. 31, 28; Mart. Brag., *De correctione rust.* 18; Ishoyahb, *Ep. can. ad Iacobum* 4; 2nd council of Mâcon, can. 1; council of Narbonne, can. 4). In the pre-Constantinian church such a proceeding would not have been possible, both because the S. rest was unknown and above all because the church had arrived at the idea that the Sabbath commandment was no longer obligatory for the New Covenant, and thus had nothing to do with the Christian S. Today it would be better to stress the original definition of S. as the day when the eucharistic liturgy is celebrated. Rest, whose socio-ethical significance must not be diminished (here the OT Sabbath legislation makes many just observations), was connected with S. not by the will of God but by imperial favour; it is not decreed that it should remain eternally bound to it.

H. Dumaine, Dimanche, *DACL* 4, 858-954; C. Callewaert, La synaxe eucharistique à Jérusalem, berceau du dimanche, *EThL* 15 (1938) 34-73; AA. VV., *Der christliche Sonntag. Probleme und Aufgaben*, Vienna 1956; J. Gaillard, Dimanche, *DSp* 3, 948-982; A. Jaubert, *La date de la Cène*, Paris 1957; AA. VV., *Der Tag des Herrn. Die Heiligung des Sonntags im Wandel der Zeit*, Vienna 1958; AA. VV., *Verlorener Sonntag?* Kirche im Volk 22, Stuttgart 1959; H. Riesenfeld, Sabbat et jour du Seigneur, *New Testament Essays*. Studies in Memory of T.W. Manson, Manchester 1959, 210-218 (= *The Sabbath and the Lord's Day, The Gospel Tradition*, Oxford 1970, 111-137); F.A. Reagan, *Dies dominica and dies solis. The beginnings of the Lord's Day in Christian Antiquity*, (Diss.), Washington 1961; O. Cullmann, Urchristentum und Gottesdienst, *Abh. z. Theol. des A.T. u. N.T.* 3 (⁴1962) 14 ff.; O. de Sousa, *Dia di Senhor. Historia, teologia espiritualidade do domingo*, Lisboa 1962; W. Rordorf, Der Sonntag. Geschichte des Rube und Gottesdiensttages im ältesten Christentum, *Abh z. Theol. des A.T. u. N.T.* 43, Zürich 1962 (Eng.: London-New York 1968; Sp.: Barcelona 1972); E. Hilgert, Jubilees and the Origin of Sunday, *Andrews University Sem. Studies* 3 (1963) 44-51; AA. VV., Le dimanche, *La MaisonD.* 83 (1965); AA. VV., *Le dimanche*, Lex Orandi 39, Paris 1965; C.S. Mosna, *Storia della domenica dalle origini fino agli inizi del V secolo. Problema delle origini e sviluppo. Culto e riposo. Aspetti pastorali e liturgici*, AG 170, Rome 1969; P.K. Jewett, *The Lord's Day: A Theological Guide to the Christian Day of Worship*, Grand Rapids 1971; W. Rordorf, *Sabbat und Sonntag in der Alten Kirche*, Traditio Christiana 2, Zürich 1972 (Fr.: Neutchâtel-Paris 1972; It.: Turin 1979); R. Staats, Ogdoas als Symbol für die Auferstehung, *VChr* 26 (1972) 29-52; G. Troxler, *Das Kirchengebot der Sonntagsmesspflicht als moral-theologisches Problem in Geschichte und Gegenwart*, Arbeiten zur prakt. Theol. 2, Fribourg 1971; S. Bacchiocchi, *From Sabbath to Sunday. A Historical Investigation of the Rise of Sunday Observance in Early Christianity*, Rome 1977; R. Staats, Die Sonntagnachtgottesdienste der christlichen Frühzeit, *ZNTW* 66 (1975) 242-263; R.T. Beckwith - W. Stott, *This is the Day: The Biblical Doctrine of the Christian Sunday*, London 1978; D.A. Carson (ed.), *From Sabbath to Lord's Day: a Biblical, Historical and Theological Investigation*, Grand Rapids 1982; W. Rordorf, Dimanche, source et plénitude du temps liturgique chrétien, *Cristianesimo nella Storia* 5 (1984) 1-9; F. Rordorf, Die theologische Bedeutung des Sonntags bei Augustin, *Der Sonntag* (Festsch. J. Baumgartner), Würzburg 1986, 30-43.

W. Rordorf

SUPERSTITIO (superstition). The notion of *superstitio* is rather ambiguous. Suetonius saw Christianity as a *superstitio nova et malefica* (*Ner.* 16, 3). Here the term designates a foreign religion. In its classical meaning the term characterizes superfluous religion, vain observance, religion vitiated by superstitious practices. In the Vulgate it corresponds to the Gk word δεισιδαιμονία (Acts 17, 22; 25, 19), meaning suspect deities and suspect practices. Among the Greek *Fathers the two terms recur with the two meanings of *idolatry (Clem. Al., *Protr.* 2; *Strom.* II, 8; VII, 4; Orig., *C. Cels.* 6, 17; Greg. Nyss., *Or. cat.* 18) and superstition or vain worship, applied to Judaism, *heresy or *paganism. Among the Latins, *Lactantius aims at the latter when he says: *Veri (Dei) cultus est, superstitio falsi* (*Div. inst.* IV, 28). It was in this sense that *Constantine and his successors repressed *superstitio*, in the same way as some Christian aberrations. This allowed the extension of the term to include survivals of paganism, divination, the formulae and practices of *magic.

J.P. Migne, *Dictionnaire des sciences occultes*, I-II, Paris 1846; *DACL* 15, 1730-1736; *EC* 11, 1574-1576; L.F. Janssen, Die Bedeutungsentwicklung von Superstitio-Superstes, *Mnemosyne* 26 (1975) 135-188; id., Superstitio and Persecution of the Christians, *VChr* 33 (1979) 131-159; D. Harmening, *Superstitio*, Berlin 1979.

A. Hamman

SUSANNA

I. In the Fathers - II. Iconography.

I. In the *Fathers. The story of S. (Dan 13, 1-65) is in the deuterocanonical part of the book of *Daniel. In the LXX it is the first supplementary chapter while in *Theodotion's version, which the Fathers used, it is the first chapter of the book of Daniel itself. In Christian literature *Tertullian, while not citing the text, is the first to mention the story of S. (*De corona* 4, 3), appealing to her as an example to Christian *women of the modest use of the veil. In *Cyprian (*Ep.* 43, 4: CSEL 3, 593) who, like Tertullian, does not cite the text, the chaste S. accused by the elders is the symbol of the church of *Carthage, whose decency and evangelical truth are threatened by *Felicissimus and his five presbyters.

The first commentator on the story was *Hippolytus (*Comm. in Dan.* I, 12, 33) who, using Theodotion's version, comments on it in a typological-allegorical sense: S. symbolizes the church, her husband Joakim Christ, the two old seducers represent the *Jews and the *pagans (the two persecuting peoples); the garden is the community of the elect, Babylon the *world. S.'s bath symbolizes *baptism: the servants, *faith and charity; the *perfumes represent the commandments of the Word; the *oil, the strength of the *Holy Spirit (*Comm. in Dan.* I, 14 and ff.). In his letter to *Origen (c.240), *Julius Africanus casts doubts on the authenticity and canonicity of the story of S., but Origen's reply upholds its complete authenticity. The story was dealt with by *Jerome in a comprehensive and philological *commentary, and by the great 4th-c. Fathers, who saw S. as a model of charity (Greg. Naz., *Poem. mor.* II, 195; Ps.-Aug., *Serm.* 112, 2), conjugal *chastity (Aug., *Serm.* 343, 1-8), silent innocence (Aug., *Serm.* 318, 2), since S. resembles Christ at his trial (Ambr., *Exp. in Lc.* X, 97; Ps.-Aug., *Serm.* 112, 1-3). In the oldest liturgical texts, S. is mentioned as an example of one saved by *God from an unjust judgment (cf. E. Lodi, *Enc. Euch. Font. Lit.*, Rome 1979, no. 306).

H. Schlosser, *LTK* 9, 1194 f.; id., RQA Ergbd. (1965); *DACL* 15, 1742-72; *RACh* 3, 575-585; A. v. Harnack, *Die Briefsammlung des Apostels Paulus und die anderen vorkonstantinische christlichen Briefsammlungen*, Leipzig 1926 (41-52 *die Sammlung der Briefe des Origenes*); Quasten I, 346-400, 429 f.; M. Lefèvre-G. Bardy, *Hippolyte, Commentaire sur Daniel*, SCh 14, 44-48, 90-127.

G. Ladocsi

II. *Iconography. Paradigm of salvation and prefiguration of Christ as innocent victim threatened with immolation, the figure of S. appears in the iconography of the Christian community relatively later (3rd-4th cc.) than other themes. Often depicted alone between two trees, praying (London, Brit.Mus., glass cup from St Severinus, Cologne; Oxford, Pusey House, glass plate) or between the two elders - always shown, however, as young men - (Rome, cem. of SS. Pietro e Marcellino: second half of 4th c.; cem. of Domitilla: first half of 4th c.; cem. Maggiore: second half of 4th c.; Leningrad Hermitage, glass cup from Podgoritza: 5th c.), S. also appears with *Daniel in the judgment scene (Rome, cem. of Callisto: second half of 3rd c.); once the depiction takes on the aspect of an anti-heretical *allegory (Rome, cem. of Pretestato, so-called arcosolium of Celerina, a *lamb threatened by two wolves: second half of 4th c.).

Esp. from the 4th c., the episodes - *temptation of S., accusation by the elders, their condemnation, absolution of S., one or both the elders executed, Daniel with a companion - which characterize the tale of S. are represented subdivided into three scenes (Rome, cem. of Priscilla, Greek Chapel; Museo delle Terme, sarc.: first third of 4th c.; Museo Pio Crist. Vat., sarc. lid: second third of 4th c.; Brescia, Mus. Crist., reliquary: 4th c.; Rome, mausoleum of Constanza, lost decoration of the dome: 4th c.) or five scenes (Gerona, church of St Felix, sarc.: first decades of 4th c.). On a *sarcophagus from Arles, and a lost one from Cahors, Daniel, judge in the scene of S.'s judgment, is matched on the other side by *Pilate, clearly alluding to S.'s parallel with Christ, the predestined victim. Finally, a singular representation: that on the glass cup of Homblières (Paris, Louvre: 4th c.), where S. appears naked between the two elders. **[Figs: 288, 289]**

P. Testini: *EC* 11, 1589 ff.; H. Schlosser: *LCI* 4, 228-231; *Wp* 334 ff.; *Ws* 2, 251 ff.; L. de Bruyne, L'imposition des mains dans l'art chrétien ancien. Contribution iconologique à l'histoire du geste, *RAC* 20 (1943) 248-252 (Le jugement de Susanne); H. Schlosser, Die Daniel-Susanna Erzählung in Bild und Literatur der christlichen Frühzeit, *ROE* Suppl. 30 (1965) 243-249; M.C. Dagens, Autour du pape Libère. L'iconographie de Suzanne et des martyrs romains sur l'arcosolium de Celerina, *MEFR Ant.* 78 (1966) 327-381; M.M. Cecchelli Trinci, Studio su Susanna nella interpretazione patristica e nell'antica iconografia cristiana, *Miscellanea G.L. Messina*, Rome n.d.

G. Santagata

SYAGRIUS. Mid 5th-c. anti-Priscillianist, member of the *conventus* of Lugo (*Gallaecia). He and *Pastor were ordained bishops against the will of *Agrestius, metropolitan of Lugo, who, acc. to *Idatius the Romanophile (cf. *Chron.* 102), was a sympathizer of *Priscillian. Acc. to C. Torres, Agrestius's opposition might have been motivated by the fact that Pastor and S. were partisans of Hermeric and faithful to the Suevophile party. *Gennadius tells us (*De vir. ill.* 65 [66]) that, during the tense and confusing situation in the ecclesiastical province of Gallaecia after the invasions, S. wrote a book on the *Trinity softening the bold terms used by the *heretics. We have a work by him: *Regulae definitionum contra haereticos prolatae*. K. Künstle also attributed to him, among other works, the pseudo-Ambrosian *Ad neophytos de symbolo* and the pseudo-Augustinian *Sermm.* 237, 238 and 239. Critics have accepted Künstle's attributions in part and with caution. We should not confuse the Gallaecian S. with the S. cited by Consentius.

Reg. definitionum: CPL 560.702-PL 13, 639-642; K. Künstle, *Antipriscilliana*, Freiburg 1902, 142-159; *Ad neophytos*: CPL 178-PLS I, 606. 1749, K. Künstle, *op.cit.*; *Serm.* 237, 238, 239: CPL 368-PL 39, 1735-2354, K. Künstle, *op.cit.*; P.B. Gams, *Die Kirchengeschichte von Spanien*, II/I, Regensburg 1864 (Graz 1956), 466-467; G. Morin, Pastor et Syagrius, deux écrivains perdus du cinquième siècle, *RBen* 10 (1983) 385-390; A.C. Vega, Un poema inédito titulado "De fide" de Agrestio obispo de Lugo, siglo V, *Boletín R. Acad. Hist.* 159 (1966) 167-209; C. Torres, *Galicia sueva*, La Coruña 1977, 76; *Patrología* III, Span. ed. Madrid 1981, 685.

E. Romero Pose

SYAGRIUS of Autun. No *Life* survives, but we have strong evidence of S.'s prestige among the great men of his time, starting with pope *Gregory I (Jaffé: *Epp.* 1438, 1491, 1743, 1747-48, etc.). He was elected bishop of *Autun (*Augustodunum*) in *c.*560. King Guntram ordered S. to accompany him to the ceremony of Lothar II's *baptism in 591. He had access to all the courts of *Gaul (Burgundy, Neustria, Austrasia, Aquitania), but esp. that of Childebert II, king of the *Franks, during the regency of his mother Brunhilde. Gregory I recommended to him the missionaries he had sent to *Britain, and wished to appoint him head of the bishops of the province of *Lyons. He took part in nearly all the *councils held in France in his time, and was asked by the *pope to organize and preside over a general council of the Gallic churches, but his death (599-600?) seems to have prevented this.

For Gregory the Great's letters to S., see also *Registrum Epistolarum*, ed. P.E. Wald-L.M. Hartmann, in MGH, *Epistol.* I and II. Cf. *EC* 11, 504; *LTK* 9, 1201; *BS* 11, 1016-1018.

L. Dattrino

SYLVESTER I, pope (314-335). Little is known for sure of S.'s quite long pontificate. He took no part in the council of *Arles, which sent him a polite *letter inviting him to communicate its synodal sentences to the bishops (Mirbt no. 239). At the council of *Nicaea (325) he was represented by two presbyters, but had no influence on the synodal debates. Nor does he appear to have been involved in *Constantine's further activity, either on the level of the universal church or on that of the Roman community (building of *churches, etc.). Only from the 5th c. did legend associate him with the emperor's *conversion and *baptism, providing the basis for the author of the so-called *donation of Constantine.

E. Ewig, Das Bild Constantins des Grossen in den ersten Jahrhunderten des abendländischen Mittelalters, *HJ* 75 (1956) 1046 (legend); *LTK* 9, 357-358; R.J. Loenertz, "Actus Sylvestri". Genèse d'une légende, *RHE* 70 (1975) 426-439; Ch. Pietri, *Roma christiana*, BEFAR 224, 2 vols., Rome 1976, 168-187 (bibl.).

B. Studer

SYMBOL - SYMBOLISM

I. In literary tradition - II. In art.

I. In literary tradition. In general, the symbols used in the Christian iconographical tradition are also attested and discussed by the written sources. But there are also many symbols attested exclusively at a literary level, often rising out of two of the early church's most common exegetical techniques, *allegory and typology. The main sources of these symbols are obviously the OT and NT *Scriptures, quickly joined by subject-matter taken from the surrounding *pagan culture. Occasionally we can make out the influence of extra-biblical Jewish traditions and *apocrypha.

Here we will briefly mention some special categories of symbols, since there is still no systematic treatment of symbols in general.

*Numbers: one of the commonest OT literary conventions is that of attributing precise values to certain "perfect" numbers (3, 4, 6, 7, 8, 10, 12, 40 . . .). In the NT, only the *Apocalypse makes much use of this kind of symbol (seven churches; seven seals; 666, the number of the beast; 12 x 12 x 1000 elect; the dimensions of the heavenly city); elsewhere they are quite rare (but cf. the "seventy times seven" of Mt 18, 22, and the "third heaven" of 2 Cor 12, 2), though there is an evident typological intention in making the number of the *apostles coincide with that of the twelve tribes of Israel (and fixing that of the disciples at 72, a multiple of 12). In subsequent Christian *tradition this type of symbolism, which borders on allegory, becomes relatively rare, perhaps because of its frequent use in *gnosticism (cf. the 30 *aeons of the *Valentinians, which become 10 for *Mark the Magus and three for the *Simonians, etc.). Notable exceptions are the *ogdoad, which represents *resurrection or eternity (with its architectonic repercussions) and some passages in *Irenaeus (correspondence between the gospels and the four cardinal points and four winds: *Adv. haer.* III, 11, 8; the seven heavens and the gifts of the Spirit: *Epid.* 9).

The letters of the Greek alphabet also had numerical value. This property was sometimes exploited by arithmological speculations analogous to those of the Rabbinic Cabbala; thus *Barnabas (*Ep.*, 9, 8) saw in *Abraham's 318 servants (cf. Gen 17, 23 and 14, 14) an image of Christ's *cross, since *iota* (10) and *eta* (8) are the initials of *Iesus*, and the shape of *tau* (300) recalls the *cross. Other interpretations of the same sort are found among the gnostics, who also used other techniques like that of adding up the numerical values of the letters of a word: e.g. *peristera*, "dove", which for the Marcosians was a symbol of the heavenly Christ, since the sum of its letters (801) was equal to that of *alpha (1) and *omega* (800).

There are few examples in the NT of the symbolic value of *names (though very common in the OT and the Semitic world generally): the best-known is the change of Simon's name, "dove", to Cephas-Peter, "stone", as well as Jesus' own name, "God saves".

Acrostics and abbreviated forms are rare: the famous ἰχθύς; that which made *Adam a symbol of the whole *world, by way of the initials of the four cardinal points (*anatole, dysis, arktos, mesembria*; Ps.-Cyprian, *De montibus Sina et Sion* 4). Two very common abbreviations, esp. in *epigraphy, are DOM=Deo Optimo Maximo (adaptation of the pagan DM=Diis Manibus?) and the *monogram of Christ formed by superimposing *chi* (or a cross) and *rho*: the christogram has two forms, decussate (combination of chi and rho) and crossed (or monogrammatic): ⳩ and ☨. The etymological and symbolic interpretations of names are collected in the onomastical lexicons. One of the most famous examples of popular etymology is the highly polemical one linking *Mani with *mania* "furious rage".

The way to symbolic interpretation of animals, plants and natural phenomena lay already open in Scripture (cf. the many gospel *parables based on sowing and fruit; or even Mt 10, 16: "be cunning as serpents and harmless as doves"; Jn 1, 29: "Jesus the *lamb of God", etc.) and was explored enthusiastically by the various exegetical schools.

The most interesting of the inventories of this kind of symbol is the *Physiologus* (3rd c.?), which gives a good 40 of them: animal, vegetable or mineral, christological or ascetic, using scriptural or sometimes pagan themes.

Material from the pagan world quickly penetrated the Christian tradition; *Clement of Rome (late 1st c.) used the theme of the *phoenix to illustrate Christ's *resurrection (1 Cor. 25-26, 1). The use of other pagan subjects is well-known: *Orpheus, the *orans*, the peacock, etc.

Probably following the example of *Justin's positive vision of the *world as the dwelling of the *logos spermatikos*, the timeliness or necessity of using subjects drawn from pagan tradition was explicitly theorized by *Clement of Alexandria (cf. *Protr.* 12, 119, 1 and *passim*). Clement was prob. the first author to introduce the terminology of pagan banquets and *mysteries into the Christian world to designate the *Eucharist.

An important characteristic of some Christian symbols is their multivalency: the lion may represent an *evangelist (*John in the *East, *Mark in the West), Christ's *incarnation and resurrection (*Physiologus*, 1), the devil (cf. 1 Pet 5, 8), royal power (for the Antiochenes), etc.

To exemplify this, we will give the interpretation of some of the best-known Christian symbols: eagle (Mark in the East, John in the West; *baptism, later *penitence in *Physiologus*); plough (*creation, *cross, Christ's two natures); axe (Christ's two natures; the last *judgment); ass (pagan peoples: cf. Mt 21, 5-7 par.); boat, ship (church; cf. *Noah's ark and the bark of Peter; the mast represents the cross); ox (*Luke the evangelist); deer (Jesus Christ, who kills the serpent or dragon, symbols of the devil: popular etymology); dove (*Holy Spirit; *virginity; virtuous life; cf. *Wing); crow (*chastity in *Physiologus*; shamelessness for the Antiochenes: cf. PG 59, 513); dragon (devil); palm (immortality, *resurrection); panther (Christ's resurrection); pelican (crucifixion of Christ, baptism and *Eucharist, born of *water and *blood from Christ's side); stone, rock (Jesus Christ the *corner-stone, rock from which water springs in the desert); tower (church).

F. Wutz, *Onomastica sacra. Untersuchungen zum Liber interpretationis nominum Hebraicorum des hl. Hieronymus*, TU 41, 1-2, Leipzig 1914-1915; H. Rahner, *Griechische Mythen in christlicher Deutung*, Zürich 1945; J. Daniélou, *Les symboles chrétiens primitifs*, Paris 1961; H. Rahner, *Symbole der Kirche. Die Ekklesiologie der Väter*, Salzburg 1964; E. Testa, *Il simbolismo dei Giudeo-Cristiani*, Jerusalem 1962; S. Sterckx-G. de Champeaux, *Introduction au monde des symboles*, La-Pierre-qui-vire 1966; A. Quacquarelli, *L'ogdoade patristica e i suoi riflessi nella liturgia e nei monumenti*, Bari 1973; R. Murray, *Symbols of Church and Kingdom: A Study in Early Syriac Tradition*, Cambridge 1975; A. Quacquarelli, *Il leone e il drago nella simbolica dell'età patristica*, Bari 1975.

S.J. Voicu

II. In art. Analogously to literary output and liturgical practice, the figurative evidence of the primitive Christian communities also shows that it left plenty of room for the use of symbolic representation. The clear function of this was to sum up visually some of the fundamental principles of Christian doctrine through a process of transposition, which would allow a coherent and immediate linking of images. This process, which is surprisingly mature even in its earliest manifestations, required the formation of a repertoire that followed a code of interpretation illustrated evidently in oral teaching (*catechesis and *homilies). So figures and scenes were elaborated which, while evolving and being enriched in meaning, remained substantially intact ideologically and never lost the necessary capacity to transmit the message. Symbolic language developed to maturity in the course of the 3rd c., when a sort of *disciplina arcani*, the natural and almost obligatory consequence of a general situation averse to the spread of Christianity, could and did become a formidable element of cohesion for the primitive communities.

Looked at formally, the figurative subjects that were used with a symbolic function are obviously taken from the heritage of the Roman world, in which - it is as well to stress - esp. in the period of *late antiquity, there was a widespread general tendency towards visual communication in terms of symbolic language, esp. in funerary iconography. It thus happened that, when convenient, the same *mythology was used to work out ideograms capable of expressing concepts of Christian faith: think of the myths of Cupid and Psyche or of *Orpheus, frescoed in the catacombs of Domitilla (*Wp* pl. 53) and Callisto (*Wp* pl. 37), i.e. in two *cemeteries belonging to the community (hence administered directly by the church hierarchy), the Four Seasons, the Cupids, the series of animals which, by universal convention, were seen as reflecting man's positive and negative qualities. Evidently it was considered that certain mythological subjects were, at least as instruments of communication, not irreconcilable with Christian doctrine.

Some of the oldest and commonest symbols - esp. those that appear in the constitutive nuclei of the Roman cemeteries (late 2nd or early 3rd c.) - refer directly to the person of Christ; this is usually so of images like the fish, anchor, lamb.

The fish, considered in relation to the Greek homologue ΙΧΘΥΣ, acrostic of Ἰ(ησοῦς) Χ(ριστὸς) Θ(εοῦ) Υ(ἱὸς) Σ(ωτήρ), seems to have been used from the start, esp. in *painting and funerary slabs, and tends to disappear with the coming of the Constantinian era. It is explicitly referred to in the famous and extremely early funerary poems of Abercius of Hierapolis (Guarducci IV, n. 1, p. 377) and Pectorius of Autun (Guarducci IV, n. 1, p. 487) (cf. *epigraphy). With a basket of *bread and a chalice of *wine - i.e. in a context evoking the saving food of the *Eucharist - the fish is frescoed at the start of the 3rd c. in a cubicle in the region of Cornelius in the cemetery of Callisto (*Wp* pl. 28). In some fishing scenes, e.g. those frescoed in the cubicles "of the Sacraments" in Callisto (*Wp* pl. 27, 2-3), the fish may represent the believer saved by Christ, the baptized person; this sense is directly reflected in the famous passage of *Tertullian's *De baptismo* (I, 3: CCL 1, 277) in which Christians are called *pisciculi* and Christ ἰχθύς.

The anchor, also a very ancient image and one often connected with the fish, was considered a symbol of *hope in eternal *life and, by its form, allusive of the *cross. The anchor-fish combination, very pregnant in content, recurs in the late 2nd and early 3rd cc. in a group of Greek and Latin inscriptions in the catacombs beneath the *basilica Apostolorum* on the Appian Way at *Rome (ICUR V, 12891, 12892, 12900). The hypothesis that the anchor represented on Christian monuments has no symbolic meaning whatever (Stumpf, Anker, *RACh* 1, 441) seems frankly contrary to the available figurative evidence.

The so-called *stellar christogram* is also a christological symbol when it is not obviously being used as a *compendium scripturae* of the name of Christ. Made of the crossed Greek initials of Ἰησοῦς and Χριστός, it is very rare in the West, but particularly common in *Asia Minor. Its use in the Christian world was probably favoured by the fact that in classical antiquity it was accepted as a general symbol of divinity and *light. This original type of christogram was replaced at the start of the 4th c. by the so-called Constantinian (or *decussate*) *monogram, made by combining the first two letters of the name XPICTOC into a single sign. Widely used, mainly at *Rome and esp. in funerary *epigraphy, its symbolic value was progressively enriched in the course of the 4th c. by the addition, at each side, of the apocalyptic letters *alpha* and *omega*. Alongside the decussate form, the other christogram called the monogrammatic cross spread from the 4th c.

Some symbols allude directly to the believer seen in relation to his earthly existence or, more properly, to the afterlife. Thus the *sheep, esp. if in a paradisiacal context, is an image of the afterlife; so too the peacock, which also represents the concept of immortality. The two symbols appear significantly together in an funerary inscription dedicated to a dead woman called Aurelia Proba (ICUR III, 8787). Images alluding to the attainment of eternal salvation include the ship and the *lighthouse, represented esp. on funerary slabs, where they often appear together.

Depictions of the crown, palm, or horse are connected with the allegorical images of life as a contest, often employed by St *Paul. The *phoenix, a mythical bird symbolizing *resurrection, appears on a palm or by itself in frescos (so-called Greek Chapel in Priscilla, perhaps the earliest Christian depiction) and on *mosaics, sometimes in apses (cf. SS. Cosma e Damiano, Rome). Palm and crown were extremely common; the crown in particular, from the Constantinian era, also came to represent *martyrdom, becoming indeed an almost constant attribution in the *iconography of martyrs on the walls of hypogea and basilicas (e.g. the well-known "procession" of saints in the nave of St Apollinare at *Ravenna).

The vine, a many-sided symbol since innate to human existence and allusive of *blood, vehicle of life and supreme token of *sacrifice, generally designates the heavenly *refrigerium*, but also assumes ecclesiological and christological value in connection with Jn 15, 1. 5: *Ego sum vitis vera, Ego sum vitis vos palmites*; analogously, the grape-cluster can represent an *imago brevis* of the vine, alluding to the fullness of the blessings of *paradise or, in reference to the episode of the spies in Canaan (Num 13, 1-26), evoking the image of the promised land.

Besides the sheep and the *lamb, the depictions most used to symbolize the *soul are the *orans* and the dove. The so-called *orans*, a male or female figure always shown in a *gesture of *expansis manibus*, has its formal antecedent in the Roman image of *pietas*: however it does not mean a request for help but, as De Bruyne pointed out (Les lois ... 12 ff.), the state of salvation attained. This explains why this gesture can be common to any image, OT or NT (*Noah, *Daniel, *Mary, a dead woman, etc.). The dove, image of the soul entered into its heavenly rest, often shown by itself, at other times sitting in a tree or holding a palm or olive branch in its beak, or again preparing to drink from a *cantharus* or pecking at a grape-cluster, can be considered as having the same meaning. The elements that accompany the dove (tree, olive, palm, cantharus, grape) are intended to characterize the afterlife as a paradisiacal place, i.e. a garden. In other contexts the dove can take on extended meanings: thus in scenes of Christ's *baptism (cf. e.g. *Wp* pll. 29, 1; 27, 3; 240, 1), where it represents the *Holy Spirit; in baptistery scenes of *Noah's ark, as a bringer of *peace, where, to the number of twelve, it symbolizes the apostolic college (WMM, pl. 88, baptistery of Albenga); finally, with a *nimbus, signifying Christ among his *apostles (hypogeum of Leo in the catacomb of Commodilla, Rome: *RAC* 33 [1957] 7-44 and 34 [1958] 5-58). **[Figs: 290, 291, 292]**

F.J. Dölger, ΙΧΘΥΣ, IV-V, Münster i. W. 1927-43; M. Guarducci, *I graffiti sotto la confessione di San Pietro in Vaticano*, I-III, Vatican City 1958; J. Daniélou, *Les symboles chrétiens primitifs*, Paris 1961; E. Testa, *Il simbolismo dei Giudeo-Cristiani*, Jerusalem 1962; P. Bruun, Symboles signes monogrammes, in *Sylloge Inscriptionum Christianarum Veterum Musei Vaticani*, 2, Helsinki-Helsingfors 1963, 73-166; as well as individual entries in *DACL*, *LCI* and *RACh*. For specific problems mentioned above, see De Bruyne, Les "lois" de l'art paléochrétien comme instrument herméneutique, *RAC* 39 (1963) 7-92; P. Testini, *Le catacombe e gli antichi cimiteri cristiani in Roma*, Bologna 1966, 255-340, 359-366; H.W. Bartsch, Das alttestamentliche Bilderverbot und die frühchristliche Verwendung des Bildes im Wort und in den Anfängen christlicher Kunst, *Symbolon* 6 (1968) 150-178; P. Testini, Il sarcofago del Tuscolo, *RAC* 52 (1976) esp. 72 ff.; M. Guarducci, *Epigrafia greca*, IV, Rome 1978; A. Perkins, *The Art of Dura-Europos*, Oxford 1973. Cf. also under *Epigraphy.

C. Carletti

SYMMACHIANS. *Judaeo-Christian sect mentioned only by Latin Christian writers. They took their origin from the Pharisees (*Ambrosiaster) or from *James (*Marius Victorinus). *Filaster's observation that the S. were libertines conflicts with their supposed Jewish background. We have no concrete information on them except that of Marius Victorinus, who wrote that they said that Jesus is *Adam and *anima generalis* (*In epist. ad Gal.* 1, 15).

A.F.J. Klijn-G.J. Reinink, *Patristic Evidence for Jewish-Christian Sects*, SNT 36, Leiden 1973, 52-54.

A.F.J. Klijn

SYMMACHUS, translator. Late 2nd-c. author of the Gk version of the Hebrew OT which, in *Origen's *Hexapla*, occupied the column after Aquila's version and before those of the LXX and *Theodotion. *Eusebius, who considers S. an *Ebionite, calls him the author of commentaries which Origen received from a certain Juliana, who got them from S. himself (*HE* VI, 17; cf. also *Dem. ev.* VII, 1. On this Juliana, see *Palladius's information in *Hist. Laus.* 64, ed. Bartelink, p. 272).

*Jerome too considers S. an Ebionite and adds that he wrote commentaries on *Matthew's gospel, on the basis of which he tried to reinforce his doctrine (*Vir. ill.* 54). But *Epiphanius calls S. a *Samaritan who went over to Judaism (*Mens.* 16).

H.J. Schoeps, *Aus frühchristlicher Zeit. Religionsgeschichtliche Untersuchungen*, Tübingen 1950, 82-119.

S. Zincone

SYMMACHUS. Otherwise unknown 5th-c. exegete, presumed author of a *Commentary on the *Song of Songs*, of which a large fragment of a *Syriac version remains. On the basis of a recurrent abbreviation in the *Catena on

Proverbs (Cod. Vat. Gr. 1802), some think S. also commented on Proverbs, but the hypothesis is disputed.

CPG 6547; C. van den Eynde, *La version syriaque du Commentaire de Grégoire de Nysse sur le Cantique des Cantiques*, Bibliothèque du Muséon 10, Louvain 1939, 77-89 (text), 104-116 (Fr. tr.); M. Faulhaber, *Hoheliéd-Proverbien-und Predigercatenen*, Vienna 1902, 90-94; G. Mercati, *Pro Symmacho: Nuove note di letteratura biblica e cristiana antica*, ST 95, Rome 1941, 91-93; M. Richard, Les fragments du commentaire de S. Hippolyte sur les Proverbes de Salomon, *Muséon* 78 (1965) 286-287 and *Muséon* 80 (1967) 356, n. 41 (= *Opera minora*, Turnhout-Leuven 1976, n. 17).

F. Scorza Barcellona

SYMMACHUS, pope (498-514). A native of *Sardinia, S. was elected by the majority, discontented with the pro-Byzantine policy of Anastasius II, while the minority chose *Lawrence. Despite *Theodoric's intervention, the dissidents remained in the community. The very synod (501) which Theodoric had intended to take a decision, declared that S., as bishop of the first see, could not be summoned before any human tribunal (MGH, *Auct. ant.* XII, 416-437). The whole *schism reflected disagreements between clergy and senate, *Rome and *Byzantium, as well as the tendency to enfranchise church affairs from outside interference. For four years S.'s role was limited. Only when Theodoric abandoned Lawrence and his pro-Byzantine followers did S. enter into the full exercise of his functions. He defended *orthodoxy against the *Henotikon*, helped *Catholics persecuted by *Arian rulers and made *Caesarius of Arles primate of *Gaul and *Spain. During his pontificate *Sigismund of Burgundy came to *Rome and (perhaps) *Clovis I received *baptism. The Symmachian *falsifications which seek to demonstrate the adage: *prima sedes a nemine iudicatur*, date from this time. They are: *Gesta de Xysti purgatione* (PLS 3, 1249-1255), *Synodi Sinuessanae de Marcellino papa* (PL 6, 11-20), *Gesta Liberii* (PL 8, 1388-1393), *Constitutum Sylvestri* (PL 8, 829-840).

CPL 1678-1682; Thiel I, 641-734; *LP* I, 44-46, 260-268; III, 87-88; W. Politz, A propos des synodes apocryphes du pape Symmaque, *RHE* 32 (1936) 81-88; A. Alessandrini, Teodorico e papa Simmaco durante lo scisma laurenziano, *Archivio romano di storia patria* 67 (1944) 152-207; G.B. Picotti, I sinodi romani nello scisma laurenziano, in *Studi in onore di G. Volpe*, II, Florence 1958, 741-786; G. Sigismondi, I sinodi simmachiani e la sede episcopale di Nocera Umbra, *Boll. di Deputaz. di Storia Pat. Umbra* 59 (1962) 5-42; H. Fuhrmann, Die Fälschungen im Mittelalter, *HZ* 197 (1963) 529-554; Ch. Pietri, Le sénat, le peuple chrétien et les parties du cirque à Rome sous le Pape Symmaque, *MEFR Ant.* 78 (1966) 123-139; P.A.B. Llewellyn, The Roman Church during the Laurentian Schism: Priests and Senators, *ChHist* 45 (1976) 417-427; G. Zecchini, I "Gesta de Xysti purgatione" e le fazioni aristocratiche a Roma alla metà del V secolo, *RSCI* 34 (1980) 60-74; S.T. Stevens, The Circle of Bishop Fulgentius, *Traditio* 38 (1982) 327-341.

B. Studer

SYMMACHUS, AURELIUS ANICIUS. Grandson of the famous orator Quintus Aurelius *Symmachus and related to the noble family of Anicii, he was *proconsul Africae* (415) and *praefectus urbis* (24 Dec 418 - early 420). He played an important part in the troubles over the election of a successor to pope *Zosimus (†26 Dec 418). Quarrels and mutual accusations led to the Roman *clergy splitting in favour of two candidates, Eulalius, *archdeacon of S. Lorenzo, and the priest Boniface. S.'s task was to keep the court of *Ravenna and in particular *Galla Placidia, who supported Boniface, informed of the uncertain course of the election. His *letters are preserved in the *Collectio Avellana* (*Epistulae* XIV-XXXIV: 28.12.418-8.4.419). S.'s action at Easter 419 was decisive: he freed the Lateran basilica, in which Eulalius and his supporters had shut themselves up, and restored it to bishop *Achilleus of Spoleto, who was at *Rome to preside over the *Easter celebrations. This episode sealed Eulalius's defeat, and there were no more obstacles to Boniface's succession.

PWK 23/1, 1158-1159; Fliche-Martin 4, 549; *PLRE* 2, 1043-1044.

M.L. Angrisani Sanfilippo

SYMMACHUS, QUINTUS AURELIUS (340-402). Played an important role in the public and religious life of the empire. A member of the crown council in 369 as *tertii ordinis*, he was named *proconsul Africae* in 373 under *Valentinian I, *praefectus urbis* in c.384 under *Valentinian II and consul with *Theodosius I in 391. *Ambrose's *De exc. fratris* I, 32 and S.'s *Ep.* I, 63 seem to allude to a kinship between the two of them. More certain is their youthful *friendship in the Roman period and a feeling of respect which never lessened (cf. Ambr., *Epp.* 17, 6; 57, 2), not even during the controversy over the *Altar of Victory. After the Altar's second removal, decreed by *Gratian in 382, the senate, through S., vainly requested its restoration. In 384, S. (*Relatio* III) presented Valentinian II with the remonstrances of his fellow-*pagans, recalling the religious heritage of the empire and asking for the abrogation of the main anti-pagan measures decreed by *Gratian in 382. S.'s protest failed to achieve the hoped-for result because of Ambrose's appeal (*Epp.* 17 and 18) to Valentinian. Ambrose returned to the same problem towards the end of 389 (*Ep.* 57, 4) after a new petition by S. The thesis of S.'s *Relatio* III and the refutation of this by Ambrose were taken up by *Prudentius in his two books *Contra Symmachum*. Behind this controversy was an encounter between two different ways of conceiving the relationship between politics and religion: for Roman tradition, whose greatest 4th-c. exponent S. was, this relationship was one of close alliance; for Christianity it was one of distinction, with the difficulties and contradictions this involved in practice.

MGH, *Auct. ant.* VI, 1; F. Paschoud, Réflexions sur l'idéal religieux de Symmaque, *Historia* 14 (1965) 215-235; F. Canfora, *Simmaco e Ambrogio o di un'antica controversia sulla tolleranza e sull'intolleranza*, Bari 1970; R. Klein, *Symmachus. Eine tragische Gestalt des ausgehenden Heidentums*, Darmstadt 1971; id., *Der Streit um den Victoriaaltar. Die dritte Relatio des Symmachus und die Briefe 17, 18 und 57 des Mailänder Bischof Ambrosius*, Darmstadt 1972; F. Zuddas Del Chicca, Rassegna di studi simmachiani, *StudRom* 20 (1972) 526-540; M. Forlin Patrucco-S. Roda, Le lettere di Simmaco ad Ambrogio, Vent'anni di rapporti amichevoli, *Ambrosius episcopus* (Atti del Congresso internazionale di studi ambrosiani), II, Milan 1976, 284-297; P. Meloni, Il tempo e la storia in Simmaco e Ambrogio, *SSR* I (1977) 105-123; id., Il rapporto fra impegno politico e fede religiosa in Simmaco e Ambrogio, *Sandalion* 1 (1978) 153-169; J.F. Matthews, The Letters of Symmachus, in (ed.) J.W. Binns, *Latin Literature of the Fourth Century*, London 1974, ch. 3; J.J. O'Donnell, The Demise of Paganism, *Traditio* 35 (1979) 69-80; F. Paschoud, Le rôle de providentialisme dans le conflit de 384 sur l'autel de la Victoire, *MH* 40 (1983) 197-206; C.R. Ruggini, Simmaco, *Otia et Negotia* di Classe, fra conservazione e rinnovamenta, in (ed.) F. Paschoud, *Symmache*, Paris 1986, 97-118. Other papers in *Colloquium genevois sur Symmache à l'occasion du mille six centième anniversaire du conflit de l'autel de Victoire* (eds. G.W. Bowersock et al.), Paris 1986.

M.G. Mara

SYMMACHUS, QUINTUS AURELIUS MEMMIUS (†525). Historian and orator, great-grandson of the famous Q. A. *Symmachus. Consul (485) and *patricius*, he was the senate's most authoritative spokesman during the reign of *Theodoric, with whom he tried to maintain friendly relations. After the death of his son-in-law *Boethius (524), he was suspected of treason and executed (525). *Ennodius lists him among the most cultivated men of his time. S.'s cultural activity was indeed considerable. His revision of Macrobius's commentary on the *Somnium Scipionis* bears the famous subscription *Aurelius Memmius Symmachus v. c. emendabam vel distinguebam meum (sc. exemplar) Ravennae cum Macrobio Plotio Eudoxio v. c.* He also compiled a *Roman History* in seven books (lost), marked by a clear Christian spirit. He was in touch with Boethius (*Cons. phil.* I, 40, 40 and II, 4, 5), Ennodius, *Cassiodorus and the Byzantine grammarian *Priscian.

PWK 23/1, 1160; *KLP* 5, 446; Schanz IV, 2, 83 f.; W. Ensslin, *Des Symmachus Historia romana als Quelle für Iordanes*, Munich 1948; A. Momigliano, Gli Anicii e la storiografia latina del VI sec. d.C., *RAL* ser. 8ª 11 (1956) 279-297; R. Klein, *Symmachus. Eine tragische Gestalt des ausgehenden Heidentums*, Darmstadt 1971; B. Luiselli, Note sulla perduta Historia Romana di Quinto Aurelio Memmio Simmaco, *Studi Urbinati* 49/1 (1975) 529-535; *PLRE* 2, 1044-1046.

M.L. Angrisani Sanfilippo

SYMPHOSIUS Scholasticus. Latin writer, perhaps 4th or 5th c., wrongly associated by synonymy with *Lactantius's lost *Symposium* (Jerome, *Vir. ill.* 80: *Habemus eius Symposium, quod adulescentulus scripsit...*). Indeed some editors, accepting this opinion, ended up including S.'s brief work on riddles in editions of Lactantius (PL 7, 289-298). Under S.'s name survives the only complete collection of ancient riddles (*Symphosii scholastici Aenigmata*). But nothing can be known from it except the author's name, and even this has been disputed. The collection consists of 100 riddles, all contained in triplets of three hexameters each. The collection is preceded by a brief verse preface, humorous in intent (v. 16: *Insanos inter sanum non esse necesse est*). These enigmatic formulations (no. 2 *Harundo*; no. 3 *Anulus cum gemma*) were very popular in the Middle Ages and certainly influenced the future course of enigmatic poetry.

CPL 1458; PL 7, 289-298; A. Riese, *Anthologia latina*, I, Leipzig 1894, 221-246; R.T. Ohl, *The enigmas of Symphosius*, Philadelphia 1928 (with Eng. tr., notes and commentary); Fr. tr.: E.F. Corpet, *Enigmes de Symphosius*, Paris 1868; Schanz IV, 2, 74-76; *PLRE* 2, 1047.

L. Dattrino

SYMPRONIANUS. Spanish *Novatianist known only by three *letters sent to him by *Pacianus of Barcelona, which allow us to reconstruct his thought. Unknown and unsolicited, S. had written asking Pacianus for explanations of the use of the word *Catholic, of *penitence and of repentance (Pacianus, *Ep.* 1). Pacianus replying, S. sent him a letter and a treatise in which he maintained that "there is no penitence after *baptism; the church cannot pardon mortal *sin; indeed the church destroys itself by readmitting sinners" (Pacianus, *Ep.* 3, 1: PL 13, 1065). He proved his thesis using *Scripture and theological argument.

CPL 784; PL 13, 1051-1082; L. Rubio Fernández, *San Paciano. Obras*, Barcelona 1958, 48-134; L. Wohleb, Bischof Pacianus von Barcelona und seine Gegner Novatianer Sympronianus, *Spanische Forschungen der Görresgesellschaft* 2 (1930) 25-35; Schanz VI, 1, 369; *Patrologia* III, 124-126.

A. Di Berardino

SYNAXIS. This word, almost unknown before the appearance of Christianity, became a technical term in the *Byzantine liturgy to indicate both the *assembly of believers and the *Eucharist, more properly the *communion. S. and its homonym "synagogue" come from the same Gk root. The first Christians, to distinguish their assemblies from those of the *Jews and, later, of the *heretics, adopted a different designation (cf. Epiph., *Pan.* 30, 18). *Cyril of Jerusalem clearly distinguishes the assemblies that took place on *Sundays and the day of the Lord's *Ascension, of which he fixes the order of readings (*Cat.* XIV, 24), from those during which he held his catechetical instructions. While *John Chrysostom does not seem to keep up this distinction (*Hom.* 29 *in Act; Hom.* 5 *in Mt.*), *Basil speaks of a visible s. and a spiritual s. of believers, who serve *God in spirit and in truth, and which he calls holy (*Hom. I in Psalmum* 28). The term also indicates the participation of the people in the celebration of the holy *mysteries and their intimate union with Christ (cf. Dion. Areop., *Hier. cael.* I, 3) and also, but at a distance from the original meaning, the form of *prayer or worship of monks and nuns (*Apophthegmata Patrum*: PG 65, 220). In the Latin monastic world it appears late, with the meaning of *collect* (cf. John Cass., *Conlat.* VIII, 16; IX, 36; *Vitae Patrum Jurensium* 52, 9; 64, 5; 130, 2), and means both the liturgical celebration whose culminating point is the Eucharist and the celebration composed of prayer and *psalmody (*Reg. S. Bened.* XVII, 17).

DACL 15, 1835-1836; Lampe 1302-1303.

E. Peretto

SYNCLETICA. The historical existence of S., an Egytian nun, is documented only by the Greek *Life* attrib., seemingly first by Nicephorus Callistus (*HE* VIII, 40), to *Athanasius. Acc. to this *Life*, S. was born at *Alexandria in *Egypt of an originally Macedonian family. Consecrated to *God, she passed her youth in *prayer and penance and, on her parents' death, retired to solitude. Her fame spread rapidly, several *women gathered around her and persuaded her to direct them spiritually. She died at an advanced age after supporting severe and painful illnesses with fortitude, to the end consoling and comforting those who attended her.

BS 11, 1209-1210.

S. Zincone

SYNCLETICA. 5th-c. *virgin and *deaconess celebrated by the poet *Sedulius who, in a *letter to the presbyter Macedonius (CSEL 10, 9), exalts S.'s humility, despite her noble birth, and warmly praises her theological knowledge.

DCB 4, 756.

S. Zincone

SYNCRETISM. This term, derived from the Greek σύγκρασις ("mixture"), designates the union and fusion of disparate elements of different origins, which come together to form either a new religious creed or a new philosophical system. This process of synthesis, whether on the religious or the philosophical level, is particularly characteristic of the Hellenistic-Roman period and one of its main centres was *Alexandria in *Egypt. Leaving aside the syncretistic religions of the Hellenistic-Roman era, we will briefly point out the syncretistic character of *gnosis, whether *pagan or Christian, and of Hellenistic and late-antique thought, from which the thought of the *Fathers cannot be detached. Pagan gnosis, which finds its greatest expression in that group of treatises known as the *Corpus Hermeticus* (cf. *Hermetism), formed in Egypt in the first centuries of our era, is a combination of *astrology, occult science, neopythagoreanism, Platonism, Stoicism and, acc. to R. Reitzenstein, Iranian religious doctrines; Christian *gnosis* shows some close analogies with the pagan (e.g. the motif - present in *Corp. Herm.* I and XIII and in the *Valentinian system: cf. *Ogdoad - of the *Himmelsreise* or journey of the gnostic soul which, having passed through the seven lower heavens, reaches the eighth heaven or *ogdoas*), and incorporates in itself occultism, astrological, neopythagorean and Platonic doctrines, more or less fantastic interpretations of the OT and NT, heterodox tendencies of late Judaism, Zoroastrianism and myths of Oriental origin (e.g. the motif of the fall of Sophia, represented in various gnostic systems).

In Hellenistic thought, the most illustrious examples of the "syncretistic" tendency are the Stoic Poseidonius (2nd c. BC) and the Academic Antiochus of Ascalon (1st c. BC). Poseidonius's commentary on Plato's *Timaeus* remained famous throughout *late antiquity and was in all probability used by authors like *Philo, *Plotinus, *Gregory of Nyssa and *Basil the Great; evident "Platonic" motifs in his thought are the image of the chariot guided by two horses, derived from *Phaedrus* (246b), the ethic of simple moderation of the passions, present in the *Republic* (IV, 423a 4-5, 431c 5-7; X, 619a 5-6), and the doctrine of the tripartition of the human *soul (though Poseidonius prefers to talk of "functions" rather than "parts" of the soul: cf. S. Lilla, *Clement of Alexandria*, Oxford 1971, 87 n. 1, 98 n. 4, 100 n. 3). The whole teaching of Antiochus of Ascalon aims to show the essential agreement between the academic and the peripatetic schools (cf. Cic., *Ac. post.* I 37 and 38) and is profoundly influenced by the "middle Stoicism" of Panaetius and Poseidonius: Antiochus may be defined as a "Stoicizing Platonist". The same syncretistic approach is observable in some exponents of the so-called *middle Platonism of the first two centuries AD, such as Plutarch and Alcinous, in whose writings the presence of Platonic, Aristotelian and Stoic elements (despite Plutarch's polemic against Stoicism) is particularly evident. A s. with close affinities to that of middle Platonism, but enriched with various neopythagorean motifs, can be seen in *Philo of Alexandria (cf. *Judaeo-Hellenism). The "syncretistic" approach is one of the main characteristics of all *neoplatonism, from *Ammonius Saccas to its last exponents. As Hierocles records in his work *On providence*, Ammonius had aimed to show the agreement between Plato and Aristotle (Photius, *Bibl.*, cod. 214) and had brought the doctrines of the two great philosophers into a single system, which had been handed down by Plotinus and by *Origen the neoplatonist (Photius, *Bibl.*, cod. 251). In ch. 14 of his *Vita Plotini*, *Porphyry (I, pp. 15-16 Bréhier) records that in his lectures Plotinus examined and commented on the works of numerous Greek philosophers (Stoics, peripatetics, Platonists, neopythagoreans), accepting many of their doctrines and following Ammonius's example in his investigation (cf. p. 15, 15-16): this last assertion proves that Ammonius, besides demonstrating the agreement between Plato and Aristotle, also subjected the main philosophical tendencies to careful examination. In fact, as E.R. Dodds and W. Theiler have observed (cf. bibl.), all Plotinus's thought reveals a vast synthesis of neopythagorean, peripatetic and Stoic, as well as Platonic and middle Platonic, doctrines. The same type of s. can also be observed in later exponents of neoplatonism such as Porphyry, Iamblichus, *Julian the Apostate, *Proclus and Damascius, though in each of these authors it may take on different nuances, being gradually enriched with new contributions such as, e.g., the cult of the *sun, that of the Chaldaean *Oracles, *magic and *theurgy.

In *patristics, s. finds its theoretical formulation (on the level of theological interpretation of history) and practical application (on the doctrinal level) in *Justin martyr and esp. in *Clement of Alexandria: ascribing the origin of Greek *philosophy to the inspiring or "seminal" action of the *logos* and the use by the Greeks of the Mosaic books (cf. *Hellenism and Christianity), they feel it legitimate to accept doctrines of various Greek philosophical schools into their thought and thus to bring about a type of s. wholly analogous to that of middle Platonism and Philo (the influence of Philo's s. on Clement is direct and marked, while it cannot be proved in the case of Justin, who probably did not know Philo and depended rather on the middle Platonism of the 2nd c. AD (cf. H. Chadwick, *Early Christian Thought and the classical Tradition*, Oxford 1966, 4). Acc. to the evidence of *Gregory the Thaumaturge (*Thanks to Origen* XIII, 150-153), Origen the Christian made his students study the works of many Greek philosophers, excluding the atheists (cf. Hellenism and Christianity): in this way he followed the example of his teacher Ammonius Saccas and of Plotinus. In Origen's thought - as in that of *Basil the Great, *Gregory of Nyssa and *Gregory Nazianzen - we can observe a synthesis of *Platonic, *Stoic and *Aristotelian-peripatetic doctrines.

R. Reitzenstein, *Poimandres*, Leipzig 1904; id., *Die hellenistischen Mysterienreligionen*, Darmstadt 1956 (reprint of the 1927 edition); id., *Studien zum antiken Synkretismus aus Iran und Griechenland*, Leipzig 1926; id., *Die Göttin Psyche in der hellenistischen und frühchristlichen Literatur*, Heidelberg 1917; H. Strache, *Der Eklektizismus des Antiochos von Askalon*, Berlin 1921; E.R. Dodds, The Parmenides of Plato and the Origin of the neoplatonic One, *CQ* 22 (1928) 129-142 (on Plotinus's dependence on neopythagoreanism in the interpretation of the Parmenides); id., Tradition and Personal Achievement in the Philosophy of

Plotinus, *JRS* 50 (1960) 2 (on the "syncretistic" character of Plotinian philosophy); F. Cumont, *Les religions orientales dans le paganisme romain*, Paris ⁴1929 (It. tr. Florence 1913); W. Theiler, Plotin und die antike Philosophie, *MH* 1 (1944) 209-225 (on the syncretistic character of Plotinian philosophy); id., *Die Vorbereitung des Neuplatonismus*, Berlin 1930 (on Plotinus's dependence particularly on Poseidonius); id., Philo von Alexandreia und der Beginn des Kaiserzeitlichen Platonismus, in *Parusia, Festgabe J. Hirschberger*, Frankfurt 1965, 199-218; H. Chadwick, *Early Christian Thought and the classical Tradition*, Oxford 1966, 5-7; S. Lilla, *Clement of Alexandria*, Oxford 1971 (on the substantial agreement between the s. of Clement and Philo and that of middle Platonism); A. Graeser, *Plotinus and the Stoics*, Leiden 1972; M. Simon, *Hercule et le Christianisme*, Paris 1955; *Paganisme, Judaïsme, Christianisme* (Mél. Marcel Simon), Paris 1978.

S. Lilla

SYNESIUS of Cyrene. Born at Cyrene in *Libya c. 370-75 of a noble *pagan family; studied at *Alexandria, where the celebrated Hypatia initiated him into *neoplatonism. A journey to Greece left him disillusioned with the cultural decadence of *Athens (*Ep.* 136). In 399 his compatriots sent him on a mission to the emp. *Arcadius at *Constantinople to try to obtain a reduction of taxes for the region of Pentapolis, impoverished by barbarian raids, earthquakes and famines.

His *Discourse on kingship*, pronounced at Constantinople in 400, is a courageous address to the emperor in which he criticizes the luxury and court life which leads to losing touch with reality. His treatise *On providence* was begun at Constantinople and finished in *Egypt: in allegorical form it describes what he saw at Constantinople, in terms of relations and conflict between *virtue and vice.

Having obtained the tax reduction, S. returned to Cyrene in 402. Between then and 405/6 he contracted a Christian *marriage at Alexandria in the presence of patriarch *Theophilus (*Ep.* 105) and defended the city from the raids of the Macheti. We do not know when he received *baptism. In 410 the *clergy and people elected him bishop of Cyrene and metropolitan of the Ptolemaïd. S. hesitated long and seriously before accepting the episcopal *ministry: his *Epp.* 105 and 110 are a sincere explanation to Theophilus of the reasons for his reservations: a fear of conflict between intellectual truth and the religion of the "vulgar"; some elements of neoplatonic philosophy that he felt unable to renounce; a certain attachment to the life of a country gentleman. Having by now three *children, he also asked to remain married. Theophilus consecrated him bishop. These were years of severe trial for the whole population on account of the continuous bloody invasions of the barbarians, as his two discourses (*Catastases*) attest, esp. the second one, of 412. We have no writings or news of him after 413.

As well as S.'s surviving works (*Discourse on kingship*, the treatise *On providence, Dion, In praise of baldness*; the treatise *On dreams, The gift*, the *Letters* (156), the *Hymns*, the two *Discourses*, two fragments of *Homilies*), the existence of other juvenile poetic works inspired by Anacreontism has been supposed (cf. E. Cavalcanti, *Studi Eunomiani*, p. 106, n. 2). Of the surviving writings, those belonging to the Christian period comprise: not many of the 156 *letters; the two *Discourses*; the two frr. of *Homilies and nine *Hymns, the tenth being certainly spurious. Of the nine *Hymns*, the first five (in the numbering of modern editors) are theological; the other four are of *prayer or celebration.

The Synesian thought of the *Hymns* - markedly of the first five - is extremely complex. In the past, critics confined themselves to calling it "a web of echoes" and it was studied mostly in terms of these echoes, taking for granted that there was little Christian in it. At worst, the idea of S. as a bishop was found repugnant. More recent studies have reproposed the problem of S.'s thought in relation to the neoplatonist sources common to him and to *Marius Victorinus (P. Hadot, *Porphyre et Victorinus*, I, 461-474). Other researches allow us to establish a relationship between events of church life which S. had to face during his episcopate and - on the level of thought - the more complex subject-matter of the *Hymns. Epp.* 142 and esp. 141 attest the presence of late Eunomian propaganda in Cyrenaica, linked to a tendency in the imperial court and supported on the ground by the prefect Andronicus, with whom S. had severe clashes in late 411 and 412. The hermit *Isidore of Pelusium supported S. in this struggle. The clergy were divided, and local interests were tinged with tones of late *Arian opposition (cf. *Ep.* 141). The trinitarian subject-matter of the *Hymns* is not unrelated to anti-Eunomian problems; in particular hymns I, II - parallel in structure - and V allow the emergence of a theological elaboration that is anti-technicalistic and anti-nominalistic. S.'s treatment of the *Trinity contains elements of Origenian Alexandrian tradition, esp. as regards generation in the trinitarian context, as well as his *theology of the *Holy Spirit, on which he brings together his thought on the divine nature and on relations within the *Trinity (cf. *Hymn* III).

CPG 3, 5630-5640 where, for each work, the main editions and studies are given.

In view of the highly complex interweaving of problems and consequently of studies touching on S., the reader is directed to the synthesis worked out in E. Cavalcanti, *Studi Eunomiani*, OCA 202, Rome 1976, 106-128. Add to this the most recent edition of the letters: Synesii Cyrenensis, *Epistolae*, rec. A. Garzya, Rome 1979; cf. Altaner, 293 f.; Quasten II, 108 ff. S.'s works are in PG 66, 1021-1756; Eng. tr.: A. Fitzgerald, 3 vols., London 1926-1930; Fr. tr: H. Druon, Paris 1878; It. tr. *Sulla provvidenza*, S. Nicolosi, Palermo 1959; *Sul regno*, A. Garzya, Naples 1973; *Inni*, (ed. and tr.) A. Dell'Era, Rome 1968. Studies: Grützmacher, *Synesios von Kyrene, ein Charakterbild aus dem Untergang des Hellenentums*, Leipzig 1913; G. Bettini, *L'attività pubblica di Sinesio di Cirene*, Udine 1938; J.C. Pando, *The Life and Times of Synesius of Cyrene as Revealed in his Work*, Washington 1940; C. Bizzochi, La tradizione storica della consacrazione episcopale di Sinesio di Cirene, *Gregorianum* 25 (1944) 130-170; id., La irregolarità della consacrazione di Sinesio come congettura?, *Gregorianum* 27 (1946) 261-299; H.I. Marrou, La "conversion" de Sinésius, *REG* 65 (1952) 474-484; id., Synesius of Cyrene and Alexandrian neoplatonism, in (ed.) A. Momigliano, *The Conflict between Paganism and Christianity in the Fourth Century*, Oxford 1963, 126-150; A.J. Bregman, Synesius of Cyrene. Early Life and Conversion to Philosophy, *California Studies in Class. Antiquity* 7 (1974) 55-88; A. Garzya, *Storia e interpretazione di testi bizantini. Saggi e ricerche*, London 1974 (repr. of 9 earlier articles on S.); G. Gellie, Repetitions on the Letters of Synesius, *Antichthon* 13 (1979) 70-102; A.J. Bregman, *Synesius of Cyrene, Philosopher-Bishop*, Berkeley 1982; S. Vollenweider, *Neuplatonische und christliche Theologie bei Synesios von Kyrene*, Göttingen 1985; J.H.W.G. Liebeschutz, Why did Synesius become bishop of Ptolemais?, *Byzantion* 56 (1986) 180-195; D. Roques, *Synésios de Cyrène et la Cyrenaïque du Bas-Empire*, Paris 1987 (full bibl.) (rev. W.H.C. Frend, *JbAC* 32 [1989] 203-206).

E. Cavalcanti

SYNOUSIASTS. From σύν (= with, together) and οὐσία, *ousia (= essence, substance). Name given to the *Apollinarists, since they asserted that, in Christ, human substance and divine substance were united so that a single substance resulted from them.

M. Simonetti

SYRACUSE

I. The city - II. Archaeology.

I. The city. S. (*Syracusae*) was a Greek colony founded in 734 BC by Archias of Corinth, who subdued the native inhabitants, the Siculi. It developed considerably under Gelon (*c.*540-478); its influence grew all through the 5th c., reaching its apogee under Dionysius the Elder (405-367), an able statesman, and Timoleon of Corinth (343-337/6). S. passed under Roman rule in 212; in 21 BC *Augustus sent a colony there, since it was becoming depopulated; it became the seat of Roman authority in Sicily. St *Paul spent three days there on his way to *Rome (Acts 28, 12). A Christian presence seems well-established in the 3rd c.: the evidence of various *cemeteries suggests the existence of a large community. Sicily had its *lapsi during the *persecution of *Decius (cf. Cypr., *Ep.* 30, 5: CSEL 3, 2, p. 553). One Marcian, a 1st-c. *martyr, is claimed as the first bishop of S.; but the sources are too late to be trustworthy. The first historical bishop is Chrestus (the name appears in various forms in codices; cf. CCL 148, index): in 314 *Constantine invited him - in a *letter preserved by *Eusebius - to go to *Arles with two presbyters to settle the question of *Donatism, and authorized him to use the *cursus publicus* (Euseb., *HE* X, 5, 21-24). Chrestus was indeed present at the *council of Arles with his deacon Florus and was the first to sign the *acts.

S. had its martyrs, but information on them is uncertain; however, *Lucy died in *Diocletian's persecution. A cemetery developed round her tomb, and her cult spread beyond Sicily. A *monastery dedicated to her existed at S. in the 6th c. (Greg. Gt, *Ep.* VIII, 36: MGH, *Epist.* I, 1, 484).

To find a name that is both historically certain and of some importance, we must wait until Eulalius (late 5th-early 6th c.), present at the council of *Rome in 502 (under pope *Symmachus) at which he took the floor to condemn a legislative text of *Odoacer's prefect Basil, issued some 20 years before, forbidding the election of a successor to pope *Simplicius (†483) without the king's consent and prohibiting the alienation of church *property (Mansi 8, 266). Eulalius also took part in the council of 503 (Mansi 8, 299). *Fulgentius of Ruspe, fleeing from the *Vandal persecution in *Africa and intending to go to the *Thebaïd, stopped at S. and spent several months there as a guest of Eulalius who, also a lover of monastic life, dissuaded him from going to *Egypt since the Christians there were not in *communion with Rome (PL 65, 128-130, where there is a brief *biography of Eulalius). Eulalius also seems to be the recipient of a *letter of *Ennodius of Pavia (3, 18: MGH, *Auct. ant.* 7, 116).

Sicily, and consequently S., passed successively under the rule of the Vandals, Odoacer and the *Goths. In 535 it was reconquered by *Belisarius and came under the rule of *Justinian, with S. as capital: barbarian rule was replaced by *Byzantine exploitation. *Constans II established his residence at S. from 663 to 668 and it became the capital city of the empire. In 878 it passed under Muslim rule.

In the time of pope *Gregory the Great two characters of some importance were bishops of S.: *Maximian and John. Maximian, a native of Sicily, had been *abbat of St Andrew's monastery at Rome, founded by Gregory before he became *pope, and Gregory's longstanding personal *friend. He was bishop of S. from 591 to 594. Gregory allowed him the pallium and made him his representative for lesser church affairs, with the job of supervising discipline all over Sicily. On his death Gregory wrote words of high praise: "we have no other Maximian" (*Ep.* V, 54: MGH, *Epist.* I, 1, 358).

Maximian was succeeded in 595 by John, *archdeacon of the church of Catania (*Catana*), suggested by Gregory himself (*Ep.* VI, 20: MGH, *Epist.* I, 1, 302) who allowed him the pallium and wrote him numerous letters, giving him various tasks administering the property that the Roman church possessed in the area of S. and authorizing him to watch over discipline in Sicily and settle various disputes. John died in 609.

PWK Suppl. 8 (1973) 815-836; *EC* 11, 718-726; *EI* 31, 869-878; *EAA* 7, 329-339; CE 13, 892-893; C. Barreca, *I primordi del cristianesimo in Siracusa*, Rome 1935; L. Giuliano, *Storia di Siracusa antica*, Milan ³1936; A. Amore, *S. Marciano di Siracusa*, Vatican City 1958; A. Pincherle, Sulle origini del cristianesimo in Sicilia, *Kokalos* 10/11 (1964-1965) 547-564.

A. Di Berardino

II. *Archaeology. Among the monumental evidence of early Christianity in Sicily - and not only there - S. has an outstanding position, due to the richness and quality of its surviving monuments, mostly funerary. These can be divided into four great groups by their technical and iconographical characteristics and their different distances from the ancient city.

A first group consists of the four great catacombs - S. Lucia, S. Maria di Gesù, Vigna Cassia and S. Giovanni - distributed from E to W along the S slope of the middle and western Acradina. The initial nucleus of the S. Lucia complex, which predates the religious peace, developed around *Lucy's tomb in four regions and on three levels, ranging from the typically archaic in region A to the use of hydraulic systems in B, to the incorporation of classical rooms in C and D. The *cemetery's long frequentation is attested by two Byzantine oratories and the presence of early mediaeval sculptural fragments.

Further W, the catacomb of S. Maria di Gesù, smaller and communicating through the arm of an old aqueduct with the contiguous one of Vigna Cassia, consists essentially of two axes at right angles, in which we can distinguish two chronological phases, the later of which (post-Constantinian) saw the use of polysarcous arcosolia (cf. *Italy: Sicily). In the cemetery of Vigna Cassia - developed around a rectangular room now open to the sky and of disputed date and function - a vast zone resulting from the multiplication of nuclei over a period extending from the 3rd c. through the 4th and including the use of pre-existent hydraulic systems contrasts with a sector postdating the peace of the church and characterized by great polysarcous arcosolia containing more than 20 burials, outstanding among which is the well-known tomb of Marcia, decorated with *paintings. A substantially rectangular scheme characterizes the excavation of the monumental catacomb of S. Giovanni, where the particular nature of the rock has allowed the opening of a long, wide E-W *decumanus* dividing the cemeterial network into two great more-or-less equal halves into which, mainly in the N sector, run *cardines* intersected in turn by a minor *decumanus*. In the S sector, stretches of shorter galleries gravitate round circular or quadrangular rooms. Among these rooms, and two more in the N sector, are those bearing the names of Adelphia (whence comes the famous Constantinian-era *sarcophagus), the Seven Virgins, Antiochia, Eusebius and the Sacred Ampulla. The cemetery's architectural structure, the residual frescos, the sarcophagus of Adelphia, all suggest an origin in the first decades of the 4th c.; a chronology confirmed by the earliest inscription found, of 349, whilst *epigraphy attests that it was used until at least 452. At a short distance from the S. Giovanni complex, in the Villa Landolina, opens the catacomb of Predio Maltese, held in the past to be part of S. Giovanni, but more recently recognized as autonomous. In the same area, the three-apsed crypt of S. Marciano, part of a group of classical monuments, is related to the figure of the martyr Marcian, presumed first bishop of S., and shows signs of use up to and beyond the Norman era.

The second group of cemeterial systems is that of the hypogea found between Predio Adorno-Avolio, the Capuchin convent and the coast of Pietralonga in E Acradina. These (the Führer, Attanasio, Brianciamore, Capuchin, Belloni-Monteforte, Troia, Trigilia, Mauceri, Russo, Bonaiuto, Fortuna and Grotticelli catacombs), pre-Constantinian in origin, are characterized by less crowded burials, which however retain the usual typological characteristics. These hypogea seem to have belonged to small groups, perhaps even *families, who, as finds, esp. pottery, show, expressed themselves with considerable freedom outside the control of the community.

The third group - rather later in date - comprises burials reusing or added to classical monuments. Opening along the street of tombs and in the rocky part of Colle Temenite surrounding the Greek theatre are numerous cruciform-plan cubiculi with arcosolia, including the one installed in the Nymphaeum and later transformed into a *church building.

We must also consider as part of the city a group of suburban hypogea, from those of Riuzzo and Manomozza at Priolo to the catacombs of Molinello in the anchorage of Augusta near the station of Megara Iblea. For the most part these are divided into cubiculi and sepulchral rooms centred on grandiose "baldacchino" tombs, being thus differentiated - though chronologically coeval - from the city proper, from which this typology is absent.

The primitive cathedral is also cemeterial in origin. Over the crypt of S. Marciano (mentioned above) rise the remains of a grandiose basilica, now dedicated to St John, but thought to be originally dedicated to the martyr bishop, with three columned aisles, semicircular apse and presbytery perhaps raised, presumably roofed over. The building has been assigned to the 6th c. and was the episcopal church until the 7th, when that was transferred to Ortygia inside the city, to the new church adapted from the temple of Athena and dedicated to the Virgin. This three-aisled church, created by reversing the orientation and opening arcades in the walls of the *cella*, shows, despite much rebuilding, esp. in the N apsidiole, evident traces of the original building.

The reuse of existing buildings was common at S., as is shown by, e.g., the church contained in the *nàos* of the temple of Diana, also in Ortygia, or the two smaller churches, one inserted in a *baths connected to the amphitheatre and today called Piscina di S. Nicolò, from the church above it, the other a hypogeum found beneath the city's railway station.

The urban architectural picture is completed by the church of S. Pietro *intra moenia*, again in Ortygia, with longitudinal plan of three aisles with square pillars, semicircular apse and barrel-vaulted roof; that of S. Martino, also with three pillared aisles and semicircular apse, is characterized by an exaggerated longitudinal scheme.

In the suburbs, the church of S. Pietro ad Baias, with three pillared aisles and a trichorate presbytery (second half of 5th c.), has the characteristic presbytery ground-plan present in other Sicilian churches (cf. *Italy), while outside S. the church of S. Focà di Priolo has, like that of S. Pietro just mentioned, the peculiarity that the perimeter walls are pierced by arches corresponding to those of the aisles, interpreted in the past as continuous openings subsequently blocked up. **[Fig: 293]**

Bibliography in: P. Testini, *Archeologia cristiana*, Bari ²1980, 272 ff., 292 f., and esp. 809 f.; O. Garana, *Le catacombe siciliane e i loro martiri*, Palermo 1961; G. Agnello, Recenti scoperte di monumenti paleocristiani nel siracusano, in *Akten des VII. Intern. Kongresses f. christliche Archäologie*, Vatican City-Berlin 1969, 309-326; id., Nuovi ritrovamenti nella catacomba di S. Maria a Siracusa, *RAC* 49 (1973) 7-31; S.L. Agnello, Siracusa sotterranea: Nuovi contributi, in *Atti del III Congresso Naz. Archeol. Cristiana*, Trieste 1974, 467-473; id., Chiese siracusane del VI secolo, *CCAB* 27 (1980) 13-26; P. Testini, La cultura artistica in Italia nella antichità, in *La cultura in Italia fra Tardo Antico e Alto Medioevo*, Rome 1981, 812-815 (cf. 810).

L. Pani Ermini - R. Giordani

SYRIA

I. The provinces; Christian origins - II. Archaeology.

I. The provinces; Christian origins. Pompey conquered S. in 65-62 BC and made it a Roman province, leaving various independent cities and numerous native princes in some way subject to Rome. Its ill-defined borders went from the Taurus to Arabia and *Egypt (Josephus, *Ant. Iud.* 12, 317-321); its E border was the Euphrates. S. was a mosaic of different political systems which continued into the 1st c. AD, while a process of annexing and transforming some territories into imperial properties went on alongside them. The borders of S. underwent radical changes in the course of time; at the end of the 1st c. they went from the Amanus chain with Commagene, annexed in AD 72 with Samosata, along the Euphrates to beyond Mt Carmel and all-'Hauran. We do not know when *Palmyra was annexed. After the Jewish war of AD 70, the province of Judaea was detached: from the time of *Hadrian it was officially called Syria Palaestina and later just Palaestina (*Palestine), with capital at *Caesarea. In 106 *Trajan turned the vassal kingdom of the Nabataeans into the Roman province of *Arabia, capital *Bostra. *Septimius Severus divided S. into two provinces: Syria Coele (or Coele-Syria) to N, capital *Antioch, after a brief interlude with Laodicea, and Syria Phoenice to S, capital Berytus (Beirut). During the 4th c. there was a further division (exact date unknown) into four provinces: Augusta Euphratensis (Commagene, Cyrrhestica, the cities along the Euphrates as far as Circesium), Augusta Libanensis (the territory of Phoenicia beyond Lebanon and Hermon), Phoenice (on the coast) and Syria Coele (region of Antioch). Hellenized in the cities, S. kept its local languages in the villages and countryside. The Roman period saw great development, both commercial and cultural; the territory of S. included great cities like *Antioch, *Seleucia, *Apamea, *Laodicea, *Damascus, Ptolemaïs, *Tyre, Sidon, Berytus, *Emesa (Homs), *Palmyra, Samosata, etc. In the mid 3rd c. it was invaded by the *Sassanids under Shapur I who occupied Antioch, destroyed *Dura Europos and deported much of the population, including bishop *Demetrian of Antioch (cf. BS 4, 550 f.).

NT writers often name S. and understand it in the sense of the territory of the Roman province (Mt 4, 24; Lk 2, 2; Acts 15, 23; 18, 18; 20, 3; Gal 1, 21). Christianity arrived there very early, from the first Christian missions (e.g. Tyre, Acts 21, 4-7; Ptolemaïs, Acts 21, 7; Damascus, Acts 9, 2-22; Sidon, Acts 27, 3; Antioch, Acts 11, 19 ff., etc.). In some centres a Christian presence is documented only from the 2nd or 3rd c. At any rate, in the 3rd c. S. was one of the most Christianized regions of the empire.

Although in antiquity the Euphrates was the geographical and political border of S., it was not its ethnic, cultural or linguistic border; in these respects S. went beyond its natural borders. Culturally and linguistically, Syriac literature embraces all writers who expressed themselves in *Syriac, even if they came from *Persia, *Adiabene, *Palestine or elsewhere. For this reason we use the expression Syrian Fathers in a broad sense, to mean not just authors from S. proper but all those who, regardless of geographical area, wrote in Syriac, and particularly writers from the territory between the Tigris and the Euphrates, i.e. Mesopotamia. This region saw the first flowering of Syriac Christian literature.

The emp. *Trajan was the first to extend direct Roman rule beyond the Euphrates: but his new provinces of Mesopotamia and Assyria were abandoned by his successor *Hadrian. Lucius Verus and *Marcus Aurelius recreated the province of Mesopotamia, which was always a theatre of war, first with the Parthians and then with the *Sassanids. *Septimius Severus enlarged its territory and created the provinces of Osrhoene, capital *Edessa, and E Mesopotamia, capital *Nisibis, which city was transferred to the Sassanids in 365. Modern Syria corresponds broadly to its ancient counterpart, though today part of the N belongs to Turkey and part of the S to Lebanon, with a small extension in Mesopotamia.

The early spread of Christianity in Mesopotamia is a much-debated historical argument since the earliest sources and traditions, which ascribe its *evangelization to Thaddaeus (*Addai) and his disciples Aggai and Mari, are not fully guaranteed. But a series of documents attests that Christianity had some strength at the end of the 2nd c., a strength that must have had quite distant roots going back to *Judaeo-Christian missionaries. In Mesopotamia from the 3rd c. the Syriac language, derived from the Aramaic spoken at Edessa, acquired more and more literary value. After the council of *Ephesus (431), the bishops of Mesopotamia, supporters of *Nestorius, set up the Nestorian church, which they detached from the patriarchate of *Antioch, establishing the see of their *catholicós at *Seleucia-Ctesiphon. Also after the council of Ephesus, the *monophysites (later called *Jacobites) became predominant in S. and N Mesopotamia. In both S. and Mesopotamia, *monasticism followed a development of its own, moulding a form of Christianity quite different from that of Western lands.

PWK 23/2 (1932) 1622-1727; EC 9, 737-740; EAA 6, 576-578; RGG³ 6, 571-581; E. Honigmann, Evêques et évêchés monophysites d'Asie Antérieure au VIᵉ s., CSCO 127, subs. 2, Louvain 1951; id., Le couvent de Bar Sauma et le patriarcat jacobite d'Antioche et de Syrie, CSCO 146, subs. 7, Louvain 1954; A. Vööbus, A History of Asceticism in the Syrian Orient, CSCO 184, subs. 14, Louvain 1958; W. Hage, Die syrisch-Jacobitische Kirche in frühislamischer Zeit, Wiesbaden 1966; J. Fiey, Jalons pour une histoire de l'Eglise en Iraq, CSCO 310, subs. 36, Louvain 1970; J.B. Segal, Edessa, the "Blessed City", Oxford 1970; A.H.M. Jones, The Cities of Eastern Roman Provinces, Oxford ²1971, 226-294; R. Murray, Symbols of Church and Kingdom, Cambridge 1975 (with intr. on Christian origins and bibl.); B. Lifshitz, Etudes sur l'histoire de la province romaine de Syrie, ANRW II.8, 3-30; J.P. Rey-Coquais, Syrie romaine de Pompée à Dioclétien, JRS 68 (1978) 44-73; E. Perettoi, Il problema degli inizi del cristianesimo in Siria, Augustinianum 19 (1979) 197-214; U. Bianchi, Questioni storico-religiose relative al cristianesimo in Siria nei secoli II-IV, ibid. 19 (1979) 41-52; W.H.C. Frend, The Rise of the Monophysite Movement, Cambridge ²1979; id., The Monks and the Survival of the East Roman Empire in the Fifth Century, P&P 54 (1972) 3-24; P. Brown, Town, Village and the Holy Man: the case of Syria, in Society and the Holy in Late Antiquity, New York 1982, 153-165; S.A. Harvey, Asceticism and Society in Crisis: John of Ephesus and the Lives of the Eastern Saints, Berkeley 1990.

A. Di Berardino

II. *Archaeology. In the first Christian centuries S. was part of the patriarchate of *Antioch and extended from the Gulf of Antalya to the Tigris, from the Taurus mountains to the E shore of the Dead Sea. This latter part formed *Arabia. The regions of S. differed in language and customs: this brief review will begin from the W, following the scheme of R. Devreesse (Patriarcat d'Antioche, Paris 1945).

1. Isauria, now in Turkey, extended along the Mediterranean coast opposite *Cyprus. Its main city was Kalykadnian Seleucia, now Silifke. Remains of the famous sanctuary of St *Thecla, N of the city, have been discovered at Merianlik. The *church, 80 m. long, was built over a grotto said to have been inhabited by the saint. It had a semicircular apse, with narthex and portico to S, and was richly decorated. On the site are remains of the church built by the emp. *Zeno in the 5th c., with rectangular plan and dome; and the ruins of a third basilical church (Guyer-Herzfeld, Mon. Asiae min. ant., II; Keil-Wilhelm, Mon. Asiae min. ant., III). Four churches have been found near Kaulidivan.

2. Diocaesarea, now Uzundia-Burdj, has remains of the church built on the site of the great temple of Jupiter, and of two other basilicas, one cemeterial, built over the tomb of the *martyr Lucius, and another dedicated to St *Stephen in the 4th c. (Keil-Wilhelm, Mon. Asiae min. ant., III, 44-79). Other ruins of churches have been found at Olba, modern Ura, with three buildings (Keil-Wilhelm, Mon. Asiae min. ant., III, 86).

S. proper is divided into two parts: prima to N, and secunda to S. In the N part we find the main monuments that have made the religious architecture of S. famous, esp. in the three mountains: Djebel Barisc, Djebel Halaka, Djebel Sem'an with the well-known sanctuary of St *Simeon Stylites. G. Tchalenko's general and particular study (cf. bibl.) and other recent studies by the Franciscans I. Peña, P. Castellana and R. Fernández (cf. bibl.) dispense us from the duty of describing these places.

3. Beroea, now Aleppo, was one of the main dioceses. Traces of the quadrilobate church of St Helena have been found in the mosque called Hallawiya. The city also had a *Judaeo-Christian community: there St *Jerome was able to consult the Aramaic Gospel called "of the Hebrews" (Bagatti, Church from Circumcision, 84).

4. Anasartha, now Khanasir, 60 km. S-E of Aleppo, still keeps its old city wall. Two great architraves survive, one near the great 5th-c. church and one further to the S-E. The church is the martyrium built by Silvanus dux Arabiae and on the architrave is sculpted the image of the Virgin and Child with two *angels carrying her up to heaven (Lassus, Sanctuaires, 252, 294). Many Christian inscriptions have been found in the region (Devreesse, Patriarcat d'Antioche, 164-167).

5. Epiphania, now Hama, preserves in the Great Mosque the Qubbat el-Khazna, made with Byzantine columns and capitals from some church, prob. that built on the *pagan temple, later turned into a mosque, the Giame el-Kebir. It was inhabited by a recluse (Peña-Castellana-Fernández, Reclus, 293-295). Excavations have uncovered vestiges of the Byzantine period with a church and a "refuge", dedicated to the Virgin and Sts *Cosmas and Damian. Inscriptions also mention the sanctuary of Sts *Sergius and Bacchus with liturgical objects offered by the *faithful (Jalabert-Mouterde, Inscriptions, V, 15-38). In this region too, some evidence of ancient Christianity remains (Devreesse, Patriarcat d'Antioche, 184-191). A bas-relief found E of Hama shows St *Simeon Stylites with a disciple bringing food (Peña-Castellana-Fernández, Stylites, 181-183).

6. Phoenice too was divided into two parts: prima along the Mediterranean, secunda further E.

7. Sidon, now Saida, preserves a church with a 4th-c. *mosaic floor with geometrical motifs. In a cistern were found several Christian *lamps with *crosses and the image of *David with his sling, now in the Museum of the Flagellation at Jerusalem.

8. Beirut. Studies of the topography of this city have ascertained that the church called Anastasis, or cathedral, rose near the present Greek Orthodox church of St George. Nearby was the law school. But it is not certain that the wall under the mediaeval church of St John the Baptist belonged to a Byzantine church: it seems more probable that it was part of the nearby *baths. An inscription of Ps 28, 3 on a reused stone in the same church suggests a Byzantine baptistery near by. A pavement mosaic and remains of an apse have been found E of the *Maronite church of St George. In the city museum are many mosaics from the region, including that from Jenat showing a youthful Good *Shepherd with many animals.

9. At Byblos, now Gebail, excavations at the foot of the acropolis have yielded the well-known bas-relief of *Orpheus surrounded by animals, some exotic, recalling the *paintings in the catacomb of Domitilla. The 1961 dig also uncovered early Christian remains.

10. Antarados, now Tartus. Reused Corinthian capitals in the N wall of the mediaeval church may be part of an older chapel. There is also a house inscription in marble with two crosses at the beginning and end of the text.

11. Heliopolis, now Baalbek, had a church in the atrium of the great temple of Jupiter, with an apse polygonal outside and circular inside. Its orientation to W, following that of the temple, was reversed in the 6th c. The church, quite well preserved, was destroyed in 1933 to investigate the original structure of the pagan temple; so today we can only see traces of the building. The circular temple of Venus, converted to the church of St Barbara, preserves a cross on the inside of the wall. The Christians say that the Great Mosque also rests on a church, and we can indeed see remains of the baptistery backing onto an arcade in the N part. Christian funerary remains are known in various grottos.

12. Salamias, now Selemye. We know a 5th-c. church dedicated to St Sergius, and a 7th-c. one dedicated to the Virgin (Devreesse, *Patriarcat d'Antioche*, 207). Inscriptions also mention an "asylum" of the martyr Cyriac and various deceased people. A wooden *encolpion bears the figure of Christ (Jalabert-Mouterde, *Inscriptions*, V, 221-231).

The E provinces of S. comprise: Euphratensis, with cities W of the Euphrates; Osrhoene, with cities E of the Euphrates; and Mesopotamia, which went as far as the Tigris, on the borders of the empire. In Euphratensis the main city, Hierapolis, now Membdj, keeps nothing of its old buildings because the stones were used for new ones. A bronze *cross or *encolpion has been found, with a figure seemingly of St *Chrysogonus, martyred under *Diocletian. At el-Bire, 18 km. S-E of Hierapolis, is a mosaic with flowers and birds, and an inscription recording an oratory of martyrs.

13. Cyrrhus is today a field of ruins. In the S-W of what was the town are traces of buildings ascribed to a pagan temple succeeded by a Christian chapel, perhaps that of the martyrs *Cosmas and Damian.

14. Sergiopolis, now Resafa, is one of the finest ruins in the Syrian *desert. Inside its great rectangular wall, often three metres high with towers, are three longitudinal-plan churches, of which one, to S-E, is the sanctuary of St *Sergius, and a fourth with central plan, called "the cathedral". The sanctuary of St Sergius is imposing, its apsidal conch well preserved. The altar is above a grotto which held the saint's body. Access to the crypt is by a double stairway (Lassus, *Sanctuaires*, XXIX, 39, 210, 300).

15. Carrhes, now Harran in Turkey, is a small village with remains of the early Christian period.

Some dioceses of Mesopotamia are now in Turkey, such as Amid, now Diyarbekr. In the citadel is a building thought to have been a Nestorian church, built using local techniques. It has two distinct parts and various rebuildings. Its date of construction is disputed, but in all probability pre-Islamic (Monneret de Villard, *Chiese della Mesopotamia*, 31). There is also a central-plan church dedicated to the Virgin (Lassus, *Sanctuaires*, XXIX, 154).

16. Ctesiphon (*Seleucia-Ctesiphon), now Takh i Kesra, is in Iran. On the Tigris, 40 km. S of Baghdad, it was the capital of the Persian empire and had several churches. Only two superimposed Nestorian ones remain from the patristic period. The lower church, longitudinal, has thick walls and columns almost touching them, the upper one has pillars and three rectangular cells like apses, the central one of which has three rectangular niches. There are many doors, all on the long sides (Monneret de Villard, *Chiese della Mesopotamia*, 9-31).

17. Al-Hirah is in Iraq. Two Nestorian churches have been excavated, though not completely. Both have thick walls with columns close to them, as at Ctesiphon. One, called XI, has a squared apse on the E with an altar at the end and two side rooms; in the middle is the bema with semicircular sedilia. The other church, called V, simpler but not completely excavated, is oriented like the other and, like it, has columns close to the walls (Monneret de Villard, *Chiese della Mesopotamia*, 32-44); we do not yet know what saints (there were several) the two churches were dedicated to. We know that the Christians were transferred to Mesopotamia on account of the wars, so it is no wonder if we find *church buildings constructed in forms proper to the host country. In this region, the *bema* probably took on a different function from what it had in S., which was to imitate liturgically the functions it had at the Holy Sepulchre in *Jerusalem (Blomme, *La liturgie*, SBF 29 [1979] 221-237). To the patristic period under the *Sassanids belong, among other things, some seals with OT scenes like the Sacrifice of *Abraham, or *Daniel in the lions' den, and NT scenes like *Mary with the Child Jesus, the Visitation and Jesus' *Entry into Jerusalem.

An example of a *monophysite church, closer to S. than the Nestorian ones, may be mentioned at Qasr Serij, 60 km. N-W of Mosul. It had a portico developed on three sides and was dedicated to St *Sergius. **[Figs: 294-300]**

S. Guyer-E. Herzfeld, *Meriamlik und Korykos. Zwei christliche Ruinenstätten des rauhen Kilikiens. Monumenta Asiae Minoris antiqua*, II, Manchester 1930; J. Keil-A. Wilhelm, *Denkmäler aus dem rauhen Kilikien. Monumenta Asiae minoris Antiqua*, III, Manchester 1931; U. Monneret de Villard, *Le chiese della Mesopotamia*, Rome 1940; R. Devreesse, *Le Patriarcat d'Antioche depuis la paix de l'Eglise jusqu'à la conquête arabe*, Paris 1945; J. Lassus, *Sanctuaires chrétiens de Syrie*, Paris 1947; G. Tchalenko, *Villages antiques de la Syrie du Nord*, I-III, Paris 1953-1958; L. Jalabert-R. Mouterde, *Inscriptions grecques et latines de la Syrie*, V, Paris 1959; B. Bagatti, *The Church from the Circumcision*, Jerusalem 1971; I. Peña-P. Castellana-R. Fernández, *Les stylites syriens*, Jerusalem 1975; iid., *Les reclus syriens*, Jerusalem 1980; J.A. Lerner, *Christian Seals in the Sassanian Period*, Nederlands Inst. te Istanbul 1977; V. Blomme, La liturgie de Jérusalem et son influence sur les Eglises nestorienne et syrienne, SBF 29 (1979) 221-237; I. Peña-P. Castellana-R. Fernández, *Les Cénobites Syriens*, Jerusalem 1983; J.H.W.G. Liebeschutz, *Antioch: City and Imperial Administration in the Later Roman Empire*, Oxford 1972; id., The defences of Syria in the sixth century, *Acten des X internat. Limes Conferenz*, Cologne-Bonn 1977, 487-499.

B. Bagatti

SYRIAC

I. Preamble - II. Language - III. Literature - IV. Liturgy - V. Syriac studies.

I. Preamble. Because of its wide geographical diffusion and lasting use, but also as a language very close to the dialect used by Jesus of Nazareth to proclaim the coming of the Kingdom of God, S. is the most important language of the Christian *East. It was the mother-tongue of the first orthodox communities and then of the *Nestorian (Assyrian or East Syrian or Chaldaean, now the Eastern church), *monophysite (*Jacobite or West Syrian, now the orthodox Syrian church), *Maronite (from the monk John *Maro [†c.410] and his disciples of the convent near *Apamea where this branch of the Antiochene church originated in the 7th c.) and *Melchite (from *malko*: king, to designate the communities dependent on the Byzantine emperor; i.e. *orthodox) communities of *Syria and Mesopotamia, i.e. the majority of the Eastern dioceses of the Roman empire, but also of *Persia. The missionary impetus of these churches took the S. language as far as India, prob. very early (the *Malabar and Malankar churches are its heirs), and even to China. For various reasons - from history to *hagiography, touching on *exegesis and history of *dogma - S. is essential to patristic studies, partly because of the contribution of Aramaic culture, partly because of translations of works by Greek *Fathers often lost in the original. We may further stress the mediating position of the Syrian churches towards their neighbour churches in *Armenia and *Georgia, and their important role in transmitting to the Arabs (and through them to the West) the cultural heritage of ancient Greece and Mesopotamia.

II. Language. This Semitic language assimilated such a number of foreign words (Akkadian, Hebrew, Persian and above all Greek) that it has been rightly called the "least Semitic of all the Semitic languages" (cf. bibl. B1). It is generally defined as a late Aramaic dialect, derived from the common N-W Semitic stock.

Derived from Edessene (attested by the earliest pagan inscriptions there), thanks to the prestige of the small kingdom of Osrhoene (capital *Edessa; Roman province after 217) S. became the *lingua franca* of the Aramaic peoples of Mesopotamia and *Syria. It prevailed as their literary language in the 4th c., by which time they had largely adhered to the Christian faith. Later, S. owed its growth and survival to their defence of their own political and cultural identity against Roman, *Byzantine, Persian and Arab pressure. Though, from the 8th c., *Arabic rapidly supplanted it as a spoken language, it remained very much alive as a literary and liturgical language. It is still used as an educated and liturgical language in some parts of E Turkey, Syria and Iraq.

The Syriac script is derived from the square Hebrew script: its earliest examples, the pagan inscriptions of the Edessa region (1st and 2nd cc. AD), show many similarities with Palmyrene cursive. From the start of the 5th c. the alphabet settled down into a sharp, clear script, called "Estrangela" (from Gk *strongoulos*: round, squat). From the 8th c. two more alphabets developed alongside it: "Serta" (Jacobites and Maronites) and "Chaldaean" (the Eastern church). Until the 8th c. the vowels were indicated by diacritical points (above and below the word). This system was subsequently developed in the East, and replaced in the West by Greek vowels.

We possess a considerable number of manuscripts in S. (but so far only one *papyrus), the oldest of which are preserved in Europe. These collections are relatively well-recorded, though their catalogues have not all been drawn up with the same criteria. The manuscripts preserved in the East are less well-known and still hold surprises for scholars of S. - e.g. recent discoveries of manuscripts on Mt *Sinai and in China. The oldest parchments (the first goes back to 411) are of capital importance, and their antiquity is all-important for the *manuscript tradition of works translated from Greek.

III. Literature

1. 2nd-4th cc. Though from the 3rd c. *Edessa was the main centre of Aramaic Christianity, its *evangelization must date from the end of the 1st c. and have been the work of *Judaeo-Christians from Jewish communities of Palestinian obedience from the small kingdom of *Adiabene (E of the

Tigris). The first attempts to translate the OT into S. were made in these communities, as is shown by the mutual relationship of some parts of the Peshitta and the *Targumîm.

It was in these Persian regions of Mesopotamia, not forgetting *Nisibis which possessed a Rabbinic school, in the syncretistic atmosphere of Edessa and in bilingual *Syria that, in the 2nd, 3rd and early 4th cc., the first works in the Syriac language were written. They were mostly connected with *gnosticism, heterodox Judaism (*Ebionites, *Elkesaïtes) and Judaeo-Christianity. Many of them have survived in other languages (*Coptic, Greek: in some cases we still hesitate to declare in favour of a Syriac original) or have been reused in works compiled later. The most famous are: the *corpus* of traditions relating to *Thomas, incl. the *Gospel of Thomas* and the *Hymn of the Pearl*; the *Gospel of Philip*; the *Odes of Solomon*; the *Pseudo-*Clementines*; *Tatian's *Diatessaron*; the *Didascalia*; as well as the *Acts of Thomas* and the *Manichaean Psalter*. The teaching of the "gnostic" astrologer *Bardesanes (†222) deserves special mention.

This literature helps to bring out the role and originality of the two main *orthodox figures of Syriac literature: *Aphraates, the Persian sage (†c.345), and Ephrem of Nisibis (*Ephrem the Syrian) (†373), in whose footsteps followed *Maruta of Maiferqat (†c.420) and a few minor late 4th-c. Edessene writers (esp. *Cyrillona). The main characteristics of their Christianity, not yet contaminated by *Hellenism, are the following: biblicism, "ascetico-*encratism" (cf. the tendencies of Tatian's *Diatessaron*), Jewish traditions (targumîm, *midrash and liturgy) but also anti-Judaism, Mesopotamian traditions, symbolism and apophatic theology, poetic form.

We meet these elements later in the great Syrian writers, e.g. *Jacob of Sarug (†521), but attenuated and often overshadowed by Greek traditions. In fact the decreasing isolation of the Persian communities, their frequent contacts with *Antioch and *Constantinople, the growing *organization and the centralization of the Syrian churches in both Persian and Roman territory, led to their becoming strongly Hellenized from the start of the 5th c. The translation of the NT known as the *vetus syriaca* (known from citations of writers and fragments contained in two very old MSS: Sinaiticus and Curetonianus), as well as probably the first translations of Greek Fathers, had already been made before this date.

2. *5th-7th cc*. If the beginnings of Syriac literature are clearly "Persian", and if Syriac Christianity gives the impression of having gravitated round small centres, with the 5th c. we see two blocks organize themselves: on one hand Persian Christianity, with *Seleucia-Ctesiphon as its administrative centre and *Edessa - later, *Nisibis and Seleucia-Ctesiphon - as its scholastic centre; on the other, the Eastern provinces of the diocese of the *East (Mesopotamia, Osrhoene, Euphratasia, Syria I and II), intellectually centred on Edessa - though this city lost the School of the Persians in 489 - and administratively more and more dependent on *Antioch. Alongside a still *orthodox output, attested by *Balaî (†c.435), the *Liber Graduum*, the *Spelunca thesaurorum* and *John of Apamea, the 5th c. saw an increase in translations, starting with that of the "standard" version of the NT (Peshitta). At the School of Edessa, particularly under the episcopates of *Rabbula (†435) and *Ibas (†457), the main Greek Fathers were translated as and when necessary. The importance of the institution of *monasticism, which developed greatly from this time, and the division of Syriac Christianity explain the existence of different translations of the same work. The great *schism of the late 5th c. led to the subsequent classification of writers according to their confession. The main ones are, on the *Nestorian-Persian side: *Narsai (†502), Barsuma or *Barsauma (†c.495), Mar *Aba (†552: founder of the School of Seleucia), then Isho`yahb I (596), *Babai the Great (†628); on the *monophysite side *Philoxenus of Mabbug (†522), Polycarp, author of a new translation of the NT (called the Philoxenian, after its inspirer), the moderate *Jacob of Sarug (†521), *Jacob Baradaeus (†578) who gave his name to the *Jacobite church, John of Tella (†538), the historian *John of Ephesus (†586), *Peter of Callinicum (†591), *Henana of Adiabene (†610), Thomas of Harkel. This last, who fled to *Alexandria on the eve of the Muslim invasion, was the author of a new translation of the NT, made on the basis of the Philoxenian, but now lost. In Alexandria at the same time, Paul of Tella completed the Syro-Hexaplar version of the OT, based on *Origen's *Hexapla*.

3. *7th-10th cc*. With the Muslim invasion (636, fall of the Persian empire; 638, occupation of Syria), the Syriac-speaking churches benefited from a certain tolerance. They took on the role of a cultural bridge between their own Greek and Aramaic (esp. philosophical and medical) heritage and the new culture. As Chabot says (p. 80), "the burning theological controversies of the previous centuries... came to an end. The main aim of *exegesis was no longer to take the texts apart in a search for dogmatic arguments; it became literal and philological. The long treatises of moral teaching gave way to hagiographical accounts, and history took on an important role in literature... Arabic made progress at the expense of Syriac, and the official language quickly became the popular idiom". This period can be defined as one of translations (and soon of retranslations from S. to *Arabic), reflections on the past, commentary and compilation. Some of the main writers of this period are: in the Nestorian church, *Theodore bar Koni, *Isaac of Nineveh (†c.680), *Martyrius Sahdona, *Timothy I (†823); in the *Jacobite church, *Jacob of Edessa (†708) who, in particular, retranslated works by *Severus of Antioch already translated by *Paul of Callinicum during Severus's lifetime, bishop *George of the Arabs (†724), Moses bar Kepha (†903).

4. *10th-16th c*. This period saw the decline of Syriac literature despite a revival in the 12th and 13th cc. We should mention some authors important for patristic studies: Dionysius bar Salibi (†1171), Michael the Syrian (†1199), bar Hebraeus (†1282) and Abdiso bar Berika (Ebedjesu) (1318). These last two wrote in S. or Arabic indifferently.

IV. *Liturgy.* We observe in liturgy the same movement already outlined in literature. A primitive period, Persian and Edessene, is characterized on one hand by its links with Judaism, shown not just in its celebrating *Easter on 14 Nisan (*quartodecimans), but also in Eucharistic blessings, baptismal rites, sanctuary architecture (cf. the well-known *bema*) and music; on the other, by its links with Mesopotamia, of which we find traces esp. in liturgical poetry: the *hymn (of Edessene origin?), the dramatic poem (*sogito*). These elements recur, often corrected, in the second period, that of Hellenization (cf. e.g. the prestige of *Theodore of Mopsuestia's sacramentary *theology in the School of the Persians). Separate liturgies were now formed in the various geographical areas, though without interrupting links and influences. Thus the Chaldaean rite is composed mainly of Edessene and Persian elements (cf. e.g. the *anaphora of *Addai and Mari); the Syrian rite unites the traditions of *Antioch, *Jerusalem (cf. the anaphora of St James) and *Edessa; the *Maronite rite depends on Edessa, though it underwent many Syrian influences. Another important factor for this period, mainly connected with the formation of the office, was the growing influence of *monasticism. The third period can be defined as that of *commentaries, compilations, repetitions and prolixity (cf. e.g. the proliferation of poetic passages, often attrib. to Ephrem and Jacob of Sarug, in the Syrian and Maronite rites).

V. Syriac studies. Though humanism saw the printing of a first grammar (1539) and a first NT (1555; cf. W. Strothmann, *Die Anfänge der syrischen Studien in Europa*, Göttingen 1971) and though E. Renaudet (1646-1720) undertook important works, only the transference to Europe of Oriental manuscripts, esp. those of the Syrian *monasteries in the *desert of Nitria, really gave an impetus to Syriac studies. From the start of the 17th c., at Rome, the work of the cultivated Maronite Assemani family revealed the riches of Syriac literature (cf. J.S. Assemani's well-known *Bibliotheca Orientalis*). Little more than a century later, the formation at London of the most prestigious collection of manuscripts set off innumerable labours and editions. The catalogue of manuscripts in the British Museum (now British Library) by W. Wright (1870-72) continues to stimulate new studies. The start of the 20th c. saw the birth of the great collections of texts: the CSCO of Louvain and the PO of Paris.

A) General: 1) Introductions: J. Assfalg-A. Krüger, *Kleines Wörterbuch des christlichen Orients*, Wiesbaden 1975; S.P. Brock-P. Robson-F. Young-S. Ashbrook, *Horizons in Semitic Studies: Articles for the Student*, University Semitics Aids 8, Univ. of Birmingham 1980; 2) Histories of literature: as well as the classical works of Wright, Duval and Chabot, see esp. A. Baumstark and I. Ortiz de Urbina; 3) Bibliographies: C. Moss, *Catalogue of Syriac Printed Books and Related Literature in the British Museum*, London 1962; S.P. Brock, Syriac Studies 1960-1970: a Classified Bibliography, *PdO* 4 (1973) 393-465; id., for the years 1970-80 (to be published in the same periodical); S.P. Brock, *Syriac Perspectives in Late Antiquity*, London 1984; H.J.W. Dryvers, *East of Antioch: Studies in early Syriac Christianity*, London 1984, Part 1; S.P. Brock, Syriac Studies 1981-1985. A classified bibliography, *PdO* 14 (1987), 289-360; with S.A. Harvey, *Holy Women of the Syrian Orient*, Berkeley 1987.
B) Language: 1) R. Köbert: *EC* 9, 734-40; 2) Grammars: as well as the classical works of Duval, Nöldeke and Brockelmann, we may consult L. Palacios, *Grammatica Syriaca*, Rome 1954; L. Costaz, *Grammaire syriaque*, Beirut 1955; H. Healey, *First Studies in Syriac*, Rome 1981; 3) Dictionaries: the most accessible are: C. Brockelmann, *Lexicon syriacum*, Berlin 1895 (repr. 1966); ed. J. Payne Smith, *Compendious Syriac Dictionary*, Oxford 1903 (often reprinted); W. Jenning, *A Lexicon to the Syriac NT*, Oxford 1926 (reprintings); L. Costaz, *Dictionnaire syriaque-français (anglais-arabe)*, Beirut 1963; the reference work remains R. Payne Smith's *Thesaurus Syriacus*, Oxford 1889-1901.
C) Auxiliary bibliography: 1) Epigraphy: cf. Ortiz de Urbina 23; and S.P. Brock's bibliography in *Annali dell'Istituto Orientale di Napoli* 38, n.s. 28 (1978) 254-271 and A. Desreumaux, *ibid*. 40, n.s. 30 (1980) 704-708; 2) Manuscripts: a) Catalogues: cf. Ortiz de Urbina 3 and S.P. Brock, above A)3; b) Palaeography: W.H.P. Hatch, *An Album of Dated Syriac Manuscripts*, Boston 1946; c) Codicology: cf. above A)3 and esp. the works of J.-M. Sauget and, by the same,

Un cas très curieux de restauration de manuscrit: le Borgia syriaque 39, ST 292, Vatican City 1981; d) Iconography: J. Leroy, *Les manuscrits syriaques à peintures*, Paris 1964.

D) Collections: 1) *Patrologia Syriaca*, PO (Paris), CSCO (Louvain), OCA, Göttinger Orient Forschung; cf. in particular the *Symposia syriaca* in OCA and the Vööbus, Graffin and Strothmann miscellanies; 2) Periodicals: *AB, Muséon, OC, OCP, OLP, OrSyr, PdO, ROC, Journal of the Syriac Academy* (Baghdad, in Arabic).

E) 1) History: cf. above A)3 and J.-M. Fiey's bibliography, *Jalons pour une histoire de l'Eglise en Iraq*, CSCO 310, sub 36 1970; W.H.C. Frend, *The Rise of the Monophysite Movement*, Cambridge 1972; general picture: A.S. Atiya, *A History of Eastern Christianity*, London 1968; 2) Bible: I. Ortiz de Urbina, *Vetus Evangelium Syrorum: Diatessaron Tatiani*, Madrid 1967; the publications of the Peshitta Institute at Leiden (ed. P.A.H. Boer); B.M. Metzger, *The Early Versions of the NT*, Oxford 1977; 3) Theology: cf. above A)2 and A)3, and R. Murray, *Symbols of Church and Kingdom*, Cambridge 1975 (with an introduction on the beginnings of Syriac Christianity and a bibliography); S.P. Brock, Two Syriac verse homilies on the binding of Isaac, *Muséon* 99 (1986) 61-129; 4) Liturgy: cf. above A)2 and A)3 and J.-M. Sauget, *Bibliographie des liturgies orientales*, 1900-1960, Rome 1962; S. Janeras, *Bibliografia sulle Liturgie Orientali*, Rome 1969 (p.m.); 5) Hagiography: cf. above A)1; S. Ashbrook and P. Peeters, *Orient et Byzance: le tréfonds oriental de l'Hagiographie byzantine*, Brussels 1950; 6) Cultural exchanges: AA. VV., Syncretismus im syrisch-persischen Kulturgebiet, in *Ab.d. Ak.d. Wiss. in Göttingen*, Phil.-hist. KL., III F. 96, Göttingen 1975; J.-M. Sauget, L'apport des traductions syriaques pour la patristique grecque, *RThPh* 110 (1978) 139-148; the 3rd Symposium Syriacum (to be published in OCA).

F. Rilliet

T

TABARKA. Thabraca, modern Tabarka (Tunisia), was a *municipium* at the start of the empire, a colony under *Antoninus Pius, but did not truly develop until the 4th c. It was part of Proconsular *Africa.

Christianity is attested at T. in the 3rd c. with bishop Victoricus, who signed the synodal letters of spring 253, 254/5 and the *Acts of the council of *Carthage of 1 Sept 256 (Cypr., *Epp.* 57; 67; *Sent. epp.* 25). After him we know Clarentius, present at the conference of Carthage of 411, and his Donatist rival Rusticianus. *Victor of Vita (*Hist. pers. Afr. prov.*, 1, 10) mentions a *monastery at T.

Two basilicas are preserved from the early Christian and Byzantine eras: the urban basilica with almost square central aisle, and baptistery attached to the N perimeter wall; a funerary basilica, doubtless a *memoria* of a *martyr unknown to us, notable for its *sarcophagi with *mosaic decoration: the deceased in prayer between two lit candles; a *chalice from which issue peacocks and doves; various animals, birds, *lambs, fish among flowers, sometimes with the deceased in an attitude of prayer (*DACL* 1, 716-719; figs. 152-155). Another mosaic is famous for its depiction of a church of African type, symbol of *Ecclesia Mater* (*DACL* 4, 2231-2234, with colour plate).

PWK 9, 1178-79; *LTK* 10, 6; *DACL* 15, 2146-69; N. Duval, L'Evêque et la cathédrale en Afrique du Nord, *Actes du XIe Cong. Internat. d'arch. chrét.*, CEFR 123, Rome 1989, 345-403.

V. Saxer

TABITHA. A symbol, like other scenes of the early Christian figurative repertoire, of faith in the *resurrection (Acts 9, 36-46), T. first appears in the late 4th-c. *sarcophagus of Fermo cathedral (*Ws* 116, 3). *Peter's miracle is depicted in two scenes: his arrival at T.'s house and the raising of T., who is not on her bed but standing, with Peter holding her hand. Other examples are in two Gallic sarcophagi: one, a late 4th- or early 5th-c. fragment in Arles Museum (*Ws* 145, 6=Benoit, *Sarc.* VII, 4), shows two moments of the episode: Peter's arrival at T.'s house (partly damaged) and the raising (with two women standing and two under T.'s bed): Peter takes the pulse of the girl lying on the bed and raises her. A similar formulation appears on the short side of a sarcophagus in the crypt of St Maximin (*Ws* 145, 5 - same period). Stylistically it has a more lively rhythm than that of Arles. The sole example in the sumptuary arts is an ivory panel in the British Museum, first decade of 5th c.

DACL 15, 2, 1947-48; *EC* 11, 1684; W.F. Volbach, *Elfenbeinarbeiten der Spätantike und des frühen Mittelalters*, Mainz 1952, pl. 38, 117; M. Sotomayor, *S. Pedro en la iconografía paleocristiana*, Granada 1962, 156-160.

A.M. Di Nino

TABULA PASCHALIS PETROCORICENSIS. An Easter table incised in uncial characters on a marble slab once on the right side of the choir altar in St Stephen's church, once the cathedral of Périgueux (*Petrocoricum*) (Dordogne), and now inserted in the E side of the S wall of the same church. The opening: "*Hoc est Pascha sine termino et numero: cum finierit a capite reincipe*", is followed by 91 dates of *Easter, giving days and months but not years, from which we deduce that the author was proposing a perpetual Easter table. Comparison with Easter tables inspired by the 19-year Alexandrian cycle popularized in the West by *Dionysius Exiguus show that the Périgueux table shows the Easter dates from 631 to 721.

A. Cordoliani, La table pascale de Périgueux, *CCM* 4 (1961) 57-60; CPL, *Addenda* 2297 b.

V. Loi

TACITUS and the Christians. This well-known Roman historian lived from AD 55 to 120. His *cursus honorum* was helped by his marriage to the daughter of Julius Agricola. He kept to himself during the reign of *Domitian, but in 97, under Nerva, was made consul. In the 20 years between Nerva and *Hadrian he wrote his historical works. Describing the fire of Rome of 19 July 64, he writes that *Nero, to deflect suspicion from himself, accused the Christians. T.'s judgment on these, in contrast to that of *Pliny the Younger, is harsh, due to an indirect and distorted knowledge of Christians and Christianity. He relates that the emperor "put forward as guilty and inflicted the most refined tortures on those whom the vulgar, hating them for their crimes, called Chrestians. The author of this sect, Chrestus, had been condemned to death by the procurator Pontius *Pilate during the reign of *Tiberius; but, repressed for the moment, the deadly *superstition broke out again, not just in Judaea, the origin of this sickness, but even in the City, where all atrocious and shameful things come together from every direction and are magnified. Those who confessed their faith were arrested first, then - on their information - a multitude of others, accused not just of having set fire to the city, but of hatred for mankind" (*Annals* XV, 44). T. goes on to describe the tortures to which the Christians were condemned. Apart from the historian's brusque judgment, his evidence confirms some fundamental facts about Jesus' life, handed down by the gospels, and the spread of Christianity at *Rome. While holding the Christians guilty of other crimes, which he carefully avoids naming, probably for lack of proof - and popular hostility could not constitute proof - he admits that they seemed to be sacrificed not to the general interest, but to the cruelty of Nero, the man truly responsible for the disaster.

The Tacitian description, limited to the essentials, attests that in the 70s of the 1st c. the Christian community of Rome had a considerable number of *members (*ingens multitudo*) and that the reason for the *persecution was not the discovery in Christianity of a threat to the state, but that Nero used *pagan hostility to clear himself of the grave charge of incendiarism. The Christian religion was not attacked as such. *Odium humani generis* turned essentially on the fact that the Roman Christians, living according to their own customs, appeared suspect. The norm of behaviour in Greco-Roman society was *philanthropia*, humanism, to which was opposed misanthropy (*odium generis humani*), which allowed them to impute to the Christians the worship of an ass, criminal rituals and incest. This was the first stage of the pagan judgment on the Christians, which T. noted and set down. Suetonius (*Vita Caesarum: Claudius* XXIX, 1; *Nero* XVI, 3) had upheld the *accusation of seditious activities linked to messianism.

H. Fuchs, Tacitus über die Christen, *VChr* 4 (1950) 69-74; J.B. Bauer, Tacitus und die Christen, *Gymnasium* 69 (1957) 495-503; J. Beaujeu, *L'incendie de Rome en 64 et les Chrétiens*, Brussels 1960; J. Daniélou-H. Marrou, *Nuova Storia della chiesa. I. Dalle origini a S. Gregorio Magno*, Turin 1970, 123-124; (Eng. tr. *The Christian Centuries*, I. *The First Six Centuries*, London 1964); K. Baus, *Storia della chiesa*. I. *Le origini*, Milan 1976, 168 f; W.H.C. Frend, *Martyrdom and Persecution in the Early Church*, Oxford 1965, ch. 6.

E. Peretto

TAIO of Saragossa. Bishop of *Saragossa (*Caesaraugusta*) after *Braulio (651-683), he wrote five books of *Sententiae*: the first two are theological and exegetical (I on *God and the OT, II on Christ and the church); the other three are mainly interested in *ethics: *virtues, *sins, last *judgment. Following a common custom of the time, the work is a *collage* of passages of other authors: besides *Augustine, it uses esp. *Gregory the Great, whom T. praises highly in a surviving *letter to *Eugenius of Toledo. On a *journey to *Rome he had transcribed in his own hand those *books of Gregory's which he lacked.

CPL 1267-1270; Díaz 205-210; PL 80, 727-990; *DHEE* 4, 2516 f.; A.C. Vega, Tajón de Zaragoza. Una obra inédita, *Ciudad de Dios* 156 (1943) 145-177; P. Martínez, El pensamiento penitencial de Tajón, *RET* 6 (1946) 185-222.

M. Simonetti

TANNAIM (literally "teachers"). Plural of *tanna*, from an Aramaic root meaning to repeat, transmit orally, hence teach. Name given to the succession of 1st- and 2nd-c. Jewish teachers starting from Hillel and Shammai. The collection of their opinions constitutes the *Mishnah*; those not included in this, the *baraitoth*, found a place in the *Tosefta* and the *Talmud*. They were also the creators of the ancient *midrashim* (called *tannaitic*) that were published by their successors, the *Amoraim.

Jewish Encyclopedia 12, 49; *EJ* 15, 798-803.

R. Le Déaut

TARGUM. Meaning "translation", it indicates the Aramaic versions of the *Bible made when that language supplanted Hebrew as the common tongue. The biblical texts (of the Torah and Prophets) were then translated from *memory in the synagogues after being read in Hebrew. Aiming to make the text immediately comprehensible to their audience, the translators did not hesitate to complete, *gloss, clarify, even paraphrase, often taking inspiration from all the sources of *midrash. So T. is not a translation in the modern sense, but a transposition reflecting the religious conceptions of a certain period, and showing in particular how the Bible was understood in ancient times. Hence its interest for the study of the NT and ancient Judaism. Targumic traditions are often attested in the *Fathers (*Origen, *Jerome, *Ephrem).

T. exist of the whole Bible (except Ezra-Nehemiah and Daniel), but their composition ranges over some ten centuries. The discovery of fragments of Lev 16 and Job 17-42 at Qumrân proves the existence of written texts in the 1st c. BC.

The official T. of the Pentateuch (*Onkelos*) and that of the Prophets (*Jonathan ben Uziel*), in literary Aramaic, go back to a basic text compiled in *Palestine early in the 2nd c. and later published in the Babylonian academies (4th-5th c.). These two versions follow the Hebrew quite closely, but also contain numerous original interpretations. Also from Palestine come two complete recensions of the T. of the Pentateuch: that of Vatican codex *Neophyti 1*, in Palestinian Aramaic close to that of the Talmud, may go back substantially to the 3rd c., while that called *of Pseudo-Jonathan* is a late (7th-c.?) compilation of original material from a Palestinian recension together with elements taken from Onkelos and the midrashim. Other recensions of the Palestinian traditions have survived in the so-called *Fragmentary Targum* (c.800 verses) and the Cairo Gheniza fragments published in 1930 by P. Kahle (whose affinity with *Neophyti* is evident).

The *Targum of the Hagiographa*, Palestinian in origin, are of more recent date (4th-8th cc.) and each of them poses particular problems. Some, like those of Esther and the Song of Songs, are so paraphrastic that they may be considered midrashim. Finally, the *Samaritan Pentateuch* has its own T., which may go back to the 3rd or 4th c.

We should point out that the exegetical traditions contained in these texts may be much older than the date when they were entrusted to writing or that of the Aramaic dialects in which they were written.

Texts: B. Walton, *Biblia Sacra Polyglotta*, London 1653-57 (with Lat. vers.); J.W. Etheridge, *The Targums of Onkelos and Jonathan ben Uzziel*, 2 vols., London 1862-65 (repr. New York 1968, tr. only); P. de Lagarde, *Prophetae Chaldaice*, Leipzig 1872 (repr. Osnabrück 1967), *Hagiographa Chaldaice*, Leipzig 1873 (repr. Osnabrück 1967); A. Berliner, *Targum Onkelos*, Berlin 1884; P. Kahle, *Masoreten des Westens*, II, Stuttgart 1930; J.F. Stenning, *The Targum of Isaiah*, Oxford 1949 (text and tr.); A. Sperber, *The Bible in Aramaic*, 4 vols., Leiden 1959-73; A. Díez Macho, *Ms. Neophyti 1*, 6 vols., Madrid-Barcelona 1968-79 (with Sp., Fr. and Eng. tr.); R. Le Déaut-J. Robert, *Targum des Chroniques*, Rome 1971 (text and tr.); D. Rieder, *Pseudo-Jonathan - Targum Jonathan ben Uziel on the Pentateuch*, Jerusalem 1974; R. Le Déaut, *Targum du Pentateuque (Neofitti et Ps. Jonathan)*, 4 vols. (SCh 245, 256, 261, 271), Paris 1978-1980 (tr. only); M.L. Klein, *The Fragment-Targums of the Pentateuch*, Rome 1980.
Studies: R. Le Déaut, *Introduction à la littérature targumique*, Rome 1966; J. Bowker, *The Targums and Rabbinic Literature*, Cambridge 1969; M. McNamara, *Targum and Testament*, Shannon 1972; id., Targums, in *Interp.Dict.Bible* (Suppl.), 856-61; A. Díez Macho, *El Targum*, Barcelona 1972 (repr. 1979); B. Grossfeld, *Bibliography of Targum Literature*, 2 vols., Cincinnati 1972, 1977; J.T. Forestell, *Targumic Traditions and the New Testament*, Missoula 1981. Cf. also E. Schürer-G. Vermes-F. Millar, *The History of the Jewish People*, I, Edinburgh 1973, 99-114.

R. Le Déaut

TARRACONENSIS. Hispano-Roman province, capital Tarraco (*Tarragona), containing the episcopal sees of Caesaraugusta (*Saragossa: 254-258), Barcino (*Barcelona: 347), Calagurris (Calahorra: 306-457), Turiaso (Tarazona: 449), Egara (*Tarrasa: 450), Ausona (516), Ampuriae (Castelló de Ampurias: 516), Gerunda (Gerona: 516), Dertosa (Tortosa: 516), Ilerda (*Lérida: 516), Osca (Huesca: 527), Urgellum (Seo de Urgel: 527), Pampilona (Pamplona: 589), Auca and Amaia (?) (foundation dates unknown). We have early evidence of the existence of Christian communities (*martyrs of Tarraco, Caesaraugusta, etc.) in the E part of the province: their close relationship with *Rome is shown by the oldest funerary monuments, i.e. the group of imported *sarcophagi, of Roman make, in the necropolises of Barcelona (three examples of Constantinian date), Saragossa (late Constantinian and Theodosian examples), Castiliscar, Tarraco and Gerona (four pre-Constantinian and two Constantinian examples). The sparse products of Rosae and Badalona have the same origin. The two Aquitanian-type sarcophagi of Ampuriae attest relations with S *Gaul as well. When Roman output dried up, local workshops arose and developed (cf. *Tarragona), with an unrelated group in La Bureba (region of Burgos), with possible iconographical links with N *Africa, and similar relations with the *East.

In this first period, artistic output clearly imitates and reflects the models coming from the imperial capital and Eastern centres, as is easy to establish for the mausoleum of Centcelles (Tarragona), a central-plan building rising on one of the thermal installations of a 4th-c. rustic villa. The cupola survives with a *mosaic cycle of hunting, *commendatio animae*, seasons and other certainly Christian motifs. It is a considerable work, comparable to the mausoleum of S. Costanza at Rome, and perhaps, since it reflects imperial motifs, to be attributed to a member of *Constantine's family.

Contacts of the province of T. with N African Christianity from the end of the 4th c. are attested by numerous mosaic tombstones: at Barcelona, San Cugat, Egara, Coscojuela de Fontova (Huesca), Alfaro (Logroño) and perhaps Ampuriae and Cornellá, as well as the Tarragona group. But the ornamental motifs of these compositions differ widely.

Though written sources attest the existence of *church buildings in the cited sees, the oldest remains are from well into the 5th c. and form a group of generally ill-known and typologically quite different buildings. To the basilica type belong the church of Santa María at Egara, with one aisle, paved in mosaic, rectangular altar, baptistery in a separate building (N Italian in form and position); that of Barcelona, with three aisles, separate octagonal baptistery, alterations of Visigothic date; that of Bobalá (Serós, Lérida) with three aisles, font in a side room, headed by lateral sacristies and rectilinear apse, much modified in the Visigothic era. The group also includes the church of Santa Cristina at Aro with a single aisle and rectilinear head; that of Porqueras (Bañolas) with polygonal apse curved on the inside; from the transitional period, the hall of San Cugat (late 6th c.) with horseshoe apse; and the memoria-cell in the necropolis of Ampuriae, with apse semicircular inside and perhaps a sacristy (modified in the Visigothic period). In the period of transition, towards the Visigothic period, arose the church of Villa Fortunatus at Fraga, cruciform with projecting rectilinear apse (the proprietor's name can be read in a secular mosaic with *Alpha and *Omega*) and rectangular baptismal font at the end.

Other remains and evidence of church buildings are: apse and altar-*mensa* at Rosae, baptistery of Santa Margherita at Ampuriae, domed octagonal baptistery at Torre del Fum (San Feliu de Guixols), fragments of altar-*mensae* at San Fust di Campsentelles, Santa María l'Antiga, San Pere di Egara, San Salvatore della Vedella and San Pere De Casserres, and of semicircular *mensa* at San Feliuet de Vilamilans (Rubi). Remains of buildings of the Visigothic period are few. There are materials of ornamental church *sculpture (Barcelona, Bobalá, Fraga) and other fragments like the altar-foot of Santes Creus and the recent discoveries at Ribarroja (Castellón). The church of Santa María de Ventas Blancas (Logroño) may date from the late 7th c. Through inscriptions (Siero-Burgos) or thanks to early mediaeval continuity (San Millán de la Cogolla, San Vincenzo del Valle), we know other church buildings.

*Liturgical furnishings of the late 6th and 7th cc. are important: e.g. the Coptic censers of Lladó and Bobalá and the small Coptic bosses and patenae (Calonge, la Grassa), later imitated in Hispania (Rosas), on the coast of T.

E. Camps-J. Ferrandis, *Art hispano-visigodo* (Historia de España, III, dir. R. Menéndez Pidal), Madrid 1940; H. Schlunk-T. Hauschild, *Informe preliminar sobre los trabajos realizados en Centcelles*, Madrid 1963; T. Hauschild, Vorbericht über die Arbeiten in Centcelles, *Madr.Mitt.* 6 (1965) 127-194; P. de Palol, Demografía y arqueología de los siglos IV-VIII, *BSEAA* (1966) 5-66; id., *Art hispanico de la época visigoda*, Barcelona 1966; *Actas I reunión nacional de Arqueología paleocristiana hispánica*, Vitoria 1966 (1967); id., *Arqueología cristiana de la España romana*, Madrid-Valladolid 1967; id., Arqueología cristiana hispánica de tiempos romanos y visigodos, *RAC* Miscell.Josi II (1967) 177-232; T. Hauschild, Untersuchungen im spätrömische Monument von Centcelles, Tarragona, *VIII Congres.Int.Arch.Crist.*, Vatican City-Barcelona 1969, 45 ff.; Pita Merce-P. de Palol, La basílica de Babalá y su mobiliario litúrgico, *ibid.*, 44 ff.; P. de Palol, Arqueología cristiana romana y visigoda, *DGEE* 1, 1972, 96-113; J. Fontaine, *Le pré-roman hispanique*, I, La pierre-qui-vire, 1973; M. Sotomayor, *Datos históricos sobre sarcófagos romano-cristianos de España*, Granada 1973; id., *Sarcófagos romano-cristianos de España*, Granada 1975; R. Puertas, *Iglesias hispánicas (s. IV al VIII). Testimonios literarios*, Madrid 1975; X. Barral, La basilique paléochrétienne et visigotique de Sant Cugat del Vallés, Barcelona, *MEFR Ant.* 86 (1974/2) 891-928; S. Alavedra, *Les ares d'altar de Sant Pere de Terrassa-Egara*, Tarrasa 1979; M. Sotomayor, Historia de la Iglesia en España. I. *La Iglesia en la España romana y visigoda* (dir. García Villoslada), Madrid 1979; H. Schlunk-T. Hauschild, *Hispania Antiqua, Die Denkmäler der frühchristlichen und westgotischen Zeit*, Mainz am Rhein 1979; *Actas II reunió d'Arqueología paleocristiana hispánica*, Montserrat 1978 (1983).

P. de Palol

TARRAGONA

I. City - II. Councils - III. Archaeology.

I. City. Ancient Tarraco was capital of Hispania Citerior from 197 BC and, after its division by *Diocletian into three provinces, capital of the province of *Tarraconensis. All through the Roman period it was a flourishing city. It was conquered by the Visigoth *Euric in 456, and occupied and destroyed by the Muslims in 714. We do not know when Christianity reached it. Its Pauline origin cannot be demonstrated, but the Apostle must have arrived there very early, given its position as a port close to *Rome and the importance of the city itself. The first valid document is the Acts of the *martyrdom of St *Fructuosus, bishop of T., and his deacons Augurius and Eulogius. These Acts take us back to 259, year of the martyrdom, and show us a flourishing Christian community with no tension with the *pagans. In 385 pope *Siricius sent a *letter in reply to one of bishop *Himerius. It seems that T. was not then a metropolitan see, since the text of the letter says clearly that the see's *authority was due to his personal seniority as a bishop (*Ep. ad Himerium* 16: PL 56, 562). But it was not long before it became a religious metropolis too. In 468 pope *Hilarus calls bishop Ascanius a metropolitan (*Ep. ad Ascanium*: PL 58, 17), a dignity recognized by the bishop of *Saragossa, who appealed to him, as metropolitan, to do as he thought fit with the bishop of Calahorra, his suffragan. Ascanius's whole behaviour confirms it, and at the council of *Toledo of 516 it was said *expressis verbis* that T. was a metropolitan see.

Bishops: St *Fructuosus (†259); Himerius (before 385); Hilary (402); Ascanius (465 ff.); John (516); Sergius (520-555); Tranquillinus (*c*.560); Stephen (†589); Arthemius (†599); Asiaticus (†599); Eusebius (610-632); Audax (*c*.633); Protasius (638-646); Faulax (645-668?); Cyprian (683-688); Vera (693); St Prosper (711?).

P. Palol de Salellas, *Terraco hispano visigoda*, Tarragona 1955; J. Blanch, *Arxiepiscologi de la santa Esgresia metropolitana i primada de Tarragona*, Tarragona 1955; D. Mansilla, Orígenes de la organización metropolitana en la iglesia española, *Hispania Sacra* 12 (1959) 262-267; E. Flórez, *ES* 24 and 25.

II. *Councils. *464*. The celebration of a provincial council at T. in 464 has been surmised by some authors (Hefele-Leclercq) from a letter sent by Ascanius and other bishops of the province of *Tarraconensis to pope St *Hilarus asking how they should behave towards Silvanus of Calahorra who was carrying out illicit *ordinations, and requesting advice on the transference of Irenaeus from *Tarrasa to the see of *Barcelona (PL 58, 15). No other evidence of this council exists.

516. Provincial council. Metropolitan John presided over seven other bishops of the province with the metropolitan of Carthago Nova and the bishop of Elvira. Its 13 canons aim at regulating the life of the *clergy (and monks: can. 11). Worth mentioning are its decision to *excommunicate *ad tempus* a bishop unjustifiably absent from a council though he had been called (can. 6), and its invitation to bishops to bring country priests and *laymen with them to a council (can. 13).

Synoden 124-126, 128-133 and *passim*.

P. de Luis

III. *Archaeology. The earliest archaeological evidence goes back to the 4th c. and is funerary. The much-studied necropolis of Francolí has yielded sepulchral rooms and burials of various types, including some in the form of sigma-shaped *mensae*, and a group of *sarcophagi and mosaic tombstones. Unlike other centres of the province, there are no imported sarcophagi of Roman make from the first half of the 4th c. Those found are of Theodosian date: fragments in the necropolises, the "Bethesda"-type sarcophagus in the cathedral and a frontal fragment. The activity of local workshops producing pagan work was interrupted in the 3rd c. and did not resume until Roman production ceased at the start of the 5th c. Close contacts with N *Africa would explain the iconographical and formal analogies between some local sarcophagi (Apóstoli, Leocadio) and those of *Carthage. The marble tombstone of the *orantes* was imported from N Africa. Local work varies considerably in ornamental conception: smooth panels with *tabula* and *alpha* and *omega*, strigilate panels with or without figured compartments. This output coincides chronologically with that of the mosaic tombstones, quite different in motifs and decorations from African ones. They form a group of ten pieces, including those of Optimus (late 4th c.), Ampelius and the Good *Shepherd, mostly grouped around or inside the necropolis church. This building, of disputed identification, shows more than one building phase. Located above the earlier necropolis, it is thought to be a memorial of the first *martyrs buried there, around which Christian tombs were arranged *ad sanctos*. We can deduce, at a certain period, a basilica form with three aisles separated by columns, free semicircular apse and square room at the end. Two irregular rooms, used as tombs, flanking the apse, and the baptistery, on the S side with rectangular pool and lateral steps, are the result of modifications. Plan and pool have parallels in *Syria and N Africa.

The city was devastated by *Euric in 468, and Visigothic rule involved a break in its fruitful relations with N Africa. The impoverishment of the Roman artistic tradition affected the vitality of the Christian community. It is practically impossible to identify and locate the churches known from the *Orational*, except that of St *Fructuosus in the arena of the amphitheatre, perhaps built in his memory on the place of his *martyrdom. It is not of martyrial type, but corresponds to early Roman basilica models. It may date from the mid 6th c. or even the 7th. It has three aisles separated by pillars and a transverse wall between the aisles themselves and the apse, which is horseshoe-shaped inside and uncertain outside. A Romanesque church above it prevents us finding out much about its structure.

Sculptural elements from the Visigothic period survive (two columns of the altar-*mensa*), as well as remains of *cancelli* (Tarragona Arch. Mus.), very baroque but still containing the element of the classical pediment, as in the workshops of *Mérida. **[Fig: 301]**

E. Camps-J. Ferrandis, *Arte hispano-visigoda* (Historia de España, III, dir. R. Menéndez Pidal), Madrid 1940; P. de Palol, *Tarraco hispano-visigoda*, Tarragona 1953; J. Vives-T. Marín, *Concilios visigóticos e Hispano-romanos*, Madrid 1963; P. de Palol, Demografía y arqueología de los siglos IV-VIII, *BSEAA* 32 (1966) 5-66; id., *Arte hispánico de la época visigoda*, Barcelona 1966; *Actas I reunión nacional de Arqueología paleocristiana hispánica*, Vitoria 1966 (1967); P. de Palol, *Arqueología cristiana de la España romana*, Madrid-Valladolid 1967; id., Arqueología cristiana hispánica de tiempos romanos y visigodos, *RAC* Miscell. Josi II (1967) 177-232; J. Vives, *Inscripciones cristianas de la España romana y visigoda*, Barcelona ²1969; P. de Palol, Arqueología cristiana romana y visigoda, *DHEE* 1, 1972, 96-113; J. Fontaine, *Le pré-roman hispanique*, I, La pierre-qui-vire, 1973; M. Sotomayor, *Datos históricos sobre sarcófagos romano-cristianos de España*, Granada 1973; id., *Sarcófagos romano-cristianos de España*, Granada 1975; R. Puertas, *Iglesias hispánicas (s. IB al VII). Testimonios literarios*, Madrid 1975; M. Sotomayor, *Historia de la Iglesia en España*, I, *La Iglesia en la España romana y visigoda* (dir. García Villoslada), Madrid 1979; H. Schlunk-Th. Hauschild, *Hispania Antiqua. Die Denkmäler der frühchristlichen und westgotischen Zeit*, Mainz am Rhein 1979; *Actas II reunió d'Arqueologia paleocristiana hispánica*, Montserrat 1978 (1981); M.O. Del Amo, *Estudio crítico de la necrópolis paleocristiana de Tarragona*, Tarragona 1979; X.B.I. Altet, L'image littéraire de la ville dans la péninsule ibérique pendant l'antiquité tardive, *Actes XI*ᵉ *Cong. internat. d'arch. chrét.*, CEFR 123, Rome 1989, 1393-1403.

P. de Palol

TARRASA (ancient *Egara*). *Council in 614 of the bishops of the province of *Tarraconensis, presided over by metropolitan Eusebius. 11 bishops were present; two more were represented. It issued one canon, confirming what had been established at the council of Huesca in 598 and extending to bishops what had been prescribed for lower *clergy on *chastity.

Mansi 10, 532; *ES* 3, 193-194; Palazzini 2, 39; *Synoden* 137.

P. de Luis

TARSICIUS or Tarcisius. The earliest evidence of T. is provided by pope *Damasus (Ferrua, *Epigr. Damas*. no. 15), who compares him to *Stephen the protomartyr because both were executed by stoning. But this tells us nothing about his membership of the hierarchy. The fact remains that he died for refusing to hand over the *Eucharist. Damasus says nothing of the date of his *martyrdom. The 7th-c. *Itineraries put his tomb close to that of *Zephyrinus. The *Passio S. Stephani* pp. (BHL 7845) makes him an acolyte of the Roman church. The Passion is of epic type, and late. The first to introduce T. into his *martyrology was Ado (15 Aug).

Vies des SS. 8, 270; *LTK* 10, 11; *BS* 12, 136-38.

V. Saxer

TARSUS. City of Cilicia Campestris, on the river Cydnos (today Tarsus Çay), of very ancient origin (known in the 9th c. BC). Capital of the kings of Cilicia and the Persian satraps: but Greek coins issued by the local mint in the 5th and 4th cc. BC show that it soon became autonomous and was deeply Hellenized. It came under the rule of Alexander and then of the Seleucids of *Syria: annexed to the Roman state by Pompey in 67 BC during his campaign against the pirates, Anthony made it a free city and conceded wide privileges to groups of its citizens. With the creation of the province of *Cilicia, T. became its capital; its internal constitution was timocratic and, among other things, imposed a tax of 500 drachmae on the exercise of civil rights, thus excluding the great mass of the population. A famous inscription recording the passsage of *Septimius Severus through the pass of the Pylae Ciliciae, not far from the city, praises it as the greatest and finest metropolis of the three provinces of Cilicia, Isauria and Lycaonia; with *Diocletian's reform it became capital of Cilicia Prima.

A lively centre of production (the local flax industry was famous), T. was also a flourishing centre of cultural life, seat of a renowned school of philosophy rivalling *Athens and *Alexandria (cf. Strabo, *Geogr.* XIV, 5, 13-14). *Paul was born there (Acts 9, 11), passed his childhood there and returned for some time after his *conversion. It was an episcopal see of great importance and had bishops from the apostolic era, starting with

Jason, a disciple of Paul (Rom 16, 21), and his successor Urban; then follow, of known names, Helenus (252-269), Athanasius (martyred 259), Clino, Lupus (314-325), Theodore (at *Nicaea), Anthony (an *Arian), *Silvanus (leader of the *homoeousians, expelled after the council of *Constantinople of 360, died perhaps 371), *Diodore (379-381), Falerius (394), Dositheus (415), Marianus, Helladius (431-434), Theodore (449-451), Pelagius (458), Nestor (489), Syncletius (530), Peter (553), Theodore (680).

W. Ruge, Tarsos, *PWK* 23, 2, 2413-2439; W.M. Ramsay, *The Cities of St. Paul*, London 1907, 85-244; D. Magie, *Roman Rule in Asia Minor. To the End of the Third Century*, Princeton 1950, 272 ff.

M. Forlin Patrucco

TATIAN. 2nd-c. Christian writer classed among the *Apologists. Of Eastern origin ("born in the land of the Assyrians" he says of himself [*Discourse to the Greeks* 42]: which could mean Mesopotamia proper or, more vaguely, *Syria), but Hellenistic by education. After all kinds of *journeys and wanderings (*Disc.* 35), he encountered the Christian *Scriptures and was converted (*Disc.* 29). He was with *Justin, whose disciple he considered himself and was considered (at least in the first period after his *conversion); with him he was involved in the controversy against the Cynic philosopher Crescens, who repaid both with mortal hatred (*Disc.* 19).

His one remaining work, the *Discourse to the Greeks* (Λόγος πρὸς Ἕλληνας), originally handed down in the compendium of the Greek Apologists put together by Aretas of Caesarea (cod. *Par. Gr.* 451: the pages containing T. were later lost, but exact copies remain), is a violent polemic against the whole of Greek culture, mingled with a defence of Christian ideas, which he thus expounds for apologetic ends. Its date is uncertain: different points of view put it between 155 (presumably quite close to his conversion) and 170 (the last, confused news of his activity). He accuses the Greeks of depending on barbarian cultures, while the one invention of the Greeks was *philosophy, which T. vilifies in the persons of its representatives, the philosophers, who fall into contradiction and deliver the most foolish judgments, though T. himself then makes wide use of conceptual instruments of philosophical (esp. *Stoic) origin.

A rhetor, expert in the most refined sophistic cavils to the point of competing with his adversaries in subtlety and polemical aggressiveness, he presents himself ostensibly as a "barbarian", rich in that wisdom which the Greeks despised, a wisdom which he identifies in the biblical tradition. In this way T. combines Eastern nationalism, Christian *faith, restless theological speculation, intransigent polemic, sarcasm and contempt for his adversaries. Judging from the *Discourse to the Greeks*, he seems a man of violent, passionate character, often obscure, aspiring to formulations of thought that try to be original, sometimes through the accumulation of arguments that are not always coherent.

Some aspects of his *theology still have an archaic (*Judaeo-Christian) character; others anticipate or echo more radical and almost gnosticizing tendencies. After a violent polemical introduction against the whole of Greek culture, not just intellectual, but especially the philosophers (chs. 1-3), T. summarily expounds the Christian concept of *God (ch. 4), the God-Logos relationship and the *creation of matter (ch. 5), the doctrine of the *resurrection, consequence of divine omnipotence (ch. 6), the creation of the angelic creatures (ch. 7), the fall of the *angels and the protoparents, which involved the introduction of *idolatry and *astrology (ch. 8). After a polemical interlude against idolatry, T. goes on to deal with demonology, pneumatology, psychology and *soteriology (chs. 9-16), according to a theological scheme that might seem strange but is in fact constant for those times - there are still abundant traces of it in *Origen's *De principiis*. This exposition is punctuated by continual polemical attacks on the various opinions of the Greeks; it goes on to condemn the use of drugs, medicine, *entertainments, *pagan moral and political behaviour, as having demonic origins (chs. 17-28). The influence of Jewish *apocryphal literature (esp. the *Book of Enoch*) is evident all through this part.

After another autobiographical interlude (chs. 29-30), T. develops two arguments: firstly, the moral superiority of the "Christian" (or "barbarian") "philosophy" over the licentiousness of the Greeks, demonstrated by the low moral level of their most famous men, their vulgarly immoral works of art (catalogue of statues), their mythological tales that teach immorality and corrupt the young (chs. 31-35). The second argument, fully developed in the final part of the *Discourse* (chs. 36-41), is the so-called "chronological argument", tending to prove that the barbarians (i.e. the Jews and consequently the Christians) are, since they go back to *Moses, prior to and so older and more authoritative than the Greeks. The former go back to Moses, the latter to Homer: on the basis of a series of comparative chronological calculations, following a path already beaten by Alexandrian-Jewish apologetes, T. concludes with the Christians' superiority to the pagans on the basis of greater antiquity. The *Discourse* closes (ch. 42) with a proclamation of faith ("I know who God is, and his creation") and a challenge: T. declares himself ready for a contest of ideas with his adversaries.

T.'s newest and most original theological positions concern the procession of the Logos, his presentation of the theme of *resurrection and his very full development of demonology. More archaic are his formulation of pneumatology and the cosmic conception of the fall of the angels. Traces of rigorous *ascesis are present in the *Discourse*, though it is hard to say whether his position is properly heretical.

Heresiological tradition (Iren., *Adv. haer.* I, 28, 1; III, 23, 8; Euseb., *HE* IV, 29, 1; Epiph., *Pan.* 46) attests that, after Justin's death, T. detached himself from the Great Church, inclining towards *encratism and accentuating his pessimism over the problem of the fall of mankind until he ended by maintaining the doctrine of the damnation of *Adam. These ideas seem to bring him close, in a certain sense, to conceptions that are not just encratite but truly *gnostic (*economy of the OT subject to Satan or to a lower god). T. wrote various other works. The one that most ensured his fame in the following centuries, esp. in the *East, was the *Diatessaron. But even before the *Discourse* he had written a treatise *On animals* (Περὶ ζῴων) which, despite its name, was probably a treatise on natural *anthropology, strongly pessimistic in tone. Another work had been dedicated to demonology (*Disc.* ch. 16). *Perfection according to the Saviour* (Περὶ τοῦ κατὰ τὸν Σωτῆρα καθαρτισμοῦ), cited and refuted by *Clement of Alexandria (*Strom.* III, 12, 81, 1 ff.), was a clearly encratite ascetic work: in it, among other things, T. interpreted *Paul's words in 1 Cor 7, 5 to mean that only abstention from *marriage united one with God, while the custom of marriage meant *communion of incontinence and *fornication with the devil. In this treatise T. proposed Christ, virgin and continent (and *poor), as a model, radically contrasting the new man (Christ, and with him the continent Christian) with the old man (Adam; damned for this). Another work, the *Problems*, dealt with the difficult questions present in the sacred texts: his disciple *Rhodo proposed in turn to provide the explanations. It may be (but is not certain) that this work of T.'s contained the *exegesis of Gen 1, 3 (*fiat lux*) understood as an entreaty by the Creator God to the Supreme God that there should be *light, an exegesis that seems to be clearly *Valentinian-gnostic in tone (Clem. Al., *Ecl. Proph.* 38, 1; Orig., *C. Cels.* VI, 51; *De orat.* 24, 5). Acc. to later (and not very certain) information, T. also substituted *water for *wine in the *Eucharist, in accordance with his encratite convictions, which proscribed the use of wine (Epiph., *Pan.* 46; Jerome, *In Amos* 2, 12). Of the NT *Scriptures, T. is said to have rejected certain of *Paul's epistles; of some at least, he corrected certain expressions (acc. to Euseb., *HE* IV, 29, 6), though he accepted the pastoral epistles. Acc. to *Epiphanius, T. returned to the *East and propagated his encratite ideas from *Antioch on the Daphne to *Cilicia, Pisidia and elsewhere. Though this last information may concern the development of encratite trends contemporary with Epiphanius himself (*loc. cit.*), the information may have a certain basis, this being in fact the region in which T.'s best-known work, the *Diatessaron*, was propagated and long remained the official gospel of the local churches.

PG 6, 803-888; E. Schwartz, *Tatiani Oratio ad Graecos*, TU 4, 1, Leipzig 1888 (with testimonies and fragments); E.J. Goodspeed, *Die ältesten Apologeten*, Göttingen 1914 (Schwartz's text, pp. 266-305, without the fragments). Translations: Ger.: R.C. Kukula, *BKV*² 12, Kempten 1913; Fr.: A. Puech, *Recherches sur le Discours aux Grecs de Tatien, suivies d'une traduction française du Discours avec notes*, Paris 1903; It.: P. Ubaldi, Turin 1931, M. Fermi, Rome 1924; Sp.: D. Ruiz Bueno, BAC 116, Madrid 1954, 572-628 (with Gk text); Eng.: M. Whittaker, *T. Oratio ad Graecos and fragments*, Oxford 1982. Studies: C.W. Steuer, *Die Gottes-und Logoslehre des T. (Diss.)*, Jena 1892; R.C. Kukula, *Tatians sogennante Apologie*, Leipzig 1900; id., "Altersbeweis" und "Künstlerkatalog" in Tatians Rede an die Griechen, Vienna 1900; A. Puech, *op.cit.*; R.M. Grant, The Date of Tatian's Oration, *HThR* 46 (1953) 99-101; id., The Heresy of Tatian, *JThS* n.s. 5 (1954) 62-68; id., Tatian's theological Method, *HThR* 51 (1958) 123-128; id., T. and The Bible, *SP* 1 (1957) 297-308; F. Bolgiani, La tradizione eresiologica sull'encratismo, I: Le notizie di Ireneo; II: La confutazione di Clemente di Alessandria, *AAT* 91 (1956-57); 96 (1961-62); A. Orbe, A propósito de Gen. 1, 3 en la exégesis de T., *Gregorianum* 42 (1961) 401-443; M. Elze, *T. und seine Theologie*, Göttingen 1960; S. Di Cristina, L'idea di dynamis nel "De mundo" e nell' "Oratio ad Graecos" di Taziano, *Augustinianum* 17 (1977) 485-504; M. Whittaker, *Tatian: Oratio ad Graecos and fragments* (Ed. and Eng. tr.), Oxford 1982; L.W. Barnard, The Heresy of Tatian, *Studies in Church History and Patristics*, AV 26, Thessalonica 1978, 181-193.

F. Bolgiani

TAUROBOLIUM. Sacrificial rite pertaining to the originally Phrygian cult of the Great Mother Cybele, officially accepted at Rome in 204 BC. The rite, first attested in AD 160 by an inscription at *Lyons, consisted of the *sacrifice of a bull, sometimes accompanied by that of a ram (*criobolium*), and could have a dual purpose, public and private. In the former case, the t. was celebrated for the benefit of the emperor (*pro salute imperatoris*), sometimes associated with other members of the imperial family, and was intended to ensure the goddess's favour for the highest representative of the state, as a basis for and guarantee of the common good. The t. performed by individual believers for their own benefit, in the forms described in *Prudentius's *Peristephanon* (X, 1006-1050) and the anon. *Carmen contra paganos* (vv. 57-62), had a cathartic purpose, with efficacy limited to a period of 20 years, after which the rite could be repeated. It consisted of a sort of "bath" of *blood which the believer received standing in a pit covered by a grating, over which the animal was killed. Widespread all over the empire in the 2nd-4th cc., the rite was particularly favoured in the Roman aristocratic circles which promoted the anti-Christian pagan renaissance of the last part of the 4th c. The title of "tauroboliate" of the Great Mother rounded off the numerous religious offices of distinguished members of the Roman aristocracy, such as Vettius Agorius *Praetextatus (CIL VI, 1778), while another member of the same circle was proclaimed, through it, to be *in aeternum renatus* (CIL VI, 510).

H. Graillot, *Le culte de Cybèle, Mère des dieux à Rome et dans l'Empire romain*, Rome 1916; R. Duthoy, *The Taurobolium. Its Evolution and Terminology*, Leiden 1969; M.J. Vermaseren, *Cybele and Attis. The Myth and the Cult*, London 1977; G. Sfameni Gasparro, *Soteriologia e aspetti mistici nel culto di Cibele e Attis*, Palermo 1979; U. Bianchi, *Mysteria Mithrae*, Rome 1979.

G. Sfameni Gasparro

TEBESSA. Theveste, modern Tébessa (Algeria), a great agricultural market and important strategic post, is one of the oldest cities of N *Africa. There was stationed the 3rd Legion Augusta at least from the time of *Vespasian (69-79) to the start of the 2nd c. T. became a *municipium* under the Flavians and a colony under *Trajan. We do not know what province it then belonged to, but from the time of *Diocletian it was part of Proconsularis. It reached its apogee in the 3rd c., as witness *Caracalla's triumphal arch (inserted as N gate into the Byzantine wall), the temple of Minerva (city museum), amphitheatre and forum. The *Vandals destroyed the walls in 429, but the *magister militum* Solomon rebuilt them after *Belisarius's reconquest (CIL 8, 16507). The walls still stand, which is one of the reasons why the urban centre has never ceased to be inhabited. The 11th-c. Arab geographer El Bekri vaunted the city's wealth.

Christianity appeared at T. in the 3rd c. Lucius was one of the bishops at the council of *Carthage, 1 Sept 256, which deliberated about the *baptism of *heretics (*Sent. epp.* 31). A marginal note to its *Acts, written after 259, call him "*confessor and *martyr". His name appears in *Mart. hier.* (18 Jan). This suggests that he died in 259. In 361-63 a Donatist *council was held at T. (Opt. of Mil., *Schism. don.* 2, 18), but despite this fact the city was never a stronghold of *Donatism. Among bishops we know Romulus (council of Carthage 345/48), Urbicus and his Donatist rival Perseverantius (conference of 411), Felix (that of 484), a bishop Palladius, known only from his early 5th-c. epitaph (CIL 8, 2011), and finally a 6th-c. Faustinus. The city's martyrological *fasti* record on 12 March 295 the martyrdom of the conscript *Maximilian, buried at Carthage (BHL 5813), and of *Crispina and her companions on 5 Dec 304 (BHL 1988ab). Over their tomb, in the reign of *Theodosius (383-395), was raised a great basilica, a place of *pilgrimage. In 883 Leo VI the Wise still numbered T. among the bishoprics of Africa.

PWK 11, 249-52; DACL 15, 1998-2028; LTK 10, 111-12; J. Christern, *Das frühchristliche Pilgerheiligtum von Tebessa*, Wiesbaden 1976; F. Kadra, Rapport sur les récentes découvertes en Algérie, *Actes XIe Cong. internat. d'arch. chrét.*, CEFR 123, Rome 1989, 1961-1974.

V. Saxer

TE DECET LAUS. More than a *hymn, a liturgical formula in the form of an *acclamation to the *glory of the *Trinity. Like the *Gloria Patri*, it is a minor *doxology, whereas the *Te Deum* is considered a major doxology. Of very ancient origin, it appears in the *Apostolic Constitutions* (VII, 48) and the *Byzantine liturgy, for the morning hours. St *Benedict introduced it into the divine Office in this form: *Te decet laus / Te decet hymnus / tibi gloria Deo Patri / Et Filio, cum Sancto Spiritu / in saecula saeculorum. Amen.* It is no longer used in the Roman liturgy.

M. Righetti, *Storia liturgica*, I, Milan ²1950, 201; cf. *EC* 11, 1861.

L. Dattrino

TE DEUM. *Hymn written in Latin in praise of the *Trinity. In the 9th c. Hincmar of Rheims (*Praedestinatione* XXIX) attributed it to *Ambrose and *Augustine, who composed it on the occasion of Augustine's *baptism, but more recent criticism rejects such a paternity. *Nicetas of Remesiana has been considered its author, interpreting what *Paulinus of Nola says about Nicetas's composition of hymns and liturgical *chants (*Carm.* 17, 90 ff.). The rhythmical prose in which the *T. D.* is composed, unusual in Latin hymnology, has led to the hypothesis of the hymn's derivation from a Greek source. The *T. D.* was recited during the morning Office, acc. to *Caesarius of Arles (*Reg. mon.* 21).

A.E. Burn, *The Hymn Te Deum and its Author*, London 1926; M. Simonetti, Studi sull'innologia popolare cristiana dei primi secoli, *Atti Acc. Nazionale Lincei, Memorie*, ser. 8, vol. 4 (1952) 478-481; *Patrologia* III, 183.

M.G. Mara

TEMPTATION

I. Temptation in general and the temptation of Jesus - II. Typology of paradise: Christ, the new Adam - III. Typology of the desert: Christ, the true Israel - IV. Temptation and Baptism - V. Temptation and Passion - VI. Ecclesial dimension of Jesus' temptation.

I. Temptation in general and the temptation of Jesus. On the subject of t. the *Fathers, both Greek and Latin, express a group of reflections that can be schematically summed up thus: *1) Types* of t. Some authors speak of t. arising out of adversity and the misfortunes of life (emblematic figure: *Job) and t. arising out of well-being and prosperity (type-figure: the rich man in the gospel). Others distinguish inner and outer t.; and others voluntary and involuntary t. The most illuminating distinction is in *Augustine, who speaks of t. as "testing" and t. as "seduction": *Alia est tentatio seductionis, alia tentatio probationis* (*Ep.* 205, 16); *2)* T. *in relation to *God*: God does not properly tempt, but he "puts to the test", to assay, purify and educate; *3)* T. *in relation to the devil*: it is to the devil that we owe t. as solicitation to *evil. This aspect is present particularly in ascetic literature, esp. apropos of monks and hermits (*Athanasius, *Nilus, *Palladius, *Maximus Confessor, *Jerome, *Cassian); *4) Behaviour* in t.: courage and patience, *prayer and *fasting, obedience and prudence (in the sense of not voluntarily exposing oneself to t.) are recommended; *5) Effects*: right behaviour in t. attests *love of God, nourishes humility and favours the progress of spiritual *life.

Alongside this range of general ascetico-spiritual meanings, the vocabulary of t. is charged with a more strictly theological or salvation-historical value, which is made explicit apropos of the t. of Jesus. For this meaning we will reserve fuller treatment.

Analysis of the synoptic account of the t. reveals, linked to its messianic character, a network of other themes closely connected with antecedent biblical history (Christ, new *Adam, new Israel, new *Moses), with other *mysteries of Jesus' life (*Baptism, Passion) and with the life of the Christian community (exemplary and redemptive value of Christ's struggle). We will examine the patristic interpretation by taking up these themes (synchronic aspect) and bringing out the contributions and tonalities of different authors (diachronic aspect).

II. Typology of *paradise: Christ, the new *Adam. The parallel, present in *Justin (*Dial.* 103), is systematically elaborated by *Irenaeus against a background of "*recapitulation" (*Adv. haer.* V, 21). Indeed, starting from the triple t. of Jesus, he reconstructs a triple t. of Adam (V, 21, 2). The link between the two events is Gen 3, 15, which is related to Christ through Gal 3, 19 and 4, 4 (V, 21, 1). Acc. to *Tertullian, Jesus by his fast "inaugurated the new man in reproof of the old" (*novum hominem inveteris sugillationem fastidiendi cibi initiabat: De ieiun.* 8, 2). For *Origen, the devil, out of jealousy, tried to make Jesus fall from the dignity of son of God, using the same tactic he had employed on the first man (cf. *Fragm. in Lc.* 56). *Hilary describes Jesus' t. on the model of Adam's (*tenens ordinem fraudis antiquae: In Mt.* 3, 5): followed in this by *Ambrose, who offers this eloquent text: "In the desert Adam, in the desert Christ; indeed he knew where to find the condemned man in order to lead him back to paradise after having freed him from his error" (*Exp. in Lc.* 4, 7). The comparison between the two Adams, present in all the 4th- and 5th-c. Fathers, is elaborated particularly by *Maximus of Turin: "The Saviour acts in such a way as to destroy crimes by following in the same tracks (*eisdem vestigiis*) through which they were committed" (*Serm.* 51a: CCL 23, 202), echoed by *Gregory the Great: "(The devil) was defeated by the second man using the same methods (*eisdem modis*) by which he boasted of having defeated the first man" (*In Ev. hom.* XVI).

III. Typology of the *desert: Christ, the true Israel. The parallel, in its precise outlines as present in Mt, is evoked only incidentally by the Fathers. The first to appeal to the contrast between the t. of the people in the desert and that of Jesus is *Tertullian, for whom the outcome of the first t. marks the transition from the old carnal people to the authentic representative of God's people, who "direct onto themselves the censure incurred by Israel" (*de figura Israelis exprobationem in ipsum retorsit*: *De bapt*. 20, 4). Another very close text is that in which he deplores Israel's incapacity to resist hunger, thus drawing down God's punishment (*Nunquam non per impatientiam delinquendo perierunt*: *De pat*. 5, 25). Acc. to *Origen, Israel failed in its historical task by rejecting the true manna which is the Word of God, in which Jesus, on the contrary, found his true nourishment (cf. *Fragm. in Mt*. 63). For both authors, Christ in the desert inaugurates the new man, bringing to fulfilment the divine pedagogy with which Israel's experience had failed to comply. To give greater prominence to absolute *faith in God and in his Word, both eliminate the term "only" in Jesus' reply, thus stressing the spiritualization of Dt 8, 2-3 in the light of Wis 16, 26. Later Fathers, while referring to Israel's experience in the desert, do not interpret it in an antithetical-typological sense, but record it under the miraculous aspect, as a time of *prodigious* nourishment (*manna and *quails), a time that, e.g., *Hilary calls "angelic" (*In Mt*. 3, 1). The text of *Maximus of Turin (*Serm*. 51) must seemingly be interpreted in the same way, while another text of Hilary in which he speaks of Jesus' "resistance to hunger" (*patientia esuritionis*), in probable dependence on the above-cited text of Tertullian - one of his teachers - could be interpreted in an antithetical-typological sense. The figures of *Moses, *Elijah, Elisha and even *John the Baptist are evoked almost unanimously by the Fathers with the same idea of miraculous nourishment.

IV. Temptation and *Baptism. The link between the two moments, in the life of Jesus and his disciples, is programmatically made all through the *tradition. Its first appearance is in *Justin (*Dial*. 103, ff.). But "the earliest sure evidence that has come down to us of this theme, dear to the meditation of primitive Christianity" (Steiner, 43), comes from the *Extracts of *Theodotus* (85), a *gnostic work, where the parallel between the tt. of the gnostic and those of Jesus is closely linked to the rite of *baptism. The connection is brought out by *Tertullian (*De bapt*. 20, 3); and particularly by *Origen, acc. to whom, after baptism, in which both Jesus and the Christian "break the hammer of the whole earth" (cf. Jer 50, 23), i.e. the devil, we must be ready for a harder struggle: "You have come to the *water of baptism: this is the beginning of *spiritual combat: here the struggle against the devil begins for you" (*Hom. in Iud*. 9, 2). The devil, insists *Hilary, specially tempts the "sanctified", i.e. the baptized, as he tempted Jesus after the Jordan: "So the Lord is tempted immediately after (*statim post*) baptism, showing by his temptation that the devil's attempts become rampant particularly when we are sanctified (*in sanctificatis*), because a victory achieved over the saints is more attractive to him" (*quia victoria ei est magis exoptata de sanctis*: *In Mt*. 3, 1). For *Cyril of Alexandria too, tt. become greater after baptism because the Spirit has received the ability to struggle; before baptism it would not have been possible to resist these assaults (*Ad reginas*, 36). *John Chrysostom warns the baptized not to be disturbed, as by something unexpected, if after baptism he experiences more violent tt. Instead of being disturbed, he should resist and support it all with courage, as something normal. He receives arms not to remain in idleness, but so that he may realize that he has become stronger than steel, and as confirmatory evidence of the treasure he has received (baptism). The *demon would not assail him if he did not see him raised to such an honour (cf. *Com. in Mt. hom*. 13). The author of the *Opus imperfectum in Matthaeum* follows the same line: compared to the *catechumen, the baptized is more exposed to t. because, escaped from the devil, become a son of God and anointed as an *athlete of the Spirit, he now belongs to Christ and like him is forced into the desert to struggle and overcome (*Hom*. 5).

V. Temptation and Passion. This aspect too the Fathers refer simultaneously to Christ and to believers. Sufficiently alluded to in *Justin (*Dial*. 125, 4-5), little stressed in *Irenaeus (but cf. *Adv. haer*. V, 23, 2), touched on by *Tertullian (*Scorp*. 15, 6), the theme emerges in *Origen. The most vigorous text is in his comment on Jer 50, 23, already mentioned apropos of Baptism: "At that moment (= in the desert) Jesus did not demolish, but merely broke the hammer of all the earth. But when he returned, then . . . it was not just broken, but demolished. And since the hammer of all the earth was first broken and then demolished, for this reason it is also broken by each one of us when we are introduced into the church and advance in faith; then it is demolished by us when we reach perfection" (*Hom. in Jer*. 20, 2). For Origen, the culmination of "perfection" is *martyrdom, the supreme trial that conforms most to the Passion of the Lord Christ. The t.-Passion parallel, rather muted in subsequent Fathers, is alluded to in *Ambrose (*In Lc*. 35) and more explicitly in *Augustine: "As teacher, he wished to be tempted in all, since we are tempted; he wished to be a martyr, since we die; he wished to rise, since we will rise" (*In Ps. 60 serm*. 2; cf. *Serm*. 284). The thought returns, almost identically, in *Maximus of Turin and *Gregory the Great, of whom we will cite the following text: "It was right that he should conquer our temptations with his temptation, as he conquered our death with his death" (*In Ev. hom*. 16). Finally, in an individual idea, we read: "He was led into the desert by the Spirit to be baptized in the fire of temptation, as later he had to be baptized in the fire of death" (*Op. imp. in Mt. hom*. 5).

VI. Ecclesial dimension of Jesus' temptation. The exemplary and salvific character of Jesus' t., already brought out in the preceding paragraphs, is the most prominent aspect of the Fathers' reflections. We must limit ourselves to the main witnesses. For *Irenaeus, Christ's victory becomes every Christian's victory. He cites to this effect Lk 10, 19, a *logion* constantly taken up by *tradition in the context of t. (cf. *Adv. haer*. V, 24, 4). The author in whom the paradigmatic and soteriological aspect of Jesus' t. is most vibrant is *Origen. Jesus' struggle is seen by him in terms of the "truer" battle that Christ fights in his church: "He was tempted so that we too could overcome thanks to his victory" (*Hom. in Lc*. 29, 3). A precious teaching is provided by *Athanasius in his *Vita Antonii*: what the Lord said in his third t., he said and did for us (37, 1 f.); he struggles *in* and *with* *Anthony (5, 7; 7, 1; 10, 2-3; 34, 1; 41, 5-6; 42, 2; 91, 3). Acc. to *Cyril of Alexandria, the Lord underwent t. "not because fasting was necessary for him, but to propose his acts to us as type and model, and to outline for us the ideal of an excellent and marvellous life" (*Com. in Lc*. 4). "We have conquered in Christ . . . . Christ, conquering, has given us the strength to conquer (*to dynasthai nikan*)" (*ibid*.). Again: "Christ first conquered for us, so that we could have *peace after having trampled and defeated the enemy (cf. Lk 10, 19). It was necessary that he should measure himself against Satan in our favour to make us partakers of his fullness . . . . Fallen in Adam, in Christ we conquer" (*Ad reginas* 36). *Theodore of Mopsuestia, lapidarily: "The Lord, having thrice prevailed over the devil, gave us the victory he had gained" (*Fragm. dogm. ex libro incarn. Filii Dei* XIII). Particularly rich and stimulating is *John Chrysostom (*Com. in Mt. hom*. 13; *Op. imp. in Mt. hom*. 5). With an original thought, *Hilary brings out Jesus' true hunger: the salvation of men: "He no longer hungered for the food of men, but for their salvation" (*Non cibum etiam hominum esuriit, sed salutem . . . esuritionem se humanae salutis habiturum*: *In Mt*. 3, 2). "The Lord hungered less for *bread than for men's salvation" (*non panem potius quam salutem hominum esuriens*: *In Mt*. 3, 3). With evident dependence on Hilary, *Ambrose: "He who could not hunger for 40 days, makes me understand that then he desired not the nourishment of the body so much as the salvation of mankind" (*se non cibum esurisse corporis, sed salutem*: *In Lc*. 4, 16). The same insistence is in *Maximus of Turin (*Esuriit enim non cibum hominum sed salutem*: *Serm*. 60, 4). On the exemplary and salvific character, Ambrose again offers us some very happy ideas: "He let himself be tempted by the devil, so that we should learn to conquer in him" (*In Lc*. 4, 4). "So we follow in his footsteps, so that from the desert we may return to paradise" (*ibid*., 4, 12). "If he had not accepted the challenge, Jesus would not have conquered for me" (*ibid*., 4, 12). "You too, learn to overcome the devil. The Spirit leads you, follow the Spirit" (*ibid*., 4, 24). The exemplarity and soteriological value of Jesus' t. attain their theologically most profound and stylistically most effective expression in *Augustine: "He was tempted first so that we should not be overcome in temptation. For him temptation was not necessary: Christ's temptation is for our teaching" (*tentatio Christi, nostra doctrina*: *In Ps. 90, serm*. 1). "If he had not overcome the tempter, how would you have learnt to fight against the tempter?" (*ibid*., *serm*. 2). But the most vigorous text is in his comment on Ps 60, where the theme is centred on the category of the "whole Christ". We will give the most important passage: "Our life in this pilgrimage cannot be exempt from trials and our progress is made through temptation. No-one can know himself unless he is tempted, nor can he be crowned unless he wins, nor can he win without fighting; but combat supposes an enemy, a test . . . . The Lord wished to *prefigure* us, who are his body, in the vissicitudes of that body in which he died, rose and ascended to heaven. In this way too the members can hope to arrive where the head has preceded them. So he *as it were transfigured us in himself* when he wished to be tempted by Satan . . . . Christ was tempted by Satan, but in Christ you too will be tempted. For Christ took flesh from you, but your salvation from himself; death from you, your life from himself; humiliation from you, your glory from himself: so also he took his temptation from you, your victory from himself. If we have been tempted in him, it will be in him that we overcome the devil. You rest your attention on the fact that Christ was tempted: why do you not also consider that he overcame? It was you who were tempted in him, but recognize that in him you are the victor. He could have kept the devil far from him, but if he had not let himself be tempted he would not have taught you to overcome when you are tempted" (*En. in Ps*. 60, 3).

K.P.Köppen, *Die Auslegung der Versuchungsgeschichte unter besonderer Berücksichtigung der Alten Kirche*, Tübingen 1961; M. Steiner, *La tentation de Jésus dans l'interprétation patristique de saint Justin à Origène*, Paris 1962; G. Leonardi, Le tentazioni di Gesù nella interpretazione patristica, *Studia Patavina* 15 (1968) 229-262; S. Raponi, *Tentazione ed esistenza cristiana. Il racconto sinottico della tentazione di Gesù alla luce della storia della salvezza nella prima letteratura patristica*, Roma 1974; id., Tentazione, *Diz. spirit. laici*, Milan 1981, 332-333; id., Cristo tentato e il cristiano. La lezione dei Padri, *Studia Moralia* 21 (1983) 209-236.

S. Raponi

TERTIUM GENUS. This term was used to designate Christians and distinguish them on the level of faith from *Jews and *Pagans. 1 Pet 2, 9-10 calls the Christians "a chosen race" (cf. Is 43, 20), "God's own people", "object of mercy" (cf. Hos 1, 8-9; 2, 1. 23-24), keeping themselves faithful, but with veiled allusions to Israel's deviation. Eph 2, 11-22 develops a not dissimilar concept with special attention to the ex-gentiles, historically saved with Israel and forming with Israel, in a single church, a unique new man (allusion to *Adam). The Ephesian text sets out the premises for the affirmation and subsequent evolution of the concept of t. g., but referred to a sole and evident reality "Christ, our *peace" (Eph 2, 14), with clear allusion to the elimination of the conflict of religion between Jews and Pagans (Eph 2, 17) and the unification of the two areas of operation ideally occupied and defended by two opposed religious components, the Jewish and the pagan. A first clear hint appears in the *Martyrdom of *Polycarp* (III, 2): the crowd awaiting the bishop's arrival in the stadium is astonished "at the courage of the holy and pious race (*genos*) of the Christians". The word "race" must be taken in the religious sense, as it must in *To *Diognetus* I, where it is asked "why this new race has come to light now and not before". The title *t. g.*, as specific to the Christians, in that they are in a different situation distinct from that of the Jews and pagans, appears in the *Kerygma of Peter* (frag. V) and is closely connected to the new worship of *God through Christ and the new covenant. *Clement of Alexandria explains how this comes about: "these are they who are gathered into a single race of people, which is saved, from the Hellenistic and the legal (Jewish) teachings. The three people are not divided chronologically, since no-one thinks of three natures, but are educated with three different Testaments by the word of a single Lord, the Lord being only one" (*Strom.* VI, 5). Not an ethnographical distinction, then, but one of religious content (cf. *To Diognetus* V-VI). *Aristides (*Apol.* II) uses the idea and the term, but the uncertainty of the textual tradition, Hellenistic as well as *Armenian and *Syriac, does not allow us to be exact. Acc. to the Syriac edition of the *Apologia*, there are four stocks of men who are distinguished by religious motives: the barbarians by Chronos and Rhea, the Greeks by Helenus, the Jews by *Abraham and the Christians by Jesus Christ. While the religious motive is clear for Christians and Barbarians, it is not so for Jews and Greeks. Acc. to Aristides, Christian particularity does not isolate believers, since Christianity is the universal religion immanent in the whole universe (*Apol.* XV-XVII) and Christians take part in history and the life of the *world (cf. *To Diognetus* VI, 2). *Irenaeus treads the same ground of the t. g. (with eyes fixed on Eph 2, 11 ff.) with much delicacy and elegance, referring to the salvific project that God has been awaiting from the beginning (*Epideixis* 8. 34).

A. von Harnack, *Missione e propagazione del cristianesimo nei primi tre secoli*, Turin 1906, 31-52; C. Vona, *L'Apologia di Aristide*. Rome 1950; P. Batiffol, *La Chiesa nascente e il cattolicesimo*, Florence 1971, 87 f.: C. Mohrmann, Tertium genus, in *Etudes sur le latin des chrétiens*, IV, Rome 1977, 195-210.

E. Peretto

TERTULLIAN

I. Life and works - II. Doctrine and problems.

I. Life and works. The little we know about T.'s life is derived from references in his own works and from *Jerome's words in *De vir. ill.* 35 (but some of Jerome's facts have been doubted; e.g. T.'s priestly *ordination). Even the dates of his birth and death remain unknown. We can confidently assert that his literary activity spanned the last years of the 2nd c. and the first two decades of the 3rd, and that his first reliably dateable works, *Ad nationes* and the *Apologeticum* (AD 197), reveal a mature spirit, an author in command of his material.

T. was a native of N *Africa, of *Carthage and - acc. to Jerome (*op. cit.*) - the son of a proconsular centurion. So he lived in a geographical area where Christianity seems from the start to have assumed special characteristics expressed in a spirit of autonomy and sometimes of particularism; a spirit that had distant roots in the fidelity with which Africans had always looked to their origins and the dissatisfaction often shown with Roman rule. So in that land Christianity found itself becoming the catalytic element of a state of ancient irritability and, in the course of time, gathered around itself forms of latent resistance to Roman power. Perhaps it was not by chance that T. protested sarcastically against the abandonment of typically Carthaginian manners and customs in favour of Roman ones (he did so in a treatise, *De pallio* [1, 2] which constitutes a real crux for critics, but which, being in some way autobiographical, well represents the enigmatic figure of its author). T. wrote under the emperors *Septimius Severus (193-211) and Antoninus *Caracalla (211-217), in a critical period for Rome on account of internal disturbances and external wars. There were violent episodes against Christians (a *persecution broke out at Carthage in 197, prob. caused by the arrival of a new proconsul). In 202 the emperor may have forbidden Jewish and Christian proselytism (but some recent critics consider this information in the *Historia Augusta* [cf. Spartianus, *Sept. Sev.* 17, 1] false). They were important years for the church, which saw the rise of strong personalities within itself: *Victor (189-99), *Zephyrinus (199-217), *Callistus (217-222), among the bishops of *Rome; *Hippolytus, *Clement of Alexandria, *Origen, among writers in Greek. T.'s life, so uncertain to reconstruct, developed within these co-ordinates. He grew up in a *pagan family and was taught in the schools of his time; he profited from study, which he continued to cultivate afterwards: his works attest his considerable knowledge in different fields - history, *philosophy, *rhetoric, languages (as well as Latin, he knew Greek well enough to compose treatises in it). As for law, the identification suggested by some between our T. and the jurist of the same *cognomen*, author of a *Liber singularis de castrensi peculio*, whose fragments are preserved in the *Digest*, has recently been denied again (cf. R. Martini, *Tertulliano giurista*, 79 ff.) and it is disputed whether the arguments of, e.g., the *Apologeticum* are those of a jurist (cf. J. Gaudemet, *Le droit romain*, 15 ff.) or rather those of an advocate. His passionate and sensual temperament led him in youth to a dissipated existence (cf. *Ad nat.* I, 10, 47; *Apol.* 15, 5; *De res.* 59, 3); he had a wife, whom he addresses in *Ad uxorem* (1, 1); at a time impossible to specify (before 197), he was converted to Christianity, perhaps attracted by the example of the *martyrs (cf. *Apol.* 50, 15; *Ad Scap.* 5, 4). This is what we know of the first or - as some critics have it, giving *Minucius Felix first place - one of the very first of the Latin-speaking Christian writers whose works can be reliably ascribed to a distinct personality (it is well-known that esp. J. Daniélou [cf. *Les origines du christianisme latin*] has proposed to identify a series of pseudo-Cyprianine treatises as the expression of an earlier *Judaeo-Christian Latin literature, against which T. reacted; the credit for having made the Christian message assimilable to the structures and ways of thought of the Latin West would then belong to Minucius Felix first and to T. next. A hypothesis that is far from being commonly accepted).

T.'s works are often divided according to content, into three categories: apologetic, doctrinal and polemical, moral and ascetic. Though a convenient, didactically clear partition, it does not seem to correspond to the complexity of his output. In the list that follows we prefer to keep to the chronological criterion: among other things, it allows us to watch the author's religious evolution, though raising difficulties where we can find no inner or outer criterion capable of placing one or another work chronologically. This is the case of *De pallio*, a bitter polemical reply aimed at those of his fellow-citizens who made fun of the writer for having doffed the Roman toga to don the philosopher's cloak: in its brevity and obscurity, it is presented now as the first, now as the last of T.'s treatises, or else assigned to an intermediate period. From 197 date the two best-known *apologetic works, *Ad nationes* in two books and the *Apologeticum*, directed at the pagans' *accusations against the new religion; in it, defence and illustration of Christian morals and doctrines alternate with an attack on the conduct and beliefs of the gentiles. The precise placing of *Ad martyras*, a brief but intense exhortation to the *brothers to face *persecution courageously, is uncertain: either at the start of 197 or in the course of that year, or in 202-203, in which case the recipients would be the martyrs known from another ancient document, the *Passio Perpetuae et Felicitatis* (see *Perpetua), a text sometimes attributed to T. as editor (the latest research opposes this hypothesis on the basis of linguistic and stylistic observations [cf. R. Braun, *Nouvelles observations*, 105 ff.]). No less problematic is the date attributed to *Adversus Iudaeos*, which presents itself as the completion of an unfinished debate between a Christian and a Jewish *proselyte, and which expounds the major points of controversy between *Jews and Christians. The work, whose authenticity has long been disputed, has been ascribed to AD 200 or, by others, before 197. *De testimonio animae*, an apologetic work appealing to the evidence of the *soul to demonstrate the existence of *God and other truths affirmed by Christian doctrine, seems to have been composed shortly before 200. Between c.200 and 206 comes an important series of moral treatises: *De spectaculis* (condemning games in circus, stadium and amphitheatre and forbidding Christians to take part in them), *De oratione* (on *prayer, esp. the *Lord's Prayer), *De patientia* (on the importance of Christian *patientia*, of which Jesus gave an example), *De paenitentia* (on the first "penitence", necessary for receiving *baptism, and the second *penitence, after the *sacrament of *initiation, which precedes ecclesiastical reconciliation), *De cultu feminarum* (on *women's *dress and ornaments and the need for modesty) in two books, *Ad uxorem* (a sort of spiritual testament in which he recommends his wife not to remarry) in two books. To the same period belong another three important works: *De baptismo*, against the *Cainite sect (on baptism, its necessity, its effects, the invalidity of that administered by heretics), *De praescriptione haereticorum* (on the right to possess and hence to interpret Holy *Scripture, reserved not to heretics but only to the church, which is its heir by legitimate transmission, having received the Scriptures from Christ through the *Apostles; on the reasons why *heresies are in error. The term *praescriptio* is understood by our author in at least two senses: with a properly legal value, and with a rhetorical and dialectical value) and finally *Adversus Hermogenem* (a defence of the Christian doctrine of *creation against those who, like the gnostic *Hermogenes, considered matter eternal).

In the treatises composed from 207 we see the ever-clearer influence of *Montanism, the Phrygian religious movement which arose in the second half of the 2nd c. and to which T. was to adhere. Writings of a doctrinal and anti-gnostic character follow each other from 207 to 212: the first four books of *Adversus Marcionem*; this is a third edition of a work already composed earlier (against *Marcion and any attempt by him to separate the *God of the OT from that of the NT); *Adversus Valentinianos* (an exposition and refutation of the doctrine of the *Valentinian *gnostics), *De anima* (on the nature, origin, development and destiny of the *soul, which is at the same time a refutation of heretical doctrines), *De carne Christi* (on the Lord's *incarnation), *De resurrectione mortuorum* (on Christ's second *parousia, on the salvation of the corporeal element, destined to rejoin the soul, on the need for *judgment and the necessity of *resurrection), book V of *Adversus Marcionem*. In this great doctrinal undertaking, T. seems to be trying to expound the essential points of the *regula fidei* in a setting that bears in mind the difficulties and objections of the heretics and, more generally, the mentality and culture of his time. Other works composed in this most intense and fertile season of his literary activity have a moral and practical character and clearly reveal the writer's Montanist tendency: this is so of *De exhortatione castitatis* (again on second *marriage, and more rigid in attitude), *De virginibus velandis* (on the need for *virgins to wear a veil not just in church, but in all public places), *De corona* (on the incompatibility between Christianity and military service), *Scorpiace* or medicine against the scorpion's sting, i.e. against the *heresy of *gnosticism, in which he exalts the value of *martyrdom, denied by the heretics. *Ad Scapulam* is like an open *letter, apologetic in nature, addressed to *Scapula, proconsul of Africa, who in *c*.200 had begun to persecute the Christians. Opinions are divided on the date of *De idolatria* (against all idolatrous practices and against any activity and occupation that came into contact with such): many historians consider it composed shortly before 212; others propose 197, or the years immediately after.

The third and last period sees T. lined up on the side of the Montanists against the Great Church. Questions have been asked about the nature of African *Montanism in relation to that of *Phrygia; and if some critics have supposed differences, others have denied them, identifying the former with the so-called Tertullianist movement (cf. D. Powell, *Tertullianists and Cataphrygians*, 33 ff.). From 212-213 on, then, come *De fuga in persecutione* (on the inadmissibility of *flight during *persecution); *Adversus Praxean* (expounding the doctrine of the *Trinity against the *patripassian *Praxeas), *De monogamia* (once more against remarriage, with more radical theses and harder tones), *De ieiunio adversus Psychicos* (a defence of the Montanist practice of *fasting and an attack on the *psychici*, or *Catholics, accused of being laxist) and *De pudicitia*, where he denies the church the right to remit *sins, reserving it to "spiritual men", that is to say *apostles and *prophets; he claims that some very grave sins (*idolatry, *fornication, murder) are not pardonable by anybody, and attacks a bishop who has expressed a contrary opinion on the last point. The religious parabola described by T. has thus reached its goal: the polemic which he set in action is at its acme, so much so that some of the ideas that we read in the last works contradict others in previous works.

*Jerome (cf. *De vir. ill.* 53) suggests a reason, hard to verify, which pushed the African writer into the orbit of Montanism; the *invidia* and *contumeliae* that the *clergy of *Rome displayed towards him in a context that we are totally ignorant of and which, if it existed, we may suppose to have concerned disciplinary questions; in this context, we should not forget Jerome's own rancour against the Roman clergy. In all probability, outer circumstances and inner compatibility led him to adhere to the Montanist movement, allowing him to develop to its extreme consequences a rigid and uncompromising ideal of life for which he had always had a propensity; an outcome certainly favoured by T.'s conception of *ethics: his ascetic and rigorist spirit was urged on by the ideas of *justice, retribution, fear, even *hope, but does not seem to have been sufficiently inspired by *love. In this sense Montanism did no more than accelerate a process begun much earlier, welling up from the deep personal experience of the man.

With *De pudicitia*, one of his last works (217-222), we lose track of him. *Cyprian, bishop of *Carthage (†258), knows his works, but does not mention him. Jerome, sheltering behind a prudent *they say*, says he lived to a decrepit old age. *Augustine (*De haeres.* 86) states that T. soon quarrelled with the Montanists and formed his own *conventiculae*, which still existed at Carthage in his own time; indeed Augustine's work helped reconcile them with the church.

Among his lost works are: *De spe fidelium*, *De paradiso*, *Adversus Apelleiacos* (against the followers of *Marcion's disciple *Apelles), *De censu animae* (on the origin of the *soul, against *Hermogenes), *De fato*, *Ad amicum philosophum* (on the difficulties of conjugal life), *De ecstasi* in seven books (in defence of Montanism, against the church and the Asiatic bishop *Apollonius). Unauthentic writings attrib. to him by tradition include *Adversus omnes haereses*, *De execrandis gentium diis*, the five-book poem entitled *Carmen adversus Marcionem*, the *Carmen ad Flavium Felicem de resurrectione mortuorum et de iudicio Domini* and other minor works.

II. Doctrine and problems. T.'s surviving works are very numerous and would seem sufficient to give a reliable interpretation of the man and the writer: but it is not so. The complexity of his writings is accompanied by the enigma of his person when, after having patiently delineated the individual treatises, scholars try to understand him overall. T.'s historical importance for the knowledge of the times in which he lived is undoubted; as is his considerable contribution to the formation of Christian theological doctrine in the West. By looking at the articles of the *regula fidei*, more than once formulated by him (cf. *De praescr. haer.* 13, 1-6; *De virg. vel.* 1, 3; *Adv. Prax.* 2, 1), we can gather the African writer's most important doctrinal contribution, beginning with the *theology of the *Trinity. He is the first to speak of *trinitas unius divinitatis, Pater et Filius et Spiritus sanctus* (*De pud.* 21, 16); he is the first to use the term *persona*: . . . *quia iam adhaerebat illi Filius secunda persona, sermo ipsius et tertia, Spiritus in sermone, ideo pluraliter pronuntiavit "faciamus" et "nostram" et "nobis"; alium quomodo accidere debeas iam professus sum, personae, non substantiae nomine, ad distinctionem, non ad divisionem* (*Adv. Prax.* 12, 3 and 6). He was the first to work out the formula of the dogma: one single substance in three persons, even if he later underwent a certain *subordinationist influence. As regards *christology, we owe to him the statement of two "substances" in one sole person (*Videmus duplicem statum, non confusum sed coniunctum in una persona, Deum et hominem Iesum - de Christo autem differo - et adeo salva est utriusque proprietas substantiae* . . . : *Adv. Prax.* 27, 11). These and many others are expressions of great interest, which yet, to be fully understood, must be both inserted in the context of the works to which they belong and also considered in the light of a historically rigorous terminological study (cf. R. Braun, *Deus Christianorum*, 141 ff.). In this way we can be present, in T.'s pages, at the birth of a systematic theological reflection (cf. J. Moingt, *Théologie Trinitaire*, I, 8 ff.), but we do not have to believe in T. as the founder of Western *theology: he borrowed his thought from more than one source, but developed and organized it more systematically. We can say something analogous of his concept of the church, called "mother" throughout his works, even in the Montanist period (cf. *Ad mart.* 1, 1; *De orat.* 2, 6; *De bapt.* 20, 5; *De pud.* 5, 14, etc.). For other aspects it is instructive to follow the way in which his idea of the church, which at first follows the line of previous *tradition, of which *Irenaeus too had been an interpreter, reflects in the last period the viewpoint of *Montanism. On the other hand it is interesting to stress the character of his *eschatological doctrine, marked, especially in some treatises, by clear opposition to *gnostic interpretations (think of the defence of the material element in man, the reasons invoked to prove the *resurrection of the *flesh, etc.), which require the development of an *anthropology based on the *Genesis account. On yet other points the content of his writings is important for understanding living tendencies in the Christian communities of his time: on the *Eucharist, primacy, *traducianism, *penitence, the virginity of *Mary, which - to oppose the *Docetist heresy - he denies *in partu* and *post partum*. T. then, is architect and witness of a thought, a milieu and an era, and much is known of him; and yet criticism, which in recent decades has dedicated an imposing number of studies to him, is far from unanimous in delineating his personality. If we could superimpose, like photographic negatives, the different portraits critics have drawn of the African writer, only a few common traits would appear clearly, while many others would not coincide, so that in the end it would be hard to identify the subject (cf. J.-C. Fredouille, *Tertullien*, 17).

We can usefully identify at least three areas on which the most lively recent discussions have turned. The first concerns T.'s attitude towards the *res publica* and *pagan society. It is often said that T. was one of the most rigid and inflexible representatives of the position which held that Christianity was irreconcilable with the *saeculum* (C. Guignebert: to the names cited here as examples others could easily be added); it is said that he proclaimed the separation of Christians from *Romanitas*, since there could be no link of solidarity between the believer and those who surrounded him (C. N. Cochrane); whence some have seen in his character a revolutionary vein, founded on *apocalyptic hopes and dominated by exaltation of *martyrdom, so much as to desire the overthrow of the society of his time (W.H.C. Frend) - a position very differently nuanced by other critics (J.M. Hornus, W. Rordorf) and completely denied by others, on the basis of an opposed thesis: T. as representative of a position aimed at preparing a harmonious understanding between Christianity and Roman *civitas* (R. Klein).

Another area, which has given material for many debates, which still continue, is that relating to the author's attitude towards ancient *philosophy and towards reason itself. Some have maintained that he radically opposed philosophy and that his anti-philosophism originated its most celebrated formulae when it developed into anti-rationalism; he rejected philosophy as such, not just certain systems or certain philosophies of his time; in short, he showed a substantial mistrust, which appears in the opposition between the irrational truth of the Gospel and the rational "truths" discovered by the human mind (C.G. Jung; A. Labhardt). Other scholars come to quite different conclusions. Starting from the very passages invoked to support his anti-rationalism (cf. e.g. *De carn. Chr.* 5, 4; *De praescr. haer.* 7, 9), the problem is reproposed with reference to other data that seem essential to an understanding of it, esp. in a literary-historical perspective (Decarié, Siniscalco, Ayers, Sider). In fact it allows us better to understand T.'s contest with the heretics, and with pagan philosophers in that they professed theories which contributed to the elaboration of heretical speculations, and, on the other hand, not to forget the anti-philosophical tradition that existed well before him (cf. J.-C. Fredouille, *Tertullien*, 337 ff.) and at the same time not to be surprised by the use he makes of philosophical concepts to construct his *theology (R. Braun, J. Moingt).

Finally a third area, the linguistic and literary, has shown itself to be the most fertile in results and the most capable of furnishing useful interpretative keys (also apropos of the writer's attitude towards philosophy and reason). Alongside original elements - present in great number in his works - this illuminates the strength of literary tradition, not just in the use of the *virtutes dicendi* or in the taste imposed by the movement of the Second Sophistic, but even in the elaboration of the structure of his writings, his search for means of argument proper to the philosophical logic dear to the ancients. In particular the study of his work in the light of his *rhetoric (as regards *elocutio* and, above all, *dispositio* and *inventio* in discourse) has contributed to a better identification of T.'s features (Siniscalco, *Ricerche*, 77 ff.; R.D. Sider, *Ancient Rhetoric*; J.-C. Fredouille, *Tertullien*, 29 ff.). The language used by T. makes a chapter of its own, of great interest and great utility: today he is no longer considered the creator of Christian *Latin, but rather the creator of theological Latin, as a structured language (cf. R. Braun, *Deus christianorum*, 10 ff.; 687 f.; Puente Sandidrian, 106 ff.). From the totality of recent bibliography, some reliable data of considerable importance on T.'s personality emerge, alongside others more vague and ambivalent. In any case, he has a place of considerable importance in the history of what was about to become the world of *late antiquity. His debt to earlier pagan and Christian authors (among the latter we should especially mention *Irenaeus) is manifest; nevertheless he is original and highly personal, as his attempts to harmonize not so much pagan culture as some elements of it (philosophical, legal, rhetorical, linguistic, etc.) with the Christian faith, and to reach new, even if sometimes imperfect and unsatisfactory, solutions, demonstrate. Certainly, on the basis of a well-grounded analysis of the texts, it is no longer possible to accept the image, often recurrent in textbooks and monographs, of T. as a man and writer determined to demonstrate in theory and practice the irreconcilability of Christianity and the pagan world. His life and writings are, in short, more consistent and complex than was thought even a few years ago.

CPL 1-36; PL 1-2; PLS 1, 29-32; CSEL 20, 47, 69, 70, 76; CCL 1-2. For critical editions of individual works, see the *Bibliographia selecta*, in CCL 1, XII-XIV (up to 1954) and the *Index bibliographicus*, in R. Braun, *Deus Christianorum*. 2nd ed., 596-98 and 725-26 (up to 1976). Furthermore, from 1975 on, *REAug* publishes annually a *Chronica Tertullianea*, whose first section is dedicated to editions (other sections give bibliographical references and critical information on publications concerning our author). G. Claesson, *Index Tertullianeus*, 3 vols., Paris 1974-1975.
Translations: Fr.: D. de Genoude, *Oeuvres*, 3 vols., Paris-Louvain 1852; Ger.: K. Kellner-G. Esser, *Private und katechetische Schriften. Apologetische, dogmatische und montanistische Schriften*, BKV² 2 vols., Kempten-Munich 1912-1916; Dutch: H.U. Meyboom, *Oud-Christelijke Geschriften in Nederlandesche vertaling*, 6 vols., Leiden 1928 ff. There are many translations of one or more works in modern languages: cf. *supra*, the references given for editions. We should, however, point out the initiative of the series *Sources Chrétiennes* which intends to make available the entire Tertullianean *oeuvre* with introd., text, Fr. tr. and notes; works so far published are: nos. 35, 46, 173, 216 f., 280 f., 310.
Editions since 1983: J.H. Waszink-J.C.M. van Winden, *Tertullianius, De Idololatria* (critical text, trans. and commentary), VChr. Suppl., Leiden 1987; C. Moreschini-J.C. Fredouille, *Tertullien; Exhortation à la Chastité*, SCh 319 (1985); M. Turcan, *Tertullien, Les Spectacles*, SCh 332 (1986); P. Mattei, *Tertullien, Le mariage unique*, SCh 343 (1988).
The critical bibliography is vast: besides CCL 1, XV-XXV and R. Braun, *Deus Christianorum*, 599-623, 726-732, cf. J. Quasten, *Patrologia*, Turin 1967, 494 ff.; B. Altaner-A. Stuiber, *Patrologie*, Freiburg i. Br. ⁸1978, 118 ff. and 149 f. Here we will confine ourselves to mentioning a small part of it, referring, though not exclusively, to that used in drawing up the present article: C. Guignebert, *T. Etudes sur les sentiments à l'égard de l'Empire et de la société civile*, Paris 1901; P. Monceaux, *Histoire littéraire de l'Afrique chrétienne*, I, Paris 1901; A. D'Alès, *La théologie de T.*, Paris 1905; C.G. Jung, *Psychologische Typen*, Zürich 1921; R. Roberts, *The Theology of T.*, London 1924; J. Lortz, *T. als Apologet*, 2 vols., Münster 1927; J. Morgan, *The Importance of T. in the Development of Christian Dogma*, London 1928; C.L. Shortt, *The Influence of Philosophy on the Mind of T.*, London 1933; C. N. Cochrane, *Christianity and Classical Culture*, New York 1940; A. Labhardt, *T. et la philosophie ou la recherche d'une "position pure"*, MH 7 (1950) 159-180; B. Nisters, *T. Seine Persönlichkeit und seine Schicksal*, Münster 1950; F. Refoulé, *T. et la philosophie*, RSR 30 (1956) 42-45; H. Finé, *Die Terminologie der Jenseitsvorstellungen*, Bonn 1958; J.M. Hornus, *Evangile et Labarum. Etude sur l'aptitude du christianisme primitif devant les problèmes de l'Etat, de la guerre et de la violence*, Geneva 1960 (Eng. ed. reviewed, Scottdale [Pennsyl.]-Kitchener [Ontario] 1980); P. Siniscalco, Il motivo razionale della resurrezione in due passi di T. (Apol. 48, 4; De res. 14, 3 ss.), AAT 95 (1960-1961) 195-221; V. Decarié, Le paradoxe de T., VChr 15 (1961) 23-31; W.H.C. Frend, *Martyrdom and Persecution in the Early Church*, Oxford 1965; J. Moingt, *Théologie Trinitaire de T. Histoire, Doctrine, Méthodes*, 4 vols., Paris 1966-1969; P. Siniscalco, *Ricerche sul "De resurrectione" di T.*, Rome 1966; R. Klein, *T. und das römische Reich*, Heidelberg 1968; W. Rordorf, T.s Beurteilung des Soldatenstandes, VChr 23 (1969) 105-141; T.D. Barnes, *T. A historical and Literary Study*, Oxford 1971; R.D. Sider, *Ancient Rhetoric and the Art of T.*, Oxford 1971; R. Martini, T. giurista e T. padre della Chiesa, SDHI 41 (1975) 79-124; J.C. Fredouille, *T. et la conversion de la culture antique*, Paris 1972; D. Powell, Tertullianists and Cataphrygians, VChr 29 (1975) 33-54; R. Braun, *Deus Christianorum. Recherches sur le vocabulaire doctrinal de T.*, Paris ²1977; J. Daniélou, *Les origines du christianisme latin*, Paris 1978; J. Gaudemet, *Le droit romain dans la littérature chrétienne occidentale du IIIᵉ au Vᵉ siècle*, Milan 1978, 15 ff.; R.H. Ayers, *Language, Logic and Reason in the Church Fathers: A Study of T., Augustine and Aquinas*, Hildesheim-New York 1979; R. Braun, Nouvelles observations sur le rédacteur de la "Passio Perpetuae", VChr 33 (1979) 105-117; P. Puente Sandidrian, T. y el latín cristiano. Revisión de las diversas posiciones, Durius 6 (1978) 93-115; R.D. Sider, Credo quia absurdum?, Class. World 73 (1980) 417-19; S. Vicastillo, *Tertuliano y la muerte del hombre*, Madrid 1980; R.F. Evans, *One and Holy; the Church in Latin Patristic Thought*, London 1972, ch. 1; C. Aziza, *Tertullien et le Judaisme*, Nice 1977; G.L. Bray, *Holiness and the Will of God. Perspectives on the Theology of Tertullian*, London 1979; R.D. Sider, Approaches to Tertullian: a study of recent scholarship, Second Century 2 (1982) 228-260; G. Hallonsten, *Satisfactio bei Tertullian*, Malmö 1984; id., *Meritum bei Tertullian*, Malmö 1985; L.I. van der Lob, Tertullian on the continued existence of things and beings, REAug 34 (1988) 14-24.

P. Siniscalco

TESTAMENT OF O. L. J. C. (TESTAMENTUM DOMINI). Various works bear this name. The best-known is a liturgical-canonical treatise preserved in *Syriac. It presents itself as a work of the *apostles reporting a conversation that Jesus had with them after his *resurrection. There we find: *1)* a description of the signs that will precede the coming of *Antichrist; *2)* some rules for the building of churches; *3)* prescriptions concerning the *ordination and duties of *clerics, the *Eucharist, *baptism and Christian life. This third part is based on the *Apostolic Tradition (as we read in the Latin, *Coptic, *Arabic and *Ethiopic collections), but the *prayers are considerably developed, while other prayers and rites are added. Some prayers of the T. are still used in the Syriac Pontifical. The same treatise exists in Arabic and Ethiopic versions. Another "Testament of O. L. J. C.", in Ethiopic, published by H. Guerrier and S. Grébaut in PO IX, 2, has nothing to do with the former and is of less interest.

I.E. Rahmani, *Testamentum Domini nostri Jesu Christi*, Mainz 1899 (Syriac and Lat. tr.); F. Nau, *La version syriaque de l'Octateuque de Clément traduite en français*, Paris 1913, 18-77.

P. Nautin

TESTIMONIA. J.R. Harris coined this term to indicate the systematic collections of citations, particularly of the Prophets, Psalms and OT generally, whose existence was recognized at the basis of some pages of the oldest Christian literature, characterized by some stereotyped citations, textual variants and tendentious interpretations (*Testimonies*, I-II, London 1916-1920). It was hard to admit that these texts came directly from the Hebrew or Greek *Bible; out of their original context, these elements seemed to evolve, to illuminate each other, to change attribution, to tend towards *midrash. The very work of selection that was behind such collections favoured certain themes, not just the person and work of Christ, but of all those who attracted controversy, reflection, worship. We have here an archaic phase of primitive *theology, even earlier than the redaction of the first gospel (K. Stendahl, *The School of St Matthew*, Uppsala 1954), earlier than *Paul, *John (C.H. Dodd, *According to the Scriptures*, London 1952; id., *Historical Tradition in the Fourth Gospel*, Cambridge 1965), the Acts of the Apostles (J. Dupont, *Etudes sur les Actes*, Paris 1967, 245-390) and, naturally, than *Barnabas, *Justin, etc. The correct exegesis of many pages of early Christian literature supposes the reconstruction of these dossiers and their history. The only surviving collections are already rather elaborate and revised on the originals (Cypr., *Ad Quirinum* [ed. R. Weber, CCL 3, Turnhout 1972] and Basil, *Moral rules* [PG 31, 653-869]).

J. Daniélou, *Etudes d'exégèse judéo-chrétienne (Les Testimonia)*, Paris 1966; E. Lupieri, *Il cielo è il mio trono. Is. 40, 12 e 66, 1 nella tradizione testimoniaria*, Rome 1980; L.W. Barnard, The Use of Testimonies in the Early Church and in the Epistle of Barnabas, in *Studies in the Apostolic Fathers and their Background*, Oxford 1966, 109-137; O. Skarsaune, *The Proof from Prophecy: a study in Justin Martyr's proof-text tradition*, Leiden 1987.

J. Gribomont

TESTIMONY - WITNESS. Very frequent terms. Their presence confers a character of truthfulness on an account. "Testimony" is the active form: coming forward and testifying as a witness (Homer, *Odyssey* XI, 235). The original meaning, linked to its etymology, suggests becoming aware of a remote reality that cannot be forgotten, but must be brought to the knowledge of others (Plato, *Symposium* 179b).

In the Greek *Bible, we should look at those cases where "testimony" recalls the meaning of confirmatory evidence recording a precise event, or that of evidence of the Law or the Covenant. The terms tent of meeting (testimony) and ark of the covenant (testimony) call to mind respectively the time when the agreement with Jahweh was made and the place where the evidence of this is preserved (cf. Ex 29, 4 ff.; 40, 2 ff.; Lev 4, 4 ff.; Num 4, 25 ff.). The OT knows no testimony expressed on the basis of unverifiable subjective conviction, but supposes personal knowledge of the event.

The NT, while considering testimony based on fact to be fundamental, allows room for subjective conviction of (religious) truth for which proofs are admitted and for which one is prepared to give one's life (cf. Mk 6, 11 and par., 13, 9; Lk 21, 13). The terms "witness" and "testimony" occur with about the same frequency: their judicial use in the condemnation of Jesus on the evidence of false witnesses (cf. Mk 14, 55. 56. 59. 63 and par.) and in that of *Stephen (Acts 6, 13; 7, 58) is important. In *Paul, the message of salvation in Christ, object of the proclamation, coincides with testimony (cf. 1 Cor 2, 1; 2 Th 1, 10); Acts, however, discounting the utterances of human testimonies, introduce the meaning of bearing witness understood as preaching Christ, entrusted to Paul in *Rome (Acts 23, 11; cf. 4, 33). The *Apostles, the disciples, especially since they are in a position to testify to the *resurrection and works of Jesus (Acts 2, 32; 3, 15; 13, 31; 22, 15; 26, 16 ff.), are charged to spread the message. The Johannine writings present testimony in the form of showing in the Baptist, in personal form in Jesus and as proclamation in the disciples.

In the fourth gospel, testimony has as its particular object, besides the documentable facts, God's self-communication in Jesus in line with the *prophets. This concept, typically Johannine, is supplemented by the testimony of the *Holy Spirit (Jn 15, 25. 27) and by that of those who have accepted the Word, thus confirming God's truth (Jn 3, 33; 4, 39). In 1 Jn 5, 10, testimony, knowledge and message of faith are identical. Here the experience of the senses does not say much, but only the involvement of *faith.

In patristic literature testimony has great importance and combines the classical meaning with the biblical one. *God testifies in His favour in *Scripture (Clem. Al., *Paed*. III, 1), in Christ's *baptism (*Const. Apost*. II, 32, 2). The Christian, faithful to the choice he has made, is a witness of Christ (*To Diognetus* 12, 6; Clem. Al., *Strom*. IV, 6; Orig., *C. Cels*. I, 8) and, urged by charity (John Chrys., *Hom. 27, 3 in Rom.*), gives his life for Christ and for the faith (Euseb., *HE* V, 21, 4; John Chrys., *Hom. 66 in Gen.*). In the texts of the *Fathers, where the abstract term is preferred, the semantic connotations do not change: God testifies in favour of Christ (Orig., *C. Cels*. VIII, 9), of the Christians (Just., *Dil*. 123-124; Tat., *Ad Graecos* 12). The latter's testimony to their faith is a thanksgiving to God (Clem. Al., *Strom*. IV, 4), made by virtue of the *love that is in them (Clem. Al., *Strom*. IV, 4. 7). The *martyrs, in giving their *blood, have borne witness to God the *Father (Athan., *C. Arianos* III, 10), to Christ (*Mart. Polyc*., 2, 25; Orig., *Mart*. 14), whom the heavenly Father has presented as his firstborn in baptism (Clem. Al., *Paed*. 16; Meth. Olym., *Symp*. VII, 1), and to the *Holy Spirit (Hipp., *In Danielem* 2, 21, 1; Athan., *De fuga* 22; Did., *De Trinitate* II, 1), and have proclaimed the superiority of the Christian religion to all others (Iren., *Adv. haer*. V, 9, 2; Clem. Al., *Strom*. IV, 5). The purity of God's testimony to Scripture, to the prophets, to Christ, to the bishops (*Const. Apost*. II, 25) is an object of contamination by the *heretics (Iren., *Adv. haer*. IV, 33, 9).

C.L. Masson, Le témoignage de Jean, *RThPh* 38 (1950) 5-63; L. Cerfaux, Témoins du Christ d'après le Livre des Actes, *Recueil L. Cerfaux* II, Gembloux 1954, 157-174; A. Vanhoye, Témoignage et vie en Dieu selon le quatrième évangile, *Christus* 16 (1955) 150-171; R. Koch, Témoignage d'après les Actes, *Masses Ouvrières* 129 (1957) 16-33; 131 (1957) 4-25; I. de la Potterie, La notion de témoignage dans Saint-Jean, *Sacra Pagina* II, Paris 1959, 193-208; H. Strathmann, μάρτυς, *GLNT* 6, 1269-1392; *Enc. Bibbia*, Turin 1971, 6, 907-910; E. Günther, Zeuge und Martyrer, *ZNTW* 47 (1956) 145-161; H.A.M. Hoppenbrouwers, *Recherches sur la terminologie du martyr de Tertullien à Lactance*, Nijmegen 1961.

E. Peretto

THALASSIUS. *Abbat of a *monastery in the Libyan *desert (c.650). *Friend of *Maximus Confessor; wrote a *Collection of maxims* (in four centuries) on the *virtues and on efforts to reach perfection.

CPG 3, 7848; PG 91, 1427-70; M. Th. Disdier, Le témoignage spirituel de Thalassius le Libyen, *REByz* 2 (1944) 79-118.

G. Ladocsi

THALASSIUS of Angers. Mid 4th-c. bishop of *Angers (*Andegavum*). On the occasion of his *ordination there, a *council was held (Mansi 7, 941), in which he took part. We have a *letter to him from *Lupus of Troyes replying to some questions of his (CPL 988) on the way to celebrate the *Christmas, *Epiphany and *Easter vigils and on the remarriage of minor *clerics.

PG 58, 66; L. Duchesne, *Fastes* II, 246-250; P.-M. Duval, *La Gaule jusqu'au milieu du Ve siècle*, Paris 1971, 808.

F. Cocchini

THEBAÏD. Southern region of *Egypt, one of the provinces into which *Diocletian divided the diocese of Egypt. It took its name from the ancient city of Thebes. In monastic texts this name is extended to designate generically the *desert on both sides of the Nile Valley, in which the monks took up residence to exercise the various types of *ascesis.

PWK 48, 1577-1582; *CE* 14, 6 f. (bibl.).

T. Orlandi

THEBAN Legion. St *Maurice of Agaunum is presented by legend as the commander of the T. L. which, for its collective refusal to *sacrifice to the gods, passes for having provided as many *martyrs as it counted soldiers, i.e. 6666. In fact in the 4th c. we know numerous legions which bore this surname: the *II Flavia Constantia*, stationed at Cusae, which dated from the time of *Diocletian, around 296-7; the *II Flavia Constantia Thebaea*, detached from the former by *Constantius, which depended on the *magister militum per Orientem*; the *II Felix Valentis Thebaea*, which probably formed a whole with the *II Valentiniana* in *Egypt; the *I Maximiana Thebaea* and the *II Diocletiana Thebaea*; the *Thebaea legiones* mentioned by *Ammianus Marcellinus; the *Thebaei*, stationed in *Italy. Of these bodies of troops, only those that bore the names of *Maximian or Diocletian were stationed in Europe, more precisely in *Thrace, and it is these that are identified with the T. L. of the *Passio mm. Acaunensium*. But the probability is that this is simply a hagiographical romance, whose topographical precision does not at all guarantee its historical accuracy.

PWK 12, 1467 (xxiv); *DACL* 10, 2701-2; *Vies des SS*. 9, 454-55; A. Dufourcq, *Etudes sur les Gesta martyrum romains*, II, Paris 1907, 15-33; *BS* 9, 193-205; *EC* 11, 1855-1857; D. van Berchem, *Le martyre de la Légion Thébane: essai sur la formation d'une legende*, Schweizerische Beiträge zur Altertumswissenschaft 8, Basle 1956; L. Dupraz, *Les Passions de S. Maurice d'Agaune*, Studia Friburgensia n.s. 27, Frieburg 1961.

V. Saxer

THECLA. Various saints of this name were venerated in antiquity: *1)* the T. of *Aquileia and Trieste is a duplicate of the saint of *Iconium; *2)* Santa Sofia, at Benevento, claims to possess the relics of the 12 children of Boniface and T., doubtless to be identified with the *martyrs of Hadrumetum in *Africa, or those of *Abitina; *3)* at Chamalikres in Auvergne, the relics of the saint of Iconium are venerated in a 7th-c. *monastery; *4)* in *Palestine, *Eusebius of Caesarea (*Mart. Palest.* 3, 1) and the *Mart. hier.* (24-25 March) mention a saint of this name; *5)* at *Rome, on the Via Labicana: this doubtless, in the *Mart. rom.* (25-26 March), is derived from a confusion with the preceding one; *6)* again at Rome, in 1963 a 3rd-c. catacomb was found on the Via Ostiense with the tomb of a martyr of this name.

G. Morin, La formation des légendes provençales. Faits et aperçus nouveaux, *RBen* 26 (1909) 24-33; Monceaux 3, 140-47; U. Fasola, Il complesso catacombale di S. Tecla, *RAC* 40 (1964) 19-50; id., La basilica sotterranea di S. Tecla, *RAC* 46 (1970) 193-288; *Vies des SS.* 9, 477-84; *LTK* 10, 18-19; *BS* 12, 172-81; G. Dagron, *Vie et Miracles de sainte Thecle*, Brussels 1978; D.R. Macdonald, *The Legend and the Apostle: the battle for Paul in Story and Canon*, I, Philadelphia 1983.

V. Saxer

THELA, *Council of.* On 24 Feb 418 some 30 bishops of *Byzacena met under the presidency of their primate, Donatianus of Thelepte, in the presence of Vincent and Fortunatian, legates to Proconsular *Africa. It is probable that they discussed problems relating to *Pelagianism, thus preparing for the plenary African council of 1 May 418. But what remains of the conciliar *Acts speaks only of the reading of the circular *letter sent by pope *Siricius to the African bishops, to communicate to them the decisions taken at the council of *Rome of Jan 386.

CCL 149, 54-65; Hfl-Lecl 2, 68-75; Palazzini 5, 298.

Ch. Munier

THEMISTIUS. 6th-c. deacon of *Alexandria and religious writer. Under patriarch *Timothy III of Alexandria (518-535) he had published his ideas on Christ's ignorance, views not shared by his bishop. Under Timothy's successor *Theodosius (535-566) he struggled so relentlessly as to consummate a *schism at *Constantinople and become the leader of the *Agnoetae. He died before 600.

Of his many works, only short fragments have survived. Their titles are: *Antirrheticus against the tome of Theodosius*, addressed to the empress *Theodora; *Treatise against *Colluthus*, adviser of the patriarch Theodosius, who had composed a book against T.; dissertation "Against those who say that in virtue of the one divine energy of Christ, his humanity had knowledge of everything"; *Apologia in twenty chapters*; *letters to the monk Carisius, the priest Marcellus and the deacon Stephen, to bishop Constantine of Laodicea, to the Salamites. He collaborated with *Conon and Eugenius in a work against *John Philoponos and, with Calonymus, in a *Defence* of St Theophobius against the Alexandrian monk Theodore, who replied with a volume criticized by T., and then answered this criticism with a work (no better identified) in three volumes.

The fragments were collected, with a precise aim, by an opponent of *monothelism. Hence they give a partial vision of T.'s thought, which, in exaggerating the deficiencies of Christ's humanity, tended to compromise the integrity of the incarnate Word, inclined as T. was to accept Severian *monophysism and monothelism.

CPG 7285-7292; PG 91, 172; *DTC* 15, 219-222; Beck 290. 391 ff.

S. Stiernon

THEOCTISTUS of Caesarea. Bishop of Palestinian *Caesarea, *friend and protector of *Origen. In agreement with *Alexander of Jerusalem, he allowed him to preach while still a *layman and answered the protests of *Demetrius of Alexandria (Euseb., *HE* VI, 19, 16-19). Later he ordained Origen priest, acc. to *Photius (*Bibl.*, cod. 118) - who confuses him with his successor *Theotecnus - and welcomed him to Caesarea after his expulsion from *Egypt. On Origen's death, he received a letter of condolence from *Dionysius of Alexandria (Photius, *Bibl.*, cod. 232, 291 B). He invited Dionysius to the council of *Antioch on the *schism of *Novatian (Euseb., *HE* VI, 46, 3). He survived the *persecutions of *Decius and *Valerian and died after 260 under *Gallienus (*ibid.* VII, 5, 1; VII, 14).

H. Crouzel

THEODELINDA (or THEODOLINDA). Daughter of Garibald, king of Bavaria, she married *Authari, king of the *Longobards (*c.*589). On Authari's death (590) she married his successor *Agilulf. She prepared the *conversion of the Longobards from *Arianism to *Catholicism and favoured better relations with the church (in particular with *Gregory the Great) though, like Agilulf, she did not adhere to the condemnation of the *Three Chapters. She obtained her husband's permission for their son Adaloald to receive Catholic *baptism. Various precious objects belonging to her are preserved, like the Iron Crown, in Monza cathedral treasury.

CPL 1719a; *EC* 11, 1925; O. Bertolini, I Papi e le Missioni fino alla metà del Secolo VIII, *La Conversione al Cristianesimo nell'Europa dell'alto Medioevo*, Settimane di Studio 14, Spoleto 1967, 327-363.

E. Malaspina

THEODORA. Eastern empress (527-548), wife of *Justinian I. Daughter of a hippodrome guard, born in *Constantinople in 497, she was an actress and courtesan before following a lover to the African Pentapolis. In the *East she came into contact with the *monophysite hierarchy, renounced her earlier conduct and returned, redeemed, to the capital, where she maintained herself by spinning wool. In 520 she met Justinian, who married her, after the abolition of restrictive laws and T.'s elevation to patrician rank (524/5). T. had an exceptional personality, a sharp mind, experience of life and an energetic character. As Augusta from 527 and her husband's co-regent, she played a great part in the leadership of the state, more than any other queen before or after her. Her name was cited in decrees and treaties; she demanded *proskynesis*, met ambassadors and exercised great influence on all political officials. She supported the monophysites. With her immense power she supported ecclesiastical and humanitarian initiatives. She appears majestic and autocratic at her husband's side in the *mosaic of S. Vitale, *Ravenna. She died 28 June 548 and was buried in the church of the Apostles at Constantinople. The contemporary historian Procopius, who expressed the interests of the senate and landowning aristocracy, distorted T.'s image, stressing her negative characteristics and also influencing subsequent biographers.

PWK 5A, 1776-1791; R. Browning, *Justinian and Theodora*, London 1971; A. Cameron, *Procopius and the Sixth Century*, London 1985; E.A. Fisher, Theodora and Antonina in the *Historia Arcana*, History and/or fiction?, *Arethusa* 11 (1978) 253-279.

J. Irmscher

THEODORE. T., a soldier, died a *martyr at *Amasea and was buried at Euchaita in *Pontus (Aukhat, Turkey), where his tomb became a place of *pilgrimages, famous from the 4th c., and where *Gregory of Nyssa preached a *homily (PG 46, 736-748) which is our earliest source of information on him. His feast was celebrated there on 17 Feb. Other churches in his honour were built at *Constantinople in 452 (*dedication 5 Nov), *Ancyra (5 Nov 518), *Asia Minor and all over the *East. At *Rome the saint is depicted in the apse *mosaic of SS. Cosma e Damiano and venerated in a church bearing his name at the foot of the Palatine (dedication 9 Nov). Evidence of his cult existed at Messina, *Ravenna, *Naples, in Latium, at Venice and Vercelli. In the 9th c., legend divided the saint into two: a soldier and a general, leading to a division of the liturgical feasts.

H. Delehaye, *Les légendes grecques des saints militaires*, Brussels 1909, 11-43, 121-202; P. Franchi de' Cavalieri, Attorno al più antico testo del Martyrium S. Theodori tironis, *ST* 22, 91-107; *Vies des SS.* 2, 150-52; *LTK* 10, 39-40; *BS* 12, 238-242.

V. Saxer

THEODORE. Bishop of Echinus in *Thessaly (first quarter of 5th c.). This city was Christianized after the victory of *Constantine. Of its church life we know little; we know only four of its bishops, of whom T. took part in the council of *Ephesus and subscribed its *Acts (Mansi 4, 1216 C; 6, 873 A).

G. Ladocsi

THEODORE. *Jacobite patriarch of *Alexandria (575-after 582), successor of *Theodosius after an interval. The circumstances of his *ordination are of interest for the history of *monophysism. *Theodosius the *archpriest, with his nephew *Theodore (of Copros) the *archdeacon, invited *Longinus, bishop of the Nobadae in *Nubia, to consecrate a new patriarch. On his way to Alexandria, in the church of St Mennas in the desert *monastery of Rhamnis, Longinus consecrated *abbat T., of Syrian origin, to the patriarchate almost against his will. *Paul, suspended patriarch of Antioch, whom Longinus found in the monastery of Mareotis, collaborated in this illegal ordination. They informed the Alexandrian Christians of the fact by *letter; the *clergy were indignant, since they already had their patriarch in the person of Peter. T. remained in his *monastery, was a humble and peaceable monk and took little notice of his consecration. After Peter's death, the Alexandrians ignored him and elected *Damian. His protectors could do nothing for him. T. was still alive in the year of Paul of Antioch's death (582).

G. Ladocsi

THEODORE, pope (642-649). T.'s father was a Greek from *Jerusalem. Succeeding *John IV, the first months of his pontificate coincided with the revolt of the *cartularius* Maurice at *Rome against the exarch of *Ravenna (*LP* I, 331 f.). T.'s election was a sign of the close relations between Rome and *Palestine which, with *Africa, was a centre of opposition to *monothelism (Bertolini, 327 f.). T. continued those relations, deposing Sergius of Jerusalem and appointing his vicar, Stephen of Dor, mainly to safeguard *orthodoxy there (Mansi 10, 891). Soon after his election he received three *letters from *Constantinople: from the emp. *Constans II; the synodical letter of *Paul, the new patriarch who had succeeded *Pyrrhus; and a letter from the consecrating bishops, addressed to John IV, who had meanwhile died. T. answered all three (Jaffé 2049 f.; 2052): from Paul he required as condition of his recognition a regular trial to depose Pyrrhus, and from all he required the condemnation of the monothelite doctrine. Meanwhile Pyrrhus had withdrawn to Africa, where in 645 he was refuted in public debate by *Maximus Confessor (Mansi 10, 709-760); from there he went to Rome, where he publicly and solemnly abjured and was treated with all honours (*LP* I, 332); but subsequently, having fled to Ravenna, he returned to monothelism, for which T. *excommunicated him (*LP* I, 332). Meanwhile T.'s relations with Paul had also become more tense; under pressure from the African bishops (Jaffé 2055), he sent a delegation to Constantinople to demand a profession of faith from Paul, who replied with a clear adhesion to monothelism (Mansi 10, 1019-1026), for which the *pope deposed him (*LP* I, 333; Mansi 10, 878) and the patriarch responded with harsh reprisals (*LP* I, 333). In 648 the emp. Constans, with the intention of overcoming the disputes, published the *Typos* (Mansi 10, 1029-1039), which was condemned together with *Heraclius's *Ekthesis* at the Lateran *council of 649. T. built various churches at Rome, in particular the chapel of St Venantius in S. Giovanni del Fonte, where he is depicted in a *mosaic. He died 14 May 649 and was succeeded by *Martin I.

CPL 1732; *Verzeichnis* 558; PL 78, 75-102; PLS 4, 1997-1999; *LP* I, 331-335; III, 94; *DTC* 15, 224-226; *CE* 14, 16; Fliche-Martin 5, 230-235, 531 f.; G. Ladner, Die Bildnisse der östlichen Päpste des 7. und 8. Jahrhunderts in römischen Mosaiken und Wandgemälden, in *Atti del V Congr. Inter. Studi Bizantini*, II, Rome 1940, 169-182 (repr. 1978); O. Bertolini, *Roma di fronte a Bisanzio e ai Longobardi*, Bologna 1941, 327-337; P. Conte, *Chiesa e Primato nelle lettere dei Papi del secolo VII*, Milan 1971, 433-442 (transmission and discussion of the correspondence; also cf. index p. 585); L. Magi, *La sede romana nella corrispondenza degli imperatori e patriarchi bizantini (VI-VII sec.)*, Rome-Louvain 1972, 212-221; A.N. Stratos, Il Patriarca Pirro, *Byzantina* 8 (1976) 9-19 (in Greek); J. Richards, *The Popes and the Papacy in the early Middle Ages, 476-752*, London 1979, ch. 2.

A. Di Berardino

THEODORE, presbyter. The priest T., twice candidate for bishop of *Rome, is prob. the same as the homonymous Roman legate at the 6th ecumenical *council (*Constantinople 680). The first time he was supported by the military party against the *archpriest Peter, the *clergy's candidate. After two months of wrangling, the two parties agreed on Conon, who was consecrated *pope 21 Oct 686. Conon dying in 687, T., now an archpriest, entered the lists against the archpriest Paschal. Once again a third candidate, the priest *Sergius, was preferred. When he gained entry to the Lateran by forcing the doors, T., unlike Paschal, desisted from any form of resistance and professed obedience to him.

LP I, 350. 368 and 371 f.; *EC* 11, 1932.

A.V. Nazzaro

THEODORE. Mid 7th-c. bishop of Pharan in *Arabia; acc. to the documents of the council of *Rome (649), he was the first representative of *monoenergism, who tried to spread his doctrine by means of the patriarch of Constantinople, *Sergius I. Some theologians consider him a *monophysite, others a Chalcedonian. The 6th ecumenical *council (*Constantinople, 681) condemned him, and local councils cite some extracts from his *letter to Sergius of Arsion, in Egypt, and his commentary on the words of the *Fathers (cf. *Theodore of Raithu).

Mansi 11, 555 ff.; *DCB* 4, 948, 35; *DTC* 15, 279-82; *LTK* 10, 44.

G. Ladocsi

THEODORE ASCIDAS. *Archbishop of Cappadocian *Caesarea, died Jan 558 at Constantinople. *Abbat of the New *Laura and a warm admirer of *Origen, T. secretly favoured *monophysism. He took part in the council of *Constantinople of 536 and remained in the capital, where he enjoyed great credit with the empress *Theodora and became an intimate of the emp. *Justinian (Evagr. Schol., *HE* IV, 38). In 537 he obtained the metropolitan see of *Cappadocia, but spent more time in the capital than in his diocese. When Justinian and patriarch *Menas condemned the *Origenists (543), T. subscribed the edict while continuing to favour them secretly. In 543 he suggested to the emperor that the best way to bring the monophysites back to unity was to condemn the so-called *Three Chapters as tainted with *Nestorianism, which Justinian did. This act aroused a vast controversy, during which pope *Vigilius, already under surveillance, prepared a sentence of deposition against T. (July 551), which was published 14 Aug 551. In 552 T. made his submission; he passed his last years in *peace.

CPG 3, 6988; Hfl-Lecl 3, 1-5. 13 f. 62 f. 88. 95; F. Diekamp, *Die origenistischen Streitigkeiten im sechsten Jahrhundert und das fünfte allgemeine Concil*, Münster 1899, 37-40, 50-52; J. Pargoire, *L'Eglise byzantine de 527 à 847*, Paris ³1923, 34-40; L. Duchesne, *L'Eglise au VIᵉ siècle*, Paris ²1925, 168-175, 206-108; *EC* 2, 151; A. Guillaumont, *Les "Kephalaia gnostica" d'Evagre le Pontique*, Paris 1962, 128-36, 173-175; W.H.C. Frend, *The Rise of the Monophysite Movement*, Cambridge 1972, 279 f.; L. Perrone, *La chiesa di Palestina e le controversie cristologiche*, Brescia 1980, 75.

G. Ladocsi

THEODORE bar Koni. We know nothing precise of the life of this *Nestorian author, seemingly from the village of Kaskar (prob. second half of 8th c.). Abdiso's *Catalogue of Syrian authors* attributes to him an *Ecclesiastical History*, ascetic instructions, *sermons of consolation for the dead and a collection of scholia. Only the latter is preserved: an encyclopaedic compilation, whose sources are not cited and which concerns *theology, philosophy, *exegesis and *apologetic, it has never yet received an unabridged edition. J.B. Chabot sums up its contents thus: "It is divided into 11 books. Books I-V contain the scholia on the OT, books VII-IX those on the NT. Book VI, which is like the introduction to the NT, contains mainly theological definitions; book VIII consists of two treatises, one against the *orthodox and the *monophysites, the other against the *Arians; book X is a dialogue between a Christian and a *pagan, who really represents the objections of the Muslims. Book XI is a treatise of *heresies, derived from St *Epiphanius but amplified with very interesting information on the *Manichees, *Mandaeans, Kantaeans and several other lesser-known Oriental sects".

CSCO, *Script. syri* 55; 56; 69; 187; 193; 194; Abdiso, *Cat. Libr. syror.* 133 (Assemani, Bibl. Orient., III 1, Rome 1725, 198-199); Duval 368-369; Baumstark 218-219; Chabot 107-108; Ortiz de Urbina 216-21; L. Brade; *Untersuchungen zum Scholienbuch des Theodoros bar Konai; die Übernahme des Erbes von Theodoros von Mopsuestia in der nestorianischen Kirche*, Wiesbaden 1975; id., Die Herkunft von Prologen in den Paulusbriefexegesen des Theodors bar Konai und Ishodad von Merv, *OC* 60 (1976) 162-171; S.H. Griffith, Chapter ten of the "Scholion": Theodore bar Kônî's Apology for Christianity, *OCP* 47 (1981) 158-188.

J.-M. Sauget

THEODORE of Aquileia. At the time when *Constantine promulgated the *Edict of Milan (313), the church of *Aquileia was ruled by T., whom the *episcopal lists make the 4th or 5th Aquileian bishop. The dates of his episcopate are disputed (308-319?), but he took part in the anti-Donatist synod of *Arles in 314, where, with his deacon Agatho, he subscribed the synodal *acts as *episcopus de civitate Aquileiensi, provincia Dalmatiae*. But his chief monument is the splendid *mosaic floor of the first *church buildings built by him in the rich and populous Adriatic metropolis immediately after the edict of freedom.

G. Cuscito, *Cristianesimo antico ad Aquileia e in Istria*, Trieste 1977, 156-168 (with bibl.).

G. Cuscito

THEODORE of Bostra. Together with *Jacob Baradaeus (542-578), T. was of decisive importance for the survival of the *monophysite (*Jacobite) church after the death of *Severus, patriarch of *Antioch (512-518, †538). When the Ghassanid Emir Hareth bar Gabala asked the empress *Theodora for two monophysite bishops for the regions of Mesopotamia, *Syria, *Palestine and *Arabia, the monophysite patriarch *Theodosius of Alexandria, in exile at *Constantinople, named (542) Jacob Baradaeus (whence "Jacobite" church) for the see of *Edessa and T., a monk of Arabia, for the province of Arabia Secunda, whose metropolis was *Bostra. T. never resided in Bostra, but in the Field (*hertha*) of the Saracens, to be located in Golan. His episcopal activity is unknown, but was certainly less than Jacob's. While at *Constantinople in 564, he approved, in his *Letter to Paul of Beth Ukkame*, the election of *Paul as patriarch of Antioch (J.B. Chabot, *Documenta ad origines monophysitarum illustrandas*, CSCO 17, 94-96: *text*; 113, 65-66: *translation*). In 567 he was in the *East, in 569 again at Constantinople, where he attacked the *tritheists and *excommunicated *Conon of Tarsus. In 570, while in the East, he was invited by *Justin II (565-578) to discussions on the faith at Constantinople, whence he managed to return to the East with a safeconduct, despite the opposition of patriarch *John III. He died soon after 570. We have two fragments of *Letters written with Jacob, one to the monks, exhorting them to guard themselves against the tritheists, the other to Paul of Antioch comforting him in his persecutions (J.B. Chabot, *op. cit.*, CSCO 17, 165 f. and 179 f.: *text*; 113, 115 f. and 125 f.: *version*).

CPG 3, 7201; cf. 7173 f.; E. Honigmann, *Evêques et évêchés monophysites d'Asie Antérieure au VIᵉ siècle*: CSCO 127, Subsidia 2, Louvain 1951, 157-164. 169. 172. 176. 185. 201, 1. 211, 5.

A. De Nicola

THEODORE of Canterbury. Greek monk originally from *Cilicia, but residing at *Rome. On 26 March 668, aged 70, he was ordained *archbishop of *Canterbury by pope *Vitalian on the recommendation of abbat *Hadrian of Nisida, who followed T. to England. Here the condition of the church had been deteriorating for some time: the lack of bishops and the inadequacy of the few who remained had harmful consequences for all aspects of religious life. With great energy T. set about suppressing abuses and filling the gaps in the episcopate. In 673 he called a *council at Hertford to give the English church a stronger disciplinary organization. He promoted studies, with Hadrian's collaboration, and introduced Gregorian *chant into England. He did not hesitate to depose *Wilfrid from his see of *York as insufficiently deferential to his *authority. His great achievement was the reorganization of the English dioceses: he divided those of Mercia and Northumbria, which were too big, managed to rein in particularistic tendencies and gave unity to the whole ecclesiastical *organization. With him, after the troubled outcome of *Augustine's mission, the church of England took on its definitive appearance. He died in 690.

CPL 1885; DCB 4, 926-932; W. Reany, *St. Theodore of Canterbury*, Saint Louis 1944; BS 12, 243 f; H. Mayr-Harting, *The Coming of Christianity to Anglo Saxon England*, London ²1977, Part 1.

M. Simonetti

THEODORE of Constantinople. Deacon and also rhetor and *synodicarius* of patriarch *Paul II of Constantinople (641-654), author of two *aporiae* (questions or difficulties) against the *orthodox doctrine of two wills in Christ. They occur among the writings of *Maximus Confessor (580-662; PG 91, 216 f.), who answered them for the presbyter Marinus (PG 91, 217-228). The first *aporia* puts ignorance and will on the same level in Christ and hence deduces that, in Christ, will must be predicated κατ' οἰκείωσιν "according to appropriation", like ignorance; Maximus demonstrates that the *aporia* is contradictory on its own terms: in fact, will means the positing of a thing, while ignorance is its destruction; so if will and ignorance have the same *raison d'être* (λόγος), it follows that there is ignorance in *God and that inanimate things have will; then he lists the consequences to *christology (in Christ there would be two ignorances and two wills; according to one's way of understanding appropriation, one would fall into *Docetism, *Apollinarism, etc.). The second *aporia* claims that the term natural will, not having been used by the *Fathers, involves a new doctrine; Maximus proves the patristic use of the expression, and asserts that the Fathers attributed will to nature.

CPG 3, 7632; Bardenhewer 5, 20; Beck 433.

A. De Nicola

THEODORE of Copros. Monophysite deacon of Copros (or Copris near *Alexandria), second half of 6th c. Perhaps the same man described by *John of Ephesus as an *archdeacon. With the presbyter and church attorney (*ecclesiekdikos*) *Theodosius (John of Ephesus speaks of an *archpriest Theodosius), he invited *Longinus, bishop for six years of the Nobadae (a people of *Nubia, near the frontier of the Byzantine empire), to come secretly to Mareotis (near Alexandria) to consecrate a patriarch for the *Jacobite church of Alexandria, whose see had been vacant for nine years: the Syrian *archimandrite *Theodore was ordained (575; John of Ephesus IV, 9, in E.W. Brooks, CSCO 105, 188 f.: *text*; 106, 140 f.: *version*). A fragment of the *Letter to Longinus* is in J.B. Chabot, *Documenta ad origines monophysitarum illustrandas* (CSCO 17, 273 f.: *text*; 103, p. 191: *version*).

CPG 3, 7225; E. Honigmann, *Evêques et évêchés monophysites d'Asie Antérieure au VIᵉ siècle*, CSCO 127, Subsidia 2, Louvain 1951, 226 f.

A. De Nicola

THEODORE of Heraclea. Bishop of Heraclea in *Thrace; one of the leaders of the anti-Nicene front set up by *Eusebius of Nicomedia in the *East, c.330. At the council of *Tyre (335) he was a member of the commission that went to Mareotis to investigate *Athanasius's actions. In 342 he took part in the commission that visited the emp. *Constans at *Milan to try, unsuccessfully, to put an end to the disputes between Easterners and Westerners. He took part in the council of *Sardica (343) and was one of the bishops whom the Westerners *excommunicated by name. In c.355 he was again working against Athanasius, who had returned to *Alexandria. His death is put around this date. *Jerome (*Vir. ill.* 90) says he was active in the field of *exegesis, with *commentaries on *Matthew, *John, Acts, *Psalms: clear, precise writings concentrating on literal interpretation. Surviving fragments on Matthew and John, handed down in the *catenae, substantially confirm this judgment, though *allegory is not completely absent (cf. frs. 45 and 124 on Mt). Some fragments published under his name (e.g. frs. 117 and 126 on Jn, 134 on Mt) are hard to consider authentic, since they relate to a later theological context.

CPG 3, 3561-67; DCB 4, 933; LTK 10, 40; J. Reuss, *Matthaeus-Kommentare aus der griechischen Kirche*, TU 61, Berlin 1957, 55-95; id., *Johannes Kommentare aus der griechischen Kirche*, TU 89, Berlin 1966, 65-176; Simonetti 596.

M. Simonetti

THEODORE of Marseilles. Living at the time of the intestine struggles which, for dynastic reasons, convulsed the city of *Marseilles (*Massilia*) after the death of the Merovingian king Lothar I in 561, T. took an active part in them after he became bishop, just as he concerned himself with the religious reordering of a diocese equally tormented and rebellious. Though imprisoned twice (the second time for supporting the regal pretensions of Gundovald, who came from *Constantinople, perhaps sent by the emp. Tiberius II: hence the hypothesis of T.'s Eastern origin) (Greg. of Tours, *Hist. Franc.* VI, 24) and long persecuted, he always miraculously escaped death (Greg. of Tours, *Hist. Franc.* VIII, 12). He took part in the synod of Mâcon (Mansi 9, 957). *Gregory the Great wrote to him advising greater leniency in the work of Christianizing the *Jews (*Ep.* 47). He died 2 Jan 594.

BS 12, 253-257.

F. Cocchini

THEODORE of Martigny. First bishop of Martigny (*Octodurus*), present with *Ambrose at the council of *Aquileia in 381 (Mansi 3, 599) and that of *Milan in 390, subscribing the epistle sent by the synod to pope *Siricius (Mansi 3, 667, 689-690). Acc. to the *letter written by *Eucherius to bishop *Salvius, T. was the source of the information on the *martyrdom of St *Maurice and his companions at Agaunum (Euch., *Ep. ad Salv.*). He had rediscovered the bodies and built a basilica in their honour (Euch., *Pass. Acaun* 7). For the miraculous events attrib. to T. and the spread of his cult, cf. BS 12, 257-258.

L. Lathion, *Théodore d'Octodure et les origines chrétiennes du Valois*, Lausanne 1961.

A. Pollastri

THEODORE of Mopsuestia. As a pupil of his fellow-Antiochene *Diodore of Tarsus, T.'s birth can be fixed around 350. If he is the same Theodore addressed by *John Chrysostom, he was a monk and fellow-pupil at the school of *Libanius whom John dissuaded from abandoning the ascetic life for the *world. A priest in 383, he was ordained bishop of *Mopsuestia (*Cilicia) in 392 and died in 428. An outstanding representative of Antiochene *exegesis and *theology, he was esteemed and admired in life, but subsequently dragged into the Nestorian controversy, being considered, like Diodore, a precursor of *Nestorius's *christology. He became the favourite target of the *monophysites and was condemned at the council of *Constantinople of 553 (question of the *Three Chapters).

This condemnation caused the disappearance of nearly all his many theological and exegetical works. The few to survive are nearly all in *Syriac translation, since his memory was greatly venerated in the Nestorian church. The 14th-c. Nestorian Ebedjesu left a catalogue of his works. Apart from three ascetic treatises (*The priesthood, To the monks, The perfect direction*) and a collection of *letters (*Book of pearls*), all lost but for fragments, T.'s work is divided between theology and exegesis. In the first field, his interests were in the *Arian controversy (*For Basil, against Eunomius*, in which he refuted *Eunomius's reply to *Basil: a description by *Photius remains [Bibl., cod. 4]; Dispute with the *Macedonians, surviving in Syriac tr., which goes back to a debate in 392 at *Anazarbus); in the *Apollinarist controversy (The assumer and the assumed, lost); and in questions relating to free will and original *sin (Against *magic, discussed by Photius [Bibl., cod. 81]; Against the defenders of original sin, also described by Photius [Bibl., 177]). This last work took up a position against *Augustine and those who maintained that original sin is inherent in our nature. Indeed T. gave a friendly welcome to Augustine's opponent *Julian of Eclanum who, condemned in the West, sought help and protection in the *East.

T.'s most important and disputed doctrinal work was The incarnation in 15 books, directed against *anomoeans and *Apollinarists and dealing with problems of *christology, *anthropology and the *Trinity. A Syriac tr. was discovered in the East in 1905, but the MS disappeared during the 1914-18 war. Copious fragments remain in Latin, Greek and Syriac, but they require delicate appraisal, since they were collected either by T.'s defenders or by his detractors.

A better knowledge of his christology, together with information relating to worship and *liturgy, can be deduced from the Catechetical homilies, a collection of 16 texts discovered in Syriac tr. in 1932. Acc. to a traditional custom, the first ten, addressed to *catechumens, deal with trinitarian and christological questions based on the Nicene *creed; the other six explain the *Lord's Prayer and the baptismal and eucharistic liturgy. Besides a trinitarian theology of the now traditional type, anti-Arian in its affirmation of the unity of divine substance in three distinct *hypostases of equal dignity, they provide a treatment of the christological question which takes account of the difficulties brought out by the Apollinarist controversy. T. is a typical representative of the Antiochene tradition: so he gives the maximum value to the humanity assumed by the divine Logos and its capacity to operate autonomously: it is a genuine subject, which T. likes to define as Son of David in correlation to the Son of God, the Logos. And it is superfluous to add that he makes normal use of the phraseology of the homo assumptus to present the union (indwelling of the Logos in the temple represented by the *body, etc.). But T. also bears well in mind the criticisms aimed by the Apollinarists against this so clearly divisionist christology, and so he also takes care to outline the union between Christ's two components, divine and human, with more attention than is usual for the Antiochenes. He denies that we can speak of two Lords and two Sons, since the two natures are united in an ineffable and eternally indissoluble way in a single *prosopon. As the union does not destroy the distinction of natures, so this distinction does not prevent the two natures being one. T.'s efforts to safeguard Christ's unity without lessening the integrity and autonomy of his human nature are evident: but it is beyond doubt that to an Alexandrian the result would appear insufficient, both in the terminology used and because the way in which the union is achieved is unclear, all the more since prosopon (=figure, external aspect) appeared less weighty than *hypostasis to indicate this union. Indeed T., esp. in his Commentary on John, deals with Christ, man and *God, as two subjects distinct from each other.

T. was above all an exegete, the most typical representative of Antiochene *exegesis in its opposition to Alexandrian allegorism. We know that he wrote *Commentaries on *Genesis, the other books of the Pentateuch, the *Psalms, the major and minor *Prophets, *Job, *Exodus, Ecclesiastes. In a letter he makes clear his thought on the *Song of Songs. As for the NT, we know of Commentaries on *Matthew, *Luke, *John, Acts, *Paul's epistles. Besides fragments, copious for chs. 1-3 of Genesis, we possess the Commentary on the minor Prophets in the original Greek, that on John in Syriac tr., that on Paul's minor epistles in *Latin translation, and fragments of that on the four major epistles. Devreesse, using the *catenae, Julian of Eclanum's Lat. tr. and other evidence, has reconstructed T.'s commentary on Psalms 1-80.

T. precedes his commentaries on a book of *Scripture with an introduction in which he describes the book's characters and discusses its author and *chronology. His historical observations are worthy of note: e.g., he has understood that the historical situations described in the Psalms are often later than the time of *David, traditionally considered the author of the whole book, and to safeguard this authority he explains that, as a prophet, David described facts and circumstances yet to come. The interpretation of the texts flows smoothly, attentive to the literal sense, with historical, linguistic, grammatical, more rarely doctrinal observations. Though attention is paid to passages whose interpretation provides difficulties, vast tracts of the commentary are simply a broad paraphrase of the biblical text. T. is particularly careful to stress that, when the sacred text gives a *number, this number has no other aim than the obvious one of determining the quantity of persons or things, in evident polemic against the Alexandrians, who were always ready to give numbers symbolic values. And when a text is expressed in emphatic and figurative language, he always explains it in the literal sense, using the figure of hyperbole, where the Alexandrians, in cases of this kind, saw difficulties in the literal interpretation and passed straight on to *allegory.

On the other hand T. does not wholly deny that various OT passages should be considered as prefigurations, typoi, of NT facts and figures: *Jonah is a figure of Christ; the liberation of the Israelites from Egypt, the legal prescriptions and the episode of the bronze serpent are typoi of Christ's death and the liberation from *sins brought by it. He specifies that an OT fact, to be considered a typos of a NT fact, must show a similarity with it, be useful in its time and express a reality inferior to the future reality (PG 66, 320 ff.). Basing themselves on this and other analogous passages, some modern scholars have sought to minimize the conflict between T.'s literalism and Alexandrian allegorism in the interpretation of the OT. But if we go from the theory to the facts, we find that T. restricts as much as possible the number of OT passages that he considers typoi of the NT: he includes only four psalms (2, 8, 44 [45], 109 [110]) and very few prophetic passages, rejecting typologies however traditional, like that which referred the "sun of righteousness" of Mal 4, 2 (3, 20 LXX) to Christ. He prefers to consider nearly all the prophetic passages usually referred to Christ and the church as having been fulfilled at the time of Zerubbabel and the Maccabees. Alone among all the ancient exegetes, he denied that the bride and groom of the Song of Songs could be considered typoi of Christ and the church, reducing the work to a song of *Solomon's love for an Egyptian princess. In fact T. tends to consider the OT as the expression of a closed *economy, that of Jewish monotheism, as opposed to *pagan polytheism and Christian trinitarianism, expressed in the NT. Besides this, his doctrine of the two ages, the present age of sin and the future age of liberation and happiness, and his view of the phase of the present age inaugurated by Christ's *incarnation as an anticipation of the future age, leads him to read the NT more as a projection of the future than as a continuation of the past. Theological presuppositions and programmatic reaction to the excessive allegorism of the Alexandrians led T. to minimize the presence of Christ in the OT.

The NT does not pose the same problems for T., since here it is the letter of the sacred text that presents Christ and the church, and so his interpretation is programmatically only literal. But for this reason a text like *John's gospel, so rich in symbolic meanings, comes out greatly impoverished from this flat, literal explanation with some doctrinal openings. By contrast the tendency, described above, to see a break rather than a continuity between the OT and the NT put T. in perfect sympathy with the thought of *Paul, unlike previous commentators, who, out of anti-*gnostic preoccupations, had sought to read Paul in harmony with the OT. In fact T. is able to understand fully and illustrate eloquently the Pauline opposition between Law and *Grace, between the slavery of the old man and the freedom of the new man.

CPG 2, 3827-3873; PG 66l CCL 88A (Julian of Eclanum's Lat. tr. of Exp. in ps.); CSCO 115-116 (Com. Jo.); PO 9, 637-667 (Disp. Mac.); H.B. Swete, Theodori episcopi Mopsuesteni in epistulas B. Pauli commentarii, I-II, Cambridge 1880-1882; R. Devreesse, Le Commentaire de Théodore de Mopsueste sur les Psaumes, ST 93, Vatican City 1933; Commentarius in XII prophetas, ed. H.N. von Sprenger, Göttingen 1977; K. Staab, Pauluskommentare aus der griechischen Kirche, Münster 1933, 113-172; R. Tonneau-R. Devreesse, Les homélies catéchétiques de Théodore de Mopsueste, ST 145, Vatican City 1949; R. Devreesse, Essai sur Théodore de Mopsueste, ST 141, Vatican City 1949; DTC 15, 235-279; Quasten II, 405-426 (bibl.); U. Wickert, Studien zu den Pauluskommentaren Theodors von Mopsuestia, Berlin 1962; R.A. Norris, Manhood and Christ, Oxford 1963; J.M. Lera, "y se hizo hombre". La economía trinitaria en las Catequesis de Teodoro de Mopsuestia, Bilbao 1977; M. Simonetti, Note sull'esegesi veterotestamentaria di Teodoro di Mopsuestia, VetChr 14 (1977) 69-102; R.P. Vaggione, Some Neglected Fragments of Theodore of Mopsuestia, Contra Eunomium, JThS n.s. 31 (1980) 403-470; D.Z. Zakaropoulos, Theodore of Mopsuestia on the Bible, New York 1989.

M. Simonetti

THEODORE of Paphos. Mid 7th-c. bishop of Paphos in *Cyprus; on 12 Dec 655 he preached the panegyric for the feast of St *Spyrid(i)on, which we possess. Spyridon, a shepherd, married with *children, was bishop of Trimethus (Cyprus); he may have been at the council of *Nicaea (325); he subscribed the *Acts of the council of *Sardica (343; Athan., Apol. c. Arian. 2: PG 25, 340b; Mansi 3, 69a); he is mentioned by *Rufinus (HE 1, 5: PL 21, 471 f.), *Socrates (HE 1, 12: PG 67, 164 f., dependent on Rufinus) and *Sozomen (HE 1, 11: PG 67, 885-889). Oral and biographical traditions (lost) circulated about him, as well as a poem in iambic verses attrib. to Triphyllios (lost). This poem was used by T.; T.'s panegyric was in turn reworked by Simeon Metaphrastes and by two anon. Lives of St Spyridon.

CPG 3, 7987; Bardenhewer 5, 138 f.; P. van den Ven, *La légende de S. Spiridon*: Bibliothèque du Muséon 33, Louvain 1933; Beck 463, cf. 456, 1; *BS* 11, 1354-1356; Altaner 247.

A. De Nicola

THEODORE of Petra. 6th-c. author of a panegyric on St *Theodosius the coenobiarch. From it we deduce that the writer was a disciple of the saint in the *monastery he had founded. The encomium, very verbose and composed according to the rhetorical canons of this *literary genre, was preached to the monks of the monastery on the first anniversary of the saint's death, i.e. 11 Jan 530, and, retouched in several points, published after 552 (death of patriarch *Peter of Jerusalem). From the heading we learn that T. was later bishop of *Petra (*Arabia).

CPG 3, 7533; H. Usener, *Der heilige Theodosios. Schriften des Theodoros und Kyrillos*, Leipzig 1890; A-J. Festugière, *Les moines d'Orient*, III, 3, Paris 1963, 83-160; Bardenhewer 5, 128 f.; Beck 406 and 409.

A. De Nicola

THEODORE of Philae. *Monophysite bishop of Philae in the Upper *Thebaïd, he and *Longinus, bishop of the Nobadae (a people living near the Roman frontier), were eminent members of the *Jacobite church of *Egypt in the mid 6th c. Consecrated by *Timothy IV (517-535) in 525/6, he ruled his diocese for *c.*50 years and died between 577 and 584. For an 18-year period he was also in charge of the Nobadae. In 575 T. sent Longinus a *mandatum* (*entolikon*) authorizing him to proceed in his name to consecrate a Jacobite patriarch for *Alexandria, whose see had been vacant for nine years (J.B. Chabot, *Documenta ad origines monophysitarum illustrandas*: CSCO 17, 274 f.: text; 103, 192: *version*). In pursuance of this mission (called by *Theodore of Copros and the presbyter *Theodosius), Longinus went, unknown to the civil authorities of *Nubia, to Mareotis (near Alexandria): passing through Philae, he invited the old bishop to accompany him, but he declined because of his advanced age and "because of the perfidy [*neklô, dolum*: version] of those who at present govern the churches" (perhaps an allusion to bishop *John of Cellae[Kellia]); however, he gave his mandate for the *ordination.

CPG 3, 7227; John of Ephesus, IV, 7, 9; R. Payne Smith, *The Third Part of the Ecclesiastical History of John Bishop of Ephesus*, Oxford 1860, 255-259; W. Brooks: CSCO 105, 186-189: text; 106, 139. 141: *version*; *Theodorus* (36), *DCB* 4, 949; J. Maspéro, Théodore de Philae, *RHR* 59 (1909) 299-317; E. Honigmann, *Evêques et évêchés monophysites d'Asie Antérieure au VIe siècle*: CSCO 127, Subsidia 2, Louvain 1951, 175. 224-227. 232 f.

A. De Nicola

THEODORE of Raithu. Monk and priest in the monastic colony at Raithu, a port city in the S-W Sinai peninsula; author of a προπαρασκευή, a propaedeutic work, part heresiological, part dialectical (philosophical notions) and *neochalcedonian in tendency, i.e. characterized by a desire to harmonize the formulae of *Chalcedon with those of *Cyril of Alexandria. Acc. to Richard, we must also attribute to T. the *De Sectis* (CPG 3, n. 6823), composed after 579, a sort of *résumé* of contemporary *theology, severely critical of the theology of *Leontius of Byzantium and the religious policy of *Justinian I. T. became *abbat of his *monastery. If he were identified with bishop *Theodore of Pharan, a city near Raithu, this would make him one of the first eminent representatives of *monoenergism - something not impossible even for a former neochalcedonian (Beck) - and the confidant of patriarch *Sergius of Constantinople, condemned with him by the council of *Constantinople of 680/1. Finally, his death would be no later than 649, prob. 638.

Editions: CPG 3, 7600 ff.; cf. 6823; PG 91, 1484-1504 (incomplete); F. Diekamp, *Analecta Patristica*, Rome 1938, 173-227; Mansi 10, 957-961; 11, 567-572 (fragments); PG 86, 1193-1268 (= *De Sectis*).
Studies: *DTC* 15/1, 282 ff.; W. Elert, *Der Ausgang der altkirchlichen Christologie*, Berlin 1957; Beck 374, 382 f., 430 f.; J. Speigl, Der Autor der Schrift "De Sectis" über die Konzilien und die Religionspolitik Justinians, *Annuarium Hist. Conc.* 2 (1970) 207-230.

B. Studer

THEODORE of Scythopolis. 6th-c. Palestinian Origenist, of the *isochristoi* faction, whose centre was the New *Laura. T. was its *hegumen before becoming *staurophylax* and metropolitan of Scythopolis, thanks to the support of *Theodore Ascidas (Cyr. of Scyth., *V. Sab.* 89). Following a reversal of *Justinian's religious policy, leading to the condemnation of *Origenism at the *council of 553, he was obliged to retract his convictions. The *Libellus de erroribus Origenianis*, drawn up at *Constantinople prob. near the end of 552, contains 12 *anathemata, directed in general against the doctrines of pre-existence of *souls and *apocatastasis, and in particular against the christological and soteriological developments of the *isochristoi*.

CPG 6993, PG 86, 232-236; F. Diekamp, *Die origenistischen Streitigkeiten im 6. Jahrhundert und das 5. allgemeine Concil*, Münster i. W. 1899, 125-129.

L. Perrone

THEODORE of Tabennesi. Superior of the *monastery of Tabennesi and, as such, head of the Pachomian monasteries from *c.*350 to his death in 368. The job was given him by *Horsiesi, who seemingly failed to put an end to the disputes that arose after *Pachomius's death. T. brought back tranquillity to the monasteries, so that after 368 Horsiesi was able to resume their leadership without difficulty. The facts of T.'s life are related in some enlarged versions of the lives of Pachomius. *Jerome translated a *letter of his into Latin with other works of Pachomius and Horsiesi. Another letter, which we have in the original, was recently discovered in *Coptic: it was written to the convents on the occasion of their annual meeting in the month of Mesor, at which they celebrated a festival of "remission" (of *sins?).

CPG 2373-2376; PG 23, 99-100; A. Boon, *Pachomiana latina*, Louvain 1932, 105 f.; ASS *Maii* III, 63-71; L. T. Lefort, CSCO 159, Louvain 1959, 37-62 (text); CSCO 160, 195, 38-62 (Fr. tr.); Quasten II, 161-162.

T. Orlandi

THEODORE of Trimethus. Bishop of Trimethus in *Cyprus, he attended the 6th ecumenical *council (*Constantinople III, 680) from the 14th session on, as representative of Epiphanius II, *archbishop of Constantia in Cyprus (Mansi 11, 584e, 603c, 613a; substitute Theodore for Basil in 625c). He spoke once, with other Cyprian bishops, in favour of two wills in Christ (*ibid.*, 596e, cf. Hfl-Lecl 3, 1, 506: reading of a discourse of *Athanasius) and subscribed the *Acts (*ibid.*, 640e, 669b, 688de). He was the author of a life of St *John Chrysostom, first published by A. Mai (*Nova Patrum Bibliotheca*, VI, 263 ff.) and reproduced in PG (47, LI-LXXXVIII). This life, better than the one attrib. to patriarch George of Alexandria (cf. Photius, *Bibl.*, cod. 96; Beck 460), cites long documents illustrating the last tormented period of the saint's life (*Letters of *Arcadius, *Honorius, pope *Innocent, *Theophilus, the Roman synod, *Eudoxia, etc.).

CPG 3, 7989; F. Halkin, *Douze récits byzantins sur saint Jean Chrysostome*: Subs. Hagiogr. 60, Brussels 1977, 7-68; Bardenhewer 5, 140; Beck 463.

A. De Nicola

THEODORE of Tyana. Bishop (from 383 to at least 404) of *Tyana, metropolis of *Cappadocia II, detached from *Caesarea at the start of St *Basil's episcopate with clamorous claims of autonomy by its bishop *Anthimus. T. was thus the metropolitan of *Gregory Nazianzen, who dedicated to him the *Philocalia* of *Origen and wrote him *letters (*Epp.* 139, 152, 157, 160-163 and certainly 122-124, but not 77 or 183). He showed sympathy towards St *John Chrysostom in his exile (Pall., *Dialogue* 132, 4).

M.M. Hauser-Meury, *Prosopographie zu den Schriften Gregors von Nazianz*, Bonn 1970, 161-162.

J. Gribomont

THEODORE Spudaeus. (On the Σπουδαῖος cf. S. Pétridès, Le monastère des Spoudaei à Jérusalem et les Spoudaei de Constantinople, *EO* 4 [1900-1901] 225-231; 7 [1904] 341-348.) T. was almost certainly the author of the *Commemoratio* (Sirmond, 81-113) and the *Hypomnesticum* (Sirmond, 251-272; BHG 2261), vital evidence of the persecutions connected with the dispute over *monothelism. The first certain information on him is from 653, when he was present at the arrival at *Constantinople of pope *Martin I, taken there after being arrested on the orders of *Constans II. Moved by the sad spectacle, T. made contact with the prisoner - we do not know how - and received from him two *letters, *Quoniam agnovi* (Jaffé 2078) and *Noscere voluit* (Jaffé 2079), which he sums up in the *Commemoratio*. This work, composed prob. towards the end of 654 (Devreesse, Le texte grec, 58) and aimed at Christians in *Rome and *Africa, illustrates Martin's vicissitudes and praises his behaviour. We possess only a Latin version of it by Anastasius the Librarian. The *Hypomnesticum*, edited by Devreesse (who at first attributed it to Theodosius Gangra, T.'s brother: Le texte grec, 49) from a single witness, *Vatic.* 1671 (10th c.), and previously known from a Latin version by Anastasius the Librarian (BHL 5844), describes the sufferings of other victims of monothelism, *Anastasius the apocrisiary, his two disciples, the brothers Theodore and Euprepius, *Maximus the monk (Confessor) and his disciple *Anastasius, all *friends of T. whom he visited and comforted in their exile. The work, which ends with a long *doxology, is grammatically and stylistically full of errors: the author is aware of this, but holds it his duty to describe the work of the martyrs and to invite the *faithful to pray for a speedy end to the persecutions.

CPG 3, 7968-7969; CPL 1734; PL 129, 591-604, 681-690; PG 90, 173-178, 193-202; R. Devreesse, Le texte grec de l'Hypomnesticum de Théodore Spoudée (le supplice, l'exil et la mort des victimes illustres du monothélisme), *AB* 53 (1935) 49-80; Beck 462-463.

A. Labate

THEODORE Syncellus (7th c.). Presbyter and *syncellus* of the "Great Church" (Hagia Sophia) of *Constantinople, a member of the embassy sent to Adrianople by the emp. *Heraclius in July 626 on a peace mission to the khan of the Avars, who were moving against the *Byzantine capital (*Chronicon Paschale*: PG 82, 1009BD). A number of codices transmit under his name (and some anonymously) a *Discourse* on the discovery at *Jerusalem, translation to Constantinople and deposition in the sanctuary of Blachernae of the clothing of the Mother of God, a discourse given on 2 July some time between 620 and 626 (BHG 1058). We also possess another Logos of his for the feast of the Liberation of Constantinople from the Avar siege in August 626.

CPG 7935-7936; A. Wenger, *L'Assomption de la T.S. Vierge dans la tradition byzantine du VIe au Xe siècle*, Paris 1955, 111-136, 294-303; Beck 545 (with erroneous reference to the 9th c.); J.L. van Dieten, *Geschichte der Patriarchen von Sergios I. bis Johannes VI. (610-715)*, Amsterdam 1972, 114-121, 174-178.

D. Stiernon

THEODORE the Lector. Ecclesiastical historian (early 6th c.). Lector of Hagia Sophia, Chalcedonian in sympathy, author of a *Historia tripartita a compilation in four books taken from the histories of *Socrates, *Sozomen and *Theodoret, going from *Constantine to *Theodosius II. Large portions of this work, important for reconstructing the textual form of its original models, were translated into Latin by the monk *Epiphanius for *Cassiodorus's *Historia ecclesiastica tripartita* (*c*.560). T. followed it up with an original *Ecclesiastical history*, also in four books, from 439 to the reign of *Justin I (518). Its dominant interest is in politico-ecclesiastical events rather than the general history of the empire, and esp. in the restricted horizon of *Constantinople, with no understanding of *Anastasius's overtures towards the *monophysites. Both works survive fragmentarily, but we can reconstruct their arrangement and individual parts thanks to an "Epitome" of church history drawn up in the 7th or 8th c.

CPG 7502-7503; PG 86, 165-228; G.C. Hansen, *Theodoros Anagnostes, Kirchengeschichte* (GCS), Berlin-DDR 1971; *PWK* 24/2 (1934) 1869-1881; G. Moravcsik, *Byzantinoturcica* I, Berlin-DDR ²1958, 519-520.

L. Perrone

THEODORET of Cyrrhus. Born at *Antioch in *Syria *c*.393; educated in Syrian monastic circles where he learnt both Christian and *classical culture. He later brought this ascetic milieu to life in his 30 chapters of *biographies of 28 monks and three *women, the *Religious history* or *Manners of ascetic life*. Some unproven traditions give him *John Chrysostom and *Theodore of Mopsuestia as teachers and *Nestorius and *John of Antioch as fellow-pupils (cf. Tillemont, vol. XV, 207 ff.).

In 423 he was elected bishop of Cyrrhus, a town near Antioch. At the start of his episcopate he seems to have opposed, in works now lost, various *heresies present in communities in his region: besides refuting the *Jews, he seems to have written against the *Arians, *Macedonians and *Marcionites. Also from this first period must be his refutation of Hellenic *paganism, *Graecarum affectionum curatio*, last flower of Christian *apologetic, preserved entire, consisting of 12 discourses in which he compares pagan and Christian answers to the fundamental philosophical questions. The years immediately preceding the council of *Ephesus were probably his most exacting and decisive: when *Cyril wrote his 12 *anathemata against *Nestorius (Nov 430), T. took on Nestorius's defence and wrote, early in 431, at John of Antioch's request, a refutation of the *Anathemata*. Since this work, *Reprehensio duodecim capitum seu anathematismorum Cyrilli*, was condemned by the 5th ecumenical *council (*Constantinople, 553), it is lost; however Cyril defended his *Anathemata* in three apologies, and the second of these: *Epistula ad Euoptium adversus impugnationem duodecim capitum a Theodoreto editam* (PG 76, 385-452), preserves T.'s text.

Here, at the most delicate moment of the controversies between the schools of *Antioch and *Alexandria, T. develops what is historically the most interesting character of his theological work: a sort of re-examination of the *tradition up to his time. Cyril is seen as dependent on *Apollinaris and as risking denying Christ's full humanity; at the same time T. points out the Arian danger of a doctrine that does not distinguish humanity from divinity in the incarnate Word.

From the same immediately pre-Ephesine years come the two books *De sancta et vivifica Trinitate* and *De incarnatione Domini*, surviving under Cyril's name (PG 75, 1147-1190, 1419-1478) and returned to T. by A. Ehrhard. They are works of great interest because it is the Athanasian tradition that is invoked to distinguish *theology from *economy, and on this basis the divisive *christology is constructed and justified: the text of Phil 2, 5-8: "Jesus Christ, though he was of divine nature . . . emptied himself, taking the form of a servant . . . " plays an important part in this argument.

This was prob. the climate and milieu in which pseudo-Athanasian works were written: it seems to me this could be the historical setting of the two pseudo-Athanasian dialogues *C. Macedonianos* (PG 28, 1291-1338).

After the council T. published five books, again attacking Cyril and the council's decisions: these are the *Pentalogus*, also lost because of their condemnation in 553. Latin fragments remain in the *Collectio Palatina* and some Greek citations in Nicetas of Heraclea's *catena on *Luke.

The moves towards union in 433, though T. was almost certainly employed in person to draw up the formulae of reconciliation, did not gain his personal adherence, since they still required subscription to the condemnation of Nestorius (cf. *Ep.* 171).

In 447 T. composed *Eranistes* (the Beggar) or *Polymorphos* (the man of many shapes): a work of great theological importance written to refute the *monophysite doctrine being propagated by *Eutyches at Constantinople, with some success in *Egypt, esp. at *Alexandria, through the support of patriarch *Dioscorus, Cyril's successor. The work, in four books, is in the form of a *dialogue between an *orthodox and a monophysite (beggar or many-shaped, because in his doctrine could be found elements picked up from previous *heresies, from *Docetism to *Apollinarism). It is a work truly typical of the time and of T.'s method of work: a time and a method of evaluation and summing-up of tradition; compilation of florilegia; ample use of sources (*Eranistes* preserves 238 passages from 88 different patristic sources). All this is important for the sense of *tradition that was developing and for the transmission of at least parts of works that we would not otherwise know; it has its limits if we bear in mind that alongside the sense of tradition was developing a tendency to manipulate earlier materials, to produce *pseudoepigrapha aimed at polemic and to use traditions in a confused way. *Eranistes* survives in a second edition enlarged after the council of *Chalcedon.

In 449, during the so-called robber-council (*latrocinium*) of *Ephesus, which forcibly imposed the absolution of Eutyches, whom pope *Leo the Great had already condemned in a dogmatic *letter, T. was deposed from his episcopal see. He was restored by the council of Chalcedon (451), at which he agreed to subscribe the condemnation of Nestorius, not out of personal interest but out of a sense of compliance to the real needs of the whole church, since a formulation had been reached which sufficiently comprehended some fundamental requirements of Antiochene *christology. He remained at the head of the church of Cyrrhus until his death, *c*.466.

T.'s *exegesis is attested in various works, in which we can distinguish two main genres: a series of continuous *commentaries: *Interpretation of the *Psalms*, commentary on the *Song of Songs, the major and minor *Prophets, Chronicles, St *Paul's Epistles; and the genre known as *Quaestiones et Responsiones*: on the *Octateuch* and the Books of Kings. *Palaeoslavonic versions remain, whole or in part, of nearly all these exegetical works; as well as *Armenian and *Georgian versions of the *Commentary on the Psalms*; and the Armenian version of that on *Ezekiel. The *Ten discourses on *Providence*, whose date is much disputed, probably post-date the council of Ephesus. From citations which T. often makes of it, we know that he also wrote a work of refutation against the attitude of the Persians towards the Christians (*Replies to the Magi*).

T. wrote much historical work: besides the *Religious history* cited, he wrote an *Ecclesiastical history*, continuing that of *Eusebius from 323 to 428, and also - in accordance with T.'s method - a precious trove of otherwise unknown sources and documentation. Finally he wrote the *Haereticorum fabularum compendium* in five books: the first four describe the *heresies from *Simon Magus to *Nestorius and *Eutyches; the fifth compares the main heretical doctrines with the teaching of the church.

Of his *Sermons, fragments remain; of his *letters, more than 200, a precious historical source for events of the time, and evidence of his pastoral work.

Two works, handed down under the name of *Justin, have been returned by various scholars to the Antiochene milieu and prob. to T.; these are an *Expositio rectae fidei* and *Quaestiones et responsiones ad orthodoxos*. Other scholars attribute to T. a treatise entitled *There is only one Son, Our Lord Jesus Christ* (PG 83, 1433-1441).

CPG 3, 6200-6288. For specific bibliography on individual works, see Quasten II, 541-559; M. Simonetti, *La letteratura cristiana antica greca e latina*, Florence 1969, 317-322; *PWK* 2, 5, 1791-1801; *DTC* 15, 299-325; A. Ehrhard, *Die Cyrill von Alexandrien zugeschriebene Schrift* Περὶ τῆς τοῦ Κυρίου ἐνανθρωπήσεως, *ein Werk des Theodoret von Cyrus*, Tübingen, 1888; E. Cavalcanti, *I Dialoghi pseudoatanasiani contro i Macedoniani*, crit. text, intr., It. tr. and comm., Turin 1983, 157-169; R.V. Sellers, *Two Ancient Christologies*, London 1940 and *The Council of Chalcedon*, London 1961; P. Canivet, *Le Monachisme syrien selon Theodoret de Cyro*, Paris 1977; G.F. Chesnut, *The First Christian Historians, Eusebius, Socrates, Sozomen, Theodoret*, Mâcon²1986; P. Canivet - A. Leroy-Molinghen, *Théodoret de Cyr; Histoire des Moines de Syrie*, SCh 234, 257, (1977-1979); *A History of the Monk in Syria*, (Eng. tr. R.M. Price), Cistercian Studies 88, Kalamazoo 1985; J.N. Guinot, *Commentaire sur Isaïe 1*, SCh 276, 295 (1980, 1982).

E. Cavalcanti

THEODORIC (or THEODERIC) (*c*.453-526). King of the Ostrogoths, of the Amaling family. As a boy he lived at *Constantinople as a hostage. Having succeeded his father as head of the Ostrogoths stationed in *Italy as *foederati* and gained influence with the emperor *Zeno, he defeated and treacherously killed *Odoacer. As king of Italy, soon recognized by Constantinople, he ruled over *Arian *Goths and *Catholic Romans, respecting Roman traditions and creating an enlightened syncretism. A highly educated man, he embellished the city. The so-called *Theodericianan Edict* is attrib. to him; epistles inspired by him are collected in *Cassiodorus's *Variae*. He enjoyed great prestige among the other Roman-barbarian sovereigns. Towards the end of his reign, T., worried by the emp. *Justinian's anti-Arian measures, sent pope *John I to Constantinople and on his return imprisoned him, suspecting him of treachery; *Boethius and his father-in-law *Symmachus also fell victim to his suspicions. T. was buried at *Ravenna in the mausoleum he had prepared; his nephew Athalaric, son of Amalasuntha, succeeded him. T. appears in the Germanic sagas under the name of Dietrich of Berne.

Editions: CPL 1805-1807; PL 72, 1117B; MGH, *Leges* V, 145-170 (Fr. Bluhme); *Edictum*: P.L. Falaschi, Milan 1966; G. Vismara (*Ius Romanum medii aevi* I, 2b aa), Milan 1967.
Studies: A. Lippold: *KLP* 5, 685-687 (with bibl.); G. Della Valle, Teoderico e Roma, *Rendiconti Acc. Arch. Lett. Napoli* 34 (1959) 119-176; W. Ensslin, *Theoderich der Grosse*, Munich ²1959; G. Astuti, Note sull'origine e attribuzione dell'Edictum Theoderici regis, in *Studi in onore di E. Volterra*, Milan 1971, V, 647-686; O. Hoelfler, Theoderich der Grosse und sein Bild in der Sage, *AAWW* 111 (1974) 349-372; E.A. Thompson, *Romans and Barbarians; the Decline of the Western Empire*, Madison 1982; J. Richards, *The Popes and the Papacy in the Early Middle Ages, 476-752*, London 1979, Part 2; T.A.S. Burns, *A History of the Ostrogoths*, Bloomington 1984; W.H.C. Frend, *The Rise of Christianity*, London 1984, ch. 22; A.H.M. Jones, The Constitutional Position of Odoacer and Theodoric, *JRS* 52 (1962) 126-30.

E. Malaspina

THEODOSIUS. *Monophysite monk. *Evagrius (*HE* II, 5) claims he was expelled from his *monastery for his sins and later incited the population of *Alexandria to revolt, for which he was whipped for subversion and then dragged through the streets of the city. T. was one of the most zealous followers of the brutal archimandrite *Barsauma. During and immediately after the council of *Chalcedon, he raised another revolt in the church of *Jerusalem and falsified pope *Leo I's *Tomus ad Flavianum*. He found a great protectress in the person of *Eudoxia, widow of *Theodosius II, who lived at Jerusalem far from the imperial court, and could thus act as "a pillar of *orthodoxy". After having roused the population, T. forced patriarch *Juvenal, just back from Chalcedon, to denounce the council's decisions; Juvenal refused but, after an attempt to kill him, fled to *Constantinople. T. was then ordained in the church of the Anastasis. For three months he used brutality to get his *ordination accepted, taking into his service malefactors who committed innumerable acts of cruelty: the governor of *Palestine, Dorotheus, obeyed him out of terror. Finally the emp. *Marcian, tired of his activities and helped by those *clergy and monks who opposed T., managed to expel him. He fled, and his hiding-place was never discovered. The rest of his life is unknown, as is the date of his death.

DCB 4, 968-21; W.H.C. Frend, *The Rise of the Monophysite Movement*, Cambridge 1972, 143-83; L. Perrone, *La chiesa di Palestina e le controversie cristologiche*, Brescia 1980, 31-139.

G. Ladocsi

THEODOSIUS, archdeacon. In *c*.530 he composed a brief description of *Palestine entitled *De situ terrae sanctae* (or *Expositio civitatis Ierusalem*), discovered in 1864. It shows a good knowledge of the buildings constructed at *Jerusalem by the emp. *Anastasius I, but not those of *Justinian, from which we deduce the date cited above. T.'s N African origin has been conjectured from the fact that in a passage of *De situ terrae sanctae* (ch. 14) the conflict between *Arians and *Catholics is seen in terms of *Vandals and Romans.

CPL 2328; PLS 4, 1456-1463; J. Gildemeister, *Theodosius de situ terrae sanctae*, Bonn 1882; CSEL 39, 137-150; CCL 175, 113-125; *LTK* 10, 50.

S. Zincone

THEODOSIUS, deacon. Lived in N *Africa, prob. in the 6th c. Translated or compiled a collection of material important for our knowledge of the history of the 4th-c. *Arian controversy (*Cod. Veron.* LX), divided into 27 parts. His sources were synodal canons, *letters, the *Historia Athanasii* (*Historia acephala*) and the records of the patriarchate of *Alexandria.

C.H. Turner, EOMIA I/625-671; P. Franchi de' Cavalieri, Una pagina di storia bizantina del sec. IV: il martirio dei Santi Notari, *AB* 64 (1946) 132-175; E. Schwartz, *Gesammelte Schriften*, III, Berlin 1959, 30-72.

G. Ladocsi

THEODOSIUS I, "the Great". Of Spanish origin, appointed Augustus for the Eastern Roman empire in 379, he died at *Milan 17 Jan 395, aged 48. His politico-military activities were particularly vigilant on two fronts: that of Illyricum, where *Gratian had given him the job of restoring peace, which he sought to achieve by establishing a *foedus* with the *Goths stationed there; and that of the West, where the continually unstable situation led him to intervene first in favour of *Valentinian II against the usurper *Maximus, then against the *Frank Arbogast, then against the Catholic *Eugenius. Of no less interest is his religious policy, which overturned that of *Valens. Partly for the sake of political unity, T. worked to obtain religious unity in the empire. To this end he followed an anti-*pagan policy and, within the church, tried to heal the breach between *Catholics and *Arians. A convinced supporter of the Nicene *creed (for his *baptism cf. Socr., *HE* V, 6), at *Thessalonica in 380 he issued an edict ordering his subjects to follow the Catholic faith of *Rome and *Alexandria (*CT* XVI, 1, 2). In the same year he obliged the Arian *Demophilus to resign from the patriarchate of *Constantinople and replaced him with the Catholic *Gregory Nazianzen, forbidding those who did not profess the Nicene faith to hold religious meetings in the city (*CT* XVI, 5, 6). The council of *Constantinople was called by him (381) and the conciliar *Fathers gave him credit for having restored *peace among Christians (Mansi 3, 557; Socr., *HE* V, 8; Soz., *HE* VII, 9), while they requested ratification of the sanctions promulgated against the *heresies variously linked to the Arian controversy (*CT* XVI, 1, 3). Between 381 and 383 came a series of imperial provisions against *pagans and *apostates (*CT* XVI, 7, 1-2). After the failure of an attempt to unify the Christians doctrinally, T. issued a series of edicts (25 July and 23 Dec 383, 21 Jan 384) recognizing only the Catholic religion and taking away the heretics' places of worship (*CT* XVI, 5, 11-13; Philost., *HE* X, 6). A law of 14 June 388 struck at the heretics again (*CT* XVI, 5, 15). *Ambrose's *Epp.* 13 and 14 to T. in the name of the bishops of *Italy who had taken part in the council of *Aquileia (381) are also part of this anti-Arian polemic. Other *letters of Ambrose to T. concern the emperor's actions and the bishop of *Milan's response. *Ep.* 40 echoes the imperial orders after the episode of Callinicum and the position taken by Ambrose (cf. also *Ep.* 41 and Paulinus of Milan, *Vita Ambr.* 22). *Ep.* 51 marks Ambrose's intervention after the massacre of *Thessalonica in 390 (on the remote causes of the massacre cf. the differing interpretations of Palanque, *Saint Ambroise*, 228 ff., and Mazzarino, *L'impero romano*, III, Bari 1973, 379 ff.): Ambrose complains of the imperial provisions directed against himself (*Ep.* 51, 2) and of a series of provisions hardly favourable to the church (*CT* XII, 1, 121; XVI, 2, 27), but above all he alludes to various actions of his own aimed at avoiding the massacre (apropos of the number of victims cf. Soz., *HE* VII, 25; Theodoret, *HE* V, 17-18; for T.'s vain attempt to cancel his previous orders cf. *CT* IX, 40, 13), and explicitly invites T. to do public *penitence. Reconciliation took place at Christmas 390 (Paul. Mil., *Vita Ambr.* 24; Aug., *De civ. Dei* V, 26, 9-16). Ambrose's *Epp.* 61 and 62 are also addressed to T.; the first to justify his recognition of Eugenius, the second to plead clemency for the defeated. How much T. became a point of reference for *orthodoxy is shown by the *Libellus precum* sent to him in 384 by *Faustinus asking him to intervene on behalf of the *Luciferians. Order, repeatedly disturbed by the monks, was restored by the provisions of 2 Sept 390 (*CT* XVI, 3, 1), 9 April 392 (*CT* XI, 36, 31) and 17 April 392 (*CT* XVI, 3, 2). No less important were T.'s actions against *paganism. On 25 May 385 (*CT* XVI, 10, 9) he renewed the prohibition against offering bloody and divinatory *sacrifices. On 24 Feb 391 he prohibited any pagan ceremony at *Rome (*CT* XVI, 10, 10) and then extended the provision to *Egypt (*CT* XVI, 10, 11). If the edict of Concordia of 391 removed all civil and political rights from *apostates (*CT* XVI, 7, 4-5), that of Constantinople of 8 Nov 392 proscribed paganism *de facto* (*CT* XVI, 10, 12). The funeral oration pronounced by Ambrose in T.'s honour (*De obitu Theod.*) is indicative of the harmony that the emperor and bishop had reached. A panegyric in T.'s honour was requested from *Paulinus of Nola by a certain *Endelechius (Paul. Nol., *Ep.* 28).

A. Guldenpennung, *Kaiser Theodosius der Grosse*, Halle 1878; J.R. Palanque, *Saint Ambroise et l'empire romain*, Paris 1933; W. Ensslin, La politica ecclesiastica dell'imperatore Teodosio agli inizi del suo governo, *NDid* 2 (1948) 5-35; M. Pavan, I cristiani e il mondo ebraico nell'età di Teodosio il Grande, *AFLPer* 3 (1965-1966) 367-550; Y.-M. Duval, L'éloge de Théodose dans la Cité de Dieu (V 26, 1). Sa place, son sens et ses sources, *RecAug* 4 (1966) 135-79; A. Lippold, *Theodosius der Grosse und seine Zeit*, Stuttgart 1968; J. Rongé, La législation de Théodose contre les hérétiques. Traduction de CTh. XVI 5, 6-24, *Epektasis*, Mélanges J. Daniélou, Paris 1972, 636 ff.; M.G. Mara, *La storia di Naboth*, L'Aquila 1975, 12-23; L. De Giovanni, *Chiesa e Stato nel Codice Teodosiano. Saggio sul libro XVI*, Naples 1980; N.Q. King, *The Emperor Theodosius and the establishment of Christianity*, London 1961.

M.G. Mara

THEODOSIUS II, emperor. Eastern emperor 408-450, born 10 April 401 at *Constantinople, died 28 July 450. Son of *Arcadius and *Eudoxia; grandson of *Theodosius I. Appointed Augustus in 402 while still an infant, he succeeded his father in 408, becoming sole sovereign of the Eastern empire. A weak personality attached to the honours of the imperial dignity, all his life he was dominated by high dignitaries and court ladies: his tutor the *praefectus praetorio* Anthemius, his sister *Pulcheria, his wife *Eudoxia Athenaïs, bishops *Atticus, *Nestorius and *Proclus, the *praepositus sacri cubiculi* Chrysaphius. His reign saw wars with the Persians, the Armenians and the great empire of the Huns, to whom, from 430 on, he had to pay huge tributes. In 424/5 T. once more gave official recognition to *Valentinian III, and hence to his dynasty, over the Western empire. The *Codex Theodosianus*, which between 429 and 438 collected the imperial edicts from 312 on, was promulgated in the name of the two emperors and enforced in both parts of the empire. T. benefited the capital by building the city wall (still surviving) and reorganizing higher education.

PWK, Suppl. 13, 961-1044. I.E. Karagiannopoulos, 'Ιστορία βυζαντινοῦ κράτους, 1, Thessalonica 1978, 231-275; P.E. Pieler, in H. Hunger, *Die hochsprachliche profane Literatur der Byzantiner*, 2, Munich 1978, 388-390 (for the *Codex Theodosianus*); editions of *CT*: T. Mommsen-P.M. Meyer, 2 vols., Berlin 1905; P. Krueger, 2 vols., Berlin 1923-26 (books 1-8 only). G.G. Archi, *Teodosio II e la sua codificazione*, Naples 1976; W.H.C. Frend, *The Rise of the Monophysite Movement*, Cambridge ²1979, ch. 1; W.E. Kaegi Jr., *Byzantium and the Decline of Rome*, Princeton 1968, ch. 1; C. Liubheid, Theodosius II and Heresy, *JEH* 16 (1965) 13-38.

J. Irmscher

THEODOSIUS of Alexandria. Disciple and *friend of the great anti-Chalcedonian theologian *Severus of Antioch and secretary of patriarch *Timothy III of Alexandria, whom he was elected to succeed in 535. But the election was contested: a hard-line anti-Chalcedonian, *Gaianus, was put up against him. With the help of the empress *Theodora, wife of *Justinian, T. prevailed, but almost at once he was called to *Constantinople (536) by the emperor, who tried to make him embrace the *Catholic party. T. refused and was detained in the capital, in gilded exile, until his death (566). Of his works, all in Greek, there remain: a few fragments in the original version; a collection of theological and canonical works in *Syriac tr.; four *homilies (prob. "cathedral") in *Coptic tr., one of which is almost certainly spurious.

W.H.C. Frend, *The Rise of the Monophysite Movement*, Cambridge 1972; T. Orlandi, Teodosio di Alessandria nella letteratura copta, *GIF* ser. II 2 (1971) 175-185.

T. Orlandi

THEODOSIUS of Alexandria. 6th-c. *monophysite presbyter and church legal consultant (ἐκκλησιέκδικος). In a collection, handed down in *Syriac tr., of the *capitula* of John Claudus, *archimandrite of the *monastery of Mar Bassus, is a *letter from T. to *Longinus, previously a presbyter in *Alexandria, now a bishop in *Nubia, exhorting him to *communion with *Theodore, bishop of Philae.

CPG 3, 7223; J.B. Chabot, *Documenta ad origines monophysitarum illustrandas*, Louvain ²1952, 273-276 (text), 191-193 (tr.); E. Honigmann, *Evêques et évêchés monophysites d'Asie Antérieure au VIᵉ siècle*, Louvain 1951, 226 f.

J. Irmscher

THEODOSIUS the Coenobiarch. Saint (feast 11 Jan). Born 423 at Mogarissus (Mogariassus, Garissus) in *Cappadocia of a pious and religious family, he was ordained a lector when still young. He visited St *Simeon Stylites near *Antioch. At *Jerusalem he put himself under the spiritual direction of the old *ascetic Longinus in the Tower of David (451), whence he went (455) to the *monastery of the Theotokos of the Kathisma (between Jerusalem and *Bethlehem), founded by the pious Ikelia; to avoid being made *hegumen, he retired to the monastery of Metopa and thence to solitude in the Grotto of the Magi (Judaean *desert N-E of Bethlehem), where he lived for 38 years in great austerity and continual *prayer. Not wishing to reject the numerous believers who wished to be his disciples, he founded a monastery there with annexes for the old, sick, mad (called *energumenoi* or "possessed") and *pilgrims. The *psalmody was in Greek, Armenian and Slavonic (Bexic), each in its own chapel: but all met in the Greek church for *liturgy and *communion; a fourth church was for penitents. The locality is still called *Dejr Dosi* (monastery of Theodosius) and the church and monastery have been rebuilt. In 494 patriarch Sallustius of Jerusalem appointed T. as *archimandrite of the *coenobia (hence his surname) and St *Sabas as archimandrite of the *laurae in his jurisdiction. Both T. and Sabas opposed *monophysism, which was promoted by the emp. *Anastasius. The emperor exiled T., but his immediate death (1 July 518) allowed T. to return to his monastery with the favour of the fervently *orthodox *Justin I.

T. died in 529 and was buried in the Grotto of the Magi, where he had begun his eremitical life. His *Life* (encomium) was written by his disciple *Theodore, later bishop of *Petra (H. Usener, *Der heilige Theodosios*, Leipzig 1890, 5-101), by *Cyril of Scythopolis (H. Usener, *op. cit.*, 105-113; (ed.) Schwartz, *Cyrillos von Skythopolis*, TU 49, 2, Leipzig 1939, 235-241) and reworked by Simeon Metaphrastes (PG 114, 469-553).

CPG 3, 7533 (Theodore of Petra), 7539 (Cyril of Scythopolis); ASS *Ianuar.*, I, 680-701; Bardenhewer 5, 128 f.; *EC* 11 1940-1942; Beck 833 (index); *LTK* 10, 48 f. (Theodosios Koinobiarches) and 49 (Theodosioskloster); *BS* 12, 290-292.

A. De Nicola

THEODOTION. Author of a Greek version of the OT written in the reign of *Commodus (180-192), prob. at *Ephesus. Information on him is inconsistent. *Irenaeus makes him a *proselyte at Ephesus (*Adv. haer.* III, 21, 2); *Jerome calls him an *Ebionite (*Vir. ill.* 54), *Epiphanius calls him a disciple of *Marcion (*Mens. et pond.* 17). T.'s translation is for the most part essentially a revision of the LXX, harmonized with the Hebrew text; but in various passages of the *Bible he seems to have done no more than retouch the anonymous Palestinian-Jewish revision of the LXX, prob. used by the authors of the NT.

His version is of singular importance esp. for the book of *Daniel, because it contains the deuterocanonical part of the book. His version constitutes the sixth column of *Origen's *Hexapla*. T.'s work was little appreciated by the Jews or by *Eusebius of Caesarea, but was much more esteemed by the early Christians, esp. by Origen.

EC 6, 1056; *Enc. de la Biblia*, Barcelona 1963, 6, 394 f.; *Bibel-Lexikon*, Einsiedeln 1968, 232; P. Kahle, *Die Kairoer Geniza*, Berlin 1962, 266-72.

G. Ladocsi

THEODOTUS and the SEVEN VIRGINS. All *martyrs at *Ancyra. Accused of being Christians during *Diocletian's *persecution, seven *virgins of advanced age were drowned in a pond: Thecusa, Alexandra, Fainê, Claudia, Euphrasia, Matrona and Julitta. Thecusa's nephew, the innkeeper T., recovered their bodies and buried them, but suffered martyrdom in his turn. T.'s martyrdom is related by a certain Nilus, who claims to be an eyewitness (BHG 1782). Delehaye thought the account a hagiographical romance; P. Franchi de' Cavalieri substantially agrees with him, but acknowledges a historical basis for the text. The saints are commemorated on 18 May.

H. Delehaye, La Passion de S. Théodote d'Ancyre, *AB* 22 (1903) 320-28; P. Franchi de' Cavalieri, Sopra alcuni Atti di martiri di Settimio Severo e di Massimo Daza, *NBAC* 10 (1904) 27-37; id., *I martìri di S. Teodoto e di S. Ariadne*, ST 6, 7-87; *Vies des SS.* 5, 358; *LTK* 9, 1343; *BS* 12, 309-12.

V. Saxer

THEODOTUS of Ancyra. Bishop of *Ancyra in *Galatia, at first *Nestorius's *friend, then his determined enemy at the council of *Ephesus (431; cf. Mansi 4, 1181b), where he was one of the delegation sent by the *orthodox part of the council to inform the emp. *Theodosius II of the actions of patriarch *John of Antioch and the *conciliabulus* held by him (Mansi 4, 1457-1464). John and the other Orientals repeatedly *excommunicated him (at Ephesus, cf. *Synodicum* 38: PG 84, 640 f.; at the synod of *Tarsus in 432). He died before 446, since his successor was consecrated by *Proclus of Constantinople (434-446; cf. Mansi 7, 452b). The iconoclast council of Constantinople (754) cited a pretended text of his against sacred *images, but it was demonstrated at the 2nd Nicene council (784) that the citation belonged to none of T.'s works, which were then listed (Mansi 13, 309-311).

Of T. there remain: *Three books against Nestorius* in *Syriac, but incomplete; *Expositio symboli Nicaeni*, aimed against the Nestorians, demonstrating that there is one sole Son of God having two natures; a fragment of the *Epistula ad Vitalem monachum* (PG 84, 814; ACO 1, 4, p. 212); six or seven *Homilies*. The two homilies *In diem nativitatis Domini* (I and II), preached at Ancyra, were read at the council of Ephesus; as was the homily (III) *In Nestorium*; all three, included in the *Acts of the council, are aimed at Nestorius (without naming him), whom they accuse among other things of renewing the error of *Photinus; also against Nestorius is the homily (IV) *In s. Deiparam et in Simeonem* (called by Nicaea II εἰς τά φῶτα); the homily (V) *In Domini nostri Iesu Christi diem natalem*, once known only in *Latin translation, was published in the original text by M. Aubineau (Une homélie de Théodote d'Ancyre sur la nativité du Seigneur, *OCP* 26 [1960] 221-250), as was the homily (VI) *In s. Deiparam et in nativitatem Domini* by M. Jugie (*Homélies mariales byzantines*, PO 19, 289-317 [introduction], 318-335 [text and Lat. tr.]), whose genuineness, however, has been doubted; finally we have in *Ethiopic a homily preached at Ephesus during the council (A. Dillmann, *Chrestomathia Aethiopica*, Leipzig 1866, 103-106). The homily *In baptisma Domini* (ed. M. Aubineau, in Διακονία πίστεως, Mélanges J.A. Aldama, Granada 1969, 6-30) is dubious. The 2nd Nicene council called T. *Cyril's *fellow-fighter* (συναγωνιστής); he clearly professes the two natures of the single *person Christ, but does not want to investigate the mystery of the manner of their union; he eloquently celebrates *Mary, praising particularly her perpetual virginity united with divine motherhood.

CPG 3, 6124-6141; PG 77, 1308-1432; Hfl-Lecl 2 (*passim*); Bardenhewer 4, 197-200; DTC 15, 328-330; G. Roschini, *Dizionario di Mariologia*, Rome 1960, 474 f.; id., *Maria SS. nella storia della salvezza*, I, Isola del Liri 1969, 314; *LTK* 10, 51; L. Cignelli, *Maria Nuova Eva*, Assisi 1966, 157-201.

A. De Nicola

THEODOTUS of Antioch. Succeeded *Alexander as bishop of *Antioch in 420. Like his predecessor he rehabilitated *John Chrysostom. He readmitted the *Apollinarists to the church (Theodoret, *HE* V, 38), condemned *Pelagius in a synod at Antioch over which he presided perhaps in 422 (Mansi 4, 296) and died in 429. *Theodoret has words of great esteem for him (*HE* V, 38; *Ep. ad Dioscorum*).

DCB 4, 983.

E. Prinzivalli

THEODOTUS of Byzantium. Called the Tanner. Towards the end of the 2nd c. he propagated the doctrine of *adoptionism (so-called) at *Rome. He proclaimed Christ to be only a man, in order to mitigate in some way the gravity of the *apostasy he had fallen into during a *persecution, since he had denied a man and not *God.

A. Hilgenfeld, *Die Ketzergeschichte des Urchristentums*, Leipzig 1884, 610-611; *DTC* 5, 2427 (cf. 8, 1255; 10, 2197 f.).

M. Simonetti

THEODOTUS of Byzantium. Called the Banker. A disciple of *Theodotus the Tanner, he devised the *Melchizedekian variant of *adoptionism.

A. Hilgenfeld, *Die Ketzergeschichte des Urchristentums*, Leipzig 1884, 611-612; *DTC* 10, 514 f.; 5, 2427; 7, 464; 8, 1256.

M. Simonetti

THEODOTUS of Laodicea. Among *Arius's supporters from the start: condemned as such, with *Eusebius of Caesarea and *Narcissus of Neronias, in the *council held at *Antioch shortly before the council of *Nicaea (325). His presence at Nicaea is unknown; but immediately afterwards *Constantine threatened to punish him if he continued to help the Arians. In 327 he was one of the bishops who, at the council of Antioch presided over by Eusebius of Caesarea, condemned and deposed *Eustathius.

DCB 4, 981; Simonetti 596.

M. Simonetti

THEODOTUS the Valentinian. Preserved among *Clement of Alexandria's works are *Extracts of the works of Theodotus and the so-called Oriental school at the time of Valentinus*. T., of whom no more is known, is given as the author of some of the *gnostic propositions cited by Clement, but others are ascribed more vaguely to the followers of *Valentinus. The character of the *Extracts* is very composite and, besides passages by Clement himself, shows derivations from more than one Valentinian source with differing tendencies. So, though the *Extracts* are very important for our knowledge of Valentinian doctrines in general and in its various ramifications, it seems difficult to infer anything precise about T. from them.

SCh 23, Paris 1948; *LTK* 10, 52; *CE* 14, 27; A.-J. Festugière, Notes sur les Extraits de Théodote de Clément d'Alexandrie, *VChr* 3 (1949) 193-207; A. Orbe, A propósito de Excerpta ex Theodoto 54, 2 (κατ' ἰδίαν), *Gregorianum* 41 (1960) 481-485; id., La trinidad maléfica (A propósito de "Excerpta ex Theodoto" 80, 3), *ibid.* 49 (1968) 726-761.

M. Simonetti

THEODULUS. Presbyter in Coele-Syria, died at the time of the emp. *Zeno (474-491). Prob. a disciple of *Theodore of Mopsuestia. Ebedjesu calls him a "Nestorian". *Gennadius says T. wrote many books, but all are lost except five groups of fragments: *a)* two fragments of the book *De consonantia Veteris et Novi Testamenti*, an anti-heretical work, prob. aimed against either the *Manichees or the *Marcionites; *b)* two fragments of the *Comm. in Jdth.*; *c)* a fragment of the *Comm. in Is.*, originally in two volumes; *d)* a fragment of the *Tract. in Ps.* and the best way of reciting them; *e)* a fragment of the *Comm. in ep. Rom.*

CPG 3, 6540-6544; *EC* 2, 1947; Bardenhewer 4, 262.

G. Ladocsi

THEOGNIS of Nicaea. One of *Arius's supporters from the start. At *Nicaea he gave way and subscribed the anti-Arian formula of faith under pressure from *Constantine but, soon after, Constantine deported and exiled him and *Eusebius of Nicomedia for continuing to support the Arians. Exiled to *Gaul, he returned in 328 and regained his episcopal see. In 335 he was one of the protagonists of the council of *Tyre and the subsequent talks with the emperor at *Constantinople, whose outcome was the condemnation, deposition and exile of *Athanasius. *Sozomen (*HE* III, 7) says that in 342 T. ordained *Macedonius bishop of Constantinople after the death of Eusebius, but this election does not seem to have been recognized as valid at the time. He must have died soon after, since he was not present at the council of *Sardica (343).

DCB 4, 989; Quasten II, 195-196; D. de Bruyne, Deux lettres inconnues de Theognius, *ZNTW* 27 (1908) 107-110; G. Bardy, *Recherches sur S. Lucien d'Antioche et son école*, Paris 1936, 210-216; Simonetti 597.

M. Simonetti

THEOGNOSTUS. Information derived from *Philip of Side makes T. head of the school of *Alexandria after *Pierius, i.e. *c.*300; but such a late date seems unacceptable, so T.'s scholarchate is usually placed between those of *Dionysius and Pierius, 265-*c.*280. His one work we know of is the *Hypotyposeis*, known through *Photius's description (*Bibl.*, cod. 106). Its seven books dealt with *God the *Father and the *creation of the *world, the Son, the *Holy Spirit, the other spiritual creatures and the Son's *incarnation. Photius appreciates T.'s literary gifts and spirit of *piety, but rebukes his erroneous opinions on the Son, the Holy Spirit and the rational creatures, which were those of *Origen. In fact, the *Hypotyposeis* appears to have been a work whose structure and content were analagous to those of Origen's *De principiis*, i.e. a comprehensive work on questions relating to God and the world. And the few surviving fragments, cited by *Athanasius and others, confirm T.'s close adherence to Origen's thought.

In two theologically very important fragments, T., perhaps writing against *Lucian of Antioch, maintains that the Son's *ousia* (essence, substance) does not derive from nothing, but from the *ousia* of the Father, as reflection derives from *light and vapour from water: it is the emanation (*apórrhoia*) of the Father's *ousia*. Therefore the Son, as *logos* and wisdom of the Father, is like him in *ousia*: a concept that would be taken up in the 4th c. by the *homoeousians.

PG 10, 235-242; Quasten I, 375-376; F. Diekamp, Ein neues Fragment aus den Hypotyposen des Alexandriners Theognostus, *ThQ* 84 (1902) 481-494; A. v. Harnack, *Die Hypotyposen des Theognost*, TU 24, 3, Leipzig 1903, 73-92; L.B. Radford, *Three Teachers of Alexandria: Theognostus, Pierius and Peter*, Cambridge 1908; G. Anesi, La notizia di Fozio sulle Hypotyposeis di Teognosto, *Augustinianum* 21 (1981) 491-516.

M. Simonetti

THEOLOGY. For the pagan Greeks the term θεολογία and the related words θεολογεῖν and θεόλογος referred principally to the mythical-religious way of speaking of the gods, their origin and relation to the *world (Plato, *Rep.* 379ᵃ). Aristotle however, rather exceptionally, used θεολογική for a part of philosophy. The Stoic tradition introduced the distinction between mythical, political and natural t., later discussed by Christian authors (Tertull., *Nat.* II, 1, 2; Euseb., *Praep. ev.* IV, 1, 1-4; Aug., *De civ. Dei* VI, 5-10). Since the mythico-religious meaning remained prevalent, we can understand that up to the 3rd c. Christian authors, while speaking of true *philosophy, spoke of θεολογία only in the negative sense of *pagan *mythology. *Origen began the Christian use of the word (*C. Cels.* 6, 18; 7, 41; *Com. Jo.* II, 34, 205: Trinity) which, thanks above all to *Eusebius (*HE* I, 1, 7; II praef. 1), became common in the 4th c. (Lampe 627 f.). In general θεολογία meant the recognition of the divinity of the Son and the *Holy Spirit or doctrine of the *Trinity, often distinguished from the doctrine of the saving *incarnation of the Word, called οἰκονομία (*economy), while sacred authors or those who spoke correctly of Christ's divinity were called "theologians" (*John, *Athanasius). T. also had some particular uses: for the *Trisagion (Cyr. of Jer., *Cat.* 23, 6); mystical t. in monastic authors; negative and affirmative t. in Ps.-*Dionysius (*Myst.* 3). Among the Latins, only *Marius Victorinus (*Ephes. prol.*) used the Greek θεολογία in a positive sense. So did *Augustine, who gave a definition of the word (*De civ. Dei* VIII, 1: *quo verbo graeco significari intelligimus de divinitate rationem sive sermonem*) that would be decisive for the introduction of *theologia* in the Middle Ages.

DTC 15/1, 341-346 (bibl.); *LTK* 10, 62 ff.; K. Bärthlein, Θεολογία, *JbAC* 15 (1972) 181-185; K.B. Skouteris, (in Greek)*On the teaching of the Fathers up to the Cappadocians*, Athens 1972.

B. Studer

THEONAS of Alexandria. Bishop of *Alexandria, saint; feast 23 Aug. Succeeded Maximus and governed from 281/2 to c.300. Appointed *Achillas to the *didaskaleion*; ordained *Peter, who succeeded him, and *Pierius (Euseb., *HE* VII, 32, 30). A *letter to *Diocletian's chamberlain Lucian, handed down under his name (PG 10, 1567-1574), is a *falsification.

E. Prinzivalli

THEOPASCHITES. Those who insisted on the distinction of the two natures in Christ (anti-*Apollinarists, *Nestorians, Chalcedonians) gave the name t. (from θεός: *God; πάσχειν: to suffer) to those who attributed suffering and death to the Word itself. After the council of *Chalcedon, the controversy mainly concerned two formulae: *qui crucifixus est pro nobis*, added to the *Trisagion, and *unus ex Trinitate passus est*; in fact this was an aspect of the doctrine of *communicatio idiomatum*, differently understood by the Alexandrians on one hand and the Antiochenes and Latins on the other, but finally accepted by all in the Alexandrian sense, earlier sanctioned by the council of *Ephesus (DS 263: a meaning also comprised in *Theotokos). From the t. properly so-called we distinguish the *patripassians who, ignoring the distinction of the two natures, confused the *persons of *Father and Son. But in a broad sense they too are called t., since they speak, in a way hardly differentiated, of the death of God (cf. Ign., *Rom.* 6, 2; *Eph.* 1, 1).

LTK 10, 83; *EC* 11, 1977 f.; A. Grillmeier, *Mit ihm und in ihm*, Freiburg 1975, *passim*.

B. Studer

THEOPHILUS. Late 2nd-c. bishop of *Caesarea in Palestine. At the time of pope *Victor (189-199), T., with bishop *Narcissus of Jerusalem, was president of the provincial *council called over the question of the date of *Easter. On the basis of their decisions, they wrote an encyclical *letter - sent also to *Rome - in which they declared that, having celebrated Easter on the *Sunday after 14 Nisan, they, like the Romans and Alexandrians, followed their apostolic *tradition (Euseb., *HE* V, 22. 25). The *Acts of this council under his name are *falsifications; a polemical work on the date of Easter, also fake, is prob. the work of a 6th- or 7th-c. British author.

EC 11, 1954; *LTK* 10, 90; *BS* 12, 399 f.; P. Nautin, *Lettres et écrivains chrétiens des IIᵉ et IIIᵉ siècles*, Paris 1961, 85-89.

G. Ladocsi

THEOPHILUS. Patriarch of *Alexandria (385-412), uncle of his successor St *Cyril. An outstanding personality, intelligent, full of energy and decision, conscious of the prestige that his position conferred on him, he was also rather inflexible and did not think overmuch about the means he adopted. Our sources (*Palladius, *Socrates, *Sozomen) are hostile to him, but *Arnobius, *Theodoret, *Leo the Great, Virgil of Thapsus and monastic circles have good words for him.

At the start of his pontificate he violently promoted the end of *paganism in *Egypt: his destruction of the Serapeion (391: the attached library was also destroyed) is famous. But T. was also a great builder of *churches. In ecclesiastical questions he was for a brief period a peacemaker (*schism of *Antioch, controversy between *Jerome and *Rufinus), but later acted in a rather inconsiderate and imprudent way. From an admirer of *Origen he became in 399 a fanatical enemy, condemned him in a synod of 401 and ceaselessly persecuted his supporters, among them the four "long brothers", Dioscorus, Ammon, Eusebius and Euthymius. When these accused T. before the emp. *Arcadius and were given *hospitality by *John Chrysostom, T. was very angry and began to intrigue against John, whom he had himself consecrated (398). Called to *Constantinople to defend himself at a synod presided over at the emperor's wish by the patriarch, he formed a coalition of John's many enemies, transformed himself through intrigue from accused into John's accuser and, at the synod of the *Oak (near *Chalcedon) in 403, deposed him. The popular riots that broke out immediately on John's departure for exile forced T. to flee precipitously from Constantinople. *Innocent I, to whom John had appealed, *excommunicated T. He has always been regarded with severity and disfavour in the Byzantine and Latin churches because of the heavy shadows that conceal his person, but enjoys considerable esteem among the Copts and Syrians, who venerate him as a saint.

Of T.'s quite numerous writings, we have only a few relics and very little in the original text. Jerome's correspondence (*Epp.* 87, 89, 90, 92, 96, 98, 100) preserves two *letters sent by T. to Jerome, one to *Epiphanius, a *synodica* to the bishops of *Palestine and *Cyprus and three *festal letters* of the years 401, 402, 404: all closely concern the question of *Origenism and, in part, *Apollinarism. The *Ep. ad Theodosium imp.*, preserved entire only in Latin, is a dedicatory letter prefacing the Easter cycle of the years 380-479 sent to the emperor for him to adopt. Of several other letters, incl. the *festals*, we have only fragments or only the recipient's name. The *Homilia in mysticam coenam* among Cyril's works (PG 77, 1016-1029), which comments on the Last Supper, is T.'s. Many *homilies survive in *Coptic under T.'s name: two might be authentic (*De crucifixione* and *De poenitentia*); others are late legendary constructions. We have fragments of other works: *De morte et iudicio* (PG 65, 200), on the woman with an *issue of blood (F. Diekamp, *Doctrina Patrum*, Münster i.W. 1907, 120 under the title *De poenitentia*), etc. Several other homilies and narrations of dubious or no authenticity have come down under his name. T. also wrote against Origen and John Chrysostom, commented on various books of *Scripture and issued ecclesiastical canons. Various *apophthegmata* of him circulated among the monks of the *desert (PG 65, 197-201).

CPG 2, 2580-2684; PG 65, 33-68; *DTC* 15, 523-530; *LTK* 10, 80; F. Rossi, *I papiri copti del Museo Egizio di Torino*, I, Turin 1887, 64-90; E.A.W. Budge, *Coptic Homilies in the Dialect of Upper Egypt*, London 1910, 65-225 (text and Eng. tr.); G. Lazzati, *Teofilo d'Alessandria*, Milan 1936; A. Favale, *Teof. d'Al. Scritti, vita e dottrina*, Turin 1958; Bardenhewer 3, 115-117; Quasten II, 102-108; Altaner 291; Fliche-Martin 4, 21-193 (*passim*).

A. De Nicola

THEOPHILUS of Antioch. This late 2nd-c. bishop of *Antioch wrote at least four works. The only surviving one, *Ad Autolycum*, provides some biographical information on T.: born not far from the Tigris and Euphrates (II, 24), a *convert to Christianity (I, 24), he lived among Christians who were an opposed and denigrated minority (III, 4); when he finished *Ad Autolycum*, the emp. *Marcus Aurelius was dead (III, 28): i.e. after 17 March 180. This Autolycus was a cultivated *pagan who did not understand how T. could become a Christian. T. sought not just to answer his objections, but also to justify his own *faith in the invisible creator *God and in the *resurrection (book I). Book II brings out the contradictions of the Greek philosophers and poets about God and the origin of the *world, and contrasts them with the *prophets, inspired by God. Here we find the first Christian *commentary on the beginning of *Genesis and (II, 15) the first example of the use of the term "*Trinity" (*trias*) to designate "God, the Word and Wisdom (*Holy Spirit)". Book III contains a chronicle of world history aiming to demonstrate that *Moses predated Homer and the earliest Greek writers.

Another of his works was directed *Against *Hermogenes. *Eusebius tells us only that it cited John's *Apocalypse (*HE* IV, 24). *Jerome specifies that it consisted of "a single book" (*Vir. ill.* 25). It has been possible to establish that it was *Tertullian's source for his *Adversus Hermogenem*. And T.'s own statement in *Ad Autolycum* (II, 28) that he has explained elsewhere that the dragon or devil was originally an *angel (cf. Rev 20, 2-3) prob. refers to this treatise; Tertullian's *Adv. Hermogenem* (11, 3) mentions this explanation.

T. also composed a work *Against *Marcion*, mentioned by Eusebius as "a work of great value" (*HE* IV, 24). T.'s undeniable influence on *Irenaeus and the *List against all the heresies* is probably explained by the reading of this treatise.

T. himself refers to a work *On history* which he had published in several books (*Ad Autol.* II, 30, 31; III, 19). It set out the genealogy of mankind after Seth, identified *Noah with Deucalion and listed the descendants of Shem, Ham and Japheth. It was prob. there that *Clement of Alexandria (*Strom.* I, 21, 142, 1) and the author of the *List* (X, 31) found, independently of each other, that the descendants of Shem, Ham and Japhet formed 72 nations (cf. *RHR* 130 [1971] 164-166). In his reference to T., Jerome claims to have "read commentaries on the Gospel and Solomon's Proverbs under his name", whose authenticity seem to him doubtful (*Vir. ill.* 25). He also specifies (*Ep.* 121, 6) that the former was a synoptic commentary on the four gospels.

Edition of *Ad Autolycum*, ed. and tr. by R.M. Grant, Oxford 1970.
Studies: P. Nautin on *Ad Aut.* III, 10-13 (creation of the world): Centre d'études des religions du livre, *In principio*, Paris 1973, 69-79; N. Zeegers-Van der Vorst on II, 18, 8-12 (creation of man): *VChr* 30 (1976) 258-267; F. Bolgiani on III, 19 (Noah's ascesis): *Forma futuri (Studi M. Pellegrino)*, Turin 1975, 295-333, and on the lost work against Hermogenes: *Paradoxos Politeia* (Misc. G. Lazzati), Milan 1979, 77-118. Cf. also W. Hartke, Über Jahrespunkte und Feste, insbesondere das Weihnachtsfest, *Akad. der Wiss., Sektion für Altertumwiss.*, 6, Berlin 1956, 13-17; J. Bentivegna, A Christianity without Christ by Theophilus of Antioch, *SP* 13 (1975) 107-130; R.M. Grant, *The Greek Apologists of the Second Century*, London 1988, chs. 16-19.

P. Nautin

THEOPHILUS of Castabala. Originally a bishop in *Palestine (Soz., *HE* IV, 24), then transferred by *Silvanus of Tarsus to Castabala in *Cilicia, where he was bishop from 362 to at least 375. In 362 he was one of the leaders of the *homoeousians and took part, with Silvanus and *Eustathius of Sebaste, in the embassy sent to *Rome by the synod of *Lampsacus. He subscribed the *Creed of *Nicaea and was accepted into *communion by pope *Liberius. In 375 he contributed to the break between his *friend Eustathius and *Basil of Caesarea. He was later one of the group of *pneumatomachi (cf. Basil, *Epp.* 130, 1; 244, 2; 245).

R. Devreesse, *Le Patriarcat d'Antioche*, Paris 1945, 157; Simonetti 393. 395. 415; *DCB* 4, 999.

J. Gribomont

THEOPHILUS the Indian, called the Thaumaturge. *Anomoean bishop, died *c.* 365. A native of Diva (modern Sokotra), taken hostage as a young man by the Romans under *Constantine; converted to Christianity, he took to the monastic life and was consecrated deacon by *Eusebius of Nicomedia. Made a bishop, he was sent by *Constantius II on missionary journeys, first to the Sabaeans, among whom he founded three churches at Taphanam, Adane (Aden) and Hormuz, then to *Ethiopia (already evangelized by *Frumentius, sent by *Athanasius in 327). Exiled a first time by Constantius for interceding in favour of the Caesar Gallus, he was recalled to cure his sick wife by his thaumaturgic powers. A partisan of *Aetius, he was exiled by the council of *Seleucia (359) to Heraclea Pontica. He returned at *Julian's accession and was present at *Eudoxius's consecration as patriarch (360) and Aetius's election to the episcopate at the Arian synod of 362.

LTK 10, 89-90; Simonetti 339. 358. 453.

E. Prinzivalli

THEOPHYLACT SIMOCATTA (first half of 7th c.): Byzantine author, of Egyptian origin, known esp. for his *Histories*, eight books on the reign of the emp. Maurice (582-602), compiled with no precise plan and with many concessions to anecdote. We also possess 84 *letters (mostly fictitious), *Quaestiones physicae* and a treatise *De praedestinatione*.

Krumbacher 247-251; H. Hunger, *Die hochsprachliche profane Literatur der Byzantiner*, I, Byzantinisches Handbuch 5, I, Munich 1978, 313-319; C. de Boor, *Theophylacti Simocattae Historiae*, Leipzig 1887; R. Hercher, *Epistolographi graeci*, Paris 1873, 763-786; G. Zanetto, La tradizione manoscritta delle Epistole di Teofilatto Simocatta, *Bollettino del Com. Prep. Ed. dei classici* 24 (1976) 64-86; L. Massa Positano, *Teofilatto Simocatta. Questioni naturali*, Naples 1965; P. Moraux, Le début inédit de l'ouvrage de Théophylacte Simocatès sur la prédestination, *Le monde grec . . . Hommages à Claire Préaux*, Brussels 1975, 234-244; L.G. Westerink, Theophylactus Simocates on Predestination, *Studi in onore di Vittorio de Falco*, Naples 1971, 535-551; C. Garton, Theophylact, On Predestination: A First Translation, *GRBS* 14 (1973) 82-102; B. Baldwin, Theophylact's Knowledge of Latin, *Byzantion* 47 (1977) 357-360.

S.J. Voicu

THEORIA. All through Greek thought, from the Presocratics to the *Corpus Hermeticum* and later neoplatonism, the terms θεωρία (*contemplation) and θεωρεῖν (to contemplate), with other analogues such as θέα, θεᾶσθαι, ὅρασις, ὁρᾶν, designate the highest form of knowledge attainable by the human mind (cf. R. Arnou's article cited in the bibliography). Already in Plato contemplation assumes an importance of the first rank (cf. R. Arnou's article and A.J. Festugière's book, cited in the bibliography; Plato's most important passages are collected in S. Lilla, *Clement of Alexandria*, Oxford 1971, 165). But it is Aristotle above all who exalts t.: for him it is the supreme end of man (*Protr.* fr. 6 Walzer [*Aristotelis dialog. fragm.*, Florence 1934, 35, 13-41]), it is closely associated with his love of knowledge (*Protr.* fr. 2 Walzer), it is the most pleasant and sublime thing (*Met.* Λ1072b 24), it is productive of pleasure and happiness (*Eth. Nic.* VII 1152b 36-1153a 1, X 1178b 28-32). All through the Platonic tradition - from Plato to *Philo, to *middle Platonism and *neoplatonism - t. appears often associated with terms taken from mystery language, such as ἐποπτεία and ἐποπτεύειν, and thus comes to take on an esoteric stamp. Greek *patristics fully accepts this conception of t.: like Plato, Philo and the exponents of middle Platonism and neoplatonism, Christian authors like *Clement of Alexandria, *Origen, *Gregory of Nyssa, *Evagrius and Ps.-*Dionysius the Areopagite see in t. the contemplation - by the human *nous* - of the transcendent realities or ideas which, for them, all exist together in the divine *logos*; they associate it intimately with terminology taken from mystery language, and hence identify it with the higher esoteric *gnosis*, though even in "mystical" authors like Gregory of Nyssa and Ps.-Dionysius, *gnosis* remains distinct from ἕνωσις with the supreme principle, which is an experience that transcends any form of intellectual knowledge (in neoplatonism too, t. and ἕνωσις remain distinct). Contemplation also plays an important role in St *Augustine and in *Cassian (cf. M. Olphe-Galliard's article, cited in the bibliography).

As E. Kihn (*ThQ* 62 [1880] 581-582) and esp. A. Vaccari (*Biblica* 1 [1920] 4-36 [= *Scritti di erudizione e di filologia*, I, Rome 1952, 101-142]) have shown, in the Antiochene school t. appears to be closely connected to the type of scriptural *exegesis that resolutely rejects the *allegory practised by the Alexandrian school, which followed the example of Philo. *Diodore of Tarsus wrote a treatise on the difference between t. and *allegoria* of which the preface to his commentary on Ps 118, preserved in codex Paris gr. Coislin. 275 and edited by L. Mariés, is a summary (cf. Vaccari, *Scritti di erudizione e di filologia*, I, 101-102, 110-111). A concise definition of t. in the Antiochene sense is given by *Julian of Eclanum (PL 21, 971B; cf. Vaccari, *op. cit.*, 111-112). This type of t. is held to by, among others, *Severian of Gabala and St *John Chrysostom (cf. Vaccari, *op. cit.*, 109-110).

R. Arnou, La contemplation chez les anciens philosophes du monde gréco-romain, *DSp* 2, 1716-1742; J. Daniélou, Mystique de la ténèbre chez Grégoire de Nysse, *DSp* 2, 1762-1787; M. Olphe-Galliard, La contemplation dans la littérature chrétienne latine, *DSp* 2, 1911-1929; R. Roques, Contemplation, extase et ténèbre chez le pseudo-Denys, *DSp* 2, 1885-1911; E. Kihn, Über θεωρία und ἀλληγορία nach den verlorenen hermeneutischen Schriften der Antiochener, *ThQ* 62 (1880) 581-582; A. Vaccari, La "teoria esegetica" della scuola antiochena, *Biblica* 1 (1920) 4-36 (= *Scritti di erudizione e di filologia*, I, Rome 1952, 101-142); J. Kroll, *Die Lehren des Hermes Trismegistos*, Münster 1914, 350-354; A.J. Festugière, *Contemplation et vie contemplative selon Platon*, Paris 1936; W. Jaeger, *Aristoteles*, Berlin ²1955, 81 ff.; id., *On the Origin and Cycle of the philosophical Ideal of Life*, Appendix II of the Eng. tr. of his *Aristoteles*, by R. Robinson, Oxford 1948, 430-431; W. Völker, *Der wahre Gnostiker nach Clemens Alexandrinus*, TU 57, Berlin 1952, 316-321, 403-411; E. Fascher, Epoptie, *RACh* 5, 973-983; S. Lilla, *Clement of Alexandria*, Oxford 1971, 163-169.

S. Lilla

THEOTECNUS of Caesarea. Bishop of *Caesarea in Palestine after *Theoctistus and the brief episcopate of Domnus (after 260), T. was a pupil of *Origen (Euseb., *HE* VII, 14). *Eusebius calls him his contemporary. He exhorted the officer Marinus to confession of the faith and *martyrdom (*ibid.*, VII, 15). With other disciples of Origen, he took part in the first *council against *Paul of Samosata, bishop of *Antioch (VII, 26) and in the last, subscribing the final *letter (VII, 30, 2). *Photius (*Bibl.*, cod. 118 and 232) confuses T. with his predecessor Theoctistus.

H. Crouzel

THEOTECNUS of Livia. Bishop of Livia in *Palestine (*Tell er-Ram* in W Jordan, near the Dead Sea), 6th-7th c. We have a *homily of his on the feast of *Mary's *assumption (15 Aug), which he is one of the first to attest explicitly. The feast's name is significant: ᾿Ανάληψις τῆς ἁγίας Θεοτόκου. The homily more than once states explicitly that Mary was assumed to heaven *body and *soul, since the former was separated from the latter only for a short time, though remaining incorrupt. Though the author refers to the traditions of the *apocrypha on Mary's death and assumption, he places the dogmatic foundation for the assertion of her assumption into heaven in her relation to the *mystery of the *incarnation: Mary's dignity as Mother of God, her perpetual virginity and exceeding sanctity require that her body, made exempt from corruption, should be assumed into heaven and reunited to her soul, so that the Virgin may enjoy the *glory of her Son. All T.'s theological arguments in favour of the dogma of the assumption recur in later Byzantine homilists.

CPG 3, 7418; A. Wenger, *L'Assomption de la très sainte Vierge dans la tradition byzantine du VIe au Xe siècle. Etudes et documents*, Paris 1955, 96-110, 271-291; id., in *Maria: Etudes sur la S. Vierge* (ed. H. du Manoir) 5, Paris 1958, 936-938; Beck 400; Altaner 555; G. Söll, *Storia dei dogmi mariani*, Rome 1981, 192-196 and 429.

A. De Nicola

THEOTIMUS. Bishop of *Tomi in *Scythia Minor, called "the philosopher", first quarter of 5th c. Originally a *pagan philosopher (Socr., *HE* VI, 12; Soz., *HE* VII, 26), soon after his *conversion he was ordained bishop. He worked fruitfully among the pagan tribes, Huns and *Goths. The barbarians called him "the God of the Romans". *Jerome still knew some of his dialogues (*Vir. ill.* 131), but all are now lost. In 403 T. protested against the condemnation of *Origen, whom many considered an *orthodox teacher. Date of death unknown: certainly before the council of *Ephesus (431), which his successor Timothy attended.

G. Ladocsi

THEOTOKOS. Title given to *Mary as Mother of *God; its origin must be sought in the Alexandrian school. *Origen, in his *Homilies on Luke*, affirms the concept of Mary's divine motherhood in analogous formulae (*Hom.* VII, 6; VIII, 4). Acc. to *Philip of Side, *Pierius of Alexandria (281-*c*.300) wrote a treatise on the T. *Peter martyr, bishop of Alexandria (†311), knew and used the title. Well attested by *Alexander of Alexandria and *Athanasius, our surest evidence comes from the troparion *Sub tuum praesidium*, where the word is cited intact. From the 4th c. on, writers and church *Fathers commonly use it as a specific title of Mary. The objections of *Nestorius, who thought it was improper, unbiblical, absent from the vocabulary of the council of *Nicaea, contaminated by *paganism - because it represented Mary as a goddess - and should be replaced by *Christotokos*, were definitively answered by the council of *Ephesus (431), which sanctioned the legitimacy of its use and its exact meaning.

Philip of Side, *Historia christiana, fragmenta*, TU, 5/2, 165 ff.; Peter Martyr, *Fragmenta*: PG 18, 517; Alexander of Alexandria, *Epistola I ad Alexandrum episcopum Constantinopolitanum* I, 12: PG 18, 568; Athanasius, *Contra Arianos, Oratio* III, 14. 19. 33; id., *Vita Antonii* 36; id., *De Incarnatione et Contra Arianos* 22; H. Rahner, Hippolytus von Rom als Zeuge der Ausdruck Theotokos, *ZKTh* 59 (1935) 73-81; G. Giamberardini, Il "Sub tuum Praesidium" e il titolo Theotokos nella tradizione egiziana, *Marianum* 31 (1969) 350-358; A. Grillmeier, *Christ in Christian Tradition*, London ²1975, 392-413; G. Söll, *Storia dei dogmi mariani*, Rome 1981, 90-91, 108-113, 152-164; A. Poós, *La "Theotokos" ad Efeso, Calcedonia e nel Vaticano II*, Rome 1981; L.I. Scipione, *Nestorio e il Concilio di Efeso*, Milan 1974; J. Meyendorff, *Christ in Christian Tradition*, New York 1975; A. Cameron, The Theotokos in Sixth Century Constantinople, *JThS* n.s. 29 (1978) 79-108; G.S. Bebis, The Virgin Mary in the Eastern Traditions, *New Catholic World* (1986) 258-263.

E. Peretto

THESSALONICA

I. City and Christianity - II. Archaeology.

I. City and Christianity. Thessalonica (now Salonica; Greek Thessaloniki) was founded in 316 BC by Cassander, who gave the city the name of his wife, sister of Alexander the Great. Its splendid geographical position ensured its rapid growth and development: immediately after the Roman conquest of Macedon (186 BC), it became capital of the second of the four autonomous districts into which the whole territory was divided; later the city became capital of the whole eparchy of *Macedonia (146 BC) with the exceptional title of "free city". All through the Roman period T. held first place on account of its excellent commercial and political position, from time to time gaining special recognition from Rome, e.g. the title of guardian of the temple (*neókoros*) - i.e. of the imperial temple - and, later, that of Roman colony. At the time of the Tetrarchy (AD 300), T. reached its greatest prosperity as the seat of Maximilianus *Galerius, who enriched it with many buildings. The language used during the period of Roman rule remained Greek, as *epigraphy testifies. Religion, once purely Hellenic, slowly declined towards a *syncretism with other foreign deities until, by the start of the 4th c., the Christian religion emerged. When the apostle *Paul reached the city in AD 50 he found the terrain suitable for the preaching of the new creed, despite the opposition of the *Jews. In the next centuries the city stood out as the religious centre to which the leaders of the surrounding churches converged to share a religious climate *par excellence*, as *archbishop Aetius of T. said at the council of *Sardica (343). Coming into contact with this environment, *Theodosius the Great (379-395) decided to embrace Christianity officially, and here in 380 he promulgated the decree recognizing the decisions of *Nicaea and condemning *Arianism. In 390 sedition broke out in T. and the emperor ordered a massacre in the circus, for which *Ambrose obliged him to do public penance.

Since T. was the centre of an important road network and a meeting-point between the Greek *East and the Latin West, it had been evangelized as early as the apostolic era. At the start of his second missionary journey, in the first months of AD 50, Paul, after preaching at *Philippi (Acts 16) where he had met opposition from the Jews, who caused him "suffering and shameful treatment" (1 Th 2, 2), went with Silas (Acts 16, 25-17, 4), also called Silvanus, and *Timothy to *Thessaly and, passing through Amphipolis and Apollonia, arrived at T. There, in the synagogue, for three consecutive days he explained the Holy *Scriptures, demonstrating Jesus' messianicity. Some Jews believed and joined Paul and Silas, as did "a great many of the devout Greeks and not a few of the leading *women" (Acts 17, 4). The Jews, jealous, incited some men to do away with Paul and Silas, but they fled by night to Beroea, whence they went to *Athens, where Paul sent for Silas and Timothy (Acts 17, 15; 18, 5), who then returned to T. (1 Th 3, 1). At T. Paul had founded the second Christian community in Europe, after Philippi. From Athens he wrote the *First letter to the Thessalonians*, after Timothy had described to him the position of the Christian community there, assailed by the Jews, but firm in the faith. The *Second letter to the Thessalonians* is of some months later. Subsequently the Christian community developed (cf. the edict of *Antoninus Pius [138-161] recorded by *Melito of Sardis in *Eusebius [*HE* IV, 26, 16]). During *Diocletian's *persecution, in 304, the sisters *Agape, Chione and Irene suffered *martyrdom there (cf. P. Franchi de' Cavalieri, *Il testo greco originale degli atti delle ss. Agape, Irene e Chione*, ST 9, Vatican City 1902, 3-19), recorded in *Mart. hier.* on 1 April, the first two in *Brev. syr.* on 2 April (*Passio graeca*, in BHG 2, 34). Agathopus and Theodulus, martyred under the emp. *Galerius (305-311), are recorded on 4 April (*Synaxarion Ecclesiae Constantinopolitanae*, 583). The martyr *Demetrius was also venerated there (BHG 2, 496-547; H. Delehaye, *Les légendes grecques des saints militaires*, Paris 1909, 107 f.). Other saints of T. were David (*c*.535) and Theodora (*c*.893). The *episcopal lists are fragmentary, so that few names are certain before the 4th c. The first bishop was Aristarchus, a disciple of Paul. Bishop Demetrius, patron of the city, died a martyr in 290 or 306 (ASS *Octobr.* IV, 50-209). The metropolitan bishop Alexander was present at the council of *Nicaea (325) and the *dedication of the basilica of the Holy Sepulchre at *Jerusalem (335) (cf. Euseb., *Vita Const.* IV, 43, 3). After 379 Macedonia, though politically joined to the *East, remained under the patriarchate of *Rome; from 382 to 389 the metropolitans of T. were apostolic vicars for E Illyricum. After a *council called at Capua in 391 under pope *Siricius (384-399) had dealt with the questions raised by *Bonosus, bishop of Naissus in Dacia Inferior, who denied *Mary's perpetual virginity and repeated the trinitarian errors of *Photinus of Sirmium, the bishops of Illyricum re-examined Bonosus's doctrines more deeply: to this end *Anisius, archbishop of T., called and presided over a council, which also looked at the validity of the *ordinations carried out by Bonosus after his deposition by the council of Capua. *Ambrose (*Ep.* 56), who watched over the council's progress and had deferred judgment on Bonosus to Anisius, wrote to Bonosus that he was obliged to obey the council's provisions. In 518 another council was held at T. From 649 T. came into conflict with Rome (Hfl-Lecl 2, 208), so that in 732 the emp. *Leo III separated Illyria from Rome. T., the land of Cyril and Methodius, was from 860 the centre for the *evangelization of the Slavs.

O. Tafrali, *Thessalonique des origines jusqu' au XIVe siècle*, Paris 1919; F. Tafel, *De Thessalonica eiusque agro dissertatio geographica*, Berlin 1939; on the first council of T. cf.: Mansi 3, 689; Jaffé, nn. 261, 299 and 303; Palazzini 5, 966 f; K.P. Donfried, The cults of Thessalonica and Thessalonian correspondence, *NTS* 31 (1985) 336-356.

E. Pelekanidou - C. Nardi

II. *Archaeology. The first Christian *church buildings were constructed at T. from the time of *Theodosius the Great, and later enlarged and splendidly decorated.

1. St George. The Rotunda, as it is called locally because of its circular plan, is contemporary with *Galerius's triumphal arch, i.e. the end of the 3rd c. or first years of the 4th. At the end of the 4th and in the mid 5th c. it was transformed into a church, with the addition of a vast circular corridor. The main entrance was moved to the W side; the apse was added and the inside ornamented with splendid *mosaics. Windows and rosettes show simple motifs, which also recur in the pavement mosaics: on a gilded or silvered ground, in various positions and at irregular intervals, are birds, baskets full of fruit, isolated fruits, rendered in glowing colours; even the frames are alive with animals, birds and fruit. But the main decoration is in the vault of the dome, divided into two sections: the first containing saints framed by architecture; the second filling the remaining surface. Among the architecture are *exedrae* enclosing altars, whose forms recall the structure of the tombs of *Petra and the temple of Dionysius at Baalbek. On the bases and plinths of these fantastic constructions are *martyrs in the attitude of *orantes*, shown frontally and full-length. The great mosaic compositions once set above this area are lost; but from what remains it is possible to reconstruct their general scheme: 22 figures filled the whole surface (now bare); running vertically were a band representing the vault of heaven, another of blue stars on a gilded ground and a rich garland of leaves and fruit. In the centre, raised by four *angels, was Christ standing and blessing. The apse is ornamented with a fresco of Christ's *ascension, divided into two zones: in the upper part Christ in a luminous aureole, flanked by two angels; in the lower part the Virgin praying between two angels, surrounded by the *apostles. The fresco is dateable to the 9th c.

2. Monì Latomou. This little church, now known as Hosios David, goes back to the mid 5th c. The building, of which only the W part still remains, was originally in the form of a Greek cross with a dome. The apse mosaic represents the vision of *Ezekiel: Christ, young and beardless, seated on the heavenly vault and surrounded by a large, transparent aureole; at the four cardinal points, the *symbols of the *evangelists holding up their gospels; at Christ's feet the four rivers of Eden: Pishon, Gihon, Tigris and Euphrates, whose waters flow into the river Chebar, which meanders solitary and blue, full of fish; in the midst of the waters, next to Christ's feet, is the personification of the river, a bearded man with his back emerging from the water; in the left corner is Ezekiel, a prisoner in the land of the Chaldaeans, on the banks of the river Chebar; in the right corner, in a cave in the rock, *Habakkuk awaits the Word of God. The author is revealed as a master, not just in the colours and general design of the work, but also in the pregnant psychological effects of the forms. A strong expressive realism in the rendering of the storm and the climate of agitation characterizes Ezekiel's vision, while a tranquil serenity and an attitude of profound meditation shine through the face and position of Habakkuk.

3. St Demetrius. The city's most important monument, and one linked with Greek national history, is the basilica of St *Demetrius, with five aisles and a transept, founded on the place of an oratory or *martyrium* early in the 5th c. The then eparch of Illyricum, Leontius (412-413), having been cured through the saint's intervention, wished in recognition to erect a temple worthy of him. The church, destroyed by a fire in the first half of the 7th c., was rebuilt almost immediately and kept its splendour in the following centuries despite being restored and sacked by Saracens, Normans, Franks and Turks. When Leontius's great basilica arose at the beginning of the 5th c., many parts of the old *baths, either because the architectural elements were indispensable or because these underground elements were closely connected with St Demetrius's imprisonment and *martyrdom, were incorporated in the new building, particularly the crypt. At the same time the old Roman buildings, with suitable additions and modifications, served as subfoundations of the apse, transept and in general the whole E part of the basilica. The central aisle, narthex and part of the altar were covered with polychrome marbles. In the central aisle, besides this decoration, was *opus sectile* ornamentation with various polychrome geometrical and floral motifs. More valuable marbles decorated the τρίβελον (triple entrance in the façade), and white veils, lined inside with green cloth, hung in the lower part between small turquoise columns. The variety of the capitals is perhaps unique: every type of early Christian capital is represented, from the Theodosian to the wavy and stylized with wind-blown foliage; the capitals of the pilasters have ornamentations of birds drinking from vases. The *cancelli*, which date from the 5th to the 9th cc., also show great variety. The pictorial decoration follows no precise programme: slabs of marble decoration were sometimes removed to make room for the depictions of frescos and mosaics, undesired compositions or images were sometimes covered over with marble. The mosaics date from the 5th to the 9th or, at most, the 10th c. The mosaic with the healing of the boys on the W wall of the S aisle, and an *angel which is all that remains of a depiction on the W wall of the N aisle, are of the 5th c. To the 7th c. belong the mosaics of the founders on the N face of the pilaster N-E of the altar, the scene of St Demetrius with two boys on the W face of the pilaster N-E of the altar, the mosaic of St Demetrius with a bishop on the E face of the S-E pilaster of the apse, the depiction of the saint with four clerics on the N section of the τρίβελον (in poor condition), the mosaic of St Demetrius now kept in the crypt chapel (one of the most important, since it relates to the rebuilding of the church in the 7th c. and bears the name of the eparch who ruled Illyricum at that time) and the mosaic of St *Sergius on the S face of the pilaster S-E of the altar. The mosaic of the Madonna with St Theodore on the S face of the N-E pilaster of the apse is assigned to the late 9th or early 10th c. Few frescos are still preserved in good condition: the best is on the S wall of the basilica and dates from the 7th c.

4. The Acheiropoieta. Built seemingly after the condemnation of *Nestorius by the 3rd ecumenical *council (AD 431), the church took its name from an icon of the Madonna and Child, later miraculously transformed into a Madonna standing and praying without a Child. This transformation being held to have occurred without the intervention of a painter, the icon was called *Acheiropoieta* ("made without human hand"). The basilica, with three aisles divided by columns standing on wide stylobates, has lost the marble that once covered all the walls, but is still one of the few early Christian churches to preserve many architectural elements (columns, Theodosian capitals, cornices, etc.). The only surviving mosaics are those on the inner borders of the column arches. The decoration of the three arches of the τρίβελον shows a great variety of colours, designs and forms on a gilded ground. The mosaics of the *Acheiropoieta* are among the most important of the 5th c.: the presence of some motifs, e.g. the lotus, reveal a technique, like the mosaics of St George, still under the influence of classical art.
[Figs: 302, 303, 304, 305]

1. General works: O. Tafrali, *Topographie de Thessalonique*, Paris 1913; Ch. Diehl-M. Le Tourneau-H. Saladin, *Les monuments chrétiens de Salonique*, Paris 1918; A. Vakalopoulos, Ἱστορία τῆς Θεσσαλονίκης *(315 a.C.-1912)*, Thessalonica 1947. *2.* Rotunda (St George): E. Dyggve, Recherches sur le palais imperial de Thessalonique, *Studia orientalia Ioanni Pedersen dicata*, Copenhagen 1953, 59 ff.; H. Thorp, Quelques remarques sur les mosaïques de l'église Saint-George à Thessalonique, *Acta of the IXth Intern. Byzant. Congr.*, Thessalonica 1953, 489-498; H. Thorp, *Mosaikkene i St. Georg-Rutunden i. Thessaloniki. Et hovedverk i tidlig-Byzantinsk Kunst*, Oslo 1963; St. Pelekanidis, *Gli affreschi paleocristiani ed i più antichi mosaici parietali di Salonicco*, Ravenna 1963. *3.* Hosios David (*Monì Latomou*): J. Papadopoulos, Une mosaïque byzantine de Salonique, *CRAI* 1927, 215-218; Ch. Diehl, *Une mosaïque byzantine de Salonique*: CRAI 1927, 256-261; A. Xyngopoulos, Τὸ καθολικὸν τῆς Μονῆς Λατόμου ἐν Θεσσαλονίκῃ καὶ τὸ ἐν αὐτῷ ψηφιδωτόν, *AD* 12 (1929) 142 ff.; P. Grumel, La mosaïque du "Dieu Sauveur" au monastère de Latomou à Salonique, *EO* 33 (1930) 157-175; C. Diehl, A propos de la mosaïque de Hosios David à Salonique, *Byzantion* 7 (1932) 333-338; St. Pelekanidis, Παλαιοχριστιανικὰ Μνημεῖα Θεσσαλονίκης Ἀχειροποίητος, Μονὴ Λατόμου, Thessalonica 1949. *4.* St Demetrius: G. and M. Sotiriou, Ἡ Βασιλικὴ τοῦ Ἁγίου Δημητρίου Θεσσαλονίκης, Athens 1952. St. Pelekanidis, Γραπτὴ παράδοση καὶ εἰκαστιχὲς Τέχνες γιὰ τὴν προσωπικότητα τοῦ Ἁγίου Δημητρίου, Thessalonica 1970. *5.* Acheiropoieta: St. Pelekanidis, Παλαιοχριστιανικὰ Μνημεῖα Θεσσαλονίκης Ἀχειροποίητος, Thessalonica 1949; A. Xyngopoulos, Concerning Acheiropoietos Thessalonica, Μακεδονικά 2 (1941-52) 472 ff; G. Gounaris, L'archéologie chrétienne en Grèce de 1974 à 1985, *Actes XI*e *Cong. internat. d'arch. chrét.*, CEFR 123, Rome 1989, 2687-2711.

E. Pelekanidou

THESSALY. When the Delphic Amphictyon was reconstituted in the time of *Augustus, the term T. acquired a political meaning. It designated the region between Thermopylae and Mt Olympus, as far as Perrebìa. Under *Hadrian (117-138) and *Antoninus Pius (138-161), T. was administratively attached to *Macedonia: both provinces were ruled by a single governor, though in particular circumstances T. had its own administrator and was called an eparchy. In the manuscript known as *Laterculus Veronensis*, which provides an image of the empire at the time of *Diocletian, T. is part of one of the 12 dioceses, that of the Moesii, and constitutes an eparchy with autonomous administration. In this period the last remaining independent state, that of the Magneti, was annexed, so that the eparchy of T. reached its greatest geographical extension. At that time the borders of T. were as follows: to N, Mt Olympus, in the area of Pirgetus; to W, the valley of the R. Titharesios (now Sarantàporos), the ridge of the Khasiá mountains and the Pindus massif, separating T. from *Epirus; to S, the Spercheios valley; and to E, the Aegean sea. Later, halfway through the 4th c., as we see from the catalogue called *Breviarium*, Macedonia became an independent diocese comprising the eparchies of Macedonia, T., *Achaia, the two Epiri, etc. This administrative order appears in the *Notitia dignitatum*, which attests that the diocese of Macedonia was composed of six eparchies, including T. In this period T. was governed by an archon (*praeses*) and belonged to the larger administrative region of the *Praefectura praetorio per Illyricum*. In 417 Paul *Orosius, in his list of the eparchies, included T. under Macedonia: "*Macedonia habet ab Oriente Aegaeum mare, a Borea Thraciam . . . a meridie Achaiam*". *Eunapius too says that Macedonia and T. were joined to *Thrace. Similarly *Zosimus says that the pass leading from T. to Nicopolis in Epirus crosses, in a N-W direction, the sources of the Peneios in the Pindus. Our main 6th-c. source is Hierocles's *Synèkdemos*, a fundamental work of political geography for the years preceding 537, supplemented by the works of Procopius and by *archaeology. Acc. to the *Synèkdemos*, the eparchy of T., ruled by a ἡγεμών, was part of the *Praefectura praetorio per Illyricum*; the cities of Diocletianopolis and Caesarea belonged to it. This means that T. extended further to the N-W than it had under Diocletian, since Diocletianopolis was near Lake Kastorià and Caesarea was 16 km S of Kozani, as A. Keramòpoulos has established. Finally, in the 7th c., the terms T. and eparchy of T. recur in many literary and historical works; the author of the *Cosmography* says that T. bordered Macedonia, and uses the expression: *Patria que dicitur Ellas Thessalie*, thus geographically associating Hellas and T.

Both literary sources and *archaeology attest that T. was prosperous until the 4th c. But in AD 376 it was subjugated by the Scythians, as Eunapius attests. Its geographical position, though marginal to the political aims of the Byzantine empire, was still such as to feel the effects of the great tumults that shook the Balkan peninsula; its importance for maritime traffic was a misfortune for it. The invasions of the Germanic tribes and Huns in the first centuries of the Christian era led to a radical modification of urban structures and of the demographic order, on which the new Christian religion, now official, had an equally great effect. The Slav and Arab invasions of the 6th c. brought destruction, as the *Chronicle of the foundation of Monemvasia* expressly says: "In another invasion all T. and the whole of Hellas were devastated". The Slav expansion in T. is also referred to in the *Miracles of St Demetrius*, which describes the 7th-c. siege of *Thessalonica. From the same source we learn that in 675-677 the besieged asked the Belegezite Slavs of Thebae Phthiotides and Demetrias to sell them grain. From the beginning of the Christian era until the 7th c. many cities flourished in the eparchy of T. Some of them are continually cited by the sources, others completely ignored by documents after the 7th c. Some still keep their old name, others have changed theirs.

One of the most important cities in T., Larissa, still keeps its old site and its original name. In the early Christian era the city reached its greatest splendour; a circuit of walls separated the acropolis from the lower quarters. The fortress of the acropolis was built in the first years of the Christian era. Judging from the diagrammatic reconstruction of the *Tabula Peutingeriana*, it must have possessed two gates. Nothing remains of the walls, and no foundations have yet been found. Pieces of columns and other architectural elements were discovered in the past on the castle hill, but no buildings until 1978. In that year the remains of a three-aisled Christian basilica of exceptional dimensions were found. It is not yet possible to date it, though investigations are not yet complete; but an interesting inscription has been recovered with the name of Achilleus, first metropolitan of Larissa, which may refer to the foundation of the basilica. In the course of excavating, two tombs of considerable interest were found, one of which must have belonged to an important person since it was found in the N section of the *church, within the altar area. Another important discovery was the *cemetery, located next to the basilica, which seems to have been reused several times. But we consider that the cemetery's first chronological phase corresponded to the foundation of the church and consisted of tombs *ad sanctos*. At the same time, remains of another basilica were excavated near a street in the city centre. Its dimensions were even greater than those of the basilica on the castle hill. This majestic edifice was, like the others so far found, in the centre of Larissa. The walls were decorated with frescos and the floors with *mosaics; the ornamental themes were geometrical. In the surviving section of the narthex, the mosaic surface is decorated with an amphora flanked by two facing peacocks, as well as geometrical motifs. The decoration is perpendicular to the axis of the church and was exactly in front of the central entrance. The remaining parts of the basilica have not yet been discovered and may be underneath the old buildings facing each other on the city's central street. The two buildings belong, in our opinion, to the time of St Achilleus, i.e. first half of 4th c.; as we learn from the synaxaria, he embellished the city of Larissa with splendid buildings rich in religious and decorative elements. Unfortunately, research on the early Christian period in Larissa is recent, and discoveries dateable to this period are modest. However, the knowledge so far acquired attests that even in the early Christian period this old Thessalian city was one of the most important centres of the Hellenic world. We believe that many things will come to light in future, enough to acquaint us with the artistic and historical aspects of the region.

Another Thessalian centre of the same period is the city of Thebae Phthiotides, today Néa Ankhíalos. Excavations have so far brought to light many remains, mostly dating from the 5th-7th cc. Classified by Hierocles's *Synèkdemos* as the third most important city, Thebae has been identified with the ancient Pirasus. George Sotiriou, its first discoverer, has uncovered the remains of both sacred and secular buildings. Six groups with structures typical of the early Christian basilica have so far been identified: *1)* basilica A, or of St *Demetrius, late 5th-early 6th c.; *2)* basilica B, called "of Elpidius", same period; *3)* basilica C, the most imposing, called "of the archpriest Peter" on account of an epigraph of restoration. The group contains buildings from various architectural periods, from the 4th c. to the first years of the 5th. It was destroyed by a fire in the 7th c.; *4)* cemeterial basilica D, early 7th c. Two more basilicas have been identified, but not yet completely excavated. A secular building not yet identified, large remains of *baths, a road of large dimensions and the episcopal palace have also been located. The episcopal see of Thebae Phthiotides was held by that *epìskopos Thebon* who subscribed the *Acts of the council of *Nicaea. The recovered monuments give an image of a rich city, whose life is described for us in inscriptions. From these we learn the professions and places of origin of its inhabitants. *Epigraphy also testifies to the great commercial activity of the mercantile port of Thebae Phthiotides. Coins discovered, mostly from *Constantinople or *Nicomedia, give a clear idea of the importance of the city's mercantile activity. Thebae, destroyed in the 7th c., is not mentioned by sources from the second half of that century: as we see from archaeological finds, the cause of destruction must have been a fire. But the city's name survived its destruction, at least until the 14th c. Unfortunately the archaeological research carried out, though great, has not yet been published. So it is not possible to attempt a conjectural reconstruction of the artistic and cultural climate of that period, despite the prodigious wealth of finds. Great areas of pavement conceal numerous capitals and other architectural elements beneath their mosaics; numerous vases await publication. All these finds can give an idea of the wealth and splendour of this city.

Other Thessalian cities were mentioned by Hierocles, centres that were thick on the ground in this period. The sources mention Demetrìas, Nèai Patrai (today Ipàti), Echìnos, Pharsalus, etc. Yet archaeological research still has nothing of importance to communicate, or even has yet to begin. One important reason for the lack of elements able to illustrate the civilization of these cities is that in some cases their destruction was total. Naturally the two principal centres, Larissa and Thebae Phthiotides, are only the most important part of the archaeological research; but fragmentary data in our possession, partly from other areas, attests the existence of cities and inhabited centres in numerous other districts of T.

Fr. Stahlin, *Das hellenische Thessalien*, Stuttgart 1924; G. Sotiriou, Αἱ χριστιανικαὶ Θῆβαι τῆς Θεσσαλίας, *AE* (1929) 1-158; Y. Bequignon, *La vallée du Spercheios dès origines au IVe siècle*, Etudes d'archéologie et de topographie, Paris 1937; E. Honigmann, *Le Synekdèmos d'Hiéroklès et l'opuscule géographique de Georges de Chypre*, Brussels 1939; G. Sotiriou, Ἀνασκαφαὶ ἐν Νέᾳ Ἀγχιάλῳ, *PAAH* 1924, 1926, 1928, 1929, 1930, 1931, 1934, 1935, 1936, 1937, 1938, 1939, 1940, 1954, 1955, 1956; A. Avramea, Ἡ Βυζαντινὴ Θεσσαλία μέχρι τοῦ 1204, Athens 1974; L. Deriziotis, Παλαιοχριστιανικόν ψηφιδωτόν δάπεδον εἰς τήν θέσιν " Παλαια " τῆς πόλεως τοῦ Βόλου, in *La Thessalie, Actes de la table-ronde*, Lyons 1975, 193-198; H. Hunger, *Tabula Imperii Byzantini (Hellas und Thessalia)*, Vienna 1976; P. Lazaridis, Ἀνασκαφαὶ Νέας Ἀγχιάλου, *PAAH* 1959; 1960, 60-66; 1961 to 1979; L. Deriziotis, Comunicazione sugli edifici protocristiani di Larissa, in *X Congr. Intern. della Società di Archeologia Cristiana*, Thessalonica 1980 (unpublished); Achillio, *DHGE* 1, 312; *BS* 1, 157-158.

L. Deriziotis

THEURGY. As E.R. Dodds (Theurgy, 55) and H. Lewy (*Chaldean*, 461) have shown, the term *theurgy* (from Gk *theos*, god + *ergos*, working) has two meanings: that of "performing divine operations" and that of "making man a god". Acc. to Iamblichus, t. is "the art of performing 'divine' operations" (*De myst*. I, 93, p. 33, 9-10 Des Places), "the sacred performance of ineffable operations, above any thought and suitable to God" (*De myst*. II, 96, 17-19, p. 96 Des Places). Acc. to Michael Psellus, "we make ourselves like God through the theurgic virtues too, and this is the most perfect likeness. . . . Being able to make man a god (*theopoiéin*), to distance him from matter and free him from the passions, in such a way that he may 'in turn' make someone else a god (*theurgéin*), is the most perfect likeness" (*De omnif. doctr*. ch. 71, 5-11, pp. 45-46 (Westerink); "he who possesses theurgic virtue . . . makes men into gods" (*op. cit*., ch. 74, 1-2, p. 47 Westerink). In these two passages Psellus simply reproduces Proclian and Iamblichean concepts (something Westerink has not noted in the *apparatus fontium* of his edition of *De omnifaria doctrina*): for both *Proclus and Iamblichus the "theurgic virtues" represent the highest degree of human elevation (cf. Marinus, *Vita Procli*, ch. 26, pp. 20 f. Boissanade).

It appears from the *Lexikon *Suidas* that the first person officially to receive the title of "theurge" was a certain Julian, a contemporary of *Marcus Aurelius and author of a work entitled *Logia di'epón*, in all probability identical with the *Chaldaean* *Oracles (cf. Dodds, *op. cit.*, 55), of which numerous fragments remain, edited first by W. Kroll (*De oraculis chaldaicis*, Breslau 1894) and more recently by E. Des Places (*Oracles Chaldaïques*, Paris 1971). It is quite probable that Julian also wrote a commentary on these Logia, perhaps entitled *Theurgiká*: Proclus, Marinus and Damascius seem to allude to it (cf. Dodds, *op. cit.*, 56). The term "theurge" appears in the text of the oracles themselves: fr. 153 Des Places (p. 103) asserts that "theurges do not fall into the flock that is subject to fate". The first aim of t., then, is the liberation of man from the iron laws of the material *world, indispensable premise of his "deification", clearly asserted e.g. in fr. 97 (p. 90 Des Places).

Dodds (*op. cit.*, 56) brings out the fact that the *Chaldaean oracles* contained instructions for *sun and fire cults, and *magic prescriptions aiming to evoke particular spirits or deities (from this point of view, the function of t. corresponds to that of the medium in our own time). We cannot here give a detailed description of the various rituals and magical practices: for these we must refer the reader to Dodds's article (*op. cit.*, 61-69) and Lewy's book (*op. cit.*, 227-257, 467-471). This emphasis on magical practices fully accords with the philosophico-religious tendencies of the 2nd c. AD: Apuleius and the *Life of* **Apollonius of Tyana* by Philostratus give us clear examples of this. The close connection between theurgic practices and the magical *papyri has been shown by S. Eitrem (La théurgie, esp. 53-58).

A scheme analogous to that observed in the *Chaldaean oracles* can be seen in *Poimandres* (cf. *Hermetism), in which the soul's *Himmelreise* concludes with its deification: the human *soul, with all passions, impurities and bad inclinations abandoned in the seven lower heavens, ascends to the eighth heaven (*ogdoatikè physis*), becomes a divine power, enters God and is deified (*Corpus Hermeticum* I, 25-26, pp. 15, 15-16, 13 Nock-Festugière; cf. also Des Places, *La religion grecque*, Paris 1969, 324, who rightly relates this motif to t.).

*neoplatonism, the dominant philosophy of *late antiquity, was, in its initial phase represented by *Plotinus, essentially immune from the influence of t. Dodds (*op. cit.*, 57 f., 60 f.) rightly emphasizes that Plotinus, *pace* certain modern scholars, was not a theurge, but a pure rationalist who showed no interest in t. and did not hide his contempt for the magical practices so fashionable in his time. Only with *Porphyry did t. enter the neoplatonist tradition, finally becoming one of the most salient features of 4th- and 5th-c. neoplatonism, that of Iamblichus, *Proclus and Damascius. The young Porphyry, even before meeting Plotinus, nourished a great interest in "*oracles", as is shown by his juvenile work *De philosophia ex oraculis haurienda*, in which he seems as yet to have no knowledge of the *Chaldaean oracles* (cf. Dodds, *op. cit.*, 58; Des Places, *Oracles caldaïques*, 18 f.; the fragments of this Porphyrian work have been collected and edited by W. Wolff, Berlin 1856). Suppressed during the period in which he was Plotinus's disciple and most under his influence - the *Letter to Anebo*, strongly critical of magical practices and occultism, dates from this phase - Porphyry's interest in "oracular wisdom" revived after his master's death: he discovered the *Chaldaean oracles*, wrote a commentary on them, made continual use of them in *De regressu animae* (cf. Dodds, *op. cit.*, 58; Des Places, *op. cit.*, 18 f. and 23 f.) and laboured to reconcile his own philosophical system with their doctrines (cf. the section on Porphyry in *neoplatonism). But in *De regressu animae*, while partially reevaluating the role of t., he does not show the unconditional admiration for it that appears in *De philosophia ex oraculis haurienda*, written some years earlier (cf. Des Places, *op. cit.*, 247): in fr. 2 Bidez of *De regressu animae*, he admits that t. can contribute to the "purification of the lower or "spiritual" part of the soul, but at the same time he specifies that it has no effect on the conversion to God of its higher or "intellectual" part, and he repeats his condemnation of the baser magical practices (cf. the section on Porphyry in *neoplatonism). As Dodds has noted (*op. cit.*, 58), in this precise demarcation of the role of t., Porphyry is still, however partially, under Plotinus's influence.

In Iamblichus, t. played a far more important role than Porphyry had assigned to it: not only did he consider the *Chaldaean oracles* a sort of sacred book and write a commentary on it (cf. E. Des Places, *Jamblique, Les mystères d'Egypte*, Paris 1966, 15; Dodds, *op. cit.*, 58), he also wrote *De mysteriis Aegyptiorum* under the name of Anebo - to whom Porphyry had addressed his letter criticizing magical practices - in reply to Porphyry's accusations. "It is not thought" says Iamblichus "that unites theurges to the gods; if it were so, what would prevent the theoretical philosophers from possessing theurgic union with the gods? But the truth is otherwise: it is the sacred performance of ineffable operations, above any thought and suitable to God, and the power of the ineffable symbols, that produce theurgic union" (*De myst.* I, 11, 96, 13-97, 2, p. 96 Des Places).

The emperor *Julian was profoundly influenced by Iamblichus (on this point cf. the section on Iamblichus and Julian in *neoplatonism). He wrote to Priscus (*Ep.* 12 ed. Bidez, *L'empereur Julian, Oeuvres complètes*, I, 2, Paris 1924, 118-119) exhorting him to "look out for everything that Iamblichus wrote about my namesake" (i.e. Iamblichus's commentary on Julian's *Chaldaean oracles*); and he confesses: "I am crazy about Iamblichus in philosophy and about my namesake in theosophy" (i.e. theurgy, cf. Dodds, *op. cit.*, 59).

Iamblichus's example was also followed by the 5th- and early 6th-c. neoplatonist school of *Athens. Olympiodorus (*In Platonis Phaedonem*, 170) numbers Syrianus, Iamblichus and Proclus among the students of t., contrasting them with the pure philosophers, represented by Plotinus and Porphyry; and Michael Psellus recalls Proclus as one who, with Iamblichus, abandoned Greek methods of reasoning and embraced the Chaldaean doctrine (on these two cf. the references in Des Places, *Oracles chaldaïques*, 24). *Proclus, like Porphyry and Iamblichus, wrote a *Commentary on the Chaldaean philosophy* (its fragments have been edited by Des Places in his edition of the *Oracles chaldaïques*, 206-212). But Psellus's judgment of Proclus is only partly true; in fact two components coexist in Proclus without clashing: the rationalist-philosophical (think of the *Elements of Theology*) and the "super-rational" and "mystico-religious", which does not hesitate to appeal to t. and its magical practices (for Proclus too, union with the supreme principle, above being and mind, cannot be achieved by the human mind, which can at best attain *contemplation of absolute being, but is something that transcends its faculties: cf. *neoplatonism). Dodds (*op. cit.*, 61) has called attention to a passage in the *Theologia Platonica* (I, 25, I, 113, 6-10 Saffrey-Westerink) in which Proclus speaks of t. thus: "other things are 'finally' saved by theurgic power, which is better than any human wisdom and science since it contains in itself the benefits of the divinatory art, the purifying forces proper to the performance of the rites and, in a word, all the effects of divine possession". We have already had occasion to note (cf. above) that, acc. to Marinus (*Vita Procli*, ch. 26), both Iamblichus and Proclus saw the "theurgic virtues" as the supreme *virtues, those able to produce in the soul the maximum degree of deification; Marinus also records (*Vita Procli*, ch. 28) Proclus's familiarity with theurgic practices and the relevant rites. Damascius, like his predecessors, considered the Chaldaean philosophy "the most mystical" (*Dub. et sol.* 111, I, 285, 1-2 Ruelle); and did not hesitate to call philosophers like *Orpheus and Plato, who had received their philosophy by divine revelation, "theurges" (*Dub. et sol.* 111, I, 285, 5. 8. 1. 13 Ruelle). For him, t. is a "sacred practice" (*Vita Is.* 227, cf. Eitrem, *op. cit.*, 50).

PWK 24, 258-270; *LTK* 10, 111; *KLP* 6, 767; *EC* 12, 28 f.; J. Bidez, Le philosophe Jamblique et son école, *REG* 32 (1919) 29-40; J. Bidez-F. Cumont, *Les mages hellénisés*, Paris 1938, I, 158-163; II, 251-263; O. Faller, Griechische Vergottung und christliche Vergöttlichung, *Gregorianum* 6 (1925) 405-435; F. Cumont, *Les religions orientales dans le paganisme romain*, Paris 1929, 115. 173-175. 294, notes 87-89; S. Eitrem, La théurgie chez les néoplatoniciens et dans les papyrus magiques, *SO* 22 (1942) 49-79; E.R. Dodds, Theurgy and its Relationship to neoplatonism, *JRS* 37 (1947) 55-69 (= *The Greeks and the Irrational*, Berkeley-Los Angeles 1956, 283-311); A.J. Festugière, *La révélation d'Hermes Trismégiste*, I: *L'astrologie et les sciences occultes*, Paris 1950, 283-308; id., Contemplation philosophique et art théurgique chez Proclus, in *Etudes de philosophie grecque*, Paris 1971, 585 ff.; id., De la religion à la magie, in *Etude d'histoire et de philosophie*, Paris 1971, 9-156; P. Boyancé, Théurgie et télestique néoplatoniciennes, *RHR* 147 (1955) 189-209; W. Theiler, Gott und Seele in kaiserzeitlichen Denken, in *Rech. sur la trad. plat.* (Entr. sur l'ant. class. III), Geneva 1955, 65-90 (= *Forschungen zum Neuplatonismus*, Berlin 1966, 104-123); H. Lewy, *Chaldean Oracles and Theurgy*, Cairo 1956; W. Beierwaltes, *Proklos. Grundzüge seiner Metaphysik*, Frankfurt/M, 1965, 328, 385-390 (this last section particularly concerns deification); A.A. Barb, La sopravvivenza delle arti magiche, in (ed.) A. Momigliano, *Il conflitto tra paganesimo e Cristianesimo nel IV secolo*, Turin 1968, 111 ff.; E. Des Places, *Jamblique. Les mystères d'Egypte*, Paris 1966; id., *La religion grecque*, Paris 1969, 300-304. 324; id., *Oracles chaldaïques*, Paris 1971; J. Finamore, *Iamblichus and the theory of the Vehicle of the Soul*, Chico 1985.

S. Lilla

THOMAS, apostle. T. is mentioned in Mk 3, 18 and par., and Jn 11, 16; 20, 24. Jn 21, 2 says he was "called Didymus" (the twin). The Hebrew root *ta'am* means "to be a double, a twin". The *Acta Thom.* 1, 1 and the legend of *Abgar (Euseb., *HE* I, 13, 11) call him Judas Thomas; Jn 14, 22 calls him "Judas, not Iscariot"; the Sinaitic and Curatonian Syriac versions read respectively "Thomas" and "Judas Thomas". The opening of the Coptic *Ev. Thom.* calls him "Didymus Judas Thomas". The addition of the surname T. is natural to distinguish him from the two other *Judas's (Lk 6, 16) among the Twelve. His identity with Judas the Lord's brother seems a later confusion rather than an early *tradition.

T. "the Twin" exercised a particular fascination over the speculations of the *gnostics. For them, the name "the Lord's twin" did not imply consanguinity, but indicated the type of the perfect gnostic. Acc. to *Pistis Sophia* 42 and 53, *Philip, T. and *Matthew (or *Matthias) were the authors of the three gospels most accredited in certain gnostic circles. NHC II includes the *Gospel of Thomas*, the *Gospel of Philip* and the *Book of Thomas the Athlete*. The latter contains the Saviour's secret words to Judas Thomas, handed down by Matthew. Then there are the Greek MSS of group A of the *Gospel of Thomas* (infancy gospel), giving "Thomas the Israelite philosopher's account" of the Lord's infancy; while the Coptic *Gospel of Thomas* is a full collection of the Lord's sayings. The *Acta Thom.* are the sole complete survival of a group of five *apocryphal Acts used by the *Manichees. Written in *Syriac at *Edessa in the first half of the 3rd c., its theme is T.'s mission to the Indies (really to Iran). *Origen (Euseb., *HE* III, 1, 1) and Ps.-Clement (*Recogn*. IX, 29) call T. the *apostle of the Parthians. *Heracleon (Clem. Al., *Strom*. IV, 71, 3) records his natural death, speaking of Lk 12, 8-11; yet *Acta Thom*. 164, 168 and the Manichaean *Lib. Ps.* say he was killed with a lance by some soldiers. Episodes of the *Acta Thom.* are developed by *Gregory of Tours (*Glor. Mart.* 31-32). The Parthian king Gundofar, via the *Acta Thom.*, ended up in Western legend as king Caspar. The *Opus imperfectum in Matthaeum* says that the *Magi were baptized by T., a theme also developed in Zouqnîn's Syriac *Monastic Chronicle* (8th c.). In the later Middle Ages, the legend of T. and the Magi kings led to that of Prester John.

R.A. Lipsius-M. Bonnet, *Acta Apostolorum Apocrypha*, Leipzig 1903 (Darmstadt 1959); A.F.J. Klijn, *The Acts of Thomas*, Leiden 1962; G. Bornkamm, *Mythos und Legende in den apocryphen Thomas-Akten*, Göttingen 1933; J. Doresse, *Les livres secrets des gnostiques d'Egypte*, II, Paris 1959; W. Schneemelcher, *N.T. Ap.*, Cambridge 1991-92; Erbetta I, 253-288, II, 305-391 (Acts); H.H. Koester, Γνῶμαι διάφοροι. The origin and nature of diversification in the history of early Christianity, *HThR* 58 (1965), 279-318; G. Garitte, Le martyre géorgien de l'apôtre Thomas, *Muséon* 83 (1970) 497-532; id., La passion arménienne de S. Tomas l'apôtre et son model grec, *Muséon* 84 (1971) 1951-195; F. Halkin, Une nouvelle recension grecque du martyre de S. Thomas l'apôtre, *AB* 89 (1971) 386; G. Quispel, The Gospel of Thomas and the New Testament, *VChr* 11 (1957) 189-207; R. McL. Wilson, *Studies in the Gospel of Thomas*, London 1960; Text: *The Gospel according to Thomas*, Coptic text translated into English by A. Guillaumont et al., Leiden 1976.

R. Trevijano

THOMAS of Edessa († before 544). Nestorian author, to whom the catalogue of Ebedjesu/'Abdiso' attributes some works. Of these, only two treatises survive, one on *Epiphany (unpublished) and one on the *Nativity. Of his life we know only that he accompanied the future Nestorian patriarch Mar *Aba to *Constantinople, whence both had to flee for refusing to subscribe the condemnation of the *Three Chapters.

Ortiz de Urbina 127; S.J. Carr, *Thomae Edesseni Tractatus de Nativate Domini nostri Jesu Christi*, Rome 1898; W.F. Macomber, *Six Explanations of the Liturgical Feasts by Cyrus of Edessa*, II. *Translation*, CSCO 356/syr. 156, Louvain 1974, viii-x.

S.J. Voicu

THRACE. Thracia was the ancient name of the E and S-E region of the Balkan peninsula. In the millennium before Christ this territory was inhabited by the Thracii, a people divided into many tribes, whom Herodotus calls the most numerous people in the world after the Indians (*Hist*. V, 3). Geographically, the territorial extent of T. changed more than once over the centuries. Its frontiers were: to N the Danube, to E the shore of the Euxine (Black) Sea, to S and S-E the Aegean Sea and the Propontis (Sea of Marmara), to W and S-W the rivers Utus (Vit) and Nestos (Mesta), which still preserve their old names, only partially changed. After the Romans had conquered *Macedonia (148 BC) and organized it as a province, they pushed on towards T., but had to fight for nearly two centuries to overcome the resistance of the Thracii, conquer their territory and transform it (AD 46) into a Roman province, comprising the region S of Mt Haemus (Stara Planina), the Euxine Sea, the Propontis, the Aegean Sea, as far as the river Nestos. While the Latin language and culture were dominant in *Moesia to the N, T., being close to the great Greek centres, was influenced in language and culture by Hellenism. Not being a frontier province after the early 1st c. AD, fewer Roman troops were stationed there and so the territory preserved its ethnic character better. The region was crossed by only two great arterial roads: the Via Diagonalis, from Singidunum (Belgrade) to *Byzantium, and the Via Egnatia, from Dyrrachium (Durazzo, Durres) to Byzantium, passing through *Macedonia, *Thessaly and Aegean Thrace. Old Greek colonies, existing for centuries on the shores of the Black Sea and Propontis, developed into opulent cities: Abdera, Amphipolis, Maronia, Perinthus, Byzantium on the Bosphorus (later famous as *Constantinople, capital of the Eastern Roman empire). Inside T. were important centres formed from the old settlements of the Thracii: Beroë became Augusta Traiana (now Stara Zagora, in a fertile plain), Pulpudeva became Philippopolis (Plovdiv), Uscudama became Hadrianopolis (Edrene, Odrin), an important *statio* on the Via Diagonalis. Being quite far from the northern frontier of the *limes danubianus*, and fortified as well, these cities were little damaged by the "barbarian" raids of Huns, *Goths, Avars, etc., and so kept their cultural wealth and their population more or less intact. Even the people of the plain did not suffer much from the invasions and it was rarely that the invaders, not stopped at the Danube frontier, managed to push in deeply enough to reach the vicinity of the Aegean Sea and the Propontis. Of the invaders, the most audacious were the Slavs, who reached the sea during the first half of the 6th c., destroying various centres, e.g. the city of Topiros at the mouth of the Nestos in 549/50. We cannot establish with any certainty when cities like Amphipolis, at the mouth of the Strymon, and *Philippi, an early Christian centre, were damaged or destroyed. Exchanges with the old Thracian population and the Greeks who inhabited that part of T. seem to have occurred not suddenly, but gradually, so that there was no break in continuity with the new inhabitants, who retained a large number of old river-names and place-names, like that of the colony of Deultum on the Black Sea (now the village of Develt), the city of Anchialos (Bulgar Achelos, now Pomorie) on the Black Sea, Messembria (now Nesebur), Agathopolis (Achtopol), Nicopolis ad Nestum (Nevrokop in S Bulgaria), Astibus (Stip, in N-E Macedonia), etc.; and the names of various rivers: Strymon (now Struma), Nestus (Mesta), Hebrus (Ibur near the source, then commonly Maritza), Tonzus (Tungia, tributary of the Hebrus), Syrmus (Strjama, Strema); and the Rhodopes (now Rodopi) mountains, etc.

Following *Diocletian's administrative reorganization, the diocese of T. was divided into six provinces, four of them already part of T.: Europa, Rhodope, Thracia, Haemimontus, and another two: *Scythia Minor and *Moesia Inferior, acc. to the evidence of the *Laterculus Veronensis*, confirmed by other historical sources, e.g. the *Notitia dignitatum*, Hierocles's *Synèkdemos*, etc. The diocese of T. was part of the *Praefectura per Orientem*: in "civil", i.e. non-military, affairs it was dependent on a *vicarius, vir spectabilis*, while in military affairs it was ruled by a *magister militum per Thracias, vir illustris*. When *Constantine the Great officially transferred the imperial capital from *Rome to Constantinople in 330, T. ceased to be a peripheral province of the empire and became the true hinterland of the capital, acquiring particular importance, both military-strategic and economic (supplying food). But this privileged position (so to speak) of T. ended by making it more attractive to trans-Danubian invaders and its situation became particularly critical in the course of the 4th and 5th cc., after the invasions of the Huns and *Goths. Penetrating into the diocese of T., the Visigoths, allied to the Ostrogoths and Huns, mercilessly devastated the territory of T. Abandoning his war against the Persians, the emperor *Valens (364-378) concentrated his army to oppose the Goths, but suffered total defeat at the battle of Hadrianople (9 Aug 378) and lost his own life. The danger of "Germanization" of the empire became particularly keen when two Gothic military chiefs - Theodoric Strabo and *Theodoric Amaling - who had received the most exceptional military honours from the government in Constantinople, betrayed their oaths as federates and took to devastating the territory of T. It was only in the last decades of the 5th c. that the territory managed to free itself from them: the former died in 481; the latter was persuaded to leave the Balkan peninsula (488) and move towards *Italy. For some decades the territory of T. had had to suffer the invasions of the Huns, firstly during the first half of the 5th c. until the death of *Attila in spring 453, and again when his successors tried to reconquer the Balkan lands. In the late 5th and early 6th cc. began the invasions of the Slavs, who crossed the Danube *en masse*.

Christianity was propagated in T. from the first centuries of the Christian era: Christian communities, grouped around *church buildings, were formed in various centres. Leaving out the Christian communities of *Scythia Minor and those of the part of T. under the rule of the Eastern Roman empire, we must concentrate on those in the territory in which the Bulgar state would eventually be established (AD 681). We are informed of the history of these Christian nuclei by some written sources and by the evidence of *epigraphy and *archaeology. Thus, in the ancient city of Pautalia (now Kjustendil, S-W Bulgaria) and in its vicinity, Greek and Latin Christian inscriptions of the 4th-6th cc. have been discovered. Interesting, among others, is a metrical inscription in fine Greek characters, somewhat damaged, whose text reflects the christological debates of the 4th c. One important Christian centre was the city of Messembria on the Black Sea, Hellenic in culture, where various Christian inscriptions have been discovered, esp. from the 6th c., but also later. From one of these inscriptions, unfortunately rather damaged, we must deduce that in this maritime centre

Christianity was spread by people from *Asia Minor. A 6th-c. Greek inscription found at Aquae calidae (region of Burgas on the Black Sea coast, in E Bulgaria) mentions a bishop, a deacon, an *oeconomos* and a vicar, evidence of a well-developed ecclesiastical *organization. In the ancient fortress of Deultum (now the village of Develt on the Black Sea coast) and in the maritime city of Sozopolis were Christian communities, from which come some Greek sepulchral inscriptions of the 5th-6th cc. Another Christian centre was the ancient city of Augusta Traiana (once Beroë, now Stara Zagora), where Christian sepulchral inscriptions, probably of the 5th-6th cc., have been found.

At Philippopolis (now Plovdiv), metropolis of T., Greek Christian inscriptions were discovered, one (perhaps 4th-c.) *gnostic in content: its author, Tatian, came from *Asia, via *Egypt; which confirms the hypothesis of religious currents from the *East. In modern Hissar (N of Philippopolis), prob. to be identified with the late-antique city of Diocletianopolis, Greek Christian inscriptions of the 5th-6th cc. have been discovered and the remains of imposing Christian *church buildings are still preserved. Zapara (not identified with any certainty, but prob. Sveti Vrac [now Sandanski] in S Bulgaria) was an episcopal see in the 6th c., acc. to the evidence of a *mosaic inscription.

The foundation of the Bulgar state in 681 had consequences for the political and cultural situation of T., which, not being occupied by the Bulgars, remained under the rule of the Byzantine empire. The politico-administrative reform undertaken by the emp. *Heraclius (610-641) and completed by his successor Constantine IV (668-685) was prompted by the situation created by the rise of the new state, a possible enemy despite the peace treaty concluded in Sept 681. The diocese of T. was transformed into a *theme*, i.e. a military-political unit under the command of a *strategos* who united civil and military power: the better to ensure defence against the threat of the new state. But the Slavs, pushing southwards, also managed to install themselves fairly peacefully in the southern regions of S-E Europe, thus changing the ethnic composition of much of the territories under the direct rule of the empire of *Byzantium. In the most critical period of the "barbarian" invasions of the Huns and Avars in the 5th and 6th cc., it was the representatives of the Christian *clergy who organized the defence of some inhabited centres and negotiated with the enemy. Thus, in the first half of the 6th c., the bishop of the city of Margus (at the mouth of the river Morava) took the decision, to save his city from the vengeance of the Huns, to open the gates of the fortress to them. During the second half of the 6th c., in the struggle of the emp. Maurice (582-602) against the Slavs and Avars, the inhabitants of the little city of Asemus, situated in the region near the Danube, had organized an army for their defence, with the permission of the emp. *Justin II (565-578). When the Constantinopolitan *taxiarch* Guentzon tried to enrol the troops of the city in his own army, he was foiled by the resistance of the local bishop. The sanctuary dedicated to the *martyr St Alexander Romanus (cf. Halkin, BHG, nos. 48-49, martyred "*sub Maximiano*") in the village of Druzipera (region of Hadrianopolis) was destroyed by the Avars before the end of the 6th c. [Figs: 306-309]

PWK, Thrake (römisch) 25, 452-472; EAA 7, 837-839; KLP 5, 777-781; V. Ivanova: *Izvestija dell'Istituto archeologico bulgaro* 9 (1937) 214-242, 305-306; D. Concev, *Hisarskite bani*, Sofia 1937; D. Concev-K. Madzarov, *Archeologija* 7/1 (1965) 16-21; V. Besevliev, *Spätgriechische und spätlateinische Inschriften aus Bulgarien*, Berlin 1964; K. Mijatev, *Architektura v. srednovekovna Bulgarija*, Sofia 1965, 7-26; V. Velkov, *Cities in Thrace and Dacia in Late Antiquity*, Amsterdam 1977; I. Dujcev, Testimonianze epigrafiche e archeologiche sul paleocristianesimo in Bulgaria, in *Atti della Pontificia Acc. di Archeol.* (forthcoming); I. Djambov, Un centre chrétien découvert récemment en Thrace, *Actes XIe Congrès internat. d'arch. chrét.*, CEFR 123, Rome 1989, 2511-2514; M. Madjarov, Dioclétianopolis: ville paléochrétienne de Thrace, *ibid.*, 2521-2537; E. Kessiakova, Une nouvelle basilique à Philippopolis, *ibid.*, 2539-2559.

I. Dujcev

THRASEAS. Bishop of Eumenia in Proconsular *Asia, *martyr, third quarter of 2nd c. Two ancient documents preserve his name: the apologist *Apollonius's polemic against *Montanus mentions him as one of the martyrs of that time (Euseb., *HE* V, 18, 14); the *letter of bishop *Polycrates of Ephesus to pope *Victor contains this phrase: "there was T. of Eumenia, bishop and martyr, who fell asleep at Smyrna" and tells us that T., like all the bishops of Asia, celebrated *Easter on 14th Nisan, i.e. against the Roman and Alexandrian or Palestinian *tradition (Euseb., *HE* V, 24, 4. 6).

ASS *Oct.* III, 7-12: *Vies des SS.* 10, 107-108; *BS* 12, 640.

G. Ladocsi

THREE CHAPTERS, Question of the. As part of his attempts to find an agreement with the *monophysites, in *c.*544 *Justinian published an edict, of which only fragments remain, condemning *post mortem* *Theodore of Mopsuestia, *Theodoret of Cyrrhus and *Ibas of Ebessa (the "Three Chapters"), whom the monophysites hated as inspirers and supporters of *Nestorius. He demanded the approval of *Rome and, faced with the prevarications of pope *Vigilius, had him brought to *Constantinople, where he arrived early in 547. Vigilius, having had time and opportunity to consult many members of the Western episcopate, put up a long resistance to the pressures on him until, 11 April 548, he sent patriarch *Menas of Constantinople the *Iudicatum*, which condemned the T. C. and reaffirmed the validity of the council of *Chalcedon. Of this document too, only fragments remain. On hearing this, the Western episcopate reacted so violently that Vigilius subsequently withdrew the document and asked for an ecumenical *council to be called. While waiting for it to meet, Vigilius had to undergo all sorts of vexations, but though his *Constitutum*, dated 14 May 553, condemned many doctrinal propositions taken from the works of Theodore of Mopsuestia, he refused to condemn his memory, or those of Theodoret and Ibas. But the council, which had begun some days earlier at Constantinople without the *pope's participation, condemned the T. C., and Vigilius confirmed the condemnation on 8 Dec 553.

Reactions in the West were negative, esp. in *Africa and Illyricum. And though, in Africa, Justinian forcibly imposed approval of the condemnation, the churches of the patriarchate of *Aquileia separated from *communion with Rome. The *schism was favoured for political reasons by the *Longobards, who invaded *Italy shortly afterward, and, despite attempts to resolve it by *Gregory the Great and other *popes, it did not end until *c.*689, following a council of bishops of the concerned regions, meeting at Pavia (*Ticinum*) by decision of king Cunipert, with the consent of pope *Sergius I.

DTC 15, 1868-1924; EC 12, 456-460; R. Devreesse, *Essai sur Théodore de Mopsueste*, ST 141, Vatican City 1948, 194-272; B. Rubin, *Das Zeitalter Justinians* I, Berlin 1960.

M. Simonetti

TIBERIANUS. Spanish *Priscillianist from *Baetica, condemned to confiscation of his goods and exiled to the Scilly Isles with *Instantius. *Jerome mentions him as the author of an apologetic work written to defend himself from the charge of *heresy (*Vir. ill.* 123).

H. Chadwick, *Priscillian of Avila. The Occult and the Charismatic in the Early Church*, Oxford 1976, 47, 144-145.

S. Zincone

TIBERIUS, emperor. Tiberius Claudius Nero, Roman emperor, ruled AD 14-37. He attempted no new territorial conquests, but had to face a very difficult political situation at Rome from movements averse to the principate and to the aristocratic Claudian family. After the failed plot of the praetorian prefect Aelius Sejanus (AD 31), T. reigned harshly, initiating a policy of union and levelling of the peoples of the empire, at the expense of the Italians. He took steps against foreign religions, expelled the Jews from Italy (Garzetti, 31) and forbade Druidical rites. He refused to have divine honours paid him. Jesus Christ lived and died under T.; *Tertullian claims that the procurator of Judaea, Pontius *Pilate, informed T. of events in *Palestine and of the conflict between official Judaism and the Christian community of *Jerusalem, and (*Apol.* V, 1, ff.) that T. proposed to the senate the *consecratio* of Christ and the legitimation of his cult, but the proposal was turned down. The emperor T. appears widely in the NT *apocrypha, which present him in positive terms, to the detriment of Pilate (cf. L. Moraldi, *Gli apocrifi del NT*, Turin 1971, II, 2013−index).

A. Garzetti, *L'impero da Tiberio agli Antonini*, Bologna 1960; M. Sordi, *Il cristianesimo e Roma*, Bologna 1965, 25-31 and 416-19 (bibl.).

L. Navarra

TICHONIUS. The *Donatist lay exegete Tichonius or Tyconius (*c.*330-*c.*390) is presented by *Augustine (cf. *Ep. Parm.* I, 1, 1; I, 2, 2; II, 21, 40; II, 22, 42; III, 3, 17-29; *De doctrin. christ.* III, 20, 42; III, 33, 46; *Quaest. in Hept.* II, 47, 102; *C. litt. Pet.* II, 83, 184; *Epp.* 41, 2; 87, 10; 92, 14; 93, 10. 44. 45; *Retract.* II, 18) as a man of penetrating intellect who, as a Donatist, defended the church's universality to the point of provoking his own condemnation at a Donatist *council of 380. T. always continued to consider himself a Donatist and did not give way in the struggle against *Parmenian, demonstrating how good and bad live together within the church until the end of *time.

Augustine used and recommended T.'s writings, but brought his own nuances to them. He particularly used the *Liber Regularum* (*De doctr. christ.* III, 30, 42-47, 56) and the *Commentary on the Apocalypse*, which he cites explicitly (*De doctr. christ.* III, 30, 42) and uses implicitly elsewhere. *Gennadius (*De vir. ill.* 18) supplements Augustine's references by assigning two more works to T.: *De bello intestino* and *Expositiones diversarum causarum*, written in 370/375 and now lost. Gennadius stresses T.'s spiritual interpretation, the corporeity of his *angels, his ideas on *millenarism and the *resurrection. Other references to T. and his works are in *Quodvultdeus (*Dim. temp.* XIII, 22), *Primasius, *Cassiodorus and Ambrosius Autpertus. Primasius (PL 68, 793-936; PLS IV, 1208-1220) gathers up and purifies T.'s *Comm.*; Cassiodorus (PL 70, 1406-1418), in few things, and Ambrosius Autpertus, in many, follow T. while bearing in mind Primasius's critical observations. *Bede (PL 93, 129-200) summarizes the *Lib. Reg.* and the *Comm.*; *Caesarius (PL 35, 2417-2452; ed. Morin, Maredsous 1942, II, 209-277) summarizes it more literally; *Beatus of Liébana and the anon. author of the *Tichonian fragments* of Turin faithfully preserve the greater part of the *commentary (lost from the 9th c.) in the form of *catenae. All praise T.'s exegetical and doctrinal subtlety, but eliminate and correct what they consider as tending to heterodoxy (cf. Ps.-Isidore: PLS IV, 851, 7-47). Beatus is an exception. Hence the fact that, though the original *Commentary* is lost, it is possible to recover it by starting from the way it was followed by the *Catholics: *Victorinus (*Jerome), Caesarius, Primasius, *Apringius, Cassiodorus, Bede, Ambrosius Autpertus, Beatus and the Turin *Fragments*.

Surviving explicit Tichonian interpretations of the *Apocalypse are few: Rev 5, 6 (Bede: PL 93, 145, 49-50); Rev 8, 7 (Bede: PL 93, 145, 55-56); Rev 8, 7 (Bede: PL 93, 156, 12-16); Rev 9, 3-11 (Cassiodorus: PL 70, 1410, 40-44); Rev 9, 16 (Primasius: PL 68, 861, 11-15; Ambr. Autp.: CCL, med. cont. 27, 359, 24-27); Rev 12, 4 (Bede: PL 93, 166, 36-39); Rev 12, 5-6 (Cassiodorus: PL 70, 1411, 25-30); Rev 14, 5 (Bede: PL 93, 174, 22-23); Rev 14, 18-19 (Bede: PL 93, 178, 6-15); Rev 15, 5 (Ambr. Autp.: CCL, med. cont. 27A, 589, 10-590, 15); Rev 16, 17 (Bede: PL 93, 181, 48-182, 10); Rev 17, 6-7 (Bede: PL 93, 183, 37-43); Rev 17, 9 (Bede: PL 93, 183, 44-48); Rev 19, 21 (Bede: PL 93, 191, 6-9); Rev 20, 13 (Bede: PL 93, 194, 11-14). The other citations are not explicit. We can identify only conjecturally the contents of *De bello intest.* and the *Exposit. divers. caus.* The *Lib. Reg.*, written in 392, is a handbook of hermeneutics: seven rules (the complete work does not survive) by which the exegete - using a certain reasoning - can extract the secrets of *Scripture on the great *revelation: Christ and the church. By these rules, the exegete can identify whether a text refers to the Lord or his body (Rule I), the bipartite body (Rule II), the promises and the Law (Rule III), species or genus, part or whole (Rule IV), times (Rule V), *recapitulation (Rule VI), the devil or his body (Rule VII). To demonstrate the validity of these rules, and to answer ecclesial problems against the Donatists and the Catholics, T. commented on the whole Apocalypse, the ultimate exhaustive revelation of the church. We see the author's interest in correcting the *schism and rediscovering a purer church by returning to the primitive *exegesis: we see the echoes of *gnostic interpretations, and the whole commentary has its roots in the *traditions present in Ps.-*Barnabas, *Irenaeus, *Hippolytus, *Tertullian, *Cyprian and in pseudo-Cyprianine treatises.

All these traditions, laid out with logic and art, reinforce his *ecclesiology: the authentic church is that which is persecuted just as its head (Christ) was persecuted. Ecclesiology and *christology always go together. Only the elect will enter the Kingdom (Church-Kingdom). The final *judgment will openly reveal hypocrisy, the worst of evils, now latent.

For T., the sum *saints+sinners* is not admissable, only the option *saints* or *sinners*. The church is universal; but T., rather than opt for an intensive Catholicity in opposition to an extensive Catholicity, admits them both. Every verse of the Apocalypse serves him to demonstrate how the holy, universal and bipartite church walks through history amidst hypocrisy and the other manifestations of *Antichrist.

Editions: CPL 709-710; *The Book of Rules of T.*, ed. F.C. Burkitt, Cambridge 1894 (repr. Kraus 1967); PL 18, 15-66; D.L. Anderson, *The Book of Rules of T. An introd. and transl. with comm.*, Louisville 1974; *T. Afri fragmenta*, ed. Amelli, Monte Cassino 1897 (=PLS 1, 622-652); *The Turin Fragments of T. Commentary on Revelation*, ed. F. Lo Bue, Cambridge 1963.
Studies: T. Hahn, *Tyconius-Studien*, Leipzig 1900 (repr. Aalen 1971); W. Bousset, *Die Offenbarung Johannis*, Göttingen ⁶1906, 53-72; Monceaux 5, 163-219; H.D. Rauh, *Das Bild des Antichrist im Mittelalter: von T. zum deutschen Symbolismus*, Munich 1973, 102-121; *Patrologia* III, 109-112; J. Haussleiter, Die Kommentare des Victorinus. T. und Hieronymus zur Ap., *ZKWL* 5 (1886) 239-257; H.L. Ramsay, Le commentaire de l'Apocalypse par Beatus de L., *RHL* 7 (1902) 419-447; A.B. Sharpe, T. and St. Augustin, *Dublin Review* 132 (1903) 64-72; A. Souter, T. Text of the Ap., *JThS* 14 (1913) 338-358; A. Pincherle, L'ecclesiologia nella controversia donatista, *Ricerche Rel.* 1 (1925) 134-148; id., Da T. a Sant'Ambrogio, *ibid.* 1 (1925) 443-446; id., Nuovi frammenti di T., *RSLR* 3 (1969) 756-757; id., Alla ricerca di T., *SSR* 2 (1978) 357-365; H. van Barkel, T., Augustinus ante Augustinum, *NTT* 19 (1930) 36-57; I.M. Gómez, El perdido comentario de T. al Ap., *Misc. Ubach*, Montserrat 1953, 387-411; K. Forster, Die ekklesiologische Bedeutung des corpus-Begriffes im Liber Reg., *MTZ* 7 (1956) 173-183; J. Ratzinger, Beobachtungen zum Kirchenbegriff des T. im Liber Reg., *REAug* 2 (1956) 173-186; L.J. van der Lof, Warum wurde T. nicht katolisch?, *ZNTW* 57 (1966) 260-282; G. Bonner, Toward a text of T., *SP* 10 (= TU 107) (1970) 9-13; P. Cazier, Le Livre des règles de T. Sa transmission du Doctr. Christ. aux Sent. d'Isidore de Séville, *REAug* 19 (1973) 241-261; L. Mezey, Egy Korai Karoling Kódextöredek, *Magyar Konyvzemle* 15, 1976; F. Scorza Barcellona, L'interpretazione dei doni dei magi nel sermone natalizio di Ps. Ottato di M., *SSR* 2 (1978) 129-149; E. Romero Pose, T. y el sermón in nat. sanct. innoc., *Gregorianum* 60 (1979) 129-149; id., Et caelum ecclesia et terra ecclesia. Exégesis ticoniana del Ap. 4, 1, *Augustinianum* 19 (1979) 469-486; id., Una nueva edición del Comentario de Beato de L., *Accad. N. dei Lincei, Boll. dei Classici* 1 (1980) 221-231; id., La Iglesia y la mujer del Ap. 12 (Exégesis ticoniana del Ap. 12, 1-2), *Compostellanum* 24 (1979) 293-307; id., *Símbolos eclesiales en el comento a Ap. 1, 13-3, 22 de T.* (diss. P.U.G. Rome 1978); *PAC* 1, 1122-1127; E. Romero Pose, Ticonio y San Agostino, *Salmanticensis* 24 (1987) 5-16; K.B. Steinhauser, *The Apocalypse, Commentary on Tyconius: A history of its reception and influence*, Berne 1987; J.S. Alexander, Tyconius' Influence on St. Augustine, A note on their use of the distinction *corporaliter/spiritualiter*, *Congresso internazionale su S. Agostino nel XVI centenario della conversione*, Studia Ephemerides Augustinianum 24, 3 vols., Rome 1987, II, 205-212; P. Bright, *The Book of Rules of Tyconius: its purpose and inner logic*, Christianity and Judaism in Antiquity 2, Notre Dame 1988; M.G. Cox, Augustine, Jerome, Tyconius and the *Lingua Punica*, *Studia Orientalia* 64 (1988) 83-106; M. Dulaey, Tyconius, *DSp* (forthcoming).

E. Romero Pose

TIME. The God of the *Jews and Christians reveals himself above all in history, which, in the divine plan, has important moments, decisive turnings that must be pointed out for their meaning to be understood. In such a perspective, t. has a fundamental importance, since the saving acts concerning humanity, as a whole and individually, have happened and do happen in t., as does man's collective and individual response. The patristic era was well aware of this dimension, which recurs in Christian literature with a variety of emphases and a wealth of images.

In briefly examining the most important traits that characterized the intuition of t. among the Christians, it is helpful to refer to the critical debate of the last decades and to draw some clues from it. O. Cullmann's thesis is well-known: that for primitive Christianity, as for Judaism and Iranian religion, the symbolic expression of t. was the line, while for Hellenism it was the circle. According to this view, the centre of history is not - as A. Schweitzer and M. Werner had it - in the *parousia and hence in the future, but in the past: Christ came, proclaimed his message and died for man's salvation, inaugurating the new, in the expectation that, with the last things, all would be perfected and the Spirit would take possession of the world of matter. This view conflicted with that of Bultmann, for whom the expectation of the end of the *world as temporally imminent belongs to mythology. The subsequent intense debate centred around three points: the actual presence of "salvation-history" in the NT writings; the propriety and the significance of the representation of "linear" t. in connection with the early Christian sources; and the legitimacy of attributing the conception of "cyclical" t. to *Hellenism. We observe that the "cyclical" notion was not general among the Greeks: indeed great historians like Herodotus, Thucydides or Polybius did not think of history in terms of cycles (Momigliani); so the Greek way of conceiving t. was not unequivocal, and we cannot contrast it too simplistically with that of the Jews. The problem is sometimes transposed by maintaining that the true contrast in the ancient world was between a doctrine that admitted generation and corruption of the *cosmos and a doctrine that made the cosmos itself eternal: or the problem is done away with altogether (making Christianity too, on the cosmological level, express a cyclical conception, as proved by a passage from II Peter [3, 5] [Mazzarino]). Some emphasize the insufficiency of the symbol of the line to indicate the history of salvation and prefer to speak of a spiral or progressive or even vertical line (tending to eternity, not to temporal length).

This very rapid glance at some of the positions of recent criticism gives us a way of assessing the complexity of the problem, also attested by the divergence of opinions, which are based on different interpretations and sometimes on different emphases of the ancient *pagan and Christian sources.

At any rate it may be possible to shed light on some few points relating to t. that seem to be shared by primitive church writers: for one thing, their clear detachment from the viewpoint of archetypes and repetition, common to "traditional" societies; in the vision they propose, there is no "return" in the strict sense, though the idea of division into periods is dear to some of them. Moreover the importance attributed to all three parts which, according to human models, constitute t. - the past, the present and the future - and, alongside this, the identification of a continuous historical thread, woven by *God, whose centre is formed by Christ's coming and whose end we do not know: t. is necessary to allow the complete growth of the body of Christ, which must reach its perfect stature (cf. H.-I. Marrou, *Teologia della storia*, 39). Hence the tension of Christian life between the Lord's first and second *parousia, the expectation of, indeed desire for, the last things, which in the majority of cases does not lead to resignation and spiritual starvation, but to a full and impassioned commitment to the *world; just as the acute and vivid feeling of the irremediable passing of t., constant in patristic writings, induces us to make good use of it: because it is in t. that man decides his destiny.

O. Cullmann, *Christus und die Zeit*, Zürich 1946; A. Momigliano, Time in Ancient Historiography, *History and Theory* 6 (1966) 1-23 (now in *Quarto contributo alla storia degli studi classici e del mondo antico*, Rome 1969, 13-41); S. Mazzarino, *Il pensiero storico classico*, II, 2, Bari 1966, 412-61, n. 555 on the intuition of time in classical historiography; H.-I. Marrou, *Théologie de l'histoire*, Paris 1968; P. Siniscalco, Roma e le concezioni cristiane del tempo e della storia nei primi secoli della nostra èra, in AA. VV., *Roma, Costantinopoli, Mosca: Atti del Semin. Internaz. di Studi Storici, Roma, 21-23 aprile 1981*, Naples 1983, 31-62.

P. Siniscalco

TIMGAD (ancient Thamugadi). Created by *Trajan in AD 100, its chessboard pattern covered an area of 60 hectares at the crossing of the roads from *Lambaesis to *Tebessa and from *Constantine to Aurès, on the highlands of *Numidia. Its meagre fortifications could not resist the attack of the Moors who rebelled against the *Vandals early in the 6th c. They destroyed the city, which recovered only partially under *Byzantine rule before disappearing under the desert sands.

The bishops known to us are: *1)* Novatus, present at the council of *Carthage of 256 (*Sent. epp.* 4); *2)* Sextus, in 320 (*Gesta apud Zenophilum*); *3)* *Optatus (388-398), *Gildo's second-in-command, champion of *Donatism and warrior-bishop; he fell with Gildo; *4)* his successor *Gaudentius, who spoke for the Donatists at the conference of Carthage of 411; *5)* Faustinianus, his *Catholic opponent; *6)* Secundus, present at the conference of Carthage of 484.

P. Monceaux, who attributes a historical basis to the *Passio Mammarii* (BHL 5205-06), sees Lawrence, Faustinianus, Ziddinus, Crispinus and Leucius as *martyrs of T., in 259.

There is a pagan cemetery 500 m. along the Lambaesis road; a Christian *cemetery 250 m. from the walls; and another on the Aurès road to the S. Outside the walls, in the W suburb, are the two episcopal complexes, Catholic to N, Donatist to S; on the road leading S, beyond the Byzantine fort, is the chapel of the patrician Gregory (c.645). Inside the city were some sanctuaries of Byzantine date. For further archaeological details, cf. under *Africa.

P. Monceaux, *Timgad chrétien*, Paris 1911-12; *PWK* 24, 1235-36; *DACL* 15, 2313-38; *LTK* 10, 197-98; P. Romanelli, *Topografia e archeologia dell'Africa romana*, Turin 1970 (cf. index); P.-A. Février, Urbanisation et urbanisme de l'Afrique romaine, *ANRW* II.10, 321-396; N. Duval, L'Évêque et la cathédrale en Afrique du Nord, *Actes XIe Cong. internat. d'arch. chrét.*, CEFR 123, Rome 1989, 345-403.

V. Saxer

TIMOTHY. *Paul's most esteemed disciple and collaborator; named 17 times in the Epistles, six times in Acts and once in Hebrews. In the *apocrypha he hardly appears: he is named in *Acta Titi* 6, *Acta Petri cum Sim.* 4 and *Apocal. Pauli* 51. Acc. to the *Passio* (BHG 1847; BHL 8294), a late text written in the period of controversy over the antiquity of the various apostolic sees, T. preached and worked miracles at *Ephesus, where the local *pagans killed him. *Eusebius too (*HE* III, 4, 5) relates the tradition that T. was the first bishop of Ephesus, and he was cited as such at the council of *Chalcedon (Mansi 7, 293). It may be that this tradition was based on 1 Tim 1, 3. Yet *Polycrates of Ephesus appeals to the tradition of *John and makes no mention of T. (Euseb., *HE* III, 24, 2-7). T.'s *circumcision by Paul (Acts 16, 3) created difficulties for the *Fathers. In general it was justified as an adaptation to Jewish prejudice for the sake of acceptance by the *Jews (Iren., *Adv. haer.* III, 13, 3; Tertull., *Adv. Marc.* I, 20, 2-3; *Praescr.* 24, 2; Clem. Al., *Strom.* VI, 124, 1; Orig., *Comm. in Mt.* 11, 8). The question was also debated between *Augustine and *Jerome (*Epp.* 56, 3-5; 112, 4-17).

ASS *Ianuar.* II, 562-569; *BS* 5, 2217 f.; *EC* 12, 107 f.; *BS* 12, 482-488.

R. Trevijano

TIMOTHY I, Nestorian. Born 727 or 728 at Hazza (near *Arbela, in Iraq) and educated by the monk Abraham bar Dasandat, he succeeded his uncle George as bishop of Beit Bagas and in 779 manoeuvred his own election as *catholicós, though it took him two years to gain recognition and make *peace with his opponents. From then until his death (823) he governed his church with relative tranquillity. He enjoyed the friendship of the Abassid caliphs and the protection of eminent people at court. T. was a learned man: apart from *Syriac, he knew Greek and *Arabic, some Persian (the language of a considerable part of his flock) and some Hebrew. He loved *books, copied them and had them copied, and encouraged translations. He understood that, to save his church from *apostasy to Islam and *Jacobite *monophysism, he needed an educated and blameless *clergy, led by good bishops and metropolitans. So he zealously dedicated himself to the instruction and training of clerics and the choice of worthy pastors. Under his catholicate the Nestorian church reached its apogee and sent missionaries to India, China, Tibet, *Persia and the Turks. Of T.'s writings, we have lost the treatise on astronomy, the book of questions and discussions with the Jacobite patriarch George of Be'eltam, the commentary on *Gregory Nazianzen. What remain are 58 of his c.200 *Letters*, very important for a knowledge of his life, work and times; a defence of Christianity given before the Caliph (*Questions and answers spoken during the meeting between Timothy and al-Mahdi, emir of the believers*); and *canonical texts (*Provisions on ecclesiastical judgments and inheritances; Synodical tomes*). In his *apologia*, T. carefully manages to defend the *dogmas of Christianity (*Trinity of *God, *Incarnation of the Word, etc.) without hurting the emir's susceptibilities or making concessions to Islam. The *Letters* often deal with dogmatic questions. T. is a moderate *Nestorian, he uses christological formulae rejected by the rigid Nestorians; he affirms (almost like a *monothelite) the unity of operations in Christ (synergism); the aim of the Incarnation is universal renewal; man is conceived according to *Aristotelian philosophy; the *souls of the dead live in a sort of lethargy until the day of *judgment; *grace is necessary; for *baptism to be valid, right *faith is required in the *minister (priest); the *Eucharist is the body of Christ; the church is a pentarchy, in which the Nestorian patriarch has first place.

Letters: CSCO 67; *Apologia*: A. Mingana, Woodbrook Studies 2, Cambridge 1928, 1-62; Canonical texts: E. Sachau, *Syr. Rechtsbücher* 2, Berlin 1908, 53-117; O. Braun, *OC* 2 (1902) 283-311; M. Jugie, *Theol. dogm. christ. orient.*, V: *Theol. dogm. nest. et monoph.*, Paris 1935; *DTC* 15, 1121-1139; *EC* 12, 111 f.; *LTK* 10, 200 f.; Duval 382; Ortiz de Urbina 215 f; W.H.C. Frend, Christianity in the Middle East: Survey down to AD 1800, in (ed.) A.J. Arberry, *Religion in the Middle East*, I, Cambridge 1969, 239-296.

A. De Nicola

TIMOTHY I of Alexandria. Brother of *Peter II of Alexandria and priest in that city; consecrated bishop on his brother's death, early in 381, he remained so until c.385. He came late to the council of *Constantinople (381) and opposed, successfully, the election of *Gregory Nazianzen as bishop of Constantinople and, unsuccessfully, the (anti-Alexandrian) third canon sanctioning the capital's second place of honour after *Rome. He promoted monastic life. We have 18 of his disciplinary canons.

CPG 2520-2530; PG 33, 1296-1308; *DCB* 4, 1029-1030.

M. Simonetti

TIMOTHY II of Alexandria (Ailouros-Aelurus). Nicknamed "the Cat" for the spareness of his physique, he was a priest of *Alexandria at the time of bishop *Dioscorus, whom he accompanied to the council of *Ephesus of 449 and to whom he remained faithful even after his deposition by the council of *Chalcedon (451); with *Peter Mongus he organized the *monophysite opposition in *Egypt. On the death of the pro-Chalcedonian emperor *Marcian (457), this opposition brought about the murder of Proterius, Dioscorus's Chalcedonian successor, by the enraged mob. T., consecrated bishop some time before by leading monophysites, firmly took power and actively favoured his party. But these activities and the stands he took against the decisions of Chalcedon and pope *Leo's *Tomus ad Flavianum* forced the emp. *Leo I to expel him from Alexandria (458) and exile him first to Paphlagonia, then to the Crimea. Only in 475, on the accession of the pro-monophysite usurper *Basiliscus, was T. recalled. Welcomed with honour at *Constantinople by the emperor, but not by patriarch *Acacius, he attended a pro-monophysite *council at Ephesus and then returned to Alexandria, where he was once more installed in the episcopal see. The subsequent anti-monophysite measures of the emp. *Zeno were too late to harm him, for he died in 477.

T.'s fundamental work was a large *Refutation of the council of Chalcedon*, surviving only in *Armenian tr., published (untranslated) at Leipzig in 1908. Two more shorter works, one against Chalcedon, one against Leo's *Tome*, survive in *Syriac tr.; the second is unpublished, while passages of the first are published in PO 13 with fragments of an *Ecclesiastical history*.

T. expounded a moderate monophysism of the type later developed by *Severus of Antioch, characterized mainly by hostility to the council of

Chalcedon and to Leo's *Tome*, which it considered *Nestorian in tendency. Following *Cyril, he maintained that the nature of the incarnate Logos is one, since the humanity assumed by it, being without a specific *hypostasis of its own and incapable of subsisting *per se*, cannot be called a nature. This humanity, however, is whole and complete and is united to the divine nature without alteration or confusion, so that Christ, incarnate and becoming man according to the *economy, is consubstantial and congenerate with men according to the *flesh and consubstantial with *God according to nature. By virtue of this union, T., like Cyril, could claim that the Logos, while remaining impassible in its divine nature, suffered on the *cross thanks to the flesh it assumed.

CPG 5475-5491; PO 13, 202-236; DCB 4, 1031-1033; F. Nau, Sur la christologie de Timothée Aelure, *ROC* 14 (1909) 99-103; J. Lebon, La christologie de Timothée Aelure, *RHE* 9 (1908) 677-702; id., La cristologie du monophysite syrien, in *CGG* I, 425 ff; R.Y. Ebied-L.R. Wickham, A Collection of Unpublished Syriac Letters of Timothy Aelurus, *JThS* n.s. 21 (1970) 321-369; id., Timothy Aelurus: Against the Definition of the Council of Chalcedon, in (eds.) C. Laga et al., *After Chalcedon* (Festsch. A. von Roey), Leuven 1985, 115-166.

M. Simonetti

TIMOTHY IV (III) of Alexandria. *Monophysite patriarch of *Alexandria 517-535, successor of Dioscorus II and an anti-Chalcedonian, therefore opposed to pope *Leo's *Tomus ad Flavianum*. As a *friend of *Severus of Antioch, he saw with apprehension the growing support for *Julian of Halicarnassus. On his death, a *schism broke out in the monophysite church of Alexandria between the two theological tendencies. We have one entire *homily by T., in *Syriac, and fragments of others.

CPG 7090-7100; Liberatus, *Breviarium* 19-20: PL 68, 1032-38; J. Maspéro, *Histoire des patriarches d'Alexandrie*, Paris 1923; S.P. Brock, A New Syriac Baptismal Ordo attributed to Timothy of Alexandria, *Muséon* 83 (1970) 367-431 (6th-c. Antiochene text); W.H.C. Frend, *The Rise of the Monophysite Movement*, Cambridge 1972; *PWK* 25, 1357 f.

A. Di Berardino

TIMOTHY of Alexandria. *Archdeacon of *Alexandria in 412, apparent author of the original Greek text of the *Historia monachorum in Aegypto* or *Liber de vitis patrum*. T. must be identified with one (the deacon) of the seven young monks who went to visit the Egyptian monks and wrote, in Greek, an account of the visit and the stories they were told. *Sozomen attributes the writing of this account to bishop *Timothy of Alexandria (†385), but it seems more probable to attribute it to the city's homonymous archdeacon. *Rufinus's Latin text of the *Historia monachorum* (cf. Ruf., *HE* 11, 4 and Jerome, *Ep.* 133) shows discrepancies that depend both on changes made by Rufinus himself and on the fact that his original may have differed from the surviving manuscripts, which allow us to surmise three Greek recensions.

Editions: PL 21, 387-462; E. Preuschen, *Palladius und Rufinus*, Giessen 1897, 1-131, (Greek text); A.J. Festugière, *Historia monachorum in Aegypto*, Brussels 1971 (Greek text in crit. ed.).
Studies: F.X. Murphy, *Rufinus of Aquileia*, Washington 1945; A.J. Festugière, La problème littéraire de l'Historia monachorum, *Hermes* 83 (1955) 257-284; id., *Les moines d'Orient*, I: *Culture ou sainteté*, Paris 1961; id., *Les moines d'Orient*, IV, 1: *Enquête sur les moines d'Egypte*, Paris 1964.

M.G. Bianco

TIMOTHY of Constantinople (6th-7th cc.). Known only by the work attrib. to him in *manuscript tradition: *De iis qui ad ecclesiam accedunt*, dedicated to the presbyter John. Besides calling him a presbyter of the "Great Church" (Hagia Sophia), some codices also call him *scaevophylax* (treasurer) of the church of the Mother of God of Chalcopratia, at *Constantinople, though others give this title to the work's recipient, i.e. John.

T.'s work distinguishes three categories (*taxeis*) of repentant heretics, who are admitted into the *Catholic church: those who must receive *baptism, those who only need *confirmation, and those who are merely required to renounce their own and all other forms of *heresy. In the first category he lists and briefly describes 26 heretical sects: from *tascodrugi* (or *ascitae* or *ascodrugitae*) to *Melchizedekians, dwelling particularly on the *Manichees. In the second category he puts *Quartodecimans, Sabbatians or *Novatianists, *Arians, *Macedonians or *pneumatomachi and *Apollinarists; in the third are the followers of *Melitius, *Nestorius and *Eutyches. Then follows - with the aim of showing the different ramifications of the heresy of the *acephali or *theopaschites - a *brevis narratio* or list of the various *monophysite factions, subdivided into the two main denominations: Theodosians (in turn divided into *agnoetae, Paulianists [*Paulinians], *tritheists, Gaianites or Julianites [*aphthartodocetists], Dioscorians, etc.), and Marcianists (*messalians, euchites, enthusiasts and Eustathians).

The work ends with a brief treatise *On the schisms of the so-called *diacrinomeni (haesitantes)*, listing the 12 sects of anti-Chalcedonians: among which we meet the various religious factions listed earlier, touching particularly on the different shades of the *severitae* or followers of *Severus.

T.'s work, though later than the 2nd council of *Constantinople (553), seems to predate *monoenergism. "Nothing allows us better than these pages to embrace the range of heterodox sects that moved at the margins and, sometimes, the heart of the Byzantine church around the year 600" (J. Pargoire, *L'Eglise byzantine de 527 à 847*, Paris ³1923, 135).

CPG 7016; PG 86, 12-74; Beck 401 f.; V. Benesevic, *Syntagma XIV titulorum sine scholiis*, St Petersburg 1906, 707-738; J. Meyendorff, *Le Christ dans la théologie byzantine*, Paris 1969, 163-167.

D. Stiernon

TIMOTHY of Jerusalem. Priest and homilist (6th c.? - whether a real person or a fictitious name is unclear) whom *manuscript tradition makes author of two *homilies for the feast of the Presentation: *Oratio in Symeonem*, preserved in Greek and a *Georgian version, and *In occursum Domini*, only in Georgian. On the basis of the stylistic peculiarities of the former (affected vocabulary, characteristic formulae to introduce biblical citations and further oratorical procedures), it is proposed (Capelle) to assign another four homilies to T. of J. These are the *sermon *In Crucem et in Transfigurationem*, attrib. by MSS to one "Timothy, priest of Antioch", and three pseudo-Athanasian discourses previously claimed for *Proclus of Constantinople (*In nativitatem Praecursoris, in Elizabeth et in Deiparam; In censum sive descriptionem S. Mariae et in Iosephum; In caecum a nativitate*). Of these, the first two have also been considered as revisions of homilies of *Amphilochius of Iconium, perhaps by the same T. of J. (Caro). But the diversity of attributions has put the author's identity in dispute. Recently some have claimed these texts as part of the larger homiletic group that can be traced back to *Leontius of Constantinople; T. of J. would be merely one of the pseudonyms used by Leontius to propagate his own work (Sachot). T. of J. has attracted theological attention on account of a phrase in the *Oratio in Symeonem* (PG 86, 245 CD), interpreted by some as the earliest patristic evidence in favour of *Mary's bodily *assumption. In fact the passage merely asserts the Virgin's immortality "up to the present time" and her translation to heaven.

CPG 7405-7410; PG 86, 237-252, 256-265; *EC* 12, 111; M. Jugie, *La mort et l'Assomption de la S. Vierge* (ST 114), Vatican City 1944; B. Capelle, Les homélies liturgiques du prétendu Timothée de Jérusalem, *EphemLiturg* 63 (1949) 5-26; R. Caro, *La Homilética Mariana Griega en el Siglo V*, Marian Library Studies 3, Dayton 1971; M. Sachot, *L'homélie pseudo-chrysostomienne sur la Transfiguration CPG 4724, BHG 1975*, Frankfurt am Main-Berne 1981.

L. Perrone

TIMOTHY Salofaciolus ("white turban"). Patriarch of *Alexandria, Chalcedonian, elected after the exile of *Timothy Ailouros (460). Despite his conciliatory spirit, he was always in difficulties from *monophysite opposition. He was relegated to a Pachomian *monastery when, in 475, Timothy Ailouros, profiting from *Basiliscus's attempted usurpation, returned to his see. On the death of Timothy Ailouros, the monophysites elected *Peter Mongus; but the emp. *Zeno restored Salofaciolus. Despite Mongus's disturbances, T. ruled until his death in 482.

Fliche-Martin 4, 416; J. Maspéro, *Histoire des patriarches d'Alexandrie*, Paris 1923.

E. Prinzivalli

TIMOTHY the Apollinarist. In the subdivisions formed among *Apollinaris's disciples, T., bishop of Berytus, appears as head of the group which, at the opposite pole to *Polemon, tried to reach agreement with the "Great Church": indeed he subscribed the canons of the council of *Constantinople in 381 (cf. Mansi 3, 568 C). Towards the end of *Constantius's reign (361) or early in that of *Valens (364), he had been to *Rome, where he met *Athanasius, and returned with an *Epistula canonica* to "bishop" Apollinaris (cf. Leontius, *Fraud.* 151: PG 86, 1976). In *c*.373, in the middle of the *schism of *Antioch, T. sought hard for an identity that would distinguish Apollinarism from both the old and the neo-Nicenes: he pronounced *anathemata against *Paulinus, *Epiphanius, *Diodore and, eventually, *Peter of Alexandria and *Basil (cf. Lietzmann, 23, 27). In 377, as bishop of Berytus, he went a second time to Rome at Apollinaris's request to defend their interests at the *council held that year (Leontius, *ibid.*), which condemned Apollinarism (Mansi 3, 427).

*Leontius (*ibid*. 138) transmits an *Ep. ad Homonium episcopum* (Lietzmann, p. 277) in which T. distances himself from the "*synousiasts", i.e. the other pole of Apollinarism: he asserts the incarnate Christ's consubstantiality with human nature, but also the distinction of the Logos; two fragments of his *Catechesis follow the same line (Lietzmann, fr. 181, pp. 278-279). We also have a fragment of his *Historia ecclesiastica*, which transmits his *Ep. ad Iovianum*, attrib. by Athanasian MSS to Athanasius (Lietzmann, fr. 182, pp. 279-283; cf. CPG 2, 2135). Finally we have an *Ep. ad Prosdocium* (Lietzmann, 283-286), a true confession of the faith of this school of Apollinarism. It is handed down among the works of pope *Julius (PL 8, 954-959). We also have *Armenian and *Syriac versions and fragments in Greek, Latin, Syriac and Armenian (see CPG 3726).

CPG 2, 3723; *DCB* 4, 1029, 3726; H. Lietzmann, *Apollinaris von Laodicea und seine Schule*, Tübingen 1904, 5, 23, 27, 153-157; frs. pp. 277-286; G. Voisin, *L'apollinarisme. Etude historique littéraire et dogmatique sur le début des controverses christologiques au IVe siècle*, Louvain 1901, 112-113.

E. Cavalcanti

TIMOTHY the Lector. Recipient of numerous *letters from *Isidore of Pelusium (360-435): *Epp.* I, 4. 7-9. 67. 79. 102. 106. 114. 121. 131. 146. 218. 322. 415. 438. 454. 494; III, 3-5. 47-49. 63. 68. 96. 126. 133; IV, 195; V, 168. 170. 274 (PG 78, 177-1646).

PWK 25, 1360.

E. Romero Pose

TIPASA

I. The city - II. Christianity - III. Hagiography - IV. Archaeology.

I. The city. In AD 39 *Caligula put an end to the Roman protectorate over *Mauretania by assassinating king Juba II at Lyons and annexing the ancient kingdom to the empire. T. then became a Roman city, a *municipium* under Latin law in AD 46 and a colony in c.145. The original city had been built on a central promontory, anciently called "temple hill", where the remains of the forum, capitol, curia and lawcourts are, as well as a modern lighthouse. In the mid 2nd c. T. experienced considerable growth, as witness the theatre, amphitheatre and the great wall surrounding the city, from Kudiat Zarur in the E to Ras Knissia in the W, a wall able to resist Firmus's siege (371-2) victoriously. In 373 *Theodosius received the submission of the Maziri, caught up in Firmus's revolt. Conquered by the *Vandals in c.430 and demolished by them in 455, it was reconquered by the *Byzantines, doubtless at the same time as Cherchel (534). After the 6th c. T. carried on in reduced circumstances, as the late reuse of some monuments shows. With the passing of time the city was abandoned.

II. Christianity. It is generally believed that there were Christians at T. in the first three decades of the 3rd c. The inscription of Rasinia Secunda, AD 238 (CIL VIII, 20856), and others, all predating the *peace of the church (*ibid.*, 20891-94), are invoked as evidence of this. At any rate the community must certainly have organized itself and created an episcopal hierarchy in the course of the 3rd c.: indeed the general opinion is that the nine *iusti priores*, whose tombs were laid out in the basilica of Alexander, were Alexander's predecessors (*ibid.*, 20903). We know, thanks to his epitaph, that Alexander was bishop of T. and built the basilica named after him (*ibid.*, 20915), seemingly in the late 4th or early 5th c. We do not know whether Renatus, whose epitaph was found in c.1914 (*Bulletin Soc. Ant. de France* 1928, 121-124), was before or after Alexander. But he is buried in the same basilica and is not named among the nine *iusti priores*. As for Potentius, mentioned in a *letter of *Leo the Great (PL 54, 646-653), he was put in charge of Mauretania by that *pope; his title of bishop of T. appears only in the inscription which he himself put in the basilica of St Salsa in memory of the work he had carried out there (CIL VIII, 20914). The last known bishop is Reparatus, present at the conference of *Carthage of 484: a marginal note in the list of bishops present says of him: *prbt* (= *peribat*).

III. *Hagiography. Other information on T. comes from literary and hagiographical texts. *Optatus of Milevis (*Schism. don.* 2, 18-19) describes the violence exercised against the *Catholic population by two bishops who came from *Numidia, supported by the prefect Athenius: "the people, Catholics by faith, were dispersed by violence and massacred; they were chased out of their own houses; the men were maltreated; the women were dragged out by force; children were massacred; mothers were compelled to abort". A century later *Victor of Vita (*Hist. pers. Afr. prov.* 3, 29-30) gives a picture of the persecution unleashed by the *Vandals. In place of Reparatus, an *Arian bishop was sent from *Carthage to T. The great majority of the population embarked and found refuge in *Spain. Some were unable to embark and had to remain behind: the count sent on this mission by king *Huneric cut off their right hands and tongues, but despite this mutilation they continued to speak. As witness of this miracle, Victor calls a deacon of T. named Reparatus, who had found safety at *Constantinople and lived at the court of the emp. *Zeno. Byzantine historians also mention the miracle, but only Victor of Vita localizes it at T.

Finally there is the Passion of St Salsa (BHL 7467). At a time after the peace of the church, but when Christianity still had few adherents (*rara fides*) in the city, this 14-year old girl had thrown into the sea from the top of the *collis Templensis* the *idol representing the dragon that was the city's protector. For this she was lynched and thrown, as sacrilegious, into the waves: her corpse was miraculously found by some Gallic mariners, who had found refuge in the port during a storm. They buried her on the cliffs to the E.

IV. *Archaeology. St Salsa's burial brings us to T.'s extremely rich archaeological heritage, localized more precisely in the city's E necropolis. At first a small funerary chapel was built above Salsa's tomb, or she may have been laid in a pre-existing chapel belonging to her family; vestiges of a banquet room and a funerary *mensa* remain. Later, after 371/3, perhaps only in c.450, a large basilica was erected alongside, in and around which were put the tombs of the faithful. Because of the work of furnishing the basilica, this *cemetery could still have been in use in the 6th c. It has been entirely uncovered, and gives an exhaustive idea of an ancient Christian necropolis. At the N edge, backing onto the W wall, was found a small late 5th-c. basilica, consecrated, as an inscription proves, to the apostles *Peter and *Paul. On the opposite side of the city, the W necropolis was developed on a large piece of ground, with no break in continuity from the Punic and pre-Roman necropolis. In it we can distinguish numerous Christian centres: 1) a large circular mausoleum (20 m. in diameter), whose function remains unknown, but which resembles a *martyrium*; 2) the basilica of Alexander, flanked to N and S by two funerary enclosures; 3) the necropolis of Matares, with four enclosures, excavated during a rescue dig in 1968-72. All these cemeteries of T. have the following common characteristics: funerary *mensae* for *refrigeria; hydraulic installations for bringing water to the *mensae* and the tombs; enclosed spaces, of varying sizes, for *families or groups. In the inner N-W angle of the city, on the Ras Knissia, arose the episcopal complex, sheer above the sea. It comprised a great five-aisled basilica with projecting apse, supported by enormous buttresses which also reinforced the rock; a square baptistery with round pool; private *baths and, doubtless, the bishop's dwelling. Of the church's liturgical layout, nothing remains. At the centre of the city, the lawcourt basilica and the courtyard of the New Temple were also transformed into a *church building, prob. in the 5th c. For further archaeological details, cf. under *Africa. **[Fig: 310]**

PWK 12, 1413-29; *LTK* 10, 202-3; *DACL* 15, 2338-2406; *EC* 12, 118-122; J. Christern, Basilika und memoria der hl. Salsa in Tipasa, *BAA* 3 (1968) 193-257; S. Lancel-M. Bouchenaki, *Tipasa de Maurétanie*, Algiers 1971; M. Bouchenaki, *Fouilles de la nécropole occidentale de Tipasa, 1968-72*, Algiers 1975; N. Duval, L'Evêque et la cathédrale en Afrique du Nord, *Actes XIe Cong. Internal d'arch chrét.*, CEFR 123, Rome 1989, 345-403.

V. Saxer

TITULI PSALMORUM. In the Hebrew *Bible many *psalms have a title, often obscure, comprising various elements: indication of belonging to a former collection, musical or liturgical rubrics, presumed occasion of composition. The LXX adds other titles, the *Syriac version gives a different series. Patristic commentators willingly turned their attention to these mysterious signs (though *Diodore of Tarsus and others rejected their authority): they were exegetical keys. *Gregory of Nyssa wrote a relatively systematic treatise on them. To these titles from the Greek Bible, the Latin psalters often added directions for Christian *prayer, of the *Vox Christi* type. There are six distinct series, some of them very old; the first is handed down with considerable variants, which suggests much use. We can draw from them many suggestions for the history of *liturgy and *piety.

All the commentaries on the Psalter; P. Salmon, *Les Tituli Psalmorum des manuscrits latins*, Rome 1959; W. Bloemendaal, *The Headings of the Psalms in the East Syrian Church*, Leiden 1960.

J. Gribomont

TITULUS. In classical and ecclesiastical terminology this term had various meanings, which derived from τλάω, *tuli* (to carry, support): *1)* in common language, it designated any detail that conveyed knowledge of the nature, value, content, use or proprietor of an object; *2)* in Roman legal language, a t. was an act establishing a right or notifying the reason for a condemnation. John's gospel (19, 19) shows how the Latin term was carried over, with this meaning, into Greek. Among the martyrs of *Lyons of 177, Attalus bore round his neck a placard with the words: Attalus the Christian (Euseb., *HE* V, 1, 44); *3)* in a *book, the t. was the inscription indicating the author and content; in legal works, particularly Digests and Decretals, tt. also served to subdivide the book; 4) in the social sphere, it indicated the offices, honours and deeds of a noble or illustrious individual or a public figure: the totality of this information constituted the person's *titulatura*, e.g. that of the emperors, as shown on inscriptions; 5) in funerary *archaeology, the t. was the inscription put over a tomb indicating whom the monument was for and who built it. This custom, present everywhere, is particularly frequent at *Rome and at *Trier. The term *titulus* indicates primarily the inscription itself (Diehl 1787, 1991, 2138B, 3581, 4742), but it can also indicate, by metonymy, the funerary monument (Diehl 177, 3500, 3574A, 3578, 3592A, 4169A). The importance of these *tituli* can be argued from the care taken to restore them when damaged. A famous example of retoration of this kind is the inscription of M.A.I. Severianus at Cherchel (Diehl 1583); in fact they constituted not just a title to property, but often regulated the honours to be given to the dead; 6) finally, in Roman ecclesiastical language, a t. is a parish church in the city, in which religious service was performed by one or more priests. Their origin goes back to the time when church *property did not exist in law, i.e. when *church buildings were the property of private individuals, whose name appeared on a tablet close to the entrance. An inscription of 377 attests the oldest t. known to us: *Cinammius Opas lector tituli Fasciole* (Diehl 1269). The Roman synods of 499 and 595 allow us to establish the list and topographical distribution of the tt. for the 5th and 6th c.: between these two dates the tt. changed name, taking on names of saints; it often happened that the original titular was transformed into a saint. The tt. were prob. instituted in the 3rd c., but it is impossible for us to identify those going back to that time with any certainty.

Forcellini, *Totius latinitatis lexicon*, 6, Prato 1875, 107-8; *DAGR* 5/1, 347; Diehl, *indices*, 416, 598-600; J.P. Kirsch, *Die römischen Titelkirchen im Altertum*, Paderborn 1918; F. Lanzoni, I titoli presbiterali di Roma antica nella storia e nella leggenda, *RAC* 2 (1925) 115-257; R. Vielliard, *Recherches sur les origines de la Rome chrétienne*, Mâcon 1941, 25-49; *LTK* 10, 209-10; cf. Ch. Pietri, *Roma christiana*, BEFAR 224, 2 vols., Rome 1976, 90-96, 461-68; P. Testini, *Archeologia cristiana*, Bari 1980.

V. Saxer

TITUS. Collaborator of *Paul, mentioned in the Pauline epistles (2 Cor 2, 13; 8, 6. 16. 23; Gal 2, 1-3). One of the pastoral epistles (Tit) is addressed to him. He is also named in 2 Tim (4, 10). Ecclesiastical *tradition makes reference to Tit 1, 5. *Eusebius (*HE* III, 4, 5) names T. among the bishops of *Crete. So do *Theodoret of Cyrrhus (PG 82, 804C), *Paulinus of Nola (*Carm.* 20) and *Const. Ap.* (VII, 46, 8). The *Chronicon Paschale* goes so far as to identify him with one of the 72 disciples of Lk 10, 1.

T. does not appear in Acts, but he does in the *apocrypha. In *Martyr. Pauli* 6 (a section of *Acta Pauli*), he and *Luke administer *baptism, near the Apostle's tomb, to the prefect and the centurion whom Paul had evangelized while in prison. The *Acta Titi*, in Greek, certainly use the *Acta Pauli*. Acc. to Ps.-Ignatius (*Philad.* 4, 4), T. died a virgin. The 5th-c. *Ep. Titi* is written in an ascetic tone, to praise the virginal life.

As with *Timothy, the *Fathers were worried about the *circumcision of T., a *pagan by birth. The majority of commentators understood Gal 2, 3 to mean that T. was uncircumcised. But *Jerome, *Pelagius and *Augustine followed *Irenaeus and *Tertullian in considering him circumcised.

Erbetta 3, 93-110; *BS* 12, 503-505; F. Halkin, La légende crétoise de saint Tite, *AB* 79 (1961) 241-246; J.N.D. Kelly, *A Commentary on the Pastoral Epistles: I Timothy II Timothy and Titus*, London 1963; M. Dibelius-H. Conzelmann, *The Pastoral Epistles: a Commentary*, Philadelphia 1972.

R. Trevijano

TITUS, emperor. Titus Flavius Vespasianus, Roman emperor 79-81, son of *Vespasian and Flavia Domitilla; as general in his father's service, he ended the Jewish war by capturing *Jerusalem (AD 70). On Vespasian's death (79) he succeeded him, developing his policy of good relations with the senate. His solicitous interventions in Campania after the eruption of Vesuvius and in Rome after a plague and a grave fire in AD 80, together with cautious action to improve the administration of justice, gained him the epithet *amor et deliciae generis humani*. He died of fever in 81, in Sabina.

At the Flavian court there had been no anti-Christian attitudes. Acc. to *Sulpicius Severus (*Chron.* II, 30), however, T. had thought to eliminate Judaism and Christianity with the destruction of the temple of Jerusalem: but it is hard to credit this report.

S.G.F. Brandon, *The Fall of Jerusalem and the Christian Church*, London 1951; H. Montefiore, Sulpicius Severus and Titus' Council of War, *Historia* 11 (1962) 156 ff.; M. Sordi, *Il cristianesimo e Roma*, Bologna 1965, 96 ff. and 423-4 (bibl); R. Furneaux, *The Roman Siege of Jerusalem*, London 1973; B.W. Jones, *The Emperor Titus*, New York 1984.

L. Navarra

TITUS of Bostra. A *letter dated 1 Aug 362, from the emp. *Julian (*Ep.* 52) to the people of *Bostra, capital of the province of *Arabia, invited the citizens to expel from their city bishop T., who - said the emperor - had accused them of being rebellious and of having refrained from excesses only because their bishop and *clergy held them back. This suggests hardship and agitation in the city of Bostra and appeals by the bishop to the emperor in support of its people. In 363 T. appears among the signatories of the document which the synod of *Antioch, called by *Meletius after *Jovian's accession in an attempt to restore a neo-Nicene unity, worked out and sent to the emperor (Socr., *HE* III, 25). Acc. to *Jerome, T. died under *Valens, i.e. before 378 (*De vir. ill.* 102).

T. wrote a work in four books *Against the *Manichees* after the death of Julian (26 June 363), whom it mentions. This work survives entire only in a *Syriac version. Of the Greek original, books I-II and part of III (chs. 1-29) remain, as well as some fragments in *Arabic. Of a *Commentary on Luke* only fragments remain, from which we gather that it must have consisted of a series of *homilies (Sickenberger, 140-245). Some fragments of this *commentary occur in a *Coptic *catena on the gospels, and a fragment in an Arabic catena. Some fragments on *Daniel could belong to the same commentary on Luke or to a homily (Sickenberger, 130-134, 246-248). The Syriac *Florilegium Edessenum anonymum* preserves four fragments of a *sermon *On Epiphany*.

An *Oratio in ramos palmarum* (PG 18, 1264-1277), of which we possess a *Georgian version shorter than the (perhaps interpolated) Greek text, is not considered authentic. Also considered spurious is the commentary on Lk 18, 2 (the unjust judge and the importunate widow), a *cento made up from commentaries of *Cyril of Alexandria, *John Chrysostom and *Isidore of Pelusium (Sickenberger, 137).

CPG 2, 3575-3581; *PWK* 25, 6, 1586-1591; *DTC* 15, 1143-1144; P. de Lagarde, *Titi Bostreni contra Manichaeos libri IV syriace*, Berlin 1859, ²1924; id., *Titi Bostreni quae ex opere contra Manichaeos edito in codice Hamburgensi seruata sunt graece*, Berlin 1859; P. Nagel, Neues griechisches Material zu Titus von Bostra, *Studia Byzantina*, Folge II, Berlin 1973, 285-350; J. Sickenberger, *Titus von Bostra, Studien zu dessen Lukashomilien*, TU 21/1, Leipzig 1901; I. Rucker, Florilegium Edessenum Anonymum, *SBAW* 5 (1933) 82-87.

E. Cavalcanti

TOBIAS

I. In the Fathers - II. Iconography.

I. In the *Fathers. The OT Wisdom book of Tobias (in the LXX Τωβιτ, Τωβειθ = "Jahweh is good"), which tells the story of the two TT., the old (Tobit) and the young (Tobias), was considered inspired, *canonical, by the church *Fathers despite uncertainty as to whether it was originally written in Hebrew. It is not included in the Hebrew canon, and Protestants consider it "apocryphal". The works of the Fathers attest its early popularity among Christians: *Polycarp cites it (Tob 4, 10=*Phil.* 10, 2); *Hermas mentions it (*Praec.* V, 11, 3). Tobit, living in the Oriental diaspora, is a hero of fidelity to God's law and *love of neighbour, the man who endangers his life for the sake of the divine precepts. Christians saw in Tobit the fulfilment of *piety and charity, i.e. the example of the *virtues characteristic of the true Christian: thus Tobias became the prototype of the Christian. His patience in blindness has great importance for the Fathers. *Pontius compares St *Cyprian to Tobit for his magnanimity even towards *pagans during the plague at *Carthage (*Vita Cypr.* 10). Cyprian himself alludes several times to the book of T. (*Test. ad Quir.* III, 1 and 6) or to the figure of Tobit, who after so many good works and magnificent acts of mercy was put to the test by losing his sight (*De bono pat.* 18; *De mort.* 18). Still acc. to Cyprian, Tobit is an example of a person always in *prayer and practising the acts of *justice, of which the *angel Raphael is a witness (*De orat.* 33). *Clement of Alexandria stresses Tobit's continual *fasting (*Strom.* VI, 102) and young Tobias's *marriage to Sarah by the advice of the angel Raphael (*Strom.* I, 123); so Tobias is considered a symbol of the *soul led by its guardian angel. Tobias's journey towards Ecbatana expresses the journey towards *beatitude, toward *God. In catacomb frescos, Tobias usually appears with the fish. Acc. to *Paulinus of Nola (*Carm.* XXVII, 25), the elder T. was depicted in the atrium of the basilica of Nola.

LTK 10, 215-217; *EC* 12, 176-180; *LCI* 4, 320-326; J. Gamberoni, *Die Auslegung des Buches Tobias in der griechisch-lateinischen Kirche der Antike und der Christenheit des Westens bis 1680*, Mainz 1969; F. Vattioni, Studi e note sul libro di Tobia, *Augustinianum* 10 (1970) 241-284; id., Tobia nello Speculum e nella Bibbia di Alcalá, *ibid.* 15 (1975) 169-200 (bibl.); J. Doignon, Tobie et le poisson dans la littérature et l'iconographie occidentale (IIIe-Ve siècle), *RHR* 190 (1976) 113-126.

L. Vanyó

II. *Iconography. The figure of T. is linked with the symbolism of the soul of the dead, which God protects by his *angels from the snares of the enemy in its journey to eternity. But he was rarely depicted in early Christian art. The oldest representation of T. with the fish is in cubicle III of Domitilla, mid 3rd c. (*Wp* 135); then come those of the hypogeum of Via Yser (*Wp* 164, 2), the catacomb of Vigna Massimo (of the Jordani), mid 4th c., where the angel appears, and in via D. Compagni (Ferrua, *Via Latina*, XVIII), *c.*320-360. T.'s meeting with the angel occurs on the so-called *Balaam sarcophagus, in the museum of S. Sebastiano, first half of 4th c. (*Rep.* 176), a unique example in plastic art: here T. emerges from the water clothed in a *colobium* and with a *petasus* on his head. The scene refers to Tob 6, 9. Finally, the 3rd-c. sarcophagus lid of Le Mas-d'Aire (*Ws* 65, 5) shows T. extracting the inside of the fish. **[Fig: 311]**

DACL 15, 2418-2420; *EC* 12, 180; *LCI* 4, 320-326, L. De Bruyne, Sarcofago cristiano con nuovi temi iconografici scoperto a S. Sebastiano sulla via Appia, *RAC* 16 (1939) 262-263, fig. 7; U.M. Fasola, Le recenti scoperte nelle catacombe sotto Villa Savoia. Il coemeterium Jordanorum ad Alexandrum, *Actas VIII Congr. Int. Arqu. Christ.*, Vatican City-Barcelona 1972, 263 ff.

A.M. Di Nino

TOLEDO

I. Christianity - II. Councils.

I. Christianity. T., ancient Toletum, was in Hispania Citerior and, after *Diocletian's division, in the province of *Carthaginensis. After suffering the invasion of the Alans (411-418), it came under the rule of the *Goths. King Athanagild (551-568) chose it as court residence; *Leovigild (568-586) made it capital of the Visigothic kingdom. With the *conversion of *Recared and his whole people, proclaimed at the 3rd *council of T. (589), the city became the official seat of all the kingdom's national, civil and ecclesiastical policy. Under the Muslims (from 712), the capital was transferred to Cordova (*Corduba*) and T. lost its political importance. Christianity must have arrived relatively late, given the city's unimportance in the early centuries, despite its position on the road from *Saragossa (*Caesaraugusta*) to *Mérida (*Emerita*). The earliest evidence comes from the *Acts of the council of *Elvira (*c.*303), signed by Melantius as bishop of T. Facts about St Leocadia, a *martyr under Diocletian, are less certain, and those on St Eugenius, presumed 1st-c. bishop, are legendary. During the *Priscillianist crisis, T., perhaps because of its central position, hosted a council against this heretical sect (*c.*400), thus gaining prestige at the expense of Cordova, the civil metropolis. At the 2nd council of T. (527), its bishop *Montanus (522-531) appeared as a metropolitan and was entrusted with the calling of a future council. Two *letters of Montanus show him aware of his metropolitan role, though he may have felt himself to be such for only a part of the province, i.e. for the region of Celtiberia and for Carpetania, as we may also deduce from the signature of bishop Euphemius at the 3rd council of T. (589). At any rate, at the proposal of king Guntheric a synod of bishops of the province officially recognized T. as metropolitan see of the provincia Carthaginensis. Two facts influenced this: T. was a royal city, and Cartagena (*Carthago Nova*) had fallen under *Byzantine rule. The 7th council of T. (646) ordered all the suffragan bishops to spend a month every year with the metropolitan so as to constitute a sort of permanent synod. The 11th council (681) recognized the metropolitan's right to consecrate all the bishops of the kingdom, elected with the monarch's agreement. Thus T. reached the height of its prestige, well-earned during the 7th c. thanks esp. to the work of its great bishops: *Eugenius (646-657), *Ildefonsus (657-667) and *Julian (680-690), men of great civil and religious authority and, furthermore, educated men who contributed greatly to the cultural splendour of the Visigothic church in that century when even monarchs were educated men. The *monasteries too, particularly that of Agali, contributed to this splendour, preparing future bishops and producing culture. Of supreme importance was the fact that T. was inexorably linked to the vast series of provincial and national *councils which constructed that symbiosis between church and state that characterized the whole Visigothic period.

Bishops: Melantius (*c.*303), Patrunus, Turibius, Quintus, Vincentius, Paulatus, Natalis, *Audentius (all 4th c.), Asturius (*c.*400), Isitius, Martin, Castinus, Campeius, Sintitius, Praumatus, Peter (all 5th c.), Celsus, *Montanus (522-531), Julian, Bacaudas, Peter II, Euphemius (589), Exuperius, Adelphius, Conantius (these last two perhaps the other way round), *Aurasius (603-615), Eladius (615-633), *Justus (633-636), Eugenius (636-646), St *Eugenius II (646-13.11.657), St *Ildefonsus (657-23.1.667), Quiricus (667-680), St *Julian (680-6.3.690), Sisibert (690-2.5.693), Felix (2.5.694-*c.*700), Guntheric (*c.*700-710), Sintheredus (†711), Urban, Sinieredus, Concordius, Cisila (745-754).

J.F. Rivera, Encumbramiento de la sede toledana durante la dominación visigods, *Hispania Sacra* 8 (1955) 1-32; D. Mansilla, Orígenes de la organización metropolitana de la Iglesia española, *ibid.* 12 (1959) 255-290; J.F. Rivera, *Los arzobispos de Toledo. Desde sus orígenes hasta fines del siglo XI*, Toledo 1973; *ES* 5 and 6; E.A. Thompson, *The Goths in Spain*, Oxford 1969.

II. *Councils. 396: the celebration of this anti-Priscillianist council is known from the 1st council of T. (400), through the final sentence passed on bishop Symposius. It is not counted in the series of Toledan councils.

I Toledo. 400, national. The date, though disputed, is probably correct. 19 bishops were present: their sees are unknown except for Patrunus (prob. *Mérida), Asturius (Toledo) and Lampius (*Barcelona). Most of its 20 disciplinary canons concern the life of the *clergy, esp. the recruitment and *ordination of clerics to meet the consequences of *Priscillianism. Pope *Innocent I sent a *letter to those present (*Ep.* 3: PL 20, 486-493). In the *Acts appears a profession of faith, undoubtedly later than the council, against all *heretics and esp. against Priscillianism.

447. The possible existence of this council, whose whereabouts is agreed, is deduced from an invitation of *Leo the Great to the bishops of the Spanish provinces (*Baetica is not named) to celebrate a general council. Of the celebration of a provincial council in *Gallaecia (the same *pope suggested this possibility) we have only hints. It drew up a formula of faith which, with that of the 1st council of T., had to be approved by the bishops of the remaining provinces.

II Toledo. 527, provincial. Presided over by *Montanus of T. The Acts were signed by four suffragan bishops and one in exile, Marcianus. Later they were signed by bishops Nebridius and *Justus (of Tarrasa and Urgel respectively). The first four canons are on the clergy, the fifth on consanguineous *marriage. Some MSS add to its Acts two letters of Montanus to the people of Palencia and their bishop Turibius. (In 580 there was an Arian synod convoked by *Leovigild, to promote the unity of *Spain under the *Arian faith).

III Toledo. 589, general. This council begins the series of Hispano-Visigothic councils. It was called by *Recared and presided over by *Leander, who may also have been its instigator, to celebrate the *conversion of the Visigothic people to *Catholicism. It was attended by 62 bishops (eight Arian) and five delegates, from the whole Visigothic kingdom: Spain and Narbonese France. After the abjuration of the Arian faith and profession of the Catholic by the king and queen, according to the style of the first four ecumenical councils, the eight Arian bishops did the same thing, followed immediately by the presbyters, deacons and nobles. The second part of the council deliberated on the themes proposed by the king. 23 disciplinary canons were promulgated, including the following novelties: to give force of law to papal letters (can. 1); to sing the *Creed at solemnities (can. 2); participation at councils of judges and tax-collectors, to be instructed by the bishops (can. 18), etc. A royal decree confirmed all the canons of this council, giving them the value of civil law. Leander preached the closing *homily. This council marked the political and religious unity of Spain, and began the close collaboration between church and state typical of that realm.

597, general. Attended by three metropolitans (*Mérida, *Narbonne and T.) and 13 bishops from various provinces. Its two canons dealt with priestly *chastity and administration of churches. Not part of the Toledan series.

610, provincial. 15 bishops subscribed and confirmed *Gundemar's decree proclaiming T. the metropolitan see of the whole province of *Carthaginensis.

IV Toledo. 633, general. St *Isidore presided over the six metropolitans of the kingdom, 56 bishops and seven representatives of other bishops. The most important of all the Toledan councils, because of the decisions contained in its 75 canons, of which cans. 1 (profession of faith with full exposition of the doctrine of the *Trinity and the *Filioque) and 2 (unification of liturgical celebrations and rites throughout the kingdom) deserve particular attention; can. 75, very long, was intended to safeguard and strengthen the monarchy.

V Toledo. 636, general. Eugenius of T. presided. Most of the 22 bishops present belonged to the provincia Carthaginensis. The fact that seven of its nine canons were about the king and his family, particularly about the succession, demonstrates king Chintila's utterly political intention in calling it.

VI Toledo. 638, general. Selva, metropolitan of Narbonensis, signed first, followed by 47 bishops (four metropolitan) and five delegates. 19 canons of political and ecclesiastical interest (mainly the latter) were promulgated. The former referred to the succession of the royal family and their protection.

VII Toledo. 646, general. Orontius, metropolitan of Mérida, presided over 30 bishops (four metropolitan) and 11 representatives. They issued six canons on ecclesiastical *discipline. Can. 6 set up a sort of permanent synod in T., whose *authority grew continually. The political import of this council appears in the first of its canons, which threatened severe penalties against rebels and traitors.

VIII Toledo. 653, general. Presided over by Orontius of Mérida. Signed by 50 bishops (three metropolitan), 12 *abbats (for the first time), two Toledan clerics and eight nobles; even the latter put their signatures. 12 quite long canons were promulgated in reply to the *tomus regius*.

IX Toledo. 655, provincial. *Eugenius of T. presided. 15 bishops (two from the province of *Tarraconensis), six abbats, one delegate and four nobles signed. Its 17 canons were on the administration of church *property.

X Toledo. 656, general. Eugenius of T. presided over 16 bishops (two metropolitan) and five representatives. They promulgated seven canons on ecclesiastical discipline; deposed Potamius of Braga, accused of *fornication, and replaced him with *Fructuosus of Dumium; and annulled the will of Recimir of Dumium, which was detrimental to the church.

XI Toledo. 675, provincial. Quiricus presided over 16 bishops, three abbats and eight representatives. The council promulgated 16 canons on church discipline. Its profession of faith remains famous.

XII Toledo. 681, general. Julian of Seville presided over 35 bishops (four metropolitan), four abbats, three delegates and 15 nobles. It was called by Ervigius to legalize his accession to the throne after the strange deposition of Wamba. The council seconded the king's wishes. Of its 13 canons, can. 6 is important: it concedes to the metropolitan of T. the right to consecrate all the bishops of the kingdom. Thus T. acquired primacy over all the sees.

XIII Toledo. 683, provincial. *Julian of T. presided. Another 47 bishops (three metropolitan), eight abbats, one *primicerius*, 29 delegates and 26 nobles signed. It promulgated 13 canons, six on political life and the others on church life.

XIV Toledo. 684, provincial. Julian of T. presided over 16 bishops, six abbats, two representatives and eight delegates of the remaining metropolitan sees. This provincial council was celebrated in place of the general council requested by Leo II, which could not be called. At the pope's wish it approved the *Acts of the 3rd council of *Constantinople, as well as the *Apologeticum* which Julian sent to the pope. The decisions taken are collected in its 12 canons.

XV Toledo. 688, general. Julian of T. presided over 60 bishops (five metropolitans), five delegates, eight abbats and 17 nobles. Called to examine Julian's *Apologeticum*, in which Benedict II had found doubtful expressions: the council, whose Acts are quite violent against *Rome, confirmed its full *orthodoxy.

XVI Toledo. 693, general. Presided over by Felix, bishop of *Seville, whom the council appointed to replace Sisibert of T., deposed after an accusation of conspiracy. 57 bishops (four metropolitan), five abbats, three delegates and 16 nobles were present. They promulgated 11 canons of civil and religious interest, following the lines of the royal *tome*.

XVII Toledo. 694, general. We do not know its participants. It promulgated eight canons, six on church discipline, the other two on the need to protect the royal family, with consequent condemnation of the *Jews, whom the king accused of conspiring against the monarchy.

XVIII Toledo. 702. We know only that there were 58 participants. From this we can deduce that it was a general council.

Synoden, passim; Palazzini 5, 317-328; J. Vives, *Concilios visigóticos e hispano-romanos*, Barcelona-Madrid 1963; R. D'Ardabal, Els concils de Toledo, in *Homenaje a J. Vincke*, I, Madrid 1962, 21-45; P.D. King, *Law and Society in the Visigothic Kingdom*, Cambridge 1972; T.C. Akeley, *Christian Initiation in Spain c.300-1100*, London 1967, ch. 5.

P. De Luis

TOMI (Constanta, Romania). Originally a colony of Miletus, on the Black Sea coast in what is now the Dobrudja, capital of its province and see of the bishop of *Scythia Minor, its contacts with previously Christianized centres in the Aegean ensured its early reception of Christianity. The *martyrologies give T. more Christian *martyrs than anywhere else in Scythia Minor (more than 60), nearly all from *Diocletian's *persecution of 303-304. A fragmentary Greek epitaph mentions a "bishop and martyr" (name unknown), prob. from *Licinius's persecution of 320-323. From the 4th c. the bishops of T. were staunch defenders of *orthodoxy and rigorously kept up relations with the major centres. In 368-9, bishop Bertanius (Betranius) energetically opposed the emp. *Valens who, passing through T., tried to impose *Arianism there.

Though the presence of the modern city of Constanta on top of the old city has prevented systematic excavation, various casual finds and some rescue digs have uncovered more early Christian monuments than in any other centre in Scythia Minor. Among the more important are nearly 50 examples of Greek *epigraphy which provide precious information on the ecclesiastical *organization and life of the Christian community in the 4th-6th cc. A cornelian gem, now in the British Museum, on which is incised the *image of Christ on the *cross flanked by the 12 *apostles and surmounted by the word ΙΧΘΥΣ has been and continues to be wrongly dated to the 2nd or 3rd c. even though the iconographical theme of Christ on the cross was avoided by the Christians until the 4th-5th cc. A block of limestone squared in the form of a *mensa*, with two massive legs and upper surface decorated with a relief border, bears on the front surface a Greek epigraph with the name "blessed Timothy", perhaps that bishop of T. who took part in the 3rd ecumenical *council at *Ephesus (431). A well-known large silver-gilt paten, now in the Leningrad Hermitage, bears a Latin epigraph with the name of bishop Paternus (491-520). One Latin inscription is the only text in Scythia Minor to mention a priest (presbyter); another mentions a sub-deacon (ὑποδιάκονος); a third a "lector of the holy universal church", and a fourth an "administrator (πραγματευτής) of the church of St John". Other funerary epigraphs bear the names of higher civil servants or mere soldiers; some give the deceased's ethnic origin or birthplace, others cite biblical or liturgical texts.

The early Christian *church buildings of T,. razed to the ground and sometimes even to their foundations, have been identified and uncovered since 1960 as the result of a vast programme of urban renewal and building carried out by the city of Constanta. Minimal remains of two basilicas, one of them - prob. the cathedral in the 5th-6th cc. - nearly 50 m. long, have been discovered in the W sector of the ancient city. Two more have been found in the port area and, recently, two more, still unpublished, near the point of the peninsula of T. Five of the six basilicas found have crypts for relics beneath the altar: that of the presumed cathedral is considered the largest cruciform crypt of any early Christian church. Curiously, no traces of a baptistery have yet been found in the perimeter of the bishop's residence. As for *paintings, the fragments of fresco with geometrical and floral designs preserved on the walls and vault of the crypt of the basilica identified under Constanta's Lyceum no. 2, at the upper end of the old port, deserve a mention. We should mention two pavement *mosaics which, if they belong to religious buildings, give an idea of this type of paving at T. in the 5th-6th c. Numerous pieces of architectural *sculpture, esp. capitals, attest the activity of several local workshops; and there is a wealth of imported and indigenous articles including metal, glass and esp. ceramic objects. **[Fig: 312]**

Em. Popescu, *Inscriptiile grecesti si latine din secolele IV-XIII descoperite in România*, Bucharest 1976; I. Barnea, *Les monuments paléochrétiens de Roumanie*, Vatican City 1977; id., *Christian Art in Romania*, I, Bucharest 1979; A. Radulescu and V. Lungu, Le Christianisme en Scythie Mineure à la lumière des dernièrs découverts archéologiques, *Actes XIe Cong. internat. d'arch. chrét.*, CEFR 123, Rome 1989, 2561-2615.

I. Barnea

TOMUS AD FLAVIANUM. Appealed to first by the presbyter *Eutyches and then by *Flavian of Constantinople, in June 449 pope *Leo I took a stand in the christological controversy that had broken out in the *East; with the help of his secretary *Prosper of Aquitaine, he worked out a dogmatic *letter (*Ep*. 28) known as the *Tomus ad Flavianum*. Working into it large citations of his own *sermons, he put forward his *christology, centred on the two natures in one *persona*. Since the Roman delegation failed to communicate this document to the members of the synod of *Ephesus (August 449) - indeed had to withdraw, protesting against the condemnation of Flavian - next year Leo entrusted his *Tome*, supplemented by a patristic florilegium, to another delegation which, after the death of *Theodosius II, was welcomed at *Constantinople. The *Tome* was translated into Greek and accepted by a synod (Oct 450). Finally, at the council of *Chalcedon (Oct 451), it was recognized as being in agreement with *Cyril's doctrine and was partly used in composing the Chalcedonian *creed. Thus, though later contested again by the *monophysites, the *Tome*, as a testament of Latin christology, entered the dogmatic heritage of both the Byzantine and, esp., the Latin church.

Editions: PL 54, 755-782; C. Silva-Tarouca, Rome 1932; ACO II/2, 1, 24-33. Studies: A. Grillmeier, *Jesus der Christus im Glauben der Kirche*, Freiburg 1979, 734-750 (bibl.) (Eng. tr. *Christ in Christian Tradition* [tr. J.S. Bowden], London 1975); H. Arens, *Die christologische Sprache Leos des Grossen. Analyse des Tomus an den Patriarchen Flavian*, Freiburg 1982;

B. Studer

TONANTIUS. A *friend of *Sidonius Apollinaris (5th c.), through whom we know him. Prefect of *Gaul (Sid., *Epp*. I, 7; VII, 12: PL 58, 458. 581), we find him cited in *Carm*. 24 (PL 58, 746), praised with his mother (*Ep*. II, 9: PL 58, 485), and recipient of some hendecasyllables (*Epp*. IX, 13; IX,

15: PL 58, 629-633. 635). Not to be confused with the homonymous bishop of *Carthago Nova* cited by *Capreolus (5th c.-CPL 398-399; PL 53, 849).

PWK 6, 2221 f. (s.v. Ferreolus); *PLRE* 1, 1123.

E. Romero Pose

TONSURE. Τριχοκουρία, in the Greek church, was the ritual cutting of a *child's hair on the seventh day after *baptism. It was done in the form of a *cross to signify the descent of God's *blessing on the head of the newly-baptized. The cutting of hair also signified the first offering to *God (Goar, *Euchol.*, p. 306).

At the time of *Gregory the Great there were *cubicularii tonsurati* at the papal court (Ewald-Hartmann [eds], *Reg. epist.* 5, 57; vol. I, 363). These were *clerics who received the t. in imitation of monks, for whom it was a symbol of total consecration to God. The *Hadrianum* Gregorian sacramentary contains a *prayer *Ad clericum faciendum* which speaks of: "*ad deponendum comam capitis sui propter amorem Christi*" (Deshusses, *Le Sacram. Grégorien*, I, 417). The same prayer mentions the *habitus religionis* of clerics, which the *Holy Spirit would have them wear for ever (cf. *Habit, monastic). Even at this early state, then, the clerical state began to assimilate monastic elements.

While monks and, later, clerics practised the t., other Christians, at least until the 5th c., wore their hair moderately long. We can see this in the catacombs, where men are depicted without t. but not with long hair. *Jerome notes: "*Perspicue demonstratur nec rasis captibus, sicut sacerdotes cultoresque Isidis atque Sarapidis nos esse debere, nec rursum comam dimittere, quod proprium luxuriosum est, barbarorumque et militantium*" (*In Ezech.* 13, 44: PL 25, 437).

J. Goar (ed.), *Euchologion sive Rituale Graecorum*, Venice 1730 (repr. Graz 1960); I. Schuster, *Liber Sacramentorum*, ed. recast and updated by C. D'Amato, Casale 1967; J. Deshusses, *Le Sacramentaire Grégorien*, Spicil. Friburg. 16, Fribourg (Switz.) 1971, 16.

A. Chupungco

TOTILA. Ostrogothic king, successor of Heraric, ruled 541-552. He combined extraordinary military qualities with a sense of justice towards the Italian people, oppressed by heavy *Byzantine taxes. A member of the Gothic aristocracy, he abandoned the cultural policy that had characterized the Amaling family and deliberately distanced himself from Roman intellectual life. After *Belisarius left *Italy, T. advanced to conquer N Italy. When that Byzantine general returned, he fought hard battles to take *Rome (548). Narses, sent against T. when Belisarius returned to *Constantinople, defeated him at Tegina, where he lost his life (552).

H.N. Roisl, *PWK* 14 (1974) 799-809; Z.V. Udalcova, La campagne de Narsès et l'écrasement de Totila, *CCAB* 17 (1971) 557-564; id., L'Italie et Byzance au VIᵉ f., *CCAB* 18 (1971) 547-555; P.A. Cusak, Some Literary Antecedents of the Totila Encounter in the Second Dialogue of Pope Gregory I, *SP* 12 (1975) 87-90; A. Cameron, *Procopius*, London 1985.

M.L. Angrisani Sanfilippo

TOULOUSE. We know five bishops of the Roman city of Tolosa: *Saturninus, martyred under *Decius in 250, Hilary, Rodanius and Silvius in the 4th c., *Exuperius in the 5th. *Jerome's *letters introduce us to: the monk and deacon Sisinnius, sent to the *East by Exuperius in 402 and 406 to take *alms to the monks of *Palestine and *Egypt and a voluminous correspondence to Jerome (*Epp.* 102, 105, 109, 119); the monks Minervius and Alexander, two closely-related ex-advocates to whom Jerome dedicated his *Commentary on Malachi* (*prol.*: CCL 76A, 902) and wrote *Ep.* 119 in answer to their exegetical problems; the priest Riparius, whom we find in the *civitas* of T. but close to the region of Comminges, who denounced *Vigilantius's propaganda to Jerome (*Ep.* 109), sending him in 406, through Sisinnius, the writings of Vigilantius collected by Riparius and his "brother" Desiderius. The most important event was the building of the basilica dedicated to the *martyr Saturninus (Saint-Sernin), begun by Silvius and finished by Exuperius, who consecrated it on 1 Nov 402: the *Passio S. Saturnini* and the *mass for the saint's translation, preserved through the *Liber mozarabicus sacramentorum*, are slightly later.

H. Crouzel, Saint Jérôme et ses amis toulousains, *BLE* 73 (1972) 125-146.

H. Crouzel

TOURS

I. Christianity - II. Council.

I. Christianity. T. is the ancient Caesarodunum, capital of the *civitas* of the Turoni; in the late empire civitas Turonorum, promoted (second half of 4th c.) to capital of the province of Lugdunensis III; Urbs Turonica in the 6th c. A late tradition cited by *Gregory of Tours attributes the foundation of the Turonic church to Gatianus, one of the seven missionaries sent in c.250 to evangelize *Gaul. Historical sources, however, ignore Gatianus and give the first bishop of T. as Litorius (337/8-370), the second on Gregory's list. The Turonic church emerges from obscurity with Litorius's successor St *Martin (372-397), whose evangelistic activity, warm charity and thaumaturgical "virtues" gave Christianity at T. a decisive impulse. Magnified by the work of his biographer, *Sulpicius Severus, the figure and work of Martin had a lasting impact as a model for all Christians enamoured of *ascesis. In T. itself, after the long and mediocre episcopate of Bricius (397-442), who tried hard to obliterate the memory of his too illustrious predecessor, the bishops began to claim for their church, now metropolis of an ecclesiastical province, the spiritual legacy handed down by Martin. Eustochius (442-458/9), *Perpetuus (458/9-488/9) - the most famous - Volusianus (488/9-495/6) and Verus (495/6-507) worked to place T. officially under Martin's patronage: from then on two annual feasts were celebrated in his memory (11 Nov: deposition; 4 July: *ordination) in a great basilica raised over his tomb. Strong in this protection, when the city, torn away from the Roman empire, fell under the rule of the *Arian Visigoths (471-507), the community of T. remained firmly attached to its Nicene faith; then, when the political situation was ripe, it gave the signal of resistance to the occupier; forming relations with the still *pagan *Franks, settled N of the Loire, and gambling on the *conversion of their king *Clovis to *Catholicism, Volusianus and Verus invoked Clovis as a liberator: both paid for this initiative by exile. It was their successor Licinius (507-519) who welcomed Clovis, meanwhile baptized, to T. in 508 after his recent triumph over the Visigoths: the Frankish king, wearing in St Martin's basilica the purple tunic and diadem sent by *Anastasius, emperor of the *East, did homage to Martin for his victory. But Clovis's successors, desiring to keep such a prestigious city under their tutelage, arrogated to themselves the right to designate the prelates of T. The see of Martin went through a long period of obscurity under the ephemeral direction of old, docile bishops: Theodore and Proculus (519-521), Dinifius (521-522), Ommatius (522-526), Leo (526), Francilio (526-529), Injuriosus (529-546), Baudinus (546-552), Gunthar (552-555). But Eufronius (556-573) and esp. *Gregory, the bishop-historian, succeeded in restoring their church's independence and prestige. In Gregory's time, despite wars and natural calamities, T. shone in its brightest splendour, now appearing as the city of Martin *par excellence*. A contribution to the success of this enterprise, began by Perpetuus and completed by Gregory, was the organization of *pilgrimages. By making known, in their own works or those of contemporary poets, the miracles that Martin continued to work at his tomb, the bishops drew pilgrims in ever-increasing numbers. By the end of the 6th c. T. was the greatest pilgrimage centre in Gaul.

In this period Martin's memory made a visible mark on the topography of T.; remodelled on the pattern of a spiritual geography, T. was, even physically, the Urbs Martini. To the E, inside the walls of the old *castrum*, rose the buildings of the episcopal complex, in particular the *ecclesia prima*, the first cathedral built by Litorius and rebuilt by Gregory. Sanctified by the relics of the martyrs of Agaunum (whence the patronage of St *Maurice, attested in the 8th c.), it was above all the venerable sanctuary in which Martin had been consecrated to the episcopate and where "*paintings" recorded the most famous miracles worked by the *confessor in life. W of the *castrum*, a first funerary basilica, erected by Litorius (St Lidoire), was eclipsed by St Martin's basilica, built by Perpetuus. It was to this vast sanctuary, adorned with "paintings", with their legends celebrating the posthumous power of the thaumaturge, that the multitude of devotees, natives of T. or pilgrims, flocked. Numerous sanctuaries were erected in the shadow of St Martin's, in the *atrium* of the basilica (two successive baptisteries) or within a radius of some hundreds of metres (basilica of Sts Peter and Paul; *monastery of St Venantius). Thus began the formation, opposite the old *castrum*, of a new town centre which later took the significant name of Martinopolis. Thus, between the two poles of the life of T. appeared the first links in a network that would later bring them together, particularly through the construction, in Gregory's time, of the monastic basilica dedicated to St *Julian of Brioude. Finally, on the right bank of the Loire, the monastery, once founded by Martin, became the Maius Monasterium (Marmoutier), the largest monastic complex in T. and the third pole of its Martinian geography. After Gregory's death and before the Carolingian renaissance, T. returned, through lack of sufficiently numerous evidence, to obscurity. Most of its bishops are for us just names handed down by a signature on a charter or conciliar *act. Pelagius, Gregory's successor, was followed in the 7th c. by: Leupacar (before 609-before 614), Egiric (616), Gwalac, Sigilaic, Leubald, Medigisilus (before 627-after 638), Latinus (650), Caregiselus, Rigobert (654), Papolenus, Chrodobert (before 668-after 672), Bertus (before 680-after 688), Peladius, Ebarcius (696/8); then, in the first half of the 8th c., we know: Ibbo (c.710), Gumtramnus, Dido, Ragambert, Audbert, Ostald, Eusebius (757-762).

Duchesne, *Fastes* II, 283-312; AA. VV., *Mémorial de l'année martinienne*, Paris 1962; *Histoire religieuse de la Touraine* (under the direction of S. Oury), Tours 1975; L. Pietri, *La ville de Tours du IVe au VIe s.: Naissance d'une cité chrétienne*, Rome 1983; P. Saint-Roch, L'Utilisation liturgique de l'espace urbain et suburbain. L'example de quatre villes de France, *Actes XIe Cong. internat. d'arch. chrét.*, CEFR 123, Rome 1989, 1103-1115.

L. Pietri

II. *Council. On 18 Nov 461 a few bishops of the province of T. and the neighbouring provinces (Bourges [*Bituriga*], Le Mans [*Cenomannis*]) met for the feast of St Martin and, on the occasion, promulgated 13 disciplinary canons on the duties of *clerics (*chastity, sobriety, stability, disinterest), consecrated *virgins and penitents. For the council of 567 cf. *Merovingian councils.

CCL 148, 142-149; Hfl-Lecl 2,898 ff.; Palazzini 5, 371.

Ch. Munier

TRACTATUS. In ecclesiastical Latin the verb *tractare*, used absolutely, meant to preach (cf. e.g. Paul. Mil., *Vita Ambrosii* 17, 18, 23); sometimes, to explain a passage of Holy *Scripture. The corresponding noun *tractatus* (or *tractatio*) was equivalent to **sermo*=*homily, preaching (cf. e.g. Aug., *Epp.* 29, 3; 224, 2), and frequently meant an explanation of and commentary on the sacred text heard during the *lectio* in the first part of the liturgical *synaxis (cf. e.g. Ambr., *Ep.* 20, 4; *De off.* I, 99-101). The term T. was also used to mean exegetical writing, *commentary, treatise relating to Holy Scripture; but we must remember that such works often arose in preparation for homiletic activity or resulted from it.

G. Bardy, Tractatus, tractare, *RecSR* 33 (1946) 211-235; Ch. Mohrmann, Praedicare-Tractare-Sermo. Essai sur la terminologie de la prédication paléochrétienne, *La MaisonD.* 39 (1954) 97-107 (repr. in *Etudes sur le latin des Chrétiens*, II, Rome 1961, 63-72).

P. Siniscalco

TRACTATUS CONTRA ARIANOS. Name usually given to a long passage from an anti-Arian work, contained in the famous *Codex Vindobonensis* 2160 with part of *Hilary's *De Trinitate*. The unknown author points out to the *Arians the difficulties deriving from their consideration of the Son as a creature of different substance from the *Father. The fragment appears to have been written before the end of the 4th c.

H.S. Sedlmayer, Der Tractatus contra Arianos in der Wiener Hilarius-Handschrift, *SAW* 146, 2 (1903) 1-18; G. Morin, *ibid.* 18-21; id., Deux fragments d'un Traité contre les Ariens attribués parfois à St. Hilaire, *RBen* 20 (1903) 125-131.

M. Simonetti

TRADITIO LEGIS ET CLAVIUM: *iconography. It is well-known that, in parallel with the great spread of Christianity, numerous other scenes came to supplement and enrich the earlier iconographical repertoire characteristic of the period between the late 3rd and first years of the 4th c. In this context, *traditio* scenes acquired a precise symbolic value: these scenes (*traditio legis* and *traditio clavium*) did not refer to episodes known from literary evidence, but were intended to recall to the observer certain precise concepts. Both developed over the last decades of the 4th c. and lasted into the 5th, i.e. in a politically uncertain period when it is no surprise to find the concept of the priority of the otherworldly over this *world confirmed in the figurative arts. The so-called *traditio legis* appears quite early in monuments of that period, with well-marked characteristics: Christ in the foreground, with unrolled *rotulus*, between the *apostles *Peter and *Paul. Thus, in *painting, we have the arcosolium of the catacomb *ad Decimum* on the Via Latina, where Peter, bearing a *cross, is receiving the scroll, while Paul acclaims. In *sculpture we could mention numerous examples, like sarcophagus Lat. 174 (late 4th c.) or the sarcophagus fragment, more or less contemporary, where Christ is represented on foot, bearded, on the mount, in a composition recalling an analagous monument at S. Sebastiano; in both cases, obviously, Peter and Paul are present.

A so far unique variant is the *mosaic of one of the two apsidioles of *Constantia's mausoleum at *Rome: here Christ has in his hands the unrolled *rotulus*, on which however we read *Dominus pacem dat*. It seems to be a *Traditio pacis*, not *legis*, a variant probably suggested by the imperial clients who commissioned the scene.

Recently A. Quacquarelli has tended to give compositions of this kind a different meaning, no longer linked to the mere transmission of the law, but to the acquisition of the Holy *Scriptures (the *rotulus*), always present in early Christian literature.

The *traditio clavium* certainly leaves less room for interpretation: in it Christ appears, handing over the keys to Peter, while Paul assists or acclaims. The composition is necessarily charged with symbolic value if we consider the choice of Peter as the foundation of the church, despite his betrayal.

Some have suggested a Roman origin for the representation, recalling that one of the first examples on which it appears is a *sarcophagus preserved in the sculpture museum of S. Sebastiano: but it was widespread, both at *Rome and elsewhere, monuments being known at *Ravenna, in *Gaul, etc., all chronologically referable to a period contemporary with the development of the so-called *traditio legis*.

The two scenes come together in the mausoleum of Constantia: a mosaic of the *traditio clavium*, damaged by poor restoration, is in the lunette diametrically opposite to the one described above.

Finally, the *traditio clavium* returns on the panels of ivory *diptychs, a fact of some interest since it well exemplifies the type of commission.

The so-called *traditio legis* and *traditio clavium* are frequent on funerary monuments, often on sarcophagi or works commissioned by a certain social class, in a period which saw the beginnings of conflict between the political and religious powers, the former - in the West - now far removed from the actual population, the latter necessarily called to take the former's place, in support of the less well-off classes and supported by the nobility and aristocracy. **[Figs: 313, 314]**

(Cf. also *Peter and *Paul); W.N. Schumacher, Dominus legem dat, *RQA* 54 (1959) 1-39, 137-202; C. Davis-Weyer, Das Traditio legis-Bild u. seine Nachfalge, *Münchener Jahrb. der Bildenden Kunst* 12 (1961) 7-45; M. Sotomayor, *S. Pedro en la iconografía paleocristiana*, Granada 1962; R. Farioli, *Pitture di epoca tarda nelle catacombe romane*, Ravenna 1963; F.W. Deichmann-G. Bovini-H. Brandenburg, *Repertorium der christlich-antiken Sarkophage*, I, *Rom und Ostia*, Wiesbaden 1967; P. Testini, *Gli apostoli Pietro e Paolo nella più antica iconografia cristiana: Studi Petriani*, Rome 1968, 105-130; U.M. Fasola, *Pietro e Paolo a Roma*, Rome 1980, 91-94.

U. Broccoli

TRADITION. When we study the problem of t. in the early church, we must remember first of all that the NT canon was not formed until the end of the 2nd c. If we except the OT, which at this time had already received its final *canonical form and so no longer posed a problem of t., but only one of interpretation, in this long interval of 200 years there was no "*Scripture" alongside "tradition": the whole of the preaching and teaching referring to Jesus and the *apostles constituted "tradition", παράδοσις.

We will try to follow briefly the stages of evolution during this period. Chronologically, the Pauline writings come first. O. Cullmann has asserted, with convincing arguments, that we find in the apostle *Paul a dual concept of t.: on the one hand, as a theologian formed in the rabbinic schools, he faithfully handed down the t. he had received; on the other, as an *apostle who had received his mission from the risen Christ, he felt himself invested with the power to transmit this t. with the *authority of the living Lord (cf. 1 Cor 11, 23 f.; 15, 1-4). The Lord Jesus was not just the object, but is at the same time the active subject of his preaching. It was because of his awareness of his apostolic authority, conferred by the Spirit, that Paul tended to distinguish clearly between the t. received and its clarification and application, which could and must change according to concrete circumstances (cf. 1 Cor 7, 10-12. 25. 40). L. Goppelt has particularly insisted on this "hierarchy" of traditions in the Apostle.

During the "pseudoepigraphical" period that followed, and which lasted from AD *c.*60 to *c.*120, the canonical gospels and the deutero-Pauline and Catholic epistles were drawn up. These different writings were characterized by the same determination to speak in the name of the apostolic authority, which faithfully transmitted t. but did not hesitate to update it freely and sovereignly according to the needs of preaching and exhortation. But this cannot prevent us seeing that, compared to the Pauline writings, the sense of hierarchy of traditions and consequently of differences in the value to be attributed to them, tends to disappear; it is significant that, e.g., the Pastoral epistles now speak only of the "deposit" which must be preserved as such (cf. 1 Tim 6, 20), and now we often find the stereotyped formula that the t. must be maintained "with nothing added and nothing taken away".

In the 2nd c. the church passed through a real crisis of t. The generation of the apostles and their disciples, the charismatic *prophets and doctors, had passed away; Christian communities were dispersed in the various provinces of the empire, unconnected by any central court of appeal. In this precarious situation they began to emphasize the apostolic *succession of bishops, which guaranteed the authenticity of t. (cf. *1 Clem.* 42-44), and they appealed to recognized authorities like the apostle Paul (cf. Polyc., *Phil.* 3,

2) and the presbyters (Papias, in Euseb., *HE* III, 39, 3-4), as well as to the *regula fidei* (cf. Arist., *Apol.* 15, 2). But, faced with the *gnostic trend that was spreading rapidly and universally at this time, in various forms, but with a clear tendency to dissolve t., the Great Church, to avoid having its own foundations put in question, had to find a firmer and more coherent position.

It is not our task here to outline the concept of t. as it appears in the various gnostic schools. Suffice it to say that the gnostics were not content to interpret the common apostolic t. in their own way, but appealed to a vast treasure of esoteric apostolic t. The gnostic writings also insist on a form of "apostolic succession" (cf. Ptol., *Letter to Flora*, = Epiph., *Pan.* 33, 7, 9). *Irenaeus, *Tertullian and *Clement of Alexandria then intervened - each in his own way and his own place - to put a brake on this movement of peaceful infiltration of *gnosis into the church. Thanks to these theologians, the gnostic wave was blocked and the church received the doctrinal pillars on which it would from now on be built. It is worth the trouble of examining more closely the respective positions of these three *Fathers.

For *Irenaeus of Lyons, the canon of NT writings, recently formed in reaction to *Marcion, became the main weapon in his battle against gnosis. But he was aware that this weapon alone was not enough. He also had to prove that this canon of the church contained the complete, authentic, apostolic t., and that this t. was correctly interpreted in the church. The first proof was furnished, in Irenaeus's eyes, by the uninterrupted *succession of bishops in the apostolic churches (cf. *Adv. haer.* III, praef.; 1, 1; 2, 1-4, 1); the second proof was given by the fact that there was perfect agreement between the *regula fidei* of the apostolic churches, as well as between their interpretations of *Scripture. Irenaeus also claimed that the bishops had received the *charisma veritatis* (*Adv. haer.* IV, 26, 2), but the opinions of specialists are divided as to the exact significance of this concept.

In his treatise *De praescriptione haereticorum*, *Tertullian apparently moves along the lines marked out by Irenaeus, but his thought is more legalistic, more marked by formal logic than that of his predecessor: acc. to Tertullian, in fact, the transmission of t. takes place within the churches that are linked by family relationships; this transmission corresponds to the administration of an inheritance left by a founder, an administration that is entrusted only to the legitimate descendants, family members. *Heretics are, on every evidence, usurpers; their claims can be rejected by means of "*praescriptiones*" (cf. *Praescr.* 19-21).

Finally, *Clement of Alexandria strove to render inoffensive the attraction exercised by gnosis on certain Christians by demonstrating that the sources of "true gnosis" were found within traditional Christianity and that one could never make sufficient progress on the arduous journey that led to perfection; Clement hoped thus to incite the best to undertake this journey (cf. *Strom.* VI, 7, 61, 1-3; 15, 124, 3-5; 131, 4-5; VII, 16, 103, 5; 104, 1-2). And the Great Church too had its "secret" apostolic t., which was unknown to the mass of believers and was transmitted from master to pupil (*Strom.* I, 1, 11, 3-12, 1; *Hypotyposeis*=Euseb., *HE* II, 1, 4).

With these replies, divergent in detail but basically convergent, gnosis was now considered a *heresy. But this did not mean that all problems were solved. On another level, it was precisely at this time that another debate arose. Despite the value of Scripture, interpreted according to the norms of the *regula fidei* and in continuity with the apostolic origins thanks to the uninterrupted succession of bishops, "grey areas" existed which could not be reclaimed simply by appealing to fixed t. What should one think, in particular, of the church's liturgical and disciplinary t.? Had it the same value as doctrinal t.? Tertullian himself, whom we have seen at work as a convinced *heresiologist, after becoming a *Montanist spoke for the first time of "traditions" in the plural and designated as such some liturgical and disciplinary customs of the church which had no biblical foundation; these customs were linked to ecclesiastical *consuetudo* and could be explained rationally (*Cor.* 3-4). At the same period, *Hippolytus brought together a series of liturgical formulations and disciplinary rules in a book to which he gave the significant title *Apostolic Tradition.

Now, we must emphasize that the church was still aware of the fact that this customary law was not to be confused with the kerygmatic and doctrinal t., which alone was normative. Two incidents of that time make this sufficiently clear: the controversy over the date of *Easter at the end of the 2nd c. (cf. Euseb., *HE* V, 24, 2-6. 13) and the controversy over the *baptism of *heretics, halfway through the 3rd c. (cf. Cypr., *Epp.* 71-75). In both cases the bishops of *Rome - *Victor and *Stephen I respectively - sought to impose the customary law of their own church on the other churches; in both cases the attempt was unsuccessful, because it was pointed out to them - by Irenaeus in the first case, and then by *Cyprian, *Firmilian and *Dionysius of Alexandria (= Euseb., *HE* VII, 3-4) - that they were going beyond the limits of their competence by threatening to *excommunicate sister churches which followed another tradition just as well-founded as that of Rome. So the church of the late 2nd and mid 3rd c. still tolerated a certain pluralism in matters of *liturgy and *discipline.

But it was not just in the sphere of liturgical customs that a "grey area" remained, but also in that of doctrine; it was here that personal theological research found a vast field of exercise, and here that theological schools developed. After still timid and very traditional beginnings in 2nd-c. *apologetic (cf. Just., *Apol.* I, 6, 2; 10, 1; 66, 1), this *theology found its first development at *Alexandria. *Origen began his systematic essay *De principiis* by saying that he wished to study the questions left open by the *regula fidei* (cf. *De princ.* I, praef. 2-3; 7-8; 10). As long as the fruits of this theological research remained the private affair of a teacher and a circle of his disciples, pluralism of opinions was no problem; but when these opinions reached the great public, finding a wide hearing and arousing opposition, it was necessary to find a general and common solution to the questions raised. The church would find itself faced with these problems in the course of the two following centuries.

The division of the church caused by the early 4th-c. disputes over the *Trinity took on such dimensions that a new tribunal of decision in matters of *dogma and *discipline had to be created. This tribunal was the ecumenical *council. The particular circumstances which, as a result of the emperor's *conversion, transformed the church of *martyrs at one blow into an institution favoured by the state, made possible the transition from the pre-Constantinian model of the provincial synod to a manifestation of the universal church. The emperor himself convoked the council, took part in its deliberations and ordered the application of the conciliar decisions. We can never overestimate the importance of the 1st council of *Nicaea. It was a theological event of the first order, and one which marked a turning-point in the evolution of the concept of t. in the church. What made this event important was not the number of bishops present, nor the consequences of imperial protection, so much as the following notable fact: for the first time, and in full knowledge of the import of its action, a council formulated a doctrine that went beyond the limits of the t. accepted in the church, whose exclusive norm had until then been the witness of Scripture and its interpretation in accordance with the *regula fidei*: now, however, the term *homoousios* was adopted in the *Creed, a term with no scriptural basis and one, moreover, which was somewhat compromised by its earlier history. The contest over this crucial problem lasted all through the 4th c.: on one side were the *Arians and their sympathizers, who cast themselves as "traditionalists"; on the other, *Athanasius and his friends, who played the role of "progressives"; against the former they had to provide proof that, in this particular case, a term which was not traditional corresponded to the "spirit" of the t. better than the traditional theological language (cf. Athan., *Ep. ad Afros* 6).

It is interesting to see that at the same time proof from patristic *argumentation made its appearance in Christian t. The *Fathers - in particular the Fathers of Nicaea - now constituted a new *authority to which it was necessary to submit. Their definition of trinitarian dogma had created a new t. that was juxtaposed with the old as its correct interpretation and elucidation. Furthermore, the bishops were no longer just the guarantors and managers of the apostolic legacy (e.g. against heretical falsifications), but they were enabled - as a conciliar assembly - to create a new t., in the name of the apostolic spirit.

Athanasius held resolutely to this point of view and defended it more than once. But it was above all *Basil the Great who reflected this new concept of t. in his definition of the doctrine of the *Holy Spirit, which found its dogmatic formulation at the council of *Constantinople in 381. In his work *On the Holy Spirit*, Basil went back to liturgical customs of his church, themselves derived from the trinitarian baptismal formula, for a basis of the Holy Spirit's *homoousia* with the *Father and the Son. *Lex orandi* became *lex credendi*: Basil tried, it is true, to show the ancient origins of the *lex orandi*, but in the final account he started from the axiom of the "*apostolicity" of all that the church, in its living reality animated by the Spirit, represented and confessed; he thought that it would be wrong to abolish it. As we see, unwritten traditions, considered *adiaphora* in the 3rd c., now received the consecration of apostolicity and became constitutive.

After Basil, the Greek Fathers frequently referred to 2 Th 2, 15 as a biblical justification of t. in its dual form, written and oral (cf. e.g. John Chrys., *In Ep. II ad Thess. hom.* 4, 2; John Dam., *De fide orth.* IV, 16). It is an astonishing fact that the Eastern churches never developed their t. further, despite the basic affirmation - expressed particularly by *Gregory Nazianzen (*Oratio* 31, 26-27) - that an evolution of *dogma was possible. In the West, the council of Trent did not fail to cite St Basil, in turn provoking a protest from the Reformers against this putting of "Tradition" and "traditions" on a level of equality (cf. H. Beintker).

In the West, evolution followed a different course, in accordance with the taste of the Latin spirit for clear and established order. On the one hand, we find texts here that attest the same esteem for the conciliar decisions and the t. of the Fathers. In particular St *Augustine - like Basil of Caesarea before him - willingly attributed an apostolic origin to ecclesiastical traditions that were universal in the church of his time (cf. *De bapt.* II, 7, 12; V, 17, 23; 23, 31; *De Gen. ad litt.* I, 10, 20). On the other hand, there was reflection on the criteria that permitted critical acceptance of t.: what was it that allowed authentic and acceptable t. to be distinguished from inauthentic and unacceptable t.? It was *Vincent of Lérins, in his *Commonitorium*, in dialogue with the *Augustinian heritage, who laid the systematic bases that would determine the concept of t. in the West. Vincent unequivocally put *Scripture above Tradition (*Comm.* 2, 1-2). Now, the correct interpretation of Scripture is not without problems, and consequently the church must have recourse to t. (*ibid.* 2, 2). But t. appears in various forms: not all its aspects are of the same value. It was in this context that Vincent pronounced his famous definition: *id teneamus, quod ubique, quod semper, quod ab omnibus creditum est* (*ibid.* 2, 3). It is important for the understanding of this definition that the three criteria evoked, *universitas, antiquitas* and *consensio*, should not be placed on the same level. When, in the actual situation of the church, a disputed question cannot be resolved by the majority of the churches in a definite sense (*ubique*), recourse must be had to t. (*semper*); but within t., conciliar decisions (*ab omnibus*) count for more than the isolated voices of Fathers. It is evident that appeal to t. may not bring the solution to all problems. For this reason, Vincent admitted the possibility of a doctrinal evolution in the church (*ibid.* 23, 28). It is interesting to observe the role he allowed, in this context, to councils (*ibid.* 23, 32). Even today, the definition of the function and authority of t. as proposed by Vincent of Lérins can still render precious service in ecumenical dialogue.

DTC 15, 1252-1350; D. van den Eynde, *Les Normes de l'Enseignement Chrétien dans la littérature patristique des trois premiers siècles*, Gembloux-Paris 1933; O. Cullmann, *La Tradition, Problème exégétique, historique et théologique*, Neuchâtel-Paris 1953 (²1968); E. Flesseman van Leer, *Tradition and Scripture in the Early Church*, Assen 1954; R.P.C. Hanson, *Origen's doctrine of Tradition*, London 1954; H.E.W. Turner, *The Pattern of Christian Truth. A Study in the Relations between Orthodoxy and Heresy in the Early Church*, London 1954; L. Goppelt, Tradition nach Paulus, *Kerygma und Dogma* 4 (1958) 213-233; J.N. Bakhuizen van den Brink, Traditio im theologischen Sinne, *VChr* 13 (1959) 65-85; M. Kremser, *Die Bedeutung des Vincenz von Lerinum für die römischkatholische Wertung der Tradition* (Thesis), Hamburg 1959; Y. Congar, *La Tradition et les traditions. Essai historique*, Paris 1960; H. Holstein, *La tradition dans l'Eglise*, Paris 1960; H. Beintker, *Die evangelische Lehre von der Heiligen Schrift und von der Tradition*, Lüneburg 1961; G.G. Blum, *Tradition und Sukzession. Studien zum Normbegriff des Apostolischen von Paulus bis Irenäus*, Berlin-Hamburg 1963; H. v. Campenhausen, *Kirchliches Amt und geistliche Vollmacht*, Tübingen ²1963; R.P.C. Hanson, *Tradition in the Early Church*, London 1962, Philadelphia 1963; H. v. Campenhausen, *Schrift und Tradition*, Zürich 1963; B. Studer, Träger der Vermittlung, *MySal* I, Einsiedeln 1965, 545-605; J. Beumer, *La Tradition orale*, Paris 1967; G.G. Blum, *Offenbarung und Überlieferung. Die dogmatische Konstitution Dei Verbum des II. Vaticanums im Lichte altkirchlicher und moderner Theologie*, Göttingen 1971; J. Pelikan, *The christian tradition. A history of the development of doctrine*, I, Chicago-London 1971; J.B. Bauer, Das Verständnis der Tradition in der Patristik, *Kairos* 20 (1978) 193-208; R.E. Person, *The Mode of Theological Decision Making at the Early Ecumenical Councils*, Basel 1978; H.J. Sieben, *Die Konzilsidee der Alten Kirche*, Paderborn-Munich-Vienna-Zürich 1979; AA. VV., La Tradition Apostolique régulatrice de la communauté ecclésiale aux premiers siècles, *L'Année canonique* 23 (1979); J.F. Kelly, (ed.) *Perspectives on Scripture and Tradition*, Notre Dame 1976; M.F. Wiles, The Patristic Appeal to Tradition, *Explorations in Theology* 4, London 1979, 41-52; W. Rordorf - A. Schneider, *L'évolution du concept de tradition dans l'Eglise ancienne*, Berne 1982.

W. Rordorf

TRADITOR (Traitor). Its predominant meaning of one who crossed over to the enemy involved a gamut of accepted meanings embracing the actions and words of one who abandons a position to which he had a precise commitment.

The fundamental meaning remained unchanged in Greek and Latin classical literature, as well as in biblical and patristic literature (Herodotus, *Histories* VIII, 30; Aeschylus, *Prometheus* 1068; Sophocles, *Philoctetes* 94; Euripides, *Orestes* 1657; Plato, *Laws* 9).

In the OT those who broke promises or plotted against their fellows could be considered traitors (cf. 1 Macc 12, 42-13, 23; 16, 15-16; 2 Macc 3, 4-6; 4, 1; 13, 21): Cain (Gen 4, 8), Delilah (Jdg 16, 19-21), Doeg (1 Sam 22, 9-10), Joab (2 Sam 3, 27), even *David (2 Sam 11, 14-17). In the NT the t. by antonomasia was *Judas Iscariot. Only Lk 6, 16 calls him this proleptically, but all the *evangelists concur in presenting his actions as a series of betrayals of Jesus (cf. Mk 3, 19; 14, 11. 18. 21. 22. 44 and par.; Jn 6, 64. 71; 12, 4; 13, 2. 11. 21; 18, 2. 5. 36). The Jews who handed over Jesus (and the prophets before him) to be put to death were traitors (Acts 7, 51-53) and, in the Christian community, so were those who handed their *friend over to the persecutors (2 Tim 3, 4; cf. Lk 21, 16-17). In early Christianity those who denounce their parents (Hermas, *Vis.* II, 2, 2; *Const. Apost.* II, 43, 4) or, driven by perverse conduct, their *brothers in faith (Orig., *In Jo.* 1, 11) are called traitors, as are *apostates (Hermas, *Sim.* VIII, 6, 4; IX, 19, 1; Orig., *In Jo.* 28, 23; *Const. Apost.* V, 14, 3; VI, 12, 1; John Chrys., *Hom. 71, 2 in Jo.*) and violators of the *disciplina arcani* (Orig., *In Jo. 1, 11* 28, 23; *Const. Apost.* III, 5, 5). Dominating all stands the figure of Judas, to whom *Athanasius compared the *Arians who betrayed the Nicene faith. Judas was dismissed from the dignity of an *apostle, not because he sacrificed to the gods, but because he was a t. (Athan., *Epist. ad Episc. Aeg. et Libiae* 21).

In *Africa during *Diocletian's *persecution, churchmen who handed the sacred *books over to the persecutors were called *traditores*. *Ordination conferred by *traditor* bishops was considered invalid by the *Donatists (Opt. of Mil., *C. Parm. don.* 1, 13; Aug., *De bapt.* 7, 2; *Ep.* 35, 4).

Enc. della Bibbia 6, Turin 1971, 971; Lampe, *ad loc.*; W.H.C. Frend, *The Donatist Church*, Oxford ³1985, ch. 1; B. Kriegbaum, *Kirche der Traditoren oder Kirche der Märtyrer?*, Innsbrucker theol. Studien 13, Innsbruck-Vienna-Tyrolia 1986.

E. Peretto

TRADUCIANISM. T. was put forward in antiquity as one of the solutions to the problem of the origin of the *soul: *1)* originated from matter which dissolves at the end of each cyclical conflagration (*Stoicism); *2)* created before each man's *birth and hence existing before its fall into the *body (*Platonism); *3)* created at the same time as the body (Christian creationism) or transmitted by generation at the same time as the body (material and spiritual t., nowadays also called generationism). In Christian authors before the 5th c., esp. in *Tertullian (*De anima*) in the West and *Apollinaris (Ps.-Athan., *C. Apoll.* 2, 8: PG 26, 1143) in the *East, t. was connected with the conception of the soul as "corporeal substance", though this corporeity was different from that which commonly came under the experience of the senses. T. drew attention in the 5th c., esp. in the West, as a result of the controversy with *Pelagianism over original *sin and its transmission in every descendant of *Adam. For the Pelagians (particularly *Julian of Eclanum), admitting the transmission of original sin involved accepting the thesis of the t. of the soul, against the established Christian doctrine of creationism. Those who believed in original sin were called, acc. to *Augustine (*Op. imp. c. Iul.* I, 6), "*traduciani*". The question was much discussed by *Jerome (*Ep.* 126; *C. Rufinum* 2, 8; 13, 28, 30), by the anon. author of the *dialogue *De origine animarum* (PL 30, 270) and by Augustine (*Epp.* 166 and 190; *De gen. ad litt.* chs. 6; 7; 10, 11-26; *De origine animae* = *Epp.* 176 and 177; *De anima et eius origine* 1, 16, 26; 4, 11, 2). Augustine, while rejecting the t. of a material soul, held the possibility of spiritual t. as well as of creationism (*Retr.* I, 1, 3; II, 45 and 56), since in neither case was faith in the transmission of original sin compromised (*Op. imp. c. Iul.* II, 168). This possibility was accepted by Augustine's disciples, e.g. *Fulgentius (*De vera praedest.* 3, 28) up to *Gregory the Great (*Ep.* 9, 52).

DTC 15, 1350-1365; *CE* 14, 230.

V. Grossi

TRAJAN, emperor. Marcus Ulpius Traianus, Roman emperor 98-117, native of Hispania *Baetica, a brave military commander, famous particularly for his conquest of *Dacia (described on the reliefs of Trajan's column) and his great public works in Rome (markets, forum, basilica Ulpia). Under his government there were grave financial difficulties. He died at Selinous in *Cilicia in 117 after a campaign against the Parthians. He was given the title *Optimus princeps*. Pliny the Younger wrote a famous *Panegyric* of him.

Under T. there was a new *persecution of the Christians, guilty of *superstitio illicita*: among its victims were *Simeon of Jerusalem and *Ignatius of Antioch. In AD 112 the proconsul of *Bithynia, *Pliny the Younger, responsible for condemning numerous Christians, asked the emperor for instructions concerning them (*Ep.* X, 96): in a rescript, T. said he should not officially seek out adherents of the new religion, but if denounced and confessed they must be punished, though never as a result of anonymous denunciations. *Tertullian criticizes his ambiguity in not wishing to seek out Christians, as innocent, while ordering them to be punished as guilty, if denounced. T.'s rescript long served to oppose Christianity, but the state's formal renunciation of any persecutory initiative (*conquirendi non sunt*) favoured, in a way, the existence of Christians in the empire. In essence, T. did not consider the Christians politically dangerous and sought, as far as possible, to tolerate or ignore them.

A. Garzetti, *L'impero da Tiberio agli Antonini*, Bologna 1960; M. Sordi, *Il cristianesimo e Roma*, Bologna 1965, 131-149 and 425-6 (bibl.); G. Lanata, *Gli atti dei martiri come documenti processuali*, Milan 1973, 58 ff; A.N. Sherwin-White, *The Letters of Pliny*, Oxford 1966, 691-713 (Pliny-Trajan correspondence); W.H.C. Frend, *Martyrdom and Persecution in the Early Church*, Oxford 1965, ch. 8; G.E.M. de St. Croix, Why were the Early Christian Persecuted?, *P&P* 26 (1963) 6-38; E.J. Bickerman, Trajan, Hadrian and the Christians, *RFIC* 96 (1968) 290-315; J. Molthagen, *Der römische Staat und die Christen im zweiten und dritten Jahrhundert*, Hypomnemata 28, Göttingen 1970; T.D. Barnes, Legislation against the Christians, *JRS* 58 (1968) 32-50; P. Keresztes, *Imperial Rome and the Christians*, 2 vols., New York-London 1989 (bibl.).

L. Navarra

TRANSITUS MARIAE. Title given generically to a group of texts handed down in various languages (Greek, Latin, *Syriac, *Coptic, *Arabic, *Ethiopic) under various titles: *Transitus Mariae, Dormitio Mariae*, etc., and with various attributions: *John the evangelist, *Melito of Sardis, *Evodius of Antioch, etc. These texts seem to be related to each other; the first versions, prob. in Greek, appeared towards the end of the 4th c. Their content turns on the death of the Virgin, and consequently on the *assumption of her *body into heaven. Not all the versions attest this belief: this difference is evidence of its rise and establishment, or otherwise, in different cultural areas.

The common nucleus is the meeting of the *apostles in the place where *Mary was dying, at *Jerusalem; then follows the description of her death, with the intervention of *angels and of Jesus himself; finally the fate of her body; assumption, semi-assumption, or mere burial, as in one important Coptic version (Ps.-Cyril of Jerusalem). Two other Coptic versions, one attrib. to *Theodosius of Alexandria, show clear anti-Chalcedonian theological characteristics, of the Severian (as opposed to Julianine) kind.

The two oldest Greek versions appear to be that attrib. to Leukios Charinos and that attrib. to Melito of Sardis. Both, however, post-date the council of *Ephesus and include the bodily assumption into heaven.

M. Jugie, *La mort et l'assomption de la Sainte Vierge*, ST 114, Vatican City 1944; Erbetta I, 2, 409-632 (418-420, bibliography: texts, versions and studies).

T. Orlandi

TRANSLATIO IMPERII. From pre-Christian times, the historical fortunes of the historiographical model known as *t. i.* can be followed in a relatively uniform way. Of Greek (Trieber) or Oriental (Swain) origin, it refers to the handing over from one people to another of political and military hegemony (the corresponding expression *translatio studii* refers to the handing over of cultural hegemony). The model was very popular among *pagan historians (e.g. Dionysius of Halicarnassus [*Ant. Rom.* I, 2, 2 ff., 3, 3], Appian [*Hist. Rom.* praef. 9 f.], Pompeius Trogus [in Justin, *Hist. Philippicarum Epit.* 41, 1, 1 ff.; 43, 1, 1 f.]), as well as among the Jews (as confirmed by Daniel chs. 1 and 2) and Christians (see, among others, passages in *Jerome [*Comm. in Dan.* 2, 38 ff.] and *Orosius [*Hist. adv. paganos* 2, 1, 1 ff.]). *T. i.* is considered by ancient writers under both its chronological aspect (in every period a single "kingdom", replacing another, assumes the leadership of the "civilized world") and its spatial aspect (many interpreters make it follow a movement from *East to West; though variants to this scheme are not lacking, with the "kingdoms" corresponding to the four cardinal points). In the first centuries of our era, Rome was commonly judged to be the heir to the power of earlier empires (Babylonian, Median, Persian, Macedonian, Carthaginian): its exaltation was tempered, esp. by Jewish and Christian authors, who emphasized its fragility and announced its coming end. Thus an interpretative model of universal history pointed to a trans-historical end, affirming, beyond human kingdoms destined to disappear, the coming of the final kingdom of God.

C. Trieber, Die Idee der vier Weltreiche, *Hermes* 27 (1892) 321-344; J.W. Swain, The Theory of the Four Monarchies. Opposition History under the Roman Empire, *CPh* 35 (1940) 1-21; W. Baumgartner, Zu den vier Reichen von Daniel 2, *ThZ* 1 (1945) 17-22; W. Goez, *Translatio Imperii. Ein Beitrag zur Geschichte des Geschichtsdenkens und der politischen Theorien in Mittelalter und in der frühen Neuzeit*, Tübingen 1958, 4-36 (for the period that interests us here); B. Gatz, *Weltalter, goldene Zeit und sinnverwandte Vorstellungen*, Hildesheim 1967, 106 f. passim.

P. Siniscalco

TRIER

I. Christianity - II. Council.

I. Christianity. Ancient Colonia Augusta Treverorum, administrative centre of the *civitas* of the Treviri, then Civitas Treverorum (4th c.), seat of the praetorian prefecture, capital of the diocese of *Gaul and the province of Belgica Prima; then Urbs Treverica, Treverus (7th c.). Its Christian origins are little known: the *episcopal list, which Duchesne considers a serious document, cites three bishops before *Agricius, who subscribed the council of *Arles in 314. The first prelate, Eucherius, may have occupied the see in the last quarter of the 3rd c., but he is not attested until the 5th c., when bishop Cyril dedicated a *memoria* to him and to Valerius, the second bishop on the list (Gauthier, *Evangélisation*, 11-16). Maternus, cited third on the list, is in fact reliably attested for Cologne (*Colonia*): his presence in the catalogue of T. may be due to the introduction of his relics. The *Vita Eucherii, Valerii et Materni* (BHL 2655), written in the 10th c., has no value.

The Christian history of T. begins in the 4th c., precisely at the moment when the city, for a century an imperial residence, experienced a new monumental and demographic development. Hence the growing role played in ecclesiastical politics by its bishops, close to the centre of power, and the notable progress of *conversions, attested particularly by *epigraphy (Gauthier, *Recueil*, I, 59; id., *Evangélisation*, 84-86). After Agricius (314), *Maximinus (335-before 351), whom *Athanasius knew during his exile at T. (335), played an important role in the negotiations that preceded the council of *Sardica (343); his successor *Paulinus, who also opposed the Arianizing policy of the emp. *Constantius, was expelled from his see for refusing, in a council at *Arles (353), to sign Athanasius's condemnation, and died in exile (358). After an obscure Bonosius (died 375), Britto, who had links with *Ithacius, the Spanish bishop and enemy of *Priscillian, seems indirectly implicated in Priscillian's trial; his successor *Felix, elected in 386 shortly before Priscillian's execution, was exposed until his death to the hostility of a whole party of bishops who condemned the actions of the emp. *Maximus. After the sacks of the city (evoked by *Salvian), there began for T. a period of tumult ending at the end of the 5th c. with its annexation by the *Franks. We know little of bishops Maurice, Legontius (perhaps, acc. to Ewig [*Trier*, 40], mentioned in a *letter of pope *Leo), Severus, a disciple of *Lupus of Troyes, who evangelized the Alamanni, Cyril, known from a funerary poem of his (Gauthier, *Recueil*, I, 59), Iamblichus, who had to leave T. and died at Chalon-sur-Saône (*Cabillonum*). After some completely obscure bishops: Emerus (?), Marus, Volusianus, Miletus, Modestus, attested in *Mart. rom.* (19 Sept) and finally Apricolus, named by *Gregory of Tours (*Vitae Patrum* VI, 3), the 6th c. was dominated by two metropolitans: *Nicetius (525-566/9), a monk who owed his episcopate to king Theodoric I, king of Austrasia, and Magneric, his successor († after 587). Both were great builders: Nicetius restored the episcopal complex and erected a *castellum* at *Mediolanus*; Magneric built St Martin's. Nicetius, an ascetic figure, celebrated as a protector of the weak, oppressed by the Frankish aristocracy, enjoyed a prestige that authorized him to write *Justinian a *letter of reprimand for the condemnation of the "*Three Chapters" and a letter to Clodoswinth, wife of the *Longobard king Alboin, an *Arian, exhorting *Clovis's granddaughter to imitate *Clothilde (CPL 1063). For the 7th c. the episcopal list mentions Guntheric, Sabaudus (council of Paris, 614), Modoald, less obscure (council of Clichy, 627), propagator of *monasticism, acc. to the 11th-c. *Gesta Treverorum*, Numerian (643/7), Basinus and Leothwine, succeeded by his son Milo (723/751), also titular of the see of *Rheims (*Remis*). The evidence of *epigraphy (esp. 4th- and 5th-c.: cf. *Recueil*, pl. VII) and, above all, monumental development show T.'s splendour and then decline: *intra muros*, the episcopal complex is among the most important of the 4th c., with two basilicas (St Peter's and, to S of it, St Mary's: cf. Kempf; and Oswald, 340 f.), a baptistery and the *monasterium S. Mariae* (of Modoald?); outside the walls, in the *cemetery area to N, Felix erected a *memoria* to his predecessor Paulinus, to which another was added before the 6th c., dedicated to St Maximinus; to N-W, St Martin's; and to S, the *memoria* of Eucharius (St Matthias's) founded by Cyril in the 5th c. (Cüppers), not counting some foundations of uncertain date attested by later texts (*Gesta Treverorum* [11th c.]: *ecclesia S. Symphoriani, S. Crucis*). **[Fig: 315]**

Duchesne, *Fastes*, III, 33-39; E. Ewig, *Trier in Merowingerreich. Civitas, Stadt, Bestim*, Trier 1954; T. K. Kempf, Untersuchungen und Beobachtungen am Trierer Dom 1961-1963, *Germania* 42 (1964) 126-141; H. Cüppers, Das Gräberfeld von St. Matthias . . . , in W. Reusch, *Frühchristliche Zeugnisse*, Trier 1965, 165-174 and *TZ* 32 (1969) 269 ff.; F. Oswald, *Vorromanische Kirchenbauten*, Munich 1971, III, 340-350; N. Gauthier, *Recueil des inscriptions chrétiennes de la Gaule*, I, Paris 1975, 1-237; id., *L'évangélisation des pays de la Moselle*, Paris 1980 (bibl.); E. Wightman, *Trier and the Treveri*, London 1970; W. Weber, *Constantinische Deckengemälde aus dem römische Palast unter dem Dom*, Trier 1984; H. Brandenburg, Zur Deutung der Deckenbilder aus der Trierer Domgyrabung, *Boreas* 8 (1985) 143-189.

Ch. Pietri

II. *Council. Despite the reservations of *Ambrose of Milan, *Martin of Tours, *Hyginus of Cordova and many other Western bishops, many Gallic bishops, meeting at T., approved the conduct of *Ithacius of Ossonuba during the case of *Priscillian, and urged the emp. *Maximus to take new measures against Priscillian's partisans. They elected *Felix as successor to Britto of T. Hence the name "Felician *schism" given to this division among the episcopate.

CCL 148, 47-48; Gaudemet 117-123; Palazzini 5, 407.

Ch. Munier

TRIFOLIUS. Presbyter (city unknown), c.520. Only his *letter to the senator Faustus is known and important. Faustus asked him about the doctrine of a certain Scythian monk, who had arrived at *Rome from *Constantinople. The substance of his demand was this: does the (*theopaschite) formula *Unus ex Trinitate passus est* correspond to the doctrine of the *Fathers or not? T. replied in the negative, invoking the *authority of the first four ecumenical *councils.

CPL 655; PL 63, 534-536; E. Schwartz, *Publizistische Sammlungen zur Acacianischen Schisma*, Munich 1934, 115-117.

G. Ladocsi

TRINITY

I. Premises - II. The ante-Nicene period - III. The Nicene dogma and its universal acceptance - IV. The legacy of the Fathers.

I. Premises. There is no doubt that the Christians were distinguished from the start from other believers, *Jews and non-Jews, by the fact that they admitted *converts to the community of Jesus Christ by baptizing them in the name of the *Father, Son and Holy Spirit (Mt 28, 19; *Didachè* 7). The origin of this form of *initiation, based on Christ's own command, is still not yet clear (apropos of this we should bear in mind the exegetical discussions on the apostolic origins of the trinitarian faith). But two things are certain. Firstly, the baptismal *creed had its roots in the Easter exerience of the primitive community, an experience itself founded on the prophetic experience of Jesus, who felt himself so united with the *Father that he could speak in his name and die according to his will for the salvation of mankind, an experience prepared by Israel's faith in the one God, creator of all things and saviour of the just, and by their *hope in the coming of the Messiah and the outpouring of the *Holy Spirit. Secondly, this baptismal creed, based on the experience of the spiritual presence of Jesus of Nazareth, confirmed by God in his *resurrection as Christ and Lord, represented a decisive influence for the whole evolution of trinitarian *dogma and *theology. As *norma normans* (cf. *Regula fidei*) it determined three aspects of this: it ensured the predominance of the Father-Son-Spirit terminology; it showed the order of these three guarantees of Christian *baptism; and it suggested that all three exist equally in the divine sphere.

In fact, the baptismal creed was continually reinterpreted during the whole patristic period, so that *orthodoxy was to a great extent assessed in relation to it. The perspective of this reinterpretation also changed continually. In a first period, ending with the council of *Nicaea (325), it was mainly *apocalyptic, *soteriological and *gnostic. In the second period (325-381), however, dominated by the debates over the Nicene Creed, the context was creationist and politico-dogmatic. Finally, in a sort of epilogue, the baptismal creed was reinterpreted under the influence of the christological disputes and of the concern, greater than ever, to preserve the faith of the *Fathers intact. But, for an evaluation of all this doctrinal evolution, it is not enough to consider the general approach of Christian theology. We must also bear in mind the other factors of the evolution, partly common to the different stages and often connected to each other: the development of Christian *exegesis, the development of *liturgy, esp. that of Christian *initiation (*baptism), which included the *anamnesis of the Easter *mystery (*Sunday - feast of *Easter), the development of Christian *spirituality, with the charismatic experiences of *conversion, *martyrdom and the community's synodal and non-synodal union, the development of theological language brought about especially by the transition from Semitic to *classical culture, and finally the communities' external development, conditioned by their growth and political circumstances. We must also always take account of the difficulties of a hermeneutic that had to distinguish properly between the experience of *faith and its expression, things which interpenetrated each other, and consider attentively the influences of the first formulae on later ones and the interpretation of the former in terms of the latter.

II. The ante-Nicene period. The evolution of trinitarian doctrine during the first three centuries was characterized mainly by the transition of the baptismal *creed from the context of *apocalyptic to that of *Hellenism. Before 150 the apocalyptic perspective, the setting in which the first Christians had confessed and understood the baptismal creed, was still present, more or less so in different areas. Christ and the Spirit, by whose means *God fulfilled his plans of salvation, were thus considered principally under the aspect of the *revelation which made manifest the correspondence of what happened in the church and what existed in heaven (cf. the exegesis of Is 6, the theme of the two paracletes, the heavenly church and the earthly church). Always prevalent, therefore, was the economic consideration (cf. *economy) which insisted on the functionality, the salvific role, of the Son and the Spirit, who lead us from this *world to the *Father. Nevertheless a certain feeling for Christ's pre-existence was not lacking, as is shown by the titles applied to him: Name - Law - *Light - Beginning (cf. Daniélou, *Histoire des doctrines chrétiennes*, I, 195-226), titles which in Judaism connoted heavenly origin, and consequently, in that apocalyptic climate, were easily referred to the risen Christ. Moreover, there were no doubts as to the divine origin of the Spirit, often called prophetic, who however was considered more as a Messianic gift than as a giver of *grace (Martín, *Espíritu Santo*).

Quite soon we see a change of theological approach, marked by greater interest in the divinity of Jesus, now more frequently called *God (cf. *Ignatius), reinterpretation of the messianic titles "Son of God" and "Son of man" (cf. Ignatius) and new *anthropological nuances in pneumatology. However, from the mid 2nd c. the influence of *Hellenism, which had already led to the facts mentioned, became massive. This appears in what is called the soteriological perspective. *Justin and the other Greek *Apologists were concerned about the salvation of all men, even those who had lived before Christ's coming (cf. *I Apol.* 46; *Ep. ad Diogn.* 9). Replying to this question of *cur tam sero?*, they developed the biblical concept of God's Logos, always present in all periods of the world. Since this soteriological universalism required the pre-existence of Christ, they had recourse to the texts of *Genesis (1, 1) and the *Wisdom books (Pr 8, 22 ff.) and consequently linked the idea of the Word still more profoundly to the idea of *creation than it had been in the apocalyptic perspective of the NT (Jn 1, 1-4; Col 1, 15; Heb 1, 1-3). On the other hand, as the Apologists, partly for apologetic motives, reread the biblical theme of the Word with God in the context of *middle Platonism, confession of the Trinity (Τριάς, a word first found in *Theophilus) took on clearly metaphysical nuances. It was understood in the Greek context of being (οὐσία), participation, divine immutability (cf. Daniélou, II, 297-353). This transition to a more philosophical consideration was favoured by the need to explain the liturgical worship of Christ in view of both biblical and philosophical monotheism (cf. Pelikan, *Emergence*, I, 173 ff.). The problems of universal salvation were accentuated in the gnostic context (cf. *Gnosticism, *Soteriology). On one hand the origin of *evil (and of the *persecutions) had to be explained without compromising either human freedom or the unicity of God the creator. On the other, the *resurrection, the salvation of the whole man, was in question. To this set of problems, approached from a still more clearly Hellenistic viewpoint, the theologians of the Great Church at first responded defensively. Against gnostic speculations about emanations, they insisted on the *mystery of God (Iren., *Adv. haer.* III, 19, 2, etc.). Equally they insisted on the reality of the *incarnation, while leaving open the question of how Jesus' true death could be fitted in with the fact of his divinity. Since these anti-gnostic disputes included a major re-evaluation of the apostolic writings, we see at the same time a much wider revival of the trinitarian perspectives of *Paul and *John (cf. esp. *Irenaeus). Since on the other hand the anti-gnostic reaction, reinforced by certain Jewish trends, led to a strongly unitarian tendency, it was necessary to defend the real distinction of Father, Son and Holy Spirit. This was done, leaning heavily on the traditional *Testimonia* of the *Bible, by *Tertullian, *Hippolytus of Rome, *Novatian and the author of *Contra Noetum*. Succeeding quite well in formulating, using technical terms (above all *persona*), the distinction of the divine persons (*distributio*), they were less happy when they expressed the substantial unity in the distinction, being unable to overcome a certain *subordinationist tendency (gradation of persons and a too close link of the persons with the act of *creation).

The taking up of positions, aroused more or less directly by the gnostic movement of the 2nd c., could not remain purely negative, nor at the level of speculations on the origin of the Son and the Holy Spirit. In fact, almost contemporary with Tertullian and Hippolytus, *Origen worked out his Christian gnosis. In this cosmic-soteriological viewpoint, following up the true aspirations of the gnostics, he seems at first sight to put the stress completely on the Logos's mediation between the one God and the multiple *world. In reality, by keeping closely to the *regula fidei* he does not just replace the soul, in other words the world, with the Holy Spirit, but puts the latter clearly on the side of the incorporeal and adorable Trinity, distinct from all creatures (*Princ.* I, 6, 4; II, 2, 2; *Com. Io.* VI, 33, 166). Indeed he does not fail to seek an explanation of the unity of the three *hypostases

(*Com. Io.* II, 10, 75), insisting on the priority of the *Father, beginning (ἀρχή) of all beginnings (*Com. Rom.* 7, 13; *Dial. Heracl.* 3). But he too fails to distinguish sufficiently between the eternal generation of the Son and the temporal creation of rational beings, not to mention the fact that he does not dare give an opinion on the procession of the *Holy Spirit. Thus he bequeathed to posterity a trinitarian theology which made clear the real distinction of Father, Son and Holy Spirit, against any modalist simplification, but which did not formulate with necessary clarity either the equality or the eternal union of the three divine persons. Nor were his *aporiae* given any solution immediately after him (cf. *Dionysius of Alexandria), but rather they led to the Arian crisis.

III. The Nicene dogma and its universal acceptance. Since the great theologians of the 3rd c., despite their untiring efforts, had failed to find a formula in which the unity and the distinction of the divine persons were counterbalanced, still less to explain the unique divinity of Father, Son and Holy Spirit, it must not surprise us to meet, at the start of the 4th c., two opposed tendencies in trinitarian theology, the "pluralistic" and the "unitarian". Later a clash between these two tendencies was inevitable. But it was not immediately frontal. The crisis broke out rather within the former tendency, defended mainly by Origen's disciples, so that we may speak of an *Origenist crisis, or a crisis of Origen's Logos-*christology. For reasons hard to define, *Arius, an Alexandrian presbyter, radicalized the Origenian positions on the priority of the Father in such a way that the Son appeared to be a creature and his origin seemed to be confused with that of the creatures. In the new situation in which the biblical doctrine of *creatio ex nihilo* had obviously made theologians more sensitive to the temporality of the created world, Arius's doctrine of Christ as the first creature could not remain unchallenged (cf. Ricken, *Nikaia als Krisis*).

Arius's bishop *Alexander, also an inheritor of the Origenian theology, put his troublesome presbyter on trial. Indeed, to break his resistance, now supported by influential *friends like *Eusebius of Caesarea, the bishop of *Alexandria allied himself, despite his Origenian preferences, with bishops of a unitarian tendency. Thus, at the council of *Nicaea, convoked in 325 by *Constantine, the first Christian emperor, as well as in the subsequent debates, the two tendencies met head on. It was no longer just a question of Christ, only-begotten or first creature, but increasingly also a question of one or three *hypostases.

The Nicene council itself, referring clearly to the baptismal creed, pronounced in favour of Christ's true divinity, insisting on his sonship (cf. the *Creed [DS 125]: only-begotten, begotten not created, through whom all things were created) and indeed expressing it in technical terms (of the Father's substance, *homoousios* with the Father). The second question of one or three hypostases remained open (cf. the *anathemata: DS 126). The problem of the Holy Spirit was not even touched on. Nor was the dogma of the Son, defined as not created but begotten, sufficiently clarified. These open questions could not remain unanswered. Consequently the bishops and theologians, with the more or less close complicity of the civil authorities, became involved in an embittered controversy over the true meaning of Christian *orthodoxy, now the basis not just of the *communion of the churches but also of the unity of the Roman empire. This struggle was at first fought out on the level of church politics; after Constantine's death (337), it became a search for a confession of faith that could unite both the defenders and the opponents of the Nicene Creed (cf. esp. DS 139 f.). From *c.*350 the debates took on a more strictly theological character. Under the leadership of *Aetius and *Eunomius, the earlier Arian positions were radicalized in a way that almost led to a *rapprochement* between the Nicene theologians, led above all by *Athanasius, and the so-called *homoeousians, under the leadership of *Basil of Ancyra. Reconciliation of the more moderate parties was obtained at the council of *Alexandria (362), which admitted both the formula of the three hypostases and that of the one *ousia* (cf. Simonetti, *Crisi ariana*, 367 f.). On the basis of this agreement, *Basil of Caesarea and the other Cappadocians worked out a logical distinction between *hypostasis and *ousia* which allowed them to arrive at the formula of one *ousia* and three hypostases (no earlier than 375). At the same time, these theologians returned to the question of the *Holy Spirit, already posed by Athanasius in *c.*360. Seeking to remain as faithful as possible to the baptismal creed, Basil eventually demonstrated in his magisterial work *De Spiritu sancto* (375) that the third person was truly part of the T., being substantially distinct from all spirits, giver of all *life and worthy of the same worship as the Father and the Son. Under the leadership of *Gregory Nazianzen and *Gregory of Nyssa, this reinterpretation of the baptismal creed was officially accepted by the council of *Constantinople (381) (DS 150). In the same period, though not involved in this synod, *Damasus of Rome and *Ambrose of Milan induced the Western churches to accept Nicene *orthodoxy, as it had been formed after 360, including the doctrine of the divinity of the Holy Spirit and the explicit exclusion of the *Sabellian positions of *Marcellus of Ancyra (cf. *Tomus Damasi*: DS 152-180, and Ambrose's *De Fide* and *De Spiritu sancto*). The same happened in the Eastern churches outside the Roman empire, though their acceptance of the Nicene belief was less influenced by Greek philosophical culture. Thus towards the end of the 4th c. all the Christian churches, except the Gothic communities, were unanimous in the baptismal creed, as reinterpreted by the councils of Nicaea and *Constantinople and as the religious basis of the unity of the Roman empire (cf. decree of the three emperors: Mirbt no. 310).

IV. The legacy of the Fathers. In the course of the disputes over the Nicene creed, along with the concept of *orthodoxy as the common faith of all the *Catholic churches and basis of the political unity of the empire, the idea was formed of the *authority of the *Fathers (cf. *Argumentation). From *Athanasius on, appeals were made to the Fathers of *Nicaea, and gradually the same authority was attributed to all the bishops and to the theologians, considered as witnesses to the true faith. In this setting, trinitarian theology itself became primarily a theology of the Fathers. This is attested in the *East by the trinitarian writings of *Cyril of Alexandria, the numerous treatises on *theology and *economy, and finally the most famous expression of the theology of the Fathers, *John Damascene's *Expositio fidei*. In the West, *Augustine intended his *De Trinitate* primarily to sum up the whole patristic legacy on this matter (*Trin.* III, pr., 1). He was followed by others, esp. *Fulgentius of Ruspe.

But this fidelity to the doctrinal legacy of the Fathers was not without a certain creativity. In the first place, esp. in the East, a further exploration is visible in the christological perspective of later *patristics. It is true that the christological controversies, which went back to the year 360, were at bottom no more than a logical consequence of the debates over the trinitarian faith. In particular it was necessary to harmonize the terminology adopted for the *mystery of the T. (οὐσία - ὑπόστασις) with that of *christology. But all these efforts to formulate belief in Jesus' true humanity and the true unity of Christ, God and Man, by means of a recognized trinitarian terminology, could not fail to have repercussions on trinitarian theology itself. By accepting the christological adage *Unus ex Trinitate passus est*, the Byzantine and Latin theologians definitively ruled out any *patripassianism. Discussing the significance of Christ's one hypostasis, as defined by the council of *Chalcedon (451), *Leontius of Byzantium and other Greek theologians supplemented Basil's concept of ὑπόστασις χαρακτηριστική with the definition of *hypostasis as καθ᾽ ἑαυτὸ εἶναι. Though convenient for the explanation of the *hypostatic union, this definition, in which hypostasis was understood as the subject of individual properties and at the same time as subsistence, created difficulties with regard to the T. It was necessary to avoid speaking of three subsistent beings, i.e. *tritheism. So the triple καθ᾽ ἑαυτὸ εἶναι, *qua* common being, was identified with the divine essence. All these further developments, made under the influence of the *Aristotelianism of the Alexandrian *neoplatonists, did not remain unknown in the West. In a very similar christological and philosophical context, it was *Boethius who made a particular contribution to the evolution of trinitarian theology, both by his definition of *persona* as *naturae rationalis individua substantia* and by his adage *est relatio quae multiplicat trinitatem*.

No less original was the contribution of *Augustine in this later period, above all in his *De Trinitate*. Written in a spirit of spiritual inquiry (cf. *Trin.* X, 19: on *exercitatio mentis*), this most theological of Augustine's works did not just formulate satisfactorily the trinitarian dogma inherited from the Fathers, but, by his doctrine of *homo interior imago trinitatis*, developed earlier trinitarian theology in a totally personal way. Starting from the one *God, Augustine tries to define and explain the *aequalitas personarum*. Leaning primarily on *Scripture, in the first (mainly dogmatic) part he presents the content of the trinitarian dogma, considered especially under the aspect of the indivisibility of all three divine persons (books I-IV); then he discusses the dogmatic terminology (*relatio, essentia, persona*, etc.) (books V-VII). After a transition (book VIII), in the second part he tries to understand the *aequalitas personarum* in the one deity, above all by developing man's inner life as an *image of the T. (books IX-XV). But it would be mistaken to see in the more speculative part only the so-called psychological theory of the T. (cf. Schmaus, *Trinitätslehre*). In it, in fact, Augustine demonstrates, with a clarity not attained before him, the correspondence between the economic T. and the immanent T. Thus, basing himself primarily on *John, he provides in particular the theological bases for the later dogmatization of the *Filioque. But his reinterpretation of the baptismal *creed (cf. his *De fide et symbolo*) was not merely decisive for this dogmatic detail, but determined the whole future of Latin trinitarian theology.

1. For bibliography: cf. the numerous suggestions in: *Estudios Trinitarios* (Publicación del Segretariado Trinitario, Salamanca) 1 (1967), and ff.; V. Venanzi, *Dogma e linguaggio trinitario nei Padri della Chiesa. Un panorama bibliografico, Augustinianum* 13 (1973) 425-453.
2. General presentations: *DTC* 15, 1545-1702; G.L. Prestige, *God in Patristic Thought*, London ²1952; L. Scheffczyk, Lebramtliche Formulierung und Dogmengeschichte der Trinität, *MySal* II, Einsiedeln 1967, 146-220 (best Catholic presentation, bibl.); J. Pelikan, *The Emergence of the Catholic Tradition*, I, Chicago 1971; B. de Margerie, *La Trinité chrétienne dans l'histoire*, Paris 1975; H. Saake, Pneuma, *PWK* Suppl. 14 (1975) 387-412; A. Grillmeier, *Jesus der Christus im Glauben der Kirche*, I, Freiburg 1979; G.C. Stead, *Divine Substance*, Oxford 1977; B. Studer, *Gott und unsere Erlösung im Glauben der Alten Kirche*, Düsseldorf 1985; R.P.C. Hanson, *The Search for the Christian Doctrine of God*, Edinburgh 1988; id., The Doctrine of the Trinity achieved in 381, *Scottish Journal of Theology* 36 (1983) 41-58; id., The Transformation of Images in Trinitarian Theology, *SP* 17 (1982), ch. 12; T.F. Torrance, *The Trinitarian Faith, The Evangelical Theology of the Ancient Catholic Church*, Edinburgh 1988; A. de Halleux, Personalisme ou essentialisme trinitaire chez les Pères cappadociens?, *RThL* 17 (1986) 129-155.
3. Ante-Nicene period: J. Lebreton, *Histoire du dogme de la Trinité*, II, *De s. Clément à s. Irénée*, Paris 1928; G. Kretschmar, *Studien zur frühchristlichen Trinitätslehre*, Tübingen 1956; A. Orbe, *Estudios Valentinianos*, 6 vols., Rome 1956-1966; J. Daniélou, *Histoire des doctrines chrétiennes avant Nicée*, 3 vols., Paris 1958-1961-1978; M. Lods, *Précis d'histoire de la théologie chrétienne du début au 4ᵉ siècle*, Neuchâtel 1966; F. Ricken, Nikaia als Krisis des altchristlichen Platonismus, *Theologie und Philosophie* 44 (1969) 321-341; J.P. Martín, *El Espíritu Santo en los orígenes del cristianismo*, Zürich 1971; W.D. Hauschild, *Gottes Geist und der Mensch*, Munich 1972; B. Studer, Der Person-Begriff in der frühen kirchenamtlichen Trinitätslehre, *Theologie und Philosophie* 57 (1982) 161-177.
4. Nicene period: A.M. Ritter, *Das Konzil von Konstantinopel und sein Symbol*, Göttingen 1965; E. Boularand, *L'hérésie d'Arius et la "foi" de Nicée*, Paris 1972; M. Simonetti, *La crisi ariana nel quarto secolo*, Rome 1975 (in bibl., earlier studies by the same author); A.M. Ritter, Arianismus, *TRE* 3, 692-719; M. Simonetti, Ancora su Homoousios a proposito di due recenti studi, *VetChr* 17 (1980) 85-98.
5. Later patristics: V. Schurr, *Die Trinitätslehre des Boethius im Lichte der skytischen Kontroversen*, Paderborn 1935; R. Roques, *L'univers dionysien*, Paris 1954 (cf. *DHGE* 14, 265-310); S. Otto, *Person und Subsistenz. Die philosophische Anthropologie des Leontius von Byzanz*, Munich 1968; D.B. Evans, *Leontius of Byzantium*, Washington 1970; L. Scipioni, *Nestorio e il concilio di Efeso*, Milan 1974; A. Basdekis, *Die Christologie des Leontius v. Jerusalem*, Münster 1974; B. Brons, *Gott und die seienden. Untersuchung zum Verhältnis von neuplatonischer Metaphysik und christlicher Tradition bei Dionysius Areopagita*, Munich 1975; B. Studer, Soteriologie in der Schrift und Patristik, *HDG* 3/2ª, Freiburg 1978, 175-223.
6. Augustine: M. Schmaus, *Die psychologische Trinitätslehre des hl. Augustinus*, Münster 1927-²1967; L. Maier, *Les missions divines selon s. Augustin*, Fribourg 1960; A. Schindler, *Wort und Analogie in Augustins Trinitätslehre*, Tübingen 1965; O. du Roy, *L'intelligence de la foi en la Trinité selon s. Augustin*, Paris 1966; J. Verhees, Heiliger Geist und Gemeinschaft bei Augustinus von Hippo, *REAug* 23 (1977) 245-264 (where cf. other studies by the same author); E. Bailleux, L'Esprit du Père et du Fils selon s. Augustin, *Rev. Thomiste* 85 (1977) 5-29 (cf. other studies by the same author); B. Studer, Augustin et la foi de Nicée, *REAug* 30 (1984); id., Credo in Deum Patrem omnipotentem, Zum Gottesbegriff des Heiligen Augustins, *Congresso Internazionale su S. Agostino nel XVI centenario della conversione*, Studia Ephemeridis Augustinianum 24, 3 vols., Rome 1987, I, 163-184; F. Genn, Amt und Trinität bei Augustinus, *ibid.*, II, 158-168; E. Bromuri, Le analogie trinitarie di S. Agostino tra psicologia e mistica, *ibid.*, II, 169-186; H.R. Drobner, *Person-exegese und christologie bei Augustin*, Leiden 1986. Cf. also the bibliographies for individual Fathers and esp. the introductions to their trinitarian works in recent editions.

B. Studer

TRISAGION. The T. - "Holy God, holy and mighty, holy and immortal, have mercy on us!" - seems to have been introduced into the Eastern *liturgies after 450. *John Damascene claims that it was revealed to St *Proclus (*De fide orth.* III, 10: PG 94, 1021). It was recited before the readings, except in the Egyptian liturgy, where it was put before the gospel. Both *Antioch and *Constantinople had a fixed formula, against attempts at modification by the addition of ὁ σταυροθεὶς δι' ἡμᾶς. The Egyptian liturgies, however, esp. that of the *Jacobites of *Ethiopia, have a formula that assimilates elements of the *Creed.

The T. was introduced into the *Gallican liturgies before the readings, as in the *Byzantine liturgy, and before the gospel, as in the Egyptian liturgies. The *expositio brevis antiquae liturgiae gallicanae* calls it simply *"Aius"* and explains why it is sung both in Greek and in Latin. The *Hispanic liturgy, however, which recited it only on solemn feasts, used Latin exclusively, with additions similar to those of *Egypt. The Roman liturgy did not accept the T. until the 11th c., and then only for the adoration of the cross on Good Friday (*Ordo Romanus* 31). The T. did not reach *Rome directly from the *East, but arrived through the Gallican liturgies.

J. Quasten (ed.), *Expositio brevis antiquae liturgiae gallicanae*, Münster 1934; id., Oriental Influence in the Gallican Liturgy, *Traditio* 1 (1943) 55-78; J. Mateos, La célébration de la parole dans la liturgie byzantine, *OCA* 191 (1971) 91-126.

A. Chupungco

TRITHEISM. During the patristic era, the charge of defending a t. (τριθεία) or of being a tritheist (τριθείτης) was made against various theological tendencies and for various reasons (Lampe 1408; Isid., *Etym.* 8, 5, 68). *Dionysius of Rome accused certain anti-*Sabellians of breaking up the monarchy into three *hypostases and thus speaking of three gods (DS 112). The Eunomians and *Macedonians accused the Cappadocians of t. because they recognized the divinity of the Son and the *Holy Spirit and failed to safeguard *God's one *ousia* (*Ep.* 38 attrib. to Basil; Greg. Nyss., *Tres dii*; Greg. Naz., *Or.* 31, 13). This criticism was made later against *Nestorius (Leontius of Jer., *Adv. Nest.* 2, 6) and, with more reason, against *John Philoponos (Ps.-Leont., *Sect.* 5, 6), because they confused nature and hypostasis too much.

LTK 10, 365 (bibl.). See also *Monophysism.

B. Studer

TROIANUS of Saintes. Bishop of Saintes (*Santonae*) at the start of the 5th c. We have a *letter from him to bishop Eumerus of Nantes, present at the *council of Orléans in 533, which dealt with a case of conscience concerning *baptism. He must have died in 432. *Gregory of Tours praises him. Feast 30 Nov. Two places bear his name.

CPL 1074; PL 67, 995; MHG *Epist.* 3, 437; CCL 157, 489; *BS* 12, 678-679 (bibl.).

A. Hamman

TROPAEUM TRAIANI (Adamclisi, district of Constanta, Romania). A city founded by the emp. *Trajan near the triumphal monument of that name (109), rebuilt from the foundations by *Constantine the Great and *Licinius (313-316), restored under *Anastasius and *Justinian, abandoned and fallen into ruin after 600; excavations directed by Gr. Tocilescu in 1891-1909 revealed five Christian basilicas of the 4th-6th cc., four inside the walls and one in the *cemetery on the slope of the hills N of the city.

The "simple" basilica (A), *c.*50 m. from the E gate, N of the *via principalis*, is well-proportioned (30 x 17.6 m.), with three aisles, tripartite narthex, semicircular apse to E, atrium to W discovered in 1971-73, as was the crypt under the presbytery (2.70 x 2.30 m.; height *c.*2.50 m., barrel vault). Attached to the S side is a five-roomed building; other annexes to N. Date: time of Anastasius (491-518).

The transept or T-shaped basilica (D), the only one of this type in *Scythia Minor, opposite the "simple" basilica and S of the *via principalis*, has an inner length of 33.80 m.: in the apse, a semicircular *synthronos*; between this and the wall of the apse, a small passage (*deambulatorium*); opposite the *synthronos*, very probably under a ciborium, an altar with a small crypt for relics; a narthex divided into three parts corresponding to the aisles of the basilica; to W, connected by two steps, an atrium with triportico and a small inner courtyard (8 x 6 m.) with brick floor. Date: 6th c.

The "marble" basilica (B), so-called because of the great quantity of marble used in its construction or in a restoration at the time of Justinian (527-565), has a triporticoed atrium to W communicating with the narthex, 0.60 m. above the level of the atrium, by two entrances corresponding to the N and S porticos. A third entrance, on the short (S) side of the narthex, was preceded by a fine *propylon*. Three stone steps linked the narthex to the aisles, on a higher level. The presbytery, beneath which no crypt has been found, was prolonged beneath the central aisle as far as the enclosure of the *cancelli*, whose foundations have been found. A small, elegant baptistery rose S of the atrium.

S of the main street, *c.*30 m. from the city's great W gate, are the ruins of the "cistern" basilica (C), so-called because it rose over a 2nd/3rd-c. Roman cistern. Dimensions of the basilica: 19.7 x 6.80 m.; towards the E end, under the presbytery, a small crypt for relics was built later (6th c.). A platform paved with stone slabs, discovered E of the basilica, was probably the inside of the atrium. **[Fig: 316]**

I. Barnea, *Les monuments paléochrétiens de Roumanie*, Vatican City 1977; id., *Christian Art in Romania*, 1, Bucharest 1979; N. Duval, *Revue archéologique*, 2, Paris 1980, 333-336.

I. Barnea

TROPHIMUS of Arles. *Innocent I claimed for *Rome the *evangelization of the West: *Italy, *Gaul, *Spain, *Africa and Sicily (PL 20, 552). He was echoed by the inhabitants of *Arles under *Patroclus (412-426), who claimed that their first bishop, T., had been sent to Arles from Rome - a claim taken up on his own account by *Zosimus (417-18). In c.450 the bishops of the province of Arles even made T. a disciple of the apostle *Peter and the teacher of three other Gallic bishops: *Paul of Narbonne, *Saturninus of Toulouse, Daphnus of Vaison. This was the starting-point of the apostolic legends. *Gregory of Tours, however, placed T. in the 3rd c. with the other six missionaries sent from Rome and did not endorse the claims of Arles, which nevertheless continued to prosper in Provence, where T. appeared accompanied by the three *Maries and a whole troop of bishops to conquer Gaul from *paganism after a bitter struggle. We may accept as true that T. was bishop of Arles around, or rather before, the middle of the 3rd c.

Duchesne, *Fastes* I, 253-54; E. Griffe, *La Gaule chrétienne à époque romaine*, 1, Paris ²1964, 104-8; *BS* 12, 665-672.

V. Saxer

TROPICI. So called by *Athanasius (*Serap.* 1, 10), refuting their doctrine for the benefit of *Serapion who had brought it to his notice, because of their way of interpreting *Scripture (according to τρόποι, figures). The t., active in *Egypt, maintained, on the basis of Amos 4, 13, Zech 1, 9 and 1 Tim 5, 21, the non-divinity of the *Holy Spirit - who would otherwise have been another Son - and hence its creaturality, though as an *angel superior to the others. As regards the doctrine of the Son, they had passed from *Arianism to *orthodoxy (*Serap.* 1, 1). Some have tried to identify the t. with the *pneumatomachi* mentioned by *Hilary (*Trin.* XII, 55-56).

A. Laminski, *Der Heilige Geist als Geist Christi und Geist der Glaübigen*, Leipzig 1969, 30 ff.; Simonetti 365-366.

F. Cocchini

TURIBIUS of Astorga. Bishop of Astorga (*Asturica*), mid 5th c. A committed anti-Priscillianist, he wrote a series of accounts of the activities of this group in his time and sent them to *Leo the Great, whose detailed replies are preserved in the Hispana *Canonical Collection (*Decret.* 61). All that remains of T. himself is a *letter to the Gallaecian bishops Idatius and Ceponius, on the *apocrypha used by the *Priscillianists. It appears from the letter, whose authenticity is not beyond doubt, that he was much-travelled. It is unfortunately impossible to establish whether his cult, perhaps local and late, is really his or has been confused with bishop Turibius of Palencia (6th c.).

CPL 564; PL 54, 693-695; L. Alonso Luengo, *Santo Toribio de Astorga*, Madrid 1939; *DHEE* 4, 2575; H. Chadwick, *Priscillian of Avila: the Occult and the Charismatic in the Early Church*, Oxford 1976, 208 ff.

M. Díaz y Díaz

TURIN

I. Christian origins - II. Council.

I. Christian origins. The Roman Augusta Taurinorum may have been built on the old Taurasia, capital of the Taurini. Christianity arrived there in the 3rd c.; the first signs of Christian presence are a group of *martyrs: Adventor, Solutor and Octavius (Max. of T., *Serm.* 12: CCL 23, 41 f.). But the first bishop was *Maximus (late 4th-early 5th c.), whose preaching gives us an idea of the importance of his community and the difficulties it met amidst a very strong *pagan presence, esp. in the countryside. Before him the territory of T. was dependent on the see of Vercelli (*Vercellae*). Maximus took part in the synod of T. in 398 (cf. *Serm.* 21 and 58); he died after 412 (cf. *Serm.* 30 and 31). Other important bishops were: *Maximus II (cf. AA.VV. *Congrès*, 502-509), who attended the synods of *Milan in 451 and *Rome in 465 and preached an address at Milan for the inauguration of the *ecclesia maior*; Victor (end of 5th c.), a *friend of *Epiphanius of Pavia; Ursinus (c.562-c.609), under whom the diocese was dismembered, despite his own opposition and that of *Gregory the Great (*Epp.* IX, 214: MGH *Epist.* II, 200; IX, 226: *ibid.* II, 217). In c.570 T. came under the rule of the *Longobards, who made it an important duchy.

EC 12, 318-319; F. Lanzoni, *Le diocesi d'Italia*, II, Faenza 1927, 1044-1050; AA. VV., *Congrès archéologique du Piémont (1971)*, Paris 1978 (important articles); G. Casiraghi, Il problema della diocesi di Torino nel Medioevo, *Bollettino storico-bibl. subalpino* 75 (1977) 405. 534; F. Bolgiani, La penetrazione del cristianesimo in Piemonte, *Atti del V Congresso naz. di archeol. cristiana*, I, Rome 1982, 37-61.

A. Di Berardino

II. *Council. Called in 398 by bishop *Simplicianus (*Ambrose's successor) at the request of the Gallic bishops, the council resolved conflicts dividing the churches of *Gaul. *Felix of Trier resigned, ending the Felician *schism. The council also settled disputes between Proculus of Marseilles and his suffragans (can. 1), the bishops of *Arles and *Vienne over the primacy (can. 2) and the Martinians of *Tours and bishop Bricius.

CCL 148, 52-60; Gaudemet 133-145; Hfl-Lecl 2, 129-134; Palazzini 5, 345-346; L. Duchesne, Le concile de Turin, *Revue Hist.* 87 (1905) 178-302; E. Griffe, *La Gaule chrétienne à l'époque romaine*, I, Paris ²1964 (cf. index); id., La date du concile de Turin (398 ou 471), *BLE* 74 (1973) 289-295.

Ch. Munier

TYANA

I. City - II. Council.

I. City. T. (now Kemerhishar) in *Cappadocia was known above all as the birthplace of the famous *Apollonius, philosopher and theurge, a sort of *pagan double of Christ, whose life was related by Philostratus in the 3rd c. An important *statio* on the main road to *Syria, this ancient city was the capital of a Hittite kingdom in the 2nd millennium BC and a flourishing commercial centre during the Persian empire. Acc. to Strabo (XII, 2, 7, 538C), it and Mazaca (later *Caesarea) were the only cities in Cappadocia worthy of the name, and it was deeply Hellenized in an essentially rural country generally little touched by *Hellenism (cf. Philostr., *Vita Apoll.* I, 4).

Bishops: Eupsychius (at *Nicaea), Theophronius (at *Antioch in 341), *Anthimus (372, enemy of *Basil), Etherius, *Theodore (†404), Callopius, Longinus, Theodore II, Eutherius (at *Ephesus), Patricius (448 at *Constantinople), Cyrus, Cyriac, Paul (536), Euphratas (553), Justin (680), Paphnutius.

PWK 7, A2, 1630-1642; A.H.M. Jones, *The Cities of the Eastern Roman Provinces*, Oxford ²1971, 177 ff., 181 ff., 185 ff.; for the bishops cf. M. Le Quien, *Oriens Christianus*, I, Paris 1740, 395 f.; P.B. Gams, *Series Episcoporum*, Graz 1957 (repr.), 440.

M. Forlin Patrucco

II. *Council. In 366 many *homoeousian bishops from *Syria and the regions of *Asia Minor near Syria, incl. *Eusebius of Caesarea (Cappadocia), *Athanasius of Ancyra, *Pelagius of Laodicea and Gregory of Nazianzus, father of the famous orator, met at T. They approved the actions of the homoeousian delegation (*Eustathius of Sebaste, *Silvanus of Tarsus, *Theophilus of Castabala) which, sent to pope *Liberius in the West for help and solidarity against the persecution of *Valens, had been persuaded to subscribe the Nicene *homoousios*. They arranged that all the Eastern bishops should receive a copy of the *letters (from Liberius and the bishops of *Gaul, *Africa and *Sicily) which the delegates had brought with them. To ratify their *communion with all the Western bishops, they proposed a new council at *Tarsus in *Cilicia, but this was never held.

Hfl-Lecl 1, 979; Simonetti 398.

M. Simonetti

TYRE

I. City and Christian origins - II. Councils - III. Archaeology.

I. City and Christian origins. Tyrus (now Sûr in Lebanon) was an old Phoenician city whose inhabitants in the 8th c. BC had founded many colonies on the coasts of the W Mediterranean, including *Carthage and *Hippo. It was conquered by Alexander the Great in 332 BC; in 64 BC Pompey assigned it to the Roman province of *Syria. *Septimius Severus, whom its inhabitants had supported in his struggle against Pescennius Niger, embellished it with monuments and raised it to the status of a colony, sending veterans there. It was an important commercial centre, esp. because of its port and its famous glass and purple factories. *Jerome calls it the finest city in Phoenicia (*Comm. in Ezech.* 26, 7; 27, 2: PL 25, 241 ff.; 246 f.). It fell under Arab rule in AD 638.

T. is named many times in the gospels, usually together with Sidon (Mt 11, 21 f.; 15, 21; Mk 3, 8; 7, 24; Lk 6, 17; 10, 13). *Paul, embarking at T. in c.57, found Christians already there (Acts 21, 4-7). Acc. to the *Clementines*, *Peter preached there (*Hom.* 3, 58 ff.: PG 2, 147 f.). Its first known bishop was Cassius, who took part in the *council of 190 on the problem of *Easter (Euseb., *HE* V, 25); the mid 3rd-c. Marinus is mentioned by *Dionysius of Alexandria (Euseb., *HE* VII, 5, 1). *Origen died at T. (Jerome, *De vir. ill.* 54; Photius, *Bibl.*, cod. 118; cf. Epiph., *Pan.* LXIV, 3, 6: PG 41, 1074); acc. to Jerome, *Methodius of Olympus was bishop there (*De vir. ill.* 83). During *Diocletian's *persecution T. had several *martyrs, incl. bishop Tyrannius, Theodoxia (Euseb., *HE* VIII, 13, 3) and Ulpianus (Euseb., *De mart. Pal.* 5, 1). *Eusebius himself was present at the torture of many Christians (*HE* VIII, 7); a letter from the emp. *Maximinus Daia, praising the inhabitants for having issued decrees against the Christians, was set up at T. (*HE* IX, 7). The cultivated priest *Dorotheus was appointed by an emperor (perhaps Diocletian) as procurator of the imperial purple-dye works, and Eusebius had "heard him in church thoughtfully expounding the *Scriptures" (*HE* VII, 32, 4). The Christian community of T. must have had considerable numbers and wealth if, after the great persecution, it was able to build a great church. In the inaugural address that Eusebius pronounced for the occasion (*HE* X, 4, 2-72), he describes the church, which he calls the finest in Phoenicia (*HE* X, 4, 1). At this period the bishop of T. was *Paulinus, previously a priest of *Antioch, to whom Eusebius dedicated book X of his *Historia ecclesiastica* (cf. X, 1, 2). Bishop Zeno was present at the council of *Nicaea in 325; he was succeeded by Paul. *Athanasius was condemned at the synod of T. in 335. Also from T. came Edesius and *Frumentius, who landed by chance in *Ethiopia and found Christians there. Frumentius, consecrated by Athanasius in c.330, was the first bishop of Ethiopia (Ruf., *HE* 9: PL 21, 478-480). In the mid 4th c. the Arian Uranius was bishop of T. (Athan., *De synod*. 12: PG 26, 702; Socr., *HE* II, 41; Epiph., *Pan*. LXXXIII, 23); his successor Zeno II, ordained by *Meletius of Antioch, took part in the synod of *Tyana in 365 (Soz., *HE* VI, 12) and that of *Constantinople in 381 (Mansi 3, 568). Photius, elected bishop in 448, was a firm opponent of *Nestorianism and took part in the *latrocinium* of *Ephesus and the council of *Chalcedon. Eusebius of T. was present at the council of *Constantinople of 553, partly at the invitation of pope *Vigilius.

PWK 26, 1876-1908; *EC* 12, 135-137; *CE* 14, 355 f.; W.B. Fleming, *The History of Tyre*, New York 1915; G. Bardy, Sur Paulin de Tyr, *RSR* 2 (1922) 35-45; N. Jidejian, *Tyre through the Ages*, Beirut 1969; A.H.M. Jones, *The Cities of Eastern Roman Provinces*, Oxford ²1971. For the bishops: P.B. Gams, *Series episcoporum*, Regensburg 1873, 434.

A. Di Berardino

II. *Councils. *335*. The anti-Nicene reaction that developed in the *East after 327 struck at *Athanasius in 335. The *Eusebians used as pretext some violence ventured by Athanasius's followers in the struggle against the *Melitian schismatics and, having *Constantine on their side, convoked a council at T. at which Athanasius was obliged by a *letter from the emperor to present himself. Some 60 Eastern bishops took part, and some Egyptian ones who came with their patriarch. Among Athanasius's enemies were *Eusebius of Caesarea, *Eusebius of Nicomedia and *Flacillus of Antioch, who presided. To obtain confirmation of the charges brought against Athanasius by the Melitians, a commission of inquiry consisting of six anti-Athanasians was sent to Mareotis. The Egyptian believers protested in vain against the partiality with which the commission carried out its work. Athanasius then fled secretly from T. to appeal directly to the emperor. In his absence the council condemned him for violence and insubordination and deposed him (Sept 335). On 30 Oct Athanasius presented himself before the emperor at *Constantinople, but Constantine confirmed the condemnation and exiled him to *Trier.

448. In the wake of the Nestorian controversy, *Theodosius II ordered *Ibas of Edessa, who had written a letter to *Maris strongly criticizing *Cyril's actions, to be judged by a council which met at T. and Berytus. Ibas denied being a Nestorian and was acquitted.

513/515. At an uncertain date between these two years, *Severus of Antioch called at T. a council of Syrian bishops of *monophysite tendency, incl. *Philoxenus of Mabbug. The council gave a clearly pro-monophysite interpretation of *Zeno's *Henoticon*. This interpretation aroused opposition both among the *Henoticon*'s more moderate supporters and among the radical monophysites, who repudiated the formula *tout court*.

518. A small council held on 16 Sept approved the anti-monophysite decisions taken the previous July by the council of *Constantinople.

335: Hfl-Lecl 1, 659-666; Simonetti 124-128. *448*: Hfl-Lecl 2, 493-498. *513/515*: Palazzini 5, 431-432; K.M. Girardet, *Kaisergericht und Bischopsgericht*, Bonn 1975. *518*: Hfl-Lecl 2, 1049; W.H.C. Frend, *The Rise of the Monophysite Movement*, Cambridge ²1979, 226-228.

M. Simonetti

III. *Archaeology. At the limit of the modern city of Sûr has been found a monumental arch with a Christian chapel, once ornamented with wall *mosaics representing the Virgin and saints. Many Christian *sarcophagi from T., though they cannot rival the *pagan ones in wealth of figures, remain important evidence of Christianity. Many Christian funerary inscriptions indicate the deceased's occupation: of two deacons, one was a carpenter and the other a goldsmith; a subdeacon sold cloth; of two cantors, one was employed at the church of the Virgin Mary, the other at the church of Old St Mary's. A consecrated *virgin also appears. One invocation is singular: "O Lord who hast tasted death, grant rest" (Jidejian, *Tyre*, 118). From T. comes a terracotta with a representation of the "*cross personified" (Bagatti, *Church from Circumcision*, 222 f.). From the territory near the city comes the well-known pavement mosaic of the church of St Christopher at Qabr Hiran, with scenes of real life, seasons, etc., impregnated with theological and liturgical significance.

N. Jidejian, *Tyre through the Ages*, Beirut 1969; B. Bagatti, *The Church from the Circumcision*, Jerusalem 1971; J.-P. Rey-Coquais, *Inscriptions grecques et latines dans les fouilles de Tyr (1963-1974)*, I: *Inscriptions de la nécropole*, Paris 1977; D. Feissel, Notes d'épigraphie chrétienne III, *BCH* 102 (1978) 545-555; D. Feinel, L'Évêque, titres et fonctions d'après les inscriptions grecques jusqu'au VIIe siècle, *Actes XIe Cong. internat. d'arch. chrét.*, CEFR 123, Rome 1989, 801-826.

B. Bagatti

U

ULFILA. Of Cappadocian origin and Christian family, born *c*.311, as a child he and his parents were taken beyond the Danube by raiding *Goths. Keeping his Christian faith, he was ordained bishop at *Constantinople in 341 by *Eusebius of Nicomedia and dedicated himself to spreading *Arian Christianity among the Goths. For seven years he worked N of the Danube; then a persecution by the still *pagan Goths forced him and his Gothic *converts to seek refuge in *Moesia, within the borders of the empire. In 360 he attended the council of *Constantinople and subscribed the *homoean (moderate Arian) formula of faith. He died in 383 while travelling to Constantinople to intercede for *Palladius of Ratiaria.

Despite recent attempts to minimize it, his evangelizing activity among the barbarians was of prime importance. To sustain it adequately on the cultural level, U. created an alphabet adapted to the language and translated the *Scriptures into Gothic, a work that proved fundamental for culture as well as religion. We know that he also wrote in Greek and Latin, but all that survives is a profession of faith. Modern scholars generally consider him a moderate Arian, on the strength of his subscription to the formula of 360; but this formula served only to support a radical Arianism (*anomoeism) directly based on that of *Eunomius. This form of doctrine was transmitted by U. to the Western Arianism active in the late 4th and early 5th cc. and to the Germanic peoples who became Christian.

CPL 689; SCh 267; *PWK* 11A, 512-531; *Patrologia* III, 88-90; J. Zeiller, *Les origines chrétiennes dans les provinces danubiennes de l'Empire romain*, Paris 1918; G. Friedrichsen, *The Gothic Version of the Gospel*, Oxford 1926; M. Simonetti, L'arianesimo di Ulfila, *RomBarb* 1 (1976) 297-323; H. Schäferdiek, Wulfila: vom Bischof von Goten zum Gotenbischof, *ZKG* 90 (1979) 253-292; M. Forlin Patrucco and S. Roda, Religione e cultura dei Goti transdanubiani nel IV-V secolo, *Augustinianum* 19 (1979) 167-187.

M. Simonetti

UNUS EX TRINITATE PASSUS EST. Basing itself on *Cyril of Alexandria's 12 *anathemata (DS 263), the *monophysite tradition introduced the christological formula *unus ex* (or *de*) *Trinitate passus est*, which, for its opponents, signified nothing but *theopaschism, of which the *dyophysites had accused Cyril and his followers. It was accepted only slowly by the Chalcedonians and, under the influence of *Dionysius Exiguus (†*c*.545) and the Scythian monks, gained acceptance in the Latin church too (DS 401).

V. Schurr, *Die Trinitäslehre des Boethius im Lichte der skytischen Kontroversen*, Paderborn 1935; A. Grillmeier, Vorbereitung des Mittelalters, *CGG* 2, 791-839.

B. Studer

URANIUS, presbyter. A disciple of *Paulinus of Nola, in 431 he wrote a *letter on the death of his teacher; it was addressed to a certain *Pacatus who wished to illustrate Paulinus's life in verse (cf. *Ep.*, par. 1: PL 53, 860).

CPL 207; PL 53, 859-866; Y.-M. Duval, *La Gaule jusqu'au milieu du Ve siècle*, Paris 1971, n. 298, pp. 700 f.; A. Pastorino, Il De obitu sancti Paolini di Uranio, *Augustinianum* 24 (1984) 115-141.

S. Zincone

URBANUS of Sicca. Early 5th-c. bishop of *Sicca Veneria, successor of *Fortunatian (Aug., *Epp.* 148, 1; 149, 3, 34). It was he who gave *Augustine information on Darius, court magistrate (Aug., *Ep.* 229, 1), who in turn praised U. to Augustine (Aug., *Ep.* 230, 1). U. found himself mixed up in the affair of *Apiarius, priest of Sicca, restored by pope *Zosimus after having been deposed in *Africa. The *pope's gesture led to mutual embassies and tensions between the see of *Rome and the African church.

DTC 15, 2307-2312; *PAC* 1, 1232 f.

E. Romero Pose

URBS BEATA JERUSALEM. *Hymn of rare beauty, of unknown authorship (6th to 8th c.). It consists of eight strophes. Its original trochaic metre was changed to iambics at the time of Urban VIII (1623-1644). It thus lost some of its effect, a little rough, but no less robust for that. It is read at Vespers of the *Dedication of the church, and celebrates the praises of the heavenly *Jerusalem, figure of the earthly church. In the restored rhythm of the new breviary, it begins: *Urbs Jerusalem beata . . .*

Dreves u. Blume, *Analecta hymnica Medii Aevi*, Leipzig 51 (1908) 110-112; A. Lentini, *Hymni instaurandi Breviarii romani*, Vatican City 1968, 143 (with comm.); J. Szövérffy, *Die Annalen der latein. Hymnendichtung*, Berlin 1964, I, 151-154; H. Rahner, *Mater Ecclesia* (Hymns of praise in the church, drawn from the first thousand years of Christian literature), Milan 1972; P.W.L. Walker, *Holy City, Holy Places? Christian attitudes to Jerusalem and the Holy Land in the Fourth Century*, Oxford 1990.

L. Dattrino

URSACIUS of Singidunum. See **VALENS of Mursa.**

URSINUS, antipope. The oldest source for the origin of the Ursinian *schism is anti-Damasian (*Coll. Avel.* 1: CSEL 55, 1-4), but substantially agrees with the facts given by *Ammianus Marcellinus and *Jerome. Immediately after the death of pope *Liberius (24 Sept 366), some priests and deacons elected *Damasus, while others in the basilica *Iulii* (S. Maria in Trastevere) elected U., who was immediately consecrated by Paul, bishop of Tibur (Tivoli). The Damasians at once besieged the basilica for three days and dragged various victims out of it. Meanwhile Damasus had installed himself in the Lateran and been consecrated. U. and others were exiled by the prefect of the city, but his followers occupied the basilica Liberiana (*basilica Sicinini*), prob. on the Esquiline, where they were besieged by the Damasians and a massacre took place (*Coll. Avel.* 1, 7: CSEL 55, 3; Amm. Marc., *Rerum gest.* XXVII, 3, 11 ff.). On 15 Sept 367 U. and his followers were allowed to return to *Rome and they immediately occupied churches, for which the prefect expelled them again (16 Nov 367); but since many of his followers remained at Rome, fights still took place, which led to the intervention of the prefect and the emperor, who forbade the Ursinians to reside in Rome or its vicinity. But the schismatic party continued to molest Damasus by means of the converted Jew *Isaac. U. and some of his followers were sent to *Gaul (*Coll. Avel.* 11 and 12: CSEL 55, 52-54); U. was later at *Milan, where he joined the *Arians (Ambr., *Ep.* 4) and whence he was exiled to Cologne in *Germany, before 378.

DCB 4, 1068-1070; *EC* 12, 923 f.; Fliche-Martin 3, 241 f.; Ch. Pietri, *Roma christiana*, BEFAR 224, 2 vols., Rome 1976, I, 408-423, 733-737 (cf. also Index: II, 1701); A. Alföldi, *A Conflict of Ideas in the Late Roman Empire: The Clash between the Senate and Valentinian I* (tr. H. Mattingly), Oxford 1952.

A. Di Berardino

URSULA and the 11,000 Virgins. Modern scholars agree to reduce the legend of the 11,000 virgins to the following historical nucleus: some *virgins martyred at Cologne (*Colonia*) in the 3rd c. were once more honoured in the 9th c. after 500 years of near-neglect. Thanks to a Passion composed at the end of the 10th c. (BHL 8427) and the finding of bones in 1106, they then became objects of great devotion until the end of the Middle Ages. Behind the legend is a distorted reading of the inscription of Clematius who, in the 4th or 5th c., restored the basilica of the virgin *martyrs on his property. Near the basilica in the 9th c. was a convent of nuns, ceded in the 10th c. to the ladies of Gerresheim and transformed in the 14th c. into a female lay foundation, dissolved in 1802. The liturgical cult of the 11,000 virgins developed from the Carolingian era: in the 8th c. there was an Office in their honour; in the 9th c. the name appears in *calendars, litanies and missals. From Cologne it spread all through Europe. Feast 21 Oct.

W. Levison, *Das Werden der Ursula-Legende*, Cologne 1928; G. de Tervarent, *La légende de S. Ursule dans la littérature et dans l'art du moyen âge*, 2 vols., Paris 1931; J. Salzbacher-V. Hopmann, *Die Legende der hl. Ursula*, Cologne 1963; *Vies des SS.* 10, 674-88; *LTK* 10, 574-75; *BS* 9, 1252-71.

V. Saxer

USURY. Although the NT does not explicitly denounce u. (Lk 6, 34 cannot be so interpreted), the *Fathers, appealing particularly to OT texts (Ex 18, 7-8; 22, 25; Lev 23, 35; Dt 15, 3; 23, 20; Ps 15, 5; Ezek 18, 4-9), held it incompatible with Christian *life (Clem. Al., *Paed.* 1, 10; *Strom.* II, 19). In condemning u. they followed a doctrine widespread in the Jewish world, to the extent that *Tertullian calls it one of the aspects indicative of the harmony between the OT and NT (*Adv. Marc.* IV, 17; Greg. Nyss., *Epist. can.* 6: PG 45, 234). *Cyprian dedicates an entire thesis of his *Testimonia* to demonstrating the illicitness of u. (*Test.* III, 48). It is condemned primarily as contrary to the *love the Christian must show to the *poor: the appearance of giving is really a mask for its true *raison d'être*: to impoverish one's neighbour (Commod., *Instruct.* 65; Lact., *Inst. div.* 6, 18; Hilary, *In* *ps.* 14, 15). For *Basil, u. does not just make the poor still poorer, it even deprives him of freedom (*Hom. II in ps.* 14). Basil's *Ep.* 188 (PG 32, 682) gives important information on the penitential practice to be applied to usurers. For *Gregory of Nyssa, u. is a *sin unknown to nature because it draws gain from inanimate things, while in nature it is only living things that provide fruit (*Hom. IV in Eccl.*; Greg. Naz., *Or.* 16, 18; Ambr., *De off.* III, 3); it adds to the numbers of the poor (*Hom. IV in Eccl.*) and the usurer is like the thief (*Contra usurarios*). An entire work of *Ambrose (*De Tobia*) is dedicated to the theme of u. Though permitted by imperial law (*De Tob.* 42; John Chrys., *Hom. LVI in Mt.*), Ambrose sees the practice of u. as possible only towards an enemy in *war (*De Tob.* 51; though even this concession is apparently only due to the interpretative difficulties of the text of Dt 23, 20-21). But in other works Ambrose's condemnation of u. is absolute (*De Nabuthe* 4, 15; *De bono mortis* 12, 56), and includes the cornering of grain and speculating on prices of agricultural products in time of famine, all considered as manoeuvres of usurers (*De off.* III, 41). The same condemnation reappears in the works of *John Chrysostom (*Hom. 41 in Gen.*; *Hom. XIII in I Cor.*; *Hom. X in I Thess.*), *Jerome (*Com. in Ez.* 18, 6), *Augustine (*Enarr. in ps. 36, serm.* 3, 6; *Serm.* 38; 86; *De bapt.* 4, 9), *Leo the Great (*Ep.* 4: PL 54, 613; *Serm.* 17), can. 20 of the council of *Elvira (Mansi 2, 9) and the 1st council of *Arles (Mansi 2, 472).

DTC 15, 2316-2333; R.P. Maloney, Early conciliar legislation on usury, *RecTh* 39 (1972) 145-157; id., The Teaching of the Fathers on Usury: an Historical Study on the Development of Christian Thinking, *VChr* 27 (1973) 241-263; R.M. Grant, *Early Christianity and Society*, London 1978.

M.G. Mara

V

VAGHARSHAPAT - ETCHMIADZIN. See of the Armenian *catholicós until the 5th c., sources tell us that the oldest *church buildings in *Armenia (with that of *Ashtishat) were built in this city by St *Gregory the Illuminator on the site of the *martyrdom of saints Gayané and Hripsimé and rebuilt respectively in 630 and 618: two domed quadrangular buildings, which still use the squinch. Sources also tell us that there was a fortified episcopal palace at V. From the point of view of art history, the most important building is the cathedral, whose present version, the result of numerous rebuildings, dates partly from the 7th c.: in it, as in a sort of architectural palimpsest, it is possible to identify older sections. The tetraconch form, adopted in the phase attrib. to Vahan Mamikonian (late 5th c.), was respected in later rebuildings, but it is improbable that the original 4th-c. building had that plan, as Khatchatrian claims (an apse belonging to the primitive phase has been found under the present building): it must rather have been a three-aisle building divided by pillars, of the same kind as the basilica of Qassaq. The well-known synod that officially marked the Armenian church's adherence to *monophysism was held at V. in 491.

A. Khatchatrian, *RBK* s.v. Etchmiadzin (with earlier bibliography); id., *Inscriptions et histoire des églises arméniennes*, Milan 1974, 94; *CE* 14, 513; *EC* 5, 181-183.

M. Falla Castelfranchi

VAISON, *Council of. Meeting at Vasio in *Gaul on 13 Nov 442 - bishop Auspicius was honorary president, but the effective president was the primate, *Hilary of Arles - this council completed the disciplinary work of the council of *Orange of 441.

CCL 148, 94-104; Hfl-Lecl 2, 454-460; Palazzini 6, 16-17.

Ch. Munier

VALENCE, *Council of. Interprovincial meeting of the bishops of *Gaul, 12 July 374, at Valentia. In their synodal letter the *Fathers handed down four disciplinary canons concerning access to *orders, *penitence of *virgines lapsae* and the case of those who had been rebaptized by *heretics.

CCL 148, 35-45; Hfl-Lecl 1, 2, 982; Palazzini 6, 18-19.

Ch. Munier

VALENS, emperor. Appointed Augustus by his brother *Valentinian I, V. was emperor of the *East from 364 to 9 Aug 378, when he was defeated and killed at Adrianople. A follower of the *Arian creed according to the formula of *Rimini, he defended it against *homoeousians and Nicenes, whom he persecuted (Soz., *HE* VI, 7: apropos of the fate of the delegates of the assembly of *Lampsacus). In 365 he promulgated an edict exiling the bishops who had returned to their sees under *Julian (Socr., *HE* IV, 2; Soz., *HE* VI, 7). In 367, at the beginning of the Gothic war, he was baptized by the Arian *Eudoxius of Constantinople; in 369, on the occasion of the appointment of the Arian *Demophilus to succeed Eudoxius, he began a new persecution of those who did not adhere to the Arian faith (Soz., *HE* VI, 14; Basil, *Ep.* 243; Greg. Naz., *Or.* 20; 25; 43). The change that marked V.'s religious policy in his last two years (in 376 he revoked the sentences of exile on the Nicenes) has been interpreted in two ways: as a sincere change of mind (Socr., *HE* IV, 35), or as a political necessity faced with the worsening military situation (Ruf., *HE* II, 13). References to V. as responsible for the empire's difficult situation occur in *Ambrose (*Exp. ev. Lc.* I, 32; II, 37; cf. J.R. Palanque, *Saint Ambroise*, 334, n. 51).

L.-S. Le Nain de Tillemont, *Histoire des empereurs*, V, Venice 1732; J.R. Palanque, *Saint Ambroise et l'empire romain*, Paris 1933.

M.G. Mara

VALENS of Mursa and URSACIUS of Singidunum. These two important spokesmen of Illyrican *Arianism first appear, still young, at the council of *Tyre (335) in the ranks of *Athanasius's enemies. Their pro-Arian tendency is thought to have been derived from contacts with *Arius during his exile in Illyricum (326-328). Though Westerners, at *Sardica (343) they sided with the anti-Nicene and anti-Athanasian Easterners, and the Western bishops arraigned them for their Arianism; Valens was also charged with having tried, some time before, to have himself transferred irregularly from Mursa to *Aquileia. The position of the two, residing in a part of the empire which was governed by *Constantius but was anti-Arian in faith, was not easy and in 347, on the occasion of Athanasius's return to *Alexandria, they were formally reconciled with him. But when Constantius was left sole emperor, the influence of the two, and esp. of Valens, who had been very close to Constantius at the battle of Mursa against the usurper Magnentius (351), gradually grew and they played a very important part in the actions aimed at aligning the Western episcopate on the anti-Nicene and anti-Athanasian line of the *East. Working in close harmony with *Germinius, bishop of Sirmium from 351, they took part in the council of *Milan (355), where Valens prevented the approval of the Nicene *Creed, which his opponents had proposed, and in 357 worked out the second formula of *Sirmium, the most openly pro-Arian of all those published in those years.

The success of the *homoeousian reaction took them by surprise and, obedient to Constantius's will, they subscribed the anti-Arian decisions of the council of *Sirmium in 358. But at the council of *Rimini (359) they headed the pro-Arian minority which, after long manoeuvres and strong in the emperor's support, imposed the formula of Rimini on the anti-Arian majority. The anti-Arian reaction that followed Constantius's death (362) forced them to a difficult defence. Yet, thanks to *Valentinian I's policy of non-intervention, they were able to keep their sees and on the whole to safeguard Arian positions in Illyricum, which were considerable thanks to their own activity. In c.365 Valens welcomed *Eunomius, who was passing through on his way to exile, and got his sentence revoked by interceding with the emp. *Valens. In 366 they sought to dissuade Germinius from his proposal to abandon the formula of Rimini in favour of a more decidedly anti-Arian one, and for this they called a small Arian council at Singidunum (Belgrade), but without result. After this we know no more of them.

CPL 682-684; M. Meslin, *Les Ariens d'Occident*, Paris 1967, 71-84; Simonetti 597; R.P.C. Hanson, *The Search for the Christian Doctrine of God*, Edinburgh 1988.

M. Simonetti

VALENTINIAN I. Roman emperor of the West 364-375; though he professed the Nicene faith, his religious policy was based on the principle of neutrality, not just for political reasons but also because he held that, as a *layman, he was not qualified to judge in questions of faith (Soz., *HE* VI, 7). In 373 he ratified the election of *Ambrose, who later reminded *Valentinian II of his father's respect for ecclesiastical autonomy and approval of his own appointment (*Ep.* 21, 2-5). In *Ep.* 17, 16 (PL 16, 1006) Ambrose imagines a speech given by V. when informed of the presence of the *Altar of Victory in the senate-house; in his funeral oration, Ambrose remembers the *faith V. demonstrated under *Julian (*De obit. Valent.* 55). A proof of the respectful distance that separated V. from Ambrose appears in Ambrose's silence over V.'s second *marriage with *Justina after repudiating his first wife.

W. Heering, *Kaiser Valentinian I*, Magdeburg 1927; R. Andreotti, Incoerenza della legislazione dell'imperatore Valentiniano I, *Nuova Rivista Storica* 15 (1931) 456-516; J. Gaudemet, *L'Eglise dans l'Empire romain*, Paris 1958; R. Soraci, *L'imperatore Valentiniano I*, Catania 1971; A. Alföldi, *A Conflict of Ideas in the Late Roman Empire. The Clash between the Senate and Valentinian I* (tr. A. Mattingly), Oxford 1952.

M.G. Mara

VALENTINIAN II. In 375, on the death of his father *Valentinian I, V., still a *child, was proclaimed Augustus by the troops of Illyricum. At first he governed under the tutelage of his mother *Justina and, after her death, of the *pagan Arbogastes, his minister and tutor; nor should we forget the political protection offered him first by *Gratian and then by *Theodosius. But the main influence on V.'s religious policy was his relationship with *Ambrose, at first difficult because of the influence of the Arianizing Justina and her personal hostility towards the bishop's intrusiveness (Ambr., *Ep.* 20, 27), then marked by a filial dependence on him. When, after his mother's death, Theodosius imposed Arbogastes on him as minister, conflict between the young emperor, by now a fervent *Catholic, and his pagan minister was inevitable and such as to cost him his life. In 392 he was found dead in his palace at *Vienne (suicide or murder by Arbogastes?). Various Ambrosian works are indicative of the first and second periods of V.'s life. In 384 the second petition to the senate over the *Altar of Victory led Ambrose to send him *Epp.* 17 and 18 against *Symmachus's request. In the field of anti-Arian polemic, in *Ep.* 21 Ambrose reminded V. of the respect his father had had for the autonomy of the church and in his *Sermo contra Auxentium* he explained to the congregation meeting in the basilica the basic reason for his refusal to hand it over to the emperor as he had requested. But in 387 it was V. himself who asked Ambrose to go to *Maximus at *Trier, to obtain Gratian's body and esp. to prevent him carrying out his project of invading *Italy (Ambr., *Ep.* 24). After Justina's death in 388, Ambrose sang V.'s return "to the door of faith" (*Exp. Luc.* IX, 32) and exonerated him from responsibility for the earlier policy (*De Joseph* 36). To V.'s *friendship with Ambrose we owe the decree of 391 proscribing pagan worship and once more refusing to restore the Altar of Victory. In 392, when the young V.'s body returned to *Milan, Ambrose mourned his death in a *letter to Theodosius (*Ep.* 53) and in a funeral oration (*De obit. Valent.*).

A. Morpurgo, *Arbogaste e l'Impero romano dal 379 al 394*, Trieste 1883; A. Solari, *La crisi dell'impero romano*. II: *Gli ultimi Valentiniani*, Rome 1933; J.R. Palanque, *Saint Ambroise et l'empire romain*, Paris 1933; R. Paribeni, *Da Diocleziano alla caduta dell'impero d'occidente*, Bologna 1941; F. H. Dudden, *Life and Times of St. Ambrose*, 2 vols., Oxford 1935.

M.G. Mara

VALENTINIAN III. Son of *Constantius III and *Galla Placidia (sister of *Honorius), V. was Roman emperor of the West 424-455. After his coronation at Rome in 425, aged six, he moved to *Ravenna where the court poet *Merobaudes drew a picture of V. and his *family in a short poem of which 23 vv. remain (cf. MGH, *Auct. ant.* 14, 3). His reign, under the tutelage of his mother, was marked by the last great crisis of the Western empire, threatened by *Vandals in N *Africa, lacerated by the revolt of the two generals *Aetius and Boniface, invaded by Huns. He was saved by Aetius's military valour and *Leo the Great's embassy to *Attila. His political unwisdom, culminating in the murder of Aetius (21 Sept 454), and his lack of moral scruple were the causes of his assassination. His religious policy, marked by Leo's influence, further suppressed *paganism, forbidding under pain of death the reopening of the temples and the offering of *sacrifice (*CJ* I, 11, 7); he issued provisions against *heretics and *Nestorians (*CJ* I, 1, 3; 5, 6), *Manichees (PL 54, 622-24), *Pelagians in S *Gaul (PL 48, 409), confirming all the privileges of the *Catholic church (*CJ* I, 2, 12). At Leo's request V. repeatedly pronounced in favour of the primacy of the bishop of *Rome (cf. Leo Gt, *Epp.* 10 and 11 in the controversy with *Hilary of Arles). In *Idatius's *Chronicle* (SCh 218, 162), references to V. are frequent.

G.A. Balducci, *La politica di Valentiniano III*, Bologna 1934; A. Solari, *La crisi dell'impero romano*, IV, Rome 1936; P. Stockmeier, *Leo I des Grossen Beurteilung der kaiserlichen Religionspolitik*, Munich 1959; W.E. Kaegi Jr., *Byzantium and the Decline of Rome*, Princeton 1968, ch. 1; S.I. Oost, *Galla Placidia Augusta*, Chicago 1968.

M.G. Mara

VALENTINUS, martyr. The *Mart. hier.* (14 Feb) mentions a *martyr at Terni (*Interamna Nahars*), whose basilica is attested in the 8th and 9th cc. (Duchesne *LP* I, 427; II, 154). On the same road, the Flaminia, where it meets *Rome, is another basilica of V., whose construction was attrib. to *Julius I (337-52), but whose titular was not called a saint (Duchesne *LP* II, 9). Pope *Theodore (642-49) was said to have restored it (*ibid.*, I, 333); in fact he finished the work begun by pope *Honorius (*ibid.*, I, 334, n. 10). All the *itineraries mention the basilica (De Rossi, *Roma sotterranea cristiana*, I, Rome 1864, 176-177). These sure facts leave three possibilities: *1)* two distinct saints, one at Terni, the other at Rome; *2)* a single V., of Terni and buried there, whose *brandea* were taken to Rome; *3)* a single Roman V., whose cult spread along the via Flaminia to Terni. The third hypothesis seems to be favoured by the *epigraphy of the Roman basilica: an inscription of 318 attests the early use of the *cemetery and some later inscriptions give the name of St V. Against this is the fact that of the oldest documents, i.e. the *Mart. hier.* and the *Liberian catalogue*, the former puts the martyr at Terni and the latter does not call V. a saint. For this reason we propose to see the Roman V. as the person who provided pope Julius with the funds for building the suburban basilica; the Passion (BHL 8450-61) will have contributed to confusing him with the V. of Terni.

Vies des SS. 2, 322-23; *LTK* 10, 998-99; *BS* 12, 896-99; A. Amore, *I martiri di Roma*, Rome 1975, 13-16.

V. Saxer

VALENTINUS. *Catholic bishop (4th-5th c.) of Baiana in *Numidia, part of an embassy to the imperial court (Aug., *Ep.* 88, 10). He attended the council of *Cirta against the *Donatists (*Ep.* 141, 1) and that of Milevis (416) against the *Pelagians (*Ep.* 176), and was present at the conference of *Carthage in 411 (*Coll. Carth.* I, 57, 99).

PAC 1, 1130-1132.

E. Romero Pose

VALENTINUS the Apollinarist. *Leontius of Byzantium (*Fraud.* 104 ff.) preserves some chapters of an *Apologia* written by the *Apollinarist V. "against those who say that we maintain that (Christ's) *body is consubstantial with *God". The cited chapters distance themselves from other Apollinarists; in particular they are aimed - acc. to the heading cited by Leontius - "against the doctrine of Timothy and his followers, and their teacher Polemon". In fact *Polemon and *Timothy took up positions opposed in practice and differentiated in doctrine: at any rate, we find in V. the explicit statement that Christ is "true Son of God σαρκοφόρος" (Lietzmann, 288, l. 20), and that this *flesh is the "clothing of the hidden *mystery" (*ibid.*, 289, l. 13). The capacity to save comes from its Union with the Spirit (*ibid.*, 290, ll. 10-12). Voisin speaks of "moderate Apollinarists" (124 ff.); but we should rather speak of elements of Antiochene Apollinarism (cf. Lietzmann, 9, 15 ff.).

CPG 2, 3732; H. Leitzmann, *Apollinaris von Laodicea und seine Schule*, TU 1, Tübingen 1904, 157; framm. pp. 287-291; G. Voisin, *L'apollinarisme. Etude historique littéraire et dogmatique sur le début des controverses christologiques au IVe siècle*, Louvain 1901, 112-113; DTC 15, 2519-2520.

E. Cavalcanti

VALENTINUS the Gnostic. This *gnostic heresiarch, of Egyptian origin, came to *Rome in *c.* 140 (cf. Iren., *Adv. haer.* III, 4, 3; Euseb., *HE* IV, 11, 1). At some point he abandoned *orthodoxy and founded a school where he propagated his doctrines. Under pope Anicetus he left Rome to go to the *East, perhaps to *Cyprus. Returning to Rome, he died there soon after 160. Of his works we possess only a few fragments handed down mainly by *Clement of Alexandria, not enough to reconstruct his teaching, even broadly. The works in the library of *Nag Hammadi showing doctrinal characteristics that can be related to the Valentinian system (*Ev. Veritatis; Tract. Tripartitus; Ev. Philippi; Ep. ad Rheginum*; etc.) are hard to ascribe to V. himself. V. wrote psalms, *homilies and *letters; *Hippolytus (*Refut.* VI, 37, 7) preserves one of his *hymns. From what remains, an image emerges of V. as a biblical theologian, influenced by *Platonism but straying little from the - as yet ill-defined - borders of Christian *orthodoxy.

However, we possess much more detailed information on the complex theological systems worked out by V.'s pupils; the *heresiologists have given us six distinct portraits of them (Iren., *Adv. haer.* I, 1-8, 11-12, 13-21; Hipp., *Refut.* VI, 29-36; Tertull., *Adv. Valent.*; Orig., *In Johannem*, esp. book XIII; Clem. Al., *Exc. ex Theodoto*; Epiph., *Pan.* 31, 58; 35, 4). The main characteristics of the Valentinian doctrinal system are: the divine *Pleroma is composed of 30 *aeons, arranged in pairs (*syzygiae*); the first four pairs are the most important and form the primordial *ogdoad*; from them all the other aeons originate. The unity of the two elements of the *syzygia* is presented as the model of the unity that was broken by *sin, symbolized by the sexual separation of Eve from *Adam, which the spiritual man must rebuild by rejoining his heavenly partner. The first origin of sin is placed in the vissicitudes of the last aeon, Sophia, who, through her immoderate desire to know the unknowable *Father, brought about the degradation of the divine element into the *world, which originated from the materialization of Sophia's ignorance or error. The process of recovery of the fallen divine element by the heavenly Saviour began at the same time; this process, which involves several stages - intervention of various divine beings; activity of the Demiurge; distinction of men into three kinds (spiritual or pneumatic, psychic, and material or hylic) - will eventually lead to the reintegration in the divine pleroma of the divine element, torn away from its mixture with matter, and the placing of the other beings according to their nature (complete salvation and full reintegration for the spirituals; partial, inferior salvation for the psychics; dissolution for the hylics).

V.'s school divided into two branches: an Italic and an Eastern (cf. Hipp., *Refut.* VI, 35). To the Italic branch belonged *Heracleon, *Ptolemy and *Florinus; to the Eastern branch *Theodotus and *Mark.

W. Völker, *Quellen zur Geschichte der christlichen Gnosis*, Tübingen 1932, 57-141; M. Simonetti, *Testi gnostici cristiani*, Bari 1970, 119-259; H. Leisegang, *Die Gnosis*, Leipzig 1924, 281-297; F.M. Sagnard, *La gnose valentinienne et le témoignage de saint Irénée*, Paris 1947; A. Orbe, *Estudios valentinianos*, 5 vols., Rome 1955-66; K. Rudolph, *Die Gnosis*, Göttingen 1977, 339-47; B. Layton (ed.), *The Rediscovery of Gnosticism*, I: *The School of Valentinus*, Leiden 1980. Bibliographical list: D.M. Scholer, *Bibliographia Gnostica*, Leiden 1971 and subsequent *Supplementa*, published annually by Scholer in the review *Novum Testamentum*; H. Jonas, *The Gnostic religion: the message of the Alien God and the beginnings of Christianity*, Boston ²1963; C.G. Stead, The Valentinian Myth of the Sophias, *JThS* n.s. 20 (1969) 75-104; G. Quispel, Origen and Valentinian Gnosis, *VChr* 28 (1974) 29-42; M. van den Broek, The Present State of Gnostic Studies, *VChr* 37 (1983) 47-71; (ed. and tr.) J.E. Ménard, *L'exposé Valentinien. Les fragments sur le Baptême et sur l'Eucharistie* (NH XI.2), Sainte Foy 1985; M. Desjardins, The Sources for Valentinian Gnosticism: A Question of Methodology,

VChr 40 (1986), 342-347; M.J. Edwards, Gnostics and Valentinians, JThS n.s. 40 (1989) 26-47; M.R. Desjardins, Sin in Valentinianism, Missoula 1990.

C. Gianotto

VALERIA. A province of the former Pannonia Inferior, reorganized in 293, its N and E borders were the Danube; its S border was just N of the Drava; to W, a N-S line over the Pannonian plain separated it from Pannonia Prima. Christianity was introduced by immigrants from the *East who settled in *Pannonia in the late 3rd and early 4th cc. The period of transition from *paganism to Christianity is well-documented from tombs, principally the *sarcophagi found at Brigetium. In one of these tombs, placed centrally to the others, was found a silver staff decorated in niello technique, late 3rd c. The staff (*lituus*) prob. belonged to an augur as symbol of his dignity, and was prob. destined to pass to the new priest at his death. That it was found among his tomb furniture suggests that he may have been the city's last augur before the coming of Christianity.

In this transitional period, Judaism too played an important part in forming the religious life of V. There was a Jewish community with a synagogue at Intercisa in the 3rd c. *Persecution of Christians, more or less documented in the other Pannonian provinces, is so far not attested in V.

The region's administrative centre and most important diocese was Sopianae (Pécs); the other diocesan sees are unknown. The bitter clashes between *Catholics and *Arians attested in Pannonia Prima must surely have had repercussions in V.

The most important Christian monuments are at Sopianae, where six funerary *cells were found around the modern cathedral: no. 1 composed of a *memoria* and a crypt, with pilaster strips outside; no. 2, similar; nos. 5 and 6, of the same type, but more modest; the others polylobate, no. 3 trichoral and no. 4 septichoral. Chronologically, no. 1, similar to that of Obud (*Aquincum*), is of the 4th c.; no. 4 resembles the plan of St Gereon at Cologne and can be dated to the 5th c. Some of these *memoriae* are painted. The best frescos are in no. 1: they have general affinities with the *paintings of the Roman catacombs, but with local characteristics. They show scenes from OT and NT (Adoration of the *Magi, *Adam and Eve, *Noah, *Daniel in the lions' den, *Jonah, Good *Shepherd). On a wall the apostles *Peter and *Paul acclaim a *monogram of Christ among the stars; on another wall is a garden of *paradise. At the centre of the vault is a christogram at whose sides are *clipei* with two busts of young men and two bearded male busts, and in the spaces between the *clipei* are fountains with peacocks or small birds. Cell no. 2 is also decorated with frescos, though of inferior quality and badly preserved: in a niche is painted a vase with grapes; above, vines with grapes; on the entrance wall, a garden of paradise. Both *memoriae* are of the 4th c. As for *church buildings, at *Aquincum* is a twin basilica without apse, but with clergy bench of the type common in the Upper Adriatic in the 5th c. At Intercisa was found a three-aisled basilica with rectilinear apse; at Gorsium, a longitudinal-plan basilica.

Rich archaeological finds have been made in V.: vases of ceramic and glass, bronze panels with Christian scenes, pieces of architectural *sculpture and some *sarcophagi.

J. Zeiller, *Les origines chrétiennes dans les provinces danubiennes de l'empire romain*, Paris 1918; T. Nagy, *History of Christianity in Pannonia up to the fall of the Roman protection of the border* (in Hungarian), Budapest 1939; F. Gerke, Die Wandmalereien der neu gefundene Grabkammer in Pécs (Fünfkirchen). Ihre Stellung in der spätrömischen Kunstgeschichte, in *Spätantike und Byzanz*, Baden-Baden 1952, 115 ff.; F. Gerke, Die Wandmalereien der Petrus-Paulus Katakombe in Pécs (Südungarn), *Neue Beiträge zur Kunstgeschichte des I Jahrhunderts*, II Halbband, Baden-Baden 1954, 147 ff.; F. Fülep, Scavi archeologici a Sopianae, *CCAB* 16 (1969) 151 ff.; Z. Kádár, Lineamenti dell'arte della Pannonia nell'epoca dell'antichità tarda paleocristiana, *ibid.*, 179 ff.; A. Mòcsy, *Pannonia and Upper Moesia*, London 1974.

N. Cambi

VALERIAN, emperor. Publius Licinius V., Roman emperor 253-260; appointed by the legions on the death of Trebonianus Gallus, he associated his son *Gallienus on the throne and entrusted him with the defence of the Western frontiers. V. was taken prisoner by the Persian Shapur I during a military campaign in the *East, and died soon after. At first favourable to the Christians, whom he even welcomed into his home (cf. Dionysius of Alexandria's *letter to Hermammon, in Euseb., *HE* VII, 10, 3), he later became hostile to them for political reasons (with V. the state officially took cognizance of the church's existence and for the first time declared it illegal). In 257 and 258 he issued two rigorous edicts against the Christians: victims of the *persecution included popes *Stephen I and *Sixtus II, *Dionysius of Alexandria, *Cyprian and *Fructuosus. *Commodian compared the seven years of V.'s reign to the seven years of the Apocalypse (*Carmen apol.*, *passim*).

J. Healy, *The Valerian Persecution. A Study of the Relations between Church and State in the Third Century*, London 1905; M. Sordi, *Il cristianeimo e Roma*, Bologna 1965, 286-311 and 441-2 (bibl.); R. Rémondon, *La crisi dell'impero romano da Marco Aurelio a Anastasio*, Milan 1975.; W.H.C. Frend, *Martyrdom and Persecution in the early Christian Church*, Oxford 1965, ch. 13; id., *The Rise of Christianity*, London 1984, 324-328; G.W. Clarke, *The Letters of St. Cyprian of Carthage*, vol. 4, ACW 47, New York 1989; C.J. Haas, Imperial Religious Policy and Valerian's persecution of the Church, AD 257-260, *ChHist* 52 (1983) 133-144; K.H.S. Schwarte, Die Christengesetze Valerians, in (ed.) W. Eck, *Religion und Gesellschaft in der römischen Kaiserzeit* (Festsch. Vittinghoff), Cologne-Vienna 1989, 103-163.

L. Navarra

VALERIAN of Aquileia. *Fortunatian, compromised by *Arianism, was succeeded in the see of *Aquileia by V. (368?-388), who inaugurated a more clearly Nicene theological and pastoral line. Under his government Aquileia acquired a prominent role among the churches of N *Italy, including *Milan, see of the *semi-Arian *Auxentius, and the churches between the Adriatic and the Danube. Perhaps it was V.'s vigorous activity on behalf of *orthodoxy, as well as its geographical position, that made Aquileia the seat of a Western *council in Sept 381. V. was able to recruit a well-trained and hardworking *clergy, recalled by St *Jerome as *quasi chorus beatorum*.

G. Cuscito, *Cristianesimo antico ad Aquileia e in Istria*, Trieste 1977, 177-179 (with bibl.).

G. Cuscito

VALERIAN of Bierzo. Spanish ascetic and writer (second half of 7th c.). Native of the region of Asturica (prov. of León), probably from the Bierzo (the area bordering *Gallaecia), after receiving the scholastic education of his time he entered the *monastery of Complutum, founded by *Fructuosus of Braga, whose ascetic footsteps he followed. Obsessed by the idea of the *demon and the dangers of life in the *world, he lived always in solitude. He passed his life in continual penance, guiding disciples and monks, who were numerous in the Bierzo.

He dedicated his writings to his *friend and perhaps disciple, *abbat Donadeus. His intransigent *ascesis saw diabolical tricks everywhere. He died after 691. Memory of him lasted there at least until the 10th c., when bishop Gennadius of Astorga began to celebrate him with Fructuosus of Braga as prototypes of ascetic life in the Bierzo.

His personality is known from his works, almost all written between 675 and 690. For the instruction of the monks whom he guided, he drew up a hagiographical compilation of some 50 texts, preserved more or less entire in various MSS of the 10th-12th cc. There we find *Jerome's life of Paul, *Hilarion and Malchus; *Possidius's *Indiculus*, the lives of St *Germanus, brief edifying narrations selected from the *Vitae Patrum*, the *Vita Fructuosi*, some narrations written by himself, and others of minor interest. The compilation should be read with two poems (*Epithamaeron*) which illustrate its objectives, describing it as a means of spiritual edification, providing an axis of orientation for monastic conversion. Small texts included in it describe *visions of *paradise seen by monks or *conversi*; one curious passage, in the form of a *letter, outlines for the monks of the Bierzo how V. came to know the *Itinerarium* of *Egeria, whom he presents as an intrepid heroine of faith, devoted to *Scripture. He also composed a treatise on monastic life, of which only fragments are preserved.

Broader in character is a book of instruction whose aim combines moral and spiritual edification - in part echoing *Isidore's *Sentences* - with refined literary techniques: compositions in which the initial words of each verse or of the whole poem begin with the same letter. V. seems also to have been interested in a particular sort of writing, spiritual texts as *centos of the Psalter.

On V.'s personality, we find much more material in his *Autobiography*, written in at least three well-defined periods, repeating previously presented themes in different perspectives. At its centre are his spiritual struggles, esp. his resistance to *temptation, which at times took on a concrete and violent form. This narration too should be included among the poems.

The *Vita Fructuosi*, though commonly attributed to V., is not his.

CPL 1276-1293; Díaz 285-303; PL 87, 421-455; R. Fernández Pousa, *San Valerio. Obras*, Madrid 1942; C.M. Aherne, *Valerio of Bierzo*, Washington 1949; M.C. Díaz y Díaz, La compilación hagiográfica de Valerio del Bierzo, *Hispania Sacra* 4 (1951) 3-25; M.C. Díaz y Díaz, *Anecdota wisigothica*, I, Salamanca 1958, 49-61, 89-116 (PLS 4, 2019-2029): M.C. Díaz y Díaz, *Códices visigóticos en la monarquía leonesa*, León 1983, 117-148.

M. Díaz y Díaz

VALERIAN of Calahorra (*Calagurris*). Early 5th-c. bishop mentioned by *Prudentius (*Perist.* 11, 2). Wrote a profession of faith, which survives. Prob. from a rich family from the region of *Saragossa (*Caesaraugusta*).

CPL 558a; PLS 1, 1045; J. Madoz, Valeriano obispo calagurritano, escritor del siglo V, *Hispania Sacra* 3 (1950) 131-7; *DHEE* 4, 2705.

M. Díaz y Díaz

VALERIAN of Cimiez. Perhaps a member of the noble family of the Valeriani, married, he later embraced the religious life, living first on the continent, then on the isle of *Lérins. We find him bishop of Cimiez, a suburb of Nice, in 439. As such he took part in the councils of *Riez (439) and *Vaison (442), aimed at strengthening church discipline. He supported *Hilary of Arles against *Leo I. In 449 he approved the election of *Ravennius, *abbat of Lérins, to the see of Arles. He subscribed the *Tomus ad Flavianum* (Leo Gt, *Ep.* 102). His name appears at the council of *Arles which composed the quarrel between the bishop of Fréjus and Lérins. He must have died soon after. A collection of 20 *homilies and a *letter *Ad monachos* remain. Riberi's restoration to V. of the homily *De dedicatione ecclesiae* is uncertain. Feast 23 July.

PLS 3, 184-188; PL 52, 691-758; *DTC* 14, 2520-2522; *BS* 12, 912-914 (bibl.); J.P. Weiss, La personnalité de V., *Annales fac. lettres de Nice*, 1970, 141-162; *Patrologia* III, 514-515; P.-M. Duval, *La Gaule jusqu'au Ve siècle*, Paris 1971, 761-762; C. Tibiletti, Valeriano di Cimiez e la teologia dei Maestri Provenzali, *Augustinianum* 22 (1982) 513-532.

A. Hamman

VANDALS. A people of Germanic stock (but originally from Scandinavia) who in the 1st c. AD lived on the Baltic between the Elbe and the Vistula. Pliny (*Nat. hist.* 4, 99) gives the name Vandili to a vast group of E Germanic tribes (Burgodiones, Varinnae, Charini, Gutones); the form Vandilii recurs in Tacitus (*Germ.* 2) as one of the group of peoples called Lugii (*ibid.*, 43, 2), mentioned by Strabo (*Geogr.* 7, 1, 3) as allies of Maroboduus, and by Tacitus (*Ann.* 12, 29 f.) as allies of the Hermunduri against Vannius (AD 50). The V. headed S towards the Danubian *limes* and are first expressly mentioned during *Marcus Aurelius's *war against the Marcomanni and Quadi (171-3). They were divided into two groups, the Silings (Ptol., *Geogr.* 2, 11, 10) - hence the name Silesia - and the Asdings (Dio Cass., *Hist.* 71, 12). In 214, under *Caracalla, we find them in dispute with the Marcomanni; in 270 the Asding V., breaking into *Pannonia, were defeated by *Aurelian and pushed back beyond the Danube; going W up the Danube and making for the Rhine border, in 279 the Siling V. together with the *Burgundians were defeated in *Rhaetia by Probus. In *c.* 335 the Asding V., beaten on the Danube by the *Goths, obtained permission from *Constantine to settle in Pannonia, under the Romans and with an obligation to provide military contingents. They remained here for some 60 years, after which the pressure of the Huns forced them to retreat to the Rhine border, where, with the Suevi, Alani and Siling V., already forced W in the 3rd c., they formed a great barbarian column which, overcoming the resistance of the *Franks (in an encounter with whom the V. lost their king Godigiselus), crossed the Rhine on 31 Dec 406 (*Orosius [*Hist.* 7, 38, 3 f.] accuses the Vandal *Stilicho of complicity), invaded *Gaul and inflicted severe devastation on it. In autumn 409 the V., Suevi and Alani, favoured by Gerontius (Spanish supporter of the usurper Maximus), crossed into *Spain, which they raided ruinously, until in 411 they made an agreement with the emp. *Honorius, who recognized the settlement of the barbarians as *foederati*: The Asdings and Suevi took *Gallaecia, the Silings *Baetica, the Alani *Lusitania and *Carthaginensis. In 416 the Visigothic king Wallia pledged himself to Honorius to liberate Spain from the other barbarians: in 418 he routed the Silings, whose king Fredbal was taken prisoner to Italy, and the Alani; the Asding king Guntheric then united the whole Vandal people and the remains of the Alani under himself, with the title "king of the Vandals and Alani". Subsequently the V. descended on S Spain (whence the name Andalusia), conquered Cartagena and *Seville (425) and thus ensured their control of all Spain. While still in Spain, the intolerance of the *Arian V. towards the *Catholic Romans, which exploded in particularly violent forms in Africa, began to manifest itself.

In 428 Guntheric was succeeded by his brother *Genseric (428-477), who in 429 led his people (about 80,000 strong, including V., Alani and disbanded Visigoths who had remained in Spain) to *Africa, attracted there by the region's wealth and fertility (the provision of *Italy itself depended on African grain) and favoured by the dissension between Boniface, governor of the province, and the imperial government (though the claim that Boniface himself invited the V. to Africa seems unfounded). The Romans were quickly routed and managed to keep possession only of the major centres such as *Carthage and *Hippo (during the Vandal siege of Hippo St *Augustine died, 28 Aug 430); subsequently an agreement was reached by which the Romans recognized the V. as *foederati* stationed in the proconsulate of *Numidia, with Hippo as capital. Thus began the African kingdom of the V., which reached the height of its power under Genseric himself. Vandal rule gradually spread: Carthage was conquered on 19 Oct 439 (which became the first year of a new Vandal dating system) and in 442 an agreement with the Western emperor recognized the independence of Vandal rule and its extension to other African regions; at the same time the extremely expert Vandal fleet scourged the coasts of Sicily and Sardinia with repeated pirate raids. The death of *Valentinian III offered Genseric an opportunity to fall on *Rome and repeat (June 455) *Alaric's sack of 410 (a certain moderation was observed towards the Christian churches, partly through the intervention of pope *Leo I). After two weeks Genseric left Rome and took back to Africa an enormous amount of booty as well as important hostages, including Valentinian's widow Eudoxia and her daughters Eudocia (who was given in marriage to Genseric's son *Huneric) and Placidia (later released with her mother). Repeated military and diplomatic attempts to stem and overthrow the Vandal power came to nothing, and in 476 the emperor of *Byzantium officially had to recognize Vandal rule over a vast area: the whole province of Africa, the *Baleari, Ischia, *Corsica, *Sardinia and Sicily (later, except for the city of Lilybaeum, ceded to *Odoacer in exchange for an annual tribute).

The Vandal settlement in Africa brought grave social and religious upheavals. Particularly in the province of Proconsularis (or Zeugitana) around Carthage, the landed proprietors were dispossessed and banished or forced to remain on their former possessions as slaves or *coloni*; but in the other regions too there were state confiscations and impositions of heavy tributes. At the same time the Arian V. made the Catholics undergo harsh persecution: innumerable acts of physical violence were committed, *orthodox worship was forbidden, the Catholic *clergy were exiled, *churches and church *property were either laid waste or confiscated and given to the Arian clergy (on the dismay of the first impact cf. Aug., *Ep.* 228; on the situation up to the conquest of Carthage in 439, cf. a group of 12 *Sermones* attrib. to bishop *Quodvultdeus; more generally on the tragedy of the Catholics under Vandal domination, cf. *Victor of Vita's *Historia persecutionis Africanae provinciae*, published in *c.* 488). Towards the end of Genseric's reign the oppression was alleviated and exiled clerics were allowed to return (475).

Genseric was succeeded by his son *Huneric (477-484) who, after a first phase of relative tolerance towards the Catholics (in 481 a bishop, *Eugenius, was reinstalled at Carthage), resumed ferocious repression: of particular gravity was the decree of 24 Jan 484 which laid down that all edicts promulgated by the emperors against *heretics would now be applied to the Catholics, with all the consequences that this involved; among other things, all who served in the army or public administration were forced to choose between conversion to *Arianism or dismissal, exile and confiscation of their goods. Under Huneric, the Mauri (Berbers) managed to extract themselves from Vandal rule.

His successor Gunthamund (484-496) took a milder attitude in religious affairs: the Catholic churches were reopened and the exiles allowed to return. He also put down a revolt of the Mauri and tried, unsuccessfully, to recover possession of Sicily. For a poem that may have been dedicated to the emp. *Zeno, he imprisoned the poet *Dracontius.

Gunthamund was followed by his brother Thrasamund (496-523), who established a policy of close collaboration with *Theodoric, whose sister Amalafrida he married in 500 as part of the project of a "grand alliance" of Arian Germans, of which the Ostrogothic king was to be the leader. Under Thrasamund, sentences of exile against the Catholic clergy were resumed (though without the former violence); among the exiles was the bishop of Ruspe, *Fulgentius, destined to become the most prestigious figure in the African episcopate of his time: for which reason Thrasamund recalled him to Carthage in 415 to attempt a compromise on the question of the *Trinity - an attempt frustrated by the intransigent *orthodoxy of Fulgentius, who was therefore obliged to return to the exile of Cagliari.

With Hilderic (523-530), son of Huneric and Eudocia, came a political turning-point: the Vandal king broke the previous understanding with the Ostrogoths (he arrested Amalafrida and slew the Goths of her entourage), established relations with the emperor of Byzantium and the Romans and restored full freedom to Catholic worship (in 525 a *council of Catholic bishops was held at Carthage). But Hilderic's pro-Roman and pro-Catholic policy aroused profound discontent among the V., which exploded on the occasion of their severe defeat by the Mauri: the king was deposed, imprisoned and replaced by Gelimer (530), Genseric's great-grandson. *Justinian made use of the occasion to send to Africa an expeditionary force (the historian Procopius of Caesarea fought in this corps and wrote an account of the campaign) commanded by *Belisarius, who, helped by Ostrogothic support in Sicily, rebellion in Sardinia and the favour of the African population, in a short time (June 533-March 534) overthrew the kingdom of the V., who thus made their final exit from history.

L. Schmidt, *Geschichte der Wandalen*, Munich 1942; H. Helbling, *Goten und Wandalen. Wandlung der historischen Realität*, Zürich 1954; Ch. Courtois, *Les Vandales et l'Afrique*, Paris 1955; F. Giunta, *Generico e la Sicilia*, Palermo 1958; H.J. Diesner, *Das Vandalenreich. Aufstieg und Untergang*, Leipzig 1966; A. Morazzani, Essai sur la puissance maritime des Vandales, BAGB (1966) 539-561; H. Schreiber, *I Vandali*, Milan 1984.

A. Marchetta

VANNES, *Council of. Meeting in Venetus in *Gaul some time between 461 and 491 to ordain a new bishop, some bishops of the province of *Tours, under the presidency of their metropolitan *Perpetuus, promulgated 16 disciplinary canons.

CCL 148, 150-158; Hfl-Lecl 2, 2, 904-906, 1382.

Ch. Munier

VATICAN: investigations under St Peter's Basilica. The archaeological investigations beneath St Peter's Basilica are the only way of proving or disproving the *tradition that the tomb of *Peter really existed in the heart of the basilica, under the papal altar.

These investigations, to which for centuries the Roman pontiffs had not dared put their hand, were ordered in 1939 by Pius XII. They developed, to all practical purposes, in two phases: 1940-1949, and from 1952 on.

Taking part in the works of the first phase, the excavation proper, were Enrico Josi, Antonio Ferrua, Engelbert Kirschbaum and the architect Bruno Maria Apollonj Ghetti. Their official account was published in 1951. The excavations brought to light the following: *a)* under the central aisle of the Basilica, a grandiose *pagan necropolis of the 2nd and 3rd cc., oriented from E to W like the famous circus of *Caligula and *Nero, and completely interred - a very serious act - in the time of *Constantine in order to create the plan of the first Basilica; *b)* under the area of the present "open Confession", an area of respect (the so-called Area *P*), confined to the W by the so-called red wall; *c)* under the papal altar (of Clement VIII: 1594), a series of superimposed monuments related to the cult of the Apostle and demonstrating its age-old continuity: altar of Callistus II (1123); altar of *Gregory the Great (590-604); Constantinian monument (after 321 or 322 and before 326); structure backing onto the "red wall", the so-called *trophaeum* of Gaius (2nd c.); signs of an original earth tomb under the "trophaeum"; *d)* a small 3rd-c. wall (the so-called "wall *g*"), perpendicular to the "red wall" and enclosed by the Constantinian masonry, whose N face was covered in a close network of Christian graffiti and inside which was a *loculus* covered by marble slabs. It was also observed that other tombs converged towards the apostolic monument in a singular way.

On this basis Pius XII announced in 1950 that Peter's tomb had been discoverd. The great news corresponded to the truth. But in its demonstration there remained various disquieting gaps which did not fail to arouse uncertainties and controversies: the name of Peter was nowhere recognized; the graffiti were almost undeciphered; the loculus of "wall *g*" was not clearly explained; relics of the *Apostle were totally absent.

A second phase followed, on which I myself worked, accompanied for some time by A. Prandi and D. Mustilli. But it fell to me alone to fill those gaps. Thus, I recognized the name of Peter both on "wall *g*" and in a mausoleum in the necropolis (mausoleum of the *Valerii*); I completely deciphered the graffiti of "wall *g*", which were revealed as an outstanding example of mystical *cryptography and a wonderful page of Christian *spirituality (with, among other things, numerous acclamations of the victory of Christ, Peter and *Mary); I traced and identified the surviving relics of Peter, which - through a set of strange but explicable happenings - had escaped the excavators of 1940-1949, and demonstrated that they had been transferred in the time of Constantine from the earth tomb under the "trophaeum of Gaius" to the specially prepared resting-place inside "wall *g*". I attained certainty in the identification of the relics in 1964.

An effective contribution to this certainty was made by the experimental sciences, above all by the anthropological research carried out by Prof. Venerando Correnti. The ancient deposition of the relics in the *loculus* of "wall *g*" was later corroborated by a precious Greek graffito incised on the inside of the *loculus* itself, on the surface of the so-called red wall: "Peter is in here". The identification of Peter's relics was announced by Paul VI in 1968 and subsequently confirmed by him several times, each time more decisively. The church's age-old tradition about Peter's tomb had received definitive confirmation from science. **[Figs: 317, 318]**

B.M. Apollonj Ghetti-A. Ferrua-E. Josi-E. Kirschbaum, *Esplorazioni sotto la Confessioni di San Pietro in Vaticano*, Vatican City 1951; A. Prandi, *La zona archeologica della Confessio Vaticana*, Vatican City 1957 (republished with additions in *Convegni del Centro di studi sulla spiritualità medievale*, 4, Todi 1963, 283-447, figs. 1-159); M. Guarducci, *I graffiti sotto la Confessione di San Pietro in Vaticano*, Vatican City 1958; ead., *Le reliquie di Pietro sotto la Confessione della Basilica Vaticana*, Vatican City 1963; ead., *Le reliquie di Pietro sotto la Confessione della Basilica Vaticana: una messa a punto*, Rome 1967; ead., *Pietro ritrovato*, Verona ²1970; ead., *Pietro fondamento della Chiesa*, Rev. Fabbrica di San Pietro, Vatican City 1977; ead., *Pietro in Vaticano*, Rome 1983; J.M.C. Toynbee - J. Ward-Perkins, *The Shrine of St. Peter and the Vatican Excavations*, London 1956; J. Carcopino, *Les Reliques de Saint Pierre à Rome*, Paris 1965; C. Smith, Pope Damasus' Baptistery in St. Peter's: A Reconsideration, *RAC* 64 (1988) 257-286.

M. Guarducci

VEDASTUS, saint. Born, acc. to his *Life* (BHL 8501-05), on the Limousin-Périgord border, but, acc. to a recent hypothesis, at Chalus; V. became a hermit near Toul (*tullo*), whose bishop ordained him priest. *Clovis, whose catechist he may have been, took him to *Rheims (*Remis*). *Remigius had him appointed bishop of Arras and later of Cambrai (*Cameracum*), with the task of bringing those regions, which had become *pagan with the arrival of the *Franks, back to Christianity. He died in 540, after 40 years as a bishop. His tomb became the centre of a powerful abbey. We owe his first *Life* to *Jonas of Bobbio (BHL 8501-03), another to Alcuin (BHL 8506). Feast 6 Feb, translation 5 Oct.

Vies des SS. 2, 135-38; *LTK* 10, 649; *BS* 12, 965-68; M. Rouche, *L'Aquitaine des Wisigoths aux Arabes*, Paris 1978, 428.

V. Saxer

VENANTIUS FORTUNATUS. Latin Christian poet, born *c.*530 at Duplavilis (Valdobbiadene). His literary formation, guided by *Arator, was much influenced by the Byzantine milieu of *Ravenna. His classical knowledge appears in his numerous borrowings from Virgil, Ovid, Lucan and, in lesser measure, Horace and other silver age poets; his biblical culture from frequent Scriptural citations, esp. from the Psalms, Isaiah and the Gospels. He shows signs of knowledge, however partial, of some *Fathers, echoing *Sedulius, *Orosius, *Caesarius of Arles and *Hilary of Poitiers. Healed of a serious eye-disease after being *anointed with *oil from the lamp that burned before the image of St *Martin of Tours in the basilica of Sts John and Paul, around 565 he left for *Gaul on a *pilgrimage of thanksgiving to his protector's tomb, but also perhaps to escape the vexations heaped on him by the authorities of Ravenna for his nonconformist attitude in the matter of the *Three Chapters. In 566 he was at Metz (*Mettis*) at the court of Sigebert I, king of Austrasia, whose marriage to Brunhilde he celebrated in song, thus acquiring a position of prestige. Having made his pilgrimage to *Tours (*Turones*), in 567 he settled at *Poitiers (*Pictavium*) at the insistence of ex-queen *Radegund who, with her daughter Agnes who had taken the veil, had retired to a *monastery. He also spent some time at the court of Chilperic, king of Neustria. Consecrated bishop of Poitiers in 597, he died there soon after 600. An important though debatable figure, he lived through some of Gaul's darkest historical moments: the intrigues of queens Fredegund and Brunhilde, the rivalries between Chilperic and Sigebert, the assassinations of Sigebert and Galaswinth. At times he kept quiet, perhaps out of prudence, at times he adulated the Merovingian sovereigns, following an accepted literary practice. His lyrical effusions over Radegund and Agnes must be understood in a spirit of spiritual transfiguration.

Works: 11 books of *Miscellanea*, *c.*300 compositions mostly in verse, on sacred and secular subjects; *Vita S. Martini*, a hexametrical reworking, with personal references, of the *biography by *Sulpicius Severus; *Vitae* in prose of Sts Hilary, *Germanus, Albinus, Paternus, Radegund, Marcellus, Amantius, Remedius, Médard, Leobinus, Maurilius. The religious lyrics contain V.'s whole Christian subject-matter: the sense of death, the cult of the *cross, devotion to *Mary guide of souls. His search for the transcendent finds singular expression in his travel poetry: a pilgrim, he always recommends the sanctuaries he encounters or aims for, understood as points of privileged encounter with the supernatural. As a hagiographer, given the panegyrical nature of the *Vitae*, he is more interested in moral edification than in historical accuracy, though he does not fail to emphasize at times his status as an eyewitness. In these biographies we notice thaumaturgical and demoniac elements, and we see a marked propensity for aristocratic circles. V. F. is remembered above all for his two *hymns *Vexilla regis prodeunt* and *Pange lingua gloriosi*, written in 568-569 for the solemn entry into Poitiers of a relic of the Cross and, because of their intensity, soon inserted into the *liturgy.

CPL 1033-1052; PL 88; *BS* 12, 985-987; D. Tardi, *Fortunat. Etude sur un dernier représentant de la poésie latine dans la Gaule mérovingienne*, Paris 1927; J. Laporte, *Le royaume de Paris dans l'oeuvre hagiographique de Fortunat*, Poitiers 1953; K. Steinmann, *Die Gelesuintha-Elegie des Venantius Fortunatus (Carm. VI, 5)*, Zürich 1975; L. Navarra, A proposito del "De navigio suo" di V.F. in rapporto alla "Mosella" di Ausonio e agli "Itinerari" di Ennodio, *SSR* 3 (1979) 79-131; B. Brennan, The career of Venantius Fortunatus, *Traditio* 41 (1985) 49-78.

L. Navarra

VENERIUS of Milan. Deacon during the episcopate of *Ambrose, he became bishop of *Milan on the death of *Simplicianus, in 400 or 401 (Paul. Mil., *Vita Ambr.* 46; Paul. Nol., *Ep.* 20, 3). He found himself mixed up in the *Origenist controversy: pope Anastasius I, in a *letter to *John of Jerusalem (400 or 401), claims to have written previously to V. on the subject of *Origen's condemnation (Anast., *Ep.* 1, 5; Jaffé nn. 281 and 282; Mansi 3, 944). In June 401 V. received from the bishops meeting in *council at *Carthage a request to assist the needs of the African *clergy (Mansi 3, 749-752). In 404 *John Chrysostom appealed to him, as well as to pope *Innocent and *Chromatius of Aquileia (Pall., *Vita Chrys.* 2), for support against the hostility of *Theophilus of Alexandria and against his own deposition; *Palladius mentions a letter of V. on the subject (*Vita Chrys.* 4); Chrysostom's *Ep.* 182, written in 406, is also addressed to V. *Ennodius dedicated one of his poems (*Carm.* II, 79) to V.

BS 12, 1009-1011 (bibl.).

A. Pollastri

VERANUS of Vence. Second son of St *Eucherius of Lyons. V. found himself on St Margaret's Isle when his father, by agreement with his wife, retired to *Lérins. There he became a disciple of *Vincent. His father addressed to him the *Formulae spiritualis intelligentiae* on biblical *exegesis. We find him bishop of Vence (*Vincium*) in 451: his brother *Salonius was bishop of Geneva (*Genava*) from 440. At this period he put his signature to the Arlesian *letter to *Leo the Great. With bishops Ceretius of Grenoble and Salonius of Geneva, he wrote to the same *pope on the subject of the *Tomus ad Flavianum* (PL 54, 887). V. is always appearing in the correspondence of pope *Hilarus up to 465 (Jaffé 555, 562); perhaps also in the letter of *Lucidus, in which case he will have lived until 474. With his brother Salonius he appears in the *episcopal list of *Lyons, in which they were inserted because of their father. Acc. to *Mart. hier.* (11 Nov), he was buried at Lyons (*Lugdunum*).

Vies des SS. 9, 346; *LTK* 10, 670; *DHGE* 15, 1315-17; *BS* 12, 1021-23.

V. Saxer

VERECUNDUS of Iunca. Neither the date nor place of the 6th-c. African writer V.'s birth are known exactly. We know for sure that he was bishop of Iunca in Byzacena and played an important role in the controversy over the *Three Chapters, during which he held firm against the imperial authorities, so much so that in 551 he was forced to flee with pope *Vigilius and take refuge at *Chalcedon, where he died in 552. To V. certainly belong: the *Commentarii super Cantica Ecclesiastica*, which comment allegorically on nine OT canticles, following *Origen and *Augustine but not without vigour and originality; the *Carmen de satisfactione paenitentiae* in 212 hexameters, which, sincerely but not without literary complacency, confesses, like *Dracontius before him, the sufferings of one who, after sad experience, rests his *hopes on divine mercy. Attribution to V. of the *Excerptiones de gestis Chalcedonensis Concilii* is doubtful; that of the *Exhortatio paenitendi* and the *Lamentum paenitentiae*, the *Carmen ad Flavium Felicem de resurrectione mortuorum* and the *Crisiados Libri III* (the work of a humanist) is unsustainable.

CPL 869-871; PLS 4, 39-234; J.B. Pitra, *Spicilegium Solesmense*, IV, Paris 1858; R. Demeulenaere, CCL 93, Turnholt 1976; H. Scheider, *Die altlateinischen biblischen Cantica*, Beuron 1938; E. Kullendorff, *Textkritische Beiträge zu Verecundus Iuncensis*, Lund 1943; O. Rousseau, La plus ancienne liste de cantiques liturgiques tirés de l'Ecriture, *RecSR* 35 (1948) 120-129; M. Schuster, Verecundus, *PWK* 27/1, 1010-1012; A. Hudson-Williams, Notes on Verecundus, *VChr* 6 (1952) 47-51; L. Brou, Etudes sur les Collectes du Psautier, I, La série africaine et l'évêque Verecundus de Junca, *SEJG* 6 (1954) 73-95; C. Magazzù, *Tecnica esegetica nei "Commentarii super cantica ecclesiastica" di Verecondo di Junca*, Messina 1983; M.G. Bianco, Verecondo di Iunca, un poeta ancora trascurato del VI secolo, in *Disiecta membra poetae*, I, Foggia 1984, 216-231; J.L. Kugel, Is there but One Song?, *Biblica* 63 (1982) 329-350.

S. Costanza

VERONA. Important Roman city of N *Italy which preserves numerous remains of that period. Being in an area of considerable strategic importance, it was often a battleground: *Decius defeated *Philip the Arab there in 249, *Constantine defeated the army of *Maxentius, *Stilicho defeated *Alaric in 452 and *Theodoric defeated *Odoacer. The emp. *Gallienus rebuilt and enlarged the walls. Christianity must have been organized in V. in the course of the 3rd c., after *Aquileia, as the Veronese *episcopal lists suggest (the 8th-c. *Veil of Classis* and the early 9th-c. *Versus de Verona* [cf. MGH, *Poet. aev. Car.* 1, 121 ff.]), acc. to which the first bishop was a certain Euprepius. But the first bishop we know for certain, from other sources, is Lucilius (or Lucillus), who took part in the council of *Sardica in 343/4 (Mansi 3, 38; Athan., *Apol. c. Ar.* 50: PG 25, 338; *Apol. ad Const.* 3: PG 25, 599). The eighth of the series is *Zeno (after 356), of African origin, the most important pastor of the ancient Veronese church, to which he gave a great impetus with his work of *evangelization and *organization, in a still largely *pagan environment with a strong Jewish presence; in his time Christian *church buildings were still few (cf. *Tract.* II, 6, 2: CCL 22, 168). Zeno also favoured the *ascetic life, since one of the first Western communities of *virgins was at V.

In 380 the bishop of V. was Syagrius, to whom *Ambrose addressed two *letters rebuking him for the procedure followed towards the virgin Indicia (Ambr., *Epp.* 5 and 6: PL 16, 929-943). In this period the Veronese church was dependent on that of *Milan, though it later passed over to depend on *Aquileia. In the 5th c. we meet bishop Petronius, the 13th of the series, to whom Lanzoni attributes two *Sermones* usually ascribed to *Petronius of Bologna (CPL 210-211: PLS 3, 141-143).

In the early Middle Ages V. was furnished with a *scriptorium*, whose first known ammanuensis was Ursicinus (517). Its Capitular *Library is very rich in ancient MSS, many of which come from other *scriptoria*. Among the most famous MSS are St *Augustine's *De civitate Dei*, *Hilary's *In Psalmos* and *De Trinitate*, the *Purple Evangelary*, the *Didascalia apostolica* (all 5th-c.), the work of *Sulpicius Severus transcribed in 517 by Ursicinus, and various sacramentaries including the *Sacramentarium Leonianum* (cf. E.A. Lowe, *Codices latini antiquiores*, IV, Oxford 1947, 32-40). For V.'s early Christian remains, cf. *Italy.

DACL 15, 2959-2962; EC 12, 1282-1299; F. Lanzoni, *Le diocesi d'Italia*, Faenza 1927, 929-934; A. Grazioli, La giurisdizione di Milano a Verona all'epoca di s. Ambrogio, *Scuola Catt.* 68 (1940) 373-379; G.B. Pighi, *Versus de Verona. Versus de Mediolano*, Bologna 1960; P.L. Zovatto, Arte paleocristiana a Verona: il cristianesimo a Verona, in *Verona e il suo territorio*, Verona 1960, 553-613; E. Lodi, Le due omelie di san Petronio vescovo di Bologna, in *Misc. Liturgica card. G. Lercaro*, II, Rome 1967, 263-301; AA. VV., *Storia della cultura veneta*, I: *Dalle origini al Trecento*, Vicenza 1976 (various articles); G.B. Pighi, *Cenni storici sulla Chiesa Veronese*, I, Verona ²1980; P. Testini et al., La Cathedrale in Italia: Schede, Verona, *Actes XIe Cong. internat. d'arch. chrét.*, CEFR 123, Rome 1989, 203-205 (bibl.).

A. Di Berardino

VERUS of Orange. V. is habitually presented as St *Eutropius's biographer and successor at Arausio, but in fact we know him only as author of the latter's *biography (BHL 2782). In the *titulus* of the text we read: *Domino sancto papae Stephano Verus*; the "*pope" Stephen in question was the bishop of *Lyons, who held the see from 501 to 515. As regards V., we would like to be more certain of his being a bishop. He may only have been a *cleric. In this case, Stephen would be Eutropius's successor and the predecessor of Florentius, who appears in 517.

Duchesne, *Fastes* I, 266; E. Griffe, *La Gaule chrétienne à l'époque romaine*, 2, Paris ²1965, 256-57.

V. Saxer

VESPASIAN, emperor. Titus Flavius V., Roman emperor AD 69-79, founder of the military dynasty of the Flavii (his sons *Titus and *Domitian were at his side immediately after his proclamation by the legions of *Egypt, *Syria and Judaea. Flavius Josephus, the Jewish historian who worked at the Flavian court, disapproving the violence of extremists against the Christians and the killing of *James (*Antiq. Iud.* XX, 200 ff.), shows us the attitude of the Flavian dynasty towards the followers of the new religion. Two fragments of *Hegesippus, cited by *Eusebius (*HE* III, 12; III, 20, 1-7), allude to the action brought by V. and his sons against *David's descendants in *Palestine and their attempt to ascertain whether the Christians were a political threat. V.'s tolerance was due to his observation of the non-existence of a Christian danger and of the great difference between Judaism and Christianity. It should be pointed out that V., observant of traditional Roman religion, did not wish in life to be considered a god. He died of fever in 79, after having reorganized the army, carried out financial reforms and increased legal and legislative activities. When his son Titus returned from the *East in 71, he and V. celebrated a triumph for the destruction of *Jerusalem.

M. Sordi, *Il cristianesimo e Roma*, Bologna 1965, 95-103.

L. Navarra

VESTIARIUS or Vestararius. Custodian of the treasury and wardrobe (*vestiarium*) of high lay and church dignitaries, or of chapters, churches and *monasteries. At *Rome, from at least the 7th c., this name was given to the official in charge of the *vestiarium* of the Lateran Patriarchate, in which, as evinced by a passage in the *Liber Pontificalis* (Duchesne *LP* I, 328 f.), were preserved the pontiff's sacred *vestments and the various *cymilia episcopii*, sacred vessels and other precious objects, gifts of emperors and high dignitaries. From the *biography of Stephen III (768-772) (Duchesne *LP* I, 470), we learn that the building contained an oratory dedicated to St *Caesarius, whose position - and consequently that of the *vestiarium* - within the Lateran Palace is unknown (Lauer, *Le Palais de Latran*, 96, n. 2). Within the province of the *vestiarium* administration and, in particular, of its head, the *v.*, also lay the preparation, carrying out and recording in special *books of the papal activities, such as foundations, construction of buildings, restorations, *donations. On the basis of this consideration, it has been hypothesized (Geertman, *More Veterum*, 34, 63) that the compilation of the *Liber Pontificalis* was done under the aegis of the *vestiarium*, and that perhaps its writers should be considered as employed by this administration. From *Ordo Romanus I* (nos. 10, 22) it seems that on festival days the *v.* had to supply, putting his own seal on them, the sacred vessels and evangelaries needed for the *mass; in *processions he rode immediately after the *pope. Only one *v.*'s name - a certain Baiolus, mentioned in a Roman inscription of uncertain date and provenance (Diehl, 1325) - is known before the pontificate of Hadrian I (772-795), during which four persons filled this office: among them the future pope Leo III, in charge until his own election as pontiff (Duchesne *LP* I, 519, n. 77).

P.L. Galletti, *Del Vestarario di Santa Romana Chiesa*, Rome 1758; Duchesne *LP* I, pp. CLXII, CCXLIII-IV; P. Lauer, *Le Palais de Latran. Etude historique et archéologique*, Paris 1911; M. Andrieu, *Les Ordines Romani du haut moyen âge*, II, Louvain 1948, 43, 70, 73; H. Geertman, *More veterum. Il Liber Pontificalis e gli edifici ecclesiastici di Roma nella tarda antichità e nell'alto medioevo*, Groningen 1975.

V. Fiocchi Nicolai

VESTMENTS, LITURGICAL. Garments worn by sacred *ministers during religious functions. If, as it seems, in ancient religions (looking particularly at the Jewish) it was quite natural to wear special clothing for particular religious circumstances, the same was not true of the pre-Nicene church. In this church, whether out of a spirit of distinction and contrast with practices and customs recorded by the old Mosaic Law or whether because of the more restrained and reserved character of primitive Christian worship, the clothes ordinarily used in Graeco-Roman civil life (cf. *Dress) were also employed for the liturgical office. On the other hand it does seem that best or cleanest and, apparently, white clothes were quickly reserved for liturgical celebrations (Clem. Al., *Strom*. IV, 22, 141, 4; *Canones Hipp*. 37 and cf. later Jerome, *Adv. Pelag*. I, 29): so it seems logical to speak, from the late 2nd or early 3rd c., of clothes reserved for liturgical use, although perfectly identical in form and cut with the clothes of everyday life. When *Origen says (*In Lev. hom*. IV, 6) that *aliis indumentis sacerdos utitur dum est in sacrificiorum ministerio, et aliis cum procedit ad populum*, he is undoubtedly speaking on two levels, one real and referring to the OT text, the other, as it were, moral and referring to the Pauline conception of ἐνδύειν. *Jerome, by choosing the word *habitus*, seems to retain the same perspective when, perhaps echoing Origen, he maintains (*In Ezech*. XIII, 44, 17 ff.: PL 25, 437): *religio divina alterum habitum habet in ministerio, alterum in usu vitaque communi*. Such testimonies, I believe, are not expressly referable, as is often claimed (cf. M. Righetti I, 585), to Christian liturgical vestments. Equally untrustworthy, because of its known late composition (5th-6th c.), is a statement in the *Liber Pontificalis* (XXIV, 3) under the time of *Stephen I (254-257): *constituit sacerdotes et levitas ut vestes sacratas in usu cotidiano non uti, nisi in ecclesia*.

At any rate a ἱερὰ στολή is mentioned in an important passage of *Theodoret of Cyrrhus (*HE* II, 23) which refers to the 330s and in addition demonstrates how, with the liberalization of worship and as a result of certain privileges and honours accorded to the Christian religion in the 4th c., part of the *clergy gave way to the temptation to employ quite precious materials in making up their own clothing, often to the point of enriching them with all kinds of ornaments and embroidery. Slightly later, we see the church's most prominent personalities showing some reluctance in this respect. Once *Augustine, e.g., had emphasized the simplicity of his manner of dress - certainly referring to that of his liturgical *ministry - he not only publicly rejected such affectations, but also disapproved of the tendency that followed from them, to make distinctions of dress within the various grades of clergy; further, he made it clear that any precious article of clothing given him for the sake of distinction would be sold and the proceeds distributed among the *poor (*Serm*. 356, 13 and cf. also *Ep*. 263, 1). *Celestine I (428) took a similar stand against such attempts at a hieratic restoration of the figure of the priest when some bishops of *Gaul introduced certain distinctions into their dress (. . . *amicti pallio et lumbos praecincti*) which, in contrast to the customs of the *Fathers, tended to give greater value, in worship, to the letter than to the spirit of Holy *Scripture (Lk 12, 35). The clergy - he maintained - must be distinguished from the people *doctrina, non veste; conversatione, non habitu; mentis puritate, non cultu* (*Ep*. 4, 1, 2: PL 50, 430 f.).

The substantial identity in style between clothes commonly used in civil life (and by clerics) and vestments reserved for liturgical acts seems to have been kept up until, esp. with the impact of barbarian manners and customs, secular fashion gradually changed; then the clergy continued to retain the old clothing for liturgical celebrations, both out of the understandable sense of sacral inalterability in which customs linked to religion are often clothed and (as Theodoret and Augustine have already indirectly testified) because of the richness with which such garments were often tailored, which logically contributed to magnifying this sacrality and hence the office of the priesthood. Indeed, that by the end of the 5th c. there was an awareness of the distance that had grown up between the two kinds of dress can be deduced from some church regulations providing for the dress of the clergy even in civil life (cf. *Stat. Eccl. Ant*. 25-26: CCL 148, 171), making it possible to preserve the old vestments. Thus, e.g., can. 5 of the council of Mâcon (581 or 583) sanctioned that *nullus clericus sagum* (a shorter tunic) *aut vestimentum vel calceamenta saecularia, nisi quae religionem deceant induere praesumat*. Finally, it is important to record that in the late 6th c. (but it may have happened earlier) we have indirect knowledge of liturgical vestments worn over ordinary clothes (can. 12, council of *Narbonne, 589).

For the *East too, can. 27 of the so-called 2nd Trullan council (*Constantinople, 691) expressly forbids the clergy to wear anything other than the prescribed dress. We may therefore conclude, in the light of what has been said, that, alongside "liturgical" vestments, the clergy were recommended to continue wearing, even in ordinary life, one or more tunics as well as a cloak (*paenula, pluviale*, etc.). The sobriety of this costume was likewise emphasized by the choice to be made of rather simple materials and the prohibition against using showy colours (can. 1, council of *Narbonne, 589; can. 16, 2nd council of Nicaea, 787). But on this cf., among others, G. Retzlaff's recent contribution, mentioned in the bibliography.

Returning to liturgical vestments proper, it must be added that, because of a particular symbolic meaning assigned to them by ancient (esp. early mediaeval) *spirituality and often inherent in their ancient use, form and *colour, such garments, with other marks of distinction discussed below, ended by becoming an integral part of the liturgical *actio*. Can. 6 of the council of Mâcon (CCL 148A, 224) recommends, e.g., the *archbishop not to celebrate *mass without a pallium; and in the 9th c. Amalarius more explicitly understands the vestments prescribed for the divine office as *exemplum bonae conversationis* for the Christian people (*Lib. off*. II, 15). So it is certainly to the "letter" of this symbolism, perfectly understood for centuries - as is shown by *homilies preached on them from time to time - that we owe the preservation of these clothes up to our own day, though with more or less slight modifications.

As far as the assigned chronological limits permit, I will now discuss the various liturgical vestments, distinguishing them, as it is customary to do, into under-garments (alb and girdle, amice, cotta, rochet) and upper garments (chasuble, cope, dalmatic and tunic). I will also deal with other articles, classified by some as *insignia*, together with other specific signs that distinguish the various grades of the ecclesiastical hierarchy and are therefore an organic part of the liturgical wardrobe. Finally, as regards Oriental-rite paraments (Greek, Syriac, Armenian, Nestorian-Chaldaean, Coptic), these are less numerous than the Latin. More precise information will be given when necessary under the corresponding Western garment, while some iconographical references are deferred for convenience to the conclusion of this article.

1. Alb: derived from the *tunica talaris et manicata* of the Romans, which, broad and long, covered the whole body to the feet. It was also commonly called *camisia* because of its origin in the (shorter) secular garment, *alba* because of its colour, *linea* because of its material, and *colobium* if the sleeves were short or altogether absent. From the earliest times of Christianity it was given a particular meaning, one of purity and incorruptibility; for which reason it was later donned by the faithful immediately after receiving *baptism (cf. under *Colour: white). As a special liturgical vestment of priests, it is recorded in the West (398?) by can. 60 of the so-called 4th council of *Carthage (CCL 148, 176), which allows its use by deacons only *tempore oblationis . . . vel lectionis*, but earlier in the *East, by *Eusebius of Caesarea (*HE* X, 4, 2) under the term ποδήρης, and by *John Chrysostom (*In Mt. hom.* 82 [83], 6), who calls it χιτωνίσκος. In the *Byzantine rite the corresponding garment is the στιχάριον, worn by priests and deacons.

2. *Cingulum*: along with the alb we must consider the *cingulum* or girdle (ζωνή). In the ancient world this was tied round the waist when appearing in public. Its liturgical use seems to be hinted at by *Celestine (cf. above) in his letter to the bishops of *Gaul restating the traditional symbolic interpretation of Lk 12, 35 (. . . *in lumborum praecinctione castitas*). Amalarius takes the same line (*Lib. off.* II, 22, 2).

3. *Amice*: corresponds to the *humerale* or *anagolaium* (cf. *Ordo Rom.* I, 34: ed. Andrieu, II, 78). In Greek we have ἀναβόλαιον, ἀνάλαβος. It is a rectangle of cloth placed over the shoulders, before the alb in the Roman practice, and then passed under the arms to tie the garments to the body and thus make movement freer. At first a privilege of the pontiff and his deacons, it was extended to all clerics in the Carolingian period. Various hypotheses have been put forward as to its origins: some mediaeval liturgists like Honorius of Autun (*Gemma animae*, I, 201) hold it to be derived from the Hebrew *ephod*; a more recent suggestion (O. Casel: *JLW* [1927] 139-141) derives it from the *amictus* used by the Romans in their sacrifices, or, more simply, from the *focale* or *sudarium* which in civil life was put around the neck to preserve it from sweat or cold (J. Braun, *Die lit. Gew.*, 45-46). More ample and convincing evidence is brought by C. Callewaert (241-248) in favour of an Egyptian monastic origin. Amalarius (*Lib. off.* II, 17), looking at the way the amice is worn (*collum undique cingimus*), sees it as symbolizing restraint in the use of the tongue (*custodia vocis, castigatio vocis*). Guillaume Durand's Pontifical (I, 11, 16: ed. Andrieu, 357) says the same, since in the *ordination of the sub-deacon the amice was put on his head: a custom also encountered in some old monastic orders.

4. *Cotta* (surplice) and *rochet*: considered to be modifications of the *camisia*, they are mentioned here to complete the essential picture of the lower vestments, even though of relatively late use. The cotta, with long, wide sleeves, was derived from the *superpelliceum*, an ample, comfortable alb used in liturgical functions above the everyday padded garments, esp. in cold northern countries. Beginning as a choir garment (replacing the camisia in lower clerics), it was gradually shortened to the knee.

The rochet is distinguished from the cotta by its narrower, shorter sleeves. It is worn directly over the cassock and is proper to prelates. It has never been considered a liturgical vestment. Anciently reaching to the feet, it was derived from the camisia or *alba romana*, *succa* or *subuta* used by mediaeval churchmen as a daily garment over which the liturgical alb was then put.

5. *Chasuble* (φελόνιον): the upper garment worn by the priest when celebrating mass. Historically the *planeta* or *casula* was derived from the *paenula* (φαινόλης), an upper garment much used in the Graeco-Roman world up to the 4th-5th cc. for sheltering esp. from rain or cold and for travelling. It was in the form of a circular cloak into which the head was inserted through an opening from which often hung a hood (*cucullus*). It too having fallen off from civil use, it remained in use by the clergy both in daily life and in the liturgical office. It initally appeared in liturgical use under the name *amphibalus* (cf. Sulp. Sev., *Dial.* I (II), 1, 5 ff.: CSEL 1, 181).

The use of the chasuble, before the spread of the cope, does not seem to have been exclusively reserved for *mass: at *Rome it was worn, though for distinct functions, by nearly all clerics; but in the *Gallican liturgy it was very probably a prerogative of bishops and priests only (cf. Ps.-Germ. Par., *Exp. ant. lit. gall.* II: PL 72, 97). For some time the subject of various ornamental motifs and depictions which made it heavier and heavier, the *planeta* underwent various transformations which gradually shortened it, first just in front, then along the arms, until it became, from the 14th-15th c. to our own day, a sort of scapular. In the Eastern rite it has mostly preserved the old form. For Amalarius (*Lib. off.* II, 19, 2 and 26, 2), the *casula* designates the *opera corporis pia* (*fames, sitis, vililiae, nuditas*, etc.) proper to each member of the clergy. For other mediaeval authors like Rupert of Deutz (*De div. off.* I, 22), it represents Christ's clothing, i.e. the church.

6. *Cope*: acc. to J. Wilpert (I, 84), it originates not so much from the *paenula* as from the lengthening of the *lacerna* (a short summer cloak, open in front, lighter and more comfortable than the *paenula*) or from the *byrrhus* (cf. Aug., *Serm.* 356, 13), hooded, heavier and suitable for winter. But acc. to others (cf. C. Callewaert, 227 f.), it is derived from the 8th/9th-c. clerical and monastic *capa* or *cappa*, worn mainly in choir when taking part in minor functions or *processions. The hood was in time reduced to a sort of shield (*clipeus*) laid out over the shoulders. It seems to be unknown before the 9th c. (Amalarius does not name it in his list of liturgical vestments); for a first symbolic meaning of it (= *sancta conversatio*), we must go back to Honorius of Autun (*Gemma animae* I, 227).

7. *Dalmatic* and *Tunic*: the dalmatic is a sort of cassock, with sleeves shorter and disproportionately wide, ornamented with two perpendicular purple *clavi*. Anciently used as an upper garment by senators and distinguished persons, it later came to form part of the paraments proper to the pontiff and was later a distinctive and honorific habit conceded by the pontiff to his deacons, perhaps already from the 4th c. (*LP* I, 171). Gradually becoming an object of privilege for the deacons of various churches, towards the 9th c. it finally began to be worn by presbyters under the casula (cf. Walafr. Str., *De reb. eccl.* 24). It is certain however that in the West from the 11th c. the dalmatic was considered the upper liturgical garment of the deacon, while bishops and presbyters continued to wear it under the chasuble, to whose various *colours it was later made to conform. Looking for its symbolic meaning, Amalarius sees in the white dalmatic, intrinsic to the *ministry of the deacon, the application of Jas 1, 27, i.e. *religio sancta et immaculata* and the purity in which those devoted to *cura proximorum* are preserved (*Lib. off.* II, 21 and 26, 2). In the Greek rite the σάκκος by which bishops were, after a certain date, distinguished from priests, for whom the φελόνιον was reserved, must be considered a sort of dalmatic.

The tunic is the *dalmatica minor* or *tunica stricta* of the subdeacon. In its original form it was without *clavi*, shorter and narrower, with long, narrow sleeves. This too is mentioned as a pontifical garment - thus we must interpret *Ordo Romanus* I, n. 34, which speaks of two dalmatics - and it was already a subdiaconal tunic in the 6th c., but it was soon abolished by *Gregory the Great (*Ep.* IX, 12) and until the 9th c. subdeacons wore the chasuble as a liturgical overgarment.

Both dalmatic and tunic, because of their original white colour, were always considered garments of happiness and, as such, suitable for liturgical days of joy and feasting, their use being forbidden in Advent and Lent, exceptions being made for *Gaudete* and *Laetare* Sundays.

We must now consider some other garments usually classified as distinctive *insignia* of the various grades of the ecclesiastical hierarchy. Some of them are first referred to by *Paulinus of Nola (*Carm.* XXIV, 783 ff.: CSEL 30, 232 f.).

8. *Maniple*: subdiaconal *insigne* derived from a sort of ceremonial napkin (*mappa*) much in use at Rome. In the Roman rite it more correctly goes under the name *mappula*; corresponding to it in the Greek rite is the *enchirion* which, like the *epigonation* (a rigid rhomboid ornament hung from the belt), was worn on the right side and only by the bishop. The *Liber Pontificalis* is probably referring to this parament when, first in the life of *Sylvester (314-335) and then in that of *Zosimus (417-418), it mentions a *pallium linostimum* carried in the left hand by deacons (I, 171; I, 223). But it did not long remain a prerogative of deacons alone. Amalarius, who calls the maniple *sudarium*, sees in it the *piae et mundae cogitationes, quibus detergimus molestias animi ex infirmitate corporis* (*Lib. off.* II, 21, 2). The 13th-c. Roman Pontifical, however, interprets it as the fruit of good works (IX, 6: ed. Andrieu, p. 336).

The *epimanikía* worn in the East at the end of each arm of the tunic resemble more the sleeves of a cotta.

9. *Stole*: acc. to G. Dix (404), this is perhaps the oldest of the *insignia*. It is a parament of the deacon. Like the maniple it is worn by him over one shoulder; but by bishops and presbyters over the neck, with the ends hanging down in front. Presbyters, if vested in an alb, wear it crossed over the breast (can. 4, synod of Braga, 675). The term *stola* became more widespread in the 9th c.: its original name was *orarium* (ὠράριον, ἐπιτραχήλιον), probably a sort of bandage to protect the mouth against wind or other things (cf. A. Pincherle, 55), but which may already have become a distinctive sign of office or dignity in civil life, if it is true that the council of *Laodicea (end of 4th c.?) forbade the use of the *orarium* to subdeacons, lectors and cantors (cann. 22. 23). For Amalarius it is an emblem of humility, a necessary *virtue for the deacon in carrying out the duties of his office; i.e. it represents Christ's yoke, the Gospel, *onus leve ac suave* (*Lib. off.* II, 20 and 26, 2).

In the East, the term apparently used by *Isidore of Pelusium to indicate the stole is ὀθόνη (*Ep.* I, 136). In describing the office of the deacon, here too the reference is to the Lord's humility. But the Greek ἐπιτραχήλιον differs from the Western stole in that it has an opening through which the head passes, and it reaches almost to the feet.

10. *Pallium*: a liturgical *insigne* reserved for the *pope and metropolitan *archbishops. It consists of a strip encircling the neck like a ring, from which two short bands of equal width hang down before and behind. The use of the pallium in the Western church goes back to the 5th c. and is probably derived from another liturgical *insigne*, the ὠμοφόριον, already widespread around the year 400 among Eastern bishops. A theory of T. Klauser (*Der Ursprung . . .* , 17-19. 25 and nn. 23-26), who saw it as derived from a secular garment, privilege of dignitaries of the *Byzantine court, extended at the time of *Constantine to the episcopal authorities, now meets with some reservations (M. Marinone, 95 f.). It is certain, however, that at

the beginning of the 5th c. *Isidore of Pelusium (*Ep.* I, 136) developed its symbolic meaning and saw it as representing the spiritual *authority and and the pastoral care of the bishop. In the West the pallium, initially adopted as an exclusively papal *insigne*, was from the 6th c. a privilege extended to metropolitans and even to some bishops.

Besides the *rationale* (*superhumerale*), a late arm-ornament with two appendages before and two behind, worn on the chasuble as a sign of distinction by some bishops in Germany, France and Poland, we must also record here some pontifical *insignia*: the mitre, the pastoral staff, the ring, the pectoral cross and other accessories such as gloves and shoes.

11. Mitre (κίδαρις): ornament of *authority of bishops in particular ceremonies. Its origin as a liturgical emblem is uncertain. Similarities can be found in the *infula* or *mitra* (μίτρα) of *pagan priests (cf. Virg., *Aen.* II, 430; IX, 616; X, 538) or in the headgear of Jewish ministers (Ex 28, 4. 40; 29, 9; 39, 28; Lev 8, 9. 13). In the Christian world we more often encounter the *mitra* as the headgear of consecrated *virgins. Thus in *Optatus of Milevis (*De schism. Don.* II, 19). *Isidore of Seville later mentions it (*Orig.* XIX, 31, 4), distinguishing it from the man's *pileus*. The old *Hispanic liturgy (*Lib. ord.* 23: ed. Férotin, chs. 66-67) knew the *inpositio* of a *mitra religiosa* in the consecration of an *abbess. But the evolution of this headgear into a liturgical emblem seems to have occurred around the 9th c. and it is probable once more that, analogously to what happened for other items of clothing, the liturgical *mitra* was derived from a hat, *camelaucium* (καμελαύκιον), much used in ordinary life, mainly in the East, by emperors and high officials. For the 8th c. the *Liber Pontificalis* (I, 390) tells of a *camelaucium* used by the *pope on solemn occasions.

In the West we see a primitive wedge-shaped form, but in the Oriental rite it assumes the form of a crown (surmounted later by a cross) which, acc. to G. Dix (407), seems to have originated in the Persian turban.

The papal *tiara* (*LP* II, 296, 22) also seems to be derived from the *camelaucium*. It may be this that the late 9th-c. *Ordo Romanus* XXXVI, 55 alludes to when it speaks of a *regnum*, in the form of a helmet of white cloth, placed on the head of the newly-elected pope. It probably began to be distinguished from the mitre when, with the exercize of temporal power by the popes, another crown was added to it, gradually becoming three (*triregnum*) in the 14th c.

12. Pastoral staff (*baculus, ferula,* *virga, cambuta*): called by *Celestine I (*Ep.* 4, 1, 2) *regimen pastorale*. It is named again by can. 28 of the synod of *Toledo of 633 and by the *Liber Ordinum*, 19 (ed. Férotin, chs. 59-60), apropos of the consecration of an *abbat. P. Radó (II, 1468) opines for a probable Oriental derivation from the liturgical use of the episcopal baton (ῥάβδος, βακτηρία). As for its symbolic meaning, Isidore (*De eccl. off.* II, 5, 12) sees represented in it the episcopal *authority and office. But as a liturgical emblem its diffusion was not universal: particularly at Rome, where its use did not prevail until very late (G. Dix, 412 and n. 3).

*13. *Ring*: Augustine seems to refer to an episcopal ring (*Ep.* 59 [217], 2 [PL 33, 227]). But it is first mentioned as the emblem of a bishop in can. 28 of the 4th council of Toledo (633), at one with Isidore (*De eccl. off.* II, 5, 12), who expressly describes it as *signum pontificalis honoris, vel signaculum secretorum*. Its use, at first reserved for bishops, was later extended to *abbats as well. But the so-called *anellus piscatorius* used by the popes to seal *breves* does not seem to have appeared until about the 13th c.

*14. Pectoral *cross* (σταυρός): often ornamented with precious stones with a relic inside, it is worn hanging round the neck mainly by bishops as their distinctive emblem. Even the earliest examples are quite late: one of the first to speak of such a thing is Innocent III (*De sacro alt. myst.* I, 53). It is only mentioned here because some have supposed a derivation from the φυλακτήρια (as well as from the *encolpia) anciently worn on the breast by Christians (cf. Jerome, *In Matth.* IV, 23, 5) and which variously enclosed relics, sacred objects, or even sentences of the gospels.

15. Gloves (*chirothecae*, χειροθήκη): again we do not know when they were first used liturgically: some suppose it to have happened by the 6th c. (*DACL* 6, 624-627), but it more certainly goes back, acc. to the evidence of some inventories, to the 9th c. The use of gloves properly so-called remains perhaps unknown in the East, but in practice they were replaced, as we have seen, by ἐπιμανίκια, embroidered cuffs.

16. Shoes: consisted of an outer part, the sandal (*campagus*), and an inner part, the sock (*udo, caliga*). Already used by the upper classes, they must long have been a liturgical emblem. *Gregory the Great, e.g., speaks of them in a *letter to bishop *John of Syracuse (*Ep.* VIII, 27). At that time their use was reserved for *popes and Roman deacons. Amalarius (*Lib. off.* II, 25, 1 ff.) considers them a symbol of preaching.

17. Finally, for the sake of completeness, we will record that the papal paraments also included: the *falda* (an ample undergarment that dragged on the ground and was belted at the hips), the *succinctorium* (worn tied on the right of the girdle), and the *fano*, an oval garment worn on the arm over the chasuble and often confused with the amice and the maniple, which were also called *fano, phano*, as happened with the *velum* (cf. Radó II, 1459 f.) on which the celebrant offered the eucharistic *bread. These papal *insignia* are in general only mentioned in the Middle Ages.

As for the oldest iconographical evidence, we will mention here only the most important examples (others are in the bibliography). In the *mosaic of S. Agnese *extra muros* at *Rome (8th c.), one of the two popes is wearing a white tunic, and over it a chasuble and pallium. St Sixtus and St Optatus, in the fresco of the crypt of pope *Cornelius (6th-7th c.) in the area of the catacomb of S. Callisto, are also wearing an ample chasuble on which the pallium stands out. St *Sylvester is represented in papal garments in the mosaic of S. Martino ai Monti (5th c.) at Rome. In the fresco of the crypt of the "Velatio" (late 3rd c.) in the Roman *cemetery of Priscilla, the left-hand group may represent a bishop with his deacon, but the matter is still much disputed (cf. C. Carletti: *VetChr* 9 (1972) 124-126). In the mosaics of S. Vitale at *Ravenna (6th c.), Ecclesius and *Maximian seem to be wearing a stole rather than a pallium. 8th-c. examples of monastic costume (cf. *Habit, monastic) are in the frescos of the Roman monastery of S. Sabina and in the oratory attached to the basilica of S. Ermete on the Salaria Vetus. Outstanding among the numerous examples of *encolpia* or pectoral *crosses is the fine example in Monza cathedral sent by *Gregory the Great to queen *Theodelinda.

Besides the articles Dress and Vestments, some historical considerations present under those headings in the *DCA* are still valid. Of a number of specific studies of vast breadth dedicated to this subject by J. Braun, we will cite only two: *Die liturgische Gewandung im Occident und Orient nach Ursprung und Entwicklung, Verwendung und Symbolik*, Freiburg i.Br. 1907 and *Die liturgische Paramente in Gegenwart und Vergangenheit*, Freiburg i.Br. ²1924, a revision of the *Handbuch der Paramentik* of which G. Alliod has prepared an Italian version (*I Paramenti sacri. Loro uso, storia e simbolismo*, Turin 1914). Certainly more rigorous, however, are the contributions of P. Battifol and C. Cecchelli (cf. *infra*). For a rather general panorama, we must also point out: G. Dix, *The Shape of the Liturgy*, London ²1945 (1975); P. Radó, *Enchiridion Liturgicum* II, Rome 1961; M. Righetti, *Manuale di Storia Liturgica*, Milan ³1964, 584-642, while, for more specific aspects, cf.: L. Eisenhofer, *Das bischöfliche Rationale. Seine Entstehung und Entiscklung*, Munich 1904; C. Callewaert, *Sacris Eruditi. Fragmenta liturgica collecta . . .*, Steenbrugge 1940; P. Salmon, *Etude sur les insignes du pontife dans le rite romain*, Rome 1955; id., La "ferula", bâton pastoral de l'Evêque de Rome, *RSR* 30 (1956) 313-27; id., Aux origines de la Crosse des Evêques, *Mélanges M. Andrieu*, Strasbourg 1957, 373-383; id., *Mitra und Stab. Die Pontifikalinsignien im römischen Ritus*, Mainz 1960; Th. Klauser, Der Ursprung der bischöflichen Insignien und Ehrenrechte, in *Gesammelte Arbeiten zur liturgiegeschichte, kirchengeschichte und christlichen Archäologie*, JbAC Erg. 3, Münster W. 1974, 195-211 (art. of 1949); B. Sirch, *Der Ursprung der bischöflichen Mitra und päpstlichen Tiara*, St. Ottilien 1975.

For Oriental and Byzantine liturgy, besides E. Trenkle *Liturgische Geräte und Gewänder der Ostkirche*, Munich 1962 and Papas Bibliografia sulle vesti religiose e liturgiche del rito bizantino, Θεολογία 52 (1981) 754-778 (in Greek) (other titles in n. 1), for a specific bibliography on the various rites and for the terminological correspondences of individual vestments the reader is referred in general to J. Assfalg-P. Krüger, *Kleines Wörterbuch des Christlichen Orients*, Wiesbaden 1975, s.v. Liturgische Gewänder. Useful for its identification of a symbolic and euchological aspect is R. Bornert, *Les Commentaires Byzantins de la Divine Liturgie du VII*ᵉ *au XV*ᵉ *siècle*, Paris 1966; C. Walter, *Art and Ritual of the Byzantine Church*, London 1982; H.J. Schultz, *The Byzantine Liturgy*, New York 1986.

The following, among others, have looked at the correct meaning of some terms: S. Bertina, 'Οθόνια ex papyrorum testimoniis linteamenta, *Studia papyrologica* 4 (1965) 27-38; A. Pincherle, Orarium and Sudarium, *RSLR* 9 (1973) 52-6; G.J.M. Bartelink, Φυλακτήριον-phylacterium, *Mélanges Chr. Mohrmann*, Utrecht-Amsterdam 1973, 25-69. For research into iconographical evidence cf., besides G. Wilpert, *Le pitture delle catacombe romane*, Rome 1903, the contribution of P. Battifol, Le costume liturgique romain, *Etudes de Liturgie et d'Archéologie chrétienne*, Paris 1919, 30-83, that of C. Cecchelli, *La vita di Roma nel Medio Evo*, I, 2: *Le arti minori e il costume*, Rome 1960, 755-840 and particularly 933-1011 and, for a more particular aspect, M. Marinone's other recent contribution, Un'antica testimonianza iconografica sull'uso della stola diaconale in Occidente, *EphemLiturg* 90 (1976) 88-99. For the East cf. N. Thierry, Le costume épiscopal byzantin du IXᵉ au XIIIᵉ s. d'après les peintures datées (miniatures, fresques), *REByz* 21 (1966) 308-15; id., Les plus anciennes représentations cappadociennes du costume épiscopal byzantin, *REByz* 34 (1976) 325 ff; M.G. Houston, *Ancient Greek, Roman and Byzantine Costume*, London ²1931, ch. 5.

On the religious habit in general, cf. the articles "Abito religioso" and "Costume dei monaci e dei religiosi" in *DIP* 1, 50-79 and 3, 204-49; M. Augé, *L'abito religioso*, Rome 1977; G. Retzlaff, Die äussere Erscheinung des Geistlichen im Altag, *Intern. kirchliche Zeit.* 69 (1979) 46-57, 88-115, 129-208; K. Michalowski, *Faras*, Warsaw 1974; id., *Faras, Die Kathedrale aus dem Wüstensand*, Zürich 1967 (illustrations of vestments worn in 10th-11thc. AD).

V. Pavan

VICES. See **VIRTUES and VICES.**

VICTOR I, pope (189?-198?). With V. the *papacy entered a new phase of its history. For the first time the interventions of the community of *Rome were crystallized into the person of its bishop and this in matters of a universal *discipline, that of the celebration of *Easter. Except for the church of *Ephesus under *Polycrates (Euseb., *HE* V, 24, 1-8), all the communities shared the Roman *tradition of celebrating Easter on a *Sunday instead of on 14 Nisan, though not all, e.g. *Irenaeus of Lyons (Euseb., *HE* V, 24, 1-18), approved of the Roman bishop's authoritarian procedure. V. was no less decisive against *adoptianism, *excommunicating *Theodotus the Elder (the Tanner). V.'s writings are not preserved. They are only mentioned by *Eusebius (*HE* V, 24, 9) and *Jerome (*Vir. ill.* 53).

LTK 10, 768 f. (bibl.); *BS* 12, 1281-1285; H. von Campenhausen, Ostertermin oder Osterfasten. Zum Verständnis des Irenäusbriefes an Viktor, *VChr* 28 (1974) 114-138; R. Cantalamessa, *La Pasqua nella Chiesa antica*, Turin 1978.

B. Studer

VICTORINUS of Petovium. We owe mainly to St *Jerome what little we know of the person and work of V., bishop of Petovium (*Vir. ill.* 74) in *Pannonia Superior (modern Ptuj in Slovenia), who lived in the second half of the 3rd c. and was a *martyr (*ibid.*) in the first years of *Diocletian's *persecution, prob. in 304. Forerunner of Latin biblical *exegesis, he dedicated the greater part of his literary activity to it, as we see from the numerous writings Jerome attributes to him: interpretations of *Genesis, *Exodus, Leviticus, Isaiah, *Ezekiel, Habakkuk, Ecclesiastes, *Song of Songs and, the sole surviving work, the *Apocalypse. And the list is incomplete: it ends with the expression *et multa alia*, and Jerome himself mentions elsewhere (*Comm. Matth., praef.* [CCL 77, 5, 96] and *Transl. Lat. homil. Origenis in Lucam, prol.* [PL 26, 231]) a *commentary by V. on *Matthew, to which the fragments published by G. Mercati from the Ambrosian MS I 101 sup. entitled *Anonymi chiliastae in Matthaeum fragmenta* may belong. The attribution to V. seems strengthened by linguistic-stylistic reasons as well as a conceptual one: its *millenarism, which closely recalls that of the *Commentary on the Apocalypse*.

Apart from the original text, preserved in the 15th-c. MS *Ottobonianus Lat.* 3288 A, the *Commentary on the Apocalypse* survives in three subsequent recensions. The first was made by Jerome at the request of one Anatolius: he radically revised the text, made additions (whose main source has been recognized as *Tichonius's commentary on the *Apocalypse*), suppressed and emended, improved the style, and above all, as an implacable enemy of millenarism, eliminated all trace of that doctrine; in compliance with this last criterion, he entirely replaced the ending, in which Rev 20-21 were expounded mostly *ad litteram*, with another strictly allegorical one composed *ex novo* by himself. Of the other two versions, the first repeats Jerome's reworking, retouching some points and citing the text of the Apocalypse more extensively, the other amalgamates all the previous versions, with some emendations and transpositions. In the light of the original version, the only one from which we can reconstruct V.'s genuine ideas on millenarism, we can say that his adherence to that doctrine was neither total nor unconditional. Following the traditional model, his millenium is characterized by the preponderance of material goods, but alongside these, which in V. are less crass than in *Papias or *Nepos or *Irenaeus, spiritual goods also coexist: "In this kingdom of those who were spoiled of their goods on account of the Lord's name, also the many killed with all kinds of wickedness and in prison . . . 'will receive their consolation', i.e. a crown and heavenly riches" (CSEL 49, 152, 16-20). Just like the millenarists, V. gives a fundamentally literal interpretation of Rev 20-21, but - and this is the most characteristic aspect of his millenarian kingdom - he does not hesitate to incorporate some allegorical explanations into the web of his literal commentary: e.g., spiritualizing the heavenly *Jerusalem as much as possible, he sees in its "gates of pearl", "the apostles", and in its "stones different in kind and colour", "the men and the most precious variety of their faith" (CSEL 49, 154, 12-14). In his exegesis, V. may be called a spiritual disciple of *Origen; in fact, even if he distances himself from Origen in adhering to millenarism, which Origen had opposed in defence of a spiritual interpretation of those biblical passages which the supporters of millenarism explained literally, he comes back to him in drawing allegorical meanings from the context of Rev 20-21, to which the millenarists applied a strictly literal exegesis. The evidence of Jerome, who informs us more than once of Origen's influence, becomes extremely important when he refers specifically to V.'s lost works, which were still available to him. From him, e.g., we learn that in the *sex alae uni et sex alae alteri* of Is 6, 2, V. saw allegorically the twelve *apostles (*Ep.* 18 A, 6 [CSEL 54, 82]), and in the *melior est puer pauper et sapiens quam rex senex et stultus* of Eccl 4, 13 he saw, following Origen, an *allegory of Christ and the devil (*Comm. Eccl.* 4, 13-16 [CCL 72, 290]).

V.'s predilection for Origenian-allegorical exegesis and his millenarist tendencies also manifestly emerge from the fragment of the *Tractatus de fabrica mundi*, attested by *Lambethana Bibliotheca Londiniensis* MS 414 (9th c.), which may be one of those "numerous other" works to which Jerome alludes without giving their titles, and which has been claimed as V.'s on the basis of particularities of language, style and thought. The author explains the seven days of *creation, dwelling with subtle considerations on the *number four and, particularly, on the virtues of the number seven, on its importance in the history of mankind and the development of the moral life. His millenarian beliefs are also evident when, on the basis of the citation, by now canonical, of Ps 89, 4 (*in oculis tuis, domine, mille anni ut dies una*), he relates the seven days of creation to the seven thousand years of the duration of the *world, and the seventh day of rest to the thousand years of Christ's kingdom, in which the Son of God will rule on earth in company with his elect.

Among V.'s works Jerome also mentions a treatise *Against all the heresies* (*Vir. ill.* 74), which Harnack has identified with the pseudo-Tertullianean work of that title (a rapid review of 32 *heresies, from *Dositheus to *Praxeas), appended in some MSS to *Tertullian's *De praescriptione haereticorum*. But this identification, which indeed its own author cooled towards in time, seems to be opposed by stylistic and chronological considerations.

Jerome's negative judgment weighs heavily on V.: while recognizing his countryman's good will and erudition, Jerome reproaches him more than once with not having full command of Latin, or at least not knowing that language as well as Greek (*Vir. ill.* 74; *Comm. Is., prol.* [CCL 73, 3]; *Ep.* 58, 10 [CSEL 54, 539]; *Transl. Lat. homil. Origenis in Lucam, prol.* [PL 26, 231]). And in *Ep.* 70, 5 (CSEL 54, 707) he does not even allow him erudition, but only *eruditionis voluntas*.

CPL 79 ff.; PL 5, 301-344; CSEL 59 (= PLS 1, 103-172); gives the four redactions of the *Commentary on the Apocalypse*. A. von Harnack, Geschichte der marcionitischen Kirchen, *Zeitschr. f. wiss. Theologie* 19 (1875) 11 ff.; J. Haussleiter, Die Kommentare des Victorinus, Ticonius und Hieronymus zur Apokalypse, *Zeitschr. f. kirchl. Wiss. u. kirchl. Leben* 7 (1886) 239 ff.; L. Atzberger, *Geschichte der christlichen Eschatologie innerhalb der vornicänischen Zeit*, Freiburg i.B. 1896, 566-573; G. Mercati, Anonymi chiliastae in Matthaeum fragmenta, *RIL*, series II, 31 (1898) 1203-1208 (ST 11, 3-49: with the text of the *fragmenta*; J. Haussleiter, *Beiträge zur Würdigung der Offenbarung des Johannes und ihres ältesten lateinischen Auslegers, Victorinus von Pettau*, Greifswald 1901; W. Macholz, *Spuren binitarischer Denkweise im Abendland seit Tertullian*, Jena 1902, 16 ff.; C.H. Turner, An Exegetical Fragment of the Third Century, *JThS* 5 (1904) 218-241; J. Barbel, *Christos Angelos*, Bonn 1941, 79-80; L. Bieler, The "Creeds" of St. Victorinus and St. Patrick, *ThS* 9 (1948) 121-124; J. Daniélou, La typologie millénariste de la semaine dans le christianisme primitif, *VChr* 2 (1948) 1-16; J. Fischer, Die Einheit der beiden Testamente bei Laktanz, Victorin von Pettau und deren Quellen, *Münch. Theol. Zeit.* 1 (1950) 96-101; J. Daniélou, *Théologie du judéo-christianisme*, Tournai 1958, 341-346; M. Simonetti, Il millenarismo in Oriente da Origene a Metodio, *Corona Gratiarum*, I, Brugge 1975, 37 ff.; C. Curti, Il regno millenario in Vittorino di Petovio, *Augustinianum* 18 (1978) 419-433.

C. Curti

VICTORINUS the Poet. Dekkers (CPL 1458-1459) attributes to one Victorinus poeta (perhaps 5th c.) two short poems: the first, entitled *De Pascha* (variants: *De cruce, De ligno vitae, *De ligno crucis*: beginning "*Est locus ex omni medius . . .* "), wrongly attrib. to *Tertullian or *Cyprian; and the other, *De Iesu Christo Deo et homine*: beginning "*Verbum Christe Dei, Patris*". This attribution is still much contested, as is the very existence of a poet named V. Acc. to Roncoroni (p. 387), it is always risky to attempt too easy attributions of the didactic compositions of the 4th and 5th cc., as well as of biblical epic poems of the same period, surviving almost anonymously. Antiquity itself was in great uncertainty, as is shown by false attributions to Tertullian, Cyprian and *Lactantius attested in the MSS. In any case, "attribution to a Victorinus is not documented by *manuscript tradition" (Roncoroni, 387, n. 21).

CPL 1458/1459; *De Iesu Christo*: PLS 3, 1135-39; *De ligno vitae*: CSEL 3, 305-308; A. Roncoroni, Ps. Cipriano. De ligno crucis, *RSLR* 12 (1976) 380-90.

L. Dattrino

VICTORIUS of Aquitaine. Mid 5th-c. Aquitanian scholar. *Gennadius calls him a *calculator scrupulosus* (*De vir. ill.* 89 [88]). Pope *Hilarus, when still *archdeacon of *Rome, wrote to him a first time to consult him on the *Easter *computus (PLS 3, 380-381). His reply consisted of an accompanying *letter and a table allowing the Easter computus to be worked out. This computus was adopted by the *council of Orléans in 541 and was widespread in *Gaul in the 7th and 8th cc., until the introduction of that of *Dionysius Exiguus. An extract exists under *Bede's name (PL 90, 712 and PLS 4, 2218). Acc. to Gennadius, V. composed a chronological table from the *creation of the world, used by *Prosper for his Chronicle and then by *Cassiodorus. He was prob. also the author of the *Prologus Paschae* (PLS 3, 381).

PLS 3, 381-426 (*Cursus Paschalis*); PL 90, 677 (*Liber calculi*, also ed. by Friedlein, Rome 1872); PLS 3, 427-441 (*Prologus Paschae*); E. Schwartz, *Christliche und jüd Ostertafeln*, Berlin 1905, 427-441; C. Jones, *Bedae Pseudepigrapha*, Ithaca 1939.

A. Hamman

VICTOR of Antioch. Presbyter (AD 500), wrongly known as author of a *commentary on *Mark; this text is part of a *catena, whose main sources are the *homilies on *Matthew by *John Chrysostom, *Origen, *Cyril of Alexandria, *Titus of Bostra and *Theodore of Heraclea. It still remains to be clarified whether the author of this compilation, who evidently used the catena on Matthew in its primitive form, was the same as the author of the OT and NT scholia going under V.'s name in various MS catenae.

CPG 3, 6529; C.F. Matthaei, Βίκτωρος πρεσβυτέρου Ἀντιοχείας κατ' ἄλλων, τινῶν ἁγίων πατέρων ἐξήγησις εἰς τὸ κατὰ Μάρκον ἅγιον εὐαγγέλιον. 2 vols., Moscow 1775; Commentaries on Luke in A. Maj, *Scriptorum veterum nova collectio e Vaticanis codicibus edita*, 9, Rome 1837, 633-693. Fragments of the exegesis on Jeremiah and Lamentations are in M. Ghislerius, *In Ieremiam Prophetam Commentarii*, Leiden 1623; individual scholia on Deuteronomy, Judges, Kings and Daniel are of little importance. 8 A, 2066; *LTK* 10, 791; *EC* 12, 1540; B. Altaner, *Patrologia*, 479; Beck 420 f.

J. Irmscher

VICTOR of Capua. Bishop of Capua 541-554 (cf. his tombstone inscription in CIL X, 4503). His major contribution to cultural history was the revision of the pre-Vulgate Latin text of *Tatian's *Diatessaron* to bring it into line with the Vulgate text of the gospels. His *Praefatio in evangelicas harmonias Ammonii* describes the circumstances that led him to revise Tatian's Latin text, a revision finished before April 546. V.'s gospel harmony is preserved in codex *Fuldense* (F) of the gospels (Fulda, Landesbibl., *Bonifatianus* I); it is of capital importance in Germanic studies, since it was the text normally used by High German translators for the various mediaeval *Diatessara*; though these may also be influenced by local variants of gospel texts, derived from liturgical or homiletic traditions: this would explain readings close to the *Vetus Latina* attested in High German *Diatessara*. V. also composed a *De Pascha* discussing the credibility of the *Easter cycle worked out by *Victor of Aquitaine: a fragment of it is handed down by *Bede (*Rat. Temp.* 51); 17 more fragments handed down under the name of *John the deacon are edited by Pitra (*Spicilegium Solesmense*, I, Paris 1852, 296-301). Six fragments of his short exegetical work *Reticulus seu de arca Noe* are edited by Pitra (*op. cit.*, 287-289). Also under V.'s name are *Scolia Veterum Patrum* (Polycarp [actually *Pacatus], *Origen, *Basil, *Didymus, *Severian), edited by Pitra (*op. cit.*, 265-277). V. also composed *Capitula de resurrectione Domini* (lost). Acc. to G. Morin (*RBen* 7 [1890] 416-423), we should attribute to V. a *letter to bishop Constantinus of Aquino, which, in some codices (e.g. the 9th-c. *Paris.* 9451), is premised to the Roman lectionary under *Jerome's name. In it the author explains the nature and function of lectionaries and claims to have composed one adapting the various biblical pericopes to the feasts of the liturgical cycle. But the *Lectionary* or *Comes Romanus* which follows the letter in the codices is not that outlined by V. and seems to be inspired by a later liturgical situation.

CPL 953a, 954, 955, 1976; PL 68, 251 f. (*Praefatio in evangelicas armonias*); PL 90, 490 (fragment of the Easter cycle); Schanz IV, 2, Munich 1920, 596 f.; *BS* 12, 1253-56; Moricca III, 2, 1529-30 (bibl.), 1559-61; *DTC* 15, 2874-76; B. Fischer, *Bibelausgaben des frühen Mittelalters*, in *La Bibbia nell'Alto Medioevo*, Settimane di Studio 10, Spoleto 1963, 545-557.

V. Loi

VICTOR of Cartenna. 5th-c. African bishop and writer mentioned only by *Gennadius (*De vir. ill.* 77), who attributes to him: a treatise *Adversus Arianos* dedicated to *Genseric (428-477), lost; a book on public *penitence, perhaps the pseudo-Ambrosian *De paenitentia* ascribed in two codices to *Victor of Tunnuna; a *consolatio*, lost, dedicated to a certain Basil on the death of his son (improbably identified with the *De consolatione in adversis* attrib. to *Basil of Caesarea: PG 31, 1687-1704); numerous *homilies collected into *books by V.'s brother, also lost. *De paenitentia* is an exhortation emphasizing the function of repentance for remission of *sins and the confidence which must be inspired by divine mercy, able to pardon even one who falls into sin again (it is disputed whether the treatise refers only to divine pardon or also to the custom of repeating confession before the bishop or a priest).

CPL 854: PL 17, 971-1004 (1059-1094); *DTC* 15, 2876 f.; Moricca III, 1, 746 f.

F. Scorza Barcellona

VICTOR of Tunnuna. 6th-c. African bishop (the name of his see is uncertain in the *manuscript tradition of his works, as in *Isidore of Seville [*Vir. ill.* 49. 50 and *Chron.* 1]). For his opposition to the condemnation of the *Three Chapters, he was exiled to *Egypt by *Justinian in 555. Recalled to *Constantinople in 564-565 in order to subscribe the condemnation, he maintained his position, for which he was shut up in a convent of that city, where he died soon after 566. Of his book of *Chronicles* from the *creation of the world to the first year of the emp. *Justin II (566), only the part relating to 443-566 remains. The work is interesting for the religious history of the 5th and 6th cc. and for its information on the author's own life. Up to 443, V.'s *Chronicles* must have depended on those of *Prosper of Aquitaine, as is asserted at the start of the long remaining fragment.

CPL 2260: PL 68, 941-962; T. Mommsen, *Chronica minora* 2 (MGH, *Auct. ant.* 11), Berlin 1894, 163-206; *DTC* 15, 2880 f.; *EC* 12, 1543; Moricca III, 2, 1364-1366; J. Fitton, The Death of Theodora, *Byzantion* 46 (1976) 119.

F. Scorza Barcellona

VICTOR of Vita. Bishop and author of a history of the *Vandal persecution in *Africa at the time of *Genseric and *Huneric. The supposed identification of this V., called bishop in the *manuscript tradition of his *History*, with bishop Victor Vitensis - mentioned in the *Notitia provinciarum et civitatem Africae* (a list of African bishops at the council of Carthage of 484) under the bishops of Byzacena (*Victor Vitensis non occurrit*) - clashes with what the author says in his *History*, that at the time of the events narrated he was a member of the *clergy of *Carthage (cf. *Hist.* II, 5. 18. 27. 40; III, 49, etc.). To this must be added the way in which he refers to the bishops, always in the third person and never including himself among their number (*Hist.* II, 53-55; III, 2. 5), and the fact that he was not, like them, exiled after the failure of the *council of 484 (*Hist.* III, 15 ff.). So the manuscript tradition must not be understood to mean that V. was bishop of Vita (*episcopus Vitensis*) during the persecution, but that Vita was his birthplace (*patriae Vitensis*) and that he, a priest of Carthage until Huneric's death, only later became a bishop (of Vita or another unknown see). It is hard to ascertain the year of his birth. That he relates the events of the landing of the Vandals (429) and the reign of Genseric (book I) in a more general way than those of the reign of Huneric (books II-III), and that for Genseric's reign he seems to use earlier sources, while for that of Huneric he appeals to his own testimony, suggests that in 429 V. was not born or was very young. This supposition is confirmed by the information that, when Valerian of Abensa was outlawed by Genseric (after 455), V. did not yet hold the priestly dignity (*Hist.* I, 40), usually assumed around the age of 30. The date of his death is uncertain. Acc. to Chifflet and Liron, V. was that bishop of Byzacena who ordained *Fulgentius of Ruspe in 507 and, soon after, was exiled by Thrasamund to *Sardinia, where he died in 510 or 512.

V.'s work - which in the archetype was entitled *Historia persecutionis Africanae provinciae temporum Geiserici et Hunerici regis Vandalorum* - begins with a prologue (whose authenticity and literary unity have been unjustly doubted), a *letter dedicated to an authoritative churchman, very probably bishop *Eugenius of Carthage. The historical narrative is divided into three books: book I contains the events of Genseric's reign from 429 (the invasion year) to 477 (Genseric's death); book II narrates the events of Huneric's reign from 477 to Feb 484 (council of Carthage); to this book is added the *Liber fidei catholicae* presented to the *Arians, to expound *Catholic doctrine, by four delegated bishops after the failure and suspension of the council of 484, and apparently written by Eugenius with the collaboration of other bishops; book III narrates the events of February to summer 484 (ch. 71, the last, describing Huneric's death, December 484, is an addition. The work ends with an invective against the barbarians and a call for help to God and men (chs. 61-70).

The *History* was written at Carthage during the persecution, before Huneric's death. At the beginning the author gives an explicit chronological reference: *sexagesimus nunc agitur annus ex eo, quo populis ille crudelis ac saevus Vandalicae gentis Africae miserabiles attigit fines* (I, 1). The dating of the work depends on the year *a quo* this *sexagesimus* must begin: whether 429, year of the Vandal invasion, or 430, when the occupation was completed, or 425, time of the first raids. Thus we have 488 (429+59), 489 (430+59) or 484 (425+59). The last date is the most probable, since it would also justify the final call for help to the Romans, which suggests an unsupportable situation of extreme peril, such as that immediately after the failure of the council of Feb 484.

V.'s *History* is an account of the persecution of the African Catholic church by the Arian Vandals, developed with the passion of a man belonging to the losing party and therefore not without sectarianism and bias. The reason for its composition is given in the prologue, addressed apparently to bishop Eugenius: to offer the recipient, as requested, an exposition of the events that occurred in Africa after the Vandal invasion, sure that he will know how to elaborate them suitably and give literary value to this still unrefined material. It is probable that Eugenius, who had come to fill the episcopal see of Carthage during the persecution, had asked V. for an account of the Vandal occupation, to be referred in turn to the Roman emperor to show the urgent need for immediate help. So the *History* was composed to inform the Romans overseas, through Eugenius, of the dramatic situation in Africa and ask the Roman emperor for help. Despite the limitations due to apologetic intent and partisan passion, despite some inexactnesses and contradictions (though not enough to impugn the overall credibility of the narrative), despite the exaggerations, fantasies and irrational interpretations of events and, in general, the naively theological interpretation of facts, and despite the distorting perspective which he mechanically places before Vandals and Romans as Arians and Catholics, V.'s work has considerable historical value because it is the only source for many of the events narrated, because it offers the modern historian important historical, political and geographical data, and because of the various documents inserted in the narrative, particularly important among which are Huneric's three edicts (II, 3-4; II, 39; III, 3-14), which shed light on barbarian-Roman law.

V.'s language is, though with certain popularizing novelties of common speech, that of the Carthaginian educated classes of the 5th c., with all the modifications that the times, the social and religious conditions of Africa and the influence of biblical literature and *liturgy had brought. Though he sometimes indulges in turgidity and prolixity, the writer succeeds in being an effective narrator, who knows how to interest and attract his reader.

CPL 798 ff.; PL 58; MGH, *Auct. ant.* 3, 1, Berlin 1879; CSEL 7, Vienna 1881. *It. tr.*: S. Constanza, Rome 1981.
Studies: A. Schönfelder, *De Victori Vitensi episcopo*, Breslavia 1899; G. Vismara, Gli editti dei re vandali, *Studi in on. di G. Scherillo*, II, n.d., 857 ff.; G. Ghedini, *Le clausole ritmiche dell'Historia persecutionis Africanae provinciae di Victor de Vita*, Milan 1927; G. Capello, Il latino di Vittore di Vita, *Atti della XXV riun. della Soc. Ital. delle Scienze a Tripoli* 25 (1937) 74-108; F. Di Capua, *Il ritmo prosaico nelle lettere dei papi e nei documenti della cancelleria romana dal IV al XIV sec.*, III, Rome 1946, ch. 5: Il ritmo prosaico nelle cancellerie dei regni romano-barbarici, 49 ff.; C. Courtois, *Victor de Vita et son oeuvre*, Alger 1954; id., *Les Vandales et l'Afrique*, Paris 1955 (repr. Aalen 1964); D. Romano, Osservazioni sul prologo alla Historia di Vittore Vitense, *AAPalser.* IV, 20 (1959-1960), I, 19-36 (= *Lett. e Storia nell'età tardoromana*, Palermo 1979, 155-172); S. Costanza, Uuandali-Arriani e Romani-Catholici nella Historia persecutionis Africanae provinciae di Vittore di Vita, *Oikoumene. Studi paleocr. in on. del Conc. Vat. II*, Catania 1964, 223 ff.; id., Considerazioni storiografiche nella Historia persecutionis Africanae provinciae di Vittore di Vita, *BStudLat* 6 (1976) 30-36; L. Alfonsi, L'Historia persecutionis Africanae provinciae di Vittore Vitense, ovvero il rifiuto di un ipocrita rinunciatorismo velleitario: "Romani" e "barbari", *SicGymn* n.s. 29 (1976) 1-18; A. Roncoroni, Vittore Vitense, Hist. III, 55-60, *ibid.*, 387-395; A. Pastorino, Osservazioni sulla "Historia persecutionis Africanae provinciae" di Vittore di Vita, in *La storiografia eccl. nella tarda antichità*, Atti Conv. di Erice, Messina 1980, 45-112; S. Costanza, Vittore di Vita e la "Historia persecutionis Africanae provinciae", *VetChr* 17 (1980) 229-268; R. Pitkäranta, *Studien zum Latein des Victor Vitensis*, Helsinki 1978 (cf. version by S. Costanza: *Orpheus* 2 [1981] 223-232).

S. Costanza

VICTRICIUS of Rouen. Acc. to *Paulinus of Nola (*Ep.* 18; cf. *Ep.* 37), V., born *c.* 340 "at the limits of the empire", renounced a military career to serve Christ, like *Martin of Tours. Ordained priest, he dedicated himself esp. to the *evangelization of the Nervi and Morini, who occupied what is now Flanders, Brabant and the Cambrésis. In 385 he became bishop of Rouen (*Rotomagus*). Like Martin, whom he visited, he gave a great impetus to the spiritual life of his community, built churches, watched over discipline, developed male and female monastic life. In *c.* 396 he went to *Britain to combat *Arianism. He wrote *de laude sanctorum* in praise of some *martyrs. In 403 he went to *Rome to exculpate himself from a charge of *Apollinarism. Died *c.* 410. Feast 7 Aug.

CPL 481; PL 20, 443-458; *De laude sanctorum*, ed. Sauvage (publ. by A. Tougard), Paris 1895; Fr. tr. by R. Herval, Rouen 1966; J. Mulders, Victricius van Rouaan, *Bijdragen Tijdschrift voor Ph. en Th.* 17 (1956) 1-25; 18 (1957) 19-40, 270-289; P. Andrieu-Guitrancourt, Essai sur Victrice, *Année canonique* 14 (1970) 1-23; E. Griffe, *La Gaule chrétienne*, I, Paris ²1964, 306-310, 383-385; *Vies des SS.* 8, 121-123; *BS* 12, 310-315 (bibl.); J. Fontaine, Victrice de Rouen et les origines du monachisme dans l'Ouest de la Gaule, in L. Musset, *Aspects du Monachisme en Normandie, IVᵉ-XVIIIᵉs.*, Paris 1982, 9-29.

A. Hamman

VIENNE

I. City and Christianity - II. Council.

I. City and Christianity. Colonia Julia Vienna in Narbonensis, later Civitas Viennensium or Vienna (in the 4th c.), metropolis of Viennensis and for some time capital of a civil diocese, one of the residences of the king of the *Burgundians in the 5th c., knew a fresh period of prosperity in the second half of the 6th c. It had one of the first Christian communities in *Gaul: in 177 Sanctus appears as deacon (Euseb., *HE* V, 1, 17 f.), but the first certainly attested bishop is Verus (council of *Arles, 314), though in a dubious list compiled by Ado of Vienne in the 9th c. he is preceded by Crescens (identified with Paul's disciple!), Zacharias and Martin. Despite its antiquity, the provincial expansion of the episcopal see was thwarted by the development of *Arles; after the obscure Justus, Dionysius, Parocos, Florentius and Lupicinus, *Simplicius tried to assert some metropolitan rights at the council of *Turin (late 4th c.), but pope *Zosimus (417-418) favoured the ambitions of *Patroclus of Arles (*Ep.* 5, *Coll. Arelat.* 5). After Claudius (. . . 441/2 . . .), Nectarius and Nicetas (. . . 449 . . .), Mamertus, despite his pastoral zeal (Rogations) and the intellectual brilliance of his brother *Claudianus (CPL 983-984), failed to expand outside the limited provincial jurisdiction allowed to his see through the arbitration of pope *Leo in 450 (*Ep.* 66; Hilary, *Ep.* 10). From the end of the 5th c. the new political geography favoured V. for a time, esp. after Hesychius, under the rule of his son, the famous *Avitus (494-518; CPL 990-996), active in Burgundian politics; after Julian (518/23-532) and Domninus, Pantagatus (538) and Hesychius (. . . 549-552 . . .) had held other posts before being bishops. But after Namatius (†559), Philip (. . . 570-573 . . .), Evantius (. . . 581-585 . . .) and Verus (586), Desiderius failed to obtain the pallium from *Gregory the Great and was exposed to the hostility of Aridius, bishop of *Lyons. The bishops of the 7th c., Domnulus, Aetherius, Clarentius, Sindulf, Ecdicus, Coaldus, Dodolenus, Bobolinus, George, Deodatus, Blidrannus, Eocald, Eobolinus and Austrobert, are not particularly famous. Wilicarius, who received the pallium from Gregory III (731-741), lived at the time of the Saracen invsion (725) and finally retired to a *monastery, acc. to Ado, the bishop and martyrologist who died in 875. Despite all these difficulties, the local mission was very active in the 4th and 6th cc., as inscriptions (esp. 6th-c.) and numerous monuments attest: in the city we count, besides the episcopal group (dedicated to the Maccabees), the temple of *Augustus and Livia consecrated to Christian worship (St Mary) and two monasteries (St André-le-Haut and St Nicetus); outside the walls, esp. on the right bank of the Rhône, are some basilicas: St Gervasius from the 4th c., the *basilica Stephani* in the 5th c., St Peter's (time of Mamertus) and some oratories; on the left bank, the 5th-c. *basilica Ferreoli.*. **[Fig: 319]**

L. Duchesne, *Fastes* I, 86-211; H. Leclercq, Vienne, *DACL* 15, 3038-3054; G. Langgärtner, *Die Gallienpolitik der Päpste*, Theophaneia 16, Bonn 1964; E. Griffe, *La Gaule chrétienne à l'époque romaine*, Paris ²1964, II, 146-189; 1965, III, 209-234, 323-325; F. Descombes, La topographie chrétienne de Vienne des origines à la fin du VIIᵉ, in *Les martyrs de Lyon (177)*, Paris 1978, 267-276; J.F. Reynaud et al., *Les Edifices funéraires et les nécropoles dans les Alpes et la Vallée du Rhone*, Actes XIᵉ Cong. internat. arch. chrét., CEFR 123, Rome 1989, 1475-1514.

Ch. Pietri

II. *Council. In about 471/5, Mamertus of V. got his colleagues to accept the Rogations and *fasts that he had instituted for the three days preceding the *Ascension of Jesus. This custom spread rapidly (Sidon. Ap., *Epp.* V, 14; VII, 1; Greg. of Tours, *Hist. Franc.* II, 34).

CCL 148, 137; Palazzini 6, 45-46; 6, 129.

Ch. Munier

VIGILANTIUS. Born at Calagurris in the *civitas* of the Convenae (Saint-Martory in Comminges, Haute-Garonne), the priest V. travelled to *Palestine in 395 distributing *alms from *Paulinus of Nola. He stayed with *Jerome at *Bethlehem, in harmony, despite some friction (Jerome, *Ep.* 58 to Paulinus). But Jerome took offence when he learnt that, on his return *journey to Nola and thence to *Gaul (Paul. Nol., *Ep.* 5 to *Sulpicius Severus), V. had described him as an *Origenist (Jerome, *Ep.* 61 to V.). In 403 Riparius, a priest of the *civitas* of *Toulouse near the region of Comminges, denounced to Jerome V.'s campaign against the cult of the *martyrs (Jerome, *Ep.* 109 to Riparius), prob. sparked off by the consecration at Toulouse of the basilica dedicated to the bishop-martyr *Saturninus. In 406, the Tolosan deacon Sisinnius brought V.'s writings to Jerome, who in one night drew up a violent *libellus, *Contra Vigilantium*: acc. to what we read there, V. condemned the cult of the martyrs, some liturgical customs, ecclesiastical *celibacy, monastic life and the alms sent to the Holy Land.

R. Lizop, *Les Convenae et les Consoranni*, Toulouse-Paris 1931, 248-258; P.-M. Duval, *La Gaule jusqu'au milieu du Ve siècle*, Paris 1971, 674 f.; H. Crouzel, Saint Jérôme et ses amis toulousains, *BLE* 73 (1972) 125-146; J.N.D. Kelly, *Jerome*, London 1975, 286-290.

H. Crouzel

VIGILIUS. To the deacon V. (cf. Gennad., *De vir. ill.* 51) is attrib. the *Regula orientalis*, composed c.420 perhaps in *Gaul. This *Monastic *Rule*, in 47 chapters (33 of which make use of *Pachomius), is on the whole very close to the long recension of *Jerome's translation of Pachomius's *Rule*.

CPL 1840; PL 103, 477-484 (= PL 50, 373-380 and PG 34, 983-990); A. Boon, *Pachomiana Latina*, Louvain 1932; C. De Clercq, L'influence de la règle de saint Pachôme en Occident, *Mél. L. Halphen*, Paris 1951, 169-176; A. Mundó, Les anciens synodes abbatiaux et les "Regulae SS. Patrum", *SA* 44 (1959) 107-125.

S. Zincone

VIGILIUS, pope (537-555). After the irregular deposition of *Silverius (537), V., of senatorial origin, got himself elected bishop of *Rome. Recognized by all after Silverius's death, he broke his promise to the empress *Theodora by confessing the faith according to the doctrine of *Chalcedon and *Leo. But *Justinian I, who in 543/4 had condemned the so-called *Three Chapters and obliged V. to come to *Constantinople, induced in him an attitude so hesitant that he lost all respect among the Western bishops, all the more since V., in his *Iudicatum* of 548, explicitly condemned the Three Chapters. V. protested against a new imperial edict *excommunicating patriarch *Menas of Constantinople. Indeed he refused to take part in the synod called by the emperor (553), justifying his position in his *Constitutum I*. Excommunicated by the synod, he was condemned to prison or exile. In two documents, including the *Constitutum II*, he revoked his refusal to condemn the Three Chapters, and was given permission to return to Rome. But he died at *Syracuse, and long remained dishonoured in the West.

CPL 1694-1697; PLS 4, 1249-1252; PL 69, 15-114; 143-178; ACO IV, II, 15-18 (PLS 4, 1250 f.) 138-168; K. Baus et alii, *Storia della Chiesa*, III, It. tr. Milan 1978, 37-43 (bibl.); J. Straub, Die Verurteilung der Drei Kapitel durch Vigilius, *Kleronomia* 2 (1970) 347-375; E. Zettl, *Die Bestätigung des 5. ökumenischen Konzils durch Papst Vigilius*, Bonn 1974; G. Every, Was Vigilius a Victim or an Ally of Justinian?, *Heythrop Journal* 20 (1979) 257-266; *PWK* Suppl. 14, 864 ff.

B. Studer

VIGILIUS of Thapsus. Took part, as bishop of Thapsus, in the dramatic talks between *Catholics and *Arians called by the *Vandal king *Huneric in Feb 484. That he subsequently fled to *Constantinople is a supposition of modern scholars. Various works have been wrongly attrib. to him, including the pseudo-Athanasian *De Trinitate*; but the *Contra Eutychetem* and the *Dialogus contra arrianos*... are certainly his, and prob. the pseudo-Augustinian *Contra Felicianum*. This brief work presents in *dialogue form the themes of the trinitarian controversy, always topical in Vandal *Africa, expounded on a purely rational basis without recourse to scriptural citations. The same material is treated in a much more vast and complete way in the *Dialogus contra arrianos, sabellianos et photinianos* in three books, where the dialogue between a Catholic and three heretical spokesmen aims to refute one *heresy by means of the other. But the main interest is anti-Arian, and in fact the work also survives in a recension shortened to two speakers, a Catholic and an Arian. In the *Dialogus*, V. mentions two more of his anti-Arian works, against Maribadus and against *Palladius of Ratiaria, both lost. The *Contra Eutychetem* in five books presents Catholic doctrine on matters of *christology as intermediate between the opposite errors of *Nestorius and *Eutyches, and defends the Chalcedonian doctrine that sees in Christ two natures, human and divine, coexistent, without confusion and without separation, in a single *person. In the last two books of the work, V. refutes, adducing various literal citations, two *monophysite works against *Chalcedon and against *Leo the Great, apparently otherwise unknown. Perhaps they should be identified with two works by *Timothy Ailouros.

CPL 806-812; PL 42, 1157-1172; 62, 95-238; Moricca III, 2, 764-773; M. Simonetti, Letteratura antimonofisita d'Occidente, *Augustinianum* 18 (1978) 505-522.

M. Simonetti

VIGILIUS of Trent. Third bishop of Trent (*Tridentum*) (385?-405), at the start of his *ministry V. received from *Ambrose a *letter accompanied by pastoral instructions (*Ep.* 17: PL 16, 1024-1036). Acc. to his *Life* (BHL 8602 ff.), written in the 6th c., he was distinguished for his missionary activity and the building of numerous churches. During his episcopate, in 397, Sisinnius, Martyrius and Alexander, his co-workers, were killed. V. himself sent an account of their *martyrdom to bishop *Simplicianus of Milan and, later, another to *John Chrysostom accompanied by relics. Still acc. to the *Life*, he himself was stoned by the *pagans, and became the patron of his diocese.

CPL 212 ff.; PL 13, 549-558; *LTK* 10, 789; *BS* 12, 1086-1088.

B. Studer

VINCENT, martyr. The oldest non-hagiographical information on St V. comes from *Paulinus of Nola (*Carm*. XIX, 164), *Prudentius (*Perist*. 4, 77-80, 179-80; 5 *passim*), and *Augustine's *sermons (*Serm*. 4, 274-277, *Caillou* 1, 47: PLS 2). These texts present V. as a deacon of bishop Valerius of Saragossa and *martyr of *Diocletian's *persecution; while the bishop survived, V. was condemned to death and executed at Valencia on 22 Jan 304.

On these facts is based his Passion (BHL 8627-37), in epic form, written before the end of the 4th c. It abounds in tortures, *visions and other customary and established features, showing how early these literary procedures were present in Spanish *hagiography. The wordplay on *Vincentius* and *vincere* certainly takes up a traditional theme, that of martyrdom as victory over the *demon, but they also inspired St Augustine's sermons and the Visigothic liturgical *prayers.

The cult of the saint spread from the 5th c. in *Spain, *Gaul and *Africa, as witness the churches erected in his honour and the diffusion of his relics. His feast, 22 Jan, appears in all the Mozarabic *calendars, the *Cal. Carth*. and that of Carmona (6th and 7th cc.), the calendar of *Polemius Silvius (448) and the *Mart. hier*.

PL 38, 1252-1268; *Miscellanea Agostiniana*, I, Rome 1930, 243-245; J. Vives, *Oracional visigótico*, I, Barcelona 1946, 157-166; *Vies des SS.* 1, 431-36; *LTK* 10, 802-3; *BS* 12, 1149-55; A. Fábrega Grau, *Pasionario hispánico*, Madrid-Barcelona 1953-55, 1, 92-107; 2, 187-96; C. García Rodríguez, *El culto de los santos en la España roma y visigoda*, Madrid 1966, 257-78.

V. Saxer

VINCENT. Presbyter of the church of *Constantinople (Jerome, *C. Joan. Hier.* 41), known to *Jerome in 380 and from then on his *friend and companion in travel (Jerome, *Ep.* 61; *C. Ruf.* 3, 22) and *ascesis. Jerome dedicated to him his *Latin translation of *Eusebius's *Chronicle*, since V. knew Greek and Latin.

A. Di Berardino

VINCENT of Capua. Bishop of Capua, perhaps the same as the Roman priest of that name who represented pope *Sylvester at *Nicaea (325). In 344 he was sent with *Euphratas of Cologne on a mission to *Antioch to mitigate the emp. *Constantius's anti-Nicene provisions, not without some result. Sent again to Constantius by pope *Liberius in 353, this time to *Arles, at the local *council there he submitted to subscribe the condemnation of *Athanasius. In 357 Liberius, exiled to *Thrace for his stand in favour of Athanasius, wrote to V., among others, asking him to work with the other bishops of Campania for the revocation of his exile.

DCB 4, 1152.

M. Simonetti

VINCENT of Lérins. Though the best-known of the Lerinese monks, we know little of his life (Gennad., *De vir. ill.* 65 [64]). He began as a soldier, then retired to *Lérins, where he was ordained priest. *Gennadius stresses his knowledge of the *Bible and of the history of *dogma. He died before 450, perhaps in 435. Feast 24 May.

V.'s famous *Commonitorium* (Memorial), which appeared under the pseudonym of Peregrinus, has sufficed to ensure his fame. The *Obiectiones Vincentianae* are lost. J. Madoz has discoverd at Ripoli the *Excerpta s. mem. Vincentii Lir.*: the first ever Augustinian florilegium.

The *Commonitorium*, which depends on *Tertullian's *De praescriptione haereticorum*, is not just a handbook, but a sort of "Discourse on Method" in *theology, to distinguish the *Catholic faith from *heresy: *Quod ubique, quod semper, quod ab omnibus* (ch. 2). The three criteria, then, are universality, antiquity and unanimity. V. applies them to some past controversies, such as *Arianism and *Donatism.

To avoid the principle of *tradition leading to theological immobilism, he enunciates the complementary rule: *Crescat... et multum vehementerque proficiat..., sed in suo dumtaxat genere, in eodem sc. dogmate, eodem sensu, eaque sententia* (ch. 23). The criterion would be taken up and consecrated by the first Vatican council.

V. had great influence from the 16th c., when Roman Catholics, Protestants, Old Catholics and Anglicans all appealed to him: above all Bossuet, later Newman. Few books have exercised such lasting influence.

Editions: *Commonitorium*: PL 50, 637-686; FP 5, Bonn 1906; R.S. Moxon, Cambridge 1915; *Excerpta*: ed. Madoz, Madrid 1940=PLS 3, 23-45; W.J. Mountain: *SEJG* 18 (1967-1968) 385-405 (MS of Novara). Fr. tr.: P. de Labriolle, Paris 1906 (repr. 1978); M. Meslin, Namur 1959; It. tr.: C. Colafemmina, Alba 1967.
Studies: J. Madoz, *El concepto de la Tradición en s. V. de Lérins*, Rome 1933; id., Un tratado desconocido de S.V., *Gregorianum* 21 (1940) 75-94; id., Los "Excerpta Vincentii Lirinensis" en la controversia adopcionista, *RET* 13 (1953) 475-483; E. Griffe, Pro Vincentio, *BLE* 62 (1961) 26-32; W. O'Connor, S. Vincent . . . und S. Augustine, *Doctor Communis* 16 (1963) 123-257; *Patrologia* III, 517-521 (bibl.); N.K. Chadwick, *Poetry and Letters in Early Christian Gaul*, London 1955; J. Pelikan, *The Emergence of Catholic Tradition (100-600)*, Chicago 1971, 319-333.

A. Hamman

VIRGA: *iconography. There appears in early Christian figurative art, with some frequency and from the earliest times, either isolated or more often as part of a context, a man holding a staff, which hermeneuts, following Wilpert (*Wp* 41 ff.), have always considered a thaumaturgical attribute and have conventionally called *virga virtutis*. In pre-Christian antiquity the *v.* signified primarily a symbolic instrument of power over animals, men and things, and secondarily a sign of didactic activity, esp. of philosophical discipline. To these meanings patristic *exegesis added that of sacramental, divine power and liberation from physical and spiritual imperfections and death (Aug., *En. in ps.* XLIV, 17-18; *Quaest. in Num.* XXXV; Ambr., *De myst.* IX, 50).

During the 3rd c. this attribute is exclusively associated with the figures of Christ, in scenes of the raising of *Lazarus, the multiplication of loaves and the miracle of *Cana, and *Moses, in the miracle of the spring. From the 4th c. we see it held either by Christ or by Moses-*Peter in the Petrine cycle, or by *Ezekiel in the raising of the dry bones. More exceptionally it is held by isolated characters, indefinable or vaguely allusive of doctrine, either in community *cemeteries (Fasola, *Le catacombe di S. Gennaro*, 68) or in private hypogea, incl. some of dubious Christianity or *orthodoxy (Boyancé, Aristote sur une peinture, 107 ff.; Himmelmann, *Das Hypogäum der Aurelier*, 24 ff.). Sometimes, in the same figurative context, two scenes appear with characters holding *vv.*, placed symmetrically to create a formal-symbolic harmony (Van Moorsel, Il miracolo della roccia, 235-236). **[Fig: 320]**

A. Hermann, Aegyptologische Marginalien zur spätantiken Ikonographie, *JbAC* 5 (1962) 60-69; L. De Bruyne, Les lois de l'art paléochrétien comme instrument herméneutique, *RAC* 39 (1963) 14 ff.; P. Van Moorsel, Il miracolo della roccia nella letteratura e nell'arte paleocristiana, *RAC* 40 (1964) 235-236; P. Boyancé, Aristote sur une peinture de la voie Latine, in *Mélanges E. Tisserant*, Vatican City 1964, 107 ff.; M. Dulaey, Le symbole de la baguette dans l'art paléochrétien, *REAug* 19 (1973) 3-38; N. Himmelmann, *Das Hypogäum der Aurelier am Viale Manzoni*, Main 1975; U.M. Fasola, *Le catacombe di S. Gennaro a Capodimonte*, Rome 1975, 68.

F. Bisconti

VIRGILIUS MARO, grammaticus. Native almost certainly of *Gaul, 6th-7th cc. It does not seem we can identify him with the person aimed at by *Ennodius in his *epigrams *De quodam stulto qui Virgilius dicebatur* (*Carm.* II, 118-122). The grammarian V. composed *epitomae* and epistles; the former are 15 condensed accounts of grammatical or more general problems, while the latter turn on the eight parts of speech. *Bede was among the first to know and use V.'s works.

CPL 1559; J. Huemer, *Virgilii Maronis grammatici opera*, BT, Leipzig 1886; *Virgilio Marone grammatico, Epitomi ed Epistole*, crit. ed. by G. Polara, tr. by L. Caruso and G. Polara, Naples 1979; G. Pesenti, Frammento bobiense di Virgilio grammatico, *BollFilolClass* 27 (1920-1921) 49-52 (ed. of a new fragment of the *Epitoma de metris*); D. Tardi, *Les epitomae de Virgile de Toulouse. Essai de traduction critique avec une bibliographie, une introduction et des notes*, Paris 1928.

S. Zincone

VIRGIN - VIRGINITY - VELATIO. By Virginity (παρθενία, εὐνουχία, ἐγκράτεια, ἀγαμία, ἀγνεία) is understood a form or kind of ascetic life (βίος), consisting of total renunciation of the exercise of *sexuality. None of the terms given above adequately express the reality to which it refers. Virginity suggests a situation of physical integrity usually applied only to the female sex. In the case of men the term "*ascetic" is often used; the term *singularitas* (Tertull., *Exhor. castit.* 1, 1; Ps.-Cypr., *De singul. cler.*) applies to the solitary life of an unmarried man. By *virginitas* *Tertullian designates a twofold physical integrity: *a nativitate*, from *birth; *a lavacro*, from *baptism (*Exhor. castit.* 1, 4). By *chastity, man, *God's *image (*ibid.* 1, 3), becomes like God (ibid. 1, 3). Like *widows,

virgins were a definite category recognized by the church. They lived in the *world; later the *monastery was the ideal place for virgins and ascetics.

Virginity had followers from the earliest Christian times. Many, men and *women of all conditions, following Christ from childhood, kept themselves spotless all their life (Just., *I Apol.* 15). Numerous Christians of both sexes grew old without marrying (ἄγαμοι) (Athenag., *Legat.* 33). Acc. to Tertullian, some live in virginal continence, remaining even in old age as pure as *children (*senes pueri*: Apol. 9, 19). *Eusebius of Caesarea relates the tradition of the two daughters of the apostle *Philip who grew old as virgins (*HE* V, 24, 2), placing them in the context of *eschatology and the *resurrection. Acc. to St *Ambrose, Christian virgins were so numerous that, if they had been subsidized like Vestals, the treasury would not have been able to meet the expense (*Ep.* 18, 12).

Virginity, as a form of life, was perpetual, not temporary like that of the Vestals, which was *non morum pudicitia, sed annorum* (Ambr., *De virginibus* I, 4, 15). Like all the *Fathers, Ambrose insists on the inner and spiritual nature of Christian virginity, embraced for *love of God. In the Vestals it is a purely physical fact: *nulla meritorum est pietas, nulla mentis integritas*, but only *carnis virginitas* (*De virginitate* 13). Their virtue is fallacious, since it is not free and not disinterested: *illic praemiis revocantur a nuptiis . . . violentia fit, ut capiantur* (*op. cit.* 13). The bishop's clear duty is to sow seeds of *chastity and stimulate love of virginity: *virginitatis studia provocare* (*op. cit.* 26). The practice of virginity, far from compromising the future of the human race, is accompanied by an increase in births (*op. cit.* 36). The virgin's veil is given only to that girl who, young in age, is mature in *faith and *virtue, reserved in conduct, guarded by her mother, firm in her resolve (*op. cit.* 39). The virgin must go out of doors only accompanied by her mother and rarely, even to go to church: living apart is a school of modesty (*Exhor. virg.* 71). Visits even to parents or *friends should be rare (*De virginitate* III, 3, 9). St *Cyprian forbids virgins any display of wealth, luxury, ornaments, participation in wedding-banquets, frequenting *baths (*De hab. virg.* 19). To realize her ideal, the virgin must overcome the resistance of her *family: *vince prius, puella, pietatem. Si vincis domum, vincis saeculum* (Ambr., *De virginibus* I, 12, 64).

1. Motivation. In Christian writers, virginity regularly has a biblical motivation: it is part of salvation-history, from which it receives its significance. The kingdom of God is already realized in the person of Christ the virgin: whoever professes virginity puts himself in harmony with the kingdom (Mt 19, 12). With Christ the last times have already begun, but are not yet complete. Virginity is one of the realities characteristic of the kingdom, at once present and future: it has already begun its reign, which, however, still coexists with *marriage. The marriage-virginity *antithesis is not philosophical in nature, but is the present-future dialectic of salvation-history. Marriage belongs to a provisional order, destined to disappear; virginity is based on the fact that we are in the end-time, in which the figure of this *world is passsing away (1 Cor 7, 29-31) and earthly values, including marriage, are relativized.

The centre of *ascesis thus becomes God, to whose plan of salvation man conforms. Just as present and future coexist, virginity is optional and marriage remains permitted; marriage belongs to the present age, while virginity is the beginning, as well as the expectation and prophecy, of the future age. Hence its superiority to marriage. It represents a "new", "messianic" ideal, essential to Christianity. In the OT a fertile marriage was a sign of God's blessing: in giving Elizabeth a son, God took away a cause of dishonour (Lk 1, 25). *Mary herself speaks of the "lowliness" of her virginal state (Lk 1, 48).

The excellence of virginity could suggest thoughts of pride, to which the Fathers called attention: "If one can persevere in chastity to the glory of the Lord, let him remain in humility. If he boasts, he is lost" (Ign., *Polyc.* 5, 2).

Nearly all Christian writers, in considering virginity, recur to the τόπος, typical of the cynic-stoic diatribe, of the inconveniences and troubles of marriage: worries about *children, the humiliating position of inferiority of the *woman, who has a *dominus* in her husband (Aug., *Confess.* IX, 9, 19). To this end they remove the eschatological significance from 1 Cor 7, 28, which refers to the tribulations linked to the end-time, not those deriving from the contingencies of married life. *Methodius of Olympus praises virginity, placing it in the context of salvation-history and ignoring considerations of this sort.

This form of Christian asceticism has value - for the Fathers - only if joined to *orthodoxy. Apropos of Vestal Virgins and continent *pagan priests, Tertullian speaks of chastity that leads to ruin (*perditricem*); only Christian chastity saves (*conservatricem*) (*Exhor. castit.* 13, 2). The asceticism of *Marcion and the *Encratites is considered by *Clement of Alexandria and Tertullian as impious and blasphemous, an expression of hatred for the *God of the OT. The chastity of these *heretics is motivated by the intention not to collaborate, by marriage, in God's creative work. Chastity, homage to God, cannot stand apart from *faith in the true God. St *Augustine teaches the same: we cannot call *veraciter pudicum* one who does not practise virtue *propter Deum verum* (*Nupt. et concup.* I, 3, 4).

In Tertullian's thought as a *Montanist, continence ensures "a great capital of sanctity; by renouncing the use of sex we acquire the *Holy Spirit". Free from woman, man has a taste for things of the spirit (*Exhor. castit.* 10, 11). Many, men and women, grow old in the virginal life, in the hope of being closer to God (Athenag., *Legat.* 33). Women who renounce first and second marriage are brides of Christ and of God (*Deo/Christo nubere*), and already belong to the family of the *angels (Tertull., *Uxor.* 1, 4; *De virg. vel.* 16, 4). The association of virginity with angelic life is common in the Fathers.

The ideal of virginity as total consecration to God is *per se* disinterested: no less than the Christian life of *contemplation, of which it is an element, it is not ordered by its nature to any practical aim of service. "The ascetic life has one sole aim: salvation" (Basil); the ascetic as such has no cure of souls (Hausherr).

Virgins, who had a reserved place in liturgical *assemblies, made public profession of their choice of life (Ps.-Ambr., *Laps. virg.* 5, 19) and received a veil: *sacro velamine tecta es* (*op. cit.* 5, 20). "To take the veil" became synonymous with making a profession of virginity (Ambr., *De virginitate* 39). St Ambrose's sister *Marcellina devoted herself to virginity before pope *Liberius in St Peter's basilica on Christmas day, perhaps in 353. The ceremony involved an exchange of clothing and an address to the newly professed woman (Ambr., *De virginibus* III, 1, 1).

Tertullian requires that virgins wear the veil, in church and out of church; if they remove it, the Spirit is offended. An unveiled girl is no longer a virgin (*De virg. vel.* 3). The veil over the head is the helmet and shield that protects virginity from *temptation, scandal, gossip and envy (*op. cit.* 15). The veil must not be small like a bonnet, but must come down to cover the shoulders, like unbound hair (*op. cit.* 17, 2).

R. Metz, *La consécration des vierges dans l'église romaine*, Paris 1954; H. Crouzel, *Virginité et mariage selon Origène*, Paris-Bruges 1963; L. Legrand, *La dottrina biblica della verginità*, It. tr. Turin 1965; S. Frank, *Anghelikòs Bíos. Begriffsanalytische und Begriffsgeschichtliche Untersuchung zum "engelgleichen Leben" im frühen Mönchtum*, Münster 1964; C. Tibiletti, *Verginità e matrimonio in antichi scrittori cristiani*, Rome ²1983; P.F. Beatrice, Continenza e matrimonio nel cristianesimo primitivo (sec. I-II), in AA. VV., *Etica sessuale e matrimonio nel cristianesimo delle origini*, Studia patr. Mediol. 5, Milan 1976, 3-68; C. Tibiletti, Ascetismo e storia della salvezza nel De habitu virginum di Cipriano, *Augustinianum* 19 (1979) 431-442. Also useful are the introductions to the eds. of Methodius of Olympus's *Symposium* (SCh 95) and Gregory of Nyssa's (SCh 119) and John Chrysostom's (SCh 125) *De virginitate*; H. von Campenhausen, Early Christian Asceticism, in *Tradition and Life in the Church* (tr. Littledale), Philadelphia 1974, 90-122; (ed.) U.Bianchi, *La Tradizione dell'Enkrateia*, Milan 1985; P. Brown, *The Body and Society: men, women and sexual renunciation in early Christianity*, New York-Columbia 1988; A. Rousselle, *Porneia: On desire and the body in Antiquity* (Eng. tr. F. Pheasant), Oxford 1988.

C. Tibiletti

VIRTUES AND VICES. Virtue was strongly recommended in antiquity. *Philo of Alexandria reconciled the formulae of Aristotle and the Stoics with the *Bible; virtues are planted and perfected in the *soul by *God's power (*Legum alleg.* I, 52. 48. 49). For the *Fathers the doctrine of virtue was an essential part of their moral and ascetic teaching. They willingly took up popular axioms, mostly inherited from *Stoic aretalogy: virtue is the sole good, vice is the sole *evil, the rest is indifferent. Virtue lies in the just mean between the two extremes (Aristotle's *metron* or *mesotes*, cf. *Eth. Nic.* 1106b-1107a, 1-3). The reward of virtue is virtue itself (*virtus sibi ipsi praemium*). The way of virtue is difficult, that of vice is easy. Virtue is one, vices are many (the *anakolouthia* of virtues: one supposes the other).

The transposition of this moralism into a Christian context was made by *Origen, who identified virtues with Christ: He *is* *Justice, Wisdom, Truth. We, on the contrary, not *are* but *have* these virtues, as we participate in Christ's life, perfecting his *image in us (cf. H. Crouzel, *Théologie de l'image de Dieu chez Origène*, Paris 1956, 239 ff.).

Because of this participation in Christ in every virtue, the Fathers did not distinguish like the scholastics between "natural" and "supernatural" virtues. However they did accept and often expounded Plato's division (*Polit.* 439a f.) of the four cardinal virtues: *1) prudence* which perfects *logistikón*, the mind; *2) courage* which is the strength of *thymoeidés*, irascible appetite against evil; *3) temperance* which resists *epithymetikón*, concupiscence; *4)* *justice* which harmonizes the exercise of the previous virtues in the right proportion. The idea dear to the Stoics was that the virtues are like a chain, one following the other. The Fathers too composed various genealogical lists of virtues (cf. e.g. Hermas, *Vis.* III, 8, 7). These lists agree in putting *faith at the beginning and charity as the end (Ign., *Eph.* 14, 1).

Emphasizing the dynamic aspect of virtue, the Stoics saw its strength in the struggle against the vices, the passions, sicknesses of the soul. Their idea of impassibility, *apátheia*, was taken up by the Eastern Fathers, but with an essentially different meaning.

With fine psychological observation the Greek Fathers (and after them the Byzantines, esp. the *hesychasts) describe the birth of the passions. They start from the conviction that malice does not belong to man's true nature, therefore it must come "from outside" by means of a wicked thought, *logismós*. "The source and beginning of all *sin is bad thoughts" (Orig., *Comm. in Mt.* 21: GCS 40, 58). The first degree of onset is *prosbolé*, suggestion, a sinful image. Then follows *syndiasmós*, a "conversation" with what was suggested, without any decision for or against. *Synkatáthesis* is the consent given to the suggestion. *Pale* is an inner struggle, the effort to free oneself from malice. For the sinner this becomes more and more difficult. Whoever consents more often falls into *aichmalosía*, captivity to *pathos*, passion (cf. e.g. John Dam., *De virt. et vit.*: PG 95, 93A). In the West, *Augustine gives three degrees: *suggestus-delectatio-consensus* (*Enarr. in Ps.* 143, 6), but he too speaks of struggle and *consuetudo* in malice.

To avoid the first suggestions, say the ascetics, is impossible to man. To avoid consent is necessary. The art of the spiritual *life consists in knowing how to avoid even "conversations" with evil.

To be able immediately, from the beginning, to oppose the wicked thought, discernment is necessary. To facilitate it, *Evagrius of Pontus presents us with a list of eight fundamental vices (*hoi genikotatoì logismoí*) which later became traditional. His teaching is condensed in chapter 6 of the *Praktikós*: "Eight in all are the fundamental thoughts which make up all malice: the first is of gluttony (*gastrimargía*), then that of fornication (*pornéia*), the third of avarice (*philargyría*), the fourth of sadness (*lypé*), the fifth of anger (*orgé*), the sixth of *accidia (*akédia*), the seventh of vainglory (*kenodoxía*), the eighth of pride (*hyperephanía*)" (SCh 171, 506 ff.).

The Latins reduced the list to seven vices, joining vainglory with pride and replacing sadness with envy (sadness at the good of others). They also turned the order round on the basis of Sir 10, 15 (*Vulg.*): *Initium omnis peccati est superbia*. The Easterners on the contrary remained faithful to Evagrius's order.

DSp 9, 955-957; DTC 15, 2739-2799; J.S. Stelzenberger, *Die Beziehungen der frühchristlichen Sittenlehre zur Ethik der Stoa*, Munich 1933, 307-354; G.E. Konstantinou, *Die Tugenlehre Gregors von Nyssa*, Würzburg 1966; W. Völker, *Scala Paradisi. Eine Studie zu Johannes Climacus und zugleich eine Vorstudie zu Symeon dem Neuen Theologen*, Wiesbaden 1968; A. Guillaumont, Introduction to Evagrius's *Praktikós*, SCh 170, 55 ff.; T. Spidlík, *La spiritualité de l'Orient chrétien. Manuel systématique*, OCA 206, Rome 1978, 255 ff. (vices), 277 ff. (virtues).

T. Spidlík

VISION. The vision of the afterlife in early Christian literature has an almost occasional origin, as an edifying *exemplum* inserted in a wider context; it then assumed its own particular features, structured according to its own rules and characteristics, and took the form of an independent *literary genre, which took root and rapidly developed in the West between the 7th and 8th cc. The first surviving example of a literary composition entirely dedicated to a vision of the afterlife is the so-called *Visio Baronti*, datable around 678-9, which takes up and elaborates the narrative pattern already consolidated by tradition and destined to characterize this genre over the centuries. It is divided into three fundamental moments: *1)* unexpected illness and apparent death of the protagonist; consternation of those present and funeral vigil; unexpected return to life; *2)* narrative of the returned man, with description of ultramondane places visited and meetings with various persons; *3)* re-entry into the body and exhortations to the readers.

The literary model constantly referred to by all authors of *visiones* in the Middle Ages, one who, until Dante, was the undisputed father of the voyage to the other world, was *Gregory the Great of *Dial.* IV, 37-38 who, explaining to the deacon Peter the consciousness of those who died in error, cites the singular experiences of some people (the monk Peter, the nobleman Stephen, a nameless soldier struck down by plague at Rome), believed dead and marvellously recalled to life, who declared that they had been in the other world and provided descriptions of it. All the fundamental motifs of the Gregorian representation of the other world recur almost unchanged in the literature of visions that takes its starting-point from here: the infernal black smoke into which souls judged guilty are thrown; the narrow bridge that crosses it and acts as *probatio* to separate the just from the unjust; the dramatic *certamen* between *angels and devils for the possession of the *soul contested between good and *evil; the dwellings of the blessed, *paradise represented as a marvellous garden of perfumed flowers.

But the genesis of this type of literature is older than Gregory, who was able to draw widely on the copious material offered by Christian and pre-Christian tradition. Enough to record the *descensus ad inferos* described in book VI of the *Aeneid*; the OT visions of *Ezekiel and *Daniel and above all those in the *Book of Enoch*; the apocryphal *Apocalypses*, in first place the *Visio Pauli*, with their fantastic descriptions of angels, devils and infernal tortures. As specific antecedents in Latin Christian literature we can cite the v. included in the *Passio Perpetuae et Felicitatis* (ch. 11) and attributed to Saturus, a prisoner awaiting *martyrdom; the so-called *dream of *Jerome* related by the author (*Ep.* 22, 30); a passage in *Sulpicius Severus's *Vita Martini* (7, 6); the widow's dream related by *Evodius of Uzalis in a *letter to St *Augustine (*Ep.* 158, 3).

That Gregory the Great should not be considered the "inventor" of this literary genre in a strict sense is demonstrated by the contemporary presence of visions of the afterlife in the historical work of his namesake *Gregory of Tours (*Hist. Franc.* 4, 33 and 7, 1): especially notable is the vision of Salvius, in which the description of the other world is set wholly in the narrative scheme already analysed apropos of the *Visio Baronti*. Before this, on the model of Gregory the Great, visions of the other world are recounted by the author of the *Vitae sanctorum Patrum Emeretensium* (before the middle of the 7th c.), the author of the *Vita Fursei* (mid 7th c.) and by *Valerius of Bierzo in his *Dicta ad beatum Donadeum* (second half of 7th c.). Subsequently, in the first half of the 8th c., a determining contribution to the genre was made by the visions related in the letters of *Boniface (*Epp.* 10 and 115) and in *Bede's *Ecclesiastical History* (V, 12-14), which develop some of the themes derived from Gregory the Great, but confer new significance on them, thanks e.g. to the introduction of *purgatory into an afterlife that had traditionally been bipartite. While the first visions contained only partial descriptions of the other world, occupying themselves prevalently either with *hell or with paradise, and in a very summary manner until the more complex elaboration of the *Visio Baronti*, with Bede there begins the systematic representation of the Christian afterlife, characterized by the hell/purgatory/paradise tripartition: to give just one example, the *Visio Wettini*, dated 824, in its first prose version and subsequent verse redaction by Walafrid Strabo.

In the course of the 9th c. there arose and flourished a particular type of v. of the afterlife, prevalently political in nature and aim, which presented the powers of this world - mostly rulers of the Carolingian dynasty - as subjected to eschatological *judgment: the eternal reward or punishment allotted to them thus became an expression of the author's political convictions (*Visio cuiusdam paupercolae mulieris, Visio Bernoldi, Visio Karoli*). At the end of the 9th c. we find a full awareness that the v. of the afterlife constituted a genuine literary genre with a consolidated tradition, as witness Hincmar of Rheims in the *Visio Bernoldi*, when, to guarantee the authenticity of his narration, he cites the authors of previous visions, Gregory the Great, Bede, Boniface and the *Visio Wettini*.

C. Fritzsche, Die lateinischen Visionen des Mittelalters bis zur Mitte des 12. Jahrhunderts, *Rom. Forsch.* 1886, II 247-79 and III 337-69; J.A. MacCulloch, *Early Christian Visions of the Other World*, Edinburgh 1912; A. Rüegg, *Die Jenseitsvorstellungen vor Dante*, 2 vols., Einsiedeln-Cologne 1945; F. Bar, *Les routes de l'autre monde: descentes aux enfers et voyages dans l'au-delà*, Paris 1946; H.R. Patch, *The Other World according to descriptions in medieval literature*, Cambridge, Mass. 1950; P. Dinzelbacher, Die Visionen des Mittelalters. Ein geschichtlicher Umriss, *ZRGG* 30 (1978) 116-28; M. Aubrun, Caractères et portée religieuse et sociale des "Visiones" en Occident du VIᵉ au XIᵉ siècle, *Cahiers de Civ. Méd.* 22, 2 (1980) 109-30; P. Dinzelbacher, *Vision und Visionsliteratur im Mittelalter*, Stuttgart 1981; M.P. Ciccarese, Alle origini della letteratura delle Visioni: il contributo di Gregorio di Tours, *SSR* 5 (1981) 251-66; ead., La Visio Baronti nella tradizione letteraria delle Visiones dell'aldilà, *RomBarb* 6 (1981/82) 25-52; ead., La più antiche rappresentazioni del purgatorio, dalla "Passio Perpetuae" alla fine del sec. IX, *RomBarb* 7 (1983); A. Louth, *The origins of the Christian Mystical Tradition from Plato to Denys*, Oxford 1981.

M.P. Ciccarese

VITAE PATRUM. After St *Benedict (chs. 42 and 73), the monks of the Middle Ages gave this name to the collections of translations of Eastern monastic writings, very varied compilations corresponding to the *Gerontikón* of the Greeks or the *Paradise* of the Syrians. In 1615 H. Rosweyde published ten books of them and an appendix. Recent labours going back to the MSS in the various languages have made it possible to follow the tracks of the translators and identify the sources. Rosweyde's books III and VII (PL 73) deserve no credit, and we should rather go back to *Paschasius's translation: cf. the study by J.G. Freire, *A versao latina por Pascásio de Dume dos Apophthegmata Patrum*, I-II, Coimbra 1971, and Freire's edition of the *Commonitiones Sanctorum Patrum*, Coimbra 1974. To make up for this, Rosweyde's books V and VI are a good translation, made by the Roman deacons Pelagius and John, of an excellent Greek collection; but Rosweyde's edition must be corrected by C.M. Batlle, *Die "Adhortationes sanctorum Patrum"*, Münster W. 1972.

Cf. G. Philippart, Vitae Patrum, *AB* 92 (1974) 353-365.

J. Gribomont

VITALIAN, pope (657-672). A native of Segni. At the time of his election, *Constans II's *Typos* of 648, opposed by *Martin I and condemned by the Lateran *council of 649, was still in force. But he sought good relations with *Constantinople: he wrote both to the emperor (Jaffé 2085) and to patriarch Peter (Mansi 11, 572), both of whom welcomed his *letters, and his name was inserted in the *diptychs. He also received Constans with great honour when he visited *Rome in 663, even though the visit ended with great quantities of bronze being pillaged and removed from the city. Under his pontificate Maurus of Ravenna obtained from the emperor, who had settled in Sicily, exemption from Roman jurisdiction, causing lively protests on V.'s part (Jaffé 209 f.) and a break between the churches of Rome and *Ravenna. Like his predecessors, V. continued to interest himself in the Anglo-Saxons (Jaffé), some of whom came to Rome on *pilgrimage; he also sent the Greek monk *Theodore, whom he consecrated himself, to be *archbishop of *Canterbury.

CPL 1735; *Verzeichnis* 576 f.; PL 87, 999-1010; PLS 4, 2090; *LP* I, 343-344; *DTC* 15, 3115-3117; *BS* 12, 1232-1235; P. Conte, *Chiesa e primato nelle lettere dei Papi del secolo VII*, Milan 1971, 456-464 (cf. index); P. Felici, San Vitaliano papa, assertore dell'unità con l'Oriente, *Oikoumenikon* 12 (1972) 394-400; B. Navarra, *S. Vitaliano papa*, Rome 1972 (cf. *RSCI* 27 (1973) 759-781); V. Monachino, I tempi e la figura del papa Vitaliano, in *Storiografia e Storia* (Miscel. E. Duprè Theseider), II, Rome 1974, 573-588.

A. Di Berardino

VITALIS the Apollinarist. Around 375, the priest V., ordained by *Meletius, appears in *Antioch at the head of an important group which adhered to the doctrine of *Apollinaris of Laodicea (cf. Soz., *HE* VI, 25: PG 67, 1357 B). He went to *Rome to submit his doctrine to pope *Damasus, who allowed him a *letter of communion, though reserving the ultimate decision to *Paulinus, bishop of the anti-Meletian party in the church of Antioch, which Rome supported (cf. *Ep. per filium meum Vitalem*: PL 13, 556-557). When, after the death of the emp. *Valens (9 Aug 378), Meletius returned from exile, the aspirants to the see of Antioch included not just Meletius and Paulinus, but also V., though he was soon set aside (Theodoret, *HE* V, 4: PG 82, 1203).

A fr. *De fide* by V. is provided by *Cyril of Alexandria in *De recta fide ad reginas* (PG 76, 1216=fr. 172 Lietzmann, p. 273). This brief text claims Christ to be τέλειος θεός and τέλειος ἄνθρωπος; it *anathematizes those who say that "Christ had a celestial body, consubstantial with *God according to the *flesh"; but at the same time it anathematizes those who say that "the Lord's flesh is consubstantial with men"; and again it condemns those who say that Christ's humanity was without a rational *soul, or that Christ's divinity, not his flesh, suffered, or who "divide" the Saviour and do not confess that the Logos and the man are one thing.

CPG 2, 3705; H. Lietzmann, *Apollinaris von Laodicea und seine Schule*, Tübingen 1904, 9, 15, 152 ff.; fr. p. 273 and E. Schwartz, ACO I, 1, 5, 67 f.; F. Diekamp, Das Glaubensbekenntnis des apollinaristischen Bischofs Vitalis von Antiochien, *Tübinger Theol. Quartalschrift* 86 (1904) 497-511; E. Amann, Mélèce d'Antioche, *DTC* 10, 527-528.

E. Cavalcanti

VIVARIUM. *Monastery N of Mt Moscius (Punta di Staletti) on the Gulf of Squillace in Calabria. Founded by *Cassiodorus in 550 after his return from *Constantinople. With the establishment of this centre of studies, Cassiodorus proposed to carry out a cultural programme adapted to the requirements of young monks. The *scriptorium* that flourished at V. was famous for its production of manuscripts of the *Scriptures. The Echternach gospels (Paris, Bibl. Nat., Lat. 9389) are a copy of an evangelary written at V. in 558. The *library of V. contained biblical texts, *commentaries on Scripture, works of history, medicine, *rhetoric, dialectic, *music and agriculture. After the destruction of the library in the course of the 7th and 8th cc., its works were dispersed and ended up on both sides of the Alps: not just at Montecassino, Bobbio, Milan, Florence, but also at Luxeuil, Aquisgrana, Corbie, Lyons, St Gall. By this means *Benedict Biscop enriched the library of Yarrow. Some manuscripts from V. have recently been discovered by Lowe (E.A. Lowe, *Codices Latini Antiquiores*, III, Oxford 1938, nn. 391-398).

Cassiodorus, *Variae* VIII, 32 and XII, 14. *PWK* 11 A 1, 495 (G. Radke); E. Josi, *EC* 12, 1564-1565; M. Cappuyns, Cassiodore, *DHGE* 11, 1349-1408; T. Klauser, *Vivarium* (R. Böhringer, eine Freundesgabe), Tübingen 1957, 337-44; P. Courcelle, *Les lettres grecques en Occident*, Paris 1948, 313-388; P. Riché, *Educazione e cultura nell'occidente barbarico dal VI all'VIII secolo*, (It. tr. G. Grimaldi), Rome 1966; G. Ludwig, *Cassiodor. Über die Entstehung der abendländischen Schule*, Frankfurt am Main 1967, 145 ff.; L. Viscido, Norme per la trascrizione del testo biblico a Vivarium, *VetChr* 15 (1978) 75-84.

M.L. Angrisani Sanfilippo

VIVENTIOLUS of Lyons. 6th-c. bishop of *Lyons (*Lugdunum*), known to us mainly from the *letters of *Avitus of Vienne (*Epp.* 19, 57, 59, 67, 69, 73). In the era of the *Burgundians, V. was one of the most eminent heads of the Lyonese church. As a monk at Condat in the Jura (modern Saint-Claude), his spiritual gifts and erudition were judged by Avitus to be worthy of a bishop: which he was in 515, after the death of his *abbat Eugendus. He played an important role in the *councils of the Burgundian kingdom, esp. that of Epaon, 517. His letter of convocation survives, but none of his other writings. An epitaph found in the 14th c. in the church of Saint-Nizier gives the day of his death: 12 July, his feast-day. The year is unknown: perhaps 524.

Ep. ad ep. prov.: PL 67, 993; MGH, *Auct. ant.* 6, 2, 165; *Ep. ad Avitum*: PL 59, 272; MGH, *ibid.* 89; A. Coville, *Recherches sur l'histoire de Lyon*, Paris 1928, 308-316; *Vies des Pères du Jura*: Sch 142, 54-55 (bibl.); *BS* 12, 1319.

A. Hamman

VOLUSIANUS. R.A.A. Volusianus, a 5th-c. *pagan intellectual, was son of the *pontifex maximus* Albinus and uncle of *Melania junior. He was proconsul of *Carthage. *Augustine admired him (*Epp.* 137, 138) just as he admired *Evodius (*Ep.* 161), and in the same way V. esteemed Augustine. Augustine exhorted him to read the *Scriptures (*Ep.* 132) and, at *Marcellinus's request (*Ep.* 136), replied to the objections (*Ep.* 136) of the pagan V. (*Ep.* 135, 2), putting forward the doctrine of Christianity (*Ep.* 137).

P. Martain, Une conversion au Ve s.: Volusien, *Revue Augustinienne* 10 (1907) 145-172; *PAC* 1, 1228; P. Brown, Aspects of the Christianisation of the Roman Aristocracy, *JRS* 51 (1961) 1-11.

E. Romero Pose

W

WAR. Christ refused to confuse his kingdom with a political régime and resort to force or w. to inaugurate it. He did not condemn the soldier's or the centurion's profession of arms. The problem for the gospel is man's inner state.

*Paul uses the image of w. and military armour to illustrate *spiritual combat against infernal powers and the carnal man's opposition to *God's service (2 Tim 2, 3; 1 Thess 5, 8; Rom 13, 12; Eph 6, 10-17).

The first Christian centuries did not worry about w. at all. *Clement of Rome admires military order (37). *Athenagoras quite favours the conquests of the Romans in *Suppl.* 37, but is more reserved in *De res. mort.* 19. *Tatian is equally severe (*Or.* 11, 1; 19, 2; 23, 12). *Tertullian's attitude, originally open (*Apol.* 30, 4), hardens, esp. in *De idol.* 19. *Lactantius rejects w. because it kills (*Div. Inst.* VI, 20, 15-17). *Clement of Alexandria sees it as a machination of the devil (*Strom.* V, 126, 5). Similarly *Cyprian (*Ad Donat.* 6).

After the *peace of *Constantine, Christians were once more employed in the army. The council of *Arles condemns deserters (can. 3). *Eusebius surprisingly presents w. as a calamity (*HE* I, 2 and V, proem.), while describing the emperor praying before each battle and bearing the emblem of the *cross in his "crusades" (*Vita Const.* II, 4; IV, 56). *Athanasius and *Ambrose praise those who defend their country; Ambrose even puts the warrior's strength among the *virtues (*De off.* I, 129). *Gregory of Tours wants Christian princes not to hesitate to make w. when necessary. In the 4th c. the church tardily remembers the *persecutions it suffered (Greg. Naz., *Or.* 2, 87) and becomes aware of a certain incompatibility, which appears in *canonical texts (*Trad. Ap.* 16; *Can. Hipp.* 14, 74). *Augustine tries to justify the profession of arms and makes a distinction between just and unjust w. (*Ep.* 138, 15), which he takes from Cicero and develops in his *City of God* (19, 7).

R.H. Bainton, The early church and the war, *HThR* 39 (1946) 189-212; E. Ryan, The Rejection of Military Service by the Early Church, *ThS* 13 (1952) 1-32; J.M. Hornus, *Evangile et Labarum. Etude sur l'attitude du christianisme primitif devant les problèmes de l'Etat, de la guerre et de la violence*, Geneva 1960; S. Windass, *Le christianisme et la violence. Etude sociologique et historique de l'attitude du christianisme à l'égard de la guerre*, Paris 1966; A. Colombo, *La problematica della guerra nel pensiero politico cristiano dal I sec. al V sec.*, Milan 1970; R. Klein, *Das Frühe Christentum in Römischen Staat*, Darmstadt 1971, 187-216; H.T. McElwain, *St. Augustine's doctrine on war in relation to earlier ecclesiastical writers (a comparative analysis)*, Rome 1972; M. Toschi, *Pace e Vangelo. La tradizione cristiana di fronte alla guerra*, Brescia 1980; A. von Harnack, *Militia Christi* (Eng. tr. D. McI. Gracie), Philadelphia 1981 (repr.).

A. Hamman

WARNAHARIUS. 7th-c. priest of Langres (*Lingonica civ.*); in reply to bishop Caraunius of Paris, who had asked him for information on the *martyrs of his city, he wrote two texts: one concerns the three brothers Speusippus, Eleusippus and Meleusippus, martyred during *Aurelian's *persecution; the other deals with Desiderius, bishop of Langres, who was killed by the *Vandals when they occupied and sacked the city. The first text is legendary in content, of the so-called "epic passion" type; the second is historically much more valuable and particularly brings out the ferocity of the barbarians.

CPL 1308-1310; PL 80, 185-200; MGH, *Ep.* 3, 457.

M. Simonetti

WASHING OF FEET. A baptismal rite practised in the *quartodeciman communities of Johannine observance, in contention with the communities of Petrine and synoptic *tradition, who baptized, like *John the Baptist, by total immersion (Jn 13, 10: he who receives *baptism by w. of f. has no need to wash his head and hands too). Through this rite they asserted that the active principle of baptismal *purification was not the physical element of *water, but the crucified Messiah's service of self-humiliation which founded the community of mutual *love.

Remembrance of the original baptismal significance of the w. of f. survived long in the Syrian *East (Aphraates, *Dem. XII: De Paschate*; Cyrillona, *Hymn on the washing of feet*, etc.), despite the normal practice of baptism by immersion. The Johannine quartodeciman mission from *Asia brought the baptismal rite of w. of f. to the West. For *Irenaeus of Lyons (*Adv. haer.* IV, 22, 1) the Saviour purified the just dead during his *descent into Hell by conferring the *sacrament of baptism on them by w. of f., a claim that presupposes his knowledge and practice of this rite within the Johannine quartodeciman tradition which he defended in his *letter to *Victor of Rome (Euseb., *HE* V, 24, 11-18).

After the sources' long silence all through the 3rd c., from the 4th c. we can directly follow the rite's evolution in the West. At *Aquileia it took place just before baptism (Chrom., *Sermo* XV); all other witnesses make it one of the post-baptismal rites. In the 4th c., understanding of its baptismal value finally reached such a state of crisis that the council of *Elvira decreed its abolition (can. 48). At *Milan, *c.*390, *Ambrose had to justify the existence of this rite to those who, appealing to its absence from Roman baptismal practice, wanted to abolish it or practise it as a mere gesture of *hospitality (*De Sacr.* III, 1, 4-7); for Ambrose the w. of f. purified the *neophyte from original *sin, while the baptismal washing purified him from personal sins (*De Myst.* 6, 32; *In psalm.* 48, 8-9). But for *Augustine the w. of f. has nothing to do with the sacrament of baptismal regeneration, but is simply the example of supreme humility left by the Lord to his disciples (*Ep.* 55, 18, 33; *Virg.* 32, 32; *In psalm.* 92, 3). In particular, the w. of f. puts into effect the *mystery of the *sacrifice of the *cross which allows the penitential remission of the inevitable post-baptismal daily sins (*In evang. Ioh.* 55-58).

The joint influence of Augustine and monastic tradition, which saw the w. of f. exclusively as an act of hospitality and humility (John Cass., *Inst.* IV, 19, 2; *Reg. Bened.* 53, 13-15, etc.), conditioned the interpretation of the post-baptismal w. of f. in the early mediaeval Gallican and Celtic churches (Caes. of Arles, *Serm.* 64, 2; 204, 3, etc.; Ps.-Max. of Turin, *Sermo de baptismo* III; *Missale Gothicum, Missale Gallicanum vetus, Missale Bobbiensis, Stowe Missal*). In some churches the post-baptismal w. of f. was put off until Tuesday or *Sunday of Octave week (Ps.-Fulg., *Sermones IV de pedibus lavandis*); it was eventually ousted by the new rite of w. of f. on Maundy Thursday and disappeared irreversibly.

The Greek non-quartodeciman tradition, from *Clement of Alexandria (*Paed.* II, 38, 1 and II, 63, 2) to the exegetes of the Antiochene school, had considered the w. of f. as the supreme example of the Master's humility and love and his preparation of the *apostles for their evangelical mission. *Origen, through an allegorical-moral reading, came to interpret it as the *mystery of purification of the feet of the *soul and of the inner man (cf. *Comm. in Joh.* XXXII). Greek *exegesis, together with the fact that Eastern churches like *Alexandria, *Jerusalem, *Antioch and *Constantinople had never practised the baptismal rite of w. of f., led to the new Maundy Thursday rite, which competed with and replaced the baptismal rite. Originating in the cathedral *liturgy of the Holy City no earlier than the 5th c. (*Georgian lectionary* in CSCO 189, 92 and 205, 100 f.), this Maundy Thursday rite spread swiftly from Jerusalem to *Byzantium (*Cod. Barber. Vat. Graec. 336* [8th c.]) and to the West, where it seems to be attested in *Spain (conc. Tolet. XVII [694], can. 3) and *Rome, in the Lateran papal liturgy, *c.*700, as appears from the *Ordo* edited by A. Chavasse in *RHE* 50 (1955) 21-35. From Rome the rite spread all through the Latin West from the end of the 8th c. on the wave of liturgical Romanization inaugurated by Pippin the Short and Charlemagne.

T. Schäfer, *Die Fusswaschung im monastischen Brauchtum und in der lateinischen Liturgie*, Beuron 1956; H. Giess, *Die Darstellung der Fusswaschung Christi in den Kunstwerken des 4.-12. Jahrhunderts*, Rome 1962; G. Richter, *Die Fusswaschung im Johannesevangelium. Geschichte ihrer Deutung*, Biblische Untersuchungen 1, Regensburg 1967; M.A. Argal Echarri, El lavatorio des los pies y el pecado original en san Ambrosio, *El pecado original, XXIX Semana España de teologia, 15-19 Sept. 1969*, Madrid 1970, 141-159; B. Kötting, Fusswaschung, *RACh* 8, 743-777; P.F. Beatrice, *Tradux peccati. Alle fonti della dottrina agostiniana del peccato originale*, Milan 1978, 181-185; id., Due nuovi testimoni della lavanda dei piedi in età patristica: Cromazio di Aquileia e Severiano di Gabala, *Ecclesia orans* (Mélanges A. Hamman) (= *Augustinianum* 20/1-2), Rome 1980, 23-54; id., *La lavanda dei piedi. Contributo alla storia delle antiche liturgie cristiane*, Rome 1983.

P.F. Beatrice

WATER. W. in *Scripture is theologically ambivalent: bearer of *life or death, sign of *blessing or curse, its abundance a sign of the Messianic kingdom or the infernal realm. Its symbolism in Christian *liturgy is derived from its liturgical use in Scripture. The *Fathers explored the significance of its use in *baptism (see Acts 8, 36 and 10, 47) through typological interpretation, seeing prefigurations in OT and NT texts. W. communicates new life and gives birth to new creatures, as in Gen 1, 2 (Tertull., *De bapt.*

3, 4; Cyr. of Jer., *Catech.* III, 4; Ambr., *De myst.* 9; *De sacram.* 3, 3; John Chrys., *Catech.* II, 25); its power to destroy *sin and the old man is anticipated in the *flood (cf. 1 Pet 3, 18-21), whose ww. prefigured baptism since, washing away men's sins, they were themselves a "baptism of the world" (Tertull., *De bapt.* 8); by them the ark was "as it were baptized without being immersed in them" (Aug., *In ev. Jo.* 7, 3); they buried *evil (Ambr., *De myst.* 11) and brought salvation to those who entered the ark, just as the ww. of baptism save only those inside the Church (Cypr., *Ep.* 69, 2). They brought destruction, but also renewal (Leo Gt., *Serm.* 60, 3). The double function that passed from the ww. of the flood to the ww. of baptism is also prefigured in the ww. of the Red Sea (cf. 1 Cor 10, 2). Baptism washes away the filth of sin, leaving the new man to sing the new song of exodus (Orig., *Hom. V, 5 in Ex.*). The ww. of the Red Sea are above all liberating, and this is the function most stressed (Tertull., *De bapt.* 9; Orig., *Hom. VI, 4 in Ex.*; Cypr., *Ep.* 69, 15; Cyr. of Jer., *Catech.* III, 5; XIX, 3; Ambr., *De myst.* 12; Greg. Nyss., *Bapt. Chr.*; Aug., 353, 4; Leo Gr., *Serm.* 55, 5), but the superiority of the ww. of baptism is seen in their ability to bring back to *life, while the ww. of the Red Sea could not prevent the Hebrews dying in the desert (Ambr., *De sacr.* 1, 22). Again, the ww. which *Moses caused to spring from the rock (Num 20, 1-11) and those of Mara sweetened by the branch thrown into them (Ex 15, 22-25) are types of the baptismal ww. sanctified by Christ the Rock (Tertull., *De bapt.* 9; Cypr., *Ep.* 63, 8) or by the wood of the *cross (Tertull., *De bapt.* 9; *Adv. Iud.* 13, 12; Ambr., *De myst.* 14; Did., *De Trin.* II, 14); this time too the typology is scriptural (1 Cor 10, 4; Jn 19, 34). Other OT ww. interpreted as types of baptism are: the w. shown miraculously by an *angel to Hagar and Ishmael dying in the desert after exhausting the w. in their skin (Gen 21, 14-19): the episode shows the need for a divine w., given the insufficiency of the w. of the synagogue (Greg. Nyss., *Bapt. Chr.*); the w. in which Naaman washed (2 Kgs 5, 1 ff.), mentioned by *Ambrose (*De myst.* 17-18) and *Gregory of Nyssa (*Bapt. Chr.*); the living w. of Ezek 47, 1-3, which gave rise in early times to the practice of baptizing in springs (*Didachè* VII, 1, 3). The w. present in some gospel episodes is also interpreted as a figure of baptismal w.: the ww. of the Jordan in which Christ was baptized are a prototype of the w. in which Christians are baptized (Hil., *In Matth.* 2, 5; Cyr. of Jer., *Catech.* XII, 15; XXI, 1; Greg. Naz., *Or.* 29, 20; Ambr., *In Luc.* 83: Jerome, *In Matth.* 1; John Chrys., *Hom. XII, 2 in Mt.*) and which are called "the origin of the Gospel" as the w. of *creation had been "the origin of the *world" (Cyr. of Jer., *Catech.* III, 5); the w. of the pool of Bethsaida (Jn 5, 1-9) (Tertull., *De bapt.* 5; Chrom., *Serm.* 14); and the w. that flowed from Christ's side (Jn 19, 34) (Cyr. of Jer., *Catech.* III, 10; Ruf., *Symb.* 21). For *Origen the w. of which Christ spoke to Nicodemus (Jn 3, 5) shows that baptism is from above, being the same w. mentioned in Ps 148, 4 (*Comm. in Rom.* V, 8). For *Basil the ww. are images of death, "receiving bodies as in a tomb" (*De Spir.* XV, 35).

Apart from baptism, w. is an indispensable element of the *Eucharist, again taking its significance from Scripture. *Justin says that the Eucharist is the totality of consecrated *bread, *wine and w. (*I Apol.* 65 and 67). The practice of tempering the wine with w. is attested by *Irenaeus (*Adv. haer.* V, 2, 3) and *Hippolytus (*Trad. Ap.* 21); *Cyprian offers Scriptural support (Pr 9, 1-5 and Mt 26, 28-29) in *Ep.* 63, 5, and gives a symbolic interpretation based on Rev 17, 15 where the ww. represent the peoples; thus the w. united to the wine "is the people united to Christ" (*Ep.* 63, 13). Other explanations are that the w. is the salvation effected by Christ's *blood (Clem. Al., *Paed.* II, 2) or a sign of his humanity (Iren., *Adv. haer.* V, 1, 3). For *Augustine the presence of w. in the *chalice is based on Jn 19, 34 (*In ev. Jo.* 120, 2), for *Theodore of Mopsuestia it is to show that the Eucharist, like baptism, commemorates and is a participation in Christ's death (*Hom.* XV). For *Ambrose the w. springing from the rock (Ex 17, 1-7) and that flowing from Christ's side are both types of the w. poured into the Eucharistic chalice (*De sacr.* V, 3, 4), but for *Cyprian "whenever the Holy Scriptures speak only of w., they presage baptism" (*Ep.* 63, 8).

Outside sacramental typology, w. provides the Fathers with other symbolic references, particularly to the Word (Orig., *Hom. XIII, 3 in Gen.*): the w. of *Cana (Jn 2, 1-11) is Scripture, tasteless until the coming of Christ to explain it (Orig., *In Jo.* XIII, 436 ff.; Aug., *In ev. Jo.* 9, 3-4); the w. of *Jacob's well (Jn 4, 6) is the OT, which Christ asks for and receives from the Samaritan woman (Orig., *In Jo.* XIII, 23-25), while the living w. which he gives represents what is "beyond what is written" (Orig., *In Jo.* XIII, 31-37; *Hom. III in Cant.* 2, 8). W. is a sign of the Spirit (Ignat., *Rom* VII, 2; Aug., *In Ps.* 103, &.2, 3); of the Church (Ambr., *Hex.* IV, 1-3); of the Jewish people (Aug., *In ev. Jo.* 17, 2); of the virginal womb that regenerates the faithful as *Mary generated Christ (Leo Gt., *Serm.* XXV, 5). Generally, w. is considered "the most beautiful of the four elements" that make up the *world (Cyr. of Jer., *Catech.* III, 5), a sign of God's *providence (Theodor., *Or. de Prov.* 2), always obedient to Him (Ambr., *Hex.* IV, 1-2).

The liturgical custom of *blessing the w. of baptism is not based on Scripture, but is attested by *tradition (Tertull., *De bapt.* 4; Hipp., *Trad. Apost.* 21; Clem. Al., *Exc. Theod.* 82; Cypr., *Ep.* 70, 1; Cyr. of Jer., *Catech.* III, 3; Basil, *De Spir.* 66; Ambr., *De Sacr.* I, 5; Greg. Nyss., *Bapt. Chr.*; John Chrys., *Catech.* II, 10). The use of holy w. outside baptism is quite late: the resemblance to Jewish, *pagan or syncretistic rites discouraged the practice (Tertull., *De bapt.* 5). The earliest formulae for blessing w. come from late 3rd-c. *Egypt (*Sacr. Serap.*) and are for exorcism and curing illness. At *Antioch, as all through the *East, w. intended for the personal use of the *faithful was blessed during the feast of *Epiphany, in memory of the sanctification of the ww. of the Jordan (John Chrys., *De bapt.* 2). In the West the earliest mention of the use of holy w. comes from pope *Vigilius (537-555) and concerns the *dedication of a church (*Ep.* I, 4). For this, so-called "Gregorian" w. was made up with salt, *wine and ashes. The washing of the hands with w. during the *Eucharist, however, is ancient (Tertull., *Apol.* 39).

DACL 4, 1680-1690; DTC 4, 1978-1984; B. Capelle, L'"aqua exorcizata" dans les rites romains de la dédicace au VI siècle, *RBen* 50 (1938) 306-308; A.V. Nazzaro, *Simbologia e poesia dell'acqua e del mare in Ambrogio di Milano*, Naples 1977.

F. Cocchini

WEDDING: Procedure and Liturgy. *Ignatius of Antioch, writing to *Polycarp (V, 2), insists that Christians should contract their unions with the approval of the bishop; the *Montanist *Tertullian in *Monog.* XIX, 1-2 - which, despite K. Ritzer, explicitly concerns "*psychici*", i.e. *Catholics - speaks of "demanding" (*postulare*) *marriage from members of the church's *clergy who will "join them" (celebrate their union) (*coniungere*). In *Ad uxorem* (II, 8, 6), the still Catholic Tertullian attributes to the church the role of *conciliator* who takes on the preliminary negotiations, and speaks of an *oblatio*, a term that in Tertullian often has a eucharistic meaning, and of a *benedictio*, without specifying whether this is imparted by the cleric. The *Letter to *Diognetus* (V, 6) says that Christians "marry like everyone else", but the context shows this to refer to the conjugal state, not to the ceremony. It is true that *archaeology attests *pagan rites adopted by Christians and preserved in later liturgies, like the coronation of bride and groom and the *dextrarum iunctio*, but the religious rites of pagan marriage could not but be replaced by Christian *prayer (Tertull., *De corona* XIII, 4). Certainly nothing permits us to generalize what Tertullian, alone in the 2nd and 3rd cc., writes, or to suppose that such rites were obligatory. In the 4th c. *Basil mentions the *blessing (*Hom. in Hexaem.* VII, 5-6), as do *John Chrysostom, *Ambrosiaster and others. Pope *Siricius (letter to *Himerius of Tarragona), supported by popular sentiment, claims indissolubility from the moment the *sponsalia*, the first act of the ceremony of marriage, are celebrated, even before the second, the *deductio in domum mariti*. The *sponsalia* comprised two simultaneous rites, *velatio* (bride and groom placed under a single veil) and *benedictio*. *Paulinus of Nola (*Carm.* XXV) describes these two rites at the wedding of the future Pelagian bishop *Julian of Eclanum. The *Eucharist is mentioned in connection with marriage in **Praedestinatus* III (5th c.).

K. Ritzer, *Le mariage dans les Eglises chrétiennes du Ier au XIe siècle*, Paris 1970; H. Crouzel, Deux textes de Tertullien concernant la procédure et les rites du mariage chrétien, *BLE* 74 (1973) 3-13.

H. Crouzel

WEEK. The Christian w. is a mixture of Jewish, astrological and Christian elements.

1. The Israelite-Jewish w., of seven days ending in the *Sabbath, remained the foundation of the Christian w. The earliest Christianity simply took it over: it too called the last day the Sabbath, the penultimate day the *parasceve* or day before the Sabbath, and the others simply the 1st, 2nd, 3rd, 4th or 5th day. So *Sunday was called "first day of the w." before receiving its new name of "Lord's day".

2. The Sabbath, in Graeco-Roman antiquity, was compared with the day of Cronos/Saturn (Jewish influence may have played a role here: Josephus, *C. Apion.* II, 31 f.; Tertull., *Nat.* I, 13; *Apol.* 16, 11). The day of Cronos/Saturn later became the first day of the "planetary w.", so called because every day was under the rule of one of the seven planets, which, according to the ancient view, revolved around the earth and influenced the course of the *world positively or negatively. Strangely, the Christian church in the West took over these pagan-astrological names: *dies Solis* became Eng. *Sunday*, Ger. *Sonntag*; *dies Lunae* became Eng. *Monday*, Ger. *Mon[d]tag*, Fr. *lundi*, It. *lunedì*; *dies Martis* became Fr. *mardi*, It. *martedì* (in Ger., from the god Thingus/Tyr/Ziu, *Dienstag*; Eng. *Tuesday*); *dies Mercurii* became Fr. *mercredi*, It. *mercoledì* (in Nordic countries it was put under Wotan: cf.

Eng. *Wednesday*; Ger. "Mittwoch" is a more recent creation); in the same way, *dies Jovis* became Fr. *jeudi*, It. *giovedì* (in Nordic countries, the god Donar or Thor gave Ger. *Donnerstag* and Eng. *Thursday*); finally, *dies Veneris* became Fr. *vendredi*, It. *venerdì* (in Nordic lands, Ger. *Freitag* and Eng. *Friday* are the day of the goddess Fr(e)ia, the Nordic Venus). In the English linguistic area, *dies Saturni* became *Saturday*. But the *order* of the w. remained the Jewish one (Sunday is the first, not the second day of the w.).

3. For Christian elements connected with the w., cf. *Sabbath, *Sunday, *Fasting (for Wednesday and Friday).

E. Schürer, Die siebentägige Woche im Gebrauche der christlichen Kirche der ersten Jahrhunderte, *ZNTW* 6 (1905) 1-66; F.H. Colson, *The Week*, Cambridge 1926; F.J. Dölger, Die Planetenwoche der griechisch-römischen Antike und der christliche Sonntag, *AC* 6 (1941) 202-238; R.L. Odom, *Sunday in Roman Paganism*, Washington 1944; G. Schreiber, Die Wochentage im Erlebnis der Ostkirche und des christlichen Abendlandes, *Wissenschaft. Abh. d. Arbeitsgemeinschaft f. Forschung d. Landes Nordrhein-Westf.* 11 (1959) 207-220; W. Rordorf, Der Sonntag. Geschichte des Ruhe-und Gottesdiensttages im ältesten Christentum, *Abh. z. Theol. d. A. T. u. N. T.* 43, Zürich, 1962, 11-44; id., Le christianisme et la semaine planétaire, *Augustinianum* 19 (1979) 189-196.

W. Rordorf

WIDOWS. The social and economic position of the widow in antiquity was precarious. The widow belonged to the world of the *poor and assisted. St *Paul exhorts young widows to remarry (1 Tim 5). Those who found themselves in need were taken in charge by the Christian community (Acts 6; *Trad. Ap.* 50). The term "widow" came to indicate a particular category, which had its own position and function in the community. The widows lived together under the direction of one of their number. So they had to possess certain requisites: to be 60 years old, only once married, outstanding as mothers of *families and for their social concern (1 Tim 5, 9-16). The term ended by losing its original meaning and just indicating an ecclesiastical function (Karlsbach), which was gradually taken over by *deaconesses. *Ignatius mentions "the virgins called widows" (*Smyrn.* 13, 1). These seem to have been *virgins accepted among the widows, who already formed religious communities. The *Didascalia* and the *Testamentum Domini* present them as nuns, with no social activity (*Didasc.* XV; *Trad. Ap.* 10, Sahidic recension; *Test.* I, 42).

J. Viteau, L'institution des diacres et des veuves, *RHE* 22 (1926) 513-536; *RACh* 3, 917-927; A. Hamman, *Vie liturgique et vie sociale*, Paris 1968, 140-143 and *passim*: G. Stählin, Das Bild der Witwe, *JbAC* 17 (1974) 5-20 (with bibl.); J. Daniélou, *The Ministry of Women in the Early Church*, Leighton Buzzard 1974; C. Osiek, The Widow as Altar: The Rise and Fall of a Symbol, *Second Century* 3 (1983) 159-169; B.B. Thurston, *The Widows - A woman's ministry in the early Church*, Minneapolis 1989.

A. Hamman

WILFRID of York. Born of a noble family in Northumbria in 634, for a time he entered a *monastery. He was at *Rome on *pilgrimage and then for some years in *Gaul: from which he learnt absolute loyalty to the *pope and contempt for the roughness and ignorance of his countrymen. Returning home, he played an important part in the triumph of Roman over Celtic customs on the observance of *Easter, at the *council of Whitby (664). Appointed bishop of *York that year, he went to Gaul for consecration. Disliked by his compatriots, and by *Theodore of Canterbury, for his haughty and angular character, he had to undergo all kinds of vexations from the local princes, who deprived him of his dignity, imprisoned him and, in the face of Rome's support for him (he twice more had to go back to Rome), sought to elude him in every way. Only in old age, in 705, did he obtain the monastery-bishoprics of Hexham and Ripon. He died in 709. The monk Eddi, his companion in travel and adversity, wrote an important *biography of him, clearly apologetic in tone (MGH, *Script. rer. merov.* 6, 193-263).

DCB 4, 1179-1185; BS 12, 1092-1094.

M. Simonetti

WILLIBRORD of Utrecht. An autograph marginal annotation by W. himself in the Echternach *calendar (*Paris. Lat.* 10837), information given by the Venerable *Bede (BHL 8395-97; *HE* 5, 13; 6. 10-11. 19) and some Acts inserted in the *Liber aureus Epternacensis* are the oldest sources of W.'s life and activity.

Born in 658, an oblate at Ripon aged six when his father Wilgilis retired as a hermit to the Humber estuary, a disciple of *Wilfrid and then of Egbert, W. set out in 690 with 11 companions to evangelize Friesland. He put himself under the protection of Pippin II, mayor of the palace, and during a visit to *Rome in 692 obtained full missionary powers from pope *Sergius I (687-702). The same *pope ordained him *archbishop of the Frisians on 21 or 22 Nov 695. W. established his cathedral at Utrecht and in 697-8 fell back on Echternach, where he installed the Benedictines, to whom he left his martyrology-calendar (*Paris. Lat.* 10837). The *martyrology is the famous *Epternacensis*, the oldest surviving exemplar of the *Mart. hier.* (early 8th c.). The hostility of Radbod, king of the Frisians, and the vicissitudes of Frankish politics after the death of Pippin II and before the rise to power of Pippin's bastard son Charles Martel explain the momentary retreat of the Frisian mission. Christianity finally conquered the land after Radbod's death (719) and thanks to the collaboration of *Boniface. W. died on 7 Nov 739 and was buried at Echternach.

Vies des SS. 11, 218-27; LTK 10, 1168; BS 12, 1113-21.

V. Saxer

WINE. In ancient Christian literature this word was used in four different contexts: *a)* moral-pastoral; *b)* liturgical; *c)* sacramentary; *d)* exegetical.

In writings of a moral nature the *Fathers defend the modest pleasure of w. against rigorist movements like those of the *Ebionites, *encratites, *Marcionites and *Manichees. The pleasure of w. is not bad or sinful, while intemperance and drunkenness are (John Chrys., *De stat.* 2, 5); w. is a gift of *God, we must use it properly (Basil, *Hom.* 14, 1); *Clement of Alexandria devotes a whole chapter of his *Pedagogue* to this question: *In what way should we participate in banquets?* (*Paed.* II, 3). *Paul recommended w. to *Timothy as a medicine (1 Tim 5, 23). In the early centuries the first of the w. was the portion of the *prophets (*Didachè* 13, 6) and, later, of the bishop (*Const. Apost.* 2, 34, 5) and the *clergy (*Const. Apost.* 7, 29, 3).

Christian texts mention w. as the material of *pagan libations (*Act. Thom.* 77; Ps.-Clem. Rom., *Hom.* 2, 15). It was an important element in the rites of some extremist sects, e.g. the rite of the gnostic mage *Mark, as described by *Irenaeus (*Adv. haer.* I, 13, 2). It was no less important in the ritual banquets of the Qumrân communities (*Community Rule* 1QS^a 2, 17-22), which were very similar to the liturgical banquets of the first Christian communities. For the *Eucharist, w. was an essential material both in *orthodox and in heterodox *liturgies, except for some rigorist sects. For liturgical use, the colour of the w. was never fixed in antiquity, but there were standards of quality (*Const. Apost.* 2); the manner of its use both in liturgy and in daily life was in accordance with the custom of the time, i.e. normally w. mixed with *water (the earliest evidence comes from *Justin [*I Apol.* 65, 5]), but there were exceptions: e.g. can. 3 of the Trullan synod of 692 condemns the Armenians, who used w. not mixed with water in their eucharistic service; to drink w. "*akratos*", not mixed, was synonymous with drunkenness (John Chrys., *Hom. in Eutrop.* 1, I).

The rigorist movements already mentioned forbade the use of w. not just in daily life, but also in the liturgy, in reference to Jesus' words in Mt 26, 29 (John Chrys., *Hom. in Mt.* 82, 2), or simply because they considered it a work of the devil (Epiph., *Haer.* 47, 1); for this reason they were named *aquarii, hydroparastae, hydrothei*. The orthodox tradition was defended against them esp. by Irenaeus (*Adv. haer.* V, 1, 3), Clement of Alexandria (*Strom.* I, 19, 96, 1), *Cyprian (*Ep.* 63, 5) and *Epiphanius (*Haer.* 30, 16).

The w. of the Eucharist was identified with Christ's *blood by Paul (1 Cor 10, 16) and by *Ignatius of Antioch (*Rom.* 7, 3). Irenaeus emphasized its real identity (*Adv. haer.* IV, 17, 5). *Cyril of Jerusalem was one of the pioneers of the doctrine of the conversion of the elements (*Catech. myst.* 1, 7; 2, 9; 5, 7).

In the field of *exegesis, many passages of *Scripture containing this word were interpreted by the Fathers in a eucharistic sense, but there are innumerable passages where w. was considered as the symbol of divine doctrine (the law and the prophets), i.e. the w. that existed before *Cana (Orig., *Comm. in Jo.*, fr. 74), or of Christ's own words (id., *op. cit.*, fr. 74); not infrequently w. metaphorically means the *Pneuma (Hipp., *Bened. Jacobi* 18), or the *grace of the *Holy Spirit (Orig., *De princip.* I, 3, 7).

E.R. Godenough, *Jewish Symbols in the Greco-Roman Period*, vol. VI, New York 1956; W. Rordorf, La vigne et le vin dans la tradition juive et chrétienne, *Ann. Univ. Neuchâtel* 1969/70, 131-146.

G. Ladocsi

WING: Flight of the Soul. In antiquity the immortal *soul, which at death evaded the *body and rose to heaven, was often compared to a bird, which rose up in flight with its wings. In harmony with Plato, esp. with his *Phaedrus* (many commentaries), the motif of the soul-bird with wings and flight was developed by the philosophical tradition in various directions: physical death, *ascesis, the transcending of the material *world

(Poseidonius) and *contemplation of spiritual reality (Beauty). Like *Philo before them, Christian authors enriched the whole picture with biblical passages (esp. Ps 54, 7 [55, 6]). They also christianized it by the addition of various motifs: *sin (*Origen), dove of the *Holy Spirit, *virginity (*Methodius, *Ambrose), virtuous *life (*love of neighbour, humility, *chastity) (*Augustine), *prayer, elevation to *God (*Gregory of Nyssa), the devil who as predator impedes the soul's flight, the death of the saints (*Gregory the Great). Esp. from the 4th c. flight was taken as a symbol for the overcoming of distance in epistolary literature.

P. Courcelle, Flügel I, *RACh* 8, 29-65 (bibl.); K. Thraede, Flügel II (epistolary motif), *RACh* 8, 65 ff.; V. Pöschl, *Bibliographie zur antiken Bildersprache*, Heidelberg 1964, 478.

B. Studer

WISDOM BOOKS

I. Job - II. Solomonic books - III. Wisdom and Ecclesiasticus.

The Christians gave the name Wisdom books, on account of their content, to the poetic books of the OT which the *Jews put in the wider group of *kethubîm*. They are: Job; Proverbs, Ecclesiastes and Song of Songs (Solomonic books); Wisdom and Ecclesiasticus or ben Sirach (deuterocanonical, hence not part of the Hebrew canon). Some would include the book of *Psalms, but we discuss this separately. See also separate articles on *Job and *Song of Songs.

I. Job. Citations of or allusions to Job occur in the NT (Mt 19, 26; Mk 10, 27; Lk 1, 52; 1 Cor 3, 19; Phil 1, 19; 1 Th 5, 22; 2 Th 2, 8; Jas 5, 11; Rev 9, 6), but less so in the earliest Christian literature (1st and 2nd cc.), which seems to have had little interest in this book. Exceptions are *Clement of Rome (*1 Clem.* 17, 3-4; 26, 3) and *Justin (*Dial.* 46, 32; 79, 4; 103, 5). The first writer to make much use of Job is *Clement of Alexandria (esp. in the *Stromata*: 15 explicit citations), who draws arguments from it for his well-known doctrine of the authentic Christian "gnosis". A little later Job makes his first appearance in Latin Christian literature with *Cyprian (*Testimonia*; *De op. et eleem.* 18, etc.). The first commentator on Job was *Origen, whose *exegesis is largely lost: there are Greek fragments in the *catenae (CPG 1, 1424; CPG 4, 213-215), Latin fragments in *Hilary of Poitiers (cf. *infra*) and perhaps also in the *commentary of the priest *Philip or Ps.-Jerome (PL 26, 619-802), which certainly contains some Origenian ideas (Altaner, 203-204). Origen explains the mystery of *evil, debated in the book of Job, by recourse to the well-known doctrine of the pre-existence of *souls and their initial fall; significantly, he sees Job not as a type of Christ but as the prototype of all Christian *martyrs. After Origen we should mention, among the Greek *Fathers, *Evagrius's *Commentary*, of which many fragments remain in the catenae (CPG 2, 2458), and that of *Athanasius (CPG 2, 2141), also surviving in fragments. Almost entirely lost is *Hilary of Poitiers's *Commentary*, whose few remains should be considered no more than a translation of Origen (*Tractatus in Job*: PL 10, 723-724; CSEL 65, 229-230). *Didymus of Alexandria's *Commentary*, as far as Job 16, 2, has recently been found among the Tura *papyri (ed. Henrichs-Hagedorn-Koenen, I-III, Bonn 1968-1973); we also know fragments of it, in a different redaction, up to Job 34, 25, through the catenae (CPG 2, 2553). Origen's doctrines of the pre-existence and fall of souls and of *apocatastasis recur in Didymus. Didymus also derives from Origen his interpretation of Job as a symbol of the just man persecuted and subjected to *temptation (not a type of Christ). The Job-Christ typology (later adopted by *Gregory the Great) appears for the first time in a *Commentary on Job* attrib. to a certain *Julian the Arian (4th c.?) (CPG 2, 2075; D. Hagedorn, *Der Hiobkommentar des Arianers Iulianus*, Berlin 1973). From *John Chrysostom we have a *sermon on Job (PG 63, 477-486), a commentary only partly published (PG 64, 503-506) and many catenistic fragments (PG 64, 506-656), whose authenticity usually needs looking into (CPG 2, 4443-4444); four more *sermones* handed down under Chrysostom's name (CPG 2, 4564) are certainly not authentic (*Proclus of Constantinople's?). With Chrysostom, ancient Christian *exegesis reaches one of its peaks of excellence, surpassed only by Gregory the Great's *Moralia in Job*. In a true osmosis of biblical *spirituality and late imperial moral philosophy, Chrysostom sees Job as the model and sum of all the *virtues of the biblical and Graeco-Roman traditions. *Theodore of Mopsuestia's (lost) commentary on Job denied the book's canonicity and held it to be the work of a *pagan. We must also mention, after Chrysostom, the commentary of *Olympiodorus (PG 93, 13-469: lengthy extracts in the catena of Nicetas; acc. to a recent hypothesis, the Latin commentary now commonly attrib. to Julian contains material belonging to Olympiodorus: cf. CPG 3, 7453 and *infra*).

Among the Latin Fathers, after Hilary (mentioned above) the first commentator is *Ambrose, who preached some sermons on Job which later became part of his *De interpellatione Job et David* (PL 14, 793-850; CSEL 32, 211-296). *Augustine's *Adnotationes in Job* (PL 34, 825-886, cf. PL 32, 635; CSEL 28/2, 509-628) have a theological flavour: against the *Pelagians, Augustine brings his exegesis into line with his well-known doctrines of the universality of *sin and limited salvation through free divine choice: Job is a just man but, conceived and born in sin, he knows that he deserves no reward for his just conduct in life. A diametrically opposite point of view to that of Augustine is expressed in the *Expositio libri Job*, commonly attrib. to *Julian of Eclanum after Vaccari's authoritative essay (*Un commento a Giobbe di Giuliano de Eclano*, Rome 1915; but cf. *contra* Stiglmayr in *ZKTh* [1919] 269-288), a work in the Antiochene mould, whose ascription to Julian would seem to be validated by its Pelagian interpretation, which has all the feeling of a reply to Augustine's exegesis (ed. PLS 1, 1573-1679; full bibl. in Altaner, 390-391, and *DSp* 8, 1224-1225). Latin commentaries also include that by Ps.-Pelagius or Ps.-Jerome (PL 23, 1407-1470), besides the one (mentioned above) by Ps.-Jerome (PL 26, 619-802) which should be restored to the priest Philip and which contains Origenian material. But the fullest and best work on Job in the whole of patristic literature is *Gregory the Great's *Moralia in Job*, not so much a work of *exegesis as a real manual of moral and ascetic *theology (ed. Gillet - De Gaudemaris, Paris 1952, SCh 32), composed in the form of *homilies preached by Gregory to the monks of *Constantinople between 579 and 585 (cf. the dedicatory *letter to *Leander of Seville). We cannot here discuss the work's rich moral and ascetic content (for which see *DSp* 6, 876 ff. and 8, 1222-1223): we can only point out, from the strictly exegetical point of view, the typology established by Gregory, which sees Job as the type of the suffering Christ and consequently also as the church (*Praef.*: Job represents the *mystery of Christ's passion, not just for having prophecied it in his speeches, but also for having prefigured it in his sufferings. And since Christ and the church are *unum*, Job, like any saint who bears Christ's *image in himself, is also a type of the church.

DSp 8, 1218-1225, with full bibl. (1224-1225) on the Greek and Latin Fathers. Further bibliographical information under the various nos. of CPG mentioned above and in Altaner, 390-391 (Julian of Eclanum) and 499-500 (Gregory the Great). For Didymus see M. Simonetti, *VetChr* 20 (1983) 357-363; for Theodore of Mopsuestia: L. Pirot, *L'oeuvre exégétique de Th. de Mops.*, Rome 1913, 131-134. Isidore of Pelusium's exegesis on Job (which does not amount to a commentary on this biblical book) has recently been studied by R. Maisano: *Koinonia* 4 (1980) 63-68. For the catenae, important particularly for the exegesis of John Chrysostom and Olympiodorus, some hundreds of fragments of whom are preserved, cf. CPG 4, 213-215 (with further bibl., also on some Fathers).

II. Solomonic books (Proverbs, Ecclesiastes, Song of Songs). Christian antiquity, following earlier Jewish tradition, unanimously attributed these three books to *Solomon. From *Origen on (*Prol. in Cant.*: GCS 33, pp. 75 ff.), they were considered as a single trilogy in which could be seen the symbol and figure of *humana doctrina*, i.e. of secular wisdom and its traditional division into *ethics, physics and metaphysics, to which the books were made to correspond: Proverbs to ethics, Ecclesiastes to physics, Song of Songs to metaphysics.

S. Leanza, La classificazione dei libri salomonici e i suoi riflessi sulla questione dei rapporti tra Bibbia e scienze profane, da Origene agli scrittori medievali, *Augustinianum* 14 (1974) 651-666; id., L'esegesi di Origene al libro dell'Ecclesiaste, *Reggio Cal.* 1975, 25-31; M. Harl, Les trois livres de Salomon et les trois parties de la philosophie dans les Prologues des Commentaires sur le Cantique des Cantiques (d'Origène aux chaînes exégétiques grecques), *Festschrift M. Richard* I, Berlin (forthcoming); W. Thiele, Sapientia Salamonis 18, 18-end, *Vetus Latina* 11/1 fasc 8, Freiburg i. Breisgau, 1986; id., Sirach (Ecclesiasticus) fasc 1, Einleitung, *Vetus Latina* 11/2, Freiburg i. Br. 1987.

1. Proverbs. Because of its moral content, the book of Proverbs was often cited and used by the Fathers for paraenetic ends (the earliest citations are in the *Epistle of *Barnabas, *Clement of Rome, *Ignatius of Antioch and *Polycarp); but not many *commentaries were expressly dedicated to the entire book or to individual sections of it. The pearl of patristic literature on Proverbs is *Basil the Great's *Homilia in principium Proverbiorum* (cf. *infra*). The book's inspiration and canonicity were never in dispute; even *Theodore of Mopsuestia did not deny its canonicity, but allowed Pr and Eccl a lower degree of inspiration (*sapientiae gratia*, rather than *prophetiae gratia*: a doctrine condemned at the 2nd council of *Constantinople in 553). The Greek Fathers' *exegesis of Proverbs is largely preserved in the *catenae (of *Procopius, unedited: the Gr. text attrib. to Procopius in PG 87, 1221-1544 is not authentic, while Corderius's partial Lat. tr. in PG 87, 1779-1800, made from a MS of the Procopian tradition, seems trustworthy; of *Polychronius, published in Lat. tr. by T. Peltanus [Anversa 1614]; other catenae, connected in some way with the Procopian tradition, like that of *Vat. Gr.* 1802, partly edited by A. Mai [*Nova Patrum Biblioth.* VII/2, 1-81]:

all, including that of Procopius, dependent on an older lost catena), so far only partly explored and not always trustworthy as regards attributions. We possess various scholia and fragments of *Hippolytus (CPG 1, 1883; PG 10, 615-630; H. Achelis, GCS 1/2, 155-167: moral *exegesis), *Origen (CPG 1, 1430; PG 13, 17-34; 17, 149-160; C. Tischendorf, *Notitia editionis codicis bibliorum Sinaitici*, Leipzig 1860, 76-122: allegorical and spiritual exegesis), *Didymus of Alexandria (CPG 2, 2552; PG 39, 1621-1646; 95, 1297B; 39, 180-182: literal explanation and spiritual application), *Evagrius of Pontus (CPG 2, 2456-2457; crit. ed. of the catenistic fragments by P. Gehin, *sub prelo*; perhaps Evagrian material under Origen's name, PG 17, 161-252; *Syriac fragments: J. Muyldermans, *Evagriana syriaca*, 133 f. and 163 f.: all Evagrius's works are profoundly influenced by the book of Proverbs, even borrowing its sententious style; cf. esp. the *Metrical sentences for monks and virgins*, a real imitation of Pr), *Gregory Nazianzen (CPG 2, 3052, 6), *Eustathius of Antioch (CPG 2, 3366; PG 18, 675-686; Pitra, *Analecta sacra* II, p. XXXVIII; M. Spanneut, *Recherches sur les écrits d'Eustathe d'Antioche*, Lille 1948, 122 f., on Pr 8, 22 *et al.*: theological and allegorico-christological exegesis), *Eusebius of Caesarea (CPG 2, 3469, 7; PG 24, 76-78), *Apollinaris of Laodicea (CPG 2, 3683; Mai, *Nova Patrum Biblioth.* VII/2, 76-80; other fragments still unedited), *Epiphanius (CPG 2, 3761, 3), *John Chrysostom (CPG 2, 4445-4446; fragments in PG 64, 659-740 and whole commentary in Cod. Patm. 161, to be published in CCG), *Cyril of Alexandria (CPG 3, 5205, 3; PG 69, 1277 f. on Pr 8, 22: literal and theological exegesis), *Olympiodorus (CPG 3, 7464; PG 93, 469-478, of doubtful authenticity: literal exegesis with moralizing tone, sometimes allegorical), Julianus Diaconus (Mai, *op. cit.*, 80). Various occasional interpretations occur in the correspondence of *Isidore of Pelusium (PG 78). *Basil of Caesarea's *Homilia in principium Proverbiorum*, on Pr 1, 1-5, has come down entire in direct tradition (CPG 2, 2856; PG 31, 385-424) as well as being much cited in the catenae: a true jewel of exegetical literature on Proverbs, in which the author, after emphasizing the sublime character of this book, most useful for education in morals, and of Ecclesiastes, which reveals the vanity of earthly things and directs us towards divine and eternal values, and after outlining the mystical interpretation of the *Song of Songs, illustrates the initial verses of Proverbs. Also under Basil's name we have another *homily on Pr 6, 4 (PG 31, 1497-1508).

Among the Latin Fathers, *Augustine comments on some passages of Proverbs in his *Sermones* (CPL 284): Pr 9, 12 (acc. to the LXX) in *Serm.* 35 (PL 38, 213-214; Pr 13, 7-8 in *Serm.* 36 (PL 38, 215-221; Pr 31, 10-31 in *Serm.* 37 (PL 38, 221-235). To *Salonius is attrib. a disputed commentary in *dialogue form on Pr and Eccl which, in the literary form of *quaestiones et responsiones*, confronts and resolves the main exegetical difficulties of the two books (CPL 499; Pr: PL 53, 967-994; Eccl: PL 53, 994-1012; ed. C. Curti, *Salonii Commentarii in Parabolas Salomonis et in Ecclesiasten*, Catania 1964). *Gregory the Great's exegesis of Pr is preserved by *Paterius in his *Liber testimoniorum . . . ex opusculis S. Gregorii* (CPL 1718; PL 79, 683 ff.; on Pr: PL 79, 895-906). *Bede wrote an *Allegorica expositio super Parabolas Salomonis* (PL 91, 937-1040), wrongly handed down among the works of Rabanus Maurus, and a *Libellus de muliere forti* (Pr 31: PL 91, 1039-1052); fragments also remain of his *Allegorica interpretatio in Proverbia Salomonis*, on chs. 7, 30, 31, 36 (PL 91, 1051-1066): for all these writings cf. CPL 1351-1352.

L. Bigot: *DTC* 13, 932; A. Barucq: *DBS* 8, 1469 and 1472-1473. For the catenae: CPG 4, 225-226. On Hippolytus: M. Richard, Les fragments du Commentaire de S. Hippolyte sur les Proverbes de Salomon, *Muséon* 78 (1965) 257-290; 79 (1966) 61-94; 80 (1967) 327-364. On Origen: M. Richard, Les fragments d'Origène sur Pr. 30, 15-31, *Epektasis*. Mélanges J. Daniélou, Paris 1972, 385-394. On Evagrius: G. Mercati, Intorno ad uno scolio creduto di Evagrio, *RBi* 14 (1914) 534-542 (= *Opere minori* 3, ST 78, 393-401); H. Urs von Balthasar, Die Hiera des Evagrius, *ZKTh* 63 (1939) 86-106, 181-206; M. Richard: *Muséon* 79 (1966) 70 f.; Quasten II, 177-178; crit. ed. of all the scholia, *sub prelo* (P. Gehin); Syriac fragments: Muyldermans, *cit. supra*. On Eustathius: M. Spanneut, *cit. supra*. On John Chrysostom: M. Richard, Le Commentaire de S. Jean Chrysostome sur les Proverbes de Salomon, Συμπόσιον. *Studies on St. John Chrysostom*, AV 18, Thessalonica 1973, 99-103, and *Muséon* 78 (1965) 257-263. On Olympiodorus: M. Faulhaber, *Hohelied- Proverbien- und Prediger- Catenen*, 1902, 89 and 113. On Isidore of Pelusium: R. Maisano, L'esegesi veterotestamentaria di Isidoro di Pelusio. I libri sapienziali, *Koinonia* 4 (1980) 39-75. On Bede: A. Vaccari: *Miscell. Geronimiana*, Rome 1920, 5-7 (other bibl. in CPL 1351-1352). For the use of Pr, and esp. of 8, 22-25, in the Arian dispute: M. Simonetti, *La crisi ariana nel IV secolo*, Rome 1975, *passim*, and *Studi sull'arianesimo*, Rome 1965, 9-87.

2. *Ecclesiastes*. The first patristic *commentary on Eccl was that of *Hippolytus, of which only a fragment remains (H. Achelis, GCS, *Hippolytus Werke* I/2, 179). We have more abundant fragments, reconstructed from the *catenae, of *Origen, who wrote on Eccl a commentary in scholia and eight *Homilies (ed. S. Leanza, *L'esegesi di Origene...*, and new fragments from Cod. Vindob. Theol. Gr. 147 [CCG 4, Suppl.]), marked as usual by an allegorical interpretation, which exercized a determining influence on later patristic *exegesis. Fragments remain (in the *Catena of *Procopius*) of *Dionysius of Alexandria's *Commentary on Ecclesiastes*, limited to the first chapters (Feltoe's ed., Cambridge 1904, is unreliable; better ed. in CCG 4, *Catena of Procopius*, and new fragments from Cod. Vindob. Theol. Gr. 147 [CCG 4, Suppl.]; the Dionysian paternity of the fr. on Eccl 12, edited by Bienert: *Kleronomia* 5 [1973] 308-314, is debatable), with a mystical-allegorical interpretation aimed mainly at overcoming the difficulties deriving from Eccl's "Epicurean" teaching. But we do possess the whole of *Gregory the Thaumaturge's *Metaphrasis in Ecclesiasten Salomonis* (PG 10, 988-1017), which some codices wrongly attribute to *Gregory Nazianzen: more than a commentary, a succinct paraphrase with prevalently moral intention, important because it introduces for the first time the hermeneutical hypothesis of *prosopopea* (constantly followed by subsequent interpreters) as an alternative to the allegorical interpretation, to explain Eccl's embarrassing "Epicureanism". The Tura *papyri recently gave us almost the whole of *Didymus of Alexandria's ample commentary (*Kommentar zum Ecclesiastes*, ed. Binder-Liesenborghs-Gronewald, Bonn 1969 ff.), whose exegesis of Eccl was previously known only from a few fragments in the catenae, not always corresponding to the Tura interpretation; however, both the Tura commentary and the catenistic fragments reveal a mainly allegorical and tropological interpretation, with frequent ecclesiological applications. We possess all eight of *Gregory of Nyssa's *Homilies on Ecclesiastes* (PG 44 and P. Alexander, Leiden 1962), also marked by mystical and moralizing exegesis. The aim of Gregory, whose interpretation aroused great interest and was much used in the catenae, was to show that *Solomon, far from exhorting to enjoyment, taught contempt of sensible goods, whose vanity he shows the reader. The "Epicurean" passages are referred, leaning on the theory of *prosopopea*, to a hypothetical hedonistic interlocutor of Solomon; or else it is supposed that Solomon himself raised the objections and then resolved them. The same apologetic intention, i.e. to defend Eccl from charges of Epicureanism, lies behind *Jerome's *Commentary on Ecclesiastes* (ed. M. Adriaen, CCL 72), another work which aroused great interest in later centuries down to the Middle Ages: a sober commentary which, with rare balance, tempers the mystical-allegorical interpretation with the philological doctrine typical of Jerome's best commentaries. The *Commentary* attrib. to *Salonius (ed. C. Curti, Catania 1964) is really dependent on Alcuin and was composed in the form of a *dialogue between an interlocutor who puts the difficulties and the writer who resolves them, following the method of *quaestiones et responsiones*. *Olympiodorus of Alexandria's *Commentarius in Ecclesiasten* (PG 93, 477-628) is clearly compilatory in character, almost a *catena under the appearance of a *commentary: it depends mainly on Origen, Dionysius of Alexandria, Gregory of Nyssa, Didymus of Alexandria, Nilus and Evagrius, and its exegesis is mostly mystical and allegorical. Of the exegesis of *Nilus and *Evagrius, just mentioned, we possess only a few fragments in catenae and other collections of a similar nature: their interpretation is allegorical and moral. The one surviving commentary of the Antiochene school is pseudo-Chrysostomic (ed. S. Leanza, CCG 4), with a polemically non-allegorical interpretation. But also Antiochene, lost in the original Greek but surviving incomplete in the *Syriac version, is *Theodore of Mopsuestia's *Commentary* (cf. CCG 4, 62), whose erroneous doctrine of the lesser degree of inspiration of the Solomonic books we have mentioned (CCG 4, 61). The patristic period is closed, among the Greeks, by *Gregory of Agrigento's *Explanatio super Ecclesiasten* (PG 98, 741-1181, crit. ed. by G. H. Ettlinger, in CCG) which, together with Jerome's, is the best patristic commentary. The commentary on Ecclesiastes - Gregory's sole surviving work - reveals the wide education of the author, who shows his familiarity with both classical (Aristotle, *Philo, Mimnermus, the Greek rhetors) and Christian authors (Basil, Gregory Nazianzen, Gregory of Nyssa, John Chrysostom, earlier commentaries on Eccl), and also manifests considerable interest in philosophical and theological problems, with particular attention to the problem of free will. Important from a more specifically exegetical point of view is the work's brief *Prologue* in which Gregory, taking the idea from Pr 30, 33 ("press milk and you will have butter"), identifies milk, in a transparent metaphor, with the literal-historical sense and butter with the spiritual sense (PG 98, 741 f.), according to a clear bipartite exegetical scheme that would seem to favour the spiritual sense. In the commentary, however, the literal interpretation is prevalent, most often alone, but sometimes followed by the spiritual; sometimes the spiritual interpretation occurs alone without the literal. The tripartite exegetical scheme (literal+allegorical+anagogical interpretation) postulated by Cataudella as

the dominant characteristic of the Agrigentine's Commentary ("this scheme - its three parts clearly distinct - is essentially the same all through the *Commentary*, which thus gives the impression of an architectural construction, clear, ordered and precise": Q. Cataudella, *Storia della Sicilia*, IV, Naples 1980, 17) is without foundation and has no correspondence with Gregory's work. In fact Gregory does not constantly and systematically use any fixed exegetical scheme, not even the bipartite one theoretically put forward in the *Prologue*, but rather lets himself be guided by the tenor of the biblical text, which in many cases lends itself better to exploration of the literal-historical sense (under this type of interpretation come the long, involved philosophical and theological examinations), but sometimes offers the opportunity, without having to deny the literal sense, of developing spiritual interpretations (whether allegorical or tropological) and is sometimes such that it requires - or at least so it seems to Gregory - the rejection of the literal sense and the adoption of the spiritual interpretation alone, according to the well-known Alexandrianizing principle that there is no literal sense whenever this is unsatisfactory or somehow unworthy of the majesty of the divine word (cf. e.g. *Comm. in Eccl.* 10, 11: "This expression must be understood anagogically and only anagogically"; other examples in S. Leanza, Sul Commentario all'Ecclesiaste . . .). As regards the overall interpretation of Ecclesiastes, at the start of his *Commentary* Gregory takes over from earlier exegetical tradition the christological interpretation which sees *Solomon as the type of Christ speaking mystically to the church (PG 98, 745 ff., deriving the allegorical meaning of the term Ἐκκλησιαστής from *Gregory of Nyssa). But this typology is not followed up in the rest of the commentary, which, as we have said, favours the literal interpretation. From earlier exegetical tradition Gregory derives the commentary's apologetic and moralizing character, which seeks to demonstrate, against those who saw in Ecclesiastes a hedonistic and Epicurean teaching, that in fact this book teaches contempt for earthly goods, as being vain, and invites us to search for spiritual and eternal goods (PG 98, 752 ff., 805-809, 829, 897). However, unlike other earlier interpreters, Gregory does not seem excessively disturbed or worried by the "Epicurean" content of Ecclesiastes (for which cf. S. Leanza, L'atteggiamento della più antica esegesi cristiana . . .). In any case, to the traditional solution of *prosopopea*, which eluded the difficulty of the more embarrassing passages by attributing hedonistic statements to Solomon's imaginary interlocutors, Gregory adds or substitutes a more attentive and appropriate consideration of the literal sense, which allows him to play down the importance of the so-called "Epicurean" passages and interpret them as an acceptable invitation to moderate use of earthly goods, in accordance with the divine plan and the necessities of nature (PG 98, 833, 873, 1072-1076, etc.). Gregory of Agrigento's strong personality, evident in his stand on the Epicurean problem, is still more manifest in the notable spirit of independence he shows towards earlier interpreters - continually, but always anonymously, called to account - and more particularly in the freedom with which he contests and refutes the *exegesis of illustrious Fathers like *Origen, *Dionysius of Alexandria, *Gregory of Nyssa, *Nilus (documentation in S. Leanza, Sul Commentario all'Ecclesiaste . . .): an attitude all the more remarkable in a period when exegesis had widely fallen into the catenistic genre and few dared contest the authority of the "*venerandi Patres*". Among the Latins, *Gregory the Great's *Quaest. sup. Eccl* 3, 18 ff. (= *Dial.* IV, 3-4: PL 77, 321 ff.) confronts the thorny passage of Ecclesiastes which asserts the identity of the human *soul's nature and destiny with those of the beasts, resolving it by the expedient of *prosopopea*. For other commentaries of lesser importance, or of which we know little, cf. Leanza, *L'Ecclesiaste nell'interpretazione dell'antico cristianesimo* (cf. bibl.). Among the *catenae, that of *Procopius (CCG 4, Suppl., ed. S. Leanza) and that of the three Fathers (*Gregory of Nyssa, *Gregory the Thaumaturge, *Maximus Confessor: ed. S. Lucà, CCG 11) have been published; the *Catena Hauniense* (ed. A. Labate) will soon be published. Other catenae, unedited, are the *Catena Barberiniana* (Cod. Vat. Barb. Gr. 388), the *Catena of *Polychronius, etc.

Comprehensive treatment: S. Leanza, *L'Ecclesiaste nell'interpretazione dell'antico cristianesimo*, Messina 1978; id., L'atteggiamento della più antica esegesi cristiana dinanzi all'epicureismo ed edonismo di Qohelet, *Orpheus* n.s. 3 (1982) 73-90. Particular studies: on Origen: S. Leanza, *L'esegesi di Origene al libro dell'Ecclesiaste*, Reggio Cal. 1975; on Dionysius of Alexandria: S. Leanza, Il Commentario sull'Eccl. di Dionigi Aless., *Scritti in onore di S. Pugliatti*, V, Milan 1978, 399-429; on Didymus: Aless., *SP* 18 (1982) 300-316; M. Simonetti: *VetChr* 20 (1983) 375-385; on Nilus: S. Lucà: *Sileno* 1 (1977) 13-39; *Biblica* 60 (1979) 237-246; *Augustinianum* 19 (1979) 287-296; on Evagrius: A. Labate, L'esegesi di Evagrio al libro dell'Ecclesiaste, *Studi in nore di A. Ardizzoni*, Rome 1978, 485-490; P. Géhin, Un nouvel inédit d'Evagre le Pontique: son commentaire sur l'Ecclésiaste, *Byzantion* 49 (1979) 188-198 (disputable conclusions; Géhin is preparing the ed. of Evagrius's frs.); on Ps.-Chrysostom and Theodore of Mopsuestia: S. Leanza, in CCG 4, 53-62; on Gregory of Agrigento: S. Gennaro, Influssi di scrittori greci nel Comm. all'Eccl. di Gregorio di Agrigento, *Misc. di Studi di Lett.crist.ant.*, III, Catania 1951, 162-184; S. Leanza, Sul Commentario all'Ecclesiaste di Gregorio di Agrigento, *VetChr* 21 (1984); on the catenae: CPG 4, 227-228; A. Labate, Nuove catene esegetiche sull'Ecclesiaste, *Antidoron*, Homm. à M. Geerard, I, Turnhout 1984, 137-159; for the eschatological interpretation of Eccl 12: *Augustinianum* 18 (1978) 191-207.

3. *Song of Songs*. In the history of early Christian *exegesis, the Song, more than any other book, boasts a constant and uninterrupted allegorical interpretative tradition, whose finest expression is in the *commentaries of *Origen, *Gregory of Nyssa and *Gregory the Great. The sole exception to this exegetical tendency was the naturalistic interpretation of *Theodore of Mopsuestia, condemned at the 2nd council of *Constantinople (553). Even before the Christians, the Song had been interpreted allegorically by the *Jews, who saw in it a nuptial *allegory of the love between God (the bridegroom) and Zion or Israel (the bride). The earliest Christian commentary on the Song is that of *Hippolytus (surviving up to Song 3, 7 in *Georgian tr.; various fragments in Greek, *Syriac, *Armenian and *Palaeoslavonic; Lat. tr. by G. Garitte [CSCO 263-264]), who gives a Christian sense to the Jewish allegorical interpretation, seeing the bridegroom as Christ and the bride as the church (only rarely and sporadically does it hint at the psychological interpretation typical of Origen, which identifies the bride as the *soul). But the most important commentator on the Song in antiquity is *Origen, whose exegesis became an obligatory reference-point for all later interpreters. Origen commented on the Song in a series of *homilies and in a big ten-book commentary. The Greek original is almost entirely lost (there is a passage on Song 1, 5 in the *Philocalia* and various scholia in the *Catena of *Procopius*), but we have the first two homilies, containing the interpretation of Song 1, 1-2, 14, in *Jerome's *Latin translation (ed. W.A. Baehrens, GCS 33, 27-60), and the beginning of the *Commentary* up to Song 2, 15 in *Rufinus's translation (ed. Baehrens, *ibid.*). Jerome judged these writings Origen's masterpiece (*Hom. in Cant.*, *Prol.*: "*Origenes cum in caeteris libris omnes vicerit, in Cantico Canticorum ipse se vicit*"). Origen derived from Hippolytus the ecclesiological allegorical interpretation, which he propounded esp. in the *Homilies*; but his great innovation was the psychological interpretation (bride=soul), almost completely absent in Hippolytus but which later, following Origen's example, became enormously popular. In the *Commentary* he gave every pericope a triple interpretation: literal, allegorical (= *ecclesiological) and tropological (= psychological). The main theme of the allegorical-ecclesiological interpretation is the contrast between the *economy of the OT and that of the NT: the bridegroom's friends are the *prophets, the daughters of Jerusalem (whom the bride addresses) are the Jewish people who did not accept Christ's message. For Origen, the church did not originate in NT times, but existed from the *creation of the *world and lived in expectation of the coming of Christ, the bridegroom; Christ's coming and his mystical union with the church mark the transition from the imperfection of the law to the perfection of *grace. In the tropological-psychological interpretation, the bride is the perfect soul who adheres mystically to Christ's divinity, while the maidens are the imperfect souls who adhere only to the incarnate Christ. The Origenian interpretation of the Song has had enormous influence, not just - as we said - on the history of *exegesis, but also on Western *mysticism and *spirituality (St Bernard, St Teresa of Avila, St John of the Cross, etc.). After Origen, we should just mention the *Commentary on the Song of Songs* by *Methodius of Olympus (lost, but attested by Jerome [*De vir. ill.* 83]; see also some exegetical ideas on the Song in Methodius's *Symposium*), that of *Athanasius (who seems to have included the interpretation of the Song and Eccl in a single work: cf. Leanza, L'Eccl. nell'interpret. dell'ant. crist., *op. cit.*, 23), a few fragments of which survive, and that of *Apollinaris of Laodicea (15 fragments in the *Catena of Procopius*: PG 87, 1545 ff.). *Gregory of Nyssa's 15 *Homilies on the Song of Songs* (as far as Song 6, 8: ed. H. Langerbeck, Leiden 1960) deserve greater attention: with Origen's *Homilies* and *Commentary* and Gregory the Great's *Expositio in Cant. Canticorum*, they are absolutely - not just in the patristic period - the finest commentary on the Song. Gregory had before him the great model of Origen, whom he praised in the dedicatory prologue to the virgin *Olympias, but from whom he intentionally departed. Among other things Gregory's exegesis, unlike Origen's, has a considerable philosophical substratum (frequent allusions to Plato's *Dialogues*, polemic against Aristotle in the doctrine of the indefinite progress of the soul, dependences on *Plotinus, etc.: cf. Langerbeck, *ed. cit., praef.*) and reveals different theological ideas (cf. Quasten II, 269). From a more strictly exegetical point of view, we can distinguish two kinds of interpretation in Gregory's *Homilies*: in the first part he exclusively follows the tropological-psychological interpretation (bride=soul), but in the second part (from *Hom.* VIII on) the allegorical-ecclesiological interpretation prevails. In all, going beyond the Origenian balance between the two interpretations, Gregory gives more importance to the psychological interpretation and the theme of the soul's mystical union

with the Logos. For him, the Song corresponds to the highest degree of spiritual life, the moment when the *soul, freed from every bond, reaches the stage of perfection, which consists in *contemplation and mystical union with *God.

Among the Latins we should record the two commentaries of *Victorinus of Petovium and *Reticius of Autun, of which nothing remains, but which were known to Jerome (*De vir. ill.* 74 and 82), while Jerome himself doubted the existence of a commentary by *Hilary of Poitiers (*De vir. ill.* 100: "*aiunt quidam scripsisse eum et in Canticum Canticorum, sed a nobis hoc opus ignoratur*"). Five surviving *Homilies on the Song of Songs* by *Gregory of Elvira (ed. J. Fraipont, CCL 69) show the influence of Hippolytus and Origen. Latin exegetes of the Song include *Ambrose, who composed no specific commentary on this book, but who commented on nearly all of it (except 10 verses!) in various works, esp. in *De virginitate, Expositio Psalmi CXVIII* and *De Isaac et anima* (a continuous reading of Ambrose's exegesis on the Song can easily be made from the mediaeval compilation of *excerpta* from Ambrose made by William of St Thierry: PL 15, 1851-1962): his interpretation is mystical and allegorical, dependent on Origen and Hippolytus. *Jerome proposed to comment on the Song in a specific work (*Prol. in Matth.*: CCL 77, 6), but he never wrote it: but he too interpreted the Song frequently, in dependence on Origen, in various places in his other exegetical works and in his *letters.

Greek Fathers of the 5th and 6th cc. who commented on the Song include: *Philo of Carpasia, whose interpretation survives in *Epiphanius's old Latin version, commissioned by *Cassiodorus, while the Greek text is known in a reduced version (recent crit. ed. of Epiphanius's version by A. Ceresa-Gastaldo: CP 6); *Nilus of Ancyra (mostly preserved in the catenae, but now also partly known in direct tradition: PG 87, 1545-1754 and S. Lucà, cited below); *Cyril of Alexandria (PG 69, 1277-1293; PG 87, 1545-1753 [= *Catena of Procopius*]); *Theodoret of Cyrrhus, a spokesman of the Antiochene school who yet rejected the naturalistic interpretation of Theodore of Mopsuestia (who saw in the Song a profane love poem of Solomon for his Egyptian bride) and accepted the allegorical-ecclesiological interpretation. It is uncertain whether *Theodore of Mopsuestia himself composed a commentary on the Song: at any rate nothing has come down under his name, not even in the catenae; the 2nd ecumenical council of *Constantinople (553) condemned his interpretation of the Song on the basis of a letter-treatise to a *friend, who had asked him to elucidate this book.

Among the Latins of the 5th c., we should mention the commentary of *Aponius (PLS 1, 799-1031) and that of *Justus of Urgel (*Explanatio mystica in Cant. Canticorum*: PL 67, 693-994), who follows the ecclesiological interpretation but already foreshadows the mediaeval mystical interpretation. But the most important of the Latin commentaries is *Gregory the Great's *Expositio in Cant. Canticorum*, which worthily closes the patristic period and inaugurates the monastic mystical exegesis of the Middle Ages (ed. P. Verbraken, CCL 144). Gregory inherits from earlier patristic *exegesis, esp. Origen and Gregory of Nyssa, the fundamental themes of the dual allegorical-ecclesiological and tropological-psychological interpretation, but favours the latter. For him too the bridegroom and bride of the Song represent Christ and the church, or else the Word and the soul: the first interpretation, the allegorical, serves the doctrinal knowledge of the church, while the tropological interpretation serves the ascetic elevation of the individual soul. Gregory's work is addressed to *praedicatores* and those who have reached the heights of mystical experience, such as *virgins; just as the Song itself is addressed to *perfecti*, those who have already been through the experience of Proverbs and Ecclesiastes (Gregory takes up and modifies the old Origenian and Hieronymian ideas, which saw adumbrated in the three books of Solomon three degrees of spiritual understanding: *incipientes, progredientes, perfecti*).

Among the Greek *catenae we should mention that of *Procopius (PG 87), that of the Three Fathers (*Gregory of Nyssa, *Nilus, *Maximus Confessor), perhaps by the same author who compiled the Catena of the Three Fathers on Ecclesiastes, that of *Polychronius, that of Ps-Eusebius, etc. There is also an Ethiopic catena.

Comprehensive treatment: *DSp* 2, 93-101; E. Cunitz, *Histoire critique de l'interprétation du Cantique des Cantiques*, Strasbourg 1834; F. Ohly, *Hohelied-Studien. Grundzüge einer Geschichte der Hoheliedauslegung des Abendlandes bis um 1200*, Wiesbaden 1958; P. Meloni, *Il profumo dell'immortalità. L'interpretazione patristica di Cant. 1, 3*, Rome 1975. For particular bibliography on the various Fathers, see the works of Ohly and Meloni. Here we will mention only: for Origen: O. Rousseau, *Origène. Homélies sur le Cantique des Cantiques*, Paris 1966 (SCh 37a); J. Chênevert, *L'Eglise dans le Commentaire d'Origène sur le Cantique des Cantiques*, Brussels-Paris-Montréal 1969; M. Simonetti, *Origene. Commento al Cantico dei Cantici*, Rome 1976; for Gregory of Nyssa: *La colombe et la ténèbre. Textes extraits des "Homélies sur le Cantique des Cantiques" de Grégoire de Nysse*: Tr. M. Canevet. Intro. and notes de J. Daniélou, Paris 1967; for Theodoret of Cyrrhus and Theodore of Mops.: L. Pirot (cited in the bibl. for Job, col. 3084), 134-137, and J.-N. Guinot, in *Orpheus* 1984; for Theodoret and Origen: M. Simonetti, in *Letterature comparate*, Studi Paratore, Bologna 1981, 919-930; for Gregory the Great: V. Recchia, *L'esegesi di Gregorio Magno al Cantico dei Cantici*, Turin 1967; for Nilus: S. Lucà, in *Atti del IV Congr. nazionale di St. Biz.*, Naples 1981 and *Augustinianum* 22 (1982) 365-403 (*Status quaestionis*, earlier bibl., ed. of the text, known in direct tradition, of Song 6, 8-8, 14). For the catenae: CPG 4, 222-224 and S. Lucà, in *Augustinianum* cited above.

III. Wisdom and Ecclesiasticus.

1. Wisdom. We possess no patristic *commentary on this book, though it is frequently cited by the *Fathers (cf. *Biblia Patristica*). Acc. to *Cassiodorus (*Instit. div. litt.* 5), *Ambrose and *Augustine preached *homilies (now lost) on Wisdom, and *Bellator (6th c.) wrote a full commentary in VIII books (*Expositio Sapientiae* - also lost). *Paterius (or Ps.-Paterius) collected and ordered the scattered *exegesis on Wis and Sir of his teacher *Gregory the Great (*Testimonia in libr. Sapientiae et Ecclesiastici*: PL 79, 917-940). The earliest commentaries we possess are quite late: Rabanus Maurus, 9th c. (PL 109, 671-762); in the Middle Ages, the *Glossa ordinaria* (PL 113, 1167ff.), Hugh of Saint-Cher (*Opera* III, Lyon 1669), etc.

L. Bigot: *DTC* 15, 742-743; R. Petraglio, *Eloge des ancêtres, Sirach 44-50* (forthcoming); Thiele, Sirach 3 (Prol. 1-4), *Vetus Latina* 11.2 (1988); S. Leanza, Sul commentaria all'Ecclesiastico di Girolamo, *Actes du Colloque de Chantilly 1986*, Paris 1988, 267-282.

2. Ecclesiasticus (ben Sirach). Even more than Wis, Sir was overlooked in the patristic period, certainly on account of its contested canonicity. What we read in the pseudo-Chrysostomic *Synopsis Scripturae Sacrae* (PG 56, 375-376) is hardly more than a summary of the book. *Paterius (or Ps.-Paterius) collected for Sir, as for Wis, the *excerpta* of *Gregory the Great's exegesis (cf. above; for Eccl: PL 79, 921-940). It is equally significant that Rabanus Maurus (9th c.), who commented on Wis, was also one of the few to comment on Sir (PL 109, 763-1126): his interpretation, allegorical, was much used by mediaeval exegetes (*Glossa ordinaria*, etc.). Otherwise, in the patristic period, there is only the 9th-c. Nestorian Isodad of Merw, whose publication is announced in CSCO. Recently R. Maisano has studied the exegesis of Sir in the letters of *Isidore of Pelusium, who did not write a commentary on it.

H. Duesberg: *DSp* 4, 61; H. Duesberg-I. Fransen, *Ecclesiastico*, Turin 1966, 19; J.M. Vosté: *RBi* 38 (1929) 382-395, 542-554 (Isodad of Merw); R. Maisano: *Koinonia* 4 (1980) 68-72 (Isidore of Pelusium); M.M. Winter, Ben Sira in Syriac?, *SP* 16.2 (1985) 121-123.

S. Leanza

WITNESS. See **TESTIMONY.**

WOMAN. It was undoubtedly the two biblical figures of Eve and *Mary who had the greatest influence on the conception of w. formulated by the *Fathers of the first centuries. But to be able to give a historical evaluation of this conception, we must not fail to take account of the weight that continued to be exercised on it by the historical and cultural context in which it was formed. It is a fact that the position of w. in the ancient world, whether Jewish, Hellenistic or Roman (yet with different - and not unimportant - nuances), was one of inferiority towards men. This could be expressed either in a reduction of the spheres in which they were allowed to act, though without denying them dignity and prestige, or in an effective consideration of subjection and existential imparity with regard to men. The tension between fidelity to that announcement of novelty brought by the gospel, which included even the way of considering w., and the influence deriving from the environment from which the NT texts sprang and in which the *Fathers lived, is easy to see in every patristic text that deals with w. But in most cases the two figures of Eve and Mary lent themselves to keeping this tension rigorously religious in scope, they being the two poles between which discussion of w. moved: Eve, representative of w. in her existential situation of inferiority, and Mary, the aim to which precisely this w. considered inferior - when not given a decidedly negative connotation - tends and with which it is called to identify itself. Besides Eve and Mary, other female characters from *Scripture (sometimes anonymous) provided the Fathers with material to draw a portrait of w. which was new to *pagan culture, in fidelity to the data of *revelation. The very fact of returning innumerable times to moral exhortations (1 Tim 2, 9-10; 1 Pet 3, 3-6) - which appear in Scripture and seem to reduce w. to an exclusively inner world, considering any outside interest or care unseemly in her - is in fact an explicitly new proposal, aimed at diversifying the Christian w. from the pagan w. (Clem. Al, *Strom.* 2, 146; *Paed.* 3, 56, 1-2; 62-68; Jerome, *Ep.* 148,

25-27; John Chrys., *Catech*. I, 34. 37. 38; Chromat., *Serm*. 35).

W.'s new position derives from certain primacies that Scripture attributes to her: above all, being the first to receive the announcement of Christ's *resurrection (Orig., *Comm. in Jo*. 13, 29, 179; Cyr. of Jer., *Catech*. XIV, 12; Jerome, *In Mat*. 28, 9; Ambr., *In Lc*. 2, 28; Aug., *In Jo*. 1; *In I Jo*. 3, 2). But she is also the first to reveal the right way to approach Christ to obtain mercy (on the basis of Mt 9, 20 ff.; Peter Chrysol., *Serm*. 34, 3). From a spiritual point of view w. is equal to man: both have a single Lord, a single Teacher, a single church (Clem. Al., *Paed*. 1, 10-11; Tertull., *Ad ux*. II, 9). But this equality is not always recognized: *John Chrysostom, after having admitted it (*Hom. 10, 3 in 2 Tim*.), seems in another text to deny it (*Hom. 5, 5 in 2 Thess*.; cf. also Aug., *Gen. ad litt*. 11, 58): she, created as man's helpmeet (Gen 2, 18), lost that dignity because of *sin (John Chrys., *De virg*. 46, 1). The original fault is always attributed to w., who is thus held to be the cause of sin (Iren., *Adv. haer*. 3, 22, 4; Tertull., *De c. fem*. 1, 1; Cyr. of Jer., *Catech*. 13, 21; Ambr., *Hex*. 5, 18; *De parad*. 14, 70; Ruf., *Symb*. 21; John Chrys., *De virg*. 46, 1; Aug., *In ps*. 48, 1, 6): every w. bears Eve in herself and must therefore do penance (Tertull., *De c. fem*. 1, 1). The debt towards man, contracted by the female sex at the moment of her creation, has been paid by Mary who gave *birth to Christ virginally (Cyr. of Jer., *Catech*. 12, 29). It is through her relation to Christ that w.'s situation is overturned (Hilary, *In Mt*. 33, 9; Greg. Gt, *Mor. Job*. XIV, 49, 57). Even in allegorical interpretation, w. has a dual value in the Fathers: positive, whenever she is interpreted as an image of the church (Jerome, *In Mt*. 15, 22; 26, 12-14; *In Mc*. 5, 34; John Chrys., *In Mt. hom*. 51); negative, or at least subordinate to man, when she is seen as the *soul while man is the spirit (Orig., *Hom. 1, 15 in Gen*.), as the *flesh that must follow the spirit (Orig., *Hom. 4, 4 in Gen*.), as the senses while man is the mind (Ambr., *De parad*. 15, 73), as synonymous with weakness (Greg. Gt, *Mor. Job* XI 49, 65). Within the church, the prophetic function is acknowledged to w. (1 Cor 11, 4-5) and the existence of *deaconesses is amply attested (1 Tim 3, 11; Rom 16, 1; Pliny, *Ep*. 10, 96, 8; Clem., *Strom*. 3, 6, 53; Orig., *Comm. Rom*. 10, 17; *Didasc. Apost*. II 26, 4-6; III 12, 1; 13, 1). But it is not clear whether the w. deacon received an *ordination apt to confer an official *ministry on her: the council of *Nicaea (can. 19) says that deaconesses belong to the *lay state, since they have not received *laying-on of hands. But the *Apostolic Constitutions* (4th c.-Syrian milieu) cite the entire rite for the ordination of deaconesses, which occurred through laying-on of hands by the bishop (*Const. Apost*. VIII, 20, 1-2). The characterization of w. in the various states of *virgin, wife, mother, *widow, is amply dealt with by the Fathers, drawing on models already known to pagan culture, as well as those of Scripture or those from the first centuries were held up as new examples of w., such as *Monica, *Nonna, *Macrina, *Melania, etc.

DACL 5, 1300-1353; *DSp* 5, 131-151; J. Daniélou, Le ministère des femmes dans l'Eglise ancienne, *La MaisonD*. 60 (1960) 70-96; R. Gryson, *Il ministero della donna nella chiesa antica*, Rome 1974; E. Giannarelli, *La tipologia femminile nella biografia e nell'autobiografia cristiana del IV secolo*, Studi Storici fasc. 127, Rome 1980; M.G. Mara, Le funzioni della donna nella Chiesa antica, *Rivista di pastorale liturgica* 19/2 (1981) 5-16; B. Witherington III, *Women in the earliest churches*, SNTS Monograph. Ser. 19, Cambridge 1988; E.A. Clark, *Women in the early Church*, Wilmington 1983; S.F. Brock-S.A. Harvey, *Holy Women of the Syrian Orient*, Berkeley 1987; J. Simpson, Women and Asceticism in the Fourth Century, *JRH* 15 (1988-89) 38-60; E.A. Clark, *Ascetic piety and women's faith: Essays on late ancient Christianity*, Lewiston 1986.

M.G. Mara

WORLD. To understand the *Fathers' cosmology we must bear in mind not just the fundamental distinction between the w. of ideas, intelligible things (νοητά), and the w. of sensible realities (αἰσθητά), but also two different tendencies: for radical *Platonism the sensible w. could not be the abode of immutable truth, while for *Stoicism the w. was permeated by *God. For the Fathers these cosmological conceptions were the substratum of the data of faith, put at the service of religion.

In relation to the w., the Christians found themselves in the same complex situation as Christ during his passage on this earth: they were not of the w. (Jn 15, 19; 19, 17), but they were in the w. (17, 11) and for the salvation of the w. In a certain sense Christian writers distinguish two "worlds": one which is the natural place of our existence and which we must love, and the other which is the enemy of God and which we must flee. Against *gnosticism they affirm the goodness and beauty of the universe, its unity, in which the *Holy Spirit, Wisdom, the Logos, *Providence provide cohesion. Western writers frequently praise the contributions of the w. to man's life; the Greek Fathers claim that the visible w. is above all a school for souls (Basil, *In Hex*. 6, 1: PG 29, 117), to bring man to the *contemplation of God. Man's responsibility for the universe derives from the fact that he is in a certain sense a microcosm, a *cosmos in miniature (cf. *Macrocosm). In the Creator's plan the mysterious union of *soul and *body has a well-defined aim: the *purification of the w. and its progressive *divinization.

On the other hand, the literature of *ascesis expresses the conviction that "the whole world lies in the power of the Evil One" (1 Jn 5, 19). Therefore abandonment of the w. is for Christians the necessary consequence of *love for God (cf. Theodoret, *Hist. rel*. 2: PG 82, 1308; 3, 1324-25).

Renunciation of the w. is, naturally, primarily an inner attitude. There is a *flight from the w. which is indispensable for all (τῆς ἁμαρτίας ἡ ἀναχώρησις): separation from *sin (Basil, *Reg. fus*. 7, 3: PG 31, 932 c), working to free oneself from all that may impede salvation. *Origen may be considered the ideological precursor of flight from the w. in the monastic sense, a flight that presupposes renunciation of whatever may retard spiritual ascent. On some specific points the gospel, ancient *dualism and common psychological experience are all in agreement. In the fervour of their exhortations, the ascetics did not always worry about terminological precision. We must bear this in mind when we find ourselves in the presence of expressions implying radical opposition between the "two worlds", so frequent in ascetic writers, though expressed in different terms: present/future worlds; visible/invisible worlds; worldly life/solitary life; worldly opinion/true gnosis. Sometimes we have the impression of being in the presence of two radically opposite tendencies: *Evagrius's contemplative tension is characterized by a total oblivion towards things, while *Basil in the *Hexaemeron* goes back to the original project of *creation, when man, contemplating the universe, was nourished by the memory of God. But in practice the *spirituality of the Fathers is not aware of two opposed "tendencies" towards the problem of the w. The authors emphasize now one, now the other aspect of the redemptive action: radical renunciation always presupposes that those who have attained perfection will reopen their eyes to discover the marvels of the visible world in "natural contemplation", a vision of the invisible God in the visible universe.

DSp 5, 1575-1605; P. Duhem, *Le système du monde. Histoire des doctrines cosmologiques de Platon à Copernic*, Paris 1914; V. Monod, *Dieu dans l'univers. Essai sur l'action exercée sur la pensée chrétienne par les grands systèmes cosmologiques depuis Aristote jusqu'à nos jours*, Paris 1933; A. Frank-Duquesne, *Cosmos et gloire. Dans quelle mesure l'univers physique a-t-il part à la chute, à la rédemption et à la gloire finales?*, Paris 1947; A.J. Festugière, *La révélation d'Hermès Trismégiste*, II, *Le Dieu cosmique*, Paris 1949; I. Hausherr, L'hésychasme. Etudes de spiritualité, *OCP* 22 (1956) 5-40, 247-285; id., *Hésychasme et prière*, OCA 176, Rome 1966, 163-237; M. Spanneut, *Le stoïcisme des Pères de L'Eglise de Clément de Rome à Clément d'Alexandrie*, Paris 1957; J. Gribomont, Le renoncement au monde dans l'idéal ascétique de saint Basile, *Irénikon* 31 (1958) 282-307, 460-475; L. Thunberg, *Microcosm and Mediator. The Theological Anthropology of Maximus the Confessor*, Lund 1965; H. Urs v. Balthasar, *Herrlichkeit. Eine theologische Aesthetik*, Einsiedeln 1965; A.I. Orbán, *Les dénominations du monde chez les premiers auteurs chrétiens*, Nijmegen 1970; T. Spidlík, Stare nel mondo o fuggire il mondo, *Vita consacrata* 13 (1977) 170-177; id., *La spiritualité de l'Orient chrétien. Manuel systématique*, OCA 206, Rome 1978, 121 ff., 199 ff.

T. Spidlík

WRITING. W., for the greater part of antiquity, meant dictating. Cicero and *Pliny the Younger dictated their works. Thanks to rich patrons, *Origen and *Jerome had at their disposal stenographers who transcribed the text dictated by the author. Arns states that "Jerome wrote none of his works with his own hand". To dictate a *letter or a work meant, in part, renouncing the chance to finish the text. Enough to examine some of Origen's and Jerome's *commentaries. The same is true of the *verba captata* and the edited text, as we see by comparing *Ambrose's *De mysteriis* with his *De sacramentis*. Jerome confessed: "I do not dictate as elegantly as I write. In the latter case I often manoeuvre the style to compose something that deserves to be read; in the former, I dictate what comes to my mind" (*Ep*. 74, 6, 2. Cf. Ambr., *Ep*. 47, 1-2).

J. de Ghellinck, *Patristique et moyen âge*, Brussels, II-III, 1947-1949; E. Arns, *La technique du livre d'après saint Jérôme*, Paris 1953.

A. Hamman

Y

YEZDEGERD I. Persian emperor (399-420) of the *Sassanid dynasty, called the "Persian Constantine". Ended the persecution begun by Shapur II in the mid 4th c. and conceded liberty to the church in *Persia, which was able to reorganize. "He ordained that all through his empire the churches destroyed by his ancestors should be magnificently rebuilt, overturned altars should be diligently repaired, those who had been tried for God by chains and blows should regain their freedom, and presbyters and rectors, with all their communities, should be without fear and without risk. And this took place at the time and on the occasion of the election of our reverend ... Mar *Isaac, bishop of *Seleucia-Ctesiphon, *catholicós and head of the bishops of all the *East" (Acts of the synod of Seleucia-Ctesiphon, in Chabot, *Sinod. Or.*, p. 18). Y. called this synod in 419 and advised the Persian bishops to remain united with the West. He also maintained good relations with the *Byzantine court; his legate was probably the bishop of Seleucia-Ctesiphon, Jahballaha.

J.B. Chabot, *Synodicon orientale*, Paris 1902, 18; I. Ortiz de Urbina: *EC* 6, 1590; id., Storia e cause dello scisma della Chiesa di Persia, *OCP* 3 (1937) 468-75.

G. Ladocsi

YORK. City in N-E England (Lat. *Eboracum*), military headquarters of Roman *Britain from the 1st to the 5th c. *Constantine was proclaimed emperor there in 306. In 314 bishop Eborius (a Celtic name) attended the council of *Arles. There is no evidence of a Christian community at Y. from the time of the Saxon invasion (mid 5th c.) to 625 when *Paulinus was consecrated bishop, but in 601 pope *Gregory I had arranged in a *letter to *Augustine of Canterbury for the bishop of Y. to have metropolitan dignity and consecrate 12 bishops. In 627 Paulinus baptized king Edwin of Northumbria at Y. In 633, when the Saxon king Edwin was overthrown, Paulinus fled and Y. ended up under the jurisdiction of the Celtic bishops of Lindisfarne. In 634 pope *Honorius sent Paulinus a pallium in response to a request made by Edwin before his death, and arranged that whenever the see of Y. or *Canterbury became vacant, the occupant of the other metropolitan see should ordain a new bishop for the vacant see and have supremacy over him. The struggle for supremacy between the two sees did not end until the 14th c. The see of Y. was restored in 664, and *Wilfrid I reintroduced Roman customs and restored the *church building. His episcopate was interrupted by disputes with the *archbishops of Canterbury and the king of Northumbria, who twice forced him to abandon his see. Y. became an archbishopric in 735.

Bishops: Paulinus 625 (until 633); vacant 633-664; *Chad 664 (until 669); Wilfrid I 669; Boisil 678; Wilfrid I (recalled) 686; Boisil (recalled) 691; John of Beverley 705; Wilfrid II 718; Egbert 732 or 734.

M. Deanesly, *The Pre-Conquest Church in England*, London 1961; C. Thomas, *Christianity in Roman Britain to 500 AD*, London 1981 (refs. to York and other sites).

E. Harbert

Z

ZACCHAEUS, antignostic. 2nd-c. bishop of Caesarea (no better identified). Named exclusively by *Praedestinatus* (I, 11; I, 13: PL 53, 591) as an opponent of *Valentinus and *Ptolemy and their disciples, whom he anathematized.

E. Prinzivalli

ZACHARIAS of Jerusalem. Patriarch of *Jerusalem 609-c.628. When the Persians conquered Jerusalem in 614, Z. was taken prisoner together with the Holy Cross and numerous believers. From this period dates a consolatory work composed by *Modestus, *abbat of the *monastery of *Theodosius, who later succeeded Z. The *consolatio* also survives in *Georgian and *Arabic versions. Z. was added to the *calendar of saints (feast 21 Feb).

CPG 3, 7825; PG 86, 2, 3219-3234; Georgian version: CSCO 202, 70-76 and 203, 46-50 (G. Garitte); Arabic version: 340, 49-53 and 98-101; 341, 33-35 and 66-68 (G. Garitte; one strand of the tradition); also 347, 143-145 and 185-188; 348, 97 f. and 126-129 (a parallel strand); *LTK* 10, 1301; *BS* 12, 1451-1453; Beck 448.

J. Irmscher

ZACHARIAS Scholasticus or the Rhetor. Born at Maiuma of *Gaza (S *Palestine) in the last third of the 5th c., of a large family (one of the five brothers was *Procopius of Gaza). From 485 to 487 he studied at *Alexandria where he made *friends with *Severus, the future *monophysite patriarch of *Antioch. After having favoured his *conversion, he joined him in autumn 487 at Berytus (Beirut) to study law (487-492), leading an austere life among the monophysite *ascetics. At least from 492 he practised as a *scholastikos* (advocate) at *Constantinople where he also reached high office, including that of legal adviser to the *comes sacri palatii*, while remaining in close touch with religious circles, indeed living in the same *monastery as his brothers, the priests Philip and Victor. After 512 he abandoned Severian monophysism. Between 527 and 536 he was promoted to the metropolitan see of Mitylene (Lesbos) and as such he spoke, in May 536, at the synod of *Constantinople which deposed patriarch *Anthimus, who was condemned with his monophysite followers. The date of Z.'s death is unknown.

Apart from *Ammonius* or the *Dispute on the *creation of the world*, which took place at Alexandria with his professor, the *pagan sophist Ammonius, and opuscula *Against the *Manichees* written in 527 or slightly earlier, Z.'s literary legacy, or that attrib. to him, is historical. The *Life of Severus of Antioch*, written in c.515 and preserved in *Syriac (PO 2, 7-115), emerges as an autobiographical document. Of the *Life of *Peter the Iberian* only a brief fragment in Syriac survives, while Z.'s authorship of the *Life of *Isaiah of Gaza* is denied by some scholars. The Greek text of Z.'s so-called *Ecclesiastical History* is lost, but a Syriac revision survives, inserted as books III-VI (450-491) of an anon. historical compilation in 12 books from the creation of the world to 569. *Evagrius is dependent on Z.'s *History*. Even in its reduced form, the work remains the most important historical source for the reign of *Marcian, *Leo I and *Zeno, whose *Henoticon* the author approves.

PG 85, 1011-1178; CPG 6995-7001; M. Minniti Colonna, *Zaccaria Scolastico. Ammonio* (introd., crit. text and tr., commentary), Naples 1973; B. Tatakis, *La philosophie byzantine*, Paris 1949, 34-37; *DTC* 15, 3676-3680; *EC* 12, 1761; *PWK* 9, A, 2, 2212-2216; W.H.C. Frend, *The Rise of the Monophysite Movement*, Cambridge ²1979, 202 ff.

D. Stiernon

ZENO. Roman emperor 474-475 and 476-491. Taracosidissa - such was his name - was born in Isauria in c.426. The emp. *Leo I, deciding to free himself from the Ostrogothic tutelage of Aspar (later assassinated, 471), called the Isaurian chief to *Constantinople and made him *magister militum per Orientem*. He took the Greek name Z. and married the emperor's eldest daughter Ariadne. The son born of this marriage ideally succeeded as *basileus* (early 474) under the name Leo II but, after his immature death (autumn 474), Z., from regent and co-emperor, became emperor. He was overthrown (Jan 475) by the plot that brought *Basiliscus, brother of Verina, wife of Leo I (who also died in 474), to the throne. There followed a disastrous interregnum of 20 months. At the end of August 476, Z. returned to power and remained on the throne for 15 years, despite other factional struggles and civil wars. To rid himself of the turbulent Ostrogothic chieftain who had helped him overthrow Basiliscus, Z. persuaded *Theodoric the Great to move against *Italy and bring down *Odoacer, previously appointed by *Byzantium *magister militum per Italiam*. Theodoric's victory opened a new page in Italian history on the morrow of the fall of the Roman empire in the West.

In the *East, the danger of disintegration came from the *monophysism that was alienating *Egypt and *Syria-*Palestine from Byzantium. To restore religious unity Z., advised by patriarch *Acacius of Constantinople, promulgated in 482, on the occasion of *Peter Mongus's accession to the see of *Alexandria, a document that aimed to reconcile the monophysites without provoking the Chalcedonians, the so-called *Henoticon*. It presented itself as an imperial *letter sent "to the most reverend bishops, clerics, monks and laymen of Alexandria, Egypt, *Libya and Pentapolis", but was really aimed at the whole empire. It proclaimed the profession of faith of *Nicaea-*Constantinople as the one orthodox *creed, according to the definition of the council of *Ephesus. It repeatedly condemned *Nestorius and *Eutyches as well as those who, at *Chalcedon or in other synods, had spoken differently of the one Christ, the one Son, and it approved the 12 *anathemata of *Cyril of Alexandria.

The imperial decree did not totally satisfy the monophysites: Peter Mongus accepted it, but his church did not; Calendio of Antioch also rejected it, while *Martyrius of Jerusalem accepted it. Above all it was condemned by pope Felix III at the Roman synod of 28 July 484, thus opening a *schism with Byzantium (the *Acacian schism) that lasted until 519.

Evagrius, *HE* III, 14; PG 86, 2620-2625; Liberatus, *Breviarium* 17; PL 68, 1022-1024; *EC* 5, 365-367; 12, 1792-1793; *Cath* 5, 605; Fliche-Martin 4, 362-373; G. Ostrogorsky, *Storia dell'impero bizantino*, Turin 1968, 54-56; *PWK* 10, A (1972) 149-213; W.H.C. Frend, *The Rise of the Monophysite Movement*, Cambridge ²1979, ch. 4; E. Dovere, L'Enotico di Zenone Isaurico, *SDHI* 54 (1988) 170-190.

D. Stiernon

ZENOBIA. Queen of *Palmyra. On the death of her husband Odenathus (267), she took command in the name of her son Vaballatus. Unlike her husband, she followed a policy of hostility towards the Roman empire, enlarging her borders in *Syria, *Egypt and *Asia Minor, as far as *Ancyra in *Galatia. *Aurelian, busy elsewhere, recognized the state of affairs (270). The caravan city of Palmyra enjoyed a period of splendour, and this Hellenistic sovereign surrounded herself with advisers like the neoplatonist philosopher Cassius Longinus and the bishop of *Antioch and *ducenarius* (finance minister) of the Palmyrene kingdom *Paul of Samosata, whom Z. protected, allowing him to keep his see even after his condemnation by the Antiochene synod of 268. Driven by ambition, Z. aimed to create an independent state and began to coin money (271). Aurelian intervened, took Palmyra at the head of his army and took Z. prisoner (272).

J.C. Février, *Essai sur l'histoire politique et économique de Palmyre*, Paris 1931, 103 ff.; J. Gagé, *La montée des Sassanides et l'heure de Palmyre*, Paris 1964, 349 ff.; *PWK* 10, A (1972) 1-8.

U. Dionisi

ZENOBIUS of Zephyrium. Bishop of Zephyrium in *Cilicia around 433. After the council of *Ephesus, the conciliatory *creed compiled by *Theodoret of Cyrrhus was approved under pressure from the emperor, in order to resolve the *Nestorian problem. It was signed by the Antiochenes, represented by *John of Antioch, by Theodoret and by the Alexandrians, represented by *Cyril of Alexandria (Hfl-Lecl 2, 1, 385-419). But Z., with other Cilician bishops, went into opposition against John of Antioch, for which he was deposed, perhaps in 435. Exiled to Tiberias, he managed to escape. The *acts of the council contain a *letter from Z. to bishop *Alexander of Hierapolis, metropolitan of the province of Euphrates, also in opposition, in which the author informs him of the sanctions adopted against bishop *Meletius of Mopsuestia (in Cilicia).

CPG 3, 6470; PG 84, 792-793; ACO I, 4, 195; *PWK* 10 A, 14 f.

J. Irmscher

ZENO of Verona. *Ambrose mentions Z. (*Ep*. I, 5, 1) as bishop of *Verona (†c.380) and local tradition puts him eighth in the *episcopal list of that city. Some MSS, also derived from local tradition, attribute to him some 90 *homilies written after 360. This collection has reached us in two parts: part

I comprises 62 texts, part II only 30. Only about 30 homilies are complete. The rest are rough drafts or summaries. It seems clear that the collection was not made by the author but after his death, prob. for liturgical requirements.

Many of the homilies are exegetical, mostly dedicated to OT passages, and their interest in OT events and characters also appears in homilies of a different sort, accompanied by a certain anti-Jewish vein. The *exegesis is the traditional typological one, mainly christological, for which *Jonah, *Jacob, *Job, etc., are seen as *typoi* of Christ. Some homilies deal with baptismal and Paschal themes, others are moral (*de continentia, de avaritia*, etc.). Few are specifically doctrinal (I, 2; II, 5), but doctrinal ideas are disseminated a bit throughout. We see some traces of *millenarism (I, 2), but considerable mastery of Nicene trinitarian *theology. Z. often polemizes against the Photinians, whose centre of propagation, *Pannonia, was not far from N *Italy. In I, 3, 19 he repeats the archaic idea that Christ entered *Mary through her ear, i.e. through the word of the announcing *angel. *Hom.* I, 39 is the only one dedicated to a *martyr, the African Arcadius: this fact, together with echoes of *Tertullian and *Lactantius, has led some to suggest an African origin for Z.

CPL 208-209; CCL 22; A. Bigelmair, *Zeno von Verona*, Münster 1904; AA. VV., *Studi Zenoniani in occasione del XVI centenario della morte di san Zeno*, Verona 1976; *Patrologia* III, 117-120 (bibl.); V. Boccardi, *L'esegesi di Zenone di Verona*, *Augustinianum* 23 (1983) 453-485; C. Truzzi, *Zeno, Gaudenzio e Cromazio: Testi e contenuti della predicazione cristiana per le Chiese di Verona Brescia e Aquileia (360-410 ca)*, Brescia 1985.

M. Simonetti

ZEPHYRINUS, pope (199-217). Acc. to *Eusebius he presided over the church of *Rome for 18 years (*HE* VI, 21); the Chronographer of 354 says he died in 217 (MGH, *Chron. min.* I, 74). The *Liber Pontificalis* (I, 139) adds that he issued two decrees on the presence of the community at *ordinations and the celebration of *mass by the bishop in the presence of the priests. Z. was buried in the *cemetery of S. Callisto, but not underground. During his episcopate *Origen came to Rome (Euseb., *HE* VI, 14, 10) and may have met him, and *Gaius wrote against the *Montanists (Euseb., *HE* VI, 20, 3). Pope *Victor, Z.'s predecessor, had condemned the *adoptianism propagated at Rome by *Theodotus "the Tanner" of Byzantium; Z. renewed the condemnation and readmitted the modalist bishop *Natalius to *communion, but as a *layman (Euseb., *HE* V, 28, 8-19). Z. is known mainly through the work of a bitter enemy of his successor *Callistus (who had been his deacon), *Hippolytus of Rome's *Philosophumena*, which expound the theological disputes current at the start of the 3rd c. (IX, 7 ff.). Modalism was being propagated at Rome by *Epigonus, *Cleomenes and *Sabellius. Z., who, acc. to Hippolytus, was a "simple man, uneducated and ignorant of ecclesiastical rules and a lover of *gifts and money" (*Philos.* IX, 11) and completely at Callistus's mercy, followed a very ambiguous line, which favoured the new *heresy; even his profession of faith, acc. to Hippolytus, was wrong. Hippolytus's personal enmity with Callistus plays an important part in these heavy judgements.

DTC 15, 3690-91; 2, 1336-37; *BS* 12, 1464-65; Quasten I, 249 f.; Fliche-Martin, cf. indexes of vol. 2; A. von Harnack, *Die älteste uns im Wortlaut bekannte dogmatische Erklärung eines Römisches Bischofs (Zephyrin bei Hippolyt, Ref. IX, 11)*, in *Kleine Schriften zur alten Kirche*, II, Leipzig 1980, 619-625; B. Capelle, *Le cas du pape Zéphyrin*, *RBen* 38 (1926) 329-330; W.H.C. Frend, *The Rise of Christianity*, London 1984, 340-345.

A. Di Berardino

ZOSIMUS, pope (417-418). Certainly of non-Roman origin, his ecclesiastical policy was aimed at reorganizing the hierarchy in *Gaul and solving the *Pelagian problem. No sooner elected *pope than he accorded *Patroclus, bishop of *Arles, a privileged position that made him primate of the seven Gallican provinces, a solution that could not last long. In the Pelagian controversy Z.'s attitude was rather ambiguous. After having readmitted *Pelagius and *Caelestius to ecclesial *communion (autumn 417), he had to give way to unanimous and vigorous pressure from the African episcopate (*Ep.* 12). Constrained by the interventions of the imperial court and by a new African synod, he finally condemned Pelagius and Caelestius openly in the *letter known as the *Tractoria* (DS 231), provoking a *schism in *Italy. Ruling for only a year, he left his successors numerous difficulties.

CPL 1644-1647; PLS 1, 796 f.; PL 20, 642-686 (Coustant); *Ep. Tractoria*: PL 20, 693 f.; *BS* 12, 1493-1497; W. Marschall, *Karthago und Rom*, Stuttgart 1971, 150-159, 166-173; O. Wermelinger, *Rom und Pelagius*, Stuttgart 1975, 134-218 (bibl.); id., *Das Pelagiusdossier in der Tractoria des Zosimus*, ZPhTh 26 (1979) 336-368; B. Studer, *Patrologia* III, 553 f.

B. Studer

ZOSIMUS. Byzantine historian (5th-6th cc.). From the *intitulatio* of his work we know only that he was *comes* and *exadvoctus fisci*, a detail also given by *Photius (*Bibl.*, cod. 98), who identified the historian with two homonymous sophists, Zosimus of Gaza (which is possible) and Zosimus of Ascalon (which is not). From the same work - the *Historia Nova* - we can deduce that the author, a convinced and enthusiastic *pagan not without literary culture, lived at least for some time at *Constantinople, where he wrote his *History* between 498 (or more precisely 502) and 518.

Starting from the Trojan War, in the first seven books Z. makes a rapid summary of the events of the Graeco-Oriental world up to the fall of the Macedonian empire; then he briefly outlines the history of the first imperial Roman era and, starting esp. from the 4th c., gradually amplifies his narrative, which is interrupted *ex abrupto* in 410, while the author proposed to go beyond that date. So the work either remained incomplete or survives in a mutilated state.

The author of the *Historia Nova* (meaning not a modern history, but an unusual historical approach) has been considered the Polybius of the late empire. His originality consists in his attributing the decadence of Rome firstly to the "mad presumption" of the monarchical system and secondly to the abandonment of traditional religion, sacrificed on the altar of Christianity.

He thus conceives his own inquiry as a programmatic overthrow of Christian *historiography, in particular the Eusebian *apologia* of the monarchical ideal and the Augustan *pax* understood as a providential preparation for the spread and triumph of Christianity, consecrated by the *conversion of *Constantine. Beyond his confusions and contradictions, Z.'s work, harshly judged by Photius, is at any rate the last monument of ancient historiography.

L. Mendelssohn, *Zosimi comitis et exadvocati fisci Historia nova*, Leipzig 1887, repr. Hildesheim 1963. Ed. with Fr. vers.: F. Paschoud, *Zosime. Histoire nouvelle*, 3 vols., Paris 1969-1979 (vol. IV = books V and VI, to be published next). It. tr. only: F. Conca, Milan 1977. M.E. Colonna, *Gli storici bizantini dal IV al XV secolo*, Naples 1956; PWK 10 A, 797-841; W. Goffart, *Zosimus, the First Historian of Rome's Fall*, American Historical Review 76 (1971) 412-441; L. Berardo, *Struttura, lacune e struttura delle lacune nell' 'Ιστορία νέα di Zosimo*, *Athenaeum* 54 (1976) 472-481; F. Paschoud, *Cinq études sur Zosime*, Paris 1975; L. Cracco Ruggini, *Zosimo, ossia il rovesciamento delle "Storie ecclesiastiche"*, *Augustinianum* 16 (1976) 23-36; F. van Ommeslaeghe, *Jean Chrysostome et le peuple de Constantinople*, AB 99 (1981) 329-349; W.E. Kaegi, Jr., *Byzantium and the Decline of Rome*, Princeton 1968, ch. 3.

D. Stiernon

ZOSIMUS. 6th-c. Palestinian *abbat and *ascetic who used the monk *Isaiah (cf. L. Regnault: *RAM* 46 [1970] 40) and was in turn used by *Dorotheus of Gaza (L. Perrone, *La Chiesa di Palestina e le controversie cristologiche*, Brescia 1980, 310). His *Alloquia* survive (PG 78, 1680-1701, incomplete; CPG 7361).

K. Kunze: *BS* 12, 1499-1500, n. 4.

J. Gribomont

SYNOPTIC TABLE

Date	Secular Events	Ecclesiastical Events	Cultural and Doctrinal Matters
40BC	37: Herod the Great, king of Judaea (39-34), occupies Jerusalem with Anthony's support.		
	31: Battle of Actium: Octavian defeats Anthony and Cleopatra.		
	30: Egypt a Roman prefecture.		
	27: On 13th Jan Octavian receives *tribunicia potestas, imperium proconsulare* and the *cognomen Augustus*: date of the beginning of the Empire.		
	18: The *leges Iuliae* on morality.		
	15: Rhaetia, Noricum and Vindelicia annexed to the Empire.		
	12: Augustus *pontifex maximus*.		
10	12-9: Campaigns in Germany (Drusus and Tiberius); conquest of Pannonia (Tiberius). Inauguration of the *ara pacis Augustae*.	Tarsus: birth of Saul (Paul) (10-5). He frequents the school of Gamaliel.	
	8: Death of Maecenas and Horace at Rome.	*c.*7-6: Bethlehem: birth of Jesus.	
	*c.*4-3: Death of Herod. Palestine split between Archelaus (deposed AD 6), Herod Antipas (4 BC-AD 39) and Philip (4 BC-AD 34). With the deposition of Archelaus Judaea passes under a Roman procurator (until AD 41).		
AD			
10	14: Death of Augustus. Tiberius emperor (14-37).		
	Organization of the praetorium. The two Rhine provinces.		
	Cappadocia and Commagene Roman provinces (17 and 18).		
20	Pontius Pilate procurator of Judaea (26-36).	27: Preaching of John the Baptist; start of Jesus' ministry.	
		28: Jesus at Jerusalem.	
		29: Herod Antipas has John killed.	
30		30: Passover at Jerusalem: death and resurrection of Jesus. Pentecost: outpouring of the Holy Spirit.	
		Pilate in difficulties with the Jews over the question of the military standards.	
		*c.*33: Election of the seven "deacons".	
		*c.*34: Martyrdom of Stephen. Dispersion of the Christian community of Jerusalem.	
		*c.*36: Conversion of Paul. Pontius Pilate leaves Jerusalem.	
	37: Death of Tiberius. C. Caligula emperor (37-41). Temple of Divus Augustus consecrated at Rome.	*c.*37: Church of Antioch. Peter preaches in Samaria and the coastal cities.	
	38: Persecution against the Jews of Alexandria.		
	39: Herod Antipas exiled; Galilee passes to Herod Agrippa.		
40	41: Assassination of Caligula. Claudius emperor (41-54). Claudius allows Herod Agrippa to reunify the kingdom, including Judaea. Edict and letter of Claudius, restoring		

Date	Secular Events	Ecclesiastical Events	Cultural and Doctrinal Events
	to the Jews their privileges and immunities (end of 41).		
	42: Mauretania a Roman province. 43-44: Provinces of Britain, Lycia and Pamphilia. The Romans found Londinium (London). 44: Death of Herod Agrippa. Palestine once more under a Roman procurator: procuratorial province (44-66). The situation remains tense. Procurators: Antonius Felix (52-60); Porcius Festus (60-62); Lucceius Albinus (62-64); Gessius Florus (64-66). 49: Expulsion of the Jews from Rome.	44: James (the Great), brother of John, beheaded at Jerusalem. Peter imprisoned. 46-48: Paul's first mission: Syrian Antioch, Cyprus, Pisidian Antioch, Lystra. . . ; return to Antioch. Famine in Palestine: help from the community of Antioch brought by Paul and Barnabas to Jerusalem.	49: Council of Jerusalem: those converted from paganism are not obliged to observe the Mosaic law. Oral preaching is put down in writing: *Aramaic Matthew*.
50	Seneca Nero's tutor. Claudius marries Agrippina Minor and adopts Nero. 52: Rome: the aqueduct of Acqua Claudia is completed. Porta Maggiore. 54: Claudius murdered. Nero emperor (54-68). Seneca counsellor.	Paul's second mission: Lystra (Timothy), Phrygia, Galatia, Macedonia, Athens. . . , Antioch (between 50 and 52). 53-58: Paul's third mission: Phrygia, Galatia, Ephesus, Macedonia (57), Corinth (57-58), Caesarea, Jerusalem (his arrest during Pentecost of 58). Taken to Caesarea: before the governor Felix. Prisoner at Caesarea (58-60). In AD60 before Festus: he appeals to Caesar; before Agrippa and Berenice. Journey to Rome (autumn 60). Paul at Rome under guard, but with a certain freedom (61-63).	51: *Epistles to the Thessalonians* from Corinth. From Ephesus: *Epistle to the Philippians* (56). *First epistle to the Corinthians* (57). *Epistle to the Galatians* (?). 57: *Second epistle to the Corinthians*. From Corinth: (*Epistle to the Galatians*?); *Epistle to the Romans*. *Epistle of James* to the Jews of the dispersion (before 49?, after 62?).
60	60: Nero has his mother Agrippina killed. Violent earthquake at Pompeii. 61: Revolt in Britain under Boudicca. 62: Nero exiles his wife Octavia and marries Poppaea. Retirement of Seneca: *Naturales quaestiones*, tragedies, *Epistulae morales ad Lucilium*. 64: Fire of Rome, lasting nine days. Accusations against Nero. Gessius Florus, as governor of the province of Judaea (64-66), provokes much resentment. Petronius: *Satyricon*. Lucan: *Pharsalia*. 65: Plot against Nero. Great men, such as Seneca, Lucan and Petronius, are eliminated. The Neopythagorean Apollonius of Tyana. 67: Insurrection in Palestine; exodus from Jerusalem; the Christians flee to Pella(?). Roman intervention: first Cestius Gallus, then Vespasian, with 60,000 soldiers. Reconquest of Galilee, then of other territories (67-68). Destruction of Qumran. 68: Revolt of the Spanish legions. Suicide of Nero. Sulpicius Galba (68-69), emperor of the senate. Otho, emperor of the praetorians, killed by Vitellius, emperor of the legions. Vespasian emperor (69-79).	62: In Jerusalem James (the Less), brother of Jesus, is stoned by order of the high priest Annas. Simeon, son of Cleophas and Mary, succeeds James. 63: Paul free. Journey to Spain? Rome: Persecution of the Christians, due to the burning of the city. Peter killed at Rome (64 or 67). c.65: Paul at Ephesus? Rome: bishop Linus (67?-79?).	*Epistles to the Colossians, to the Ephesians, to Philemon* (61-63). *First epistle of Peter* (c.64?); *Gospel of Mark* (?). *First epistle to Timothy* (?); *Epistle to Titus* (?). *Greek gospel of Matthew* (?); *Gospel of Luke* and *Acts of the Apostles* (?). *Second epistle to Timothy* (?). *Epistle to the Hebrews* (?).
70	70: Siege and conquest of Jerusalem by Titus; destruction of the Temple	Some of the Judaeo-Christians return to Jerusalem.	*Epistle of Jude* (?); *Second epistle of Peter* (?); *IV Esdras* (?).

Date	Secular Events	Ecclesiastical Events	Cultural and Doctrinal Matters
	(29 Aug). Judaea a Roman province: Caesarea a Roman colony. Vespasian's *lex de imperio*. The rabbis of Jabne (Jamnia). 71: Triumph of Titus. 72: Foundation of Flavia Neapolis (Nablus). Pliny: *historia Naturalis* (*c.*77).Flavius Josephus:*Bellum Iudaicum*. Construction of the Colosseum (Flavian Amphitheatre) (75-80). Martial: epigrams for its inauguration.		
80	79: Titus emperor (79-81). Eruption of Vesuvius. Death of Pliny the Elder. Domitian emperor (81-96). Rome: Arch of Titus. 83: *Agri decumates*. 84: Britain subdued as far as Caledonia. 85-88: Moesian and Thracian campaigns.	Rome: bishop Anacletus (Cletus) (79?-88?).	*Didachè* (late 1st c.?) (Syria / Palestine?): anonymous compilation (by a Judaeo-Christian) from various sources, deriving from the living tradition of ecclesial communities; teachings useful for the edification of the converted.
90	Campaigns against Suevi and Sarmatians (89-97). 93: Persecution against intellectuals at Rome. Epictetus teaches at Rome (from 94, at Nicopolis in Epirus). Flavius Clemens, cousin of Domitian, executed on a charge of "atheism" (Christianity?). His wife Domitilla exiled to Pandataria. Eusebius calls her a Christian (*HE* III, 18, 4). 93-94: Flavius Josephus:*Antiquitates*. 96: Acilius Glabrio executed for the same reason. Plot of Domitia. Nerva emperor of the senate (96-98). He adopts the Spaniard Trajan (adoptive emperors).	Rome: bishop Clement (88?-97?; 92-101 according to Eusebius's chronology). Persecution of Christians in various parts of the Empire. *c.*95: John relegated to Patmos.	Nicolaïtans: practise a certain doctrinal and moral laxity. *Apocalypse* (95?). *Letter of Clement* to the Christians of Corinth, where a sedition has broken out against the responsible members of the community. Important document for our knowledge of Roman theology and liturgy at the end of the 1st century and the Roman church's awareness of its right to intervene in other communities. *Gospel and epistles of John* (?). Docetism: a doctrine that devalues Christ's humanity.
100	Trajan emperor (98-117). Tacitus: *Germania*. Dacian campaigns (101-106). Dacia and Arabia provinces.	Rome: bishop Evaristus (97?-106?), who is certainly Clement's successor. John's death at Ephesus (*c.*100). Rome: bishop Alexander I (105?-115?). *c.*107: Jerusalem: martyrdom of Simeon.	
110	110: Pliny the Younger (61-113) Governor of Bithynia: *Letters*. Juvenal: *Satires*. Campaigns in Armenia. Provinces of Mesopotamia and Assyria. Plutarch of Chaeronea: opuscula, *Parallel lives* (23 pairs of characters). Ostia: Trajan's Port (100-112). Rome: Trajan's Forum (112-113). 114: Benevento, Arch of Trajan. Revolt of the Jews in Cyprus, Cyrenaica and Egypt. Many dead. 117: Death of Trajan at Selinous in Cilicia. Emperor Hadrian (117-138) and Sabina.	*c.*110: Ignatius's journey to Rome. His martyrdom at Rome (112?). In Bithynia: Pliny persecutes the Christians. Pliny's letter to Trajan and the latter's rescript. Rome: bishop Sixtus I (115-125?).	The *Seven letters of Ignatius of Antioch*, written, during his journey to Rome, to the communities of Ephesus, Magnesia, Rome, Philadelphia, Smyrna, Tralles and to Polycarp. Against Docetism; importance of the bishop. *Letter of Ps.-Barnabas*. A treatise in epistolary form, anti-Jewish. Allegorical interpretation of the OT. The true covenant is that of Jesus, sealed in the heart of those who hope and believe in him (datable from the end of the 1st c. to *c.*130).

Date	Secular Events	Ecclesiastical Events	Cultural and Doctrinal Events
120	Abandonment of the provinces of Assyria and Mesopotamia. Suetonius: *Lives of the Caesars*. Rome: rebuilding of the Pantheon. Tivoli: Villa Adriana (125-135). Ptolemy writes his *Geography*. Hadrian's Wall in Britain. Hadrian's journey to the East (Greece, Asia Minor, Egypt) (128-132).	125: Hadrian's rescript to Minucius Fundanus, governor of Asia. Rome: bishop Telesphorus (125?-136?). Gnosticism spreads: Basilides at Alexandria; Saturnilus (or Saturninus) in Syria.	Quadratus, the first apologist (123/129), perhaps of Athens or Asia Minor, consigns his work, now lost, to Hadrian.
130	Jurists: S. Pomponius and Salvius Julianus (c.100 - c.169). (*Edictum perpetuum*: c.130). Arrian of Nicomedia publishes the *Enchiridion* (*Handbook*) and *Diatribes* of Epictetus. Insurrection of the Jews in Palestine (Simon bar Kochba = S. bar Kosiba). 135: Jerusalem rebuilt and called Aelia Capitolina. On the site of the Jewish temple is a temple dedicated to Jupiter. Rome: Mausoleum of Hadrian (Castel S. Angelo) (132-139). Death of Hadrian. Antoninus Pius emperor (138-161); Faustina Major. Britain: Antonine Wall.	Irenaeus born in Asia Minor (130/140). Rome: bishop Hyginus (136?-140?).	Papias of Hierapolis: *Explanation of the Lord's sentences*,, (post 130), dedicated to exegesis of the Lord's sayings; only fragments survive. Ebionites: various Judaeo-Christian sects, which consider Jesus a mere man, do not accept Paul, and follow the Jewish Law. A *Gospel of the Ebionites* is cited by Origen. Aristo of Pella: *Dispute between Jason and Papiscus*. A lost work: it purported to be an account of a discussion between a Christian and a Jew about the OT prophecies and their fulfilment in Christ. *Kerygma Petri* (apocryphal).
140		Rome: bishop Pius I (140?-155?). Marcion comes to Rome from Sinope (Pontus), whence he was expelled by his father, bishop of that city. 144: Marcion, excluded from the Roman community, founds his own church. Valentinus, an Egyptian come to Rome around 140, founds a school: he develops an original system which synthesizes the scattered elements of gnostic ideology.	Hermas: *Shepherd*: announcement of a day of pardon, once only, for all those who repent of their sins committed after baptism. From here spreads the doctrine that penitence is possible only once in a lifetime. Marcion distinguishes between the God of the OT, a just judge, but wrathful and voluble, and the God and Father of Jesus Christ, a good and benign God. Rejection of the OT; acceptance only of Paul, emended, and Luke's gospel.
150	Apuleius of Madaura: *Apologia*; *Metamorphoses* (*Golden Ass*): book XII is the story of the author's initiation into the mysteries of Isis. During this period the Germanic peoples move southwards.	150-161: *Apologies* of Justin martyr. Construction, in the Vatican necropolis at Rome, of a shrine over the earth tomb of the Apostle Peter (c.150). Birth of Clement of Alexandria (c.150). Aristides addresses to the emperor Antoninus Pius a defence of Christianity presented, through Platonic concepts, as the true philosophy; his *Apology*, in 17 chapters, is the oldest preserved in the original Greek. Polycarp, bishop of Smyrna, martyred c.155. Rome: pope Anicetus (155?-166?).	*Second letter of Clement*: a homily, the oldest, penitential in character, of unknown author (Syria? Egypt?). (The *Epistulae ad Virgines* were attributed to Clement of the late 4th c.) The *Odes of Solomon*: (post 150): 42 poetic compositions, which comment on the baptismal and Easter liturgy of a Judaeo-Christian community. Justin is the first Christian to make use of Aristotelian categories, to use philosophical terminology in Christian thought and to reconcile faith and reason. He identifies Christ with Plato's World-Soul. Besides the *Apologies*, he wrote the *Dialogue with Trypho*. Polycarp is author of the *Letter to the Philippians*, which survives in Latin translation. The facts of his martyrdom are related in the so-called *Martyrium Polycarpi*, which attests for the first time the cult of martyrs and relics.

Date	Secular Events	Ecclesiastical Events	Cultural and Doctrinal Matters
160	161: Death of Antoninus Pius. Emperor Marcus Aurelius (161-180) and Faustina. He takes his brother Lucius Verus as partner in the empire. War with the Parthians (163-166), against Vologeses. Occupation of Armenia and Mesopotamia. Bubonic plague first starts to spread (post 160). Lucian of Samosata: satirical works, dialogues, *Historia Vera*. Aulus Gellius: *Noctes Atticae*. Appian the historian. Invasion of barbarians as far as Aquileia. Campaigns in Germany, Pannonia (167-175). Death of Lucius Verus (169).	Start of Montanus's activity in Phrygia (155-160): founder of Montanism, rejected by the Great Church. *c*.162: Birth of Tertullian. Rome: martyrdom of Justin (163-167). Rome: pope Soter (166?-175?).	Montanus claims to be the spokesman of the Holy Spirit and that the Paraclete promised by Jn 14, 26; 16, 7 is incarnate in his person.
170	Jurists: Scaevola and Gaius. Spread of Oriental cults. Marcus Aurelius: *To himself*. 175: Revolt in the East by Avidius Cassius, governor of Syria. Barbarians accepted into the *auxilia*. The plague continues. Depopulation and economic crisis.	Tatian active. *Address to the Greeks* and *Diatessaron* (*c*.170). Pantaenus at Alexandria. Rome: pope Eleutherius (175?-189). 177: Athenagoras the Athenian addresses to Marcus Aurelius and Commodus an apology entitled *Embassy for the Christians*. Irenaeus of Smyrna elected bishop at Lyons, whence he will work to evangelize Gaul. *c*.177: Martyrs of Lyons (among them Pothinus and Blandina). Celsus writes his *True speech* against the Christians (*c*.180).	In the apology entitled *Discourse to the Greeks* Tatian polemizes violently against the whole of Greek culture, and defends Christian ideas. Athenagoras refutes the three traditional accusations of atheism, incest and cannibalism; for the first time he offers a rational proof of the uniqueness of God. The first polemical work against the Christians, which survives thanks to Origen's criticism in *Contra Celsum*.
180	Equestrian statue of Marcus Aurelius. Death of Apuleius. Column of Marcus Aurelius (180-193). 180: Marcus Aurelius dies at Vindobona (Vienna) (17th Mar). Emperor Aurelius Commodus (180-193), who concludes peace with the barbarians. Quasi-monarchical and anti-senatorial government. Aelius Aristides and Maximus of Tyre.	180: The execution of the so-called Scillitan Martyrs takes place at Carthage. Theophilus of Antioch writes three books *Ad Autolycum* (180-183). Irenaeus of Lyons writes *Adversus haereses* (*c*.180-*c*.185). 185: Birth of Origen. Pope Victor (189-199) intervenes in the Easter controversy with the Churches of Asia Minor.	The *Acts* of this martyrdom constitute the first written document of early Christian literature in Latin. A defence of Christianity in which the term "trinity", designating the three divine persons, first appears. This is the first Christian commentary on the beginning of Genesis. The existence of the invisible God is argued from his visible works in creation. A work in five books, surviving in Latin translation, a precious document for our knowledge of the various forms of gnosis and the development of Catholic doctrine. Irenaeus maintains that the true gnosis is that which has been handed down by the Church. The Churches of Rome and Alexandria celebrate Easter on the Sunday after the first full moon of spring; but the Churches of Asia Minor on the 14th of the month Nisan.

Date	Secular Events	Ecclesiastical Events	Cultural and Doctrinal Events
190	Death of Lucian of Samosata at Athens (after 190). 193: Assassination of Commodus. Emperor Pertinax, in turn eliminated on 28th March. Didius Julianus at Rome (soon killed); replacement of the praetorians by the legionaries. Pescennius Niger at Byzantium; Clodius Albinus in Britain; 194: Septimius Severus (193-211) at Carnuntum in Pannonia, defeats Pescennius Niger at Issus. 197: Victory over Albinus at Lyons (Lugdunum). With Caracalla, war against the Parthians as far as Seleucia. Osrhoene and Mesopotamia Roman provinces (198-200); Nisibis a colony. Reform of the army, which acquires greater importance, and devaluation of the currency. Caracalla Augustus.	At Rome Theodotus of Byzantium propagates "dynamic monarchianism" or adoptianism, a heresy that interprets God's unity by denying Christ's divine nature (c.190). Clement teaches and writes at Alexandria, perhaps as Pantaenus's successor. First Latin translation of the Bible, perhaps made by a Christian community in Africa (c.190). 197: Tertullian composes his *Apologeticum* and *Ad nationes*. Construction of the Christian catacombs in the suburban area of Rome. Their administration is organized according to the form of the *collegia funeratica* (c.199). Rome: pope Zephyrinus (199-217).	Clement of Alexandria's works (*Protrepticus*, *Paedagogus*, *Stromata*) aim essentially to base the nexus between faith and philosophy on the search for an "orthodox gnosis". The date constitutes a useful "terminus post quem" to date contemporary Christian literary works in Latin, beginning with Minucius Felix's *Octavius*. Tertullian's apologetic writings articulate the defence of Christianity according to the models and style of classical Latin rhetoric.
200		Minucius Felix writes his dialogue *Octavius* (c.200). Veneration of relics, arising out of the cult of the martyrs, enters ecclesiastical practice (end of 2nd c.). Origen is called by bishop Demetrius to direct the catechetical school of Alexandria. 202: Septimius Severus issues an edict against Jews and Christians, forbidding any form of proselytism. 203: *Passio Perpetuae et Felicitatis*, archetype of acts of Christian martyrs. Birth of Plotinus in Egypt (Lycopolis or Lykon) (c.205).	*Octavius*: a dialogue between three friends, at Ostia: Caecilius, a pagan; Octavius and Minucius, Christians. Evocation of their conversation and of Caecilius's conversion. Origen (author of *Contra Celsum* and the treatise *De principiis*) contributes considerably to the doctrinal systemization of early Christianity, through neoplatonic concepts and making use of biblical exegesis. The text of the *Passio* contains typically apocalyptic Judaeo-Christian elements.
	Rome: Arch of Septimius Severus. Wars in Britain (208-210). Accentuation of absolute rule: oppressive rule; importance of the military in politics. Influence at court of Julia Domna, wife of Septimius Severus, and of Emesa. Spread of		
210	*collegia*. 211: Death of Septimius Severus at Eboracum (York). Emperors: Caracalla (211-217) and Geta, murdered in 212. 212: The *Constitutio Antoniniana*: extension of Roman citizenship to all the inhabitants of the Empire. Oppressive rule; devaluation. Rome: Baths of Caracalla. Ammonius Saccas the Neoplatonist; Papinian the jurist. c.216: Birth of Mani. Parthian war (216-217). Assassination of Caracalla at Carrhae in Mesopotamia by Macrinus (217-218), deposed by the army. Varius Avitus, called Elagabalus (Heliogabalus), emperor (218-222), son of Julia Soemia and nephew of Septimius Severus.	Tertullian adheres to Montanism (c.213). 215: Clement of Alexandria finishes the *Stromata*. Rome: pope Callistus I (217-222). Sabellius at Rome.	210: In *De anima*, Tertullian maintains a certain corporeality of the soul and traducianism. He becomes more rigorist, rejecting disciplinary positions that he had earlier accepted. A sort of miscellany in which he confronts, as in a work for private use for pedagogical expositions, many problems in a Christian perspective. For Sabellius and his followers, there is only one God, manifesting himself as Father in the OT, Son in the incarnation and Spirit poured out on the Apostles (three modes, modalism).

Date	Secular Events	Ecclesiastical Events	Cultural and Doctrinal Matters
	Importance of Syrian women at court: Julia Domna, Julia Soemia and Julia Mamaea. Adoption of Severus Alexander, Elagabalus' cousin, son of Julia Mamaea.	The emperor Elagabalus's establishes policy of religious tolerance and of syncretism.	
220		*c.*220: Sabellius is condemned by Callistus I.	
	Assassination of Elagabalus and his mother. Severus Alexander emperor (222-235). Religious tolerance. In Persia the Sassanid dynasty (226-652) begins: Ardashir I (226-242). Ulpian the jurist. Dio Cassius the historian.	*c.*222: Death of Bardesanes, gnostic thinker born in 154. *c.*222: Death of Tertullian. Rome: pope Urban I (222-230).	Bardesanes's doctrine is known to us thanks to his disciple Philip's *Book of the Laws of the Nations.* The Christian "house" at Dura Europos.
230		230: Origen leaves Alexandria for Caesarea. Invited to Greece (*c.* 231). He teaches, writes and preaches in church as a simple presbyter. Rome: pope Pontian (230-235).	Origen is perhaps the most prolific author of antiquity. Main works: the *Hexapla*, a synopsis of the OT, in Hebrew and translations; many *exegetical works* and *homilies*; *De principiis*: the first essay of Christian theological reflection taking the *regula fidei* as starting-point, leaning on Scripture and reason; *Contra Celsum* (248): refuting in eight books Celsus's criticisms in *True Speech* (*c.*180) of Christianity: the most important apologetic work of Christian antiquity.
	War against the Persians (231-232). Birth of Porphyry (232/33 - 305?), disciple of Plotinus and prestigious and original representative of the neoplatonist tradition.		
		Origen's disciple Dionysius directs the catechetical school of Alexandria (late 2nd c. - *c.*265).	Dionysius writes a treatise in epistolary form, *On nature*, against the atomistic philosophies of Democritus and Epicurus, and numerous *letters*, dogmatic or disciplinary in content, important among which are those against Sabellianism.
	235: Murder of Severus Alexander. Maximinus Thrax emperor (235-238). Period of military anarchy (235-238). 238: Antonius Gordianus emperor in Africa, together with his son Gordian II: the latter is killed and his father commits suicide. Assassination of Maximinus Thrax at Aquileia. Joint emperors: Calvinus Balbinus and C. Pupienus Maximus; assassinated. Emperor Gordian III (late 238 - 241). Sarcophagus of Balbinus.	235: Persecution of Maximinus Thrax. Pope Pontian is deported to Sardinia, where he dies (235); succeeded by Antherus (235-236) and then Fabian (236-250). After 235 Hippolytus (not the same as the Roman priest deported to Sardinia in 235), a Greek-speaking Roman writer, writes *Refutation (Elenchos) of all the heresies*, also called *Philosophoumena*.	The *Elenchos* aims to be a refutation of the heresies: the author sees in each of them the reflowering of a pagan philosophical school.
240	Mani, friend of Shapur, founds Manichaeism. Plotinus (204-270), disciple of Ammonius Saccas and friend of Gordian III and then of Gallienus. 242: War on the Danube; then on the Persian front at Resaina. 244: Assassination of Gordian III; Julius Verus Philippus, called the Arab, emperor (244-249). Peace with Shapur. Wars against the Goths in Dacia. 248: Millenium of Rome. Philip defeated by G. Messius Decius near Verona. Decius emperor (249-251).	247: Dionysius becomes bishop at Alexandria. 249: Cyprian bishop of Carthage. Persecution of Decius, who issues a general edict (late 249) ordering Roman citizens to participate in a *supplicatio*. Many "lapsi".	Catacombs of Domitilla: chapel of the Good Shepherd.

Date	Secular Events	Ecclesiastical Events	Cultural and Doctrinal Events
250	251: Decius defeated by the Goths at Abrittus (June 251). His son Hostilian succeeds him; Trebonianus Gallus, acclaimed emperor, adopts him. Hostilian dies of plague, which is widespread in many parts of the Empire. The Goths in Asia; Shapur reaches Antioch. Trebonianus killed by M. Aemilius Aemilianus, emperor (253), in turn defeated by P. Licinius Valerianus, emperor (253-259). 256: Destruction of Dura Europos; Invasion in the North by Franks and Alamanni, stopped by Gallienus. But in 258 they break into numerous regions of the West. The Alamanni are defeated at Milan (259). Shapur's Sassanids in the East invade the Empire more than once, reaching Cappadocia and Antioch; deportations.	250: Pope Fabian martyred. Only after 14 months does Cornelius succeed him (251-253). Novatian, a Roman priest, opposes him and is elected antipope, giving rise to the schismatic church of the Novatianists. Cornelius confronts the question of Novatian at a Roman synod. Question of the "lapsi": Cyprian's *De lapsis*. Pope Lucius, hardly elected, is exiled; succeeded by Stephen I (254-257), who dies a martyr. Death of Origen, in prison (254?). Councils of Carthage (255/256) on the rebaptism of heretics. Persecution of Valerian (257/258). Rome: death of pope Sixtus II (257-258) and the Roman deacon Lawrence; Dionysius then becomes pope (259-268). 258: Death of Cyprian at Carthage.	In *De Trinitate* Novatian opposes the monarchians. He does not use the term "trinitas" and does not call the Holy Spirit either God or person. Decius's persecution makes the problem of penitence acute. Cyprian's solution: before receiving laying on of hands and being readmitted to the Eucharist, the "lapsi" must submit to a suitable penance and make public confession of their fault. Cyprian, in accord with a part of African tradition, confers baptism again on those baptized outside the Church. Pope Stephen opposes this practice and doctrine, maintaining that laying on of hands alone is sufficient. Catacomb of Priscilla on the Via Salaria with paintings of the Good Shepherd and the Madonna and Child.
260	260: Valerian is defeated near Edessa and taken prisoner by the Sassanids. Gallienus sole emperor. Military reforms. Porphyry Plotinus's pupil: the *Enneads*. 267: Invasion of Greece. Gaul and Spain independent under Posthumus († 268); rebellions in the East; Septimius Odenathus, *corrector totius Orientis* for defence against the Parthians and Goths, resident at Palmyra, assassinated in 267; succeeded by Zenobia, with their son Vaballatus, who enlarges his dominion. 268: Further barbarian invasions. Rebellion of Aureolus at Milan. Assassination of Gallienus. M. Aurelius Claudius, later called Gothicus, emperor (268-270). Wars with the barbarians in the Balkans. Epidemic of plague.	260: The emperor Gallienus issues an edict of religious tolerance and restores confiscated goods. According to some calculations there are about six million Christians in the Empire. Death of Gregory the Thaumaturge (after 264), evangelizer of Pontus. Felix I (269-274) answers the letter of the members of the council of Antioch (268), which had condemned Paul of Samosata.	260: Synod of Rome under pope Dionysius. Some Christians of Pentapolis have denounced the bishop of Alexandria, Dionysius, for christological errors. The council condemns both Sabellianism and tritheism (the question of the two Dionysii). Paul of Samosata maintains a monarchianism of adoptianist type. The Logos is just a *dynamis*, an operative faculty of God, and not the Son of God, which the man Jesus, in whom the Logos has taken up its dwelling, is.
270	270: Claudius Gothicus dies at Sirmium of the epidemic. Emperor L. Domitius Aurelianus (270-275); revolt of the *Bacaudae* in Gaul; Aurelian defeats the barbarians who are invading Italy. The Aurelian Walls at Rome. Abandonment of Dacia. Victory over Zenobia: reunification of East (272/273) and West (274). The solar cult becomes official in the Empire: feast 25th December. Trades become hereditary. 272: Death of Shapur I. Rome: temple of *Sol Invictus*. Porphyry: *Contra Christianos*. Death of Mani (274/277).	Anthony begins his ascetic life in Egypt	

Date	Secular Events	Ecclesiastical Events	Cultural and Doctrinal Matters
	275: Assassination of Aurelian near Byzantium. Emperor M. Claudius Tacitus (275-276), killed near Tyana. Florianus emperor, then Probus (276-282). 277-278: Wars against the barbarians in central Europe (Franks, Alamanni, Burgundians, Senoni, Vandals).	Rome: pope Eutychian (275-283).	
280	282: Assassination of Probus. Emperor Carus (282-283): occupies Seleucia-Ctesiphon. His sons emperors: Carinus in the West and Numerian in the East (†284). Emperor C. Aurelius Valerius Diocletianus (284-305). 285: Civil war; assassination of Carinus. Maximian, with the title of Caesar, in Gaul to put down the revolt of the Bacaudae. 286: Rebellion of Carausius in Britain. Maximian given the title Augustus. 289: Diocletian *Jovius*, residence at Nicomedia; Maximian *Herculius*, residence at Milan.	Rome: pope Caius (or Gaius) (283-296).	
290	Constantius Chlorus against Carausius. 293: Tetrarchy with the two Caesars: Constantius Chlorus for the West, Galerius Valerius Maximianus for the East. Constantius Chlorus defeats the rebels in Britain. Galerius against Narsetes's Sassanids (296-298): reconquest of Mesopotamia. Peace treaty favourable to the Romans. 296: Repression of an insurrection in Egypt by Diocletian. 297: Persecution against the Manichees. 298: Rome: start of the construction of the Baths of Diocletian. Administrative reforms: 12 dioceses ruled by *vicarii* and 87 provinces. Distinction in provinces between civil power (*praesides*) and military power (*duces*); financial and fiscal reforms. The *consilium principis* virtually replaces the senate.	Birth of Pachomius (*c*.292). Rome: pope Marcellinus (293-304). Conversion of Arnobius. Conversion of Tiridates, king of Armenia. Evangelization of Armenia by Gregory the Illuminator.	Arnobius writes *Adversus Nationes*: a refutation of the worship and rites of the pagan religion.
300	301: Edict on prices: fixed price of products and services. 302: Diocletian visits Rome. Death of Porphyry (305?). 305: Abdication of Diocletian and Maximian (1st May). Augusti:	Conversion of Lactantius (*c*.300). Persecution of Diocletian: four edicts (303/304). Numerous martyrs, among them Agnes, Sebastian, Cosmas and Damian, Maurice, Genesius, etc. The problem of the "traditores".	300: Paintings in the hypogeum of the Acilii in the catacomb of Priscilla and paintings in the catacomb of Peter and Marcellinus, on the Via Labicana at Rome. Transformation of a house on the Coelian into a "domus ecclesiae" (later SS. Giovanni e Paolo).

Date	Secular Events	Ecclesiastical Events	Cultural and Doctrinal Events
	Galerius (East) and Constantius Chlorus (West). Caesars: Maximinus Daia (East) and Flavius Severus (West).		
	306: War among the successors of Diocletian and Maximian. Death of Constantius Chlorus; his son Constantine acclaimed Augustus. Rebellion at Rome of Maxentius, son of Maximian.		
	307: Death of Severus.		
	308: Meeting of Carnuntum: West: Licinius Augustus, Constantine Caesar; East: Galerius Augustus, Maximinus Daia Caesar. Start of the basilica of Maxentius.	308: Some monks join Anthony in the eremitical life. Hilarion visits him and returns to Gaza. Rome: pope Marcellus I (308-309); then Eusebius (309-310).	
310	310: Death of Maximian. 311: Death of Galerius.	Council of Elvira in Spain (before 314), important for our knowledge of the Spanish church of that period and of the spread of Christianity. Rome: pope Miltiades (311-314). Anthony abbot goes to Alexandria to encourage the martyrs. 311: Edict of religious freedom by Galerius. Death of Methodius of Olympius, martyr at Euboea.	310: Methodius criticizes Origen's doctrine of the preexistence of souls in *De resurrectione*. His main work, the *Symposium,* is a dialogue on virginity.
	312: Constantine defeats Maxentius at the Milvian Bridge (28 Oct) and is sole emperor of the West.	312: Lucian of Antioch, teacher of Arius and of Eusebius of Nicomedia, dies a martyr at Nicomedia (7th Jan). In the course of that year a series of measures in favour of the Catholic Church introduced by Constantine. Start of the Donatist schism in Africa. Caecilian is opposed at Carthage by Maiorinus, who will be succeeded by Donatus.	Lucian has been considered the founder of the school of Antioch, literalist in approach, in contrast to Alexandrian allegorism: but little information is available. Lactantius writes his *Divinae institutiones* (306-313), in six books; between 318 and 321 he composes *De mortibus persecutorum* to testify to future generations the vengeful justice of God, who punishes all the main persecutors of the Christians.
	313: Licinius defeats Maximinus Daia and is sole emperor of the East.	313: The so-called edict of Milan (9 Jan) issued by Constantine and Licinius: the document provides freedom of worship and restitution of the goods confiscated from the Christian communities. The clergy are exempt from *munera*. Rome: council about the Donatist question; bishops from Italy and Gaul take part. Eusebius bishop of Caesarea in Palestine.	Eusebius of Caesarea's two apologetic works: *Praeparatio evangelica* and *Demonstratio evangelica*, the former addressed to the pagans, the latter to the Jews, are also aimed at Porphyry (312-320).
		314: Arles; council on the Donatist question, convened by the emperor Constantine. Rome: pope Sylvester (314-335). Lactantius at Trier to educate Constantine's son Crispus. Condemnation and persecution of the Donatists (315/316).	Under the presidency of Crestus of Syracuse, in the presence of Caecilian of Carthage and his accusers, the council of Arles endorses the decisions of the council of Rome of 313. It confronts various other questions: the date of Easter, baptism, ordinations, etc.

Date	Secular Events	Ecclesiastical Events	Cultural and Doctrinal Matters
	Rome: construction of the Arch of Constantine.	316: *Manumissio in ecclesia* permitted. 318: *Audientia episcopalis* conceded (23 Jun). Pachomius (*c*.292-347), a converted Egyptian soldier, founds his first monastery, initiating coenobitic monasticism. He founds the monasteries of Tabennisi (*c*.320) and Pbou (*c*.328).	
320		320: Abolition of the Augustan laws against celibacy. Saint Nino evangelizes Georgia. 321: Law of festive repose for tribunals and in cities. The Church can receive donations.	Rome: the Palace of Maxentius is made over to the pope: basilica Lateranense (after 320). In 324 begins the construction of St Peter's basilica, dedicated in 340.
	323-324: Constantine defeats Licinius at Adrianople and at Chrysopolis. Start of the construction of the new capital (Constantinople).	323: Alexandria: synod against Arius, born *c*.260, who had adhered to the Melitian schism.	323: Eusebius of Caesarea's *Ecclesiastical History*: though unrevised, it includes numerous otherwise unknown documents; its apologetic intention is evident.
		325: Nicaea (now Iznik): the first great (ecumenical) council convoked by the emperor Constantine. Condemnation of Arius and redaction of a formula of faith (Nicene creed). Prohibition of gladiatorial games.	The polemical originality of the "faith of Nicaea" lies in these words: *of the Father's substance* and *true God from true God, generated not created, consubstantial (*homoousios*) with the Father*. The "faith of Nicaea" will remain the dogmatic rule invoked by all the other ecumenical councils of the early Church.
	326: Construction of the imperial basilica.	326: Edict against heretics, who are excluded from the benefits of the clergy.	
		328: Athanasius succeeds Alexander as bishop of Alexandria. Constantine asks Eusebius for 50 copies of Holy Scripture. Frumentius, ordained by Athanasius, evangelizes Ethiopia. Conversion of king Ezana (perhaps later).	Helena, Constantine's mother, promotes the construction of various sacred buildings at Jerusalem.
330	330: Inauguration of the new capital Constantinople, the New Rome, where Constantine's mother Helena, dies (*c*.330).	330: Athanasius goes to the Thebaïd. Macarius at Scete. First monasteries in Palestine. Constantine protects the Arian bishops. At Rome the feast of Christmas is celebrated on 25th December. 331: Decurions cannot enter the clergy.	Juvencus composes the first Christian poetical work: *Evangeliorum libri IV*, a poetic transposition of part of Matthew's gospel, in paraphrastic form. "Dogmatic sarcophagus" of the Vatican Museums, with stories from the Old and New Testament.
	332: Goths defeated. In Armenia Tiridates disappears: territory contested between Romans and Sassanids. 336: Start of a war against the Persians.	335: Council (*latrocinium*) of Tyre; condemnation of Athanasius, exiled to Gaul, where he spreads knowledge of monastic life. Consecration of the church of the Anastasis at Jerusalem. Rome: death of pope Sylvester; 336: Mark succeeds Sylvester; then Julius I (337-352). Death of Arius (*c*.336). Carthage: Donatist council, with more than 270 bishops taking part.	Mosaics of the Constantinian villa of Antioch (Louvre).
	337: Death of Constantine (22nd May), aged 63, baptized by Eusebius of Nicomedia. His sons become emperors: Constantine II (337-340):	337: Pachomius settles at Pbou; Anthony visits Alexandria and the desert of Nitria; 338: Cellae founded.	

Date	Secular Events	Ecclesiastical Events	Cultural and Doctrinal Events
	Spain, Gaul, Britain; Constans I (337-350): Italy, Illyricum, Africa; Constantius II (337-361): Asia, Syria, Egypt.	The emperors Constantine and Constans support Athanasius; Constantius supports the Arians. 339: Persia;. persecution of the Christians by Shapur II.	
340	340: Constan kills his brother Constans at Aquileia. The Empire divided between the two remaining brothers. War against the Franks in Gaul and the Picts in Britain.	c.340: Death of Eusebius of Caesarea. 340: Athanasius at Rome. Pachomius's sister Miriam founds a female monastery. 341: Synod of Antioch (*in Encaeniis*). Ulfila (c.311-383), consecrated bishop at Constantinople by Eusebius of Nicomedia, starts to propagate Arian Christianity among the Goths. 342: Exemption of the clergy from curial contributions. 343: Council of Sardica (Sophia), according to the wishes of the emperor. 344: Macedonius bishop of Constantinople: will give his name to Macedonianism. Death of Aphraates the Syrian (c.345), first Father of the Syrian church, known as the Persian Sage. 347: Death of Pachomius. Birth of Jerome and Rufinus (c.347). Persecution of the Donatists by the emperor Constans. Firmicus Maternus born in Sicily; lives mostly at Syracuse. Cyril of Jerusalem's *Catecheses*, preached as a simple priest or immediately after his election as bishop (c.348). 348: Council of Carthage against the Donatists.	The council of Sardica is summoned to settle disputes between Eastern and Western bishops, headed by Ossius, over the Arian controversy. It is continued by the Westerners alone, who approve various disciplinary canons and a doctrinal document without the *homoousios*. Rome: construction of S. Agnese and S. Costanza. Mosaics of S. Costanza. Firmicus writes *De errore profanarum religionum*. In it he opposes the paganism of his time. He reminds the emperors (Constans and Constantius) of their grave duty to destroy the pagan religion. These are a series of 24 catecheses: one figures as introduction (procatechesis); 18, addressed to those who are to be baptized at Easter, turn on the detailed exposition of the creed in use at Jerusalem; five (mystagogical) are addressed to the baptized.
350	350: Death of Constans, killed by Magnentius, who in Gaul proclaims himself Augustus, in turn defeated by Constantius II in 353. 352: Rome: Arch of Janus. Invasion of Franks and Alamanni in Gaul. Appearance of the Huns in Russia. 354: Gallus beheaded at Milan.	350: Ulfila translates the Bible into Gothic, a work that proves to be of great importance not just religiously but also culturally. 351: Horsiesi has himself replaced by Theodore as head of the Pachomian community. Rome: pope Liberius (352-366). 354: Birth of Augustine to Monica and Patricius at Thagaste.	*Chronographer of 354*: Furius Dionysius Philocalus, calligrapher to pope Damasus.

Date	Secular Events	Ecclesiastical Events	Cultural and Doctrinal Matters
	355: Julian appointed Caesar.	355: Death of Donatus, who gives his name to Donatism. Council of Milan: condemnation of Athanasius. Bishops are exempted from civil jurisdiction. c.355: Death of Anthony in Egypt. c.355: Conversion of Marius Victorinus at Rome. 356: Liberius exiled by Constantius to Thrace and replaced by Felix II. Closure of pagan temples and cessation of worship.	Donatus's theologiy is centred on the "remnant". He accepts that the church should be universal, but thinks that present time it is a small body of the saved, surrounded by false Christians. Marius Victorinus writes numerous works both before and after his conversion. He knows Platonic philosophy much better than Christian theology; while defending the Nicene *homoousios*, he conceives the Trinity as a double dyad.
	357: Julian fights against Franks and Alamanni.	357: Athanasius writes his *Vita Antonii*. Sirmium: Arian synod (*blasphemia sirmiensis*). Basil, having visited Palestine and Egypt, retires to Annesi. He will propagate coenobitic life, also writing monastic rules. 358: Liberius resumes his place as pope. Another synod at Sirmium (358/359), where the homoeousian doctrine triumphs.	The *Vita Antonii* is an important document for making Egyptian monastic life known, particularly in the West. Between 350 and 360 Hilary of Poitiers's *De Trinitate* reworks the traditional data in an original way; it affirms the unity of nature and distinction of persons in the Father and the Son. It speaks little of the Holy Spirit, a theme still not discussed.
	359: Shapur II resumes hostilities against Rome.	359: Rimini; council willed by Constantius, with 400 bishops, presided over by Restitutus of Carthage; at it the Nicene creed is confirmed; the bishops, constrained by the emperor, accept a formula of pro-Arian compromise.	359: Rome; sarcophagus of the 42-year-old Junius Bassus, *praefectus urbi*, a masterpiece of the classicistic Christian sculpture of the 4th c.
360	360: Refusal of the army of Gaul to go to Constantius's help against the Sassanids. Julian acclaimed Augustus.	360: Constantinople: homoean council called by the emperor Constantius. c.360: Julian abandons Christianity.	Macedonius of Constantinople, is considered among the initiators, c.360, of the question of the Holy Spirit, because with other homoeousians he refused to recognize its divine character.
	361: Death of Constantius (3 Nov) of sickness, in Cilicia, aged 44. Julian aged 31 is sole emperor (361-363). Declares equality of cults; wishes to restore paganism. 363: Death of Julian in battle against the Sassanids (26 Jun). Jovian emperor (363-364): makes peace by renouncing part of the territory, including Nisibis. 364: Death of Jovian. Emperor Flavius Valentinianus (364-375): establishes himself at Milan, and associates his brother Valens in the Empire (364-378) for the East, with seat at Constantinople. Valentinian appoints his son Gratian, aged eight, Augustus (367-383) and moves to Trier.	361: Martin, born in Pannonia (316/317), founds the community of Ligugé: the first in Europe. Apollinaris (310-390) bishop of Laodicea. 362: Alexandria; council of Egyptian bishops, willed by Athanasius, with the participation of Eusebius of Vercelli. 364: The goods donated to pagan temples pass to the State; the rich cannot enter the clergy. 365: First council in Armenia at Ashtishat. Rome: pope Damasus I (366-384). 367: Death of Hilary of Poitiers. Epiphanius bishop of Salamis in Cyprus. 368: Death of Theodore; Horsiesi returns to be head of the Pachomian community.	Apollinaris, enunciating Christ's unity as his hypostasis, conceives the composite being of the Word made flesh as a substantial integration of the flesh with the Word. He therefore excludes reason (*nous*), or the higher soul, from Christ's being, since it is a subject capable of self-determination.
370	Two historical works dedicated to the emperor Valens: Rufus Festus's	In 370 Basil (c.339-379) is bishop of Caesarea in Cappadocia, one of the	Basil of Caesarea's *Great Asceticon*: it comprises the *Great Rules* and the

Date	Secular Events	Ecclesiastical Events	Cultural and Doctrinal Events
	Breviarium and Eutropius's *Breviarium ab urbe condita*.	three great Cappadocian Fathers; his brother Gregory of Nyssa, bishop in 371, and Gregory Nazianzen, bishop in c.372.	*Little Rules*, i.e. 400 questions on ascetic and monastic life.
		371: Death of Lucifer of Cagliari and Eusebius of Vercelli (371/372).	The paintings of the catacomb of Via Latina (c.370) Beginnings of Priscillianism in Spain (c.370). Priscillian preaches a very rigid ascetic doctrine; he admits the existence of canonical books other than those of the official canon; he distinguishes between the God of the OT and that of the NT; he affirms the unreality of Christ's humanity.
	372: Theodosius the Elder restores Roman rule in Britain.	372: Martin bishop at Tours. Manichaean assemblies forbidden.	
		373: Death of Athanasius of Alexandria; succeeded by Peter II, who is forced to flee to Rome; Death of Ephrem the Syrian, considered the most important writer of the Syrian church. Rebaptism prohibited (373: *Cod. Theod.*, XVI, 6, 1).	Ephrem opposes the biblico-Semitic tradition and its symbols to the influence of Greek philosophy. Founder of the school of Edessa.
		374: Ambrose bishop of Milan. Council of Rome with pope Damasus: condemnation of Eustathius of Sebaste and Apollinaris of Laodicea (Apollinarism). Jerome goes to Palestine (c.374). Council of Valence in Gaul on disciplinary subjects.	
	375: Death of Valentinian. The West divided between his two sons: Gratian and Valentinian II (375-392) (Illyricum). Gratian condemns Theodosius the Elder to death.	Rufinus and Melania in Egypt (373-375). c.376: Melania on the Mount of Olives.	375: Basil of Caesarea's *De Spiritu Sancto*.
		377: The clergy are exempted from personal public duties.	
	378: Defeat and death of Valens at Adrianople by the Huns, Ostrogoths and Alans. Gratian appoints Theodosius, son of Theodosius the Elder, Augustus for the East (379-395); Theodosius fights against Sarmatians and Goths. Invasion of Pannonia by Huns and Alans. Gratian and Theodosius: edict in favour of the Christians (379/380).	379: Death of Basil of Caesarea. Peter II returns to Alexandria. Heretics cannot hold assemblies. Latin becomes the liturgical language at Rome.	
380	380: Edict of Thessalonica (*Cod. Theod.*, XVI, 1, 2): Catholic Christianity the religion of the Empire (28th Feb).	380: Theodosius makes Christianity the religion of the Empire with the edict of Thessalonica. The faith of Damasus of Rome and Peter of Alexandria is orthodox. Those who are not in communion with them are heretics. Council of Saragossa against Priscillian. Timothy patriarch of Alexandria.	Gregory Nazianzen preaches at Constantinople (summer 380) his five *Theological orations*, a coherent whole on the spiritual conditions of knowledge of God and on the Trinity.
	381: Proscription of pagan cults by Theodosius: next year the Altar of Victory is removed from the senate. Ammianus Marcellinus: *Rerum gestarum libri XXXI*.	381: Assemblies of heretics interdicted; apostate Christians lose the right to make wills, as do Manichees; Easter amnesty. Enumeration of the orthodox Eastern bishops (381: *Cod. theod.*, XVI, 1, 3). Prohibition of sacrifices. Council of Constantinople (2nd ecumenical council): the doctrine of the Holy Spirit is defined and Constantinople is declared to have second place after Rome. Council of Aquileia.	From 381 to 385 Gregory of Nyssa's writings against Eunomius, aimed at refuting Eunomius and defending Basil and the trinitarian doctrine.

Date	Secular Events	Ecclesiastical Events	Cultural and Doctrinal Matters
	382: The Visigoths are allowed to settle within the Empire.	382: Numerous measures in favour of the Catholic Church and against pagans, heretics, apostates and Manichees.	
	383: Rebellion of Magnus Maximus; death of Gratian at Paris; Magnus Maximus recognized as Augustus of the West.	383: Death of Ulfila: born c.311, ordained bishop in 341 by Eusebius of Nicomedia, Arian evangelizer of the Goths. Evagrius of Pontus in Palestine (382) and in Egypt at Scete and Cellae (383-385). 383: Augustine at Rome. Death of Damasus; Siricius succeeds him (384-399).	
		385: Priscillian executed at Trier; the first time that a heretic is condemned to death. Jerome leaves Rome for the East.	c.385: Jerome's *Vulgate*, a new Latin version of the Bible. Gregory of Nyssa's *Great Catechesis*: (c.385), a doctrinal compendium for teachers who, in their instructions, have need of a "summa". The author is heir to a tradition of baptismal initiation on the one God in three persons, on the creation, the fall, the incarnation, etc.
		386: John bishop of Jerusalem. The affair of the basilicas at Milan between Ambrose and the empress Justina. Milan: conversion of Augustine of Hippo. 387: Augustine of Hippo is baptized by Ambrose (24 April). Composition of the philosophical dialogues at Cassiciacum. Death of Monica at Ostia. 388: Augustine at Rome. Some anti-Manichaean works.	386: The Ambrosian hymns, creating a new literary genre, are composed mainly during the question of the basilicas. His liturgical hymnology spreads in the West. 386: Rome: church of St Paul's outside the Walls, founded by the emperor Valentinian; 384-389: basilica of S. Pudenziana, founded by pope Siricius. c.386: John Chrysostom's treatise *On the priesthood* is in the form of a dialogue between the author and his friend Basil, otherwise unknown.
	388: Theodosius captures Magnus Maximus near Aquileia, and has him killed.		
390		389: Death of Gregory Nazianzen. Baptism of Paulinus of Nola. 390: Monks are forbidden to reside in cities. Tension between Theodosius and Ambrose over the massacre of Thessalonica. 391: Edict of Theodosius prohibiting pagan worship under any form. Augustine ordained priest at Hippo; Aurelius becomes bishop of Carthage.	Ambrose of Milan's *De mysteriis* and *De sacramentis* (390-391).
	392: Valentinian II killed by Arbogastes; the rebel Eugenius, emperor in Gaul, favours a pagan reaction. Stilicho defeats Alans, Goths and Huns on the Danube. At Constantinople an obelisk is erected in the Hippodrome.	392: Theodore bishop of Mopsuestia in Cilicia. Interdiction on destroying synagogues (392, 393 and 397). Right of asylum in churches.	392: Augustine has already commented on the first 32 psalms; the work is finished in 416 (or after 422 for psalm 118 [119]). A vast work, rich in spiritual doctrine. Some comentaries are preached, others dictated.
		393: The Origenist question. Epiphanius visits Jerusalem and insists on the condemnation of Origen; Jerome rails against Origen. Plenary council of the African provinces at Hippo. Augustine, a	393: At Olympia the last Olympiad is celebrated. Abolition of the Olympic Games.

Date	Secular Events	Ecclesiastical Events	Cultural and Doctrinal Events
		simple priest, preaches the opening address, which will become *De fide et symbolo*.	
	394: Victory of Theodosius, on the river Frigidus, over Eugenius.	394: Jerome composes *De viris illustribus*. Council of Constantinople; present is Gregory of Nyssa, who dies in 395. c.394: Death of John of Lycopolis.	
	395: Death of Theodosius at Milan. The Empire divided between his two sons. West: Honorius (395-423) (tutor Stilicho, *magister militum*); East: Arcadius (395-408) (prefect Rufinus). The division will remain permanent. Alaric invades the Balkan peninsula. 396: Alaric in Greece: sack of Athens.	Diodore of Tarsus, founder of the exegetical school of Antioch, dies before 394. 395: Days of pagan celebrations are no longer holidays. 396: Paulinus of Nola begins composition of the *Carmina natalicia*. Augustine becomes bishop of Hippo (395 or 396). The pagan clergy are deprived of all privileges.	Diodore reacts against what he considers the excessive allegorism of the Alexandrians, in the name of a mainly literal appreciation of the sacred text. Paulinus's 14 *Carmina natalicia* are written to celebrate St Felix of Nola, his protector.
	397: Africa: rebellion of Maurus Gildo. 398: Poetic compositions of Claudius Claudianus.	397: Council of Carthage at which the *Breviarium Hipponense* is approved. Death of Martin of Tours. Death of Ambrose of Milan. 398: John Chrysostom elected bishop of Constantinople to succeed Nectarius. Ninian evangelizes Scotland.	Of the council's acts only fragments survive, but their substance has passed into the *Breviarium Hipponense*, a collection of 39 disciplinary canons, composed by the bishops of Byzacena in 397 and approved by the council of Carthage in the same year.
	399: Persia: Bahram dies; Iezdegerd I (Iazdgart) king.	399: Rome; pope Anastasius I (399-401). Death of Evagrius of Pontus (346-399), Origenist and important spiritual writer. Palladius and John Cassian leave Egypt. Alexandria: Theophilus holds a council to condemn Origen. Prohibition of pagan feasts: to preserve temples (*Cod. Theod.*, XVI, 10, 18; XVI, 10, 15); those in the countryside must be pulled down (*Cod. Theod.*, XVI, 10, 116).	Augustine's *Confessions*. An autobiographical, but also philosophical, theological and mystical work, as well as rich in poetry. Begun after 397 and finished around 400, it is divided into two parts: the first (I-IX) takes the story up to his conversion; the second (X-XIII), a later addition, is a self-reflection of the writer.
400	400: Ostrogoths in Pannonia. Eudoxia appointed Augusta of the East. Consular diptychs: many are in the British Museum. 402-403: Alaric invades Italy: defeated by Stilicho at Pollentia and Verona. 403: Honorius settles his residence at Ravenna.	Honoratus of Arles founds the monastery of Lérins. Rome: pope Innocent I (401-417). 401: Council of Toledo. 402: Confiscation of the heretics' places of worship. 404: Constantinople: John Chrysostom is exiled. Death of Paula; Jerome translates the rule of Pachomius into Latin. 405: Imperial edict against the Donatists (*Cod. Theod.*, XVI, 5, 8). Rebaptism prohibited. Anti-Donatist measures (405 ff.).	Church of S. Lorenzo at Milan. Rome: the mosaics of the church of S. Pudenziana. Augustine preaches at Carthage. 404-405: Prudentius publishes his collection of poems. In the *praefatio* he describes his intention: "that my sinful soul may compose hymns all day long and not pass a night without singing to the Lord".

Date	Secular Events	Ecclesiastical Events	Cultural and Doctrinal Matters
	406: Ostrogoths defeated at Fiesole. 407: Invasion of Gaul. Picts and Saxons in Britain: Constantine III (407-411) proclaimed Augustus in Britain. 408: Death of Arcadius in the East; Theodosius II emperor (408-450). Assassination of Stilicho by order of Honorius. Rome: the senate deposes Honorius and appoints Attalus. 409: The Alans in Lusitania, the Suevi in Gallaecia, the Vandals in Baetica.	407: Death of John Chrysostom. 408: *Audientia episcopalis* confirmed. 409: No public entertainments on Sundays.	
410	410: Siege and sack of Rome by Alaric, who reaches Calabria, where he dies; his successor Ataulf leads the Visigoths to Gaul (412).	410: Council of Seleucia-Ctesiphon. Death of Rufinus of Aquileia, the translator of many Greek works, including Origen's *De principiis* and Eusebius's *Historia Ecclesiastica.*, into Latin. Death of Melania at Jerusalem. Lérins becomes a monastic centre. Synesius of Cyrene becomes bishop (410) of Ptolemaïs in Libya; dies after 413.	Synesius's works on Christian subjects are few; important are his nine hymns, of which five are theological and four are prayers.
	411: Gaul; defeat of Constantine III.	411: Augustine's journeys to Carthage and Cirta. Again at Carthage. Conference (*Collatio*) of Carthage between Catholics and Donatists in the *thermae Gargilianae*, under the presidency of the imperial tribune, the *notarius* Marcellinus, with 285 Donatist and 286 Catholic bishops taking part. Ecclesiastical goods are exempted from tax. Caelestius the Pelagian condemned at Carthage (411/412). 412: Cyril patriarch of Alexandria. Edicts against the Donatists (412 and 414): whoever does not convert will be subject to taxes.	Augustine' begins *De civitate Dei* (books I-V in 413). Augustine's *Tractatus in Iohannem*: part preached and part dictated. A commentary of pastoral type, but rich in theological, philosophical and spiritual doctrine (date disputed, but before 420). The Pelagian doctrine maintains that Adam would have died even without sin; his sin harmed only himself, and so the original fault is not inherited. 412-413: Church of St Demetrius at Thessalonica.
	413: Burgundians in Gaul. 414: Ataulf marries Galla Placidia, sister of Honorius. 415: In the East power is in the hands of Pulcheria. Ataulf succeeded by Wallia.	415: Murder of Hypatia at Alexandria by the Christians. Pagan worship not subsidized; its buildings pass to the Catholic Church (*Cod. Theod.*, XVI, 10, 20). Pagans excluded from the army and from public functions. Synod of Diospolis against Pelagius.	
		416: Orosius in Africa; he brings St Stephen's relics with him. At the request of various African bishops, pope Innocent condemns Pelagius and Caelestius. 417: Death of Eustochium. John of Jerusalem dies. Rome: pope Zosimus (417-418); at Zosimus's death, some of the clergy elect Eulalius and some Boniface I (418-422).	Orosius's *Historiae* are the oldest Christian universal history: they go from the flood to 417. The work intends to rebut the criticisms of pagans who claimed to see in the sack of Rome by Alaric (410) the consequence of the abandonment of the traditional cults, definitively prohibited by Theodosius in 391.
	418: Aquitaine; Theodoric I king of the Visigoths.	419: Julian of Eclanum's literary activity begins: Augustine's replies.	Palladius of Helenopolis writes his *Historia Lausiaca* (419-420), a

Date	Secular Events	Ecclesiastical Events	Cultural and Doctrinal Events
		Persecution in Persia by the Sassanids.	collection of profiles of ascetics, mainly from Egypt and, to a lesser extent, Palestine. He relies both on personal recollections and on accounts by others, giving room to the legendary element.
420	Persia: king Bahram V.	420: Death of Jerome. Rome: pope Celestine I (422-432).	Around 420 Augustine finishes *De Trinitate*, his principal dogmatic work and very influential. Its most original aspects are: the doctrine of the divine relations, the "psychological" explanation, the doctrine of the personal properties of the Holy Spirit, the ideas on the relationship between the trinitarian mystery and the life of grace.
	423: Death of Honorius. Theodosius II recognizes as emperor Valentinian III, aged four (423-455), son of Galla Placidia and Constantius. His mother holds actual power.	423: Death of Sulpicius Severus. 425: Juvenal bishop of Jerusalem. John Cassian (*c*.360-*c*.432) active at Marseilles. Around 426 he writes his *Collationes*. Reconfirmation of the privileges of the Church and the clergy. Honoratus of Lérins becomes bishop of Arles.	
	427: Rebellion of Boniface, governor in Africa.		In 426 Augustine finishes *De civitate Dei*, his masterpiece, an apologetic and dogmatic work: Augustine answers the accusations of the pagans and expounds the Christian doctrine of the beginnings, course and eternal destinies of the two cities, founded on two loves, of self and of God, mingled in the historical process, separate in the eternal dwelling.
	428: Aetius defeats the Franks. Armenia under the Sassanids.	428: Enumeration of prohibited sects and punishments by the emperor. Nestorius becomes patriarch of Constantinople. He opens the question of Mary as *theotokos*, mother of God. Cyril of Alexandria's reaction. Consecration of the Laura of Euthymius.	
	429: The Vandals invade Africa.		
430	430: Defeat of Boniface in Africa; siege of Hippo by the Vandals, while St Augustine is still living. Martianus Capella writer.	430: Death of Augustine at Hippo (28th Aug). Right of asylum in churches (430 and 432).	
	431: The Franks in Gaul.	431: Council of Ephesus, convened by the emperor Theodosius II: 3rd ecumenical council.	The council, taking place amid tumult, defines Mary as *theotokos*, i.e. mother of God, since mother of Christ, God and man, thereby consecrating this title.
		432: Rome: pope Sixtus III (432-440). Sedulius the poet. St Patrick arrives in northern Ireland. 433: Agreement between Cyril of Alexandria and John of Antioch on the Nestorian question. Devastation of Scete (*c*.434). 435: Pagan temples pass to the church. 438: John Chrysostom's body brought back to Constantinople.	Rome, 432-440: basilica of S. Maria Maggiore with its mosaics on the virgin.
	434: Ravenna: Aetius appointed Patrician.		
	439: Genseric takes Carthage.	439: Death of Melania *junior*. Death of Mesrob (439/440).	Mesrob invents the Armenian alphabet, a necessity for translating Scripture and liturgical books.
440	440: The Vandals invade Sicily.	Rome: pope Leo I, the Great (440-461).	

Date	Secular Events	Ecclesiastical Events	Cultural and Doctrinal Matters
	442: Agreement between Aetius and the Vandals. 444-446: Attila in the Balkan peninsula.	443: Eudocia retires to Jerusalem. 444: Death of Cyril of Alexandria; Dioscorus succeeds him. 447: Theodoret of Cyrrhus denounces the doctrines of the monk Eutyches in his three books of *Eranistes*. 448: Eutyches condemned by a synod held in the capital by Flavian, patriarch of Constantinople (22nd Nov). Porphyry's writings against the Christians are burned.	*Eranistes* (*The Beggar*) is a work of great theological import, in four books, in the form of a dialogue between an orthodox and a monophysite (a beggar who scrapes together elements of previous heresies).
	449: Saxons and Angles, led by Hengist and Horsa, land in Britain.	449: Leo the Great writes his *Tomus ad Flavianum*, an important document on the christological question (13th Jun). "Latrocinium Ephesinum": council organized by Dioscorus, bishop of Alexandria: it rehabilitates Eutyches and condemns Flavian and Theodoret.	Literary activity of Leo the Great with homilies and letters. He affirms the primacy of the apostolic see.
450	450: Theodosius II, emperor of the East, dies. Emperor Marcian (450-457) and Pulcheria. Ravenna: death of Galla Placidia. Her Mausoleum (*c.*450).	450: Death of Nestorius and of Eucherius of Lyons. Death of Peter Chrysologus.	
	451: Aetius defeats Attila at the Catalaunian Fields (near Châlons-sur-Marne).	451: Council of Chalcedon (4th ecumenical), with 500 bishops taking part: it abolishes the acts of the "latrocinium Ephesinum" and adopts the doctrine of the *Tomus ad Flavianum*. The see of Jerusalem is recognized as a patriarchate. Further prohibition of pagan worship. Genoveffa encourages the inhabitants of Paris during the siege.	The council of Chalcedon confesses "one and the same Christ, Son, Lord, Only-begotten, without confusion, without change, without division, without separation, since the difference between the natures is not at all taken away by the union, but rather the properties of each are safeguarded and united in a single person and a single hypostasis".
	Attila in Italy (452 or 453); meeting with pope Leo the Great. The Venetian lagoon is populated by refugees. 453: Constantinople: death of Pulcheria. Theodoric II king of the Visigoths in Aquitaine. Valentinian III kills Aetius (453 or 454) and withdraws to Rome. 455: Valentinian III dies at Rome. Emperor Petronius Maximus. Sack of Rome by the Vandals; emperor Avitus. 456: The Vandals occupy all the African provinces and the islands (Sicily, Sardinia and Corsica).	452: Peter the Iberian, bishop of Maiuma. Death of Quodvultdeus (before *c.*454) at Carthage. 454: Death of Dioscorus at Gangra. Severinus's pastoral and charitable activity in Noricum.	Ravenna: construction of the Cathedral baptistery (Neonian or of the Orthodox); its mosaics.
	Constantinople: emperor Leo I (457-474). In the West: Majorian (457-461).	457: Timothy Aelurus (the Cat) consecrated bishop of Alexandria by Peter the Iberian; 458: Timothy Aelurus is expelled from Alexandria for his anti-Chalcedonian position, where he returns only in 475. The consecration of nuns is permitted only for women who have reached the age of 40. 459: Death of Simeon Stylites the Elder in Syria. Death of Juvenal of Jerusalem.	

Date	Secular Events	Ecclesiastical Events	Cultural and Doctrinal Events
460		460: Death of Eudocia.	
	West: emperor Libius Severus (461-465).	461: Priestly ordination cannot be imposed. Rome: death of Leo the Great; Hilarus (or Hilarius) succeeds him (461-468).	
		Death of Theodoret of Cyrrhus (c.466).	Theodoret is author of *Graecarum affectionum curatio*, the last patristic apology, preserved in its entirety, comparing the pagan and Christian replies to fundamental philosophical questions.
	West: emperor A. Procopius (467-472); then Olybrius (472).	Rome: pope Simplicius (468-483).	
470	470: Third sack of Rome.	Death of Salvian of Marseilles (after 469).	Salvian's principal work is *De gubernatione Dei*, in eight books, the last of which is incomplete: it develops the theme of divine providence, which includes the "government" and "judgment" of the world, referring to the situation of the Empire, convulsed by disasters.
	Theodoric, king of the Ostrogoths, obtains territories on the Danube from Leo I.	Peter the Fuller first monophysite patriarch of Antioch, at various times (471, 465-477, 485-488), initiator of some liturgical novelties. 473: Death of Euthymius. Council of Arles (470-475) condemns Lucidus's predestinationist doctrine.	
	474: Constantinople; death of Leo I. Emperors: first Leo II and then Zeno the Isaurian (474-491). 475: Brief interregnum of the usurper Basiliscus. Ravenna: general Orestes proclaims his son Romulus Augustulus emperor. Odoacer defeats Orestes, deposes Romulus Augustulus and sends the imperial insignia to Zeno.		
	477: Death of Genseric. Theodoric occupies the south part of the Balkan peninsula.	478: Death of Timothy Aelurus. Persecution in Africa by the Vandals. Birth of Boethius at Rome, between 475 and 480, of the family of the Anicii.	Timothy Aelurus's most important work is the *Refutation of the council of Chalcedon*, where he expounds a moderate monophysism. Milan: chapel and dome of S. Vittore in Ciel d'Oro.
480	481: Death of Childeric, king of the Franks; succeeded by Clovis, who unifies nearly all Gaul.	482: *Henotikon* of the emperor Zeno the Isaurian, edict of union, drawn up under the influence of Acacius, patriarch of Constantinople. Death of Timothy Salofaciolus at Alexandria. Patriarch Peter Mongus, monophysite.	The *Henotikon* is an equivocal formula of faith, hence unacceptable both to the Chalcedonians and to the monophysites. It condemns both Nestorius and Eutyches, with ample concessions to monophysism. It omits any reference to the two natures in Christ. It ignores the importance of the council of Chalcedon.
	483: Odoacer extends his rule over Dalmatia.	483: Sabas founds the Great Laura at Mar Saba. Rome: pope Felix III (II) (483-492). He condemns the *Henotikon* and Acacius. Acacian schism, which lasts from 484 to 519.	
	484: Death of Euric, king of the Visigoths; death of Huneric, king of the Vandals, succeeded by Thrasamund. 486: Clovis's victory over the Gallo-Roman leader Syagrius initiates the Frankish kingdom of the Merovingians.	484: Death of Sidonius Apollinaris: he leaves 147 letters, useful for our knowledge of the times. 486: Council of Seleucia-Ctesiphon, celebrated at the time of Barsauma: it attests the transition of the Persian church to Nestorianism. Victor of Vita, priest of the church of Carthage, writes his *Historia persecutionis Africanae provinciae* between 484 and 489.	Victor of Vita's work is an account of the persecution of the African Catholic church by the Arian Vandals, written with the passion of a man belonging to the losing party. The language is that of educated Carthaginian circles.

Date	Secular Events	Ecclesiastical Events	Cultural and Doctrinal Matters
	489: Theodoric, consul of the emperor Zeno, enters Italy with the Ostrogoths.	488: Death of Peter the Fuller. 489: Zeno closes the school of Edessa. Birth of Cassiodorus (between 485 and 490).	
490		c.490: Avitus bishop of Vienne in Gaul, author of *De spiritalis historiae gestis*. 490: Athanasius II patriarch of Alexandria.	
	491: Death of the emperor Zeno; succeeded by Anastasius I (491-518). 493: Ravenna, after three years of siege, is conquered by Theodoric, who marries a sister of Clovis, king of the Franks, and in 497 receives the purple from the emperor Anastasius; in Gaul, Clovis marries the Burgundian Clothilde and later is converted to Catholicism.	491: Death of Peter the Iberian, monophysite. The church of Armenia goes over to monophysism following the council of Vagharshapat. Rome: pope Gelasius (492-496). Rome: pope Anastasius II (496-498). Death of Faustus of Riez (490/495). Rome: pope Symmachus (498-514): Laurentian schism, which divides the Roman church for several years (until 506). Numerous anonymous writings and mutual accusations. King Theodoric supports pope Symmachus. 498: Flavian patriarch of Antioch.	c.495: Ravenna: baptistery of the Arians.
500	Priscian of Caesarea writes the *Institutiones Grammaticae*; the poet Blossius Aemilius Dracontius; Fabius Planciades. 506: *Lex romana Visigothorum*.	c.500: Clovis receives baptism from Remigius of Rheims: a conversion important for the history of Gaul. 504: Foundation of the New Laura in Palestine. 506: Council of Agde, presided over by Caesarius of Arles, important for religious policy towards Alaric, king of the Visigoths. It also marks the transition from the Roman to the Merovingian councils, from the Gallo-Roman to the Gallo-Frankish church. Pope Symmachus appoints Caesarius primate of Gaul and Spain.	c.500: Definitive redaction of the Babylonian Talmud, which comments on the Mishnah in a more extensive way. It is presented as the concise and hardly retouched minutes of academic disputes. c.505: Ravenna: S. Apollinare Nuovo. *Apophthegms of the Desert Fathers*: collections that bring together reflections and anecdotes, fruit of spiritual experience in the desert. There are various collections, whose beginnings go back to the 4th century.
	507: Wars in Gaul between Visigoths and Franks; 509: intervention of Theodoric, who enlarges his dominion.		
510			
	511: Death of Clovis; the Frankish kingdom is divided between his four sons: Theoderic I (Austrasia), Chlodomir (Orléans), Childebert (Paris), Lothar (Soissons). 513: Theodoric's kingdom attains great extension (Italy, Provence, Pannonia, Rhaetia, Noricum). Rebellion of Vitalian in the Byzantine empire. 515: Amalasuntha, sister of Theodoric, marries Eutharic, Gothic king of Spain. 517: *Lex romana Burgundiorum*.	511: The emperor Anastasius deposes the patriarch Marcian. 512: Severus patriarch of Antioch; deposition of Flavian. 513: M. Felix Ennodius bishop of Pavia. Rome: pope Hormisdas (514-523). 516: The Burgundian king Sigismund, a Catholic, takes power and favours his people's transition to Catholicism. 517: Council of Epaon, Burgundy, with Avitus of Vienne. Timothy IV (III) monophysite patriarch of Alexandria (517-535).	

Date	Secular Events	Ecclesiastical Events	Cultural and Doctrinal Events
	Justin I emperor of the East (518-527).		Ravenna: S. Apollinare Nuovo (518-519).
	519: Foundation of the kingdom of Wessex by Cerdic.	519: With the approval by the emperor Justin and John II of Constantinople, of pope Hormisdas's *libellus*, the Acacian schism is ended.	
520	520: Cosmas Indicopleustes visits Ethiopia.	520: Death of Jacob of Sarug.	Jacob of Sarug, moderate monophysite and fertile poet of rhythmical homilies (*Mêmre*) is inspired by the biblical tradition.
	War between the Franks and the Burgundians (523/524).	523: Death of Philoxenus of Mabbug, pioneer of Syrian monophysism. Rome: pope John I (523-526); dies in prison; Felix IV (III) succeeds him (526-530).	
	524: Imprisonment of Boethius: *De consolatione philosophiae*; 525: Boethius' execution. Justinian marries Theodora; elected emperor (527-565). Under him Italy, Africa and part of Spain are reconquered by the Byzantines. 526: Death of Theodoric; Athalaric, son of Amalasuntha, succeeds him. Theodoric's mausoleum at Ravenna. Spain: Amalaric king of the Visigoths.	Death of Boethius at Pavia, on the orders of king Theodoric. Execution of Boethius's father-in-law Symmachus.	
	527: War between Justin and the Sassanids.	527: 2nd council of Toledo. Council of Dvin (527?) in Armenia.	The *Regula Magistri*, which has considerable influence on monastic life, as well as on the *Rule* of St Benedict (pre-530).
	529: First redaction of the *Codex Justinianus* under the direction of Tribonian.	529: Benedict of Nurcia founds the monastery of Montecassino, after the twelve in the Aniene valley, among them that of Subiaco. Council of Orange.	Benedict's *Rule*, a prologue and 73 chapters, sums up the spiritual journey of return to God through obedience, under Christ's guidance. Important council for the theology of grace: it publishes 25 *capitula* against so-called semipelagianism.
530	530: Belisarius, Justinian's general, defeats the Sassanids at Dara; he is defeated the following year.	530: Rome; pope Dioscorus, designated by his predecessor. Part of the clergy chooses Boniface II, who succeeds him (530-532). Then follow John II (533-535), Agapitus I (535-536), Silverius (536-537) and Vigilius (537-555).	
	531: Activity of Aeneas of Gaza. Spain: death of Amalaric. The Frankish kings extend their dominion. Peace between Justinian and the Sassanids; 532: Fire and destruction at Constantinople following revolt of Nika (= Victory).	532: Death of St Sabas in Palestine.	532-538: the church of Hagia Sophia at Constantinople; Ravenna (532-547): S. Vitale. The first mention of the works of Ps.-Dionysius the Areopagite dates from 532; we must place the *Celestial hierarchy*, the *Ecclesiastical hierarchy* and the *Divine names* after 482.
	533: Belisarius in Italy; 534: Belisarius reconquers Vandal Africa; death of Athalaric. Constantinople: publication of the *Codex Justinianus*, the *Digests*, the *Institutiones* and the *Novellae*. Invasion of Italy by the Byzantines (534/539) and wars against Witiges, Ostrogothic king; 537: Witiges besieges Rome.	533: Death of Fulgentius of Ruspe.	Fulgentius's works confront the trinitarian and christological problem and the soteriological problem (grace and free will) against the semipelagians.
		538: Severus of Antioch dies in Egypt. Birth of Gregory of Tours.	Immense literary work of Severus of Antioch in defence of moderate monophysism, in polemic against

Date	Secular Events	Ecclesiastical Events	Cultural and Doctrinal Matters
			both radical monophysism and Chalcedonianism. However his doctrine is very close to that of Chalcedon.
540	540: The Byzantines occupy Ravenna. Advance of Chosroes I in the East. Cassiodorus retires to Vivarium, in Calabria. 541: Totila king of the Ostrogoths.	c.540: Death of Dionysius Exiguus. 542: Death of Ceasarius of Arles. Jacob Baradaeus, elected bishop (542/543), by his consecration of numerous bishops and priests organizes the "Jacobite" monophysite church. 543: Edict against Origen and Origenism. 544: Justinian's edict of the "Three Chapters", in which Theodore of Mopsuestia, Theodoret of Cyrrhus and Ibas of Edessa are condemned. Birth of Leander of Seville (c.545) at Cartagena, in Spain. 547: Pope Vigilius is abducted by Justinian.	Many of Dionysius Exiguus's works are translations from Greek. But his fame is due to his work as a canonist, in particular the *Codex canonum ecclesiasticorum*, for which he may be considered the founder of canon law. Dionysius substituted the Christian era for that of Diocletian: Christ was born in 753 *ab urbe condita*. Gildas the Wise's *De excidio Britanniae* (c.547), describing the drama of Roman Britain abandoned by the Roman legions at the end of the 4th c. and devastated. Gildas is the first Celtic writer in Latin, after the abandonment of Britain by Rome.
	548: Spain: death of the Visigothic king Theudis. 549: Totila conquers Rome. The writer and poet Flavius Cresconius Corippus.		549: St Catherine's monastery on Mt Sinai. Ravenna: S. Vitale consecrated.
550		c.550: Cassiodorus founds the monastery of Vivarium in Calabria, where, alongside manual work, ample room is given to the study of sacred and secular texts, their transcripton and translation from Greek. 552: Death of Verecundus of Iunca.	Principal works of Cassiodorus: *Variae* in 12 books, *Historia Gothorum* and *Institutiones divinarum litterarum*. c.550: Cosmas Indicopleustes's *Topographia christiana*: a treatise on cosmography.
	552: The Byzantine general Narses in Italy. Death of Totila; Teia succeeds him. 553: Spain: Athanagild defeats king Agila. 554: *Pragmatica Sanctio*: Justinian extends imperial legislation to Italy. 555: Austrasia: death of king Theodebald, whose widow is married by Lothar I, king of Soissons, who annexes his kingdom. 558: Lothar I reunites the kingdom of the Franks.	553: 5th ecumenical council, Constantinople II. 555: The Origenists are expelled from the New Laura. c.555: Death of Romanus Melodus. Rome: pope Pelagius I (556-561).	Literary activity of Cyril of Scythopolis (556 f.): he writes biographies of celebrated Palestinian monks, based on archival documents, or accounts of other monks, or personal knowledge.
560	England: Ethelbert becomes king of Kent (560-616); Ceawlin king of Wessex (560-593). The Frankish kingdom is divided among the sons of Lothar I, who dies in 561. 562: Peace between Chosroes's Sassanids and the Byzantines. 565: Death of Justinian; Justin II emperor (565-578). 567: Spain: death of Athanagild: Liuva (capital Narbonne) and	Rome: pope John III (561-574). 561: Paul of Antioch (or of Beth Ukkame, or the Black), patriarch of Antioch.	567: Consecration of the cathedral of Nantes.

Date	Secular Events	Ecclesiastical Events	Cultural and Doctrinal Events
	Leovigild (Toledo) become kings. Alboin, king of the Longobards, invades Italy, which he divides into duchies (568/569). Leovigild sole king of the Visigoths.		569: Radegund founds the monastery of the Holy Cross at Poitiers. For the occasion, Venantius Fortunatus composes *Pange Lingua* and *Vexilla Regis*.
570	c.570: Birth of Mohammed. 571: Britain: conquest of the Saxon Cuthwulf. 572: Death of Alboin; Clephi succeeds him. Struggles between Lothar's sons in Gaul (573; 575).	572: Gregory, future pope, *praefectus urbi*. The Suevi are converted to Catholicism. 575: Rome; pope Benedict I (575-579). Gregory founds the monastery of St Andrew on the Coelian. 576: Death of Germanus, bishop of Paris.	
	577: Ceawlin of Wessex extends his realm. 578: Death of Justin II; Tiberius II emperor (578-582). The Byzantine general Maurice defeats Chosroes. 579: Death of Chosroes.	578: Death of Jacob Baradaeus: from him the "Jacobite" Church of Antioch received its name and also its structure and propagation. Rome: pope Pelagius II (579-590). Death of Martin of Braga (after 579). Gregory at Constantinople (579-585).	Martin of Braga is author of *De correctione rusticorum*.
580	582: Spain; rebellion of Hermenegild, a convert to Catholicism, against his father Leovigild, an Arian. Maurice, eastern emperor (582-602). 584: Italy: Authari king of the Langobards. Death of Chilperic, king of Neustria; Lothar II succeeds him. England: war between Cutha and the Britains. 585: Spain; Hermenegild, taken prisoner, is killed. 586: Death of Leovigild; succeeded by Recared, who is converted to Catholicism. In England Cridda founds the kingdom of Mercia. 587: Death of Radegund, wife of king Lothar I.	580: Death of Cassiodorus. 585: Columbanus, in Gaul, founds various monasteries including that of Luxeuil. 587: Recared becomes a Catholic. The patriarch of Constantinople assumes the title of "ecumenical". 589: 3rd council of Toledo, convoked by Recared and presided over by Leander of Seville: it celebrates the conversion to Catholicism of the Visigothic people.	Beginning of the *Moralia in Iob* in the form of conversations addressed to his brother monks, and continued partly *dictando*.
590	590: Plague in Italy. Death of Authari at Pavia. 591: Agilulf marries Authari's widow Theodelinda and succeeds him. 592: Death of Ceawlin; Ceol (Ceolric) succeeds him. The Langobards threaten Rome, defended by pope Gregory the Great.	Rome: Gregory I, the Great, pope (590-604). c.590: The greater part of Armenia passes under the rule of the Byzantines, who try to impose the council of Chalcedon.	Gregory the Great's liturgical reform. 593: His *Homiliae in Evangelium*: a collection of 40 homilies on as many gospel passages; his *Homiliae in Hiezechihelem*: preached between 593 and 594. Other important works of Gregory are the *Regula pastoralis* and the *Dialogues*.

Date	Secular Events	Ecclesiastical Events	Cultural and Doctrinal Matters
	595: Austrasia: death of king Childebert II; his two sons succeed him.		
	597: Conversion of king Ethelbert of Kent : he allows the missionary Augustine the palace of Canterbury. Callicus hexarch of Ravenna (597-603).	597: Augustine goes from Rome to England, beginning the evangelization of the kingdom of Kent. Venantius Fortunatus consecrated bishop of Poitiers. Ethelbert, king of Kent, converted.	Venantius is author of 11 books of *Miscellanea*, some 300 compositions, mostly in verse, on both sacred and secular subjects.
600	Saebert king of Essex (600-617). 601: Spain; death of Recared; Liuva II succeeds him. Scotland: the kingdoms of the Picts and Scots are formed.		
	602: Phocas emperor of the East (602-610). 603: Agilulf has his son Adaloald baptized. Smaragdus exarch of Ravenna (603-711).	The synod of "Augustine's Oak" (Worcester) (603?).	The synod was the outcome of an attempt by Augustine to get Roman ecclesiastical traditions adopted, in particular the Roman way of celebrating Easter.
	604: Ethelbert, king of Kent, promulgates a code of laws.	Rome: pope Sabinian (604-606); then Boniface III (607) and Boniface IV (608-615), who turns the Pantheon into a Christian church.	
	605/6: Defeat of Byzantines at Edessa and invasion of Eastern provinces.	606: At Grado a Catholic patriarch (Candidian); at Aquileia a schismatic (John).	
	609: The Slavs in the Balkan peninsula. Death of Witteric, king of the Visigoths of Spain;		
610	610: Gundemar succeeds Witteric. Death of Phocas; Heraclius succeeds him as emperor of the East (610-641).	610: Foundation of Westminster Abbey.	
	612: Spain; Sisebut becomes king; he also has literary aspirations. 613: Invasion of Friuli by Avars. France: murder of Brunhilde; Lothar II reunifies the kingdom. Defeat of the Byzantines near Antioch.		
	614: *Edictum Clotarii*: to regulate administrative and political life (614). The Slav advance into the Balkans continues: city names change. Conquest of Jerusalem (614).	614: Council of Paris, convened by Lothar II: important for legislation regarding the clergy. St Gall's missionary activity among Suevi, Helvetii and Alamanni. Columbanus founds the monastery of Bobbio and dies in 615.	Columbanus left various writings of a monastic character which reveal not just an ecclesiastical, but also a classical culture: in particular the *Regula monachorum* and the *Regula coenobialis*.
	615: Constantinople threatened from the East by the Sassanids and from the North by the Slavs. 616: Death of Agilulf; succeeded by his son Adaloald (615-626) under the regency of his mother Theodolinda. England: Eadbald (616-640) succeeds Ethelbert. Edwin, king of Northumbria (616-632), conquers Bernicia and Deira.	Rome: pope Adeodatus I (Deusdedit) (615-618); succeeded by Boniface V, (619-625). Death at Rome of John Moschus (619 or 634).	John Moschus, with the collaboration of Sophronius, publishes the *Pratum spirituale*, an important collection of monastic narratives.
620	621: Spain: death of Sisebut; After Recared II, Suintila becomes king in the same year.		Sisebut is author of the *Carmen de luna*, a short poem describing a lunar eclipse.

Date	Secular Events	Ecclesiastical Events	Cultural and Doctrinal Events
	Mohammed chooses Medina as refuge for his followers, perhaps about 70, persecuted at Mecca: the start of the Muslim era (Hejira: 622).		
	622-23: Eastern advance of Heraclius. 623-39: Dagobert king of Austrasia.		
	624-26: Wars of Mohammed's followers against the inhabitants of Mecca and the Bedouin.	625: Rome: pope Honorius I (625-38). His intense pastoral activity to get ecclesiastical discipline observed. He intervenes unhappily in the monothelite question, for which he is condemned as a heretic by the 6th ecumenical council.	
	626: Ariovald (Arioald) king of the Langobards (626-636). Siege of Constantinople.		
	627: Earpwald, king of East Anglia (627-631), is converted. Baptism of king Edwin.	627: Edwin of Northumbria receives baptism with all his household at York. Paulinus of York has full liberty to evangelize.	
	628 or 629: Dagobert also obtains Neustria and Burgundy.		
630	630: Heraclius defeats the Sassanids and brings back the relic of the Cross to Jerusalem. 631: Spain: Sisenand is king. 632: Dagobert also obtains Aquitaine. Death of Mohammed, succeeded by Abu Bakr, who concentrates both civil and religious power in himself.		
	633: Wars among Saxons and Britons. The Muslims conquer central Arabia. Britain: Oswald occupies Northumbria and calls missionaries to his territories. Abu Bakr succeeded by the caliph Omar, who in a few years will conquer Iran, Syria and Mesopotamia.	633: Sophronius of Jerusalem opposes the doctrine of Sergius of Constantinople. The gospel is spread in East Anglia its king Earpwald being a Catholic. 633: 4th council of Toledo, presided over by Isidore of Seville.	Sergius, patriarch of Constantinople, intending to reconcile Chalcedonians and monophysites, thinks to set aside the concept of "nature", instead making use of that of "energy": Christ's energy or activity derives from his personality (hypostasis), which is one, and not from his two natures. So there is a single energy, a single will, in Christ.
	636: Rothari king of the Longobards (636-652); Chintila king in Spain.	636: Nestorian Christianity spreads in China. Death of Isidore of Seville.	Isidore of Seville's principal work: the *Etymologiae*, a vast encyclopaedia of sacred and secular culture.
	637: The Arabs put an end to the Sassanid kingdom.		
	638: The caliph Omar occupies Jerusalem.	638: The emperor Heraclius publishes the *Ekthesis*, inspired by Sergius. The Arabs in Egypt: the monophysites free from Byzantine imperial opresion (638-640). Rome: pope Severinus elected in 638. But he has to wait nearly 20 months for the emperor Heraclius's confirmation. Death of Sophronius of Jerusalem.	The *Ekthesis* affirms the perfect harmony, in Christ, between the divine and the human will, which therefore form a single will, while the two natures of the one person of the incarnate Logos remain unconfused.
	639: Death of Dagobert.		
640	640: Tula king in Spain. 641: Constantine II emperor of the East (641-668). Rothari wars against the Byzantines in Italy. The Slavs in Istria and Dalmatia. 642: The Arabs occupy Alexandria. Oswy king of Bernicia and Deira (642-651); of Northumbria (651-670). 643: Edict of Rothari: Longobard legislation. The Arabs in Egypt and Cyrenaica.	Rome: pope John IV (640-642); succeeded by Theodore I (642-649), who struggles against monothelism; then Martin I (649-655) who condemns both Heraclius's *Ekthesis* and Constans II's *Typos* (648) at a Roman council (649). He is arrested, brought a prisoner to Constantinople and then deported to the Crimea.	Constans II's *Typos* forbids any discussion of Christ's will and energy.

Date	Secular Events	Ecclesiastical Events	Cultural and Doctrinal Matters
650	651: Rebellion of Olympius, hexarch of Ravenna. Death of Oswin, king of Deira; Oswiu, king of Bernicia, unites the two kingdoms. 652: Death of Rothari; succeeded by Rodoald. 653: Aribert succeeds Rodoald (653-661). Spain: Receswinth is king (653-672) and publishes the *Liber iudiciorum*. Oswiu conquers Mercia.	653: Conversion of the Longobards. 654: Rome; pope Eugenius I (654-657). Elected during the exile of pope Martin († 655) in the Crimea. 657: Pope Martin succeeded by Vitalian (657-672).	
660	661: Death of Aribert. Conflicts among the Longobards. The Muslims divide into Sunni (orthodox) and Shia, followers of the caliph Alì († 661), who stress the caliphs' role as religious leader. 662: Grimoald king of the Longobards (662-671).	662: Death of Maximus Confessor, after the mutilation of his tongue and his right hand, the parts of his body with which he had opposed monothelism and the imperial edict. 664: Council of Whitby: convoked by king Oswin to resolve the conflict over ecclesiastical traditions between the Celtic and Anglo-Roman rites.	Maximus Confessor's numerous works include the *Opuscula theologica et polemica*, the *400 capita de caritate*, the *Mystagogia*.
	668: Death of Constans II; Constantine IV Pogonatus succeeds him as emperor (668-685).	667: Death of Ildefonsus of Toledo. 668: Theodore of Canterbury is consecrated bishop and reorganizes the church in England.	
670	671: Bertharid king of the Longobards (671-688). Wamba (672-680) king of the Visigoths. Death of the first Angle poet, Caedmon. 673: France: death of Lothar III, king of Neustria; Childeric II (673-675?) becomes king of Austrasia, Neustria and Burgundy. Dagobert II (675-679), king of Austrasia; Pippin of Héristal mayor of the palace.	672: Rome; pope Adeodatus II (672-676), then Donus (676-678) and Agatho (678-681), who condemns monothelism at a Roman council (680); with him ends the autocephaly of Ravenna.	
680	680: Spain; king Ervig (680-687). 685: Justinian II emperor of the East (685-695). Peace treaty with the Arabs. 687: Pippin sole mayor of the Frankish kingdom. Egica king of the Visigoths (687-701). 688: Cunipert king of the Longobards (688-700).	680/681: Council of Constantinople, 6th ecumenical, also called "in Trullo" or Trullan. Conversion of the Croats. 682: Rome; pope Leo II (682-683) has to wait some time for the Eastern emperor's permission before being consecrated. Succeeded by Benedict II (684-685), John V (685-686), Conon (686-687) and Sergius I (687-701), who refuses to approve the Quinisext council (or "in Trullo") of 692. Antipope Paschal II (687-692). Because of the Arab invasions, the Maronite monks emigrate to the mountains of Lebanon.	The council condemns the doctrines of monothelism and monoenergism and their principal supporters, living and dead; it confirms that in Christ are two inseparable wills and two inseparable energies.
690	690: The Byzantines are defeated by the Arabs near Sebastopol (Armenia) and lose this territory.	690: Willibrord begins the evangelization of Friesland.	

Date	Secular Events	Ecclesiastical Events	Cultural and Doctrinal Events
	Erection of the mosque of Omar at Jerusalem.

695: Deposition of Justinian II: his nose is cut off and he is exiled to Cherson; Leontius emperor (695-698).

Around this time the Venetians first elect a Doge (*dux*).
697-698: The Arabs conquer the area of Carthage.
698: Tiberius III Apsimar emperor of the East (698-705). | 692: Quinisext council (in Trullo). | The council "in Trullo" of 692 is important for the disciplinary aspects of the Eastern church: 102 canons touching on all aspects of religious life and the clergy, e.g. can. 13 relating to married deacons and priests. Can. 36 affirms that the see of Constantinople is second after Rome and endowed with equal authority. |
| 700 | 701: Witiza king of the Visigoths (701-710); Aribert II of the Longobards.

705: Justinian II restored as emperor of the East (705-711).
Construction of the great mosque of Damascus. | 700: Rupert bishop at Salzburg. Rome: pope John VI (701-705), John VII (705-707), Sisinnius (708), who lasts only 20 days, and Constantine (708-715).

In Malabar, in India, Nestorian bishops.

708: Death of Jacob, bishop of Edessa. | Jacob of Edessa translates many works from Greek to Syriac; he also revises the Syriac version (Peshitta) of the OT. |
| 710 | 710: The Arabs occupy Persia.
711: Philippicus Bardanes emperor (711-713): deposed and blinded; Anastasius II (Arthemius) appointed (713-715).
The Arabs invade Spain, conquering Toledo. Roderic is the last Visigothic king (710-711).

712: Liutprand king of the Longobards (712-744).
714: Death of Pippin of Héristal.
715: Charles Martel, Pippin's illegitimate son, mayor of the palace of Austrasia. Theodosius III emperor (715-717).

717: Leo III the Isaurian (717-741). Pelagius king of the Asturias (717-737).

719: Charles Martel, by now the dominant power in the Frankish kingdom, has Chilperic II appointed king (719-720). | 715: Rome; pope Gregory II (715-731). He opposes the iconoclasm of Leo III the Isaurian; rebuilds the abbey of Montecassino; favours St Boniface, who goes to Rome in 718.

719: Boniface continues his missionary activity in central Germany and Friesland. | |
| 720 | Incursions of Arabs into France (720; 721-725).

725: Wessex: Eadbert king (725-748).

727: Mercia: Penda king (727-755). | 723: Liutprand acquires and transfers the bones of St Augustine from Sardinia to Pavia.

725: Leo the Isaurian begins an intense political activity against the cult of images.

727: The Longobard king Liutprand occupies Sutri, which he then gives to the pope. This is regarded as the birth of the Papal States.
Pirminus in Alsace. | |
| 730 | | 731: Rome; pope Gregory III (731-741): condemns the iconoclasts at a council of 731.

735: Death of Bede at Jarrow. | Bede's *Historia ecclesiastica gentis anglorum* goes up to 731. |

Date	Secular Events	Ecclesiastical Events	Cultural and Doctrinal Matters
740	738: The Arabs invade France; battle of Poitiers against Charles Martel. Charles Martel expels the Arabs from Provence. Asturias: king Alfonso I the Catholic (738-757). 741: Constantine Copronymus emperor of the East (741-775). Death of Charles Martel: his kingdom is divided among his sons.	741: Rome; pope Zacharias (741-752): recognizes Pippin the Short as king of the Franks. Abbey of Sturm, founded by Sturm, a disciple of Boniface. Death of John Damascene (*c.*749).	Damascene's most important work is the *Fount of Gnosis*, in three parts: one is on philosophy, the second on the heresies and the last on the orthodox faith.

MAPS

1. Palestine at the time of Jesus
2. Egypt at the time of the Council of Nicaea (325)
3. Africa: mid 3rd c.
4. Asia Minor: early 4th c.
5. Italy: episcopal sees of the mid 3rd c.
6. Gaul at the time of the Emperor Constantine
7. Iberian Peninsula: Roman era
8. Mauretania Tingitana
9. Mauretania Sitifensis and Caesariensis: 4th-5th cc.
10. Britain: 4th c.
11. Germany (Roman)
12. Palestine III (Sinai): 6th c.
13. Palestine (I and II) and Arabia: 6th c.
14. Palestinian Monasteries
15. Lower Egypt: 4th c.
16. Upper Egypt: 4th c.
17. Tripolitana
18. African Provinces: 4th-5th cc.
19. Phoenicia and Syria: 6th c.
20. Libya: 6th c.
21. Nubia in the early Christian age
22. Ethiopia
23. Crete
24. Cyprus: autocephalous church (Council of Ephesus 431)
25. Armenia and Surrounding Regions: 6th c.
26. Central Provinces of Asia Minor: 6th c.
27. Isauria and Cilicia: 6th c.
28. Western Provinces of Asia Minor: 6th c.
29. Moesia and Thrace: 6th c.
30. Southern Phrygia Pacatiana
31. Dioceses of Macedonia (and Crete)
32. Northern Italy: end of 6th c.
33. Southern Italy and the Islands: end of 6th c.
34. Central Italy: end of 6th c.
35. Noricum
36. Dalmatia: 6th c.
37. Pannonia
38. Iberian Peninsula: Suevian and Visigothic era
39. Gaul: end of 6th c.
40. Central Europe: at the time of Charlemagne
41. Mesopotamia: 7th c.
42. Persia and Surrounding Regions: 6th c.
43. Ireland (Hibernia)
44. British Isles: end of 9th c.

1. PALESTINE at the time of Jesus

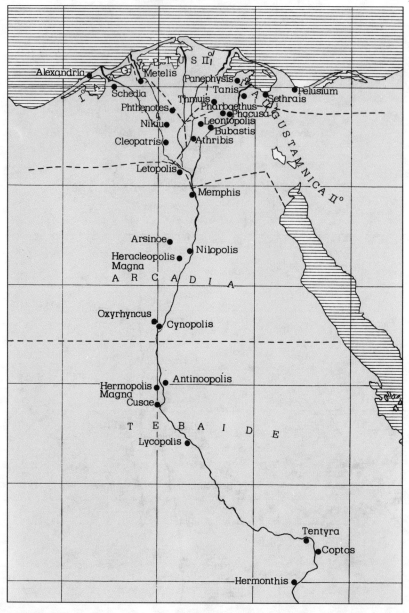

2. EGYPT at the time of the Council of Nicaea (325) Scale 1 : 4 500 000

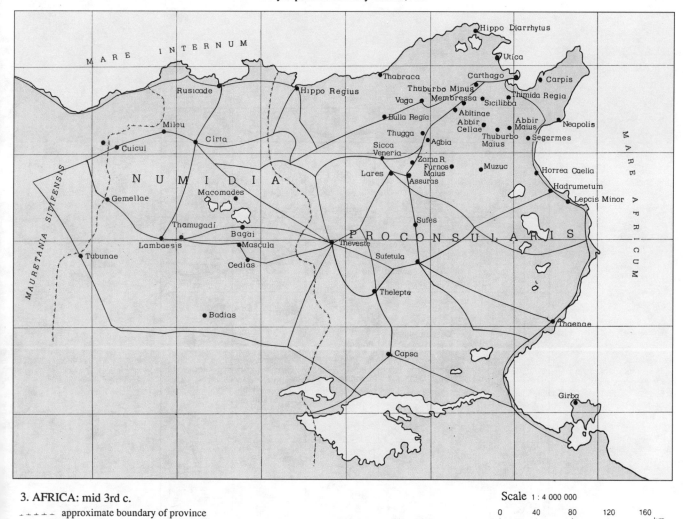

3. AFRICA: mid 3rd c.

- - - - approximate boundary of province

Scale 1 : 4 000 000

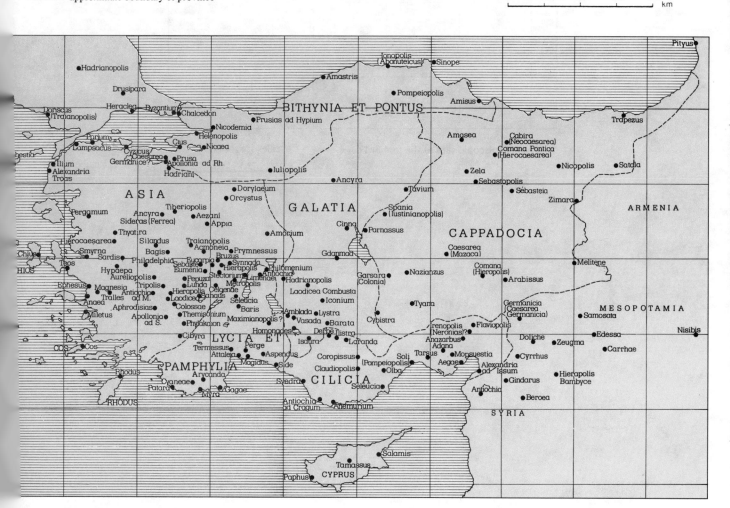

4. ASIA MINOR: early 4th c.

Scale 1 : 3 800 000

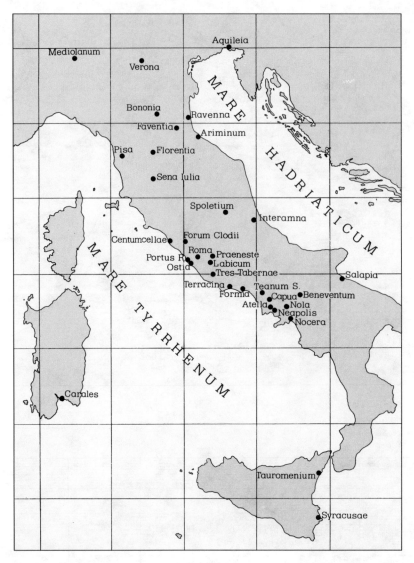

5. ITALY: episcopal sees of the mid 3rd c.

Scale 1:4 200 000

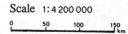

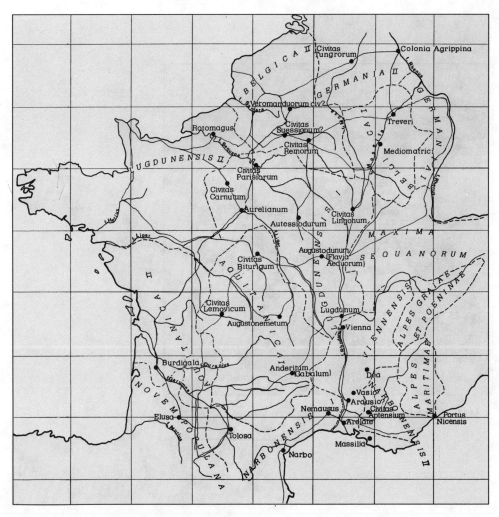

6. GAUL at the time of the Emperor Constantine
• episcopal sees
Scale 1 : 8 000 000

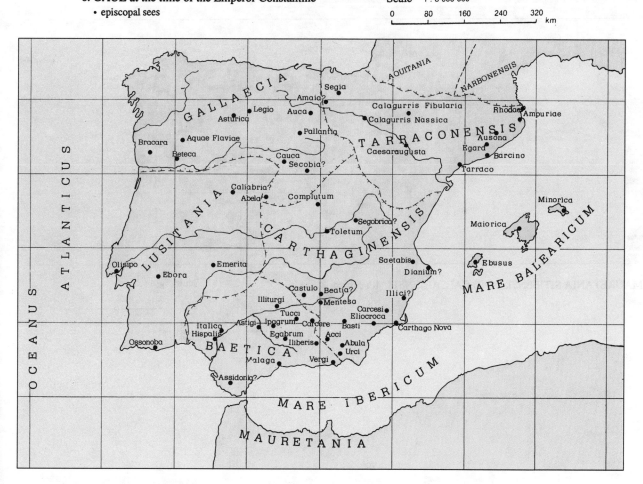

7. IBERIAN PENINSULA: Roman era
Scale 1 : 9 000 000

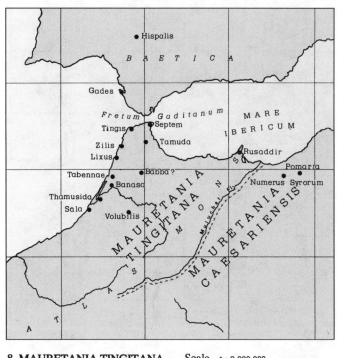

8. MAURETANIA TINGITANA   Scale 1 : 9 000 000

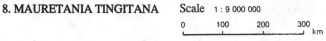

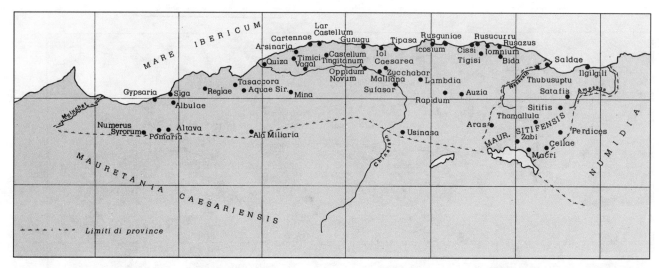

9. MAURETANIA SITIFENSIS AND CAESARIENSIS: 4th-5th cc.   Scale 1 : 6 000 000

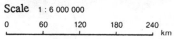

10. BRITAIN: 4th c.

11. GERMANY (ROMAN)

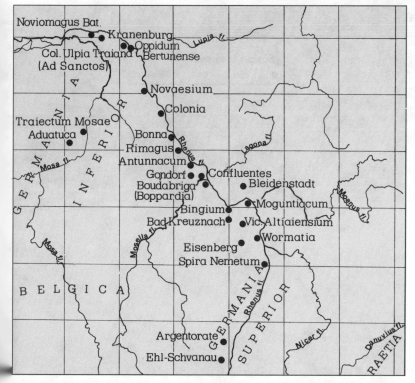

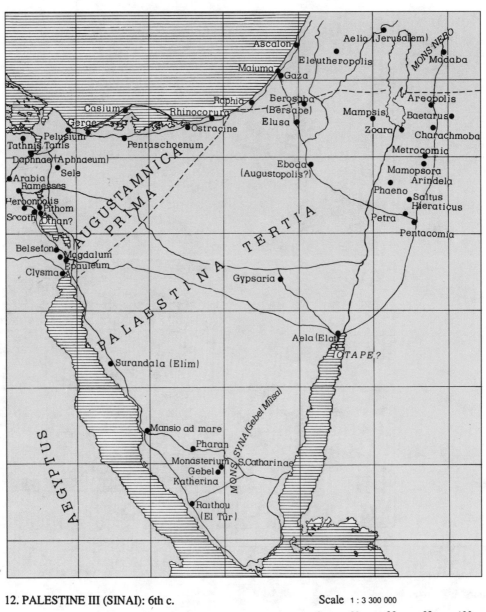

12. PALESTINE III (SINAI): 6th c.

13. PALESTINE (I AND II) AND ARABIA: 6th c.

Scale 1 : 2 000 000

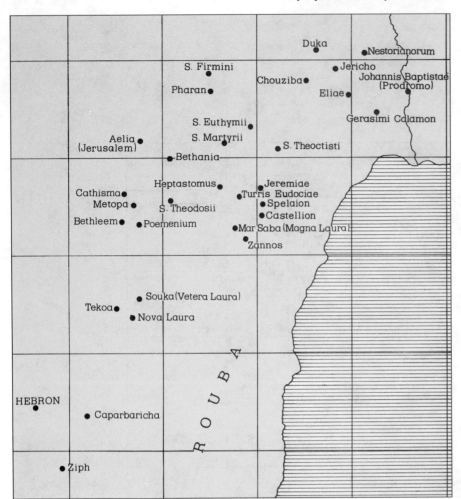

14.

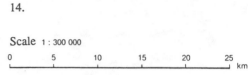

- Palestinian monasteries

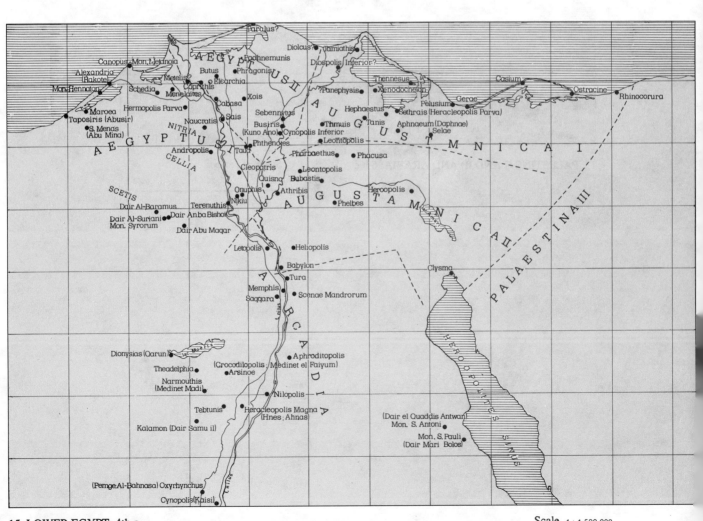

15. LOWER EGYPT: 4th c.

Scale 1 : 1 500 000

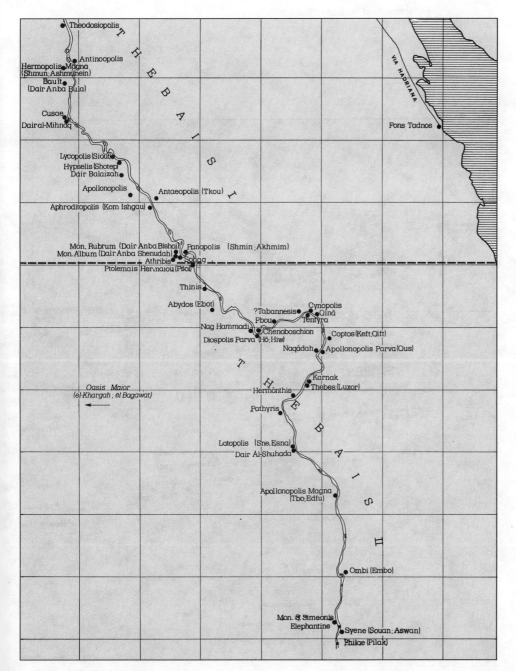

16. UPPER EGYPT: 4th c.   Scale 1 : 1 500 000

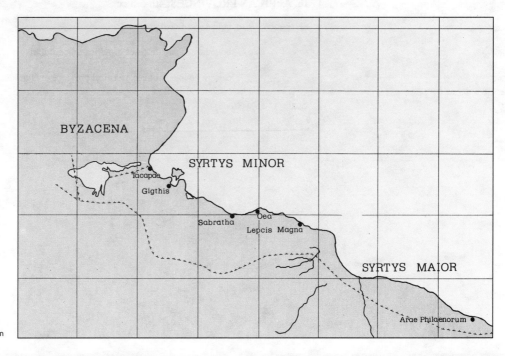

17. TRIPOLITANA

Scale 1 : 9 000 000

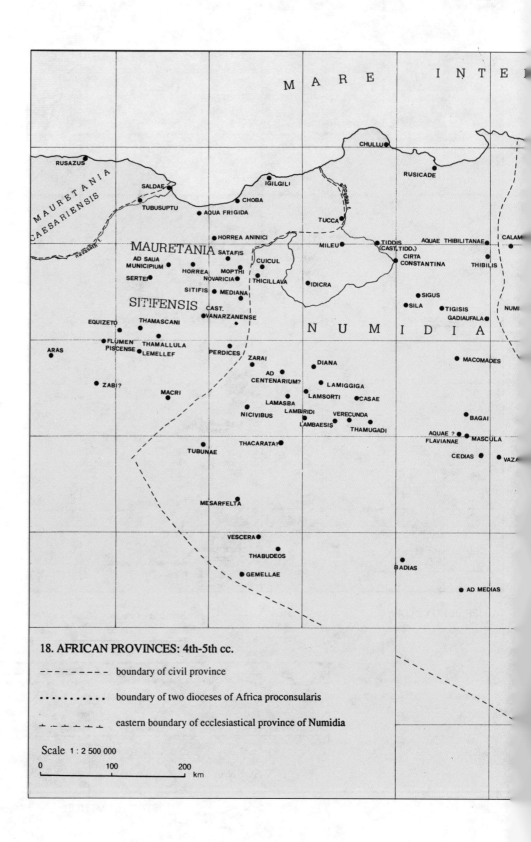

18. AFRICAN PROVINCES: 4th-5th cc.

- - - - - - - - boundary of civil province
· · · · · · · · · · boundary of two dioceses of Africa proconsularis
- - - - - - eastern boundary of ecclesiastical province of Numidia

Scale 1 : 2 500 000

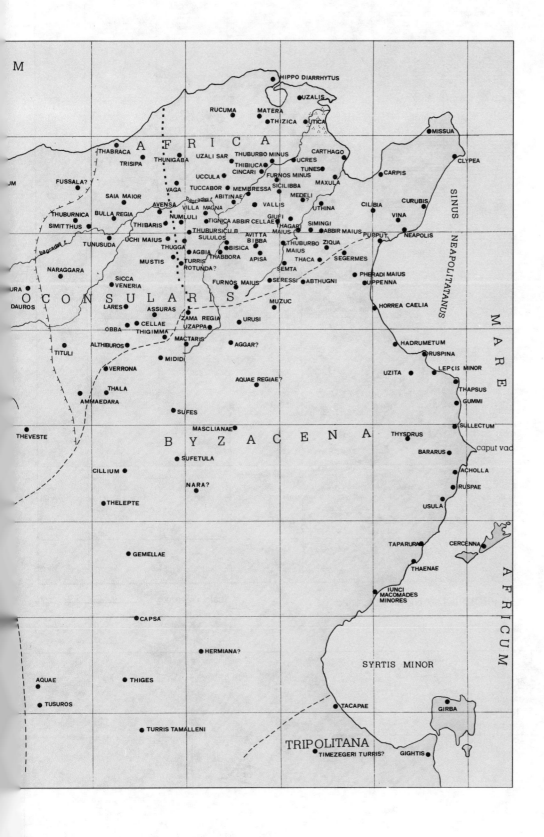

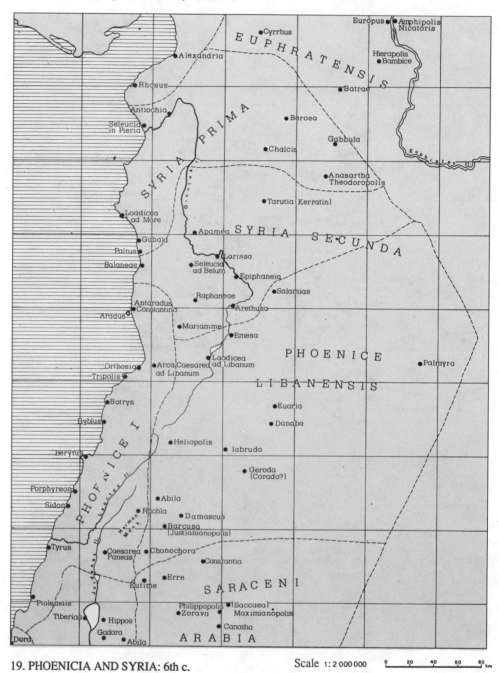

19. PHOENICIA AND SYRIA: 6th c.　　Scale 1:2 000 000

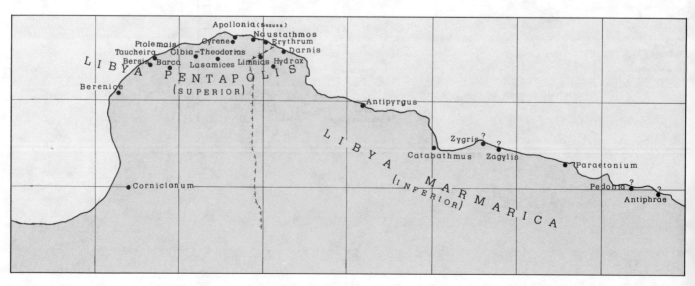

20. LIBYA: 6th c.　　Scale 1:1 650 000

21. NUBIA in the early Christian age

Scale 1 : 7 000 000

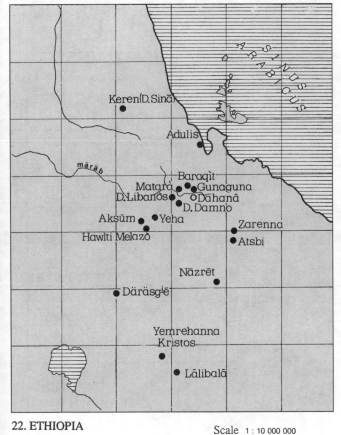

22. ETHIOPIA

Scale 1 : 10 000 000

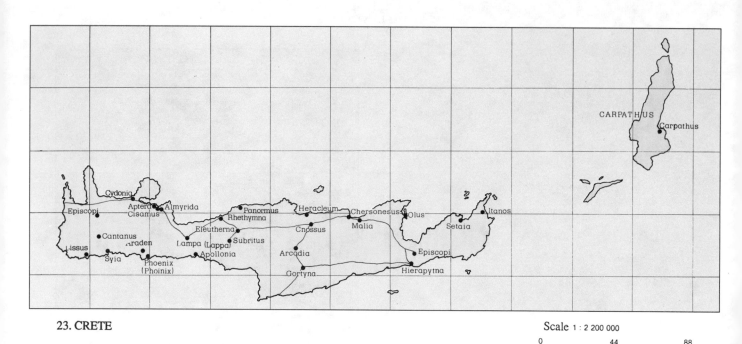

23. CRETE — Scale 1 : 2 200 000

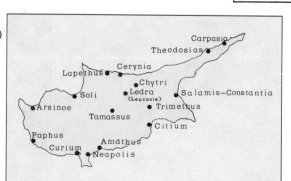

24. CYPRUS: autocephalous church (Council of Ephesus 431) — Scale 1 : 3 500 000

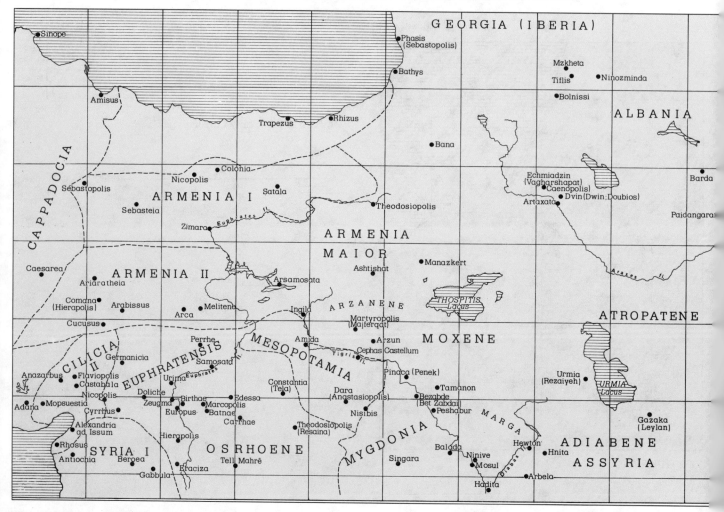

25. ARMENIA AND SURROUNDING REGIONS: 6th c. — Scale 1 : 4 000 000

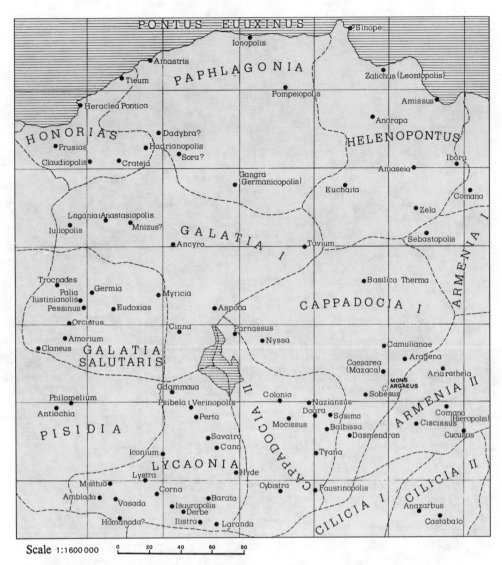

26. CENTRAL PROVINCES OF ASIA MINOR: 6th c.

27. ISAURIA AND CILICIA: 6th c.

28. WESTERN PROVINCES OF ASIA MINOR: 6th c.

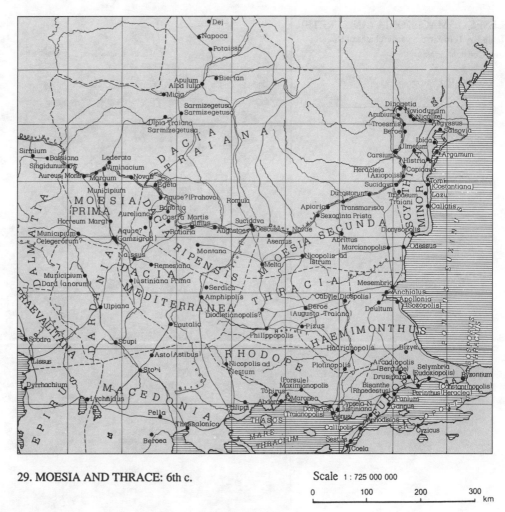

29. MOESIA AND THRACE: 6th c.

Scale 1 : 725 000 000

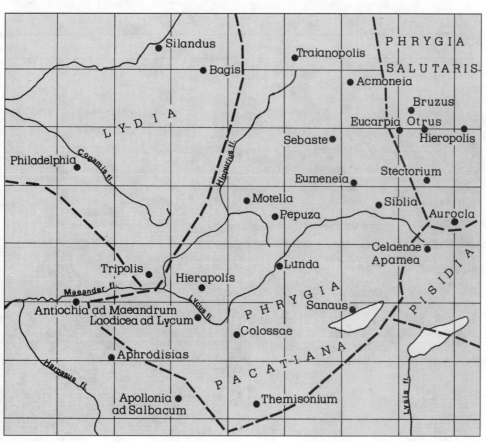

30. SOUTHERN PHRYGIA PACATIANA

Scale 1 : 1 500 000

31. DIOCESES OF MACEDONIA (AND CRETE)

••••• boundary between the patriarchates of Rome and Constantinople

Scale 1 : 4 500 000

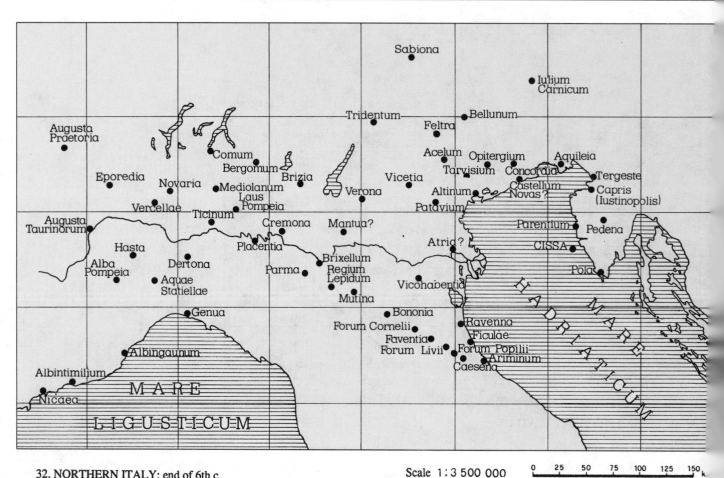

32. NORTHERN ITALY: end of 6th c.

Scale 1 : 3 500 000

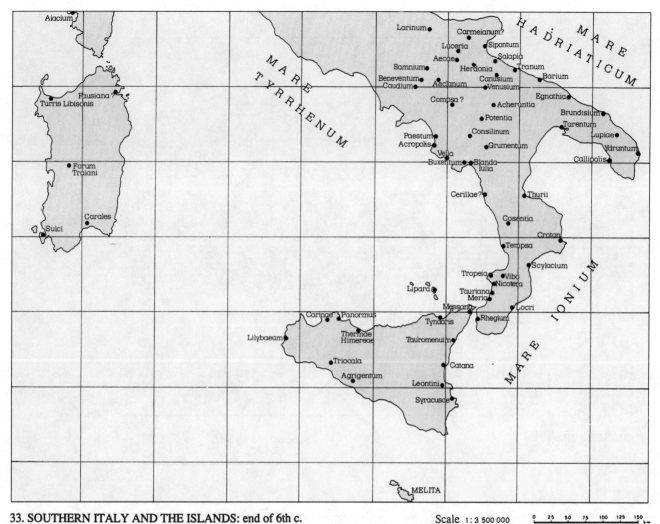

33. SOUTHERN ITALY AND THE ISLANDS: end of 6th c. Scale 1 : 3 500 000

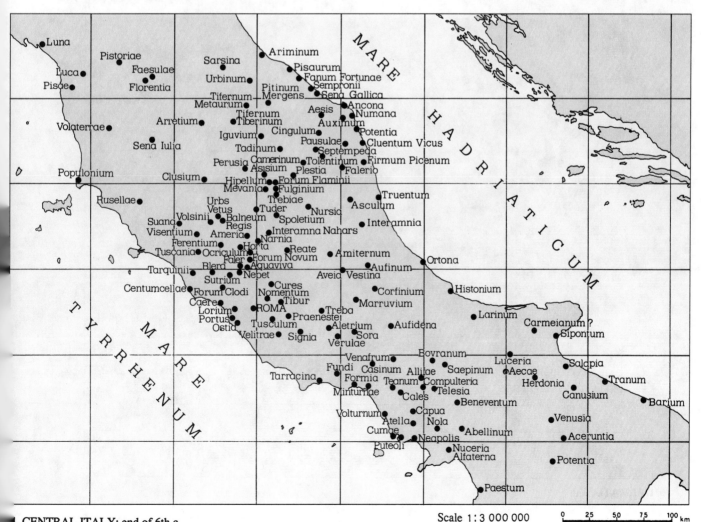

34. CENTRAL ITALY: end of 6th c. Scale 1 : 3 000 000

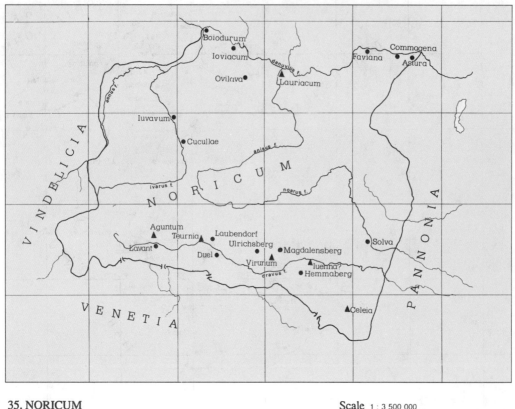

35. NORICUM

▲ episcopal sees

Scale 1 : 3 500 000

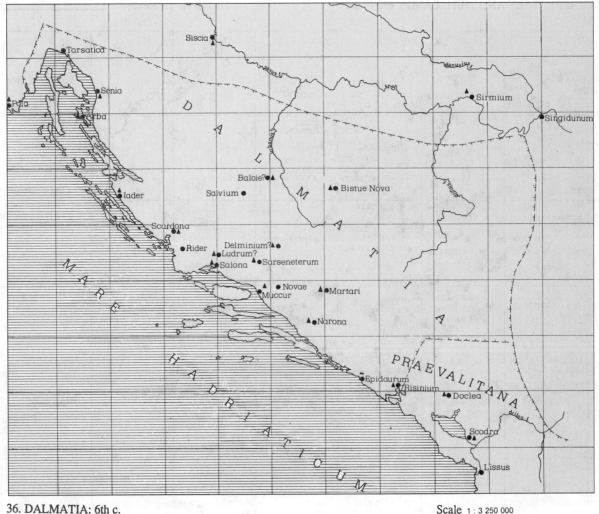

36. DALMATIA: 6th c.

▲ episcopal sees

Scale 1 : 3 250 000

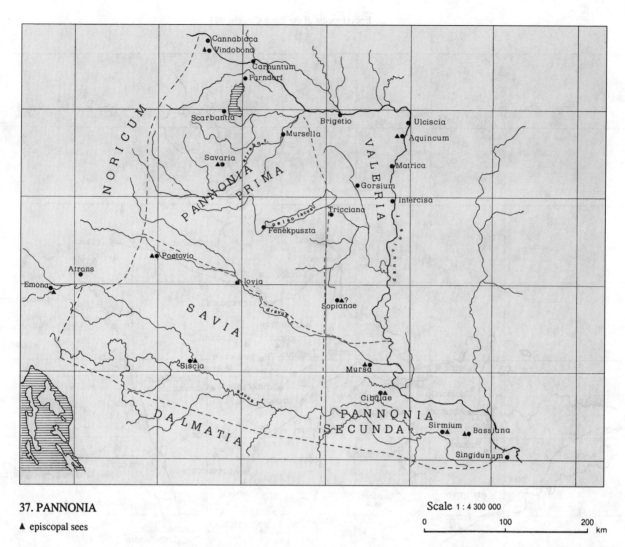

37. PANNONIA

▲ episcopal sees

Scale 1 : 4 300 000

38. IBERIAN PENINSULA: Suevian and Visigothic era

- - - - - - provincial boundary
──── boundary during Suevian era
· · · · · · · · boundary during Visigothic era
─ · ─ · ─ Byzantine territory

Scale 1 : 10 000 000

39. GAUL: end of 6th c.
◉ episcopal cities

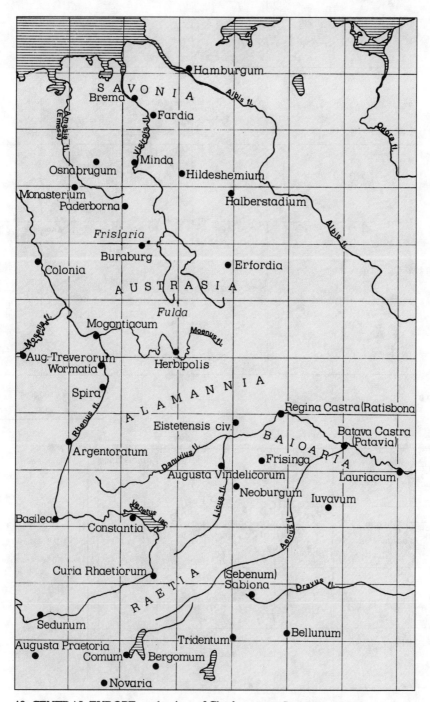

40. CENTRAL EUROPE: at the time of Charlemagne Scale 1 : 5 000 000

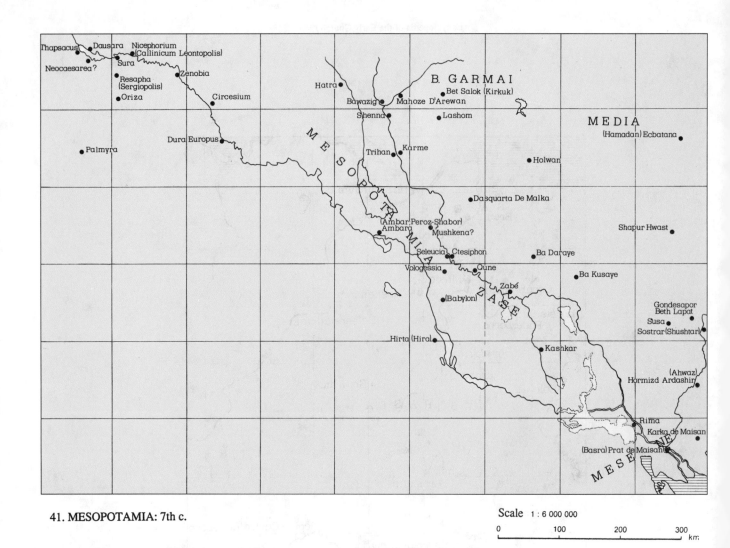

41. MESOPOTAMIA: 7th c.

Scale 1:6 000 000

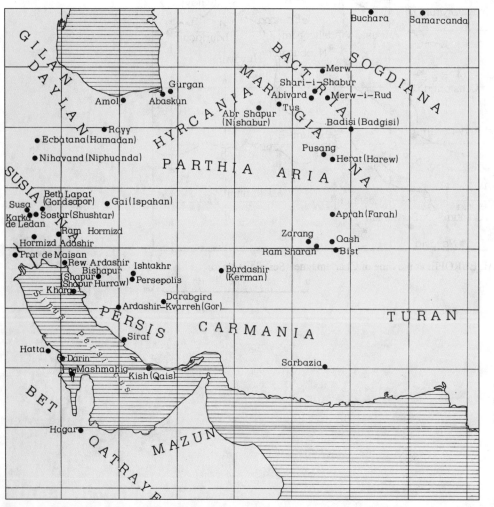

42. PERSIA AND SURROUNDING REGIONS: 6th c.

Scale 1:12 000 000

43. IRELAND (HIBERNIA)

Scale 1 : 3 000 000

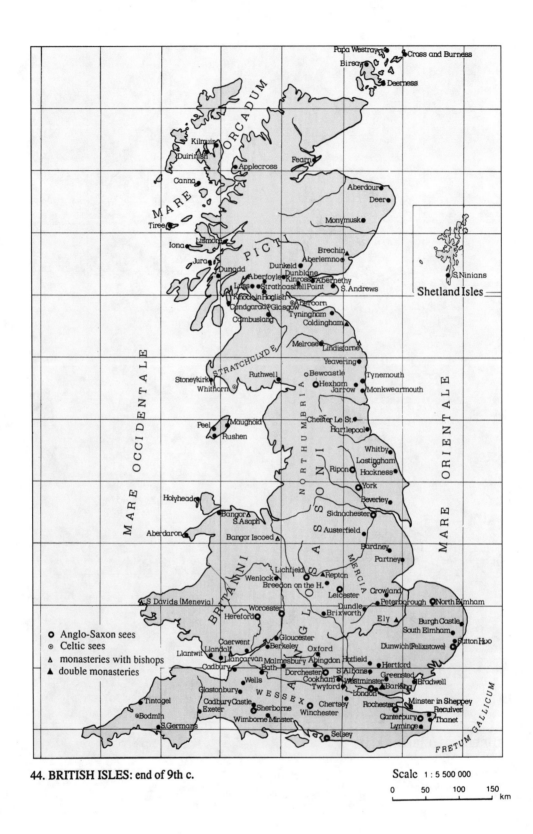

44. BRITISH ISLES: end of 9th c.

ILLUSTRATIONS

Fig.
1 Aaron
  Rome, wooden door panel
2 Abdon and Sennen
  Rome, fresco
3 Abel
  Ravenna, mosaic
4 Cain and Abel, Rome, fresco
5 Abraham
  Rome, fresco
6 Abrasax
  London, amulet
7 Achaia
  Epidaurus, plan of basilica
8 Adam and Eve
  Gargaresc, fresco
9 Ad Bestias
  Karlsruhe, terracotta plate
10 Africa
   Djemila, main complex of Christian quarter
11 Panoramic view of model
12 La Skhira, basilica
13 Sbeitla, plan showing Christian buildings
14 Sbeitla, cathedral complex
15 Agape
   Rome, fresco
16 Agnes
   Rome, gold-glass
17 Vatican City, gold-glass
18 Agnus Dei
   Rome, fresco
19 Albania (Caucasus)
   Lekit, plan of tetraconch church
20 Ambrose
   Milan, mosaic
21 Ananias and Sapphira
   Brescia, ivory reliquary
22 Angel
   Rome, fresco
23 Annunciation
   Ravenna, sarcophagus
24 Vatican City, silk cloth
25 Antioch in Syria
   New York, Antioch chalice
26 Antioch, pavement mosaic
27 Kausiye, plan of martyrium of St Babylas
28 Apamea
   Plan of episcopal complex
29 Ambulatory of tetraconch church
30 Apostle
   Barletta, funerary relief
31 Aquileia
   Episcopal complex
32 Basilica of the Beligna
33 Arabia
   Ezra, martyrium of St George
34 Ezra, plan of martyrium of St George
35 Madaba, pavement mosaic
36 Gerasa, cathedral and apse of martyrium of St Theodore
37 Gerasa, plan of the city
38 Mount Nebo, plan of monastery complex of the prophet Moses
39 Mount Nebo, Basilica of the prophet Moses
40 Arles
   Plan of the city
41 sarcophagus
42 Armenia
   Yerevan, Ecmiadzin Gospels
43 Odzoun, cathedral
44 Odzoun, cathedral: interior
45 Ererouk, plan of basilica
46 Zvart'noc, plan of church of St Gregory the Illuminator
47 Zvart'noc, arcade fragment
48 Ascension
   Munich, ivory tablet
49 Asia Minor
   Alahan Monastir, plan of monastic complex
50 Alahan Monastir, domed basilica
51 Alahan Monastir, cruciform baptismal pool
52 Hierapolis, plan of martyrium of St Philip
53 Sardis, Christian church

Fig.
54 Athens
   Stoa of Hadrian
55 Stoa of Hadrian, mosaic pavement
56 Plan of the Basilica of St Leonidas
57 Augustine
   Rome, fresco
58 Baetica
   Malaga, Basilica of Vega del Mar
59 Baleari
   Majorca, basilica of S. Peretó
60 Baptism
   Rome, wall fresco
61 Walesby, font
62 Bethlehem
   Church of the Nativity
63 Plan of the Church of the Nativity
64 Bostra
   Plan of the Church of Sts Sergius, Bacchus and Leontius
65 Britain
   Lullingstone, christogram
66 Hinton St Mary, mosaic pavement
67a-b Casket of St Cuthbert
68 Callatis
   Reconstruction of Syrian-type basilica
69 Cana, wedding feast of
   Milan, five-part gospel cover
70 Carthage
   Basilica of Damous el Karita
71 Basilica Maiorum Arearum
72 Cathedra
   Rome, Cimitero Maggiore catacomb
73 Cemetery
   Amiternum, catacomb of S. Vittorino
74 Cimitile, plan of monumental complex
75 Naples, catacombs of S. Gennaro
76 Lower catacomb
77 Mosaic portrait
78 Rome, painted cubicle
79 Plan of Basilica of S. Alessandro
80 Rome, catacomb of the Jordani
81, 82, 83 Details of Basilica of S. Alessandro
84, 85 Plan of Catacombs of SS. Pietro e Marcellino
86 Rome, Catacomb of Priscilla
87 Cilicia
   Korykos, extra-muros church
88 Korykos, extra-muros church, detail
89 Meriamlik, plan of martyrium of St Thecla
90 Constantinople
   Plan of church of St John of Studios
91 Apse of church of St John of Studios
92 Pillar capital
93 Plan of church of Hagia Sophia
94 Corinth
   Lechaion, plan of Basilica of St Leonidas
95 baptistery of Basilica of St Leonidas
96 Corsica
   Mariana, plan of Christian basilica
97 Cross
   Florence, Gospel of Rabbula
98 Rome, wooden door panel
99 Crucifix
   Isle of Man, incised stone fragment
100 Cyprus
    Kiti, mosaic
101 New York, Silver plate
102 Dacia Traiana
    Sucidava, plan of one-aisled basilica
103 Damasus
    Rome, Damasian inscription
104 Examples of Filocalian alphabet
105 Daniel
    Rome, Niche fresco
106 Ecija, detail of sarcophagus
107 Diptych
    Monza, ivory diptych of Stilicho
108 Dura Europos
    Plan of domus ecclesiae
109 Dvin
    Reconstruction of main quarter
110 Edessa
    Fragment of mosaic pavement

Fig.
111 Egypt
    Abu Mina, plan of cruciform church
112 Baptistery of cruciform church
113 Plan of East church
114 Sohag, niche decoration
115 Plan of church of monastery of Shenoute
116 Qusur Al-Rubayyat, plan
117 Elijah
    Milan, sarcophagus of Stilicho
118 Emesa
    Damascus, funerary mosaic
119 Entry of Jesus into Jerusalem
    Rossano Calabro, Codex Purpureus
120 Ephesus
    Selçuk, basilica of St John the Evangelist
121 Baptistery of basilica of St John
122 Baptistery of basilica of St John
123 Baptistery of church of St Mary
124 Epigraphy, Christian
    Rome, fragmentary inscription
125 Sepulchral inscription
126 Thessalonica, dome mosaic
127 Chios, mosaic pavement
128 Rome, catacomb of Priscilla, devotional graffiti
129 Epirus
    Nicopolis, mosaic head
130 Dodona, plan of basilica
131 Eucharist
    Istanbul, Stûma paten
132 Evangelists
    Rome, apse mosaic
133 Fiery Furnace
    Egnazia, clay lamp
134 Flood
    Rome, fresco of catacomb cubicle vault
135 Fossor
    Rome, fresco on catacomb cubicle wall
136 Gabriel
    Ravenna, cathedra backrest
137 Gammadia
    Ravenna, mosaic
138 Gaul
    St-Maximin-la-Ste-Baume, crypt engraving
139 Clermont, sarcophagus
140 Montpellier, sarcophagus
141 Georgia
    Bolnissi, pilaster capital
142 Plan of the basilica
143 Mzcheta-Djvari, church of the Holy Cross: exterior
144 Plan of the church of the Holy Cross
145 Germany
    Cologne, Basilica of St Gereon
146 Gestures
    Rome, cubicle vault
147 Habakkuk
    Vatican City, "dogmatic" sarcophagus
148 Hadrumetum
    Sousse, catacomb of Hermes
149 Healing of the Blind Man
    Ravenna, wall mosaic
150 Hippo
    Basilica and baptistery
151 Histria
    Plan of some basilicas
152 Issue of Blood
    Milan, sarcophagus
153 Italy
    Aosta, complex of S. Orso
154 Castelseprio, fresco
155 Nocera Superiore, plan
156 Florence, mosaic pavement and plan
157 Egnazia, plan of excavation
158 Teramo, Byzantine triforium
159 Castelbordino, plan of basilica
160 Cornus, excavations
161 Jacob
    Vienna, miniature codex of Genesis
162 Jacob's Ladder
    Rome, cubicle fresco
163 Jerusalem
    Madaba, pavement mosaic

Fig.
164 **Plan of the Holy Sepulchre complex**
165 Plan of basilica of the Eleona
166 **Jesus Christ**
   Rome, sarcophagus
167 Rome, "Lot" sarcophagus
168 Vatican City, sarcophagus
169 Rome, cubicle fresco
170 Milan, sarcophagus: front
171 Milan, sarcophagus: rear
172 **Job**
   Rome, cubicle fresco
173 **John the Baptist**
   Ravenna, dome mosaic
174 **Jonah**
   Cleveland, marble sculpture
175 Rome, "Jonah" sarcophagus
176 **Joseph (St)**
   Ravenna, backrest of Cathedra of Maximian
177 **Joshua**
   Rome, detail of mosaic
178 **Judas**
   Ravenna, wall mosaic
179 London, ivory diptych
180 **Judgment**
   Rome, cubicle fresco
181 **Kiss**
   Castellammare di Stabia, ivory comb(?)
182 **Lamp**
   Rome, clay lamp
183 Aquileia, bronze chandelier
184 **Laying on of Hands**
   Rome, cubicle fresco
185 **Lazarus**
   Paris, reliquary lid
186 **Leper**
   London, ivory diptych
187 **Libya**
   Apollonia, plan of western basilica
188 Plan of the palace of the Byzantine *dux*
189 East cathedral
190 Cyrene, plan of cathedral
191 **Liturgical Furniture**
   Mildenhall, three silver spoons
192 Water Newton, silver cup
193 Traprain Law, a chalice (?)
194 Mobile lamp
195 Riha paten
196 Stûma paten
197 Vase from Homs
198 Censer from Syracuse
199 Reliquary cross
200 Reliquary cross
201 Capsella from Grado
202 **Lot**
   Rome, cubicle fresco
203 **Lusitania**
   Idanha-a-Velha, plan
204 Torre de Palma, plan of basilica complex
205 **Macedonia**
   Bargala, plan of basilica
206 Dion, plan of basilica B
207 **Magi**
   Rome, fresco in "Greek Chapel"
208 Traprain Law, silver flask
209 **Manna**
   Rome, wooden door panel
210 **Manuscript Tradition**
   London, Lindisfarne Gospels
211 Vatican Library, Vat. gr. 1809
212 Vat. gr. 2200
213 **Marriage**
   Nicosia, paten
214 **Marseilles**
   St Victor, plan of martyrium
215 **Martyrdom, inscriptions of**
   Niculitel, martyrial crypt
216 **Martyrdom, scenes of**
   Rome, column of ciborium
217 **Mary**
   Parenzo, apse vault
218 Rome, fresco in unnderground basilica
219 Rome, cubicle fresco

Fig.
220 **Medical Lecture**
   Rome, catacomb fresco
221 **Merida**
   Casa Herrera, basilica
222 **Milan**
   Plan of Basilica of S. Simpliciano
223 Plan of Basilica of S. Lorenzo
224 Baptistery of S. Tecla
225 **Monogram**
   Grado, Basilica of S. Eufemia
226 Naples, dome mosaic
227 **Mosaic**
   Mount Nebo, pavement mosaic
228 Ravenna, Mausoleum of Galla Placidia
229 **Moses**
   Rome, cubicle fresco
230 **Mzcheta-Sveti Tzkhoveli**
   Church of the Holy Cross
231 **Naples**
   Episcopal complex
232 **Nativity**
   Milan, five-part Gospel cover
233 **Nazareth**
   Plan of Church of Annunciation
234 **Nimbus**
   Vatican Cityn, Necropolis vault
235 **Nisibis**
   Baptistery: exterior
236 Baptistery: decoration detail
237 **Noah**
   Rome, catacomb fresco
238 **Noricum**
   Teurnia, cemeterial basilica
239 **Orans**
   Rome, catacomb fresco
240 **Orpheus**
   Rome, catacomb fresco
241 **Painting**
   Rome, catacomb fresco
242 Rome, catacomb fresco
243 Rome, ceiling painting
244 **Palestine I**
   El-Gib, plan of church
245 **Palestine II**
   Et-Tabgha, plan of church
246 mosaic pavement
247 **Palestine III**
   Shivta, plan of church
248 **Paradise**
   Rome, catacomb painting
249 **Paralytic**
   Dura Europos, fresco
250 **Parapetasma**
   Vatican City, sarcophagus lid
251 **Parentium**
   Plan of Euphrasian basilica
252 **Paul, apostle**
   Rome, cubicle niche painting
253 **Perpetua and Felicity**
   Ravenna, medaillons
254 **Peter**
   Vatican City, sarcophagus front
255 **Philippi**
   Plan and ruins of basilica B
256 **Phoenix**
   Rome, fresco of "Greek Chapel"
257 **Pilate**
   Vatican City, "Passion" sarcophagus
258 **Pious Women**
   Milan, ivory diptych
259 **Portrait**
   Rome, sarcophagus of "two brothers"
260 **Prophet**
   Rome, catacomb fresco
261 **Quails, miracle of**
   Rome, wooden door panel
262a,b,c **Ravenna**
   Classe, excavations
263 **Rome**
   Basilica Apostolorum
264 S. Pancrazio
265 Basilica of SS. Giovanni e Paolo

Fig.
266 Santa Maria in Cosmedin
267 Santa Balbina
268 Basilica of S. Vitale
269 Monastery of Tre Fontane
270 **Salona**
   Plan of Manastirine
271 **Samson**
   Rome, catacomb painting
272 **Sarcophagi**
   Rome, Sarcophagus of S. Maria Antica
273 Vatican City, Sarcophagus of Junius Bassus
274 **Sardica**
   Plan of the church of St Sophia
275 **Sermon on the Mount**
   Rome, Sarcophagus front
276 **Shepherd, Good**
   Rome,fresco of cubicle of "Velata"
277 Vatican City, "Bath-type" sarcophagus
278 Ravenna, Mausoleum of Galla Placidia
279 **Side**
   Plan of episcopal complex
280 **Sinai**
   Mount Sinai, St Catherine's monastery
281 Plan of church of St Catherine's monastery
282 apse mosaic
283 **Sirmium**
   Plano f city
284 **Soldierscasting lots for Jesus' clothing**
   Rome, catacomb painting
285 **Stobi**
   Plan of the city
286 Plan of "Central Basilica"
287 Baptistery of episcopal church
288 **Susanna**
   Rome, wall fresco
289 Rome, fresco of Arcosolium of Celerina
290 **Symbolism**
   Aquileia, mosaic pavement
291 Albenga, christigram
292 Rome, catacomb painting of fish
293 **Syracuse**
   Plan of Pupillo hypogea
294 **Syria**
   Resafa-Sergiopolis, plan of city
295 Inside of Church A
296 Qal'at Seman, complex of St Simeon Stylites
297 Qal'at Seman, central zone
298 Huarte, planof ecclesial complex
299 Qirkbiza, plan of church
300 Inside of the church
301 **Tarraconensis**
   Centcelles, Roman villa
302 **Thessalonica**
   Plan of the city
303 Plan of church of Acheiropoietis
304 Plan of martyrium of St Demetrius
305 Church of St George, mosaic
306 **Thrace**
   Mesembria, plan of episcopal church
307 View of episcopal church
308 Perustica, plan of "Red Church"
309 View of "Red Church"
310 **Tipasa**
   Basilica
311 **Tobias**
   Le Mas D'Aire, sarcophagus lid
312 **Tomi**
   Plan of the Great Basilica
313 **Traditio Clavium**
   Rome, catacomb fresco
314 **Traditio Legis**
   Vatican City, sarcophagus front
315 **Trier**
   Plan of cemeterial area of city
316 **Tropaeum Traiana**
   Plan of Romano-Byzantine fortress
317 **Vatican**
   Plans of St Peter's basilica
318 Vatican necropolis
319 **Vienne**
   Plan of St Peter's basilica
320 **Virga**
   Naples, catacomb fresco

Fig. 1 **AARON**
ROME. Basilica of S. Sabina. Wooden door. Detail of the last panel of the third column, with Aaron (first half of 5th c.).

The entire panel refers to the OT episodes of Moses and the exodus from Egypt. A., depicted bottom left, bearded and with flowing hair, is the protagonist of two of the three scenes narrated *boustrophedon*, from bottom to top: with the *virga* he directs the two great serpents towards Pharaoh (in imperial dress), terrifying him; his very gesture raises the waters of the Red Sea which engulf Pharaoh's chariot and his army. Top right, God's hand indicates the column of fire; top left, the angel places himself at the head of the army of Israel. While the two figures below seem rather rigid, the image of the powerful flow of the waves which have already overturned Pharaoh's servant, and against whose impetus the horses fight, is rendered particularly effectively.

Fig. 2 **ABDON AND SENNEN**
ROME. Catacomb of Ponziano. Room of the Baptistery: coronation of Abdon and Sennen (6th c.).

Framed by pyramidal shrubs and called by their names: to the left, *Milis*, in soldier's tunic and chlamys, praying and haloed; to the right, *Bincentius*, in cassock-like tunic and *paenula*. In the centre, above, the Eternal with cruciform nimbus uses both hands to crown the two martyrs: A. to left and S. to right; both bearded and in Persian dress, in sleeved tunic with *lacerna* closed on the breast by a round brooch (the only known example in cemetery painting of this type of clothing) and *pileus* (or Phrygian cap). On the band that frames the fresco, above: *De donis d(e)i et s(an)c(to)r(um) m(artyrum) Abdon [et Sennen Gaudiosus fecit]*.

Fig. 3 **ABEL**

RAVENNA. Mosaic of the basilica of S. Apollinare in Classe (6th c.).

In the centre Melchizedek, with caption, places his hands on the altar on which are two loaves and a chalice; to his right A. advances, clothed in light skins, offering a lamb; to his left Abraham, in tunic and pallium, pushes forward his son Isaac to offer him to God. Neutral ground between two curtains; above, the divine hand among the clouds.

Fig. 4 **ABEL AND CAIN**

ROME. Catacomb of Via Latina. Fresco in an arcosolium of a cubicle (second half of 4th c.).

To the left sit Eve and Adam, dejected, dressed in animal skins; to the right A., in tunic and pallium, offers a lamb; C., in exomis tunic, bears a bundle of ears of grain.

Fig. 5 ABRAHAM

ROME. Catacomb of S. Callisto. Fresco in the so-called "Chapel of the Sacraments" (first half of 3rd c.).

In the centre A. in tunic, in attitude of *expansis manibus*; to his right Isaac in similar attitude and attire; to his left the ram, a tree and a bundle of wood. The gesture of the two characters expresses the salvation that has occurred.

Fig. 6 ABRASAX
LONDON. British Museum. Amulet.

On one side the lion, with solar significance and also understood as the constellation: between his paws is a star, and above his body the moon. On the other side the term A., a name given by Basilides, founder of Egyptian gnosticism, to his supreme deity. The sum of the numerical values of the Greek letters (or "psephy") of this name gives 365, a figure corresponding to the days of the solar year.

Fig. 7 ACHAIA
EPIDAURUS. Plan of the basilica (*c*.400).

The church, with five aisles preceded by a vast atrium, has a five-part transept and ends in a single apse. It rises near an enclosure sacred to Asclepios.

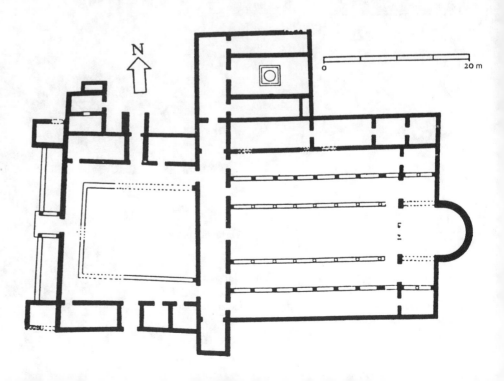

Fig. 8 **ADAM AND EVE**
GARGARESC (Tripoli). Fresco in the hypogeum of Adam and Eve (mid 4th c.).

A. and E., naked beside the tree around whose trunk is wound the serpent, following Gen 3, 1-13; the scene signifies the concept of *felix culpa*, which made the redemption necessary.

Fig. 9 **AD BESTIAS**
KARLSRUHE. Badisches Landesmuseum. Terracotta plate (4th c.).

The condemned man, bearded, with thick hair and a loincloth, is tied to a stake, while his feet rest on a small pedestal. To the right of the space, which represents the arena, a lion approaches threateningly.

Fig. 10 **AFRICA**
DJEMILA. The main complex of the city's "Christian quarter".

It consists of the double episcopal basilica. These show a system of communicating double crypts. Between the two churches is inserted the circular baptistery. This is still preserved entire and has a pool surmounted by a canopy. 5th century.

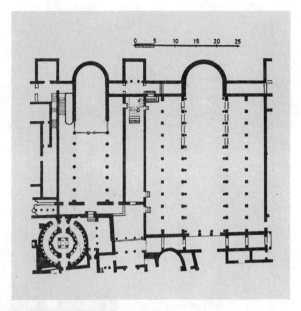

Fig. 11 Panoramic view (model).

Fig. 12
LA SKHIRA. Tunisia. Basilica.

A great five-aisled religious building erected in the 4th c., in a zone previously occupied by burials. Three main rebuilding phases in the 6th c. have been identified. The history of this basilica involved the creation of two opposed apses, a baptistery with three aisles and cruciform pool, and a wealth of mosaic decoration, particularly in the baptistery.

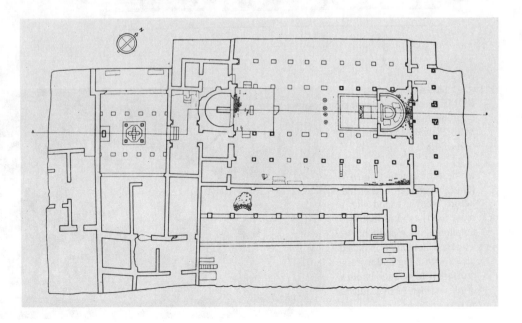

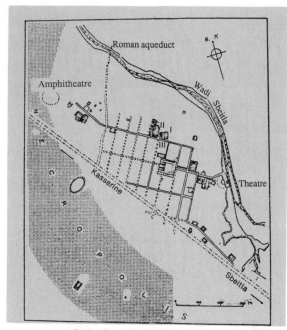

Positions of the churches of Sufetula, now Sbeitla
I: So-called basilica of Bellator
II: So-called basilica of Vitalis
III: Oratory of Jocundus, primitive baptistery

Fig. 13 **AFRICA**
SBEITLA. Plan of the ancient city of Sufetula with the Christian buildings marked.

Among these the churches with two opposed apses, a frequent phenomenon in Africa and then in Spain, I and II, respectively called "of Bellator" and "of Vitalis", and the city's first and second cathedral.

Fig. 14 SBEITLA. The cathedral complex.

It dates from the 5th-6th cc. The baptistery of the first, smaller episcopal church was transformed into a *martyrium* to receive the remains of bishop Jocundus, a martyr of the Vandal persecution. At the same time the new, larger cathedral, in turn provided with a baptistery, was built by the priest Vitalis.

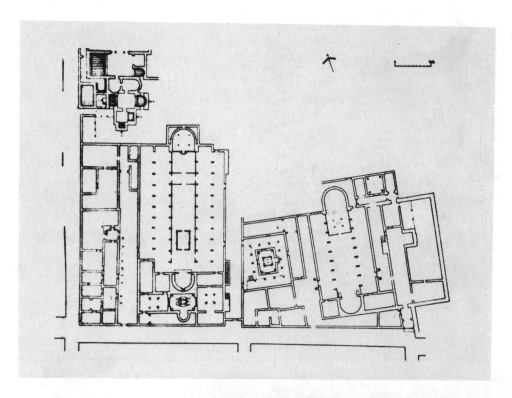

Fig. 15 **AGAPE**
ROME. Fresco in a lunette of an arcosolium in the catacomb of SS. Pietro e Marcellino (first half of 4th c.).

In the centre: four figures in tunics banquet around a table; before them a tripod with a roast fowl, a small loaf, two amphorae, a jug and a bottle. To left: a handmaid in belted tunic, with a jug and a glass, proffers wine obtained from a drink-warmer. Above: *Sabina mixes*.

Fig. 16 **AGNES**
ROME. Catacombs of Panfilo. Gold-glass (4th c.).

The martyr A., with caption *Acnnes*, haloed and richly dressed, in attitude of *expansis manibus*; beside her two doves on columns, two scrolls, two stars.

Fig. 17 VATICAN CITY. Museo Sacro della Biblioteca Vaticana. Gold-glass (4th c.).

In the centre A. (captioned *Annes*), sumptuously dressed, head covered by a *palla*, with two stars beside her, prays between the princes of the Apostles, indicated by the captions *Petrus* and *Paulus* and dressed in tunic and pallium.

Fig. 18 **AGNUS DEI**
ROME. Basilica of SS. Cosma e Damiano. Detail of the lower band of the apse: *Agnus Dei* (first half of 6th c.).

The mystical Lamb is on the mount from which flow the four rivers of Paradise; beside it, two of the twelve sheep representing the Apostles, which advance - six from one side, six from the other - from the symbolic gates of the cities of Bethlehem and Jerusalem. Above the Lamb flows the river Jordan, which surrounds the land. The rendering of the animals' fleeces by scaly stripes is purely conventional, anti-naturalistic.

Fig. 19 **ALBANIA (CAUCASUS)**
LEKIT (near). Plan of the tetraconch church (7th c.?).

The building, part of a vast complex, shows close analogies with the tetraconch church of Zvart'noc (mid 7th c.) and is built of river stones, a masonry technique peculiar to this region.

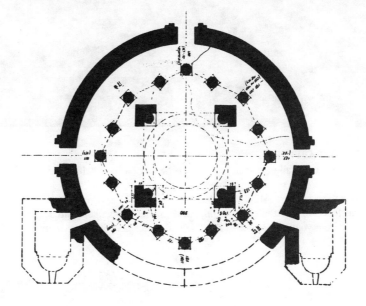

Fig. 20 **AMBROSE**
MILAN. Basilica of S. Ambrogio. Chapel of S. Vittore in Ciel d'oro. Detail of St Ambrose (second half of 5th c.).

The saint is at the base of the dome, on an intense blue background: it seems a very expressive portrait, with his triangular face framed by a short beard, large ears, long nose, melancholy eyes and full mouth. A different official iconography of A. would be propagated only in the following centuries.

Fig. 21 **ANANIAS AND SAPPHIRA**
BRESCIA. Museo Civico. Detail of rear side of ivory reliquary (second half of 4th c.).

To the left St Peter, bearded, with short curly hair and features by now established, wearing tunic and pallium, sits on a *subsellium* with his feet on a footstool. Before him stands S., her hair gathered up on her head; at her feet is the sack with the money. To the right, four youthful figures in short, belted tunics take away the inert body of A., stiffened into an unnatural position.

Fig. 22 **ANGEL**
ROME. Catacomb of Via Latina. Fresco on an arcosolium of a cubicle ([c.] mid 4th c.).

The a. is barring the way to Balaam on his ass (Num 12, 21-23). The divine messenger, still without wings in accordance with the older iconography, brandishes a short sword in his right hand, while his left is covered by his pallium; he is bearded, with great eyes, as is the prophet. Their faces are turned not towards each other, but towards the spectator.

Fig. 23 **ANNUNCIATION**
RAVENNA. Quadrarco di Braccioforte. Right short side of the sarcophagus "of the prophet Elisha" (or Pignatta sarcophagus) (late 4th c.).

To the left sits Our Lady wrapped in tunic and very light palla, lifting a great spindle with her left hand; at her feet is a wicker basket full of wool. To the right the angel, with wings unfolded, wearing tunic and belted overgarment, makes the gesture of announcement.

Fig. 24 VATICAN CITY. Museo Pio Cristiano (from the Chàpel of the Sancta Sanctorum). Silk cloth (7th-8th c.).

Inside a circular frame with inner floral decoration, the winged, haloed angel in tunic, pallium and sandals approaches from the right, raising one hand in the gesture of announcement, while holding a *baculus* with the other. Our Lady, seated on a jewelled throne, is spinning, as in the account of pseudo-Matthew (IX, 2).

Fig. 25 **ANTIOCH IN SYRIA**
NEW YORK. Metropolitan Museum. The Cloisters Collection: so-called "Antioch chalice", silver (Syria, first half of 6th c.).

On the chalice, an object probably belonging to the Church of Antioch (perhaps a lamp, according to the most recent proposals), the seated figure of Christ appears twice, and those of ten Apostles. The two Apostles seated on a *cathedra*, low down, have been identified with Peter and Paul. Identification of the scenes, set among a close network of vine-leaves and grape-clusters, with birds, is disputed: they are probably the Risen Christ and Christ Teaching.
The "chalice" was found at the start of the 19th c., with other liturgical objects, near Antioch.

Fig. 26 ANTIOCH. Mosaic museum. Pavement mosaic of the *pistikòn* of the *martyrium* of St Babylas (5th c.).

The mosaic inscription, of 420-429, hence later than the baptistery of the *martyrium*, records bishop Theodotus, who with the priest Athanasius commissioned the mosaic, which was executed under the deacon and *paramonarius* Akkiba.

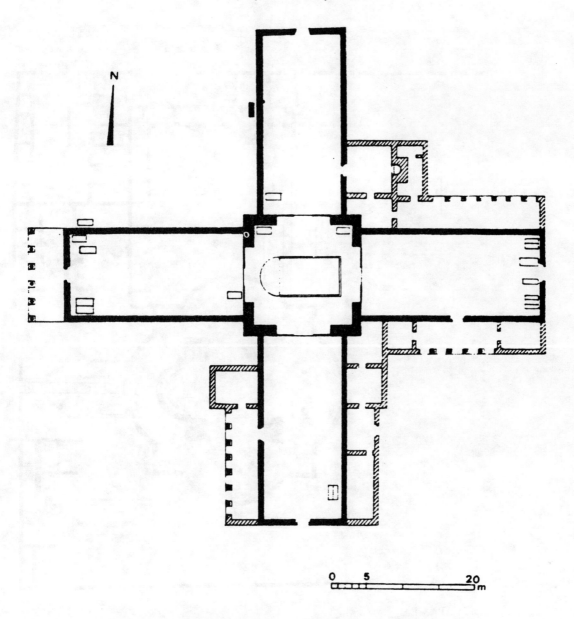

Fig. 27 **ANTIOCH IN SYRIA**
KAUSIYE. Plan of the *martyrium* of St Babylas (381).

The building, in a free Greek cross plan, stood in a suburb of Antioch and was dedicated to the martyr Babylas, whose body was translated there in 381 (perhaps the earliest known translation).

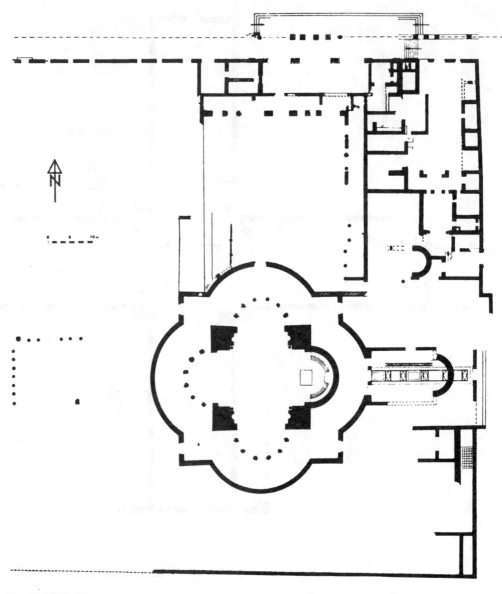

Fig. 28 **APAMEA**
Plan of the episcopal complex (5th and 6th c.).

The church has a double tetraconch plan (a quatrefoil with ambulatory) peculiar to the religious architecture of Syria, where in the majority of cases this type of building had the functions of an episcopal church. Beginning probably as a *martyrium*, according to Balty's hypothesis it became the cathedral of Apamea only in the 6th c., under bishop Paul (before 540), to whom some dated inscriptions, found *in situ*, attribute the rebuilding of the edifice, and in particular the paving.

Fig. 29
Ambulatory of the tetraconch church: fragment of pavement in *opus sectile* (6th c.).

Fig. 30 **APOSTLE**
BARLETTA (Bari). Museo Civico. Funerary relief in three fragments (early 5th c.).

The sculpture, of Constantinopolitan origin and coming perhaps from Canosa, must have belonged to the facing of a masonry tomb or to a monumental sarcophagus. Christ and the Apostles, animated by a gentle movement and in unusual attitudes, are identified by captions (whole names or abbreviations). From the left: Matthew, Simon, Andrew, Thomas, Paul, Christ, John, Peter, Philip, Luke, Mark, Bartholemew. Here as elsewhere Paul and the two evangelists are included among the twelve. All are in tunic and pallium and hold scrolls in their hands. At Christ's feet, a small female figure whose attitude resembles that of the woman with an issue of blood may also be identifiable as a deceased woman.

Fig. 31 **AQUILEIA**
The episcopal complex.

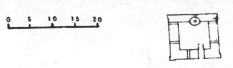

The work of bishop Theodore († 319?), it was essentially divided into two halls. It is "the oldest architectural evidence of an organized church in Northern Italy", equipped with a baptistery, whose exact position has recently been identified, adjacent to the South hall. The North room was probably the original seat of the assembly of the faithful, while the South one functioned as a catechumeneum. However, between the late 4th and early 5th cc., both halls were replaced by two basilical constructions. In particular the South hall was linked with a new baptistery.

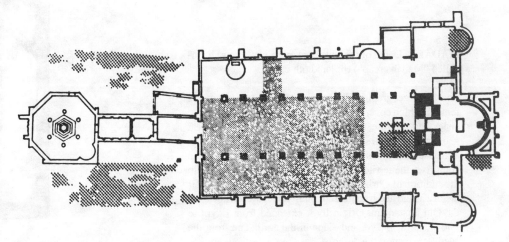

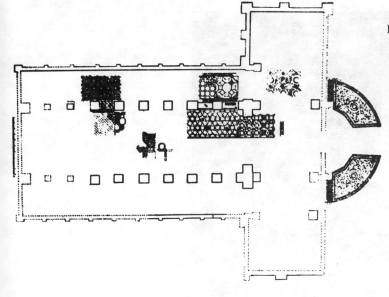

Fig. 32 Basilica of the Beligna or of the "fondo Tullio".

Situated to the south, outside the old city walls, it was originally dedicated to the cult of the martyrs. Between the 5th and 6th cc. there rose alongside it a monastery dedicated to St Martin, a dedication perhaps referable to the anti-Arian reaction, championed at the turn of the 5th c. by bishop Chromatius.

Fig. 33 **ARABIA**
EZRA. *Martyrium* of St George: exterior.

Fig. 34 EZRA. Plan of the *martyrium* of St George (515).

The church, an octagon inscribed in a square, with projecting apse, has a wooden dome supported by eight pillars. The masonry is in ashlar of local basalt, a building technique peculiar to Southern Syria.

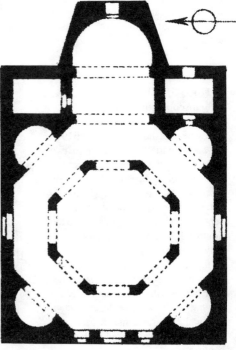

Fig. 35
MADABA. Detail of the pavement mosaic of a church (so-called "Madaba map") ([*c*.] mid 6th c.).

The great "geographical" mosaic of which ample fragments remain, which covered the pavement of a church with three aisles and an apse, is an important document of biblical geography, though seen in the light of the Christian reality, at whose centre was the "holy city of Jerusalem". The map comprises the territory of the biblical Twelve Tribes, with the addition of Christian sites, whose inclusion attests their historical and geographical continuity with the territories of the Old Testament. Originally it extended from Egypt to Phoenicia, with Tyre and Sidon in the north, and from the Mediterranean to the desert.

Fig. 36 **ARABIA**
GERASA. The cathedral and apse of the *martyrium* of St Theodore (late 4th - late 5th c.).

The grandiose complex rose near the south side of the temple of Artemis, and was connected to the *cardo maximus*.
The three-aisled cathedral church was built perhaps at the close of the 4th c.: an atrium with a miraculous fountain at the centre, still existing, linked it with the later church of St Theodore, a *martyrium* built between 494 and 496 on the same axis as the cathedral. The complex also includes numerous rooms for the clergy and a small baptistery.

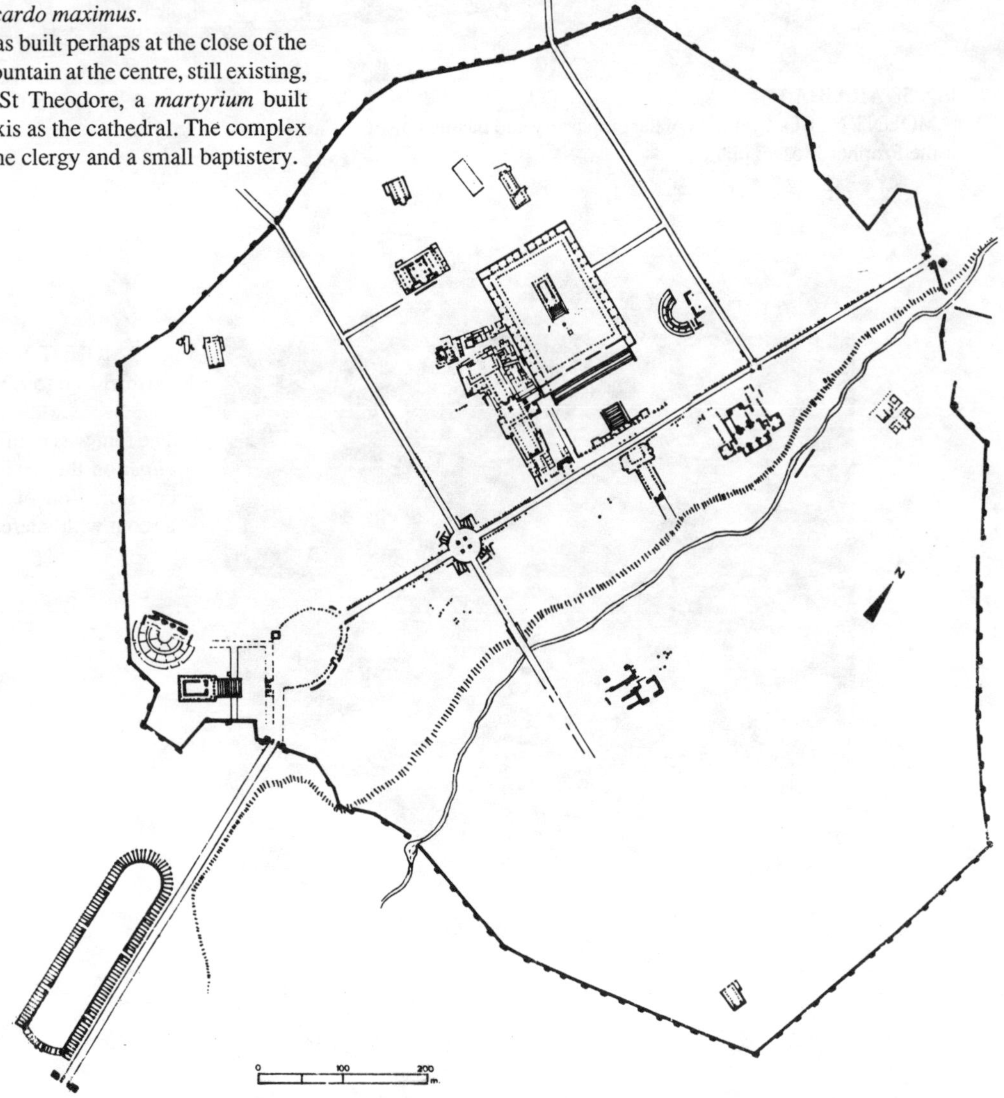

Fig. 37
GERASA. Plan of the city.

The plan of the old city, of Hellenistic origin but greatly developed in Roman times, shows a colonnaded *cardo* and two *decumani*, between which were distributed the city's monuments, first pagan and then, from the 4th c., the numerous Christian buildings. With the Christianization of the city, the grandiose pagan temples were closed but not destroyed, and in the same area rose some of the most important Christian religious complexes. So the Christian city occupied the same space as the pagan one, and reused its stones, sculptures and monuments: cases of cities built *ex novo* are very rare in this period.

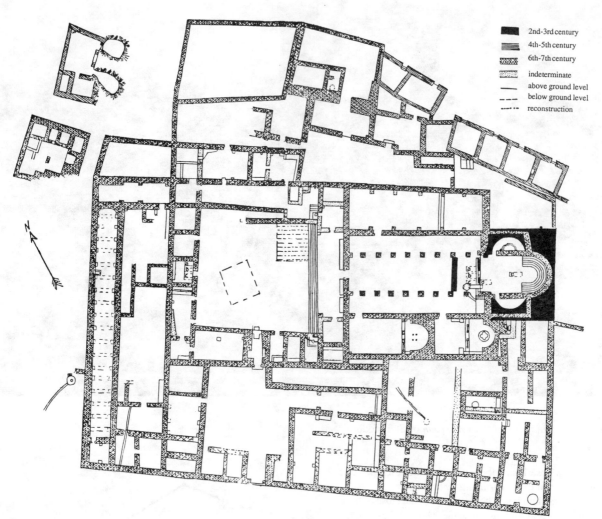

Fig. 38 **ARABIA**
MOUNT NEBO. Complex of the monastery and basilica of the Prophet Moses: plan.

Fig. 39 MOUNT NEBO. Basilica of the Prophet Moses: baptismal font (597).

The font was built in 597, as the inscription incised in *clipei* on the outside attests, and replaced an earlier cruciform font of 531, situated on the opposite side, in a room with interesting pavement mosaics.

Fig. 40 **ARLES**
Plan of the city.

The *civitas romana* of Arles had an episcopal see reliably attested from the mid 3rd c. The local martyr known to us from tradition is St Genesius. Among other places of worship, the plan shows the site of the primitive cathedral, within the walls near the S-E corner (transformed into the women's monastery of St Caesarius), and the site of the second cathedral of St Trophimus, which existed in the mid 5th c., but under the name of St Stephen. It was near the city centre, almost opposite the cryptoporticos. The plan also shows the two important cemetery areas of Trinquetaille, where St Genesius died, and Les Alyscamps, where he was buried, as was St Trophimus, later transferred to the pre-Romanesque cathedral.

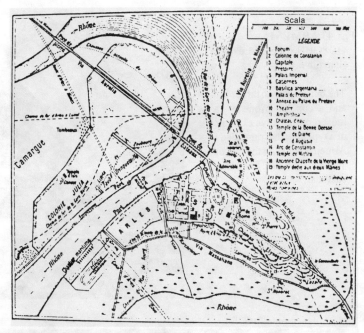

Fig. 41 Columned sarcophagus with biblical scenes and an *orans* at the centre of the lower level (second half of 4th c.).

Fig. 42 **ARMENIA**
YEREVAN. Library. Ečmiadzin Gospels, of 989 (Matedanaran 2374). Adoration of the Magi ([*c.*] 6th-7th c.).

Executed in the monastery of Noravank', the Ečmiadzin Gospels contain, at the end, four older miniature pages, dated to the 6th-7th cc., together with the fine ivory binding showing the Virgin and Child. Originally attributed to Syria, these miniatures are now seen to have close parallels with contemporary Armenian painting.

Fig. 43 **ARMENIA**
ODZOUN. Cathedral: exterior (late 6th c.).

The building, with three aisles and a cupola, is one of the most representative examples of a typology that prevailed in Armenia between the late 6th and the 7th c., i.e. the domed basilica.

Fig. 44 ODZOUN. Cathedral: interior; detail of the cupola.

The cupola with drum, always over a square space, is set on four corner squinches, the so-called "Armenian squinch" that preceded the pendentive, whose use prevailed in the region in the 7th c.

Fig. 45
EREROUK. Plan of the basilica (5th c.).

With three aisles and enclosed semi-circular apse, the basilica of Ererouk shows greater affinity than any other Armenian building with Syrian early Christian architecture, including the presence, unusual in Armenia, of two towers on the façade and the so-called "Syrian band".

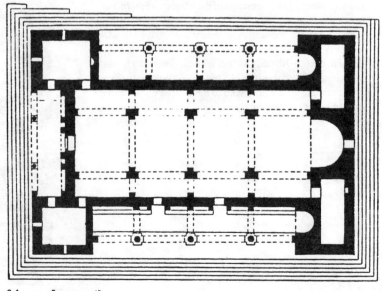

Fig. 46 **ARMENIA**
ZVART'NOC. Plan of the church of St Gregory the Illuminator (mid 7th c.).

This is a theophanic *martyrium* rising over the place where, according to legend, Gregory the Illuminator, the evangelizer of Armenia, encountered the angels. The church, a tetraconch with ambulatory, is attached to an episcopal palace. Built under the episcopate of the pro-Greek Armenian catholicos Nersetes III the Builder (641-661), it shows considerable typological affinities with the great Syrian tetraconchs. It must further be said that in Armenia we know other similar buildings, all attributed to the patronage of Nersetes III.

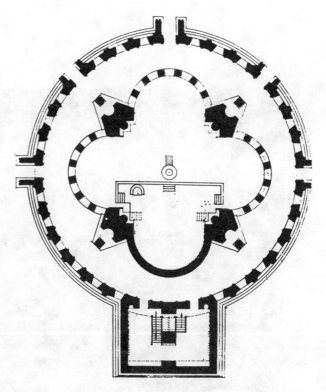

Fig. 47 ZVART'NOC. Church of St Gregory the Illuminator: fragment of arcade with human figure.

The fragment, a sort of enclosing pendentive, like others found in this building and at Dvin, a human figure, must be referred to the system of arcades, with functions pre-eminently bearing rather than decorative, which animated the building's circular external wall.

Fig. 48 **ASCENSION**
MUNICH. Bayerisches Nationalmuseum. Ivory tablet (5th c.).

Bottom left, the angel sits on a rocky spur, in tunic and pallium; with his right hand he makes the gesture of speech towards the three Maries who arrive in dejected attitudes, dressed in tunic and *palla*; behind the angel is the tomb on which lean two Roman soldiers; in the background, a great tree on which sits a blackbird; to right the haloed Christ, in tunic and pallium and with a scroll in his left hand, ascends to heaven, holding the divine hand which comes out of a cloud; at his feet two Apostles show evident surprise.

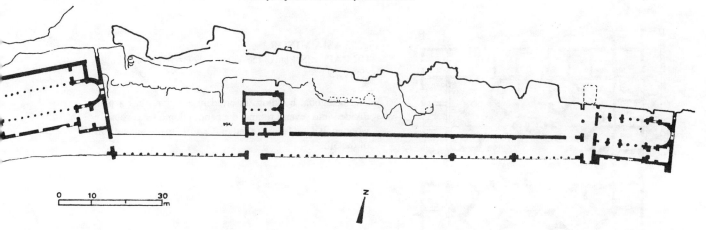

Fig. 49 **ASIA MINOR**
ALAHAN MONASTIR. Plan of the monastic complex (late 5th c.).

Situated on the side of a hill, the complex, reached by a gate whose richly decorated posts remain, consists of two churches and a two-aisled baptistery.

Fig. 50 ALAHAN MONASTIR. Domed basilica: exterior (6th c.).

Fig. 51 ALAHAN MONASTIR. Baptistery of the monastic complex: detail of the cruciform baptismal pool.

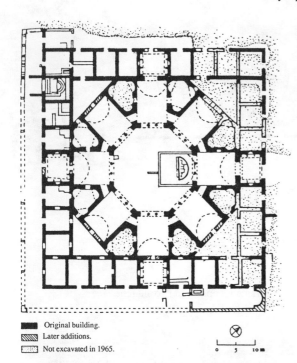

Fig. 52 **ASIA MINOR**
HIERAPOLIS. Plan of the octagonal *martyrium* of St Philip (5th c.).

The octagonal building, inserted in a square, has a plan divided into several rooms: the central dome was probably wooden. It dates from the end of the 4th c. to the second half of the 5th.

■ Original building.
▨ Later additions.
▦ Not excavated in 1965.

Fig. 53 SARDIS. Christian church situated in the S-E corner of the temple of Artemis (late 4th c.).

The edict of Milan on the closure of pagan temples (390) led to the phenomenon, documented even earlier, of the transformation of pagan temples into Christian churches, which intensified in this period. This is the case of the small single-aisled building, situated in the temple of Artemis, of around the end of the 4th c.

Fig. 54 **ATHENS**
Stoa of Hadrian (late 4th - 5th c.).

The building is in the form of a double tetraconch, of the type attested in the Balkans in the early Christian era. Once interpreted as a civil construction, it is now considered a place of worship, perhaps the city cathedral. The early dating rests partly on the stylistic analysis of the pavement mosaics in geometrical and floral designs that decorate it.

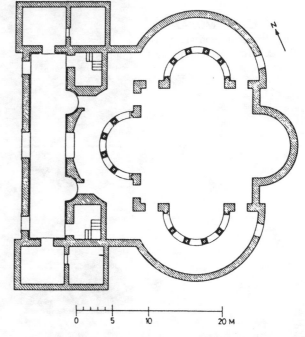

Fig. 55 **ATHENS**
Tetraconch in the Stoa of Hadrian: detail of the mosaic pavement.

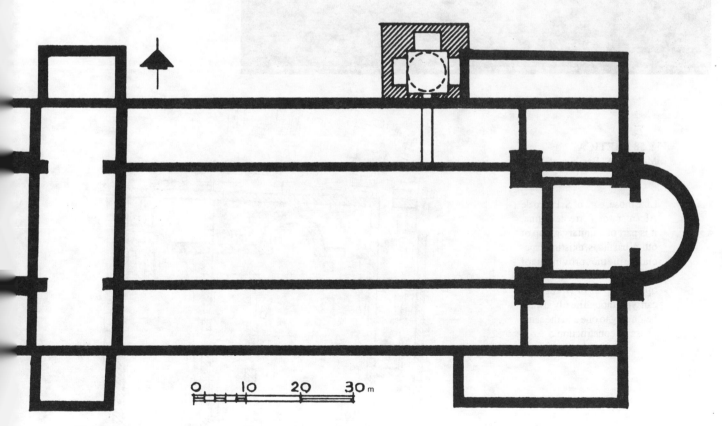

Fig. 56
Plan of the basilica of St Leonidas (Ilisos) ([c.] mid 6th c.).

Fig. 57 **AUGUSTINE**
ROME. Library of the Sancta Sanctorum in the Lateran. Fresco (late 6th c.).

The earliest image of the saint, with the characteristics of a portrait. A. sits in the attitude of a teacher, on a great folding chair with high backrest and large arms. His expression is suffering, his features marked. His appearance is that of an old man, with sparse grey hair and a short beard that frames his face. He wears a *tunica clavata* and pallium; his sandalled feet rest on a high footstool. On a lectern is an open *codex* with schematic lines of writing; A. lays his right hand on it, while his left clasps a scroll. It is more the depiction of a Doctor of the Church than of a saintly bishop. Beneath the scene, which is silhouetted on a red-brown background, a painted inscription says among other things that the saint "treated everything in the Latin language, proclaiming mystical sentences in a voice of thunder".

Fig. 58 **BAETICA**
MALAGA. Basilica of Vega del Mar.

Like those, e.g., of S. Pedro de Alcantara or Torre de Palma, it is part of a unitary group of 6th-c. basilicas, existing especially in the provinces of Baetica and Lusitania, with opposed apses of North African type, though nearly always belonging to one and the same phase of construction.

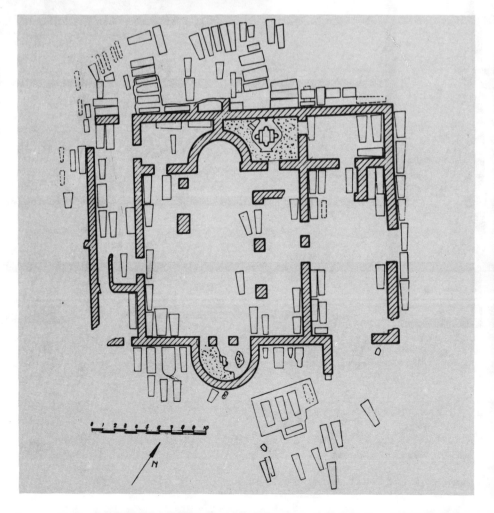

Fig. 59 **BALEARI**
MAJORCA. Basilica of S. Peretò (5th-6th c.).

One of the island's three known basilicas. Its plan is of Syrian type mediated by North Africa, with three aisles (in which are the remains of late mosaics) and tripartite head, and concluded by a baptistery. The baptistery has two baptismal pools of the same importance, in contrast to what we generally meet in other Spanish examples, in which the second pool is smaller.

Fig. 60 **BAPTISM**
ROME. Wall fresco of a cubicle of the catacomb of S. Callisto (first half of 3rd c.).

A man in tunic and pallium lays his hands on the head of a naked figure, of smaller dimensions, with its arms along its body, immersed in water up to the calves.

Fig. 61 **BAPTISM**
WALESBY, near Market Rasen, Lincolnshire. Font (4th c.). Photography by Lincoln City Libraries, Museum and Art Gallery.

The complete font measured about 1 metre in diameter and 50 cm. in depth; it shows a scene distributed probably on three horizontal panels separated by four columns. The right-hand panel shows three standing male figures, in mantle and tunic, a scene probably repeated in the lost left-hand panel. The central panel represents a naked woman between two more women richly veiled and draped. According to the most convincing interpretation, it shows the act of baptism and the point of view is that of the officiating cleric or bishop, who is close to the font, opposite the candidate, who is receiving baptism inside the font, accompanied by her godmothers.

Fig. 63 **BETHLEHEM**
Plan of the church of the Nativity (4th and 5th cc.).

Built by Constantine († 337) on the traditional site of the Cave of the Nativity, the building originally had a five-aisle plan, characterized by an unusual apsidal octagon at the end. Later, with the reconstruction carried out by Justinian (527-565), the presbyterial zone was modified, with the addition of three apses.

Fig. 62 **BETHLEHEM.** Church of the Nativity: right apse (6th c.).

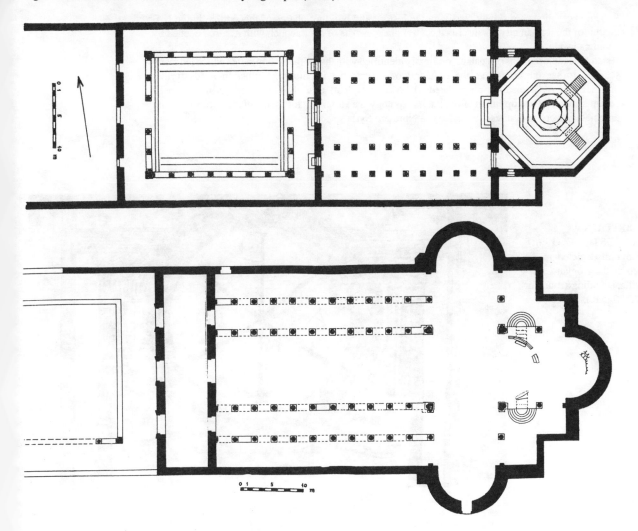

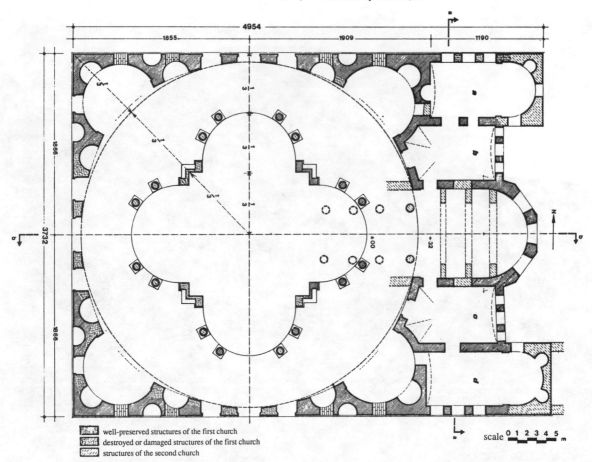

Fig. 64 **BOSTRA**
Plan of the church of Sts Sergius, Bacchus and Leontius (512-513).

Recent excavations are slowly clarifying the history and building phases of this edifice in the N-E quarter of the Roman city: its inscribed central plan consists of a tetraconch with *exedrae* opened by columns inside.

It stands near the remains of a palace, originally identified with that of the bishop, a fact which led many scholars to see the church as the cathedral of Bostra: but its liturgical function needs further clarification. A Greek inscription dated 512-513 (now lost) described the work as having been commissioned by bishop Julian of Bostra. Its typology is a variant on the theme of the tetraconch with ambulatory, very widespread, especially in Northern Syria.

Fig. 65 **BRITAIN**
LULLINGSTONE (Kent). Christogram depicted in the portico of the Roman Christian villa, including a room for worship (4th c.).

Fig. 66 **BRITAIN**
HINTON ST MARY (Dorset). Roman Christian villa including a room for worship: mosaic pavement (detail).

Perhaps the first representation of Christ in Britain (late 4th - 5th c.).

Fig. 67a, 67b Casket of St Cuthbert: the two extremities.

St Cuthbert's casket, the *levis theca* (Bede, *Vita Cuthberthi*, ch. 42) in which the saint was laid in 698, is the one carved wood object of any importance to survive from Anglo-Saxon England. On the outside of the lid is incised a figure of Christ with the symbols of the evangelists at the four corners; at the narrower end of the coffin are the archangels Michael and Gabriel, and at the wider end a representation of the Virgin and Child.

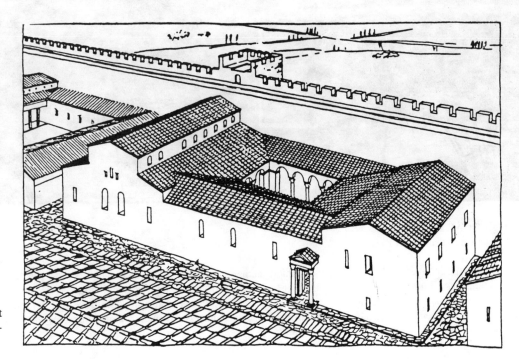

Fig. 68 **CALLATIS**
Reconstruction of the front view of the "Syrian-type" basilica (5th-6th c.).

Fig. 69 **CANA, wedding-feast of**
MILAN. Cathedral Treasury. Five-part gospel cover (rear side) (second half of 5th c.).

In the central panel the jewelled cross on Mount Paradise is framed by an architrave upheld by columns; the symbols and busts of the evangelists Luke and John inside wreaths are set at the four corners. In the upper level, horizontally, is the adoration of the Magi. To the left, gospel episodes: healing of the blind man, miracle of the paralytic, raising of Lazarus. To the right, Christ on the globe between two saints, the Last Supper, the widow's mite. Below, the wedding-feast of Cana: Christ Apollonian and beardless, surrounded by the Apostles, in front of three great jars lays the *virga* on one of them, working the miracle and turning into wine the water poured out by the slave on the right. The scene appears more articulate here than in the more "summary" iconography of the sarcophagi.

Fig. 70 **CARTHAGE**
Basilica of Damous el Karita.

Cemeterial complex near Carthage, divided among numerous buildings. The fulcrum consists of a basilica of unusual dimensions, of nine aisles (or eleven if we look at the apse from East to West), with porticos, *trichorae* and apsed oratories. It also includes within itself an underground rotunda, which Delattre proposes to see as the tomb of St Cyprian (5th c.). At the end of the semicircular atrium and connected with it is a trifoliate oratory, probable *memoria martyrum*, which was built earlier than the great complex.

Fig. 71 **CARTHAGE**
Basilica Maiorum Arearum.

Cemeterial complex dedicated to St Perpetua and her companions. The basilica had several aisles. The altar was probably situated above a small apsed crypt in which the martyrs' relics were preserved (5th c.).

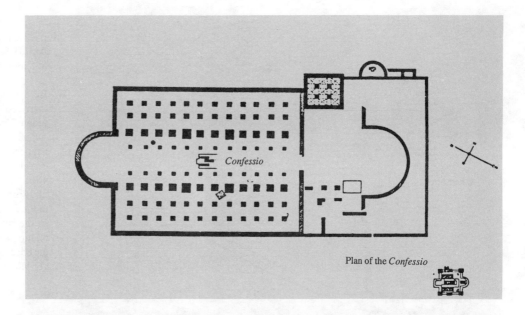

Plan of the *Confessio*

Fig. 72 **CATHEDRA**
ROME. Catacomb called "Cimitero Maggiore". Cathedra in tufa (4th c.).

Cathedra completely dug in the tufa and provided with back and arms, probable symbol of the invisible presence of the deceased at the funeral meal celebrated in his honour.

Fig. 73 **CEMETERY**
AMITERNUM (Abruzzo). Catacomb of S. Vittorino.

Here was buried the priest Victorinus, martyred, according to his Passion, at the time of Nerva, on the Via Salaria, at the *Aquae Cotiliae*. Perhaps, it more probably happened in the 3rd or early 4th c. The cemeterial area in question predates the 4th c. The martyr's tomb was transformed into a monument by a bishop Quodvultdeus in the 5th c.

In the plan we also see the catacomb's relation to the basilica, first mentioned in the 8th c. and later dedicated to the archangel Michael.

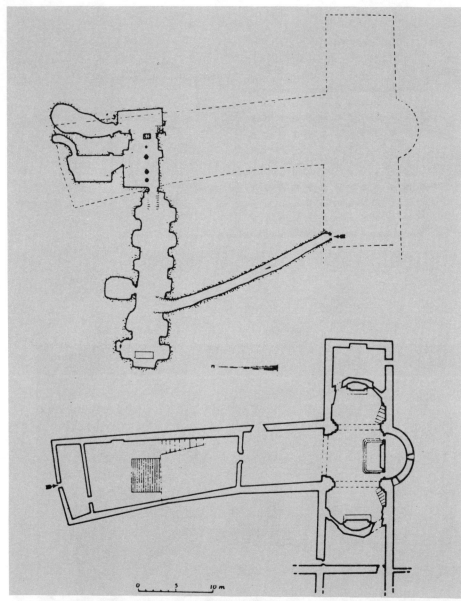

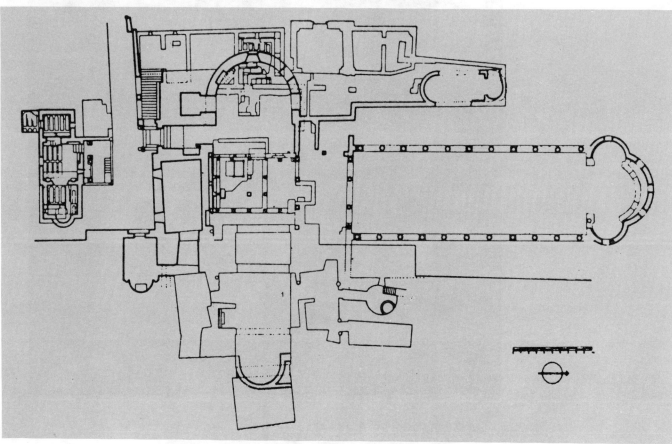

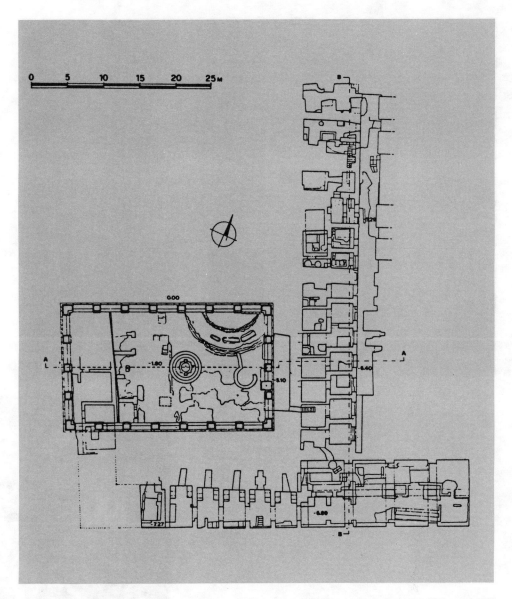

Fig. 75 **CEMETERY**
NAPLES. Catacombs of S. Gennaro.

A complex, which was always accessible, of two hypogea and cemeterial basilicas, originating from numerous primitive nuclei. Among the latter the most imposing and splendid is the so-called "lower vestibule" (late 2nd - early 3rd c.). Another important nucleus can be identified in the room which later (second half of 3rd c.) held the tomb of bishop Agrippinus. The whole of the lower catacomb seems older than the upper one. The latter preserves the city's earliest examples of pictorial art, apart from a rich mosaic decoration. In this area (upper catacomb) was recently discovered the crypt (5th c.) of the Neapolitan bishops. The relics of St Januarius were translated there in the 5th c., and the earliest depiction of the martyr dates from the same period.

◀ Fig. 74
CIMITILE (Nola). Plan of the monumental complex.

Its fulcrum was the tomb of the martyr Felix, organized by Paulinus of Nola.

Fig. 76 **CEMETERY**
Detail of the lower catacomb.

Fig. 77 Detail of an arcosolium with mosaic portrait of local bishop.

Fig. 78 ROME. Catacomb of Panfilo. Detail of the painted cubicle (4th c.).

The cubicle (in other words, a sepulchral chamber) is entirely decorated with frescos, both on the walls and the ceiling. The space is subdivided by wide red bands, inside which are festoons and flowers, panels in imitation marble, biblical characters (the Good Shepherd at the centre of the vault). The most common types of tombs in the cubicle are arcosolia and loculi.

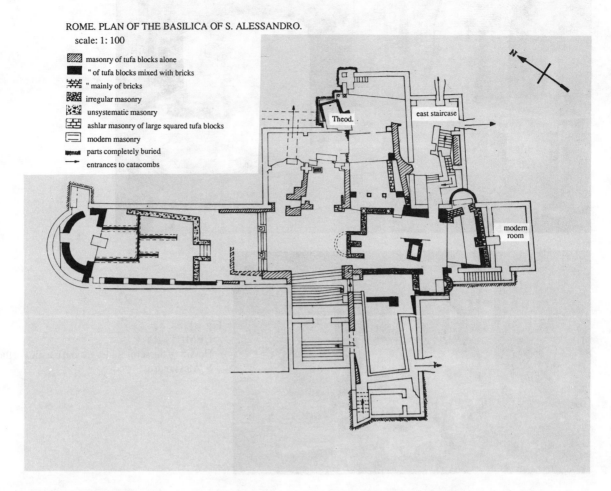

Fig. 79 **CEMETERY**

ROME. Catacomb of S. Alessandro. Plan of the entire complex dedicated in particular to the martyrs Alexander, Eventius and Theodulus.

The plan shows the South sector of the complex, dedicated to the memory of Alexander and Eventius, who were buried together. The cemetery was at the 7th mile of the Via Nomentana and was in the diocese of Ficulea and Nomentum. The arrangement of the remains of the two martyrs of the South sector was the work of bishop Ursus (*Christi signifer Ursus*), who probably ruled that diocese at the start of the 5th c.

Fig. 80

ROME. Catacomb of the Jordani.

Cubicle with arcosolia in the primitive region. Above note modern restorations; at the bottom the development of a later gallery.

Fig. 81, 82, 83
CEMETERY
Details of the semi-underground basilica of the catacomb of S. Alessandro.

Fig. 86
ROME. Catacomb of Priscilla. A gallery on the first floor.

The most common types of tombs in the underground cemeteries are loculi, dug horizontally in the tufa walls and closed by marble slabs, tiles or heterogeneous filling materials, sometimes with a coat of plaster. On the right is an arcosolium, a tomb surmounted by an arched (or rectangular) niche.

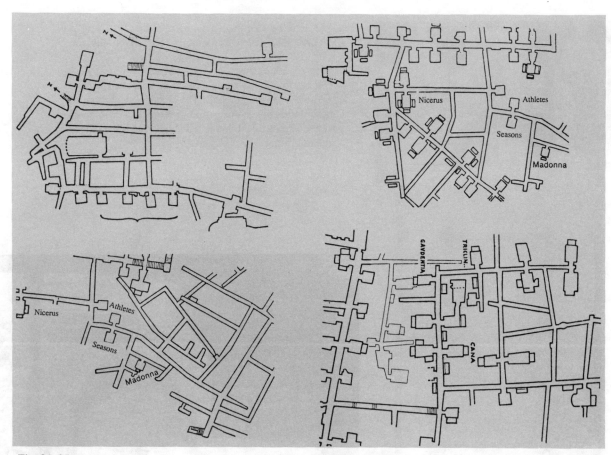

Fig. 84, 85
CEMETERY
ROME. Catacomb of SS. Pietro e Marcellino (Via Labicana): regions between mid 3rd and mid 4th c. Plan of some nuclei of the vast hypogeum, whose surface area contains a large circular basilica and the mausoleum of Constantine's mother Helena.
Region X. Comprises, among other things, the cubicle of Binkentia whose figured decoration is clearly allusive to the *refrigerium*.
Region Y. Among the best known of this region is the cubicle of Nicerus, probably of the last quarter of the 3rd c.
Region Z. Perhaps the most important in the catacomb, especially for the cubicles called "of the Madonna and the Magi" and "of the Seasons".
Region of the Agapai. So called because banquet scenes predominate there.

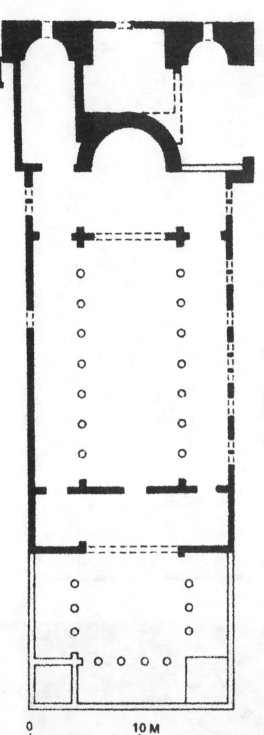

Fig. 87 **CILICIA**
KORYKOS. So-called "extra-muros" church (5th-6th c.).

This three-aisled church is characterized by a presbytery which has two apsed rooms projecting near the apse and closed off by a straight wall. The architectural school of Cilicia, flourishing in this period, is distinguished by some peculiarities both in the architectural field and in the refined sculptural fittings.

Fig. 88
KORYKOS. So-called "extra-muros" church: detail of the internal fittings.

Fig. 89 MERIAMLIK. Plan of the *martyrium* of St Thecla (late 5th c.).

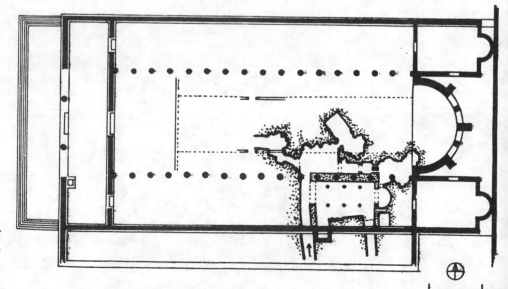

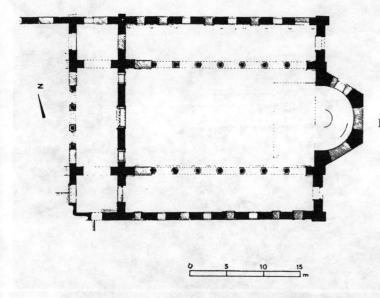

Fig. 90 **CONSTANTINOPLE**
Plan of the monastic church of St John of Studios (450).

The building was founded by the patrician Studios. It had three aisles, preceded by an atrium, and an apse polygonal outside and semicircular inside. Above the side aisles were galleries, reached from the narthex. Inside, note the rich paving in *opus sectile* and the architectural sculpture mainly marked by the use of the precious marble *verde antico*. Near the area of the apse opened a small cruciform crypt, analogous in form and function to that of the contemporary church of St Maria Chalcoprateia.

Fig. 91 Apse of the church of St John of Studios: exterior.

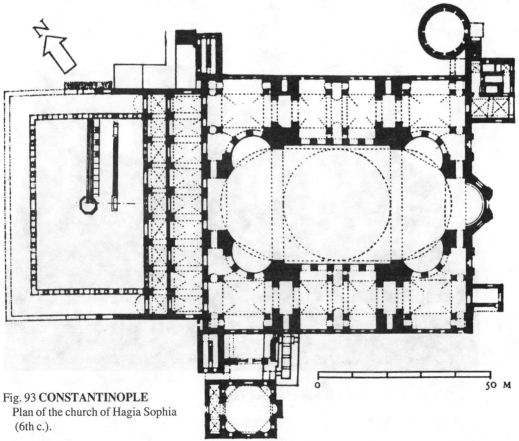

Fig. 93 **CONSTANTINOPLE**
Plan of the church of Hagia Sophia (6th c.).

The church was built under Justinian (527-565) after 532, the year when the previous building was destroyed during the Nika revolt. The present building was in fact preceded by two phases: the oldest dates from 360, when it was begun under Constantius II (337-361). Its plan is unknown, but it would seem that the *skeuophylakion*, which still exists, should be attributed to this phase. Destroyed by a fire in 408, the church was rebuilt under Theodosius II in 415: of this building, part of the trabeation decorated with twelve lambs, to be referred to the broken tympanum of the main façade, still remains *in situ*. The rebuilding under Justinian, entrusted to the Asia Minor "architects" Anthemius of Tralles and Isidore of Miletus, aimed to express, in the grandiosity and richness of the project, exalted by the plan which happily combined the basilica with the dome, and in the rich decoration, the new theocratic ideas of power.

Fig. 92 **CONSTANTINOPLE**
Archaeological Museum. Pillar capital from the church of St Polyeuctes (524-527).

The church of St Polyeuctes was founded by princess Juliana Anicia near the aqueduct of Valens. Its exact plan is unknown, since excavations have uncovered only the framework of the building, perhaps domed.
The church's extraordinary architectural decoration is almost a *unicum* even in the variegated picture of Constantinopolitan sculpture of the period, showing notable affinities, both in tectonics and in iconographical motifs, with Sassanid art.

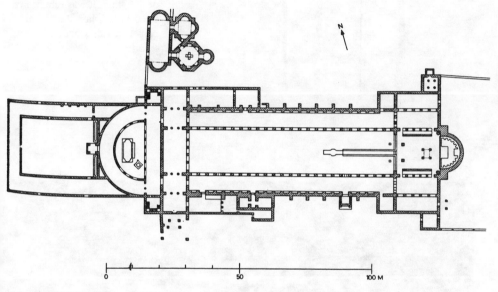

Fig. 94 **CORINTH**
LECHAION (near Corinth). Plan of the basilica of St Leonidas (5th-6th c.), situated in the port.

The construction of this vast church, one of the largest in Greece, begun around the middle of the 5th c., was completed in the reign of Justinian (518-527). The basilica had three aisles with a five-part transept and numerous attached rooms. The central zone of the presbytery was covered by a dome, probably of wood, resting on a canopy consisting of four great pillars.

Fig. 95 LECHAION (near Corinth). View of the baptistery of the basilica of St Leonidas.

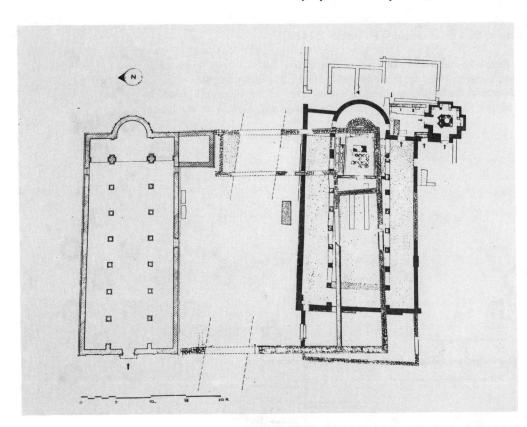

Fig. 96 **CORSICA**
MARIANA (Lucciana). Plan of the early Christian basilica and baptistery.

The ancient Corsican city of Mariana was founded as a Roman colony in the 1st c. BC by Marius, on still older beginnings. The Christian origins of the site and the chronological order of the bishops are not easy to follow. But they are attested by the basilica with its baptistery, whose first phase dates from the end of the 4th c. It was rebuilt in the 5th c., when the mosaic carpet was also renewed.

Fig. 97 **CROSS**
FLORENCE. Biblioteca Laurenziana. So-called Gospel of Rabbula: Crucifixion (second half of 6th c.).

The full-page miniature shows Christ at the centre of the scene, bearded and haloed, with the *colobium* as in the Oriental iconography, crucified with four nails. He is flanked by the two thieves, their loins girt with the *subligaculum*. To the left Mary and John, to the right the pious women in a group, at the foot of the cross three soldiers contending for the Lord's clothing. The lancebearer and spongebearer are beside Christ, not yet dead, framed in the background by the sun and moon, between two rocky crags. Below, the Resurrection: from the open tomb issue rays of light that terrify the soldiers; on the left, the angel speaks to the pious women; on the right, the women beside the Risen Christ.

Fig. 98 **CROSS**
ROME. Basilica of S. Sabina. Top left panel of the wooden door, with Crucifixion (first half of 5th c.).

Christ is flanked by the two thieves, in smaller dimensions: the crosses are hidden by their outstretched arms. They wear only a *subligaculum* around their loins; in the background, a wall with three *tympana* divided by pilaster strips frames the scene and alludes to the city wall of Jerusalem. This is one of the oldest depictions of the Crucified Christ: bearded, eyes open, he looks to the right towards the condemned, at the moment before death.

Fig. 99 **CRUCIFIX**
Fragment of Crucifixion scene incised on stone. The Manx Museum, Isle of Man.

This fragment belongs to an altar frontal. It was found in Calf of Man, the islet near the Isle of Man (Great Britain). 8th c.; height 76.5 cm.

Fig. 100 **CYPRUS**
KITI. Church of the Panaghia Angeloktistos: apse vault: Virgin and Child between Michael and Gabriel (mosaic) ([c.] early 7th c.).

Fig. 101 NEW YORK. Metropolitan Museum. Silver plate with Samuel anointing David (7th c.).

Coming from Cyprus, the well-known series of liturgical plates with stories of David was produced probably at Constantinople under Heraclius between 613 and 629-630, as the seals impressed on the back indicate. They were probably executed to commemorate the victory over the Persians by Heraclius, here compared to the biblical hero.

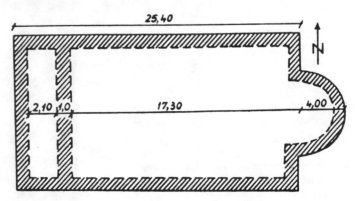

Fig. 102 **DACIA TRAIANA**
SUCIDAVA. Plan of the little one-aisled basilica (6th c.?).

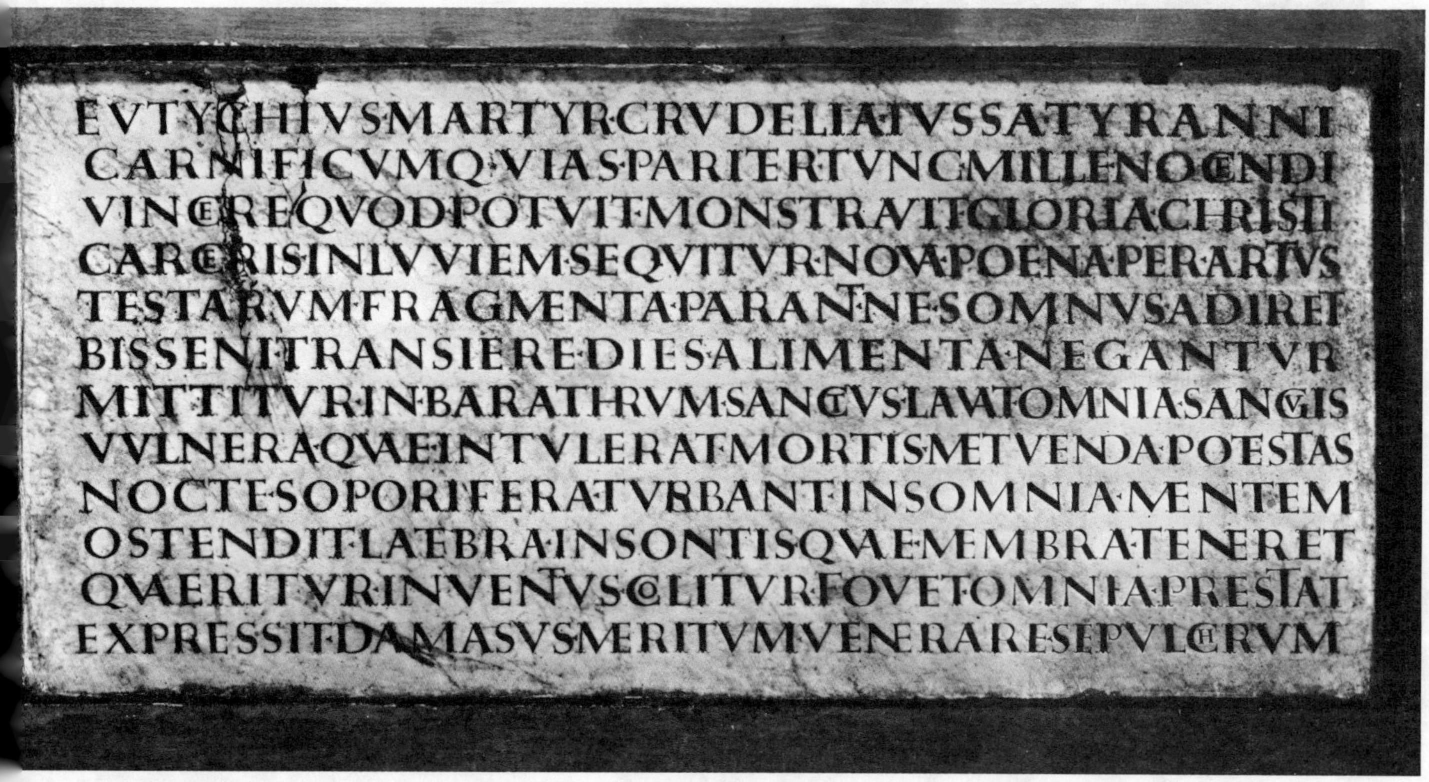

Fig. 103 **DAMASUS**
ROME. Basilica of S. Sebastiano. Inscription of the martyr Eutychius (Damasian, 366-384).

The only stone incised by the pontiff's orders to have survived entire and unbroken (only the triangular punctuation marks are arbitrary 17th-c. additions). It is also considered one of the finest examples of Filocalian script. The twelve hexameters of text describe the martyrdom of Eutychius - an otherwise unknown character - first imprisoned, then subjected to torture of various kinds, which caused his death probably under Diocletian. According to Damasus's account, his tomb was found after a revelation in a dream; it is now up to the faithful, after having read the poem, to venerate his memory.

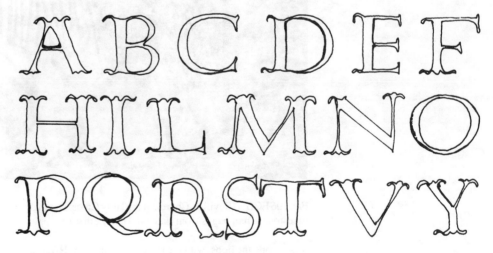

Fig. 104
Examples of Filocalian alphabet.

The particular characters used for many Damasian inscriptions were invented for the purpose by the calligrapher Furius Dionysius Filocalus. A beautiful squared capital, with strong *chiaroscuro* effects created by triangular grooves; the letters are inscribable in squares or rectangles and always furnished with curved apexes, the P's and R's have their upper eyelets open. On each slab a preliminary *ordinatio* of the texts was made, arranged with extreme regularity; to make all the lines end in the same place, expedients were often used such as contractions, linking, insertion of small letters inside larger ones. It has been calculated that some 20 Damasian epigraphs were incised in Filocalian characters.

Fig. 105 **DANIEL**
ROME. Catacomb of Via Anapo. Niche fresco. D. among the lions (late 3rd - early 4th c.).

The prophet D. is a model of wisdom and hope in God, who freed him from death, and so he expresses hope in salvation and resurrection from death. The iconography of D. condemned to the lions' den was widespread from the 3rd c.

Fig. 106 ECIJA (Seville). Church of the Holy Cross. Detail of limestone sarcophagus front (late 5th - early 6th c.).

D. among the lions, with the Greek caption ΔΑΝΙΗΛ, is at the right extremity of the front: praying, in short unbelted tunic and mantle, his hair is short and curly. At his sides the lions sit on their hind paws, with heads turned towards the prophet. The den is indicated by a band that passes behind D.'s back and delimits a rectangular space.

Fig. 107 **DIPTYCH**
MONZA. Cathedral Treasury. Ivory diptych of Stilicho (5th c.).

The consul Stilicho, *magister militum*, is on the right tablet in military costume, with embroidered chlamys buckled at the right shoulder by a large fibula and short, sleeved tunic belted at the waist; he has a sword at his side, while on his round, bossed shield appears the double portrait of the emperors of East and West. In his right hand he holds a lance. On the left tablet is his wife Serena, daughter of the emperor Theodosius I, with her young son Eucherius: she is richly dressed and jewelled and in her left hand she holds a *mappula*, while her raised right hand bears a flower. Eucherius has the dignity of an adult, in tunic and chlamys, and probably with a scroll in his left hand. In the background is symbolic architecture. The faces stand out by their genuine individual characterization.

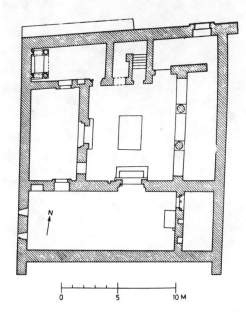

Fig. 108 **DURA EUROPOS**
Plan of the *domus ecclesiae*.

A rare example of a 3rd-c. *domus ecclesiae*. In the house, which is arranged around a courtyard, we can distinguish among the various rooms the baptistery, with baptismal pool and canopy and decorated with frescos, and a hall for the liturgical synaxis. The city of D. E. was destroyed by the Sassanids in 256 and the *domus* must date from some decades earlier, as a readaptation of a pre-existing private house.

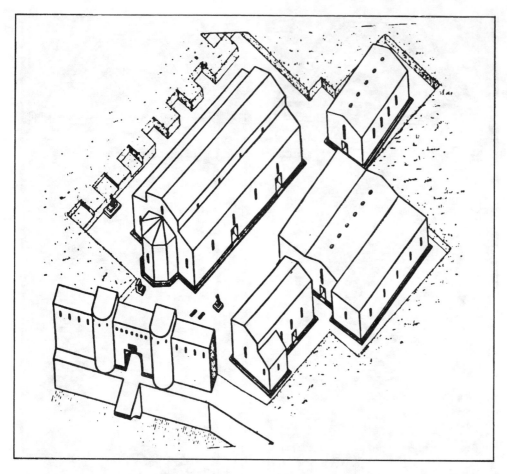

Fig. 109 **DVIN**
Reconstruction of the main quarter in the 5th c.

Fig. 110 **EDESSA**
(Urfa). Fragment of pavement mosaic (5th-6th c.).

E. was an ancient city and metropolitan see of Osrhoene. The sources (the Chronicle of Edessa and the Stylite Chronicle of Joshua) attest the antiquity of the Christian presence and of numerous church buildings present in the city from the start of the 3rd c. Through such sources, in particular the mediaeval ones, it is possible to reconstruct the topography of the Christian city.

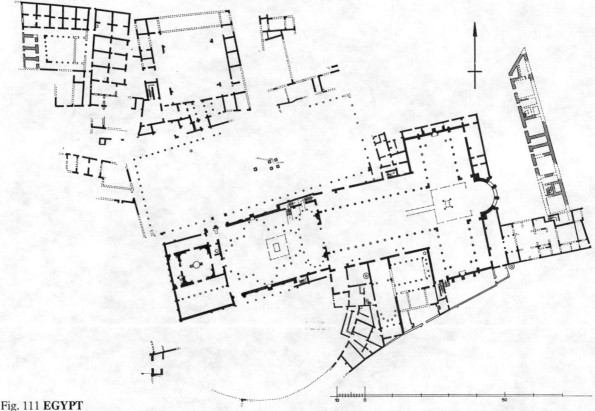

Fig. 111 **EGYPT**
ABU MINA. Plan of the cruciform church (late 5th c.) and the buildings belonging to the great pilgrimage centre.

The present phase of the great cruciform basilica, whose original construction is attributed by the sources to the emperor Arcadius (395-408), must date, following discoveries that emerged after the recent excavations, largely from the end of the 5th c. To imperial patronage - Arcadius and then Zeno (474-491) - must probably be attributed the particular typology, unusual in the region, and the great octagonal baptistery, placed in continuation of the church's main axis, both elements that recur, with cogent analogies, in other important contemporary complexes, the cruciform basilica that arose over the Marneion of Gaza and the church of St John at Ephesus (Selçuk).

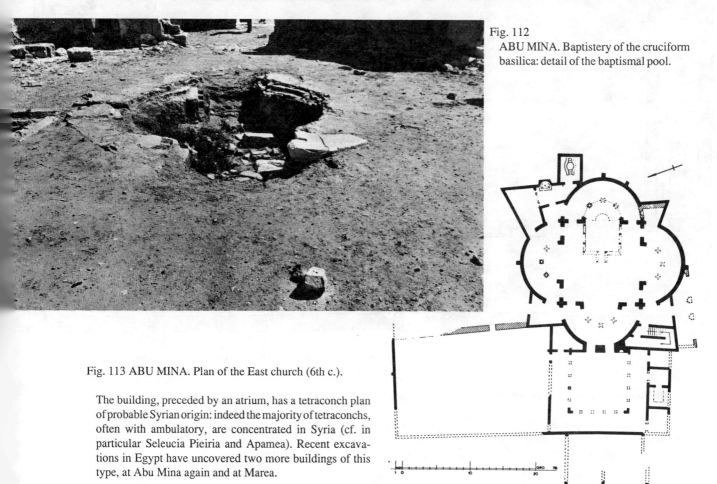

Fig. 112
ABU MINA. Baptistery of the cruciform basilica: detail of the baptismal pool.

Fig. 113 ABU MINA. Plan of the East church (6th c.).

The building, preceded by an atrium, has a tetraconch plan of probable Syrian origin: indeed the majority of tetraconchs, often with ambulatory, are concentrated in Syria (cf. in particular Seleucia Pieiria and Apamea). Recent excavations in Egypt have uncovered two more buildings of this type, at Abu Mina again and at Marea.

Fig. 114 **EGYPT**
SOHAG (near). Church of the monastery of Shenoute: detail of the decoration of a niche.

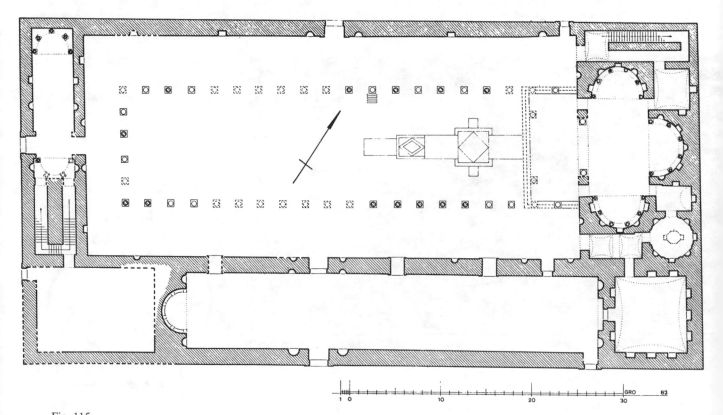

Fig. 115
SOHAG (near). Plan of the church of the monastery of Shenoute (White Convent) (c.440).

The building, of three aisles with columns, preceded by a narrow narthex, is flanked by a long hall running parallel to the south side: characteristic is the accentuation of the longitudinal axis and the triconch termination of the apse. The complex, consisting of several buildings and enclosed in high walls, also contained a baptistery, located behind the apse.

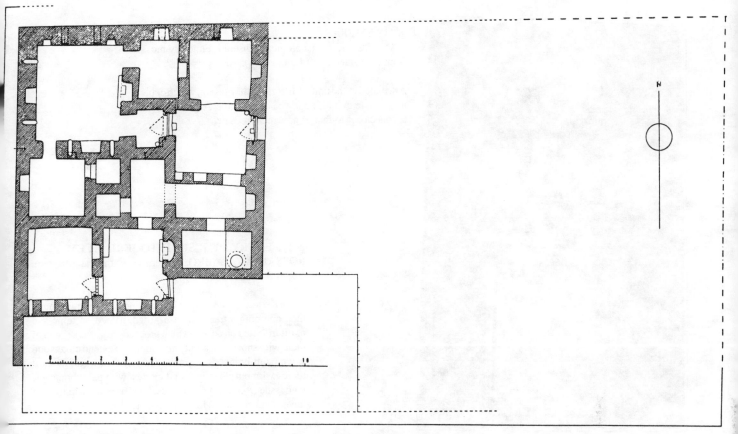

Fig. 116 **EGYPT**
QUSUR AL-RUBAYYAT, Kum K167 (Kellia), 5th/6th-c. phase (drawing by G. Castel).

As time passed the dwellings of the Egyptian anchorites, originally isolated, approached one another to constitute genuine monastic settlements, however modest and low-key compared to the great Egyptian monasteries. Between the 5th and 6th centuries some *kellia* comprised, as well as dormitories for the monks, rooms for prayer and visits, kitchens, etc.

Fig. 117 **ELIJAH**
MILAN. Basilica of S. Ambrogio. Short side of the sarcophagus "of Stilicho" (late 4th c.).

E., of youthful appearance, in tunic and pallium, drives a quadriga and hands his symbolic cloak to Elisha, who accepts it with veiled hands. To right and to left, in the background, the busts of two *filii prophetarum* in the niches of the "city gates" that characterize the scene of the whole sarcophagus. Below, under the horses, Adam and Eve beside the tree of sin, in smaller scale. To right, Noah's Ark amidst the waves of the sea.

Fig. 118 **EMESA**
DAMASCUS. Archaeological Museum. Funerary mosaic from the catacomb of Emesa (6th c.?).

Attributed to the time of Justinian, the mosaic, which depicts two people in tunics, may be older (4th c.?). The technique seems closer to that of pavement mosaic.

Fig. 119 **ENTRY OF JESUS INTO JERUSALEM**
ROSSANO CALABRO (Cosenza). Cathedral Museum. *Codex Purpureus* (6th c.).

From the left: two Apostles in tunic and pallium; two young men climbing a tree; Christ bearded, haloed, in tunic and pallium seated on an ass, which proceeds towards a group of men, with short tunics and *paenulae*, who extend cloaks and wave palm branches; a small group of four boys, in short tunics with purple stripes and circles, with palm branches, who issue from the gate of Jerusalem; from the windows of the city other people look out celebrating.

Fig. 120 **EPHESUS**
SELÇUK (near Ephesus). View of the basilica of St John the Evangelist (527-548).

Fig. 121 **EPHESUS**
SELÇUK (near Ephesus). Baptistery of the basilica of St John ([c.] first half of 5th c.).

Divided into several rooms, the vast complex of the baptistery of St John's, together with the small octagon recently found, perhaps the *secreton* mentioned in an inscription, should probably be ascribed to the pre-Justinianean building: this would also seem to be indicated by the masonry structure in striped work, different from that employed in the construction of the vast 6th-c. building.

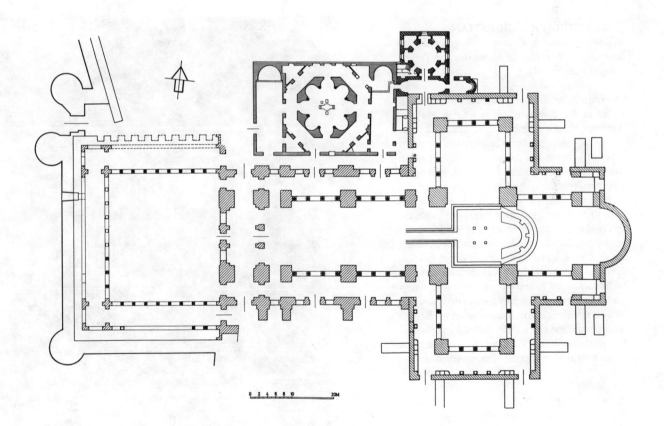

Fig. 122
SELÇUK (near Ephesus). Plan of the church of St John the Evangelist. Justinianean phase.

The grandiose building, which Justinian and Theodora, whose monograms remain on the capitals, had rebuilt from the foundations on the site where the original church stood between the late 4th and mid 5th cc., had a Latin-cross plan with six domes. The earlier building had a similar plan and was of considerable dimensions, as has emerged from excavations.

Fig. 123 **EPHESUS**
Baptistery of the episcopal church of St Mary: interior (before 400).

The octagonal baptistery, situated along the north side of the church, was covered within by marble slabs decorated with crosses: the pool has been excavated at the centre of the octagon.

Fig. 124 **EPIGRAPHY, CHRISTIAN**
ROME. Museo Pio Cristiano. Fragmentary inscription of Abercius of Hierapolis, in Turkey (second half of 2nd c.).

The oldest Christian inscription so far known. It refers to the bishop of Hierapolis in Phrygia, Abercius, already known from a manuscript of the Byzantine hagiographer Metaphrastes. This text was partly repeated in the later stele of Alexander, dated 216. The content is heavily symbolic: Abercius proclaims himself "a disciple of the pious pastor" and speaks of a mission to Rome, and then to Syria and the river Euphrates, with the spiritual assistance of St Paul. It alludes to the strength of faith, to the Eucharistic food (already we find the association of the fish with Christ); it specifies that the epigraph was dictated when the bishop was aged 72 and invites "those who understand these things" to pray for the deceased. Finally it threatens a fine, to devolve on the Roman treasury and the city of Hierapolis, for any violators of the tomb. His journey to Rome took place at the time of the emperor Marcus Aurelius.

Fig. 125 **EPIGRAPHY, CHRISTIAN**
ROME. Catacomb of Panfilo. Sepulchral inscription of *Aproniane* (4th c.).

Her parents dedicate the stone to their sweet daughter *Aproniane*, who died aged five years and four months. The dogmatic content of the text emerges from the last line: "*Aproniane*, you believed in God, you will live in Christ". In this formulation the expression has no other exact comparisons.

Fig. 126 THESSALONICA. Church of St George. Detail of the dome mosaic: St Onesiphorus (5th c.).

Lower band: in the form of a monumental calendar, against a background of a complex architectural scenario consisting of niches, arcades and apsidal conchs, the saints, in an attitude of prayer, are each identified by their name written alongside, with the addition of the Church's monthly feasts in their honour. the architecture recalls the fourth Pompeian style, while a very close link with Hellenism appears in the figures. St Onesiphorus is young, beardless, with thick hair in wavy locks streaked with gold around his oval face; in an attitude of prayer, he wears tunic and chlamys with sumptuous *tablion*, in the Byzantine manner.

Fig. 127 **EPIGRAPHY, CHRISTIAN**
CHIOS (Greek islands). Mastikhari. Church of St John. Votive inscription of mosaic pavement ([*c.*] mid 6th c.).

Eustochiané, shipowner (an important role held by a woman), makes the offering together with the crew of her boat, whose name, Maria, is also specified; they provided to have the whole portico covered in mosaic.

Fig. 128 **EPIGRAPHY, CHRISTIAN**
ROME. Catacomb of Priscilla. Devotional graffiti in the region of the martyr Crescention.

The graffiti of pilgrims are often an effective clue to identifying a venerated tomb. Writings in various hands, at different times, intersecting each other, in many cases present serious difficulties of interpretation. In this stretch of plastered wall we read the name of the martyr (SALBA ME DOMNE CRESCENTIONE MEAM LVCE) and, in a complex text in cursive, a reference to one Priscilla. Besides christograms and names of the faithful, in other graffiti the date of the funeral commemoration is recorded.

Fig. 129 **EPIRUS**
NICOPOLIS. Museum. Detail of a mosaic head in the ambo of basilica B (early 6th c.).

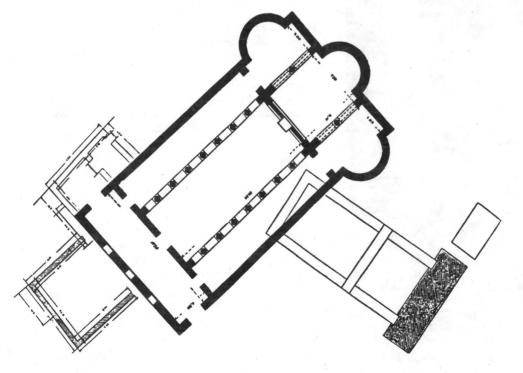

Fig. 130 DODONA. Plan of the basilica (6th c.).

The basilica, of three aisles preceded by a narthex, has a presbytery with three apses arranged to form a sort of triconch: this belongs to an earlier phase, to which the aisles and narthex were added in the 6th c.

Fig. 131 **EUCHARIST**
ISTANBUL. Archaeological Museum. So-called Stûma paten (second half of 6th c.).

Found in 1908 at Stûma in Syria, the paten is in partly-gilt silver. Communion is administered to the apostles by Christ, who appears "doubled" in the act of holding out, to the left, the chalice and, to the right, the bread to the two groups of people at the extremities of the scene. The raised altar on which the rite is celebrated is covered by a cloth falling in fine folds and surmounted by a sort of crescent-shaped ciborium, from which hangs a lamp. The inscription which runs along the outer border faithfully reflects a prayer of leave-taking taken from the liturgy of the Syrian Jacobites.

Fig. 132 **EVANGELISTS**
ROME. Apse mosaic of the basilica of S. Pudenziana (5th c.).

Above Christ enthroned, surrounded by the Apostles and the two personified *ecclesiae*, are the symbolic representations of the four evangelists in the form of the tetramorph, following Ezek 1, 4-14; Rev 4, 6-8. From the left: the angel (Matthew), the lion (Mark), the ox (Luke) and the eagle (John). In the background the fantastic architecture of the heavenly Jerusalem; in the centre of the composition, behind Christ, the jewelled cross, triumphal symbol, on Mount Paradise.

Fig. 133 **FIERY FURNACE**
EGNAZIA (Bari). Early Christian basilica. Clay lamp (4th-5th c.).

The lamp, of "Mediterranean" type, has a shoulder with plant decorations; in the central disc, between the two *infundibula*, is the scene of the three young Hebrews' refusal to worship the idol. From left to right: king Nebuchadnezzar sitting on a seat, the bust of the idol atop a column, a soldier in small dimensions and the three young men, in Oriental dress, with Phrygian cap and wide belted tunic.

Fig. 134 **FLOOD**
ROME. Catacomb of Via Latina. Fresco of cubicle vault ([*c.*] mid 4th c.).

God leans out from a window (*catarracta*) of heaven and pours water on earth as if from a sponge. On the right, two people and trees are submerged. The scene, which seems to be inspired by Gen 7, 10-24, is a *unicum*.

Fig. 135 **FOSSOR**
ROME. Catacomb of SS. Marcellino e Pietro. Fresco on the entrance wall of a cubicle (first half of 4th c.).

The f., in short tunic and headgear, is drawn from behind while delivering a blow of his *dolabra* to a tufa spur; on the right a lantern, lit and hanging from a hook, defines the underground scene.

Fig. 136 **GABRIEL**
RAVENNA. Museo del Arcivescovado. So-called *cathedra* of Maximian. Detail of the front part of the backrest: Annunciation (mid 5th c.).

The virgin, in sleeved tunic, wrapped in a cloak, head veiled by a *palla*, sits on a high wicker seat with footstool: beneath a tympanate roof, she holds a distaff and spindle, while at her feet is the basket of wool, as described by the apocryphal narratives. Gabriel, who addresses her, is youthful in appearance, beardless and with thick curls. In tunic, winged, with his right hand he makes the gesture of speech, and in his left he holds the *baculus*, attribute of wayfarers.

Fig. 137 **GAMMADIA**
RAVENNA. Basilica of S. Apollinare Nuovo. Figure of prophet on the right wall (first half of 6th c.).

The prophet, old, bearded and half-bald, haloed, wears tunic and pallium, on whose hem is the gammadion Γ. In his left hand he holds a closed scroll, on which are indicated letters of an undecipherable symbolic script.

Fig. 138 **GAUL**
SAINT-MAXIMIN-LA-SAINTE-BAUME.
Abraham's sacrifice of Isaac in the basilica of St Mary Magdalen, from the crypt of an early Christian mausoleum (6th c.).

Of all the episodes of Abraham's life, the sacrifice attained greatest popularity in that it served to express God's saving work.

Fig. 139 **GAUL** CLERMONT. Cathedral. Sarcophagus front (first half of 4th c.).

From the left, the miracle of the Petrine spring with the two standing soldiers (one in the background); the healing of a blind man, who leans on a staff and is presented to Christ by a bearded Apostle; the deceased praying between Peter and Paul; the miracle of the woman with an issue of blood, and finally the raising of Lazarus, with Christ seen from the back and the sister Martha standing.

Fig. 140 MONTPELLIER. Sarcophagus front with trees (second half of 4th c.).

From the left, Moses removing his sandals; the miracle of multiplication of loaves, with Christ laying his hand on one of them and touching two more on the ground with his *virga*; a female *orans* with a *capsa* of *volumina* in the background (the deceased or, according to another interpretation, the *Ecclesia*); the miracle of the paralytic; Daniel among the lions; the raising of the son of the widow of Nain. Among the branches of the trees, in the central zone, two birds.

Fig. 141 **GEORGIA**
BOLNISSI. Sion: pilaster capital.

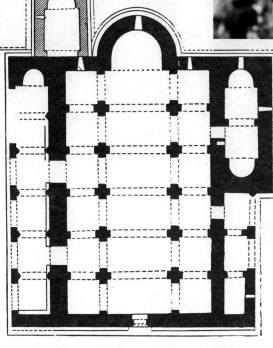

Fig. 142 BOLNISSI. Sion: plan of the basilica (late 5th c.).

The basilica, of three aisles with pillars, and with a small portico attached to the south side, was built - as an inscription, one of the oldest in Georgian, specifies - between 478 and 492-93. In plan and elevation it shows characteristics typical of early Christian building in the Caucasus, like a lean-to façade and cross pillars. Its architectural decoration is of interest, with motifs of Sassanid tradition.

Fig. 143 **GEORGIA**
MZCHETA-DJVARI. Church of the Holy Cross, exterior: the apsidal area.

Fig. 144
MZCHETA-DJVARI. Plan of the church of the Holy Cross (586/87 - 604/05).

Built by prince Stephen, depicted more than once in the external bas-reliefs, on the site where, according to tradition, stood a cross erected by St Nino, the church has a tetraconch plan with corner rooms, and a dome partly supported by the perimeter walls. This typology would have immense popularity in the region, especially in the 7th c.

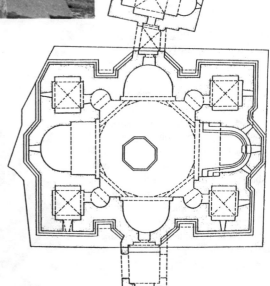

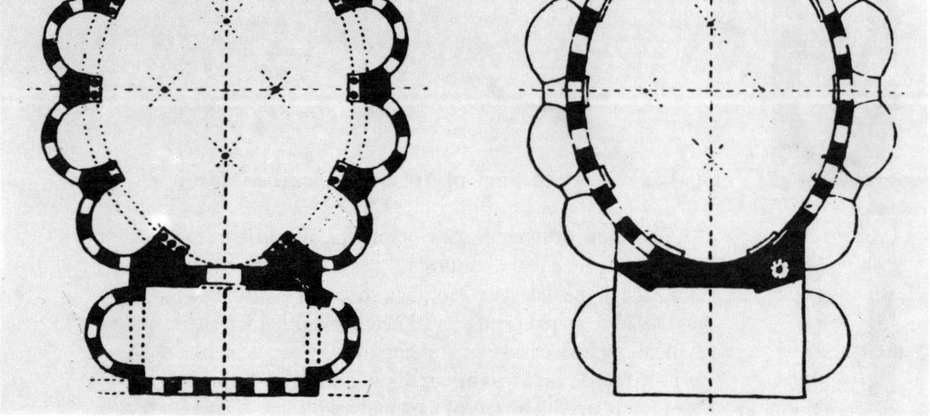

Fig. 145 **GERMANY**
COLOGNE. Basilica of St Gereon ([*c.*] mid 4th c.).

A singular construction with an elliptical central plan, with double and triple columns punctuating the niches and the apse, and polylobate, with "forceps" atrium, lightened by numerous openings. Typologically this church is related to late imperial Roman examples, to other Oriental ones, and finally to Christian buildings such as the mausoleums of Pécs (Pannonia) and Toulouse (France).

Fig. 146 **GESTURES**
ROME. Catacomb of SS. Marcellino e Pietro. Cubicle vault (late 4th c.).

The g. of acclamation is particularly explicit in this painting: in the upper part Christ, haloed and bearded, sits enthroned, with the apocalyptic letters A and Ω at each side of him. Around him Peter and Paul, their features already well defined, acclaim with a raised arm and open hand. The same attitude characterizes, below, the four martyrs identified by captions (Gorgonius, Peter, Marcellinus and Tiburtius), who face the *Agnus Dei*, also haloed, standing on Mount Paradise, from which flow the four rivers of Paradise (Gihon, Pishon, Tigris, Euphrates). The background of the scene is strewn with festoons and flowers.

Fig. 147 **HABAKKUK**
VATICAN CITY. Museo Pio Cristiano. So-called "dogmatic" sarcophagus (325-350).

At the centre of the lower level, Daniel among the lions; to his left H., in short sleeved tunic, offers a basket of loaves signed with the cross (Dan 14, 31-39), an evident allusion to the saving food. To the right: the Petrine trilogy (prediction of his denial, arrest and miracle of the spring). To the left: the offering of gifts by the Magi (the first of them points to the star), the miracle of the healing of the blind man. In the upper level, from the left: the creation of Adam and Eve by the divine Triad, the handing over to them of the symbol of work, a *clipeus* with unfinished portraits of the deceased couple supported by two winged genii, the miracle of Cana, the multiplication of loaves and the raising of Lazarus.

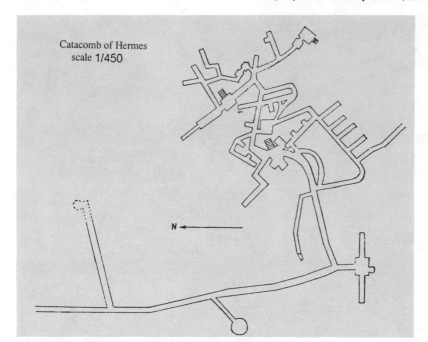

Fig. 148 **HADRUMETUM**
SOUSSE (Hadrumetum). So-called catacomb of Hermes.

One of the underground funerary nuclei of this city, which experienced a considerable development of Christianity between the second half of the 3rd and the 4th c. Like the catacomb of the Good Shepherd, ours is connected with a pagan hypogeum. The catacombs of Hadrumetum have yielded a large number of mosaic tomb slabs, largely preserved in the Bardo Museum, Tunis.

Fig. 149 **HEALING OF THE BLIND MAN**
RAVENNA. Wall mosaic in S. Apollinare Nuovo (6th c.).

From the left: the two blind men in long tunic and *paenula*, leaning on a staff; Christ in tunic and sacred pallium, with crossed halo, places his fingers on the eyes of one of them; to the right a person (Apostle?) in tunic and pallium shows wonder.

Fig. 150 **HIPPO**
Basilica and baptistery.

These form a religious complex, whose origins are traditionally made to go back to St Augustine. At the centre of this church are preserved the mosaics of the peristyle and *tablinum* of an earlier Roman house. The annexes too are carpeted with similarly-derived mosaics. The ornamentation of the paving is altered by the presence of mosaic-covered tombs.

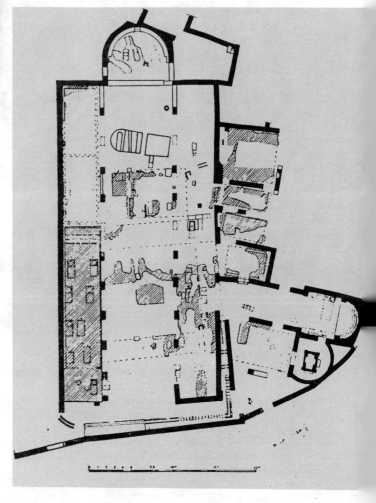

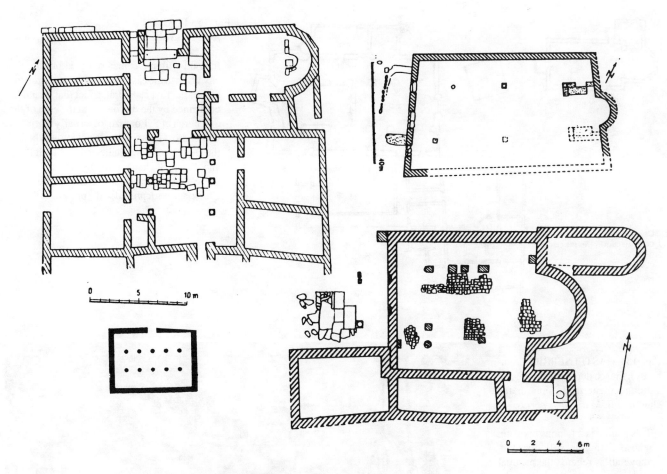

Fig. 151 **HISTRIA**
(Romania). Plan of some basilicas (5th-7th cc.).

Basilicas discovered during the excavations begun in 1914 and still in progress. The most important, that discovered in 1950, some 50 metres east of part of the side of the enclosure, has small dimensions (21 x 12.5 m.), a semicircular apse to the east and a simple, not divided, narthex.

Fig. 152 **ISSUE OF BLOOD**
MILAN. Church of S. Celso. Short side of sarcophagus (5th c.).

The depiction is particularly faithful to the gospel passage (Mt 9, 20-22; Mk 5, 25-34; Lk 7, 43-48). The veiled woman bends down to touch the hem of the pallium of a youthful-looking Christ, who turns brusquely. No other character, no element of landscape appear in the essential representation.

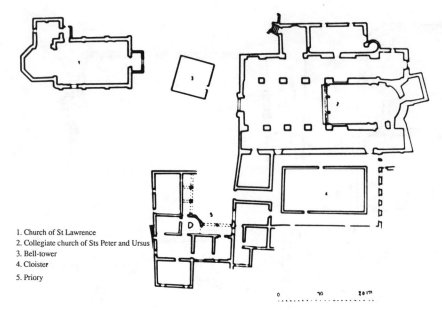

Fig. 153 **ITALY**
AOSTA (Valle d'Aosta). Complex of S. Orso.

It stood within a cemeterial area outside the city walls to the east. It may also have been the seat of the first cathedral, a fact that would be confirmed by the discovery, in the area of the ex-church of S. Lorenzo, of an early Christian Greek-cross church. We do not know whether the church building, dedicated first to St Peter and then to St Ursus, certainly earlier than the 9th c., was erected on the site of the early Christian one or preceded the present building.

1. Church of St Lawrence
2. Collegiate church of Sts Peter and Ursus
3. Bell-tower
4. Cloister
5. Priory

Fig. 154 CASTELSEPRIO (Lombardy). Fresco depicting the angel's announcement to Joseph in the church of S. Maria *foris portas*.

This church is a famous building, whose dating is still disputed, above all as regards the pictorial cycle, variously dated from the 6th c. on.

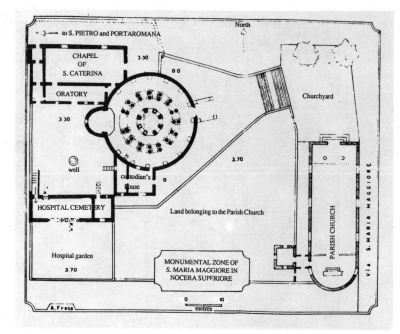

Fig. 155
NOCERA SUPERIORE (or DEI PAGANI) (Campania).

An imposing construction, probably baptismal in character, with a circular plan, comparable to buildings of the same type in Central and Southern Italy (from Constantina's mausoleum in Rome to the basilica of S. Stefano in the same City, from the baptistery of Canosa di Puglia to the Rotunda of S. Angelo at Perugia). The relationship of this building to the church building has not yet been clarified (6th c.).

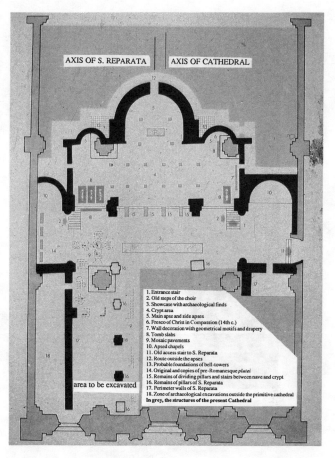

Fig. 156a, 156b
ITALY
FLORENCE (Tuscany). Church of S. Reparata. Detail of the pavement mosaic and plan.

Discovered in the early 1960s, the original position of the city's episcopal church covered part of the area of the present Cathedral. The primitive church building (4th-5th c.) had a single apse and a basilica plan with three aisles. Inside it have been recovered numerous stretches of the ancient mosaic pavement, which shows typical figurative motifs: stylized flowers, cruciform and geometrical figures. In the 8th c. - probably - two apses and two side-chapels were attached to the original nucleus. The entire body of the basilica was finally torn down in 1375 to make way for the present Cathedral.

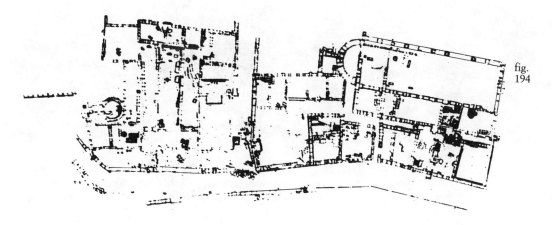

Fig. 157 EGNAZIA (Puglia). Plan of the excavations west of Via Traiana, marking the second basilica discovered on this site during the campaigns of 1969-1970.

The first basilica was uncovered in 1912-1913 and later recognized as being built over two older church buildings, the first of which perhaps belongs to the 4th c. The second basilica can be ascribed to the mid 5th c.: its mosaic pavement has a white background with ornamental and geometrical motifs.

Fig. 158 **ITALY**
TERAMO (Abruzzo). The so-called Byzantine "triforium".

This element was probably part of the entrance to the early mediaeval cathedral of Teramo, structures of which, belonging to a three-aisled basilica, have recently been discovered.

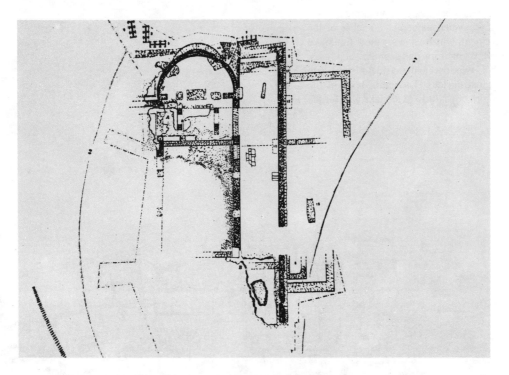

Fig. 159 CASALBORDINO (Abruzzo). Plan of the recently excavated basilica, probably corresponding to the celebrated monastery of St Stephen *rivo maris*, of which a chronicle exists.

The archaeological testing of this settlement dates the foundation of the church to the early Christian period, as also documented by the mosaic decoration. The monastic settlement was probably later, in the early mediaeval period.

Fig. 160a, 160b **ITALY**
CORNUS (Sardinia).

The excavations of Cornus, begun in 1955, are still in progress and have uncovered a cemeterial area, probably beginning in the 3rd c., and an imposing religious complex with two basilicas and a baptistery. The lesser of the two church buildings was then turned into a baptistery. The history of this complex seems to be identifiable with that of the Sardinian diocese of Senafer, whose bishop was present at the council of Carthage of 484.

Fig. 161 **JACOB**
VIENNA. National Bibliothek. Miniatured codex of Genesis (6th c.).

J. crosses his hands and places them on the heads of Ephraim and Manasseh in sign of blessing, as in Gen 48, 14. On his right assist: Rachel in tunic and *palla* worn on her head, and Joseph who tries to change round his father's hands to respect Manasseh's primogeniture.

Fig. 162 **JACOB'S LADDER**
ROME. Catacomb of Via Latina. Right arcosolium of a cubicle ([c.] mid 4th c.).

Jacob, old and bearded, is half lying on a pile of stones with his eyes open. The dream (Gen 28, 10-13) is rendered by the depiction of two young men in smaller dimensions, in tunic and pallium - the angels - who ascend a runged ladder set obliquely across much of the picture.

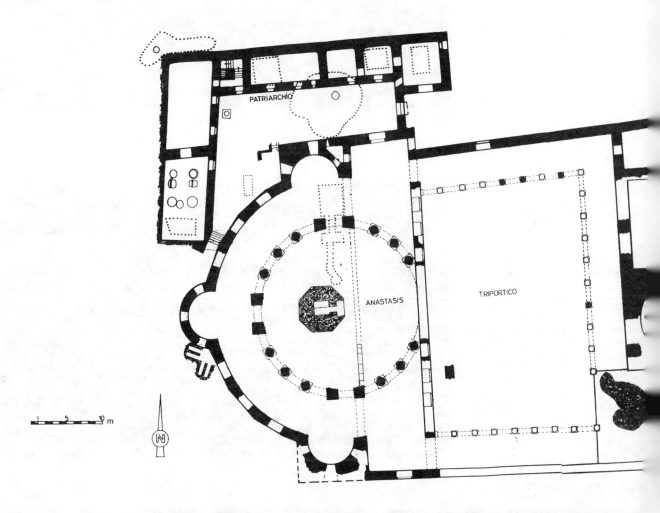

Fig. 163 **JERUSALEM**
MADABA (Arabia). Detail of the pavement mosaic with representation of the city of Jerusalem ([c.] mid 6th c.).

Fig. 164 **JERUSALEM**
Plan of the Holy Sepulchre complex (Constantinian phase, 4th c.).

Built under the patronage of the emperor Constantine († 337) and his mother Helena, at the time of bishop Macarius of Jerusalem, the complex is divided into two main blocks, the Rotunda of the Anastasis and the five-aisled church, the two being separated by a vast atrium. The complex, of which we have a detailed inscription by bishop Eusebius of Caesarea, who took part in its consecration, was dedicated to the Saviour. In the 12th c. it underwent radical rebuilding at the hands of the Crusaders, who destroyed the five-aisled church, replacing it by the so-called "Canons' Choir", which still exists.

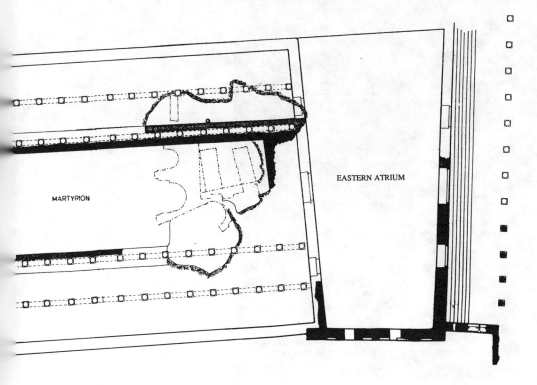

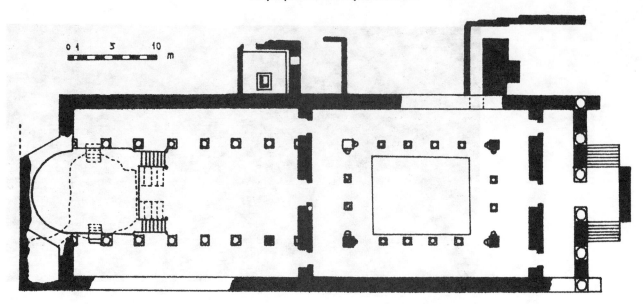

Fig. 165 **JERUSALEM**
Plan of the basilica of the Eleona (4th c.), later rebuilt after the Persian invasion of 614.

Fig. 166 **JESUS CHRIST**
ROME. Museum of S. Sebastiano. Detail of sarcophagus front (first half of 4th c.).

At the centre J. C., shown full-face, in tunic and pallium, with youthful, beardless face, holds a scroll in his left hand and places his right on a basket of loaves, held by one of the two Apostles at his sides. The miracle of multiplication of loaves is further indicated by five more wicker baskets full of loaves marked with a cross, at the feet of the ternary group.

Fig. 167 **JESUS CHRIST**
ROME. Catacomb of S. Sebastiano. Detail of the "Lot" sarcophagus (mid 4th c.).

In the upper level is the scene of Peter's denial; in the centre J. C. in tunic and pallium, youthful in appearance, holds a scroll in his left hand and makes the gesture of speech with his right, turning towards the Apostle who holds one hand to his chin, in sign of sorrow; below is the cock and to the right a person who follows the event. Rare traces of red colouring remain.

Fig. 168 **JESUS CHRIST**
VATICAN CITY. Vatican Grottos. Detail of the front of the sarcophagus of Junius Bassus (359).

J. C., young and "Apollonian", enthroned, has in his left hand a half-unrolled scroll and is flanked by Peter and Paul. His feet rest on a cloth supported by a bearded person, image of *Caelus* and the heavenly vault. In the lower part of the panel is Christ's entry into Jerusalem.

Fig. 169
ROME. Catacomb of Commodilla. Detail of fresco on the vault of the so-called "Cubicle of Leo" (last decades of 4th c.).

At the centre of the coffered ceiling, inside a frame, is the bust of J. C. bearded and haloed, between the apocalyptic letters A and ω, with a hieratic expression foreshadowing the subsequent early Byzantine representations.

Fig. 170 **JESUS CHRIST**
MILAN. Basilica of S. Ambrogio. Detail of sarcophagus front (late 4th c.).

J. C. teaching is enthroned among the apostolic college; youthful in appearance, beardless, with long hair in ringlets, in his left hand he has an open *codex*; in the background, a rich architrave and two palms. At his feet two deceased with hands veiled in sign of deference.

Fig. 171 MILAN. Basilica of S. Ambrogio. Detail of rear side of sarcophagus (late 4th c.).

J. C. standing, in a niche surmounted by an arch, bearded, with long flowing hair, in tunic and pallium, consigns the open scroll to Peter, who carries a cross; on his left is Paul and on both sides the other Apostles acclaiming. In the background the crenellated walls of the heavenly city and at J. C.'s feet two believers in adoration; in the lower socle the *Agnus Dei* among the twelve lambs issuing from the gates of the two cities.

Fig. 172 **JOB**
ROME. Catacomb of Via Latina. Left wall of a cubicle arch ([*c*.] mid 4th c.).

J., in short exomis tunic, leans his right hand on one of the three rocks on which he sits, while his left hand is abandoned along his leg, clearly covered in sores. At his shoulders his wife, wearing just a *tunica clavata*, pushes bread to him with a long stick to protect herself from contagion, but does not hold her nose, as in other similar scenes.

Fig. 173 **JOHN THE BAPTIST**
RAVENNA. Baptistery of the Arians. Dome mosaic (early 6th c.).

On a background of gold *tessellae*, within a *clipeus* with stylized leaf decoration, the naked Christ is immersed to the waist in water; on him descends the dove of the Holy Spirit, while the Baptist places his right hand on his head. John is in a raised place, on a rock, with short tunic of spotted skin, girt at the waist and sleeveless; he has flowing hair, thick beard and holds a *baculus*. On the left: the personification of the river Jordan, with a reed in his right hand and at his shoulders an amphora symbolizing the source of the river. Around them is the procession of the twelve Apostles, facing the empty throne (so-called "etimasìa"). All are haloed, bear in their veiled hands the crowns of their martyrdom and wear tunic and pallium. Ten of them have gammadia on the hems of their overgarments (mainly Z and H).

Fig. 174 **JONAH**
CLEVELAND (USA). The Cleveland Museum of Art. Marble sculpture: Jonah thrown up by the sea-monster (second half of 3rd c.).

The *cetus* is a monster with great fins, serpentine tail, leonine paws, elongated muzzle with curved tusks. J.'s head and bust emerge from its jaws; he is bearded, with thick hair and deep-set eyes. This marble was probably worked in Asia Minor and was perhaps part - like other sculptures of the same J. cycle - of a baptismal context.

Fig. 175 **JONAH**
ROME. Museo Pio Cristiano. Front of "Jonah" sarcophagus. (*c*.300).

The front is divided into two levels by a discontinuous break. The three scenes of the J. cycle occupy the greater part of the space: to the left, the prophet is thrown into the sea by the sailors (above the ship's sail are the bust-personifications of the sun and moon; the sun has puffed-up hair and wears a rayed and indented crown); immediately next to him the sea-monster, with dragon's head and serpentine body, is about to swallow him. The same monster throws him up, in accordance with the biblical account, on the beach of Nineveh. Top right, J. is resting under the booth, gently lying down. Also linked with the theme of water are the figures of Noah in the ark welcoming the dove (in smaller proportions) and the fisherman with the line, next to whom are another small character in *subligaculum* and a goose. In the bottom left-hand corner, two more fishermen, one in short tunic, the other in loincloth, exchange a basket full of fish. In the upper level, the raising of Lazarus, the miracle of the spring and arrest of Peter, unusual in this formulation, also interpreted as the invitation to flee from prison by some believers.

Fig. 176 **JOSEPH (St)**
RAVENNA. Museo del Arcivescovado. Detail of the front part of the backrest of the so-called *Cathedra* of Maximian: the trial of the bitter waters (mid 5th c.).

The episode is taken from the apocryphal narratives: J. and Mary are forced by the Chief Scribe to drink the water of trial to demonstrate their innocence. Mary holds in her right hand an elliptical container with the liquid that emerges from the spring at her feet, at the centre of the panel; J., bearded, holds a *baculus* in his left hand, with his right he perhaps hints at a rebuke to Mary, while the winged angel who faces him attests the Virgin's innocence and convinces him of his error. In the background a wall and the tympanate prospect of a building.

Fig. 177 **JOSHUA**
ROME. Basilica of S. Maria Maggiore. Detail of mosaic of the central aisle (first half of 5th c.).

Above, J., depicted in large dimensions, from the front and in military costume, stops the sun with his raised right hand, as in Jos 10, 12-13. Below, a lively and disordered mass of soldiers are at his sides, in a bold striving for perspective effect.

Fig. 178 **JUDAS**
RAVENNA. Wall mosaic in S. Apollinare Nuovo (6th c.).

In the centre Christ, bearded and with crossed halo, receives the kiss from J., who approaches from the left, also bearded and in tunic and pallium. At his back a squad of armed men, the first of whom has his sword unsheathed. To the right, the group of dejected apostles: Peter, white-haired, is about to draw his sword. The three in first place have the letter Γ on their pallium.

Fig. 179 LONDON. British Museum. Ivory diptych (7th-8th c.).

In the same panel in which Christ's crucifixion is represented, J. is depicted to the left, hanging from a tree.

Fig. 180 **JUDGMENT**
ROME. Catacomb of Pretestato. Fresco on the vault of a cubicle (first half of 4th c.).

At the centre of the medallion a shepherd, in exomis tunic and leg-bands, with a *virga* drives a wild ass and a pig, symbols of evil, away from the flock on his right. Earliest example of the last judgment.

Fig. 181 **KISS**
CASTELLAMMARE DI STABIA (Naples). Cathedral. Ivory comb (?) (6th c.).

Peter and Paul, whose features are well defined, kiss before parting to undergo martyrdom, to symbolize the unity and harmony of the Church. To the left, Paul is partly bald, with thick beard, and wears tunic and pallium, as does Peter, with round head, perhaps the great tonsure, hair covering his forehead, short beard and deep-set eyes. This is the oldest example of this scene and the only one known in this formulation.

Fig. 182 **LAMP**
ROME. Catacomb of Commodilla. Clay lamp (4th c.).

In the central disc, between two *infundibula*, a monogrammatic cross decorated with braided motifs, on which are set two birds in symmetrical position, with their beaks turned upward. They should be doves.

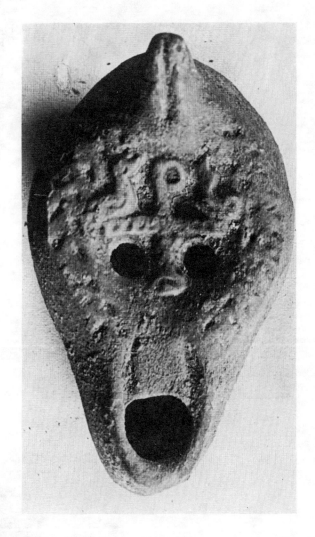

Fig. 183 AQUILEIA. Museo Archeologico Nazionale. Bronze chandelier (4th c.).

The great bronze chandelier consists of a circular crown decorated with bows *"à jour"* with symbolic figures set in them; from the left: the phoenix on the date-palm, the apocalyptic letters and the *Agnus Dei*.

Fig. 184 **LAYING ON OF HANDS**
ROME. Catacomb of Priscilla (3rd c.).

In this fresco, Christ lays his hand on the head of a person of reduced dimensions, perhaps a blind man. The gesture of laying on of hands (or hand) has many values in the development of the sacred rites: exorcism, baptism, reconciliation in the penitential act, confirmation, blessing, ordination to some ministry.

Fig. 185 **LAZARUS**
PARIS. Louvre Museum. Reliquary lid in silver gilt with biblical scenes: the raising of Lazarus (early 5th c.).

L. wrapped in bandages is inside a domed building supported by columns; on the right Christ performs the miracle with a *virga*. At his feet, a kneeling figure wrapped in a cloak: Martha, L.'s sister. A palm bounds the scene to the right. The flat relief and clearly delineated outlines further simplify this composition, already reduced to the essential elements.

Fig. 186 **LEPER**
LONDON. Victoria and Albert Museum. Detail of an ivory diptych (5th c.).

The scene consists of a ternary group in the foreground: to the left Christ, in tunic and pallium, with crossed halo and sandals, makes the gesture of speech towards the l., naked, with dotted traces of his disease on his skin, leaning on a stick and accompanied by a standing woman in tunic and *palla*, with a scroll between her hands, as in the iconography of Job and his wife; in the background two buildings and a tree.

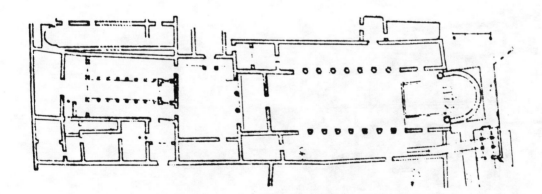

Fig. 187 **LIBYA**
APOLLONIA (Cyrenaica). Plan of the western basilica.

The plan makes clear the readaptation of a previous church, whose orientation has been reversed. For this reason the apse seems to be incorporated and inappropriately inscribed in the city walls, near which the church had been built. An inner staircase in the north aisle constituted a further direct link with the church building. This basilica also came to incorporate the area of the baptistery, which was provided with a furnace to heat the water and had, beyond the atrium, a series of numerous other rooms around a second atrium (5th - early 6th c.).

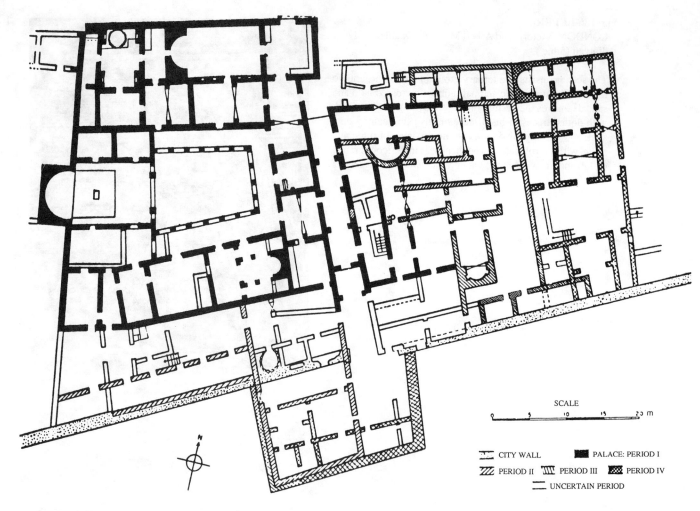

Fig. 188 **LIBYA**
APPOLLONIA (Cyrenaica). Plan of the palace of the Byzantine *dux* (first half of 6th c.).

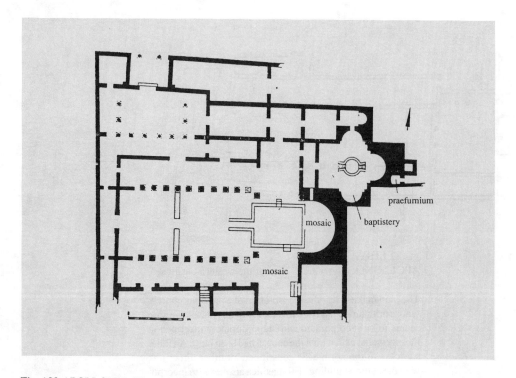

Fig. 189 APOLLONIA (Cyrenaica). East cathedral.

One of the city's oldest basilicas ([*c.*] first half of 5th c.). Perfectly oriented, it has three aisles with side entrances and a transept which distinguishes it from the other basilicas of the area. The baptistery, inserted in place of one of the pastophories, must be connected to a second phase of the building (6th c.), which also involved a second mosaic pavement.

Fig. 190 **LIBYA**
CYRENE. Plan of the cathedral (5th and 6th cc.).

This was a three-aisled building with opposed apses (the western apse was added in the 6th c.), a peculiarity of the religious building of North Africa, whose functions are still not very clear. The case of Cyrene is one of the rare examples of North African influence, given that Cyrenaica, like its Church, generally moved in the orbit of Alexandria and Egypt.

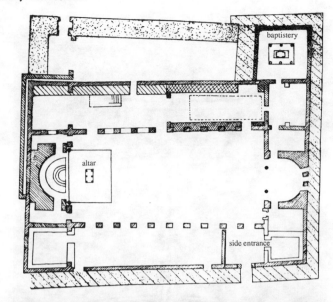

Fig. 191 **LITURGICAL FURNITURE**
MILDENHALL. British Museum photography.

Three silver spoons, with the christogram between Alpha and Omega, part of the treasure discovered in 1942 and hidden around 360. There are two more spoons with Christian inscriptions: one with *Pappitedo vivas* and the other with *Pascentia vivas*.

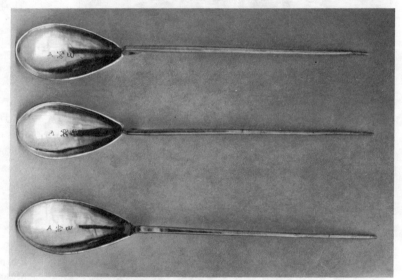

Fig. 192 WATER NEWTON. British Museum photography.

Silver cup from the Water Newton treasure, with the christogram between Alpha and Omega. This is a Christian silver treasure in a particular sense, in that it may have been the property of a church or a Christian community. It was found in 1975 at Durobrivae near Chesterton, Huntingdonshire. The treasure comprises some thirty objects.

Fig. 193 **LITURGICAL FURNITURE**
TRAPRAIN LAW. National Museum of Antiquities of Scotland.

From the Traprain Law treasure: perhaps a chalice, 4th century (height 10.4 cm).

Fig. 194 1. Lamps.
Example of mobile lamp preserved in the West Berlin Museum (6th-7th c.).

2. Patens.
These were great plates for the distribution of communion:
Fig. 195 a) Riha paten.

Fig. 196 b) Stûma paten.

LITURGICAL FURNITURE
3. Reliquaries:

Fig. 197 Vase from Homs.

Fig. 198 Censer from Syracuse.

Reliquary crosses of the emperor Justin.

Fig. 199

Fig. 200

Fig. 201 *Capsella* from Grado.

Fig. 202 **LOT**
ROME. Catacomb of Via Latina. Inner sector of an arcosolium of a cubicle ([*c.*] mid 4th c.).

L., bearded and with the appearance of an old person, flees from Sodom (Gen 19, 15-26); he is in tunic and pallium and holds his two daughters, dressed in *dalmatica*, by the hand. In the background are mountainous outlines with trees and bushes, while to the right his wife, in smaller dimensions, makes a gesture of disappointment with her hand, but is already transformed into a statue of salt. Behind her the city walls, from which come tongues of fire.

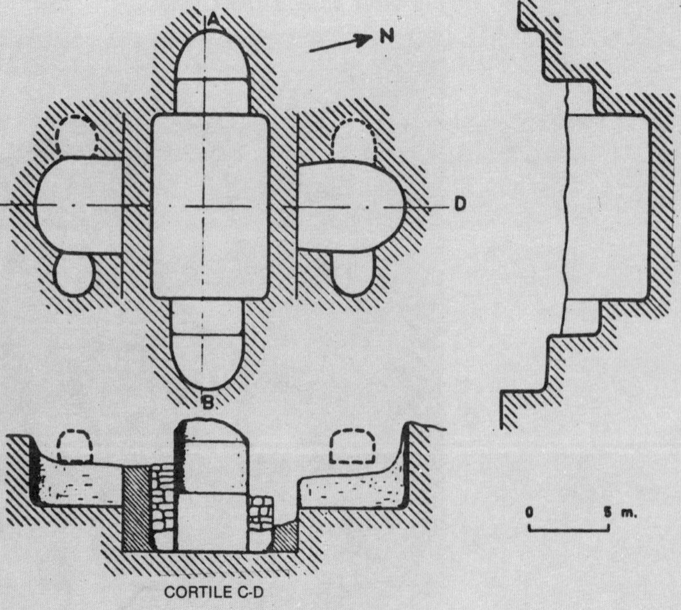

Fig. 203 **LUSITANIA**
IDANHA-A-VELHA (diocese of Egitania [?]), founded during the Suevian domination. S. Almeida. Plan.

Following a pattern typical of Portugal, as well as of the province of Baetica, the baptistery has more than one pool. These buildings in turn are part of religious complexes that arose within great rustic villas (another example is the baptistery of Torre de Palma: see fig. 210). Usually however the pools were not of the same importance or size. The smaller ones seem truly subordinate to the larger ones. The dating of the baptistery oscillates between the 5th and 6th cc.

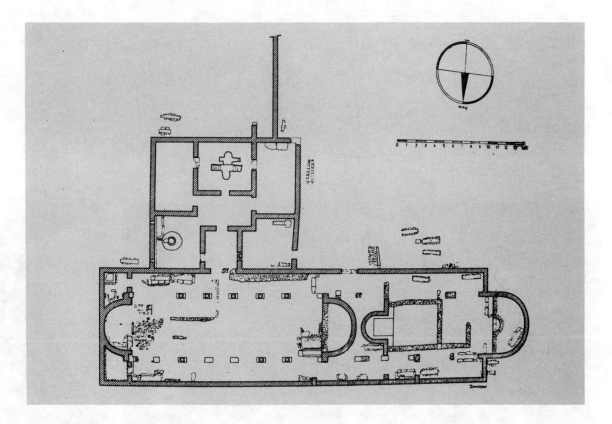

Fig. 204 **LUSITANIA**
TORRE DE PALMA. Basilica complex within a great rustic villa: church with double apse.

A typology with numerous examples in the Iberian peninsula. It is adjacent to a great baptistery with double pool and to various annexes, of which the most important is a sort of oratory, also with double apse, forming a continuation of the main church. 5th-6th c.

Fig. 205 **MACEDONIA**
BARGALA. Plan of the basilica (6th c.).

The building, preceded by an inner and outer narthex, with *tribelon*, has three aisles with a single semicircular projecting apse. On the north side is a baptistery, which contains a cruciform baptismal pool, built of brick.

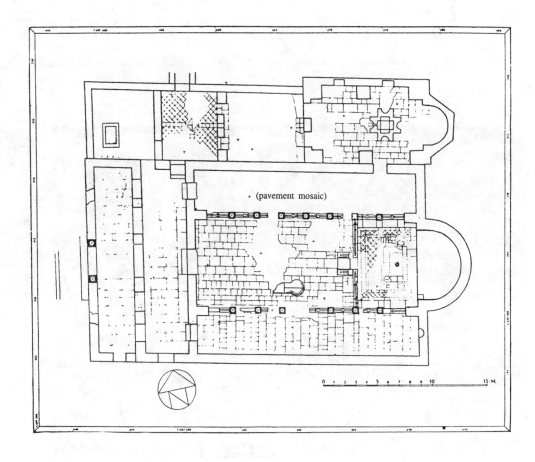

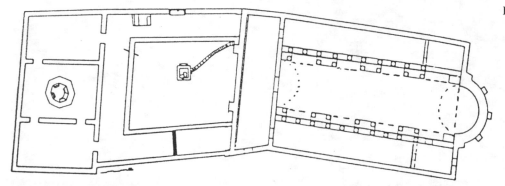

Fig. 206 **MACEDONIA**
DION. Plan of basilica B (mid 5th c.), rebuilt in the mid 6th c.

Fig. 207 **MAGI (the three kings)**
ROME. Catacomb of Priscilla. Fresco of the so-called "Greek Chapel" (mid 3rd c.).

Mary sits on a *cathedra* and holds the Child in her arms; the three characters approach them bearing gifts in their hands.

Fig. 208 Figured silver flask (height 21 cm.) from the 4th-c. treasure of Traprain Law. National Museum of Antiquities of Scotland.

In the lower part we see the scene of the Adoration of the M. The treasure was discovered in 1919 inside a fortified hill 32 km. east of modern Edinburgh and comprises 150 different objects. It was buried at the end of the 5th c.

Fig. 209 **MANNA**
ROME. Basilica of S. Sabina. Panel of the wooden door (first half of 5th c.).

At the centre three women, in tunic and *palla*, sitting on seats around a circular table, eat a flat loaf prepared with manna, kneaded with a rolling-pin by a figure on the left and collected in a container by a final figure to the right of the scene, which is rich in movement.

Fig. 210 **MANUSCRIPT TRADITION**

The Lindisfarne Gospels (f. 139): detail of the largest page-heading decoration at the start of Luke's gospel (*c*.698). British Library.

The Gospels, written in Latin at the monastery of Lindisfarne on Holy Island, near the Northumberland coast, constitute one of the greatest works of miniature. They are probably connected with the translation of the body of St Cuthbert (died 687) in 698. The Latin text of the gospels is that of St Jerome's Vulgate.

Fig. 211 **MANUSCRIPT TRADITION**

Vat. gr. 1809 (second half of 10th c.). St Maximus Confessor: f. 213r, col. 1, end of the commentary on the Lord's Prayer; col. 2, opusc. 1, *ad Marinum*. Vatican Library.

Italo-Greek codex of the "Nilene" school, written partly in brachygraphic script, visible in col. 2 of f. 213. On the script of the codex, apart from P. Canart, *Codices Vaticani graeci. Codices 1745-1962*, Vat. Apost. Lib. 1970, pp. 176-177, cf. S. Lilla, *Il testo tachigrafico del De divinis nominibus*, Studi e Testi 263, Vatican City 1970, and N.P. Chionides - S. Lilla, *La brachigrafia italo-byzantina*, Studi e Testi 290, Vatican City 1981 (note by S. Lilla).

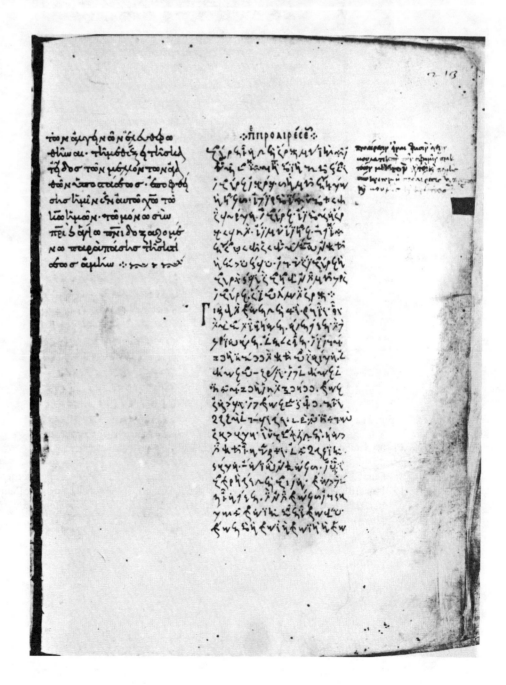

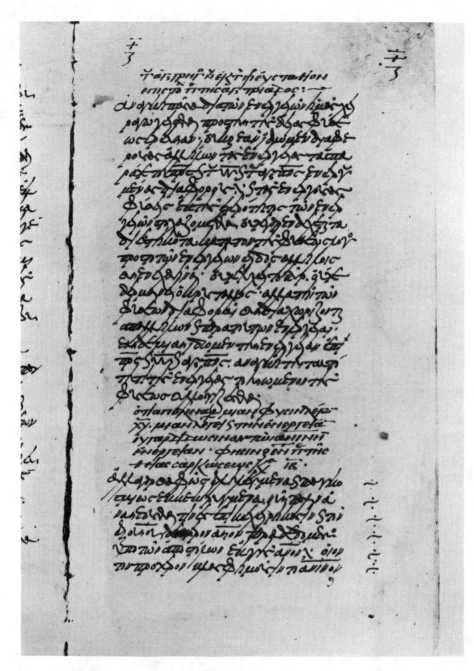

Fig. 212 **MANUSCRIPT TRADITION**
Vat. gr. 2200 (8th-9th c.), p. 121: *Doctrina Patrum de incarnatione Verbi* (cf. article "Aristotelianism"), ch. XII. Vatican Library.

Cursive book-script, ancestor of book-minuscule, rather than "Damascene" (term proposed by P. Maas) it should be called "hagiopolitan" (cf. E. Follieri, Tommaso di Damasco e l'antica minuscola libraria greca, *Atti dell' Acc. Naz. dei Lincei, Rendiconti, Classe di sc. morali, str. e fil.*, serie 8, 29 [1974],15-16) (note by S. Lilla).

Fig. 213 **MARRIAGE**
NICOSIA (Cyprus). Museum of Antiquities. Paten with the wedding of David (613-630).

Against a background of architecture is arranged a group of figures in heavily modelled forms: David and his wife Bathsheba, in the presence of the Priest, haloed and standing on a footstool, hold out their right hands united in the act of *dexterarum iunctio*, culminating moment of the rite. David and the Priest wear the chlamys, the bride a rather clinging sleeved tunic and a *lacerna* clasped on her breast, while in her left hand she holds a *mappula*. At the sides of the scene, two musicians in short, belted tunic, also with chlamys. In the foreground, on the earth, three objects probably connected with the rite as symbols of good luck: a basket (perhaps containing gold coins) and two full bags.

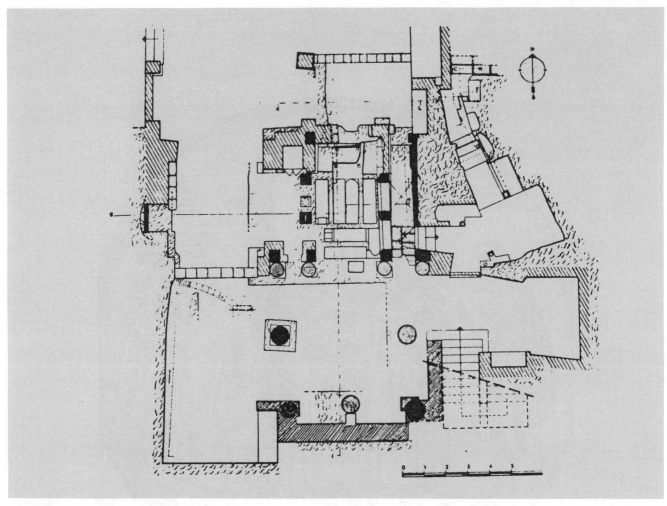

Fig. 214 **MARSEILLES**
St Victor. Rock-cut and "catacombal" *martyrium*.

It was founded by St Cassian at the start of the 5th c., in the neighbourhood of the rock-cut tomb of St Victor. The saint's grotto is in turn prolonged into a gallery of catacombal type.

Fig. 215 **MARTYRDOM, inscriptions of**
NICULITEL (Romania). Inscription painted in red on the south wall of the martyrial crypt (late 4th - early 5th c.).

The epigraph reveals the names of the martyrs laid here, mentioned by the Martyrology of Jerome: Zoticus, Attalus, Kamàsis and Philip. In the first line is a monogrammatic cross. They were local martyrs, victims probably of Diocletian's persecution.

Fig. 216 **MARTYRDOM, scenes of**
ROME. Basilica of SS. Nereo e Achilleo in the catacomb of Domitilla. Column of ciborium (late 4th - early 5th c.).

The martyr Achilleus, defined by the caption *ACILLEVS*, in unbelted tunic, with hands tied behind his back, is receiving a blow from a *gladium* from a soldier: in the background is incised a *vexillum* which recalls the crown of martyrdom.

Fig. 217 **MARY**
PARENZO. Apse vault of the Euphrasian basilica (mid 6th c.).

M. is at the centre of the apse, on a jewelled throne with footstool and ample cushion, regal attributes that contrast with the simplicity of her clothing; on her knees the Child with crossed halo, in *tunica clavata* and pallium. Beside her are two angelic guards with brown wings and the wayfarer's *baculus*, while a group of figures with their names written next to them proceed along the grassy and flowery background. On the left, from the centre: St Maurus, patron of the Parentine church; bishop Euphrasius, in purple *casula* and with the model of the rebuilt basilica; archdeacon Claudius in white *dalmatica* and, lower down, a boy with two candles (or two scrolls): Euphrasius, son of the archdeacon. On the right: three saints in tunic and pallium (perhaps the protectors of the city). Eleutherius, Proiectus and Elpidius. The image of M. *Theotokos* is higher than the others, symbol of the dignity conferred on her by the council of Ephesus of 431. From the sky streaked with gaily coloured clouds, the hand of the Eternal reaches down to crown her.

Fig. 218 **MARY**
ROME. Catacomb of Commodilla. Fresco in the little underground Basilica (6th c.).

At the centre is the Madonna seated with the Child on a jewelled throne, her feet on a footstool; to the right is St Felix with white hair and beard, golden halo, *tunica clavata*, pallium with an upside-down Γ on one hem, and sandals; to the left St Adauctus, youthful in appearance, dressed like his colleague, places his right hand, with a gesture of introduction, on the shoulder of the deceased Turtura. She is dressed in a tunic and with *palla* on her head, on her veiled hands are two *rotuli*. The figures follow the rigid forms of Byzantine art, except for the portrait of Turtura, which is very expressive and intense.

Fig. 219 ROME. Catacomb of SS. Marcellino e Pietro. Fresco in the arcosolium of a cubicle (late 3rd - early 4th c.).

M. seated on a high-backed *cathedra* has the Child in her arms. Beside her two Magi, in the usual Oriental dress, offer gifts on trays.

Fig. 220 **MEDICAL LECTURE (so-called)**
ROME. Catacomb of Via Latina. Fresco of an arcosolium (second half of 4th c.).

At the centre of the group is a bearded figure, clothed in cynic's pallium, of grave aspect, who seems to point to a naked, supine man with an opening in his abdomen, towards whom another figure is pointing a *virga*. It may be a scene of surgery, philosophy or, less probably, resurrection.

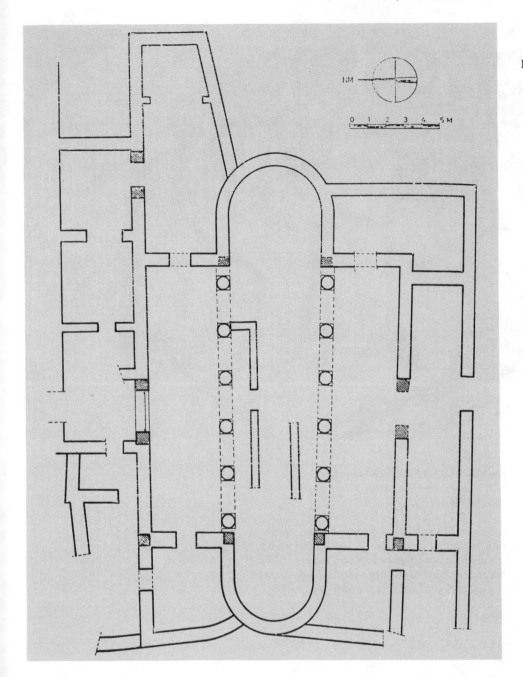

Fig. 221 **MERIDA**
CASA HERRERA. Basilica (6th c.).

Like numerous other church buildings of the same period in the provinces of Baetica and Lusitania, this sacred building has opposed apses. If its model of reference should be sought in Christian Africa, these Iberian examples are distinguished by the fact that they all belong to a single historical moment.

Fig. 222 **MILAN**
Plan of the basilica of S. Simpliciano, originally called the basilica Virginum.

Probably begun by St Ambrose, the basilica was later named for his successor. Investigations begun in 1944 have revealed the early Christian cruciform-plan structure, still in large part substantially recognizable, despite Romanesque interference.

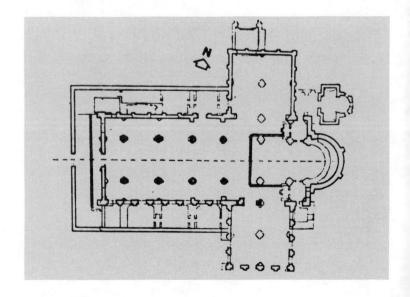

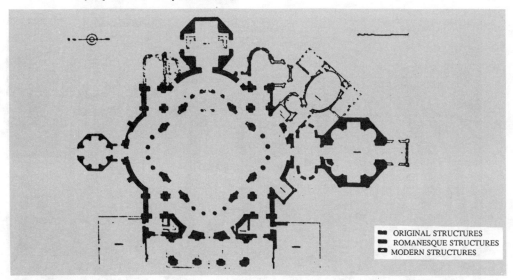

Fig. 223 **MILAN**
Basilica of S. Lorenzo (probably from the last decades of the 4th c.), an elaborate central plan with attached chapels.

With it are associated those dedicated to St Hippolytus and St Lawrence and the so-called "Chapel of St Aquilinus", ornamented with an interesting christological mosaic, probably to be considered a baptistery. On account of its position outside the walls, the basilica's function has been variously interpreted and there is still no agreement on whether or not it should be considered the Palatine church. It seems we can rule out any connection with the Portian basilica, known for Arian controversies of the Ambrosian period.

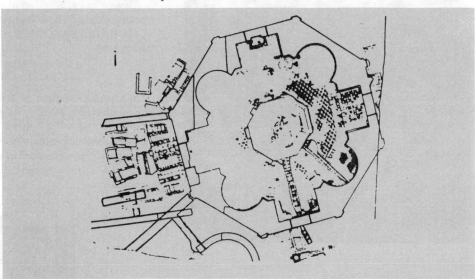

Fig. 224a, 224b

Baptistery of the "new" basilica, later S. Tecla. Plan and detail of the pool.

The grandiose five-aisled cathedral (c. mid 4th c.), whose remains can be observed in the north part of the churchyard of the present cathedral, was destroyed in the second half of the 15th c. The baptistery, situated behind the apse of the church, was linked to it by a covered passage. Its octagonal form and dimensions are similar to the chapel of S. Aquilino at S. Lorenzo.
On the sides of the octagon, inside, four rectilinear niches alternate with four semicircular ones; the pool too is octagonal.

Fig. 225 **MONOGRAM**
GRADO (Gorizia). Basilica of S. Eufemia. Detail of the mosaic pavement of the so-called "Mausoleum of Elias" (579).

The monogram of the (schismatic) patriarch Elias (571-586) is repeated twice: here and in the so-called *salutatorium* (much damaged). The m. contains all the letters of the two words HELIAS EPISCOPVS, in black *tessellae* on a white background.

Fig. 226 NAPLES. Baptistery of S. Giovanni *in Fonte*. Dome mosaic (detail) (early 5th c.).

The monogrammatic cross, allusive to Christ, appears here in triumphal form: surrounded by a band of plant and animal motifs, on a stellar background, with a crown of beribboned laurel above it.

Fig. 227 **MOSAIC**
MOUNT NEBO (Jordan). Church of Sts Lot and Procopius. Detail of the pavement mosaic of the central aisle (first half of 6th c.).

The first of the two panels of the central aisle, it is subdivided into six rows of vine-shoot volutes issuing from four tufts of acanthus: inside them are hunting, harvest and pastoral scenes, including a pair of partridges, a shepherd with his dog and flock, a youth killing a bear with a lance, the grape-harvest, a fox eating the grapes, the harvest transported on an ass and in baskets, two people treading the grapes accompanied by a fluteplayer. Lower down, a dog is urged on by a hunter against a hare and an archer is firing an arrow against a lion. The mosaic was commissioned by bishop John and is one of the best-known examples of mosaic work in the Jordanian area.

Fig. 228 **MOSAIC**
RAVENNA. Mosaics of the Mausoleum of Galla Placidia (5th c.).

The mosaic decoration covers the inner surface of the mausoleum and shows all the characteristics of Byzantine art, associating Oriental taste - especially in the ornamentation and the partition motifs - with the Hellenistic-Roman tradition in the rendering of the figures. In the lunette of the middle niche: St Lawrence, in long white *tunica clavata*, with a golden cross in his right hand and an open book in his left, as he approaches the gridiron; to the left a small cabinet contains the Holy Scriptures. Above, two Apostles in tunic and pallium acclaim and are separated by a small fountain approached by two doves. Geometrical and plant motifs decorate the extra spaces.

Fig. 229 **MOSES**
ROME. Catacomb of Via Latina. Arcosolium of a cubicle ([*c.*] mid 4th c.).

M. is in the act of unfastening his sandals (Ex 3, 1-6). Of youthful appearance, in tunic and pallium, he has his left foot on two rocks. Above, to the left, on a red background the divine hand is partly preserved, holding a scroll.

Fig. 230 **MZCHETA - SVETI TZKHOVELI**
Church of the Holy Cross. Bas-relief with depiction of Christ and the donor, prince Stephen (586/7-604/5).

Fig. 231 **NAPLES**
Episcopal complex.

The cathedral church of Naples, later S. Restituta, was built by Constantine, perhaps over a previous Christian structure. It was given a square-plan baptistery, now at the back, beyond the apse of the right aisle, but originally among the front parts of the basilica, which we think underwent a total reversal of orientation. The great problem of the Neapolitan complex is connected with the identification of the zone where the "second cathedral" (Stephania: 5th c.) stood and of the new Vincentian spring, a complex later replaced by the Angevin cathedral (13th c.).

Fig. 232 **NATIVITY**
MILAN. Cathedral Treasury. Gospel-cover in five parts (second half of 5th c.).

Scenes of Christ's life are framed by the wreathed symbols and portraits of the evangelists Matthew and Luke; in the centre the haloed, jewelled Lamb inside a beribboned garland. In the upper panel the N.: following western iconography the Child, wrapped in swaddling bands, is beneath the roof, which also houses the ox, while the ass, to the left, seems to put its head out of a brick building. To the side: Mary, eyes turned towards Jesus' pallet, sits on a rise in the ground; in the same pose Joseph, to the left, in exomis tunic, with saw in hand, is seen in realistic accents. To the left of the panel: the Annunciation (at the spring), the Magi pointing to the star and Christ's Baptism. To the right: Mary's Presentation in the Temple, Christ among the doctors and the Entry into Jerusalem, while at the bottom the Massacre of the Innocents spreads out horizontally.

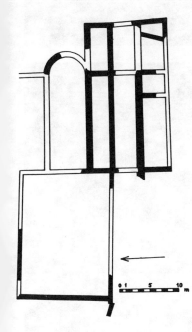

Fig. 233 **NAZARETH**
Plan of the church of the Annunciation ([*c.*] early 5th c.).

Preceded by a large atrium, the church, of which only the foundations are largely preserved, was connected with a monastery, situated along the south side. The presence of crosses in the mosaic pavement would concur to date the complex to a period before 427, when a law of the Theodosian Code forbade the representation of crosses in pavements.

Fig. 234 **NIMBUS**
VATICAN CITY. Necropolis beneath St Peter's basilica. Vault of the so-called "Mausoleum of the Julii" ([*c.*] mid 3rd c.).

The oldest known Christian mosaic, in small squares of opaque glassy paste. At the centre of the vault is Christ-Helios on the hurrying quadriga, with a great radial nimbus. This motif from the pagan iconographical repertoire, eschatologically interpreted, celebrates Christ's triumph, glorified in the empyrean, with the radial attribute of the Sun-god: Christ *Sol Invictus*, light of salvation. The mosaic's christological significance is also confirmed by the vine-shoots that clothe the entire yellow-grounded vault, starting from the corners and going towards the centre, which recall Christ's words: *Ego sum vitis vera*.

Fig. 235 **NISIBIS**
Baptistery: exterior.

The original building was divided into three rooms: a central domed room flanked by two apsed rooms, of which only that on the left still remains. A Greek inscription, now disappeared, records bishop Volageos of Nisibis, who built the baptistery in 359. Inside, a richly decorated cornice runs along the walls and marks the openings. Note the presence of a crypt, probably open at a more recent time, with a sarcophagus which, according to tradition, contained the body of the protobishop of Nisibis, Mar Yakub († 338).

Fig. 236 Baptistery: detail of the internal decoration.

Fig. 237 **NOAH**
ROME. Fresco of an arcosolium in the catacomb of Via Latina (second half of 4th c.).

Noah bearded, praying and in *tunica clavata*, stands in the ark, which comes up to his knees and is rendered as a trunk with a lock.

Fig. 238 **NORICUM**
TEURNIA (Tiburnia). S. Peter im Holz. Cemeterial basilica.

Teurnia was one of the three most important Roman cities of Noricum and the region's major episcopal see, existing before the middle of the 5th c. The basilica has a single aisle, with apsed pastophories, a flat end and an apsed form of modest height enclosing the presbyterial area, characteristic of the churches of Noricum. There are two main phases of the building: the first belongs to the 5th c., before 472, year of the Gothic siege; the second (6th c.) involved substantially the addition of the narthex and lateral corridors.

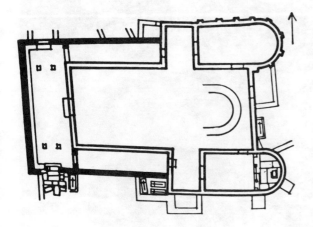

Fig. 239 **ORANS**
ROME. Catacomb called "Cimitero Maggiore". Fresco of the lunette of an arcosolium (first half of 4th c.).

At the centre a female figure raises her arms in the gesture of an *orans*, to signify the happiness of the life of Paradise; the woman, in sleeved tunic, has a "turban" headdress. To the left a shepherd, in exomis tunic of skin and leg-bands, sits and milks a sheep; to the right another shepherd, similarly dressed, is standing with a sheep on his shoulders. The bucolic-paradisiac scene is completed by a *mulcra* and a sheep to the right, as well as two trees beside the *orans*.

Fig. 240 **ORPHEUS**
ROME. Catacomb of SS. Pietro e Marcellino. Fresco in a lunette of an arcosolium (first half of 4th c.).

In the centre O. seated, dressed in Oriental style, with Phrygian cap, short tunic, *bracae* and long cloak, sounds the lyre with a large plectrum. Beside him trees and birds seem enchanted by his music, sweet as the words of Christ.

Fig. 241 **PAINTING**
ROME. Catacomb of SS. Marcellino e Pietro. Fresco of the entrance wall of the so-called "Cubicle of Orpheus" (first half of 4th c.).

The white ground of the entrance wall is divided by wide red bands into five fields. At the top, in a crescent-shaped space, Orpheus with his lyre among the animals; below, in four rectangular fields: to the left Moses striking the rock and the miracle of the paralytic, to the right the healing of the woman with an issue of blood and Noah in the ark. The figures, isolated and free from the superstructures of landscape and context, are uniquely accompanied by the key elements of interpretation and show a rediscovered classicity in volumes and forms, typical of Constantinian art.

Fig. 242 ROME. Catacomb of Pretestato. Fresco in the so-called "Cubicle of the *Coronatio*" (first half of 3rd c.).

The fresco, executed in an impressionistic style, through lively and primary colour-matching, shows a unique scene: at the centre a character in tunic and pallium with a crown of thorns on his head; behind him a bird on a tree; to the left two characters, in short tunic and chlamys, proceed towards the crowned man, one of them holding out a cane. The scene has been referred to Christ's passion and has been considered among those depictions that appeared at the beginnings of Christian art and then ceased to be depicted until the Middle Ages.

Fig. 243 **PAINTING**
ROME. Catacomb of S. Callisto (Area "of Lucina"). Ceiling of the so-called "Cubicle Y" (early 3rd c.).

Into a complex system obtained with red-green lines are inserted various figures: in the central medallion, Daniel praying among the lions; in the diagonal fields, busts of seasons, figures of *orantes* and of the good shepherd alternate on plant bases. The expedient of dividing the space by means of segments, on one hand simplifies the ancient decorative system of the Pompeian tradition; on the other, it announces an illusionistic style, consistent with the new wholly symbolic figurative language.

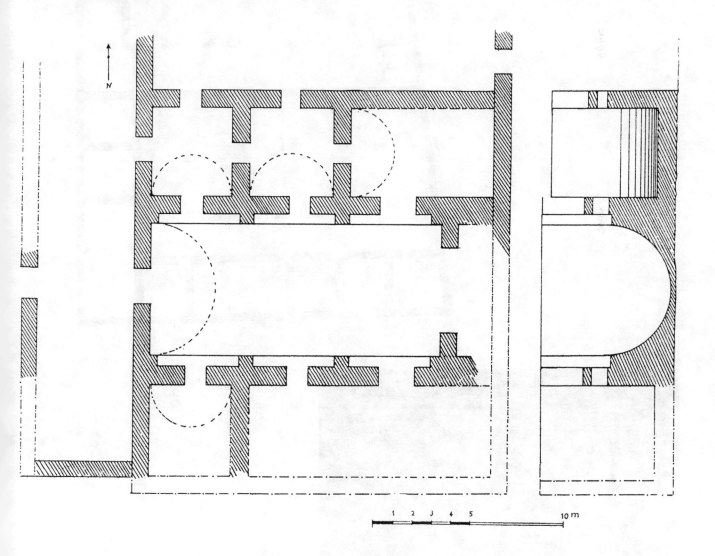

Fig. 244 **PALESTINE I**
EL-GIB (Gabaon/Gibeon). Plan of the Persian-style church.

It differs greatly from the Palestinian church buildings of Greco-Latin type. It seems logical to see the builders as individuals who came from Persia, in the Sassanid period, around the 5th-6th c.

Fig. 245 **PALESTINE II**
ET-TABGHA. Plan of the church of the Multiplication of Loaves and Fishes ([*c.*] mid 5th c.).

Preceded by an atrium of irregular form, and part of a complex with several rooms, the church has three aisles divided by columns, with transept; the semicircular apse, included inside a straight wall, is flanked by pastophories. It is a memorial church, linked to the episode of the multiplication of the loaves and fishes.

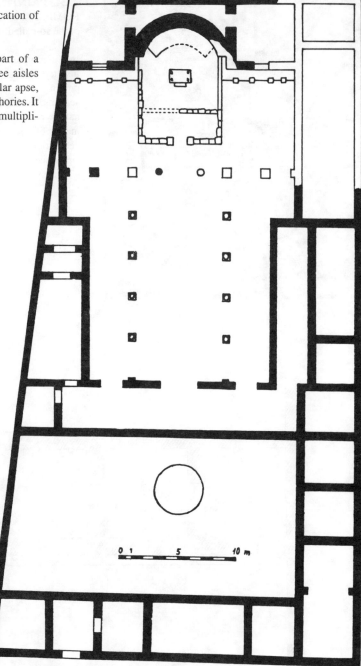

Fig. 246 **PALESTINE II**
ET-TABGHA. Church of the Multiplication of Loaves and Fishes: detail of the pavement mosaic.

Fig. 247 **PALESTINE III**
SHIVTA. Plan of the so-called "North Church" ([c.] early 6th c.).

Dedicated to St George, the church, with three aisles terminating in as many enclosed apses, is preceded by a vast atrium flanked by rooms. Along the south side an apsed room preserves a cruciform baptismal pool with steps.

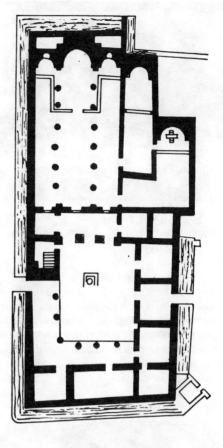

Fig. 248 **PARADISE**
ROME. Catacomb of Domitilla. Arcosolium called "of Veneranda" ([c.] mid 4th c.).

The deceased Veneranda appears in the attitude of an orans, *expansis manibus*; she is introduced into a paradisiac setting, rendered as a garden, by the martyr Petronilla (also identified by a caption). At the far right an open *codex* with seals may allude to the Apocalypse, while next to it is a *capsa* full of scrolls (of Holy Scripture), with the lid near by.

Fig. 249 **PARALYTIC**
DURA EUROPOS. *Domus ecclesiae*. Fresco on panel, preserved in Yale University Museum, New Haven (USA) (mid 3rd c.).

The p. lies to the right on his bed, in short, sleeved tunic, while Christ, in the middle ground, works the miracle by extending his right hand. On the left the man, now healed, departs with his bed on his shoulders, following the iconography that would always be used from now on.

Fig. 250 **PARAPETASMA**
VATICAN CITY. Museo Pio Cristiano. Detail of sarcophagus lid (320-330).

Two figures in tunic and pallium, bearded (perhaps two Apostles) hold the unfolded p. at the shoulders of the dead woman, who in tunic and *palla* is in the attitude of an *orans*. Her face is unfinished and the drapery is sketchily rendered, just by grooves. At the left of the lid are Adam and Eve beside the tree with the serpent wound round it; they have rather heavy bodies, as does the Cupid who supports the left side of the *tabula*.

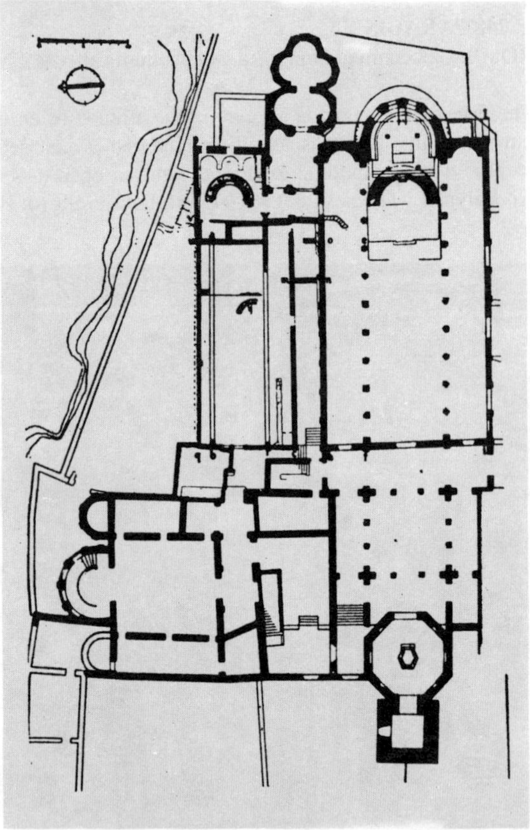

Fig. 251 **PARENTIUM**
(Istria). Plan of the Euphrasian basilica (6th c.).

A complex furnished with atrium, baptistery, bishop's palace and *trichora*, built over two earlier religious buildings of the 4th and first half of 5th cc.

Fig. 252 **PAUL, APOSTLE**
ROME. Catacomb of Domitilla. Detail of the rear niche of the cubicle called "of Diogenes" (late 3rd c.).

One of the earliest images in which P.'s features are individualized: bald, long face, pointed beard, thick eyebrows, aquiline nose. The eyes appear sunken in this painting; the expression intense. P., in tunic and pallium, has a scroll in his hands, while at his feet is an open *capsa*, in which we can make out five more scrolls.

Fig. 253a, 253b **PERPETUA AND FELICITY**
RAVENNA. Archiepiscopal Chapel. Detail of medallions with saints: Perpetua and Felicity (first half of 6th c.).

Busts of men and women saints are inserted in the *clipei* that frame the cross-vault of the chapel, distinguished by a subtle individual characterization. Perpetua, traditionally of high rank, has fine features, with her hair in small, schematically-rendered locks that frame her veiled head, and wears a rich garment. By contrast Felicity, her companion of lower rank, is more modest in appearance, simply wrapped in a dark *palla* and characterized by less noble features.

Fig. 254 **PETER**
VATICAN CITY. Museo Pio Cristiano. Sarcophagus front with Petrine cycle (325-350).

The episodes of the Petrine cycle - Denial, Arrest, Miracle of the spring - are flanked by numerous OT and NT scenes, while at the centre is the *clipeus* with the busts of the two deceased spouses, facially characterized. From top left: the raising of Lazarus, the multiplication of loaves, the sacrifice of Abraham, the healing of the blind man, the denial (below is the cock, "key to the reading of the scene"), the handing of the symbols of labour to Adam and Eve. Below, under the shell, the animated episode of Jonah, thrown into the sea and at rest under the booth; to his left, the wedding-feast of Cana, the woman with an issue of blood, Moses unfastening his sandals; to the right, Daniel among the lions with the bearded, wingless angel next to him, the arrest and the miracle of the spring, with the *bibentes* in military dress (chlamys and *pileus pannonicus*). P.'s portrait is already well defined: short, curly beard, wavy hair, mature appearance.

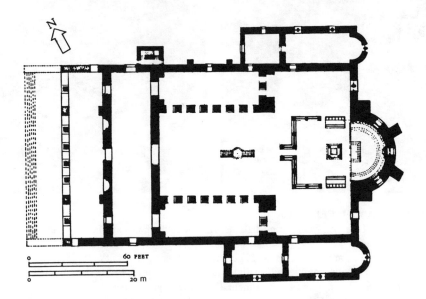

Fig. 255a, 255b
PHILIPPI
Plan and ruins of basilica B (built shortly before 540).

This was a great domed basilica, situated in the heart of the Roman city. It had three aisles with a semicircular apse and was preceded by a narthex. Note some lateral service rooms.

Fig. 256 **PHOENIX**
ROME. Catacomb of Priscilla. Fresco of the so-called "Greek Chapel" (mid 3rd c.).

The phoenix, haloed and rayed, is surrounded by the flames that release it from its nest, to indicate the violent moment of death, which, however, is a prelude to that of rebirth.

Fig. 257 **PILATE**
VATICAN CITY. Museo Pio Cristiano. So-called "Passion" sarcophagus (mid 4th c.).

The front is intercalated by twisted columns, which alternately support triangular pediments and architraves. In the centre arch is the *Anastasis* with the *Crux invicta*, surmounted by a laurel wreath, inside which is the monogram of Christ; beside it are two doves, while below sit the two guards of the sepulchre, in the traditional chlamys and helmet. To the left: Simon of Cyrene with the cross; Christ's coronation, with the symbol of humiliation (the thorns) replaced by that of triumph (laurel); to the right, the Lord accompanied by a soldier, in the presence of P. He, seated on a *sella curulis*, is thoughtful, with his hand under his chin; he is assisted by two *camilli* in exomis tunics, one of whom pours water into a *cantharus*, set on a tall base with animal supports. This alludes to P.'s symbolic gesture of washing his hands. In chlamys and short tunic, he has his head encircled by a diadem; he turns away from Christ: P. incarnates the ephemeral and earthly imperial authority, in contrast to the perennial nature of the divine magisterium. P.'s judgment also alludes to the last act of Christ's earthly life; so the scene often appears in the so-called "Passion" sarcophagi.

Fig. 258 **PIOUS WOMEN**
MILAN. Museum of Castello Sforzesco. Tablet of ivory diptych (5th c.).

In the upper section the two soldiers with *pileus pannonicus*, chlamys and lance sleep in unnatural positions before the sepulchre; at the upper corners are the symbols of the evangelists Luke and Matthew. In the lower panel an angel in tunic and pallium, haloed and wingless, sits on a rock, announcing the Resurrection to the two pious women, in tunic and *palla*, stretching out towards him (one is on her knees). At their shoulders can be seen the masonry façade of the sepulchre, with decorated architrave and half-open door, in turn ornamented with biblical scenes: from the top, the raising of Lazarus, Zacchaeus on the sycamore, the miracle of the *mulier inclinata* (or of the Canaanite woman).

Fig. 259 **PORTRAIT**
ROME. Front of the sarcophagus "of the two brothers" (*c*.340).

In the centre of the upper level, within the shell, the busts of the deceased - the two brothers - recognizable as such by their facial similarity. The sarcophagus must originally have been prepared for a married couple, but was subsequently adapted for the inclusion of a second male figure, that on our left, whose bust is in fact female. The sparse hair and the beards of both are rendered with small, soft, tidy locks; beneath the high forehead the eyes are deep-set and "brought alive" by drilled holes; the thoughtful, distant look gives them an attitude of noble and detached dignity. On either side of the *clipeus* we see: top left, the raising of Lazarus, the denial of Peter, Moses receiving the tables of the law; top right, the sacrifice of Abraham and a thoughtful Pilate washing his hands. Below, from left, the miracle of the spring and the arrest of Peter; Daniel naked among the lions; beneath the shell the less common scene of Peter reading between the two soldiers; the healing of the blind man and the multiplication of the loaves. This sarcophagus, with that of Junius Bassus, is considered the finest expression of the so-called "beautiful style".

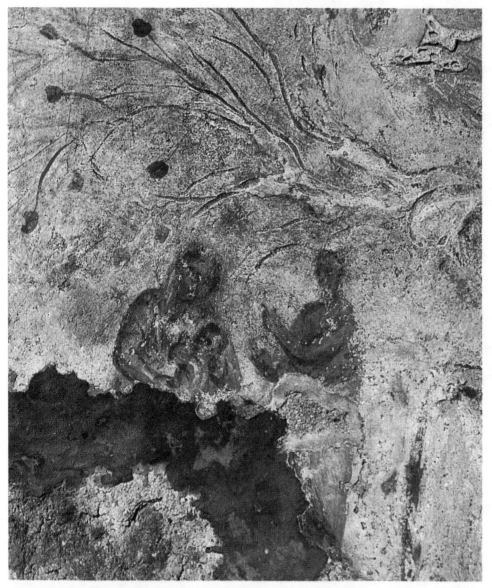

Fig. 260 **PROPHET**
ROME. Catacomb of Priscilla. Fresco of the Madonna and Child with Prophet (early 3rd c.).

The Virgin is seated on a *cathedra* with the Child on her lap: her head covered by a *palla*, she looks towards her Son. The Prophet, identified with Balaam, the Oracle "who sees the star come forth from Jacob and the sceptre from Israel" (Num 24, 15-17), is beside them in the act of pointing to the star. But the scene could also refer to the passage of Isaiah foretelling the coming of the Lord, light of salvation (Is 7, 14; 60, 1). This group of the Madonna and Child is the earliest in a cemeterial depiction.

Fig. 261 **QUAILS, miracle of the**
ROME. Basilica of S. Sabina. Panel of the wooden door (first half of 5th c.).

Three characters in tunic and pallium sit on seats around a small circular table, covered by a round, fringed *mappa*, on which is set a plate of roast quails, as in Ex 16, 13 and Num 11, 13-14.

Fig. 262a, 262b, 262c
RAVENNA
CLASSE (excavations 1963-1965).

a) Petrian basilica. Situated some 350 metres N-E of that of S. Severo. The building of this church was completed by bishop Neon (449-475). Begun by St Peter Chrysologus (429-449), it underwent a further restoration at the hands of king Astulf, after having been ruined by an earthquake in the second quarter of the 8th c.

b) Basilica Probi. Situated some 200 metres from the Basilica of S. Apollinare in the direction of Cervia. Furnished with an atrium of particular length, it presents a sort of singular transept. Within it were buried numerous bishops of Ravenna, especially by Maximian in the middle of the 6th c.

c) Basilica, discovered in 1965 2 km south of S. Apollinare in Classe, of uncertain identification (St Eleucadius, St Sergius the Arian, St Demetrius?).

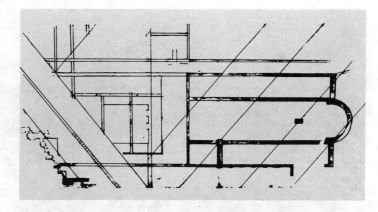

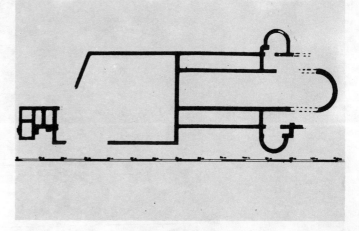

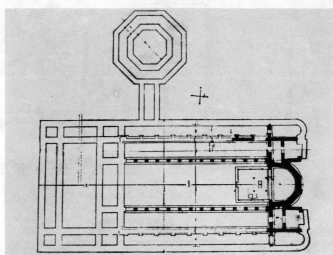

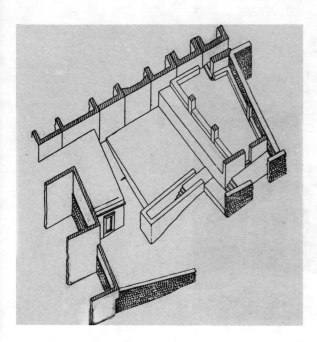

Fig. 263 **ROME**
Basilica Apostolorum: detail of the zone of the *triclia*.

A religious centre on the Via Appia, related to a temporary translation of the bodies of Peter and Paul from their original graves. The zone of the *triclia* corresponds to the place where *refrigeria* were performed. In the 4th c. the area was included in the perimeter of a great basilica, which followed the typology of the Roman circiform churches of the period (SS. Marcellino e Pietro, S. Lorenzo *fuori le mura*, S. Agnese *fuori le mura*).

Fig. 264 **ROME**
S. Pancrazio. Cemeterial basilica of the Via Aurelia Antica.

The primitive construction was promoted by bishop Symmachus at the start of the 6th c. A second phase, to which the present structures correspond, should be put under the pontificate of Honorius I (first half of 7th c.), to which we must also assign the construction of the semicircular crypt. Beneath the building run two of the three regions of a catacomb.

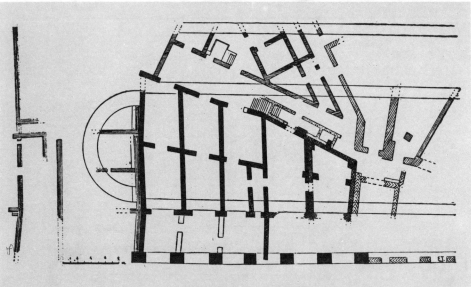

Fig. 265a, 265b Basilica of SS. Giovanni e Paolo. Plan and façade.

A Roman complex whose peculiarities make it a *unicum*. In it we can find the traces of a 3rd-c. *domus ecclesiae*, later related to the history of two *tituli* and the original positioning of the city's sole urban *martyrium*, followed by the establishment of an early Christian basilica, probably promoted by the senator Pammachius († 410).

Fig. 266 **ROME**
Santa Maria in Cosmedin.

One of the *diaconiae*, probably the most important, near the Tiber, together with S. Teodoro and S. Giorgio al Velabro. Its origins may go back to the 6th c., but it was wholly rebuilt by pope Hadrian I at the end of the 8th c.

Fig. 267 Santa Balbina.

A basilica on the "Via Nova", probably built in an *aula* of the Aventine property of the patrician Cilo. It may correspond to the *Titulus Tigridae*, whose presbyters were present at the Roman synod of 499. It has a single aisle.

Fig. 268 **ROME**
Basilica of S. Vitale.

It was built through the munificence of the pious matron Vestina under the pontificate of Innocent I (401-417). It kept the name of its benefactress for a long time in conjunction with its intitulation to the non-Roman martyrs Gervasius and Protasius and Vitalis. It was an important *titulus* of the Quirinal, sited on the *Vicus Longus*.

Characteristic of this church is the open façade, or absence of doors, a peculiarity also encountered in other basilicas of the same period, e.g. that of SS. Giovanni e Paolo.

Fig. 269 Monastery of the Tre Fontane.

Probably one of the oldest Oriental monasteries in Rome, though the origins of its foundation still remain obscure and may go back to an initiative of pope Gregory the Great.

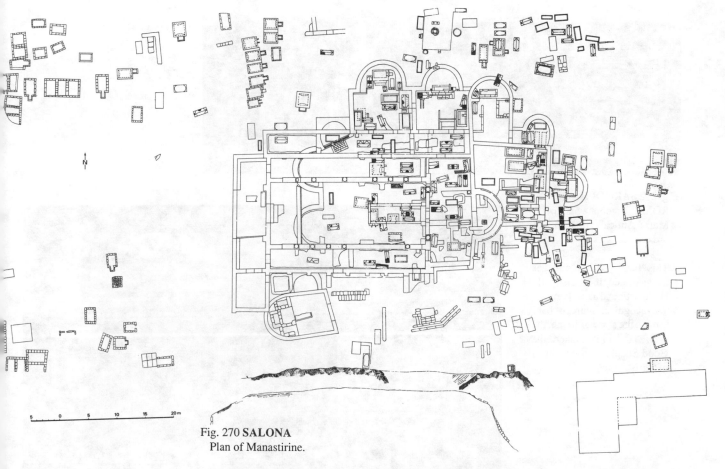

Fig. 270 **SALONA**
Plan of Manastirine.

A complex some hundred metres north of the city walls. It was the Christian cemeterial zone where, at the start of the 4th c., the body of the martyr Domnius was laid. The cemetery was destroyed towards the end of the 4th c., but at the start of the 5th a three-aisled basilica was built there, which enclosed Domnius's memorial in its presbytery. A narthex was added to the church in the course of the 6th c.

Fig. 271 **SAMSON**
ROME. Catacomb of Via Latina. End arcosolium of a cubicle ([c.] mid 4th c.).

S. kills the Philistines with the jawbone of an ass (Jdg 15, 14-16). The bearded figure of S., in tunic and pallium, is standing and brandishes his unusual weapon in his right arm. On the left the fleeing Philistines, in short tunics: some already lie dead on the ground, some are wounded in the head, some flee terrified. To the right, stairs precede a temple set on a high podium, with two columns, a half-open curtain at the front and two windows at the side.

Fig. 272 **SARCOPHAGI**
ROME. Sarcophagus "of S. Maria Antica" (first half of 3rd c.).

This "bath-type" sarcophagus is decorated in a distinctly Hellenistic and narrative taste. We recognize: figures of fishermen, the story of Jonah, the *orans*, the philosopher, the Good Shepherd, the baptism of Christ.

Fig. 273 VATICAN CITY. Vatican Grottos. Sarcophagus of Junius Bassus (359).

The lid is partly lost. The front is divided into two levels and shows a distinction of scenes between the columns. In the upper level, from the left: the sacrifice of Abraham, the capture of Peter, the *traditio legis*, the capture of Christ, Christ before Pilate; in the lower level: Job with his wife, Adam and Eve, Christ's entry into Jerusalem, Daniel (falsely completed by the restorer) in the lions' den, capture of Paul. In the triangles over the columns are depicted miracles allegorically performed by little lambs. Stylistically and structurally, note the typical canons of the art of the "Constantinian Renaissance".

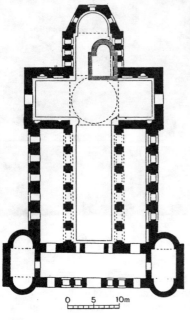

Fig. 274 **SARDICA**
Plan of the church of St Sophia (6th c.).

St Sophia, the city's episcopal church, has a plan that unites the three-aisled basilica pattern with the cruciform plan. A narthex flanked by a sort of towers precedes the building, while a rather projecting presbytery ends in an apse circular inside and polygonal outside. The church stands on a site on which various church buildings followed each other, the earliest of which, of limited dimensions, had a single aisle, with mosaic pavements.

Fig. 275 **SERMON ON THE MOUNT**
ROME. Museo Nazionale Romano. Detail of sarcophagus front (early 4th c.).

At the centre of the lower level Christ is seated on a rise, facing us; he wears the cynic philosopher's pallium, which leaves part of his chest uncovered; in his left hand he holds a half-unrolled scroll, with his right hand he makes the gesture of speech; his face, very intense in expression, is framed by his beard and unkempt hair; at his feet six seated persons, seen from behind, with faces naïvely turned upwards, represent the Apostles as *imago brevis* or the crowd. The scene seems replicated in the upper level, among other depictions all inspired by the New Testament. The slab retains traces of colouring.

Fig. 276 **SHEPHERD, GOOD**
ROME. Fresco of the vault of the so-called cubicle of the "Velata" in the catacombs of Priscilla (second half of 3rd c.).

In the central medallion of the ceiling a shepherd, in exomis tunic with delicate purple striping and leg-bands, holds a kid on his shoulders; at the sides two sheep and two shrubs, on which are two birds in symmetrical positions.

Fig. 277 **SHEPHERD, GOOD**
VATICAN CITY. Museo Pio Cristiano. "Bath-type" sarcophagus from the Via Salaria (mid 3rd c.).

At the centre of the front the figures of the G.S., in exomis tunic and leg-bands with a ram on his shoulders, and the *orans*, in tunic and *palla*, between trees and sheep; to the left a teaching scene with a seated man in tunic and pallium, his reading interrupted by two standing persons similarly dressed, and a sundial, symbol of doctrine and the ineluctable passing of time; to the right a seated matron in tunic and *palla*, with a scroll in her left hand, is accompanied by another female figure. At the sides two great rams define the front. The various figures allude symbolically to doctrine, philosophy, *pietas* and philanthropy, concepts very widespread in the culture of the Gallienic era, which saw the rise and spread of Christianity. The forms, marked by a renewed classicism, respect volumes and gestures in a naturalistic manner.

Fig. 278 **SHEPHERD, GOOD**
RAVENNA. Mausoleum of Galla Placidia.

Mosaic depicting Christ as G. S., with processional cross in hand and head surrounded by a halo, seated on a mountain amid six sheep, arranged symmetrically and all turning towards him. We have here a complete assimilation of pastoral allegory to a typically christological theme. The mosaic dates from the years *c*.424-425.

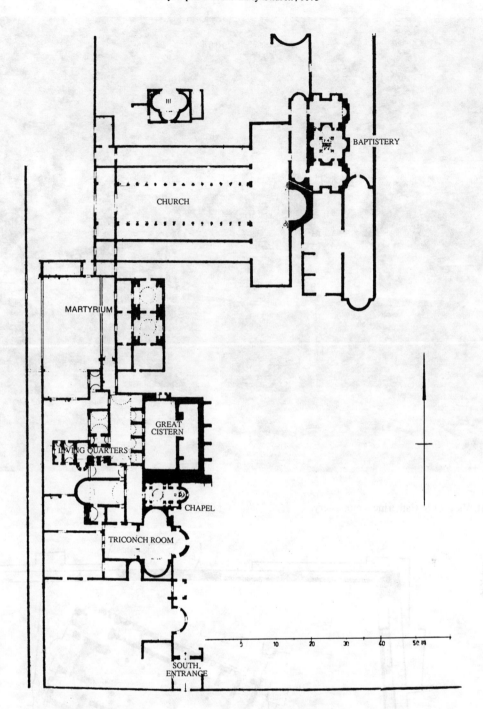

Fig. 279 **SIDE**
Plan of the episcopal complex and the episcopal church (4th-5th c.?).

This complex, one of the largest in Asia Minor, comprises various church buildings, including the episcopal church, a baptistery and rooms suited for ecclesiastical service, as well as the bishop's residence.

Fig. 280 **SINAI**
MOUNT SINAI. View of St Catherine's monastery.

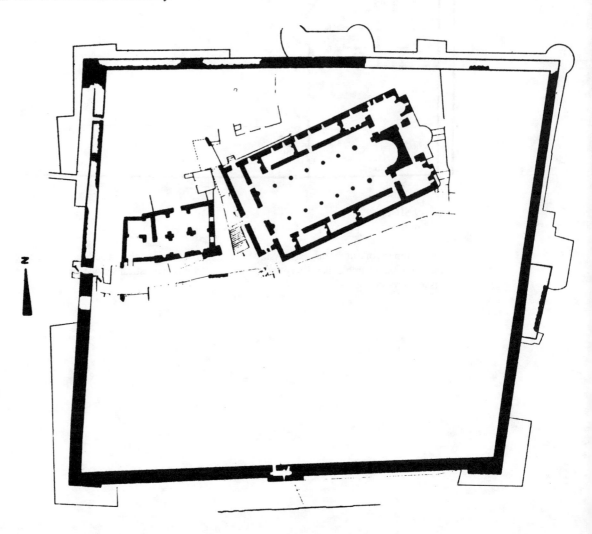

Fig. 282 **SINAI**
MOUNT SINAI. Church of St Catherine's monastery: apse mosaic (6th c.).

In the vault appears the iconographical theme of Christ's Transfiguration, documented in at least two other cases in the 6th c. Around it *clipei* depict Apostles and Prophets, together with the images of two local authorities, the hegumen of the monastery, Longinus, and the deacon John, then still living, distinguished by a sort of square halo (the earliest example known): this element, here similar to a cloth, recalls the drapery placed behind the heads of Palmyrene funerary portraits.

On the arch of the apse appears the episode of Moses and the Burning Bush. Rather than to Constantinople, as critics would have it, the execution of the mosaic was probably entrusted to local workshops. Indeed the style is very far from that of the rare pictorial documents produced in the capital in the same period. Here the characters involved participate dramatically in the event unfolding before their eyes: the expressive faces marked by thick contours, the very poses of each figure are very far from the refined abstract images, so permeated by Hellenistic culture, of Constantinopolitan painting of the pre-iconoclastic era.

 Fig. 281 **SINAI**
MOUNT SINAI. St Catherine's monastery: plan of the church (548-565).

The fortified monastery stood near mount Sinai, at the mouth of a narrow valley, on the place where, at least from the 4th c., a flourishing community of monks had established itself. The present complex was rebuilt by Justinian after the death of Theodora († 548), as some Greek inscriptions indicate. The church, built by the local architect Stephen of Aila (Eilat) and originally dedicated to the *Theotokos*, has three aisles with rooms along the sides: behind the apse, which is enclosed, is the later chapel of the Burning Bush, which preserves an interesting aniconic mosaic in the vault of the apse.

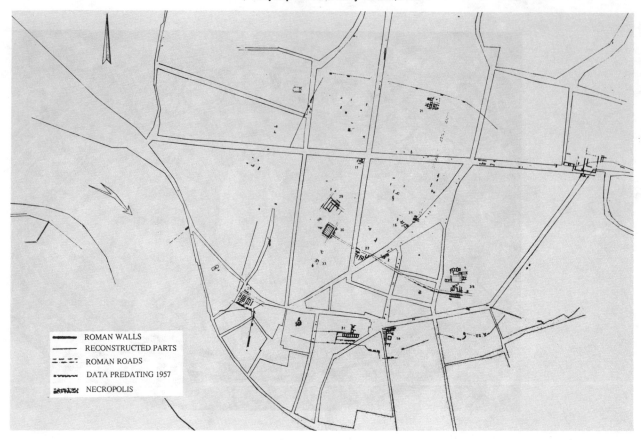

Fig. 283 **SIRMIUM**
Plan of the city.

S. became an imperial residence under the tetrarchy and was an important centre in the 4th and 5th cc. The areas of the necropolis are in grey: they date mainly from the early period of Christianity in the city, which numbered among its martyrs Irenaeus, Anastasius and Demetrius and which, after a period of tenacious Arianism, became a very important bastion of orthodoxy.

Fig. 284 **SOLDIERS WHO CAST LOTS FOR JESUS' CLOTHING**
ROME. Catacomb of Via Latina. Arcosolium of the left wall of a cubicle (second half of 4th c.).

Two Roman soldiers wear short tunic and chlamys, and on their heads the *pileus pannonicus* (or cylindrical cap); at their shoulders two bossed shields are held upright by supports. The soldier on the right holds two spears with his left hand, while his right hand activates the gaming instrument, formed from an amphora revolving on a pole held in a wooden frame, from which issue coloured discs. In the background Christ's sepulchre is in the form of a small temple (similar to that of Lazarus) with steps in front; on the side wall we note windows. The picture was damaged by unknown people after its discovery.

Fig. 285 **STOBI** (Macedonia). Plan of the city.

Situated at the confluence of the rivers Crna (Erigon) and Vardar (Axius), some 160 km north of Thessalonica, it was an episcopal see and capital of Macedonia Salutaris and subsequently of Macedonia Secunda.

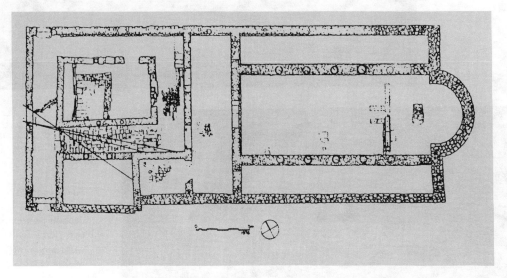

Fig. 286
Plan of the so-called "Central Basilica" (second half of 5th c.).

Situated in the N-E part of the Roman city, near the walls. Like the North Basilica and the episcopal basilica, sited in the S-W zone, it was conditioned by the pre-existence of earlier buildings, in this case belonging to a synagogue, of which clear traces of persistence can be identified, especially in the singular configuration of the atrium.

Fig. 287 **STOBI**
Baptistery of the episcopal church, with tetraconch plan inscribed in a square.

The episcopal basilica is of particular importance on account of the quantity and quality of the mosaics.

Fig. 288 **SUSANNA**
ROME. Wall fresco of the so-called "Greek Chapel" of the catacombs of Priscilla (mid 3rd c.).

To the right is S. in tunic and *palla*; beside her, set symmetrically, are the two elders placing their hands on her head in an act of accusation. To the left Daniel and S. praying after the recognition of her innocence.

Fig. 289 ROME. Fresco of the so-called "Arcosolium of Celerina" in the catacombs of Pretestato (4th c.).

At the centre S. is depicted in the form of a lamb with the caption *Susanna*; beside her two fierce wolves, with the word *senioris* (!), signify the wicked elders.

Fig. 290 **SYMBOLISM**
AQUILEIA. South Theodorian basilica. Detail of the mosaic pavement: the so-called "Eucharistic Victory" (second decade of 4th c.).

Fig. 291 ALBENGA. Baptistery. Vault of the central niche with christogram (late 5th c.).

At the centre of the primitive hall, the panel bears the traces of the foundations of an altar or an ambo superimposed at a later date. The scene is a *unicum*, the object of various interpretations: Victory over the passions, because of the attributes of the crown and the palm; symbol of Baptism, fruit of struggle for the acceptance of faith and of victory; allegorical representation of the Eucharist, because of the presence of symbolic baskets with bread and grapes, or wine. This is the "classical" Victory, iconographically of pagan origin: a blonde girl, with heavenly wings, in sleeveless tunic belted at the waist, her right hand raising a laurel crown and her left holding a palm. The image has been considered the ideal conclusion of the whole allegorical discourse developed by the other scenes of the vast mosaic pavement.

On an intense blue ground, scattered with white stars, the great Constantinian monogram of gilded *tessellae* is flanked by the apocalyptic letters A and ω, and is surrounded by a triple halo of graded tonality, from lighter to darker blue, which probably alludes to the Trinity.

Fig. 292 **SYMBOLISM**
ROME. Catacomb of S. Callisto. So-called "Chapel of the Sacraments" in the region of Lucina: Eucharistic fish (first decades of 3rd c.).

Among the most ancient and allusive symbols, the fish recalls the Saviour on account of its literary references and because it figuratively translates the celebrated acrostic ιχσυς, formed by the initials of the Greek words: "Jesus Christ son of God Saviour"; joined with the basket of loaves and the cup of wine, it takes on a Eucharistic, as well as christological, significance. This is one of the oldest examples of cemetery painting.

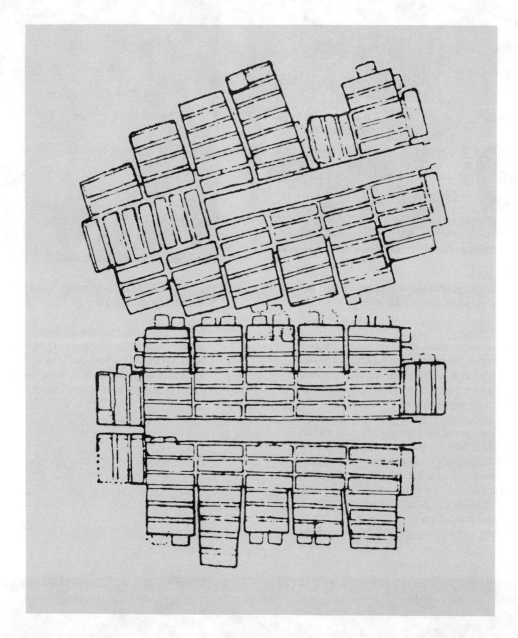

Fig. 293 **SYRACUSE**
Plan of the Pupillo hypogea.

These are part of the group of numerous small underground cemeteries which supplied Syracuse, apart from the great nuclei of S. Giovanni, S. Lucia, S. Maria and Vigna Cassia, along the coastal strip of the eastern Acradina. For these hypogea the Christian phase is dated between the second half of the 4th and the 6th c.

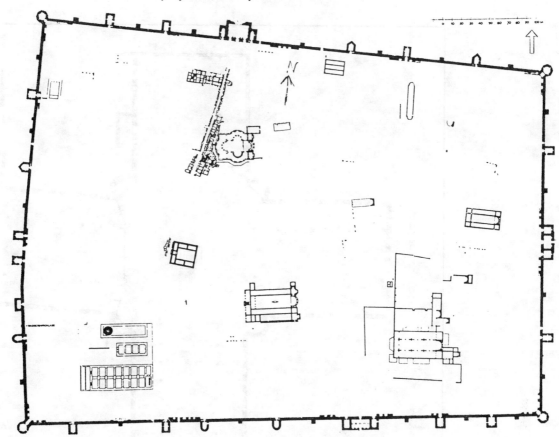

Fig. 294 **SYRIA**
RESAFA-SERGIOPOLIS. Plan of the city.

An extraordinarily well-preserved city wall encloses the monuments, roads and houses of the old caravan city of Resafa, which at the time of Anastasius I (491-518) took the name of Sergiopolis, linking it to the cult of Sts Sergius and Bacchus, but in particular to St Sergius. Campaigns of excavation have discovered a plethora of monuments, specifically the so-called "Church A" (Holy Cross), a vast three-aisled building, of 559, with a *bema* at the centre of the central aisle, the tetraconch, perhaps the city's episcopal church, and the *martyrium* of St Sergius, also with three aisles.

Fig. 295
RESAFA-SERGIOPOLIS. Inside of the so-called "Church A" (Holy Cross).

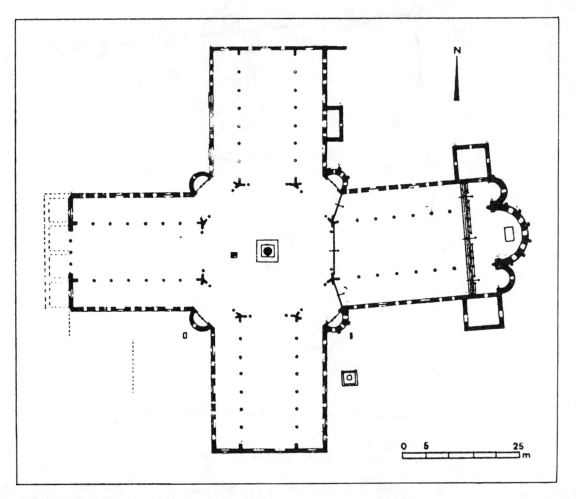

Fig. 297 **SYRIA**
QAL'AT SEMAN. Complex of St Simeon Stylites: central zone.

Fig. 296 **SYRIA**
QAL'AT SEMAN. Plan of the complex of St Simeon Stylites (late 5th c.).

The complex is centred on the column on which St Simeon Stylites († 459) lived for many years. The main nucleus consists of a vast octagon, now uncovered, which forms the point of departure for four three-aisled arms, of which the eastern one constitutes the basilica proper, with three projecting apses. The plan happily unites two forms already attested *ab antiquo* in the Syro-Palestinian area, the octagon and the free Greek cross.

Hostel for pilgrims, dwellings for the clergy and various other rooms stand near the building, while to the south is situated an octagonal baptistery, with pool inserted in the eastern apse. Remains of pavements in *opus sectile*, as well as the fine architectural decoration, contribute to emphasize the role played by this grandiose building project, probably under imperial patronage - Zeno (474-491) - in the range of early Christian religious building in Syria.

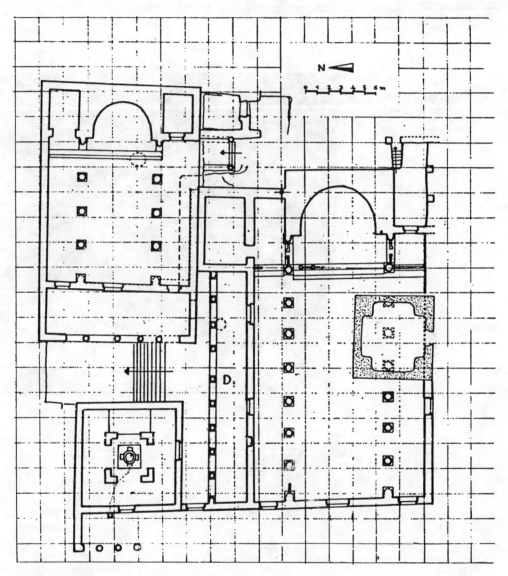

Fig. 298 **SYRIA**
HUARTE. Plan of the ecclesial complex.

This important complex consists of two churches and a baptistery. The south basilica (B) or basilica of Photius, of 483, as specified by an inscription, is divided into three aisles by columns, with enclosed apse: a well-preserved mosaic pavement, with geometrical, plant and animal motifs, covers the pavement of the aisles and the apsidal zone of the church. During excavation traces were found of a previous mosaic pavement, at a lower level, datable to the middle of the 5th c., evidence of the existence of an earlier phase preceding the present building.

The north basilica (A), or *Michaelion*, has three aisles with an enclosed apse, preceded by a narthex, more or less contemporary with the south basilica: the pavement mosaics are interesting, with animal scenes and panels with an offering scene and a representation of Adam. In the continuation of the *Michaelion*, and separated from it by set of steps, stands the quadrangular baptistery, a typology peculiar to Syria, with round pool at the centre.

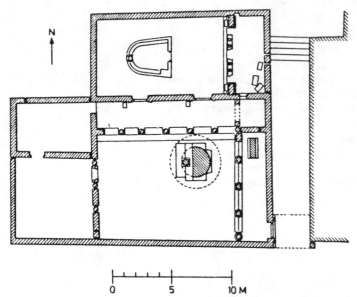

Fig. 299 **SYRIA**
QIRKBIZA. Plan of the church (early 4th c.).

The building, one of the oldest in Syria, of limited dimensions, has a rectangular plan with no apse: inside, at the centre, is a *bema* in which a sculpted seat is visible. The entrance is on the south side, where there is an atrium containing tombs, and a small *martyrium*.

Fig. 300 QIRKBIZA. Inside of the church.

Fig. 301 **TARRACONENSIS**
CENTCELLES (Tarragona). Roman villa.

In this villa a sumptuous mausoleum was installed (second half of 4th c.). The supposition that the monument belonged to an imperial personage (Constans?) is still a point of debate. From a typological point of view it is related to the numerous funerary complexes which include, among other things, examples of central-plan (Constantina's mausoleum at Rome) and rectangular-plan buildings (mausoleum of La Alberca [Murcia]).

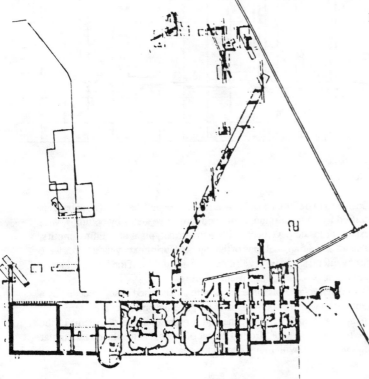

Fig. 302 **THESSALONICA**
Plan of the city.

Crossed by a tributary of the Via Egnatia, the great artery which ended at Constantinople, T. was at the centre of an important road network and a meeting-point between East and West. Its Christian community went back to the apostolic age and remained linked to the church of Rome until the start of the 8th c. In late antiquity the Christian city occupied the site of the Roman one, so that pagan buildings themselves were in part transformed into churches. This was the case, for example, of the church of St George, which was harboured in the Rotunda of Galerius, while the palace of the same Galerius probably became the residence of the Byzantine eparch.

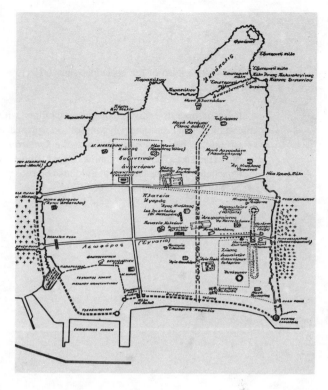

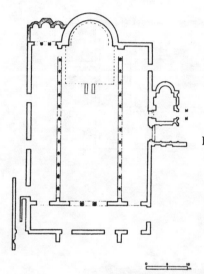

Fig. 303 Plan of the church of the *Acheiropoietos* ([*c.*] mid 5th c.).

This is a three-aisled basilica with semicircular apse and attached baptistery, dedicated to the Virgin, of whom it contained a precious and venerated icon.

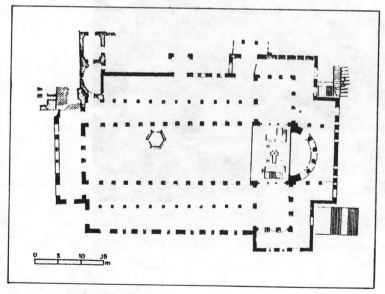

Fig. 304 **THESSALONICA**
Plan of the so-called *martyrium* of St Demetrius ([*c.*] between middle and second half of 5th c.).

After the transfer of the Prefecture of Illyricum, in 442, from Sirmium to Thessalonica, the cult of St Demetrius, originating in Sirmium, arrived at Thessalonica, giving rise to the legend that made Demetrius martyred in this city. So the *pseudo-martyrium*, with cruciform plan and five aisles, was built over a crypt that contained a venerated ciborium, later linked to a miraculous oil, but which in reality never contained a relic of the saint.

Fig. 305 Church of St George: the so-called "Rotunda". Detail of the mosaic decoration of the vault ([*c.*] late 5th c.).

Fig. 306 **THRACE**
MESEMBRIA. Plan of the episcopal church ([*c.*] second half of 5th c.).

A church of three aisles divided by masonry pillars, with its apse semicircular inside and polygonal outside, preceded by an atrium and provided with galleries: the *synthronon* is visible in the apse. The building shows clear affinities with the Constantinopolitan churches of St John of Studios and St Mary Chalcoprateia, both of *c.*450.

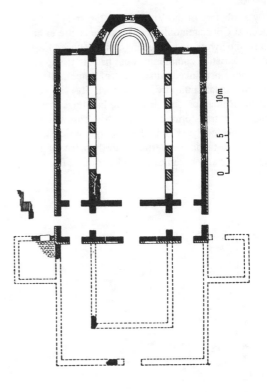

Fig. 307 MESEMBRIA. View of the episcopal church.

Fig. 308 **THRACE**
PERUSTICA (near Philippopolis). Plan of the so-called "Red Church" (6th c.).

The building, whose original purpose is unknown, now isolated, has a tetraconch plan, of the type widespread in the Balkans (Athens, Adrianople, Lin Ohrid), which differs, partly because of the outlets in the elevation, from the group of Syrian tetraconchs. A small baptistery with pool still *in situ* is also part of the complex.

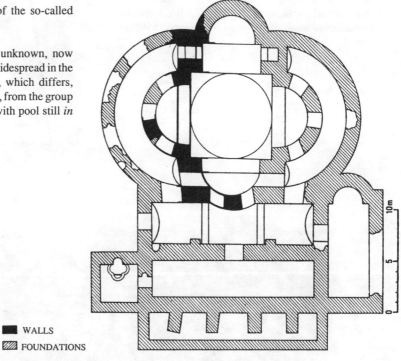

■ WALLS
▨ FOUNDATIONS

Fig. 309 PERUSTICA. View of the so-called "Red Church".

Fig. 310 **TIPASA**
Basilica.

This great basilica, one of the largest in Africa, was built within the walls in the west part of the city. It has one of the largest number of aisles of any basilica, seven, divided by pillars.

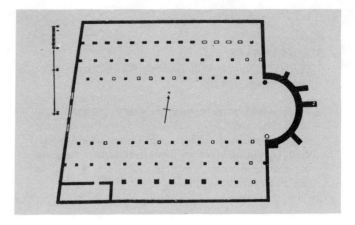

Fig. 311 **TOBIAS**
LE MAS D'AIRE (France). Detail of a sarcophagus lid (early 4th c.).

At the right of the lid the little T., naked, stretches out towards a great fish. On the rest of the lid are depicted: the sacrifice of Abraham, the miracle of the paralytic and the story of Jonah; on the front, from the left: the raising of Lazarus, Daniel among the lions, the Good Shepherd between two deceased, Adam and Eve, a baptism scene.

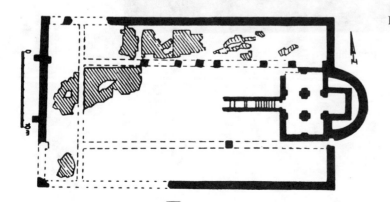

Fig. 312 **TOMI**
Plan of the "Great Basilica" (5th-6th c.).

The building, with three aisles and a single semicircular apse, contains a crypt reached by a staircase: it is one of the largest in the region, and probably functioned as the episcopal church.

Fig. 313 **TRADITIO CLAVIUM**
ROME. Catacomb of Commodilla. Fresco in the so-called "little basilica with *traditio clavium*" (first third of 6th c.).

The consignment of the keys, a *unicum* for cemeterial painting, recalls the Pelagian mosaic on the triumphal arch of S. Lorenzo *fuori le mura*. At the centre Christ, on the blue globe, beardless and youthful in appearance, has the jewelled codex on his knees and with his right hand makes the *traditio* to Peter, who has his hands veiled by his pallium; to his left Paul approaches bearing a bundle of six scrolls. Beside the Apostles appear the local martyrs: to the left Felix, who bears a crown and is flanked by the protomartyr St Stephen praying; to the right Adauctus, Felix's companion, also with a crown, and beside him Merita, now almost completely disappeared. Beneath, in line with the last two saints, two date-palms.

Fig. 314 **TRADITIO LEGIS**
VATICAN CITY. Vatican Grottos. Sarcophagus front with *traditio legis* (mid 4th c.).

At the centre, joined for the first time with Passion scenes, a *traditio legis*: Christ young, beardless, with round boyish face, framed by long curls, is seen as symbol of the eternal immutability of his divine essence. He is flanked by two Apostles and sits with his feet resting on the wind-blown cloak of a young man who personifies the *caelus*, in heavenly glory. With his left hand he hands over to Peter, who approaches with his hands veiled by an *orarium*, the open scroll of the law. The entire context is connected with the episodes that flank the central scene, punctuated by columns decorated with vine-shoots and grape-clusters: to the left, the sacrifice of Abraham, Peter led to martyrdom accompanied by a cross-bearer and Paul acclaiming; to the right, Christ before Pilate. The Lord has an almost feminine appearance; he is not yet the severe, bearded judge of later depictions.

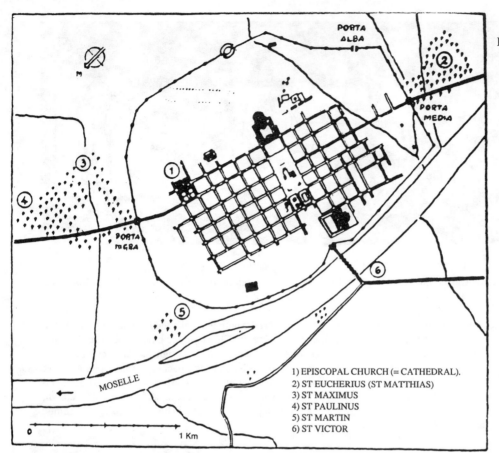

Fig. 315 **TRIER**
Plan showing in particular the cemeterial areas of the city.

1) EPISCOPAL CHURCH (= CATHEDRAL).
2) ST EUCHERIUS (ST MATTHIAS)
3) ST MAXIMUS
4) ST PAULINUS
5) ST MARTIN
6) ST VICTOR

Fig. 316 **TROPAEUM TRAIANI**
Plan of the Romano-Byzantine fortress.

Abandoned towards the start of the 7th c., Trajan's city preserves the remains of four urban Christian basilicas, and one cemeterial one, built in the period between the 4th and 6th cc.

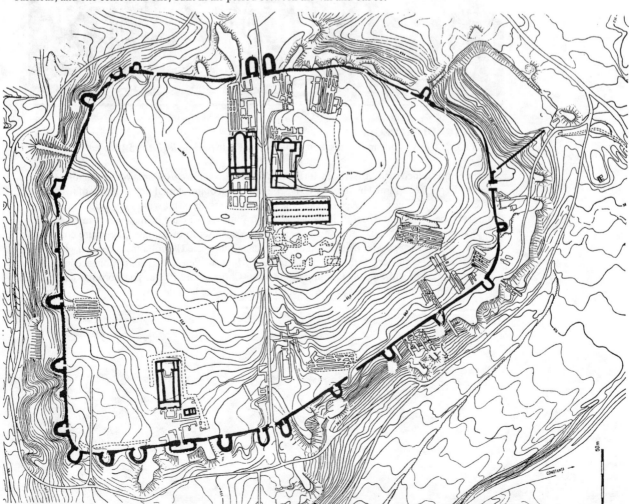

Fig. 317a, 317b
VATICAN
Plans of St Peter's basilica.

Built by the emperor Constantine, the basilical complex consisted of a staircase which gave access to the vestibule and from there to an atrium with lateral porticos, in whose centre was a fountain for ritual ablutions, and thence to the portico. From there five doors gave access to the five-aisled basilica. Its essential structure remained unchanged for over a thousand years.

A - Central aisle
B - Transept
C - Apse with altar of the *Confessio*
D - Side aisles
E - Side aisles
F - Portico in front of the five entrance doors
G-I - Lateral porticos
H - Atrium
K - Vestibule
L - Flat area at the summit of the staircase

Fig. 318 Vatican necropolis. Area near Peter's tomb.

Alignment of the mausoleums from East to West. Observe how the latter have respected the area of field P affected by the accommodation of the Apostle's tomb.

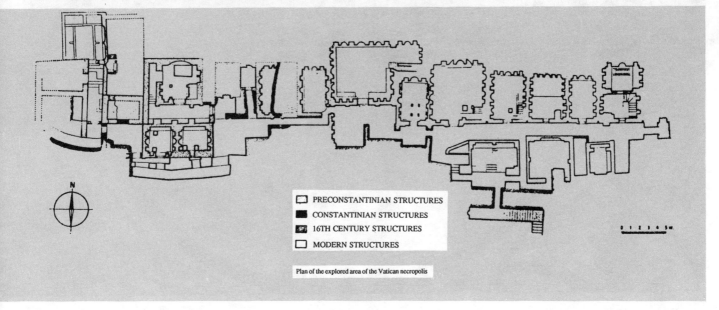

PRECONSTANTINIAN STRUCTURES
CONSTANTINIAN STRUCTURES
16TH CENTURY STRUCTURES
MODERN STRUCTURES

Plan of the explored area of the Vatican necropolis

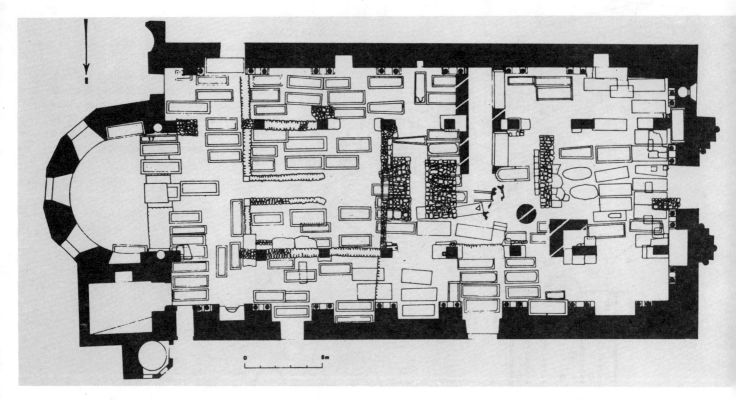

Fig. 319 **VIENNE**
Plan of St Peter's basilica.

For a long time this city disputed with Arles the primacy over the southern regions of Gaul. Excavations made inside St Peter's church between 1860 and 1864 have brought to light structures and tombs from the early Christian era. The basilica had a funerary character.

Fig. 320 **VIRGA**
NAPLES. Frescoed vault of a cubicle of the catacomb of S. Gennaro, (4th c.).

An isolated person, in tunic and pallium, is shown inside a *clipeus*, among foliage, holding in his right hand the thaumaturgical *virga* that, for the Christians, signified the sacramental power, divine, of liberation from physical imperfections and death.

INDEX

INDEX

Aaron, **1**, 41, 123, 504, 561, 562, 575, 616, 635, 752, *1*
Aba, Mar, **1**, 3, 583, 674, 810, 837
Aba I, Mar, **1**, 203, 216, 663
Abas, 439
Abbat, Abbess, **1**, 27, 119, 130, 184, 320, 371, 411, 566, 740, 866
Abda, Mar, 106
Abd el-Aziz, 438
Abd-al Malik, 48
Abdias, Ps.-, 1, 676, Passion of, 429
Abdias of Babylon, **1**
Abdiso of Nisibis, 67, 106, 110, 537
Abdiso bar Berika (Ebedjesu), 810
Abdon and Sennen, **1**, 631, *2*
Abel, **1**, 2, 71, 100, 294, 504, 550, 652, 756, 771, 777, *3*
Abelard, 178
Abelites (Abeloitae), **2**
Abellinum (Avellino), 424
Abercius, Epitaph of, 52, 232, 263, 280, 293, 426, 598, 674, 778, 803, Life of, 110
Aberdaron, 131
Aberlemno cemetery, Angus, 132, Slab-cross, 132
Abgar, **2**
Abgar V "Uchama" ("The Black"), 2, 7, 10, 79, 263, 836
Abgar IX the Great, 2, 110, 263
Abila, 636
Abitina, 246, 533, Martyrs of, **2**, 13, 137
Ablabius, **2**, 150
Abortion, **2**, 122, 318, 739, 780
Abraham, **2**, 11, 19, 32, 40, 111, 148, 151, 199, 205, 294, 299, 309, 315, 316, 317, 320, 341, 349, 389, 391, 398, 400, 407, 425, 434, 451, 455, 461, 462, 471, 504, 506, 525, 550, 573, 575, 649, 683, 687, 688, 730, 751, 756, 791, 798, 802, 809, 818, *5*, Abraham's bosom, 192, 731, *see also* Paradise
Abraham bar Dasandat, 840
Abraham bar Kili, 445
Abraham of Albatanzi, **3**, 81
Abraham of Beth Rabban, **3**, 374
Abraham of Beyt Qiduna (or Qidunaya), **3**
Abraham of Clermont, **4**
Abraham of Ephesus (Abrahamios), **4**, 216, 640
Abraham of Harran (Mesopotamia), **4**
Abraham of Kashkar, **4**, 106, 271
Abraham of Nisibis, 106
Abraham of Pbou, **4**
Abraham the Confessor, **4**
Abrasax, **4**, 217, 377, 517, *6*
Absamya, 214
Abu Girga, 268
Abula, 148
Abu Mina, 265, 267, 553
Abundius (or Abundantius) of Como, **4**, 420, 735, 767
Abu Sarbut, 66
Acacian schism, **4**, 5, 29, 36, 40, 105, 125, 142, 197, 241, 251, 268, 272, 325, 339, 366, 399, 438, 569, 570, 642, 645, 678, 744, 884
Acacius of Aleppo, 729
Acacius of Beroea, **4**, 107, 609, 744
Acacius of Caesarea, **4**, 5, 49, 58, 78, 95, 215, 310, 395, 546, 550, 658, 665, 727, 767, 785
Acacius of Constantinople, **4**, 5, 6, 11, 47, 49, 58, 78, 107, 112, 113, 118, 138, 197, 220, 275, 296, 305, 325, 330, 339, 343, 432, 438, 444, 447, 527, 550, 569, 642, 645, 657, 678, 743, 744, 780, 783, 840, 884

Acacius of Seleucia, **5**
Academy, 53
Acathistus, **5**, 26, 346
Acaunensian martyrs, Passion of, 295, 347, 544, 754, 821
Acci, 148
Accidia, **5**, 469, 781, 872
Acclamations, Liturgical, **5**, 6, 160, 228, 232, 498, 499, 541, 744, 793, 816
Accusations against the Christians, **6**, 39, 60, 82, 95, 99, 134, 167, 209, 271, 326, 369, 389, 517, 671, 672, 674, 812, 818
Acephali, **6**, 446, 569, 638, 678, 775, 841
Acesius, **7**
Achaia, **7**, 38, 96, 106, 172, 201, 225, 356, 510, 515, 670, 834, *7*
Acheiropoietai, **7**, 834
Achillas, Arian priest, **7**
Achillas, bishop of Alexandria, **7**, 23, 551
Achilleus, bishop of Spoleto, **7**, 730, 804, 831
Achilleus and Nereus, *see* Nereus and Achilleus
Achilleus of Larissa, 835
Acholius of Thessalonica, **8**, 41, 218, 744
Achtarak, 80
Acilius Glabrio, **8**, 245, 711
Acilius Severus, **8**
Acoemetae ("sleepless ones"), **8**, 22, 138, 195, 417, 438, 500, 570, 679, 708
Acoluthia, **8**, 135
Acolyte, 6, 232
Acquaviva, 422
Acta Proconsularia, 672
Actistetae, **8**
Acts, 13, 57, 64, 65, 67, 200, 739
Acts, Passions, Legends, 191, 209, 533, 707, *see also* Martyrology
Acts and Canons, Conciliar, **8**, 104, 244, 318
Adaima, 266
Adaloald, 17
Adam and Eve, **2**, 9, 10, 44, 45, 46, 71, 107, 133, 135, 137, 163, 166, 183, 198, 226, 227, 242, 243, 255, 257, 260, 261, 271, 293, 299, 309, 318, 321, 331, 340, 349, 353, 357, 358, 359, 389, 414, 415, 433, 452, 458, 461, 462, 463, 465, 504, 517, 519, 525, 528, 542, 549, 598, 612, 621, 629, 630, 631, 668, 670, 683, 686, 701, 716, 717, 718, 730, 755, 775, 781, 788, 791, 802, 803, 815, 816, 818, 849, 859, 860, *8*, Book of Adam, 9, Life of Adam, 56, 718, Life of Adam and Eve, 9, 56, Teachings of Adam, 446, *see also* Apocalypse of Adam
Adamantius, **10**, 229
Adamitae (Adamiani), **10**
Adamnan, **10**, 118, 130, 186, 411, 412, 426, 762
Adane (Aden), 832
Adauctus and Felix, *see* Felix and Adauctus
Ad Autolycum, 60, 166, 176, 448, 831
Ad Bestias, **10**, 186, *9*
Addai or Addaeus, 2, **10**, 11, 48, 79, 263, 598, 637, 674, 808, Doctrine of 2, 10, 263, Anaphora of Addai and Mari, 10, 33, 34, 277, 496, 501, 810
Addas, 11, 519
Ad Deum, **11**
Ad Donatum, 121, 211, 212
Adelphius, 778
Adeodatus, **11**, 743
Adeodatus I (Deusdedit), 645
Adeodatus II, 645, 743
Adi ibn Zaid, 65, 472
Adiabene, **11**, 67, 106, 263, 271, 674, 809

Adimantus the Manichaean, **11**, Against the disciples of, 11
Ad Metalla, **11**, 188
Ad Novatianum, **11**, 473, 603, 784
Ado of Vienne, 537, 582, 663, 869
Adoptianism, -ists, **11**, 53, 83, 84, 109, 164, 303, 507, 566, 583, 694, 621, 640, 660, 663, 685, 686, 741, 830, 867, 885
Adra, 106
Adrianople, 32
Ad Sanctos, **11**, 157, 171, 212, 347, 535, 753, 814, 835
Adulis, 289
Adultery, 8, **12**, 161, 243, 270, 273, 318, 328, 409, 556, 701, 781, 793
Adventor, Solutor and Octavius, 854
Advent, 186, 493, 865
Aed of Sletty, 576
Aedesius, 289, 307
Aegina, 7
Aelianus, 351
Aelius Aristides, 121
Aelius Lampridius, 388
Aelius Paulinus, 321
Aemilianus, 487, 672, Life of, 127
Aemilius Leto, 188
Aeneas of Gaza, **12**, 589, 704
Aeon (AIΩN), **12**, 27, 109, 163, 204, 226, 239, 327, 328, 352, 353, 397, 519, 526, 542, 603, 610, 698, 717, 724, 752, 769, 771, 802, 859
Aerius, **13**, 303
Aeschylus, 639
Aetherius, 513
Aetius Flavius, **13**
Aetius of Antioch, **13**, 35, 37, 40, 42, 49, 77, 95, 96, 134, 162, 195, 295, 296, 305, 334, 357, 447, 480, 487, 555, 683, 684, 783, 832, 852, 859
Aetius of Thessalonica, 833
Africa, 2, 8, 11, **13**, 17, 18, 24, 27, 29, 35, 53, 55, 60, 69, 78, 79, 96, 98, 104, 105, 107, 109, 128, 142, 143, 148, 151, 155, 156, 167, 170, 171, 173, 174, 175, 177, 185, 188, 192, 202, 212, 225, 228, 230, 233, 246, 247, 248, 249, 282, 307, 315, 334, 369, 377, 386, 400, 410, 450, 464, 486, 490, 497, 498, 517, 532, 535, 541, 552, 554, 561, 572, 604, 607, 613, 616, 617, 618, 631, 641, 645, 651, 653, 660, 665, 666, 667, 673, 679, 731, 741, 742, 756, 761, 762, 765, 768, 778, 790, 795, 812, 814, 816, 822, 823, 826, 838, 849, 853, 859, 861, 868, 870, *10-14*,liturgy, 497
Agapè, **16**, 48, 63, 86, 90, 91, 123, 148, 173, 247, 274, 292, 297, 506, 518, 541, 699, 707, 731, 741, 800, *15*
Agape, Chionia and Irene, **16**, 35, 533, 534, 833
Agapeti (fem. agapetae), **16**, 284, 318, 440
Agapetus I, 43, 150, 241, 365, 486, 553, 555, 642, 645, 743, 769
Agapetus of Constantinople, **16**, 655
Agapetus of Synnada, 684
Agapius, 32, 66, 158, 372
Agatha, **16**
Agathangelos, **16**, 80, 81, 83, 84, 368, 478
Agatho, 16, 119, 197, 200, 446, 645, 743, 823
Agathodorus, 17
Agathonicus of Tarsus, **17**
Agathonicus of Tarsus, Ps.-, 96
Agathonice and companions, **17**, 533, 534
Agathopus and Theodulus, 833
Agaune, 544

Agde, C. of (506), **17**, 138, 188, 335, 360, 582
Agelius of Constantinople, **17**
Agen (Aginnum), 685
Aggai, 10, 674, 808
Agilulf, **17**, 103, 187, 254, 307, 366, 506, 774, 822
Agnellus, **17**
Agnes, **17**, 615, 741, *16-17*, Passion of, 270
Agnoetae, **17**, 185, 343, 794, 822, 841
Agnus Dei, **17**, 542, 768, *18*
Agraecius of Sens, **18**, 754
Agraphon, **18**
Agrestius, **18**, 652, 801
Agricius of Antibes, 555
Agricius of Trier, **18**, 348, 850
Agrigento (Agrigentum), 425
Agrippa I, **18**, 429, 432, 671
Agrippa II, **18**
Agrippa Castor, **18**
Agrippinus of Carthage, 13, **18**, 26, 109, 146, 205, 212
Agriustia, **18**, 96, 380
Ahay, 537
Ahenny (co. Tipperary), 132, crosses, 132
Aidan of Lindisfarne, **18**, 129, 130, 159
Aidesia, 31
Aila, 636
Aileran the Wise, **19**, 412
Aimoin of Fleury, 526
Ain Ghorab, 228
Ain Karem, 434
Ain Zara, 16
Aipgitir Crábaid, 412
Aisle, 170
Aistulf, 505
Aithalla of Edessa, **19**, 263
Aix (Aquae), 654
Akiva (or Akiba), rabbi, **19**, 110, 563, 786
Aksum, **19**, 289, 290, 307, 315, 472
Alaethia, 278
Alahan, 87
Alamanni, 183, 187, 198, 336, 347, 349
Ala Milaria (Benian), 247
Alan of Lille, 178
Aland, Barbara, 523-24, 545-46, 570-71, 702, 712-13
Alans, 349
Alaric I, 13, 17, **19**, 67, 96, 201, 283, 334, 356, 397, 409, 416, 521, 559, 612, 624, 666, 729, 770, 795, 861, 863
Alaric II, 183, 356
Alb, 864
Al-Bagawat, 268
Alba Pompeia (Alba), 421
Alban of Verulamium, **19**, 128, 129
Albania, 403, 556
Albania of Caucasus, **19**, 439, *19*
Albenga (Albingaunum), 421, 573
Albina, 550
Albintimilium (Ventimiglia), 421
Al-Bîrunî, 519
Alboin, 505
Alcalà de los Gazules, 106
Alcalá de Henares, 790
Alcalà la Real, 107
Alcaudete, 106, 148
Alcibiades of Apamea, 269
Alcinous, 204, 558, 587, 602, 690, 693
Alcisone, 283
Alconétar (Cáceres), 512
Alcuin, 43, 101, 117, 118, 161, 242, 278, 394, 655, 706, 753, 862
Alcuin, Ps.-, 412
Aldfrid of Northumbria, 10, 19
Aldhelm, **19**, 154, 202, 411

Volume I contains pages 1-578; Volume II contains pages 579-1100 and the index

Alemanni, 459
Aleppo, see Beroea
Alerion, 32
Alessandria, 421
Alethia, 651
Alexander, Ps.-, 586
Alexander I, pope, 644
Alexander Romanus, 838
Alexander Severus, 22, 126, 271, 371, 434, 460, 672
Alexander of Alexandria, 7, 20, 21, 23, 24, 48, 63, 64, 72, 76, 90, 93, 142, 164, 185, 207, 301, 343, 507, 538, 595, 743, 833, 852
Alexander of Antaradus, 196
Alexander of Antioch, 20, 50
Alexander of Apamea, 20, 53
Alexander of Aphrodisias, 20, 74, 75, 585, 587, 601
Alexander of Cappadocia, 582, 688
Alexander of Constantinople, 21, 302, 344
Alexander of Cyprus, 21
Alexander of Hierapolis, 20, 21, 372, 444, 546, 550, 652, 884
Alexander of Jerusalem, 21, 84, 120, 180, 223, 225, 432, 486, 619, 639, 822
Alexander of Latium, 143
Alexander of Thessalonica, 7, 20, 21
Alexander of Tipasa, 14
Alexander the Acoemete, 8, 22, 401, 499, 514, 597, 729
Alexander the Great, 63, 184
Alexander the Novatianist, 337
Alexandria, 1, 6, 7, 11, 12, 20, 22, 25, 28, 29, 31, 32, 34, 36, 37, 44, 47, 48, 50, 54, 56, 58, 64, 66, 71, 72, 73, 75, 76, 82, 84, 89, 93, 105, 111, 113, 119, 124, 138, 140, 142, 147, 148, 150, 151, 152, 163, 178, 180, 185, 187, 189, 195, 196, 200, 202, 214, 216, 219, 223, 224, 225, 235, 236, 237, 238, 257, 258, 264, 265, 266, 282, 284, 289, 290, 291, 293, 294, 298, 300, 301, 303, 307, 310, 315, 340, 343, 360, 409, 417, 432, 438, 448, 449, 453, 455, 466, 480, 484, 486, 487, 495, 496, 499, 505, 507, 528, 538, 551, 564, 568, 569, 594, 596, 604, 614, 618, 632, 638, 639, 641, 642, 643, 648, 662, 672, 677, 678, 679, 682, 687, 691, 699, 704, 705, 713, 722, 741, 742, 743, 752, 760, 761, 763, 768, 769, 773, 792, 805, 806, 814, 822, 826, 827, 828, 830, 831, 841, 848, 875, 884, C. of (231/232), 23, C. of (c.305), 23, C. of (c.323), 23, C. of (c.324), 23, 185, C. of (338), 23, C. of (346), 23, C. of (362), 24, 58, 78, 92, 195, 302, 401, 487, 508, 626, 852, C. of (363), 24, C. of (c.370), 24, C. of (399), 24, liturgy, 34, 45, 50, 257, 496, Alexandrian-Ethiopian rite, 290
Alexius, 4, 25, 535
Alfred the Great, 518
Algasia, 126
Al-Hareth bar Gabala, 662
Al-Hirah, 216, 809
Alicula, 253
Alijezares (Murcia), 148
Aliki, 174
Allegory, 9, 23, 25, 28, 30, 37, 46, 50, 55, 67, 73, 89, 98, 106, 111, 114, 139, 152, 183, 187, 188, 214, 223, 224, 228, 281, 294, 295, 300, 301, 309, 310, 311, 312, 331, 340, 341, 363, 365, 379, 382, 395, 455, 457, 504, 511, 529, 548, 560, 578, 598, 621, 683, 686, 703, 716, 721, 754, 763, 772, 786, 801, 802, 824, 825, 832, 867, 879, 880, see also Typology
Alleluia, 5, 6
Allogenes, 71
Almachius, 25
Almyris Apokoronou, 209
Alms - Almsgiving, 15, 25, 151, 192, 211, 319, 351, 372, 461, 493, 507, 517, 519, 548, 555, 570, 609, 703, 737, 781, see also Assistance and Charity
Al-Nakhla, 576
Alogi, see Gaius and the Alogi
Alpes Cottiae, 421
Alpha and Omega, 26, 133, 209, 210, 279, 292, 334, 349, 435, 471, 554, 573, 597, 606, 607, 675, 802, 803, 813, 814
Alphabetical Poems, 5, 26, 332, 362, 765
Altar, 123, 171
Altar of Victory, 26, 28, 124, 218, 325, 360, 552, 804, 858, 859
Altava, 14
Altercatio De Anima, 26
Altercatio Heracliani laici cum Germinius episcopo, 26, 27
Altercationes, 27
Altinum, 371, 678
Altus prosator, 186
Alypius, 25, 27, 147, 148, 225, 488, 702, 739, 746, 774, 797
Ama or Amma, 27
Amalafrida, 861
Amalaric, 183
Amalarius of Metz, 750
Amalasuntha, 357
Amalfi, 38
Amandus, 27
Amandus of Bordeaux, 27
Amante Simoni, Clorinda, 322-23
Amaraz, 19
Amasea of Pontus, 27, 91, 122, 305, 702, 822
Amastris, 27, 637
Amatorium Canticum, 27
Amatus, 213
Amaya, 148
Ambo, 7, 17, 172
Ambrose, Ps.-, 676
Ambrose, disciple of Didymus, 28
Ambrose, friend of Origen, 28
Ambrose of Milan, 1, 2, 9, 11, 17, 25, 26, 27, 28, 29, 30, 33, 35, 41, 42, 50, 52, 54, 55, 56, 57, 60, 63, 64, 67, 78, 92, 98, 105, 107, 116, 118, 123, 125, 126, 127, 149, 151, 152, 159, 160, 161, 166, 167, 168, 173, 177, 178, 183, 187, 190, 192, 196, 202, 205, 218, 220, 224, 225, 232, 238, 242, 243, 255, 257, 261, 269, 277, 279, 286, 288, 293, 296, 298, 299, 303, 306, 307, 308, 311, 312, 320, 322, 327, 330, 335, 336, 337, 340, 341, 347, 350, 359, 360, 361, 370, 371, 380, 396, 397, 398, 400, 401, 419, 420, 431, 448, 450, 452, 454, 457, 464, 475, 478, 483, 493, 494, 495, 497, 498, 499, 501, 503, 504, 511, 521, 528, 532, 533, 538, 540, 546, 547, 548, 549, 559, 560, 562, 577, 584, 589, 592, 604, 609, 612, 615, 616, 617, 620, 637, 643, 646, 649, 655, 656, 658, 660, 665, 666, 672, 676, 682, 697, 702, 704, 708, 709, 712, 716, 718, 719, 721, 722, 723, 725, 731, 734, 737, 738, 742, 744, 745, 749, 750, 758, 765, 770, 776, 780, 782, 787, 788, 796, 804, 816, 817, 824, 828, 833, 850, 852, 854, 857, 858, 863, 870, 871, 875, 876, 878, 881, 882, 884, 20, Life of, 121, 660
Ambrosian Liturgy, 29, 155, 228, 334, 335, 370, 464, 490, 493, 497, 499, 500, 797
Ambrosiaster, 30, 152, 243, 311, 312, 328, 359, 382, 406, 417, 521, 542, 562, 615, 658, 665, 727, 803, 876
Ambrosius Autpertus, 55, 839
Amen, 6, 34
Amice, 865
Amid (now Diyarbekr), 809
Amida, 30
Amiternum (now San Vittorino), 423
Ammaidara (Haidra), 14
Ammianus Marcellinus, 30, 31, 92, 282, 302, 348, 351, 486, 592
Ammonas, 31
Ammonius (Hermiae of Alexandria), 8, 31, 63, 124, 298, 446, 519, 590, 687, 785, 805, 884
Ammonius Saccas, 31, 74, 75, 373, 374, 558, 585, 586, 587, 589, 603, 606, 619, 624, 639, 691, 692, 695, 699, 805
Ammon, 31, 515
Ammon of Adrianople, 32
Amoraim, 32, 563, 812
Ampelum (Zlatna), 217
Amphilochia, 234
Amphilochius, Ps.-, 678
Amphilochius of Iconium, 32, 35, 114, 115, 142, 143, 361, 362, 403, 484, 611, 686, 724, 778, 841
Amphion of Epiphania, 32
Amphion of Nicomedia, 32
Amphipolis, 515
Amphitheatre, 273, 296
Amphitryon, 32
Ampuriae (Castelló de Ampurias), 813
Amra Coluimchille, 413
Amulets, 32, 232, 271, 291, 333, 371, 377, 517, 759
Amwas, 187
Anachoresis, 231
Anacletus, 96, 644
Anaemius, 28, 33, 78
Anamnesis, 10, 33, 34, 50, 292, 293, 336, 496, 532, 851
Ananias, 432
Ananias and Sapphira, 33, 21
Ananias of Sirak, 81, 575
Anaphora, 10, 33, 82, 113, 135, 265, 277, 290, 293, 335, 383, 490, 495, 498, 499, 541, 679, of St Basil, see Basil, Liturgy of, of St James, 33, 34, 50, 135, 429, 496, 810, of St John Chrysostom, 34, 50, 135, 441, 442, of St Justin, 34, of St Mark, 34, 35, 265, 496, of St Peter Sarar, 496, of Timothy of Alexandria, 34, of the Twelve Apostles, 135
Anasartha (now Khanasir), 808
Anastasia, 16, 35
Anastasiana, 35, 244
Anastasius, secretary to Eudoxia, 35
Anastasius, martyr, 35, 36
Anastasius, emperor, 5, 6, 27, 36, 134, 197, 251, 268, 272, 329, 399, 438, 439, 447, 516, 569, 642, 645, 662, 670, 742, 743, 746, 827, 828, 829, 846, 853
Anastasius I, pope, 641, 645, 863
Anastasius I of Antioch, 36, 49, 140, 233, 283, 297, 339, 363, 582
Anastasius II, pope, 645, 804
Anastasius II of Antioch, 36, 142, 237, 346
Anastasius Magundat, 36, 343
Anastasius of Alexandria, 219
Anastasius of Jerusalem, 268, 547
Anastasius of Thessalonica, 36, 642
Anastasius the Apocrisiary, 36, 826
Anastasius the Librarian, 4, 36, 346, 540, 826
Anastasius the Sinaite, 35, 36, 37, 52, 56, 76, 180, 311, 318, 437, 582, 639, 655, 682, 728
Anathalon, 559
Anathema, 37, 38, 65, 112, 159, 205, 207, 214, 218, 275, 304, 314, 342, 444, 465, 484, 530, 538, 546, 550, 569, 570, 585, 594, 623, 659, 685, 726, 737, 747, 783, 798, 826, 827, 841, 852, 884
Anatolius, Ps.-, 188
Anatolius of Constantinople, 37, 196, 271, 455, 458, 484, 523, 532, 752, 760
Anatolius of Laodicea, 37, 189, 272, 736
Anaxyrides, 253
Anazarbus, 37, 62, 95, 175, 546, 825
Anchor, 803
Anchoratus, 200
Anchorite, -ism, 4, 6, 31, 37, 44, 48, 84, 112, 115, 150, 184, 230, 264, 266, 276, 327, 330, 371, 379, 500, 567
Ancona, 105, 422, 423
Ancoratus, 282, 637
Ancyra, 37, 52, 85, 87, 95, 113, 245, 296, 333, 596, 597, 822, 829, 884, C. of (314), 16, 32, 137, 142, 154, 162, 205, 237, 333, 522, 617, C. of (c.375), 333, C. of (358), 77, 113, 154, 333, 343, 395, 783
Andrew, apostle, 7, 24, 27, 38, 61, 112, 194, 282, 284, 344, 619, 643, 680, 764, 775, Acts of, 38, 112, Acts of Andrew and Matthias in the land of the ant..., 543
Andrew, priest, 37
Andrew of Caesarea, 38, 55, 137, 311, 610
Andrew of Crete, 38, 135, 160, 209, 343, 538
Andrew of Samosata, 38, 72, 214, 444, 546, 729
Angels, 3, 6, 12, 28, 38, 39, 41, 53, 54, 80, 86, 90, 107, 108, 111, 113, 116, 122, 128, 132, 135, 145, 158, 163, 186, 190, 195, 202, 203, 213, 219, 223, 226, 227, 239, 240, 244, 251, 255, 262, 269, 288, 290, 315, 323, 333, 340, 341, 345, 352, 356, 368, 583, 591, 597, 610, 614, 615, 621, 622, 626, 631, 636, 649, 651, 664, 683, 689, 696, 699, 710, 715, 717, 718, 720, 745, 749, 758, 774, 776, 780, 781, 808, 815, 831, 834, 839, 843, 844, 850, 854, 872, 876, 885, 22
Angers (Andigava), 40, C. of (453), 40, 482, 821
Angoulême Sacramentary, 224
Angrisani Sanfilippo, Maria Luisa, 30-31, 103, 119, 137, 149-50, 211, 334, 356-57, 451, 459-60, 505, 505-6, 580-81, 600, 610, 745, 804, 846, 873-74
Ani, 81
Anianus of Orléans, 40, 166, 241
Anicetus, 145, 258, 644, 701, 740
Anicia, Juliana, Roman noblewoman, 40
Anicia, Juliana, widow, 40, 88, 225
Anicii, 149, 365, 666, 804
Anima Mundi, 12, 40, 73, 74, 75, 209, 353, 455, 463, 557, 558, 586, 588, 590, 592, 601, 606, 614, 639, 698, 699, 717, 796
Anisius of Thessalonica, 41, 125, 782, 833
Annals of Ulster, 412, 413
Annals of the Four Masters, 412
Annesi, 114
Annexes, 174
Annianus, 41
Annianus of Celeda, 41
Annius Bassus, 2
Annunciation, 4, 5, 38, 39, 41, 43, 81, 82, 135, 138, 322, 333, 451, 466, 511, 538, 539, 631, 712, 23
Anointing, 611, 670, 710, 751, see also Oil
Anomoeans, 4, 13, 42, 49, 62, 77, 78, 114, 195, 295, 296, 301, 343, 395, 441, 683, 825, 832, Against the Anomoeans, 339
Anonymous Apollinarist, 42
Anonymous Apollinarists, 42
Anonymous Arians, 43
Anonymous of Lyons, 537
Anonymous of Melk, 654
Anonymus (or Antoninus) Placentinus, 43, 379, 426
Anonymus Gallus, 43
Anonymus Valesianus, 43
Ansedunum, 92
Anselm, 101, 683, 758
Anselm of Laon, 153, 352
Antalya, 89
Antarados (now Tartus), 808
Antelias, 154
Anthedon, 339
Anthemius, 778, 829
Antherus, 315, 644, 741
Anthimus of Constantinople, 135, 197,

401, 553, 596, 645, 662, 677, 678, 749, 769, 826, 884
Anthimus of Nicomedia, **43**
Anthimus of Trebizond, **43**
Anthimus of Tyana, **43**, 114, 143, 854
Anthimus of the Via Salaria, 43
Anthologia Latina, **43**, 119, 144, 256, 612
Anthologia Palatina, **44**, 664
Anthologia Salmasiana, 279
Anthony, abbat, 22, 31, 37, **44**, 84, 93, 116, 132, 184, 224, 230, 264, 291, 319, 327, 380, 514, 521, 532, 552, 567, 664, 688, 702, 768, 817, Life of, 44, 81, 84, 93, 94, 121, 140, 227, 230, 269, 273, 431, 475, 597, 746, 793, 799, 817
Anthony, hagiographer, **44**
Anthony, poet, **44**
Anthony of Baghdad, 65
Anthony of Choziba, **44**
Anthony of Tarsus, 185
Anthropology, 12, 23, 34, **44**, 99, 110, 113, 117, 123, 183, 223, 242, 251, 255, 285, 340, 352, 405, 412, 414, 415, 429, 553, 557, 584, 614, 621, 664, 665, 670, 672, 718, 719, 746, 767, 786, 794, 815, 851 819, 825
Anthropomorphism, 17, **46**, 82, 90, 94, 96, 149, 309, 311, 323, 328, 354, 355, 369, 414, 621, 622, 666, 733
Antichrist, **46**, 55, 188, 218, 248, 269, 372, 415, 614, 625, 733, 820, 839
Antidicomarianitae, **47**
Antifons, 729
Antigonus, bishop of Magina, 360
Antigonus of Caristus, 121
Antikeimenon, 459
Antilegomenon, **47**, 58
Antinopolis, 264
Antioch, 13, 18, 20, 22, 23, 30, 36, 37, 44, **47**, 54, 56, 58, 69, 71, 72, 78, 84, 86, 89, 90, 91, 92, 102, 106, 110, 111, 114, 115, 135, 142, 148, 151, 152, 159, 163, 164, 169, 174, 186, 187, 195, 202, 211, 214, 219, 223, 225, 235, 236, 245, 246, 251, 257, 263, 264, 276, 284, 289, 293, 294, 297, 298, 300, 301, 302, 303, 307, 308, 310, 315, 325, 340, 343, 360, 363, 364, 369, 382, 409, 428, 432, 440, 441, 448, 456, 466, 480, 484, 490, 493, 495, 496, 499, 507, 510, 533, 541, 550, 552, 553, 561, 562, 569, 572, 594, 596, 617, 618, 619, 637, 642, 643, 657, 660, 662, 663, 673, 679, 682, 684, 685, 688, 704, 715, 720, 722, 746, 755, 760, 761, 763, 768, 773, 779, 807, 808, 810, 827, 830, 831, 832, 855, 870, 875, 876, *25-27*, C. of (251), 371, C. of (264), 301, 368, 371, 400, C. of (268), 48, 246, 324, 396, 400, 518, C. of (272), 371, C. of (324-325), 20, 48, 77, 237, 299, C. of (327), 48, 77, 84, 196, 303, C. of (338), 48, C. of (339), 302, C. of (340), 303, C. of (341: in Encaenis), 49, 77, 92, 93, 142, 162, 205, 207, 271, 303, 325, 337, 460, 472, 507, 522, 527, 546, 617, 757, 759, 854, C. of (357), 5, 49, 295, C. of (360/361), 49, 343, C. of (c.362), 49, C. of (363), 49, 78, 95, 215, 305, 395, 550, 665, 843, C. of (379), 49, 59, 550, C. of (c.380), 49, 556, C. of (383), 49, C. of (389), 49, C. of (397), 363, C. of (c. 422), 830, C. of (424), 49, C. of (432), 49, C. of (435), 49, C. of (440), C. of (445), 49, C. of (447/448), 49, C. of (471), 49, C. of (478), 49, C. of (482), 49, C. of (485), 49, C. of (508-509: conciliabulus), 49, C. of (c. 538), 276, C. of (542), 49, 276, C. of (565), 49, Schism of, 4, 20, 28, 48, 49, 236, 325, 550, 641, 660, 665, 678, 749, 831, 841, liturgy, *see* Syro-occidental (Antiochene) rite
Antioch Chalice, 51
Antioch in Caria, C. of, **47**
Antiochene creed/formula, 77, 207, 767, 783, 853

Antiochicus, 459
Antiochus Strategius, **52**
Antiochus of Ascalon, 53, 557, 585, 587, 602, 797, 805
Antiochus of Ptolemais, **51**, 609
Antipater of Bostra, 35, **52**
Antiphonaries, 503
Antipraedicamenta, 124
Antiquity, **52**, 166, 358, *see also* Novelty
Antisemitism, **52**, 436
Antithesis, **52**, 53, 226, 329, 359, 397, 408, 415, 523, 578, 721, 731, 734, 736, 871
Antitrinitarianism, **53**
Antonine column, 253
Antoninus, hagiographer, 53
Antoninus, poet, 53
Antoninus Honoratus, 193
Antoninus of Aleppo, 769
Antoninus of Ephesus, 32, 196
Antoninus Pius, **53**, 60, 72, 113, 724, 812, 833, 834
Anulinus, 2, **53**, 322
Anversa, 27
Anz, 67
Aonès, 296
Aosta (Augusta Praetoria), 360, 421
Apa, *see* Abbat, Amma
Apamea, **53**, 171, 174, 527, 572, 599, 662, 678, 702, 808, 809, *28, 29*
Apamene, 172
Apatheia, **53**, 83, 92, 161, 231, 243, 355, 398, 443, 444, 455, 547, 585, 587, 589, 592, 691, 693, 694, 725, 789, 792, 872
Apelles, **54**, 244, 384, 621, 732, 736, 776, 819
Aphra, 347
Aphraates, 10, 16, **54**, 81, 235, 269, 276, 312, 395, 429, 436, 452, 472, 520, 562, 583, 649, 674, 708, 765, 782, 810
Aphrodisia, 662
Aphrodisias, 87, 88
Aphthartodocetism, 8, 36, 37, 49, **54**, 302, 305, 333, 439, 458, 569, 773, 841, Against the Aphthatrodocetists, 444
Aphu of Oxyrhynchus, Life of, 96
Apianus, 203
Apiarius of Sicca, **55**, 142, 146, 147, 148, 154, 605, 641, 645, 778, 856
Apion, **55**
Apocalypse, 593, 596, 606, 610, 671, 712, 733, 763, 798, 802, 831, 839, 867, of Adam, 9, 771, of Dositheus, 251, of Elijah, 56, of Esdras, 56, 57, of James, 56, of John, 1st, 120, of St John the Baptis, of Mary, Bartholomew, Daniel and Zechariah, 56, of Moses, 9, 56, 257, 718, of Paul, 39, 56, of Peter, 47, 56, 57, of Stephen, 5, 56, of Thomas, 56, of Zephaniah, 56
Apocalyptic, 19, 23, 26, 38, 46, 47, **55**, 56, 57, 64, 90, 111, 117, 158, 168, 188, 219, 226, 236, 238, 252, 271, 275, 279, 291, 295, 306, 311, 314, 333, 342, 353, 383, 431, 435, 449, 463, 489, 503, 511, 560, 577, 671, 684, 718, 820, 851
Apocatastasis, 622, 623, 693, 718, 725, 733, 826, 878
Apocorius, 138
Apocrisarius (or apocrisiarius), **56**, 619
Apocriticus, 514
Apocrypha, 9, 18, 31, 39, 41, 47, 55, **56**, 62, 63, 64, 66, 84, 91, 139, 141, 176, 181, 199, 223, 281, 309, 345, 379, 412, 446, 451, 452, 474, 537, 539, 558, 647, 656, 680, 707, 712, 714, 723, 762, 763, 767, 791, 802, 815, 833, 837, 838, 840, 843, 854
Apodemius, 126
Apokritikòs, 233
Apollinaris, senior, **58**
Apollinaris of Hierapolis, **58**, 60, 258, 332, 436, 728, 729
Apollinaris of Laodicea, 9, 28, 35, 38, 46, 50, **58**, 94, 164, 215, 218, 233, 250, 256,

310, 318, 326, 343, 362, 364, 368, 401, 419, 438, 442, 449, 459, 524, 560, 569, 573, 594, 650, 658, 660, 700, 704, 715, 723, 734, 744, 761, 790, 827, 841, 849, 873, 879, 880
Apollinaris of Ravenna, **59**
Apollinaris of Valence, **59**, 342
Apollinarists, 24, 29, 42, 49, 50, 58, 59, 165, 196, 208, 236, 275, 296, 315, 318, 362, 379, 407, 408, 410, 538, 594, 806, 824, 825, 827, 830, 831, 841, 859, 869
Apollonia, **59**, 171, 487, 488
Apollonius, **59**, 160, 472, 533, 534, 551, 552, 563, 672, 741, 819, 838, 854
Apollonius of Antiphrae, 487
Apollonius of Ephesus, **60**
Apollonius of Tyana, **60**, 300, 380, 434, Life of, 121, 779
Apollonius the Sophist, 379
Apollos, 59
Apologeticum, 60, 117, 271, 725, 818
Apologia, 264, 551
Apologists, Apologetic, 2, 9, 39, 45, 46, 52, 56, 57, 59, **60**, 61, 72, 73, 82, 93, 96, 118, 134, 144, 155, 160, 163, 168, 176, 177, 179, 180, 209, 226, 233, 237, 242, 244, 262, 271, 284, 285, 286, 288, 289, 300, 308, 309, 313, 316, 340, 341, 354, 355, 372, 378, 389, 399, 436, 457, 463, 469, 472, 483, 517, 533, 537, 558, 560, 561, 562, 563, 594, 620, 624, 650, 658, 660, 684, 700, 704, 705, 710, 715, 722, 723, 727, 734, 741, 744, 761, 767, 790, 798, 818, 823, 827, 841, 848, 849, 873, 879, 880
Aponius, **60**, 787, 881
Apophthegmata, 251, 475, 514, 567, 597, 655, 793
Apophthegmata Patrum, **60**, 83, 220, 271, 401, 417, 652
Apostasy, 7, 8, 47, **61**, 79, 92, 144, 208, 211, 212, 223, 246, 260, 261, 307, 369, 396, 414, 415, 432, 439, 473, 484, 530, 620, 667, 623, 828, 829, 830, 840, 849
Apostle, 2, 24, 38, 57, **61**, 63, 98, 103, 109, 111, 132, 135, 141, 160, 181, 189, 190, 192, 202, 207, 218, 221, 224, 225, 232, 233, 234, 235, 241, 242, 250, 259, 261, 262, 263, 274, 283, 290, 292, 294, 306, 307, 313, 316, 320, 326, 334, 336, 355, 361, 365, 370, 375, 376, 380, 407, 421, 429, 432, 434, 448, 449, 452, 456, 457, 467, 471, 477, 487, 492, 493, 495, 510, 525, 527, 541, 543, 561, 573, 575, 579, 615, 616, 617, 630, 631, 640, 641, 656, 657, 659, 669, 676, 680, 705, 706, 715, 723, 733, 742, 746, 749, 755, 746, 749, 756, 762, 771, 777, 798, 802, 803, 819, 820, 821, 834, 845, 847, 849, 850, 867, 875, *30*, Apocryphal acts of the apostles, 1, 61, 179, 234, 271, 320
Apostoleion, 64, 224, 656
Apostolic Canons, **62**, 63, 85, 106, 142, 154, 198, 237, 259, 460, 556
Apostolic Constitutions, 33, 42, 50, 56, 57, **62**, 63, 104, 123, 189, 198, 221, 264, 352, 370, 404, 410, 460, 494, 495, 496, 500, 504, 541, 561, 609, 614, 723, 751, 793, 816, 882
Apostolic creed, 323, 640
Apostolic succession, *see* Succession, Apostolic
Apostolic Tradition, 16, 33, 34, 35, 42, 58, 62, **63**, 103, 107, 123, 143, 182, 189, 190, 191, 198, 231, 245, 252, 258, 259, 277, 371, 385, 469, 475, 492, 494, 495, 497, 498, 499, 502, 541, 561, 577, 611, 639, 640, 643, 723, 740, 741, 744, 788, 793, 800, 820, 848
Apostolici, 13, **63**, 566, 703
Apostolicity, **63**, 72, 99, 258, 284, 640, 643, 644, 730, 734, 741, 746, 798, 848
Apostolos, 136
Apotactites, 13, 38, 686
Apotheosis, **63**, 721

Apringius of Beja, 791
Apringius of Chalcis, 20
Apse, 170
Apsorus, 217
Apthugni, 168
Apuleius, 121, 558
Apulia and Calabria, 424
Apulum (Alba Iulia), 217
Apa Phoebammon, 689
Aqaba, 636
Aquae Iasse (Varazdinske toplice), 759
Aquae Statiellae (Acqui Terme), 421
Aquae calidae, 838
Aquariani, 542
Aquarii, **64**, 877
Aquas Salvias, 36
Aquila, 122, 740
Aquila and Priscilla, 178
Aquileia, 16, 26, 28, 29, 33, 53, **64**, 69, 70, 78, 85, 96, 109, 166, 167, 169, 170, 172, 173, 174, 187, 193, 299, 323, 360, 377, 419, 420, 421, 430, 435, 450, 485, 495, 497, 549, 571, 572, 600, 618, 642, 645, 659, 667, 675, 685, 705, 743, 759, 778, 822, 823, 838, 860, 863, 875, C. of (381), 64, 116, 218, 305, 324, 371, 465, 530, 546, 637, 749, 765, 824, 828, *31, 32*, liturgy, 166, 497, 499
Aquincum, **65**, 860
Aquisgranae, 791, C. of (836), 118
Arab invasions, 22, 85, 247, 433, 446, 488, 497, 527, 547, 553, 568, 570, 605, 755, 835
Arabia, Arabs, 13, 23, 48, 52, **65**, 67, 86, 119, 165, 186, 200, 208, 216, 225, 257, 263, 375, 380, 428, 438, 457, 472, 567, 583, 607, 635, 674, 763, 782, 807, 808, 823, 843, *33-39*
Arabian council, 65
Arabian heresy, 65, **67**, 620
Arabic, 605, 609, 674, 687, 748, 769, 782, 791, 797, 809, 810, 820, 843, 850, 884
Arabic language and literature, 2, 9, 21, 38, 42, 44, 52, 57, 60, 61, 63, 65, 113, 115, 119, 143, 179, 198, 199, 200, 211, 216, 234, 235, 251, 291, 345, 383, 394, 416, 429, 442, 446, 472, 496, 540, 564
Arabici, 65
Arabissos, 772
Aramaic, 65, 567, 584
Arator, **67**, 476, 651, 862
Arba, 217
Arbela, 11, **67**, Chronicle of, **67**
Arbogastes, 28, 103, 296, 828
Arcadius, 19, 24, 26, 41, **67**, 83, 88, 96, 138, 193, 194, 196, 295, 324, 334, 337, 339, 397, 537, 553, 607, 609, 611, 704, 725, 745, 760, 806, 826, 829, 831
Arcadius of Cyprus, **67**
Arcavica, 148
Archaeology, Christian, 11, 48, **68**, 138, 153, 156, 158, 172, 203, 209, 221, 278, 281, 338, 518, 552, 564, 581, 598, 600, 637, 644, 654, 656, 757, 788, 790, 814, 833, 835, 837, 842, 843, 855, 876
Archaeus, **71**
Archangels, 39, 40
Archbishop, 4, **71**, 137, 730, 864, 865
Archdeacon, **71**, Roman, **71**
Archelaus of Carcara, **71**, 115, 371, Acts of, 11, 71, 113, 233, 519, 542,
Archidamus, **71**
Archimandrite, 1, 6, **71**, 112, 241, 439
Architas, Ps.-, 586
Archpriest, **71**
Arcontici, **71**, 610
Arculf, 10
Ardagh (co. Limerick), 133, chalice, 133
Ardwall Island, 130
Areopolis, 636
Aretas, **71**, 238, 815
Arezzo (Arretium), 264, 422
Arganum, 765
Argumentation, Patristic, **72**, 655, 687, 848

Ariamirus, 127
Arians, -ism, 1, 5, 7, 13, 14, 17, 20, 23, 24, 26, 27, 28, 29, 32, 33, 37, 39, 43, 44, 48, 49, 50, 53, 58, 59, 62, 64, 71, 72, 76, 83, 84, 92, 93, 96, 103, 105, 113, 114, 121, 122, 124, 134, 138, 141, 143, 145, 151, 158, 160, 166, 173, 187, 192, 193, 194, 195, 205, 207, 208, 214, 215, 218, 228, 233, 236, 238, 243, 244, 250, 251, 252, 256, 271, 273, 275, 276, 278, 282, 292, 296, 297, 298, 299, 300, 303, 307, 318, 321, 324, 325, 326, 329, 331, 338, 339, 341, 343, 347, 348, 352, 354, 355, 356, 357, 360, 361, 363, 365, 377, 379, 381, 382, 391, 395, 396, 401, 403, 405, 417, 418, 438, 450, 464, 449, 475, 478, 480, 482, 487, 488, 506, 508, 511, 516, 524, 525, 526, 527, 531, 538, 541, 546, 548, 550, 551, 559, 585, 594, 595, 601, 613, 614, 620, 623, 626, 634, 637, 641, 651, 652, 660, 662, 680, 682, 683, 684, 685, 700, 706, 722, 730, 734, 736, 737, 742, 743, 749, 753, 755, 757, 758, 766, 770, 771, 774, 779, 783, 785, 790, 791, 799, 804, 822, 824, 823, 827, 828, 833, 841, 844, 845, 847, 848, 849, 854, 856, 858, 860, 861, 868, 869, 870, Against the Arians, 94, 331, 868
Arianus, 200
Aridius of Lyons, 869
Arindella, 636
Ariomanitae, **72**
Aristarchus, 740
Aristeas, Letter of, **72**
Aristenes, 142
Aristides, 54, 60, 61, **72**, 96, 111, 168, 176, 264, 315, 683, 703, 796, 818
Aristion, **73**, 647
Aristo of Pella, **73**, 233, 436
Aristobulus, 455
Aristocles, 20
Aristotelianism, 20, 31, 59, **73**, 101, 125, 316, 372, 446, 463, 547, 558, 584, 589, 603, 621, 624, 663, 683, 690, 691, 694, 695, 775, 781, 796, 805, 840, 852, *see also* Philosophy and the Fathers
Aristotle, 3, 12, 32, 81, 82, 92, 124, 161, 204, 233, 398, 489, 553, 601, 602, 626, 639, 648, 689, 692, 695, 713, 769, 805, 879
Aristoxenus of Tarentum, 121
Arius, 7, 20, 21, 23, 35, 37, 38, 42, 50, 64, 65, 72, 76, 84, 90, 92, 93, 95, 105, 141, 164, 185, 192, 207, 208, 224, 298, 299, 301, 303, 305, 323, 355, 358, 391, 396, 434, 507, 514, 522, 525, 526, 566, 582, 595, 596, 626, 633, 657, 661, 682, 705, 734, 735, 797, 830, 852, 858, Against, 697
Arius Didymus, 75
Arius of Petra, 487
Arles (Arelate), 17, 62, **79**, 102, 138, 154, 188, 205, 260, 306, 335, 337, 338, 339, 347, 369, 381, 396, 429, 452, 456, 479, 575, 730, 861, C. of (314), 8, 18, 64, 79, 104, 109, 126, 129, 137, 138, 169, 191, 225, 243, 247, 249, 250, 270, 274, 322, 338, 348, 419, 420, 513, 530, 554, 559, 595, 613, 641, 728, 733, 735, 757, 790, 802, 806, 823, 850, 857, 869, 875, *40, 41*, C. of (353), 77, 79, 93, 348, 485, 559, 609, 661, 736, 850, 870, C. of (437), 508, C. of (449-461), 79, C. of (453), 548, 730, 861, C. of (c.463), 79, 304, C. of (c.470-475), 79, 304, 321, 480, 513, 653, 665, 767, C. of (524), 212, 555, C. of (554), 556, C. of (c.463), 79, so-called 2nd council, 79, 142
Armagh, 130, 411, 653, Book of, 412, 576
Armenia, 3, 4, 16, 19, 43, 48, **79**, 83, 84, 96, 112, 114, 153, 209, 255, 257, 307, 344, 368, 375, 403, 433, 439, 457, 478, 496, 550, 567, 582, 593, 597, 673, 674, 752, 809, 858, 877, *42-47*, Language

and Literature, 2, 9, 10, 16, 19, 31, 38, 41, 42, 52, 60, 61, 66, 73, **81**, 113, 115, 152, 166, 234, 257, 263, 269, 299, 305, 306, 312, 315, 320, 345, 383, 394, 413, 429, 439, 442, 460, 467, 468, 472, 501, 518, 520, 534, 537, 540, 556, 584, 599, 655, 660, 683, 752, 775, 785, 786, 818, 827, 840, 842, 880
Armentarius, 737
Arnobius of Sicca, 60, **82**, 168, 177, 470, 563, 612, 614, 628, 629, 655, 696, 760, 778, 831
Arnobius the Younger, **82**, 543, 706
Arnulf of Metz, 338
Aro, 813
Aroutch, 80
Arsaces of Armenia, 2
Arsacius, **83**, 437, 584
Arsenius of Hypselis, **83**, 439
Arsenius the Great, 67, **83**, 93, 551
Artasat, C. of (449), 482
Artaxata, 4, **83**
Artemision, 87
Artemon or Artemas, 11, **83**, 560, 566
Artemonites, 83
Arthemius, **83**, Passion of, 684
Artik, 80
Artotyritae, **83**, 542
Arvernum, 778
As-Safi'i, 339
Asarbus, **83**, 712
Ascalon, 635
Ascanius, 814
Ascension, 33, 43, **83**, 132, 135, 258, 269, 290, 298, 364, 410, 493, 540, 546, 669, 671, 712, 731, 738, 805, 834, 869, *48*
Ascesis, Ascetic, 4, 9, 29, 39, 44, 45, 48, 54, 63, **83**, 84, 93, 98, 100, 101, 103, 114, 115, 116, 121, 122, 124, 149, 150, 176, 181, 184, 186, 200, 212, 221, 223, 224, 225, 226, 227, 228, 230, 231, 233, 242, 253, 262, 264, 269, 282, 284, 287, 289, 298, 303, 306, 318, 319, 320, 321, 330, 338, 342, 353, 358, 359, 363, 367, 371, 380, 388, 391, 396, 401, 406, 411, 417, 432, 440, 445, 475, 500, 508, 516, 521, 523, 527, 528, 531, 532, 536, 557, 559, 560, 566, 567, 611, 621, 659, 661, 665, 670, 674, 683, 688, 707, 718, 721, 725, 746, 754, 755, 758, 760, 767, 775, 779, 786, 787, 790, 792, 797, 799, 815, 821, 829, 846, 860, 863, 870, 871, 877, 882, 884
Asceterium, **84**, 295, 396
Asceticon, 114, 115, 475, 556, 746
Ascitae, **84**, 841
Asclepas of Gaza, 757
Asclepias of Antioch, **84**
Asclepiodotus or Asclepias, **84**
Asclepius, **84**
Ascodrugitae, 84
Asculum (Ascoli Piceno), 423
Ashaqlûn, 519
Ashburnham Pentateuch, 1, 2, 10, 452, 575, 631, 730
Ashmûnein, 267
Ashtishat, **84**, 858, C. of (449), 315
Asia, 22, 47, 59, **85**, 89, 150, 195, 274, 275, 313, 314, 512, 523, 533, 619, 671, 686, 701, *49-53*
Asia Minor, 26, 27, 38, 47, 48, 51, 63, 69, 77, **85**, 109, 111, 114, 122, 137, 143, 155, 158, 164, 170, 171, 172, 173, 174, 175, 236, 245, 263, 266, 267, 275, 279, 280, 303, 307, 314, 333, 343, 445, 448, 512, 564, 566, 567, 604, 618, 657, 660, 670, 702, 727, 728, 764, 797, 803, 822, 854
Asiatic culture, 23, 45, 46, 50, **89**, 257, 276, 303, 340, 733
Asidonia, 106
Asilus, 769
Asot I, 344
Aspasius, 145, 556

Asprenus, 580
Assembly, 63, **90**, 91, 123, 152, 181, 232, 246, 277, 293, 325, 335, 394, 410, 474, 490, 494, 498, 499, 500, 501, 540, 541, 542, 561, 609, 665, 699, 707, 711, 715, 722, 741, 745, 784, 792, 794, 805, 872
Assistance and Charity, works of, 16, 25, 48, 71, **90**, 104, 338, 371, 432, 441, 554, 555, 609, 731, 740, 741, 742, 746, 794, *see also* Alms - Almsgiving
Assos, 87
Assumption, **91**, 538, 539, 832, 841, 850
Assyria, 11
Asterius of Amasea, 27, **91**, 141, 163, 296, 407, 477, 536
Asterius of Ansedunum, **92**
Asterius of Arabia, **92**, 487
Asterius the Sophist, 76, **91**, 92, 185, 300, 449, 598, 722
Asterius Turcius Rufus, **92**, 766
Astorga-León (Asturica), 127, 334, 790, 854
Astrology, **92**, 106, 110, 227, 263, 269, 287, 301, 324, 378, 517, 542, 595, 612, 614, 634, 712, 769, 805, 815, Against the astronomers, the astrologers and fate, 236
Asylum, right of, **92**, 147, 338, 555, 556, 784
Ataraxia, **92**, 455
Atarbius of Neocaesarea, **93**, 114, 623
Ataulf, 198, 334
Athalaric, 67, 357, 438, 761, 828
Athanagild, 162, 791, 844
Athanaric, 357
Athanasians, 454
Athanasian creed, 208
Athanasius, Ps.-, 9, 16, 58, 449, 720
Athanasius Gammal, 95
Athanasius of Alexandria, 1, 9, 20, 22, 23, 24, 31, 35, 37, 38, 44, 45, 48, 58, 72, 77, 78, 79, 81, 83, 84, 92, **93**, 113, 114, 120, 121, 129, 140, 142, 149, 160, 164, 168, 175, 188, 191, 194, 195, 200, 207, 208, 215, 224, 227, 228, 230, 231, 235, 238, 243, 250, 257, 262, 269, 278, 289, 290, 296, 298, 299, 300, 301, 302, 303, 308, 318, 319, 323, 325, 327, 328, 331, 338, 343, 347, 358, 359, 362, 363, 365, 388, 392, 396, 405, 406, 417, 430, 434, 439, 449, 460, 480, 485, 487, 494, 496, 500, 508, 509, 511, 516, 520, 521, 522, 525, 538, 543, 546, 548, 550, 551, 559, 561, 567, 577, 582, 585, 595, 600, 616, 626, 628, 634, 640, 641, 643, 657, 658, 660, 661, 662, 663, 664, 678, 680, 700, 705, 719, 722, 734, 735, 736, 742, 743, 746, 757, 763, 768, 771, 776, 782, 785, 788, 789, 790, 797, 799, 805, 816, 817, 824, 830, 831, 832, 833, 841, 848, 849, 850, 852, 854, 855, 858, 870, 875, 878, 880
Athanasius of Anazarbus, **95**, 329
Athanasius of Ancyra, **95**, 854
Athanasius of Balad, 428
Athanasius of Pena, 49
Athanasius of Perrhes, 196
Athaulf, 110, 356, 770
Atheism, 672
Athenagoras, 2, 23, 60, 72, 73, 84, 93, **95**, 96, 121, 161, 163, 168, 176, 235, 236, 373, 384, 391, 524, 533, 603, 614, 690, 691, 695, 732, 760, 785, 796, 875
Athenaïs, *see* Eudoxia
Athertodorus, **95**, 368
Athenogenes, 84
Athens, 7, 22, 31, 95, 114, 141, 162, 174, 236, 297, 369, 382, 552, 590, 592, 596, 619, 704, 713, 714, 725, 727, 760, 814, 833, 836, *54-56*
Athleta Christi, **96**
Athlete, 122, 287, 817
Atrium, 170
Attala, 451
Attalus, 64, 842

Atticus, 37, 74, 83, **96**, 295, 437, 558, 594, 694, 752, 784, 829
Attila, 40, **96**, 189, 301, 342, 357, 511, 555, 559, 743, 757, 837, 859
Aubenya (Majorca), 107
Audacht Moraind, 412
Audentius, **96**
Audiani, 19, 46, **96**, 235, 281, 355
Audientia Episcopalis, **96**, 98, 104, 193, 408, 520
Audius, 96
Augsburg (Augusta Vindelicorum), 347
Augurius, 814
Augusta Bagiennorum (Bene Vagenna), 421
Augusta Traiana (once Beroë, now Stara Zagora), 838
Augustalis, 18, **96**, 189
Augustine of Aquileia, **96**
Augustine of Canterbury, **96**, 129, 131, 143, 513, 551, 676, 883, Gospels, 131, 294, 631
Augustine of Hippo, 1, 2, 3, 6, 9, 10, 11, 13, 14, 15, 18, 25, 26, 27, 28, 30, 32, 39, 40, 41, 45, 46, 47, 52, 53, 55, 56, 57, 60, 63, 65, 67, 72, 74, 82, 83, 84, 90, 92, **96**, 101, 102, 104, 105, 107, 108, 109, 112, 116, 118, 119, 121, 123, 124, 127, 128, 129, 131, 132, 138, 141, 145, 146, 147, 149, 151, 152, 153, 154, 159, 160, 161, 162, 165, 167, 168, 175, 177, 178, 179, 182, 183, 185, 186, 187, 191, 193, 202, 205, 208, 209, 211, 212, 221, 222, 223, 225, 226, 227, 228, 232, 233, 241, 242, 243, 244, 246, 247, 248, 249, 250, 255, 257, 273, 277, 282, 285, 286, 288, 293, 296, 298, 306, 308, 311, 312, 313, 317, 318, 320, 321, 322, 324, 326, 327, 328, 330, 331, 332, 336, 337, 339, 340, 341, 342, 351, 354, 355, 356, 357, 361, 366, 369, 370, 372, 373, 375, 376, 379, 381, 383, 389, 392, 394, 395, 396, 397, 398, 399, 405, 406, 407, 408, 409, 410, 414, 430, 433, 437, 449, 450, 451, 452, 453, 454, 457, 458, 462, 463, 465, 469, 472, 474, 476, 477, 478, 482, 483, 486, 488, 490, 492, 493, 494, 501, 503, 504, 505, 506, 507, 511, 516, 517, 519, 520, 521, 524, 526, 527, 528, 532, 533, 538, 542, 543, 545, 546, 549, 552, 553, 555, 559, 560, 562, 567, 568, 576, 577, 578, 584, 585, 587, 588, 589, 592, 600, 602, 603, 605, 606, 607, 609, 611, 612, 613, 614, 615, 616, 624, 625, 637, 642, 645, 646, 648, 649, 650, 651, 652, 653, 655, 656, 658, 660, 664, 665, 666, 667, 669, 670, 675, 676, 679, 683, 689, 692, 697, 701, 702, 703, 704, 705, 708, 709, 710, 711, 713, 714, 717, 719, 720, 722, 723, 725, 727, 728, 729, 731, 733, 734, 735, 736, 737, 738, 739, 742, 743, 745, 746, 747, 748, 749, 750, 751, 757, 758, 759, 760, 762, 763, 765, 766, 767, 769, 770, 774, 776, 777, 778, 780, 781, 782, 784, 785, 788, 789, 793, 794, 795, 812, 816, 817, 819, 824, 825, 831, 832, 838, 840, 843, 849, 852, 856, 857, 861, 863, 864, 866, 870, 872, 873, 874, 875, 876, 878, 879, 881, *57*, City of God, 6, 98, 99, 100, 101, 168, 177, 211, 226, 229, 373, 389, 521, 624, 704, 720, 863, 875, De Trinitate, 99, 102, 121, 125, 235, 296, 317, 381, 449, 553, 603, 696, 725, 747, 847, 852, 863, Life of, 96, 121
Augustine, Ps.-, 412
Augustine's Oak, Synod of (c.603), 129
Augustinianism, **99**, 101, 118, 285, 308, 331, 359, 408, 418, 479, 508, 530, 616, 664, 709, 717, 731, 734, 745, 767, 770
Augustinus Hibernicus, **102**, 412
Augustus, 63, 79, **102**, 145, 208, 333, 510, 607, 700, 702, 729, 774, 778, 790, 806, 834
Aulus Gellius, 179

Aunacarius of Auxerre, **102**, 556
Aupsichius, 96
Aurasius of Toledo, **102**
Aurelia Proba, 803
Aurelia Zosima, 7
Aurelian, emperor, **102**, 160, 163, 190, 217, 246, 336, 338, 356, 565, 637, 663, 756, 799, 861, 875, 884
Aurelian of Arles, **102**
Aurelian of Carthage, **102**, 142, 146, 147, 205, 225, 247, 383, 742, 782
Aureobindus, 187
Ausona, 813
Ausonius, 27, **102**, 126, 338, 347, 360, 483, 530, 660, 661, 721, 799
Ausonius of Bordeaux, 799
Auspicius of Toul, **103**
Austrebert of Pavilly, 532
Austremonius, 238
Authari, 17, **103**, 366, 505, 506, 822
Authority in the Church, 8, 38, 72, 79, 83, 93, 96, 98, **103**, 114, 125, 132, 137, 139, 141, 147, 153, 154, 159, 162, 181, 182, 197, 211, 212, 223, 233, 242, 244, 246, 248, 258, 260, 309, 315, 320, 329, 338, 339, 376, 388, 397, 400, 409, 410, 419, 430, 432, 466, 469, 479, 487, 495, 510, 554, 567, 570, 603, 613, 616, 617, 640, 641, 642, 651, 657, 670, 676, 710, 712, 713, 730, 733, 738, 739, 741, 742, 744, 746, 763, 766, 781, 782, 794, 798, 814, 824, 845, 847, 848, 851, 852, 866
Autun (Augustodunum), **104**, 280, 337, 480, 604, 733, 760, 802
Auvergne, 4
Auxanon, 785
Auxentius of Durostorum, **105**
Auxentius of Milan, 24, 28, 78, 94, **105**, 218, 238, 323, 329, 374, 381, 382, 523, 559, 743, 860
Auxentius of Mopsuestia, 684
Auxerre (Autessiodurum), 346, C. of (561-605), 556, C. of (573), 517
Auxilius, 411, 616
Auximum (Osimo), 422
Avenches (Aventicum), 525
Avila, 712
Aviti, 127
Avitus, emperor, 59, 134
Avitus of Braga, **105**, 134, 321
Avitus of Vienne, 59, 67, **105**, 144, 231, 249, 270, 329, 339, 393, 399, 555, 651, 778, 779, 869, 874
Awgin, 296
Axido, 175
Axiopolis, **105**, 535, 565, 764, 765
Axum, *see* Aksum
Ayasoluk, 85

Ba'eda Maryam, 291
Bab, 43
Babaway, 112, 118
Babai bar Nesibnaye (the Less), **106**
Babai of Seleucia-Ctesiphon, 47
Babai the Great, **106**, 598, 810
Babylas of Antioch, 50, 51, 86, **106**, 223, 315, 672, 682, 684
Babylon, 1, 60
Bacchus and Sergius, *see* Sergius and Bacchus
Bacchyllus, **106**
Bachiarius, **106**, 137, 790
Baetica, **106**, 148, 270, 386, 399, 418, 774, 775, 790, 838, 861, *58*
Bagadius, 32, 372
Bagadius of Bostra, 32, 196, 372
Bagai, C. of (394), 208, 247, 321, 711
Bagaran, 80
Bagatti, Bellarmino, 53, 66-67, 120, 127, 138, 175, 213, 218, 255, 263, 271, 433-34, 584, 585, 598, 635-36, 637, 680, 808-9, 855
Bahrâm I, 519
Baile Chuind, 413

Bailén, 107
Bakur, 344
Balaam, 40, **107**, 517, 596
Balai, **107**, 472, 583, 810
Baleari, **107**, 148, 174, 191, 572, 774, 795, 861, *59*
Balsamon, 142
Balteus, 254
Balthasar Bebel, 68
Bana, 19, 344
Bandolier, 254
Bangor, 131, 186, 411, Antiphonary, 155, 412, 497, 765, Chronicle, 413
Baptism, 11, 13, 18, 27, 28, 29, 42, 52, 54, 62, 63, 64, 76, 80, 81, 90, 96, 98, 100, **107**, 111, 114, 122, 123, 126, 127, 133, 134, 137, 138, 141, 142, 145, 146, 147, 150, 151, 153, 154, 162, 163, 173, 174, 175, 176, 180, 183, 189, 198, 199, 200, 205, 206, 211, 212, 213, 215, 219, 221, 225, 227, 228, 229, 232, 233, 234, 237, 238, 243, 245, 247, 250, 258, 260, 261, 269, 271, 273, 274, 276, 280, 282, 285, 292, 309, 315, 316, 317, 318, 319, 321, 324, 325, 332, 334, 340, 341, 351, 357, 358, 359, 360, 361, 363, 364, 370, 372, 377, 382, 383, 386, 390, 394, 396, 397, 398, 399, 403, 404, 405, 406, 408, 409, 410, 415, 429, 431, 433, 437, 441, 442, 447, 461, 462, 463, 465, 466, 467, 469, 473, 477, 479, 489, 491, 492, 493, 494, 495, 502, 503, 504, 506, 517, 518, 580, 585, 595, 596, 598, 603, 604, 609, 611, 613, 616, 617, 621, 622, 628, 629, 640, 646, 650, 651, 666, 667, 668, 669, 672, 675, 710, 712, 714, 719, 732, 734, 741, 744, 745, 749, 750, 751, 755, 758, 767, 775, 780, 781, 782, 792, 794, 800, 801, 802, 803, 805, 816, 817, 819, 820, 821, 822, 828, 840, 841, 843, 846, 848, 851, 853, 864, 871, 875, 876, *60, 61*, On baptism, 98, 107, 114, 115, 150, 190, 312, 494, 616, 775, 803, 819
Baptismal catechesis, 441
Baptismal creed, 852
Baptistery, 173
Barbarus Scaligeri, 109
Barbelioti, 110
Barbelo, 109, 771
Barbelognostics, 610
Barberini euchologion, 224
Barcelona (Barcino), 110, 604, 611, 628, 660, 728, 791, 813, 814, 844, C. of (540), 110, 386, 497, C. of (599), 110, 548
Bardesanes, of Edessa, 10, 110, 143, 179, 229, 233, 263, 276, 370, 634, 674, 680, 810
Barhadbsabba, 110
Bar Hebraeus, 214, 238, 428, 518, 810
Barium (Bari), 424
Bar-Jesus, 213
Bar Kokhebas (Cochba, Koziba), 73, **110**
Barlaam, **110**, 115, 739
Barlaam and Josaphat, Life of, 73, **111**
Barletta relief, 62
Barnabas, 9, 21, 48, **111**, 213, 279, 316, 343, 398, 432, 456, 543, 559, 561, 618, 657, 723, 763, 802, 821, 839, Acts and martyrdom of, at Cyprus, 111, Barnabas, Ps.-, Epistle of, 2, 9, 22, 47, 53, 56, **111**, 150, 163, 176, 234, 309, 312, 358, 436, 454, 483, 648, 706, 782, 792, 878, Gospel of, 111
Barnea, Jon, 105, 140, 217, 390, 605, 764-65, 845, 853
Baronius, C., 4, 68, 654
Barsanuphius, **112**, 251, 442, 567
Barsauma, **112**, 118, 119, 583, 598, 674, 766, 810, 828
Barsauma of Nisibis, 5, **112**
Bartholemew, 65, 79, **112**, 680, 780, Acts of, 112
Baruch, Book of, 56, 465

Basel (Basileia), 347
Basil, archimandrite, **112**
Basil I, emperor, 686
Basil and Emmelia, **113**
Basileiad, 113, 114
Basileias of Caesarea, **115**
Basilica, 7, 14, 15, 169, 209, 217, 265
Basilica Apostolorum, 169, 174
Basilica of SS. Cosma e Damiano, 435
Basilideans, 113, 163, 264, 493
Basilides, 4, 18, 23, 57, 75, **113**, 204, 282, 308, 316, 352, 353, 373, 384, 419, 603, 610, 732, 758
Basilides of Léon and Astorga, 146
Basiliscus, 5, **113**, 197, 220, 275, 357, 432, 479, 569, 642, 679, 780, 840, 841, 884
Basil of Ancyra, 13, 77, **113**, 115, 195, 243, 268, 281, 343, 395, 526, 719, 776, 783, 852
Basil of Caesarea in Cappadocia (Basil the Great), 2, 4, 9, 25, 27, 29, 32, 35, 37, 39, 43, 45, 49, 50, 56, 57, 59, 72, 75, 78, 81, 84, 91, 93, 95, 96, 108, 110, **113**, 114, 115, 116, 119, 135, 137, 142, 143, 152, 168, 177, 178, 184, 188, 191, 192, 195, 200, 205, 207, 215, 218, 230, 233, 235, 237, 243, 244, 246, 250, 252, 262, 269, 273, 276, 286, 287, 296, 299, 302, 303, 305, 306, 308, 310, 312, 315, 317, 319, 328, 346, 351, 356, 358, 361, 362, 363, 369, 372, 373, 380, 382, 392, 395, 386, 398, 403, 405, 407, 409, 417, 449, 452, 457, 459, 475, 484, 495, 499, 501, 504, 511, 514, 516, 518, 522, 528, 532, 538, 550, 556, 561, 566, 567, 575, 577, 584, 587, 597, 608, 633, 641, 643, 648, 649, 655, 660, 668, 678, 685, 686, 694, 703, 714, 716, 723, 734, 737, 742, 746, 747, 749, 750, 751, 759, 760, 761, 763, 775, 781, 785, 792, 796, 805, 826, 832, 841, 848, 849, 852, 857, 868, 876, 878, 879, 882, Liturgy of, 34, 113, **115**, 496
Basil of Cilicia, **115**, 389
Basil of Seleucia, **115**, 312, 410, 436, 450, 795
Bassano, 422
Bassianus of Ephesus, **115**, 275
Bassianus of Lodi, **116**
Basti, 148
Bathild, 532
Baths, 14, 47, 85, 87, **116**, 144, 158, 169, 173, 194, 213, 265, 423, 425, 681, 725, 757, 759, 807, 842, 871
Batna, 16
Battikha, Aboud, 363
Batz (Brittany), 663
Baucalis, 20
Baudonivia, **116**, 729
Baukalis, 76
Bauzanum (Bolzano), 420
Bawit, 267, 268, 631
Beast Canon Tables, 131
Beatitude, 39, 99, 101, 116, 117, 186, 229, 261, 270, 285, 295, 351, 406, 489, 615, 621, 649, 665, 725, 731, 733, 751, 752, 792, 843
Beatitudes, **117**, 221, 364, 461, 703, 708, 792
Beatrice, Pier Franco, 121, 181, 183, 223, 233-34, 599, 704, 775-76, 780-81, 875
Beatus Rhenanus, 655
Beatus of Liébana, 839
Beccanus, 211
Bede, the Venerable, 10, 19, 26, 36, 42, 55, 60, 67, 96, **117**, 119, 129, 132, 137, 166, 167, 186, 188, 211, 221, 248, 272, 294, 296, 312, 367, 389, 393, 411, 426, 465, 493, 503, 511, 537, 538, 576 597, 605, 654, 655, 669, 727, 753, 839, 868, 871, 873, 877, 879
Behnam, Mar, 417
Beijing, 496
Beisan, 572

Beit Adrai, C. of (485), 112, **118**
Beit Bagas, 840
Beit Lapat, C. of (484), 112, **118**
Beit Ras, 636
Beja, 555
Belial, 47, 226
Belisarius, **119**, 145, 357, 464, 779, 806, 816, 846, 862
Belisarius Scholasticus, 119
Bellarmine, R., 654
Bellator, 119, 881
Bellum vandalicum, 202
Belmonte, 424
Bema, 172
Benedict I, 642, 645
Benedict II, **458**, 646, 743, 845
Benedict Biscop, 117, **119**, 129, 132, 446, 873
Benedict of Nurcia, 55, **119**, 121, 160, 230, 233, 250, 268, 307, 319, 322, 366, 367, 369, 394, 395, 447, 486, 500, 526, 531, 743, 746, 749, 759, 816, 873
Benedictus qui venit, 6
Benevento, 26, 112, 424, 430, 497, 822
Benjamin, 200, 265
Benjamin of Alexandria, **119**
Berenice, 18, 682
Berenice and Prosdoce, 50, 665
Bergomum (Bergamo), 421
Berig, 356
Berja, 106
Bernard of Clairvaux, 101, 394, 452, 623
Berny-Rivière, C. of (580), 556
Beroea (Aleppo), 4, 107, 228, 303, 515, 808
Bertanius (Betranius), 845
Bertcrann of Bordeaux, 637
Bertha, 96, 129, 143, 286
Bertoli, 69
Bertrand, bishop of Bordeaux, 556
Bérulle, De, 101
Beryllus of Bostra, 65, 119, 126, 127, 620
Berytus (Beirut), 58, 196, 301, 303, 403, 773, 808, 841, 855
Besançon (Vesontium), 250, 346, C. of, 120
Besa, 120, 200
Beth'Abe, 271
Beth Alpha, 3, 572
Beth Arbaye, 110
Beth 'Aynata, 106
Beth Gabba, 417
Beth Govrin, 450
Bethlehem, 69, 120, 149, 160, 169, 174, 209, 255, 264, 315, 345, 431, 475, 500, 659, 688, 869, *62, 63*
Beth Rabban, 3
Beth Yerak, 635
Beth Zabday, 106
Bethany, 433
Bewcastle, Cumbria, 132
Beyt Qiduna, 4
Béziers (Biterris), 581, 747, 766, C. of (356), 77, 120, 736, 758
Bet Qoqa, 11
Biagius, 120
Bianchi, Ugo, 717-18
Bianco, Maria Grazia, 6-7, 7, 27, 44, 84, 113, 154, 161, 184, 221, 230-231, 296, 304, 346, 371, 485, 500-1, 521, 567-68, 660, 664, 737-738, 841
Bibiana, 688
Bible, 9, 12, 28, 40, 46, 56, 65, 84, 93, 106, 120, 123, 127, 141, 150, 152, 177, 187, 203, 205, 212, 225, 226, 227, 232, 244, 247, 255, 286, 291, 300, 307, 309, 337, 345, 352, 355, 356, 357, 358, 359, 379, 395, 397, 405, 408, 412, 429, 431, 464, 472, 474, 478, 479, 489, 495, 498, 503, 539, 548, 552, 558, 576, 577, 579, 604, 606, 620, 621, 626, 629, 647, 650, 656, 658, 663, 666, 678, 687, 688, 703, 714, 725, 727, 733, 734, 735, 736, 746, 750, 751, 752, 754, 760, 762, 763, 764, 766,

Volume I contains pages 1-578; Volume II contains pages 579-1100 and the index

769, 782, 788, 794, 796, 812, 813, 821, 829, 842, 851, 870, 872, see also Scripture
Bierzo, 860
Bigastrum, 148
Binbirkilise, 87
Bingham, Joseph, 68
Binitarianism, 121
Biography and Autobiography, 4, 22, 44, 116, 117, 119, 121, 139, 200, 211, 233, 377, 381, 489, 510, 644, 705, 798, 799
Birr, C. of (697), 411
Birten (oppidum Bertunensium), 347, 349
Birth, 2, 9, 29, 94, **121**, 260, 351, 376, 408, 441, 466, 511, 537, 583, 666, 670, 707, 709, 775, 781, 789, 849, 871, 882
Bisconti, Fabrizio, 3, 328-29, 370-71, 428, 429, 453, 482, 535-36, 607-8, 625, 675, 685, 871
Bishops, 704
Bithynia and Pontus, 2, 85, 87, 122, 165, 195, 512, 551, 594, 596, 671, 699, 702, see also Pontus
Blandina, **122**, 552
Blasphemia Sirmiensis, 77, 381, 783
Blastus, **122**, 728
Blesilla, **122**, 192, 303
Blessing, 17, 50, **123**, 127, 136, 141, 190, 224, 234, 265, 277, 297, 351, 370, 386, 477, 490, 493, 494, 500, 502, 504, 528, 541, 580, 617, 679, 709, 722, 749, 846, 875, 876
Blood, 1, 2, 18, **123**, 188, 189, 226, 232, 257, 277, 293, 313, 358, 404, 415, 453, 463, 470, 504, 531, 532, 542, 564, 671, 673, 676, 699, 731, 750, 780, 802, 803, 816, 821, 876, 877
Bobbio, 186, 307, 451, 486, 633
Boccea, 422
Body, 3, 8, 9, 12, 17, 18, 32, 42, 44, 45, 46, 54, 65, 67, 73, 74, 75, 76, 83, 84, 90, 91, 92, 94, 95, 99, 110, 117, **123**, 127, 145, 163, 164, 181, 183, 188, 190, 191, 215, 219, 231, 233, 243, 244, 251, 255, 260, 261, 276, 277, 285, 287, 288, 293, 305, 316, 318, 319, 326, 340, 353, 366, 372, 381, 395, 405, 406, 408, 413, 441, 442, 458, 463, 465, 469, 470, 481, 492, 494, 504, 516, 518, 519, 523, 532, 557, 569, 577, 587, 589, 592, 601, 603, 604, 614, 620, 621, 622, 623, 649, 653, 662, 670, 675, 678, 690, 691, 693, 694, 696, 720, 725, 732, 733, 751, 773, 775, 780, 781, 789, 796, 832, 849, 859, 877
Boethius, 101, **124**, 149, 177, 233, 241, 272, 357, 576, 592, 642, 675, 698, 736, 747, 804, 828, 852
Bogomils, 556, 623, 633, 660
Boldetti, Marcantonio, 68
Bolgiani, Franco, 234, 271-72, 390-93, 460-61, 815
Bollandists, 644
Bolnissi, 345
Bolsena, 169
Bonaria, 425
Bonaventure, 101
Boniface (Wynfrith), **125**, 129, 162, 175, 330, 334, 349, 394, 643, 646, 676, 873, 877
Boniface I, 7, 47, 532, 618, 641, 645, 654, 670, 730, 741, 742, 743, 804
Boniface II, 101, 241, 281, 438, 642, 645, 743, 744, 749, 782, 794
Boniface III, 645
Boniface IV, 187, 551, 645
Boniface V, 396, 465, 645, 743
Boniface of Senafer (Cornus?), 757
Bonifacius Primicerius, 125, 189
Bonifaziu, 203
Bonn (Bonna), 347, 349
Bononia (Bologna), 421, 680
Bonorva, 425
Bonosus, 28, 41, 96, **125**, 430, 555, 559, 776, 833

Bonus secundicerius, 125
Book, 90, 120, **125**, 130, 264, 335, 394, 407, 409, 411, 428, 478, 631, 633, 647, 669, 724, 761, 763, 812, 842, 849, 864
Book of Life, **126**
Book of the Laws of the Countries, 110, 233, 263, 634, 674, 680
Book of the Tower, 674
Book of the Union, 106
Boppard (Boudabriga), 348
Borboriani, 126, 740
Bordeaux (Burdigala), 27, 102, 126, 225, 237, 238, 756, 799, C. of (384), 126, 410, 547, 712, C. of (385), 126, C. of (662-675), 556
Bordeaux pilgrim, see Itinerarium Burdigalense
Bordonali, Felicia, 149
Borrovadus, 130
Bosio, Antonio, 68, 158, 654
Bossuet, J.B., 101
Bostra, 32, 52, 66, 119, 126, 174, 209, 428, 769, 794, 807, 823, 843, C. of (c.238-244), 65, 127, 64
Botosana (district of Suceava), 217
Bottari, G.G., 68
Bourges (Bituriga), 18, 27, 847
Brac islands (Povlja), 217
Bracae, 253, 254
Braga (Bracara), 105, **127**, 139, 255, 334, 386, 531, 790, 791, C. of (561), 127, 226, 386, 531, C. of (563), 160, C. of (572), 127, 497, C. of (675), 127, 324
Brahmins, 184, 185
Braidotti, Cecilia, 134, 329-30
Brandenburg, Hugo, 68-70
Braulio of Saragossa, 127, 145, 296, 331, 418, 743, 755, 791, 812
Bravona, 203
Bread, 8, 33, 34, 82, 83, 117, 123, 127, 141, 186, 220, 222, 223, 231, 258, 262, 277, 280, 285, 292, 294, 296, 319, 383, 436, 491, 504, 506, 541, 542, 550, 564, 577, 609, 611, 744, 749, 750, 786, 800, 803, 817, 866, 876
Breedon-on-the-Hill, 132
Brefi, 221
Bregenz, 334
Brendan, 128, 411, Voyage of, 128
Brescia (Brixia), 324, 420, 456, reliquary, 33, 62, 219, 221, 269, 429, 450, 477, 575, 650
Bressanone, 149
Breviarium Hipponense, 8, **128**, 142, 147, 383
Breviarius de Hierosolyma, **128**, 426
Breviarum in psalmos, **128**
Brezzi, 644
Bricius, 846
Brigid of Kildare, **128**, 186, 411, Life of, 412
Brioude, 458
Britain, 20, 96, 119, 125, 126, **128**, 178, 188, 189, 193, 198, 279, 307, 346, 351, 366, 369, 397, 411, 446, 476, 486, 497, 511, 545, 547, 551, 567 633, 642, 676, 743, 747, 768, 784, 802, 869, 883, 65-67
Britain and Ireland, **129**
Brother, 6, 90, **134**, 230, 279, 287, 355, 376, 494, 506, 548, 552, 714, 793, 849
Brundisium (Brindisi), 112, 424
Brunhilde, 162, 346, 802, 862
Bruno of Segni, 186
Brutinus, Ps.-, 586
Buddha, Buddhism, 111, 161, 230, 519
Bulic, F., 69
Bulla Regia, 15, 173
Bulwark of the cross, 291
Buonarotti, 69
Burchard of Würzburg, 393
Burgh Castle, Suffolk, 131
Burgos, C. of (1080), 497
Burguillos, 106

Burgundians, 105, 124, 134, 138, 142, 183, 187, 329, 349, 505, 525, 544, 555, 642, 653, 779, 861, 869, 874
Burry Holms, 131
Busento (Cosenza), 19
Byblos (now Gebail), 808
Byrrus, 253
Byzacena, 13, 128, 134, 137, 142, 146, 147, 170, 247, 369, 822, C. of (393), 134, 545
Byzantium, Byzantines, 1, 17, 21, 38, 48, 80, 112, 136, 139, 163, 193, 194, 274, 292, 343, 344, 366, 375, 377, 382, 418, 453, 505, 545, 565, 619, 642, 643, 666, 667, 674, 708, 730, 742, 743, 757, 779, 791, 804, 806, 809, 837, 838, 842, 846, 861, 875, 884, liturgy, 5, 6, 8, 15, 27, 34, 50, 82, 83, 113, 120, 127, 134, 160, 232, 276, 293, 346, 369, 478, 490, 496, 501, 541, 797, 805, 816, 853, 865
Bznunik, 4

Cabarsussi, 143, C. of (393), 545
Cadbury, 131, Castle, 131
Cadiz, 106
Cadwalla, 768
Cadwallon, 662
Caecilian, 53, 79, **137**, 185, 211, 246, 248, 250, 321, 332, 350, 508, 517, 554, 560, 641, 742, 743, 765, 779
Caedmon, **137**
Caelestians, **137**
Caelestius, 96, 98, 102, **137**, 147, 241, 359, 375, 447, 525, 641, 660, 665, 666, 742, 744, 746, 782, 885
Caelestis urbs Jerusalem, 90
Caena Cypriani, 137
Caerwent (Venta Silurum), 129
Caesarea in Cappadocia, 38, 80, 109, 113, 120, **137**, 307, 372, 496, 547, 640, 703, 823, 826, C. of (314), 137, 480, Acts of, Ps.-, 188
Caesarea in Mauretania (Cherchel), 13, 15, **138**, 156, 543
Caesarea in Palestine, 28, 36, 43, 53, 71, 85, 95, 113, 114, **138**, 143, 202, 215, 235, 269, 270, 299, 305, 307, 339, 363, 368, 380, 430, 448, 486, 544, 618, 619, 673, 761, 807, 822, 832
Caesar Gallus, 192
Caesarius, Ps.-, 138, 634
Caesarius of Arles, 6, 17, 55, 67, 79, 101, 102, 105, **138**, 208, 212, 233, 250, 298, 307, 311, 312, 322, 335, 338, 356, 359, 361, 393, 394, 399, 410, 483, 490, 497, 500, 503, 517, 555, 609, 645, 663, 667, 669, 702, 709, 725, 743, 746, 758, 761, 766, 767, 770, 781, 794, 804, 816, 839, Life of, 270, 761
Caesarius of Nazianzus, **138**
Caesarius of Nazianzus, Ps.-, 234
Caesarius the African, 339
Caesaropapism, **139**, 339, 618, 701
Cahors (Cadurcum), 231
Cain, 1, 71, 139, 309, 771, and Abel, 2, 9, 4
Cáin Adomnáin, 411, 412
Cainites, **139**, 384, 771, 819
Caius or Gaius, 645, 741
Calabria, 150
Calahorra (Calagurris), 790, 813
Calama, 14, 705
Calcagnini Carletti, Daniella, 1, 9-10, 33, 274, 369, 549, 555, 575, 579
Calcea, 253
Calcidius, **139**
Calcio, 421
Caldonius, 139, 146
Calendar, 4, 51, **140**, 200, 229, 338, 356, 370, 429, 438, 535, 536, 671
Calendio, 49, 50, 684, 884
Caligula, 18, **140**, 178, 682, 862
Callatis, 140, 765, 68
Callinicum, 28, 52, 95, 521, 662, C. of (567), 570

Callinicus, monk, **140**
Callinicus of Pelusium, **140**
Callinicus I of Constantinople, 120, 121, **140**, 175, 337, 366, 401
Callistus I, pope, **140**, 145, 153, 154, 156, 181, 189, 229, 269, 384, 385, 522, 536, 598, 604, 631, 640, 644, 654, 706, 741, 748, 757, 784, 818, 885
Calvary, 433
Calvin, John, 101
Calvisius Taurus, 558
Calydon, 169
Cambi, Nenad, 217-18, 600, 638-39, 753, 759, 783, 860
Camelaucus, 253
Campagi, 254
Canakkale, 88
Canart, Paul, 125-26, 520-21, 632-33
Cana, Wedding-Feast of, 24, 38, 119, **140**, 163, 282, 290, 294, 457, 493, 504, 529, 539, 540, 723, 740, 749, 755, 871, 876, 877, 69
Cancellus, 172
Candida Casa, 130, 597
Candidus the Arian, 141
Candidus the Valentinian, 141
Candle, Paschal, 123, 141, 232, 386, 493
Candlemas, 493
Canne, 424
Canonical, 8, 39, 47, 55, 56, 57, 58, 66, **141**, 200, 205, 306, 474, 710, 712, 763, 843, 847
Canonical collections, 13, 48, 62, 63, **141**, 146, 147, 148, 205, 237, 342, 382, 418, 439, 475, 479, 585, 609, 655, 669, 791, 793, 875
Canons, 8, 193, in causa Apiarii, 142, of Athanasius, 200, of Basil, 200, of Hippolytus, 122, 143, 191, 200, of the Apostles, 200
Canosa (Canusium), 424, 749
Canterbury (Durovernum Cantiacorum), 96, 119, 129, 131, 143, 231, 286, 369, 397, 465, 476, 486, 743, 824, 883
Cantianilla, 64
Cantianus, 64
Cantius, 64
Canusia (Canosa), 424
Caorle, 189
Capernaum, 169, 635
Capitolias, 636
Capitula, 180, 717
Cappadocia, 7, 21, 43, 82, 85, 87, 92, 105, 109, 114, 135, 137, **143**, 163, 167, 175, 180, 195, 203, 230, 237, 282, 290, 293, 304, 307, 356, 371, 493, 518, 561, 608, 635, 703, 746, 768, 823, 826, 854, C. of (372), 143
Capreolus, **143**, 728
Capsa, 13, 143
Capua, 3, 424, 497, 868, 870, C. of (391/392), 41, 125, 559, 833
Caracalla, 20, 60, 110, **143**, 176, 253, 356
Carales (Cagliari), 35, 425, 459, 728, 757
Caraunius of Paris, 875
Carausius, 198
Carcara, 71
Carcesi, 148
Carentius, 348
Caricin Grad, 174
Caricus, 58
Carisius, 822
Carletti, Carlo, 40, 61-62, 227-28, 254-55, 306, 323, 333, 403, 557, 802-3
Carmen adversus marcionitas, **144**, 819
Carmen adversus paganos, **144**
Carmen ad Flavium Felicem de resurrectione mortuorum, **144**, 819, 863
Carmen ad quendam senatorem, **144**
Carmen de Christi Iesu beneficiis, 270, 747
Carmen de laudibus Domini, 144
Carmen de passione Domini, 144
Carmen de Providentia, **144**, 278, 381

Numbers in bold refer to the main entry on the subject. *Numbers in italics refer to illustration numbers (vol II, pages 951-1094)*

Carmen Paschale, 92, 119, 178, 312, 651, 766
Carmona, 107
Carnsore, 130
Carpentras (Carpentorate) (527), 555
Carpocrates, 7, 10, **145**, 180, 281, 353, 384, 434, 467, 521
Carpophorus, M. Aurelius, **145**
Carpus, 17
Carrhes (now Harran), 809
Cartagena (Carthago Nova), 148, 368, 386, 456, 790, 844, C. of (254/255), 530
Carterius, 145, 213
Carthage, 2, 11, 13, 14, 15, 18, 29, 40, 69, 83, 90, 91, 92, 98, 102, 121, 137, 143, **145**, 151, 170, 171, 173, 177, 202, 225, 228, 246, 247, 248, 250, 252, 270, 296, 321, 350, 360, 487, 517, 521, 537, 541, 544, 545, 547, 551, 554, 572, 618, 641, 643, 651, 652, 653, 670, 672, 688, 702, 706, 711, 712, 714, 725, 728, 732, 742, 752, 756, 761, 762, 765, 768, 779, 782, 790, 792, 801, 814, 843, 854, 861, 865, 868, *70, 71*, C. of (c.220), 109, 146, C. of (251), 146, 211, 212, 328, 743, C. of (252), 146, C. of (255: 1st baptismal), 109, 146, Cc. of (256: 2nd and 3rd baptismal), 109, 146, 193, 229, 382, 470, 482, 584, 672, 812, 816, 840, C. of (312), 321, 332, C. of (336), 247, 250, C. of (345/348), 8, 142, 146, 360, C. of (390), 142, 146, 617, C. of (392), 545, 711, C. of (394), 146, C. of (397), 128, 146, 274, 319, 383, 499, 541, 609, 617, C. of (398), 142, C. of (399), 147, C. of (401), 147, 863, C. of (403), 147, 705, C. of (404), 147, C. of (405), 147, C. of (407), 147, 705, 739, C. of (411), 137, 138, 143, 247, 248, 270, 328, 337, 351, 359, 516, 521, 544, 545, 600, 607, 613, 665, 680, 705, 711, 746, 774, 812, 840, 859, C. of (412), 660, C. of (416), 328, 744, 782, C. of (417), 147, C. of (418), 102, 605, 665, 666, C. of (419), 142, 148, 237, 605, 705, 759, C. of (c.424), 148, C. of (484), 138, 544, 601, 757, 840, 842, 868, C. of (525), 148, C. of (536), 148
Carthaginensis, 107, **148**, 368, 386, 387, 790, 844, 861
Cara cognato, 153
Casae Nigrae, 250
Casa Herrera, 174, 554
Casalbordino, 423
Casaranello, 573
Caspar, 644
Caspari, C. P., 654
Cassia, 16
Cassian, John, 5, 17, 37, 46, 48, 61, 72, 119, **149**, 184, 185, 269, 282, 295, 306, 319, 338, 339, 369, 406, 483, 500, 531, 549, 567, 621, 646, 707, 708, 746, 764, 767, 781, 793, 796, 816, 832
Cassian of Imola, 149
Cassiciacum, 11, 27, 96, 98, 584
Cassiodorus, Flavius Magnus Aurelius, 3, 55, 67, 119, 124, 149, 152, 166, 177, 180, 203, 235, 237, 270, 272, 282, 312, 366, 388, 389, 475, 486, 503, 521, 576, 595, 649, 655, 658, 716, 723, 760, 761, 785, 790, 804, 827, 839, 868, 873, 881
Cassius Longinus, 884
Castabala (now Budrum), 175, 832
Castelseprio, 631
Castelvecchio Subequo, 423
Castillo de Locubín, 106
Castricianus, 559
Castulo, 148
Catacomb, 155, 535, 629, of Aproniano, 450, of the Aurelii, 9630, 771, of Balbina, 645, 778, of Bassilla, 329, of Bonaria, 450, of Calepodio, 157, 536, 631, of Cartagena, 148, of Ciriaca, 715, of Commodilla, 62, 156, 202, 294, 321, 540, 555, 631, 659, 705, 803, of Cyriaca, 107, 294, 520, of Domitilla, 156, 219, 245, 246, 253, 269, 281, 450, 536, 693, 597, 615, 625, 680, 705, 715, 801, 803, 808, of Felicita, 685, of S. Gaudioso, 581, of Grottaferrata, 685, of the Jordani, 520, of Maximus, 321, of Nunziatella, 482, of Panfilo, 599, of Pontianus, 1, 447, 688, of Porta Maggiore, 169, of Predio Maltese, 807, of Pretestato, 419, 489, 536, 629, 668, 685, 801, of S. Priscilla, 8, 41, 70, 107, 108, 154, 156, 219, 294, 323, 429, 434, 477, 521, 568, 575, 625, 630, 682, 685, 715, 801, 803, 866, of SS. Marcellino e Pietro, 3, 9, 10, 41, 62, 107, 108, 140, 157, 169, 187, 294, 306, 325, 333, 419, 434, 477, 482, 597, 625, 630, 668, 715, 801, of SS. Marco e Marcelliano, 306, 323, 475, 597, of S. Callisto 3, 69, 70, 108, 140, 154, 156, 210, 212, 219, 281, 294, 315, 329, 447, 450, 477, 554, 575, 597, 615, 625, 629, 630, 631, 644, 650, 668, 685, 702, 705, 787, 801, 803, 866, 885, of S. Gennaro, 10, 70, 156, 450, 476, 536, 580, 581, 630, 631, 728, of S. Giovanni, 807, of S. Hippolyto, 157, of S. Lorenzo, 169, of S. Maria di Gesù, 807, of S. Massimino, 379, of S. Maria in Stelle, 274, 313, 583, 630, of S. Sebastiano, 68, 107, 155, 169, 281, 552, 650, 676, 785, of S. Severo, 202, 349, of S. Tecla, 157, 536, 559, 715, of Trebius Justus, 705, of S. Valentino, 210, 476, of S. Vittorino, 457, of Via Anopo, 3, 62, 158, 329, of Via D. Compagni, 2, 3, 9, 10, 41, 107, 187, 202, 269, 327, 329, 428, 429, 477, 506, 549, 575, 630, 675, 684, 715, 755, 771, 786, 844, of Via della Lungara, 108, of Via Latina, 10, 156, 225, 280, 452, 597, of Via Manzoni, 612, of Via Nomentana, 192, of Via Salaria Antiqua, 122, of Via Yser, 844, of Vibia, 457, of Vigna Cassia, 274, 313, 807, of Vigna Massimo, 844, of Vigna Randanini, 625
Catania (Catana), 425, 807
Cataphrygians, 37, 84, 150, 384, 542, 736
Catechesis, 57, 83, 91, 99, 107, 108, 120, 150, 151, 190, 200, 215, 222, 225, 233, 235, 238, 245, 264, 277, 293, 338, 360, 370, 394, 395, 398, 403, 409, 434, 441, 474, 478, 493, 494, 503, 504, 506, 527, 537, 552, 619, 628, 651, 656, 668, 708, 736, 749, 750, 761, 780, 792, 802
Catechumen, -ate, 6, 42, 50, 79, 90, 107, 108, 123, 128, 150, **151**, 170, 173, 174, 177, 207, 215, 227, 232, 235, 242, 257, 258, 273, 290, 296, 317, 334, 364, 370, 383, 410, 463, 467, 477, 490, 494, 506, 509, 510, 540, 541, 585, 664, 668, 707, 708, 727, 728, 741, 744, 745, 751, 780, 792, 817
Categories, 76, 124, 769
Catenae, Biblical, 4, 55, **152**, 214, 311, 352, 380, 655, 722, 880
Cathars, 63, 556623, 634, 686
Cathedra, 153, 157, 172,.174, 260, 262, 333, 432, 452, 593, 600, 651, 676, 764, 72, Petri, 153, 229, 261, 437, 641, 671, 676, 741, of Maximian, 41, 274, 313, 447, 539, 583
Cathemerinon, 401, 721
Catherine of Alexandria, **153**
Catholic, Catholicism, 2, 6, 7, 14, 17, 18, 20, 27, 28, 29, 49, 50, 53, 63, 91, 98, 101, 105, 109, 134, 147, **153**, 163, 165, 205, 212, 218, 224, 226, 228, 244, 247, 262, 309, 310, 320, 328, 329, 331, 351, 357, 360, 377, 383, 395, 405, 406, 418, 436, 462, 473, 495, 545, 601, 626, 642, 643, 651, 706, 711, 755, 774, 777, 791, 794, 798, 804, 819, 822, 828, 844, 846, 852, 860, 870
Catholicós, 1, 3, 19, 27, 48, **153**, 766, 858
Cathwulf, 720
Cavalcanti, Elena, 42, 43, 58, 438, 459, 460, 521, 611, 700, 827-28, 841-42, 843, 859, 873
Cavalletti, Sofia, 19, 52, 110, 222, 435-36, 454, 716, 762
Cebarsussa (393), 247, 711
Cecchelli Trinci, Margherita, 252-54, 351, 486, 490-91
Cecropius of Nicomedia, 154
Cedd, 159
Ceillier, R., 654
Celebra,Iuda, in sanctorum apostolorum, 211, 412
Celeia, 600
Celeres, 143
Celestine I, 24, 55, 99, 148, **154**, 155, 196, 214, 275, 314, 325, 369, 411, 444, 530, 538, 545, 594, 604, 636, 641, 645, 717, 742, 743, 744, 864, 865, 866
Celibacy of the Clergy, 5, 16, 17, 48, 54, 84, 118, 150, **154**, 161, 288, 382, 562, 596, 646, 776, 869
Cell, 6, 130, 131, **154**, 156, 184, 217, 220, 264, 266, 566, 581
Cellae, 731
Celsus, 6, 46, 73, **155**, 167, 192, 205, 226, 275, 285, 314, 372, 373, 508, 551, 558, 564, 620, 672, 674, 690, 692, 763, True Doctrine (Alethes Logos), 167, 620, 672, see also Contra Celsum
Celsus and Nazarius, see Nazarius and Celsus
Celtic church, 18, Liturgy, 10, 96, 129, 143, **155**, 187, 188, 497, 645
Cemetery, 7, 17, 50, 64, 65, 70, 79, 85, 92, 107, 129, 133, 140, 145, **155**, 169, 189, 201, 209, 210, 213, 217, 218, 219, 221, 228, 232, 279, 315, 326, 328, 334, 336, 338, 347, 348, 350, 370, 377, 390, 403, 420, 421, 422, 423, 424, 425, 434, 450, 475, 520, 581, 600, 617, 630, 640, 651, 656, 690, 714, 728, 741, 753, 806, 807, 835, 840, 842, 850, 853, 859, 871, *73-86*, see also Catacomb
Cenchreae, 201, 221
Ceneda (Vittorio Veneto), 420
Centcelles, Mausoleum of, 450, 599, 813
Cento, 118, **158**, 165, 721, 843
Centumcellae (now Civitavecchia), 422
Ceolfrith, 10, 131
Cerdo, **158**, 384, 644
Cerealis, **158**
Ceresa Gastaldo, Aldo, 547-48
Cerethius of Grenoble, 753, 863
Cericus and Julitta, 175
Cerinthians, 158, 176
Cerinthus, 26, 55, 158, 333, 384, 448, 560
Cerne, Book of, 132
Ceylon (Sri Lanka), 307
Chad, **159**, 883
Chaeremon, 683
Chalcedon, 8, 36, 85, **159**, 218, 356, 641, 688, 742, 747, C. of (451: 4th ecumenical), 5, 22, 23, 24, 36, 37, 48, 49, 52, 59, 80, 84, 85, 109, 112, 113, 115, 122, 136, 142, 159, 194, 195, 196, 200, 201, 205, 207, 208, 214, 223, 233, 237, 241, 245, 256, 275, 289, 296, 301, 303, 305, 314, 325, 327, 330, 333, 342, 346, 350, 400, 403, 416, 428, 432, 433, 438, 444, 458, 464, 466, 475, 479, 480, 484, 487, 496, 523, 527, 538, 548, 550, 556, 559, 566, 569, 582, 593, 594, 596, 608, 617, 618, 619, 641, 642, 652, 655, 679, 684, 686, 699, 714, 725, 742, 767, 773, 780, 784, 827, 828, 831, 838, 840, 845, 852, 855, 884
Chalcedonian formula/creed, 47, 115, 165, 205, 315, 344, 358, 401, 402, 432, 443, 444, 446, 530, 584, 585, 734, 788, 826, 870
Chalcedonians, 831
Chaldaean oracles, 12, 587, 588, 589, 590, 591, 614, 805, 836
Chaldaean rite, 496, 810
Chaldaeans, 10, 50, 153, 517
Chalice, 33, 89, 123, 130, 133, **159**, 277, 292, 294, 323, 349, 351, 406, 434, 491, 541, 668, 812, 876
Châlon-sur-Saône (Cabillonum), 325, C. of (579), 556, C. of (647-653), 360, 556
Chamalikres, 822
Chant, 6, 17, 28, 29, 50, 136, **160**, 224, 274, 290, 325, 335, 386, 387, 400, 446, 472, 492, 497, 498, 504, 532, 541, 576, 685, 722, 745, 816, 824
Charachmoba, 636
Charibert, 162, 556
Charisms, 35, 63, 103, 104, **160**, 163, 189, 366, 390, 445, 477, 513, 527, 528, 533, 561, 609, 622, 715, 763, 780, of truth, 561
Charito, **160**, 304, 454
Charlemagne, 43, 142, 208, 242, 246, 278, 324, 334, 366, 497, 503, 505, 616, 646, 653, 720, 745, 791, 875
Charles Martel, 330, 505, 700, 735, 877
Charles the Bald, 153
Chastity, 63, 103, 146, **161**, 176, 198, 222, 271, 284, 288, 333, 377, 383, 485, 517, 528, 538, 604, 621, 665, 666, 685, 732, 775, 792, 801, 802, 814, 844, 847, 871, 878
Chasuble, 865
Chelidonius, 120
Chenoboskion, 57
Cherubim, 39, 40
Chester-le-Street, 131
Chesterton, Huntingdonshire, 133
Childebert I, 102, 329, 349, 482, 555, 556, 663, 735
Childebert II, 366, 556, 802
Childeric, 183, 329, 342
Childeric II, 105, 556
Child, 43, 121, 147, 156, **161**, 174, 232, 273, 317, 318, 319, 320, 383, 411, 440, 441, 528, 566, 656, 666, 718, 739, 744, 745, 760, 776, 777, 781, 792, 806, 846, 871
Chiliasm, 560
Chilperic I, 162, 346, 556, 653, 862
China, 263, 519, 674, 809, 840
Chindaswinth, 296
Chios, 417
Chiusi, 422
Chlamys, 253, 254
Chlodomir, 183, 329
Chorbishop, 48, 107, 114, 143, **162**, 617
Chosroes I, 3, **162**, 464, 551, 598
Chosroes II Abharvez, 36, 65, 83, 106, **162**, 263, 564
Chouhoud el Batin, 2
Choziba, 44
Chrestus, 806
Chrism, 42
Christ, see Jesus Christ
Christe eleison, 6
Christians, 47, 53, 72, **163**, 176, 376, 456, 510, 699, 755
Christmas, 4, 17, 35, 50, 52, 82, **163**, 186, 229, 257, 274, 282, 298, 337, 351, 364, 410, 433, 493, 494, 504, 511, 671, 712, 738, 740, 742, 797, 800, 821
Christodorus of Thebes, 44
Christogram, 107, see also Epigraphy, Cryptography
Christology, 16, 19, 20, 23, 29, 30, 34, 35, 36, 44, 45, 46, 50, 54, 55, 58, 59, 60, 72, 90, 92, 94, 99, 101, 106, 109, 113, 124, 136, 151, 159, **163**, 167, 193, 196, 201, 202, 205, 208, 214, 215, 220, 223, 231, 234, 242, 243, 244, 251, 256, 275, 282, 286, 296, 303, 304, 306, 307, 315, 318, 321, 322, 325, 331, 357, 362, 363, 374, 378, 379, 382, 401, 402, 405, 408, 429, 434, 436, 442, 443, 446, 454, 465, 479,

480, 481, 483, 496, 524, 536, 537, 538, 547, 551, 568, 569, 585, 593, 594, 604, 638, 646, 675, 684, 686, 699, 708, 713, 716, 723, 725, 742, 758, 786, 819, 824, 825, 827, 839, 845, 852, 870
Christopher, **165**
Christotokos, 833
Christus patiens, 158, 165
Chromatius of Aquileia, 29, 64, 83, 96, 151, **166**, 189, 371, 430, 497, 499, 543, 655, 723, 863
Chronicle, 431, 460, 870, of Arbela, 11, of Edessa, 214, of Séert, 106, of Zuqnin, 517, of the foundation of Monemvasia, 835
Chronicon, 81, **166**, 299, 389, 404, 428, 705
Chronicon Horosii, 166
Chronicon paschale, 43, **166**, 258, 539, 682, 843
Chronographer of 354, 140, 166, 228, 315, 493, 644, 675, 682, 885
Chronograph of Ravenna, 426
Chronography, Chronology, **166**, 237, 343, 388, 389, 418, 639, 644, 656, 718, 728, 766, 799, 825
Chrysanthius, 296
Chrysaphius, 304, 305, 829
Chrysippus, 53, 74, 91, **167**, 560
Chrysogonus, 16, 35, 64, **167**, 809
Chrysokava, 213
Chrysopolis, 35
Chupungco, Anscar, 6, 17-18, 141, 160, 185, 232-33, 277-78, 313-14, 410, 490, 738-39, 846, 853
Church and Empire, 102, 167, 193, 250, 271, 299, 435, 626, 641, 673, 701, 744
Church buildings, 3, 7, 11, 14, 15, 19, 24, 40, 47, 51, 64, 66, 70, 80, 82, 85, 90, 106, 107, 122, 127, 129, 168, 189, 201, 209, 217, 246, 255, 257, 265, 266, 283, 338, 344, 348, 371, 377, 390, 403, 406, 422, 423, 425, 433, 434, 435, 447, 465, 487, 488, 513, 541, 549, 552, 554, 571, 573, 581, 600, 605, 617, 629, 630, 635, 636, 638, 651, 656, 676, 681, 714, 725, 749, 753, 757, 759, 764, 765, 771, 778, 783, 795, 802, 807, 809, 813, 823, 831, 833, 837, 838, 842, 843, 845, 858, 860, 863, 883
Church Island, Valentia, 130
Church of Hagia Sophia, 16, 89, 135, 172, 174, 195, 197, 209, 213, 292, 305, 343, 346, 374, 439, 464, 488, 491, 553, 573, 664, 682, 725, 827, of S. Agata dei Goti, 377, of S. Agnese, 174, 574, 656, 866, of S. Ambrogio, 221, of S. Angelo in Formis, 255, of S. Apollinare Nuovo, 17, 40, 59, 141, 169, 170, 186, 202, 209, 219, 228, 294, 306, 377, 419, 450, 456, 464, 477, 509, 531, 557, 573, 650, 671, 688, 689, 715, 730, 803, of S. Apollinare in Classe, 2, 59, 269, 294, 464, 550, 573, 730, 756, of S. Aquilino, 62, 202, 269, of SS. Apostoli, 175, of S. Balbinae, 656, of S. Carrotja, 107, of S. Clemente, 25, 218, of SS. Cosma e Damiano, 51, 66, 292, 574, 685, 803, 822, of S. Crisogono, 631, of S. Croce, 175, of S. Eulalia de Campo, 296, of S. Eusebio, 377, 656, of S. Francesca Romana, 202, of S. Gavino at Porto Torres, 425, of S. Giorgio, 581, of S. Giovanni Battista, 377, of S. Giovanni Evangelista, 170, of S. Giovanni Maggiore, 377, 581, of S. Giovanni al Sepolcro, 424, of S. Giovanni in Fonte, 219, 447, 573, 581, 774, 823, of SS. Giovanni e Paolo, 170, 209, 447, 536, 630, 638, 656, of S. Lorenzo, 174, 574, of S. Lorenzo Maggiore, 377, 581, of S. Lorenzo at Aosta, 175, of S. Lorenzo fuori le mura, 476, 656, 667, of S. Lucia, 807, of S. Maria Antiqua, 108, 221, 450, 540, 631,
764, 769, of S. Maria Maggiore, 1, 3, 40, 41, 170, 218, 292, 306, 333, 379, 428, 447, 452, 453, 471, 485, 506, 529, 540, 550, 573, 575, 659, 710, 728, 741, 784, of S. Maria Maggiore alla Pietra Santa, 581, of S. Maria Nova, 632, of S. Maria dell'Anima, 3, of S. Maria de Quintanilla de las Viñas, 148, of S. Maria di Castelseprio, 292, of S. Maria in Porto Fuori, 756, of S. Maria in Trastevere, 154, 202, 489, 540, 632, 656, of S. Martino ai Monti, 491, 787, of S. Martino in Ciel d'Oro, 59, of S. Michele in Affricisco, 464, of S. Paulo fuori le Mura, 69, 169, 435, 646, 656, of S. Peretó, 107, 174, of S. Pietro in Ciel d'oro, 124, of S. Pietro in Vaticano, 169, 171, 656, of S. Pietro in Vincoli, 67, of S. Prisco, 292, of S. Pudenziana, 62, 218, 292, 435, 675, of S. Saba, 52, 204, 491, 631, of S. Sabina, 1, 141, 154, 210, 219, 253, 269, 369, 435, 520, 540, 573, 575, 675, 689, 866, of S. Salvatore, 162, of S. Silvestro in capite, 688, of SS. Silvestri et Martini, 656, of S. Simpliciano, 175, of S. Stefano Rotondo, 795, of S. Teodoro, 574, of S. Venanzio, 574, of S. Vitale, 2, 40, 174, 219, 253, 294, 306, 333, 435, 464, 491, 550, 568, 573, 575, 597, 656, 705, 715, 730, 822, 866, of the Anastasis, 195, 224, 409, 433, 500, of the Blachernae, 195, 197, 209, 827, of the Dormition, 89, of the Holy Sepulchre, 62, 224, 302, 433, 809, 833, of the Nativity, 209
Chur (Curia Rhaetiorum), 347
Ciacconio, Alfonso, 68
Ciarán of Clonamacnois, 324, 411
Cibalae, 639
Ciccarese, Maria Pia, 483, 872-73
Cicero, 74, 75, 82, 121, 177, 179, 233, 286, 288, 363, 469, 470, 483, 490, 602, 650, 696, 697, 720, 796, 875, 882
Cilicia, 32, 37, 62, 63, 89, 109, 143, 175, 203, 258, 371, 403, 512, 599, 618, 657, 720, 814, 824, 87-89, C. of (423), 175
Cimiez, 861
Cimitero Maggiore, 270, 536
Cimitile (Nola), 9, 423, 424, 450, 630, 661
Cingulum, 865
Circumcellions, **175**, 228, 248, 249, 250, 516, 545, 567, 613, 651
Circumcision, 54, 111, **176**, 290, 321, 402, 432, 454, 456, 580, 604, 716, 755, 840, 843, of Jesus, 38, 82
Circus, 273, 296
Cirga, 89
Cirta, 13, 14, 157, 168, 247, 328, 351, 382, 607, 714, 779, C. of (305), 765, C. of (c.414), 176, 247, 337, 859
Citizenship, Roman, 144, **176**, 178, 669, 701
Cittanova, 420
Ciudadela, 107
Cixala, 405
Clarentius, 812
Clarus of Mascula, 561, 584
Classical culture and Christianity, 72, **176**, 247, 272, 338, 342, 367, 397, 459, 470, 717, 791, 827, 851
Claudian, 67, **178**, 202, 212, 335, 555, 599, 766
Claudianus, 869
Claudianus Mamertus, 29, **178**, 278, 295, 321, 339, 592
Claudius, 10, 18, **178**, 194, 264, 348, 403, 419, 512, 593, 607, 676, 687, 740
Claudius II Gothicus, **178**, 356, 594
Claudius Callistus Hilarius, 30
Clausium, 7
Clausula, **179**, 213, 646, 723, 738
Cleanthes, 53, 74
Clemens, 245
Clemens Contemptus, 103
Clement I, pope, 9, 56, 57, 72, **179**, 181, 191, 269, 284, 286, 309, 311, 326, 449, 453, 469, 472, 476, 483, 543, 616, 644, 657, 664, 675, 709, 710, 802, 875, 878, 587, 616, 626, 639, 644, 657, 671, 672, 675, 689, 691, 693, 694, 695, 696, 706, 709, 710, 724, 732, 734, 740, 792, 802, 875, 878
Clement VIII, 101
Clement X, 27
Clement of Alexandria, 2, 9, 10, 21, 22, 23, 32, 47, 53, 55, 56, 57, 62, 64, 72, 74, 75, 82, 84, 92, 96, 108, 113, 117, 121, 122, 134, 150, 151, 161, 164, 166, 168, 177, 178, **179**, 189, 205, 206, 221, 226, 232, 235, 240, 242, 243, 244, 252, 257, 258, 259, 260, 264, 273, 281, 282, 284, 286, 287, 293, 304, 310, 312, 316, 319, 320, 326, 330, 355, 356, 358, 361, 373, 376, 391, 395, 398, 401, 406, 409, 429, 434, 437, 448, 449, 455, 457, 460, 462, 465, 469, 472, 475, 486, 489, 492, 493, 499, 500, 526, 528, 534, 537, 552, 558, 561, 577, 596, 603, 606, 610, 614, 615, 625, 629, 640, 648, 658, 665, 680, 682, 683, 691, 703, 707, 709, 719, 720, 723, 725, 732, 737, 750, 760, 761, 775, 781, 784, 785, 788, 792, 795, 796, 798, 802, 805, 815, 818, 830, 832, 843, 848, 859, 871, 875, 877, 878
Clement of Ochrida, 633
Clementines, Ps.-, 45, 56, 57, 74, 110, 163, **179**, 251, 259, 318, 337, 454, 657, 674, 676, 746, 785, 790, 810, 854
Cleomenes, 181, 278, 566, 653, 885
Clephi, 103, 505
Clergy, 128, 134, 139, 146, 147, 154, 162, 169, 172, 173, 175, **181**, 189, 193, 197, 200, 211, 219, 221, 224, 231, 232, 242, 246, 249, 260, 270, 286, 289, 315, 329, 338, 350, 351, 360, 367, 369, 383, 400, 404, 408, 411, 440, 447, 473, 477, 483, 495, 528, 595, 600, 630, 640, 642, 648, 667, 705, 708, 710, 714, 722, 739, 741, 742, 753, 755, 790, 793, 794, 814, 820, 821, 838, 840, 844, 846, 847, 861, 864, 876, 877
Clermont-Ferrand (Arverna), 4, 238, 338, C. of (535), 555
Cletus, 617, 644
Clibanon, 344
Clichy, C. of (626-627), 517, 556, 850, C. of (657), 360
Clinicus, 152, 182, 183, 315
Clipeus, 254
Clodius and Domnina, 175
Clodoswinth, 506, 850
Clonard (Cluain Iraird), 19, 130, 411, 324
Clonfert, 128, 211, 411
Clonmacnoise (county Offaly), 130, 411
Clotaire (Lothar) I, 162, 556, 729, 824
Clotaire (Lothar) II, 231, 268, 325, 329, 330, 556, 802
Clotilde, **183**, 329, 342, 653, 676, 850
Clovesho (746)
Clovis (Chlodovic) I, 59, 105, 124, **183**, 199, 286, 307, 329, 338, 342, 349, 356, 365, 555, 653, 676, 700, 732, 735, 804, 846, 850, 862
Clovis II, 556
Cluana (now Porto Civitanova), 422
Cluj-Napoca, 217
Clusium (Chiusi), 422
Coats of skins, 124, **183**, 704, 775, 776
Cocchini, Francesca, 2, 2-3, 8, 10, 13, 17, 47, 54, 64, 67, 83, 84, 150-51, 159-60, 186, 233, 281, 297-98, 402, 431-33, 471, 521, 574-75, 651-52, 660, 662, 669, 776, 821, 824, 854, 875-76
Codex, 125, 647, Alexandrinus, 181, 352, Amiatinus, 131, Apiarii causae, 146, Barberinus, 501, Bezae, 647, canonum ecclesiasticorum, 237, canonum ecclesiae africanae, 142, Fuldensis, 234,
497, Grandior, 131, Gregorianus, 183, Hermogenianus, 183, Justinianus, 183, 184, 203, 464, 648, Laurentianus, 793, Marchalianus, 632, Purpureus Rossanensis, 274, 294, 313, 371, 456, 477, 688, 863, Salmasianus, 612, Sinaiticus, 111, 647, 782, Sinopensis, 631, Theodosianus, 183, 184, 191, 532, 648, 784, 785, 829, Vaticanus, 647, Vercellensis, 302, Vindobonensis, 847, of Cosmas Indicopleustes, 1, 369, 575, of Rabbula, 1, 109, 210, 220, 228, 274, 294, 313, 369, 435, 437, 447, 450, 456, 540, 583, 631, 729, 786, see also Book
Coelfrid, 117
Coenobites, 27, 83, 112, 119, 149, 150, **184**, 217, 230, 264, 266, 327, 350, 500, 567, 746, 628
Coenobium, 1, 135, **184**, 281, 475, 483, 486, 829
Cofar Aden, 65
Cogitosus, 130
Colarbasius, 384
Colchester (Camulodunum), 129
Collatio Alexandri et Dindimi, **184**
Collatio cum Donatistis, 147, **185**, 247
Collatio Legum, **185**
Collect, **185**, 386, 541, 805
Collectio Avellana, 804
Collectio Palatina, 446, 525
Collegno, 421
Colle Temenite, 807
Collucianists, **185**, 301, 507
Colluthus, 23, **185**, 343, 185, Against, 185
Collyridiani, **186**, 538
Colmán mac Lénéni, 18, 129, 413
Colman moccu Béognai, 412
Colobium, 253
Cologne (Colonia Agrippina), 347, 348, 349, 772, 850, 856, C. of (346), 348, 546, 771, Cologne Mani Codex, 269
Colosseum, **186**
Colour, **186**, 864, 865
Columba, 10, 130, **186**, 307, 324, 411, 412, 486, 762, Life of, 412, Cathach, 131
Columbanus, 43, **186**, 230, 250, 258, 268, 307, 334, 338, 349, 411, 412, 451, 497, 669, 747
Columbanus, Ps.-, 412
Comacina, 420
Comgall, 411
Comitas, 467
Commentaries, Biblical, 28, 30, 152, **187**, 205, 282, 311, 393, 394, 398, 431, 504, 506, 613, 652, 666, 707, 715, 716, 722, 727, 736, 764, 810, 825, 831, 847, 873
Commodian, 26, 55, 56, 57, 144, **187**, 560, 614, 860
Commodilla, 675
Commodus, 32, 54, 60, 95, 145, **188**, 189, 333, 522
Commonitorium, 208, 483, 525, 624, 723, 870
Communion, 6, 9, 13, 17, 20, 21, 33, 48, 49, 50, 61, 63, 78, 82, 90, 93, 101, 103, 104, 114, 123, 128, 135, 137, 148, 159, 185, **188**, 196, 218, 224, 230, 234, 236, 241, 243, 246, 248, 250, 258, 260, 261, 262, 270, 274, 277, 282, 283, 284, 287, 292, 293, 302, 305, 309, 319, 320, 330, 348, 358, 364, 371, 376, 392, 399, 400, 410, 413, 468, 479, 484, 491, 494, 496, 504, 508, 527, 541, 542, 550, 567, 571, 583, 590, 626, 637, 640, 643, 667, 691, 695, 738, 742, 743, 752, 758, 763, 784, 800, 805, 815, 829, 852
Complutum, 148, 860
Computi, Apocryphal, **188**
Computus, Ecclesiastical, 18, 96, **188**, 237, 258, 343, 380, 384, 385, 409, 411, 412, 557, 583, 590, 626, 637, 640, 643, 667, 691, 695, 738, 742, 743, 752, 758, 763, 784, 800, 805, 812, 815, 829, 852, 868

Computus Carthaginensis a (455), 18, 96
Computus de Pascha, 189
Comum (Como), 4, 420
Conantius, 791
Concordia, **189**
Concordius, 62
Concubinage, 189
Condat, 874
Confessions, 96, 99, 121, 272, 553, 568, 705, 708
Confessor, 21, 160, 176, **189**, 191, 208, 212, 223, 232, 246, 248, 270, 400, 531, 546, 583, 603, 707, 752
Confirmation, 63, 108, 151, **189**, 215, 223, 335, 473, 477, 491, 504, 751, 841
Conlationes, 149, 227, 234, 295
Conon, 89, **190**, 646, 672, 743, 822, 824
Cononite, 190
Conscience, Freedom of, **190**, 376, 701
Conscientious objection, 79, **191**, 545, 664
Consilinum (Sala Consilina), 424
Consolatio, **191**, 398
Consolatio philosophiae, 124, 233, 736
Constans I, 2, 49, 77, 93, 146, **192**, 193, 198, 207, 228, 250, 268, 295, 324, 360, 485, 516, 525, 559, 582, 662, 714, 757, 824
Constans II, 36, 96, **192**, 439, 530, 547, 568, 645, 659, 725, 806, 823, 826, 873
Constantia, 17, **192**, 301, 302, 407, 434, mausoleum of, 659, 847
Constantia, Flavia Julia, **192**
Constantine, cartophylax, **193**
Constantine, pope, 646
Constantine (Cirta), *see* Cirta
Constantine I the Great, 2, 3, 7, 14, 21, 23, 27, 47, 48, 53, 62, 76, 79, 89, 91, 93, 96, 116, 120, 122, 134, 136, 137, 139, 141, 144, 153, 168, 169, 173, 175, 176, 184, 186, 189, 191, 192, 193, 194, 198, 207, 217, 246, 263, 269, 274, 279, 296, 298, 299, 321, 338, 347, 335, 371, 407, 409, 432, 439, 471, 473, 475, 476, 488, 491, 514, 520, 522, 531, 545, 546, 554, 559, 560, 595, 596, 614, 618, 626, 641, 647, 653, 656, 662, 673, 676, 679, 688, 702, 721, 733, 734, 741, 742, 743, 755, 757, 777, 788, 800, 801, 802, 806, 813, 822, 823, 830, 837, 840, 852, 853, 855, 861, 863, 883
Constantine II, 93, 192, 193, 198, 347
Constantine IV Pogonatus, **197**, 520, 568
Constantine V Copronymus, **197**, 407
Constantine Porphyrogenitus, 679, 798
Constantine of Assiut, 219
Constantine of Laodicea, 193, 822
Constantine of Preslav, 634
Constantine of Siout, 194
Constantine the Philosopher (Cyril), 633, 634, Life of, 634
Constantinople, 2, 3, 4, 5, 6, 7, 8, 16, 17, 20, 21, 22, 24, 25, 32, 35, 36, 37, 38, 43, 44, 50, 56, 58, 69, 71, 83, 84, 85, 87, 96, 110, 112, 114, 116, 135, 136, 137, 138, 140, 143, 148, 149, 154, 159, 163, 169, 170, 172, 174, 183, 184, 186, 193, 194, 200, 203, 208, 209, 213, 216, 218, 224, 228, 230, 241, 255, 264, 265, 267, 273, 274, 281, 282, 284, 292, 293, 297, 301, 302, 305, 325, 330, 333, 336, 339, 346, 350, 361, 366, 371, 374, 375, 409, 410, 426, 433, 438, 439, 447, 458, 466, 479, 484, 491, 495, 496, 499, 515, 516, 518, 521, 523, 533, 549, 550, 565, 566, 567, 573, 584, 594, 596, 618, 619, 626, 641, 642, 643, 645, 652, 656, 659, 662, 663, 669, 678, 679, 681, 682, 688, 709, 725, 742, 744, 756, 761, 767, 768, 769, 771, 784, 795, 810, 822, 823, 826, 828, 837, 838, 870, 873, 875, 886, *90-93*, C. of (335/336), 195, 302, C. of (360), 13, 49, 58, 78, 95, 113, 195, 215, 268, 296, 305, 343, 381, 395, 472, 516, 521, 525, 585, 758, 779, 787, 856, C. of (381: 2nd ecumenical), 4, 5, 8, 22, 32, 37, 47, 49, 59, 62, 64, 67, 78, 85, 93, 114, 142, 159, 164, 194, 195, 196, 205, 207, 215, 218, 236, 257, 268, 275, 277, 324, 325, 361, 364, 372, 390, 392, 396, 401, 487, 516, 538, 549, 550, 569, 585, 758, 779, 787, 856, 585, 595, 618, 619, 620, 641, 660, 665, 700, 702, 716, 742, 744, 759, 828, 840, 841, 848, 852, 855, C. of (382), 196, 215, 364, 376, 759, C.of (394), 32, 37, 62, 196, 364, 372, C. of (399), 275, C. of (c.400), 196, C. of (404), 196, C. of (426), 556, C. of (431), 196, C. of (c.443), 196, C. of (c.458), 197, C. of (448), 1, 115, 196, 301, 327, C. of (449), 196, 327, C. of (449/450), 196, C. of (450), 196, C. of (457), 196, C. of (477), 197, C. of (482), 197, C. of (492), 197, C. of (496), 197, C. of (497/498), 197, C. of (518), 197, 432, C. of (530/531), 197, C. of (532), 238, C. of (533: 5th ecumenical), 6, 205, 570, 623, C. of (536), 122, 197, 366, 401, 428, 662, 823, 884, C. of (543), 197, 553, C. of (553), 165, 187, 193, 197, 305, 306, 366, 402, 403, 432, 487, 538, 570, 608, 642, 667, 747, 824, 827, 841, 855, 878, 880, 881, C. of (c.588 and c.592), 197, C. of (638), 197, C. of (639), 197, C. of (680), 38, 205, 318, 397, 514, 559, 642, 659, 726, 768, 823, 826, C. of (680/681: 6th ecumenical, in Trullo I), 165, 197, 568, 643, 645, 826, C. of (691), 142, 864, C. of (692: Quinisext, in Trullo II), 62, 63, 116, 142, 197, 407, 484, 646, 768, 877, C. of (754), 197, 830, C. of (786), 197, C. of (869), 71
Constantinopolitan formula/creed, 78, 159, 324, 853
Constantius I Chlorus, 2, 5, 13, 24, 38, 49, 65, 77, 79, 92, 93, 113, 120, 189, 192, 193, 195, 198, 236, 238, 268, 289, 295, 298, 302, 305, 324, 328, 333, 338, 343, 347, 348, 356, 371, 395, 407, 459, 485, 508, 525, 545, 548, 559, 582, 596, 656, 661, 662, 732, 737, 766, 767, 771, 783, 787, 850, 870
Constantius II, 21, 26, 47, 168, 192, 193, 195, 198, 302, 347, 607, 674, 681, 712, 736, 768, 832
Constantius III, 198, 612, 859
Constitution of Sirmond (333), 96
Contemplation, 100, 116, 198, 204, 231, 233, 306, 362, 367, 373, 394, 396, 406, 414, 441, 443, 444, 455, 469, 495, 525, 553, 558, 567, 577, 587, 589, 590, 622, 649, 660, 670, 690, 692, 694, 696, 706, 707, 719, 720, 725, 832, 872, 878, 881, 882
Continence, 13, 223, 236, 251, 271, 318, 360
Continentes, **198**
Contra Celsum, 28, 167, 177, 300, 373, 398, 620, 640, 664, 692
Contumeliosus, bishop of Riez, 555, 645
Conversion, 11, 16, 19, 27, 36, 44, 47, 48, 61, 78, 90, 93, 98, 111, 129, 131, 147, 150, 151, 181, 183, 193, **199**, 205, 211, 218, 226, 230, 231, 234, 235, 245, 263, 272, 278, 289, 307, 313, 316, 329, 340, 355, 357, 358, 359, 361, 365, 388, 398, 399, 405, 411, 418, 436, 449, 456, 461, 462, 469, 506, 510, 517, 531, 535, 536, 548, 551, 568, 576, 578, 585, 624, 628, 633, 659, 672, 683, 702, 710, 721, 734, 740, 741, 742, 743, 755, 758, 791, 792, 802, 814, 822, 833, 844, 848, 850, 851
Cope, 865
Coptic, **199**, art, 557, church, 428, 689, literature, 2, 5, 9, 11, 17, 18, 20, 22, 23, 24, 32, 44, 57, 60, 61, 63, 66, 69, 84, 93, 94, 96, 109, 113, 115, 119, 120, 140, 142, 143, 152, 154, 181, 194, 199, 211, 219, 220, 225, 257, 265, 266, 267, 290, 292, 352, 353, 378, 380, 383, 392, 394, 400, 431, 438, 445, 446, 461, 472, 477, 491, 496, 499, 514, 519, 540, 551, 569, 579, 605, 609, 612, 618, 626, 628, 648, 673, 676, 680, 687, 689, 699, 722, 724, 746, 755, 759, 763, 772, 775, 776, 797, 810, 820, 826, 829, 831, 836, 843, 850, liturgy, 34, 224, 276, 496,
Copts, 33, 66, 249
Corbie, 633
Cordova (Corduba), 106, 148, 464, 774, 790, 844
Coren, 4
Corfinium, 423
Coricius, 201
Coricus, 175
Corinth, 7, 84, 96, 106, 156, 171, 181, **201**, 238, 259, 355, 376, 398, 404, 456, 483, 551, 640, 644, 664, 670, 671, 710, 740, 764, *94, 95*
Corippus, Flavius Cresconius, **202**, 761
Cormons, 774
Cornelius I, 25, 138, 139, 146, **202**, 212, 238, 315, 328, 371, 456, 603, 604, 617, 640, 644, 646, 667, 672, 714, 741, 743, 866
Cornelius Nepos, 121
Cornelius the Centurion, **202**
Corner-stone, 111, **202**, 336, 802, *see also* Gammadion
Cornus, 425
Cornutus, 683
Corona, 253
Corpus, 120, 139, 203, 655, Einsiedlense, 279, Hermeticum, 12, 240, 320, 378, 579, 603, 805, 832, Laureshamense, 279, Virdunense, 279
Corsica, **203**, 757, 861, *96*
Cosmas, 167
Cosmas and Damian, **203**, 808, 809
Cosmas Indicopleustes, 1, **203**, 517, 583, 634, 683
Cosmas Melodus, 748
Cosmas of Maiuma, 135, **204**
Cosmas the monk, **204**
Cosmas Vestitor, **204**
Cosmocrator, 204, 435
Cosmos, 40, 45, 46, 92, 124, 136, 181, **204**, 255, 285, 320, 327, 352, 353, 363, 378, 409, 415, 516, 558, 683, 685, 690, 708, 720, 729, 733, 752, 792, 796, 839, 882
Cotopitae, 175
Cotta, 865
Cotton Bible, 10, 452, 599, 631
Cotyaeum, 216
Councils, 1, 8, 18, 36, 100, 103, 104, 120, 141, 173, **205**, 242, 245, 309, 346, 376, 408, 409, 472, 484, 528, 739, 743, 746, 784, 793, 848, 851
Cranaeum, 7, 201
Creation, 9, 11, 21, 26, 33, 35, 39, 42, 44, 45, 46, 77, 84, 94, 98, 99, 100, 105, 109, 123, 144, 151, 161, 163, 164, 166, 183, 205, 207, 222, 225, 251, 255, 262, 285, 293, 300, 308, 309, 327, 340, 353, 355, 356, 364, 380, 391, 395, 405, 413, 429, 442, 446, 448, 458, 469, 494, 504, 518, 525, 528, 546, 577, 603, 622, 623, 634, 664, 666, 670, 695, 717, 718, 719, 734, 749, 767, 773, 787, 792, 800, 802, 815, 830, 851, 867, 876, 880, 882
Creeds and confessions of faith, 17, 49, 63, 65, 98, 151, 152, 196, 206, 207, 208, 215, 226, 229, 242, 244, 258, 285, 293, 296, 313, 354, 361, 364, 374, 376, 391, 392, 398, 409, 493, 499, 537, 538, 595, 654, 678, 679, 750, 848, 851, 852, 884
Cremona, 421
Crepacore, 424
Crescens, 159, 193, 869
Crescentius, 37
Cresconius, 14, 142, **208**, 248
Crete, 38, **208**, 515, 657, 843, C. of (667), 209, Cretan martyrs, 38
Crikvina, 217
Crimen laesae romanae religionis, 209
Crimen maiestatis immunitae, 209
Crimen religionis, 209
Crimthann, 186
Crispina, 13, 14, 15, **209**, 816
Crispinus, Crispinianus, **209**, 536
Crispinus of Calama, 248
Crispus, 347
Cross, Crucifix, 17, 21, 38, 40, 41, 53, 63, 64, 66, 89, 105, 108, 109, 111, 130, 132, 135, 138, 140, 151, 162, 170, 175, 209, 210, 213, 217, 218, 221, 226, 227, 231, 245, 255, 258, 259, 271, 272, 276, 279, 289, 290, 291, 292, 316, 322, 323, 336, 344, 345, 350, 358, 369, 371, 387, 390, 398, 401, 407, 408, 409, 415, 416, 423, 433, 434, 435, 442, 463, 471, 476, 491, 492, 517, 518, 554, 557, 564, 573, 574, 578, 598, 605, 607, 636, 637, 651, 654, 660, 670, 675, 680, 688, 707, 752, 753, 756, 764, 765, 778, 786, 787, 788, 802, 803, 808, 809, 841, 845, 846, 847, 855, 862, 875, 876, *97, 98*
Crossing of the Red Sea, 108
Crouzel, Henri, 12, 21, 27, 28, 126, 141, 154, 189, 199, 225, 235, 236, 243-44, 314, 328, 368, 374, 405-7, 409, 438, 480, 528-29, 619-24, 638, 682, 682-83, 732-33, 736, 793, 822, 832, 846, 869-70, 876
Cryptography, **210**, 280, 281, 403, 862, *see also* Epigraphy, Christogram
Cubicularius, **210**
Cú-Chuimne of Iona, 412
Cucuphas, 110
Cuenca, 148
Cuicul (Djemila), 14, 15
Culcianus, 534
Culdees, 762
Culross, 771
Cumae, 424
Cumanin Camii, 88
Cummineus Longus (Cummian; Cummíne Fota), **211**, 412, 669
Cuneo, 421
Cunimund, 505
Cunipert, 838
Cursus, 179
Curti, Carmelo, 152-53, 299-301, 339-40, 340, 727-28, 753-54, 867
Cusanus, 101
Cuscito, Guiseppe, 64-65, 97, 189, 465-66, 544-45, 545, 651, 729-30, 773-74, 823, 860
Cuthbert, 129, 130, 131, 132, 133, **211**, Life of, 118, 211, Coffin, 132, Pectoral cross, 133
Cutzupitae, 175
Cybele, 144, 815
Cyclades, 685
Cynics, 508
Cyprian, Ps.-, 802
Cyprian of Antioch, 211, 212
Cyprian of Carthage, 9, 11, 13, 14, 16, 18, 25, 27, 52, 53, 56, 57, 60, 63, 64, 72, 79, 82, 91, 102, 103, 104, 109, 121, 134, 139, 140, 142, 143, 144, 145, 146, 151, 153, 154, 156, 168, 177, 182, 186, 188, 189, 191, 202, 206, 207, 211, 222, 223, 228, 229, 231, 232, 238, 241, 247, 250, 260, 269, 273, 292, 293, 295, 296, 298, 305, 309, 311, 312, 315, 320, 321, 324, 326, 334, 336, 337, 359, 376, 386, 398, 437, 453, 472, 473, 478, 484, 490, 494, 495, 497, 499, 506, 507, 513, 517, 523, 532, 534, 541, 544, 554, 560, 561, 563, 577, 583, 600, 603, 604, 605, 609, 611, 612, 615, 616, 619, 643, 644, 646, 649, 651, 655, 656, 664, 667, 668, 672, 682, 702, 703, 707, 712, 714, 716, 725, 738, 749, 750, 751, 752, 758, 759, 763, 776,

777, 782, 784, 785, 790, 792, 794, 798, 801, 819, 839, 843, 848, 856, 860, 867, 871, 875, 876, 877, 878, Acta proconsularis S. Cypriani, 155, Acts of, 228, Life of, 121, 784, 790
Cyprian of Toulon, **212**
Cyprian the Gaul, 137, 312
Cyprian the Poet, **212**, 213, 231, 651
Cyprian the presbyter, **212**, 213
Cyprianus Gallus, **212**
Cyprus, 21, 24, 47, 48, 67, 73, 111, **213**, 333, 336, 518, 618, 793, 825, 826, *100, 101*
Cyrenaica, 47, 170, 171, 172
Cyrene, 208, 487, 806
Cyriac, **213**, 484, 540, Life of, 66
Cyriac of Nisibis, 598
Cyriac of the laura of Sonka, 215
Cyrillona (Qurillona), 214, 472, 810
Cyril of Alexandria, 1, 5, 20, 21, 22, 23, 24, 35, 37, 38, 46, 49, 50, 59, 72, 81, 96, 112, 116, 142, 154, 159, 164, 167, 187, 188, 189, 196, 197, 200, 208, **214**, 218, 233, 236, 237, 241, 243, 251, 256, 262, 274, 275, 281, 291, 295, 298, 303, 304, 310, 312, 313, 314, 318, 320, 324, 325, 342, 346, 358, 379, 401, 402, 405, 406, 407, 429, 444, 449, 450, 452, 466, 472, 481, 496, 500, 511, 538, 545, 546, 548, 550, 552, 561, 569, 570, 577, 593, 594, 596, 603, 641, 643, 644, 646, 648, 652, 655, 658, 663, 677, 679, 695, 696, 698, 709, 713, 715, 716, 723, 729, 734, 742, 744, 773, 776, 784, 788, 795, 798, 817, 826, 827, 830, 831, 841, 843, 845, 852, 856, 868, 873, 879, 881 884, 886
Cyril of Jerusalem, 5, 39, 42, 47, 57, 72, 90, 107, 151, 153, 176, 186, 190, 195, 200, **215**, 232, 242, 244, 269, 277, 305, 306, 312, 346, 380, 392, 398, 406, 433, 490, 494, 504, 520, 546, 560, 562, 567, 577, 595, 634, 649, 665, 725, 749, 750, 751, 805, 877
Cyril of Scythopolis, 66, 167, **215**, 304, 350, 379, 447, 480, 523, 540, 660, 677, 748, 829
Cyrrhus, 203, 809, 827
Cyrus Muqauqas, 119, 775
Cyrus of Alexandria, 112, 197, **216**, 263, 568
Cyrus of Constantinople, 194
Cyrus of Edessa, **216**
Cyrus of Panopolis, **216**
Cyrus of Phasis, 768
Cyzicus, 44, **218**, 340, 784, C, of (c.376), 268

Daburieh, 636
Dachios of Berenice, 487
Dacia, 217, 565, 595, 756, 849, *102*
Dadiso, 4, 106
Dado of St Oer, 338
Dagobert, 231, 238, 268, 330, 338, 349, 664
Dallán Forgaill, 413
Dalmais, Irénée-Henri, 33, 264-65, 494
Dalmatia, 169, 171, 172, 178, 217, 339, 610, 639, 645, 753, 759
Dalmatic, 865
Dalmatius, 83, 218, 784
Dalmatius of Cyzicus, 218
Dalriada, 130
Damascius, 31, 121, 603, 696, 805
Damascus, 38, 65, 127, 218, 657, 808
Damasus, 4, 8, 17, 30, 48, 49, 59, 63, 78, 114, 116, 142, 145, 153, 157, 166, 195, 196, 218, 223, 232, 249, 250, 261, 278, 281, 282, 302, 305, 321, 348, 351, 382, 416, 430, 431, 496, 502, 508, 522, 535, 552, 593, 618, 641, 643, 645, 646, 678, 682, 706, 712, 727, 738, 741, 742, 743, 744, 745, 749, 755, 771, 814, 852, 856, 873, *103*, *104*, see also Epigrams, Damasian

Damian of Alexandria, 194, 200, 219, 505, 679, 689, 746, 823
Damit el-Alya, 67
Damous el Karita, 171
Dance, 219, 290, 469
Daniel, 3, 16, 40, 55, 95, 148, 161, 187, 219, 294, 310, 323, 342, 369, 379, 383, 425, 431, 450, 471, 477, 615, 629, 631, 668, 684, 688, 702, 704, 715, 716, 733, 763, 764, 801, 803, 809, 829, 860, 873, *105, 106*
Daniel of Constantinople, 797
Daniel of Scete, 220
Daphnè, 47, 106
Daphnus of Vaison, 854
Daras, 112
Dasius, 105, 220
Dassmann, Ernst, 108-9, 294, 346-50, 657-58, 668
Dateus of Milan, 162
Dativus of Bades, 584
Dattrino, Lorenzo, 11, 44, 137, 144, 211, 224, 224-25, 225, 228, 229, 231, 284, 294, 294-95, 301, 302, 317, 322, 323, 337, 352, 396, 409, 419-20, 446, 475, 485, 615, 628, 652, 662, 679, 711, 747, 802, 804, 816, 856, 867
David, 37, 41, 127, 132, 151, 161, 200, 213, 220, 230, 255, 315, 324, 330, 433, 451, 461, 529, 537, 562, 604, 625, 635, 722, 723, 779, 786, 808, 825, 849, 863
David of Wales, **221**
David the Armenian, **221**
David the philosopher, 81
Dayr al-Gub, 417
Dag Pazari, 87
Dar el Kous, 15, 778
Deacon, *see* Diaconate
Deaconess, 48, 91, 221, 235, 304, 555, 699, 740, 805, 877, 882
Dead, Cult of the, 11, 13, 16, 140, 153, 157, 186, 221, 241, 294, 532, 554, 612, 630, 725, 729, 731, 752
De Aleatoribus, **222**
De Bernardi Ferrero, Daria, 85-88
De Bono Pudicitiae, **222**
Decalogue, 222, 235, 288, 336, 406, 415, 748, 781
De Centesima, Sexagesima, Tricesima, **223**
Decentius of Gubbio, 67, 223, 232, 532, 616, 738
Decius, 1, 13, 17, 21, 79, 85, 106, 137, 139, 145, 151, 168, 178, 190, 202, 208, 211, 212, 223, 238, 261, 275, 307, 314, 315, 321, 324, 326, 338, 356, 368, 432, 473, 484, 507, 603, 604, 620, 640, 664, 667, 672, 712, 737, 741, 758, 772, 794, 806, 822, 846, 863
Decretum Gelasianum, 2, 9, 58, 62, 105, 111, 180, 223, 379, 474, 484, 486, 619, 712, 744, 763, 766
Dedication, Council of the, 23, 224, 302, 305, 434, 833
Dedication of churches, 19, 49, 135, 171, 224, 350, 494, 502, 535, 536, 856, 876
Dedimia, 116
Defensor, 224, 367
Defensor, monk, 224
Defensor civitatis, 224, 338
Defensor ecclesiae, 147, 224
Defensor of Ligugé, 655
Degile, 87
Deir Abu Hennis, 141, 268, 447, 452
Deir Sharqi, 218
Deir al-ahmar, 267
Deir el-Adas, 127
Deir er-Rawat, 120
Deir-Balizeh, 265
Deir-Balizeh papyrus, 34, 496, 732
Dej (Transylvania), 217
De Jesu Christo Deo et Homine, **224**, 876
Dejr Dosi, 829
De Ligno Crucis, **224**, 867

De Luis, Pio, 110, 127, 255, 270, 350, 400, 483, 554, 755, 774-75, 790-91, 813-14, 844-45
Delphinus of Bordeaux, 27, 126, 225, 660
Demetrian, 246
Demetrian of Antioch, 225, 246, 808
Demetrias, 40, 225
Demetrius of Alexandria, 21, 23, 225, 235, 619, 822
Demetrius of Antioch, Ps.-, 225
Demetrius of Philippi, 410
Demetrius of Thessalonica, 783
Demetrius the martyr, 225, 515, 833, 834
De Miraculis S. Stephani, **225**, 308, 774
Demoniac, 38, 227, 466
Demon, 38, 39, 44, 53, 56, 82, 94, 106, 107, 122, 124, 149, 180, 220, 222, 226, 227, 230, 231, 232, 247, 255, 322, 326, 372, 404, 415, 444, 463, 517, 519, 542, 557, 560, 578, 591, 610, 614, 621, 622, 626, 664, 689, 696, 699, 720, 731, 751, 758, 775, 781, 782, 786, 788, 792, 817, 860, 870
Demophilus of Constantinople, 228, 305, 525, 828, 858
De mortibus persecutorum, 60, 388, 469, 470, 673, 720
Demosthenes, 115
De mysteriis, 190, 312, 382, 494, 562, 882
Dendritae, 228
Denia (Alicante), 148
De Nicola, Angelo, 36, 36-37, 37, 51-52, 52, 92, 112, 119, 140, 160-61, 203, 238, 299, 320, 437, 438, 444, 447, 518, 557, 638, 652, 664, 678, 681, 685, 686, 740, 774, 794, 794-95, 823-24, 824, 825-26, 826, 829, 829-30, 831, 832-33, 840
Deo Laudes, 176, 228, 248
Deogratias, 228
De Palol, Pedro, 106-7, 107, 148-49, 334, 512, 554-55, 813, 814
Depositio episcoporum, 140, 166, 228, 229, 486, 521, 522, 532, 536, 537
Depositio Martyrum, 1, 17, 140, 153, 163, 166, 212, 228, 229, 475, 493, 532, 536, 537, 552, 742
De principiis, 177, 229, 236, 310, 609, 620, 621, 622, 623, 624, 692, 746, 830, 848
Dereagzi, 89
Der el-Kahf, 67
Deriziotis, Lazaros, 834-35
Der Jauni, 67
Dermech, 14, 15, 174, 762
De Rossi, G. B., 158
Derry, 186, 307, 411
Derrynavlan (co. Tipperary), 133
Dertona (Tortona), 421
Dertosa (Tortosa), 813
Descent into Hell, 33, 36, 57, 137, 226, 229, 258, 372, 437, 523, 622, 731, 875
Desert, 4, 37, 44, 48, 83, 84, 149, 184, 230, 264, 266, 269, 319, 327, 440, 454, 477, 567, 636, 816, 821
Desiderius, bishop of Langres, 875
Desiderius of Cahors, 231, 338, 664, 846
De Simone, Russell J., 109, 222, 229, 315-17, 319, 462-64, 604, 683, 739
De Trinitate, ps.-Athanasian, **231**
Deusdedit, 231
Deus Tuorum Militum, 231
Deuterius, 67
Devil, 226, 230, 817
De Vogüé, M., 69
Devotion, 57, 158, 226, 227, 231, 245, 261, 278, 288, 290, 296, 363, 492, 510, 511, 621, 687, 792
Dexter, 30, 232
Diaconate, 79, 90, 91, 123, 182, 221, 232, 235, 259, 561, 616, 617, 740, 800
Diacrinomeni, 233, 841
Diadochus of Photice, 233, 283, 296, 507, 524, 707, 793
Dialectica, 76

Dialogue, 34, 105, 119, 233, 366, 514, 557, 563, 579, 827, 879
Dialogue on the Life of St John Chrysostom, 233, 374, 611, 634, 636, 704, with Aelian, 368, with Heraclides, 233, 620, with Trypho the Jew, 9, 60, 121, 233, 312, 354, 431, 436, 463, 584, 690
Diamartyria, 179
Diamper (1599), 518
Dianium, 148
Diatessaron, 19, 60, 81, 234, 263, 276, 345, 448, 489, 543, 729, 763, 810, 815, 868
Díaz y Díaz, Manuel C., 102, 296, 304, 330-31, 382, 458-59, 465, 488, 509, 531, 542, 571, 669-70, 714, 728, 774, 854, 860-61
Diban, 66
Dictinius, 191, 234
Didachè, 2, 35, 47, 53, 61, 62, 107, 111, 117, 134, 150, 160, 182, 190, 234, 252, 255, 259, 264, 292, 293, 306, 316, 319, 358, 400, 472, 494, 495, 500, 521, 541, 560, 570, 577, 609, 707, 715, 723, 792, 800
Didascalia, 42, 90, 200, 315, 318, 490, 561, 714, 723, 751, 793, 810, 863, 877
Didascalia Apostolorum, 62, 153, 207, 221, 235, 259, 291, 494, 495, 496, 500, 541
Didaskaleion, 23, 150, 187, 235, 238, 264, 638, 681, 687, 761, 831
Didaskalos, 235
Didyma, 87
Didymus the Blind of Alexandria, 22, 23, 29, 56, 183, 187, 199, 202, 214, 235, 310, 312, 336, 358, 392, 431, 437, 449, 496, 514, 543, 550, 648, 656, 658, 716, 719, 723, 746, 761, 776, 782, 878, 879
Dierna, 217
Digamy, 154, 236
Dijon, 59
Dindimus, 184
Dindimus de Bragmanibus, 185
Di Nino, Antonella, 187, 429, 456, 506, 520, 598-99, 755, 785, 786, 812, 844
Dinogetia, 764, 765
Diocaesarea (Ouzoundja-Bourdj), 635, 798, 808
Dio Cassius, 8
Dio Chrysostom, 230
Dioceses, 617
Diocletian, 13, 16, 22, 23, 27, 43, 44, 47, 48, 65, 82, 85, 90, 107, 122, 127, 129, 136, 143, 148, 162, 167, 169, 176, 183, 186, 193, 198, 200, 201, 217, 220, 236, 270, 274, 299, 313, 328, 333, 338, 346, 351, 368, 369, 380, 388, 419, 420, 421, 432, 434, 473, 486, 488, 512, 521, 522, 543, 545, 559, 596, 600, 605, 607, 618, 626, 638, 673, 686, 687, 700, 702, 735, 739, 741, 757, 764, 771, 774, 779, 790, 806, 809, 813, 814, 816, 821, 829, 833, 834, 837, 844, 845, 849, 854, 867, 870
Diodore of Tarsus, 5, 45, 49, 50, 56, 112, 164, 195, 203, 214, 236, 310, 312, 315, 325, 436, 440, 507, 658, 684, 713, 723, 727, 744, 824, 832, 841, 842
Diogenes, 459
Diogenes Laertius, 75, 121, 558
Diogenes of Apollonia, 639
Diogenianus of Heraclea, 379
Diognetus, to, 9, 52, 60, 73, 155, 176, 199, 237, 358, 483, 657, 720, 792, 796, 818, 876
Diomede, 175
Dionisi, Ulpia, 2, 7, 17, 18, 19, 53, 67, 96-97, 122, 232, 249-50, 250, 281, 295, 324-25, 325, 327, 328, 342, 397, 454, 682, 743-44, 757, 884
Dionysiana, 142, 237
Dionysius, Rusticus and Eleutherius, Passion, 581, 663

Dionysius bar Salibi, 159, 333, 383, 385, 428, 496, 810
Dionysius Exiguus, 62, 125, 142, 146, 167, 188, 189, 237, 364, 438, 473, 475, 585, 764, 812, 856, 868
Dionysius Thrax, 81
Dionysius of Alexandria, 23, 55, 57, 59, 72, 91, 94, 109, 142, 159, 164, 207, 212, 235, 237, 238, 298, 301, 315, 336, 371, 396, 473, 487, 504, 533, 560, 566, 593, 640, 641, 656, 663, 672, 741, 743, 797, 822, 848, 854, 860, 879, 880
Dionysius of Byzantium, 159, 168
Dionysius of Corinth, 91, 181, 208, 238, 637, 641, 644, 675, 681, 688, 725, 740, 775, 800
Dionysius of Milan, 30, 238, 497, 559
Dionysius of Paris, 238, 651
Dionysius of Rome, 238, 518, 617, 640, 643, 644, 646, 741, 743, 853
Dionysius of Tellmahre, 238, 428
Dionysius of Tellmahre, Ps.-, 445
Dionysius the Areopagite, Ps.-, 12, 34, 39, 40, 73, 74, 76, 96, 135, 205, 206, 238, 241, 242, 243, 312, 355, 356, 358, 364, 405, 446, 475, 562, 577, 580, 587-92, 603, 615, 634, 655, 679, 679, 689, 692, 693, 696, 713, 723, 735, 750, 769, 792, 793, 831, 832
Dionysio-Hadriana, 142
Dion, 515, 806
Diophantus, 484
Dioscorus, deacon, 240
Dioscorus, martyr, 240
Dioscorus of Alexandria, 22, 24, 37, 112, 115, 159, 195, 200, 240, 241, 275, 304, 325, 337, 416, 466, 487, 496, 514, 569, 642, 645, 758, 776, 827, 840
Diospolis (Lydda), 7, 36, 507, 635, C. of (415), 98, 137, 241, 438, 665, 744, 782
Diospontus, 27
Diptychs, 5, 10, 48, 49, 126, 141, 187, 197, 214, 222, 241, 253, 274, 313, 330, 349, 386, 452, 482, 484, 532, 535, 540, 541, 550, 619, 632, 689, 705, 738, 758, 784, 847, 873, *107*, of Boethius, 477
Disciplina arcani, 151, 242, 395, 409, 463, 802, 849
Discipline, 61, 63, 132, 141, 152, 205, 242, 260, 288, 309, 342, 360, 368, 376, 492, 495, 626, 845, 848, 86
Dissertatio Maximini contra Ambrosium, 105, 546, 637
Dittochaeon, 270, 278, 721
Divinization, 9, 163, 242, 293, 355, 358, 378, 396, 406, 414, 462, 481, 489, 693, 731, 734, 788, 882
Divorce, 27, 62, 116, 146, 147, 236, 243, 318, 701, 739
Djebel Maqlub, 543
Djebel Mousa, 782
Djemila, 172, 174
Dobrudja, 765
Docetism, 29, 54, 113, 122, 151, 158, 163, 244, 303, 316, 326, 404, 408, 523, 537, 538, 569, 621, 660, 819, 824, 827
Doctrina Apostolorum, 111, 609
Doctrina Patrum, 244, 274, 313, 339, 340, 638
Doctrina Patrum de Incarnatione Verbi, 35, 37, 76, 459, 655
Dogma, 12, 35, 37, 53, 59, 91, 99, 100, 104, 120, 136, 151, 179, 187, 244, 262, 315, 317, 346, 366, 393, 462, 481, 496, 538, 620, 621, 623, 656, 679, 700, 727, 782, 788, 796, 809, 840, 848, 851, 870
Dogmatic Epistle, 47
Domitian, emperor, 8, 181, 245, 326, 419, 671, 779, 812, 863
Domitian of Ancyra, 245, 677
Domitilla, 8, 246, 326, 659, 671, 680, 741, 844
Domnina, 50
Domninus III, 36

Domnus, 49, 102, 196, 225, **246**, 275, 416
Donadeus, 860
Donatianus of Thelepte, 822
Donation of Constantine, 246, 318, 653, 802
Donations to the church, 17, 168, 246, 318, 338, 348, 600, 653, 667, 714, 735, 741, 864
Donatism, 2, 13, 14, 27, 53, 55, 72, 79, 84, 98, 100, 102, 109, 137, 143, 145, 146, 147, 151, 173, 175, 176, 185, 193, 205, 208, 218, 223, 228, 246, 250, 260, 261, 270, 288, 308, 318, 321, 325, 328, 332, 348, 350, 351, 360, 376, 377, 383, 397, 482, 485, 495, 508, 516, 517, 521, 524, 542, 544, 554, 560, 595, 604, 607, 612, 613, 617, 641, 645, 651, 666, 679, 711, 713, 716, 733, 738, 742, 743, 759, 766, 779, 790, 806, 816, 838, 840, 870
Donatist Martyrs, **249**, 533
Donatists, Anonymous, **249**
Donatulus, 143
Donatus Aelius, 249
Donatus of Bagai, 524
Donatus of Besançon, 250
Donatus of Carthage, 13, 14, 137, 246, 248, 249, 250, 360, 470, 517, 545, 613, 651, 679, 711, 738, 743, 760
Donus, 645, 743
Doorkeeper, 232
Dor, 635
Dorothea, **250**
Dorotheus, presbyter, **250**
Dorotheus of Constantinople, 525
Dorotheus of Gaza, 61, 112, **251**, 379, 442, 567, 828, 854, 885
Dorotheus of Marcianopolis, **251**, 545
Dorotheus of Thessalonica, **251**
Dorylaeum, 301
Dos Hermanas, 107
Dositheus of Samaria, 112, **251**, 384
Double Creation, **251**, 358, 378, 717, 719
Double procession, 392
Dougga, 172
Dove, 749, 803
Doxology, 12, 22, 33, 34, 35, 72, 94, 232, **252**, 293, 335, 355, 361, 495, 498, 499, 685, 706, 738, 752, 793, 816, 827
Dracontius Blossius Aemilius, 67, 144, 252, 296, 651, 761, 861, 863
Dream of Jerome, 873
Dream of the Rood, 132
Dress, 40, 108, 241, 252, 322, 351, 369, 435, 493, 540, 625, 705, 711, 739, 756, 793, 819, 864
Drobeta - Turnu Severin, 217
Dropsy, 254
Drosis, 50
Druim Cett (575), 411
Drunkenness, **255**
Druzipera, 838
Dualism, 22, 45, 54, 56, 83, 98, 113, 123, 163, 205, 226, 234, 244, 251, 255, 281, 289, 308, 352, 354, 358, 378, 390, 397, 405, 470, 518, 609, 614, 633, 660, 703, 709, 717, 718, 737, 775, 796, 882
Duchesne, L., 654
Dujcev, Ivan, 194-95, 565-66, 633-34, 756-57, 837-38
Duka, 160
Dula, 175
Dulcitius, 16, 337
Dumezio, 283
Dumium, 127, 255, 330, 531, 652
Dunkeld, 130
Duns Scotus, 101
Du Nuwas, 71
Dura Europos, 1, 3, 9, 70, 169, 210, 221, 234, 255, 315, 406, 407, 429, 433, 452, 540, 572, 575, 625, 629, 650, *108*
Durazzo, 36, 283
Durostorum, 13, 105, 220
Durrow, 130, 186, 307, 411, 672, 677, 689, 714, 715, 778, 786, 808, Book of, 131, 141

Duval, Noël, 168-75, 535
Dvin, 80, 81, 83, 209, 255, *109*, C. of (506), 496, C. of (527), 80, C. of (554-555), 593
Dynamius Patricius, 256, 548
Dyophysism, 159, 165, 208, 256, 275, 304, 325, 438, 480, 481, 568, 569, 570, 773, 856
Dyrrachium, 283
Dzveli-Chouamta, 344
Dzveli-Gavazi, 344

Eadbald of Kent, 465, 551, 662
Eagle, 802
East, 1, 5, 6, 7, 13, 14, 23, 24, 29, **30**, 37, 38, 40, 42, 43, 47, 48, 49, 50, 54, 57, 59, 63, 69, 77, 78, 83, 85, 90, 93, 107, 109, 116, 119, 126, 135, 141, 142, 148, 150, 151, 153, 154, 159, 160, 162, 163, 164, 166, 167, 168, 171, 173, 174, 180, 182, 183, 187, 188, 190, 193, 195, 198, 206, 207, 208, 210, 212, 219, 224, 228, 230, 233, 234, 250, 257, 262, 275, 282, 293, 306, 310, 314, 317, 319, 323, 334, 345, 357, 366, 369, 372, 386, 393, 408, 409, 410, 428, 430, 444, 448, 450, 452, 458, 462, 472, 479, 499, 501, 507, 522, 524, 532, 535, 538, 539, 552, 560, 566, 567, 568, 569, 572, 583, 604, 617, 641, 642, 652, 653, 655, 658, 668, 673, 675, 676, 686, 707, 712, 714, 733, 734, 742, 743, 746, 749, 757, 760, 761, 766, 776, 780, 792, 798, 802, 809, 810, 815, 822, 833, 849, 852, 855, 864, 866, 875, 876, 884
Easter, 6, 13, 17, 20, 41, 50, 58, 71, 79, 83, 90, 91, 94, 96, 108, 125, 128, 129, 135, 141, 147, 150, 151, 152, 160, 166, 187, 188, 200, 205, 210, 211, 214, 225, 229, 235, 237, 257, 263, 264, 274, 281, 298, 300, 312, 319, 337, 341, 351, 364, 380, 384, 399, 405, 408, 411, 412, 413, 438, 439, 476, 482, 492, 494, 495, 501, 504, 510, 542, 546, 551, 557, 582, 595, 606, 609, 637, 640, 644, 646, 652, 660, 669, 671, 701, 714, 720, 728, 730, 732, 738, 742, 743, 744, 749, 762, 768, 786, 788, 800, 810, 812, 821, 831, 838, 848, 851, 854, 867, 868, 877
Easter controversy, 604, 701, 741, *see also* Quartodecimans
Eauze (Elusa), 636, C. of (551), 556
Ebedjesu, 236
Ebedjesus bar Berika, 654
Eberigisil, 348, 349
Ebionites, 64, 66, 163, 176, 179, 220, 258, 269, 281, 319, 384, 454, 542, 659, 803, 810, 829, 877, Gospel of the, 57, 489
Eboda, 636
Ebroin, 480
Ecclesiastes, 187, 362, 374, 383, 879
Ecclesiasticus (ben Sirach), 455, 881
Ecclesiology, 34, 38, 55, 99, 100, 101, 153, 182, 210, 248, 259, 262, 342, 357, 408, 460, 484, 529, 577, 628, 643, 646, 651, 675, 742, 786, 839, 880
Echternach, 877, calendar, 877, gospels, 131, 873
Ecija, 106, 219
Eclectus, 188
Economy, 33, 34, 35, 53, 92, 135, 163, 204, 206, 237, 259, 262, 276, 285, 289, 293, 316, 340, 357, 359, 361, 382, 391, 408, 409, 414, 415, 437, 442, 443, 452, 463, 467, 509, 510, 514, 548, 551, 574, 577, 622, 684, 715, 718, 720, 730, 733, 734, 746, 751, 775, 788, 815, 825, 827, 831, 841, 851, 852, 880
Ecstasy, 56, 116, 160, 219, 255, 262, 293, 467, 547, 560, 568, 570, 577, 614, 621, 649, 786, 792
Eddi, 877
Edessa, 1, 2, 4, 5, 7, 10, 19, 25, 48, 49, 51, 79, 95, 110, 112, 118, 144, 181, 190, 203, 234, 263, 274, 276, 307, 340, 403,

416, 428, 429, 515, 583, 637, 674, 729, 763, 809, 810, 823, *110*
Edict of Milan, 193, 217, 263, 407, 459, 488, 541, 559, 617, 673, 681, 714, 823
Edwin of Northumbria, 129, 397, 662, 883
Egabrum, 106
Egara, *see* Tarrasa
Egbert, 118, 877
Egeria, 2, 6, 85, 90, 134, 141, 151, 152, 257, 263, 282, 409, 410, 426, 433, 491, 493, 501, 504, 567, 688, 709, 770, 782, 793, 797, 860, *see also* Itinerarium Egeriae
Eghvard, 80
Egidius, 556, 735
Egnathia, 424
Egypt, 1, 3, 4, 6, 13, 17, 20, 21, 22, 23, 29, 31, 32, 34, 37, 39, 48, 49, 50, 54, 61, 69, 77, 78, 83, 84, 85, 92, 93, 95, 114, 119, 125, 135, 140, 149, 154, 162, 163, 164, 165, 167, 170, 171, 172, 181, 184, 193, 194, 196, 199, 200, 216, 219, 220, 223, 224, 225, 230, 232, 234, 258, 264, 268, 271, 279, 282, 289, 307, 331, 353, 380, 392, 430, 450, 452, 458, 484, 486, 488, 493, 496, 551, 554, 561, 566, 567, 568, 569, 599, 605, 618, 631, 635, 640, 643, 647, 670, 672, 673, 677, 704, 722, 730, 748, 761, 768, 773, 780, 787, 797, 821, 827, 829, 831, 840, 854, 876, 884, *111-116*, liturgy, 34, 264, 768, 853
Egyptians, Gospel of the, 57, 181, 271, 579, 719, 771, 775
Einsiedeln Itinerary, 427
Eitha, 769
Ekthesis, 49, 77, 197, 207, 228, 268, 295, 439, 516, 522, 530, 559, 725
Ekthesis makrostichos, 685
El Asnam, 14, 15, 169
El Bagawat, 3, 10, 437, 450, 536, 575, 599, 631, 730
El Bekri, 816
El Faiyum, 519
El Germo, 106
El-Habis, 636
Elenchos, 384
El-Qoweilbe, 636
Eleusis, 96
Eleusius of Cyzicus, 195, 268, 395, 516, 767
Eleutheris, 7
Eleutherius, 2, 413, 571, 644, 646, 741
Eleutheropolis-Beit Gebrin, 635
Elias, 432, 748
Elias of Jerusalem, 268, 569
Elias of Nisibis, 429
Elicroca, 148
Eligius of Noyon, 268, 338, 394, Life of, 394
Elijah, 47, 55, 66, 84, 200, 230, 268, 319, 715, 817, *117*
Elipandus, 117
Elipa, 106
Elis, 7
Elisaeus, 4, 19, 81, 269, 482
Elkesaïtes, 259, 269, 519, 674, 810
Elnon, 27
Elo, 148
Elounda, 209
Elpidius, layman, 270
Elpidius Rusticus, 3, 270, 350
Elpidius of Huesca, 465
Elvira, C. of (c.303), 37, 106, 148, 153, 154, 270, 274, 400, 406, 409, 528, 554, 629, 669, 739, 748, 774, 776, 784, 790, 791, 844, 857, 875
Elxai, 269
Ember days, 493
Emerentiana, 270
Emeritus, 249, 270
Emerius of Saintes, 556
Emesa (now Homs), 4, 26, 270, 301, 584, 663, 799, 808, *118*

Emilia Hilaria, 103
Emmelia, 184
Emperor, Cult of the, 64, 199, 209, 271, 596, 672
Enaniso, 271
Encolpion, 271, 452, 491, 809, 866
Encratism, 13, 38, 48, 49, 57, 58, 63, 64, 84, 121, 161, 181, 223, 230, 251, 263, 271, 318, 319, 390, 448, 528, 542, 576, 674, 686, 688, 703, 704, 712, 717, 718, 719, 763, 775, 792, 810, 815, 871, 877
Encyclica, 780
Encyclopaedia, 98, 124, 272, 418, 461, 576
Endelechius, 272, 829
Enfide (now Affile), 119
Enhypostasis, 272, 481
Enigmas of Solomon, 465
Ennead, 587, 698
En Ngila, 16
Ennodius, 67, 121, 270, 272, 278, 281, 399, 421, 476, 650, 702, 761, 767, 804, 806, 863, 871
Enoch, 39, 47, 55, 410, 718, Book of, 56, 57, 291, 815, 873
Entertainments, 10, 147, 182, 186, 194, 219, 226, 273, 604, 794, 815
Entry of Jesus into Jerusalem, 38, 274, 323, 466, 493, 518, 631, 712, 809, *119*
Eobulus of Lystra, 274, 313
Epaenetus, 740
Epaon, 105, C. of (517), 17, 59, 134, 409, 555, 874
Epetion, 217
Ephesus, 4, 69, 71, 85, 86, 87, 88, 89, 115, 158, 174, 200, 258, 274, 307, 313, 356, 374, 401, 404, 538, 552, 619, 701, 761, 772, 831, 840, 867, *120-123*, C. of (400), 275, 314, C. of (431: 3rd ecumenical), 5, 19, 20, 21, 22, 36, 37, 48, 49, 59, 72, 112, 137, 142, 143, 154, 159, 165, 196, 197, 200, 205, 208, 214, 218, 223, 241, 251, 275, 292, 301, 304, 314, 315, 320, 324, 325, 402, 416, 432, 444, 452, 466, 487, 496, 525, 538, 539, 545, 546, 550, 552, 556, 569, 573, 594, 596, 618, 641, 663, 666, 670, 677, 679, 729, 732, 742, 774, 782, 808, 822, 829, 830, 833, 845, 854, 884, C. of (434), 275, **C.** of (447), 275, C. of (449: latrocinium Ephesinum), 36, 37, 112, 115, 159, 208, 241, 257, 301, 303, 305, 325, 380, 403, 461, 466, 487, 548, 569, 648, 744, 827, 840, 845, 855, C. of (c.475), 275
Ephrem of Antioch, 6, 35, 49, 245, 276, 339, 439, 440, 445, 585, 655, 769
Ephrem of Edessa, 160
Ephrem the Syrian, 1, 2, 3, 9, 16, 41, 42, 50, 54, 81, 110, 200, 214, 234, 263, 269, 274, 276, 282, 312, 315, 328, 370, 395, 416, 429, 452, 472, 496, 504, 520, 524, 543, 556, 562, 567, 583, 598, 611, 634, 649, 674, 715, 716, 723, 740, 764, 782, 810, 813
Epiclesis, 580, 738, 768
Epictetus, 796
Epictetus and Astio, 764
Epictetus of Corinth, 75, 83, 92, 278, 326, 334
Epicureans, 74, 82
Epicurus, 74, 92, 689, 691
Epidaurus, 7
Epigonus, 181, 278, 566, 653, 706, 885..
Epigrams, 44, 270, 278, 281, 508, 682, 717, 721, Damasian, 281, 535, 721
Epigraphy, Christian, 19, 24, 25, 29, 43, 51, 64, 68, 70, 87, 96, 104, 134, 156, 210, 211, 217, 218, 221, 222, 228, 249, 272, 278, 279, 293, 307, 317, 321, 337, 350, 370, 377, 421, 422, 424, 425, 489, 499, 513, 529, 535, 544, 564, 566, 568, 585, 599, 605, 615, 647, 649, 651, 655, 656, 665, 681, 682, 686, 724, 725, 729, 731, 741, 752, 753, 765, 772, 790, 802,

803, 807, 833, 835, 837, 845, 850, *124-128*
Epimanikía, 865
Epiphania (now Hama), 32, 808
Epiphanius, Archdeacon, 282
Epiphanius, Ps.-, 122
Epiphanius II, 826
Epiphanius Scholasticus, **235**, 282, 388, 6389, 683
Epiphanius of Constantinople, 43, 197, 281, 744, 794
Epiphanius of Cyprus, 520
Epiphanius of Pavia, 272, 281, 307, 476, 854
Epiphanius of Salamis, 4, 9, 10, 13, 16, 23, 38, 46, 47, 56, 57, 63, 66, 71, 72, 75, 83, 84, 91, 96, 110, 126, 154, 158, 159, 164, 183, 185, 186, 199, 200, 232, 235, 236, 244, 251, 259, 269, 272, 281, 306, 315, 324, 333, 353, 355, 372, 373, 375, 380, 384, 407, 430, 436, 438, 448, 454, 522, 537, 538, 556, 562, 571, 584, 596, 612, 623, 634, 637, 660, 680, 683, 685, 686, 689, 693, 724, 728, 744, 749, 767, 771, 785, 790, 803, 815, 823, 827, 829, 841, 877, 879, Life of, 273
Epiphanius the Latin, 282
Epiphany, 50, 82, 135, 163, 247, 264, 274, 282, 290, 298, 364, 405, 410, 493, 494, 504, 546, 625, 671, 679, 709, 821, 837, 876
Epirus, 7, 38, 69, 282, 514, 834, Vetus, 233, *129, 130*
Episcopal Lists, 283
Epistula Apostolorum, 150, 158, 206, 723
Epistulae Austrasicae, **284**
Epistula Fermetis ad Hadranum Imperatoren, **284**
Epistula Titi, **284**
Epitaphs, 278, 535
Equilium (Iesolo), 420
Equitius, 147
Eraclea, 420
Eranistes (the Beggar) or Polymorphos (the man of many shapes), 233, 827
Erasmus, 612, 624, 655
Erechtheum, 96
Ererouk, 19, 80, *45*
Erotapokriseis, 536
Eros of Arles, 241
Erustes, 148
Ervigius, 458, 845
Es Fornás, 107
Es-Salt, 636
Esau, 139
Esbous (now Esban), 66
Eschatology, 9, 16, 23, 33, 34, 38, 44, 46, 55, 61, 64, 90, 99, 100, 108, 116, 123, 136, 159, 166, 167, 182, 229, 234, 242, 269, 284, 291, 292, 306, 309, 311, 312, 341, 352, 353, 390, 412, 434, 453, 463, 489, 510, **5**29, 542, 560, 563, 625, 651, 706, 707, 717, 719, 720, 725, 732, 733, 749, 750, 791, 792, 794, 800, 819, 871
Eskender, 291
Essenes, 161, 220, 703
Et-Tabga, 572, 635
Etchmiadzin, 80, 154
Ethelbert, 96, 129, 143, 286
Ethelburga, 662
Ethics, 92, 177, 180, 199, 204, 234, 264, 286, 352, 520, 691, 693, 719, 727, 812, 819, 878
Ethiopia, 9, 19, 33, 60, 61, 63, 66, 95, 113, 142, 179, 200, 211, 220, 230, 235, 289, 291, 307, 324, 331, 361, 383, 394, 428, 446, 472, 496, 540, 567, 763, 832, 853, church, 34, 57, 490, liturgy, 34, 225, 290, 496
Ethiopic book of revelations, 791
Ethiopic texts, 609, 687, 718, 723, 724, 732, 755, 774, 820, 830, 850, 881
Etimasia, 291, 573
Etruria, 422

Etymologiae, 50, 127, 272, 418, 783
Eubulus of Lystra, 35
Eucharist, 1, 6, 16, 29, 33, 34, 37, 63, 64, 79, 90, 116, 120, 123, 127, 136, 141, 151, 155, 159, 188, 208, 214, 215, 220, 221, 222, 224, 232, 234, 239, 242, 243, 247, 248, 255, 257, 258, 261, 264, 276, 279, 280, 285, 289, 290, 292, 296, 305, 312, 318, 334, 335, 351, 355, 358, 361, 383, 387, 390, 396, 408, 409, 440, 470, 473, 475, 490, 493, 494, 495, 496, 500, 502, 504, 506, 528, 532, 540, 541, 542, 548, 550, 554, 561, 562, 577, 583, 612, 616, 617, 622, 630, 664, 667, 668, 707, 710, 712, 714, 715, 722, 725, 738, 744, 745, 748, 749, 750, 751, 752, 764, 792, 798, 800, 802, 803, 805, 814, 815, 819, 820, 840, 876, 877, *131*
Eucharisticos, 530, 651
Eucharistos, 121, 441, 463, 467, 661
Eucheria, 294
Eucherius, Ps.-, 294, 426
Eucherius, poet, 294
Eucherius of Lyons, 149, 178, 294, 294, 295, 311, 312, 347, 483, 543, 544, 615, 680, 688, 700, 725, 727, 747, 753, 754, 761, 767, 824, 834, 863
Euchites, 22
Eucrocia, 126
Eudorus of Alexandria, 558
Eudoxia Athenaïs, 24, 35, 67, 83, 158, 194, 211, 216, 295, 304, 339, 350, 440, 607, 725, 795, 826, 828, 829
Eudoxiana, 339
Eudoxius, 5, 13, 49, 77, 78, 195, 228, 295, 296, 302, 305, 343, 395, 550, 783, 832
Eudoxius of Constantinople, 858
Eugenius, monk, **296**
Eugenius I, **296**, 530, 642, 643, 645
Eugenius II, 497
Eugenius Flavius, 26, 28, 144, **296**, 325, 351, 828
Eugenius of Ancyra, **296**
Eugenius of Carthage, 198, **296**, 861, 868
Eugenius of Constantinople, 822
Eugenius of Toledo, 202, **296**, 458, 743, 791, 812, 844, 845
Eugippius, 121, **296**, 567, 600, 652, 655, 735, 747, 772
Eugnostos, 579
Eulalia of Barcelona, **297**
Eulalia of Mérida, 110, 297, 536, 554, 582, Passion, 296
Eulalius, antipope, 645, 646, 730, 804, 806
Eulalius of Antioch, 8, 22, 297, 362
Eulogia, 127, 172, 351
Eulogius of Alexandria, 6, 24, 251, **297**, 487, 787, 814
Eulogy, 192, **297**
Eumenius the thaumaturge, 208
Eumerus of Nantes, 853
Eunapius, 684, 760, 834
Eunapius of Sardis, 121, **297**
Eunomius of Beroea, 700
Eunomius of Cyzicus, 35, 42, 49, 62, 75, 77, 78, 179, 195, 295, **297**, 303, 355, 364, 365, 392, 449, 658, 683, 684, 783, 824, 852, 856, 858, Against, 235, 296
Eunomius the Arian, 162
Euphemia, 536
Euphemia of Chalcedon, **297**
Euphemius of Constantinople, 36, 197, 242, **297**, 569
Euphrasius, 568
Euphratas of Cologne, **298**, 348, 772, 870
Euphronius of Antioch, 105, 296, **298**, 343
Euplus of Catania, **298**, 533, 534
Euprepius, **298**
Euric, **298**, 320, 338, 356, 778, 813, 814
Euripides, 639, Life of, 121
Eusebian canons, 126, 298, 300
Eusebians, 5, 13, 23, 49, 93, 113, 215,

268, 295, 298, 302, 325, 365, 395, 439, 460, 509, 516, 522, 525, 527, 551, 582, 662, 783, 855
Eusebius, Ps.-, 881, 883
Eusebius Gallicanus, 295, **298**, 321
Eusebius of Alexandria, **298**
Eusebius of Alexandria, Ps.-, **298**, 436
Eusebius of Bologna, 65, **299**
Eusebius of Caesarea in Cappadocia, **299**, 854
Eusebius of Caesaera in Palestine, 2, 5, 17, 18, 20, 21, 23, 29, 31, 32, 37, 38, 41, 43, 47, 48, 49, 52, 54, 55, 57, 58, 60, 62, 65, 67, 72, 73, 75, 77, 80, 81, 83, 84, 93, 95, 102, 106, 110, 111, 112, 113, 114, 115, 118, 119, 121, 126, 138, 139, 141, 150, 159, 166, 168, 180, 188, 192, 193, 207, 219, 224, 225, 235, 238, 244, 245, 246, 250, 251, 258, 259, 262, 263, 269, 271, 272, 284, 298, **299**, 301, 302, 303, 306, 310, 312, 315, 328, 333, 336, 337, 340, 355, 371, 373, 380, 388, 395, 396, 404, 407, 424, 428, 434, 436, 449, 522, 536, 562, 582, 583, 595, 596, 603, 605, 607, 614, 619, 626, 629, 638, 639, 640, 644, 645, 646, 647, 649, 650, 654, 655, 656, 658, 663, 665, 672, 673, 680, 682, 683, 684, 687, 688, 693, 698, 704, 705, 715, 716, 720, 722, 725, 727, 760, 763, 764, 768, 773, 775, 779, 785, 788, 790, 798, 803, 822, 829, 830, 831, 832, 840, 843, 852, 854, 855, 865, 871, 875, 879, 885, 886
Eusebius of Cremona, **301**, 623
Eusebius of Dorylaeum, 196, 275, **301**, 305, 325, 569
Eusebius of Emesa, 81, 115, 236, 269, **301**, 310, 312, 325, 395, 436, 658, 739
Eusebius of Heraclea, 301
Eusebius of Laodicea, 298, **301**
Eusebius of Milan, **301**
Eusebius of Nicomedia, 20, 21, 37, 48, 76, 77, 84, 122, 141, 185, 298, 299, **301**, 303, 325, 337, 434, 460, 507, 522, 525, 526, 567, 595, 662, 682, 705, 743, 824, 830, 832, 855
Eusebius of Qennesrin, 729
Eusebius of Samosata, 49, 114, **302**
Eusebius of Thessalonica, **302**
Eusebius of Valentinopolis, 32
Eusebius of Vercelli, 24, 77, 105, 231, 281, 300, **302**, 305, 350, 381, 508, 548, 559, 567, 617, 657, 757
Eustathians, 58
Eustathius, 582, 660, 662, 663, 678, 830
Eustathius, monk, 302
Eustathius of Antioch, 13, 20, 35, 46, 48, 50, 52, 58, 72, 77, 90, 93, 164, 184, 296, 298, 299, 302, 303, 310, 364, 451, 475, 507, 566, 623, 879
Eustathius of Berytus, 303
Eustathius of Epiphania, 303
Eustathius of Marseilles, 335, 497
Eustathius of Salonica, 797
Eustathius of Sebaste, 13, 114, 195, 250, 303, 337, 514, 528, 550, 556, 567, 744, 767, 775, 832, 854
Eustochium, 122, 303, 430, 521, 659, 688, 787
Eustochius of Tours, 40, 482
Eustorgius of Milan, 30, 303, 497, 559
Eustrathius of Constantinople, 303, 794
Euteknios, 513
Euthalius of Sulci, 304
Eutharic Cillica, 150
Euthemon of Smyrna, 672
Eutherius of Tyana, 196, 304, 545
Euthymius the Great, 5, 52, 167, 215, 304, 475, 567, 677, Life of, 167
Eutropia, 688
Eutropius of Orange, 304
Eutropius of Saintes, 304
Eutropius of Valence, 304, 488, 628, 785, 774, 790, 863, 774

Numbers in bold refer to the main entry on the subject. *Numbers in italics refer to illustration numbers (vol II, pages 951-1094)*

Eutropius the Presbyter, 304
Eutyches, 1, 37, 105, 112, 165, 196, 241, 275, 301, 304, 325, 375, 461, 466, 556, 559, 569, 678, 738, 767, 769, 827, 841, 845, 870, 884, Against Eutyches, 125, 342
Eutychians, 47, 428, 741, Against, 105
Eutychianus, 89, 229, 645
Eutychia, 16
Eutychius of Constantinople, 27, 197, 303, 305, 366, 439, 664
Eutychius of Eleutheropolis, 305
Euzoius of Antioch, 13, 49, 78, 296, 305, 339, 343, 550
Euzoius of Caesarea, 305
Evagrius, 230, 305, 417, 514
Evagrius, priest, 218
Evagrius of Antioch, 28, 30, 48, 49, 230, 305, 430, 431, 475, 597, 660, 684, 707, 725, 746, 775, 785, 788, 792, 828, 832, 878, 879, 882, 884
Evagrius of Constantinople, 305
Evagrius of Epiphania (Scholasticus), 303, 305, 363, 389
Evagrius of Kellia, 22
Evagrius of Pontus (Ponticus), 5, 31, 46, 96, 106, 115, 122, 227, 262, 306, 342, 362, 405, 478, 481, 485, 507, 514, 549, 567, 577, 623, 636, 723, 781, 785, 792, 794, 872, 879
Evagrius of Soli, 732
Evangelarion, 136
Evangelicus, 764
Evangelists, 19, 55, 62, 87, 123, 132, 290, 291, 292, 306, 345, 389, 430, 435, 447, 509, 597, 755, 802, 834, 849, *132*
Evangelius the presbyter, 41
Evangelization, 2, 11, 27, 47, 48, 50, 65, 79, 85, 117, 130, 154, 221, 225, 263, 289, 307, 330, 334, 337, 338, 344, 348, 349, 357, 366, 394, 411, 429, 432, 454, 472, 510, 531, 543, 551, 552, 582, 598, 600, 605, 625, 636, 642, 653, 674, 714, 762, 768, 799, 808, 809, 833, 853, 863, 869
Evaristus, 617, 644
Evelpius, 138, 156
Evergetinon, 655
Eve, 71, 261, 341, 463, 504, 781, 881, *see also* Adam and Eve
Evil, 6, 25, 32, 39, 40, 53, 71, 98, 99, 110, 113, 186, 204, 220, 221, 226, 227, 238, 239, 240, 248, 251, 255, 271, 281, 287, 288, 308, 315, 319, 328, 353, 354, 355, 372, 378, 399, 419, 457, 465, 470, 519, 546, 557, 587, 589, 592, 598, 612, 614, 625, 666, 670, 695, 698, 703, 709, 716, 717, 718, 720, 725, 731, 767, 775, 776, 778, 781, 785, 786, 788, 816, 851, 872, 876, 878
Evodius of Antioch, 308, 850
Evodius of Uzalis, 174, 225, 308, 383, 520, 774, 873, 874
Example, 308, 309, 312, 337, 396, 400
Excommunication, 8, 20, 28, 37, 52, 58, 79, 104, 109, 110, 127, 140, 148, 212, 251, 309, 318, 365, 376, 400, 444, 550, 609, 667, 743, 759, 793, 848
Exegesis, Patristic, 23, 24, 27, 28, 30, 40, 46, 50, 54, 55, 60, 78, 89, 92, 94, 101, 111, 114, 117, 118, 150, 152, 154, 166, 189, 203, 213, 219, 220, 223, 229, 236, 248, 260, 261, 276, 286, 295, 298, 299, 300, 301, 309, 328, 340, 345, 355, 363, 367, 373, 374, 379, 381, 384, 393, 394, 395, 398, 403, 408, 413, 415, 417, 437, 441, 449, 451, 458, 463, 470, 472, 504, 506, 509, 511, 514, 526, 549, 583, 613, 620, 621, 624, 629, 650, 657, 658, 661, 668, 678, 682, 683, 686, 701, 716, 719, 722, 723, 727, 736, 737, 746, 754, 763, 786, 809, 810, 815, 823, 824, 827, 832, 839, 851, 863, 867, 871, 875, 877, 878, 880, 881, 885

Exodus, 105, 137, 187, 276, 311, 314, 337, 382, 470, 492, 504, 575, 620, 631, 746, 782, 825, 867
Exorcism, 42, 122, 151, 152, 227, 232, 242, 477, 615, 876
Expiation, 313, 731, 752, 758
Exultet, **313**, 335, 493
Exuperius of Toulouse, 58, **314**, 742, 758, 846
Ezana, 289, **314**, 472
Ezekiel, 89, 219, 306, **314**, 366, 433, 511, Apocrypha of, 57
Eznik of Kolb, 81, **314**, 472, 620, 702, 715, 716, 827, 834, 867, 871, 873
Ezra (now Zorava), 67, 209

Fabian, 157, 202, 223, 229, **315**, 470, 603, 617, 644, 672, 702, 741, 743
Fabiola, 91, 312, **315**, 609
Fabius of Antioch, 106, 138, 202, 225, **315**, 324
Facundus of Hermiane, 246, **315**, 475, 576
Faesulae (Fiesole), 422
Faeth Fiada, 412, 413
Fahey, Patrick, 123, 540-42, 712, 765
Faith, 3, 6, 30, 54, 60, 74, 90, 94, 96, 98, 99, 100, 101, 107, 114, 120, 123, 127, 140, 141, 145, 151, 166, 168, 179, 186, 188, 191, 199, 205, 219, 220, 223, 226, 230, 243, 244, 245, 260, 261, 279, 282, 287, 288, 292, 315, 317, 319, 326, 328, 341, 347, 355, 357, 360, 375, 377, 378, 396, 397, 398, 399, 405, 408, 419, 436, 477, 478, 483, 494, 506, 510, 517, 524, 526, 531, 537, 551, 560, 575, 579, 598, 616, 621, 629, 640, 657, 659, 661, 664, 689, 703, 707, 708, 709, 721, 732, 733, 734, 737, 742, 745, 751, 754, 758, 760, 763, 767, 769, 781, 788, 792, 794, 801, 815, 817, 821, 840, 851, 871, 872
Faithful, 134, 242, 280, 290, **317**, 326, 336, 388, 390, 406, 410, 467, 493, 540, 615, 664, 667, 708, 710, 759
Faiyum papyrus, 232
Fall, 9, 10
Falla Castelfranchi, Marina, 11, 19, 24-25, 67, 79-81, 83, 84, 88-89, 209, 255-56, 344-45, 487-88, 578, 858
Falsification, 42, 57, 59, 302, 317, 370, 723, 784, 804
Famagusta, 213
Family, 91, 104, 161, 162, 221, 287, 318, 319, 337, 441, 469, 528, 552, 566, 580, 596, 667, 688, 739, 760, 784, 787, 793, 842, 871, 877
Fanum Fortunae (Fano), 423
Farina, Rafaele, 139
Farne Island, 130, 133, 211
Fasir, 175
Fasola, Umberto, 155-58
Fastidiosus, 318
Fastidius, 318
Fasting, 3, 13, 17, 25, 63, 84, 111, 117, 152, 160, 210, 219, 223, 227, 231, 234, 235, 258, 290, 318, 319, 337, 383, 442, 450, 454, 461, 466, 473, 492, 493, 519, 548, 570, 609, 667, 669, 671, 707, 728, 742, 744, 748, 781, 792, 793, 816, 819, 843, 869, 877
Father (name of God), 3, 4, 5, 9, 11, 13, 20, 26, 27, 29, 33, 35, 36, 39, 42, 46, 49, 54, 59, 64, 71, 76, 77, 78, 89, 92, 94, 99, 108, 109, 113, 114, 120, 121, 123, 141, 150, 158, 159, 163, 164, 168, 185, 195, 205, 207, 215, 233, 245, 252, 255, 260, 287, 296, 300, 303, 324, 352, 354, 358, 362, 381, 390, 391, 395, 396, 402, 405, 406, 415, 430, 448, 449, 463, 467, 468, 470, 507, 510, 516, 519, 523, 524, 525, 526, 528, 553, 566, 595, 606, 621, 622, 653, 657, 734, 735, 737, 748, 757, 771, 780, 783, 797, 831, 848, 851
Fathers of the Church, 1, 2, 8, 17, 18, 24,

25, 38, 40, 46, 56, 57, 60, 61, 72, 98, 103, 104, 108, 116, 123, 128, 152, 160, 161, 176, 200, 219, 223, 227, 228, 234, 320, 393, 494, 652, 654, 852
Fausta, 192
Faustianus, 637
Faustinus, 30, 55, **320**, 508, 522, 828
Faustus, priest, 43
Faustus of Byzantium, 81, 83, 84, **320**, 478, 593
Faustus of Milevis, 11
Faustus of Riez, 43, 105, 178, 223, 298, **320**, 331, 393, 480, 483, 508, 513, 548, 653, 660, 669, 761, 747, 767, 796
Faustus of Thimida Regia, 584
Faustus the Manichaean, 38, 98, **321**
Favorita, 425
Feinan, 636
Felician of Musti, 252, **321**, 545, Against, 870
Felician schism, 126, 321, 322, 851
Felicianus, 761
Felicissimus of Carthage, 139, 146, **321**, 328, 605, 801
Felicity and Perpetua, *see* Perpetua and Felicity
Felix, bishop of Seville, 845
Felix, exorcist, 18
Felix, manichee, 98, **321**
Felix I, 644, 741
Felix II, antipope, 168, 218, 485, 646, 742
Felix III, 5, 49, 330, 339, 365, 447, 569, 610, 642, 645, 743, 744, 884
Felix IV (III), 645, 743
Felix and Adauctus, 156, **321**, 555, 631
Felix and Fortunatus, 64, 420
Felix and Nabor, *see* Nabor and Felix
Felix of Acci, 270
Felix of Apthugni, 137, 246, 247, 250, **321**, 351, 613, 766
Felix of Bagai, 584
Felix of Bamacorra, 584
Felix of Gerona, 582
Felix of Nola, 3, 192, **322**, 407, 661, Life of, 118
Felix of Thibiuca, 13, **322**
Felix of Trier, 79, **322**, 348, 559, 597, 850, 851, 854
Feltria (Feltre), 420
Feminalia, 253
Femoralia, 253
Fénékpuszta, 638
Fenelon, 101
Feriana, 15
Ferrandus of Carthage, 134, 142, **322**, 331, 497
Ferreolus of Uzès, **322**, 747, 761
Festus, **322**, 430, 432, 657
Février, Paul-Albert, 14-16
Fibula, 219, 253, 254, **322**, 390, 759
Ficino, 101
Fidelis, 554, 663
Fidentia (Fidenza), 421
Fidentius, 383
Fides Hieronymi, **323**
Fides Nicaena, **323**
Fiducia, **323**, 355, 397, 398
Fiery furnace, 3, 16, 40, 138, 219, **323**, 349, 471, 615, 668, 755, 763, *133*
Filaster, 64, 83, 84, 105, 158, 159, 186, 282, **323**, 337, 375, 384, 653, 655, 728, 767, 771, 803
Filioque, 208, **324**, 392, 652, 844, 852
Filoramo, Giovanni, 71, 284-86, 352-54, 371-72, 377-78, 457, 579, 649-50, 724
Finalmarina, 421
Finnian of Clonard, **324**, 411
Finnian of Moville, 324
Find Fili, 413
Fiocchi Nicolai, Vincenzo, 210-11, 271, 327, 489, 629-32, 684-85, 786, 864
Firmicus Maternus, 198, **324**, 409, 592
Firmilian of Caesarea in Cappadocia, 72, 109, 137, 143, 212, 315, **324**, 371, 403, 620, 640, 663, 794, 848

Firminus of Trieste, 773
Firmus of Caesarea in Cappadocia, 248, **324**, 771
Fish, 803
Flaccilla, 320, **324**, 351, 364, 607
Flaccillus of Antioch, 49, **325**, 300, 855
Flaccus, 682
Flavia Neapolis, 635
Flavian, martyr, **325**, 571
Flavian I of Antioch, 4, 49, 50, 51, 196, 225, 236, **325**, 440, 584, 643, 660, 684, 744, 778
Flavian II of Antioch, 49, 268, **325**
Flavian of Chalon-sur-Saône, **325**
Flavian of Constantinople, 112, 159, 241, 275, 303, 305, **325**, 380, 403, 461, 466, 569, 678, 845
Flavian of Philippi, **325**, 744
Flavianopolis, 175
Flavianus Virius Nicomachus, 144, 296, **325**, 742
Flavias (now Kars el Bazar), 175
Flavius Anastasius, 729
Flavius Clemens, 8, 246, **326**
Flavius Constantinus, 397
Flavius Philostratus, 60
Flavius Vopiscus, 388
Flesh, 9, 34, 44, 45, 59, 84, 94, 111, 116, 117, 123, 127, 128, 161, 164, 165, 220, 233, 244, 261, 288, 319, 322, **326**, 341, 353, 361, 391, 395, 404, 408, 414, 415, 438, 441, 448, 460, 470, 489, 523, 528, 557, 560, 596, 604, 613, 614, 621, 651, 657, 685, 708, 732, 733, 734, 758, 781, 792, 796, 841, 859, 873, 882
Flight, 212, 238, **326**, 603, 819, from persecution, **326**, 570, from the world, 83, 230, **327**, 587, 697, 882, of the Soul, 71, 877
Flodoard, 540
Flood, 105, 151, 188, **327**, 341, 415, 453, 504, 525, 598, 624, 631, 668, 712, 749, 876, *134*
Florentia (Firenze, Florence), 422
Florentinus of Sardis, 196
Florentius, civil servant, 327
Florentius of Strasburg, **327**
Florian, **328**, 600
Florilegia, 76, 311, 444
Florilegium Edessenum Anonymum, 327, 843
Florinus, **328**, 859
Florus, 806
Florus and Usward, 584
Florus of Hadrumetum, **328**
Florus of Lyons, 393, 537, 655
Focale, 254
Folgado, Segundo, 116-17, 308, 709
Fontaine, Jacques, 400-1, 405, 418-19, 531, 721-22, 799
Font, 123
Footwear, 253
Forcona (now Civita di Bagno), 423
Forlin Patrucco, Marcella, 36, 85, 143, 159, 176, 183-84, 184, 192, 193, 198, 236, 263, 274-75, 333, 371, 403, 488, 512, 520, 545, 546, 594, 608, 637, 669, 686, 702-3, 711, 814-15, 854
Formosus, 616
Fornells, 107
Fornication, 243, 270, 328, 364, 528, 596, 719, 780, 781, 815, 819, 845
Fortunatian of Aquileia, **328**, 497, 543, 778, 856, 860
Fortunatian of Sicca Veneria, **328**
Fortunatus, 605, 712
Fortunatus, deacon, 102
Fortunatus, martyr, 64
Fortunatus of Carthage, 193, **328**
Fortunatus the Manichaean, 98, **328**
Forty Martyrs of Nokalakevi, 344
Forty martyrs of Sebaste, 110, 328
Forum Clodii, 422
Forum Traiani, 425

Fossor, 157, **328**, 555, 630, 742, *135*
Fraga, 813
Fragmenta Arriana, **329**, 637
Frampton, 130
Francis de Sales, 101
Francolí, 814
Franks, 103, 104, 105, 124, 134, 138, 142, 162, 183, 198, 199, 254, 307, 329, 330, 335, 336, 338, 342, 349, 356, 365, 459, 473, 505, 525, 555, 582, 633, 642, 646, 653, 663, 667, 680, 700, 743, 779, 802, 828, 846, 850, 861, 862, Franco-Roman rite, 387
Frank's casket, 132
Fravitta, 330
Fredbal, 861
Fredegar, Ps.-, **330**
Fredegund, 162, 365, 556, 735
Freer codex, 632
Fréjus (Forum Iulii), 582
Frend, W.H.C., 137, 175-76, 185, 208, 228, 246-49, 250, 270, 332, 337, 351, 360, 613, 651, 671-74, 679-80, 711
Friendship, 46, 90, 128, 191, 222, 287, 296, **330**, 368, 441, 467, 469, 483, 580, 661, 787, 871, 793, 661
Frigitil, 307
Fritigern, 356
Friuli, 420
Frogan, 102
Fronto of Cirta, **330**
Fructuosus, Augurius and Eulogius, 534
Fructuosus of Braga, **330**, 747, 791, 814, 860, Life of, 860
Fructuosus of Dumium, 845
Fructuosus of Tarragona, **331**, 386, 752, 790, 814, 860
Frumentius, 289, 290, 307, 315, **331**, 496, 832, 855
Fulda, 125, 486
Fulgentius, Fabius Planciades, **331**
Fulgentius of Astigi (Ecija), **331**, 418
Fulgentius of Ruspe, 26, 101, 102, 116, 218, 321, 322, **331**, 392, 406, 507, 655, 702, 709, 757, 806, 849, 852, 861, 868
Fulgentius the Donatist, **332**, Against, 246
Fulginium (Foligno), 423
Fulminata, **332**
Fulvo, Andrea, 68
Funerals, 221
Fursa (Fursey), 131, 154
Fussala, 383
Fustat (Cairo), 200

Gabriel, 40, 132, 163, 167, 213, 271, 290, **333**, 447, 539, 557, *136*
Gabriel of Katar, 106
Gadara, 636
Gaiane, 80
Gaianus, 54, **333**, 458, 829
Gainas, 67
Gaius (and the Alogi), 55, 159, **333**, 448, 657, 675, 712, 885
Gaius of Paraetonium, 487
Galaswinth, 162, 862
Galatia, 37, 52, 85, 87, 109, 143, 197, 307, **333**, 336, 371, 403, 456, 585
Galen, **333**, 508, 558, 584, 672, 693, 769
Galerius, 43, 168, 193, 198, 236, 263, **333**, 388, 544, 546, 641, 673, 714, 833
Galilaeans, Christians, **333**
Galilee, 334
Gall, **334**, 348, 349
Gallaecia (Galicia), 18, 106, 127, 148, 255, 263, **334**, 404, 531, 652, 714, 742, 790, 801, 844, 860, 861
Galla Placidia, 2, 198, **334**, 366, 397, 605, 607, 804, 859, Mausoleum of, 202, 306, 334, 435, 476, 536, 573, 729, 778
Gallicae, 254
Gallican liturgy, 6, 123, 141, 155, 185, 22, 224, **334**, 335, 338, 346, 385, 475, 493, 500, 615, 616, 745, 853, 865, Gallican-Gothic rite, 497

Gallienus, 156, 178, 237, **336**, 640, 673, 822, 860, 863
Gallus, 106, 187
Gamaliel II, 436, 657
Gammadia, 202, 209, **336**, *137*
Gangra, 27, 237, 303, **336**, 528, 567, 609, C. of (c.340), 8, 142, 205, 337
Garibald of Bavaria, 103
Garrucci, R., 69
Gasr Elbia (ancient Olbia), 489, 675
Gatianus, 23
Gaudentius of Brescia, 150, 277, 312, 328, **337**, 393, 494, 649, 840
Gaudentius of Novara, **337**, 421
Gaudentius of Thamugadi, 247
Gaudentius the Donatist, **337**
Gaudete Sunday, 186
Gaul, 8, 13, 17, 28, 30, 43, 53, 59, 71, 78, 79, 90, 102, 104, 126, 134, 142, 154, 167, 170, 174, 178, 184, 191, 192, 193, 198, 205, 218, 224, 230, 238, 246, 249, 274, 278, 282, 307, 324, 328, 329, 334, 336, **337**, 342, 346, 351, 356, 359, 365, 380, 381, 393, 397, 399, 411, 453, 486, 493, 512, 530, 545, 552, 559, 567, 571, 572, 579, 581, 601, 604, 617, 618, 633, 641, 642, 645, 653, 676, 700, 712, 714, 742, 743, 761, 767, 784, 802, 804, 846, 847, 853, 854, 861, 864, 865, 868, 869, 870, 885, *138-40*
Gayané and Hripsimé, 858
Gaza, 12, 24, 84, 112, 141, 201, **339**, 417, 527, 625, 635, 662, 710, 713, 884, C. of (541), 339
Ge'ez, 290, 291, 307
Gelasius I, pope, 5, 6, 36, 62, 103, 104, 168, 123, 197, 203, 223, 246, 277, 320, **339**, 396, 410, 447, 486, 493, 496, 499, 618, 642, 643, 645, 646, 678, 714, 738, 742, 745, Sacramentary, 42, 232, 493, 497, 498, 500, 502, 611, 738, 745, 749
Gelasius of Caesarea, 138, 305, **339**, 340, 389
Gelasius of Cyzicus, **339**, 340, 389
Gelasius of Mar Saba, 677
Gelimer, 861
Gelsi, Danielle, 33-35, 134-36
Gemes, 689
Geminianus, **340**
Geminius, **340**
Genesis, 9, 65, 67, 98, 99, 105, 110, 137, 180, 187, 205, 212, 235, 251, 276, 295, 310, 311, 318, **340**, 342, 353, 374, 378, 380, 382, 409, 446, 465, 504, 514, 525, 530, 557, 620, 634, 652, 682, 697, 718, 727, 746, 764, 798, 819, 825, 831, 851, 867
Genesius of Arles, 79, **341**, Passion, 79
Genesius of Rome, 342
Geneva, 174
Gennadius, Ps.-, 702
Gennadius of Astorga, 860
Gennadius of Constantinople, 142, 197, 312, **342**, 450, 523, 536, 658
Gennadius of Marseilles, 84, 96, 121, 142, 149, 178, 339, **342**, 374, 389, 405, 418, 497, 516, 525, 530, 548, 576, 611, 652, 654, 660, 680, 702, 753, 754, 772, 793, 801, 830, 838, 867, 868, 870
Genoveffa of Paris, 307, **342**, 346, 676
Genseric, 78, 96, 145, 228, 248, 326, **342**, 396, 400, 581, 605, 705, 728, 757, 861, 868
Gentiles, **342**
Genua (Genova, Genoa), 29, 421
George, martyr, 38, **343**, 786, Acts of Martyrdom of, 106
George I, catholicós, 416
George Arsas, 768
George Choiroboskos, **343**
George Grammaticus, **343**
George Pachimera, 76
George Scholarius, 76
George Syncellus, 41, 639

George of Amastris, 27
George of Arethusa, 298
George of Be'eltam, 840
George of Cappadocia, 93
George of Constantinople, 197
George of Laodicea, 58, 77, 113, 301, **343**, 395, 785
George of Nicomedia, 346
George of Pisidia, **343**
George of Ptolemais, 487
George of Resh'aina, 547
George of Rusafah, 71
George of Sykeon, **343**
George of the Arabs, 110, 343, 428, 810
George the Monk, 339, **343**
Georgia, 3, 19, 81, 153, 292, **344**, 403, 425, 567, 578, 597, 679, 809, *141-144*
Georgian Lectionary, 134
Georgian language and literature, 9, 21, 31, 38, 44, 52, 60, 61, 66, 83, 89, 112, 113, 234, 257, 334, **345**, 383, 394, 447, 472, 481, 501, 527, 556, 599, 677, 782, 786, 794, 797, 827, 841, 843, 880, 884
Gepidae, 179, 505
Gerasa, 66, 171, 174, 572, 599, *36*
Gerhard, J., 654
Gerizim, 36, 69
Germanus of Auxerre, 120, 129, 130, 307, 342, **346**, 653, 678
Germanus of Constantinople, 135, 136, 204, **346**, 407, 479, 511, 538
Germanus of Paris, **346**, 497, 503, 556
Germany, 17, 64, 125, 126, 129, 174, 338, **346**, 486, 567, 633, 646, 676, 689, *145*
Germinius of Sirmium, 26, 33, 77, 350, 737, 783, 858
Gerona (Gerunda), 315, **350**, 790, 813, C. of (517), 386
Gerontius, 679
Gerontius, monk, 121, **350**
Gerontius of Nicomedia, **350**
Gerson, Jean, 655
Gertrude of Nivelles, 532
Gervasius and Protasius, 28, 224, 347, **350**, 370, 521, 559, 579, 584, 656
Gesta apud Zenophilum, 14, 193, 533, 607, 779, 840
Gesta collationis Carthaginensis, 322, 533
Gesta de Xysti purgatione, 784, 804
Gesta purgationis Caeciliani et Felicis, 322, **350**
Gestures, 62, 109, 255, 274, 292, 294, 323, 333, **351**, 370, 378, 409, 457, 477, 552, 607, 615, 650, 706, 707, 745, 751, 770, 803, *146*, see also Liturgical Gestures
Geta, 143
Gethsemane, 433
Getica, 150
Gharandal, 636
Ghent, 27
Ghor es-Safi, 636
Gianotto, Claudio, 54, 109-10, 158, 251, 328, 374, 375-76, 526, 546, 550, 612, 680, 771, 859-60
Gift, 91, 123, 296, **351**, 409, 491
Gildas-de-Rhuys, 351
Gildas the Wise, 19, **351**, 412
Gildo, 101, 146, 147, 247, 248, 351, 545, 613, 660, 795, 840
Giordani, Roberto, 377, 419, 422-26, 434-435, 764, 807
Gisulf II, 774
Giuntella, Anna Maria, 140-41, 471-72, 529, 597, 615, 777-78
Glamis (Glamis, Angus), 132
Glaphyra, 214, 312
Glastonbury Abbey, 131, Tor, 131
Glaucias, 113
Glendalough, 130
Gloria, 232, **352**, 541, 555, 556, 685
Gloria in excelsis Deo, 62, 160, 252, **352**
Gloria Patri, 816
Glory, 6, 35, 76, 91, 116, 122, 123, 135, 186, 192, 233, 242, 247, 252, 261, 269, 293, 316, 340, **352**, 359, 360, 397, 399, 401, 414, 441, 461, 462, 489, 506, 510, 532, 533, 649, 651, 669, 707, 716, 720, 734, 735, 792, 816, 833
Gloss, Glossary, 84, 152, 184, 187, **352**, 356, 412, 418, 600, 714, 715, 716, 812
Glossolalia, 499
Gloves, 866
Gnosis, 45, 71, 161, 163, 204, 205, 210, 259, 271, **352**, 378, 413, 448, 449, 465, 523, 537, 551, 553, 566, 612, 619, 674, 724, 733, 734, 763, 798, 805, 848
Gnostic, 2, 4, 9, 18, 22, 26, 39, 53, 55, 56, 57, 63, 71, 84, 92, 109, 110, 111, 112, 113, 123, 126, 127, 129, 145, 151, 158, 160, 161, 180, 199, 205, 207, 220, 226, 242, 251, 255, 262, 264, 281, 282, 285, 287, 289, 293, 309, 316, 320, 326, 328, 333, 350, 355, 377, 390, 429, 430, 434, 448, 460, 465, 518, 543, 550, 566, 579, 587, 596, 609, 612, 614, 628, 653, 657, 658, 670, 680, 698, 700, 707, 712, 717, 732, 734, 735, 758, 763, 766, 771, 775, 777, 780, 781, 786, 788, 796, 798, 815, 817, 819, 830, 837, 838, 839, 848, 851
Gnosticism, 10, 12, 23, 41, 44, 53, 55, 56, 61, 82, 181, 205, 243, 255, 308, **352**, 371, 375, 405, 485, 603, 610, 639, 656, 709, 796, 802, 810, 819, 851, 882
Goar, J., 349, 654
God, 2, 3, 5, 6, 9, 11, 12, 13, 16, 20, 22, 25, 33, 35, 36, 38, 39, 44, 45, 46, 52, 53, 54, 55, 56, 62, 71, 72, 73, 74, 75, 76, 77, 78, 82, 83, 84, 94, 95, 96, 98, 99, 101, 103, 107, 110, 111, 113, 116, 119, 120, 121, 123, 125, 127, 139, 145, 150, 158, 161, 163, 167, 168, 175, 181, 186, 191, 192, 198, 199, 201, 202, 203, 205, 206, 215, 220, 222, 226, 227, 229, 230, 232, 239, 240, 242, 243, 244, 245, 248, 252, 255, 256, 260, 261, 276, 279, 280, 281, 284, 289, 301, 308, 309, 312, 315, 316, 319, 320, 323, 327, 328, 330, 336, 340, 351, **352**, 354, 359, 362, 365, 368, 372, 378, 391, 392, 395, 396, 397, 402, 404, 405, 406, 408, 413, 414, 415, 416, 443, 445, 447, 448, 455, 457, 458, 461, 462, 463, 467, 469, 470, 473, 485, 489, 492, 494, 495, 500, 506, 509, 510, 516, 523, 526, 547, 553, 566, 577, 589, 595, 604, 609, 614, 621, 626, 651, 652, 653, 657, 664, 668, 670, 683, 684, 689, 690, 691, 692, 694, 695, 698, 699, 701, 706, 707, 708, 709, 710, 712, 717, 718, 719, 720, 721, 725, 730, 731, 732, 733, 734, 737, 748, 750, 751, 752, 758, 762, 765, 767, 769, 771, 773, 775, 776, 781, 786, 788, 789, 792, 796, 797, 800, 801, 815, 816, 818, 819, 821, 824, 830, 831, 839, 841, 843, 846, 851, 852, 871, 872, 877, 878, 881, 882
Godomar, 134
Gojam, 290
Golgotha, 433
Golindukh, 794
Gomon, 22
Gondeshapur, 225
Good Friday, 57
Gordian III, 672
Gordianus, 365
Gordius, 137
Gorgonia, 356, 361
Gortyna, 208, 209, 681
Gospel of Truth, 448, 579
Gothic Literature, 322, 323, **356**
Goths, 7, 17, 19, 40, 86, 113, 119, 138, 150, 194, 195, 199, 201, 217, 223, 254, 283, 307, 336, **356**, 360, 363, 366, 368, 438, 444, 451, 478, 505, 506, 515, 521, 525, 548, 549, 550, 560, 565, 594, 596, 638, 642, 679, 681, 741, 742, 746, 753, 757, 770, 806, 828, 833, 837, 844, 856, 861

Numbers in bold refer to the main entry on the subject. *Numbers in italics refer to illustration numbers (vol II, pages 951-1094)*

Grace, 46, 72, 82, 90, 98, 99, 100, 101, 102, 108, 126, 138, 147, 154, 160, 195, 199, 212, 213, 225, 241, 260, 261, 262, 269, 277, 287, 288, 289, 293, 308, 311, 316, 321, 328, 331, 355, **357**, 362, 369, 370, 376, 391, 399, 405, 406, 408, 409, 430, 440, 443, 454, 456, 461, 462, 466, 477, 479, 507, 508, 528, 529, 542, 555, 561, 562, 585, 604, 616, 621, 646, 653, 657, 658, 665, 668, 678, 709, 717, 720, 745, 751, 767, 781, 782, 788, 792, 793, 825, 840, 851, 877, 880
Graces, 538
Grado, 170, 173, 174, 420, 572, 774
Grammaticus, 469
Granada, 106
Gratian, 26, 28, 64, 78, 102, 168, 195, 218, 325, 347, 348, **360**, 404, 407, 410, 416, 426, 547, 641, 655, 702, 712, 804, 828, 858
Gratus, bishop of Carthage, **360**
Gratus, saints, 143, 146, 250, **360**
Gravedona, 420
Greaves, 254
Grech, Prospero, 518
Greece, 1, 19, 23
Greek, Christian, **360**, 472, 474, 494, 498, 656, 736
Greensted, Essex, 131
Gregentius, **361**
Gregentius, Ps.-, 437
Grégoire, Réginald, 117-18, 324, 393-94, 394-95, 476, 551, 769
Gregorianum, 123
Gregory I, the Great 6, 16, 17, 30, 36, 39, 53, 55, 60, 79, 96, 101, 118, 119, 121, 142, 143, 187, 197, 211, 213, 224, 232, 233, 242, 247, 262, 269, 286, 293, 296, 302, 305, 311, 312, 329, 335, 359, **365**, 372, 374, 377, 397, 407, 410, 418, 420, 432, 437, 439, 442, 465, 466, 476, 477, 478, 484, 486, 488, 493, 495, 496, 497, 499, 502, 503, 506, 509, 511, 513, 532, 541, 545, 551, 562, 567, 593, 616, 637, 638, 642, 645, 652, 654, 655, 662, 676, 679, 685, 716, 738, 743, 744, 745, 747, 749, 757, 759, 763, 773, 787, 793, 798, 807, 812, 816, 817, 822, 824, 838, 846, 849, 854, 865, 866, 869, 872, 878, 879, 880, 881, Gregorian Sacramentaries, 366, 497, 500, 502, 745, 846
Gregory II, 125, 407, 479, 646
Gregory III, 125, 407, 479, 646
Gregory Nazianzen, 9, 16, 38, 43, 44, 49, 57, 59, 76, 81, 84, 96, 113, 114, 115, 121, 124, 126, 135, 138, 142, 143, 149, 158, 160, 163, 164, 165, 177, 178, 192, 194, 195, 196, 199, 200, 204, 207, 211, 226, 230, 232, 245, 269, 278, 282, 299, 302, 306, 312, 325, 344, 351, 356, **361**, 363, 364, 372, 382, 392, 402, 405, 406, 408, 410, 417, 428, 430, 431, 450, 459, 475, 483, 484, 511, 518, 538, 549, 550, 561, 584, 599, 611, 614, 633, 634, 648, 649, 655, 662, 665, 670, 678, 682, 695, 709, 714, 719, 723, 725, 731, 744, 746, 749, 760, 761, 785, 790, 805, 826, 828, 840, 848, 852, 854, 879
Gregory of Agrigento, **362**, 480, 879, 880
Gregory of Akritas, 208
Gregory of Albania of Caucasus, 19
Gregory of Antioch, 197, 305, 363
Gregory of Caesarea, 137, 480
Gregory of Elvira, 55, 311, 312, 323, 363, 453, 508, 669, 723, 787, 790, 881
Gregory of Khandzta, 344
Gregory of Nyssa, 9, 25, 27, 45, 49, 54, 59, 74, 75, 76, 83, 91, 95, 113, 114, 115, 116, 121, 122, 140, 142, 143, 151, 152, 161, 163, 164, 177, 183, 186, 192, 195, 200, 205, 207, 226, 233, 237, 240, 243, 244, 252, 255, 257, 262, 269, 276, 277, 282, 296, 308, 310, 312, 324, 325, 328, 346, 355, 356, 358, 361, 362, **363**, 368,
372, 392, 396, 398, 402, 405, 406, 407, 410, 449, 475, 477, 493, 514, 516, 517, 520, 536, 538, 556, 561, 575, 576, 577, 578, 587, 603, 608, 611, 648, 649, 652, 655, 656, 658, 665, 678, 682, 683, 688, 690, 691, 693, 694, 699, 703, 709, 717, 719, 723, 725, 731, 749, 776, 784, 786, 788, 792, 793, 795, 805, 822, 832, 842, 852, 857, 876, 879, 880
Gregory of Nyssa, Ps.-, 436
Gregory of Rimini, 101
Gregory of Tours, 1, 4, 27, 40, 104, 106, 121, 140, 143, 162, 168, 203, 238, 256, 284, 296, 322, 329, 335, 339, 347, 348, 349, 365, 366, 389, 417, 458, 483, 493, 497, 499, 500, 513, 530, 532, 533, 556, 595, 638, 663, 671, 687, 735, 758, 769, 771, 772, 837, 846, 853, 854, 873, 875
Gregory the Cappadocian, 49, 93, 105, 302, **365**, 420
Gregory the Illuminator, 16, 19, 80, 83, 84, 307, **368**, 480, 496, 556, 752, 858
Gregory the Presbyter, 368
Gregory the Thaumaturge, 27, 72, 93, 95, 96, 122, 137, 138, 142, 233, 235, 272, 355, 368, 373, 405, 516, 620, 629, 650, 663, 693, 703, 761, 785, 805, 879, 880, Life of, 200, 364
Gribomont, Jean, 8, 10, 16, 31, 37, 52, 83-84, 93, 95, 113, 114-15, 138, 140, 167, 251, 302, 302-3, 303, 304, 305, 306, 315, 350, 361-62, 363-65, 379, 380, 388, 400, 417, 430-31, 454, 476, 480, 514, 516, 527, 549, 556-57, 566-67, 583, 584, 593, 597, 599, 600, 628, 646, 662, 663, 678, 680-81, 682, 684, 714, 716, 722-23, 746, 746-47, 762-64, 775, 779-80, 821, 826, 832, 842, 873, 885
Grisbrooke, W. Jardine, 501-2
Grosseto, 422
Grossi, Vittorino, 37, 44-46, 61, 137, 209-10, 242, 309, 318-19, 320, 376-77, 408, 458, 525, 553, 625-26, 665-66, 710-11, 732, 732-33, 759, 789-90, 849
Grossmann, Peter, 265-66
Guadix, 772
Guarrazar, 491
Gubba, 238
Gubbio, 223
Guentzon, 838
Gundemar, 102, **368**, 844
Gundioc, 134
Gundobad, 124, 134, 183
Gundobald, 281
Gundofar, 837
Gundovald, 824
Gunthamund, 252, 400, 861
Guntheric, 774, 844, 861
Gunthiges Baza, 451
Guntram, 105, 162, 186, 556, 637, 802
Guntrud, 505
Gymnasium of Zeusippus, 44

Habakkuk, 219, 294, 306, **369**, 834, 867, *147*
Habit, Monastic, 252, **369**, 846
Habr Hiram, 51
Hades, 371
Hadiab, 663
Hadita, 11
Hadrian, 60, 72, 73, 85, 95, 111, 113, 120, 183, 284, **369**, 403, 432, 594, 671, 688, 727, 807, 812, 834
Hadrian, exegete, **369**
Hadrian I, pope, 142, 246, 366, 502, Sacramentary, 502
Hadriana, 142
Hadrian of Canterbury, 19, **369**
Hadrian of Nicomedia, **369**
Hadrian of Nisida, 824
Hadrianopolis, 27
Hadrumetum (Sousse), 14, 99, 328, 359, **369**, 666, 711, *148*, C. of (397), 146
Hagiography, 5, 118, 121, 138, 139, 154,
160, 161, 186, 200, 211, 215, 227, 232, 233, 245, 264, 278, 281, 291, 296, 338, 345, 361, 365, **370**, 389, 394, 412, 426, 475, 489, 520, 532, 533, 535, 537, 553, 596, 634, 638, 648, 705, 723, 724, 762, 765, 770, 787, 791, 793, 795, 799, 809, 842, 862, 870
Haidar-Pacha, 165
Haidra, 15, 170, 173
Hairstyles, 253
Halawe, 218
Hallawiya, 808
Haller, 644
Halmyris (Salmorus), 764
Hamartigenia, 651, 721, 781
Hamman, Adalbert, 2, 6, 10-11, 16, 18, 25-26, 90, 102, 103, 105, 106, 107-8, 134, 151-52, 155, 176, 186, 188, 190, 191, 203, 228, 231-32, 246, 263-64, 292-93, 304, 305, 317-18, 320, 322, 330, 333, 381, 409, 410-11, 416, 427, 465, 474-75, 480, 482, 511, 517, 544, 551, 551-52, 559-60, 576, 577, 585, 600, 609, 612, 614-15, 653, 654-56, 664-65, 685, 699, 706-7, 708-9, 715, 717, 723-24, 724-25, 725, 730, 731, 747, 751, 766, 767-68, 768, 780, 784, 787, 792-93, 794, 801, 853, 861, 867-68, 869, 870-71, 874, 875, 877, 882
Hannibalianus, 192
Hanson, Richard, 206-8
Harbert, Edward, 18-19, 128-29, 143, 159, 221, 762, 883
Hârith ibn Amr, 65
Hârith ibn Gabala (Hareth bar Gabala), 428, 823
Harmonius, **370**
Harnack, A., 654
Harrân, 4, 65
Hartlepool, 131
Hassan Ibn an-Nu'man, 145
Hasta (Asti), 421
Hatfield, Synod of (680), 129, 446
Haugia el Hafi, 636
Hauran, 66
Hauwarte, 218
Hazza, 11
Healing of the Man Born Blind, **370**, 755, *149*
Hebdomas, 204
Hebrews, Gospel of the, 57
Hecataeus, Ps.-, 455
Hedybia, 126
Heep, 778
Hegemonius, 71, 113, 233, **371**
Hegesippus, 181, 245, 251, 284, **371**, 375, 451, 561, 640, 644, 779, 790, 798, 863
Hegesippus, Ps.-, 371, 676
Hegumen, 1, 8, 71, 112, **371**, 566
Helena, 89, 189, 193, 195, 210, 258, **371**, 688, 741
Helenopolis, 636
Helenopontus, 27
Helenus of Tarsus, 48, 315, 324, **371**
Heliand, 234
Heliodorus, 230, **371**, 430, 593
Heliodorus of Sozusa, 487
Heliogabalus, 22, 26, 166, 271, **371**, 799
Heliopolis (now Baalbek), 809
Hell, 3, 94, 208, 229, 285, 308, 352, 353, **371**, 415, 575, 693, 873
Helladius, hermit, 149
Helladius of Caesarea, **372**
Helladius of Tarsus, 196, **372**, 545, 785
Hellenism and Christianity, 23, 39, 45, 46, 163, 177, 205, 306, 336, 341, 354, 355, **372**, 391, 415, 582, 628, 686, 760, 775, 785, 805, 810, 839, 851, 854
Hellón (Albacete), 148
Helmet, 254
Helmwald, 118
Helpidius, 747
Helvidians, 47
Helvidius, 47, 145, **374**, 405, 538 776, Against Helvidius, 374, 431
Henana of Adiabene, 110, **374**, 598, 674, 810
Henaniso, 417
Henchir bou Said, 228
Henchir Faraoun, 15
Henchir Seffan, 15
Henoticon, 5, 6, 47, 49, 112, 118, 197, 268, 325, 432, 438, 487, 516, 569, 570, 585, 593, 610, 642, 678, 679, 684, 723, 744, 773, 780, 804, 855, 884
Henos sômatos, 20, 301
Henry of Gand, 654
Henry Savile, 655
Hephaisteion (Theseion), 96
Heptateuchos, 212, 312, 651
Heraclas of Alexandria, 23, 235, 238, **374**, 460, 619
Heraclea, 174, 301
Heraclea (Perinthus) in Thrace, 194
Heraclea Lynkestis, 174
Heracleon, 187, 309, 354, 373, **374**, 384, 448, 724, 837, 859
Heraclian of Chalcedon, 26, 27, 71, 159, **374**
Heraclidas of Nyssa, **374**, 636, 768
Heraclides, 647
Heraclides of Cyprus, 207, 213, **375**
Heraclides of Ephesus, 275, 314, 514
Heraclides of Oxyrhynchus, 508
Heraclitus, **375**, 639, 658
Heraclius, antipope, 646
Heraclius, emperor, 14, 35, 36, 80, 95, 163, 192, 194, 215, 216, 263, 274, 313, 330, 343, **375**, 383, 433, 439, 447, 530, 564, 598, 645, 768, 797, 827, 838
Heraclius of Hippo, **375**
Heraclius of Paris, 482
Herbauge, 27
Herculaneum, 210
Herennius, 624
Heresiology, 37, 110, 245, 282, 342, 343, 352, 353, 384, 579, 859, 848
Heresy, Heretics, 3, 6, 7, 11, 13, 17, 18, 22, 27, 28, 30, 33, 37, 47, 48, 49, 56, 58, 59, 63, 67, 75, 79, 80, 96, 98, 109, 112, 114, 123, 125, 126, 127, 128, 140, 145, 146, 147, 150, 151, 158, 160, 163, 165, 185, 193, 196, 197, 198, 205, 212, 214, 217, 219, 222, 226, 227, 229, 231, 233, 237, 242, 243, 244, 245, 248, 251, 255, 256, 258, 259, 261, 271, 276, 278, 281, 282, 284, 299, 309, 315, 320, 321, 324, 330, 340, 342, 346, 354, 360, 372, 373, **375**, 376, 378, 384, 388, 389, 391, 397, 398, 401, 403, 408, 410, 432, 436, 437, 446, 463, 464, 467, 468, 473, 479, 484, 495, 507, 508, 513, 516, 523, 528, 531, 536, 537, 555, 560, 566, 571, 583, 595, 613, 616, 619, 620, 623, 622, 625, 628, 629, 634, 640, 645, 646, 658, 664, 665, 668, 672, 684, 686, 687, 691, 693, 703, 706, 712, 713, 714, 721, 726, 728, 730, 732, 734, 736, 739, 740, 741, 745, 754, 758, 759, 764, 767, 771, 773, 775, 798, 799, 801, 805, 816, 819, 821, 827, 828, 838, 841, 848, 858, 859, 861, 870, 871, Against all the heresies, 342, 375, 819, 867, Against the heresies, 1, 138, 264, 281, 375, 413, 475
Heretical Monuments, **377**
Heribald, 669
Hermacoras, 64
Hermaius, 208
Hermas, Shepherd of, 25, 39, 47, 56, 61, 63, 121, 153, 160, 162, 163, 186, 222, 235, 243, 245, 259, 264, 287, 291, 316, 376, **377**, 391, 398, 454, 457, 472, 475, 519, 561, 609, 614, 615, 644, 648, 654, 667, 703, 715, 734, 740, 781, 782, 785, 792, 843
Hermenegild, **377**, 418, 478, 482, 542, 730, 774, 791
Hermes Trismegistus, 82, 92, 378, 695

Hermetism, 45, 82, 92, 199, **377**, 579, 603, 610, 805, 836, *see also* Corpus Hermeticum
Hermias, 31, 60, 74, 75, **378**, 689, 691
Hermippus the peripatetic, 121
Hermogenes, **378**, 429, 819, Against, 378, 819, 831
Hermopolis Magna, 265
Herod the Great, 5, 18, 138, 282, 323, **378**, 410, 431, 466, 510, 687, 755
Herodians, 384
Herodotus, 83
Hertford (672/673), 129, 824
Heruli, 179, 505
Hesychasm, Hesychast, 231, **379**, 442, 514, 556, 684, 708, 792, 872
Hesychius, Alexandrian exegete, **379**
Hesychius, exegete, **379**
Hesychius, lexicographer, 352, **379**
Hesychius of Alexandria, 352, **379**
Hesychius of Jerusalem, 81, 91, 312, **379**, 389, 450, 562, 634, 716, 723, 727
Hesychius of Miletus, 166, **379**, 654
Hesychius of Salona, **379**, 753
Hesychius the Sinaite, **379**, 708
Hesychius the homologete, 27
Heterodoxy, 625
Hexaemeron, 55, 114, 205, 287, 310, 340, 343, 363, **380**, 428, 475, 634, 736, 754, 792, 882
Hexapla, 310, **380**, 431, 478, 620, 716, 722, 764, 803, 810, 829
Hexham, 877
Heylin, Johann, 655
Hieracas of Leontopolis, 199, **380**
Hierapolis (Mabbug), 86, 87, 112, 142, 175, 647, 797
Hieratikon, 135
Hierius, 459
Hierocles Sossianus, 12, **380**, 586, 589, 603, 693, 805, Against Hierocles, 300, 380
Hierolyte, 404
Hierosolymitana, 496
Hierotheus, 239, 794, Book of, 623, 794
Hilarianus (Quintus Julius), **380**
Hilarianus Hilarius, 30
Hilarion of Gaza, **380**, 431, 532, Life of, 121, 787
Hilarus or Hilarius, 30, 79, **380**, 480, 486, 582, 642, 645, 646, 742, 780, 814, 863, 867, Life of, 381, 396
Hilary, Gallic poet, **381**
Hilary, presbyter, **381**
Hilary of Arles, 79, 120, 225, 295, **381**, 393, 396, 483, 511, 582, 615, 700, 737, 744, 753, 754, 767, 799, 858, 859, 861
Hilary of Pavia, 30
Hilary of Poitiers, 9, 26, 46, 57, 77, 78, 79, 105, 120, 121, 149, 154, 168, 177, 187, 225, 243, 261, 269, 293, 295, 300, 302, 311, 312, 323, 324, 338, 340, 350, 359, **381**, 401, 437, 449, 452, 453, 478, 497, 501, 522, 531, 543, 562, 567, 592, 612, 626, 649, 655, 656, 696, 700, 708, 709, 721, 722, 723, 736, 758, 788, 816, 817, 847, 863, 878, 881
Hilary the Luciferian, **382**
Hilda, 129, 137
Hildebert of Lavardin, 540, 729
Hilderic, 861
Himerius, 596, 714, 814
Himerius, sophist, **382**
Himerius of Nicomedia, 545
Himerius of Tarragona, 104, **382**, 742, 782, 876
Hincmar, 732
Hind, 65
Hinduism, 110, 185, 230
Hinton St Mary, Dorset, 130
Hippo, 2, 14, 15, 16, 27, 83, 98, 101, 142, 146, 147, 249, 328, 342, **382**, 486, 504, 546, 562, 572, 607, 653, 705, 713, 854, 861, *150*, C. of (393), 128, 146, 319, 383, 499, 541, C. of (427), 383

Hippodrome, 273, 296
Hippolytus, 36, 38, 42, 47, 55, 56, 57, 63, 75, 83, 103, 108, 109, 113, 121, 123, 134, 140, 145, 150, 157, 158, 159, 164, 166, 183, 187, 188, 189, 190, 205, 206, 219, 220, 232, 233, 244, 252, 258, 269, 278, 284, 285, 310, 311, 312, 315, 319, 333, 341, 373, 375, 376, 380, **383**, 385, 391, 410, 424, 436, 448, 465, 477, 485, 490, 492, 494, 498, 500, 501, 502, 532, 538, 543, 546, 558, 560, 561, 596, 598, 604, 611, 612, 615, 616, 619, 629, 640, 644, 647, 654, 655, 672, 689, 691, 693, 702, 706, 715, 716, 722, 728, 733, 738, 741, 744, 757, 764, 771, 786, 793, 798, 800, 801, 818, 839, 848, 851, 876, 879, 880, 885
Hippolytus, Ps.-, 60
Hippolytus, Statue of, 63, 153, 188, **384**, 385
Hippolytus, antipope, 646
Hippolytus of Bostra, **385**
Hippos-Sussita, 53, 636
Hîra, 65
Hispalis, 148
Hispana, 142, 147, 270
Hispana Canonical Collection, 714, 854
Hispanic liturgy, 83, 155, 228, 334, 335, 346, **385**, 405, 410, 459, 475, 485, 490, 493, 497, 499, 500, 582, 714, 853, 866, Hispano-Visigothic rite, 497
Hisperica Famina, 20, **388**, 412
Hisperism, 20
Hissar, 838
Historia acephala, 23, **388**, 828
Historia Augusta, 22, 121, 168, 223, **388**, 768
Historia Francorum, 284, 365, 389, 533, 671
Historia Gothorum, 150, 389, 451
Historia monachorum in Aegypto, 31, 61, 271, 374, **388**, 443, 445, 514, 567, 636, 646, 746, 841
Historia tripartita, 282, **388**, 389, 827
Historiography, Christian, 118, 166, 121, 166, 245, 299, **388**, 412, 451, 625, 644, 790, 885, of the papacy, 644, 841
Histria, 390, 765, *151*
Hodegos, 37
Hohenburg, 610
Hoischügel, 600
Holy Spirit, 5, 9, 24, 28, 31, 35, 36, 39, 42, 46, 52, 53, 54, 72, 76, 77, 78, 90, 94, 99, 100, 101, 103, 108, 114, 115, 117, 120, 121, 128, 151, 159, 160, 163, 164, 190, 195, 198, 205, 208, 215, 220, 221, 222, 223, 229, 231, 232, 243, 245, 247, 248, 250, 252, 257, 259, 260, 261, 262, 265, 268, 269, 277, 278, 282, 291, 293, 300, 303, 317, 321, 324, 340, 352, 354, 356, 358, 359, 361, 362, 363, 364, 372, 381, **390**, 395, 396, 398, 399, 403, 405, 406, 414, 430, 431, 449, 462, 467, 470, 473, 489, 490, 492, 493, 496, 509, 510, 516, 521, 526, 528, 537, 538, 543, 550, 553, 561, 562, 570, 573, 575, 595, 604, 606, 611, 616, 621, 622, 626, 652, 657, 669, 670, 696, 699, 700, 706, 707, 708, 715, 734, 744, 748, 750, 763, 767, 769, 780, 781, 786, 788, 792, 797, 798, 800, 801, 802, 803, 806, 821, 830, 831, 846, 848, 851, 852, 853, 854, 872, 877, 878, 882
Homiliary, 387, **393**, 394, 503, 655, 745, 769
Homily, 12, 50, 90, 181, 187, 200, 232, 273, 292, 298, 309, 311, 345, 393, **394**, 410, 436, 437, 475, 489, 498, 503, 504, 506, 507, 541, 668, 708, 715, 716, 727, 740, 744, 745, 764, 769, 802, 847
Homoeans, 4, 47, 49, 58, 78, 94, 113, 195, 215, 218, **395**, 550, 585, 767, 856
Homoeousians, 4, 5, 13, 47, 49, 77, 78, 94, 95, 113, 154, 195, 215, 268, 296, 301, 305, 343, 381, **395**, 396, 472, 507, 516, 521, 526, 527, 550, 582, 585, 657, 663,

767, 778, 779, 783, 787, 830, 832, 852, 854, 858
Homoiosis Theô, 45, **395**, 455 589, 592, 691, 694
Homoiousios, 77, 268, 305, 381, **395**, 396, 472, 516, 585, 737, 767, 783
Homoousians, 78, 381, **396**, 665
Homoousios, 5, 20, 47, 49, 62, 77, 78, 94, 95, 114, 129, 165, 207, 208, 215, 237, 268, 299, 302, 305, 350, 363, 381, 395, **396**, 516, 526, 538, 550, 566, 569, 585, 595, 604, 626, 652, 663, 685, 737, 757, 767, 778, 779, 783, 848, 852, 854
Honings, Bonifacio, 2
Honoratus Antoninus, **396**
Honoratus of Aquae Sirenses, 544
Honoratus of Bourges, 555
Honoratus of Lérins and Arles, 149, 295, 381, **396**, 483, 548, 705, 753, 767
Honoratus of Marseilles, 381, **396**
Honoratus of Vercelli, **396**
Honorius, Flavius, 26, 28, 67, 83, 96, 102, 147, 176, 178, 185, 194, 198, 224, 247, 273, 296, 324, 334, 337, 396, **397**, 419, 516, 521, 559, 607, 612, 613, 643, 645, 665, 729, 730, 743, 744, 768, 795, 826, 859, 861, 883, 638, 642, 645
Honorius, pope, 25, 59, 127, 197, 369, **397**, 433, 509, 530, 532, 568
Honorius III, 476
Honorius Scholasticus, **397**
Honorius of Autun, 654, 865
Hope, 56, 116, 219, 243, 260, 278, 280, 294, 309, 316, 323, 372, **397**, 461, 462, 560, 622, 668, 677, 748, 755, 788, 792, 803, 819, 851, 863
Hormisdas, 5, 36, 48, 104, 223, 237, 241, 251, 272, 281, **399**, 438, 443, 464, 476, 528, 553, 570, 642, 645, 742, 774, 779
Hormuz, 65, 832
Horrea (Hergla), 13
Horsiesi, 200, 400, 826
Hortensius, 75
Hosae, 253
Hosanna in excelsis, 6
Hospitality, Christian, 79, 91, 160, 230, 234, 260, **400**, 688, 707, 875
Hripsimè, 368
Huesca (598), 400
Hummidius Quadratus, 188
Hunaim ibn Ishaq, 66
Huneric, 78, 138, 143, 158, 193, 296, **400**, 544, 601, 757, 842, 861, 868, 870
Huns, 13, 96, 134, 217, 447
Hunterston brooch, 133
Hyacinth, 27
Hyginus of Cordova, 96, 121, 126, 158, **400**, 644, 850
Hylas, 252
Hymation, 253
Hymenaeus of Jerusalem, 48, **400**, 432
Hymn, Hymnology, 1, 3, 5, 25, 28, 29, 56, 106, 107, 110, 115, 118, 135, 137, 155, 160, 162, 222, 231, 261, 274, 276, 280, 291, 325, 335, 352, 370, 374, 387, **400**, 412, 465, 496, 497, 498, 499, 500, 501, 503, 504, 541, 595, 685, 707, 721, 722, 740, 765, 772, 810, 816, 856
Hypapante, 135, 493, 539
Hypatia, 7, 214, 806
Hypatius, 8, 22, 37, 140, 175, 238, **401**, Life of, 121, 140
Hypatius of Ephesus, 238, 339, **401**
Hyperchius, **401**
Hypogeum, see Catacomb, Cemetery
Hypostasis, 5, 12, 20, 23, 24, 42, 45, 49, 53, 58, 59, 76, 77, 78, 93, 159, 164, 165, 190, 196, 207, 214, 215, 233, 237, 268, 272, 274, 276, 296, 308, 313, 324, 325, 353, 358, 364, 374, 391, 395, **401**, 402, 443, 446, 455, 481, 519, 568, 569, 585, 586, 587, 588, 590, 594, 603, 606, 626, 663, 684, 690, 694, 695, 698, 716, 749, 757, 773, 825, 841, 851, 852, 853

Hypostatic union, 36, 59, **402**, 418, 443, 481, 538, 622, 675, 852
Hypotyposeis, 115, 180, 691, 830
Hypselis (Shotep), 83
Hypsistarians, 361, **402**

Iader, 217, 584
Iamblichus, 74, 75, 296, 363, 587, 589, 591, 603, 614, 695, 805, 835, 836
Iamona (Minorca), 107
Iao, 4
Ibas of Edessa, 5, 49, 112, 159, 197, 263, 275, 303, **403**, 570, 583, 667, 729, 810, 838, 855
Iberia, 344, **403**, 578
Ibligum (Invillino), 420
Ibn Qutaiba, 65
Icklingham, Suffolk, 129, 133
Iconium, 822
Iconoclasm, 646, 708
Iconography, Iconology, 26, 53, 68, 70, 81, 90, 96, 108, 132, 136, 149, 153, 202, 210, 228, 241, 252, 267, 285, 290, 291, 323, 329, 333, 336, 345, 349, 378, **403**, 419, 476, 477, 489, 509, 549, 550, 574, 607, 612, 630, 649, 655, 676, 686, 687, 705, 708, 731, 735, 755, 756, 774, 803
Icons, 632
Idanha-a-Velha (Beira Baixa), 512
Idatius (Hydatius, Ydacius) of Mérida, 166, 167, 400, **404**, 426, 554, 652, 711, 791, 859
Idatius and Ceponius, 854
Idatius the Romanophile, 801
Idolatry, 39, 43, 46, 60, 93, 102, 111, 167, 168, 181, 191, 226, 237, 270, 273, 327, 351, 355, **404**, 406, 407, 408, 415, 434, 445, 469, 528, 530, 596, 625, 689, 706, 736, 751, 757, 760, 780, 781, 801, 815, 819
Idolothytes, 16, **404**, 596
Idols, 4, 22, 23, 39, 82, 147, 226, 270, 273, 296, 402, 404, 428, 469, 473, 555, 596, 700, 721, 742, 764
Ignatius, Ps.-, 843
Ignatius of Antioch, 16, 35, 50, 51, 57, 84, 90, 92, 93, 123, 153, 160, 163, 186, 209, 244, 259, 262, 275, 287, 292, 308, 314, 316, 327, 358, 376, 390, 398, **404**, 408, 448, 454, 483, 515, 528, 537, 561, 615, 640, 649, 657, 665, 675, 686, 701, 706, 724, 732, 762, 788, 792, 798, 849, 851, 876, 877, 878
Ikalto, 345
Ildefonsus of Toledo, 55, 385, 386, 387, 389, **405**, 459, 497, 654, 728, 791, 844
Iliberis, 363
Ilice, 148
Illa del Rei, 107
Illumination, 39, 99, 101, 141, 239, 245, 313, 319, 358, 370, **405**, 457, 489, 519, 767, 788
Illus, 113
Illyricum, 197, 208, 217, 218
Image, 1, 5, 7, 9, 12, 39, 44, 45, 46, 52, 54, 81, 89, 90, 92, 99, 100, 101, 135, 136, 163, 164, 168, 192, 197, 207, 220, 232, 239, 241, 242, 243, 251, 262, 268, 288, 292, 301, 309, 340, 346, 353, 356, 358, 373, 395, 404, **405**, 413, 414, 434, 435, 443, 444, 462, 479, 481, 491, 496, 507, 510, 523, 526, 532, 533, 540, 553, 557, 573, 578, 582, 585, 587, 597, 598, 602, 603, 621, 622, 629, 632, 633, 634, 685, 687, 694, 704, 706, 708, 717, 719, 741, 749, 750, 764, 767, 778, 786, 789, 794, 830, 845, 852, 871, 872, 878
Immunities and Privileges of Clerics, 193, 198, 397, **407**, 556, 739
Impiety, 52, 73, 121, 271, **408**, 578, 629, 687, 691, 703
Improperia, **408**
Incarnation, 5, 6, 29, 33, 36, 41, 45, 57, 59, 81, 94, 112, 118, 123, 127, 136, 144,

Numbers in bold refer to the main entry on the subject. *Numbers in italics refer to illustration numbers (vol II, pages 951-1094)*

155, 163, 167, 191, 200, 202, 205, 206, 208, 219, 221, 224, 227, 231, 237, 239, 243, 244, 259, 260, 262, 275, 276, 282, 290, 293, 300, 317, 326, 336, 340, 355, 358, 359, 361, 364, 392, 396, 402, 406, 407, **408**, 414, 415, 416, 417, 438, 443, 448, 449, 452, 458, 467, 479, 511, 537, 539, 577, 583, 604, 621, 622, 629, 638, 651, 654, 660, 663, 670, 678, 683, 687, 720, 730, 734, 735, 749, 769, 786, 788, 798, 802, 819, 825, 830, 831, 833, 840, 851

Incense, 33, 61, 110, 221, 223, 271, **409**, 517
Incest, **409**, 528, 556
Inchleraun, 130
Index Oleorum, **409**, 420, 427, 680
India, 48, 60, 110, 112, 203, 307, 428, 496, 517, 519, 639, 674, 809, 840
Indicia, 559, 863
Indiculus de gratia Dei, 495, 708
Indition, **409**
Inebriation, 293, *see also* Drunkenness
Infensor et defensor, 616
Ingobert, 349
Inishmurray, 130
Initiation, Christian, 42, 62, 123, 150, 151, 190, 239, 242, 293, 342, 364, 370, **409**, 468, 477, 494, 497, 503, 504, 536, 541, 576, 577, 611, 667, 745, 749, 750, 751, 851
Innisfallen, 128
Innocent I, 20, 40, 41, 42, 48, 58, 102, 104, 142, 147, 149, 154, 218, 223, 225, 232, 315, **409**, 486, 491, 496, 515, 532, 595, 604, 616, 617, 636, 641, 643, 645, 646, 665, 704, 705, 738, 741, 742, 744, 758, 782, 826, 831, 844, 853, 863
Innocent of Maronea, **410**
Innocents, Feast of the, 163, 378, **410**, 466, 493, 536, 613
Inscription of Sertoria, 771
Instantius, 126, 410, 712, 838
Instituta regularia divinae legis, 461, 663
Interamnia Nahars (Terni), 423
Intercession, 10, 34, 265, 293, **410**, 496, 541, 752
Intercisa, 860
Interpolation, 19, 22, 31, 34, 179, 317, 335, **410**, 614, 725
Ioca Monachorum, 411
Iona, 18, 130, 132, 186, 307, **411**, 426, 486, 762, Chronicle, 412
Ipagro, 106
Ireland, 10, 19, 20, 126, 128, 130, 131, 143, 154, 155, 159, 184, 186, 189, 211, 258, 307, 324, 346, 351, 388, **411**, 497, 567, 633, 636, 653, 720, 747, 762, 765, *see also* Britain and Ireland
Irenaeus, 4, 9, 29, 38, 39, 41, 45, 47, 55, 56, 57, 61, 63, 64, 71, 72, 81, 90, 93, 107, 110, 113, 121, 123, 127, 145, 150, 158, 160, 163, 164, 183, 198, 204, 205, 206, 226, 242, 243, 245, 258, 259, 262, 264, 269, 271, 277, 281, 284, 285, 293, 306, 307, 309, 312, 313, 315, 316, 319, 320, 324, 328, 337, 341, 347, 353, 354, 355, 358, 370, 372, 375, 376, 383, 391, 395, 398, 404, 405, 406, 408, **413**, 421, 448, 449, 453, 454, 457, 467, 469, 475, 482, 483, 492, 495, 497, 506, 509, 510, 512, 517, 526, 537, 551, 553, 560, 561, 577, 596, 606, 609, 612, 616, 626, 629, 640, 644, 647, 649, 655, 656, 657, 682, 701, 707, 713, 715, 718, 719, 720, 730, 731, 732, 734, 735, 740, 743, 758, 762, 763, 769, 771, 775, 781, 788, 789, 790, 792, 796, 797, 798, 802, 816, 817, 818, 819, 820, 829, 832, 839, 843, 875, 876, 877
Irenaeus of Sirmium, **416**, 638
Irenaeus of Tyre, 372, **416**
Irene, 197
Irenim, 228

Irmscher, Johannes, 16, 38, 113, 115, 136, 162-63, 204, 216, 375, 397, 401, 464, 646-65, 473, 479, 522-23, 679, 682, 702, 713, 724, 725, 732, 822, 829, 868, 884
Isaac, 3, 30, 40, 148, 218, 294, 315, 341, 428, 433, 537, 683, 766, 856, 883
Isaac of Amida, **416**
Isaac of Antioch, **416**
Isaac of Edessa, **416**
Isaac of Nineveh, **416**, 445, 725, 810
Isaac of Persia, **416**
Isaac of Qalamun, 755
Isaac of Scete, **417**
Isaac of Tsurtav, 345
Isaac the Jew, **416**
Isaac the Persian, **417**
Isacius of Jerusalem, 484
Isaiah, 89, 236, 567, 855, Ascension of, 39, 57, 71, 410
Isaiah of Gaza, 31, 679, Life of, 884
Isaiah of Scete, 251, **417**
Isauria, 88, 808
Ischyras, 140, 551, 561
Iserninus, 411
Ishmael, 341
Ishobokht of Rev-Ardashir, **417**
Ishoyahb I, 11, 271, 374, 810
Ishoyahb III, 417
Ishozeka, 106
Isidore of Alexandria, **417**, 623
Isidore of Chios, **417**
Isidore of Pelusium, 56, **417**, 470, 496, 634, 806, 842, 843, 865, 866, 879, 881
Isidore of Seville, 42, 48, 52, 55, 71, 106, 118, 127, 144, 158, 161, 166, 167, 169, 176, 177, 182, 186, 232, 233, 269, 272, 277, 311, 312, 331, 342, 367, 386, 389, 405, 412, **418**, 465, 467, 478, 486, 497, 507, 542, 576, 606, 654, 655, 660, 702, 711, 722, 727, 736, 747, 770, 771, 774, 775, 783, 791, 844, 866, 868, Life of, 121
Isidore the Gnostic, **419**
Isis, 144
Iskan, 344
Islam, 48, 86, 208, 230, 289, 428, 674
Isna, 268
Isochristi, 623
Isocrates, 121, 469
Isodad of Merw, 881
Isola Sacra, 489
Israel, bishop, 19
Issue of blood, 407, **419**, 429, 466, 668, 831, *152*
Istanbul, 194
Istria, 419, *see also* Histria
Itacius, 126
Italy, 2, 7, 13, 19, 27, 28, 29, 30, 43, 69, 96, 104, 126, 142, 150, 167, 169, 170, 171, 173, 179, 180, 192, 193, 218, 224, 230, 242, 246, 249, 307, 336, 356, 386, **419**, 497, 545, 559, 565, 600, 604, 610, 617, 618, 641, 642, 667, 671, 678, 705, 714, 730, 741, 742, 743, 747, 756, 761, 772, 780, 784, 837, 838, 853, 863, 884, 885, *153-160*
Ithacius of Ossonuba, 404, 426, 711, 850
Ithamar, 397
Itineraries, 1, 128, 158, 209, 264, 278, 370, **426**, 427, 688
Itinerarium Burdigalense, 426, **427**, 688
Itinerarium/Peregrinatio Egeriae, 257, 496, 500, 501, *see also* Egeria
Itinerarium/Peregrinatio ad loca sancta, 263
Iulianum, C., 99
Iunca, 15, 172, 174, 175, 863, C. of (523), 702
Iuvavum (Salzburg), 600, 747
Ivo of Chartres, 62
Ixworth, 133
Izla, 4, 106, 296

Jaber, 66
Jacob, abbat, 688
Jacob, patriarch, 56, 111, 261, 276, 341, 351, **428**, 429, 442, 454, 477, 517, 573, 683, 687, 876, 885, *161*
Jacob Baradaeus, 263, **428**, 444, 445, 496, 570, 619, 662, 810, 823
Jacob of Edessa, 343, 416, **428**, 439, 609, 662, 810
Jacob of Hodata, 106
Jacob of Nisibis, 296, **429**, 598
Jacob of Sarug, 214, 312, **429**, 440, 452, 772, 794, 810
Jacob of Tsurtav, 345
Jacobite, 4, 11, 24, 50, 95, 107, 175, 238, 263, 416, 428, **429**, 445, 496, 570, 619, 626, 662, 674, 679, 768, 808, 809, 810, 823, 824, 826, 840, 853, Against the Jacobites, 442
Jacob's Ladder, 429, *162*
Jahballaha, 883
Jairus, **429**, 466
Jaldabaoth, 353
James of Cyrrhestica, 527
James of Jerusalem, 561
James the Great, 38, 176, **429**, 430, 432, 454, 456, 493, 532, 657, 671, 790, 803, 863, Protogospel, 41, 57, 364, 451, 647, 648, 723
James the Less, 62, **430**, 432, 493, 561, 779, 780
Januarian, **430**
Januarius and Companions, **430**, 438, 581, 772
Januarius of Cagliari, 757
Jarrow, 117, 119, 129, 131, 486
Jason, 27, 73
Java, 496
Jebel Nefusa, 16
Jebel Oust, 14
Jeremiah, 55
Jerez de la Frontera, 106
Jericho, 44, 160, 453, 635
Jerome, 5, 16, 18, 19, 27, 28, 30, 37, 41, 47, 55, 56, 57, 59, 60, 61, 65, 71, 72, 73, 82, 91, 92, 95, 105, 113, 116, 117, 118, 121, 122, 126, 141, 145, 149, 152, 154, 157, 159, 161, 166, 177, 182, 184, 186, 187, 192, 199, 212, 213, 217, 218, 225, 228, 230, 232, 233, 235, 236, 241, 242, 243, 244, 247, 249, 250, 257, 261, 269, 282, 298, 299, 300, 301, 302, 303, 305, 306, 311, 312, 315, 318, 319, 320, 323, 328, 340, 342, 347, 351, 361, 362, 371, 374, 380, 389, 395, 404, 405, 406, 419, **430**, 432, 438, 450, 452, 453, 463, 469, 470, 472, 474, 475, 478, 483, 496, 500, 503, 507, 508, 509, 511, 518, 521, 527, 528, 528, 543, 550, 560, 562, 563, 567, 584, 593, 600, 609, 612, 613, 616, 619, 620, 623, 624, 637, 639, 645, 649, 654, 655, 658, 659, 660, 662, 664, 666, 669, 680, 682, 685, 686, 687, 688, 704, 708, 712, 714, 716, 718, 719, 722, 723, 727, 733, 734, 736, 738, 744, 745, 746, 747, 759, 760, 763, 767, 775, 776, 785, 786, 787, 790, 794, 795, 796, 801, 803, 808, 813, 816, 818, 819, 824, 826, 829, 831, 833, 840, 843, 846, 849, 850, 854, 857, 864, 867, 869, 870, 879, 880, 881, 882
Jerome, Ps.-, 412, 878
Jerome of Jerusalem, **431**, 437
Jerusalem, 4, 5, 6, 18, 21, 33, 36, 38, 44, 47, 50, 52, 56, 62, 66, 69, 71, 73, 82, 91, 105, 111, 120, 122, 128, 134, 138, 141, 151, 152, 154, 167, 169, 174, 180, 186, 188, 203, 208, 210, 213, 215, 220, 224, 225, 232, 257, 259, 264, 268, 271, 279, 280, 284, 289, 290, 302, 305, 307, 369, 371, 375, 400, 409, 410, 427, 429, 430, **431**, 436, 438, 439, 456, 466, 484, 486, 490, 493, 495, 496, 499, 501, 504, 510, 518, 538, 560, 561, 564, 570, 610, 618, 619, 624, 625, 635, 638, 649, 657, 671,

682, 688, 698, 703, 708, 709, 712, 715, 722, 733, 756, 762, 763, 768, 779, 780, 787, 797, 809, 828, 838, 843, 856, 863, 867, 875, *163-165*, C. of (49), 432, C. of (335), *see* Dedication, C. of the, C. of (346), 546, C. of (357), 215, C. of (536), 678, liturgy, 345, 433, 500, 541, 688, 810
Jesus Christ, 7, 81, 87, 241, **434**, *166-171*
Jewish worship, 35, 47, 52, 60, 73, 237, 293, **435**, 437, 499, 503, 533, 715, 722, 749, 800
Jews, 10, 17, 24, 27, 30, 36, 47, 52, 53, 54, 57, 58, 61, 72, 73, 111, 120, 155, 160, 162, 163, 176, 178, 187, 188, 191, 201, 202, 219, 226, 245, 260, 261, 270, 276, 287, 300, 309, 316, 319, 320, 341, 354, 376, 382, 397, 398, 404, 405, 407, 408, 419, 429, 430, 431, 436, 438, 439, 495, 496, 499, 501, 504, 510, 518, 538, 560, 561, 564, 570, 582, 604, 614, 628, 635, 657, 659, 671, 678, 687, 688, 702, 710, 716, 720, 728, 734, 735, 748, 751, 755, 763, 768, 776, 780, 783, 791, 801, 805, 818, 824, 827, 833, 839, 840, 845, 851, 878, 880
Jews and Christians, 255, **436**, 550, 674, 774, 818, 863
Joannitae, **437**
Job, 28, 62, 66, 152, 219, 227, 233, 235, 310, 351, 374, 431, **437**, 477, 482, 611, 668, 680, 702, 756, 816, 825, 878, 885, *172*
Jobius, **438**
Jobius the Apollinarian, **438**
John, bishop of Lappa, 209
John I, 124, 125, 189, **438**, 581, 642, 645, 743, 642, 749, 828
John I of Naples, 430, **438**
John II, 197, 401, 410, **438**, 555, 642, 645, 656, 743
John II of Constantinople (the Cappadocian), 190, 242, 399, **438**
John II of Jerusalem, 46, 151, 204, 215, 241, 312, 407, 430, 433, **438**, 442, 494, 496, 507, 562, 623, 660, 704, 744, 746, 863, Against, 438
John III, pope, 642, 645, 743
John (III) of Alexandria, **438**
John III of Constantinople, 824
John IV, pope, **439**, 568, 642, 645, 725, 743
John IV of Constantinople, 197, 397, 667
John IV of Jerusalem, **439**
John IV the Faster, 200, **439**, 442, 771, Life of, 685
John V, 646, 743
John VI, pope, 646
John VI of Constantinople, **439**
John VII, 646
John VIII, 30, 62, 153
John Arkaph, 83, **439**, 551
John bar Aphtonia, 439
John bar Kursos, **440**
John bar Penkayè, 445
John Cadonatus, 49
John Cassian, see Cassian, John
John Chrysostom, 1, 4, 9, 16, 25, 32, 33, 39, 40, 41, 47, 48, 50, 51, 52, 56, 81, 83, 96, 102, 106, 107, 108, 110, 116, 121, 135, 149, 151, 152, 160, 163, 166, 178, 186, 187, 194, 196, 200, 204, 214, 220, 225, 232, 233, 236, 242, 243, 251, 262, 269, 273, 275, 277, 282, 286, 287, 291, 292, 293, 295, 306, 310, 312, 325, 337, 350, 358, 362, 372, 374, 379, 398, 406, 407, 409, 417, 430, 436, 437, **440**, 443, 449, 450, 452, 472, 475, 477, 483, 484, 490, 494, 496, 499, 500, 511, 514, 517, 521, 527, 528, 537, 543, 552, 561, 576, 577, 594, 597, 609, 611, 612, 616, 623, 633, 634, 636, 643, 646, 648, 655, 656, 658, 665, 670, 676, 681, 682, 686, 703, 704, 709, 713, 716, 719, 723, 737, 739,

751, 760, 761, 763, 768, 772, 776, 782, 784, 790, 793, 796, 805, 817, 824, 826, 827, 830, 831, 832, 843, 857, 863, 865, 868, 870, 876, 878, 879, 882, Life of, 437
John Chrysostom, Ps.-, 436, **442**, 716 723
John Climacus (or Scholasticus), 24, 37, 61, 62, 142, 231, 232, 306, 312, 379, 439, **442**, 443, 446, 473, 567, 585, 634, 662, 684, 708, 792
John Colobos, 83
John Damascene, 5, 6, 52, 63, 67, 73, 76, 84, 111, 135, 160, 202, 204, 235, 256, 359, 376, 405, 407, 437, **442**, 445, 521, 533, 538, 633, 654, 696, 707, 735, 748, 771, 775, 794, 852, 853
John Diacrinomenos, 389, **443**
John Grammaticus, 593, 769, Against, 773
John Khozibita, 138
John Lateran, 169
John Malalas, 102, 166, 303, 439, **443**, 446, 465, 472
John Mandakuni, 81, **443**
John Mark, 111, 213
John Maxentius, 165, **443**, 585, 764
John "Mediocris" of Naples, **443**, 723
John Moschus (Moschos), 4, 24, 61, 204, 363, **443**, 447, 540, 634, 787, 797
John Philoponos, 31, 36, 190, 439, 444, **446**, 519, 570, 794, 822, 853
John Primicerius, **447**
John Saba, 445
John Scholasticus, *see* John Climacus
John Scotus Eriugena, 139, 364, 475
John Silentiarius, 215, **447**
John Stobaeus, 795
John Talaias, **447**, 678
John Troglita, 202
John of Antioch, 5, 20, 21, 38, 49, 166, 214, 215, 238, 275, 304, **444**, 545, 546, 548, 550, 552, 594, 596, 641, 663, 677, 679, 713, 729, 744, 784, 827, 829, 884
John of Apamea, 49, **445**, 794, 810
John of Arbela, 11
John of Ashmunein, 219
John of Berytus, **444**
John of Beth Susan, 416
John of Biclaro, **444**, 548, 747, 791
John of Bostra, 501
John of Caesarea, 138, 165, **444**, 585, 662, 704, 773
John of Cak'nut, 443
John of Carpathus, **444**
John of Cellae, **444**, 826
John of Dalyatha, **445**
John of Elche, 399
John of Ephesus, 389, 440, **445**, 448, 505, 605, 810, 824
John of Euboea, **445**
John of Fécamp, 106
John of Germanicia, 546
John of Lycopolis, **445**, 485
John of Lydia, 588
John of Maiuma (John Rufus), 350, **445**, 679, 699
John of Nikiu, 24, **446**
John of Ocun, 81
John of Paralos, 219, **446**
John of Parentium, 773
John of Ravenna, 502, 780
John of Rhodes, 83
John of Rila, 634
John of Saint Martin, **446**
John of Scythopolis, 73, 115, **446**, 585
John of Seville, 26
John of Stonyhurst, Gospel of, 131
John of Syracuse, 807, 866
John of Tarragona, 350
John of Tella, 810
John of Thessalonica, **446**
John of Tomi, 443, **446**
John of Tritenheim, called Trithemius, 654

John of Zagylis, 487
John of the Cross, 101
John of the Goths, 27
John the Almsgiver, 444, **447**, 564, 618, Life of, 481
John the Baptist, 38, 66, 84, 108, 135, 199, 230, 235, 251, 269, 333, 374, 401, 446, **447**, 466, 510, 518, 536, 562, 575, 635, 661, 716, 764, 817, 875, *173*
John the Deacon, 6, 152, **448**, 545, 868
John the Egyptian, **448**
John the Evangelist, Apocryphon of, 12, 109, 448, 579, 603, 771, Gospel of, 26, 38, 55, 85, 86, 91, 123, 158, 163, 176, 187, 204, 205, 214, 226, 235, 245, 257, 258, 275, 284, 293, 306, 310, 313, 314, 316, 324, 356, 370, 374, 390, 391, 429, **448**, 472, 475, 489, 493, 511, 552, 599, 619, 620, 647, 650, 657, 671, 680, 698, 706, 712, 723, 724, 727, 743, 751, 752, 755, 763, 777, 802, 821, 824, 825, 831, 840, 842, 850, 851, 852, Acts of, 41, 271
John the Exarch, 634
John the presbyter, 73, **449**, 647, 701
Jonah, 64, 66, 217, 231, 309, 323, 349, 351, **449**, 489, 572, 573, 615, 629, 630, 635, 668, 755, 763, 778, 825, 860, 885, *174, 175*
Jonas Khelasvili, 345
Jonas of Bobbio, 187, 397, **451**, 862
Jordanis (Jordanes), 150, 389, **451**
Josaphat and Barlaam, Life of, *see* Barlaam and Josaphat, Life of
Joseph, 47, 57, 107, 161, 309, 326, **451**, 452, 528, 539, 583, 675, 687, 710, *176*
Joseph Barsabbas, 73
Joseph Hazzaya (Joseph-'Abdiso' Hazzaya, or Hazzaia), 106, 238, 445, 598
Joseph Melodus, 25
Joseph of Syracuse, 135
Joseph of Thessalonica, 135
Joseph the Hebrew, **452**
Josephus, Flavius, 72, 121, 161, 430
Joshua, 141, 309, **453**, 573, 575, *171*
Joshua Scroll, 453, 675
Journeys, 57, 98, 147, 176, 179, 264, 284, 383, 388, 400, 426, **453**, 509, 510, 688
Jovia (modern Ludbreg), 759
Jovian, 24, 49, 58, 93, 113, 276, 305, **454**, 598, 680
Jovinian, 116, 145, 340, 405, **454**, 559, 638, 666, 776, Against, 454
Jovinian, monk, 744
Jovinus, 134, **454**
Juansher, 19
Juba II, 138
Jubaianus, 109, 146, 211, 229
Jubeiha, 66
Jubilees, Book of, 56, 291, 718
Judaea, Desert of, 432, **454**, 567
Judaeo-Christianity, 16, 22, 26, 31, 35, 38, 39, 47, 54, 55, 57, 61, 66, 67, 73, 91, 108, 109, 155, 159, 163, 169, 179, 180, 210, 222, 223, 226, 227, 229, 233, 234, 258, 259, 264, 269, 276, 293, 309, 316, 360, 391, 395, 430, 436, 450, **454**, 470, 494, 519, 584, 609, 635, 640, 647, 657, 658, 671, 674, 703, 706, 740, 748, 763, 773, 779, 780, 781, 792, 803, 808, 809, 810, 815, 818
Judaeo-Hellenism, 23, 25, 180, 181, 240, 257, 315, 360, 372, 380, **455**, 558, 585, 650, 682, 689, 763, 805
Judah, Rabbi, 563
Judaism, 6, 13, 38, 53, 55, 72, 83, 84, 111, 155, 161, 163, 168, 198, 199, 210, 235, 245, 257, 271, 309, 313, 315, 320, 340, 342, 355, 360, 361, 376, 388, 390, 397, 408, 436, 459, 461, 477, 489, 517, 537, 560, 572, 576, 577, 649, 669, 671, 672, 674, 701, 705, 716, 717, 718, 722, 733, 739, 762, 763, 764, 801, 805, 810, 813, 838, 839, 843, 851, 860, *see also* Jewish worship

Judaizers, 30, 208, 432, **456**, 523, 659, 723, 748
Judas Iscariot, 780, 849, *178, 179*
Judas Thaddaeus, **456**, Gospel of, 57, 139
Judas son of James, 780
Judas the Galilaean, 333
Judgment, 34, 126, 144, 179, 188, 192, 220, 227, 259, 260, 269, 285, 286, 287, 288, 291, 341, 353, 372, 380, 453, **457**, 462, 463, 467, 469, 560, 614, 651, 668, 703, 721, 725, 733, 735, 778, 781, 796, 802, 812, 819, 839, 840, 873, *180*
Julia Castaphronia, 103
Julia Domna, 60, 143, 768, 799
Julia Dryadia, 103
Julia Mammaea, 22
Julian, anomoean, 62
Julian, martyr, 175
Julianists, 54, 428
Julian of Atramitium, 401
Julian of Brioude, **458**, 846
Julian of Cos, **458**
Julian of Eclanum, 72, 82, 99, 175, 311, 320, 359, 447, **458**, 538, 661, 665, 666, 706, 716, 723, 776, 784, 825, 832, 849, 876, 878
Julian of Halicarnassus, 8, 49, 54, 200, 333, 439, **458**, 460, 480, 569, 570, 662, 773, 841
Julian of Seville, 845
Julian of Toledo, 311, 335, 367, 387, 405, **458**, 582, 791, 844, 845
Julian Sabas, 782
Julian the Apollinarist, **459**, 700
Julian the Apostate, 13, 24, 26, 30, 47, 51, 58, 59, 78, 83, 91, 93, 96, 106, 120, 138, 168, 192, 198, 214, 219, 222, 236, 247, 268, 296, 299, 302, 325, 329, 334, 339, 356, 381, 387, 406, 407, **459**, 484, 497, 508, 517, 518, 590, 651, 681, 695, 714, 738, 760, 763, 799, 805, 832, 836, 843, Against the Galilaeans, 214, 459
Julian the Arian, 62, 404, **460**, 878
Julian the Cappadocian, 714
Julianus Argentarius, 730
Julianus Diaconus, 879
Julianus Valens, 64
Julitta, 137
Julius I, 42, 49, 71, 77, 93, 229, 278, 302, 325, **460**, 485, 522, 546, 626, 641, 645, 646, 741, 742, 743, 757, 842, 859
Julius II, 774
Julius Africanus, 22, 56, 72, 110, 166, 299, 373, 374, 451, **460**, 620, 724, 790, 801
Julius Capitolinus, 388
Julius Cassianus, **460**, 717, 718, 719, 775
Julius Rufinianus, 52
Julius of Aqfahs, **461**, 534
Julius of Puteoli, **461**
Junilius, 3, 311, **461**, 663
Junius Bassus, 169
Jung Codex, 374
Justice, 1, 3, 25, 98, 117, 126, 138, 164, 179, 192, 226, 239, 248, 286, 287, 288, 289, 318, 351, 355, 357, 359, 372, 397, 399, 408, 433, 457, **461**, 462, 505, 516, 564, 629, 670, 703, 736, 781, 788, 819, 843, 872
Justification, 100, 101, 288, 316, 357, **462**, 526, 657, 658, 709, 718, 758, 782
Justin martyr, 6, 9, 23, 35, 39, 41, 45, 55, 56, 57, 60, 63, 72, 73, 84, 89, 90, 107, 110, 111, 121, 134, 151, 155, 160, 163, 168, 176, 178, 180, 188, 190, 197, 205, 206, 209, 226, 232, 233, 236, 242, 243, 244, 245, 251, 252, 262, 269, 285, 287, 292, 306, 309, 312, 316, 320, 325, 333, 354, 372, 373, 375, 376, 391, 394, 395, 408, 410, 415, 432, 436, 438, 440, 448, 449, 453, 454, 455, 457, 462, **464**, 472, 489, 494, 501, 504, 517, 524, 526, 532, 533, 534, 537, 541, 543, 551, 552, 560, 563, 564, 570, 577, 578, 609, 612, 614,
615, 640, 644, 657, 662, 665, 672, 683, 687, 689, 690, 693, 695, 709, 713, 715, 718, 724, 727, 729, 731, 732, 734, 735, 741, 744, 749, 750, 760, 761, 762, 785, 792, 796, 797, 800, 802, 805, 815, 816, 817, 821, 827, 851, 876, 877, 878
Justin, Ps.-, **464**, 691, 727
Justin I, 36, 37, 124, 193, 195, 201, 399, 438, **464**, 642, 684, 829
Justin II, 36, 202, 305, 363, **464**, 491, 570, 598, 729, 824, 838
Justina, 28, 33, 105, **464**, 521, 858
Justina of Padua, **464**
Justinian I, 4, 5, 16, 21, 36, 43, 48, 49, 51, 54, 56, 67, 85, 86, 96, 116, 119, 120, 136, 139, 140, 142, 154, 162, 165, 168, 176, 183, 184, 194, 195, 197, 201, 203, 217, 219, 224, 244, 245, 251, 267, 271, 283, 305, 306, 315, 339, 345, 352, 357, 379, 399, 410, 432, 433, 438, 439, 445, **464**, 473, 484, 487, 506, 514, 531, 553, 566, 570, 571, 573, 585, 592, 596, 597, 598, 605, 607, 620, 622, 623, 638, 642, 645, 662, 663, 664, 666, 678, 702, 705, 709, 730, 738, 743, 748, 757, 761, 773, 778, 779, 782, 806, 822, 823, 826, 828, 829, 838, 850, 853, 861, 868, 870
Justinian II, 140, 194, 197
Justinian of Valencia, 465, 791
Justin the Gnostic, **465**
Justus, **465**, 551
Justus and Pastor of Complutum, 582
Justus of Canterbury, **465**
Justus of Toledo, **465**
Justus of Trieste, **465**
Justus of Urgel, **465**, 466, 497, 787, 791, 844, 881
Juvenal, **466**
Juvenal of Jerusalem, 159, 163, 275, 314, 350, 423, 432, **466**, 548, 618, 635, 677, 744, 795, 828
Juvencus, 202, 311, **466**, 651, 721, 766, 790

Kairouan, 13, 14, 16
Kaisun, 238
Kalliana, 517
Kalykadnian Seleucia (now Silifke), 808
Kamulia, 7
Kandys, 253
Kannengiesser, Charles, 9, 20, 21, 58-59, 195-96, 268-69, 276, 301-2, 522, 595, 609-10, 687
Kantaeans, 823
Kaoussié, 174
Kaphar-Gamala, 105
Karm al-Ahbariya, 268
Karm el-Arabis, 271
Karmoutz, 24
Kefer Kama, 636
Kekkut, 638
Kelibia (Cape Bon), 14, 15, 16, 174, 572
Kellia, 22, 265, 266, 268, 306
Kells, 130, 132, Book of, 130, 131, 132, 133, 412
Kentigern, 762, 771
Kephalaia, 52, 216, 234, Gnostica, 306, 623
Kepht (Koptos), 689
Kerak, 636
Kerygma, 35, 150, 206, 244, 262, 316, 401, **467**, 521, 523, 647, 760, Iohannis, 446, Petri, 57, 176, 179, 675, 818
Kestoi, 460
Keturah, 3
Kevin, 130
Khalasa, 636
Khenchela, 15, 228
Khirbet es-Suq, 66
Khirbet Hâss, 51
Kherbet-Oum-el-Ahdam, 676, 762
Khirbet Samra, 66
Khosrovidukht', 368
Kilcullen, 411

Numbers in bold refer to the main entry on the subject. *Numbers in italics refer to illustration numbers (vol II, pages 951-1094)*

Kildare, 130, 411, Chronicle, 413
Killashee, 411
Kiss, 33, 50, 90, 223, 231, 290, 386, 410, 467, 491, 494, 492, 541, 665, *181*
Kiti, 213
Klapavica, 217
Klasis, 16
Klijn, A.F.J., 159, 258-59, 269-70, 454, 584, 803
Knik' Hawatoy (Seal of faith), 81, 385, **467**, 655
Knossos, 208, 688
Knowledge of God, 355, 358, 696, 733
Koblenz (Confluentes), 348
Koha, 766
Koiné, **467**, 472, 509, *see also* Greek, Christian
Koinonia, 188, 242, 467, 500, 567, 643
Kokanâya, 51
Kollwitz, J., 69
Kore Kosmou, 378
Koriatha, 66
Koriun, 81, **468**, 482, 556
Korycos (now Narli Kuyu), 175
Krithénion, 22
Kum, 19
Kumi, 403
Kune, 2
Kurion, Cyprus, 213
Kuzlü Kule, 175
Kyriainè, 175

La Alberca (Murcia), 148
Labarum, 607
Labate, Antonio, 21, 67-68, 204, 303, 343, 346, 360-61, 374-75, 375, 379, 439, 445, 593, 596, 610, 639, 640, 683-84, 768, 785, 790, 797, 798, 826-27
Labbé, Ph., 654
Labour, 22, 119, 150, 151, 266, 287, 381, **469**, 478, 549, 556, 567, 703, 708, 717, 736, 746, 772
Labubna son of Sennak, 10
Lacerna, 253, 254
La Cocosa, 106
Lactantius, 8, 55, 56, 57, 60, 82, 90, 121, 122, 144, 177, 191, 263, 272, 286, 288, 311, 347, 351, 355, 372, 388, 406, 453, **469**, 560, 563, 609, 614, 640, 649, 655, 656, 665, 696, 720, 721, 752, 758, 760, 769, 785, 788, 801, 804, 875, Divinae institutiones, 60, 82, 177, 469, 696
Lactantius, Ps.-, 685
Ladocsi, Gaspar, 18, 63, 65, 161-62, 257, 298, 302, 340, 371, 375, 379, 505, 521-22, 564, 637, 662, 665, 670, 677, 702, 704, 725, 779, 801, 821, 822-23, 823, 828, 829, 830, 831, 833, 838, 851, 877, 883
Laetare Sunday, 186
Laidcend mac Bairchedo, 413
Laidcenn mac Baith Bannaig of Clonfertmulloe, 412
Laïty, *see* Lay
Lake Triton (Chott el Djérid), 13
Lalibelà, 289, 290
Lamb, 1, 2, 10, 17, 62, 64, 123, 221, 257, 258, 291, 293, 294, 306, 312, 380, 407, 408, 420, 423, 436, 447, 453, **470**, 492, 503, 504, 573, 600, 675, 715, 728, 764, 777, 801, 802, 803, 812
Lambaesis, 13, 193, **470**, 607, 840
Lambert, **471**
Lamp, 1, 3, 64, 68, 70, 105, 123, 141, 157, 173, 217, 218, 219, 232, 323, 369, 409, 450, **471**, 477, 489, 491, 518, 535, 599, 600, 611, 639, 664, 753, 759, 765, 808, *182, 183*
Lampsacus, C. of (364), 78, 114, 303, **472**, 832, 858
Landulf the Wise, 389
Langres, 875
Languages of the Fathers, **472**
Laodicea (Latakieh), 37, 58, 85, 193, 232, 237, 258, 301, 307, 343, 665, 670, 686, 808, 865, C. of (c.380), 8, 32, 116, 141, 142, 152, 205, 274, **472**, 617, 759, C. of (481), 473
Laoghaire, 411
Laon (Laudunum), 633
La Peñuela, 106
Lapithos, 213
Lapsi, Problem of the, 11, 13, 23, 61, 145, 146, 189, 202, 212, 248, 250, 261, 270, 321, 467, **473**, 483, 484, 507, 522, 523, 551, 585, 603, 604, 628, 640, 645, 668, 677, 741, 743, 759, 806
Larissa, 835
La Skhira, 15
Las Tamujas de Malpica de Tajo (Toledo), 148
La Vega del Mar, 106
Las Vegas de Puebla Nueva (Toledo), 148
Lastingham, Yorkshire, 159
Late antiquity, Late antique, 25, 52, 51, 70, 88, 121, 126, 191, 194, 199, 224, 266, 267, 273, 296, 352, 371, 454, **473**, 560, 579, 585, 629, 642, 681, 687, 723, 795, 803, 820, 836
Lateran councils, *see* Rome
Laterculum, 295
Laterculus, 96, 754, Veronensis, 834, 837
Latin, Christian, 9, 361, 394, 472, **474**, 498, 604, 646, 656, 736, 759, 763, 769, 820
Latin Translations of Greek Texts, 2, 10, 26, 38, 41, 44, 60, 63, 180, 229, 235, 237, 257, 264, 282, 301, 302, 311, 339, 346, 362, 371, 372, 378, 413, 442, 472, **474**, 494, 496, 508, 567, 620, 655, 658, 693, 696, 701, 746, 764, 785, 786, 790, 799, 870, 880
Latium and Campania, 423
Latopolis, 628
Latrocinium, *see* Ephesus, C. of (449)
Latronianus, **475**, 712
Latrun, 488
Laudes Eunomiae, **475**
Laumellum (Lomello), 421
Laura, 44, 135, 160, 167, 215, 265, 266, 304, 363, 454, 447, **475**, 480, 797, 823, 826, 829
Laus Pompeia (Lodi Vecchio), 420
Laus tibi, Christe, 6
Lausanne (Lausannum), 525
Lausiac History, 61, 121, 271, 303, 374, 388, 567, 636
Lavant, 600
Lavenant, René, 54, 110, 263, 416, 583, 598
Lawrence, antipope, 36, 365, 399, **476**, 521, 642, 646, 652, 678, 730, 744, 804, Lawrentian schism, 149, 476, 645, 744
Lawrence, bishop of Novae, **476**
Lawrence, Roman martyr, 17, **475**, 536, 573, 574, 741, 784
Lawrence of Canterbury, **476**, 551
Lawrence of Milan, 29, 67, 272, 421, **476**
Lawrence of Novara, 337
Laying on of hands, 13, 42, 103, 108, 151, 152, 183, 190, 221, 223, 229, 232, 473, **477**, 482, 551, 561, 562, 604, 616, 667, 668, 745, 751, 882, *184*, *see also* Orders, ordination
Lay, Layman, Laïty, 17, 21, 48, 79, 91, 112, 116, 123, 146, 182, 188, 208, 211, 224, 225, 248, 249, 259, 268, 338, 367, 393, 400, 407, 412, 473, **476**, 500, 530, 551, 555, 556, 609, 616, 619, 647, 668, 710, 714, 737, 741, 754, 761, 785, 791, 792, 793, 794, 822, 858, 882
Layos, 148
Lazarus, 38, 79, 180, 200, 330, 349, 433, 466, 471, **477**, 480, 518, 530, 629, 706, 755, 871, *185*
Lazarus of Pharp, 4, 81, 320, **477**, 482
Lazio, 423
Lazzarus of Aix, 241
Leander of Seville, 106, 307, 331, 366, 367, 377, 386, 418, **478**, 485, 488, 497, 499, 730, 743, 747, 774, 775, 791, 844, 878
Leanza, Sandro, 305-6, 525, 453, 878-81
Lebbaeus, 487
Le Deaut, Roger, 32, 558-59, 563-64, 812-13
Le Kef, 15
Lechaeum, 7, 201
Lectio divina, 119, **478**, 549, 650, 661, 708, 721
Lector (reader), 232, 504
Lécuyer, Joseph, 311-13, 560-62
Le Blant, G., 692
Lekit, 19, 81
Le Mans (Cenomannis), 847
Le Mas d'Aire, 3
Lemellef, 15
Lemsa, 16
Leningrad Bede, 132
Lent, 5, 6, 50, 135, 150, 151, 152, 186, 227, 257, 298, 319, 351, 493, 504, 548, 660, 745, 792, 865
Lentini, 425
Leo I, Eastern emperor, 44, 52, 113, 220, 233, 407, **479**, 487, 677, 679, 840, 884, 594, 618, 643, 645, 646, 652, 678, 684, 702, 717, 725, 730, 738, 741, 744, 745, 747, 758, 767, 828, 845, 861
Leo I the Great, 4, 5, 6, 25, 35, 36, 37, 47, 72, 96, 101, 104, 120, 122, 149, 159, 160, 165, 177, 182, 185, 189, 196, 197, 208, 211, 218, 225, 242, 243, 244, 245, 249, 257, 262, 275, 301, 305, 319, 325, 330, 339, 342, 380, 381, 394, 410, 438, 458, 461, **479**, 484, 487, 495, 499, 503, 514, 520, 522, 523, 543, 559, 562, 569, 576, 615, 616, 641, 669, 676, 705, 731, 742, 743, 750, 753, 758, 765, 769, 770, 773, 788, 798, 827, 831, 840, 841, 842, 844, 850, 854, 857, 859, 863, 869, 870, sacramentary, 496, 500, 502, 545, 745, 863
Leo II, 397, **479**, 645, 743
Leo III, pope, 864
Leo III the Isaurian, 208, 346, 407, **479**, 540, 646, 743, 833
Leo VI the Wise, 143, 816
Leo IX, 62
Leo XIII, 124
Leo of Bourges, **482**
Leo of Isauria, rhetor, 713
Leo of Narbonne, 298
Leo of Sens, **482**
Leo the Armenian, 407
Leocadia, 844
Leodegar of Autun, 105, 338, **480**, 700
Leonard, **480**
León-Astorga (Asturica), 334
Leonidas, 201
Leonides, 162, **480**, 552, 619, 768
Leontian martyrs, 482
Leontius, bishop of Fréjus, 149
Leontius of Ancyra, 37
Leontius of Antioch, 13, 185, **480**
Leontius of Apamea, 487
Leontius of Arabissus, **480**
Leontius of Arles, 79, 380, **480**
Leontius of Bordeaux, 556
Leontius of Byzantium, 42, 52, 54, 56, 233, 244, 245, 272, 339, 438, **480**, 481, 638, 747, 826, 852, 859
Leontius of Byzantium, Ps.-, 437
Leontius of Caesarea in Cappadocia, 80, **480**
Leontius of Constantinople, 442, **481**, 773, 841
Leontius of Jerusalem, 165, 244, 272, 480, **481**, 570, 585
Leontius of Magnesia, 275, 314
Leontius of Neapolis, 437, 447, **481**
Leontius of St Saba, 362, 480
Leontius of Tripoli, **481**, 684
Leontius the Armenian, historian, **482**
Leontius the Armenian, martyr, **482**
Leontocephalus, 564
Leontopolis, 85
Leovigild, 148, 356, 377, 418, 444, 478, **482**, 542, 554, 663, 730, 774, 791, 844
Lepcis Magna, 15, 169, **482**
Leper, 370, **482**, *186*
Leporius, 165, **482**
Lérida (Ilerda), **483**, 791, 813
Lérins, 79, 138, 154, 208, 295, 307, 338, 396, **483**, 567, 582, 653, 704, 746, 753, 754, 761, 767
Lesteb, 66
Letter, 2, 56, 79, 106, 120, 146, 196, 203, 224, 260, 252, 286, 315, 318, 351, 370, 400, 477, **483**, 484, 504, 619, 725, 827, 882
Letter of the church of Lyons and Vienne, 96, 189, 337, 413, **483**, 497, 512, 532, 533
Letters of Communion, **484**, of presentation, 400, of the popes, 646
Leucius Carinus, 91, **484**
Leudast of Tours, 556
Leukios Charinos, 850
Leviathan, 612
Libanius, 47, 305, 440, **484**
Libellatici, 61, 146, 223
Libellus, 61, 249, 335, 399, 473, **484**, 499, 507, 508, 530, 546, 745, 746
Liber Diurnus, 367, **485**, 497
Liber Genealogus, 247, **485**
Liber Graduum, **485**, 717, 723, 810
Liber Monstrorum, **485**
Liber orationum Psalmographus, 386, 387, **485**
Liber Pontificalis, 17, 120, 141, 211, 218, 225, 246, 284, 314, 315, 380, 396, 471, 476, **486**, 496, 500, 521, 554, 580, 617, 644, 667, 678, 784, 864, 865, 866, 885
Liber Regularum, 55, 311, 838
Liberatus of Byzacena, 148
Liberian catalogue, 166, 486
Liberius, 77, 78, 218, 228, 302, 328, **485**, 486, 508, 559, 585, 641, 643, 645, 646, 737, 741, 742, 778, 779, 832, 854, 856, 870, 872
Libosus of Vaga, 584
Libraries, Christian, 5, 8, 14, 21, 138, 150, 200, 255, 272, 299, 305, 366, 380, 384, 393, 404, 418, 423, 430, 459, **486**, 476, 519, 579, 638, 655, 722, 759, 761, 873
Libya, 13, 20, 23, 169, 224, 264, **486**, 489, 618, 748, *187-190*
Licentius, **488**, 739
Lichfield, 159
Licinianus of Cartagena, **488**, 774, 791
Licinius, 21, 47, 166, 192, 193, 263, 302, 470, **488**, 515, 559, 673, 682, 764, 845, 853
Lièges, 471
Life, 42, 44, 52, 76, 100, 116, 120, 122, 123, 127, 128, 151, 159, 161, 164, 181, 186, 192, 199, 206, 209, 210, 220, 225, 239, 242, 257, 279, 281, 284, 286, 288, 309, 316, 319, 327, 340, 351, 358, 361, 391, 392, 397, 398, 406, 414, 415, 461, 462, 467, 470, **489**, 492, 526, 577, 610, 614, 621, 625, 649, 666, 670, 687, 703, 707, 708, 709, 718, 725, 732, 737, 749, 758, 763, 767, 776, 780, 781, 786, 792, 798, 803, 816, 872, 875, 876, 878
Life of Christ (Monotessaron), 234
Light, Symbolism of, 32, 34, 39, 42, 53, 94, 108, 109, 116, 164, 187, 192, 204, 206, 210, 226, 239, 240, 255, 261, 262, 279, 294, 308, 340, 352, 353, 359, 397, 401, 405, 470, **489**, 493, 510, 518, 576, 577, 621, 626, 649, 670, 685, 692, 718, 725, 733, 783, 787, 800, 803, 815, 830, 851
Lighthouse, 441, **489**, 675, 787, 803
Ligugé, 224, 531

Liguria and Aemilia, 420
Lilla, Salvatore, 12-13, 20-21, 31, 31-32, 53-54, 73-76, 92-93, 204, 204-5, 238-40, 372-74, 454-56, 465, 517, 557-58, 584-85, 585-93, 601-3, 606, 610-11, 624, 639, 639-40, 689-98, 699, 713, 785, 805-6, 832, 835-36
Lilybaeum, 652
Limoges (Lemovicum), 747
Lincoln (Lindum), 129
Lindisfarne, 18, 129, 130, 211, 486, 762, Gospels, 131, 132, 133, 633
Linus, 644
Lipari islands, 112
Lisbon (Olisipo), 706
Literary genres, 56, 57, 66, 74, 121, 135, 150, 165, 166, 192, 201, 233, 276, 278, 284, 291, 317, 345, 351, 370, 378, 379, 426, 436, 455, 460, 461, 483, **489**, 506, 533, 535, 558, 563, 579, 582, 597, 629, 646, 650, 655, 661, 699, 701, 705, 717, 718, 722, 725, 747, 769, 826, 872
Litorius, 846
Littaeus of Gemellae, 584
Liturgical books, 29, 107, 370, 484
Liturgical Dialogue, **490**
Liturgical furniture, 133, 218, 290, 323, 334, 471, **490**, 607, 764, 813, *191-201*
Liturgical gestures, 490, **492**
Liturgical language, 498, 541
Liturgical year, 29, 50, 82, 123, 135, 298, 335, 351, 387, 393, 394, 479, **492**, 494, 497, 503, 548, 678, 744
Liturgy, 6, 8, 29, 33, 37, 39, 57, 63, 68, 71, 82, 90, 101, 106, 107, 110, 118, 120, 132, 134, 135, 141, 150, 151, 152, 155, 159, 160, 170, 186, 187, 188, 190, 199, 215, 218, 226, 230, 232, 238, 241, 245, 247, 248, 249, 257, 258, 264, 274, 285, 290, 291, 292, 293, 307, 309, 313, 315, 337, 338, 345, 352, 355, 360, 370, 374, 385, 394, 401, 403, 408, 409, 410, 418, 428, 432, 454, 467, 472, 474, 475, 478, 479, 485, **493**, 494, 503, 504, 506, 528, 540, 541, 547, 548, 549, 562, 567, 577, 583, 595, 609, 614, 616, 629, 633, 646, 656, 664, 665, 668, 671, 674, 678, 696, 708, 709, 710, 721, 722, 725, 727, 728, 731, 736, 741, 742, 746, 750, 752, 764, 766, 770, 787, 788, 790, 792, 793, 797, 800, 810, 825, 829, 842, 848, 851, 853, 862, 869, 875, 877, Liturgy of the hours, 499, *see also* Acclamations, Anaphora, Eucharist, Mass
Liturgy and Bible, **503**
Liutprand, **505**, 653
Liutulf of Mainz, 774
Livia, 832
Llantwit Major, 131
Lmbat, 80
Lodève (Luteva), 581
Lodi, 116
Logos, 10, 40, 50, 163, 165, 181, 448, 449
Logos-theology, 164, 507, 566, 748, 797, 852
Loi, Vincenzo, 18, 41, 71, 97, 121-22, 125, 188-89, 237, 258, 326, 330, 380, 409, 464-70, 474, 485, 498, 548, 657, 700, 720, 720-21, 728, 752, 812, 868
Loja, 107
London (Londinium), 96, 129
Long Brothers, 282, 623, 831
Longinus, **505**, 588, 704, 822, 824, 826, 829
Longobards (Lombards), 29, 78, 103, 158, 189, 253, 254, 307, 322, 366, 420, 421, 422, 424, 464, 486, **505**, 525, 642, 643, 645, 645, 646, 653, 667, 743, 745, 774, 822, 850
Lord's Prayer, 32, 123, 127, 151, 152, 242, 252, 258, 260, 320, 364, 398, 409, 436, 490, 493, 494, 496, 498, 503, **506**, 541, 620, 678, 706, 707, 708, 781, 819, 825

Lorica, 254
Lorium, 422
Loros, 253
Los Palacios, 106
Lothar, *see* Clotaire
Lot, 231, 400, **506**, 525, 785, *202*
Louis II, 26
Love, 2, 16, 34, 84, 100, 101, 116, 117, 119, 145, 199, 211, 239, 242, 255, 262, 279, 286, 287, 288, 289, 294, 330, 358, 398, 399, 443, 444, 462, 467, 523, 547, 578, 621, 626, 657, 661, 687, 709, 734, 737, 788, 792, 819, 821, 856, 875, of God and neighbour, 54, 90, 99, 101, 287, 288, 408, 441, **506**, 687, 816, 843, 871, 878, 882
Lucania and Brutium, 424
Luccreth moccu Cérai, 413
Lucernarium, 63
Lucian of Antioch, 23, 47, 49, 50, 53, 77, 76, 92, 185, 207, 301, 310, 384, 433, 480, **507**, 525, 596, 673, 795, 830
Lucian of Caphargamala, **507**
Lucian of Carthage, **507**
Lucian of Samosata, 134, 155, 400, **508**, 578
Lucidus the presbyter, 79, 321, 480, **508**, 653, 863
Lucifer of Cagliari, 48, 49, 77, 92, 93, 168, 302, 381, 431, **508**, 559, 660, 757
Luciferians, 151, 218, 233, 318, 320, 363, 410, 508, 522, 626, 828
Lucilla, 137, 188, 246, 249, 250, **508**, 517
Lucinius Baeticus, **509**
Lucius I, pope, 644, 646
Lucius and Ptolemy, 724
Lucius of Adrianople, 13, 77, **509**, 551
Lucius of Alexandria, **509**, 678, 680
Lucius of Carthage, 325
Lucius of Castra Galbae, 584
Lucius of Membressa, 584
Lucius of Theveste, 584
Lucretius, 82
Luculentius, **509**
Lucullanum, 747
Luca (Lucca), 422
Lucy of Syracuse, **509**, 806, 807
Ludolf of Saxony, 234
Lugo, 18, 127, 801
Luke, evangelist, 7, 28, 38, 61, 62, 67, 191, 214, 217, 228, 292, 300, 306, 311, 390, 401, 431, 451, 461, 503, 506, **509**, 518, 523, 537, 540, 620, 631, 647, 657, 671, 687, 723, 737, 740, 746, 755, 762, 802, 825, 843
Luke the Stylite, 797
Lullingstone, Kent, 130
Lupus of Troyes, 103, 130, 307, 483, **511**, 767
Lusitania, 399, **512**, 554, 774, 790, 861, *203*, *204*
Luther, 101
Luxeuil (Luxovium), 187, 307, 334, 338, 451, 633
Luxorius, 761
Lycaonia, 112, 175
Lycia, 85, 87, **512**, 596, 673
Lyons (Lugdunum), 64, 95, 105, 134, 178, 249, 280, 295, 326, 337, 421, 465, 483, 499, **512**, 533, 551, 552, 653, 672, 762, 802, 815, 863, 874, C. of (470-475), 508, 513, 767, C. of (518-23), 59, 555, C. of (567-570), 556, C. of (583), 325, Martyrs of, 10, 122, 280, 293, 358, 483, 512, 524, 571, 649, 706, 752, 842, *see also* Letter of the church of Lyons and Vienne
Lyre of the Virgin, 291
Lystra, 274, 313
Lytrankomo, 213

Maastricht (Mosae Traiectum), 27, 349, 471
Macarius, Johannes, 68

Macarius, Ps.-, 115, 140, 303, 364, 507, 718
Macarius/Simeon, **514**, 556, 567
Macarius I of Jerusalem, 160, 432, **514**, 546
Macarius Magnetes, 704
Macarius of Antioch, 197, **514**
Macarius of Egypt, *see* Macarius of Scete
Macarius of Magnesia, 127, 183, 233, **514**
Macarius of Scete, 22, 31, 417, 496, **514**, 567, 723, 759
Macarius of Tkow, **514**
Macedonia, 7, 8, 19, 69, 172, 174, 192, 283, **514**, 515, 565, 657, 681, 795, 833, 834, 837, *205*, *206*
Macedonians, 17, 21, 32, 78, 114, 195, 236, 268, 296, 302, 303, 321, 324, 329, 392, 395, 516, 521, 595, 700, 767, 827, 841, 853
Macedonius, bishop of Mopsuestia, **516**
Macedonius, vicarius Africae, **516**
Macedonius of Constantinople, 21, 78, 195, 268, 296, 302, 303, 392, **516**, 521
Macedonius II of Constantinople, 6, 36, 197, 242, 337, 458, **516**, 569
Machaeron, 66
Mâcon (Matisco), C. of (581-583), 102, 325, 493, 556, 609, 824, 864, C. of (585), 102, 325, 530, 556 609, 637, 800
MacRegol Gospels, 132
Macrina junior, 54, 113, 114, 184, 192, 232, 363, **516**, 678, 882, Life of, 121, 364, senior, 114, 363, **516**
Macrinus, 348
Macrobius, 592, 713, 799
Macrobius, Donatist bishop of Hippo, 249, **516**
Macrobius, Donatist bishop of Rome, **516**
Macrocosm, 40, 45, **516**, 602, 683, 882
Mactar, 14, 15, 169, 171
Madaba, 66, 572
Madaura, 14
Madensehir, 87
Maeldubh, 19
Magen, 635
Maghreb, 14, 15, 16
Magi, 5, 57, 81, 92, 107, 133, 138, 163, 227, 232, 253, 282, 322, 323, 378, 410, 452, 466, 493, **517**, 539, 540, 582, 583, 715, 723, 837, 860, *207*, *208*
Magic, 4, 8, 74, 82, 92, 107, 120, 124, 126, 144, 145, 160, 210, 218, 226, 227, 232, 242, 291, 350, 377, 378, 444, 460, **517**, 553, 579, 582, 671, 672, 712, 749, 780, 801, 805, 825, 836
Magical papyri, 836
Magnentius, 192, 198, 607, 663, 712
Magnus, 109
Magnus Maximus, 360
Mahón, 774
Mai, Angelo, 654
Maiferqat, 537
Mainz (Moguntiacum), 348, 486
Maiorianus, 778
Maiorinus, 137, 246, 250, 508, **517**, 766, 779
Maiuma of Ascalon, 635
Maiuma of Gaza, 204, 339, 344, 635
Majorca, 107
Majsan, 217
Malabar, 48, 307, **517**, 674, 809, rite, 50, 496
Malabarese, 10
Malaga, 106, 107
Malankar rite, 50
Malaspina, Elena, 119, 183, 186-87, 286, 296, 297, 298, 334, 340, 381, 400, 411-13, 447, 576, 637, 653, 765, 822, 828
Malchion of Antioch, 48, **518**, 663, 760
Malchus, Life of, 787
Malingrey, Anne-Marie, 440-42
Malkîkarib, 65
Mallosus, 347, 349
Malmesbury, 19, Itinerary, 427

Malta (Melita), 157, 424, **518**, 550
Mamasios, 518
Mamas (or Mammas), 137, 337, **518**
Mambre Vercanol, 81, **518**
Mamertus of Vienne, 79
Mammarius, Passion of, 840
Mampbis-Kurnub, 174
Mamre, 3
Manda d'Haijê, 518
Mandaeans, Mandaeism, 9, **518**, 670, 823
Mandyas, 253
Mani, 11, 98, 113, 263, 269, 276, 321, 354, 371, **519**, 674, 802
Manichaean Psalter, 810
Manichaeism, Manichaeans, 2, 11, 22, 27, 38, 39, 48, 57, 63, 64, 84, 96, 98, 113, 128, 147, 151, 161, 191, 205, 219, 226, 236, 255, 264, 269, 281, 288, 289, 301, 308, 315, 319, 321, 328, 339, 343, 354, 359, 379, 399, 404, 405, 407, 410, 418, 426, 443, 458, 484, **519**, 524, 528, 542, 568, 574, 584, 629, 633, 645, 660, 665, 673, 674, 680, 703, 709, 712, 718, 734, 740, 744, 746, 758, 765, 775, 781, 823, 830, 837, 841, 859, 877, Against the Manichees, 21, 235, 374, 511, 843, 884, 768
Maniple, 865
Manna, 128, 293, 294, 504, **520**, 575, 750, 817, *209*
Mansi, 654
Mansuetus of Milan, **520**
Mantua (Mantova), 421
Manumissio in ecclesia, 147, 193, **520**, 739, 784
Manuscript tradition, 82, 152, 394, **520**, 724, 772, 809, *210*, *211*, *212*
Mappula, 253, 865
Mara, Maria Grazia, 18, 26, 28-29, 43, 56-58, 102, 125, 282, 330, 360, 437, 464, 467, 467-68, 475, 521, 542, 595, 658-59, 660, 703, 715-16, 736-37, 745, 758, 804, 816, 828-29, 856-57, 858-59, 881-82
Maranatha, 232, **521**, 793
Marano (591), 773
Marathonius of Nicomedia, 195, **521**, 567
Marcella, 28, 430, **521**
Marcellians, 196
Marcellianum (now S. Giovanni in Fonte), 424
Marcellina, Carpocratian, 145, **521**
Marcellina, sister of Ambrose, 28, **521**, 787, 872, Life of, 521
Marcellinus, pope, 521, **522**, 645, 741, 782, 874
Marcellinus Comes, 166, **522**
Marcellinus Flavius, 185, 247, **521**
Marcellinus the Luciferan, 218, 320, 508, **522**
Marcellus, 741
Marcellus, Ps.-, 676
Marcellus, martyr, 13
Marcellus I, pope, 8, 77, 521, **522**, 617, 640, 645
Marcellus of Ancyra, 5, 11, 37, 43, 53, 77, 90, 92, 93, 113, 195, 207, 215, 296, 299, 300, 302, 363, 365, 449, 460, **522**, 559, 566, 641, 685, 743, 749, 757, 783, 852, Against, 113, 522
Marcellus of Apamea, 49
Marchetta, Antonio, 861-62
Marchi, P., 69
Marcia, 188, 189, **522**
Marcian, emperor, 4, 159, 201, 241, 327, 408, 432, 479, 484, **522**, 569, 725, 779, 828, 840, 884
Marciana, 138, 544, Passion, 534
Marcian of Arles, 79, 212, 337, **523**, 794
Marcian of Gaza, 201
Marcian of Lampsacus, 516
Marcian the Ascetic, **523**
Marcian the Presbyter, **523**
Marcion, 54, 58, 144, 158, 198, 226, 229,

244, 263, 276, 285, 287, 308, 309, 319, 353, 372, 384, 398, 449, 457, 510, **523**, 564, 640, 641, 644, 657, 681, 714, 731, 732, 741, 746, 763, 766, 819, 829, 848, 871, Against, 312, 725, 819, 832, Gospel of, 57
Marcionite Prologues, **524**, 526
Marcionites, Marcionist heresy, 10, 46, 110, 121, 163, 205, 237, 301, 315, 319, 326, 462, 509, 510, 523, 524, 528, 574, 621, 622, 644, 658, 736, 674, 775, 777, 827, 830, 877
Marcosians, 526, 802
Marculf, **524**
Marculus, **524**
Marcus, pope, 645
Marcus Aurelius, 17, 53, 58, 60, 92, 95, 96, 155, 166, 167, 188, 280, 326, 330, 332, 333, 373, 421, 426, 472, 508, **524**, 531, 551, 560, 672, 727, 796, 808, 831, 835, 861
Marcus Diadochus, **524**
Mareotis, 824
Margus, 838
Mari of Beth Qardu, 112
Mari (or Maris) of Rewardasir 10, 11, 403, 598, 674, 808, 855
Marialba de León, Mausoleum of 334
Mariana, 203
Marian devotions, 712
Marian feasts, 539, 709
Marianus and James, 13, 470, 524, 533, 607, Passion, 193, 470, 607
Maries, The Three, **524**, 854
Marinone, Mariangela, 41, 452, 475-76, 539-40, 583, 650, 709-10, 715
Marinus, martyr, **525**, 592
Marinus of Constantinople, 175, **525**
Mariolata, 7
Maris of Rewardashir, *see* Mari
Maris of Chalcedon, 159, 185, 197, **525**, 757
Marius Claudius Victorius, 212, 224, 278, **525**, 530
Marius Mercator, **525**, 660, 746
Marius Victorinus, 141, 177, 205, 311, 324, 355, 359, 392, 405, **525**, 589, 592, 614, 651, 658, 683, 689, 696, 697, 704, 714, 734, 760, 780, 803, 806, 831
Marius Victorinus, Ps.-, 520
Marius of Avenches, 166, **525**
Mark, disciple of Valentinus, **526**
Mark, poet, **526**
Mark, pope, 742
Mark of Arethusa, 78, 395, **527**, 783
Mark of Scete, 782
Mark of Zylgra, 487
Mark the Deacon, **527**, 704
Mark the Evangelist, 22, 24, 50, 62, 64, 73, 111, 180, 200, 213, 228, 264, 306, 374, 384, 431, 451, 472, 487, 496, 509, 510, **527**, 542, 543, 551, 596, 610, 647, 740, 755, 802, 868, 877, Apocryphal Gospel of, 180
Mark the Gnostic, 542, 551, 610, 859
Mark the Hermit, 251, **527**
Mark the Magus, 802
Marmoutier, 772
Marneion, 339
Maro, John, 496, **527**, 809
Maronite church of St George, 808
Maronites, 66, 107, 228, 429, 496, **527**, 809, liturgy, 34, 527, 810
Maro the anchorite, **527**
Marriage, 1, 3, 10, 12, 17, 63, 84, 98, 99, 100, 112, 121, 123, 140, 154, 161, 176, 181, 189, 198, 220, 223, 236, 243, 244, 261, 271, 280, 281, 285, 287, 288, 317, 318, 328, 335, 337, 351, 356, 359, 377, 404, 409, 416, 417, 428, 429, 436, 451, 452, 454, 458, 466, 473, 477, 485, 493, 523, **528**, 555, 570, 585, 622, 665, 669, 701, 712, 717, 718, 719, 725, 738, 739, 742, 751, 758, 773, 775, 776, 784, 786,

787, 790, 792, 793, 796, 806, 815, 819, 843, 844, 858, 871, 876, *213*
Marsala (Lilybaeum), 425
Marseilles (Massilia), 99, 101, 149, 154, 193, 335, 337, 339, 342, 396, **530**, 576, 597, 754, 756, 767, 772, 824, *214*, C. of (533), 555
Martha, 79
Martial and Basilides, **530**, 554
Martial of Limoges, 238
Martial of Mérida, 146
Martias and Marsianus, 71
Martigny (Octodurus), 824
Martin I, pope, 192, 304, **530**, 547, 568, 640, 642, 645, 646, 659, 743, 768, 823, 826, 873
Martina, **530**
Martin of Braga (or of Dumium), 127, 142, 188, 255, 497, **530**, 595, 652, 689, 791
Martin of Lampsacus, 195
Martin of Liébana, 117
Martin of Tours, 121, 126, 191, 322, 338, 347, 369, 486, **530**, 531, 532, 552, 559, 567, 618, 708, 712, 743, 746, 752, 846, 850, 869, Life of, 121, 531, 562, 651, 664, 793, 799, 862, 873
Martinianus, 559
Martinian of Forum Traiani (Fordongianus), 757
Martos, 106, 107
Martvili cross, 345
Martyrdom, 1, 2, 11, 13, 16, 17, 35, 60, 80, 96, 105, 110, 117, 120, 121, 140, 143, 165, 168, 186, 189, 190, 191, 192, 193, 201, 209, 211, 212, 217, 218, 223, 225, 230, 231, 245, 247, 248, 249, 261, 273, 287, 289, 292, 293, 296, 319, 326, 337, 347, 357, 358, 368, 396, 404, 416, 429, 432, 435, 447, 461, 476, 483, 494, 506, 507, **531**, 542, 554, 570, 621, 640, 652, 659, 668, 670, 671, 673, 675, 680, 707, 710, 736, 756, 762, 781, 784, 792, 795, 803, 817, 819, 820, 833, 834, 851, *215*, *216*, of Andrew, 41, 682, of Dasius, 534 of Paul, 676, of Peter, 701, Pionius, 534, of Polycarp, 532, 533, 724, 818
Martyria, 24, 174, 573
Martyrium Pilati, 687
Martyrium S. Apollonii, 785
Martyrium S. Justini et sociorum, 463
Martyrius (Sahdona), **536**, 810
Martyrius of Antioch, 342, **536**, 679
Martyrius of Antioch, Ps.-, 437
Martyrius of Jerusalem, 350, 432, 445, **536**, 569, 884
Martyrology, 17, 37, 118, 140, 165, 180, 202, 338, 356, 370, 536, 676, 770, 877
Martyrs, 1, 6, 8, 14, 16, 17, 19, 21, 24, 25, 28, 40, 50, 51, 64, 66, 71, 86, 90, 93, 102, 103, 104, 106, 110, 120, 122, 128, 137, 151, 156, 168, 170, 171, 175, 176, 186, 189, 191, 200, 208, 209, 219, 220, 222, 223, 224, 227, 228, 232, 238, 241, 246, 252, 269, 270, 271, 274, 278, 281, 298, 299, 336, 337, 338, 350, 360, 365, 368, 372, 388, 389, 404, 407, 409, 410, 419, 420, 421, 422, 426, 433, 457, 473, 487, 489, 508, 522, 525, 531, **536**, 544, 547, 552, 554, 560, 581, 583, 584, 604, 605, 615, 630, 651, 657, 659, 665, 668, 670, 701, 705, 721, 724, 725, 731, 740, 741, 742, 744, 752, 762, 764, 779, 785, 790, 793, 814, 818, 821, 822, 834, 848, 854, 869, 878
Marusinac, 148
Maruta of Maiferqat, 416, **537**, 674, 757, 766, 810
Mary, 5, 7, 29, 34, 38, 39, 41, 47, 57, 91, 107, 108, 112, 125, 135, 141, 161, 165, 186, 202, 204, 220, 221, 251, 256, 275, 280, 289, 290, 301, 312, 314, 322, 326, 333, 341, 344, 346, 374, 379, 386, 392, 408, 433, 442, 447, 448, 451, 452, 454,

463, 493, 509, 524, 528, 536, **537**, 548, 549, 552, 564, 573, 577, 582, 583, 584, 594, 597, 599, 615, 622, 631, 654, 664, 675, 685, 710, 713, 715, 723, 730, 735, 740, 744, 776, 781, 784, 787, 797, 803, 809, 819, 830, 832, 833, 841, 850, 862, 871, 876, 881, 885, *217*, *218*, *219*, feasts, *see* Marian feasts
Mary Magdalene, 79, 275, 314, 370, 453, 524
Mary the Egyptian, **540**, Life of, 787
Mascazel, 351, 660
Mass, 6, 8, 17, 50, 57, 102, 123, 133, 154, 155, 160, 185, 198, 215, 222, 223, 224, 232, 239, 290, 293, 335, 386, 393, 411, 484, 490, 493, 498, 532, **540**, 619, 679, 725, 736, 738, 744, 745, 768, 864, 865, 885, *see also* Anaphora, Eucharist, Liturgy
Massa, 90, **542**
Massa damnationis, 709
Massa perditionis, 430, 767
Massona of Mérida, **542**, 554, 663, 791
Mastara, 80
Materano, 424
Maternus, 348
Mathematici, 92, **542**
Matronea, 173
Mattai, Mar, **543**
Matthew, Ps.-, 41
Matthew the Evangelist, Gospel of, 30, 57, 106, 166, 187, 214, 232, 236, 258, 301, 306, 311, 320, 382, 383, 431, 451, 461, 466, 503, 506, 510, 511, 527, **543**, 613, 620, 647, 680, 703, 723, 727, 746, 755, 766, 778, 803, 824, 825, 837, 867, 868
Matthias, 62, 113, 181, **543**, 837, Gospel of, 543
Maundy Thursday, 71
Mauretania, 13, 109, 138, 175, 228, 248, 250, **543**, 605, 651, 653, 738, 771, 790, 842
Maurice, saint, **544**, 773, 821, 824, 846
Maurists, **544**, 612, 644, 655, 774
Maurus of Parentium, **544**, 651
Maurus of Ravenna, 873
Mavia, 576
Maxentius, 193, 522, **545**, 554, 645, 673, 863
Maximian, emperor, 22, 102, 134, 169, 185, 186, 192, 193, 208, 236, 251, 270, 295, 321, 333, 366, **545**, 573, 663, 673, 711, 713, 730, 807, 866, cathedra, *see* Cathedra of Maximian
Maximian of Constantinople, 5, 281, **545**
Maximian of Pola, 730
Maximian of Ravenna, 420, **545**
Maximian of Syracuse, 366, **545**
Maximian the Donatist, **545**
Maximianists, 143, 208, **545**, 613, 738
Maximianopolis (now Shaqqa), 67
Maximianus, 247
Maximianus Herculeus, 615
Maximilian, 11, 13, 534, **545**, 816, Passion, 212, 545
Maximilianszelle (Pongau), 747
Maximilla and Priscilla (or Prisca), 60, 150, 570, **545**
Maximus Daia, 193, 250, **546**, 673, 705, 854
Maximinus (or Maximian) of Anazarbus, 37, **546**
Maximinus of Trier, 348, **546**, 850
Maximinus the Arian, 329, 393, 410, 503, **546**
Maximus Thrax, 28, 324, 356, **546**, 672, 741
Maximus, Ps.-, 655
Maximus, anti-gnostic, **546**
Maximus, bishop of Jerusalem, 432, **546**
Maximus, emperor, 28, 77, 126, 322, **547**, 400, 426, 464, 547, 660, 687, 712, 828, 850, 851, 859

Maximus, martyr, **547**
Maximus II, 854
Maximus Confessor, 35, 36, 39, 73, 76, 112, 192, 206, 242, 293, 304, 306, 358, 364, 374, 379, 443, 446, 507, 530, **547**, 568, 618, 639, 642, 655, 696, 708, 725, 727, 734, 735, 788, 792, 793, 816, 821, 823, 824, 826, 880
Maximus Planudis, 44
Maximus of Alexandria, 518
Maximus of Antioch, **548**
Maximus of Constantinople, 678
Maximus of Madaura, 607
Maximus of Riez, 321, 483, **548**, Life of, 256
Maximus of Saragossa, **548**
Maximus of Turin, 261, 421, 452, 546, **547**, 549, 552, 617, 676, 751, 816, 817, 854
Maximus of Turin, **548**
Maximus of Tyre, 558
Maximus of Zagylis, 487
Maximus the Cynic, 196, **549**, 744
Mazdaism, 1, 4, 162
Mazzoleni, Danilo, 278-79, 450-51, 568, 580, 599, 650, 659, 704-5
Medea, 252
Medical Lecture, **549**, 630, *220*
Medina Sidonia, 107
Meditation, 28, 100, 118, 135, 285, 360, 401, 478, 499, 548, **549**, 552, 577, 669, 781, 793
Medula, 210
Mees, Michael, 179-81
Megalopsychia of Antioch, 572
Meggen, 14
Meigle, Perthshire, 132
Meir, Rabbi, 563
Melania junior, 350, 388, 430, 433, 481, **550**, Life of, 121, senior, 306, 549, 550, 688
Melchites, 24, 48, 50, 66, 126, 200, 265, 444, **550**, 568, 570, 618, 755, 809, rite, 496
Melchizedek, 2, 9, 11, 128, 293, 294, 341, 391, 504, **550**, 750, 751, 771
Melchizedekians, 384, **550**, 830, 841
Meletian schism, 48, 302, 677
Meletius of Antioch, 4, 5, 47, 48, 49, 51, 58, 72, 78, 93, 106, 114, 142, 195, 218, 219, 236, 250, 302, 303, 305, 325, 343, 361, 364, 379, 395, 398, 440, 508, **550**, 559, 660, 665, 843, 855, 873
Meletius of Mopsuestia, 37, **550**, 884
Melitene, 5, 30, 337, 673, C. of, **550**
Melitian schism, 76, 77, 83, 93, 439, 551, 855
Melitius of Lycopolis, 7, 20, 22, 23, 140, 439, **551**, 677, 841
Melito, Ps.-, 614
Melito of Sardis, 9, 20, 47, 55, 57, 60, 90, 102, 168, 199, 245, 256, 257, 258, 264, 276, 307, 312, 320, 408, 426, 524, **551**, 560, 647, 657, 665, 671, 728, 732, 760, 764, 781, 833, 850
Mellitus, 465, **551**
Meloni, Pietro, 117, 126, 220-21, 255, 319, 670, 786-87
Melrose, 211
Members of the Church, 259, 316, 474, 477, 496, **551**, 793
Memnon of Ephesus, 115, 275, 314, 444, **552**, 663
Memoria, **552**, Apostolorum, 552, 568, S. Stephani, 375
Memory, 158, 180, 223, 235, 364, 389, 394, 398, 399, 406, 440, 478, 535, 549, **552**, 563, 687, 744, 760, 764, 812
Menaia, 135
Menander, 353, 384, **553**, 758
Menander, Ps.-, 455
Menas of Constantinople, 197, 268, 305, **553**, 662, 677, 678, 743, 823, 838, 870
Menas of Pshati, 200

Mennas, 24, **553**, 688
Mensa, 14, 107, 148, 171, 173, 213, 334, 390, 425, 450, 518, 535, **554**, 573, 731, 753, 764, 813, 814, 842, 845
Mensurius of Carthage, 137, 246, 250, 517, **554**, 765
Mentesa, 148
Mercurius, 438
Merianlik, 808
Merida (Emerita), 148, 296, 334, 386, 512, 530, 542, **554**, 663, 790, 791, 814, 844, *221*, C. of (c.650), 554, C. of (666), 554
Merita, **555**
Merobaudes, 13, **555**, 859
Merosaba, 112
Merovingian councils, 17, 212, 338, **555**, 582, 651
Mesembria, 36
Meskiana, 15
Mesopotamia, 22, 48, 54, 184, 203, 257, 263, 264, 823
Mesrob, 80, 81, 315, 468, 472, 482, **556**, 752, Life of, 81
Messalians, Messalianism, 8, 22, 32, 49, 183, 196, 272, 298, 358, 364, 392, 445, 485, 514, 527, **556**, 567, 597, 598, 635, 674, 684, 708, 723, 775, 776, 778, 781, 784, 792, 841
Metaphrastes, 829
Meteorologica, 75
Methodius of Olympus, 10, 38, 53, 55, 57, 90, 183, 229, 233, 261, 269, 293, 308, 310, 315, 380, 449, 477, **557**, 560, 606, 622, 623, 633, 634, 693, 704, 729, 733, 776, 854, 871, 878, 880
Methonis, 7
Metrodorus, **557**
Metropolitans, 71
Metrum in Genesim, 651
Metz (Mediomatrici), 337
Metz (Mettis), 348
Michael, Archangel, 40, 132, 163, 213, 271, 290, 333, **557**
Michael II, 407
Michael Psellus, 76, 713, 835
Michael Syncellus, 748
Michael of Antioch, 772
Michael of Ephesus, 20, 76.
Michael the Syrian, 95, 119, 145, 238, 363, 428, 445, 496, 810
Microcosm, 516
Middle Platonism, 32, 45, 73, 74, 75, 163, 181, 204, 240, 285, 320, 354, 355, 363, 373, 405, 448, 455, 463, **557**, 585, 586, 587, 590, 601, 606, 621, 689, 690, 691, 693, 696, 697, 795, 797, 805, 832, 851
Middle Stoicism, 53
Midrash, 276, 309, 395, **558**, 563, 674, 810, 812, 821
Migetius of Narbonne, 556
Migliarina, 421
Mikaelites, 289
Milan (Mediolanum), 11, 24, 26, 27, 28, 29, 30, 62, 67, 69, 77, 78, 94, 98, 105, 147, 149, 160, 169, 173, 174, 175, 186, 187, 202, 218, 223, 224, 232, 236, 263, 264, 269, 274, 301, 303, 307, 336, 347, 348, 350, 370, 377, 381, 386, 396, 419, 420, 421, 427, 476, 490, 495, 497, 499, 527, 532, **559**, 573, 579, 583, 600, 604, 618, 642, 643, 656, 660, 669, 689, 697, 715, 722, 729, 731, 738, 739, 742, 756, 767, 774, 780, 860, 863, 875, *222*, *223*, *224*, C. of (345), 522, 685, C. of (347 or 349), 559, C. of (355), 77, 93, 238, 302, 382, 485, 508, 559, 657, 736, 758, 858, C. of (380), 559, C. of (386), 559, C. of (389), 116, C. of (390), 340, 559, C. of (392), 125, 749, C. of (393), 559, C. of (451), 360, 549, 559, 735, 767, 854, diptych, 456, 540, 688, sarcophagus, 756
Mildenhall, Suffolk, 133, treasure, 133

Miletus, 86
Milevis, 774, C. of (402), 147, C. of (416), 499, 605, 705, 744, 774, 782, 859
Military costume, 254
Militia, **559**
Militia Christi, 740
Militia spiritalis, 182
Millennarism, 23, 55, 64, 90, 138, 159, 166, 191, 285, 286, 299, 338, 342, 372, 380, 391, 433, 453, 454, 463, 470, 557, **560**, 593, 621, 651, 732, 733, 839, 867, 885
Milreu-Estoi (Faro), 512
Miltiades, apologist, 60, 250, 436, 521, 524, **560**
Miltiades, pope, 137, 247, **560**, 641, 645, 713, 741, 742, 743
Mime, 273
Minervius and Alexander, 846
Minim, 436
Ministries, 6, 34, 103, 109, 182, 221, 232, 235, 239, 247, 259, 260, 270, 278, 284, 326, 312, 335, 441, 477, 479, 490, 494, 501, 512, 541, 542, 548, **560**, 567, 613, 615, 616, 642, 644, 655, 682, 710, 711, 715, 740, 742, 744, 751, 794, 798, 840, 864, 865, 882, *see also* Clergy
Minorca, 105, 107, 624, *see also* Baleari
Minucius Felix, 2, 60, 82, 155, 168, 177, 211, 233, 330, 406, 472, **562**, 672, 696, 703, 785, 796, 818
Mirabella Roberti, Mario, 64, 420-22, 559
Mirocles, 559
Misael, 563
Mishnah, 32, 436, 490, **563**, 812
Mishneqah, 66
Missal, 335, 497, 502
Mithras, Mithraism, 12, 65, 82, 128, 169, 347, **564**, 799, 800
Mitre, 866
Mitrophanes, 194
Mljet (Polace), 217
Mocianus scholasticus, 576
Moderatus the Neopythagorean, 691
Modestus, 36, 115, 433, **564**, 884
Modestus of Jerusalem, 447, 538, **564**
Modicia (Monza), 420
Moesia (Mysia, Misia), 13, 19, 105, 199, 223, 307, 356, **565**, 764, 837
Molanus, Johannes, 68
Monaci Castagno, Adele, 13, 145, 160, 553, 758
Monarchians, Monarchianism, 11, 48, 53, 77, 90, 113, 164, 195, 300, 301, 303, 320, 391, 395, 396, 507, **566**, 653, 663, 675, 685, 706, 712, 713, 714, 716, 748, 757, 797
Monasteries, Double, **566**
Monastery, 1, 3, 4, 11, 16, 24, 27, 35, 36, 44, 51, 54, 65, 70, 71, 79, 84, 96, 98, 105, 106, 112, 118, 122, 128, 130, 131, 137, 147, 148, 150, 162, 184, 186, 195, 200, 213, 217, 218, 224, 264, 266, 268, 274, 279, 296, 304, 307, 308, 324, 330, 338, 344, 348, 366, 397, 400, 411, 423, 432, 475, 518, 527, 530, **566**, 628, 633, 655, 659, 666, 689, 704, 714, 729, 759, 761, 772, 794, 806, 864, 871
Monastic constitutions, 115
Monasticism, 1, 22, 31, 39, 40, 48, 52, 84, 94, 112, 114, 119, 122, 123, 131, 132, 136, 149, 154, 160, 161, 184, 199, 230, 247, 251, 264, 290, 296, 304, 319, 320, 337, 338, 344, 366, 379, 411, 417, 430, 431, 468, 477, 478, 487, 496, 499, 514, 521, 531, 536, 544, 556, **566**, 597, 636, 666, 674, 708, 722, 723, 746, 748, 755, 761, 776, 781, 791, 796, 799, 808, 810, 850
Monastic liturgy, 500
Monazontes, 567
Monica, 11, 97, 222, 308, 532, **568**, 584, 653, 882
Monkwearmouth, 131

Monoenergism, 165, 197, 433, 530, 547, **568**, 642, 768, 823, 826, 841
Monogram, 66, 89, 193, 209, 217, 228, 232, 279, 322, 334, 336, 421, 423, 435, 471, 491, 518, 554, **568**, 572, 597, 607, 638, 639, 656, 675, 771, 802, 803, 860, *225, 226*
Monophysites, Monphysism, 5, 6, 8, 11, 17, 19, 23, 24, 35, 36, 37, 43, 47, 48, 49, 50, 51, 54, 55, 59, 65, 82, 95, 105, 112, 118, 119, 126, 135, 159, 165, 185, 193, 195, 196, 197, 208, 233, 238, 249, 256, 263, 268, 272, 273, 275, 289, 297, 301, 303, 305, 307, 315, 325, 331, 333, 339, 342, 343, 344, 346, 350, 362, 363, 374, 375, 379, 401, 402, 407, 416, 428, 429, 432, 435, 439, 440, 443, 445, 446, 447, 448, 464, 465, 473, 480, 481, 487, 496, 505, 543, 547, 548, 550, 553, 563, 564, 567, **568**, 569, 585, 598, 619, 638, 642, 659, 662, 674, 678, 679, 684, 723, 744, 747, 748, 768, 769, 773, 779, 780, 787, 794, 808, 809, 810, 822, 823, 824, 826, 827, 828, 829, 838, 840, 841, 845, 855, 856, 858, 870, 884, Against, 481
Monothelites, Monothelism, 35, 36, 37, 38, 43, 67, 129, 165, 192, 195, 197, 216, 346, 397, 402, 438, 439, 443, 444, 514, 520, 527, 530, 547, 559, **568**, 570, 642, 645, 659, 726, 732, 743, 787, 822, 823, 826, 840
Mons Mirabilis, 174
Montanists, Montanism, 60, 151, 160, 161, 191, 205, 221, 255, 319, 333, 337, 326, 391, 403, 406, 513, 528, 531, 542, 545, 560, **570**, 603, 604, 644, 665, 667, 686, 710, 712, 715, 716, 768, 781, 819, 798, 848, 872, 876
Montanus, 13, 58, 60, 150, 160, 262, 325, 513, 545, **570**, 640, 713, 741, 838, 844
Montanus, Lucius and Companions, 533, **571**
Montanus of Toledo, **571**
Montecassino, 119, 447, 646, 747
Monte della Casetta, 422
Monte Gargano, 557
Monteu da Po, 421
Monymusk reliquary, 133
Monza, 447
Mopsuestia (now Misis), 89, 175, 447, 599, 755, 824, C. of, **571**
Morin, Jean, 654
Morinus of Alexandria, 188
Morlupo, 422
Mosaic, 1, 3, 14, 15, 16, 40, 41, 51, 53, 62, 64, 66, 70, 85, 86, 87, 88, 89, 106, 107, 108, 120, 127, 129, 138, 141, 148, 156, 157, 170, 173, 175, 187, 202, 203, 209, 213, 218, 228, 253, 255, 263, 268, 269, 271, 278, 283, 292, 294, 306, 315, 327, 333, 334, 336, 345, 349, 369, 377, 379, 419, 420, 421, 422, 423, 424, 425, 428, 433, 434, 435, 447, 450, 452, 453, 456, 471, 476, 477, 487, 488, 489, 506, 515, 520, 529, 535, 536, 539, 540, 550, 554, 557, 559, **571**, 575, 581, 597, 599, 600, 625, 631, 635, 636, 638, 650, 651, 659, 664, 667, 675, 677, 681, 685, 688, 689, 705, 710, 715, 721, 725, 728, 729, 730, 753, 755, 756, 757, 771, 778, 784, 795, 803, 808, 812, 813, 822, 823, 834, 835, 838, 845, 847, 855, 866, *227, 228*
Moselkern, 350
Moses, 1, 37, 41, 56, 66, 108, 111, 133, 159, 163, 212, 214, 221, 222, 230, 259, 261, 288, 299, 309, 312, 315, 319, 349, 356, 373, 446, 453, 460, 464, 465, 529, 562, **573**, 574, 578, 606, 612, 636, 652, 677, 687, 691, 693, 697, 724, 748, 756, 762, 763, 771, 777, 782, 815, 816, 817, 831, 871, 876, 229, Against Moses and the other prophets, 11, Assumption of, 56, 57, Blessings of, 312, Life of, 312, 364

Moses, bishop, 65
Moses bar Kepha, 428, 810
Moses of Eghvard, 80
Moses of Khorene, 81, 83, 518, **575**
Moses of Outi, 19
Moses the Saracen, **576**
Mos Maiorum, 576
Moylough Townland (co. Sligo), 133, belt-reliquary, 133
Moyne (co. Mayo), 130
Mren, 81
Msihazeka, 3
Msuffin, 228
Mughtasilah, 269
Muhatt el-Urdi, 635
Muirchú, **576**, 653
Mukawer, 66
Mummolinus, 268
Munier, Charles, 8-9, 17, 40, 79, 90-91, 97, 103-4, 120, 126, 134, 141-43, 146-48, 176, 205, 246, 271, 318, 383, 400, 407-8, 436-37, 469, 513, 597, 615, 651, 714, 737, 739, 793-94, 822, 847, 850-51, 854, 858, 862, 869
Muratori, L.A., 69, 654
Muratorian canon, 153, 222, 306, 448, 509, 526
Muratorian fragment, 38, 58, 644
Murbach, 689
Mursa, 198, 639
Musaeus of Marseilles, 335, 339, 393, 497, 530, **576**
Musanus, **576**
Music, 124, 219, 220, 272, 274, 387, 446, 478, **576**, 629, 736, 873
Mutawakkil, 686
Mutianus of Antioch, **576**
Mutina (Modena), 340, 421
Mutius of Amphipolis, 515
Myron, 42
Mystagogia, 547
Mystagogical catecheses, 611, 750
Mystagogy, 136, 150, 152, 258, 264, 409, **576**
Mystery, 11, 30, 34, 43, 46, 54, 99, 100, 101, 108, 123, 128, 136, 151, 186, 188, 198, 199, 219, 231, 237, 240, 242, 243, 245, 252, 258, 260, 261, 276, 282, 291, 292, 316, 319, 328, 335, 336, 340, 341, 352, 355, 358, 362, 374, 399, 408, 414, 438, 441, 446, 458, 467, 470, 495, 499, 506, 528, 537, 548, 557, 564, 576, **577**, 621, 622, 652, 693, 706, 708, 709, 731, 733, 734, 750, 751, 786, 788, 792, 802, 805, 816, 833, 851, 852, 859, 875, 878
Mystical Body, **577**
Mystical theology, 239
Mysticism, 46, 101, 240, 352, 363, 418, 529, 531, 557, **577**, 770, 792, 793, 880
Mythology, 46, 73, 109, 156, 181, 213, 226, 252, 273, 285, 324, 352, 372, 485, 557, 572, **578**, 591, 599, 677, 736, 760, 771, 803, 831
Mzcheta-Djvari, 344
Mzcheta - Sveti Tzkhoveli, 344, 345, **578**, *230*

Naas, Naassenes, 57, 139, 465, 612, *see also* Ophites
Nabataea, 126
Nablus, 635
Nabor and Felix, **579**
Nag Hammadi, 12, 22, 71, 109, 199, 251, 271, 287, 352, 378, 448, 478, **579**, 628, 648, 656, 674, 680, 771, 775, 859
Nagran, 71
Nain, 261, 509, 511, **579**
Najrân, 65
Naldini, Mario, 32-33, 165-66, 264, 647-48, 664
Name, 355, 436, 470, 506, **579**, 606, 615
Names, 223, 239, 279, 294, 320, 356, 431, 568, **580**, 622, 656, 802
Namrâel, 519

Nantes (Namnetes), 676
Naples, 69, 71, 140, 141, 152, 155, 156, 157, 202, 306, 336, 370, 377, 419, 423, 430, 435, 438, 450, 476, 497, 536, 573, **580**, 630, 631, 743, 772, 774, 778, 822, *231*
Narbolia, 425
Narbonne (Narbona), 249, 338, 386, **581**, 654, 663, 747, C. of (589), 556, 582, 800, 864
Narcissus of Jerusalem, 21, 432, **582**, 831
Narcissus of Neronias, 48, **582**, 757, 830
Nardi, Carlo, 37, 65, 118, 118-19, 126-27, 175, 213, 224, 241, 337, 339, 472-73, 550-51, 571, 585, 833
Narnia (Narni), 423
Narona, 217
Narratio, **582**, 649
Narratio de Rebus Armeniae, 81, **582**
Narratio de Rebus Persicis, **582**
Narrationes de Caede Monachorum in Monte Sinai, **583**, 597
Narsai or Nerses, 3, 112, 357, 429, 562, **583**, 598, 674, 810
Narthex, 170
Natale, 96, 140, 222, 229, 279, 322, 532, 546, **583**, 731, 744
Natale Petri de cathedra, 153
Natalius, 84, **583**, 885
Nathanael, 112
Nativity, 135, 282, 452, 489, 494, 539, **583**, 837, *232*
Naucratius, 114, **584**
Nautin, Pierre, 62, 62-63, 63, 95, 119-20, 143, 198, 235, 235-36, 238, 259, 305, 315, 324, 377, 380, 383-85, 404-5, 483-84, 701, 820, 831-32
Navarra, Leandro, 22, 53, 60, 102, 113, 115, 116, 140, 143-44, 144, 162, 178, 178-79, 186, 188, 245, 270, 272-73, 369, 371, 478, 524, 546, 555-56, 593, 730-31, 838, 843, 849-50, 860, 862-63
Navigius, 97, 568, **584**, 653
Nazarenes, 66, 163, Gospel of the, 57
Nazareth, **584**, *233*
Nazarius and Celsus, 28, 421, 559, **584**, 656
Nazianzus, 361
Nazoraei, **584**
Nazzano, 422
Nazzaro, Antonio, 212-13, 278, 369, 650-51, 704, 823
Nea Anchialos (Thebae Phthiotides), 174, 835
Neapolis, 62
Nebo, 66, 488
Nebridius, 351, 465, **584**, 844
Nechepsos, 92
Nechtan, 762
Nectarius of Constantinople, 17, 59, 83, 142, 196, 350, 549, **584**, 611
Nectarius of Tarsus, 135
Negev, 576
Nemesianus and Companions, **584**
Nemesianus of Thubunae, 584
Nemesius of Emesa, 31, 76, 408, **584**, 606, 695, 699
Nendrum, 130
Neocaesarea, 93, 122, 368, 585, 702, C. of (314-325), 32, 137, 142, 154, 205, 237, 409, **585**, 617
Neochalcedonians, Neochalcedonianism, 115, 165, 276, 362, 402, 443, 444, 446, 465, 481, 570, **585**, 593, 773, 826
Neonas of Seleucia, **585**
Neophytes, 108, 122, 134, 150, 182, 183, 491, 493, 494, 506, **585**, 617, 629, 660, 750, 875
Neoplatonism, 12, 21, 31, 39, 53, 54, 73, 74, 75, 76, 96, 121, 139, 177, 178, 181, 183, 205, 206, 239, 240, 296, 308, 327, 336, 339, 340, 355, 359, 362, 373, 392, 396, 401, 405, 445, 455, 459, 485, 525, 557, 558, 584, **585**, 601, 603, 606, 614,
619, 639, 658, 670, 683, 689, 690, 691, 692, 693, 696, 697, 713, 734, 751, 780, 781, 788, 789, 795, 799, 806, 832, 836, 852
Neopythagoreanism, 60, 455, 585, 587, 589, 590, 602, 692, 805
Nepet (Nepi), 422
Nephalius, 585, **593**
Nepos, **593**, 867
Nepotianus, 182, **593**
Nereus and Achilleus, 563, **593**, 680, Passion of, 246, 680
Nero, 18, 47, 55, 60, 122, 178, 186, 188, 370, 419, 432, **593**, 614, 657, 671, 675, 688, 702, 740, 812, 862, Life of, 6
Nersetes, 80, **593**
Nersetes II, 80
Nersetes III the Builder, 80, 344
Nersetes the Great, 320
Nerva, 30, 60
Nesazio, 420
Nestorians, Nestorianism, 1, 5, 4, 10, 11, 34, 38, 43, 47, 48, 49, 50, 55, 66, 106, 112, 115, 118, 127, 153, 164, 197, 203, 216, 233, 236, 263, 271, 272, 305, 307, 315, 331, 374, 379, 407, 416, 429, 443, 444, 445, 446, 473, 525, 527, 536, 552, 583, 593, **594**, 598, 619, 626, 658, 663, 674, 684, 706, 766, 769, 776, 809, 810, 823, 831, 840, 841, 855, 859, 884, rite, 496, Against, 481, Against, Nestorians and Eutychians, 480
Nestorius, 5, 11, 20, 21, 23, 24, 37, 38, 48, 49, 50, 72, 105, 112, 135, 149, 154, 159, 165, 196, 208, 214, 233, 236, 251, 263, 275, 281, 295, 301, 303, 304, 314, 325, 372, 375, 401, 403, 410, 416, 436, 442, 444, 466, 482, 484, 501, 525, 538, 545, 550, 556, 569, 570, **594**, 641, 643, 652, 663, 677, 679, 713, 716, 729, 732, 740, 744, 773, 774, 784, 808, 824, 827, 829, 833, 834, 838, 841, 853, 870, 884, Against Nestorius, 214, 342, Liturgy, 1
Nestorius of Constantinople, 742
Neunheuser, Burkhard, 41-42, 189-90, 492, 611
Neuss (Novaesium), 349
Nicaea, 40, 85, 87, 89, 96, 159, 169, 186, 194, 200, 205, 207, 407, **594**, 596, C. of (325: 1st ecumenical), 5, 7, 19, 20, 21, 22, 32, 42, 47, 48, 49, 50, 64, 67, 72, 77, 79, 90, 93, 95, 106, 109, 122, 136, 137, 138, 142, 143, 148, 162, 166, 188, 191, 193, 196, 200, 205, 207, 223, 237, 245, 257, 263, 271, 275, 276, 298, 299, 302, 303, 305, 307, 314, 320, 323, 325, 344, 357, 403, 417, 429, 432, 460, 487, 496, 512, 514, 522, 525, 538, 546, 551, 556, 566, 582, 585, 595, 596, 598, 604, 617, 618, 620, 626, 637, 641, 643, 646, 657, 660, 662, 663, 669, 705, 720, 728, 742, 743, 759, 783, 788, 793, 802, 825, 830, 833, 835, 848, 851, 852, 854, 855, 870, 882, C. of (784), 830, C. of (787: 7th ecumenical), 39, 52, 122, 182, 192, 193, 205, 346, 407, 442, 566, 830, 864
Nicene creed/formula, 23, 63, 77, 78, 93, 128, 143, 159, 202, 205, 207, 228, 275, 314, 323, 391, 396, 403, 479, 538, 569, 585, 643, 757, 778, 828, 858
Nicene-Constantinopolitan creed, 324, 584
Nicephorus Callistus, 95, 308, 356, 659, 724, 785, 805
Nicephorus of Constantinople, 407
Nicephorus Uranos, 67
Niceta Paphlagonius, 346
Niceta Syncellus, 346
Nicetas Acominatus, 684
Nicetas of Aquileia, 244
Nicetas of Heraclea, 827
Nicetas of Remesiana, 30, 323, 398, **595**, 661, 816
Nicetius of Trier, 348, 513, **595**, 850

Nicholas of Ancyra, **596**
Nicholas of Cusa, 397
Nicholas of Damascus, 121
Nicholas of Myra, 38, 135, **596**
Nicodemus, Gospel of, 9, 57, 517, 687
Nicodemus the Hagiorite, 684
Nicolaïtans, 26, 384, 556, **596**
Nicolaus, 596
Nicomachean Ethics, 75, 76
Nicomachus Flavianus, 706
Nicomedia, 32, 43, 85, 122, 154, 220, 236, 301, 307, 350, 356, 444, 469, 484, 521, 594, **596**, 639, 673
Nicopius, 780
Nicopolis, 283, 572
Nicopolis-Emmaus, 635
Nicostratus, 558
Niculitel, 605
Nigrinus, 148
Nigg cemetery, Ross-shire, 132
Nika revolt, 194, 464, 488
Nilus of Ancyra, 37, 84, 262, 306, 333, 369, 407, 577, 583, **597**, 787, 792, 816, 879, 880, 881
Nimbus, 40, 292, 345, 435, 540, **597**, 675, 685, 689, 705, 759, 778, 803, *234*
Nîmes (Nemausus), 581, 766, C. of (394), 79, 338, 400, 597
Ninian, 130, 132, **597**, 762
Ninotsminda, 344
Nino, 344, 578, **597**, Life of, 345
Nisibis, 1, 3, 4, 30, 48, 65, 106, 110, 112, 216, 263, 271, 307, 374, 403, 429, 454, 583, **585**, 663, 674, 808, 810, *235*, *236*
Nitria, 264, 567, 759
Noah, 3, 89, 108, 161, 175, 219, 229, 247, 276, 293, 309, 327, 341, 349, 351, 363, 415, 429, 453, 477, 504, 525, 575, **598**, 615, 668, 802, 803, 832, 860, ark, 572, 598, 621, *237*
Nobadae, 505
Nocent, Adrian, 615-16, 749-51
Nocera Umbra, 423
Noetus of Smyrna, 53, 90, 278, 384, 566, **599**, 653, 706, 749, Against, 121, 488, 851
Nola, 27, 493, 843
Noli, 421
Nomina Sacra, **599**, 647
Nonantula, 126
Nonna, 361, **599**, 882
Nonnberg, 747
Nonnus, Ps.-, 599
Nonnus of Panopolis, **599**, 650
Norea, 771
Noricum, 70, 307, 572, *238*
Noris, 101
North Elmham, Norfolk, 131
Nostrianus of Naples, 728
Notae de Textu Evangeliorum, **600**
Notarii, 247, 521, **600**, 711, 794
Notitia Dignitatum, 189, **600**, 601, 834, 837
Notitia Galliarum, 349, **601**
Notitia Provinciarum et Civitatum Africae, **601**, 607, 868
Nous, 20, 40, 44, 45, 59, 73, 75, 163, 164, 198, 251, 306, 340, 353, 373, 378, 396, 406, 516, 553, 557, 569, 586, 587, 588, 590, 592, **601**, 639, 692, 696, 698, 699, 717, 789, 832
Novalja (island of Pag), 218
Novara (Novaria), 337, 421
Novatian, 109, 121, 161, 164, 202, 207, 212, 222, 238, 298, 311, 315, 321, 324, 337, 448, 472, 473, **603**, 605, 628, 640, 646, 733, 741, 743, 769, 797, 822, 851
Novatianists, Novatianism, 7, 17, 63, 146, 154, 175, 222, 236, 238, 260, 324, 328, 337, 371, 376, 377, 450, 473, 528, **604**, 628, 668, 686, 720, 804, 841
Novatus, 321, 328, **605**, 640
Novatus of Carthage, 604, **605**
Novatus the Catholic, **605**

Novellae, 122, 142, 183
Novelty, 52, 155, 166, *see also* Antiquity
Noviodunum, **605**, 764, 765
Noyon (Noviomagus), 268
Nubia, 200, **605**, 824
Nuceria Alfaterna (Nocera Inferiore), 424
Numbers, 20, 50, 55, 56, 202, 214, 223, 228, 310, 336, 366, 415, 455, 526, **605**, 624, 802, 825, 867
Numenius of Apamea, 183, 455, 585, 587, 589, 602, **606**, 692, 693, 695
Numidia, 13, 14, 18, 109, 147, 175, 176, 193, 228, 246, 247, 248, 249, 543, **607**, 612, 613, 653, 680, 743, 765, 774, 840, 842, 859, 861
Numismatics, 19, **607**
Nundinarius, 779
Nyssa, 43, 137, 363, **608**

Oak, Synod of the, 5, 51, 196, 214, 295, 350, 440, 514, 537, **609**, 636, 772, 831
Oblations of the Faithful, 90, 91, 170, 246, 259, 277, 491, **609**
Oceanus, **609**
Octateuch of Clement, 62, 104, 198, 428, 440, **609**
Octavius, 10, 60, 233, 563, 672, 696
Odenathus of Palmyra, 336
Odes of Solomon, 9, 176, 263, 316, **609**, 674, 718, 792, 810
Odilia, **610**
Odoacer, 43, 149, 281, 339, 342, 357, 400, 476, 486, 505, 565, **610**, 645, 730, 780, 806, 828, 861, 863, 884
Odysseus, 578
Odzoun, 80
Oecumenius, 38, 55, 311, 352, **610**
Offa, 143
Ogdoas - Ogdoad, 71, 204, 220, 274, 309, 328, 336, 354, 598, 606, **610**, 749, 800, 802, 805, 859
Ohrid, 174
Oil, 33, 41, 42, 108, 123, 172, 174, 190, 219, 221, 232, 265, 296, 332, 409, 420, 427, 491, **611**, 651, 749, 751, 801, *see also* Annointing
Oktoichos, 135
Olba, modern Ura, 808
Olbia (Fausiana?), 425, 488
Olisipo (Lisbon), 334, 512
Olteni (district of Ilfov), 217
Olybrius, 40
Olybrius Probus, 225
Olympia, 2
Olympia(s), 441, **611**, 880, Life of, 374, 768
Olympiodorus of Alexandria, 31, 35, **611**, 716, 878, 879
Olympiodorus of Thebes, 790
Olympius, **611**
Olympus, **612**
Omega, *see* Alpha and Omega
Onomasticon, 300, 431, 654
Onuphrius, **612**, 646
Opellius Macrinus, 144
Opelt, Ilona, 144, 187-88, 272, 466, 525, 555, **612**, 712
Opera Omnia, 203, **612**, 655
Ophites, 57, 110, 139, 161, 353, 384, 542, **612**, 771, *see also* Naas, Naassenes
Opitergium (Oderzo), 420
Optatian, 210, **612**, 721
Optatus of Milevis, 15, 145, 163, 175, 228, 249, 282, 322, 350, 410, 493, 497, 524, 607, **612**, 651, 796, 842, 866
Optatus of Thamugadi, 247, 351, 545, 607, **613**, 711, 840
Opus Imperfectum in Matthaeum, 78, 442, 517, 543, 546, **613**, 817, 837
Opus Paschale, 312, 766
Oracles, 51, 55, 75, 82, 543, 587, **614**, 722, 734, 769, 787
Orange (Arausio), 304, C. of (441), 99, 188, 295, 381, 548, **615**, 753, 858, C. of

(529), 101, 138, 359, 555, 645, 717, 767, 782
Orans, 130, 219, 323, 434, 477, 536, 540, 555, 574, 599, **615**, 629, 650, 668, 675, 705, 707, 755, 778, 802, 803, 834, *239*
Orarium, 253
Oratio Cypriani, **615**
Orbe, Antonio, 413-16
Orders, ordination, 23, 71, 79, 103, 110, 123, 142, 147, 148, 153, 154, 182, 183, 185, 225, 232, 235, 259, 260, 276, 278, 289, 335, 361, 363, 383, 399, 407, 439, 467, 473, 477, 494, 502, 504, 555, 556, 560, 561, 562, 609, **615**, 616, 622, 645, 737, 739, 745, 750, 755, 793, 794, 800, 814, 818, 820, 833, 844, 849, 858, 865, 882, 885, *see also* Clergy, Diaconate, Laying on of hands, Ministry
Ordines Romani, 211, 222, 393, 394, 492, 493, 502, **616**, 745
Organization, Ecclesiastical, 22, 47, 91, 103, 105, 132, 143, 155, 200, 202, 219, 260, 264, 315, 336, 337, 347, 349, 397, 411, 438, 472, 483, 495, 512, 537, 566, 581, 600, 601, **617**, 715, 740, 742, 810, 824, 838, 845, 863
Orientalis, 126
Orientation, 170, 257, *see also* East
Orientius, 11, 144, **619**
Origen, 5, 7, 9, 10, 21, 22, 23, 24, 27, 29, 32, 37, 39, 45, 46, 47, 49, 50, 52, 54, 55, 56, 57, 60, 65, 67, 72, 73, 74, 76, 82, 84, 90, 93, 95, 96, 102, 107, 111, 113, 114, 116, 117, 119, 120, 121, 128, 138, 139, 141, 150, 155, 160, 161, 162, 163, 164, 167, 168, 176, 177, 178, 180, 183, 187, 190, 191, 199, 202, 204, 205, 206, 207, 214, 219, 221, 223, 225, 226, 227, 229, 230, 232, 233, 235, 236, 238, 242, 243, 244, 245, 251, 252, 255, 256, 257, 258, 260, 261, 262, 264, 269, 272, 283, 285, 286, 287, 299, 300, 303, 305, 306, 307, 308, 310, 311, 312, 313, 315, 317, 318, 319, 320, 323, 324, 327, 341, 354, 355, 356, 358, 361, 363, 364, 365, 368, 372, 373, 374, 375, 380, 381, 382, 391, 395, 396, 398, 404, 405, 406, 408, 409, 410, 414, 415, 430, 431, 432, 437, 448, 449, 453, 455, 457, 460, 467, 470, 475, 477, 478, 480, 486, 492, 495, 499, 500, 506, 510, 511, 514, 517, 520, 526, 528, 530, 537, 543, 550, 552, 558, 560, 561, 566, 575, 577, 578, 584, 587, 593, 598, 603, 606, 609, 612, 614, 615, **619**, 623, 624, 626, 628, 629, 638, 639, 640, 647, 648, 649, 655, 656, 658, 663, 664, 672, 677, 678, 682, 683, 689, 691, 692, 694, 695, 696, 697, 698, 704, 705, 707, 708, 709, 713, 715, 716, 717, 718, 719, 720, 722, 723, 725, 727, 748, 750, 751, 760, 761, 763, 764, 765, 768, 775, 781, 785, 786, 788, 789, 792, 793, 797, 798, 801, 803, 805, 810, 813, 815, 816, 817, 818, 822, 823, 826, 829, 830, 831, 832, 833, 837, 848, 851, 854, 863, 864, 867, 868, 872, 875, 876, 878, 879, 880, 882, 885, 886, Address of thanks to Origen, 235, 368, 373, 619, 620, 693, 761, 805
Origenist controversy, 180, 215, 311, 372, 431, 432, 438, 475, 521, 567, 620, 636, 852, 660, 863
Origenists, Origenism, 38, 49, 112, 114, 127, 197, 215, 225, 227, 245, 276, 282, 310, 339, 343, 358, 362, 372, 374, 430, 465, 480, 507, 514, 547, 553, 567, 593, 597, 609, 622, 623, 648, 666, 674, 677, 687, 733, 746, 748, 759, 776, 794, 823, 826, 831, 869
Origen the Neoplatonist, 32, 586, **624**, 805
Original sin, 98, 100, 101, 147, 175, 359, 462, 539, 611, 623, 666, 733, 775, 776, 781, 825, 849, *see also* Sin
Orlandi, Tito, 4, 17, 22, 44, 97, 119, 120, 194, 199, 201, 219, 225, 297, 438-39,

445-46, 446, 514, 612, 678, 689, 699, 724, 746, 755, 759, 776, 821, 826, 829, 850
Orlea, 217
Orléans (Aurelianum), 40, 329, C. of (454), 102, C. of (511), 183, 409, 517, 555, C. of (533), 482, 555, 853, C. of (538), 409, 482, 555, C. of (541), 212, 556, 867, C. of (549), 212, 556
Orontius, 41, 554, 845
Orosius, 102, 105, 127, 134, 389, 397, 432, 507, **624**, 628, 665, 687, 712, 774, 791, 795, 834, 850
Orpheus, 434, 572, 578, **625**, 630, 691, 802, 803, 808, 836, *240*, Orphism, 82, 106
Orsera, 420
Orthodoxy, 1, 10, 27, 29, 32, 36, 37, 38, 49, 50, 58, 59, 64, 66, 72, 77, 91, 101, 112, 123, 136, 153, 160, 162, 165, 177, 192, 198, 205, 207, 208, 213, 218, 223, 224, 233, 238, 242, 244, 267, 274, 282, 284, 296, 306, 307, 320, 322, 325, 347, 357, 364, 366, 372, 376, 383, 391, 392, 405, 407, 429, 432, 436, 439, 441, 443, 447, 448, 470, 479, 484, 495, 499, 508, 533, 538, 546, 549, 551, 560, 564, 569, 570, 595, 600, 609, 619, 623, **625**, 638, 645, 656, 654, 659, 664, 665, 684, 686, 687, 688, 698, 718, 727, 730, 732, 738, 744, 748, 758, 763, 767, 771, 775, 778, 779, 788, 790, 804, 823, 828, 845, 851, 852, 859, 860, 861, 871, 877
Orthodoxy Sunday, 135
Ortygia, 807
Osborn, Eric, 286-89
Osca (Huesca), 813
Osenovo, 631
Osidius Geta, 158
Osius, 139
Osk Vank, 345
Osrhoene, 2, 48, 263, 808, 809
Osseni, 269
Ossius of Cordova, 23, 48, 77, 106, 110, 139, 246, 270, 347, 460, 546, 554, 595, **626**, 757, 790
Ostanes, 82
Ostia, 178, 423
Ostiarius, 232
Ostiense, 211
Ostrogoths, 119, 179, 283, 335, 356, 451, 464, 565, *see also* Goths
Oswald, 18, 130
Oswin, 18, 119, 129
Oswiu, 159
Otacilia Severa, 682
Othmar, 334
Otreius of Melitene, 304
Otricoli, 423
Ousia, 49, 58, 59, 77, 78, 93, 196, 207, 214, 256, 305, 308, 355, 391, 392, 395, 396, 472, 558, 566, 589, 602, 603, **626**, 654, 663, 692, 737, 796, 806, 830, 852, 853
Oxama, 148
Oxyrhynchus, 24, 203, 264, 519, 647, papyri, 18

Pacatus, **628**, 856
Pachino, 425
Pachomius, 22, 31, 119, 184, 199, 230, 264, 266, 296, 319, 369, 400, 472, 499, 500, 521, 567, **628**, 746, 761, 826, Rule of, 870, Life of, 237
Pacianus of Barcelona, 30, 110, 232, 304, **628**, 790, 804
Pact of Union, 214
Pact of mercy, 291
Padula, 424
Padua (Patavium), 420, 464
Paedagogus, 74, 177, 180, 252, 401, 552
Paenula, 253, 254
Paestum, 424
Pagans, Paganism, 1, 2, 4, 6, 7, 8, 10, 16,

22, 25, 26, 27, 28, 30, 32, 39, 44, 46, 47, 48, 49, 52, 53, 60, 61, 72, 83, 90, 92, 96, 116, 134, 136, 138, 141, 147, 155, 157, 163, 166, 167, 168, 169, 171, 177, 181, 194, 198, 199, 202, 205, 208, 211, 212, 214, 218, 219, 222, 226, 228, 231, 232, 234, 236, 237, 242, 245, 256, 260, 268, 270, 271, 279, 285, 288, 294, 298, 300, 307, 309, 315, 316, 320, 322, 325, 336, 339, 340, 342, 347, 351, 352, 354, 358, 360, 367, 372, 375, 376, 378, 381, 382, 384, 389, 397, 399, 402, 404, 406, 408, 409, 411, 412, 415, 432, 433, 434, 436, 449, 457, 459, 460, 463, 469, 473, 476, 493, 508, 517, 527, 528, 531, 533, 548, 551, 552, 577, 578, 582, 583, 595, 604, 612, 614, 624, 625, **628**, 629, 657, 659, 665, 672, 678, 684, 685, 687, 689, 702, 704, 706, 710, 713, 715, 717, 721, 727, 728, 731, 732, 734, 736, 739, 740, 741, 745, 751, 752, 757, 763, 767, 774-76, 780, 790, 792, 800, 801, 802, 803, 804, 805, 808, 812, 814, 815, 818, 820, 823, 825, 827, 828, 831, 833, 839, 840, 843, 854, 856, 859, 862, 866, 871, 876, 877, Against the pagans, 778
Pahlavi, 417, 518, **629**
Paideia, 61, 115, 167, 230, 367, **629**, 683
Painter, Kenneth, 129-34
Painting, 2, 3, 9, 24, 41, 62, 68, 70, 81, 88, 89, 107, 108, 130, 138, 141, 156, 158, 187, 213, 217, 219, 221, 230, 255, 267, 269, 274, 278, 290, 313, 315, 323, 327, 329, 345, 370, 378, 406, 419, 425, 428, 429, 434, 435, 437, 450, 452, 457, 476, 477, 482, 520, 536, 540, 554, 555, 581, 598, 612, **629**, 636, 650, 659, 668, 675, 677, 684, 685, 687, 689, 705, 715, 753, 756, 757, 759, 764, 778, 785, 786, 795, 803, 807, 808, 845, 846, 847, 860, *241, 242, 243*
Palaeography, 126, 278, 599, **632**, 647, 682, 724
Palaeoslavonic, 2, 9, 16, 38, 60, 113, 115, 211, 257, 383, 442, 498, 557, **633**, 763, 786, 827, 880
Palazzo Lancellotti, 187
Palchonius of Braga, 105
Palencia (Pallantia), 148, 791
Paleotti, Gabriel, 68
Palermo (Panormus), 425
Palestine, 1, 3, 6, 7, 23, 24, 37, 39, 43, 48, 52, 60, 65, 66, 69, 89, 109, 134, 138, 160, 164, 167, 172, 173, 180, 187, 205, 215, 225, 230, 241, 264, 268, 271, 282, 304, 307, 310, 339, 371, 426, 427, 432, 447, 450, 454, 472, 486, 510, 562, 567, 568, 599, 618, 631, **635**, 688, 755, 797, 807, 823, 863, 869, 884, *244, 245, 246, 247*
Palla, 253
Palladius, evangelist of Ireland, 154, 155, 411, **636**
Palladius of Helenopolis, 17, 61, 84, 121, 184, 233, 235, 271, 282, 303, 306, 374, 561, 611, 616, 634, **636**, 660, 664, 704, 746, 771, 785, 790, 803, 816, 831
Palladius of Ratiaria, 64, 116, 153, 299, 329, **637**, 765, 856
Palladius of Saintes, **637**
Palladius of Suedri, **637**
Pallas, Demetrios, 7, 201-2, 283, 514-16
Palliolum, 253
Palliolus, 253
Pallium, 253, 865
Palmas, 27, 637
Palmyra, 22, 102, 126, 210, **637**, 807, 808, 884
Paludamentum, 253, 254
Palut, 10, 48, 263, **637**
Pammachius, 91, 192, 303, 454, 609, **637**, 746
Pamphilus, martyr, 673
Pamphilus of Caesarea, 23, 138, 225, 235,

299, 300, 620, 623, **638**, 639, 687, 746, 789
Pamphilus of Jerusalem, **638**
Pamphylia, 63, 85, 86, 512
Pampilona (Pamplona), 813
Panaetius, 53, 720, 805
Panajir Dagh, 772
Panarion, 4, 75, 282, 384, 436
Pancras, **638**
Pandecta, 52
Pange lingua, 729, 862
Pani, Giancarlo, 40, 225, 449-50
Pani Ermini, Letizia, 422-26, 457-58, 581, 728, 807
Pannonia, 19, 26, 65, 96, 329, 356, 450, 565, 572, 600, **638**, 759, 783, 860, 861, 885
Panodorus, 166, **639**
Panopolis, 216
Pantaenus, 603, **639**, 658, 761
Pantaleon, 513, **639**
Pantheism, 98, 243, 306, 378, 623, **639**
Pantocrator, 708
Pantoleon, **640**
Pantomime, 273
Panvino, Onofrio, 68
Papacy, 7, 9, 104, 218, 245, 330, 497, **640**, 653, 667, 704, 782, 867, *see also* Pope
Papas Bar Aggai, **646**, 674
Paphlagonia, 27, 122
Paphnutius, **646**
Paphos, 213, 772
Papias of Hierapolis, 55, 61, 73, 285, 306, 449, 526, 560, **647**, 657, 675, 680, 681, 701, 732, 867
Papiscus, 27, 73
Pappus, 736
Papylus, 17
Papyrus, 12, 32, 120, 125, 181, 232, 264, 265, 632, 633, **647**, 656, 724, 763, 793, 797
Parabalani, 648
Parable, 1, 55, 56, 199, 223, 243, 250, 260, 261, 276, 282, 286, 372, 377, 394, 466, 506, 509, 510, 543, 605, **648**, 749, 778, 802
Paradise, 9, 10, 91, 135, 136, 183, 186, 257, 262, 276, 285, 292, 306, 319, 358, 398, 414, 457, 465, 525, 536, 572, 615, **649**, 670, 675, 685, 688, 707, 712, 718, 719, 731, 775, 788, 792, 803, 816, 860, 873, *248*
Paralytic, Healing of the, 38, 108, 255, 466, 629, **650**, 668, *249*
Parapetasma, **650**, 705, *250*
Paraphrase, Biblical, 599, **650**
Parasceva, Passion of, 445
Parchment, 647
Parchor, 419
Parentalia, 103
Parentium (Parenzo-Porec), 41, 173, 174, 202, 336, 420, 464, 509, 539, 544, 545, 568, 572, 573, **651**, 671, 705, *251*
Paris (Parisius), 174, 183, 307, 329, 337, 342, C. of (361), 381, **651**, C. of (552), 556, C. of (556-573), 556, C. of (573), 102, 556, 637, C. of (577), 556, C. of (614), 330, 349, 530, 556, 850
Paris Bible, 631
Parishes, *see* Organization, Ecclesiastical
Parma, 421
Parmenian, 98, 247, 248, 249, 545, 607, 613, **651**, 738, 838, Against, 613
Parmenides, 239, 586, 587, 589, 590, 591, 603, 692, 697
Paros, 209
Parousia, 34, 36, 47, 56, 167, 210, 226, 242, 285, 286, 291, 361, 380, 397, 408, 437, 457, 463, 467, 493, 506, 510, 532, 573, 577, 621, **651**, 706, 732, 734, 819, 839, 840
Parrhesia, **652**
Parthemius, **652**, 779
Parthenius, 67

Numbers in bold refer to the main entry on the subject. *Numbers in italics refer to illustration numbers (vol II, pages 951-1094)*

Parthenius of Constantinople, **652**
Parthenon, 96
Pascentius the Arian, **652**
Paschal I, 646, 706
Paschal II, 122
Paschasinus of Lilybaeum, 159, **652**
Paschasius of Dumium, **652**
Paschasius of Rome, 61, 255, **652**, 772
Paschoud, François, 624-25
Pasquato, Ottorino, 47-48, 49, 50, 51, 116, 219, 273-74, 617, 618-19
Passau (Batava Castra), 347
Passover, 210, 257, 312, 408, 436, 470, 541, 728
Pastor, 18, **652**, 801
Pastoral staff, 866
Patapius, 38
Patarini, 634
Paterikon, 60, 220, 271, **652**
Paterius, 366, **652**, 655, 879, 881
Paternus, 18, 127, 534, 651
Paternus of Tomi, 845
Patiens of Lyons, 513, **653**
Patmowt'iwn Hayoc', 16
Patras, 7, 38
Patria Constantinopolis, 379
Patriarchates, 618
Patriciani, **653**
Patricius, 97, 113, 568, 584, **653**
Patrick, 121, 130, 186, 307, 346, 411, 636, **653**, 743, 765, Synod of, 411
Patrimony of St Peter, 366, 367, 397, 642, **653**, 743
Patripassian, 140, 164, 181, 278, 511, 566
Patristics, *see* Patrology
Patroclus of Arles, 191, 530, 643, 853, **654**, 869, 885
Patrology, 12, 66, 68, 101, 606, 646, 647, **654**, 724
Patron, -age, 171, 535, **656**, 676
Patrophilus of Pithyonta, 344
Patrophilus of Scythopolis, 215, 301, 302, **657**
Paul, 12, 17, 18, 25, 44, 45, 47, 52, 53, 57, 61, 62, 64, 67, 69, 81, 84, 87, 91, 95, 96, 98, 111, 120, 132, 138, 141, 150, 151, 154, 156, 159, 160, 161, 163, 167, 175, 176, 178, 181, 190, 201, 204, 205, 208, 213, 217, 218, 221, 226, 229, 235, 238, 252, 258, 260, 272, 283, 284, 287, 292, 296, 306, 308, 309, 310, 311, 313, 316, 318, 319, 320, 322, 326, 333, 336, 337, 338, 356, 358, 359, 373, 374, 375, 376, 380, 390, 397, 399, 404, 407, 414, 419, 430, 431, 432, 435, 453, 454, 456, 457, 461, 462, 463, 467, 469, 471, 472, 490, 492, 493, 498, 503, 504, 509, 510, 515, 516, 517, 518, 521, 523, 526, 528, 532, 536, 540, 546, 551, 552, 559, 561, 562, 573, 574, 577, 581, 593, 596, 597, 609, 615, 617, 630, 640, 641, 643, 647, 649, **657**, 664, 666, 670, 671, 675, 676, 677, 678, 681, 685, 688, 703, 705, 706, 707, 718, 740, 748, 749, 756, 762, 763, 767, 773, 775, 777, 786, 789, 790, 798, 806, 814, 815, 821, 825, 827, 833, 840, 842, 843, 847, 851, 854, 860, 875, 877, 252, Acts of, 657, 723, 843, Acts of Peter and Paul, 676, Acts of Paul and Thecla, 659, Life of, 121, 230, 664, Passion of Peter and Paul, 676
Paul I, 676, 680
Paul II, **659**, 726, 823, 824
Paul III of Constantinople, 140
Paul VI, 232
Paula, 122, 161, 184, 192, 303, 430, 500, 521, 637, **659**, 688
Paul Helladicus, **660**
Paulicians, 80, 136, 407, 633, **660**
Paulina, 192, 303, 688
Paulinians, 182, **660**, 663, 841
Paulinianus, **660**
Paulinism, 657
Paulinus of Antioch, 20, 24, 48, 49, 51, 58, 59, 78, 93, 114, 168, 195, 196, 218, 250, 282, 298, 305, 325, 430, 431, 508, 550, 559, **660**, 678, 744, 841, 873
Paulinus of Aquileia, 26, 655
Paulinus of Béziers, 278
Paulinus of Bordeaux, **660**
Paulinus of Ireland, 551
Paulinus of Milan, 29, 102, 121, 137, 497, **660**, 705, 765, 828
Paulinus of Nola, 1, 3, 11, 27, 28, 29, 44, 110, 126, 177, 182, 191, 192, 225, 272, 278, 282, 292, 295, 312, 315, 322, 401, 407, 426, 438, 453, 483, 488, 497, 501, 528, 529, 549, 595, 628, 645, 651, **660**, 661, 669, 676, 721, 735, 739, 766, 774, 780, 793, 799, 816, 829, 843, 856, 865, 869, 870, 876
Paulinus of Périgueux, 144, 651, **661**, 799
Paulinus of Pella, 121, 126, 530, 651, **661**
Paulinus of Trier, **661**, 850
Paulinus of Tyre, 297, 302, **661**, 855
Paulinus of York, 129, 131, 397, 465, **662**, 883
Paul of Alexandria, 339
Paul of Alexandria [gnostic], 23
Paul of Antioch, 462, 768, 823, 824
Paul of Apamea, **662**
Paul of Aphrodisia, **662**
Paul of Callinicum, 428, **662**, 769
Paul of Canopus, 570, **662**
Paul of Cirta, 673
Paul of Concordia, **662**
Paul of Constantinople, 21, 84, 195, 302, 388, 516, 546, 655, **662**
Paul of Edessa, 428, **662**
Paul of Emesa, 163, **663**
Paul of Ephesus, 197
Paul of León, **663**
Paul of Mérida, **663**
Paul of Narbonne, 238, 581, **663**, 854
Paul of Nisibis (or Paul the Persian), 3, 374, 598, **663**
Paul of Obba, 584
Paul of Samosata, 11, 16, 23, 47, 48, 50, 53, 95, 102, 164, 207, 237, 238, 246, 301, 303, 324, 368, 371, 396, 400, 507, 518, 522, 559, 566, 594, 644, 660, **663**, 685, 704, 832, 884
Paul of Tella, 810
Paul of Thebes, 230, 291, **664**
Paul of Verdun, **664**
Paul Silentiarius, 278, 465, **664**
Paul the Black, 428
Paul the Deacon, 103, 189, 389, 393, 540, 597, 772, 773
Paul the Donatist, 778
Paul the Simple, **664**
Paul Warnefried, 502
Paulus Evergetinos, 61
Pautalia (now Kjustendil), 837
Pavan, Enzo, 186, 296, 599-600, 638, 735, 772, 864-66
Pavia (Ticinum), 67, 124, 281, 838
Pbou (now Faw al-Qibli), 4, 31, 265, 628
Peace, 53, 96, 116, 117, 123, 147, 181, 212, 224, 239, 260, 279, 287, 288, 294, 307, 309, 319, 336, 361, 398, 411, 434, 467, 473, 479, 484, 489, 528, 629, 649, **664**, 710, 748, 778, 803, 817, 818, 828, 875
Peacock, 803
Pécs, 631
Pectoral cross, 866
Pectorius, 104, 232, 280, 293, 337, 803
Pedagogue, 434, 877
Pelagia, 13, 175, **665**
Pelagian controversy, 9, 72, 100, 225, 462
Pelagians, Pelagianism, 9, 27, 41, 46, 72, 96, 98, 100, 101, 102, 105, 145, 147, 154, 175, 191, 213, 225, 233, 269, 288, 318, 320, 321, 328, 331, 339, 346, 381, 399, 409, 458, 462, 475, 479, 511, 521, 525, 624, 641, 645, 653, **665**, 709, 717, 719, 744, 745, 767, 774, 782, 784, 799, 822, 849, 859, 878, 885, 886

Pelagius, 30, 40, 49, 58, 61, 72, 79, 96, 98, 102, 129, 137, 147, 225, 241, 261, 311, 339, 357, 359, 375, 430, 432, 438, 447, 475, 507, 538, 624, 641, 658, 660, 662, **665**, 714, 742, 743, 744, 746, 767, 776, 781, 796, 830, 843, 885, Against Pelagius, 342
Pelagius, Ps.-, 878
Pelagius I, 642, 645, **666**, 667, 773
Pelagius II, 102, 197, 224, 366, 475, 486, 642, 645, **667**, 685, 743, 773
Pelagius of Ecija, 775
Pelagius of Laodicea, **665**, 854
Pelekanidou, Elli, 681-82, 833-34
Pella, 259, 432, 436, 584, 635, 779
Pelland, Gilles, 340-44
Pellegrino, Michele (Card.), 60, 176-78, 494-95, 548-49, 754
Pellikan, Johann, 655
Pelusium, 140, 417
Penance, *see* Penitence
Penda, 397
Penitence, 29, 42, 61, 105, 106, 110, 115, 146, 151, 181, 182, 183, 186, 187, 211, 212, 242, 261, 288, 309, 322, 350, 361, 377, 379, 382, 383, 409, 410, 411, 439, 473, 477, 484, 492, 493, 494, 522, 555, 560, 583, 585, 595, 598, 603, 604, 622, 628, 660, **667**, 709, 725, 741, 745, 750, 752, 758, 759, 765, 780, 781, 790, 792, 799, 802, 804, 819, 828, 858, 868
Penitential books, 223, 669
Penitents, 17, 50, 71, 540
Pentalogus, 827
Pentapolis, 20, 23, 224, 238, 264, 486, 806
Pentecost, 17, 34, 50, 122, 135, 186, 200, 208, 222, 229, 258, 290, 298, 319, 364, 374, 394, 481, 493, 494, 504, 540, 561, 575, 669, 671, 738, 749, 780
Pentecostarion, 135
Peplon, 253
Pepuza, 571
Pepuziani, 83
Pepysian Gospel Harmony, 234
Perati, 612
Peregrinatio, *see* Pilgrimage, Itinerary
Peregrinatio Egeriae, *see* Egeria, Itinerarium Egeriae
Peregrinus, 106, 508, **669**, 723
Peregrinus of Auxerre, **670**
Perennes, 59, 534
Peretto, Eli, 5, 8, 26, 38, 47, 52, 63-64, 91, 120-21, 141, 150, 153, 163, 223-24, 229-30, 326, 333-34, 342-43, 352, 404, 408, 429-30, 430, 449, 451-52, 456, 456-57, 461-62, 467, 484, 509-11, 539, 543, 596, 687, 723, 755, 771, 780, 797, 798, 805, 812, 818, 821, 827, 833, 849
Perfection according to the Saviour, 815
Perfume, 409, **670**, 786, 801
Perga, 86
Pergamum, 17, 85, 87
Pergola, 172
Pergola, Philippe, 203
Perichoresis, 392, 443, **670**
Perigenes, **670**
Périgueux (Petrocoricum), 661, 812
Periodeuta, **670**
Peripatetics, 53, 74, 75
Peristephanon, 721, 816
Perizoma, 253
Perones, 254
Peroz, 112, 118, 119, 518
Perpetua and Felicity, 10, 13, 14, 145, 186, 321, 532, 533, 552, **670**, 752, 818, 253, Passion of, 13, 150, 151, 228, 497, 524, 533, 534, 571, 670, 672, 768, 818, 873
Perpetuus of Tours, 140, 661, **671**, 676, 846, 862
Perrona Scottorum, 154
Perrone, Lorenzo, 439, 480, 480-81, 679, 826, 841

Persecution, 1, 13, 14, 16, 20, 22, 23, 28, 39, 43, 44, 47, 48, 54, 55, 60, 79, 82, 84, 92, 93, 95, 96, 102, 106, 117, 122, 129, 139, 143, 145, 151, 155, 156, 157, 167, 176, 177, 180, 181, 184, 188, 189, 191, 193, 200, 202, 208, 211, 212, 213, 217, 219, 220, 223, 225, 230, 232, 236, 237, 238, 245, 246, 247, 248, 260, 261, 263, 270, 271, 275, 279, 280, 295, 299, 301, 307, 314, 315, 321, 324, 326, 329, 333, 336, 337, 338, 342, 347, 368, 370, 380, 388, 398, 404, 419, 420, 430, 432, 433, 434, 473, 480, 483, 484, 486, 507, 512, 521, 522, 524, 546, 551, 554, 559, 571, 584, 585, 593, 596, 603, 604, 605, 607, 614, 619, 620, 625, 626, 640, 649, 657, 664, 667, 670, **671**, 676, 684, 687, 689, 699, 701, 705, 712, 737, 739, 741, 745, 748, 752, 757, 758, 759, 764, 768, 772, 777, 779, 790, 792, 794, 806, 812, 818, 819, 829, 830, 833, 845, 849, 854, 860, 867, 870, 875, 883
Persia, 2, 47, 48, 54, 118, 230, 263, 276, 307, 416, 446, 454, 457, 531, 536, 567, 594, 598, 619, **674**, 757, 766, 780, 797, 809, 840, 883
Person, 121, 125, 163, 164, 191, 205, 207, 208, 215, 239, 240, 245, 260, 336, 352, 356, 364, 374, 381, 391, 392, 406, 479, 483, 546, 547, 568
Persona, 29, 78, 163, 164, 245, 276, 446, 526, 569, 594, 604, 621, 622, 626, 652, 670, **675**, 679, 685, 695, 706, 716, 743, 747, 788, 797, 819, 830, 831, 845, 851, 870
Personifications, **675**
Perteu, 203
Perusia (Perugia), 422
Pesher, 187
Pesilta, 428
Pessinus, 333
Petasus, 253
Peter, abbat, 148
Peter, apostle, 8, 17, 22, 24, 27, 30, 33, 38, 47, 57, 61, 64, 67, 70, 81, 104, 108, 113, 129, 138, 153, 156, 179, 200, 210, 212, 229, 252, 255, 260, 261, 262, 282, 284, 292, 306, 319, 326, 337, 348, 351, 374, 380, 407, 409, 419, 429, 432, 433, 435, 439, 449, 454, 456, 457, 460, 466, 467, 471, 479, 486, 493, 509, 510, 517, 526, 532, 534, 549, 551, 552, 562, 573, 574, 575, 580, 593, 597, 598, 617, 618, 630, 632, 640, 641, 643, 644, 654, 657, 659, 671, **675**, 680, 688, 703, 705, 711, 725, 730, 740, 742, 755, 756, 775, 777, 780, 798, 812, 842, 847, 854, 860, 862, 871, 254, Acts of, 675, 680, Acts of Peter and Paul, 676, Acts of Peter and the twelve apostles, 579, Gospel of, 57, Passion of Peter and Paul, 676, *see also* Kerygma Petri, Cathedra Petri
Peter, bishop of Myra, **677**
Peter, bishop of Parembolae, **677**
Peter, metropolitan of Gortyna, 209
Peter, patriarch of Jerusalem, 49, 245, 339, 432, 480, **677**
Peter I of Alexandria, 20, 23, 76, 142, 200, 208, 218, 380, 551, 623, 634, 673, **677**, 687, 724, 831, 833
Peter II of Alexandria, 196, 302, 509, 521, 549, **678**, 840, 841
Peter II of Sebaste, **678**
Peter Apselamos or Balsamos, **678**
Peter Chrysologus, 29, 59, 149, 151, 282, 395, 398, 410, 419, 452, 497, 499, 511, 669, **678**, 708, 730, 749
Peter Damian, 774
Peter Lombard, 101, 152
Peter Mongus, 5, 6, 330, 447, 487, 536, 569, 645, **678**, 744, 840, 841, 884
Peter of Altinum, **678**
Peter of Apamea, 662, **678**
Peter of Béziers, 330

Peter of Damascus, 238
Peter of Hippo Diarrhytus, 584
Peter of Sebaste, 113, 195, 364
Peter of Trajanopolis, **679**
Peter the Fuller, 5, 47, 49, 51, 197, 416, 439, 445, 473, 536, 569, **679**, 684
Peter the Iberian, 344, 445, 481, **679**, Life of, 446, 884
Peter the neo-martyr, 208
Peter the Patrician, **679**
Petilian, 98, 185, 193, 208, 248, 270, 337, 607, **679**
Petitiones Arrianorum, **680**
Petosiris, 92
Petovium, 28, 64, 867
Petra, 65, 92, 635, **680**, 826
Petrarch, 101
Petrine succession, *see* Papacy
Petronilla, 680, 705
Petronius of Bologna, 421, **680**, 863
Peyia, 213
Phantasiastae, 54, 81
Phaona (now Mismiyeh), 67
Pharan, 304, 636
Pharisees, 384
Phasis (Sebastopolis), 216
Philadelphia, 87
Philadelphia-Amman, 66
Philae, 826
Philalethes, 380, 662, 773, Apology of, 773
Philantia and companion, 27
Phileas of Thmuis, 534, **680**, Acts of, 534
Philebus, 74, 398, 558, 601, 602, 606
Philetus, 429
Philiatra di Triphilia, 7
Philip, apostle, 86, 87, 110, 112, 138, 223, 264, 449, **680**, 682, 755, 837, 871, Gospel of, 281, 448, 451, 579, 810, 837
Philip, disciple of Bardesanes, **680**
Philip, priest, **680**
Philippa, 16
Philippi, 171, 172, 174, 209, 515, 566, **681**, 701, 833, 837, *255*
Philippicus, 439
Philip of Gortyna, **681**
Philip of Side, 23, 235, 389, 582, **681**, 687, 784, 785, 830, 833
Philippopolis (now Plovdiv), 838, C. of (343), 207
Philippopolis (now Shohba), 67
Philipps Sacramentary, 224
Philip the Arab, 65, 315, 356, 604, 672, **682**, 863
Philip the Deacon, **680**, 682, 780
Philo, historiographer, **682**
Philo of Alexandria, 8, 22, 25, 28, 40, 44, 45, 46, 53, 72, 74, 75, 81, 111, 161, 180, 181, 183, 187, 204, 205, 230, 240, 251, 255, 309, 310, 312, 315, 340, 341, 363, 373, 376, 380, 395, 406, 455, 463, 516, 529, 558, 566, 574, 575, 576, 578, 585, 586, 587, 597, 601, 602, 603, 626, 639, 670, **682**, 689, 691, 693, 716, 717, 718, 727, 762, 763, 789, 805, 832, 872, 878, 879
Philo of Carpasia, **683**, 786, 881
Philo the Elder, 650
Philocalia, 114, 361, 620, 655, **682**, 684, 708, 761, 763, 786, 826, 880
Philocalus (or Filocalus), 166, 218, 228, 278, 281, 535, **682**
Philodemus, 40
Philogonius of Antioch, 47, **682**
Philomelium, 533
Philomena, 54, **682**
Philosophoumena, 153, 154, 156, 181, 375, 383, 885
Philosophy, 12, 22, 23, 25, 37, 54, 60, 82, 93, 98, 115, 154, 164, 176, 177, 180, 199, 205, 242, 243, 244, 247, 272, 276, 287, 288, 313, 319, 355, 356, 363, 367, 372, 373, 374, 378, 379, 390, 399, 404, 408, 434, 460, 461, 470, 506, 517, 526, 528, 551, 558, 566, 577, 601, 620, 623, 629, **683**, 689, 691, 710, 725, 734, 739, 741, 760, 761, 763, 773, 785, 795, 805, 815, 818, 820, 823, 831

Philostorgius, 62, 92, 228, 389, **683**
Philostratus, 121
Philotheus, **684**, 792
Philotheus Kokkinos, **684**
Philotheus the Sinaite, 708
Philoxenus of Mabbug, 59, 71, 263, 312, 325, 337, 445, 569, 577, **684**, 794, 810, 855
Phinehas, **684**
Phocas of Sinope, 17, 27, 36, 213, 274, 366, 375, 645, **685**, 703
Phocilis, Ps.-, 455
Phoebadius, 626, **685**, 706
Phoebadius of Agen, 225, **685**, 737
Phoebe, 221
Phoenix, 181, 469, 573, 597, **685**, 732, 802, 803, 808, *256*
Phos hilaron, 252, 401, **685**
Photinians, 96, 196
Photinus of Constantinople, 53, 113, 154, 228, 303, 323, 350, 522, **685**
Photinus of Sirmium, 11, 207, 559, 566, 582, **685**, 783, 830, 833
Photinus the Manichee, 663
Photius, 5, 31, 58, 62, 71, 76, 115, 180, 201, 203, 225, 233, 234, 238, 295, 297, 299, 343, 346, 352, 393, 438, 484, 619, 639, 654, 684, **686**, 687, 691, 778, 790, 794, 798, 830
Phrygia, 63, 85, 150, 195, 272, 280, 307, 377, 403, 404, 512, 533, **686**, 720, 774, 819
Phrygium, 253
Physiologus, 291, **686**, 802
Pianabella (Ostia), 423
Piazza Armerina, 169, 186
Piazza S. Giovanni, 202
Pico della Mirandola, 624
Pierius of Alexandria, 23, 235, 638, 658, **687**, 715, 830, 831, 833
Pietri, Charles, 104-5, 218-19, 337-39, 735, 740-43, 850, 869
Pietri, Luce, 846-47
Pietro de' Natali, 772
Piety, 23, 32, 58, 93, 98, 192, 231, 288, 320, 378, 395, 409, 478, 492, 495, 629, 685, **687**, 708, 746, 761, 763, 767, 775, 793, 830, 842, 843
Pilate, 57, 138, 435, 466, 671, 676, **687**, 755, 756, 801, 812, 838, *257*, Acts of, 57, 451, 687, Death of, 687
Pileus, 253
Pilgrimage, 10, 15, 21, 43, 70, 86, 90, 102, 109, 113, 128, 130, 134, 151, 157, 159, 168, 172, 174, 176, 186, 220, 232, 257, 265, 274, 278, 289, 327, 339, 410, 417, 423, 426, 432, 433, 434, 438, 458, 475, 500, 527, 532, 541, 553, 635, 638, 643, 656, 659, 668, 670, 676, **687**, 688, 712, 741, 768, 793, 799, 816, 822, 829, 846, 862, 873, 877
Pimeniola, 381
Pimenius, **688**
Pimenius of Asidonia, 106
Pinell, Jordi, 155, 334-36, 385-88, 485-86
Pinytus, **688**
Pinytus of Cnossus, 775
Pionius, 672, 688, Acts of, 134, 534
Pious women, 221, 255, 430, 539, 540, 629, **689**, *258*
Pippin, 505
Pippin II, 330, 471, 877
Pippin the Short, 43, 167, 616, 646, 653, 680, 700, 875
Pirminius, **689**
Pisae (Pisa), 422
Pisaurum (Pesaro), 423
Pisenthius of Kepht, **689**
Pispir, 31

Pistis Sophia, 57, 354, 609, 680, 837
Piso of Darnis, 487
Pitra, J.B., 654
Pius I, 644, 740
Pius II, 38
Pizzani, Ubaldo, 124-25, 272, 331, 576
Placentia (Piacenza), 421, 749
Placidia, 555
Planudean Anthology, 44
Plato, 7, 12, 31, 32, 40, 41, 53, 73, 74, 75, 82, 92, 123, 161, 180, 187, 204, 230, 233, 239, 251, 288, 373, 398, 405, 415, 455, 463, 507, 557, 558, 577, 578, 585, 586, 587, 589, 590, 591, 601, 606, 626, 639, 689, 690, 691, 693, 695, 697, 713, 717, 720, 725, 805, 832, 872, 877, 880, Against Plato on God and the gods, 236
Platonism and the Fathers, 22, 23, 45, 73, 90, 93, 98, 99, 114, 116, 124, 154, 176, 177, 181, 205, 209, 239, 243, 244, 286, 288, 310, 320, 352, 355, 362, 363, 365, 372, 373, 395, 398, 399, 448, 455, 463, 470, 526, 528, 558, 621, 624, 626, **689**, 694, 697, 789, 805, 832, 849, 859, 882
Pleguina, 118
Pleroma, 12, 41, 109, 204, 262, 352, 353, 610, **698**, 724, 733, 776, 859
Plerophoriae, 445, **699**
Pliny the Younger, 16, 90, 122, 161, 167, 221, 242, 271, 531, 551, 596, 671, **699**, 800, 812, 849, 882
Plotinus, 12, 20, 28, 32, 40, 53, 74, 75, 82, 93, 205, 206, 230, 240, 296, 311, 327, 352, 355, 363, 396, 553, 558, 585, 586, 587, 588, 589, 590, 591, 603, 606, 624, 639, 689, 691, 692, 693, 694, 695, 696, 697, **699**, 704, 717, 805, 836, 880, Life of, 121, 696, 699, 805
Plusianus of Lycopolis, 83
Plutarch of Athens, 590
Plutarch of Chaeronea, 40, 121, 558, 592, 602, 690, 695
Pneuma, 34, 41, 44, 45, 163, 204, 251, 391, 561, 621, **699**, 733, 796, 877
Pneumatius and Fidelis, 192
Pneumatomachi, 62, 72, 78, 114, 151, 195, 218, 361, 364, 381, 392, 403, 516, **700**, 742, 832, 841, 854
Poimandres, 378
Poitiers, 102, 116, 339, 381, 497, 518, **700**, 729, 862
Pola (Pula), 420
Polemius Silvius, 282, 295, 531, **700**, 754
Polemon the Apollinarist, 59, 459, **700**, 841, 859
Polianus of Milevis, 584
Policastro Busentino, 424
Polientus, 219
Politics, 700, *see also* Church and Empire
Politiko, 213
Pollard, T. Evan, 448-49
Pollastri, Alessandra, 30, 120, 127-28, 185, 298, 325, 350, 416-17, 484, 506, 508, 524, 545, 549-50, 576, 637-38, 659, 745, 747, 751-52, 754, 774, 776-77, 787, 824, 863
Pollentia in Piemonte, 19
Pollentius, **701**
Pollio, 638
Polochronius, 284
Polycarp of Ephesus, 258
Polycarp of Smyrna, 10, 53, 61, 63, 117, 140, 153, 252, 284, 293, 316, 320, 326, 328, 398, 404, 413, 447, 457, 467, 475, 533, 552, 561, 640, 644, 647, 657, 672, 688, **701**, 707, 728, 740, 792, 798, 843, 876, 878
Polychronius of Apamea, 702, 878, 881
Polycrates of Ephesus, 61, 258, 551, 640, 680, 682, 701, 728, 743, 838, 840, 867
Polyeuctos, 66
Pomerius Julianus, 233, 339, **702**
Pompeo Ugonio, 68
Pomponia Graecina, **702**

Pomponius, 145
Pontianus, bishop, **702**
Pontianus, pope, 315, 384, 385, 546, 640, 644, 646, 672, **702**, 741, 743, 757
Ponticianus, **702**
Pontifex Maximus, 103, 168, 211, 360, 503, **702**, 705
Pontius, 58, 121, 211, **702**, 784, 843
Pontus, 13, 27, 85, 87, 93, 95, 114, 115, 122, 137, 143, 163, 195, 205, 238, 523, 567, 619, 637, 662, 685, 699, **702**, 822, *see also* Bithynia and Pontus
Pontus Polemoniacus, 585
Poor, 3, 13, 16, 17, 22, 25, 47, 48, 63, 91, 96, 98, 113, 117, 155, 173, 180, 182, 184, 185, 192, 194, 211, 225, 246, 258, 315, 319, 338, 346, 364, 367, 369, 377, 441, 447, 478, 485, 511, 514, 527, 531, 541, 550, 551, 552, 555, 556, 562, 566, 567, 597, 609, 611, 628, 637, 638, 648, 653, 664, 679, **703**, 704, 714, 715, 725, 731, 736, 737, 741, 754, 768, 775, 794, 815, 856, 877
Pope, 99, 104, 132, 139, 141, 147, 153, 155, 168, 197, 210, 218, 221, 223, 237, 241, 253, 260, 261, 262, 278, 281, 366, 408, 413, 479, 523, 617, 618, 619, 640, 656, 676, 702, **704**, 705, 739, 741, 742, 744, 771, 791, 864, 865, 866, 877, *see also* Papacy
Poppo, 64
Porcarius of Lérins, 700, **704**
Porcuna, 107
Porolissum (Moigrad, Transylvania), 217
Porphyrius, 681
Porphyrius, bishop of Gaza, **704**
Porphyrius of Antioch, 681, **704**
Porphyry, 31, 32, 53, 59, 75, 81, 121, 124, 183, 226, 296, 310, 311, 363, 373, 584, 585, 587, 588, 589, 590, 591, 603, 614, 624, 689, 692, 695, 696, 697, 699, **704**, 733, 763, 769, 799, 805, 836, Against Porphyry, 300
Porta Romana, 224
Portofino, 421
Porto Torres (Turris Libisonis), 425
Porto Venere (San Pietro), 421
Portrait, 659, 677, **704**, *259*
Portugal, **790**
Portus (Porto), 156, 423, 424, 771
Poseidonius, 53, 74, 455, 805
Poseidonius of Apamea, 683
Posidippus, 82
Possidius, 96, 121, 128, 147, 224, 383, 612, 698, **705**
Potaissa gem (intaglio), 217
Potamiaena and Basilides, 532, **705**, 752
Potamion of Heraclea, **705**
Potamius, 90, **706**, 790
Pothinus, 413, 497, 512, **706**
Pottern, Wiltshire, 131
Poverty, *see* Poor
Praedestinatus, 82, 375, 644, 666, **706**, 876
Praeneste (Palestrina), 423
Praetextatus of Assuras, 545
Praetextatus of Rouen, 338, 556
Praetextatus Vettius Agorius, **706**, 816
Pragmateia, 519
Pratum spirituale, 61, 363, 540, 634
Praxeas, 53, 384, 566, 571, 653, **706**, 741, 819, Against, 685, 706, 819
Praxedis, 476, **706**, 711, 725
Prayer, 6, 8, 22, 25, 29, 33, 34, 46, 50, 62, 63, 84, 90, 100, 101, 102, 113, 119, 121, 135, 140, 144, 147, 151, 152, 154, 155, 185, 219, 221, 224, 225, 226, 227, 230, 231, 232, 234, 239, 241, 252, 255, 257, 259, 261, 262, 266, 277, 287, 290, 292, 309, 319, 335, 351, 358, 359, 372, 377, 379, 383, 386, 401, 404, 406, 409, 410, 411, 433, 442, 444, 446, 450, 461, 478, 490, 492, 493, 494, 498, 499, 500, 501, 506, 510, 514, 517, 519, 521, 528, 541,

Numbers in bold refer to the main entry on the subject. *Numbers in italics refer to illustration numbers (vol II, pages 951-1094)*

548, 549, 553, 556, 560, 562, 566, 577, 606, 609, 616, 621, 652, 664, 665, 667, 676, 688, **706**, 710, 722, 723, 744, 745, 746, 750, 754, 758, 767, 781, 792, 793, 797, 800, 805, 816, 819, 820, 829, 842, 843, 846, 870, 876, 878
Predestination, 46, 82, 99, 100, 101, 206, 261, 285, 308, 332, 357, 359, 414, 508, 621, 658, 666, 706, **709**, 767
Presbeia, 95
Presbyterium, 172
Presentation in the Temple, 82, 290, 452, 539, 540, **709**
Pricoco, Salvatore, 1, 295, 342, 369, 381, 396, 483, 548, 702, 759-62, 778-79
Priene, 87, 88
Priesthood of Believers, 463, 562, **710**
Primasius of Hadrumentum, 55, 102, 311, 461, 658, 706, **711**, 839, 663
Primian of Carthage, 247, 545, 321, **711**
Primianist, 270
Primicerius, 447, 521, 600, **711**
Primuliacum, 407, 799
Principia, 521
Prinzivalli, Emannuela, 7, 8, 10, 18, 28, 31, 32, 33, 41, 43, 43-44, 55, 55-56, 71, 83, 84, 92, 95, 106, 112, 126, 139, 140, 145, 154, 181, 190, 216, 218, 225, 235, 241, 246, 250, 268, 278, 281, 296, 299, 308, 333, 370, 371, 372, 378, 385, 400, 416, 439, 446, 509, 516, 521, 548, 612, 662, 678, 681, 683, 773, 775, 793, 830, 831, 832, 841, 844
Prisc(ill)a, 150, 570, 706, **711**, 740
Priscian, **711**, 760, 804
Priscillian, 28, 53, 56, 79, 83, 96, 106, 126, 218, 270, 322, 338, 347, 376, 392, 400, 404, 410, 475, 531, 547, 658, **711**, 714, 742, 758, 799, 801, 850, 851
Priscillianists, Priscillianism, 18, 38, 57, 106, 126, 127, 145, 147, 191, 219, 226, 234, 249, 284, 319, 360, 363, 400, 426, 479, 531, 554, 571, 597, 652, 658, **711**, 714, 742, 755, 758, 775, 776, 790, 799, 801, 838, 844, 850, 851, 854
Priscus of Lyons, 513, 556
Privatus of Lambaesis, 13, 315, 328, 470, **712**
Proba Faltonia, 40, 158, 225, 232, 492, 708, **712**
Probatus of Berenice, 487
Probus, 175
Procession, 174, 224, 274, 369, **712**, 722, 745, 864, 865
Proclus, montanist, 333, 682, **712**
Proclus Lycius Diadochus, 12, 31, 35, 39, 96, 112, 187, 206, 218, 238, 239, 240, 437, 442, 446, 482, 586, 587, 588, 589, 590, 591, 603, 614, 689, 696, 697, 698, **713**, 805, 835, 836, 853, Against Proclus on the eternity of the world, 446
Proclus of Constantinople, 36, 49, 115, 196, 200, 237, 444, 539, **713**, 752, 784, 829, 830, 841, 878
Proclus of Cyzicus, 545
Procopius of Caesarea, **713**, 862
Procopius of Gaza, 12, 138, 152, 183, 201, 202, 270, 311, 465, 683, **713**, 716, 786, 881, 884
Procula, 126
Proculianus, **713**
Proculus of Marseilles, 559, 854
Profuturus, **714**
Profuturus of Braga, 127, 386, **714**, 738
Profuturus of Constantine, 193
Prohaeresius, 96, 296, 382, **714**, 760
Prois, 339
Prologues to the Biblical Books, **714**
Prologus Paschae, 868
Promotus, 556
Property, Ecclesiastical, 17, 53, 91, 98, 139, 154, 198, 246, 330, 338, 367, 407, 479, 555, 556, 610, 642, 653, 672, 679, **714**, 736, 741, 780, 793, 806, 843, 845, 861

Prophecy, 160, 259, 269, 309, 316, 513, 557, **715**
Prophet, 55, 62, 71, 103, 168, 180, 192, 199, 219, 220, 230, 232, 234, 235, 251, 259, 261, 292, 310, 311, 336, 355, 356, 367, 377, 401, 432, 447, 449, 450, 464, 499, 510, 541, 560, 561, 604, 611, **715**, 716, 720, 722, 729, 734, 763, 764, 798, 819, 821, 825, 827, 831, 847, 877, 880, 260
Prophetiae ex Omnibus Libris Collectae, **716**
Proselyte, 107, 199, 223, 342, 360, 436, 456, 672, **716**, 818
Prosopon, 50, 159, 164, 268, 303, 325, 391, 594, 654, 675, **716**, 749, 773, 825
Prosper of Aquitaine, 72, 101, 145, 154, 166, 225, 229, 359, 381, 410, 411, 495, 525, 655, 702, 705, 708, **717**, 745, 767, 845, 868
Protadius, 555
Protasius and Gervasius, *see* Gervasius and Protasius
Protasius of Tarragona, 386
Proterius of Alexandria, 24, 196, 487
Protoctetus, 546
Protology, 44, 46, 205, 341, **717**
Protonice, 10
Protopaschites, 258, **720**
Protrepticus, 74, 75, 76, 177, 180, 258
Provence, 96, 704, councils, 17
Proverbia Graecorum, **720**
Proverbs, 235, 374, 878
Providence, 17, 38, 60, 73, 74, 75, 76, 118, 125, 204, 205, 212, 227, 236, 239, 262, 320, 346, 398, 469, 504, 543, 563, 624, 639, 664, 689, 691, 695, 716, **720**, 729, 754, 790, 796, 876, 882
Pruba, 403
Prudentius, 17, 26, 67, 110, 144, 149, 157, 160, 177, 202, 211, 212, 278, 282, 296, 331, 341, 401, 407, 426, 453, 475, 477, 491, 536, 542, 651, 661, 663, 676, **721**, 739, 766, 781, 790, 804, 816, 861, 870
Psalms, Psalmody, 50, 82, 90, 92, 94, 99, 102, 110, 114, 150, 152, 160, 187, 214, 222, 232, 235, 236, 255, 261, 274, 293, 295, 310, 311, 312, 343, 352, 364, 374, 382, 394, 398, 399, 401, 405, 431, 478, 485, 494, 497, 498, 499, 504, 577, 595, 620, 652, 678, 708, **722**, 727, 745, 762, 764, 787, 793, 815, 824, 825, 827, 829, 842, 878
Psalms of Solomon, 56
Psalmus responsorius, 648, **723**
Psathyrianoi, 525
Pseudoepigraphy, 56, 63, 66, 317, **723**, 827
Psotes of Psoi, **724**
Psychius, 792
Psychomachia, **721**, 781
Ptghni, 80
Ptolemaïs, 51, 487
Ptolemy and Lucius, **724**
Ptolemy Philadelphus, 72
Ptolemy the Gnostic, 309, 354, 373, 374, 384, 483, 551, **724**, 763, 859, 884
Publication, 187, 394, 483, 654, 655, **724**
Publicia, 14
Publicola, 550
Publius, 96, 524, **725**
Pudens and Pudentiana, 706, 711, **725**
Pulcheria (Aelia P. Augusta), 8, 22, 195, 241, 295, 324, 522, 523, 569, **725**, 744, 767, 829
Pulpitum, 172
Purgatory, 285, 286, **725**, 873
Purification, 3, 39, 42, 53, 84, 100, 123, 144, 239, 259, 313, 319, 365, 380, 405, 409, 443, 445, 564, 575, 578, 589, 693, 707, **725**, 751, 775, 792, 836, 875, 882
Purple Codex of Rossano, *see* Codex Purpureus Rossanensis
Purpurius of Limata, 765

Puteoli (Pozzuoli), 419
Pyrrhus of Constantinople, 197, 439, 547, 659, **725**, 823
Pythagoras, 690, 695
Pythagoreanism, 75, 470, 557
Pyx of S. Ambrogio, 450

Qabr Hiran, 855
Qalaat Seman, 174
Qalamun, 755
Qalb Lozé, 170
Qasr Serij, 809
Qasr el Lebia, 488
Qasr el-Jehud, 447
Qassaq, 80
Qatoûra, 51
Qennesrîn, or Qal'at Semon (Chalcis), 107, 220, 238, 439
Qirq-Bize, 169
Qona, 263
Quacquarelli, Antonio, 52-53, 179, 191-92, 202, 336, 552-53, 605-6, 629, 735-36
Quadratus, 60, 72, 96, 168, 201, **727**
Quadratum populi, 170
Quadrivium, 272
Quaestiones Veteris et Novi Testamenti, 30, 312, **727**
Quaestiones et Responsiones, 37, 138, 187, 234, 310, 385, 655, **727**, 753, 827, 879, ad orthodoxos, 234, 727, 827
Quails, Miracle of the, 1, 133, 520, 575, **728**, 817, 261
Quartodecimans, 90, 106, 210, 257, 258, 281, 333, 384, 408, 492, 551, 557, 626, 701, **728**, 743, 810, 841, 875, *see also* Easter controversy
Qubbat el-Khazna, 808
Quesnelliana, 142
Quicumque vult, 208
Quietus of Uruc, 584
Quintasius of Carales, 143, **728**, 757
Quintilian, 52, 179, 650
Quintilliani, 83
Quintillus, 36
Quintus, 109
Quintus Aurelius Memmius Symmachus, 124
Quintus Julius Hilarianus, 18
Quiricus of Barcelona, 297, **728**, 845
Quirinus of Siscia, 638, 759
Quo vadis?, 676
Quodvultdeus, 14, 228, 229, 312, 376, **728**, 839, 861
Quweismeh, 66

Rabanus Maurus, 152, 669, 753, 879, 881
Rabba, 636
Rabban Sliba, 4, 356
Rabbi, 762
Rabbula of Edessa, 19, 22, 49, 234, 263, 403, 524, **729**, 810, *see also* Codex of Rabbula
Rabbula of Samosata, 444
Radagaisus, 795
Radbod, 877
Radegund, 116, 329, 338, 513, 518, 532, 700, **729**, 862, Life of, 116
Raffaele Fabretti, 68
Rahab, 453
Raineri, Osvaldo, 289-91
Raithu, 44, 826
Raphael, 40, 132, 290, 557, 843
Raponi, Santino, 816-18
Ras el Hilal, 488
Rasinia Secunda, 842
Rationes Seminales, **729**, 733
Ravenna, 7, 13, 17, 24, 29, 51, 59, 62, 67, 69, 70, 96, 108, 109, 141, 147, 151, 169, 170, 171, 172, 174, 183, 185, 186, 202, 203, 209, 213, 219, 228, 253, 269, 292, 294, 297, 306, 333, 334, 336, 357, 377, 419, 435, 447, 450, 452, 456, 464, 476, 477, 479, 491, 495, 497, 505, 509, 511, 531, 536, 539, 545, 550, 555, 557, 559, 568, 572, 573, 575, 583, 597, 610, 615, 618, 642, 643, 645, 650, 653, 667, 671, 675, 678, 688, 689, 705, 715, 726, 753, 756, 768, 769, 774, 778, 803, 804, 822, 823, 828, 847, 859, 862, 866, 873, *262a*, *262b, 262c*, liturgy, 497, 499
Ravenna Scroll, 678
Ravennius of Arles, 79, 582, **730**, 747, 861
Reate (Rieti), 423
Rebecca, 261, 341, **730**
Recapitulation, 55, 262, 313, 340, 358, 467, 537, 719, **730**, 731, 733, 816, 839
Recared, 148, 356, 357, 400, 418, 444, 478, 482, 556, **730**, 774, 791, 844
Recchia, Vincenzo, 242, 365-68, 749, 783
Receswinth, 148, 331, 356, 554, 791
Rechila, 783
Recimir of Dumium, 845
Recognitiones, 74, 75, 179, 534, 746
Reculver cross, 132
Redemption, 8, 10, 29, 34, 53, 56, 84, 94, 98, 100, 101, 123, 164, 173, 205, 206, 224, 226, 231, 285, 293, 341, 352, 353, 357, 364, 398, 406, 415, 434, 447, 452, 453, 459, 462, 463, 477, 523, 525, 538, 547, 657, 666, 668, 719, **731**, 755, 767, 780, 781, 782, 788, 789, 792
Refâdi, 51
Refrigerium, 16, 64, 153, 157, 192, 222, 232, 285, 425, 532, 544, 552, 554, 676, **731**, 803, 842
Regensburg (Regina Castra), 347
Reginus, **732**
Regula fidei, 63, 103, 206, 244, 358, 376, 620, 621, 623, **732**, 734, 750, 751, 819, 848, 851
Regula pastoralis, 367, 442, 562, 793
Reichenau, 126, 689
Relics, 171, 224, 225, 228, 232, 248, 271, 274, 278, 295, 296, 308, 338, 347, 350, 365, 370, 383, 385, 387, 417, 433, 439, 491, 502, 507, 513, 531, 532, 533, 535, 536, 537, 538, 552, 554, 564, 582, 616, 624, 636, 637, 643, 656, 660, 670, 676, 688, 704, 712, 742, 747, 755, 765, 774, 791, 795, 845, 846, 850, 853, 866, 870
Remagen (Rigomagus), 349
Remigius of Rheims, 183, 307, 329, 349, 482, **732**, 735, 862
Renanus, 612
Renatus, 842
Renaudot, 654
Reparatus, 842
Resafa, 174, 688, 768, 688, 768
Restitutus (or Restutus), **732**, 737
Resurrection of the Dead, 3, 11, 12, 17, 18, 23, 26, 33, 36, 45, 52, 54, 56, 57, 60, 64, 65, 67, 71, 84, 90, 91, 95, 100, 108, 116, 117, 135, 144, 155, 158, 165, 181, 186, 190, 191, 203, 204, 207, 210, 217, 219, 220, 221, 225, 226, 229, 232, 233, 235, 242, 243, 257, 258, 286, 290, 293, 300, 309, 313, 316, 317, 319, 323, 328, 336, 340, 341, 343, 353, 366, 372, 380, 382, 390, 395, 397, 398, 399, 404, 408, 433, 436, 442, 450, 458, 461, 462, 463, 466, 467, 477, 492, 494, 504, 510, 515, 516, 521, 523, 531, 532, 543, 551, 553, 557, 560, 570, 575, 604, 629, 634, 638, 649, 651, 657, 662, 670, 685, 693, 707, 710, 712, 718, 719, **732**, 749, 750, 773, 775, 778, 786, 788, 789, 792, 796, 800, 802, 803, 812, 815, 819, 820, 821, 831, 839, 851, 871, 882
Reticius of Autun, 104, 311, **733**, 787, 881
Reticulus seu de arca Noe, 868
Retractations, 96, 612
Revelation, 39, 54, 55, 56, 58, 198, 219, 222, 233, 237, 245, 255, 269, 308, 309, 317, 355, 356, 357, 358, 365, 377, 378, 398, 405, 406, 408, 413, 415, 429, 463, 489, 506, 507, 509, 557, 571, 577, 620,

621, 625, 666, 683, 696, 697, 698, 707, 715, 720, 727, 732, **733**, 762, 763, 775, 780, 788, 851, 881
Revello, 421
Revardashir, 417, 517
Rhaetia, 330, 347, 349, 559, 572, 618, **735**, 861
Rheims (Remis), 329, 337, 454, 732, **735**, 850
Rhetoric, 12, 21, 23, 27, 28, 52, 58, 82, 91, 114, 177, 179, 181, 208, 232, 245, 272, 309, 321, 366, 367, 381, 382, 396, 411, 479, 483, 552, 553, 575, 576, 582, 605, 625, 629, 648, 650, 732, 734, **735**, 754, 758, 760, 767, 818, 820, 873
Rhipsime, 80
Rhodanius of Toulouse, 381, **736**
Rhodo, 54, **736**, 815
Ribarroja (Castellón), 813
Rich, Riches, 16, 47, 48, 114, 117, 180, 182, 192, 287, 377, 441, 552, 703, **736**
Richborough (Rutupiae), 129
Ricimer, 134, 479, 778
Riegel, A, 69
Riez, 320, C. of (439), 338, 381, 548, **737**, 861
Riggi, Calogero, 11, 21-22, 27-28, 102-3, 233, 282, 320, 328, 519, 557, 637, 765
Rignano, 422
Riha, 491
Rihab, 66
Rila, 634
Rilliet, Frédéric, 67, 276-77, 343-44, 543, 674-75, 684, 766, 809-11
Rima (now Rimet el-Lohf), 67
Rimini, 24, 94, 464, 787, C. of (359), 78, 129, 195, 207, 223, 228, 350, 360, 381, 395, 526, 559, 651, 685, 706, 732, **737**, 758, 766, 771, 783, 858, Formula of, 363, 382, 395, 472, 508, 732, 858
Ring, 108, 232, 350, 390, 529, 568, **737**, 866
Riparius, 846
Ripon, 132, 211, 877
Ris'ayna (Theodosiopolis), 769
Ritual, 503
Riva San Vitale, 421
Rivenhall, Essex, 131
Rizinice, 217
Robba, 247
Rochester, Kent, 465
Rochet, 865
Rod, 140
Rodrigo di Cerrato, 405
Rogatists, 738
Rogatus, 14, 651, **738**
Roman canon, 2, 33, 63, 277, 278, 294, 335, 386, 497, 541, 700, **738**, 745
Roman creed, 205
Romanianus, 162, 488, **739**
Roman law and Christianity, 168, 286, 412, 451, 520, 528, 620, **739**
Romanus, **739**
Romanus Melodus, 5, 17, 36, 50, 115, 160, 175, 271, 276, 465, **740**
Romanus of Roso, **740**
Rome, 2, 3, 4, 5, 7, 8, 9, 11, 13, 14, 17, 19, 23, 25, 27, 29, 30, 35, 36, 47, 48, 49, 54, 55, 56, 59, 60, 64, 67, 71, 72, 77, 78, 79, 83, 84, 91, 93, 96, 98, 102, 104, 105, 106, 109, 112, 114, 116, 119, 120, 122, 132, 136, 139, 140, 141, 142, 144, 145, 146, 147, 149, 150, 152, 153, 154, 155, 157, 158, 159, 160, 162, 163, 166, 167, 168, 170, 178, 179, 180, 181, 187, 188, 189, 192, 193, 194, 195, 196, 197, 200, 201, 203, 205, 206, 207, 208, 209, 211, 212, 218, 219, 223, 224, 225, 227, 228, 229, 237, 238, 245, 246, 248, 249, 250, 258, 259, 260, 261, 262, 267, 269, 273, 274, 277, 278, 279, 280, 281, 283, 284, 292, 293, 297, 307, 311, 315, 319, 323, 324, 325, 333, 336, 337, 339, 342, 348, 351, 353, 356, 360, 361, 363, 369, 370,

371, 376, 377, 386, 389, 393, 397, 399, 409, 410, 419, 423, 426, 427, 429, 430, 431, 435, 439, 447, 450, 452, 453, 458, 464, 472, 475, 476, 477, 482, 484, 486, 490, 493, 495, 496, 499, 500, 501, 502, 506, 509, 515, 522, 523, 530, 532, 535, 540, 549, 550, 551, 552, 553, 560, 570, 573, 574, 575, 579, 593, 596, 599, 600, 611, 613, 614, 617, 618, 619, 624, 629, 630, 631, 639, 640, 641, 642, 643, 653, 654, 656, 657, 666, 669, 671, 673, 675, 676, 678, 679, 680, 684, 685, 688, 689, 701, 702, 704, 705, 706, 709, 710, 712, 714, 726, 728, **740**, 749, 755, 757, 761, 763, 764, 769, 771, 772, 778, 779, 784, 785, 786, 790, 791, 795, 798, 800, 803, 804, 812, 813, 814, 819, 821, 822, 823, 824, 826, 828, 829, 830, 831, 833, 837, 838, 840, 843, 845, 846, 847, 848, 853, 856, 859, 861, 864, 865, 866, 867, 873, 875, 885, *263*, *264*, *265*, *266*, *267*, *268*, *269*, C. of (251), 211, C. of (311), 595, C. of (313), 79, 104, 321, 419, 559, 560, C. of (340/341), 49, 460, 522, 757, C. of (368/372), 116, 218, C. of (377), 218, C. of (378), 218, 249, 407, C. of (382), 8, 28, 33, 59, 218, 223, 360, 549, 660, C. of (384), 645, C. of (386), 249, 822, C. of (390), 749, C. of (430), 154, 444, 594, C. of (449), 301, C. of (465), 380, 549, 854, C. of (484), 884, C. of (499), 154, 476, 794, 843, C. of (501), 476, C. of (502), 476, 806, C. of (531), 281, 749, C. of (595), 843, C. of (607), 642, C. of (610), 551, C. of (641), 439, C. of (649), 530, 547, 568, 659, 743, 823, 873, C. of (680), 645, C. of (729), 646, liturgy, 6, 17, 62, 82, 83, 129, 132, 141, 155, 160, 181, 185, 223, 224, 257, 334, 335, 366, 410, 446, 490, 493, 496, 499, 616, 744, 745, 762, 797, 816, Romano-Frankish liturgy, 334, 397
Romero Pose, Eugenio, 55, 84, 97, 117, 145, 184-85, 191, 249, 321, 322, 328, 331, 368, 465, 485, 543, 605, 611, 652, 652-53, 663, 670-71, 701, 738, 774, 801, 838-39, 842, 845-846, 856, 859, 874
Romula, 217
Romulea, 252
Rordorf, Willy, 235, 257-58, 531-32, 748, 800-1, 847-49, 876-77
Rosamund, 505
Rossano, 631, *see also* Codex Purpureus Rossanensis
Rossi, G.B. De, 654
Rosvita, 16
Rothari, 505, 506, **745**
Rothbury in Northumbria, 132
Rotunda of Galerius, 88, 833
Rouen (Rotomagus), 338, 512, 676, 869
Rouillard, Philippe, 499-500, 744-45
Ruadhan of Lorrha, 133
Ruben of Dairinis, 412
Rufiananes, 22
Rufina, 210
Rufiniani, 140
Rufinianus, 465, 551
Rufinus, **745**
Rufinus (the Syrian), 431, **746**
Rufinus Flavius, **745**
Rufinus of Aquileia, 10, 61, 63, 67, 141, 149, 151, 166, 179, 184, 229, 235, 244, 250, 271, 289, 299, 301, 306, 318, 323, 337, 339, 340, 344, 345, 364, 371, 388, 389, 406, 410, 430, 431, 475, 478, 497, 549, 550, 559, 567, 597, 609, 620, 623, 638, 645, 646, 655, 658, 664, 666, 693, 717, 724, 736, **746**, 775, 784, 785, 786, 790, 793, 825, 831, 841, 880
Rufus of Shotep, 219, **746**
Rufus of Thessalonica, 96, 670
Rules, Monastic, 4, 8, 44, 79, 96, 101, 112, 116, 118, 119, 138, 139, 150, 160, 187,

211, 230, 234, 250, 266, 268, 286, 287, 297, 319, 322, 323, 331, 335, 338, 366, 394, 396, 451, 478, 483, 499, 526, 628, 652, 708, **746**, 761, 791, 793, 870, of Aurelian, 761, of St Benedict, 1, 6, 184, 410, 478, 480, 560, 567, 708, 747, 761, 770, of the four fathers, 1, 500, of St. Pachomius, 291, of the master, 1, 500, 567, 761
Rupert of Deutz, 865
Rupert of Salzburg, **747**
Ruricius of Limoges, 320, 338, 702, **747**, 766
Rusguniae, 15
Ruspe, 331
Rusticiana, 124
Rusticianus (or Rusticanus), 516, 812
Rusticus, 582, **747**
Rusticus, Roman deacon, **747**
Rusticus, presbyter, **747**
Rusticus of Narbonne, 126, 154, 189, 231, 372, 417, 546, **747**
Ruthwell, Dumfriesshire, 132
Ruthwell and Bewcastle crosses, 132
Rutilius Namatianus, 6, 795
Rylands Papyrus, 448

Saamasas (Lugo), 334
Sabaeans, 832
Sabaoth, 71
Sabas of Iskhan, 344, 797
Sabbath, 50, 54, 111, 117, 223, 251, 258, 290, 321, 360, 402, 404, 433, 504, 604, 610, 649, 701, **748**, 755, 800, 876, 877
Sabbatians, 841
Sabellians, Sabellianism, 23, 76, 77, 96, 164, 196, 207, 215, 238, 245, 296, 299, 300, 301, 320, 348, 363, 381, 395, 445, 448, 487, 507, 511, 526, 566, 585, 640, 653, 662, 716, 737, 743, **748**, 852
Sabellius, 53, 140, 196, 237, 323, 487, 566, 653, **748**, 885
Saben, 149
Sabina, **749**
Sabines and Peligni, 423
Sabinianus, 366, 645, 743
Sabinus of Canosa, **749**
Sabinus of Heraclea, **749**, 785, 790
Sabinus of Piacenza, 65, **749**
Sabinus of Seville, 774
Sabratha, 15, 169, 174
Sabris, 106
Sabriso', 374, 517, 518
Saccophori, 686
Sacramentary, 502, 725, of Drogo of Metz, 224, of Gello, 224, of Pippin the Short, 502
Sacraments, 8, 37, 41, 63, 83, 90, 99, 100, 104, 108, 109, 123, 127, 136, 150, 152, 180, 215, 239, 243, 246, 247, 257, 258, 260, 261, 262, 273, 276, 277, 280, 288, 290, 292, 293, 332, 334, 335, 342, 344, 353, 357, 364, 370, 374, 387, 396, 405, 409, 414, 441, 479, 493, 494, 502, 504, 528, 529, 548, 560, 562, 577, 613, 616, 651, 655, 660, 665, 679, 710, 745, **749**, 751, 754, 776, 780, 784, 819, 875
Sacramentum, 708, **750**, 751
Sacra Parallela, 180, 235, 442, 655, 775
Sacrificati, 146, 223
Sacrifice, 1, 2, 3, 6, 26, 33, 82, 94, 110, 111, 123, 164, 165, 171, 226, 227, 232, 237, 241, 273, 277, 288, 292, 294, 296, 313, 321, 337, 341, 402, 404, 407, 408, 409, 436, 441, 462, 473, 477, 491, 494, 500, 504, 528, 541, 550, 562, 583, 622, 670, 688, 706, 710, 721, **751**, 758, 772, 788, 803, 816, 828, 859, 875
Sadducees, 384
Saeculum, **752**
Saetabis, 148
Saffaracus of Paris, 556
Sagalassos, 86
Sagion, 253

Sagittarius, 556, 645
Sagochlamys, 253
Sagona, 203
Sagum, 254
Sahak the Great, 5, 80, 84, 482, 496, 556, **752**
Sahdona, 536
Sahel el-Jaulan, 66
St Albans (Verulamium), 128, 129
Saint-Bertrand-de-Comminges, 170
Saintes (Santonae), C. of (561-567), 556, 853
St Gall, 126, 486
Saint-Jean de Losne, C. of (673-675), 556
St Mary's (previously St Bertelin's), Stafford, 131
St Michael's, Thetford, 131
St Ninian's Isle, Shetland Islands, 133
Saints, Intercession of, 7, 197, **752**
Salamias (now Selemye), 809
Salamina (now Salamis), 111, 213, 281
Salemi, 425
Salic Law, 286
Sallustius of Seville, 399
Salome, 524
Salona, 69, 169, 172, 174, 186, 217, 236, 377, 379, 424, **753**, *270*, C. of (530 & 533), 753, 759
Salonius of Geneva, 295, 483, 556, 645, 727, **753**, 754, 767, 863, 879
Salsa, 14, 544
Salvian of Marseilles, 177, 338, 339, 348, 349, 393, 483, 530, 656, 753, **754**, 784
Salvianus, 584, 640
Salvina, 351
Salvius, 295, 754, 824
Salvius of Membressa, 545
Salzburg, 747
Samaria, 432
Samaritan Pentateuch, 813
Samaritans, 24, 138, 251, 803, 748, **755**, 780, 803
Samon, 165
Samosata, 38
Samothrace, 515
Sampsaei, 269
Sams ad- Din (Oasis al-Harga), 265
Samson, 89, 175, 187, 755, *271*
Samuel of Qalamun, 755
Sancti Notarii, 663
Sanctus, 6
Sanctus, deacon of Vienne, 512
Sangallensis, 617
Santagata, Giuhina, 1, 107, 219-20, 221, 447-48, 477, 671, 801
Santra, 121
Sapaudus, 178
Saqqarah, 3, 267, 268, 631
Sarabaites, 184, 230
Saraballa, 253
Saracens, 704
Saragossa, 127, 296, 493, 548, 688, 712, **755**, 790, 791, 812, 813, 814, 844, C. of (380), 96, 126, 145, 270, 282, 400, 404, 426, 755, C. of (385), 225, C. of (592), 755, C. of (691), 755
Sarcophagi, Early Christian, 2, 3, 33, 51, 53, 62, 64, 68, 69, 70, 79, 105, 107, 108, 132, 138, 141, 148, 149, 157, 171, 172, 187, 189, 210, 217, 219, 221, 267, 269, 274, 294, 313, 315, 323, 334, 349, 369, 370, 378, 419, 420, 421, 422, 423, 424, 425, 428, 429, 434, 435, 437, 447, 450, 452, 456, 471, 477, 489, 491, 520, 528, 530, 536, 559, 575, 578, 583, 596, 597, 600, 615, 625, 635, 636, 639, 650, 659, 668, 671, 675, 676, 677, 680, 681, 685, 687, 705, 728, 753, **755**, 759, 771, 778, 801, 807, 812, 813, 814, 855, 860, *272*, *273*, Sarc. of Adelphia, 807, of Agricius, 349, Albani, 650, of St Ambrose, 3, 269, 583, of St. Andrew, 132, of Apt, 306, Balaam, 10, Bethesda, 274, 313, 370, Borghese, 269, 756, of Concordius, 306,

Numbers in bold refer to the main entry on the subject. *Numbers in italics refer to illustration numbers (vol II, pages 951-1094)*

of Elisha the prophet, 41, 452, of S. Engracia, 419, of Flavius Gorgonius, 221, of Frende, 334, of Hydria Tertulla, 456, of Isacius, 756, of Ithacius, 334, of Jonah, 450, 477, 598, 675, of Junius Bassus, 435, 447, 688, 756, of La Puebla Nueva, 148, Lot, 10, 506, 785, of S. Marcello, 3, of Paulinus, 349, Pignatta, 40, 41, 333, 756, Plotinus, 650, of Praetextatus, 370, of Qarara, 268, Ravenna, 41, Sarigüzel, 40, Servannes, 1, 456, 575, 689, of St Trophimus, 583, Tusculum, 292, of the two brothers, 219, 435, 477, 688, 705, 756, Verona, 456

Sardanapalus, 75

Sardica, 125, 565, **756**, C. of (343), 5, 23, 71, 77, 84, 92, 93, 110, 113, 121, 140, 142, 148, 153, 182, 237, 250, 268, 295, 298, 338, 343, 348, 349, 360, 365, 382, 419, 420, 430, 460, 487, 509, 513, 522, 525, 546, 554, 566, 582, 600, 616, 617, 626, 641, 681, 730, 742, 757, 768, 774, 793, 824, 825, 833, 850, 858, 863, *274*

Sardinia, 425, 653, **757**, 861

Sardis, 87

Sargis, 428

Sasima, 43, 114, 143

Sassanids, 80, 106, 169, 417, 537, 637, 674, **757**, 766, 808, 809

Satan, 46, 53, 71, 152, 220, 226, 255, 463, 622, 718, 817

Satisfactio, 252, **758**

Satornilus (or Saturninus), 90, 198, 238, 338, 384, 612, **758**, Acts of, 248, Passion, 814

Saturninus of Arles, 120, 381, 382, 651, **758**

Saturninus of Thugga, 584

Saturninus of Toulouse, **758**, 846, 854, 869

Satyrus, 13, 29, 121, 192, 279, 398, 521, **758**

Sauget, Joseph-Marie, 1, 3-4, 4, 12, 16, 22, 83, 106, 107, 110, 112, 214, 215-16, 220, 271, 371, 374, 379-80, 401, 403, 428-29, 439-440, 443, 445, 445-46, 447, 527, 537, 546, 564-65, 662, 679, 688, 704, 729, 748, 769, 794, 823

Saul, 41, 200, 562

Savaria (Sombathely), 638

Savia, 638, **759**

Saxer, Victor, 2, 4, 8, 10, 11, 11-12, 13-14, 16, 17, 18, 19, 25, 27, 35, 55, 59, 59-60, 71, 79, 83, 102, 120, 122, 138, 139, 140, 143, 145-46, 149, 153, 154, 163, 165, 167, 183, 189, 192-93, 202, 203, 209, 211, 211-12, 212, 220, 221-23, 222, 223, 224, 225, 225-26, 228, 228-29, 229, 231, 238, 239-40, 246, 250, 268, 270, 282, 296, 297, 298, 308, 321, 322, 325, 327, 328, 331, 332, 341-42, 342, 343, 346, 350, 356, 360, 369, 370, 375, 377, 382-83, 417, 426-27, 430, 440, 458, 461, 464, 466, 470-71, 475, 481-82, 482, 492-94, 507, 509, 512-13, 518, 522, 523, 524, 524-25, 530, 531, 532-35, 536-37, 540, 543-44, 546, 547, 552, 553, 554, 571, 579, 581-82, 583, 584, 593, 595-596, 596, 597, 607, 610, 616-17, 638, 639, 653, 654, 656, 663, 665, 669, 670, 671, 675-77, 678, 680, 682, 685, 688, 700, 702, 704, 705, 706, 711, 712, 724, 725, 728, 729, 732, 733, 739, 747, 749, 752-53, 758, 762, 765, 768-69, 771, 771-72, 778, 779, 787, 795, 812, 814, 816, 821-22, 822, 829, 840, 842-43, 853-54, 856, 859, 862, 863, 870, 877

Sazana, 289

Sbeitla, 14, 15, 172, 174, 269

Scala Paradisi, 61, 312, 442, 634

Scapula, 60, 143, 759, 819

Scete (Wadi-el-Natrûn), Desert of, 4, 22, 31, 83, 119, 149, 200, 220, 264, 267, 271, 514, 567, 755, **759**

Schism, Schismatic, 6, 13, 17, 20, 22, 29, 32, 43, 63, 79, 96, 102, 109, 114, 136, 176, 185, 195, 195, 208, 211, 212, 246, 247, 248, 321, 324, 350, 376, 385, 393, 400, 404, 480, 533, 551, 595, 603, 607, 610, 613, 619, 629, 641, 652, 665, 711, 741, 743, 748, 749, **759**, 804, 810, 822, 838, 831, 839, 854, 856

Scholastica, **759**

Scholasticus, **759**

School, 22, 58, 138, 143, 150, 161, 162, 177, 180, 235, 299, 335, 339, 373, 403, 411, 412, 446, 619, 633, **759**, 763, 774, 791

Scillitani, 188, 228, 534, 552, 600, **762**, Acts of, 13, 14, 209, 370, 472, 479, 534

Sclua of Narbonne, 330

Scorpiace, 819

Scorza Barcellona, Francesco, 20, 60, 73, 111, 111-12, 138, 185, 185-86, 281, 301, 330, 143, 371, 409, 410, 444, 448, 482, 508-9, 516, 517, 524, 527, 545, 554, 582-83, 612-13, 652, 670, 702, 774, 779, 803-4, 868

Scotland, 18, 129, 130, 131, 132, 186, 486, 653, **762**, *see also* Britain and Ireland

Scribe, **762**

Scripture, 5, 10, 18, 19, 23, 25, 28, 30, 32, 38, 39, 46, 47, 53, 54, 55, 57, 58, 64, 72, 77, 78, 81, 83, 94, 96, 98, 99, 100, 102, 103, 104, 111, 114, 116, 118, 119, 120, 125, 126, 128, 139, 145, 150, 151, 152, 155, 158, 160, 180, 181, 187, 188, 207, 222, 223, 226, 228, 229, 230, 234, 239, 243, 244, 245, 246, 251, 260, 269, 274, 276, 288, 291, 295, 298, 300, 301, 303, 309, 310, 311, 315, 317, 319, 320, 322, 323, 326, 327, 330, 333, 336, 356, 360, 362, 365, 367, 368, 369, 372, 373, 379, 380, 384, 388, 390, 392, 394, 395, 396, 400, 403, 404, 406, 407, 411, 413, 415, 428, 435, 437, 448, 451, 455, 458, 461, 462, 468, 472, 474, 489, 492, 495, 496, 498, 499, 504, 508, 511, 517, 523, 527, 533, 537, 538, 539, 541, 542, 548, 549, 552, 556, 558, 563, 570, 577, 604, 606, 613, 620, 621, 622, 629, 633, 638, 647, 650, 652, 655, 666, 683, 691, 693, 696, 701, 704, 707, 708, 720, 727, 732, 734, 736, 737, 739, 740, 746, 752, 760, **762**, 773, 783, 792, 793, 802, 805, 815, 819, 821, 833, 839, 847, 848, 849, 852, 854, 856, 860, 864, 873, 875, 877, 881, *see also* Bible

Sculpture, 3, 9, 24, 51, 69, 81, 87, 88, 148, 157, 217, 263, 266, 269, 290, 323, 345, 370, 390, 420, 422, 450, 488, 554, 575, 579, 659, 677, 681, **764**, 765, 813, 845, 860

Scyllacium (Calabria), 149

Scythia, 38, 105, 140, 237, 565

Scythopolis (now Beysan), 215, 635, 826

Sebaeos, 81

Sebaste, 13, 447, 635, 702

Sebastian, 741, **765**

Secrecy, Sacramental, **765**

Secundianus of Singidunum, 14, 64, 299, 637, **765**

Secundinus, 98, 411, **765**

Secundus of Ptolemais, 595

Secundus of Tigisis, 176, 246, 384, 554, **765**, 779

Secundus of Trent, 389

Secundus the Pentapolitan, 487

Sedatus of Nîmes, 393, **766**

Sedilia, 172

Sedulius, 26, 67, 92, 119, 160, 202, 311, 312, 651, **766**, 805

Sedulius Scotus, 720

Seekirchen (Wallersee), 747

Segienus, 211

Segobriga (Cuenca), 148

Segontia, 148

Segovia, 148

Seir Kieran, 130

Sejanus, 687

Seleucia in Isauria, 115, C. of (359), 5, 78, 113, 207, 215, 268, 296, 305, 343, 381, 516, 527, 550, 657, 688, 737, **766**, 779, 783, 787, 832

Seleucia-Ctesiphon, 1, 3, 71, 112, 216, 305, 416, 417, 537, 550, 646, 674, 757, **766**, 808, 809, 810, 883, C. of (399-400), 766, C. of (410), 416, 757, 766, C. of (421), 19, C. of (486), 5, 11, 118, 119, 766, C. of (605), 110

Selva, metropolitan of Narbonensis, 844

Semiarians, 93, 94, 196, 207, 395, 546, 550, **767**, 860

Semipelagians, 138, 178, 273, 321, 331, 359, 381, 483, 495, 555, 645, 665, 709, 717, **767**

Senator of Milan, **767**

Seneca, 177, 184, 373, 397, 398, 483, 531, 593, 796

Seneca and Paul, Correspondence of, **767**

Sens (Senonas), 18, 337, 512, 556

Sephronus, 148

Seppelt, 644

Septimius Severus, 20, 21, 47, 60, 65, 84, 110, 143, 168, 180, 194, 223, 225, 388, 480, 607, 619, 637, 672, **768**, 807, 808, 814, 818, 854

Septuagint, 22, 72, 326, 360, 380, 431, 455, 620, 763, Hexaplar, 380

Seraphim, 39

Serapion of Antioch, 10, 58

Serapion of Thmuis, 23, 58, 265, 410, 501, 520, 577, 611, 750, **768**, 845, Euchologion, 34, 123, 136, 252, 265, 293, 494, 496, 500, 541, 611, 768

Serapion of Zarzma, 344

Serapis, 799

Serenus of Marseilles, 407, 530

Sergia, **768**

Sergiopolis (now Resafa), 809

Sergius, pope, 17, 197, 343, 539, 542, 643, 646, 709, 743, **768**, 809, 823, 838, 877

Sergius II, 787

Sergius and Bacchus, 632, 688, **768**, 808

Sergius Grammaticus, 662, **769**, Against, 773

Sergius Magister, 27

Sergius of Arsion, 823

Sergius of Constantinople, 5, 197, 216, 397, 547, 568, 570, **768**, 787, 826

Sergius of Jerusalem, 823

Sergius of Ris'ayna, **769**

Sergius of Tarragona, 483

Sergius of Tella, 662

Sergius Paulus, 663

Seridos, 112, 251

Seripando, 101

Sermo, 99, 138, 286, 298, 393, 394, 410, 503, 604, 725, 727, **769**, 792, 847

Sermo Arrianorum, 637, **770**

Sermon on the Mount, 25, 117, 461, 466, 738, **770**, *275*

Serpent, 612

Serra de Paraetonium, 487

Serre, 515

Sertei, 16

Servanus, **771**

Servatius of Tongres, 349, **771**

Servus Dei, **771**

Seth, Sethians, 9, 71, 251, 353, 384, 579, 612, **771**

Sétif (Sitifis), 13, 14, 15, 16, 543, 572, 676, **771**

Seven Sleepers of Ephesus, 85, 89, 275, 314, **772**

Severian of Gabala, 51, 81, 203, 282, 312, 436, 442, 561, 609, 616, 634, 658, **772**, 832

Severians, 37, 410, 542, 569, 773

Severin, Hans-Georg, 266-68

Severinus, pope, 645

Severinus of Cologne, **772**

Severinus of Noricum, 307, 347, 348, 743, **772**, Life of, 121, 600, 735, 772

Severus, encratite, **773**

Severus ibn al-Muqaffa, 66

Severus of Antioch, 6, 17, 36, 43, 47, 58, 59, 106, 110, 126, 160, 165, 193, 197, 200, 233, 268, 272, 276, 303, 312, 325, 328, 333, 339, 343, 364, 419, 428, 429, 432, 444, 458, 481, 501, 520, 527, 562, 563, 569, 570, 593, 609, 610, 611, 638, 655, 662, 678, 679, 684, 730, 739, 740, 747, 769, 772, **773**, 810, 823, 829, 840, 841, 855, 884, Against, 438, 446, Life of, 95, 444, 884

Severus of Aquileia, **773**

Severus of Malaga, 488, **774**

Severus of Milevis, **774**

Severus of Minorca, **774**, 791

Severus of Naples, 581, **774**

Severus of Ravenna, **774**

Severus of Synnada, **774**

Severus of Trieste, 773

Severus Sabokt, 428, 439

Seville, 106, 126, 233, 377, 386, 418, 486, 497, 499, **774**, 791, 861, C. of (590), 148, 775, C. of (619), 106, 331, 775, C. of (624), 775

Sextus, 287, 655, **775**, Sentences of, 272, 579, 704, 775, 784

Sexuality, 244, 252, 717, 719, **775**, 871

Sfameni Gasparro, Giulia, 4, 40-41, 92, 123, 251-52, 255, 489, 518, 542-43, 564, 614, 625, 718-20, 799-800, 815-16

Sfax, 16

Shabuhragan, 519

Shahrbarâz, 36, 433

Shapur I, 118, 225, 519, 757, 860

Shapur II, 11, 54, 106, 637, 674, 757, 883

Shechem, 174

Sheep, 10, 42, 53, 243, 306, 606, 668, 771, **776**, 778, 803

Shekînah, 26

Shellal, 635

Shem, 9

Shemona, Gurya and Habib, 263

Shenoute, 4, 119, 120, 200, 265, 267, 472, **776**

Shepherd, the Good, 1, 52, 64, 138, 152, 164, 217, 221, 255, 280, 349, 406, 434, 572, 573, 578, 600, 615, 625, 629, 668, 675, 755, 764, 771, **776**, 788, 808, 814, 860, *276, 277, 278*

Sherborne, 19

Shiroè, 106

Shivat, 174

Shoes, 866

Shotep, 746

Shushanik, 345

Siagu, 173

Sibylline oracles, 609, 614

Sicca Veneria (El Kef, Tunisia), 15, 55, 82, 148, 641, **778**, 856

Sicily, 7, 157, 169, 170, 186, 342, 424, 685, 743, 853, C. of, **778**

Side, 32, 86, 174, 364, 512, 556, *279*, C. of (c.390), 567, **778**

Sidi Khrebish, 488

Sidon (now Saida), 268, 808, 854, C. of (511), 325

Sidonius Apollinaris, 4, 18, 40, 59, 60, 67, 103, 178, 270, 298, 304, 307, 320, 335, 338, 401, 497, 499, 511, 513, 592, 617, 653, 671, 725, 732, 747, **778**, 845

Sigebert I, 162, 556, 735, 862

Sigebert of Gembloux, 654

Sigismund, 105, 134, 544, 555, **779**, 804

Sigistheus, 652, **779**

Silchester (Calleva Atrebatum), 129

Siligo, 425

Silistria, 89
Silvanus dux Arabiae, 808
Silvanus of Calahorra, 814
Silvanus of Cirta, 176, 351, 765, **779**, 782
Silvanus of Tarsus, 767, **779**, 832, 854
Silverius, 119, 642, 645, 743, **779**, 870
Simeon, 432
Simeon Metaphrastes, 3, 43, 83, 161, 518, 793, 825
Simeon Petritsi, 238
Simeon Stylites the Elder, 44, 51, 175, 220, 688, 723, 808, 829, **779**, 797
Simeon Stylites the Younger, 67, 439, 688, **779**, 797, Life of, 67
Simeon of Bet Arsam, 5, 71
Simeon of Emesa, Life of, 481
Simeon of Jerusalem, 849
Simeon of Mesopotamia, 514
Simeon of Nisibis, 374
Simeon the New Theologian, 708
Simon, apostle, 334, **779**, 780
Simonetti, Manlio, 4, 4-5, 5, 11, 13, 17, 19, 19-20, 20, 21, 22-24, 25, 26-27, 37, 38, 42, 43, 47, 48-49, 49-50, 50-51, 64, 72, 76-78, 84, 89-90, 92, 95, 97, 108, 113, 121, 125, 127, 139, 141, 154, 158, 159, 163-65, 185, 187, 195, 196-98, 214-15, 215, 228, 236-37, 238, 245-46, 251, 268, 276-78, 284, 295-96, 297, 298, 299, 301, 303, 304, 305, 309-11, 315, 318, 320, 320-21, 322, 323-24, 325, 329, 331-32, 337, 343, 350, 351, 363, 365, 369, 381-82, 388, 395, 396, 403, 303, 411, 434, 443, 444, 446, 446-47, 451, 458, 460, 461, 466, 472, 480, 482, 484, 507, 508, 509, 521, 522, 524, 525, 525-26, 527, 546, 550, 551, 560, 566, 568, 569-70, 582, 585, 594, 599, 609, 613-14, 626, 628, 636, 637, 653-54, 657, 660, 661-62, 662-63, 663, 677-78, 680, 682, 685, 685-86, 689, 700, 706, 711, 711-12, 713, 737, 747, 748-49, 749, 757, 765, 766-67, 770, 773, 778, 779, 783, 787, 794, 797, 806, 812, 824, 824-25, 830, 838, 840-41, 847, 854, 855, 856, 858, 870, 875, 877, 884-85
Simonians, 158, 353, 776, 780, 802
Simon Magus, 158, 179, 251, 353, 375, 384, 517, 553, 676, **780**, 827
Simon of Cyrene, 113, 435, 487
Simon of Zayté, 598
Simplicianus of Milan, 28, 29, 497, 658, 660, 697, **780**, 854, 863, 870
Simplicius, pope, 20, 31, 120, 446, 447, 519, 642, 645, 714, 741, 742, 774, **780**, 806, 869
Simplicius of Vienne, 559, **780**
Simpronianus, 628
Sin, 2, 5, 9, 10, 25, 29, 39, 41, 94, 101, 105, 107, 108, 117, 122, 123, 127, 128, 137, 145, 151, 163, 164, 181, 183, 186, 188, 190, 191, 192, 201, 220, 222, 226, 227, 242, 243, 244, 247, 252, 259, 260, 262, 277, 287, 288, 304, 308, 309, 312, 313, 316, 318, 319, 321, 326, 327, 328, 330, 331, 337, 340, 341, 353, 357, 359, 360, 372, 377, 379, 396, 398, 399, 404, 406, 414, 415, 433, 436, 450, 457, 458, 461, 465, 470, 476, 489, 492, 504, 506, 507, 516, 523, 525, 528, 531, 535, 538, 542, 547, 548, 575, 583, 598, 603, 604, 609, 612, 621, 622, 624, 634, 650, 658, 659, 664, 665, 667, 668, 669, 670, 684, 694, 701, 703, 707, 709, 717, 720, 725, 731, 733, 740, 741, 745, 748, 751, 752, 758, 765, 767, 775, **780**, 784, 785, 788, 789, 792, 793, 805, 812, 819, 825, 849, 857, 859, 872, 875, 876, 878, 882, *see also* Original sin
Sinai, 4, 16, 36, 37, 43, 56, 65, 84, 170, 222, 228, 264, 276, 312, 380, 433, 540, 562, 567, 575, 578, 583, 597, 632, 636, 669, 684, **782**, 786, *280, 281, 282*

Sines, 555
Singidunum (Belgrade), 765, 858
Siniscalco, Paolo, 56, 72-73, 82, 166-67, 167-68, 190-91, 199, 209, 223, 282, 307-8, 326-27, 336, 378, 388, 388-90, 453-54, 454, 472, 489-90, 508, 545, 562-63, 582, 628, 700-1, 768, 818-20, 839-40, 847, 850
Sion of Bolnissi, 344
Siout, 194
Sipontum (Siponto), 424
Siret el Rheim, 487
Siricius, 41, 104, 125, 142, 154, 182, 237, 282, 382, 430, 493, 536, 547, 559, 593, 641, 645, 669, 741, 742, 744, **782**, 814, 822, 824, 833, 876
Sirmium, 26, 28, 33, 35, 350, 360, 416, 515, 547, 600, 618, 638, 639, 685, **783**, *283*, C. of (347), 783, C. of (351), 113, 154, 207, 228, 295, 582, 685, 783, C. of (357), 77, 783, C. of (358/359), 77, 113, 207, 343, 395, 783, 858, C. of (377/378), 783, Formula of (351), 207, 516, Formula of (357), 49, 207, 296, 350, 395, 626, 685, 706, 783, 790, 858, Formula of (358), 281, Formula of (359), 5, 78, 395, 527, 732, 737
Sis, 154
Siscia (modern Sisak), 759
Sisebut of Toledo, 387, 418, **783**
Sisenand, 330
Sisinnius, 713
Sisinnius, Martyrius and Alexander, 870
Sisinnius, archbishop of Constantinople, 196, 218, **784**
Sisinnius, novatianist, **784**
Sisinnius, pope, 646
Sisinnius of Toulouse, 869
Sisoes the Theban, 552
Sixtus I, 644, 741
Sixtus II, 96, 212, 229, 237, 238, 644, 670, 741, 758, 775, **784**, 860
Sixtus III, 36, 304, 379, 452, 540, 641, 645, 670, 676, 710, 741, 742, **784**
Sixtus of Siena, 47
Siyagha, 66
Skaramangion, 253
Skaranicon, 253
Skellig Michael, 130
Skiàdion, 253
Slaves, Slavery, 17, 47, 53, 140, 148, 156, 162, 175, 176, 182, 192, 212, 337, 411, 441, 464, 469, 473, 476, 520, 551, 556, 566, 596, 600, 672, 701, 736, 737, 739, 745, **784**, 775
Smaragdus, 366, 655, 774
Smbat Bagratuni, 255
Smile of the Apostles, 446
Smyrna, 73, 160, 284, 533, 672, 688
Soccus, 254
Socotra, 517
Socrates, 13, 23, 24, 47, 58, 84, 96, 235, 282, 340, 496, 626, **785**
Socrates Scholasticus, 388, 389, 644, 749, **785**, 790, 793, 825, 827, 831
Sodom, **785**, and Gomorrah, 309, 506, 525
Sogdian, 61, **785**
Soissons (Suessiones), 209, C. of (853), 616
Soldiers Casting Lots for Jesus' Clothing, **786**, *284*
Solea, 173, 253
Solignac, 268
Soliloquia, 233
Soll, Georg, 537-39
Solomon, 32, 138, 291, 377, 433, 631, **786**, 825, 878, 879, 880
Somnium Scipionis, 697
Son, 5, 13, 16, 20, 29, 34, 36, 39, 42, 49, 53, 59, 64, 76, 77, 78, 89, 92, 94, 99, 108, 114, 121, 141, 150, 163, 165, 195, 199, 205, 207, 215, 223, 233, 243, 252, 260, 293, 296, 300, 303, 324, 340, 352,

354, 362, 381, 390, 391, 395, 396, 402, 405, 418, 429, 430, 437, 448, 449, 463, 467, 468, 524, 525, 526, 553, 566, 595, 621, 622, 653, 670, 684, 685, 700, 734, 735, 737, 749, 757, 780, 783, 788, 797, 830, 831, 848, 851, 853, 854
Song of Songs, 25, 28, 53, 152, 187, 214, 255, 282, 293, 310, 363, 364, 374, 375, 383, 398, 431, 466, 505, 506, 538, 557, 597, 620, 621, 634, 651, 652, 670, 683, 713, 722, 733, 746, **786**, 793, 803, 813, 825, 827, 867, 878, 879, 880
Sophia, 353
Sophia of Jesus Christ, 352, 579
Sophia Prunikos, 110
Sophist, 589
Sophocles, 82
Sophronius, **787**
Sophronius of Jerusalem, 24, 135, 433, 443, 540, 568, 619, 655, 768, **787**
Sophronius of Pompeiopolis, **787**
Sophronius the Sophist, 447
Sopianae (Pécs), 860
Sorrina Nova, 422
Sosigenes, 20
Soter, 181, 238, 258, 644, 646, **787**
Sotericus of Caesarea in Cappadocia, 374
Soteriology, 46, 98, 99, 100, 101, 116, 164, 210, 226, 229, 245, 258, 262, 291, 308, 313, 331, 352, 357, 456, 479, 481, 537, 542, 684, 730, 731, **788**, 815, 851
Soteris, 638
Sotomayor, Manuel, 755-56
Soul, Human, 2, 9, 10, 12, 23, 25, 26, 32, 40, 42, 44, 45, 46, 54, 56, 59, 60, 65, 67, 73, 75, 76, 82, 90, 91, 92, 93, 94, 95, 98, 99, 110, 113, 123, 161, 163, 164, 178, 183, 208, 215, 221, 229, 233, 236, 242, 251, 255, 260, 261, 276, 280, 285, 287, 288, 301, 308, 316, 317, 318, 319, 321, 323, 340, 353, 356, 367, 372, 378, 380, 392, 395, 396, 399, 405, 408, 414, 415, 419, 433, 443, 453, 455, 457, 459, 462, 463, 465, 481, 489, 516, 518, 519, 523, 526, 528, 529, 552, 553, 556, 557, 569, 577, 584, 587, 589, 590, 601, 606, 610, 614, 615, 620, 621, 623, 638, 659, 666, 670, 675, 678, 683, 686, 690, 691, 693, 694, 695, 696, 698, 699, 707, 717, 719, 721, 725, 732, 733, 751, 773, 781, 786, **789**, 792, 793, 796, 803, 805, 818, 819, 832, 836, 843, 849, 872, 873, 875, 877, 880, 881, 882
Source of Gnosis, 442
Sozomen, 6, 11, 23, 24, 47, 56, 57, 58, 65, 84, 244, 282, 388, 389, 496, 509, 538, 611, 614, 646, 736, 749, 764, 765, **790**, 793, 825, 827, 830, 831, 841
Sozopolis, 838
Spain, 127, 230, 572, 582, 599, 604, 617, 618, 633, 641, 642, 657, 660, 663, 669, 676, 710, 711, 712, 714, 730, 742, 743, 747, 756, 761, 774, 780, 783, 784, **790**, 804, 842, 853, 861, 870, 875
Sparta, 7
Spatha, 254
Spelunca Thesaurorum, **791**, 810
Speratus, 513
Spes, 423, 791
Speusippus, 586, 601
Speusippus, Eleusippus and Meleusippus, 875
Speyer (Spira Nemetum), 348
Spidlík, Thomas, 5-6, 71, 123-24, 198, 228, 262-63, 442, 443-44, 489, 529-30, 577-78, 652, 707-8, 725, 787, 792, 797, 872, 882
Spinelli, Mario, 40, 92, 102, 301, 339, 476-77, 506-7, 582, 644-46, 759, 774
Spiritual combat, 223, 312, 621, 708, **792**, 817, 875
Spirituality, 44, 54, 84, 98, 100, 101, 111, 112, 119, 124, 140, 161, 168, 182, 230, 244, 272, 280, 282, 287, 311, 338, 339,

355, 358, 381, 394, 398, 405, 430, 434, 441, 443, 445, 459, 462, 478, 483, 489, 492, 510, 514, 521, 531, 536, 562, 567, 573, 620, 621, 624, 661, 682, 683, 708, 717, 719, 722, 723, 776, 778, 786, 787, 788, **792**, 851, 862, 864, 878, 880, 882
Split, 217
Spoleto (Spoletium), 7, 8, 423, 791
Sponsa Christi, 787, **793**
Spyridion, **793**, 825, Life of, 481
Statuta ecclesiae antiqua, 17, 142, 147, 182, 342, 530, 714, **793**
Stead, Christopher George, 93-95, 626-27
Stegaurach, 772
Stele of Mesha, 66
Stenography, 185, 394, **794**
Stephania, 581
Stephanites, 289
Stephen I, pope, 640, 643, 644, 646, 672, **794**, 848, 860, 864
Stephen II, pope, 643, 653
Stephen II of Hierapolis-Mabbug, **794**
Stephen baru Sudayle, 445, 623, **794**
Stephen Gobar, **794**
Stephen Melodus, 748
Stephen of Antioch, 49, 473, 757
Stephen of Bostra, **794**
Stephen of Dor, 530, 823
Stephen of Ephesus, 115
Stephen of Larissa, 197, 281, 744, 749, **794**
Stephen of Ptolemaïs, 487
Stephen of Salona, 237
Stephen the protomartyr, 105, 117, 175, 213, 225, 295, 308, 383, 432, 433, 493, 507, 509, 533, 536, 575, 624, 671, 707, 741, 748, 774, **795**, 814, 821, 822, Passion of, 814
Stephen the Thaumaturge, 748
Stiernon, Daniel, 4, 5, 27, 35, 35-36, 36, 37, 43, 44, 109, 122, 137, 138, 166, 175, 193, 201, 208-9, 213-14, 241, 274, 304, 305, 454, 374, 431, 438, 439, 444, 446, 479-80, 486-87, 516, 523, 545, 546, 549, 550, 552, 553, 584, 596, 611, 659, 660, 662, 663, 725-26, 768, 784, 822, 827, 841, 884
Stiffel, 772
Stilicho, 19, 28, 178, 334, 356, 397, 559, 612, 745, **795**, 861, 863
Stipanska, 217
Stobi, 171, 174, 515, **795**, *285, 286, 287*
Stoicism and the Fathers, 25, 40, 45, 53, 74, 75, 83, 90, 92, 96, 114, 116, 154, 161, 176, 178, 180, 191, 204, 205, 209, 226, 230, 240, 244, 262, 276, 287, 288, 316, 326, 354, 363, 372, 390, 395, 398, 401, 405, 448, 455, 463, 470, 504, 516, 517, 528, 557, 584, 585, 587, 601, 604, 621, 626, 639, 683, 689, 690, 695, 698, 699, 716, 720, 729, 737, 767, 781, 789, **795**, 805, 815, 849, 872, 882
Stowe missal, 133, 155
Strasbourg (Argentorate), 348
Strategius (or Strategus), 52, **797**
Strato of Lampsacus, 12
Strato of Sardis, 44
Straulesti (Bucharest), 217
Strbinci, 639
Streanaeshalch, 137
Stromata, 74, 76, 166, 177, 180, 373, 719, 775
Strophium, 253
Studer, Basilio, 26, 27, 38-40, 46, 53, 60, 71, 72, 82, 111, 140, 154, 166, 198, 202, 205-6, 226-27, 237, 242-43, 244, 244-45, 256, 262, 272, 308-9, 313, 324, 328, 351, 352, 354-56, 357-60, 371, 395-96, 398-99, 401-2, 405, 408-9, 409-10, 442-43, 458, 460, 461, 479, 482-83, 485, 530, 560, 576, 576-77, 585, 640-44, 646, 652, 666-67, 670, 675, 678, 680, 704, 706, 716, 729, 730, 731, 733-35,

Numbers in bold refer to the main entry on the subject. *Numbers in italics refer to illustration numbers (vol II, pages 951-1094)*

758, 782-83, 788-89, 794, 802, 804, 826, 831, 845, 851-53, 856, 867, 870, 877-78, 885
Stuma, 491
Stylites, Stylitism, 48, 228, 500, 688, 779, **797**
Subiaco, 119
Subintroductae, 16
Subordinationism, 43, 50, 76, 92, 114, 164, 168, 215, 223, 252, 276, 282, 298, 301, 341, 391, 395, 405, 463, 507, 622, 743, **797**, 819, 851
Successus, 215, 237, **798**
Succession, apostolic, 48, 63, 103, 141, 260, 284, 299, 316, 495, 617, 619, **798**, 847
Successus of Abbir Germaniciana, 584
Sucidava (Celei - Corabia, district of Olt), 217, 765
Sudarium, 253
Suetonius, 6, 121, 812
Suetonius Tranquillus, 121
Suevi, 187, 505, 554, 774
Sufetula (Sbeitla), 14
Suhag, 265
Suidas (or Sudas), 95, 115, 236, 303, 352, 654, 684, **798**, 835
Suintila, 418, 783, 791
Suka, 160
Sulpicius Severus, 121, 126, 166, 233, 270, 338, 369, 407, 531, 562, 618, 661, 736, 752, 755, 758, 793, **799**, 843, 846, 862, 863, 873
Sun, Cult of the, 26, 65, 102, 143, 163, 193, 257, 271, 273, 296, 371, 434, 546, 564, 590, 607, 685, 706, 750, **799**, 800, 836
Sunday, 17, 50, 71, 90, 116, 135, 147, 152, 186, 188, 210, 230, 232, 257, 258, 274, 292, 319, 394, 404, 454, 473, 475, 482, 492, 493, 504, 518, 537, 541, 548, 556, 557, 562, 595, 625, 637, 640, 669, 699, 701, 728, 739, 741, 744, 748, 784, **800**, 805, 831, 851, 867, 875, 876, 877
Sunna, 542
Supagvlane, 19
Superstitio, 6, 32, 73, 167, 226, 231, 241, 257, 264, 295, 365, 376, 473, 517, 548, 552, 671, 702, **801**, 812, 849
Susanna, 161, 219, 350, 351, 460, 536, 615, 630, 675, **801**, *288*, *289*
Sutrium (Sutri), 422
Sutton Hoo, Suffolk, 133, ship-burial, 133
Suwayfiyeh, 66
Sub-deacon, 233
Syagrius, 18, 183, 652, **801**, 863
Syagrius of Autun, 105, **802**
Sybilline Oracles, 56, 57, 410, 455, 469
Syias, 209
Sykada, 213
Syllogismoi, 54
Sylvanus of Cirta, 247
Sylvanus of Tarsus, 195, 236
Sylvester, 153, 246, 476, 521, 641, 645, 741, 742, **802**, 865, 866, 870, Acts of, 82
Sylvia, 687
Symbols, Symbolism, 4, 10, 17, 18, 19, 20, 25, 26, 32, 41, 50, 53, 54, 55, 56, 62, 89, 111, 116, 132, 136, 156, 173, 186, 187, 202, 209, 210, 213, 217, 219, 232, 239, 240, 260, 271, 276, 279, 285, 291, 294, 306, 309, 310, 336, 341, 345, 346, 377, 390, 395, 398, 404, 405, 409, 420, 423, 434, 435, 437, 470, 471, 489, 491, 511, 528, 544, 550, 554, 571, 573, 597, 598, 599, 605, 607, 611, 612, 615, 625, 629, 648, 650, 668, 685, 686, 689, 695, 737, 749, 750, 755, 771, 776, 786, 778, 786, **802**, 834, 844, 878, *290*, *291*, *292*
Symbolum Apostolorum, 63
Symmachians, 149, 803
Symmachus, 16, 26, 36, 59, 79, 116, 124, 142, 149, 237, 241, 325, 357, 360, 365, 380, 389, 397, 399, 431, 455, 476, 521, 552, 567, 706, 745, **803**

Symmachus, exegete, 767, **803**
Symmachus, pope, 638, 642, 645, 646, 652, 678, 730, 742, 744, 784, **804**, 806, 858
Symmachus, Aurelius Anicius, 645, **804**
Symmachus, Quintus Aurelius, 742, 774, **804**
Symmachus, Quintus Aurelius Memmius, **804**, 828
Symmikta Zetemata, 584, 695
Symphonia, 71
Symphosius Scholasticus, 761, **804**, Aenigmata, 804
Symposium of the ten virgins, 233
Symposius of Astorga, 127, 234, 844
Sympronianus, **804**
Synagogè, 62
Synagogè of 50 Titles, 439
Synaxis, 2, 50, 90, 171, 173, 241, 493, 499, 699, 741, 759, **805**, 847
Syncletica (deaconess), **805**
Syncletica (nun), **805**
Syncretism, 9, 23, 32, 73, 201, 226, 349, 350, 352, 361, 412, 455, 485, 519, 534, 557, 585, 587, 589, 599, 602, 639, 674, 775, 800, **805**, 833
Syneisaktoi, 16
Synekdemos, 136, 208, 835, 837
Synesius of Cyrene, 96, 154, 320, 401, 487, 496, **806**
Synnada, 774
Synonyms, 418
Synopsis Scripturae Sacrae, 406, 881
Synousiasts, **806**, 841
Syntagma, 375, 384, 463
Syntagmation, 13, 92
Synthronos, 172
Syracuse, 156, 157, 192, 274, 313, 381, 425, 457, 509, 666, **806**, *293*
Syria, 2, 4, 6, 23, 33, 37, 38, 43, 47, 48, 49, 53, 58, 60, 65, 66, 69, 70, 80, 84, 91, 95, 96, 107, 108, 111, 114, 126, 135, 138, 152, 154, 163, 164, 165, 167, 169, 170, 171, 173, 174, 175, 180, 181, 188, 193, 205, 210, 219, 223, 224, 225, 230, 234, 258, 263, 264, 265, 266, 279, 280, 282, 296, 310, 448, 454, 472, 478, 494, 541, 566, 567, 568, 569, 585, 604, 609, 618, 619, 631, 635, 637, 662, 670, 673, 684, 688, 720, 757, 758, 761, 763, 768, 797, **807**, 809, 810, 814, 815, 823, 854, 884, *294*, *295*, *296*, *297*, *298*, *299*, *300*
Syriac Christian theology, 700
Syriac Didascalia, 91, 258, 668
Syriac Literature, 1, 2, 3, 5, 9, 10, 19, 20, 21, 23, 31, 32, 42, 43, 44, 47, 48, 50, 51, 54, 56, 60, 62, 65, 66, 71, 73, 82, 94, 110, 113, 114, 115, 142, 152, 179, 181, 187, 193, 198, 204, 206, 211, 214, 234, 235, 240, 257, 259, 262, 263, 269, 271, 276, 298, 299, 300, 306, 312, 327, 330, 345, 364, 368, 370, 380, 389, 394, 403, 404, 413, 416, 417, 428, 429, 438, 439, 440, 442, 444, 445, 446, 448, 449, 452, 458, 467, 472, 481, 482, 485, 496, 499, 504, 505, 511, 514, 517, 518, 519, 520, 521, 523, 527, 536, 537, 540, 547, 551, 556, 561, 562, 567, 593, 594, 597, 599, 609, 610, 623, 629, 640, 654, 655, 662, 679, 680, 699, 713, 722, 727, 729, 739, 759, 763, 769, 770, 775, 779, 782, 785, 786, 791, 794, 798, 803, 808, **809**, 818, 820, 824, 825, 829, 830, 837, 840, 841, 842, 850, 879, liturgy, 34, 80, 293, 562, 810, Syro-occidental (Antiochene) rite, 6, 33, 34, 45, 50, 82, 496, 500, 550, 810, Syro-oriental rite, 34, 50, 496, 500
Syrianus, 590

Tabarka, 13, 15, 16, 55, 148, 382, 450, 705, **812**
Tabitha, 25, **812**
Tabula Paschalis Petrocoricensis, **812**
Tabula peutingeriana, 426, 675, 835

Tacitus, 6, 30, 121, 167, 370, 531, 593, 671, 674, 675, 702, 740, **812**
Tafha, 67
Taggia, 421
Taio of Saragossa, 465, 728, 755, 791, **812**
Takla Haymanot, 291
Talich, 81
Talinn, 81
Talleleus, 175
Talmud, 32, 309, 563, **812**
Tamallula, 13
Tannaim, 32, 563, **812**
Taphanam, 832
Taphar, 361
Taprobane (Ceylon, Sri Lanka), 203
Taracus and Andronicus, 175, Acts of, 534
Tarasius, 197
Tarcisius, 232
Tarentum (Taranto), 424
Targum, 54, 276, 395, 674, 763, 810, **812**, of the Hagiographa, 813
Taron, 84
Tarquinii (Tarquinia), 422
Tarraconensis, 107, 110, 148, 191, 356, 755, 790, **813**, 814, 845, *301*
Tarragona (Tarraco), 69, 110, 157, 331, 377, 382, 386, 497, 597, 721, 752, 755, 756, 790, 791, **813**, C. of (464), 814, C. of (516), 386, 814
Tarrasa (ancient Egara), **814**, C. of (614), 400, 548, 813
Tarsicius or Tarcisius, **814**
Tarsus, 60, 111, 175, 190, 236, 307, 657, **814**, C. of (432), 550, 830
Tarvisium (Treviso), 420
Tascodrugitae, 84
Tatian, 9, 54, 56, 60, 63, 64, 73, 82, 92, 121, 163, 168, 176, 178, 198, 205, 234, 243, 244, 263, 271, 319, 373, 378, 384, 391, 448, 460, 463, 472, 489, 614, 644, 683, 689, 690, 691, 693, 709, 717, 718, 719, 732, 736, 761, 775, 785, 810, **815**, 868, 875
Tato, 505
Taurobolium, 409, **815**
Tavium, 333
Tayk, 81
Tberkjot (Ar. Farshut), 4
Tbilisi, 344, 578
Tebessa, 14, 15, 170, 171, 174, 209, 545, 572, **816**, 840
Te Decet Laus, **816**
Te Deum, 160, 595, **816**
Teias, 357
Tekor, 19, 80
Telesphorus, 156, 644
Tell Hassan, 635
Temes, 334
Templon, 172
Temptation, 9, 100, 124, 137, 145, 161, 226, 282, 312, 399, 414, 446, 454, 506, 523, 528, 538, 614, 621, 731, 781, 792, 801, **816**, 860, 872, 878
Tenes, 16
Teramo (Interamnia Praetuttianorum), 423
Terapius of Bulla, 146, 584
Terapon, 38
Terebus, 677
Terentius, 115
Teresa of Avila, 101
Terni (Interamna Nahars), 859
Tertium Genus, 61, **818**
Tertius, 145
Tertullian, 2, 6, 9, 13, 14, 16, 25, 27, 38, 42, 45, 46, 52, 53, 55, 56, 57, 60, 61, 63, 72, 79, 82, 90, 92, 103, 107, 108, 111, 121, 123, 128, 140, 143, 144, 145, 150, 151, 154, 155, 156, 158, 160, 161, 163, 164, 168, 177, 178, 186, 188, 189, 190, 191, 198, 205, 206, 209, 211, 212, 221, 222, 223, 227, 231, 232, 242, 243, 244,

245, 247, 250, 252, 256, 257, 258, 260, 262, 271, 273, 278, 283, 285, 286, 287, 305, 308, 309, 311, 312, 313, 318, 319, 320, 323, 326, 331, 332, 347, 355, 359, 372, 373, 375, 376, 378, 391, 398, 400, 405, 406, 408, 409, 410, 433, 434, 448, 449, 452, 453, 457, 469, 470, 472, 473, 474, 477, 478, 492, 494, 495, 497, 498, 499, 504, 506, 507, 510, 517, 528, 529, 532, 533, 537, 541, 544, 551, 552, 560, 561, 563, 564, 566, 571, 576, 583, 585, 593, 603, 611, 612, 614, 615, 628, 629, 640, 643, 649, 654, 655, 656, 657, 665, 667, 668, 669, 670, 671, 672, 675, 683, 685, 687, 689, 698, 699, 703, 706, 707, 710, 712, 716, 719, 723, 724, 725, 731, 732, 734, 737, 749, 750, 752, 758, 759, 760, 768, 769, 775, 776, 778, 781, 782, 785, 788, 789, 790, 792, 793, 796, 797, 798, 801, 803, 816, 817, **818**, 831, 838, 839, 843, 848, 849, 851, 856, 867, 871, 875, 876
Tertullian, Ps.-, 158, 251, 771
Testament of O. L. J. C. (Testamentum Domini), 34, 63, 104, 609, 723, **820**, 877
Testament of the 12 Patriarchs, 39, 56, 57, 410, 718, 781
Testimonia, 72, 150, 211, 276, 305, 309, 312, 324, 354, 436, 437, 472, 655, 763, 788, **821**, 851, 856, 881
Testimony, 316, **821**
Tetrabiblos, 92
Teurnia (St Peter im Holz), 600
Thaddaeus, 2, 10, 62, 79
Thagaste, 27, 29, 97
Thala, 15
Thalassius, 40, 112, 511, 728, **821**
Thalassius of Angers, **821**
Thalia, 20, 76
Thamugadi, *see* Timgad
Thapsus, 870
Tharros, 425
Thasos, 515
Thebaïd, 4, 20, 23, 83, 149, 184, 264, 445, 486, 521, 664, 673, 768, **821**, 826
Theban Legion, 295, 338, 347, 360, **821**
Thebes, 7
Thecla, 89, 369, 536, 688, 808, **822**
Thela, C. of (418), 134, **822**
Themistius, 17, 67, 185, 760, 785, **822**
Theoctistus of Caesarea, 21, 23, 225, 315, 525, 619, **822**, 832
Theodahat, 357
Theodebert, 329
Theodelinda (or Theodolinda), 17, 103, 187, 307, 366, 409, 420, 427, 447, 506, 783, **822**, 866
Theodericion Edict, 828
Theodo, 747
Theodora, 27, 43, 85, 88, 119, 192, 200, 428, 464, 505, 573, 597, 605, 662, 678, 679, **822**, 823, 829, 870
Theodore, archimandrite, 824
Theodore, bishop of Darnis, 487
Theodore, bishop of Echinus, **822**
Theodore, bishop of Merv, 769
Theodore, bishop of Pharan, **823**
Theodore, Jacobite patriarch of Alexandria, 505, 662, **822**
Theodore, martyr, 407, 536, **822**
Theodore, metropolitan of Libya Pentapolis, 487
Theodore, pope, 642, 645, 659, 726, **823**, 859
Theodore, presbyter, **823**
Theodore Abu Qurra, 66, 437
Theodore and Theophanes, 748
Theodore Ascidas, 137, 245, 677, **823**, 826
Theodore bar Koni, 71, 520, 810, **823**
Theodore Daphnopate, 442
Theodore Graptus, 346
Theodore of Antinoë, 593

Theodore of Aqileia, 64, 419, **823**
Theodore of Bostra, **823**
Theodore of Canterbury, 119, 129, 132, 143, 159, 369, 446, **824**, 873, 877
Theodore of Constantinople, **824**
Theodore of Copros, 822, **824**, 826
Theodore of Edessa, 748
Theodore of Fréjus, 79
Theodore of Heraclea, 302, 449, 543, **824**, 868
Theodore of Marseilles, **824**
Theodore of Martigny, **824**
Theodore of Mopsuestia, 5, 18, 42, 49, 50, 107, 108, 112, 151, 154, 164, 175, 187, 197, 203, 214, 216, 236, 243, 293, 310, 311, 312, 315, 325, 342, 374, 397, 430, 444, 449, 450, 475, 480, 482, 484, 494, 500, 501, 504, 528, 543, 561, 570, 571, 583, 594, 598, 616, 656, 658, 663, 667, 684, 702, 713, 715, 716, 719, 723, 729, 750, 776, 781, 787, 788, 810, 817, **824**, 827, 830, 838, 876, 878, 879, 880, 881
Theodore of Paphos, 793, **825**
Theodore of Petra, **826**, 829
Theodore of Pharan, 768, **826**
Theodore of Philae, **826**, 829
Theodore of Raithu, **826**
Theodore of Scythopolis, **826**
Theodore of Sykeon, 343, Life of, 213
Theodore of Tabennesi, 31, 44, 199, 628, **826**
Theodore of Trimethus, **826**
Theodore of Tyana, 682, **826**, 854
Theodore of Tyre, 27
Theodore Prodromos, 142
Theodore Spudaeus, **826**
Theodore Syncellus, **827**
Theodore the Lector, 65, 251, 282, 388, 389, **827**
Theodore the Studite, 6, 83, 135, 251, 407
Theodoret of Cyrrhus, 1, 4, 6, 21, 24, 31, 38, 44, 50, 57, 58, 61, 64, 71, 75, 121, 154, 159, 185, 197, 203, 214, 233, 243, 275, 310, 312, 325, 339, 340, 351, 352, 372, 375, 388, 389, 403, 407, 416, 429, 436, 444, 450, 472, 481, 507, 520, 521, 523, 524, 527, 550, 562, 567, 569, 570, 603, 634, 640, 644, 646, 655, 658, 665, 667, 676, 695, 698, 700, 704, 713, 716, 723, 727, 728, 745, 760, 764, 779, 783, 787, 797, **827**, 830, 831, 838, 843, 864, 881, 884
Theodoric II, 127, 298
Theodoric Strabo, 113, 565
Theodoric the Great, 43, 62, 67, 124, 149, 169, 183, 186, 241, 252, 254, 273, 281, 298, 329, 339, 357, 438, 464, 476, 521, 565, 610, 642, 645, 678, 730, 744, 749, 761, 804, **828**, 837, 861, 863, 884, 886
Theodosians, 37
Theodosius, deacon, **828**
Theodosius, ecclesiekdokos, 824
Theodosius, suffragan bishop of Larissa, 795
Theodosius, monophysite monk, **828**
Theodosius I, "the Great", 7, 8, 17, 19, 24, 26, 28, 47, 49, 51, 52, 59, 62, 63, 64, 67, 78, 85, 88, 92, 105, 135, 168, 175, 185, 194, 195, 196, 200, 215, 247, 268, 272, 283, 296, 320, 324, 325, 333, 334, 339, 351, 356, 360, 363, 397, 408, 432, 444, 445, 447, 472, 484, 491, 505, 508, 521, 522, 537, 547, 549, 559, 564, 584, 595, 607, 611, 614, 624, 637, 641, 741, 745, 760, 771, 795, 804, 816, 822, **828**, 833, 842, 858
Theodosius II, 4, 5, 49, 67, 85, 92, 96, 102, 112, 154, 159, 184, 194, 209, 214, 216, 218, 241, 251, 275, 295, 304, 305, 314, 325, 327, 407, 416, 444, 447, 522, 550, 552, 569, 594, 596, 597, 607, 641, 648, 681, 704, 713, 725, 744, 752, 767, **829**, 855
Theodosius of Alexandria, 428, 448, 662, 769, 822, 823, **829**, 850

Theodosius of Auxerre, 482
Theodosius of Ephesus, 197
Theodosius of Gangra, 36
Theodosius the coenobiarch, 826, **829**
Theodosius the Younger, 790
Theodoti, 90
Theodotion, 199, 360, 455, 620, 716, 801, 803, **829**
Theodotus and the Seven Virgins, **829**
Theodotus of Ancyra, 22, 26, 35, 58, 180, 384, 650, **829**, 859
Theodotus of Antioch, 22, 784, **830**
Theodotus of Byzantium (the Banker), 11, 84, 583, **830**
Theodotus of Byzantium (the Tanner), 11, 84, 566, 583, 640, 741, **830**, 867, 885
Theodotus of Laodicea, 48, 300, 582, **830**
Theodotus of Nicopolis, 114
Theodotus the Valentinian, 9, **830**
Theodulus, **830**
Theognis of Nicaea, 77, 185, 329, 594, **830**
Theognius, 216
Theognostus, 23, 235, **830**
Theology, 12, 13, 20, 21, 22, 23, 25, 27, 29, 33, 34, 35, 39, 42, 46, 50, 54, 59, 63, 74, 75, 82, 90, 93, 94, 98, 99, 100, 101, 107, 108, 109, 110, 113, 118, 123, 135, 138, 139, 141, 161, 162, 168, 177, 181, 182, 199, 200, 205, 206, 207, 208, 211, 218, 226, 227, 233, 239, 242, 243, 244, 245, 246, 248, 249, 250, 257, 262, 288, 293, 298, 299, 300, 301, 305, 307, 308, 313, 315, 316, 317, 324, 328, 331, 341, 352, 355, 356, 357, 359, 363, 365, 366, 368, 373, 379, 381, 390, 391, 394, 395, 396, 397, 398, 401, 405, 407, 408, 409, 413, 415, 418, 429, 430, 436, 441, 442, 443, 444, 446, 459, 463, 478, 479, 481, 486, 499, 500, 504, 510, 521, 526, 528, 530, 533, 544, 548, 551, 555, 562, 576, 577, 580, 583, 584, 593, 595, 603, 604, 606, 609, 614, 616, 620, 621, 623, 626, 640, 646, 654, 656, 657, 658, 661, 672, 675, 678, 682, 683, 684, 685, 687, 689, 695, 696, 699, 707, 708, 709, 713, 716, 717, 731, 733, 734, 738, 740, 741, 749, 751, 761, 763, 764, 767, 769, 773, 775, 776, 782, 786, 788, 791, 792, 794, 797, 806, 815, 819, 820, 823, 824, 826, 827, **831**, 848, 851, 852, 870, 878, 885, 886
Theolyte, 404
Theonas of Alexandria, 24, **831**
Theonas of Marmarica, 487, 595
Theopaschite controversy, 438
Theopaschite formula, 237, 315, 401
Theopaschites, 51, 125, 322, 400, 416, 438, 443, 446, 679, 684, 713, 740, **831**, 841, 851, 856
Theophanes Graptós, 135
Theophanes the Chronicler, 251
Theophania, 93
Theophany, 300, 735
Theophilus Bosphoritanus, 357
Theophilus of Alexandria, 22, 23, 24, 31, 41, 46, 48, 50, 51, 56, 59, 90, 138, 149, 155, 163, 166, 168, 189, 196, 199, 200, 205, 214, 264, 275, 277, 282, 306, 309, 314, 355, 364, 372, 375, 407, 409, 417, 430, 432, 438, 440, 457, 496, 537, 543, 567, 609, 614, 623, 636, 704, 776, 806, 826, **831**, 851, 863
Theophilus of Antioch, 9, 57, 60, 90, 122, 176, 259, 273, 299, 378, 380, 391, 406, 448, 543, 583, 732, 760, 785, 796, **831**
Theophilus of Caesarea, 188, 432, 582, **831**
Theophilus of Castabala, **832**, 854
Theophilus the Indian, 49, 65, **832**
Theophrastus, 12, 74, 233
Theophronius of Tyana, 49
Theophylact Simocatta, **832**
Theoria, 198, 236, 364, 443, 489, 558, 577, 601, 693, 695, 725, **832**

Theotecnus of Caesarea in Palestine, 37, 48, 663, 822, **832**
Theotecnus of Livia, 538, **832**
Theotimus, **833**
Theotokos, 51, 72, 165, 213, 251, 275, 301, 314, 364, 402, 433, 444, 446, 452, 538, 539, 540, 552, 594, 599, 622, 652, 713, 744, 797, 831, **833**
Thessalonica, 7, 8, 16, 28, 36, 41, 71, 88, 169, 172, 186, 195, 201, 203, 225, 251, 281, 283, 292, 302, 306, 315, 369, 409, 434, 446, 479, 515, 540, 566, 573, 631, 670, 705, 742, 745, 784, 795, 828, **833**, 835, *302, 303, 304, 305*
Thessaly, 69, 197, 515, 795, 822, **833**, 834, 837
Theudila, 783
Theurgy, 587, 589, 614, 805, **835**
Theveste, 15
Thibilis, 15
Thmuis, 768
Thomas, 10, 57, 62, 66, 290, 517, 519, 674, 680, 775, 810, 836, Acts of, 674, 810, **837**, Gospel of, 18, 57, 161, 181, 263, 271, 430, 448, 567, 674, 718, 775, 810, 837
Thomas, Mar, 518
Thomas Aquinas, 99
Thomas of Edessa, 216, **837**
Thomas of Harkel, 810
Thomas of Heraclea, 439
Thrace, 32, 85, 113, 195, 198, 356, 514, 523, 565, 566, 619, 631, 681, 756, 834, **837**, *306, 307, 308, 309*
Thrasamund, 78, 144, 203, 252, 296, 757, 861, **868**
Thraseas, 838
Three Chapters, 8, 29, 102, 187, 197, 203, 223, 236, 246, 305, 315, 322, 366, 403, 416, 432, 484, 545, 553, 570, 571, 623, 642, 645, 666, 667, 677, 679, 702, 711, 730, 743, 747, 773, 822, 823, 824, 837, 838, 850, 862, 863, **868**, 870
Thrones, 39
Thuburgo Maius, 14
Thubursicu Numidarum, 14
Thuccabor, 211
Thurificati, 61, 146, 223
Thuringians, 183
Thusum, 36
Tiara, 253, 866
Tiberianus, 83, 712, **838**
Tiberias, 635
Tiberius, 10, 85, 92, 137, 140, 143, 453, 512, 593, 687, 812, **838**
Tibet, 519
Tibiletti, Carlo, 198-99, 529, 766, 767, 795-97, 871-72
Tichonius, 30, 55, 64, 117, 161, 176, 248, 286, 311, 613, 651, 711, 733, **838**, 867
Ticineto, 421
Ticinum (Pavia), 421
Tiddis, 15
Tigisis (now Ain el Bordj), 16, 765
Tigzirt, 15
Tillemont, L.S. Le Nain de, 654
Timaeus of Locris, 12
Time, 12, 45, 55, 56, 94, 99, 101, 118, 151, 167, 205, 239, 240, 262, 285, 308, 327, 340, 379, 389, 399, 409, 489, 500, 509, 510, 528, 553, 577, 586, 588, 601, 621, 622, 651, 692, 698, 733, 752, 778, **838**, 839
Timgad (ancient Thamugadi), 14, 15, 16, 157, 337, 351, 572, 607, 613, 799, **840**
Timothy, 59, 120, 142, 195, 438, 496, 504, 561, 562, 740, 833, **840**, 843, 877
Timothy I, 445, 810, **840**
Timothy I of Alexandria, **840**
Timothy II of Alexandria (Ailouros-Aelurus), 24, 197, 200, 275, 337, 479, 569, 655, 678, 679, **840**, 841, 870
Timothy IV (III) of Alexandria, 333, 342, 822, 826, 829, **841**

Timothy Salofaciolus, 6, 447, 678, **841**
Timothy of Alexandria, **841**
Timothy of Berytus, 59
Timothy of Constantinople, 251, **841**
Timothy of Jerusalem, 91, **841**
Timothy the Apollinarist, 700, 785, **841**, 859
Timothy the Lector, **842**
Tingis (Tangier), 13, 543
Tintagel, 131
Tipasa, 14, 15, 16, 53, 148, 157, 169, 170, 174, 544, 731, **842**, *310*
Tirechán, 653
Tiridates I, 83
Tiridates II, 83, 368
Tiridates III, 80, 83
Titoe, abbat, 688
Titulus, 157, 168, 218, 223, 232, 500, 522, 617, 656, 742, 745, **842**
Titulus Praxidae, 706
Tituli Psalmorum, 364, **842**
Tituli urbani, 211
Titus, 38, 186, 208, 217, 246, 326, 432, 614, 682, **843**, 863, Acts of, 843
Titus, emperor, **843**
Titus of Bostra, 511, 519, **843**, 868
Tkow, 514
Tobias, 161, **843**, *311*
Toga, 253
Tokra (Taucheira), 487
Toledan creed, 96, 652
Toledo (Toletum), 96, 102, 142, 148, 331, 368, 377, 386, 405, 418, 458, 497, 555, 611, 783, 791, **844**, C. of (396), 844, C. of (400), 96, 127, 189, 234, 323, 554, 704, 844, C. of (440), 755, C. of (447), 844, C. of (516), 814, C. of (527), 571, 761, 844, C. of (580), 482, C. of (589), 106, 148, 208, 418, 478, 542, 554, 582, 730, 774, 791, 844, C. of (597), 148, 542, 582, 844, C. of (610), 331, 844, C. of (633), 774, 844, 866, C. of (633-638), 148, 386, 387, 418, 497, 521, 582, C. of (636), 148, 844, C. of (638), 127, 844, C. of (646), 844, C. of (653), 296, 405, 845, C. of (655), 296, 405, 845, C. of (656), 255, 296, 386, 728, 845, C. of (675), 125, 148, 845, C. of (681), 845, C. of (683), 582, 845, C. of (684), 455, 646, 845, C. of (688), 845, C. of (693), 845, C. of (694), 845, C. of (702), 845
Tolentinum (Tolentino), 422
Tomi, 85, 140, 565, 764, 765, 833, **845**, *312*
Tomus ad Flavianum, 37, 159, 165, 196, 197, 208, 275, 318, 325, 400, 461, 479, 548, 559, 569, 594, 618, 684, 730, 742, 753, 773, 828, 840, 841, **845**, 861, 863
Tonantius, 143, **845**
Tonsure, 20, 182, 187, 224, 369, 411, 476, **846**
Topics, 124
Torcello, 420
Torre del Fum, 813
Totila, 119, 254, 357, 464, 505, 666, 749, **846**
Toulouse (Tolosa), 183, 238, 315, 338, 356, 581, 736, 756, 758, **846**, 869
Tours (Turones), 17, 27, 102, 126, 183, 187, 238, 279, 284, 337, 339, 365, 512, 531, 671, 676, **846**, 854, 862, C. of (461), 847, C. of (563), 493, C. of (567), 556
Tours Pentateuch, 187, 506, 785
Toul (Tullo), 103, 348
Tovina (c.527), 765
Tractatus, 394, 436, 716, 723, 770, **847**
Tractatus Contra Arianos, **847**
Tractoria, 96, 458, 665, 666, 885
Traditio Legis et Clavium, 62, 64, 202, 219, 351, 435, 466, 676, 677, **847**, *313, 314*
Tradition, 23, 33, 39, 45, 47, 55, 57, 61, 62, 63, 72, 99, 103, 104, 114, 120, 127,

163, 177, 181, 182, 200, 205, 206, 223, 226, 234, 244, 252, 257, 260, 264, 284, 285, 309, 312, 320, 326, 336, 337, 346, 347, 355, 358, 359, 361, 364, 365, 388, 390, 395, 397, 398, 411, 415, 441, 457, 470, 478, 479, 495, 499, 510, 526, 530, 537, 538, 539, 543, 550, 553, 571, 593, 604, 609, 621, 623, 639, 640, 644, 674, 685, 701, 703, 718, 719, 720, 732, 733, 734, 740, 744, 746, 748, 760, 762, 763, 768, 769, 773, 779, 781, 782, 792, 793, 794, 798, 800, 802, 817, 819, 827, 831, 838, 843, **847**, 862, 870, 875, 876
Traditions of Matthias, 543
Traditor, 61, 79, 137, 145, 176, 246, 248, 250, 321, 332, 351, 516, 521, 533, 613, 651, 765, 766, 779, **849**
Traducianism, 26, 41, 98, 137, 538, 622, 666, 782, 789, 819, **849**
Trajan, 11, 65, 122, 126, 143, 156, 217, 271, 284, 369, 565, 615, 671, 699, 729, 756, 779, 782, 807, 808, 816, 840, **849**, 853
Trajanopolis, 679
Transept, 171
Transitus Mariae, 57, 91, **850**
Translatio Imperii, **850**
Tranum (Trani), 424
Trapè, Agostino, 11, 27, 97-101, 101-2, 462, 568, 781-82
Traprain Law treasure, 133
Trasamund, 331
Treasure of St Ninian's Isle, 133
Treasury of Guarrazar, 148
Trebellius Pollio, 388
Trebizond (Trapezous), 36, 43
Trebonianus Gallus, 223
Tremetuschia, 213
Trent (Tridentum), 870, C. of, 101, 120, 751, 770
Trevijano, Ramón, 46-47, 61, 63, 72, 112, 179, 306, 526-27, 680, 682, 687, 798, 836-37, 840, 843
Triacca, Achille, 29-30, 495-98, 498-99
Tricenius, 678
Trier (Treveri), 18, 28, 69, 83, 102, 169, 174, 186, 230, 322, 329, 337, 347, 349, 360, 400, 404, 430, 546, 547, 559, 595, 702, 712, 799, 843, **850**, *315*, Council of, 850
Trieste (Tergeste), 420
Trifolius, 851
Triform Protennoia, 771
Triginta Capita contra Severum, 480
Triglia, 87
Trimethus, 793, 825, 826
Trinity, 2, 9, 11, 17, 23, 24, 29, 34, 35, 39, 42, 43, 45, 46, 53, 54, 58, 76, 89, 93, 94, 95, 99, 100, 101, 108, 121, 124, 128, 162, 179, 181, 190, 191, 196, 198, 201, 202, 207, 231, 235, 237, 238, 239, 240, 244, 245, 260, 261, 268, 276, 277, 289, 291, 293, 296, 300, 301, 317, 322, 323, 331, 336, 354, 358, 362, 364, 381, 382, 385, 390, 391, 392, 395, 396, 401, 402, 403, 406, 412, 416, 417, 439, 442, 443, 446, 453, 462, 467, 479, 489, 495, 499, 525, 526, 547, 566, 570, 578, 583, 599, 603, 604, 621, 623, 626, 634, 638, 640, 646, 649, 663, 670, 675, 684, 687, 695, 706, 712, 713, 716, 738, 743, 749, 751, 783, 786, 797, 801, 806, 816, 819, 825, 831, 840, 844, 848, **851**, 861
Triphyllios, 793, 825
Tripoli, 15
Tripolitana, 13
Trisagion, 36, 40, 49, 51, 232, 416, 439, 440, 496, 498, 569, 615, 679, 793, 831, **853**
Tritheism, 36, 53, 190, 237, 276, 362, 392, 446, 570, 638, 640, 641, 662, 679, 743, 824, 841, 852, **853**
Trocondus, 113
Troianus of Saintes, **853**

Tropaeum Traiana, 765, **853**, *316*
Tropeia (Tropea), 424
Trophimus of Arles, 79, 238, 581, 654, 663, 743, **853**
Tropici, **854**
Troyes (Trecas), 511
Tsitsernavank, 80
Tucci, 106
Tuna al-Gabal, 267
Tunic, 253, 865
Tunis, 14
Tunisia, 170
Tura, 235, papyri, 187, 310, 647, 716, 723, 878, 879
Tur Alpheph, 543
Turbo, 371
Turfan, 519
Turiaso (Tarazona), 813
Turibius, 484, 571
Turibius of Astorga, **854**
Turibius of Palencia, 844, **854**
Turin (Augusta Taurinorum), 193, 421, 548, **854**, C. of (398), 79, 338, 530, 780, 854, 869, Holy Shroud of, 7
Turis, Mar, 536
Tusculanae, 75
Tyana, 43, 143, 304, 608, 826, **854**, C. of (366), 47, 78, 95, 778, 779, 854, 855
Tychicus, 199
Tynemouth, 131
Typology, 1, 25, 30, 11, 176, 236, 293, 312, 341, 453, 470, 493, 504, 530, 548, 551, 557, 568, 575, 598, 621, 669, 750, 751, 786, 792, 801, 802, 816, 876, 878, *see also* Allegory
Tyre, 23, 36, 93, 94, 196, 224, 251, 289, 303, 307, 416, 673, 685, 752, 808, **854**, C. of (335), 77, 83, 140, 195, 224, 302, 325, 347, 432, 434, 439, 516, 522, 525, 546, 551, 582, 657, 662, 705, 824, 830, 855, 858, C. of (367), 665, C. of (448), 403, 855, C. of (515), 769, 855, C. of (518), 678, 855

Ui Néill, 186
Ukhtanès, 3
Ulfila, 78, 105, 195, 199, 302, 307, 329, 356, 357, 546, 566, 637, 770, **856**
Ulmerings, 356
Ulmetum, 765
Ulpian of Antioch, 22, 714
Ulpia Traiana (Sarmizegetuza), 217
Ulrichsberg, 600
Ulster, 128
Umbo, 254
Umbria and Picenum, 422
Umm el-Gemal, 66
Umm Gerar, 635
Umm Harteyn, 218
Umm Qeis, 636
Ungaro Testini, Luerezia, 291-92, 571-74, 770-71
Unus ex Trinitate crucifixus, 570
Unus ex Trinitate passus est, 165, 315, 322, 400, 443, 444, 446, 585, 713, 747, 831, 851, 852, **856**
Uranius, 49, 430, 438, 628, **856**
Urban I, 617, 644, 741
Urbanus, 148
Urbs Beata Jerusalem, **856**
Urci, 148
Urgellum (Seo de Urgel), 813
Uriel, 132
Ursacius of Singidunum, 77, 120, 559, 783, **858**
Ursicinus, 30, 863
Ursinus, antipope, 30, 218, 249, 348, 416, 646, 742, **856**
Ursula and the 11,000 Virgins, 347, **856**
Ursus, 729
Usury, 79, 270, 364, 737, 793, **856**
Usward of Saint-Germain-des-Prés, 537
Utica, 542
Utrecht, 877

Vagharshapat - Etchmiadzin, 80, **858**, C. of (491), 19, 80
Vahan Mamikonian, 80, 255, 478, **858**
Vaison, C. of (442), 381, 410, 548, 609, 753, **858**, 861, C. of (529), 555, 761
Valdebert, 187
Val d'Olona, 421
Valence, 59, C. of (374), 154, 338, 465, **858**, C. of (375), 338, C. of (529), 212, C. of (585), 325
Valencia (Valentia), 688, 791
Valens, emperor, 5, 17, 30, 43, 47, 49, 78, 93, 114, 120, 137, 143, 194, 215, 218, 224, 228, 236, 263, 268, 296, 299, 305, 339, 350, 356, 360, 363, 440, 509, 543, 550, 565, 665, 678, 737, 742, 783, 828, 837, 845, 854, **858**, 873
Valens of Mursa, 77, 559, **858**
Valentinian I, 26, 28, 78, 92, 105, 183, 192, 204, 218, 224, 243, 264, 271, 305, 316, 328, 343, 347, 360, 375, 391, 407, 460, 464, 526, 706, **858**
Valentinian II, 26, 28, 65, 88, 347, 464, 547, 559, 660, 782, 828, **858**
Valentinian III, 665, 704, 739, 829, **859**, 861
Valentinians, 10, 12, 45, 163, 205, 242, 244, 309, 353, 396, 398, 413, 415, 560, 579, 603, 606, 610, 612, 619, 621, 622, 657, 680, 771, 776, 802, 805, 815, 819, 859, Against the Valentinians, 819
Valentinus, 23, 229, 280, 308, 328, 352, 353, 373, 374, 384, 423, 438, 569, 644, 717, 724, 732, 741, 769, 830, **859**, 884
Valentinus, catholic bishop, **859**
Valentinus the Apollinarist, **859**
Valentinus the Gnostic, 610, 629, **859**
Valentio, 366
Valenza, 148
Valeria, **860**
Valeria Maria, 729
Valerian, 64, 145, 156, 166, 178, 193, 237, 301, 307, 326, 336, 338, 497, 571, 584, 607, 667, 672, 676, 741, 757, 784, 794, 822, **860**
Valeriana and Victoria, 383
Valerian of Aquileia, **860**
Valeria Ripensis, 65
Valerian of Bierzo, **860**
Valerian of Calahorra, **861**
Valerian of Cimiez, 393, **861**
Valerius of Bierzo, 331, 383, 791, **860**, 873
Valerius of Saragossa, 870
Vandals, 13, 14, 55, 78, 113, 134, 142, 143, 144, 145, 146, 148, 158, 193, 202, 228, 247, 248, 249, 283, 315, 326, 331, 342, 349, 356, 369, 382, 383, 396, 400, 418, 464, 479, 523, 543, 554, 607, 610, 641, 652, 664, 705, 741, 742, 743, 761, 774, 779, 791, 795, 806, 816, 828, 840, 842, 859, **861**, 868, 869, 870, 875
Vanesbroeck, Michel, 3, 5, 19, 36, 60-61, 65-66, 95, 106, 220, 251, 345, 372, 410, 416, 417, 428, 429, 517-18, 576, 609, 646
Vannes, 671, C. of (461-91), 17, **862**
Vanyó, Lazló, 1-2, 219, 579-80, 598, 647, 698, 699-700, 843
Vardan I, 269
Vardan II, 269
Variae, 149, 828
Varro, 82, 121
Vatican, 38, 70, 156, 169, 171, 210, 211, 434, 450, 573, 597, 675, 741, **862**, *317, 318*
Vatican II, 6, 99, 120, 171, 502
Vedastus, **862**
Vegas de Pedroza (Segovia), 148
Vegesela, 524
Vejer de la Frontera, 107
Velatio, 871
Velletri, 10
Venantius, bishop, 217

Venantius Fortunatus, 19, 26, 64, 116, 144, 160, 162, 202, 278, 304, 339, 341, 346, 347, 349, 401, 426, 464, 497, 595, 700, 729, 732, 758, 799, **862**
Venantius of Thinissa, 584
Vence (Vincium), **863**
Venerius of Milan, 147, **863**
Venetia and Histria, 419, 420
Ventaroli, 424
Ventrale, 253
Venusia (Venosa), 424
Vera Cruz de Marmelar (Alto Alentejo), 512
Veranus of Vence, 295, 753, 767, **863**
Vercelli (Vercellae), 28, 302, 337, 396, 420, 854
Verdun (Viredunum), 348, 644
Verecundus of Iunca, 144, **863**
Verina, 113, 479
Verona, 19, 126, 193, 223, 356, 420, 583, 630, **863**, 884
Verona palimpsest, 107
Verona sacramentary, 484, 496, 499, 616
Veronica, Veil of, 7
Verulus of Rusiccade, 584
Verus, 846
Verus of Orange, **863**
Vescera (Biskra), 13
Vespasian, 143, 186, 245, 246, 614, 779, 816, 843, **863**
Vestiarius or Vestararius, **864**
Vestibule, 170
Vestments, Liturgical, 82, 108, 123, 186, 233, 252, 500, 502, **864**
Vetus Gallica, 142
Vetus Latina, 38, 472, 524, 752, 868
Vexilla regis prodeunt, 729, 862
Vice, *see* Virtues and Vices
Vicetia (Vicenza), 420
Vicovaro, 119
Victor, bishop of Constantine, 193
Victor, priest, 146
Victor I, 11, 106, 205, 222, 225, 258, 328, 413, 472, 492, 512, 551, 563, 626, 640, 643, 644, 646, 701, 728, 741, 743, 762, 768, 818, 831, 838, 848, **867**, 875, 885
Victorianus, 14
Victori(n)us of Aquitaine, 557, **867**
Victorinus of Munatiana, 711
Victorinus of Petovium, 14, 55, 90, 117, 121, 207, 224, 225, 311, 431, 543, 560, 638, 647, 715, 733, 787, **867**, 881
Victorinus the Poet, **867**
Victorius of Le Mans, 482
Victor of Antioch, 527, 716, **868**
Victor of Aquitaine, 188, 189, **868**
Victor of Assuras, 584
Victor of Capua, 152, 234, 497, **868**
Victor of Cartenna, 393, **868**
Victor of Garba, 250
Victor of Limoges, 380
Victor of Octavu, 584
Victor of Ravenna, 116
Victor of Tunnuna, 166, 444, 548, **868**
Victor of Turin, 307
Victor of Vita, 14, 233, 296, 400, 497, 671, 778, 812, 842, 861, **868**
Victricius of Rouen, 104, 191, 477, 552, 742, **869**
Vienna (Vindobona), 638
Vienna Dioscorides, 88
Vienna Genesis, 10, 428, 452, 453, 506, 599, 631, 730
Vienne (Vienna), 79, 105, 134, 178, 268, 326, 337, 339, 483, 533, 597, 654, 676, 712, 854, **869**, *319*, C. of (c.471/5), 869
Vigilantius, 776, 846, **869**, Against, 869
Vigilius, 27, 67, 79, 102, 119, 127, 150, 154, 197, 305, 315, 366, 386, 420, 438, 451, 486, 496, 545, 553, 623, 642, 643, 645, 662, 666, 679, 714, 738, 743, 747, 779, 823, 838, 855, 863, **870**, 876
Vigilius, deacon, **870**
Vigilius of Thapsus, **870**

Vigilius of Trent, **870**
Vilicus, bishop of Paris, 342
Villa Ludovisi, 187
Villanueva de Lorenzana, 334
Viminacium, 192
Vincent, 581, 660, 753, 767, 863, **870**
Vincent, priest of Constantinople, **870**
Vincent of Capua, 298, 348, **870**
Vincent of Ibiza, 488
Vincent of Lérins, 72, 208, 244, 245, 320, 483, 576, 655, 849, **870**
Vincent of Saragossa, 582, 774
Vindemialis, 143
Vindemius of Cyssa, 773
Vindicta Salvatoris, 687
Virga, 315, 477, 536, 549, 575, 579, 778, 866, **871**, *320*
Virgilius Maro, **871**
Virgilius of Arles, 483
Virgil of Thapsus, 831
Virgins, Virginity, 2, 16, 17, 29, 39, 43, 47, 48, 52, 84, 91, 94, 99, 100, 110, 103, 120, 122, 123, 125, 130, 146, 154, 161, 162, 173, 181, 184, 192, 199, 200, 218, 221, 223, 225, 232, 251, 264, 271, 276, 281, 284, 288, 303, 306, 347, 357, 358, 368, 383, 399, 411, 431, 439, 440, 452, 454, 467, 477, 515, 521, 528, 529, 532, 534, 538, 557, 559, 561, 621, 666, 682, 685, 718, 719, 722, 725, 738, 746, 774, 775, 776, 783, 787, 790, 792, 793, 802, 819, 829, 847, 856, 863, 866, **871**, 877, 878, 881, 882
Virtues and vices, 1, 3, 19, 29, 54, 74, 83, 94, 100, 117, 144, 145, 161, 177, 184, 198, 220, 222, 251, 255, 260, 286, 287, 288, 316, 318, 367, 372, 394, 401, 406, 417, 437, 442, 455, 462, 517, 529, 547, 558, 562, 587, 590, 597, 610, 621, 629, 652, 655, 687, 691, 693, 699, 702, 703, 719, 721, 737, 771, 780, 781, 786, 788, 792, 806, 812, 821, 836, 843, 865, **871**, 872, 875, 878
Visigothic rite, 385
Visigoths, 17, 19, 96, 127, 142, 167, 179, 183, 198, 297, 298, 307, 335, 338, 356, 387, 418, 451, 464, 465, 478, 482, 554, 565, 730, 774, 778, 783, 791, 795, 813, 837, 846, 861
Vision, 13, 56, 71, 262, 367, 412, 531, 533, 571, 649, 670, 725, 860, 870, **872**
Vitae Columbae, 186
Vitae Patrum, 61, 119, 121, 237, 255, 443, 448, 532, 652, 655, 860, **873**
Vitalian, 143, 209, 369, 399, 446, 642, 645, 743, 824, **873**
Vitalis and Agricola, 656
Vitalis of Sulci (S. Antioco), 757
Vitalis the Apollinarist, 14, 29, 59, 143, 218, **873**
Vitiges, 357
Vivarium, 126, 131, 150, 272, 282, 451, 486, 576, 655, **873**
Viventiolus of Lyons, 555, **874**
Vladimir, 356
Vogel, Cyril, 477, 486, 502-3, 667-68
Vogt, Hermann J., 259-62, 473, 604, 617-18, 688

Voicu, Sever J., 1, 3, 4, 16-17, 19, 32, 41, 81, 82, 91-92, 115, 153-54, 216, 221, 244, 269, 276, 298, 298-99, 303-4, 314, 324, 325, 327-28, 331, 342, 356, 362-63, 368, 394, 416, 428, 437, 442, 443, 445, 467, 468, 477-78, 480, 482, 514, 518, 527, 536, 550, 556, 575-76, 576, 582, 593, 605, 629, 660, 686, 752, 766, 772, 785, 791, 794, 802, 832, 837
Volsinii (Bolsena), 422
Volusianus, 846, 874
Volusianus and Fortunatus, 530
Vranje, 600
Vrthanes K'erdol, 81
Vulcacius Gallicanus, 388
Vulci, 422
Vulgate, 131, 524, 600, 649, 746, 750, 764, 769, 801

Waccho, 505
Wadi ed-Defali, 636
Wadi-el-Kelt, 44
Walafrid Strabo, 518, 873
Wales, 131
Walesby, 133, font, 133
Walid I, 218
Wallia, 356, 861
Wamba, 458
War, 191, 288, 411, 664, 857, **875**
Warnaharius, 875
Washing of feet, 190, 325, **875**
Water, 34, 42, 64, 82, 107, 108, 111, 122, 123, 159, 173, 190, 221, 222, 229, 278, 280, 292, 296, 312, 316, 319, 341, 390, 409, 465, 490, 493, 504, 518, 525, 541, 542, 564, 575, 611, 649, 651, 679, 744, 749, 751, 781, 800, 802, 815, 817, **875**, 877
Water Newton treasure, 133
Wearmouth, 119, 129, 446
Wedding: Procedure and Liturgy, 786, **876**
Week, 82, 90, 96, 135, 188, 319, 494, 504, 625, 744, 800, **876**
Weli El-Khader, 636
Wendelin, 349
Western liturgical books, 502
Wharram Percy, Yorkshire, 131
Whitby, 131, 137, Synod of (664), 129, 143, 187, 211, 762, 877
White Convent, 265
White Monastery (Deir el Abiad), 200, 776
Whithorn, 130, 597, 762
Wicthed, 118
Widows, 48, 91, 123, 130, 173, 199, 218, 221, 232, 244, 259, 288, 304, 319, 360, 383, 476, 521, 528, 561, 605, 617, 710, 741, 759, 793, 871, 882, **877**
Widow's son, 579
Wilfrid I, 883
Wilfrid of York, 129, 132, 159, 824, **877**
William of Malmesbury, 427
William of Saint-Thierry, 623, 881
Willibald, 125
Willibrord of Utrecht, 162, 768, **877**, Life of, 394

Wilton, 133
Windisch (Vindonissa), 347
Wine, 8, 34, 52, 64, 82, 84, 128, 159, 222, 223, 255, 271, 277, 280, 292, 294, 319, 383, 436, 473, 490, 491, 541, 542, 550, 609, 611, 744, 749, 773, 786, 803, 815, 876, **877**
Wing, 262, 601, 690, 694, 802, **877**
Wings of the soul, 506
Wisdom books, 44, 152, 242, 371, 405, 431, 455, 504, 611, 669, 752, 787, 803, 815, 867, **877**
Wisdom of Jesus Christ, 610
Wiseman, James R., 795
Witeric, 368
Witiges, 451
Witimer of Orense, 531
Witness, 189, 191, 531, 543, 821
Woman, 10, 12, 91, 116, 153, 154, 160, 173, 198, 221, 243, 253, 259, 268, 273, 281, 284, 296, 318, 328, 333, 341, 411, 441, 467, 500, 509, 514, 524, 528, 532, 542, 551, 556, 566, 567, 580, 597, 689, 718, 736, 737, 773, 775, 792, 801, 805, 819, 833, 871, **881**
Worcester, C. of (c.603), 129
Work, *see* Labour
World, 2, 11, 12, 34, 37, 40, 52, 53, 54, 56, 75, 82, 84, 92, 94, 110, 113, 114, 125, 145, 155, 158, 163, 164, 166, 204, 205, 219, 226, 227, 230, 231, 247, 250, 251, 252, 255, 259, 285, 287, 288, 291, 306, 308, 309, 310, 327, 352, 354, 357, 358, 367, 373, 378, 389, 397, 405, 415, 440, 455, 457, 465, 477, 485, 506, 510, 518, 523, 553, 556, 560, 587, 590, 606, 610, 621, 633, 683, 689, 690, 698, 707, 708, 709, 712, 717, 720, 729, 734, 746, 752, 753, 754, 758, 761, 775, 776, 780, 782, 788, 789, 792, 797, 801, 802, 818, 830, 831, 836, 839, 840, 847, 859, 860, 867, 871, 876, 877, 880, **882**
Worms (Wormatia), 134, 348
Writing, 241, 794, **882**
Wrmonoc, 663
Wurzburg Lectionary, 59

Xanten (Colonia Ulpia Traiana), 347, 349
Xenophon, 649

Yahya ibn Adi, 66
Yahya ibn Yahya of Antioch, 66
Yakto mosaic, 51
Yaqût, 65
Yarnold, Edward, 503-5
Yarrow, 873
Ydruntum (Otranto), 424
Yeavering, 131
Yecla (Murcia), 148
Yerevan diptych, 228
Yezdegerd I, 416, 537, 674, 752, 757, **883**
Yezdegerd II, 4, 482
Ymnum Sancti Patrici, 765
Yonan the Slave, Mar, 11
York, 96, 118, 129, 132, 159, 662, 762, 824, 877, **883**

Zacchaeus, **884**, 886
Zacharias, pope, 433, 523, 564, 643, 646
Zacharias of Chrysopolis, 234
Zacharias of Jerusalem, **884**
Zacharias of Shkou, 200
Zacharias Scholasticus or the Rhetor, 43, 330, 389, 417, 444, 563, 593, **884**
Zahara, 107
Zakai, Mar, 95
Zangara, Vincenza, 58, 73, 560, 736
Zapara, 838
Zarathustra, 564
Zar'a Ya'eqob, 291
Zebboc, 14
Zebed, 65
Zebennus, 106
Zegani, 19
Zenaide and Philonilla, 175
Zeno, 5, 6, 36, 47, 49, 85, 88, 109, 112, 113, 118, 197, 246, 251, 252, 263, 296, 325, 357, 400, 403, 416, 420, 432, 438, 447, 473, 479, 553, 565, 569, 585, 610, 678, 684, 743, 744, 780, 808, 828, 840, 841, 855, 863, **884**
Zeno of Kurion, 732
Zeno of Seville, 780
Zeno of Verona, 108, 138, 311, 312, 398, 450, 494, 538, 719, **884**
Zeno the Isaurian, 679
Zenobia, 102, 246, 637, 663, **884**
Zenobius of Zephyrium, 120, **884**
Zenonis, 113
Zenophilus, 779
Zephyrinus, 11, 83, 84, 140, 156, 181, 375, 583, 640, 644, 741, 768, 818, **885**
Zincone, Sergio, 40, 43, 83, 119, 213, 224, 234, 256, 270, 284, 303, 333, 374, 380, 382, 397, 400, 404, 410, 417-18, 426, 430, 438, 448, 475, 520, 526, 547, 578, 593, 636-37, 648-49, 660, 688-89, 720, 720, 740, 759, 772, 780, 784, 803, 805, 828, 838, 856, 870, 871
Zion, 433
Ziporis, 635
Zoara, 636
Zoilus, 339
Zokom, 65
Zonara, 142
Zopyrus of Barca, 487
Zorita de los Canes (Guadalajara), 148
Zoroaster, 82, 557
Zoroastrians, -ism, 162, 255, 315, 519, 564, 718, 805
Zosimus, abbat, **885**
Zosimus, historian, **885**
Zosimus, pope, 55, 79, 96, 98, 102, 137, 141, 142, 147, 148, 175, 182, 314, 338, 379, 458, 617, 641, 643, 645, 654, 660, 665, 666, 742, 744, 745, 780, 782, 804, 834, 854, 856, 865, 869, **885**
Zosimus of Ascalon, **885**
Zosimus of Gaza, **885**
Zostrianus, 771
Zoui, 15
Zouqnîn, 837
Zvart'noc, 19, 80, 81, 256, 344

Numbers in bold refer to the main entry on the subject. *Numbers in italics refer to illustration numbers (vol II, pages 951-1094)*